Pearson New International Edition

Organic Chemistry
Paula Y. Bruice
Seventh Edition

Pearson Education Limited
Edinburgh Gate
Harlow
Essex CM20 2JE
England and Associated Companies throughout the world

Visit us on the World Wide Web at: www.pearsoned.co.uk

© Pearson Education Limited 2014

ISBN 10: 1-292-02436-4
ISBN 13: 978-1-292-02436-3

British Library Cataloguing-in-Publication Data
A catalogue record for this book is available from the British Library

Printed in the United States of America

PEARSON CUSTOM L

Table of Contents

Remembering General Chemistry: Electronic Structure and Bonding

From Chapter 1 of *Organic Chemistry*, Seventh Edition. Paula Yurkanis Bruice. Copyright © 2014 by Pearson Education, Inc.

Remembering General Chemistry: Electronic Structure and Bonding

To stay alive, early humans must have been able to distinguish between the different kinds of materials in their world. "You can live on roots and berries," they might have said, "but you can't eat dirt. You can stay warm by burning tree branches, but you can't burn rocks."

By the early eighteenth century, scientists thought they had grasped the nature of that difference, and in 1807 Jöns Jakob Berzelius gave names to the two kinds of materials. Compounds derived from living organisms were believed to contain an immeasurable vital force—the essence of life. These he called "organic." Compounds derived from minerals—those lacking the vital force—were "inorganic."

Because chemists could not create life in the laboratory, they assumed they could not create compounds that had a vital force. Since this was their mind-set, you can imagine how surprised chemists were in 1828 when Friedrich Wöhler produced urea—a compound known to be excreted by mammals—by heating ammonium cyanate, an inorganic mineral.

$$\overset{+}{N}H_4 \, \overset{-}{O}CN \quad \xrightarrow{\text{heat}} \quad \underset{\substack{\text{urea} \\ \text{an "organic" compound}}}{H_2N \overset{\overset{\displaystyle O}{\|}}{C} NH_2}$$

ammonium cyanate
an inorganic mineral

For the first time, an "organic" compound had been obtained from something other than a living organism and certainly without the aid of any kind of vital force. Chemists, therefore, needed a new definition for "organic compounds." **Organic compounds** are now defined as *compounds that contain carbon.*

Organic compounds are compounds that contain carbon.

Why is an entire branch of chemistry devoted to the study of carbon-containing compounds? We study organic chemistry because just about all of the molecules that make life possible and that make us who we are—proteins, enzymes, vitamins, lipids, carbohydrates, DNA, RNA—are organic compounds. Thus the chemical reactions that take place in living systems, including our own bodies, are reactions of organic compounds. Most of the compounds found in nature—those that we rely on for all of our food, for some of our clothing (cotton, wool, silk), and for energy (natural gas, petroleum)—are organic as well.

Organic compounds are not limited, however, to those found in nature. Chemists have learned to synthesize millions of organic compounds never found in nature, including synthetic fabrics, plastics, synthetic rubber, and even things like compact discs and Super Glue. And most importantly, almost all of our commonly prescribed drugs are synthetic organic compounds.

Some synthetic organic compounds prevent shortages of naturally occurring products. For example, it has been estimated that if synthetic materials—nylon, polyester, Lycra—were not available for clothing, then all of the arable land in the United States would have to be used for the production of cotton and wool just to provide enough material to clothe us. Other synthetic organic compounds provide us with materials we would not have—Teflon, Plexiglas, Kevlar—if we had only naturally occurring organic compounds. Currently, there are about 16 million known organic compounds, and many more are possible that we cannot even imagine today.

What makes carbon so special? Why are there so many carbon-containing compounds? The answer lies in carbon's position in the periodic table. Carbon is in the center of the second row of elements. We will see that the atoms to the left of carbon have a tendency to give up electrons, whereas the atoms to the right have a tendency to accept electrons (Section 3).

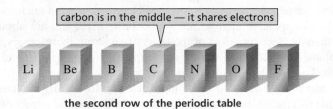

carbon is in the middle — it shares electrons

Li Be B C N O F

the second row of the periodic table

Because carbon is in the middle, it neither readily gives up nor readily accepts electrons. Instead, it shares electrons. Carbon can share electrons with several different kinds of atoms, and it can share electrons with other carbon atoms. Consequently, carbon is able to form millions of stable compounds with a wide range of chemical properties simply by sharing electrons.

Natural Organic Compounds Versus Synthetic Organic Compounds

It is a popular belief that natural substances—those made in nature—are superior to synthetic ones—those made in the laboratory. Yet when a chemist synthesizes a compound, such as penicillin or morphine, the compound is exactly the same in all respects as the compound synthesized in nature. Sometimes chemists can even improve on nature. For example, chemists have synthesized analogues of penicillin that do not produce the allergic responses that a significant fraction of the population experiences from naturally produced penicillin, or that do not have the bacterial resistance of the naturally produced antibiotic. Chemists have also synthesized analogues of morphine—compounds with structures similar to but not identical to that of morphine—that have pain-killing effects like morphine but, unlike morphine, are not habit forming.

A field of poppies growing in Afghanistan. Most commercial morphine is obtained from opium, the juice extracted from this species of poppy. Morphine is the starting material for the synthesis of heroin. One of the side products formed in the synthesis has an extremely pungent odor; dogs used by drug enforcement agencies are trained to recognize this odor. Nearly three-quarters of the world's supply of heroin comes from the poppy fields of Afghanistan.

When we study organic chemistry, we learn how organic compounds react. Organic compounds consist of atoms held together by bonds. When an organic compound reacts, some of these bonds break and some new bonds form. *Bonds form when two atoms share electrons, and bonds break when two atoms no longer share electrons.*

How readily a bond forms and how easily it breaks depend on the particular electrons that are shared, which depend, in turn, on the atoms to which the electrons belong. So, if we are going to start our study of organic chemistry at the beginning, we must start with an understanding of the structure of an atom—what electrons an atom has and where they are located.

1 THE STRUCTURE OF AN ATOM

An atom consists of a tiny dense nucleus surrounded by electrons that are spread throughout a relatively large volume of space around the nucleus called an electron cloud. The nucleus contains **positively charged protons** and **uncharged neutrons,** so it is positively charged. The **electrons** are **negatively charged.** The amount of positive charge on a proton equals the amount of negative charge on an electron. Therefore, the number of protons and the number of electrons in an uncharged atom must be the same.

Electrons move continuously. Like anything that moves, electrons have kinetic energy, and this energy is what counteracts the attractive force of the positively charged protons that would otherwise pull the negatively charged electrons into the nucleus.

Protons and neutrons have approximately the same mass and are about 1800 times more massive than an electron. Most of the *mass* of an atom, therefore, is in its nucleus. Most of the *volume* of an atom, however, is occupied by its electrons, and this is where our focus will be because it is the electrons that form chemical bonds.

The **atomic number** of an atom is the number of protons in its nucleus. The atomic number is unique to a particular element. For example, the atomic number of carbon is 6, which means that all uncharged carbon atoms have six protons and six electrons. Atoms can gain electrons and thereby become negatively charged, or they can lose electrons and become positively charged, but the number of protons in an atom of a particular element never changes.

Although *all carbon atoms have the same atomic number,* they do not all have the same mass number because they do not all have the same number of neutrons. The **mass number** of an atom is the sum of its protons and neutrons. For example, 98.89% of all carbon atoms have six neutrons—giving them a mass number of 12—and 1.11% have seven neutrons—giving them a mass number of 13. These two different kinds of carbon atoms (^{12}C and ^{13}C) are called **isotopes.**

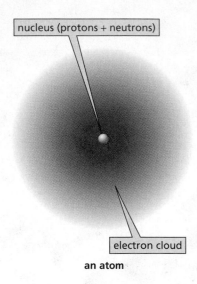

The nucleus contains positively charged protons and uncharged neutrons.

The electrons are negatively charged.

nucleus (protons + neutrons)

electron cloud

an atom

atomic number = the number of protons in the nucleus

mass number = the number of protons + the number of neutrons

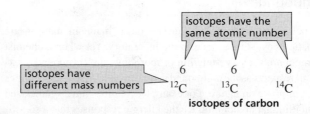

isotopes have the same atomic number

isotopes have different mass numbers

^{12}C ^{13}C ^{14}C

6   6   6

isotopes of carbon

Carbon also contains a trace amount of ^{14}C, which has six protons and eight neutrons. This isotope of carbon is radioactive, decaying with a half-life of 5730 years. (The *half-life* is the time it takes for one-half of the nuclei to decay.) As long as a plant or animal is alive, it takes in as much ^{14}C as it excretes or exhales. When it dies, however, it no longer takes in ^{14}C, so the ^{14}C in the organism slowly decreases. Therefore, the age of a substance derived from a living organism can be determined by its ^{14}C content.

The **atomic weight** of an element is the average mass of its atoms. Because an *atomic mass unit (amu)* is defined as exactly 1/12 of the mass of ^{12}C, the atomic mass of ^{12}C is 12.0000 amu; the atomic mass of ^{13}C is 13.0034 amu. Therefore, the atomic weight of carbon is 12.011 amu because $(0.9889 \times 12.0000) + (0.0111 \times 13.0034) = 12.011$.

atomic weight = the average mass of the atoms in the element

molecular weight = the sum of the atomic weights of all the atoms in the molecule

The bronze sculpture of Albert Einstein, on the grounds of the National Academy of Sciences in Washington, D.C., measures 21 feet from the top of the head to the tip of the feet and weighs 7000 pounds. In his left hand, Einstein holds the mathematical equations that represent his three most important contributions to science: the photoelectric effect, the equivalency of energy and matter, and the theory of relativity. At his feet is a map of the sky.

2 HOW THE ELECTRONS IN AN ATOM ARE DISTRIBUTED

For a long time, electrons were perceived to be particles—infinitesimal "planets" that orbit the nucleus of an atom. In 1924, however, a French physicist named Louis de Broglie showed that electrons also have wavelike properties. He did this by combining a formula developed by Albert Einstein that relates mass and energy with a formula developed by Max Planck that relates frequency and energy. The realization that electrons have wavelike properties spurred physicists to propose a mathematical concept known as quantum mechanics.

Quantum mechanics uses the same mathematical equations that describe the wave motion of a guitar string to characterize the motion of an electron around a nucleus. The version of quantum mechanics most useful to chemists was proposed by Erwin Schrödinger in 1926.

According to Schrödinger, the electrons in an atom can be thought of as occupying a set of concentric shells that surround the nucleus. The first shell is the one closest to the nucleus. The second shell lies farther from the nucleus. The third and higher numbered shells lie even farther out.

Each shell contains subshells known as **atomic orbitals.** *Each atomic orbital has a characteristic shape and energy and occupies a characteristic volume of space.*

The first shell consists only of an *s* atomic orbital; the second shell consists of *s* and *p* atomic orbitals; the third shell consists of *s*, *p*, and *d* atomic orbitals; and the fourth and higher shells consist of *s, p, d,* and *f* atomic orbitals (Table 1).

Table 1	Distribution of Electrons in the First Four Shells That Surround the Nucleus			
	First shell	**Second shell**	**Third shell**	**Fourth shell**
Atomic orbitals	*s*	*s, p*	*s, p, d*	*s, p, d, f*
Number of atomic orbitals	1	1, 3	1, 3, 5	1, 3, 5, 7
Maximum number of electrons	2	8	18	32

Each shell contains one *s* orbital. Each second and higher shell—in addition to its *s* orbital—contains three *degenerate p* orbitals. **Degenerate orbitals** are orbitals that have the same energy. The third and higher shells—in addition to their *s* and *p* orbitals—contain five degenerate *d* orbitals, and the fourth and higher shells also contain seven degenerate *f* orbitals.

Because a maximum of two electrons can coexist in an atomic orbital (see the Pauli exclusion principle), the first shell, with only one atomic orbital, can contain no more than two electrons (Table 1). The second shell, with four atomic orbitals—one *s* and three *p*—can have a total of eight electrons. Eighteen electrons can occupy the nine atomic orbitals—one *s*, three *p*, and five *d*—of the third shell, and 32 electrons can occupy the 16 atomic orbitals of the fourth shell. In studying organic chemistry, we will be concerned primarily with atoms that have electrons only in the first and second shells.

Degenerate orbitals are orbitals that have the same energy.

The **ground-state electronic configuration** of an atom describes the orbitals occupied by the atom's electrons when they are all in the available orbitals with the lowest energy. If energy is applied to an atom in the ground state, one or more electrons can jump into a higher-energy orbital. The atom then would be in an excited state and have an **excited-state electronic configuration.**

The ground-state electronic configurations of the smallest atoms are shown in Table 2. (Each arrow—whether pointing up or down—represents one electron.)

Atom	Name of element	Atomic number	$1s$	$2s$	$2p_x$	$2p_y$	$2p_z$	$3s$
H	Hydrogen	1	↑					
He	Helium	2	↑↓					
Li	Lithium	3	↑↓	↑				
Be	Beryllium	4	↑↓	↑↓				
B	Boron	5	↑↓	↑↓	↑			
C	Carbon	6	↑↓	↑↓	↑	↑		
N	Nitrogen	7	↑↓	↑↓	↑	↑	↑	
O	Oxygen	8	↑↓	↑↓	↑↓	↑	↑	
F	Fluorine	9	↑↓	↑↓	↑↓	↑↓	↑	
Ne	Neon	10	↑↓	↑↓	↑↓	↑↓	↑↓	
Na	Sodium	11	↑↓	↑↓	↑↓	↑↓	↑↓	↑

Table 2 The Electronic Configurations of the Smallest Atoms

The following three rules specify which orbitals an atom's electrons occupy:

1. The **aufbau principle** (*aufbau* is German for "building up") tells us the first thing we need to know to be able to assign electrons to the various atomic orbitals. According to this principle:

An electron always goes into the available orbital with the lowest energy.

It is important to remember that the closer the atomic orbital is to the nucleus, the lower is its energy.

Because a $1s$ orbital is closer to the nucleus, it is lower in energy than a $2s$ orbital, which is lower in energy—and closer to the nucleus—than a $3s$ orbital. When comparing atomic orbitals in the same shell, we see that an s orbital is lower in energy than a p orbital, and a p orbital is lower in energy than a d orbital.

Relative energies of atomic orbitals:

lowest energy ▶ $1s < 2s < 2p < 3s < 3p < 3d$ ◀ highest energy

2. The **Pauli exclusion principle** states that
   a. **no more than two electrons can occupy each atomic orbital, and**
   b. **the two electrons must be of opposite spin.**

It is called an exclusion principle because it limits the number of electrons that can occupy any particular shell. (Notice in Table 2 that spin in one direction is designated by ↑, and spin in the opposite direction by ↓.)

From these first two rules, we can assign electrons to atomic orbitals for atoms that contain one, two, three, four, or five electrons. The single electron of a hydrogen atom occupies a $1s$ orbital, the second electron of a helium atom fills the $1s$ orbital, the third electron of a lithium atom occupies a $2s$ orbital, the fourth electron of a beryllium atom fills the $2s$ orbital, and the fifth electron of a boron atom occupies one of the $2p$ orbitals. (The subscripts x, y, and z distinguish the three $2p$ orbitals.) Because the three p orbitals are degenerate, the electron can be put into any one of them. Before we can discuss atoms containing six or more electrons, we need to define Hund's rule.

3. Hund's rule states that

when there are two or more atomic orbitals with the same energy, an electron will occupy an empty orbital before it will pair up with another electron.

In this way, electron repulsion is minimized.

The sixth electron of a carbon atom, therefore, goes into an empty $2p$ orbital, rather than pairing up with the electron already occupying a $2p$ orbital (see Table 2). There is one more empty $2p$ orbital, so that is where nitrogen's seventh electron goes. The eighth electron of an oxygen atom pairs up with an electron occupying a $2p$ orbital rather than going into the higher-energy $3s$ orbital.

The locations of the electrons in the remaining elements can be assigned using these three rules.

The electrons in inner shells (those below the outermost shell) are called **core electrons.** Core electrons do not participate in chemical bonding. The electrons in the outermost shell are called **valence electrons.**

Carbon has two core electrons and four valence electrons (Table 2). Lithium and sodium each have one valence electron. If you examine the periodic table, you will see that lithium and sodium are in the same column. Elements in the same column of the periodic table have the same number of valence electrons. Because the number of valence electrons is the major factor determining an element's chemical properties, elements in the same column of the periodic table have similar chemical properties. Thus, the chemical behavior of an element depends on its electronic configuration.

> Core electrons are electrons
> in inner shells.

> Valence electrons are electrons
> in the outermost shell.

> The chemical behavior of
> an element depends on its
> electronic configuration.

PROBLEM 3♦

How many valence electrons do the following atoms have?

a. boron   **b.** nitrogen   **c.** oxygen   **d.** fluorine

PROBLEM 4♦

a. Write electronic configurations for chlorine (atomic number 17), bromine (atomic number 35), and iodine (atomic number 53).

b. How many valence electrons do chlorine, bromine, and iodine have?

PROBLEM 5

Look at the relative positions of each pair of atoms listed here in the periodic table. How many core electrons does each have? How many valence electrons does each have?

a. carbon and silicon   **c.** nitrogen and phosphorus
b. oxygen and sulfur   **d.** magnesium and calcium

---

3   IONIC AND COVALENT BONDS

Now that you know about the electronic configuration of atoms, let's now look at why atoms come together to form bonds. In explaining why atoms form bonds, G. N. Lewis proposed that

> *an atom is most stable if its outer shell is either filled or contains*
> *eight electrons, and it has no electrons of higher energy.*

According to Lewis's theory, an atom will give up, accept, or share electrons in order to achieve a filled outer shell or an outer shell that contains eight electrons. This theory has come to be called the **octet rule** (even though hydrogen needs only two electrons to achieve a filled outer shell).

Lithium (Li) has a single electron in its 2*s* orbital. If it loses this electron, the lithium atom ends up with a filled outer shell—a stable configuration. Lithium, therefore, loses an electron relatively easily. Sodium (Na) has a single electron in its 3*s* orbital, so it too loses an electron easily.

Each of the elements in the first column of the periodic table readily loses an electron because each has a single electron in its outermost shell.

When we draw the electrons around an atom, as in the following equations, core electrons are not shown; only valence electrons are shown because only valence electrons are used in bonding. Each valence electron is shown as a dot. When the single valence electron of lithium or sodium is removed, the species that is formed is called an ion because it carries a charge.

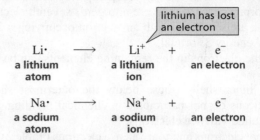

Fluorine has seven valence electrons (Table 2). Consequently, it readily acquires an electron in order to have an outer shell of eight electrons, thereby forming F^-, a fluoride ion. Elements in the same column as fluorine (such as chlorine, bromine, and iodine) also need only one electron to have an outer shell of eight, so they, too, readily acquire an electron.

Elements (such as fluorine and chlorine) that readily acquire an electron are said to be electronegative.

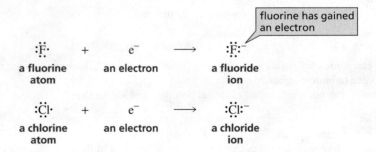

A hydrogen atom has one valence electron. Therefore, it can achieve a completely empty shell by losing an electron or a filled outer shell by gaining an electron.

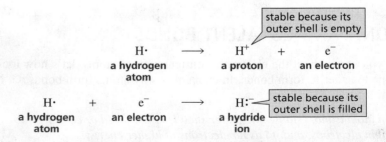

Loss of its sole electron results in a positively charged **hydrogen ion.** A positively charged hydrogen ion is called a **proton** because when a hydrogen atom loses its valence electron, only the hydrogen nucleus—which consists of a single proton—remains. When a hydrogen atom gains an electron, a negatively charged hydrogen ion—called a **hydride ion**—is formed.

PROBLEM 6♦

a. Find potassium (K) in the periodic table and predict how many valence electrons it has.

b. What orbital does the unpaired electron occupy?

Ionic Bonds Are Formed by the Attraction Between Ions of Opposite Charge

We have just seen that sodium gives up an electron easily and chlorine readily acquires an electron, both in order to achieve a filled outer shell. Therefore, when sodium metal and chlorine gas are mixed, each sodium atom transfers an electron to a chlorine atom, and crystalline sodium chloride (table salt) is the result. The positively charged sodium ions and negatively charged chloride ions are held together by the attraction of opposite charges (Figure 1).

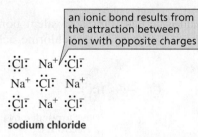

an ionic bond results from the attraction between ions with opposite charges

$:\ddot{Cl}:^-$ Na^+ $:\ddot{Cl}:^-$
Na^+ $:\ddot{Cl}:^-$ Na^+
$:\ddot{Cl}:^-$ Na^+ $:\ddot{Cl}:^-$

sodium chloride

a.

b.
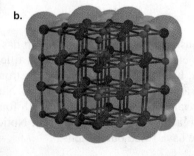

An ionic bond results from the attraction between ions of opposite charge.

▲ **Figure 1**

(a) Crystalline sodium chloride.

(b) The electron-rich chloride ions are red, and the electron-poor sodium ions are blue. Each chloride ion is surrounded by six sodium ions, and each sodium ion is surrounded by six chloride ions. Ignore the sticks holding the balls together; they are there only to keep the model from falling apart.

A **bond** is an attractive force between two ions or between two atoms. Attractive forces between opposite charges are called **electrostatic attractions.** A bond formed as a result of the electrostatic attraction between ions of opposite charge is called an **ionic bond.**

Sodium chloride is an example of an ionic compound. **Ionic compounds** are formed when an element on the left side of the periodic table *transfers* one or more electrons to an element on the right side of the periodic table.

Covalent Bonds Are Formed by Sharing a Pair of Electrons

Instead of giving up or acquiring electrons to achieve a filled outer shell, an atom can achieve a filled outer shell by sharing a pair of electrons. For example, two fluorine atoms can each attain a filled second shell by sharing their unpaired valence electrons.

Salar de Uyuni in Bolivia—the largest deposit of natural lithium in the world.

Lithium salts are used clinically. Lithium chloride (Li^+Cl^-) is an antidepressant, lithium bromide (Li^+Br^-) is a sedative, and lithium carbonate ($Li_2^+CO_3^{2-}$) is used to stabilize mood swings in people who suffer from bipolar disorder. Scientists do not yet know why lithium salts have these therapeutic effects.

A bond formed as a result of sharing electrons between two nuclei is called a **covalent bond.** A covalent bond is commonly shown by a solid line rather than as a pair of dots.

A covalent bond is formed when two atoms share a pair of electrons.

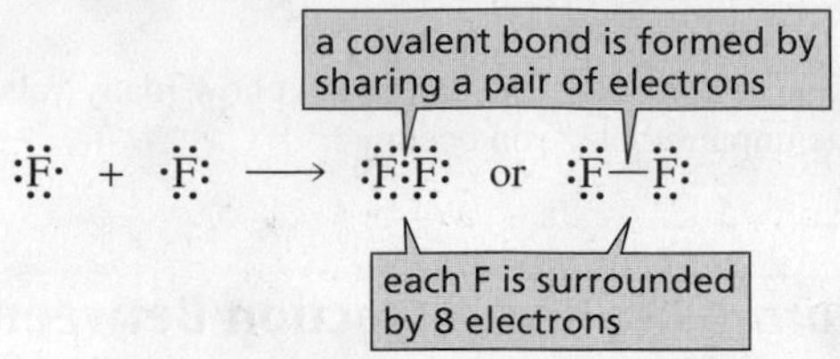

$$:\ddot{F}\cdot \ +\ \cdot\ddot{F}:\ \longrightarrow\ :\ddot{F}\!:\!\ddot{F}:\ \text{ or }\ :\ddot{F}\!-\!\ddot{F}:$$

> a covalent bond is formed by sharing a pair of electrons

> each F is surrounded by 8 electrons

Two hydrogen atoms can form a covalent bond by sharing electrons. As a result of covalent bonding, each hydrogen acquires a stable, filled first shell.

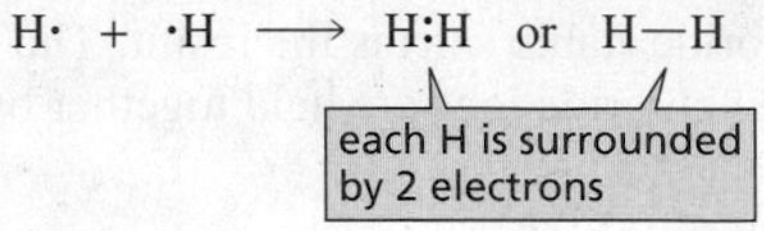

$$H\cdot \ +\ \cdot H\ \longrightarrow\ H\!:\!H\ \text{ or }\ H\!-\!H$$

> each H is surrounded by 2 electrons

Similarly, hydrogen and chlorine can form a covalent bond by sharing electrons. In doing so, hydrogen fills its only shell, and chlorine achieves an outer shell of eight electrons.

$$H\cdot \ +\ \cdot\ddot{C}l:\ \longrightarrow\ H\!:\!\ddot{C}l:\ \text{ or }\ H\!-\!\ddot{C}l:$$

> H is surrounded by 2 electrons
> Cl is surrounded by 8 electrons

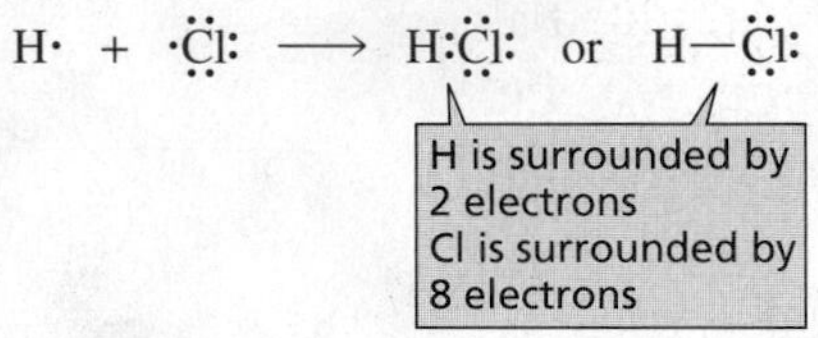

We just saw that because hydrogen has one valence electron and chlorine has seven valence electrons, each can achieve a filled outer shell by forming one covalent bond. Oxygen, however, has six valence electrons, so it needs to form two covalent bonds to achieve an outer shell of eight electrons. Nitrogen, with five valence electrons, must form three covalent bonds, and carbon, with four valence electrons, must form four covalent bonds to achieve a filled outer shell. Notice that all the atoms in water, ammonia, and methane have filled outer shells.

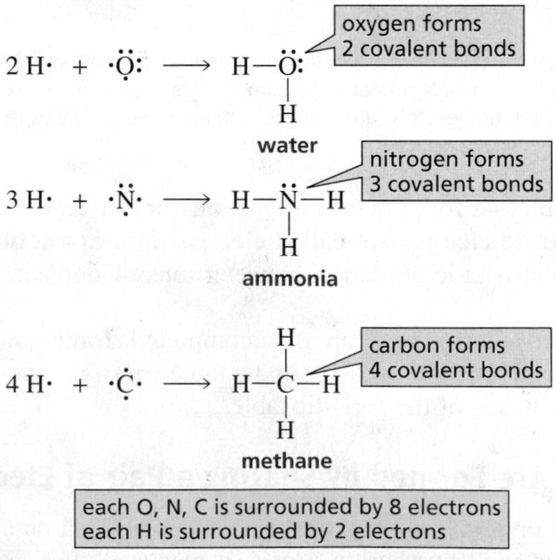

$$2\,H\cdot \ +\ \cdot\ddot{O}:\ \longrightarrow\ H\!-\!\underset{\underset{\displaystyle H}{|}}{\ddot{O}}:$$

> oxygen forms 2 covalent bonds

water

$$3\,H\cdot \ +\ \cdot\ddot{N}\cdot\ \longrightarrow\ H\!-\!\underset{\underset{\displaystyle H}{|}}{\ddot{N}}\!-\!H$$

> nitrogen forms 3 covalent bonds

ammonia

$$4\,H\cdot \ +\ \cdot\dot{\underset{\displaystyle .}{C}}\cdot\ \longrightarrow\ H\!-\!\underset{\underset{\displaystyle H}{|}}{\overset{\overset{\displaystyle H}{|}}{C}}\!-\!H$$

> carbon forms 4 covalent bonds

methane

> each O, N, C is surrounded by 8 electrons
> each H is surrounded by 2 electrons

Polar Covalent Bonds

The atoms that share the bonding electrons in the F—F and H—H covalent bonds are identical. Therefore, they share the electrons equally; that is, each electron spends as much time in the vicinity of one atom as in the other. Such a bond is called a **nonpolar covalent bond.**

In contrast, the bonding electrons in hydrogen chloride, water, and ammonia are more attracted to one atom than to another because the atoms that share the electrons in these molecules are different and have different electronegativities. **Electronegativity** is a measure of the ability of an atom to pull the bonding electrons toward itself.

The bonding electrons in hydrogen chloride, water, and ammonia are more attracted to the atom with the greater electronegativity. The bonds in these compounds are **polar covalent bonds.**

The electronegativities of some of the elements are shown in Table 3. Notice that electronegativity increases from left to right across a row of the periodic table and from bottom to top in any of the columns.

> A nonpolar covalent bond is a covalent bond between atoms with the same electronegativity.
>
> A polar covalent bond is a covalent bond between atoms with different electronegativities.

Table 3   The Electronegativities of Selected Elements[a]

1A	2A	1B	2B	3A	4A	5A	6A	7A
H 2.1								
Li 1.0	Be 1.5			B 2.0	C 2.5	N 3.0	O 3.5	F 4.0
Na 0.9	Mg 1.2			Al 1.5	Si 1.8	P 2.1	S 2.5	Cl 3.0
K 0.8	Ca 1.0							Br 2.8
								I 2.5

increasing electronegativity →

↑ increasing electronegativity

[a]Electronegativity values are relative, not absolute. As a result, there are several scales of electronegativities. The electronegativities listed here are from the scale devised by Linus Pauling.

A polar covalent bond has a slight positive charge on one end and a slight negative charge on the other. Polarity in a covalent bond is indicated by the symbols δ^+ and δ^-, which denote partial positive and partial negative charges. The negative end of the bond is the end that has the more electronegative atom. The greater the difference in electronegativity between the bonded atoms, the more polar the bond will be.

$$\overset{\delta^+}{H}-\overset{\delta^-}{\ddot{C}l}: \qquad \overset{\delta^+}{H}-\overset{\delta^-}{\underset{\underset{\delta^+}{H}}{\ddot{O}}}: \qquad \overset{\delta^+}{H}-\overset{\delta^-}{\underset{\underset{\delta^+}{H}}{N}}-\overset{\delta^+}{H}$$

The direction of bond polarity can be indicated with an arrow. By convention, chemists draw the arrow so that it points in the direction in which the electrons are pulled. Thus, the head of the arrow is at the negative end of the bond; a short perpendicular line near the tail of the arrow marks the positive end of the bond. (Physicists draw the arrow in the opposite direction.)

$$H-\ddot{C}l: \quad \boxed{\text{the negative end of the bond}}$$

You can think of ionic bonds and nonpolar covalent bonds as being at the opposite ends of a continuum of bond types. All bonds fall somewhere on this line. At one end is an ionic bond—a bond in which no electrons are shared. At the other end is a nonpolar covalent bond—a bond in which the electrons are shared equally. Polar covalent bonds fall somewhere in between.

The greater the difference in electronegativity between the atoms forming the bond, the closer the bond is to the ionic end of the continuum.

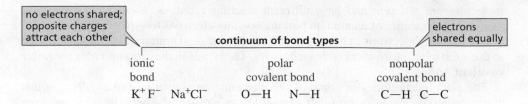

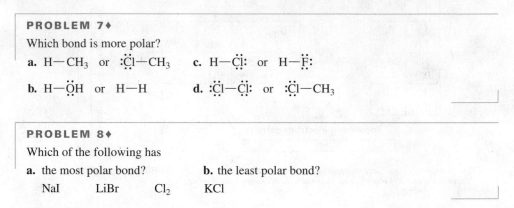

C—H bonds are relatively nonpolar, because carbon and hydrogen have similar electronegativities (electronegativity difference = 0.4; see Table 3); N—H bonds are more polar (electronegativity difference = 0.9), but not as polar as O—H bonds (electronegativity difference = 1.4). Even closer to the ionic end of the continuum is the bond between sodium and chloride ions (electronegativity difference = 2.1), but sodium chloride is not as ionic as potassium fluoride (electronegativity difference = 3.2).

PROBLEM 7♦

Which bond is more polar?

a. H—CH$_3$ or :C̈l—CH$_3$ **c.** H—C̈l: or H—F̈:

b. H—ÖH or H—H   **d.** :C̈l—C̈l: or :C̈l—CH$_3$

PROBLEM 8♦

Which of the following has

a. the most polar bond?   **b.** the least polar bond?

   NaI   LiBr   Cl$_2$ KCl

A polar bond has a **dipole**—it has a negative end and a positive end. The size of the dipole is indicated by the dipole moment, symbolized by the Greek letter μ. The **dipole moment** of a bond is equal to the magnitude of the charge on either atom (either the partial positive charge or the partial negative charge, because they have the same magnitude) times the distance between the two charges:

dipole moment of a bond = μ = size of the charge × the distance between the charges

A dipole moment is reported in a unit called a **debye (D)** (pronounced de-bye). Because the charge on an electron is 4.80×10^{-10} electrostatic units (esu) and the distance between charges in a polar bond will have units of 10^{-8} cm, the product of charge and distance will have units of 10^{-18} esu cm; so 1.0 D = 1.0×10^{-18} esu cm. Thus, a dipole moment of 1.5×10^{-18} esu cm can be more simply stated as 1.5 D. The dipole moments of some bonds commonly found in organic compounds are listed in Table 4.

When a molecule has only one covalent bond, the molecule has a dipole moment, which is identical to the dipole moment of the bond. For example, the dipole moment of hydrogen chloride (HCl) is 1.1 D because the dipole moment of the H—Cl bond is 1.1 D.

The dipole moment of a molecule with more than one covalent bond depends on the dipole moments of all the bonds in the molecule and the geometry of the molecule. We will examine the dipole moments of molecules with more than one covalent bond in Section 16 after you learn about the geometry of molecules.

Table 4 The Dipole Moments of Some Commonly Encountered Bonds

Bond	Dipole moment (D)	Bond	Dipole moment (D)
H—C	0.4	C—C	0
H—N	1.3	C—N	0.2
H—O	1.5	C—O	0.7
H—F	1.7	C—F	1.6
H—Cl	1.1	C—Cl	1.5
H—Br	0.8	C—Br	1.4
H—I	0.4	C—I	1.2

PROBLEM 9 Solved

Use the symbols δ^+ and δ^- to show the direction of the polarity of the indicated bond in

$$H_3C—OH$$

Solution The indicated bond is between carbon and oxygen. According to Table 3, the electronegativity of carbon is 2.5 and the electronegativity of oxygen is 3.5. Because oxygen is more electronegative than carbon, oxygen has a partial negative charge and carbon has a partial positive charge.

$$\overset{\delta+}{H_3C}—\overset{\delta-}{OH}$$

PROBLEM 10♦

Use the symbols δ^+ and δ^- to show the direction of the polarity of the indicated bond in each of the following compounds:

a. HO—H   **c.** $H_3C—NH_2$ **e.** HO—Br   **g.** I—Cl

b. F—Br   **d.** $H_3C—Cl$ **f.** $H_3C—Li$ **h.** $H_2N—OH$

PROBLEM 11 Solved

Determine the partial negative charge on the fluorine atom in a C—F bond. The bond length is 1.39 Å* and the bond dipole moment is 1.60 D.

Solution If fluorine had a full negative charge, the dipole moment would be

(charge on an electron in esu units) (length of the bond in centimeters)

$$(4.80 \times 10^{-10}\, esu)\,(1.39 \times 10^{-8}\, cm) = 6.67 \times 10^{-18}\, esu\ cm = 6.67\ D$$

Knowing that the dipole moment is 1.60 D, we can calculate that the partial negative charge on the fluorine atom is about 0.24 of a full charge:

$$\frac{1.60}{6.67} = 0.24$$

PROBLEM 12

Explain why HCl has a smaller dipole moment than HF, even though the H—Cl bond is longer than the H—F bond.

---

*The angstrom (Å) is not a Système International (SI) unit. Those who prefer SI units can convert Å into picometers (pm): 1 Å = 100 pm. Because the angstrom continues to be used by many organic chemists, we will use angstroms in this text. Dipole moment calculations require the bond length to be in centimeters: 1 Å = 10^{-8} cm.

Understanding bond polarity is critical to understanding how organic reactions occur, because a central rule governing the reactivity of organic compounds is that *electron-rich atoms or molecules are attracted to electron-deficient atoms or molecules*. **Electrostatic potential maps** (often called simply potential maps) are models that show how charge is distributed in the molecule under the map. The potential maps for LiH, H_2, and HF are shown here.

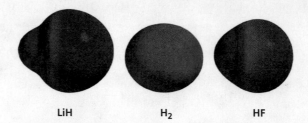

LiH H_2 HF

The colors on a potential map indicate the relative charge distribution in the molecule and therefore the degree to which the molecule or an atom in a molecule attracts other species. Red, signifying the most negative electrostatic potential, is used for regions that attract electron-deficient species most strongly. Blue is used for areas with the most positive electrostatic potential—regions that attract electron-rich species most strongly. Other colors indicate intermediate levels of attraction.

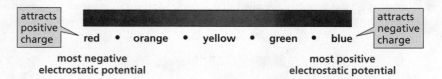

The potential map for LiH shows that the hydrogen atom (red) is more electron-rich than the lithium atom (blue). By comparing the three maps, we can tell that the hydrogen in LiH is more electron-rich than a hydrogen in H_2, whereas the hydrogen in HF is less electron-rich than a hydrogen in H_2.

Because a potential map roughly marks the "edge" of the molecule's electron cloud, the map tells us something about the relative size and shape of the molecule. A given kind of atom can have different sizes in different molecules, because the size of an atom in a potential map depends on its electron density. For example, the negatively charged hydrogen in LiH is bigger than a neutral hydrogen in H_2, which is bigger than the positively charged hydrogen in HF.

PROBLEM 13♦

After examining the potential maps for LiH, HF, and H_2, answer the following questions:

a. Which compounds are polar?

b. Why does LiH have the largest hydrogen?

c. Which compound has the hydrogen that would be most apt to attract a negatively charged molecule?

---

4   HOW THE STRUCTURE OF A COMPOUND IS REPRESENTED

First we will see how compounds are drawn using Lewis structures. Then we will look at the kinds of structures that are used more commonly for organic compounds.

Lewis Structures

The chemical symbols we have been using, in which the valence electrons are represented as dots or solid lines, are called **Lewis structures.** Lewis structures show us which atoms are bonded together and tell us whether any atoms possess *lone-pair electrons* or have a *formal charge*, two concepts described below. The Lewis structures for H_2O, H_3O^+, HO^-, and H_2O_2 are shown here.

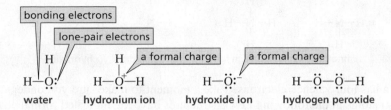

Notice that the atoms in Lewis structures are always lined up linearly or at right angles. Therefore, they do not tell us anything about the bond angles in the actual molecule.

When you draw a Lewis structure, make sure that hydrogen atoms are surrounded by two electrons and that C, O, N, and halogen (F, Cl, Br, I) atoms are surrounded by eight electrons, in accordance with the octet rule. Valence electrons not used in bonding are called **nonbonding electrons** or **lone-pair electrons.**

Once you have the atoms and the electrons in place, you must examine each atom to see whether a formal charge should be assigned to it. A **formal charge** is the *difference* between the number of valence electrons an atom has when it is not bonded to any other atoms and the number it "owns" when it is bonded. An atom "owns" all of its lone-pair electrons and half of its bonding (shared) electrons.

formal charge = number of valence electrons

− (number of lone-pair electrons + 1/2 number of bonding electrons)

For example, an oxygen atom has six valence electrons (Table 2). In water (H_2O), oxygen "owns" six electrons (four lone-pair electrons and half of the four bonding electrons). Because the number of electrons it "owns" is equal to the number of its valence electrons ($6 - 6 = 0$), the oxygen atom in water does not have a formal charge.

The oxygen atom in the hydronium ion (H_3O^+) "owns" five electrons: two lone-pair electrons plus three (half of six) bonding electrons. Because the number of electrons oxygen "owns" is one less than the number of its valence electrons ($6 - 5 = 1$), its formal charge is $+1$.

The oxygen atom in the hydroxide ion (HO^-) "owns" seven electrons: six lone-pair electrons plus one (half of two) bonding electron. Because oxygen "owns" one more electron than the number of its valence electrons ($6 - 7 = -1$), its formal charge is -1.

Lone-pair electrons are valence electrons that do not form bonds.

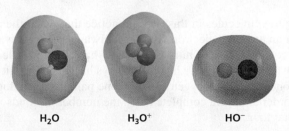

H_2O H_3O^+ HO^-

PROBLEM 14♦

An atom with a formal charge does not necessarily have more or less electron density than the atoms in the molecule without formal charges. We can see this by examining the potential maps for H_2O, H_3O^+, and HO^-.

a. Which atom bears the formal negative charge in the hydroxide ion?

b. Which atom has the greater electron density in the hydroxide ion?

c. Which atom bears the formal positive charge in the hydronium ion?

d. Which atom has the least electron density in the hydronium ion?

Nitrogen has five valence electrons (Table 2). Prove to yourself that the appropriate formal charges have been assigned to the nitrogen atoms in the following Lewis structures:

$$\overset{\overset{\textstyle H}{|}}{H-\overset{\cdot\cdot}{N}-H} \qquad H-\overset{\overset{\textstyle H}{|}}{\overset{+}{N}}-H \qquad H-\overset{\overset{\textstyle H}{|}}{\overset{\cdot\cdot}{N}}:^{-} \qquad H-\overset{\overset{\textstyle |}{N}}{\underset{H}{}}-\overset{}{\underset{H}{N}}-H$$

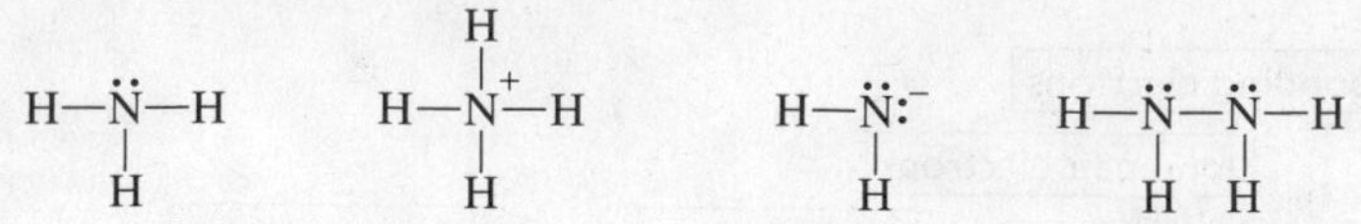

ammonia   ammonium ion   amide anion   hydrazine

Carbon has four valence electrons. Take a moment to make sure you understand why the carbon atoms in the following Lewis structures have the indicated formal charges:

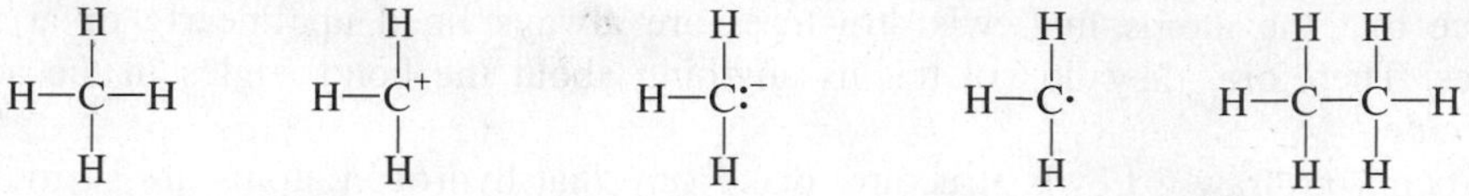

$$\overset{\overset{\textstyle H}{|}}{H-\overset{|}{C}-H}_{H} \qquad H-\overset{+}{\underset{H}{\overset{|}{C}}} \qquad H-\overset{\cdot\cdot}{\underset{H}{\overset{|}{C}}}^{-} \qquad H-\overset{\cdot}{\underset{H}{\overset{|}{C}}} \qquad H-\overset{|}{\underset{H}{C}}-\overset{|}{\underset{H}{C}}-H$$

methane   methyl cation   methyl anion   methyl radical   ethane
         a carbocation   a carbanion

A species containing a positively charged carbon is called a **carbocation,** and a species containing a negatively charged carbon is called a **carbanion.** (Note that a *cation* is a positively charged ion and an *anion* is a negatively charged ion.) A species containing an atom with a single unpaired electron is called a **radical** (often called a **free radical).**

Hydrogen has one valence electron, and each halogen (F, Cl, Br, I) has seven valence electrons, so the following species have the indicated formal charges:

$$H^{+} \qquad H:^{-} \qquad H\cdot \qquad :\overset{\cdot\cdot}{\underset{\cdot\cdot}{Br}}:^{-} \qquad :\overset{\cdot\cdot}{\underset{\cdot\cdot}{Br}}\cdot \qquad :\overset{\cdot\cdot}{\underset{\cdot\cdot}{Br}}-\overset{\cdot\cdot}{\underset{\cdot\cdot}{Br}}: \qquad :\overset{\cdot\cdot}{\underset{\cdot\cdot}{Cl}}-\overset{\cdot\cdot}{\underset{\cdot\cdot}{Cl}}:$$

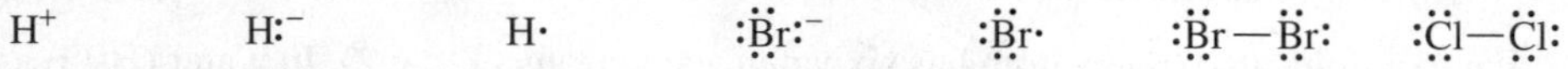

hydrogen ion   hydride ion   hydrogen radical   bromide ion   bromine radical   bromine   chlorine

---

PROBLEM 15

Give each atom the appropriate formal charge:

a. $CH_3-\overset{\cdot\cdot}{\underset{H}{O}}-CH_3$ **b.** $H-\overset{\cdot\cdot}{\underset{H}{C}}-H$ **c.** $CH_3-\overset{\overset{\textstyle CH_3}{|}}{\underset{CH_3}{N}}-CH_3$ **d.** $H-\overset{\overset{\textstyle H}{|}}{\underset{H}{N}}-\overset{\overset{\textstyle H}{|}}{\underset{H}{B}}-H$

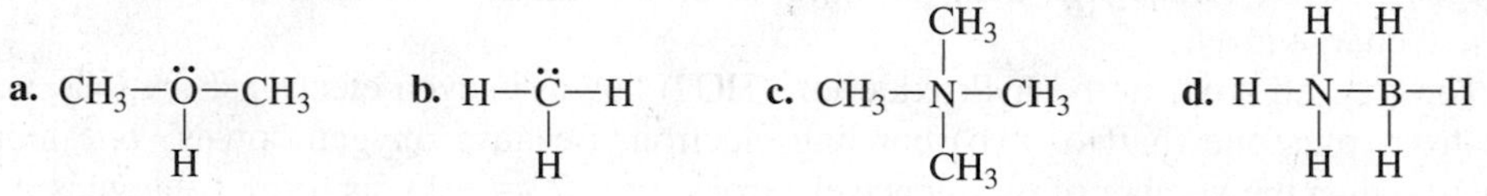

While studying the molecules in this section, notice that when the atoms do not bear a formal charge or an unpaired electron, hydrogen always has *one* covalent bond, carbon always has *four* covalent bonds, nitrogen always has *three* covalent bonds, oxygen always has *two* covalent bonds, and a halogen always has *one* covalent bond. Also notice that nitrogen has one lone pair, oxygen has two lone pairs, and a halogen has three lone pairs, because in order to have a complete octet, the number of bonds and the number of lone pairs must total four.

$$H- \qquad -\overset{|}{\underset{|}{C}}- \qquad -\overset{\cdot\cdot}{\underset{|}{N}}- \qquad :\overset{\cdot\cdot}{O}- \qquad \begin{array}{l} :\overset{\cdot\cdot}{F}- \quad :\overset{\cdot\cdot}{\underset{\cdot\cdot}{Br}}- \\ :\overset{\cdot\cdot}{\underset{\cdot\cdot}{Cl}}- \quad :\overset{\cdot\cdot}{\underset{\cdot\cdot}{I}}- \end{array}$$

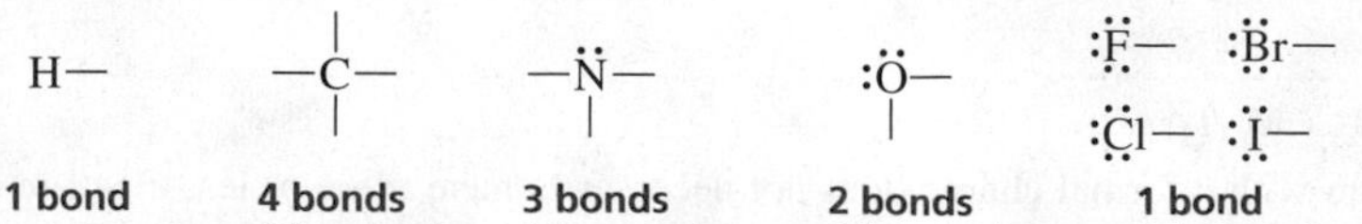

1 bond	4 bonds	3 bonds	2 bonds	1 bond
	0 lone pairs	1 lone pair	2 lone pairs	3 lone pairs

# of bonds +
# of lone pairs $\dfrac{}{4}$

Atoms that have more bonds or fewer bonds than the number required for a neutral atom will have either a formal charge or an unpaired electron. These numbers are very important to remember when you are first drawing structures of organic compounds because they provide a quick way to recognize when you have made a mistake.

   Each atom in the following Lewis structures has a filled outer shell. Notice that since none of the molecules has a formal charge or an unpaired electron, H forms 1 bond, C forms 4 bonds, N forms 3 bonds, O forms 2 bonds, and Br forms 1 bond. Notice, too, that each N has 1 lone pair, each O has 2 lone pairs, and Br has 3 lone pairs.

| 2 covalent bonds holding 2 atoms together is called a double bond |

$$
\underset{\underset{H}{|}}{\overset{\overset{H}{|}}{H-C-\ddot{\underset{\cdot\cdot}{Br}}:}} \qquad
\underset{\underset{H}{|}\;\;\underset{H}{|}}{\overset{\overset{H}{|}\;\;\overset{H}{|}}{H-C-\ddot{\underset{\cdot\cdot}{O}}-C-H}} \qquad
\underset{}{\overset{\overset{:\ddot{O}:}{\|}}{H-C-\ddot{\underset{\cdot\cdot}{O}}-H}} \qquad
\underset{\underset{H}{|}\;\underset{H}{|}}{\overset{\overset{H}{|}}{H-C-\ddot{N}-H}} \qquad
:N\equiv N:
$$

| 3 covalent bonds holding 2 atoms together is called a triple bond |

PROBLEM-SOLVING STRATEGY

Drawing Lewis Structures

a. Draw the Lewis structure for CH_4O.   **b.** Draw the Lewis structure for HNO_2.

a. 1. Determine the total number of valence electrons (4 for C, 1 for each H, and 6 for O adds up to $4 + 4 + 6 = 14$ valence electrons).

2. Distribute the atoms, remembering that C forms four bonds, O forms two bonds and each H forms one bond. Always put the hydrogens on the outside of the molecule since H can form only one bond.

$$
\underset{\underset{H}{|}}{\overset{\overset{H}{|}}{H-C-O-H}}
$$

3. Use the total number of valence electrons to form bonds and fill octets with lone-pair electrons.

| lone-pair electrons |

$$
\underset{\underset{H}{|}}{\overset{\overset{H}{|}}{H-C-\ddot{\underset{\cdot\cdot}{O}}-H}}
$$

| bonding electrons |

10 bonding electrons
 4 lone-pair electrons
_____
14 valence electrons

4. Assign a formal charge to any atom whose number of valence electrons is not equal to the number of its lone-pair electrons plus one-half of its bonding electrons. (None of the atoms in CH_4O has a formal charge.)

b. 1. Determine the total number of valence electrons (1 for H, 5 for N, and 6 for each O adds up to $1 + 5 + 12 = 18$ valence electrons).

2. Distribute the atoms putting the hydrogen on the outside of the molecule. If a species has two or more oxygen atoms, avoid oxygen–oxygen single bonds. These are weak bonds, and few compounds have them.

$$
H-O-N-O
$$

3. Use the total number of valence electrons to form bonds and fill octets with lone-pair electrons.

$$
H-\ddot{\underset{\cdot\cdot}{O}}-N-\ddot{\underset{\cdot\cdot}{O}}:
$$

6 bonding electrons
12 lone-pair electrons
_____
18 valence electrons

4. If, after all the electrons have been assigned, an atom (other than hydrogen) does not have a complete octet, use a lone pair to form a double bond to that atom.

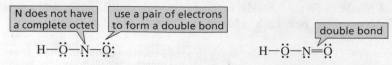

| N does not have a complete octet | use a pair of electrons to form a double bond | | double bond |

$$H-\ddot{O}-\ddot{O}: \qquad\qquad H-\ddot{O}-N=\ddot{O}$$

18 electrons have been assigned   **using one of oxygen's lone pairs to form a double bond gives N a complete octet**

5. Assign a formal charge to any atom whose number of valence electrons is not equal to the number of its lone-pair electrons plus one-half of its bonding electrons. (None of the atoms in HNO_2 has a formal charge.)

Now use the strategy you have just learned to solve Problem 16.

PROBLEM 16 Solved

Draw the Lewis structure for each of the following:

a. NO_3^- **c.** $^-C_2H_5$ **e.** $CH_3\overset{+}{N}H_3$ **g.** HCO_3^-

b. NO_2^+ **d.** $^+C_2H_5$ **f.** $NaOH$ **h.** H_2CO

Solution to 16a The total number of valence electrons is 23 (5 for N, and 6 for each of the three Os). Because the species has one negative charge, we must add 1 to the number of valence electrons, for a total of 24. The only way we can arrange one N and three Os and avoid O—O single bonds is to place the three Os around the N. We then use the 24 electrons to form bonds and fill octets with lone-pair electrons.

$$:\ddot{O}:$$
$$:\ddot{O}-N-\ddot{O}:$$

| incomplete octet |

All 24 electrons have been assigned, but N does not have a complete octet. We complete N's octet by using one of oxygen's lone pairs to form a double bond. (It does not make a difference which oxygen atom we choose.) When we check each atom to see whether it has a formal charge, we find that two of the Os are negatively charged and the N is positively charged, for an overall charge of -1.

$$:O:$$
$$\|$$
$$^-:\ddot{O}-\overset{+}{N}-\ddot{O}:^-$$

Solution to 16b The total number of valence electrons is 17 (5 for N and 6 for each of the two Os). Because the species has one positive charge, we must subtract 1 from the number of valence electrons, for a total of 16. The 16 electrons are used to form bonds and fill octets with lone-pair electrons.

| incomplete octet |

$$:\ddot{O}-N-\ddot{O}:$$

Two double bonds are necessary to complete N's octet. We find that the N has a formal charge of $+1$.

$$\ddot{O}=\overset{+}{N}=\ddot{O}$$

PROBLEM 17♦

a. Draw two Lewis structures for C_2H_6O.

b. Draw three Lewis structures for C_3H_8O.

(*Hint:* The two Lewis structures in part **a** are **constitutional isomers**—molecules that have the same atoms, but differ in the way the atoms are connected. The three Lewis structures in part **b** are also constitutional isomers.)

Kekulé Structures and Condensed Structures

Kekulé structures are like Lewis structures except lone pairs are normally omitted. Structures are often further simplified by omitting some (or all) of the covalent bonds and listing atoms bonded to a particular carbon (or nitrogen or oxygen) next to it (with a subscript if there is more than one of a particular atom). Lone-pair electrons are usually not shown, unless they are needed to draw attention to some chemical property of the molecule. These structures are called **condensed structures.** Compare the **condensed** structures shown here with the Lewis structures shown earlier in this section.

$$CH_3Br \qquad CH_3OCH_3 \qquad HCO_2H \qquad CH_3NH_2 \qquad N_2$$

(Although lone pairs are not shown, you should remember that neutral nitrogen, oxygen, and halogen atoms always have them: one pair for nitrogen, two pairs for oxygen, and three pairs for a halogen.)

You can find examples of Kekulé and condensed structures and the conventions commonly used to create condensed structures in Table 5. Notice that since none of the molecules in Table 5 has a formal charge or an unpaired electron, each C has four bonds, each N has three bonds, each O has two bonds, and each H or halogen has one bond.

Table 5 Kekulé Structures and Condensed Structures

Atoms bonded to a carbon are shown to the right of the carbon. Atoms other than H can be shown hanging from the carbon.

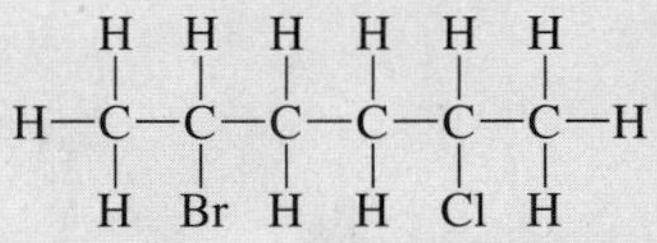

$$\text{H—C—C—C—C—C—C—H} \quad \text{or} \quad CH_3CHBrCH_2CH_2CHClCH_3 \quad \text{or} \quad CH_3CHCH_2CH_2CHCH_3$$

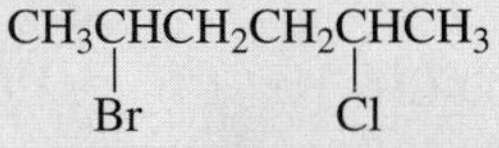

Repeating CH_2 groups can be shown in parentheses.

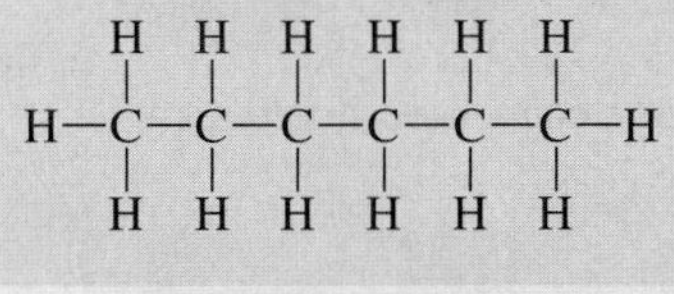

$$\text{H—C—C—C—C—C—C—H} \quad \text{or} \quad CH_3CH_2CH_2CH_2CH_2CH_3 \quad \text{or} \quad CH_3(CH_2)_4CH_3$$

Groups bonded to a carbon can be shown (in parentheses) to the right of the carbon, or hanging from the carbon.

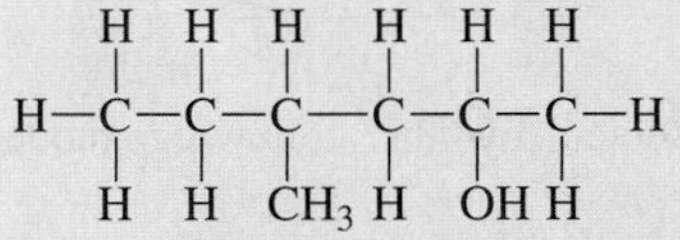

$$\text{H—C—C—C—C—C—C—H} \quad \text{or} \quad CH_3CH_2CH(CH_3)CH_2CH(OH)CH_3 \quad \text{or} \quad CH_3CH_2CHCH_2CHCH_3$$

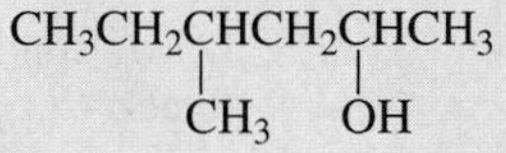

A single group bonded to the far-right carbon is not put in parentheses.

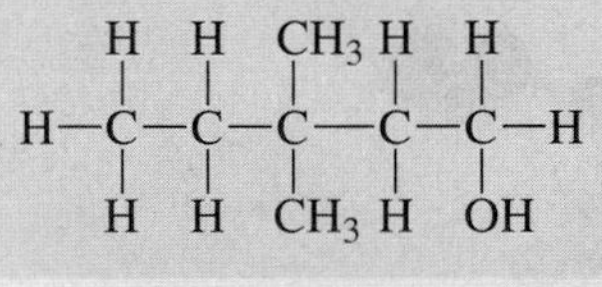

$$\text{H—C—C—C—C—C—H} \quad \text{or} \quad CH_3CH_2C(CH_3)_2CH_2CH_2OH \quad \text{or} \quad CH_3CH_2CCH_2CH_2OH$$

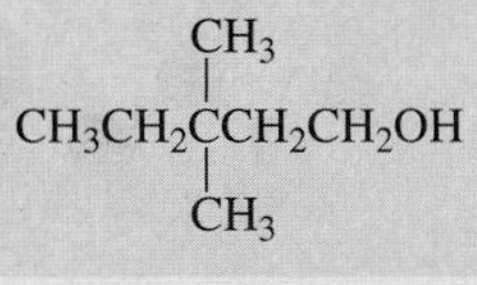

Two or more identical groups on the right side of the molecule can be shown in parentheses or hanging from a carbon.

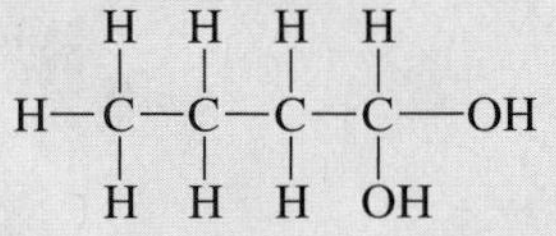

$$\text{H—C—C—C—C—OH} \quad \text{or} \quad CH_3CH_2CH_2CH(OH)_2$$

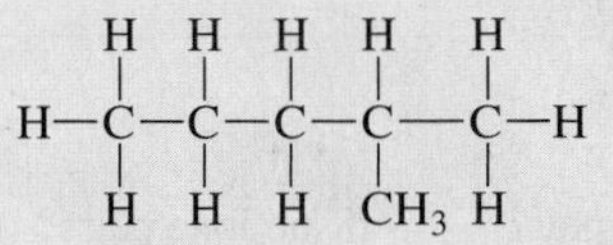

$$\text{H—C—C—C—C—C—H} \quad \text{or} \quad CH_3CH_2CH_2CH(CH_3)_2$$

(Continued)

Table 5 (*Continued*)

Two or more identical groups bonded to the "first" atom on the left can be shown (in parentheses) to the left of that atom, or hanging from the atom.

$$H-\overset{\overset{\displaystyle H}{|}}{\underset{\underset{\displaystyle H}{|}}{C}}-\overset{\overset{\displaystyle }{|}}{\underset{\underset{\displaystyle CH_3}{|}}{N}}-\overset{\overset{\displaystyle H}{|}}{\underset{\underset{\displaystyle H}{|}}{C}}-\overset{\overset{\displaystyle H}{|}}{\underset{\underset{\displaystyle H}{|}}{C}}-\overset{\overset{\displaystyle H}{|}}{\underset{\underset{\displaystyle H}{|}}{C}}-H$$ or $(CH_3)_2NCH_2CH_2CH_3$ or $CH_3NCH_2CH_2CH_3$
CH_3

$$H-\overset{\overset{\displaystyle H}{|}}{\underset{\underset{\displaystyle H}{|}}{C}}-\overset{\overset{\displaystyle H}{|}}{\underset{\underset{\displaystyle CH_3}{|}}{C}}-\overset{\overset{\displaystyle H}{|}}{\underset{\underset{\displaystyle H}{|}}{C}}-\overset{\overset{\displaystyle H}{|}}{\underset{\underset{\displaystyle H}{|}}{C}}-\overset{\overset{\displaystyle H}{|}}{\underset{\underset{\displaystyle H}{|}}{C}}-H$$ or $(CH_3)_2CHCH_2CH_2CH_3$ or $CH_3CHCH_2CH_2CH_3$
CH_3

An oxygen doubly bonded to a carbon can be shown hanging from the carbon or to the right of the carbon.

$$CH_3CH_2\overset{\overset{\displaystyle O}{\|}}{C}CH_3$$ or $CH_3CH_2COCH_3$ or $CH_3CH_2C(=O)CH_3$

$$CH_3CH_2CH_2\overset{\overset{\displaystyle O}{\|}}{C}H$$ or $CH_3CH_2CH_2CHO$ or $CH_3CH_2CH_2CH=O$

$$CH_3CH_2\overset{\overset{\displaystyle O}{\|}}{C}OH$$ or $CH_3CH_2CO_2H$ or CH_3CH_2COOH

$$CH_3CH_2\overset{\overset{\displaystyle O}{\|}}{C}OCH_3$$ or $CH_3CH_2CO_2CH_3$ or $CH_3CH_2COOCH_3$

PROBLEM 18♦

Draw the lone-pair electrons that are not shown in the following condensed structures:

a. $CH_3CH_2NH_2$ c. CH_3CH_2OH e. CH_3CH_2Cl

b. CH_3NHCH_3 d. CH_3OCH_3 f. $HONH_2$

PROBLEM 19♦

Draw condensed structures for the compounds represented by the following models (black = C, gray = H, red = O, blue = N, and green = Cl):

a.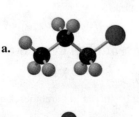

c.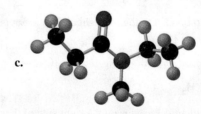

b.

d.

PROBLEM 20♦

Which of the atoms in the molecular models in Problem 19 have

a. three lone pairs?   b. two lone pairs?   c. one lone pair?   d. no lone pairs?

PROBLEM 21

Expand the following condensed structures to show the covalent bonds and lone-pair electrons:

a. $CH_3NH(CH_2)_2CH_3$ **b.** $(CH_3)_2CHCl$ **c.** $(CH_3)_3CBr$ **d.** $(CH_3)_3C(CH_2)_3CHO$

5   ATOMIC ORBITALS

We have seen that electrons are distributed into different atomic orbitals (Table 2). An atomic orbital is a three-dimensional region around the nucleus where an electron is most likely to be found. Because the **Heisenberg uncertainty principle** states that both the precise location and the exact momentum of an atomic particle cannot be simultaneously determined, we can never say precisely where an electron is—we can only describe its probable location.

Mathematical calculations indicate that an *s* atomic orbital is a sphere with the nucleus at its center, and experimental evidence supports this theory. Thus, when we say that an electron occupies a 1*s* orbital, we mean that there is a greater than 90% probability that the electron is in the space defined by the sphere.

> An atomic orbital is the three-dimensional region around the nucleus where an electron is most likely to be found.

Because the second shell lies farther from the nucleus than the first shell (Section 2), the average distance from the nucleus is greater for an electron in a 2*s* orbital than it is for an electron in a 1*s* orbital. A 2*s* orbital, therefore, is represented by a larger sphere. Because of the greater size of a 2*s* orbital, its average electron density is less than the average electron density of a 1*s* orbital.

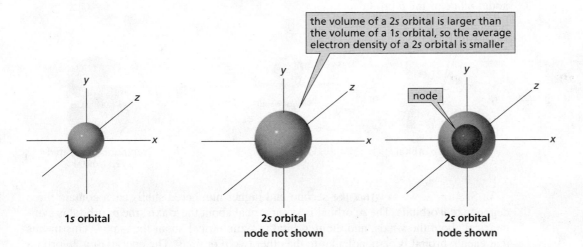

the volume of a 2s orbital is larger than the volume of a 1s orbital, so the average electron density of a 2s orbital is smaller

node

1s orbital

2s orbital
node not shown

2s orbital
node shown

An electron in a 1*s* orbital can be anywhere within the orbital, but a 2*s* orbital has a region where the probability of finding an electron falls to zero. This is called a **node,** or, more precisely, a **radial node,** since this absence of electron density lies at one set distance from the nucleus. As a result, a 2*s* electron can be found anywhere within the 2*s* orbital—including the region of space defined by the 1*s* orbital—except at the node.

To understand why nodes occur, you need to remember that electrons have both particle-like and wave-like properties. A node is a consequence of the wave-like properties of an electron.

> Electrons have particle-like and wave-like properties.

There are two types of waves: traveling waves and standing waves. Traveling waves move through space. Light is an example of a traveling wave. A standing wave, on the other hand, is confined to a limited space. The vibrating string of a guitar is a standing wave—the string moves up and down, but does not travel through space. The amplitude is $(+)$ in the region above where the guitar string is at rest and it is $(-)$ in the region below where the guitar string is at rest—the two regions are said to have opposite phases. The region where the guitar string has no transverse displacement (zero amplitude) is a node.

An electron behaves like a standing wave but, unlike the wave created by a vibrating guitar string, it is three-dimensional. This means that the node of a 2*s* orbital is actually a spherical surface within the 2*s* orbital. Because the electron wave has zero amplitude at the node, there is zero probability of finding an electron at the node.

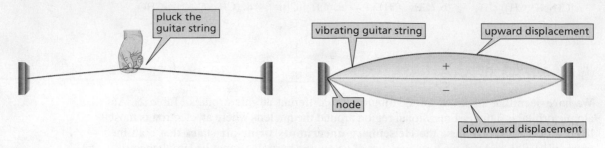

There is zero probability of finding an electron at a node.

Unlike *s* orbitals, which resemble spheres, *p* orbitals have two lobes. Generally, the lobes are depicted as teardrop shaped, but computer-generated representations reveal that they are shaped more like doorknobs (as shown on the right below). Like the vibrating guitar string, the lobes have opposite phases, which can be designated by plus (+) and minus (−), or by two different colors. (Notice that in this context, + and − indicate the phase of the orbital; they do not indicate charge.) The node of the *p* orbital is a plane—called a **nodal plane**—that passes through the center of the nucleus, between its two lobes. There is zero probability of finding an electron in the nodal plane of the *p* orbital.

2*p* orbital **2*p* orbital** **computer-generated 2*p* orbital**

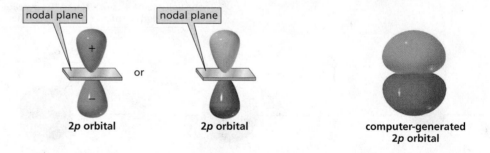

In Section 2, we saw that the second and higher numbered shells each contain three degenerate *p* orbitals. The p_x orbital is symmetrical about the *x*-axis, the p_y orbital is symmetrical about the *y*-axis, and the p_z orbital is symmetrical about the *z*-axis. This means that each *p* orbital is perpendicular to the other two *p* orbitals. The energy of a 2*p* orbital is slightly greater than that of a 2*s* orbital because the average location of an electron in a 2*p* orbital is farther away from the nucleus.

$2p_x$ orbital **$2p_y$ orbital** **$2p_z$ orbital**

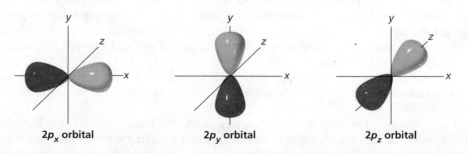

PROBLEM 22

Draw a picture of

a. a 3*s* orbital. **b.** a 4*s* orbital. **c.** a 3*p* orbital.

6 AN INTRODUCTION TO MOLECULAR ORBITAL THEORY

How do atoms form covalent bonds in order to form molecules? The Lewis model, which shows atoms attaining a complete octet by sharing electrons, tells only part of the story. A drawback of the model is that it treats electrons like particles and does not take into account their wavelike properties.

Molecular orbital (MO) theory combines the tendency of atoms to fill their octets by sharing electrons (the Lewis model) with their wavelike properties, assigning electrons to a volume of space called an orbital. According to MO theory, covalent bonds result when atomic orbitals combine to form *molecular orbitals*. Like an atomic orbital, which describes the volume of space around an atom's nucleus where an electron is likely to be found, a **molecular orbital** describes the volume of space around a molecule where an electron is likely to be found. And, like atomic orbitals, molecular orbitals, too, have specific sizes, shapes, and energies.

> *An atomic orbital surrounds an atom.*
>
> *A molecular orbital surrounds a molecule.*

Let's look first at the bonding in a hydrogen molecule (H_2). Imagine a meeting of two separate H atoms. As one atom with its $1s$ atomic orbital approaches the other with its $1s$ atomic orbital, the orbitals begin to overlap. The atoms continue to move closer, and the amount of overlap increases until the orbitals combine to form a molecular orbital. The covalent bond that is formed when the two s orbitals overlap is called a **sigma (σ)** bond. A σ bond is cylindrically symmetrical—the electrons in the bond are symmetrically distributed about an imaginary line connecting the nuclei of the two atoms joined by the bond.

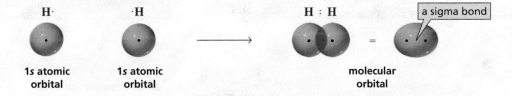

As the two orbitals begin to overlap, energy is released because the electron in each atom is attracted to its own nucleus *and* to the nucleus of the other atom (Figure 2).

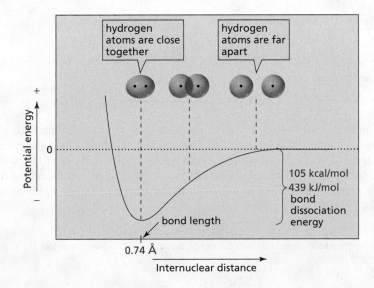

◀ **Figure 2**

The change in energy that occurs as two $1s$ atomic orbitals approach each other. The internuclear distance at minimum potential energy is the length of the H—H covalent bond.

The attraction of the two negatively charged electrons for the two positively charged nuclei is what holds the two H atoms together. The more the orbitals overlap, the more the energy decreases, until the atoms are so close that their positively charged nuclei begin to repel each other. This repulsion causes a large increase in energy. Figure 2 shows that minimum energy (maximum stability) is achieved when the nuclei are a particular distance apart. This distance is the **bond length** of the new covalent bond; the bond length of the H—H bond is 0.74 Å.

Minimum energy corresponds to maximum stability.

As Figure 2 shows, energy is released when a covalent bond forms. We see that when the H—H bond forms, 105 kcal/mol (or 439 kJ/mol)* of energy is released. Breaking the bond requires precisely the same amount of energy. Thus, the **bond dissociation energy**—a measure of bond strength—is the energy required to break a bond, or the energy released when a bond is formed. Every covalent bond has a characteristic bond length and bond dissociation energy.

Orbitals are conserved. In other words, the number of molecular orbitals formed must equal the number of atomic orbitals combined. In describing the formation of an H—H bond, we combined two atomic orbitals but discussed only one molecular orbital. Where is the other molecular orbital? As you will see shortly, it is there; it just doesn't contain any electrons.

It is the wavelike properties of the electrons that cause two atomic orbitals to form two molecular orbitals. Two atomic orbitals can combine in an additive (constructive) manner, just as two light waves or two sound waves can reinforce each other (Figure 3a). The constructive combination of two *s* atomic orbitals is called a σ **(sigma) bonding molecular orbital.**

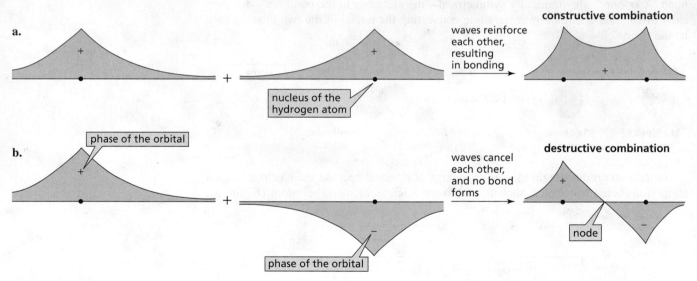

▲ **Figure 3**

(a) Waves with the same phase (+/+) overlap constructively, reinforcing each other and resulting in bonding.

(b) Waves with opposite phases (+/−) overlap destructively, canceling each other and forming a node.

Atomic orbitals can also combine in a destructive way, canceling each other. The cancellation is similar to the darkness that results when two light waves cancel each other or to the silence that results when two sound waves cancel each other (Figure 3b). The destructive combination of two *s* atomic orbitals is called a σ^* **antibonding molecular orbital.** An antibonding orbital is indicated by an asterisk (*), which chemists read as "star." Thus, σ^* is read as "sigma star."

---

*Joules are the Système International (SI) units for energy, although many chemists use calories (1 kcal = 4.184 kJ). We will use both in this text.

The σ bonding molecular orbital and the σ^* antibonding molecular orbital are shown in the molecular orbital (MO) diagram in Figure 4. In an MO diagram, the energies of both the atomic orbitals and the molecular orbitals are represented as horizontal lines, with the bottom line being the lowest energy level and the top line the highest energy level.

When two atomic orbitals overlap, two molecular orbitals are formed—one lower in energy and one higher in energy than the atomic orbitals.

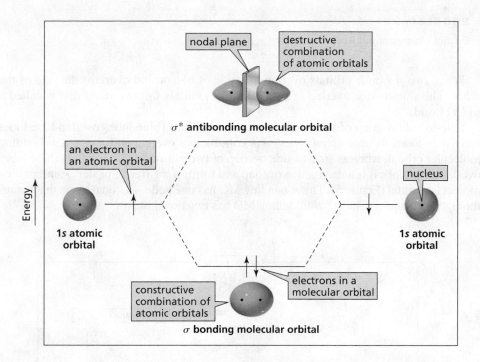

◀ **Figure 4**

Atomic orbitals of H· and molecular orbitals of H_2. Before covalent bond formation, each electron is in an atomic orbital. After covalent bond formation, both electrons are in the bonding MO. The antibonding MO is empty.

The electrons in the bonding molecular orbital are most likely be found between the nuclei, where they can more easily attract both nuclei simultaneously. This increased electron density between the nuclei is what binds the atoms together (Figure 3a). Any electrons in the antibonding molecular orbital are most likely to be found anywhere except between the nuclei, because there is a nodal plane between the nuclei (Figure 3b). As a result, any electrons in the antibonding orbital leave the positively charged nuclei more exposed to one another than do electrons in the bonding molecular orbital. Thus, electrons that occupy this orbital detract from, rather than assist in, the formation of a bond between the atoms.

Electrons in a bonding MO assist in bonding.

Electrons in an antibonding MO detract from bonding.

The MO diagram shows that the bonding molecular orbital is lower in energy, and therefore more stable, than the individual atomic orbitals. This is because the more nuclei an electron senses, the more stable it is. The antibonding molecular orbital, with less electron density between the nuclei, is less stable—and therefore higher in energy—than the atomic orbitals.

Electrons are assigned to the molecular orbitals using the same rules used to assign electrons to atomic orbitals: electrons always occupy available orbitals with the lowest energy (the aufbau principle), and no more than two electrons can occupy a molecular orbital (the Pauli exclusion principle). Thus, the two electrons of the H—H bond occupy the lower-energy bonding molecular orbital (the σ bonding MO in Figure 4), where they are attracted to both positively charged nuclei. It is this electrostatic attraction that gives a covalent bond its strength. We can conclude, therefore, that the strength of the covalent bond increases as the overlap of the atomic orbitals increases.

Covalent bond strength increases as atomic orbital overlap increases.

The MO diagram in Figure 4 allows us to predict that H_2^+ would not be as stable as H_2 because H_2^+ has only one electron in the bonding molecular orbital. Using the same diagram, we can also predict that He_2 does not exist, because the four electrons of He_2 (two from each He atom) would fill the lower energy bonding MO *and* the higher energy antibonding MO. *The two electrons in the antibonding MO would cancel the advantage to bonding that is gained by the two electrons in the bonding MO.*

PROBLEM 23♦

Predict whether or not He_2^+ exists.

When two *p* atomic orbitals overlap, the side of one orbital overlaps the side of the other. The side-to-side overlap of two parallel *p* orbitals forms a bond that is called a **pi (π) bond**.

Side-to-side overlap of two in-phase *p* atomic orbitals (blue lobes overlap blue lobes and green lobes overlap green lobes) is a constructive overlap and forms a π bonding molecular orbital, whereas side-to-side overlap of two out-of-phase *p* orbitals (blue lobes overlap green lobes) is a destructive overlap and forms a π* (read "pi star") antibonding molecular orbital (Figure 5). The π bonding MO has one node—a nodal plane that passes through both nuclei. The π* antibonding MO has two nodal planes.

Side-to-side overlap of two *p* orbitals forms a π bond. All other covalent bonds in organic molecules are σ bonds.

In-phase overlap forms a bonding MO; out-of-phase overlap forms an antibonding MO.

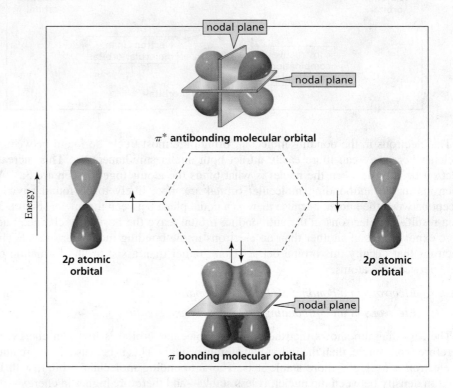

▶ **Figure 5**

Side-to-side overlap of two parallel *p* atomic orbitals forms a π bonding molecular orbital and a π* antibonding molecular orbital.

Now let's look at the MO diagram for side-to-side overlap of a *p* atomic orbital of carbon with a *p* atomic orbital of oxygen. In this case, the orbitals are the same type, but they belong to different kinds of atoms. When these two *p* atomic orbitals combine to form molecular orbitals, the atomic orbital of the more electronegative atom (oxygen) contributes more to the bonding MO, and the atomic orbital of the less electronegative atom (carbon) contributes more to the antibonding MO (Figure 6).

This means that when we put electrons in the bonding MO, they will be more apt to be around oxygen than around carbon. Thus, both Lewis theory and molecular orbital theory

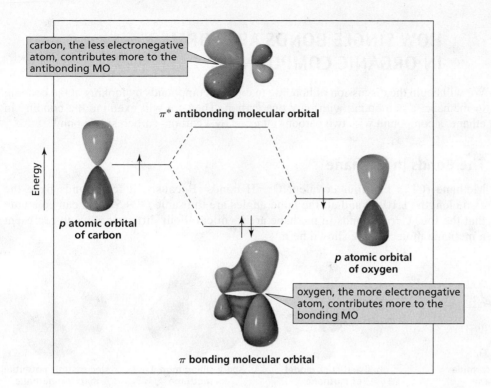

Side-to-side overlap of a p atomic orbital of carbon with a p atomic orbital of oxygen forms a π bonding molecular orbital and a π^* antibonding molecular orbital. The atomic orbital of oxygen contributes more to the bonding MO, because it is closer in energy to the bonding MO. The atomic orbital of carbon contributes more to the antibonding MO, because it is closer in energy to the antibonding MO.

tell us that the electrons shared by carbon and oxygen are not shared equally—the carbon atom of a carbon–oxygen bond has a partial positive charge and the oxygen atom has a partial negative charge.

To learn even more about the bonds in a molecule, organic chemists turn to the **valence-shell electron-pair repulsion (VSEPR) model**—a model for the prediction of molecular geometry based on the minimization of electron repulsion between regions of electron density around an atom. In other words, atoms share electrons by overlapping their atomic orbitals and, because electron pairs repel each other, the bonding electrons and lone-pair electrons around an atom are positioned as far apart as possible. Thus, a Lewis structure gives us a first approximation of the structure of a simple molecule, and VSEPR gives us a first glance at the shape of the molecule.

Because organic chemists generally think of chemical reactions in terms of the changes that occur in the bonds of the reacting molecules, the VSEPR model often provides the easiest way to visualize chemical change. However, the model is inadequate for some molecules because it does not allow for antibonding molecular orbitals. We will use both the MO and the VSEPR models in this text. Our choice will depend on which model provides the best description of the molecule under discussion. We will use the VSEPR model in Sections 7–13.

PROBLEM 24♦

Indicate the kind of molecular orbital (σ, σ^*, π, or π^*) that results when atomic orbitals are combined as indicated:

a.

b.

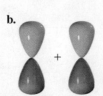

c.

d.

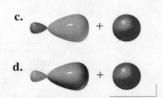

7 | HOW SINGLE BONDS ARE FORMED IN ORGANIC COMPOUNDS

We will begin the discussion of bonding in organic compounds by looking at the bonding in methane, a compound with only one carbon. Then we will examine the bonding in ethane, a compound with two carbons attached by a carbon–carbon single bond.

The Bonds in Methane

Methane (CH_4) has four covalent C—H bonds. Because all four bonds have the same length (1.10 Å) and all the bond angles are the same (109.5°), we can conclude that the four C—H bonds in methane are identical. Four different ways to represent a methane molecule are shown here.

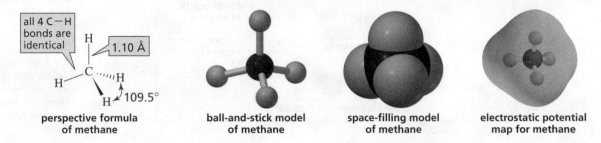

perspective formula of methane   **ball-and-stick model of methane**   **space-filling model of methane**   **electrostatic potential map for methane**

In a **perspective formula,** bonds in the plane of the paper are drawn as solid lines (and they must be adjacent to one another), a bond protruding out of the plane of the paper toward the viewer is drawn as a solid wedge, and one projecting back from the plane of the paper away from the viewer is drawn as a hatched wedge.

The potential map of methane shows that neither carbon nor hydrogen carries much of a charge: there are neither red areas, representing partially negatively charged atoms, nor blue areas, representing partially positively charged atoms. (Compare this map with the potential map for water later in this chapter.) The absence of partially charged atoms can be explained by the similar electronegativities of carbon and hydrogen, which cause them to share their bonding electrons relatively equally. Methane, therefore, is a **nonpolar molecule.**

You may be surprised to learn that carbon forms four covalent bonds, since you know that carbon has only two unpaired valence electrons (Table 2). But if carbon formed only two covalent bonds, it would not complete its octet. We need, therefore, to come up with an explanation that accounts for the observation that carbon forms four covalent bonds.

If one of the electrons in carbon's $2s$ orbital were promoted into its empty $2p$ orbital, then carbon would have four unpaired valence electrons (in which case four covalent bonds could be formed).

The blue colors of Uranus and Neptune are caused by the presence of methane, a colorless and odorless gas, in their atmospheres. Natural gas—called a fossil fuel because it is formed from the decomposition of plant and animal material in the Earth's crust—is approximately 75% methane.

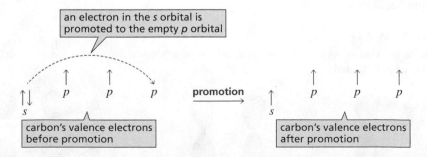

an electron in the *s* orbital is promoted to the empty *p* orbital

carbon's valence electrons before promotion

promotion

carbon's valence electrons after promotion

However, we have seen that the four C—H bonds in methane are identical. How can they be identical if carbon uses an *s* orbital and three *p* orbitals to form these four bonds? Wouldn't the bond formed with the *s* orbital be different from the three bonds formed with p orbitals? The four C—H bonds are identical because carbon uses hybrid atomic orbitals.

Hybrid orbitals are mixed orbitals that result from combining atomic orbitals. The concept of combining atomic orbitals, called **hybridization,** was first proposed by Linus Pauling in 1931.

If the one s and three p orbitals of the second shell are all combined and then apportioned into four equal orbitals, each of the four resulting orbitals will be one part s and three parts p. This type of mixed orbital is called an sp^3 (read "s-p-three," not "s-p-cubed") orbital. (The superscript 3 means that three p orbitals were mixed with one s orbital—the superscript 1 on the s is implied—to form the four hybrid orbitals.) Each sp^3 orbital has 25% s character and 75% p character. The four sp^3 orbitals are degenerate—that is, they all have the same energy.

Hybrid orbitals result from combining atomic orbitals.

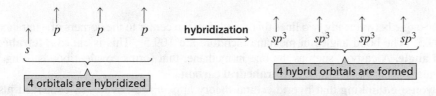

Like a p orbital, an sp^3 orbital has two lobes. The lobes differ in size, however, because the s orbital adds to one lobe of the p orbital and subtracts from the other lobe of the p orbital (Figure 7).

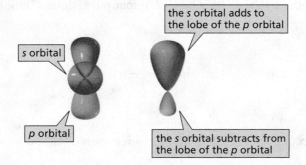

◀ Figure 7

The s orbital adds to one lobe of the p orbital and subtracts from the other. The result is a hybrid orbital with two lobes that differ in size.

The larger lobe of the sp^3 orbital is used to form covalent bonds. The stability of an sp^3 orbital reflects its composition; it is more stable than a p orbital, but not as stable as an s orbital (Figure 8.) (To simplify the orbital dipictions that follow, the phases of the orbitals will not be shown.)

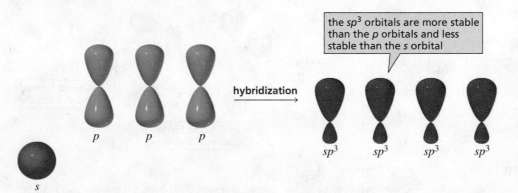

◀ Figure 8

An s orbital and three p orbitals hybridize to form four sp^3 orbitals. An sp^3 orbital is more stable (lower in energy) than a p orbital, but less stable (higher in energy) than an s orbital.

The four sp^3 orbitals adopt a spatial arrangement that keeps them as far away from each other as possible. They do this because electrons repel each other, and moving as far from each other as possible minimizes the repulsion. (See the description of the VSEPR model earlier in this chapter.)

When four sp^3 orbitals move as far from each other as possible, they point toward the corners of a regular tetrahedron—a pyramid with four faces, each an equilateral triangle

Electron pairs stay as far from each other as possible.

(Figure 9a). Each of the four C—H bonds in methane is formed from the overlap of an sp^3 orbital of carbon with the s orbital of a hydrogen (Figure 9b). This explains why the four C—H bonds are identical.

▶ Figure 9

a. The four sp^3 orbitals are directed toward the corners of a tetrahedron, causing each bond angle to be 109.5°. This arrangement allows the four orbitals to be as far apart as possible.

b. An orbital picture of methane, showing the overlap of each sp^3 orbital of carbon with the s orbital of a hydrogen. (For clarity, the smaller lobes of the sp^3 orbitals are not shown.)

The angle between any two lines that point from the center to the corners of a tetrahedron is 109.5°. The bond angles in methane therefore are 109.5°. This is called a **tetrahedral bond angle.** A carbon, such as the one in methane, that forms covalent bonds using four equivalent sp^3 orbitals is called a **tetrahedral carbon.**

If you are thinking that hybrid orbital theory appears to have been contrived just to make things fit, then you are right. Nevertheless, it gives us a very good picture of the bonding in organic compounds.

The Bonds in Ethane

Each carbon in ethane (CH_3CH_3) is bonded to four other atoms. Thus, both carbons are tetrahedral.

$$
\begin{array}{c}
\text{H} \quad \text{H} \\
| \quad\;\; | \\
\text{H}-\text{C}-\text{C}-\text{H} \\
| \quad\;\; | \\
\text{H} \quad \text{H}
\end{array}
$$

ethane

One bond connecting two atoms is called a **single bond.** All the bonds in ethane are single bonds.

Each carbon uses four sp^3 orbitals to form the four covalent bonds (Figure 10). One sp^3 orbital of one carbon of ethane overlaps an sp^3 orbital of the other carbon to form the C—C bond.

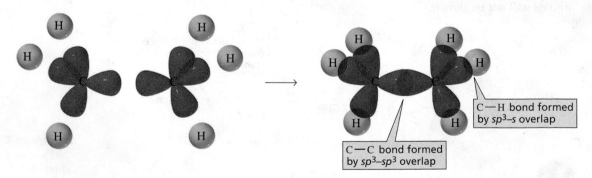

▲ Figure 10

An orbital picture of ethane. The C—C bond is formed by sp^3–sp^3 overlap, and each C—H bond is formed by sp^3–s overlap. (The smaller lobes of the sp^3 orbitals are not shown.) As a result, both carbons are tetrahedral and all bond angles are ~109.5°.

The three remaining sp^3 orbitals of each carbon overlap the s orbital of a hydrogen to form a C—H bond. Thus, the C—C bond is formed by sp^3–sp^3 overlap, and each C—H bond is formed by sp^3–s overlap. Each of the bond angles in ethane is nearly the tetrahedral bond angle of 109.5°, and the length of the C—C bond is 1.54 Å. The potential map shows that ethane, like methane, is a nonpolar molecule.

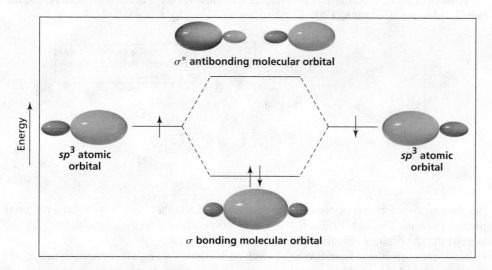

H 109.6° H

H⟍C—C⟍H

H 1.54 Å H

**perspective formula
of ethane**

**ball-and-stick model
of ethane**

**space-filling model
of ethane**

**electrostatic potential
map for ethane**

The MO diagram illustrating the overlap of an sp^3 orbital of one carbon with an sp^3 orbital of another carbon shows that the two sp^3 orbitals overlap end-on (Figure 11). End-on overlap forms a cylindrically symmetrical bond—a sigma (σ) bond (Section 6). *All single bonds found in organic compounds are sigma bonds.* Thus, all the bonds in methane and ethane are sigma (σ) bonds.

σ^* **antibonding molecular orbital**

Energy

sp^3 **atomic
orbital**

sp^3 **atomic
orbital**

σ **bonding molecular orbital**

Sigma bonds are
cylindrically symmetrical.

All single bonds found in organic
compounds are sigma bonds.

◀ **Figure 11**

End-on overlap of two sp^3 orbitals
to form a σ bonding molecular orbital
and a σ^* antibonding molecular orbital.

Notice in Figure 11 that that the electron density of the σ bonding MO is concentrated between the nuclei. This causes the back lobes (the nonoverlapping green lobes) to be quite small.

PROBLEM 25♦

What orbitals are used to form the 10 sigma bonds in propane ($CH_3CH_2CH_3$)?

PROBLEM 26

Explain why a σ bond formed by overlap of an s orbital with an sp^3 orbital of carbon is stronger than a σ bond formed by overlap of an s orbital with a p orbital of carbon.

8 HOW A DOUBLE BOND IS FORMED: THE BONDS IN ETHENE

Each of the carbon atoms in ethene (also called ethylene) forms four bonds, but each carbon is bonded to only three atoms:

H⟍   ⟋H
   C=C
H⟋   ⟍H

**ethene
ethylene**

To bond to three atoms, each carbon hybridizes three atomic orbitals: an s orbital and two of the p orbitals. Because three orbitals are hybridized, three hybrid orbitals are formed. These are called sp^2 orbitals. After hybridization, each carbon atom has three degenerate sp^2 orbitals and one unhybridized p orbital:

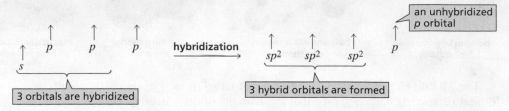

an unhybridized p orbital

hybridization

3 orbitals are hybridized

3 hybrid orbitals are formed

To minimize electron repulsion, the three sp^2 orbitals need to get as far from each other as possible. Therefore, the axes of the three orbitals lie in a plane, directed toward the corners of an equilateral triangle with the carbon nucleus at the center. As a result, the bond angles are all close to 120° (Figure 12a).

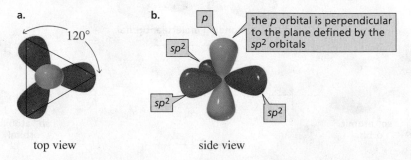

a.

120°

b.

p

the p orbital is perpendicular to the plane defined by the sp^2 orbitals

sp^2

sp^2

sp^2

top view   side view

▶ **Figure 12**

a. The three degenerate sp^2 orbitals lie in a plane, oriented 120° from each other. (The smaller lobes of the sp^2 orbitals are not shown.)

b. The unhybridized p orbital is perpendicular to this plane.

Because an sp^2 carbon is bonded to three atoms that define a plane, it is called a **trigonal planar carbon.** The unhybridized p orbital is perpendicular to the plane defined by the axes of the sp^2 orbitals (Figure 12b).

The carbons in ethene form two bonds with each other. Two bonds connecting two atoms is called a **double bond.** The two carbon–carbon bonds in the double bond are not identical. One of them results from the overlap of an sp^2 orbital of one carbon with an sp^2 orbital of the other carbon; this is a sigma (σ) bond because it is cylindrically symmetrical (Figure 13a). Each carbon uses its other two sp^2 orbitals to overlap the s orbital of a hydrogen to form the C—H bonds. The second carbon–carbon bond results from side-to-side overlap of the two unhybridized p orbitals. Side-to-side overlap of p orbitals forms a pi (π) bond (Figure 13b). Thus, one of the bonds in a double bond is a σ bond, and the other is a π bond. All the C—H bonds are σ bonds. (Note that all single bonds in organic compounds are σ bonds.)

A double bond consists of one σ bond and one π bond.

For maximum overlap to occur, the two p orbitals that overlap to form the π bond must be parallel to each other (Figure 13b). This forces the triangle formed by one carbon and two hydrogens to lie in the same plane as the triangle formed by the other carbon and two hydrogens. As a result, all six atoms of ethene lie in the same plane, and the electrons in the p orbitals occupy a volume of space above and below the plane (Figure 14).

The potential map for ethene shows that it is a nonpolar molecule with a slight accumulation of negative charge (the pale orange area) above the two carbons. (If you could turn the potential map over, you would fine a similar accumulation of negative charge on the other side.)

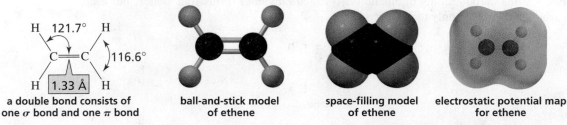

H   121.7°   H

C=C   116.6°

H   1.33 Å   H

a double bond consists of one σ bond and one π bond

ball-and-stick model of ethene

space-filling model of ethene

electrostatic potential map for ethene

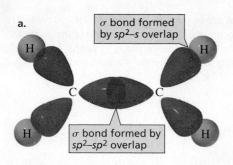

a.

b.

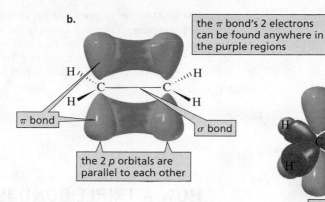

σ bond formed by sp²–s overlap

σ bond formed by sp²–sp² overlap

the π bond's 2 electrons can be found anywhere in the purple regions

π bond

σ bond

the 2 p orbitals are parallel to each other

▲ **Figure 13**

a. One C—C bond in ethene is a σ bond formed by sp²–sp² overlap, and the C—H bonds are σ bonds formed by sp²–s overlap.

b. The second C—C bond is a π bond formed by side-to-side overlap of a p orbital of one carbon with a p orbital of the other carbon. The two p orbitals are parallel to each other.

π bond   σ bond

▲ **Figure 14**

The two carbons and four hydrogens lie in the same plane. Perpendicular to that plane are the two parallel p orbitals. This results in an accumulation of electron density above and below the plane containing the two carbons and four hydrogens.

Four electrons hold the carbons together in a carbon–carbon double bond but only two electrons hold the carbons together in a carbon–carbon single bond. Since more electrons hold the carbons together, a carbon–carbon double bond is stronger (174 kcal/mol or 728 kJ/mol) and shorter (1.33 Å) than a carbon–carbon single bond (90 kcal/mol or 377 kJ/mol, and 1.54 Å).

Diamond, Graphite, Graphene, and Fullerenes: Substances that Contain Only Carbon Atoms

The difference that hybridization can make is illustrated by diamond and graphite. Diamond is the hardest of all substances, whereas graphite is a slippery, soft solid most familiar to us as the "lead" in pencils. Both materials, in spite of their very different physical properties, contain only carbon atoms. The two substances differ solely in the hybridization of the carbon atoms.

Diamond consists of a rigid three-dimensional network of carbon atoms, with each carbon bonded to four others via sp³ orbitals.

The carbon atoms in graphite, on the other hand, are sp² hybridized, so each bonds to only three other carbons. This trigonal planar arrangement causes the atoms in graphite to lie in flat, layered sheets. Since there are no covalent bonds between the sheets, they can shear off from neighboring sheets.

Diamond and graphite have been known since ancient times—but a third substance found in nature that contains only carbon atoms was discovered just 8 years ago. Graphene is a one-atom-thick planar sheet of graphite. It is the thinnest and lightest material known. It is transparent and can be bent, stacked, or rolled. It is harder than diamond and it conducts electricity better than copper. In 2010, the Nobel Prize in Physics was given to Andre Geim and Konstantin Novoselov of the University of Manchester for their ground-breaking experiments on graphene.

Fullerenes are also naturally occurring compounds that contain only carbon. Like graphite and graphene, fullerenes consist solely of sp² carbons, but instead of forming planar sheets, the carbons join to form spherical structures. (Fullerenes are discussed in Section 9.)

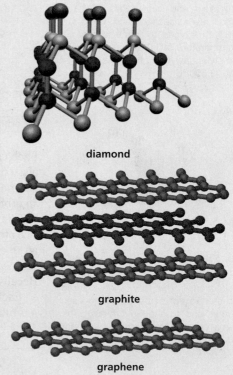

diamond

graphite

graphene

PROBLEM 27 Solved

Do the sp² carbons and the indicated sp³ carbons have to lie in the same plane?

a. H_3C —— H / $C=C$ / H —— CH_3

b. H_3C —— H / $C=C$ / H —— CH_2CH_3

Solution The two sp^2 carbons and the atoms that are bonded to each of the sp^2 carbons all lie in the same plane. The other atoms in the molecule will not lie in the same plane as these six atoms. By putting stars on the six atoms that do lie in the same plane, you will be able to see if the indicated atoms lie in the same plane. They are in the same plane in part **a.** but they are not necessarily in the same plane in part **b.**

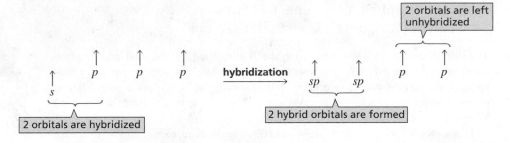

9 HOW A TRIPLE BOND IS FORMED: THE BONDS IN ETHYNE

Each of the carbon atoms in ethyne (also called acetylene) forms four bonds, but each carbon is bonded to only two atoms—a hydrogen and another carbon:

$$H{-}C{\equiv}C{-}H$$

ethyne
acetylene

In order to bond to two atoms, each carbon hybridizes two atomic orbitals—an s and a p. Two degenerate sp orbitals result.

Oxyacetylene torches are used to weld and cut metals. The torch uses acetylene and mixes it with oxygen to increase the temperature of the flame. An acetylene/oxygen flame burns at ~3,500 °C.

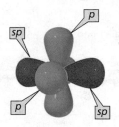

▲ **Figure 15**

The two sp orbitals point in opposite directions. The two unhybridized p orbitals are perpendicular to each other and to the sp orbitals. (The smaller lobes of the sp orbitals are not shown.)

Each carbon atom in ethyne, therefore, has two sp orbitals and two unhybridized p orbitals. To minimize electron repulsion, the two sp orbitals point in opposite directions. The two unhybridized p orbitals are perpendicular to each other and are perpendicular to the sp orbitals (Figure 15).

The two carbons in ethyne are held together by three bonds. Three bonds connecting two atoms is called a **triple bond.** One of the sp orbitals of one carbon in ethyne overlaps an sp orbital of the other carbon to form a carbon–carbon σ bond. The other sp orbital of each carbon overlaps the s orbital of a hydrogen to form a C—H σ bond (Figure 16a). Because the two sp orbitals point in opposite directions, the bond angles are 180°.

Each of the unhybridized p orbitals engages in side-to-side overlap with a parallel p orbital on the other carbon, resulting in the formation of two π bonds (Figure 16b).

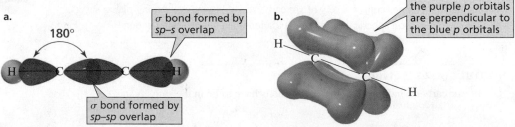

▲ **Figure 16**

a. The C—C σ bond in ethyne is formed by sp–sp overlap, and the C—H bonds are formed by sp–s overlap. The carbon atoms and the atoms bonded to them form a straight line.
b. The two carbon–carbon π bonds are formed by side-to-side overlap of the two p orbitals of one carbon with the two p orbitals of the other carbon.

Thus, a triple bond consists of one σ bond and two π bonds. Because the two unhybridized p orbitals on each carbon are perpendicular to each other, they create regions of high electron density above and below *and* in front of and back of the internuclear axis of the molecule (Figure 17).

The overall result can be seen in the potential map for ethyne—the negative charge accumulates in a cylinder that wraps around the egg-shaped molecule.

A triple bond consists of one σ bond and two π bonds.

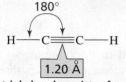

a triple bond consists of one
σ bond and two π bonds

ball-and-stick model
of ethyne

space-filling model
of ethyne

electrostatic potential map
for ethyne

The two carbon atoms in a triple bond are held together by six electrons, so a triple bond is stronger (231 kcal/mol or 967 kJ/mol) and shorter (1.20 Å) than a double bond (174 kcal/mol or 728 kJ/mol, and 1.33 Å).

▲ **Figure 17**
The triple bond has an electron dense region above and below and in front of and in back of the internuclear axis of the molecule.

PROBLEM 28

Put a number in each of the blanks:

a. ___ s orbital and ___ p orbitals form ____ sp^3 orbitals.
b. ___ s orbital and ___ p orbitals form ____ sp^2 orbitals.
c. ___ s orbital and ___ p orbitals form ____ sp orbitals.

PROBLEM 29 Solved

For each of the given species:

a. Draw its Lewis structure.
b. Describe the orbitals used by each carbon atom in bonding and indicate the approximate bond angles.

   1. H_2CO **2.** CCl_4 **3.** CH_3CO_2H **4.** HCN

Solution to 29a1 Our first attempt at a Lewis structure (drawing the atoms with the hydrogens on the outside of the molecule) shows that carbon is the only atom that does not form the needed number of bonds.

$$H-C-O-H$$

If we place a double bond between carbon and oxygen and move the H from O to C (which still keeps the Hs on the outside of the molecule) then all the atoms end up with the correct number of bonds. Lone-pair electrons are used to give oxygen a filled outer shell. When we check to see if any atom needs to be assigned a formal charge, we find that none of them does.

$$\begin{array}{c} \ddot{\text{O}}: \\ \parallel \\ H-C-H \end{array}$$

Solution to 29b1 Because carbon forms a double bond, we know that it uses sp^2 orbitals (as it does in ethene) to bond to the two hydrogens and the oxygen. It uses its "left-over" p orbital to form the second bond to oxygen. Because carbon is sp^2 hybridized, the bond angles are approximately 120°.

$$\begin{array}{c} \cdot\ddot{\text{O}}\cdot \\ 120°\nearrow\parallel\nwarrow120° \\ \text{C} \\ H\swarrow\searrow H \\ 120° \end{array}$$

10 THE BONDS IN THE METHYL CATION, THE METHYL RADICAL, AND THE METHYL ANION

Not all carbon atoms form four bonds. A carbon with a positive charge, a negative charge, or an unpaired electron forms only three bonds. Now we will see the orbitals that carbon uses when it forms three bonds.

The Methyl Cation ($^+CH_3$)

The carbon in $^+CH_3$ is sp^2 hybridized.

The positively charged carbon in the methyl cation is bonded to three atoms, so it hybridizes three orbitals—an s orbital and two p orbitals. Therefore, it forms its three covalent bonds using sp^2 orbitals. Its unhybridized p orbital remains empty. The positively charged carbon, and the three atoms bonded to it, lie in a plane. The unhybridized p orbital stands perpendicular to the plane.

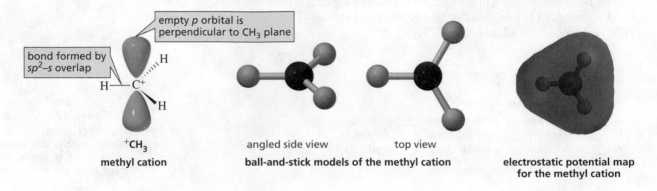

empty p orbital is perpendicular to CH_3 plane

bond formed by sp^2–s overlap

$^+CH_3$
methyl cation

angled side view   top view
ball-and-stick models of the methyl cation

electrostatic potential map for the methyl cation

The Methyl Radical ($\cdot CH_3$)

The carbon in $\cdot CH_3$ is sp^2 hybridized.

The carbon atom in the methyl radical is also sp^2 hybridized. The methyl radical, though, has one more electron than the methyl cation. That electron is unpaired and it resides in the p orbital, with half of the electron density in each lobe. Although the methyl cation and the methyl radical have similar ball-and-stick models, the potential maps are quite different because of the additional electron in the methyl radical.

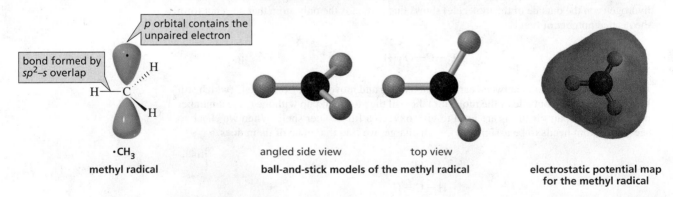

p orbital contains the unpaired electron

bond formed by sp^2–s overlap

$\cdot CH_3$
methyl radical

angled side view   top view
ball-and-stick models of the methyl radical

electrostatic potential map for the methyl radical

The Methyl Anion ($^-:CH_3$)

The carbon in $:CH_3$ is sp^3 hybridized.

The negatively charged carbon in the methyl anion has three pairs of bonding electrons and one lone pair. Four pairs of electrons are farthest apart when the four orbitals containing the bonding and lone-pair electrons point toward the corners of a tetrahedron. Thus, a negatively charged carbon is sp^3 hybridized. In the methyl anion, three of carbon's sp^3 orbitals each overlap the s orbital of a hydrogen, and the fourth sp^3 orbital holds the lone pair.

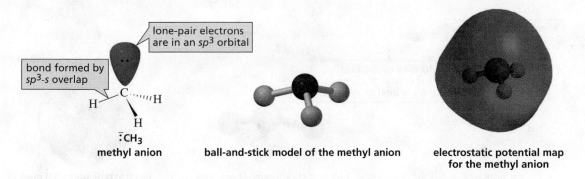

methyl anion

ball-and-stick model of the methyl anion

electrostatic potential map for the methyl anion

Take a moment to compare the potential maps for the methyl cation, the methyl radical, and the methyl anion.

11 THE BONDS IN AMMONIA AND IN THE AMMONIUM ION

The nitrogen atom in ammonia (NH_3) forms three covalent bonds. Nitrogen's electronic configuration shows that it has three unpaired valence electrons (Table 2), so it does not need to promote an electron to form the three covalent bonds required to achieve an outer shell of eight electrons—that is, to complete its octet.

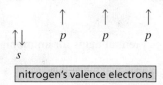

nitrogen's valence electrons

However, this simple picture presents a problem. If nitrogen uses p orbitals to form the three N—H bonds, as predicted by its electronic configuration, then we would expect bond angles of about 90° because the three p orbitals are at right angles to each other. But the experimentally observed bond angles in NH_3 are 107.3°.

The observed bond angles can be explained if we assume that nitrogen uses hybrid orbitals to form covalent bonds—just as carbon does. The s orbital and three p orbitals hybridize to form four degenerate sp^3 orbitals.

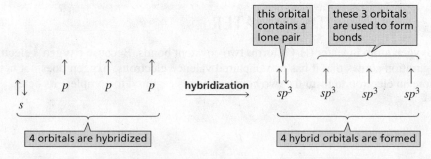

The bond angles in a molecule indicate which orbitals are used in bond formation.

Each of the three N—H bonds in NH_3 is formed from the overlap of an sp^3 orbital of nitrogen with the s orbital of a hydrogen. The lone pair occupies the fourth sp^3 orbital. The observed bond angle (107.3°) is a little smaller than the tetrahedral bond angle (109.5°) because of the lone pair. A lone pair is more diffuse than a bonding pair that is shared by two nuclei and relatively confined between them. Consequently, a lone pair exerts more electron repulsion, causing the N—H bonds to squeeze closer together, which decreases the bond angle.

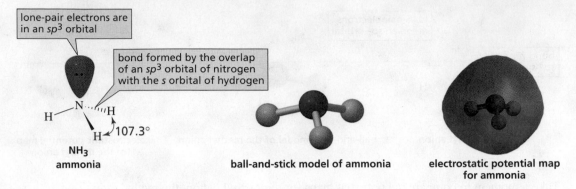

NH₃
ammonia

ball-and-stick model of ammonia

electrostatic potential map for ammonia

Because the ammonium ion ($^+NH_4$) has four identical N—H bonds and no lone pairs, all the bond angles are 109.5°, just like the bond angles in methane.

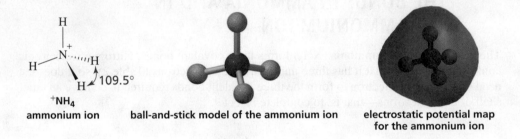

$^+NH_4$
ammonium ion

ball-and-stick model of the ammonium ion

electrostatic potential map for the ammonium ion

Take a moment to compare the potential maps for ammonia and the ammonium ion.

PROBLEM 30♦

Predict the approximate bond angles in
a. the methyl cation.   **b.** the methyl radical.   **c.** the methyl carbanion.

PROBLEM 31♦

According to the potential map for the ammonium ion, which atom has the greatest electron density?

---

12   THE BONDS IN WATER

The oxygen atom in water (H_2O) forms two covalent bonds. Because oxygen's electronic configuration shows that it has two unpaired valence electrons, oxygen does not need to promote an electron to form the two covalent bonds required to complete its octet.

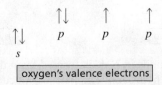

oxygen's valence electrons

The experimentally observed bond angle in H_2O is 104.5°. The bond angle indicates that oxygen, like carbon and nitrogen, uses hybrid orbitals when it forms covalent bonds.

Also like carbon and nitrogen, the one *s* and three *p* orbitals hybridize to form four degenerate *sp*³ orbitals:

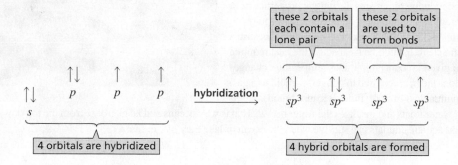

4 orbitals are hybridized

4 hybrid orbitals are formed

Each of the two O—H bonds is formed by the overlap of an *sp*³ orbital of oxygen with the *s* orbital of a hydrogen. A lone pair occupies each of the two remaining *sp*³ orbitals.

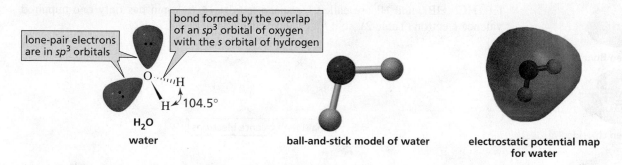

The bond angle in water (104.5°) is even smaller than the bond angles in NH_3 (107.3°) because oxygen has two relatively diffuse lone pairs, whereas nitrogen has only one.

PROBLEM 32♦

Compare the potential maps for methane, ammonia, and water. Which is the most polar molecule? Which is the least polar?

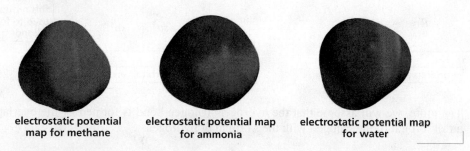

electrostatic potential map for methane

electrostatic potential map for ammonia

electrostatic potential map for water

PROBLEM 33 Solved

The bond angles in H_3O^+ are less than _____ and greater than _____.

Solution The carbon atom in CH_4 has no lone pairs; its bond angles are 109.5°. The oxygen atom in H_3O^+ has one lone pair. A lone pair is more diffuse than a bonding pair, so the O—H bonds squeeze together to minimize electron repulsion. However, they do not squeeze as closely together as they do in water (104.5°), where oxygen has two lone pairs. Therefore, the bond angles in H_3O^+ are less than 109.5° and greater than 104.5°.

Water—A Unique Compound

Water is the most abundant compound found in living organisms. Its unique properties have allowed life to originate and evolve. For example, its high heat of fusion (the heat required to convert a solid to a liquid) protects organisms from freezing at low temperatures because a lot of heat must be removed from water to freeze it. Its high heat capacity (the heat required to raise the temperature of a substance by a given amount) minimizes temperature changes in organisms, and its high heat of vaporization (the heat required to convert a liquid to a gas) allows animals to cool themselves with a minimal loss of body fluid. Because liquid water is denser than ice, ice formed on the surface of water floats and insulates the water below. That is why oceans and lakes freeze from the top down (not from the bottom up), and why plants and aquatic animals can survive when the ocean or lake they live in freezes.

13 THE BOND IN A HYDROGEN HALIDE

hydrogen fluoride

hydrogen chloride

hydrogen bromide

hydrogen iodide

HF, HCl, HBr, and HI are called hydrogen halides. A halogen has only one unpaired valence electron (Table 2), so it forms only one covalent bond.

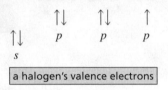

a halogen's valence electrons

Bond angles will not help us determine the orbitals that form the hydrogen halide bond, as they did with other molecules, because hydrogen halides have only one bond and therefore no bond angles. We do know, however, that a halogen's three lone pairs are energetically identical and that lone-pair electrons position themselves to minimize electron repulsion. Both of these observations suggest that the halogen's three lone pairs are in hybrid orbitals.

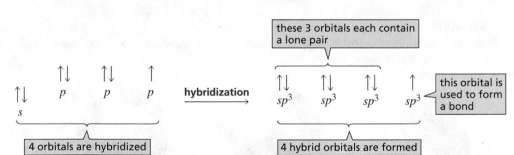

these 3 orbitals each contain a lone pair

this orbital is used to form a bond

4 orbitals are hybridized

4 hybrid orbitals are formed

Therefore, we will assume that the hydrogen–halogen bond is formed by the overlap of an sp^3 orbital of the halogen with the s orbital of hydrogen.

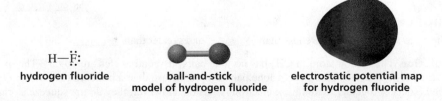

H—F̈:
hydrogen fluoride

ball-and-stick model of hydrogen fluoride

electrostatic potential map for hydrogen fluoride

In the case of fluorine, the sp^3 orbital used in bond formation belongs to the second shell of electrons. In chlorine, the sp^3 orbital belongs to the third shell. Because the average distance from the nucleus is greater for an electron in the third shell than it is for an electron in the second shell, the average electron density is less in a $3sp^3$ orbital than it is in a $2sp^3$ orbital. This means that the electron density in the region where the s orbital of hydrogen overlaps the sp^3 orbital of the halogen decreases as the size of the halogen increases (Figure 18). Therefore, the hydrogen–halogen bond becomes longer and weaker as the size (atomic weight) of the halogen increases (Table 6).

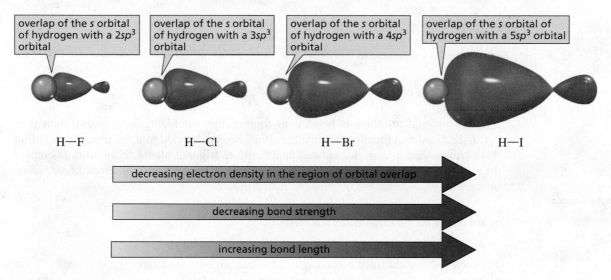

| overlap of the s orbital of hydrogen with a $2sp^3$ orbital | overlap of the s orbital of hydrogen with a $3sp^3$ orbital | overlap of the s orbital of hydrogen with a $4sp^3$ orbital | overlap of the s orbital of hydrogen with a $5sp^3$ orbital |

H—F   H—Cl   H—Br   H—I

decreasing electron density in the region of orbital overlap

decreasing bond strength

increasing bond length

▲ **Figure 18**
There is greater electron density in the region of overlap of the s orbital of hydrogen with a $2sp^3$ orbital than in the region of overlap of the s orbital of hydrogen with a $3sp^3$ orbital, which is greater than in the region of overlap of the s orbital of hydrogen with a $4sp^3$ orbital.

Table 6	Hydrogen–Halogen Bond Lengths and Bond Strengths		
Hydrogen halide	**Bond length (Å)**	**Bond strength**	
		(kcal/mol)	**(kJ/mol)**
H—F	0.917	136	571
H—Cl	1.275	103	432
H—Br	1.415	87	366
H—I	1.609	71	298

The hydrogen–halogen bond becomes longer and weaker as the size of the halogen increases.

PROBLEM 34♦

a. Predict the relative lengths and strengths of the bonds in Cl_2 and Br_2.

b. Predict the relative lengths and strengths of the carbon–halogen bonds in CH_3F, CH_3Cl, and CH_3Br.

PROBLEM 35♦

a. Which bond would be longer?

b. Which bond would be stronger?

    1. C—Cl or C—I   **2.** C—C or C—Cl   **3.** H—Cl or H—F

14   HYBRIDIZATION AND MOLECULAR GEOMETRY

The *hybridization* of an atom determines the *arrangement in space* of the bonds around the atom, and that this arrangement in space determines the *bond angle*.

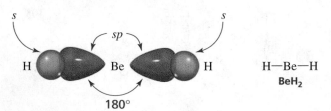

number of atoms C is bonded to:	4	3	2
hybridization of C:	sp^3	sp^2	sp
bond angles:	109.5°	120°	180°
molecular geometry of C:	tetrahedral	trigonal planar	linear

For example, if an atom is bonded to four groups (including lone pairs), then it is sp^3 hybridized and therefore tetrahedral. If an atom is bonded to three groups (including lone pairs), then it is sp^2 hybridized and therefore trigonal planar. If an atom is bonded to two groups (including lone pairs), then it is sp hybridized and therefore linear. *Thus, molecular geometry is determined by hybridization.*

Molecular geometry is determined by hybridization.

PROBLEM-SOLVING STRATEGY

Predicting the Orbitals and Bond Angles Used in Bonding

Describe the orbitals used in bonding and the bond angles in the following compounds:

a. BeH_2 **b.** BH_3

a. Beryllium (Be) does not have any unpaired valence electrons (Table 2). Therefore, it cannot form any bonds unless it promotes an electron. After promotion of an electron from the *s* orbital to the empty *p* orbital and hybridization of the two orbitals that now contain an unpaired electron, two *sp* orbitals result.

Each *sp* orbital of beryllium overlaps the *s* orbital of a hydrogen. To minimize electron repulsion, the two *sp* orbitals point in opposite directions, resulting in a bond angle of 180°.

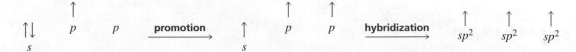

b. Without promotion, boron (B) could form only one bond because it has only one unpaired valence electron (Table 2). Promotion gives it three unpaired electrons. When the three orbitals that contain an unpaired electron (one *s* orbital and two *p* orbitals) are hybridized, three sp^2 orbitals result.

Each sp^2 orbital of boron overlaps the *s* orbital of a hydrogen. When the three sp^2 orbitals orient themselves to get as far away from each other as possible, the resulting bond angles are 120°.

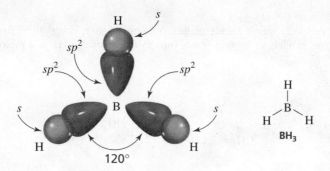

Now use the strategy you have just learned to solve Problem 36.

PROBLEM 36

Describe the orbitals used in bonding and the bond angles in the following compounds:

a. CCl_4 **b.** CO_2 **c.** HCOOH   **d.** N_2 **e.** BF_3

---

15   SUMMARY: HYBRIDIZATION, BOND LENGTHS, BOND STRENGTHS, AND BOND ANGLES

We have seen that all *single bonds* are σ bonds, all double bonds are composed of one σ bond and one π bond, and all triple bonds are composed of one σ bond and two π bonds.

| one σ bond | one σ bond one π bond | one σ bond two π bonds |

Bond order describes the number of covalent bonds shared by two atoms. A single bond has a bond order of one, a double bond has a bond order of two, and a triple bond has a bond order of three.

The easiest way to determine the hybridization of carbon, nitrogen, or oxygen is to count the number of π bonds it forms. If it forms no π bond, it is sp^3 hybridized; if it forms one π bond, it is sp^2 hybridized; and if it forms two π bonds, it is sp hybridized. The exceptions are carbocations and carbon radicals, which are sp^2 hybridized—not because they form a π bond, but because they have an empty or a half-filled p orbital (Section 10).

The hybridization of a C, N, or O is $sp^{(3\ \text{minus the number of}\ \pi\ \text{bonds})}$.

$CH_3-\ddot{N}H_2$

$\begin{array}{c} CH_3 \\ \diagdown \\ C=\ddot{N}-\ddot{N}H_2 \\ \diagup \\ CH_3 \end{array}$

$CH_3-C\equiv N:$

$CH_3-\ddot{O}H$

$\begin{array}{c} \ddot{O} \;\leftarrow sp^2 \\ \parallel \\ C \\ CH_3 \diagup \diagdown CH_3 \end{array}$

$:\!\ddot{O}=C=\ddot{O}:$

sp^3 sp^3 sp^3 sp^2 sp^2 sp^3 sp^3 sp sp sp^3 sp^3 sp^3 sp^2 sp^3 sp^2 sp sp^2

PROBLEM 37 Solved

In what orbitals are the lone pairs in each of the following molecules?

a. $CH_3\ddot{O}H$ **b.** $\begin{array}{c} \ddot{O} \\ \parallel \\ C \\ CH_3 \diagup \diagdown CH_3 \end{array}$ **c.** $CH_3C\equiv N:$

Solution

a. Oxygen forms only single bonds in this compound, so it is sp^3 hybridized. It uses two of its four sp^3 orbitals to form σ bonds (one to C and one to H), and the other two for its lone pairs.

$$
\text{a. } CH_3\overset{sp^3}{\underset{sp^3}{\ddot{O}}}H \qquad \text{b. } \overset{sp^2 \longrightarrow \cdot\ddot{O}\cdot \longleftarrow sp^2}{\underset{CH_3 \qquad CH_3}{C}} \qquad \text{c. } CH_3C\equiv N\overset{sp}{:}
$$

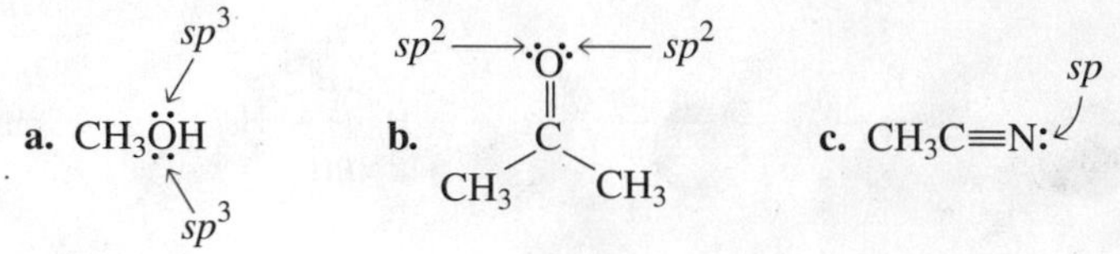

b. Oxygen forms a double bond in this compound, so it is sp^2 hybridized. It uses one of its three sp^2 orbitals to form the σ bond to C, and the other two for its lone pairs.

c. Nitrogen forms a triple bond in this compound, so it is sp hybridized. It uses one of the sp orbitals to form the σ bond to C and the other one for its lone pair.

In comparing the lengths and strengths of carbon–carbon single, double, and triple bonds, we see that the carbon–carbon bond gets shorter and stronger as the number of bonds holding the two carbon atoms together increases (Table 7). As a result, triple bonds are shorter and stronger than double bonds, which are shorter and stronger than single bonds.

The shorter the bond, the stronger it is.

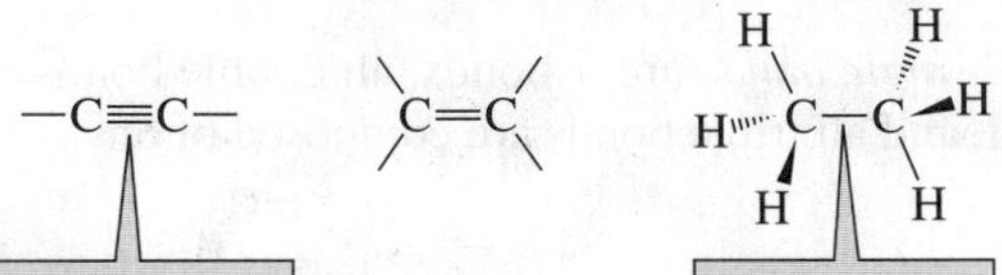

strongest bond
shortest bond

weakest bond
longest bond

bond strength increases as bond length decreases

Table 7 Comparison of the Bond Angles and the Lengths and Strengths of the Carbon–Carbon and Carbon–Hydrogen Bonds in Ethane, Ethene, and Ethyne

Molecule	Hybridization of carbon	Bond angles	Length of C—C bond (Å)	Strength of C—C bond (kcal/mol) (kJ/mol)		Length of C—H bond (Å)	Strength of C—H bond (kcal/mol) (kJ/mol)	
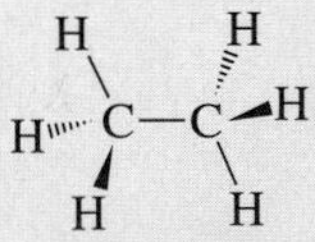 ethane	sp^3	109.5°	1.54	90.2	377	1.10	101.1	423
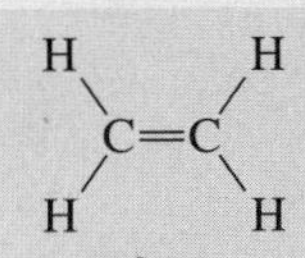 ethene	sp^2	120°	1.33	174.5	730	1.08	110.7	463
ethyne	sp	180°	1.20	230.4	964	1.06	133.3	558

A C—H σ bond is shorter than a C—C σ bond (Table 7) because the s orbital of hydrogen is closer to the nucleus than is the sp^3 orbital of carbon. Consequently, the nuclei are closer together in a bond formed by sp^3–s overlap than they are in a bond formed by sp^3–sp^3 overlap. In addition to being shorter, a C—H σ bond is stronger than a C—C σ bond because there is greater electron density in the region of overlap of an sp^3 orbital with an s orbital than in the region of overlap of an sp^3 orbital with an sp^3 orbital.

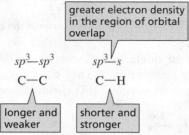

the greater the electron density in the region of orbital
overlap, the stronger and shorter the bond

The length and strength of a C—H bond both depend on the hybridization of the carbon to which the hydrogen is attached. The more s character in the orbital used by carbon to form the bond, the shorter and stronger is the bond—again because an s orbital is closer to the nucleus than is a p orbital. Thus, a C—H bond formed by an sp carbon (50% s) is shorter and stronger than a C—H bond formed by an sp^2 carbon (33.3% s), which in turn is shorter and stronger than a C—H bond formed by an sp^3 carbon (25% s).

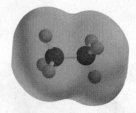

Ethane

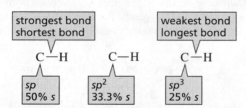

bond strength increases as bond length decreases

Ethene

A double bond (a σ bond plus a π bond) is stronger (174 kcal/mol) than a single bond (a σ bond; 90 kcal/mol), but it is not twice as strong, so we can conclude that the π bond of a double bond is weaker than the σ bond. We expect the π bond to be weaker than the σ bond because the side-to-side overlap that forms a π bond is less effective for bonding than the end-on overlap that forms a σ bond (Section 6).

The strength of a C—C σ bond given in Table 7 (90 kcal/mol) is for a bond formed by sp^3–sp^3 overlap. A C—C σ bond formed by sp^2–sp^2 overlap is expected to be stronger, however, because of the greater s character in the overlapping sp^2 orbitals; it has been estimated to be ~112 kcal/mol. We can conclude then, that the strength of the π bond of ethene is about 62 kcal/mol (174 − 112 = 62).

Ethyne

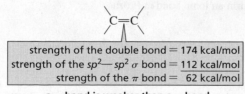

a π bond is weaker than a σ bond

The bond angle, too, depends on the orbital used by carbon to form the bond. The greater the amount of s character in the orbital, the larger the bond angle. For example, sp^3 carbons have bond angles of 109.5°, sp^2 carbons have bond angles of 120°, and sp carbons have bond angles of 180°.

$$
\underset{\substack{sp^3 \\ 25\%\ s}}{\overset{\displaystyle H\overset{109.5°}{\,}H}{\overset{\displaystyle H\cdots C - C - H}{\underset{\displaystyle H\quad H}{}}}}
\qquad
\underset{\substack{sp^2 \\ 33.3\%\ s}}{\overset{\displaystyle H\overset{120°}{\,}H}{\overset{\displaystyle C=C}{\underset{\displaystyle H\quad H}{}}}}
\qquad
\underset{\substack{sp \\ 50\%\ s}}{\overset{\displaystyle 180°}{H-C\equiv C-H}}
$$

The more *s* character in the orbital, the larger the bond angle.

bond angle increases as *s* character in the orbital increases

You may wonder how an electron "knows" what orbital it should go into. In fact, electrons know nothing about orbitals. They simply occupy the space around atoms in the most stable arrangement possible. It is chemists who use the concept of orbitals to explain this arrangement.

PROBLEM 38♦

Which of the bonds in a carbon–oxygen double bond has more effective orbital–orbital overlap, the σ bond or the π bond?

PROBLEM 39♦

Would you expect a C—C σ bond formed by sp^2–sp^2 overlap to be stronger or weaker than a C—C σ bond formed by sp^3–sp^3 overlap?

PROBLEM 40

Caffeine is a natural insecticide, found in the seeds and leaves of certain plants, where it kills insects that feed on the plant. Caffeine is extracted for human consumption from beans of the coffee plant, from Kola nuts, and from the leaves of tea plants. Because it stimulates the central nervous system, it temporarily prevents drowsiness. Add caffeine's missing lone pairs to its structure.

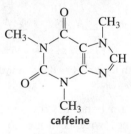

caffeine

coffee beans

PROBLEM 41

a. What is the hybridization of each of the carbon atoms in the following compound?

$$
\underset{\displaystyle CH_3}{\overset{\displaystyle CH_3CHCH=CHCH_2C\equiv CCH_3}{\big|}}
$$

b. What is the hybridization of each of the atoms in Demerol and Prozac?

c. Which two atoms form an ionic bond in Prozac?

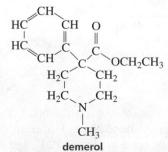

demerol
used to treat
moderate to severe pain

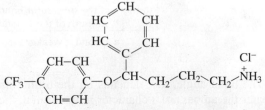

Prozac®
used to treat depression, obsessive-compulsive disorder, some eating disorders, and panic attacks

PROBLEM-SOLVING STRATEGY

Predicting Bond Angles

Predict the approximate bond angle of the C—N—H bond in $(CH_3)_2NH$.

First we need to determine the hybridization of the central atom (the N). Because the nitrogen atom forms only single bonds, we know it is sp^3 hybridized. Next we look to see if there are lone pairs that will affect the bond angle. An uncharged nitrogen has one lone pair. Based on these observations, we can predict that the C—N—H bond angle will be about 107.3°, the same as the H—N—H bond angle in NH_3, which is another compound with an sp^3 nitrogen and one lone pair.

Now use the strategy you have just learned to solve Problem 42.

PROBLEM 42♦

Predict the approximate bond angles for

a. the C—N—C bond angle in $(CH_3)_2\overset{+}{N}H_2$.   **c.** the H—C—N bond angle in $(CH_3)_2NH$.

b. the C—N—H bond angle in $CH_3CH_2NH_2$.   **d.** the H—C—O bond angle in CH_3OCH_3.

16   THE DIPOLE MOMENTS OF MOLECULES

In Section 3, we saw that if a molecule has one covalent bond, then the dipole moment of the molecule is identical to the dipole moment of the bond. When molecules have more than one covalent bond, the geometry of the molecule must be taken into account because both the *magnitude* and the *direction* of the individual bond dipole moments (the vector sum) determine the overall dipole moment of the molecule.

> *The dipole moment depends on the magnitude of the individual bond dipoles and the direction of the individual bond dipoles.*

Because the direction of the bond dipoles have to be taken into account, totally symmetrical molecules have no dipole moment. In carbon dioxide (CO_2), for example, the carbon is bonded to two atoms, so it uses sp orbitals to form the two C—O σ bonds. The remaining two p orbitals on the carbon form the two C—O π bonds. The sp orbitals form a bond angle of 180°, which causes the individual carbon–oxygen bond dipole moments to cancel each other. Carbon dioxide therefore has a dipole moment of 0 D.

the 2 bond dipole moments cancel because they are identical and point in opposite directions

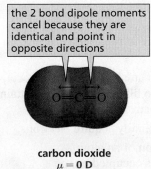

the bond dipole moments cancel because all 4 are identical and project symmetrically out from carbon

carbon dioxide
$\mu = 0\ D$

carbon tetrachloride
$\mu = 0\ D$

Another symmetrical molecule is carbon tetrachloride (CCl_4). The four atoms bonded to the sp^3 carbon atom are identical and project symmetrically out from the carbon atom. Thus, as with CO_2, the symmetry of the molecule causes the bond dipole moments to cancel. Therefore, methane also has no dipole moment.

The dipole moment of chloromethane (CH_3Cl) is greater (1.87 D) than the dipole moment of its C—Cl bond (1.5 D) because the C—H dipoles are oriented so that they reinforce the dipole of the C—Cl bond. In other words, all the electrons are pulled in the same general direction.

the 4 bond dipole moments point in the same general direction

chloromethane	water	ammonia
$\mu = 1.87$ D	$\mu = 1.85$ D	$\mu = 1.47$ D

The dipole moment of water (1.85 D) is greater than the dipole moment of a single O—H bond (1.5 D) because the dipoles of the two O—H bonds reinforce each other; the lone-pair electrons also contribute to the dipole moment. Similarly, the dipole moment of ammonia (1.47 D) is greater than the dipole moment of a single N—H bond (1.3 D).

PROBLEM 43♦

If the dipole moment of CH_3F is 1.847 D and the dipole moment of CD_3F is 1.858 D, which is more electronegative, hydrogen or deuterium?

PROBLEM 44

Account for the difference in the shape and color of the potential maps for ammonia and the ammonium ion in Section 11.

PROBLEM 45♦

Which of the following molecules would you expect to have a dipole moment of zero? To answer parts **g** and **h,** you may need to review the Problem Solving Strategy earlier in this chapter.

a. CH_3CH_3 c. CH_2Cl_2 e. $H_2C{=}CH_2$ g. $BeCl_2$

b. $H_2C{=}O$ d. NH_3 f. $H_2C{=}CHBr$ h. BF_3

SOME IMPORTANT THINGS TO REMEMBER

- **Organic compounds** are compounds that contain carbon.
- The **atomic number** of an atom is the number of protons in its nucleus (or the number of electrons that surrounds the neutral atom).
- The **mass number** of an atom is the sum of its protons and neutrons.
- **Isotopes** have the same atomic number, but different mass numbers.
- **Atomic weight** is the average mass of the atoms in the element.
- **Molecular weight** is the sum of the atomic weights of all the atoms in the molecule.
- An **atomic orbital** tells us the volume of space around the nucleus where an electron is most likely to be found.
- The closer the atomic orbital is to the nucleus, the lower is its energy.
- **Minimum energy** corresponds to **maximum stability.**
- **Degenerate orbitals** have the same energy.

- Electrons are assigned to orbitals (atomic or molecular) following the **aufbau principle,** the **Pauli exclusion principle,** and **Hund's rule.**
- An atom is most stable if its outer shell is either filled or contains eight electrons, and if it has no electrons of higher energy.
- The **octet rule** states that an atom will give up, accept, or share electrons in order to fill its outer shell or attain an outer shell with eight electrons.
- **Electronegative** elements readily acquire electrons.
- The **electronic configuration** of an atom describes the atomic orbitals occupied by the atom's electrons.
- A **proton** is a positively charged hydrogen ion; a **hydride ion** is a negatively charged hydrogen ion.
- Attractive forces between opposite charges are called **electrostatic attractions.**
- An **ionic bond** results from the electrostatic attraction between ions with opposite charges.

- A **covalent bond** is formed when two atoms share a pair of electrons.

- A **polar covalent bond** is a covalent bond between atoms with different **electronegativities.**

- The greater the difference in electronegativity between the atoms forming the bond, the closer the bond is to the ionic end of the continuum.

- A polar covalent bond has a **dipole** (a positive end and a negative end), measured by a **dipole moment.**

- The **dipole moment** of a bond is equal to the **size of the charge × the distance between the charges.**

- The **dipole moment** of a molecule depends on the magnitude and direction of all the bond dipole moments.

- **Core electrons** are electrons in inner shells. **Valence electrons** are electrons in the outermost shell. **Lone-pair electrons** are valence electrons that do not form bonds.

- **formal charge** = # of valence electrons – # of electrons the atom has to itself (all the lone-pair electrons and one-half the bonding electrons)

- **Lewis structures** indicate which atoms are bonded together and show **lone pairs** and **formal charges.**

- When the atom is neutral: C forms 4 bonds, N forms 3 bonds, O forms 2 bonds, and H or a halogen forms 1 bond.

- When the atom is neutral: N has 1 lone pair, O has 2 lone pairs, and a halogen has 3 lone pairs.

- A **carbocation** has a positively charged carbon, a **carbanion** has a negatively charged carbon, and a **radical** has an unpaired electron.

- According to **molecular orbital (MO) theory,** covalent bonds result when atomic orbitals combine to form **molecular orbitals.**

- Atomic orbitals combine to give a **bonding MO** and a higher energy **antibonding MO.**

- Electrons in a bonding MO assist in bonding. Electrons in an antibonding MO detract from bonding.

- There is zero probability of finding an electron at a **node.**

- Cylindrically symmetrical bonds are called **sigma** (σ) bonds; side-to-side overlap of parallel p orbitals forms a **pi (π) bond**.

- Bond strength is measured by the **bond dissociation energy**; a σ bond is stronger than a π bond.

- To be able to form four bonds, carbon has to promote an electron from an s orbital to an empty p orbital.

- C, N, O, and the halogens form bonds using **hybrid orbitals.**

- The **hybridization** of C, N, or O depends on the number of π bonds the atom forms: no π bonds = sp^3, one π bond = sp^2, and two π bonds = sp. Exceptions are carbocations and carbon radicals, which are sp^2.

- All **single bonds** in organic compounds are σ bonds.

- A **double bond** consists of one σ bond and one π bond; a **triple bond** consists of one σ bond and two π bonds.

- The greater the electron density in the region of orbital overlap, the shorter and stronger the bond.

- Triple bonds are shorter and stronger than double bonds, which are shorter and stronger than single bonds. The shorter the bond, the stronger it is.

- **Molecular geometry** is determined by **hybridization:** sp^3 is tetrahedral, sp^2 is trigonal planar, and sp is linear.

- Bonding pairs and lone-pair electrons around an atom stay as far apart as possible.

- The more s character in the orbital used to form a bond, the shorter and stronger the bond and the larger the bond angle.

PROBLEMS

46. Draw a Lewis structure for each of the following species:
    a. H_2CO_3 **b.** CO_3^{2-} **c.** CH_2O **d.** CO_2

47. a. Which of the following has a nonpolar covalent bond?
    b. Which of the following has a bond closest to the ionic end of the bond spectrum?

    CH_3NH_2 CH_3CH_3 CH_3F CH_3OH

48. What is the hybridization of all the atoms (other than hydrogen) in each of the following species? What are the bond angles around each atom?
    a. NH_3 **c.** $^-CH_3$ **e.** $^+NH_4$ **g.** HCN   **i.** H_3O^+
    b. BH_3 **d.** $\cdot CH_3$ **f.** $^+CH_3$ **h.** $C(CH_3)_4$ **j.** $H_2C{=}O$

49. Draw the condensed structure of a compound that contains only carbon and hydrogen atoms and that has
    a. three sp^3 hybridized carbons.
    b. one sp^3 hybridized carbon and two sp^2 hybridized carbons.
    c. two sp^3 hybridized carbons and two sp hybridized carbons.

50. Predict the approximate bond angles:
    a. the $C-N-C$ bond angle in $(CH_3)_2\overset{+}{N}H_2$
    b. the $C-O-H$ bond angle in CH_3OH
    c. the $C-N-H$ bond angle in $(CH_3)_2NH$
    d. the $C-N-C$ bond angle in $(CH_3)_2NH$

51. Draw the ground-state electronic configuration for the following:
    a. Mg **b.** Ca^{2+} **c.** Ar **d.** Mg^{2+}

52. Draw a Lewis structure for each of the following species:
    a. CH_3NH_2 **b.** HNO_2 **c.** N_2H_4 **d.** NH_2O^-

53. What is the hybridization of each of the carbon and oxygen atoms in vitamin C?

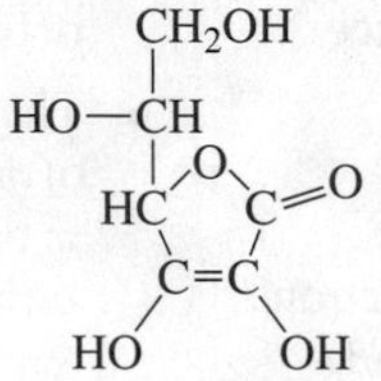

vitamin C

54. List the bonds in order from most polar to least polar.
    a. $C-O, C-F, C-N$ **b.** $C-Cl, C-I, C-Br$ **c.** $H-O, H-N, H-C$ **d.** $C-H, C-C, C-N$

55. Draw the Lewis structure for each of the following compounds:
    a. CH_3CHO **b.** CH_3OCH_3 **c.** CH_3COOH

56. What is the hybridization of the indicated atom in each of the following molecules?

    a. $CH_3CH=CH_2$ **b.** $CH_3\overset{O}{\underset{||}{C}}CH_3$ **c.** CH_3CH_2OH **d.** $CH_3C\equiv N$ **e.** $CH_3CH=NCH_3$ **f.** $CH_3OCH_2CH_3$

57. Predict the approximate bond angles for the following:
    a. the $H-C-H$ bond angle in $H_2C=O$
    b. the $F-B-F$ bond angle in $^-BF_4$
    c. the $C-C-N$ bond angle in $CH_3C\equiv N$
    d. the $C-C-N$ bond angle in $CH_3CH_2NH_2$

58. Show the direction of the dipole moment in each of the following bonds (use the electronegativities given in Table 3):
    a. H_3C-Br **b.** H_3C-Li **c.** $HO-NH_2$ **d.** $I-Br$ **e.** H_3C-OH **f.** $(CH_3)_2N-H$

59. Draw the missing lone-pair electrons and assign the missing formal charges for the following.

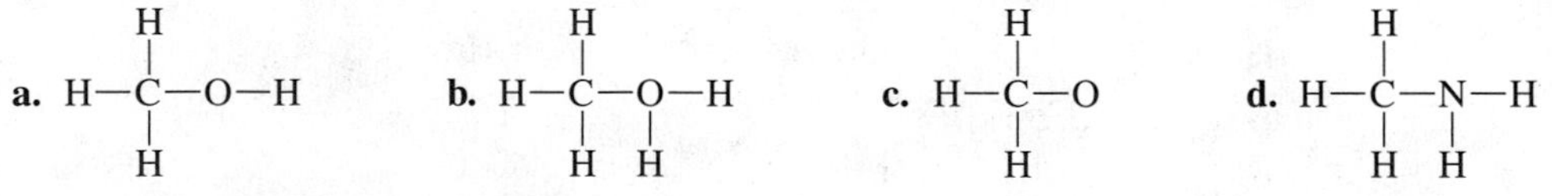

60. a. Which of the indicated bonds in each molecule is shorter?
    b. Indicate the hybridization of the C, O, and N atoms in each of the molecules.

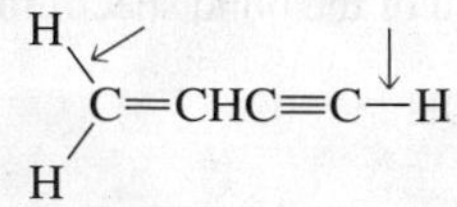

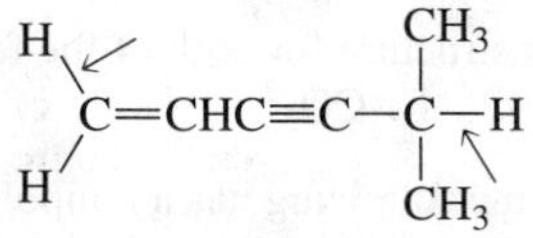

61. For each of the following molecules, indicate the hybridization of each carbon and give the approximate values of all the bond angles:

a. $CH_3C\equiv CH$ 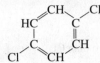 b. $CH_3CH=CH_2$ c. $CH_3CH_2CH_3$ d. $CH_2=CH-CH=CH_2$

62. Draw the Lewis structure for each of the following compounds:
a. $(CH_3)_3COH$ b. $CH_3CH(OH)CH_2CN$ c. $(CH_3)_2CHCH(CH_3)CH_2C(CH_3)_3$

63. Rank the following compounds from highest dipole moment to lowest dipole moment:

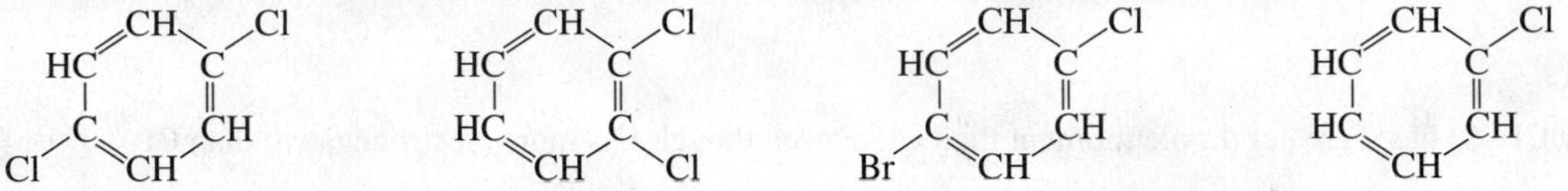

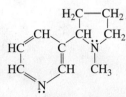

64. In which orbitals are the lone pairs in nicotine?

nicotine

**nicotine increases the concentration of dopamine
in the brain; the release of dopamine makes a
person feel good—the reason why nicotine
is addictive**

65. Indicate the formal charge on each carbon that has one. All lone pairs are shown.

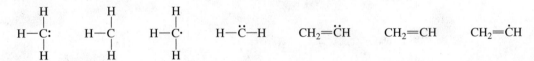

66. Do the sp^2 carbons and the indicated sp^3 carbons lie in the same plane?

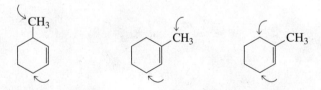

67. a. Which of the species have bond angles of 109.5°?
b. Which of the species have bond angles of 120°?

H_2O H_3O^+ $^+CH_3$ BF_3 NH_3 $^+NH_4$ $^-CH_3$

68. Which compound has a longer C—Cl bond?

CH_3CH_2Cl
**at one time it was used as a
refrigerant, an anesthetic, and a
propellant for aerosol sprays**

$CH_2=CHCl$
**used as the starting material
for the synthesis of a plastic that
is used to make bottles, flooring,
and clear packaging for food**

69. Which compound has a larger dipole moment, $CHCl_3$ or CH_2Cl_2?

70. The following compound has two isomers. One isomer has a dipole moment of 0 D, whereas the other has a dipole moment of 2.95 D. Propose structures for the two isomers that are consistent with these data.

$ClCH=CHCl$

71. Explain why the following compound is not stable:

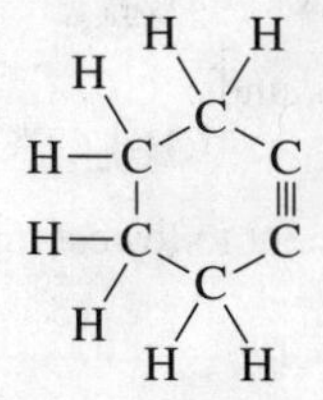

72. Explain why CH_3Cl has a greater dipole moment than CH_3F even though F is more electronegative than Cl.

73. Draw a Lewis structure for each of the following species:

    a. $CH_3N_2^+$ **b.** CH_2N_2 **c.** N_3^- **d.** N_2O (arranged NNO)

GLOSSARY

antibonding molecular orbital   a molecular orbital that results when two atomic orbitals with opposite signs interact. Electrons in an antibonding orbital decrease bond strength.

atomic number   the number of protons (or electrons) that the neutral atom has.

atomic orbital   an orbital associated with an atom.

atomic weight   the average mass of the atoms in the naturally occurring element.

aufbau principle   states that an electron will always go into that orbital with the lowest available energy.

bond length   the internuclear distance between two atoms at minimum energy (maximum stability).

bond order   the number of covalent bonds shared by two atoms.

carbanion   a compound containing a negatively charged carbon.

carbocation   a species containing a positively charged carbon.

constitutional isomers (structural isomers)   molecules that have the same molecular formula but differ in the way their atoms are connected.

covalent bond   a bond created as a result of sharing electrons.

degenerate orbitals   orbitals that have the same energy.

dipole moment (μ)   a measure of the separation of charge in a bond or in a molecule.

dissociation energy   the amount of energy required to break a bond, or the amount of energy released when a bond is formed.

double bond   a σ bond and a π bond between two atoms.

electronegativity   tendency of an atom to pull electrons toward itself.

electrostatic attraction   attractive force between opposite charges.

excited-state electronic configuration   the electronic configuration that results when an electron in the ground-state electronic configuration has been moved to a higher energy orbital.

formal charge   the number of valence electrons – (the number of non-bonding electrons + 1/2 the number of bonding electrons).

ground-state electronic configuration   a description of which orbitals the electrons of an atom or molecule occupy when all of the electrons of atoms are in their lowest-energy orbitals.

Heisenberg uncertainty principle   states that both the precise location and the momentum of an atomic particle cannot be simultaneously determined.

Hund's rule   states that when there are degenerate orbitals, an electron will occupy an empty orbital before it will pair up with another electron.

hybrid orbital   an orbital formed by mixing (hybridizing) orbitals.

hydride ion   a negatively charged hydrogen.

hydrogen ion (proton)   a positively charged hydrogen.

ionic bond   a bond formed through the attraction of two ions of opposite charges.

isotopes   atoms with the same number of protons, but different numbers of neutrons.

Kekulé structure   a model that represents the bonds between atoms as lines.

Lewis structure   a model that represents the bonds between atoms as lines or dots and the valence electrons as dots.

lone-pair electrons (nonbonding electrons)   valence electrons not used in bonding.

mass number   the number of protons plus the number of neutrons in an atom.

molecular orbital   an orbital associated with a molecule.

molecular orbital theory   describes a model in which the electrons occupy orbitals as they do in atoms, but with the orbitals extending over the entire molecule.

node   that part of an orbital in which there is zero probability of finding an electron.

nonbonding electrons (lone-pair electrons)   valence electrons not used in bonding.

nonpolar covalent bond   a bond formed between two atoms that share the bonding electrons equally.

octet rule   states that an atom will give up, accept, or share electrons in order to achieve a filled shell. Because a filled second shell contains eight electrons, this is known as the octet rule.

orbital hybridization   mixing of orbitals.

organic compound   a compound that contains carbon.

Pauli exclusion principle   states that no more than two electrons can occupy an orbital and that the two electrons must have opposite spin.

perspective formula   a method of representing the spatial arrangement of groups bonded to an asymmetric center. Two bonds are drawn in the plane of the paper; a solid wedge is used to depict a bond that projects out of the plane of the paper toward the viewer, and a hatched wedge is used to represent a bond that projects back from the plane of the paper away from the viewer.

pi (π) bond   a bond formed as a result of side-to-side overlap of p orbitals.

polar covalent bond   a covalent bond between atoms of different electronegativites.

proton   a positively charged hydrogen (H^+); a positively charged particle in an atomic nucleus.

radical   an atom or a molecule with an unpaired electron.

sigma (σ) bond   a bond with a cylindrically symmetrical distribution of electrons.

single bond   a σ bond.

tetrahedral bond angle   the bond angle ($109.5°$) formed by adjacent bonds of an sp^3 hybridized carbon.

tetrahedral carbon   an sp^3 hybridized carbon; a carbon that forms covalent bonds by using four sp^3 hybridized orbitals.

trigonal planar carbon   an sp^2 hybridized carbon.

triple bond   a σ bond plus two π bonds.

valence electron   an electron in an unfilled shell.

valence shell electron-pair repulsion (VSEPR) model   combines the concept of atomic orbitals with the concept of shared electron pairs and the minimization of electron pair repulsion.

PHOTO CREDITS

Photo credits are listed in order of appearance.

Ocean/Corbis; USDA/Stephane Alix; Paula Bruice; (left) Blaz Kure/Shutterstock; (bot.) modestlife/Fotolia; JPL/NASA; An Post; terekhov igor/Shutterstock; NASA/Kathryn Hansen; Fellks/Shutterstock

TEXT CREDITS

Text credits are listed in order of appearance.

(Table 3) The electronegativities listed here are from the scale devised by Linus Pauling

Acids and Bases: Central to Understanding Organic Chemistry

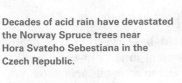

Decades of acid rain have devastated the Norway Spruce trees near Hora Svateho Sebestiana in the Czech Republic.

The chemistry you will learn in this chapter explains such things as the cause of acid rain and why it destroys monuments and plants, why exercise increases the rate of breathing, how Fosamax prevents bones from being nibbled away, and why blood has to be buffered and how that buffering is accomplished. Acids and bases play an important role in organic chemistry. What you learn about them in this chapter will be useful in almost every aspect of organic chemistry. The importance of organic acids and bases you learn is particularly important to understanding how and why organic compounds react.

It is hard to believe now, but at one time chemists characterized compounds by tasting them. Early chemists called any compound that tasted sour an acid (from *acidus,* Latin for "sour"). Some familiar acids are citric acid (found in lemons and other citrus fruits), acetic acid (found in vinegar), and hydrochloric acid (found in stomach acid—the sour taste associated with vomiting).

Compounds that neutralize acids, thereby destroying their acidic properties, were called bases, or alkaline compounds. Glass cleaners and solutions designed to unclog drains are familiar alkaline solutions.

1   AN INTRODUCTION TO ACIDS AND BASES

We will look at two definitions for the terms *acid* and *base,* the Brønsted–Lowry definitions and the Lewis definitions.

According to Brønsted and Lowry, an **acid** is a species that loses a proton, and a **base** is a species that gains a proton. (*Positively* charged hydrogen ions are called protons.) For example, in the reaction shown next, hydrogen chloride (HCl) is an acid because it loses a proton, and water is a base because it gains a proton. In the

From Chapter 2 of *Organic Chemistry*, Seventh Edition. Paula Yurkanis Bruice. Copyright © 2014 by Pearson Education, Inc. All rights reserved.

55

reverse reaction, H_3O^+ is an acid because it loses a proton, and Cl^- is a base because it gains a proton.

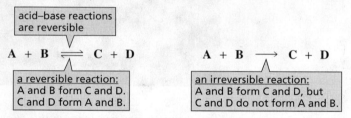

Water can accept a proton because it has two lone pairs, either of which can form a covalent bond with the proton, and Cl^- can accept a proton because any one of its lone pairs can form a covalent bond with a proton. Thus, according to the Brønsted–Lowry definitions:

Any species that has a hydrogen can potentially act as an acid.

Any species that has a lone pair can potentially act as a base.

The reaction of an acid with a base is called an **acid–base reaction** or a **proton transfer reaction.** Both an acid and a base must be present in an acid–base reaction, because an acid cannot lose a proton unless a base is present to accept it. *Acid–base reactions are reversible.* Two half-headed arrows are used to designate reversible reactions. In Section 5 we will see how we can determine whether reactants or products are favored when the reaction has reached equilibrium.

Acid–base reactions are reversible.

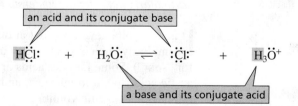

When an acid loses a proton, the resulting species without the proton is called the **conjugate base** of the acid. Thus, Cl^- is the conjugate base of HCl, and H_2O is the conjugate base of H_3O^+. When a base gains a proton, the resulting species with the proton is called the **conjugate acid** of the base. Thus, HCl is the conjugate acid of Cl^-, and H_3O^+ is the conjugate acid of H_2O.

A conjugate base is formed by removing a proton from an acid.

A conjugate acid is formed by adding a proton to a base.

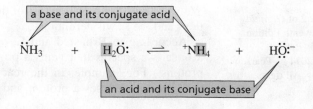

Another example of an acid–base reaction is the reaction between ammonia and water: ammonia (NH_3) is a base because it gains a proton, and water is an acid because it loses a proton. In the reverse reaction, ammonium ion ($^+NH_4$) is an acid because it loses a proton, and hydroxide ion (HO^-) is a base because it gains a proton. Thus, HO^- is the conjugate base of H_2O, $^+NH_4$ is the conjugate acid of NH_3, NH_3 is the conjugate base of $^+NH_4$, and H_2O is the conjugate acid of HO^-.

Notice that in the first of these two reactions water is a base, and in the second it is an acid. Water can behave as a base because it has a lone pair, and it can behave as an acid because it has a proton that it can lose. In Section 4, we will see how we can predict that water is a base in the first reaction and is an acid in the second reaction.

Acidity is a measure of the tendency of a compound to lose a proton, whereas **basicity** is a measure of a compound's affinity for a proton. A strong acid is one that has a strong tendency to lose a proton. This means that its conjugate base must be weak because it has little affinity for the proton. A weak acid has little tendency to lose its proton, indicating that its conjugate base is strong because it has a high affinity for the proton. Thus, the following important relationship exists between an acid and its conjugate base:

> A strong base has a high affinity for a proton.

> A weak base has a low affinity for a proton.

The stronger the acid, the weaker its conjugate base.

For example, HBr is a stronger acid than HCl, so Br^- is a weaker base than Cl^-.

PROBLEM 1♦

Which of the following are *not* acids?

CH_3COOH CO_2 HNO_2 $HCOOH$ CCl_4

PROBLEM 2♦

Draw the products of the acid–base reaction when

a. HCl is the acid and NH_3 is the base.   **b.** H_2O is the acid and $^-NH_2$ is the base.

PROBLEM 3♦

a. What is the conjugate acid of each of the following?
   1. NH_3 **2.** Cl^- **3.** HO^- **4.** H_2O

b. What is the conjugate base of each of the following?
   1. NH_3 **2.** HBr   **3.** HNO_3 **4.** H_2O

---

2 pK_a AND pH

When a strong acid such as hydrogen chloride is dissolved in water, almost all the molecules dissociate (break into ions), which means that the *products* are favored at equilibrium—the equilibrium lies to the right. When a much weaker acid, such as acetic acid, is dissolved in water, very few molecules dissociate, so the *reactants* are favored at equilibrium—the equilibrium lies to the left. A longer arrow is drawn toward the species favored at equilibrium.

$$H\ddot{C}l: \quad + \quad H_2\ddot{O}: \quad \rightleftharpoons \quad H_3\ddot{O}^+ \quad + \quad :\ddot{C}l:^-$$

hydrogen
chloride

$$\underset{\text{acetic acid}}{CH_3-\overset{\overset{\displaystyle :O:}{\|}}{C}-\ddot{O}H} \quad + \quad H_2\ddot{O}: \quad \rightleftharpoons \quad H_3\ddot{O}^+ \quad + \quad CH_3-\overset{\overset{\displaystyle :O:}{\|}}{C}-\ddot{O}:^-$$

The degree to which an acid (HA) dissociates in an aqueous solution is indicated by the **equilibrium constant** of the reaction, K_{eq}. Brackets are used to indicate the concentrations of the reactants and products (in moles/liter).

$$HA \quad + \quad H_2O \quad \rightleftharpoons \quad H_3O^+ \quad + \quad A^-$$

$$K_{eq} = \frac{[H_3O^+][A^-]}{[H_2O][HA]}$$

The degree to which an acid (HA) dissociates is normally determined in a dilute solution, so the concentration of water remains essentially constant. Combining the two constants

(K_{eq} and H_2O) allows the equilibrium expression to be rewritten using a new equilibrium constant, K_a, called the **acid dissociation constant.**

$$K_a = \frac{[H_3O^+][A^-]}{[HA]} = K_{eq}[H_2O]$$

The acid dissociation constant is the equilibrium constant multiplied by the molar concentration of water (55.5 M).

The stronger the acid, the more readily it loses a proton.

The larger the acid dissociation constant, the stronger is the acid—that is, the greater is its tendency to lose a proton. Hydrogen chloride, with an acid dissociation constant of 10^7, is a stronger acid than acetic acid, with an acid dissociation constant of 1.74×10^{-5}. For convenience, the strength of an acid is generally indicated by its **pK_a** value rather than its K_a value, where

$$pK_a = -\log K_a$$

The stronger the acid, the smaller its pK_a value.

The pK_a of hydrogen chloride is -7 and the pK_a of acetic acid, a much weaker acid, is 4.76. Notice that the stronger the acid, the smaller its pK_a value.

very strong acids	$pK_a < 1$
moderately strong acids	$pK_a = 1-3$
weak acids	$pK_a = 3-5$
very weak acids	$pK_a = 5-15$
extremely weak acids	$pK_a > 15$

Unless otherwise stated, the pK_a values given in this text indicate the strength of the acid *in water*.

The concentration of protons in a solution is indicated by **pH**. This concentration can be written as either $[H^+]$ or, because a proton in water is solvated, as $[H_3O^+]$.

$$pH = -\log[H^+]$$

The pH values of some commonly encountered solutions are shown in the margin. Because pH values decrease as the acidity of the solution increases, we see that lemon juice is more acidic than coffee, and rain is more acidic than milk. Solutions with pH values less than 7 are acidic, whereas those with pH values greater than 7 are basic. The pH of a solution can be changed simply by adding acid or base to the solution.

Do not confuse pH and pK_a. The pH scale is used to describe the acidity of a *solution*, whereas the pK_a indicates the tendency of a compound to lose its proton. Thus, the pK_a is characteristic of a particular compound, much like a melting point or a boiling point.

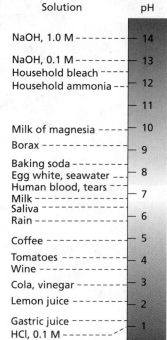

Solution	pH
NaOH, 1.0 M	14
NaOH, 0.1 M	13
Household bleach	
Household ammonia	12
	11
Milk of magnesia	10
Borax	9
Baking soda	8
Egg white, seawater	
Human blood, tears	7
Milk	
Saliva	6
Rain	
Coffee	5
Tomatoes	4
Wine	
Cola, vinegar	3
Lemon juice	2
Gastric juice	1
HCl, 0.1 M	
HCl, 1.0 M	0

PROBLEM 4♦

a. Which is a stronger acid, one with a pK_a of 5.2 or one with a pK_a of 5.8?

b. Which is a stronger acid, one with an acid dissociation constant of 3.4×10^{-3} or one with an acid dissociation constant of 2.1×10^{-4}?

PROBLEM 5♦

An acid has a K_a of 4.53×10^{-6} in water. What is its K_{eq} for reaction with water in a dilute solution? ($[H_2O] = 55.5$ M)

PROBLEM-SOLVING STRATEGY

Determining K_a from pK_a

Vitamin C has a pK_a value of 4.17. What is its K_a value?

You will need a calculator to answer this question. Remember that $pK_a = -\log K_a$.

Press the key labeled 10^x; then enter the negative value of the pK_a and press =.

You should find that vitamin C has a K_a value of 6.8×10^{-5}.

Now use the strategy you have just learned to solve Problem 6.

Acid Rain

Rain is mildly acidic (pH = 5.5) because water reacts with the CO_2 in the air to form carbonic acid (a weak acid with a pK_a value of 6.4).

$$CO_2 + H_2O \rightleftharpoons H_2CO_3$$
carbonic acid

In some parts of the world, rain has been found to be much more acidic (pH values as low as 4.3). This so-called acid rain is formed where sulfur dioxide and nitrogen oxides are produced, because water reacts with these gases to form strong acids—sulfuric acid ($pK_a = -5.0$) and nitric acid ($pK_a = -1.3$). Burning fossil fuels for the generation of electric power is the factor most responsible for forming these acid-producing gases.

Acid rain has many deleterious effects. It can destroy aquatic life in lakes and streams; it can make soil so acidic that crops cannot grow and forests can be destroyed; and it can cause the deterioration of paint and building materials, including monuments and statues that are part of our cultural heritage. Marble—a form of calcium

photo taken in 1935   photo taken in 2012

Statue of George Washington in Washington Square Park, in Greenwich Village, New York.

carbonate—decays because protons react with CO_3^{2-} to form carbonic acid, which decomposes to CO_2 and H_2O (the reverse of the reaction shown above on the left).

$$CO_3^{2-} \overset{H^+}{\rightleftharpoons} HCO_3^- \overset{H^+}{\rightleftharpoons} H_2CO_3 \rightleftharpoons CO_2 + H_2O$$

PROBLEM 6♦

Butyric acid, the compound responsible for the unpleasant odor and taste of sour milk, has a pK_a value of 4.82. What is its K_a value? Is it a stronger acid or a weaker acid than vitamin C?

PROBLEM 7

Antacids are compounds that neutralize stomach acid. Write the equations that show how Milk of Magnesia, Alka-Seltzer, and Tums remove excess acid.

a. Milk of Magnesia: $Mg(OH)_2$
b. Alka-Seltzer: $KHCO_3$ and $NaHCO_3$
c. Tums: $CaCO_3$

PROBLEM 8♦

Are the following body fluids acidic or basic?

a. bile (pH = 8.4)   **b.** urine (pH = 5.9)   **c.** spinal fluid (pH = 7.4)

---

3 ORGANIC ACIDS AND BASES

The most common organic acids are carboxylic acids—compounds that have a COOH group. Acetic acid and formic acid are examples of carboxylic acids. Carboxylic acids have pK_a values ranging from about 3 to 5, so they are weak acids. The pK_a values of a wide variety of organic compounds are listed in the table titled *"pK_a Values"* in the appendix at the end of this text.

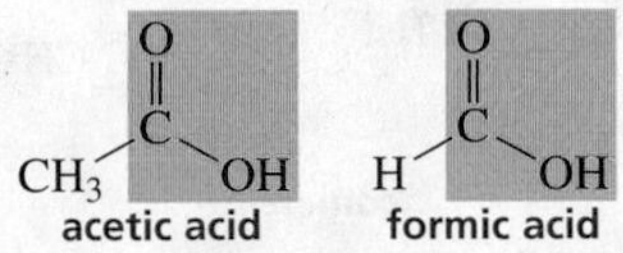

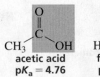

acetic acid
$pK_a = 4.76$

formic acid
$pK_a = 3.75$

Alcohols—compounds that have an OH group—are much weaker acids than carboxylic acids, with pK_a values close to 16. Methyl alcohol and ethyl alcohol are examples of alcohols. We will see why carboxylic acids are stronger acids than alcohols in Section 8.

$$CH_3OH \qquad CH_3CH_2OH$$
methyl alcohol   ethyl alcohol
$pK_a = 15.5$ $pK_a = 15.9$

Amines are compounds that result from replacing one or more of the hydrogens bonded to ammonia with a carbon-containing subsitutent. Amines and ammonia have such high pK_a values that they rarely behave as acids—they are much more likely to act as bases. In fact, they are the most common organic bases. We will see why alcohols are stronger acids than amines in Section 6.

$$CH_3NH_2 \qquad NH_3$$
methylamine   ammonia
$pK_a = 40$ $pK_a = 36$

We can assess the strength of a base by considering the strength of its conjugate acid—remembering that *the stronger the acid, the weaker its conjugate base*. For example, based on their pK_a values, protonated methylamine (10.7) is a stronger acid than protonated ethylamine (11.0), which means that methylamine is a weaker base than ethylamine. (A protonated compound is a compound that has gained an additional proton.) Notice that the pK_a values of protonated amines are about 11.

$$CH_3\overset{+}{N}H_3 \qquad CH_3CH_2\overset{+}{N}H_3$$
protonated methylamine   protonated ethylamine
$pK_a = 10.7$ $pK_a = 11.0$

Protonated alcohols and protonated carboxylic acids are very strong acids. For example, protonated methyl alcohol has a pK_a of -2.5, protonated ethyl alcohol has a pK_a of -2.4, and protonated acetic acid has a pK_a of -6.1.

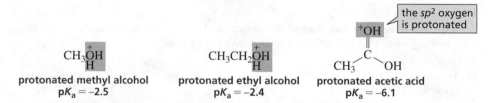

the sp^2 oxygen is protonated

$$CH_3\overset{+}{O}H \qquad\qquad CH_3CH_2\overset{+}{O}H$$
$$H \qquad\qquad\qquadH$$

protonated methyl alcohol   protonated ethyl alcohol   protonated acetic acid
$pK_a = -2.5$ $pK_a = -2.4$ $pK_a = -6.1$

Notice that it is the sp^2 oxygen of the carboxylic acid that is protonated (meaning that it acquires the proton).

Poisonous Amines

hemlock

Exposure to poisonous plants is responsible for an average of 63,000 calls each year to poison control centers. Hemlock is an example of a plant known for its toxicity. It contains eight different poisonous amines—the most abundant primary one is coniine, a neurotoxin that disrupts the central nervous system. Ingesting even a small amount can be fatal because it causes respiratory paralysis, which results in oxygen deprivation to the brain and heart. A poisoned person can recover if artificial respiration is applied until the drug can be flushed from the system. A drink made of hemlock was used to put Socrates to death in 399 BC; he was condemned for failing to acknowledge the gods that natives of the city of Athens worshipped.

coniine

We saw in Section 1 that water can behave both as an acid and as a base. An alcohol, too, can behave as an acid and lose a proton, or it can behave as a base and gain a proton.

the lone-pair electrons form a
new bond between O and H

new bond

the bond breaks
and the bonding
electrons end up
on O

$CH_3\ddot{O}-H$ + $H-\ddot{O}:$ ⇌ $CH_3\ddot{O}:$ + $H-\ddot{O}-H$

an acid

A curved arrow points
from the electron donor
to the electron acceptor.

$CH_3\ddot{O}-H$ + $H-\overset{+}{\ddot{O}}-H$ ⇌ $CH_3\overset{+}{\ddot{O}}-H$ + $H-\ddot{O}-H$

a base

H

H   new bond

bond breaks

Chemists frequently use curved arrows to indicate the bonds that are broken and formed as reactants are converted into products. They are called *curved arrows* to distinguish them from the *straight* arrows used to link reactants with products in the equation for a chemical reaction. Each curved arrow with a two-barbed arrowhead signifies the movement of two electrons. The arrow always points *from* the electron donor *to* the electron acceptor.

In an acid–base reaction, one of the arrows is drawn *from* a lone pair on the base *to* the proton of the acid. A second arrow is drawn *from* the electrons that the proton shared *to* the atom on which they are left behind. As a result, the curved arrows let you follow the electrons to see what bond is broken and what bond is formed in the reaction.

A carboxylic acid also can behave as an acid (lose a proton) or as a base (gain a proton).

$\ddot{O}:$

$CH_3-C-\ddot{O}-H$ + $H-\ddot{O}:$ ⇌ $CH_3-C-\ddot{O}:$ + $H-\ddot{O}-H$

an acid   bond breaks

new bond

new bond

$\ddot{O}:$

$CH_3-C-\ddot{O}-H$ + $H-\overset{+}{\ddot{O}}-H$ ⇌ $CH_3-C-\overset{+}{\ddot{O}}-H$ + $H-\ddot{O}-H$

a base   bond breaks   H

new bond

Similarily, an amine can behave as an acid (lose a proton) or as a base (gain a proton).

new bond

$CH_3\ddot{N}H$ + $H-\ddot{O}:$ ⇌ $CH_3\overset{-}{\ddot{N}}H$ + $H-\ddot{O}-H$

bond
breaks

H

an acid

H   new bond

$CH_3\ddot{N}H$ + $H-\overset{+}{\ddot{O}}-H$ ⇌ $CH_3\overset{+}{\ddot{N}}H$ + $H-\ddot{O}-H$

H   H

a base   bond breaks   H

It is important to know the approximate pK_a values of the various classes of compounds we have looked at. An easy way to remember them is in units of five, as shown in Table 1. (R is used when the particular carboxylic acid, alcohol, or amine is not specified.) Protonated alcohols, protonated carboxylic acids, and protonated water have pK_a values less than 0, carboxylic acids have pK_a values of ~5, protonated amines have pK_a values of ~10, and alcohols and water have pK_a values of ~15.

Table 1	Approximate pK_a Values		
$pK_a < 0$	$pK_a \sim 5$	$pK_a \sim 10$	$pK_a \sim 15$
$R\overset{+}{O}H_2$ a protonated alcohol	R—C(=O)—OH a carboxylic acid	$R\overset{+}{N}H_3$ a protonated amine	ROH an alcohol
$\overset{+}{O}H$ R—C—OH a protonated carboxylic acid			H_2O water
H_3O^+ protonated water			

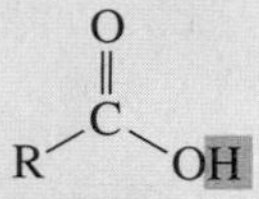

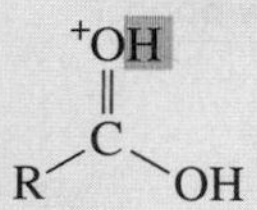

You need to remember these approximate pK_a values because they will be very important when you learn about the reactions of organic compounds.

PROBLEM 9♦

Draw the conjugate acid of each of the following:

a. CH_3CH_2OH **b.** $CH_3CH_2O^-$ **c.** 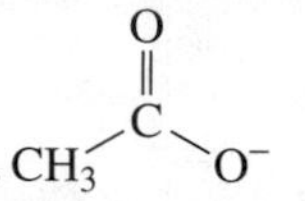 $CH_3\overset{\displaystyle O}{\underset{}{C}}O^-$ **d.** $CH_3CH_2NH_2$ **e.** 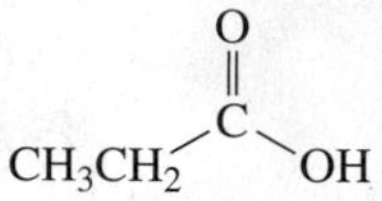 $CH_3CH_2\overset{\displaystyle O}{\underset{}{C}}OH$

PROBLEM 10

a. Write an equation showing CH_3OH reacting as an acid with NH_3 and an equation showing it reacting as a base with HCl.
b. Write an equation showing NH_3 reacting as an acid with CH_3O^- and an equation showing it reacting as a base with HBr.

PROBLEM 11♦

Estimate the pK_a values of the following compounds:

$CH_3CH_2CH_2NH_2$ $CH_3CH_2CH_2OH$ CH_3CH_2COOH $CH_3CH_2CH_2\overset{+}{N}H_3$

PROBLEM-SOLVING STRATEGY

Which atom of the following compound is more apt to be protonated?

$$HOCH_2CH_2CH_2NH_2$$

Solution One way to solve this problem is to look at the pK_a values of the conjugate acids of the groups, remembering that the weaker acid will have the stronger conjugate base. The stronger base will be the one more apt to be protonated.

$pK_a < 0$ $pK_a \sim 10$

$H\overset{+}{O}CH_2CH_2CH_2\overset{+}{N}H_3$ $HOCH_2CH_2CH_2NH_2 \xrightarrow{\text{HCl}} HOCH_2CH_2\overset{+}{N}H_3\ Cl^-$
      |
      H

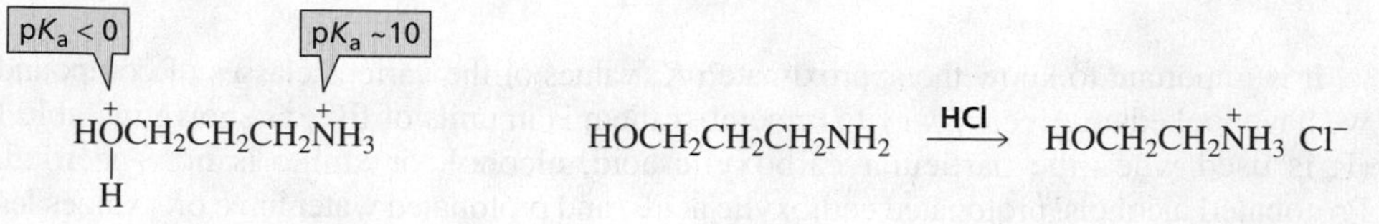

The conjugate acids have pK_a values of ~0 and ~10. Because the $^+NH_3$ group is the weaker acid, the NH_2 group is the stronger base, so it is the group more apt to be protonated.

Now use the skill you have just learned to solve Problem 12.

PROBLEM 12◆

a. Which is a stronger base, CH_3COO^- or $HCOO^-$? (The pK_a of CH_3COOH is 4.8; the pK_a of $HCOOH$ is 3.8.)

b. Which is a stronger base, HO^- or $^-NH_2$? (The pK_a of H_2O is 15.7; the pK_a of NH_3 is 36.)

c. Which is a stronger base, H_2O or CH_3OH? (The pK_a of H_3O^+ is -1.7; the pK_a of $CH_3\overset{+}{O}H_2$ is -2.5.)

PROBLEM 13◆

Using the pK_a values in Section 3, rank the following species in order from strongest base to weakest base:

$$CH_3NH_2 \quad CH_3\bar{N}H \quad CH_3OH \quad CH_3O^- \quad CH_3\overset{\displaystyle O}{\overset{\|}{C}}O^-$$

4   HOW TO PREDICT THE OUTCOME OF AN ACID–BASE REACTION

Now let's see how we can predict that water will behave as a base when it reacts with HCl (the first reaction in Section 1) but as an acid when it reacts with NH_3 (the second reaction in Section 1). To determine which of two reactants will be the acid, we need to compare their pK_a values.

For the reaction of H_2O with HCl: the reactants are water ($pK_a = 15.7$) and HCl ($pK_a = -7$). Because HCl is the stronger acid (it has the lower pK_a value), it will be the reactant that loses a proton. Therefore, HCl is the acid and water is the base in this reaction.

this is the acid				this is the acid
	reactants		reactants	
HCl	+ H_2O		NH_3 +	H_2O
$pK_a = -7$	$pK_a = 15.7$		$pK_a = 36$	$pK_a = 15.7$

For the reaction of H_2O with NH_3: the reactants are water ($pK_a = 15.7$) and NH_3 ($pK_a = 36$). Because water is the stronger acid (it has the lower pK_a value), it will be the reactant that loses a proton. Therefore, water is the acid and ammonia is the base in this reaction.

PROBLEM 14◆

Does methanol behave as an acid or a base when it reacts with methylamine?

5   HOW TO DETERMINE THE POSITION OF EQUILIBRIUM

To determine the position of equilibrium for an acid–base reaction (that is, to determine whether reactants or products are favored), we need to compare the pK_a value of the acid on the left of the equilibrium arrows with the pK_a value of the acid on the right of the arrows. The equilibrium favors *formation* of the weaker acid (the one with the higher pK_a value). In other words, the equilibrium lies toward the weaker acid.

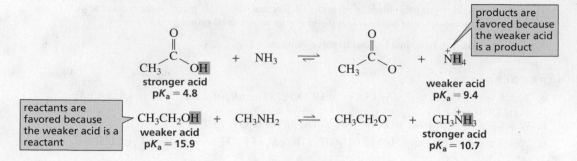

In an acid–base reaction, the equilibrium favors the formation of the weaker acid.

Since the equilibrium favors formation of the weaker acid, we can say that *an acid–base reaction will favor products if the conjugate acid of the base that gains the proton is a weaker acid than the acid that loses a proton in the first place.*

The precise value of the equilibrium constant can be calculated from the following equation:

$$pK_{eq} = pK_a(\text{reactant acid}) - pK_a(\text{product acid})$$

Thus, the equilibrium constant for the reaction of acetic acid with ammonia is 4.0×10^4.

$$pK_{eq} = 4.8 - 9.4 = -4.6$$
$$K_{eq} = 10^{4.6} = 4.0 \times 10^4$$

and the equilibrium constant for the reaction of ethyl alcohol with methylamine is 6.3×10^{-6}.

$$pK_{eq} = 15.9 - 10.7 = 5.2$$
$$K_{eq} = 10^{-5.2} = 6.3 \times 10^{-6}$$

PROBLEM 15

a. For each of the acid–base reactions in Section 3, compare the pK_a values of the acids on either side of the equilibrium arrows to prove that the equilibrium lies in the direction indicated. (The pK_a values you need can be found in Section 3 or in Problem 12.)
b. Do the same for the acid–base reactions in Section 1.

PROBLEM 16

Ethyne has a pK_a value of 25, water has a pK_a value of 15.7, and ammonia (NH_3) has a pK_a value of 36. Draw the equation, showing equilibrium arrows that indicate whether reactants or products are favored, for the acid–base reaction of ethyne with

a. HO^-.   **b.** $^-NH_2$.
c. Which would be a better base to use if you wanted to remove a proton from ethyne, HO^- or $^-NH_2$?

PROBLEM 17♦

Which of the following bases can remove a proton from acetic acid in a reaction that favors products?

$$HO^- \qquad CH_3NH_2 \qquad HC{\equiv}C^- \qquad CH_3OH \qquad H_2O \qquad Cl^-$$

PROBLEM 18♦

Calculate the equilibrium constant for the acid–base reaction between the reactants in each of the following pairs:

a. $HCl + H_2O$ **b.** $CH_3COOH + H_2O$ **c.** $CH_3NH_2 + H_2O$ **d.** $CH_3\overset{+}{N}H_3 + H_2O$

6 HOW THE STRUCTURE OF AN ACID AFFECTS ITS pKₐ VALUE

The strength of an acid is determined by the stability of the conjugate base that forms when the acid loses its proton: the more stable the base, the stronger its conjugate acid.

A stable base readily bears the electrons it formerly shared with a proton. In other words, stable bases are weak bases—they do not share their electrons well. Thus we can say either:

> *The weaker the base, the stronger its conjugate acid or the*
> *more stable the base, the stronger its conjugate acid.*

Stable bases are weak bases.

The more stable the base, the stronger its conjugate acid.

The weaker the base, the stronger its conjugate acid.

Electronegativity

Two factors that affect the stability of a base are its *electronegativity* and its *size*.

The atoms in the second row of the periodic table are all *similar in size*, but they have very *different electronegativities*, which increase across the row from left to right. Of the atoms shown, carbon is the least electronegative and fluorine is the most electronegative.

relative electronegativities: $C < N < O < F$

most electronegative

If we look at the acids formed by attaching hydrogens to these elements, we see that the most acidic compound is the one that has its hydrogen attached to the most electronegative atom. Thus, HF is the strongest acid and methane is the weakest acid.

> *When the atoms are similar in size, the strongest acid has its*
> *hydrogen attached to the most electronegative atom.*

relative acidities: $CH_4 < NH_3 < H_2O < HF$

strongest acid

If we look at the stabilities of the conjugate bases of these acids, we find that they too increase from left to right, because the more electronegative the atom, the better it can bear its negative charge. Thus, *the strongest acid has the most stable (weakest) conjugate base.*

relative stabilities: $^-CH_3 < \ ^-NH_2 < HO^- < F^-$

most stable

When atoms are similar in size, the strongest acid has its hydrogen attached to the most electronegative atom.

The effect that the electronegativity of the atom bonded to a hydrogen has on the compound's acidity can be appreciated when the pKₐ values of alcohols and amines are compared. Because oxygen is more electronegative than nitrogen, an alcohol is more acidic than an amine.

CH_3OH
methyl alcohol
pKₐ = 15.5

CH_3NH_2
methylamine
pKₐ = 40

Again, because oxygen is more electronegative than nitrogen, a protonated alcohol is more acidic than a protonated amine.

$CH_3\overset{+}{O}H_2$
protonated methyl alcohol
pKₐ = –2.5

$CH_3\overset{+}{N}H_3$
protonated methylamine
pKₐ = 10.7

PROBLEM 19◆

List the ions ($^-CH_3$, $^-NH_2$, HO^-, and F^-) in order from most basic to least basic.

Hybridization

The hybridization of an atom affects the acidity of a hydrogen bonded to it because hybridization affects electronegativity: an sp hybridized atom is more electronegative than the same atom that is sp^2 hybridized, which is more electronegative than the same atom that is sp^3 hybridized.

An sp carbon is more electronegative than an sp^2 carbon, which is more electronegative than an sp^3 carbon.

relative electronegativities

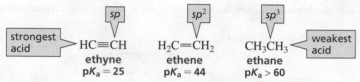

Because the electronegativity of carbon atoms follows the order $sp > sp^2 > sp^3$, ethyne is a stronger acid than ethene, and ethene is a stronger acid than ethane. Again, the most acidic compound is the one with its hydrogen attached to the most electronegative atom.

sp	sp^2	sp^3
strongest acid HC≡CH	$H_2C=CH_2$	CH_3CH_3 **weakest acid**
ethyne	**ethene**	**ethane**
$pK_a = 25$	$pK_a = 44$	$pK_a > 60$

Why does the hybridization of the atom affect its electronegativity? Electronegativity is a measure of the ability of an atom to pull the bonding electrons toward itself. Thus, the most electronegative atom will be the one with its bonding electrons closest to the nucleus. The average distance of a $2s$ electron from the nucleus is less than the average distance of a $2p$ electron from the nucleus. Therefore, an sp hybridized atom with 50% s character is the most electronegative, an sp^2 hybridized atom (33.3% s character) is next, and an sp^3 hybridized atom (25% s character) is the least electronegative.

Pulling the electrons closer to the nucleus stabilizes the carbanion. Once again we see that the stronger the acid, the more stable (the weaker) is its conjugate base. Notice that the electrostatic potential maps show that the strongest base (the least stable) is the most electron-rich (the most red).

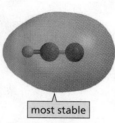

most stable

HC≡C$^-$

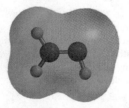

$H_2C=\overset{..}{C}H$

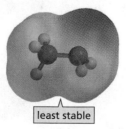

least stable

$CH_3\overset{..}{C}H_2$

PROBLEM 20◆

List the carbanions shown in the margin in order from most basic to least basic.

PROBLEM 21◆

Which is a stronger acid?

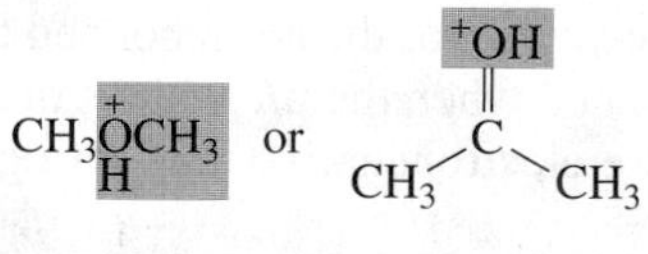

$CH_3\overset{+}{\underset{H}{O}}CH_3$ or $CH_3\overset{\overset{+OH}{\|}}{C}CH_3$

PROBLEM 22

a. Draw the products of the following reactions:

A HC≡CH + $CH_3\overset{..}{C}H_2$ ⇌

B $H_2C=CH_2$ + HC≡C$^-$ ⇌

C CH_3CH_3 + $H_2C=\overset{..}{C}H$ ⇌

b. Which of the reactions favor formation of the products?

Size

When comparing atoms that are very different in size, the *size* of the atom is more important than its *electronegativity* in determining how well it bears its negative charge. For example, as we proceed down a column in the periodic table, the atoms get larger and the *stability* of the anions *increases* even though the electronegativity of the atoms *decreases*. Because the stability of the bases increases going down the column, the strength of their conjugate acids *increases*. Thus, HI is the strongest acid of the hydrogen halides (that is, I^- is the weakest, most stable base), even though iodine is the least electronegative of the halogens.

When atoms are very different in size, the strongest acid will have its hydrogen attached to the largest atom.

relative size: $F^- < Cl^- < Br^- < I^-$

largest

relative acidities: $HF < HCl < HBr < HI$

strongest acid

Size overrides electronegativity when determining relative acidities.

When atoms are very different in size, the strongest acid will have its hydrogen attached to the largest atom.

Why does the size of an atom have such a significant effect on stability that it more than overcomes any difference in electronegativity? The valence electrons of F^- are in a $2sp^3$ orbital, the valence electrons of Cl^- are in a $3sp^3$ orbital, those of Br^- are in a $4sp^3$ orbital, and those of I^- are in a $5sp^3$ orbital. The volume of space occupied by a $3sp^3$ orbital is significantly larger than the volume of space occupied by a $2sp^3$ orbital because a $3sp^3$ orbital extends out farther from the nucleus. Because its negative charge is spread over a larger volume of space, Cl^- is more stable than F^-.

Thus, as a halide ion increases in size (going down the column of the periodic table), its stability increases because its negative charge is spread over a larger volume of space (its electron density decreases). As a result, HI is the strongest acid of the hydrogen halides because I^- is the most stable halide ion, even though iodine is the least electronegative of the halogens (see Table 2). The potential maps illustrate the large difference in size of the hydrogen halides.

Table 2 The pK_a Values of Some Simple Acids			
CH_4	NH_3	H_2O	HF
pK_a = 60	pK_a = 36	pK_a = 15.7	pK_a = 3.2
		H_2S	HCl
		pK_a = 7.0	pK_a = −7
			HBr
			pK_a = −9
			HI
			pK_a = −10

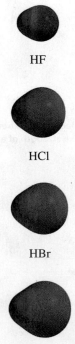

HF

HCl

HBr

HI

In summary, atomic size does not change much as we move from left to right across a row of the periodic table, so the atoms' orbitals have approximately the same volume. Thus, electronegativity determines the stability of the base and, therefore, the acidity of its conjugate acid. Atomic size increases as we move down a column of the periodic table, so the volume of the orbitals increases and, therefore, their electron density decreases. The electron density of an orbital is more important than electronegativity in determining the stability of a base and, therefore, the acidity of its conjugate acid.

CH₃O⁻

CH₃S⁻

PROBLEM 23◆

List the halide ions (F⁻, Cl⁻, Br⁻, and I⁻) in order from strongest base to weakest base.

PROBLEM 24◆

a. Which is more electronegative, oxygen or sulfur?
b. Which is a stronger acid, H_2O or H_2S?
c. Which is a stronger acid, CH_3OH or CH_3SH?

PROBLEM 25◆

Which is a stronger acid?

a. HCl or HBr

b. $CH_3CH_2CH_2\overset{+}{N}H_3$ or $CH_3CH_2CH_2\overset{+}{O}H_2$

c. 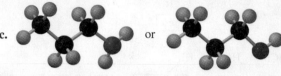 or

black = C, gray = H, blue = N, red = O

d. $CH_3CH_2CH_2OH$ or $CH_3CH_2CH_2SH$

PROBLEM 26◆

a. Which of the halide ions (F⁻, Cl⁻, Br⁻, and I⁻) is the most stable base?
b. Which is the least stable base?

PROBLEM 27◆

Which is a stronger base?

a. H_2O or HO^- **b.** H_2O or NH_3 **c.** $CH_3\overset{O}{\overset{\|}{C}}O^-$ or CH_3O^- **d.** CH_3O^- or CH_3S^-

---

7 HOW SUBSTITUENTS AFFECT THE STRENGTH OF AN ACID

Although the acidic proton of each of the following carboxylic acids is attached to the same atom (an oxygen), the four compounds have different pK_a values:

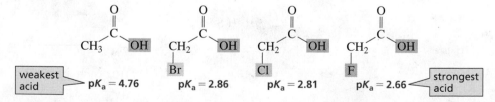

Inductive electron withdrawal increases the strength of an acid.

The different pK_a values indicate that there must be another factor that affects acidity other than the nature of the atom to which the hydrogen is bonded.

From the pK_a values of the four carboxylic acids, we see that replacing one of the hydrogens of the CH_3 group with a halogen increases the acidity of the compound. (The term for replacing an atom in a compound is *substitution*, and the new atom is called a *substituent*.) The halogen is more electronegative than the hydrogen it has replaced, so the halogen pulls the bonding electrons toward itself more than a hydrogen would. Pulling electrons through sigma (σ) bonds is called **inductive electron withdrawal.**

If we look at the conjugate base of a carboxylic acid, we see that inductive electron withdrawal *decreases the electron density* about the oxygen that bears the negative charge, thereby stabilizing it. And we know that stabilizing a base increases the acidity of its conjugate acid.

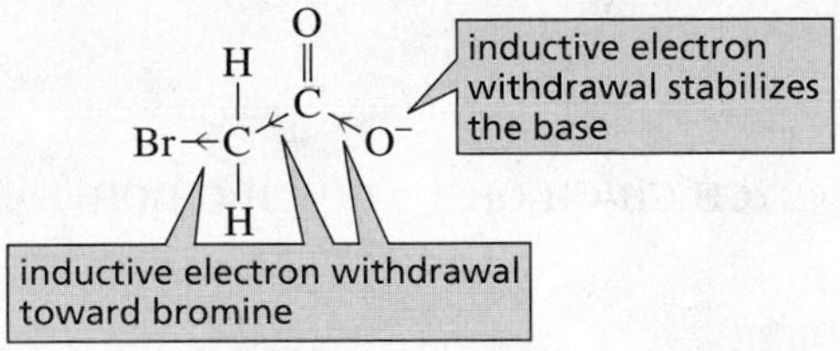

The pK_a values of the four carboxylic acids shown on the previous page decrease (become more acidic) as the electron-withdrawing ability (electronegativity) of the halogen increases. Thus, the fluoro-substituted compound is the strongest acid because its conjugate base is the most stabilized (is the weakest).

The effect a substituent has on the acidity of a compound decreases as the distance between the substituent and the acidic proton increases.

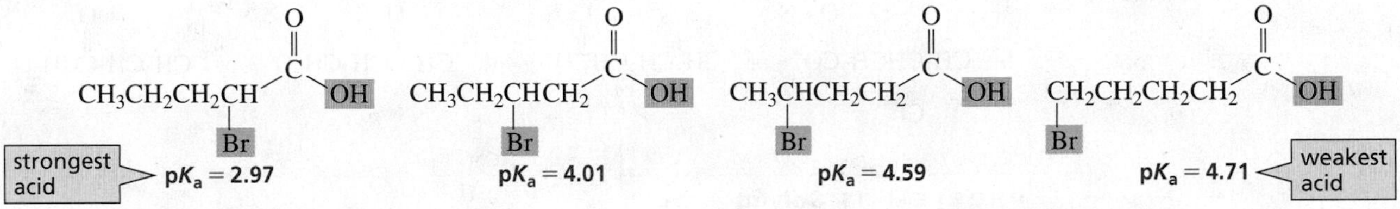

PROBLEM-SOLVING STRATEGY

Determining Relative Acid Strength from Structure

a. Which is a stronger acid?

$$CH_3CHCH_2OH \quad \text{or} \quad CH_3CHCH_2OH$$
$$\quad\;\; | \qquad\qquad\qquad\qquad | $$
$$\quad\;\; F \qquad\qquad\qquad\quad\; Br$$

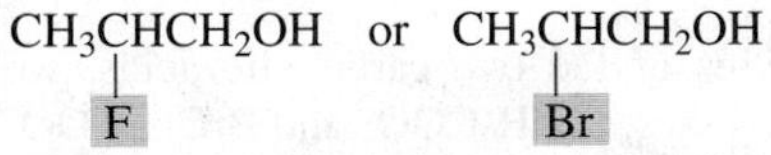

When you are asked to compare two items, pay attention to where they differ, ignore where they are the same. These two compounds differ only in the halogen that is attached to the middle carbon. Because fluorine is more electronegative than bromine, there is greater inductive electron withdrawal from oxygen in the fluorinated compound. The fluorinated compound, therefore, will have the more stable conjugate base, so it will be the stronger acid.

b. Which is a stronger acid?

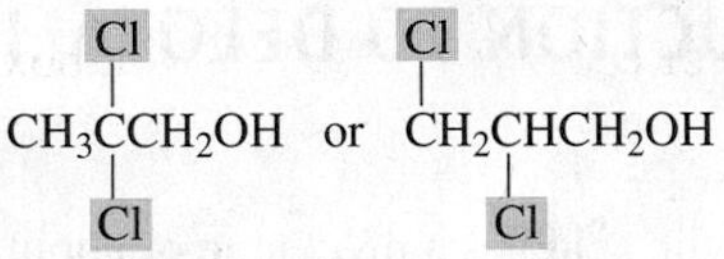

$$\qquad\qquad\; Cl \qquad\quad\; Cl$$
$$\qquad\qquad\; | \qquad\qquad | $$
$$CH_3CCH_2OH \quad \text{or} \quad CH_2CHCH_2OH$$
$$\qquad\qquad\; | \qquad\qquad\qquad\; | $$
$$\qquad\qquad\; Cl \qquad\qquad\qquad Cl$$

These two compounds differ in the location of one of the chlorines. Because the second chlorine in the compound on the left is closer to the O—H bond than is the second chlorine in the compound on the right, the compound on the left is more effective at withdrawing electrons from the oxygen. Thus, the compound on the left will have the more stable conjugate base, so it will be the stronger acid.

Now use the strategy you have just learned to solve Problem 28.

PROBLEM 28♦

Which is a stronger acid?

a. $CH_3OCH_2CH_2OH$ or $CH_3CH_2CH_2CH_2OH$

b. $CH_3CH_2CH_2\overset{+}{N}H_3$ or $CH_3CH_2CH_2\overset{+}{O}H_2$

c. $CH_3OCH_2CH_2CH_2OH$ or $CH_3CH_2OCH_2CH_2OH$

d. $CH_3\overset{O}{\overset{\|}{C}}CH_2OH$ or $CH_3CH_2\overset{O}{\overset{\|}{C}}OH$

PROBLEM 29♦

List the following compounds in order from strongest acid to weakest acid:

$$CH_3CHCH_2OH \qquad CH_3CH_2CH_2OH \qquad CH_2CH_2CH_2OH \qquad CH_3CHCH_2OH$$
$$\;\;\;\;\;|\qquad\qquad\qquad\qquad\qquad\qquad\qquad\qquad\;\;\;|\qquad\qquad\qquad\qquad\;\;\;|$$
$$\;\;\;\;\;F\qquad\qquad\qquad\qquad\qquad\qquad\qquad\qquad\;\;\;Cl\qquad\qquad\qquad\qquad\;\;\;Cl$$

PROBLEM 30♦

Which is a stronger base?

a. $CH_3\overset{\overset{\textstyle O}{\|}}{C}HCO^-$ or $CH_3\overset{\overset{\textstyle O}{\|}}{C}HCO^-$
$\quad\;\;\;|\qquad\qquad\qquad\qquad\;\;|$
$\quad\;\;\;Br\qquad\qquad\qquad\qquad\;F$

c. $BrCH_2CH_2\overset{\overset{\textstyle O}{\|}}{C}O^-$ or $CH_3CH_2\overset{\overset{\textstyle O}{\|}}{C}O^-$

b. $CH_3CHCH_2\overset{\overset{\textstyle O}{\|}}{C}O^-$ or $CH_3CH_2CH\overset{\overset{\textstyle O}{\|}}{C}O^-$
$\quad\;\;\;\;\;\;\;|\qquad\qquad\qquad\qquad\qquad\;\;|$
$\quad\;\;\;\;\;\;\;Cl\qquad\qquad\qquad\qquad\qquad\;Cl$

d. $CH_3\overset{\overset{\textstyle O}{\|}}{C}CH_2CH_2O^-$ or $CH_3CH_2\overset{\overset{\textstyle O}{\|}}{C}CH_2O^-$

PROBLEM 31 Solved

If HCl is a weaker acid than HBr, why is $ClCH_2COOH$ a stronger acid than $BrCH_2COOH$?

Solution To compare the acidities of HCl and HBr, we need to compare the stabilities of their conjugate bases, Cl^- and Br^-. (Notice that an H—Cl bond breaks in one compound and an H—Br bond breaks in the other.) Because we know that size is more important than electronegativity in determining stability, we know that Br^- is more stable than Cl^-. Therefore, HBr is a stronger acid than HCl.

In comparing the acidities of the two carboxylic acids, we again need to compare the stabilities of their conjugate bases, $ClCH_2COO^-$ and $BrCH_2COO^-$. (Notice that an O—H bond breaks in both compounds.) The only way the conjugate bases differ is in the electronegativity of the atom that is drawing electrons away from the negatively charged oxygen. Because Cl is more electronegative than Br, Cl exerts greater inductive electron withdrawal. Thus, it has a greater stabilizing effect on the base that is formed when the proton leaves, so the chloro-substituted compound is the stronger acid.

8 AN INTRODUCTION TO DELOCALIZED ELECTRONS

We have seen that a carboxylic acid has a pK_a value of about 5, whereas the pK_a value of an alcohol is about 15. Because a carboxylic acid is a much stronger acid than an alcohol, we know that the conjugate base of a carboxylic acid is considerably more stable than the conjugate base of an alcohol.

$$CH_3CH_2O{-}H \qquad\qquad CH_3{-}\overset{\overset{\textstyle O}{\|}}{C}{-}O{-}H \quad \boxed{\text{stronger acid}}$$
$$pK_a = 15.9 \qquad\qquad\qquad pK_a = 4.76$$

Two factors cause the conjugate base of a carboxylic acid to be more stable than the conjugate base of an alcohol. First, the conjugate base of a carboxylic acid has a doubly bonded oxygen where the conjugate base of an alcohol has two hydrogens. Inductive electron withdrawal by this electronegative oxygen decreases the electron density of the negatively charged oxygen, thereby stabilizing it and increasing the acidity of the conjugate acid.

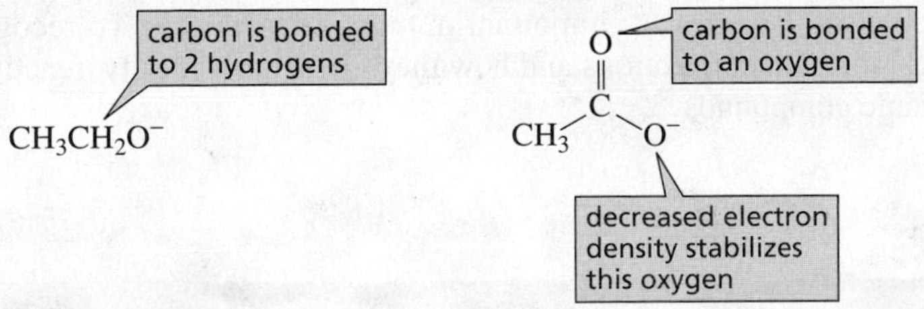

carbon is bonded
to 2 hydrogens

carbon is bonded
to an oxygen

$CH_3CH_2O^-$

O

CH_3 C O^-

decreased electron
density stabilizes
this oxygen

The second factor that causes the conjugate base of the carboxylic acid to be more stable than the conjugate base of the alcohol is *electron delocalization*. When an alcohol loses a proton, the negative charge resides on its single oxygen atom. These electrons are said to be *localized* because they belong to only *one* atom. In contrast, when a carboxylic acid loses a proton, the negative charge is shared by both oxygens. These electrons are *delocalized*.

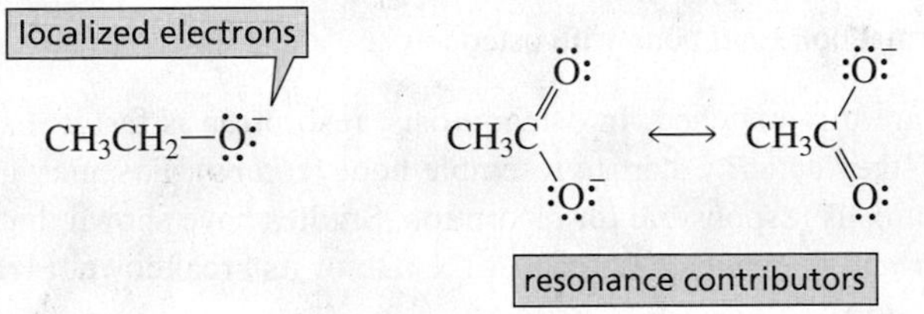

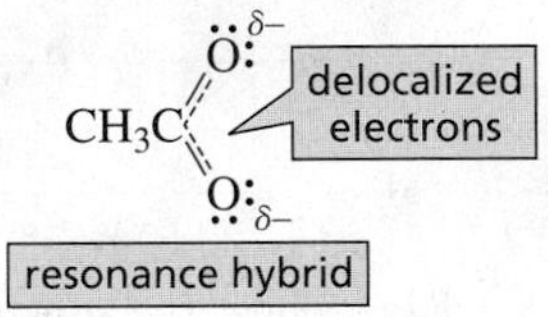

localized electrons

$CH_3CH_2-\overset{\cdot\cdot}{\underset{\cdot\cdot}{O}}:^-$

CH_3C $\overset{\overset{\cdot\cdot}{O}:}{\underset{:\overset{\cdot\cdot}{O}:^-}{}}$ ⟷ CH_3C $\overset{:\overset{\cdot\cdot}{O}:^-}{\underset{\overset{\cdot\cdot}{O}:}{}}$

resonance contributors

CH_3C $\overset{\overset{\cdot\cdot}{O}:^{\delta-}}{\underset{\overset{\cdot\cdot}{O}:_{\delta-}}{}}$

delocalized
electrons

resonance hybrid

The two structures shown for the conjugate base of the carboxylic acid are called **resonance contributors.** Neither resonance contributor alone represents the actual structure of the conjugate base. Instead, the actual structure—called a **resonance hybrid**—is a composite of the two resonance contributors. The double-headed arrow between the two resonance contributors is used to indicate that the actual structure is a hybrid.

the negative charge is
localized in one oxygen

the negative charge is
shared by two oxygens

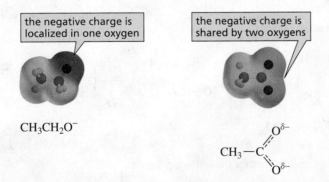

$CH_3CH_2O^-$

CH_3-C $\overset{O^{\delta-}}{\underset{O^{\delta-}}{}}$

**Delocalized electrons are
shared by more than two atoms.**

Notice that the two resonance contributors differ only in the location of their π electrons and lone-pair electrons—all the atoms stay in the same place. In the resonance hybrid, an electron pair is spread over *two oxygens and a carbon*. The negative charge is shared equally by the two oxygens, and both carbon–oxygen bonds are the same length—they are not as long as a single bond, but they are longer than a double bond. A resonance hybrid can be drawn by using dotted lines to show the delocalized electrons.

Thus, the combination of inductive electron withdrawal and the ability of two atoms to share the negative charge makes the conjugate base of the carboxylic acid more stable than the conjugate base of the alcohol.

Delocalized electrons are very important in organic chemistry. To recognize when a compound has delocalized electrons and how they affect the stability, reactivity, and pK_a values of organic compounds.

Fosamax Prevents Bones from Being Nibbled Away

Fosamax is used to treat osteoporosis, a condition characterized by decreased bone density. Under normal conditions, the rate of bone forma-

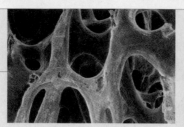
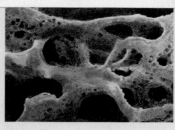

photo of normal bone and bone with osteoporosis

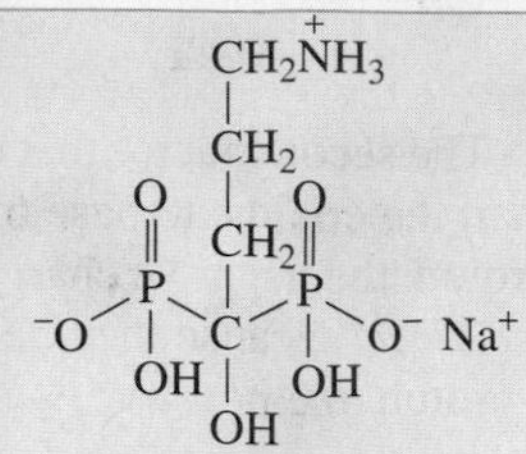

sodium alendronate
(generic form of Fosamax)

tion and the rate of bone resorption (breakdown) are carefully matched. In osteoporosis, resorption is faster than formation, so bone is nibbled away, causing bones to become fragile (they actually start to resemble honeycombs). Fosamax goes specifically to the sites of bone resorption and inhibits the activity of cells responsible for resorption. Studies have shown that normal bone is then formed on top of Fosamax, and the rate of bone formation becomes faster that the rate of its breakdown. (Trade name labels in this text are green.)

PROBLEM 32

Fosamax has six acidic groups. The structure of the active form of the drug is shown in the box. (Notice that the phosphorus atom in Fosamax and the sulfur atom in Problem 33 can be surrounded by more than eight electrons since P and S are below the second row of the periodic table.)

a. The OH groups bonded to phosphorus are the strongest acids of the six groups. Why?
b. Which of the remaining four groups is the weakest acid?

PROBLEM 33♦

Which is a stronger acid? Why?

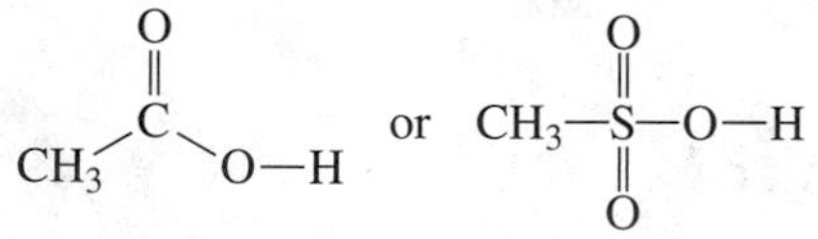

$$CH_3-\overset{\displaystyle O}{\overset{\|}{C}}-O-H \quad \text{or} \quad CH_3-\overset{\displaystyle O}{\underset{\displaystyle O}{\overset{\|}{\underset{\|}{S}}}}-O-H$$

PROBLEM 34♦

Draw resonance contributors for the following species:

a.

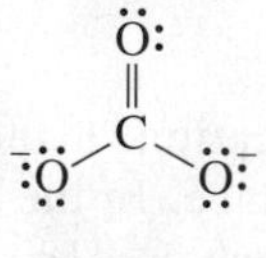

b.

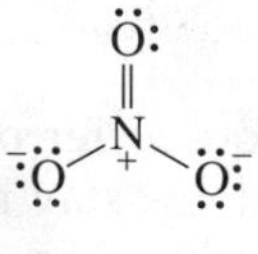

9 A SUMMARY OF THE FACTORS THAT DETERMINE ACID STRENGTH

We have seen that the strength of an acid depends on five factors: the *size* of the atom to which the hydrogen is attached, the *electronegativity* of the atom to which the hydrogen is attached, the *hybridization* of the atom to which the hydrogen is attached, *inductive electron withdrawal*, and *electron delocalization*. All five factors affect acidity by affecting the stability of the conjugate base.

1. **Size:** As the atom attached to the hydrogen increases in size (going down a column of the periodic table), the strength of the acid increases.

2. **Electronegativity:** As the atom attached to the hydrogen increases in electronegativity (going from left to right across a row of the periodic table), the strength of the acid increases.

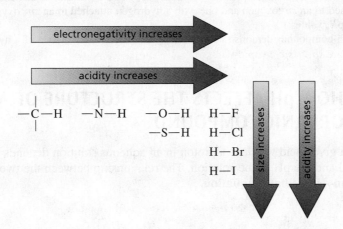

3. **Hybridization:** The relative electronegativities of an atom change with hybridization as follows: are $sp > sp^2 > sp^3$. Because an sp carbon is the most electronegative, a hydrogen attached to an sp carbon is the most acidic, and a hydrogen attached to an sp^3 carbon is the least acidic.

$$\boxed{\text{strongest acid}} \quad HC \equiv CH \; > \; H_2C = CH_2 \; > \; CH_3CH_3 \quad \boxed{\text{weakest acid}}$$

$$\underset{sp}{} \qquad \underset{sp^2}{} \qquad \underset{sp^3}{}$$

4. **Inductive electron withdrawal:** An electron-withdrawing group increases the strength of an acid. As the electronegativity of the electron-withdrawing group increases and moves closer to the acidic hydrogen, the strength of the acid increases.

$$\boxed{\text{strongest acid}} \; \underset{F}{CH_3CHCH_2OH} > \underset{Cl}{CH_3CHCH_2OH} > \underset{Br}{CH_3CHCH_2OH} > CH_3CH_2CH_2OH \; \boxed{\text{weakest acid}}$$

$$\boxed{\text{strongest acid}} \; \underset{F}{CH_3CHCH_2OH} > \underset{F}{CH_2CH_2CH_2OH} > CH_3CH_2CH_2OH \; \boxed{\text{weakest acid}}$$

5. **Electron delocalization:** An acid whose conjugate base has delocalized electrons is more acidic than a similar acid whose conjugate base has only localized electrons.

PROBLEM 35♦

Using the table of pK_a values given in the appendix at the end of this text, answer the following:

a. Which is the most acidic organic compound in the table?
b. Which is the least acidic organic compound in the table?
c. Which is the most acidic carboxylic acid in the table?
d. Which is more electronegative, an sp^3 oxygen or an sp^2 oxygen? (*Hint:* Pick a compound in given in the table titled "pKa values" in the appendix at the end of this text with a hydrogen attached to an sp^2 oxygen and one with a hydrogen attached to an sp^3 oxygen, and compare their pK_a values.)
e. Which compounds demonstrate that the relative electronegativities of a hybridized nitrogen atom are $sp > sp^2 > sp^3$?

10 HOW pH AFFECTS THE STRUCTURE OF AN ORGANIC COMPOUND

Whether a given acid will lose a proton in an aqueous solution depends on both the pK_a of the acid and the pH of the solution. The relationship between the two is given by the **Henderson–Hasselbalch equation.**

the Henderson–Hasselbalch equation

$$pK_a = pH + \log\frac{[HA]}{[A^-]}$$

This is an extremely useful equation because it tells us whether a compound will exist in its acidic form (with its proton retained) or in its basic form (with its proton removed) at a particular pH. Knowing this will be important when we are assessing the reactivity of organic compounds.

The Henderson–Hasselbalch equation tells us that

- when the pH of a solution equals the pK_a of the compound that undergoes dissociation, the concentration of the compound in its acidic form (HA) will equal the concentration of the compound in its basic form (A^-) (because log 1=0).
- if the pH of the solution is less than the pK_a of the compound, the compound will exist primarily in its acidic form.
- if the pH of the solution is greater than the pK_a of the compound, the compound will exist primarily in its basic form.

A compound will exist primarily in its acidic form (HA) if the pH of the solution is less than the compound's pK_a value.

In other words, *compounds exist primarily in their acidic forms in solutions that are more acidic than their* pK_a *values and primarily in their basic forms in solutions that are more basic than their* pK_a *values.*

A compound will exist primarily in its basic form (A^-) if the pH of the solution is greater than the compound's pK_a value.

PROBLEM-SOLVING STRATEGY

Determining the Structure at a Particular pH

Write the form in which the following compounds will predominate in a solution at pH 5.5:

a. CH_3CH_2OH (pK_a = 15.9) **b.** $CH_3CH_2\overset{+}{O}H_2$ (pK_a = −2.5) **c.** $CH_3\overset{+}{N}H_3$ (pK_a = 11.0)

To answer this kind of question, we need to compare the pH of the solution with the pK_a value of the compound's dissociable proton.

a. The pH of the solution is more acidic (5.5) than the pK_a value of the compound (15.9). Therefore, the compound will exist primarily as CH_3CH_2OH (with its proton).
b. The pH of the solution is more basic (5.5) than the pK_a value of the compound (−2.5). Therefore, the compound will exist primarily as CH_3CH_2OH (without its proton).
c. The pH of the solution is more acidic (5.5) than the pK_a value of the compound (11.0). Therefore, the compound will exist primarily as $CH_3\overset{+}{N}H_3$ (with its proton).

Now use the skill you have just learned to solve Problem 36.

PROBLEM 36♦

For each of the following compounds (shown in their acidic forms) write the form that will predominate in a solution at pH = 5.5:

a. CH_3COOH ($pK_a = 4.76$)

b. $CH_3CH_2\overset{+}{N}H_3$ ($pK_a = 11.0$)

c. H_3O^+ ($pK_a = -1.7$)

d. HBr ($pK_a = -9$)

e. $^+NH_4$ ($pK_a = 9.4$)

f. $HC\equiv N$ ($pK_a = 9.1$)

g. HNO_2 ($pK_a = 3.4$)

h. HNO_3 ($pK_a = -1.3$)

PROBLEM 37♦

As long as the pH is not less than _____, at least 50% of a protonated amine with a pK_a value of 10.4 will be in its neutral, nonprotonated form.

PROBLEM 38♦ Solved

a. Indicate whether a carboxylic acid (RCOOH) with a pK_a value of 4.5 will have more charged molecules or more neutral molecules in a solution with the following pH:

1. pH = 1   **3.** pH = 5   **5.** pH = 10

2. pH = 3   **4.** pH = 7   **6.** pH = 13

b. Answer the same question for a protonated amine ($R\overset{+}{N}H_3$) with a pK_a value of 9.

c. Answer the same question for an alcohol (ROH) with a pK_a value of 15.

> **You are what you're in:** a compound will be mostly in the acidic form in an acidic solution (pH < pK_a) and mostly in the basic form in a basic solution (pH > pK_a).

Solution to 38a 1. First determine whether the compound is charged or neutral in its acidic form and charged or neutral in its basic form: a carboxylic acid is neutral in its acidic form (RCOOH) and charged in its basic form ($RCOO^-$). Then compare the pH and pK_a values and remember that if the pH of the solution is less than the pK_a value of the compound, then more molecules will be in the acidic form, but if the pH is greater than the pK_a value of the compound, then more molecules will be in the basic form. Therefore, at pH = 1 and 3, there will be more neutral molecules, and at pH = 5, 7, 10, and 13, there will be more charged molecules.

PROBLEM 39

A naturally occurring amino acid such as alanine has a group that is a carboxylic acid and a group that is a protonated amine. The pK_a values of the two groups are shown.

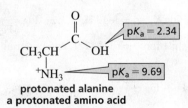

protonated alanine
a protonated amino acid

a. If the pK_a value of carboxylic acid such as acetic acid is about 5 (see Table 1), then why is the pK_a value of the carboxylic acid group of alanine so much lower?

b. Draw the structure of alanine in a solution at pH = 0.

c. Draw the structure of alanine in a solution at physiological pH (pH 7.4).

d. Draw the structure of alanine in a solution at pH = 12.

e. Is there a pH at which alanine will be uncharged (that is, neither group will have a charge)?

f. At what pH will alanine have no net charge (that is, the amount of negative charge will be the same as the amount of positive charge)?

If we know the pH of the solution and the pK_a of the compound, the Henderson–Hasselbalch equation makes it possible to calculate precisely how much of the compound will be in its acidic form and how much will be in its basic form.

For example, when a compound with a pK_a of 5.2 is in a solution of pH 5.2, half the compound will be in the acidic form and the other half will be in the basic form (Figure 1). If the pH is one unit less than the pK_a of the compound (pH = 4.2), then there will be 10 times

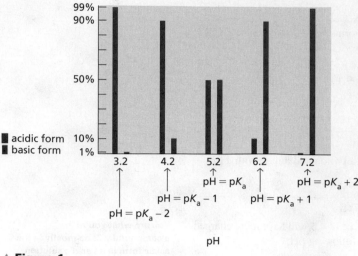

▲ Figure 1

The relative amounts of a compound with a pK_a of 5.2 in the acidic and basic forms at different pH values.

more compound present in the acidic form than in the basic form (because log 10 = 1).

$$5.2 = 4.2 + \log \frac{[HA]}{[A^-]}$$

$$1.0 = \log \frac{[HA]}{[A^-]} = \log \frac{10}{1}$$

If the pH is two units less than the pK_a of the compound (pH = 3.2), then there will be 100 times more compound present in the acidic form than in the basic form (because log 100 = 2).

Now consider pH values that are greater than the pK_a value: if the pH is 6.2, then there will be 10 times more compound in the basic form than in the acidic form, and at pH = 7.2, there will be 100 times more compound present in the basic form than in the acidic form.

Derivation of the Henderson–Hasselbalch Equation

The Henderson–Hasselbalch equation can be derived from the expression that defines the acid dissociation constant:

$$K_a = \frac{[H_3O^+][A^-]}{[HA]}$$

Take the logarithms of both sides of the equation and remember that when expressions are multiplied, their logs are added. Thus, we obtain

$$\log K_a = \log [H_3O^+] + \log \frac{[A^-]}{[HA]}$$

Multiplying both sides of the equation by −1 gives us

$$-\log K_a = -\log [H_3O^+] - \log \frac{[A^-]}{[HA]}$$

Substituting pK_a for −log K_a, pH for −log $[H_3O^+]$, and inverting the fraction (which means the sign of its log changes), we get

$$pK_a = pH + \log \frac{[HA]}{[A^-]} \quad \text{or} \quad pH = pK_a - \log \frac{[HA]}{[A^-]}$$

PROBLEM 40◆ Solved

a. At what pH will the concentration of a compound with a pK_a = 8.4 be 100 times greater in its basic form than in its acidic form?

b. At what pH will the concentration of a compound with a pK_a = 3.7 be 10 times greater in its acidic form than in its basic form?

c. At what pH will the concentration of a compound with a pK_a = 8.4 be 100 times greater in its acidic form than in its basic form?

d. At what pH will 50% of a compound with a pK_a = 7.3 be in its basic form?

e. At what pH will the concentration of a compound with a pK_a = 4.6 be 10 times greater in its basic form than in its acidic form?

Solution to 40a If the concentration in the basic form is 100 times greater than the concentration in the acidic form, then the Henderson–Hasselbalch equation becomes

$$pH = pK_a - \log 1/100$$
$$pH = 8.4 - \log 0.01$$
$$pH = 8.4 + 2.0$$
$$pH = 10.4$$

There is a faster way to get the answer: if 100 times more compound is present in the basic form than in the acidic form, then the pH will be two units more basic than the pK_a value. Thus, $pH = 8.4 + 2.0 = 10.4$.

Solution to 40b If 10 times more compound is present in the acidic form than in the basic form, then the pH will be one unit more acidic than the pK_a value. Thus, $pH = 3.7 - 1.0 = 2.7$.

PROBLEM 41♦

For each of the following compounds, indicate the pH at which

a. 50% of the compound will be in a form that possesses a charge.
b. more than 99% of the compound will be in a form that possesses a charge.

    1. CH_3CH_2COOH ($pK_a = 4.9$)   **2.** $CH_3\overset{+}{N}H_3$ ($pK_a = 10.7$)

The Henderson–Hasselbalch equation can be very useful in the laboratory for separating compounds in a mixture. Water and diethyl ether are barely soluble in each other, so they form two layers when combined. Diethyl ether is less dense than water, so the ether layer will lie above the water layer. Charged compounds will be more soluble in water than in diethyl ether, whereas uncharged compounds will be more soluble in diethyl ether than in water.

Two compounds, such as a carboxylic acid (RCOOH) with a $pK_a = 5.0$ and a protonated amine ($R\overset{+}{N}H_3$) with a $pK_a = 10.0$, dissolved in a mixture of water and diethyl ether, can be separated by adjusting the pH of the water layer. For example, if the pH of the water layer is 2, then the carboxylic acid and the protonated amine will both be in their acidic forms because the pH of the water is less than the pK_a values of both compounds. The acidic form of a carboxylic acid is not charged, whereas the acidic form of an amine is charged. Therefore, the uncharged carboxylic acid will dissolve in the ether layer, and the positively charged protonated amine will dissolve in the water layer.

    acidic form   **basic form**

$$RCOOH \rightleftharpoons RCOO^- + H^+$$
$$R\overset{+}{N}H_3 \rightleftharpoons RNH_2 + H^+$$

For the most effective separation, the pH of the water layer should be at least two units away from the pK_a values of the compounds being separated. Then the relative amounts of the compounds in their acidic and basic forms will be at least 100:1 (Figure 1).

Aspirin Must Be in Its Basic Form to Be Physiologically Active

Aspirin has been used to treat fever, mild pain, and inflammation since it first became commercially available in 1899. It was the first drug to be tested clinically before it was marketed. Currently one of the most widely used drugs in the world, aspirin is one of a group of over-the-counter drugs known as NSAIDs (nonsteroidal anti-inflammatory drugs).

Aspirin is a carboxylic acid. The carboxylic acid group must be in its basic form to be physiologically active.

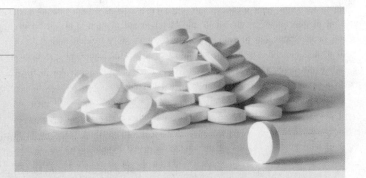

    acidic form   **Aspirin**   **basic form**

The carboxylic acid group has a pK_a value of ~5. Therefore, it will be in its acidic form while it is in the stomach (pH = 1–2.5). The uncharged acidic form can pass easily through membranes, whereas the negatively charged basic form cannot. In the environment of the cell (pH 7.4), the drug will be in its active basic form and will therefore be able to carry out the reaction that reduces fever, pain, and inflammation.

The undesirable side effects of aspirin (ulcers, stomach bleeding) led to the development of other NSAIDs . Aspirin also has been linked to the development of Reye's syndrome, a rare but serious disease that affects children who are recovering from a viral infection such as a cold, the flu, or chicken pox. Therefore, it is now recommended that aspirin not be given to anyone under the age of 16 who has a fever-producing illness.

PROBLEM 42♦ Solved

Given that $C_6H_{11}COOH$ has a $pK_a = 4.8$ and $C_6H_{11}\overset{+}{N}H_3$ has a $pK_a = 10.7$, answer the following:

a. What pH would you make the water layer in order to cause both compounds to dissolve in it?
b. What pH would you make the water layer in order to cause the acid to dissolve in the water layer and the amine to dissolve in the ether layer?
c. What pH would you make the water layer in order to cause the acid to dissolve in the ether layer and the amine to dissolve in the water layer?

Solution to 42a A compound has to be charged in order to dissolve in the water layer. The carboxylic acid will be charged in its basic form—it will be a carboxylate ion. For > 99% of the carboxylic acid to be in its basic form, the pH must be two units *greater* than the pK_a of the compound. Thus, the water should have a pH > 6.8. The amine will be charged in its acidic form—it will be an ammonium ion. For > 99% of the amine to be in its acidic form, the pH must be two units *less* than the pK_a value of the ammonium ion. Thus, the water should have a pH < 8.7. Both compounds will dissolve in the water layer if its pH is 6.8–8.7. A pH in the middle of the range (for example, pH = 7.7) would be a good choice.

11   BUFFER SOLUTIONS

A solution of a weak acid (HA) and its conjugate base (A^-) is called a **buffer solution.** A buffer solution will maintain nearly constant pH when small amounts of acid or base are added to it, because the weak acid can give a proton to any HO^- added to the solution, and its conjugate base can accept any H^+ that is added to the solution.

can give an H^+ to HO^-

$$HA \;+\; HO^- \longrightarrow A^- \;+\; H_2O$$

$$A^- \;+\; H_3O^+ \longrightarrow HA \;+\; H_2O$$

can accept an H^+ from H_3O^+

PROBLEM 43♦

Write the equation that shows how a buffer made by dissolving CH_3COOH and $CH_3COO^-Na^+$ in water prevents the pH of a solution from changing appreciably when

a. a small amount of H^+ is added to the solution.
b. a small amount of HO^- is added to the solution.

PROBLEM 44 Solved

You are planning to carry out a reaction that produces hydroxide ion. In order for the reaction to take place at a constant pH, it will be buffered at pH = 4.2. Would it be better to use a formic acid/formate buffer or an acetic acid/acetate buffer? (Note: the pK_a of formic acid = 3.75 and the pK_a of acetic acid = 4.76.)

Solution Constant pH will be maintained because the hydroxide ion produced in the reaction will remove a proton from the acidic form of the buffer. Thus the better choice of buffer is the one that has the highest concentration of buffer in the acidic form at pH = 4.2. Because formic acid's pK_a is 3.75, the majority of the buffer will be in the basic form at pH = 4.2. Acetic acid, with pK_a = 4.76, will have more buffer in the acidic form than in the basic form. Thus, it would be better to use acetic acid/acetate buffer for your reaction.

Blood: A Buffered Solution

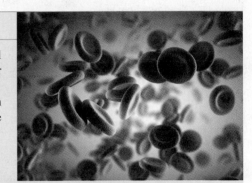

Blood is the fluid that transports oxygen to all the cells of the human body. The normal pH of human blood is ~7.4. Death will result if this pH decreases to less than ~6.8 or increases to greater than ~8.0 for even a few seconds.

Oxygen is carried to cells by a protein in the blood called hemoglobin (HbH^+). When hemoglobin binds O_2, hemoglobin loses a proton, which would make the blood more acidic if it did not contain a buffer to maintain its pH.

$$HbH^+ + O_2 \rightleftharpoons HbO_2 + H^+$$

A carbonic acid/bicarbonate $\left(H_2CO_3/HCO_3^-\right)$ buffer controls the pH of blood. An important feature of this buffer is that carbonic acid decomposes to CO_2 and H_2O, as shown below:

$$\underset{\textbf{bicarbonate}}{HCO_3^-} + H^+ \rightleftharpoons \underset{\textbf{carbonic acid}}{H_2CO_3} \rightleftharpoons CO_2 + H_2O$$

During exercise our metabolism speeds up, producing large amounts of CO_2. The increased concentration of CO_2 shifts the equilibrium between carbonic acid and bicarbonate to the left, which increases the concentration of H^+. Significant amounts of lactic acid are also produced during exercise, which further increases the concentration of H^+. Receptors in the brain respond to the increased concentration of H^+ by triggering a reflex that increases the rate of breathing. Hemoglobin then releases more oxygen to the cells and more CO_2 is eliminated by exhalation. Both processes decrease the concentration of H^+ in the blood by shifting the equilibrium of the top reaction to the left and the equilibrium of the bottom reaction to the right.

Thus, any disorder that decreases the rate and depth of ventilation, such as emphysema, will decrease the pH of the blood—a condition called acidosis. In contrast, any excessive increase in the rate and depth of ventilation, as with hyperventilation due to anxiety, will increase the pH of blood—a condition called alkalosis.

12   LEWIS ACIDS AND BASES

In 1923, G. N. Lewis offered new definitions for the terms *acid* and *base*. He defined an acid as *a species that accepts a share in an electron pair* and a base as *a species that donates a share in an electron pair*.

All Brønsted-Lowry (proton-donating) acids fit the Lewis definition because all proton-donating acids lose a proton and the proton accepts a share in an electron pair.

(Curved arrows show where the electrons start from and where they end up.)

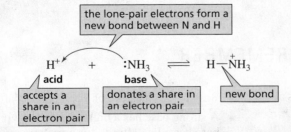

Lewis acids, however, are not limited to compounds that lose protons. Compounds such as aluminum chloride ($AlCl_3$), ferric bromide ($FeBr_3$), and borane (BH_3) are acids according to the Lewis definition because they have unfilled valence orbitals that can accept a share in an electron pair. These compounds react with a compound that has a lone pair, just as a proton reacts with a compound that has a lone pair.

aluminum trichloride dimethyl ether
a Lewis acid   a Lewis base

Lewis acid: Need two from you.

Lewis base: Have pair, will share.

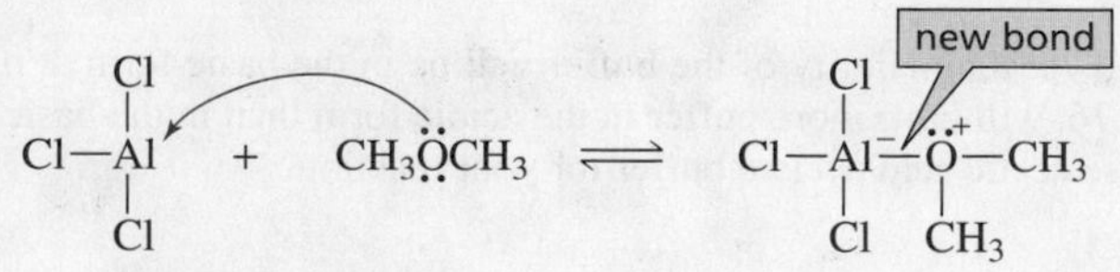

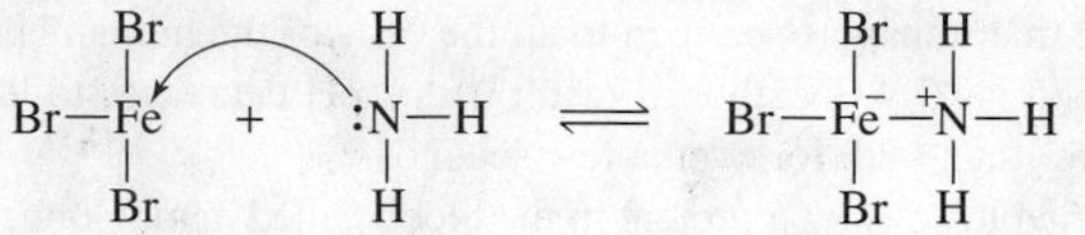

ferric bromide   ammonia
a Lewis acid   a Lewis base

Thus, the Lewis definition of an acid includes all proton-donating compounds and some additional compounds that do not have protons. Throughout this text, the term *acid* is used to mean a proton-donating acid, and the term **Lewis acid** is used to refer to non-proton-donating acids such as $AlCl_3$ or BF_3. Lewis acids are also called **electrophiles**. In the reactions just shown, H^+, aluminum trichloride, and borane are electrophiles.

All bases are **Lewis bases** because they all have a pair of electrons that they can share, either with a proton or with an atom such as aluminum, boron, or carbon. Throughout this text, the term *base* is used to mean a compound that shares its lone pair with a proton, and the term **nucleophile** is used for a compound that shares its lone pair with an atom other than a proton. In the three reactions just shown, ammonia is a base in the first reaction, and dimethyl ether and ammonia are nucleophiles in the next two reactions.

Most organic reactions involve *the reaction of an electrophile with a nucleophile*. In other words, they involve the reaction of a Lewis acid with a Lewis base.

PROBLEM 45

Draw the products of the following reactions. Use curved arrows to show where the pair of electrons starts from and where it ends up.

a. $ZnCl_2$ + $CH_3\ddot{O}H$ $\rightleftharpoons$ **b.** $FeBr_3$ + $:\ddot{Br}:^-$ $\rightleftharpoons$ **c.** $AlCl_3$ + $:\ddot{Cl}:^-$ $\rightleftharpoons$

PROBLEM 46

What products are formed when each of the following reacts with HO^-?

a. CH_3OH
b. $^+NH_4$ **d.** BF_3 **f.** $FeBr_3$ **h.** CH_3COOH
c. $CH_3\overset{+}{N}H_3$ **e.** $^+CH_3$ **g.** $AlCl_3$

SOME IMPORTANT THINGS TO REMEMBER

- An **acid** is a species that donates a proton; a **base** is a species that accepts a proton.

- A **Lewis acid** is a species that accepts a share in an electron pair; a **Lewis base** is a species that donates a share in an electron pair.

- **Acidity** is a measure of the tendency of a compound to lose a proton.

- **Basicity** is a measure of a compound's affinity for a proton.

- A strong base has a high affinity for a proton; a weak base has a low affinity for a proton.

- The stronger the acid, the weaker its conjugate base.

- The strength of an acid is given by the **acid dissociation constant** (K_a).

- The stronger the acid, the smaller its pK_a value.

- Approximate pK_a values are as follows: protonated alcohols, protonated carboxylic acids, protonated water < 0; carboxylic acids ~5; protonated amines ~10; alcohols and water ~15.

- The **pH** of a solution indicates the concentration of protons in the solution; the smaller the pH, the more acidic the solution.

- In **acid–base reactions,** the equilibrium favors formation of the weaker acid.

- Curved arrows indicate the bonds that are broken and formed as reactants are converted into products.

- The strength of an acid is determined by the stability of its conjugate base: the more stable (weaker) the base, the stronger its conjugate acid.

- When atoms are similar in size, the strongest acid will have its hydrogen attached to the more electronegative atom.

- When atoms are very different in size, the strongest acid will have its hydrogen attached to the larger atom.

- Hybridization affects acidity because an *sp* hybridized atom is more electronegative than an sp^2 hybridized atom, which is more electronegative than an sp^3 hybridized atom.

- **Inductive electron withdrawal** increases acidity: the more electronegative the electron-withdrawing group and the closer it is to the acidic hydrogen, the stronger is the acid.

- **Delocalized electrons** (electrons that are shared by more than two atoms) stabilize a compound.

- A **resonance hybrid** is a composite of the **resonance contributors,** structures that differ only in the location of their π electrons and lone-pair electrons.

- The **Henderson–Hasselbalch equation** gives the relationship between pK_a and pH: a compound exists primarily in its acidic form (with its proton) in solutions more acidic than its pK_a value and primarily in its basic form (without its proton) in solutions more basic than its pK_a value.

- A **buffer solution** contains both a weak acid and its conjugate base.

- In this text, the term *acid* is used to mean a proton-donating acid, the term **Lewis acid** is used to refer to non-proton-donating acids such as $AlCl_3$ or BF_3, and the term **electrophile** refers to both proton-donating and non-proton-donating acids.

- In this text, the term *base* is used to mean a compound that shares its lone pair with a proton, and the term **nucleophile** is used for a compound that shares its lone pair with an atom other than a proton.

PROBLEMS

47. a. List the following alcohols in order from strongest acid to weakest acid:

CCl_3CH_2OH CH_2ClCH_2OH $CHCl_2CH_2OH$
$K_a = 5.75 \times 10^{-13}$ $K_a = 1.29 \times 10^{-13}$ $K_a = 4.90 \times 10^{-13}$

  b. Explain the relative acidities.

48. Which is a stronger base?

  a. HS^- or HO^- **c.** CH_3OH or CH_3O^- **e.** CH_3COO^- or CF_3COO^-
  b. CH_3O^- or $CH_3\overset{-}{N}H$ **d.** Cl^- or Br^- **f.** $CH_3CHClCOO^-$ or $CH_3CHBrCOO^-$

49. Draw curved arrows to show where the electrons start from and where they end up in the following reactions:

  a. $\overset{..}{N}H_3 + H-\overset{..}{\underset{..}{C}}l: \; \rightleftharpoons \; {}^+NH_4 + :\overset{..}{\underset{..}{C}}l:^-$

  c. (structures with curved-arrow reaction)

  b. $H_2\overset{..}{O}: + FeBr_3 \; \rightleftharpoons \; H_2\overset{..}{O}{}^+\!-\overset{-}{F}eBr_3$

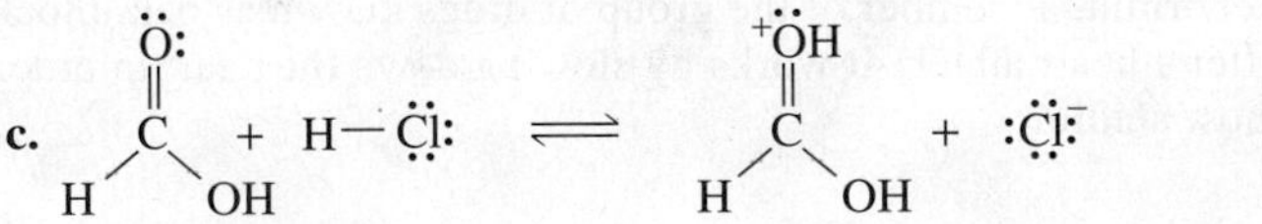

50. a. List the following carboxylic acids in order from strongest acid to weakest acid:

$CH_3CH_2CH_2COOH$ $CH_3CH_2CHCOOH$ $ClCH_2CH_2CH_2COOH$ CH_3CHCH_2COOH
$K_a = 1.52 \times 10^{-5}$ $\overset{|}{Cl}$ $K_a = 2.96 \times 10^{-5}$ $\overset{|}{Cl}$
                    $K_a = 1.39 \times 10^{-3}$ $K_a = 8.9 \times 10^{-5}$

  b. How does the presence of an electronegative substituent such as Cl affect the acidity of a carboxylic acid?
  c. How does the location of the substituent affect the acidity of the carboxylic acid?

51. Draw the products of the following reactions:

a. $CH_3\ddot{O}CH_3 + BF_3 \longrightarrow$ b. $CH_3\ddot{O}CH_3 + H—Cl \longrightarrow$ c. $CH_3\ddot{N}H_2 + AlCl_3 \longrightarrow$

52. For the following compound,
 a. draw its conjugate acid.   b. draw its conjugate base.

$$HOCH_2CH_2CH_2NH_2$$

53. List the following compounds in order from strongest acid to weakest acid:

CH_3CH_2OH $CH_3CH_2NH_2$ CH_3CH_2SH $CH_3CH_2CH_3$

54. For each of the following compounds, draw the form in which it will predominate at pH = 3, pH = 6, pH = 10, and pH = 14:
 a. CH_3COOH $pK_a = 4.8$ b. $CH_3CH_2\overset{+}{N}H_3$ $pK_a = 11.0$ c. CF_3CH_2OH $pK_a = 12.4$

55. Give the products of the following acid–base reactions, and indicate whether reactants or products are favored at equilibrium (use the pK_a values that are given in Section 3):

a. $CH_3\overset{O}{\overset{\|}{C}}OH + CH_3O^- \rightleftharpoons$ c. $CH_3\overset{O}{\overset{\|}{C}}OH + CH_3NH_2 \rightleftharpoons$

b. $CH_3CH_2OH + {}^-NH_2 \rightleftharpoons$ d. $CH_3CH_2OH + HCl \rightleftharpoons$

56. a. List the following alcohols in order from strongest acid to weakest acid.
 b. Explain the relative acidities.

$CH_2{=}CHCH_2OH$ $CH_3CH_2CH_2OH$ $HC{\equiv}CCH_2OH$

57. For each compound, indicate the atom that is most apt to be protonated.

a. $CH_3—\underset{\overset{|}{OH}}{CH}—CH_2NH_2$ b. $CH_3—\underset{\overset{|}{NH_2}}{\overset{\overset{|}{CH_3}}{C}}—OH$ c. $CH_3—\underset{\overset{|}{NH_2}}{\overset{\overset{|}{CH_3}}{C}}—CH_2OH$

58. a. Given the K_a values, estimate the pK_a value of each of the following acids without using a calculator (that is, is it between 3 and 4, between 9 and 10, and so on?):
  1. nitrous acid (HNO_2), $K_a = 4.0 \times 10^{-4}$
  2. nitric acid (HNO_3), $K_a = 22$
  3. bicarbonate (HCO_3^-), $K_a = 6.3 \times 10^{-11}$
  4. hydrogen cyanide (HCN), $K_a = 7.9 \times 10^{-10}$
  5. formic acid (HCOOH), $K_a = 2.0 \times 10^{-4}$

 b. Determine the exact pK_a values, using a calculator.
 c. Which is the strongest acid?

59. A single bond between two carbons with different hybridizations has a small dipole. What is the direction of the dipole in the indicated bonds?

a. $CH_3\overset{\downarrow}{—}CH{=}CH_2$ b. $CH_3\overset{\downarrow}{—}C{\equiv}CH$

60. Tenormin, a member of the group of drugs known as beta-blockers, is used to treat high blood pressure and improve survival after a heart attack. It works by slowing down the heart in order to reduce its workload. Which hydrogen in Tenormin is the most acidic?

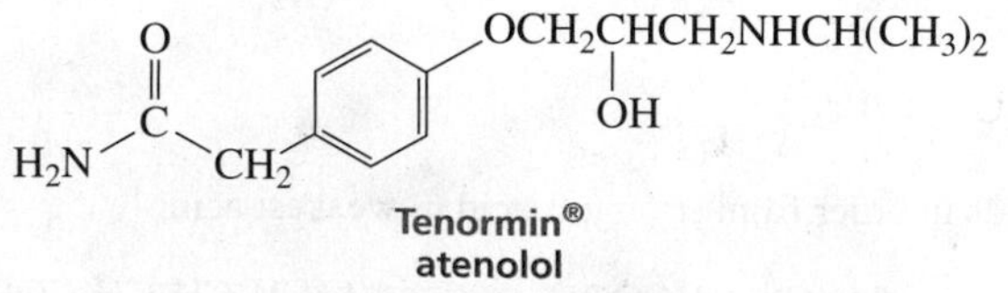

$$H_2N—\overset{O}{\overset{\|}{C}}—CH_2—\underset{\underset{\text{atenolol}}{\text{Tenormin}^®}}{\boxed{}}—OCH_2\underset{\overset{|}{OH}}{CH}CH_2NHCH(CH_3)_2$$

61. From which acids can HO^- remove a proton in a reaction that favors product formation?

CH_3COOH $CH_3CH_2NH_2$ $CH_3CH_2\overset{+}{N}H_3$ $CH_3C{\equiv}CH$
    A   **B**   **C**   **D**

62. a. For each of the following pairs of reactions, indicate which one has the more favorable equilibrium constant (that is, which one most favors products):

1. $CH_3CH_2OH + NH_3 \rightleftharpoons CH_3CH_2O^- + \overset{+}{N}H_4$

   or

   $CH_3OH + NH_3 \rightleftharpoons CH_3O^- + \overset{+}{N}H_4$

2. $CH_3CH_2OH + NH_3 \rightleftharpoons CH_3CH_2O^- + \overset{+}{N}H_4$

   or

   $CH_3CH_2OH + CH_3NH_2 \rightleftharpoons CH_3CH_2O^- + CH_3\overset{+}{N}H_3$

b. Which of the four reactions has the most favorable equilibrium constant?

63. You are planning to carry out a reaction that produces protons. The reaction will be buffered at pH = 10.5. Would it be better to use a protonated methylamine/methylamine buffer or a protonated ethylamine/ethylamine buffer? (pK_a of protonated methylamine = 10.7; pK_a of protonated ethylamine = 11.0)

64. Which is a stronger acid?

a. $CH_2{=}CHCOOH$ or CH_3CH_2COOH

c. $CH_2{=}CHCOOH$ or $HC{\equiv}CCOOH$

b.

d.

65. Citrus fruits are rich in citric acid, a compound with three COOH groups. Explain the following:

a. The first pK_a (for the COOH group in the center of the molecule) is lower than the pK_a of acetic acid.

b. The third pK_a is greater than the pK_a of acetic acid.

$$pK_a = 4.5 \quad HO-\overset{\overset{O}{\|}}{C}-CH_2-\overset{\overset{OH}{|}}{\underset{\underset{\underset{OH}{|}}{\overset{\|}{C}}}{C}}-CH_2-\overset{\overset{O}{\|}}{C}-OH \quad pK_a = 5.8$$

$$pK_a = 3.1$$

66. Given that pH + pOH = 14 and that the concentration of water in a solution of water is 55.5 M, show that the pK_a of water is 15.7. (*Hint:* pOH = $-\log [HO^-]$)

67. How could you separate a mixture of the following compounds? The reagents available to you are water, ether, 1.0 M HCl, and 1.0 M NaOH. (*Hint:* See Problem 42.)

COOH	$\overset{+}{N}H_3Cl^-$	OH	Cl	$\overset{+}{N}H_3Cl^-$
$pK_a = 4.17$	$pK_a = 4.60$	$pK_a = 9.95$		$pK_a = 10.66$

68. Carbonic acid has a pK_a of 6.1 at physiological temperature. Is the carbonic acid/bicarbonate buffer system that maintains the pH of the blood at 7.4 better at neutralizing excess acid or excess base?

69. a. If an acid with a pK_a of 5.3 is in an aqueous solution of pH 5.7, what percentage of the acid is present in its acidic form?

b. At what pH will 80% of the acid exist in its acidic form?

70. Calculate the pH values of the following solutions:

a. a 1.0 M solution of acetic acid ($pK_a = 4.76$)

b. a 0.1 M solution of protonated methylamine ($pK_a = 10.7$)

c. a solution containing 0.3 M HCOOH and 0.1 M HCOO$^-$ (pK_a of HCOOH $= 3.76$)

TUTORIAL

Enhanced by MasteringChemistry®

ACIDS AND BASES

This tutorial is designed to give you practice working problems based on some of the concepts you learned in this chapter. Most of the concepts are given here without explanation because full explanations can be found in this chapter.

An Acid and Its Conjugate Base

An acid is a species that can lose a proton (the Brønsted–Lowry definition). When an acid loses a proton (H^+), it forms its conjugate base. When the proton comes off the acid, the conjugate base retains the electron pair that attached the proton to the acid.

acid — conjugate base — a proton

Often, the lone pairs and bonding electrons are not shown.

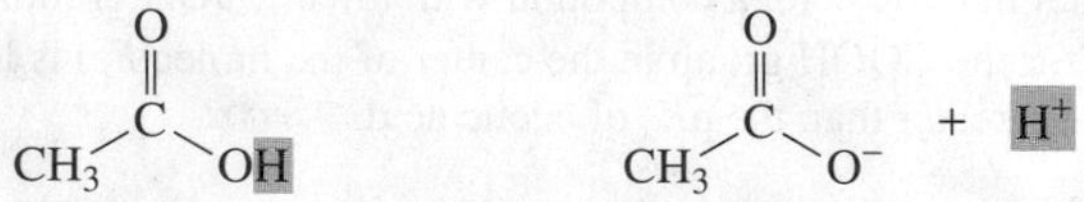

acid — conjugate base

$$CH_3\overset{+}{O}H_2 \qquad CH_3OH + H^+$$

acid — conjugate base

Notice that a neutral acid forms a negatively charged conjugate base, whereas a positively charged acid forms a neutral conjugate base. (The difference in charge *decreases* by one because the acid *loses* H^+.)

PROBLEM 1 Draw the conjugate base of each of the following acids:

a. CH_3OH **b.** $CH_3\overset{+}{N}H_3$ **c.** CH_3NH_2 **d.** H_3O^+ **e.** H_2O

A Base and Its Conjugate Acid

A base is a species that can gain a proton (the Brønsted–Lowry definition). When a base gains a proton (H^+), it forms its conjugate acid. In order to gain a proton, a base must have a lone pair that it can use to form a new bond with the proton.

base — conjugate acid

Often, the lone pairs and bonding electrons are not shown.

$$CH_3O^- + H^+ \qquad CH_3OH$$

base — conjugate acid

$$CH_3NH_2 + H^+ \qquad CH_3\overset{+}{N}H_3$$

base — conjugate acid

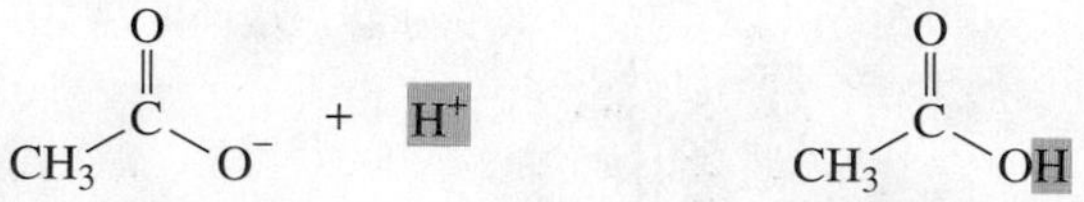

base — conjugate acid

Notice that a negatively charged base forms a neutral conjugate acid, whereas a neutral base forms a positively charged conjugate acid. (The difference in charge *increases* by one because the compound *gains* H^+.)

PROBLEM 2 Draw the conjugate acid of each of the following bases:

a. H_2O b. HO^- c. CH_3OH d. NH_3 e. Cl^-

Acid–Base Reactions

An acid cannot lose a proton unless a base is present to accept the proton. Therefore, an acid always reacts with a base. The reaction of an acid with a base is called an acid–base reaction or a proton transfer reaction. Acid–base reactions are reversible reactions.

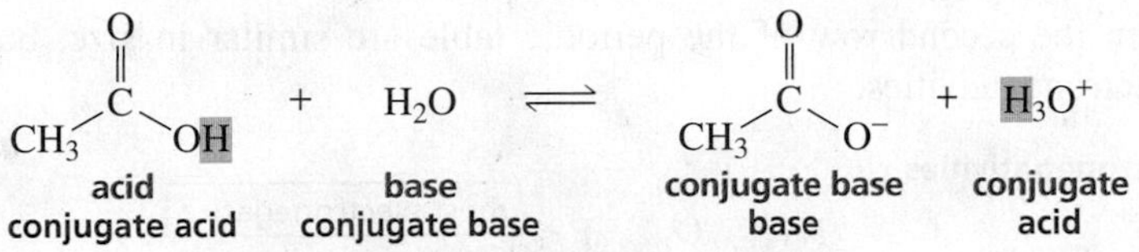

acid   **base**
conjugate acid **conjugate base**   **conjugate base**   **conjugate acid**

Notice that an acid reacts with a base in the forward direction (blue labels) and an acid reacts with a base in the reverse direction (red labels).

The Products of an Acid–Base Reaction

Both CH_3COOH and H_2O in the preceding reaction have protons that can be lost (that is, both can act as acids), and both have lone pairs that can form a bond with a proton (that is, both can act as bases). How do we know which reactant will lose a proton and which will gain a proton? We can determine this by comparing the pK_a values of the two reactants; these values are 4.8 for CH_3COOH and 15.7 for H_2O. The stronger acid (the one with the lower pK_a value) will be the one that acts as an acid (it will lose a proton). The other reactant will act as a base (it will gain a proton).

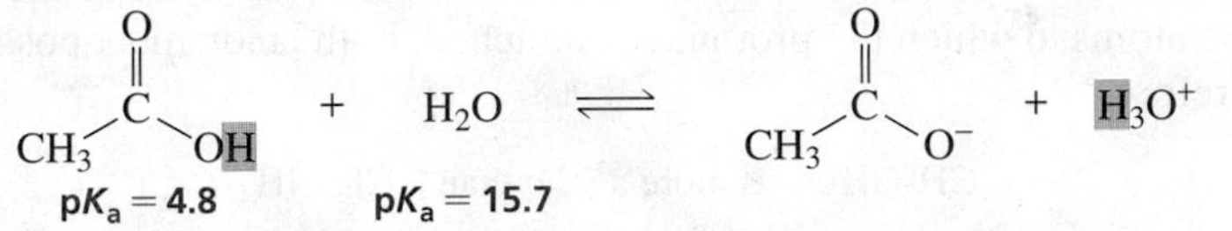

$pK_a = 4.8$ $pK_a = 15.7$

PROBLEM 3 Draw the products of the following acid–base reactions:

a. $CH_3\overset{+}{N}H_3$ + H_2O c. $CH_3\overset{+}{N}H_3$ + HO^-

b. HBr + CH_3OH d. CH_3NH_2 + CH_3OH

The Position of Equilibrium

Whether an acid–base reaction favors formation of the products or formation of the reactants can be determined by comparing the pK_a value of the acid that loses a proton in the forward direction with the pK_a value of the acid that loses a proton in the reverse direction. The equilibrium will favor the reaction of the stronger acid to form the weaker acid. The following reaction favors formation of the reactants, because $CH_3\overset{+}{O}H_2$ is a stronger acid than CH_3COOH.

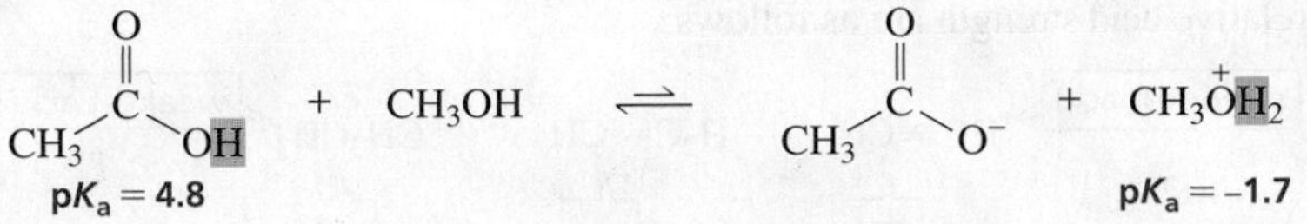

$pK_a = 4.8$ $pK_a = -1.7$

The next reaction favors formation of the products, because HCl is a stronger acid than $CH_3\overset{+}{N}H_3$.

$$HCl \quad + \quad CH_3NH_2 \quad \rightleftharpoons \quad Cl^- \quad + \quad CH_3\overset{+}{N}H_3$$
$$pK_a = -7 \qquad\qquad\qquad\qquad\qquad\qquad pK_a = 10.7$$

PROBLEM 4 Which of the reactions in Problem 3 favor formation of the reactants, and which favor formation of the products? (The pK_a values can be found in Sections 3 and 6.)

Relative Acid Strengths When the Proton Is Attached to Atoms Similar in Size

The atoms in the second row of the periodic table are similar in size, but they have different electronegativities.

relative electronegativities

$$C \; < \; N \; < \; O \; < \; F \overset{\text{most electronegative}}{}$$

When acids have protons attached to atoms similar in size, the strongest acid is the one with the proton attached to the more electronegative atom. The relative acid strengths are as follows:

$$\overset{\text{strongest acid}}{} HF \; > \; H_2O \; > \; NH_3 \; > \; CH_4 \overset{\text{weakest acid}}{}$$

A positively charged atom is more electronegative than the same atom when it is neutral. Therefore,

$$CH_3\overset{+}{N}H_3 \quad \text{is more acidic than} \quad CH_3NH_2$$

$$CH_3\overset{+}{O}H_2 \quad \text{is more acidic than} \quad CH_3OH$$

When the relative strengths of two acids are determined by comparing the electronegativities of the atoms to which the protons are attached, both acids must possess the same charge. Therefore,

$$CH_3\overset{+}{O}H_2 \quad \text{is more acidic than} \quad CH_3\overset{+}{N}H_3$$

$$CH_3OH \quad \text{is more acidic than} \quad CH_3NH_2$$

PROBLEM 5 Which is the stronger acid?

a. CH_3OH or CH_3CH_3 \qquad **c.** CH_3NH_2 or HF

b. CH_3OH or HF \qquad\qquad **d.** CH_3NH_2 or CH_3OH

The Effect of Hybridization on Acidity

The electronegativity of an atom depends on its hybridization.

$$\overset{\text{most electronegative}}{} sp \; > \; sp^2 \; > \; sp^3$$

Once again, the stronger acid will have its proton attached to the more electronegative atom. Thus, the relative acid strength are as follows:

$$\overset{\text{strongest acid}}{} HC\equiv CH \; > \; H_2C=CH_2 \; > \; CH_3CH_3 \overset{\text{weakest acid}}{}$$
$$\quad\quad sp \quad\quad\quad\quad sp^2 \quad\quad\quad\quad sp^3$$

PROBLEM 6 Which is the stronger acid?

a. CH_3CH_3 or $HC\equiv CH$ **b.** $H_2C\!=\!CH_2$ or $HC\equiv CH$ **c.** $H_2C\!=\!CH_2$ or CH_3CH_3

Relative Acid Strengths When the Proton Is Attached to Atoms Very Different in Size

The atoms in a column of the periodic table become considerably larger as you go down the column.

largest halide ion $\quad I^- > Br^- > Cl^- > F^-$ smallest halide ion

When comparing two acids with protons attached to atoms that are very different in size, the stronger acid is the one attached to the larger atom. Thus, the relative acid strengths are as follows:

strongest acid $\quad HI > HBr > HCl > HF$ weakest acid

PROBLEM 7 Which is the stronger acid? (*Hint:* You can use the periodic table at the back of this book.)

a. HCl or HBr   **b.** CH_3OH or CH_3SH **c.** HF or HCl   **d.** H_2S or H_2O

The Effect of Inductive Electron Withdrawal on Acidity

Replacing a hydrogen with an electronegative substituent—one that pulls bonding electrons towards itself—increases the strength of the acid.

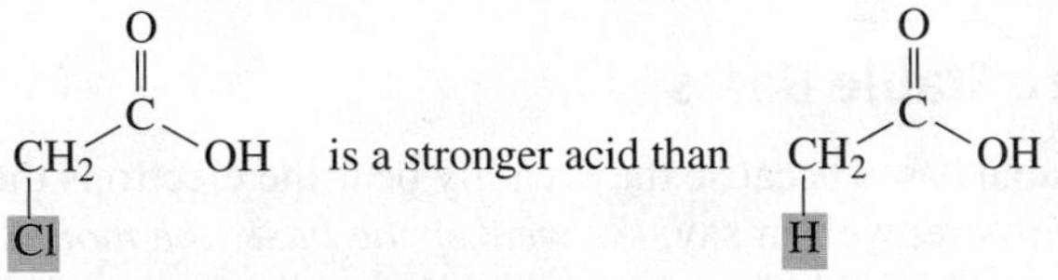

is a stronger acid than

The halogens have the following relative electronegativities:

most electronegative $\quad F > Cl > Br > I$ least electronegative

The more electronegative the substituent that replaces a hydrogen, the stronger the acid. Thus, the relative acid strength are as follows:

strongest acid $\quad CH_3CHCH_2OH > CH_3CHCH_2OH > CH_3CHCH_2OH > CH_3CHCH_2OH$ weakest acid
$\qquad\qquad\quad\ \ \ F \qquad\qquad\quad Cl \qquad\qquad\quad Br \qquad\qquad\quad I$

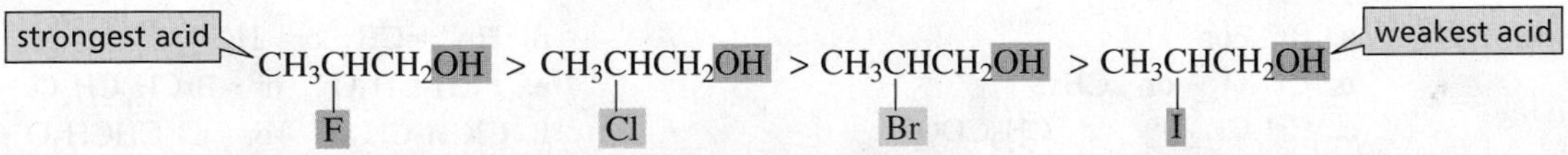

The closer the electronegative substituent is to the group that loses a proton, the stronger the acid will be. Thus, the relative acid strength are as follows:

strongest acid $\quad CH_3CH_2CHCOOH > CH_3CHCH_2COOH > CH_2CH_2CH_2COOH$ weakest acid
$\qquad\qquad\qquad\ \ Cl \qquad\qquad\qquad\quad Cl \qquad\qquad\qquad Cl$

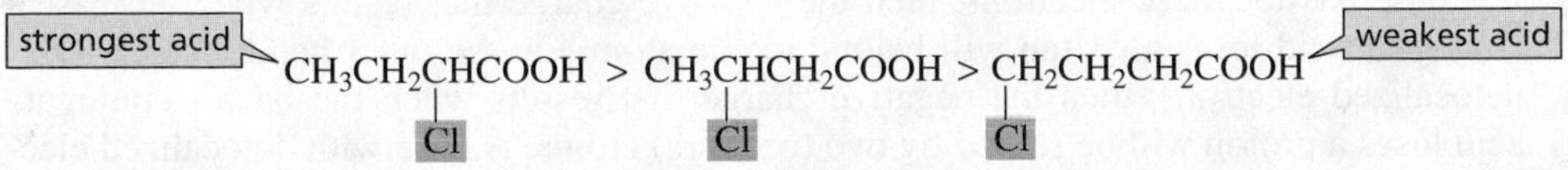

PROBLEM 8 Which is the stronger acid?

a. $ClCH_2CH_2OH$ or FCH_2CH_2OH

c. $CH_3CH_2OCH_2OH$ or $CH_3OCH_2CH_2OH$

b. $\underset{\underset{Br}{|}}{CH_2}\overset{\overset{Br}{|}}{CH}CH_2OH$ or $CH_3\overset{\overset{Br}{|}}{\underset{\underset{Br}{|}}{C}}CH_2OH$

d. $CH_3\overset{\overset{O}{\|}}{C}CH_2CH_2OH$ or $CH_3CH_2\overset{\overset{O}{\|}}{C}CH_2OH$

Relative Base Strengths

Strong bases readily share their electrons with a proton. In other words, the conjugate acid of a *strong* base is a *weak* acid because it does not readily lose a proton. This allows us to say, *the stronger the base, the weaker its conjugate acid* (or *the stronger the acid, the weaker its conjugate base*).

For example, which is the stronger base?

a. CH_3O^- or CH_3NH^-?   b. $HC\equiv C^-$ or $CH_3\bar{C}H_2$

In order to answer the question, first compare their conjugate acids:

a. CH_3OH is a stronger acid than CH_3NH_2 (because O is more electronegative than N). Since the stronger acid has the weaker conjugate base, $CH_3\bar{N}H$ is a stronger base than CH_3O^-.

b. $HC\equiv CH$ is a stronger acid than CH_3CH_3 (an *sp* hybridized atom is more electronegative than an sp^3 hybridized atom). Therefore, $CH_3\bar{C}H_2$ is a stronger base.

PROBLEM 9 Which is the stronger base?

a. Br^- or I^-

b. CH_3O^- or CH_3S^-

c. $CH_3CH_2O^-$ or CH_3COO^-

d. $H_2C=\bar{C}H$ or $HC\equiv C^-$

e. $FCH_2CH_2O^-$ or $BrCH_2CH_2O^-$

f. $ClCH_2CH_2O^-$ or $Cl_2CHCH_2O^-$

Weak Bases Are Stable Bases

Weak bases are stable bases because they readily bear the electrons they formerly shared with a proton. Therefore, we can say, *the weaker the base, the more stable it is*. We can also say, *the stronger the acid, the more stable (the weaker) its conjugate base*.

For example, which is a more stable base, Cl^- or Br^-?

In order to determine this, first compare their conjugate acids:
HBr is a stronger acid than HCl (because Br is larger than Cl). Therefore, Br^- is a more stable (weaker) base.

PROBLEM 10 Which is the more stable base?

a. Br^- or I^-

b. CH_3O^- or CH_3S^-

c. $CH_3CH_2O^-$ or CH_3COO^-

d. $H_2C=\bar{C}H$ or $HC\equiv C^-$

e. $FCH_2CH_2O^-$ or $BrCH_2CH_2O^-$

f. $ClCH_2CH_2O^-$ or $Cl_2CHCH_2O^-$

Electron Delocalization Stabilizes a Base

If a base has localized electrons, then the negative charge that results when the base's conjugate acid loses a proton will belong to one atom. On the other hand, if a base has delocalized electrons, then the negative charge that results when the base's conjugate acid loses a proton will be shared by two (or more) atoms. A base with delocalized electrons is more stable than a similar base with localized electrons.

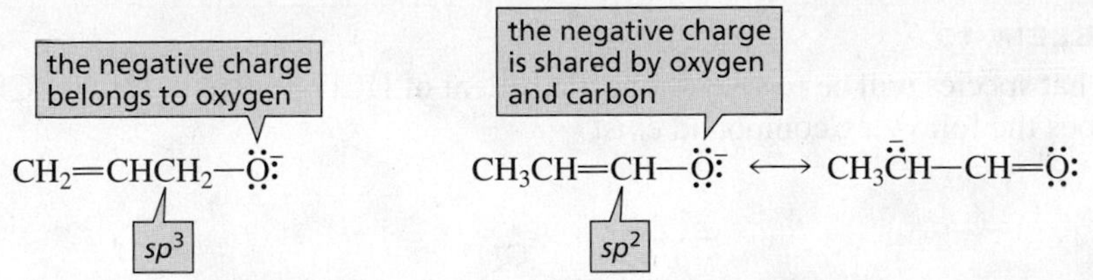

the negative charge belongs to oxygen

the negative charge is shared by oxygen and carbon

$$CH_2=CHCH_2-\ddot{O}:^- \qquad\qquad CH_3CH=CH-\ddot{O}:^- \longleftrightarrow CH_3\ddot{C}H-CH=\ddot{O}:$$

sp^3 sp^2

How do we know whether a base has delocalized atoms? If the electrons left behind, when the base's conjugate acid loses a proton, are on an atom bonded to an sp^3 carbon, then the electrons will belong to only one atom—that is, the electrons will be *localized*. If the electrons left behind, when the base's conjugate acid loses a proton, are on an atom bonded to an sp^2 carbon, then the electrons will be *delocalized*.

PROBLEM 11 Which is a more stable base?

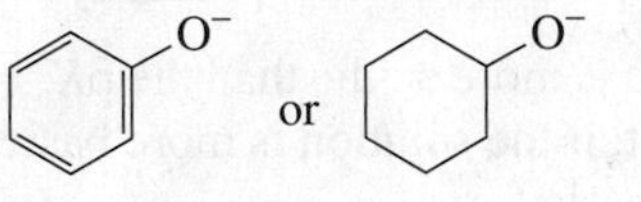

or

Remembering that the more stable (weaker) base has the stronger conjugate acid, solve Problem 12.

PROBLEM 12 Which is the stronger acid?

a.   or   b.   or

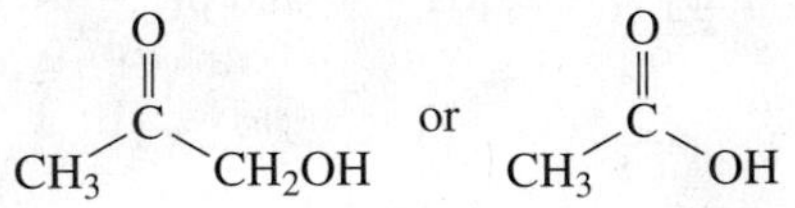

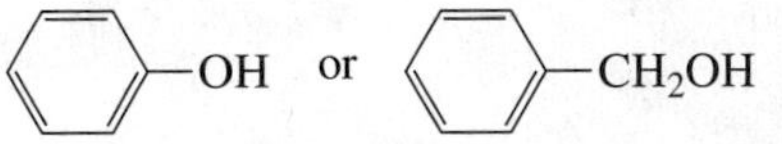

Compounds with More Than One Acidic Group

If a compound has two acidic groups, then a base will remove a proton from the more acidic of the two groups first. If a second equivalent of base is added, then the base will remove a proton from the less acidic group.

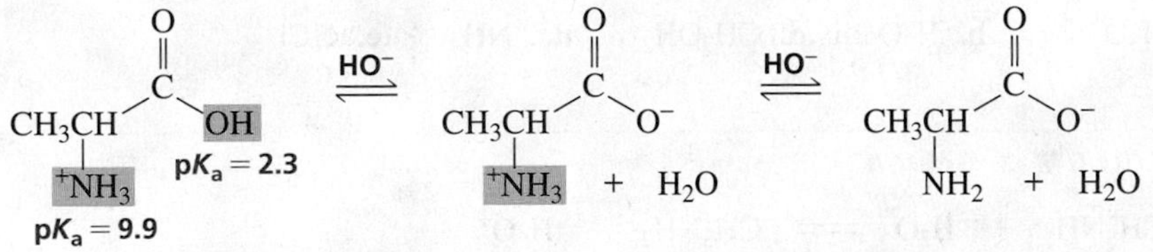

Similarly, if a compound has two basic groups, then an acid will protonate the more basic of the two groups first. If a second equivalent of acid is added, then the acid will protonate the less basic group.

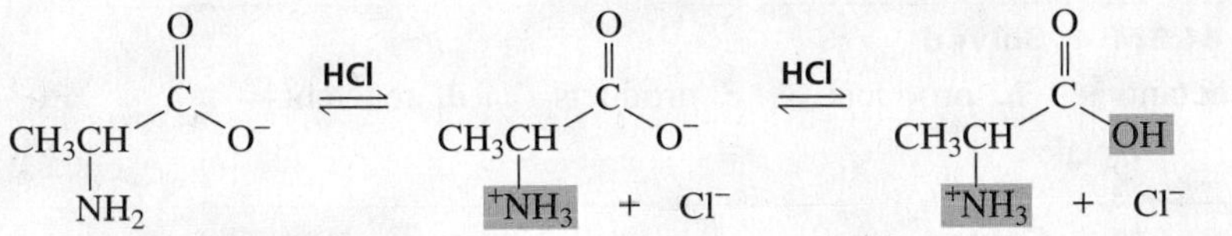

PROBLEM 13

a. What species will be formed if one equivalent of HCl is added to $HOCH_2CH_2NH_2$?

b. Does the following compound exist?

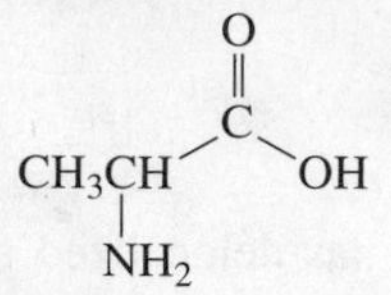

$$\underset{\underset{NH_2}{|}}{CH_3CH}\overset{\overset{O}{\|}}{C}OH$$

The Effect of pH on Structure

Whether an acid will be in its acidic form (with its proton) or its basic form (without its proton) depends on the pK_a value of the acid and the pH of the solution:

- If $pH < pK_a$, then the compound will exist primarily in its acidic form.

- If $pH > pK_a$, then the compound will exist primarily in its basic form.

In other words, if the solution is more acidic than the pK_a value of the acid, the compound will be in its acidic form. But, if the solution is more basic than the pK_a value of the acid, the compound will be in its basic form.

PROBLEM 14

a. Draw the structure of CH_3COOH ($pK_a = 4.7$) at pH = 2, pH = 7, and pH = 10.

b. Draw the structure of CH_3OH ($pK_a = 15.5$) at pH = 2, pH = 7, and pH = 10.

c. Draw the structure of $CH_3\overset{+}{N}H_3$ ($pK_a = 10.7$) at pH = 2, pH = 7, and pH = 14.

ANSWERS TO PROBLEMS ON ACIDS AND BASES

PROBLEM 1 Solved

a. CH_3O^- b. CH_3NH_2 c. $CH_3\bar{N}H$ d. H_2O e. HO^-

PROBLEM 2 Solved

a. H_3O^+ b. H_2O c. $CH_3\overset{+}{O}H_2$ d. $^+NH_4$ e. HCl

---

PROBLEM 3 Solved

a. $CH_3\overset{+}{N}H_3$ + H_2O $\rightleftharpoons$ CH_3NH_2 + H_3O^+

b. HBr + CH_3OH $\rightleftharpoons$ Br^- + $CH_3\overset{+}{O}H_2$

c. $CH_3\overset{+}{N}H_3$ + HO^- $\rightleftharpoons$ CH_3NH_2 + H_2O

d. CH_3NH_2 + CH_3OH $\rightleftharpoons$ $CH_3\overset{+}{N}H_3$ + CH_3O^-

---

PROBLEM 4 Solved

a. reactants   b. products   c. products   d. reactants

PROBLEM 5 Solved

a. CH_3OH b. HF c. HF d. CH_3OH

PROBLEM 6 Solved

a. $HC \equiv CH$ **b.** $HC \equiv CH$ **c.** $H_2C = CH_2$

PROBLEM 7 Solved

a. HBr **b.** CH_3SH **c.** HCl **d.** H_2S

PROBLEM 8 Solved

a. FCH_2CH_2OH **b.** $CH_3\overset{\displaystyle Br}{\underset{\displaystyle Br}{C}}CH_2OH$ **c.** $CH_3CH_2OCH_2OH$ **d.** $CH_3CH_2\overset{\displaystyle O}{C}CH_2OH$

PROBLEM 9 Solved

a. Br^- **c.** $CH_3CH_2O^-$ **e.** $BrCH_2CH_2O^-$
b. CH_3O^- **d.** $H_2C = CH^-$ **f.** $ClCH_2CH_2O^-$

PROBLEM 10 Solved

a. I^- **c.** CH_3COO^- **e.** $FCH_2CH_2O^-$
b. CH_3S^- **d.** $HC \equiv C^-$ **f.** $Cl_2CHCH_2O^-$

PROBLEM 11 Solved   **PROBLEM 12 Solved**

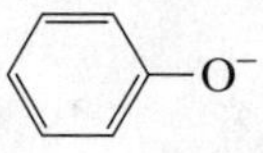

a. $CH_3\overset{\displaystyle O}{C}OH$ **b.** phenol-OH structure

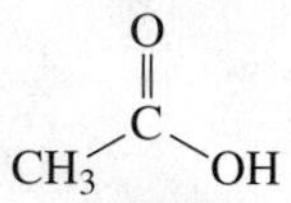

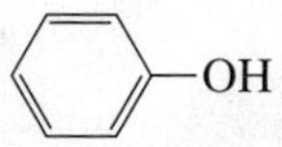

PROBLEM 13 Solved

a. $HOCH_2CH_2\overset{+}{N}H_3$

b. The compound does not exist. For it to be formed, a base would have to be able to remove a proton from a group with a $pK_a = 9.9$ more readily than it would remove a proton from a group with a $pK_a = 2.3$. This is not possible, because the lower the pK_a, the stronger the acid—that is, the more readily the group loses a proton. In other words, a weak acid cannot lose a proton more readily than a strong acid can.

PROBLEM 14 Solved

a. CH_3COOH at pH = 2, because pH $< pK_a$
     CH_3COO^- at pH = 7 and 10, because pH $> pK_a$

b. CH_3OH at pH = 2, 7, and 10, because pH $< pK_a$

c. $CH_3\overset{+}{N}H_3$ at pH = 2 and 7, because pH $< pK_a$
     CH_3NH_2 at pH = 14, because pH $> pK_a$

MasteringChemistry®
for Organic Chemistry

MasteringChemistry tutorials guide you through the toughest topics in chemistry with self-paced tutorials that provide individualized coaching. These assignable, in-depth tutorials are designed to coach you with hints and feedback specific to your individual misconceptions. For additional practice on Acids and Bases, go to MasteringChemistry where the following tutorials are available:

• Acids and Bases: Equilibrium Basics
• Acids and Bases: Factors Influencing Acid Strength
• Acids and Bases: pH Influence on Acid and Base Structure

GLOSSARY

acid (Brønsted) a substance that donates a proton.
acid–base reaction a reaction in which an acid donates a proton to a base or accepts a share in a base's electrons.
acid dissociation constant a measure of the degree to which an acid dissociates in solution.
base[1] a substance that accepts a proton.
basicity the tendency of a compound to share its electrons with a proton.
buffer a weak acid and its conjugate base.
conjugate acid a species accepts a proton to form its conjugate acid.
conjugate base a species loses a proton to form its conjugate base.
electrophile an electron-deficient atom or molecule.
equilibrium constant the ratio of products to reactants at equilibrium or the ratio of the rate constants for the forward and reverse reactions.
Henderson–Hasselbalch equation $pK_a = pH + \log[HA]/[A^-]$
inductive electron withdrawal withdrawal of electrons through a σ bond.
Lewis acid a substance that accepts an electron pair.
Lewis base a substance that donates an electron pair.
nucleophile an electron-rich atom or molecule.
pH the pH scale is used to describe the acidity of a solution ($pH = -\log[H^+]$).
proton transfer reaction a reaction in which a proton is transferred from an acid to a base.
resonance contributor (resonance structure, contributing resonance structure) a structure with localized electrons that approximates the true structure of a compound with delocalized electrons.
resonance hybrid the actual structure of a compound with delocalized electrons; it is represented by two or more structures with localized electrons.

PHOTO CREDITS

Photo credits are listed in order of appearance.

Michael Fernahl/iStockphoto; (left) NYC Parks Photo Archive/Fundamental Photographs; (rt.) Brian Mickelson; USDA-NRCS PLANTS Database/William & Wilma Follette; (left) SPL/Photo Researchers, Inc.; (rt.) Professor Pietro M. Motta/ Photo Researchers, Inc.; (top) Richard Megna/Fundamental Photographs; (bot.) Ravi/Shutterstock ; Sashkin/Shutterstock

An Introduction to Organic Compounds

From Chapter 3 of *Organic Chemistry*, Seventh Edition. Paula Yurkanis Bruice. Copyright © 2014 by Pearson Education, Inc.

An Introduction to Organic Compounds

Nomenclature, Physical Properties, and Representation of Structure

cholesterol crystals

The material in this chapter explains why drugs with similar physiological effects have similar structures, how high cholesterol is treated clinically, why fish is served with lemon, how the octane number of gasoline is determined, and why starch (a component of many of the foods we eat) and cellulose (the structural material of plants) have such different physical properties, even though both are composed only of glucose.

The presentation of organic chemistry in this text is organized according to how organic compounds react. When a compound undergoes a reaction, a new compound is synthesized. In other words, while you are learning how organic compounds react, you are simultaneously learning how to synthesize organic compounds.

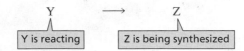

Later in this chapter, we will be looking at the structures and physical properties of alkanes, alkyl halides, ethers, alcohols, and amines. As we learn about the structures, physical properties, and reactions of organic compounds and the products the reactions form, we will need to be able to refer to the compounds by name. Therefore, we will begin the study of organic chemistry by learning how to name these five families of compounds.

First we will see how *alkanes* are named because their names form the basis for the names of all other organic compounds. **Alkanes** are composed of only carbon atoms and hydrogen atoms and contain only *single bonds.* Compounds that contain only carbon and hydrogen are called **hydrocarbons.** Thus, an alkane is a hydrocarbon that has only single bonds.

Alkanes in which the carbons form a continuous chain with no branches are called **straight-chain alkanes.** The names of the four smallest straight-chain alkanes have historical roots, but the others are based on Greek numbers. It is important that you learn the names of at least the first 10 straight-chain alkanes in Table 1.

Table 1 Nomenclature and Physical Properties of Straight-Chain Alkanes

Number of carbons	Molecular formula	Name	Condensed structure	Boiling point (°C)	Melting point (°C)	Density[a] (g/mL)
1	CH_4	methane	CH_4	−167.7	−182.5	
2	C_2H_6	ethane	CH_3CH_3	−88.6	−183.3	
3	C_3H_8	propane	$CH_3CH_2CH_3$	−42.1	−187.7	
4	C_4H_{10}	butane	$CH_3CH_2CH_2CH_3$	−0.5	−138.3	
5	C_5H_{12}	pentane	$CH_3(CH_2)_3CH_3$	36.1	−129.8	0.5572
6	C_6H_{14}	hexane	$CH_3(CH_2)_4CH_3$	68.7	−95.3	0.6603
7	C_7H_{16}	heptane	$CH_3(CH_2)_5CH_3$	98.4	−90.6	0.6837
8	C_8H_{18}	octane	$CH_3(CH_2)_6CH_3$	125.7	−56.8	0.7026
9	C_9H_{20}	nonane	$CH_3(CH_2)_7CH_3$	150.8	−53.5	0.7177
10	$C_{10}H_{22}$	decane	$CH_3(CH_2)_8CH_3$	174.0	−29.7	0.7299
11	$C_{11}H_{24}$	undecane	$CH_3(CH_2)_9CH_3$	195.8	−25.6	0.7402
12	$C_{12}H_{26}$	dodecane	$CH_3(CH_2)_{10}CH_3$	216.3	−9.6	0.7487
13	$C_{13}H_{28}$	tridecane	$CH_3(CH_2)_{11}CH_3$	235.4	−5.5	0.7546
⋮	⋮	⋮	⋮	⋮	⋮	⋮
20	$C_{20}H_{42}$	eicosane	$CH_3(CH_2)_{18}CH_3$	343.0	36.8	0.7886
21	$C_{21}H_{44}$	heneicosane	$CH_3(CH_2)_{19}CH_3$	356.5	40.5	0.7917
⋮	⋮	⋮	⋮	⋮	⋮	⋮
30	$C_{30}H_{62}$	triacontane	$CH_3(CH_2)_{28}CH_3$	449.7	65.8	0.8097

[a]Density is temperature dependent. The densities given are those determined at 20 °C ($d^{20°}$).

The family of alkanes shown in the table is an example of a homologous series. A **homologous series** (*homos* is Greek for "the same as") is a family of compounds in which each member differs from the one before it in the series by one **methylene (CH_2) group.** The members of a homologous series are called **homologs.**

homologs

$CH_3CH_2CH_3$ $CH_3CH_2CH_2CH_3$

If you look at the relative numbers of carbons and hydrogens in the alkanes listed in Table 1, you will see that the general molecular formula for an alkane is C_nH_{2n+2}, where *n* is any integer. So, if an alkane has one carbon, it must have four hydrogens; if it has two carbons, it must have six hydrogens; and so on.

Carbon forms four covalent bonds and hydrogen forms only one covalent bond. This means that there is only one possible structure for an alkane with molecular formula CH_4 (methane) and only one possible structure for an alkane with molecular formula C_2H_6 (ethane). There is also only one possible structure for an alkane with molecular formula C_3H_8 (propane).

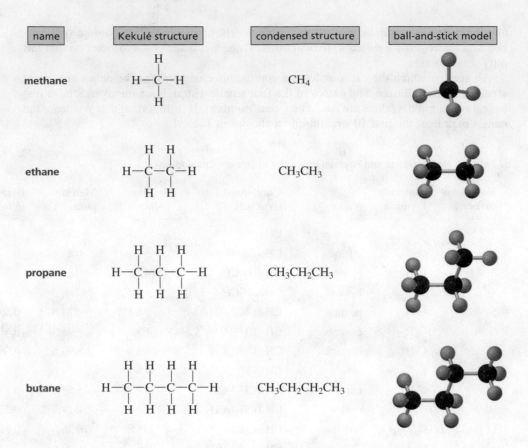

name
Kekulé structure
condensed structure
ball-and-stick model
methane
CH_4
ethane
CH_3CH_3
propane
$CH_3CH_2CH_3$
butane
$CH_3CH_2CH_2CH_3$

However, there are two possible structures for an alkane with molecular formula C_4H_{10}. In addition to butane—a straight-chain alkane—there is a branched butane called isobutane. Both of these structures fulfill the requirement that each carbon forms four bonds and each hydrogen forms one bond.

$CH_3CH_2CH_2CH_3$
butane
CH_3CHCH_3
CH_3
isobutane
constitutional isomers
CH_3CH-
CH_3
an "iso" structural unit

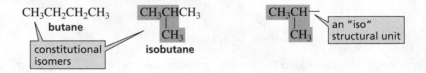

Compounds such as butane and isobutane that have the same molecular formula but differ in the way the atoms are connected are called **constitutional isomers**—their molecules have different constitutions. In fact, isobutane got its name because it is an *iso*mer of butane. The structural unit consisting of *a carbon bonded to a hydrogen and two CH_3 groups,* which occurs in isobutane, has come to be called "iso." Thus, the name *isobutane* tells you that the compound is a four-carbon alkane with an iso structural unit.

There are three alkanes with molecular formula C_5H_{12}. You have already learned how to name two of them. Pentane is the straight-chain alkane. Isopentane, as its name indicates, has an iso structural unit and five carbons. We cannot name the other branched-chain alkane without defining a name for a new structural unit. (For now, ignore the names written in blue.)

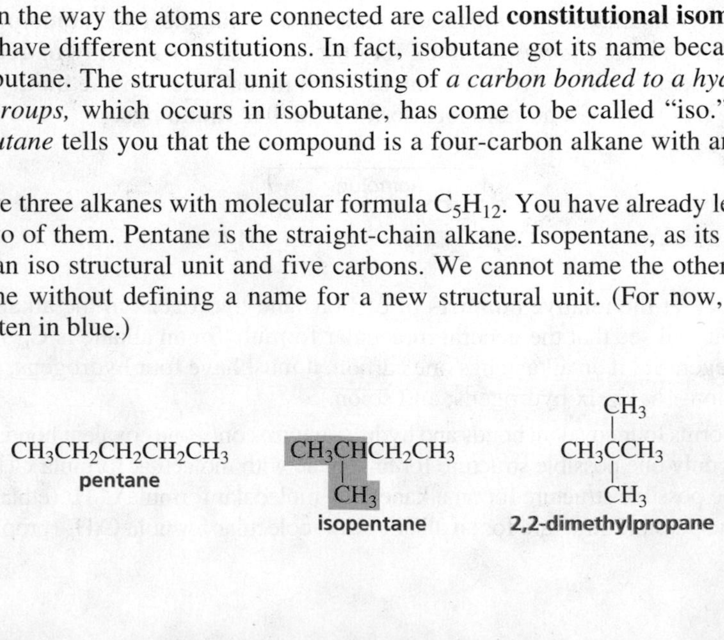

$CH_3CH_2CH_2CH_2CH_3$
pentane
$CH_3CHCH_2CH_3$
CH_3
isopentane
CH_3
CH_3CCH_3
CH_3
2,2-dimethylpropane

There are five constitutional isomers with molecular formula C_6H_{14}. Again, we are able to name only two of them, unless we define new structural units.

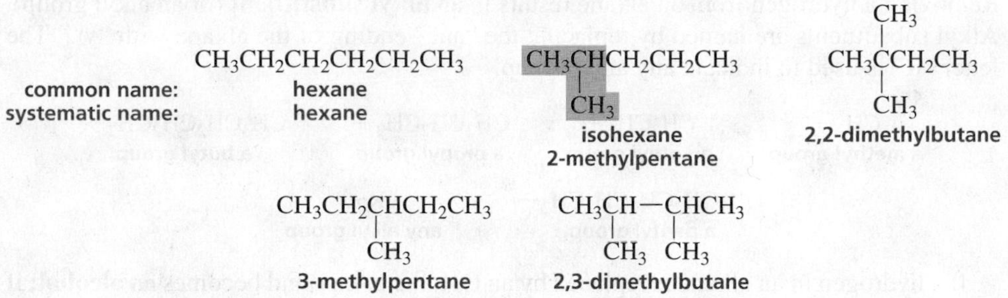

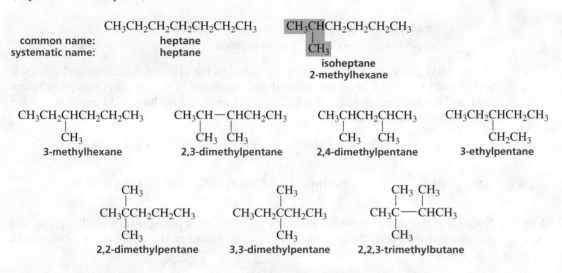

common name: hexane
systematic name: hexane

CH₃CH₂CH₂CH₂CH₂CH₃

CH₃CHCH₂CH₂CH₃
|
CH₃
isohexane
2-methylpentane

CH₃CCH₂CH₃ with CH₃ above and CH₃ below
2,2-dimethylbutane

CH₃CH₂CHCH₂CH₃
|
CH₃
3-methylpentane

CH₃CH—CHCH₃
| |
CH₃ CH₃
2,3-dimethylbutane

There are nine alkanes with molecular formula C_7H_{16}. We can name only two of them (heptane and isoheptane).

common name: heptane
systematic name: heptane

CH₃CH₂CH₂CH₂CH₂CH₂CH₃

CH₃CHCH₂CH₂CH₂CH₃
|
CH₃
isoheptane
2-methylhexane

CH₃CH₂CHCH₂CH₂CH₃
|
CH₃
3-methylhexane

CH₃CH—CHCH₂CH₃
| |
CH₃ CH₃
2,3-dimethylpentane

CH₃CHCH₂CHCH₃
| |
CH₃   CH₃
2,4-dimethylpentane

CH₃CH₂CHCH₂CH₃
|
CH₂CH₃
3-ethylpentane

CH₃CCH₂CH₂CH₃ with CH₃ above and CH₃ below
2,2-dimethylpentane

CH₃CH₂CCH₂CH₃ with CH₃ above and CH₃ below
3,3-dimethylpentane

CH₃C—CHCH₃ with CH₃ CH₃ above and CH₃ below
2,2,3-trimethylbutane

The number of constitutional isomers increases rapidly as the number of carbons in an alkane increases. For example, there are 75 alkanes with molecular formula $C_{10}H_{22}$ and 4347 alkanes with molecular formula $C_{15}H_{32}$. To avoid having to memorize the names of thousands of structural units, chemists have devised rules for creating systematic names that describe the compound's structure. That way, only the rules have to be learned. Because the name describes the structure, these rules make it possible to deduce the structure of a compound from its name.

This method of nomenclature is called **systematic nomenclature.** It is also called **IUPAC nomenclature** because it was designed by a commission of the International Union of Pure and Applied Chemistry (abbreviated IUPAC and pronounced "eye-you-pack") in 1892.

The IUPAC rules have been continually revised by the commission since then. A name such as *isobutane*—a nonsystematic name—is called a **common name.** When both names are shown in this text, common names will be shown in red and systematic (IUPAC) names in blue. Before we can understand how a systematic name for an alkane is constructed, we must learn how to name alkyl substituents.

PROBLEM 1♦

a. How many hydrogens does an alkane with 17 carbons have?
b. How many carbons does an alkane with 74 hydrogens have?

PROBLEM 2

Draw the structures of octane and isooctane.

---

1 HOW ALKYL SUBSTITUENTS ARE NAMED

Removing a hydrogen from an alkane results in an **alkyl substituent** (or an alkyl group). Alkyl substituents are named by replacing the "ane" ending of the alkane with "yl." The letter "R" is used to indicate any alkyl group.

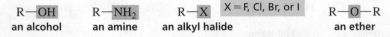

CH_3-	CH_3CH_2-	$CH_3CH_2CH_2-$	$CH_3CH_2CH_2CH_2-$
a methyl group	**an ethyl group**	**a propyl group**	**a butyl group**

$CH_3CH_2CH_2CH_2CH_2-$ $R-$
a pentyl group   **any alkyl group**

If a hydrogen in an alkane is replaced by an OH, the compound becomes an **alcohol;** if it is replaced by an NH_2, the compound becomes an **amine;** if it is replaced by a halogen, the compound becomes an **alkyl halide;** and if it is replaced by an OR, the compound becomes an **ether.**

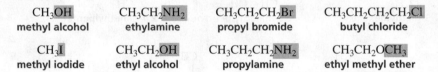

$R-OH$	$R-NH_2$	$R-X$	X = F, Cl, Br, or I	$R-O-R$
an alcohol	**an amine**	**an alkyl halide**		**an ether**

The alkyl group name followed by the name of the class of the compound (alcohol, amine, etc.) yields the common name of the compound. The two alkyl groups in ethers are listed in alphabetical order. The following examples show how alkyl group names are used to build common names:

CH_3OH	$CH_3CH_2NH_2$	$CH_3CH_2CH_2Br$	$CH_3CH_2CH_2CH_2Cl$
methyl alcohol	**ethylamine**	**propyl bromide**	**butyl chloride**
CH_3I	CH_3CH_2OH	$CH_3CH_2CH_2NH_2$	$CH_3CH_2OCH_3$
methyl iodide	**ethyl alcohol**	**propylamine**	**ethyl methyl ether**

Notice that for most compounds there is a space between the name of the alkyl group and the name of the class of compound. For amines, however, the entire name is written as one word.

PROBLEM 3♦

Name each of the following:

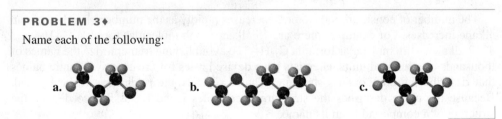

a.   b.   c.

There are two alkyl groups—the propyl group and the isopropyl group— that have three carbons. A propyl group is obtained when a hydrogen is removed from *a primary carbon* of propane. A **primary carbon** is a carbon bonded to only one other carbon. An isopropyl group is obtained when a hydrogen is removed from the *secondary carbon* of propane. A **secondary carbon** is a carbon bonded to two other carbons. Notice that an isopropyl group, as its name indicates, has its three carbon atoms arranged as an iso structural unit—that is, a carbon bonded to a hydrogen and to two CH_3 groups.

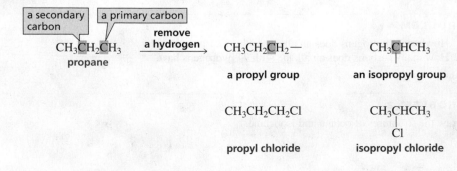

a secondary carbon	a primary carbon			
$CH_3CH_2CH_3$	remove a hydrogen →	$CH_3CH_2CH_2-$		CH_3CHCH_3
propane		**a propyl group**		**an isopropyl group**

$CH_3CH_2CH_2Cl$ CH_3CHCH_3
                                    |
                                    Cl
propyl chloride   **isopropyl chloride**

methyl alcohol

methyl chloride

methylamine

Molecular structures can be drawn in different ways. For example, isopropyl chloride is drawn below in two different ways. Both representations depict the same compound. Although the two-dimensional representations may appear at first to be different (the methyl groups are placed at opposite ends in one structure and at right angles in the other), the structures are identical because carbon is tetrahedral. The four groups bonded to the central carbon—a hydrogen, a chlorine, and two methyl groups—point to the corners of a tetrahedron. If you rotate the three-dimensional model on the right 90° in a clockwise direction, you should be able to see that the two models are the same.

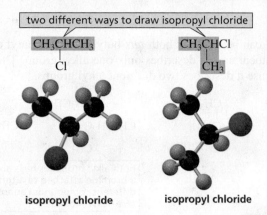

isopropyl chloride   isopropyl chloride

Build models of the two representations of isopropyl chloride to see that they represent the same compound.

There are four alkyl groups that have four carbons. Two of them, the butyl and isobutyl groups, have a hydrogen removed from a primary carbon. A *sec*-butyl group has a hydrogen removed from a secondary carbon (*sec*-, sometimes abbreviated *s*-, stands for secondary), and a *tert*-butyl group has a hydrogen removed from a tertiary carbon (*tert*-, often abbreviated *t*-, stands for tertiary). A **tertiary carbon** is a carbon that is bonded to three other carbons. Notice that the isobutyl group is the only one with an iso structural unit.

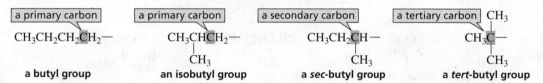

a butyl group an isobutyl group a *sec*-butyl group a *tert*-butyl group

The names of straight-chain alkyl groups often have the prefix "*n*" (for "normal") to emphasize that the carbons are in an unbranched chain. If a name does not have a prefix such as "*n*", "iso," "*sec*," or "*tert*," we assume that the carbons are in an unbranched chain.

A primary carbon is bonded to one carbon, a secondary carbon is bonded to two carbons, and a tertiary carbon is bonded to three carbons.

$CH_3CH_2CH_2CH_2Br$ $CH_3CH_2CH_2CH_2CH_2F$
butyl bromide   **pentyl fluoride**
or   **or**
***n*-butyl bromide**   ***n*-pentyl fluoride**

Like the carbons, the hydrogens in a molecule are also referred to as primary, secondary, and tertiary. **Primary hydrogens** are attached to a primary carbon, **secondary hydrogens** are attached to a secondary carbon, and **tertiary hydrogens** are attached to a tertiary carbon.

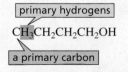 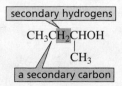 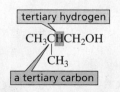

Primary hydrogens are attached to a primary carbon, secondary hydrogens to a secondary carbon, and tertiary hydrogens to a tertiary carbon.

A chemical name must specify one compound only. The prefix "*sec*," therefore, can be used only for *sec*-butyl compounds. The name "*sec*-pentyl" cannot be used because pentane has two different secondary carbons. Thus, removing a hydrogen from a secondary carbon of pentane produces one of two different alkyl groups, depending on which hydrogen is removed. As a result, *sec*-pentyl chloride would specify two different alkyl chlorides, so it is *not* a correct name.

A name must specify one compound only.

> Both alkyl halides have five carbon atoms with a chlorine attached to a secondary carbon, but two compounds cannot be named *sec*-pentyl chloride.

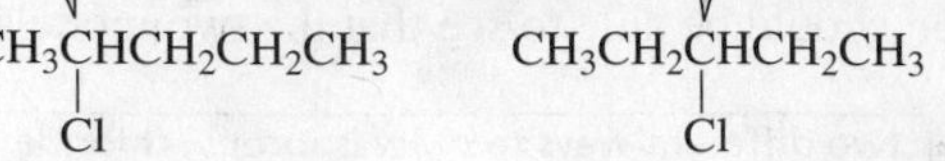

$$CH_3CHCH_2CH_2CH_3 \qquad CH_3CH_2CHCH_2CH_3$$
$$\hspace{0.5cm} | \hspace{5cm} |$$
$$\hspace{0.5cm} Cl \hspace{5cm} Cl$$

The prefix "*tert*" can be used for both *tert*-butyl and *tert*-pentyl compounds because each of these substituent names describes only one alkyl group. The name "*tert*-hexyl" cannot be used because it describes two different alkyl groups.

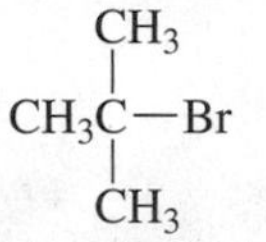

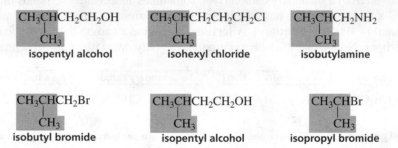

tert-butyl bromide 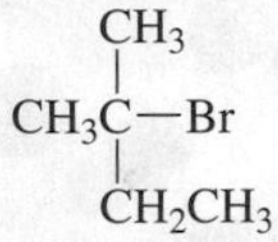 **tert-pentyl bromide**

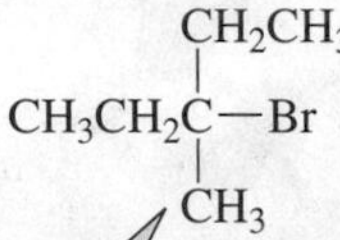

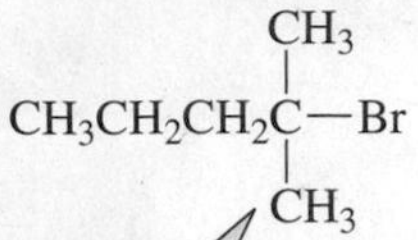

> Both alkyl bromides have six carbon atoms with a bromine attached to a tertiary carbon, but two different compounds cannot be named *tert*-hexyl bromide.

Notice in the following structures that whenever the prefix "iso" is used, the iso structural unit is at one end of the molecule, and any group replacing a hydrogen is at the other end:

$$CH_3CHCH_2CH_2OH \qquad CH_3CHCH_2CH_2CH_2Cl \qquad CH_3CHCH_2NH_2$$
$$\hspace{0.5cm} | \hspace{3cm} | \hspace{3cm} |$$
$$\hspace{0.5cm} CH_3 \hspace{3cm} CH_3 \hspace{3cm} CH_3$$

isopentyl alcohol **isohexyl chloride** **isobutylamine**

$$CH_3CHCH_2Br \qquad CH_3CHCH_2CH_2OH \qquad CH_3CHBr$$
$$\hspace{0.5cm} | \hspace{3cm} | \hspace{3cm} |$$
$$\hspace{0.5cm} CH_3 \hspace{3cm} CH_3 \hspace{3cm} CH_3$$

isobutyl bromide **isopentyl alcohol** **isopropyl bromide**

Notice that an iso group has a methyl group on the next-to-the-last carbon in the chain. Also notice that all isoalkyl compounds have the substituent (OH, Cl, NH_2, etc.) on a primary carbon, except for isopropyl, which has the substituent on a secondary carbon. Thus, the isopropyl group could have been called a *sec*-propyl group. Either name would have been appropriate because the group has an iso structural unit and a hydrogen has been removed from a secondary carbon. Chemists decided to call it isopropyl, however, which means that "*sec*" is used only for *sec*-butyl.

Alkyl group names are used so frequently that you need to learn them. Some of the most common alkyl group names are compiled in Table 2.

Table 2	Names of Some Common Alkyl Groups				
methyl	CH_3-	isobutyl	CH_3CHCH_2-	pentyl	$CH_3CH_2CH_2CH_2CH_2-$
ethyl	CH_3CH_2-		$\qquad\;\; | \atop CH_3$	isopentyl	$CH_3CHCH_2CH_2-$
propyl	$CH_3CH_2CH_2-$	*sec*-butyl	CH_3CH_2CH-		$\qquad | \atop CH_3$
isopropyl	$CH_3CH- \atop \quad\; | \atop \quad CH_3$		$\qquad\quad | \atop \qquad CH_3$		
			$\qquad\quad CH_3$	hexyl	$CH_3CH_2CH_2CH_2CH_2CH_2-$
butyl	$CH_3CH_2CH_2CH_2-$	*tert*-butyl	$CH_3C- \atop \quad\; | \atop \quad CH_3$	isohexyl	$CH_3CHCH_2CH_2CH_2- \atop \qquad | \atop \qquad CH_3$

PROBLEM 4♦

Draw the structure and give the systematic name of a compound with molecular formula C_5H_{12} that has

a. one tertiary carbon.

b. no secondary carbons.

PROBLEM 5♦

Draw the structures and name the four constitutional isomers with molecular formula C_4H_9Br.

PROBLEM 6♦

Which of the following statements can be used to prove that carbon is tetrahedral?

a. Methyl bromide does not have constitutional isomers.
b. Tetrachloromethane does not have a dipole moment.
c. Dibromomethane does not have constitutional isomers.

PROBLEM 7♦

Draw the structure for each of the following compounds:

a. isopropyl alcohol
b. isopentyl fluoride

c. *sec*-butyl iodide
d. *tert*-pentyl alcohol

e. *tert*-butylamine
f. *n*-octyl bromide

PROBLEM 8♦

Name the following compounds:

a. $CH_3OCH_2CH_3$

c. $CH_3CH_2CHNH_2$
     $\quad\quad\quad\; | $
     $\quad\quad\quad CH_3$

e. CH_3CHCH_2Br
     $\quad\; | $
     $\quad CH_3$

b. $CH_3OCH_2CH_2CH_3$

d. $CH_3CH_2CH_2CH_2OH$

f. CH_3CH_2CHCl
     $\quad\quad\quad\; | $
     $\quad\quad\quad CH_3$

2   THE NOMENCLATURE OF ALKANES

The systematic name of an alkane is obtained using the following rules:

1. Determine the number of carbons in the longest continuous carbon chain. This chain is called the **parent hydrocarbon.** The name that indicates the number of carbons in the parent hydrocarbon becomes the alkane's "last name." For example, a parent hydrocarbon with eight carbons would be called *octane*. The longest continuous chain is not always in a straight line; sometimes you have to "turn a corner" to obtain the longest continuous chain.

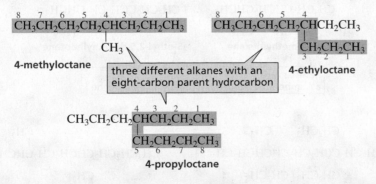

three different alkanes with an eight-carbon parent hydrocarbon

4-methyloctane

4-ethyloctane

4-propyloctane

First, determine the number of carbons in the longest continuous chain.

2. The name of any alkyl substituent that hangs off the parent hydrocarbon is placed in front of the name of the parent hydrocarbon, together with a number to designate the carbon to which the alkyl substituent is attached. The carbons in the parent chain are numbered in the direction that gives the substituent as low a number as possible. The substituent's name and the name of the parent hydrocarbon are joined into one word, preceded by a hyphen that connects the substituent's number with its name.

$$\overset{1}{C}H_3\overset{2}{C}H\overset{3}{C}H_2\overset{4}{C}H_2\overset{5}{C}H_3 \qquad \overset{6}{C}H_3\overset{5}{C}H_2\overset{4}{C}H_2\overset{3}{C}H_2\overset{2}{C}HCH_2\overset{1}{C}H_3 \qquad \overset{1}{C}H_3\overset{2}{C}H_2\overset{3}{C}H_2\overset{4}{C}HCH_2\overset{5}{C}H_2\overset{6}{C}H_2\overset{7}{C}H_2\overset{8}{C}H_3$$

Number the chain in the direction that gives the substituent as low a number as possible.

CH_3	CH_2CH_3	$CHCH_2CH_3$
2-methylpentane	**3-ethylhexane**	**4-propyloctane**
not	not	not
4-methylpentane	**4-ethylhexane**	**5-propyloctane**

Only systematic names have numbers; common names *never* contain numbers.

$$CH_3$$
$$CH_3CHCH_2CH_2CH_3$$

common name:	isohexane
systematic name:	2-methylpentane

Numbers are used only for systematic names, never for common names.

3. If more than one substituent is attached to the parent hydrocarbon, the chain is numbered in the direction that will produce a name containing the lowest of the possible numbers. The substituents are listed in alphabetical order, with each substituent preceded by the appropriate number. In the following example, the correct name contains a 3 as its lowest number, whereas the incorrect name contains a 4 as its lowest number:

$$CH_3CH_2CHCH_2CHCH_2CH_2CH_3$$
$$\qquad\quad CH_3 \quad\;\; CH_2CH_3$$

5-ethyl-3-methyloctane
not
4-ethyl-6-methyloctane
because 3 < 4

Substituents are listed in alphabetical order.

If two or more substituents are the same, the prefixes "di," "tri," and "tetra" are used to indicate how many identical substituents the compound has. The numbers indicating the locations of the identical substituents are listed together, separated by commas. There are no spaces on either side of a comma. There must be as many numbers in a name as there are substituents. The prefixes "di," "tri," "tetra," "*sec*," and "*tert*" are ignored in alphabetizing substituents.

A number and a word are separated by a hyphen; numbers are separated by a comma.

"di," "tri," "tetra," "*sec*," and "*tert*" are ignored in alphabetizing substituents.

$$CH_3CH_2CHCH_2CHCH_3 \qquad\qquad \overset{\displaystyle CH_2CH_3}{CH_3CH_2CCH_2CH_2CHCH_3}$$
$$\qquad CH_3 \quad\; CH_3 \qquad\qquad\qquad\qquad CH_3 \quad\;\; CH_3$$

2,4-dimethylhexane   **5-ethyl-2,5-dimethylheptane**

numbers are separated by a comma;
a number and a word are separated by a hyphen

$$\overset{\displaystyle CH_2CH_3 \quad\; CH_3}{CH_3CH_2CCH_2CH_2CHCHCH_2CH_2CH_3} \qquad\qquad \overset{\displaystyle CH_3}{CH_3CH_2CH_2CHCH_2CH_2CHCH_3}$$
$$\qquad CH_2CH_3 \;\; CH_2CH_3 \qquad\qquad\qquad\qquad\quad CH_3CH_2$$

3,3,6-triethyl-7-methyldecane   **5-ethyl-2-methyloctane**

4. When numbering in either direction leads to the same lowest number for one of the substituents, the chain is numbered in the direction that gives the lowest possible number to one of the remaining substituents.

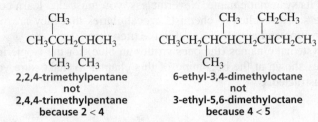

2,2,4-trimethylpentane
not
2,4,4-trimethylpentane
because 2 < 4

6-ethyl-3,4-dimethyloctane
not
3-ethyl-5,6-dimethyloctane
because 4 < 5

5. If the same substituent numbers are obtained in both directions, the first group listed receives the lower number.

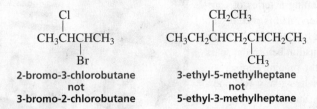

2-bromo-3-chlorobutane
not
3-bromo-2-chlorobutane

3-ethyl-5-methylheptane
not
5-ethyl-3-methylheptane

Only if the same set of numbers is obtained in both directions does the first group listed get the lower number.

6. Systematic names for branched substituents are obtained by numbering the alkyl substituent starting at the carbon attached to the parent hydrocarbon. This means that the carbon attached to the parent hydrocarbon is always the number-1 carbon of the substituent. In a compound such as 4-(1-methylethyl)octane, the substituent name is in parentheses; the number inside the parentheses indicates a position on the substituent, whereas the number outside the parentheses indicates a position on the parent hydrocarbon. (If a prefix such as "di" is part of a branch name, it *is* included in the alphabetization.)

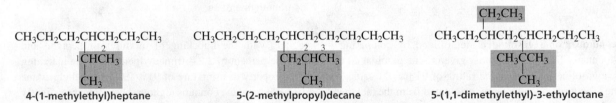

4-(1-methylethyl)heptane

5-(2-methylpropyl)decane

5-(1,1-dimethylethyl)-3-ethyloctane

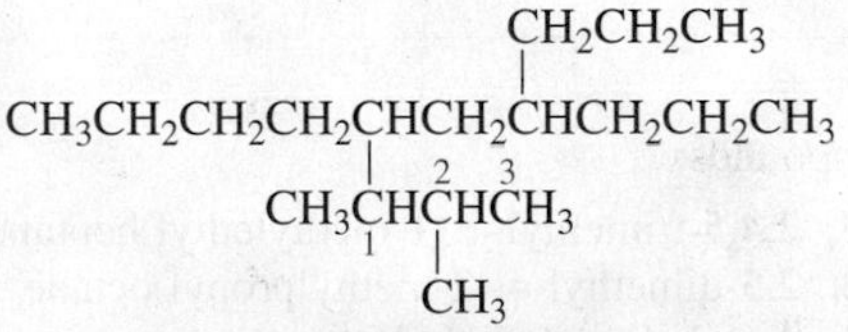

6-(1,2-dimethylpropyl)-4-propyldecane

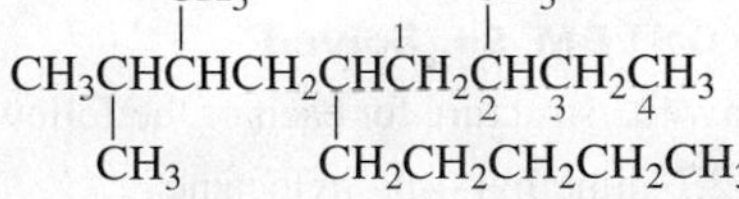

2,3-dimethyl-5-(2-methylbutyl)decane

7. If a compound has two or more chains of the same length, the parent hydrocarbon is the chain with the greatest number of substituents.

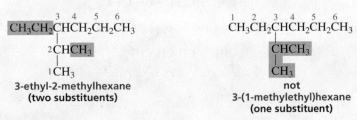

3-ethyl-2-methylhexane
(two substituents)

not
3-(1-methylethyl)hexane
(one substituent)

In the case of two hydrocarbon chains with the same number of carbons, choose the one with the most substituents.

These rules will allow you to name thousands of alkanes, and eventually you will learn the additional rules necessary to name many other kinds of compounds. The rules are important for looking up a compound in the scientific literature, because it usually will be listed by its systematic name. Nevertheless, you must also learn common names because they are so entrenched in chemists' vocabularies that they are widely used in scientific conversation and are often found in the literature.

Look at the systematic names (the ones written in blue) for the isomeric hexanes and isomeric heptanes shown at the beginning of this chapter to make sure you understand how they are constructed.

How Is the Octane Number of Gasoline Determined?

The gasoline engines used in most cars operate by creating a series of carefully timed, controlled explosions. In the engine cylinders, fuel is mixed with air, compressed, and then ignited by a spark. When the fuel used is too easily ignited, then the heat of compression can initiate combustion before the spark plug fires. A pinging or knocking may then be heard in the running engine.

As the quality of the fuel improves, the engine is less likely to knock. The quality of a fuel is indicated by its octane number. Straight-chain hydrocarbons have low octane numbers and make poor fuels. Heptane, for example, with an arbitrarily assigned octane number of 0, causes engines to knock badly. Branched-chain alkanes have more hydrogens bonded to primary carbons. These are the bonds that require the most energy to break and, therefore, make combustion more difficult to initiate, thereby reducing knocking. 2,2,4-Trimethylpentane, for example, does not cause knocking and has arbitrarily been assigned an octane number of 100.

$$CH_3CH_2CH_2CH_2CH_2CH_2CH_3$$
heptane
octane number = 0

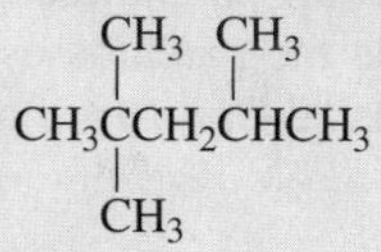

$$CH_3CCH_2CHCH_3$$
with CH₃ CH₃ above and CH₃ below
2,2,4-trimethylpentane
octane number = 100

The octane number of a gasoline is determined by comparing its knocking with the knocking of mixtures of heptane and 2,2,4-trimethylpentane. The octane number given to the gasoline corresponds to the percent of 2,2,4-trimethylpentane in the matching mixture. Thus, a gasoline with an octane rating of 91 has the same "knocking" property as a mixture of 91% 2,2,4-trimethylpentane and 9% heptane. The term *octane number* originated from the fact that 2,2,4-trimethylpentane contains eight carbons. Because slightly different methods are used to determine the octane number, gasoline in Canada and the United States will have an octane number that is 4 to 5 points less than the same gasoline in Europe and Australia.

PROBLEM 9♦ Solved

Draw the structure for each of the following compounds:

a. 2,2-dimethyl-4-propyloctane
b. 2,3-dimethylhexane
c. 4,4-diethyldecane

d. 2,4,5-trimethyl-4-(1-methylethyl)heptane
e. 2,5-dimethyl-4-(2-methylpropyl)octane
f. 4-(1,1-dimethylethyl)octane

Solution to 9a The parent (last) name is *octane,* so the longest continuous chain has eight carbons. Now draw the parent chain and number it.

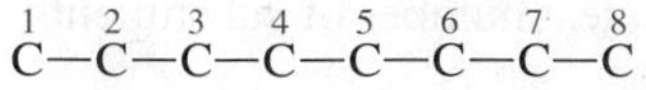

$$\overset{1}{C}-\overset{2}{C}-\overset{3}{C}-\overset{4}{C}-\overset{5}{C}-\overset{6}{C}-\overset{7}{C}-\overset{8}{C}$$

Put the substituents (two methyl groups and a propyl group) on the appropriate carbons.

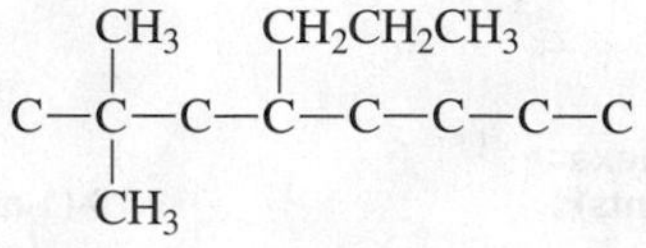

$$C-\underset{CH_3}{\overset{CH_3}{C}}-C-\overset{CH_2CH_2CH_3}{C}-C-C-C-C$$

Add the appropriate number of hydrogens so each carbon is bonded to four atoms.

$$CH_3-\overset{\overset{\displaystyle CH_3}{|}}{\underset{\underset{\displaystyle CH_3}{|}}{C}}-CH_2-\overset{\overset{\displaystyle CH_2CH_2CH_3}{|}}{CH}-CH_2-CH_2-CH_2-CH_3$$

PROBLEM 10 Solved

a. Draw the 18 isomeric octanes.
b. Give each isomer its systematic name.
c. How many isomers have common names?
d. Which isomers contain an isopropyl group?
e. Which isomers contain a *sec*-butyl group?
f. Which isomers contain a *tert*-butyl group?

Solution to 10a Start with the isomer with an eight-carbon continuous chain. Then draw isomers with a seven-carbon continuous chain plus one methyl group. Next, draw isomers with a six-carbon continuous chain plus two methyl groups or one ethyl group. Then draw isomers with a five-carbon continuous chain plus three methyl groups or one methyl group and one ethyl group. Finally, draw a four-carbon continuous chain with four methyl groups. (Your answers to Problem 10b will tell you whether you have drawn duplicate structures, because if two structures have the same systematic name, they represent the same compound.)

PROBLEM 11♦

What is each compound's systematic name?

a. $CH_3CH_2\overset{\overset{\displaystyle CH_3}{|}}{CH}CH_2\overset{\overset{\displaystyle CH_3}{|}}{\underset{\underset{\displaystyle CH_3}{|}}{C}}CH_3$

b. $CH_3CH_2C(CH_3)_3$

c. $CH_3CH_2CH_2\overset{\overset{}{|}}{\underset{\underset{\displaystyle CH_3CHCH_2CH_3}{|}}{CH}}CH_2CH_2CH_3$

d. $CH_3\overset{\overset{\displaystyle CH_3}{|}}{CH}CH_2CH_2\overset{\overset{\displaystyle CH_3}{|}}{\underset{\underset{\displaystyle CH_3}{|}}{C}}CH_3$

e. $CH_3CH_2C(CH_2CH_3)_2CH(CH_3)CH(CH_2CH_2CH_3)_2$

f. $CH_3\overset{\overset{\displaystyle CH_3}{|}}{\underset{\underset{\displaystyle CH_2CH_2CH_3}{|}}{C}}\overset{\overset{\displaystyle CH_2CH_2CH_3}{|}}{CH}CH_2CH_3$

g. $CH_3CH_2C(CH_2CH_3)_2CH_2CH_2CH_3$

h. $CH_3CH_2CH_2CH_2\overset{\overset{}{|}}{\underset{\underset{\displaystyle CH(CH_3)_2}{|}}{CH}}CH_2CH_2CH_3$

i. $CH_3\overset{\overset{}{|}}{\underset{\underset{\displaystyle CH_2CH_3}{|}}{CH}}CH_2CH_2\overset{\overset{\displaystyle CH_3}{|}}{CH}CH_3$

PROBLEM 12♦

Draw the structure and give the systematic name of a compound with molecular formula C_5H_{12} that has

a. only primary and secondary hydrogens.
b. only primary hydrogens.

c. one tertiary hydrogen.
d. two secondary hydrogens.

---

3 THE NOMENCLATURE OF CYCLOALKANES • SKELETAL STRUCTURES

Cycloalkanes are alkanes with their carbon atoms arranged in a ring. Because of the ring, a cycloalkane has two fewer hydrogens than an acyclic (noncyclic) alkane with the same number of carbons. This means that the general molecular formula for a cycloalkane is C_nH_{2n}. Cycloalkanes are named by adding the prefix "cyclo" to the alkane name that signifies the number of carbons in the ring.

cyclopropane   **cyclobutane**   **cyclopentane**   **cyclohexane**

Cycloalkanes are almost always written as **skeletal structures.** Skeletal structures show the carbon–carbon bonds as lines, but do not show the carbons or the hydrogens bonded to carbons. Each vertex in a skeletal structure represents a carbon, and each carbon is understood to be bonded to the appropriate number of hydrogens to give the carbon four bonds.

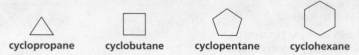

cyclopropane   cyclobutane   cyclopentane   cyclohexane

Acyclic molecules can also be represented by skeletal structures. In skeletal structures of acyclic molecules, the carbon chains are represented by zigzag lines. Again, each vertex represents a carbon, and carbons are assumed to be present where a line begins or ends.

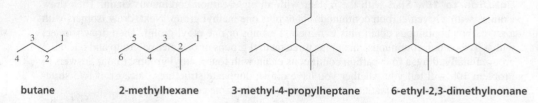

butane   2-methylhexane   3-methyl-4-propylheptane   6-ethyl-2,3-dimethylnonane

The rules for naming cycloalkanes resemble the rules for naming acyclic alkanes:

1. In a cycloalkane with an attached alkyl substituent, the ring is the parent hydrocarbon unless the substituent has more carbons than the ring. In that case, the substituent is the parent hydrocarbon and the ring is named as a substituent. *There is no need to number the position of a single substituent on a ring.*

If there is only one substituent on a ring, do not give that substituent a number.

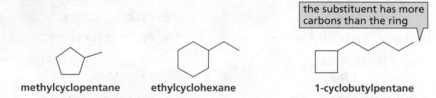

the substituent has more carbons than the ring

methylcyclopentane   ethylcyclohexane   1-cyclobutylpentane

2. If the ring has two different substituents, they are listed in *alphabetical order* and the number-1 position is given to the substituent listed first.

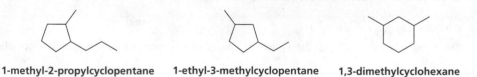

1-methyl-2-propylcyclopentane   1-ethyl-3-methylcyclopentane   1,3-dimethylcyclohexane

3. If there are more than two substituents on the ring, they are listed in alphabetical order, and the substituent given the number-1 position is the one that results in a second substituent getting as low a number as possible. If two substituents have the same low numbers, the ring is numbered—either clockwise or counterclockwise—in the direction that gives the third substituent the lowest possible number.

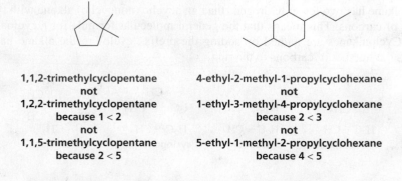

1,1,2-trimethylcyclopentane
not
1,2,2-trimethylcyclopentane
because 1 < 2
not
1,1,5-trimethylcyclopentane
because 2 < 5

4-ethyl-2-methyl-1-propylcyclohexane
not
1-ethyl-3-methyl-4-propylcyclohexane
because 2 < 3
not
5-ethyl-1-methyl-2-propylcyclohexane
because 4 < 5

Skeletal structures can be drawn for compounds other than alkanes. Atoms other than carbon are shown, and hydrogens bonded to atoms other than carbon are also shown.

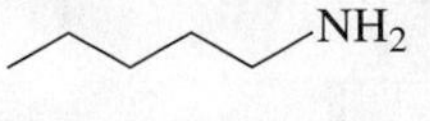

pentylamine 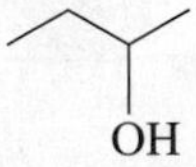   *sec*-butylalcohol 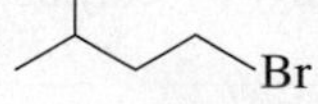   isopentyl bromide

---

PROBLEM-SOLVING STRATEGY

Interpreting a Skeletal Structure

How many hydrogens are attached to each of the indicated carbons in cholesterol?

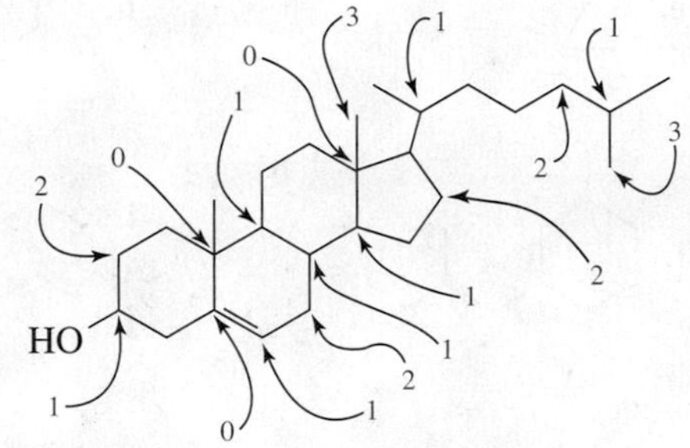

cholesterol

None of the carbons in the compound have a charge, so each needs to be bonded to four atoms. Thus, if the carbon has only one bond showing, it must be attached to three hydrogens that are not shown; if the carbon has two bonds showing, it must be attached to two hydrogens that are not shown, and so on. Check each of the answers (shown in red) to see that this is so.

Now use the strategy you have just learned to solve Problem 13.

PROBLEM 13

How many hydrogens are attached to each of the indicated carbons in morphine?

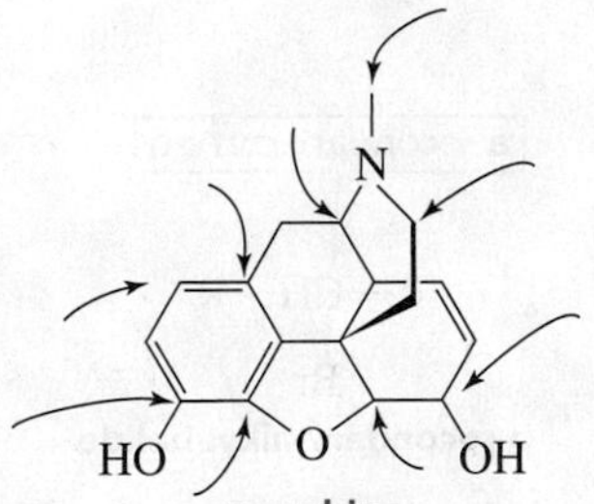

morphine

PROBLEM 14♦

Convert the following condensed structures into skeletal structures:

a. $CH_3CH_2CH_2CH_2CH_2CH_2OH$

b. $CH_3CH_2CH_2CH_2CH_2CH_3$

c. $CH_3CH_2\overset{\displaystyle CH_3}{\underset{|}{C}H}CH_2\overset{\displaystyle CH_3}{\underset{|}{C}H}CH_2CH_3$

d. $CH_3CH_2CH_2CH_2OCH_3$

e. $CH_3CH_2NHCH_2CH_2CH_3$

f. $CH_3\overset{\displaystyle CH_3}{\underset{|}{C}H}CH_2CH_2\underset{\underset{\displaystyle Br}{|}}{C}HCH_3$

Condensed structures show atoms but show few, if any, bonds, whereas skeletal structures show bonds but show few, if any, atoms.

PROBLEM 15

Convert the structures in Problem 11 into skeletal structures.

PROBLEM 16

Draw a condensed and a skeletal structure for the each of following:

a. 3,4-diethyl-2-methylheptane **b.** 2,2,5-trimethylhexane

PROBLEM 17♦

What is each compound's systematic name?

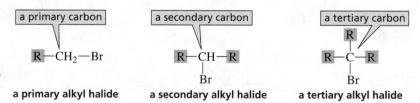

4 THE NOMENCLATURE OF ALKYL HALIDES

An **alkyl halide** is a compound in which a hydrogen of an alkane has been replaced by a halogen. Alkyl halides are classified as primary, secondary, or tertiary, depending on the carbon to which the halogen is attached. **Primary alkyl halides** have a halogen attached to a primary carbon, **secondary alkyl halides** have a halogen attached to a secondary carbon, and **tertiary alkyl halides** have a halogen attached to a tertiary carbon (Section 1). The lone-pair electrons on the halogens are generally not shown unless they are needed to draw your attention to some chemical property of the atom.

> The carbon to which the halogen is attached determines whether an alkyl halide is primary, secondary, or tertiary.

a primary carbon	a secondary carbon	a tertiary carbon
		R
R—CH₂—Br	R—CH—R	R—C—R
	Br	Br
a primary alkyl halide	a secondary alkyl halide	a tertiary alkyl halide

The common names of alkyl halides consist of the name of the alkyl group, followed by the name of the halogen—with the "ine" ending of the halogen name (fluorine, chlorine, bromine, and iodine) replaced by "ide" (fluoride, chloride, bromide, and iodide).

	CH_3Cl	CH_3CH_2F	CH_3CHI CH_3	CH_3CH_2CHBr CH_3
common name:	methyl chloride	ethyl fluoride		
systematic name:	chloromethane	fluoroethane		
			isopropyl iodide 2-iodopropane	*sec*-butyl bromide 2-bromobutane

In the IUPAC system, alkyl halides are named as substituted alkanes. The prefixes for the halogens end with "o" (that is, fluoro, chloro, bromo, and iodo). Therefore, alkyl halides are also called haloalkanes. Notice that although a name must specify only one compound, a compound can have more than one name.

> A compound can have more than one name, but a name must specify only one compound.

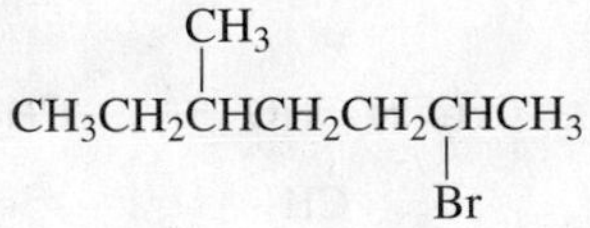

CH₃
|
CH₃CH₂CHCH₂CH₂CHCH₃
|
Br

2-bromo-5-methylheptane

CH_3F
methyl fluoride

1-chloro-6,6-dimethylheptane

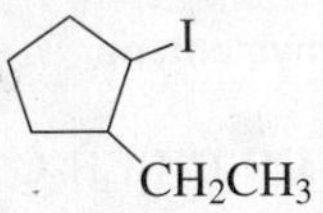

CH_3Cl
methyl chloride

1-ethyl-2-iodocyclopentane

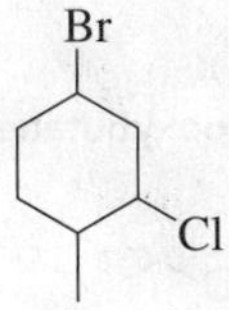

4-bromo-2-chloro-1-methylcyclohexane

CH_3Br
methyl bromide

PROBLEM 18♦

Give two names for each of the following alkyl halides and indicate whether each is primary, secondary, or tertiary:

a. 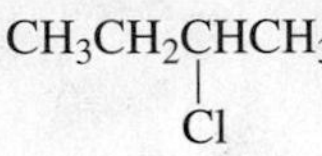 CH₃CH₂CHCH₃
　　　　　|
　　　　　Cl

b. (cyclohexane)—Br

c. 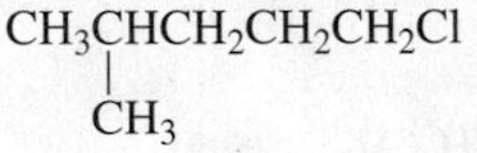 CH₃CHCH₂CH₂CH₂Cl
　　　　|
　　　　CH₃

d. 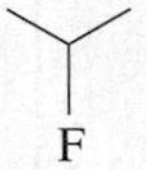 (isopropyl)—F

PROBLEM 19

Draw the structures and provide systematic names for parts **a, b,** and **c** by substituting a chlorine for a hydrogen of methylcyclohexane:

a. a primary alkyl halide　　**b.** a tertiary alkyl halide　　**c.** three secondary alkyl halides

CH_3I
methyl iodide

---

5 THE NOMENCLATURE OF ETHERS

An **ether** is a compound in which an oxygen is bonded to two alkyl substituents. In a **symmetrical ether,** the alkyl substituents are identical. In an **unsymmetrical ether,** the alkyl substituents are different.

R—O—R　　　　R—O—R′
a symmetrical ether　　**an unsymmetrical ether**

The common name of an ether consists of the names of the two alkyl substituents (in alphabetical order), followed by the word "ether." The smallest ethers are almost always named by their common names.

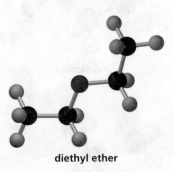

dimethyl ether

CH₃OCH₂CH₃
ethyl methyl ether

CH₃CH₂OCH₂CH₃
diethyl ether
often called ethyl ether

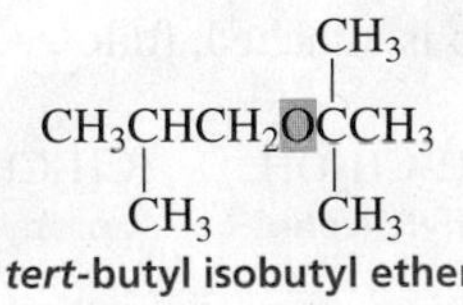

　　　　　　　　CH₃
　　　　　　　　|
CH₃CHCH₂OCCH₃
　　|　　　　|
　　CH₃　　CH₃
***tert*-butyl isobutyl ether**

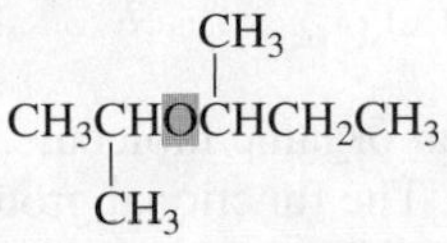

CH₃CHOCHCH₂CH₃
　|　　　|
　CH₃　CH₃
***sec*-butyl isopropyl ether**

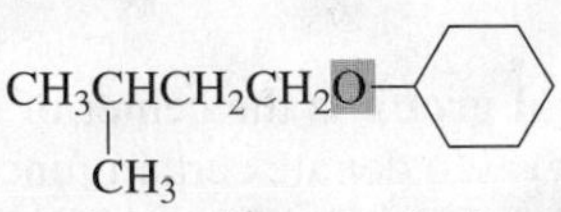

CH₃CHCH₂CH₂O—(cyclohexyl)
　|
　CH₃
cyclohexyl isopentyl ether

diethyl ether

The IUPAC system names an ether as an alkane with an RO substituent. The substituents are named by replacing the "yl" ending in the name of the alkyl substituent with "oxy."

Chemists sometimes neglect the prefix "di" when they name symmetrical ethers. Try not to do this.

CH_3O-
methoxy

CH_3CH_2O-
ethoxy

CH_3CHO-
$\quad\quad|$
$\quad CH_3$
isopropoxy

CH_3CH_2CHO-
$\quad\quad\quad|$
$\quad\quad CH_3$
sec-butoxy

$\quad CH_3$
$\quad\ |$
CH_3CO-
$\quad\ |$
$\quad CH_3$
tert-butoxy

$CH_3CHCH_2CH_3$
$\quad\ |$
$\quad OCH_3$
2-methoxybutane

$CH_3CH_2CHCH_2CH_2OCH_2CH_3$
$\quad\quad\quad|$
$\quad\quad\ CH_3$
1-ethoxy-3-methylpentane

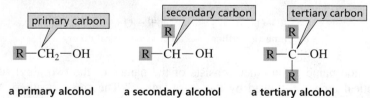

1-butoxy-2,3-dimethylpentane

$CH_3CHOCH_2CH_2CH_2CH_2OCHCH_3$
$\quad\ |\quad\quad\quad\quad\quad\quad\quad\ |$
$\quad CH_3\quad\quad\quad\quad\quad\quad\ CH_3$
1,4-diisopropoxybutane

PROBLEM 20♦

a. What is each ether's systematic name?

  1. $CH_3OCH_2CH_3$

  2. $CH_3CH_2OCH_2CH_3$

  3. $CH_3CH_2CH_2CH_2CHCH_2CH_2CH_3$
  $\quad\quad\quad\quad\quad\quad\ |$
  $\quad\quad\quad\quad\quad\ OCH_3$

  4. $CH_3CHOCH_2CH_2CHCH_3$
  $\quad\quad\ |\quad\quad\quad\quad\ |$
  $\quad\quad CH_3\quad\quad\ CH_3$

  5.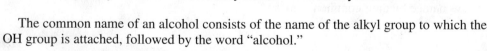

  6.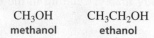

b. Do all of these ethers have common names?

c. What are their common names?

---

6 | THE NOMENCLATURE OF ALCOHOLS

An **alcohol** is a compound in which a hydrogen of an alkane has been replaced by an OH group. Alcohols are classified as **primary, secondary,** or **tertiary,** depending on whether the OH group is attached to a primary, secondary, or tertiary carbon—just like the way alkyl halides are classified.

$R-CH_2-OH$
a primary alcohol
primary carbon

$R-CH-OH$
$\quad\ |$
$\quad R$
a secondary alcohol
secondary carbon

$\quad\ R$
$\quad\ |$
$R-C-OH$
$\quad\ |$
$\quad\ R$
a tertiary alcohol
tertiary carbon

The common name of an alcohol consists of the name of the alkyl group to which the OH group is attached, followed by the word "alcohol."

CH_3CH_2OH
ethyl alcohol

$CH_3CH_2CH_2OH$
propyl alcohol

CH_3CHOH
$\quad\ |$
$\quad CH_3$
isopropyl alcohol

CH_3CHCH_2OH
$\quad\ |$
$\quad CH_3$
isobutyl alcohol

The **functional group** is the center of reactivity in an organic molecule. The IUPAC system uses *suffixes* to denote certain functional groups. The functional group of an alcohol is the OH group, which is denoted by the sufffix "ol." Thus, the systematic name of an alcohol is obtained by replacing the "e" at the end of the name of the parent hydrocarbon with "ol."

CH_3OH
methanol

CH_3CH_2OH
ethanol

methyl alcohol

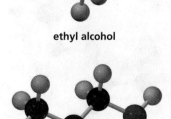
ethyl alcohol

propyl alcohol

The carbon to which the OH group is attached determines whether an alcohol is primary, secondary, or tertiary.

When necessary, the position of the functional group is indicated by a number immediately preceding the name of the parent hydrocarbon or immediately preceding the suffix. The most recently approved IUPAC names are those with the number immediately preceding the suffix. However, the chemical community has been slow to adopt this change, so the names most likely to appear in the literature, on reagent bottles, and on standardized tests are those with the number preceding the name of the parent hydrocarbon. They will also be the ones that appear most often in this text.

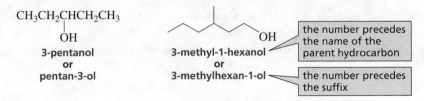

The following rules are used to name a compound that has a functional group suffix:

1. The parent hydrocarbon is the longest continuous chain *containing the functional group*.

2. The parent hydrocarbon is numbered in the direction that gives the *functional group suffix the lowest possible number*.

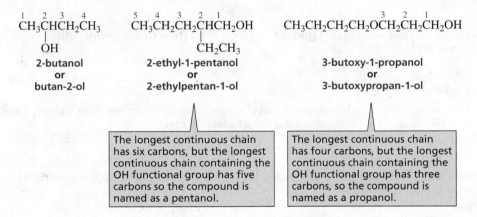

3. If there are two OH groups, the suffix "diol" is added to the name of the parent hydrocarbon.

CH₃CHCH₂CHCH₂CH₃
  | |
  OH   OH

2,4-hexanediol
or
hexane-2,4-diol

         OH
         |
    ────────

         OH

4-methyl-2,3-pentanediol
or
4-methylpentane-2,3-diol

4. If there is a functional group suffix and a substituent, the functional group suffix gets the lowest possible number.

When there is only a substituent, the substituent gets the lowest possible number.

When there is only a functional group suffix, the functional group suffix gets the lowest possible number.

When there is both a functional group suffix and a substituent, the functional group suffix gets the lowest possible number.

$\overset{1}{H}OC\overset{2}{H}_2C\overset{3}{H}_2CH_2Br$

3-bromo-1-propanol

$\overset{4}{Cl}C\overset{3}{H}_2C\overset{2}{H}_2\overset{1}{CH}CH_3$
                    |
                    OH

4-chloro-2-butanol

        CH₃
        |
$\overset{5}{C}H_3\overset{4}{C}\overset{3}{C}H_2\overset{2}{C}H\overset{1}{C}H_3$
        | |
        CH₃ OH

4,4-dimethyl-2-pentanol

5. If counting in either direction gives the same number for the functional group suffix, then the chain is numbered in the direction that gives a substituent the lowest possible number. Notice that a number is not needed to designate the

position of a functional group suffix in a cyclic compound, because it is assumed to be at the 1 position.

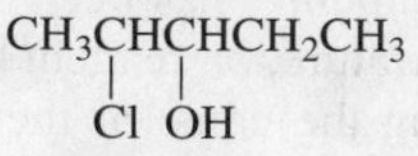

CH₃CHCHCH₂CH₃   CH₃CH₂CH₂CHCH₂CHCH₃
    | |
   Cl OH   OH   CH₃

2-chloro-3-pentanol
not
4-chloro-3-pentanol

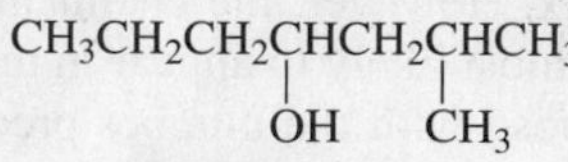

2-methyl-4-heptanol
not
6-methyl-4-heptanol

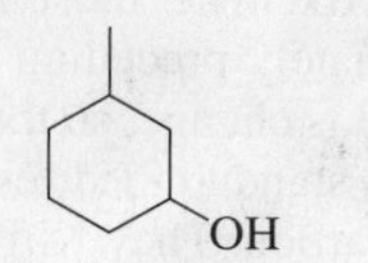

3-methylcyclohexanol
not
5-methylcyclohexanol

6. If there is more than one substituent, the substituents are listed in alphabetical order.

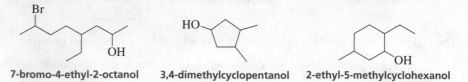

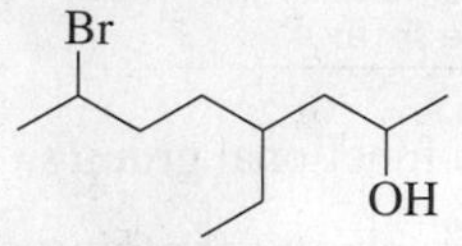

7-bromo-4-ethyl-2-octanol   **3,4-dimethylcyclopentanol**   **2-ethyl-5-methylcyclohexanol**

Remember that the name of a substituent is stated *before* the name of the parent hydrocarbon, and the functional group suffix is stated *after* the name of the parent hydrocarbon.

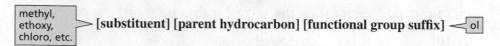

methyl, ethoxy, chloro, etc. ⟶ **[substituent] [parent hydrocarbon] [functional group suffix]** ⟵ ol

---

PROBLEM 21

Draw the structures of a homologous series of alcohols that have from one to six carbons, and then give each of them a common name and a systematic name.

PROBLEM 22♦

Give each of the following a systematic name, and indicate whether each is a primary, secondary, or tertiary alcohol:

a. CH₃CH₂CH₂CH₂CH₂OH

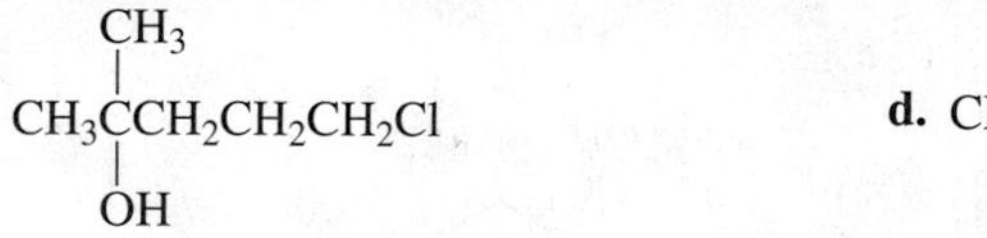

c. 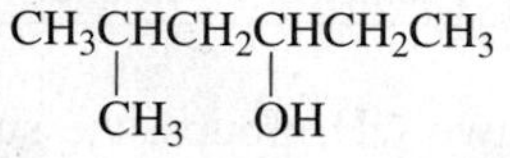 CH₃CHCH₂CHCH₂CH₃
          | |
          CH₃ OH

        CH₃
        |
b. CH₃CCH₂CH₂CH₂Cl
        |
        OH

d. 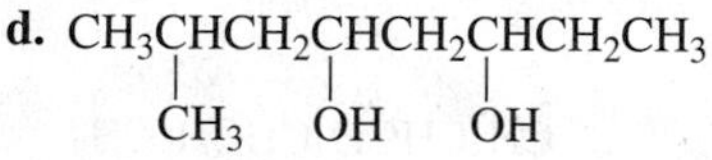 CH₃CHCH₂CHCH₂CHCH₂CH₃
          | | |
          CH₃ OH OH

PROBLEM 23♦

Write the structures of all the tertiary alcohols with molecular formula C₆H₁₄O and give each a systematic name.

PROBLEM 24♦

Give each of the following a systematic name, and indicate whether each is a primary, secondary, or tertiary alcohol:

a.

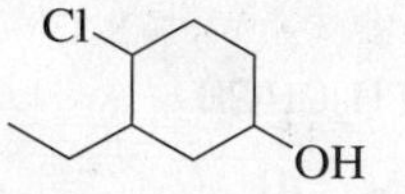

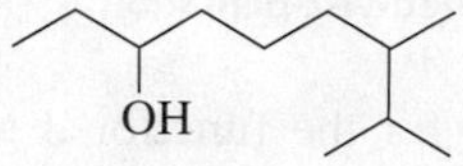

c.

b.

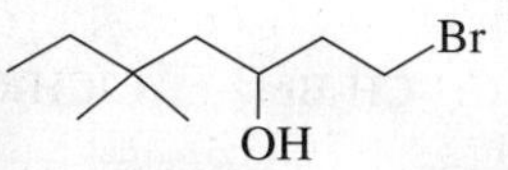

d.

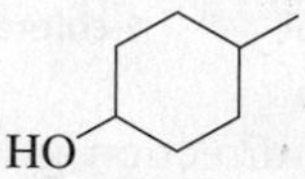

7 THE NOMENCLATURE OF AMINES

An **amine** is a compound in which one or more hydrogens of ammonia have been replaced by alkyl groups. Amines are classified as **primary, secondary,** and **tertiary,** depending on how many alkyl groups are attached to the nitrogen. Primary amines have one alkyl group attached to the nitrogen, secondary amines have two, and tertiary amines have three.

NH₃   R—NH₂   R—NH   R—N—R
ammonia   a primary amine   a secondary amine   a tertiary amine

> The number of alkyl groups attached to the nitrogen determines whether an amine is primary, secondary, or tertiary.

Be sure to note that the number of alkyl groups *attached to the nitrogen* determines whether an amine is primary, secondary, or tertiary. In contrast, whether the X (halogen) or OH group is *attached to a primary, secondary, or tertiary carbon* determines whether an alkyl halide or alcohol is primary, secondary, or tertiary (Sections 4 and 6).

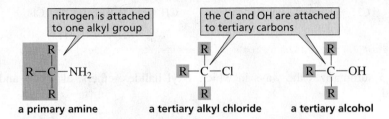

nitrogen is attached to one alkyl group — a primary amine

the Cl and OH are attached to tertiary carbons — a tertiary alkyl chloride — a tertiary alcohol

The common name of an amine consists of the names of the alkyl groups bonded to the nitrogen, in alphabetical order, followed by "amine." The entire name is written as one word (unlike the common names of alcohols, ethers, and alkyl halides, in which "alcohol," "ether," and "halide" are separate words).

CH₃NH₂   CH₃NHCH₂CH₂CH₃   CH₃CH₂NHCH₂CH₃
methylamine   methylpropylamine   diethylamine

CH₃   CH₃   CH₃
CH₃NCH₃   CH₃NCH₂CH₂CH₂CH₃   CH₃CH₂NCH₂CH₂CH₃
trimethylamine   butyldimethylamine   ethylmethylpropylamine

The IUPAC system uses the suffix "amine" to denote the amine functional group. The "e" at the end of the name of the parent hydrocarbon is replaced by "amine"—similar to the way alcohols are named. Also similar to the way alcohols are named, a number identifies the carbon to which the nitrogen is attached, and the number can appear before the name of the parent hydrocarbon or before "amine." The name of any alkyl group bonded to nitrogen is preceded by an "*N*" (in italics) to indicate that the group is bonded to a nitrogen rather than to a carbon.

⁴CH₃³CH₂²CH₂¹CH₂NH₂   ¹CH₃²CH₂³CH⁴CH₂⁵CH₂⁶CH₃   ³CH₃²CH₂¹CH₂NCH₂CH₃
1-butanamine | |
or   NHCH₂CH₃   CH₃
butan-1-amine

N-ethyl-3-hexanamine   *N*-ethyl-*N*-methyl-1-propanamine
or   or
N-ethylhexan-3-amine   *N*-ethyl-*N*-methylpropan-1-amine

The substituents—regardless of whether they are attached to the nitrogen or to the parent hydrocarbon—are listed in alphabetical order, and then a number or an "*N*" is assigned to each one. The chain is numbered in the direction that gives the functional group suffix the lowest number.

$$\overset{4}{C}H_3\overset{3}{C}H\overset{2}{C}H_2\overset{1}{C}H_2NHCH_3$$
$$\underset{Cl}{|}$$

3-chloro-*N*-methyl-1-butanamine

$$\overset{1}{C}H_3\overset{2}{C}H_2\overset{3}{C}H\overset{4}{C}H_2\overset{5}{C}H\overset{6}{C}H_3$$
with CH_3 branch and $NHCH_2CH_3$

N-ethyl-5-methyl-3-hexanamine

4-bromo-*N*,*N*-dimethyl-2-pentanamine

2-ethyl-*N*-propylcyclohexanamine

Nitrogen compounds with four alkyl groups attached to the nitrogen—thereby giving the nitrogen a positive formal charge—are called **quaternary ammonium salts.** Their names consist of the names of the alkyl groups in alphabetical order, followed by "ammonium" (all in one word), and then the name of the accompanying anion as a separate word.

$$CH_3\overset{CH_3}{\underset{CH_3}{\overset{|}{-}N^{\pm}-}}CH_3 \quad HO^-$$

tetramethylammonium hydroxide

$$CH_3CH_2CH_2\overset{CH_3}{\underset{CH_2CH_3}{\overset{|}{-}N^{\pm}-}}CH_3 \quad Cl^-$$

ethyldimethylpropylammonium chloride

Table 3 summarizes the ways in which alkyl halides, ethers, alcohols, and amines are named.

Table 3	Summary of Nomenclature	
	Systematic name	**Common name**
Alkyl halide	substituted alkane	alkyl group attached to halogen plus *halide*
	CH_3Br bromomethane	CH_3Br methyl bromide
	CH_3CH_2Cl chloroethane	CH_3CH_2Cl ethyl chloride
Ether	substituted alkane	alkyl groups attached to oxygen, plus *ether*
	CH_3OCH_3 methoxymethane	CH_3OCH_3 dimethyl ether
	$CH_3CH_2OCH_3$ methoxyethane	$CH_3CH_2OCH_3$ ethyl methyl ether
Alcohol	functional group suffix is *ol*	alkyl group attached to OH, plus *alcohol*
	CH_3OH methanol	CH_3OH methyl alcohol
	CH_3CH_2OH ethanol	CH_3CH_2OH ethyl alcohol
Amine	functional group suffix is *amine*	alkyl groups attached to *N*, plus *amine*
	$CH_3CH_2NH_2$ ethanamine	$CH_3CH_2NH_2$ ethylamine
	$CH_3CH_2CH_2NHCH_3$ *N*-methyl-1-propanamine	$CH_3CH_2CH_2NHCH_3$ methylpropylamine

Bad-Smelling Compounds

Amines are responsible for some of nature's unpleasant odors. Amines with relatively small alkyl groups have a fishy smell. For example, fermented shark, a traditional dish in Iceland, smells exactly like triethylamine. Fish is often served with lemon, because the citric acid in lemon protonates the amine, thereby converting it to its better smelling acidic form.

The citric acid in lemon juice decreases the fishy taste.

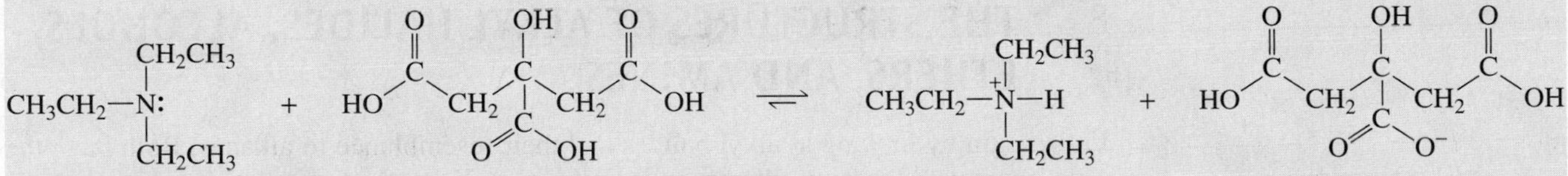

The reaction scheme shows:

triethylamine (bad smelling) + **citric acid** ⇌ **protonated triethylamine** (better smelling) + **conjugate base of citric acid**

The amines putrescine and cadaverine are poisonous compounds formed when amino acids are degraded in the body. Because these amines are excreted as quickly as possible, their odors may be detected in the urine and breath. Putrescine and cadaverine are also responsible for the odor of decaying flesh.

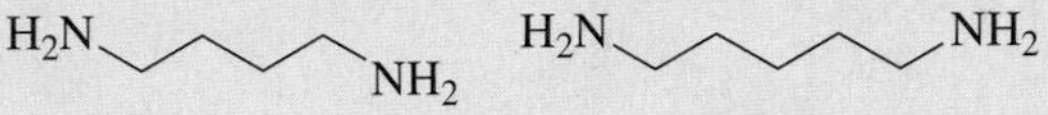

1,4-butanediamine
putrescine

1,5-pentanediamine
cadaverine

PROBLEM 25♦

Are the following compounds primary, secondary, or tertiary?

a. 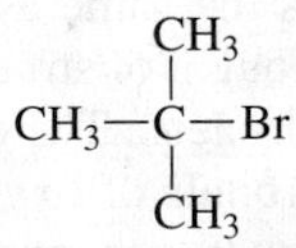 $CH_3-\overset{\displaystyle CH_3}{\underset{\displaystyle CH_3}{C}}-Br$

b. $CH_3-\overset{\displaystyle CH_3}{\underset{\displaystyle CH_3}{C}}-OH$

c. $CH_3-\overset{\displaystyle CH_3}{\underset{\displaystyle CH_3}{C}}-NH_2$

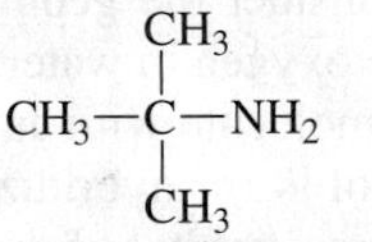

PROBLEM 26♦

Give a systematic name and a common name (if it has one) for each of the following amines and indicate whether each is a primary, secondary, or tertiary amine:

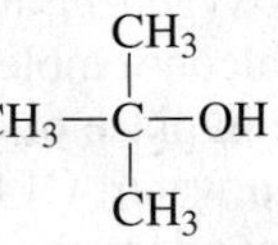

a. $CH_3CH_2CH_2CH_2CH_2CH_2NH_2$

b. 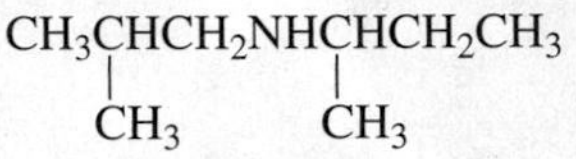 $CH_3\underset{\displaystyle CH_3}{CH}CH_2NH\underset{\displaystyle CH_3}{CH}CH_2CH_3$

c. $(CH_3CH_2)_2NCH_3$

d. $CH_3CH_2CH_2NHCH_2CH_2CH_2CH_3$

e. 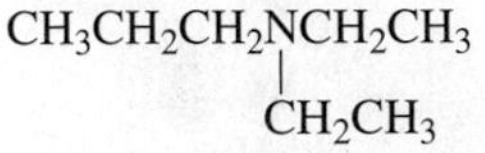 $CH_3CH_2CH_2N\underset{\displaystyle CH_2CH_3}{CH_2CH_3}$

f. 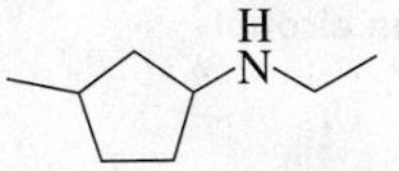 (cyclopentane with N-ethyl amine substituent)

PROBLEM 27♦

Draw the structure for each of the following amines:

a. 2-methyl-*N*-propyl-1-propanamine
b. *N*-ethylethanamine
c. 5-methyl-1-hexanamine
d. methyldipropylamine
e. *N,N*-dimethyl-3-pentanamine
f. cyclohexylethylmethylamine

PROBLEM 28♦

For each of the following, give the systematic name and the common name (if it has one), and then indicate whether it is a primary, secondary, or tertiary amine:

a. 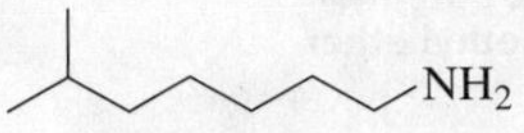 (structure with NH₂)
b. 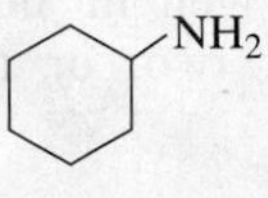 (cyclohexane with NH₂)
c. 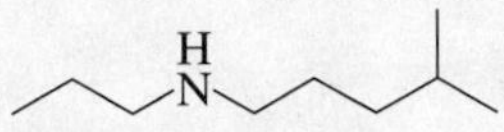 (structure with NH)
d. 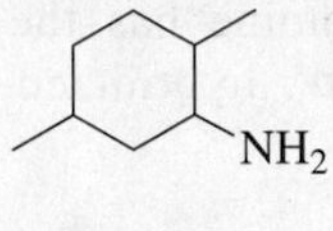 (cyclohexane with NH₂)

8 THE STRUCTURES OF ALKYL HALIDES, ALCOHOLS, ETHERS, AND AMINES

Let's begin by looking at alkyl halides and their resemblance to alkanes. Both have the same geometry; the only difference is that a C—X bond of an alkyl halide (where X denotes a halogen) has replaced a C—H bond of an alkane.

The C—X bond of an alkyl halide is formed from the overlap of an sp^3 orbital of carbon with an sp^3 orbital of the halogen. Fluorine uses a $2sp^3$ orbital to overlap with a $2sp^3$ orbital of carbon, chlorine uses a $3sp^3$ orbital, bromine a $4sp^3$ orbital, and iodine a $5sp^3$ orbital. Thus, the C—X bond becomes longer and weaker as the size of the halogen increases because the electron density of the orbital decreases with increasing volume.

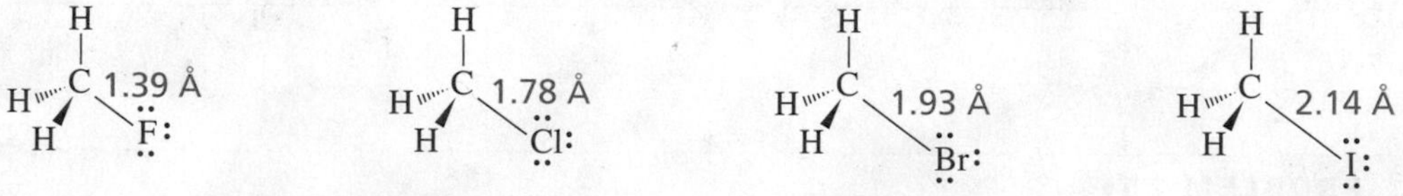

Now let's consider the geometry of the oxygen in an alcohol; it is the same as the geometry of the oxygen in water. In fact, an alcohol molecule can be thought of structurally as a water molecule with an alkyl group in place of one of the hydrogens. The oxygen in an alcohol is sp^3 hybridized, as it is in water. Of the four sp^3 orbitals of oxygen, one overlaps an sp^3 orbital of a carbon, one overlaps the s orbital of a hydrogen, and the other two each contain a lone pair.

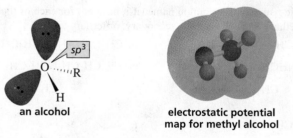

an alcohol

electrostatic potential
map for methyl alcohol

The oxygen in an ether also has the same geometry as the oxygen in water. An ether molecule can be thought of structurally as a water molecule with alkyl groups in place of both hydrogens.

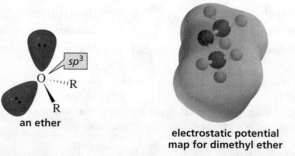

an ether

electrostatic potential
map for dimethyl ether

The nitrogen in an amine has the same geometry as the nitrogen in ammonia. The nitrogen is sp^3 hybridized as in ammonia, with one, two, or three of the hydrogens replaced by alkyl groups. Remember that the number of hydrogens replaced by alkyl groups determines whether the amine is primary, secondary, or tertiary (Section 7).

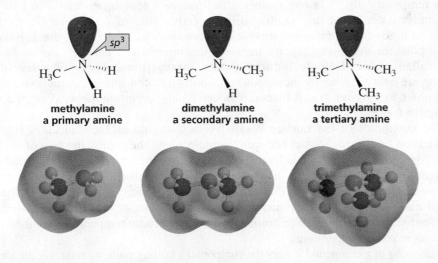

methylamine
a primary amine

dimethylamine
a secondary amine

trimethylamine
a tertiary amine

PROBLEM 29♦

Predict the approximate size of the following bond angles.

a. the C—O—C bond angle in an ether
b. the C—N—C bond angle in a secondary amine
c. the C—O—H bond angle in an alcohol
d. the C—N—C bond angle in a quaternary ammonium salt

---

9   THE PHYSICAL PROPERTIES OF ALKANES, ALKYL HALIDES, ALCOHOLS, ETHERS, AND AMINES

Now we will look at the physical properties of the five families of compounds whose names and structures we have just examined. Comparing the physical properties of several families of compounds makes learning them a little easier than if the physical properties of each family were presented separately.

Boiling Points

The **boiling point (bp)** of a compound is the temperature at which the liquid form becomes a gas (vaporizes). In order for a compound to vaporize, the forces that hold the individual molecules close to each other in the liquid must be overcome. Thus, the boiling point of a compound depends on the strength of the attractive forces between the individual molecules. If the molecules are held together by strong forces, a lot of energy will be needed to pull the molecules away from each other and the compound will have a high boiling point. On the other hand, if the molecules are held together by weak forces, only a small amount of energy will be needed to pull the molecules away from each other and the compound will have a low boiling point.

Van Der Waals Forces

Alkanes contain only carbons and hydrogens. Because the electronegativities of carbon and hydrogen are similar, the bonds in alkanes are nonpolar—there are no significant partial charges on any of the atoms. Alkanes, therefore, are neutral, nonpolar molecules, so the attractive forces between them are relatively weak. The nonpolar nature of alkanes gives them their oily feel.

However, it is only the average charge distribution over the alkane molecule that is neutral. Electrons move continuously, and at any instant the electron density on one side

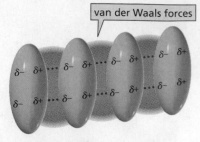

van der Waals forces

▲ **Figure 1**
Van der Waals forces, the weakest of all the attractive forces, are induced-dipole–induced-dipole interactions.

of a molecule can be slightly greater than that on the other side, causing the molecule to have a temporary dipole. Note that a molecule with a dipole has a negative end and a positive end.

A temporary dipole in one molecule can induce a temporary dipole in a nearby molecule. As a result, the (temporarily) negative side of one molecule ends up adjacent to the (temporarily) positive side of another, as shown in Figure 1. Because the dipoles in the molecules are induced, the interactions between the molecules are called **induced-dipole–induced-dipole interactions.** The molecules of an alkane are held together by these induced-dipole–induced-dipole interactions, which are known as **van der Waals forces.** Van der Waals forces are the weakest of all the attractive forces.

The magnitude of the van der Waals forces that hold alkane molecules together depends on the area of contact between the molecules. The greater the area of contact, the stronger the van der Waals forces and the greater the amount of energy needed to overcome them. If you look at the boiling points of the alkanes listed in Table 1, you will see that they increase as their molecular weight increases, because each additional methylene (CH_2) group increases the area of contact between the molecules. The four smallest alkanes have boiling points below room temperature, so they exist as gases at room temperature.

Branching in a compound lowers the compound's boiling point by reducing the area of contact. If you think of *unbranched* pentane as a cigar and *branched* 2,2-dimethylpropane as a tennis ball, you can see that branching decreases the area of contact between molecules— that is, two cigars make contact over a greater surface area than do two tennis balls. Thus, if two alkanes have the same molecular weight, the more highly branched alkane will have a lower boiling point.

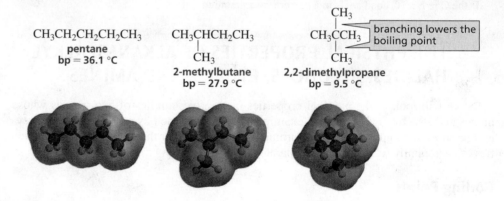

$$CH_3CH_2CH_2CH_2CH_3$$
pentane
bp = 36.1 °C

$$CH_3CHCH_2CH_3$$
$$|$$
$$CH_3$$
2-methylbutane
bp = 27.9 °C

$$CH_3CCH_3$$ (with CH_3 above and CH_3 below)
branching lowers the boiling point
2,2-dimethylpropane
bp = 9.5 °C

PROBLEM 30♦

What is the smallest alkane that is a liquid at room temperature (which is about 25 °C)?

Dipole–Dipole Interactions

The boiling points of a series of ethers, alkyl halides, alcohols, or amines also increase with increasing molecular weight because of the increase in van der Waals forces. The appendix on Physical Properties (or the relevant title) can be found in the Study Area of Mastering Chemistry. (See Appendix: "Physical Properties of Organic Compounds.") The boiling points of these compounds, however, are also affected by the polar C — Z bond. Note that the C — Z bond is polar because nitrogen, oxygen, and the halogens are more electronegative than the carbon to which they are attached.

$$R-\overset{|}{\underset{|}{C}}-\overset{\delta+ \ \delta-}{Z} \qquad Z = N, O, F, Cl, or Br$$

polar bond

Molecules with polar bonds are attracted to one another because they can align themselves in such a way that the positive end of one dipole is adjacent to the negative end of another dipole. These electrostatic attractive forces, called **dipole–dipole interactions,** are stronger than van der Waals forces, but not as strong as ionic or covalent bonds.

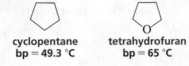
dipole–dipole interaction

Ethers generally have higher boiling points than alkanes of comparable molecular weight (Table 4), because *both* van der Waals forces *and* dipole–dipole interactions must be overcome for an ether to boil.

cyclopentane
bp = 49.3 °C

tetrahydrofuran
bp = 65 °C

The boiling point of a compound depends on the strength of the attraction between the individual molecules.

Table 4 Comparative Boiling Points (°C)			
Alkanes	**Ethers**	**Alcohols**	**Amines**
$CH_3CH_2CH_3$	CH_3OCH_3	CH_3CH_2OH	$CH_3CH_2NH_2$
–42.1	–23.7	78	16.6
$CH_3CH_2CH_2CH_3$	$CH_3OCH_2CH_3$	$CH_3CH_2CH_2OH$	$CH_3CH_2CH_2NH_2$
–0.5	10.8	97.4	47.8

For an alkyl halide to boil, both van der Waals forces and dipole–dipole interactions must be overcome. As the halogen atom increases in size, both of these interactions become stronger. A larger electron cloud means that the van der Waals contact area is greater. A larger electron cloud also means that the cloud's polarizability is greater. **Polarizability** indicates how readily an electron cloud can be distorted to create a strong induced dipole. The larger the atom, the more loosely it holds the electrons in its outermost shell, and the more they can be distorted. Therefore, an alkyl fluoride has a lower boiling point than an alkyl chloride with the same alkyl group. Similarly, alkyl chlorides have lower boiling points than analogous alkyl bromides, which have lower boiling points than analogous alkyl iodides (Table 5).

Table 5 Comparative Boiling Points of Alkanes and Alkyl Halides (°C)					
—Y	**H**	**F**	**Cl**	**Br**	**I**
CH_3—Y	–161.7	–78.4	–24.2	3.6	42.4
CH_3CH_2—Y	–88.6	–37.7	12.3	38.4	72.3
$CH_3CH_2CH_2$—Y	–42.1	–2.5	46.6	71.0	102.5
$CH_3CH_2CH_2CH_2$—Y	–0.5	32.5	78.4	101.6	130.5
$CH_3CH_2CH_2CH_2CH_2$—Y	36.1	62.8	107.8	129.6	157.0

More extensive tables of physical properties can be found in Appendix: "Physical Properties of Organic Compounds."

Hydrogen Bonds

Alcohols have much higher boiling points than ethers with similar molecular weights (Table 4) because, in addition to van der Waals forces and the dipole–dipole interactions of the polar C—O bond, alcohols can form **hydrogen bonds.** A hydrogen bond is a special kind of dipole–dipole interaction that occurs between a hydrogen that is attached to an oxygen, nitrogen, or fluorine and a lone pair of an oxygen, nitrogen, or fluorine in another molecule.

The length of the covalent bond between oxygen and hydrogen is 0.96 Å, whereas a hydrogen bond between an oxygen of one molecule and a hydrogen of another molecule is almost twice as long (1.69–1.79 Å). Thus, a hydrogen bond is not as strong as an O—H covalent bond, but it is stronger than other dipole–dipole interactions. The strongest hydrogen bonds are linear—the two electronegative atoms and the hydrogen between them lie on a straight line.

Hydrogen bonds are stronger than other dipole–dipole interactions, which are stronger than van der Waals forces.

Although each individual hydrogen bond is weak, requiring only about 5 kcal/mol (or 21 kJ/mol) to break, there are many such bonds holding alcohol molecules together. The extra energy required to break these hydrogen bonds is why alcohols have much higher boiling points than ethers with similar molecular weights.

The boiling point of water illustrates the dramatic effect that hydrogen bonding has on boiling points. Water has a molecular weight of 18 and a boiling point of 100 °C. The alkane nearest in size is methane, with a molecular weight of 16 and a boiling point of −167.7 °C.

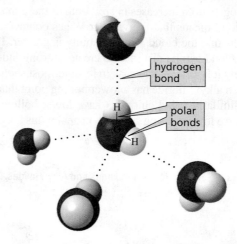

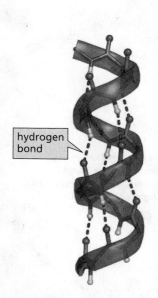

▲ **Figure 2**

Hydrogen bonds hold a segment of a protein chain in a helical structure. Notice that each hydrogen bond forms between a lone pair on oxygen (red) and a hydrogen (white) that is attached to a nitrogen (blue).

Primary and secondary amines also form hydrogen bonds, so they have higher boiling points than ethers with similar molecular weights. Nitrogen is not as electronegative as oxygen, however, so the hydrogen bonds between amine molecules are weaker than those between alcohol molecules. An amine, therefore, has a lower boiling point than an alcohol with a similar molecular weight (Table 4).

Because primary amines have stronger dipole–dipole interactions than secondary amines, hydrogen bonding is more significant in primary amines than in secondary amines. Tertiary amines cannot form hydrogen bonds between their own molecules because they do not have a hydrogen attached to the nitrogen. Consequently, when we compare amines with the same molecular weight, the primary amine has a higher boiling point than the secondary amine and the secondary amine has a higher boiling point than the tertiary amine.

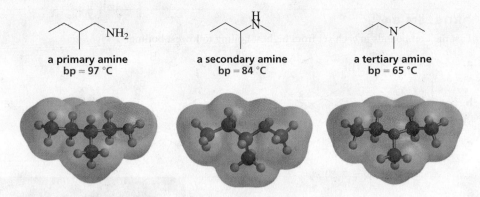

a primary amine
bp = 97 °C

a secondary amine
bp = 84 °C

a tertiary amine
bp = 65 °C

Hydrogen bonds play a crucial role in biology, including holding proteins chains in the correct three-dimensional shape (Figure 2) and making it possible for DNA to copy all its hereditary information (Figure 3).

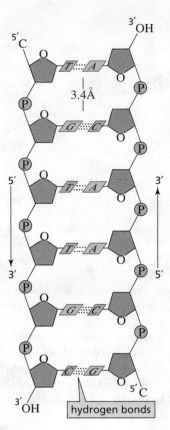

▲ **Figure 3**

DNA has two strands that run in opposite directions. The phosphates (P) and the sugars (five-membered rings) are on the outside, and the bases (A, G, T, and C) are on the inside. The two strands are held together by hydrogen bonding between the bases. A always pairs with T (using two hydrogen bonds), and G always pairs with C (using three hydrogen bonds).

PROBLEM-SOLVING STRATEGY

Predicting Hydrogen Bonding

a. Which of the following compounds will form hydrogen bonds between its molecules?

   1. $CH_3CH_2CH_2OH$ **2.** $CH_3CH_2CH_2F$ **3.** $CH_3OCH_2CH_3$

b. Which of these compounds will form hydrogen bonds with a solvent such as ethanol?

To solve this type of question, start by defining the kind of compound that will do what is being asked.

a. A hydrogen bond forms when a hydrogen attached to an O, N, or F of one molecule interacts with a lone pair on an O, N, or F of another molecule. Therefore, a compound that will form hydrogen bonds with itself must have a hydrogen attached to an O, N, or F. Only compound 1 will be able to form hydrogen bonds with itself.

b. Ethanol has an H attached to an O, so it will be able to form hydrogen bonds with a compound that has a lone pair on an O, N, or F. All three compounds will be able to form hydrogen bonds with ethanol.

Now use the strategy you have just learned to solve Problem 31.

PROBLEM 31♦

a. Which of the following compounds will form hydrogen bonds between its molecules?

   1. $CH_3CH_2OCH_2CH_2OH$ **3.** $CH_3CH_2CH_2CH_2Br$ **5.** $CH_3CH_2CH_2COOH$

   2. $CH_3CH_2N(CH_3)_2$ **4.** $CH_3CH_2CH_2NHCH_3$ **6.** $CH_3CH_2CH_2CH_2F$

b. Which of the preceding compounds will form hydrogen bonds with a solvent such as ethanol?

PROBLEM 32

Explain why

a. H_2O (100 °C) has a higher boiling point than CH_3OH (65 °C).
b. H_2O (100 °C) has a higher boiling point than NH_3 (−33 °C).
c. H_2O (100 °C) has a higher boiling point than HF (20 °C).
d. HF (20 °C) has a higher boiling point than NH_3 (−33 °C).

PROBLEM 33♦

List the following compounds from highest boiling to lowest boiling:

PROBLEM 34

List the compounds in each set from highest boiling to lowest boiling:

a.

b.

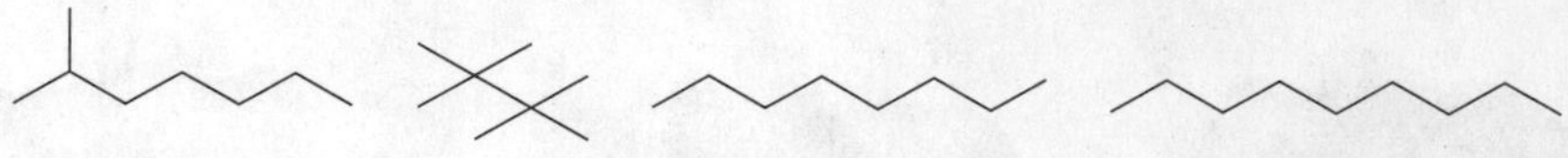

c.

Drugs Bind to Their Receptors

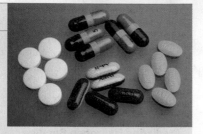

Many drugs exert their physiological effects by binding to specific sites, called *receptors,* on the surface of certain cells. A drug binds to a receptor using the same kinds of bonding interactions—van der Waals interactions, dipole–dipole interactions, hydrogen bonding—that molecules use to bind to each other.

The most important factor in the interaction between a drug and its receptor is a snug fit. Therefore, drugs with similar shapes and properties, which causes them to bind to the same receptor, have similar physiological effects. For example, each of the compounds shown here has a nonpolar, planar, six-membered ring and substituents with similar polarities. They all have anti-inflammatory activity and are known as NSAIDs (non-steroidal anti-inflammatory agents).

Salicylic acid has been used for the relief of fever and arthritic pain since 500 B.C. In 1897, acetylsalicylic acid (known by brand names such as Bayer Aspirin, Bufferin, Anacin, Ecotrin, and Ascriptin) was found to be a more potent anti-inflammatory agent and less irritating to the stomach; it became commercially available in 1899.

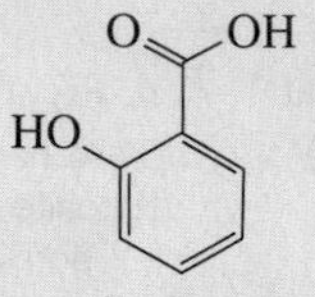

salicylic acid

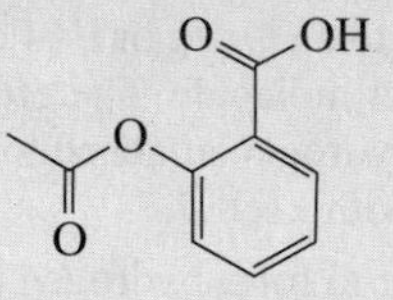

acetylsalicylic acid

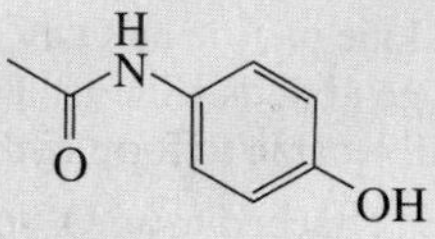

acetaminophen
Tylenol®

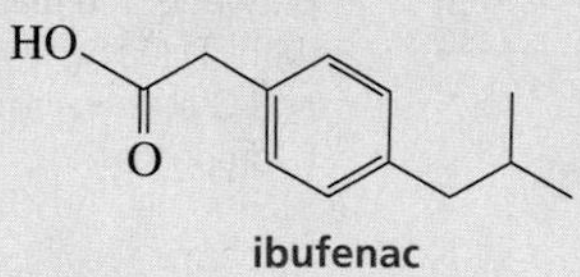

ibufenac

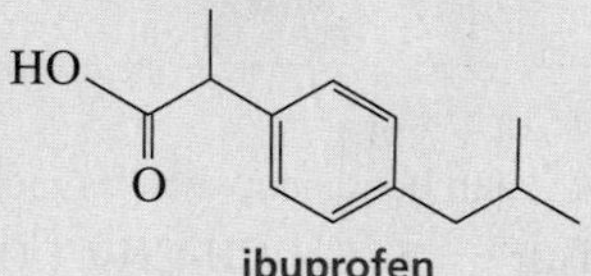

ibuprofen
Advil®

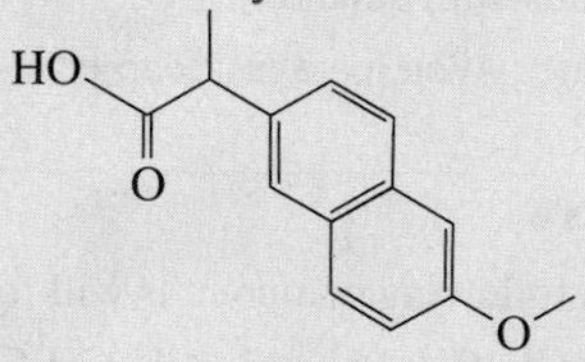

naproxen
Aleve®

Changing the substituents and their relative positions on the ring produced acetaminophen (Tylenol), which was introduced in 1955. It became a widely used drug because it causes no gastric irritation. However, its effective dose is not far from its toxic dose. Subsequently, ibufenac emerged; adding a methyl group to ibufenac produced ibuprofen (Advil), which is a much safer drug. Naproxen (Aleve), which has twice the potency of ibuprofen, was introduced in 1976.

Melting Points

The **melting point (mp)** of a compound is the temperature at which its solid form is converted into a liquid. The melting points of the alkanes listed in Table 1 show that they increase (with a few exceptions) as the molecular weight increases. The increase in melting point is less regular than the increase in boiling point because, in addition to the intermolecular attractions we just considered, the melting point is influenced by the **packing** (that is, the arrangement, including the closeness and compactness, of the molecules) in the crystal lattice. The tighter the fit, the more energy is required to break the lattice and melt the compound.

The melting points of straight-chain alkanes with an even number of carbons fall on a smooth curve (the red line in Figure 4). The melting points of straight-chain alkanes with an odd number of carbons also fall on a smooth curve (the green line). The two curves do not overlap, however, because alkanes with an odd number of carbons pack less tightly than alkanes with an even number of carbons.

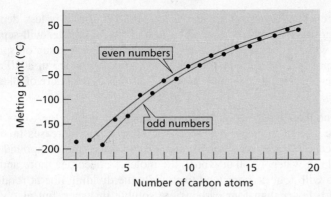

◀ **Figure 4**
Straight-chain alkanes with an even number of carbons fall on a melting-point curve that is higher than the melting-point curve for straight-chain alkanes with an odd number of carbons.

Alkanes with an odd number of carbons pack less tightly because the molecules (each a zigzag chain with its ends tilted the same way) can lie next to each other with a methyl group on the end of one facing and repelling the methyl group on the end of the other, thus increasing the average distance between the chains. Consequently, they have weaker intermolecular attractions and correspondingly lower melting points.

odd number of carbons　　　**even number of carbons**

Solubility

The general rule that governs **solubility** is "like dissolves like." In other words,

> *polar compounds dissolve in polar solvents,*
> *and nonpolar compounds dissolve in nonpolar solvents.*

"Polar dissolves polar" because a polar solvent, such as water, has partial charges that can interact with the partial charges on a polar compound. The negative poles of the solvent molecules surround the positive pole of the polar compound, and the positive poles of the solvent molecules surround the negative pole of the polar compound. The clustering of the solvent molecules around the polar molecules separates them from each other, which is what makes them dissolve. The interaction between solvent molecules and solute molecules (molecules dissolved in a solvent) is called **solvation.**

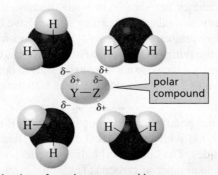

solvation of a polar compound by water

"Like dissolves like."

Because nonpolar compounds have no charge, polar solvents are not attracted to them. In order for a nonpolar molecule to dissolve in a polar solvent such as water, the nonpolar molecule would have to push the water molecules apart, disrupting their hydrogen bonding. Hydrogen bonding, however, is strong enough to exclude the nonpolar compound. On the other hand, nonpolar solutes dissolve in nonpolar solvents because the van der Waals interactions between solvent molecules and solute molecules are about the same as those between solvent molecules and those between solute molecules.

Alkanes

Alkanes are nonpolar, so they are soluble in nonpolar solvents and insoluble in polar solvents such as water. The densities of alkanes increase with increasing molecular weight

(Table 1), but even a 30-carbon alkane ($d^{20°} = 0.8097$ g/mL) is less dense than water ($d^{20°} = 1.00$ g/mL). Therefore, a mixture of an alkane and water will separate into two distinct layers, with the less dense alkane floating on top. The Alaskan oil spill in 1989, the Gulf War oil spill in 1991, and the oil spill in the Gulf of Mexico in 2010 are large-scale examples of this phenomenon because crude oil is primarily a mixture of alkanes.

Alcohols

Is an alcohol nonpolar, due to its alkyl group, or is it polar, due to its OH group? It depends on the size of the alkyl group. As the alkyl group increases in size, becoming a more significant fraction of the entire alcohol molecule, the compound becomes less and less soluble in water. In other words, the molecule becomes more and more like an alkane. Groups with four carbons tend to straddle the dividing line at room temperature, so alcohols with fewer than four carbons are soluble in water, but alcohols with more than four carbons are insoluble in water. Thus, an OH group can drag about three or four carbons into solution in water.

The four-carbon dividing line is only an approximate guide because the solubility of an alcohol also depends on the structure of the alkyl group. Alcohols with branched alkyl groups are more soluble in water than alcohols with unbranched alkyl groups with the same number of carbons, because branching minimizes the contact surface of the nonpolar portion of the molecule. Thus, *tert*-butyl alcohol is more soluble than *n*-butyl alcohol in water.

Ethers

The oxygen of an ether, like the oxygen of an alcohol, can drag only about three carbons into solution in water (Table 6). Diethyl ether—an ether with four carbons—is not fully soluble in water.

Smoke billows from a controlled burn of spilled oil off the Louisiana coast in the Gulf of Mexico.

Table 6 Solubilities of Ethers in Water

2 Cs	CH_3OCH_3	soluble
3 Cs	$CH_3OCH_2CH_3$	soluble
4 Cs	$CH_3CH_2OCH_2CH_3$	slightly soluble (10 g/100 g H_2O)
5 Cs	$CH_3CH_2OCH_2CH_2CH_3$	minimally soluble (1.0 g/100 g H_2O)
6 Cs	$CH_3CH_2CH_2OCH_2CH_2CH_3$	insoluble (0.25 g/100 g H_2O)

Amines

Low-molecular-weight amines are soluble in water because amines can form hydrogen bonds with water. Primary, secondary, and tertiary amines have a lone pair they use to form a hydrogen bond. Primary amines are more soluble than secondary amines with the same number of carbons, because primary amines have two hydrogens that can engage in hydrogen bonding with water. Tertiary amines do not have hydrogens to donate for hydrogen bonds, so they are less soluble in water than are secondary amines with the same number of carbons.

Alkyl Halides

Alkyl halides have some polar character, but only alkyl fluorides have an atom that can form a hydrogen bond with water. Alkyl fluorides, therefore, are the most water soluble of the alkyl halides. The other alkyl halides are less soluble in water than ethers or alcohols with the same number of carbons (Table 7).

Table 7 Solubilities of Alkyl Halides in Water

CH_3F	CH_3Cl	CH_3Br	CH_3I
very soluble	soluble	slightly soluble	slightly soluble
CH_3CH_2F	CH_3CH_2Cl	CH_3CH_2Br	CH_3CH_2I
soluble	slightly soluble	slightly soluble	slightly soluble
$CH_3CH_2CH_2F$	$CH_3CH_2CH_2Cl$	$CH_3CH_2CH_2Br$	$CH_3CH_2CH_2I$
slightly soluble	slightly soluble	slightly soluble	slightly soluble
$CH_3CH_2CH_2CH_2F$	$CH_3CH_2CH_2CH_2Cl$	$CH_3CH_2CH_2CH_2Br$	$CH_3CH_2CH_2CH_2I$
insoluble	insoluble	insoluble	insoluble

Cell Membranes

Cell membranes demonstrate how nonpolar molecules are attracted to other nonpolar molecules, whereas polar molecules are attracted to other polar molecules. All cells are enclosed by a membrane that prevent the aqueous (polar) contents of the cell from pouring out into the aqueous fluid that surrounds the cell. The membrane consists of two layers of phospholipid molecules—called a lipid bilayer. A phospholipid molecule has a polar head and two long nonpolar hydrocarbon tails. The phospholipids are arranged so that the nonpolar tails meet in the center of the membrane. The polar heads are on both the outside surface and the inside surface, where they face the polar solutions on the outside and inside of the cell. Nonpolar cholesterol molecules are found between the tails in order to keep the nonpolar tails from moving around too much. The structure of cholesterol is shown and discussed in Section 15.

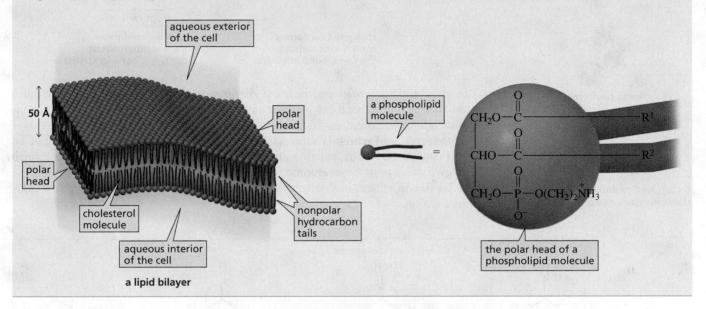

a lipid bilayer

PROBLEM 35♦

Rank the following compounds in each set from most soluble to least soluble in water:

a.

b. $CH_3CH_2CH_2OH$ $CH_3CH_2CH_2CH_2Cl$ $CH_3CH_2CH_2CH_2OH$ $HOCH_2CH_2CH_2OH$

PROBLEM 36♦

In which solvent would cyclohexane have the lowest solubility, 1-pentanol, diethyl ether, ethanol, or hexane?

---

10 ROTATION OCCURS ABOUT CARBON–CARBON SINGLE BONDS

A carbon–carbon single bond (a σ bond) is formed when an sp^3 orbital of one carbon overlaps an sp^3 orbital of another carbon. Figure 5 shows that rotation about a carbon–carbon single bond can occur without any change in the amount of orbital overlap. The different spatial arrangements of the atoms that result from rotation about a single bond are called **conformers** or **conformational isomers.**

Chemists commonly use *Newman projections* to represent the three-dimensional structures that result from rotation about a σ bond. A **Newman projection** assumes that the viewer is looking along the longitudinal axis of a particular C—C bond. The carbon in front is represented by a point (where three lines are seen to intersect), and the carbon

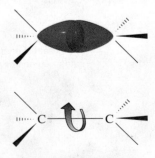

▲ **Figure 5**

A carbon–carbon single bond is formed by the overlap of cylindrically symmetrical **sp**^{3} orbitals, so rotation about the bond can occur without changing the amount of orbital overlap.

staggered conformer

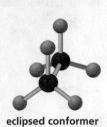

eclipsed conformer

at the back is represented by a circle. The three lines emanating from each of the carbons represent its other three bonds. (Compare the three-dimensional structures shown in the margin with the two-dimensional Newman projections.)

$$H_3C-CH_3$$
ethane

Newman
projections

staggered conformer
from rotation about
the C—C bond in ethane

eclipsed conformer
from rotation about
the C—C bond in ethane

A staggered conformer is more stable than an eclipsed conformer.

The *staggered conformer* and *eclipsed conformers* represent two extremes because rotation about a C—C bond can produce an infinite number (a continuum) of conformers between the two extremes.

A **staggered conformer** is more stable, and therefore lower in energy, than an **eclipsed conformer.** Thus, rotation about a C—C bond is not completely free since an energy barrier must be overcome when rotation occurs (Figure 6). However, the energy barrier in ethane is small enough (2.9 kcal/mol or 12 kJ/mol) to allow continuous rotation.

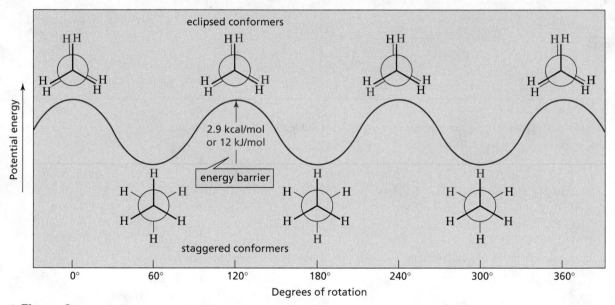

▲ Figure 6
The potential energies of all the conformers of ethane obtained in one complete 360° rotation about the C—C bond. Notice that staggered conformers are at energy minima, whereas eclipsed conformers are at energy maxima.

A molecule's conformation changes from staggered to eclipsed millions of times per second at room temperature. As a result, the conformers cannot be separated from each other. At any one time, approximately 99% of the ethane molecules will be in a staggered conformation because of the staggered conformer's greater stability, leaving only 1% in less stable conformations. The investigation of the various conformers of a compound and their relative stabilities is called **conformational analysis.**

Why is a staggered conformer more stable than an eclipsed conformer? The major contributions to the energy difference are stabilizing interactions between the C—H σ bonding molecular orbital on one carbon and the C—H $\sigma*$ antibonding molecular orbital on the other carbon: the electrons in the filled bonding MO move partially into the unoccupied antibonding MO. Only in a staggered conformation are the two orbitals parallel, so staggered conformers maximize these stabilizing interactions. The delocalization of electrons by the overlap of a σ orbital with an empty orbital is called **hyperconjugation.**

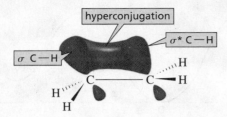

Butane has three carbon–carbon single bonds, and rotation can occur about each of them.

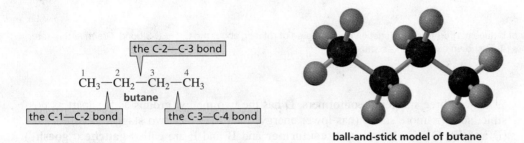

ball-and-stick model of butane

The Newman projections that follow show the staggered and eclipsed conformers that result from rotation about the C-1—C-2 bond. Notice that the carbon with the lower number is placed in the foreground in a Newman projection.

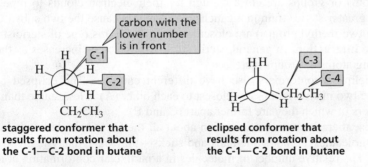

staggered conformer that results from rotation about the C-1—C-2 bond in butane

eclipsed conformer that results from rotation about the C-1—C-2 bond in butane

Although the staggered conformers that result from rotation about the C-1—C-2 bond in butane all have the same energy, the staggered conformers that result from rotation about the C-2—C-3 bond do not. The staggered and eclipsed conformers that result from rotation about the C-2—C-3 bond in butane are

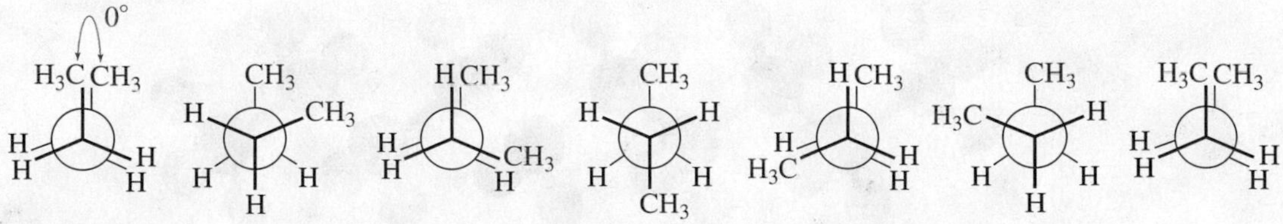

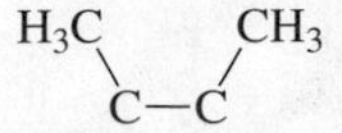

The relative energies of the conformers are shown in Figure 7. The letters in the figure correspond to the letters that identify the above structures. The degree of rotation of each conformer is identified by the dihedral angle—the angle formed in a Newman projection by a bond on the front carbon and a bond on the back carbon. For example, the conformer (A) in which one methyl group stands directly in front of the other has a dihedral angle of 0°, whereas the conformer (D) in which the methyl groups are opposite each other has a dihedral angle of 180°.

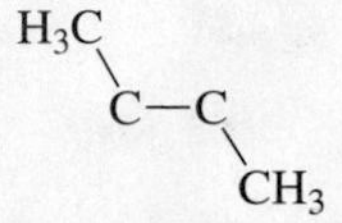

127

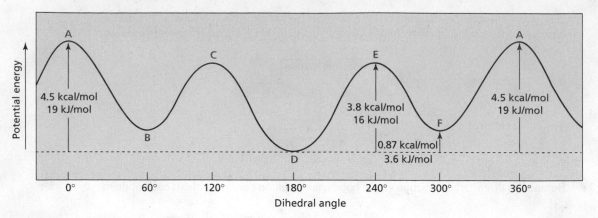

▲ **Figure 7**
Potential energy of butane conformers as a function of the degree of rotation about the C-2—C-3 bond. Green letters refer to the conformers (A–F) shown earlier in this section.

Of the three staggered conformers, D has the two methyl groups as far apart as possible, so D is more stable (has lower energy) than the other two staggered conformers (B and F); D is called the **anti conformer** and B and F are called **gauche ("goesh") conformers.** (*Anti* is Greek for "opposite of"; *gauche* is French for "left.") The two gauche conformers have the same energy.

The anti and gauche conformers have different energies because of steric strain. **Steric strain** is the strain experienced by a molecule (that is, the additional energy it possesses) when atoms or groups are close enough for their electron clouds to repel each other. There is greater steric strain in a gauche conformer because the two substituents (in this case, the two methyl groups) are closer to each other. This type of steric strain is called a **gauche interaction.** In general, steric strain in molecules increases as the size of the interacting atoms or groups increases.

The eclipsed conformers also have different energies. The eclipsed conformer in which the two methyl groups are closest to each other (A) is less stable than the eclipsed conformers in which they are farther apart (C and E).

Because there is continuous rotation about all the C—C single bonds in a molecule, organic molecules are not static balls and sticks—they have many interconvertible conformers. The relative number of molecules in a particular conformation at any one time depends on its stability—the more stable it is, the greater the fraction of molecules that will be in that conformation. Most molecules, therefore, are in staggered conformations at any given instant, and there are more anti conformers than gauche conformers. The preference for the staggered conformation gives carbon chains a tendency to adopt zigzag arrangements, as seen in the ball-and-stick model of decane.

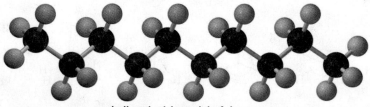

ball-and-stick model of decane

PROBLEM 37

a. Draw all the staggered and eclipsed conformers that result from rotation about the C-2—C-3 bond of pentane.

b. Draw a potential-energy diagram for rotation about the C-2—C-3 bond of pentane through 360°, starting with the least stable conformer.

PROBLEM 38

Convert the following Newman projections to skeletal structures and name them.

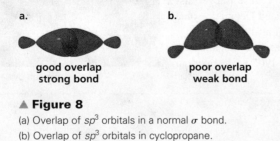

PROBLEM 39♦

Using Newman projections, draw the most stable conformer for each of the following:

a. 3-methylpentane, viewed along the C-2 — C-3 bond
b. 3-methylhexane, viewed along the C-3 — C-4 bond
c. 3,3-dimethylhexane, viewed along the C-3 — C-4 bond

11 SOME CYCLOALKANES HAVE ANGLE STRAIN

Ideally, an sp^3 carbon has bond angles of 109.5°. In 1885, the German chemist Adolf von Baeyer, believing that all cyclic compounds were planar, proposed that the stability of a cycloalkane could be predicted by determining the difference between this ideal bond angle and the bond angle in the planar cycloalkane. For example, the bond angles in cyclopropane are 60°, representing a 49.5° deviation from 109.5°. According to Baeyer, this deviation causes **angle strain,** which decreases cyclopropane's stability.

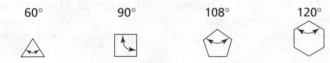

the bond angles of planar cyclic hydrocarbons

The angle strain in a cyclopropane can be understood by looking at the overlap of the orbitals that form the σ bonds (Figure 8). Normal σ bonds are formed by the overlap of two sp^3 orbitals that point directly at each other. In cyclopropane, the overlapping orbitals cannot point directly at each other, so the amount of overlap between them is less than in a normal C — C bond. Decreasing the amount of overlap weakens the C — C bonds, and this weakness is what is known as angle strain.

a.   **b.**

good overlap   poor overlap
strong bond   weak bond

▲ **Figure 8**
(a) Overlap of sp^3 orbitals in a normal σ bond.
(b) Overlap of sp^3 orbitals in cyclopropane.

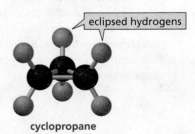

eclipsed hydrogens

cyclopropane

In addition to the angle strain of the C — C bonds, all the adjacent C — H bonds in cyclopropane are eclipsed rather than staggered, making it even more unstable.

If cyclobutane were planar, the bond angles would have to be compressed from 109.5° to 90°. Planar cyclobutane would therefore have less angle strain than cyclopropane because the bond angles in cyclobutane would be only 19.5° (not 49.5°) less than the ideal bond angle. It would, however, have eight pairs of eclipsed hydrogens, compared with six pairs in cyclopropane. Because of the eclipsed hydrogens, we will see that cyclobutane is not planar.

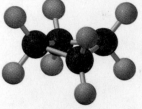

cyclobutane

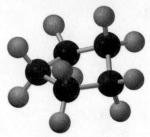

cyclopentane

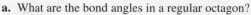

PROBLEM 40♦

The bond angles in a regular polygon with n sides are equal to

$$180° - \frac{360°}{n}$$

a. What are the bond angles in a regular octagon?
b. What are the bond angles in a regular nonagon?

Baeyer predicted that cyclopentane would be the most stable of the cycloalkanes because its bond angles ($108°$) are closest to the ideal tetrahedral bond angle. He also predicted that cyclohexane, with bond angles of $120°$, would be less stable than cyclopentane and that the stability of cycloalkanes would continue to decrease as the number of sides in the cycloalkanes increased beyond six.

Contrary to what Baeyer predicted, however, cyclohexane is more stable than cyclopentane. Furthermore, cyclic compounds do not become less and less stable as the number of sides increases beyond six. The mistake Baeyer made was to assume that all cyclic molecules are planar.

Because three points define a plane, the carbons of cyclopropane must lie in a plane. The other cycloalkanes, however, twist and bend out of a planar arrangement in order to attain a structure that maximizes their stability by minimizing ring strain and the number of eclipsed hydrogens. For example, one of the methylene groups of cyclobutane is at an angle of about $25°$ from the plane defined by the other three carbons.

If cyclopentane were planar, as Baeyer had predicted, it would have essentially no angle strain, but it would have 10 pairs of eclipsed hydrogens. Therefore, cyclopentane puckers, allowing some of the hydrogens to become nearly staggered. However, in the process, the molecule acquires some angle strain. The puckered form of cyclopentane is called the *envelope conformation,* because the shape of the ring resembles a squarish envelope with the flap up.

Von Baeyer, Barbituric Acid, and Blue Jeans

Johann Friedrich Wilhelm Adolf von Baeyer (1835–1917) was a professor of chemistry at the University of Strasbourg and later at the University of Munich. In 1864, he discovered barbituric acid—the first of a group of sedatives known as barbiturates—and named it after a woman named Barbara. Who Barbara was is not certain. Some say she was his girlfriend, but because Baeyer discovered barbituric acid in the same year that Prussia defeated Denmark, some believe he named it after Saint Barbara, the patron saint of artillerymen.

Baeyer was also the first to synthesize indigo, the dye used in the manufacture of blue jeans. He received the Nobel Prize in Chemistry in 1905 for his work in synthetic organic chemistry.

indigo dye

PROBLEM-SOLVING STRATEGY

Calculating the Strain Energy of a Cycloalkane

If we assume that cyclohexane is completely free of strain, then we can use the **heat of formation**—the heat given off when a compound is formed from its elements under standard conditions—to calculate the strain energy of the other cycloalkanes. Taking the heat of formation of cyclohexane (–29.5 kcal/mol in Table 8) and dividing by 6 for its six CH_2 groups gives us a value of –4.92 kcal/mol for a "strainless" CH_2 group. With this value, we can calculate the heat of formation of any other "strainless" cycloalkane by multiplying the number of CH_2 groups in its ring by -4.92 kcal/mol. The strain in the compound is the difference between its actual heat of formation and its calculated "strainless" heat of formation (Table 8). For example, cyclopentane has an actual heat of formation of -18.4 kcal/mol and a "strainless" heat of formation of $(5)(-4.92) = -24.6$ kcal/mol. Therefore, cyclopentane has a strain energy of 6.2 kcal/mol, because $[-18.4 - (-24.6) = 6.2]$. Dividing the strain energy by the number of CH_2 groups in the cyclic compound gives the strain energy per CH_2 group for that compound.

Now use the strategy you have just learned to solve Problem 41.

Table 8	Heats of Formation and Strain Energies of Cycloalkanes		
	Heat of formation (kcal/mol)	**"Strainless" heat of formation (kcal/mol)**	**Strain energy (kcal/mol)**	**Strain energy per CH_2 group (kcal/mol)**
cyclopropane	+12.7	−14.6	27.3	9.1
cyclobutane	+6.8	−19.7	26.5	6.6
cyclopentane	−18.4	−24.6	6.2	1.2
cyclohexane	−29.5	−29.5	0	0
cycloheptane	−28.2	−34.4	6.2	0.9
cyclooctane	−29.7	−39.4	9.7	1.2
cyclononane	−31.7	−44.3	12.6	1.4
cyclodecane	−36.9	−49.2	12.3	1.2
cycloundecane	−42.9	−54.1	11.2	1.0

PROBLEM 41

Verify the strain energy shown in Table 8 for cycloheptane.

PROBLEM 42♦

The effectiveness of a barbiturate as a sedative is related to its ability to penetrate the nonpolar membrane of a cell. Which of the following barbiturates would you expect to be the more effective sedative?

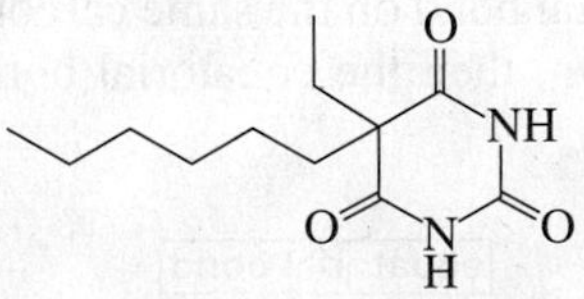

hexethal 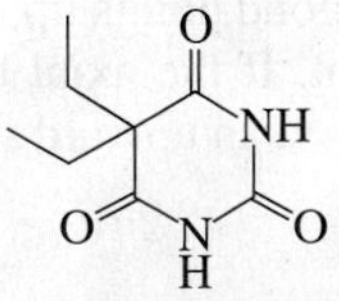   barbital

12   CONFORMERS OF CYCLOHEXANE

The cyclic compounds most commonly found in nature contain six-membered rings because carbon rings of that size can exist in a conformation—called a *chair conformer*—that is almost completely free of strain. All the bond angles in a **chair conformer** are 111° (which is very close to the ideal tetrahedral bond angle of 109.5°) and all the adjacent bonds are staggered (Figure 9).

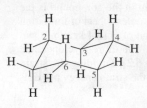

**chair conformer of
cyclohexane**

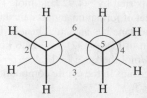

**Newman projection of
the chair conformer
looking down the
C-1—C-2 and C-5—C-4 bonds**

**ball-and-stick model of
the chair conformer**

▶ **Figure 9**

The chair conformer of cyclohexane, a Newman projection of
the chair conformer showing that
all the bonds are staggered, and a
ball-and-stick model.

The chair conformer is so important that you should learn how to draw it:

1. Draw two parallel lines of the same length, slanted upward.

2. Connect the tops of the lines with a V whose left side is slightly longer than its right side. Connect the bottoms of the lines with an inverted V. (The bottom-left and top-right lines should be parallel; the top-left and bottom-right lines should be parallel.) This completes the framework of the six-membered ring.

3. Each carbon has an axial bond and an equatorial bond. The **axial bonds** (red lines) are vertical and alternate above and below the ring. The axial bond on one of the uppermost carbons is up, the next is down, the next is up, and so on.

4. The **equatorial bonds** (red lines with blue balls) point outward from the ring. Because the bond angles are greater than 90°, the equatorial bonds are on a slant. If the axial bond points up, then the equatorial bond on the same carbon is on a downward slant. If the axial bond points down, then the equatorial bond on the same carbon is on an upward slant.

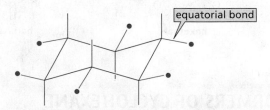

Notice that each equatorial bond is parallel to two ring bonds (red lines) one bond away.

Remember that in this depiction, cyclohexane is viewed edge-on. The lower bonds of the ring are in front and the upper bonds are in back.

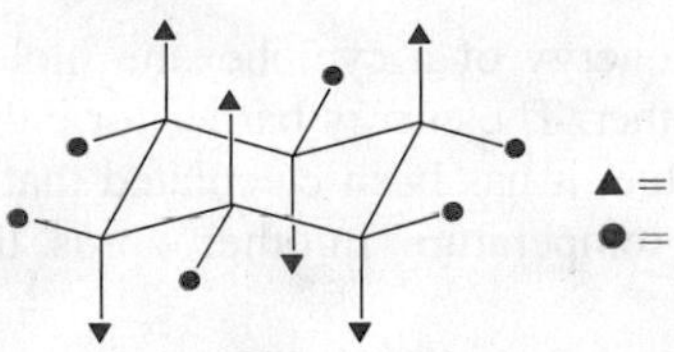

▲ = axial bond
● = equatorial bond

PROBLEM 43

Draw 1,2,3,4,5,6-hexachlorocyclohexane with

a. all the chloro groups in axial positions.
b. all the chloro groups in equatorial positions.

Cyclohexane rapidly interconverts between two stable chair conformers because of the ease of rotation about its C—C bonds. This interconversion is called **ring flip** (Figure 10). When the two chair conformers interconvert, bonds that are equatorial in one chair conformer become axial in the other chair conformer, and bonds that are axial become equatorial.

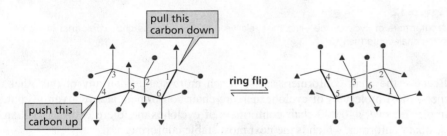

pull this carbon down

ring flip

push this carbon up

Bonds that are equatorial in one chair conformer are axial in the other chair conformer.

◀ **Figure 10**

Ring flip causes equatorial bonds to become axial bonds and axial bonds to become equatorial bonds.

Cyclohexane can also exist as a **boat conformer,** shown in Figure 11. Like the chair conformer, the boat conformer is free of angle strain. However, the boat conformer is not as stable because some of the C—H bonds are eclipsed. The boat conformer is further destabilized by the close proximity of the **flagpole hydrogens**—the hydrogens at the "bow" and "stern" of the boat—which cause steric strain.

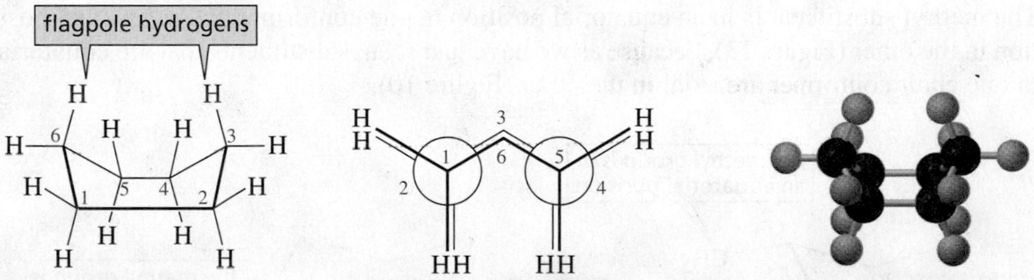

flagpole hydrogens

boat conformer of cyclohexane

Newman projection of the boat conformer

ball-and-stick model of the boat conformer

◀ **Figure 11**

The boat conformer of cyclohexane, a Newman projection of the boat conformer showing that some of the C—H bonds are eclipsed, and a ball-and-stick model.

The conformers that cyclohexane assumes when interconverting from one chair conformer to the other are shown in Figure 12. To convert from the boat conformer to a chair conformer, one of the two topmost carbons of the boat conformer must be pulled down so that it becomes the bottommost carbon of the chair conformer. When the carbon is pulled down just a little, the **twist-boat conformer** is obtained, which is more stable than the boat conformer because the flagpole hydrogens have moved away from each other, thus relieving some steric strain. When the carbon is pulled

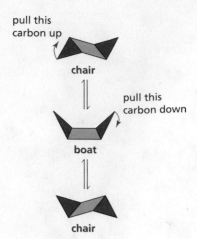

pull this
carbon up

chair

pull this
carbon down

boat

chair

Build a model of cyclohexane.
Convert it from one chair conformer
to the other by pulling the topmost
carbon down and pushing the
bottommost carbon up.

down to the point where it is in the same plane as the sides of the boat, the very unstable **half-chair conformer** is obtained. Pulling the carbon down farther produces the *chair conformer*.

Figure 12 shows the relative energy of a cyclohexane molecule as it interconverts from one chair conformer to the other. The energy barrier for interconversion is 12.1 kcal/mol (50.6 kJ/mol). Using this value, it has been calculated that cyclohexane undergoes 10^5 ring flips per second at room temperature. In other words, the two chair conformers are in rapid equilibrium.

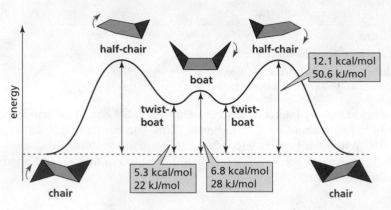

half-chair

boat

half-chair

12.1 kcal/mol
50.6 kJ/mol

energy

twist-
boat

twist-
boat

chair

5.3 kcal/mol
22 kJ/mol

6.8 kcal/mol
28 kJ/mol

chair

▲ **Figure 12**

The conformers of cyclohexane—and their relative energies—as one chair conformer interconverts to the other chair conformer.

Because the chair conformers are so much more stable than any of the other conformers, most molecules of cyclohexane are chair conformers at any given instant. For example, for every 10,000 chair conformers of cyclohexane, there is no more than one twist-boat conformer, which is the next most stable conformer.

13   CONFORMERS OF MONOSUBSTITUTED CYCLOHEXANES

Unlike cyclohexane, which has two equivalent chair conformers, the two chair conformers of a monosubstituted cyclohexane (such as methylcyclohexane) are not equivalent. The methyl substituent is in an equatorial position in one conformer and in an axial position in the other (Figure 13), because as we have just seen, substituents that are equatorial in one chair conformer are axial in the other (Figure 10).

▶ **Figure 13**

A substituent is in an equatorial position in one chair conformer and in an axial position in the other. The conformer with the substituent in the equatorial position is more stable.

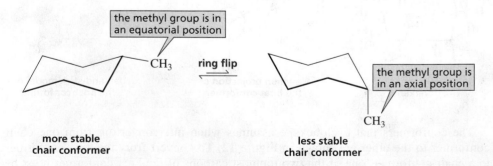

the methyl group is in
an equatorial position

CH_3

ring flip

the methyl group is
in an axial position

CH_3

**more stable
chair conformer**

**less stable
chair conformer**

The chair conformer with the methyl substituent in an equatorial position is the more stable of the two conformers because a substituent has more room and, therefore, fewer steric interactions when it is in an equatorial position. This can be understood by looking at Figure 14a, which shows that a methyl group in an equatorial position extends into space, away from the rest of the molecule.

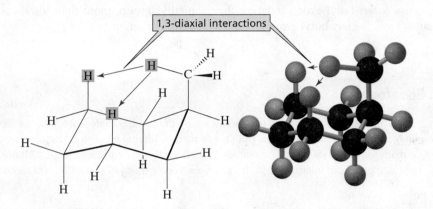

a.

equatorial substituent points into space

b.

axial substituent is parallel to 2 Hs

◀ **Figure 14**

Newman projections of methylcyclohexane:

a. the methyl substituent is equatorial

b. the methyl substituent is axial

In contrast, any axial substituent will be relatively close to the axial substituents on the other two carbons on the same side of the ring because all three axial bonds are parallel to each other, (Figure 14b). Because the interacting axial substituents are in 1,3-positions relative to each other, these unfavorable steric interactions are called **1,3-diaxial interactions.**

Build a model of methylcyclohexane so you can see that a substituent has more room if it is in an equatorial position than if it is in an axial position.

1,3-diaxial interactions

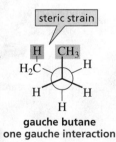

A gauche conformer of butane and the axially substituted conformer of methylcyclohexane are compared in Figure 15. Notice that the gauche interaction in butane is the same as a 1,3-diaxial interaction in methylcyclohexane.

steric strain

steric strain

gauche butane
one gauche interaction

axial methylcyclohexane
two 1,3-diaxial interactions

▲ **Figure 15**

Butane has one gauche interaction between a methyl group and a hydrogen, whereas methylcyclohexane has two 1,3-diaxial interactions between a methyl group and a hydrogen.

In Section 10, we saw that the gauche interaction between the methyl groups of butane causes a gauche conformer to be 0.87 kcal/mol (3.6 kJ/mol) less stable than the anti conformer. Because there are two such interactions in the chair conformer of methylcyclohexane when the methyl group is in an axial position, this conformer is 1.74 kcal/mol (7.2 kJ/mol) less stable than the chair conformer with the methyl group in an equatorial position.

Because of the difference in stability of the two chair conformers, a sample of methylcyclohexane (or any other monosubstituted cyclohexane) will, at any point in time, contain more chair conformers with the substituent in an equatorial position than with the substituent in an axial position. The relative amounts of the two chair conformers depend on the substituent (Table 9).

The substituent with the greater bulk in the vicinity of the 1,3-diaxial hydrogens will have a greater preference for an equatorial position because it will have stronger 1,3-diaxial interactions. For example, the experimental equilibrium constant (K_{eq}) for the

Table 9

Equilibrium Constants for Several Monosubstituted Cyclohexanes at 25 °C

Substituent	$K_{eq} = \dfrac{[\text{equatorial}]}{[\text{axial}]}$
H	1
CH_3	18
CH_3CH_2	21
CH_3CH with CH_3	35
CH_3C with CH_3, CH_3	4800
CN	1.4
F	1.5
Cl	2.4
Br	2.2
I	2.2
HO	5.4

conformers of methylcyclohexane (Table 9) indicates that 95% of methylcyclohexane molecules have the methyl group in an equatorial position at 25 °C:

$$K_{eq} = \frac{[\text{equatorial conformer}]}{[\text{axial conformer}]} = \frac{18}{1}$$

$$\% \text{ of equatorial conformer} = \frac{[\text{equatorial conformer}]}{[\text{equatorial conformer}] + [\text{axial conformer}]} \times 100$$

$$\% \text{ of equatorial conformer} = \frac{18}{18 + 1} \times 100 = 95\%$$

In *tert*-butylcyclohexane, where the 1,3-diaxial interactions are even more destabilizing because a *tert*-butyl group is larger than a methyl group, more than 99.9% of the molecules have the *tert*-butyl group in an equatorial position.

Starch and Cellulose—Axial and Equatorial

Polysaccharides are compounds formed by linking many sugar molecules together. Two of the most common naturally occurring polysaccharides are amylose (an important component of starch) and cellulose. Both are formed by linking glucose molecules together. Starch, a water-soluble compound, is found in many of the foods we eat—potatoes, rice, flour, beans, corn, and peas. Cellulose, a water-insoluble compound, is the major structural component of plants. Cotton, for example, is composed of about 90% cellulose and wood is about 50% cellulose.

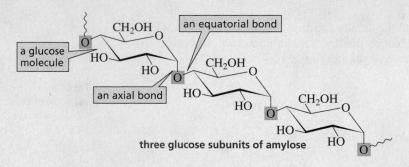

three glucose subunits of amylose

foods rich in starch

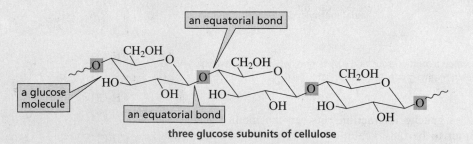

three glucose subunits of cellulose

cotton plant and cotton towel

How can two compounds with such different physical properties both be formed by linking together glucose molecules? If you examine their structures, you will see that the linkages in the two polysaccharides are different. In starch, an oxygen on an *axial* bond of one glucose is linked to an equatorial bond of another glucose, whereas in cellulose, an oxygen on an *equatorial* bond of one glucose is linked to an equatorial bond of another glucose. The axial bonds cause starch to form a helix that promotes hydrogen bonding with water molecules—as a result, starch is soluble in water. The equatorial bonds cause cellulose to form linear arrays that are held together by intermolecular hydrogen bonds, so it cannot form hydrogen bonds with water—as a result, cellulose is not soluble in water.

Mammals have digestive enzymes that can break the axial linkages in starch but not the equatorial linkages in cellulose. Grazing animals have bacteria in their digestive tracts that possess the enzyme that can break the equatorial bonds, so cows and horses can eat hay to meet their nutritional need for glucose.

PROBLEM 44♦

The chair conformer of fluorocyclohexane is 0.25 kcal/mol more stable when the fluoro substituent is in an equatorial position than when it is in an axial position. How much more stable is the anti conformer than a gauche conformer of 1-fluoropropane considering rotation about the C-1-C-2 bond?

PROBLEM 45♦

Using the data in Table 9, calculate the percentage of molecules of cyclohexanol that have the OH group in an equatorial position at 25 °C.

---

14   CONFORMERS OF DISUBSTITUTED CYCLOHEXANES

If a cyclohexane ring has two substituents, we must take both substituents into account when predicting which of the two chair conformers is more stable. Let's use 1,4-dimethylcyclohexane as an example.

First of all, note that there are two different dimethylcyclohexanes. One has both methyl substituents on the *same side* of the cyclohexane ring (both point downward)—it is called the **cis isomer** (*cis* is Latin for "on this side"). The other has the two methyl substituents on *opposite sides* of the ring (one points upward and one points downward)—it is called the **trans isomer** (*trans* is Latin for "across").

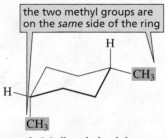

cis-**1,4-dimethylcyclohexane**

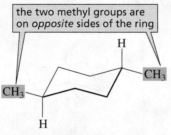

trans-**1,4-dimethylcyclohexane**

The cis isomer of a disubstituted cyclic compound has its substituents on the same side of the ring.

The trans isomer of a disubstituted cyclic compound has its substituents on opposite sides of the ring.

cis-1,4-Dimethylcyclohexane and *trans*-1,4-dimethylcyclohexane are examples of **cis–trans isomers** or **geometric isomers.** Geometric isomers have the same atoms, and the atoms are linked in the same order, but they have different spatial arrangements. The cis and trans isomers are different compounds with different melting and boiling points, so they can be separated from one another.

PROBLEM-SOLVING STRATEGY

Differentiating Cis–Trans Isomers

Does the cis isomer or the trans isomer of 1,2-dimethylcyclohexane have one methyl group in an equatorial position and the other in an axial position?

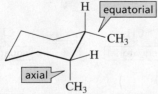

Is this the cis isomer or the trans isomer?

To solve this kind of problem, we need to determine whether the two substituents are on the same side of the ring (cis) or on opposite sides of the ring (trans). If the bonds bearing the substituents are both pointing upward or both pointing downward, then the compound is the cis isomer; if one bond is pointing upward and the other downward, then the compound is the trans isomer. Because the conformer in question has both methyl groups attached to downward-pointing bonds, it is the cis isomer.

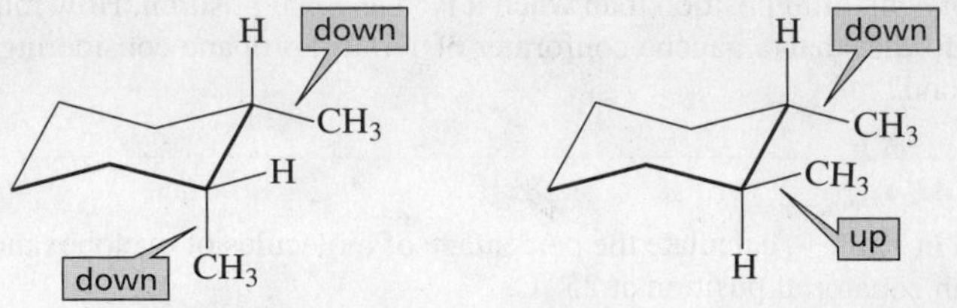

the cis isomer   **the trans isomer**

The isomer that is the most misleading when drawn in two dimensions is a *trans*-1,2-disubstituted isomer. At first glance, the methyl groups of *trans*-1,2-dimethylcyclohexane (on the right in the preceding image) appear to be on the same side of the ring, so you might think the compound is the cis isomer. Closer inspection shows, however, that one bond points upward and the other downward, so we know that it is the trans isomer. Alternatively, if you look at the two axial hydrogens, they are clearly trans (one points straight up and the other straight down), so the methyl groups must also be trans.

Now use the strategy you have just learned to solve Problem 46.

PROBLEM 46♦

Is each of the following a cis isomer or a trans isomer?

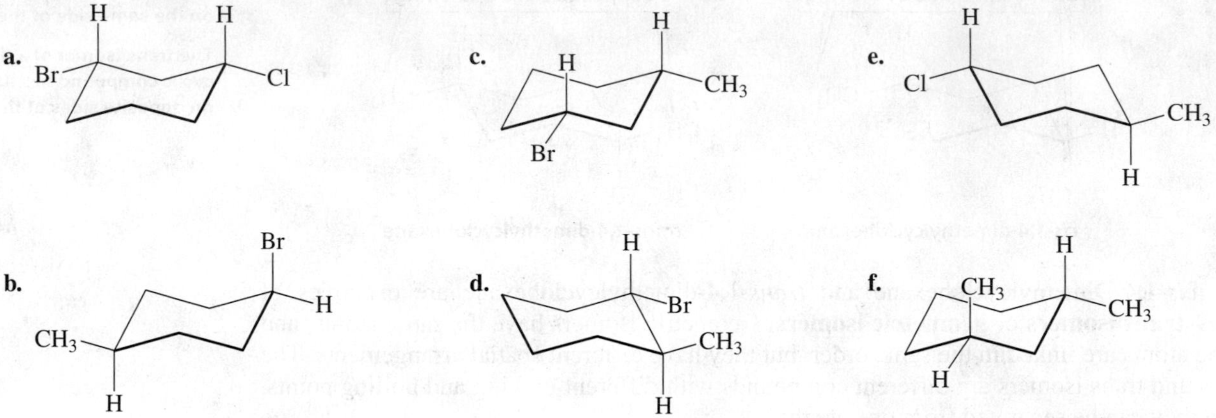

Every compound with a cyclohexane ring has two chair conformers; thus, both the cis isomer and the trans isomer of a disubstituted cyclohexane have two chair conformers. Let's compare the structures of the two chair conformers of *cis*-1,4-dimethylcyclohexane to see if we can predict any difference in their stabilities.

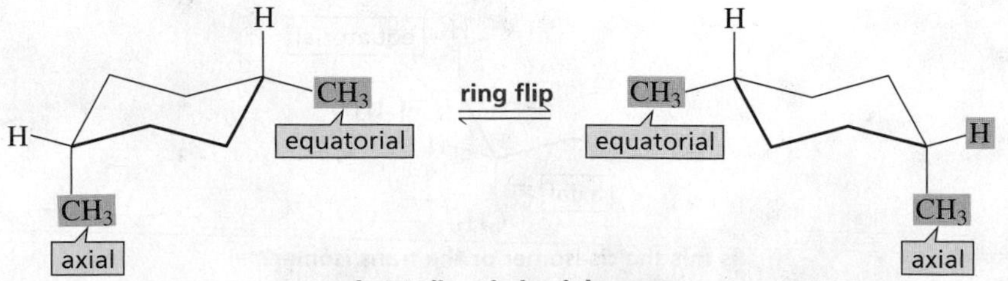

cis-1,4-dimethylcyclohexane

The conformer shown on the left has one methyl group in an equatorial position and one methyl group in an axial position. The conformer on the right also has one methyl group in an equatorial position and one methyl group in an axial position. Therefore, both chair conformers are equally stable.

In contrast, the two chair conformers of *trans*-1,4-dimethylcyclohexane have different stabilities because one has both methyl substituents in equatorial positions and the other has both methyl groups in axial positions. The conformer with both substituents in equatorial positions is more stable.

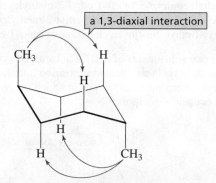

trans-1,4-dimethylcyclohexane

The chair conformer with both substituents in axial positions has four 1,3-diaxial interactions, causing it to be about 4×0.87 kcal/mol $= 3.5$ kcal/mol (14.6 kJ/mol) less stable than the chair conformer with both methyl groups in equatorial positions. Thus, almost all the molecules of *trans*-1,4-dimethylcyclohexane will be chair conformers with both substituents in equatorial positions.

this chair conformer has four 1,3-diaxial interactions

Now let's look at the geometric isomers of 1-*tert*-butyl-3-methylcyclohexane. Both substituents of the cis isomer are in equatorial positions in one chair conformer and both are in axial positions in the other. The conformer with both substituents in equatorial positions is more stable.

cis-1-tert-butyl-3-methylcyclohexane

Both chair conformers of the trans isomer have one substituent in an equatorial position and the other in an axial position. Because the *tert*-butyl group is larger than the methyl group, the 1,3-diaxial interactions will be stronger when the *tert*-butyl group is in an axial position. Therefore, the conformer with the *tert*-butyl group in an equatorial position is more stable.

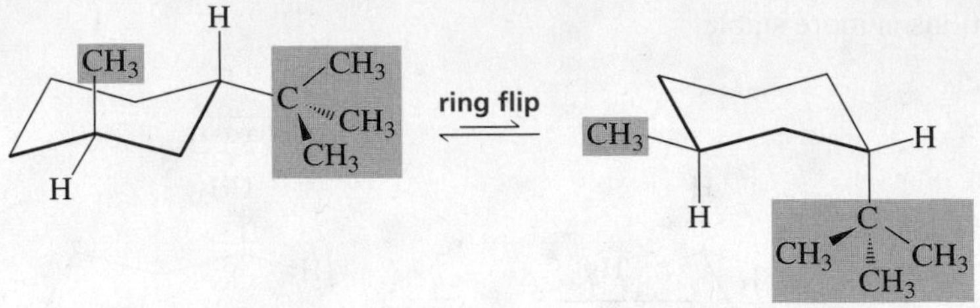

more stable   less stable

trans-1-*tert*-butyl-3-methylcyclohexane

PROBLEM 47♦

Which will have a higher percentage of the diequatorial-substituted conformer compared with the diaxial-substituted conformer, *trans*-1,4-dimethylcyclohexane or *cis*-1-*tert*-butyl-3-methyl-cyclohexane?

PROBLEM 48 Solved

a. Draw the more stable chair conformer of *cis*-1-ethyl-2-methylcyclohexane.
b. Draw the more stable chair conformer of *trans*-1-ethyl-2-methylcyclohexane.
c. Which is more stable, *cis*-1-ethyl-2-methylcyclohexane or *trans*-1-ethyl-2-methylcyclohexane?

Solution to 48a If the two substituents of a 1,2-disubstituted cyclohexane are to be on the same side of the ring, one must be in an equatorial position and the other must be in an axial position. The more stable chair conformer is the one in which the larger of the two substituents (the ethyl group) is in the equatorial position.

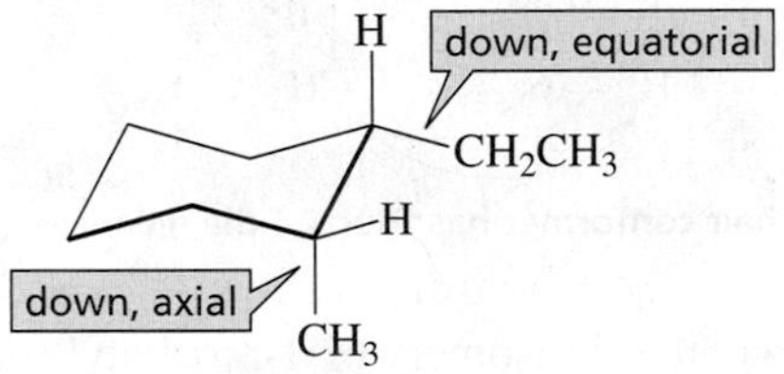

PROBLEM 49♦

For each of the following disubstituted cyclohexanes, indicate whether the substituents in the two chair conformers will be both equatorial in one chair conformer and both axial in the other *or* one equatorial and one axial in each of the chair conformers:

a. *cis*-1,2-   **b.** *trans*-1,2-   **c.** *cis*-1,3-   **d.** *trans*-1,3-   **e.** *cis*-1,4-   **f.** *trans*-1,4-

---

PROBLEM 50 Solved

a. Draw Newman projections of the two conformers of *cis*-1,3-dimethylcyclohexane.
b. Which of the conformers would predominate at equilibrium?
c. Draw Newman projections of the two conformers of the trans isomer.
d. Which of the conformers would predominate at equilibrium?

Solution to 50a Draw two staggered conformers adjacent to one other, connect them, and number the carbons so you know where to put the substituents.

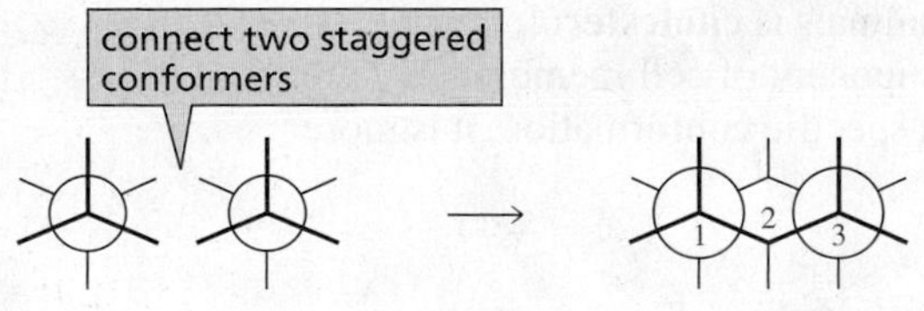

connect two staggered conformers

$\rightarrow$

Attach the methyl groups to the 1 and 3 carbons. Because you are drawing the cis isomer, the two substituents should both be on upward-pointing bonds or both be on downward-pointing bonds.

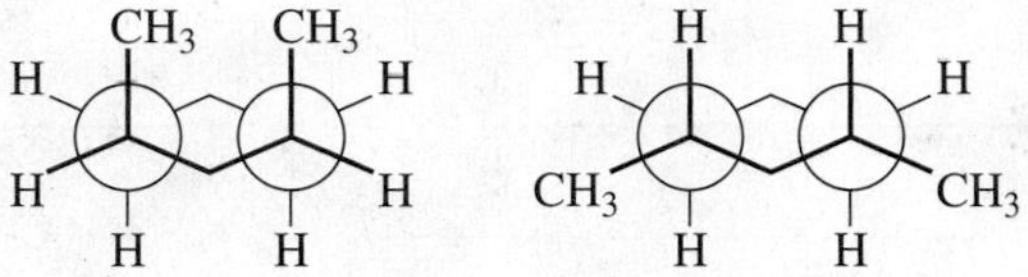

Solution to 50b The conformer on the right would predominate at equilibrium because it is more stable—both methyl groups are on equatorial bonds.

PROBLEM 51♦

a. Calculate the energy difference between the two chair conformers of *trans*-1,4-dimethyl-cyclohexane.

b. What is the energy difference between the two chair conformers of *cis*-1,4-dimethyl-cyclohexane?

---

15 FUSED CYCLOHEXANE RINGS

When two cyclohexane rings are fused—**fused rings** share two adjacent carbons—one ring can be considered to be a pair of substituents bonded to the other ring. As with any disubstituted cyclohexane, the two substituents can be either cis or trans. The trans isomer (in which one substituent bond points upward and the other downward) has both substituents in the equatorial position. The cis isomer has one substituent in the equatorial position and one in the axial position. **Trans-fused** rings, therefore, are more stable than **cis-fused** rings.

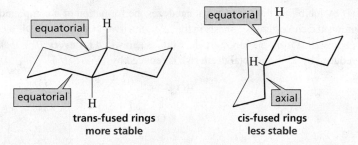

trans-fused rings
more stable

cis-fused rings
less stable

Hormones are chemical messengers—organic compounds synthesized in glands and delivered by the bloodstream to target tissues in order to stimulate or inhibit some process. Many hormones are **steroids.** Steroids have four rings designated here by A, B, C, and D. The B, C, and D rings are all trans fused, and in most naturally occurring steroids, the A and B rings are also trans fused.

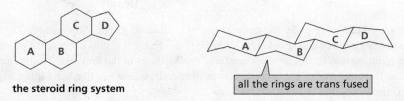

the steroid ring system

all the rings are trans fused

The most abundant member of the steroid family in animals is **cholesterol,** the precursor of all other steroids. Cholesterol is an important component of cell membranes. (See the box in this section.) Because its rings are locked in a specific conformation, it is more rigid than other membrane components.

cholesterol

Cholesterol and Heart Disease

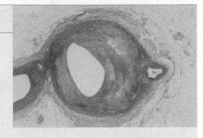

cholesterol (brown) blocking an artery

Cholesterol is probably the best-known steroid because of the widely publicized correlation between cholesterol levels in the blood and heart disease. Cholesterol is synthesized in the liver and is present in almost all body tissues. It is also found in many foods, but we do not require cholesterol in our diet because the body can synthesize all we need. A diet high in cholesterol can lead to high levels of cholesterol in the bloodstream, and the excess can accumulate on the walls of arteries, restricting the flow of blood. This disease of the circulatory system is known as *atherosclerosis* and is a primary cause of heart disease.

Cholesterol travels through the bloodstream packaged in particles that are classified according to their density. Low-density lipoprotein (LDL) particles transport cholesterol from the liver to other tissues. Receptors on the surfaces of cells bind LDL particles, allowing them to be brought into the cell so it can use the cholesterol. High-density lipoprotein (HDL) is a cholesterol scavenger, removing cholesterol from the surfaces of membranes and delivering it back to the liver, where it is converted into bile acids. LDL is the so-called "bad" cholesterol, whereas HDL is the "good" cholesterol. The more cholesterol we eat, the less the body synthesizes. But this does not mean that dietary cholesterol has no effect on the total amount of cholesterol in the bloodstream, because dietary cholesterol inhibits the synthesis of the LDL receptors. So the more cholesterol we eat, the less the body synthesizes, but also the less the body can get rid of by transporting it to target cells.

How High Cholesterol Is Treated Clinically

Statins are drugs that reduce serum cholesterol levels by inhibiting the enzyme that catalyzes the formation of a compound needed for the synthesis of cholesterol. As a consequence of diminished cholesterol synthesis in the liver, the liver forms more LDL receptors—the receptors that help clear LDL (the so-called "bad" cholesterol) from the bloodstream. Studies show that for every 10% that cholesterol is reduced, deaths from coronary heart disease are reduced by 15% and total death risk is reduced by 11%.

lovastatin
Mevacor®

simvastatin
Zocor®

atorvastatin
Lipitor®

Lovastatin and simvastatin are natural statins used clinically under the trade names Mevacor and Zocor. Atorvastatin (Lipitor), a synthetic statin, is the most popular statin. It has greater potency and lasts longer in the body than natural statins because the products of its breakdown are as active as the parent drug in reducing cholesterol levels. Therefore, smaller doses of the drug may be administered. In addition, Lipitor is less polar than lovastatin and simvastatin, so it persists longer in liver cells, where it is needed. Lipitor has been one of the most widely prescribed drugs in the United States for the past several years.

SOME IMPORTANT THINGS TO REMEMBER

- **Alkanes** are **hydrocarbons** that contain only single bonds. Their general molecular formula is C_nH_{2n+2}.

- **Constitutional isomers** have the same molecular formula, but their atoms are linked differently.

- Alkanes are named by determining the number of carbons in their **parent hydrocarbon.** Substituents are listed as prefixes in alphabetical order, with a number to designate their position on the chain.

- When there is only a substituent, the substituent gets the lower of the possible numbers; when there is only a functional group suffix, the functional group suffix gets the lower of the possible numbers; when there is both a functional group suffix and a substituent, the functional group suffix gets the lower of the possible numbers.

- A **functional group** is a center of reactivity in a molecule.

- **Alkyl halides** and **ethers** are named as substituted alkanes. **Alcohols** and **amines** are named using a functional group suffix.

- **Systematic names** can contain numbers; **common names** never do.

- A compound can have more than one name, but a name must specify only one compound.

- Whether alkyl halides or alcohols are **primary, secondary,** or **tertiary** depends on whether the X (halogen) or OH group is attached to a primary, secondary, or tertiary carbon.

- Whether amines are **primary, secondary,** or **tertiary** depends on the number of alkyl groups attached to the nitrogen.

- Compounds with four alkyl groups attached to a nitrogen are called **quaternary ammonium salts.**

- The oxygen of an alcohol or an ether has the same geometry as the oxygen of water; the nitrogen of an amine has the same geometry as the nitrogen of ammonia.

- The boiling point of a compound increases as the attractive forces between its molecules—**van der Waals forces, dipole–dipole interactions,** and **hydrogen bonds**—increase.

- Hydrogen bonds are stronger than other dipole–dipole interactions, which are stronger than van der Waals forces.

- A **hydrogen bond** is an interaction between a hydrogen bonded to an O, N, or F and a lone pair of an O, N, or F in another molecule.

- In a series of **homologs,** the boiling point increases with increasing molecular weight. Branching lowers the boiling point.

- **Polarizability** indicates the ease with which an electron cloud can be distorted. Larger atoms are more polarizable.

- Polar compounds dissolve in polar solvents; nonpolar compounds dissolve in nonpolar solvents.

- **Solvation** is the interaction between a solvent and a molecule or an ion dissolved in that solvent.

- The oxygen of an alcohol or an ether can drag three or four carbons into solution in water.

- Rotation about a C—C bond results in staggered and eclipsed conformers that rapidly interconvert.

- **Conformers** are different conformations of the same compound. They cannot be separated.

- A **staggered conformer** is more stable than an **eclipsed conformer** because of **hyperconjugation.**

- The **anti conformer** is more stable than a **gauche conformer** because of **steric strain,** which is repulsion between the electron clouds of atoms or groups.

- A **gauche interaction** causes steric strain in a gauche conformer.

- Five- and six-membered rings are more stable than three- and four-membered rings because of the **angle strain** that results when bond angles deviate markedly from the ideal bond angle of $109.5°$.

- Cyclohexane rapidly interconverts between two stable chair conformers—this is called **ring flip.**

- Bonds that are **axial** in one chair conformer are **equatorial** in the other and vice versa.

- A chair conformer with an equatorial substituent has less steric strain and is, therefore, more stable than a chair conformer with an axial substituent.

- An axial substituent experiences unfavorable **1,3-diaxial interactions.**

- Cis and trans isomers (**geometric isomers**) are different compounds and can be separated.

- A **cis isomer** has its two substituents on the same side of the ring; a **trans isomer** has its substituents on opposite sides of the ring.

- The more stable conformer of a disubstituted cyclohexane has its larger substituent on an equatorial bond.

PROBLEMS

52. Draw a condensed structure and a skeletal structure for each of the following compounds:

a. *sec*-butyl *tert*-butyl ether
b. isoheptyl alcohol
c. *sec*-butylamine
d. isopentyl bromide

e. 5-(1,2-dimethylpropyl)nonane
f. triethylamine
g. 4-(1,1-dimethylethyl)heptane
h. 5,5-dibromo-2-methyloctane

i. 3-ethoxy-2-methylhexane
j. 5-(1,2-dimethylpropyl)nonane
k. 3,4-dimethyloctane
l. 4-(1-methylethyl)nonane

53. List the following compounds from highest boiling to lowest boiling:

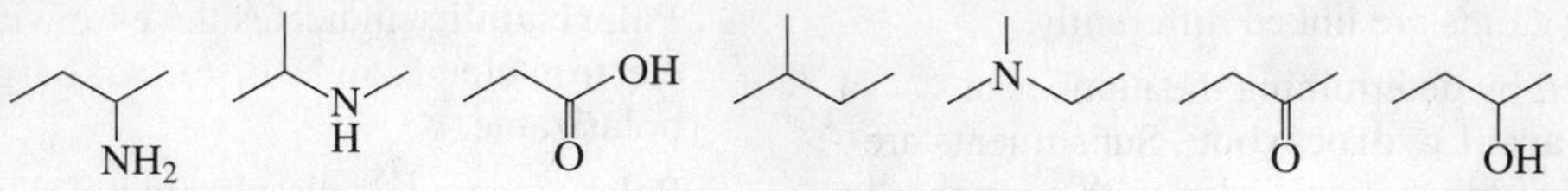

54. a. What is each compound's systematic name?

b. Draw a skeletal structure for each condensed structure given, and draw a condensed structure for each skeletal structure given in the following list:

1. $(CH_3)_3CCH_2CH_2CH_2CH(CH_3)_2$

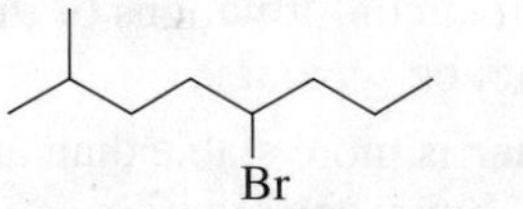

5. $BrCH_2CH_2CH_2CH_2CH_2NHCH_2CH_3$

9.

2.

6.

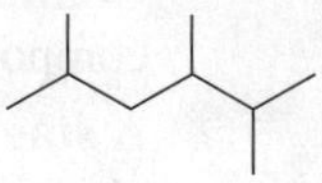

10.

3. $CH_3CHCH_2CHCH_2CH_3$
   $\quad\ \ |\qquad\quad |$
   $\quad CH_3\ \ OH$

4. $(CH_3CH_2)_4C$

7. $CH_3CH_2CHOCH_2CH_3$
   $\qquad\qquad\ |$
   $\qquad\quad CH_2CH_2CH_2CH_3$

8. $CH_3OCH_2CH_2CH_2OCH_3$

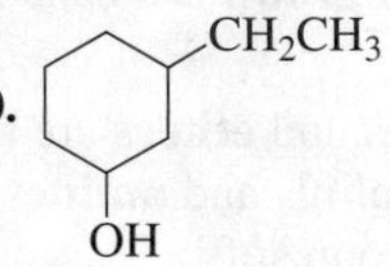

11.

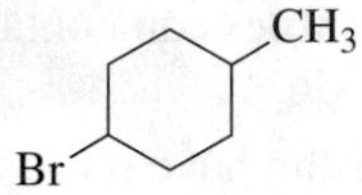

55. Which of the following represents a cis isomer?

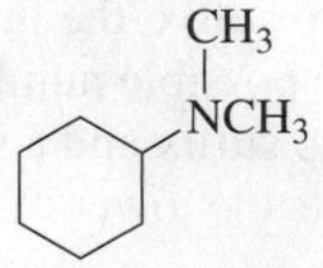

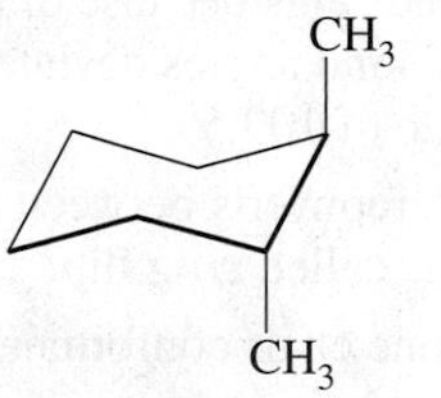

A 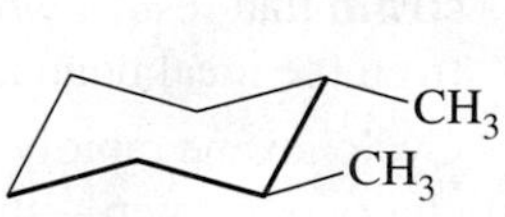   B 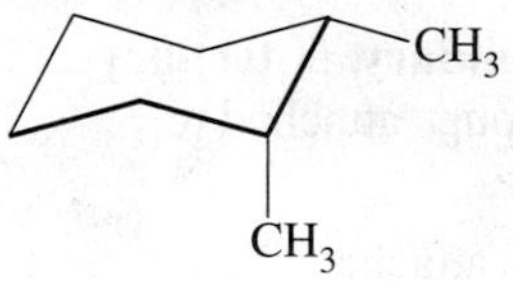   C 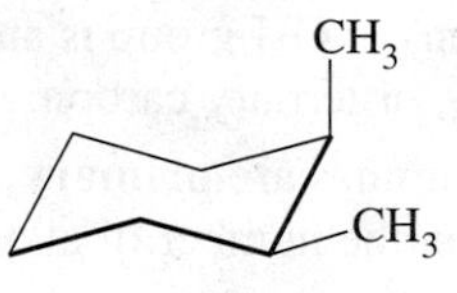   D

56. a. How many primary carbons does each of the following compounds have?
b. How many secondary carbons does each one have?
c. How many tertiary carbons does each one have?

1.

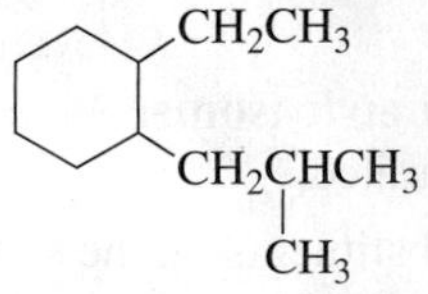

2.

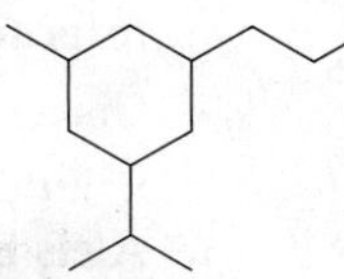

57. Which of the following conformers of isobutyl chloride is the most stable?

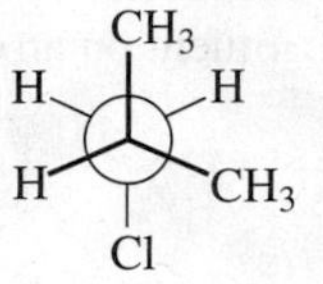

A 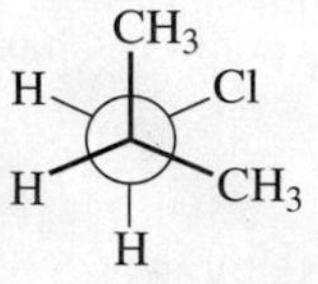   B 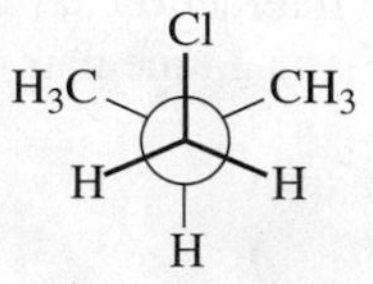   C

58. Draw the structural formula for an alkane that has
   a. six carbons, all secondary.
   b. eight carbons and only primary hydrogens.
   c. seven carbons with two isopropyl groups.

59. What is each compound's systematic name?

a. $CH_3CH_2CHCH_3$
        |
        NH_2

b. $CH_3CH_2CHCH_3$
        |
        Cl

c. $CH_3CH_2CHNHCH_2CH_3$
        |
        CH_3

d. $CH_3CH_2CH_2OCH_2CH_3$

e. $CH_3CHCH_2CH_2CH_3$
        |
        CH_3

f. CH_3CHNH_2
        |
        CH_3

g. CH_3
        |
        CH_3CBr
        |
        CH_2CH_3

h. $CH_3CHCH_2CH_2CH_2OH$
        |
        CH_3

i. [cyclopentane with Br]

j. [cyclohexane with OH]

60. Which has
   a. the higher boiling point: 1-bromopentane or 1-bromohexane?
   b. the higher boiling point: pentyl chloride or isopentyl chloride?
   c. the greater solubility in water: 1-butanol or 1-pentanol?
   d. the higher boiling point: 1-hexanol or 1-methoxypentane?
   e. the higher melting point: hexane or isohexane?
   f. the higher boiling point: 1-chloropentane or 1-pentanol?
   g. the higher boiling point: 1-bromopentane or 1-chloropentane?
   h. the higher boiling point: diethyl ether or butyl alcohol?
   i. the greater density: heptane or octane?
   j. the higher boiling point: isopentyl alcohol or isopentylamine?
   k. the higher boiling point: hexylamine or dipropylamine?

61. a. Draw Newman projections of the two conformers of *cis*-1,3-dimethylcyclohexane.
   b. Which of the conformers would predominate at equilibrium?
   c. Draw Newman projections of the two conformers of the trans isomer.
   d. Which of the conformers would predominate at equilibrium?

62. Ansaid and Motrin belong to the group of drugs known as nonsteroidal anti-inflammatory drugs (NSAIDs). Both are only slightly soluble in water, but one is a little more soluble than the other. Which of the drugs has the greater solubility in water?

[structures: **Ansaid®** and **Motrin®**]

63. Draw a picture of the hydrogen bonding in methanol.

64. A student was given the structural formulas of several compounds and was asked to give them systematic names. How many did the student name correctly? Correct those that are misnamed.
   a. 4-bromo-3-pentanol
   b. 2,2-dimethyl-4-ethylheptane
   c. 5-methylcyclohexanol
   d. 1,1-dimethyl-2-cyclohexanol
   e. 5-(2,2-dimethylethyl)nonane
   f. isopentyl bromide
   g. 3,3-dichlorooctane
   h. 5-ethyl-2-methylhexane
   i. 1-bromo-4-pentanol
   j. 3-isopropyloctane
   k. 2-methyl-2-isopropylheptane
   l. 2-methyl-*N*,*N*-dimethyl-4-hexanamine

65. Which of the following conformers has the highest energy (is the least stable)?

[cyclohexane conformers A, B, C with CH₃ and Cl substituents]

A   B   C

66. Give the systematic names for all alkanes with molecular formula C_7H_{16} that do not have any secondary hydrogens.

67. Draw skeletal structures for the following:
   a. 5-ethyl-2-methyloctane
   b. 1,3-dimethylcyclohexane
   c. 2,3,3,4-tetramethylheptane
   d. propylcyclopentane
   e. 2-methyl-4-(1-methylethyl)octane
   f. 2,6-dimethyl-4-(2-methylpropyl)decane

68. For rotation about the C-3 — C-4 bond of 2-methylhexane, do the following:
   a. Draw the Newman projection of the most stable conformer.
   b. Draw the Newman projection of the least stable conformer.
   c. About which other carbon–carbon bonds may rotation occur?
   d. How many of the carbon–carbon bonds in the compound have staggered conformers that are all equally stable?

69. Draw all the isomers that have molecular formula $C_5H_{11}Br$. (*Hint:* There are eight.)
   a. Give the systematic name for each of the isomers.
   b. Give a common name for each isomer that has a common name.
   c. How many of the isomers are primary alkyl halides?
   d. How many of the isomers are secondary alkyl halides?
   e. How many of the isomers are tertiary alkyl halides?

70. What is each compound's systematic name?

   a.   **e.**   **i.**   NH₂

   b. OH   **f.** Cl   **j.**

   c.   **g.** O

   d. OH   **h.**

71. Draw the two chair conformers for each of the following, and indicate which conformer is more stable:
   a. *cis*-1-ethyl-3-methylcyclohexane
   b. *trans*-1-ethyl-2-isopropylcyclohexane
   c. *trans*-1-ethyl-2-methylcyclohexane
   d. *cis*-1,2-diethylcyclohexane
   e. *cis*-1-ethyl-3-isopropylcyclohexane
   f. *cis*-1-ethyl-4-isopropylcyclohexane

72. Why are lower molecular weight alcohols more soluble in water than higher molecular weight alcohols?

73. **a.** Draw a potential energy diagram for rotation about the C — C bond of 1,2-dichloroethane through 360°, starting with the least stable conformer. The anti conformer is 1.2 kcal/mol more stable than a gauche conformer. A gauche conformer has two energy barriers, 5.2 kcal/mol and 9.3 kcal/mole.
   b. Draw the conformer that would be present in greatest concentration.
   c. How much more stable is the most stable staggered conformer than the most stable eclipsed conformer?
   d. How much more stable is the most stable staggered conformer than the least stable eclipsed conformer?

74. For each of the following compounds, is the cis isomer or the trans isomer more stable?

   a.   **b.**   **c.**

75. How many ethers have molecular formula $C_5H_{12}O$? Draw their structures and give each a systematic name. What are their common names?

76. Draw the most stable conformer of the following molecule. (A solid wedge points out of the plane of the paper toward the viewer. A hatched wedge points back from the plane of the paper away from the viewer.)

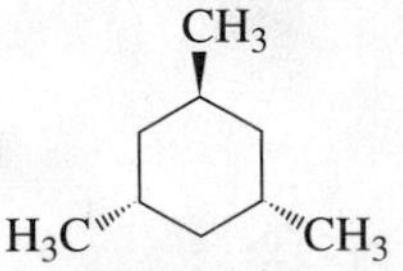

77. What is each compound's systematic name?

a. 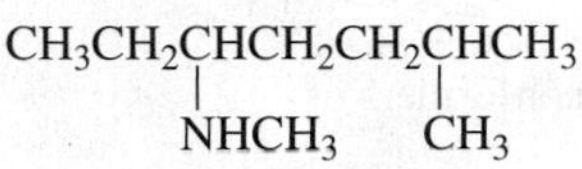 $CH_3CH_2CHCH_2CH_2CHCH_3$
    with $NHCH_3$ and CH_3 substituents

b. 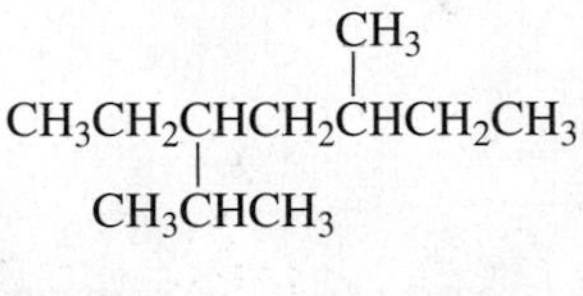 $CH_3CH_2CHCH_2CHCH_2CH_3$
    with CH_3 and CH_3CHCH_3 substituents

c. 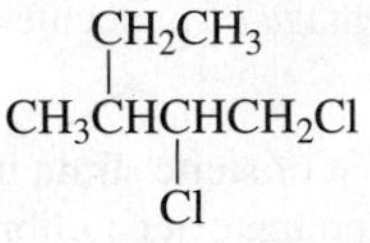 $CH_3CHCHCH_2Cl$
    with CH_2CH_3 and Cl substituents

d. 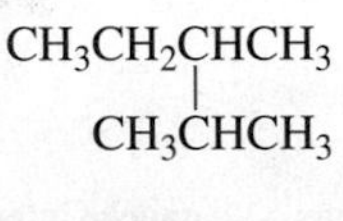 $CH_3CH_2CHCH_3$
    with CH_3CHCH_3 substituent

e.

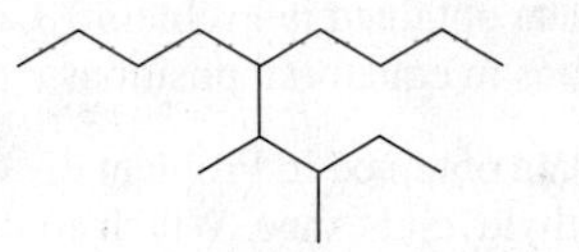

f.

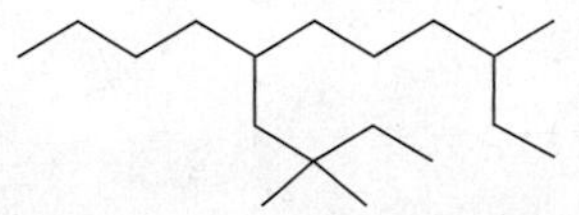

78. Calculate the energy difference between the two chair conformers of *trans*-1,2-dimethylcyclohexane.

79. The most stable form of glucose (blood sugar) is a six-membered ring in a chair conformation with its five substituents all in equatorial positions. Draw the most stable conformer of glucose by putting the OH groups on the appropriate bonds in the structure on the right.

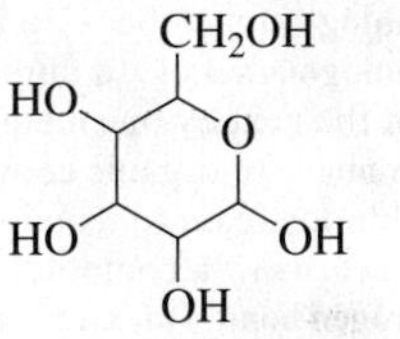

glucose

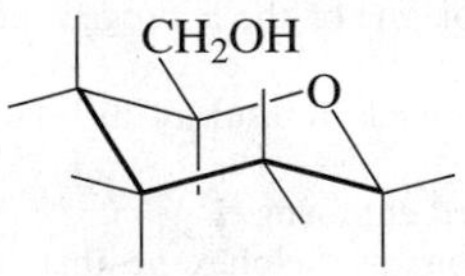

80. What is each compound's systematic name?

a. 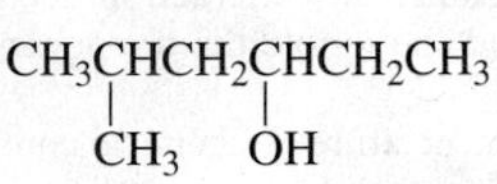 $CH_3CHCH_2CHCH_2CH_3$
    with CH_3 and OH substituents

b.

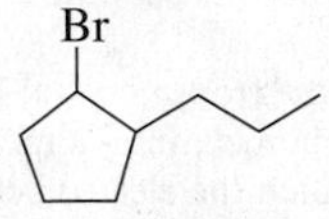

c. 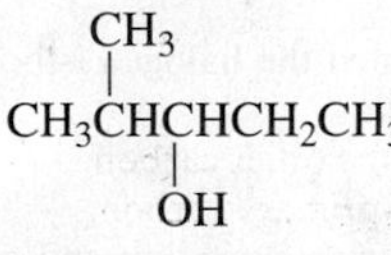 $CH_3CHCHCH_2CH_3$
    with CH_3 and OH substituents

d. 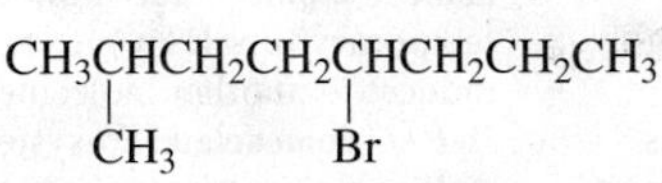 $CH_3CHCH_2CH_2CHCH_2CH_2CH_3$
    with CH_3 and Br substituents

e.

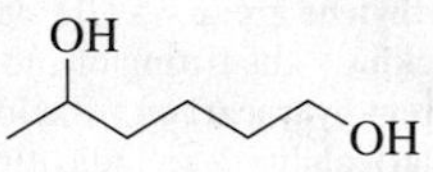

f.

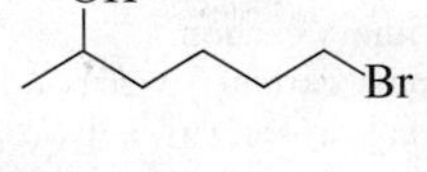

g.

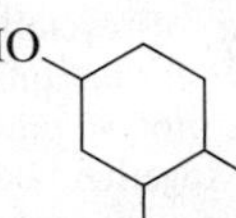

h.

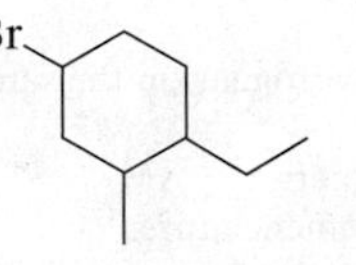

81. Explain the following:
   a. 1-Hexanol has a higher boiling point than 3-hexanol.
   b. Diethyl ether has very limited solubility in water, but tetrahydrofuran is completely soluble.

tetrahydrofuran

82. One of the chair conformers of *cis*-1,3-dimethylcyclohexane is 5.4 kcal/mol (23 kJ/mol) less stable than the other. How much steric strain does a 1,3-diaxial interaction between two methyl groups introduce into the conformer?

83. Bromine is a larger atom than chlorine, but the equilibrium constants in Table 9 indicate that a chloro substituent has a greater preference for the equatorial position than a bromo substituent does. Suggest an explanation for this fact.

84. Name the following compounds:

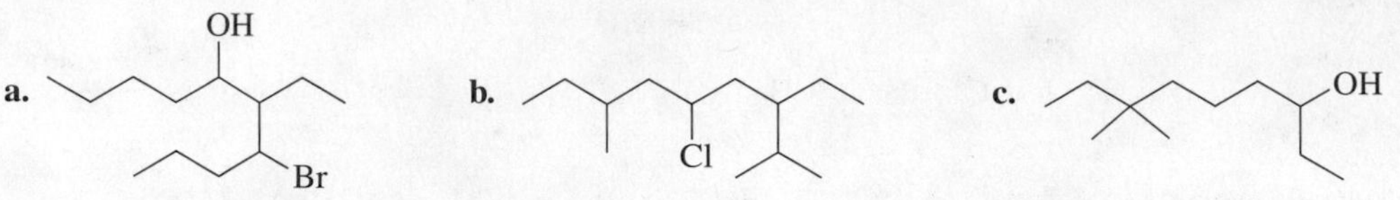

85. Using the data obtained in Problem 78, calculate the percentage of molecules of *trans*-1,2-dimethylcyclohexane that will have both methyl groups in equatorial positions.

86. Using the data obtained in Problem 82, calculate the amount of steric strain in each of the chair conformers of 1,1,3-trimethylcyclohexane. Which conformer would predominate at equilibrium?

GLOSSARY

alcohol a compound with an OH group in place of one of the hydrogens of an alkane; (ROH).

alkane a hydrocarbon that contains only single bonds.

alkyl halide a compound with a halogen in place of one of the hydrogens of an alkane.

alkyl substituent (alkyl group) formed by removing a hydrogen from an alkane.

amine a compound with a nitrogen in place of one of the hydrogens of an alkane; (RNH_2, R_2NH, R_3N)

angle strain the strain introduced into a molecule as a result of its bond angles being distorted from their ideal values.

anti conformer the most stable of the staggered conformers.

axial bond a bond of the chair conformation of cyclohexane that is perpendicular to the plane in which the chair is drawn (an up–down bond).

boat conformation the conformation of cyclohexane that roughly resembles a boat.

boiling point the temperature at which the vapor pressure from a liquid equals the atmospheric pressure.

chair conformation the conformation of cyclohexane that roughly resembles a chair. It is the most stable conformation of cyclohexane.

cholesterol a steroid that is the precursor of all other animal steroids.

cis fused two cyclohexane rings fused together such that if the second ring were considered to be two substituents of the first ring, one substituent would be in an axial position and the other would be in an equatorial position.

cis isomer the isomer with the hydrogens on the same side of the double bond or cyclic structure.

cis–trans isomers geometric isomers.

common name nonsystematic nomenclature.

conformational analysis the investigation of the various conformations of a compound and their relative stabilities.

conformers different conformations of a molecule.

constitutional isomers (structural isomers) molecules that have the same molecular formula but differ in the way their atoms are connected.

cycloalkane an alkane with its carbon chain arranged in a closed ring.

1,3-diaxial interaction the interaction between an axial substituent and the other two axial substituents on the same side of the cyclohexane ring.

dipole–dipole interaction an interaction between the dipole of one molecule and the dipole of another.

eclipsed conformation a conformation in which the bonds on adjacent carbons are aligned as viewed looking down the carbon–carbon bond.

equatorial bond a bond of the chair conformer of cyclohexane that juts out from the ring in approximately the same plane that contains the chair.

ether a compound containing an oxygen bonded to two carbons (ROR).

flagpole hydrogens (transannular hydrogens) the two hydrogens in the boat conformation of cyclohexane that are closest to each other.

functional group the center of reactivity in a molecule.

gauche conformer a staggered conformer in which the largest substituents are gauche to each other.

gauche interaction the interaction between two atoms or groups that are gauche to each other.

geometric isomers cis–trans (or *E,Z*) isomers.

half-chair conformation the least stable conformation of cyclohexane.

heat of formation the heat given off when a compound is formed from its elements under standard conditions.

homolog a member of a homologous series.

homologous series a family of compounds in which each member differs from the next by one methylene group.

hormone an organic compound synthesized in a gland and delivered by the bloodstream to its target tissue.

hydrocarbon a compound that contains only carbon and hydrogen.

hydrogen bond an unusually strong dipole–dipole attraction (5 kcal/mol) between a hydrogen bonded to O, N, or F and the nonbonding electrons of an O, N, or F of another molecule.

hyperconjugation delocalization of electrons by overlap of carbon–hydrogen or carbon–carbon s bonds with an empty *p* orbital.

induced-dipole–induced-dipole interaction an interaction between a temporary dipole in one molecule and the dipole the temporary dipole induces in another molecule.

IUPAC nomenclature systematic nomenclature of chemical compounds.

melting point the temperature at which a solid becomes a liquid.

methylene group a CH_2 group.

packing the fitting of individual molecules into a frozen crystal lattice.

parent hydrocarbon the longest continuous carbon chain in a molecule.

polarizability an indication of the ease with which the electron cloud of an atom can be distorted.

primary alkyl halide an alkyl halide in which the halogen is bonded to a primary carbon.

primary carbon a carbon bonded to only one other carbon.

primary hydrogen a hydrogen bonded to a primary carbon.

quaternary ammonium salt a quaternary ammonium ion and an anion ($R_4N^+ X^-$).

secondary alkyl halide an alkyl halide in which the halogen is bonded to a secondary carbon.

secondary carbon a carbon bonded to two other carbons.

secondary hydrogen a hydrogen bonded to a secondary carbon.

skeletal structure shows the carbon–carbon bonds as lines and does not show the carbon–hydrogen bonds.

solvation the interaction between a solvent and another molecule (or ion).

staggered conformation a conformation in which the bonds on one carbon bisect the bond angle on the adjacent carbon when viewed looking down the carbon–carbon bond.

steric strain (van der Waals strain, van der Waals repulsion) the repulsion between the electron cloud of an atom or a group of atoms and the electron cloud of another atom or group of atoms.

steroid a class of compounds that contains a steroid ring system.

straight-chain alkane (normal alkane) an alkane in which the carbons form a continuous chain with no branches.

symmetrical ether an ether with two identical substituents bonded to the oxygen.

systematic nomenclature nomenclature based on structure.

tertiary alkyl halide an alkyl halide in which the halogen is bonded to a tertiary carbon.

tertiary carbon a carbon bonded to three other carbons.

tertiary hydrogen a hydrogen bonded to a tertiary carbon.

trans fused two cyclohexane rings fused together such that if the second ring were considered to be two substituents of the first ring, both substituents would be in equatorial positions.

trans isomer the isomer with the hydrogens on opposite sides of the double bond or cyclic structure. the isomer with identical substituents on opposite sides of the double bond.

twist-boat conformation (skew-boat conformation) a conformation of cyclohexane.

unsymmetrical ether an ether with two different substituents bonded to the oxygen.

van der Waals forces (London forces) induced-dipole–induced-dipole interactions.

PHOTO CREDITS

Photo credits are listed in order of appearance.

Charles Gellis/Photo Researchers, Inc.; Karin Hildebrand Lau/Shutterstock; VadiCo/Shutterstock; Pearson Education/Eric Schrader; Reuters/Sean Gardner; Anerila/Dreamstime; Karen Struthers/Fotolia; (top) Custom Medical Stock Photo/Newscom; (bot.) Joy Brown/Shutterstock

Isomers: The Arrangement of Atoms in Space

In this chapter we will see why interchanging two groups bonded to a carbon can have a profound effect on the physiological properties of a compound. For example, interchanging a hydrogen and a methyl group converts the active ingredient in Vicks vapor inhaler to meth-amphetamine, the street drug known as speed. The same change converts the active ingredient in Aleve, a common drug for pain, to a compound that is highly toxic to the liver.

We will now turn our attention to **isomers**—compounds with the same molecular formula but different structures. Isomers fall into two main classes: *constitutional isomers* and *stereoisomers.*

Constitutional isomers differ in the way their atoms are connected. For example, ethanol and dimethyl ether are constitutional isomers because they both have molecular formula C_2H_6O, but their atoms are connected differently (the oxygen in ethanol is bonded to a carbon and to a hydrogen, whereas the oxygen in dimethyl ether is bonded to two carbons).

constitutional isomers

CH_3CH_2OH and CH_3OCH_3
ethanol   dimethyl ether

$CH_3CH_2CH_2CH_2Cl$ and $CH_3CH_2\overset{\overset{\displaystyle Cl}{|}}{C}HCH_3$
1-chlorobutane   2-chlorobutane

$CH_3CH_2CH_2CH_2CH_3$ and $CH_3\overset{\overset{\displaystyle CH_3}{|}}{C}HCH_2CH_3$
pentane   isopentane

$CH_3\overset{\overset{\displaystyle O}{||}}{C}CH_3$ and $CH_3CH_2\overset{\overset{\displaystyle O}{||}}{C}H$
acetone   propionaldehyde

From Chapter 4 of *Organic Chemistry*, Seventh Edition. Paula Yurkanis Bruice.

Unlike constitutional isomers, the atoms in stereoisomers are connected in the same way. **Stereoisomers** (also called **configurational isomers**) differ in the way their atoms are arranged in space. Like constitutional isomers, stereoisomers can be separated because they are different compounds; they can interconvert only if bonds are broken. There are two kinds of stereoisomers: *cis–trans isomers* and isomers that contain *asymmetric centers.*

PROBLEM 1♦

a. Draw three constitutional isomers with molecular formula C_3H_8O.
b. How many constitutional isomers can you draw for $C_4H_{10}O$?

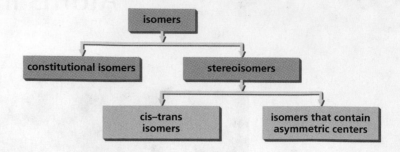

1 CIS–TRANS ISOMERS RESULT FROM RESTRICTED ROTATION

The first type of stereoisomers we will look at are **cis–trans isomers** (also called **geometric isomers**). These isomers result from restricted rotation. Restricted rotation can be caused either by a *cyclic structure* or by a *double bond.*

As a result of restricted rotation about the bonds in a ring, Cyclic compounds with two substituents bonded to different carbons have cis and trans isomers. *The cis isomer has its substituents on the same side of the ring; the trans isomer has its substituents on opposite sides of the ring.* (A solid wedge represents a bond that points out of the plane of the paper toward the viewer and a hatched wedge represents a bond that points into the plane of the paper away from the viewer.)

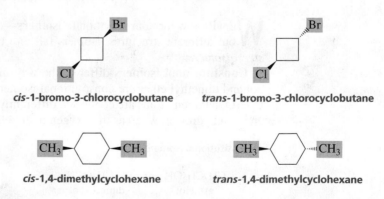

cis-1-bromo-3-chlorocyclobutane

trans-1-bromo-3-chlorocyclobutane

cis-1,4-dimethylcyclohexane

trans-1,4-dimethylcyclohexane

PROBLEM 2♦

Draw the cis and trans isomers for the following:

a. 1-bromo-4-chlorocyclohexane
b. 1-ethyl-3-methylcyclobutane

Compounds with carbon-carbon double bonds can also have cis and trans isomers. The structure of the smallest compound with a carbon–carbon double bond (ethene) shows that the double bond was composed of a σ bond and a π bond. The π bond was formed by side-to-side overlap of two parallel *p* orbitals—one from each carbon. Other compounds with carbon–carbon double bonds have similar structures.

Rotation about a double bond does not readily occur, because it can happen only if the π bond breaks—that is, only if the p orbitals are no longer parallel (Figure 1). Consequently, the energy barrier to rotation about a carbon–carbon double bond is much greater (about 62 kcal/mol or 259 kJ/mol) than the energy barrier to rotation about a carbon–carbon single bond, which is only about 2.9 kcal/mol or 12 kJ/mol.

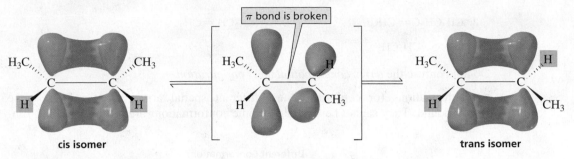

▲ Figure 1

Rotation about the carbon–carbon double bond breaks the π bond.

Because of the high energy barrier to rotation about a carbon–carbon double bond, a compound with a carbon–carbon double bond can exist in two distinct forms—the hydrogens bonded to the sp^2 carbons can be on the same side of the double bond or on opposite sides of the double bond.

The compound with the hydrogens on the same side of the double bond is called the **cis isomer;** the compound with the hydrogens on opposite sides of the double bond is called the **trans isomer.** Notice that the cis and trans isomers have the same molecular formula and the same bonds but have different *configurations*—they differ in the way their atoms are oriented in space.

Cis and trans isomers can be separated from each other because they are different compounds with different physical properties—for example, they have different boiling points and different dipole moments.

the cis isomer	the trans isomer	the cis isomer	the trans isomer
bp = 3.7 °C	bp = 0.9 °C	bp = 60.3 °C	bp = 47.5 °C
μ = 0.33 D	μ = 0 D	μ = 2.95 D	μ = 0 D

Notice that the trans isomers, unlike the cis isomers, have dipole moments (μ) of zero because the dipole moments of their individual bonds cancel.

If one of the sp^2 carbons is attached to two identical substituents, then the compound cannot have cis and trans isomers.

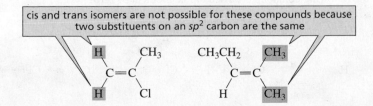

cis and trans isomers are not possible for these compounds because two substituents on an sp^2 carbon are the same

Build models to see why cis and trans isomers are not possible if one of the sp^2 carbons is attached to two identical substituents.

153

PROBLEM 3

a. Which of the following compounds can exist as cis–trans isomers?

b. For those compounds that can exist as cis and trans isomers, draw and label the isomers.

1. $CH_3CH=CHCH_2CH_2CH_3$
3. $CH_3CH=CHCH_3$

2. $CH_3CH_2C=CHCH_3$
   $\quad\quad\quad |$
   $\quad\quad CH_2CH_3$
4. $CH_3CH_2CH=CH_2$

Do not confuse the terms *conformation* and *configuration*.

- Conformations (or conformers) are different spatial arrangements of the same compound. They cannot be separated. Some conformations are more stable than others.

Different Conformations

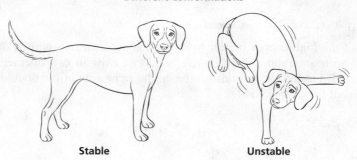

Stable   Unstable

- Compounds with different configurations are different compounds (for example, cis and trans isomers). They can be separated from each other. Bonds have to be broken to interconvert compounds with different configurations.

Different Configurations

PROBLEM 4

Draw skeletal structures for all the compounds in Problem 3, including any cis–trans isomers.

PROBLEM 5♦

Draw three compounds with molecular formula C_5H_{10} that have carbon–carbon double bonds but do not have cis–trans isomers.

PROBLEM 6♦

Which of the following compounds have a dipole moment of zero?

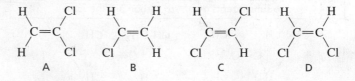

Cis–Trans Interconversion in Vision

Our ability to see depends in part on an interconversion of cis and trans isomers that takes place in our eyes. A protein called opsin binds to *cis*-retinal (formed from vitamin A) in photoreceptor cells (called rod cells) in the retina to form rhodopsin. When rhodopsin absorbs light, a double bond interconverts between the cis and trans configurations, triggering a nerve impulse that plays an important role in vision. *trans*-Retinal is then released from opsin. *trans*-Retinal isomerizes back to *cis*-retinal and another cycle begins. To trigger the nerve impulse, a group of about 500 rod cells must register five to seven rhodopsin isomerizations per cell within a few tenths of a second.

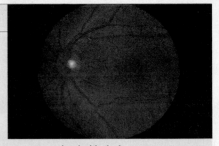

view inside the human eye

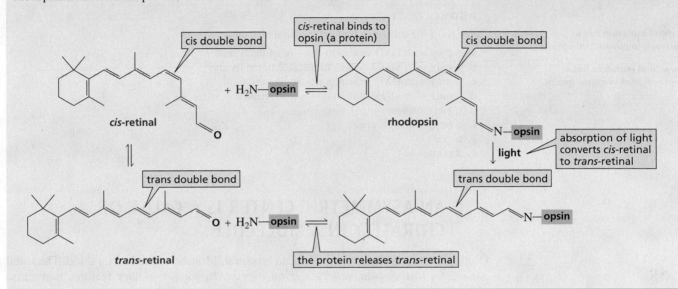

2 A CHIRAL OBJECT HAS A NONSUPERIMPOSABLE MIRROR IMAGE

Why can't you put your right shoe on your left foot? Why can't you put your right glove on your left hand? It is because hands, feet, gloves, and shoes have right-handed and left-handed forms. An object with a right-handed and a left-handed form is said to be **chiral** (ky-ral), a word derived from the Greek word *cheir,* which means "hand."

A chiral object has a *nonsuperimposable mirror image.* In other words, its mirror image *is not the same* as an image of the object itself. A hand is chiral because when you look at your right hand in a mirror, you see a left hand, not a right hand (Figure 2a).

chiral objects

right hand   left hand

◀ **Figure 2a**

A chiral object is not the same as its mirror image—they are nonsuperimposable.

In contrast, a chair is not chiral; the reflection of the chair in the mirror looks the same as the chair itself. Objects that are not chiral are said to be **achiral.** An achiral object has a *superimposable mirror image* (Figure 2b).

achiral objects

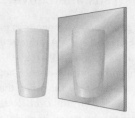

▶ **Figure 2b**

An achiral object is the same as its mirror image—they are superimposable.

A chiral molecule has a nonsuperimposable mirror image.

An achiral molecule has a superimposable mirror image.

PROBLEM 7♦

Which of the following objects are chiral?

a. a mug with DAD written to one side of the handle
b. a mug with MOM written to one side of the handle
c. a mug with DAD written opposite the handle
d. a mug with MOM written opposite the handle
e. a wheelbarrow
f. a remote control device
g. a nail
h. a screw

3 AN ASYMMETRIC CENTER IS A CAUSE OF CHIRALITY IN A MOLECULE

Objects are not the only things that can be chiral. Molecules can be chiral too. The usual *cause of chirality in a molecule is an asymmetric center*. (Other features that cause chirality are relatively uncommon and beyond the scope of this text, but you can see one example in Problem 86.)

An **asymmetric center** (also called a chiral center) is an atom bonded to four different groups. Each of the following compounds has an asymmetric center that is indicated by a star.

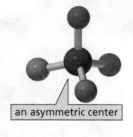

an asymmetric center

A molecule with an asymmetric center is chiral.

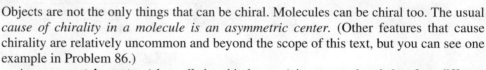

| C is bonded to H, OH, propyl, butyl | C is bonded to H, Br, ethyl, methyl | C is bonded to H, methyl, ethyl, isobutyl |

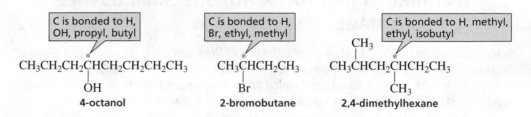

CH₃CH₂CH₂ĊHCH₂CH₂CH₂CH₃
　　　　　　|
　　　　　　OH
4-octanol

CH₃ĊHCH₂CH₃
　　　|
　　　Br
2-bromobutane

　　　　　CH₃
　　　　　|
CH₃ĊHCH₂ĊHCH₂CH₃
　　　　　　|
　　　　　　CH₃
2,4-dimethylhexane

PROBLEM 8♦

Which of the following compounds has an asymmetric center?

a. CH₃CH₂CHCH₃
　　　　|
　　　　Cl

b. CH₃CH₂CHCH₃
　　　　|
　　　　CH₃

c. CH₃CH₂CCH₂CH₂CH₃
　　　　|
　　　CH₃
　　　|
　　　Br

d. CH₃CH₂OH

e. CH₃CH₂CHCH₂CH₃
　　　　　|
　　　　　Br

f. CH₂=CHCHCH₃
　　　　　|
　　　　　NH₂

PROBLEM 9 **Solved**

Tetracycline is called a broad-spectrum antibiotic because it is active against a wide variety of bacteria. How many asymmetric centers does tetracycline have?

Solution Because an asymmetric center must have four different groups attached to it, only sp^3 carbons can be asymmetric centers. Therefore, we start by locating all the sp^3 carbons in tetracycline. (They are numbered in red.) Tetracycline has nine sp^3 carbons. Four of them (1, 2, 5, and 8) are not asymmetric centers because they are not bonded to four different groups. Tetracycline, therefore, has five asymmetric centers (3, 4, 6, 7, and 9).

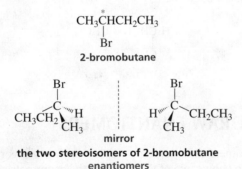

tetracycline

4 | ISOMERS WITH ONE ASYMMETRIC CENTER

A compound with one asymmetric center, such as 2-bromobutane, can exist as two stereoisomers. The two stereoisomers are analogous to a left and a right hand. If we imagine a mirror between the two stereoisomers, we can see they are mirror images of each other. Moreover, they are nonsuperimposable mirror images, which makes them different molecules.

$$CH_3\overset{*}{C}HCH_2CH_3$$
$$|$$
$$Br$$

2-bromobutane

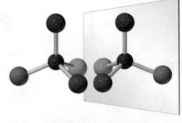

nonsuperimposable
mirror images

Br
|
CH_3CH_2—C—''''H
|
CH_3

mirror

Br
|
H''''—C—CH_2CH_3
|
CH_3

the two stereoisomers of 2-bromobutane
enantiomers

Molecules that are nonsuperimposable mirror images of each other are called **enantiomers** (from the Greek *enantion,* which means "opposite"). Thus, the two stereoisomers of 2-bromobutane are enantiomers.

A molecule that has a *nonsuperimposable* mirror image, like an object that has a *nonsuperimposable* mirror image, is *chiral* (Figure 3a). Therefore, each member of a pair of enantiomers is chiral. A molecule that has a *superimposable* mirror image, like an object that has a *superimposable* mirror image, is *achiral* (Figure 3b). Notice that chirality is a property of an entire object or an entire molecule.

NOTE TO THE STUDENT

Prove to yourself that the two stereoisomers of 2-bromobutane are not identical by building ball-and-stick models to represent them and then trying to superimpose one on the other. The tutorial for this chapter tells you what other models you should build as you go through this material.

a.

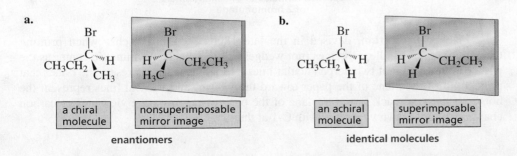

b.

◀ **Figure 3**
(a) A chiral molecule has a nonsuperimposable mirror image.
(b) An achiral molecule has a superimposable mirror image. To see that the achiral molecule is superimposable on its mirror image, mentally rotate the molecule clockwise.

5 ASYMMETRIC CENTERS AND STEREOCENTERS

An asymmetric center is also called a **stereocenter** (or a **stereogenic center**), but they do not mean quite the same thing. A stereocenter is an atom at which the interchange of two groups produces a stereoisomer. Thus, stereocenters include both (1) *asymmetric centers,* where the interchange of two groups produces an enantiomer, and (2) the sp^2 carbons of an alkene or the sp^3 carbons of a cyclic compound, where the interchange of two groups converts a cis isomer to a trans isomer or vice versa. This means that although *all asymmetric centers are stereocenters,* not all stereocenters are asymmetric centers.

PROBLEM 11

a. How many asymmetric centers does the following compound have?
b. How many stereocenters does it have?

$$CH_3CHCH = CHCH_3$$
$$|$$
$$Cl$$

6 HOW TO DRAW ENANTIOMERS

Chemists draw enantiomers using either *perspective formulas* or *Fischer projections.* A **perspective formula** shows two of the bonds to the asymmetric center in the plane of the paper, one bond as a solid wedge protruding forward out of the paper, and the fourth bond as a hatched wedge extending behind the paper. The solid wedge and the hatched wedge must be adjacent to one another. When you draw the first enantiomer, the four groups bonded to the asymmetric center can be placed around it in any order. You can then draw the second enantiomer by drawing the mirror image of the first enantiomer.

A solid wedge represents a bond that extends out of the plane of the paper toward the viewer.

A hatched wedge represents a bond that points back from the plane of the paper away from the viewer.

When you draw a perspective formula, make sure that the two bonds in the plane of the paper are adjacent to one another; neither the solid wedge nor the hatched wedge should be drawn between them.

**perspective formulas of the enantiomers
of 2-bromobutane**

A **Fischer projection,** devised in the late 1800s by Emil Fischer when printing techniques could handle only lines (not wedges), represents an asymmetric center as the point of intersection of two perpendicular lines. Horizontal lines represent the bonds that project out of the plane of the paper toward the viewer, and vertical lines represent the bonds that extend back from the plane of the paper away from the viewer. The carbon chain is usually drawn vertically, with C-1 at the top.

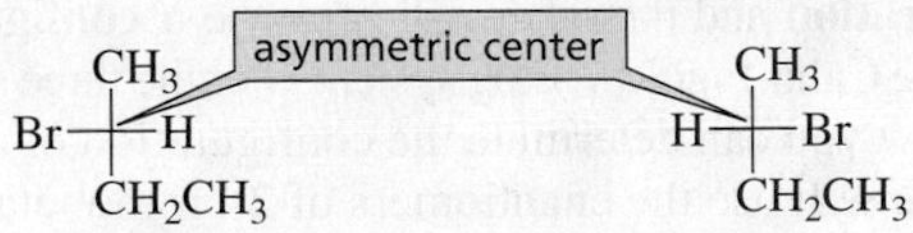

$$CH_3 \quad \boxed{\text{asymmetric center}} \quad CH_3$$

Br——H   H——Br

$$CH_2CH_3 \qquad\qquad CH_2CH_3$$

**Fischer projections of the enantiomers
of 2-bromobutane**

In a Fischer projection, horizontal lines project out of the plane of the paper toward the viewer, and vertical lines extend back from the plane of the paper away from the viewer.

When you draw an enantiomer using a Fischer projection, you can put the four atoms or groups bonded to the asymmetric center in any order around that center. You can then draw the second enantiomer by interchanging two of the atoms or groups. It does not matter which two you interchange. (Make models to convince yourself that this is true.) It is best to interchange the groups on the two horizontal bonds, because then the enantiomers look like mirror images on your paper.

Whether you are drawing perspective formulas or Fischer projections, interchanging two atoms or groups will produce the other enantiomer. Interchanging two atoms or groups a second time brings you back to the original molecule.

PROBLEM 12

Draw enantiomers for each of the following using

a. perspective formulas.
b. Fischer projections.

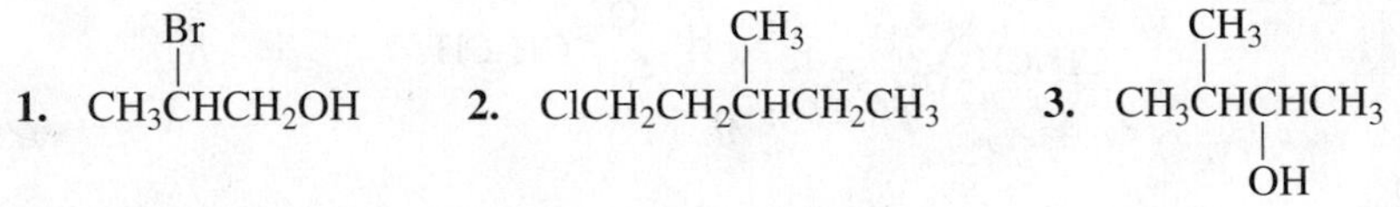

      Br CH_3 CH_3

1. CH_3CHCH_2OH **2.** $ClCH_2CH_2CHCH_2CH_3$ **3.** $CH_3CHCHCH_3$

                                                             OH

PROBLEM 13 Solved

Do the following structures represent identical compounds or a pair of enantiomers?

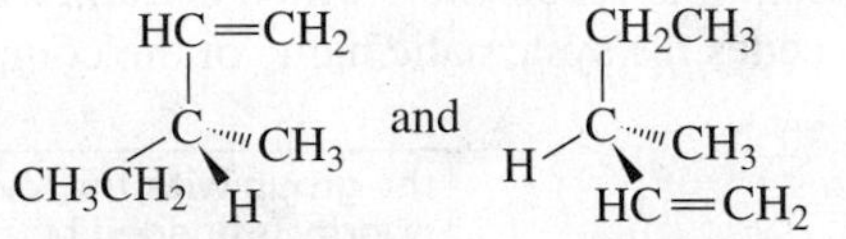

$$HC{=}CH_2 \qquad\qquad CH_2CH_3$$

          | |

        C‴‴CH_3 and   H—C‴‴CH_3

$$CH_3CH_2\ \ H \qquad\qquad HC{=}CH_2$$

Solution Interchanging two atoms or groups attached to an asymmetric center produces an enantiomer. Interchanging two atoms or groups a second time brings you back to the original compound. Because groups have to be interchanged twice to get from one structure to the other, the two structures represent identical compounds.

$$HC{=}CH_2 \qquad\qquad\qquad H \qquad\qquad\qquad CH_2CH_3$$

       | $\xrightarrow[\text{HC=CH}_2 \text{ and H}]{\text{interchange}}$ | $\xrightarrow[\text{ethyl and H}]{\text{interchange}}$ |

    C‴‴CH_3 C‴‴CH_3 H—C‴‴CH_3

$$CH_3CH_2\ \ H \qquad\qquad CH_3CH_2\ \ HC{=}CH_2 \qquad\qquad HC{=}CH_2$$

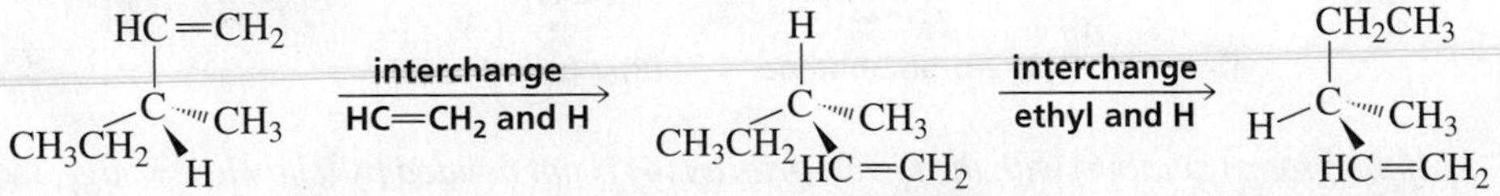

In Section 7 you will learn another way to determine if two structures represent identical compounds or enantiomers.

---

7   NAMING ENANTIOMERS BY THE *R,S* SYSTEM

How do we name the different stereoisomers of a compound like 2-bromobutane so that we know which one we are talking about? We need a system of nomenclature that indicates the arrangement of the atoms or groups around the asymmetric center. Chemists use the letters *R* and *S* for this purpose. For any pair of enantiomers with one asymmetric center, one member will have the **R configuration** and the other will have the **S configuration.** This system of nomenclature is called the Cahn-Ingold-Prelog system after the three scientists who devised it.

First, let's look at how you can determine the configuration of a compound drawn as a perspective formula. We will use the enantiomers of 2-bromobutane as an example.

the enantiomers of 2-bromobutane

1. Rank the groups (or atoms) bonded to the asymmetric center in order of priority. The atomic numbers of the *atoms* directly attached to the asymmetric center determine the relative priorities. The higher the atomic number of the atom, the higher the priority. If there is a tie, you need to consider the atoms to which the tied atoms are attached. For example, the Cs of the methyl and ethyl group tie. The C of the methyl group is attached to H, H, and H; the C of the ethyl group is attached to C, H, and H. So the ethyl group has a higher priority than the methyl group. Therefore, bromine has the highest priority (1), the ethyl group has the second highest priority (2), the methyl group has the third highest priority (3), and hydrogen has the lowest priority (4).

> The greater the atomic number of the atom directly attached to the asymmetric center, the higher the priority of the substituent.

> If the atoms attached to the asymmetric center are the same, the atoms attached to those atoms are compared.

2. *If the group (or atom) with the lowest priority (4) is bonded by a hatched wedge,* draw an arrow from the group (or atom) with the highest priority (1) to the one with the second highest priority (2), and then to the one with the third highest priority (3). If the arrow points clockwise, the compound has the *R* configuration (*R* is for *rectus,* which is Latin for "right"). If the arrow points counterclockwise, then the compound has the *S* configuration (*S* is for *sinister,* which is Latin for "left"). The letter *R* or *S* (in parentheses) precedes the systematic name of the compound.

> Clockwise specifies R if the lowest priority substituent is on a hatched wedge.

> Counterclockwise specifies S if the lowest priority substituent is on a hatched wedge.

right turn

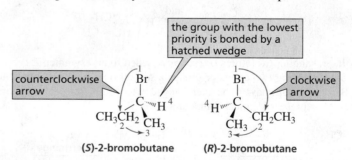

the group with the lowest priority is bonded by a hatched wedge

counterclockwise arrow

clockwise arrow

(S)-2-bromobutane   **(R)-2-bromobutane**

3. *If the group (or atom) with the lowest priority (4) is not bonded by a hatched wedge,* then interchange group 4 with the group that is bonded by a hatched wedge. Then proceed as in step 2—namely, draw an arrow from (1) to (2) to (3). Since the arrow points clockwise, the compound with the interchanged groups has the *R* configuration. Therefore, the original compound, before the groups were interchanged, has the *S* configuration.

left turn

If you forget which direction corresponds to which configuration, imagine driving a car and turning the steering wheel clockwise to make a right turn or turning it counterclockwise to make a left turn.

interchange
CH₃ and H

clockwise arrow

the group with the lowest priority is not bonded by a hatched wedge

this molecule has the *R* configuration; therefore, the molecule had the *S* configuration before the groups were interchanged

PROBLEM 14♦

Assign relative priorities to the groups or atoms in each of the following sets:

a. —CH₂OH   —CH₃   —H   —CH₂CH₂OH
b. —CH₂Br   —OH   —CH₃   —CH₂OH
c. —CH(CH₃)₂   —CH₂CH₂Br   —Cl   —CH₂CH₂CH₂Br

d. —CH=CH₂   —CH₂CH₃   ⬡   —CH₃

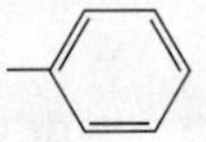

PROBLEM 15♦ Solved

Do the following compounds have the *R* or the *S* configuration?

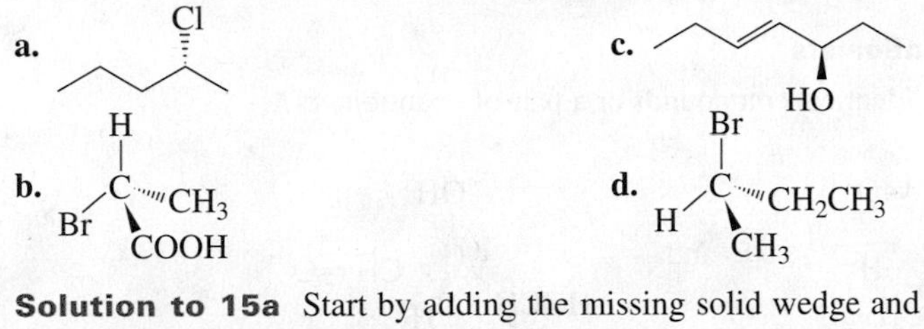

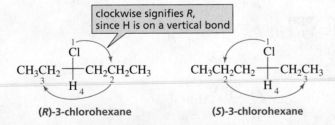

Solution to 15a Start by adding the missing solid wedge and the H to which it is bonded. The solid wedge can be drawn either to the right or to the left of the hatched wedge. (Note that the solid and hatched wedges must be adjacent.)

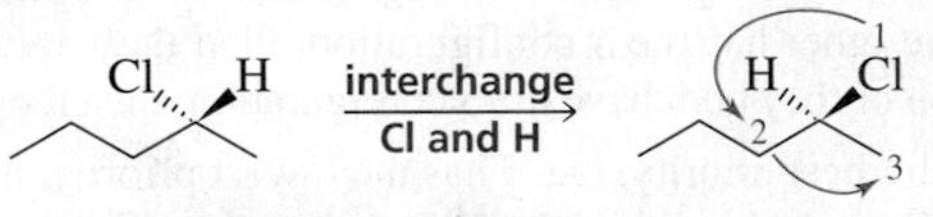

Because the group with the lowest priority is not on the hatched wedge, interchange the Cl and H so that H is on the hatched wedge. An arrow drawn from (1) to (2) to (3) indicates that the compound has the *S* configuration. Therefore, the compound before the pair was interchanged had the *R* configuration.

Now let's see how you can determine the configuration of a compound drawn as a Fischer projection.

1. Rank the groups (or atoms) that are bonded to the asymmetric center in order of priority.

2. Draw an arrow from (1) to (2) to (3). If the arrow points clockwise, then the enantiomer has the *R* configuration; if it points counterclockwise, then the enantiomer has the *S* configuration, *provided that the group with the lowest priority (4) is on a vertical bond.*

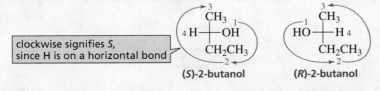

3. If the group (or atom) with the lowest priority is on a *horizontal bond,* the answer you get from the direction of the arrow will be the opposite of the correct answer. For example, if the arrow points clockwise, suggesting the *R* configuration, then the compound actually has the *S* configuration; if the arrow points counterclockwise, suggesting the *S* configuration, then it actually has the *R* configuration. In the following example, the group with the lowest priority is on a horizontal bond, so clockwise signifies the *S* configuration.

Clockwise specifies *R* if the lowest priority substituent is on a vertical bond.

Clockwise specifies *S* if the lowest priority substituent is on a horizontal bond.

If you assume that a clockwise arrow indicates the *R* configuration, then you get a VERy good answer if the group with the lowest priority is on a VERtical bond, but a HORribly bad answer if the group with the lowest priority is on a HORizontal bond.

NOTE TO THE STUDENT

When comparing two Fischer projections to see if they are the same or different, never rotate one 90° or flip it "front-to-back," because that is a quick way to get a wrong answer. A Fischer projection can be rotated 180° in the plane of the paper, but that is the only way you can move it without risking an incorrect answer.

PROBLEM 16♦

What is the configuration of each of the following compounds?

a. CH_3CH_2—$\overset{\displaystyle CH(CH_3)_2}{\underset{\displaystyle CH_3}{|}}$—$CH_2Br$

b. HO—$\overset{\displaystyle CH_2CH_2CH_3}{\underset{\displaystyle CH_2OH}{|}}$—$H$

c. CH_3—$\overset{\displaystyle Br}{\underset{\displaystyle CH_2CH_3}{|}}$—$H$

d. CH_3—$\overset{\displaystyle CH_2CH_2CH_3}{\underset{\displaystyle CH_2CH_3}{|}}$—$CH_2CH_2CH_3$

PROBLEM-SOLVING STRATEGY

Recognizing Pairs of Enantiomers

Do the structures represent identical compounds or a pair of enantiomers?

and

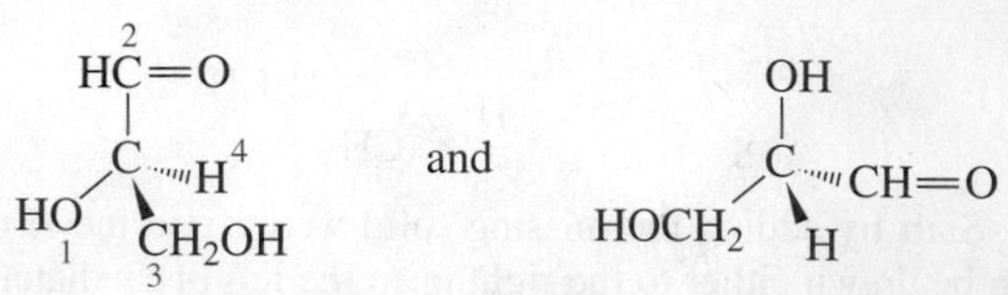

The easiest way to answer this question is to determine their configurations. If one has the *R* configuration and the other has the *S* configuration, then they are enantiomers. If they both have the *R* configuration or they both have the *S* configuration, then they are identical compounds.

The OH group has the highest priority, the H has the lowest priority, and the Cs of the other two groups tie. The CH═O group has a higher priority than the CH$_2$OH group, because if an atom is doubly bonded to another atom, the priority system treats it as if it were singly bonded to two of those atoms. Thus, the C of the CH═O group is considered to be bonded to O, O, H, whereas the C of the CH$_2$OH group is considered to be bonded to O, H, H. An O cancels in each group, leaving O, H in the CH═O group and H, H in the CH$_2$OH group.

Because the structure on the left has the *R* configuration and the structure on the right has the *S* configuration, these two structures represent a pair of enantiomers.

Now use the strategy you have just learned to solve Problem 17.

PROBLEM 17♦

Do the following structures represent identical compounds or a pair of enantiomers?

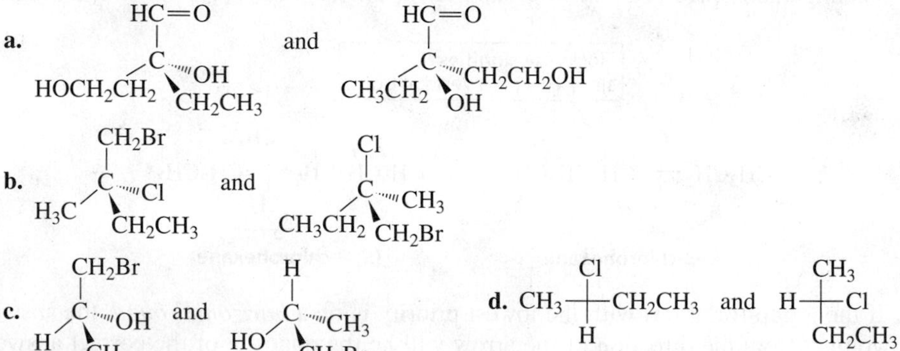

a.

and

b.

and

c.

and

d. CH_3—$\overset{\displaystyle Cl}{\underset{\displaystyle H}{|}}$—$CH_2CH_3$ and H—$\overset{\displaystyle CH_3}{\underset{\displaystyle CH_2CH_3}{|}}$—$Cl$

PROBLEM-SOLVING STRATEGY

Drawing an Enantiomer with a Desired Configuration

(*S*)-Alanine is a naturally occurring amino acid. Draw its structure using a perspective formula.

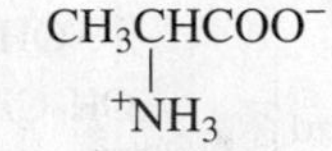

$$CH_3CHCOO^-$$
$$\underset{\displaystyle +NH_3}{|}$$

alanine

First draw the bonds about the asymmetric center. (Remember that the solid wedge and the hatched wedge must be adjacent to one another.)

Put the group with the lowest priority on the hatched wedge. Put the group with the highest priority on any remaining bond.

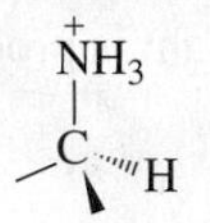

Because you have been asked to draw the *S* enantiomer, draw an arrow counterclockwise from the group with the highest priority to the next available bond and put the group with the second highest priority on that bond.

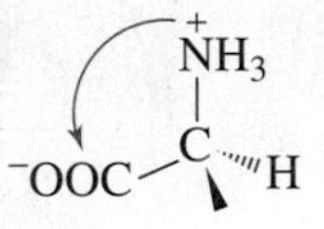

Put the remaining substituent (the one with the third highest priority) on the last available bond.

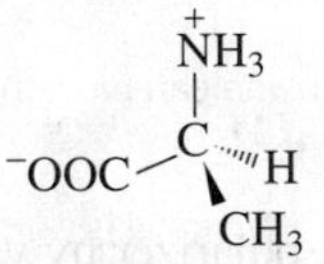

Now use the strategy you have just learned to solve Problem 18.

PROBLEM 18

Draw a perspective formula for each of the following:

a. (*S*)-2-chlorobutane

b. (*R*)-1,2-dibromobutane

PROBLEM 19 Solved

Convert each Fischer projection to a perspective formula.

a.

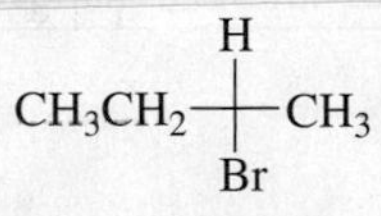

$$CH_3CH_2 \overset{\displaystyle H}{\underset{\displaystyle Br}{\vrule width0pt height1em}} CH_3$$

b.

$$H \overset{\displaystyle COO^-}{\underset{\displaystyle CH_2CH_3}{\vrule width0pt height1em}} CH_3$$

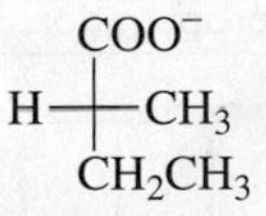

Solution to 19a First determine the configuration of the Fischer projection: it has the *R* configuration. Then draw the perspective formula with that configuration by following the steps described in the preceding problem-solving strategy.

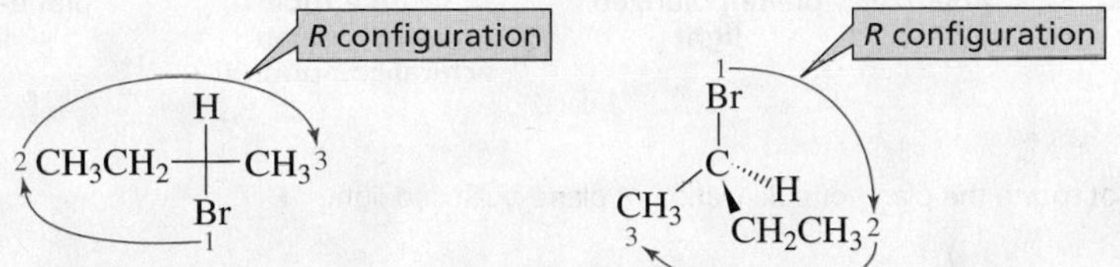

8 / CHIRAL COMPOUNDS ARE OPTICALLY ACTIVE

Enantiomers share many of the same properties, including the same boiling points, the same melting points, and the same solubilities. In fact, all the physical properties of enantiomers are the same except those that stem from how groups bonded to the asymmetric center are arranged in space. One property that enantiomers do not share is the way they interact with plane-polarized light.

Normal light, such as that coming from a light bulb or the sun, consists of rays that oscillate in all directions. In contrast, all the rays in a beam of **plane-polarized light** oscillate in a single plane. Plane-polarized light is produced by passing normal light through a polarizer (Figure 4).

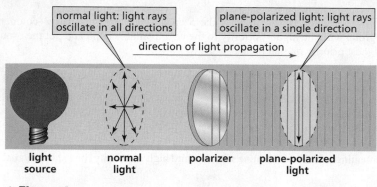

normal light: light rays oscillate in all directions

plane-polarized light: light rays oscillate in a single direction

direction of light propagation

light source   normal light   polarizer   plane-polarized light

▲ **Figure 4**

Only light oscillating in a single plane can pass through a polarizer.

You can experience the effect of a polarizer by wearing a pair of polarized sunglasses. Polarized sunglasses allow only light oscillating in a single plane to pass through, which is why they block reflections (glare) more effectively than nonpolarized sunglasses do.

In 1815, the physicist Jean-Baptiste Biot discovered that certain naturally occurring organic compounds are able to rotate the **plane of polarization** of plane-polarized light. He noted that some compounds rotated it clockwise and some rotated it counterclockwise. He proposed that the ability to rotate the plane of polarization of plane-polarized light was due to some asymmetry in the molecules. It was later determined that the asymmetry was associated with compounds having one or more asymmetric centers.

When plane-polarized light passes through a solution of achiral molecules, the light emerges from the solution with its plane of polarization unchanged (Figure 5).

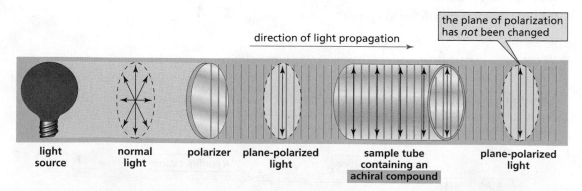

direction of light propagation

the plane of polarization has *not* been changed

light source   normal light   polarizer   plane-polarized light   sample tube containing an achiral compound   plane-polarized light

▲ **Figure 5**

An achiral compound does not rotate the plane of polarization of plane-polarized light.

On the other hand, when plane-polarized light passes through a solution of chiral molecules, the light emerges with its plane of polarization rotated either clockwise or counterclockwise (Figure 6). If one enantiomer rotates it clockwise, its mirror image will rotate it exactly the same amount counterclockwise.

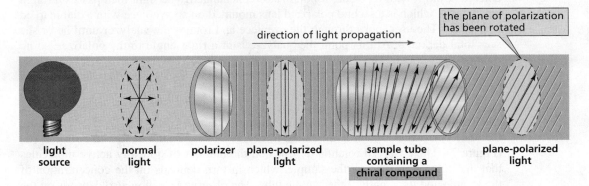

direction of light propagation

the plane of polarization has been rotated

| light source | normal light | polarizer | plane-polarized light | sample tube containing a chiral compound | plane-polarized light |

▲ **Figure 6**

A chiral compound rotates the plane of polarization of plane-polarized light.

A compound that rotates the plane of polarization of plane-polarized light is said to be **optically active.** In other words, chiral compounds are optically active, and achiral compounds are **optically inactive.**

If an optically active compound rotates the plane of polarization clockwise, then the compound is said to be **dextrorotatory,** which can be indicated in the compound's name by the prefix (+). If it rotates the plane of polarization counterclockwise, then it is said to be **levorotatory,** which can be indicated by (−). *Dextro* and *levo* are Latin prefixes for "to the right" and "to the left," respectively. Sometimes lowercase *d* and *l* are used instead of (+) and (−).

Do not confuse (+) and (−) with *R* and *S*. The (+) and (−) symbols indicate the direction in which an optically active compound rotates the plane of polarization of plane-polarized light, whereas *R* and *S* indicate the arrangement of the groups about an asymmetric center. Some compounds with the *R* configuration are (+) and some are (−). Likewise, some compounds with the *S* configuration are (+) and some are (−).

For example, (*S*)-lactic acid and (*S*)-sodium lactate both have an *S* configuration, but (*S*)-lactic acid is dextrorotatory whereas (*S*)-sodium lactate is levorotatory. When we know which direction an optically active compound rotates the plane of polarization, we can incorporate (+) or (−) into its name.

> An achiral compound does not rotate the plane of polarization of plane-polarized light.
>
> A chiral compound rotates the plane of polarization of plane-polarized light.

When light is filtered through two polarizers (polarized lenses) at a 90° angle to one another, none of the light passes through.

CH₃ · C—H · HO · COOH

(S)-(+)-lactic acid

CH₃ · C—H · HO · COO⁻ Na⁺

(S)-(−)-sodium lactate

We can tell by looking at the structure of a compound whether it has the *R* or the *S* configuration, but the only way we can tell whether a compound is dextrorotatory (+) or levorotatory (−) is to put the compound in a polarimeter. This is an instrument that measures the direction and the amount the plane of polarization of plane-polarized light is rotated.

PROBLEM 20♦

a. Is (*R*)-lactic acid dextrorotatory or levorotatory?
b. Is (*R*)-sodium lactate dextrorotatory or levorotatory?

9 HOW SPECIFIC ROTATION IS MEASURED

Figure 7 provides a simplified description of how a **polarimeter** functions. The amount of rotation caused by an optically active compound will vary with the wavelength of the light being used, so the light source must produce monochromatic (single-wavelength) light. Most polarimeters use light from a sodium arc (called the sodium D-line; wavelength = 589 nm).

The monochromatic light passes through a polarized lens and emerges as plane-polarized light, which then passes through a sample tube. If the tube is empty, the light emerges from it with its plane of polarization unchanged. The light then passes through an analyzer, which is a second polarized lens mounted on an eyepiece with a dial marked in degrees. The user looks through the eyepiece and rotates the analyzer until he or she sees total darkness. At this point the analyzer is at a right angle to the polarizer, so no light passes through. This analyzer setting corresponds to zero rotation.

The sample to be measured is then placed in the sample tube. If the sample is optically active, it will rotate the plane of polarization. The analyzer, therefore, will no longer block all the light, so some light will reach the user's eye. The user now rotates the analyzer again until no light passes through. The amount the analyzer is rotated can be read from the dial. This value, which is measured in degrees, is called the **observed rotation** (α) (Figure 7). The observed rotation depends on the number of optically active molecules that the light encounters in the sample, which in turn depends on the concentration of the sample and the length of the sample tube. The observed rotation also depends on the temperature and the wavelength of the light source.

direction of light propagation

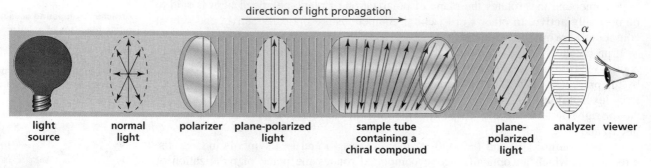

| light source | normal light | polarizer | plane-polarized light | sample tube containing a chiral compound | plane-polarized light | analyzer | viewer |

▲ **Figure 7**

A schematic drawing of a polarimeter.

Each optically active compound has a characteristic specific rotation. A compound's **specific rotation** is the rotation caused by a solution of 1.0 g of the compound per milliliter of solution in a sample tube 1.0 decimeter long at a specified temperature and wavelength.* The specific rotation can be calculated from the observed rotation using the following formula,

$$[\alpha]_{\lambda}^{T} = \frac{\alpha}{l \times c}$$

where $[\alpha]$ is the specific rotation, T is temperature in °C, λ is the wavelength of the incident light (when the sodium D-line is used, λ is indicated as D), α is the observed rotation, l is the length of the sample tube in decimeters, and c is the concentration of the sample in grams per milliliter of solution.

If one enantiomer has a specific rotation of +5.75, the specific rotation of the other enantiomer must be −5.75, because the mirror-image molecule rotates the plane of polarization the same amount but in the opposite direction. The specific rotations of some common compounds are listed in Table 1.

(R)-2-methyl-1-butanol (S)-2-methyl-1-butanol

$[\alpha]_{D}^{20\,°C} = +5.75$ $[\alpha]_{D}^{20\,°C} = -5.75$

*Unlike observed rotation, which is measured in degrees, specific rotation has units of 10^{-1} deg cm^2 g^{-1}. In this text, values of specific rotation will be given without units.

Table 1 Specific Rotation of Some Naturally Occurring Compounds

Cholesterol	−31.5	Penicillin V	+233
Cocaine	−16	Progesterone	+172
Codeine	−136	Sucrose (table sugar)	+66.5
Morphine	−132	Testosterone	+109

A mixture of equal amounts of two enantiomers—such as (R)-$(-)$-lactic acid and (S)-$(+)$-lactic acid—is called a **racemic mixture** or a **racemate.** Racemic mixtures are optically inactive because for every molecule in a racemic mixture that rotates the plane of polarization in one direction, there is a mirror-image molecule that rotates the plane in the opposite direction. As a result, the light emerges from a racemic mixture with its plane of polarization unchanged. The symbol $(\pm)$ is used to specify a racemic mixture. Thus, $(\pm)$-2-bromobutane indicates a mixture of 50% $(+)$-2-bromobutane and 50% $(-)$-2-bromobutane.

PROBLEM 21♦

The observed rotation of 2.0 g of a compound in 50 mL of solution in a polarimeter tube 20-cm long is $+13.4°$. What is the specific rotation of the compound?

PROBLEM 22♦

(S)-$(+)$-Monosodium glutamate (MSG) is a flavor enhancer used in many foods. Some people have an allergic reaction to MSG (including headache, chest pain, and an overall feeling of weakness). "Fast food" often contains substantial amounts of MSG, which is widely used in Chinese food as well. (S)-$(+)$-MSG has a specific rotation of $+24$.

$$\begin{array}{c} COO^- \ Na^+ \\ | \\ C\cdots\!\!\text{\tiny{''''}}H \\ {}^-OOCCH_2CH_2 \quad \overset{+}{N}H_3 \end{array}$$

(S)-(+)-monosodium glutamate

a. What is the specific rotation of (R)-$(-)$-monosodium glutamate?
b. What is the specific rotation of a racemic mixture of MSG?

---

10   ENANTIOMERIC EXCESS

Whether a particular sample of a compound consists of a single enantiomer, a racemic mixture, or a mixture of enantiomers in unequal amounts can be determined by its **observed specific rotation,** which is the specific rotation of the sample.

For example, if a sample of (S)-$(+)$-2-bromobutane is **enantiomerically pure** (meaning only one enantiomer is present), it will have an *observed specific rotation* of $+23.1$ because its *specific rotation* is $+23.1$. If, however, the sample of 2-bromobutane is a racemic mixture, it will have an observed specific rotation of 0. If the observed specific rotation is positive but less than $+23.1$, we will know that the sample is a mixture of enantiomers and that the mixture contains more of the S enantiomer than the R enantiomer, because the S enantiomer is dextrorotatory.

The **enantiomeric excess (ee),** also called the optical purity, tells us how much of an excess of one enantiomer is in the mixture. It can be calculated from the observed specific rotation:

$$\text{enantiomeric excess} = \frac{\text{observed specific rotation}}{\text{specific rotation of the pure enantiomer}} \times 100\%$$

For example, if the sample of 2-bromobutane has an observed specific rotation of +9.2, then the enantiomeric excess is 40%. In other words, the excess of one of the enantiomers comprises 40% of the mixture.

$$\text{enantiomeric excess} = \frac{+9.2}{+23.1} \times 100\% = 40\%$$

If the mixture has a 40% enantiomeric excess, 40% of the mixture is excess S enantiomer and 60% is a racemic mixture. Half of the racemic mixture plus the amount of excess S enantiomer equals the amount of the S enantiomer present in the mixture. Therefore, 70% of the mixture is the S-enantiomer $[(1/2 \times 60) + 40]$ and 30% is the R enantiomer.

PROBLEM 23♦

(+)-Mandelic acid has a specific rotation of +158. What would be the observed specific rotation of each of the following mixtures?

a. 50% (−)-mandelic acid and 50% (+)-mandelic acid
b. 25% (−)-mandelic acid and 75% (+)-mandelic acid
c. 75% (−)-mandelic acid and 25% (+)-mandelic acid

PROBLEM 24♦

Naproxen, a nonsteroidal anti-inflammatory drug that is the active ingredient in Aleve, has a specific rotation of +66. One commercial preparation results in a mixture with a 97% enantiomeric excess.

a. Does naproxen have the R or the S configuration?
b. What percent of each enantiomer is obtained from the commercial preparation?

PROBLEM 25 Solved

A solution prepared by mixing 10 mL of a 0.10 M solution of the R enantiomer of a compound and 30 mL of a 0.10 M solution of the S enantiomer was found to have an observed specific rotation of +4.8. What is the specific rotation of each of the enantiomers? (*Hint:* mL $\times$ M = millimole, abbreviated as mmol)

Solution One mmol (10 mL $\times$ 0.10 M) of the R enantiomer is mixed with 3 mmol (30 mL $\times$ 0.10 M) of the S enantiomer; 1 mmol of the R enantiomer plus 1 mmol of the S enantiomer will form 2 mmol of a racemic mixture, so there will be 2 mmol of S enantiomer left over. Because 2 out of 4 mmol is excess S enantiomer, the solution has a 50% enantiomeric excess. Knowing the enantiomeric excess and the observed specific rotation allows us to calculate the specific rotation.

$$\text{enantiomeric excess} = \frac{\text{observed specific rotation}}{\text{specific rotation of the pure enantiomer}} \times 100\%$$

$$50\% = \frac{+4.8}{x} \times 100\%$$

$$\frac{50}{100} = \frac{+4.8}{x}$$

$$\frac{1}{2} = \frac{+4.8}{x}$$

$$x = 2(+4.8)$$

$$x = 9.6$$

The S enantiomer has a specific rotation of +9.6, so the R enantiomer has a specific rotation of −9.6.

11 COMPOUNDS WITH MORE THAN ONE ASYMMETRIC CENTER

Many organic compounds have more than one asymmetric center. The more asymmetric centers a compound has, the more stereoisomers it can have. If we know the number of asymmetric centers, we can calculate the *maximum* number of stereoisomers for that compound: *a compound can have a maximum of 2^n stereoisomers, where* n *equals the number of asymmetric centers* (provided it does not also have stereocenters that would cause it to have cis–trans isomers; see Problem 26).

For example, 3-chloro-2-butanol has two asymmetric centers. Therefore, it can have a maximum of four ($2^2 = 4$) stereoisomers. The four stereoisomers are shown here, both as perspective formulas and as Fischer projections.

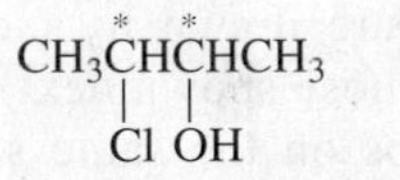

$$CH_3\overset{*}{C}H\overset{*}{C}HCH_3$$
$$|\ \ \ |$$
$$Cl\ \ OH$$

3-chloro-2-butanol

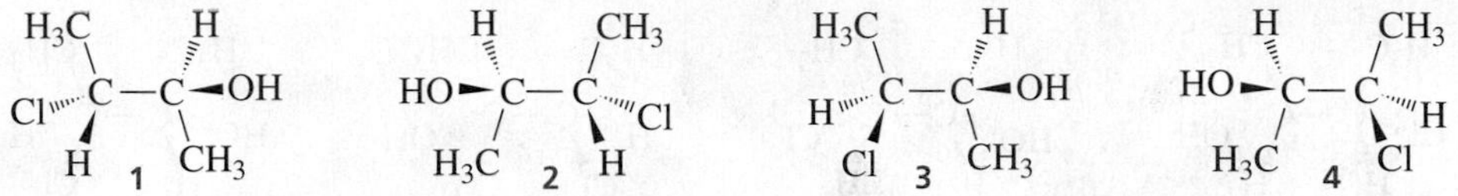

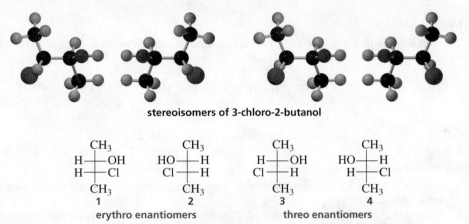

perspective formulas of the stereoisomers of 3-chloro-2-butanol (staggered)

erythro enantiomers   threo enantiomers

stereoisomers of 3-chloro-2-butanol

Fischer projections of the stereoisomers of 3-chloro-2-butanol (eclipsed)

erythro enantiomers   threo enantiomers

The four stereoisomers of 3-chloro-2-butanol consist of two pairs of enantiomers. Stereoisomers **1** and **2** are nonsuperimposable mirror images. They, therefore, are enantiomers. Stereoisomers **3** and **4** are also enantiomers. Stereoisomers **1** and **3** are not identical, and they are not mirror images. Such stereoisomers are called **diastereomers.** Stereoisomers **1** and **4, 2** and **3,** and **2** and **4** are also pairs of diastereomers.

Notice that in a pair of diastereomers, the configuration of one of the asymmetric centers is the same in both but the configuration of the other asymmetric center is different. *Diastereomers are stereoisomers that are not enantiomers.* (Note that cis–trans isomers are also considered to be diastereomers, because they are stereoisomers but they are not enantiomers.)

Note that enantiomers have *identical physical properties.* They also have *identical chemical properties*—that is, they react with a given achiral reagent at the same rate. Diastereomers, on the other hand, have *different physical properties* (meaning different melting points, boiling points, solubilities, specific rotations, and so on) and *different chemical properties*—that is, they react with a given achiral reagent at different rates.

When Fischer projections are drawn for stereoisomers with two adjacent asymmetric centers (such as those for 3-chloro-2-butanol), the enantiomers with the hydrogens on the same side of the carbon chain are called the **erythro enantiomers** (see Problem 48), whereas those with the hydrogens on opposite sides are called the **threo enantiomers.** Therefore, **1** and **2** are the erythro enantiomers of 3-chloro-2-butanol (the hydrogens are on the same side), whereas **3** and **4** are the threo enantiomers.

In each of the Fischer projections shown above, the horizontal bonds project out of the paper toward the viewer and the vertical bonds extend behind the paper away from the viewer. Groups can rotate freely about the carbon–carbon single bonds, but Fischer projections show the stereoisomers in their eclipsed conformations.

Diastereomers are stereoisomers that are not enantiomers.

Because a Fischer projection does not show the three-dimensional structure of the molecule, and because it represents the molecule in a relatively unstable eclipsed conformation, most chemists prefer to use perspective formulas. Perspective formulas (those shown in the first group of images in this section) show the molecule's three-dimensional structure in a stable, staggered conformation, so they provide a more accurate representation of structure.

When perspective formulas are drawn to show the stereoisomers in their less stable eclipsed conformations (those shown next), we can easily see that the erythro enantiomers have similar groups on the same side. We will use both perspective formulas and Fischer projections to depict the arrangement of groups bonded to an asymmetric center.

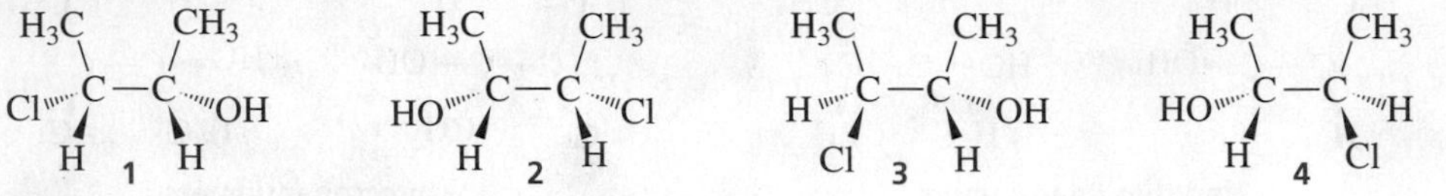

erythro enantiomers   threo enantiomers

perspective formulas of the stereoisomers of 3-chloro-2-butanol (eclipsed)

PROBLEM 26

The following compound has only one asymmetric center. Why then does it have four stereoisomers?

$$CH_3CH_2\overset{*}{C}HCH_2CH=CHCH_3$$
$$|$$
$$Br$$

PROBLEM 27♦

Are the following statements correct?

a. A compound can have a maximum of 2^n stereoisomers, where n equals the number of stereocenters.

b. A compound can have a maximum of 2^n stereoisomers, where n equals the number of asymmetric centers.

PROBLEM 28♦

a. Stereoisomers with two asymmetric centers are called ___ if the configuration of both asymmetric centers in one stereoisomer is the opposite of the configuration of the asymmetric centers in the other stereoisomer.

b. Stereoisomers with two asymmetric centers are called ___ if the configuration of both asymmetric centers in one stereoisomer is the same as the configuration of the asymmetric centers in the other stereoisomer.

c. Stereoisomers with two asymmetric centers are called ___ if one of the asymmetric centers has the same configuration in both stereoisomers and the other asymmetric center has the opposite configuration in the two stereoisomers.

PROBLEM 29♦

The stereoisomer of cholesterol found in nature is shown here.

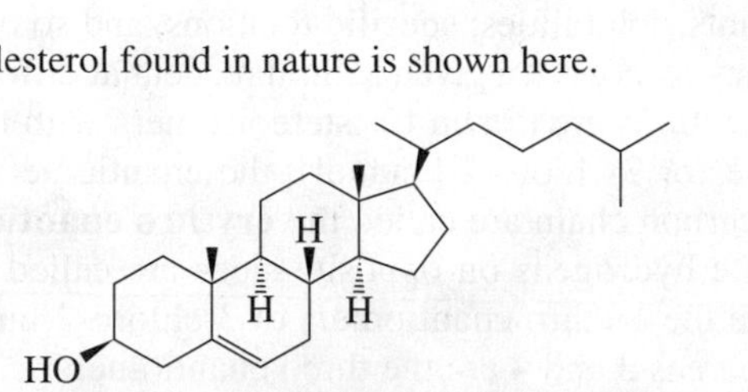

cholesterol

a. How many asymmetric centers does cholesterol have?

b. What is the maximum number of stereoisomers that cholesterol can have?

Draw the stereoisomers of the following amino acids. Indicate pairs of enantiomers and pairs of diastereomers.

a. $CH_3CHCH_2{-}CHCOO^-$
  $\quad\quad\quad\mid\quad\quad\quad\mid$
  $\quad\quad\quad CH_3\quad\ ^+NH_3$
  leucine

b. $CH_3CH_2CH{-}CHCOO^-$
  $\quad\quad\quad\quad\ \mid\quad\ \ \mid$
  $\quad\quad\quad\quad CH_3\ ^+NH_3$
  isoleucine

12 STEREOISOMERS OF CYCLIC COMPOUNDS

1-Bromo-2-methylcyclopentane also has two asymmetric centers and four stereoisomers. Because the compound is cyclic, the substituents can be either cis or trans. Enantiomers can be drawn for both the cis isomer and for the trans isomer. Each of the four stereoisomers is chiral.

Br CH₃ CH₃ Br   Br CH₃ CH₃ Br

cis-1-bromo-2-methylcyclopentane
enantiomers

trans-1-bromo-2-methylcyclopentane
enantiomers

1-Bromo-3-methylcyclohexane also has two asymmetric centers. The carbon that is bonded to a Br and an H is also bonded to two different carbon-containing groups ($-CH_2CH(CH_3)CH_2CH_2CH_2-$ and $-CH_2CH_2CH_2CH(CH_3)CH_2-$), so it is an asymmetric center. The carbon that is bonded to a CH_3 and a H is also bonded to two different carbon-containing groups, so it too is an asymmetric center.

these two groups are different Br
asymmetric center   asymmetric center
          CH₃

Because the compound has two asymmetric centers, it has four stereoisomers. Enantiomers can be drawn for both the cis isomer and for the trans isomer. Each of the four stereoisomers is chiral.

Br   Br   Br   Br

    CH₃ CH₃   CH₃ CH₃

cis-1-bromo-3-methylcyclohexane
enantiomers

trans-1-bromo-3-methylcyclohexane
enantiomers

1-Bromo-3-methylcyclobutane does not have any asymmetric centers. The C-1 carbon has a Br and an H attached to it, but its other two groups [$-CH_2CH(CH_3)CH_2-$] are identical; C-3 has a CH_3 and a H attached to it, but its other two groups [$-CH_2CH(Br)CH_2-$] are identical. Because the compound does not have a carbon with four different groups attached to it, it has only two stereoisomers, the cis isomer and the trans isomer. Both stereoisomers are achiral.

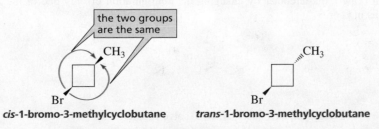

the two groups are the same

cis-1-bromo-3-methylcyclobutane   **trans-1-bromo-3-methylcyclobutane**

1-Bromo-4-methylcyclohexane also has no asymmetric centers. Therefore, the compound has only two stereoisomers, the cis isomer and the trans isomer. Both stereoisomers are achiral.

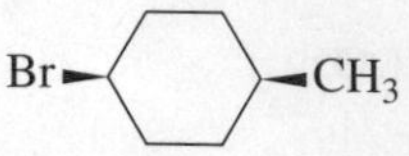

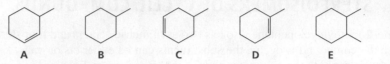

cis-1-bromo-4-methylcyclohexane    *trans*-1-bromo-4-methylcyclohexane

---

PROBLEM 31♦

Which of the following compounds has one or more asymmetric centers?

A   **B**   **C**   **D**   **E**

PROBLEM 32

Draw all possible stereoisomers for each of the following:

a. 2-chloro-3-hexanol   **c.** 2,3-dichloropentane
b. 2-bromo-4-chlorohexane   **d.** 1,3-dibromopentane

PROBLEM 33

Draw the stereoisomers of 2-methylcyclohexanol.

PROBLEM 34♦

Of all the possible cyclooctanes that have one chloro substituent and one methyl substituent, which ones do *not* have any asymmetric centers?

PROBLEM-SOLVING STRATEGY

Drawing Enantiomers and Diastereomers

Draw an enantiomer and a diastereomer for the following compound:

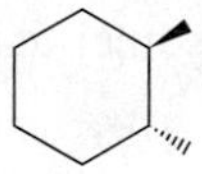

You can draw an enantiomer in one of two ways. You can change the configuration of all the asymmetric centers by changing all wedges to dashes and all dashes to wedges as in **A.** Or you can draw a mirror image of the compound as in **B.** Notice that since **A** and **B** are each an enantiomer of the given compound, **A** and **B** are identical. (You can see they are identical if you rotate **B** 180° clockwise.)

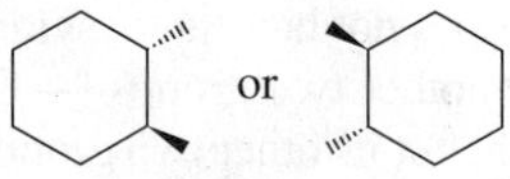

A   or   **B**

You can draw a diastereomer by changing the configuration of only one of the asymmetric centers as in **C** or **D.**

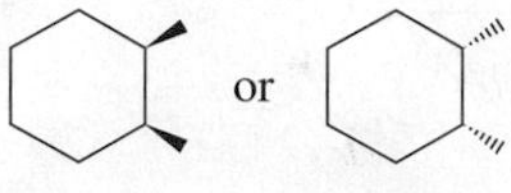

C   or   **D**

Now use the strategy you have just learned to solve Problem 35.

PROBLEM 35

Draw a diastereomer for each of the following compounds:

a. 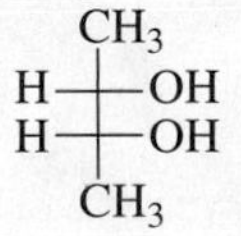
$$CH_3$$
H——OH
H——OH
$$CH_3$$

c. 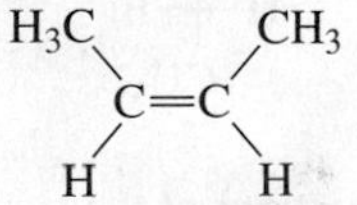
H₃C   CH₃
   \   /
    C═C
   /   \
  H   H

b. 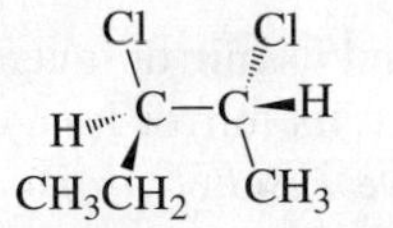
Cl   Cl
 | |
H⸝⸝⸝C—C◀H
 | |
CH₃CH₂ CH₃

d. 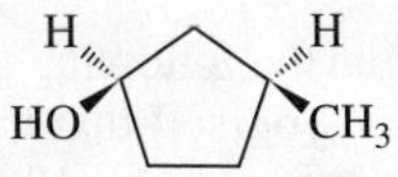
H‚‚‚   ‚‚‚H
HO⟨   ⟩CH₃

PROBLEM 36♦

Indicate whether each of the structures in the second row is an enantiomer of, is a diastereomer of, or is identical to the structure in the top row.

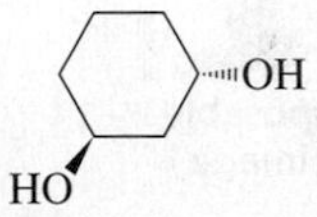

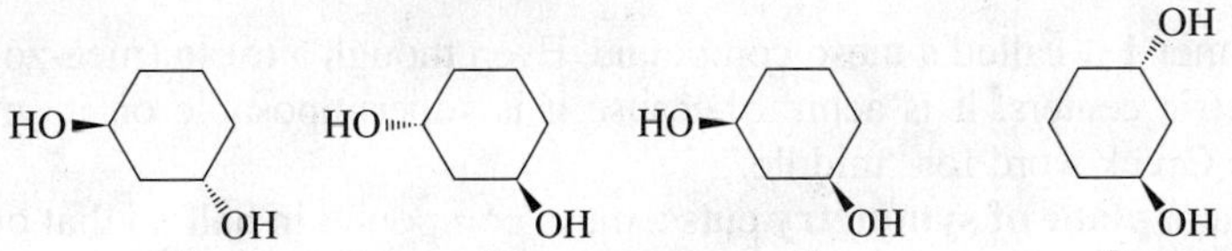

‚‚‚OH

HO‚‚‚

HO►   ‚‚‚OH
   A

HO‚‚‚   ‚‚‚OH
   B

HO►   ‚‚‚OH
   C

OH
       ‚‚‚OH
   D

13 MESO COMPOUNDS HAVE ASYMMETRIC CENTERS BUT ARE OPTICALLY INACTIVE

In the examples we have just seen, all the compounds with two asymmetric centers had four stereoisomers. However, some compounds with two asymmetric centers have only three stereoisomers.

The following is an example of a compound with two asymmetric centers that has only three stereoisomers.

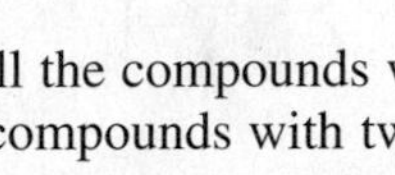

$$CH_3CHCHCH_3$$
| |
Br   Br

2,3-dibromobutane

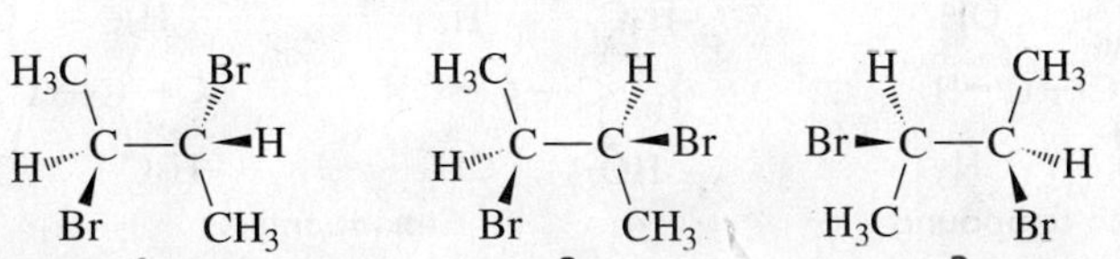

H₃C   Br
   \   /
H⸝⸝C—C◀H
   /   \
  Br   CH₃
    1

H₃C   H
   \   /
H⸝⸝C—C◀Br
   /   \
  Br   CH₃
    2

H   CH₃
  \   /
Br►C—C‚‚‚H
  /   \
 H₃C   Br
    3

perspective formulas of the stereoisomers of 2,3-dibromobutane (staggered)

The "missing" stereoisomer is the mirror image of **1** because **1** and its mirror image are the same molecule. This can be seen more clearly when the perspective formulas are drawn in eclipsed conformations or when Fischer projections are used.

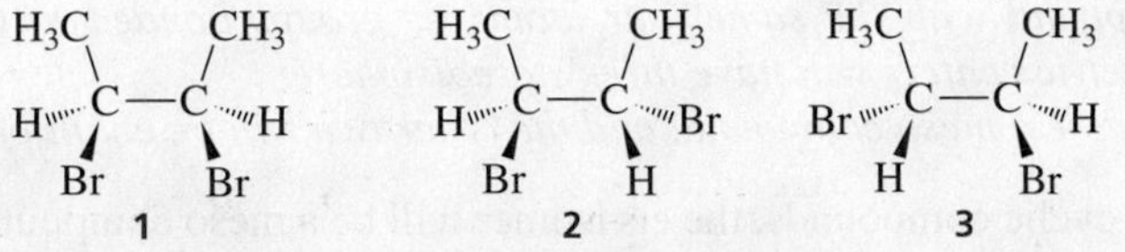

H₃C   CH₃
   \   /
H⸝⸝C—C‚‚‚H
   /   \
  Br   Br
    1

H₃C   CH₃
   \   /
H⸝⸝C—C‚‚‚Br
   /   \
  Br   H
    2

H₃C   CH₃
   \   /
Br‚‚C—C‚‚‚H
   /   \
  H   Br
    3

perspective formulas of the stereoisomers of 2,3-dibromobutane (eclipsed)

$$\begin{array}{ccc} CH_3 & CH_3 & CH_3 \\ H\!-\!\!\!\!|\!-\!Br & H\!-\!\!\!\!|\!-\!Br & Br\!-\!\!\!\!|\!-\!H \\ H\!-\!\!\!\!|\!-\!Br & Br\!-\!\!\!\!|\!-\!H & H\!-\!\!\!\!|\!-\!Br \\ CH_3 & CH_3 & CH_3 \\ \mathbf{1} & \mathbf{2} & \mathbf{3} \end{array}$$

Fischer projections of the stereoisomers of 2,3-dibromobutane (eclipsed)

It is easy to see from the following perspective formulas that **1** and its mirror image are identical. To convince yourself that the Fischer projection of **1** and its mirror image are identical, rotate the mirror image 180°. (*Remember, you can move Fischer projections only by rotating them 180° in the plane of the paper.*)

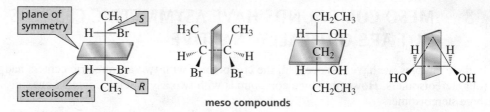

superimposable mirror image   superimposable mirror image

Stereoisomer **1** is called a meso compound. Even though a **meso** (mee-zo) **compound** has asymmetric centers, it is achiral because it is superimposable on its mirror image. *Mesos* is the Greek word for "middle."

Notice that a **plane of symmetry** cuts a meso compound in half so that one half is the mirror image of the other half. As an example, look at the first compound in the following depiction. If the top half of the molecule rotates polarized light to the right, then the bottom half will rotate the light the same amount to the left. Thus, they will cancel each other and the compound will not be optically active.

> A meso compound has two or more asymmetric centers and a plane of symmetry.
>
> Meso compounds are not optically active.
>
> If a compound has a plane of symmetry, it will not be optically active even though it has asymmetric centers.

meso compounds

It is easy to recognize when a compound with two asymmetric centers has a stereoisomer that is a meso compound because the four atoms or groups bonded to one asymmetric center are identical to the four atoms or groups bonded to the other asymmetric center. For example, both of the asymmetric centers in the following compound are bonded to an H, OH, CH$_3$, and CH(OH)CH$_3$.

a meso compound   enantiomers

a meso compound   enantiomers

A compound with the same four atoms or groups bonded to two different asymmetric centers will have three stereoisomers:
one will be a meso compound, and the other two will be enantiomers.

In the case of cyclic compounds, the cis isomer will be a meso compound and the trans isomer will be a pair of enantiomers.

cis-**1,3-dimethylcyclopentane**
a meso compound

trans-**1,3-dimethylcyclopentane**
enantiomers

cis-**1,2-dibromocyclohexane**
a meso compound

trans-**1,2-dibromocyclohexane**
enantiomers

In the perspective formula just shown, *cis*-1,2-dibromocyclohexane appears to have a plane of symmetry. Note, however, that cyclohexane is not a planar hexagon but exists preferentially in the chair conformation. The chair conformation does not have a plane of symmetry, but the much less stable boat conformation does have one. So, is *cis*-1,2-dibromocyclohexane a meso compound? The answer is yes. If a compound has a conformer with a plane of symmetry, the compound is achiral, even if the conformer with the plane of symmetry is not the most stable conformer.

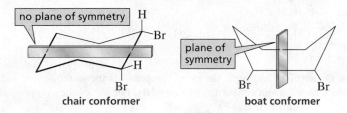

chair conformer   boat conformer

This rule holds for acyclic compounds as well. We have just seen that 2,3-dibromobutane is an achiral meso compound because it has a plane of symmetry. To see its plane of symmetry, however, we had to look at a relatively unstable eclipsed conformer. The more stable staggered conformer does not have a plane of symmetry. 2,3-Dibromobutane is still a meso compound, however, because it has a conformer with a plane of symmetry.

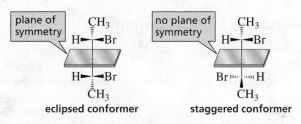

eclipsed conformer   staggered conformer

PROBLEM-SOLVING STRATEGY

Recognizing Whether a Compound Has a Stereoisomer That Is a Meso Compound

Which of the following compounds has a stereoisomer that is a meso compound?

A 2,3-dimethylbutane
B 3,4-dimethylhexane
C 2-bromo-3-methylpentane
D 1,3-dimethylcyclohexane
E 1,4-dimethylcyclohexane
F 1,2-dimethylcyclohexane
G 3,4-diethylhexane
H 1-bromo-2-methylcyclohexane

Check each compound to see if it has the necessary requirements for having a stereoisomer that is a meso compound. That is, does it have two asymmetric centers, and if so, do they each have the same four substituents attached to them?

Compounds **A, E,** and **G** do *not* have a stereoisomer that is a meso compound because they do not have any asymmetric centers.

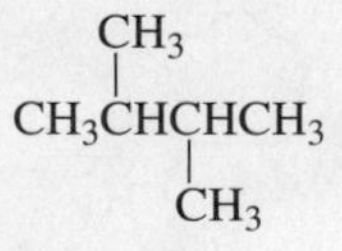

$$CH_3CHCHCH_3$$

with CH_3 groups

A

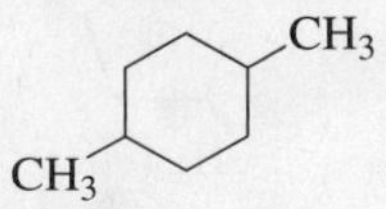

E

$$CH_3CH_2CHCHCH_2CH_3$$

with CH_2CH_3 groups

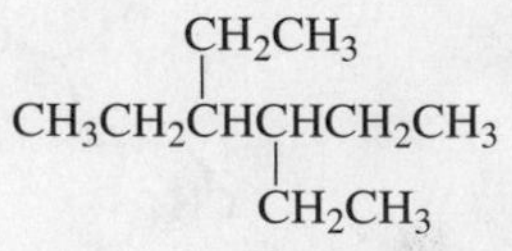

G

Compounds **C** and **H** have two asymmetric centers. They do *not* have a stereoisomer that is a meso compound, however, because the two asymmetric centers in each compound are *not* bonded to the same four substituents.

$$CH_3CHCHCH_2CH_3$$

with Br and CH_3

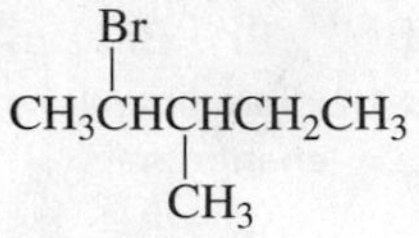

C

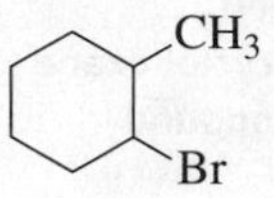

H

Compounds **B**, **D**, and **F** have two asymmetric centers, and the two asymmetric centers in each compound are bonded to the same four substituents. Therefore, these compounds have a stereoisomer that is a meso compound.

$$CH_3CH_2CHCHCH_2CH_3$$

with CH_3 groups

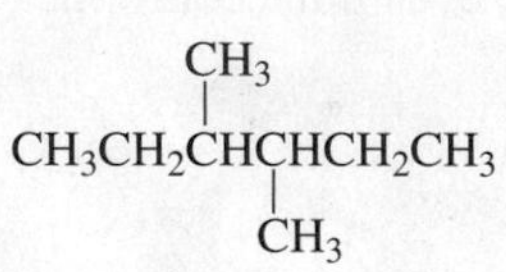

B

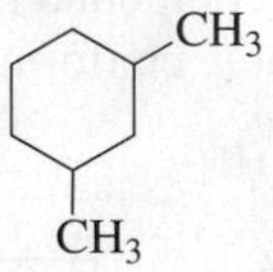

D

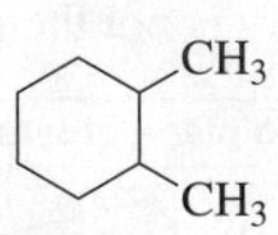

F

In the case of the acyclic compound, the meso compound is the stereoisomer with a plane of symmetry when drawn as an eclipsed conformer (**B**). For the cyclic compounds, the meso compound is the cis isomer (**D** and **F**).

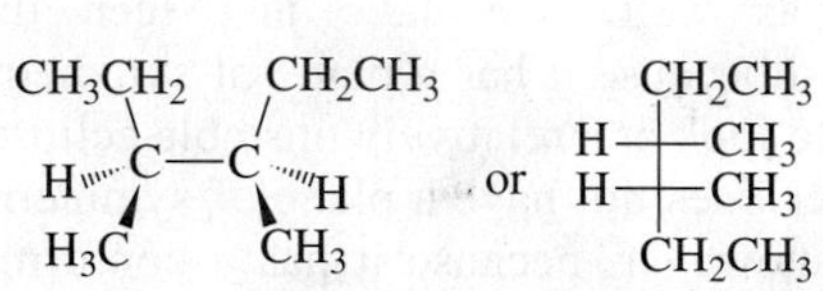

B or

H——CH₃
H——CH₃
with CH_2CH_3 groups

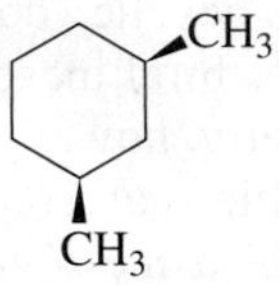

D

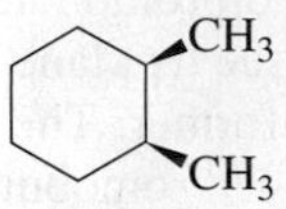

F

Now use the strategy you have just learned to solve Problem 37.

PROBLEM 37♦

Which of the following compounds has a stereoisomer that is a meso compound?

A 2,4-dibromohexane
B 2,4-dibromopentane
C 2,4-dimethylpentane

D 1,3-dichlorocyclohexane
E 1,4-dichlorocyclohexane
F 1,2-dichlorocyclobutane

PROBLEM 38 Solved

Which of the following are optically active?

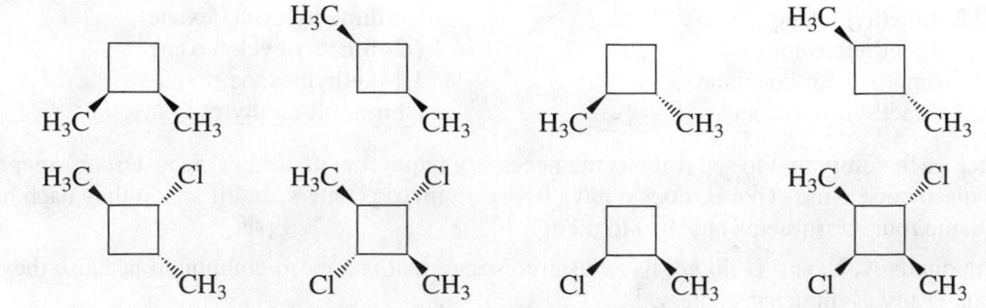

Solution In the *top row*, only the *third* compound is optically active. The first compound has a plane of symmetry, and an optically active compound cannot have a plane of symmetry; the second and fourth compounds do not have any asymmetric centers and each has a plane of symmetry. In the *bottom row*, the *first* and *third* compounds are optically active. The second and fourth compounds have a plane of symmetry.

PROBLEM 39

Draw all the stereoisomers for each of the following:

a. 1-bromo-2-methylbutane

b. 1-chloro-3-methylpentane

c. 2-methyl-1-propanol

d. 2-bromo-1-butanol

e. 3-chloro-3-methylpentane

f. 3-bromo-2-butanol

g. 3,4-dichlorohexane

h. 2,4-dichloropentane

i. 2,4-dichloroheptane

j. 1,2-dichlorocyclobutane

k. 1,3-dichlorocyclohexane

l. 1,4-dichlorocyclohexane

m. 1-bromo-2-chlorocyclobutane

n. 1-bromo-3-chlorocyclobutane

---

14   HOW TO NAME ISOMERS WITH MORE THAN ONE ASYMMETRIC CENTER

If a compound has more than one asymmetric center, the steps used to determine whether a given asymmetric center has the R or the S configuration must be applied to each of the asymmetric centers individually. As an example, let's name one of the stereoisomers of 3-bromo-2-butanol.

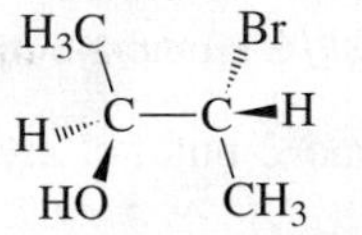

a stereoisomer of 3-bromo-2-butanol

First, we will determine the configuration at C-2. The OH has the highest priority (1), the C-3 carbon (the C attached to Br, C, H) is next (2), then comes CH_3 (3), and H has the lowest priority (4). Because the group with the lowest priority is bonded by a hatched wedge, we can immediately draw an arrow from (1) to (2) to (3). The arrow points counterclockwise, so the configuration at C-2 is S.

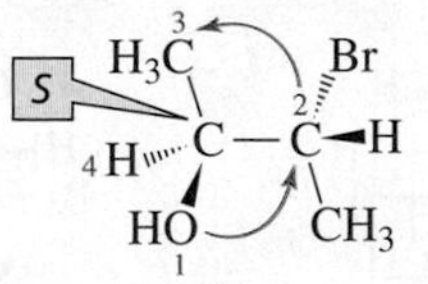

Now we need to determine the configuration at C-3. Because the group with the lowest priority (H) is not bonded by a hatched wedge, we must put it there by interchanging it with the group that is bonded by the hatched wedge.

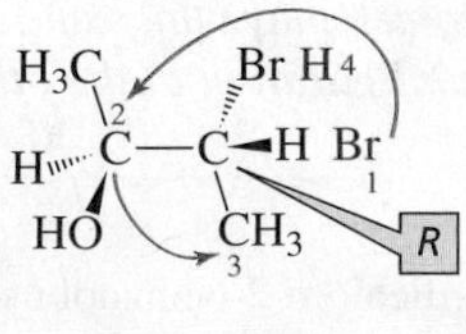

The arrow going from (1) (Br) to (2) the C-2 carbon (the C attached to O, C, H) to (3) (the methyl group) points counterclockwise, indicating that C-3 has the S configuration. However, because we interchanged two groups before we drew the arrow, C-3 in the compound before the groups were interchanged has the R configuration. Thus, the isomer is named (2S,3R)-3-bromo-2-butanol.

$$\underset{\textbf{(2S,3R)-3-bromo-2-butanol}}{\overset{\displaystyle \boxed{S}\quad\boxed{R}}{H_3C\quad Br}}$$

When Fischer projections are used, the procedure is similar. Just apply the steps that you learned for a Fischer projection with one asymmetric center to each asymmetric center individually. At C-2, the arrow from (1) to (2) to (3) points clockwise, suggesting an R configuration. However, the group with the lowest priority is on a horizontal bond, so C-2 has the S configuration instead (Section 7).

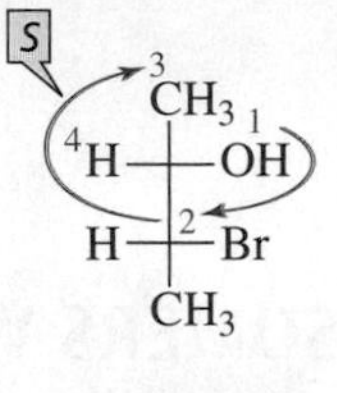

Repeating these steps for C-3 identifies that asymmetric center as having the R configuration. Thus, the isomer is named (2S,3R)-3-bromo-2-butanol.

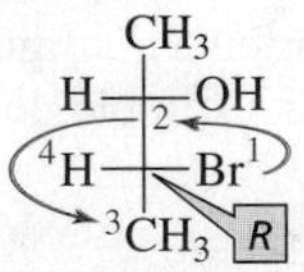

(2S,3R)-3-bromo-2-butanol

The four stereoisomers of 3-bromo-2-butanol are named as shown here. Take a few minutes to verify the names.

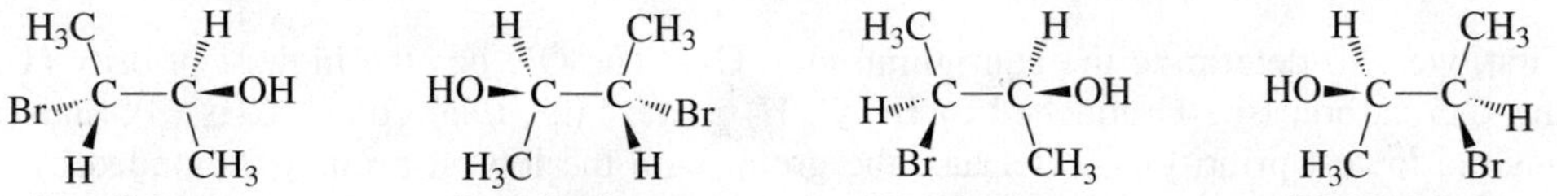

(2S,3R)-3-bromo-
2-butanol

(2R,3S)-3-bromo-
2-butanol

(2S,3S)-3-bromo-
2-butanol

(2R,3R)-3-bromo-
2-butanol

perspective formulas of the stereoisomers of 3-bromo-2-butanol

(2S,3R)-3-bromo-
2-butanol

(2R,3S)-3-bromo-
2-butanol

(2S,3S)-3-bromo-
2-butanol

(2R,3R)-3-bromo-
2-butanol

Fischer projections of the stereoisomers of 3-bromo-2-butanol

Notice that enantiomers have the opposite configuration at both asymmetric centers, whereas diastereomers have the same configuration at one asymmetric center and the opposite configuration at the other asymmetric center.

PROBLEM 40◆

Draw the four stereoisomers of 1,3-dichloro-2-pentanol using

a. Fischer projections.

b. perspective formulas.

PROBLEM 41

Name the isomers you drew in Problem 40.

Tartaric acid has three stereoisomers because each of its two asymmetric centers has the same set of four substituents. The meso compound and the pair of enantiomers are named as shown.

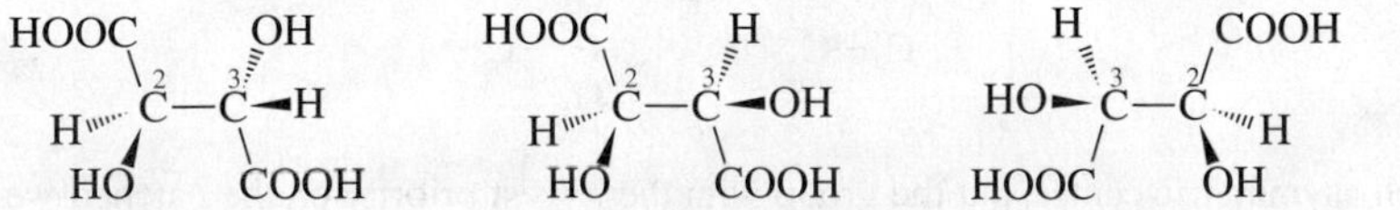

perspective formulas of the stereoisomers of tartaric acid

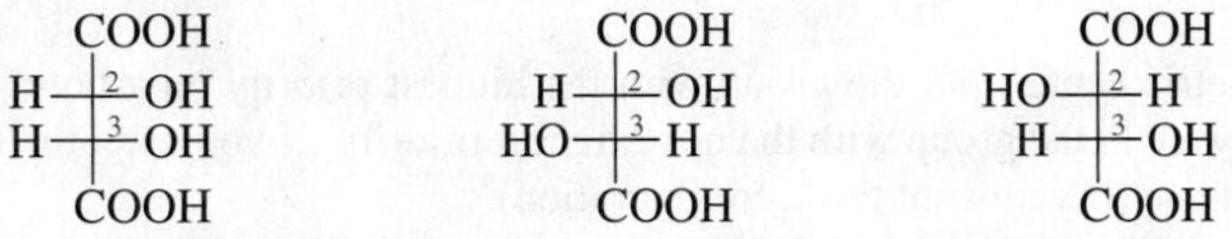

(2R,3S)-tartaric acid
a meso compound

(2R,3R)-tartaric acid

(2S,3S)-tartaric acid

enantiomers

(2R,3S)-tartaric acid
a meso compound

(2R,3R)-tartaric acid

(2S,3S)-tartaric acid

enantiomers

Fischer projections of the stereoisomers of tartaric acid

The physical properties of the three stereoisomers of tartaric acid are listed in Table 2. The meso compound and either of the enantiomers are diastereomers. Notice that the physical properties of enantiomers are the same, whereas the physical properties of diastereomers are different.

Table 2	Physical Properties of the Stereoisomers of Tartaric Acid		
	Melting point, °C	Specific rotation	Solubility, g/100 g H_2O at 15 °C
(2R,3R)-(+)-Tartaric acid	171	+11.98	139
(2S,3S)-(−)-Tartaric acid	171	−11.98	139
(2R,3S)-Tartaric acid (meso)	146	0	125
(±)-Tartaric acid	206	0	139

PROBLEM 42♦

Chloramphenicol is a broad-spectrum antibiotic that is particularly useful against typhoid fever. What is the configuration of each of its asymmetric centers?

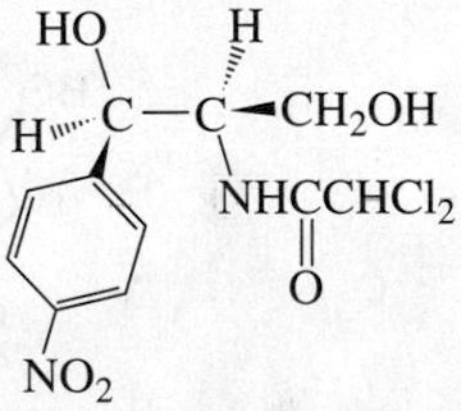

chloramphenicol

PROBLEM-SOLVING STRATEGY

Drawing a Perspective Formula for a Compound with Two Asymmetric Centers

Draw a perspective formula for (2S,3R)-3-chloro-2-pentanol.

First write a condensed structure for the compound so you know what groups are attached to the asymmetric centers.

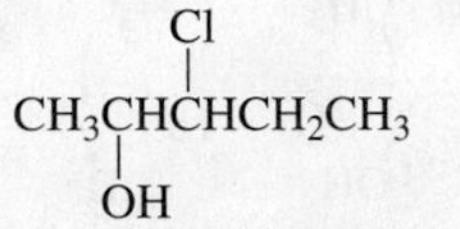

3-chloro-2-pentanol

Now draw the bonds that are in the plane of the paper, and then add the wedges. Remember that the solid and hatched wedges must be adjacent, and the hatched wedge is above the solid wedge.

At each asymmetric center, put the group with the lowest priority on the hatched wedge.

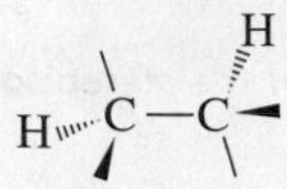

At each asymmetric center, put the group with the highest priority on a bond so that an arrow will point clockwise to the group with the next highest priority (if you want the *R* configuration), or counterclockwise (if you want the *S* configuration).

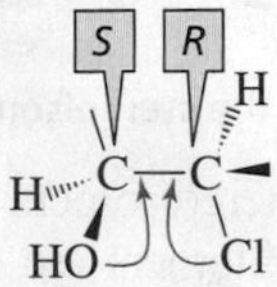

Put the remaining substituents on the last available bonds.

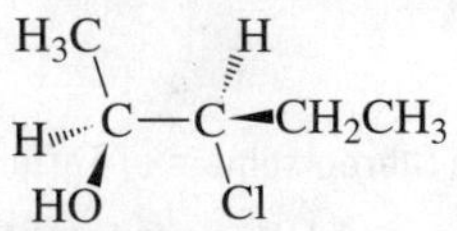

(2S,3R)-3-chloro-2-pentanol

Now use the strategy you have just learned to solve Problem 43.

PROBLEM 43

Draw a perspective formula for each of the following:

a. (*S*)-3-chloro-1-pentanol

b. (2*R*,3*R*)-2,3-dibromopentane

c. (2*S*,3*R*)-3-methyl-2-pentanol

d. (*R*)-1,2-dibromobutane

PROBLEM 44♦

Name each of the following:

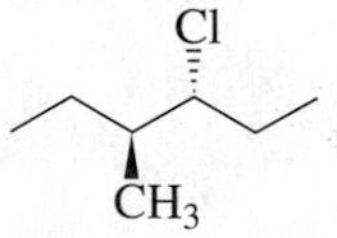

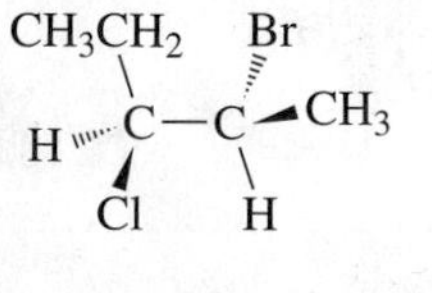

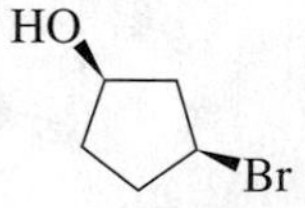

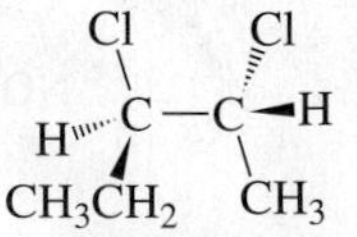

PROBLEM 45♦

Threonine, an amino acid, has four stereoisomers. The stereoisomer found in nature is (2*S*,3*R*)-threonine. Which of the following structures represents the naturally occurring amino acid?

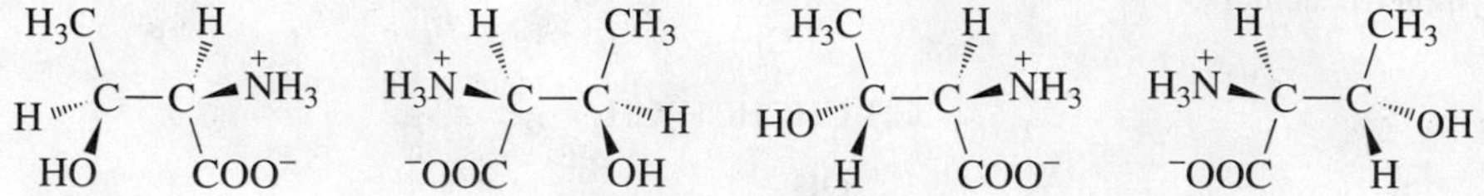

stereosiomers of threonine

PROBLEM 46♦ Solved

Convert each Fischer projection to a perspective formula.

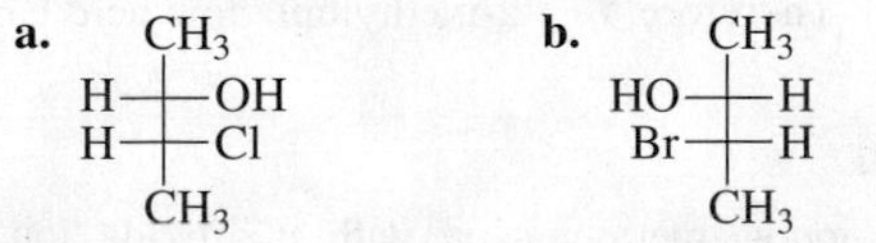

a.
CH₃
H——OH
H——Cl
CH₃

b.
CH₃
HO——H
Br——H
CH₃

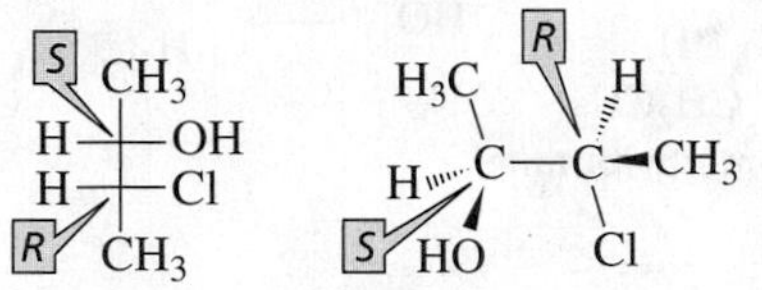

NOTE TO THE STUDENT

For practice interconverting between Newman projections, Fischer projections, perspective formulas, and skeletal structures, see the tutorial at the end of this chapter.

Solution to 46a Determine the configuration of the two asymmetric centers, and then draw the perspective formula with the same configurations, following the steps in the Problem-Solving Strategy earlier in this section.

PROBLEM 47 Solved

Convert each perspective formula to a skeletal structure.

a.
HO   H
H⋯C—C◀Br
H₃C   CH₃

b.
Cl   H
H₃C⋯C—C◀OH
H   CH₂CH₃

Solution to 47a Determine the configuration of the two asymmetric centers in the perspective formula. Now draw the skeletal structure, making sure the asymmetric centers have the same configurations they had in the perspective formulas.

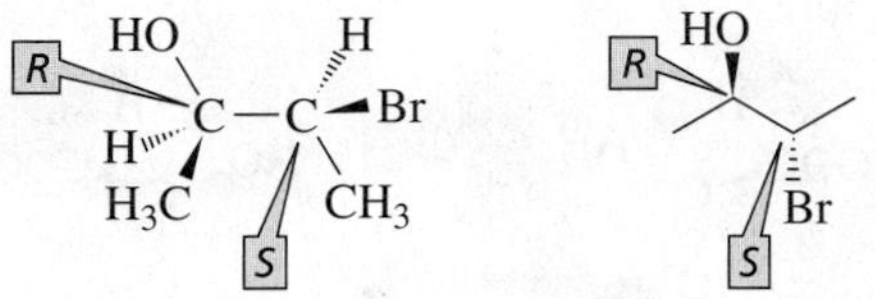

PROBLEM 48♦

The following compound has two asymmetric centers and four stereoisomers. Two of these are D-erythrose and D-threose, which are naturally occurring sugars. The configuration of D-erythrose is (2R,3R), and the configuration of D-threose is (2S,3R).

a. Which structure represents D-erythrose? **b.** Which represents D-threose?

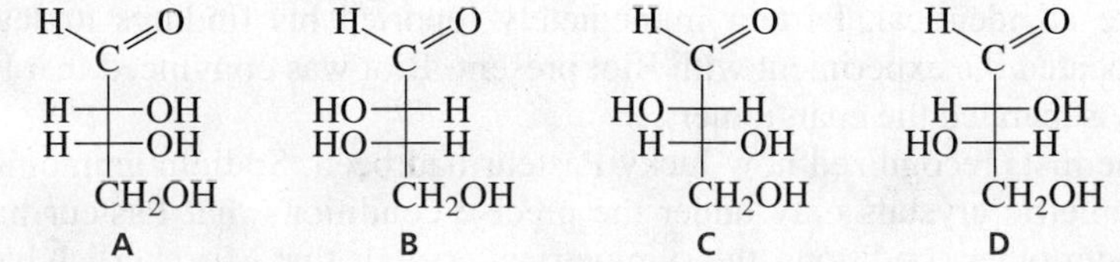

PROBLEM 49 Solved

(S)-(−)-2-Methyl-1-butanol can be converted to (+)-2-methylbutanoic acid without breaking any of the bonds to the asymmetric center. What is the configuration of (−)-2-methylbutanoic acid?

(S)-(−)-2-methyl-1-butanol   (+)-2-methylbutanoic acid

Solution We know that (+)-2-methylbutanoic acid has the configuration shown here because it was formed from (S)-(−)-2-methyl-1-butanol without breaking any bonds to the asymmetric center. From its structure, we can determine that (+)-2-methylbutanoic acid has the *S* configuration. Therefore, (−)-2-methylbutanoic acid has the *R* configuration.

PROBLEM 50♦

The reaction of (*R*)-1-iodo-2-methylbutane with hydroxide ion forms an alcohol without breaking any bonds to the asymmetric center. The alcohol rotates the plane of polarization of plane polarized light counterclockwise. What is the configuration of (+)-2-methyl-1-butanol?

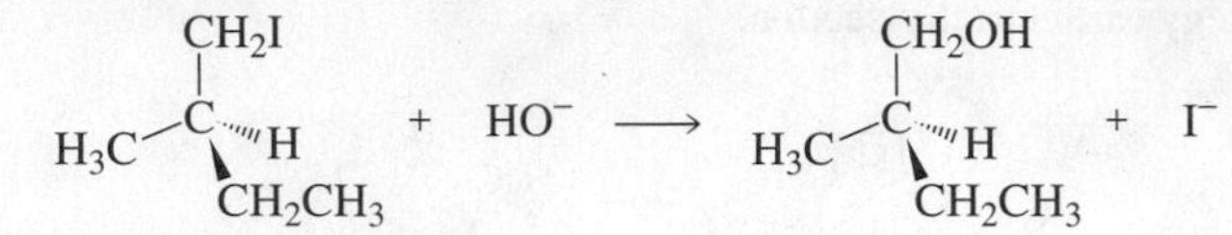

$$CH_2I \qquad\qquad CH_2OH$$

$$H_3C-C\cdots H \;+\; HO^- \longrightarrow H_3C-C\cdots H \;+\; I^-$$

$$CH_2CH_3 \qquad\qquad CH_2CH_3$$

(*R*)-1-iodo-2-methylbutane

---

15 HOW ENANTIOMERS CAN BE SEPARATED

Enantiomers cannot be separated by the usual separation techniques such as fractional distillation or crystallization because their identical boiling points and solubilities cause them to distill or crystallize simultaneously.

Louis Pasteur was the first to succeed in separating a pair of enantiomers. While working with crystals of sodium ammonium tartrate, he noted that the crystals were not identical—some were "right-handed" and some were "left-handed." After painstakingly separating the two kinds of crystals with a pair of tweezers, he found that a solution of the right-handed crystals rotated the plane of polarization of plane-polarized light clockwise, whereas a solution of the left-handed crystals rotated it counterclockwise.

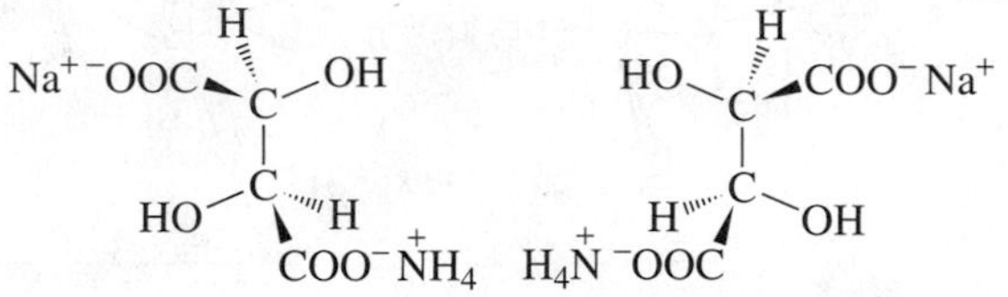

The French chemist and microbiologist Louis Pasteur (1822–1895) was the first to demonstrate that microbes cause specific diseases. Asked by the French wine industry to find out why wine often went sour while aging, he showed that the microorganisms that cause grape juice to ferment, producing wine, also cause wine to become sour. Gently heating the wine after fermentation, a process called pasteurization, kills the organisms so they cannot sour the wine.

$$Na^+{}^-OOC \diagdown C \diagup OH \qquad\qquad HO \diagdown C \diagup COO^-Na^+$$

$$HO \diagdown C \diagup H \qquad\qquad H \diagdown C \diagup OH$$

$$COO^-\overset{+}{N}H_4 \qquad\qquad H_4\overset{+}{N}{}^-OOC$$

left-handed crystals   right-handed crystals
sodium ammonium tartrate

Pasteur, only 26 years old at the time and unknown in scientific circles, was concerned about the accuracy of his observations because a few years earlier, Eilhardt Mitscherlich, a well-known German organic chemist, had reported that crystals of sodium ammonium tartrate were all identical. Pasteur immediately reported his findings to Jean-Baptiste Biot and repeated the experiment with Biot present. Biot was convinced that Pasteur had successfully separated the enantiomers.

Later, chemists recognized how lucky Pasteur had been. Sodium ammonium tartrate forms asymmetric crystals only under the precise conditions that Pasteur happened to employ. Under other conditions, the symmetrical crystals that Mitscherlich had obtained are formed. But, to quote Pasteur, "Chance favors the prepared mind."

Pasteur's experiment gave rise to a new chemical term. Tartaric acid is obtained from grapes, so it was also called racemic acid (*racemus* is Latin for "a bunch of grapes"). This is how a mixture of equal amounts of enantiomers came to be known as a **racemic mixture** (Section 9). Separation of enantiomers is called the **resolution of a racemic mixture.**

Separating enantiomers by hand, as Pasteur did, is not a universally useful method because few compounds form asymmetric crystals. Until relatively recently, separating enantiomers was a very tedious process. Fortunately, enantiomers can now be separated relatively easily by a technique called **chromatography.**

In this method, the mixture to be separated is dissolved in a solvent and the solution is passed through a column packed with a chiral material that adsorbs organic compounds. The two enantiomers will move through the column at different rates because they will have different affinities for the chiral material—just as a right hand prefers a right-hand glove to a left-hand glove—so one enantiomer will emerge from the column before the other. Because it is now so much easier to separate enantiomers, many drugs are being sold as single enantiomers rather than as racemic mixtures (see the box "Chiral Drugs" later in this section).

The chiral material used in chromatography is one example of a **chiral probe**, something capable of distinguishing between enantiomers. A polarimeter is another example of a chiral probe.

Crystals of potassium hydrogen tartrate, a naturally occurring salt found in wines. Most fruits produce citric acid, but grapes produce large quantities of tartaric acid instead. Potassium hydrogen tartrate, also called cream of tartar, is used in place of vinegar or lemon juice in some recipes.

Chiral Drugs

Until relatively recently, most drugs with one or more asymmetric centers have been marketed as racemic mixtures because of the difficulty of synthesizing single enantiomers and the high cost of separating enantiomers. In 1992, however, the Food and Drug Administration (FDA) issued a policy statement encouraging drug companies to use recent advances in synthesis and separation techniques to develop single-enantiomer drugs. Now most new drugs sold are single enantiomers. Drug companies have been able to extend their patents by marketing a single enantiomer of a drug that was previously available only as a racemate.

If a drug is sold as a racemate, the FDA requires both enantiomers to be tested because drugs bind to receptors and, since receptors are chiral, the enantiomers of a drug can bind to different receptors. Therefore, enantiomers can have similar or very different physiological properties. Examples are numerous. Testing has shown that (S)-(+)-ketamine is four times more potent an anesthetic than (R)-(−)-ketamine, and the disturbing side effects are apparently associated only with the (R)-(−)-enantiomer. Only the S isomer of the beta-blocker propranolol shows activity; the R isomer is inactive. The S isomer of Prozac, an antidepressant, is better at blocking serotonin but is used up faster than the R isomer. The activity of ibuprofen, the popular analgesic marketed as Advil, Nuprin, and Motrin, resides primarily in the (S)-(+)-enantiomer. Heroin addicts can be maintained with (−)-α-acetylmethadol for a 72-hour period compared to 24 hours with racemic methadone. This means less frequent visits to an outpatient clinic, because a single dose can keep an addict stable through an entire weekend.

Prescribing a single enantiomer spares the patient from having to metabolize the less potent enantiomer and decreases the chance of unwanted drug interactions. Drugs that could not be given as racemates because of the toxicity of one of the enantiomers can now be used. For example, (S)-penicillamine can be used to treat Wilson's disease even though (R)-penicillamine causes blindness.

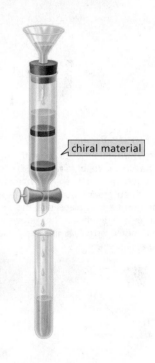

the active ingredient
in Vicks Vapor Inhaler®

methamphetamine
"speed"

PROBLEM 51♦

Limonene exists as two different stereoisomers. The R enantiomer is found in oranges and the S enantiomer is found in lemons. Which of the following molecules is found in oranges?

(+)-limonene   (−)-limonene

16 NITROGEN AND PHOSPHORUS ATOMS CAN BE ASYMMETRIC CENTERS

Atoms other than carbon can be asymmetric centers. Any atom that has four different groups or atoms attached to it is an asymmetric center. For example, the following pairs of compounds, with nitrogen and phosphorus asymmetric centers, are enantiomers.

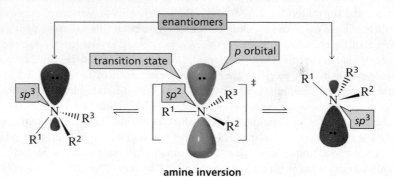

If one of the four "groups" attached to nitrogen is a lone pair, the enantiomers cannot be separated because they interconvert rapidly at room temperature. This rapid interconversion is called **amine inversion** (Figure 8). The lone pair is necessary for inversion: quaternary ammonium ions—ions with four bonds to nitrogen and hence no lone pair—do not invert. One way to picture amine inversion is to think of an umbrella that turns inside out in a windstorm.

▶ **Figure 8**

Amine inversion takes place through a transition state in which the sp^3 nitrogen becomes an sp^2 nitrogen. The three groups bonded to the sp^2 nitrogen lie in a plane and the lone pair is in a p orbital. The "inverted" and "noninverted" amine molecules are enantiomers.

amine inversion

The energy required for amine inversion is approximately 6 kcal/mol (or 25 kJ/mol), about twice the amount of energy required for rotation about a carbon–carbon single bond, but still small enough to allow the enantiomers to interconvert rapidly at room temperature. As a result, the enantiomers cannot be separated.

PROBLEM 52

Explain why compound **A** has two stereoisomers, but compounds **B** and **C** exist as single compounds.

SOME IMPORTANT THINGS TO REMEMBER

- **Stereochemistry** is the field of chemistry that deals with the structures of molecules in three dimensions.

- **Isomers** are compounds with the same molecular formula but different structures.

- **Constitutional isomers** differ in the way their atoms are connected.

- **Stereoisomers** differ in the way their atoms are arranged in space.

- There are two kinds of stereoisomers: **cis–trans isomers** and isomers that contain **asymmetric centers**.

- Because rotation about the bonds in a cyclic compound is restricted, disubstituted cyclic compounds exist as **cis-trans isomers**. The **cis isomer** has the substituents on the same side of the ring; **the trans isomer** has the substituents on opposite sides of the ring.

- Because rotation about a double bond is restricted, an alkene can exist as **cis–trans isomers**. The **cis isomer** has its hydrogens on the same side of the double bond; the **trans isomer** has its hydrogens on opposite sides of the double bond.

- A **chiral** molecule has a nonsuperimposable mirror image; an **achiral** molecule has a superimposable mirror image.

- An **asymmetric center** is an atom bonded to four different atoms or groups.

- **Enantiomers** are nonsuperimposable mirror images.

- **Diastereomers** are stereoisomers that are not enantiomers.

- Enantiomers have identical physical and chemical properties; diastereomers have different physical and chemical properties.

- An achiral reagent reacts identically with both enantiomers; a chiral reagent reacts differently with each enantiomer.

- A **racemic mixture** is a mixture of equal amounts of two enantiomers.

- The letters R and S indicate the **configuration** about an asymmetric center.

- If one member of a pair of stereoisomers has the R configuration and the other has the S configuration, they are enantiomers; if they both have the R configuration or both have the S configuration, they are identical.

- Chiral compounds are **optically active**; achiral compounds are **optically inactive**.

- If one enantiomer rotates the plane of polarization clockwise ($+$), its mirror image will rotate it the same amount counterclockwise ($-$).

- Each optically active compound has a characteristic **specific rotation**.

- A **racemic mixture**, indicated by ($\pm$), is optically inactive.

- In the case of compounds with two asymmetric centers, enantiomers have the opposite configuration at both asymmetric centers; diastereomers have the same configuration at one asymmetric center and the opposite configuration at the other asymmetric center.

- A **meso compound** has two or more asymmetric centers and a plane of symmetry; it is optically inactive.

- A compound with the same four groups bonded to two different asymmetric centers will have three stereoisomers—namely, a meso compound and a pair of enantiomers.

- Atoms other than carbons (such as N and P) can be asymmetric centers if they are bonded to four different atoms or groups.

PROBLEMS

53. Disregarding stereoisomers, draw the structures of all compounds with molecular formula C_5H_{10}. Which ones can exist as stereoisomers?

54. Draw all possible stereoisomers for each of the following. Indicate if no stereoisomers are possible.
  a. 1-bromo-2-chlorocyclohexane
  b. 2-bromo-4-methylpentane
  c. 1,2-dichlorocyclohexane
  d. 2-bromo-4-chloropentane
  e. 1-bromo-4-chlorocyclohexane
  f. 1,2-dimethylcyclopropane
  g. 4-bromo-2-pentene
  h. 3,3-dimethylpentane
  i. 1-bromo-2-chlorocyclobutane
  j. 1-bromo-3-chlorocyclobutane

55. Which of the following has an asymmetric center?

$CHBr_2Cl$ $BHFCl$ CH_3CHCl_2 $CHFBrCl$ $BeHCl$

56. Name the following compounds using *R,S* designations:

a. 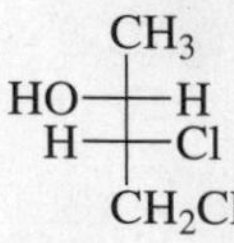
```
      CH3
HO ——— H
H ——— Cl
      CH2CH3
```

b. 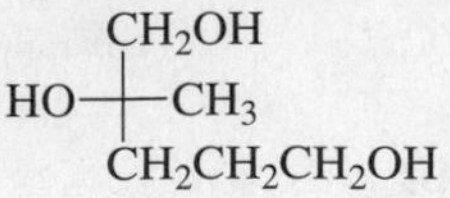
```
      CH2OH
HO ——— CH3
      CH2CH2CH2OH
```

c. 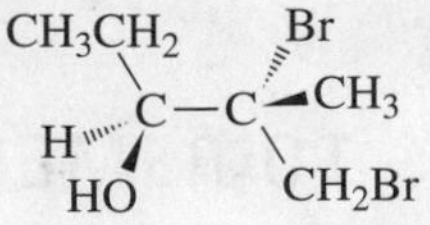
```
CH3CH2    Br
     \   /
H·····C—C····CH3
     /       \
   HO        CH2Br
```

57. Mevacor is used clinically to lower serum cholesterol levels. How many asymmetric centers does Mevacor have?

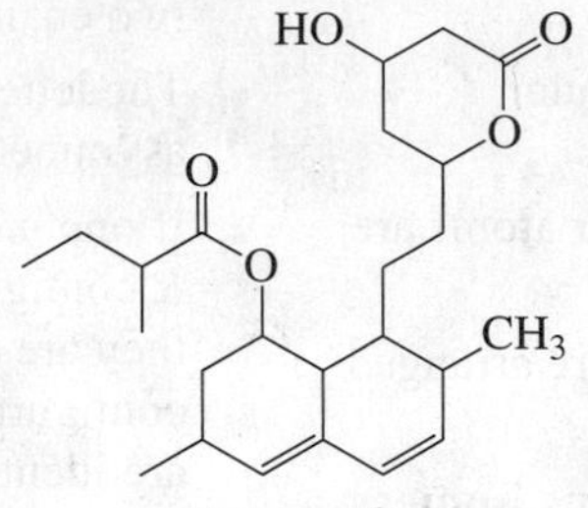

Mevacor®

58. Are the following pairs identical, enantiomers, diastereomers, or constitutional isomers?

a.

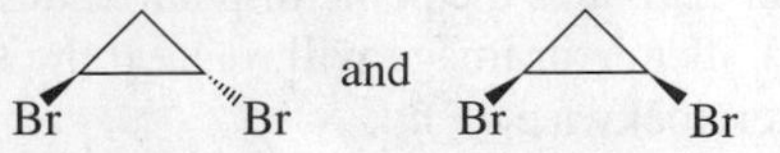

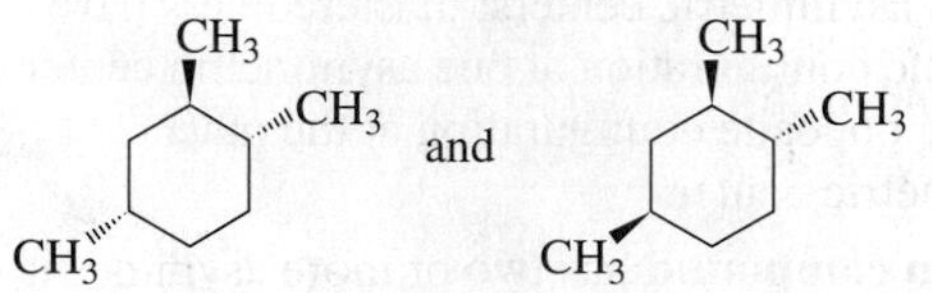

e. 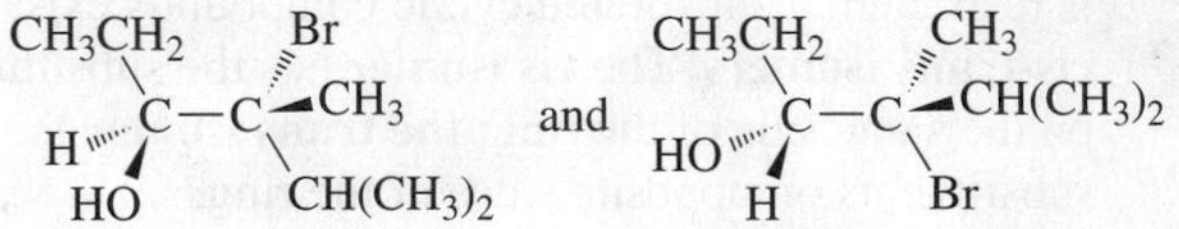
```
CH3CH2    Br                    CH3CH2    CH3
     \   /                           \   /
H·····C—C····CH3      and      HO·····C—C····CH(CH3)2
     /       \                       /       \
   HO       CH(CH3)2                H        Br
```

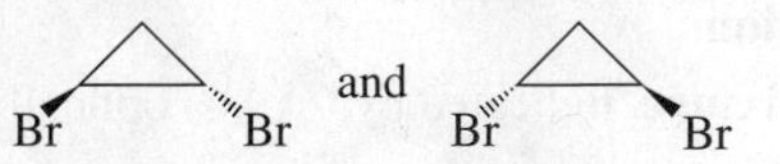

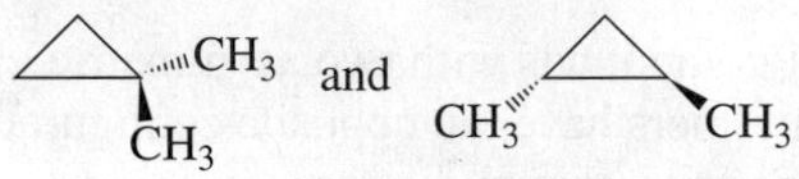

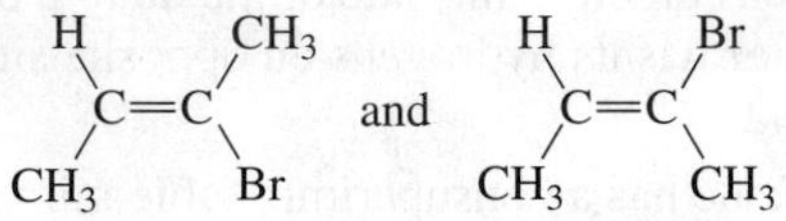

b.

f.

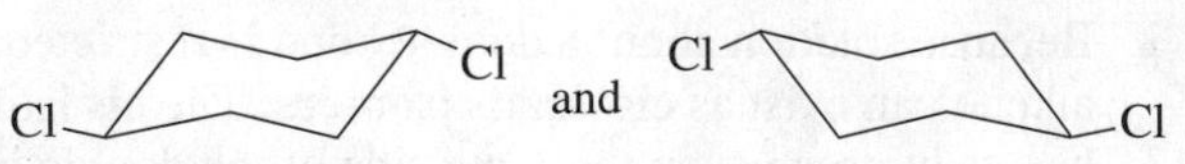

c.

g.
```
 H      CH3            H      Br
  \    /                \    /
   C=C          and      C=C
  /    \                /    \
CH3    Br             CH3    CH3
```

d.
```
    CH3                   CH3
     \···CH3               \···CH3
   (ring)      and       (ring)
CH3···/                CH3···/
```

h.
```
Br      CH3            H      CH3
  \    /                \    /
   C=C          and      C=C
  /    \                /    \
 H     CH3            Br     CH3
```

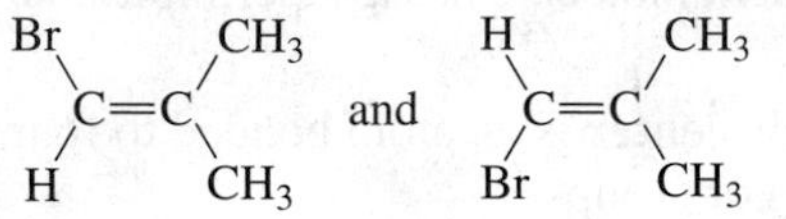

59. Which of the following are optically active?

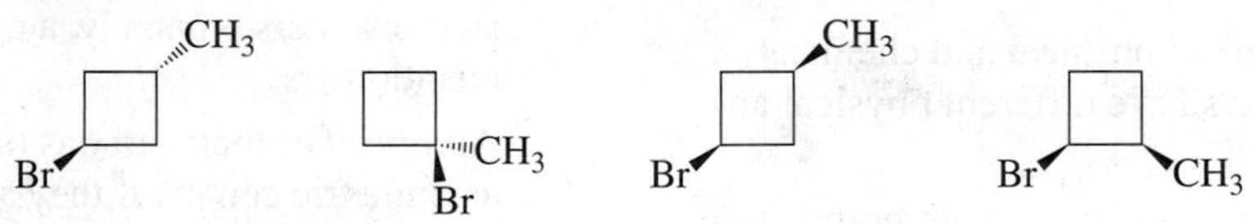

60. For many centuries, the Chinese have used extracts from a group of herbs known as ephedra to treat asthma. A compound named ephedrine has been isolated from these herbs and found to be a potent dilator of air passages in the lungs.

a. How many stereoisomers does ephedrine have?

b. The stereoisomer shown here is the one that is pharmacologically active. What is the configuration of each of the asymmetric centers?

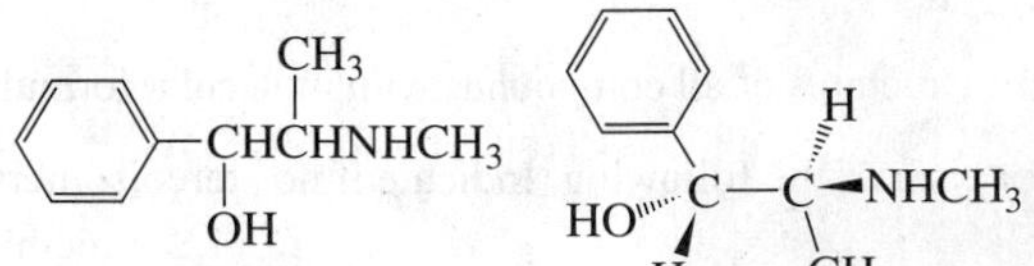

```
              CH3
               |
  [phenyl]—CHCHNHCH3
              |
             OH
         ephedrine
```

```
                     H
                     |
[phenyl]   HO·····C—C◄NHCH3
                  |   |
                  H   CH3
```

61. Name each of the following:

a.

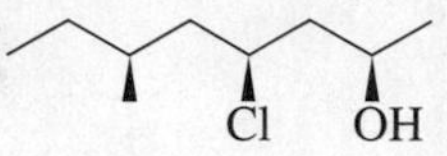

b.

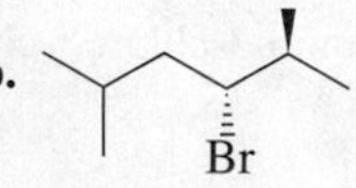

62. Which of the following has an achiral stereoisomer?

   a. 2,3-dichlorobutane
   b. 2,3-dichloropentane
   c. 2,3-dichloro-2,3-dimethylbutane
   d. 1,3-dichlorocyclopentane
   e. 1,3-dibromocyclobutane

   f. 2,4-dibromopentane
   g. 2,3-dibromopentane
   h. 1,4-dimethylcyclohexane
   i. 1,2-dimethylcyclopentane
   j. 1,2-dimethylcyclobutane

63. Are the following pairs identical, enantiomers, diastereomers, or constitutional isomers?

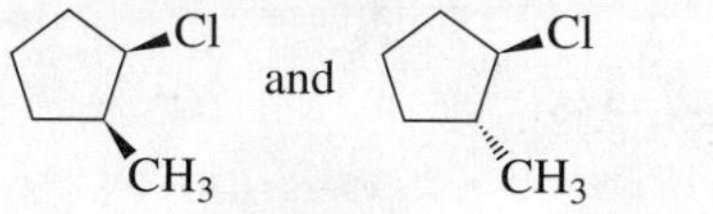

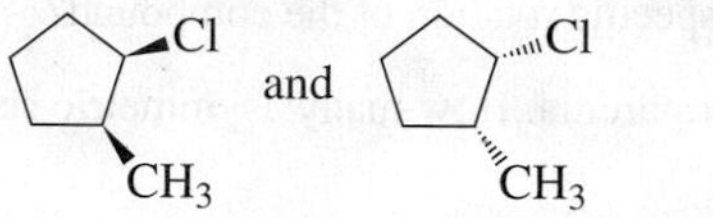

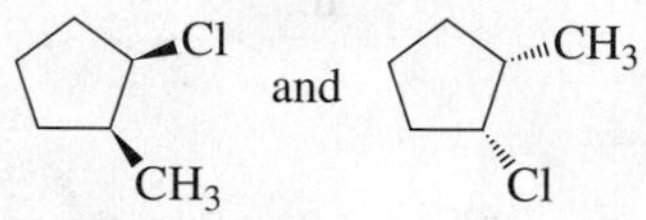

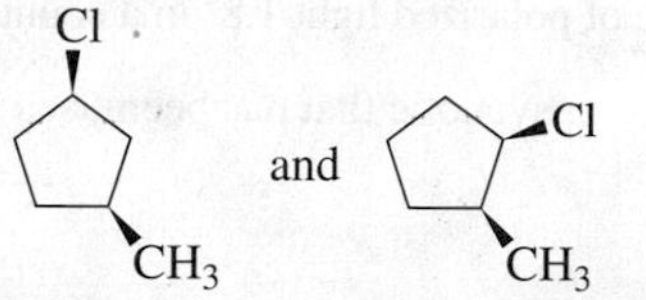

64. Citrate synthase, one of the enzymes in the series of enzyme-catalyzed reactions known as the citric acid cycle, catalyzes the synthesis of citric acid from oxaloacetic acid and acetyl-CoA. If the synthesis is carried out with acetyl-CoA that contains radioactive carbon (^{14}C) in the indicated position, the isomer shown here is obtained. (If two isotopes—atoms with the same atomic number, but different mass numbers—are being compared, the one with the greater mass number has the higher priority.)

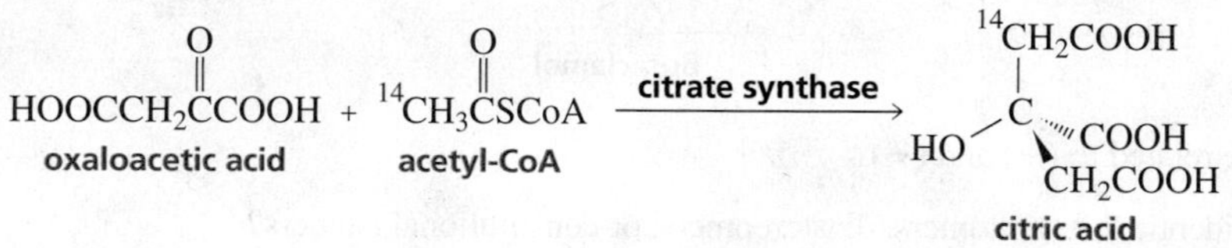

   a. Which stereoisomer of citric acid is synthesized, R or S?
   b. Why is the other stereoisomer not obtained?
   c. If the acetyl-CoA used in the synthesis contains ^{12}C instead of ^{14}C, will the product of the reaction be chiral or achiral?

65. Are the following pairs identical, enantiomers, diastereomers, or constitutional isomers?

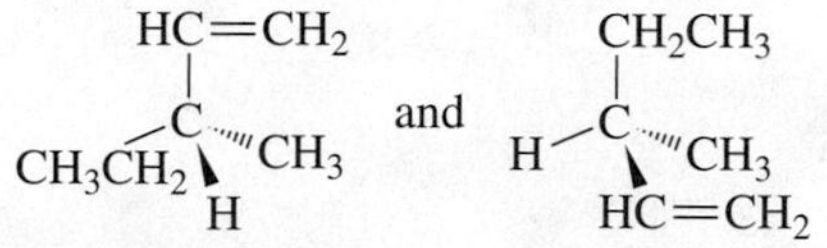

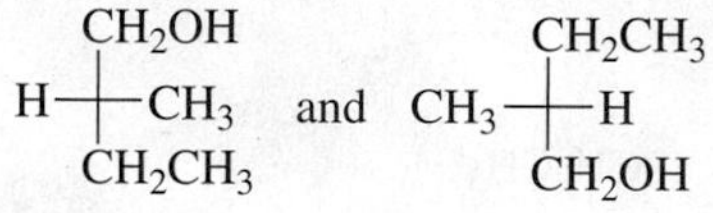

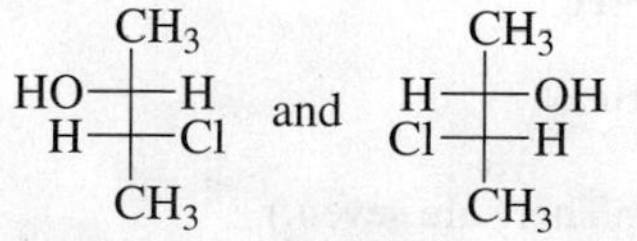

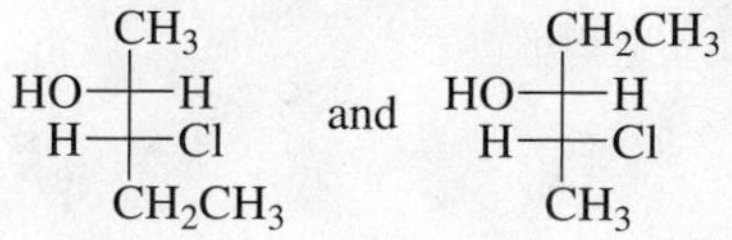

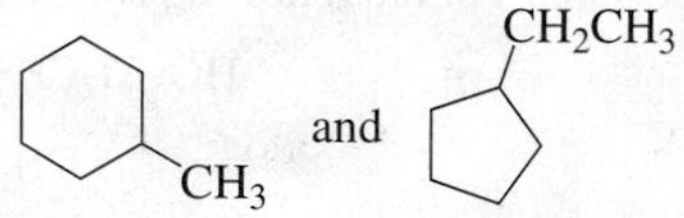

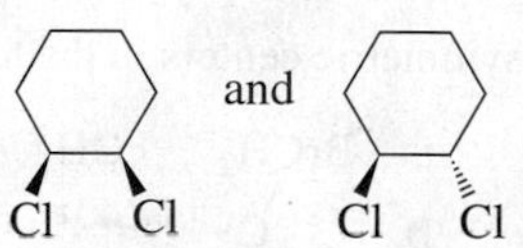

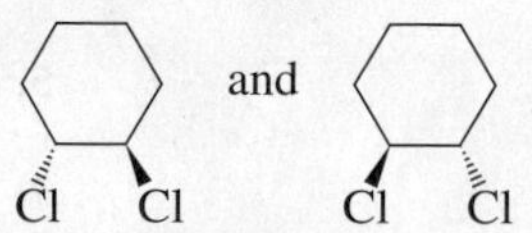

66. The specific rotation of (R)-$(+)$-glyceraldehyde is $+8.7$. If the observed specific rotation of a mixture of (R)-glyceraldehyde and (S)-glyceraldehyde is $+1.4$, what percent of glyceraldehyde is present as the R enantiomer?

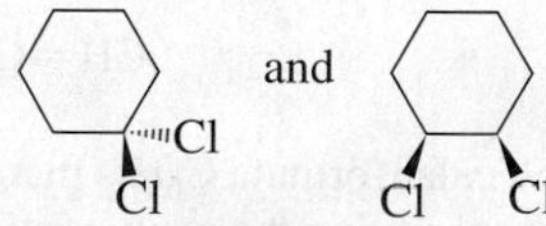

67. Indicate whether each of the following structures is (R)-2-chlorobutane or (S)-2-chlorobutane.

a.

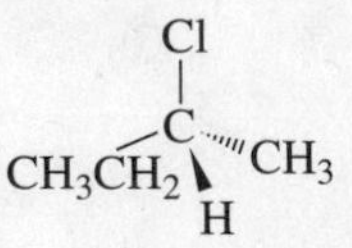

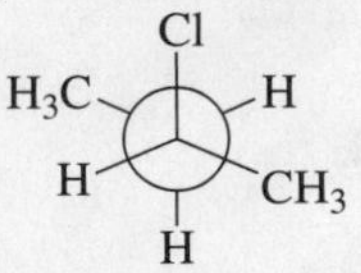

b.

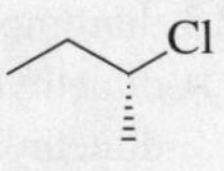

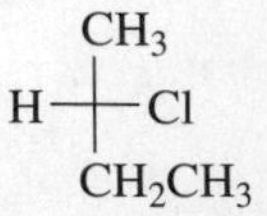

c.

d.

e.

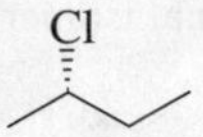

f.

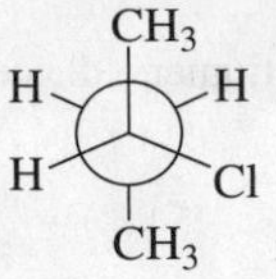

68. A solution of an unknown compound (3.0 g of the compound in 20 mL of solution), when placed in a polarimeter tube 2.0 dm long, was found to rotate the plane of polarized light 1.8° in a counterclockwise direction. What is the specific rotation of the compound?

69. Butaclamol is a potent antipsychotic that has been used clinically in the treatment of schizophrenia. How many asymmetric centers does it have?

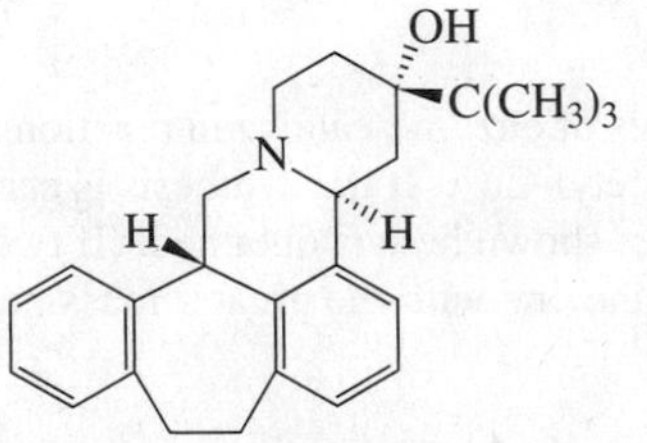

butaclamol

70. Explain how R and S are related to (+) and (−).

71. Are the following pairs identical, enantiomers, diastereomers, or constitutional isomers?

a. 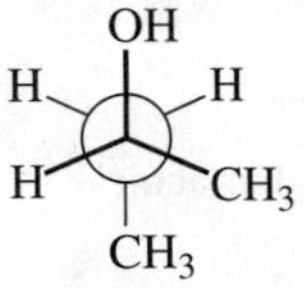 and

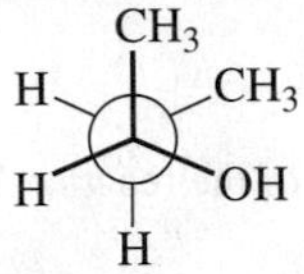

b. 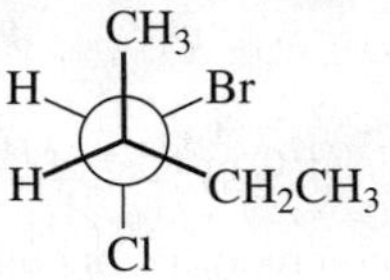 and

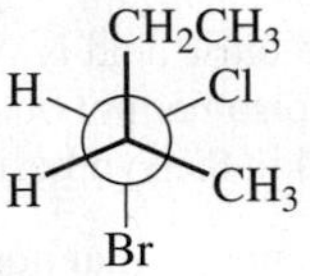

72. a. Draw all possible stereoisomers of the following compound:

$$HOCH_2CH-CH-CHCH_2OH$$
$$\quad\quad\quad |\quad\quad |\quad\quad |$$
$$\quad\quad\quad OH\quad OH\quad OH$$

b. Which stereoisomers are optically inactive?

73. What is the configuration of the asymmetric centers in the following structures?

a.

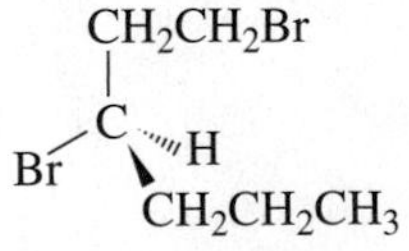

b.

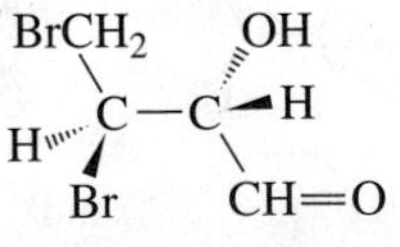

c.

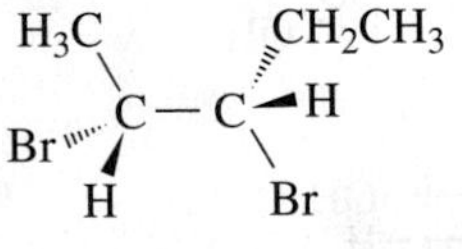

74. a. Draw all the isomers with molecular formula C_6H_{12} that contain a cyclobutane ring. (*Hint:* There are seven.)
b. Name the compounds without specifying the configuration of any asymmetric centers.
c. Identify:
  1. constitutional isomers
  2. stereoisomers
  3. cis–trans isomers
  4. chiral compounds
  5. achiral compounds
  6. meso compounds
  7. enantiomers
  8. diastereomers

75. A compound has a specific rotation of −39.0. A solution of the compound (0.187 g/mL) has an observed rotation of −6.52° when placed in a polarimeter tube 10 cm long. What is the percent of each enantiomer in the solution?

76. Are the following pairs identical, enantiomers, diastereomers, or constitutional isomers?

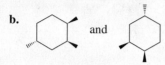

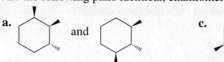

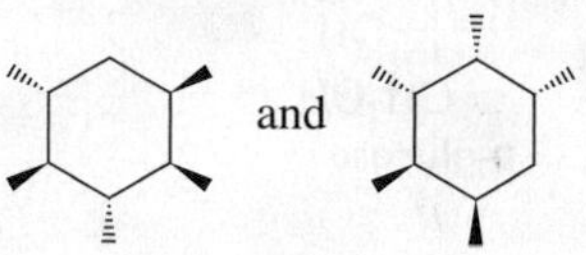

77. Draw structures for each of the following:
   a. (*S*)-1-bromo-1-chlorobutane
   b. (2*R*,3*R*)-2,3-dichloropentane
   c. an achiral isomer of 1,2-dimethylcyclohexane
   d. a chiral isomer of 1,2-dibromocyclobutane
   e. two achiral isomers of 3,4,5-trimethylheptane

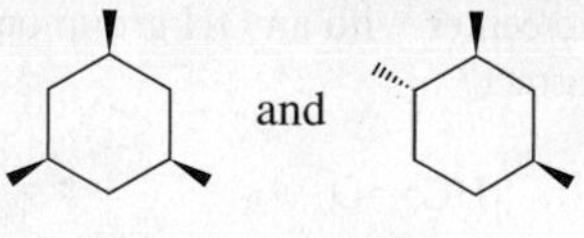

78. Explain why the enantiomers of 1,2-dimethylaziridine can be separated, even though one of the "groups" attached to nitrogen is a lone pair.

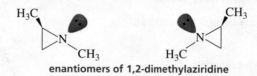

enantiomers of 1,2-dimethylaziridine

79. A sample of (*S*)-(+)-lactic acid was found to have an enantiomeric excess of 72%. How much *R* isomer is present in the sample?

80. Indicate whether each of the structures in the second row is an enantiomer of, a diastereomer of, or is identical to the structure in the top row.

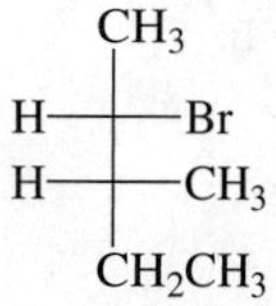

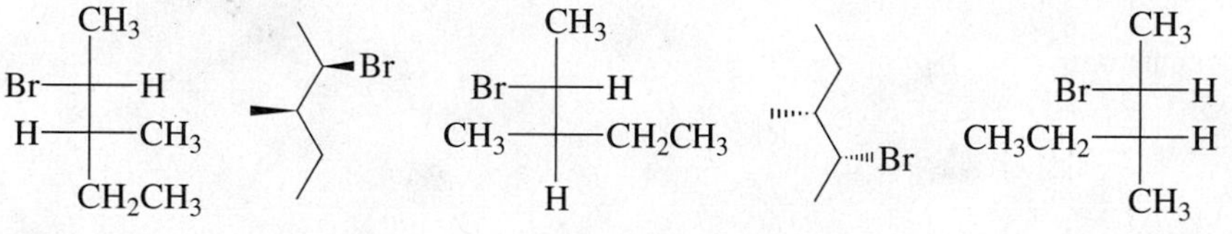

$$
\begin{array}{c}
CH_3 \\
H \!-\!\!|\!-\! Br \\
H \!-\!\!|\!-\! CH_3 \\
CH_2CH_3
\end{array}
$$

$$
\begin{array}{ccccc}
CH_3 & & CH_3 & & CH_3 \\
Br\!-\!|\!-\!H & \;Br & Br\!-\!|\!-\!H & & Br\!-\!|\!-\!H \\
H\!-\!|\!-\!CH_3 & & CH_3\!-\!|\!-\!CH_2CH_3 & \;Br & CH_3CH_2\!-\!|\!-\!H \\
CH_2CH_3 & & H & & CH_3 \\
A & B & C & D & E
\end{array}
$$

81. a. Using the wedge-and-dash notation, draw the nine stereoisomers of 1,2,3,4,5,6-hexachlorocyclohexane.
   b. From the nine stereoisomers, identify one pair of enantiomers.
   c. Draw the most stable conformation of the most stable stereoisomer.

82. Tamiflu is used for the prevention and treatment of flu. What is the configuration of each of its asymmetric centers?

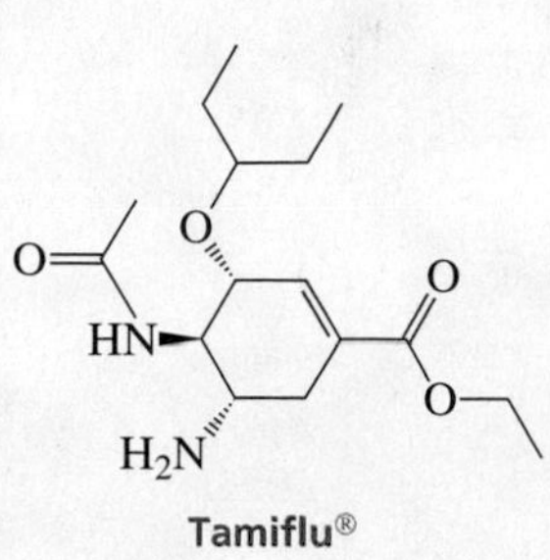

Tamiflu®

83. A student decided that the configuration of the asymmetric centers in a sugar such as D-glucose could be determined rapidly by simply assigning the *R* configuration to an asymmetric center with an OH group on the right and the *S* configuration to an asymmetric center with an OH group on the left. Is he correct?

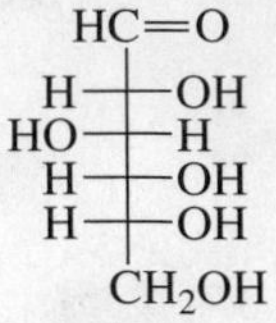

$$
\begin{array}{c}
\text{HC}{=}\text{O} \\
\text{H}{-\!\!-}\text{OH} \\
\text{HO}{-\!\!-}\text{H} \\
\text{H}{-\!\!-}\text{OH} \\
\text{H}{-\!\!-}\text{OH} \\
\text{CH}_2\text{OH}
\end{array}
$$

D-glucose

84. What the configuration of each of the asymmetric centers in the following compounds?

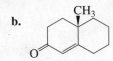

a.

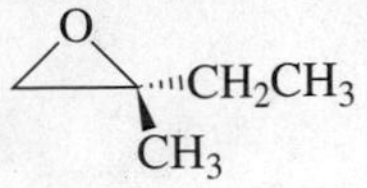

b.

c.

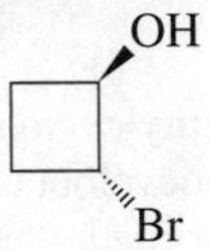

85. a. Draw the two chair conformers for each of the stereoisomers of *trans*-1-*tert*-butyl-3-methylcyclohexane.
  b. For each pair, indicate which conformer is more stable.

86. a. Do the following compounds have any asymmetric centers?
  1. $CH_2{=}C{=}CH_2$
  2. $CH_3CH{=}C{=}CHCH_3$
  b. Are the compounds chiral? (*Hint:* Make models.)

87. Is the following compound optically active?

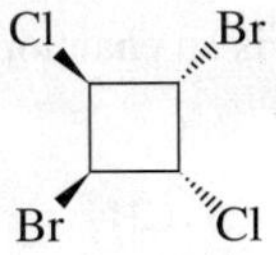

INTERCONVERTING STRUCTURAL REPRESENTATIONS

Enhanced by
MasteringChemistry®

If you have access to a set of molecular models, converting between *perspective formulas, Fischer projections,* and *skeletal structures* is rather straightforward. If, however, you are interconverting these three-dimensional structures on a two-dimensional piece of paper, it is easy to make a mistake, particularly if you are not good at visualizing structures in three dimensions. Fortunately, there is a relatively foolproof method for these interconversions. All you need to know is how to determine whether an asymmetric center has the *R* or the *S* configuration (Sections 7 and 14). Look at the following examples to learn how easy it is to interconvert the various structural representations.

1. **Converting a Fischer projection to a perspective formula.** First determine the configuration of the asymmetric center. After you find that it has the *R* configuration, draw the perspective formula with that configuration. When you draw the perspective formula, start by putting the group with the lowest priority (in this case, H) on the hatched wedge.

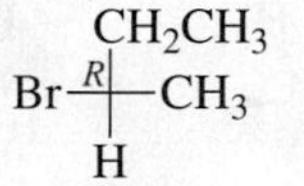

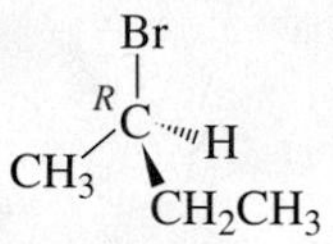

2. **Converting a perspective formula to a skeletal structure.** After you find that the asymmetric center in the perspective formula has the *R* configuration, you can draw the skeletal structure whose asymmetric center also has the *R* configuration.

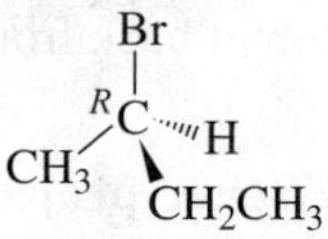

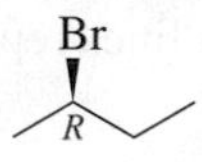

3. **Converting a skeletal structure to a Fischer projection.** Finding that C-2 has the *S* configuration and C-3 has the *R* configuration, you can draw the Fischer projection with the same configuration at C-2 and C-3.

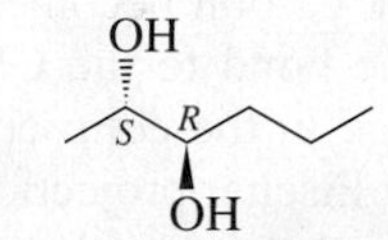

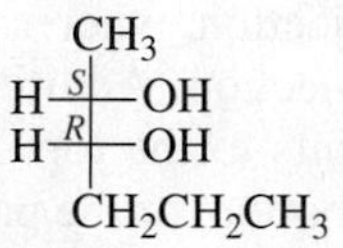

4. **Converting a perspective formula to a Fischer projection.** Determine the configuration of the asymmetric centers in the perspective formula, and then draw the Fischer projection using the same configuration for each of its asymmetric centers.

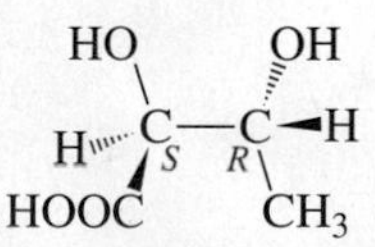

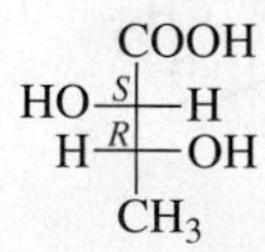

5. **Converting a Fischer projection to a skeletal structure.** Determine the configuration of the asymmetric centers in the Fischer projection, and then draw the skeletal structures using the same configuration for each of its asymmetric centers.

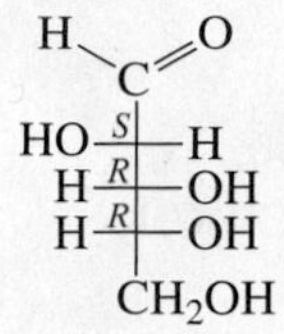

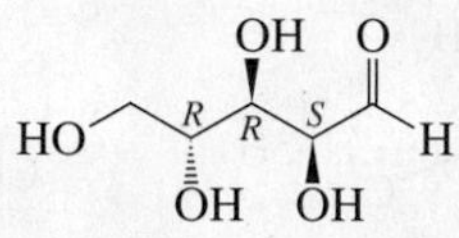

6. **Converting a Fischer projection to a perspective formula.** Determine the configuration of each of the two asymmetric centers in the Fischer projection. Draw the three solid lines of the perspective formula. Add a solid and hatched wedge at each carbon (making sure the solid wedge is below the hatched wedge), and then add the lowest priority group on each carbon to the hatched wedge. Now add the other two groups to each carbon in a way that gives the desired configuration at each asymmetric carbon.

7. **Converting a skeletal structure to a perspective formula.** Determine the configurations of the two asymmetric centers in the skeletal structure. Continue as described in Example 6 to arrive at the desired perspective formula.

Converting from a *Newman projection* to (or from) a Fischer projection, a perspective formula, or a skeletal structure is relatively straightforward, if you remember that a Fischer projection represents an eclipsed conformer. (Recall that in a Fischer projection, the horizontal lines represent bonds that point out of the plane of the paper toward the viewer, and the vertical lines represent bonds that point back from the plane of the paper away from the viewer.)

8. **Converting a Newman projection to a Fischer projection.** First convert the staggered conformer of the Newman projection to an eclipsed conformer by rotating either the front or back carbon. (Here the carbon in front is not moved while the back carbon is rotated counterclockwise.) Now you can draw the Fischer projection by pulling down the bond attached to the CH_3 group on the back carbon. As a result, the front carbon in the Newman projection becomes the top cross in the Fischer projection, whereas the back carbon becomes the next cross down in the Fischer projection. Additionally, the bond to the CH_3 group on the back carbon, which points to the top of the page in the eclipsed Newman projection, now points to the bottom of the page in the Fischer projection. Next, add the other atoms to their bonds, keeping the atoms on the right side of the Newman projection on the right side of the Fischer projection as well, and keeping the atoms on the left side of the Newman projection on the left side of the Fischer projection.

9. **Converting a Newman projection to a perspective formula.** First convert the staggered conformer to an eclipsed conformer, and then convert the eclipsed conformer to a Fischer projection as you did in Example 8. Once you have the Fischer projection, you can convert it into a perspective formula as described in Examples 6.

10. **Converting a perspective formula to a Newman projection.** Determine the configurations of the asymmetric centers in the perspective formula, and then draw the Fischer projection with its asymmetric centers in the same configuration. Because the Fischer projection is in an eclipsed conformation, you can draw the Newman projection in the eclipsed conformation by making the downward pointing bond in the Fischer projection the upward pointing bond on the back carbon of the Newman projection. Now add the other atoms to their bonds, keeping the atoms on the right side of the Fischer projection on the right side of the Newman projection as well, and keeping the atoms on the left side on the Fischer projection on the left side of the Newman projection. Next, rotate the Newman projection so it is in a staggered conformation. (Here the back carbon is rotated clockwise but it could have been rotated counterclockwise.)

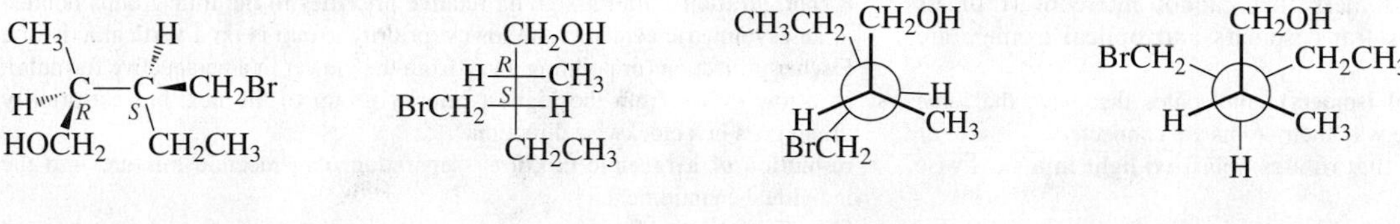

MasteringChemistry®
for Organic Chemistry

MasteringChemistry tutorials guide you through the toughest topics in chemistry with self-paced tutorials that provide individualized coaching. These assignable, in-depth tutorials are designed to coach you with hints and feedback specific to your individual misconceptions. For additional practice on Interconverting Structural Representations, go to MasteringChemistry where the following tutorials are available:

- Interconverting Structural Representations: Interpreting Fischer Projections

- Interconverting Structural Representations: Fischer Projections with Multiple Stereocenters

- Interconverting Structural Representations: Interpreting Newman Projections

GLOSSARY

achiral (optically inactive) an achiral molecule has a conformation identical to (superimposable upon) its mirror image.

amine inversion the configuration of an sp^3 hybridized nitrogen with a nonbonding pair of electrons that rapidly turns inside out.

asymmetric center an atom bonded to four different atoms or groups.

chiral (optically active) a chiral molecule has a nonsuperimposable mirror image.

chromatography a separation technique in which the mixture to be separated is dissolved in a solvent and the solvent is passed through a column packed with an absorbent stationary phase.

cis isomer the isomer with the hydrogens on the same side of the double bond or cyclic structure.

cis–trans isomers geometric isomers.

configurational isomers stereoisomers that cannot interconvert unless a covalent bond is broken. Cis–trans isomers and optical isomers are configurational isomers.

constitutional isomers (structural isomers) molecules that have the same molecular formula but differ in the way their atoms are connected.

dextrorotatory the enantiomer that rotates polarized light in a clockwise direction.

diastereomer a configurational stereoisomer that is not an enantiomer.

enantiomerically pure containing only one enantiomer.

enantiomeric excess (optical purity) how much excess of one enantiomer is present in a mixture of a pair of enantiomers.

enantiomers nonsuperimposable mirror-image molecules.

erythro enantiomers the pair of enantiomers with similar groups on the same side as drawn in a Fischer projection.

Fischer projection a method of representing the spatial arrangement of groups bonded to an asymmetric center. The asymmetric center is the point of intersection of two perpendicular lines; the horizontal lines represent bonds that project out of the plane of the paper toward the viewer, and the vertical lines represent bonds that point back from the plane of the paper away from the viewer.

geometric isomers cis–trans (or E,Z) isomers.

isomers nonidentical compounds with the same molecular formula.

levorotatory the enantiomer that rotates polarized light in a counterclockwise direction.

meso compound a compound that contains asymmetric centers and a plane of symmetry.

observed rotation the amount of rotation observed in a polarimeter.

optically active rotates the plane of polarized light.

optically inactive does not rotate the plane of polarized light.

perspective formula a method of representing the spatial arrangement of groups bonded to an asymmetric center. Two bonds are drawn in the plane of the paper; a solid wedge is used to depict a bond that projects out of the plane of the paper toward the viewer, and a hatched wedge is used to represent a bond that projects back from the plane of the paper away from the viewer.

plane of symmetry an imaginary plane that bisects a molecule into mirror images.

polarimeter an instrument that measures the rotation of polarized light.

racemic mixture (racemate, racemic modification) a mixture of equal amounts of a pair of enantiomers.

R configuration after assigning relative priorities to the four groups bonded to an asymmetric center, if the lowest priority group is on a vertical axis in a Fischer projection (or pointing away from the viewer in a perspective formula), an arrow drawn from the highest priority group to the next-highest-priority group goes in a clockwise direction.

resolution of a racemic mixture separation of a racemic mixture into the individual enantiomers.

S configuration after assigning relative priorities to the four groups bonded to an asymmetric center, if the lowest priority group is on a vertical axis in a Fischer projection (or pointing away from the viewer in a perspective formula), an arrow drawn from the highest priority group to the next-highest priority group goes in a counterclockwise direction.

specific rotation the amount of rotation that will be caused by a compound with a concentration of 1.0 g/mL in a sample tube 1.0 dm long.

stereogenic center (stereocenter) an atom at which the interchange of two substituents produces a stereoisomer.

stereoisomers isomers that differ in the way their atoms are arranged in space.

threo enantiomers the pair of enantiomers with similar groups on opposite sides when drawn in a Fischer projection.

trans isomer the isomer with the hydrogens on opposite sides of the double bond or cyclic structure. the isomer with identical substituents on opposite sides of the double bond.

PHOTO CREDITS

Photo credits are listed in order of appearance.

Alkenes

From Chapter 5 of *Organic Chemistry*, Seventh Edition. Paula Yurkanis Bruice. Copyright © 2014 by Pearson Education, Inc.
All rights reserved.

Alkenes

Structure, Nomenclature, and an Introduction to Reactivity • Thermodynamics and Kinetics

the gothic tower of Glasgow University overlooking the River Kelvin

Some of the things you will learn about in this chapter are how the Kelvin temperature scale got its name, why carrots and flamingos are orange, how insect populations can be controlled, and how compounds in biological systems recognize each other.

Alkanes are hydrocarbons that contain only carbon–carbon *single* bonds. Now we will take a look at **alkenes,** hydrocarbons that contain a carbon–carbon *double* bond.

Alkenes play many important roles in biology. For example, ethene ($H_2C=CH_2$), the smallest alkene, is a plant hormone—a compound that controls growth and other changes in the plant's tissues. Among other things, ethene affects seed germination, flower maturation, and fruit ripening. Many of the flavors and fragrances produced by plants also belong to the alkene family.

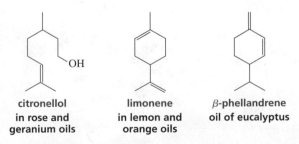

citronellol
in rose and
geranium oils

limonene
in lemon and
orange oils

β-phellandrene
oil of eucalyptus

We will begin our study of alkenes by looking at their structures and how they are named. Then we will examine a reaction of an alkene, paying close attention to the steps by which the reaction occurs and the energy changes that accompany them. You will see that some of the discussion in this chapter revolves around concepts with which you are

Tomatoes are shipped green so they will arrive unspoiled. Ripening starts when they are exposed to ethene.

already familiar, while some of the information is new and will broaden your foundation of knowledge.

Pheromones

Insects communicate by releasing pheromones—chemical substances that other insects of the same species detect with their antennae. Many of the sex, alarm, and trail pheromones are alkenes or are synthesized from alkenes. Interfering with an insect's ability to send or receive chemical signals is an environmentally safe way to control insect populations. For example, traps containing synthetic sex attractants have been used to capture such crop-destroying insects as the gypsy moth and the boll weevil.

muscalure
sex attractant of the housefly

disparlure
sex attractant of the gypsy moth

---

1 MOLECULAR FORMULAS AND THE DEGREE OF UNSATURATION

The general molecular formula for a noncyclic alkane is C_nH_{2n+2}. (Noncyclic compounds are also called **acyclic** compounds because the prefix "a" is Greek for "non" or "not.") the general molecular formula for a cyclic alkane is C_nH_{2n} because the cyclic structure reduces the number of hydrogens by two.

The general molecular formula for an *acyclic alkene* is also C_nH_{2n} because the double bond means the alkene has two fewer hydrogens than an alkane with the same number of carbons. Thus, the general molecular formula for a *cyclic alkene* must be C_nH_{2n-2}.

$CH_3CH_2CH_2CH_2CH_3$	$CH_3CH_2CH_2CH=CH_2$	a cyclic alkane	a cyclic alkene
an alkane	an alkene	C_5H_{10}	C_5H_8
C_5H_{12}	C_5H_{10}	C_nH_{2n}	C_nH_{2n-2}
C_nH_{2n+2}	C_nH_{2n}		

Thus, *the general molecular formula for a hydrocarbon is C_nH_{2n+2} minus two hydrogens for every π bond or ring in the molecule.*

The total number of π bonds and rings is called the compound's **degree of unsaturation.** Thus, C_8H_{14}, which has four fewer hydrogens than an acyclic alkane with eight carbons ($C_nH_{2n+2} = C_8H_{10}$), has two degrees of unsaturation. So we know that the sum of the compound's π bonds and rings is two.

several compounds with molecular formula C_8H_{14}

$CH_3CH=CH(CH_2)_3CH=CH_2$ $CH_3(CH_2)_5C\equiv CH$

Because *alkanes* contain the maximum number of C—H bonds possible—that is, they are saturated with hydrogen—they are called **saturated hydrocarbons.** In contrast, *alkenes* are called **unsaturated hydrocarbons** because they have fewer than the maximum number of hydrogens.

$CH_3CH_2CH_2CH_3$ $CH_3CH=CHCH_3$
an alkane   **an alkene**
a saturated hydrocarbon **an unsaturated hydrocarbon**

The general molecular formula for a hydrocarbon is C_nH_{2n+2} minus two hydrogens for every π bond or ring present in the molecule.

197

PROBLEM 1♦ Solved

What is the molecular formula for each of the following?

a. a 5-carbon hydrocarbon with one π bond and 1 ring
b. a 4-carbon hydrocarbon with two π bonds and no rings
c. a 10-carbon hydrocarbon with one π bond and 2 rings

Solution to 1a For a 5-carbon hydrocarbon with no π bonds and no rings, $C_nH_{2n+2} = C_5H_{12}$. A 5-carbon hydrocarbon with one π bond and one ring (that is, two degrees of unsaturation) has four fewer hydrogens, because two hydrogens are subtracted for every π bond or ring in the molecule. Therefore, the molecular formula is C_5H_8.

PROBLEM 2♦ Solved

Determine the degree of unsaturation for the hydrocarbons with the following molecular formulas:

a. $C_{10}H_{16}$ **b.** $C_{20}H_{34}$ **c.** C_8H_{16} **d.** $C_{12}H_{20}$ **e.** $C_{40}H_{56}$

Solution to 2a For a 10-carbon hydrocarbon with no π bonds and no rings, $C_nH_{2n+2} = C_{10}H_{22}$. A 10-carbon compound with molecular formula $C_{10}H_{16}$ has six fewer hydrogens, so the degree of unsaturation is $6/2 = 3$.

PROBLEM 3

Determine the degree of unsaturation, and then draw possible structures, for compounds with the following molecular formulas:

a. C_3H_6 **b.** C_3H_4 **c.** C_4H_6

PROBLEM 4♦

Several studies have shown that β-carotene, a precursor of vitamin A, may play a role in preventing cancer. β-Carotene has a molecular formula of $C_{40}H_{56}$ and it contains two rings and no triple bonds. How many double bonds does it have?

β-Carotene is an orange colored compound found in carrots, apricots, and flamingo feathers.

2   THE NOMENCLATURE OF ALKENES

The IUPAC system uses a suffix to denote certain functional groups. The double bond is the functional group of an alkene; its presence is denoted by the suffix "ene." Therefore, the systematic (IUPAC) name of an alkene is obtained by replacing the "ane" ending of the corresponding alkane with "ene." For example, a two-carbon alkene is called ethene, and a three-carbon alkene is called propene. Ethene also is frequently called by its common name: ethylene.

	$H_2C{=}CH_2$	$CH_3CH{=}CH_2$	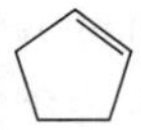 cyclopentene	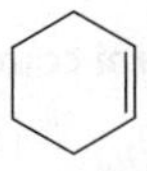 cyclohexene
systematic name:	ethene	propene		
common name:	ethylene	propylene		

Most alkene names need a number to indicate the position of the double bond. (The four names above do not because there is no ambiguity.) The IUPAC rules apply to alkenes as well:

1. The longest continuous chain containing the functional group (in this case, the carbon–carbon double bond) is numbered in the direction that gives the functional group suffix the lowest possible number. For example, 1-butene signifies that the double bond is between the first and second carbons of butene; 2-hexene signifies that the double bond is between the second and third carbons of hexene.

$$\overset{4}{C}H_3\overset{3}{C}H_2\overset{2}{C}H=\overset{1}{C}H_2$$
1-butene

$$\overset{1}{C}H_3\overset{2}{C}H=\overset{3}{C}H\overset{4}{C}H_3$$
2-butene

$$\overset{1}{C}H_3\overset{2}{C}H=\overset{3}{C}H\overset{4}{C}H_2\overset{5}{C}H_2\overset{6}{C}H_3$$
2-hexene

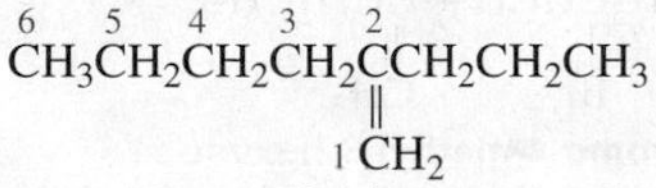

$$\overset{6}{C}H_3\overset{5}{C}H_2\overset{4}{C}H_2\overset{3}{C}H_2\overset{2}{C}CH_2CH_2CH_3$$
$$\overset{\|}{\underset{1}{C}H_2}$$
2-propyl-1-hexene

the longest continuous chain has eight carbons but the longest continuous chain containing the functional group has six carbons, so the parent name of the compound is hexene

Number the longest continuous chain containing the functional group in the direction that gives the functional group suffix the lowest possible number.

Notice that 1-butene does not have a common name. You might be tempted to call it "butylene," which is analogous to "propylene" for propene. Butylene, however, is not an appropriate name because it could signify either 1-butene or 2-butene, and a name must be unambiguous.

The stereoisomers of an alkene are named using a *cis* or *trans* prefix (in italics).

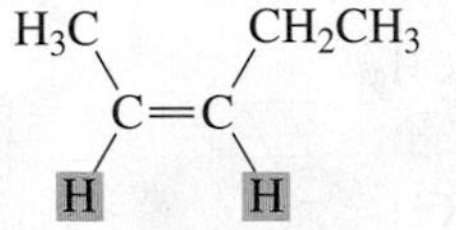

cis-2-pentene 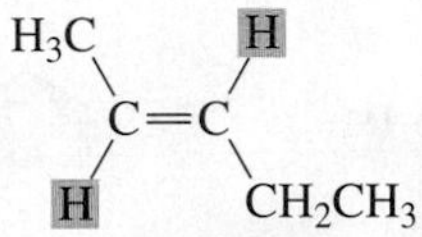   *trans*-2-pentene

2. For a compound with two double bonds, the "ne" ending of the corresponding alkane is replaced with "diene."

$$\overset{1}{C}H_3\overset{2}{C}H=\overset{3}{C}H-\overset{4}{C}H=\overset{5}{C}H\overset{6}{C}H_2\overset{7}{C}H_3$$
2,4-heptadiene

$$\overset{5}{C}H_3\overset{4}{C}H=\overset{3}{C}H-\overset{2}{C}H=\overset{1}{C}H_2$$
1,3-pentadiene

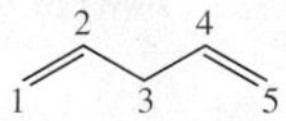

1,4-pentadiene

3. The name of a substituent is stated before the name of the longest continuous chain that contains the functional group, together with a number to designate the carbon to which the substituent is attached. Notice that *if a compound's name contains both a functional group suffix and a substituent, the functional group suffix gets the lowest possible number.*

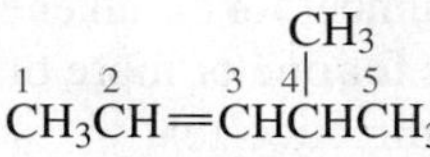

4-methyl-2-pentene

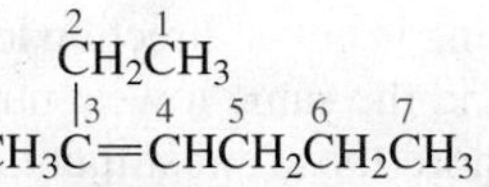

3-methyl-3-heptene

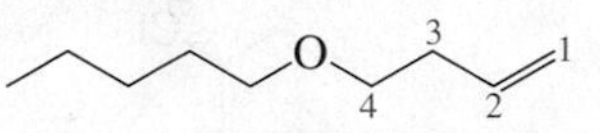

4-pentoxy-1-butene

4-methyl-1,3-pentadiene

When there are both a functional group suffix and a substituent, the functional group suffix gets the lowest possible number.

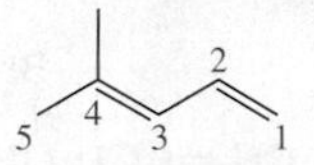

4. If a chain has more than one substituent, the substituents are stated in alphabetical order. Then the appropriate number is assigned to each substituent.

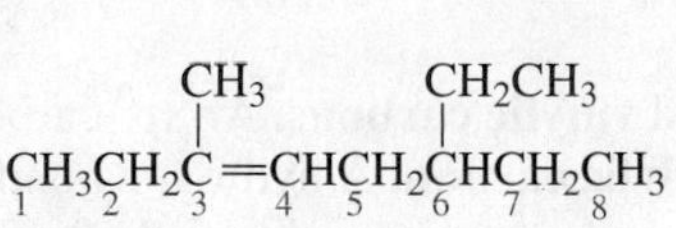

6-ethyl-3-methyl-3-octene

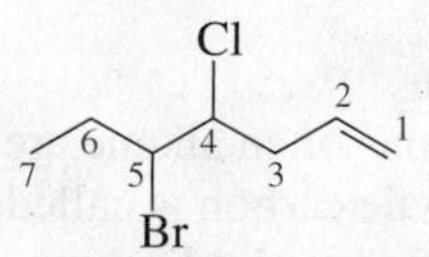

5-bromo-4-chloro-1-heptene

Substituents are stated in alphabetical order.

5. If counting in either direction results in the same number for the alkene functional group suffix, the correct name is the one containing the lowest substituent number. For example, the compound shown next on the left is a 4-octene whether the longest continuous chain is numbered from left to right or from right to left. If you number from left to right, then the substituents are at positions 4 and 7, but if you

199

number from right to left, they are at positions 2 and 5. Of those four substituent numbers, 2 is the lowest, so the compound is named 2,5-dimethyl-4-octene.

A substituent receives the lowest possible number only if there is no functional group suffix or if the same number for the functional group suffix is obtained in both directions.

$$CH_3CH_2CH_2C\!=\!CHCH_2CHCH_3$$
$$\qquad\qquad\quad | \qquad\qquad\ |$$
$$\qquad\qquad CH_3 \qquad\quad CH_3$$

2,5-dimethyl-4-octene
not
4,7-dimethyl-4-octene
because 2 < 4

$$CH_3CHCH\!=\!CCH_2CH_3$$
$$\qquad\ | \qquad\qquad\ |$$
$$\qquad Br \qquad\quad CH_3$$

2-bromo-4-methyl-3-hexene
not
5-bromo-3-methyl-3-hexene
because 2 < 3

6. A number is not needed to denote the position of the double bond in a cyclic alkene because the ring is always numbered so that the double bond is between carbons 1 and 2. To assign numbers to any substituents, count around the ring in the direction (clockwise or counterclockwise) that puts the lowest number into the name.

3-ethylcyclopentene   **4,5-dimethylcyclohexene**   **4-ethyl-3-methylcyclohexene**

Notice that 1,6-dichlorocyclohexene is *not* called 2,3-dichlorocyclohexene because the former has the lowest substituent number (1), even though it does not have the lowest sum of substituent numbers (1 + 6 = 7 versus 2 + 3 = 5).

1,6-dichlorocyclohexene
not
2,3-dichlorocyclohexene
because 1 < 2

5-ethyl-1-methylcyclohexene
not
4-ethyl-2-methylcyclohexene
because 1 < 2

7. If counting in either direction leads to the same number for the alkene functional group suffix and the same lowest number or numbers for one or more of the substituents, then ignore those substituents and choose the direction that gives the lowest number to one of the remaining substituents.

$$\qquad\qquad\qquad\qquad Br$$
$$\qquad\qquad\qquad\qquad\ |$$
$$CH_3CHCH_2CH\!=\!CCH_2CHCH_3$$
$$\quad\ | \qquad\qquad\qquad\qquad\ |$$
$$\ CH_3 \qquad\qquad\qquad CH_2CH_3$$

2-bromo-4-ethyl-7-methyl-4-octene
not
7-bromo-5-ethyl-2-methyl-4-octene
because 4 < 5

6-bromo-3-chloro-4-methylcyclohexene
not
3-bromo-6-chloro-5-methylcyclohexene
because 4 < 5

The sp^2 carbons of an alkene are called **vinylic carbons**. An sp^3 carbon that is adjacent to a vinylic carbon is called an **allylic carbon**. A hydrogen bonded to a vinylic carbon is called a **vinylic hydrogen**, and a hydrogen bonded to an allylic carbon is called an **allylic hydrogen**.

vinylic carbons

$$RCH_2\!-\!CH\!=\!CH\!-\!CH_2R$$

allylic carbons

PROBLEM 5

a. How many vinylic hydrogens does the preceding compound have?
b. How many allylic hydrogens does it have?

Two groups containing a carbon–carbon double bond are used in common names—the **vinyl group** and the **allyl group.** The vinyl group is the smallest possible group that contains a vinylic carbon and the allyl group is the smallest possible group that contains an allylic carbon. When "vinyl" or "allyl" is used in a name, the substituent must be attached to the vinylic or allylic carbon, respectively.

$CH_2=CH-$ $CH_2=CHCH_2-$
the vinyl group   **the allyl group**

$CH_2=CHCl$ $CH_2=CHCH_2Br$
common name: **vinyl chloride**   **allyl bromide**
systematic name: **chloroethene**   **3-bromopropene**

Notice how these groups and some others can be used as substituent names in systematic nomenclature.

  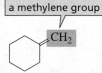 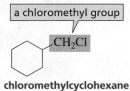

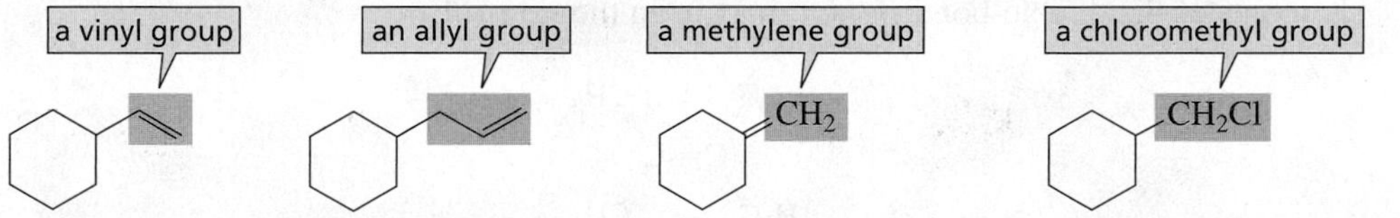

vinylcyclohexane   **allylcyclohexane**   **methylenecyclohexane**   **chloromethylcyclohexane**

PROBLEM 6♦

Draw the structure for each of the following:

a. 3,3-dimethylcyclopentene   **c.** ethyl vinyl ether
b. 6-bromo-2,3-dimethyl-2-hexene   **d.** allyl alcohol

PROBLEM 7♦

What is each compound's systematic name?

a. $CH_3CHCH=CHCH_3$
     $|$
     CH_3

b. $CH_3CH_2C=CCHCH_3$ with CH_3 above and CH_3 Cl below

c. cyclopentene with Br

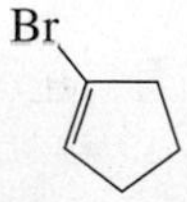

d. $BrCH_2CH_2CH=CCH_3$ with CH_2CH_3

e. cyclohexene with CH_3 and CH_3

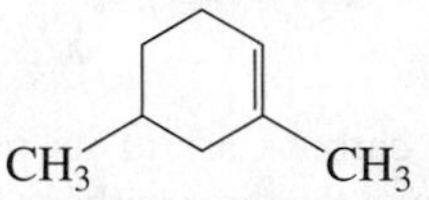

f. $CH_3CH=CHOCH_2CH_2CH_2CH_3$

g. 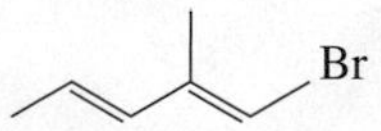 structure with Br

h. 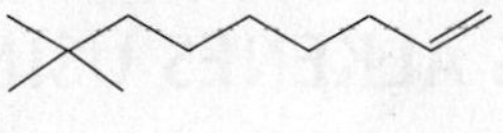 structure

3   THE STRUCTURE OF ALKENES

Alkenes have structures similar to that of ethene, the smallest alkene. Each double-bonded carbon of an alkene has three sp^2 orbitals. Each of these orbitals overlaps an orbital of another atom to form a σ bond, one of which is one of the bonds in the double bond. Thus, the σ bond of the double bond is formed by the overlap of an sp^2 orbital of one carbon with an sp^2 orbital of the other carbon, and the other bond of the double bond is a π bond formed from side-to-side overlap of the remaining p orbital on each of the sp^2 carbons.

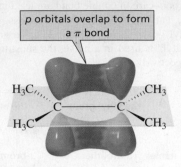

Because three points determine a plane, each sp^2 carbon and the two atoms singly bonded to it lie in a plane. In order to achieve maximum orbital–orbital overlap, the two p orbitals must be parallel to each other. For the two p orbitals to be parallel, all six atoms of the double-bond system must be in the same plane.

**the six carbon atoms
are in the same plane**

PROBLEM 8♦ Solved

How many carbons are in the planar double-bond system in each of the following compounds?

a. 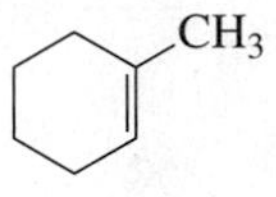   b. 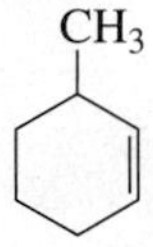   c. 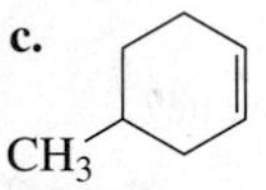   d.

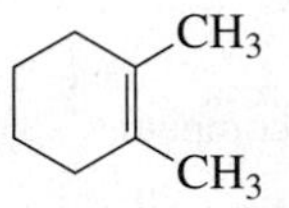

Solution to 8a Five carbons are in its planar double-bond system: the two sp^2 carbons (indicated by blue dots) and the three carbons bonded to the sp^2 carbons (indicated by red dots).

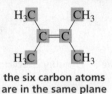

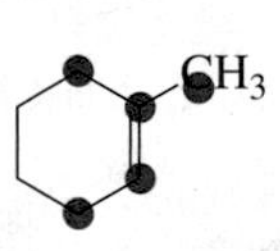

---

4 NAMING ALKENES USING THE *E,Z* SYSTEM

We have seen that as long as each of the sp^2 carbons of an alkene is bonded to one hydrogen, we can use the terms *cis* and *trans* to designate the geometric isomers of an alkene: *if the hydrogens are on the same side of the double bond, it is the cis isomer; if the hydrogens are on opposite sides of the double bond, it is the trans isomer.*

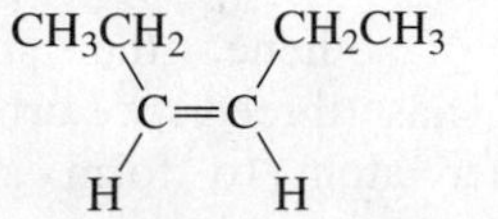

cis-2-hexene 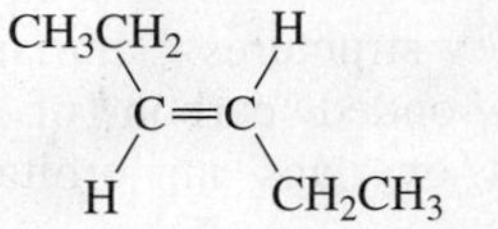   *trans*-2-hexene

But how do we designate the isomers of a compound such as 1-bromo-2-chloropropene?

$$Br \quad Cl$$
$$C=C$$
$$H \quad CH_3$$

$$Br \quad CH_3$$
$$C=C$$
$$H \quad Cl$$

Which isomer is cis and which is trans?

The *E,Z* system of nomenclature was devised for alkenes that do not have a hydrogen attached to each of the sp^2 carbons.*

To name an isomer by the *E,Z* system, we first determine the relative priorities of the two groups bonded to one of the sp^2 carbons and then the relative priorities of the two groups bonded to the other sp^2 carbon. (The rules for assigning relative priorities are explained below.)

low priority   low priority
$$C=C$$
high priority   high priority

the *Z* isomer has the high priority groups on the *same* side of the double bond

low priority   high priority
$$C=C$$
high priority   low priority

the *E* isomer has the high priority groups on *opposite* sides of the double bond

The *Z* isomer has the high-priority groups on the same side.

The *E* isomer has the high-priority groups on opposite sides.

If the two high-priority groups (one from each carbon) are on the same side of the double bond, the isomer is the *Z* isomer (*Z* is for *zusammen*, German for "together"). If the high-priority groups are on opposite sides of the double bond, the isomer is the *E* isomer (*E* is for *entgegen*, German for "opposite").

The relative priorities of the two groups bonded to an sp^2 carbon are determined using the following rules:

1. The relative priorities depend on the atomic numbers of the atoms bonded directly to the sp^2 carbon. The greater the atomic number, the higher the priority. This should remind you of the way that relative priorities are determined for *R* and *S* isomers because the system of priorities was originally devised for the *R,S* system and was later adopted for the *E,Z* system.

   For example, in the isomer below on the left, the sp^2 carbon on the left is bonded to a Br and to an H; Br has a greater atomic number than H, so **Br** has the higher priority.

high priority   high priority

$$Br \quad Br \quad Cl \quad Cl$$
$$C=C$$
$$H \quad H \quad CH_3 \quad C$$

the *Z* isomer

$$Br \quad CH_3$$
$$C=C$$
$$H \quad Cl$$

the *E* isomer

The greater the atomic number of the atom bonded to the sp^2 carbon, the higher the priority of the substituent.

   The sp^2 carbon on the right is bonded to a Cl and to a C; Cl has the greater atomic number, so **Cl** has the higher priority. (Notice that you use the atomic number of C, not the mass of the CH_3 group, because the priorities are based on the atomic numbers of atoms, *not* on the masses of groups.)

   Thus, the isomer on the left has the high-priority groups (Br and Cl) on the same side of the double bond, so it is the **Z isomer**. (Zee groups are on Zee Zame Zide.) The isomer on the right has the high-priority groups on opposite sides of the double bond, so it is the **E isomer**.

---

*The IUPAC prefers the *E* and *Z* designations because they can be used for all alkene isomers. Many chemists, however, continue to use the "cis" and "trans" designations for simple molecules.

If the atoms attached to the sp^2 carbon are the same, the atoms attached to the tied atoms are compared; the one with the greater atomic number belongs to the group with the higher priority.

2. If the two atoms attached to an sp^2 carbon are the same (there is a tie), then consider the atomic numbers of the atoms that are attached to the "tied" atoms.

For example, in the isomer shown next on the left, both atoms bonded to the sp^2 carbon on the left are carbons (in a CH_2Cl group and a CH_2CH_2Cl group), so there is a tie.

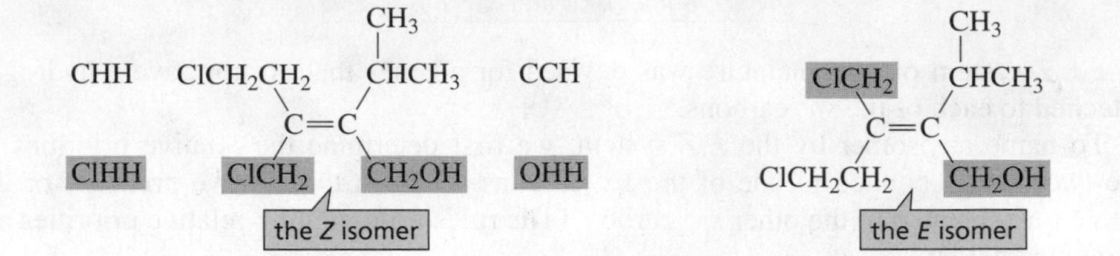

CHH   ClCH$_2$CH$_2$ $\overset{\displaystyle CH_3}{\underset{}{CHCH_3}}$ CCH $\overset{\displaystyle CH_3}{\underset{}{ClCH_2}}$ CHCH$_3$

$$C=C$$

ClHH   ClCH$_2$ CH$_2$OH   OHH   ClCH$_2$CH$_2$ CH$_2$OH

the Z isomer   the E isomer

The C of the CH_2Cl group is bonded to **Cl, H, H,** and the C of the CH_2CH_2Cl group is bonded to **C, H, H.** Cl has a greater atomic number than C, so the CH_2Cl group has the higher priority.

Both atoms attached to the sp^2 carbon on the right are Cs (in a CH_2OH group and a $CH(CH_3)_2$ group), so there is a tie on this side as well. The C of the CH_2OH group is bonded to **O, H, H,** and the C of the $CH(CH_3)_2$ group is bonded to **C, C, H.** Of these six atoms, O has the greatest atomic number, so CH_2OH has the higher priority. (Note that you do not add the atomic numbers—you take the single atom with the greatest atomic number.) The E and Z isomers are as shown above.

3. If an atom is doubly bonded to another atom, the priority system treats it as if it were singly bonded to two of those atoms. If an atom is triply bonded to another atom, the priority system treats it as if it were singly bonded to three of those atoms.

For example, in the isomer shown next on the left, the sp^2 carbon on the left is bonded to a CH_2CH_2OH group and to a $CH_2C\equiv CH$ group:

If an atom is doubly bonded to another atom, treat it as if it were singly bonded to two of those atoms.

If an atom is triply bonded to another atom, treat it as if it were singly bonded to three of those atoms.

HHO   HOCH$_2$CH$_2$ CH$=$CH$_2$ HCC   HOCH$_2$CH$_2$ CH$_2$CH$_3$

$$C=C$$

CCC   HC$\equiv$CCH$_2$ CH$_2$CH$_3$ CHH   HC$\equiv$CCH$_2$ CH$=$CH$_2$

the Z isomer   the E isomer

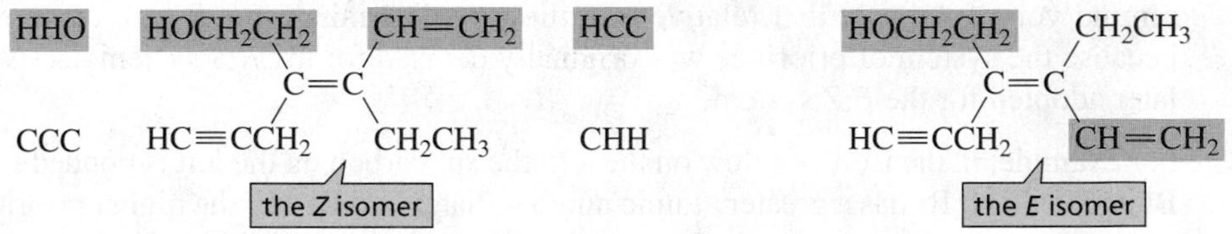

Cancel atoms that are identical in the two groups; use the remaining atoms to determine the group with the higher priority.

Because the atoms bonded to the sp^2 carbon are both carbons, there is a tie. Each of the carbons is bonded to **C, H, H,** so there is another tie. We turn our attention to the groups attached to the CH_2 groups to break the tie. One of these groups is CH_2OH, and the other is $C\equiv CH$; the C of the CH_2OH group is bonded to **H, H, O;** the triple-bonded C is considered to be bonded to **C, C, C.** Of the six atoms, O has the greatest atomic number, so CH_2OH has the higher priority.

Both atoms bonded to the sp^2 carbon on the right are Cs, so they are tied. The first carbon of the CH_2CH_3 group is bonded to **C, H, H;** the first carbon of the $CH=CH_2$ group is bonded to an H and doubly bonded to a C, so it is considered to be bonded to **H, C, C.** One C cancels in each of the two groups, leaving **H** and **H** in the CH_2CH_3 group and **H** and **C** in the $CH=CH_2$ group. C has a greater atomic number than H, so $CH=CH_2$ has the higher priority.

4. If two isotopes (atoms with the same atomic number, but different mass numbers) are being compared, the mass number is used to determine the relative priorities.

For example, in the isomer shown next on the left, the sp^2 carbon on the left is bonded to a deuterium (D) and to a hydrogen (H): D and H have the same atomic number, but D has a greater mass number, so **D** has the higher priority.

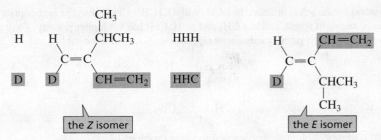

The Cs that are attached to the sp^2 carbon on the right are *both* bonded to **C, C, H,** so we must go to the next set of atoms to break the tie. The second carbon of the $CH(CH_3)_2$ group is bonded to **H, H, H,** whereas the second carbon of the $CH=CH_2$ group is bonded to **H, H, C.** (*Notice that to get the third atom, you go back along the double bond.*) Therefore, $CH=CH_2$ has the higher priority.

If atoms have the same atomic number but different mass numbers, the atom with the greater mass number has the higher priority.

PROBLEM 9♦

Assign relative priorities to each set of substituents:

a. —Br   —I   —OH   —CH₃
b. —CH₂CH₂OH   —OH   —CH₂Cl   —CH=CH₂

PROBLEM 10♦

Tamoxifen slows the growth of some breast tumors by binding to estrogen receptors. Is tamoxifen an *E* or a *Z* isomer?

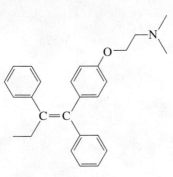

tamoxifen

PROBLEM 11

Draw and label the *E* and *Z* isomers for each of the following:

a. $CH_3CH_2CH=CHCH_3$

c. $CH_3CH_2CH_2CH_2$
   |
   $CH_3CH_2C=CCH_2Cl$
   |
   CH_3CHCH_3

b. $CH_3CH_2C=CHCH_2CH_3$
   |
   Cl

d. $HOCH_2CH_2C=CC\equiv CH$
   | |
   $O=CH$ $C(CH_3)_3$

PROBLEM 12

Draw skeletal structures for each pair of isomers in Problem 11.

PROBLEM 13♦

Name each of the following:

a.  [skeletal structure]   b. 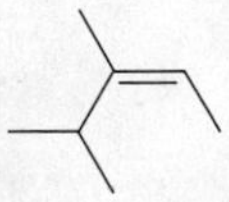 [skeletal structure]   c. 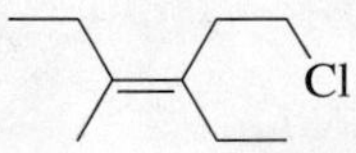 [skeletal structure with Cl]

PROBLEM-SOLVING STRATEGY

Drawing *E,Z* Structures

Draw the structure of (*E*)-1-bromo-2-methyl-2-butene.

First draw the compound without specifying the isomer so you can see what substituents are bonded to the sp^2 carbons. Then determine the relative priorities of the two groups bonded to each of the sp^2 carbons.

$$CH_3$$
$$|$$
$$BrCH_2C=CHCH_3$$

205

The sp^2 carbon on the left is attached to a CH_3 and a CH_2Br: CH_2Br has the higher priority. The sp^2 carbon on the right is attached to a CH_3 and an H: CH_3 has the higher priority. To draw the *E* isomer, put the two high-priority substituents on opposite sides of the double bond.

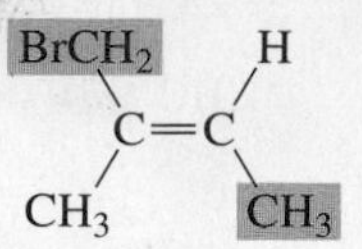

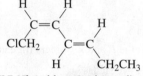

Now use the strategy you have just learned to solve Problem 14.

PROBLEM 14

Draw the structure of (*Z*)-3-isopropyl-2-heptene.

PROBLEM-SOLVING STRATEGY

Drawing Isomers for Compounds with Two Double Bonds

How many isomers does the following compound have?

$$ClCH_2CH=CHCH=CHCH_2CH_3$$

It has four isomers because each of its double bonds can have either the *E* or the *Z* configuration. Thus, there are *E-E*, *Z-Z*, *E-Z*, and *Z-E* isomers.

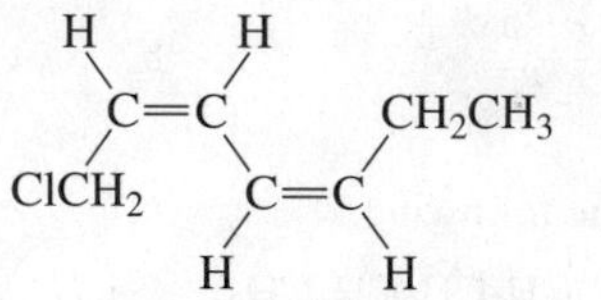

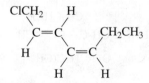

(2*Z*,4*Z*)-1-chloro-2,4-heptadiene   **(2*Z*,4*E*)-1-chloro-2,4-heptadiene**

(2*E*,4*Z*)-1-chloro-2,4-heptadiene 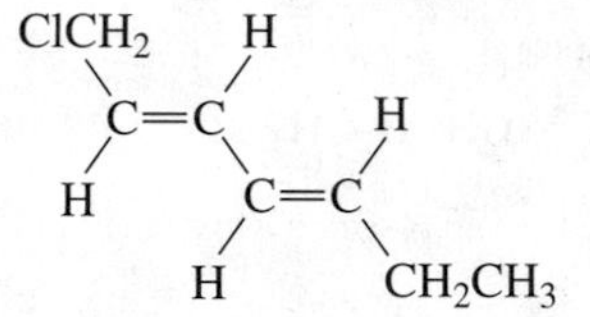   **(2*E*,4*E*)-1-chloro-2,4-heptadiene**

Now use the strategy you have just learned to solve Problem 15.

PROBLEM 15

Draw the isomers for the following compounds, and then name each one:

a. 2-methyl-2,4-hexadiene   **b.** 2,4-heptadiene   **c.** 1,3-pentadiene

---

5 HOW AN ORGANIC COMPOUND REACTS DEPENDS ON ITS FUNCTIONAL GROUP

There are many millions of organic compounds (and more being made each year). If you had to memorize how each of them reacts, studying organic chemistry would not be a very pleasant experience. Fortunately, organic compounds can be divided into families, and all the members of a family react in the same way.

The family that an organic compound belongs to is determined by its functional group. The **functional group** determines the kinds of reactions a compound will undergo. You are already familiar with the functional group of an alkene: the carbon–carbon double bond. All compounds with a carbon–carbon double bond react in the same way, whether the compound is a small molecule like ethene or a large molecule like cholesterol.

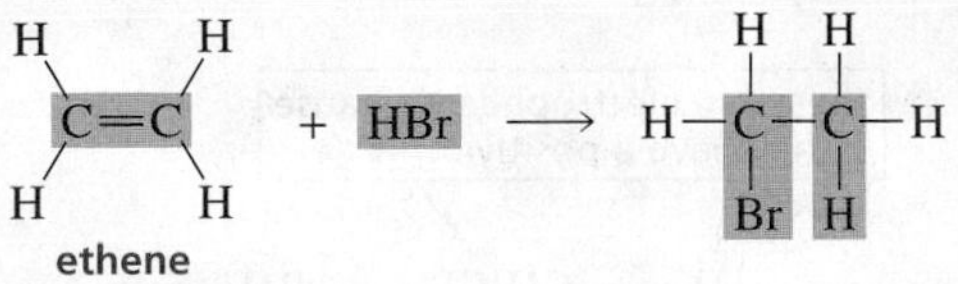

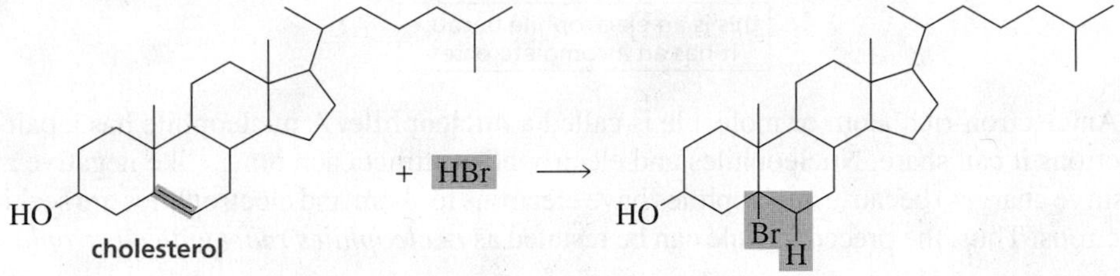

What makes learning organic chemistry even easier is that all the families of organic compounds can be placed in one of four groups, and all the families in a group react in similar ways. We will start our study of reactions by looking at the reactions of alkenes, a family that belongs to the first of the four groups.

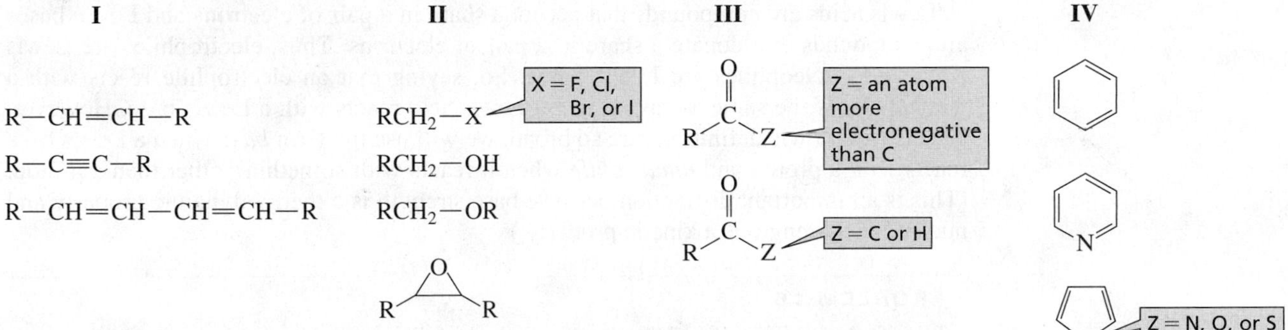

6 HOW ALKENES REACT • CURVED ARROWS SHOW THE FLOW OF ELECTRONS

When you study the reactions of a particular functional group, you need to understand *why* the functional group reacts the way it does. It is not enough to look at the two reactions shown in Section 5 and see that the carbon–carbon double bond reacts with HBr to form a product in which the H and Br atoms have taken the place of the π bond. You need to understand why the reaction occurs. If you understand the reason for each functional group's reactivity, you will reach the point where you can look at an organic compound and be able to predict the kind of reactions it will undergo.

In essence, organic chemistry is all about the interaction between electron-rich atoms or molecules and electron-deficient atoms or molecules. These are the forces that make chemical reactions happen. So each time you encounter a new functional group, remember that the reactions it undergoes can be explained by a very simple rule:

> *Electron-deficient atoms or molecules are attracted to electron-rich atoms or molecules.*

Therefore, to understand how a functional group reacts, you must first learn to recognize electron-deficient and electron-rich atoms and molecules.

An electron-deficient atom or molecule is called an **electrophile.** Literally, "electrophile" means "electron loving" (*phile* is the Greek suffix for "loving"). An electrophile looks for a pair of electrons. It is easy to recognize an electrophile—it has either a positive charge or an incomplete octet that can accept electrons.

Electron-deficient atoms or molecules are attracted to electron-rich atoms or molecules.

these are electrophiles because they have a positive charge

$$H^+ \quad CH_3\overset{+}{C}H_2 \quad BH_3$$

this is an electrophile because it has an incomplete octet

An electron-rich atom or molecule is called a **nucleophile.** A nucleophile has a pair of electrons it can share. Nucleophiles and electrophiles attract each other (like negative and positive charges) because nucleophiles have electrons to share and electrophiles are seeking electrons. Thus, the preceding rule can be restated as *nucleophiles react with electrophiles.*

A nucleophile reacts with an electrophile.

$$H\ddot{O}:^- \quad :\ddot{C}l:^- \quad CH_3\ddot{N}H_2 \quad H_2\ddot{O}:$$

these are nucleophiles because they have a pair of electrons to share

Lewis acids are compounds that accept a share in a pair of electrons and Lewis bases are compounds that donate a share in a pair of electrons. Thus, electrophiles are Lewis acids and nucleophiles are Lewis bases. So, saying that an electrophile reacts with a nucleophile is the same as saying that a Lewis acid reacts with a Lewis base. However, because the Lewis definitions are so broad, we will use the term *base* when a Lewis base reacts with a proton and *nucleophile* when it reacts with something other than a proton. (This is an important distinction because base strength is a thermodynamic property and nucleophile strength is a kinetic property.)

PROBLEM 16

Identify the nucleophile and the electrophile in the following acid–base reactions:

a. $AlCl_3 + NH_3 \rightleftharpoons Cl_3\overset{-}{Al}-\overset{+}{N}H_3$

b. $H-Br + HO^- \rightleftharpoons Br^- + H_2O$

Let's now see how the rule "nucleophiles react with electrophiles" allows us to predict the characteristic reaction of an alkene. The π bond of an alkene consists of a cloud of electrons above and below the σ bond. As a result of this cloud of electrons, an alkene is an electron-rich molecule—it is a nucleophile. (Notice the relatively electron-rich pale orange area in the electrostatic potential maps for *cis-* and *trans-*2-butene earlier in the chapter.) Also, a π bond is weaker than a σ bond. The π bond, therefore, is the bond that is most easily broken when an alkene undergoes a reaction. For these reasons, we can predict that an alkene will react with an electrophile and, in the process, the π bond will break.

Thus, if a reagent such as hydrogen bromide is added to an alkene, the alkene (a nucleophile) will react with the partially positively charged hydrogen (an electrophile) of hydrogen bromide; the product of the reaction will be a carbocation. In the second step of the reaction, the positively charged carbocation (an electrophile) will react with the negatively charged bromide ion (a nucleophile) to form an alkyl halide.

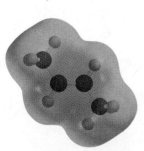

***trans*-2-butene**

***cis*-2-butene**

electrophile   nucleophile

$$CH_3CH=CHCH_3 + \overset{\delta+}{H}-\overset{\delta-}{Br} \longrightarrow CH_3\overset{+}{CH}-CHCH_3 + Br^- \longrightarrow CH_3CH-CHCH_3$$

nucleophile   electrophile   a carbocation   2-bromobutane an alkyl halide

The step-by-step description of the process by which reactants (in this case, alkene + HBr) are changed into products (an alkyl halide) is called the **mechanism of the reaction**. To help us understand a mechanism, curved arrows are drawn to show how the electrons move as new covalent bonds are formed and existing covalent bonds are broken. Each arrow represents the simultaneous movement of two electrons (an electron pair) from an electron-rich center (at the tail of the arrow) toward an electron-deficient center (at the point of the arrow). In this way, the arrows show which bonds are formed and which bonds are broken.

For the reaction of 2-butene with HBr, an arrow is drawn to show that the two electrons of the π bond of the alkene are attracted to the partially positively charged hydrogen of HBr. The hydrogen is not immediately free to accept this pair of electrons because it is already bonded to a bromine, and hydrogen can be bonded to only one atom at a time. However, as the π electrons of the alkene move toward the hydrogen, the H—Br bond breaks, with bromine keeping the bonding electrons. Notice that the π electrons are pulled away from one sp^2 carbon, but remain attached to the other. Thus, the two electrons that originally formed the π bond now form a new σ bond between carbon and the hydrogen from HBr. The product is positively charged, because the sp^2 carbon that did not form the new bond with hydrogen has lost a share in an electron pair (the electrons of the π bond).

> The mechanism of a reaction describes the step-by-step process by which reactants are changed into products.

> Curved arrows show the movement of the electrons; they are drawn from an electron-rich center to an electron-deficient center. (They are never drawn the other way around.)

> An arrowhead with two barbs signifies the movement of two electrons.

In the second step of the reaction, a lone pair on the negatively charged bromide ion forms a bond with the positively charged carbon of the carbocation. Notice that in both steps of the reaction, *a nucleophile reacts with an electrophile.*

> A curved arrow indicates where the electrons start from and where they end up.

Solely from the knowledge that a nucleophile reacts with an electrophile and a π bond is the weakest bond in an alkene, we have been able to predict that the product of the reaction of 2-butene and HBr is 2-bromobutane. The overall reaction involves the addition of 1 mole of HBr to 1 mole of the alkene. The reaction, therefore, is called an **addition reaction**. Because the first step of the reaction is the addition of an electrophile (H^+) to the alkene, the reaction is more precisely called an **electrophilic addition reaction**.

Electrophilic addition reactions are the characteristic reactions of alkenes.

At this point, you may think it would be easier just to memorize the fact that 2-bromobutane is the product of the reaction, without trying to understand the mechanism that explains why 2-bromobutane is the product. However, you will soon be encountering a great many reactions, and you will not be able to memorize them all. *It will be a lot easier to learn a few mechanisms that are based on similar rules than to try to memorize thousands of reactions.* And if you understand the mechanism of each reaction, the unifying principles of organic chemistry will soon be clear to you, making mastery of the material much easier and a lot more fun.

> Alkenes undergo electrophilic addition reactions.

PROBLEM 17♦

Which of the following are electrophiles, and which are nucleophiles?

$$H^- \qquad CH_3O^- \qquad CH_3C\equiv CH \qquad CH_3\overset{+}{C}HCH_3 \qquad NH_3$$

A Few Words About Curved Arrows

1. An arrow is used to show both the bond that forms and the bond that breaks. Draw the arrows so that they point in the direction of the electron flow; the arrows should never go against the flow. This means that *an arrow will point away from a negatively charged atom or toward a positively charged atom.*

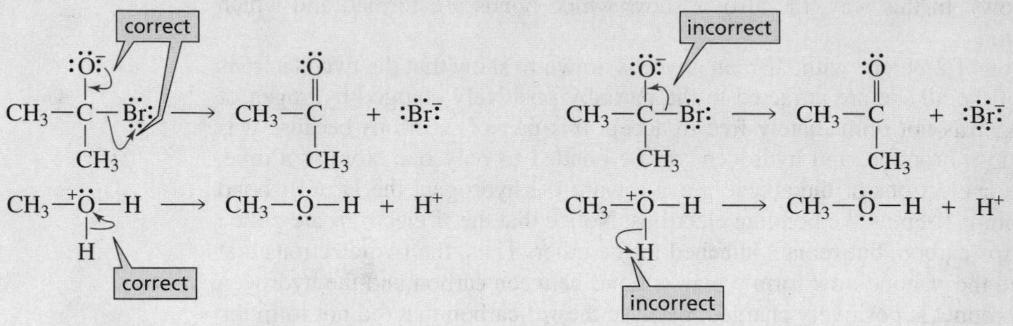

2. Curved arrows are meant to indicate the movement of electrons. *Never use a curved arrow to indicate the movement of an atom.* For example, do not use an arrow as a lasso to remove a proton, as shown in the equation on the right:

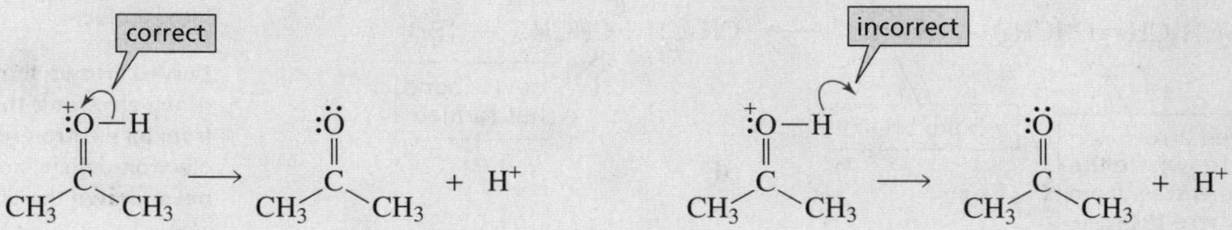

3. *The head of a curved arrow always points at an atom or at a bond.* Never draw the head of the arrow pointing out into space.

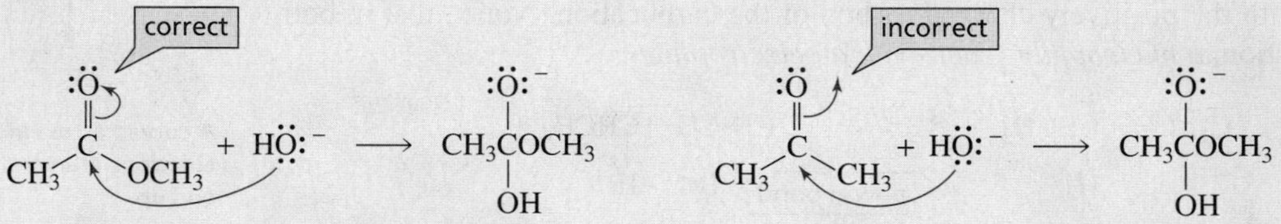

4. *A curved arrow starts at an electron source; it does not start at an atom.* In the following example, the arrow starts at the electron-rich π bond, not at a carbon atom:

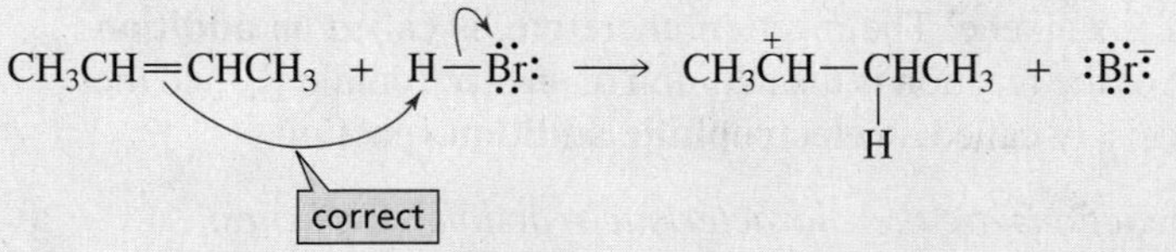

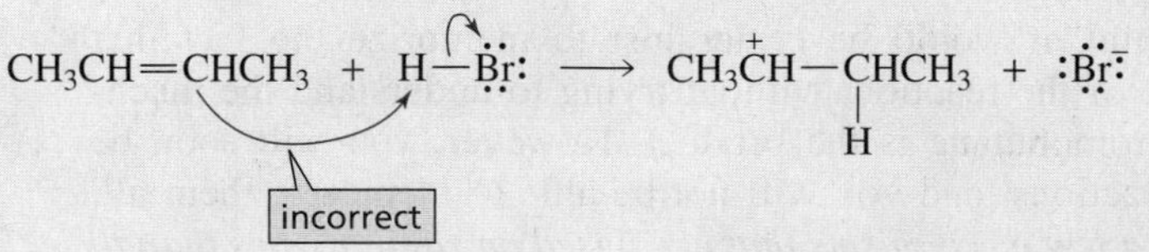

PROBLEM 18

Draw the consequence of following the incorrect arrows in Part 1 of the box on this page, "A Few Words About Curved Arrows." What is wrong with the structures that you obtain?

PROBLEM 19 Solved

Use curved arrows to show the movement of electrons in each of the following reaction steps. (*Hint:* Look at the reactants and look at the products, and then draw the arrows to convert the reactants into products.)

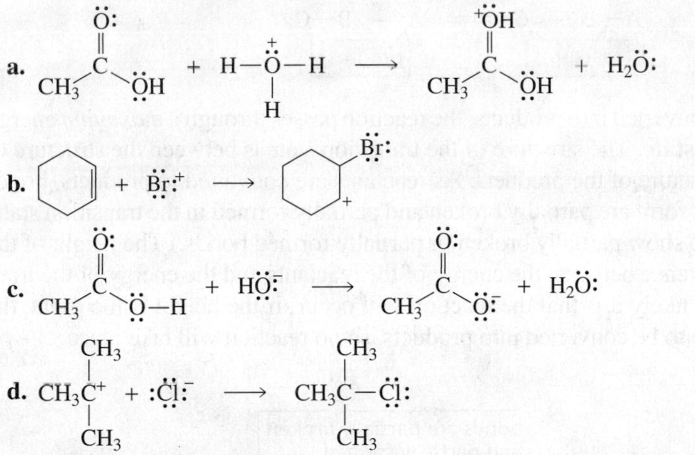

a. CH₃C(=O)OH + H–O–H → CH₃C(=O⁺H)OH + H₂Ö:

b. (cyclohexene) + Br:⁺ → (cyclohexyl-Br)⁺

c. CH₃C(=O)O–H + HÖ:⁻ → CH₃C(=O)O:⁻ + H₂Ö:

d. CH₃C⁺(CH₃) + :Cl:⁻ → CH₃C(CH₃)–Cl:

NOTE TO THE STUDENT

It is critically important that you learn how to draw curved arrows. Be sure to do the tutorial later in the chapter. It should take no more than 15 minutes, yet it can make an enormous difference to your success in this course.

Solution to 19a The double-bonded oxygen gains a proton; H₃O⁺ loses a proton with oxygen retaining the electrons it shared with the proton. Notice that the oxygen that gained a proton became positively charged, and the oxygen that lost a proton is no longer charged.

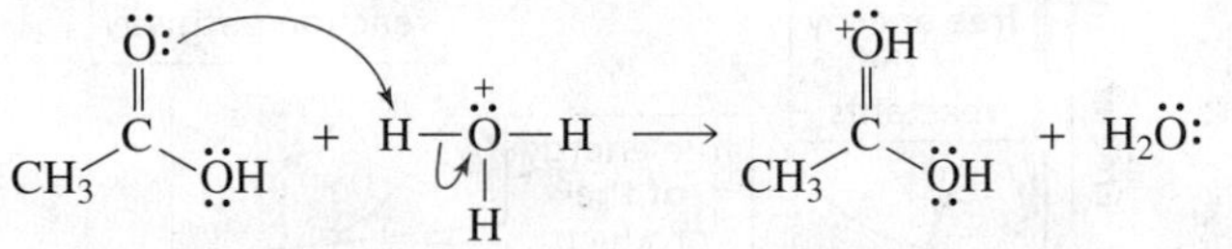

PROBLEM 20

For each of the reactions in Problem 19, indicate which reactant is the nucleophile and which is the electrophile.

7 THERMODYNAMICS AND KINETICS

To understand the energy changes that take place in a reaction such as the addition of HBr to an alkene, you need to understand some of the basic concepts of *thermodynamics,* which describes a reaction at equilibrium, and *kinetics,* which explains the rates of chemical reactions.

Consider a reaction in which Y is converted to Z: the *thermodynamics* of the reaction tells us the relative amounts of reactants (Y) and products (Z) present when the reaction has reached equilibrium, whereas the *kinetics* of the reaction tells us how fast Y is converted into Z.

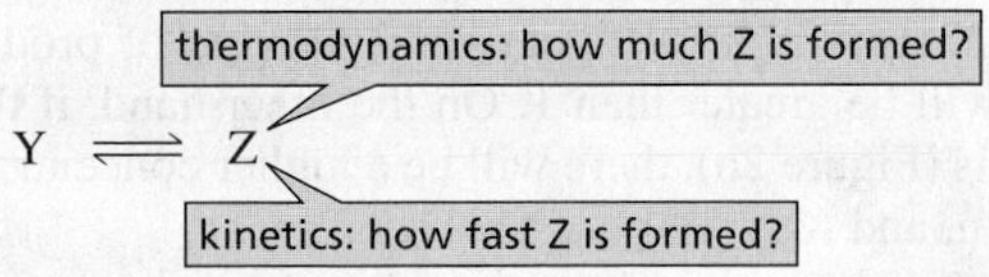

thermodynamics: how much Z is formed?

Y ⇌ Z

kinetics: how fast Z is formed?

A Reaction Coordinate Diagram Describes the Reaction Pathway

The mechanism of a reaction, as we have shown, describes the steps known to occur as reactants are converted into products. A **reaction coordinate diagram** shows the energy changes that take place in each of these steps.

In a reaction coordinate diagram, the total energy of all species is plotted against the progress of the reaction. A reaction progresses from left to right as written in a chemical equation, so the energy of the reactants is plotted on the left-hand side of the *x*-axis, and the energy of the products is plotted on the right-hand side. A typical reaction coordinate diagram is shown in Figure 1. It describes the reaction of A—B with C to form A and B—C. Remember that *the more stable the species, the lower its energy.*

$$A - B \ + \ C \ \rightleftharpoons \ A \ + \ B - C$$

reactants   products

As the reactants are converted into products, the reaction passes through a *maximum* energy state called a **transition state.** The structure of the transition state is between the structure of the reactants and the structure of the products. As reactants are converted to products, bonds that break and bonds that form are partially broken and partially formed in the transition state. (Dashed lines are used to show partially broken or partially formed bonds.) The height of the transition state (the difference between the energy of the reactants and the energy of the transition state) tells us how likely it is that the reaction will occur; if the height is too great, the reactants will not be able to be converted into products, so no reaction will take place.

The more stable the species, the lower its energy.

▶ **Figure 1**

A reaction coordinate diagram, which shows the energy changes that take place as the reaction progresses from reactants to products. The dashed lines in the transition state indicate bonds that are partially formed or partially broken.

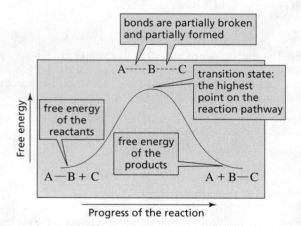

Thermodynamics: How Much Product Is Formed?

Thermodynamics is the field of chemistry that describes the properties of a system at equilibrium. The relative concentrations of reactants and products at equilibrium can be expressed by an equilibrium constant, K_{eq}.

$$m\,A \ + \ n\,B \ \rightleftharpoons \ s\,C \ + \ t\,D$$

$$K_{eq} = \frac{[\textbf{products}]}{[\textbf{reactants}]} = \frac{[\textbf{C}]^s\,[\textbf{D}]^t}{[\textbf{A}]^m\,[\textbf{B}]^n}$$

The relative concentrations of products and reactants at equilibrium depend on their relative stabilities: *the more stable the compound, the greater its concentration at equilibrium.* Thus, if the products are more stable (have a lower free energy) than the reactants (Figure 2a), there will be a higher concentration of products than reactants at equilibrium, and K_{eq} will be greater than 1. On the other hand, if the reactants are more stable than the products (Figure 2b), there will be a higher concentration of reactants than products at equilibrium, and K_{eq} will be less than 1.

Now you can understand why the strength of an acid is determined by the stability of its conjugate base—as the base becomes more stable, the equilibrium constant (K_a) for its formation becomes larger, and the larger the K_a, the stronger the acid.

The difference between the free energy of the products and the free energy of the reactants under standard conditions is called the **Gibbs free-energy change,** or $\Delta G°$. The symbol ° indicates standard conditions, which means that all species are at a concentration of 1 M, a temperature of 25 °C, and a pressure of 1 atm.

$$\Delta G° = (\textbf{free energy of the products}) - (\textbf{free energy of the reactants})$$

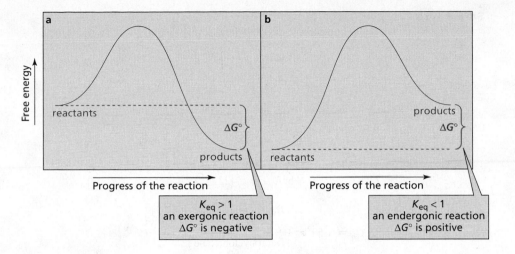

◀ **Figure 2**

Reaction coordinate diagrams for

(a) a reaction in which the products are more stable than the reactants (an exergonic reaction) and

(b) a reaction in which the products are less stable than the reactants (an endergonic reaction).

From this equation, we can see that $\Delta G°$ will be negative if the products have a lower free energy (are more stable) than the reactants. In other words, the reaction will release more energy than it consumes; such a reaction is called an **exergonic reaction** (Figure 2a).

If the products have a higher free energy (are less stable) than the reactants, $\Delta G°$ will be positive, and the reaction will consume more energy than it releases; such a reaction is called an **endergonic reaction** (Figure 2b).

(Notice that the terms *exergonic* and *endergonic* refer to whether the reaction has a negative $\Delta G°$ or a positive $\Delta G°$. Do not confuse these terms with *exothermic* and *endothermic,* which we will define later in this section.)

A successful reaction is one in which the products are favored at equilibrium (that is, the products are more stable than the reactants). Whether reactants or products are favored at equilibrium can be indicated by the equilibrium constant (K_{eq}) or by the change in free energy ($\Delta G°$). These two quantities are related by the equation

$$\Delta G° = -RT \ln K_{eq}$$

where R is the gas constant (1.986×10^{-3} kcal mol^{-1} K^{-1} or 8.341×10^{-3} kJ mol^{-1} K^{-1})* and T is the temperature in kelvins. (The Kelvin scale avoids negative temperatures by assigning 0 K to -273 °C, the lowest temperature known. Thus, because K = °C + 273, 25 °C = 298 K.)

A small change in $\Delta G°$ corresponds to a large change in K_{eq} and, therefore, a large change in the amount of product obtained at equilibrium (see Problem 44).

The more stable the compound, the greater its concentration at equilibrium.

When products are favored at equilibrium, $\Delta G°$ is negative and K_{eq} is greater than 1.

When reactants are favored at equilibrium, $\Delta G°$ is positive and K_{eq} is less than 1.

PROBLEM 21♦

a. Which of the monosubstituted cyclohexanes in Table 9 has a negative $\Delta G°$ for the conversion of an axial-substituted chair conformer to an equatorial-substituted chair conformer?

b. Which monosubstituted cyclohexane has the most negative $\Delta G°$?

c. Which monosubstituted cyclohexane has the greatest preference for an equatorial position?

d. Calculate $\Delta G°$ for the conversion of "axial" methylcyclohexane to "equatorial" methylcyclohexane at 25 °C.

PROBLEM 22 Solved

a. The $\Delta G°$ for conversion of "axial" fluorocyclohexane to "equatorial" fluorocyclohexane at 25 °C is -0.25 kcal/mol. Calculate the percentage of fluorocyclohexane molecules that have the fluoro substituent in an equatorial position at equilibrium.

b. Do the same calculation for isopropylcyclohexane (its $\Delta G°$ value at 25 °C is -2.1 kcal/mol).

c. Why is the percentage of molecules with the substituent in an equatorial position greater for isopropylcyclohexane?

---

*1 kcal = 4.184 kJ

William Thomson (1824–1907) was born in Belfast, Northern Ireland. He was a professor of natural philosophy at the University of Glasgow, Scotland. For developing the Kelvin scale of absolute temperature and other important work in mathematical physics, he was given the title Baron Kelvin, which allowed him to be called **Lord Kelvin.** The name comes from the river Kelvin that flows by the University of Glasgow. His statue is in the botanic gardens that are adjacent to The Queen's University of Belfast.

If an equilibrium is disturbed, the system will adjust to offset the disturbance.

Solution to 22a

$$\text{fluorocyclohexane} \rightleftharpoons \text{fluorocyclohexane}$$
$$\textbf{axial} \qquad\qquad \textbf{equatorial}$$

$$\Delta G° = -0.25 \text{ kcal/mol at } 25\ °C$$

$$\Delta G° = -RT \ln K_{eq}$$

$$-0.25\ \frac{\text{kcal}}{\text{mol}} = -1.986 \times 10^{-3}\ \frac{\text{kcal}}{\text{mol K}} \times 298\ K \times \ln K_{eq}$$

$$\ln K_{eq} = 0.422$$

$$K_{eq} = 1.53 = \frac{\left[\begin{array}{c}\text{fluorocyclohexane}\\ \textbf{equatorial}\end{array}\right]}{\left[\begin{array}{c}\text{fluorocyclohexane}\\ \textbf{axial}\end{array}\right]} = \frac{1.53}{1}$$

Now we must determine the percentage of the total that is equatorial, as follows:

$$\frac{\left[\begin{array}{c}\text{fluorocyclohexane}\\ \textbf{equatorial}\end{array}\right]}{\left[\begin{array}{c}\text{fluorocyclohexane}\\ \textbf{equatorial}\end{array}\right] + \left[\begin{array}{c}\text{fluorocyclohexane}\\ \textbf{axial}\end{array}\right]} = \frac{1.53}{1.53 + 1} = \frac{1.53}{2.53} = 0.60 \text{ or } 60\%$$

Fortunately, there are ways to increase the amount of product formed in a reaction. **Le Châtelier's principle** states that *if an equilibrium is disturbed, then the system will adjust to offset the disturbance.* In other words, if the concentration of C or D is decreased, then A and B will react to form more C and D in order to maintain the value of the equilibrium constant. (The value of a constant must be maintained—that is why it is called a *constant.*)

$$A + B \rightleftharpoons C + D$$

$$K_{eq} = \frac{[C][D]}{[A][B]}$$

Thus, if a product crystallizes out of solution as it is formed, or if it can be distilled off as a liquid or driven off as a gas, the reactants will continue to react to replace the departing product in order to maintain the relative concentrations of products and reactants (that is, to maintain the value of the equilibrium constant). More products can also be formed if the equilibrium is disturbed by increasing the concentration of one or more of the reactants.

The Gibbs standard free-energy change ($G°$) has an enthalpy ($H°$) component and an entropy ($\Delta S°$) component (T is the temperature in kelvins):

$$\Delta G° = \Delta H° - T\Delta S°$$

The **enthalpy** term ($\Delta H°$) is the heat given off or the heat consumed during the course of a reaction. Heat is given off when bonds are formed, and heat is consumed when bonds are broken. Thus, $\Delta H°$ is a measure of the energy of the bond-making and bond-breaking processes that occur as reactants are converted into products.

$$\Delta H° = \left(\text{heat required to break bonds}\right) - \left(\text{heat released from breaking bonds}\right)$$

If the bonds that are formed in a reaction are stronger than the bonds that are broken, more energy will be released in the bond-forming process than will be consumed in the bond-breaking process, and $\Delta H°$ will be negative. A reaction with a negative $\Delta H°$ is called an **exothermic reaction.** If the bonds that are formed are weaker than those that are broken, $\Delta H°$ will be positive. A reaction with a positive $\Delta H°$ is called an **endothermic reaction.**

Entropy ($\Delta S°$) is a measure of the freedom of motion in a system. Restricting the freedom of motion of a molecule decreases its entropy. For example, in a reaction in which two molecules come together to form a single molecule, the entropy of the product will be less than the entropy of the reactants because two separate molecules can move in ways that are not possible when they are bound together in a single molecule. In such a reaction, $\Delta S°$ will be negative. In a reaction in which a single molecule is cleaved into two separate molecules, the products will have greater freedom of motion than the reactant, and $\Delta S°$ will be positive.

$$\Delta S° = \left(\text{freedom of motion of the products}\right) - \left(\text{freedom of motion of the reactants}\right)$$

Entropy is a measure of the freedom of motion in a system.

PROBLEM 23

a. For which reaction in each set will $\Delta S°$ be more significant?
b. For which reaction will $\Delta S°$ be positive?
  1. $A \rightleftharpoons B$ or $A + B \rightleftharpoons C$
  2. $A + B \rightleftharpoons C$ or $A + B \rightleftharpoons C + D$

A reaction with a negative $\Delta G°$ has a favorable equilibrium constant ($K_{eq} > 1$); that is, the reaction is favored as written from left to right because the products are more stable than the reactants. If you examine the expression for the Gibbs standard free-energy change, you will find that negative values of $\Delta H°$ and positive values of $\Delta S°$ contribute to make $\Delta G°$ negative. In other words, *the formation of products with stronger bonds and greater freedom of motion causes $\Delta G°$ to be negative*. Notice that the entropy term is temperature dependent and, therefore, becomes more important as the temperature increases. As a result, a reaction with a positive $\Delta S°$ may be endergonic at low temperatures, but exergonic at high temperatures.

The formation of products with stronger bonds and greater freedom of motion causes $\Delta G°$ to be negative.

PROBLEM 24♦

a. For a reaction with $\Delta H° = -12$ kcal/mol and $\Delta S° = 0.01$ kcal mol^{-1} K^{-1}, calculate the $\Delta G°$ and the equilibrium constant at: (**1.**) 30 °C and (**2.**) 150 °C.
b. How does $\Delta G°$ change as T increases?
c. How does K_{eq} change as T increases?

Values of $\Delta H°$ can be calculated from bond dissociation energies, as shown in the following example. The bond dissociation energy is indicated by the special term *DH*. (Note that the bond dissociation energy of the π bond of ethene is estimated to be 62 kcal/mol).

$$\underset{\text{bonds being broken}}{\underbrace{\begin{matrix} H \\ | \\ C=C \\ | \\ H \end{matrix} \quad \begin{matrix} H \\ | \\ \\ | \\ H \end{matrix}} + H-Br} \longrightarrow \underset{\text{bonds being formed}}{\underbrace{\begin{matrix} H & H \\ | & | \\ H-C-C-H \\ | & | \\ H & Br \end{matrix}}}$$

π bond of ethene	*DH*	=	62 kcal/mol	C—H	*DH* = 101 kcal/mol
H—Br	*DH*	=	88 kcal/mol	C—Br	*DH* = 71 kcal/mol
		DH_{total}	= 150 kcal/mol		DH_{total} = 172 kcal/mol

$\Delta H°$ for the reaction $= DH$ for bonds being broken $- DH$ for bonds being formed

$$= 150 \text{ kcal/mol} - 172 \text{ kcal/mol}$$

$$= -22 \text{ kcal/mol}$$

Because values of $\Delta H°$ are relatively easy to calculate, organic chemists frequently evaluate reactions in terms of that quantity alone, since $\Delta S°$ values cannot be calculated but must be experimentally determined. However, the entropy term can be ignored only if the reaction involves just a small change in entropy, because then the $T\Delta S°$ term will be small and the value of $\Delta H°$ will be very close to the value of $\Delta G°$. Remember, though, that many organic reactions occur with a significant change in entropy or occur at high temperatures and so have significant $T\Delta S°$ terms; ignoring the entropy term in such cases can lead to a wrong conclusion. It is permissible to use $\Delta H°$ values to *approximate* whether a reaction has a favorable equilibrium constant, but if a precise answer is needed, $\Delta G°$ values must be used.

Table 1 Experimental Bond Dissociation Enthalpies Y—Z → Y + Z					
Bond	*DH*		**Bond**	*DH*	
	(kcal/mol)	**(kJ/mol)**		**(kcal/mol)**	**(kJ/mol)**
CH_3—H	105.0	439	H—H	104.2	436
CH_3CH_2—H	101.1	423	F—F	37.7	158
			Cl—Cl	58.0	243
$(CH_3)_2CH$—H	98.6	413	Br—Br	53.5	224
$(CH_3)_3C$—H	96.5	404	I—I	51.0	214
			H—F	136.3	570
CH_3—CH_3	90.2	377	H—Cl	103.2	432
CH_3CH_2—CH_3	89.1	373	H—Br	87.5	366
$(CH_3)_2CH$—CH_3	88.6	371	H—I	71.3	298
$(CH_3)_3C$—CH_3	87.5	366			
			CH_3—F	114.8	480
H_2C=CH_2	174.5	730	CH_3—Cl	83.6	350
HC≡CH	230.4	964	CH_3CH_2—Cl	84.7	355
			$(CH_3)_3C$—Br	70.7	296
H_2C=CH—H	110.7	463	CH_3—I	57.1	239
HC≡C—H	133.3	558	CH_3CH_2—I	56.3	235
			$(CH_3)_2CH$—Cl	85.1	356
HO—H	118.8	497	$(CH_3)_3C$—Cl	84.8	355
CH_3O—H	104.2	436	CH_3—Br	70.3	294
CH_3—OH	92.1	386	CH_3CH_2—Br	70.5	295
			$(CH_3)_2CH$—Br	72.0	301

S. J. Blanksby and G. B. Ellison, *Acc. Chem. Res.,* **2003,** *36,* 255.

For example, because $\Delta H°$ for the addition of HBr to ethene is significantly negative (-22 kcal/mol), we can assume that $\Delta G°$ is also negative. However, if the value of $\Delta H°$ were close to zero, we could no longer assume that $\Delta H°$ and $\Delta G°$ have the same sign. (Keep in mind that two conditions must be met to justify using $\Delta H°$ values to predict $\Delta G°$ values. First, the entropy change must be small; second, the reaction must take place in the gas phase.)

> When $\Delta G°$ values are used to construct reaction coordinate diagrams,
> the y-axis represents free energy; when $\Delta H°$ values are used,
> the y-axis represents potential energy.

When reactions occur in solution, which is the case for the vast majority of organic reactions, the solvent molecules can solvate the reagents and the products. Solvation can have a large effect on both $\Delta H°$ and $\Delta S°$.

For example, in a reaction of a solvated polar reagent, the $\Delta H°$ for breaking the dipole–dipole interactions between the solvent and the reagent has to be taken into account, and in a reaction with solvated polar product, the $\Delta H°$ for forming the dipole–dipole interactions between the solvent and the product has to be taken into account. In addition, solvation can greatly reduce the freedom of motion of the molecules, thereby affecting $\Delta S°$.

PROBLEM 25

a. Use the bond dissociation energies in Table 1 to calculate the $\Delta H°$ value for the addition of HCl to ethene.
b. Calculate the $\Delta H°$ value for the addition of H_2 to ethene.
c. Are the reactions exothermic or endothermic?
d. Do you expect the reactions to be exergonic or endergonic?

Kinetics: How Fast Is the Product Formed?

Knowing that a reaction is exergonic will not tell you how fast the reaction occurs, because $\Delta G°$ describes only the difference between the stability of the reactants and the stability of the products. It does not indicate anything about the energy barrier of the reaction, which is the energy "hill" that has to be climbed for the reactants to be converted into products. **Kinetics** is the field of chemistry that studies the rates of chemical reactions and the factors that affect those rates.

The energy barrier of a reaction (indicated in Figure 3 by $\Delta G^{\ddagger}$) is called the **free energy of activation.** It is the difference between the free energy of the transition state and the free energy of the reactants:

$$\Delta G^{\ddagger} = (\textbf{free energy of the transition state}) - (\textbf{free energy of the reactants})$$

As $\Delta G^{\ddagger}$ decreases, the rate of the reaction increases. Thus, *anything that makes the reactants less stable or makes the transition state more stable will make the reaction go faster.*

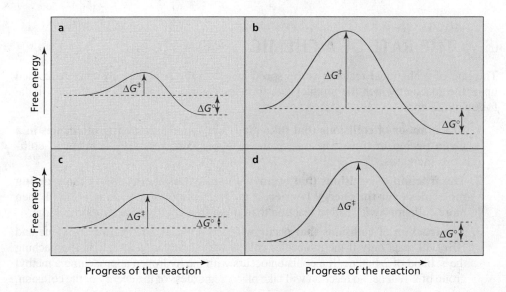

The rate of the reaction *increases* as the height of the energy barrier *decreases*.

◀ **Figure 3**
Reaction coordinate diagrams (drawn on the same scale) for
(a) a fast exergonic reaction and
(b) a slow exergonic reaction.
(c) a fast endergonic reaction and
(d) a slow endergonic reaction.

Like $\Delta G°$, $\Delta G^{\ddagger}$ has both an enthalpy component and an entropy component. Notice that any quantity that refers to the transition state is represented by a double-dagger superscript ($\ddagger$).

$$\Delta G^{\ddagger} = \Delta H^{\ddagger} - T\Delta S^{\ddagger}$$
$$\Delta H^{\ddagger} = (\textbf{enthalpy of the transition state}) - (\textbf{enthalpy of the reactants})$$
$$\Delta S^{\ddagger} = (\textbf{entropy of the transition state}) - (\textbf{entropy of the reactants})$$

Some exergonic reactions have small free energies of activation and therefore can take place at room temperature (Figure 3a). In contrast, some exergonic reactions have free energies of activation that are so large that the reaction cannot take place unless energy is supplied in addition to that provided by the existing thermal conditions (Figure 3b). Endergonic reactions can also have either small free energies of activation, as in Figure 3c, or large free energies of activation, as in Figure 3d.

Let's now look at the difference between thermodynamic stability and kinetic stability.

Thermodynamic stability is indicated by $\Delta G°$. If $\Delta G°$ is negative, then the product is *thermodynamically stable* compared with the reactant; if $\Delta G°$ is positive, then the product is *thermodynamically unstable* compared with the reactant.

Kinetic stability is indicated by $\Delta G^{\ddagger}$. If $\Delta G^{\ddagger}$ is large, then the reactant is *kinetically stable* because it reacts slowly. If $\Delta G^{\ddagger}$ is small, then the reactant is *kinetically unstable*—it reacts rapidly. Similarly, if $\Delta G^{\ddagger}$ for the reverse reaction is large, then the product is kinetically stable, but if it is small, then the product is kinetically unstable.

Generally, when chemists use the term *stability,* they are referring to thermodynamic stability.

PROBLEM 26♦

a. Which of the reactions in Figure 3 has a product that is thermodynamically stable compared with the reactant?
b. Which of the reactions in Figure 3 has the most kinetically stable product?
c. Which of the reactions in Figure 3 has the least kinetically stable product?

PROBLEM 27

Draw a reaction coordinate diagram for a reaction in which

a. the product is thermodynamically unstable and kinetically unstable.
b. the product is thermodynamically unstable and kinetically stable.

8 THE RATE OF A CHEMICAL REACTION

The rate of a chemical reaction is the speed at which the reacting substances are used up or the speed at which the products are formed. The rate of a reaction depends on the following factors:

1. **The number of collisions that take place between the reacting molecules in a given period of time.** The rate of the reaction increases as the number of collisions increases.

2. **The fraction of collisions that occur with sufficient energy to get the reacting molecules over the energy barrier.** If the free energy of activation is small, then more collisions will lead to reaction than if the free energy of activation is large.

3. **The fraction of collisions that occur with the proper orientation.** 2-Butene and HBr will react only if the molecules collide with the hydrogen of HBr approaching the π bond of 2-butene. If a collision occurs with the hydrogen approaching a methyl group of 2-butene, no reaction will take place, regardless of the energy of the collision.

$$\text{rate of a reaction} = \left(\begin{array}{c}\text{number of collisions}\\\text{per unit of time}\end{array}\right) \times \left(\begin{array}{c}\text{fraction with}\\\text{sufficient energy}\end{array}\right) \times \left(\begin{array}{c}\text{fraction with}\\\text{proper orientation}\end{array}\right)$$

- Increasing the concentration of the reactants increases the rate of a reaction because it increases the number of collisions that occur in a given period of time.

- Increasing the temperature at which the reaction is carried out also increases the rate of a reaction because it increases the kinetic energy of the molecules, which increases both the frequency of collisions (molecules that are moving faster collide more frequently) and the number of collisions that have sufficient energy to get the reacting molecules over the energy barrier (Figure 4).

- The rate of a reaction can also be increased by a catalyst (Section 11).

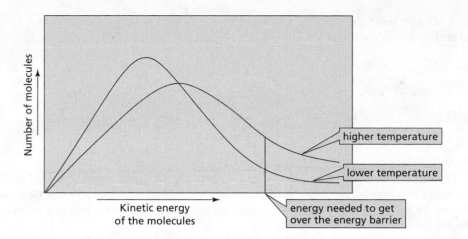

◀ **Figure 4**

Boltzmann distribution curves *(the number of molecules as a function of energy)* at two different temperatures. The curve shows the distribution of molecules with a particular kinetic energy. The energy of most molecules cluster about an average, but there are some with much lower and some with much higher energy. At a higher temperature (the red line), there will be more molecules with sufficient energy to get over the energy barrier.

9 THE DIFFERENCE BETWEEN THE RATE OF A REACTION AND THE RATE CONSTANT FOR A REACTION

For a reaction in which reactant A is converted into product B, *the rate of the reaction* is proportional to the concentration of A. If the concentration of A is doubled, the rate of the reaction will double; if the concentration of A is tripled, the rate of the reaction will triple; and so on. Because the rate of this reaction is proportional to the concentration of only *one* reactant, it is called a **first-order reaction.**

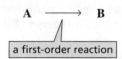

a first-order reaction

When you know the relationship between the rate of a reaction and the concentration of the reactants, you can write a **rate law** for the reaction.

$$\text{rate} \propto [A]$$

We can replace the proportionality symbol ($\propto$) with an equal sign if we use a proportionality constant k, which is called a **rate constant.** The rate constant of a first-order reaction is called a **first-order rate constant.**

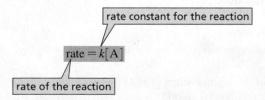

rate constant for the reaction

$$\text{rate} = k[A]$$

rate of the reaction

A reaction whose rate depends on the concentrations of *two* reactants, A and B, is called a **second-order reaction.** If the concentration of either A or B is doubled, then the rate of the reaction will double; if the concentrations of both A and B are doubled, then the rate of the reaction will quadruple; and so on.

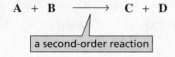

a second-order reaction

Ludwig Eduard Boltzmann (1844–1906) was born in Vienna, Austria. He received a Ph.D. from the University of Vienna in 1866; his dissertation was on the kinetic theory of gases. Boltzmann started his career as a professor of mathematical physics at the University of Graz. In 1872 he met Henriette van Aigentier, an aspiring teacher of mathematics and physics, who had not been allowed to audit lectures at the university because they were not open to women. Boltzmann encouraged her to appeal, which she did with success. In 1873 Boltzmann returned to the University of Vienna as a professor of mathematics. He and Henriette were married in 1876 and had five children. He went back to the University of Graz in 1887 to become its president. Boltzmann was subject to alternating moods of high elation and severe depression, which he attributed to having been born during the night between Mardi Gras and Ash Wednesday.

Now we can write the rate law for the reaction. In this case, the rate constant k is a **second-order rate constant.**

| rate constant for the reaction |

$$\text{rate} = k[\text{A}][\text{B}]$$

| rate of the reaction |

Do not confuse the *rate constant* for a reaction (k) with the *rate* of a reaction.

The *rate constant* tells us how easy it is to reach the transition state (how easy it is to get over the energy barrier). Low-energy barriers are associated with large rate constants (Figure 3a and c), whereas high-energy barriers are associated with small rate constants (Figure 3b and d).

The *rate* of a reaction is a measure of the amount of product that is formed per unit of time. The preceding equations show that the *rate is the product of the rate constant and the reactant concentration(s), so reaction rates depend on concentration, whereas rate constants are independent of concentration.* Therefore, when we compare two reactions to see which one occurs more readily, we must compare their rate constants and not their concentration-dependent rates of reaction. (How rate constants are determined is explained in Appendix titled "Derivations of Rate Laws.")

Although rate constants are independent of concentration, they are dependent on temperature. The **Arrhenius equation** relates the rate constant of a reaction to the experimental energy of activation and to the temperature at which the reaction is carried out:

$$k = Ae^{-E_a/RT}$$

In this expression, k is the rate constant, A is the frequency factor (which represents the fraction of collisions that occurs with the proper orientation for reaction) and $e^{-E_a/RT}$ is the fraction of collisions with the minimum energy (E_a) needed for reaction. (R is the gas constant, T is the temperature in kelvins, and E_a is the experimental energy of activation, which is an approximate value of the activation energy; see "The Difference Between $\Delta G^{\ddagger}$ and E_a."

A good rule of thumb is that an increase of 10 °C in temperature will double the rate constant for a reaction and, therefore, double the rate of the reaction.

Taking the logarithm of both sides of the Arrhenius equation, we obtain

$$\ln k = \ln A - \frac{E_a}{RT}$$

Problem 59 shows how this form of the equation is used to calculate values of E_a, $\Delta G^{\ddagger}$, $\Delta H^{\ddagger}$, and $\Delta S^{\ddagger}$ for a reaction.

PROBLEM 28 Solved

At 30 °C, the second-order rate constant for the reaction of methyl chloride and HO^- is $1.0 \times 10^{-5}\,M^{-1}s^{-1}$.

a. What is the rate of the reaction when $[CH_3Cl] = 0.10\,M$ and $[HO^-] = 0.10\,M$?
b. If the concentration of methyl chloride is decreased to 0.010 M, what will be the effect on

    1. the *rate* of the reaction?   **2.** the *rate constant* for the reaction?

Solution to 28a The rate of the reaction is given by

$$\text{rate} = k[\text{methyl chloride}][HO^-]$$

Substituting the given rate constant and reactant concentrations yields

$$\text{rate} = 1.0 \times 10^{-5}\,M^{-1}s^{-1}[0.10\,M][0.10\,M] = 1.0 \times 10^{-7}\,M s^{-1}$$

The Difference Between $\Delta G^{\ddagger}$ and E_a

The difference between the *free energy of activation* ($\Delta G^{\ddagger}$) and the *experimental energy of activation* (E_a) in the Arrhenius equation, is the entropy component. The free energy of activation has both an enthalpy component and an entropy component ($\Delta G^{\ddagger} = \Delta H^{\ddagger} - T\Delta S^{\ddagger}$), whereas the experimental energy of activation has only an enthalpy component ($E_a = \Delta H^{\ddagger} + RT$), since the entropy component is implicit in the A term of the Arrhenius equation. Therefore, E_a gives only an approximate energy barrier to a reaction. The true energy barrier is given by $\Delta G^{\ddagger}$, because the energy barriers to most reactions depend on changes in both enthalpy and entropy.

PROBLEM 29◆

The rate constant for a reaction can be increased by _____ the stability of the reactant or by _____ the stability of the transition state.

PROBLEM 30◆

From the Arrhenius equation, predict how

a. increasing the experimental activation energy will affect the rate constant for a reaction.
b. increasing the temperature will affect the rate constant for a reaction.

The next question to consider is, how are the rate constants for a reaction related to the equilibrium constant? At equilibrium, the rate of the forward reaction must be equal to the rate of the reverse reaction because the amounts of reactants and products are no longer changing:

$$A \underset{k_{-1}}{\overset{k_1}{\rightleftharpoons}} B$$

forward rate = reverse rate

$$k_1[A] = k_{-1}[B]$$

Therefore,

$$K_{eq} = \frac{k_1}{k_{-1}} = \frac{[B]}{[A]}$$

From this equation, we see that the equilibrium constant for a reaction can be determined from the relative rate constants for the forward and reverse reactions *or* from the relative concentrations of the products and reactants at equilibrium. For example, we can say that the reaction shown in Figure 2a has an equilibrium constant significantly greater than one because the rate constant for the forward reaction is greater than that for the reverse reaction *or* because the products are more stable than the reactants.

PROBLEM 31◆

a. Which reaction has a greater equilibrium constant, one with a rate constant of $1 \times 10^{-3} \sec^{-1}$ for the forward reaction and a rate constant of $1 \times 10^{-5} \sec^{-1}$ for the reverse reaction, or one with a rate constant of $1 \times 10^{-2} \sec^{-1}$ for the forward reaction and a rate constant of $1 \times 10^{-3} \sec^{-1}$ for the reverse reaction?

b. If both reactions start with a reactant concentration of 1.0 M, which reaction will form the most product in a given period of time?

NOTE TO THE STUDENT

For additional information see Appendix: "Derivations of Rate Laws."

10 A REACTION COORDINATE DIAGRAM DESCRIBES THE ENERGY CHANGES THAT TAKE PLACE DURING A REACTION

We have seen that the addition of HBr to 2-butene is a two-step process (Section 6). In each step, the reactants pass through a transition state as they are converted into products. The structure of the transition state for each of the steps is shown here in brackets.

‡ symbolizes the transition state

$$CH_3CH=CHCH_3 + HBr \longrightarrow \left[CH_3\overset{\delta+}{C}H\text{---}CHCH_3 \right]^{‡} \longrightarrow CH_3\overset{+}{C}HCH_2CH_3 + Br^-$$

H — partially formed bond

δ^-Br — partially broken bond

transition state

$$CH_3\overset{+}{C}HCH_2CH_3 + Br^- \longrightarrow \left[CH_3\overset{\delta+}{C}HCH_2CH_3 \right]^{‡} \longrightarrow CH_3CHCH_2CH_3$$

δ^-Br

Br

transition state

Notice that the bonds that break and the bonds that form during the course of the reaction are partially broken and partially formed in the transition state, as indicated by dashed lines. And atoms that either become charged or lose their charge during the course of the reaction are partially charged in the transition state. (Transition states are always shown in brackets with a double-dagger superscript.)

A reaction coordinate diagram can be drawn for each step of a reaction (Figure 5). In the first step of the addition reaction, the alkene is converted into a carbocation that is higher in energy (less stable) than the reactants. The first step, therefore, is endergonic ($\Delta G°$ is > 0). In the second step, the carbocation reacts with a nucleophile to form a product that is lower in energy (more stable) than the carbocation reactant. This step, therefore, is exergonic ($\Delta G°$ is < 0).

▶ **Figure 5**

Reaction coordinate diagrams for the two steps in the addition of HBr to 2-butene:

(a) the first step (formation of the carbocation)

(b) the second step (formation of the alkyl halide)

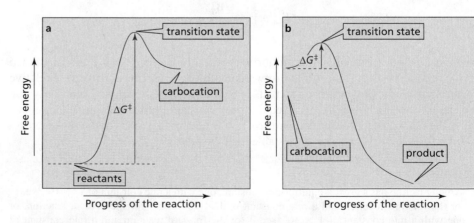

Because the products of the first step are the reactants for the second step, we can hook the two reaction coordinate diagrams together to obtain the reaction coordinate diagram that describes the pathway for the overall reaction (Figure 6). The $\Delta G°$ for the overall reaction is the difference between the free energy of the final products and the free energy of the initial reactants. Figure 6 shows that the overall reaction is exergonic ($\Delta G°$ is negative).

A chemical species that is a product of one step of a reaction and a reactant for the next step is called an **intermediate.** The carbocation intermediate formed in this reaction is too unstable to be isolated, but some reactions have more stable intermediates that can be isolated.

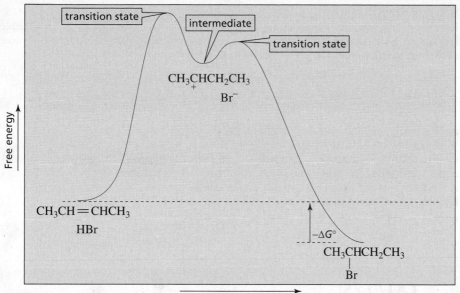

◀ **Figure 6**

Reaction coordinate diagram for the addition of HBr to 2-butene to form 2-bromobutane.

Transition states have partially formed bonds. Intermediates have fully formed bonds.

Transition states, in contrast, represent the highest-energy structures that are involved in the reaction. They exist only fleetingly and they can never be isolated (Figure 6).

Do not confuse transition states with intermediates. *Transition states have partially formed bonds, whereas intermediates have fully formed bonds.*

Figure 6 shows that the free energy of activation for the first step of the reaction is greater than the free energy of activation for the second step. In other words, the rate constant for the first step is smaller than the rate constant for the second step. This is what we would expect, since covalent bonds have to be broken in the first step, whereas no bonds are broken in the second step.

If a reaction has two or more steps, the step that has its transition state *at the highest point on the reaction coordinate* is called the **rate-determining step** or **rate-limiting step.** The rate-determining step controls the overall rate of the reaction, because the overall rate cannot exceed the rate of the rate-determining step. Thus, the rate-determining step for the reaction of 2-butene with HBr is the first step—the addition of the electrophile (the proton) to the alkene to form the carbocation.

PROBLEM 32

Draw a reaction coordinate diagram for a two-step reaction in which the first step is endergonic, the second step is exergonic, and the overall reaction is endergonic. Label the reactants, products, intermediates, and transition states.

PROBLEM 33◆

a. Which step in the reaction coordinate diagram has the greatest free energy of activation in the forward direction?

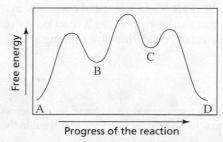

b. Is the first-formed intermediate more apt to revert to reactants or go on to form products?
c. Which step is the rate-determining step of the reaction?

PROBLEM 34♦

Draw a reaction coordinate diagram for the following reaction in which C is the most stable and B the least stable of the three species and the transition state going from A to B is more stable than the transition state going from B to C:

$$A \; \underset{k_{-1}}{\overset{k_1}{\rightleftharpoons}} \; B \; \underset{k_{-2}}{\overset{k_2}{\rightleftharpoons}} \; C$$

a. How many intermediates are there?
b. How many transition states are there?
c. Which step has the greater rate constant in the forward direction?
d. Which step has the greater rate constant in the reverse direction?
e. Of the four steps, which has the greatest rate constant?
f. Which is the rate-determining step in the forward direction?
g. Which is the rate-determining step in the reverse direction?

11   CATALYSIS

A **catalyst** increases the rate of a reaction by giving the reactants a new pathway to follow—one with a smaller $\Delta G^{\ddagger}$. In other words, a catalyst decreases the energy barrier that has to be overcome in the process of converting the reactants into products (Figure 7).

▲ **Figure 7**
A catalyst provides a pathway with a lower energy barrier but it does not change the energy of the starting point (the reactants) or the energy of the end point (the products).

> **A catalyst gives the reagents a new pathway with a lower "energy hill."**

If a catalyst is going to make a reaction go faster, it must participate in the reaction, but it is not consumed or changed during the reaction. Because the catalyst is not used up, only a small amount of it is needed to catalyze the reaction (typically, 1 to 10% of the number of moles of reactant). Notice in Figure 7 that the stability of the reactants and products is the same in both the catalyzed and uncatalyzed reactions. In other words, a catalyst does not change the relative concentrations of products and reactants when the system reaches equilibrium. Therefore, it does not change the *amount* of product formed; it changes only the *rate* at which it is formed.

> **A catalyst does not change the *amount* of product formed; it changes only the *rate* at which it is formed.**

The most common catalysts are acids, bases, and nucleophiles. Acids catalyze a reaction by giving a proton to a reactant; bases catalyze a reaction by removing a proton from a reactant, and nucleophiles catalyze reactions by forming a new covalent bond with the reactant.

PROBLEM 35♦

Which of the following parameters would be different for a reaction carried out in the presence of a catalyst, compared with the same reaction carried out in the absence of a catalyst?

$$\Delta G^{\circ}, \; \Delta H^{\ddagger}, \; E_{a}, \; \Delta S^{\ddagger}, \; \Delta H^{\circ}, \; K_{eq}, \; \Delta G^{\ddagger}, \; \Delta S^{\circ}, \; k$$

12 CATALYSIS BY ENZYMES

Essentially all reactions that occur in biological systems are reactions of organic compounds. These reactions almost always require a catalyst. Most biological catalysts are proteins called **enzymes.** Each biological reaction is catalyzed by a different enzyme.

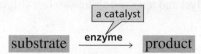

The reactant of an enzyme-catalyzed reaction is called a **substrate.** The enzyme binds the substrate in a pocket of the enzyme called the **active site.** All the bond-making and bond-breaking steps of the reaction occur while the substrate is bound to the active site.

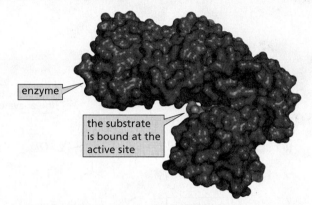

Unlike nonbiological catalysts, enzymes are specific for the substrate whose reaction they catalyze. All enzymes, however, do not have the same degree of specificity. Some are specific for a single compound and will not tolerate even the slightest variation in structure, whereas some catalyze the reaction of a family of compounds with related structures. The specificity of an enzyme for its substrate is an example of the phenomenon known as **molecular recognition**—the ability of one molecule to recognize another molecule as a result of intermolecular interactions.

The specificity of an enzyme for its substrate results from the particular amino acid side chains that reside at the active site. The side chains bind the substrate to the active site using hydrogen bonds, van der Waals forces, and dipole–dipole interactions—the same intermolecular interactions that hold molecules together.

Cell walls consist of thousands of six-membered ring molecules linked by oxygen atoms. Lysozyme is an enzyme that cleaves bacterial cell walls by breaking the bond that holds the six-membered rings together. Figure 8 shows a portion of lysozyme's active

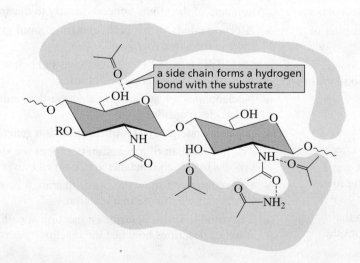

◀ **Figure 8**

The side chains at the active site of the enzyme hold the substrate in the precise position necessary for reaction.

site and some of the side chains that bind the substrate (the cell wall) in a precise location at the active site.

In addition to the side chains that bind the substrate to the active site, there are also side chains at the active site that are responsible for catalyzing the reaction. These side chains can be acids, bases, or nucleophiles—the same kinds of species that catalyze nonbiological reactions (Section 11). For example, lysozyme has two catalytic groups at its active site, an acid catalyst and a nucleophilic catalyst (Figure 9).

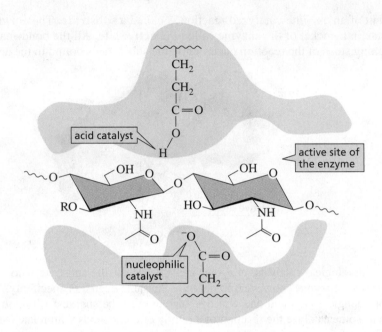

▶ **Figure 9**
Two side chains at the active site of lysozyme are catalysts for the reaction that breaks the bond holding the six-membered rings together.

SOME IMPORTANT THINGS TO REMEMBER

- An **alkene** is a hydrocarbon that contains a double bond. Because alkenes contain fewer than the maximum number of hydrogens, they are called **unsaturated hydrocarbons.**

- The double bond is the **functional group,** or center of reactivity, of an alkene.

- The general molecular formula for a hydrocarbon is C_nH_{2n+2}, minus two hydrogens for every π bond or ring in the molecule.

- The number of π bonds plus the number of rings is called the **degree of unsaturation.**

- The **functional group suffix** of an alkene is "ene."

- When there are both a functional group suffix and a substituent, the functional group suffix gets the lowest possible number.

- The **Z isomer** has the high-priority groups on the same side of the double bond; the **E isomer** has the high-priority groups on opposite sides of the double bond. The relative priorities depend on the atomic numbers of the atoms bonded directly to the sp^2 carbon.

- All compounds with a particular **functional group** react in the same way.

- Due to the cloud of electrons above and below its π bond, an alkene is a **nucleophile.**

- Nucleophiles are attracted to electron-deficient species, called **electrophiles.**

- Alkenes undergo **electrophilic addition reactions.**

- The **mechanism of a reaction** describes the step-by-step process by which reactants are changed into products.

- **Curved arrows** show the bonds that are formed and the bonds that are broken in a reaction.

- **Thermodynamics** describes a reaction at equilibrium; **kinetics** describes how fast the reaction occurs.

- A **reaction coordinate diagram** shows the energy changes that take place during the course of a reaction.

- The more stable a species, the lower its energy.

- As reactants are converted into products, a reaction passes through a maximum-energy **transition state.**

- An **intermediate** is a product of one step of a reaction and a reactant of the next step.

- Transition states have partially formed bonds; intermediates have fully formed bonds.

- The **rate-determining step** has its transition state at the highest point on the reaction coordinate.

- The equilibrium constant, K_{eq}, gives the relative concentrations of reactants and products at equilibrium.

- The more stable the product relative to the reactant, the greater is its concentration at equilibrium and the greater the K_{eq}.

- **Le Châtelier's principle** states that if an equilibrium is disturbed, the system will adjust to offset the disturbance.

- If the products are more stable than reactants, then K_{eq} is > 1, $\Delta G°$ is negative, and the reaction is **exergonic.**

- If the reactants are more stable than products, then K_{eq} is < 1, $\Delta G°$ is positive, and the reaction is **endergonic.**

- $\Delta G°$ is the **Gibbs free-energy change,** with $\Delta G° = \Delta H° - T\Delta S°$.

- $\Delta H°$ is the change in **enthalpy,** which is the heat given off or consumed as a result of bond making and bond breaking.

- An **exothermic reaction** has a negative $\Delta H°$; an **endothermic reaction** has a positive $\Delta H°$.

- $\Delta S°$ is the change in **entropy,** which is the change in the freedom of motion of the system.

- The formation of products with stronger bonds and greater freedom of motion causes $\Delta G°$ to be negative.

- $\Delta G°$ and K_{eq} are related by the formula $\Delta G° = -RT \ln K_{eq}$.

- The free energy of activation, $\Delta G^{\ddagger}$, is the energy barrier of a reaction. It is the difference between the free energy of the reactants and the free energy of the transition state.

- The rate of the reaction increases as $\Delta G^{\ddagger}$ decreases.

- Anything that makes the reactant more stable or makes the transition state less stable increases the rate constant for the reaction.

- **Kinetic stability** is given by $\Delta G^{\ddagger}$, **thermodynamic stability** by $\Delta G°$.

- The **rate** of a reaction depends on the concentration of the reactants, the temperature, and the rate constant.

- The **rate constant** for a reaction indicates how easy it is for the reactants to reach the transition state.

- A **first-order reaction** depends on the concentration of one reactant; a **second-order reaction** depends on the concentration of two reactants.

- A **catalyst** decreases the energy barrier that has to be overcome in the process of converting the reactants into products.

- A catalyst is neither consumed nor changed during the reaction.

- A catalyst does not change the *amount* of product formed, it changes only the *rate* at which the product is formed.

- Most biological catalysts are proteins called **enzymes.**

- **Molecular recognition** is the ability of one molecule to recognize another molecule.

PROBLEMS

36. What is each compound's systematic name?

a. $CH_3CH_2CHCH=CHCH_2CH_2CHCH_3$
   with Br below first CH and Br below the second CHCH₃ carbon

b.
$$\underset{CH_3CH_2}{\overset{H_3C}{>}}C=C\underset{CH_2CH_2CHCH_3}{\overset{CH_2CH_3}{<}}$$
with CH₃ below the CHCH₃

c. cyclopentene with CH₃ and CH₃ substituents

d.
$$\underset{H_3C}{\overset{H_3C}{>}}C=C\underset{CH_2CH_2CH_2CH_3}{\overset{CH_2CH_3}{<}}$$

e. cyclohexene with CH₃ substituent

f. cyclohexene with CH₂CH₃ and CH₃ substituents

37. Draw the structure of a hydrocarbon that has six carbon atoms and
   a. three vinylic hydrogens and two allylic hydrogens.
   b. three vinylic hydrogens and one allylic hydrogen.
   c. three vinylic hydrogens and no allylic hydrogens.

38. Draw the condensed structure for each of the following:
  a. (*Z*)-1,3,5-tribromo-2-pentene   **c.** (*E*)-1,2-dibromo-3-isopropyl-2-hexene   **e.** 1,2-dimethylcyclopentene
  b. (*Z*)-3-methyl-2-heptene   **d.** vinyl bromide   **f.** diallylamine

39. Draw the skeletal structures for the compounds in Problem 38.

40. **a.** Draw the condensed structures and give the systematic names for all the alkenes with molecular formula C_6H_{12}, ignoring cis–trans isomers. (*Hint:* There are 13.)
  b. Which of the compounds have *E* and *Z* isomers?

41. Name the following:

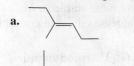

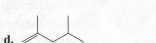

  a.   **c.**   Br   **e.**

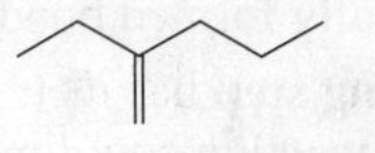

  b.   **d.**   **f.**

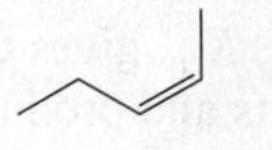

42. Draw curved arrows to show the flow of electrons responsible for the conversion of the reactants into the products:

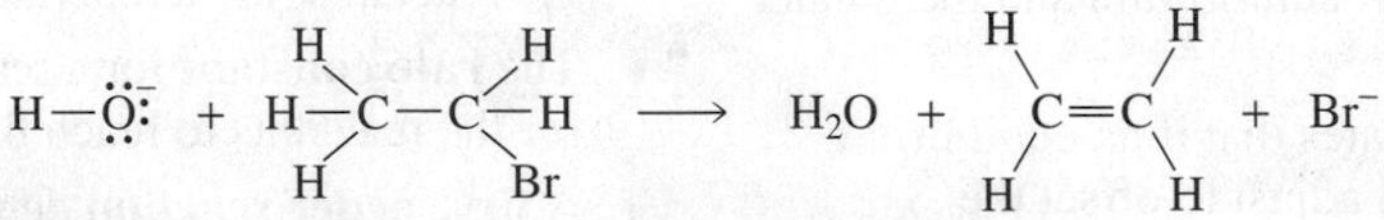

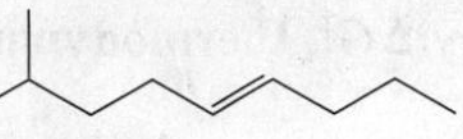

$$ H-\overset{\cdot\cdot}{\underset{\cdot\cdot}{O}}{:}^- + \ H-\overset{\overset{\displaystyle H}{|}}{\underset{\underset{\displaystyle H}{|}}{C}}-\overset{\overset{\displaystyle H}{|}}{\underset{\underset{\displaystyle Br}{|}}{C}}-H \longrightarrow H_2O + \ \overset{\overset{\displaystyle H}{|}}{\underset{\underset{\displaystyle H}{|}}{C}}=\overset{\overset{\displaystyle H}{|}}{\underset{\underset{\displaystyle H}{|}}{C} + Br^- $$

43. Draw the skeletal structure of 6-*sec*-butyl-7-isopropyl-3,3-dimethyldecane.

44. In a reaction in which reactant A is in equilibrium with product B at 25 °C, what are the relative amounts of A and B present at equilibrium if $\Delta G°$ at 25 °C is
  a. 2.72 kcal/mol?   **b.** 0.65 kcal/mol?   **c.** −2.72 kcal/mol?   **d.** −0.65 kcal/mol?

45. Which bond is stronger? Briefly explain why.
  a. CH_3-Cl or CH_3-Br **b.** $I-Br$ or $Br-Br$

46. Do the following compounds have the *E* or the *Z* configuration?

a. 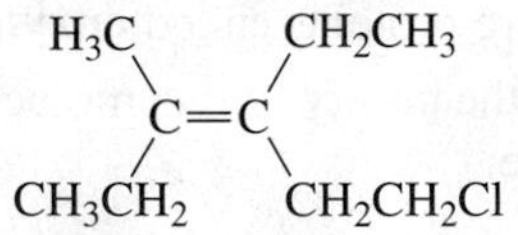
$$ \underset{CH_3CH_2}{\overset{H_3C}{}}C=C\underset{CH_2CH_2Cl}{\overset{CH_2CH_3}{}} $$

c.
$$ \underset{Br}{\overset{H_3C}{}}C=C\underset{CH_2CH_2CH_2CH_3}{\overset{CH_2Br}{}} $$

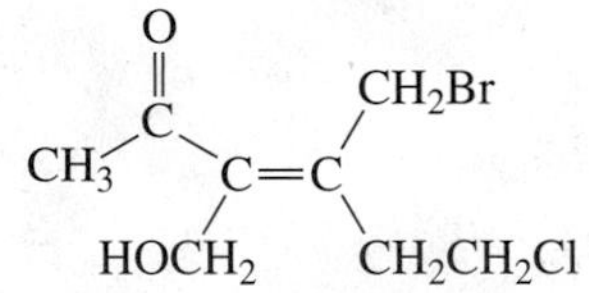

e. 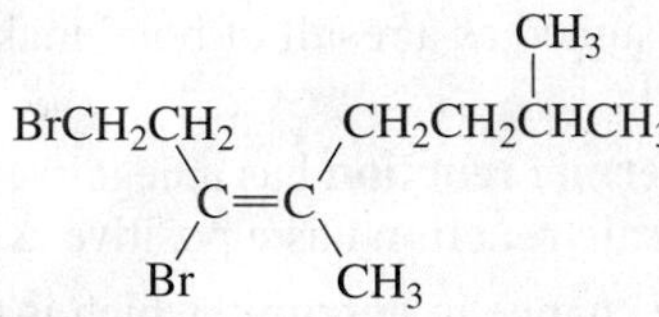
$$ \underset{Br}{\overset{BrCH_2CH_2}{}}C=C\underset{CH_3}{\overset{CH_2CH_2\overset{CH_3}{\underset{|}{CH}}CH_3}{}} $$

b.
$$ \underset{HC\equiv CCH_2}{\overset{CH_3CH_2CH_2}{}}C=C\underset{CH_2CH=CH_2}{\overset{CH(CH_3)_2}{}} $$

d.
$$ CH_3\overset{\overset{\displaystyle O}{\|}}{C}\underset{HOCH_2}{\overset{}{}}C=C\underset{CH_2CH_2Cl}{\overset{CH_2Br}{}} $$

f. 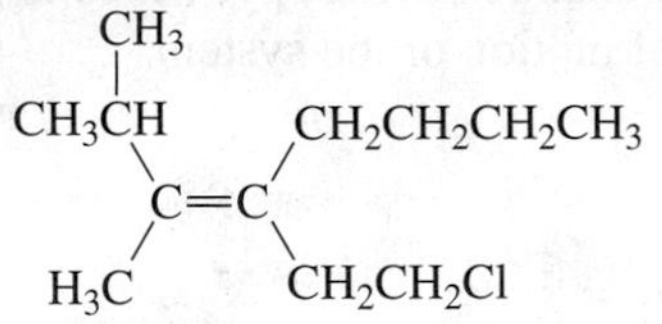
$$ \underset{H_3C}{\overset{CH_3\overset{CH_3}{\underset{|}{CH}}}{}}C=C\underset{CH_2CH_2Cl}{\overset{CH_2CH_2CH_2CH_3}{}} $$

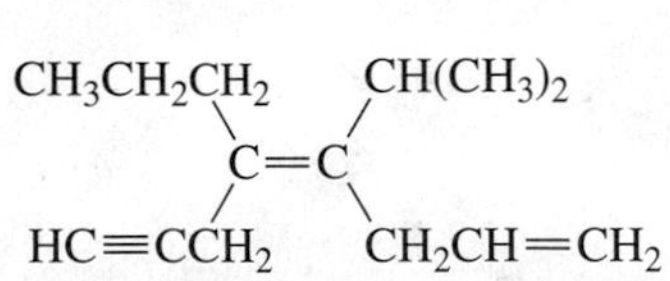

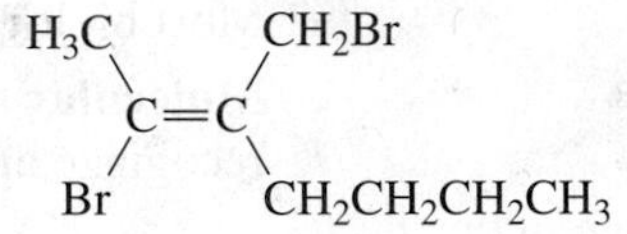

47. Squalene, a hydrocarbon with molecular formula $C_{30}H_{50}$, is obtained from shark liver. (*Squalus* is Latin for "shark.") If squalene is an acyclic compound, how many π bonds does it have?

48. Assign relative priorities to each set of substituents:
  a. $-CH_2CH_2CH_3$ $-CH(CH_3)_2$ $-CH=CH_2$ $-CH_3$
  b. $-CH_2NH_2$ $-NH_2$ $-OH$ $-CH_2OH$
  c. $-C(=O)CH_3$ $-CH=CH_2$ $-Cl$ $-C\equiv N$

49. Draw the geometric isomers for the following compounds and name each isomer:
  a. 2-methyl-2,4-hexadiene   **b.** 1,5-heptadiene   **c.** 1,4-pentadiene

50. By following the curved red arrows, draw the product(s) of each of the following reaction steps:

  a. 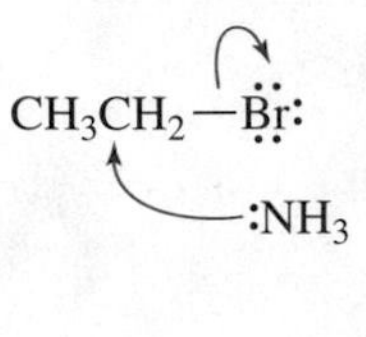 $CH_3CH_2-\overset{\cdot\cdot}{\underset{\cdot\cdot}{Br}}{:}$
      :NH_3

  b. 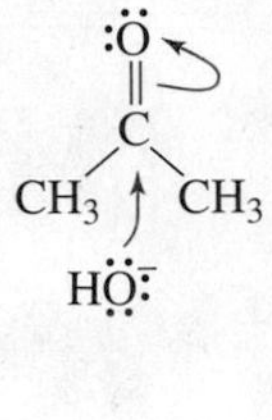
$$ CH_3\overset{\overset{\displaystyle :\overset{\cdot\cdot}{O}}{\|}}{C}CH_3 $$
    HO:⁻

  c. $CH_3\overset{CH_3}{\underset{}{C}}=\overset{CH_3}{\underset{}{C}}CH_3$
    $H-\overset{\cdot\cdot}{\underset{\cdot\cdot}{Cl}}{:}$

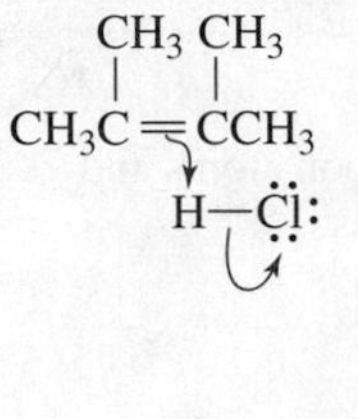

51. How many of the following names are correct? Correct the incorrect names.
- **a.** 3-pentene
- **b.** 2-octene
- **c.** 2-vinylpentane
- **d.** 1-ethyl-1-pentene
- **e.** 5-ethylcyclohexene
- **f.** 5-chloro-3-hexene
- **g.** 2-ethyl-2-butene
- **h.** (*E*)-2-methyl-1-hexene
- **i.** 2-methylcyclopentene

52. Draw structures for the following compounds:
- **a.** (2*E*,4*E*)-1-chloro-3-methyl-2,4-hexadiene
- **b.** (3*Z*,5*E*)-4-methyl-3,5-nonadiene
- **c.** (3*Z*,5*Z*)-4,5-dimethyl-3,5-nonadiene
- **d.** (3*E*,5*E*)-2,5-dibromo-3,5-octadiene

53. Given the reaction coordinate diagram for the reaction of **A** to form **G**, answer the following questions:

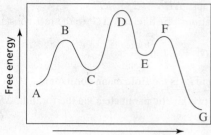

Progress of the reaction

- **a.** How many intermediates are formed in the reaction?
- **b.** Which letters represent transition states?
- **c.** What is the fastest step in the reaction?
- **d.** Which is more stable, A or G?
- **e.** Does A or E form faster from C?
- **f.** What is the reactant of the rate-determining step?
- **g.** Is the first step of the reaction exergonic or endergonic?
- **h.** Is the overall reaction exergonic or endergonic?
- **i.** Which is the more stable intermediate?
- **j.** Which step in the forward direction has the largest rate constant?
- **k.** Which step in the reverse direction has the smallest rate constant?

54. a. Which of the following reactions will have the larger $\Delta S°$ value?
  b. Will the $\Delta S°$ value be positive or negative?

A (cyclohexane with Br) $+ \ HO^- \longrightarrow$ (cyclohexane with OH) $+ \ Br^-$ **B** (cyclohexane with Br) $+ \ HO^- \longrightarrow$ (cyclohexene) $+ \ H_2O \ + \ Br^-$

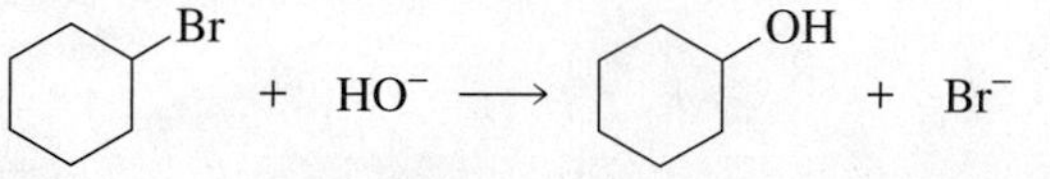

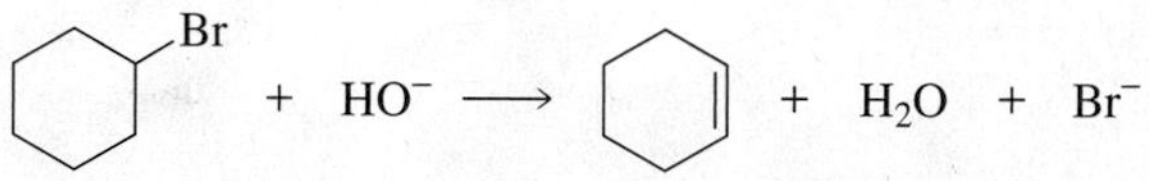

55. a. What is the equilibrium constant for a reaction that is carried out at 25 °C (298 K) with $\Delta H° = 20$ kcal/mol and $\Delta S° = 5.0 \times 10^{-2}$ kcal mol^{-1}K^{-1} ?
  b. What is the equilibrium constant for the same reaction carried out at 125 °C?

56. Using curved arrows, show the mechanism of the following reaction:

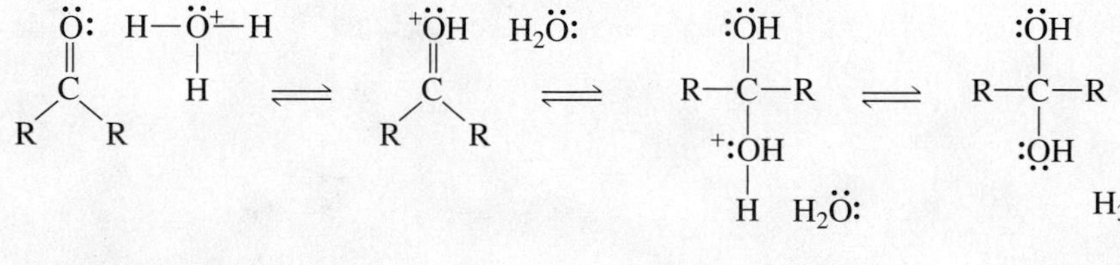

57. For a reaction carried out at 25 °C with an equilibrium constant of 1×10^{-3}, in order to increase the equilibrium constant by a factor of 10:
- **a.** How much must $\Delta G°$ change?
- **b.** How much must $\Delta H°$ change if $\Delta S° = 0$ kcal mol^{-1}K^{-1}?
- **c.** How much must $\Delta S°$ change if $\Delta H° = 0$ kcal mol^{-1}?

58. Given that the twist-boat conformer of cyclohexane is 5.3 kcal/mol higher in free energy than the chair conformer, calculate the percentage of twist-boat conformers present in a sample of cyclohexane at 25 °C.

Calculating Kinetic Parameters

After obtaining rate constants at several temperatures, you can calculate E_a, $\Delta H^{\ddagger}$, $\Delta G^{\ddagger}$, and $\Delta S^{\ddagger}$ for a reaction as follows:
- The Arrhenius equation allows E_a to be obtained from the slope of a plot of $\ln k$ versus $1/T$, because

$$\ln k_2 - \ln k_1 = -E_a / R\left(\frac{1}{T_2} - \frac{1}{T_1}\right)$$

- You can determine $\Delta H^{\ddagger}$ at a given temperature from E_a because $\Delta H^{\ddagger} = E_a - RT$.
- You can determine $\Delta G^{\ddagger}$, in kJ/mol, from the following equation, which relates $\Delta G^{\ddagger}$ to the rate constant at a given temperature:

$$-\Delta G^{\ddagger} = RT \ln \frac{kh}{Tk_B}$$

In this equation, h is Planck's constant (6.62608×10^{-34} J s) and k_B is the Boltzmann constant (1.38066×10^{-23} J K^{-1})

- You can determine the entropy of activation from the other two kinetic parameters via the formula $\Delta S^{\ddagger} = (\Delta H^{\ddagger} - \Delta G^{\ddagger})/T$.

Use this information to answer Problem 59.

59. From the following rate constants, determined at five temperatures, calculate the experimental energy of activation and $\Delta G^{\ddagger}$, $\Delta H^{\ddagger}$, and $\Delta S^{\ddagger}$ for the reaction at 30 °C:

Temperature	Observed rate constant
31.0 °C	2.11×10^{-5} s^{-1}
40.0 °C	4.44×10^{-5} s^{-1}
51.5 °C	1.16×10^{-4} s^{-1}
59.8 °C	2.10×10^{-4} s^{-1}
69.2 °C	4.34×10^{-4} s^{-1}

AN EXERCISE IN DRAWING CURVED ARROWS: PUSHING ELECTRONS

This is an extension of what you learned about drawing curved arrows. Working through these problems will take only a little of your time. It will be time well spent, however, because curved arrows are used throughout the text and it is important that you are comfortable with this notation.

Chemists use curved arrows to show how electrons move as covalent bonds break and/or new covalent bonds form. The tail of the arrow is positioned at the point where the electrons are in the reactant, and the head of the arrow points to where these same electrons end up in the product.

In the following reaction step, the bond between bromine and a carbon of the cyclohexane ring breaks and both electrons in the bond end up with bromine in the product. Thus, **the arrow starts at the electrons that carbon and bromine share in the reactant,** and **the head of the arrow points at bromine** because this is where the two electrons end up in the product.

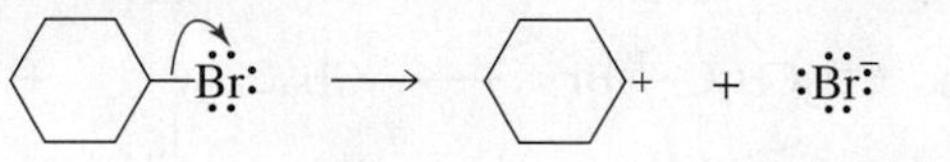

Notice that the carbon of the cyclohexane ring is positively charged in the product. This is because it has lost the two electrons it was sharing with bromine. The bromine is negatively charged in the product because it has gained the electrons that it shared with carbon in the reactant. The fact that two electrons move in this example is indicated by the two barbs on the arrowhead.

Notice that the arrow *always* starts at a bond or at a lone pair. It does *not* start at a negative charge. (And since an arrow starts at a pair of electrons, it would never start at a positive charge!)

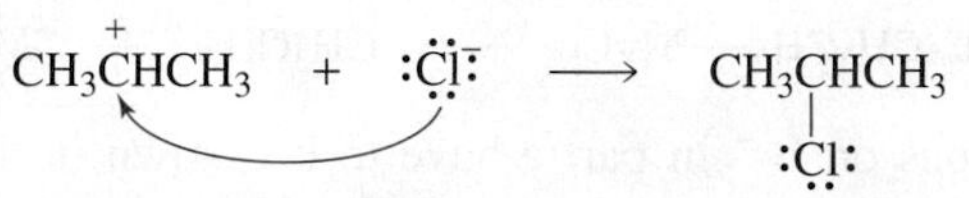

$$CH_3\overset{+}{C}HCH_3 \; + \; :\ddot{C}\ddot{l}:^- \; \longrightarrow \; CH_3CHCH_3$$
$$\underset{:\ddot{C}\ddot{l}:}{|}$$

In the following reaction step, a bond is being formed between the oxygen of water and a carbon of the other reactant. The arrow starts at one of the lone pairs of the oxygen and points at the atom (the carbon) that will share the electrons in the product. The oxygen in the product is positively charged, because the electrons that oxygen had to itself in the reactant are now being shared with carbon. The carbon that was positively charged in the reactant is not charged in the product, because it has gained a share in a pair of electrons.

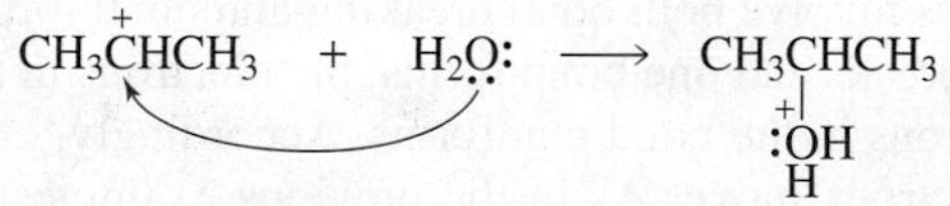

$$CH_3\overset{+}{C}HCH_3 \; + \; H_2\ddot{O}: \; \longrightarrow \; CH_3CHCH_3$$
$$\underset{H}{\overset{+|}{:\ddot{O}H}}$$

PROBLEM 1 Draw curved arrows to show the movement of the electrons in the following reaction steps. (The answers to all problems appear immediately after Problem 10.)

a.
$$CH_3CH_2\overset{\overset{\displaystyle CH_3}{|}}{\underset{\underset{\displaystyle CH_3}{|}}{C}}{-}\ddot{B}\ddot{r}: \; \longrightarrow \; CH_3CH_2\overset{\overset{\displaystyle CH_3}{|}}{\underset{\underset{\displaystyle CH_3}{|}}{C}}{}^+ \; + \; :\ddot{B}\ddot{r}:^-$$

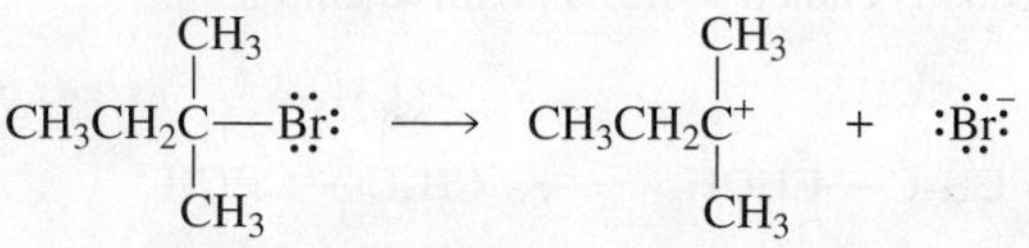

b.

c. 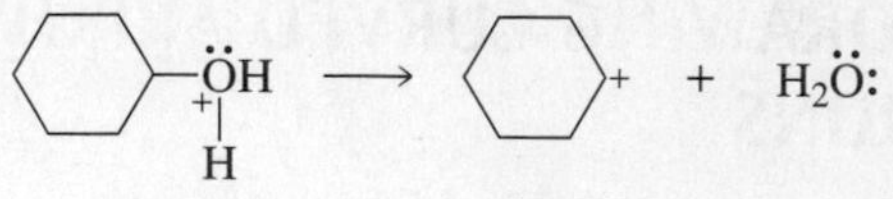   (cyclohexyl)$-\overset{..}{\underset{+\ |\ H}{O}}H$ $\longrightarrow$ (cyclohexyl)$+$ + + $H_2\overset{..}{\underset{..}{O}}:$

d. CH_3CH_2-MgBr $\longrightarrow$ $CH_3\overset{..}{C}H_2$ + ^+MgBr

e. $CH_3CH_2\overset{+}{C}HCH_3$ + $:\overset{..}{\underset{..}{Br}}:^-$ $\longrightarrow$ $CH_3CH_2\underset{\underset{:\overset{..}{\underset{..}{Br}}:}{|}}{C}HCH_3$

Frequently, chemists do not show the lone-pair electrons when they write reactions. Problem 2 shows the same reaction steps you just saw in Problem 1, except that the lone pairs are not shown.

PROBLEM 2 Draw curved arrows to show the movement of the electrons in the following reaction steps:

a. $CH_3CH_2\underset{\underset{CH_3}{|}}{\overset{\overset{CH_3}{|}}{C}}-Br$ $\longrightarrow$ $CH_3CH_2\underset{\underset{CH_3}{|}}{\overset{\overset{CH_3}{|}}{C^+}}$ + Br^-

b. 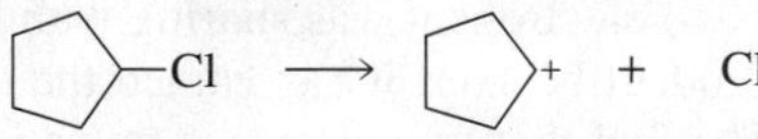   (cyclopentyl)$-Cl$ $\longrightarrow$ (cyclopentyl)$+$ + + Cl^-

c. 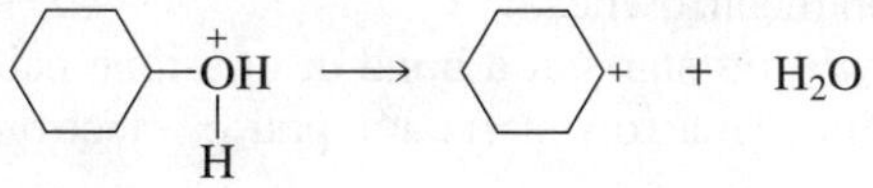   (cyclohexyl)$-\overset{+}{\underset{|\ H}{O}}H$ $\longrightarrow$ (cyclohexyl)$+$ + + H_2O

d. CH_3CH_2-MgBr $\longrightarrow$ $CH_3\overset{..}{C}H_2$ + ^+MgBr

The lone-pair electrons on Br^- in part **e** have to be shown in the reactant, because an arrow can start only at a bond or at a lone pair. The lone pair electrons on Br in the product do not have to be shown. It is never wrong to shown lone pairs, but the only time they have to be shown is when an arrow is going to start from the lone pair.

e. $CH_3CH_2\overset{+}{C}HCH_3$ + $:\overset{..}{\underset{..}{Br}}:^-$ $\longrightarrow$ $CH_3CH_2\underset{\underset{Br}{|}}{C}HCH_3$

   Many reaction steps involve both bond breaking and bond formation. In the following examples, one bond breaks and one bond forms; the electrons in the bond that breaks are the same as the electrons in the bond that forms. Accordingly, only one arrow is needed to show how the electrons move. As in the previous examples, the arrow starts at the point where the electrons are in the reactant, and the head of the arrow points to where these same electrons end up in the product (between the CH_3 carbon and the carbon that was previously positively charged in the first example, and between the carbons in the next example). Notice that the atom that loses a share in a pair of electrons (C in the first example, H in the second) ends up with a positive charge.

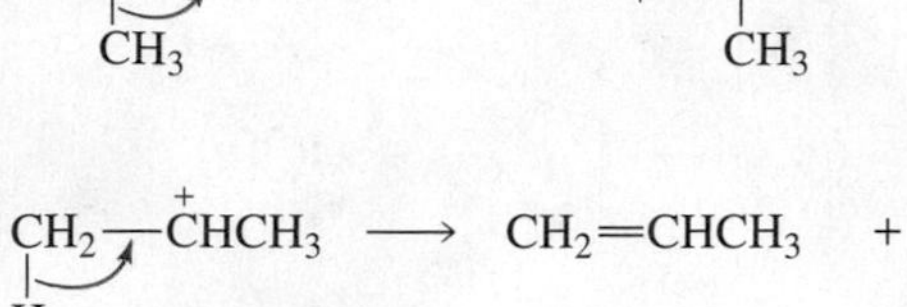

$CH_3\underset{\underset{CH_3}{|}}{\overset{\overset{CH_3}{|}}{C}}-\overset{+}{C}HCH_3$ $\longrightarrow$ $CH_3\underset{\underset{CH_3}{|}}{\overset{\overset{CH_3}{|}}{\underset{+}{C}}}-CHCH_3$

$\underset{\underset{H}{|}}{CH_2}-\overset{+}{C}HCH_3$ $\longrightarrow$ $CH_2{=}CHCH_3$ + H^+

Frequently, the electrons in the bond that breaks are not the same as the electrons in the bond that forms. In such cases, two arrows are needed to show the movement of the electrons—one to show the bond that forms and one to show the bond that breaks. In each of the following examples, look at the arrows that illustrate how the electrons move. Notice how the movement of the electrons allows you to determine both the structure of the products and the charges on the products.

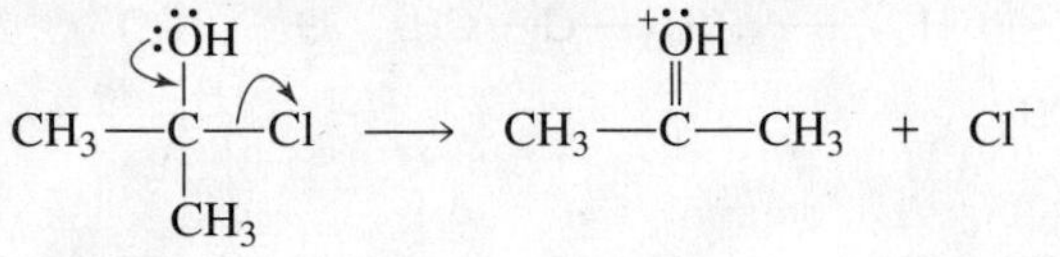

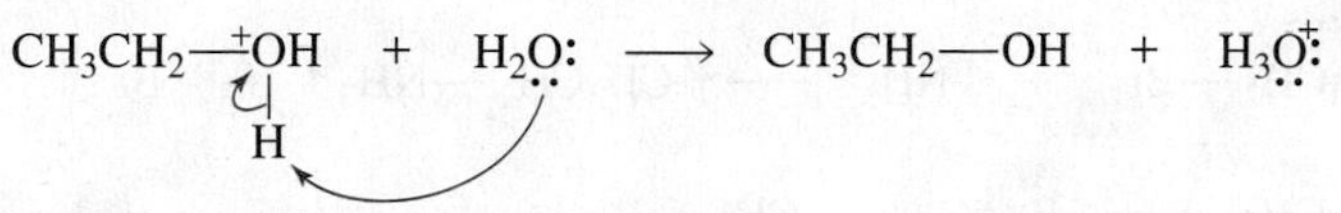

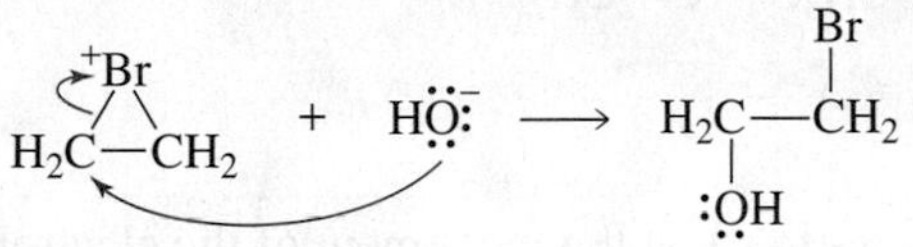

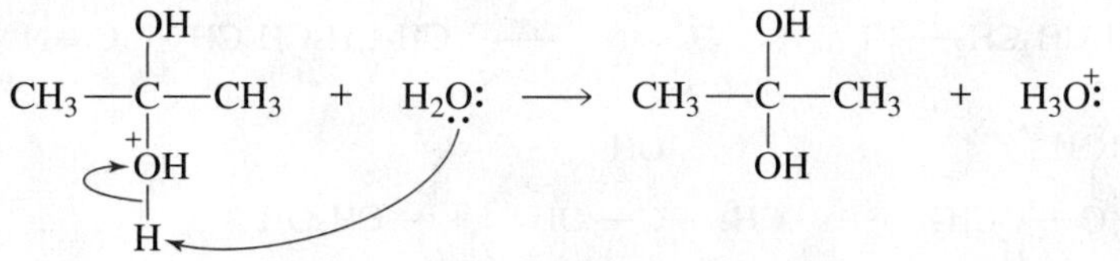

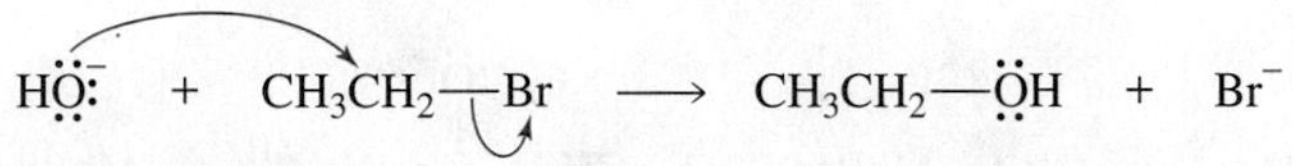

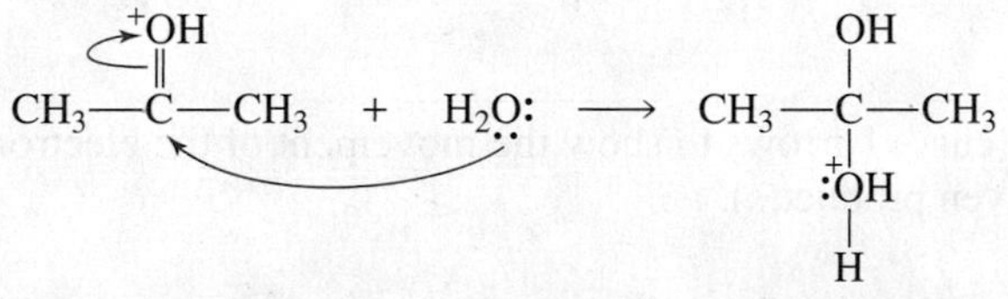

$$CH_3-\overset{\displaystyle :\ddot{O}H}{\underset{\displaystyle CH_3}{\overset{\displaystyle |}{C}}}-Cl \longrightarrow CH_3-\overset{\displaystyle \overset{+}{\ddot{O}H}}{\overset{\displaystyle \|}{C}}-CH_3 \ + \ Cl^-$$

$$CH_3CH_2-\overset{+}{\underset{H}{\overset{\displaystyle |}{\ddot{O}H}}} \ + \ H_2\ddot{O}: \ \longrightarrow \ CH_3CH_2-OH \ + \ H_3\overset{+}{\ddot{O}}:$$

$$\underset{H_2C-CH_2}{\overset{\overset{+}{Br}}{\triangle}} \ + \ H\ddot{O}:^- \ \longrightarrow \ \underset{\displaystyle :\ddot{O}H}{\overset{\displaystyle Br}{\underset{\displaystyle |}{H_2C-CH_2}}}$$

$$:\ddot{Br}:^- \ + \ CH_3-\overset{+}{\underset{\displaystyle }{\overset{\displaystyle H}{\ddot{O}H}}} \ \longrightarrow \ CH_3-\ddot{Br}: \ + \ H_2O$$

$$CH_3-\overset{\displaystyle OH}{\underset{\displaystyle \overset{+}{\underset{H}{\overset{|}{OH}}}}{\overset{|}{C}}}-CH_3 \ + \ H_2\ddot{O}: \ \longrightarrow \ CH_3-\overset{\displaystyle OH}{\underset{\displaystyle OH}{\overset{\displaystyle |}{\underset{\displaystyle |}{C}}}}-CH_3 \ + \ H_3\overset{+}{O}:$$

$$H\ddot{O}:^- \ + \ CH_3CH_2-Br \ \longrightarrow \ CH_3CH_2-\ddot{O}H \ + \ Br^-$$

$$CH_3-\overset{\overset{+}{OH}}{\overset{\|}{C}}-CH_3 \ + \ H_2\ddot{O}: \ \longrightarrow \ CH_3-\overset{\displaystyle OH}{\underset{\displaystyle \overset{+}{\underset{H}{\overset{|}{\ddot{O}H}}}}{\overset{|}{C}}}-CH_3$$

In the next reaction, two bonds break and one bond forms; two arrows are needed to show the movement of the electrons.

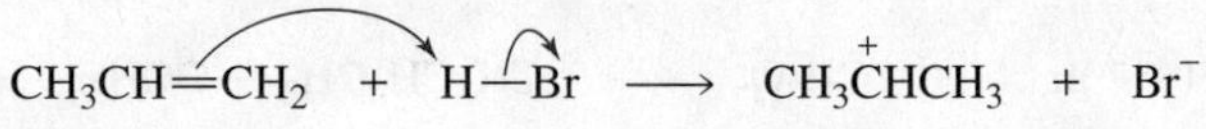

$$CH_3CH=CH_2 \ + \ H-Br \ \longrightarrow \ CH_3\overset{+}{C}HCH_3 \ + \ Br^-$$

In the next reaction, two bonds break and two bonds form; three arrows are needed to show the movement of the electrons.

$$\underset{}{\overset{H}{\triangle}}-Br \ + \ H_2\ddot{O}: \ \longrightarrow \ \bigcirc \ + \ Br^- \ + \ H_3\overset{+}{O}:$$

233

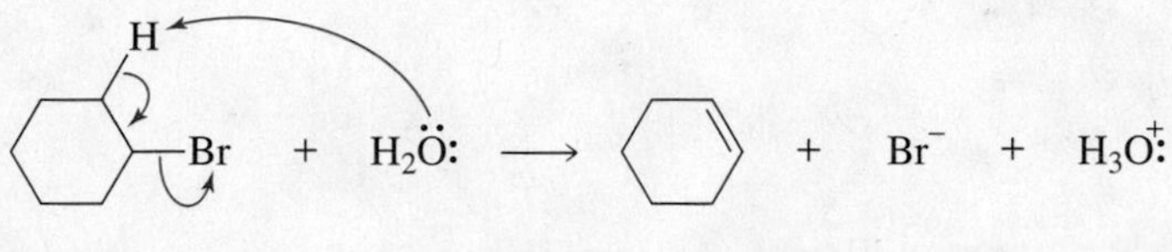

PROBLEM 3 Draw curved arrows to show the movement of the electrons that result in the formation of the given product(s). (*Hint:* Look at the structure of the product to see what bonds need to be formed and broken in order to arrive at the structure of the desired product.)

a. 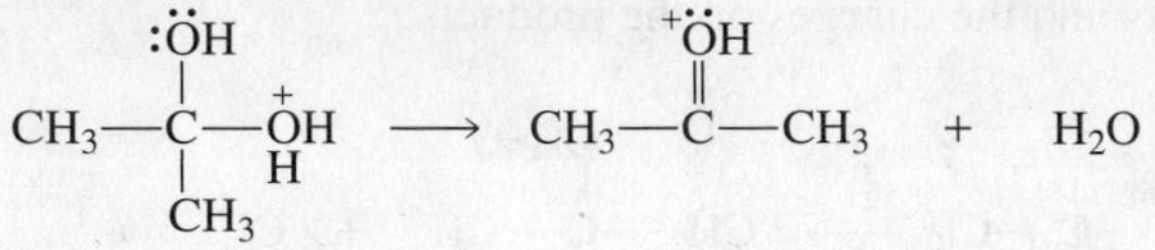 $CH_3-\overset{:\overset{..}{O}H}{\underset{\underset{CH_3}{|}}{\overset{|}{C}}}-\overset{+}{O}H \longrightarrow CH_3-\overset{\overset{+}{\overset{..}{O}H}}{\overset{||}{C}}-CH_3 + H_2O$

b. $CH_3CH_2CH{=}CH_2 + H-Cl \longrightarrow CH_3CH_2\overset{+}{C}H-CH_3 + Cl^-$

c. $CH_3CH_2-Br + \overset{..}{N}H_3 \longrightarrow CH_3CH_2-\overset{+}{N}H_3 + Br^-$

d. 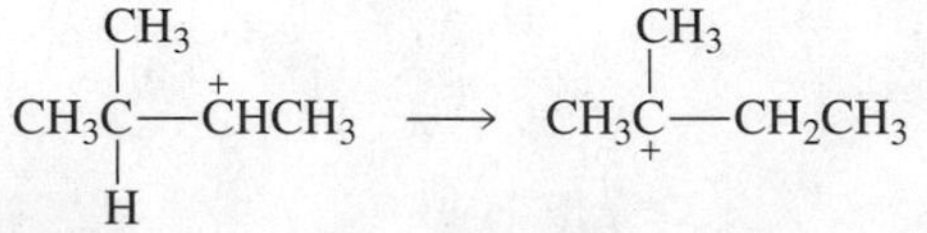 $CH_3\overset{\overset{CH_3}{|}}{\underset{\underset{H}{|}}{C}}-\overset{+}{C}HCH_3 \longrightarrow CH_3\overset{\overset{CH_3}{|}}{\underset{+}{C}}-CH_2CH_3$

PROBLEM 4 Draw curved arrows to show the movement of the electrons that result in formation of the given product(s).

a. $CH_3CH{=}CHCH_3 + H-\overset{\overset{+}{O}}{\underset{\underset{H}{|}}{}}-H \longrightarrow CH_3\overset{+}{C}H-CH_2CH_3 + H_2O$

b. $CH_3CH_2CH_2CH_2-Cl + \ ^-\overset{..}{C}{\equiv}N \longrightarrow CH_3CH_2CH_2CH_2-C{\equiv}N + Cl^-$

c. 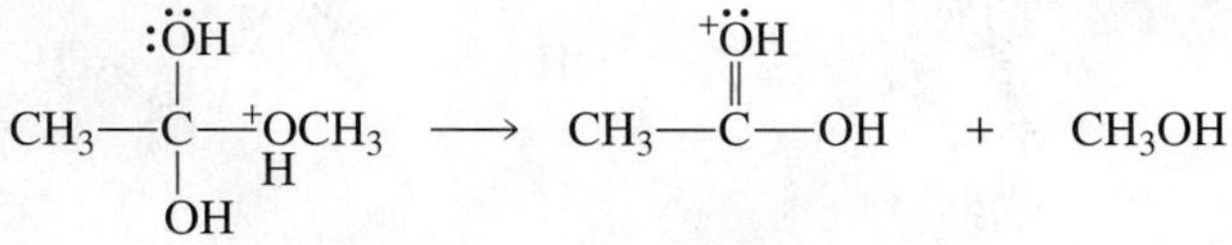 $CH_3-\overset{:\overset{..}{O}H}{\underset{\underset{OH}{|}}{\overset{|}{C}}}-\overset{+}{O}CH_3 \longrightarrow CH_3-\overset{\overset{+}{\overset{..}{O}H}}{\overset{||}{C}}-OH + CH_3OH$

d. 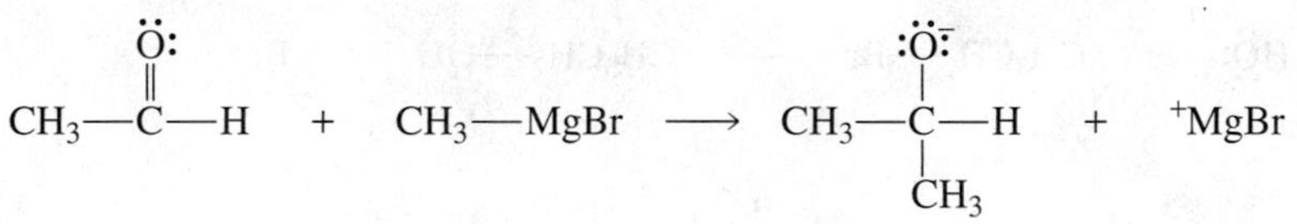 $CH_3-\overset{\overset{..}{O}:}{\overset{||}{C}}-H + CH_3-MgBr \longrightarrow CH_3-\overset{\overset{:\overset{..}{O}:^-}{|}}{\underset{\underset{CH_3}{|}}{C}}-H + \ ^+MgBr$

PROBLEM 5 Draw curved arrows to show the movement of the electrons that result in formation of the given product(s).

a. 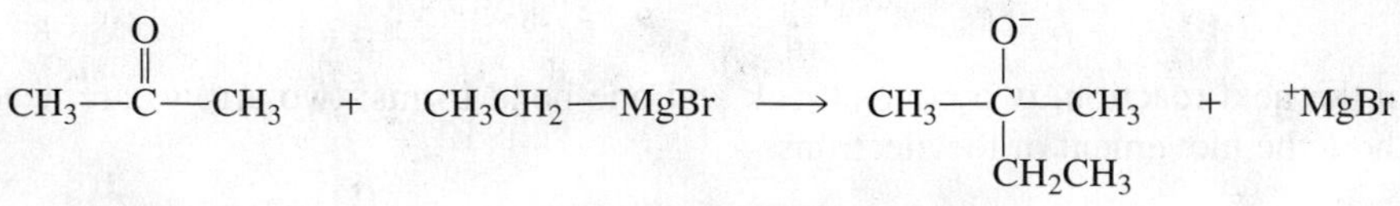 $CH_3-\overset{\overset{O}{||}}{C}-CH_3 + CH_3CH_2-MgBr \longrightarrow CH_3-\overset{\overset{O^-}{|}}{\underset{\underset{CH_2CH_3}{|}}{C}}-CH_3 + \ ^+MgBr$

b. $CH_3CH_2CH_2-Br + CH_3\overset{..}{\underset{..}{O}}:^- \longrightarrow CH_3CH_2CH_2-OCH_3 + Br^-$

c.

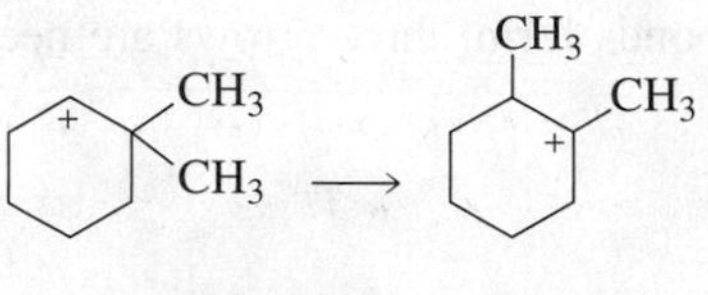

d. 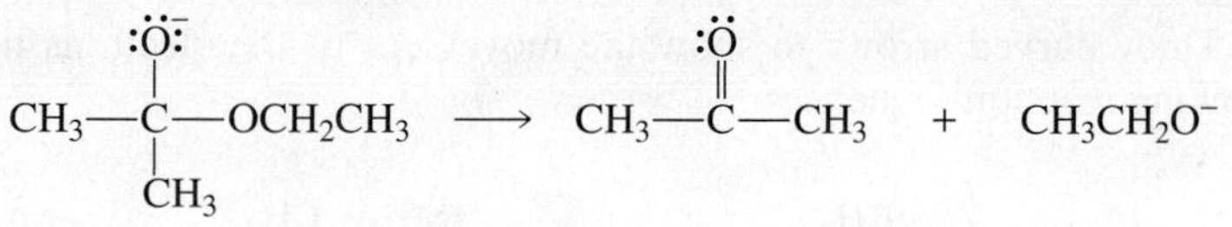 $CH_3-\overset{\overset{\displaystyle :\ddot{O}:^-}{|}}{\underset{\underset{\displaystyle CH_3}{|}}{C}}-OCH_2CH_3 \longrightarrow CH_3-\overset{\overset{\displaystyle :\ddot{O}:}{\|}}{C}-CH_3 + CH_3CH_2O^-$

PROBLEM 6 Draw curved arrows to show the movement of the electrons that result in formation of the given product(s).

a. 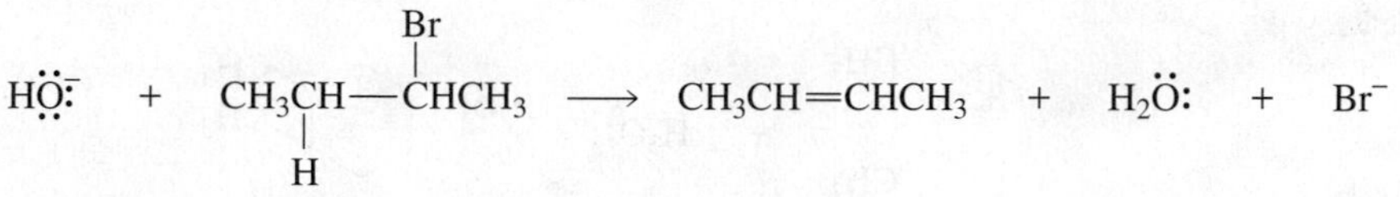 $H\ddot{O}:^- + CH_3\overset{\overset{\displaystyle Br}{|}}{\underset{\underset{\displaystyle H}{|}}{C}H-CHCH_3} \longrightarrow CH_3CH=CHCH_3 + H_2\ddot{O}: + Br^-$

b. $CH_3CH_2C\equiv C-H + :\dot{N}H_2 \longrightarrow CH_3CH_2C\equiv C:^- + \dot{N}H_3$

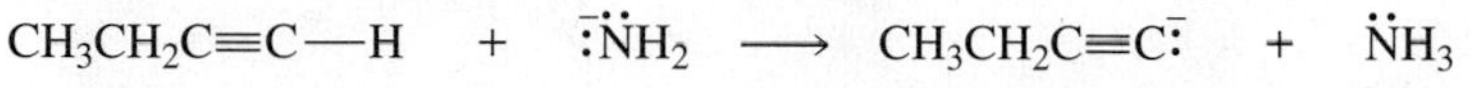

c. $CH_3\overset{\overset{\displaystyle CH_3}{|}}{\underset{\underset{\displaystyle CH_3}{|}}{\overset{+}{C}}-CHCH_2CH_3} \longrightarrow CH_3\overset{\overset{\displaystyle CH_3}{|}}{\underset{\underset{\displaystyle CH_3}{|}}{\overset{+}{C}}-CHCH_2CH_3}$

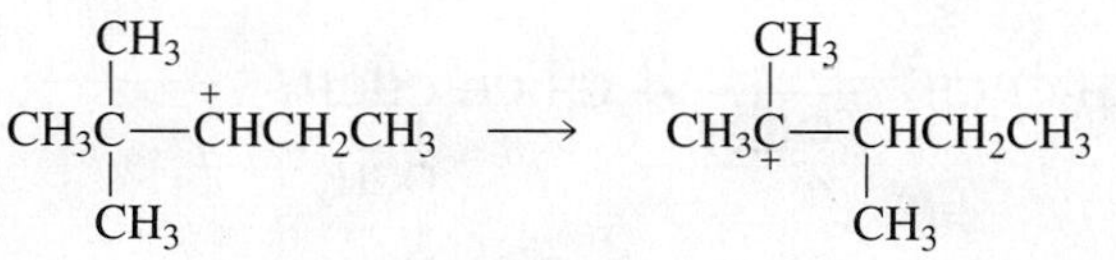

d. 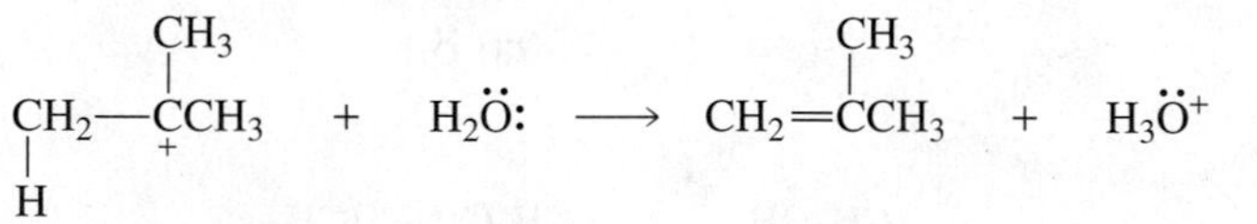 $\overset{\overset{\displaystyle CH_3}{|}}{\underset{\underset{\displaystyle H}{|}}{CH_2-\overset{+}{C}CH_3}} + H_2\ddot{O}: \longrightarrow CH_2=\overset{\overset{\displaystyle CH_3}{|}}{C}CH_3 + H_3\ddot{O}^+$

PROBLEM 7 Draw curved arrows to show the movement of the electrons that result in formation of the given product(s).

a. 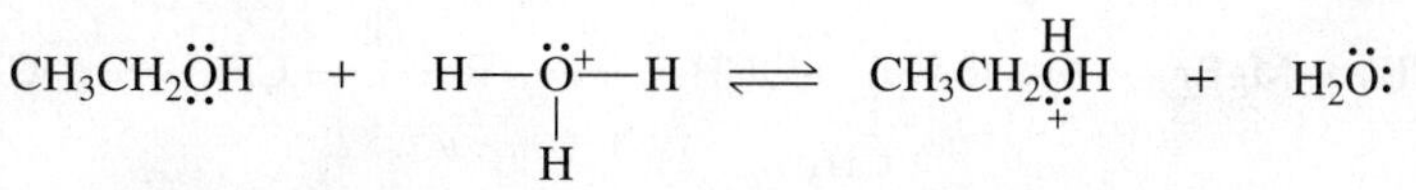 $CH_3CH_2\ddot{O}H + H-\overset{\overset{\displaystyle +}{\underset{\underset{\displaystyle H}{|}}{\ddot{O}}}-H \rightleftharpoons CH_3CH_2\overset{\overset{\displaystyle H}{|}}{\underset{\displaystyle +}{\ddot{O}H}} + H_2\ddot{O}:$

b. 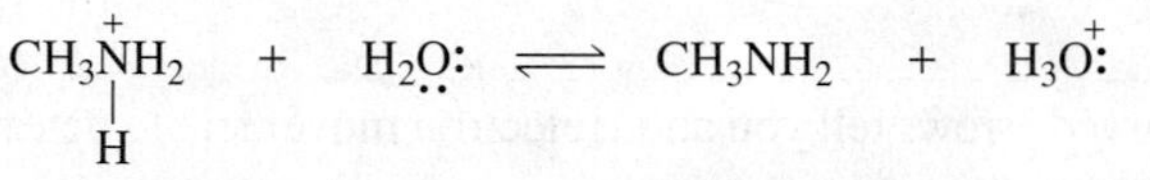 $CH_3\overset{\overset{\displaystyle +}{\underset{\underset{\displaystyle H}{|}}{N}H_2}} + H_2O: \rightleftharpoons CH_3NH_2 + H_3\overset{+}{\ddot{O}}:$

PROBLEM 8 Draw curved arrows to show the movement of the electrons in each step of the following reaction sequences. (*Hint:* You can tell how to draw the arrows for each step by looking at the products that are formed in that step as a result of the movement of electrons.)

a. 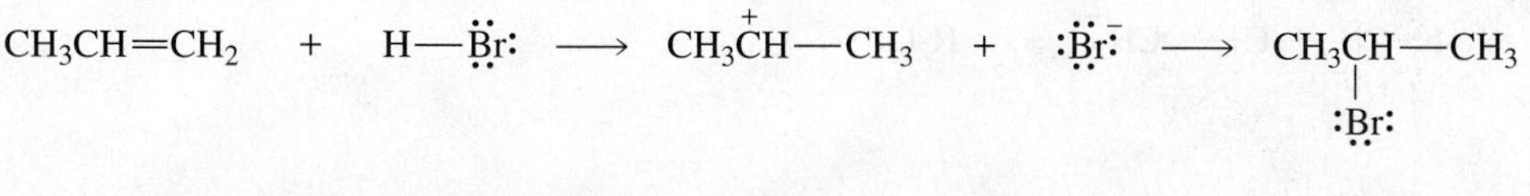 $CH_3CH=CH_2 + H-\ddot{B}r: \longrightarrow CH_3\overset{+}{C}H-CH_3 + :\ddot{B}r:^- \longrightarrow CH_3\overset{\overset{\displaystyle }{\underset{\underset{\displaystyle :\ddot{B}r:}{|}}{C}H-CH_3}}$

b. $CH_3\overset{\overset{\displaystyle CH_3}{|}}{\underset{\underset{\displaystyle CH_3}{|}}{C}-Cl} \rightleftharpoons CH_3\overset{\overset{\displaystyle CH_3}{|}}{\underset{\underset{\displaystyle CH_3}{|}}{\overset{+}{C}}} + Cl^- \overset{\dot{N}H_3}{\longrightarrow} CH_3\overset{\overset{\displaystyle CH_3}{|}}{\underset{\underset{\displaystyle CH_3}{|}}{C}-\overset{+}{N}H_3}$

c. $CH_3-\overset{\overset{\displaystyle \ddot{O}:}{\|}}{C}-Cl + H\ddot{O}:^- \longrightarrow CH_3-\overset{\overset{\displaystyle :\ddot{O}:^-}{|}}{\underset{\underset{\displaystyle :\ddot{O}H}{|}}{C}-Cl} \longrightarrow CH_3-\overset{\overset{\displaystyle \ddot{O}:}{\|}}{C}-OH + Cl^-$

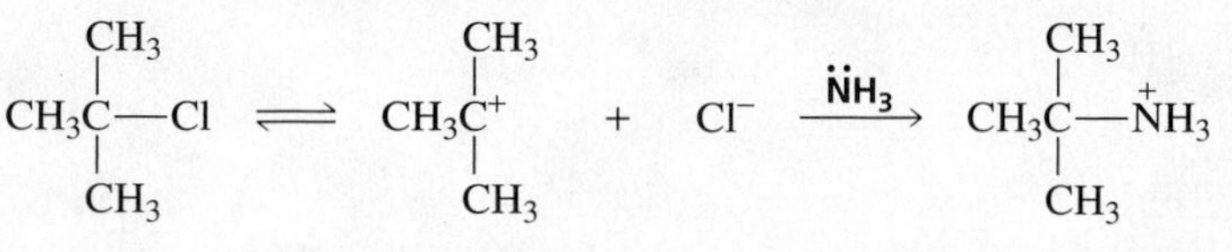

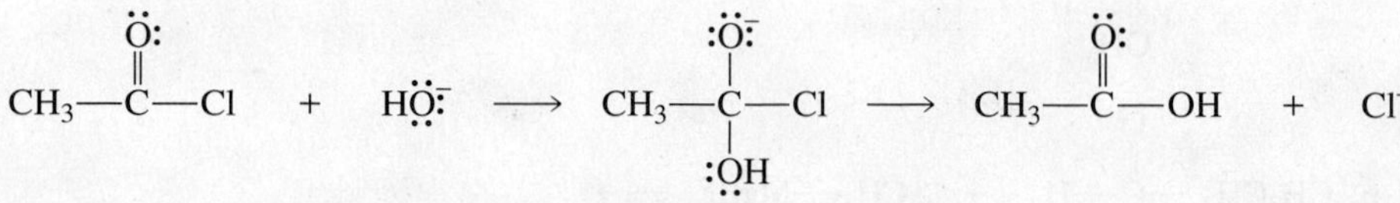

PROBLEM 9 Draw curved arrows to show the movement of the electrons in each step of the following reaction sequences.

a.

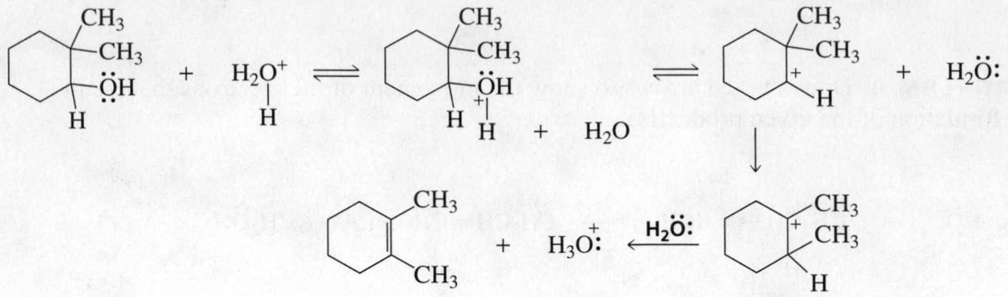

b. $CH_3CH_2CH{=}CH_2$ $\xrightarrow{CH_3\overset{+}{O}H}$ $CH_3CH_2\overset{+}{C}HCH_3$ $\xrightarrow{CH_3\ddot{O}H}$ $CH_3CH_2CHCH_3$

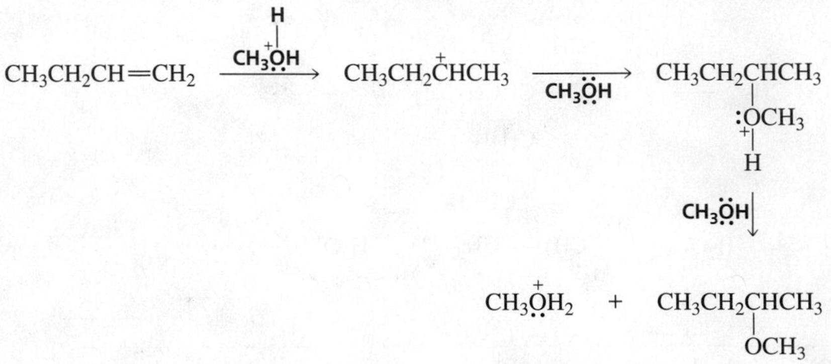

c.

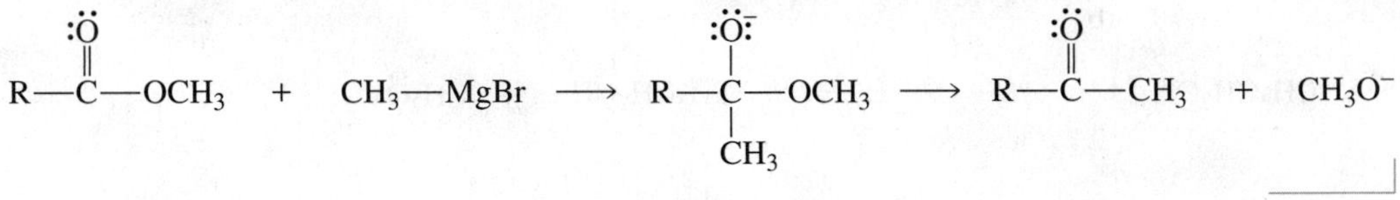

PROBLEM 10 Use what the curved arrows tell you about electron movement to determine the product of each reaction step.

a. $CH_3CH_2\ddot{O}{:}^-$ + $CH_3{-}Br$ $\longrightarrow$

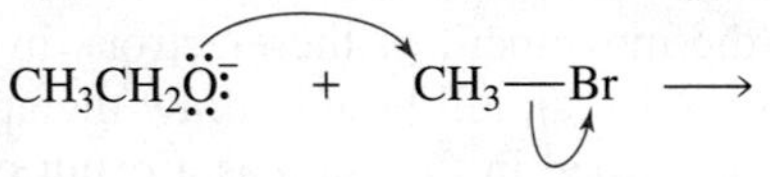

b. $CH_3{-}\overset{+\ddot{O}H}{\underset{\|}{C}}{-}OCH_3$ + $H_2\ddot{O}{:}$ $\longrightarrow$

c. $H\ddot{O}{:}^-$ + $CH_3CH_2CH{-}CH_2{-}Br$ $\longrightarrow$

with H

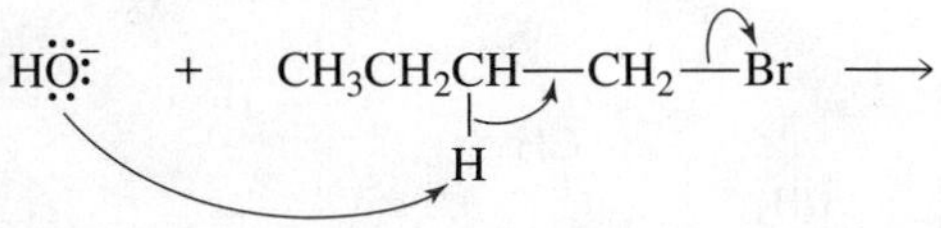

d. $CH_3CH_2{-}\overset{:\ddot{O}{:}^-}{\underset{OH}{C}}{-}NH_2$ $\longrightarrow$

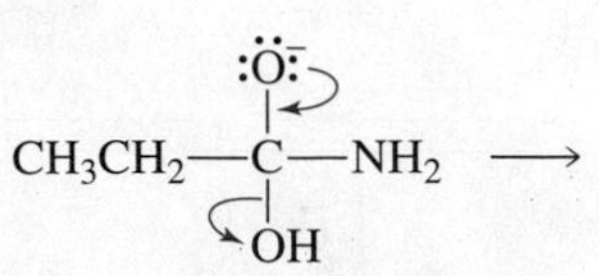

e. $CH_3CH_2{-}\overset{O}{\underset{\|}{C}}{-}H$ + $CH_3{-}MgBr$ $\longrightarrow$

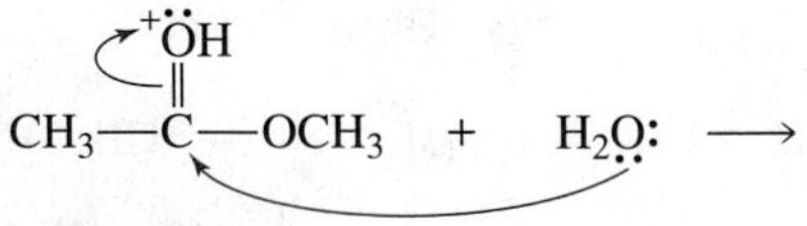

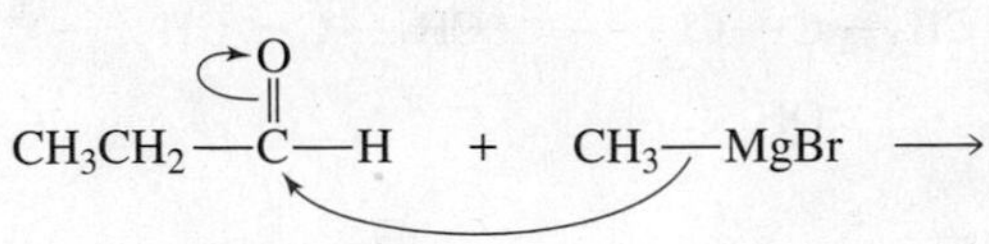

f. 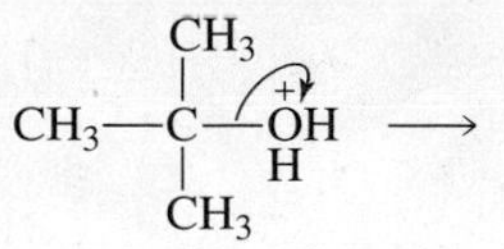 $\overset{\text{CH}_3}{\underset{\text{CH}_3}{\text{CH}_3-\text{C}-\overset{+}{\text{O}}\text{H}}}\quad\longrightarrow$

g. 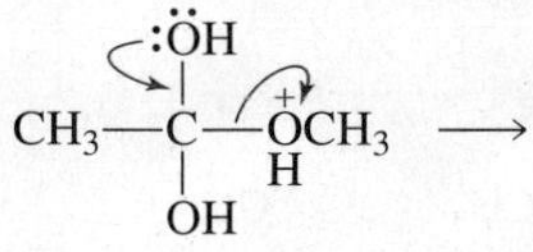 $\overset{\text{:}\ddot{\text{O}}\text{H}}{\underset{\text{OH}}{\text{CH}_3-\text{C}-\overset{+}{\text{O}}\text{CH}_3}}\quad\longrightarrow$

ANSWERS TO PROBLEMS ON DRAWING CURVED ARROWS

PROBLEM 1 Solved

a. 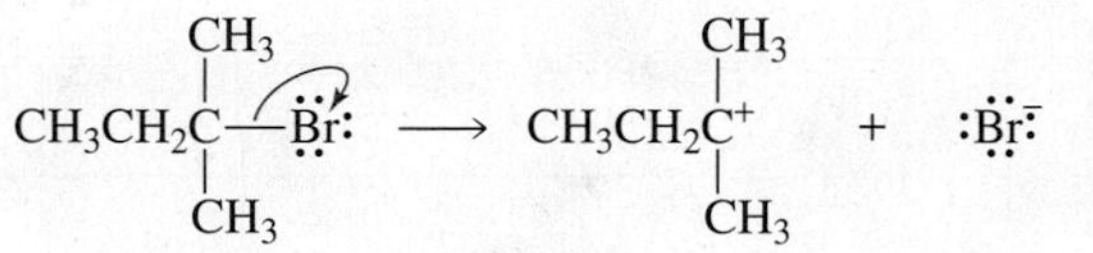 $\overset{\text{CH}_3}{\underset{\text{CH}_3}{\text{CH}_3\text{CH}_2\text{C}-\ddot{\text{B}}\ddot{\text{r}}\text{:}}}\quad\longrightarrow\quad\overset{\text{CH}_3}{\underset{\text{CH}_3}{\text{CH}_3\text{CH}_2\text{C}^+}}\quad+\quad\text{:}\ddot{\text{B}}\ddot{\text{r}}\text{:}^-$

b. ⬠—$\ddot{\text{C}}\ddot{\text{l}}$: $\longrightarrow$ ⬠+ + :$\ddot{\text{C}}\ddot{\text{l}}$:$^-$

c. ⬡—$\overset{+}{\ddot{\text{O}}}\text{H}$ below H $\longrightarrow$ ⬡+ + $\text{H}_2\ddot{\text{O}}$:

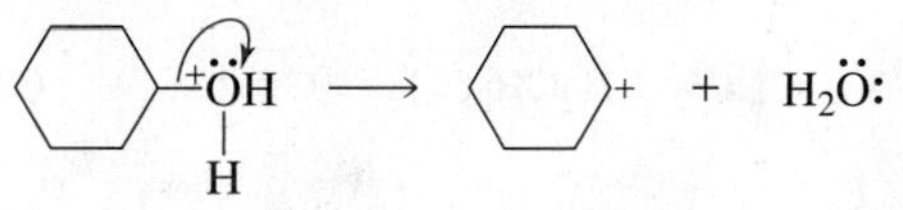

d. 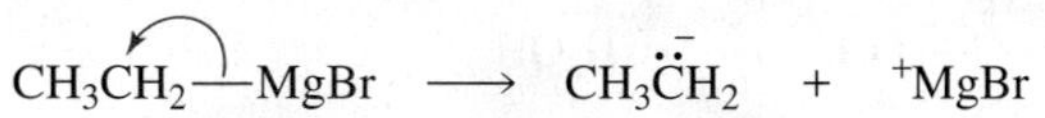 $\text{CH}_3\text{CH}_2-\text{MgBr}\quad\longrightarrow\quad\text{CH}_3\ddot{\text{C}}\text{H}_2^-\quad+\quad{}^+\text{MgBr}$

e. 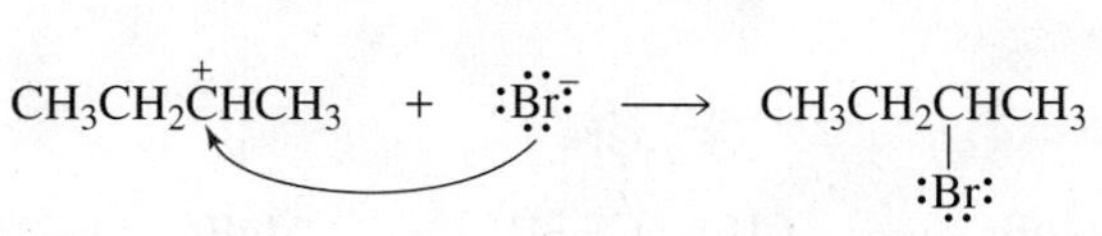 $\text{CH}_3\text{CH}_2\overset{+}{\text{C}}\text{HCH}_3\quad+\quad\text{:}\ddot{\text{B}}\ddot{\text{r}}\text{:}^-\quad\longrightarrow\quad\underset{\underset{\text{:Br:}}{|}}{\text{CH}_3\text{CH}_2\text{CHCH}_3}$

PROBLEM 2 Solved

a. 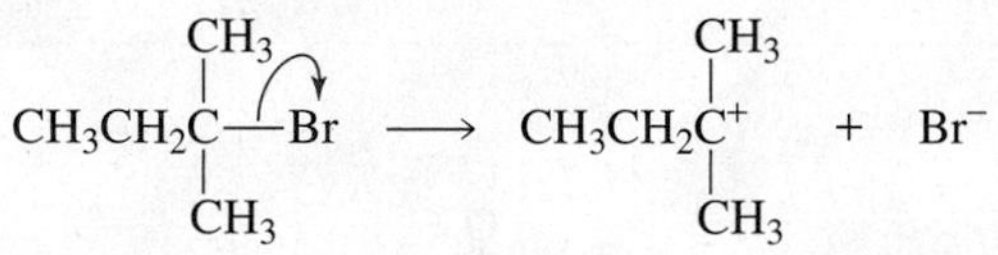 $\overset{\text{CH}_3}{\underset{\text{CH}_3}{\text{CH}_3\text{CH}_2\text{C}-\text{Br}}}\quad\longrightarrow\quad\overset{\text{CH}_3}{\underset{\text{CH}_3}{\text{CH}_3\text{CH}_2\text{C}^+}}\quad+\quad\text{Br}^-$

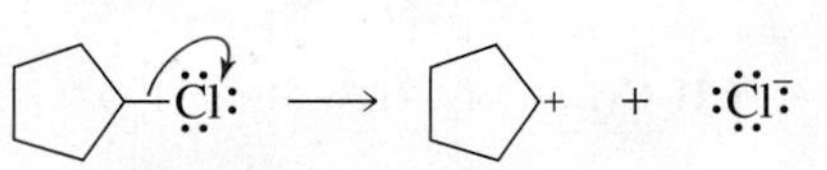

b. ⬠—Cl $\longrightarrow$ ⬠+ + Cl$^-$

c. ⬡—$\overset{+}{\text{OH}}$ below H $\longrightarrow$ ⬡+ + H_2O

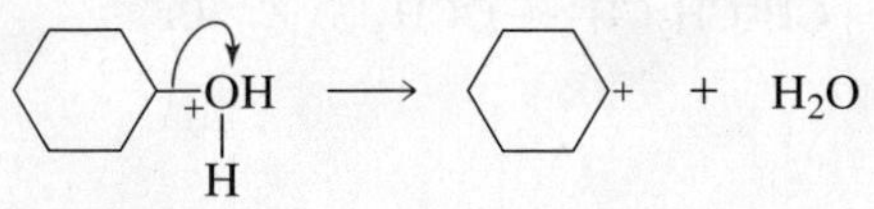

d. 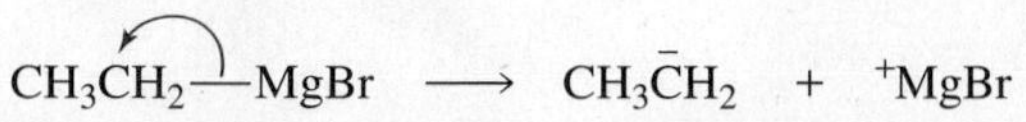 $\text{CH}_3\text{CH}_2-\text{MgBr}\quad\longrightarrow\quad\text{CH}_3\ddot{\text{C}}\text{H}_2^-\quad+\quad{}^+\text{MgBr}$

e. 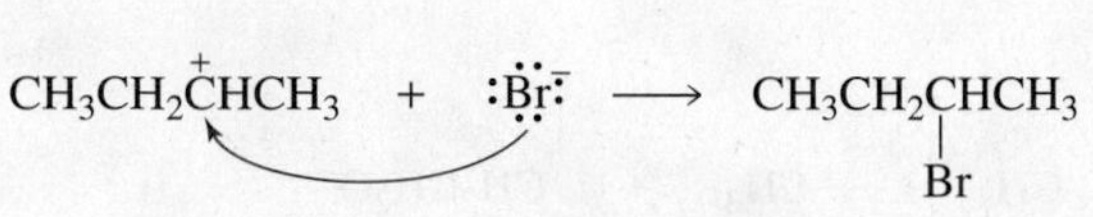 $\text{CH}_3\text{CH}_2\overset{+}{\text{C}}\text{HCH}_3\quad+\quad\text{:}\ddot{\text{B}}\ddot{\text{r}}\text{:}^-\quad\longrightarrow\quad\underset{\underset{\text{Br}}{|}}{\text{CH}_3\text{CH}_2\text{CHCH}_3}$

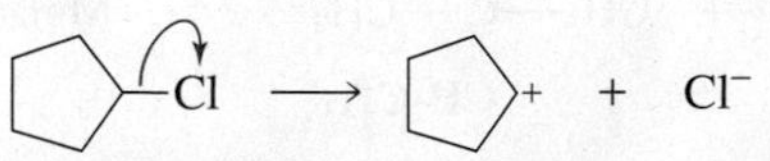

PROBLEM 3 Solved

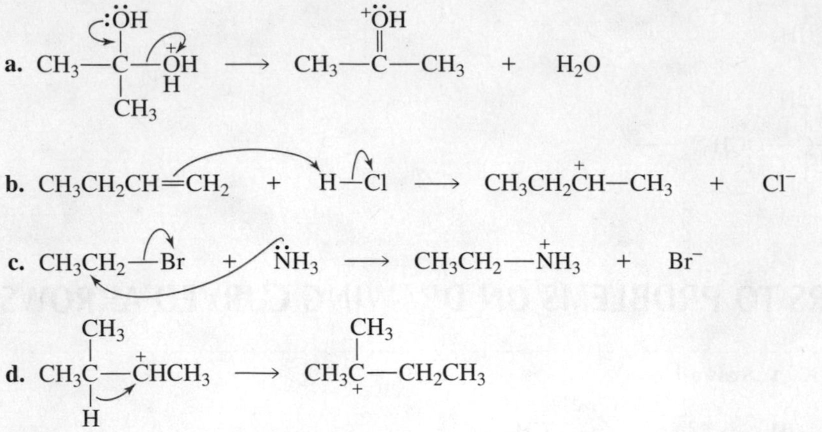

a.

$$CH_3-\overset{:\ddot{O}H}{\underset{\underset{CH_3}{|}}{\overset{|}{C}}}-\overset{+}{\underset{H}{\overset{|}{O}H}} \longrightarrow CH_3-\overset{+\ddot{O}H}{\overset{\|}{C}}-CH_3 \ + \ H_2O$$

b. $CH_3CH_2CH{=}CH_2 \ + \ H{-}Cl \longrightarrow CH_3CH_2\overset{+}{C}H{-}CH_3 \ + \ Cl^-$

c. $CH_3CH_2{-}Br \ + \ \ddot{N}H_3 \longrightarrow CH_3CH_2{-}\overset{+}{N}H_3 \ + \ Br^-$

d. $CH_3\overset{CH_3}{\underset{H}{\overset{|}{\underset{|}{C}}}}{-}\overset{+}{C}HCH_3 \longrightarrow CH_3\overset{CH_3}{\underset{+}{\overset{|}{C}}}{-}CH_2CH_3$

PROBLEM 4 Solved

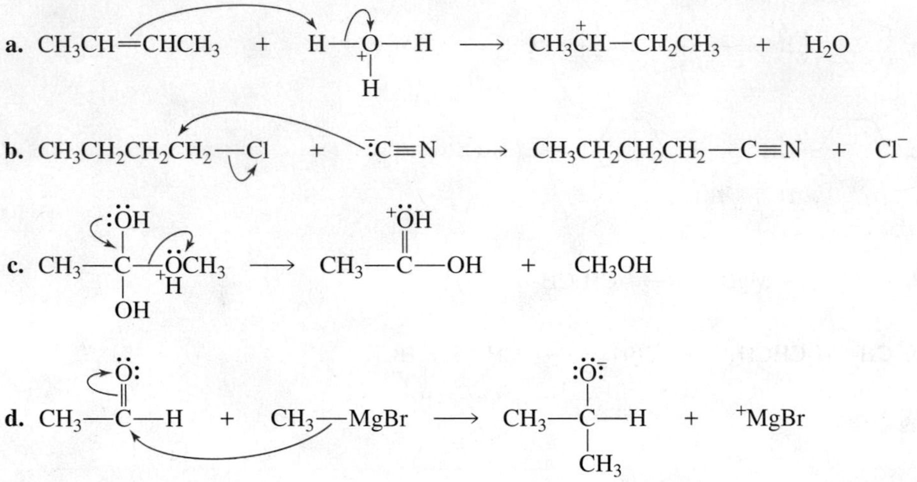

a. $CH_3CH{=}CHCH_3 \ + \ H{-}\overset{+}{\underset{H}{\overset{|}{O}}}{-}H \longrightarrow CH_3\overset{+}{C}H{-}CH_2CH_3 \ + \ H_2O$

b. $CH_3CH_2CH_2CH_2{-}Cl \ + \ \ ^-{:}C{\equiv}N \longrightarrow CH_3CH_2CH_2CH_2{-}C{\equiv}N \ + \ Cl^-$

c. $CH_3-\overset{:\ddot{O}H}{\underset{\underset{OH}{|}}{\overset{|}{C}}}-\overset{+\ddot{O}CH_3}{\underset{H}{}} \longrightarrow CH_3-\overset{+\ddot{O}H}{\overset{\|}{C}}-OH \ + \ CH_3OH$

d. $CH_3-\overset{\ddot{O}:}{\overset{\|}{C}}-H \ + \ CH_3{-}MgBr \longrightarrow CH_3-\overset{:\ddot{O}^-}{\underset{CH_3}{\overset{|}{C}}}-H \ + \ ^+MgBr$

PROBLEM 5 Solved

a. $CH_3-\overset{O}{\overset{\|}{C}}-CH_3 \ + \ CH_3CH_2{-}MgBr \longrightarrow CH_3-\overset{O^-}{\underset{CH_2CH_3}{\overset{|}{C}}}-CH_3 \ + \ ^+MgBr$

b. $CH_3CH_2CH_2{-}Br \ + \ CH_3\ddot{O}{:}^- \longrightarrow CH_3CH_2CH_2{-}OCH_3 \ + \ Br^-$

c.

d. $CH_3-\overset{:\ddot{O}^-}{\underset{CH_3}{\overset{|}{C}}}-OCH_2CH_3 \longrightarrow CH_3-\overset{O}{\overset{\|}{C}}-CH_3 \ + \ CH_3CH_2O^-$

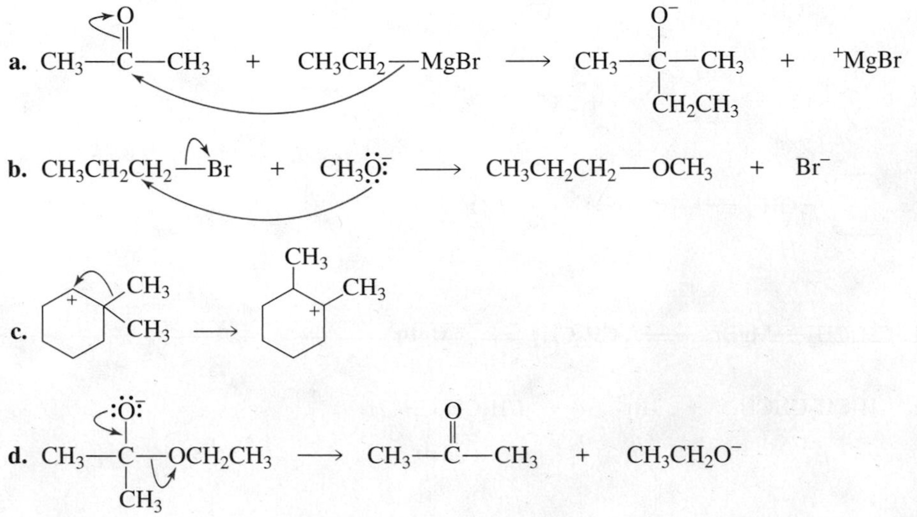

PROBLEM 6 Solved

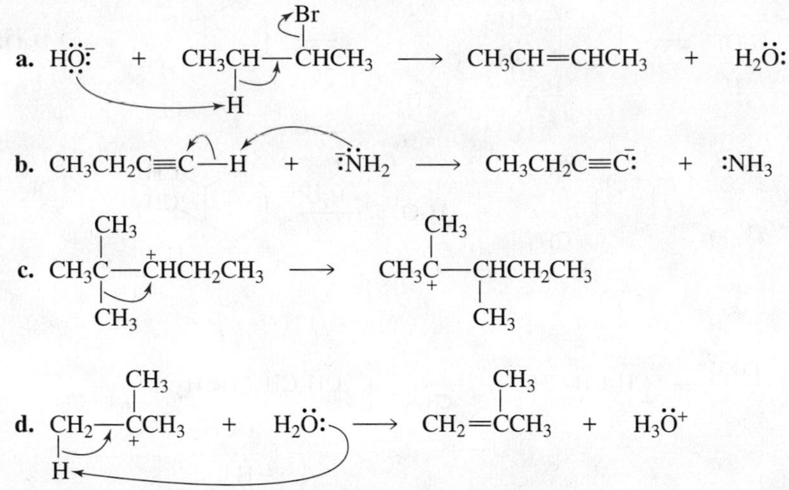

PROBLEM 7 Solved

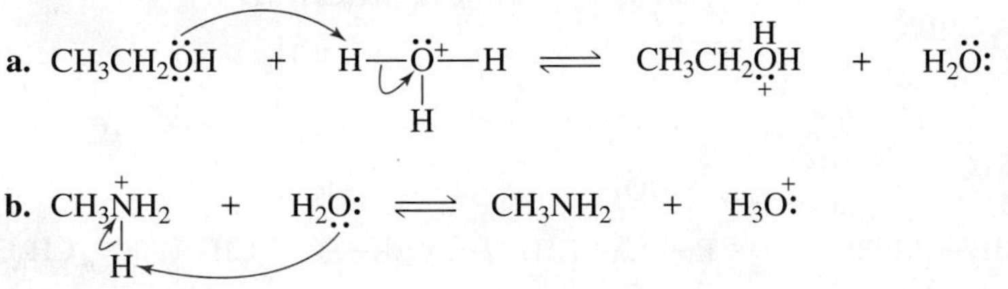

PROBLEM 8 Solved

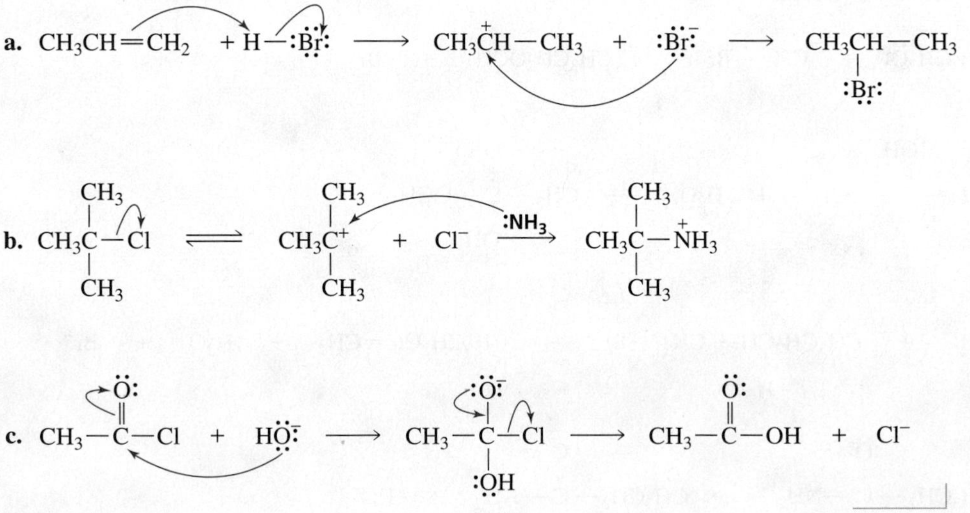

PROBLEM 9 Solved

a.

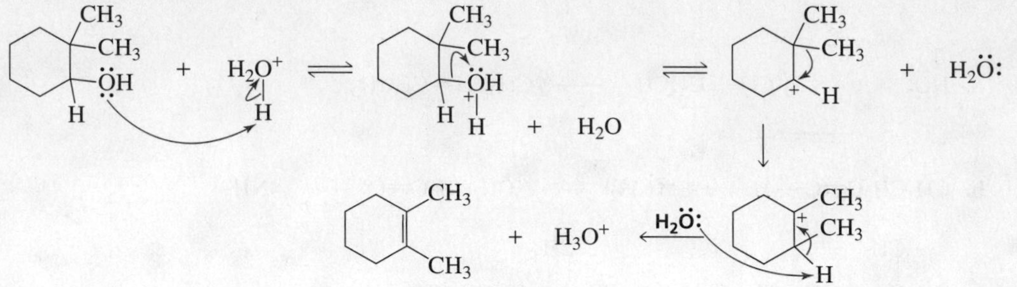

b. 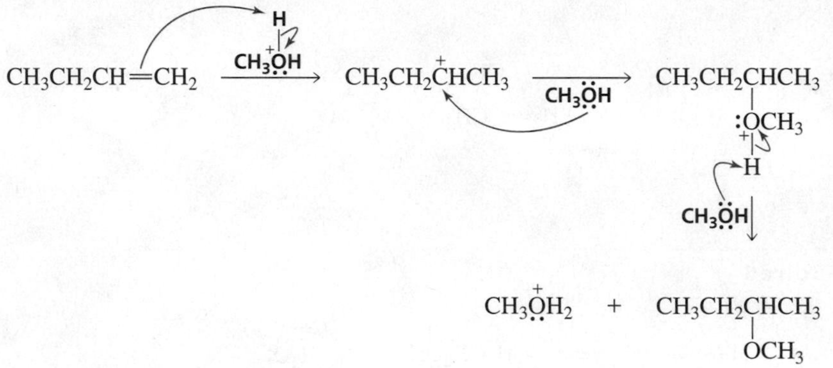 $CH_3CH_2CH=CH_2$ $\xrightarrow{CH_3\ddot{O}H}$ $CH_3CH_2\overset{+}{C}HCH_3$ $\xrightarrow{CH_3\ddot{O}H}$ $CH_3CH_2CHCH_3$

$CH_3\overset{+}{\ddot{O}}H_2$ + $CH_3CH_2CHCH_3$

$\qquad\qquad\qquad\qquad\qquad\quad OCH_3$

c. $R-\overset{:\ddot{O}:}{\underset{}{C}}-OCH_3$ + CH_3-MgBr $\longrightarrow$ $R-\overset{:\ddot{O}:^-}{\underset{CH_3}{C}}-CH_3$ $\longrightarrow$ $R-\overset{:\ddot{O}:}{\underset{}{C}}-CH_3$ + CH_3O^-

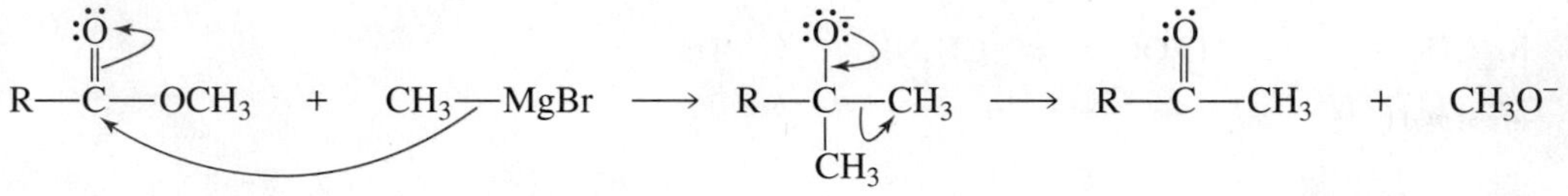

PROBLEM 10 Solved

a. 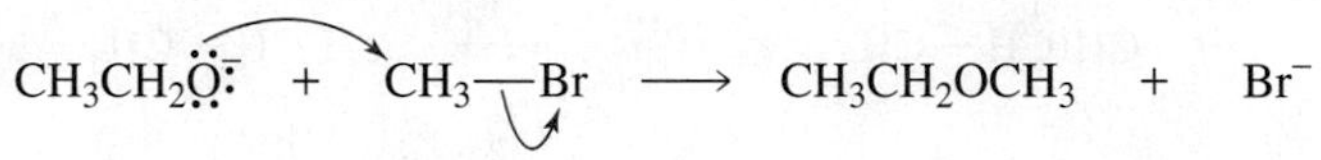 $CH_3CH_2\ddot{\ddot{O}}:^-$ + CH_3-Br $\longrightarrow$ $CH_3CH_2OCH_3$ + Br^-

b. 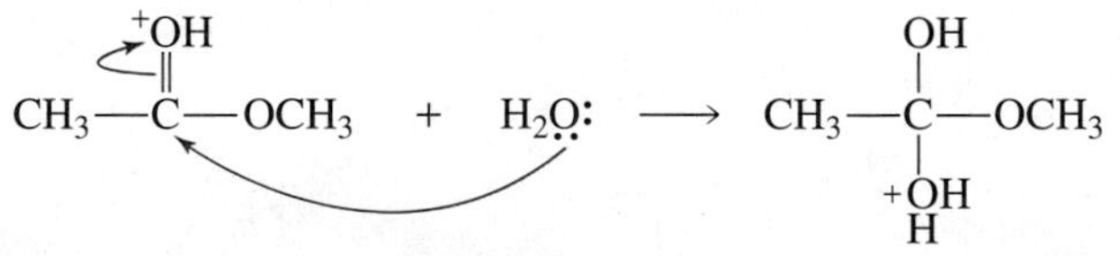 $CH_3-\overset{+OH}{\underset{}{\overset{\|}{C}}}-OCH_3$ + $H_2\ddot{O}:$ $\longrightarrow$ $CH_3-\overset{OH}{\underset{\overset{+OH}{\underset{H}{}}}{C}}-OCH_3$

c. $H\ddot{O}:^-$ + $CH_3CH_2CH-CH_2-Br$ $\longrightarrow$ $CH_3CH_2CH=CH_2$ + H_2O + Br^-

$\qquad\qquad\qquad\qquad\qquad H$

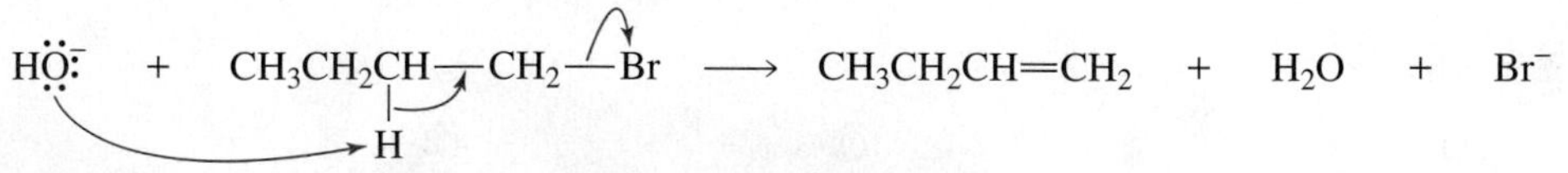

d. 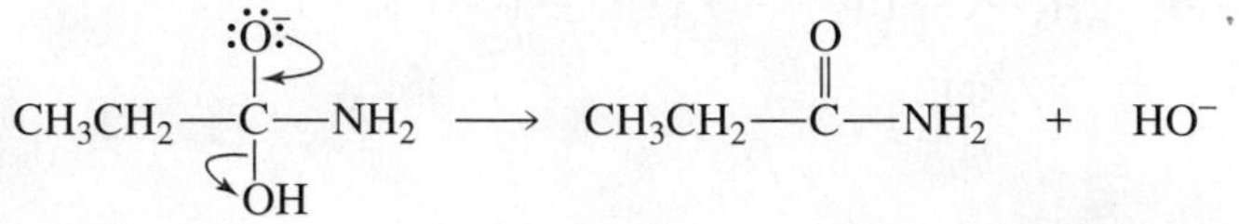 $CH_3CH_2-\overset{:\ddot{O}:^-}{\underset{\overset{}{OH}}{C}}-NH_2$ $\longrightarrow$ $CH_3CH_2-\overset{O}{\underset{}{\overset{\|}{C}}}-NH_2$ + HO^-

e. $CH_3CH_2-\overset{\overset{\displaystyle O}{\|}}{C}-H$ + CH_3-MgBr ⟶ $CH_3CH_2-\overset{\overset{\displaystyle O^-}{|}}{\underset{\underset{\displaystyle CH_3}{|}}{C}}-H$ + ^+MgBr

f. $CH_3-\overset{\overset{\displaystyle CH_3}{|}}{\underset{\underset{\displaystyle CH_3}{|}}{C}}-\overset{\overset{\displaystyle +}{}}{\underset{\underset{\displaystyle H}{}}{O}}H$ ⟶ $CH_3-\overset{\overset{\displaystyle CH_3}{|}}{\underset{\underset{\displaystyle CH_3}{|}}{C^+}}$ + H_2O

g. $CH_3-\overset{\overset{\displaystyle :\ddot{O}H}{|}}{\underset{\underset{\displaystyle OH}{|}}{C}}-\overset{\overset{\displaystyle +}{}}{\underset{\underset{\displaystyle H}{}}{O}}CH_3$ ⟶ $CH_3-\overset{\overset{\displaystyle +\ddot{O}H}{\|}}{C}-OH$ + CH_3OH

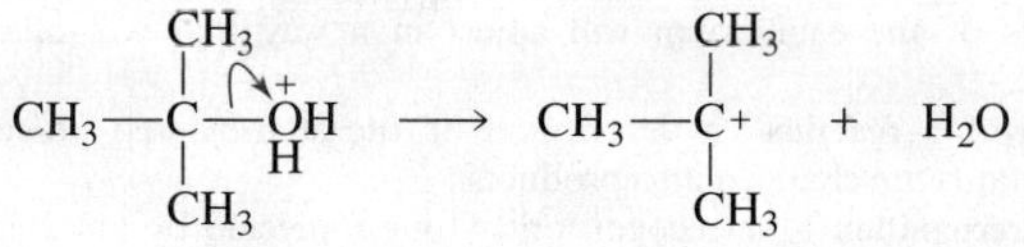

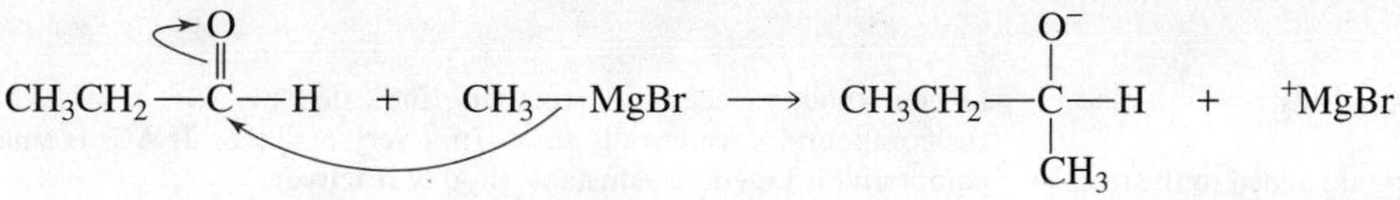

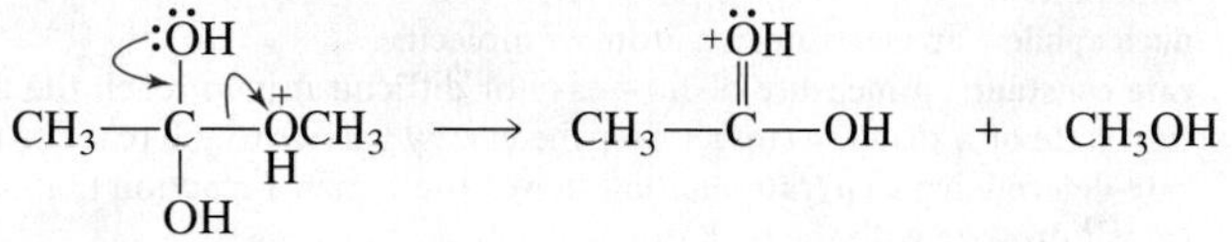

GLOSSARY

active site a pocket or cleft in an enzyme where the substrate is bound.

acyclic noncyclic.

addition reaction a reaction in which atoms or groups are added to the reactant.

alkene a hydrocarbon that contains a double bond.

allyl group $CH_2{=}CHCH_2{-}$

allylic carbon an sp^3 carbon adjacent to a vinylic carbon.

Arrhenius equation relates the rate constant of a reaction to the energy of activation and to the temperature at which the reaction is carried out ($k = Ae^{-E_a/RT}$).

catalyst a species that increases the rate at which a reaction occurs without being consumed in the reaction. Because it does not change the equilibrium constant of the reaction, it does not change the amount of product that is formed.

E isomer the isomer with the high-priority groups on opposite sides of the double bond.

electrophile an electron-deficient atom or molecule.

electrophilic addition reaction an addition reaction in which the first species that adds to the reactant is an electrophile.

endergonic reaction a reaction with a positive $\Delta G°$.

endothermic reaction a reaction with a positive $\Delta H°$.

enthalpy the heat given off ($-\Delta H°$) or the heat absorbed ($+\Delta H°$) during the course of a reaction.

entropy a measure of the freedom of motion in a system.

enzyme a protein that is a catalyst.

exergonic reaction a reaction with a negative $\Delta G°$.

exothermic reaction a reaction with a negative $\Delta H°$.

first-order rate constant the rate constant of a first-order reaction.

first-order reaction (unimolecular reaction) a reaction whose rate depends on the concentration of one reactant.

free energy of activation ($\Delta G^{\ddagger}$) the true energy barrier to a reaction.

functional group the center of reactivity in a molecule.

Gibbs standard free-energy change ($\Delta G°$) the difference between the free-energy content of the products and the free-energy content of the reactants at equilibrium under standard conditions (1 M, 25 °C, 1 atm).

intermediate a species formed during a reaction and that is not the final product of the reaction.

kinetics the field of chemistry that deals with the rates of chemical reactions.

kinetic stability chemical reactivity, indicated by $\Delta G^{\ddagger}$. If $\Delta G^{\ddagger}$ is large, the compound is kinetically stable (not very reactive). If $\Delta G^{\ddagger}$ is small, the compound is kinetically unstable (highly reactive).

Le Châtelier's principle states that if an equilibrium is disturbed, the components of the equilibrium will adjust in a way that will offset the disturbance.

mechanism of a reaction a description of the step-by-step process by which reactants are changed into products.

molecular recognition the recognition of one molecule by another as a result of specific interactions; for example, the specificity of an enzyme for its substrate.

nucleophile an electron-rich atom or molecule.

rate constant a measure of how easy or difficult it is to reach the transition state of a reaction (to get over the energy barrier to the reaction).

rate-determining step (rate-limiting step) the step in a reaction that has the transition state with the highest energy.

reaction coordinate diagram describes the energy changes that take place during the course of a reaction.

saturated hydrocarbon a hydrocarbon that is completely saturated (i.e., contains no double or triple bonds) with hydrogen.

second-order rate constant the rate constant of a second-order reaction.

second-order reaction (bimolecular reaction) a reaction whose rate depends on the concentration of two reactants.

substrate the reactant of an enzyme-catalyzed reaction.

thermodynamics the field of chemistry that describes the properties of a system at equilibrium.

thermodynamic stability is indicated by $\Delta G°$. If $\Delta G°$ is negative, the products are more stable than the reactants. If $\Delta G°$ is positive, the reactants are more stable than the products.

transition state the highest point on a hill in a reaction coordinate diagram. In the transition state, bonds in the reactant that will break are partially broken and bonds in the product that will form are partially formed.

unsaturated hydrocarbon a hydrocarbon that contains one or more double or triple bonds.

vinyl group $CH_2{=}CH{-}$

vinylic carbon a carbon in a carbon–carbon double bond.

Z isomer the isomer with the high-priority groups on the same side of the double bond.

PHOTO CREDITS

Photo credits are listed in order of appearance.

The Reactions of Alkenes •
The Stereochemistry of
Addition Reactions

From Chapter 6 of *Organic Chemistry*, Seventh Edition. Paula Yurkanis Bruice. Copyright © 2014 by Pearson Education, Inc.

The Reactions of Alkenes • The Stereochemistry of Addition Reactions

pesticides being sprayed on a crop

Two questions we will consider in this chapter: Which are more harmful, the pesticides farmers spray on our food or the pesticides plants make to ward off predators? And, how do trans fats get into our food?

Organic compounds can be divided into families and all the members of a family react in the same way. One family consists of compounds with carbon–carbon double bonds—compounds known as **alkenes.**

The π bond of a double bond is weak, so it is easily broken. This allows alkenes to undergo addition reactions. An alkene is a nucleophile, so the first species it reacts with is an electrophile. Therefore, we can more precisely say that alkenes undergo *electrophilic* addition reactions.

When an alkene undergoes an electrophilic addition reaction with HBr, the first step is a relatively slow addition of a proton (an electrophile) to the alkene (a nucleophile). A **carbocation intermediate** (an electrophile) is formed, which then reacts rapidly with a bromide ion (a nucleophile) to form an alkyl halide. Notice that each step involves the reaction of an electrophile with a nucleophile. *The overall reaction is the addition of an electrophile to one of the sp^2 carbons of the alkene and the addition of a nucleophile to the other sp^2 carbon.*

A curved arrow with a two-barbed arrowhead signifies the movement of two electrons. The arrow always points *from* the electron donor *to* the electron acceptor.

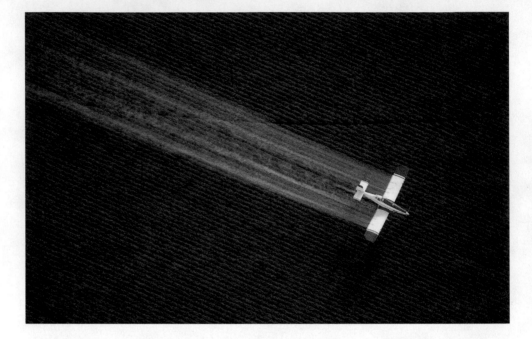

the electrophile adds to an *sp*^{2} carbon of the alkene

carbocation intermediate

the nucleophile adds to the carbocation

In this chapter, you will see that alkenes react with a wide variety of electrophiles. You will also see that some of the reactions form carbocation intermediates (like the one formed when an alkene reacts with HBr), some form other kinds of intermediates, and some form no intermediate at all. At first glance, the reactions covered in this chapter may appear to be quite different, but you will see that they all occur by similar mechanisms.

As you study each reaction, look for the feature that all alkene reactions have in common: *the relatively loosely held π electrons of the carbon–carbon double bond are attracted to an electrophile. Thus, each reaction starts with the addition of an electrophile to one of the sp^2 carbons of the alkene and concludes with the addition of a nucleophile to the other sp^2 carbon.* The end result is that the π bond breaks and the electrophile and nucleophile form new σ bonds with the sp^2 carbons. Notice that the sp^2 carbons in the reactant become sp^3 carbons in the product.

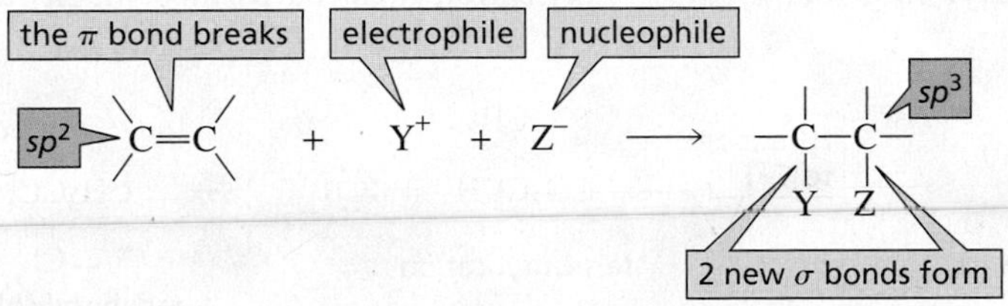

This reactivity makes it possible to synthesize a wide variety of compounds from alkenes. For example, we will see that alkyl halides, alcohols, ethers, epoxides, alkanes, aldehydes, and ketones can all be synthesized from alkenes by electrophilic addition reactions. The particular product obtained depends only on the *electrophile* and the *nucleophile* used in the addition reaction.

1 THE ADDITION OF A HYDROGEN HALIDE TO AN ALKENE

If the electrophilic reagent that adds to an alkene is a hydrogen halide (HF, HCl, HBr, or HI), the product of the reaction will be an alkyl halide:

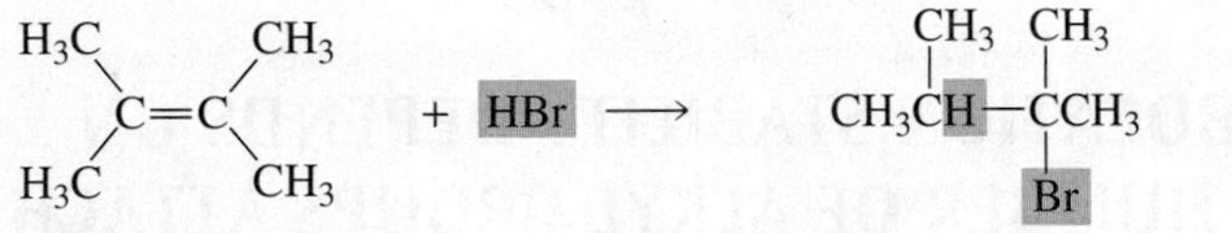

2,3-dimethyl-2-butene   **2-bromo-2,3-dimethylbutane**

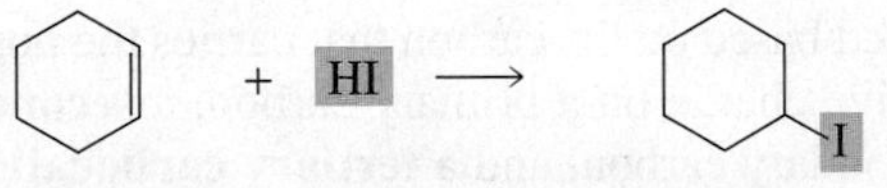

cyclohexene   **iodocyclohexane**

Because the alkenes in the preceding reactions have the same substituents on both sp^2 carbons, it is easy to predict the product of the reaction: the electrophile (H$^+$) adds to either one of the sp^2 carbons and the nucleophile (X$^-$) adds to the other sp^2 carbon. It does not matter which sp^2 carbon the electrophile adds to because the same product will be obtained in either case.

But what happens if the alkene does *not* have the same substituents on both sp^2 carbons? Which sp^2 carbon gets the hydrogen? For example, does the following reaction form *tert*-butyl chloride or isobutyl chloride?

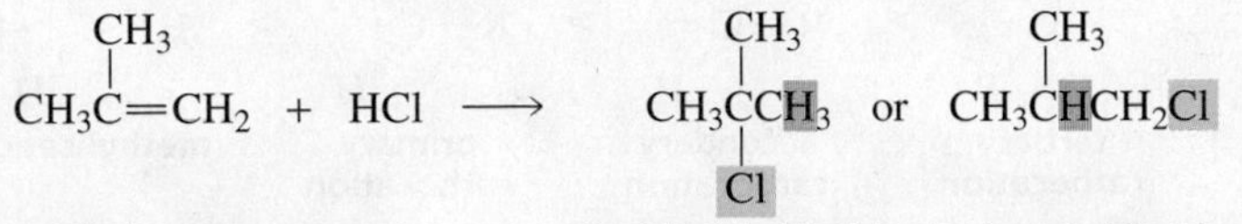

2-methylpropene   **tert-butyl chloride   isobutyl chloride**

245

To answer this question, we need to carry out this reaction, isolate the products, and identify them. When we do, we find that the only product is *tert*-butyl chloride. If we can find out why it is the only product, then we can use this knowledge to predict the products of other alkene reactions. To do this, we need to look again at the **mechanism of the reaction**.

The first step of the reaction—the addition of H$^+$ to an sp^2 carbon to form either the *tert*-butyl cation or the isobutyl cation—is the slow rate-determining step. If there is any difference in the rate of formation of these two carbocations, then the one that is formed faster will be the predominant product of the first step. Moreover, the particular carbocation formed in the first step determines the final product of the reaction. That is, if the *tert*-butyl cation is formed, it will react rapidly with Cl$^-$ to form *tert*-butyl chloride. On the other hand, if the isobutyl cation is formed, it will react rapidly with Cl$^-$ to form isobutyl chloride. Since we know that the only product of the reaction is *tert*-butyl chloride, the *tert*-butyl cation must be formed much faster than the isobutyl cation.

> The carbocation's positive charge is on the sp^2 carbon that does *not* become attached to the proton.

Why is the *tert*-butyl cation formed faster? To answer this question, we need to look at two things: (1) the factors that affect the stability of a carbocation, and (2) how its stability affects the rate at which it is formed.

PROBLEM 1

Draw the mechanism for the reaction of cyclohexene with HCl.

---

2 CARBOCATION STABILITY DEPENDS ON THE NUMBER OF ALKYL GROUPS ATTACHED TO THE POSITIVELY CHARGED CARBON

Carbocations are classified based on the carbon that carries the positive charge: a **primary carbocation** has a positive charge on a primary carbon, a **secondary carbocation** has a positive charge on a secondary carbon, and a **tertiary carbocation** has a positive charge on a tertiary carbon.

Tertiary carbocations are more stable than secondary carbocations, and secondary carbocations are more stable than primary carbocations. Thus, we see that the stability of a carbocation increases as the number of alkyl substituents attached to the positively charged carbon increases. These are relative stabilities, however, because carbocations are rarely stable enough to isolate.

relative stabilities of carbocations

> The more alkyl groups attached to the positively charged carbon, the more stable the carbocation.

Alkyl groups stabilize carbocations because they decrease the concentration of positive charge on the carbon. Notice that the blue area in the following electrostatic potential maps (representing positive charge) is the least intense for the most stable *tert*-butyl cation (a tertiary carbocation) and the most intense for the least stable methyl cation.

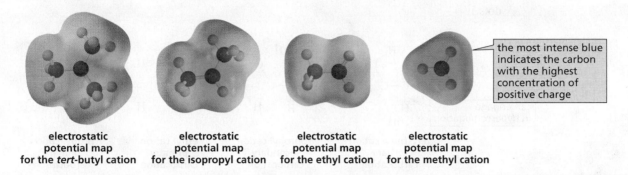

the most intense blue indicates the carbon with the highest concentration of positive charge

electrostatic potential map for the *tert*-butyl cation

electrostatic potential map for the isopropyl cation

electrostatic potential map for the ethyl cation

electrostatic potential map for the methyl cation

Carbocation stability: 3° > 2° > 1°

How do alkyl groups decrease the concentration of positive charge on the carbon? The positive charge on a carbon signifies an empty *p* orbital. Figure 1 shows that in the ethyl cation, the orbital of an adjacent C—H σ bond (the orange orbital) can overlap the empty *p* orbital (the purple orbital). This movement of electrons from a σ bond orbital toward the vacant *p* orbital decreases the charge on the sp^2 carbon and causes a partial positive charge to develop on the two atoms bonded by the overlapping σ bond orbital (the H and the C). With three atoms sharing the positive charge, the carbocation is stabilized because a charged species is more stable if its charge is dispersed over more than one atom. In contrast, the positive charge in the methyl cation is concentrated solely on one atom.

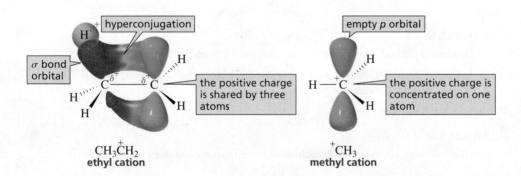

hyperconjugation

σ bond orbital

the positive charge is shared by three atoms

empty *p* orbital

the positive charge is concentrated on one atom

$CH_3\overset{+}{C}H_2$
ethyl cation

$\overset{+}{C}H_3$
methyl cation

◀ **Figure 1**

Stabilization of a carbocation by hyperconjugation. In the ethyl cation, the electrons of an adjacent C—H σ bond orbital are delocalized into the empty *p* orbital. Hyperconjugation cannot occur in the methyl cation.

Delocalization of electrons by the overlap of a σ bond orbital with an empty orbital on an adjacent carbon is called **hyperconjugation**. The molecular orbital diagram in Figure 2 is another way of depicting the stabilization achieved by the overlap of a filled C—H σ bond orbital with an empty *p* orbital.

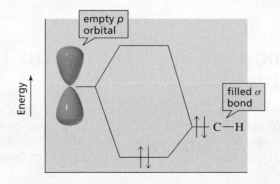

empty *p* orbital

Energy

filled σ bond

C—H

◀ **Figure 2**

A molecular orbital diagram showing the stabilization achieved by overlapping the electrons of a C—H bond with an empty *p* orbital.

247

Hyperconjugation occurs only if the orbital of the σ bond and the empty p orbital have the proper orientation. The proper orientation is easily achieved, though, because there is free rotation about the carbon–carbon σ bond. Notice that the σ bond orbitals that can overlap the empty p orbital are those *attached to an atom that is attached to the positively charged carbon.* In the *tert*-butyl cation, nine σ bond orbitals can potentially overlap the empty p orbital of the positively charged carbon. (The nine σ bonds are indicated by red dots.)

tert-butyl cation	isopropyl cation	ethyl cation	propyl cation
tertiary carbocation	secondary carbocation	primary carbocation	primary carbocation

The isopropyl cation has six such orbitals, whereas the ethyl and propyl cations each have three. Therefore, hyperconjugation stabilizes the tertiary carbocation more than the secondary carbocation, and it stabilizes the secondary carbocation more than either of the primary carbocations. Notice that both C—H and C—C σ bond orbitals can overlap the empty p orbital.

PROBLEM 2♦

a. How many σ bond orbitals are available for overlap with the vacant p orbital in the methyl cation?

b. Which is more stable, a methyl cation or an ethyl cation? Why?

PROBLEM 3♦

a. How many σ bond orbitals are available for overlap with the vacant p orbital in

    1. the isobutyl cation?   **2.** the *n*-butyl cation?   **3.** the *sec*-butyl cation?

b. Which of the carbocations in part **a** is most stable?

PROBLEM 4♦

List the following carbocations in each set in order from most stable to least stable:

$$\underset{\quad\ \ \overset{\displaystyle CH_3}{|}}{\text{a. } CH_3CH_2\overset{+}{C}CH_3}\qquad CH_3CH_2\overset{+}{C}HCH_3 \qquad CH_3CH_2CH_2\overset{+}{C}H_2$$

$$\underset{\ \ \underset{\displaystyle Cl}{|}}{\text{b. } CH_3\overset{+}{C}HCH_2\overset{+}{C}H_2}\qquad \underset{\underset{\displaystyle CH_3}{|}}{CH_3CHCH_2\overset{+}{C}H_2} \qquad \underset{\underset{\displaystyle F}{|}}{CH_3CHCH_2\overset{+}{C}H_2}$$

---

3   WHAT DOES THE STRUCTURE OF THE TRANSITION STATE LOOK LIKE?

The rate of a reaction is determined by the free energy of activation, which is the difference between the free energy of the transition state and the free energy of the reactant. Therefore, to understand how the stability of a carbocation affects the rate at which it is formed, we need to know something about the transition state of the reaction.

The structure structure of the transition state resembles both the structure of the reactants and the structure of the products. But which does it resemble more closely? Is its structure more like that of the reactants (I in the following reaction scheme), or that of the products (III), or is its structure about halfway between that of the reactants and the products (II)?

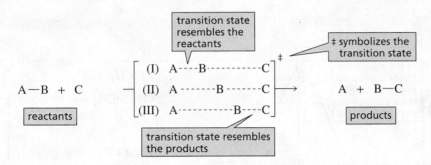

This question is answered by the **Hammond postulate,** which says that *the transition state is more similar in structure to the species to which it is more similar in energy.*

Thus, in an exergonic reaction, the energy of the transition state is more similar to the energy of the reactants (Figure 3, curve I), so its structure will be more similar to that of the reactants. In an endergonic reaction (curve III), the energy of the transition state is more similar to the energy of the products, so its structure will be more similar to that of the products. Only when the reactants and products have identical energies (curve II) would we expect the structure of the transition state to be exactly halfway between that of the reactants and that of the products.

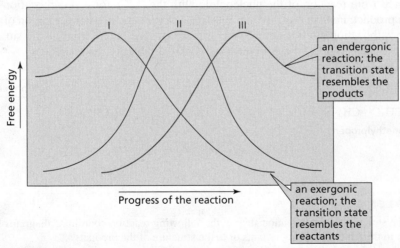

The transition state is more similar in structure to the species to which it is more similar in energy.

In an exergonic reaction ($\Delta G° < 0$), the transition state is more similar in energy to the reactants, so its structure resembles that of the reactants.

In an endergonic reaction ($\Delta G° > 0$), the transition state is more similar in energy to the products, so its structure resembles that of the products.

▲ **Figure 3**

Reaction coordinate diagrams for reactions with an early transition state (I), a midway transition state (II), and a late transition state (III).

Because the formation of a carbocation is an endergonic reaction (Figure 4), the structure of the transition state resembles the structure of the carbocation product. This means that the transition state has a significant amount of positive charge on a carbon. The same factors that stabilize the positively charged carbocation stabilize the partially positively charged transition state. Therefore, the transition state leading to the *tert*-butyl cation (a partially charged tertiary cation) is more stable (lower in energy) than the transition state leading to the isobutyl cation (a partially charged primary cation). Thus, the *tert*-butyl cation, with a smaller energy of activation, is formed faster than the isobutyl cation.

In an electrophilic addition reaction,
the more stable carbocation is formed more rapidly.

Now we know why the *tert*-butyl cation is formed faster than the isobutyl cation when 2-methylpropene reacts with HCl.

Notice that because the amount of positive charge in the transition state is less than the amount of positive charge in the product, the difference in the stabilities of the two transition states in Figure 4 is less than the difference in the stabilities of the two carbocation products.

The more stable carbocation is formed more rapidly.

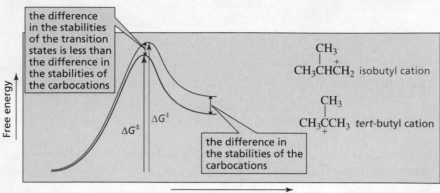

► **Figure 4**

Reaction coordinate diagram for the addition of H⁺ to 2-methylpropene to form the primary isobutyl cation and the tertiary *tert*-butyl cation.

The relative rates of formation of the two carbocations determine the relative amounts of products formed, because formation of the carbocation is the rate-limiting step of the reaction. If the difference in the rates is small, both products will be formed, but the major product will be the one formed from reaction of the nucleophile with the faster formed carbocation. If the difference in the rates is sufficiently large, however, the product formed from reaction of the nucleophile with the faster formed carbocation will be the only product. For example, in the following reaction, the rates of formation of the two possible carbocation intermediates—one primary and the other tertiary—are sufficiently different to cause *tert*-butyl chloride to be the only product of the reaction.

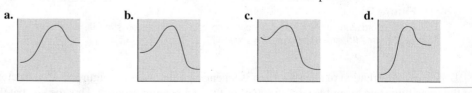

PROBLEM 5♦

Will the structure of the transition state in the following reaction coordinate diagrams be more similar to the structure of the reactants or to the structure of the products?

a.　　　　**b.**　　　　**c.**　　　　**d.**

4　**ELECTROPHILIC ADDITION REACTIONS ARE REGIOSELECTIVE**

Now that we know that the major product of an electrophilic addition reaction is the one obtained by adding the electrophile to the sp^2 carbon that results in formation of the more stable carbocation, we can predict the major product of the reaction of an unsymmetrical alkene.

For example, in the following reaction, the proton can add to C-1 to form a secondary carbocation or it can add to C-2 to form a primary carbocation. Because the secondary carbocation is more stable, it is formed more rapidly. (Primary carbocations are so unstable that they form only with great difficulty.) As a result, the only product is 2-chloropropane.

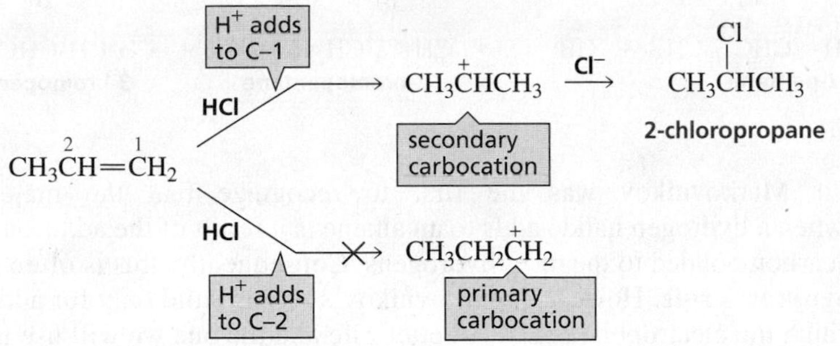

The sp^2 carbon that does not get the proton is the one that is positively charged in the carbocation intermediate.

Two products are formed in both of the following reactions but the major product is the one that results from reaction of the nucleophile with the faster formed tertiary carbocation.

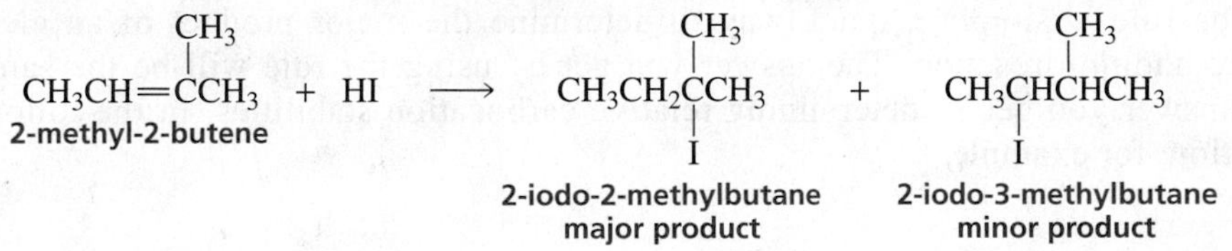

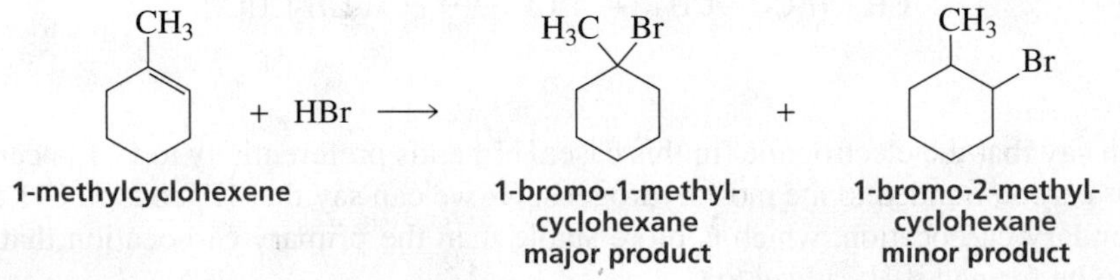

The two products of the preceding reactions are *constitutional isomers*. That is, they have the same molecular formula, but differ in how their atoms are connected. A reaction in which two or more constitutional isomers could be obtained as products, but one of them predominates, is called a **regioselective reaction.**

There are degrees of **regioselectivity**: a reaction can be *moderately regioselective, highly regioselective,* or *completely regioselective.* For example, the addition of HCl to 2-methylpropene (where the two possible carbocations are tertiary and primary) is more highly regioselective than the addition of HCl to 2-methyl-2-butene (where the two possible carbocations are tertiary and secondary), because the two carbocations formed in the latter reaction are closer in stability. In a completely regioselective reaction, only one of the possible products is formed.

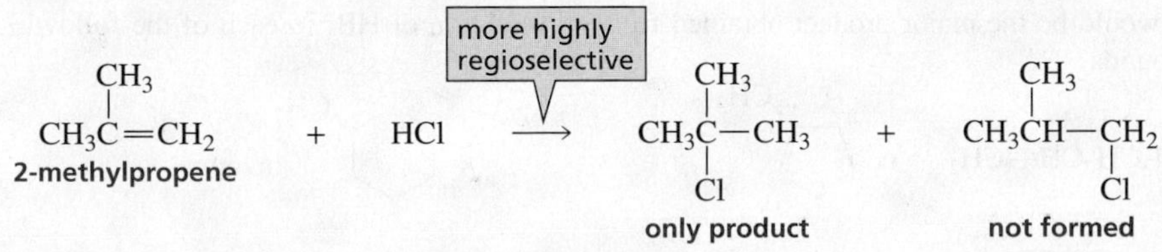

Regioselectivity is the preferential formation of one constitutional isomer over another.

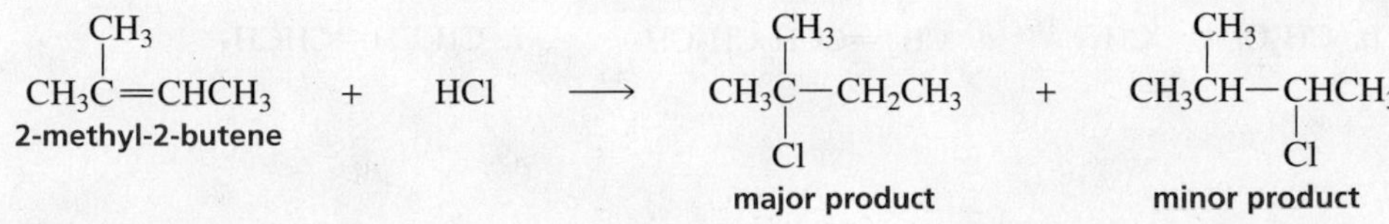

The following reaction is not regioselective. Because the addition of H$^+$ to either of the sp^2 carbons produces a secondary carbocation, both carbocations are formed at about the same rate. Therefore, approximately equal amounts of the two alkyl halides are obtained.

$$CH_3CH=CHCH_2CH_3 + HBr \longrightarrow \underset{\substack{\text{2-bromopentane} \\ \text{50\%}}}{CH_3\overset{\overset{\displaystyle Br}{|}}{C}HCH_2CH_2CH_3} + \underset{\substack{\text{3-bromopentane} \\ \text{50\%}}}{CH_3CH_2\overset{\overset{\displaystyle Br}{|}}{C}HCH_2CH_3}$$

2-pentene

The electrophile adds to the *sp*^{2} carbon bonded to the most hydrogens.

Vladimir Markovnikov was the first to recognize that the major product obtained when a hydrogen halide adds to an alkene is a result of the addition of the H$^+$ to the sp^2 carbon bonded to the most hydrogens. Consequently, this is often referred to as **Markovnikov's rule.** However, Markovnikov's rule is valid only for addition reactions in which the electrophile is H$^+$. A better rule and the one we will use in this text, is one that applies to all electrophilic addition reactions of alkenes:

> The *electrophile adds to the* sp^2 *carbon bonded to the most hydrogens* (that is, to the less substituted sp^2 carbon).

The rule is simply a quick way to determine the major product of an electrophilic addition reaction. The answer you get by using the rule will be the same as the answer you get by determining relative carbocation stabilities. In the following reaction, for example,

$$\overset{2}{C}H_3CH_2\overset{}{C}H=\overset{1}{C}H_2 + HCl \longrightarrow CH_3CH_2\overset{\overset{\displaystyle}{}}{C}HCH_3$$
$$\qquad\qquad\qquad\qquad\qquad\qquad\quad | \\ \qquad\qquad\qquad\qquad\qquad\qquad\ Cl$$

we can say that the electrophile (in this case, H$^+$) adds preferentially to C-1 because it is the sp^2 carbon bonded to the most hydrogens. Or we can say that H$^+$ adds to C-1 to form a secondary carbocation, which is more stable than the primary carbocation that would have to be formed if H$^+$ added to C-2.

The foregoing examples illustrate the way organic reactions are typically written. The reactants are written to the left of the reaction arrow, and the products are written to the right of the arrow. Any conditions that need to be stipulated, such as the solvent, the temperature, or a catalyst, are written above or below the arrow. Sometimes only the organic (carbon-containing) reagent is written to the left of the arrow, and any other reagents are written above or below the arrow.

$$CH_3CH_2CH=CH_2 \xrightarrow{\textbf{HCl}} CH_3CH_2\overset{\overset{\displaystyle}{}}{C}HCH_3$$
$$\qquad\qquad\qquad\qquad\qquad\qquad\qquad\quad | \\ \qquad\qquad\qquad\qquad\qquad\qquad\qquad\ Cl$$

PROBLEM 6♦

What would be the major product obtained from the addition of HBr to each of the following compounds?

a. $CH_3CH_2CH=CH_2$

c. (cyclopentene with CH_3 substituent)

e. (cyclohexane ring with $=CH_2$)

b. $CH_3CH=\overset{\overset{\displaystyle CH_3}{|}}{C}CH_3$

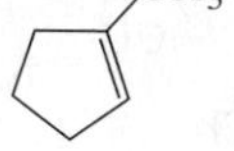

d. $CH_2=\overset{\overset{\displaystyle CH_3}{|}}{C}CH_2CH_2CH_3$

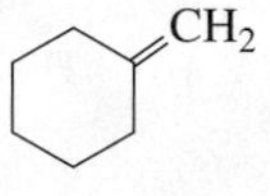

f. $CH_3CH=CHCH_3$

PROBLEM-SOLVING STRATEGY

Planning the Synthesis of an Alkyl Halide

a. What alkene should be used to synthesize 3-bromohexane?

$$? \quad + \quad HBr \longrightarrow CH_3CH_2\underset{\displaystyle Br}{\underset{|}{C}}HCH_2CH_2CH_3$$

3-bromohexane

The best way to answer this kind of question is to begin by listing all the alkenes that could be used. Because you want to synthesize an alkyl halide that has a bromo substituent at the 3-position, the alkene should have an sp^2 carbon at that position. Two alkenes fit the description: 2-hexene and 3-hexene.

$$\underset{\textbf{2-hexene}}{CH_3CH{=}CHCH_2CH_2CH_3} \qquad \underset{\textbf{3-hexene}}{CH_3CH_2CH{=}CHCH_2CH_3}$$

Because there are two possibilities, we next need to decide whether there is any advantage to using one over the other. The addition of H^+ to 2-hexene forms two different secondary carbocations. Because the carbocations have the same stability, approximately equal amounts of each will be formed. Therefore, half the product will be the desired 3-bromohexane and half will be 2-bromohexane.

$$\underset{\textbf{2-hexene}}{CH_3CH{=}CHCH_2CH_2CH_3} \xrightarrow{HBr} CH_3CH_2\overset{+}{C}HCH_2CH_2CH_3 \text{ (secondary carbocation)} \xrightarrow{Br^-} \underset{\textbf{3-bromohexane}}{CH_3CH_2\underset{\displaystyle Br}{\underset{|}{C}}HCH_2CH_2CH_3}$$

$$\underset{\textbf{2-hexene}}{CH_3CH{=}CHCH_2CH_2CH_3} \xrightarrow{HBr} CH_3\overset{+}{C}HCH_2CH_2CH_2CH_3 \text{ (secondary carbocation)} \xrightarrow{Br^-} \underset{\textbf{2-bromohexane}}{CH_3\underset{\displaystyle Br}{\underset{|}{C}}HCH_2CH_2CH_2CH_3}$$

The addition of H^+ to either of the sp^2 carbons of 3-hexene, on the other hand, forms the same carbocation because the alkene is symmetrical. Therefore, all the product (not just half) will be the desired 3-bromohexane. Thus, 3-hexene should be used for the synthesis of the desired compound.

$$\underset{\textbf{3-hexene}}{CH_3CH_2CH{=}CHCH_2CH_3} \xrightarrow{HBr} CH_3CH_2\overset{+}{C}HCH_2CH_2CH_3 \text{ (only one carbocation is formed)} \xrightarrow{Br^-} \underset{\textbf{3-bromohexane}}{CH_3CH_2\underset{\displaystyle Br}{\underset{|}{C}}HCH_2CH_2CH_3}$$

b. What alkene should be used to synthesize 2-bromopentane?

$$? \quad + \quad HBr \longrightarrow CH_3\underset{\displaystyle Br}{\underset{|}{C}}HCH_2CH_2CH_3$$

2-bromopentane

Either 1-pentene or 2-pentene could be used because both have an sp^2 carbon at the 2-position.

$$\underset{\textbf{1-pentene}}{CH_2{=}CHCH_2CH_2CH_3} \qquad \underset{\textbf{2-pentene}}{CH_3CH{=}CHCH_2CH_3}$$

When H^+ adds to 1-pentene, a secondary and a primary carbocation could be formed. The primary carbocation is so unstable, however, that little, if any, will be formed. Thus, 2-bromopentane will be the only product of the reaction.

$$CH_2{=}CHCH_2CH_2CH_3 \quad \text{1-pentene}$$

$$\xrightarrow{\text{HBr}} CH_3\overset{+}{C}HCH_2CH_2CH_3 \xrightarrow{\text{Br}^-} CH_3\underset{|}{C}HCH_2CH_2CH_3$$
$$\underset{\text{Br}}{} \quad \text{2-bromopentane}$$

$$\xrightarrow{\text{HBr}} \xcancel{\quad} \overset{+}{C}H_2CH_2CH_2CH_2CH_3$$

When H^+ adds to 2-pentene, on the other hand, two different secondary carbocations can be formed. Because they have the same stability, they will be formed in approximately equal amounts. Thus, only about half of the product will be 2-bromopentane. The other half will be 3-bromopentane.

$$CH_3CH{=}CHCH_2CH_3 \quad \text{2-pentene}$$

$$\xrightarrow{\text{HBr}} CH_3\overset{+}{C}HCH_2CH_2CH_3 \xrightarrow{\text{Br}^-} CH_3\underset{|}{C}HCH_2CH_2CH_3$$
$$\underset{\text{Br}}{} \quad \text{2-bromopentane}$$

$$\xrightarrow{\text{HBr}} CH_3CH_2\overset{+}{C}HCH_2CH_3 \xrightarrow{\text{Br}^-} CH_3CH_2\underset{|}{C}HCH_2CH_3$$
$$\underset{\text{Br}}{} \quad \text{3-bromopentane}$$

Because all the alkyl halide formed from 1-pentene is the desired product, but only half the alkyl halide formed from 2-pentene is the desired product, 1-pentene is the best alkene to use for the synthesis.

Now use the strategy you have just learned to solve Problem 7.

PROBLEM 7♦

What alkene should be used to synthesize each of the following alkyl bromides?

a. $CH_3\underset{\underset{\text{Br}}{|}}{\overset{\overset{\text{CH}_3}{|}}{C}}CH_3$ **b.** (cyclohexyl)$-CH_2\underset{\underset{\text{Br}}{|}}{C}HCH_3$ **c.** (cyclohexyl)$-\underset{\underset{\text{Br}}{|}}{\overset{\overset{\text{CH}_3}{|}}{C}}CH_3$ **d.** (cyclohexyl with CH_2CH_3 and Br)

PROBLEM 8♦

To which compound is the addition of HBr more highly regioselective?

a. $CH_3CH_2\underset{\underset{\text{CH}_3}{|}}{C}{=}CH_2$ or $CH_3\underset{\overset{\text{CH}_3}{|}}{C}{=}CHCH_3$ **b.** (cyclohexane with $=CH_2$) or (cyclohexene with CH_3)

5   THE ADDITION OF WATER TO AN ALKENE

An alkene does not react with water, because there is no electrophile present to start the reaction by adding to the alkene. The O—H bonds of water are too strong—water is too weakly acidic—to allow the hydrogen to act as an electrophile.

$$CH_3CH{=}CH_2 \quad + \quad H_2O \quad \longrightarrow \quad \text{no reaction}$$

If an acid (the acid used most often is H_2SO_4) is added to the solution, then a reaction will occur because the acid provides the electrophile. The product of the reaction is an *alcohol*.

The addition of water to a molecule is called **hydration,** so we can say that an alkene will be *hydrated* in the presence of water and acid.

H_2SO_4 ($pK_a = -5$) is a strong acid, so it dissociates almost completely. The acid that participates in the reaction, therefore, is most apt to be a hydrated proton. This is called a hydronium ion, which we will write as H_3O^+.

$$H_2SO_4 + H_2O \rightleftharpoons H_3O^+ + HSO_4^-$$
hydronium ion

When you look at the *mechanism for the acid-catalyzed addition of water to an alkene*, notice that the first two steps are the same (except for the nucleophile employed) as the two steps of the *mechanism for the addition of a hydrogen halide to an alkene.*

MECHANISM FOR THE ACID-CATALYZED ADDITION OF WATER TO AN ALKENE

- H^+ (an electrophile) adds to the sp^2 carbon of the alkene (a nucleophile) that is bonded to the most hydrogens.

- H_2O (a nucleophile) adds to the carbocation (an electrophile), forming a protonated alcohol.

- The protonated alcohol loses a proton because the pH of the solution is greater than the pK_a of the protonated alcohol. (Note that protonated alcohols are very strong acids.)

Thus, the overall reaction is the addition of an electrophile to the sp^2 carbon bonded to the most hydrogens and addition of a nucleophile to the other sp^2 carbon.

The addition of the electrophile to the alkene is relatively slow, whereas the subsequent addition of the nucleophile to the carbocation occurs rapidly. The reaction of the carbocation with a nucleophile is so fast that the carbocation combines with whatever nucleophile it collides with first. Notice that there are two nucleophiles in solution, water and HSO_4^- (the conjugate base of the acid used to start the reaction).[*] Because the concentration of water is much greater than the concentration of HSO_4^-, the carbocation is much more likely to collide with water. The final product of the addition reaction, therefore, is an alcohol.

H_2SO_4 catalyzes the hydration reaction. A catalyst increases the rate of a reaction but is not consumed during the course of the reaction. Thus,

Do not memorize the products obtained from the reactions of alkenes. Instead, for each reaction, ask yourself, "What is the electrophile?" and "What nucleophile is present in the greatest concentration?"

---

[*]HO^- cannot be a nucleophile in this reaction because there is no appreciable concentration of HO^- in an acidic solution. For example, at pH = 4 the concentration of HO^- is 1×10^{-10} M.

the proton adds to the alkene in the first step, but is returned to the reaction mixture in the final step. Overall, then, a proton is not consumed. Because the catalyst employed in the hydration of an alkene is an acid, hydration is called an **acid-catalyzed reaction.**

Note that catalysts increase the reaction rate by decreasing the free energy of activation, but they do *not* affect the equilibrium constant of the reaction. In other words, a catalyst increases the *rate* at which a product is formed but does not affect the *amount* of product formed when the reaction has reached equilibrium.

PROBLEM 9♦

The pK_a of a protonated alcohol is about -2.5, and the pK_a of an alcohol is about 15. Therefore, as long as the pH of the solution is greater than _____ and less than _____, more than 50% of 2-propanol (the product of the reaction earlier in this chapter) will be in its neutral, nonprotonated form.

PROBLEM 10♦

Answer the following questions about the mechanism for the acid-catalyzed hydration of an alkene:

a. How many transition states are there?
b. How many intermediates are there?
c. Which step in the forward direction has the smallest rate constant?

PROBLEM 11♦

What is the major product obtained from the acid-catalyzed hydration of each of the following alkenes?

a. $CH_3CH_2CH_2CH{=}CH_2$

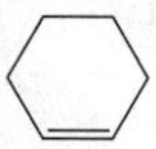

c. $CH_3CH_2CH_2CH{=}CHCH_3$

b. (cyclohexene structure)

d. 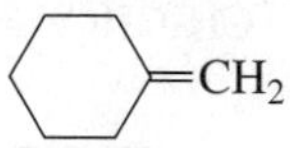 (cyclohexane with $={=}CH_2$)

---

6   THE ADDITION OF AN ALCOHOL TO AN ALKENE

Alcohols react with alkenes in the same way that water does, so this reaction, too, requires an acid catalyst. The product of the reaction is an *ether.*

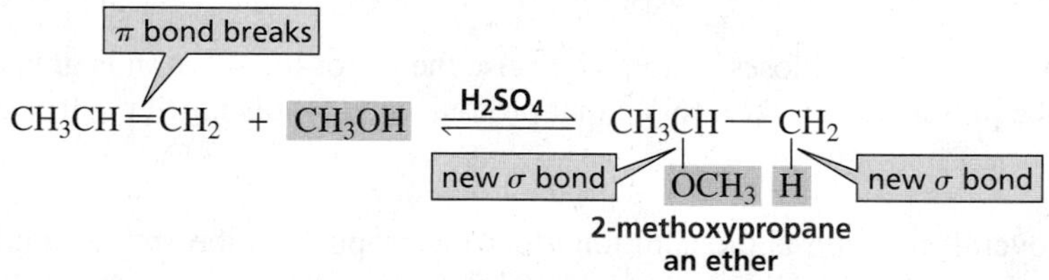

$$CH_3CH{=}CH_2 + CH_3OH \xrightarrow{H_2SO_4} CH_3CH-CH_2$$

π bond breaks

new σ bond   OCH₃ H   new σ bond

2-methoxypropane
an ether

MECHANISM FOR THE ACID-CATALYZED ADDITION OF AN ALCOHOL TO AN ALKENE

$$CH_3CH{=}CH_2 + H{-}\overset{+}{O}CH_3 \underset{\text{slow}}{\rightleftharpoons} CH_3\overset{+}{C}HCH_3 + CH_3\overset{..}{O}H \underset{\text{fast}}{\rightleftharpoons} CH_3CHCH_3 \overset{\text{fast}}{\underset{}{\rightleftharpoons}} CH_3CHCH_3 + CH_3\overset{..+}{O}H$$

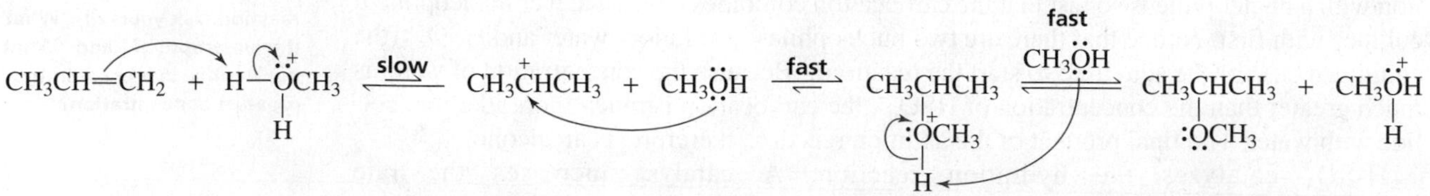

- H^+ (the electrophile) adds to the sp^2 carbon bonded to the most hydrogens.
- CH_3OH (the nucleophile) adds to the carbocation, forming a protonated ether.
- The protonated ether loses a proton, because the pH of the solution is greater than the pK_a of the protonated ether ($pK_a \sim -3.6$).

The *mechanism for the acid-catalyzed addition of an alcohol* is the same as the *mechanism for the acid-catalyzed addition of water*. The only difference in the two reactions is ROH is the nucleophile instead of H_2O.

PROBLEM 12

a. What is the major product of each of the following reactions?

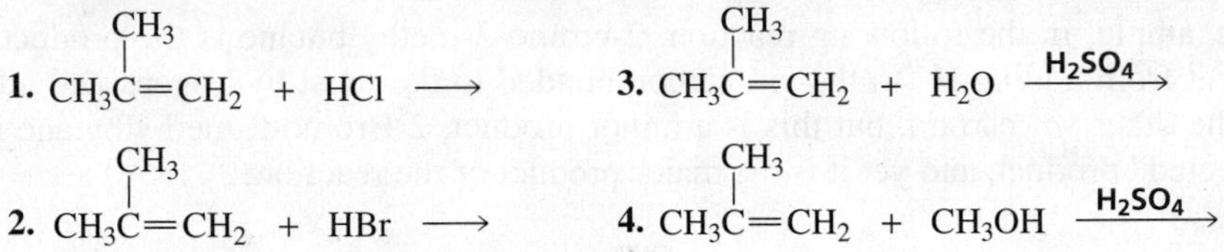

1. $\underset{\underset{CH_3}{|}}{CH_3C}=CH_2 \;+\; HCl \longrightarrow$

3. $\underset{\underset{CH_3}{|}}{CH_3C}=CH_2 \;+\; H_2O \xrightarrow{\;H_2SO_4\;}$

2. $\underset{\underset{CH_3}{|}}{CH_3C}=CH_2 \;+\; HBr \longrightarrow$

4. $\underset{\underset{CH_3}{|}}{CH_3C}=CH_2 \;+\; CH_3OH \xrightarrow{\;H_2SO_4\;}$

b. What do all the reactions have in common?

c. How do all the reactions differ?

PROBLEM 13 Solved

How could the following compound be prepared, using an alkene as one of the starting materials?

$$\underset{\underset{CH_3}{|}}{CH_3CHOCHCH_2CH_3}$$
(with CH_3 above and CH_3 below)

$$\overset{CH_3}{\overset{|}{CH_3CHOCHCH_2CH_3}}\underset{\underset{CH_3}{|}}{}$$

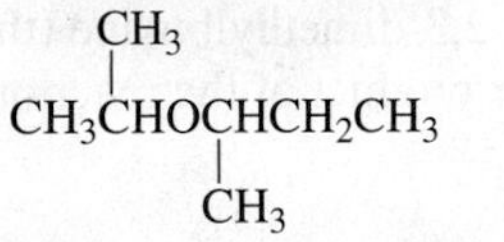

Solution The desired compound—called a **target molecule**—can be prepared by the acid-catalyzed addition of an alcohol to an alkene. There are three different combinations of an alkene and an alcohol that could be used for the synthesis of this particular ether.

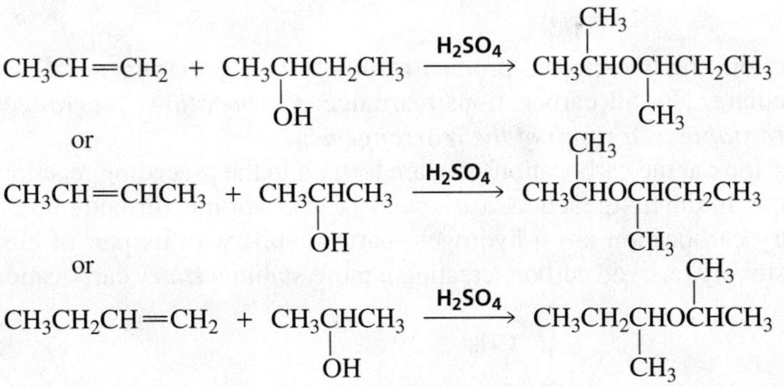

$$CH_3CH=CH_2 \;+\; \underset{\underset{OH}{|}}{CH_3CHCH_2CH_3} \xrightarrow{\;H_2SO_4\;} \overset{CH_3}{\overset{|}{CH_3CHOCHCH_2CH_3}}\underset{\underset{CH_3}{|}}{}$$

or

$$CH_3CH=CHCH_3 \;+\; \underset{\underset{OH}{|}}{CH_3CHCH_3} \xrightarrow{\;H_2SO_4\;} \overset{CH_3}{\overset{|}{CH_3CHOCHCH_2CH_3}}\underset{\underset{CH_3}{|}}{}$$

or

$$CH_3CH_2CH=CH_2 \;+\; \underset{\underset{OH}{|}}{CH_3CHCH_3} \xrightarrow{\;H_2SO_4\;} \overset{CH_3}{\overset{|}{CH_3CH_2CHOCHCH_3}}\underset{\underset{CH_3}{|}}{}$$

PROBLEM 14

How could the following compounds be prepared, using an alkene as one of the starting materials?

a. $\langle\!\!\!\bigcirc\!\!\!\rangle\text{—OCH}_3$

b. $\underset{\underset{CH_3}{|}}{\overset{\overset{CH_3}{|}}{CH_3OCCH_3}}$

c. cyclopentane ring —OH

d. $\underset{\underset{OH}{|}}{CH_3CHCH_2CH_3}$

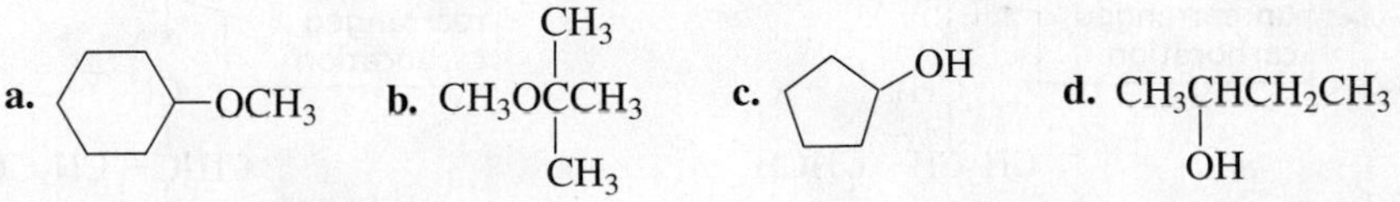

PROBLEM 15

Propose a mechanism for the following reaction (remember to use curved arrows to show the movement of electrons from the nucleophile to the electrophile):

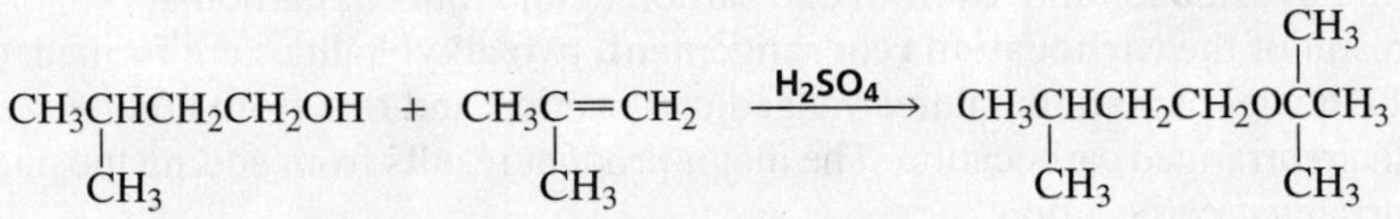

$$\underset{\underset{CH_3}{|}}{CH_3CHCH_2CH_2OH} \;+\; \underset{\underset{CH_3}{|}}{CH_3C}=CH_2 \xrightarrow{\;H_2SO_4\;} \underset{\underset{CH_3}{|}}{CH_3CHCH_2CH_2O}\underset{\underset{CH_3}{|}}{\overset{\overset{CH_3}{|}}{CCH_3}}$$

7 A CARBOCATION WILL REARRANGE IF IT CAN FORM A MORE STABLE CARBOCATION

Some electrophilic addition reactions give products that are not what you would get by adding an electrophile to the sp^2 carbon bonded to the most hydrogens and a nucleophile to the other sp^2 carbon.

For example, in the following reaction, 2-bromo-3-methylbutane is the product you would get from adding H^+ to the sp^2 carbon bonded to the most hydrogens and adding Br^- to the other sp^2 carbon, but this is a minor product. 2-Bromo-2-methylbutane is an "unexpected" product, and yet it is the major product of the reaction.

$$CH_3CHCH{=}CH_2 \ + \ HBr \longrightarrow CH_3CHCHCH_3 \ + \ CH_3CCH_2CH_3$$

3-methyl-1-butene

2-bromo-3-methylbutane
minor product

2-bromo-2-methylbutane
major product

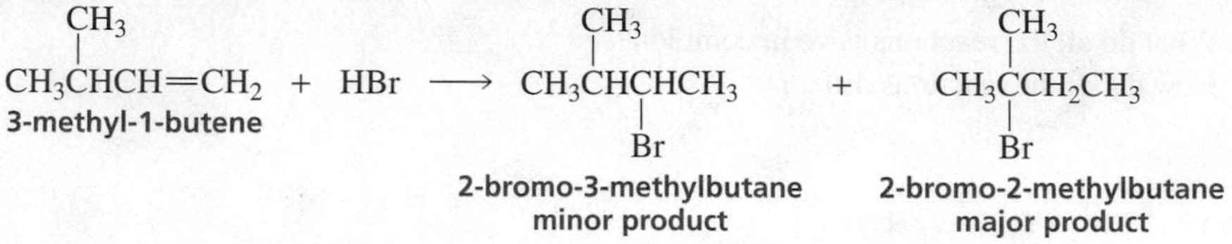

In another example, the following reaction forms both 3-chloro-2,2-dimethylbutane (the expected product) and 2-chloro-2,3-dimethylbutane (the unexpected product). Again, the unexpected product is the major product of the reaction.

$$CH_3C{-}CH{=}CH_2 \ + \ HCl \longrightarrow CH_3C{-}CHCH_3 \ + \ CH_3C{-}CHCH_3$$

3,3-dimethyl-1-butene

3-chloro-2,2-dimethylbutane
minor product

2-chloro-2,3-dimethylbutane
major product

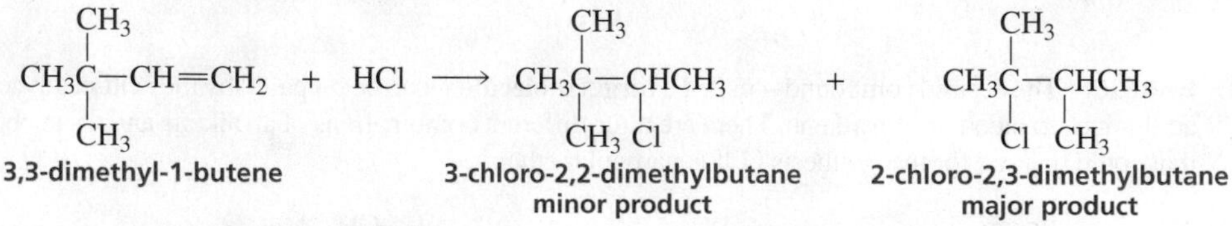

In each reaction, the unexpected product results from a *rearrangement* of the carbocation intermediate. Not all carbocations rearrange. *Carbocations rearrange only if they become more stable as a result of the rearrangement.*

Let's now look at the carbocations that are formed in the preceding reactions to see why they rearrange. In the first reaction, a *secondary* carbocation is formed initially. However, the secondary carbocation has a hydrogen that can shift with its pair of electrons to the adjacent positively charged carbon, creating a more stable *tertiary* carbocation.

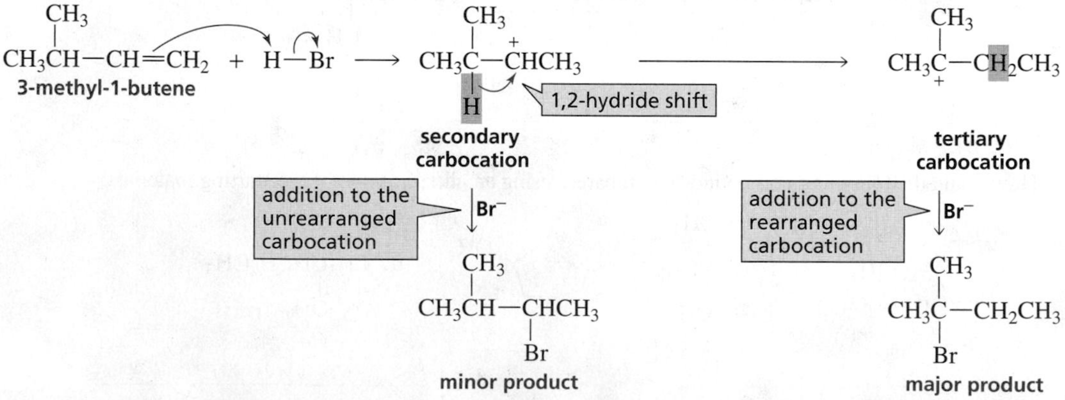

Carbocations rearrange if they become more stable as a result of the rearrangement.

$$CH_3CH{-}CH{=}CH_2 \ + \ H{-}Br \longrightarrow CH_3C{-}CHCH_3 \xrightarrow{\text{1,2-hydride shift}} CH_3C{-}CH_2CH_3$$

3-methyl-1-butene

secondary carbocation

tertiary carbocation

addition to the unrearranged carbocation | Br^-

addition to the rearranged carbocation | Br^-

$$CH_3CH{-}CHCH_3$$
Br
minor product

$$CH_3C{-}CH_2CH_3$$
Br
major product

Because a hydrogen shifts with its pair of electrons, the rearrangement is called a hydride shift. (Recall that $H{:}^-$ is a hydride ion.) More specifically, it is called a **1,2-hydride shift** because the hydride ion moves from one carbon to an *adjacent* carbon.

As a result of the **carbocation rearrangement,** two alkyl halides are formed, one from adding the nucleophile to the unrearranged carbocation and one from adding the nucleophile to the rearranged carbocation. The major product results from adding the nucleophile to the rearranged carbocation.

In the second reaction, again a *secondary* carbocation is formed initially. Then one of the methyl groups, with its pair of electrons, shifts to the adjacent positively charged carbon to form a more stable *tertiary* carbocation. This kind of rearrangement is called a **1,2-methyl shift**—the methyl group moves with its electrons from one carbon to an *adjacent* carbon. Again, the major product is the one formed by adding the nucleophile to the rearranged carbocation.

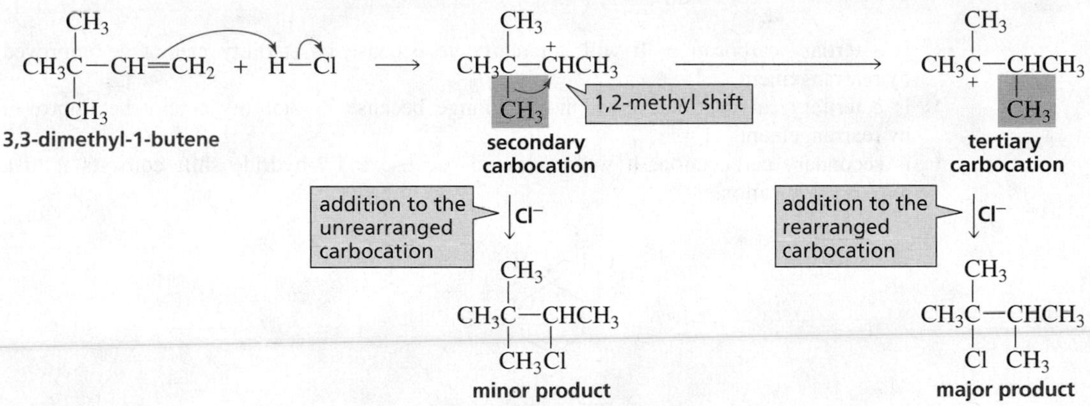

If a rearrangement does not lead to a more stable carbocation, then it typically does not occur. For example, the following reaction forms a secondary carbocation. A 1,2-hydride shift would form a different secondary carbocation. Since both carbocations are equally stable, there is no energetic advantage to the rearrangement. Consequently, the rearrangement does not occur, and only one alkyl halide is formed.

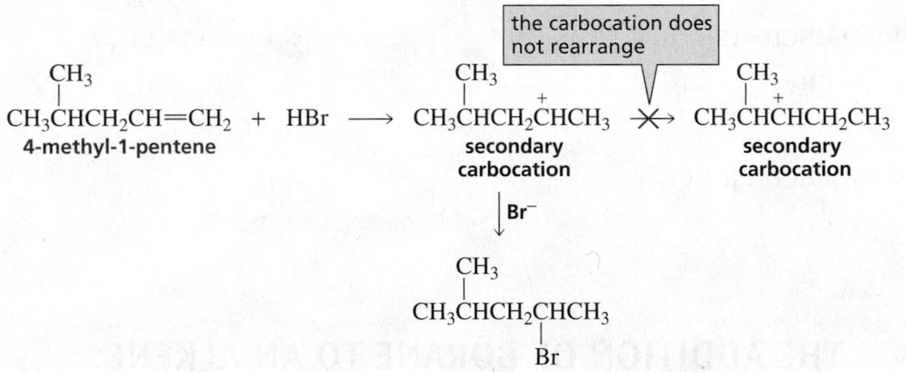

Whenever a reaction forms a carbocation intermediate, check to see if it will rearrange to a more stable carbocation.

Keep in mind that *whenever a reaction leads to the formation of a carbocation, you must check its structure to see if it will rearrange.*

PROBLEM 16 Solved

Which of the following carbocations would be expected to rearrange?

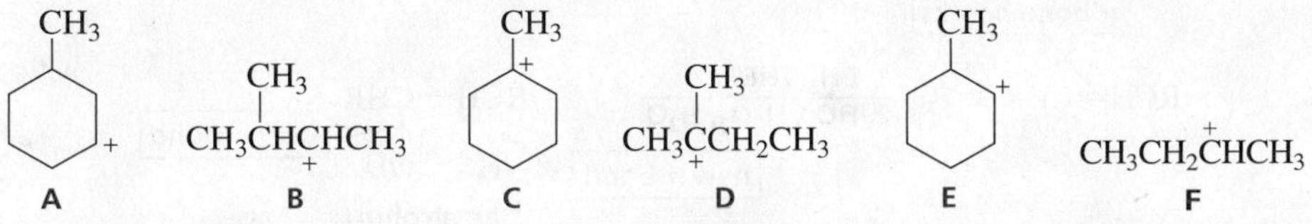

Solution

A is secondary carbocation. It will not rearrange because a 1,2-hydride shift would convert it to a different secondary carbocation, so there is no energetic advantage to the rearrangement.

B is a secondary carbocation. It will rearrange because a 1,2-hydride shift converts it to a tertiary carbocation.

$$CH_3\overset{CH_3}{\underset{H}{\overset{+}{C}-CHCH_3}} \longrightarrow CH_3\overset{CH_3}{\underset{+}{C}CH_2CH_3}$$

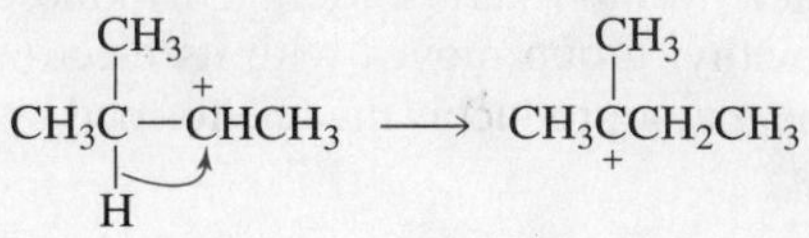

C is a tertiary carbocation. It will not rearrange because its stability cannot be improved by rearrangement.

D is a tertiary carbocation. It will not rearrange because its stability cannot be improved by rearrangement.

E is secondary carbocation. It will rearrange because a 1,2-hydride shift converts it to a tertiary carbocation.

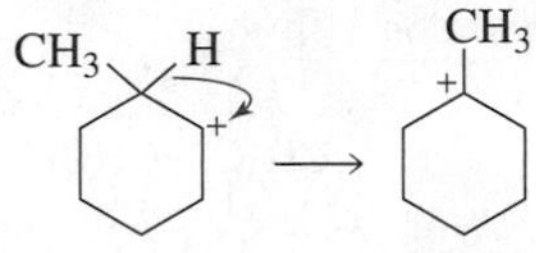

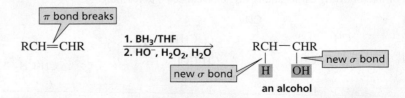

F is a secondary carbocation. It will not rearrange because rearrangement would form another secondary carbocation.

PROBLEM 17♦

What is the major product obtained from the reaction of HBr with each of the following?

a. $CH_3\overset{}{\underset{CH_3}{CHCH}}=CH_2$

c. 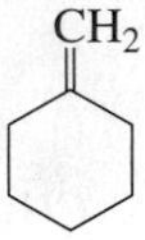 (structure: methylenecyclohexane, CH_2 exocyclic)

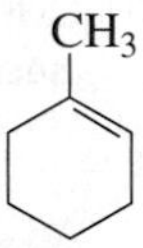

e. $CH_2=CHCCH_3$ with CH_3 groups (2-methyl-3-methyl)

b. $CH_3\overset{}{\underset{CH_3}{CHCH_2CH}}=CH_2$

d. (structure: 1-methylcyclohexene, CH_3)

f. 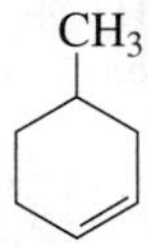 (structure: methylcyclohexene, CH_3)

---

8 THE ADDITION OF BORANE TO AN ALKENE: HYDROBORATION–OXIDATION

Another way to convert an alkene to an alcohol is by two successive reactions known as **hydroboration–oxidation.**

Hydroboration–Oxidation

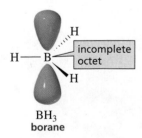

BH₃
borane

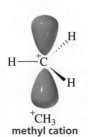

⁺CH₃
methyl cation

π bond breaks

$$RCH=CHR \xrightarrow[\text{2. } HO^-, H_2O_2, H_2O]{\text{1. } BH_3/THF} RCH-CHR$$

new σ bond   H   OH   new σ bond

an alcohol

An atom or molecule does not have to have a positive charge to be an electrophile. Borane (BH₃) is an electrophile because boron has an incomplete octet. Boron uses sp^2 orbitals to form bonds with hydrogen, so has an empty p orbital that can accept a share in a pair of electrons. Thus, BH₃ is the electrophile that adds to the alkene and the nucleophile is an H$^-$ bonded to boron. When the addition reaction is over, an aqueous solution of sodium hydroxide and hydrogen peroxide is added to the reaction mixture. The product of the reaction is an alcohol. The numbers 1 and 2 in front of the reagents above and below the reaction arrow indicate two sequential reactions; the second set of reagents is not added until reaction with the first set is over.

Borane and Diborane

Borane exists primarily as a colorless gas called diborane. Diborane is a **dimer**—a molecule formed by joining two identical molecules. Because boron does not have a complete octet—it is surrounded by only six electrons—it has a strong tendency to acquire an additional electron pair. In the dimer, therefore, two boron atoms share the two electrons in a hydrogen–boron bond by means of unusual half-bonds. These hydrogen–boron bonds are shown as dotted lines to indicate that they consist of fewer than the normal two electrons.

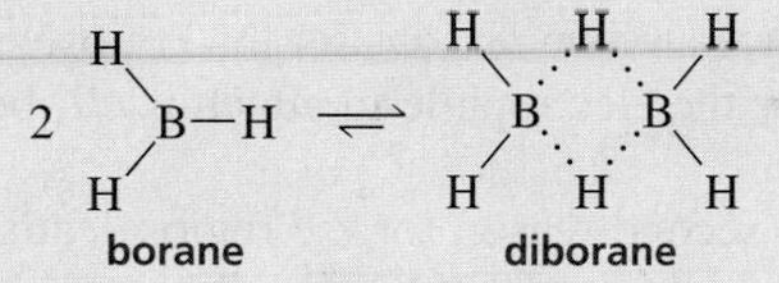

Diborane (B_2H_6) is a flammable, toxic, and explosive gas. A complex—prepared by dissolving diborane in THF—is a more convenient and less dangerous reagent. One of the lone pairs of the O atom in THF provides the two electrons that boron needs to complete its octet. The borane–THF complex is used as the source of BH_3 for hydroboration.

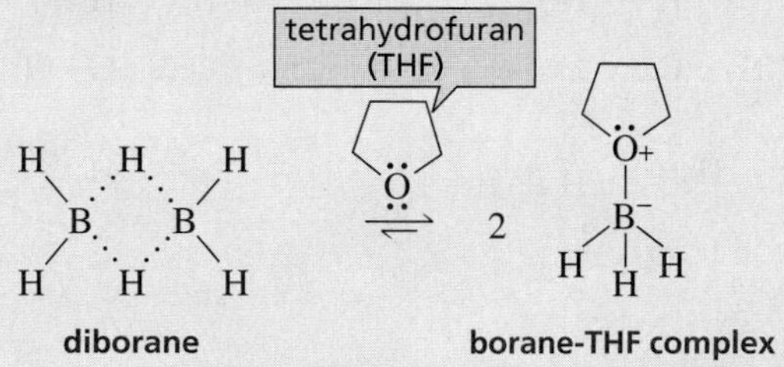

As the following reactions show, the alcohol formed from hydroboration–oxidation of an alkene has the H and OH groups switched compared to the alcohol formed from the acid-catalyzed addition of water.

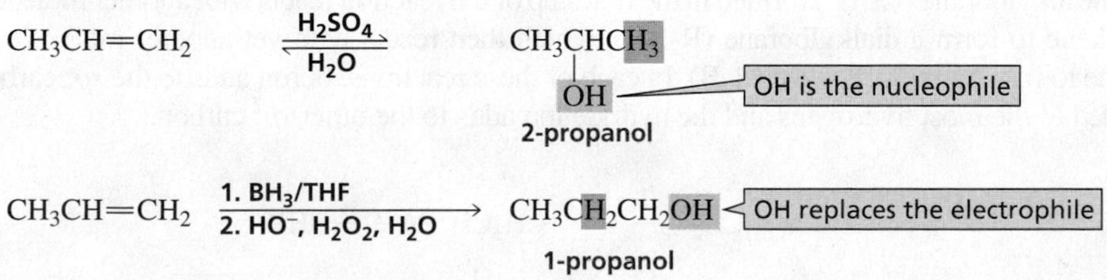

In both of the preceding reactions, *the electrophile adds to the sp^2 carbon bonded to the most hydrogens.* In the *acid-catalyzed addition of water,* H$^+$ is the electrophile and H_2O is the nucleophile; whereas in *hydroboration–oxidation,* we will see that BH_3 is the electrophile (with HO subsequently taking its place) and H$^-$ is the nucleophile.

To understand why the hydroboration–oxidation of propene forms 1-propanol and not 2-propanol, we must look at the mechanism for hydroboration, the first of the two successive reactions known as hydroboration–oxidation.

In the acid-catalyzed addition of water, H$^+$ is the electrophile; in hydroboration–oxidation, H$^-$ is the nucleophile.

MECHANISM FOR HYDROBORATION WITH BH₃

$$CH_3CH = CH_2 \longrightarrow CH_3CH - CH_2$$

nucleophile

H—BH₂

electrophile

H   BH₂

an alkylborane

- As boron (an electrophile) accepts the π electrons from the alkene and forms a bond with one sp^2 carbon, it gives a hydride ion (a nucleophile) to the other sp^2 carbon.

Hydroboration is an example of a concerted reaction. In a **concerted reaction,** all the bond-making and bond-breaking processes occur in the same step (all the events occur "in concert"). Because both the boron and the hydride ion are added to the alkene in a single step, no intermediate is formed.

Boron, like the other electrophiles we have looked at, adds to the sp^2 carbon bonded to the most hydrogens. There are two reasons for the regioselectivity. First, there is more room at this sp^2 carbon for the electrophile to attach itself, because it is the less substituted sp^2 carbon.

We can understand the second reason for the regioselectivity if we examine the two possible transition states, which show that the C—B bond has formed to a greater extent than has the C—H bond. Consequently, the sp^2 carbon that does *not* become attached to boron has a partial positive charge.

addition of BH₃   addition of HBr

| more stable transition state | less stable transition state | more stable transition state | less stable transition state |

Addition of boron to the sp^2 carbon bonded to the most hydrogens forms a more stable transition state because the partial positive charge is on a secondary carbon. In contrast, if boron had added to the other sp^2 carbon, the partial positive charge would be on a primary carbon. Thus, BH₃ and an electrophile such as H⁺ add to the sp^2 carbon bonded to the most hydrogens for the same reason—in order to form the more stable transition state.

The alkylborane (RBH₂) formed in the first step of the reaction reacts with another molecule of alkene to form a dialkylborane (R₂BH), which then reacts with yet another molecule of alkene to form a trialkylborane (R₃B). In each of these reactions, boron adds to the sp^2 carbon bonded to the most hydrogens and the hydride ion adds to the other sp^2 carbon.

$$CH_3CH = CH_2 \longrightarrow CH_3CH - CH_2 - BHR$$

H—BHR

an alkylborane

H

a dialkylborane

$$CH_3CH = CH_2 \longrightarrow CH_3CH - CH_2 - BR_2$$

H—BR₂

a dialkylborane

H

a trialkylborane

The alkylborane (RBH_2) is bulkier than BH_3 because R is larger than H. The dialkylborane (R_2BH) is even bulkier than the alkylborane. Therefore, the alkylborane and dialkylborane will have an even stronger preference for addition to the less substituted sp^2 carbon (that is, the one bonded to the most hydrogens).

Only one hydride ion is needed for hydroboration. As a result, a reagent with only one hydrogen attached to boron, such as 9-BBN, is often used instead of BH_3. Since 9-BBN has two relatively bulky R groups, it has a stronger preference for the less substituted sp^2 carbon than BH_3 has. In mechanisms, we will write this compound as R_2BH. The mechanisms for the addition of R_2BH and BH_3 to an alkene are the same.

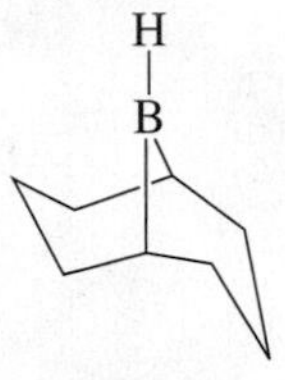

MECHANISM FOR HYDROBORATION WITH R₂BH

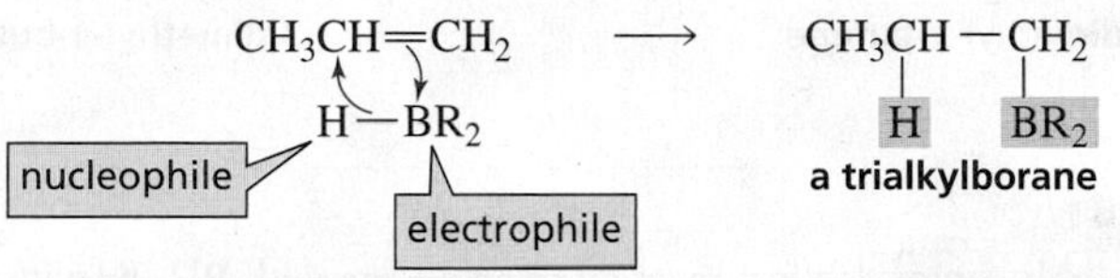

9-BBN
9-borabicyclo[3.3.1]nonane
R₂BH

When the hydroboration reaction is over, an aqueous solution of sodium hydroxide and hydrogen peroxide (HOOH) is added to the reaction mixture in order to replace BR_2 with an OH group. This is an oxidation reaction. An **oxidation reaction** decreases the number of C—H bonds or increases the number of C—O, C—N, or C—X bonds in a compound (where X is a halogen). Therefore, the overall reaction is called hydroboration–oxidation.

> An oxidation reaction increases the number of C—O, C—N, or C—X bonds and/or decreases the number of C—H bonds.

$$CH_3CH_2CH_2-\overline{B}R_2 \xrightarrow{HO^-, H_2O_2, H_2O} CH_3CH_2CH_2-OH + (HO)_2\overline{B}R_2$$

MECHANISM FOR THE OXIDATION REACTION

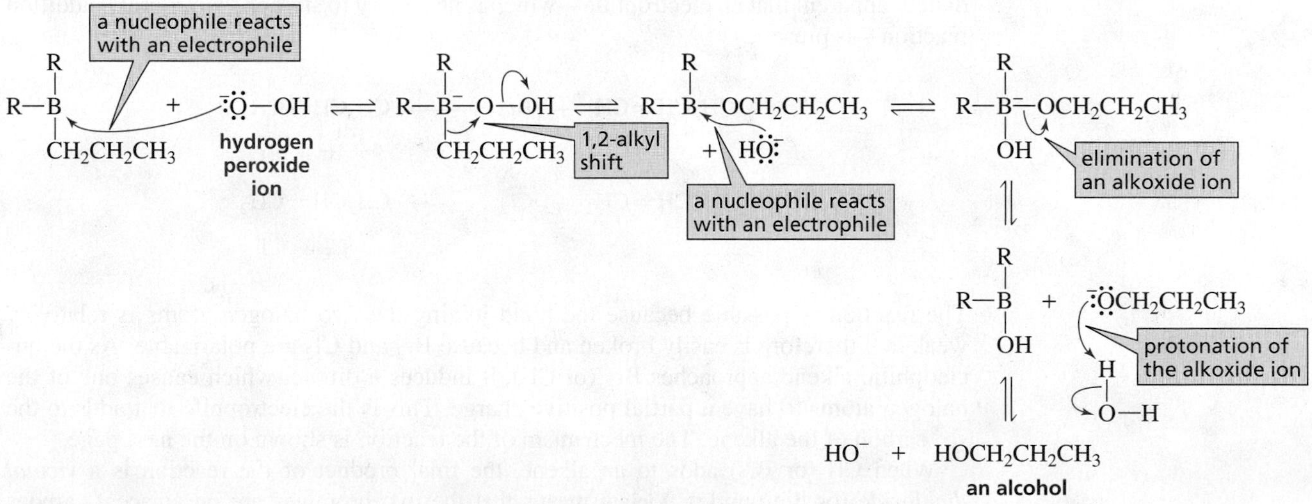

The mechanism for the oxidation reaction shows the following:

- A hydrogen peroxide ion (a nucleophile) shares a pair of electrons with the boron of $R_2BCH_2CH_2CH_3$ (an electrophile).
- A 1,2-alkyl shift displaces a hydroxide ion, allowing boron to no longer be negatively charged.
- A hydroxide ion shares a pair of electrons with the boron of $R_2BOCH_2CH_2CH_3$.
- An alkoxide ion is eliminated, allowing boron to no longer be negatively charged.
- Protonating the alkoxide ion forms the alcohol.

263

Notice that the OH group ends up attached to the sp^2 carbon bonded to the most hydrogens because it replaces boron, which was the electrophile in the hydroboration reaction.

Because carbocation intermediates are not formed in the hydroboration reaction, carbocation rearrangements do not occur.

$$\underset{\textbf{3-methyl-1-butene}}{CH_3\overset{\overset{\displaystyle CH_3}{|}}{C}HCH=CH_2} \quad \xrightarrow[\textbf{2. HO}^-\textbf{, H}_2\textbf{O}_2\textbf{, H}_2\textbf{O}]{\textbf{1. BH}_3\textbf{/THF}} \quad \underset{\textbf{3-methyl-1-butanol}}{CH_3\overset{\overset{\displaystyle CH_3}{|}}{C}HCH_2CH_2OH}$$

$$\underset{\textbf{3,3-dimethyl-1-butene}}{CH_3\overset{\overset{\displaystyle CH_3}{|}}{\underset{\underset{\displaystyle CH_3}{|}}{C}}CH=CH_2} \quad \xrightarrow[\textbf{2. HO}^-\textbf{, H}_2\textbf{O}_2\textbf{, H}_2\textbf{O}]{\textbf{1. BH}_3\textbf{/THF}} \quad \underset{\textbf{3,3-dimethyl-1-butanol}}{CH_3\overset{\overset{\displaystyle CH_3}{|}}{\underset{\underset{\displaystyle CH_3}{|}}{C}}CH_2CH_2OH}$$

PROBLEM 18♦

Which is more highly regioselective, reaction of an alkene with BH_3 or with 9-BBN?

PROBLEM 19♦

What would be the major product obtained from hydroboration–oxidation of the following alkenes?

a. 2-methyl-2-butene　　**b.** 1-methylcyclohexene

9　THE ADDITION OF A HALOGEN TO AN ALKENE

The halogens Br_2 and Cl_2 add to alkenes. This may surprise you because it is not immediately apparent that an electrophile—which is necessary to start an electrophilic addition reaction—is present.

$$CH_3CH=CH_2 \;+\; \boxed{Br_2} \;\longrightarrow\; CH_3\underset{\underset{\boxed{Br}}{|}}{C}H-\underset{\underset{\boxed{Br}}{|}}{C}H_2$$

$$CH_3CH=CH_2 \;+\; \boxed{Cl_2} \;\longrightarrow\; CH_3\underset{\underset{\boxed{Cl}}{|}}{C}H-\underset{\underset{\boxed{Cl}}{|}}{C}H_2$$

The reaction is possible because the bond joining the two halogen atoms is relatively weak and therefore is easily broken and because Br_2 and Cl_2 are polarizable. As the nucleophilic alkene approaches Br_2 (or Cl_2), it induces a dipole, which causes one of the halogen atoms to have a partial positive charge. This is the electrophile that adds to the sp^2 carbon of the alkene. The mechanism of the reaction is shown on the next page.

When Cl_2 (or Br_2) adds to an alkene, the final product of the reaction is a *vicinal dichloride* (or dibromide). **Vicinal** means that the two bromines are on adjacent carbons (*vicinus* is the Latin word for "near"). Reactions of alkenes with Br_2 or Cl_2 are generally carried out by mixing the alkene and the halogen in an inert solvent, such as dichloromethane (CH_2Cl_2), which readily dissolves both reactants, but does not participate in the reaction.

$$\underset{\textbf{2-methylpropene}}{CH_3\overset{\overset{\displaystyle CH_3}{|}}{C}=CH_2} \;+\; \boxed{Cl_2} \quad \xrightarrow{\textbf{CH}_2\textbf{Cl}_2} \quad \underset{\substack{\textbf{1,2-dichloro-2-methylpropane}\\\textbf{a vicinal dichloride}}}{CH_3\overset{\overset{\displaystyle CH_3}{|}}{\underset{\underset{\boxed{Cl}}{|}}{C}}CH_2\boxed{Cl}}$$

MECHANISM FOR THE ADDITION OF BROMINE TO AN ALKENE

cyclic bromonium ion formed from the reaction of Br₂ with cis-2-butene

- As the π electrons of the alkene approach a molecule of Br_2, one of the bromines accepts those electrons and releases the shared electrons to the other bromine, which leaves as a bromide ion. Because bromine's electron cloud is close enough to the other sp^2 carbon to form a bond, a cyclic bromonium ion intermediate is formed rather than a carbocation intermediate.

- The cyclic bromonium ion intermediate is unstable because of the strain in the three-membered ring and the positively charged bromine, which withdraws electrons strongly from the ring carbons (Figure 5). Therefore, the cyclic bromonium ion reacts rapidly with a nucleophile (Br^-). The mechanism for the addition of Cl_2 is the same as the mechanism for the addition of Br_2.

▲ **Figure 5**

The cyclic bromonium ion intermediate formed from the reaction of Br_2 with *cis*-2-butene shows that the electron-deficient region (the blue area) encompasses the carbons, even though the formal positive charge is on the bromine.

Because a carbocation is not formed when Br_2 or Cl_2 adds to an alkene, carbocation rearrangements do not occur in these reactions.

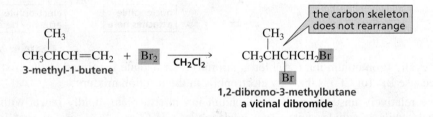

3-methyl-1-butene

the carbon skeleton does not rearrange

**1,2-dibromo-3-methylbutane
a vicinal dibromide**

PROBLEM 20◆

What would be the product of the preceding reaction if HBr were used in place of Br_2?

PROBLEM 21

a. How does the first step in the reaction of propene with Br_2 differ from the first step in the reaction of propene with HBr?

b. To understand why Br^- attacks a carbon of the bromonium ion rather than the positively charged bromine, draw the product that would be obtained if Br^- *did* attack bromine.

Although F_2 and I_2 are halogens, they are not used as reagents in electrophilic addition reactions. Fluorine reacts explosively with alkenes, so the reaction with F_2 is not useful for synthesizing new compounds. The addition of I_2 to an alkene, on the other hand, is thermodynamically unfavorable. A vicinal diiodide is unstable at room temperature, so it decomposes back to the alkene and I_2.

$$CH_3CH=CHCH_3 + I_2 \underset{CH_2Cl_2}{\rightleftharpoons} \underset{\underset{I \quad I}{\mid \quad \mid}}{CH_3CHCHCH_3}$$

unstable

If H_2O is present in the reaction mixture, the major product of the reaction will be a vicinal halohydrin (or, more specifically, a bromohydrin or a chlorohydrin). A **halohydrin** is an organic molecule that contains both a halogen and an OH group. In a vicinal halohydrin, the halogen and the OH group are bonded to adjacent carbons.

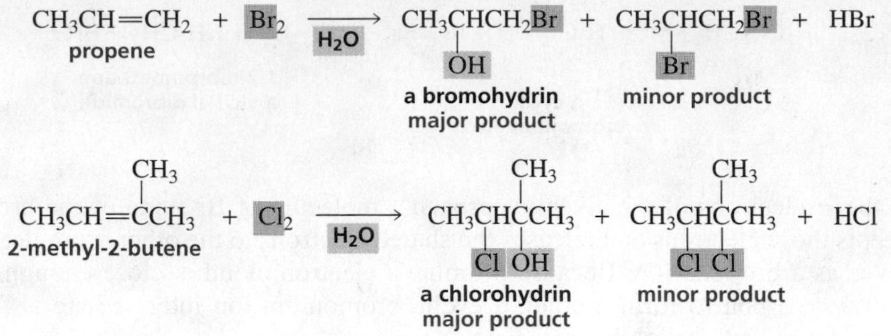

$$CH_3CH=CH_2 + Br_2 \xrightarrow{H_2O} CH_3CHCH_2Br + CH_3CHCH_2Br + HBr$$

propene   a bromohydrin (OH)   minor product (Br)

major product

$$CH_3CH=CCH_3 (CH_3) + Cl_2 \xrightarrow{H_2O} CH_3CHCCH_3 (CH_3, Cl, OH) + CH_3CHCCH_3 (CH_3, Cl, Cl) + HCl$$

2-methyl-2-butene   a chlorohydrin   minor product

major product

MECHANISM FOR HALOHYDRIN FORMATION

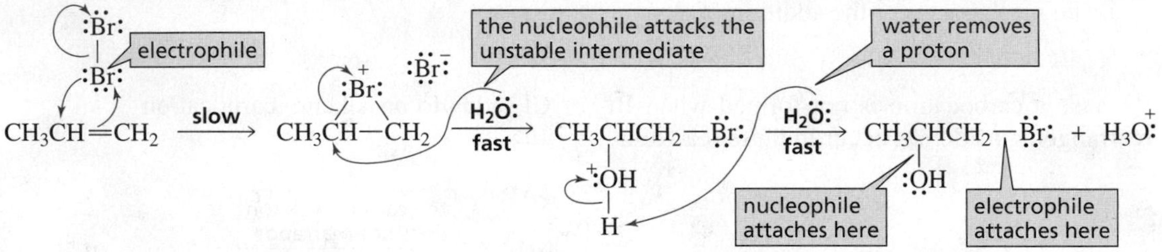

- A cyclic bromonium ion (or chloronium ion) intermediate is formed in the first step because Br^+ (or Cl^+) is the only electrophile in the reaction mixture.
- The relatively unstable cyclic bromonium ion intermediate rapidly reacts with any nucleophile it collides with. Two nucleophiles—H_2O and Br^-—are present in the solution, but the concentration of H_2O far exceeds that of Br^-. Consequently, the bromonium ion is much more likely to collide with water than with Br^-.
- The protonated halohydrin is a strong acid, so it readily loses a proton.

We can understand the regioselectivity—why the electrophile (Br^+) ends up attached to the sp^2 carbon bonded to the most hydrogens—if we examine the two possible transition states. They show that the C—Br bond has broken to a greater extent than the C—O bond has formed. As a result, there is a partial positive charge on the carbon that is attacked by the nucleophile.

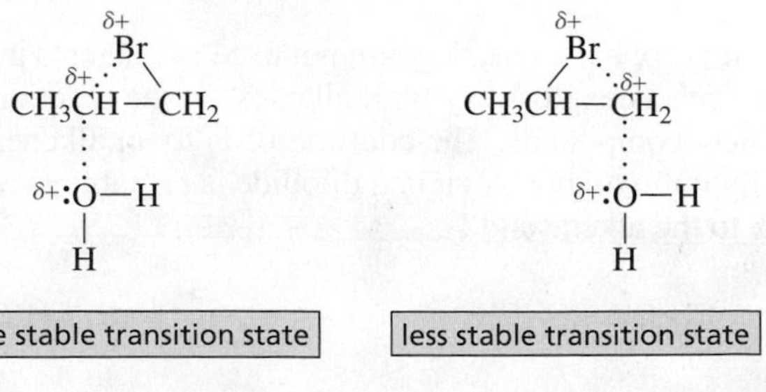

more stable transition state   less stable transition state

Water attacks the more substituted ring carbon because it leads to the more stable transition state since the partial positive charge is on a secondary rather than a primary carbon. Therefore, this reaction, too, follows the general rule for electrophilic addition reactions: *the electrophile* (here, Br^+) *adds to the* sp^2 *carbon that is bonded to the most hydrogens and the nucleophile* (H_2O) *adds to the other* sp^2 *carbon.*

When nucleophiles other than H_2O are added to the reaction mixture (for example, CH_3OH or Cl^-), they, too, change the product of the reaction. Because the concentration of the added nucleophile will be greater than the concentration of the halide ion generated from Br_2 or Cl_2, the added nucleophile will be the one most likely to participate in the second step of the reaction.

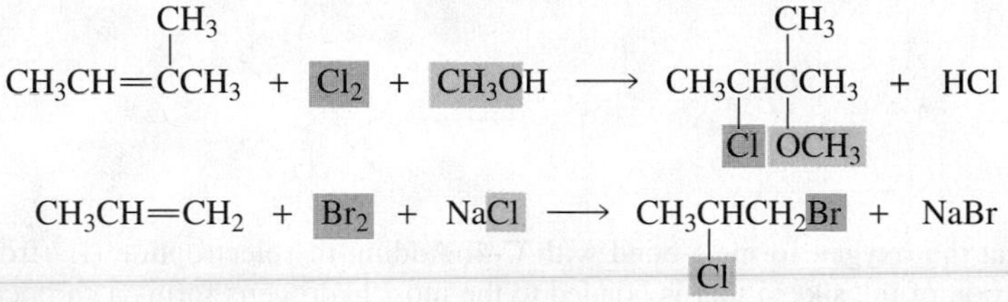

$$CH_3CH= \overset{\overset{\displaystyle CH_3}{|}}{C}CH_3 \; + \; Cl_2 \; + \; CH_3OH \; \longrightarrow \; CH_3\overset{\overset{\displaystyle CH_3}{|}}{C}\underset{\underset{\displaystyle Cl \;\; OCH_3}{|\;\;|}}{C}HCCH_3 \; + \; HCl$$

$$CH_3CH=CH_2 \; + \; Br_2 \; + \; NaCl \; \longrightarrow \; CH_3\underset{\underset{\displaystyle Cl}{|}}{C}HCH_2Br \; + \; NaBr$$

Ions such as Na^+ and K^+ cannot form covalent bonds, so they do not react with organic compounds. They serve only as counterions to negatively charged species, so their presence generally is ignored in writing chemical reactions.

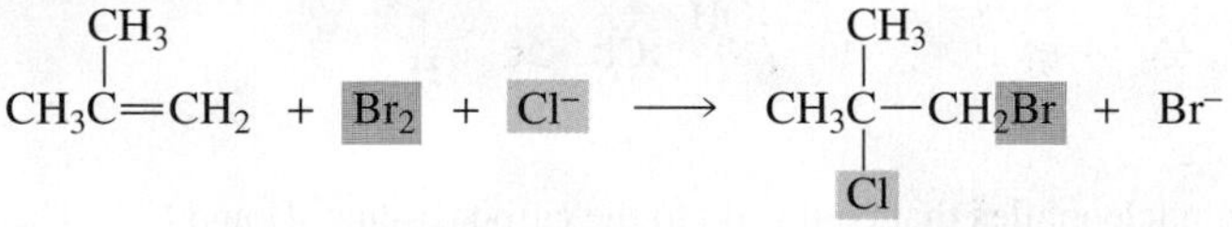

$$CH_3\overset{\overset{\displaystyle CH_3}{|}}{C}=CH_2 \; + \; Br_2 \; + \; Cl^- \; \longrightarrow \; CH_3\underset{\underset{\displaystyle Cl}{|}}{\overset{\overset{\displaystyle CH_3}{|}}{C}}-CH_2Br \; + \; Br^-$$

PROBLEM 22

Why are Na^+ and K^+ unable to form covalent bonds?

PROBLEM 23

Each of the following reactions has two nucleophiles that could add to the intermediate fromed by the reaction of the alkene with an electrophile. What will be the major product of each reaction?

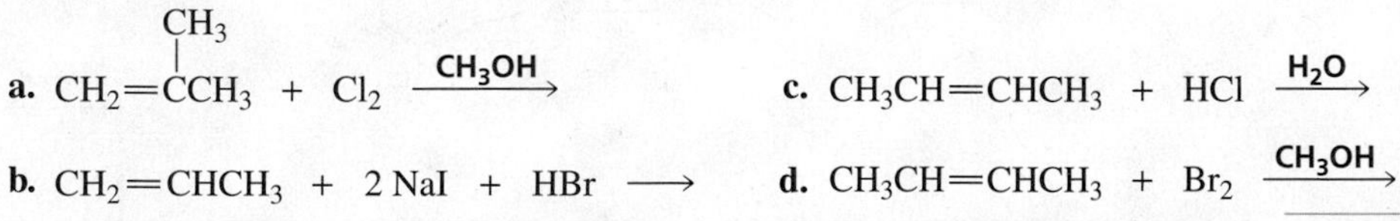

a. $CH_2= \overset{\overset{\displaystyle CH_3}{|}}{C}CH_3 \; + \; Cl_2 \; \xrightarrow{\text{CH}_3\text{OH}}$

c. $CH_3CH=CHCH_3 \; + \; HCl \; \xrightarrow{\text{H}_2\text{O}}$

b. $CH_2=CHCH_3 \; + \; 2\,NaI \; + \; HBr \; \longrightarrow$

d. $CH_3CH=CHCH_3 \; + \; Br_2 \; \xrightarrow{\text{CH}_3\text{OH}}$

PROBLEM 24

What is the product of the addition of I—Cl to 1-butene? (*Hint:* Chlorine is more electronegative than iodine.)

PROBLEM 25♦

What would be the major product obtained from the reaction of Br_2 with 1-butene if the reaction were carried out in

a. dichloromethane?   **b.** water?   **c.** ethyl alcohol?   **d.** methyl alcohol?

PROBLEM-SOLVING STRATEGY

Proposing a Mechanism

Propose a reasonable mechanism for the following reaction:

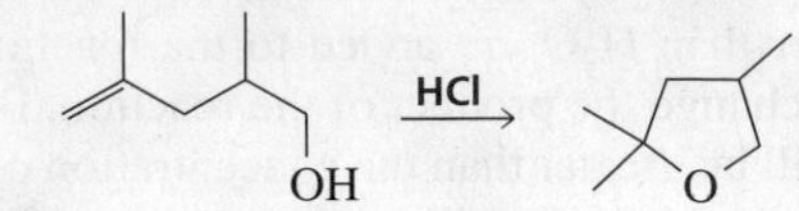

When forming a cyclic compound, start by numbering the atoms in the reactant and the product. You will then be able to see the atoms that become attached to each other in the cyclic product.

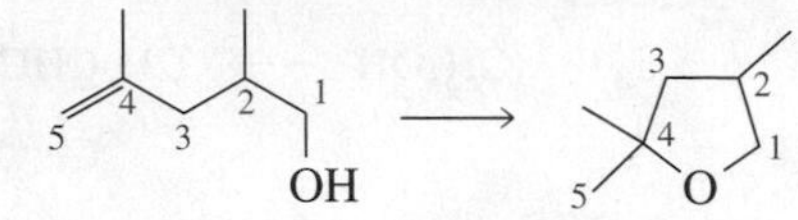

We see that the oxygen forms a bond with C-4. Adding the electrophile (H^+ from HCl) to the sp^2 carbon of the alkene that is bonded to the most hydrogens forms a carbocation (with the positive charge on C-4) that will react a nucleophile.

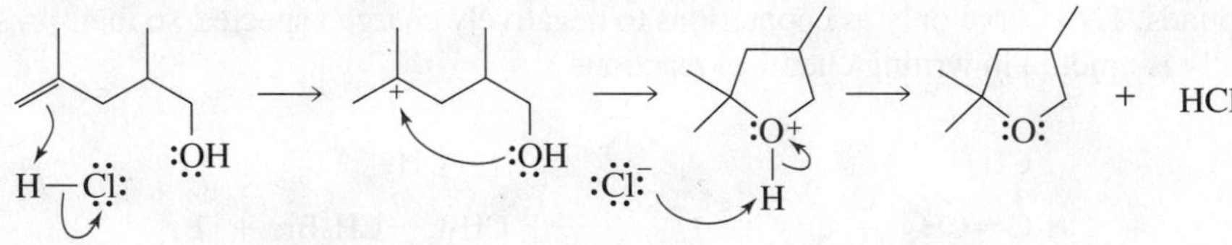

There are two nucleophiles that could add to the carbocation, OH and Cl^-. The nucleophile present in greatest concentration is the OH group, so it adds to the carbocation to from the five-membered ring. (There is little Cl^- present because HCl is a catalyst so only a small amount is needed because it is regenerated.)

Now use the strategy you have just learned to solve Problem 26.

PROBLEM 26

Propose a mechanism for the following reaction:

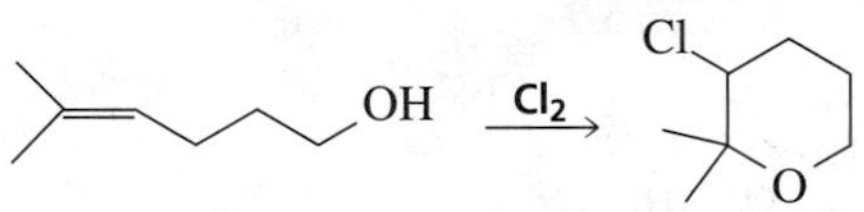

10   THE ADDITION OF A PEROXYACID TO AN ALKENE

An alkene can be converted to an *epoxide* by a peroxyacid. An **epoxide** is an ether in which the oxygen is incorporated into a three-membered ring, and a **peroxyacid** is a carboxylic acid with an extra oxygen. The overall reaction amounts to the transfer of an oxygen from the peroxyacid to the alkene. It is an oxidation reaction because it increases the number of C—O bonds.

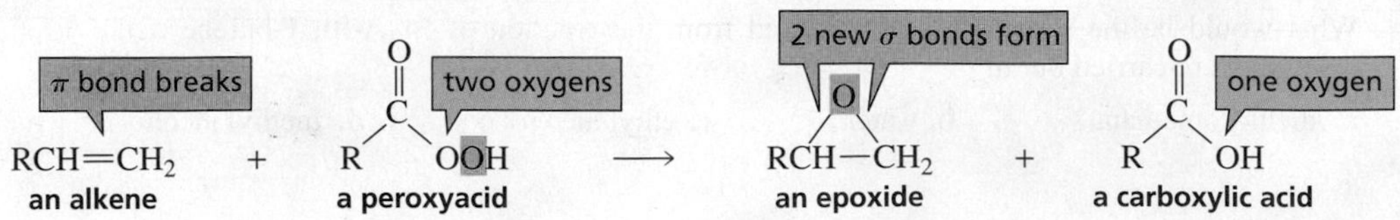

Note that an O—O bond is weak and is therefore easily broken.

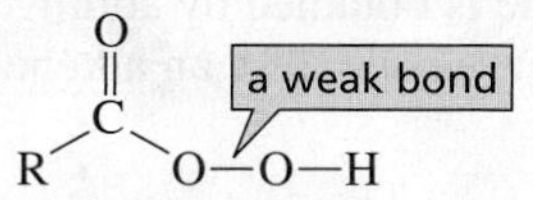

a peroxyacid

The peroxyacid commonly used for epoxidation is MCPBA (*meta*-chloroperoxy benzoic acid).

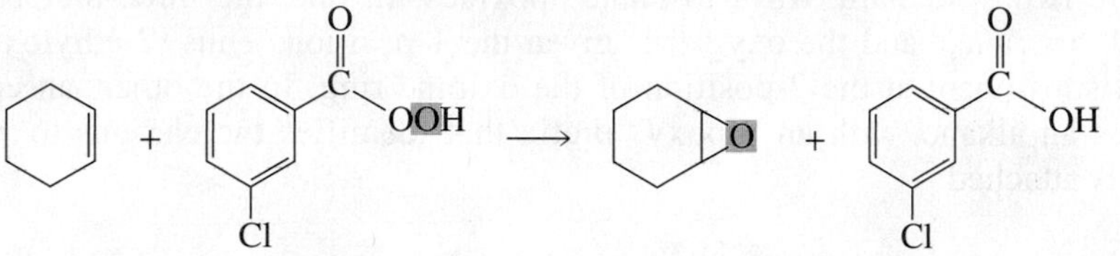

MCPBA

MECHANISM FOR THE EPOXIDATION OF AN ALKENE

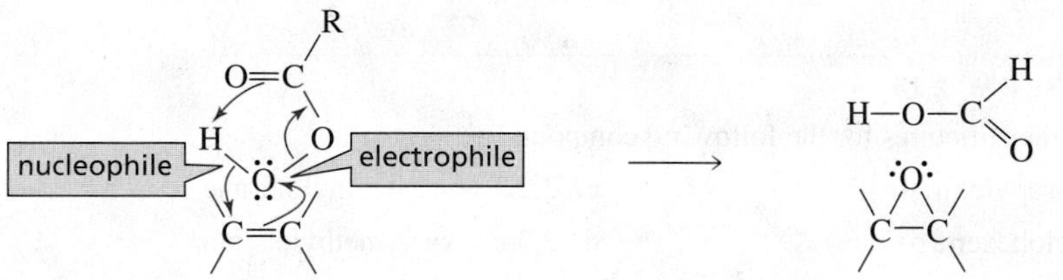

- An oxygen atom of the OOH group of the peroxyacid is electron deficient and is therefore an electrophile. It accepts a pair of electrons from the π bond of the alkene, which causes the weak O—O bond of the peroxyacid to break.

- The electrons from the O—O bond are delocalized, causing the π bond of the C=O group to break and pick up a proton.

- The electrons left behind (the nucleophile) as the O—H bond breaks add to the other sp^2 carbon of the alkene.

Thus, the oxygen atom is both the electrophile and the nucleophile. This is another example of a concerted reaction—that is, all bond-making and bond-breaking occur in the same step.

The mechanism for the addition of oxygen to a double bond to form an epoxide is analogous to the mechanism we have just seen for the addition of bromine to a double bond to form a cyclic intermediate (Section 9).

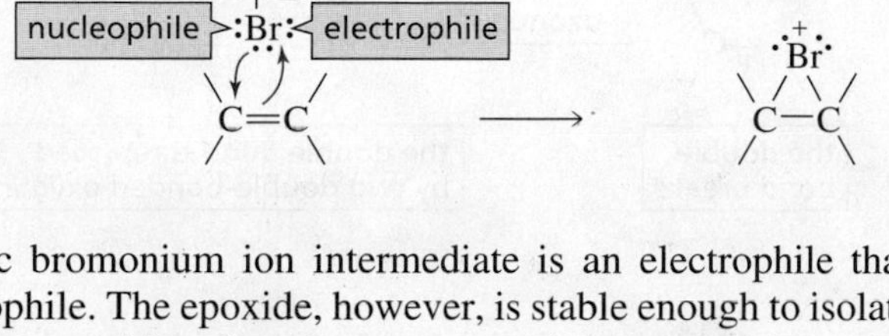

The unstable cyclic bromonium ion intermediate is an electrophile that subsequently reacts with a nucleophile. The epoxide, however, is stable enough to isolate because none of the ring atoms has a positive charge.

269

Nomenclature of Epoxides

The common name of an epoxide is obtained by adding "oxide" to the common name of the alkene; the oxygen is where the π bond of an alkene would be. The simplest epoxide is ethylene oxide.

$$H_2C{=}CH_2$$
ethylene

$$H_2C{-}CH_2$$ (with O bridging)
ethylene oxide

$$H_2C{=}CHCH_3$$
propylene

$$H_2C{-}CHCH_3$$ (with O bridging)
propylene oxide

There are two systematic ways to name epoxides. In one, the three-membered ring is called "oxirane," and the oxygen is given the 1-position. Thus, 2-ethyloxirane has an ethyl substituent at the 2-position of the oxirane ring. In the other, an epoxide is named as an alkane, with an "epoxy" prefix that identifies the carbons to which the oxygen is attached.

$$H_2C{-}CHCH_2CH_3$$ (with O bridging)
2-ethyloxirane
1,2-epoxybutane

$$CH_3CH{-}CHCH_3$$ (with O bridging)
2,3-dimethyloxirane
2,3-epoxybutane

2,2-dimethyloxirane
1,2-epoxy-2-methylpropane

PROBLEM 27♦

Draw the structures for the following compounds:

a. 2-propyloxirane

c. 2,2,3,3-tetramethyloxirane

b. cyclohexene oxide

d. 2,3-epoxy-2-methylpentane

PROBLEM 28♦

What alkene would you treat with a peroxyacid in order to obtain each of the following epoxides?

a. (cyclohexene oxide structure)

b. $H_2C{-}CHCH_2CH_3$ (with O bridging)

---

11 THE ADDITION OF OZONE TO AN ALKENE: OZONOLYSIS

When an alkene is treated with ozone (O_3) at a low temperature, both the σ and π bonds of the double bond break and the carbons that were doubly bonded to each other are now doubly bonded to oxygens instead. This is an oxidation reaction—called **ozonolysis**—because the number of C—O bonds increases.

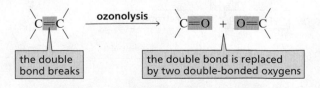

Ozonolysis is an example of **oxidative cleavage**—an oxidation reaction that cleaves the reactant into pieces (*lysis* is Greek for "breaking down").

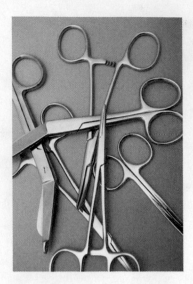

Ethylene oxide is used for the gaseous sterilization of disposable medical instruments. When the 2'-deoxyguanosine groups of a microorganism's DNA react with ethylene oxide, they are alkylated, and the microorganism can no longer reproduce.

To determine the product of ozonolysis, replace C═C with C═O O═C.

MECHANISM FOR OZONIDE FORMATION

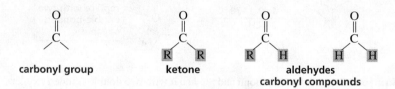

- The electrophile (an oxygen at one end of the ozone molecule) adds to one of the sp^2 carbons, and a nucleophile (the oxygen at the other end) adds to the other sp^2 carbon. The product is a **molozonide.**

- The molozonide is unstable because it has two O—O bonds; it immediately rearranges to a more stable **ozonide.**

ozone

Ozone, a major constituent of smog, is a health hazard at ground level, increasing the risk of death from lung or heart disease. In the stratosphere, however, a layer of ozone shields the Earth from harmful solar radiation.

Because ozonides are explosive, they are not isolated. Instead, they are immediately converted to ketones and/or aldehydes by dimethyl sulfide (CH_3SCH_3) or zinc in acetic acid (CH_3CO_2H). A **ketone** ("key-tone") has two alkyl groups bonded to a **carbonyl group,** and an **aldehyde** has an alkyl group and a hydrogen (or has two hydrogens) bonded to a carbonyl group.

carbonyl group   **ketone**   **aldehydes**
carbonyl compounds

The product is a ketone if the sp^2 carbon of the alkene is bonded to two carbon-containing substituents; the product is an aldehyde if one or both of the substituents bonded to the sp^2 carbon is a hydrogen.

ozonide → **a ketone** + **an aldehyde**

The following are additional examples of the oxidative cleavage of alkenes by ozonolysis. (Many organic reactions that need to be carried out in the cold are done conveniently at $-78°C$, because that is the temperature of a mixture of dry ice—solid CO_2—in acetone.)

1979

2008

The loss of stratospheric ozone over Antarctica—called the "ozone hole." The lowest ozone density is represented by dark blue.

break the double bond → **replace with two carbonyl groups**

1. O_3, -78 °C
2. $(CH_3)_2S$

an aldehyde   **a ketone**

Alkenes are oxidized to ketones and to aldehydes.

1. O_3, -78 °C
2. Zn, CH_3CO_2H

a ketone   **a ketone**

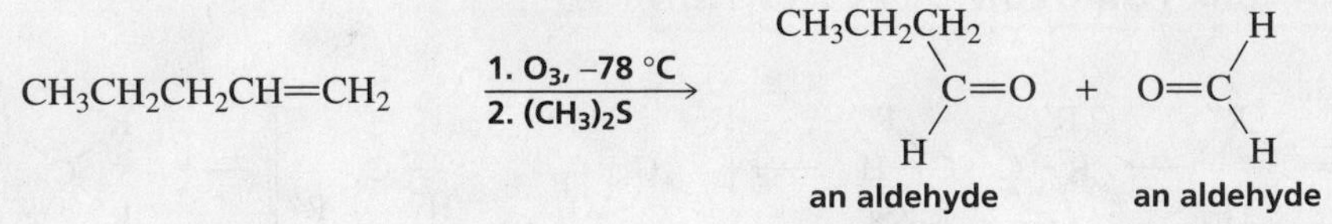

$$CH_3CH_2CH_2CH=CH_2 \xrightarrow[\text{2. (CH}_3\text{)}_2\text{S}]{\text{1. O}_3, -78\ °C}$$

$$\begin{array}{c} CH_3CH_2CH_2 \\ \diagdown \\ C=O \\ \diagup \\ H \end{array} \ + \ \begin{array}{c} H \\ \diagup \\ O=C \\ \diagdown \\ H \end{array}$$

an aldehyde   **an aldehyde**

PROBLEM-SOLVING STRATEGY

Determing the Products of Oxidative Cleavage

What products would you would expect to obtain when the following compounds react with ozone and then with dimethylsulfide?

a. 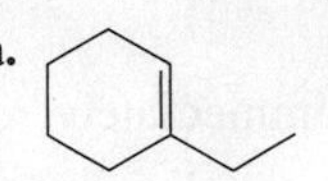   **b.** $CH_3CH_2CH=CHCH_2CH_3$ **c.**

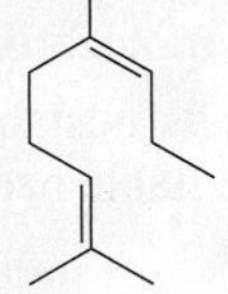

Solution to a Break the double bond and replace it with two double-bonded oxygens.

break the double bond

replace with two double-bonded oxygens

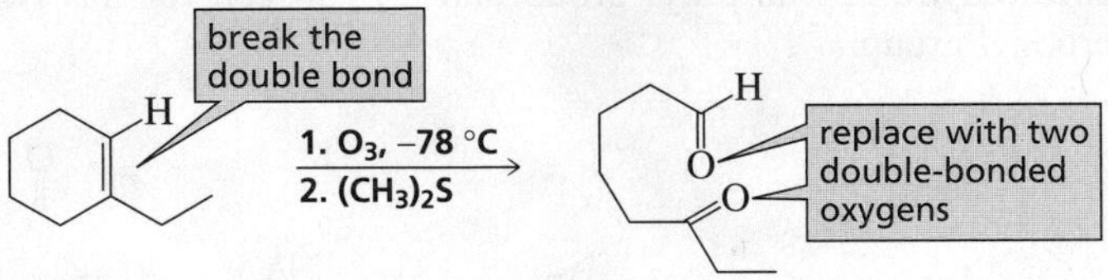

Solution to b Break the double bond and replace it with two double-bonded oxygens. Because the alkene is symmetrical, only one product is formed.

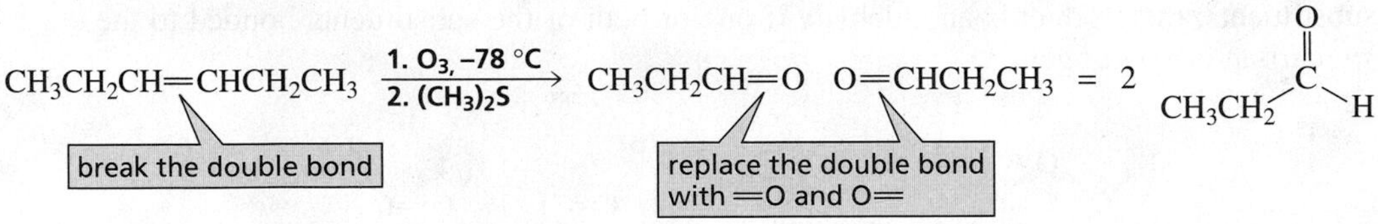

$$CH_3CH_2CH=CHCH_2CH_3 \xrightarrow[\text{2. (CH}_3\text{)}_2\text{S}]{\text{1. O}_3, -78\ °C} CH_3CH_2CH=O \ \ O=CHCH_2CH_3 = 2 \ \begin{array}{c} O \\ \parallel \\ C \\ CH_3CH_2 \diagup \ \diagdown H \end{array}$$

break the double bond

replace the double bond with =O and O=

Solution to c Since the reactant has two double bonds, each one must be replaced with two double-bonded oxygens.

break the double bond

break the double bond

replace each double bond with two double-bonded oxygens

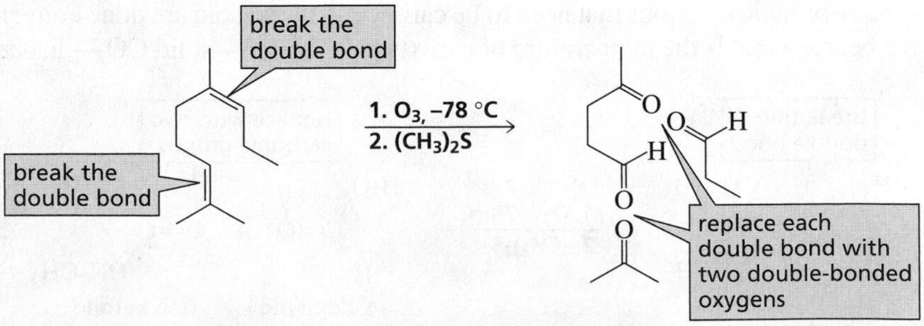

Now use the strategy you have just learned to solve Problem 29.

PROBLEM 29

What products are formed when the following compound react with ozone and then with dimethyl sulfide?

a. 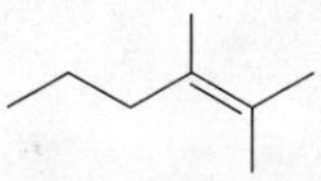 c. 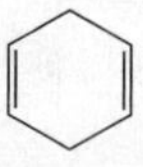 e.

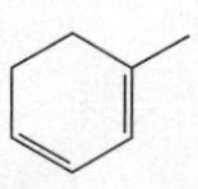

b. 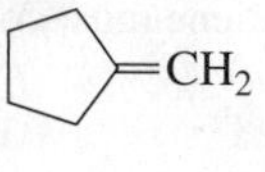 d. 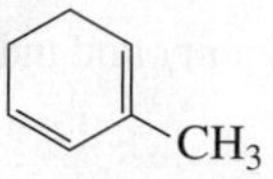 f.

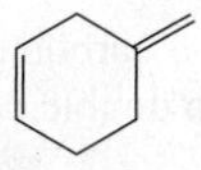

Ozonolysis can be used to determine the structure of an unknown alkene. If we know what carbonyl compounds are formed by ozonolysis, we can mentally work backward to deduce the structure of the alkene. In other words, delete the $=O$ and $O=$ and join the carbons by a double bond. (Recall that working backward is indicated by an open arrow.)

join these two carbons by a double bond

working backward is indicated by an open arrow

$$H_3C \diagdown C=O + O=C \diagup CH_2CH_2CH_3 \quad \Longrightarrow \quad H_3C \diagdown C=C \diagup CH_2CH_2CH_3$$
$$H_3C \diagup \qquad \diagdown H \qquad\qquad H_3C \diagup \qquad \diagdown H$$

ozonolysis products

alkene that underwent ozonolysis

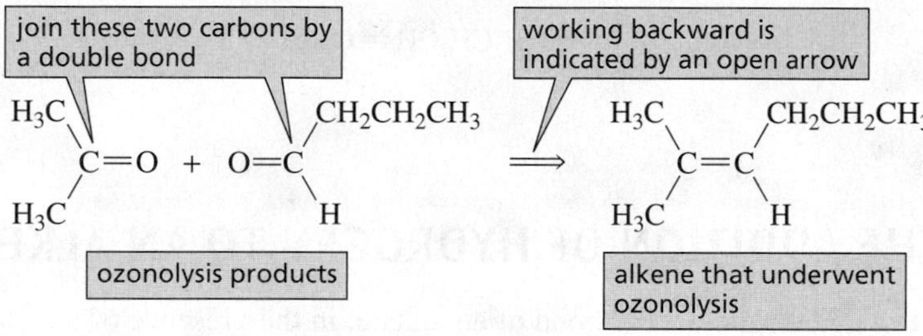

PROBLEM 30♦

a. What alkene would give only a ketone with three carbons as a product of oxidative cleavage?
b. What alkenes would give only an aldehyde with four carbons as a product of oxidative cleavage?

PROBLEM 31♦

What aspect of the structure of the alkene does ozonolysis not tell you?

PROBLEM 32 Solved

a. The following product was obtained from the ozonolysis of an alkene, followed by treatment with dimethyl sulfide. What is the structure of the alkene?

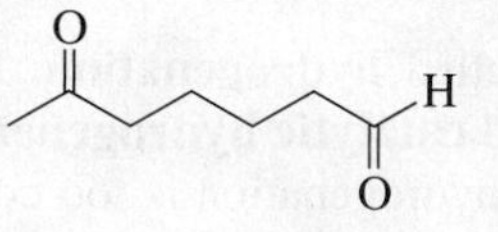

b. The following products were obtained from the oxidative cleavage of a diene. What is the structure of the diene?

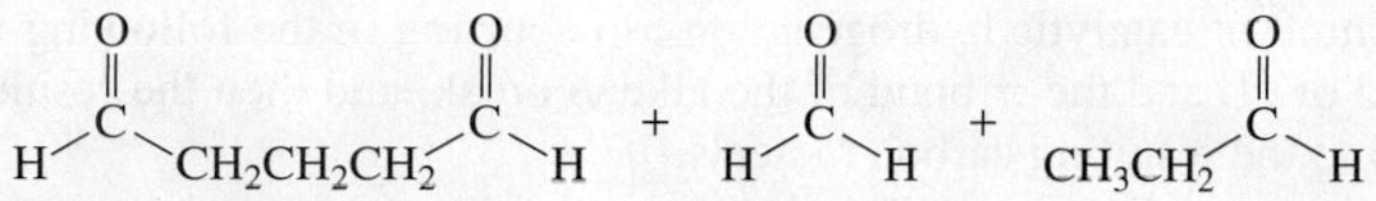

$$\underset{H}{\overset{O}{\overset{\|}{C}}} \underset{CH_2CH_2CH_2}{\overset{O}{\overset{\|}{C}}} \underset{H}{} \quad + \quad \underset{H}{\overset{O}{\overset{\|}{C}}} \underset{H}{} \quad + \quad \underset{CH_3CH_2}{\overset{O}{\overset{\|}{C}}} \underset{H}{}$$

273

Solution to 32a Because only one product is obtained, the reactant must be a cyclic alkene. Numbering the product shows that the carbonyl groups are at C-1 and C-6, so the double bonds must be between C-1 and C-6.

Solution to 32b The five-carbon dicarbonyl compound indicates that the diene must contain five carbons flanked by two double bonds.

One of the other two carbonyl compounds obtained from ozonolysis has one carbon, and the other has three carbons. Therefore, one carbon has to be added to one end of the diene, and three carbons have to be added to the other end.

$$CH_2\!=\!CHCH_2CH_2CH_2CH\!=\!CHCH_2CH_3$$

12   THE ADDITION OF HYDROGEN TO AN ALKENE

Hydrogen (H_2) adds to the double bond of an alkene, in the presence of a metal catalyst, to form an alkane. The most common metal catalyst is palladium, which is used as a powder adsorbed on charcoal to maximize its surface area; it is referred to as "palladium on carbon" and is abbreviated as Pd/C. The metal catalyst is required to weaken the very strong H—H bond.

The addition of hydrogen to a compound is a reduction reaction. A **reduction reaction** increases the number of C—H bonds and/or decreases the number of C—O, C—N, or C—X bonds (where X is a halogen).

$$CH_3CH\!=\!CHCH_3 + H_2 \xrightarrow{\text{Pd/C}} CH_3CH_2CH_2CH_3$$

2-butene   butane

> **A reduction reaction increases the number of C—H bonds.**

1-methylcyclohexene   methylcyclohexane

The addition of hydrogen is called **hydrogenation.** Because hydrogenation reactions require a catalyst, they are called **catalytic hydrogenations.**

The mechanism for catalytic hydrogenation is too complex to be easily described. We know that hydrogen is adsorbed on the surface of the metal and that the alkene complexes with the metal by overlapping its p orbitals with the vacant orbitals of the metal. All the bond-breaking and bond-forming events occur on the surface of the metal. As the alkane product forms, it diffuses away from the metal surface (Figure 6).

We can think of catalytic hydrogenation as occurring in the following way: both the H—H bond of H_2 and the π bond of the alkene break, and then the resulting hydrogen radicals add to the resulting carbon radicals.

$$CH_3CH\!=\!CHCH_3 \longrightarrow CH_3\dot{C}H\!-\!\dot{C}HCH_3 \longrightarrow CH_3CH\!-\!CHCH_3$$

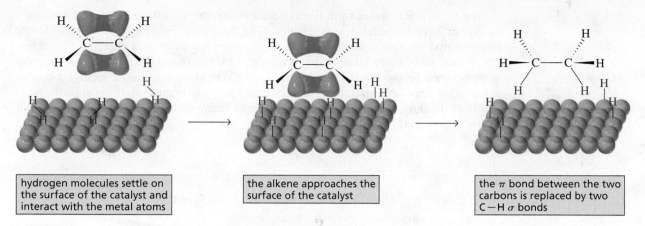

| hydrogen molecules settle on the surface of the catalyst and interact with the metal atoms | the alkene approaches the surface of the catalyst | the π bond between the two carbons is replaced by two C—H σ bonds |

▲ **Figure 6**

Catalytic hydrogenenation of an alkene to form an alkane.

The heat released in a hydrogenation reaction is called the **heat of hydrogenation.** It is customary to give it a positive value. Hydrogenation reactions, however, are exothermic (they have negative $\Delta H°$ values), so *the heat of hydrogenation is the value of $\Delta H°$ without the negative sign.*

				$\Delta H°$	
			heat of hydrogenation	kcal/mol	kJ/mol

CH₃C=CHCH₃ + H₂ $\xrightarrow{\text{Pd/C}}$ CH₃CHCH₂CH₃   26.9 kcal/mol   −26.9   −113
2-methyl-2-butene

CH₂=CCH₂CH₃ + H₂ $\xrightarrow{\text{Pd/C}}$ CH₃CHCH₂CH₃   28.5 kcal/mol   −28.5   −119
2-methyl-1-butene

CH₃CHCH=CH₂ + H₂ $\xrightarrow{\text{Pd/C}}$ CH₃CHCH₂CH₃   30.3 kcal/mol   −30.3   −127
3-methyl-1-butene

the product of each of the 3 reactions is 2-methylbutane

Because we do not know the precise mechanism for catalytic hydrogenation, we cannot draw its reaction coordinate diagram. However, we can draw a diagram like that in Figure 7, which shows the relative energies of the reactants and products for the three catalytic hydrogenation reactions just shown.

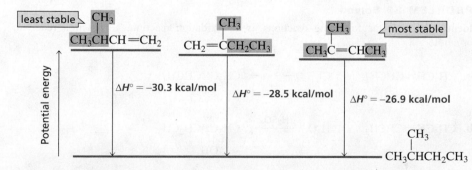

The most stable alkene has the smallest heat of hydrogenation.

▲ **Figure 7**

The relative energies (stabilities) of three alkenes that can be catalytically hydrogenated to 2-methylbutane. The most stable alkene has the smallest heat of hydrogenation.

The three reactions all form the same alkane product, so the energy of the *product* in Figure 7 is the same for each reaction. The three reactions, however, have different heats of hydrogenation, so the three *reactants* must have different energies. For example, 3-methyl-1-butene releases the most heat, so it must have the most energy to begin with (it must be the *least* stable of the three alkenes). In contrast, 2-methyl-2-butene releases the least heat, so it must have the least energy to begin with (it must be the *most* stable of the three alkenes). Notice that *the most stable compound has the smallest heat of hydrogenation.*

Trans Fats

Oils are liquids at room temperature because their fatty acid components contain several carbon–carbon double bonds, which makes it difficult for them to pack closely together. In contrast, the fatty acid components of fats have fewer double bonds so they can pack together more closely. Because of their many double bonds, oils are said to be polyunsaturated.

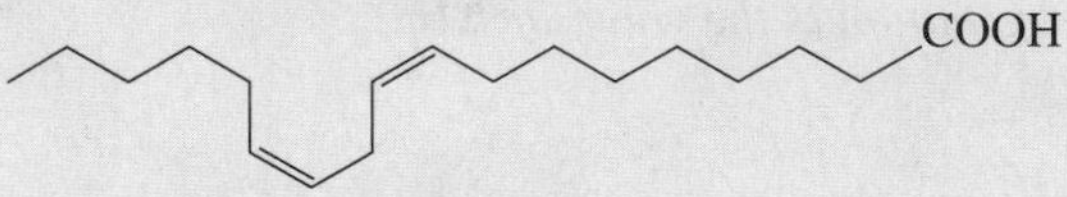

COOH

linoleic acid
an 18-carbon fatty acid with two cis double bonds

Some or all of the double bonds in oils can be reduced by catalytic hydrogenation. For example, margarine and shortening are prepared by hydrogenating vegetable oils, such as soybean oil and safflower oil, until they have the desired creamy, solid consistency of butter.

All the double bonds in naturally occurring fats and oils have the cis configuration. The catalyst used in the hydrogenation process also catalyzes cis-trans isomerization, forming what is known as a trans fat.

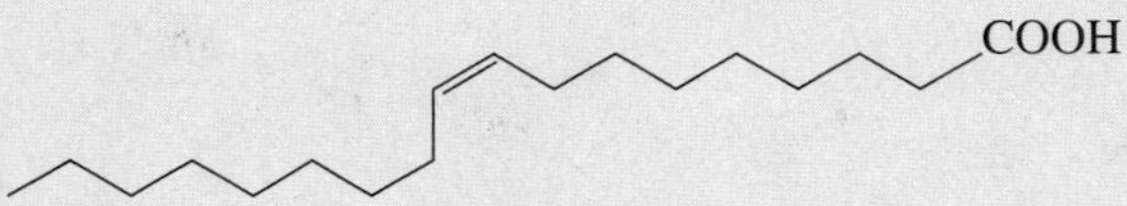

COOH

oleic acid
an 18-carbon fatty acid with one cis double bond
before being heated

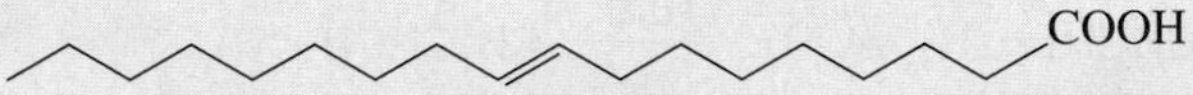

COOH

elaidic acid
an 18-carbon fatty acid with one trans double bond
after being heated

Trans fats are a health concern because they increase LDL, the so-called "bad" cholesterol. Epidemiological studies have shown that an increase in the daily intake of trans fats significantly increases the incidence of cardiovascular disease.

PROBLEM 33 Solved

Identify each of the following reactions as an oxidation reaction, a reduction reaction, or neither.

a. $CH_3CH{=}CHCH_3$ + Cl_2 $\longrightarrow$ $CH_3CHCHCH_3$
$\qquad\qquad\qquad\qquad\qquad\qquad\qquad\quad$ | |
$\qquad\qquad\qquad\qquad\qquad\qquad\qquad$ Cl Cl

b. $CH_3CH{=}CHCH_3$ + H_2O $\xrightarrow{\text{H}_2\text{SO}_4}$ $CH_3CHCH_2CH_3$
$\qquad\qquad\qquad\qquad\qquad\qquad\qquad\qquad\qquad$ |
$\qquad\qquad\qquad\qquad\qquad\qquad\qquad\qquad\quad$ OH

Solution to 33a This is an oxidation reaction because the number of C—Cl bonds increases. (Recall that an oxidation reaction decreases the number of C—H bonds and/or increases the number of C—O, C—N, or C—X bonds [where X = a halogen], whereas a reduction reaction increases the number of C—H bonds and/or decreases the number of C—O, C—N, or C—X bonds.)

Solution to 33b The product has both a new C—O bond (signifying an oxidation) and a new C—H bond (signifying a reduction). Thus, the two cancel each other; the reaction is neither an oxidation nor a reduction.

PROBLEM-SOLVING STRATEGY

Choosing the Reactant for a Synthesis

What alkene would you start with if you wanted to synthesize methylcyclohexane?

You need to choose an alkene with the same number of carbons, attached in the same way, as those in the desired product. Several alkenes could be used for this synthesis, because the double bond can be located anywhere in the molecule.

Now use the strategy you have just learned to solve Problem 34.

PROBLEM 34

What alkene would you start with if you wanted to synthesize

a. pentane?   **b.** ethylcyclopentane?

PROBLEM 35

How many different alkenes can be hydrogenated to form

a. butane?   **b.** 3-methylpentane?   **c.** hexane?

---

13   THE RELATIVE STABILITIES OF ALKENES

If you look at the structures of the three alkene reactants in Figure 7, you will see that the stability of an alkene increases as the number of alkyl substituents bonded to the sp^2 carbons increases.

For example, the most stable alkene in Figure 7 has two alkyl substituents bonded to one sp^2 carbon and one alkyl substituent bonded to the other sp^2 carbon, for a total of three alkyl substituents (three methyl groups) bonded to its two sp^2 carbons. The alkene of intermediate stability has two alkyl substituents (a methyl group and an ethyl group) bonded to its sp^2 carbons, and the least stable of the three alkenes has only one alkyl substituent (an isopropyl group) bonded to an sp^2 carbon.

relative stabilities of alkyl-substituted alkenes

The more alkyl substituents bonded to the sp^2 carbons, the more stable the alkene.

We can therefore make the following statement: *the stability of an alkene increases as the number of alkyl substituents bonded to its* sp^2 *carbons increases.* (Some students find it easier to understand this concept from the point of view of the hydrogens bonded to the sp^2 carbons—namely, *the stability of an alkene increases as the number of hydrogens bonded to its* sp^2 *carbons decreases.*)

Alkyl substituents stabilize both alkenes *and* carbocations.

PROBLEM 36♦

The same alkane is obtained from the catalytic hydrogenation of both alkene **A** and alkene **B**. The heat of hydrogenation of alkene **A** is 29.8 kcal/mol, and the heat of hydrogenation of alkene **B** is 31.4 kcal/mol. Which alkene is more stable?

PROBLEM 37♦

a. Which of the following compounds is the most stable?

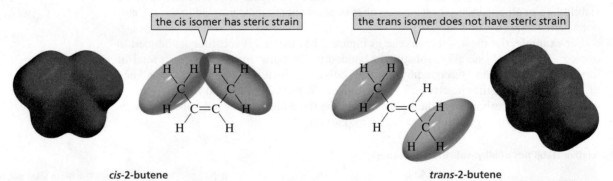

b. Which is the least stable?
c. Which has the smallest heat of hydrogenation?

Both *trans*-2-butene and *cis*-2-butene have two alkyl substituents bonded to their sp^2 carbons, but *trans*-2-butene has a smaller heat of hydrogenation. This means that the trans isomer, in which the large substituents are farther apart, is more stable than the cis isomer.

	heat of hydrogenation	$\Delta H°$ kcal/mol	kJ/mol

$$H_3C, H \quad C=C \quad H, CH_3 \quad + \quad H_2 \quad \xrightarrow{Pd/C} \quad CH_3CH_2CH_2CH_3 \qquad 27.6 \qquad -27.6 \qquad -115$$

***trans*-2-butene**

$$H_3C, CH_3 \quad C=C \quad H, H \quad + \quad H_2 \quad \xrightarrow{Pd/C} \quad CH_3CH_2CH_2CH_3 \qquad 28.6 \qquad -28.6 \qquad -120$$

***cis*-2-butene**

When large substituents are on the same side of the double bond, as in a cis isomer, their electron clouds can interfere with each other, causing steric strain in the molecule. Steric strain makes a compound less stable. When the large substituents are on opposite sides of the double bond, as in a trans isomer, their electron clouds cannot interact, so there is no destabilizing steric strain.

the cis isomer has steric strain

the trans isomer does not have steric strain

***cis*-2-butene**   ***trans*-2-butene**

The heat of hydrogenation of *cis*-2-butene, in which the two alkyl substituents are on the *same side* of the double bond, is similar to that of 2-methylpropene, in which the two alkyl substituents are on the *same carbon*. The three dialkyl-substituted alkenes are all *less* stable than a trialkyl-substituted alkene, and they are all *more* stable than a monoalkyl-substituted alkene.

relative stabilities of dialkyl-substituted alkenes

$$H_3C, H \quad C=C \quad H, CH_3 \qquad > \qquad H_3C, CH_3 \quad C=C \quad H, H \qquad \sim \qquad H_3C, H \quad C=C \quad H_3C, H$$

alkyl substituents are on *opposite sides* of the double bond	alkyl substituents are on the *same side* of the double bond	alkyl substituents are on the same sp^2 carbon

PROBLEM 38♦

Rank the following compounds in order from most stable to least stable:

trans-3-hexene, *cis*-3-hexene, *cis*-2,5-dimethyl-3-hexene, *cis*-3,4-dimethyl-3-hexene

14 REGIOSELECTIVE, STEREOSELECTIVE, AND STEREOSPECIFIC REACTIONS

When we looked at the electrophilic addition reactions that alkenes undergo, we examined the step-by-step process by which each reaction occurs (the mechanism of the reaction), and we determined what products are formed. However, we did not consider the stereochemistry of the reactions.

Stereochemistry is the field of chemistry that deals with the structures of molecules in three dimensions. When we study the stereochemistry of a reaction, we are concerned with the following questions:

1. If the *product* of a reaction can exist as two or more stereoisomers, does the reaction produce a single stereoisomer, a set of particular stereoisomers, or all possible stereoisomers?

2. If the *reactant* can exist as two or more stereoisomers, do all stereoisomers of the reactant form the same stereoisomers of the product, or does each stereoisomer of the reactant form a different stereoisomer or set of stereoisomers of the product?

Before we examine the stereochemistry of electrophilic addition reactions, we need to become familiar with some terms used in describing the stereochemistry of a reaction.

We have seen that a **regioselective** reaction is one in which two *constitutional isomers* can be obtained as products, but more of one is obtained than the other (Section 4). In other words, a regioselective reaction selects for a particular constitutional isomer. Recall that a reaction can be *moderately regioselective*, *highly regioselective*, or *completely regioselective* depending on the relative amounts of the constitutional isomers formed in the reaction.

a regioselective reaction

constitutional isomers

A ⟶ B + C

more B is formed than C

> A regioselective reaction forms more of one constitutional isomer than of another.

Stereoselective is a similar term, but it refers to the preferential formation of a *stereoisomer* rather than a *constitutional isomer*. A stereoselective reaction forms one stereoisomer preferentially over another. In other words, it selects for a particular stereoisomer. Depending on the degree of preference for a particular stereoisomer, a reaction can be described as being *moderately stereoselective*, *highly stereoselective*, or *completely stereoselective*.

a stereoselective reaction

stereoisomers

A ⟶ B + C

more B is formed than C

> A stereoselective reaction forms more of one stereoisomer than of another.

A reaction is **stereospecific** if the *reactant* can exist as stereoisomers and each stereoisomer of the reactant forms a different stereoisomer or a different set of stereoisomers of the product.

stereospecific reactions

> In a stereospecific reaction, each stereoisomer forms a different stereoisomeric product or a different set of stereoisomeric products.

In the preceding reaction, stereoisomer A forms stereoisomer B but does not form D, so the reaction is stereospecific. Because a stereospecific reaction does not form all possible stereoisomers of the product, *all stereospecific reactions are also stereoselective. However, not all stereoselective reactions are stereospecific*, because there are stereoselective reactions with reactants that do not have stereoisomers.

> **A stereospecific reaction is also stereoselective, but a stereoselective reaction is not necessarily stereospecific.**

PROBLEM 39♦

What characteristics must the reactant of a stereospecific reaction have?

---

15 THE STEREOCHEMISTRY OF ELECTROPHILIC ADDITION REACTIONS OF ALKENES

You should be familiar with stereoisomers *and* with electrophilic addition reactions, so we can combine the two topics and look at the stereochemistry of these reactions. In other words, we will look at the stereoisomers that are formed in the electrophilic addition reactions that you learned about in this chapter.

We have seen that when an alkene reacts with an electrophilic reagent such as HBr, the major product of the addition reaction is the one obtained by adding the electrophile (H^+) to the sp^2 carbon bonded to the most hydrogens and adding the nucleophile (Br^-) to the other sp^2 carbon (Section 4). Thus, the major product obtained from the following reaction is 2-bromopropane. This product does not have an asymmetric center, so it does not have stereoisomers. Therefore, we do not have to be concerned with the stereochemistry of the reaction.

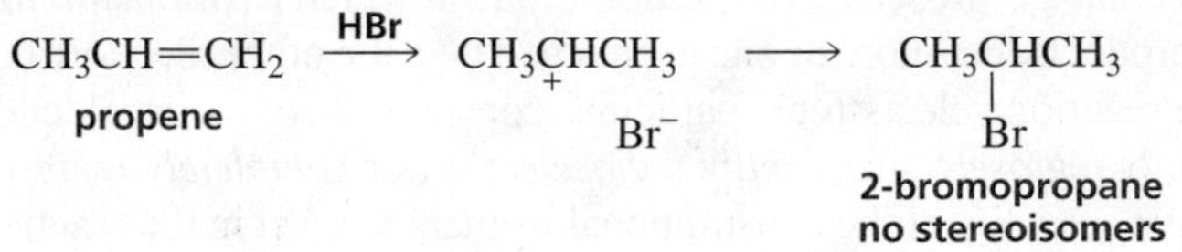

$$CH_3CH{=}CH_2 \xrightarrow{\text{HBr}} CH_3\overset{+}{C}HCH_3 \quad \xrightarrow{\quad} \quad CH_3\underset{|}{\overset{}{C}}HCH_3$$

propene · Br⁻ · Br · **2-bromopropane** · **no stereoisomers**

The following reaction forms a product with an asymmetric center, so we have to be concerned with the stereochemistry of this reaction. What is the configuration of the product? In other words, do we get the *R* enantiomer, the *S* enantiomer, or both enantiomers?

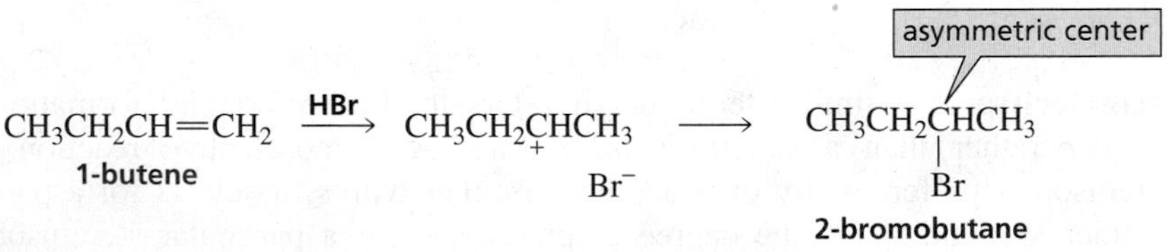

$$CH_3CH_2CH{=}CH_2 \xrightarrow{\text{HBr}} CH_3CH_2\overset{+}{C}HCH_3 \quad \xrightarrow{\quad} \quad CH_3CH_2\underset{|}{\overset{}{C}}HCH_3$$

1-butene · Br⁻ · Br · **2-bromobutane** · asymmetric center

We will begin our discussion of the stereochemistry of electrophilic addition reactions by looking at reactions that form a product with one asymmetric center. Then we will look at reactions that form a product with two asymmetric centers.

Addition Reactions that Form a Product with One Asymmetric Center

When a reactant that does *not* have an asymmetric center undergoes a reaction that forms a product with *one* asymmetric center, the product will always be a racemic mixture. For example, the reaction of 1-butene with HBr that we just looked at forms identical amounts of (*R*)-2-bromobutane and (*S*)-2-bromobutane. Thus, the reaction is not stereoselective because it does not select for a particular stereoisomer. Why is this so?

Because the products of the reaction are enantiomers, the transition states that lead to the products are also enantiomers. Thus, the two transition states have the same stability, so the two products will be formed at the same rate. The product, therefore, is a racemic mixture (Figure 8).

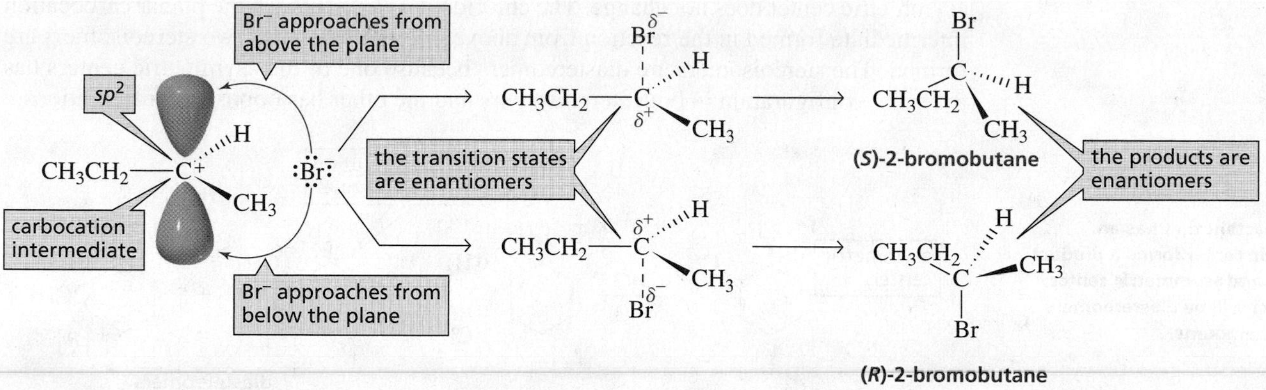

▲ Figure 8

The three groups bonded to the sp^2 carbon of the carbocation intermediate lie in a plane. When the bromide ion approaches the intermediate from above the plane, one enantiomer is formed; when it approaches from below the plane, the other enantiomer is formed. Because the bromide ion has equal access to both sides of the plane, identical amounts of the R and S enantiomers are formed.

A racemic mixture is formed by any reaction that forms a product with an asymmetric center from a reactant that does not have an asymmetric center. The products of the following reactions, therefore, are racemic mixtures.

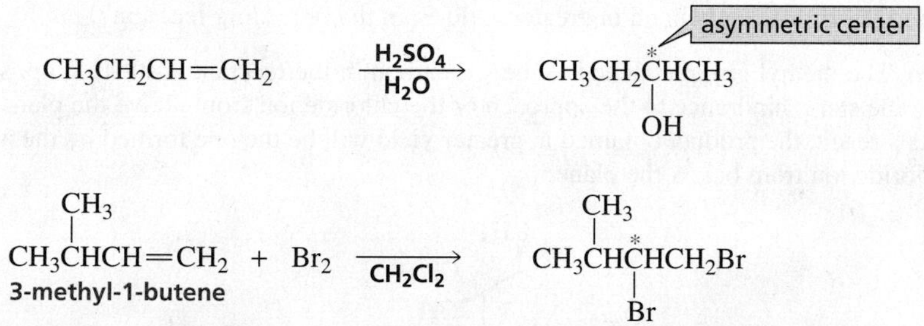

When a reactant that does not have an asymmetric center forms a product with *one* asymmetric center, the product will always be a racemic mixture.

PROBLEM 40♦

a. Is the reaction of 2-butene with HBr regioselective?
b. Is it stereoselective?
c. Is it stereospecific?
d. Is the reaction of 1-butene with HBr regioselective?
e. Is it stereoselective?
f. Is it stereospecific?

PROBLEM 41

What stereoisomers are obtained from each of the following reactions?

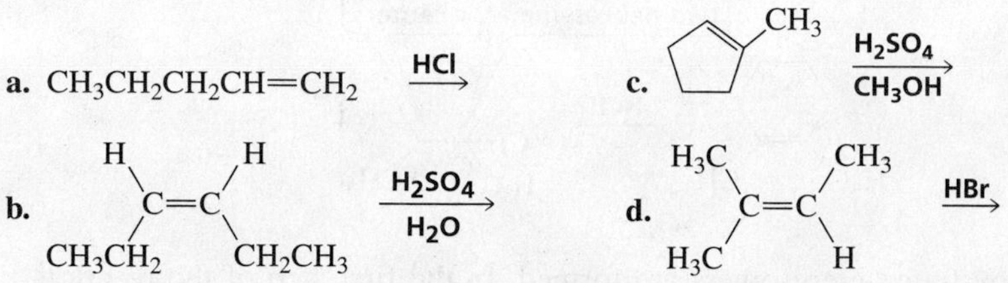

If an addition reaction creates an asymmetric center in a compound that already has an asymmetric center (and the reactant is a single enantiomer of that compound), a pair of diastereomers will be formed.

For example, let's look at the following reaction. Because none of the bonds to the asymmetric center in the reactant is broken during the reaction, the configuration of this asymmetric center does not change. The chloride ion can approach the planar carbocation intermediate formed in the reaction from above or from below, so two stereoisomers are formed. The stereoisomers are diastereomers because one of the asymmetric centers has the same configuration in both stereoisomers and the other has opposite configurations.

When a reactant that has an asymmetric center forms a product with *a second* asymmetric center, the product will be diastereomers in unequal amounts.

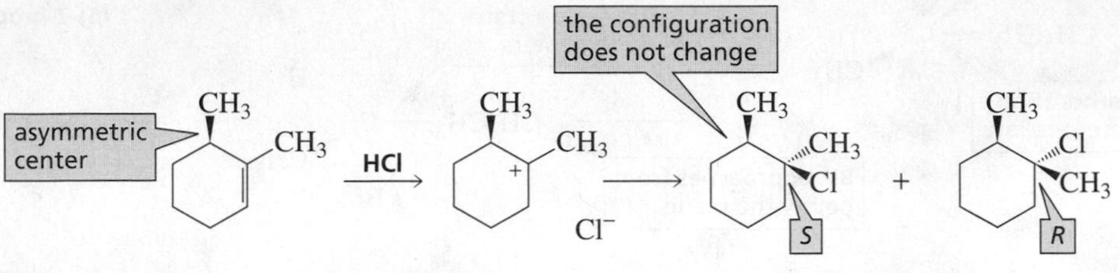

Because the products of the preceding reaction are diastereomers, the transition states that lead to them are also diastereomers. The two transition states, therefore, will not have the same stability, so they will be formed at different rates. Thus, the reaction is stereoselective—it forms more of one stereoisomer than of the other. The reaction is also stereospecific; the alkene with an asymmetric center in the S configuration forms a different pair of diastereomers than the alkene with an asymmetric center in the R configuration.

PROBLEM 42 Solved

Which product would be obtained in greater yield from the preceding reaction?

Solution The methyl group attached to the sp^3 carbon in the reactant is sticking up, so it will provide some steric hindrance to the approach of the chloride ion from above the planar carbocation. As a result, the product obtained in greater yield will be the one formed by the approach of the chloride ion from below the plane.

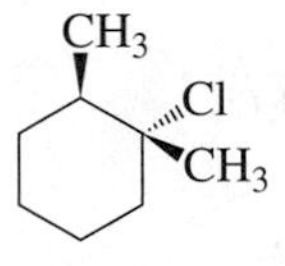

Addition Reactions that Form Products with Two Asymmetric Centers

When a reactant that does not have an asymmetric center undergoes a reaction that forms a product with *two* asymmetric centers, the stereoisomers that are formed depend on the mechanism of the reaction.

Addition Reactions that Form a Carbocation Intermediate When an addition reaction that results in a product with *two new asymmetric centers* forms a carbocation intermediate, *four stereoisomers* are formed.

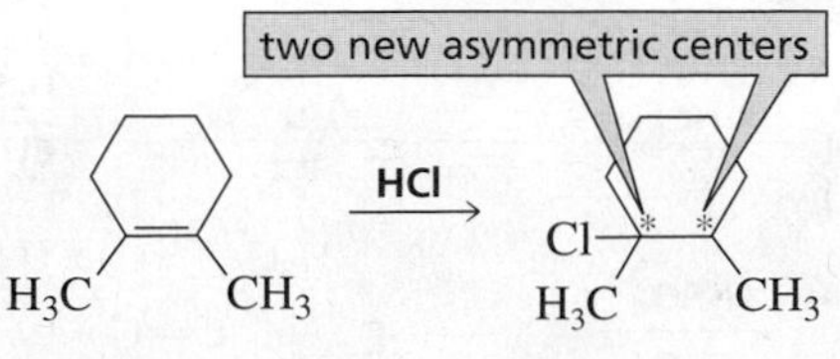

Now let's see why four stereoisomers are formed. In the first step of the reaction, the proton can approach the plane containing the double-bonded carbons of the alkene from either above or below to form the carbocation. Once the carbocation is formed, the chloride ion can approach the positively charged carbon from above or below. Thus, the addition of the proton and the chloride ion can be described as above-above (both adding from above), above-below, below-above, and below-below. As a result, four stereoisomers are obtained as products.

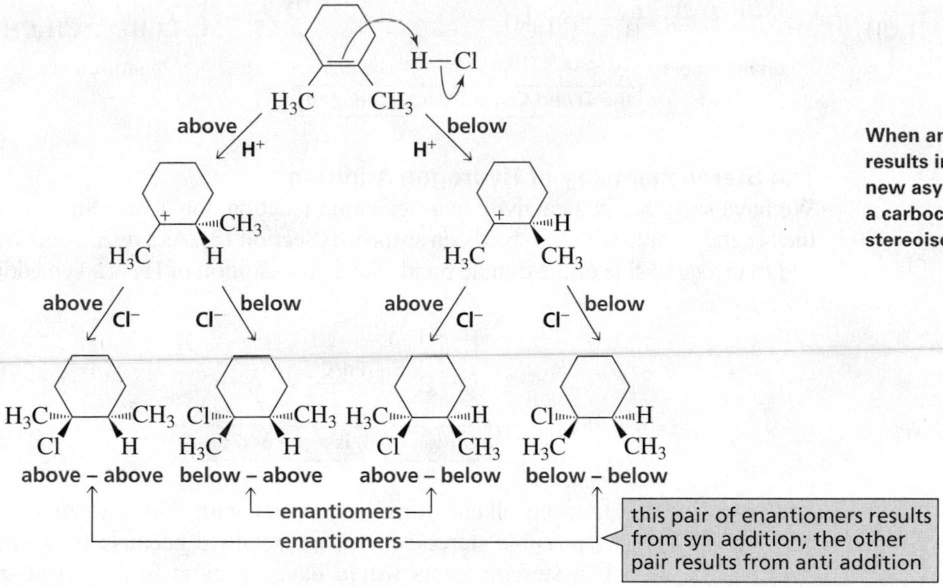

When an addition reaction that results in a product with two new asymmetric centers forms a carbocation intermediate, four stereoisomers are obtained.

When two substituents add to the same side of a double bond, the addition is called **syn addition.** When two substituents add to opposite sides of a double bond, the addition is called **anti addition.** Both syn and anti additions occur in alkene addition reactions that form a carbocation intermediate. Equal amounts of the four stereoisomers are obtained so the reaction is not stereoselective and, because the four stereoisomers formed by the cis alkene are identical to the four stereoisomers formed by the trans alkene, the reaction is also not stereospecific.

PROBLEM 43 Solved

Using perspective formulas, draw the products obtained from the reaction of *(Z)*-3,4-dimethyl-3-hexene with HCl.

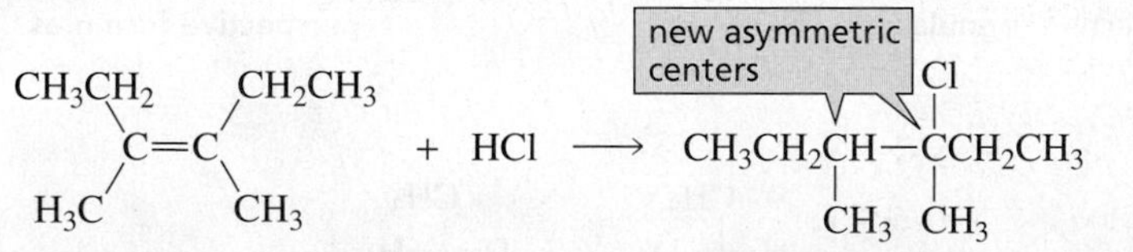

(*Z*)-3,4-dimethyl-3-hexene **3-chloro-3,4-dimethylhexane**

Because the reaction forms a carbocation intermediate on the way to forming a product with two new asymmetric centers, four stereoisomers are formed. First draw the bonds about the asymmetric centers. (Remember that the lines in the plane must be adjacent and so must the solid and hatched wedges, and the hatched wedge is above the solid wedge.) Attach the groups to the bonds in any order, then draw the mirror image of the structure you just drew.

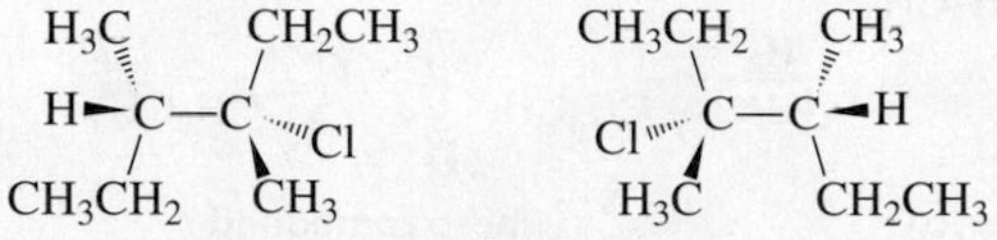

enantiomers

Next, draw a third structure by interchanging any two groups on either one of the first two structures. Then draw the mirror image of the third structure. (Notice that the two structures on the left are the same as the two structures on the bottom of the previous page.)

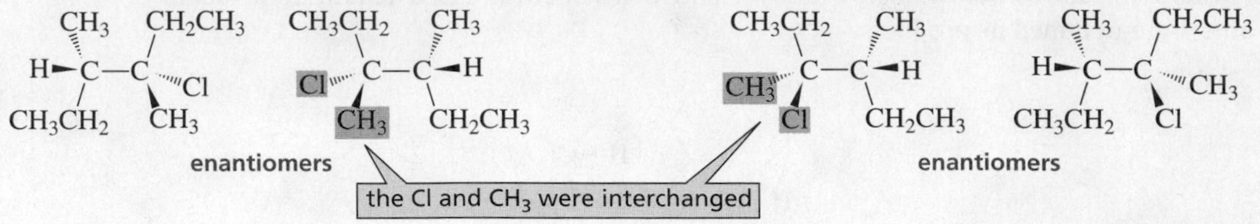

The Stereochemistry of Hydrogen Addition

We have seen that in a catalytic hydrogenation reaction, the alkene sits on the surface of a metal catalyst onto which H_2 has been absorbed (Section 12). As a result, both hydrogen atoms add to the same side of the double bond. Thus, the addition of H_2 is a syn addition reaction.

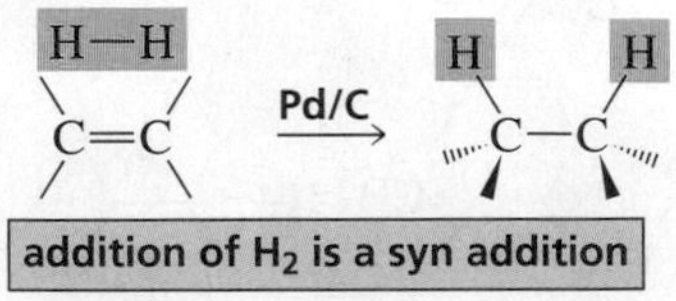

If addition of H_2 to an alkene forms a product with two asymmetric centers, then only two of the four possible stereoisomers are obtained because only syn addition can occur. (The other two stereoisomers would have to come from anti addition.) One of the two stereoisomers results from addition of both hydrogens from below the plane of the double bond, and the other results from addition of both hydrogens from above the plane, because the alkene can sit on the metal catalyst with either side facing up.

The particular pair of stereoisomers that is formed depends on whether the reactant is a *cis* alkene or a *trans* alkene. Syn addition of H_2 to a *cis* alkene forms only the erythro enantiomers. (Note that the erythro enantiomers are the ones with the hydrogens on the same side of the carbon chain in the eclipsed conformers.)

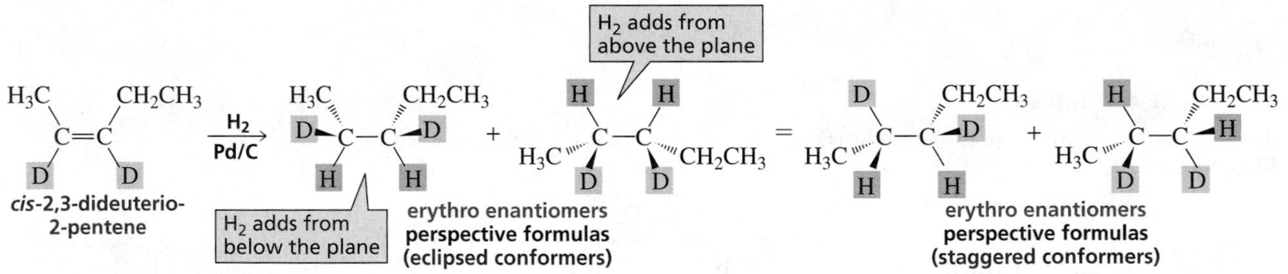

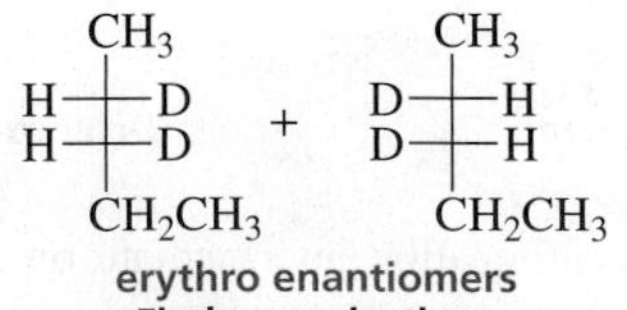

erythro enantiomers
Fischer projections

If each of the two asymmetric centers in the product is bonded to the same four substituents, a meso compound will be formed instead of the erythro enantiomers.

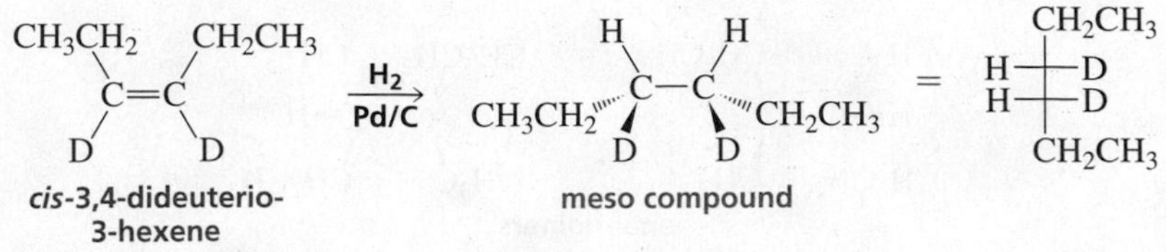

In contrast, syn addition of H_2 to a *trans* alkene forms only the threo enantiomers. Thus, the addition of hydrogen is a stereospecific reaction—the products obtained from addition to the cis isomer are different from the products obtained from addition to the trans isomer. It is also a stereoselective reaction because only two of the four stereoisomers are formed.

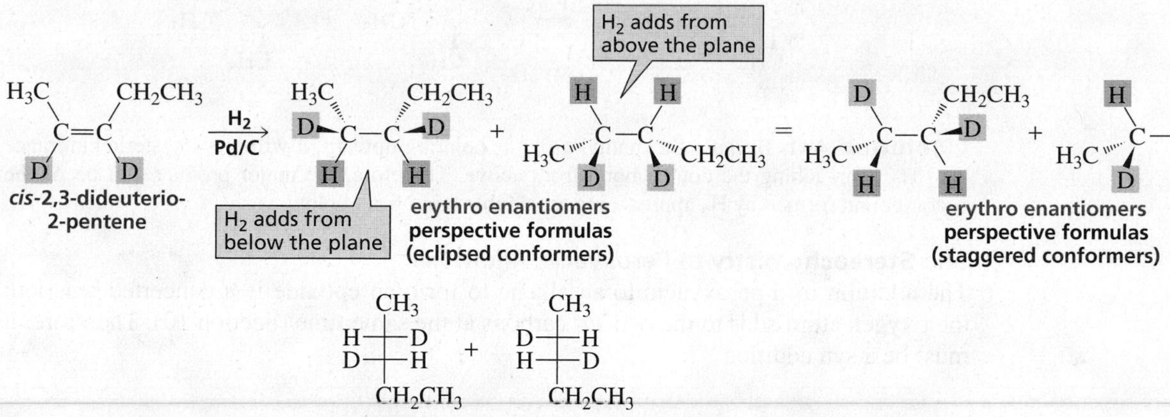

threo enantiomers
Fischer projections

If the reactant is cyclic, the addition of H_2 will form the cis enantiomers since the two hydrogens add to the same side of the double bond. You can see that the two products are enantiomers (nonsuperimposable mirror images) if you turn one of them upside down.

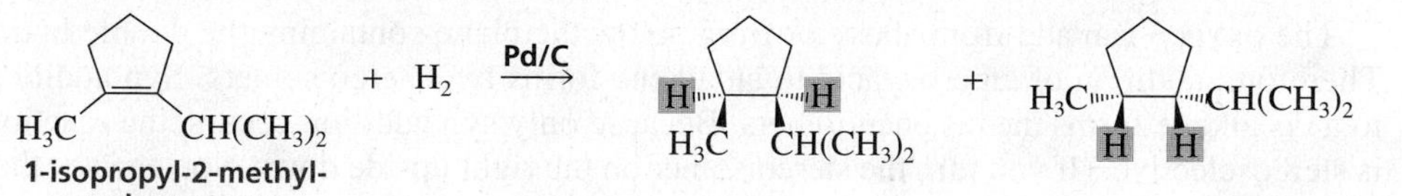

1-isopropyl-2-methyl-cyclopentene

Each of the two asymmetric centers in the product of the following reaction is bonded to the same four substituents. Therefore, syn addition forms a meso compound.

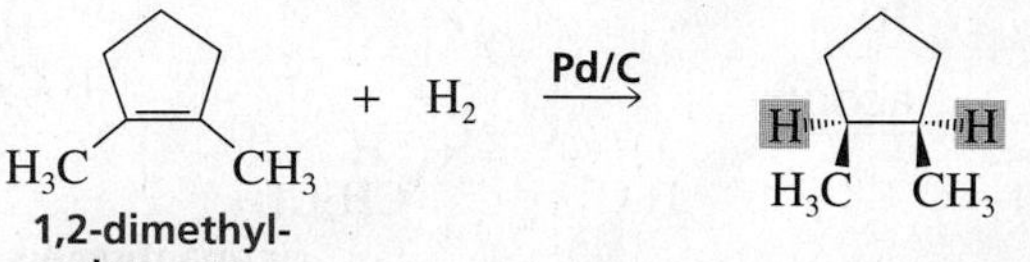

1,2-dimethyl-cyclopentene

Cyclic Alkenes

Cyclic alkenes with fewer than seven carbons in the ring, such as cyclopentene and cyclohexene, can exist only in the cis configuration because they do not have enough carbons to form a trans double bond. Therefore, it is not necessary to use the cis designation with their names. If the ring has seven or more carbons, however, then both cis and trans isomers are possible, so the configuration of the compound must be specified in its name.

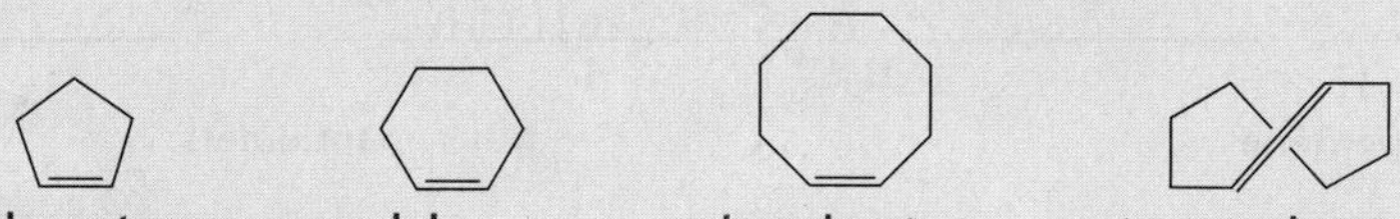

cyclopentene **cyclohexene** ***cis*-cyclooctene** ***trans*-cyclooctene**

PROBLEM 44 Solved

a. What stereoisomers are formed in the following reaction?
b. Which stereoisomer is formed in greater yield?

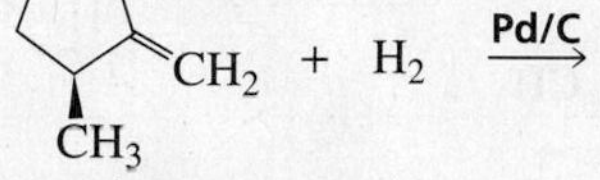

Solution 44a The reaction forms two stereoisomers because H_2 can approach the plane of the double bond from above and below. The stereoisomers are diastereomers because the reactant has an asymmetric center and the product has a new second asymmetric center.

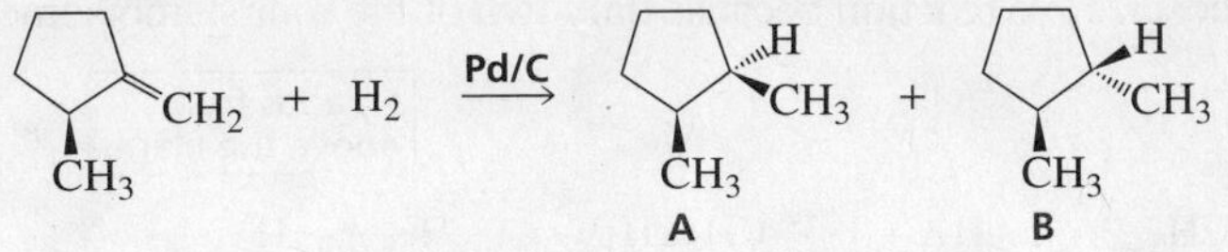

Solution 44b Because the methyl group is pointing upward, it will provide steric hindrance to H_2 approaching the double bond from above. Therefore, the major product will be **A**, the compound formed by H_2 approaching the double bond from below.

The Stereochemistry of Peroxyacid Addition

The addition of a peroxyacid to an alkene to form an epoxide is a concerted reaction: the oxygen atom adds to the two sp^2 carbons at the same time (Section 10). Therefore, it must be a syn addition.

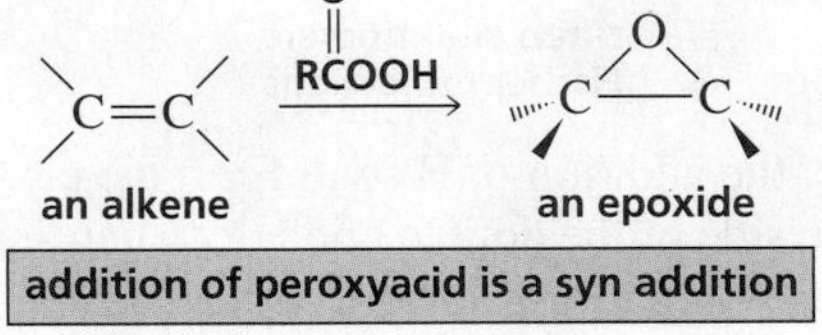

an alkene   an epoxide

addition of peroxyacid is a syn addition

The oxygen can add from above or from below the plane containing the double bond. Therefore, addition of a peroxyacid to an alkene forms two stereoisomers. Syn addition to a cis alkene forms the cis enantiomers. Because only syn addition occurs, the reaction is stereoselective. (If you turn the stereoisomer on the right upside down, you can see that it is the mirror image of the stereoisomer on the left.)

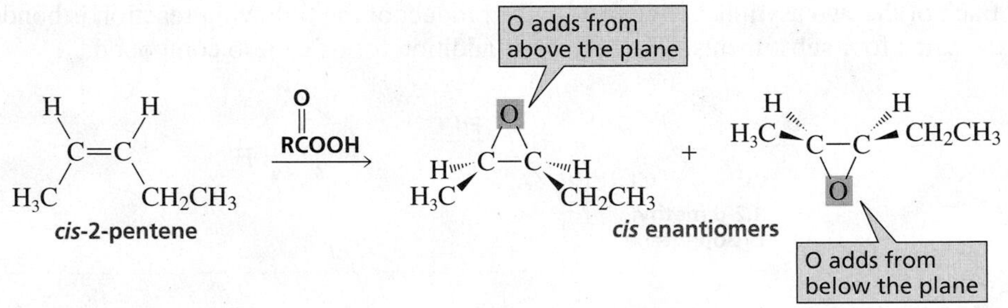

cis-2-pentene   *cis* enantiomers

Syn addition to a trans alkene forms the trans enantiomers.

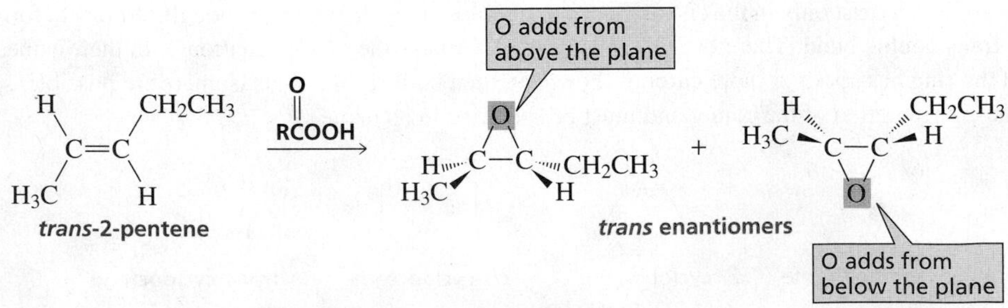

trans-2-pentene   *trans* enantiomers

Addition of a peroxyacid to *cis*-2-butene forms a meso compound since both asymmetric centers are attached to the same four groups.

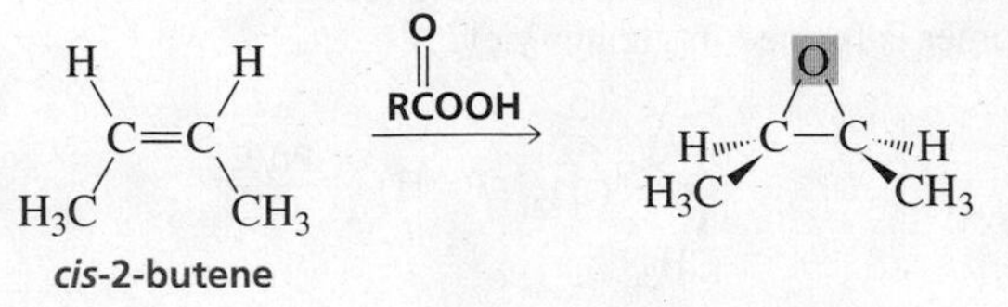

cis-2-butene

PROBLEM 45♦

a. What alkene is required to synthesize each of the following compounds?

1.

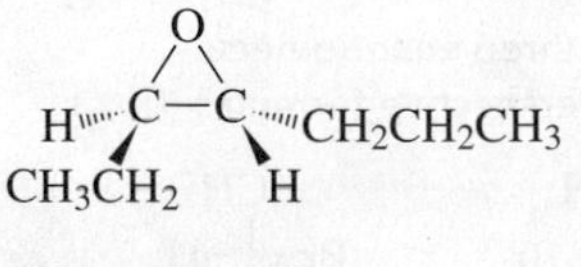

2.

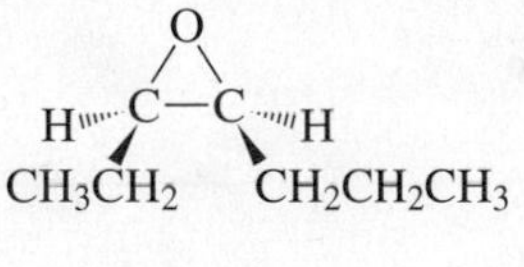

b. What other epoxide would be formed?

The Stereochemistry of Hydroboration–Oxidation

The addition of borane (or R_2BH) to an alkene is also a concerted reaction (Section 8). The boron and the hydride ion add to the two sp^2 carbons of the double bond at the same time. Because the two species add simultaneously, they must add to the same side of the double bond—that is, it is a syn addition.

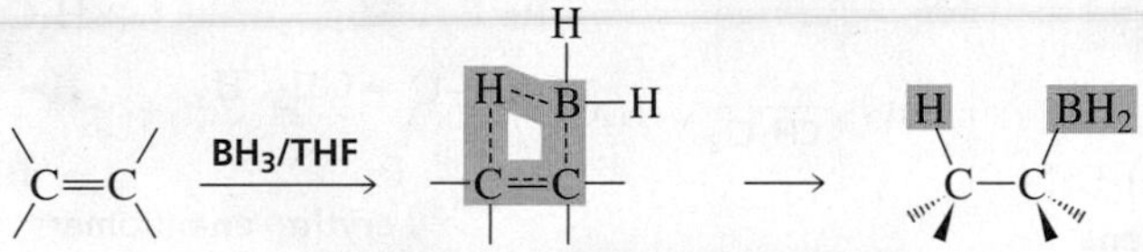

addition of borane is a syn addition

When the resulting alkylborane is oxidized by reaction with hydrogen peroxide and hydroxide ion, the OH group ends up in the same position as the boron group it replaces. Consequently, the overall reaction, called hydroboration–oxidation, amounts to a syn addition of water to a carbon–carbon double bond.

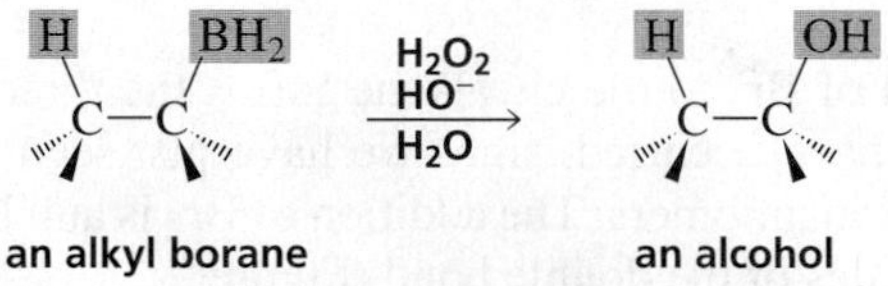

an alkyl borane   an alcohol

hydroboration–oxidation is overall a syn addition of water

Because only syn addition occurs, hydroboration–oxidation is stereoselective—only two of the four possible stereoisomers are formed. If the reactant is cyclic, syn addition forms only the enantiomers that have the added groups on the same side of the ring.

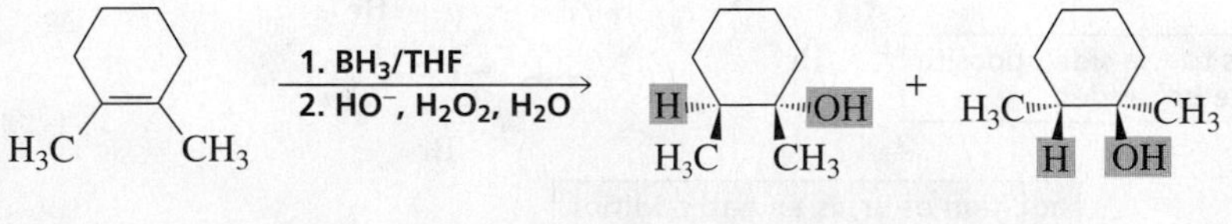

PROBLEM 46♦

What stereoisomers are obtained from hydroboration–oxidation of the following compounds?

a. cyclohexene
b. 1-ethylcyclohexene
c. *cis*-2-butene
d. (Z)-3,4-dimethyl-3-hexene

Addition Reactions that Form a Cyclic Bromonium or Chloronium Ion Intermediate

If two asymmetric centers are created from an addition reaction that forms a bromonium (or chloronium) ion intermediate, only one pair of enantiomers will be formed. For example, the addition of Br_2 to the cis alkene forms only the threo enantiomers.

287

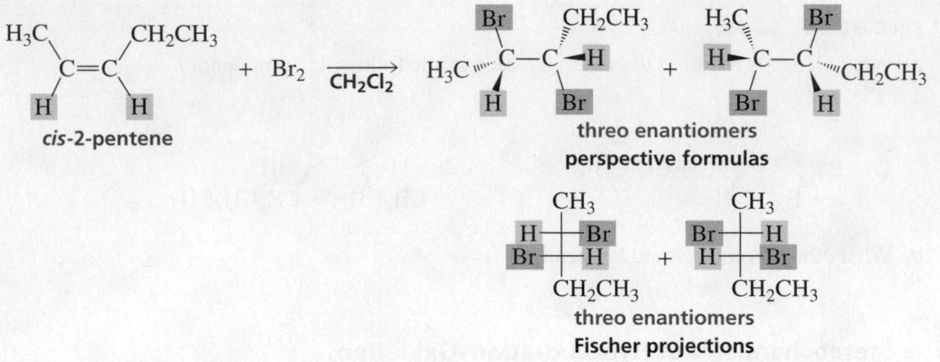

threo enantiomers
perspective formulas

threo enantiomers
Fischer projections

If you have trouble determining
the configuration of a product,
make a model.

Similarly, the addition of Br_2 to the trans alkene forms only the erythro enantiomers. Because the cis and trans isomers form different products, the reaction is stereospecific as well as stereoselective.

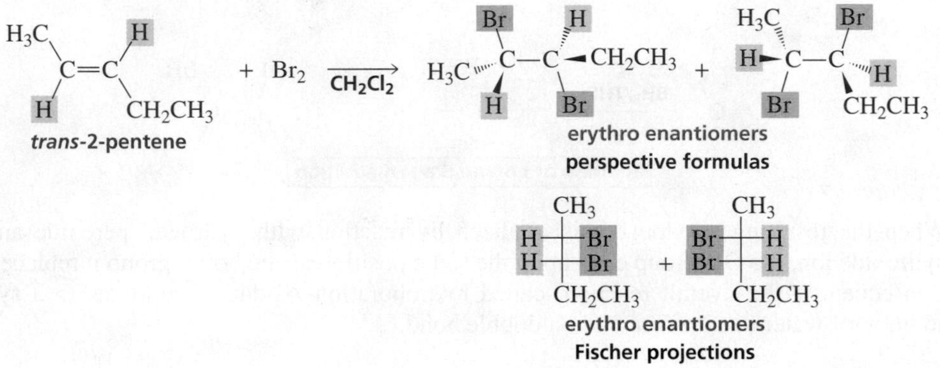

erythro enantiomers
perspective formulas

erythro enantiomers
Fischer projections

Because the addition of Br_2 to the cis alkene forms the threo enantiomers, we know that anti addition must have occurred, since we have just seen that syn addition would have formed the erythro enantiomers. The addition of Br_2 is anti because the two bromine atoms add to opposite sides of the double bond (Figure 9).

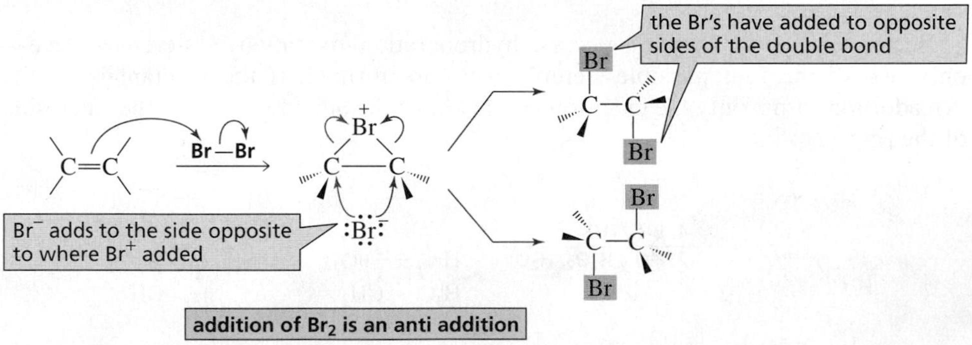

▲ **Figure 9**

A cyclic bromonium ion is formed in the first step of the reaction (Section 9). Because the bromine atom blocks one side of the intermediate, Br^- must approach from the opposite side (following either the green arrows *or* the red arrows). Thus, only anti addition of Br_2 can occur (the two bromine atoms add to opposite sides of the double bond), so only two of the four possible stereoisomers are formed.

If the two asymmetric centers in the product each have the same four substituents, the erythro isomers are identical and constitute a meso compound. Therefore, addition of Br_2 to *trans*-2-butene forms a meso compound.

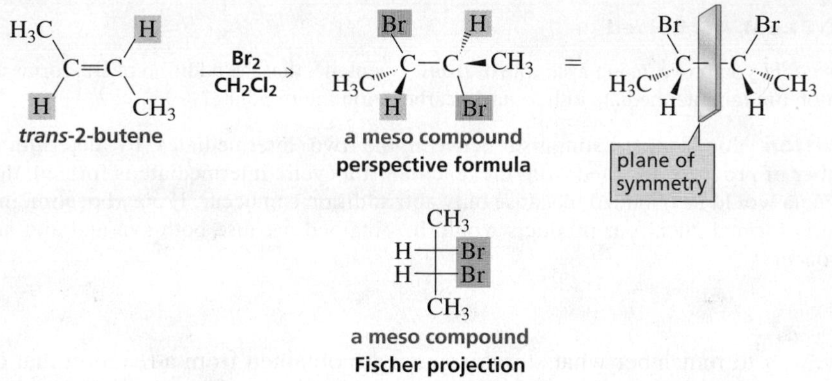

a meso compound
perspective formula

$$
\begin{array}{c}
CH_3 \\
H \!-\!\!|\!-\! Br \\
H \!-\!\!|\!-\! Br \\
CH_3
\end{array}
$$

a meso compound
Fischer projection

Because only anti addition occurs, addition of Br_2 to cyclohexene forms only the enantiomers that have the bromines on opposite sides of the ring.

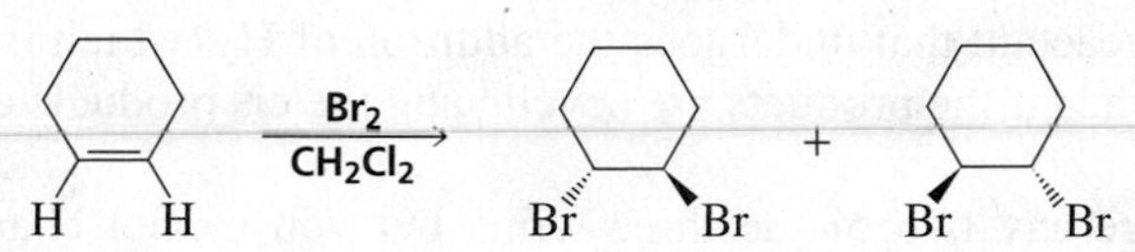

The stereochemistry of the products obtained from electrophilic addition reactions to alkenes is summarized in Table 1.

Table 1 Stereochemistry of Alkene Addition Reactions

Reaction	Type of addition	Stereoisomers formed
Addition reactions that create one asymmetric center in the product		1. If the reactant does not have an asymmetric center, then a racemic mixture is formed.
		2. If the reactant has an asymmetric center, then unequal amounts of a pair of diastereomers are formed.
Addition reactions that create two asymmetric centers in the product		
Addition of reagents that form a carbocation intermediate	syn and anti	Four stereoisomers are formed; the cis and trans isomers each form the same products.
Addition of H_2	syn	cis → erythro or cis enantiomers*
Addition of a peroxyacid	syn	
Addition of BH_3 or BHR_2	syn	trans → threo or trans enantiomers
Addition of Br_2, $Br_2 + H_2O$, $Br_2 + ROH$ (any reaction that forms a cyclic bromonium or chloronium ion intermediate)	anti	cis → threo or trans enantiomers
		trans → erythro or cis enantiomers*

* If the two asymmetric centers have the same substituents, a meso compound will be formed instead of the pair of erythro or cis enantiomers.

PROBLEM 47

The reaction of 2-ethyl-1-pentene with Br_2, with $H_2 + Pd/C$, or with R_2BH/THF followed by aqueous $HO^- + H_2O_2$ leads to a racemic mixture. Explain why a racemic mixture is obtained in each case.

PROBLEM 48 Solved

How could you prove, using a sample of *trans*-2-pentene, that the addition of Br_2 forms a cyclic bromonium ion intermediate rather than a carbocation intermediate?

Solution You could distinguish between the two intermediates by determining the number of products obtained from the reaction. If a cyclic intermediate is formed, then two products would be obtained, because only anti addition can occur. If a carbocation intermediate is formed, then four products would be obtained because both syn and anti addition can occur.

One way to remember what stereoisomers are obtained from a reaction that creates a product with two asymmetric centers is the mnemonic **CIS-SYN-(ERYTHRO or CIS).** (The third term is "erythro" if the product is acyclic, and "cis" if it is cyclic.) The three terms are easy to remember because they all mean "on the same side." Thus, if you have a cis reactant that undergoes the addition of H_2 (which is syn), the erythro products are obtained if the products are acyclic and the cis products are obtained if the product is cyclic.

You can change any two of the three terms but you cannot change just one. For example, **TRANS-ANTI-(ERYTHRO or CIS)**, and **CIS-ANTI-(THREO or TRANS)** are allowed because in each case two terms were changed. In other words, anti addition to a trans alkene forms the erythro (or cis) enantiomers, and anti addition to a cis alkene forms the threo (or trans) enantiomers. **TRANS-SYN-(ERYTHRO or CIS)** is *not* allowed, because only one term was changed. Thus, syn addition to a trans alkene does *not* form the erythro (or cis) enantiomers.

This mnemonic will work for all reactions that have products with structures that can be described by erythro and threo or by cis and trans.

PROBLEM-SOLVING STRATEGY

Predicting the Stereoisomers Obtained from the Addition Reactions of Alkenes

What stereoisomers are obtained from the following reactions?

a. 1-butene + H_2O + H_2SO_4 **c.** *cis*-3-heptene + Br_2
b. cyclohexene + HBr   **d.** *trans*-3-hexene + Br_2

Start by drawing the product without regard to its configuration to check whether the reaction has created any asymmetric centers. Then determine the stereoisomers of the products, paying attention to the configuration (if any) of the reactant, how many asymmetric centers are formed, and the mechanism of the reaction. Let's start with part **a.**

a. $CH_3CH_2\underset{\overset{|}{OH}}{C}HCH_3$

The product has one asymmetric center, so equal amounts of the *R* and *S* enantiomers will be obtained.

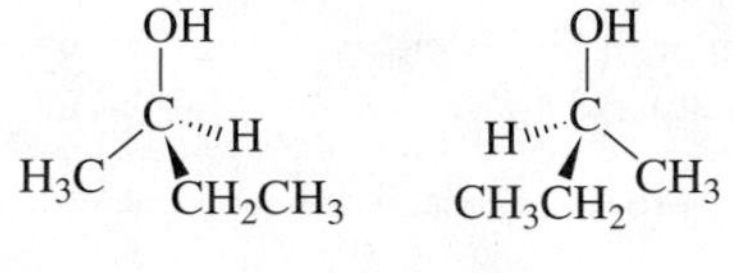

b.

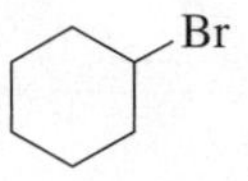

The product does not have an asymmetric center, so it has no stereoisomers.

c. CH₃CH₂CHCHCH₂CH₂CH₃
       | |
       Br Br

Two asymmetric centers have been created in the product. Because the reactant is cis and the addition of Br_2 is anti, the threo enantiomers are formed.

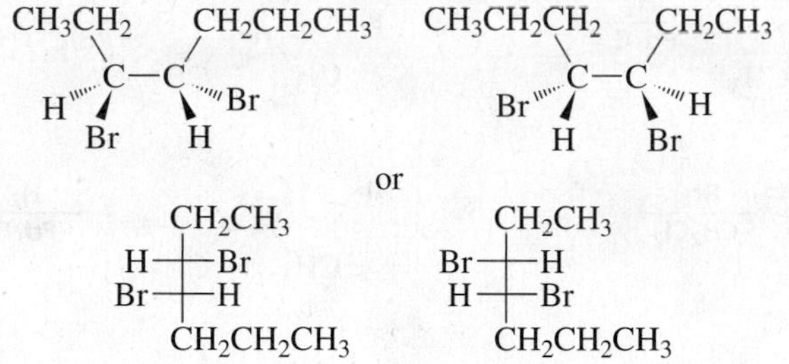

or

d. CH₃CH₂CHCHCH₂CH₃
       | |
       Br Br

Two asymmetric centers have been created in the product. Because the reactant is trans and the addition of Br_2 is anti, one would expect the erythro enantiomers. However, the two asymmetric centers are bonded to the same four groups, so the erythro product is a meso compound. Thus, only one stereoisomer is formed.

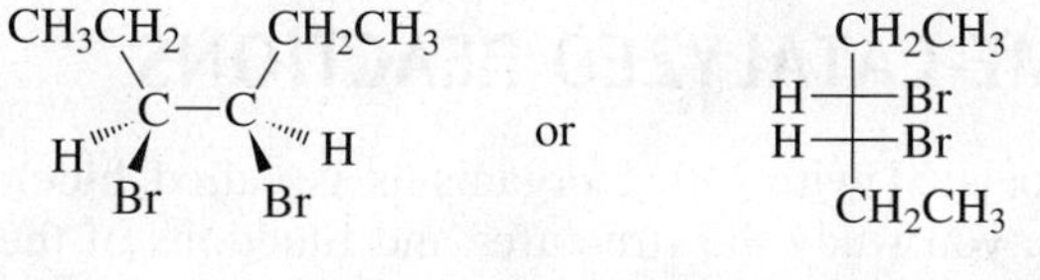

or

Now use the strategy you have just learned to solve Problem 49.

PROBLEM 49

What stereoisomers are obtained from the following reactions?

a. *trans*-2-butene + HBr
b. *(Z)*-3-methyl-2-pentene + HBr
c. *(E)*-3-methyl-2-pentene + HBr
d. *cis*-3-hexene + HBr
e. *cis*-2-pentene + Br_2
f. 1-hexene + Br_2

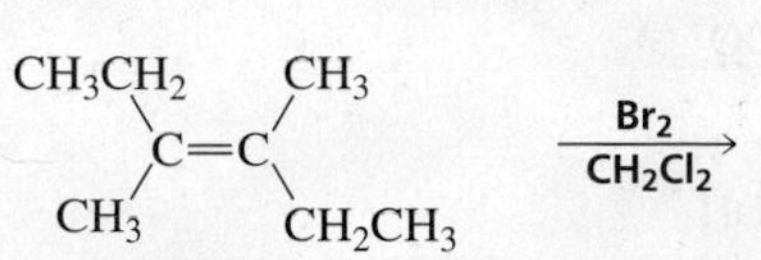

PROBLEM 50

When Br_2 adds to an alkene that has different substituents attached to each of the two sp^2 carbons, such as *cis*-2-heptene, identical amounts of the two threo enantiomers are obtained even though Br^- is more likely to attack the less sterically hindered carbon of the bromonium ion. Explain why identical amounts of the two enantiomers are obtained.

PROBLEM 51

a. What products would be obtained from the addition of Br_2 to cyclohexene if H_2O were added to the reaction mixture?
b. Propose a mechanism for the reaction.

PROBLEM 52

What stereoisomers would you expect to obtain from each of the following reactions?

291

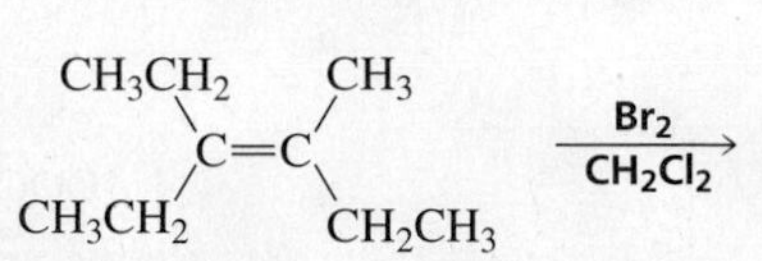

c.

$$\underset{CH_3}{\overset{CH_3CH_2}{>}} C = C \underset{CH_2CH_3}{\overset{CH_3}{<}} \xrightarrow[\text{Pd/C}]{H_2}$$

f.

$$\xrightarrow[\text{CH}_2\text{Cl}_2]{Br_2}$$

d.

$$\underset{CH_3}{\overset{CH_3CH_2}{>}} C = C \underset{CH_3}{\overset{H}{<}} \xrightarrow[\text{Pd/C}]{H_2}$$

g.

$$\xrightarrow[\text{Pd/C}]{H_2}$$

e.

$$\xrightarrow[\text{CH}_2\text{Cl}_2]{Br_2}$$

h.

$$\xrightarrow[\text{Pd/C}]{H_2}$$

PROBLEM 53♦

a. What is the major product obtained from the reaction of propene and Br_2 plus excess Cl^-?

b. Indicate the relative amounts of the stereoisomers that are obtained.

---

16 THE STEREOCHEMISTRY OF ENZYME-CATALYZED REACTIONS

The chemistry associated with living organisms is called **biochemistry.** When you study biochemistry, you study the structures and functions of the molecules found in the biological world, and the reactions involved in the synthesis and degradation of these molecules. Because the compounds in living organisms are organic compounds, it is not surprising that many of the reactions encountered in organic chemistry also occur in cells.

Cells do not contain molecules such as Cl_2, HBr, or BH_3, so you would not expect to find the addition of such reagents to alkenes in biological systems. However, cells do contain water, so some alkenes found in biological systems undergo the addition of water.

Reactions that occur in biological systems are catalyzed by proteins called **enzymes**. When an enzyme catalyzes a reaction that forms a product with an asymmetric center, only one stereoisomer is formed because enzyme-catalyzed reactions are *completely stereoselective.*

For example, the enzyme fumarase catalyzes the addition of water to fumarate to form malate, a compound with one asymmetric center.

> **When an enzyme catalyzes a reaction that forms a product with an asymmetric center, only one stereoisomer is formed.**

$$\underset{\text{-OOC}}{\overset{H}{>}} C = C \underset{H}{\overset{COO^-}{<}} \; + \; H_2O \xrightarrow{\text{fumarase}} \;\; ^-OOCCH_2\overset{\text{asymmetric center}}{\underset{OH}{CHCOO^-}}$$

fumarate   malate

However, the reaction forms only (*S*)-malate; the *R* enantiomer is not formed.

$$^-OOCCH_2 \overset{COO^-}{\underset{OH}{\overset{|}{C}{\cdots}H}}$$

(*S*)-malate

An enzyme-catalyzed reaction forms only one stereoisomer because an enzyme's binding site can restrict delivery of the reagents to only one side of the reactant.

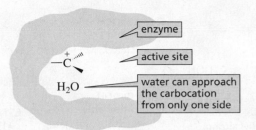

Enzyme-catalyzed reactions are also stereospecific; an enzyme typically catalyzes the reaction of only one stereoisomer. For example, fumarase catalyzes the addition of water to fumarate (the trans isomer just shown) but not to maleate (the cis isomer).

$$^-OOC \underset{H}{\overset{}{\diagdown}} C = C \overset{COO^-}{\underset{H}{\diagup}} \quad + \quad H_2O \xrightarrow{\text{fumarase}} \text{no reaction}$$

maleate

An enzyme is able to differentiate between the two stereoisomers because only one of them has the structure that allows it to fit into the enzyme's binding site where the reaction takes place.

Chiral Catalysts

The problem of having to separate enantiomers can be avoided if a synthesis is carried out that forms one of the enantiomers preferentially. **Chiral catalysts** are being developed for the synthesis of one enantiomer in great excess over the other. For example, the catalytic hydrogenation of 2-ethyl-1-pentene forms equal amounts of two enantiomers because H_2 can be delivered equally easily to both faces of the double bond.

$$\underset{CH_3CH_2CH_2}{\overset{CH_3CH_2}{\diagdown}} C = CH_2 \xrightarrow[\text{Pd/C}]{H_2}$$

(R)-3-methylhexane
50%

+

(S)-3-methylhexane
50%

If, however, the metal catalyst is complexed to a chiral organic molecule, then H_2 will be delivered to only one face of the double bond. One such chiral catalyst—using Ru(II) as the metal and BINAP (2,2'-bis[diphenylphosphino]-1,1'-binaphthyl) as the chiral molecule—has been used to synthesize (S)-naproxen, the active ingredient in Aleve and in several other over-the-counter nonsteroidal anti-inflammatory drugs, in greater than 98% enantiomeric excess. This is an example of an **enantioselective reaction**—a reaction that forms more of one enantiomer than another. This is a medically important enantioselective reaction, because a racemic mixture of naproxen cannot be given to a patient since (R)-naproxen is highly toxic to the liver.

$$\xrightarrow[\text{(R)-BINAP–Ru(II)}]{H_2}$$

(S)-naproxen
>98% ee

The Sharpless epoxidation is another example of an enantioselective reaction achieved by using a chiral catalyst. In this reaction, a peroxide (*tert*-butylhydroperoxide) delivers its oxygen to only one side of the double bond of an allylic alcohol, in the presence of a metal catalyst (titanium(IV) isopropoxide) and a chiral organic molecule (diethyl tartarate; DET).

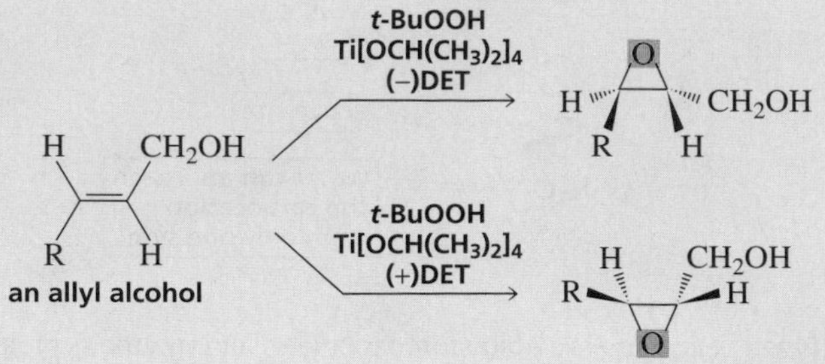

The structure of the epoxide depends on which isomer of diethyl tartarate is used.

PROBLEM 54♦

What minimum percent of naproxen is obtained as the *S* enantiomer in the synthesis shown in the preceding box?

---

17 ENANTIOMERS CAN BE DISTINGUISHED BY BIOLOGICAL MOLECULES

Enzymes and receptors can tell the difference between enantiomers because enzymes and receptors are proteins, and proteins are chiral molecules.

Enzymes

An achiral reagent, such as hydroxide ion, cannot distinguish between enantiomers. Thus, it reacts with (*R*)-2-bromobutane at the same rate that it reacts with (*S*)-2-bromobutane.

Because an enzyme is *chiral,* not only can it distinguish between cis–trans isomers, such as maleate and fumarate (Section 16), but it can also distinguish between enantiomers and catalyze the reaction of only one of them.

Chemists can use an enzyme's ability to distinguish between enantiomers to separate them. For example, the enzyme D-amino acid oxidase catalyzes only the oxidation of the *R* enantiomer but leaves the *S* enantiomer unchanged. The oxidized product of the enzyme-catalyzed reaction can be easily separated from the unreacted enantiomer because they are different compounds.

An achiral reagent reacts identically with both enantiomers. A sock, which is achiral, fits on either foot.

A chiral reagent reacts differently with each enantiomer. A shoe, which is chiral, fits on only one foot.

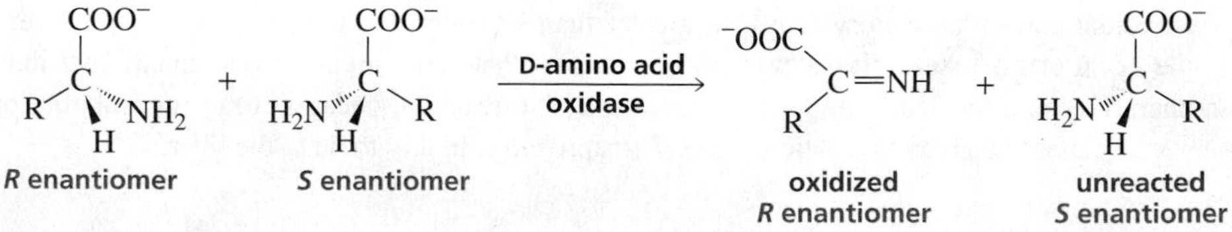

An enzyme is able to differentiate between enantiomers and between cis and trans isomers, because its binding site is chiral. Therefore, the enzyme will bind only the stereoisomer whose substituents are in the correct positions to interact with the substituents in the chiral binding site. For example, in Figure 10, the enzyme binds the *R* enantiomer but not the *S* enantiomer because the *S* enantiomer does not have its substituents in the proper positions to bind efficiently to the enzyme.

> *Like a right-handed glove, which fits only the right hand, an enzyme forms only one stereoisomer and reacts with only one stereoisomer.*

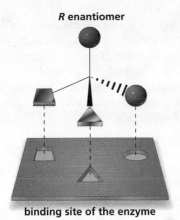

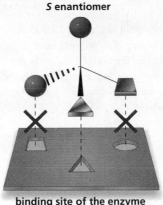

R enantiomer

S enantiomer

binding site of the enzyme

binding site of the enzyme

◀ **Figure 10**
Schematic diagram showing why only one enantiomer is bound by an enzyme. One enantiomer fits into the binding site and one does not.

PROBLEM 55♦

a. What would be the product of the reaction of fumarate and H_2O if H_2SO_4 were used as a catalyst instead of fumarase?

b. What would be the product of the reaction of maleate and H_2O if H_2SO_4 were used as a catalyst instead of fumarase?

Receptors

Like an enzyme, a **receptor** is a protein, so it too is chiral. Therefore, it will bind one enantiomer better than the other, just like an enzyme binds one enantiomer better than another.

The fact that a receptor typically recognizes only one enantiomer explains why enantiomers can have different physiological properties. For example, receptors located on the exteriors of nerve cells in the nose are able to perceive and differentiate the estimated 10,000 smells to which the cells are exposed. The reason that (R)-$(-)$-carvone (found in spearmint oil) and (S)-$(+)$-carvone (the main constituent of caraway seed oil) have such different odors is that each enantiomer fits into a different receptor.

(R)-(−)-carvone
smells like
spearmint

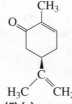

(R)-(−)-carvone

$[\alpha]_D^{20\,°C} = -62.5$

(S)-(+)-carvone

$[\alpha]_D^{20\,°C} = +62.5$

(S)-(+)-carvone
smells like
caraway seeds

Dr. Frances O. Kelsey receives the President's medal for Distinguished Federal Civilian Service from President John F. Kennedy in 1962 for preventing the sale of thalidomide. Kelsey was born in British Columbia in 1914. She received a B.Sc. in 1934 and a M.Sc. in pharmacology in 1936 from McGill University. In 1938, she received a Ph.D. and an M.D. from the University of Chicago, where she became a member of the faculty. She married a fellow faculty member and they had two daughters. She joined the FDA in 1960 and worked there until 2005, when she retired at the age of 90. Each year the FDA selects a staff member to receive the Dr. Frances O. Kelsey Award for Excellence and Courage in Protecting Public Health.

The Enantiomers of Thalidomide

Thalidomide was developed in West Germany and was first marketed (as Contergan) in 1957 for insomnia, tension, and morning sickness during pregnancy. At that time it was available in more than 40 countries but had not been approved for use in the United States because Frances O. Kelsey, a physician for the Food and Drug Administration (FDA), had insisted upon additional tests to explain a British study that had found nervous system side effects.

The (+)-isomer of thalidomide has stronger sedative properties, but the commercial drug was a racemic mixture. No one knew that the (−)-isomer is a teratogen—a compound that causes congenital deformations—until women who had been given the drug during the first three months of pregnancy gave birth to babies with a wide variety of defects, with deformed limbs being the most common. By the time the danger was recognized and the drug withdrawn from the market on November 27, 1961, about 10,000 children had been damaged. It was eventually determined that the (+)-isomer also has mild teratogenic activity and that each of the enantiomers can racemize (interconvert) in the body. Thus, it is not clear whether the birth defects would have been less severe if the women had been given only the (+)-isomer. Because thalidomide damaged fast growing cells in the developing fetus, it has recently has been approved—with restrictions and with tight controls—for the eradication of certain kinds of cancer cells.

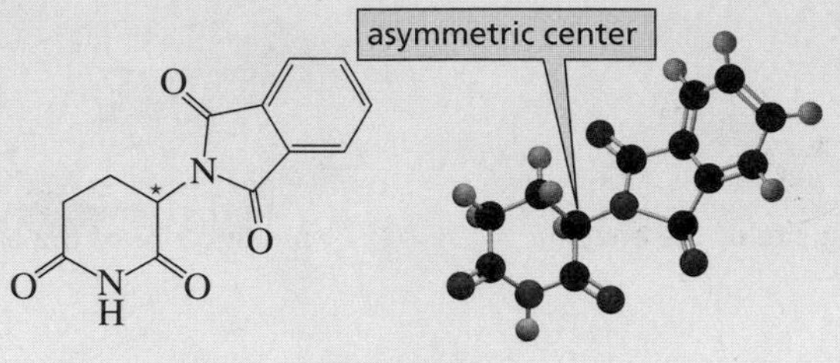

thalidomide

18 | REACTIONS AND SYNTHESIS

This chapter has focused on the reactions of alkenes. You have seen why alkenes react, the kinds of reagents with which they react, the mechanisms by which the reactions occur, and the products that are formed. Keep in mind, however, that when you are studying reactions, you are simultaneously studying synthesis. When you learn that compound **A** reacts with a certain reagent to form compound **B**, you are learning not only about the reactivity of **A**, but also about one way that compound **B** can be synthesized.

$$\mathbf{A} \longrightarrow \mathbf{B}$$

A reacts;
B is synthesized

For example, you have seen that many different reagents can add to alkenes and that compounds such as alkyl halides, vicinal dihalides, halohydrins, alcohols, ethers, epoxides, alkanes, aldehydes, and ketones are synthesized as a result.

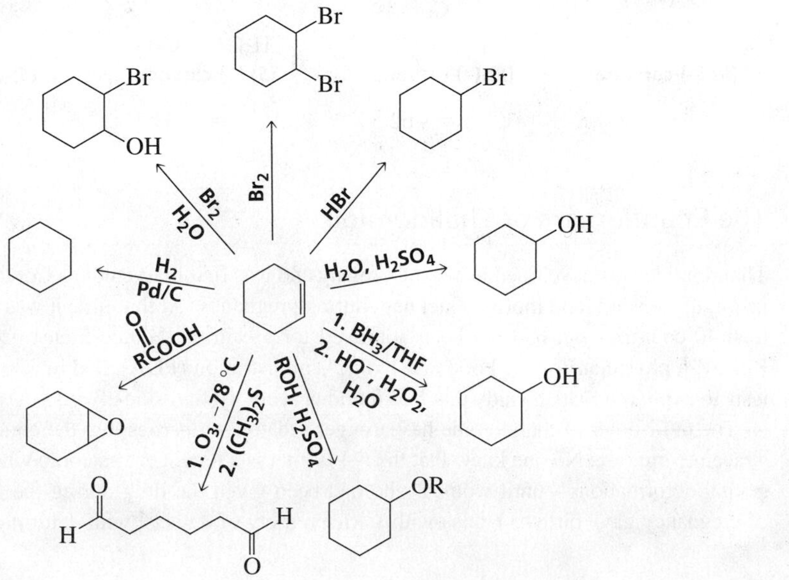

Although you have seen how alkenes react and have learned about the kinds of compounds that are synthesized when alkenes undergo reactions, you have not yet seen how alkenes are synthesized. The reactions of alkenes involve the *addition* of atoms (or groups of atoms) to the two sp^2 carbons of the double bond. Reactions that synthesize alkenes are exactly the opposite; they involve the *elimination* of atoms (or groups of atoms) from two adjacent sp^3 carbons.

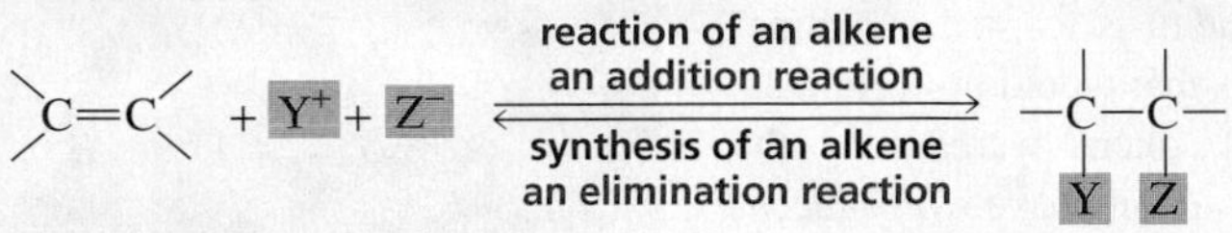

You will learn how alkenes are synthesized when you study compounds that undergo elimination reactions. The various reactions that can be used to synthesize alkenes are listed in Appendix: "Summary of Methods Used to Synthesize a Particular Functional Group."

PROBLEM 56 Solved

Show how each of the following compounds can be synthesized from an alkene:

a. 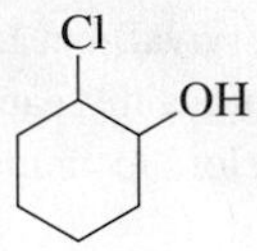   **b.**

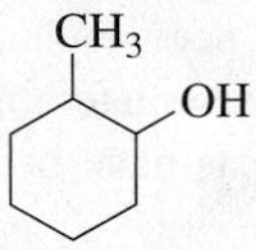

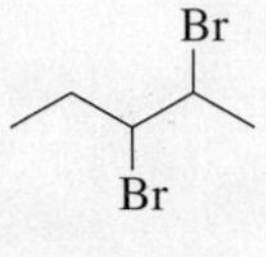

Solution to 56a The only alkene that can be used for this synthesis is cyclohexene. To get the desired substituents on the ring, cyclohexene must react with Cl_2 in an aqueous solution so that water will be the nucleophile.

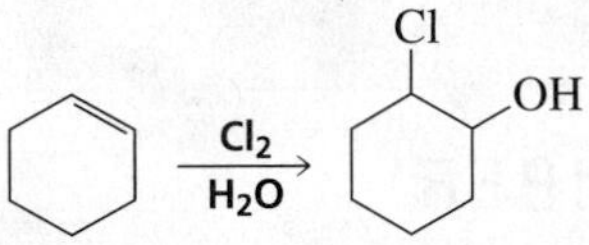

Solution to 56b The alkene that should be used here is 1-methylcyclohexene. To get the substituents in the desired locations, the electrophile must be R_2BH or BH_3, with HO taking boron's place in the subsequent oxidation reaction.

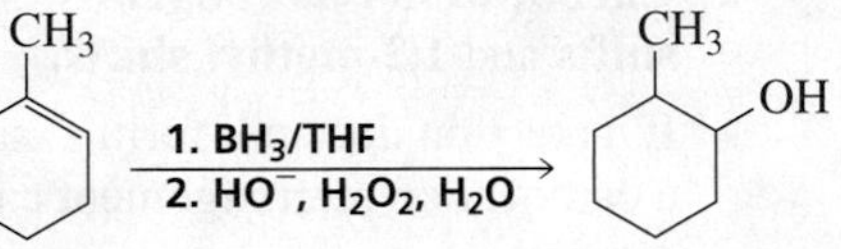

PROBLEM 57♦

Explain why 3-methylcyclohexene should *not* be used as the starting material in Problem 56b.

PROBLEM 58

Show how each of the following compounds can be synthesized from an alkene:

a. 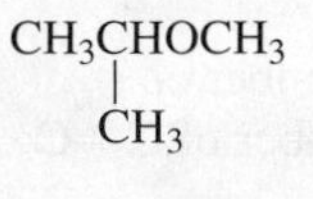 CH_3CHOCH_3
　　　　|
　　　CH_3

b.

c.

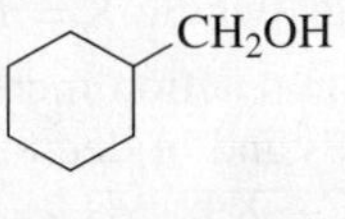

d.

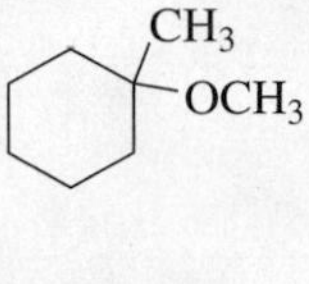

e.

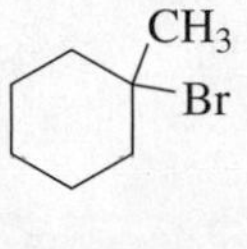

f.

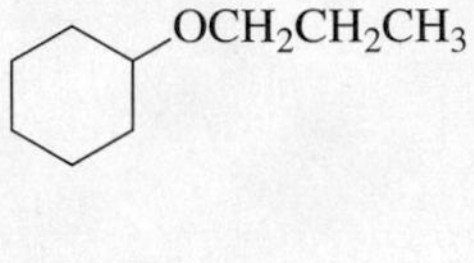

Which Are More Harmful, Natural Pesticides or Synthetic Pesticides?

Learning to synthesize new compounds is an important part of organic chemistry. Long before chemists learned to synthesize compounds that would protect plants from predators, plants were doing the job themselves. Plants have every incentive to synthesize pesticides. When you cannot run, you need to find another way to protect yourself. But which pesticides are more harmful, those synthesized by chemists or those synthesized by plants? Unfortunately, we do not know because while federal laws require all human-made pesticides to be tested for any adverse effects, they do not require plant-made pesticides to be tested. Besides, risk evaluations of chemicals are usually done on rats, and something that is harmful to a rat may or may not be harmful to a human. Furthermore, when rats are tested, they are exposed to much higher concentrations of the chemical than would be experienced by a human, and some chemicals are harmful only at high doses. For example, we all need sodium chloride for survival, but high concentrations are poisonous; and, although we associate alfalfa sprouts with healthy eating, monkeys fed very large amounts of alfalfa sprouts have been found to develop an immune system disorder.

SOME IMPORTANT THINGS TO REMEMBER

- **Alkenes** undergo **electrophilic addition reactions.** These reactions start with the addition of an *electrophile* to the sp^2 carbon bonded to the most hydrogens and end with the addition of a nucleophile to the other sp^2 carbon.

- A curved arrow always points from the electron donor to the electron acceptor.

- The addition of hydrogen halides and the acid-catalyzed addition of water and alcohols form **carbocation intermediates.**

- **Tertiary carbocations** are more stable than **secondary carbocations,** which are more stable than **primary carbocations.**

- The more stable carbocation is formed more rapidly.

- The **Hammond postulate** states that a transition state is more similar in structure to the species to which it is closer in energy.

- **Regioselectivity** is the preferential formation of one **constitutional isomer** over another.

- A carbocation will rearrange if it becomes more stable as a result of the rearrangement.

- **Carbocation rearrangements** occur by **1,2-hydride shifts** and **1,2-methyl shifts.**

- If a reaction does not form a carbocation intermediate, a carbocation rearrangement cannot occur.

- The addition of Br_2 or Cl_2 forms an intermediate with a three-membered ring that reacts with nucleophiles.

- **Ozonolysis** forms an intermediate with a five-membered ring.

- **Hydroboration, epoxidation,** and **catalytic hydrogenation** do not form an intermediate.

- An **oxidation** reaction decreases the number of C—H bonds and/or increases the number of C—O, C—N, or C—X bonds (where X = a halogen).

- A **reduction** reaction increases the number of C—H bonds and/or decreases the number of C—O, C—N, or C—X bonds.

- To determine the product of oxidative cleavage, replace C=C with C=O O=C.
- A **ketone** has two alkyl groups bonded to a carbonyl C=O group; an **aldehyde** has one hydrogen (or has two hydrogens) bonded to a carbonyl group.
- **Catalytic hydrogenation** reduces alkenes to alkanes.
- The **heat of hydrogenation** is the heat released in a hydrogenation reaction. It is the $\Delta H°$ value without the negative sign.
- The **most stable alkene** has the **smallest heat of hydrogenation.**
- The stability of an alkene increases as the number of alkyl substituents bonded to its sp^2 carbons increases.
- **Trans alkenes** are more stable than **cis alkenes** because of steric strain.
- A **regioselective** reaction selects for a particular constitutional isomer; a **stereoselective** reaction selects for a particular stereoisomer; an **enantioselective** reaction selects for a particular enantiomer.
- A reaction is **stereospecific** if the reactant can exist as stereoisomers and each stereoisomer forms a different stereoisomer or a different set of stereoisomers.

- When a reactant that does not have an asymmetric center forms a product with one asymmetric center, the product will be a racemic mixture.
- When a reactant that has an asymmetric center forms a product with *a second* asymmetric center, the product will be diastereomers in unequal amounts.
- In **syn addition** the substituents add to the same side of a double bond; in **anti addition** they add to opposite sides.
- Both syn and anti addition occur in electrophilic addition reactions that form a carbocation intermediate.
- The addition of H_2 or a peroxyacid to an alkene is a syn addition reaction; hydroboration–oxidation is overall a syn addition of water.
- The addition of Br_2 or Cl_2 is an anti addition reaction.
- An enzyme-catalyzed reaction forms only one stereoisomer, and an enzyme typically catalyzes the reaction of only one stereoisomer.
- An achiral reagent reacts identically with both geometric isomers or with enantiomers.
- A chiral reagent reacts differently with each geometric isomer of with each enantiomer.

SUMMARY OF REACTIONS

As you review the electrophilic addition reactions of alkenes, keep in mind that the first step in each of them is the addition of an electrophile to the sp^2 carbon bonded to the most hydrogens.

1. Addition of hydrogen halides: H^+ is the electrophile; the halide ion is the nucleophile (Sections 1 and Sections 4). Addition is syn and anti.

$$RCH=CH_2 \ + \ HX \ \longrightarrow \ RCHCH_3$$
$$X$$

HX = HF, HCl, HBr, HI

2. Acid-catalyzed addition of water and alcohols: H^+ is the electrophile; water or an alcohol is the nucleophile (Sections 5 and 6). Addition is syn and anti.

$$RCH=CH_2 \ + \ H_2O \ \underset{}{\overset{H_2SO_4}{\rightleftharpoons}} \ RCHCH_3$$
$$OH$$

$$RCH=CH_2 \ + \ CH_3OH \ \underset{}{\overset{H_2SO_4}{\rightleftharpoons}} \ RCHCH_3$$
$$OCH_3$$

3. Hydroboration–oxidation: BH_3 is the electrophile and H^- is the nucleophile; boron is subsequently replaced by OH (Section 9). Addition is syn.

$$RCH=CH_2 \ \xrightarrow[\text{2. HO}^-, \ H_2O_2, \ H_2O]{\text{1. BH}_3/\text{THF}} \ RCH_2CH_2OH$$

4. Addition of halogen: Br^+ or Cl^+ is the electrophile; Br^-, Cl^- (or water) is the nucleophile (Section 9). Addition is anti.

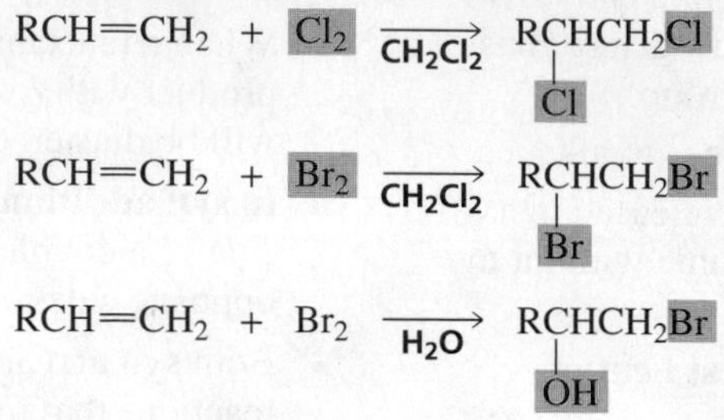

$$RCH{=}CH_2 \ + \ Cl_2 \ \xrightarrow{CH_2Cl_2} \ RCHCH_2Cl$$
$$\underset{|}{Cl}$$

$$RCH{=}CH_2 \ + \ Br_2 \ \xrightarrow{CH_2Cl_2} \ RCHCH_2Br$$
$$\underset{|}{Br}$$

$$RCH{=}CH_2 \ + \ Br_2 \ \xrightarrow{H_2O} \ RCHCH_2Br$$
$$\underset{|}{OH}$$

5. Addition of a peroxyacid: O is both the electrophile and the nucleophile (Section 10). Addition is syn.

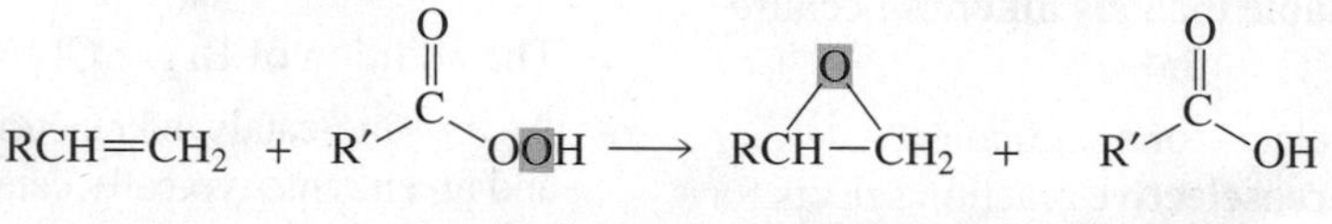

$$RCH{=}CH_2 \ + \ R'\overset{O}{\underset{}{C}}OOH \ \longrightarrow \ RCH{-}CH_2 \ + \ R'\overset{O}{\underset{}{C}}OH$$

6. Ozonolysis of alkenes: one O is the electrophile and another O is the nucleophile (Section 11).

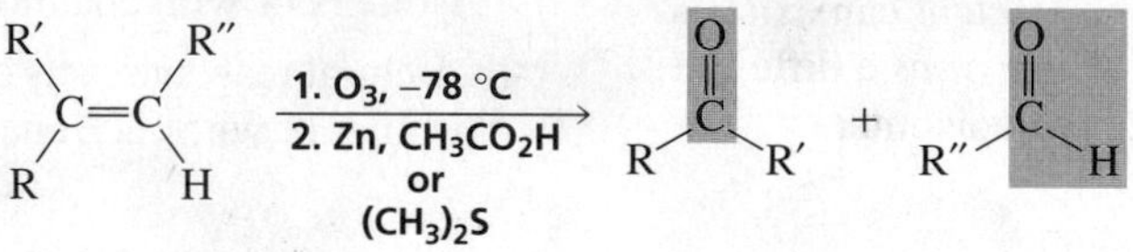

$$\underset{R \quad H}{\overset{R' \quad R''}{C{=}C}} \ \xrightarrow[\substack{2.\ \mathbf{Zn,\ CH_3CO_2H} \\ or \\ (CH_3)_2S}]{1.\ \mathbf{O_3,\ -78\ °C}} \ \underset{R \quad R'}{\overset{O}{C}} \ + \ \underset{R'' \quad H}{\overset{O}{C}}$$

7. Addition of hydrogen (Section 12): Addition is syn.

$$RCH{=}CH_2 \ + \ H_2 \ \xrightarrow{Pd/C} \ RCH_2CH_3$$

PROBLEMS

59. What is the major product of each of the following reactions?

a. cyclohexene with CH_2CH_3 substituent + HBr $\longrightarrow$

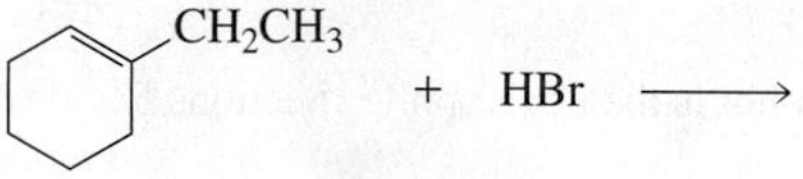

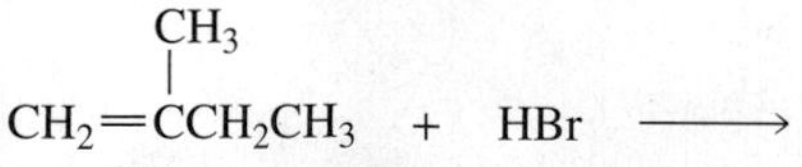

c. cyclohexane with $CH{=}CH_2$ substituent + HBr $\longrightarrow$

b. $CH_2{=}\underset{\underset{CH_3}{|}}{C}CH_2CH_3$ + HBr $\longrightarrow$

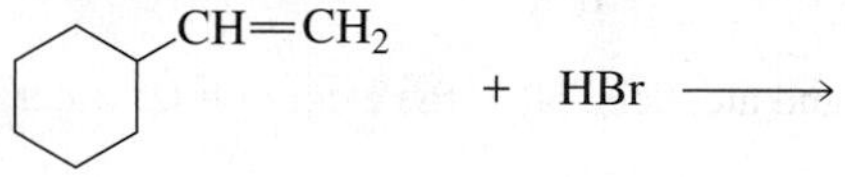

d. $CH_3CH_2\underset{\underset{CH_3}{|}}{\overset{\overset{CH_3}{|}}{C}}CH{=}CH_2$ + HBr $\longrightarrow$

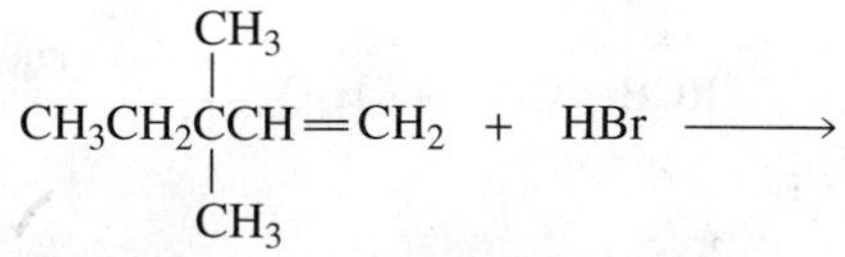

60. Identify the electrophile and the nucleophile in each of the following reaction steps, and then draw curved arrows to illustrate the bond-making and bond-breaking processes.

a. $CH_3\overset{+}{C}HCH_3$ + $:\ddot{C}l:^-$ $\longrightarrow$ CH_3CHCH_3
$$\underset{:\ddot{C}l:}{|}$$

b. $CH_3CH{=}CH_2$ + H$-$Br $\longrightarrow$ $CH_3\overset{+}{C}H{-}CH_3$ + Br^-

c. $CH_3CH{=}CH_2$ + BH_3 $\longrightarrow$ $CH_3CH_2{-}CH_2BH_2$

61. What will be the major product of the reaction of 2-methyl-2-butene with each of the following reagents?

a. HBr	**e.** H_2/Pd	**i.** Br_2/H_2O
b. HI	**f.** MCPBA (a peroxyacid)	**j.** Br_2/CH_3OH
c. Cl_2/CH_2Cl_2	**g.** H_2O + trace H_2SO_4	**k.** BH_3/THF, followed by H_2O_2, HO^-, H_2O
d. O_3, $-78\ °C$, followed by $(CH_3)_2S$	**h.** Br_2/CH_2Cl_2	

62. Give two names for each of the following:

a.

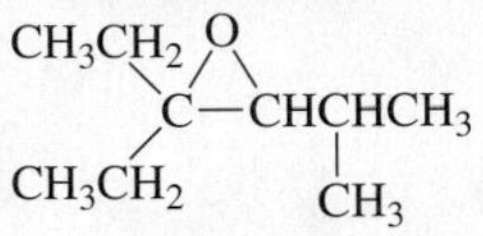

$$CH_3CH_2\ \ \ O$$
$$C-CHCHCH_3$$
$$CH_3CH_2\ \ \ \ CH_3$$

b.

$$H_3C\ \ \ O$$
$$C-CHCH_2CH_3$$
$$H_3C$$

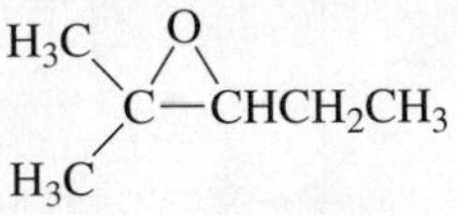

63.
a. Which is the most stable: 3,4-dimethyl-2-hexene, 2,3-dimethyl-2-hexene, or 4,5-dimethyl-2-hexene?
b. Which compound would you expect to have the largest heat of hydrogenation?
c. Which compound would you expect to have the smallest heat of hydrogenation?

64. What reagents are needed to synthesize the following alcohols?

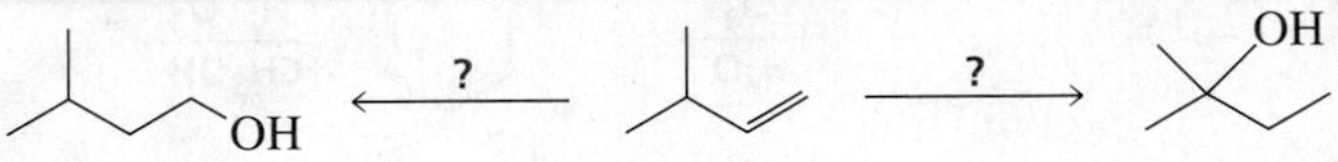

65. Identify each of the following reactions as an oxidation reaction, a reduction reaction, or neither.

a.
$$CH_3CH{=}CHCH_3\ +\ Cl_2\ \longrightarrow\ CH_3CHCHCH_3$$
$$\ Cl\ \ Cl$$

b.
$$CH_3CH{=}CHCH_3\ +\ H_2O\ \xrightarrow{H_2SO_4}\ CH_3CHCH_2CH_3$$
$$\ OH$$

66. What are the products of the following reactions? Indicate whether each reaction is an oxidation or a reduction.

a.
$$CH_3CH_2\overset{\displaystyle CH_3}{C}{=}CHCH_2CH_3\ \xrightarrow[\text{2. Zn, }CH_3CO_2H]{\text{1. }O_3,\ -78\ °C}$$

b.
$$CH_3CHCH{=}CHCH_3\ \xrightarrow{\ \ \underset{Pd/C}{H_2}\ \ }$$
$$\ \ \ \ CH_3$$

c.
$\xrightarrow[\text{2. }(CH_3)_2S]{\text{1. }O_3,\ -78\ °C}$

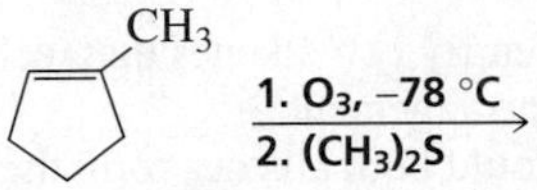

67. When 3-methyl-1-butene reacts with HBr, two alkyl halides are formed: 2-bromo-3-methylbutane and 2-bromo-2-methylbutane. Propose a mechanism that explains the formation of these two products.

68. Draw the structures for all the alkenes with molecular formula C_6H_{12}. Use those structures to answer the following questions:
a. Which of the compounds is the most stable?
b. Which of the compounds is the least stable?

69. Draw curved arrows to show the flow of electrons responsible for the conversion of the following reactants into products.

a. 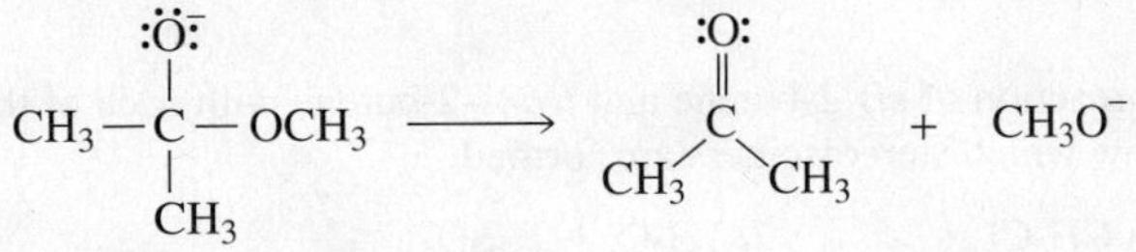
$$CH_3-\overset{\displaystyle :\ddot{O}{:}^-}{\underset{\displaystyle CH_3}{C}}-OCH_3\ \longrightarrow\ \underset{CH_3\ \ \ \ CH_3}{\overset{:O:}{C}}\ +\ CH_3O^-$$

b. $CH_3C{\equiv}C-H\ +\ :\ddot{N}H_2\ \longrightarrow\ CH_3C{\equiv}C^-\ +\ \dot{N}H_3$

c. $CH_3CH_2-Br\ +\ CH_3\ddot{\underset{..}{O}}{:}\ \longrightarrow\ CH_3CH_2-\ddot{\underset{..}{O}}CH_3\ +\ Br^-$

70. What reagents are needed to carry out the following syntheses?

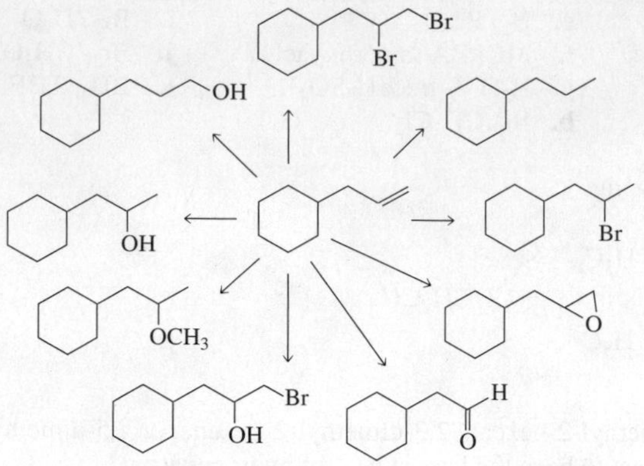

71. What is the major product of each of the following reactions?

a. 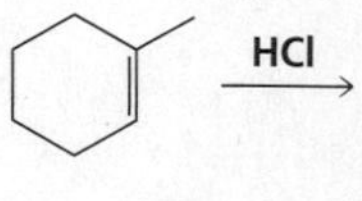 $\xrightarrow{\text{HCl}}$

c. 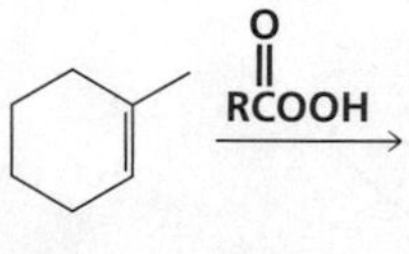 $\xrightarrow{\text{RCOOH}}$

e. 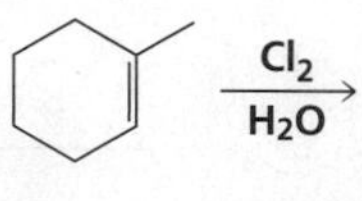 $\xrightarrow[\text{H}_2\text{O}]{\text{Cl}_2}$

g. 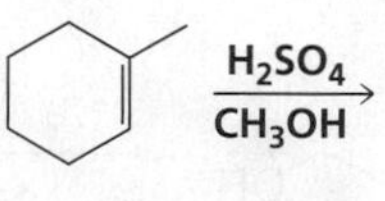 $\xrightarrow[\text{CH}_3\text{OH}]{\text{H}_2\text{SO}_4}$

b. 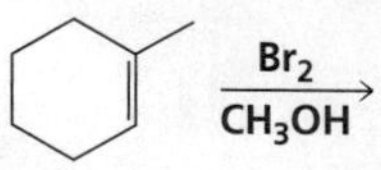 $\xrightarrow[\text{CH}_3\text{OH}]{\text{Br}_2}$

d. 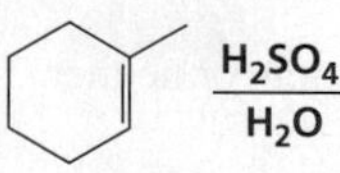 $\xrightarrow[\text{H}_2\text{O}]{\text{H}_2\text{SO}_4}$

f. 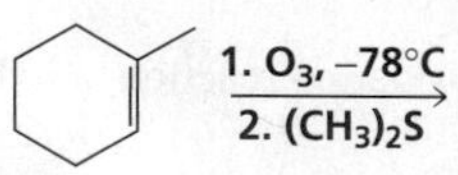 $\xrightarrow[\text{2. (CH}_3)_2\text{S}]{\text{1. O}_3,\,-78°\text{C}}$

h. $\xrightarrow[\text{CH}_2\text{Cl}_2]{\text{Cl}_2}$

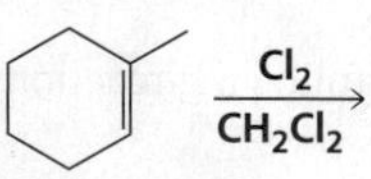

72. Using any alkene and any other reagents, how would you prepare the following compounds?

a.

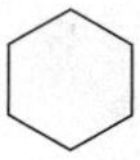

c. 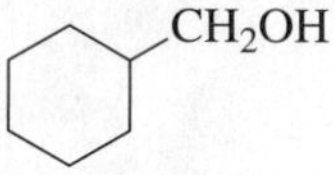 (cyclohexane with CH₂OH)

e. 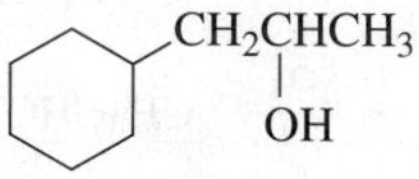 (cyclohexane with CH₂CHCH₃ / OH)

b. $CH_3CH_2CH_2CHCH_3$ with Cl

d. $CH_3CH_2CHCHCH_2CH_3$ with $Br\ OH$

f. $CH_3CH_2CHCHCH_2CH_3$ with $Br\ Cl$

73. a. Identify two alkenes that react with HBr to form 1-bromo-1-methylcyclohexane without undergoing a carbocation rearrangement.

b. Would both alkenes form the same alkyl halide if DBr were used instead of HBr? (D is an isotope of H, so D^+ reacts like H^+.)

74. Which is more stable?

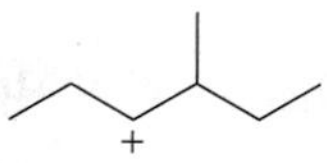

a. $\overset{CH_3}{\underset{+}{CH_3CCH_3}}$ or $CH_3\overset{+}{C}HCH_2CH_3$

c. or

e. $CH_3\overset{CH_3}{C}=CHCH_2CH_3$ or $CH_3CH=CH\overset{CH_3}{C}HCH_3$

b. $CH_3\overset{+}{C}HCH_3$ or $CH_3\overset{+}{C}HCH_2Cl$

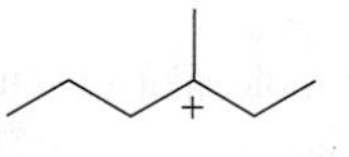

d. 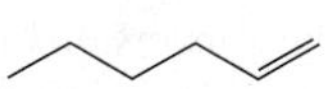 or

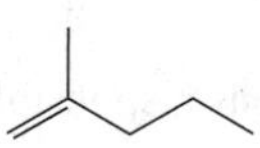

f. 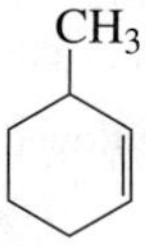 or

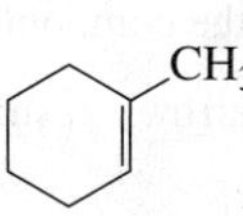

75. a. Draw the product or products that would be obtained from the reaction of *cis*-2-butene and *trans*-2-butene with each of the following reagents. If a product can exist as stereoisomers, show which stereoisomers are formed.

1. HCl
2. BH₃/THF followed by HO⁻, H₂O₂, H₂O
3. a peroxyacid

4. Br₂ in CH₂Cl₂
5. Br₂ + H₂O
6. H₂ + Pd/C

7. H₂O + H₂SO₄
8. CH₃OH + H₂SO₄

b. With which reagents do the two alkenes react to form different products?

76. The second-order rate constant (in units of $M^{-1}s^{-1}$) for acid-catalyzed hydration at 25 °C is given for each of the following alkenes:

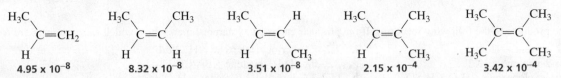

4.95 x 10^{-8} 8.32 x 10^{-8} 3.51 x 10^{-8} 2.15 x 10^{-4} 3.42 x 10^{-4}

   a. Calculate the relative rates of hydration of the alkenes. (*Hint:* Divide each rate constant by the smallest rate constant of the series: 3.51×10^{-8}.)

   b. Why does (*Z*)-2-butene react faster than (*E*)-2-butene?

   c. Why does 2-methyl-2-butene react faster than (*Z*)-2-butene?

   d. Why does 2,3-dimethyl-2-butene react faster than 2-methyl-2-butene?

77. Which compound has the greater dipole moment?

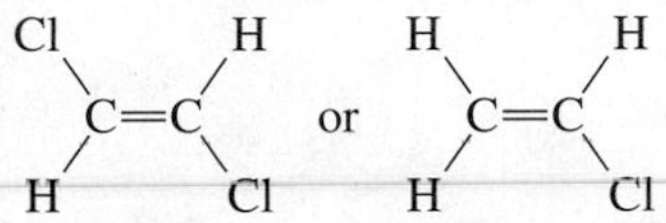

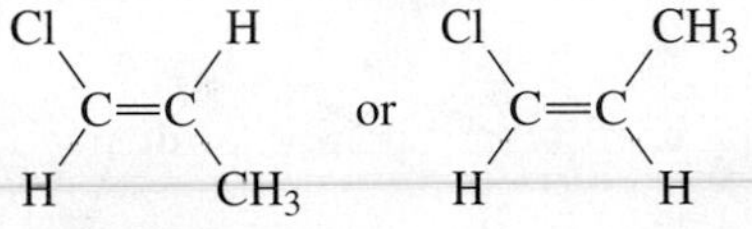

78. Draw the products of the following reactions. If the products can exist as stereoisomers, show which stereoisomers are formed.

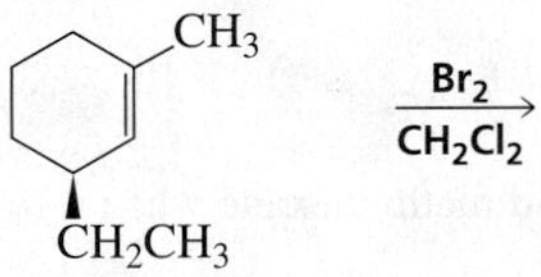

   b. $CH_3CH_2CH_2CH{=}CH_2$ + H_2O $\xrightarrow{H_2SO_4}$

79. A student was about to turn in the products he had obtained from the reaction of HI with 3,3,3-trifluoropropene when he realized that the labels had fallen off his flasks and he did not know which label belonged to which flask. His friend reminded him of the rule that says the electrophile adds to the sp^2 carbon bonded to the most hydrogens. In other words, he should label the flask containing the most product 1,1,1-trifluoro-2-iodopropane and label the flask containing the least amount of product 1,1,1-trifluoro-3-iodopropane. Should he follow his friend's advice?

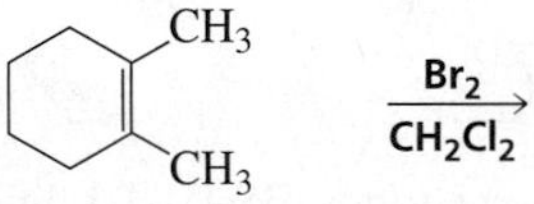

80. a. Propose a mechanism for the following reaction (show all curved arrows):

$$CH_3CH_2CH{=}CH_2 \;+\; CH_3OH \;\xrightarrow{H_2SO_4}\; CH_3CH_2CHCH_3$$
$$| $$
$$OCH_3$$

   b. Which step is the rate-determining step?
   e. What is the electrophile in the second step?

   c. What is the electrophile in the first step?
   f. What is the nucleophile in the second step?

   d. What is the nucleophile in the first step?

81. Draw the products, including their configurations, obtained from the reaction of 1-ethylcyclohexene with the following reagents:

   a. HBr   **b.** H_2, Pd/C   **c.** R_2BH/THF followed by HO^-, H_2O_2, H_2O **d.** Br_2/CH_2Cl_2

82. Which stereoisomer of 3-hexene forms a meso compound when it reacts with Br_2?

83. Which stereoisomer of 3-hexene forms (3*S*,4*S*)-4-bromo-3-hexanol and (3*R*,4*R*)-4-bromo-3-hexanol when it reacts with Br_2 and H_2O?

84. Propose a mechanism for each of the following reactions:

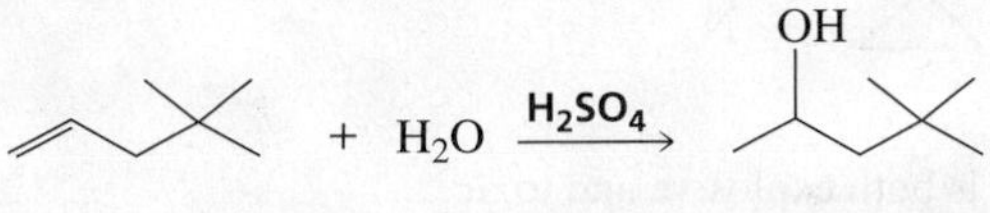

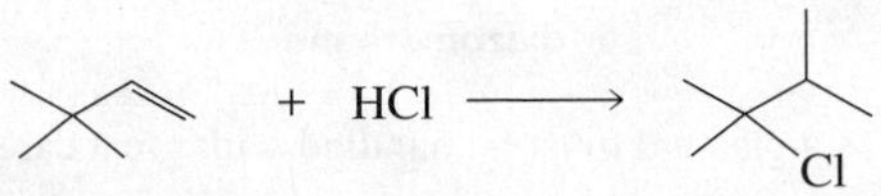

85. What is the major product of each of the following reactions?

    a. $HOCH_2CH_2CH_2CH{=}CH_2$ + Br_2 $\xrightarrow{CH_2Cl_2}$ **b.** $HOCH_2CH_2CH_2CH_2CH{=}CH_2$ + Br_2 $\xrightarrow{CH_2Cl_2}$

86. Draw the products of the following reactions. If the products can exist as stereoisomers, show which stereoisomers are formed.

    a. *cis*-2-pentene + HCl

    b. *trans*-2-pentene + HCl

    c. 1-ethylcyclohexene + H_2O + H_2SO_4

    d. 2,3-dimethyl-3-hexene + H_2, Pd/C

    e. 1,2-dimethylcyclohexene + HCl

    f. 1,2-dideuteriocyclohexene + H_2, Pd/C

    g. 3,3-dimethyl-1-pentene + Br_2/CH_2Cl_2

    h. (*E*)-3,4-dimethyl-3-heptene + H_2, Pd/C

    i. (*Z*)-3,4-dimethyl-3-heptene + H_2, Pd/C

    j. 1-chloro-2-ethylcyclohexene + H_2, Pd/C

87. a. What product is obtained from the reaction of HCl with 1-butene? With 2-butene?

    b. Which of the two reactions has the greater free energy of activation?

    c. Which compound reacts more rapidly with HCl, (*Z*)-2-butene or (*E*)-2-butene?

88. What would be the major product of the reaction of each of the following with HBr?

    a.    **b.**   **c.**   **d.**

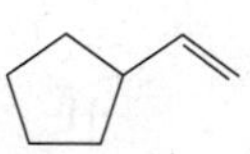

89. For each compound, show the products obtained from ozonolysis, followed by treatment with dimethyl sulfide.

    a. 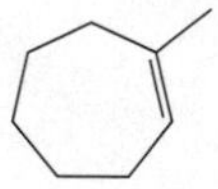   **b.** 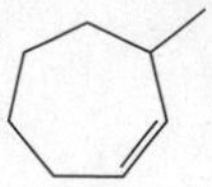    **c.** 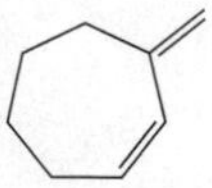 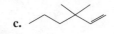   **d.** 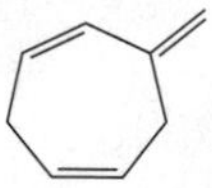   **e.**

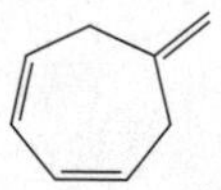

90. Which stereoisomer of 3,4-dimethyl-3-hexene forms (3*S*,4*S*)-3,4-dimethylhexane and (3*R*,4*R*)-3,4-dimethylhexane when it reacts with H_2, Pd/C?

91. a. How many alkenes could you treat with H_2, Pd/C in order to prepare methylcyclopentane?

    b. Which of the alkenes is the most stable?

    c. Which of the alkenes has the smallest heat of hydrogenation?

92. Draw the products of the following reactions. If the products can exist as stereoisomers, show what stereoisomers are formed.

    a. *cis*-2-pentene + Br_2/CH_2Cl_2

    b. *trans*-2-pentene + Br_2/CH_2Cl_2

    c. 1-butene + HCl

    d. methylcyclohexene + HBr

    e. *trans*-3-hexene + Br_2/CH_2Cl_2

    f. *cis*-3-hexene + Br_2/CH_2Cl_2

    g. 3,3-dimethyl-1-pentene + HBr

    h. *cis*-2-butene + HBr

    i. (*Z*)-2,3-dichloro-2-butene + H_2, Pd/C

    j. (*E*)-2,3-dichloro-2-butene + H_2, Pd/C

    k. (*Z*)-3,4-dimethyl-3-hexene + H_2, Pd/C

    l. (*E*)-3,4-dimethyl-3-hexene + H_2, Pd/C

93. Of the possible products shown for the following reaction, are there any that would not be formed?

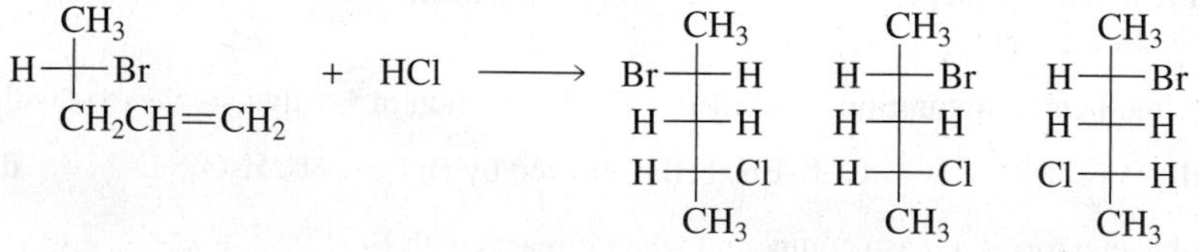

$$\begin{array}{c} CH_3 \\ H{-}\!\!\!\!-Br \\ CH_2CH{=}CH_2 \end{array} + HCl \longrightarrow \begin{array}{c} CH_3 \\ Br{-}\!\!\!\!-H \\ H{-}\!\!\!\!-H \\ H{-}\!\!\!\!-Cl \\ CH_3 \end{array} \quad \begin{array}{c} CH_3 \\ H{-}\!\!\!\!-Br \\ H{-}\!\!\!\!-H \\ H{-}\!\!\!\!-Cl \\ CH_3 \end{array} \quad \begin{array}{c} CH_3 \\ H{-}\!\!\!\!-Br \\ H{-}\!\!\!\!-H \\ Cl{-}\!\!\!\!-H \\ CH_3 \end{array}$$

94. The reaction of an alkene with diazomethane forms a cylcopropane ring. Propose a mechanism for the reaction. (*Hint:* It is a concerted reaction.)

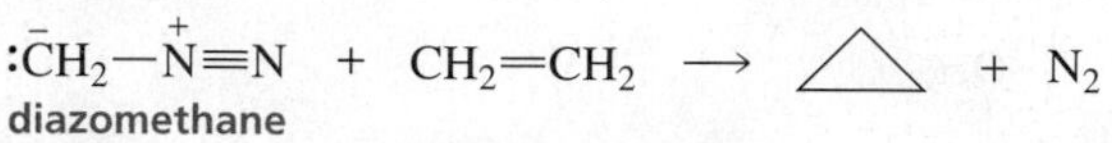

$$:\!\bar{C}H_2{-}\overset{+}{N}{\equiv}N + CH_2{=}CH_2 \longrightarrow \triangle + N_2$$

diazomethane

Note: Diazomethane is a gas that must be handled with great care because it is both explosive and toxic.

95. Two chemists at Dupont found that ICH_2ZnI is better than diazomethane at converting a $C{=}C$ bond to a cylcopropane ring. Propose a mechanism for the reaction, now known as the *Simmons–Smith reaction* in their honor.

96. a. Dichlorocarbene can be generated by heating chloroform with HO^-. Propose a mechanism for the reaction.

$$CHCl_3 \; + \; HO^- \; \xrightarrow{\Delta} \; Cl_2C\!: \; + \; H_2O \; + \; Cl^-$$

chloroform   dichlorocarbene

b. Dichlorocarbene can also be generated by heating sodium trichloroacetate. Propose a mechanism for the reaction.

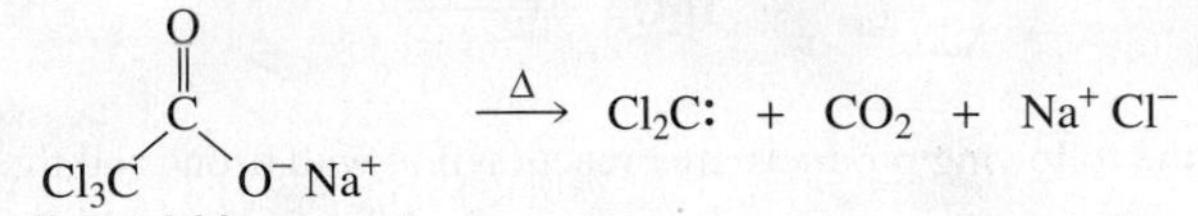

$$\xrightarrow{\Delta} \; Cl_2C\!: \; + \; CO_2 \; + \; Na^+\,Cl^-$$

sodium trichloroacetate

97. What product would be obtained from the reaction of dichlorocarbene with cyclopentene?

98. What alkene gives the product shown after reaction first with ozone and then with dimethyl sulfide

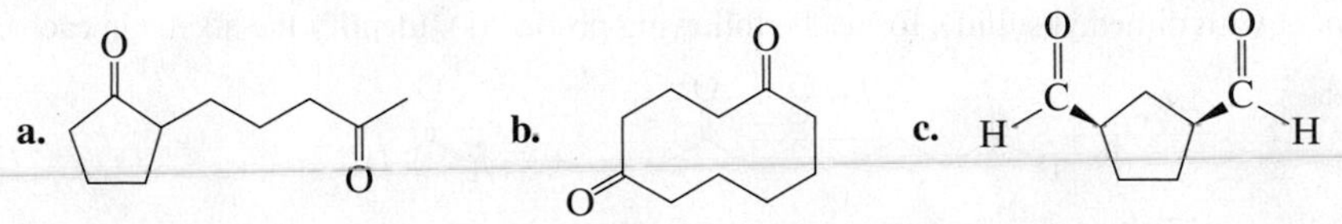

a.   **b.**   **c.**

99. Draw the products of the following reactions including their configurations:

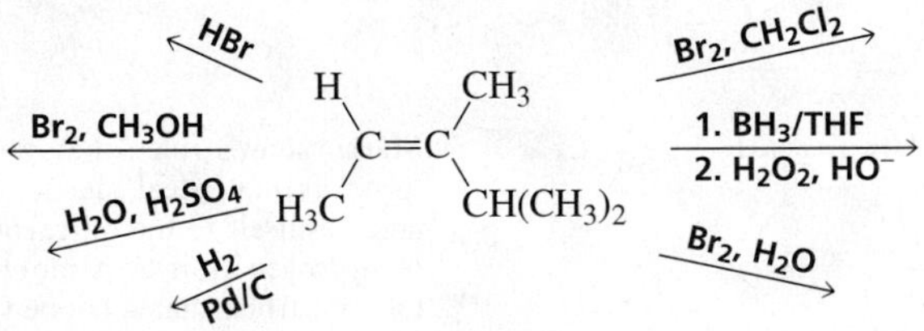

100. a. Propose a mechanism for the following reaction:

$$+ \; HBr \; \longrightarrow$$

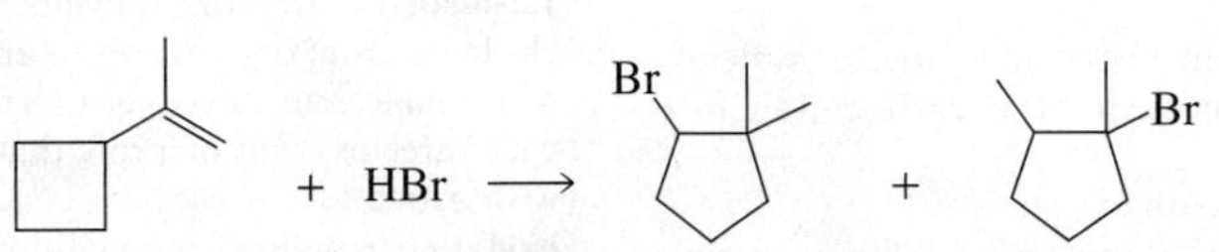

b. Is the initially formed carbocation primary, secondary, or tertiary?
c. Is the rearranged carbocation primary, secondary, or tertiary?
d. Why does the rearrangement occur?

101. Which compound would you expect to be hydrated more rapidly?

102. When the following compound is hydrated in the presence of acid, the unreacted alkene is found to have retained the deuterium atoms. What does this tell you about the mechanism for hydration?

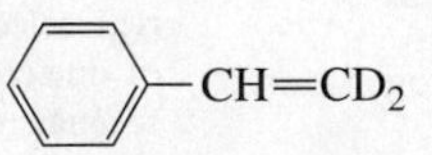

103. When fumarate reacts with D_2O in the presence of the enzyme fumarase, only one isomer of the product is formed, as shown here. Is the enzyme catalyzing a syn or an anti addition of D_2O?

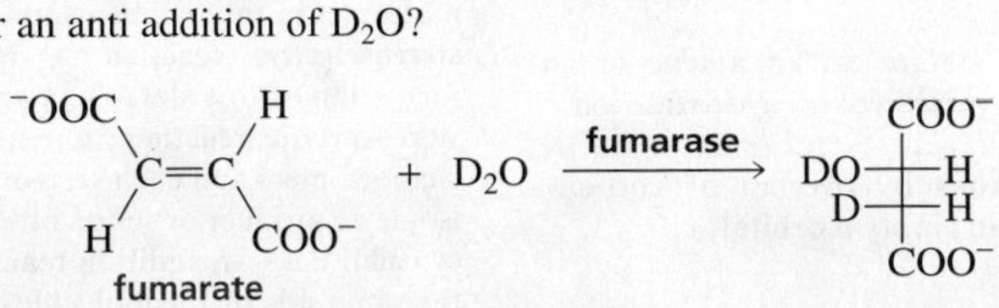

fumarate

104. When (S)-(+)-1-chloro-2-methylbutane reacts with chlorine, one of the products is (−)-1,4-dichloro-2-methylbutane. Does this product have the R or the S configuration?

105. Propose a mechanism for the following reaction:

$$H_3C \quad CH=CH_2$$

$$+ \ H_2O \xrightarrow{H_2SO_4}$$

106. What hydrocarbon would form the following products after reaction first with ozone and then with dimethyl sulfide?

107. Ozonolysis of an alkene, followed by treatment with dimethyl sulfide, forms the following product(s). Identify the alkene in each case.

a.

b.

GLOSSARY

acid-catalyzed reaction a reaction catalyzed by an acid.

aldehyde

alkene a hydrocarbon that contains a double bond.

anti addition an addition reaction in which two substituents are added at opposite sides of the molecule.

biochemistry (biological chemistry) the chemistry of biological systems.

carbocation rearrangement the rearrangement of a carbocation to a more stable carbocation.

carbonyl group a carbon doubly bonded to an oxygen.

catalytic hydrogenation the addition of hydrogen to a double or a triple bond with the aid of a metal catalyst.

concerted reaction a reaction in which all the bond-making and bond-breaking processes occur in one step.

dimer a molecule formed by the joining together of two identical molecules.

enantioselective reaction a reaction that forms an excess of one enantiomer.

enzyme a protein that is a catalyst.

epoxide (oxirane) an ether in which the oxygen is incorporated into a three-membered ring.

halohydrin an organic molecule that contains a halogen and an OH group on adjacent carbons.

Hammond postulate states that the transition state will be more similar in structure to the species (reactants or products) that it is closer to energetically.

heat of hydrogenation the heat ($-\Delta H°$) released in a hydrogenation reaction.

hydration addition of water to a compound.

1,2-hydride shift the movement of a hydride ion from one carbon to an adjacent carbon.

hydroboration–oxidation the addition of borane to an alkene or an alkyne, followed by reaction with hydrogen peroxide and hydroxide ion.

hydrogenation addition of hydrogen.

hyperconjugation delocalization of electrons by overlap of carbon–hydrogen or carbon–carbon σ bonds with an empty p orbital.

ketone

Markovnikov's rule the actual rule is "When a hydrogen halide adds to an asymmetrical alkene, the addition occurs such that the halogen attaches itself to the sp^2 carbon of the alkene bearing the lowest number of hydrogen atoms." A more universal rule is "The electrophile adds to the sp^2 carbon that is bonded to the greater number of hydrogens."

mechanism of a reaction a description of the step-by-step process by which reactants are changed into products.

1,2-methyl shift the movement of a methyl group with its bonding electrons from one carbon to an adjacent carbon.

molozonide an unstable intermediate containing a five-membered ring with three oxygens in a row that is formed from the reaction of an alkene with ozone.

oxidation reaction a reaction in which the number of C—H bonds decreases or the number of C—O, C—N, or C—X (X = a halogen) increases.

oxidative cleavage an oxidation reaction that cuts the reactant into two or more pieces.

ozonide the five-membered-ring compound formed as a result of the rearrangement of a molozonide.

ozonolysis reaction of a carbon–carbon double or triple bond with ozone.

peroxyacid a carboxylic acid with an OOH group instead of an OH group.

primary carbocation a carbocation with the positive charge on a primary carbon.

receptor site the site at which a drug binds in order to exert its physiological effect.

reduction reaction a reaction in which the number of CiH bonds increases or the number of C i O, C i N, or C i X (X = a halogen) decreases.

regioselective reaction a reaction that leads to the preferential formation of one constitutional isomer over another.

secondary carbocation a carbocation with the positive charge on a secondary carbon.

stereochemistry the field of chemistry that deals with the structures of molecules in three dimensions.

stereoselective reaction a reaction that leads to the preferential formation of one stereoisomer over another.

stereospecific reaction a reaction in which the reactant can exist as stereoisomers and each stereoisomeric reactant leads to a different stereo isomeric product or set of products.

syn addition an addition reaction in which two substituents are added to the same side of the molecule.

tertiary carbocation a carbocation with the positive charge on a tertiary carbon.

PHOTO CREDITS

Credits are listed in order of appearance.

B Brown/Shutterstock; GWImages/Shutterstock; NASA; Taushia Jackson/Shutterstock; (ctr.) National Institute of Health; (bot. left) epantha/Fotolia; (bot. rt.) Maksim Striganov/Shutterstock; Posh/Fotolia

The Reactions of Alkynes

An Introduction to Multistep Synthesis

Currently, in order to protect our environment, the challenge facing organic chemists is to design syntheses that use reactants and generate products that cause little or no toxicity to the environment. Preventing pollution at the molecular level is known as green chemistry. You will have your first opportunity to design multistep syntheses in this chapter.

An **alkyne** is a hydrocarbon that contains a carbon–carbon triple bond. Only relatively few alkynes are found in nature. Examples include capillin, which has fungicidal activity, and ichthyothereol, a convulsant used by the indigenous people of the Amazon for poisoned arrowheads.

capillin
a fungicide

ichthyothereol
a convulsant

an enediyne
a class of anti-cancer drug

A class of naturally occurring compounds called enediynes has been found to have powerful anticancer properties because they are able to cleave DNA. All enediynes have a nine- or ten-membered ring that contains two triple bonds separated by a double bond. One of the first enediynes approved for clinical use is used to treat acute myeloid leukemia. Several others are currently in clinical trials (see the box that appears later in this section).

Other drugs on the market that contain an alkyne functional group are not naturally occurring compounds; they exist only because chemists have been able to synthesize them. Their trade names are shown in green. Trade names are always capitalized; only

From Chapter 7 of *Organic Chemistry*, Seventh Edition. Paula Yurkanis Bruice. Copyright © 2014 by Pearson Education, Inc. All rights reserved.

the company that holds the patent for a product can use the product's trade name for commercial purposes.

Sinovial®

parsalmide
an analgesic

Supirdyl®

pargyline
an antihypertensive

Synthetic Alkynes Are Used to Treat Parkinson's Disease

Parkinson's disease is a degenerative condition characterized by tremors. It is caused by the destruction of cells in the *substantia nigra,* a crescent-shaped region in the midbrain. These are the cells that release dopamine, the neurotransmitter that plays an important role in movement, muscle control, and balance. A neurotransmitter is a compound used to communicate between brain cells.

Dopamine is synthesized from tyrosine (one of the 20 common amino acids). Ideally, Parkinson's disease could be treated by giving the patient dopamine. Unfortunately, dopamine is not polar enough to cross the blood–brain barrier. Therefore, L-DOPA, its immediate precursor, is the drug of choice, but it ceases to control the disease's symptoms after it has been used for a while.

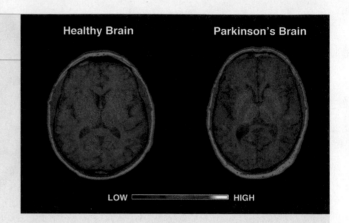

Healthy Brain **Parkinson's Brain**

LOW ▭▭▭▭▭ HIGH

tyrosine

tyrosine
hydroxylase

L-DOPA

amino acid
decarboxylase

dopamine

monoamine
oxidase

Two drugs, each containing a C≡CH group, have been developed that inhibit this enzyme, thus preventing the oxidation of dopamine and thereby increasing its availability in the brain. Both drugs have structures similar to that of dopamine, so they are able to bind to the enzyme's active site. (Note that enzymes recognize their substrates by their shape.) Because these drugs form covalent bonds with groups at the enzyme's active site, they become permanently attached to the active site, thus preventing the enzyme from binding dopamine. Patients on these drugs continue to take L-DOPA, but now this drug can be taken at longer intervals and it can control the disease's symptoms for a longer period of time.

selegiline
Eldepryl®

rasagiline
Azilect®

Selegiline was approved by the FDA first, but one of the compounds to which it is metabolized has a structure similar to that of methamphetamine (the street drug known as "speed"). So, some patients taking the drug experience psychiatric and cardiac effects. These side effects have not been found in patients taking rasagiline.

Notice that the name of most enzymes ends in "ase," preceded by an indication of what reaction the enzyme catalyzes. Thus, tyrosine hydroxylase puts an OH group on tyrosine, amino acid decarboxylase removes a carboxyl (COO⁻) group from an amino acid (or, in this case, from a compound similar to an amino acid), and monoamine oxidase oxidizes an amine.

Why Are Drugs So Expensive?

The average cost of launching a new drug is $1.2 billion. The manufacturer has to recover this cost quickly because the patent has to be filed as soon as the drug is first discovered. Although a patent is good for 20 years, it takes an average of 12 years to bring a drug to market after its initial discovery, so the patent protects the discoverer of the drug for an average of 8 years. It is only during the eight years of patent protection that drug sales can provide the income needed to cover the initial costs as well as to pay for research on new drugs.

Why does it cost so much to develop a new drug? First of all, the Food and Drug Administration (FDA) has high standards that must be met before a drug is approved for a particular use. An important factor leading to the high price of many drugs is the low rate of success in progressing from the initial concept to an approved product. In fact, only 1 or 2 of every 100 compounds tested become lead compounds. A lead compound is a compound that shows promise of becoming a drug. Chemists modify the structure of a lead compound to see if doing so improves its likelihood of becoming a drug. For every 100 structural modifications of a lead compound, only one is worthy of further study. For every 10,000 compounds evaluated in animal studies, only 10 will get to clinical trials.

Clinical trials consist of three phases. Phase I evaluates the effectiveness, safety, side effects, and dosage levels in up to 100 healthy volunteers; phase II investigates the effectiveness, safety, and side effects in 100 to 500 volunteers who have the condition the drug is meant to treat; and phase III establishes the effectiveness and appropriate dosage of the drug and monitors adverse reactions in several thousand volunteer patients. For every 10 compounds that enter clinical trials, only 1 satisfies the increasingly stringent requirements to become a marketable drug.

1 THE NOMENCLATURE OF ALKYNES

Because of its triple bond, an alkyne has four fewer hydrogens than an alkane with the same number of carbons. Therefore, while the general molecular formula for an acyclic alkane is C_nH_{2n+2}, the general molecular formula for an acyclic alkyne is C_nH_{2n-2} and that for a cyclic alkyne is C_nH_{2n-4}.

The systematic name of an alkyne is obtained by replacing the "ane" ending of the alkane name with "yne." Analogous to the way compounds with other functional groups are named, the longest continuous chain containing the carbon–carbon triple bond is numbered in the direction that gives the functional group suffix as low a number as possible. If the triple bond is at the end of the chain, the alkyne is classified as a **terminal alkyne.** Alkynes with triple bonds located elsewhere along the chain are **internal alkynes.**

**1-hexyne
a terminal alkyne**

**3-hexyne
an internal alkyne**

		a terminal alkyne	an internal alkyne	
	HC≡CH	CH₃CH₂C≡CH	CH₃C≡CCH₂CH₃	CH₃CHC≡CCH₃
Systematic:	ethyne	1-butyne	2-pentyne	4-methyl-2-hexyne
Common:	acetylene	ethylacetylene	ethylmethylacetylene	*sec*-butylmethyl-acetylene

In common nomenclature, alkynes are named as *substituted acetylenes.* The common name is obtained by stating the names of the alkyl groups (in alphabetical order) that have replaced the hydrogens of acetylene. Acetylene is an unfortunate common name for an alkyne because its "ene" ending is characteristic of a double bond rather than a triple bond.

If counting from either direction leads to the same number for the functional group suffix, the correct systematic name is the one that contains the lowest substituent number. If the compound contains more than one substituent, the substituents are listed in alphabetical order.

Cl Br
CH₃CHCHC≡CCH₂CH₂CH₃
1 2 3 4 5 6 7 8

**3-bromo-2-chloro-4-octyne
not 6-bromo-7-chloro-4-octyne
because 2 < 6**

CH₃
CH₃CHC≡CCH₂CH₂Br
6 5 4 3 2 1

**1-bromo-5-methyl-3-hexyne
not 6-bromo-2-methyl-3-hexyne
because 1 < 2**

A substituent receives the lowest possible number only if there is no functional group suffix, or if counting from either direction leads to the same number for the functional group suffix.

311

Synthetic Alkynes Are Used for Birth Control

Estradiol and progesterone are naturally occurring female hormones. Because of their ring structures, they are classified as steroids. Estradiol is responsible for the development of secondary sex characteristics in women—it affects body shape, fat deposition, bones, and joints. Progesterone is critical for the continuation of pregnancy.

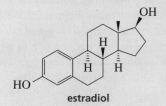

estradiol

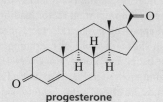

progesterone

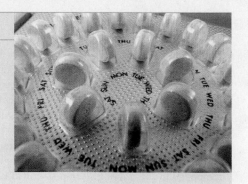

The four compounds shown next are synthetic steroids that are used for birth control; each contains an alkyne functional group. Most birth control pills contain ethinyl estradiol (a compound structurally similar to estradiol) and a compound structurally similar to progesterone (such as norethindrone). Ethinyl estradiol prevents ovulation, whereas norethindrone makes it difficult for a fertilized egg to attach to the wall of the uterus.

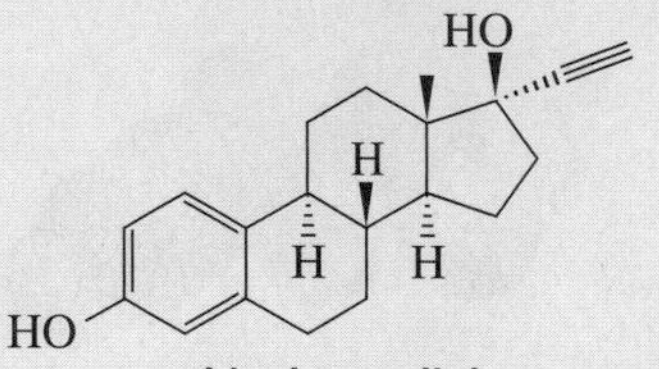

ethinyl estradiol

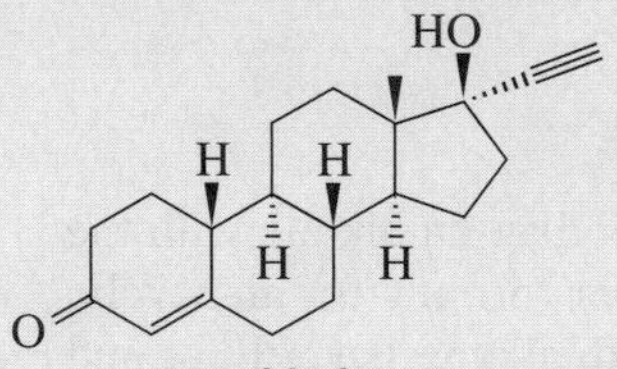

norethindrone
Aygestin®

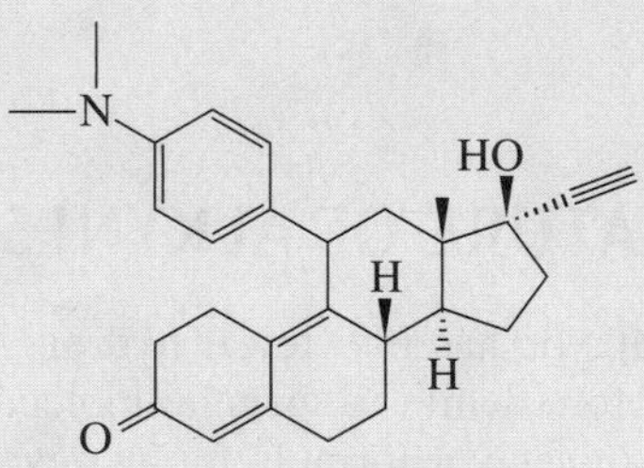

mifepristone
RU-486
Mefegyne®

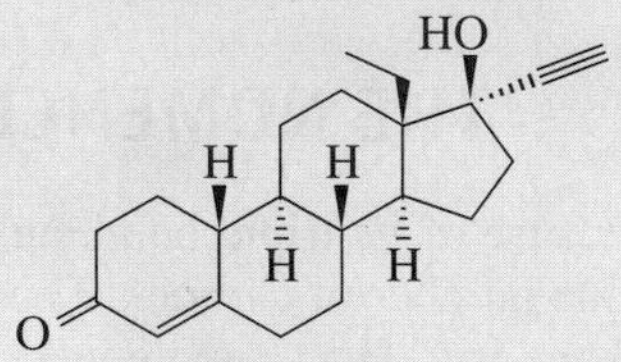

levonorgestrel
Norplant®

Mifepristone and levonorgestrel are also synthetic steroids that contain an alkyne functional group. Mifepristone, also known as RU-486, induces an abortion if taken early in pregnancy. Its name comes from Roussel-Uclaf, the French pharmaceutical company where it was first synthesized, and from an arbitrary lab serial number. Levonorgestrel is an emergency contraceptive pill. It prevents pregnancy if taken within a few days of conception.

PROBLEM 1♦

What is the molecular formula for a monocyclic hydrocarbon with 14 carbons and 2 triple bonds?

PROBLEM 2♦

Draw the structure for each of the following:

a. 1-chloro-3-hexyne

b. cyclooctyne

c. isopropylacetylene

d. *sec*-butylisobutylacetylene

e. 4,4-dimethyl-1-pentyne

f. dimethylacetylene

PROBLEM 3

Draw the structures and give the common and systematic names for the seven alkynes with molecular formula C_6H_{10}.

PROBLEM 4♦

Name the following:

a. $BrCH_2CH_2C{\equiv}CCH_3$

b. $CH_3CH_2CHC{\equiv}CCH_2CHCH_3$
   $\quad\quad\quad\quad | \quad\quad\quad\quad\quad |$
   $\quad\quad\quad\quad Br \quad\quad\quad\quad\;\; Cl$

c. $CH_3OCH_2C{\equiv}CCH_2CH_3$

d. $CH_3CH_2CHC{\equiv}CH$
   $\quad\quad\quad\quad\quad |$
   $\quad\quad\quad\quad\quad CH_2CH_2CH_3$

Name the following:

a.   b.   c.

---

2 | HOW TO NAME A COMPOUND THAT HAS MORE THAN ONE FUNCTIONAL GROUP

The rules for naming compounds with two triple bonds, using the ending "diyne," are similar to the rules for naming compounds with two double bonds.

$$CH_2=C=CH_2$$

systematic: propadiene
common: allene

$$\overset{6}{C}H_3\overset{5}{C}H=\overset{4}{C}H\overset{3}{C}H_2\overset{2}{C}=\overset{1}{C}H_2 \quad (CH_3 \text{ on } C2)$$
2-methyl-1,4-hexadiene
or
2-methylhexa-1,4-diene

$$CH_3\overset{}{C}HC\equiv CCH_2C\equiv CH \quad (CH_3 \text{ branch})$$
6-methyl-1,4-heptadiyne
or
6-methylhepta-1,4-diyne

To name an alkene in which the second functional group is not another double bond but has a functional group suffix, find the longest continuous chain containing both functional groups and put both suffixes at the end of the name. Put the "ene" ending first, with the terminal "e" omitted to avoid two adjacent vowels.

The number indicating the location of the first-stated functional group is usually placed before the name of the parent chain. The number indicating the location of the second-stated functional group is placed immediately before the suffix for that functional group. If the two functional groups are a *double bond* and a *triple bond,* number the chain in the direction that produces a name containing the lower number. Thus, in the following examples, the lower number is given to the alkyne suffix in the compound on the left and to the alkene suffix in the compound on the right.

$$\overset{7}{C}H_3\overset{6}{C}H=\overset{5}{C}H\overset{4}{C}H_2\overset{3}{C}H_2\overset{2}{C}\equiv\overset{1}{C}H$$
5-hepten-1-yne
not 2-hepten-6-yne
because 1 < 2

$$\overset{1}{C}H_2=\overset{2}{C}H\overset{3}{C}H_2\overset{4}{C}H_2\overset{5}{C}\equiv\overset{6}{C}\overset{7}{C}H_3$$
1-hepten-5-yne
not 6-hepten-2-yne
because 1 < 2

$$CH_2CH_2CH_2CH_3$$
$$\overset{1}{C}H_2=\overset{2}{C}H\overset{3}{C}H\overset{4}{C}\equiv\overset{5}{C}\overset{6}{C}H_3$$
3-butyl-1-hexen-4-yne

> the longest continuous chain has 8 carbons, but the 8-carbon chain does not contain both functional groups; therefore, the compound is named as a hexenyne because the longest continuous chain containing both functional groups has 6 carbons

If the same low number is obtained in both directions, number the chain in the direction that gives the double bond the lower number.

$$\overset{1}{C}H_3\overset{2}{C}H=\overset{3}{C}H\overset{4}{C}\equiv\overset{5}{C}\overset{6}{C}H_3$$
2-hexen-4-yne
not 4-hexen-2-yne

$$H\overset{6}{C}\equiv\overset{5}{C}\overset{4}{C}H_2\overset{3}{C}H_2\overset{2}{C}H=\overset{1}{C}H_2$$
1-hexen-5-yne
not 5-hexen-1-yne

When the functional groups are a double bond and a triple bond, the chain containing both groups is numbered in the direction that produces the name containing the lowest possible number, regardless of which functional group gets the lower number.

If there is a tie between a double bond and a triple bond, the double bond gets the lower number.

313

If the second functional group suffix has a higher priority than the alkene suffix, number the chain in the direction that assigns the lower number to the functional group with the higher-priority suffix. (The relative priorities of functional group suffixes are shown in Table 1.) The higher priority functional group is assumed to be at the 1-position in cyclic compounds.

Table 1	Priorities of Functional Group Suffixes

highest priority — C=O > OH > NH$_2$ > C=C = C≡C — lowest priority

the double bond is given priority over a triple bond only when there is a tie

Number the chain so that the lowest possible number is given to the functional group with the higher priority.

CH$_2$=CHCH$_2$OH
2-propen-1-ol
not **1-propen-3-ol**

CH$_3$C=CHCH$_2$CH$_2$OH (with CH$_3$ above)
4-methyl-3-penten-1-ol

CH$_2$=CHCH$_2$CH$_2$CH$_2$CHCH$_3$ (with NH$_2$ above)
6-hepten-2-amine

CH$_3$CH$_2$CH=CHCHCH$_3$ (with OH below)
3-hexen-2-ol

6-methyl-2-cyclohexenol

3-cyclohexenamine

How a Banana Slug Knows What to Eat

Many species of mushrooms synthesize 1-octen-3-ol, a repellent that drives off predatory slugs. Such mushrooms can be recognized by small bite marks on their caps, where the slug started to nibble before the volatile compound was released. People are not put off by the release of this compound because to them it just smells like a mushroom. 1-Octen-3-ol also has antibacterial properties that may protect the mushroom from organisms that would otherwise invade the wound made by the slug. Not surprisingly, the species of mushroom that banana slugs commonly eat cannot synthesize 1-octen-3-ol.

1-octen-3-ol

PROBLEM 6♦

Name the following:

a. CH$_2$=CHCH$_2$C≡CCH$_2$CH$_3$

b. CH$_3$CH=CCH$_2$CH=CH$_2$ (with CH$_3$ above)

c. CH$_3$CH$_2$CH=CCH$_2$CH$_2$C≡CH (with CH=CH$_2$ below)

d. HOCH$_2$CH$_2$C≡CH

e. CH$_3$CH=CHCH=CHCH=CH$_2$

f. CH$_3$CH=CCH$_2$CHCH$_2$OH (with CH$_3$ CH$_3$ above)

PROBLEM 7♦

Name the following:

a. (structure with OH)

b. (structure)

c. (structure)

3   THE PHYSICAL PROPERTIES OF UNSATURATED HYDROCARBONS

All hydrocarbons—alkanes, alkenes, and alkynes—have similar physical properties. They are all insoluble in water but soluble in nonpolar solvents. They are less dense than water and, like other homologous series, have boiling points that increase with increasing molecular weight (Table 2). Alkynes are more linear than alkenes, and a triple bond is more polarizable than a double bond. These two features cause an alkyne to have stronger van der Waals interactions and, therefore, a higher boiling point than an alkene with the same number of carbons.

Table 2   Boiling Points of the Smallest Hydrocarbons

	bp (°C)		bp (°C)		bp (°C)
CH_3CH_3 **ethane**	−88.6	$H_2C=CH_2$ **ethene**	−104	$HC\equiv CH$ **ethyne**	−84
$CH_3CH_2CH_3$ **propane**	−42.1	$CH_3CH=CH_2$ **propene**	−47	$CH_3C\equiv CH$ **propyne**	−23
$CH_3CH_2CH_2CH_3$ **butane**	−0.5	$CH_3CH_2CH=CH_2$ **1-butene**	−6.5	$CH_3CH_2C\equiv CH$ **1-butyne**	8
$CH_3(CH_2)_3CH_3$ **pentane**	36.1	$CH_3CH_2CH_2CH=CH_2$ **1-pentene**	30	$CH_3CH_2CH_2C\equiv CH$ **1-pentyne**	39
$CH_3(CH_2)_4CH_3$ **hexane**	68.7	$CH_3CH_2CH_2CH_2CH=CH_2$ **1-hexene**	63.5	$CH_3CH_2CH_2CH_2C\equiv CH$ **1-hexyne**	71
		$CH_3CH=CHCH_3$ ***cis*-2-butene**	3.7	$CH_3C\equiv CCH_3$ **2-butyne**	27
		$CH_3CH=CHCH_3$ ***trans*-2-butene**	0.9	$CH_3CH_2C\equiv CCH_3$ **2-pentyne**	55

PROBLEM 8♦

In Table 2, what is the smallest alkane, the smallest terminal alkene, and the smallest terminal alkyne that are liquids at room temperature, which is generally taken to be 20 °C to 25 °C?

PROBLEM 9♦

Why does *cis*-2-butene have a higher boiling point than *trans*-2-butene?

4   THE STRUCTURE OF ALKYNES

Each carbon in the structure of ethyne is *sp* hybridized. As a result, each carbon has two *sp* orbitals and two *p* orbitals. One *sp* orbital overlaps the *s* orbital of a hydrogen, and the other overlaps an *sp* orbital of the other carbon. (The small lobes of the *sp* orbitals are not shown.) Because the *sp* orbitals are oriented as far from each other as possible to minimize electron repulsion, ethyne is a linear molecule with bond angles of 180°.

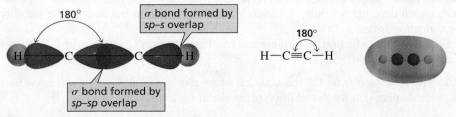

180°

σ bond formed by *sp–s* overlap

180°

$H-C\equiv C-H$

σ bond formed by *sp–sp* overlap

Other alkynes have structures similar to that of ethyne. Note that the triple bond is formed by each of the two *p* orbitals on one *sp* carbon overlapping the parallel *p* orbital on the other *sp* carbon to form two π bonds (Figure 1). The end result can be thought of as a cylinder of electrons wrapped around the σ bond.

▶ Figure 1

(a) Each of the two π bonds of a triple bond is formed by side-to-side overlap of a *p* orbital of one carbon with a parallel *p* orbital of the adjacent carbon.

(b) The electrostatic potential map for 2-butyne shows the cylinder of electrons wrapped around the σ bond.

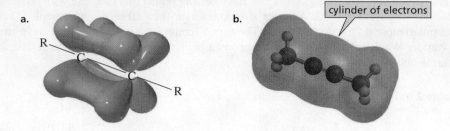

Also note that a carbon–carbon triple bond is shorter and stronger than a carbon–carbon double bond, which in turn, is shorter and stronger than a carbon–carbon single bond, and that a π bond is weaker than a σ bond.

A triple bond is composed of a σ bond and two π bonds.

Alkyl groups stabilize alkynes, just as they stabilize alkenes and carbocations. Internal alkynes, therefore, are more stable than terminal alkynes.

PROBLEM 10♦

What orbitals are used to form the carbon–carbon σ bond between the highlighted carbons?

a. $CH_3CH=CHCH_3$ **d.** $CH_3C\equiv CCH_3$ **g.** $CH_3CH=CHCH_2CH_3$

b. $CH_3CH=CHCH_3$ **e.** $CH_3C\equiv CCH_3$ **h.** $CH_3C\equiv CCH_2CH_3$

c. $CH_3CH=C=CH_2$ **f.** $CH_2=CHCH=CH_2$ **i.** $CH_2=CHC\equiv CH$

---

5 | ALKYNES ARE LESS REACTIVE THAN ALKENES

The cloud of electrons completely surrounding the σ bond makes an alkyne an electron-rich molecule. Alkynes therefore are nucleophiles, so they react with electrophiles. Thus alkynes, like alkenes, undergo *electrophilic addition reactions* because of their relatively weak π bonds. The same electrophilic reagents that add to alkenes also add to alkynes. For example, the addition of hydrogen chloride to an alkyne forms a chloro-substituted alkene.

Alkynes are less reactive than alkenes in electrophilic addition reactions.

$$CH_3C\equiv CCH_3 \xrightarrow{\text{HCl}} CH_3C=CHCH_3$$
$$\underset{Cl}{|}$$

An alkyne is *less* reactive than an alkene. This might at first seem surprising since an alkyne is less stable than an alkene (Figure 2). Note, however, that reactivity depends on $\Delta G^{\ddagger}$, which depends not only on the stability of the *reactant* but also on the stability of the *transition state*.

For an alkyne to be both less stable and less reactive than an alkene, the following two conditions must hold:

1. The transition state for the addition of an electrophile to an alkyne must be less stable than that for the addition of an electrophile to an alkene.

2. The difference in the stabilities of the transition states must be greater than the difference in the stabilities of the reactants, so that $\Delta G^{\ddagger}_{alkyne} > \Delta G^{\ddagger}_{alkene}$ (Figure 2).

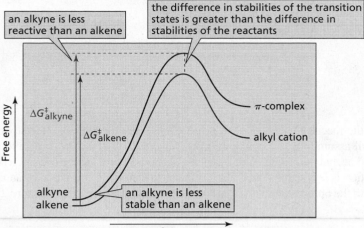

◀ **Figure 2**

Comparison of the free energies of activation for the addition of an electrophile to an alkyne and to an alkene. The $\Delta G^{\ddagger}$ for the reaction of an alkyne is greater than that for the reaction of an alkene, indicating that an alkyne is less reactive than an alkene.

Why is the transition state for the addition of an electrophile such as HCl to an alkyne less stable than that for the addition of an electrophile to an alkene?

The energy of the transition state is closer to that of the product than to that of the reactant. Therefore, the structure of the transition state resembles the structure of the product (see the Hammond postulate). The product is a positively charged intermediate—namely, *an alkyl cation* when a proton adds to an alkene and a *vinylic cation* when a proton adds to an alkyne. A **vinylic cation** has a positive charge on a vinylic carbon.

$$RCH\!=\!CH_2 \ + \ H\!-\!Cl \ \longrightarrow \ \overset{+}{R}CH\!-\!CH_3 \ + \ Cl^- \qquad RC\!\equiv\!CH \ + \ H\!-\!Cl \ \longrightarrow \ R\overset{+}{C}\!=\!CH_2 \ + \ Cl^-$$

an alkyl cation **a vinylic cation**

A vinylic cation is less stable than a *similarly substituted* alkyl cation because a vinylic cation has a positive charge on an *sp* carbon. An *sp* carbon is more electronegative than the sp^2 carbon of an alkyl cation and is, therefore, less able to bear a positive charge. By "similarly substituted," we mean that a primary vinylic cation is less stable than a primary alkyl cation, and a secondary vinylic cation is less stable than a secondary alkyl cation.

relative stabilities of carbocations

$$\boxed{\text{most stable}} \rightarrow \ \overset{R}{\underset{R}{R\!-\!\overset{+}{C}}} \ > \ \overset{R}{\underset{H}{R\!-\!\overset{+}{C}}} \ > \ RCH\!=\!\overset{+}{C}\!-\!R \ \approx \ \overset{H}{\underset{H}{R\!-\!\overset{+}{C}}} \ > \ RCH\!=\!\overset{+}{C}\!-\!H \ \approx \ \overset{H}{\underset{H}{H\!-\!\overset{+}{C}}} \ \leftarrow \boxed{\text{least stable}}$$

| a tertiary carbocation | a secondary carbocation | a secondary vinylic cation | a primary carbocation | a primary vinylic cation | the methyl cation |

A primary carbocation is too unstable to form. Since a primary vinylic cation would be even less stable, it not expected to form, either. Some chemists think that the intermediate formed when a proton adds to an alkyne is a **π-complex** rather than a vinylic cation.

a π-complex

Support for the intermediate existing as a π-complex comes from the observation that many (but not all) alkyne addition reactions are stereoselective. For example, the following reaction forms only (Z)-2-chloro-2-butene, which means that only anti addition of H and Cl occurs.

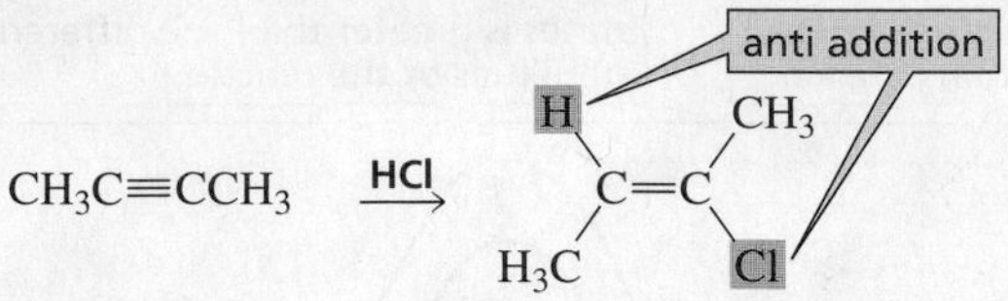

$$CH_3C\equiv CCH_3 \xrightarrow{HCl}$$

anti addition

2-butyne   (Z)-2-chloro-2-butene

Clearly, the mechanism of the addition reaction is not completely understood. For now, then, we will assume that the reaction forms a π-complex. It is more stable than a vinylic cation would be, but it is not as stable as an alkyl cation. Therefore, the transition state leading to its formation is less stable than the transition state leading to formation of an alkyl cation, agreeing with the observation that alkynes are less reactive than alkenes (Figure 2).

PROBLEM 11 Solved

Under what circumstances can you assume that the less stable of two compounds will be the more reactive compound?

Solution For the less stable compound to be the more reactive compound, the less stable compound must have the more stable transition state, or the difference in the stabilities of the reactants must be greater than the difference in the stabilities of the transition states.

---

6   THE ADDITION OF HYDROGEN HALIDES AND THE ADDITION OF HALOGENS TO AN ALKYNE

The product of the electrophilic addition reaction of an alkyne with HCl is an alkene. Therefore, a second addition reaction can occur if excess hydrogen halide is present. The second addition—like other addition reactions to alkenes—is regioselective: the H^+ adds to the *sp* carbon that is bonded to the most hydrogens.

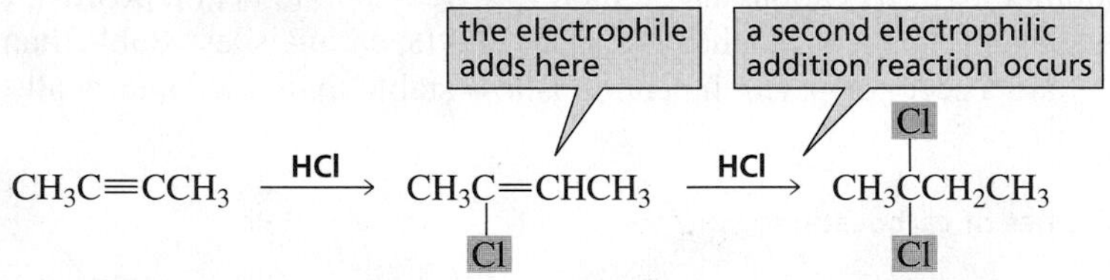

the electrophile adds here | a second electrophilic addition reaction occurs

$$CH_3C\equiv CCH_3 \xrightarrow{HCl} CH_3C=CHCH_3 \xrightarrow{HCl} CH_3CCH_2CH_3$$

Cl   Cl

a geminal
dihalide

The product of the second addition reaction is a **geminal dihalide**, a molecule with two halogens on the same carbon. "Geminal" comes from *geminus,* which is Latin for "twin."

If the alkyne is a *terminal* alkyne, the first electrophilic addition reaction is also regioselective: the H^+ adds to the less substituted *sp* carbon (that is, the one bonded to the hydrogen).

the electrophile adds here

The electrophile adds to the *sp* carbon that is bonded to the hydrogen.

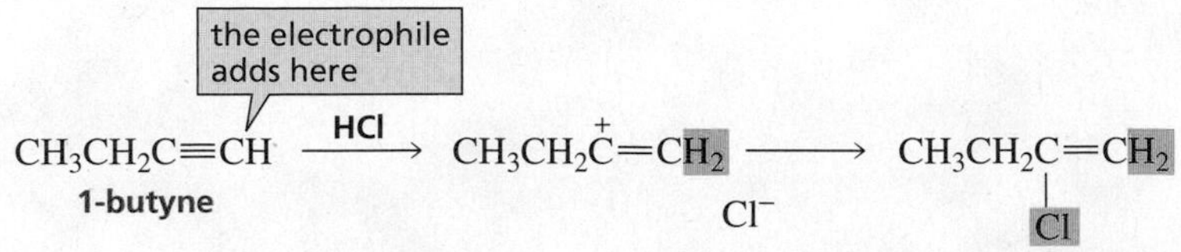

$$CH_3CH_2C\equiv CH \xrightarrow{HCl} CH_3CH_2\overset{+}{C}=CH_2 \longrightarrow CH_3CH_2C=CH_2$$

1-butyne Cl^- Cl

2-bromo-1-butene
a halo-substituted alkene

The mechanism for the addition of a hydrogen halide to an alkyne is similar to the mechanism for the addition of a hydrogen halide to an alkene. The only difference is the intermediate: an alkyne forms a π-complex, whereas an alkene forms a carbocation.

MECHANISM FOR ELECTROPHILIC ADDITION OF A HYDROGEN HALIDE TO AN ALKYNE

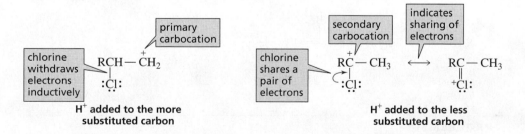

- The alkyne (a nucleophile) reacts with an electrophile to form a π-complex.
- Chloride ion adds to the intermediate, forming a halo-substituted alkene.

The regioselectivity of electrophilic addition to an alkyne can be explained easily. Of the two possible transition states for the reaction, the one with a partial positive charge on the more substituted (secondary) carbon is more stable.

To see why the second electrophilic addition reaction (the addition to the halo-substituted alkene) is also regioselective, let's look at the two possible carbocation intermediates for that reaction.

If H^+ adds to the more substituted sp^2 carbon, a *primary carbocation* would be formed; the chlorine atom further *decreases the stability* of the carbocation by withdrawing electrons inductively through the σ bond, which increases the concentration of positive charge on the carbon.

$$RC=CH_2 + HCl \longrightarrow RC-CH_3$$

a halo-substituted alkene

In contrast, adding H^+ to the less substituted sp^2 carbon (the one bonded to the most hydrogens) forms a more stable *secondary carbon*. In addition, the chlorine atom *stabilizes* the carbocation by donating a share in a lone pair to the positively charged carbon. In this way, the positive charge is shared by carbon and chlorine (Figure 3).

The addition of a hydrogen halide to an alkyne can be stopped after the addition of one equivalent of hydrogen halide because, although an alkyne is less reactive than an alkene, an alkyne is more reactive than the halo-substituted alkene that is the reactant for the second addition reaction.

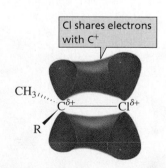

▲ **Figure 3**

Chlorine shares the positive charge with carbon by overlapping one of its orbitals that contains a lone pair with the empty *p* orbital of the positively charged carbon.

relative reactivity

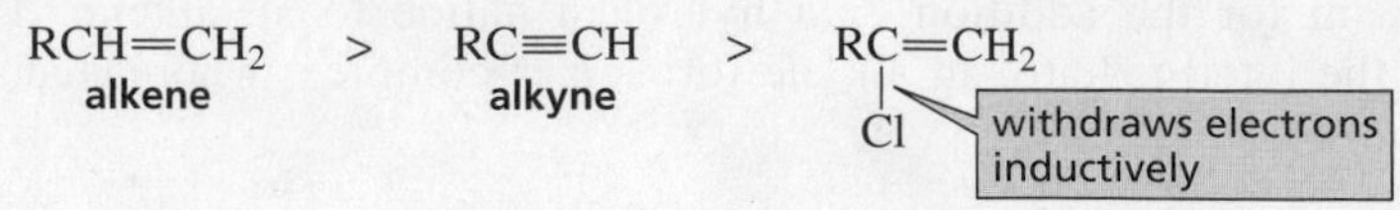

$$RCH=CH_2 \quad > \quad RC\equiv CH \quad > \quad RC=CH_2$$

alkene   alkyne

Cl — withdraws electrons inductively

halo-substituted alkene

The halo-substituted alkene is less reactive than an alkyne because the halo-substituent withdraws electrons inductively (through the σ bond), thereby decreasing the nucleophilic character of the double bond.

Addition of excess hydrogen halide to an *internal* alkyne forms two geminal dihalides, because the initial addition of H^+ can occur with equal ease to either of the *sp* carbons.

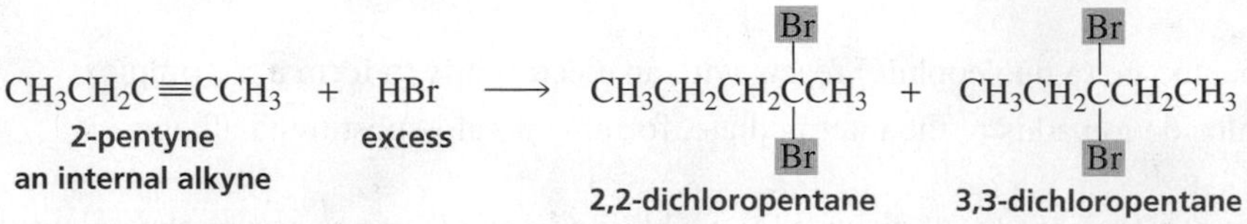

$$CH_3CH_2C\equiv CCH_3 + HBr \longrightarrow CH_3CH_2CH_2\overset{Br}{\underset{Br}{C}}CH_3 + CH_3CH_2\overset{Br}{\underset{Br}{C}}CH_2CH_3$$

2-pentyne   excess

an internal alkyne   2,2-dichloropentane   3,3-dichloropentane

Notice, however, that if the internal alkyne is symmetrical, only one geminal dihalide is formed.

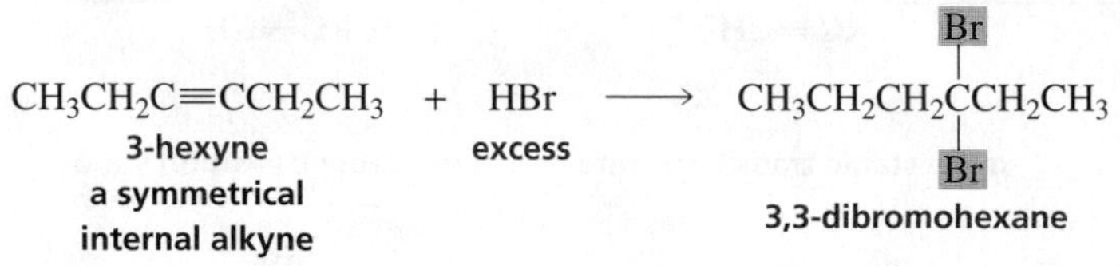

$$CH_3CH_2C\equiv CCH_2CH_3 + HBr \longrightarrow CH_3CH_2CH_2\overset{Br}{\underset{Br}{C}}CH_2CH_3$$

3-hexyne   excess

a symmetrical   3,3-dibromohexane

internal alkyne

The halogens Cl_2 and Br_2 also add to alkynes. In the presence of excess halogen, a second addition reaction occurs. The mechanism of the reaction is exactly the same as the mechanism for the addition of Cl_2 or Br_2 to an alkene.

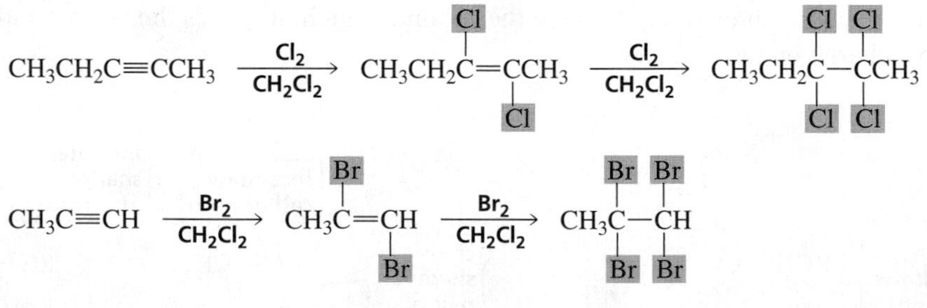

$$CH_3CH_2C\equiv CCH_3 \xrightarrow[CH_2Cl_2]{Cl_2} CH_3CH_2\overset{Cl}{C}=\overset{}{\underset{Cl}{C}}CH_3 \xrightarrow[CH_2Cl_2]{Cl_2} CH_3CH_2\overset{Cl}{\underset{Cl}{C}}-\overset{Cl}{\underset{Cl}{C}}CH_3$$

$$CH_3C\equiv CH \xrightarrow[CH_2Cl_2]{Br_2} CH_3\overset{Br}{C}=\overset{}{\underset{Br}{C}}H \xrightarrow[CH_2Cl_2]{Br_2} CH_3\overset{Br}{\underset{Br}{C}}-\overset{Br}{\underset{Br}{C}}H$$

PROBLEM 12◆

What is the major product of each of the following reactions?

a. $HC\equiv CCH_3 \xrightarrow{HBr}$

b. $HC\equiv CCH_3 \xrightarrow[HBr]{excess}$

c. $CH_3C\equiv CCH_3 \xrightarrow[CH_2Cl_2]{Br_2}$

d. $HC\equiv CCH_3 \xrightarrow[CH_2Cl_2]{excess\ Br_2}$

e. $CH_3C\equiv CCH_3 \xrightarrow[HBr]{excess}$

f. $CH_3C\equiv CCH_2CH_3 \xrightarrow[HBr]{excess}$

PROBLEM 13

Drawing on what you know about the stereochemistry of alkene addition reactions:

a. write the mechanism for the reaction of 2-butyne with one equivalent of Br_2.
b. predict the configuration of the product of the reaction.

7   THE ADDITION OF WATER TO AN ALKYNE

Alkenes undergo the acid-catalyzed addition of water. The product of the electrophilic addition reaction is an alcohol.

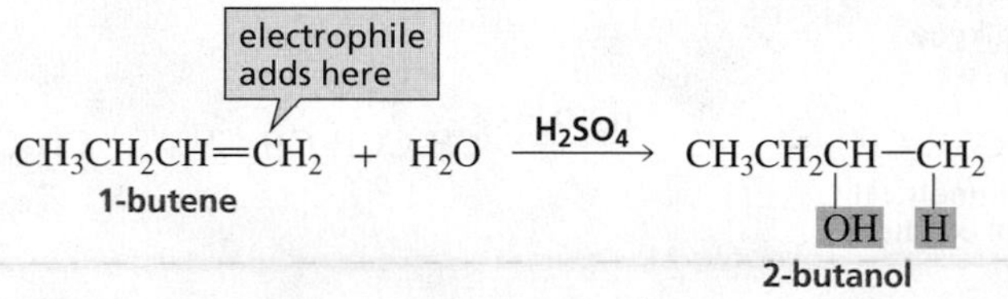

$$CH_3CH_2CH{=}CH_2 \ + \ H_2O \ \xrightarrow{H_2SO_4} \ CH_3CH_2CH{-}CH_2$$

electrophile adds here

1-butene

OH H

2-butanol

Alkynes also undergo the acid-catalyzed addition of water.

electrophile adds here

$$CH_3CH_2C{\equiv}CH \ + \ H_2O \ \xrightarrow{H_2SO_4} \ CH_3CH_2\overset{OH}{\underset{}{C{=}CH_2}} \ \rightleftharpoons \ CH_3CH_2\overset{O}{\underset{}{C{-}CH_3}}$$

an enol   a ketone

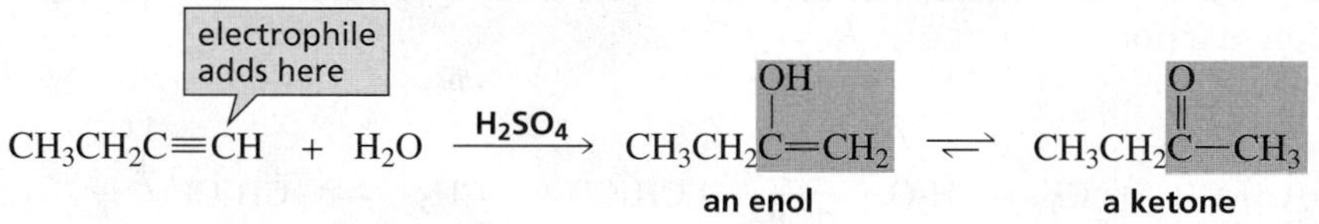

The initial product of the reaction is an *enol*. An **enol** has a carbon–carbon double bond with an OH group bonded to one of the sp^2 carbons. (The suffix "ene" signifies the double bond, and "ol" signifies the OH group. When the two suffixes are joined, the second *e* of "ene" is dropped to avoid two consecutive vowels, but the word is pronounced as if the second *e* were still there: "ene-ol.") The enol immediately rearranges to a *ketone*.

A ketone and an enol differ only in the location of a double bond and a hydrogen. A ketone and its corresponding enol are called **keto–enol tautomers. Tautomers** ("taw-toe-mers") are constitutional isomers that are in rapid equilibrium. The keto tautomer predominates in solution, because it is usually much more stable than the enol tautomer. Interconversion of the tautomers is called **keto–enol interconversion** or **tautomerization.**

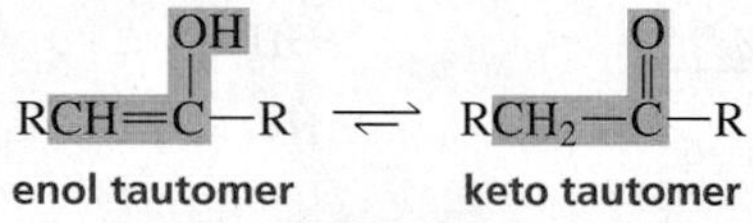

$$RCH{=}\overset{OH}{\underset{}{C}}{-}R \ \rightleftharpoons \ RCH_2{-}\overset{O}{\underset{}{C}}{-}R$$

enol tautomer   keto tautomer

tautomerization

The mechanism for the conversion of an enol to a ketone under acidic conditions is shown next. In Section 8, we will see that the reaction can also be catalyzed by bases.

MECHANISM FOR ACID-CATALYZED KETO–ENOL INTERCONVERSION

$$RCH{=}\overset{:\ddot{O}{-}H}{\underset{}{C}}{-}R \ \rightleftharpoons \ RCH_2{-}\overset{+}{\underset{}{\overset{\ddot{O}}{C}}}{-}R \ \rightleftharpoons \ RCH_2{-}\overset{O}{\underset{}{C}}{-}R \ + \ H_3O^+$$

enol

ketone

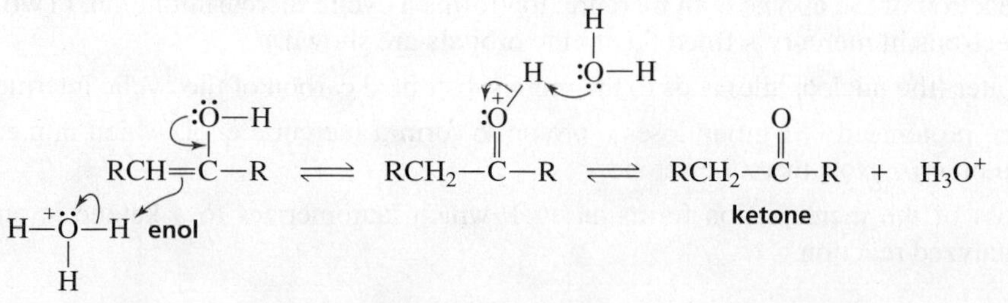

- A π bond forms between carbon and oxygen and, as the π bond between the two carbons breaks, carbon picks up a proton.
- Water removes a proton from the protonated carbonyl group.

The addition of water to a symmetrical internal alkyne forms a single ketone as a product. But if the alkyne is not symmetrical, then two ketones are formed because the initial addition of the proton can occur to either of the *sp* carbons.

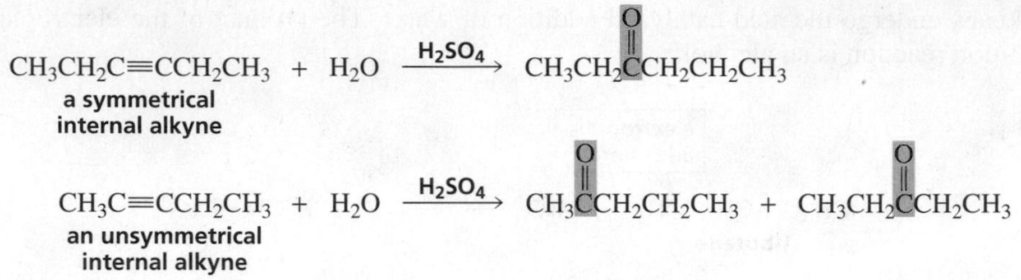

$$CH_3CH_2C{\equiv}CCH_2CH_3 \ + \ H_2O \ \xrightarrow{\text{H}_2\text{SO}_4} \ CH_3CH_2\overset{\overset{\displaystyle O}{\|}}{C}CH_2CH_2CH_3$$

a symmetrical internal alkyne

$$CH_3C{\equiv}CCH_2CH_3 \ + \ H_2O \ \xrightarrow{\text{H}_2\text{SO}_4} \ CH_3\overset{\overset{\displaystyle O}{\|}}{C}CH_2CH_2CH_3 \ + \ CH_3CH_2\overset{\overset{\displaystyle O}{\|}}{C}CH_2CH_3$$

an unsymmetrical internal alkyne

Terminal alkynes are less reactive than internal alkynes toward the addition of water. The addition of water to a terminal alkyne will occur if mercuric ion (Hg^{2+}) is added to the acidic mixture. The mercuric ion is a catalyst—it increases the rate of the addition reaction.

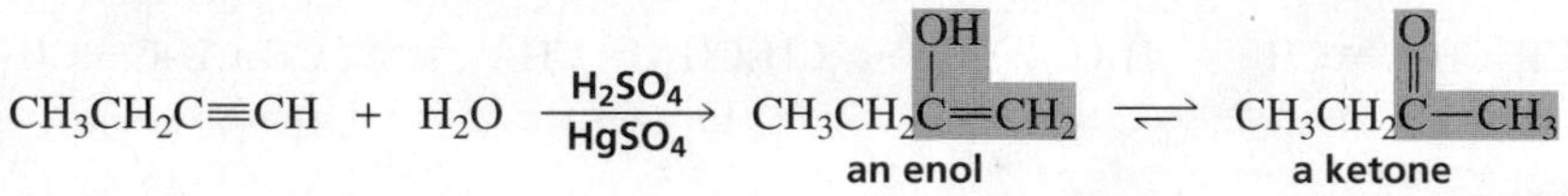

$$CH_3CH_2C{\equiv}CH \ + \ H_2O \ \xrightarrow[\text{HgSO}_4]{\text{H}_2\text{SO}_4} \ CH_3CH_2\overset{\overset{\displaystyle OH}{|}}{C}{=}CH_2 \ \rightleftharpoons \ CH_3CH_2\overset{\overset{\displaystyle O}{\|}}{C}{-}CH_3$$

an enol   **a ketone**

The mechanism for the *mercuric-ion-catalyzed hydration* of an alkyne is shown next. The intermediate formed in the first step should remind you of the cyclic bromonium ion formed as an intermediate in the addition of Br_2 to an alkene.

MECHANISM FOR THE MERCURIC-ION-CATALYZED HYDRATION OF AN ALKYNE

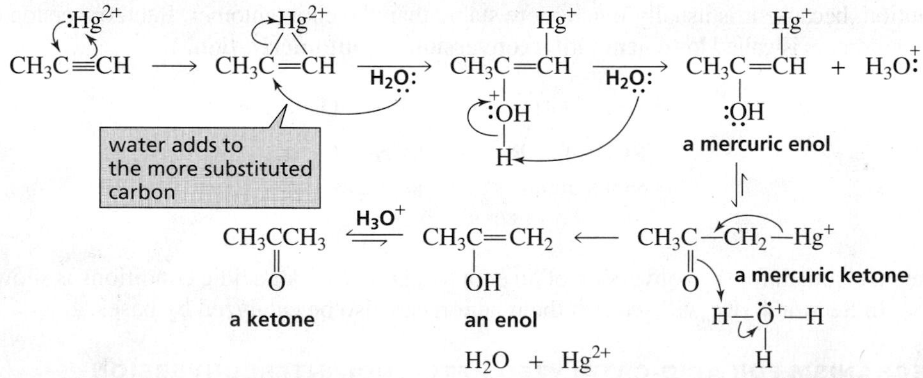

- Reaction of the alkyne with mercuric ion forms a cyclic mercurinium ion. (Two of the electrons in mercury's filled 5*d* atomic orbitals are shown.)
- Water (the nucleophile) adds to the more substituted carbon of the cyclic intermediate.
- The protonated OH group loses a proton to form a mercuric enol, which immediately tautomerizes to a mercuric ketone.
- Loss of the mercuric ion forms an enol, which tautomerizes to a ketone in an acid-catalyzed reaction.

PROBLEM 14♦

What ketones would be formed from the acid-catalyzed hydration of 3-heptyne?

PROBLEM 15♦

Which alkyne would be the best one to use for the synthesis of each of the following ketones?

          O   O   O

a. CH_3CCH_3 **b.** $CH_3CH_2CCH_2CH_2CH_3$ **c.** CH_3C—⬡

PROBLEM 16♦

Draw all the enol tautomers for each of the ketones in Problem 15.

---

8   THE ADDITION OF BORANE TO AN ALKYNE: HYDROBORATION–OXIDATION

BH_3 or R_2BH (in THF) adds to alkynes in the same way it adds to alkenes. That is, boron is the electrophile and H^- is the nucleophile. When the addition reaction is over, aqueous sodium hydroxide and hydrogen peroxide are added to the reaction mixture. The end result, as in the case of alkenes, is replacement of the boron by an OH group. The resulting enol immediately rearranges to a ketone.

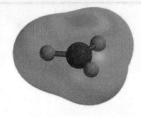

BH_3

$$CH_3C{\equiv}CCH_3 \; + \; R_2BH \; \xrightarrow{\text{THF}} \; \underset{\substack{\text{nucleophile} \\ \\ \text{boron-substituted alkene}}}{\overset{H_3C \qquad CH_3}{\underset{H \qquad \overset{|}{\underset{R}{B-R}}}{C{=}C}}} \; \xrightarrow[H_2O]{HO^-, H_2O_2} \; \underset{\text{an enol}}{\overset{H_3C \qquad CH_3}{\underset{H \qquad OH}{C{=}C}}} \; \rightleftharpoons \; \underset{\text{a ketone}}{CH_3CH_2\overset{O}{\overset{||}{C}}CH_3}$$

electrophile

A carbonyl compound will be the product of hydroboration–oxidation only if a second molecule of BH_3 or R_2BH does not add to the π-bond of the boron-substituted alkene. In the case of internal alkynes, the substituents on the boron-substituted alkene prevent the approach of the second boron-containing molecule. In the case of terminal alkynes, however, there is an H instead of a bulky alkyl group on the carbon that the second molecule adds to, so there is less steric hindrance toward the second addition reaction. Therefore, either BH_3 or R_2BH can be used with internal alkenes, but the more sterically hindered R_2BH should be used with terminal alkynes.

Boron, with its electron-seeking empty orbital, is the electrophile. When it reacts with a terminal alkyne, it, like other electrophiles, adds preferentially to the less substituted sp carbon (the one bonded to the hydrogen). Since the boron-containing group is subsequently replaced by an OH group, hydroboration–oxidation of a terminal alkyne forms an *aldehyde* (the carbonyl group is on the terminal carbon), whereas the mercuric-ion-catalyzed addition of water to a terminal alkyne forms a *ketone* (the carbonyl group is *not* on the terminal carbon).

9-BBN
9-borabicyclo[3.3.1]nonane
R_2BH

$$CH_3C{\equiv}CH \begin{array}{c} \xrightarrow[\text{2. } HO^-, H_2O_2, H_2O]{\text{1. } R_2BH/THF} \quad CH_3CH{=}\overset{OH}{\overset{|}{CH}} \rightleftharpoons CH_3CH_2\overset{O}{\overset{||}{CH}} \\ \\ \xrightarrow[\text{HgSO}_4]{H_2O, H_2SO_4} \quad CH_3\overset{OH}{\overset{|}{C}}{=}CH_2 \rightleftharpoons CH_3\overset{O}{\overset{||}{C}}CH_3 \end{array}$$

an aldehyde

a ketone

Hydroboration–oxidation of a terminal alkyne forms an aldehyde.

Acid catalyzed addition of water (in the presence of mercuric ion) to a terminal alkyne forms a ketone.

323

The mechanism for the conversion of an enol to a ketone under the basic conditions of the oxidation reaction that follows hydroboration is shown next.

MECHANISM FOR BASE-CATALYZED KETO–ENOL INTERCONVERSION

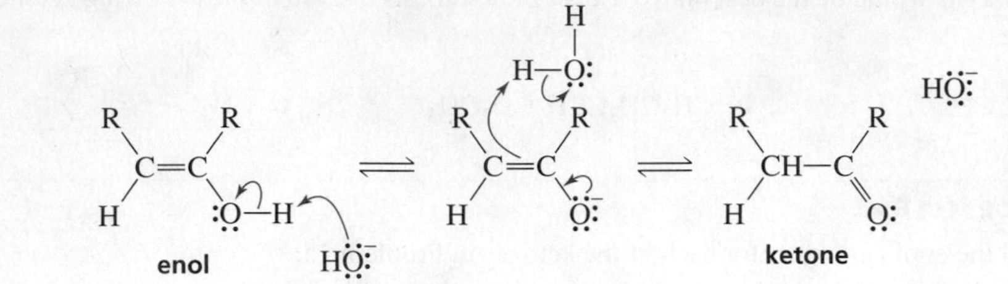

- A base removes a proton from the enol.
- A π bond forms between oxygen and carbon and, as the π bond between the two carbons breaks, carbon picks up a proton from water.

PROBLEM 17

For each of the following alkynes, draw the products of (1) the acid-catalyzed addition of water (mercuric ion is added for part a) and (2) hydroboration–oxidation:

a. 1-butyne   **b.** 2-butyne   **c.** 2-pentyne

PROBLEM 18♦

There is only one alkyne that forms an aldehyde when it undergoes the mercuric-ion-catalyzed addition of water. Identify the alkyne.

---

9   THE ADDITION OF HYDROGEN TO AN ALKYNE

Alkynes can be reduced by catalytic hydrogenation just as alkenes can. The initial product of hydrogenation is an alkene, but it is difficult to stop the reaction at this stage because of hydrogen's strong tendency to add to alkenes in the presence of these efficient metal catalysts. The final product of the hydrogenation reaction, therefore, is an alkane.

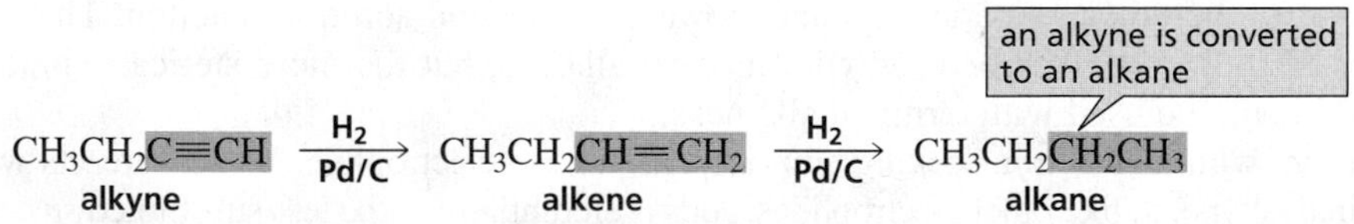

$(CH_3COO^-)_2Pb^{2+}$
lead(II) acetate

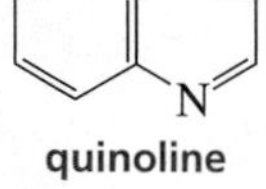

quinoline

an alkyne is converted to an alkane

$$CH_3CH_2C{\equiv}CH \xrightarrow[\text{Pd/C}]{H_2} CH_3CH_2CH{=}CH_2 \xrightarrow[\text{Pd/C}]{H_2} CH_3CH_2CH_2CH_3$$
alkyne   **alkene**   **alkane**

The reaction can be stopped at the alkene stage if a "poisoned" (partially deactivated) metal catalyst is used. The most common partially deactivated metal catalyst is known as Lindlar catalyst (Figure 4).

$$CH_3CH_2C{\equiv}CH + H_2 \xrightarrow[\text{catalyst}]{\text{Lindlar}} CH_3CH_2CH{=}CH_2$$

Because the alkyne sits on the surface of the metal catalyst and the hydrogens are delivered to the triple bond from the surface of the catalyst, both hydrogens are delivered to the same side of the double bond. In other words, syn addition of hydrogen occurs. Syn addition of H_2 to an internal alkyne forms a *cis alkene*.

▲ **Figure 4**
Lindlar catalyst is prepared by precipitating palladium on calcium carbonate and treating it with lead(II) acetate and quinoline. This treatment modifies the surface of palladium, making it much more effective at catalyzing the addition of hydrogen to a triple bond than to a double bond.

syn addition has occurred

$$CH_3CH_2C\equiv CCH_3 + H_2 \xrightarrow[\text{catalyst}]{\text{Lindlar}}$$

2-pentyne

cis-2-pentene

Internal alkynes can be converted to *trans alkenes* using sodium (or lithium) in liquid ammonia. The reaction stops at the alkene stage because sodium (or lithium) reacts more rapidly with triple bonds than with double bonds. This reaction is called a **dissolving metal reduction.** Ammonia is a gas at room temperature (bp = $-33\,°C$), so it is kept in the liquid state by cooling the reaction flask in a dry ice/acetone mixture, which has a temperature of $-78\,°C$.

$$CH_3C\equiv CCH_3 \xrightarrow[\substack{\text{NH}_3\text{(liq)} \\ -78\,°C}]{\text{Na or Li}}$$

2-butyne

trans-2-butene

As sodium dissolves in liquid ammonia, it forms a deep blue solution of dissolved Na^+ and electrons.

MECHANISM FOR THE CONVERSION OF AN ALKYNE TO A TRANS ALKENE

sodium gives up an *s* electron

sodium gives up an *s* electron

a strong base

$$CH_3-C\equiv C-CH_3 + Na\cdot \longrightarrow$$

a radical anion

a strong base

$+ Na^+$

a vinylic radical

$+\ ^-NH_2$

a vinylic anion

$+ Na^+$

a trans alkene

$+\ ^-NH_2$

The steps in the mechanism for the conversion of an internal alkyne to a trans alkene are:

- The single electron from the *s* orbital of sodium is transferred to an *sp* carbon of the alkyne. This forms a **radical anion**—a species with a negative charge and an unpaired electron.

  Notice that the movement of a single electron is represented by an arrowhead with a single barb. (Note that sodium has a strong tendency to lose the single electron in its outer-shell *s* orbital.)

- The radical anion is such a strong base that it can remove a proton from ammonia. This results in the formation of a **vinylic radical**—the unpaired electron is on a vinylic carbon.

- Another single-electron transfer from sodium to the vinylic radical forms a **vinylic anion.**

- The vinylic anion is also a strong base; it removes a proton from another molecule of ammonia, forming the trans alkene.

Both the radical anion and the vinylic anion can have either the cis or the trans configuration. The cis and trans radical anions are in equilibrium, and so are the cis and trans vinylic anions. In each case, the equilibrium favors the trans isomer. The greater stability of the trans isomers is what causes the product to be a trans alkene (Figure 5).

An arrowhead with a double barb signifies the movement of two electrons.

An arrowhead with a single barb signifies the movement of one electron.

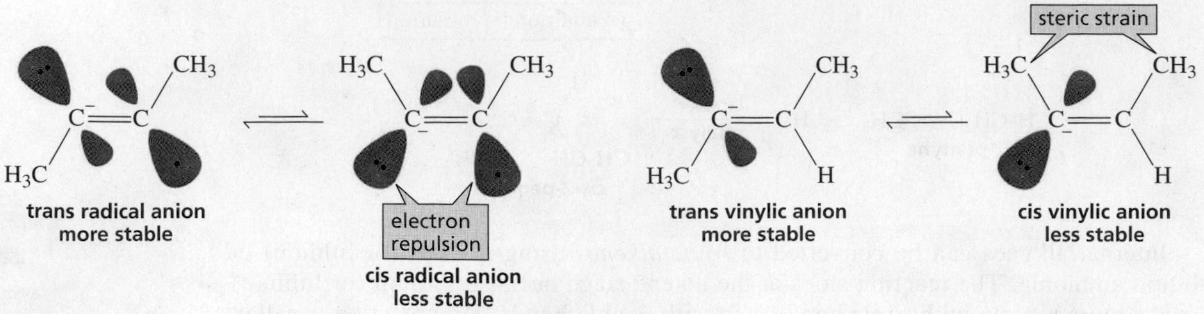

▲ Figure 5
The *trans radical anion* is more stable than the *cis radical anion,* because the nonbonded electrons are farther apart in the trans isomer. The *trans vinylic anion* is more stable than the *cis vinylic anion,* because the relatively bulky alkyl groups are farther apart in the trans isomer.

PROBLEM 19♦

Describe the alkyne you would start with and the reagents you would use if you wanted to synthesize

a. pentane. **b.** *cis*-2-butene. **c.** *trans*-2-pentene. **d.** 1-hexene.

PROBLEM 20

What are products of the following reactions?

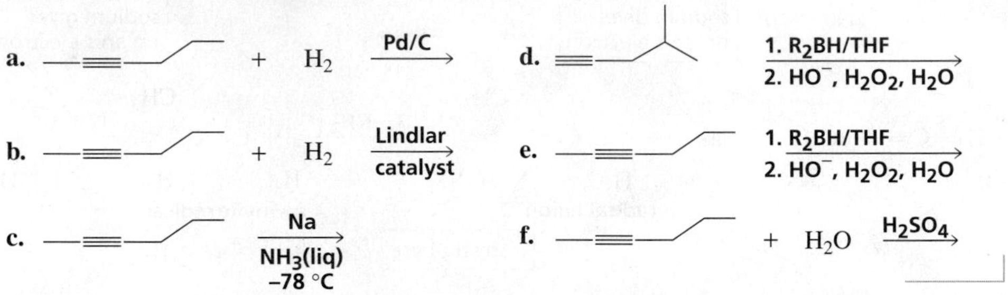

10 A HYDROGEN BONDED TO AN *sp* CARBON IS "ACIDIC"

An *sp* carbon is more electronegative than an sp^2 carbon, which is more electronegative than an sp^3 carbon.

relative electronegativities of carbon atoms

most electronegative → $sp > sp^2 > sp^3$ ← least electronegative

Because the most acidic compound is the one with its hydrogen attached to the most electronegative atom (when the atoms are the same size), ethyne is a stronger acid than ethene, and ethene is a stronger acid than ethane.

An *sp* carbon is more electronegative than an sp^2 carbon, which is more electronegative than an sp^3 carbon.

HC≡CH	$H_2C{=}CH_2$	CH_3CH_3
ethyne	**ethene**	**ethane**
$pK_a = 25$	$pK_a = 44$	$pK_a > 60$

Also, if we want to remove a proton from an acid (in a reaction that strongly favors products), we must use a base that is stronger than the base that is formed when the proton is removed.

For example, $^-NH_2$ is a stronger base than the acetylide ion that is formed, because NH_3 ($pK_a = 36$) is a weaker acid than a terminal alkyne ($pK_a = 25$). Recall that the weaker acid has the stronger conjugate base. Therefore, an amide ion ($^-NH_2$) can be used to remove a proton from a terminal alkyne to prepare an **acetylide ion.**

$$RC{\equiv}CH \ + \ ^-NH_2 \ \rightleftharpoons \ RC{\equiv}C^- \ + \ NH_3$$

	amide ion		acetylide ion		NH₃
stronger acid	stronger base		weaker base		weaker acid

The stronger the acid, the weaker its conjugate base.

In contrast, if hydroxide ion were used as the base, the reaction would strongly favor reactants because hydroxide ion is a much weaker base than the acetylide ion that would be formed.

$$RC{\equiv}CH \ + \ HO^- \ \rightleftharpoons \ RC{\equiv}C^- \ + \ H_2O$$

weaker acid · · · weaker base · · · stronger base · · · stronger acid

To remove a proton from an acid in a reaction that favors products, the base that removes the proton must be stronger than the base that is formed.

An amide ion ($^-NH_2$) cannot remove a hydrogen bonded to an sp^2 or an sp^3 carbon. Only a hydrogen bonded to an sp carbon is sufficiently acidic to be removed by an amide ion. Consequently, a hydrogen bonded to an sp carbon sometimes is referred to as an "acidic" hydrogen. Be careful not to misinterpret what is meant when we say that a hydrogen bonded to an sp carbon is "acidic." It is more acidic than most other carbon-bound hydrogens, but it is much less acidic ($pK_a = 25$) than a hydrogen of a water molecule, and we know that water is only a very weakly acidic compound ($pK_a = 15.7$).

relative acid strengths

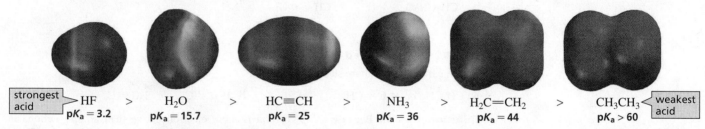

strongest acid	HF	>	H_2O	>	$HC{\equiv}CH$	>	NH_3	>	$H_2C{=}CH_2$	>	CH_3CH_3	weakest acid
	$pK_a = 3.2$		$pK_a = 15.7$		$pK_a = 25$		$pK_a = 36$		$pK_a = 44$		$pK_a > 60$	

Sodium Amide and Sodium in Ammonia

Take care not to confuse the compound sodium amide ($Na^+ \ ^-NH_2$), also called sodamide, with a mixture of sodium (Na) in liquid ammonia. Sodium amide is the strong base used to remove a proton from a terminal alkyne. Sodium in liquid ammonia is the source of electrons and protons, respectively, used to convert an internal alkyne to a trans alkene.

PROBLEM 21♦

Explain why sodium amide cannot be used to form a carbanion from an alkane in a reaction that favors products.

PROBLEM 22♦

Any base whose conjugate acid has a pK_a greater than _____ can remove a proton from a terminal alkyne to form an acetylide ion (in a reaction that favors products).

PROBLEM-SOLVING STRATEGY

Comparing the Acidities of Compounds

a. List the following compounds in order from strongest acid to weakest acid:

$$CH_3CH_2\overset{+}{N}H_3 \quad CH_3CH{=}\overset{+}{N}H_2 \quad CH_3C{\equiv}\overset{+}{N}H$$

To compare the acidities of a group of compounds, first look at how the compounds differ. The three compounds below differ in the hybridization of the nitrogen to which the acidic hydrogen is attached. Now, think about hybridization and electronegativity. For example, an sp hybridized atom is more electronegative than an sp^2 hybridized atom, which is more electronegative than an sp^3 hybridized atom. Also, the more electronegative the atom to which a hydrogen is attached, the more acidic the hydrogen. Now you can answer the question.

$$\text{relative acidities} \quad CH_3C\equiv\overset{+}{N}H \;>\; CH_3CH=\overset{+}{N}H_2 \;>\; CH_3CH_2\overset{+}{N}H_3$$

b. Draw the conjugate bases of these compounds and list them in order from strongest base to weakest base.

Removing a proton from each of the compounds provides their conjugate bases. The stronger the acid, the weaker its conjugate base, so we can use the relative acid strengths obtained in part **a** to determine that the order of decreasing basicity is as follows:

$$\text{relative basicities} \quad CH_3CH_2NH_2 \;>\; CH_3CH=NH \;>\; CH_3C\equiv N$$

Now use the strategy you have just learned to solve Problem 23.

PROBLEM 23♦

List the following in order from strongest base to weakest base:

a. $CH_3CH_2CH=\overset{-}{C}H$ $\qquad$ $CH_3CH_2C\equiv\overset{-}{C}$ $\qquad$ $CH_3CH_2CH_2\overset{-}{C}H_2$

b. $CH_3CH_2\overset{-}{O}$ $\qquad$ $\overset{-}{F}$ $\qquad$ $CH_3C\equiv\overset{-}{C}$ $\qquad$ $\overset{-}{N}H_2$

PROBLEM 24 Solved

Which carbocation is more stable?

a. $CH_3\overset{+}{C}H_2$ or $H_2C=\overset{+}{C}H$ $\qquad\qquad\qquad$ b. $H_2C=\overset{+}{C}H$ or $HC\equiv\overset{+}{C}$

Solution to 24a Because an sp^2 carbon is more electronegative than an sp^3 carbon, an sp^2 carbon with a positive charge is less stable than an sp^3 carbon with a positive charge. Thus, the ethyl carbocation is more stable.

---

11 SYNTHESIS USING ACETYLIDE IONS

Reactions that form carbon–carbon bonds are important in the synthesis of organic compounds because, without such reactions, we could not convert compounds with small carbon skeletons into compounds with larger carbon skeletons.

One reaction that forms a carbon–carbon bond is the reaction of an acetylide ion with an alkyl halide. Only primary alkyl halides or methyl halides should be used in this reaction.

an alkylation reaction puts an alkyl group on a reactant

$$CH_3CH_2C\equiv\overset{-}{C} \;+\; CH_3CH_2CH_2Br \;\longrightarrow\; CH_3CH_2C\equiv CCH_2CH_2CH_3 \;+\; Br^-$$
$$\text{3-heptyne}$$

The mechanism for this reaction is well understood. Bromine is more electronegative than carbon, and as a result, the electrons in the C—Br bond are not shared equally by the two atoms: there is a partial positive charge on carbon and a partial negative charge on bromine.

$$CH_3CH_2C\equiv\ddot{C}^- \ + \ CH_3CH_2CH_2\overset{\delta+}{-}\overset{\delta-}{Br} \ \longrightarrow \ CH_3CH_2C\equiv CCH_2CH_2CH_3 \ + \ Br^-$$

The negatively charged acetylide ion (a nucleophile) is attracted to the partially positively charged carbon (an electrophile) of the alkyl halide. As the electrons of the acetylide ion approach the carbon to form the new C—C bond, they push out the bromine and its bonding electrons because carbon can bond to no more than four atoms at a time.

The previous reaction is an example of an *alkylation reaction*. An **alkylation reaction** attaches an alkyl group to a species. Then you will understand why the reaction works best with primary alkyl halides and methyl halides.

We can convert terminal alkynes into internal alkynes of any desired chain length, simply by choosing an alkyl halide with the appropriate structure. Just count the number of carbons in the terminal alkyne and the number of carbons in the product to see how many carbons are needed in the alkyl halide.

$$CH_3CH_2CH_2C\equiv CH \ \xrightarrow{\ NaNH_2\ } \ CH_3CH_2CH_2C\equiv C^- \ \xrightarrow{\ CH_3CH_2Br\ } \ CH_3CH_2CH_2C\equiv CCH_2CH_3$$

1-pentyne   **3-heptyne**

PROBLEM 25 Solved

A chemist wants to synthesize 3-heptyne but cannot find any 1-pentyne, the starting material used in the synthesis just described. How else can 3-heptyne be synthesized?

Solution The *sp* carbons of 3-heptyne are bonded to an *ethyl* group and to a *propyl* group. Therefore, to synthesize 3-heptyne, the acetylide ion of 1-pentyne can react with an *ethyl halide* or the acetylide ion of 1-butyne can react with a *propyl halide*. Since 1-pentyne is not available, the chemist should use 1-butyne and a propyl halide.

$$CH_3CH_2C\equiv CH \ \xrightarrow[\text{2. } CH_3CH_2CH_2Cl]{\text{1. } NaNH_2} \ CH_3CH_2C\equiv CCH_2CH_2CH_3$$

1-butyne   **3-heptyne**

(Note that the numbers 1 and 2 in front of the reagents above and below the reaction arrow indicate two sequential reactions; the second reagent is not added until reaction with the first reagent is completely over.)

---

12 AN INTRODUCTION TO MULTISTEP SYNTHESIS

DESIGNING A
SYNTHESIS:

For each reaction that has been discussed so far, we have seen *why* the reaction occurs, *how* it occurs, and the *products* that are formed. A good way for you to review these reactions is to design syntheses, because when you design a synthesis, you have to be able to recall many of the reactions you have learned. And learning how to design syntheses is a very important part of organic chemistry.

Synthetic chemists consider *time, cost,* and *yield* in designing syntheses. In the interest of time, a well-designed synthesis will consist of as few steps (sequential reactions) as possible, and each of those steps will be a reaction that is easy to carry out. If two chemists in a pharmaceutical company were each asked to prepare a new drug, and one synthesized the drug in three simple steps while the other used six difficult steps, which chemist do you think will get a promotion? The costs of the starting materials must also be taken into consideration. Moreover, each step in the synthesis should provide the greatest possible yield of the desired product. The more reactant needed to synthesize one gram of product, the more expensive the product is to produce. Sometimes, a synthesis involving several steps is preferred because the starting materials are inexpensive, the reactions are easy to carry out, and the yield of each step is high. Such a synthesis is better than one with fewer steps if those steps require expensive starting materials and consist of reactions that are more difficult to run or give lower yields. At this point in your chemical education, however, you are not yet familiar with the costs of different chemicals or the difficulties encountered in carrying out specific reactions. So, for the time being, when you design a synthesis, just focus on finding the route with the fewest steps.

The following examples will give you an idea of the type of thinking required to design a successful synthesis. Problems of this kind will appear repeatedly throughout the book, because solving them is both fun and is a good way to learn organic chemistry.

Example 1. Starting with 1-butyne, how could you make the ketone shown here? You can use any reagents you need.

$$\underset{\textbf{1-butyne}}{CH_3CH_2C{\equiv}CH} \xrightarrow{?} CH_3CH_2\overset{O}{\overset{\|}{C}}CH_2CH_2CH_3$$

Many chemists find that the easiest way to design a synthesis is to work backward. Instead of looking at the reactant and deciding how to do the first step of the synthesis, look at the product and decide how to do the last step.

The product of the synthesis is a ketone. Now you need to remember all the reactions that you have learned that form a ketone. (You may find it helpful to consult Appendix: "Summary of Methods Used to Synthesize a Particular Functional Group," which lists the methods that can be used to synthesize a particular functional group.) We will use the acid-catalyzed addition of water. (You also could use hydroboration–oxidation.) If the alkyne used in the reaction has identical substituents on both *sp* carbons, only one ketone will be obtained. Thus, 3-hexyne is the alkyne that should be used for the synthesis of the desired ketone.

As the number of reactions that you know increases, you may find it helpful to consult Appendix: "Summary of Methods Used to Synthesize a Particular Functional Group" when designing syntheses; it lists the methods that can be used to synthesize each functional group.

$$\underset{\textbf{3-hexyne}}{CH_3CH_2C{\equiv}CCH_2CH_3} \xrightarrow[H_2SO_4]{H_2O} CH_3CH_2\overset{OH}{\overset{|}{C}}{=}CHCH_2CH_3 \rightleftharpoons CH_3CH_2\overset{O}{\overset{\|}{C}}CH_2CH_2CH_3$$

3-Hexyne can be obtained from the starting material (1-butyne) by removing the proton from its *sp* carbon, followed by alkylation. To produce the desired six-carbon product, a two-carbon alkyl halide must be used in the alkylation reaction.

$$\underset{\textbf{1-butyne}}{CH_3CH_2C{\equiv}CH} \xrightarrow[\text{2. } CH_3CH_2Br]{\text{1. } NaNH_2} \underset{\textbf{3-hexyne}}{CH_3CH_2C{\equiv}CCH_2CH_3}$$

Designing a synthesis by working backward from product to reactant is not simply a technique taught to organic chemistry students. It is used so frequently by experienced synthetic chemists that it has been given a name: **retrosynthetic analysis.** Chemists use open arrows when they write retrosynthetic analyses, to indicate they are working backward. Typically, the reagents needed to carry out each step are not specified until the reaction is written in the forward direction. For example, the ketone synthesis just discussed can be arrived at by the following retrosynthetic analysis.

retrosynthetic analysis

$$CH_3CH_2\overset{O}{\overset{\|}{C}}CH_2CH_2CH_3 \Longrightarrow CH_3CH_2C{\equiv}CCH_2CH_3 \Longrightarrow CH_3CH_2C{\equiv}CH$$

Once the complete sequence of reactions has been worked out by retrosynthetic analysis, the synthetic scheme can be written by reversing the steps and including the reagents required for each step.

synthesis

$$CH_3CH_2C{\equiv}CH \xrightarrow[\text{2. } CH_3CH_2Br]{\text{1. } NaNH_2} CH_3CH_2C{\equiv}CCH_2CH_3 \xrightarrow[H_2SO_4]{H_2O} CH_3CH_2\overset{O}{\overset{\|}{C}}CH_2CH_2CH_3$$

Example 2. Starting with ethyne, how could you make 2-bromopentane?

$$HC\equiv CH \xrightarrow{\ ?\ } \underset{\underset{\text{2-bromopentane}}{\overset{\displaystyle |}{\underset{\text{Br}}{}}}}{CH_3CH_2CH_2CHCH_3}$$

ethyne

2-Bromopentane can be prepared from 1-pentene, which can be prepared from 1-pentyne. 1-Pentyne can be prepared from ethyne and an alkyl halide with three carbons.

retrosynthetic analysis

$$\underset{\overset{\displaystyle |}{Br}}{CH_3CH_2CH_2CHCH_3} \Longrightarrow CH_3CH_2CH_2CH{=}CH_2 \Longrightarrow CH_3CH_2CH_2C{\equiv}CH \Longrightarrow HC{\equiv}CH$$

Now we can write the synthetic scheme:

synthesis

$$HC{\equiv}CH \xrightarrow[\text{2. } CH_3CH_2CH_2Br]{\text{1. } NaNH_2} CH_3CH_2CH_2C{\equiv}CH \xrightarrow[\substack{\text{Lindlar} \\ \text{catalyst}}]{H_2} CH_3CH_2CH_2CH{=}CH_2 \xrightarrow{HBr} \underset{\overset{\displaystyle |}{Br}}{CH_3CH_2CH_2CHCH_3}$$

Example 3. How could 2,6-dimethylheptane be prepared from an alkyne and an alkyl halide? (The prime in R′ signifies that R and R′ can be different alkyl groups.)

$$RC{\equiv}CH + R'Br \xrightarrow{\ ?\ } \underset{\underset{\text{2,6-dimethylheptane}}{\overset{\displaystyle |}{CH_3} \qquad\qquad \overset{\displaystyle |}{CH_3}}}{CH_3CHCH_2CH_2CH_2CHCH_3}$$

2,6-Dimethyl-3-heptyne is the only alkyne that will form the desired alkane upon hydrogenation. (If you don't believe this, try to draw another one.) This alkyne can be prepared in two different ways: from the reaction of an acetylide ion with a primary alkyl halide (isobutyl bromide), or from the reaction of an acetylide ion with a secondary alkyl halide (isopropyl bromide).

retrosynthetic analysis

$$\underset{\overset{\displaystyle |}{CH_3} \qquad\quad \overset{\displaystyle |}{CH_3}}{CH_3CHCH_2CH_2CH_2CHCH_3} \Longrightarrow \underset{\underset{\text{2,6-dimethyl-3-heptyne}}{\overset{\displaystyle |}{CH_3} \qquad\quad \overset{\displaystyle |}{CH_3}}}{CH_3CHCH_2C{\equiv}CCHCH_3}$$

$$\underset{\underset{\text{isobutyl bromide}}{\overset{\displaystyle |}{CH_3}}}{CH_3CHCH_2Br} + \underset{\overset{\displaystyle |}{CH_3}}{HC{\equiv}CCHCH_3} \quad \text{or} \quad \underset{\underset{\text{isopropyl bromide}}{\overset{\displaystyle |}{CH_3}}}{CH_3CHBr} + \underset{\overset{\displaystyle |}{CH_3}}{HC{\equiv}CCH_2CHCH_3}$$

Because we know that the reaction of an acetylide ion with an alkyl halide works best with primary alkyl halides and methyl halides, we should choose the synthesis that requires a primary alkyl halide.

synthesis

$$\underset{\overset{\displaystyle |}{CH_3}}{CH_3CHC{\equiv}CH} \xrightarrow[\text{2. } CH_3CHCH_2Br]{\text{1. } NaNH_2 \atop \overset{\displaystyle |}{CH_3}} \underset{\overset{\displaystyle |}{CH_3} \qquad\quad \overset{\displaystyle |}{CH_3}}{CH_3CHC{\equiv}CCH_2CHCH_3} \xrightarrow{H_2 \atop Pd/C} \underset{\overset{\displaystyle |}{CH_3} \qquad\qquad \overset{\displaystyle |}{CH_3}}{CH_3CHCH_2CH_2CH_2CHCH_3}$$

331

Example 4. How could you carry out the following synthesis using the given starting material?

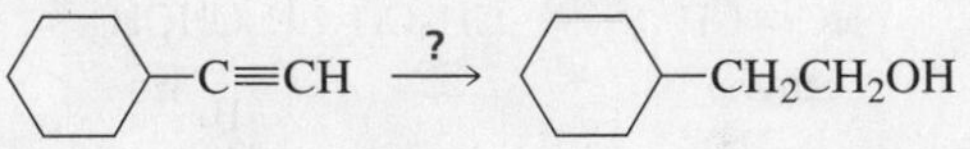

An alcohol can be prepared from an alkene, and an alkene can be prepared from an alkyne.

retrosynthetic analysis

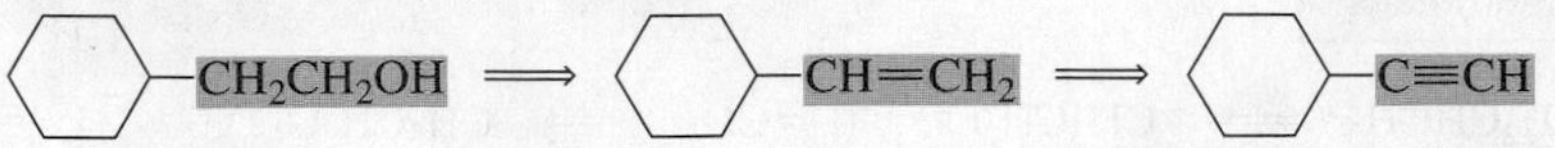

You can use either of the two methods that you have learned that convert an alkyne into an alkene, because the desired alkene does not have cis–trans isomers. Hydroboration–oxidation must be used to convert the alkene into the desired alcohol because the acid-catalyzed addition of water would form a different alcohol.

synthesis

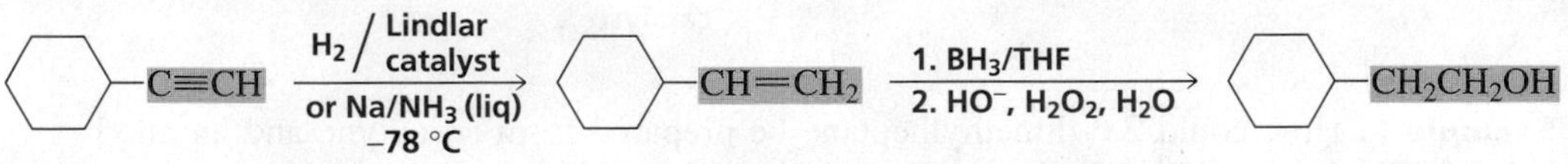

Example 5. How could you prepare (*E*)-2-pentene from ethyne?

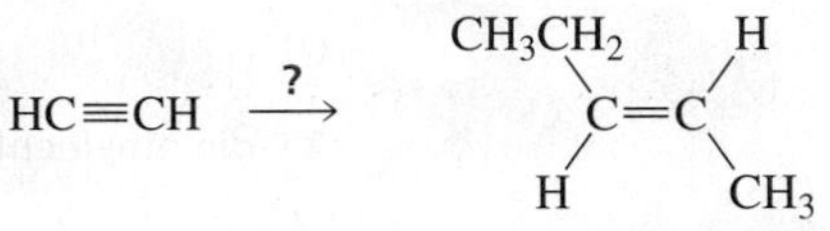

(*E*)-2-pentene

A trans alkene can be prepared from the reaction of an internal alkyne with sodium and liquid ammonia. The alkyne needed to synthesize the desired alkene can be prepared from 1-butyne and a methyl halide. 1-Butyne can be prepared from ethyne and an ethyl halide.

retrosynthetic analysis

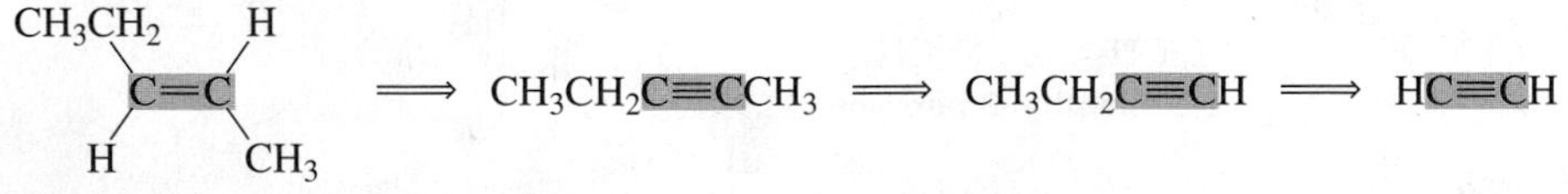

synthesis

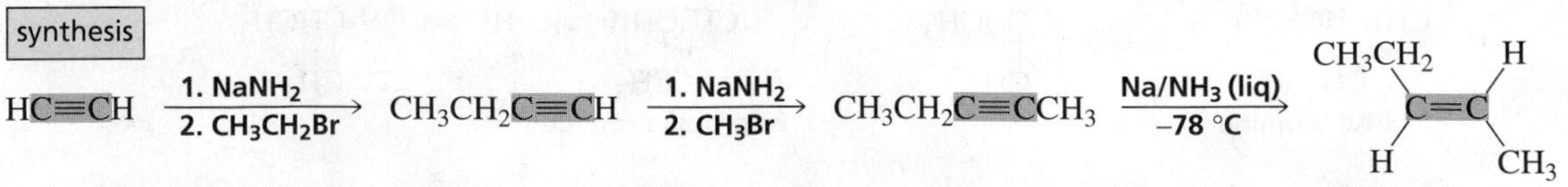

Example 6. How could you prepare *cis*-2,3-diethyloxirane from ethyne?

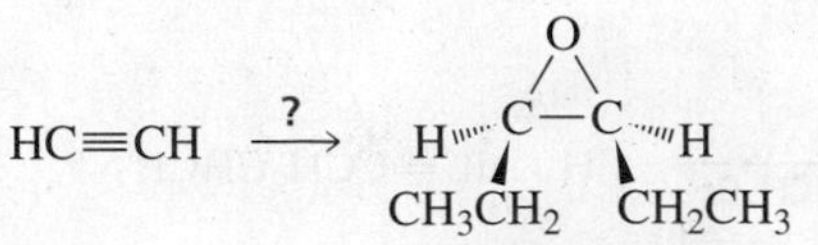

cis-2,3-diethyloxirane

A cis epoxide can be prepared from a cis alkene and a peroxyacid. The cis alkene that should be used is *cis*-3-hexene, which can be obtained from 3-hexyne. 3-Hexyne can be prepared from 1-butyne and an ethyl halide, and 1-butyne can be prepared from ethyne and an ethyl halide.

retrosynthetic analysis

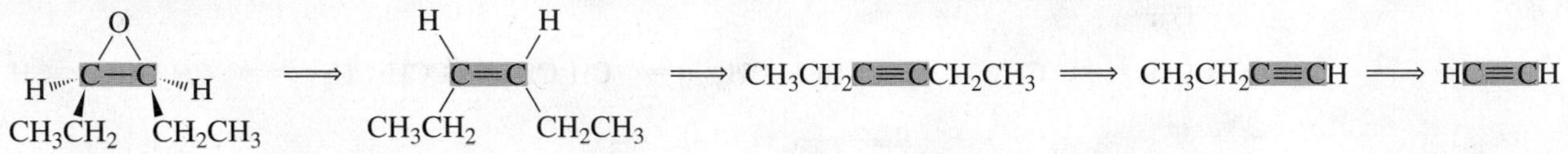
O
H
C—C
H
CH₃CH₂ CH₂CH₃
H H
C=C
CH₃CH₂ CH₂CH₃
CH₃CH₂C≡CCH₂CH₃
CH₃CH₂C≡CH
HC≡CH

The alkyne must be converted to the alkene using H_2 and Lindlar catalyst so that the desired cis alkene is obtained.

synthesis

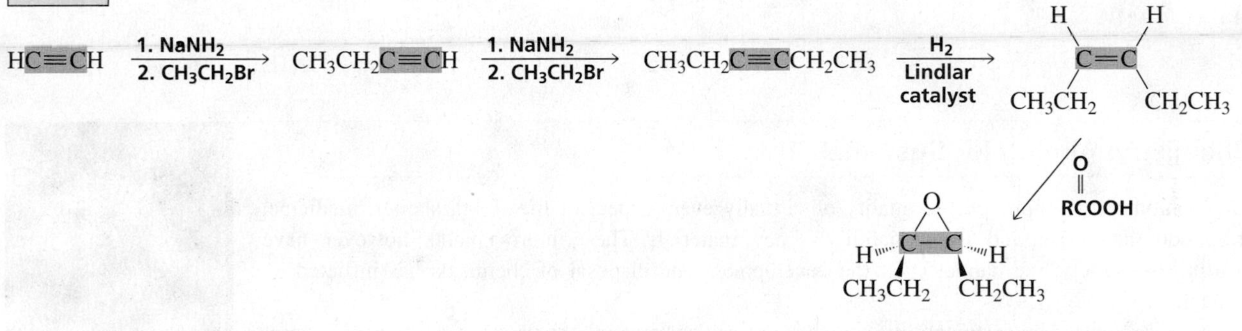
HC≡CH
1. NaNH₂
2. CH₃CH₂Br
CH₃CH₂C≡CH
1. NaNH₂
2. CH₃CH₂Br
CH₃CH₂C≡CCH₂CH₃
H₂
Lindlar
catalyst
H H
C=C
CH₃CH₂ CH₂CH₃
O
RCOOH
O
H
C—C
H
CH₃CH₂ CH₂CH₃

Example 7. How could you prepare 3,3-dibromohexane from reagents that contain no more than two carbons?

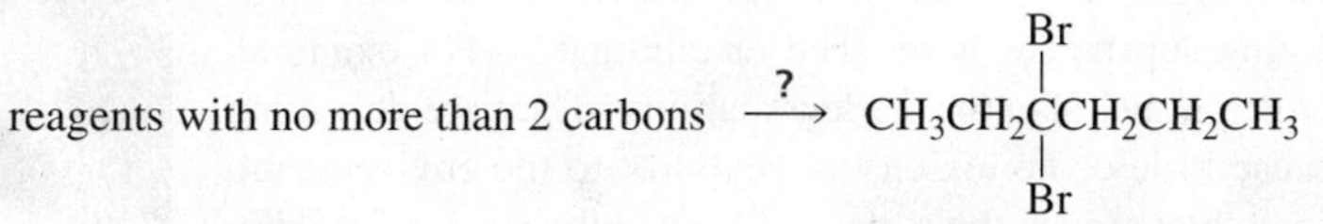
reagents with no more than 2 carbons
?
Br
CH₃CH₂CCH₂CH₂CH₃
Br

3,3-dibromohexane

A geminal dibromide can be prepared from an alkyne. 3-Hexyne is the alkyne of choice because it will form one geminal dibromide, whereas 2-hexyne would form two different geminal dibromides. 3-Hexyne can be prepared from 1-butyne and ethyl bromide, and 1-butyne can be prepared from ethyne and ethyl bromide.

retrosynthetic analysis

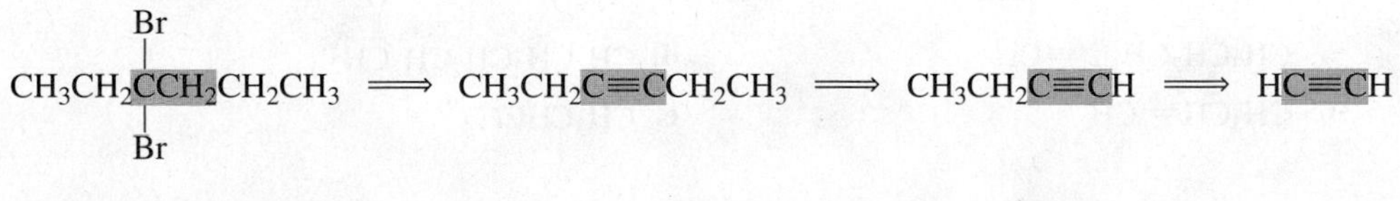
Br
CH₃CH₂CCH₂CH₂CH₃
Br
CH₃CH₂C≡CCH₂CH₃
CH₃CH₂C≡CH
HC≡CH

synthesis

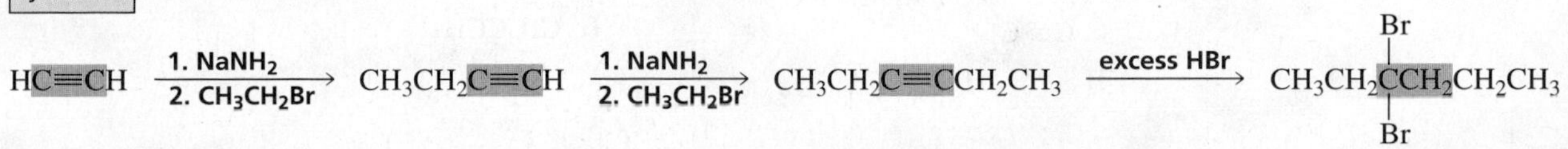
HC≡CH
1. NaNH₂
2. CH₃CH₂Br
CH₃CH₂C≡CH
1. NaNH₂
2. CH₃CH₂Br
CH₃CH₂C≡CCH₂CH₃
excess HBr
Br
CH₃CH₂CCH₂CH₂CH₃
Br

Example 8. How could you prepare a three-carbon aldehyde from 1-butyne?

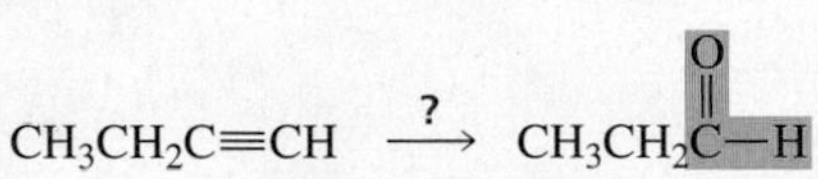
CH₃CH₂C≡CH
?
O
CH₃CH₂C—H

Because the desired compound has fewer carbons than the starting material, we know that a cleavage reaction must occur. The desired aldehyde can be prepared by ozonolysis (an oxidative cleavage) of 3-hexene. 3-Hexene can be prepared from 3-hexyne, which can be prepared from 1-butyne and ethyl bromide.

retrosynthetic analysis

$$\text{CH}_3\text{CH}_2\overset{\overset{\displaystyle O}{\|}}{\text{C}}\text{H} \implies \text{CH}_3\text{CH}_2\text{CH}=\text{CHCH}_2\text{CH}_3 \implies \text{CH}_3\text{CH}_2\text{C}\equiv\text{CCH}_2\text{CH}_3 \implies \text{CH}_3\text{CH}_2\text{C}\equiv\text{CH}$$

synthesis

$$\text{CH}_3\text{CH}_2\text{C}\equiv\text{CH} \xrightarrow[\text{2. CH}_3\text{CH}_2\text{Br}]{\text{1. NaNH}_2} \text{CH}_3\text{CH}_2\text{C}\equiv\text{CCH}_2\text{CH}_3 \xrightarrow[\substack{\text{Lindlar}\\ \text{catalyst}}]{\text{H}_2} \text{CH}_3\text{CH}_2\text{CH}=\text{CHCH}_2\text{CH}_3 \xrightarrow[\text{2. (CH}_3\text{)}_2\text{S}]{\text{1. O}_3,\, -78\,°\text{C}} \text{CH}_3\text{CH}_2\overset{\overset{\displaystyle O}{\|}}{\text{C}}\text{H}$$

Green Chemistry: Aiming for Sustainability

Chemical innovations have improved the quality of virtually every aspect of life: food, shelter, medicine, transportation, communication, and the availability of new materials. These improvements, however, have come with a price—namely, the damage that the development and disposal of chemicals has inflicted on the environment.

Chemists now are focused on sustainability, which is defined as "meeting the needs of the current generation without sacrificing the ability to meet the needs of future generations." One way to achieve sustainability is through the use of green chemistry.

Green chemistry is pollution prevention at the molecular level. It involves the design of chemical products and processes so that the generation of polluting substances is reduced or eliminated. For example, chemists are now creating products not only for function, but also for biodegradability. They are designing syntheses that use and generate substances that cause little or no toxicity to health or to the environment. Green chemical syntheses can be cost effective since they reduce the expense for such things as waste disposal, regulatory compliance, and liability. Applying the principles of green chemistry can help us achieve a sustainable future.

PROBLEM 26

How could the following compounds be synthesized from acetylene?

a. $\text{CH}_3\text{CH}_2\text{CH}_2\text{C}\equiv\text{CH}$

b. $\text{CH}_3\text{CH}=\text{CH}_2$

c.
$$\underset{\underset{\displaystyle H}{\big|}}{\overset{\overset{\displaystyle H_3C}{\big|}}{\text{C}}}=\underset{\underset{\displaystyle H}{\big|}}{\overset{\overset{\displaystyle CH_3}{\big|}}{\text{C}}}$$

d. $\text{CH}_3\text{CH}_2\text{CH}_2\text{CH}_2\overset{\overset{\displaystyle O}{\|}}{\text{C}}\text{H}$

e. $\text{CH}_3\underset{\underset{\displaystyle Br}{\big|}}{\text{C}}\text{HCH}_3$

f. $\text{CH}_3\underset{\underset{\displaystyle Cl}{\big|}}{\overset{\overset{\displaystyle Cl}{\big|}}{\text{C}}}\text{CH}_3$

SOME IMPORTANT THINGS TO REMEMBER

- An **alkyne** is a hydrocarbon that contains a carbon–carbon triple bond. The functional group suffix of an alkyne is "yne."

- A **terminal alkyne** has the triple bond at the end of the chain; an **internal alkyne** has the triple bond located elsewhere along the chain.

- Alkynes undergo electrophilic addition reactions. The same reagents that add to alkenes also add to alkynes.

- Alkynes are less reactive than alkenes.

- If excess reagent is available, alkynes undergo a second addition reaction with hydrogen halides and halogens because the product of the first reaction is an alkene.

- The product of the reaction of an alkyne with water under acidic conditions is an **enol,** which immediately rearranges to a ketone. Terminal alkynes require a mercuric ion catalyst.

- The ketone and enol are called **keto–enol tautomers;** they differ in the location of a double bond and a hydrogen. The keto tautomer usually predominates at equilibrium.

- Interconversion of the tautomers is called **tautomerization** or **keto–enol interconversion.**

- Hydroboration–oxidation of an internal alkyne forms a ketone; hydroboration–oxidation of a terminal alkyne forms an aldehyde. Either BH_3 or R_2BH can be used for internal alkynes; R_2BH should be used for terminal alkynes.

- Catalytic hydrogenation of an alkyne forms an alkane.

- Catalytic hydrogenation with Lindlar catalyst converts an internal alkyne to a *cis alkene.*

- Sodium in liquid ammonia converts an internal alkyne to a *trans alkene.*

- Electronegativity decreases in the order $sp > sp^2 > sp^3$, so ethyne is a stronger acid than ethene, and ethene is a stronger acid than ethane.

- An amide ion removes a proton from an *sp* carbon of a terminal alkyne to form an **acetylide ion.**

- An acetylide ion can undergo an alkylation reaction with a methyl halide or a primary alkyl halide to form an alkyne.

- An **alkylation reaction** attaches an alkyl group to a species.

- Designing a synthesis by working backward is called **retrosynthetic analysis.**

SUMMARY OF REACTIONS

1. Electrophilic addition reactions
   a. Addition of hydrogen halides (H^+ is the electrophile; Section 6).

$$RC\equiv CH \xrightarrow{\text{HX}} \underset{X}{RC}=CH_2 \xrightarrow{\text{excess HX}} \underset{X}{\overset{X}{RC}}-CH_3$$

   HX = HF, HCl, HBr, HI

   b. Addition of halogens (Section 6). The mechanism is the same as that for the reaction of alkenes with halogens.

$$RC\equiv CH \xrightarrow[\text{CH}_2\text{Cl}_2]{\text{Cl}_2} \underset{Cl}{\overset{Cl}{RC}}=CH \xrightarrow[\text{CH}_2\text{Cl}_2]{\text{Cl}_2} \underset{Cl\ Cl}{\overset{Cl\ Cl}{RC}}-CH$$

$$RC\equiv CR' \xrightarrow[\text{CH}_2\text{Cl}_2]{\text{Br}_2} \underset{Br}{\overset{Br}{RC}}=CR' \xrightarrow[\text{CH}_2\text{Cl}_2]{\text{Br}_2} \underset{Br\ Br}{\overset{Br\ Br}{RC}}-CR'$$

c. Acid-catalyzed addition of water and hydroboration–oxidation (Sections 7 and 8). Except for the formation of a π-complex in the acid-catalyzed reaction, the mechanisms are the same as those for the reaction of alkenes with these same reagents.

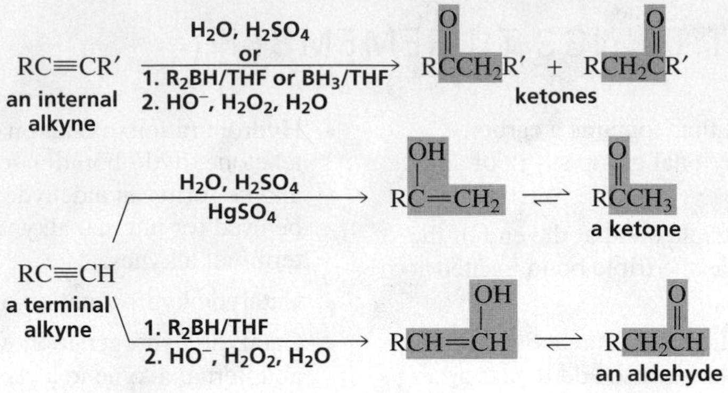

2. Addition of hydrogen (Section 9).

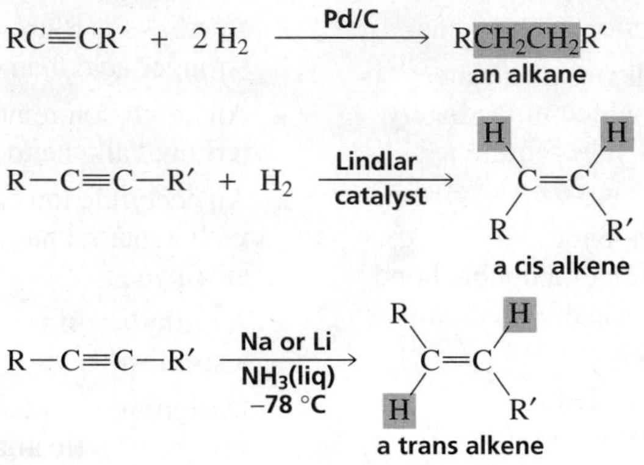

3. Removal of a proton from a terminal alkyne, followed by alkylation (Sections 10 and 11).

$$RC{\equiv}CH \xrightarrow{NaNH_2} RC{\equiv}C^- \xrightarrow{R'CH_2Br} RC{\equiv}CCH_2R'$$

PROBLEMS

27. What is the major product obtained from the reaction of each of the following compounds with excess HCl?

a. $CH_3CH_2C{\equiv}CH$ **b.** $CH_3CH_2C{\equiv}CCH_2CH_3$ **c.** $CH_3CH_2C{\equiv}CCH_2CH_2CH_3$

28. Draw a structure for each of the following:

a. 2-hexyne
b. 5-ethyl-3-octyne
c. methylacetylene
d. vinylacetylene
e. methoxyethyne
f. *sec*-butyl-*tert*-butylacetylene
g. 1-bromo-1-pentyne
h. 5-methyl-2-cyclohexenol
i. diethylacetylene
j. di-*tert*-butylacetylene
k. cyclopentylacetylene
l. 5,6-dimethyl-2-heptyne

29. A student was given the structural formulas of several compounds and was asked to give them systematic names. How many did she name correctly? Correct those that are misnamed.

a. 4-ethyl-2-pentyne **b.** 1-bromo-4-heptyne **c.** 2-methyl-3-hexyne **d.** 3-pentyne

30. Identify the electrophile and the nucleophile in each of the following reaction steps. Then draw curved arrows to illustrate the bond-making and bond-breaking processes.

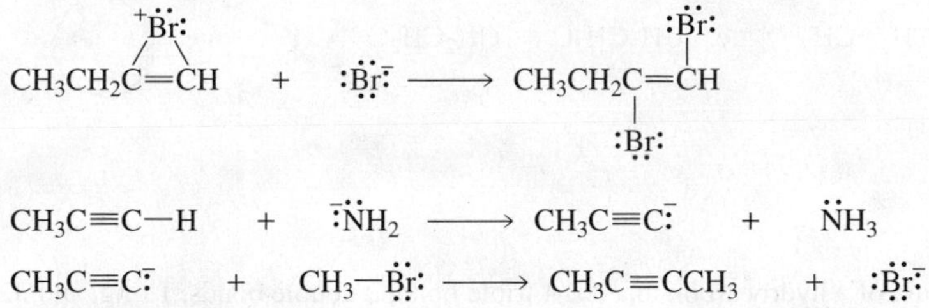

$$CH_3CH_2\overset{+\ddot{B}r:}{C}=CH \quad + \quad :\ddot{B}r:^- \quad \longrightarrow \quad CH_3CH_2\overset{:\ddot{B}r:}{\underset{:\ddot{B}r:}{C}}=CH$$

$$CH_3C\equiv C-H \quad + \quad :\ddot{N}H_2 \quad \longrightarrow \quad CH_3C\equiv C:^- \quad + \quad \ddot{N}H_3$$

$$CH_3C\equiv C:^- \quad + \quad CH_3-\ddot{B}r: \quad \longrightarrow \quad CH_3C\equiv CCH_3 \quad + \quad :\ddot{B}r:^-$$

31. What is each compound's systematic name?

a. $CH_3C\equiv CCH_2\overset{\underset{|}{Br}}{C}HCH_3$

b. $CH_3C\equiv CCH_2\overset{\underset{|}{CH_2CH_2CH_3}}{C}HCH_3$

c. $CH_3C\equiv CCH_2\overset{\overset{CH_3}{|}}{\underset{\underset{CH_3}{|}}{C}}CH_3$

d. $CH_3CHCH_2C\equiv CCHCH_3$ with Cl on first CH and CH_3 on last CH

e. 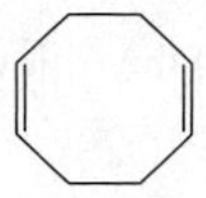 (cyclooctyne structure)

f. 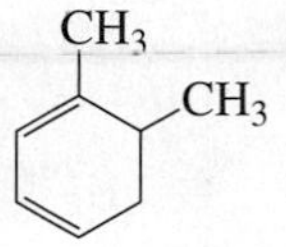 (substituted cyclohexene with two CH_3 groups)

32. What reagents could be used to carry out the following syntheses?

$$RCH_2CH_3$$
$$RCH=CH_2$$
$$R\overset{\underset{Br}{|}}{C}H=CH_2 \text{ (RCHCH}_3\text{, Br)}$$
$$R\overset{\underset{Br}{|}}{C}CH_3$$
$$RC=CH_2 \text{ (Br)}$$

$$RC\equiv CH$$

$$\underset{O}{\overset{\diagup \backslash}{RCH-CH_2}}$$
$$R\overset{\underset{Br}{|}}{C}HCH_3$$
$$R\overset{O}{\overset{||}{C}}CH_3$$
$$RCH_2\overset{O}{\overset{||}{C}}H$$

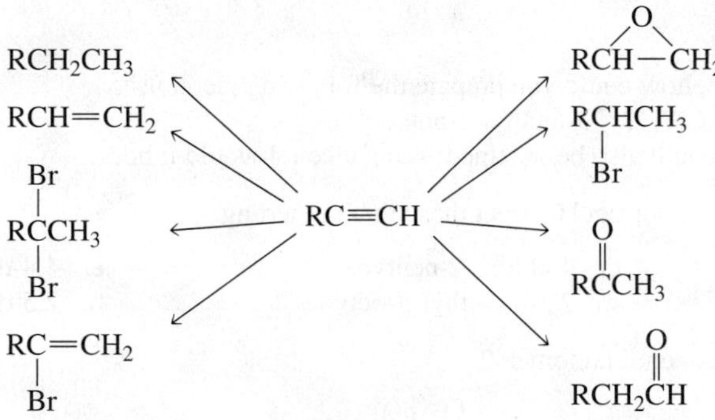

33. Draw the structures and give the common and systematic names for alkynes with molecular formula C_7H_{12}. (*Hint:* There are 14.)

34. Explain why the following reaction results in the product shown:

$$CH_3CH_2CH_2C\equiv CH \xrightarrow[\text{H}_2\text{O}]{\text{Br}_2} CH_3CH_2CH_2\overset{O}{\overset{||}{C}}CH_2Br$$

35. How could the following compounds be synthesized, starting with a hydrocarbon that has the same number of carbons as the desired product?

a. $CH_3CH_2CH_2CH_2\overset{O}{\overset{||}{C}}H$

b. $CH_3CH_2CH_2CH_2OH$

c. $CH_3CH_2CH_2\overset{O}{\overset{||}{C}}CH_2CH_2CH_2CH_3$

36. What reagents would you use for the following syntheses?

a. (*Z*)-3-hexene from 3-hexyne

b. (*E*)-3-hexene from 3-hexyne

c. hexane from 3-hexyne

37. What will be the major product of the reaction of 1 mol of propyne with each of the following reagents?

a. HBr (1 mol)
b. HBr (2 mol)
c. Br_2 (1 mol)/CH_2Cl_2
d. Br_2 (2 mol)/CH_2Cl_2

e. aqueous H_2SO_4, $HgSO_4$
f. R_2BH in THF followed by H_2O_2/HO^-/H_2O
g. excess H_2, Pd/C

h. H_2/Lindlar catalyst
i. sodium amide
j. the product of part **i** followed by 1-chloropropane

38. What is each compound's systematic name?

a. $CH_3C\equiv CCH_2CH_2CH_2CH=CH_2$ **c.** $CH_3CH_2C\equiv CCH_2CH_2C\equiv CH$ **e.**

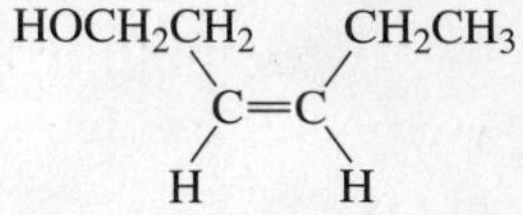

b.

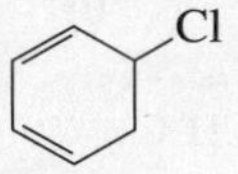

d.

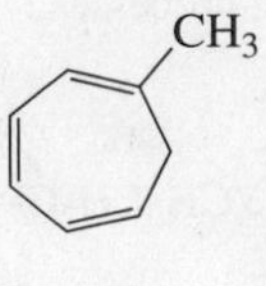

39. What is the molecular formula of a hydrocarbon that has 1 triple bond, 2 double bonds, 1 ring, and 32 carbons?

40. Answer Problem 37, parts **a–h** using 2-butyne as the starting material instead of propyne.

41. What are the products of the following reactions?

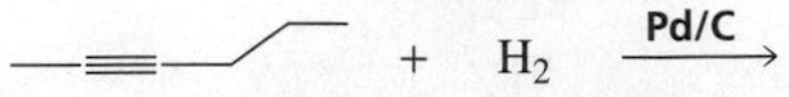

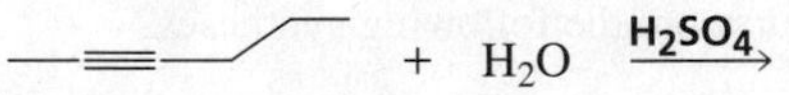

42. a. Starting with 3-methyl-1-butyne, how could you prepare the following alcohols?
    1. 2-methyl-2-butanol   **2.** 3-methyl-1-butanol
  b. In each case, a second alcohol would also be obtained. What alcohol would it be?

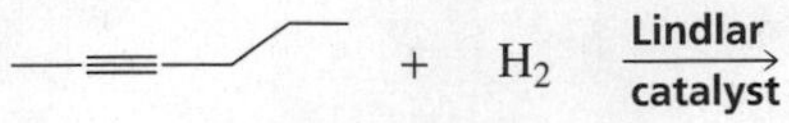

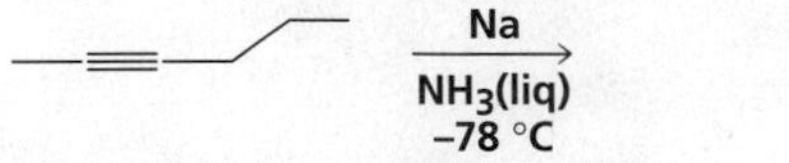

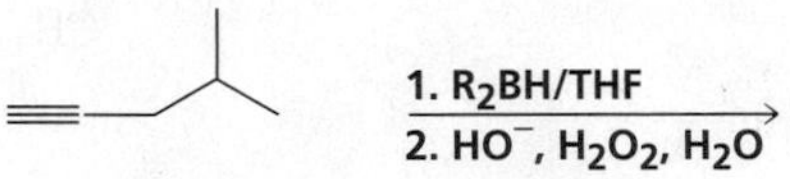

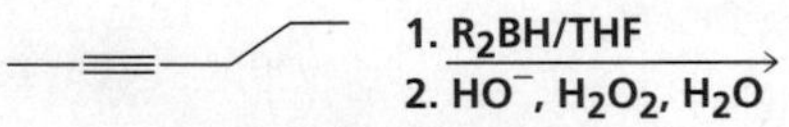

43. How many of the following names are correct? Correct those that are wrong.

a. 4-heptyne   **c.** 4-chloro-2-pentyne   **e.** 4,4-dimethyl-2-pentyne
b. 2-ethyl-3-hexyne   **d.** 2,3-dimethyl-5-octyne   **f.** 2,5-dimethyl-3-hexyne

44. Which of the following pairs are keto–enol tautomers?

a. $CH_3CH_2CH=CHCH_2OH$ and $CH_3CH_2CH_2CH_2\overset{\overset{\displaystyle O}{\|}}{C}H$

d. $CH_3CH_2CH_2CH=CHOH$ and $CH_3CH_2CH_2\overset{\overset{\displaystyle O}{\|}}{C}CH_3$

b. $CH_3\overset{\overset{\displaystyle OH}{|}}{C}HCH_3$ and $CH_3\overset{\overset{\displaystyle O}{\|}}{C}CH_3$

e. $CH_3CH_2CH_2\overset{\overset{\displaystyle OH}{|}}{C}=CH_2$ and $CH_3CH_2CH_2\overset{\overset{\displaystyle O}{\|}}{C}CH_3$

c. $CH_3CH_2CH=CHOH$ and $CH_3CH_2CH_2\overset{\overset{\displaystyle O}{\|}}{C}H$

45. How can the following compounds be prepared using ethyne as the starting material?

a. $CH_3\overset{\overset{\displaystyle O}{\|}}{C}H$ **b.** $CH_3CH_2\overset{\underset{\displaystyle Br}{|}}{C}HCH_2Br$ **c.** $CH_3\overset{\overset{\displaystyle O}{\|}}{C}CH_3$ **d.** 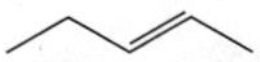   **e.** 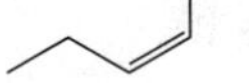   **f.**

46. Do the equilibria of the following acid–base reactions lie to the right or the left? (The pK_a of H_2O_2 is 11.6.)

$$HOOH \;+\; HO^- \;\rightleftharpoons\; HOO^- \;+\; H_2O$$

$$RC\equiv CH \;+\; HOO^- \;\rightleftharpoons\; RC\equiv C^- \;+\; HOOH$$

47. What stereoisomers are obtained when 2-butyne undergoes each of the following reaction sequences?

  a. 1. H_2/Lindlar catalyst 2. Br_2/CH_2Cl_2 **b.** 1. Na/NH_3(liq), $-78\,°C$ 2. Br_2/CH_2Cl_2 **c.** 1. Cl_2/CH_2Cl_2 2. Br_2/CH_2Cl_2

48. Draw the keto tautomer for each of the following:

a. CH$_3$CH=$\overset{\overset{\displaystyle OH}{|}}{C}CH_3$

b. CH$_3$CH$_2$CH$_2$$\overset{\overset{\displaystyle OH}{|}}{C}$=CH$_2$

c. [cyclohexene ring]—OH

d. [cyclohexane ring]=CHOH

49. Show how each of the following compounds could be prepared using the given starting material, any necessary inorganic reagents, and any necessary organic compound that has no more than four carbons:

a. HC≡CH $\longrightarrow$ CH$_3$CH$_2$CH$_2$CH$_2$$\overset{\overset{\displaystyle O}{||}}{C}CH_3$

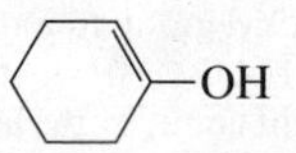

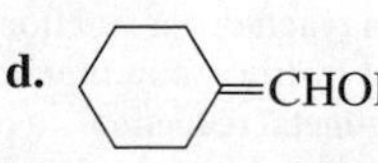

d. [cyclohexane]—C≡CH $\longrightarrow$ [cyclohexane]—CH$_2$$\overset{\overset{\displaystyle O}{||}}{C}$H

b. HC≡CH $\longrightarrow$ CH$_3$CH$_2$$\overset{\overset{\displaystyle}{\underset{\underset{\displaystyle Br}{|}}{C}}}HCH_3$

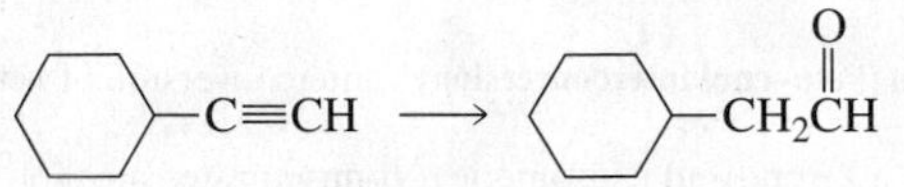

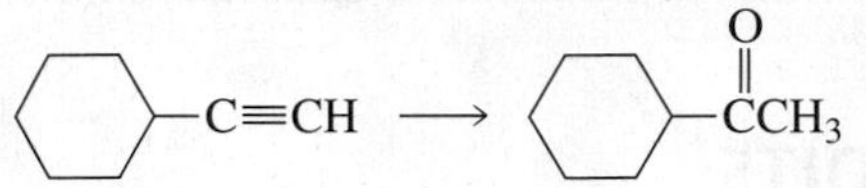

e. [cyclohexane]—C≡CH $\longrightarrow$ [cyclohexane]—$\overset{\overset{\displaystyle O}{||}}{C}CH_3$

c. HC≡CH $\longrightarrow$ CH$_3$CH$_2$CH$_2$$\overset{\overset{\displaystyle}{\underset{\underset{\displaystyle OH}{|}}{C}}}HCH_3$

f. [benzene]—C≡CCH$_3$ $\longrightarrow$ [benzene attached to C=C with H's and CH$_3$]

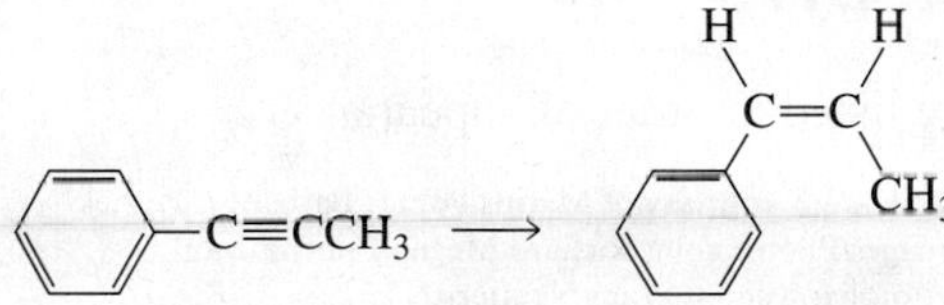

50. A chemist was planning to synthesize 3-octyne by adding 1-bromobutane to the product obtained from the reaction of 1-butyne with sodium amide. Unfortunately, however, he had forgotten to order 1-butyne. How else can he prepare 3-octyne?

51. a. Explain why a single pure product is obtained from hydroboration–oxidation of 2-butyne, whereas two products are obtained from hydroboration–oxidation of 2-pentyne.
b. Name two other internal alkynes that will yield only one product upon hydroboration–oxidation.

52. What stereoisomers are obtained from the following reactions?

a. CH$_3$CH$_2$C≡CCH$_2$CH$_3$ $\xrightarrow{\text{1. Na, NH}_3\text{(liq), }-78\ °C \atop \text{2. D}_2\text{, Pd/C}}$

b. CH$_3$CH$_2$C≡CCH$_2$CH$_3$ $\xrightarrow{\text{1. H}_2\text{/Lindlar catalyst} \atop \text{2. D}_2\text{, Pd/C}}$

53. α-Farnesene is a dodecatetraene found in the waxy coating of apple skins. What is its systematic name? Include E and Z where necessary to indicate the configuration of the double bonds.

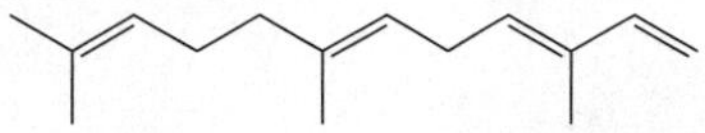

α-farnesene

54. Show how the following compounds could be synthesized starting with ethyne:

a. *cis*-2-octene
b. *trans*-3-heptene

55. Explain why, in hydroboration–oxidation, HO$^-$ and HOOH cannot be added until after the hydroboration reaction is over.

56. Starting with ethyne, describe how the following compounds could be synthesized:

a. (2*S*,3*R*)- 4-bromo-3-hexanol and (2*R*,3*S*)- 4-bromo-3-hexanol
b. (2*R*,3*R*)- 4-bromo-3-hexanol and (2*S*,3*S*)- 4-bromo-3-hexanol

57. What are the products of the following reactions?

a. [alkyne] + Cl$_2$ (excess) $\xrightarrow{\text{CH}_2\text{Cl}_2}$

e. [alkyne] + H$_2$ $\xrightarrow{\text{Pd/C}}$

b. [alkyne] + Cl$_2$ $\xrightarrow{\text{CH}_2\text{Cl}_2}$

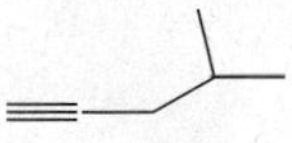

f. [alkyne] + H$_2$ $\xrightarrow{\text{Lindlar catalyst}}$

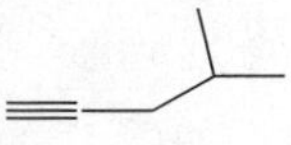

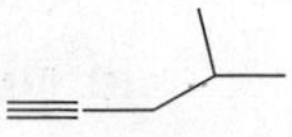

c. [alkyne] + H$_2$O $\xrightarrow{\text{H}_2\text{SO}_4 \atop \text{HgSO}_4}$

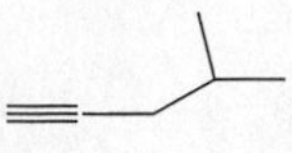

g. [alkyne] $\xrightarrow{\text{Na} \atop \text{NH}_3\text{(liq), }-78\ °C}$

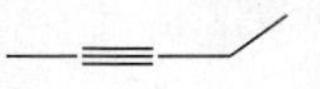

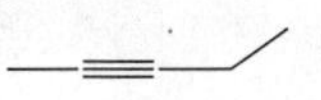

d. [alkyne] + $\xrightarrow{\text{1. R}_2\text{BH/THF} \atop \text{2. HO}^-,\ \text{H}_2\text{O}_2,\ \text{H}_2\text{O}}$

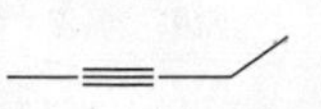

h. [alkyne] $\xrightarrow{\text{1. R}_2\text{BH, THF} \atop \text{2. HO}^-,\ \text{H}_2\text{O}_2,\ \text{H}_2\text{O}}$

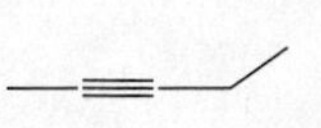

GLOSSARY

alkylation reaction a reaction that adds an alkyl group to a reactant.

alkyne a hydrocarbon that contains a triple bond.

dissolving-metal reduction a reduction brought about by the use of sodium or lithium metal dissolved in liquid ammonia.

geminal dihalide a compound with two halogen atoms bonded to the same carbon.

internal alkyne an alkyne with the triple bond not at the end of the carbon chain.

keto–enol tautomerism (keto–enol interconversion) interconversion of keto and enol tautomers.

keto–enol tautomers a ketone and its isomeric α,β-unsaturated alcohol.

pi-complex a complex formed between an electrophile and a triple bond.

radical anion a species with a negative charge and an unpaired electron.

retrosynthesis (retrosynthetic analysis) working backwards (on paper) from the target molecule to available starting materials.

tautomers rapidly equilibrating isomers that differ in the location of their bonding electrons.

terminal alkyne an alkyne with the triple bond at the end of the carbon chain.

vinylic cation a compound with a positive charge on a vinylic carbon.

vinylic radical a compound with an unpaired electron on a vinylic carbon.

PHOTO CREDITS

Photo credits are listed in order of appearance.

arquiplay77/Fotolia; Image courtesy of Marios Politis; Barbara J. Petrick/Shutterstock; Jupiterimages/Photos.com; Richard Megna/Fundamental Photographs; Infinity/Fotolia; Xuanlu Wang/Shutterstock

Delocalized Electrons and Their Effect on Stability, pK_a, and the Products of a Reaction

From Chapter 8 of *Organic Chemistry*, Seventh Edition. Paula Yurkanis Bruice. Copyright © 2014 by Pearson Education, Inc. All rights reserved.

Delocalized Electrons and Their Effect on Stability, pK_a, and the Products of a Reaction

Kekulé's Dream

Delocalized electrons play a very important role in organic chemistry. This chapter shows you how delocalized electrons are depicted. Then you will see how they affect things that are now familiar to you, such as pK_a values, the stability of carbocations, and the products formed from electrophilic addition reactions.

Electrons that are restricted to a particular region are called **localized electrons.** Localized electrons either belong to a single atom or are shared by two atoms.

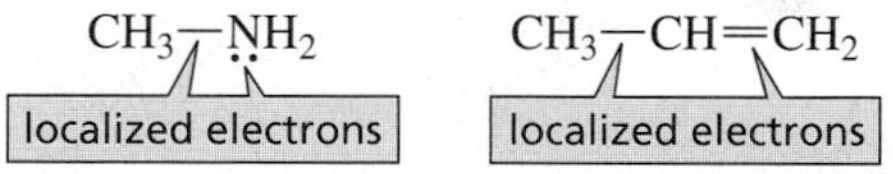

$$CH_3-\overset{\cdot\cdot}{N}H_2 \qquad CH_3-CH=CH_2$$

localized electrons   localized electrons

Many organic compounds have *delocalized* electrons. **Delocalized electrons** are shared by three or more atoms. The two electrons represented by the π bond of the COO^- group are shared by three atoms—the carbon and both oxygens. The dashed lines in the chemical structure shown here indicate that the two electrons are delocalized over three atoms.

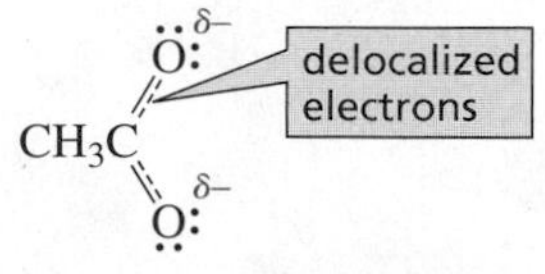

$$CH_3C\underset{\overset{\cdot\cdot}{\underset{\cdot\cdot}{O:}}^{\delta^-}}{\overset{\overset{\cdot\cdot}{O:}^{\delta^-}}{\Big\langle}}$$

delocalized electrons

In this chapter, you will learn how to recognize compounds that have delocalized electrons and how to draw structures that represent the electron distribution in molecules with delocalized electrons. You will also be introduced to some of the special characteristics of compounds that have delocalized electrons. You will then be able to understand some of the wide-ranging effects that delocalized electrons have on the reactions and properties of organic compounds. We begin by looking at benzene, a compound whose structure is ideal to illustrate the concept of delocalized electrons.

1 DELOCALIZED ELECTRONS EXPLAIN BENZENE'S STRUCTURE

Because early organic chemists did not know about delocalized electrons, they were puzzled by benzene's structure. They knew that benzene had a molecular formula of C_6H_6, that it was an unusually stable compound, and that it did not undergo the addition reactions characteristic of alkenes. They also knew that when a different atom was substituted for any one of benzene's hydrogens, only *one* product was obtained and, when the substituted product underwent a second substitution, *three* products were obtained.

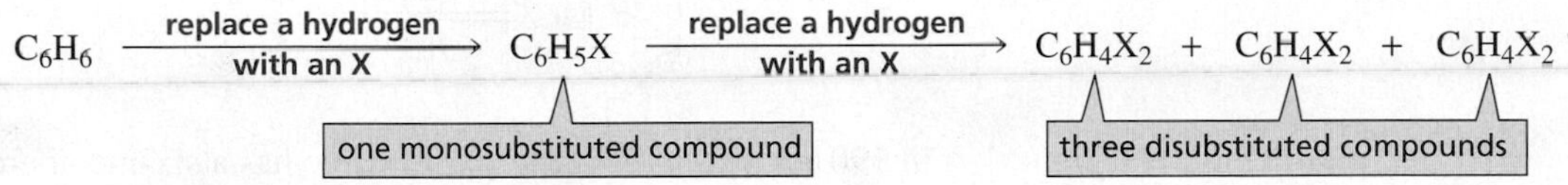

$$C_6H_6 \xrightarrow[\text{with an X}]{\text{replace a hydrogen}} C_6H_5X \xrightarrow[\text{with an X}]{\text{replace a hydrogen}} C_6H_4X_2 + C_6H_4X_2 + C_6H_4X_2$$

one monosubstituted compound   three disubstituted compounds

What kind of structure would you predict for benzene if you knew only what the early chemists knew? The molecular formula (C_6H_6) tells us that benzene has eight fewer hydrogens than an acyclic alkane with six carbons ($C_nH_{2n+2} = C_6H_{14}$). Benzene, therefore, has a degree of unsaturation of four. In other words, the total number of rings and π bonds in benzene is four.

Because only one product is obtained regardless of which of the six hydrogens of benzene is replaced with another atom, we know that all the hydrogens must be identical. Two structures with a degree of unsaturation of four and six identical hydrogens are shown here:

For every two hydrogens that are missing from the general molecular formula, C_nH_{2n+2}, a hydrocarbon has either a π bond or a ring.

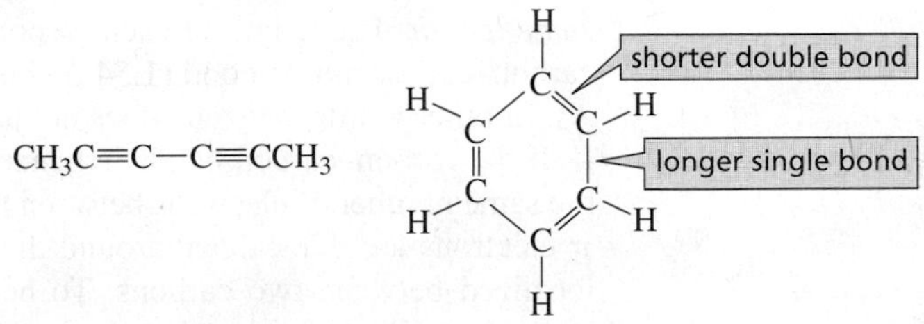

$$CH_3C{\equiv}C-C{\equiv}CCH_3$$

shorter double bond

longer single bond

Neither of these structures, however, is consistent with the observation that three compounds are obtained if a second hydrogen is replaced with another atom. The acyclic structure yields only two disubstituted products.

$$CH_3C{\equiv}C-C{\equiv}CCH_3 \xrightarrow[\text{with Br's}]{\text{replace 2 H's}} CH_3C{\equiv}C-C{\equiv}CCHBr \text{ and } BrCH_2C{\equiv}C-C{\equiv}CCH_2Br$$
$$\qquad\qquad\qquad\qquad\qquad\qquad\qquad\qquad |$$
$$\qquad\qquad\qquad\qquad\qquad\qquad\qquad\quad Br$$

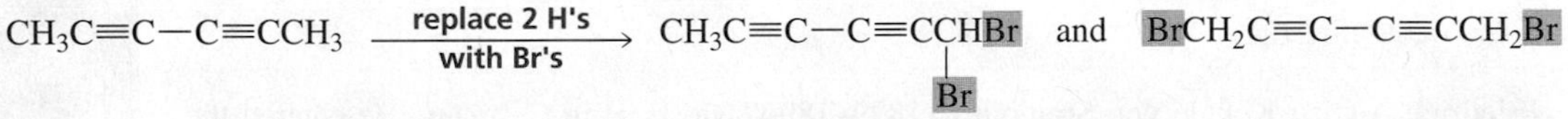

The cyclic structure, with alternating single and slightly shorter double bonds, yields four disubstituted products—a 1,3-disubstituted product, a 1,4-disubstituted product, and two 1,2-disubstituted products—because the two substituents can be placed either on two adjacent carbons joined by a single bond or on two adjacent carbons joined by a double bond.

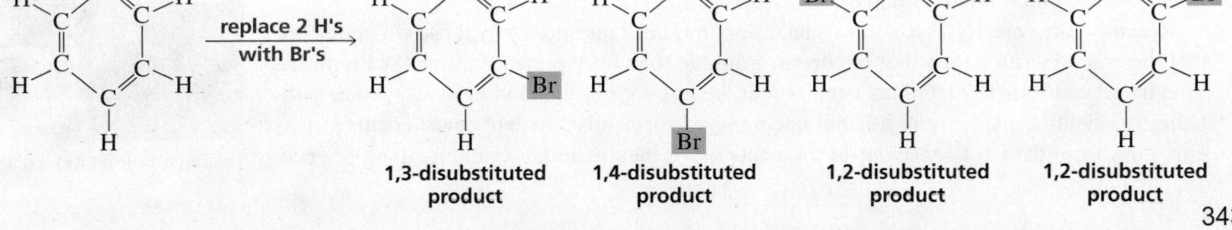

| 1,3-disubstituted product | 1,4-disubstituted product | 1,2-disubstituted product | 1,2-disubstituted product |

343

In 1865, the German chemist Friedrich Kekulé suggested a way of resolving this dilemma. He proposed that benzene was not a single compound, but a mixture of two compounds in rapid equilibrium.

rapid
equilibrium

shorter double bond

longer single bond

Kekulé's proposal explained why only three disubstituted products are obtained. According to Kekulé, there actually *are* four disubstituted products, but the two 1,2-disubstituted products interconvert too rapidly to be distinguished and separated from each other.

rapid
equilibrium

In 1901, it was confirmed that benzene has a six-membered ring, when Paul Sabatier found that catalytic hydrogenation of benzene produced cyclohexane.

benzene $\xrightarrow[\text{150–250 °C, 25 atm}]{\text{H}_2\text{, Ni}}$

cyclohexane

Controversy over the structure of benzene continued until the 1930s, when the new techniques of X-ray and electron diffraction produced a surprising result: they showed that *benzene is a planar molecule and that the six carbon–carbon bonds all have the same length.* The length of each carbon–carbon bond is 1.39 Å, which is shorter than a carbon–carbon single bond (1.54 Å) but longer than a carbon–carbon double bond (1.33 Å). In other words, benzene does not have alternating single and double bonds.

If the carbon–carbon bonds in benzene all have the same length, they must also have the same number of electrons between the carbons. This can be true, however, only if the π electrons are delocalized around the ring, rather than each pair of π electrons being localized between two carbons. To better understand the concept of delocalized electrons, we will now take a close look at the bonding in benzene.

Kekulé's Dream

Friedrich August Kekulé von Stradonitz (1829–1896) was born in Germany. He entered the University of Giessen to study architecture but switched to chemistry after taking a course in the subject. He was a professor of chemistry at the University of Heidelberg, at the University of Ghent in Belgium, and then at the University of Bonn. In 1890, he gave an extemporaneous speech at the twenty-fifth-anniversary celebration of his first paper on the cyclic structure of benzene. In this speech, he claimed that he had arrived at the structures as a result of dozing off in front of a fire while working on a textbook. He dreamed of chains of carbon atoms twisting and turning in a snakelike motion, when suddenly the head of one snake seized hold of its own tail and formed a spinning ring.

Recently, the veracity of Kekulé's snake story has been questioned by those who point out that there is no written record of the dream from the time he experienced it in 1861 until the time he related it in 1890. Others counter that dreams are not the kind of evidence one publishes in scientific papers, and it is not uncommon for scientists to experience creative ideas emerging from their subconscious at moments when they were not thinking about science.

Friedrich August Kekulé von Stradonitz

Also, Kekulé warned against publishing dreams when he said, "Let us learn to dream, and perhaps then we shall learn the truth. But let us also beware not to publish our dreams until they have been examined by the wakened mind." In 1895, he was made a nobleman by Emperor William II of Germany. This allowed him to add "von Stradonitz" to his name. Kekulé's students received three of the first five Nobel Prizes in Chemistry: van't Hoff in 1901, Fischer in 1902, and Baeyer in 1905.

PROBLEM 1♦

a. The following compounds have the same molecular formula as benzene. How many monosubstituted products would each have?
  1. $HC\equiv CC\equiv CCH_2CH_3$ **2.** $CH_2=CHC\equiv CCH=CH_2$

b. How many disubstituted products would each of the preceding compounds have? (Do not include stereoisomers.)

c. How many disubstituted products would each of the compounds have if stereoisomers are included?

PROBLEM 2

Between 1865 and 1890, other possible structures were proposed for benzene such as those shown here. Considering what nineteenth-century chemists knew about benzene, which is a better proposal for benzene's structure, Dewar benzene or Ladenburg benzene? Why?

Dewar benzene   Ladenburg benzene

2 THE BONDING IN BENZENE

Each of benzene's six carbons is sp^2 hybridized. An sp^2 carbon has bond angles of 120°, which is identical to the size of the angles in a planar hexagon. Thus, benzene is a planar molecule (Figure 1a). Because benzene is planar, the six p orbitals are parallel (Figure 1b), and are close enough for each p orbital to overlap the p orbital on either side of it (Figure 1c).

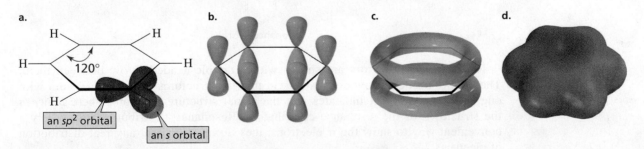

a.

H   H

H   H

120°

H

an sp^2 orbital

an s orbital

b.   c.   d.

▲ Figure 1

(a) Each of the carbons in benzene uses two sp^2 orbitals to bond to two other carbons; its third sp^2 orbital overlaps the s orbital of a hydrogen.

(b) Each carbon has a p orbital at right angles to the sp^2 orbitals. The parallel p orbitals are close enough for side-to-side overlap, so each p orbital overlaps the p orbitals on *both* adjacent carbons.

(c) The overlapping p orbitals form a continuous doughnut-shaped cloud of electrons above the plane of the benzene ring, and they form another doughnut-shaped cloud of electrons below it.

(d) The electrostatic potential map shows that all the carbon–carbon bonds have the same electron density.

Each of the six π electrons of benzene, therefore, is localized neither on a single carbon nor in a bond between two carbons (as in an alkene). Instead, each π electron is shared by all six carbons. In other words, the six π electrons are delocalized—they roam freely within the doughnut-shaped clouds that lie over and under the planar ring of carbons (Figure 1c and d). Benzene is often drawn as a hexagon containing either dashed lines or a circle to symbolize the six delocalized π electrons.

This type of representation makes it clear that there are no double bonds in benzene. Kekulé's structure was very nearly correct. The actual structure of benzene is Kekulé's structure with delocalized electrons.

3 RESONANCE CONTRIBUTORS AND THE RESONANCE HYBRID

A disadvantage to using dashed lines (or a circle) to represent delocalized electrons is that they do not tell us how many π electrons they represent. For example, the dashed lines inside the hexagon just shown indicate that the π electrons are shared equally by all six carbons and that all the carbon–carbon bonds have the same length, but they do not show how many π electrons are in the ring. Consequently, chemists prefer to use structures that portray the electrons as localized (and therefore show the number of π electrons), even though the electrons in the compound's actual structure are delocalized.

The *approximate* structure with localized electrons is called a **resonance contributor,** a **resonance structure,** or a **contributing resonance structure.** The *actual* structure with delocalized electrons is called a **resonance hybrid.** We can easily see that there are six π electrons in each of benzene's resonance contributors.

Electron delocalization is shown by double-headed arrows (⟷), whereas equilibrium is shown by two arrows pointing in opposite directions (⇌).

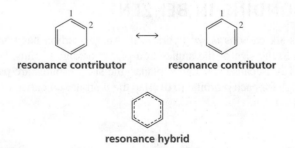

resonance contributor   resonance contributor

resonance hybrid

Resonance contributors are shown with a double-headed arrow between them. The double-headed arrow does *not* mean that the structures are in equilibrium with one another. Rather, it indicates that the actual structure lies somewhere *between* the structures of the resonance contributors. Resonance contributors are merely a convenient way to show the π electrons; they do not represent any real distribution of electrons.

The following analogy illustrates the difference between resonance contributors and the resonance hybrid. Imagine that you are trying to describe to a friend what a rhinoceros looks like. You might tell your friend that a rhinoceros looks like a cross between a unicorn and a dragon. Like resonance contributors, the unicorn and the dragon do not really exist. Furthermore, like resonance contributors, they are not in equilibrium: a rhinoceros does not change back and forth between the two forms, looking like a unicorn one minute and a dragon the next. The unicorn and dragon are simply ways to describe what the actual animal—the rhinoceros—looks like. *Resonance contributors, like unicorns and dragons, are imaginary. Only the resonance hybrid, like the rhinoceros, is real.*

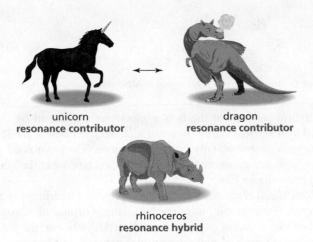

unicorn
resonance contributor

dragon
resonance contributor

rhinoceros
resonance hybrid

Electron delocalization is most effective if all the atoms sharing the delocalized electrons lie in the same plane, so that their p orbitals can maximally overlap.

For example, the electrostatic potential map shows that cyclooctatetraene is tub-shaped, not planar—its sp^2 carbons have bond angles of 120°, whereas a planar eight-membered ring would have bond angles of 135°. Because the ring is not planar, a p orbital can overlap with one adjacent p orbital, but it can have little overlap with the other adjacent p orbital. As a result, the eight π electrons are localized in four double bonds and not delocalized over the entire eight-membered ring. Thus, the carbon–carbon bonds do not all have the same length.

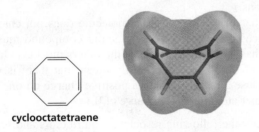

cyclooctatetraene

4 HOW TO DRAW RESONANCE CONTRIBUTORS

We have seen that an organic compound with delocalized electrons is generally represented as a structure with localized electrons to let us know how many π electrons it has. For example, nitroethane is usually represented with a nitrogen–oxygen double bond and a nitrogen–oxygen single bond.

$$CH_3CH_2 - \overset{+}{N} \overset{:\ddot{O}:}{\underset{:\ddot{O}:^-}{\Big\|}}$$

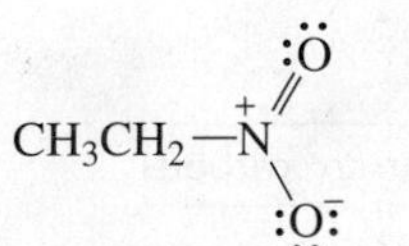

nitroethane

However, the two nitrogen–oxygen bonds in nitroethane actually have the same length. A more accurate description of the molecule's structure is obtained by drawing the two resonance contributors. Both resonance contributors show the compound with a nitrogen–oxygen double bond and a nitrogen–oxygen single bond; they indicate that the electrons are delocalized by depicting the double bond in one contributor as a single bond in the other.

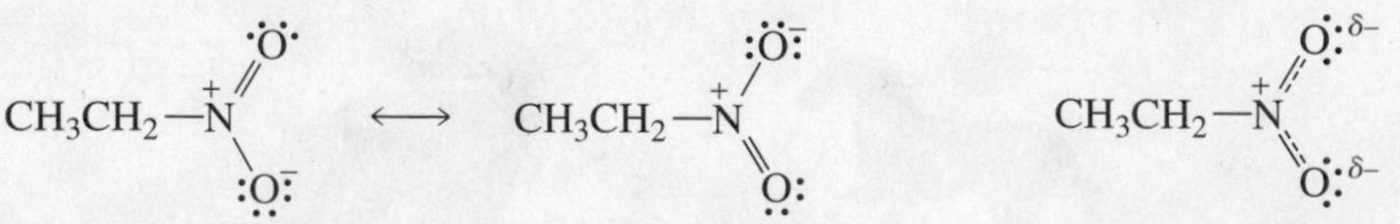

$$
\underset{\substack{\text{resonance contributor}}}{CH_3CH_2-\overset{+}{N}\!\!\underset{:\overset{..}{O}:^-}{\overset{\overset{..}{O}:}{\|}}} \quad\longleftrightarrow\quad \underset{\substack{\text{resonance contributor}}}{CH_3CH_2-\overset{+}{N}\!\!\underset{\overset{..}{O}:}{\overset{:\overset{..}{O}:^-}{\|}}} \quad\quad \underset{\substack{\text{resonance hybrid}}}{CH_3CH_2-\overset{+}{N}\!\!\underset{\overset{..}{O}:^{\delta-}}{\overset{\overset{..}{O}:^{\delta-}}{\diagup}}}
$$

The resonance hybrid shows that the two π electrons are shared by three atoms. The resonance hybrid also shows that the two nitrogen–oxygen bonds are identical and that the negative charge is shared equally by both oxygens. Thus, we need to visualize and mentally average both resonance contributors to appreciate what the actual molecule—the resonance hybrid—looks like.

Notice that *delocalized electrons result from a* p *orbital overlapping the* p *orbitals of two adjacent atoms.* For example, in nitroethane, the *p* orbital of nitrogen overlaps the *p* orbital of each of two adjacent oxygens; in the carboxylate ion, the *p* orbital of carbon overlaps the *p* orbital of each of two adjacent oxygens; and in benzene, the *p* orbital of carbon overlaps the *p* orbital of each of two adjacent carbons.

Delocalized electrons result from a *p* orbital overlapping the *p* orbitals of two adjacent atoms.

Rules for Drawing Resonance Contributors

To draw a set of resonance contributors for a molecule, first draw a Lewis structure. This is the first resonance contributor. Then, following the rules listed next, move electrons to generate the next resonance contributor.

1. Only electrons move. Atoms never move.
2. Only π electrons (electrons in π bonds) and lone-pair electrons can move. (Never move σ electrons.)
3. The total number of electrons in the molecule does not change. Therefore, each of the resonance contributors for a particular compound must have the same net charge. If one has a net charge of 0, all the others must also have net charges of 0. (A net charge of 0 does not necessarily mean that there is no charge on any of the atoms, because a molecule with a positive charge on one atom and a negative charge on another atom has a net charge of 0.)

Notice, as you study the following resonance contributors and practice drawing them, that electrons (π electrons or lone pairs) are always moved toward an sp^2 or sp atom. (An sp^2 carbon is either a positively charged carbon or a double-bonded carbon and an sp carbon is generally a triple-bonded carbon). Electrons cannot be moved toward an sp^3 carbon because an sp^3 carbon has a complete octet and does not have a π bond that can break, so it cannot accommodate any more electrons.

The carbocation shown next has delocalized electrons. To draw its resonance contributor, *move the π electrons toward an* sp^2 *carbon.* The curved arrow shows you how to draw the second contributor. The tail of the curved arrow shows where the electrons start from, and the head shows where the electrons end up. The resonance hybrid shows that the π electrons are shared by three carbons, and the positive charge is shared by two carbons.

To draw resonance contributors, move only π electrons or lone pairs toward an sp^2 (or sp) carbon.

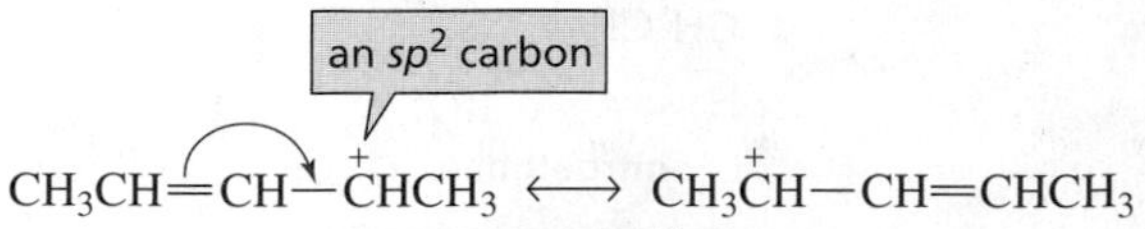

$$
\boxed{\text{an } sp^2 \text{ carbon}}
$$

$$
\underset{\substack{\text{resonance contributors}}}{CH_3CH=CH-\overset{+}{C}HCH_3 \quad\longleftrightarrow\quad CH_3\overset{+}{C}H-CH=CHCH_3}
$$

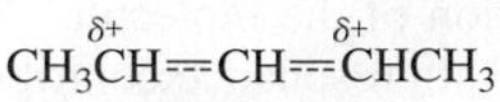

$$
\underset{\substack{\text{resonance hybrid}}}{CH_3\overset{\delta+}{C}H=\!\!=\!\!=CH=\!\!=\!\!=\overset{\delta+}{C}HCH_3}
$$

Let's compare this carbocation with a similar compound in which all the electrons are localized. The π electrons in the carbocation shown next cannot move, because the carbon they would move toward is an sp^3 carbon, and sp^3 carbons cannot accept any more electrons.

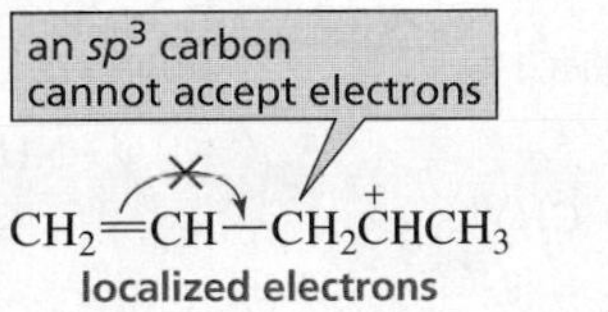

an sp^3 carbon
cannot accept electrons

$$CH_2 = CH - CH_2 \overset{+}{C}HCH_3$$
localized electrons

In the next example, π *electrons again move toward an* sp^2 *carbon.* The resonance hybrid shows that the π electrons are shared by five carbons, and positive charge is shared by three carbons.

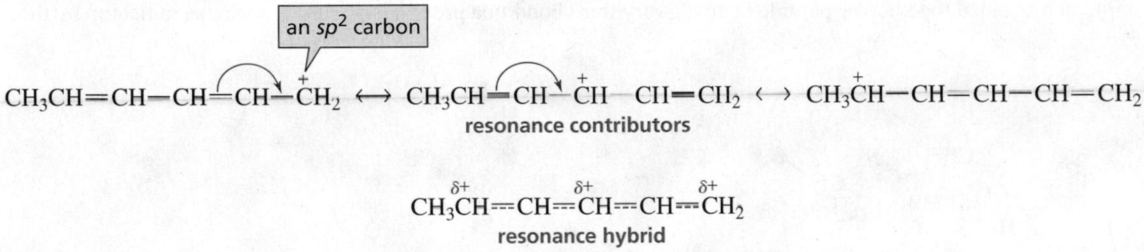

an sp^2 carbon

$$CH_3CH = CH - CH = CH - \overset{+}{C}H_2 \longleftrightarrow CH_3CH = CH - \overset{+}{C}H \quad CH = CH_2 \longleftrightarrow CH_3\overset{+}{C}H - CH = CH - CH = CH_2$$
resonance contributors

$$CH_3\overset{\delta+}{CH} = CH = \overset{\delta+}{CH} = CH = \overset{\delta+}{CH_2}$$
resonance hybrid

The resonance contributor for the next compound is obtained by *moving lone-pair electrons toward an* sp^2 *carbon.* The sp^2 carbon can accommodate the new electrons by breaking a π bond. The lone-pair electrons in the compound on the far right are not delocalized because they would have to move toward an sp^3 carbon.

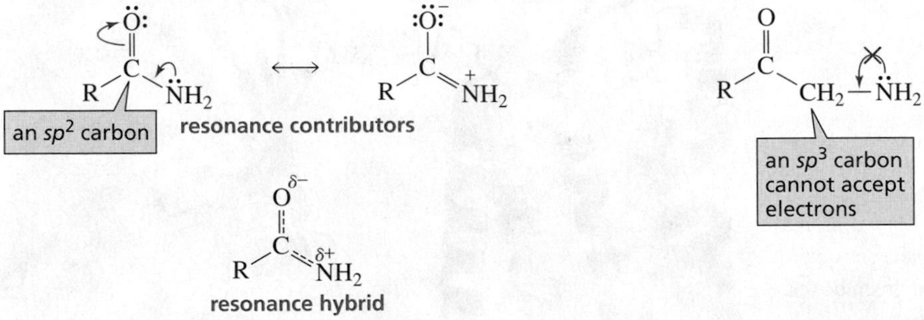

an sp^2 carbon   **resonance contributors**

an sp^3 carbon
cannot accept electrons

resonance hybrid

In the next example, *lone-pair electrons move toward an* sp *carbon.*

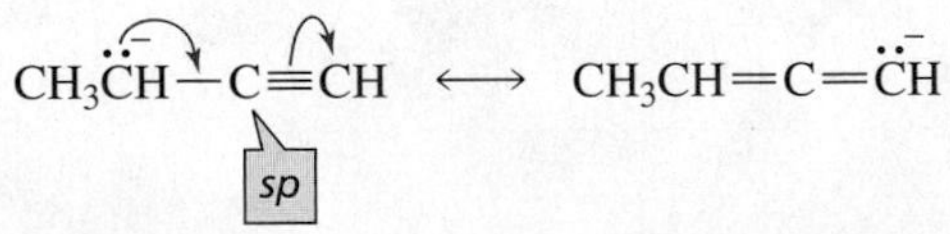

$$CH_3\overset{..}{\overset{-}{C}}H - C \equiv CH \longleftrightarrow CH_3CH = C = \overset{..}{C}H$$

sp

The resonance contributor for the next compound is also obtained by *moving electrons toward an* sp *carbon.*

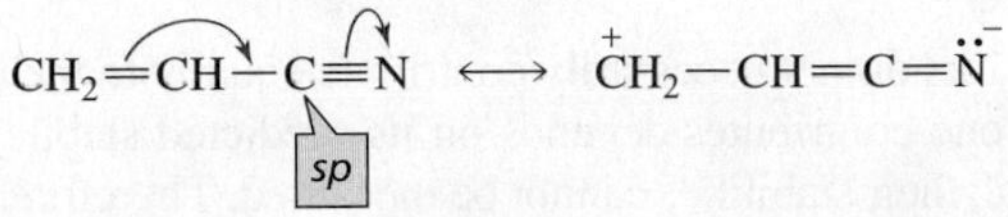

$$CH_2 = CH - C \equiv N \longleftrightarrow \overset{+}{C}H_2 - CH = C = \overset{..}{\overset{-}{N}}$$

sp

PROBLEM 3

a. Which of the following compounds have delocalized electrons?

1. $CH_2{=}CHCH_2CH{=}CH_2$

2. $CH_3CH{=}CHCH{=}CHCH_2^+$

3. $CH_3CH_2\ddot{N}HCH_2CH{=}CH_2$

4.

5.

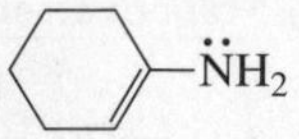

6. (structure with $\ddot{O}$)

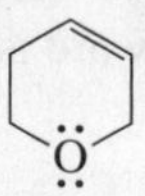

7. ($-CH_2\ddot{N}H_2$)

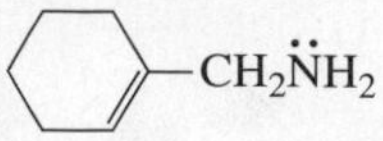

b. Draw the resonance contributors for the compounds that have delocalized electrons.

Electron Delocalization Affects the Three-Dimensional Shape of Proteins

A protein consists of amino acids joined together by peptide bonds. Every third bond in a protein is a peptide bond, as indicated by the red arrows.

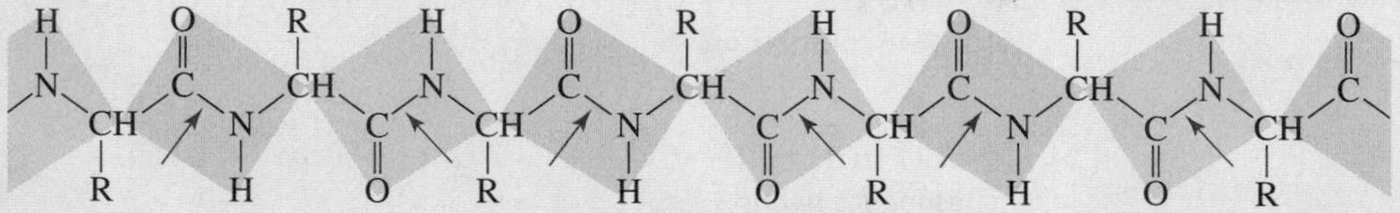

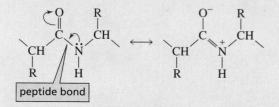

a segment of a protein

A resonance contributor can be drawn for a peptide bond by moving the lone pair on nitrogen toward the sp^2 carbon.

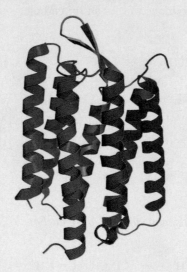

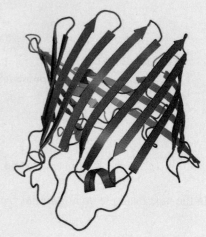

Because of the partial double-bond character of the peptide bond, the carbon and nitrogen atoms and the two atoms bonded to each of them are held rigidly in a plane, as represented in the protein segment by the blue and green boxes. Despite the rigid orientation of the peptide bond, the single bonds in the protein chain are free to rotate. Because of this, the chain is free to fold into a myriad of complex and highly intricate shapes. (Two conceptual representations of proteins are shown here.)

5 THE PREDICTED STABILITIES OF RESONANCE CONTRIBUTORS

All resonance contributors do not necessarily contribute equally to the resonance hybrid. The degree to which each one contributes depends on its predicted stability. Because resonance contributors are not real, their stabilities cannot be measured. Therefore, the stabilities of resonance contributors have to be predicted based on molecular features found in real molecules.

The greater the predicted stability of a resonance contributor,
the more it contributes to the structure of the resonance hybrid.

The more the resonance contributor contributes to the structure of the resonance hybrid, the more similar the contributor is to the real molecule.

The examples that follow illustrate these points.

The two resonance contributors for a carboxylic acid are shown in Figure 2. **B** has two features that make it less stable than **A**: one of the oxygens has a positive charge—not a stable situation for an electronegative atom—and the structure has separated charges. A molecule with **separated charges** has a positive charge and a negative charge that can be neutralized by the movement of electrons. Resonance contributors with separated charges are relatively unstable (relatively high in energy) because energy is required to keep the opposite charges separated. **A**, therefore, is predicted to be more stable than **B**. Consequently, **A** makes a greater contribution to the resonance hybrid, so the resonance hybrid looks more like **A** than like **B**.

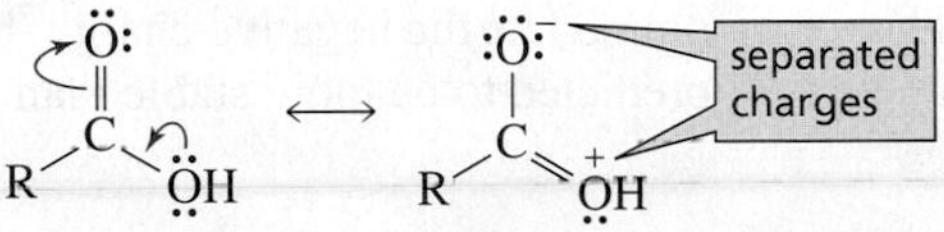

a carboxylic acid

◄ Figure 2

B is less stable than **A** because **B** has separated charges and one of its oxygens has a positive charge.

The two resonance contributors for a carboxylate ion are shown in Figure 3. **C** and **D** are equally stable, so they contribute equally to the resonance hybrid.

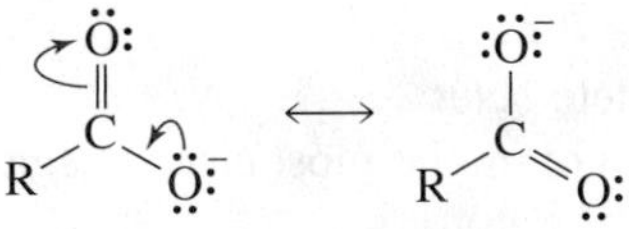

a carboxylate ion

◄ Figure 3

C and **D** are predicted to be equally stable. Therefore, they will contribute equally to the resonance hybrid.

When electrons can be moved in more than one direction, the most stable resonance contributor is obtained by moving them *toward* the more electronegative atom. For example, **G** in Figure 4 results from moving the π electrons (indicated by the red arrows in **F**) toward oxygen—the most electronegative atom in the molecule. **E** results from moving the π electrons (indicated by the blue arrows in **F**) away from oxygen.

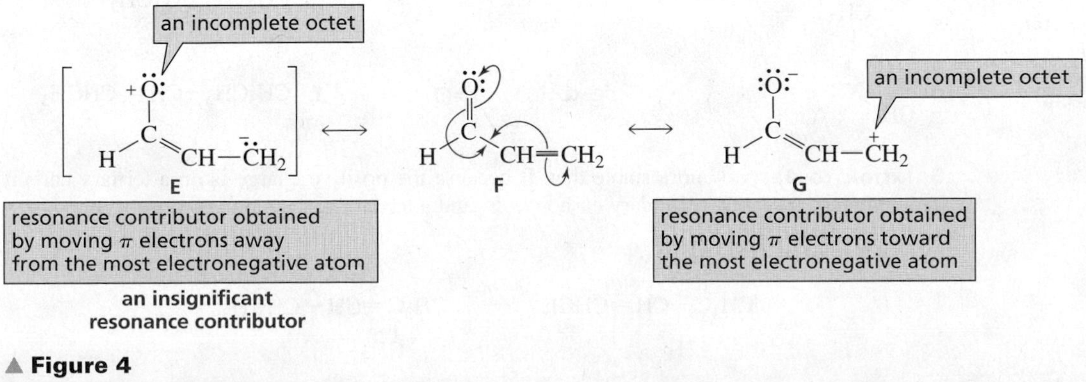

resonance contributor obtained by moving π electrons away from the most electronegative atom

an insignificant resonance contributor

resonance contributor obtained by moving π electrons toward the most electronegative atom

▲ Figure 4

G will make only a small contribution to the resonance hybrid because it has separated charges and a carbon with an incomplete octet. **E** also has separated charges and has an oxygen with an incomplete octet; its predicted stability is even less than that of **G** because **E** also has a positive charge on oxygen. Its contribution to the resonance hybrid is so insignificant that we do not need to include **E** as one of the resonance contributors. The resonance hybrid, therefore, looks very much like **F**.

The only time you need to show a resonance contributor that is obtained by moving electrons *away* from the most electronegative atom is when that is the only way the electrons can be moved. In other words, movement of electrons away from the most electronegative atom is better than no movement at all because electron delocalization makes a molecule more stable (as we will see in Section 6). For example, the only resonance contributor that can be drawn for the molecule in Figure 5 results from movement of the electrons away from oxygen.

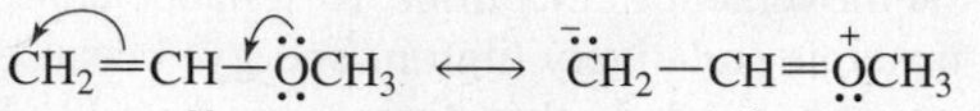

▶ **Figure 5**

I is predicted to be relatively unstable because it has separated charges and its oxygen has a positive charge. Therefore, the structure of the resonance hybrid is similar to **H**, with only a small contribution from **I**.

$$CH_2\!=\!CH\!-\!\overset{..}{\underset{..}{O}}CH_3 \quad\longleftrightarrow\quad \overset{..}{\underset{..}{C}}H_2\!-\!CH\!=\!\overset{+}{\underset{..}{O}}CH_3$$

$$\qquad\qquad H \qquad\qquad\qquad\qquad\qquad I$$

Let's now see which of the resonance contributors shown in Figure 6 has a greater predicted stability. **J** has a negative charge on carbon, whereas **K** has a negative charge on oxygen; oxygen can better accommodate the negative charge (because it is more electronegative than carbon), so **K** is predicted to be more stable than **J**.

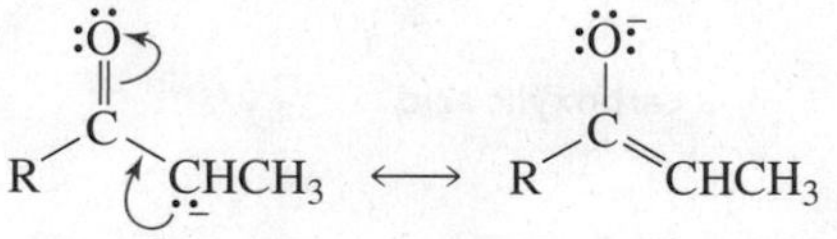

▶ **Figure 6**

The resonance hybrid more closely resembles **K**; that is, the resonance hybrid has a greater concentration of negative charge on the oxygen than on the carbon.

$$\underset{R}{\overset{\displaystyle :\overset{..}{O}:}{\underset{\displaystyle J}{\overset{\displaystyle \|}{C}}}}\!\!\diagdown\!\!\underset{..}{C}HCH_3 \quad\longleftrightarrow\quad \underset{R}{\overset{\displaystyle :\overset{..}{O}:^-}{\underset{\displaystyle K}{\overset{\displaystyle \|}{C}}}}\!\!\diagup\!\!CHCH_3$$

We can summarize the features that decrease the predicted stability of a resonance contributor as follows:

1. an atom with an incomplete octet
2. a negative charge that is not on the most electronegative atom or a positive charge that is on an electronegative atom
3. charge separation

When we compare the relative stabilities of resonance contributors, an atom with an incomplete octet (feature 1 in the preceding list) generally makes a structure more unstable than feature 2 or feature 3 do.

PROBLEM 4 Solved

Draw resonance contributors for each of the following species and rank them in order of decreasing contribution to the hybrid:

The greater the predicted stability of a resonance contributor, the more it contributes to the structure of the resonance hybrid.

The more the resonance contributor contributes to the structure of the hybrid, the more similar the contributor is to the real molecule.

a. 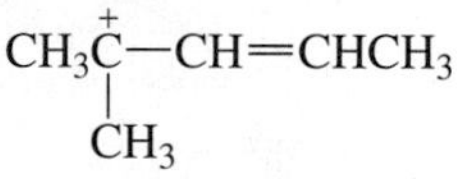 $CH_3\overset{+}{C}\!-\!CH\!=\!CHCH_3$
  $\qquad|$
  $\quad CH_3$

b. $\underset{CH_3}{\overset{\displaystyle O}{\overset{\displaystyle \|}{C}}}\diagdown OCH_3$

c. a cyclohexanone ring with $\overset{..}{\underset{..}{}}$ = O

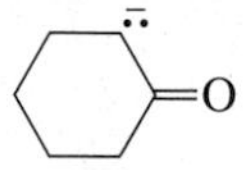

d. 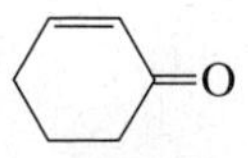 a cyclohexenone ring = O

e. 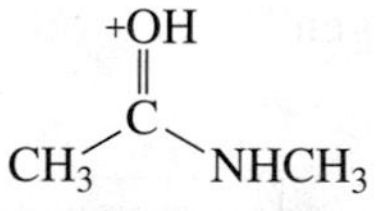 $\underset{CH_3}{\overset{\displaystyle +OH}{\overset{\displaystyle \|}{C}}}\diagdown NHCH_3$

f. $CH_3\overset{+}{C}H\!-\!CH\!=\!CHCH_3$

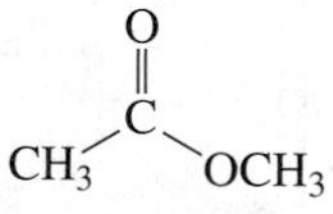

Solution to 4a **A** is more stable than **B** because the positive charge is on a tertiary carbon in **A**, whereas it is on a secondary carbon in **B**, and a tertiary carbocation is more stable than a secondary carbocation.

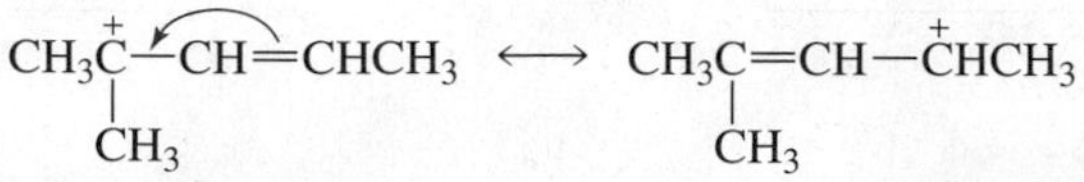

$$CH_3\overset{+}{C}\!\diagdown\!CH\!=\!CHCH_3 \quad\longleftrightarrow\quad CH_3C\!=\!CH\!-\!\overset{+}{C}HCH_3$$
$$\quad|\qquad\qquad\qquad\qquad\qquad\qquad |$$
$$\ CH_3 \qquad\qquad\qquad\qquad\qquad\quad CH_3$$
$$\qquad A \qquad\qquad\qquad\qquad\qquad\qquad B$$

NOTE TO THE STUDENT

The tutorial at the end of this chapter will give you additional practice drawing resonance contributors and predicting their relative stabilities.

PROBLEM 5

Draw the resonance hybrid for each of the species in Problem 4.

6 DELOCALIZATION ENERGY IS THE ADDITIONAL STABILITY DELOCALIZED ELECTRONS GIVE TO A COMPOUND

Delocalized electrons stabilize a compound. The extra stability a compound gains from having delocalized electrons is called the **delocalization energy. Electron delocalization** is also called **resonance,** so delocalization energy is also called **resonance energy.** Because delocalized electrons increase the stability of a compound, we can conclude that *a resonance hybrid is more stable than any of its resonance contributors is predicted to be.*

The delocalization energy associated with a compound that has delocalized electrons depends on the number *and* the predicted stability of the resonance contributors.

> *The greater the number of relatively stable resonance contributors, the greater is the delocalization energy.*

For example, the delocalization energy of a carboxylate ion with two relatively stable resonance contributors is significantly greater than the delocalization energy of a carboxylic acid with only one relatively stable resonance contributor.

| | | | |
| relatively stable | relatively unstable | relatively stable | relatively stable |

resonance contributors of a carboxylic acid

resonance contributors of a carboxylate ion

The delocalization energy is a measure of how much more stable a compound with delocalized electrons is than it would be if its electrons were localized.

The greater the number of relatively stable resonance contributors, the greater the delocalization energy.

Notice that it is the number of *relatively stable* resonance contributors—not the total number of resonance contributors—that is important in determining the delocalization energy.

For example, the delocalization energy of a carboxylate ion with two relatively stable resonance contributors is greater than the delocalization energy of the following compound with three resonance contributors since only one of its resonance contributors is relatively stable:

$$\bar{C}H_2-CH=CH-\overset{+}{C}H_2 \longleftrightarrow CH_2=CH-CH=CH_2 \longleftrightarrow \overset{+}{C}H_2-CH=CH-\bar{C}H_2$$

relatively unstable   relatively stable   relatively unstable

> *The more nearly equivalent the structures of the resonance contributors, the greater the delocalization energy.*

For example, the carbonate dianion is particularly stable because it has three equivalent resonance contributors.

The more nearly equivalent the structures of the resonance contributors, the greater the delocalization energy.

We can now summarize what we know about resonance contributors:

- The greater the predicted stability of a resonance contributor, the more it contributes to the resonance hybrid.
- The greater the number of relatively stable resonance contributors and the more nearly equivalent their structures, the greater is the delocalization energy.

PROBLEM 6♦

a. Predict the relative bond lengths of the three carbon–oxygen bonds in the carbonate ion ($CO_3{}^{2-}$).

b. What would you expect the charge to be on each oxygen?

PROBLEM-SOLVING STRATEGY

Determining Relative Stabilities

Which carbocation is more stable?

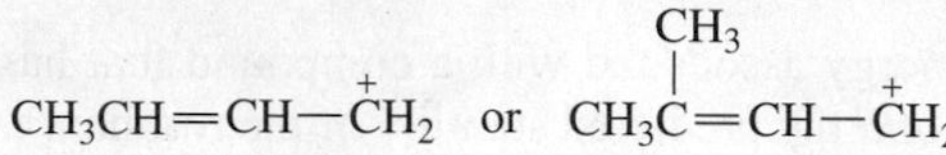

$$CH_3CH=CH-\overset{+}{C}H_2 \quad \text{or} \quad CH_3\overset{\overset{\displaystyle CH_3}{|}}{C}=CH-\overset{+}{C}H_2$$

Start by drawing the resonance contributors for each carbocation.

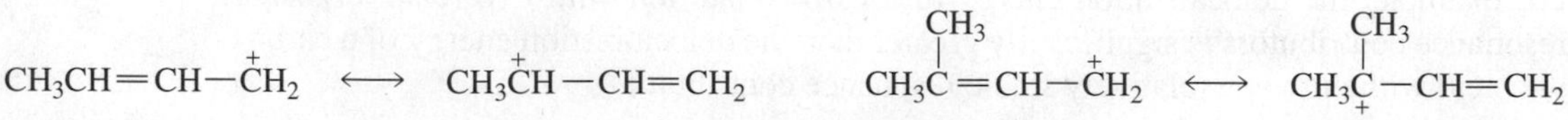

$$CH_3CH=CH-\overset{+}{C}H_2 \longleftrightarrow CH_3\overset{+}{C}H-CH=CH_2 \qquad CH_3\overset{\overset{\displaystyle CH_3}{|}}{C}=CH-\overset{+}{C}H_2 \longleftrightarrow CH_3\underset{+}{\overset{\overset{\displaystyle CH_3}{|}}{C}}-CH=CH_2$$

Now look at how the two sets of resonance contributors differ and think about how those differences affect the relative stabilities of the two resonance hybrids.

Each carbocation has two resonance contributors. The positive charge of the carbocation on the left is shared by a primary carbon and a secondary carbon. The positive charge of the carbocation on the right is shared by a primary carbon and a tertiary carbon. Because a tertiary carbon is more stable than a secondary carbon, the carbocation on the right is more stable.

Now use the strategy you have just learned to solve Problem 7.

PROBLEM 7♦

Which species in each pair is more stable?

a. 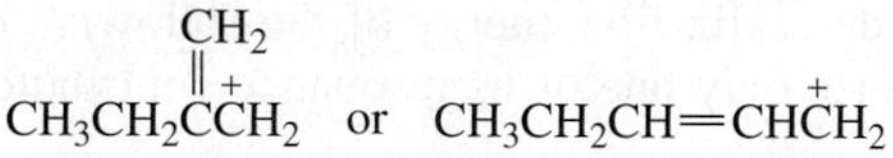 $CH_3CH_2\overset{\overset{\displaystyle CH_2}{\|}}{\underset{+}{C}}H_2$ or $CH_3CH_2CH=\overset{+}{C}HCH_2$

b. $\overset{\overset{\displaystyle O}{\|}}{\underset{CH_3}{C}}{\diagdown}_{CH=CH_2}$ or $\overset{\overset{\displaystyle O}{\|}}{\underset{CH_3}{C}}{\diagdown}_{CH=CHCH_3}$

c. $CH_3\overset{\overset{\displaystyle O^-}{|}}{C}HCH=CH_2$ or $CH_3\overset{\overset{\displaystyle O^-}{|}}{C}=CHCH_3$

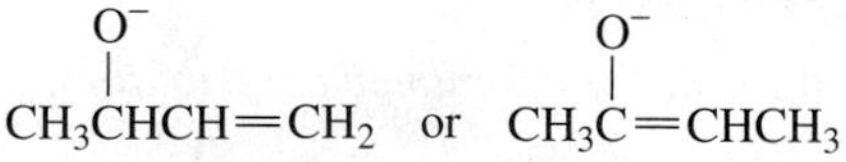

d. 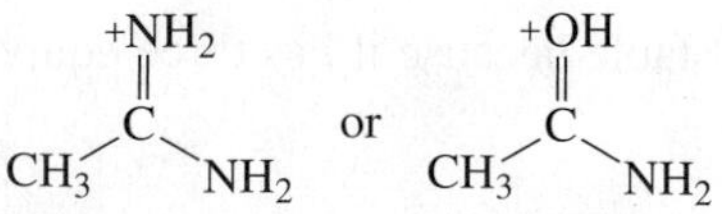 $\overset{\overset{\displaystyle +NH_2}{\|}}{\underset{CH_3}{C}}{\diagdown}_{NH_2}$ or $\overset{\overset{\displaystyle +OH}{\|}}{\underset{CH_3}{C}}{\diagdown}_{NH_2}$

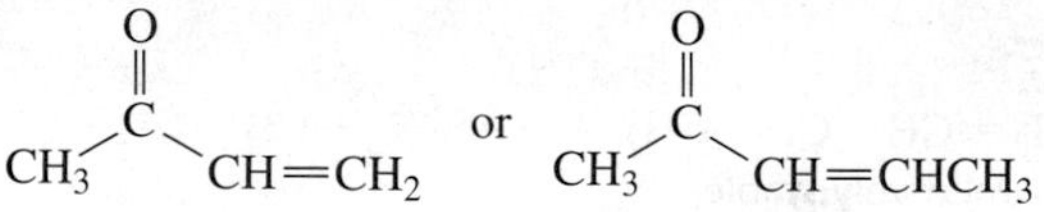

PROBLEM 8♦

Which of the following species has the greatest delocalization energy?

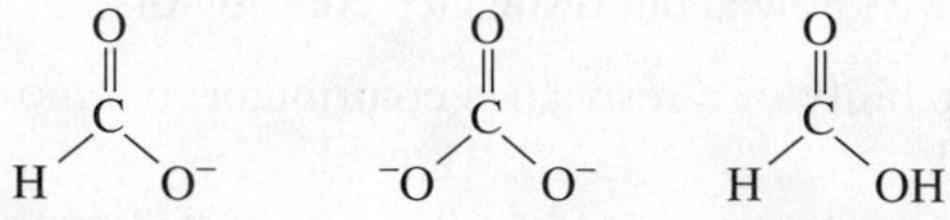

$$\overset{\overset{\displaystyle O}{\|}}{\underset{H\quad O^-}{C}} \qquad \overset{\overset{\displaystyle O}{\|}}{\underset{{}^-O\quad O^-}{C}} \qquad \overset{\overset{\displaystyle O}{\|}}{\underset{H\quad OH}{C}}$$

7 BENZENE IS AN AROMATIC COMPOUND

The two resonance contributors of benzene are identical, so we expect benzene to have a relatively large delocalization energy (Section 6).

Aromatic compounds are particularly stable.

However, the heat of hydrogenation data shown in Figure 7 indicate that benzene's delocalization energy is even larger than that expected for a compound with two equivalent resonance contributors (36 kcal/mol). Compounds with large delocalization energies, such as benzene, are called **aromatic compounds**.

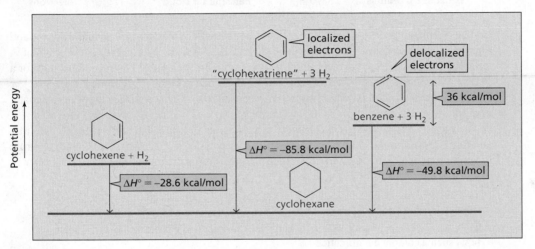

▲ **Figure 7**

Cyclohexene, a compound with one double bond with *localized* π electrons, has an experimental Δ*H*° = −28.6 kcal/mol for its reaction with H$_2$ to form cyclohexane. Therefore, the Δ*H*° of "cyclohexatriene," an unknown hypothetical compound with three double bonds with *localized* π electrons, would be three times that value (Δ*H*° = 3 × −28.6 = −85.8) for the same reaction. Benzene, which has three double bonds with *delocalized* π electrons, has an experimental Δ*H*° = −49.8 kcal/mol for its reaction with H$_2$ to form cyclohexane. The difference in the energies of "cyclohexatriene" and benzene (36 kcal/mol) is the delocalization energy of benzene—the extra stability benzene has as a result of having delocalized electrons.

Because of its large delocalization energy, benzene is an extremely stable compound. Therefore, it does not undergo the electrophilic addition reactions that are characteristic of alkenes except under extreme conditions. (Notice the conditions that Sabatier had to use in order to reduce benzene's double bonds earlier in this chapter.) Now we can understand why benzene's unusual stability puzzled nineteenth-century chemists, who did not know about delocalized electrons (Section 1).

8 THE TWO CRITERIA FOR AROMATICITY

How can we tell whether a compound is aromatic by looking at its structure? In other words, what structural features do aromatic compounds have in common? To be classified as aromatic, a compound must meet both of the following criteria:

1. *It must have an uninterrupted cyclic cloud of π electrons* (called a π cloud) *above and below the plane of the molecule.* Let's look a little more closely at what this means:

   - For the π cloud to be cyclic, *the molecule must be cyclic.*
   - For the π cloud to be uninterrupted, *every atom in the ring must have a* p *orbital.*
   - For the π cloud to form, each *p* orbital must overlap the *p* orbitals on either side of it. As a result, *the molecule must be planar.*

355

2. *The π cloud must contain an odd number of pairs of π electrons.*

Thus, benzene is an aromatic compound because it is cyclic and planar, every carbon in the ring has a *p* orbital, and the π cloud contains *three* pairs of π electrons (Figure 1).

For a compound to be aromatic, it must be cyclic and planar, and it must have an uninterrupted cloud of π electrons. The π cloud must contain an odd number of pairs of π electrons.

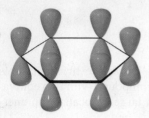

benzene's *p* orbitals

benzene's π cloud

benzene has 3 pairs of π electrons

The German physicist Erich Hückel was the first to recognize that an aromatic compound must have an odd number of pairs of π electrons. In 1931, he described this requirement in what has come to be known as **Hückel's rule,** or the **4n + 2 rule.** The rule states that for a planar, cyclic compound to be aromatic, its uninterrupted π cloud must contain $(4n + 2)$ π electrons, where *n* is any whole number. According to Hückel's rule, then, an aromatic compound must have 2 $(n = 0)$, 6 $(n = 1)$, 10 $(n = 2)$, 14 $(n = 3)$, 18 $(n = 4)$, and so on π electrons. Because there are two electrons in a pair, Hückel's rule requires that an aromatic compound have 1, 3, 5, 7, 9 and so on pairs of π electrons. Thus, Hückel's rule is a mathematical way of saying that an aromatic compound must have an *odd* number of pairs of π electrons.

PROBLEM 9♦

a. What is the value of *n* in Hückel's rule when a compound has nine pairs of π electrons?
b. Is such a compound aromatic?

9 | APPLYING THE CRITERIA FOR AROMATICITY

Cyclobutadiene has two pairs of π electrons, and cyclooctatetraene has four pairs of π electrons. These compounds, therefore, are *not* aromatic because they have an *even* number of pairs of π electrons. There is an additional reason why cyclooctatetraene is not aromatic—it is not planar, it is tub-shaped (see the earlier section on resonance contributors and resonance hybrid). Because cyclobutadiene and cyclooctatetraene are not aromatic, they do not have the unusual stability of aromatic compounds.

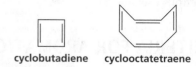

cyclobutadiene   cyclooctatetraene

Now let's look at some other compounds and determine whether they are aromatic. Cyclopentadiene is not aromatic because it does not have an uninterrupted ring of *p* orbital-bearing atoms. One of its ring atoms is *sp*³ hybridized, and only *sp*² and *sp* carbons have *p* orbitals. Therefore, cyclopentadiene does not fulfill the first criterion for aromaticity.

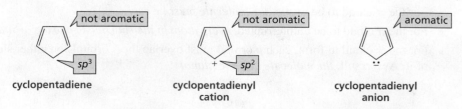

cyclopentadiene   cyclopentadienyl cation   cyclopentadienyl anion

The cyclopentadienyl cation is not aromatic either, because although it has an uninterrupted ring of *p* orbital-bearing atoms, its π cloud has an even number (two) pairs of π electrons. The cyclopentadienyl anion, on the other hand, is aromatic: it has an uninterrupted ring of *p* orbital-bearing atoms, and the π cloud contains an odd number (three) pairs of delocalized π electrons.

How do we know that the cyclopentadienyl anion's lone-pair electrons are π electrons? There is an easy way to determine this: if a lone pair can be used to form a π bond in the ring of a resonance contributor of the compound, then the lone-pair electrons are π electrons.

lone pair forms a π bond

resonance contributors of the cyclopentadienyl anion

When drawing resonance contributors, remember that only electrons move; atoms never move.

resonance hybrid

The resonance hybrid shows that all the carbons in the cyclopentadienyl anion are equivalent. Each carbon has exactly one-fifth of the negative charge associated with the anion.

The criteria that determine whether a monocyclic hydrocarbon is aromatic can also be used to determine whether a polycyclic hydrocarbon is aromatic. Naphthalene (five pairs of π electrons), phenanthrene (seven pairs of π electrons), and chrysene (nine pairs of π electrons) are aromatic.

naphthalene   **phenanthrene**   **chrysene**

Buckyballs

We have seen that diamond, graphite, and graphene are forms of pure carbon. Another form was discovered unexpectedly in 1985, while scientists were conducting experiments designed to understand how long-chain molecules are formed in outer space. R. E. Smalley, R. F. Curl, Jr., and H. W. Kroto shared the 1996 Nobel Prize in Chemistry for discovering this new form of carbon. They named the substance *buckminsterfullerene* (often shortened to *fullerene*) because its structure reminded them of the geodesic domes popularized by R. Buckminster Fuller, an American architect and philosopher. Buckminsterfullerene's nickname is "buckyball."

Consisting of a hollow cluster of 60 carbons, fullerene is the most symmetrical large molecule known. Like graphite and graphene, fullerene has only *sp²* carbons, but instead of being arranged in layers, the carbons are arranged in rings that fit together like the seams of a soccer ball. Each molecule has 32 interlocking rings (20 hexagons and 12 pentagons). At first glance, fullerene appears to be aromatic because of its benzene-like rings. However, the curvature of the ball prevents the molecule from fulfilling the first criterion for aromaticity—it must be planar. Therefore, fullerene is not aromatic.

Buckyballs have extraordinary chemical and physical properties. For example, they are exceedingly rugged, as shown by their ability to survive the extreme temperatures of outer space. Because they are essentially hollow cages, they can be manipulated to make new materials. For example, when a buckyball is "doped" by inserting potassium or cesium into its cavity, it becomes an excellent organic superconductor. These molecules are now being studied for use in many other applications, including the development of new polymers, catalysts, and drug-delivery systems. The discovery of buckyballs is a strong reminder of the technological advances that can be achieved as a result of basic research.

a geodesic dome.

C_{60}
buckminsterfullerene
"buckyball"

PROBLEM 10♦

Which compound in each set is aromatic? Explain your choice.

a.

cyclopropene   cyclopropenyl cation   cyclopropenyl anion

b.

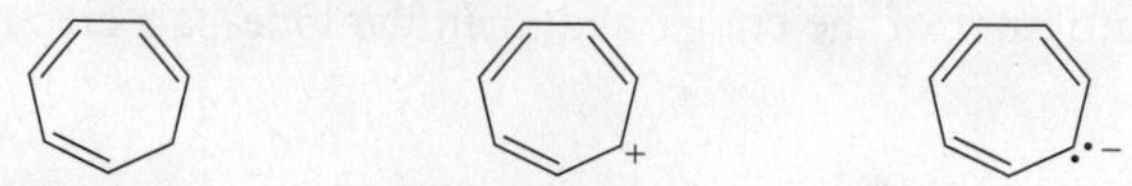

cycloheptatriene   cycloheptatrienyl cation   cycloheptatrienyl anion

---

PROBLEM 11 Solved

The pK_a of cyclopentane is > 60, which is about what is expected for a hydrogen that is bonded to an sp^3 carbon. Explain why cyclopentadiene is a much stronger acid (pK_a of 15), even though it too involves the loss of a proton from an sp^3 carbon.

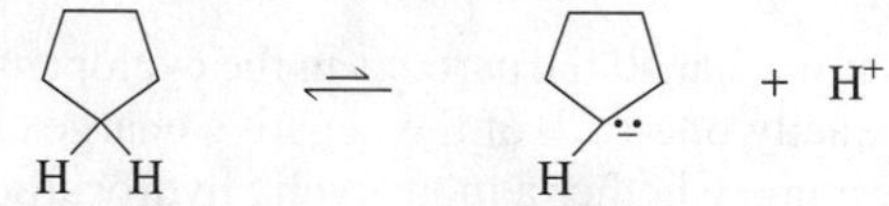

H   H
cyclopentane
p$K_a > 60$ H
cyclopentyl anion $+ \ H^+$

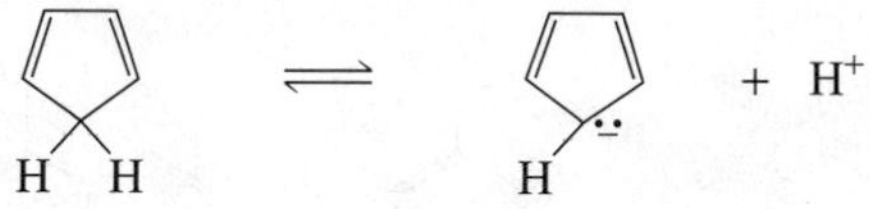

H   H
cyclopentadiene
p$K_a = 15$ H
cyclopentadienyl anion $+ \ H^+$

Solution to 11a To answer this question, we must look at the stabilities of the anions that are formed when the compounds lose a proton. (The strength of an acid is determined by the stability of its conjugate base: the more stable the base, the stronger its conjugate acid.) All the electrons in the cyclopentyl anion are localized. In contrast, the cyclopentadienyl anion is aromatic. As a result of its aromaticity, the cyclopentadienyl anion is an unusually stable carbanion, causing its conjugate acid to be an unusually strong acid compared to other compounds with hydrogens attached to $sp3$ carbons.

PROBLEM 12♦

Which of the following are aromatic?

a.   c. 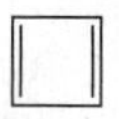   e.

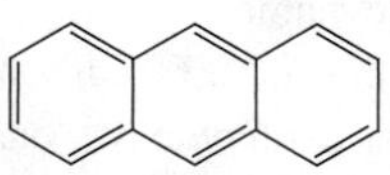

b. 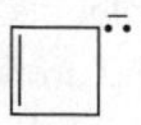   d. 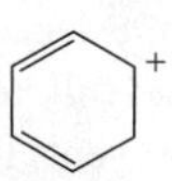   f. $CH_2{=}CHCH{=}CHCH{=}CH_2$

PROBLEM-SOLVING STRATEGY

Analyzing Electron Distribution in Compounds

Which of the following compounds has the greater dipole moment?

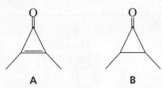

The dipole moment of these compounds results from the unequal sharing of electrons by carbon and oxygen. The dipole moment increases as the electron sharing becomes more unequal. So now the question becomes, which compound has a greater negative charge on its oxygen? To find out, we need to draw the structures with a negative charge on oxygen and determine their relative stabilities.

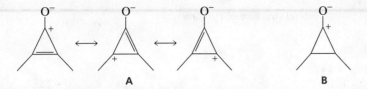

When the charges are separated, we see that charge-separated A has three resonance contributors, all of which are aromatic; charge-separated **B** has no resonance contributors. Therefore, **A** will have a greater concentration of negative charge on its oxygen, which will give **A** a greater dipole moment.

Now use the strategy you have just learned to solve Problem 13.

PROBLEM 13

a. In what direction is the dipole moment in fulvene? Explain.
b. In what direction is the dipole moment in calicene? Explain.

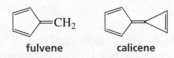

fulvene   calicene

10   AROMATIC HETEROCYCLIC COMPOUNDS

A compound does not have to be a hydrocarbon to be aromatic. Many *heterocyclic compounds* are aromatic. A **heterocyclic compound** is a cyclic compound in which one or more of the ring atoms is an atom other than carbon. The atom that is not carbon is called a **heteroatom.** The name comes from the Greek word *heteros,* which means "different." The most common heteroatoms encountered in organic compounds are N, O, and S.

heterocyclic compounds

pyridine   pyrrole   furan   thiophene

Pyridine is an aromatic heterocyclic compound. Each of the six ring atoms of pyridine is sp^2 hybridized, which means that each has a *p* orbital, and the molecule contains three

benzene

pyridine

pyrrole

pairs of π electrons. Do not be confused by the lone-pair electrons on the nitrogen—they are not π electrons. Recall that lone-pair electrons are π electrons only if they can be used to form a π bond in the ring of a resonance contributor. Because nitrogen is sp^2 hybridized, it has three sp^2 orbitals and a p orbital. The p orbital is used to form the π bond. Two of nitrogen's sp^2 orbitals overlap the sp^2 orbitals of adjacent carbons, and its third sp^2 orbital contains the lone pair.

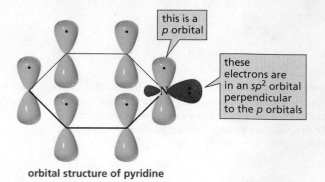

this is a
p orbital

these electrons are in an sp^2 orbital perpendicular to the p orbitals

orbital structure of pyridine

PROBLEM 14♦

What orbital do the lone-pair electrons occupy in each of the following compounds?

a. $CH_3CH_2\ddot{N}H_2$ 　　**b.** ⬡—$CH=\ddot{N}CH_2CH_3$ 　　**c.** $CH_3CH_2C\equiv N:$

The resonance contributors show that lone-pair electrons of pyrrole form a π bond in the ring of a resonance contributor; thus, they are π electrons.

lone pair
forms a
π bond

resonance contributors of pyrrole

The nitrogen atom of pyrrole is sp^2 hybridized. Thus, it has three sp^2 orbitals and a p orbital. It uses its three sp^2 orbitals to bond to two carbons and one hydrogen. The lone-pair electrons are in the p orbital that overlaps the p orbitals of adjacent carbons. Pyrrole, therefore, has three pairs of π electrons and is aromatic.

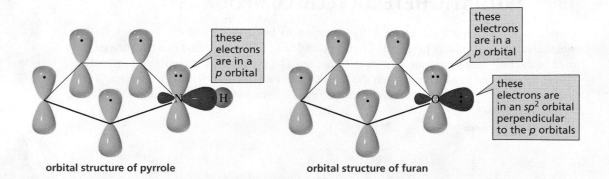

these electrons are in a p orbital

these electrons are in a p orbital

these electrons are in an sp^2 orbital perpendicular to the p orbitals

orbital structure of pyrrole　　　**orbital structure of furan**

Similarly, furan and thiophene are aromatic compounds. Both the oxygen in furan and the sulfur in thiophene are sp^2 hybridized and have one lone pair in an sp^2 orbital. The orbital picture of furan shows that the second lone pair is in a p orbital that overlaps the p orbitals of adjacent carbons, forming a π bond. Thus, they are π electrons.

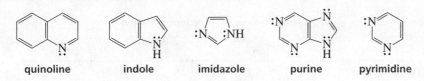

resonance contributors of furan

lone pair forms
a π bond

Quinoline, indole, imidazole, purine, and pyrimidine are other examples of heterocyclic aromatic compounds.

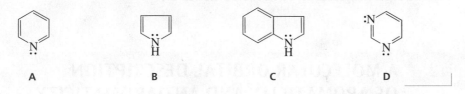

quinoline   indole   imidazole   purine   pyrimidine

PROBLEM 15♦

What orbitals contain the electrons represented as lone pairs in the structures of quinoline, indole, imidazole, purine, and pyrimidine?

PROBLEM 16♦

Which of the following compounds could be protonated without destroying its aromaticity?

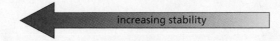

A   B   C   D

PROBLEM 17

Refer to the electrostatic potential maps earlier in this section to answer the following questions:

a. Why is the bottom part of the electrostatic potential map of pyrrole blue?
b. Why is the bottom part of the electrostatic potential map of pyridine red?
c. Why is the center of the electrostatic potential map of benzene more red than the center of the electrostatic potential map of pyridine?

11   ANTIAROMATICITY

An aromatic compound is *more stable* than a cyclic compound with localized electrons, whereas an **antiaromatic compound** is *less stable* than a cyclic compound with localized electrons. *Aromaticity is characterized by stability, whereas antiaromaticity is characterized by instability.*

relative stabilities

aromatic compound > cyclic compound with localized electrons > antiaromatic compound

◄ increasing stability

Antiaromatic compounds
are highly unstable.

A compound is classified as being antiaromatic if it fulfills the first criterion for aromaticity but does not fulfill the second. In other words, it must be a planar, cyclic compound with an uninterrupted ring of *p* orbital-bearing atoms, and the π cloud must contain an *even* number of pairs of π electrons. Hückel would say that the π cloud must contain $4n$ π electrons, where n is any whole number—a mathematical way of saying that the cloud must contain an *even* number of pairs of π electrons.

Cyclobutadiene is a planar, cyclic molecule with two pairs of π electrons. Hence, it is expected to be antiaromatic and highly unstable. In fact, it is too unstable to be isolated, although it has been trapped at very cold temperatures. The cyclopentadienyl cation also has two pairs of π electrons, so we can predict that it, too, is antiaromatic and unstable.

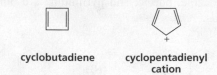

cyclobutadiene   cyclopentadienyl
cation

PROBLEM 18♦

a. Predict the relative pK_a values of cyclopentadiene and cycloheptatriene.
b. Predict the relative pK_a values of cyclopropene and cyclopropane.

PROBLEM 19♦

Which is more soluble in water, 3-bromocyclopropene or bromocyclopropane?

PROBLEM 20♦

Which of the compounds in Problem 12 are antiaromatic?

---

12   A MOLECULAR ORBITAL DESCRIPTION OF AROMATICITY AND ANTIAROMATICITY

Why are compounds with π clouds highly stable (aromatic) if the π cloud contains an odd number of pairs of π electrons but highly unstable (antiaromatic) if the π cloud contains an even number of pairs of π electrons? To answer this question, we must turn to molecular orbital theory.

The relative energies of the π molecular orbitals of a compound that has a π cloud can be determined—without having to use any math—by first drawing the cyclic compound with one of its vertices pointed down. The relative energies of the π molecular orbitals correspond to the relative levels of the vertices (Figure 8).

Aromatic compounds are stable because they have filled bonding π molecular orbitals.

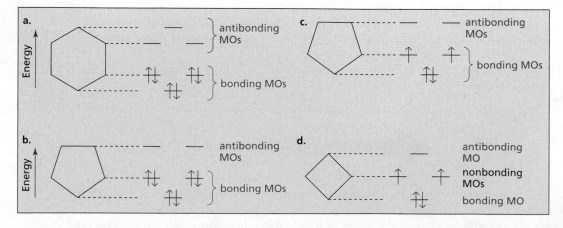

▲ **Figure 8**

The distribution of electrons in the π molecular orbitals of **(a)** benzene, **(b)** the cyclopentadienyl anion, **(c)** the cyclopentadienyl cation, and **(d)** cyclobutadiene.

Notice that the number of π molecular orbitals is the same as the number of atoms in the ring because each ring atom contributes a p orbital. (Note that orbitals are conserved.)

Molecular orbitals below the midpoint of the cyclic structure are bonding molecular orbitals, those above the midpoint are antibonding molecular orbitals, and any at the midpoint are nonbonding molecular orbitals. This simple scheme is sometimes called a Frost device (or a Frost circle) in honor of Arthur A. Frost, the scientist who devised it.

The six π electrons of benzene occupy its three bonding π molecular orbitals, and the six π electrons of the cyclopentadienyl anion occupy its three bonding π molecular orbitals. Notice that there is always an odd number of bonding molecular orbitals because one corresponds to the lowest vertex and the others come in degenerate pairs. Consequently, aromatic compounds—such as benzene and the cyclopentadienyl anion, with their odd number of pairs of π electrons—have completely filled bonding orbitals and do not have electrons in either nonbonding or antibonding orbitals. This is what gives aromatic molecules their stability.

Antiaromatic compounds have an even number of pairs of π electrons. Therefore, either they are unable to fill their bonding orbitals (cylopentadienyl cation) or they have a pair of π electrons left over after the bonding orbitals are filled (cyclobutadiene). Hund's rule states that when there are two or more atomic orbitals with the same energy, an electron will occupy an empty orbital before it will pair up with another electron. It requires that these two electrons go into two different degenerate orbitals.

PROBLEM 21◆

How many bonding, nonbonding, and antibonding π molecular orbitals does cyclobutadiene have? In which molecular orbitals are the π electrons?

PROBLEM 22◆

Can a radical be aromatic?

PROBLEM 23

Following the instructions for drawing the π molecular orbital energy levels of the compounds shown in Figure 8, draw the π molecular orbital energy levels for the cycloheptatrienyl cation, the cycloheptatrienyl anion, and the cyclopropenyl cation. For each compound, show the distribution of the π electrons. Which of the compounds are aromatic? Which are antiaromatic?

---

13 MORE EXAMPLES THAT SHOW HOW DELOCALIZED ELECTRONS INCREASE STABILITY

We will now look at two more examples that illustrate the extra stability a molecule acquires as a result of having delocalized electrons.

Stability of Dienes

Dienes are hydrocarbons with two double bonds.

- **Isolated dienes** have isolated double bonds; **isolated double bonds** are separated by more than one single bond.

- **Conjugated dienes** have conjugated double bonds; **conjugated double bonds** are separated by one single bond.

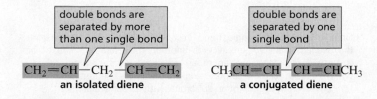

double bonds are separated by more than one single bond

double bonds are separated by one single bond

$CH_2{=}CH{-}CH_2{-}CH{=}CH_2$ $CH_3CH{=}CH{-}CH{=}CHCH_3$
an isolated diene   **a conjugated diene**

The relative stabilities of alkenes can be determined by their heats of hydrogenation. The most stable alkene has the smallest heat of hydrogenation; it gives off the least heat when it is hydrogenated because it has less energy to begin with.

The heat of hydrogenation of 1,3-pentadiene (a conjugated diene) is smaller than that of 1,4-pentadiene (an isolated diene). *A conjugated diene, therefore, is more stable than an isolated diene.*

	Heat of hydrogenation	$\Delta H°$ (kcal/mol)	(kJ/mol)
$CH_2=CH-CH_2-CH=CH_2$ + 2 H_2 $\xrightarrow{Pd/C}$ $CH_3CH_2CH_2CH_2CH_3$ **1,4-pentadiene an isolated diene**	60.2 kcal/mol	−60.2	−252
$CH_2=CH-CH=CHCH_3$ + 2 H_2 $\xrightarrow{Pd/C}$ $CH_3CH_2CH_2CH_2CH_3$ **1,3-pentadiene a conjugated diene**	54.1 kcal/mol	−54.1	−226

Why is a conjugated diene more stable than an isolated diene? Two factors contribute to the difference. The first is *electron delocalization*. The π electrons in each of the double bonds of an isolated diene are *localized* between two carbons. In contrast, the π electrons in a conjugated diene are *delocalized* (Figure 9) and electron delocalization stabilizes a compound.

The most stable alkene has the smallest heat of hydrogenation.

An increase in delocalization energy means an increase in stability.

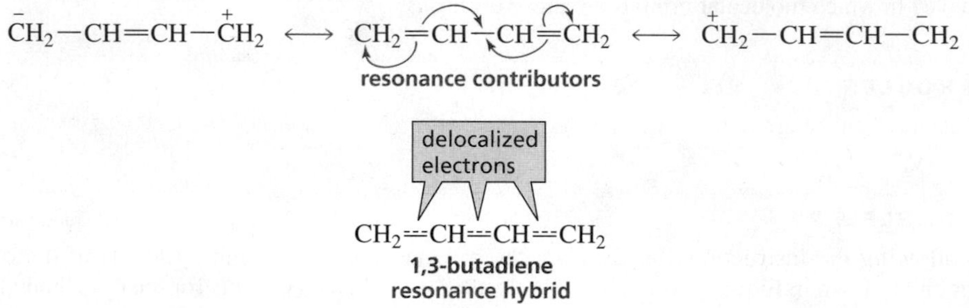

$$\bar{C}H_2-CH=CH-\overset{+}{C}H_2 \longleftrightarrow CH_2=CH-CH=CH_2 \longleftrightarrow \overset{+}{C}H_2-CH=CH-\bar{C}H_2$$

resonance contributors

delocalized electrons

$$CH_2\text{---}CH\text{---}CH\text{---}CH_2$$

1,3-butadiene resonance hybrid

▲ **Figure 9**

The resonance hybrid shows that the single bond in a conjugated diene such as 1,3-butadiene is not a pure single bond, but has partial double-bond character due to electron delocalization. Notice that because the compound does *not* have an electronegative atom that would determine the direction in which the electrons move, they can move both to the left (indicated by the blue arrows) and to the right (indicated by the red arrows).

The *hybridization of the orbitals* forming the carbon–carbon single bonds also causes a conjugated diene to be more stable than an isolated diene. Because a $2s$ electron is closer to the nucleus, on average, than a $2p$ electron, a bond formed by sp^2-sp^2 overlap is *shorter* and *stronger* than one formed by sp^3-sp^2 overlap, since an sp^2 orbital has more s character than an sp^3 orbital (Table 1). Thus, one of the single bonds in a conjugated diene is a stronger single bond than those in an isolated diene, and stronger bonds cause a compound to be more stable (Figure 10).

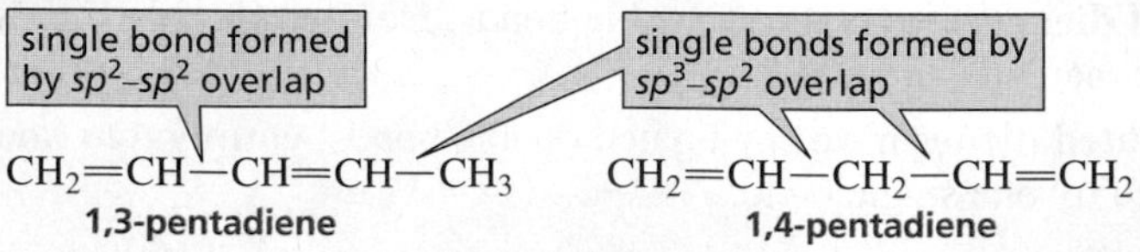

single bond formed by sp^2-sp^2 overlap

single bonds formed by sp^3-sp^2 overlap

$$CH_2=CH-CH=CH-CH_3 \qquad CH_2=CH-CH_2-CH=CH_2$$
1,3-pentadiene **1,4-pentadiene**

▲ **Figure 10**

One carbon–carbon single bond in 1,3-pentadiene is formed from the overlap of an sp^2 orbital with another sp^2 orbital and the other is formed from the overlap of an sp^3 orbital with an sp^2 orbital, whereas both carbon–carbon single bonds in 1,4-pentadiene are formed from the overlap of an sp^3 orbital with an sp^2 orbital.

Table 1	Dependence of the Length of a Carbon–Carbon Single Bond on the Hybridization of the Orbitals Used in Its Formation	
Compound	**Hybridization**	**Bond length ($\overset{\circ}{A}$)**
$H_3C—CH_3$	$sp^3–sp^3$	1.54
$H_3C—\overset{\overset{\displaystyle H}{\textstyle\vert}}{C}=CH_2$	$sp^3–sp^2$	1.50
$H_2C=\overset{\overset{\displaystyle H}{\textstyle\vert}}{C}—\overset{\overset{\displaystyle H}{\textstyle\vert}}{C}=CH_2$	$sp^2–sp^2$	1.47
$H_3C—C{\equiv}CH$	$sp^3–sp$	1.46
$H_2C=\overset{\overset{\displaystyle H}{\textstyle\vert}}{C}—C{\equiv}CH$	$sp^2–sp$	1.43
$HC{\equiv}C—C{\equiv}CH$	$sp–sp$	1.37

Allenes are compounds that have **cumulated double bonds.** These are double bonds that are adjacent to one another. The cumulated double bonds give allenes an unusual geometry because the central carbon is *sp* hybridized. The *sp* carbon forms two π bonds by overlapping one of its *p* orbitals with a *p* orbital of an adjacent sp^2 carbon and overlapping its second *p* orbital with a *p* orbital of the other adjacent sp^2 carbon (Figure 11a).

a.   **b.**

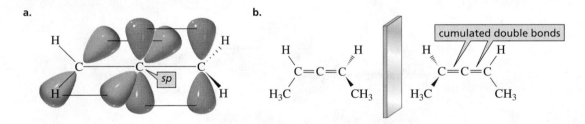

▲ **Figure 11**
(a) The two *p* orbitals on the central carbon are perpendicular, so the plane containing one H—C—H group is perpendicular to the plane containing the other H—C—H group, causing allene to be a nonplanar molecule.
(b) 2,3-Pentadiene has a nonsuperimposable mirror image, making it a chiral molecule even though it does not have an asymmetric center.

Organic Compounds That Conduct Electricity

In order for an organic compound to conduct electricity, its electrons must be delocalized so they can move through the compound just like electrons can move along a copper wire. The first organic compound able to conduct electricity was prepared by hooking together a large number of acetylene molecules in order to form a compound called polyacetylene—a process called **polymerization.** Polyacetylene is an example of a **polymer**—a large molecule made by linking together many small molecules. (Polymer chemistry is an extremely important field of organic chemistry.)

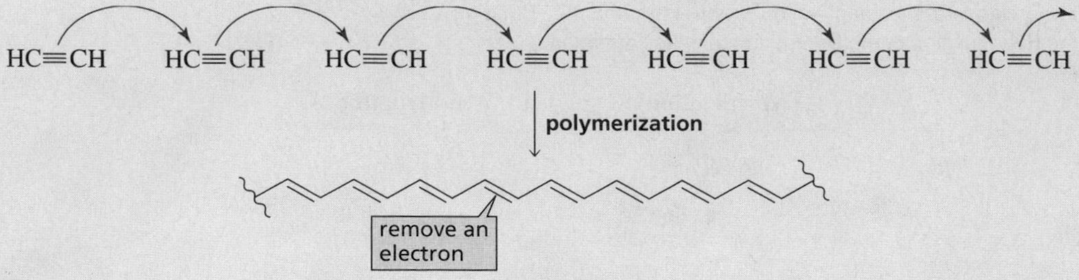

polyacetylene

The electrons in polyacetylene are not able to move along the length of the chain easily enough to conduct electricity well. However, if electrons are removed from or added to the chain (a process called "doping"), then the electrons can move down the chain and the polymer (with some refinements) can conduct electricity as well as copper can.

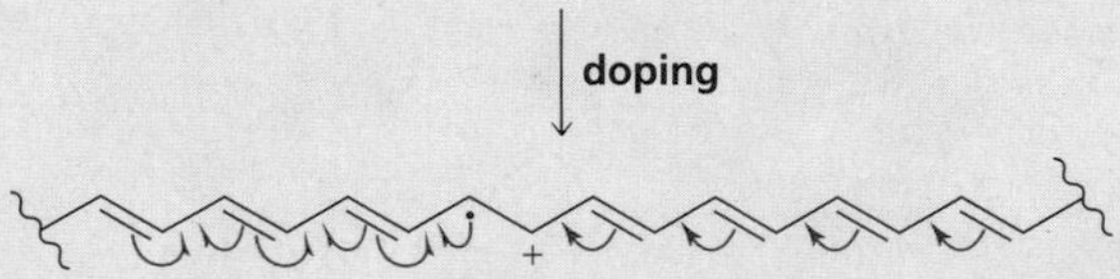

Polyacetylene is very sensitive to air and moisture, which limits its technological applications. However, many other conducting polymers have been developed (three of which are shown here) that have found many practical uses. Notice that all these conducting polymers have a chain of conjugated double bonds.

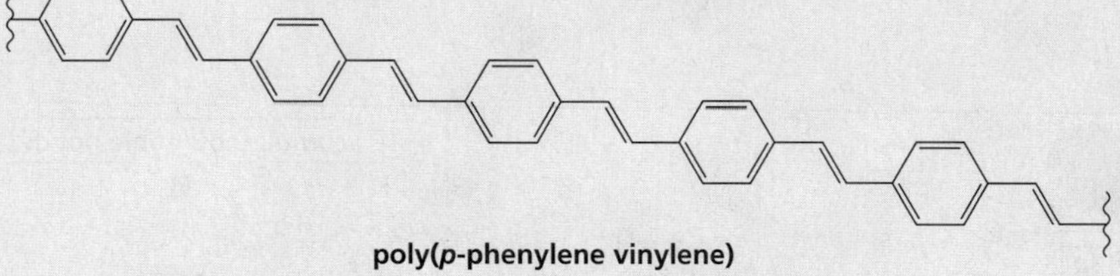

poly(*p*-phenylene vinylene)

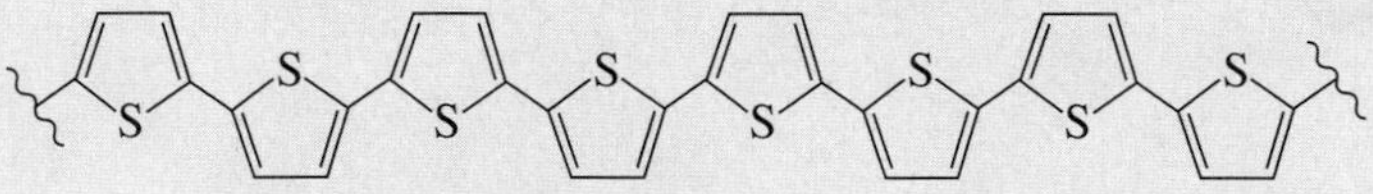

polythiophene

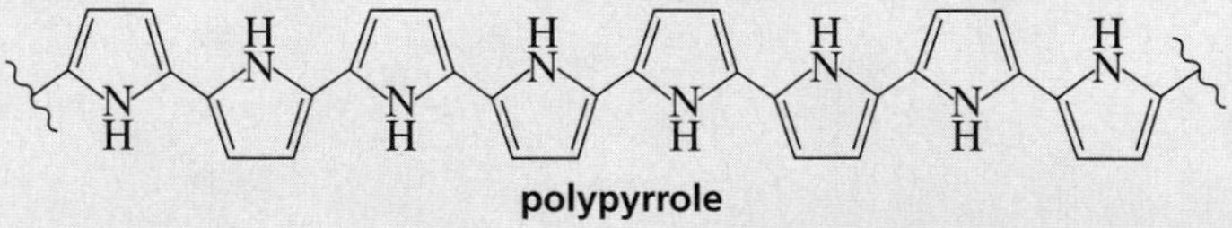

polypyrrole

An important property of conducting polymers is that they are very light. As a result, they are used to coat airplanes to prevent lightning from damaging the interior of the aircraft. The build up of static electricity can be prevented by coating an insulator with a thin coating of conducting polymer. Conducting polymers are also used in LED (light emitting diode) displays. LEDs emit light in response to an electric current—a process known as electroluminescence. LEDs are used for full color displays in flat-screen TVs, cell phones, and the instrument panels in cars and airplanes. Continued research should lead to many more applications for conducting polymers. One such area is the development of "smart structures," such as golf clubs that will adapt to the golfer's swing. Smart skis (which do not vibrate while skiing) have already been created.

Alan Heeger (University of California, Santa Barbara), Alan MacDiarmid (University of Pennsylvania), and Hideki Shirakawa (University of Tsukuba, Japan) shared the Nobel Prize in Chemistry in 2000 for their work on conducting polymers.

PROBLEM 24♦

The heat of hydrogenation of 2,3-pentadiene, a cumulated diene, is 70.5 kcal/mol. What are the relative stabilities of cumulated, conjugated, and isolated dienes?

PROBLEM 25♦

Name the following dienes and rank them in order from most stable to least stable. (Alkyl groups stabilize dienes in the same way that they stabilize alkenes.)

$$CH_3CH=CHCH=CHCH_3 \qquad CH_2=CHCH_2CH=CH_2$$

$$\overset{\displaystyle CH_3}{\underset{\displaystyle |}{CH_3C}}=CHCH=\overset{\displaystyle CH_3}{\underset{\displaystyle |}{CCH_3}} \qquad CH_3CH=CHCH=CH_2$$

Stability of Allylic and Benzylic Cations

Now we will look at allylic and benzylic cations. These are carbocations that have delocalized electrons and are therefore more stable than similar carbocations with localized electrons.

- An **allylic cation** is a carbocation with a positive charge on an allylic carbon; an **allylic carbon** is a carbon adjacent to an sp^2 carbon of an alkene.
- A **benzylic cation** is a carbocation with a positive charge on a benzylic carbon; a **benzylic carbon** is a carbon adjacent to an sp^2 carbon of a benzene ring.

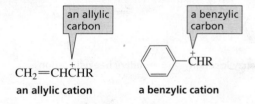

allyl cation

The *allyl cation* is an unsubstituted allylic cation, and the *benzyl cation* is an unsubstituted benzylic cation.

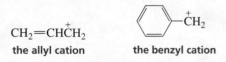

An allylic cation has two resonance contributors. The positive charge is not localized on a single carbon but is shared by two carbons.

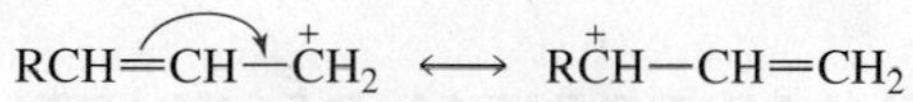

$$RCH=CH\overset{+}{-}CH_2 \quad \longleftrightarrow \quad R\overset{+}{C}H-CH=CH_2$$
an allylic cation

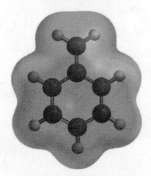

benzyl cation

A benzylic cation has five resonance contributors. Notice that the positive charge is shared by four carbons.

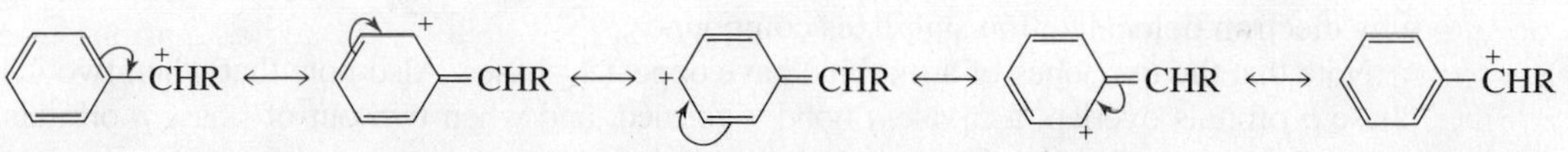

a benzylic cation

Because the allyl and benzyl cations have delocalized electrons, they are more stable than other primary carbocations in solution.

relative stabilities of carbocations

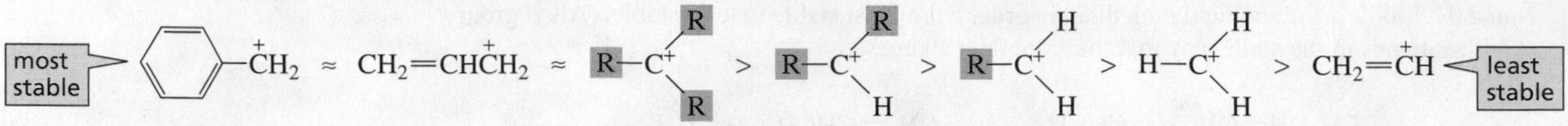

| benzyl cation | allyl cation | a tertiary carbocation | a secondary carbocation | a primary carbocation | methyl cation | vinyl cation |

Not all allylic and benzylic cations have the same stability. Just as a tertiary alkyl carbocation is more stable than a secondary alkyl carbocation, a tertiary allylic cation is more stable than a secondary allylic cation, which in turn is more stable than the (primary) allyl cation. Similarly, a tertiary benzylic cation is more stable than a secondary benzylic cation, which is more stable than the (primary) benzyl cation.

relative stabilities

tertiary allylic cation   secondary allylic cation   allyl cation

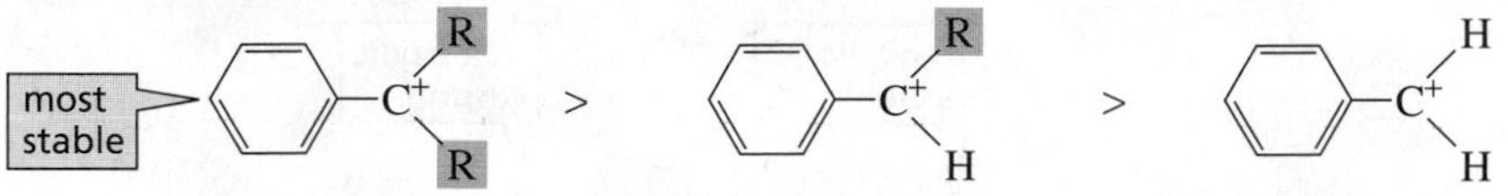

tertiary benzylic cation   secondary benzylic cation   benzyl cation

---

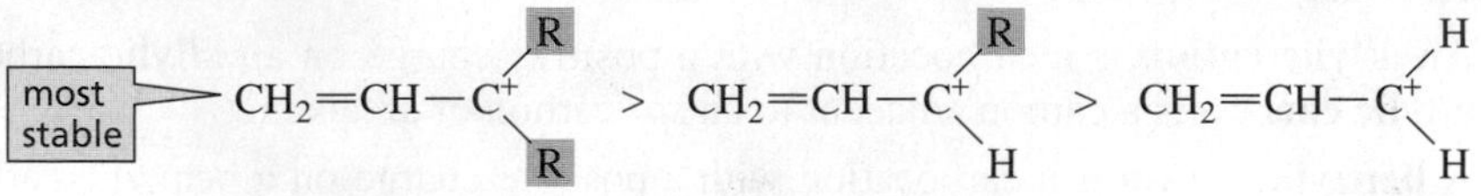

PROBLEM 26♦

Which carbocation in each pair is more stable?

a. 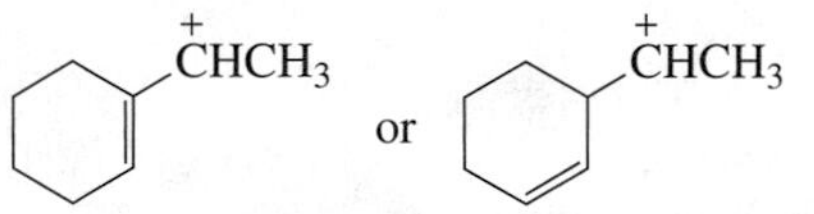   or   c. 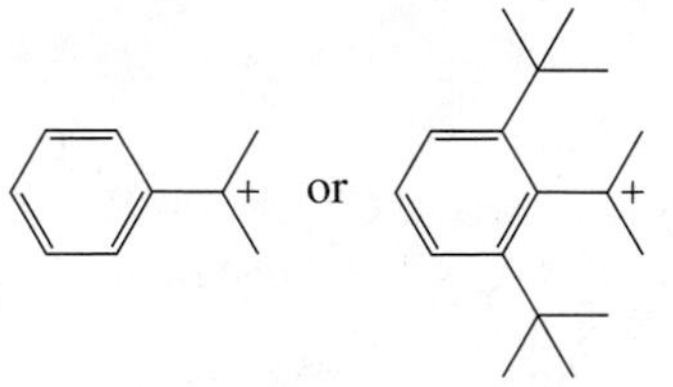   or

b. $CH_3\overset{+}{O}CH_2$ or $CH_3\overset{+}{N}HCH_2$

---

14 A MOLECULAR ORBITAL DESCRIPTION OF STABILITY

So far in our discussion we have used resonance contributors to explain why compounds are stabilized by electron delocalization. Molecular orbital (MO) theory can also explain why electron delocalization stabilizes compounds.

Note that the two lobes of a p orbital have opposite phases. Also note that when two in-phase p orbitals overlap, a covalent bond is formed, and when two out-of-phase p orbitals overlap, they cancel each other and produce a node between the two nuclei.

Let's start by discussing how the MOs of ethene are constructed. The two p atomic orbitals can be either in-phase or out-of-phase. (The different phases are indicated by different colors.) Notice that the number of orbitals is conserved—the number of molecular orbitals equals the number of atomic orbitals that produced the molecular orbitals. Thus, the two p atomic orbitals of ethene overlap to produce two MOs (Figure 12). Side-to-side overlap of in-phase p orbitals (lobes of the same color) produces a π **bonding molecular orbital,** designated ψ_1 (the Greek letter *psi*). Side-to-side overlap of out-of-phase p orbitals produces a π^* **antibonding molecular orbital,** ψ_2.

The overlap of in-phase orbitals holds atoms together; it is a bonding interaction.

The overlap of out-of-phase orbitals pulls atoms apart; it is an antibonding interaction.

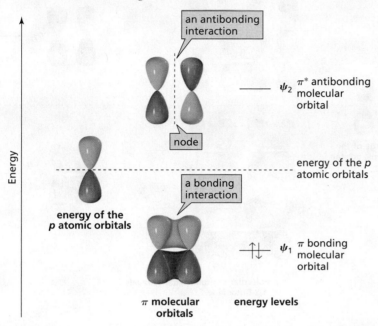

p orbitals overlap to form
π bonding and π^* antibonding
molecular orbitals.

▲ **Figure 12**

An MO diagram of ethene. The π bonding MO is lower in energy than the p atomic orbitals, and it encompasses both carbons. In other words, each electron in the bonding MO spreads over both carbons. The π^* antibonding MO is higher in energy than the p atomic orbitals; it has a node between the lobes of opposite phases.

The π electrons are placed in molecular orbitals according to the same rules that govern the placement of electrons in atomic orbitals—namely, the aufbau principle (orbitals are filled in order of increasing energy), the Pauli exclusion principle (each orbital can hold two electrons of opposite spin), and Hund's rule (an electron will occupy an empty degenerate orbital before it will pair up with an electron already present in an orbital). Thus, ethene's two π electrons are both in ψ_1.

1,3-Butadiene

The four π electrons in 1,3-butadiene are delocalized over four carbons.

$$\ddot{C}H_2-CH=CH-\overset{+}{C}H_2 \quad \longleftrightarrow \quad CH_2=CH-CH=CH_2 \quad \longleftrightarrow \quad \overset{+}{C}H_2-CH=CH-\ddot{C}H_2$$

1,3-butadiene
resonance contributors

$$CH_2{=\!=\!=}CH{=\!=\!=}CH{=\!=\!=}CH_2$$

resonance hybrid

Each of the four carbons contributes one p atomic orbital, and the four p atomic orbitals combine to produce four MOs: ψ_1, ψ_2, ψ_3, and ψ_4 (Figure 13). Thus, we see that a molecular orbital results from the **linear combination of atomic orbitals (LCAO).** Half of the MOs are π-bonding MOs (ψ_1 and ψ_2), and the other half are π^* antibonding MOs (ψ_3 and ψ_4). The energies of the bonding and antibonding MOs are symmetrically distributed above and below the energy of the p atomic orbitals.

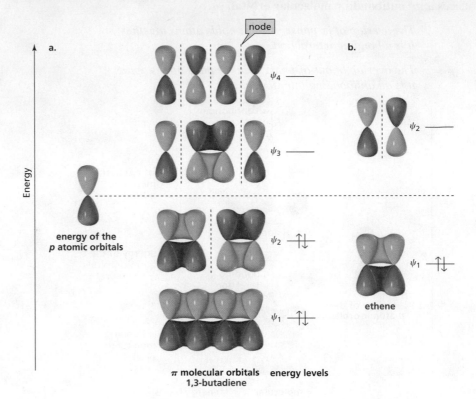

▶ Figure 13

(a) Four p atomic orbitals of 1,3-butadiene overlap to produce four MOs; two are π MOs and two are π^* MOs.

(b) Two p atomic orbitals of ethene overlap to produce two MOs; one is a π MO and the other is a π^* MO.

Notice that the average energy of the four electrons in 1,3-butadiene is lower than the energy of the two electrons in ethene, because ¥1 of ethene is closer to ¥2 of 1,3-butadiene than to ¥1 of 1,3-butadiene. This difference is the delocalization energy. In other words, 1,3-butadiene is stabilized by electron delocalization. Also notice that both compounds have filled bonding MOs.

Figure 13 shows that *as the MOs increase in energy, the number of nodes within them increases and the number of bonding interactions decreases.*

For example, 1,3-butadiene's lowest-energy MO (ψ_1) has no nodes between the nuclei (it has only the node that bisects the orbitals), and it has three bonding interactions; ψ_2 has one node between the nuclei and two bonding interactions (for a net of one bonding interaction); ψ_3 has two nodes between the nuclei and one bonding interaction (for a net of one antibonding interaction); and ψ_4 has three nodes between the nuclei and no bonding interactions (for a net of three antibonding interactions). The four π electrons of 1,3-butadiene reside in the bonding MOs (ψ_1 and ψ_2).

1,3-Butadiene's lowest-energy MO (ψ_1) is the most stable because it has three bonding interactions; its two electrons are delocalized over all four nuclei, thus encompassing all four carbons. The MO next lowest in energy (ψ_2) is also a bonding MO because it has one more bonding interaction than antibonding interaction; it is not as strongly bonding as ψ_1. Overall, ψ_3 is an antibonding MO, because it has one more antibonding interaction than bonding interaction. It is not as strongly antibonding as ψ_4 that has no bonding interactions and three antibonding interactions.

The two bonding MOs in Figure 13a show that the greatest π electron density in a compound with two double bonds joined by one single bond is between C-1 and C-2 and between C-3 and C-4, but there is some π electron density between C-2 and C-3—just as the resonance contributors show. The MOs also show why 1,3-butadiene is most stable in a planar conformation: if 1,3-butadiene weren't planar, there would be little or no overlap between the p orbitals on C-2 and C-3.

Both ψ_1 and ψ_3 are **symmetric molecular orbitals;** they have a plane of symmetry, so one half is the mirror image of the other half. In contrast, ψ_2 and ψ_4 are **antisymmetric;** they do not have a plane of symmetry (but would have one, if one half of the MO were turned upside down). Notice in Figure 13 that as the MOs increase in energy, they alternate between symmetric and antisymmetric.

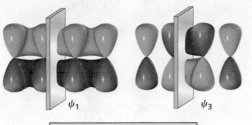
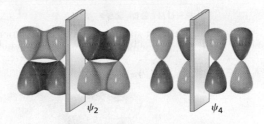

| symmetric molecular orbitals | antisymmetric molecular orbitals |

The highest-energy molecular orbital of 1,3-butadiene that contains electrons is ψ_2. Therefore, ψ_2 is called the **highest occupied molecular orbital (HOMO).** The lowest-energy molecular orbital of 1,3-butadiene that does not contain electrons is ψ_3; ψ_3 is called the **lowest unoccupied molecular orbital (LUMO).**

The MO description of 1,3-butadiene shown in Figure 13 represents the electronic configuration of the molecule in its *ground state*. If the molecule absorbs light of an appropriate wavelength, the light will promote an electron from its HOMO to its LUMO (from ψ_2 to ψ_3). The molecule will then be in an *excited* state. The excitation of an electron from the HOMO to the LUMO is the basis of ultraviolet and visible spectroscopy.

HOMO = the highest occupied molecular orbital

LUMO = the lowest unoccupied molecular orbital

PROBLEM 27♦

What is the total number of nodes in the ψ_3 and ψ_4 MOs of 1,3-butadiene?

PROBLEM 28♦

Answer the following questions for the π MOs of 1,3-butadiene:

a. Which are π bonding MOs and which are π^* antibonding MOs?
b. Which MOs are symmetric and which are antisymmetric?
c. Which MO is the HOMO and which is the LUMO in the ground state?
d. Which MO is the HOMO and which is the LUMO in the excited state?
e. What is the relationship between the HOMO and the LUMO and symmetric and antisymmetric orbitals?

1,4-Pentadiene

Now let's look at look at 1,4-pentadiene. 1,4-Pentadiene, like 1,3-butadiene, has four π electrons. However, unlike the two pairs of π electrons in 1,3-butadiene that are *delocalized,* the two pairs of π electrons in 1,4-pentadiene are *localized*—completely separate from one another.

the CH$_2$ group causes the π electrons to be localized

the π electrons are delocalized

CH_2=$CHCH_2CH$=CH_2
1,4-pentadiene

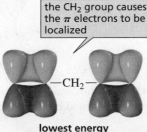

$-CH_2-$

lowest energy
π molecular orbital of
1,4-pentadiene

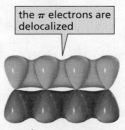

lowest energy
π molecular orbital of
1,3-butadiene

CH_2=CH—CH=CH_2
1,3-butadiene

Because its electrons are localized, the lowest energy MO of 1,4-pentadiene has the same energy as the lowest MO of ethene, a compound with one pair of localized π electrons. We have now seen that molecular orbital theory and resonance contributors are two different ways to show that the π electrons in 1,3-butadiene are delocalized and that electron delocalization stabilizes a compound.

PROBLEM 29♦

The most stable MO of 1,3,5-hexatriene and the most stable MO of benzene are shown here. Which compound is more stable? Why?

most stable MO of 1,3,5-hexatriene

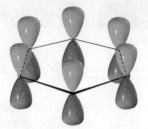

most stable MO of benzene

---

15 HOW DELOCALIZED ELECTRONS AFFECT pK_a VALUES

A carboxylic acid is a stronger acid than an alcohol, because the carboxylate ion (the conjugate base of the carboxylic acid) is a more stable (weaker) base than the alkoxide ion (the conjugate base of an alcohol). Recall that the more stable the base, the stronger its conjugate acid.

A nearby electronegative atom stabilizes an anion by inductive electron withdrawal.

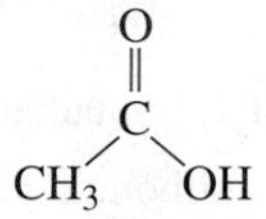

$$
\begin{array}{cc}
\underset{\substack{\text{acetic acid}\\ \text{p}K_a = 4.76}}{\overset{\displaystyle \text{O} \atop \displaystyle \|}{\underset{\text{CH}_3}{\text{C}}\text{OH}}} &
\underset{\substack{\text{ethanol}\\ \text{p}K_a = 15.9}}{\text{CH}_3\text{CH}_2\text{OH}}
\end{array}
$$

We also saw that the greater stability of the carboxylate ion is attributable to two factors—inductive electron withdrawal and electron delocalization. That is, the double-bonded oxygen stabilizes the carboxylate ion by decreasing the electron density of the negatively charged oxygen by inductive *electron withdrawal* and by an increase in *delocalization energy*.

Although both the carboxylic acid and the carboxylate ion have delocalized electrons, the delocalization energy of the carboxylate ion is greater than that of the carboxylic acid because the ion has two equivalent resonance contributors that are predicted to be relatively stable, whereas the carboxylic acid has only one (Section 6). Therefore, loss of a proton from a carboxylic acid is accompanied by an increase in delocalization energy—in other words, an increase in stability.

acetic acid

acetate ion

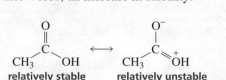

relatively stable   relatively unstable

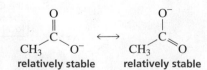

relatively stable   relatively stable

In contrast, all the electrons in both an alcohol and its conjugate base are localized, so loss of a proton from an alcohol is not accompanied by an increase in delocalization energy.

$$\underset{\text{ethanol}}{\text{CH}_3\text{CH}_2\text{OH}} \rightleftharpoons \underset{\text{ethoxide ion}}{\text{CH}_3\text{CH}_2\text{O}^-} + \text{H}^+$$

The same two factors that cause a carboxylic acid to be a stronger acid than an alcohol also cause phenol to be a stronger acid than an alcohol such as cyclohexanol—namely, the stabilization of phenol's conjugate base both by inductive *electron withdrawal* and by an increase in *delocalization energy*.

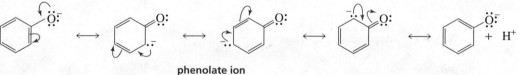

phenol	**cyclohexanol**	**ethanol**
pK_a = 10	pK_a = 16	pK_a = 16

The OH group of phenol is attached to an sp^2 carbon, which is more electronegative than the sp^3 carbon to which the OH group of cyclohexanol is attached. The greater *inductive electron withdrawal* by the more electronegative sp^2 carbon stabilizes the conjugate base by decreasing the electron density of its negatively charged oxygen.

While both phenol and the phenolate ion have delocalized electrons, the delocalization energy of the phenolate ion is greater than that of phenol because three of phenol's resonance contributors have separated charges as well as a positive charge on an oxygen. The loss of a proton from phenol, therefore, is accompanied by an increase in delocalization energy.

phenol

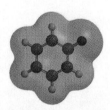

phenolate ion

phenol

phenolate ion

In contrast, the conjugate base of cyclohexanol does *not* have any delocalized electrons to stabilize it.

cyclohexanol   **cyclohexanolate ion**

Phenol is a weaker acid than a carboxylic acid because electron withdrawal by the sp^2 carbon in the phenolate ion is not as great as electron withdrawal by the oxygen in the carboxylate ion. In addition, the increased delocalization energy when a proton is lost is not as great in a phenolate ion as in a carboxylate ion, where the negative charge is shared equally by two oxygens.

These same two factors explain why protonated aniline is a stronger acid than protonated cyclohexylamine.

protonated aniline	**protonated cyclohexylamine**
pK_a = 4.60	pK_a = 11.2

First, aniline's nitrogen is attached to an sp^2 carbon, whereas cyclohexylamine's nitrogen is attached to a less electronegative sp^3 carbon. Second, the nitrogen atom of protonated aniline lacks a lone pair that can be delocalized. However, when the nitrogen loses a proton, the lone pair that formerly held the proton can be delocalized. Loss of a proton, therefore, is accompanied by an increase in delocalization energy.

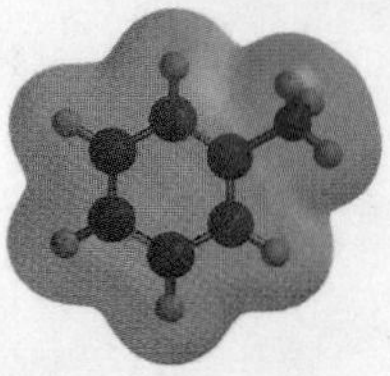

protonated aniline

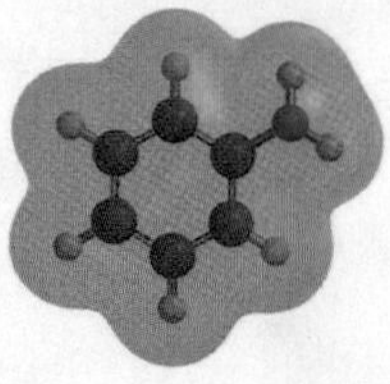

aniline

protonated aniline

aniline

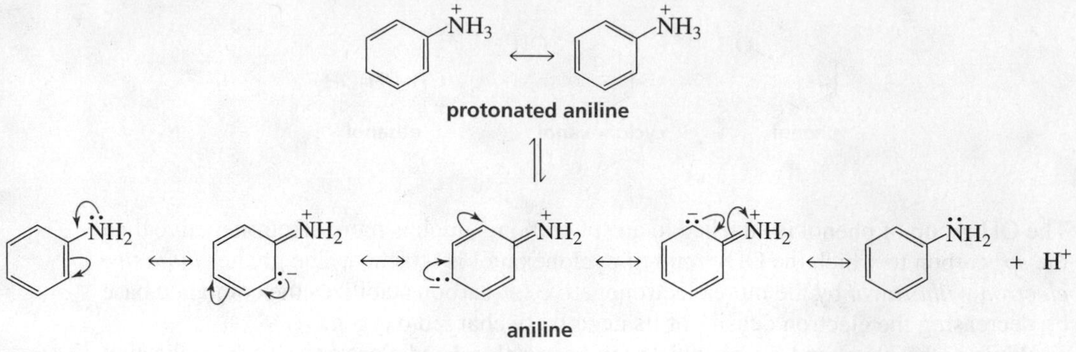

In contrast, cyclohexylamine does not have any delocalized electrons in either the acidic or the basic form to stabilize it.

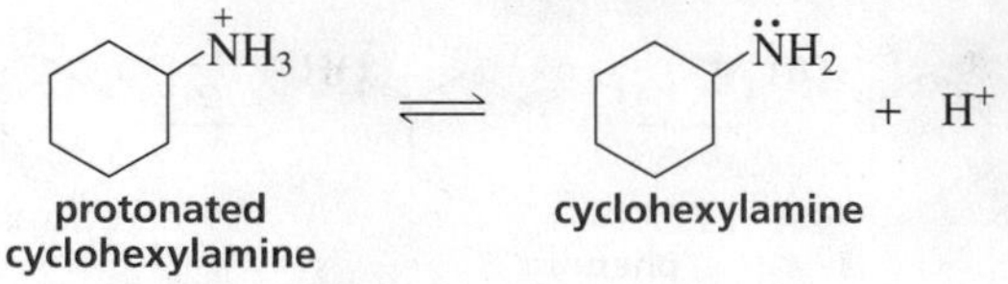

protonated
cyclohexylamine

cyclohexylamine

We can now add phenol and protonated aniline to the list of organic compounds whose approximate pK_a values you should know (Table 2).

Table 2	Approximate pK_a Values		
pK_a < 0	p$K_a \approx 5$	p$K_a \approx 10$	p$K_a \approx 15$
R$\overset{+}{O}$H with H	O=C(R)(OH)	R$\overset{+}{N}$H$_3$	ROH
$\overset{+OH}{\underset{R\ \ OH}{C}}$	$\overset{+}{N}H_3$ (aniline)	OH (phenol)	H$_2$O
H$_3$O$^+$			

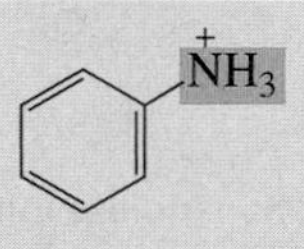

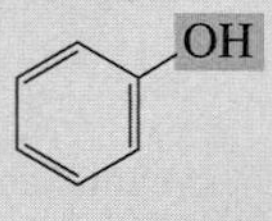

PROBLEM-SOLVING STRATEGY

Determining Relative Acidities

Which is a stronger acid?

$$CH_3CH_2OH \quad \text{or} \quad CH_2=CHOH$$
ethyl alcohol   vinyl alcohol

The strength of an acid depends on the stability of its conjugate base. We have seen that bases are stabilized by electron-withdrawing substituents and by electron delocalization. So you can answer the question by comparing the stabilities of the two conjugate bases and remembering that the more stable base will have the more acidic conjugate acid.

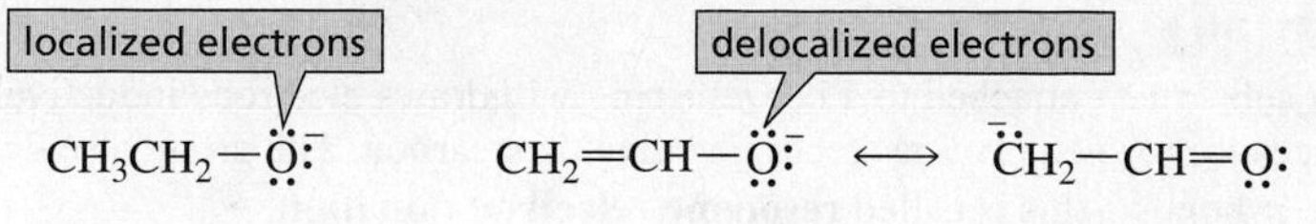

All the electrons in ethanol's conjugate base are localized. However, vinyl alcohol's conjugate base is stabilized by electron delocalization. In addition, the oxygen in vinyl alcohol is attached to an sp^2 carbon, which is more electronegative than the sp^3 carbon to which the oxygen in ethanol is attached. As a result, vinyl alcohol is a stronger acid than ethanol.

Now use the strategy you have just learned to solve Problem 30.

PROBLEM 30◆

Which member of each pair is the stronger acid?

a.

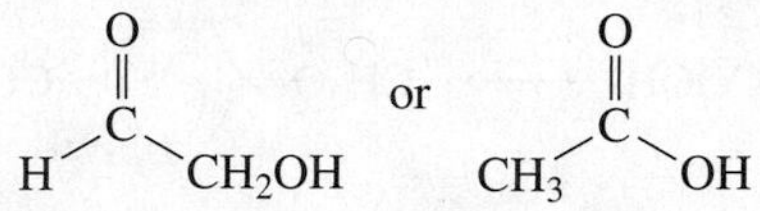

b. $CH_3CH=CHCH_2OH$ or $CH_3CH=CHOH$

c. $CH_3CH_2CH_2\overset{+}{N}H_3$ or $CH_3CH=CH\overset{+}{N}H_3$

PROBLEM 31◆

Which member of each pair is the stronger base?

a. ethylamine or aniline

b. ethylamine or ethoxide ion

c. phenolate ion or ethoxide ion

PROBLEM 32◆

Rank the following compounds in order from strongest acid to weakest acid:

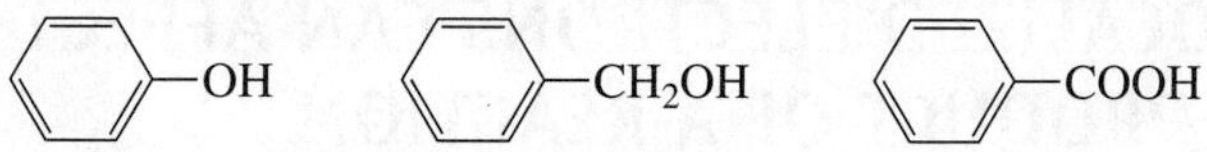

PROBLEM 33 Solved

Which of the following is the stronger acid?

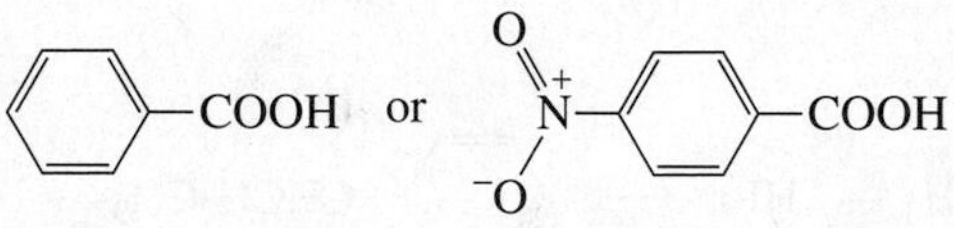

Solution The nitro-substituted compound is the stronger acid because the nitro substituent withdraws electrons both inductively (through the σ bonds) and by resonance (through the π bonds). Withdrawal of electrons through π bonds is called **resonance electron withdrawal.** We have seen that electron-withdrawing substituents increase the acidity of a compound by stabilizing its conjugate base.

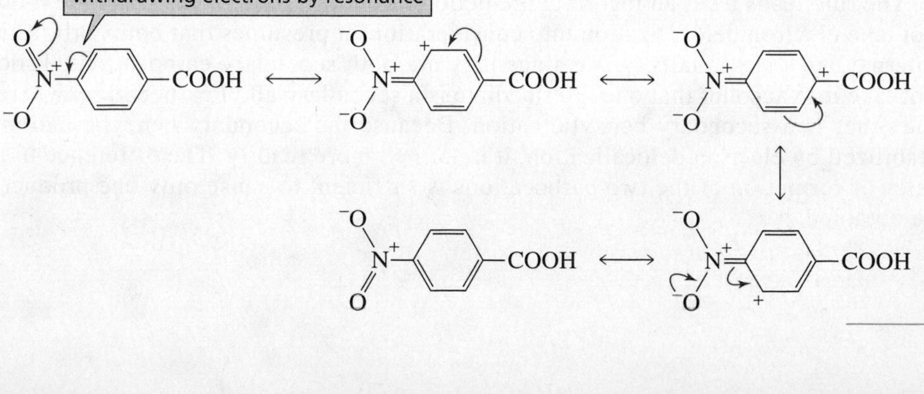

375

PROBLEM 34♦

A methoxy substituent attached to a benzene ring withdraws electrons inductively (through the σ bonds) because oxygen is more electronegative than carbon. The group also donates electrons through the π bonds—this is called **resonance electron donation.**

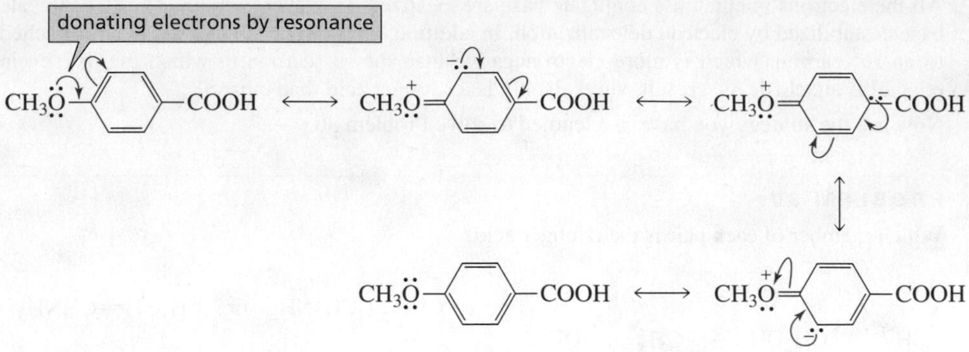

From the pK_a values of the unsubstituted and methoxy-substituted carboxylic acids, predict whether inductive electron withdrawal or resonance electron donation is a more important effect for a CH_3O^- group.

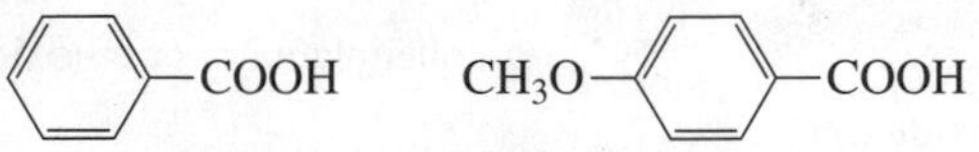

pK_a = 4.20	pK_a = 4.47

16 DELOCALIZED ELECTRONS CAN AFFECT THE PRODUCT OF A REACTION

Our ability to correctly predict the product of an organic reaction often depends on recognizing when organic molecules have delocalized electrons. For example, the alkene in the following reaction has the same number of hydrogens on both of its sp^2 carbons:

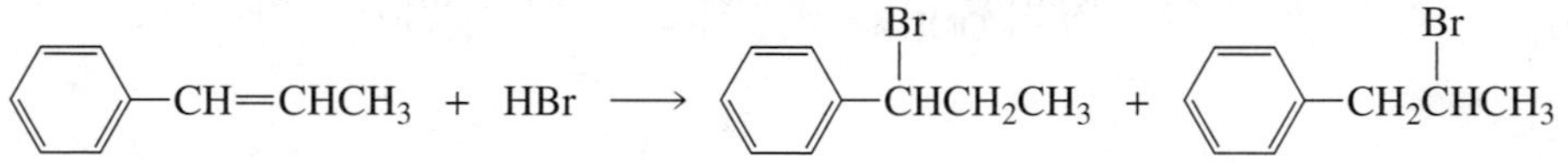

Therefore, the rule that tells us to add the electrophile to the sp^2 carbon bonded to the most hydrogens predicts that approximately equal amounts of the two products will be formed. When the reaction is carried out, however, only one of the products is obtained. (Notice that the stability of the benzene ring prevents its double bonds from undergoing electrophilic addition reactions.)

The rule leads us to an incorrect prediction of the reaction product because it does not take electron delocalization into consideration. It presumes that both carbocation intermediates are equally stable since they are both secondary carbocations. It does not take into account that one intermediate is a secondary alkyl carbocation, whereas the other is a secondary benzylic cation. Because the secondary benzylic cation is stabilized by electron delocalization, it is formed more readily. The difference in the rates of formation of the two carbocations is sufficient to cause only one product to be obtained.

$$\text{a secondary benzylic cation} \qquad \text{a secondary alkyl carbocation}$$

Let this example serve as a warning. The rule that states that the electrophile adds to the sp^2 carbon bonded to the most hydrogens cannot be applied to reactions that form carbocations that can be stabilized by electron delocalization. In such cases, you must look at the relative stabilities of the individual carbocations to predict the major product of the reaction.

PROBLEM 35♦

What is the major product obtained from the addition of HBr to the following compound?

$$\text{CH}_2\text{CH}=\text{CH}_2$$

PROBLEM 36 Solved

Predict the sites on each of the following compounds where protonation can occur.

a. $\text{CH}_3\text{CH}=\text{CHOCH}_3 + \text{H}^+$ **b.** $\bigcirc\text{—N}\bigcirc + \text{H}^+$

Solution to 36a The resonance contributors reveal that there are two sites that can be protonated: the lone pair on oxygen and the lone pair on carbon.

sites of protonation

$$\text{CH}_3\overset{\frown}{\text{CH}}=\text{CH}\overset{\frown}{\ddot{\text{O}}}\text{CH}_3 \longleftrightarrow \text{CH}_3\overset{-}{\text{CH}}-\text{CH}=\overset{+}{\ddot{\text{O}}}\text{CH}_3 \qquad \text{CH}_3\text{CH}=\text{CH}\ddot{\text{O}}\text{CH}_3$$
$$\text{resonance contributors}$$

---

17   REACTIONS OF DIENES

For another example of how delocalized electrons can affect the product of a reaction, we will compare the products formed when *isolated dienes* (dienes that have only localized electrons) undergo electrophilic addition reactions to the products formed when *conjugated dienes* (dienes that have delocalized electrons) undergo the same reactions.

$$\text{CH}_2=\text{CHCH}_2\text{CH}_2\text{CH}=\text{CH}_2 \qquad \text{CH}_3\text{CH}=\text{CH}-\text{CH}=\text{CHCH}_3$$
$$\textbf{an isolated diene} \qquad\qquad \textbf{a conjugated diene}$$

Reactions of Isolated Dienes

The reactions of *dienes with isolated double bonds* are just like the reactions of alkenes. If an excess of the electrophilic reagent is present, two independent electrophilic addition reactions will occur. In each reaction, *the electrophile adds to the sp^2 carbon bonded to the most hydrogens*.

$$\text{CH}_2=\text{CHCH}_2\text{CH}_2\text{CH}=\text{CH}_2 \;+\; \text{HBr} \;\longrightarrow\; \text{CH}_3\text{CHCH}_2\text{CH}_2\text{CHCH}_3$$
$$\textbf{1,5-hexadiene} \qquad\qquad \textbf{excess} \qquad\qquad\quad \underset{\text{Br}}{|}\qquad\quad\underset{\text{Br}}{|}$$

The reaction proceeds exactly as we would predict from our knowledge of the mechanism for the reaction of alkenes with electrophilic reagents.

MECHANISM FOR THE REACTION OF AN ISOLATED DIENE WITH EXCESS HBr

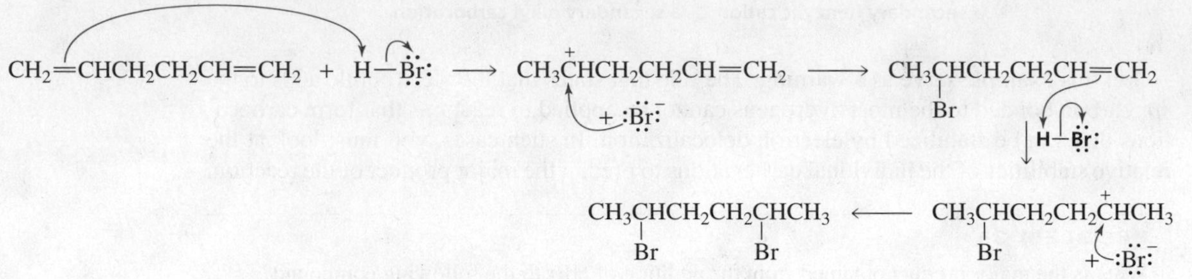

$$CH_2=CHCH_2CH_2CH=CH_2 \ + \ H-\overset{..}{\underset{..}{Br}}: \ \longrightarrow \ CH_3\overset{+}{C}HCH_2CH_2CH=CH_2 \ \longrightarrow \ CH_3CHCH_2CH_2CH=CH_2$$

$$+ :\overset{..}{\underset{..}{Br}}:^-$$

$$\underset{Br}{|}$$

$$\downarrow H-\overset{..}{\underset{..}{Br}}:$$

$$CH_3\underset{Br}{\overset{|}{C}}HCH_2CH_2\underset{Br}{\overset{|}{C}}HCH_3 \ \longleftarrow \ CH_3\underset{Br}{\overset{|}{C}}HCH_2CH_2\overset{+}{C}HCH_3$$

$$+ :\overset{..}{\underset{..}{Br}}:^-$$

- The electrophile (H^+) adds to the sp^2 carbon bonded to the most hydrogens in order to form the more stable carbocation.
- The bromide ion adds to the carbocation.
- Because there is an excess of the electrophilic reagent, there is enough reagent to add to the other double bond; again the H^+ adds to the sp^2 carbon bonded to the most hydrogens.
- The bromide ion adds to the carbocation.

If there is only enough electrophilic reagent to add to one of the double bonds, it will add preferentially to the more reactive one. For example, in the following reaction, addition of HCl to the double bond on the left forms a secondary carbocation, whereas addition to the double bond on the right forms a tertiary carbocation. Because the tertiary carbocation is more stable and is therefore formed faster, the major product of the reaction will be 5-chloro-5-methyl-1-hexene in the presence of a limited amount of HCl.

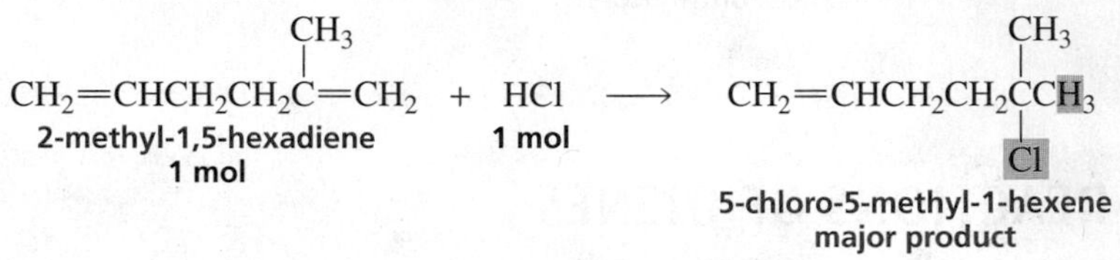

$$CH_2=CHCH_2CH_2\overset{\overset{\displaystyle CH_3}{|}}{C}=CH_2 \ + \ HCl \ \longrightarrow \ CH_2=CHCH_2CH_2\overset{\overset{\displaystyle CH_3}{|}}{\underset{\underset{\displaystyle Cl}{|}}{C}}CH_3$$

2-methyl-1,5-hexadiene
1 mol

1 mol (HCl)

5-chloro-5-methyl-1-hexene
major product

PROBLEM 37♦

What is the major product of each of the following reactions, assuming that one equivalent of each reagent is used in each reaction?

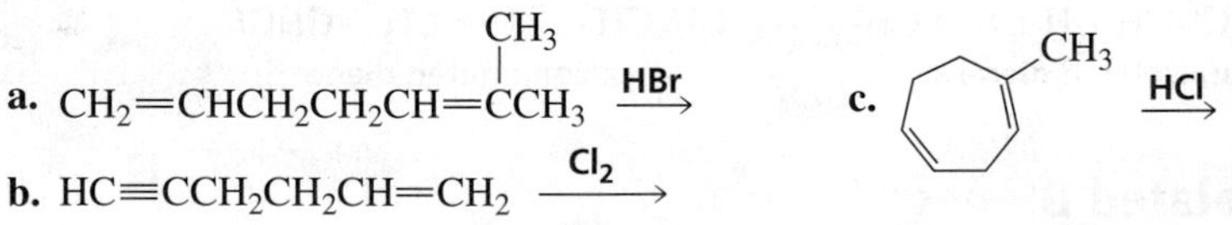

a. $CH_2=CHCH_2CH_2CH=\overset{\overset{\displaystyle CH_3}{|}}{C}CH_3 \ \xrightarrow{HBr}$

b. $HC\equiv CCH_2CH_2CH=CH_2 \ \xrightarrow{Cl_2}$

c. (cycloheptadiene with CH₃) $\xrightarrow{HCl}$

PROBLEM 38♦

Which of the double bonds in zingiberene, the compound responsible for the aroma of ginger, is the most reactive in an electrophilic addition reaction with HBr?

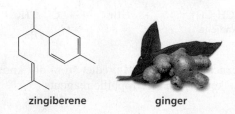

zingiberene   **ginger**

Reactions of Conjugated Dienes

When a diene with *conjugated double bonds,* such as 1,3-butadiene, reacts with a limited amount of electrophilic reagent so that addition can occur at only one of the double bonds, two addition products are formed. One is a **1,2-addition product**—the result of addition at the 1- and 2-positions. The other is a **1,4-addition product**—the result of addition at the 1- and 4-positions. **1,2-Addition** is called **direct addition,** and 1,4-addition is called **conjugate addition.**

$$\overset{1}{CH_2}=\overset{2}{CH}-\overset{3}{CH}=\overset{4}{CH_2} \;+\; Cl_2 \longrightarrow CH_2-CH-CH=CH_2 \;+\; CH_2-CH=CH-CH_2$$

1,3-butadiene   1 mol
1 mol

Cl Cl	Cl Cl
3,4-dichloro-1-butene	**1,4-dichloro-2-butene**
1,2-addition product	**1,4-addition product**

An isolated diene undergoes only 1,2-addition.

A conjugated diene undergoes both 1,2- and 1,4-addition.

$$CH_2=CH-CH=CH_2 \;+\; HBr \longrightarrow CH_3CH-CH=CH_2 \;+\; CH_3-CH=CH-CH_2$$

1,3-butadiene   1 mol
1 mol

Br   Br

3-bromo-1-butene   **1-bromo-2-butene**
1,2-addition product   **1,4-addition product**

Based on your knowledge of how electrophilic reagents add to double bonds, you would expect the 1,2-addition product. However, the 1,4-addition product is surprising, because the reagent did not add to adjacent carbons and a double bond changed its position.

When we talk about addition at the 1- and 2-positions or at the 1- and 4-positions, the numbers refer to the four carbons of the conjugated system. Thus, the carbon in the 1-position is one of the sp^2 carbons at the end of the conjugated system—it is not necessarily the first carbon in the molecule.

$$R-\overset{1}{CH}=\overset{2}{CH}-\overset{3}{CH}=\overset{4}{CH}-R$$

the conjugated system

For example, the 1- and 4-positions in the conjugated system of 2,4-hexadiene are actually C-2 and C-5.

$$CH_3\overset{1}{CH}=CH-CH=\overset{4}{CH}CH_3 \xrightarrow{Br_2} CH_3\overset{1}{CH}-\overset{2}{CH}-CH=CHCH_3 \;+\; CH_3\overset{1}{CH}-CH=CH-\overset{4}{CH}CH_3$$

2,4-hexadiene

Br Br   Br   Br

4,5-dibromo-2-hexene   **2,5-dibromo-3-hexene**
1,2-addition product   **1,4-addition product**

To understand why an electrophilic addition reaction to a conjugated diene forms both 1,2-addition and 1,4-addition products, we need to look at the mechanism of the reaction.

MECHANISM FOR THE REACTION OF A CONJUGATED DIENE WITH HBr

$$CH_2=CH-CH=CH_2 \;+\; H-\ddot{\underset{..}{Br}}\!: \longrightarrow CH_3-\overset{+}{CH}-CH=CH_2 \longleftrightarrow CH_3-CH=CH-\overset{+}{CH_2}$$

1,3-butadiene

$+\; :\!\ddot{\underset{..}{Br}}\!:^-$ an allylic cation $+\; :\!\ddot{\underset{..}{Br}}\!:^-$

$$CH_3-CH-CH=CH_2 \;+\; CH_3-CH=CH-CH_2$$

Br   Br

3-bromo-1-butene   **1-bromo-2-butene**
1,2-addition product   **1,4-addition product**

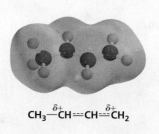

$$CH_3-\overset{\delta+}{CH}\!=\!=\!CH\!=\!=\overset{\delta+}{CH_2}$$

- The proton adds to C-1, forming an allylic cation. The allylic cation has delocalized electrons.

- The resonance contributors of the allylic cation show that the positive charge is shared by C-2 and C-4. Consequently, the bromide ion can add to either C-2 or C-4 to form the 1,2-addition product or the 1,4-addition product, respectively.

Notice that, in the first step of the reaction, adding H$^+$ to C-1 is the same as adding it to C-4 because 1,3-butadiene is symmetrical.

As we look at more examples, notice that the first step in all electrophilic additions to conjugated dienes is the addition of the electrophile to one of the sp^2 carbons *at the end of the conjugated system*. This is the only way to form a carbocation that is stabilized by electron delocalization. If the electrophile were to add to one of the internal sp^2 carbons, the resulting carbocation would not have delocalized electrons.

PROBLEM 39♦

What are the products of the following reactions, assuming that one equivalent of each reagent is used in each reaction?

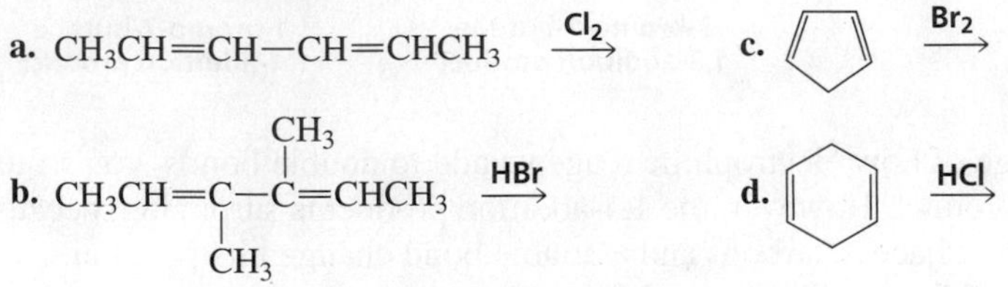

a. CH$_3$CH=CH—CH=CHCH$_3$ $\xrightarrow{\text{Cl}_2}$ **c.** (cyclopentadiene) $\xrightarrow{\text{Br}_2}$

b.
$$\begin{array}{c} \text{CH}_3 \\ | \\ \text{CH}_3\text{CH}=\text{C}-\text{C}=\text{CHCH}_3 \\ | \\ \text{CH}_3 \end{array} \xrightarrow{\text{HBr}}$$
    d. (cyclohexadiene) $\xrightarrow{\text{HCl}}$

PROBLEM 40

What stereoisomers are obtained from the two reactions shown earlier in this section?

If the conjugated diene is not symmetrical, the major products of the reaction are those obtained by adding the electrophile to whichever sp^2 carbon at the end of the conjugated system results in formation of the more stable carbocation.

For example, in the following reaction, the proton adds preferentially to C-1 because the positive charge on the resulting carbocation is shared by a tertiary allylic and a primary allylic carbon. Adding the proton to C-4 would form a carbocation in which the positive charge is shared by a secondary allylic and a primary allylic carbon.

$$\overset{\text{CH}_3}{\underset{\text{2-methyl-1,3-butadiene}}{\overset{1}{\text{CH}_2}=\overset{|}{\text{C}}-\text{CH}=\overset{4}{\text{CH}_2}}} + \text{HBr} \longrightarrow \underset{\substack{\text{Br} \\ \text{3-bromo-3-methyl-} \\ \text{1-butene}}}{\overset{\text{CH}_3}{\text{CH}_3-\overset{|}{\underset{|}{\text{C}}}-\text{CH}=\text{CH}_2}} + \underset{\substack{\text{Br} \\ \text{1-bromo-3-methyl-} \\ \text{2-butene}}}{\overset{\text{CH}_3}{\text{CH}_3-\overset{|}{\text{C}}=\text{CH}-\overset{|}{\text{CH}_2}}}$$

$$\underset{\textbf{carbocation formed by adding H}^+ \textbf{ to C-1}}{\overset{\text{CH}_3}{\text{CH}_3\overset{|}{\underset{+}{\text{C}}}-\text{CH}=\text{CH}_2} \longleftrightarrow \overset{\text{CH}_3}{\text{CH}_3\overset{|}{\text{C}}=\text{CH}-\underset{+}{\text{CH}_2}}} \qquad \underset{\textbf{carbocation formed by adding H}^+ \textbf{ to C-4}}{\overset{\text{CH}_3}{\text{CH}_2=\overset{|}{\text{C}}\underset{+}{\frown}\text{CHCH}_3} \longleftrightarrow \overset{\text{CH}_3}{\underset{+}{\text{CH}_2}-\overset{|}{\text{C}}=\text{CHCH}_3}}$$

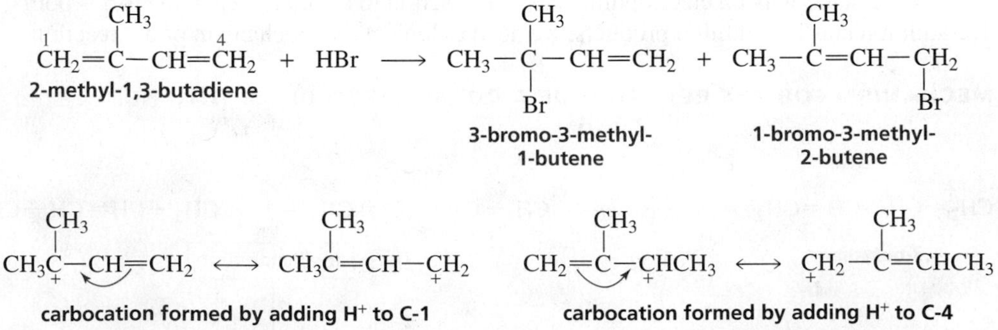

Because addition to C-1 forms the more stable carbocation, the major products of the reaction are the ones shown.

PROBLEM 41

What products would be obtained from the reaction of 1,3,5-hexatriene with one equivalent of HBr? Disregard stereoisomers.

PROBLEM 42♦

What are the products of the following reactions, assuming that one equivalent of each reagent is used in each reaction? Disregard stereoisomers.

a. 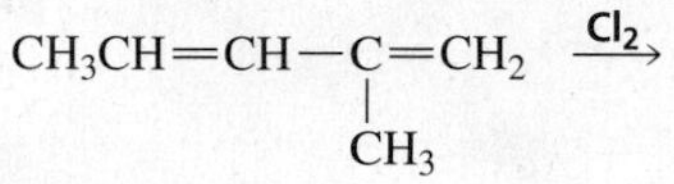 CH$_3$CH=CH—C=CH$_2$ $\xrightarrow{\text{Cl}_2}$
        |
        CH$_3$

b. 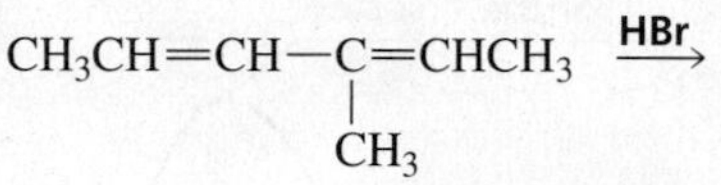 CH$_3$CH=CH—C=CHCH$_3$ $\xrightarrow{\text{HBr}}$
        |
        CH$_3$

18 THERMODYNAMIC VERSUS KINETIC CONTROL

When a conjugated diene undergoes an electrophilic addition reaction, two factors—the *structure of the reactant* and the *temperature* at which the reaction is carried out—determine whether the 1,2-addition product or the 1,4-addition product will be the major product of the reaction.

When a reaction produces more than one product, *the more rapidly formed product* is called the **kinetic product,** and *the more stable product* is called the **thermodynamic product.** Reactions that produce the kinetic product as the major product are said to be **kinetically controlled;** reactions that produce the thermodynamic product as the major product are said to be **thermodynamically controlled.**

In many organic reactions, the more rapidly formed product is also the more stable product. Electrophilic addition to 1,3-butadiene is an example of a reaction in which the kinetic product and the thermodynamic product are *not* the same: the 1,2-addition product is the kinetic product, and the 1,4-addition product is the thermodynamic product.

CH$_2$=CHCH=CH$_2$ + HBr ⟶ CH$_3$CHCH=CH$_2$ + CH$_3$CH=CHCH$_2$
1,3-butadiene | |
                                     Br   Br

1,2-addition product   **1,4-addition product**
kinetic product   **thermodynamic product**

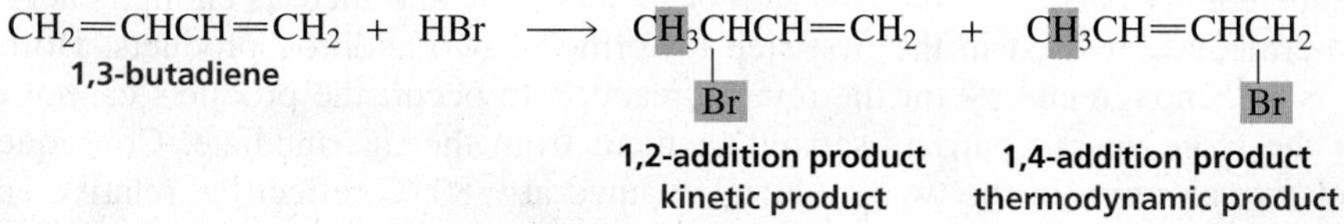

The kinetic product is the more rapidly formed product.

The thermodynamic product is the more stable product.

For a reaction, such as the one just shown, in which the kinetic and thermodynamic products are not the same, the product that predominates depends on the conditions under which the reaction is carried out. If the reaction is carried out under conditions that are sufficiently mild (at a low temperature) to cause the reaction to be *irreversible*, the major product will be the *kinetic product*—that is, the faster-formed product.

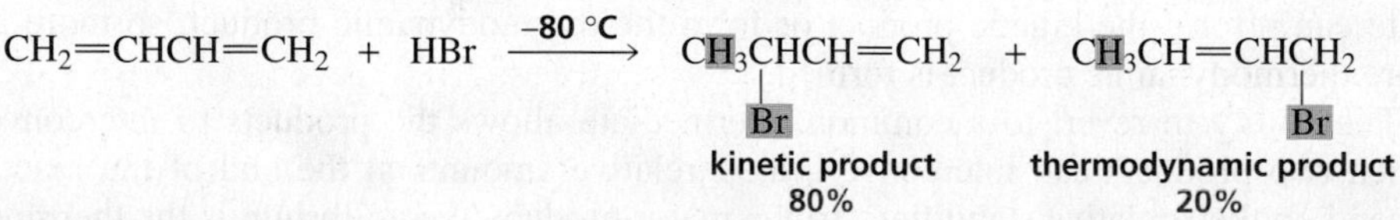

CH$_2$=CHCH=CH$_2$ + HBr $\xrightarrow{-80\,°C}$ CH$_3$CHCH=CH$_2$ + CH$_3$CH=CHCH$_2$
                                        | |
                                        Br   Br
                              kinetic product   **thermodynamic product**
                                    80%   **20%**

The kinetic product predominates when the reaction is irreversible.

If, on the other hand, the reaction is carried out under conditions that are sufficiently vigorous (at a higher temperature) to cause the reaction to be *reversible,* the major product will be the *thermodynamic product*—that is, the more stable product.

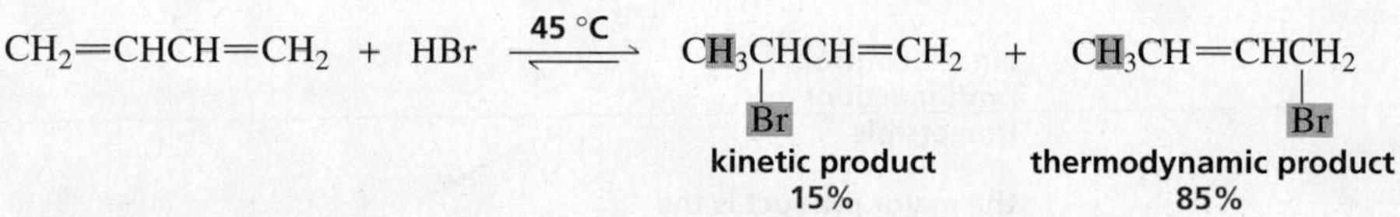

CH$_2$=CHCH=CH$_2$ + HBr $\xrightleftharpoons{45\,°C}$ CH$_3$CHCH=CH$_2$ + CH$_3$CH=CHCH$_2$
                                        | |
                                        Br   Br
                              kinetic product   **thermodynamic product**
                                    15%   **85%**

The thermodynamic product predominates when the reaction is reversible.

A reaction coordinate diagram helps explain why different products predominate under different reaction conditions (Figure 14). The first step of the addition reaction is the same whether the 1,2-addition product or the 1,4-addition product is ultimately formed: a proton adds to C-1. The second step of the reaction is the one that determines whether the nucleophile (Br⁻) adds to C-2 or to C-4.

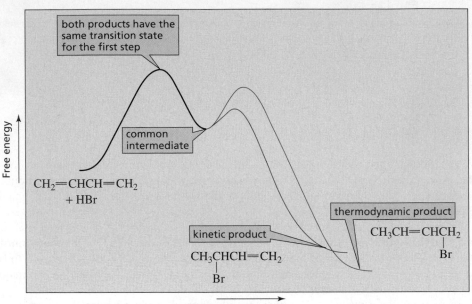

▶ **Figure 14**

A reaction coordinate diagram for the addition of HBr to 1,3-butadiene.

At low temperatures (−80 °C), there is enough energy for the reactants to overcome the energy barrier for the first step of the reaction, and there is enough energy for the intermediate formed in the first step to form the two addition products. However, there is not enough energy for the reverse reaction to occur: the products cannot overcome the large energy barriers separating them from the intermediate. Consequently, the relative amounts of the two products obtained at −80 °C reflect the relative energy barriers for the second step of the reaction. So the major product is the more rapidly formed (kinetic) product.

In contrast, at 45 °C, there is enough energy for one or more of the products to revert back to the intermediate. The intermediate is called a **common intermediate** because it is an intermediate that both products have in common. Even if both products can revert back to the common intermediate, it is easier for the 1,2-addition product to do so. Each time the kinetic product reverts back to the common intermediate, the common intermediate can reform the kinetic product or form the thermodynamic product, so more and more thermodynamic product is formed.

The ability to revert to a common intermediate allows the products to interconvert. When two products can interconvert, their relative amounts at the end of the reaction depend on their relative stabilities. So the major product at equilibrium is the thermodynamic product.

In summary, a reaction that is *irreversible* under the conditions employed in the experiment will be *kinetically controlled*. When a reaction is under *kinetic control*, the relative amounts of the products *depend on the rates* at which they are formed.

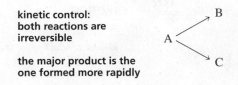

When sufficient energy is available to make the *reaction reversible, it will be thermodynamically controlled.* When a reaction is under *thermodynamic control,* the relative amounts of the products *depend on their stabilities.*

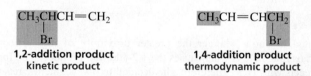

thermodynamic control: one or both reactions are reversible

the major product is the one that is more stable

For each reaction that is irreversible under mild conditions and reversible under more vigorous conditions, there is a temperature at which the change from irreversible to reversible occurs. The temperature at which a reaction changes from being kinetically controlled to being thermodynamically controlled depends on the reaction.

For example, the reaction of 1,3-butadiene with HCl remains under kinetic control at 45 °C, even though the reaction of 1,3-butadiene with HBr is under thermodynamic control at that temperature. Because a C—Cl bond is stronger than a C—Br bond, a higher temperature is required for the products containing a C—Cl bond to undergo the reverse reaction. (Remember, thermodynamic control is achieved only when there is sufficient energy to reverse one or both of the reactions.)

For the reaction of 1,3-butadiene with one equivalent of HBr, why is the 1,4-addition product the thermodynamic product? In other words, why is it the more stable product? We know that alkene stability is determined by the number of alkyl groups bonded to its sp^2 carbons—the greater number of alkyl groups, the greater its stability. The two products formed from the reaction of 1,3-butadiene with HBr have different stabilities since the 1,2-addition product has one alkyl group bonded to its sp^2 carbons, whereas the 1,4-product has two alkyl groups bonded to its sp^2 carbons. Thus, the 1,4-addition product is the more stable (i.e., the thermodynamic) product.

$$CH_3CHCH{=}CH_2 \qquad\qquad CH_3CH{=}CHCH_2$$
$$\;\;\;|\phantom{CHCH{=}CH_2} \qquad\qquad\qquad\qquad\quad |$$
$$\;\;Br \qquad\qquad\qquad\qquad\qquad\qquad Br$$

1,2-addition product
kinetic product

1,4-addition product
thermodynamic product

The next question we need to answer is why is the 1,2-addition product formed faster? For many years, chemists thought it was because the transition state for formation of the 1,2-addition product resembles the resonance contributor in which the positive charge is on a secondary allylic carbon, whereas the transition state for formation of the 1,4-addition product resembles the resonance contributor in which the positive charge is on a less stable primary allylic carbon.

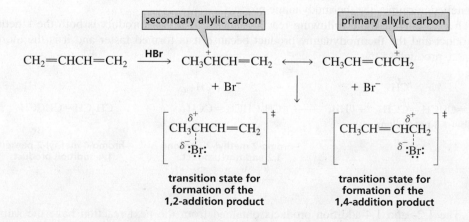

secondary allylic carbon

primary allylic carbon

$$CH_2{=}CHCH{=}CH_2 \xrightarrow{\;HBr\;} CH_3\overset{+}{C}HCH{=}CH_2 \longleftrightarrow CH_3CH{=}CH\overset{+}{C}H_2$$

$$+\; Br^- \qquad\qquad\qquad +\; Br^-$$

$$\left[\begin{array}{c} \overset{\delta^+}{CH_3CHCH{=}CH_2} \\ | \\ \underset{\delta^-}{:}\!\ddot{B}\ddot{r}: \end{array} \right]^{\ddagger} \qquad \left[\begin{array}{c} CH_3CH{=}CH\overset{\delta^+}{CH_2} \\ | \\ \underset{\delta^-}{:}\!\ddot{B}\ddot{r}: \end{array} \right]^{\ddagger}$$

transition state for formation of the 1,2-addition product

transition state for formation of the 1,4-addition product

If this explanation were correct, the kinetically controlled reaction of 1,3-pentadiene with DCl would form equal amounts of the 1,2- and 1,4-addition products, because their transition states would both have the same stability (both would have a partial positive charge on a secondary allylic carbon). However, the kinetically controlled reaction forms mainly the 1,2-addition product. Why is the 1,2-addition product formed faster?

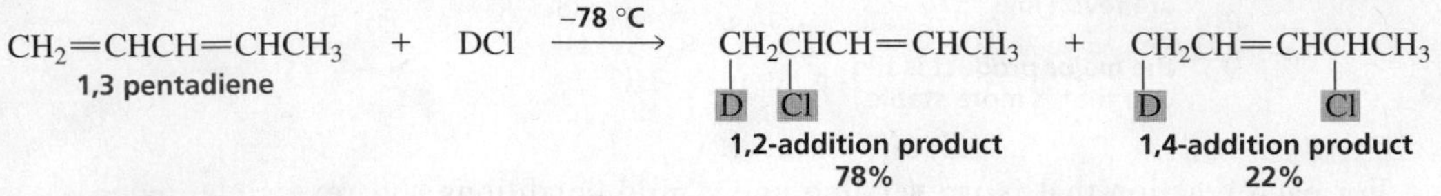

$$CH_2=CHCH=CHCH_3 \quad + \quad DCl \quad \xrightarrow{-78\ °C} \quad CH_2CHCH=CHCH_3 \quad + \quad CH_2CH=CHCHCH_3$$

1,3 pentadiene   D   Cl   D   Cl

1,2-addition product   **1,4-addition product**
       78%   **22%**

The answer is that after D^+ adds to the double bond, the chloride ion can form a bond with C-2 faster than it can form a bond with C-4 simply because it is closer to C-2 than to C-4. So it is a *proximity effect* that causes the 1,2-addition product to be formed faster. A **proximity effect** is an effect caused by one species being close to another.

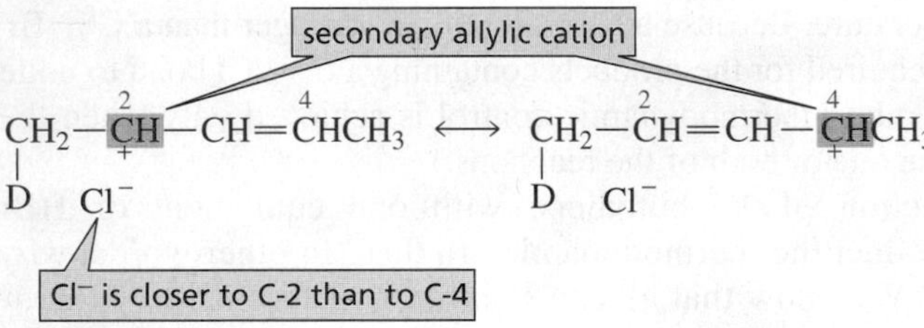

secondary allylic cation

$$CH_2-CH-CH=CHCH_3 \quad \longleftrightarrow \quad CH_2-CH=CH-CHCH_3$$

   D Cl^- D Cl^-

Cl^- is closer to C-2 than to C-4

PROBLEM 43♦

a. Why does deuterium add to C-1 rather than to C-4 in the preceding reaction?
b. Why was DCl rather than HCl used in the reaction?

PROBLEM 44♦

A student wanted to know whether the greater proximity of the nucleophile to the C-2 carbon in the transition state is what causes the 1,2-addition product to be formed faster when 1,3-butadiene reacts with HCl. Therefore, she decided to investigate the reaction of 2-methyl-1,3-cyclohexadiene with HCl. Her friend told her that she should use 1-methyl-1,3-cyclohexadiene instead. Should she follow her friend's advice?

Because a proximity effect is what causes the 1,2-addition product to be formed faster, we can assume that the 1,2-addition product is always the kinetic product for the reactions of conjugated dienes. Do *not* assume, however, that the 1,4-addition product is *always* the thermodynamic product. The structure of the conjugated diene is what ultimately determines the thermodynamic product.

For example, in the following reaction, the 1,2-addition product is both the kinetic product and the thermodynamic product because it is formed faster and it is the more stable product.

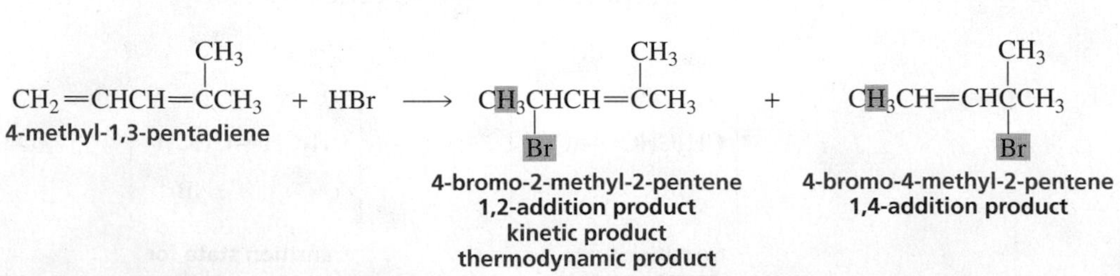

$$CH_2=CHCH=CCH_3 \quad + \quad HBr \quad \longrightarrow \quad CH_3CHCH=CCH_3 \quad + \quad CH_3CH=CHCCH_3$$

            CH_3 CH_3 CH_3

4-methyl-1,3-pentadiene   Br   Br

                    4-bromo-2-methyl-2-pentene   **4-bromo-4-methyl-2-pentene**
                        1,2-addition product   **1,4-addition product**
                           kinetic product
                      thermodynamic product

The 1,2- and 1,4-addition products obtained from the next reaction have the same stability because both have the same number of alkyl groups bonded to their sp^2 carbons.

Thus, neither is the thermodynamic product. Approximately equal amounts of both products will be formed under conditions that cause the addition reaction to be reversible. The 1,2-addition product is the kinetic product.

$$CH_3CH=CHCH=CHCH_3 \xrightarrow{\text{HCl}} CH_3CH_2CHCH=CHCH_3 \quad + \quad CH_3CH_2CH=CHCHCH_3$$

2,4-hexadiene

4-chloro-2-hexene
1,2-addition product
kinetic product

2-chloro-3-hexene
1,4-addition product

the products have the same stability

PROBLEM 45♦

a. When HBr adds to a conjugated diene, what is the rate-determining step?
b. When HBr adds to a conjugated diene, what is the product-determining step?

PROBLEM 46 Solved

What are the major 1,2- and 1,4-addition products of the following reactions? For each reaction, indicate the kinetic and the thermodynamic products.

a. 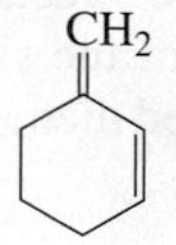 + HCl ⟶

c. 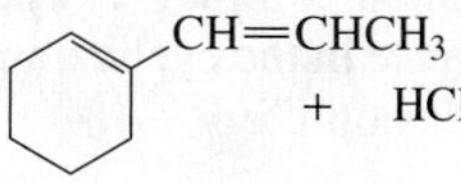 + HCl ⟶

b. 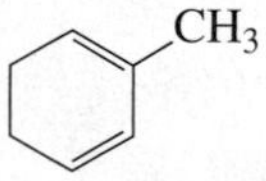 + HCl ⟶

Solution to 46a First we need to determine which of the sp^2 carbons at the ends of the conjugated system gets the proton. The proton will preferentially add to the sp^2 carbon indicated below, because the resulting carbocation shares its positive charge with a tertiary allylic and a secondary allylic carbon. If the proton were to add to the sp^2 carbon at the other end of the conjugated system, the carbocation that would be formed would be less stable because its positive charge would be shared by a primary allylic and a secondary allylic carbon. Therefore, the major products are the ones shown here. 3-Chloro-3-methylcyclohexene, the 1,2-addition product, is the kinetic product because of the chloride ion's proximity to C-2. 3-Chloro-1-methylcyclohexene, the 1,4-addition product, is the thermodynamic product because its more highly substituted double bond makes it more stable.

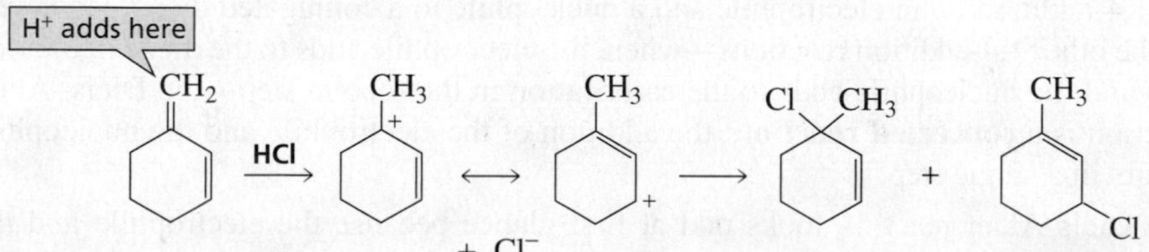

3-chloro-3-methylcyclohexene
kinetic product

3-chloro-1-methylcyclohexene
thermodynamic product

PROBLEM 47

Identify the kinetic and thermodynamic products of the following reaction:

$\xrightarrow{\text{HCl}}$

19 THE DIELS–ALDER REACTION IS A 1,4-ADDITION REACTION

Reactions that create new carbon–carbon bonds are very important to synthetic organic chemists because it is only through such reactions that small carbon skeletons can be converted into larger ones. The Diels–Alder reaction is a particularly important reaction because it creates *two* new carbon–carbon bonds and in the process forms a cyclic compound. In recognition of the importance of this reaction to synthetic organic chemistry, Otto Diels and Kurt Alder shared the Nobel Prize in Chemistry in 1950.

In a **Diels–Alder reaction,** a conjugated diene reacts with a compound containing a carbon–carbon double bond. The latter compound is called a **dienophile** because it "loves a diene." (Note that Δ signifies heat.)

CH$_2$=CH—CH=CH$_2$ + CH$_2$=CH—R $\xrightarrow{\Delta}$

conjugated diene   **dienophile**

The Diels–Alder reaction is a pericyclic reaction. A **pericyclic reaction** is a reaction that takes by a cyclic shift of electrons. It is also a cycloaddition reaction. A **cycloaddition reaction** is a reaction in which two reactants form a cyclic product. More precisely, the Diels–Alder reaction is a **[4 + 2] cycloaddition reaction** because *four* of the six π electrons that participate in the cyclic transition state come from the conjugated diene and *two* come from the dienophile.

MECHANISM FOR THE DIELS-ALDER REACTION

conjugated diene   **dienophile**   **transition state**
four π electrons   **two π electrons**   **six π electrons**

new σ bond
new double bond
new σ bond

- Although this reaction may not look like any reaction you have seen before, it is simply the 1,4-addition of an electrophile and a nucleophile to a conjugated diene. However, unlike other 1,4-addition reactions—where the electrophile adds to the diene in the first step and the nucleophile adds to the carbocation in the second step—the Diels–Alder reaction is a **concerted reaction:** the addition of the electrophile and the nucleophile occurs in a single step.

The Diels-Alder reaction looks odd at first glance because the electrophile and the nucleophile that add to the conjugated diene are the adjacent sp^2 carbons of a double bond. As with other 1,4-addition reactions, the double bond in the product is between the 2- and 3-positions of the diene's conjugated system.

the nucleophile adds to C-4

the electrophile adds to C-1

a 1,4-addition reaction

The reactivity of the dienophile is increased if an electron-withdrawing group is attached to one of its sp^2 carbons. An electron-withdrawing group, such as a carbonyl group (C=O) or a cyano group (C≡N), withdraws electrons from the dienophile's double bond. This puts a partial positive charge on the sp^2 carbon that the π electrons of the conjugated diene add to. Thus, the electron-withdrawing group makes the dienophile a better electrophile (Figure 15).

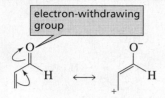

resonance contributors of the dienophile   **resonance hybrid**

A wide variety of cyclic compounds can be obtained by varying the structures of the conjugated diene and the dienophile.

Compounds containing carbon–carbon triple bonds can also be used as dienophiles in Diels–Alder reactions to prepare compounds with two isolated double bonds.

If the dienophile has two carbon–carbon double bonds, two successive Diels–Alder reactions can occur if excess diene is available.

▲ **Figure 15**
These electrostatic potential maps show that an electron-withdrawing substituent makes the bottom sp^2 carbon a better electrophile.

PROBLEM 48♦

What are the products of the following reactions?

a. $CH_2=CH-CH=CH_2$ + $CH_3\overset{\overset{O}{\|}}{C}-C\equiv C-\overset{\overset{O}{\|}}{C}CH_3$ $\xrightarrow{\Delta}$

b. $CH_2=CH-CH=CH_2$ + $HC\equiv C-C\equiv N$ $\xrightarrow{\Delta}$

c. $CH_2=\overset{\overset{CH_3}{|}}{C}-\overset{\overset{CH_3}{|}}{C}=CH_2$ + (maleic anhydride) $\longrightarrow$

d. $CH_3\overset{\overset{CH_3}{|}}{C}=CH-CH=\overset{\overset{CH_3}{|}}{C}CH_3$ + (cyclohexenedione) $\longrightarrow$

Predicting the Product When Both Reagents Are Unsymmetrically Substituted

In each of the preceding Diels–Alder reactions, only one product is formed (disregarding stereoisomers) because at least one of the reacting molecules is symmetrically substituted. If both the diene and the dienophile are unsymmetrically substituted, however, two products are possible. The products are constitutional isomers.

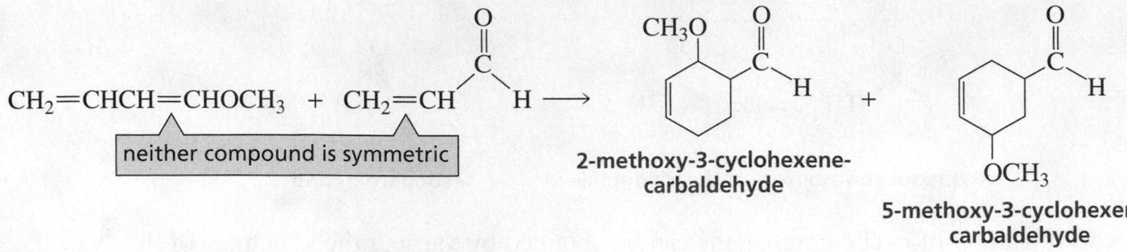

2-methoxy-3-cyclohexene-carbaldehyde

5-methoxy-3-cyclohexene-carbaldehyde

Two products are possible because the reactants can align in two different ways. (To determine the structure of the second product, don't change the position of one of the reactants and turn the other upside down.)

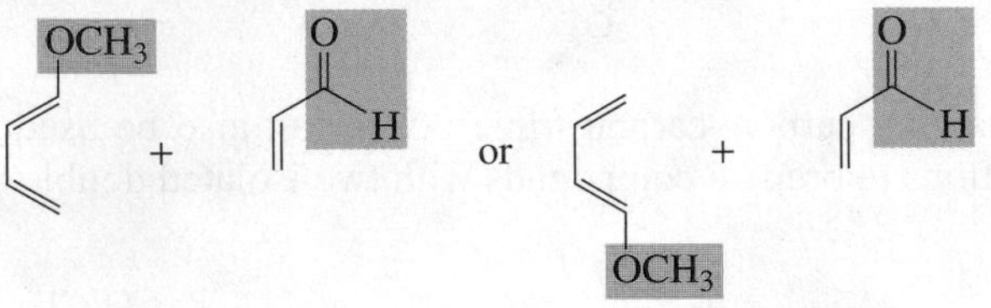

The product that will be formed in greater yield depends on the charge distribution in each of the reactants. To determine the charge distribution, we need to draw the resonance contributors of the reactants. The methoxy group of the diene *donates electrons by resonance* (see Problem 34). As a result, its terminal carbon has a partial negative charge. The carbonyl group of the dienophile, on the other hand, *withdraws electrons by resonance* (see Problem 33), so its terminal carbon has a partial positive charge.

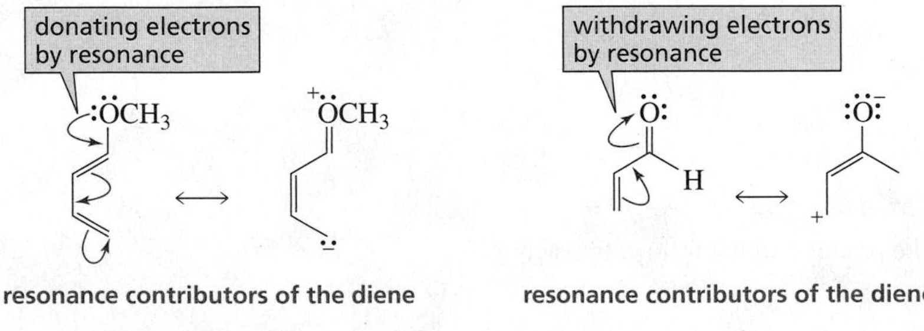

resonance contributors of the diene   resonance contributors of the dienophile

The partially positively charged carbon of the dienophile will bond preferentially to the partially negatively charged carbon of the diene. Therefore, 2-methoxy-3-cyclohexene-carbaldehyde will be the major product.

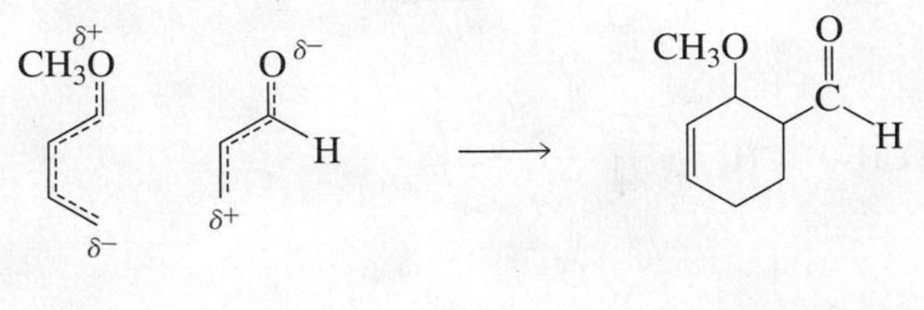

PROBLEM 49◆

What would be the major product if the methoxy substituent in the preceding reaction were bonded to C-2 of the diene rather than to C-1?

PROBLEM 50

Write a general rule that can be used to predict the major product of a Diels–Alder reaction between an alkene with an electron-withdrawing substituent and a diene with a substituent that can donate electrons by resonance depending on the location of the substituent on the diene.

PROBLEM 51◆

What two products are formed from each of the following reactions?

a. $CH_2{=}CH{-}CH{=}CH{-}CH_3$ + $HC{\equiv}C{-}C{\equiv}N$ $\xrightarrow{\Delta}$

b. $CH_2{=}CH{-}\underset{\underset{CH_3}{|}}{C}{=}CH_2$ + $HC{\equiv}C{-}C{\equiv}N$ $\xrightarrow{\Delta}$

Conformations of the Diene

We have seen that a conjugated diene, such as 1,3-butadiene, is most stable in a *planar* conformation (Section 14). However, there are two different planar conformations: an *s*-cis conformation and an *s*-trans conformation. (Note that a conformation results from rotation about single bonds.)

In an ***s*-cis conformation**, the double bonds are cis about the single bond (*s* = single), whereas they are trans about the single bond in an *s*-trans conformation. An ***s*-trans conformation** is a little more stable (by 2.3 kcal/mol or 9.6 kJ/mol) because the close proximity of the hydrogens in the *s*-cis conformation causes some steric strain. The rotational barrier between the *s*-cis and *s*-trans conformations is low enough (4.9 kcal/mol or 20.5 kJ/mol) to allow the conformations to interconvert rapidly at room temperature.

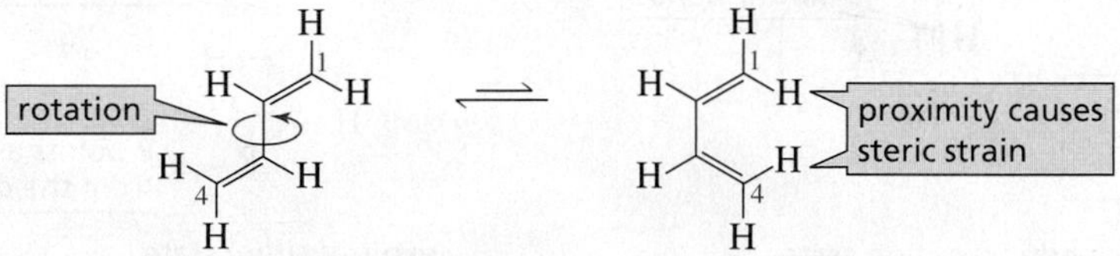

s-trans conformation   **s-cis conformation**

In order to participate in a Diels–Alder reaction, the conjugated diene must be in an *s*-cis conformation because C-1 and C-4 in an *s*-trans conformation are too far apart to react with the dienophile in a concerted reaction. Therefore, a conjugated diene that is locked in an *s*-trans conformation cannot undergo a Diels–Alder reaction because it cannot achieve the required *cis*-conformation.

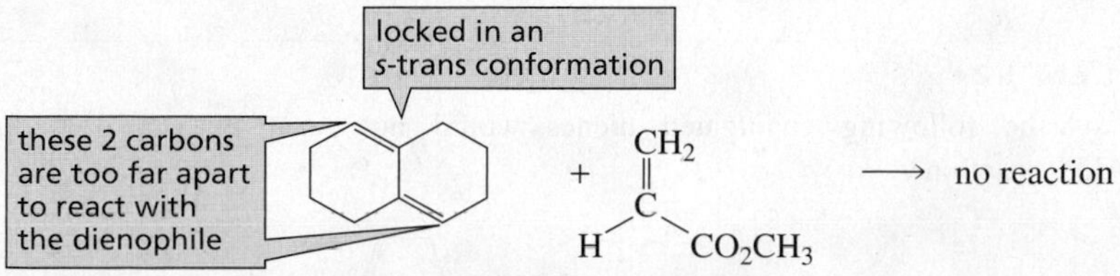

A conjugated diene that is locked in an *s*-cis conformation, such as 1,3-cyclopentadiene, is highly reactive in a Diels–Alder reaction. When the diene is a cyclic compound, the product of a Diels–Alder reaction is a **bridged bicyclic compound** (a compound that contains two rings that share two nonadjacent carbons).

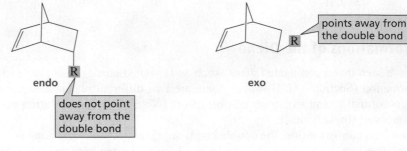

1,3-cyclopentadiene

bridged bicyclic compounds

74%   26%

bridged bicyclic rings
2 rings share 2 nonadjacent carbons

fused bicyclic rings
2 rings share 2 adjacent carbons

There are two possible configurations for substituted bridged bicyclic compounds, because the substituent (R) can either point away from the double bond (the **exo** configuration) or not point away from the double bond (the **endo** configuration). Figure 16 shows why both endo and exo products are formed.

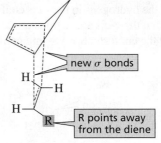

endo   exo

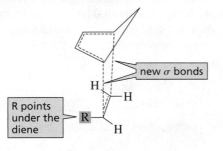

endo transition state   exo transition state

▶ **Figure 16**

The transition states for formation of the endo and exo products show that two products are formed because the dienophile can line up in two different ways. The substituent (R) can point either under the diene (endo) or away from the diene (exo).

When the dienophile has a substituent with π electrons, more of the endo product is formed.

The endo product is formed faster when the dienophile has a substituent with π electrons. Recent studies suggest that the increased rate of endo product formation is due to interaction between the π electrons of the substituent and the π electrons of the ring, which stabilizes the transition state. A substituent in the exo position cannot engage in such stabilizing interactions.

PROBLEM 52♦

Which of the following conjugated dienes would not react with a dienophile in a Diels–Alder reaction?

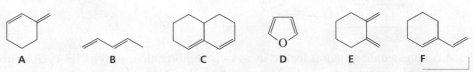

A   B   C   D   E   F

PROBLEM 53

What are the products of the following reactions?

a. 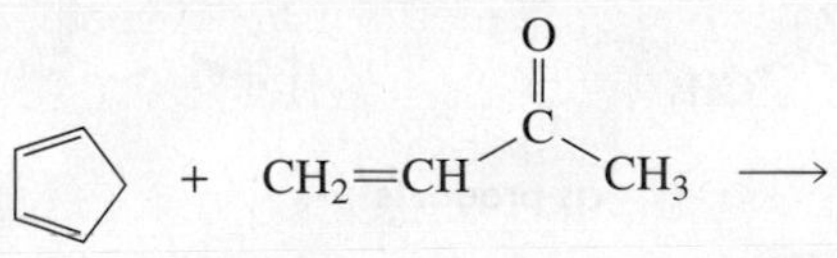 (cyclopentadiene) + $CH_2{=}CH{-}\overset{\overset{O}{\|}}{C}{-}CH_3 \longrightarrow$ **b.** 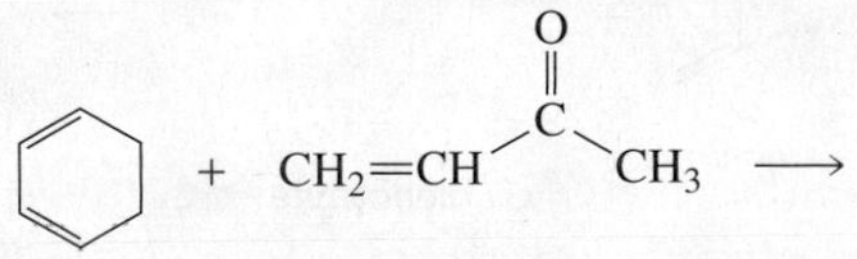 (1,3-cyclohexadiene) + $CH_2{=}CH{-}\overset{\overset{O}{\|}}{C}{-}CH_3 \longrightarrow$

PROBLEM 54 Solved

List the following dienes in order from most reactive to least reactive in a Diels–Alder reaction:

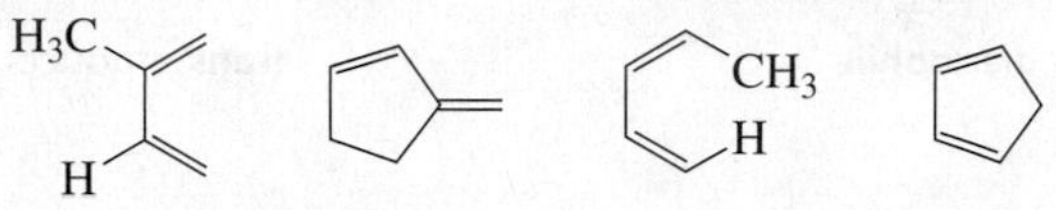

Solution The most reactive diene has its double bonds locked in an *s*-cis conformation, whereas the least reactive diene has its double bonds locked in an *s*-trans conformation. The other two compounds are of intermediate reactivity because they can exist in both *s*-cis and *s*-trans conformations. 1,3-Pentadiene is less apt to be in the required *s*-cis conformation because of steric interference between the hydrogen and the methyl group, so it is less reactive than 2-methyl-1,3-butadiene.

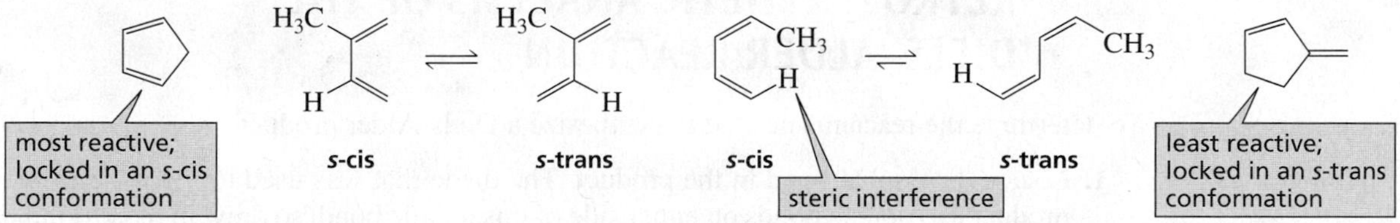

| most reactive; locked in an *s*-cis conformation | *s*-cis | *s*-trans | *s*-cis | steric interference | *s*-trans | least reactive; locked in an *s*-trans conformation |

Thus, the four dienes have the following order of reactivity:

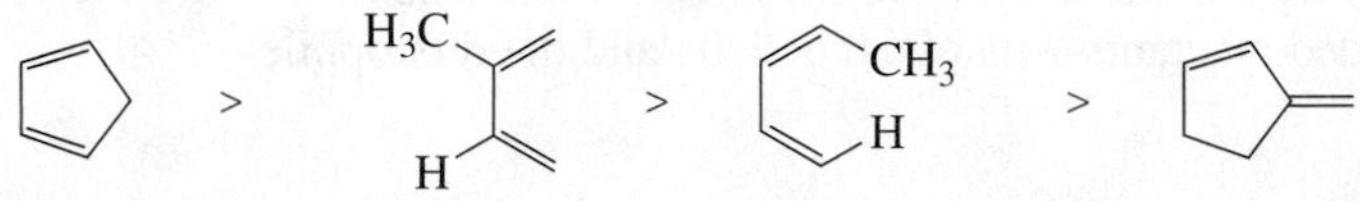

The Stereochemistry of the Diels–Alder Reaction

As with all the other reactions we have seen, if a Diels–Alder reaction creates a product with an asymmetric center, the product will be a racemic mixture.

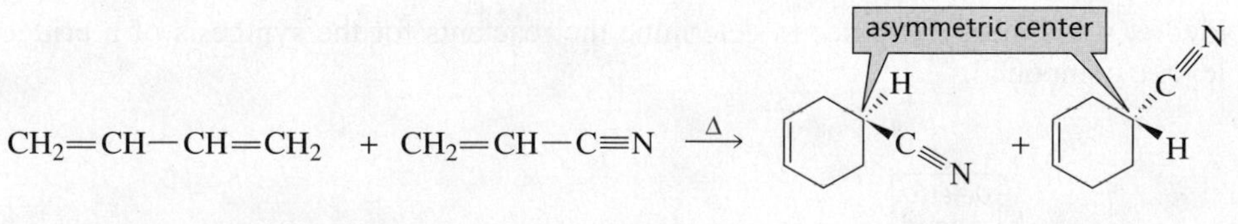

asymmetric center

$CH_2{=}CH{-}CH{=}CH_2 \quad + \quad CH_2{=}CH{-}C{\equiv}N \quad \overset{\Delta}{\longrightarrow}$

The Diels–Alder reaction is a syn addition reaction. One face of the diene adds to one face of the dienophile. Therefore, if the substituents in the *dienophile* are cis, then they will be cis in the product; if the substituents in the *dienophile* are trans, then they will be trans in the product. Because each of the following reactions forms a product with two new asymmetric centers, a pair of enantiomers is formed.

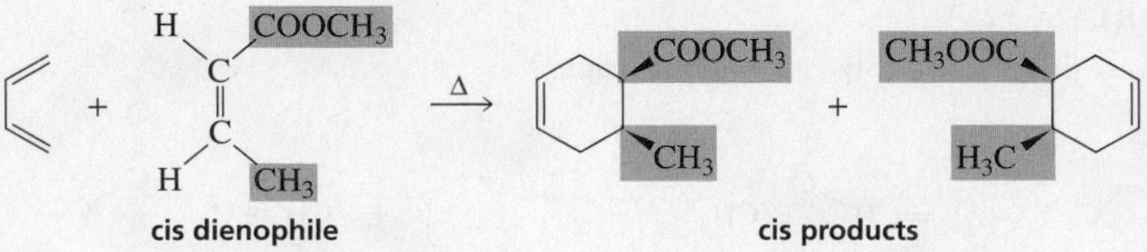

cis dienophile → **cis products**

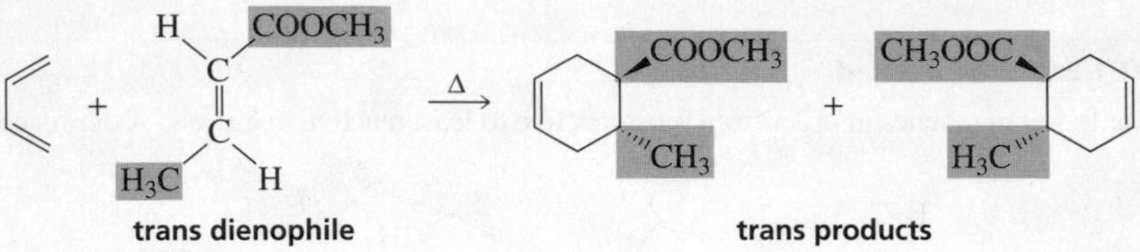

trans dienophile → **trans products**

PROBLEM 55♦

Explain why the following compounds are not optically active:

a. the product obtained from the reaction of 1,3-butadiene with *cis*-1,2-dichloroethene

b. the product obtained from the reaction of 1,3-butadiene with *trans*-1,2-dichloroethene

---

20 RETROSYNTHETIC ANALYSIS OF THE DIELS–ALDER REACTION

To determine the reactants needed to synthesize a Diels-Alder product:

1. Locate the double bond in the product. The diene that was used to form the cyclic product had double bonds on either side of this double bond, so draw in those double bonds and remove the original double bond.

2. The new σ bonds are now on either side of the double bonds. Deleting these σ bonds and putting a π bond between the two carbons whose σ bonds were deleted gives the needed reactants—that is, the diene and the dienophile.

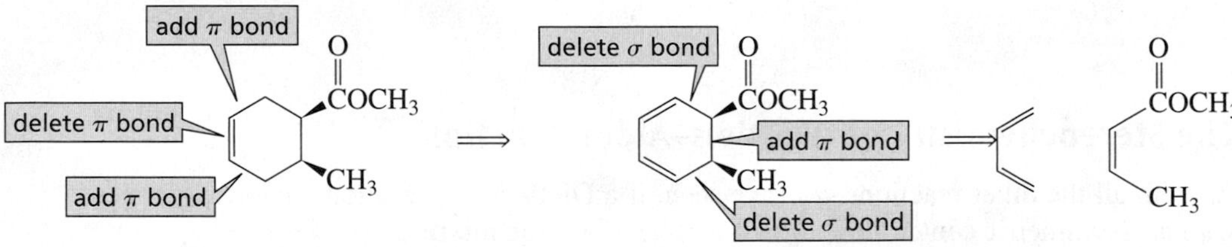

Now let's use these two rules to determine the reactants for the synthesis of a bridged bicyclic compound.

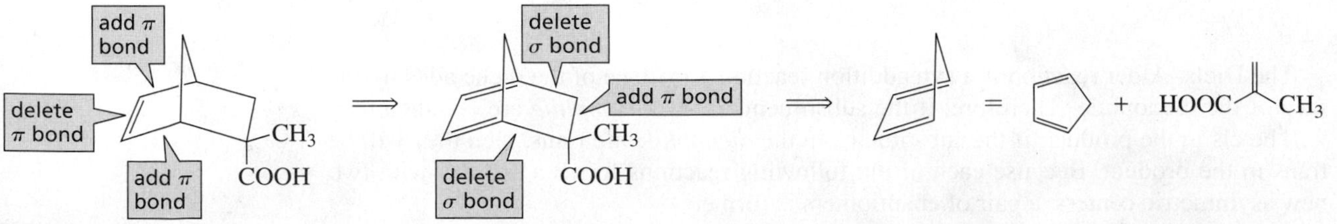

PROBLEM 56♦

What diene and what dienophile should be used to synthesize the following?

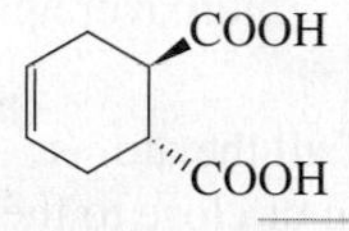

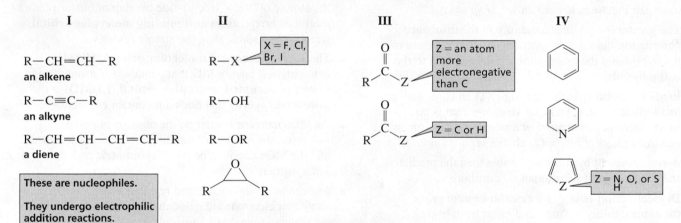

21   ORGANIZING WHAT WE KNOW ABOUT THE REACTIONS OF ORGANIC COMPOUNDS

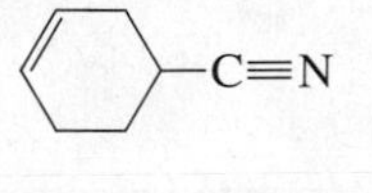

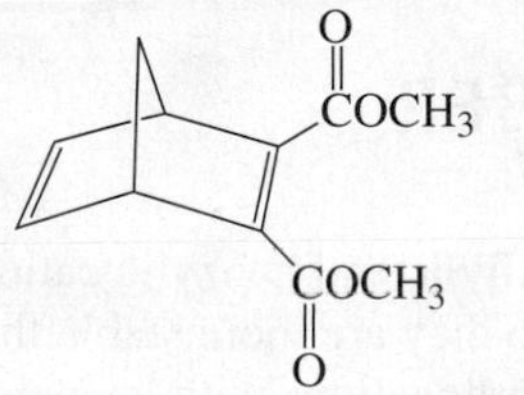

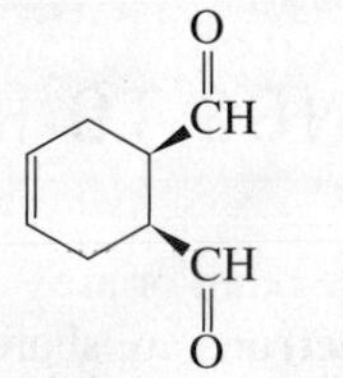

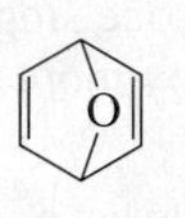

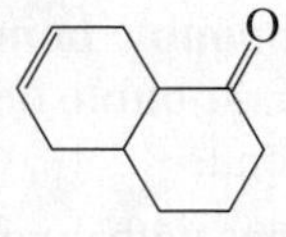

Organic compounds can be classified into families and all the members of a family react in the same way. Also, each family can be put into one of four groups, and that all the families in a group react in similar ways. Let's discuss the first group.

I	II	III	IV

R—CH=CH—R
an alkene

R—C≡C—R
an alkyne

R—CH=CH—CH=CH—R
a diene

> These are nucleophiles.
>
> They undergo electrophilic addition reactions.

R—X   X = F, Cl, Br, I

R—OH

R—OR

All the families in the first group are nucleophiles, because of their electron-rich carbon–carbon double or triple bonds. And because double and triple bonds have relatively weak π bonds, the families in this group undergo addition reactions. Since the first species that reacts with a nucleophile is an electrophile, the reactions that the families in this group undergo are more precisely called *electrophilic addition reactions*.

- Alkenes have one π bond so they undergo one electrophilic addition reaction.

- Alkynes have two π bonds so they can undergo two electrophilic addition reactions. However, if the first addition reaction forms an enol, the enol immediately rearranges to a ketone (or to an aldehyde), so a second addition reaction cannot occur.

- If the double bonds of a diene are isolated, they react just like alkenes. If, however, the double bonds are conjugated, they undergo both 1,2- and 1,4-addition reactions, because the carbocation intermediate has delocalized electrons.

SOME IMPORTANT THINGS TO REMEMBER

- **Localized electrons** belong to a single atom or are shared by two atoms. **Delocalized electrons** are shared by more than two atoms.

- Delocalized electrons result when a *p* orbital overlaps the *p* orbitals of two adjacent atoms.

- Electron delocalization occurs only if all the atoms sharing the delocalized electrons lie in or close to the same plane.

- The six π electrons of benzene are shared by all six carbons. Thus, benzene is a planar molecule with six delocalized π electrons.

- **Resonance contributors**—structures with localized electrons—approximate the structure of a compound that has delocalized electrons: the **resonance hybrid.**

- To draw resonance contributors, move π electrons or lone-pair electrons toward an sp^2 or sp atom.

- The greater the predicted stability of the resonance contributor, the more it contributes to the structure of the hybrid and the more similar its structure is to the real molecule.

- Predicted stability is decreased by (1) an atom with an incomplete octet, (2) a negative charge that is not on the most electronegative atom or a positive charge on an electronegative atom, or (3) charge separation.

- **A resonance hybrid** is more stable than the predicted stability of any of its resonance contributors.

- **Delocalization energy** (or **resonance energy**) is the extra stability a compound gains from having delocalized electrons. It tells us how much more stable a compound with delocalized electrons is than it would be if all its electrons were localized.

- The greater the number of relatively stable resonance contributors and the more nearly equivalent they are, the greater the delocalization energy.

- An **aromatic compound** has an uninterrupted cyclic cloud of π electrons that contains an *odd number of pairs* of π electrons.

- An **antiaromatic compound** has an uninterrupted cyclic cloud of π electrons that contains an *even number of pairs* of π electrons.

- Aromatic compounds are very stable; antiaromatic compounds are very unstable.

- A **heterocyclic compound** is a cyclic compound in which one or more of the ring atoms is a **heteroatom**—that is, an atom other than carbon.

- Pyridine, pyrrole, furan, and thiophene are aromatic heterocyclic compounds.

- Allylic and benzylic cations have delocalized electrons, so they are more stable than similarly substituted carbocations with localized electrons.

- **Conjugated double bonds** are separated by one single bond. **Isolated double bonds** are separated by more than one single bond.

- Because dienes with conjugated double bonds have delocalized electrons, they are more stable than dienes with isolated double bonds.

- A **molecular orbital** results from the **linear combination of atomic orbitals (LCAO).**

- The number of molecular orbitals equals the number of atomic orbitals that produced them.

- Side-to-side overlap of in-phase *p* orbitals produces a **bonding molecular orbital,** which is more stable than the atomic orbitals. Side-to-side overlap of out-of-phase *p* orbitals produces an **antibonding molecular orbital,** which is less stable than the atomic orbitals.

- The **highest occupied molecular orbital (HOMO)** is the highest-energy MO that contains electrons. The **lowest unoccupied molecular orbital (LUMO)** is the lowest-energy MO that does not contain electrons.

- As MOs increase in energy, the number of nodes increases, the number of bonding interactions decreases, and the MOs alternate between **symmetric** and **antisymmetric.**

- Molecular orbital theory and resonance contributors both show that electrons are delocalized and that electron delocalization makes a molecule more stable.

- A carboxylic acid and a phenol are more acidic than an alcohol, and a protonated aniline is more acidic than a protonated amine because inductive electron withdrawal stabilizes their conjugate bases and the *loss of a proton is accompanied by an increase in delocalization energy.*

- Donation of electrons through π bonds is called **resonance electron donation;** withdrawal of electrons through π bonds is called **resonance electron withdrawal.**

- An isolated diene, like an alkene, undergoes only 1,2-addition. If there is only enough electrophilic reagent to add to one of the double bonds, it will add preferentially to the one that forms the more stable carbocation.

- A conjugated diene reacts with one equivalent of an electrophilic reagent to form a **1,2-addition product** and a **1,4-addition product.** The first step is addition of the electrophile to one of the sp^2 carbons at the end of the conjugated system.

- The most rapidly formed product is the **kinetic product;** the most stable product is the **thermodynamic product.**

- If both reactions are irreversible, the major product will be the kinetic product; if one or both of the reactions are reversible, the major product will be the thermodynamic product.

- When a reaction is **kinetically controlled,** the relative amounts of the products depend on the rates at which they are formed; when a reaction is **thermodynamically controlled,** the relative amounts of the products depend on their stabilities.

- A **common intermediate** is an intermediate that both products have in common.

- In electrophilic addition to a conjugated diene, the 1,2-product is always the kinetic product; either the 1,2- or the 1,4-product can be the thermodynamic product, depending on their structures.

- In a **Diels–Alder reaction,** a **conjugated diene** reacts with a **dienophile** to form a cyclic compound; in this concerted [**4 + 2**] **cycloaddition reaction,** two new σ bonds and a π bond are formed at the expense of two π bonds.

- The reactivity of the **dienophile** is increased by electron-withdrawing groups attached to an sp^2 carbon.

- If both the diene and the dienophile are unsymmetrically substituted, two products are possible because the reactants can be aligned in two different ways.

- The conjugated diene must be in an *s*-**cis conformation** for a Diels-Alder reaction.

- The Diels–Alder reaction is a syn addition reaction.

- In **bridged bicyclic compounds,** a substituent can be **endo** or **exo;** endo is favored if the dienophile's substituent has π electrons.

SUMMARY OF REACTIONS

1. In the presence of excess electrophilic reagent, both double bonds of an *isolated diene* will undergo electrophilic addition (Section 7).

$$CH_2=CHCH_2CH_2\overset{\overset{\displaystyle CH_3}{|}}{C}=CH_2 \ + \ \underset{\textbf{excess}}{HBr} \ \longrightarrow \ CH_3CHCH_2CH_2\overset{\overset{\displaystyle CH_3}{|}}{\underset{\underset{\displaystyle Br}{|}}{C}}CH_3$$

In the presence of just one equivalent of electrophilic reagent, only the most reactive double bond of an *isolated diene* will undergo electrophilic addition.

$$CH_2=CHCH_2CH_2\overset{\overset{\displaystyle CH_3}{|}}{C}=CH_2 \ + \ HBr \ \longrightarrow \ CH_2=CHCH_2CH_2\overset{\overset{\displaystyle CH_3}{|}}{\underset{\underset{\displaystyle Br}{|}}{C}}CH_3$$

2. Conjugated dienes undergo 1,2- and 1,4-addition in the presence of one equivalent of an electrophilic reagent (Section 7).

$$RCH=CHCH=CHR \ + \ HBr \ \longrightarrow \ \underset{\textbf{1,2-addition product}}{RCH_2\overset{\underset{\underset{\displaystyle Br}{|}}{}}{CH}CH=CHR} \ + \ \underset{\textbf{1,4-addition product}}{RCH_2CH=CH\overset{\underset{\underset{\displaystyle Br}{|}}{}}{CH}R}$$

3. Conjugated dienes undergo 1,4-addition with a dienophile (a Diels–Alder reaction; see Section 9).

$$CH_2=CH-CH=CH_2 \ + \ \underset{CH_2=CH}{\overset{\overset{\displaystyle O}{\|}}{\underset{}{C}}}R \ \longrightarrow$$

PROBLEMS

57. Which of the following have delocalized electrons?

a. CH₂=CH—C(=O)—CH₃ *(structure with C=O)*

f.

k. *(cyclohexene structure)*

b. CH₃CH=CHOCH₂CH₃

g. CH₃ĊCH₂CH=CH₂ (with CH₃ substituent)

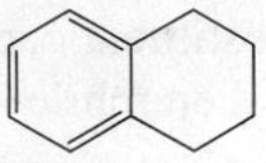

l. *(cyclohexadiene structure)*

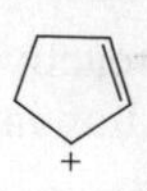

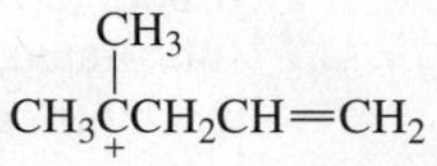

c. CH₃CH=CHCH=ĊHCH₂

h. CH₂=CHCH₂CH=CH₂

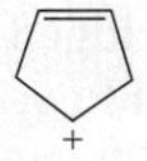

m. CH₃CH₂ĊHCH=CH₂

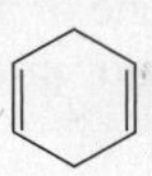

d. CH₃ĊHCH₂CH=CH₂

i. CH₃CH₂NHCH₂CH=CHCH₃

n. CH₃CH₂NHCH=CHCH₃

e.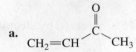

j. *(cyclopentene cation structure)*

o. *(bicyclic structure)*

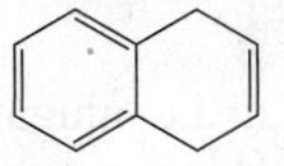

58. a. Draw resonance contributors for the following species, showing all the lone pairs:
   1. CH₂N₂   **2.** N₂O   **3.** NO₂⁻
   b. For each species, indicate the most stable resonance contributor.

59. What is the major product of each of the following reactions? Assume that there is an equivalent amount of each reagent.

a. 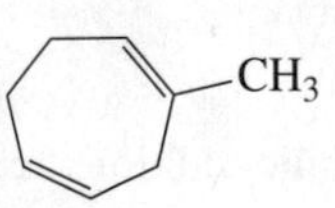 *(cycloheptadiene with CH₃)* + HBr ⟶

b. 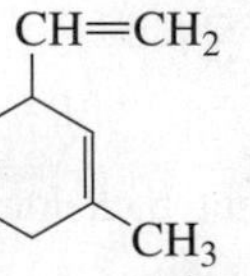 *(cyclohexene with CH=CH₂ and CH₃ substituents)* + HBr ⟶

60. Draw resonance contributors for the following ions:

a. 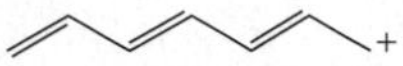 *(diene cation)*   **b.** 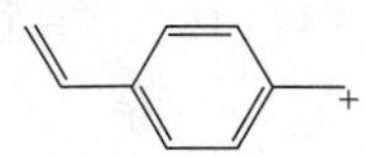 *(methylbenzyl cation)*   **c.** 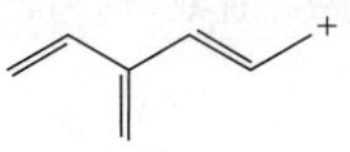 *(branched diene cation)*   **d.** 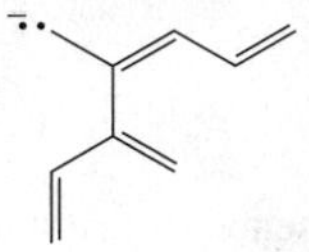 *(diene with lone pair)*

61. Draw all the products of the following reaction:

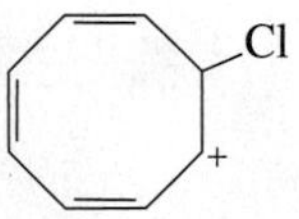

(cyclooctatriene with Cl and + charge) + HO⁻ ⟶

62. Are the following pairs of structures resonance contributors or different compounds?

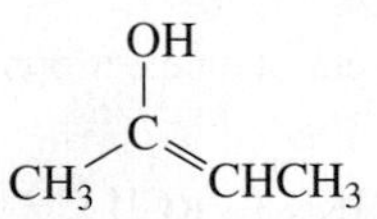

a. 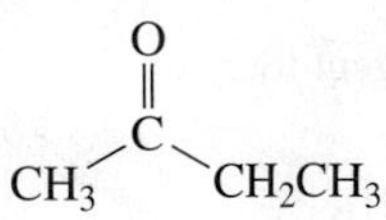 CH₃—C(=O)—CH₂CH₃ and CH₃—C(OH)=CHCH₃

d. 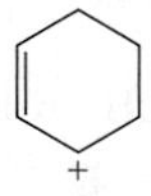 *(cyclohexadiene cation)* and *(cyclohexadiene cation)*

b. CH₃ĊHCH=CHCH₃ and CH₃CH=CHCH₂ĊH₂

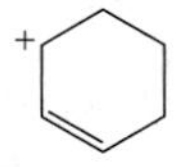

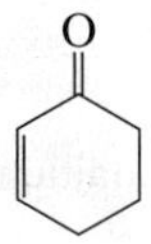

e. *(cyclohexenone)* and *(cyclohexenone)*

c. CH₃CH=CHĊHCH=CH₂ and CH₃ĊHCH=CHCH=CH₂

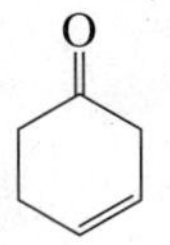

63. a. How many linear dienes have molecular formula C₆H₁₀? (Disregard cis–trans isomers.)
   b. How many of the linear dienes in part **a** are conjugated dienes?
   c. How many are isolated dienes?

64. a. Draw resonance contributors for the following species. Do not include structures that are so unstable that their contributions to the resonance hybrid would be negligible. Indicate which are major contributors and which are minor contributors to the resonance hybrid.

1. $CH_3CH=CHOCH_3$

2.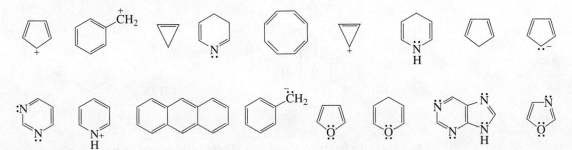

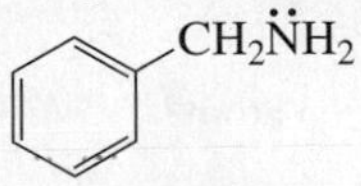

3.

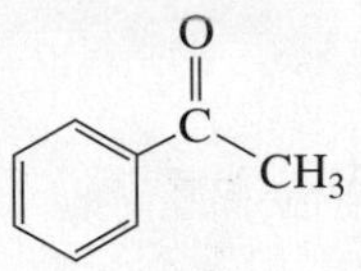

4. 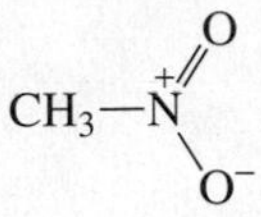 $CH_3—\overset{+}{N}$

5. 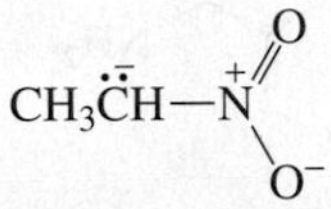 $CH_3\ddot{\overset{-}{C}}H—\overset{+}{N}$

6. $CH_3CH=CH\overset{+}{C}H_2$

7.

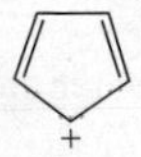

8. 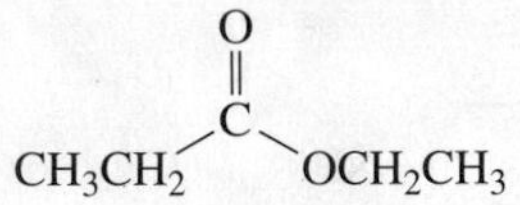 CH_3CH_2 OCH_2CH_3

9. $CH_3CH=CHCH=CH\overset{+}{C}H_2$

10. 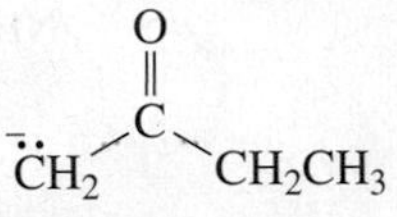 $\overset{-}{C}H_2$ CH_2CH_3

11. $CH_3\ddot{\overset{-}{C}}HCN$

12.

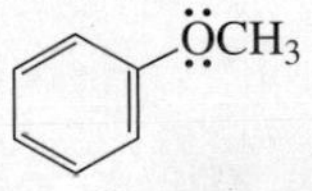

13.

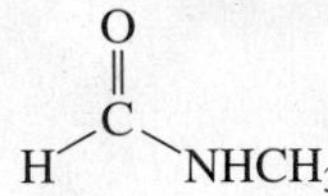

14.

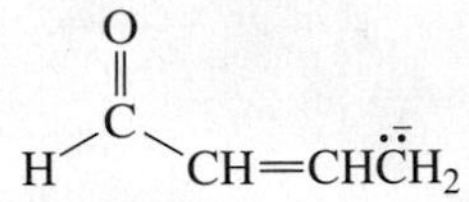

15.

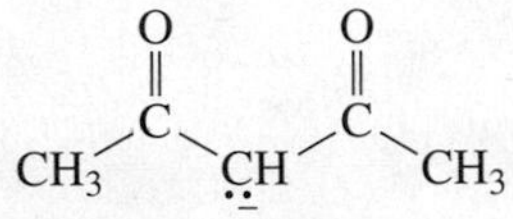

b. Do any of the species have resonance contributors that all contribute equally to the resonance hybrid?

65. Which ion in each of the following pairs is more stable and why?

a. 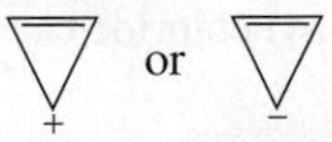 △ or ▽   **b.** 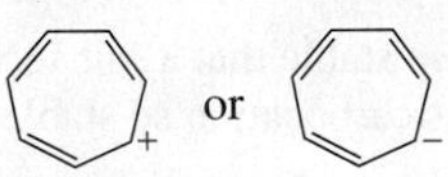 or   **c.** 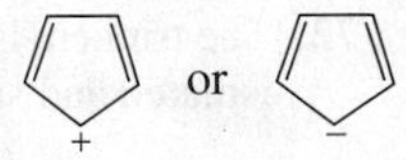 or   **d.** 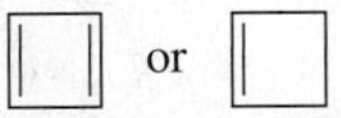 □ or □

66. Which compound would you expect to have the greater heat of hydrogenation, 1,2-pentadiene or 1,4-pentadiene?

67. Which resonance contributor in each pair makes the greater contribution to the resonance hybrid?

a. $CH_3\overset{+}{C}HCH=CH_2$ or $CH_3CH=\overset{+}{C}HCH_2$

b. 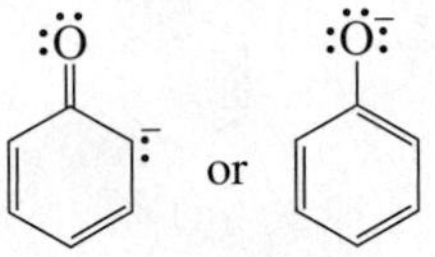 :Ö or :Ö:

c. or

d. $\overset{+}{C}HCH_2CH_3$ or $CH_2\overset{+}{C}HCH_3$

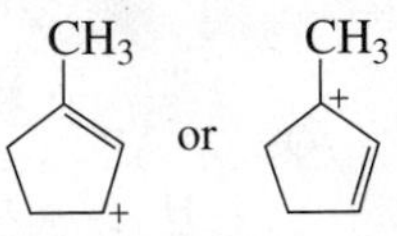

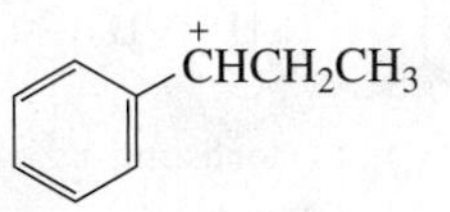

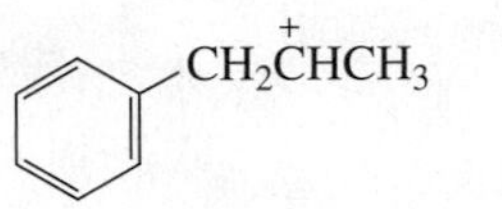

68. Classify the following species as aromatic, nonaromatic, or antiaromatic:

69. a. Which oxygen atom has the greater electron density?

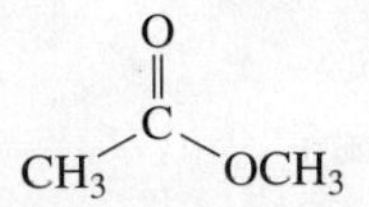

CH_3 OCH_3

b. Which compound has the greater electron density on its nitrogen atom?

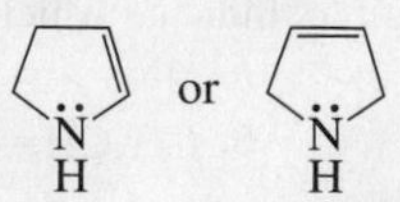

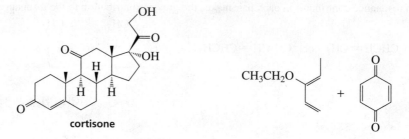

c. Which compound has the greater electron density on its oxygen atom?

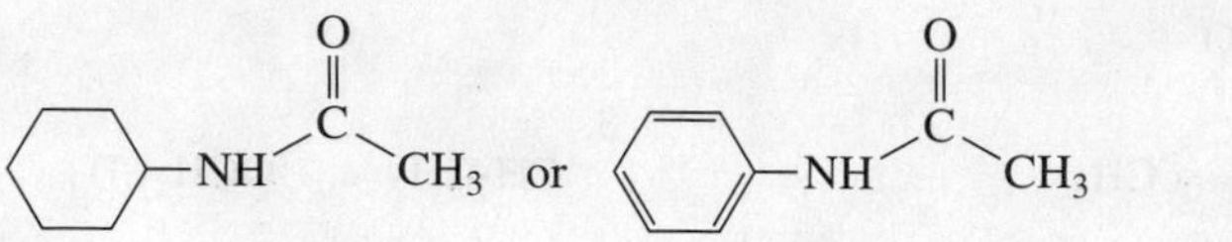

70. Which compound is the strongest base?

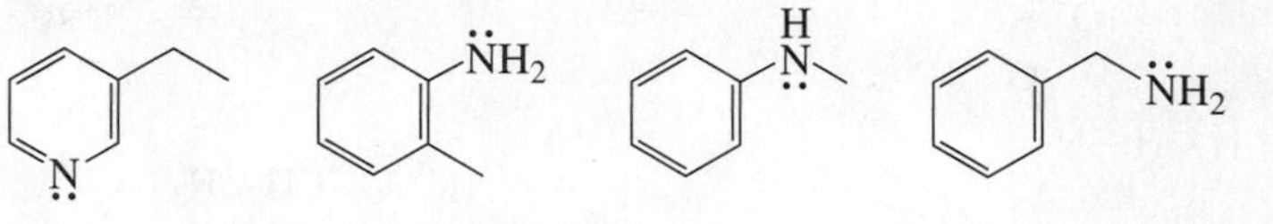

71. Which can lose a proton more readily, a methyl group bonded to cyclohexane or a methyl group bonded to benzene?

72. The triphenylmethyl cation is so stable that a salt such as triphenylmethyl chloride can be isolated and stored. Why is this carbocation so stable?

triphenylmethyl chloride

73. a. The B ring of cortisone, a steroid, is formed by a Diels–Alder reaction using the two reactants shown here. What is the product of this reaction?

cortisone

b. The C ring of estrone (a steroid) is formed by a Diels–Alder reaction using the two reactants shown here. What is the product of this reaction?

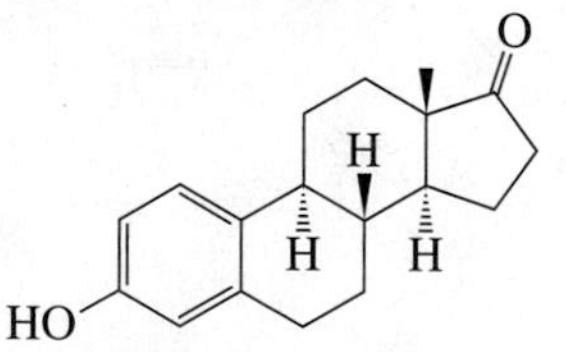

estrone

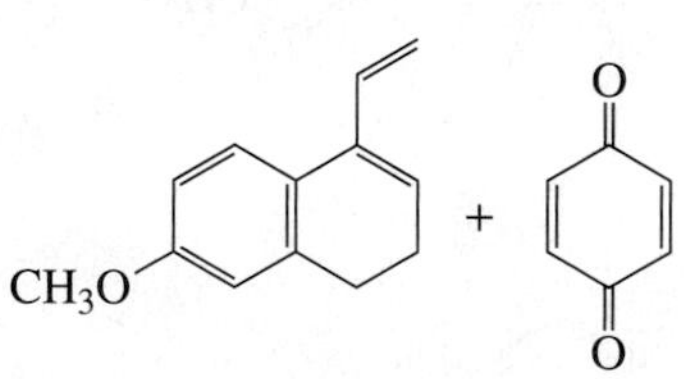

74. Answer the following questions and explain the reason for each answer:

a. Which compound is a stronger acid?

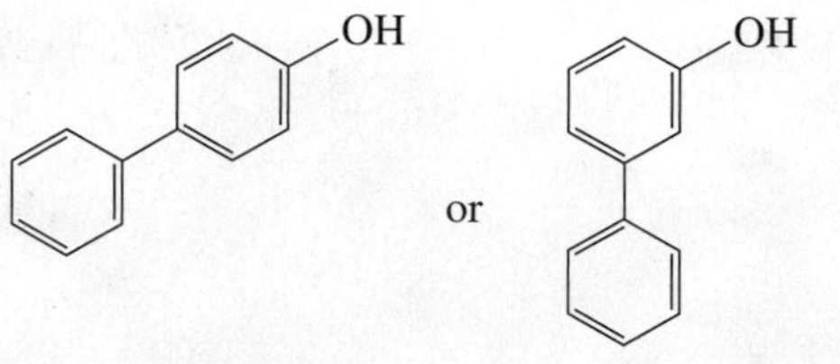

or

b. Which compound is a stronger base?

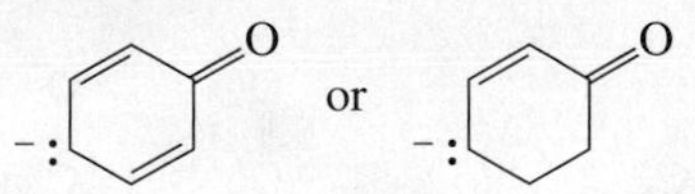

or

75. Draw the resonance contributors for the following anion and rank them in order from most stable to least stable.

$$CH_3CH_2\ddot{O}\underset{\displaystyle \overset{\displaystyle :\ddot{O}:}{\|}}{C}\ \ddot{C}H-C\equiv N:$$

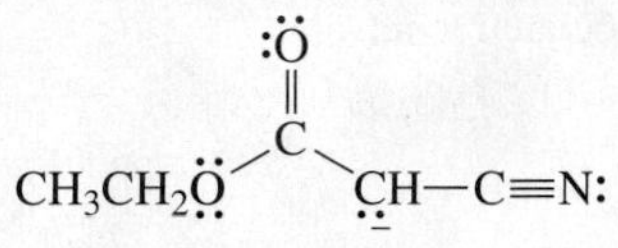

76. Rank the following compounds in order from most stable to least stable:

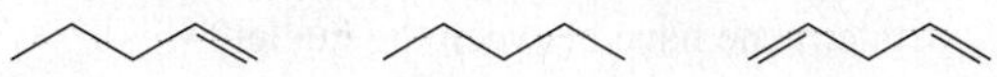

77. Which species in each pair is more stable?

a.

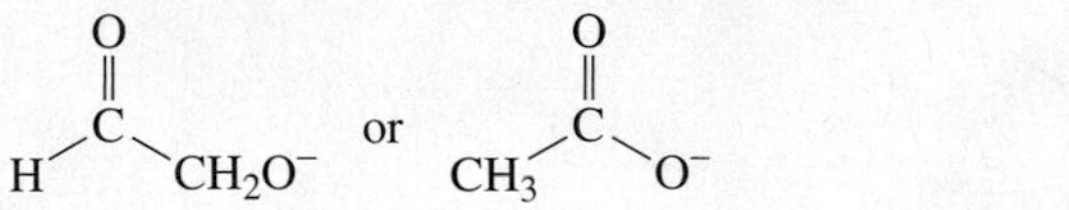

$$\underset{H}{\overset{O}{\|}}\overset{}{C}\ CH_2O^-$$ or $$\underset{CH_3}{\overset{O}{\|}}\overset{}{C}\ O^-$$

c. $$CH_3\ \underset{}{\overset{O}{\|}}C\ \ddot{C}HCH_2\ \underset{}{\overset{O}{\|}}C\ H$$ or $$CH_3\ \underset{}{\overset{O}{\|}}C\ \ddot{C}H\ \underset{}{\overset{O}{\|}}C\ CH_3$$

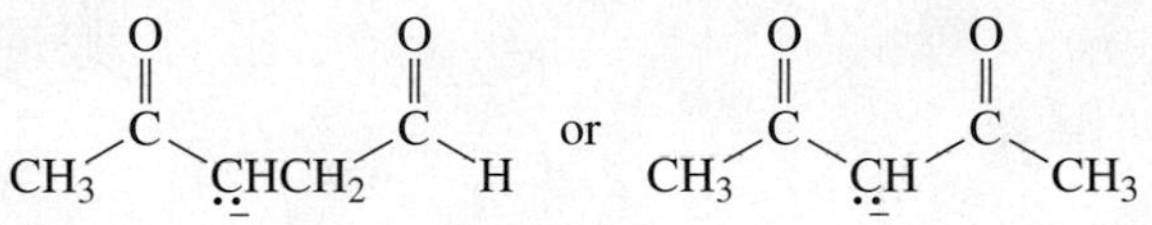

b.

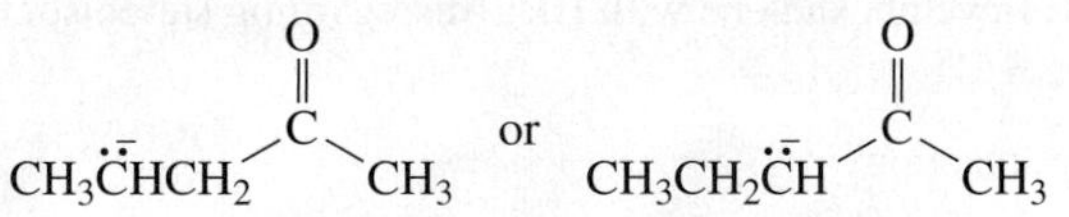

$$CH_3\ddot{C}HCH_2\ \underset{CH_3}{\overset{O}{\|}}C$$ or $$CH_3CH_2\ddot{C}H\ \underset{CH_3}{\overset{O}{\|}}C$$

d. 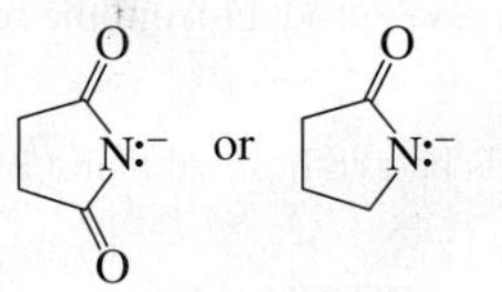 or

78. Which species in each of the pairs in Problem 77 is the stronger base?

79. Purine is a heterocyclic compound with four nitrogen atoms.

a. Which nitrogen is most apt to be protonated? **b.** Which nitrogen is least apt to be protonated?

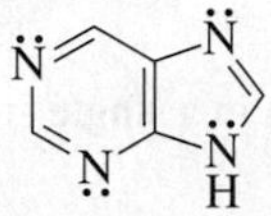

purine

80. Which of the following compounds is the strongest acid?

81. Why is the delocalization energy of pyrrole (21 kcal/mol) greater than that of furan (16 kcal/mol)?

pyrrole   **furan**

82. Rank the indicated hydrogen in the following compounds in order from most acidic to least acidic:

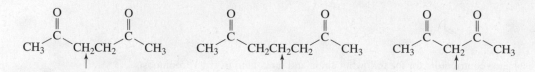

83. Answer the following questions for the π molecular orbitals (MOs) of 1,3,5,7-octatetraene:

  a. How many MOs does the compound have?
  b. Which are the bonding MOs and which are the antibonding MOs?
  c. Which MOs are symmetric and which are antisymmetric?
  d. Which MO is the HOMO and which is the LUMO in the ground state?
  e. Which MO is the HOMO and which is the LUMO in the excited state?
  f. What is the relationship between HOMO and LUMO and symmetric and antisymmetric orbitals?
  g. How many nodes does the highest-energy MO of 1,3,5,7-octatetraene have between the nuclei?

84. How could you synthesize the following compound from starting materials containing no more than six carbons? (*Hint:* A 1,6-diketone can be synthesized by oxidative cleavage of a 1,2-disubstituted cyclohexene.)

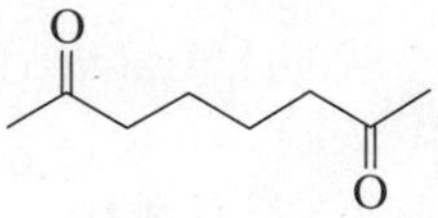

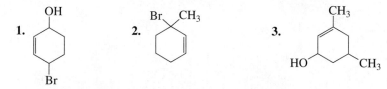

85. A student obtained two products from the reaction of 1,3-cyclohexadiene with Br_2 (disregarding stereoisomers). His lab partner was surprised when he obtained only one product from the reaction of 1,3-cyclohexadiene with HBr (disregarding stereoisomers). Account for these results.

86. How could the following compounds be synthesized using a Diels–Alder reaction?

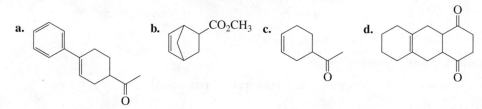

87. a. How could each of the following compounds be prepared from a hydrocarbon in a single step?

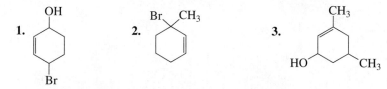

  b. What other organic compound would be obtained from each synthesis?

88. Draw the products obtained from the reaction of 1 equiv HBr with 1 equiv 1,3,5-hexatriene.

  a. Which product(s) will predominate if the reaction is under kinetic control?
  b. Which product(s) will predominate if the reaction is under thermodynamic control?

89. How would the following substituents affect the rate of a Diels–Alder reaction?

  a. an electron-donating substituent in the diene
  b. an electron-donating substituent in the dienophile
  c. an electron-withdrawing substituent in the diene

90. Draw the major products obtained from the reaction of one equivalent of HCl with the following compounds. For each reaction, indicate the kinetic and thermodynamic products.

  a. 2,3-dimethyl-1,3-pentadiene   **b.** 2,4-dimethyl-1,3-pentadiene

91. The acid dissociation constant (K_a) for loss of a proton from cyclohexanol is 1×10^{-16}.

a. Draw an energy diagram for loss of a proton from cyclohexanol.

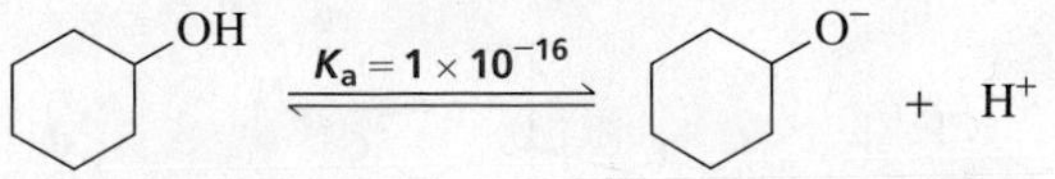

b. Draw the resonance contributors for phenol.
c. Draw the resonance contributors for the phenolate ion.
d. On the same plot with the energy diagram for loss of a proton from cyclohexanol, draw an energy diagram for loss of a proton from phenol.

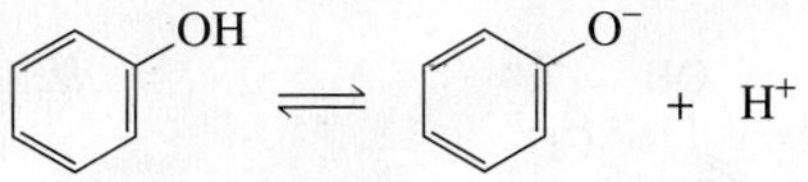

e. Which has a greater K_a, cyclohexanol or phenol?
f. Which is a stronger acid, cyclohexanol or phenol?

92. Protonated cyclohexylamine has a $K_a = 1 \times 10^{-11}$. Using the same sequence of steps as in Problem 91, determine which is a stronger base, cyclohexylamine or aniline.

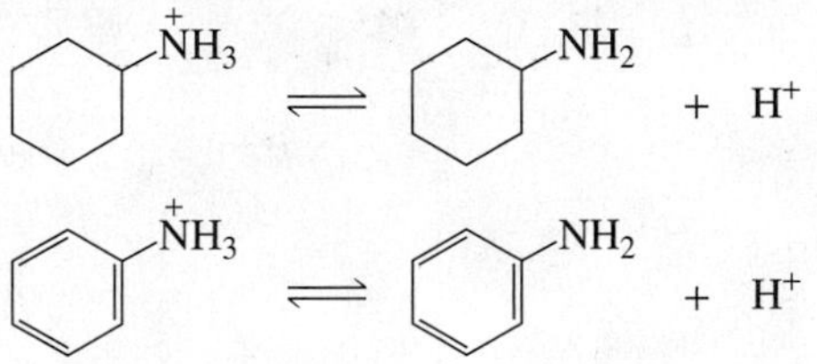

93. Draw the product or products that would be obtained from each of the following reactions:

a.
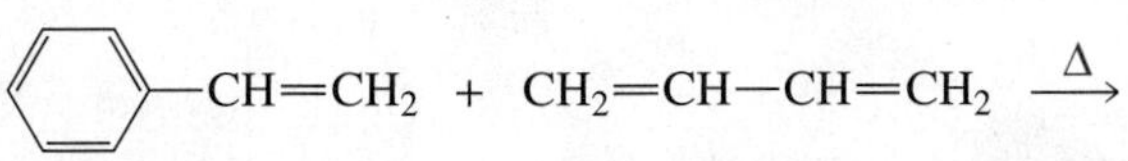

b.
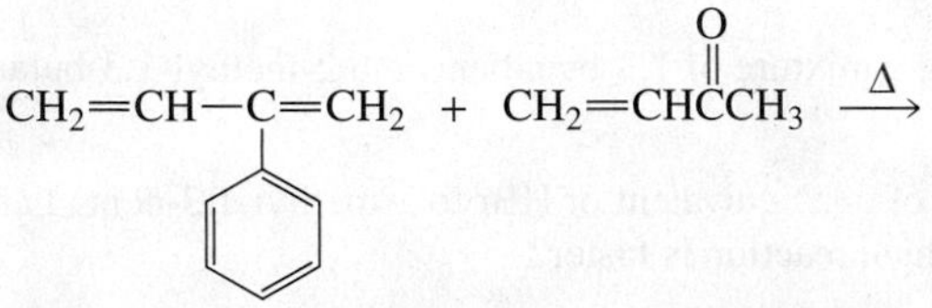

c.
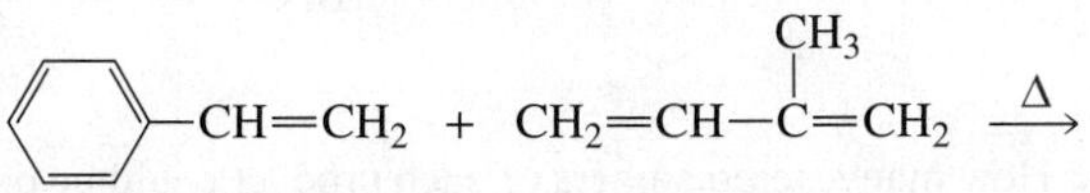

94. What two sets of a conjugated diene and a dienophile could be used to prepare the following compound?

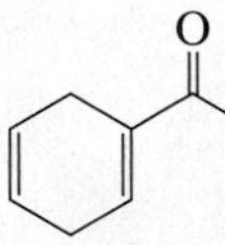

95. a. Which dienophile in each pair is more reactive in a Diels–Alder reaction?

1. $CH_2{=}CH\overset{O}{\overset{\|}{C}}H$ or $CH_2{=}CHCH_2\overset{O}{\overset{\|}{C}}H$ **2.** $CH_2{=}CH\overset{O}{\overset{\|}{C}}H$ or $CH_2{=}CHCH_3$

b. Which diene is more reactive in a Diels–Alder reaction?

$CH_2{=}CHCH{=}CHOCH_3$ or $CH_2{=}CHCH{=}CHCH_2OCH_3$

96. Cyclopentadiene can react with itself in a Diels–Alder reaction. Draw the endo and exo products.

97. Which diene and which dienophile could be used to prepare each of the following?

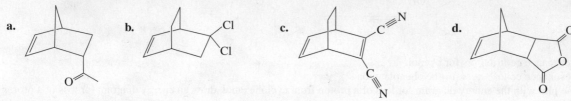

a. **b.** **c.** **d.**

98. a. Propose a mechanism for the following reaction:

b. What is the product of the following reaction?

99. a. What are the products of the following reaction?

b. How many stereoisomers of each product could be obtained?

100. As many as 18 different Diels–Alder products can be obtained by heating a mixture of 1,3-butadiene and 2-methyl-1,3-butadiene. Identify the products.

101. On a single graph, draw the reaction coordinate diagram for the addition of one equivalent of HBr to 2-methyl-1,3-pentadiene and for the addition of one equivalent of HBr to 2-methyl-1,4-pentadiene. Which reaction is faster?

102. While attempting to recrystallize maleic anhydride, a student dissolved it in freshly distilled cyclopentadiene rather than in freshly distilled cyclopentane. Was his recrystallization successful?

maleic anhydride

103. The following equilibrium is driven to the right if the reaction is carried out in the presence of maleic anhydride (see Problem 102). What is the function of maleic anhydride?

104. In 1935, J. Bredt, a German chemist, proposed that a bicycloalkene could not have a double bond at a bridgehead carbon unless one of the rings contains at least eight carbons. This is known as Bredt's rule. Explain why there cannot be a double bond at this position.

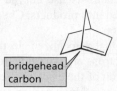

bridgehead carbon

105. The experiment shown next and discussed in Section 18 shows that the proximity of the chloride ion to C-2 in the transition state causes the 1,2-addition product to be formed faster than the 1,4-addition product.

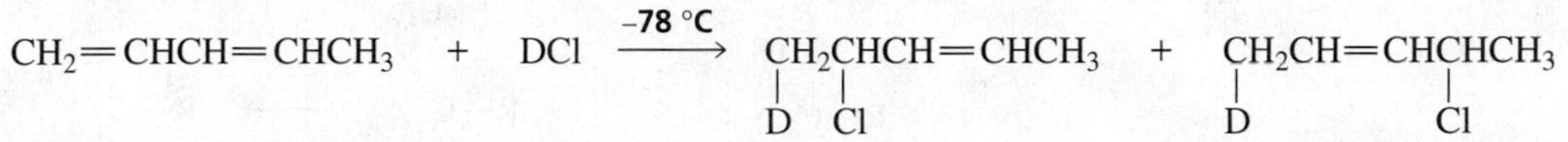

$$CH_2{=}CHCH{=}CHCH_3 \ + \ DCl \ \xrightarrow{-78\,°C} \ \underset{\underset{D \ \ Cl}{|\ \ |}}{CH_2CHCH{=}CHCH_3} \ + \ \underset{\underset{D \qquad Cl}{|\qquad |}}{CH_2CH{=}CHCHCH_3}$$

a. Why was it important for the investigators to know that the preceding reaction was being carried out under kinetic control?

b. How could the investigators know that the reaction was being carried out under kinetic control?

106. The product of the following reaction has a fused bicyclic ring. What is its structure?

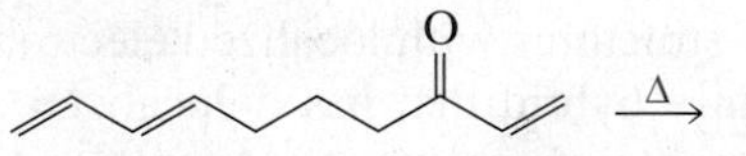

107. Draw the resonance contributors of the cyclooctatrienyl dianion.

a. Which of the resonance contributors is the least stable?

b. Which of the resonance contributors makes the smallest contribution to the hybrid?

108. Investigation has shown that cyclobutadiene is actually a rectangular molecule rather than a square molecule, and that there are two different 1,2-dideuterio-1,3-cyclobutadienes. Explain the reason for these unexpected observations.

cyclobutadiene

DRAWING RESONANCE CONTRIBUTORS

We have seen that chemists use curved arrows to show how electrons move when reactants are converted into products. Chemists also use curved arrows when they draw resonance contributors.

We have also seen that delocalized electrons are electrons that are shared by more than two atoms. When electrons are shared by more than two atoms, we cannot use solid lines to represent the location of the electrons accurately. For example, in the carboxylate ion, a pair of electrons is shared by a carbon and two oxygens. We show the pair of delocalized electrons by a dotted line spread over the three atoms. We have seen that this structure is called a **resonance hybrid.** The resonance hybrid shows that the negative charge is shared by the two oxygens.

**carboxylate ion
resonance hybrid**   **resonance contributors**

Chemists do not like to use dotted lines when drawing structures because, unlike a solid line that represents two electrons, the dotted lines do not specify the number of electrons they represent. Therefore, chemists use structures with localized electrons (indicated by solid lines) to approximate the resonance hybrid that has delocalized electrons (indicated by dotted lines). These approximate structures are called **resonance contributors.** Curved arrows are used to show the movement of electrons in going from one resonance contributor to the next.

RULES FOR DRAWING RESONANCE CONTRIBUTORS

Now we will look at three simple rules for interconverting resonance contributors:

1. Only electrons move; atoms *never* move.
2. The only electrons that can move are π electrons (electrons in π bonds) and lone-pair electrons.
3. Electrons are always moved toward an sp^2 or sp hybridized atom. An sp^2 carbon is a positively charged carbon or a double-bonded carbon; an sp carbon is a triply bonded carbon.

Moving π Electrons Toward an sp^2 Carbon That Is a Positively Charged Carbon

In the following example, π electrons are moved toward a positively-charged carbon. Because the atom does not have a complete octet of electrons, it can accept the electrons. The carbon that is positively charged in the first resonance contributor is neutral in the second resonance contributor because it has received electrons. The carbon in the first resonance contributor that loses its share of the π electrons is positively charged in the second resonance contributor.

$$CH_3CH = CH - \overset{+}{C}HCH_3 \longleftrightarrow CH_3\overset{+}{C}H - CH = CHCH_3$$

We see that the following carbocation has three resonance contributors.

$$CH_2=CH-CH=CH-\overset{+}{C}HCH_3 \longleftrightarrow CH_2=CH-\overset{+}{C}H-CH=CHCH_3$$

$$\updownarrow$$

$$\overset{+}{C}H_2-CH=CH-CH=CHCH_3$$

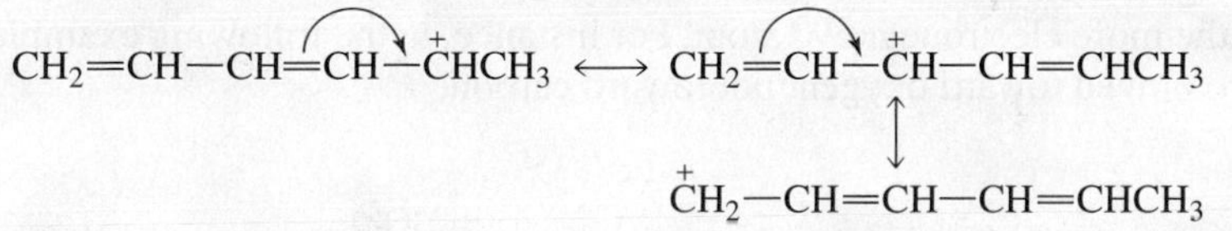

Notice that in going from one resonance contributor to the next, the total number of electrons in the structure does not change. Therefore, each of the resonance contributors must have the same net charge.

PROBLEM 1 Draw the resonance contributors for the following carbocation (the answers can be found immediately after Problem 12):

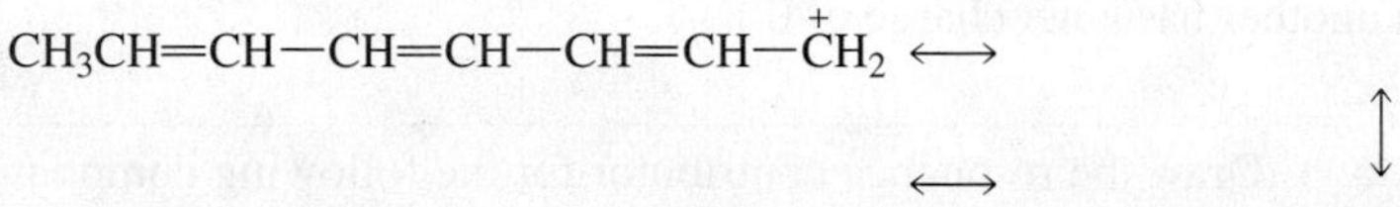

$$CH_3CH=CH-CH=CH-CH=CH-\overset{+}{C}H_2 \longleftrightarrow$$

$$\longleftrightarrow \qquad\qquad \updownarrow$$

PROBLEM 2 Draw the resonance contributors for the following carbocation:

$$\longleftrightarrow \qquad \longleftrightarrow \qquad \longleftrightarrow \qquad \longleftrightarrow$$

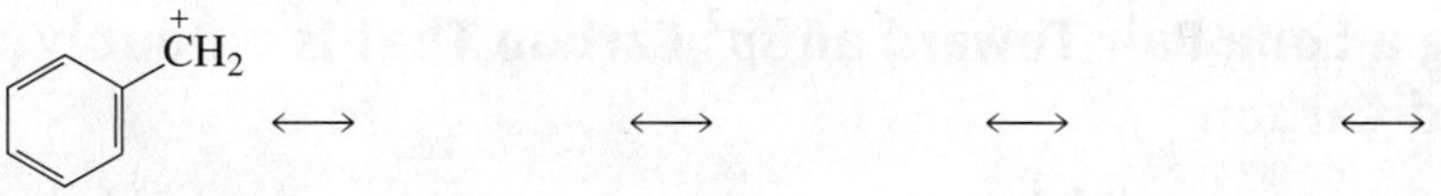

Moving π Electrons Toward an sp^2 Carbon That Is a Doubly Bonded Carbon

In the following example, π electrons are moved toward a doubly bonded carbon. The atom to which the electrons are moved can accept them because the π bond can break.

PROBLEM 3 Draw the resonance contributor for the following compound:

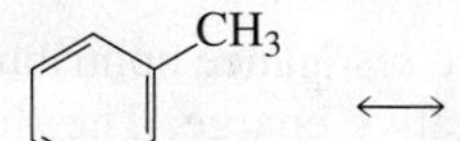

$$\longleftrightarrow$$

In the next example, the electrons can be moved equally easily to the left (indicated by the red arrows) or to the right (indicated by the blue arrows). When comparing the charges on the resonance contributors, we see that the charges on each of the end carbons cancel, so there is no charge on any of the carbons in the resonance hybrid.

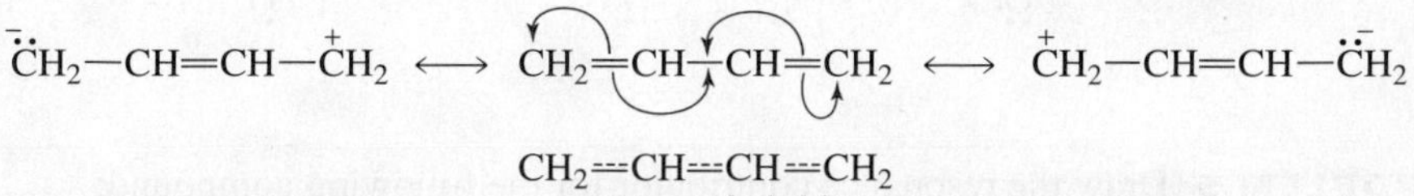

$$\overset{..}{\overset{-}{C}}H_2-CH=CH-\overset{+}{C}H_2 \longleftrightarrow CH_2=CH-CH=CH_2 \longleftrightarrow \overset{+}{C}H_2-CH=CH-\overset{..}{\overset{-}{C}}H_2$$

$$CH_2\text{---}CH\text{---}CH\text{---}CH_2$$
resonance hybrid

405

When electrons can be moved in either of two directions and there is a difference in the electronegativity of the atoms to which they can be moved, always move the electrons toward the more electronegative atom. For instance, in the following example, the electrons are moved toward oxygen, not toward carbon.

Notice that the first resonance contributor has a charge of 0. Since the number of electrons in the molecule does not change, the other resonance contributor must have a net charge of 0. (A net charge of 0 does not mean that there is no charge on any of the atoms; a resonance contributor with a positive charge on one atom and a negative charge on another has a net charge of 0.)

PROBLEM 4 Draw the resonance contributor for the following compound:

$$CH_2{=}CH{-}C{\equiv}N \longleftrightarrow$$

Moving a Lone Pair Toward an sp^2 Carbon That Is a Doubly Bonded Carbon

In the following examples, lone-pair electrons are moved toward a doubly bonded carbon. Notice that the arrow starts at a pair of electrons, not at a negative charge. In the first example, each of the resonance contributors has a charge of -1; in the second example, each of the resonance contributors has no charge or net charge of 0.

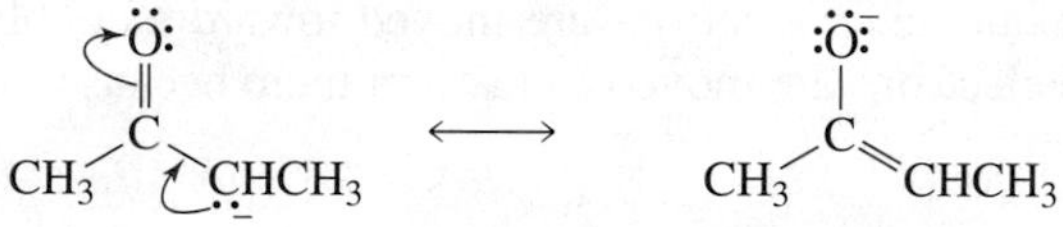

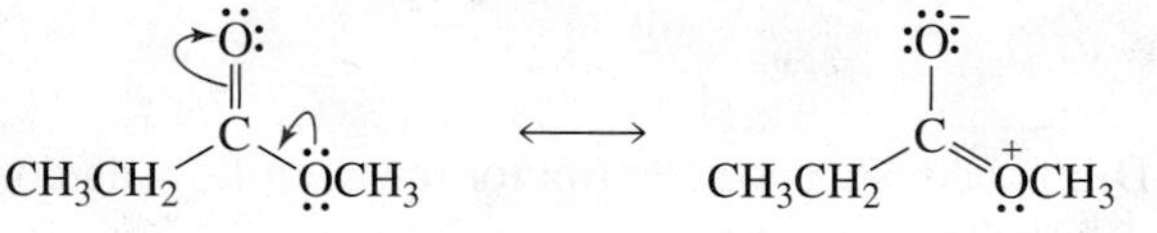

The following species has three resonance contributors. Notice again that the arrow starts at a lone pair, not at a negative charge. The three oxygen atoms share the two negative charges. Therefore, each oxygen atom in the hybrid has two-thirds of a negative charge.

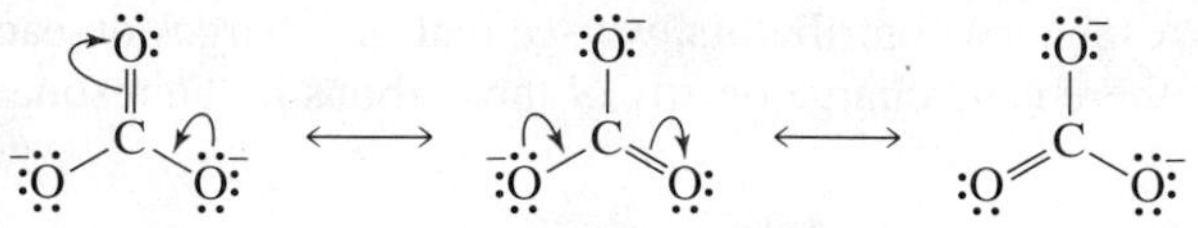

PROBLEM 5 Draw the resonance contributor for the following compound:

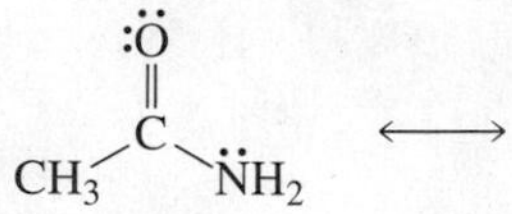

Notice in the next example that the lone pair of electrons moves away from the most electronegative atom in the molecule. This is the only way electron delocalization can occur (and any electron delocalization is better than none). The π electrons cannot move toward the oxygen because the oxygen atom has a complete octet (it is sp^3 hybridized). Recall that electrons can move only toward an sp^2 or sp hybridized atom.

$$CH_3CH=CH-\ddot{O}CH_3 \longleftrightarrow CH_3\ddot{C}H-CH=\overset{+}{O}CH_3$$

The compound in the next example has five resonance contributors. To get to the second resonance contributor, a lone pair on nitrogen is moved toward an sp^2 carbon. Notice that the first and fifth resonance contributors are not the same; they are similar to the two resonance contributors of benzene.

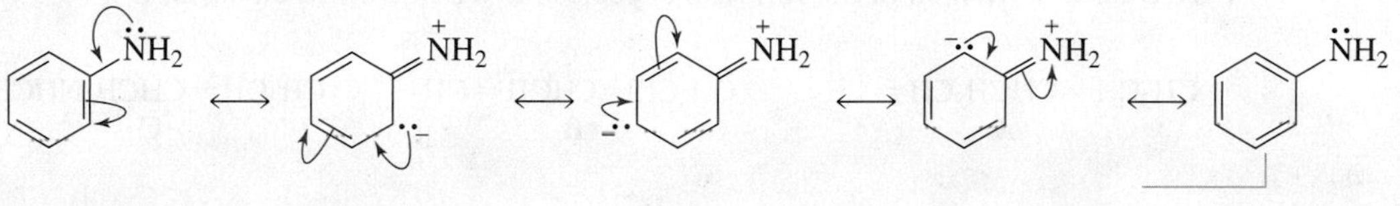

PROBLEM 6 Draw the resonance contributors for the following compound:

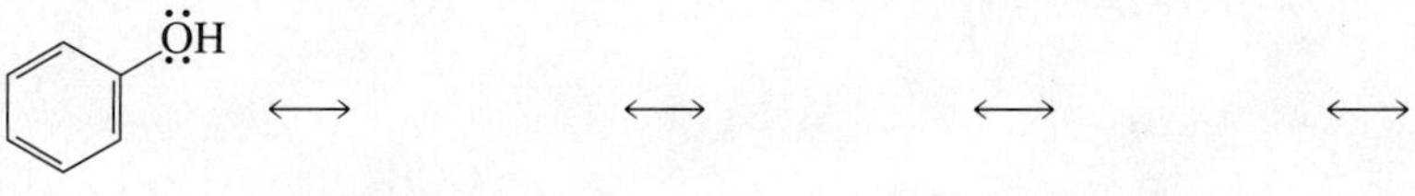

The following species do not have delocalized electrons. Electrons cannot be moved toward an sp^3 hybridized atom because an sp^3 hybridized atom has a complete octet and it does not have a π bond that can break, so it cannot accept any more electrons.

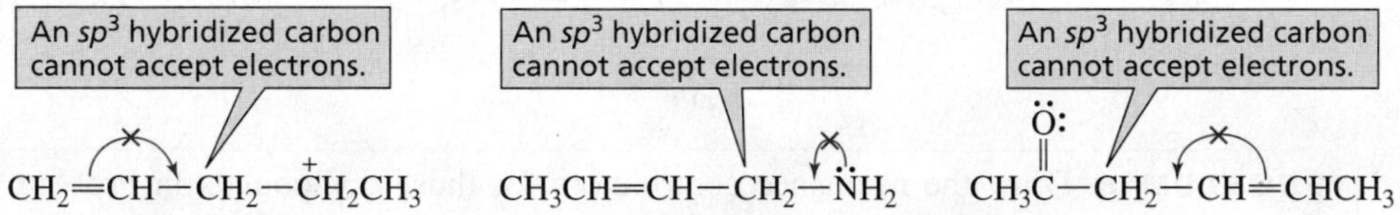

Notice the difference in the resonance contributors for the next two examples. In the first example, electrons *move into* the benzene ring. That is, a lone pair on the atom attached to the ring moves toward an sp^2 carbon.

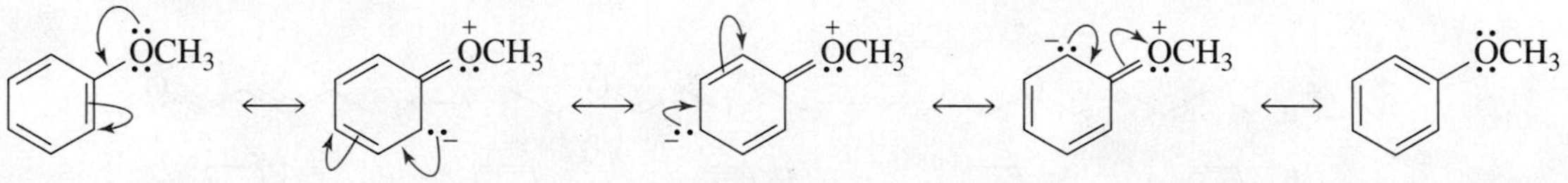

In the next example, electrons *move out* of the benzene ring. First, a π bond moves toward an sp^2 carbon. The electron movement is toward the oxygen since oxygen is more electronegative than carbon. Then, to draw the other resonance contributors, a π bond moves toward a positive charge.

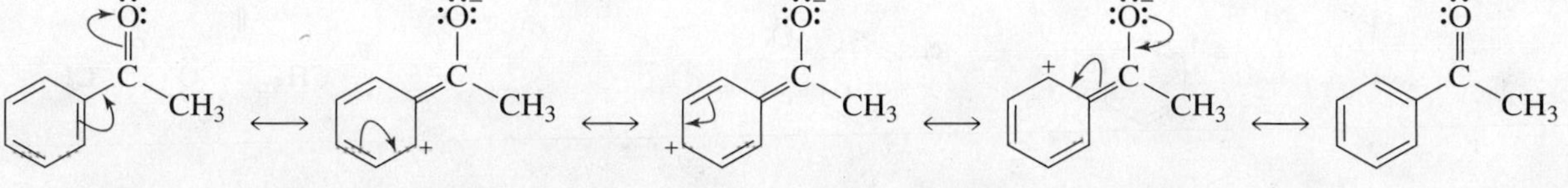

In the next two examples, the atom attached to the ring has neither a lone pair nor a π bond. Therefore, the substituent can neither donate electrons into the ring nor accept electrons from the ring. Thus, these compounds have only two resonance contributors—the ones that are similar to the two resonance contributors of benzene.

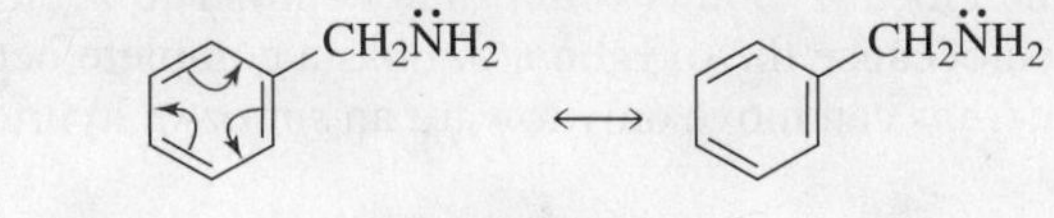

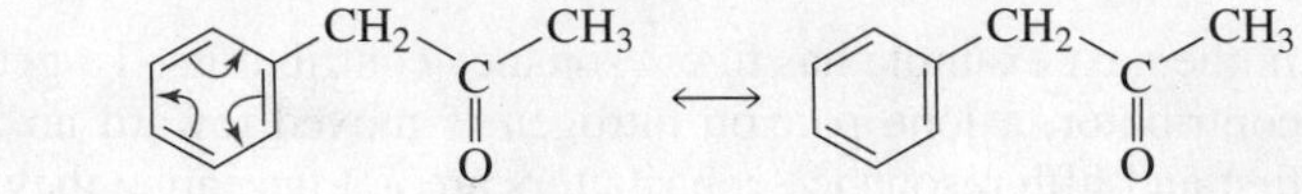

PROBLEM 7 Which of the following species have delocalized electrons?

$$CH_3CH=CHCH_2CH=CH_2 \qquad CH_3CH=CHCH=CH_2 \qquad CH_3CH=CHCH_2\overset{..}{N}HCH_3$$
$$\text{A} \qquad\qquad\qquad \text{B} \qquad\qquad\qquad \text{C}$$

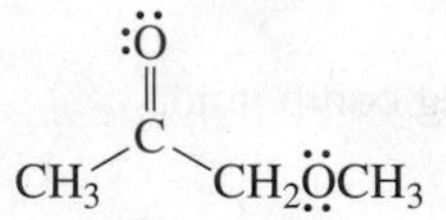

D 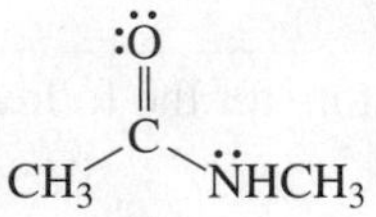 · E 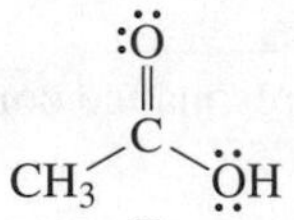 · F 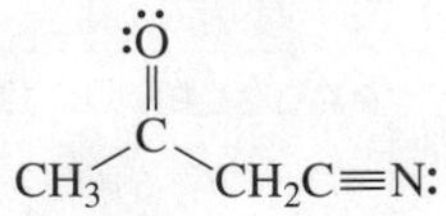 · G

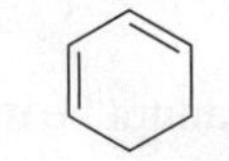

H 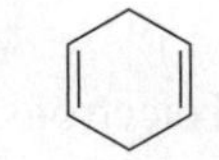 · I 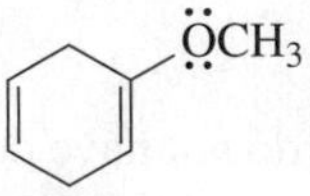 · J 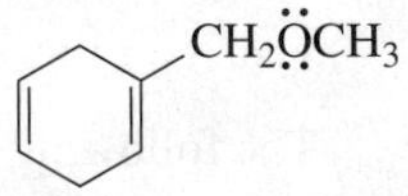 · K

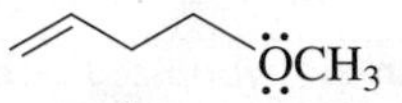

L 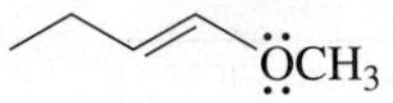 · M  · N 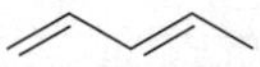 · O

PROBLEM 8 Draw the resonance contributors for those compounds in Problem 7 that have delocalized electrons.

PROBLEM 9 Draw curved arrows to show how one resonance contributor leads to the next one.

a.

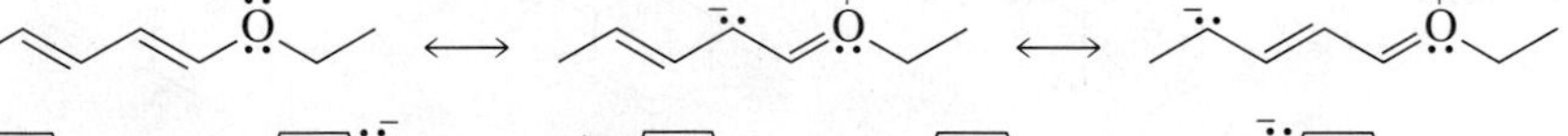

b.

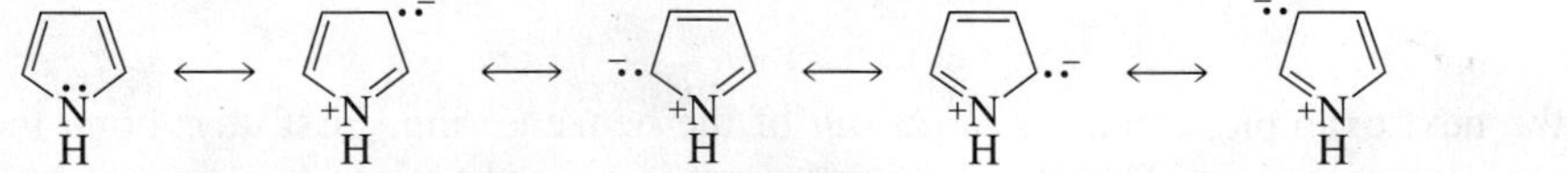

PROBLEM 10 Draw the resonance contributors for each of the following species:

a.

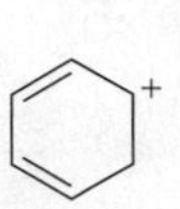

b.

c.

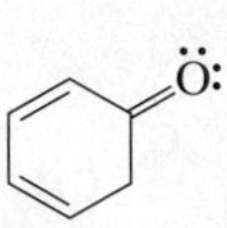

d. $CH_2=CH-CH=CH-\overset{..}{N}H_2$

e.

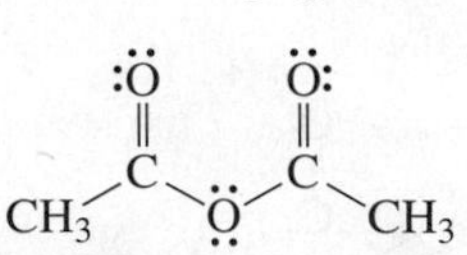

f.

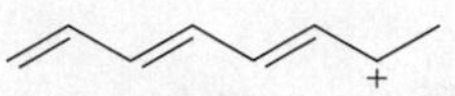

PROBLEM 11 Draw the resonance contributors for each of the following species:

a.

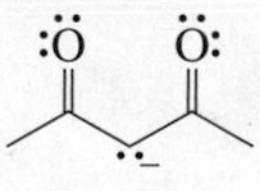

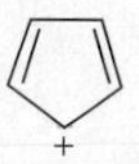

b.

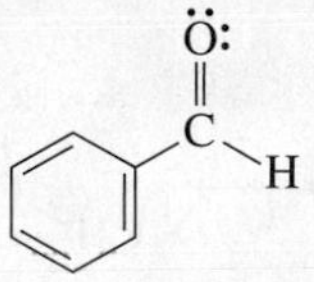

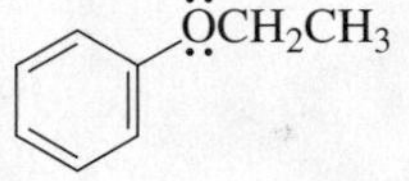

c.

d.

e.

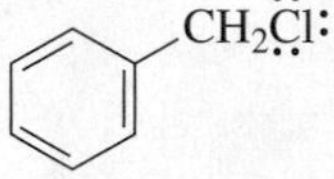

f.

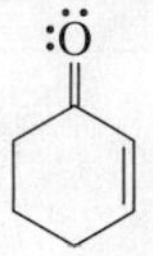

PROBLEM 12 Draw the resonance contributors for each of the following species:

a.

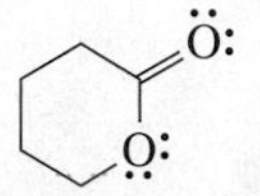

b.

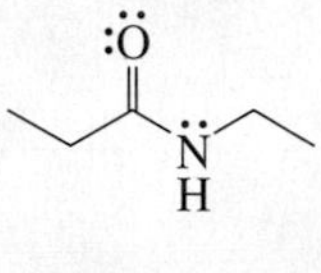

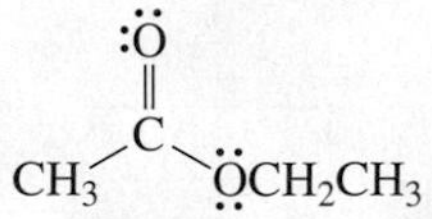

c.

d.

e.

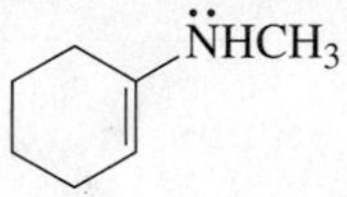

f.

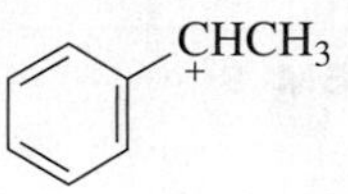

ANSWERS TO PROBLEMS ON DRAWING RESONANCE CONTRIBUTORS

PROBLEM 1

$$CH_3CH{=}CH{-}CH{=}CH{-}CH{=}CH{-}\overset{+}{C}H_2 \longleftrightarrow CH_3CH{=}CH{-}CH{=}CH{-}\overset{+}{C}H{-}CH{=}CH_2$$

$$CH_3\overset{+}{C}H{-}CH{=}CH{-}CH{=}CH{-}CH{=}CH_2 \longleftrightarrow CH_3CH{=}CH{-}\overset{+}{C}H{-}CH{=}CH{-}CH{=}CH_2$$

PROBLEM 2

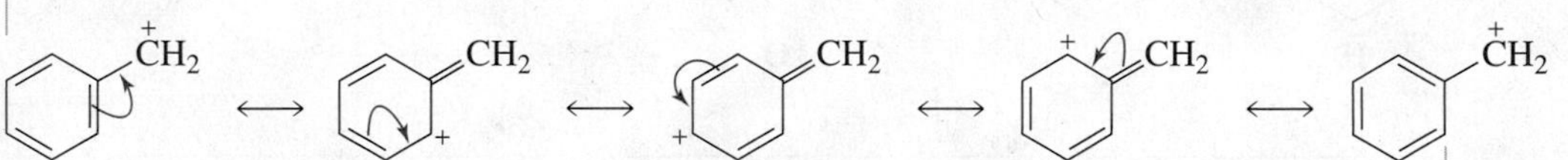

PROBLEM 3

PROBLEM 4

$$CH_2{=}CH{-}C{\equiv}N{:} \longleftrightarrow \overset{+}{C}H_2{=}CH{-}C{\equiv}\ddot{N}{:}^-$$

PROBLEM 5

PROBLEM 6

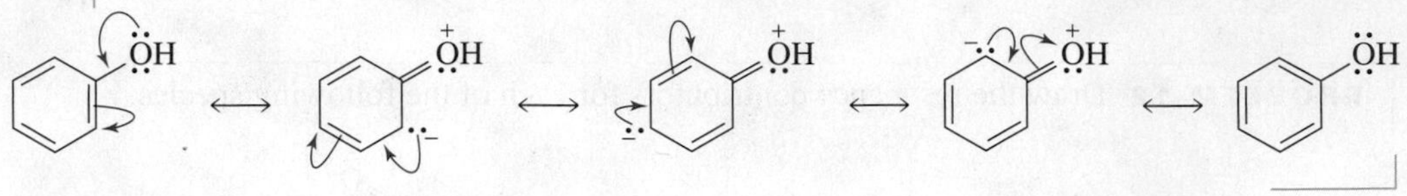

PROBLEM 7

B, E, F, H, J, M, and O

PROBLEM 8

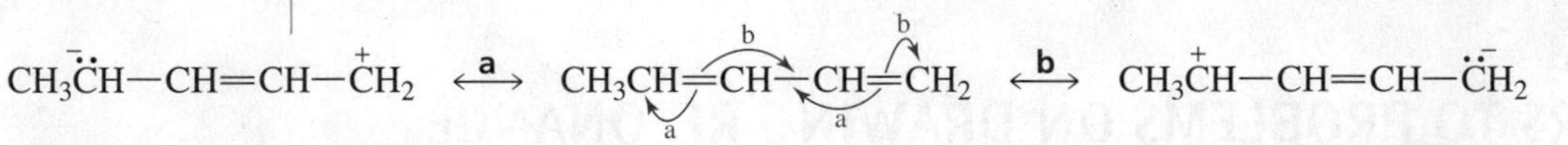

B

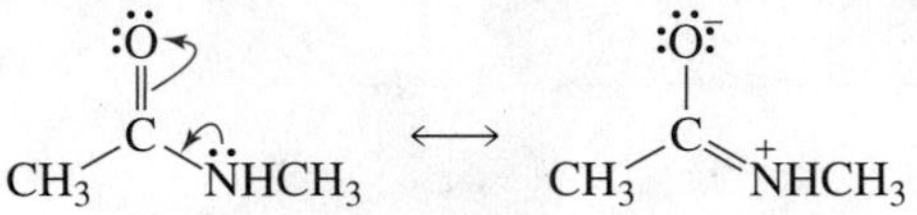

E

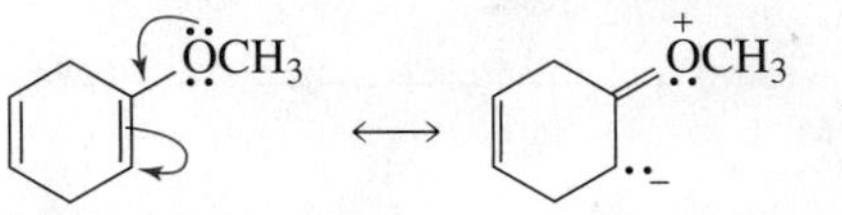

J

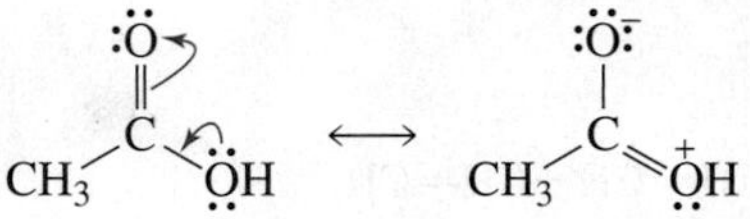

F

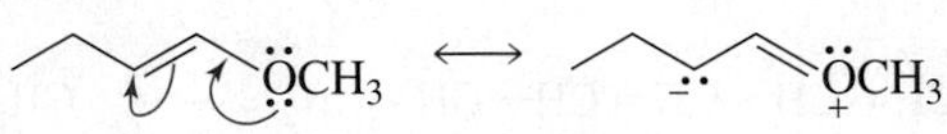

M

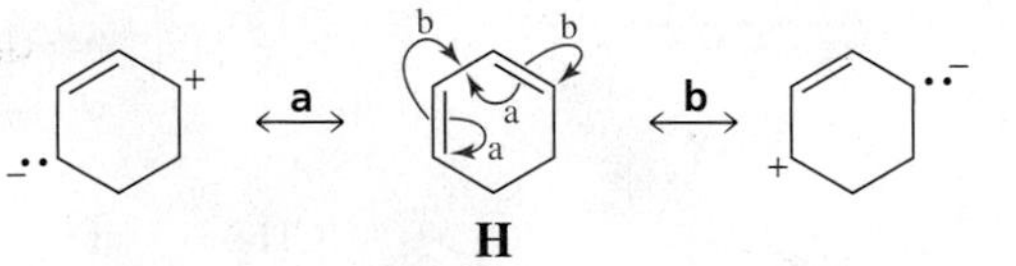

H

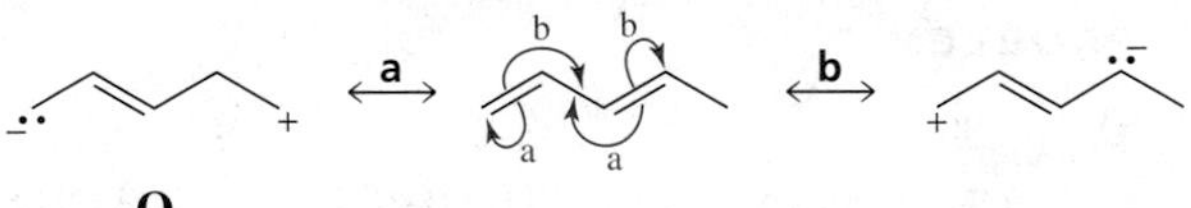

O

PROBLEM 9

a.

b.

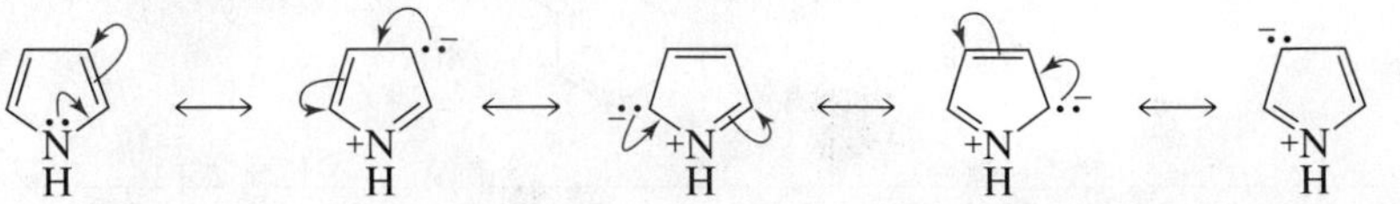

PROBLEM 10

a.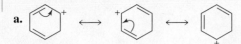

b. Notice in the following example that the electrons can move either clockwise or counterclockwise:

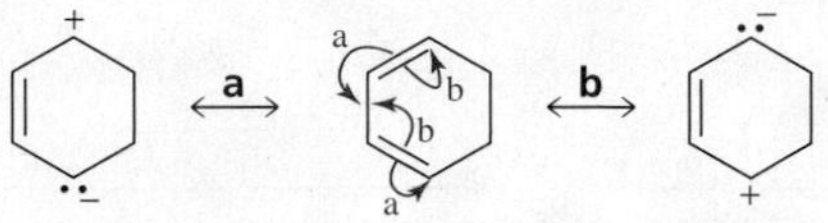

c.

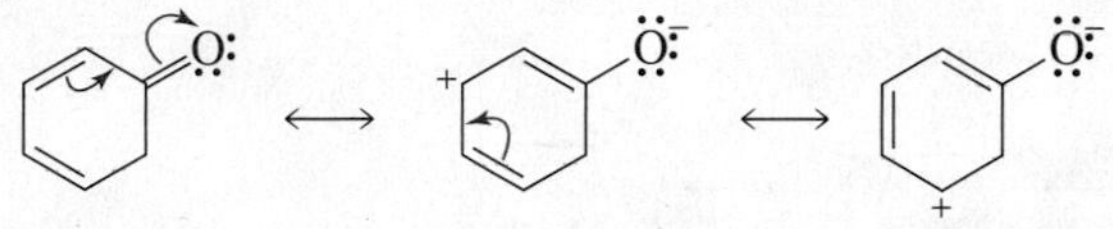

d. 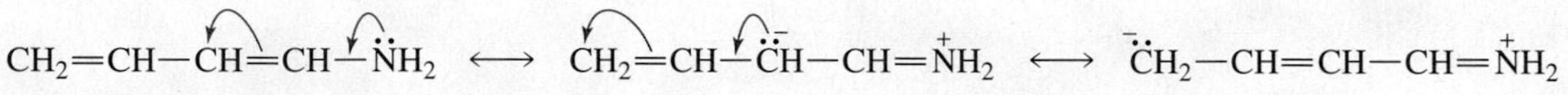 $CH_2=CH-CH=CH-\overset{..}{N}H_2 \longleftrightarrow CH_2=CH-\overset{..}{C}H-CH=\overset{+}{N}H_2 \longleftrightarrow \overset{-}{\overset{..}{C}}H_2-CH=CH-CH=\overset{+}{N}H_2$

e. Notice in the following example that the lone-pair electrons can be moved toward either of the two sp^2 carbons:

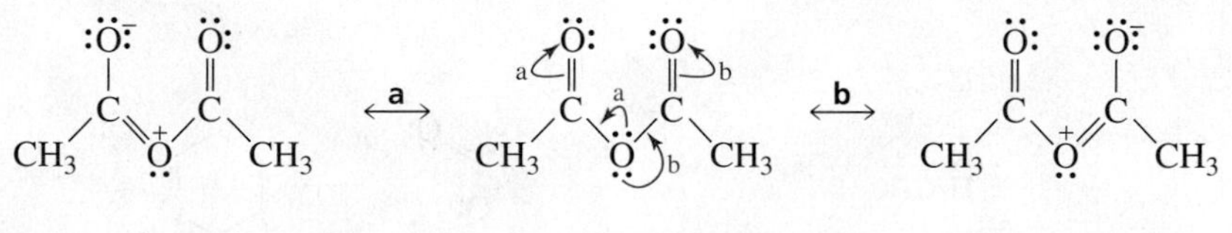

f.

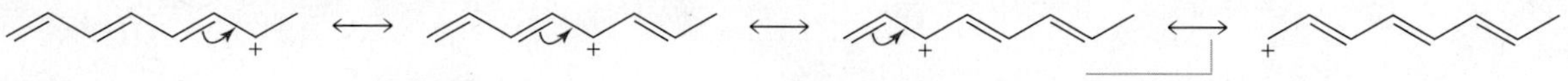

PROBLEM 11

a.

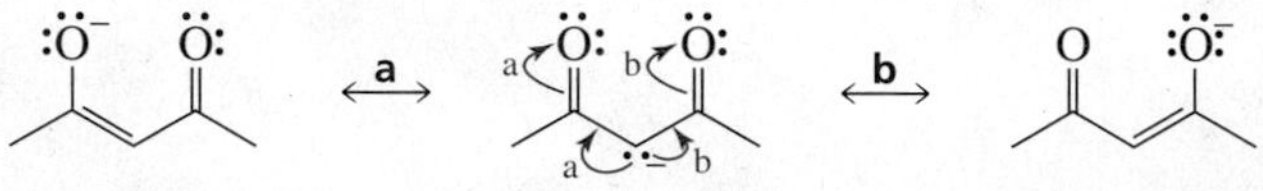

b.

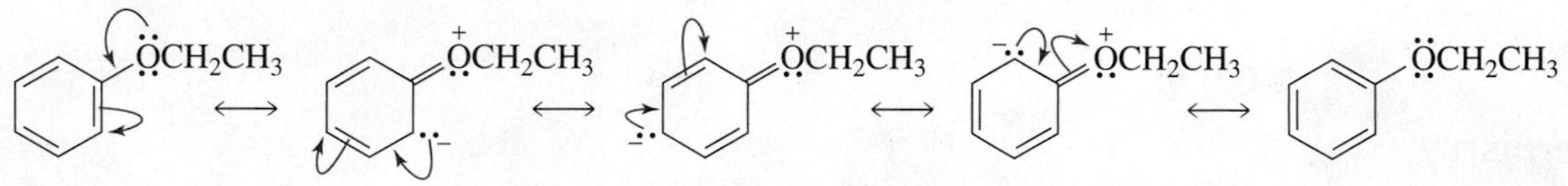

c.

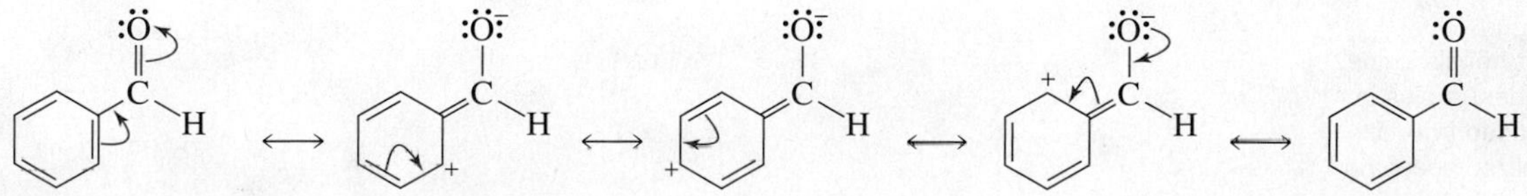

d.

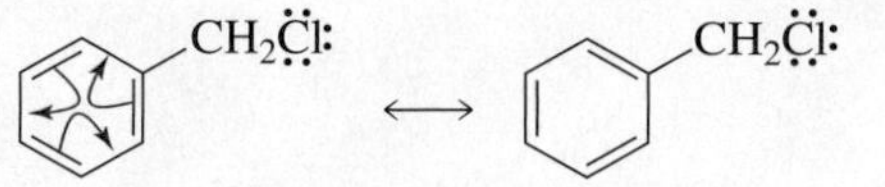

e.

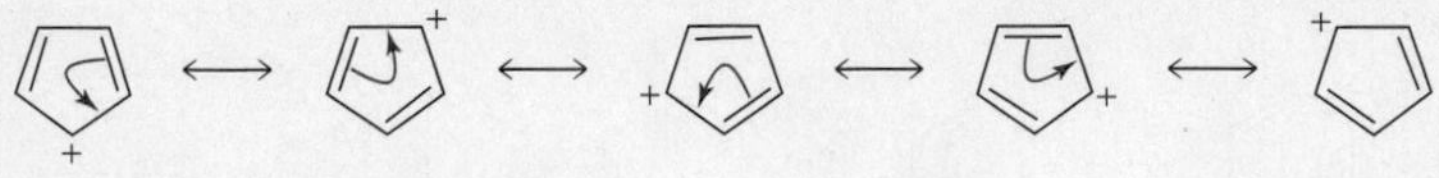

f.

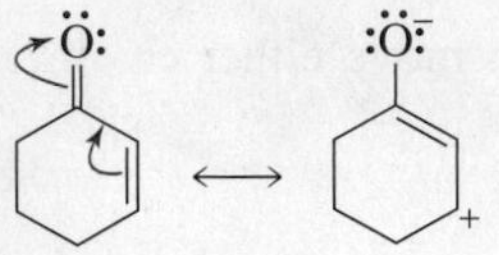

PROBLEM 12

a.

b. 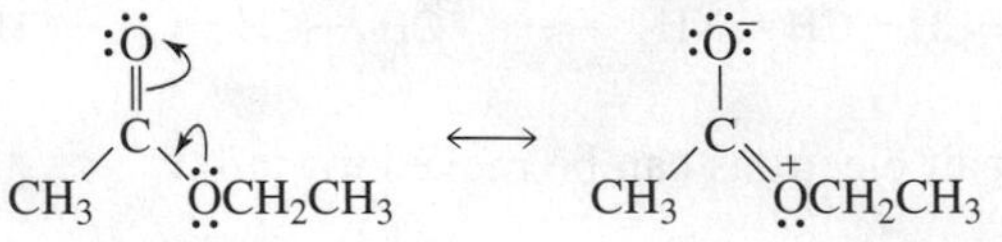   **d.**

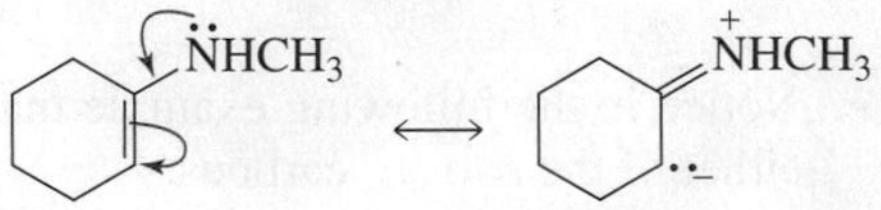

c. 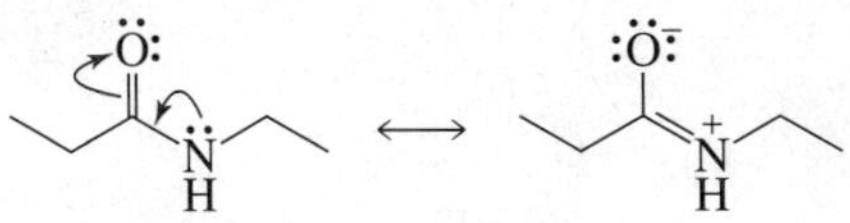   **e.**

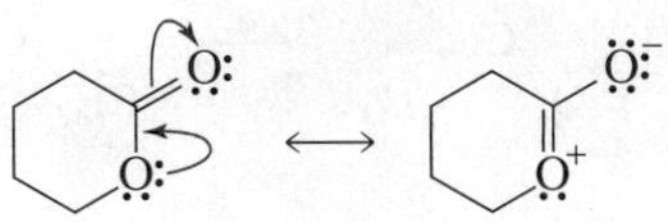

f.

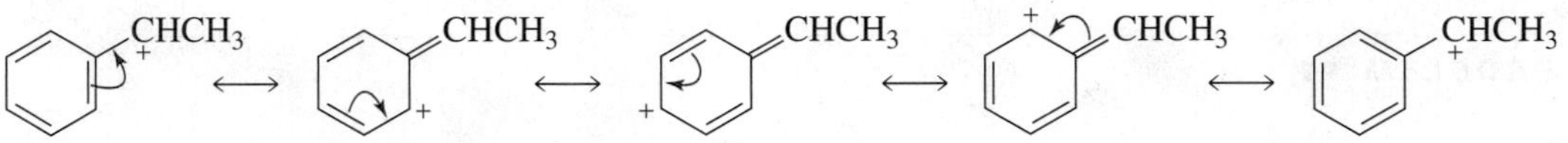

MasteringChemistry®
for Organic Chemistry

MasteringChemistry tutorials guide you through the toughest topics in chemistry with self-paced tutorials that provide individualized coaching. These assignable, in-depth tutorials are designed to coach you with hints and feedback specific to your individual misconceptions. For additional practice on Drawing Resonance Contributors, go to MasteringChemistry where the following tutorials are available:

- Drawing Resonance Contributors: Moving π Electrons
- Drawing Resonance Contributors: Predicting Contributor Structure
- Drawing Resonance Contributors: Substituted Benzene Compounds

GLOSSARY

1,2-addition (direct addition) addition to the 1- and 2-positions of a conjugated system.

1,4-addition (conjugate addition) addition to the 1- and 4-positions of a conjugated system.

allene a compound with two adjacent double bonds.

allylic carbon an sp^3 carbon adjacent to a vinylic carbon.

allylic cation a species with a positive charge on an allylic carbon.

antiaromatic a cyclic and planar compound with an uninterrupted ring of p orbital-bearing atoms containing an even number of pairs of π electrons.

antibonding molecular orbital a molecular orbital that results when two atomic orbitals with opposite signs interact. Electrons in an antibonding orbital decrease bond strength.

aromatic a cyclic and planar compound with an uninterrupted ring of p orbital-bearing atoms containing an odd number of pairs of π electrons.

benzylic carbon an sp^3 hybridized carbon bonded to a benzene ring.

benzylic cation a compound with a positive charge on a benzylic carbon.

bonding molecular orbital a molecular orbital that results when two in-phase atomic orbitals interact. Electrons in a bonding orbital increase bond strength.

bridged bicyclic compound a bicyclic compound in which rings share two nonadjacent carbons.

s-cis conformation the conformation in which two double bonds are on the same side of a single bond.

common intermediate an intermediate that two compounds have in common.

concerted reaction a reaction in which all the bond-making and bond-breaking processes occur in one step.

conjugate addition 1,4-addition to an $\alpha,\beta,$-unsaturated carbonyl compound.

conjugated double bonds double bonds separated by one single bond.

contributing resonance structure (resonance contributor, resonance structure) a structure with localized electrons that approximates the structure of a compound with delocalized electrons.

cumulated double bonds double bonds that are adjacent to one other.

cycloaddition reaction a reaction in which two π-bond-containing molecules react to form a cyclic compound.

[4 + 2] cycloaddition reaction a cycloaddition reaction in which four π electrons come from one reactant and two π electrons come from the other reactant.

delocalization energy (resonance energy) the extra stability a compound achieves as a result of having delocalized electrons.

delocalized electrons electrons that are shared by more than two atoms.

Diels–Alder reaction a [4 + 2] cycloaddition reaction.

diene a hydrocarbon with two double bonds.

dienophile an alkene that reacts with a diene in a Diels–Alder reaction.

direct addition 1,2-addition.

endo a substituent is endo if it is closer to the longer or more unsaturated bridge.

exo a substituent is exo if it is closer to the shorter or more saturated bridge.

heteroatom an atom other than carbon or hydrogen.

heterocyclic compound (heterocycle) a cyclic compound in which one or more of the atoms of the ring are heteroatoms.

highest occupied molecular orbital (HOMO) the molecular orbital of highest energy that contains an electron.

Hückel's rule states that, for a compound to be aromatic, its cloud of electrons must contain $(4n + 2)\pi$ electrons, where n is an integer. This is the same as saying that the electron cloud must contain an odd number of pairs of π electrons.

isolated double bonds double bonds separated by more than one single bond.

kinetic control when a reaction is under kinetic control, the relative amounts of the products depend on the rates at which they are formed.

kinetic product the product that is formed the fastest.

linear combination of atomic orbitals (LCAO) the combination of atomic orbitals to produce a molecular orbital.

localized electrons electrons that are restricted to a particular locality.

lowest unoccupied molecular orbital (LUMO) the molecular orbital of lowest energy that does not contain an electron.

pericyclic reaction a concerted reaction that takes place as the result of a cyclic rearrangement of electrons.

polymer a large molecule made by linking monomers together.

polymerization the process of linking up monomers to form a polymer.

proximity effect an effect caused by one species being close to another.

resonance a compound with delocalized electrons is said to have resonance.

resonance contributor (resonance structure, contributing resonance structure) a structure with localized electrons that approximates the true structure of a compound with delocalized electrons.

resonance electron donation donation of electrons through p orbital overlap with neighboring π bonds.

resonance energy (delocalization energy) the extra stability associated with a compound as a result of its having delocalized electrons.

resonance hybrid the actual structure of a compound with delocalized electrons; it is represented by two or more structures with localized electrons.

separated charges a positive and a negative charge that can be neutralized by the movement of electrons.

symmetric molecular orbital a molecular orbital in which the left half is a mirror image of the right half.

thermodynamic control when a reaction is under thermodynamic control, the relative amounts of the products depend on their stabilities.

thermodynamic product the most stable product.

s-trans conformation a conformation in which two double bonds are on opposite sides of a single bond.

PHOTO CREDITS

Photo credits are listed in order of appearance.

TEXT CREDITS

Text credits are listed in order of appearance.

Substitution Reactions of Alkyl Halides

From Chapter 9 of *Organic Chemistry*, Seventh Edition. Paula Yurkanis Bruice. Copyright © 2014 by Pearson Education, Inc.

Substitution Reactions of Alkyl Halides

This cartoon was published in *Time Magazine* on June 30, 1947.

In this Chapter you will see how the widespread use of DDT gave birth to the environmental movement. You will also learn why life is based on carbon rather than on silicon, even though silicon is just below carbon in the periodic table and is far more abundant than carbon in the earth's crust.

II

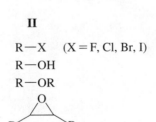

R—X   (X = F, Cl, Br, I)
R—OH
R—OR

Organic compounds can be placed in one of four groups, and all the families in a group react in similar ways. This chapter begins our discussion of the families of compounds in Group II.

Notice that all the families in Group II have an electronegative atom or an electron-withdrawing group attached to an sp^3 carbon. This atom or group creates a polar bond that allows the compound to undergo substitution and/or elimination reactions.

$$\underset{\text{a polar bond}}{\overset{\delta+\quad\quad\delta-}{RCH_2-X}}$$

An electronegative atom or an electron-withdrawing group

*In a **substitution reaction**, the electronegative atom or electron-withdrawing group is replaced by another atom or group.*

*In an **elimination reaction**, the electronegative atom or electron-withdrawing group is eliminated, along with a hydrogen from an adjacent carbon.*

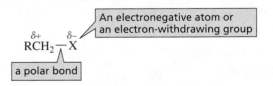

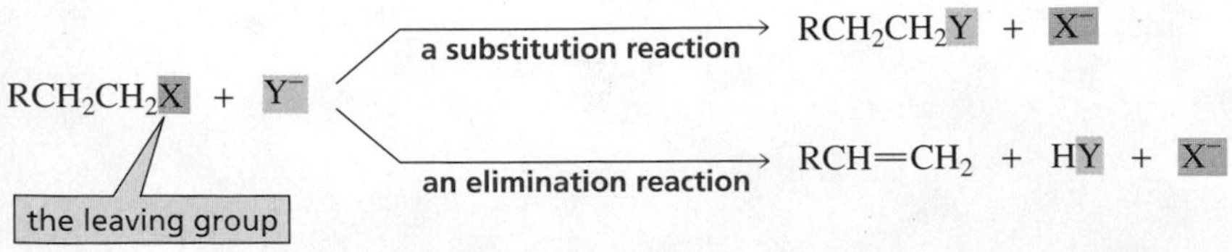

$$RCH_2CH_2X + Y^- \quad \xrightarrow{\text{a substitution reaction}} \quad RCH_2CH_2Y + X^-$$

$$\xrightarrow{\text{an elimination reaction}} \quad RCH=CH_2 + HY + X^-$$

the leaving group

The atom or group that is *substituted* or *eliminated* is called a **leaving group.**

This chapter focuses on the *substitution reactions* of alkyl halides—compounds in which the leaving group is a halide ion (F⁻, Cl⁻, Br⁻, or I⁻). This chapter assumes an understanding of the nomenclature of alkyl halides. The *elimination reactions* of alkyl halides and the factors that determine whether substitution or elimination will prevail when an alkyl halide undergoes a reaction are beyond the scope of this chapter.

alkyl halides

R—F	R—Cl	R—Br	R—I
an alkyl fluoride	**an alkyl chloride**	**an alkyl bromide**	**an alkyl iodide**

Alkyl halides are a good family of compounds with which to start the study of substitution and elimination reactions because they have relatively good leaving groups; that is, the halide ions are easily displaced.

Substitution reactions are important in organic chemistry because they make it possible to convert readily available alkyl halides into a wide variety of other compounds. Substitution reactions are also important in the cells of plants and animals. We will see, however, that because cells exist in predominantly aqueous environments and alkyl halides are insoluble in water, biological systems use compounds in which the group that is replaced is more polar than a halogen and, therefore, more water soluble.

DDT: A Synthetic Organohalide That Kills Disease-Spreading Insects

Alkyl halides have been used as insecticides since 1939, when it was discovered that DDT (first synthesized in 1874) has a high toxicity to insects and a relatively low toxicity to mammals. DDT was used widely in World War II to control typhus and malaria in both the military and civilian populations. It saved millions of lives, but no one realized at that time that, because it is a very stable compound, it is resistant to biodegradation. In addition, DDT and DDE, a compound formed as a result of elimination of HCl from DDT, are not water soluble. Therefore, they accumulate in the fatty tissues of birds and fish and can be passed up the food chain. Most adults have a low concentration of DDT or DDE in their bodies.

In 1962, Rachel Carson, a marine biologist, published *Silent Spring,* where she pointed out the environmental impacts of the widespread use of DDT. The book was widely read, so it brought the problem of environmental pollution to the attention of the general public for the first time. Consequently, its publication was an important event in the birth of the environmental movement. Because of the concern it raised, DDT was banned in the United States in 1972. In 2004, the Stockholm Convention banned the worldwide use of DDT except for the control of malaria in countries where the disease is a major health problem.

DDT

PROBLEM 1

Draw the structure of DDE.

PROBLEM 2

Methoxychlor is an insecticide that was intended to take DDT's place because it is not as soluble in fatty tissues and is more readily biodegradable. It, too, can accumulate in the environment, however, so its use was also banned—in 2002 in the European Union and in 2003 in the United States. Why is methoxychlor less soluble in fatty tissues than DDT?

methoxychlor

1   THE MECHANISM FOR AN S$_N$2 REACTION

You will see that there are two different mechanisms by which a substitution reaction can take place. As you would expect, each of these mechanisms involves the *reaction of a nucleophile with an electrophile.* In both mechanisms, the nucleophile replaces the leaving group, so the substitution reaction is more precisely called a **nucleophilic substitution reaction.**

You might be wondering how these mechanisms are determined. A mechanism describes the step-by-step process by which reactants are converted into products. It is a theory that fits the accumulated experimental evidence pertaining to the reaction. Thus, *mechanisms are determined experimentally.* They are not something that chemists make up in an attempt to explain how a reaction occurs.

Experimental Evidence for the Mechanism for an S$_N$2 Reaction

We can learn a great deal about a reaction's mechanism by studying its **kinetics**—the factors that affect the rate of the reaction.

For example, the rate of the following nucleophilic substitution reaction depends on the concentrations of both reactants. If the concentration of the alkyl halide (bromomethane) is doubled, the rate of the reaction doubles. Likewise, if the concentration of the nucleophile (hydroxide ion) is doubled, the rate of the reaction doubles. If the concentrations of both reactants are doubled, the rate of the reaction quadruples.

$$CH_3Br + HO^- \longrightarrow CH_3OH + Br^-$$
$$\text{bromomethane} \qquad\qquad \text{methanol}$$

Because we know the relationship between the rate of the reaction and the concentration of the reactants, we can write a **rate law** for the reaction:

$$\text{rate} \ \propto \ [\text{alkyl halide}][\text{nucleophile}]$$

The proportionality sign ($\propto$) can be replaced by an equals sign and a proportionality constant (k). This is a **second-order reaction** because its rate depends linearly on the concentration of each of the two reactants.

$$\text{rate} = k [\text{alkyl halide}][\text{nucleophile}]$$

the rate constant

The **rate law** tells us which molecules are involved in the transition state of the rate-determining step of the reaction. Thus, the rate law for the reaction of bromomethane with hydroxide ion tells us that *both* bromomethane and hydroxide ion are involved in the rate-determining transition state.

The proportionality constant is called a **rate constant.** The magnitude of the rate constant for a particular reaction indicates how difficult it is for the reactants to overcome the energy barrier of the reaction—that is, how hard it is to reach the transition state. The larger the rate constant, the lower is the energy barrier and, therefore, the easier it is for the reactants to reach the transition state (see Figure 3 later in this section).

PROBLEM 3♦

How would the rate of the reaction between bromomethane and hydroxide ion be affected if the following changes in concentration are made?

a. The concentration of the alkyl halide is not changed and the concentration of the nucleophile is tripled.

b. The concentration of the alkyl halide is cut in half and the concentration of the nucleophile is not changed.

c. The concentration of the alkyl halide is cut in half and the concentration of the nucleophile is doubled.

The reaction of bromomethane with hydroxide ion is an example of an **S_N2 reaction,** where "S" stands for substitution, "N" for nucleophilic, and "2" for bimolecular. **Bimolecular** means that two molecules are involved in the transition state of the rate-determining step. In 1937, Edward Hughes and Christopher Ingold proposed a mechanism for an S_N2 reaction. They based their mechanism on the following three pieces of *experimental evidence*:

1. The rate of the substitution reaction depends on the concentration of the alkyl halide *and* on the concentration of the nucleophile, indicating that both reactants are involved in the transition state of the rate-determining step.

2. As the alkyl group becomes larger or as the hydrogens of bromomethane are successively replaced with methyl groups, the rate of the substitution reaction with a given nucleophile becomes slower.

Relative Rates of an S_N2 Reaction

$$CH_3-\boxed{Br} \;>\; CH_3CH_2-\boxed{Br} \;>\; CH_3CH_2CH_2-\boxed{Br} \;>\; CH_3\overset{\displaystyle |}{\underset{\displaystyle |}{CH}}-\boxed{Br} \;>\; CH_3\overset{\displaystyle CH_3}{\underset{\displaystyle CH_3}{\overset{|}{\underset{|}{C}}}}-\boxed{Br}$$

3. The substitution reaction of an alkyl halide in which the halogen is bonded to an asymmetric center leads to the formation of only one stereoisomer, and the configuration of the asymmetric center in the product is inverted relative to its configuration in the reacting alkyl halide.

the configuration is inverted relative to that of the reactant

(S)-2-bromobutane + HO^- ⟶ (R)-2-butanol + Br^-

The Mechanism for an S_N2 Reaction

Using the preceding evidence, Hughes and Ingold proposed that an S_N2 reaction is a *concerted* reaction (that is, it takes place in a single step), so no intermediates are formed.

MECHANISM FOR THE S_N2 REACTION OF AN ALKYL HALIDE

electrophile

$H\ddot{O}{:}^- \;+\; CH_3\overset{}{-}\ddot{\underset{..}{Br}}{:} \;\longrightarrow\; CH_3-OH \;+\; {:}\ddot{\underset{..}{Br}}{:}^-$

nucleophile   leaving group

An S_N2 reaction is a one-step (concerted) reaction.

- The nucleophile attacks the back side of the carbon (the electrophile) that bears the leaving group and displaces it.

A productive collision is one that leads to the formation of the product. A productive collision in an S_N2 reaction requires the nucleophile to hit the carbon on the side opposite the side that is bonded to the leaving group. Therefore, the carbon is said to undergo **back-side attack.**

Why must the nucleophile attack from the back side? The simplest explanation is that the leaving group blocks the approach of the nucleophile to the front side of the molecule.

Molecular orbital theory explains why back-side attack is required. To form a bond, the HOMO (the highest occupied molecular orbital) of one species must interact with the LUMO (the lowest unoccupied molecular orbital) of the other.

Therefore, when the nucleophile approaches the alkyl halide to form a new bond, the nonbonding molecular orbital of the nucleophile (its HOMO) must interact with the empty σ^* antibonding molecular orbital associated with the C—Br bond (its LUMO).

Figure 1a shows that in a back-side attack, a bonding interaction (the interacting lobes are both green) occurs between the nucleophile and the larger lobe of the σ^* antibonding MO. But when the nucleophile approaches the front side of the carbon (Figure 1b), both a bonding and an antibonding interaction occur, so the two cancel each other and no bond forms. Therefore, an S_N2 reaction can be successful only if the nucleophile approaches the sp^3 carbon from its back side.

A nucleophile attacks the back side of the carbon that is bonded to the leaving group.

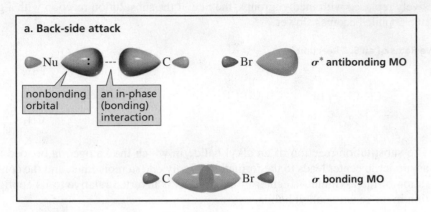

a. Back-side attack

nonbonding orbital

an in-phase (bonding) interaction

σ^* antibonding MO

σ bonding MO

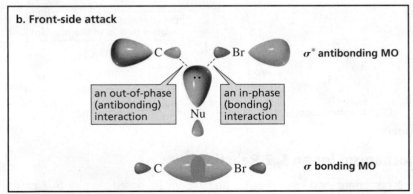

b. Front-side attack

an out-of-phase (antibonding) interaction

an in-phase (bonding) interaction

Nu

σ^* antibonding MO

σ bonding MO

▶ **Figure 1**

(a) Back-side attack results in a bonding interaction between the HOMO (the nonbonding orbital) of the nucleophile and the LUMO (the σ^* antibonding orbital) of C—Br.

(b) Front-side attack results in both a bonding and an antibonding interaction that cancel each other.

How the Mechanism Accounts for the Experimental Evidence

How does Hughes and Ingold's mechanism account for the three pieces of experimental evidence? The mechanism shows that the alkyl halide and the nucleophile are both in the transition state of the one-step reaction. Therefore, increasing the concentration of either of them makes their collision more probable, so the rate of the reaction will depend on the concentration of both, exactly as observed.

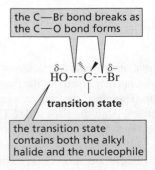

the C—Br bond breaks as the C—O bond forms

$$\overset{\delta-}{HO}\text{---}C\text{---}\overset{\delta-}{Br}$$

transition state

the transition state contains both the alkyl halide and the nucleophile

Bulky substituents attached to the carbon that undergoes back-side attack will decrease the nucleophile's access to the back side of the carbon and will therefore decrease the rate of the reaction (Figure 2). This explains why, as the size of the alkyl group increases or as methyl groups are substituted for the hydrogens in bromomethane, the rate of the substitution reaction decreases.

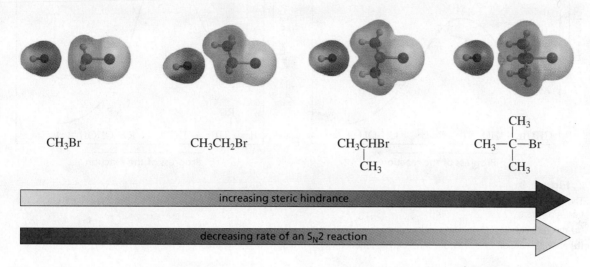

CH₃Br   CH₃CH₂Br $\underset{\text{CH}_3}{\text{CH}_3\text{CHBr}}$ $\underset{\text{CH}_3}{\overset{\text{CH}_3}{\text{CH}_3\text{—C—Br}}}$

increasing steric hindrance

decreasing rate of an S$_N$2 reaction

▲ **Figure 2**
The approach of HO⁻ (shown by the red ball) to the back sides of the carbon of methyl bromide, a primary alkyl bromide, a secondary alkyl bromide, and a tertiary alkyl bromide. Increasing the bulk of the substituents bonded to the carbon that is undergoing nucleophilic attack decreases access to the back side of the carbon, thereby decreasing the rate of the S$_N$2 reaction.

Steric effects are effects caused by the fact that groups occupy a certain volume of space. A steric effect that decreases reactivity is called **steric hindrance.** This occurs when groups are in the way at a reaction site. It is steric hindrance that causes alkyl halides to have the following relative reactivities in an S$_N$2 reaction because primary alkyl halides are usually less sterically hindered than secondary alkyl halides, and secondary alkyl halides are less sterically hindered than tertiary alkyl halides.

relative reactivities of alkyl halides in an S$_N$2 reaction

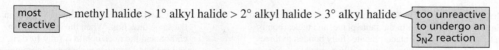

most reactive > methyl halide > 1° alkyl halide > 2° alkyl halide > 3° alkyl halide < too unreactive to undergo an S$_N$2 reaction

The three alkyl groups of a tertiary alkyl halide make it impossible for the nucleophile to come within bonding distance of the tertiary carbon, so tertiary alkyl halides are unable to undergo S$_N$2 reactions.

The rate of an S$_N$2 reaction depends not only on the *number* of alkyl groups attached to the carbon that is undergoing nucleophilic attack, but also on their size. For example, bromoethane and 1-bromopropane are both primary alkyl halides, but bromoethane is more than twice as reactive in an S$_N$2 reaction, because the bulkier alkyl group on the carbon undergoing nucleophilic attack in 1-bromopropane provides more steric hindrance to back-side attack (Figure 3).

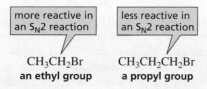

more reactive in an S$_N$2 reaction   less reactive in an S$_N$2 reaction

CH₃CH₂Br   CH₃CH₂CH₂Br
an ethyl group   **a propyl group**

The relative lack of steric hindrance causes methyl halides and primary alkyl halides to be the most reactive alkyl halides in S$_N$2 reactions.

Tertiary alkyl halides cannot undergo S$_N$2 reactions.

$$\underset{\text{CH}_3}{\overset{\text{CH}_3}{\text{CH}_3\text{CCH}_2\text{Br}}}$$

1-bromo-2,2-dimethylpropane

Although this is a primary alkyl halide, it undergoes S$_N$2 reactions very slowly because its single alkyl group is unusually bulky.

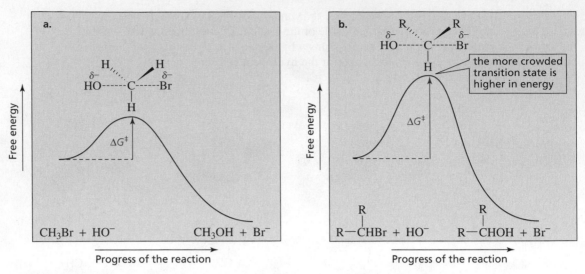

▲ **Figure 3**

The reaction coordinate diagrams show that steric hindrance decreases the rate of the reaction by increasing the energy of the transition state:

(a) the S_N2 reaction of *unhindered* bromomethane with hydroxide ion

(b) an S_N2 reaction of a *sterically hindered* secondary alkyl bromide with hydroxide ion

An S_N2 reaction takes place with inversion of configuration.

Figure 4 illustrates the third piece of experimental evidence used by Hughes and Ingold to arrive at their proposed mechanism—namely, the inversion of configuration at the carbon undergoing substitution. This **inversion of configuration** is called a *Walden inversion*. It was named for Paul Walden, who first discovered that the configuration of a compound becomes inverted in an S_N2 reaction.

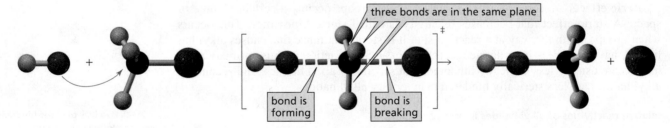

As the nucleophile approaches the back side of the carbon, the C—H bonds begin to move away from the nucleophile and its attacking electrons.

The C—H bonds in the transition state are all in the same plane, and the carbon is pentacoordinate (fully bonded to three atoms and partially bonded to two) rather than tetrahedral.

The C—H bonds have continued to move in the same direction. When the bond between carbon and the nucleophile is fully formed, and the bond between carbon and bromine is completely broken, carbon is once again tetrahedral.

▲ **Figure 4**

The reaction between hydroxide ion and bromomethane, showing that the carbon at which substitution occurs in an S_N2 reaction inverts its configuration, just like an umbrella tends to invert in a windstorm.

PROBLEM 4♦

Does increasing the energy barrier for an S_N2 reaction increase or decrease the magnitude of the rate constant for the reaction?

Because an S_N2 reaction takes place with inversion of configuration, only one substitution product is formed when an alkyl halide whose halogen atom is bonded to an asymmetric center undergoes an S_N2 reaction. The configuration of that product is inverted relative to the configuration of the alkyl halide. For example, the substitution product obtained from the reaction of hydroxide ion with (*R*)-2-bromopentane is (*S*)-2-pentanol. Thus, the mechanism proposed by Hughes and Ingold also accounts for the observed configuration of the product.

CH₃
|
CH₃CH₂—C⟍⟍H   +   HO⁻   ⟶
|
Br

(R)-2-bromobutane

CH₃
|
H⟍⟍C⟍CH₂CH₃   +   Br⁻
|
HO

(S)-2-butanol

> the configuration of the product is inverted relative to the configuration of the reactant

> If the leaving group is attached to an asymmetric center, an S_N2 reaction forms only the stereoisomer with the inverted configuration.

PROBLEM 5♦

Arrange the following alkyl bromides in order from most reactive to least reactive in an S_N2 reaction: 1-bromo-2-methylbutane, 1-bromo-3-methylbutane, 2-bromo-2-methylbutane, and 1-bromopentane.

> To draw the inverted product of an S_N2 reaction, draw the mirror image of the reactant and replace the halogen with the nucleophile.

PROBLEM 6♦ Solved

Draw the products obtained from the S_N2 reaction of

a. 2-bromobutane and methoxide ion.

c. (S)-3-chlorohexane and hydroxide ion.

b. (R)-2-bromobutane and methoxide ion.

d. 3-iodopentane and hydroxide ion.

Solution to 6a The product is 2-methoxybutane. Because the reaction is an S_N2 reaction, we know that the configuration of the product is inverted relative to the configuration of the reactant. The configuration of the reactant is not specified, however, so we cannot specify the configuration of the product. In other words, because we also do not know if the reactant is *R* or *S* or a mixture of the two, we also do not know if the product is *R* or *S* or a mixture of the two.

> the configuration is not specified

CH₃CHCH₂CH₃   +   CH₃O⁻   ⟶   CH₃CHCH₂CH₃   +   Br⁻
| |
Br   OCH₃

PROBLEM 7 Solved

Draw the substitution products that will be formed from the following S_N2 reactions:

a. *cis*-1-bromo-4-methylcyclohexane and hydroxide ion

b. *trans*-1-iodo-4-ethylcyclohexane and methoxide ion

c. *cis*-1-chloro-3-methylcyclobutane and ethoxide ion

Solution to 7a Only the trans product is obtained in this S_N2 reaction because the carbon bonded to the leaving group is attacked by the nucleophile on its back side.

> the configuration of this carbon has been inverted

CH₃▹⟨   ⟩◁Br   +   HO⁻   ⟶   CH₃▹⟨   ⟩⟍⟍⟍OH   +   Br⁻

cis-1-bromo-4-methylcyclohexane   *trans*-4-methylcyclohexanol

---

2   FACTORS THAT AFFECT S_N2 REACTIONS

We will now look at how the nature of the leaving group and the nature of the nucleophile affect an S_N2 reaction.

The Leaving Group in an S$_N$2 Reaction

If an alkyl iodide, an alkyl bromide, an alkyl chloride, and an alkyl fluoride with the same alkyl group were allowed to react with the same nucleophile under the same conditions, we would find that the alkyl iodide is the most reactive and the alkyl fluoride is the least reactive.

	relative rates of reaction	pK_a values of HX
HO$^-$ + RCH$_2$I $\longrightarrow$ RCH$_2$OH + I$^-$	30,000	−10
HO$^-$ + RCH$_2$Br $\longrightarrow$ RCH$_2$OH + Br$^-$	10,000	−9
HO$^-$ + RCH$_2$Cl $\longrightarrow$ RCH$_2$OH + Cl$^-$	200	−7
HO$^-$ + RCH$_2$F $\longrightarrow$ RCH$_2$OH + F$^-$	1	3.2

The only difference between these four reactions is the nature of the leaving group. From the relative reaction rates, we see that iodide ion is the best leaving group and fluoride ion is the worst. This brings us to an important rule in organic chemistry that you will encounter frequently: when comparing bases of the same type, *the weaker the basicity of a group, the better is its leaving propensity.*

The reason leaving propensity depends on basicity (even though the former is a kinetic concept and the latter is a thermodynamic concept) is that *weak bases are stable bases;* they readily bear the electrons they formerly shared with a proton. Therefore, they do not share their electrons well. Thus a weak base is not bonded as strongly to the carbon as a strong base would be, and a weaker bond is more readily broken.

Iodide ion is the weakest base of the halide ions (it has the strongest conjugate acid) and fluoride ion is the strongest base (it has the weakest conjugate acid) Therefore, when comparing alkyl halides with the same alkyl group, we find that the alkyl iodide is the most reactive and the alkyl fluoride is the least reactive. In fact, the fluoride ion is such a strong base that alkyl fluorides essentially do not undergo S$_N$2 reactions.

relative reactivities of alkyl halides in an S$_N$2 reaction

The weaker the base, the better it is as a leaving group.

Stable bases are weak bases.

most reactive $\longrightarrow$ RI > RBr > RCl > RF $\longleftarrow$ too unreactive to undergo an S$_N$2 reaction

At the beginning of this chapter, we saw that the polar carbon–halogen bond causes alkyl halides to undergo substitution reactions. Carbon and iodine, however, have the same electronegativity. Why, then, does an alkyl iodide undergo a substitution reaction?

We know that larger atoms are more polarizable than smaller atoms. (Polarizability is a measure of how easily an atom's electron cloud can be distorted.) The high polarizability of the large iodine atom causes the C—I bond to react as if it were polar, even though, on the basis of the electronegativities of the carbon and iodine atoms, the bond is nonpolar.

PROBLEM 8♦

Which alkyl halide would you expect to be more reactive in an S$_N$2 reaction with a given nucleophile? In each case, you can assume that both alkyl halides have the same stability.

a. [structure] or [structure]

b. [structure] or [structure]

c. [structure] or [structure]

d. [structure] or [structure]

The Nucleophile in an S$_N$2 Reaction

When we talk about atoms or molecules that have lone-pair electrons, sometimes we call them bases and sometimes we call them nucleophiles (Table 1). What is the difference between a base and a nucleophile?

Table 1	Common Nucleophiles/Bases			
HO$^-$	RO$^-$	H$_2$O	ROH	RCOO$^-$
HS$^-$	RS$^-$	H$_2$S	RSH	
$^-$NH$_2$	RNH$^-$	NH$_3$	RNH$_2$	
$^-$C≡N	RC≡C$^-$			
Cl$^-$	Br$^-$	I$^-$		

Basicity is a measure of how well a compound (a **base**) shares its lone pair with a proton. The stronger the base, the better it shares its electrons. Basicity is measured by an *equilibrium constant* (the acid dissociation constant, K_a) that indicates the tendency of the conjugate acid of the base to lose a proton.

Nucleophilicity is a measure of how readily a compound (a **nucleophile**) is able to attack an electron-deficient atom. It is measured by a *rate constant* (k). In the case of an S$_N$2 reaction, nucleophilicity is a measure of how readily the nucleophile attacks an sp^3 carbon bonded to a leaving group.

Because the nucleophile attacks an sp^3 carbon in the rate-determining step of an S$_N$2 reaction, the rate of the reaction will depend on the strength of the nucleophile: the better the nucleophile, the faster the rate of the S$_N$2 reaction.

If the attacking atoms are the same size, *stronger bases are better nucleophiles.* For example, comparing attacking atoms in the first row of the periodic table (so they are the same size), the amide ion is both the strongest base and the best nucleophile. Notice that bases are described as being strong or weak, whereas nucleophiles are described as being good or poor.

relative base strengths and relative nucleophilicities

strongest base → $^-$NH$_2$ > HO$^-$ > F$^-$

best nucleophile

A species with a negative charge is a stronger base *and* a better nucleophile than a species that has the same attacking atom but is neutral. Thus, HO$^-$ is a stronger base and a better nucleophile than H$_2$O.

stronger base, better nucleophile		weaker base, poorer nucleophile
HO$^-$	>	H$_2$O
CH$_3$O$^-$	>	CH$_3$OH
$^-$NH$_2$	>	NH$_3$
CH$_3$CH$_2$NH$^-$	>	CH$_3$CH$_2$NH$_2$

If, however, the attacking atoms of the nucleophiles are *very different in size,* another factor comes into play: the polarizability of the atom. Because the electrons are farther away in the larger atom, they are not held as tightly and can, therefore, move more freely toward a positive charge. As a result, the electrons are able to overlap the orbital of carbon from farther away, as shown in Figure 5. This results in a greater degree of bonding in the transition state, which makes the transition state more stable.

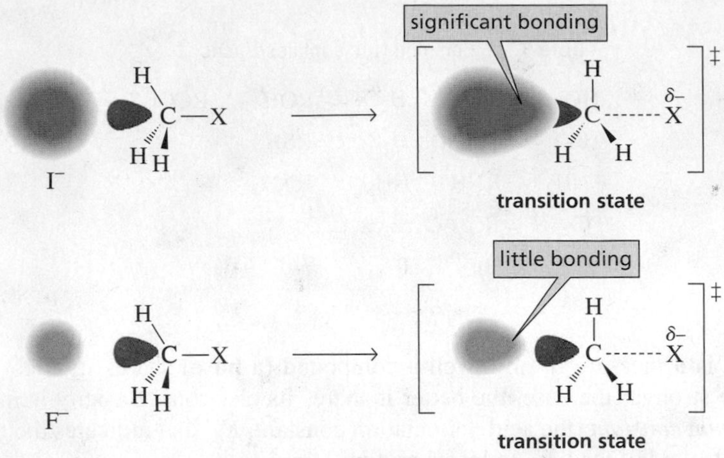

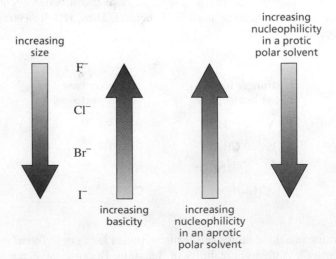

▶ **Figure 5**

An iodide ion is larger and more polarizable than a fluoride ion. Therefore, when an iodide ion approaches a carbon, the relatively loosely held electrons of the ion can overlap the orbital of carbon from farther away. The tightly bound electrons of the fluoride ion cannot start to overlap the orbital of carbon until the reactants are closer together.

Now the question becomes, does the greater polarizability that helps the larger atoms to be better nucleophiles make up for the decreased basicity that causes them to be poorer nucleophiles? The answer depends on the solvent.

If the reaction is carried out in an **aprotic polar solvent**—meaning the polar solvent molecules *do not have* a hydrogen bonded to an oxygen or to a nitrogen—the direct relationship between basicity and nucleophilicity is maintained: the strongest bases are still the best nucleophiles. In other words, the greater polarizability of the larger atoms does not make up for their decreased basicity. *Therefore, iodide ion, the weakest base, is the poorest nucleophile of the halide ions in an aprotic polar solvent.*

If, however, the reaction is carried out in a **protic solvent**—meaning the polar solvent molecules *have* a hydrogen bonded to an oxygen or to a nitrogen—the relationship between basicity and nucleophilicity becomes inverted (Figure 6). The largest atom (the most polarizable one) is the best nucleophile even though it is the weakest base. *Therefore, iodide ion, the weakest base, is the best nucleophile of the halide ions in a protic solvent.* (For a list of protic and aprotic solvents, see Table 5 later in this chapter in Section 7.)

▶ **Figure 6**

The strongest bases are the best nucleophiles except when (1) the bases differ in size *and* when (2) they are used in a protic polar solvent. Both conditions are necessary for the strongest base not to be the best nucleophile.

PROBLEM 9♦

Indicate whether each of the following solvents is protic or aprotic:

a. chloroform ($CHCl_3$)

b. diethyl ether

c. acetic acid

d. hexane

> An aprotic solvent does not contain a hydrogen bonded to either an oxygen or a nitrogen.
>
> A protic solvent contains a hydrogen bonded to an oxygen or a nitrogen.

Why Is the Nucleophilicity Affected by the Solvent?

Why, in a protic solvent, is the smallest atom the poorest nucleophile even though it is the strongest base? *How does a protic solvent make strong bases less nucleophilic?*

When a negatively charged species is placed in a protic solvent, the ion becomes solvated. Protic solvents are hydrogen bond donors, so the solvent molecules arrange themselves with their partially positively charged hydrogens pointing toward the negatively charged species. The interaction between the ion and the dipole of the protic solvent is called an **ion–dipole interaction.**

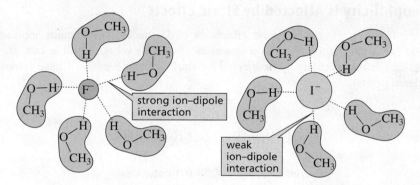

Because the solvent shields the nucleophile, at least one of the ion–dipole interactions must be broken before the nucleophile can participate in an S_N2 reaction. Weak bases interact weakly with protic solvents, whereas strong bases interact strongly because they are better at sharing their electrons. It is easier, therefore, to break the ion–dipole interactions between an iodide ion (a weak base) and the solvent than between a fluoride ion (a stronger base) and the solvent. In a protic solvent, therefore, an iodide ion, even though it is a weaker base, is a better nucleophile than a fluoride ion (Table 2).

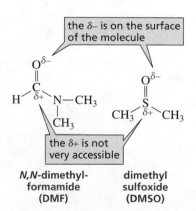

*N,N-*dimethyl-formamide (DMF)

dimethyl sulfoxide (DMSO)

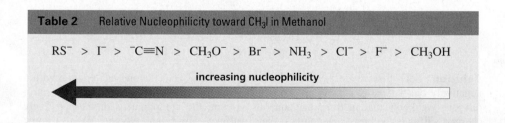

Table 2	Relative Nucleophilicity toward CH_3I in Methanol

$$RS^- > I^- > {}^-C{\equiv}N > CH_3O^- > Br^- > NH_3 > Cl^- > F^- > CH_3OH$$

increasing nucleophilicity

An **aprotic polar solvent** does not have any hydrogens with partial positive charges to form ion–dipole interactions. The molecules of an aprotic polar solvent (such as DMF or DMSO) have a partial negative charge on their surface that can solvate cations, but the partial positive charge is on the *inside* of the molecule, and therefore less accessible to solvate anions. Fluoride ion, therefore, is a good nucleophile in DMSO and a poor nucleophile in water.

Fluoride ion would be an even better nucleophile in a *nonpolar solvent* (such as hexane) because there would not be any ion–dipole interactions between the ion and the nonpolar solvent. Ionic compounds, however, are insoluble in most nonpolar solvents, but they dissolve in aprotic polar solvents. Fluoride ion is also a good nucleophile in the gas phase, where there are no solvent molecules.

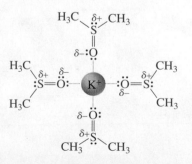

> DMSO can solvate a cation better than it can solvate an anion

PROBLEM 10♦

a. Which is a stronger base, RO^- or RS^-?
b. Which is a better nucleophile in an aqueous solution?
c. Which is a better nucleophile in DMSO?

PROBLEM 11♦

Which is a better nucleophile?

a. Br^- or Cl^- in H_2O
b. Br^- or Cl^- in DMSO
c. CH_3O^- or CH_3OH in H_2O
d. CH_3O^- or CH_3OH in DMSO

e. HO^- or $^-NH_2$ in H_2O
f. HO^- or $^-NH_2$ in DMSO
g. I^- or Br^- in H_2O
h. I^- or Br^- in DMSO

Nucleophilicity Is Affected by Steric Effects

Nucleophilicity is *affected* by steric effects. A bulky nucleophile cannot approach the back side of a carbon as easily as a less sterically hindered nucleophile can. Basicity, on the other hand, is relatively *unaffected* by steric effects because a base removes an unhindered proton.

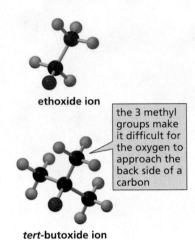

ethoxide ion

the 3 methyl groups make it difficult for the oxygen to approach the back side of a carbon

***tert*-butoxide ion**

$$CH_3CH_2{-}O^- \qquad CH_3{-}\underset{\underset{CH_3}{|}}{\overset{\overset{CH_3}{|}}{C}}{-}O^-$$

ethoxide ion
better nucleophile

***tert*-butoxide ion**
stronger base

Therefore, *tert*-butoxide ion, with its three methyl groups, is a poorer nucleophile than ethoxide ion, even though *tert*-butoxide ion is a stronger base (pK_a of *tert*-butanol = 18) than ethoxide ion (pK_a of ethanol = 16).

PROBLEM 12 Solved

List the following species in order from best nucleophile to poorest nucleophile in an aqueous solution.

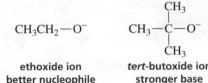

Solution Let's first divide the nucleophiles into groups. There is one nucleophile with a negatively charged sulfur, three with negatively charged oxygens, and one with a neutral oxygen. We know that in the polar aqueous solvent, the compound with the negatively charged sulfur is the best nucleophile because sulfur is larger than oxygen. We also know that the poorest nucleophile is the one with the neutral oxygen. To complete the problem, we need to rank the three nucleophiles with negatively charged oxygens, which we can do by looking at the pK_a values of their conjugate acids. A carboxylic acid is a stronger acid than phenol, which is a stronger acid than water. Because water is the weakest acid, its conjugate base is the strongest base and the best nucleophile. Thus, the relative nucleophilicities are:

$$CH_3S^- \;>\; HO^- \;>\; \text{C}_6H_5{-}O^- \;>\; CH_3\overset{\overset{O}{\|}}{C}O^- \;>\; CH_3OH$$

PROBLEM 13♦

Which reaction in each of the following pairs occurs faster:

a. $CH_3CH_2Br + H_2O$ or $CH_3CH_2Br + HO^-$

b. $CH_3\overset{\underset{|}{CH_3}}{C}HCH_2Br + HO^-$ or $CH_3CH_2\overset{\underset{|}{CH_3}}{C}HBr + HO^-$

c. $CH_3CH_2Cl + CH_3O^-$ or $CH_3CH_2Cl + CH_3S^-$
               (in ethanol)

d. $CH_3CH_2Cl + I^-$ or $CH_3CH_2Br + I^-$

Many different kinds of nucleophiles can react with alkyl halides. Therefore, a wide variety of organic compounds can be synthesized by means of S_N2 reactions. Notice that the last reaction is the reaction of an alkyl halide with an acetylide ion. This reaction can also be used to create longer carbon chains.

$$CH_3CH_2Cl + HO^- \longrightarrow CH_3CH_2OH + Cl^-$$
$$\text{an alcohol}$$

$$CH_3CH_2Br + HS^- \longrightarrow CH_3CH_2SH + Br^-$$
$$\text{a thiol}$$

$$CH_3CH_2I + RO^- \longrightarrow CH_3CH_2OR + I^-$$
$$\text{an ether}$$

$$CH_3CH_2Br + RS^- \longrightarrow CH_3CH_2SR + Br^-$$
$$\text{a thioether}$$

$$CH_3CH_2Cl + {}^-NH_2 \longrightarrow CH_3CH_2NH_2 + Cl^-$$
$$\text{a primary amine}$$

$$CH_3CH_2I + {}^-C\equiv N \longrightarrow CH_3CH_2C\equiv N + I^-$$
$$\text{a nitrile}$$

$$CH_3CH_2Br + {}^-C\equiv CR \longrightarrow CH_3CH_2C\equiv CR + Br^-$$
$$\text{an alkyne}$$

At first glance, it may seem that the reverse of each of these reactions satisfies the requirements for a nucleophilic substitution reaction. For example, the reverse of the first reaction would be the reaction of chloride ion (a nucleophile) with ethanol (with an HO^- leaving group). But ethanol and chloride ion do *not* react.

Why doesn't the reverse reaction take place? We can answer this question by comparing the leaving propensity of Cl^- and HO^-. Comparing leaving propensities means comparing basicities. Because HCl is a much stronger acid than H_2O, Cl^- is a much weaker base than HO^-; because it is a weaker base, Cl^- is a better leaving group. Consequently, HO^- can displace Cl^- (a good leaving group) in the forward reaction, but Cl^- cannot displace HO^- (a poor leaving group) in the reverse reaction.

PROBLEM 14♦

What is the product of the reaction of bromoethane with each of the following nucleophiles?

a. $CH_3CH_2CH_2O^-$ **b.** $CH_3C\equiv C^-$ **c.** $(CH_3)_3N$ **d.** $CH_3CH_2S^-$

PROBLEM 15 Solved

What product is obtained when ethylamine reacts with excess methyl iodide in a basic solution of potassium carbonate?

Solution Ethylamine and methyl iodide undergo an S_N2 reaction. The product of the reaction is a secondary amine that is predominantly in its basic (neutral) form because the pH of the basic solution is greater than the pK_a of the protonated amine. The secondary amine can undergo an S_N2 reaction with another equivalent of methyl iodide, forming a tertiary amine. The tertiary amine can react with methyl iodide in yet another S_N2 reaction. The final product of the reaction is a quaternary ammonium iodide.

$$CH_3CH_2\overset{..}{N}H_2 + CH_3\!-\!I \longrightarrow CH_3CH_2\overset{+}{N}H_2CH_3 \; I^- \xrightleftharpoons{K_2CO_3} CH_3CH_2\overset{..}{N}HCH_3$$

$$\downarrow CH_3\!-\!I$$

$$\underset{\underset{CH_3 \; I^-}{|}}{\overset{\overset{CH_3}{|}}{CH_3CH_2\overset{+}{N}CH_3}} \xleftarrow{CH_3\!-\!I} \underset{\underset{CH_3}{|}}{CH_3CH_2\overset{..}{N}CH_3} \xrightleftharpoons{K_2CO_3} CH_3CH_2\overset{+}{N}HCH_3 \; I^- \atop {\underset{CH_3}{|}}$$

The reaction of an amine with sufficient methyl iodide to convert the amine into a quaternary ammonium iodide is called **exhaustive methylation.**

PROBLEM 16

a. Explain why the reaction of an alkyl halide with ammonia gives a low yield of primary amine.

b. Explain why a much better yield of primary amine is obtained from the reaction of an alkyl halide with azide ion ($^-N_3$), followed by catalytic hydrogenation. (*Hint:* An alkyl azide is not nucleophilic.)

$$CH_3CH_2CH_2Br \xrightarrow{^-N_3} CH_3CH_2CH_2N\!=\!\overset{+}{N}\!=\!N^- \xrightarrow[\text{Pd/C}]{H_2} CH_3CH_2CH_2NH_2 + N_2$$
$$\text{an alkyl azide}$$

Why Are Living Organisms Composed of Carbon Instead of Silicon?

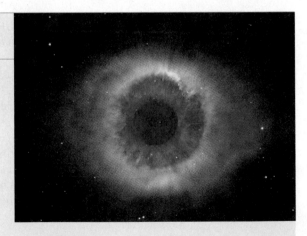

There are two reasons living organisms are composed primarily of carbon, oxygen, nitrogen, and hydrogen: the *fitness* of these elements for specific roles in life processes and their *availability* in the environment. Fitness apparently was more important than availability because carbon rather than silicon became the fundamental building block of living organisms despite the fact that silicon, which is just below carbon in the periodic table, is more than 140 times more abundant than carbon in the earth's crust.

   Why are carbon, oxygen, nitrogen, and hydrogen so well suited for the roles they play in living organisms? First and foremost, they are among the smallest atoms that form covalent bonds and they can form multiple bonds. Because of these factors, they form strong bonds (which means the molecules containing them are stable). The compounds that make up living organisms must be stable and therefore slow to react if the organisms are to survive.

   Silicon has almost twice the diameter of carbon, so silicon forms longer and weaker bonds. Consequently, an S_N2 reaction at silicon would occur much more rapidly than an S_N2 reaction at carbon. Moreover, silicon has another problem. The end product of carbon metabolism is CO_2. The analogous product of silicon metabolism would be SiO_2. But unlike carbon, which is doubly bonded to oxygen in CO_2, silicon is only singly bonded to oxygen in SiO_2. Therefore, silicon dioxide molecules polymerize to form quartz (sand). It is hard to imagine that life could exist, much less proliferate, if animals exhaled sand instead of CO_2!

	Abundance (atoms/100 atoms)	
Element	In living organisms	In Earth's crust
H	49	0.22
C	25	0.19
O	25	47
N	0.3	0.1
Si	0.03	28

3 THE MECHANISM FOR AN S$_N$1 REACTION

Given our understanding of S$_N$2 reactions, we would expect the rate of the following reaction to be very slow because water is a poor nucleophile and the alkyl halide is sterically hindered to back-side attack.

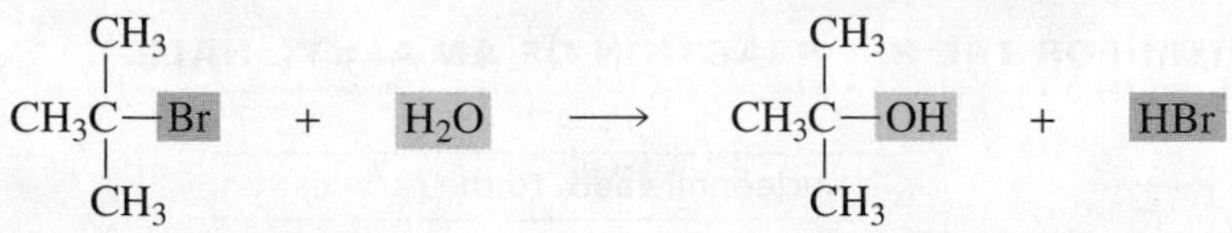

$$
\underset{\text{2-bromo-2-methylpropane}}{\underset{\overset{|}{CH_3}}{\overset{\overset{CH_3}{|}}{CH_3C}}\!\!-\!\!Br} \;+\; H_2O \;\longrightarrow\; \underset{\text{2-methyl-2-propanol}}{\underset{\overset{|}{CH_3}}{\overset{\overset{CH_3}{|}}{CH_3C}}\!\!-\!\!OH} \;+\; HBr
$$

It turns out, however, that the reaction is surprisingly fast. In fact, it is over 1 million times faster than the reaction of bromomethane (a compound with no steric hindrance) with water. The reaction, therefore, must be taking place by a mechanism different from that of an S$_N$2 reaction.

Experimental Evidence for the Mechanism of an S$_N$1 Reaction

In order to determine the mechanism of a reaction, we need to find out what factors affect the rate of the reaction, and we need to know the configuration of the products of the reaction.

Doubling the concentration of the alkyl halide doubles the rate of the reaction, but changing the concentration of the nucleophile has no effect on its rate. This knowledge allows us to write a rate law for the reaction:

$$\text{rate} = k\,[\,\text{alkyl halide}\,]$$

The rate of the reaction depends linearly on the concentration of only one reactant, so the reaction is a **first-order reaction**.

Because the rate law for the reaction of 2-bromo-2-methylpropane with water differs from the rate law for the reaction of bromomethane with hydroxide ion (Section 1), the two reactions must have different mechanisms.

We have seen that the reaction between bromomethane and hydroxide ion is an S$_N$2 reaction. The reaction between 2-bromo-2-methylpropane and water is an **S$_N$1 reaction,** where "S" stands for *substitution,* "N" stands for *nucleophilic,* and "1" stands for *unimolecular.* **Unimolecular** means that only one molecule is involved in the transition state of the rate-determining step. The mechanism for an S$_N$1 reaction is based on the following experimental evidence:

1. The rate law shows that the rate of the reaction depends only on the concentration of the alkyl halide, so only the alkyl halide is involved in the transition state of the rate-determining step.

2. Tertiary alkyl halides undergo S$_N$1 reactions but methyl halides and primary alkyl halides do not. A recent investigation of Hughes and Ingold's data showed that secondary alkyl halides also do not undergo S$_N$1 reactions.* Thus, methyl halides, primary alkyl halides, and secondary alkyl halides undergo only S$_N$2 reactions.

3. The substitution reaction of an alkyl halide in which the halogen is bonded to an asymmetric center forms two stereoisomers: one with the same relative configuration as that of the reacting alkyl halide, and the other with the inverted configuration.

*Murphy, T.J. *J. Chem. Ed.* **2009,** *86,* 519-524.

The Mechanism for an S_N1 Reaction

Unlike an S_N2 reaction, where the leaving group departs and the nucleophile approaches *at the same time,* the leaving group in an S_N1 reaction departs *before* the nucleophile approaches.

MECHANISM FOR THE S_N1 REACTION OF AN ALKYL HALIDE

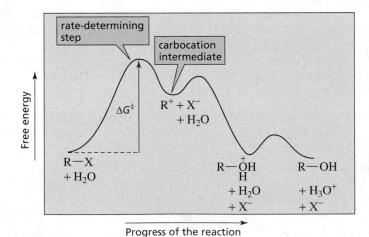

- In the first step, the carbon–halogen bond breaks and the previously shared pair of electrons stays with the halogen. As a result, a carbocation intermediate is formed.
- In the second step, the nucleophile reacts rapidly with the carbocation (an electrophile) to form a protonated alcohol.
- Whether the alcohol product will exist in its protonated (acidic) form or neutral (basic) form depends on the pH of the solution. At pH = 7, the alcohol will exist predominantly in its neutral form.

An S_N1 reaction is a two-step reaction.

Because the rate of an S_N1 reaction depends only on the concentration of the alkyl halide, the first step must be the slow (rate-determining) step (Figure 7). The nucleophile is not involved in the rate-determining step, so its concentration has no effect on the rate of the reaction.

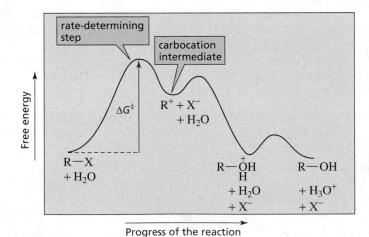

▶ **Figure 7**

The reaction coordinate diagram for an S_N1 reaction shows why increasing the rate of the second step will not make an S_N1 reaction go any faster.

How the Mechanism Accounts for the Experimental Evidence

How does the mechanism for an S_N1 reaction account for the three pieces of experimental evidence?

First, because the alkyl halide is the only species that participates in the rate-determining step, the mechanism agrees with the observation that the rate of the reaction depends only on the concentration of the alkyl halide; it does not depend on the concentration of the nucleophile.

Second, the mechanism shows that a carbocation is formed in the rate-determining step. This explains why tertiary alkyl halides undergo S_N1 reactions, but primary and secondary alkyl halides do not. Tertiary carbocations are more stable than primary and secondary carbocations and, therefore, are the most easily formed. (In Section 5 we will see that allylic and benzylic halides undergo S_N1 reactions, because they too form relatively stable carbocations.)

Third, the positively charged carbon of the carbocation intermediate is sp^2 hybridized, which means the three bonds connected to it are in the same plane (Figure 8). In the second step of the S_N1 reaction, the nucleophile can approach the carbocation from either side of the plane, so some of the product will have the same configuration as the reacting alkyl halide and some will have an inverted configuration.

Carbocation stability: 3° > 2° > 1°

Tertiary alkyl halides undergo S_N1 reactions. Primary and secondary alkyl halides undergo S_N2 reactions.

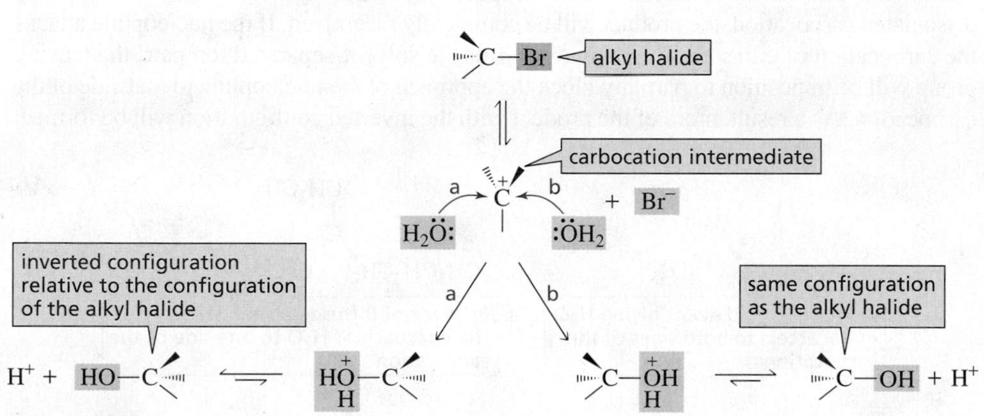

▲ Figure 8

If the nucleophile adds to the *opposite side* of the carbon from which the leaving group departs (labeled a), then the product will have the *inverted* configuration relative to the configuration of the alkyl halide.

If the nucleophile adds to the side of the carbon from which the leaving group departs (labeled b), then the product will have the *same* relative configuration as that of the reacting alkyl halide.

We can now understand why an S_N1 reaction of an alkyl halide in which the leaving group is attached to an asymmetric center forms two stereoisomers: addition of the nucleophile to one side of the planar carbocation intermediate forms one stereoisomer, and addition to the other side produces the other stereoisomer. Thus, the product is a pair of enantiomers.

An S_N1 reaction takes place with inversion and retention of configuration.

If the leaving group is attached to an asymmetric center, an S_N1 reaction forms a pair of enantiomers.

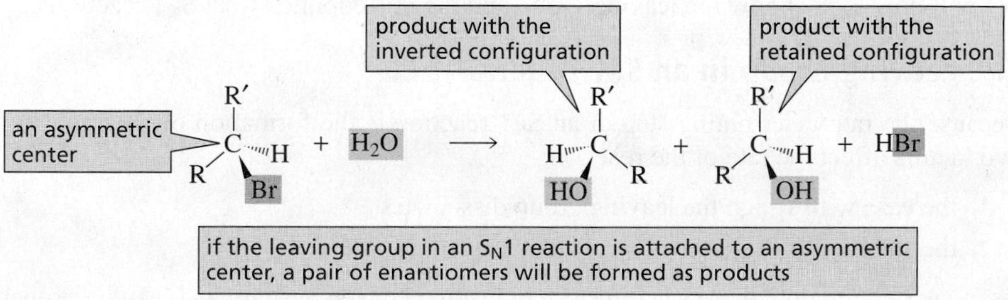

Although you probably expect that equal amounts of both products will be formed in an S_N1 reaction, a greater amount of the product with the inverted configuration is obtained in most cases. Typically, 50 to 70% of the product of an S_N1 reaction is the inverted product. If the reaction does lead to equal amounts of the two stereoisomers, the reaction is said to take place with **complete racemization.** When more of the inverted product is formed, the reaction is said to take place with **partial racemization.**

Why does an S_N1 reaction generally form more inverted product? Dissociation of the alkyl halide initially results in the formation of an **intimate ion pair.** In an intimate ion pair, the bond between the carbon and the leaving group has broken, but the cation and anion remain next to each other. When they move slightly farther apart, they become a *solvent-separated ion pair,* meaning an ion pair with one or more solvent molecules between the cation and the anion. As the ions separate further, they become dissociated ions.

R—X $\longrightarrow$ R$^+$ X$^-$ $\longrightarrow$ R$^+$ [solvent] X$^-$ $\longrightarrow$ R$^+$ X$^-$

| undissociated molecule | intimate ion pair | solvent-separated ion pair | dissociated ions |

The nucleophile can attack any of these four species. If the nucleophile attacks only the fully dissociated carbocation, the product will be completely racemized. If the nucleophile attacks the carbocation of either the intimate ion pair or the solvent-separated ion pair, the leaving group will be in position to partially block the approach of the nucleophile to that side of the carbocation. As a result, more of the product with the inverted configuration will be formed.

An S_N1 reaction takes place with racemization.

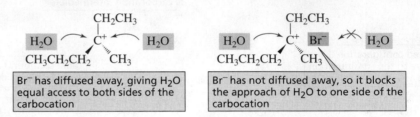

Br$^-$ has diffused away, giving H$_2$O equal access to both sides of the carbocation

Br$^-$ has not diffused away, so it blocks the approach of H$_2$O to one side of the carbocation

(Notice that if the nucleophile attacks the undissociated molecule, the reaction is an S_N2 reaction and all of the product will have the inverted configuration.)

PROBLEM 17♦

Draw the substitution products that will be formed from the following S_N1 reactions:

a. 3-bromo-3-methylpentane and methanol
b. 3-chloro-3-methylhexane and methanol

---

4   FACTORS THAT AFFECT S_N1 REACTIONS

We will now look at how the leaving group and the nucleophile affect S_N1 reactions.

The Leaving Group in an S_N1 Reaction

Because the rate-determining step of an S_N1 reaction is the formation of a carbocation, two factors affect the rate of the reaction:

  1. the ease with which the leaving group dissociates

  2. the stability of the carbocation that is formed

As in an S_N2 reaction, there is a direct relationship between basicity and leaving propensity in an S_N1 reaction: the weaker the base, the less tightly it is bonded to the carbon and the more easily the carbon–halogen bond can be broken. As a result, comparing alkyl halides with the same alkyl group, an alkyl iodide is the most reactive and an alkyl fluoride is the least reactive in both S_N1 and S_N2 reactions.

relative reactivities of alkyl halides in an S_N1 reaction

most reactive ▷ RI > RBr > RCl > RF ◁ least reactive

The Nucleophile in an S$_N$1 Reaction

Because the nucleophile does not participate in an S$_N$1 reaction until *after* the rate-determining step, the reactivity of the nucleophile has no effect on the rate of an S$_N$1 reaction.

In most S$_N$1 reactions, the solvent is the nucleophile. For example, in the following reaction methanol is both the nucleophile and the solvent. Reaction with a solvent is called **solvolysis.**

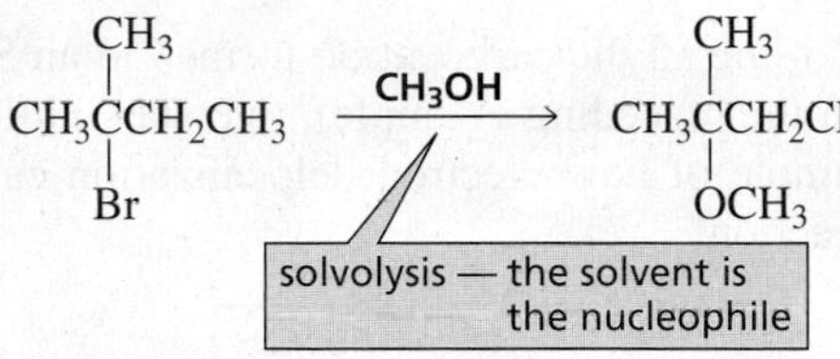

solvolysis — the solvent is the nucleophile

PROBLEM 18♦

Arrange the following alkyl halides in order from most reactive to least reactive in an S$_N$1 reaction: 2-bromo-2-methylpentane, 2-chloro-2-methylpentane, 3-chloropentane, and 2-iodo-2-methylpentane.

---

5 BENZYLIC HALIDES, ALLYLIC HALIDES, VINYLIC HALIDES, AND ARYL HALIDES

Up to this point, our discussion of substitution reactions has been limited to methyl halides and primary, secondary, and tertiary alkyl halides. But what about benzylic, allylic, vinylic, and aryl halides?

Let's first consider benzylic and allylic halides. Unless they are tertiary, these halides readily undergo S$_N$2 reactions. Tertiary benzylic and tertiary allylic halides, like tertiary alkyl halides, do *not* undergo S$_N$2 reactions because of steric hindrance.

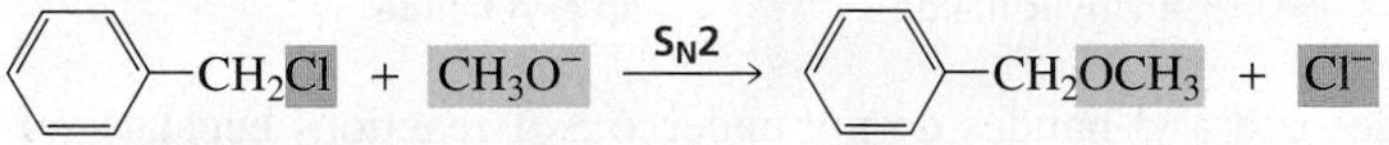

benzyl chloride   benzyl methyl ether

$CH_3CH{=}CHCH_2Br$ + HO^- $\xrightarrow{S_N2}$ $CH_3CH{=}CHCH_2OH$ + Br^-
1-bromo-2-butene   **2-buten-1-ol**
an allylic halide

Notice that a benzene ring attached to a methylene group is called a benzyl group. A substituent that consists of just a benzene ring is called a phenyl group.

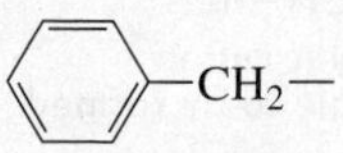

a benzyl group 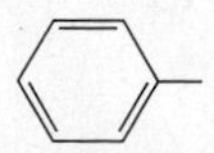   **a phenyl group**

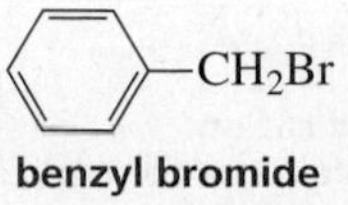

benzyl bromide 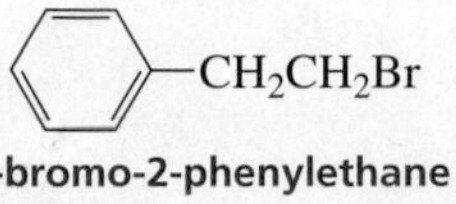   **1-bromo-2-phenylethane**

Benzylic and allylic halides readily undergo S$_N$1 reactions as well, because they form carbocations that are stabilized by electron delocalization.

Benzylic and allylic halides undergo S$_N$1 and S$_N$2 reactions.

435

$$\text{C}_6\text{H}_5\text{—CH}_2\text{Cl} \xrightleftharpoons{\text{S}_\text{N}1} \text{C}_6\text{H}_5\text{—}\overset{+}{\text{CH}}_2 \xrightarrow{\text{CH}_3\text{OH}} \text{C}_6\text{H}_5\text{—CH}_2\text{OCH}_3 + \text{H}^+$$
$$+ \text{Cl}^-$$

$$\text{CH}_2\text{=CHCH}_2\text{Br} \xrightleftharpoons{\text{S}_\text{N}1} \text{CH}_2\text{=CH}\overset{+}{\text{CH}}_2 \longleftrightarrow \overset{+}{\text{CH}}_2\text{CH=CH}_2 \xrightarrow{\text{H}_2\text{O}} \text{CH}_2\text{=CHCH}_2\text{OH} + \text{H}^+$$
$$+ \text{Br}^-$$

resonance contributors
are mirror images

If the two resonance contributors of the allylic carbocation formed in an $\text{S}_\text{N}1$ reaction are not mirror images (as they are in the preceding example), two substitution products will be formed. This is another example of how electron delocalization can affect the nature of the products formed in a reaction.

resonance contributors
are not mirror images

$$\text{CH}_3\text{CH=CHCH}_2\text{Br} \xrightleftharpoons{\text{S}_\text{N}1} \text{CH}_3\text{CH=CH}\overset{+}{\text{CH}}_2 \longleftrightarrow \text{CH}_3\overset{+}{\text{CH}}\text{CH=CH}_2 + \boxed{\text{Br}^-}$$

$$\downarrow \text{H}_2\text{O} \qquad\qquad\qquad \downarrow \text{H}_2\text{O}$$

$$\text{CH}_3\text{CH=CHCH}_2\text{OH} \qquad\qquad \text{CH}_3\text{CHCH=CH}_2$$
$$+ \text{H}^+ \qquad\qquad\qquad \text{OH} + \text{H}^+$$

Vinylic halides and aryl halides (compounds in which the halogen is attached to a benzene ring) do not undergo $\text{S}_\text{N}2$ or $\text{S}_\text{N}1$ reactions. They do not undergo $\text{S}_\text{N}2$ reactions because, as the nucleophile approaches the back side of the sp^2 carbon, it is repelled by the π electrons of the double bond or the benzene ring.

a nucleophile is repelled
by the π electron cloud

a vinylic halide   an aryl halide

Vinylic halides and aryl halides do not undergo $\text{S}_\text{N}1$ reactions because vinylic and aryl cations are even more unstable than primary carbocations. The positive charge on a vinylic or aryl cation would be on an sp carbon—sp carbons are more electronegative than the sp^2 carbons that carry the positive charge of alkyl carbocations, so sp carbons are more resistant to becoming positively charged. In addition, a ring carbon cannot form the 180° bond angles required for sp hybridization.

Vinylic and aryl halides do not undergo $\text{S}_\text{N}1$ or $\text{S}_\text{N}2$ reactions.

sp^2 carbon sp carbon

$$\text{RCH=CH—Cl} \quad\not\longrightarrow\quad \text{RCH=}\overset{+}{\text{CH}} + \text{Cl}^-$$
vinylic cation
too unstable to be formed

sp^2 carbon sp carbon

$$\text{C}_6\text{H}_5\text{—Br} \quad\not\longrightarrow\quad \text{C}_6\text{H}_5^+ + \text{Br}^-$$
aryl cation
too unstable to be formed

PROBLEM-SOLVING STRATEGY

Predicting Relative Reactivities

Which compound would you expect to be more reactive in an S_N1 solvolysis reaction?

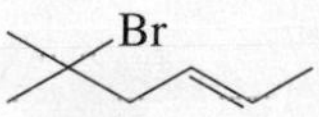

When asked to determine the relative reactivities of two compounds, we need to compare the $\Delta G^{\ddagger}$ values of their rate-determining steps. The faster-reacting compound will have the *smaller* $\Delta G^{\ddagger}$ value—that is, the smaller difference between its free energy and the free energy of its rate-determining transition state.

Both alkyl halides have approximately the same stability, so the difference in their reaction rates will be due primarily to the difference in the stabilities of the transition states of their rate-determining steps. The rate-determining step is carbocation formation. This is an endothermic reaction, so the transition state will resemble the carbocation more than it will resemble the reactant. Therefore, the compound that forms the more stable carbocation will be the one that has the faster rate of solvolysis.

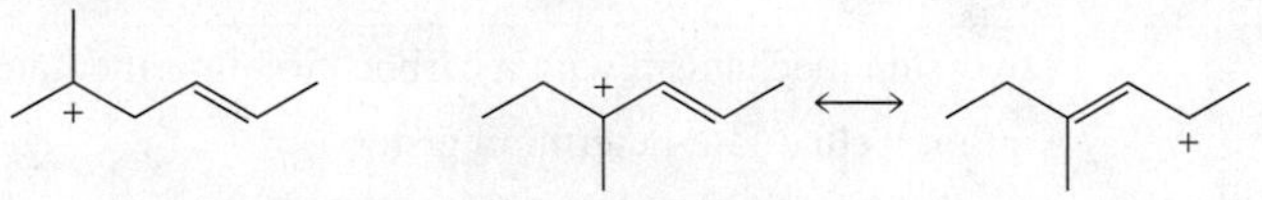

Unlike the carbocation formed by the alkyl halide on the left (which does not have delocalized electrons), the carbocation formed by the alkyl halide on the right is stabilized by electron delocalization. Thus, the alkyl halide on the right will undergo solvolysis more rapidly.

Now use the strategy you have just learned to solve Problems 19 and 20.

PROBLEM 19

Which alkyl halide would you expect to be more reactive in an S_N1 solvolysis reaction?

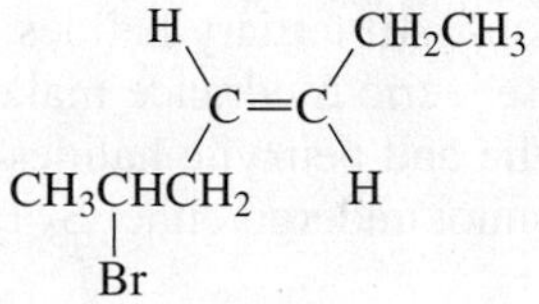

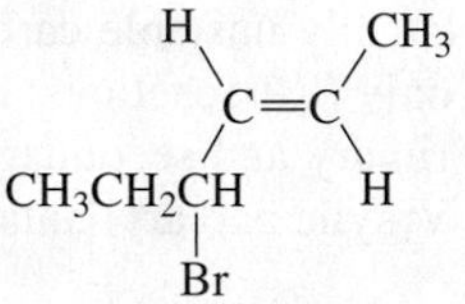

PROBLEM 20◆

Which alkyl halide would you expect to be more reactive in an S_N1 reaction? In each case, you can assume that both alkyl halides have the same stability.

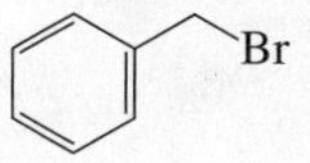

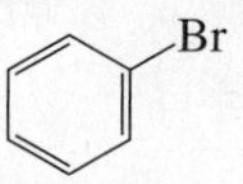

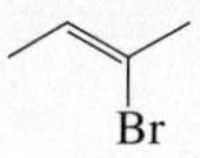

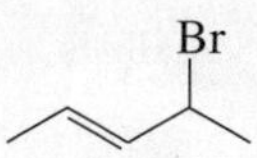

PROBLEM 21◆

For the pairs of alkyl halides in Problem 20, which would be more reactive in an S_N2 reaction?

PROBLEM 22♦

Two substitution products result from the reaction of 3-chloro-3-methyl-1-butene with sodium acetate ($CH_3COO^- Na^+$) in acetic acid under conditions that favor an S_N1 reaction. Identify the products.

---

6 COMPETITION BETWEEN S_N2 AND S_N1 REACTIONS

The characteristics of S_N2 and S_N1 reactions are compared in Table 3. Remember that the "2" in "S_N2" and the "1" in "S_N1" refer to the molecularity of the reaction (the number of molecules involved in the transition state of the rate-determining step), and *not* to the number of steps in the mechanism. In fact, the opposite is true: an S_N2 reaction proceeds by a *one*-step concerted mechanism, whereas an S_N1 reaction proceeds by a *two*-step mechanism with a carbocation intermediate.

Table 3 Comparison of S_N2 and S_N1 Reactions

S_N2	S_N1
a one-step mechanism	a two-step mechanism with a carbocation intermediate
a bimolecular rate-determining step	a unimolecular rate-determining step
the rate is controlled by steric hindrance	the rate is controlled by stability of the carbocation
product has the inverted configuration relative to that of the reactant	products have both the retained and inverted configurations relative to that of the reactant
the leaving group: $I^- > Br^- > Cl^- > F^-$	the leaving group: $I^- > Br^- > Cl^- > F^-$
the better the nucleophile, the faster the rate of the reaction	the strength of the nucleophile does not affect the rate of the reaction

Table 4 summarizes the reactivity of alkyl halides in S_N2 and S_N1 reactions. Primary alkyl halides, secondary alkyl halides, and methyl halides undergo only S_N2 reactions because of their relatively unstable carbocations. All tertiary halides (alkyl, allylic, and benzylic) undergo only S_N1 reactions, because steric hindrance makes them unreactive in S_N2 reactions. Primary and secondary allylic and benzylic halides undergo both S_N1 and S_N2 reactions. Vinylic and aryl halides cannot undergo either S_N1 or S_N2 reactions.

Table 4 Summary of the Reactivity of Alkyl Halides in Nucleophilic Substitution Reactions

1° alkyl, 2° alkyl, and methyl halides	S_N2 only
3° alkyl, 3° allylic, and 3° benzylic halides	S_N1 only
1° allylic and benzylic and 2° allylic and benzylic halides	S_N1 and S_N2
vinylic and aryl halides	neither S_N1 nor S_N2

Only primary and secondary allylic and benzylic halides undergo *both* S_N1 *and* S_N2 reactions. The conditions under which the reaction is carried out determine which reaction predominates. Therefore, we have some control over which reaction takes place.

What conditions favor an S_N2 reaction? This is an important question to synthetic chemists because an S_N2 reaction forms a single substitution product, whereas an S_N1 reaction can form a pair of enantiomers if the halogen is attached to an asymmetric center, and it can form constitutional isomers if the reactant is an allylic halide. In other words, an S_N2 reaction is a synthetic chemist's friend, but an S_N1 reaction can be a synthetic chemist's nightmare, so it is important to understand the conditions that will help avoid S_N1 reactions.

When a compound can undergo both S_N1 and S_N2 reactions, three conditions determine which reaction will predominate:

1. the *concentration* of the nucleophile
2. the *reactivity* of the nucleophile
3. the *solvent* in which the reaction is carried out

To understand how the concentration and the reactivity of the nucleophile affect whether an S_N1 or an S_N2 reaction predominates, we must examine the rate laws for the two reactions (Sections 1 and 3). The rate constants have been given subscripts that indicate the reaction order.

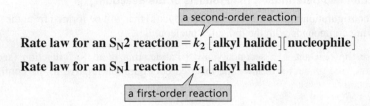

Rate law for an S_N2 reaction $= k_2$ [**alkyl halide**][**nucleophile**]

Rate law for an S_N1 reaction $= k_1$ [**alkyl halide**]

The rate law for the reaction of a compound—for example, an allylic halide—that can undergo both S_N2 and S_N1 reactions simultaneously is the sum of the individual rate laws.

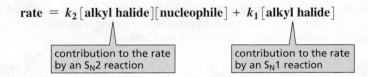

$$\text{rate} = k_2 [\textbf{alkyl halide}][\textbf{nucleophile}] + k_1 [\textbf{alkyl halide}]$$

From the rate law, you can see that increasing the *concentration* of the nucleophile increases the rate of the S_N2 reaction but has no effect on the rate of the S_N1 reaction. Therefore, when both reactions occur simultaneously, increasing the concentration of the nucleophile increases the fraction of the reaction that takes place by the S_N2 pathway.

The slow (and only) step of an S_N2 reaction is attack of the nucleophile on the allylic halide. Increasing the strength of the nucleophile increases the rate of an S_N2 reaction by increasing the value of the rate constant (k_2), because a more reactive nucleophile is better able to displace the leaving group.

The slow step of an S_N1 reaction is the dissociation of the allylic halide. The carbocation formed in the slow step reacts rapidly in a second step with any nucleophile present in the reaction mixture. Therefore, increasing the strength of the nucleophile has no effect on the rate of an S_N1 reaction. A good nucleophile, therefore, favors an S_N2 reaction over an S_N1 reaction. A poor nucleophile favors an S_N1 reaction, not by increasing the rate of the S_N1 reaction itself, but by decreasing the rate of the competing S_N2 reaction. In summary:

- An S_N2 reaction is favored by a high concentration of a good nucleophile.
- An S_N1 reaction is favored by a poor nucleophile.

Look back at the S_N1 reactions of allylic and benzylic halides in Section 5 and notice that each one has a poor nucleophile (H_2O, CH_3OH), whereas all the S_N2 reactions of these compounds have good nucleophiles (HO^-, CH_3O^-). In other words, a poor nucleophile encourages the S_N1 reaction, whereas a good nucleophile encourages the S_N2 reaction.

In Section 7, we will look at the third factor that influences whether an S_N2 or an S_N1 reaction will predominate—namely, the solvent in which the reaction is carried out.

PROBLEM 23

Draw the products obtained from each of the following reactions and show their configurations

a. under conditions that favor an S_N2 reaction.

b. under conditions that favor an S_N1 reaction.

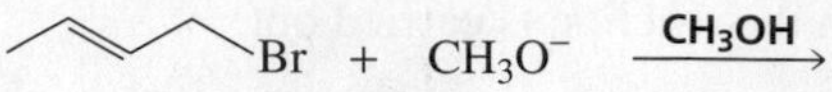

$$\text{(allyl chain)}\text{Br} + CH_3O^- \xrightarrow{CH_3OH}$$

PROBLEM-SOLVING STRATEGY

Predicting Whether a Nucleophilic Substitution Reaction Will Be an S_N1 Reaction or an S_N2 Reaction and Determining the Products of the Reaction

Draw the configuration(s) of the substitution product(s) that will be formed from the reactions of the following compounds with the indicated nucleophile:

a. Because the reactant is a secondary alkyl halide, this is an S_N2 reaction (see Table 4). Therefore, the product will have the inverted configuration relative to the configuration of the reactant.

One way to draw the inverted product is to draw the mirror image of the reacting alkyl halide and then put the nucleophile in the same location as the leaving group.

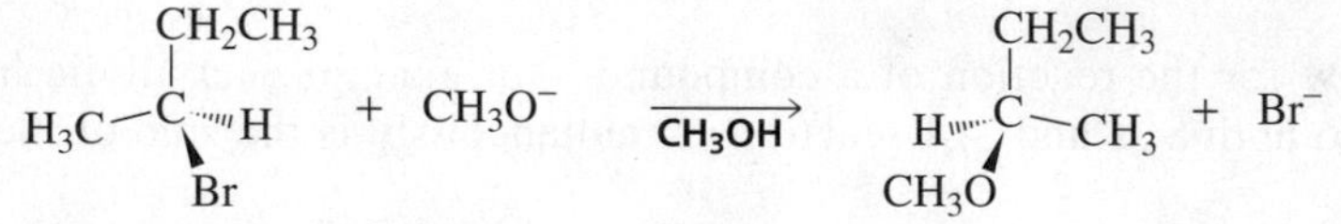

$$H_3C-\overset{CH_2CH_3}{\underset{Br}{\overset{|}{C}}}{\cdots}H + CH_3O^- \xrightarrow{CH_3OH} H{\cdots}\overset{CH_2CH_3}{\underset{CH_3O}{\overset{|}{C}}}-CH_3 + Br^-$$

b. Because the reactant is a tertiary alkyl halide, this is an S_N1 reaction. Therefore, we will obtain two substitution products, one with the retained configuration and one with the inverted configuration, relative to the configuration of the reactant.

$$CH_3CH_2CH_2-\overset{CH_2CH_3}{\underset{Br}{\overset{|}{C}}}{\cdots}CH_3 + CH_3OH \longrightarrow CH_3CH_2CH_2-\overset{CH_2CH_3}{\underset{OCH_3}{\overset{|}{C}}}{\cdots}CH_3 + CH_3{\cdots}\overset{CH_2CH_3}{\underset{CH_3O}{\overset{|}{C}}}CH_2CH_2CH_3$$

$$+ Br^-$$

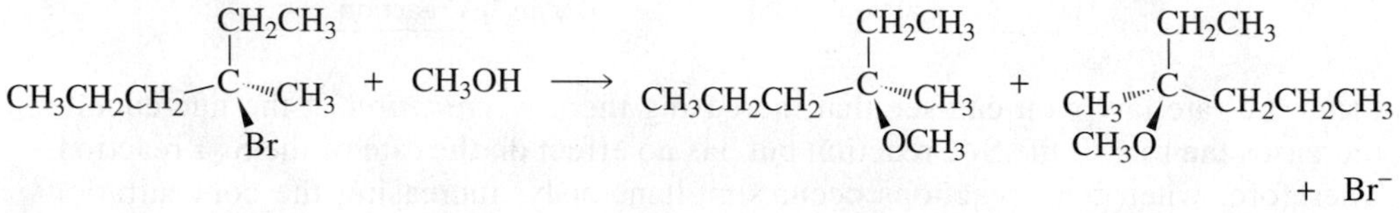

c. Because the reactant is a tertiary alkyl halide, this is an S_N1 reaction. The product does not have an asymmetric center, so it does not have stereoisomers. Thus, only one product is formed.

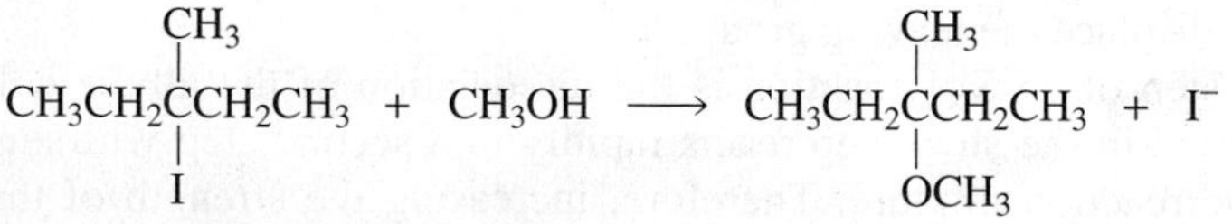

$$CH_3CH_2\overset{CH_3}{\underset{I}{\overset{|}{C}}}CH_2CH_3 + CH_3OH \longrightarrow CH_3CH_2\overset{CH_3}{\underset{OCH_3}{\overset{|}{C}}}CH_2CH_3 + I^-$$

d. Because the reactant is a secondary alkyl halide, this is an S_N2 reaction. Therefore, the configuration of the product will be inverted relative to the configuration of the reactant. But, since the configuration of the reactant is not indicated, we do not know the configuration of the product.

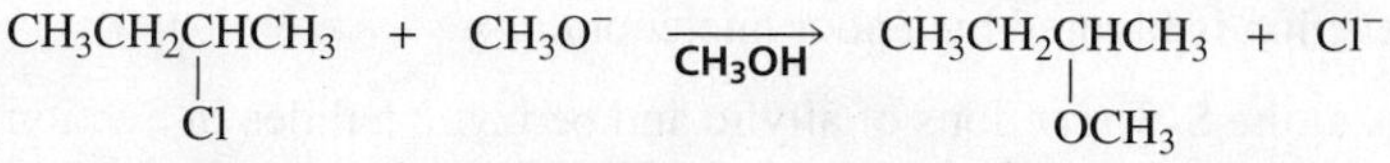

$$CH_3CH_2\overset{}{\underset{Cl}{\overset{|}{C}}}HCH_3 + CH_3O^- \xrightarrow{CH_3OH} CH_3CH_2\overset{}{\underset{OCH_3}{\overset{|}{C}}}HCH_3 + Cl^-$$

e. The reactant is a secondary benzylic halide so it can undergo both S_N1 and S_N2 reactions. Because a high concentration of a good nucleophile is employed, we know the reaction here is an S_N2 reaction. Therefore, only one substitution product is formed—the one with the configuration that is inverted relative to that of the reactant.

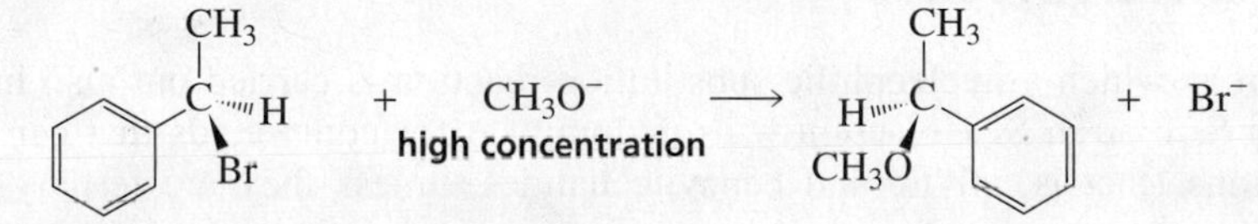

f. The reactant is a secondary benzylic halide so it can undergo both S_N1 and S_N2 reactions. Because a poor nucleophile is employed, we know the reaction here is an S_N1 reaction. Therefore, two substitution products are formed, one with the retained configuration and one with the inverted configuration, relative to the configuration of the reactant.

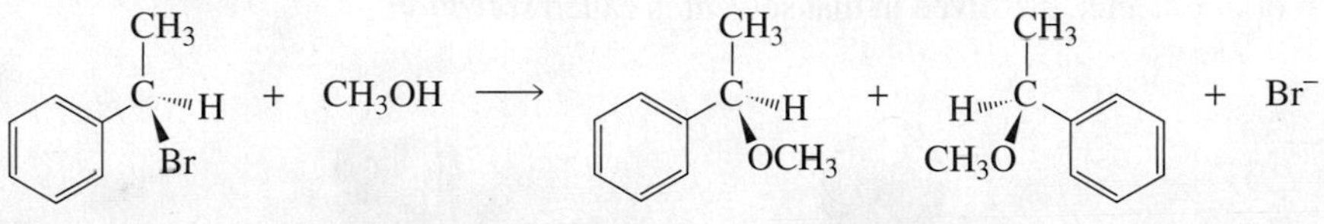

Now use the strategy you have just learned to solve Problem 24.

PROBLEM 24

Draw the configuration(s) of the substitution product(s) that will be formed from the reactions of the following compounds with the indicated nucleophile:

a.

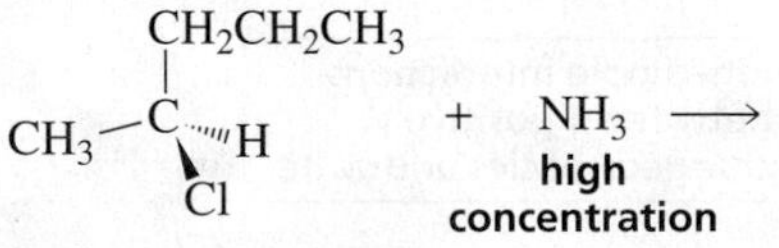

b.

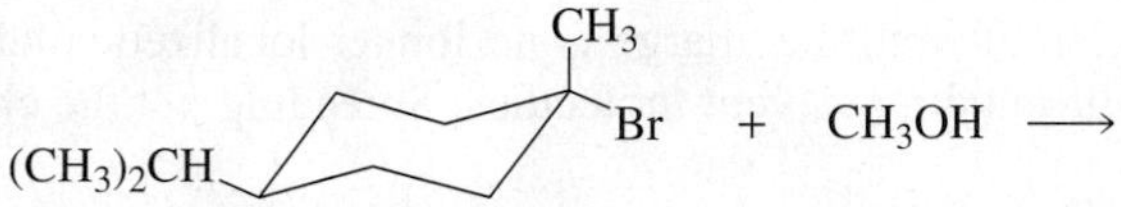

c.

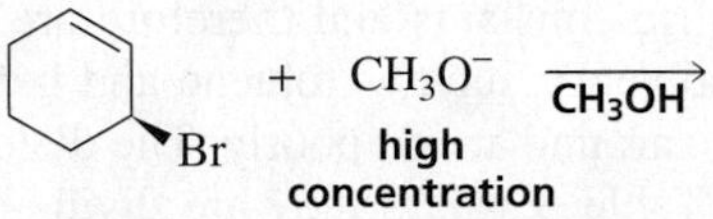

d.

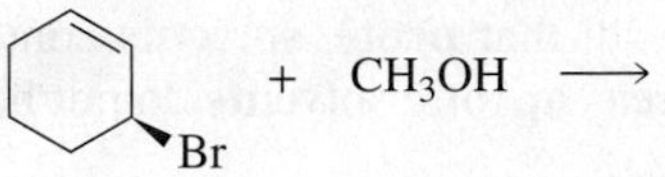

PROBLEM 25♦

Which of the following reactions will go faster if the concentration of the nucleophile is increased?

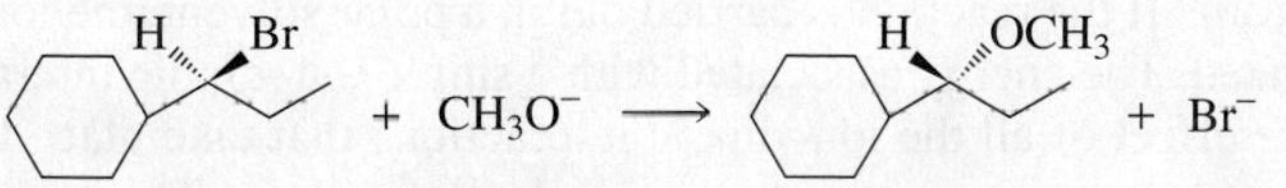

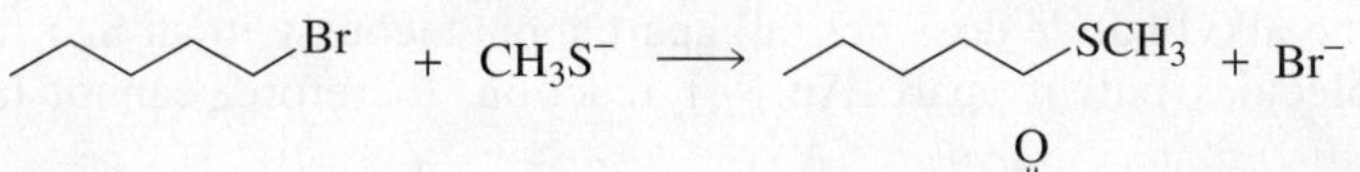

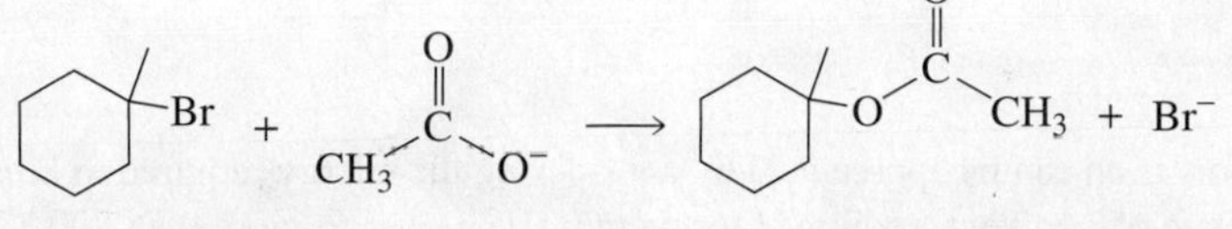

7   THE ROLE OF THE SOLVENT IN S_N1 AND S_N2 REACTIONS

The solvent in which a nucleophilic substitution reaction is carried out also influences whether an S_N1 or an S_N2 reaction will predominate for compounds that can undergo both reactions (that is, allylic and benzylic halides, unless they are tertiary). Before we can understand how a particular solvent favors one reaction over another, we must understand how solvents stabilize organic molecules.

The **dielectric constant** of a solvent is a measure of how well the solvent can insulate opposite charges from one another. Solvent molecules insulate a charge by clustering around it, so that the positive poles of solvent molecules surround negative charges while the negative poles of solvent molecules surround positive charges. The interaction between a solvent and an ion or a molecule dissolved in that solvent is called *solvation*.

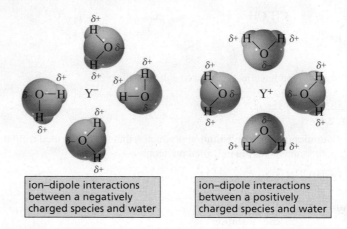

ion–dipole interactions between a negatively charged species and water

ion–dipole interactions between a positively charged species and water

When an ion interacts with a polar solvent, the charge is no longer localized solely on the ion, but is spread out to the surrounding solvent molecules. Spreading out the charge stabilizes the charged species.

Polar solvents, such as water, have high dielectric constants and therefore are very good at insulating (solvating) charges. Nonpolar solvents, such as toluene and hexane, have low dielectric constants and insulate the charge around an ion poorly. The dielectric constants of some common solvents are listed in Table 5 where they are divided into two groups: protic solvents and aprotic solvents. Recall that **protic solvents** contain a hydrogen bonded to an oxygen or to a nitrogen, whereas **aprotic solvents** do not have a hydrogen bonded to an oxygen or to a nitrogen.

The stabilization of charges by solvent interaction plays an important role in organic reactions. For example, when an alkyl halide undergoes an S_N1 reaction, the first step is dissociation of the carbon–halogen bond to form a carbocation and a halide ion. Energy is required to break the bond, but with no bonds being formed, where does the energy come from? If the reaction is carried out in a polar solvent, the ions that are produced are solvated. The energy associated with a single ion–dipole interaction is small, but the additive effect of all the ion–dipole interactions that take place when a solvent stabilizes a charged species represents a great deal of energy. These ion–dipole interactions provide much of the energy necessary for dissociation of the carbon–halogen bond. So, the alkyl halide does not fall apart spontaneously in an S_N1 reaction—polar solvent molecules pull it apart. An S_N1 reaction, therefore, cannot take place in a

Solvation Effects

The tremendous amount of energy provided by solvation can be appreciated by considering the energy required to break the crystal lattice of sodium chloride (table salt). In the absence of a solvent, sodium chloride must be heated to more than $800\,^\circ C$ to overcome the forces that hold the oppositely charged ions together. However, sodium chloride readily dissolves in water at room temperature because solvation of the Na^+ and Cl^- ions by water molecules provides the energy necessary to separate the ions.

Table 5 The Dielectric Constants and Boiling Points of Some Common Solvents

Solvent	Structure	Abbreviation	Dielectric constant (ε, at 25 °C)	Boiling point (°C)
Protic solvents				
Water	H_2O	—	79	100
Formic acid	HCOOH	—	59	100.6
Methanol	CH_3OH	MeOH	33	64.7
Ethanol	CH_3CH_2OH	EtOH	25	78.3
tert-Butyl alcohol	$(CH_3)_3COH$	*tert*-BuOH	11	82.3
Acetic acid	CH_3COOH	HOAc	6	117.9
Aprotic solvents				
Dimethyl sulfoxide	$(CH_3)_2SO$	DMSO	47	189
Acetonitrile	CH_3CN	MeCN	38	81.6
Dimethylformamide	$(CH_3)_2NCHO$	DMF	37	153
Hexamethylphosphoric acid triamide	$[(CH_3)_2N]_3PO$	HMPA	30	233
Acetone	$(CH_3)_2CO$	Me_2CO	21	56.3
Dichloromethane	CH_2Cl_2	—	9.1	40
Tetrahydrofuran		THF	7.6	66
Ethyl acetate	$CH_3COOCH_2CH_3$	EtOAc	6	77.1
Diethyl ether	$CH_3CH_2OCH_2CH_3$	Et_2O	4.3	34.6
Toluene		—	2.4	110.6
Hexane	$CH_3(CH_2)_4CH_3$	—	1.9	68.7

nonpolar solvent. It also cannot take place in the gas phase, where there are no solvent molecules and, consequently, no solvation effects.

How a Solvent Affects Reaction Rates in General

How increasing the polarity of the solvent (that is, increasing its dielectric constant) will affect the rate of most chemical reactions depends *only* on whether or not a reactant in the rate-limiting step is charged:

> *If a reactant in the rate-determining step is charged, increasing the polarity of the solvent will* decrease *the rate of the reaction.*

> *If none of the reactants in the rate-determining step is charged, increasing the polarity of the solvent will* increase *the rate of the reaction.*

Now let's see why this is true. The rate of a reaction depends on the difference between the free energy of the reactants and the free energy of the transition state of the rate-determining step. We can predict, therefore, how increasing the polarity of the solvent will affect the rate of a reaction simply by looking at the reactants and the transition state of the rate-determining step to see which will be more stabilized by a more polar solvent.

The greater or the more concentrated the charge on a molecule, the stronger will be its interactions with a polar solvent and the more the charge will be stabilized. Therefore, if the size or concentration of the charge on the reactants is greater than that on the transition state, then a polar solvent will stabilize the reactants more than it will stabilize the transition state. Therefore, *increasing the polarity of the solvent* will increase the difference in energy ($\Delta G^{\ddagger}$) between the transition state and the reactants, which *will decrease the rate of the reaction,* as shown in Figure 9.

Increasing the polarity of the solvent will decrease the rate of the reaction if a reactant in the rate-determining step is charged.

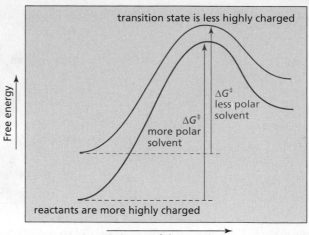

▶ **Figure 9**

The charge on the reactants is greater than the charge on the transition state. As a result, increasing the polarity of the solvent increases the stability of the reactants more than the stability of the transition state, so the reaction will be slower.

On the other hand, if the size of the charge on the transition state is greater than the size of the charge on the reactants, then a polar solvent will stabilize the transition state more than it will stabilize the reactants. Therefore, *increasing the polarity of the solvent* will decrease the difference in energy ($\Delta G^{\ddagger}$) between the transition state and the reactants, which *will increase the rate of the reaction,* as shown in Figure 10.

Increasing the polarity of the solvent will increase the rate of the reaction if none of the reactants in the rate-determining step is charged.

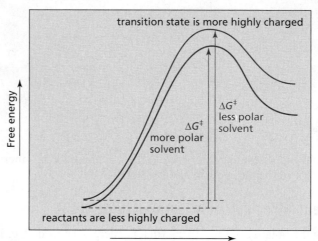

▶ **Figure 10**

The charge on the transition state is greater than the charge on the reactants. As a result, increasing the polarity of the solvent increases the stability of the transition state more than the stability of the reactants, so the reaction will be faster.

How a Solvent Affects the Rate of an S_N1 Reaction

Now let's look at specific reactions, beginning with an S_N1 reaction of an alkyl halide. The alkyl halide, which is the only reactant in the rate-determining step of an S_N1 reaction, is a neutral molecule with a small dipole moment. The rate-determining transition state has greater partial charges because as the carbon–halogen bond breaks, the carbon becomes more positive and the halogen becomes more negative. Since the partial charges on the transition state are greater than the partial charges on the reactant, increasing the polarity of the solvent will stabilize the transition state more than the reactant, which will increase the rate of the S_N1 reaction (Figure 10 and Table 6).

rate-determining step of an S_N1 reaction

the charge on the transition state is greater than the charge on the reactants

Table 6	The Effect of the Polarity of the Solvent on the Rate of Reaction of 2-Bromo-2-Methylpropane (a Tertiary Alkyl Halide) in an S_N1 Reaction	
Solvent		**Relative rate**
100% water		1200
80% water/20% ethanol		400
50% water/50% ethanol		60
20% water/80% ethanol		10
100% ethanol		1

Increasing polarity ↑

Compounds other than alkyl halides can undergo S_N1 reactions. As long as the compound undergoing an S_N1 reaction is neutral, increasing the polarity of the solvent will *increase* the rate of the S_N1 reaction because the polar solvent will stabilize the dispersed charges on the transition state more than it will stabilize the relatively neutral reactant (Figure 10).

If, however, the compound undergoing an S_N1 reaction is charged, increasing the polarity of the solvent will *decrease* the rate of the reaction because the more polar solvent will stabilize the full charge on the reactant to a greater extent than it will stabilize the dispersed charge on the transition state (Figure 9).

How a Solvent Affects the Rate of an S_N2 Reaction

How increasing the polarity of the solvent affects the rate of an S_N2 reaction depends on whether the reactants are charged or neutral, just as it does in an S_N1 reaction.

Most S_N2 reactions of alkyl halides occur between a neutral alkyl halide and a charged nucleophile. Increasing the polarity of a solvent will have a strong stabilizing effect on the negatively charged nucleophile. The transition state also has a negative charge, but that charge is dispersed over two atoms. Consequently, the interactions between the solvent and the transition state are not as strong as those between the solvent and the fully charged nucleophile. Therefore, increasing the polarity of the solvent will stabilize the nucleophile more than it will stabilize the transition state, so the reaction will be slower (Figure 9).

$$HO^- \;+\; {}^{\prime\prime\prime\prime\prime}C-Br \longrightarrow \left[\overset{\delta-}{HO} - - - C - - - \overset{\delta-}{Br} \right]^{\ddagger} \longrightarrow HO-C^{\prime\prime\prime\prime} \;+\; Br^-$$

reactants   **transition state**   **products**

If, however, the S_N2 reaction occurs between an alkyl halide and a neutral nucleophile, such as an amine, the partial charges on the transition state will be larger than the partial charges on the reactants, so increasing the polarity of the solvent will increase the rate of the reaction (Figure 10).

In summary, the way a change in the polarity of the solvent affects the rate of a reaction does not depend on the mechanism of the reaction. It depends *only* on whether or not a reactant in the rate-determining step is charged.

Because a polar solvent decreases the rate of an S_N2 reaction when the nucleophile is negatively charged, we would like to carry out such a reaction in a nonpolar solvent. However, negatively charged nucleophiles will not dissolve in nonpolar solvents, such as toluene and hexane. Instead, an aprotic polar solvent is used. Because they are not hydrogen bond donors, they are less effective than polar protic solvents, such as water and alcohols, at solvating negative charges; indeed, we have seen that the aprotic polar solvents DMSO and DMF solvate negative charges very poorly (Section 2).

Thus, the rate of an S_N2 reaction with a negatively charged nucleophile will be greater in an aprotic polar solvent than in a protic polar solvent. Consequently, an aprotic polar solvent is the solvent of choice for an S_N2 reaction in which the nucleophile is negatively charged, whereas a protic polar solvent is used if the nucleophile is a neutral molecule.

We have now seen that when an alkyl halide can undergo both S_N2 and S_N1 reactions, the S_N2 reaction will be favored by a high concentration of a good, negatively charged, nucleophile in an aprotic polar solvent or by a high concentration of a good, neutral, nucleophile in a protic polar solvent, whereas the S_N1 reaction will be favored by a poor, neutral, nucleophile in a protic polar solvent.

Environmental Adaptation

The microorganism *Xanthobacter* has learned to use alkyl halides that reach the ground as industrial pollutants as a source of carbon. The microorganism synthesizes an enzyme that uses the alkyl halide as a starting material to produce other carbon-containing compounds that it needs. This enzyme has several nonpolar groups at its active site (the pocket in the enzyme where the reaction it catalyzes takes place). The first step of the enzyme-catalyzed reaction is an S_N2 reaction with a charged nucleophile. The nonpolar groups on the surface of the enzyme provide the nonpolar environment needed to maximize the rate of the reaction.

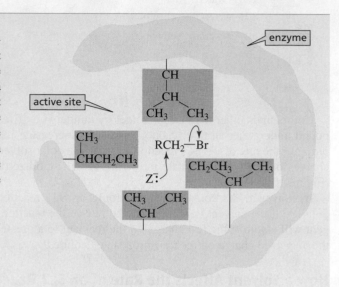

PROBLEM 26◆

Amines are good nucleophiles, even though they are neutral molecules. How would the rate of an S_N2 reaction between an amine and an alkyl halide be affected if the polarity of the solvent is increased?

An S_N2 reaction of an alkyl halide is favored by a high concentration of a good (negatively charged) nucleophile in an aprotic polar solvent or by a high concentration of a good (neutral) nucleophile in a protic polar solvent.

PROBLEM 27◆

How will the rate of each of the following S_N2 reactions change if it is carried out in a more polar solvent?

a. $CH_3CH_2CH_2CH_2Br + HO^- \longrightarrow CH_3CH_2CH_2CH_2OH + Br^-$

b. $CH_3\overset{+}{S}CH_3 + NH_3 \longrightarrow CH_3\overset{+}{N}H_3 + CH_3SCH_3$
    $\quad\,\, |$
    $\quad CH_3$

c. $CH_3CH_2I + NH_3 \longrightarrow CH_3CH_2\overset{+}{N}H_3\, I^-$

An S_N1 reaction of an alkyl halide is favored by a poor nucleophile in a protic polar solvent.

PROBLEM 28◆

Which reaction in each of the following pairs will take place more rapidly?

a. $CH_3Br + HO^- \longrightarrow CH_3OH + Br^-$
    $CH_3Br + H_2O \longrightarrow CH_3OH + HBr$

b. $CH_3I + HO^- \longrightarrow CH_3OH + I^-$
    $CH_3Cl + HO^- \longrightarrow CH_3OH + Cl^-$

c. $CH_3Br + NH_3 \longrightarrow CH_3\overset{+}{N}H_3 + Br^-$
    $CH_3Br + H_2O \longrightarrow CH_3OH + Br^-$

d. $CH_3Br + HO^- \xrightarrow{\textbf{DMSO}} CH_3OH + Br^-$
    $CH_3Br + HO^- \xrightarrow{\textbf{EtOH}} CH_3OH + Br^-$

e. $CH_3Br + NH_3 \xrightarrow{\textbf{Et}_2\textbf{O}} CH_3\overset{+}{N}H_3 + Br^-$
    $CH_3Br + NH_3 \xrightarrow{\textbf{EtOH}} CH_3\overset{+}{N}H_3 + Br^-$

PROBLEM 29 Solved

Most of the pK_a values given in this text have been determined in water. How would the pK_a values of carboxylic acids, alcohols, ammonium ions $(R\overset{+}{N}H_3)$, phenol, and an anilinium ion $(C_6H_5\overset{+}{N}H_3)$ change if they were determined in a solvent less polar than water, such as 50% water/50% dioxane?

Solution A pK_a is the negative logarithm of an equilibrium constant, K_a. Because we are determining how decreasing the polarity of a solvent affects an equilibrium constant, we must look at how decreasing the polarity of the solvent affects the stability of the reactants and products.

$$K_a = \frac{[B^-][H^+]}{[HB]} \qquad K_a = \frac{[B][H^+]}{[HB^+]}$$

a neutral acid   a positively charged acid

> A neutral acid will be a weaker acid in a solvent that is less polar than water.

Carboxylic acids, alcohols, and phenol are neutral in their acidic forms (HB) and charged in their basic forms (B^-). A polar protic solvent will stabilize B^- and H^+ more than it will stabilize HB, thereby increasing K_a. Therefore, K_a will be larger (a stronger acid) in water than in a less polar solvent, so the K_a values of carboxylic acids, alcohols, and phenol will be smaller (they will be weaker acids) and, therefore, their pK_a values will be larger in a less polar solvent.

Ammonium ions and an anilinium ion are charged in their acidic forms (HB^+) and neutral in their basic forms (B). A polar solvent will stabilize HB^+ and H^+ more than it will stabilize B. Because HB^+ is stabilized slightly more than H^+, K_a will be smaller (a weaker acid) in water than in a less polar solvent, so the pK_a values of ammonium ions and an anilinium ion will be smaller (they will be stronger acids) in a less polar solvent.

PROBLEM 30♦

Would you expect acetate ion $(CH_3CO_2^-)$ to be a better nucleophile in an S_N2 reaction with an alkyl halide carried out in methanol or in dimethyl sulfoxide?

PROBLEM 31♦

Under which of the following reaction conditions would (R)-1-chloro-1-phenylethane form the most (R)-1-phenyl-1-ethanol: HO^- in 50% water and 50% ethanol or HO^- in 100% ethanol?

8 INTERMOLECULAR VERSUS INTRAMOLECULAR REACTIONS

A molecule with two functional groups is called a **bifunctional molecule.** If the two functional groups are able to react with each other, then two kinds of reactions can occur—an *intermolecular* reaction and an *intramolecular* reaction. To understand the difference, let's look at a molecule with two functional groups that can react in an S_N2 reaction—namely, a good nucleophile (such as an alkoxide ion) and an alkyl halide.

If the alkoxide ion of one molecule displaces the bromide ion of a second molecule, then the reaction is an intermolecular reaction. *Inter* is Latin for "between," so an **intermolecular reaction** takes place between two molecules. If the product of this reaction subsequently reacts with a third bifunctional molecule (and then a fourth, and so on), a polymer will be formed. A polymer is a large molecule formed by linking together repeating units of small molecules.

an intermolecular reaction

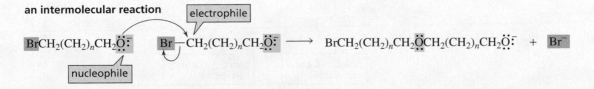

447

Alternatively, if the alkoxide ion of a molecule displaces the bromide ion of the *same* molecule (thereby forming a cyclic compound), then the reaction is an intramolecular reaction. *Intra* is Latin for "within," so an **intramolecular reaction** takes place within a single molecule.

an intramolecular reaction

Which reaction is more likely to occur, the intermolecular reaction or the intramolecular reaction? The answer depends on the *concentration* of the bifunctional molecule and the *size of the ring* that would be formed in the intramolecular reaction.

The intramolecular reaction has an advantage: the reacting groups are tethered together, so they do not have to diffuse through the solvent to find a group with which to react. Therefore, a low concentration of reactant favors an intramolecular reaction because the two functional groups have a better chance of finding each other if they are in the same molecule. A high concentration of reactant helps compensate for the advantage gained by tethering, thereby increasing the likelihood of an intermolecular reaction.

How much of an advantage an intramolecular reaction has over an intermolecular reaction also depends on the size of the ring that is formed—that is, on the length of the tether. If the intramolecular reaction forms a five- or six-membered ring, then it will be favored over the intermolecular reaction because five- and six-membered rings are stable and, therefore, easily formed. (Numbering the atoms in the reactant can help you determine the size of the ring in the product.)

Three- and four-membered rings are strained, which makes them less stable than five- and six-membered rings and so less easily formed. Therefore, the higher activation energy for formation of three- and four-membered rings cancels some of the advantage gained by tethering.

Although three-membered rings have more strain than four-membered rings, three-membered ring compounds are generally formed more easily. For a cyclic ether of any size to form, the nucleophilic oxygen atom must be oriented so that it can attack the back side of the carbon bonded to the halogen. Rotation about a C—C bond can produce conformers in which the groups point away from one another and are not able to react. A molecule that forms a three-membered ring ether has only one C—C bond that can rotate, whereas a molecule that forms a four-membered ring has two C—C bonds that can rotate. As a result, the molecule that forms the three-membered ring is more apt to have its reacting groups in the conformation required for reaction.

The likelihood of the reacting groups finding each other decreases sharply when the groups are in compounds that would form seven-membered and larger rings. Therefore, the intramolecular reaction becomes less favored as the ring size increases beyond six members.

PROBLEM 32♦

After a proton is removed from the OH group, which compound in each pair would form a cyclic ether more rapidly?

a. 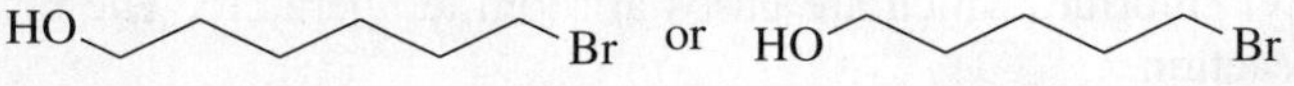 HO~~~~~Br   or   HO~~~~Br

b. 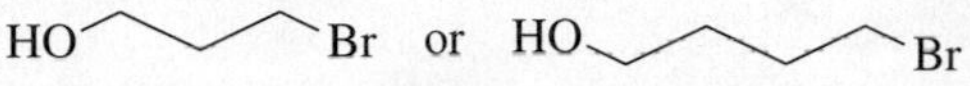 HO~~Br   or   HO~~~~Br

c. 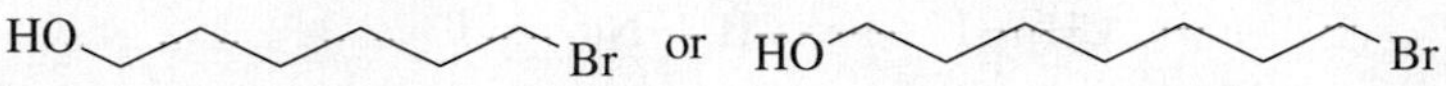 HO~~~~~~Br   or   HO~~~~~~~Br

---

PROBLEM-SOLVING STRATEGY

Investigating How Stereochemistry Affects Reactivity

Which of the following compounds will form an epoxide as a result of reacting with sodium hydride (NaH)? (*Hint:* H$^-$ is a strong base.)

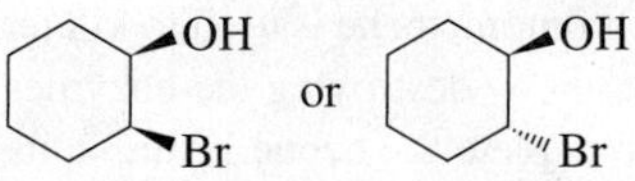

Hydride ion will remove a proton from the OH group, forming a good nucleophile that can react with the secondary alkyl halide in an S$_N$2 reaction to form an epoxide. An S$_N$2 reaction requires back-side attack. Only when the alkoxide ion and Br are on opposite sides of the cyclohexane ring will the alkoxide ion be able to attack the back side of the carbon that is attached to Br. Therefore, only the trans isomer will be able to form an epoxide.

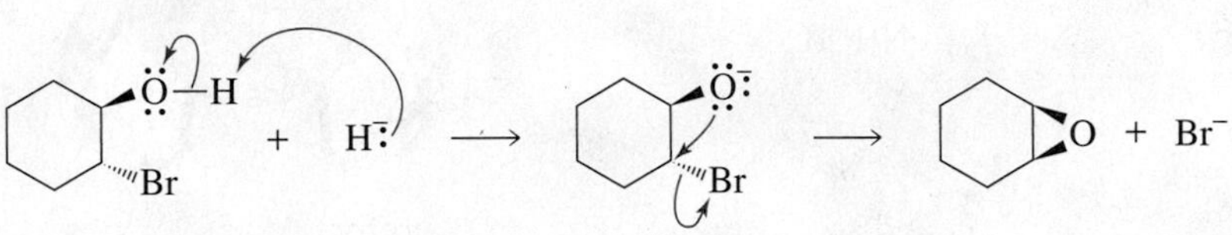

Now use the strategy you have just learned to solve Problem 33.

PROBLEM 33

Draw the products of the following reactions:

a. 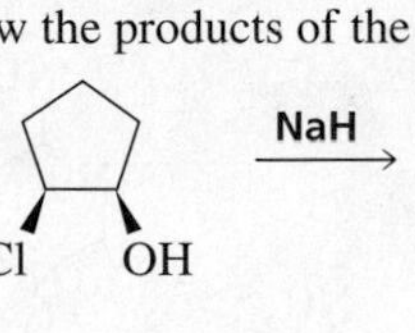 (cyclopentane with Cl and OH) $\xrightarrow{\text{NaH}}$

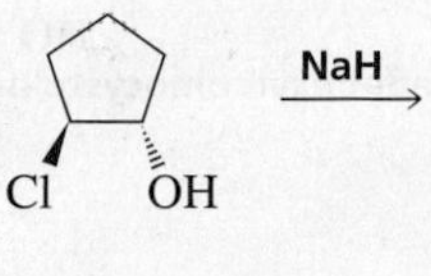

c. BrCH$_2$CH$_2$CH$_2$CH$_2$CH$_2$OH $\xrightarrow{\text{NaH}}$

d. 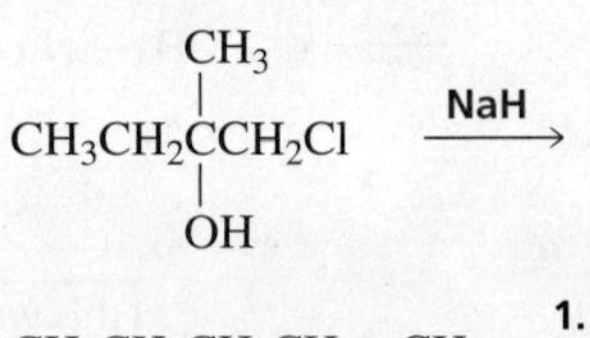 CH$_3$CH$_2\overset{\overset{\displaystyle CH_3}{|}}{\underset{\underset{\displaystyle OH}{|}}{C}}CH_2$Cl $\xrightarrow{\text{NaH}}$

b. (cyclopentane with Cl and OH) $\xrightarrow{\text{NaH}}$

e. CH$_3$CH$_2$CH$_2$CH=CH$_2$ $\xrightarrow[\text{2. NaH}]{\text{1. Cl}_2,\ \text{H}_2\text{O}}$

9 | METHYLATING AGENTS USED BY CHEMISTS VERSUS THOSE USED BY CELLS

If an organic chemist wanted to put a methyl group on a nucleophile, methyl iodide would most likely be used as the methylating agent. Of the methyl halides, methyl iodide has the most easily displaced leaving group because I^- is the weakest base of the halide ions. In addition, methyl iodide is a liquid at room temperature, so it is easier to handle than methyl bromide or methyl chloride, which are gases at room temperature. The reaction would be a simple S_N2 reaction.

$$\ddot{Nu} \; + \; CH_3{-}I \; \longrightarrow \; CH_3{-}Nu \; + \; I^-$$

Methyl halides, however, are not available in a living cell. Because they are only slightly soluble in water, they are not found in the predominantly aqueous environments of biological systems. Instead, cells use *S*-adenosylmethionine (SAM; also called AdoMet), a water-soluble compound, as a methylating agent.

Eradicating Termites

Alkyl halides can be very toxic to biological organisms. For example, bromomethane is used to kill termites and other pests. Bromomethane works by methylating the NH_2 and SH groups of enzymes, thereby destroying the enzymes' ability to catalyze necessary biological reactions. Unfortunately, bromomethane has been found to deplete the ozone layer, so its production has recently been banned in developed countries; developing countries will have until 2015 to phase out its use.

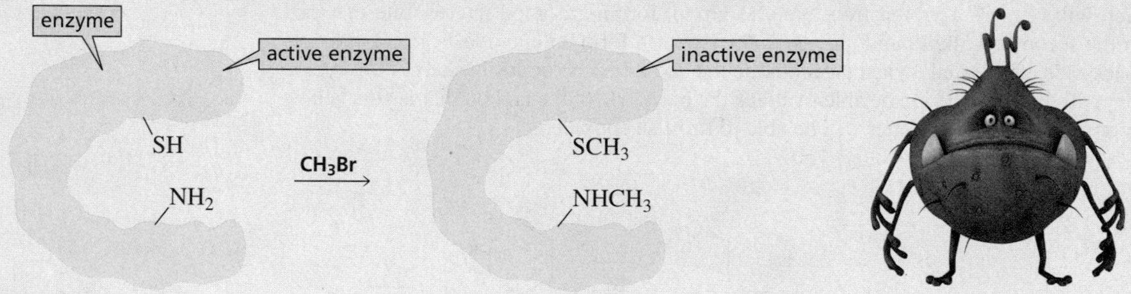

Although SAM is a much larger and more complicated looking molecule than methyl iodide, it performs the same function—namely, it transfers a methyl group to a nucleophile. Biological molecules are typically more complex than the molecules chemists use because of the need for molecular recognition.

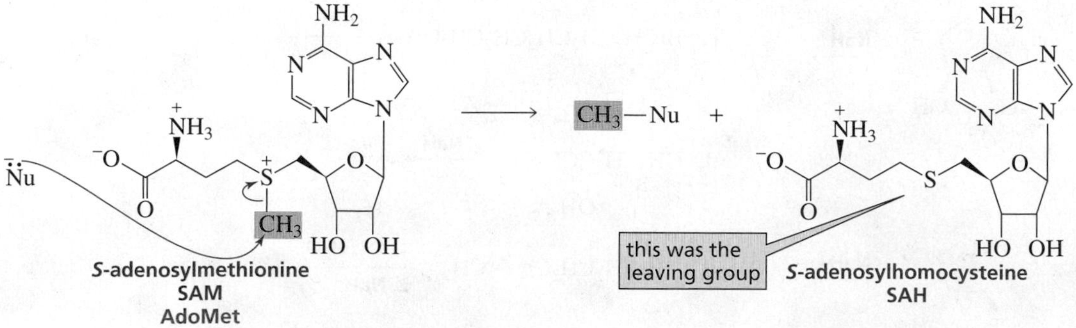

Notice that the methyl group of SAM is attached to a positively charged sulfur, which can readily accept the electrons left behind when the methyl group is transferred. In other words, the methyl group is attached to a very good leaving group, allowing biological methylation to take place at a reasonable rate.

A specific example of a biological methylation reaction that uses SAM is the conversion of noradrenaline (norepinephrine) to adrenaline (epinephrine). The reaction uses SAM to provide the methyl group. Noradrenaline and adrenaline are hormones that stimulate the breakdown of glycogen—the body's primary fuel source. You may have felt this "adrenaline rush" when preparing for a challenging activity. Adrenaline is about six times more potent than noradrenaline. This methylation reaction, therefore, is very important physiologically.

experiencing an adrenaline rush

The reaction scheme shows norepinephrine/noradrenaline and SAM converting to epinephrine/adrenaline + H^+.

norepinephrine
noradrenaline

epinephrine
adrenaline

The conversion of phosphatidylethanolamine, a component of cell membranes, into phosphatidylcholine, another cell membrane component, requires three methylations by three equivalents of SAM.

$$\text{phosphatidylethanolamine} + 3\,\text{SAM} \longrightarrow \text{phosphatidylcholine} + 3\,\text{SAH}$$

phosphatidylethanolamine

phosphatidylcholine

S-Adenosylmethionine: A Natural Antidepressant

Marketed under the name SAMe (pronounced Sammy), S-adenosylmethionine is sold in many health food and drug stores as a treatment for depression and arthritis. Although SAMe has been used clinically in Europe for more than two decades, it has not been rigorously evaluated in the United States and therefore has not been approved by the FDA. It can be sold, however, because the FDA does not prohibit the sale of most naturally occurring substances as long as the marketer does not make therapeutic claims.

SAMe has also been found to be effective in the treatment of liver diseases, such as those caused by alcohol and the hepatitis C virus. The attenuation of injury to the liver is accompanied by an increase in the concentration of glutathione in the liver. Glutathione is an important biological antioxidant. SAM is required for the biosynthesis of cysteine, one of the 20 most common naturally occurring amino acids, which is required for the biosynthesis of glutathione.

PROBLEM 34

Propose a mechanism for the following reaction:

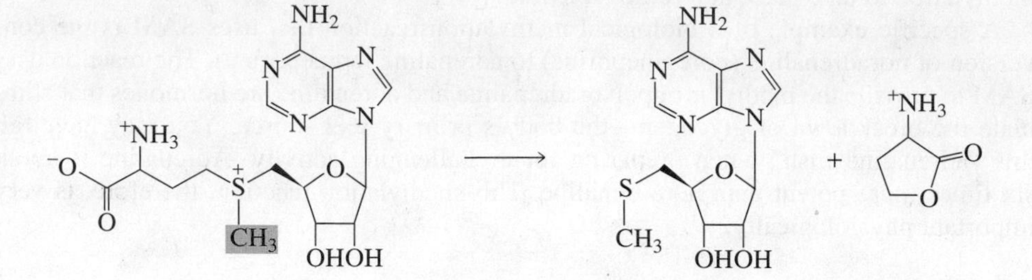

SOME IMPORTANT THINGS TO REMEMBER

- Alkyl halides undergo two kinds of **nucleophilic substitution reactions:** S_N2 and S_N1. In both reactions, a nucleophile substitutes for a halogen.

- An **S_N2 reaction** is **bimolecular:** both the alkyl halide and the nucleophile are involved in the transition state of the rate-limiting step, so the rate of the reaction depends on the concentration of both of them.

- An S_N2 reaction has a one-step mechanism: the nucleophile attacks the back side of the carbon that is attached to the halogen.

- The rate of an S_N2 reaction decreases as the size of the groups at the back side of the carbon undergoing attack increases. Therefore, the relative reactivities of alkyl halides in an S_N2 reaction are 1° > 2° > 3°. Tertiary alkyl halides cannot undergo S_N2 reactions.

- An S_N2 reaction takes place with **inversion of configuration.**

- An S_N1 reaction is **unimolecular;** only the alkyl halide is involved in the transition state of the rate-limiting step, so the rate of the reaction is dependent only on the concentration the alkyl halide.

- An S_N1 reaction has a two-step mechanism: the halogen departs in the first step, forming a carbocation intermediate that is attacked by a nucleophile in the second step. Most S_N1 reactions are **solvolysis** reactions, meaning the solvent is the nucleophile.

- The rate of an S_N1 reaction depends on the ease of carbocation formation. Tertiary alkyl halides and all allylic and benzylic halides are the only ones that form relatively stable carbocations, so they are the only ones that undergo S_N1 reactions.

- An S_N1 reaction takes place with **racemization.**

- The relative reactivities of alkyl halides that differ only in the halogen atom are RI > RBr > RCl > RF in *both* S_N2 and S_N1 reactions.

- **Basicity** is a measure of how well a compound shares its lone pair with a proton; **nucleophilicity** is a measure of how readily a species with a lone pair is able to attack an electron-deficient atom.

- In general, a stronger base is a better nucleophile. However, if the attacking atoms are very different in size *and* the reaction is carried out in a protic solvent, the stronger bases are poorer nucleophiles because of **ion–dipole interactions** between the ion and the solvent.

- Primary alkyl halides, secondary alkyl halides, and methyl halides undergo only S_N2 reactions.

- Tertiary alkyl halides undergo only S_N1 reactions.

- Vinylic and aryl halides undergo neither S_N2 nor S_N1 reactions.

- Benzylic and allylic halides (unless they are tertiary) undergo both S_N1 and S_N2 reactions. The S_N2 reaction is favored by a high concentration of a good nucleophile. The S_N1 reaction is favored by a poor nucleophile.

- **Protic solvents** (H_2O, ROH) have a hydrogen attached to an O or an N; **aprotic solvents** (DMF, DMSO) do not have a hydrogen attached to an O or an N.

- The **dielectric constant** of a solvent indicates how well the solvent insulates opposite charges from one another.

- Increasing the polarity of the solvent (that is, increasing its dielectric constant) will decrease the rate of the reaction if one or more reactants in the rate-determining step are charged, and it will increase the rate of the reaction if none of the reactants in the rate-determining step is charged.

- If the two functional groups of a **bifunctional molecule** can react with each other, both **intermolecular** (between two molecules) and **intramolecular** (within one molecule) **reactions** can occur. The reaction that is more likely to occur depends on the concentration of the bifunctional molecule and the size of the ring that would be formed in the intramolecular reaction.

SUMMARY OF REACTIONS

1. The S$_N$2 reaction has a one-step mechanism:

Relative reactivities: CH$_3$X > 1° > 2° > 3°. Tertiary alkyl halides cannot undergo S$_N$2 reactions. Only the inverted product is formed.

2. The S$_N$1 reaction reaction has a two-step mechanism with a carbocation intermediate.

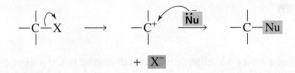

Reactivity: only tertiary alkyl halides and allylic and benzylic halides undergo S$_N$1 reactions. Products with both inverted and retained configurations are formed.

PROBLEMS

35. What product is formed when 1-bromopropane reacts with each of the following nucleophiles?
a. HO$^-$
b. $^-$NH$_2$
c. CH$_3$SH
d. HS$^-$
e. CH$_3$O$^-$
f. CH$_3$NH$_2$

36. Which member of each pair is a better nucleophile in methanol?
a. H$_2$O or HO$^-$
b. NH$_3$ or H$_2$O
c. H$_2$O or H$_2$S
d. HO$^-$ or HS$^-$
e. I$^-$ or Br$^-$
f. Cl$^-$ or Br$^-$

37. Which member in each pair in Problem 34 is a better leaving group?

38. What nucleophiles would react with 1-iodobutane to prepare the following compounds?
a. $\diagdown\diagup\diagdown$OH
d. $\diagdown\diagup\diagdown$SH
b. $\diagdown\diagup\diagdown$O$\diagup$
e. $\diagdown\diagup\diagdown$S$\diagdown$
g. $\diagdown\diagup\diagdown\diagup$N
c. $\diagdown\diagup\diagdown$N$\diagdown$ H
f. $\diagdown\diagup\diagdown$O$\diagup$ O
h. $\diagdown\diagup\diagdown\equiv\diagdown$

39. Explain how the following changes would affect the rate of the reaction of 1-bromobutane with methoxide ion in DMF.
a. The concentration of both the alkyl halide and the nucleophile are tripled.
b. The solvent is changed to ethanol.
c. The nucleophile is changed to ethanol.
d. The alkyl halide is changed to 1-chlorobutane.
e. The alkyl halide is changed to 2-bromobutane.

40. Explain how the following changes would affect the rate of the reaction of 2-bromo-2-methylbutane with methanol:
a. The alkyl halide is changed to 2-chloro-2-methylbutane.
b. The alkyl halide is changed to 2-chloro-3-methylbutane.

41. Starting with cyclohexene, how could the following compounds be prepared?
a. methoxycyclohexane
b. cyclohexylmethylamine
c. dicyclohexyl ether

42. Rank the following species in order from best nucleophile to poorest nucleophile.
a. CH$_3$CO$^-$, CH$_3$CH$_2$S$^-$, CH$_3$CH$_2$O$^-$ in methanol
c. H$_2$O and NH$_3$ in methanol
b. (ring)—O$^-$ and (ring)—O$^-$ in DMSO
d. Br$^-$, Cl$^-$, I$^-$ in methanol

453

43. The pK_a of acetic acid in water is 4.76. What effect would a decrease in the polarity of the solvent have on the pK_a? Why?

44. a. Identify the substitution products that form when 2-bromo-2-methylpropane is dissolved in a mixture of 80% ethanol and 20% water.

   b. Explain why the same products are obtained when 2-chloro-2-methylpropane is dissolved in a mixture of 80% ethanol and 20% water.

45. Draw the substitution products for each of the following reactions; if the products can exist as stereoisomers, show what stereoisomers are obtained:

   a. (R)-2-bromopentane + CH_3O^-
   b. (R)-3-bromo-3-methylheptane + CH_3OH
   c. benzyl chloride + CH_3CH_2OH

   d. allyl chloride + CH_3OH
   e. 1-bromo-2-butene + CH_3O^-
   f. 1-bromo-2-butene + CH_3OH

46. Draw the products obtained from the solvolysis of each of the following compounds in ethanol:

   a. 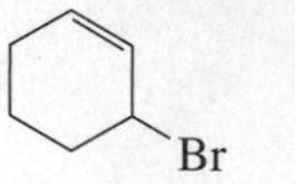   **b.** 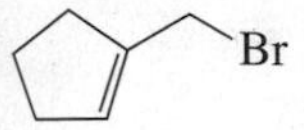   **c.**

47. Would you expect methoxide ion to be a better nucleophile if it were dissolved in CH_3OH or if it were dissolved in DMSO? Why?

48. Which reaction in each of the following pairs will take place more rapidly?

   a.

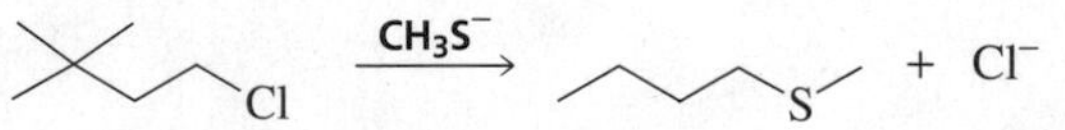

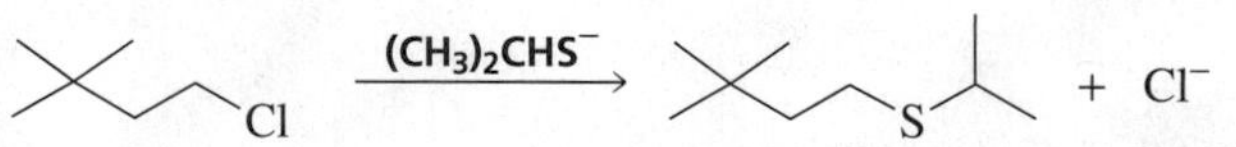

   b.

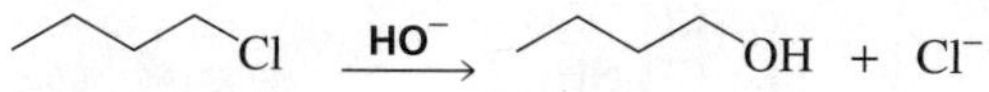

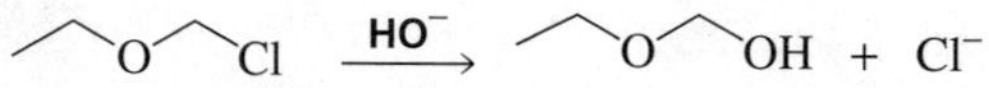

   c.

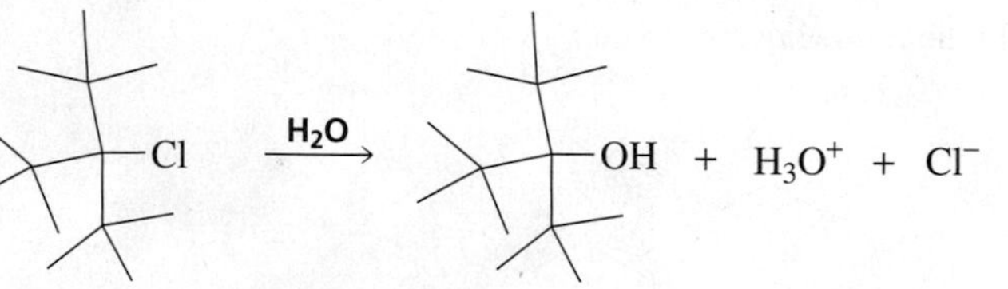

   d. $(CH_3)_3CBr \xrightarrow{H_2O} (CH_3)_3COH + H_3O^+ + Br^-$

   $(CH_3)_3CBr \xrightarrow{CH_3CH_2OH} (CH_3)_3COCH_2CH_3 + CH_3CH_2\overset{+}{O}H_2 + Br^-$

49. Alkylbenzyldimethyl ammonium chloride is a leave-on skin antiseptic used to treat such things as cuts and cold sores. It is also the antiseptic in many hand sanitizers. It is actually a mixture of compounds that differ in the number of carbons (any even number between 8 and 18) in the alkyl group. Show three different sets of reagents (each set composed of an alkyl chloride and an amine) that can be used to synthesize the alkylbenzyldimethyl ammonium chloride shown here.

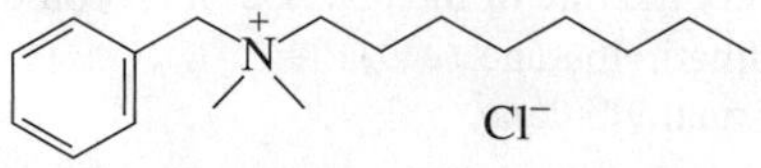

50. Fill in the blanks in the following chemical equations:

   a. 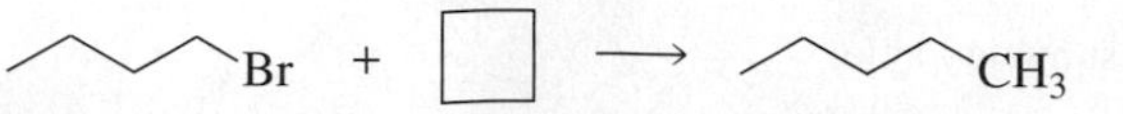   **c.**

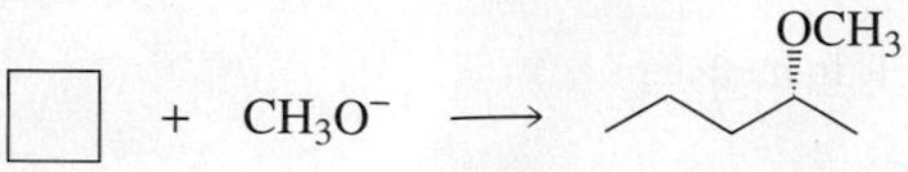

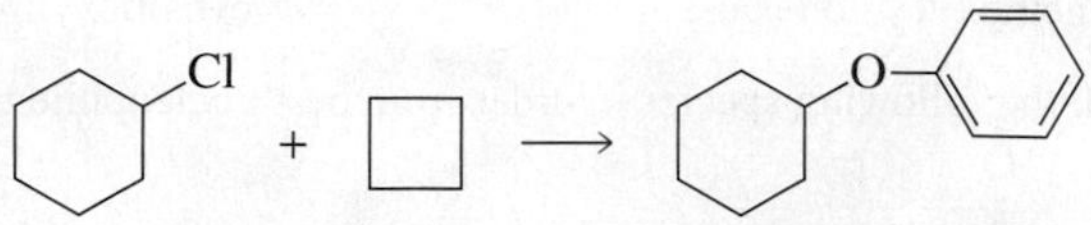

   b.   **d.**

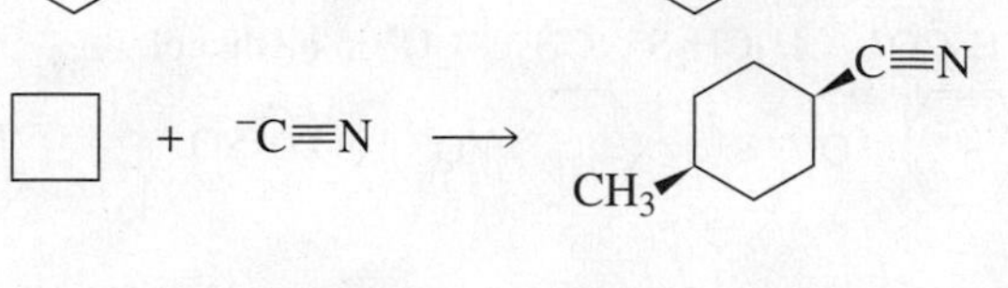

51. In Section 9, we saw that *S*-adenosylmethionine (SAM) methylates the nitrogen atom of noradrenaline to form adrenaline, a more potent hormone. If SAM methylates an OH group attached to the benzene ring instead, it completely destroys noradrenaline's activity. Show the mechanism for the methylation of the OH group by SAM.

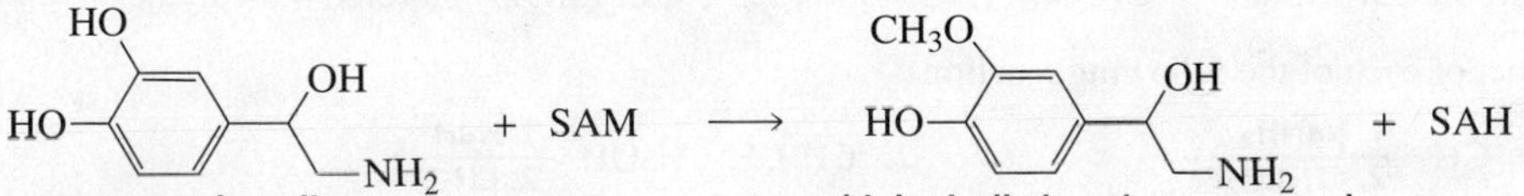

noradrenaline   a biologically inactive compound

52. *tert*-Butyl chloride undergoes solvolysis in both acetic acid and formic acid. Solvolysis occurs 5000 times faster in one of these two solvents than in the other. In which solvent is solvolysis faster? Explain your answer. (*Hint:* See Table 5.)

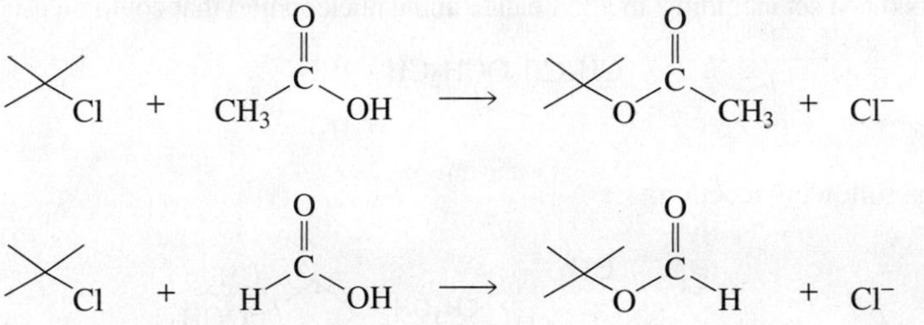

53. What substitution products are obtained when each of the following compounds is added to a solution of sodium acetate in acetic acid?
   a. 2-chloro-2-methyl-3-hexene   **b.** 3-bromo-1-methylcyclohexene

54. Show how each of the following compounds can be synthesized from the given starting materials:

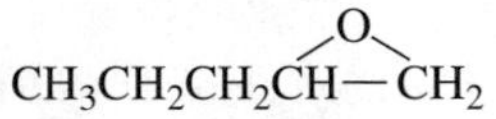

55. In which solvent—ethanol or diethyl ether—would the equilibrium for the following S_N2 reaction lie farther to the right?

$$CH_3SCH_3 + CH_3Br \rightleftharpoons CH_3\overset{\underset{\displaystyle CH_3}{|}}{\underset{+}{S}}CH_3 + Br^-$$

56. The rate of the reaction of methyl iodide with quinuclidine was measured in nitrobenzene, and then the rate of the reaction of methyl iodide with triethylamine was measured in the same solvent. The concentration of the reagents was the same in both experiments.
   a. Which reaction had the larger rate constant?
   b. The same experiment was done using isopropyl iodide instead of methyl iodide. Which reaction had the larger rate constant?
   c. Which alkyl halide has the larger $k_{quinuclidine}/k_{triethylamine}$ ratio?

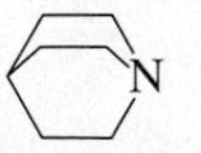

quinuclidine   triethylamine

57. Two bromoethers are obtained from the reaction of the following alkyl dihalide with methanol. Draw the structures of the ethers.

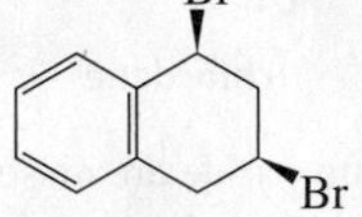

58. The rate constant of an intramolecular reaction depends on the size of the ring (*n*) that is formed. Explain the relative rates of formation of the cyclic secondary ammonium ions.

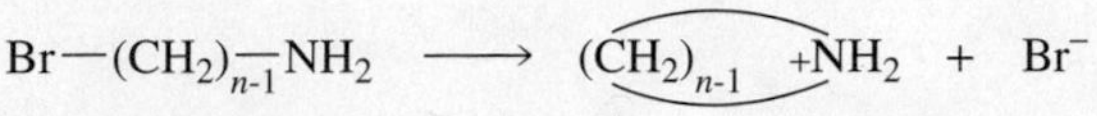

$$Br-(CH_2)_{n-1}-NH_2 \longrightarrow (CH_2)_{n-1} \overset{+}{N}H_2 + Br^-$$

$n =$	3	4	5	6	7	10
relative rate:	1×10^{-1}	2×10^{-3}	100	1.7	3×10^{-3}	1×10^{-8}

59. Draw the substitution products for each of the following reactions, assuming that all the reactions are carried out under S_N2 conditions. If the products can exist as stereoisomers, show what stereoisomers are formed:

a. (3S,4S)-3-bromo-4-methylhexane + CH_3O^-

b. (3S,4R)-3-bromo-4-methylhexane + CH_3O^-

c. (3R,4R)-3-bromo-4-methylhexane + CH_3O^-

d. (3R,4S)-3-bromo-4-methylhexane + CH_3O^-

60. a. Draw the product of each of the following reactions:

1. $CH_3CH_2C{\equiv}CH$ $\xrightarrow[\text{2. } CH_3CH_2Br]{\text{1. } NaNH_2}$

2. $CH_3CH_2CH_2OH$ $\xrightarrow[\text{2. } CH_3Br]{\text{1. } NaH}$

b. Give two sets of reactants (each set including an alkyl halide and a nucleophile) that could be used to synthesize the following alkyne:

$$CH_3CH_2C{\equiv}CCH_2CH_2CH_2CH_3$$

c. Give two sets of reactants (each set including an alkyl halide and a nucleophile) that could be used to synthesize the following ether:

$$CH_3CH_2OCH_2CH_2\underset{\underset{\displaystyle CH_3}{|}}{C}HCH_3$$

61. Propose a mechanism for the following reaction:

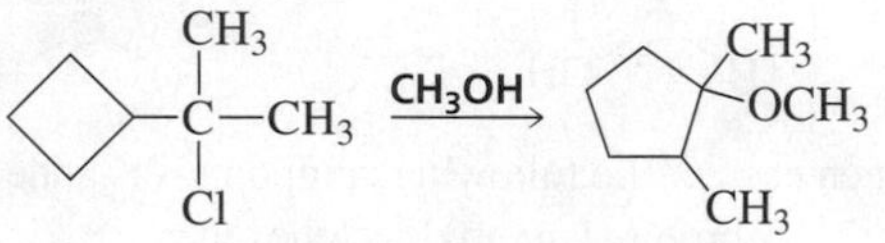

62. Explain why tetrahydrofuran can solvate a positively charged species better than diethyl ether can.

tetrahydrofuran $CH_3CH_2OCH_2CH_3$

diethyl ether

63. The reaction of an alkyl chloride with potassium iodide is generally carried out in acetone to maximize the amount of alkyl iodide that is formed. Why does the solvent increase the yield of alkyl iodide? (*Hint:* Potassium iodide is soluble in acetone, but potassium chloride is not.)

64 a. Propose a mechanism for the following reaction.

b. Explain why two products are formed.

c. Explain why methanol substitutes for only one of the bromines.

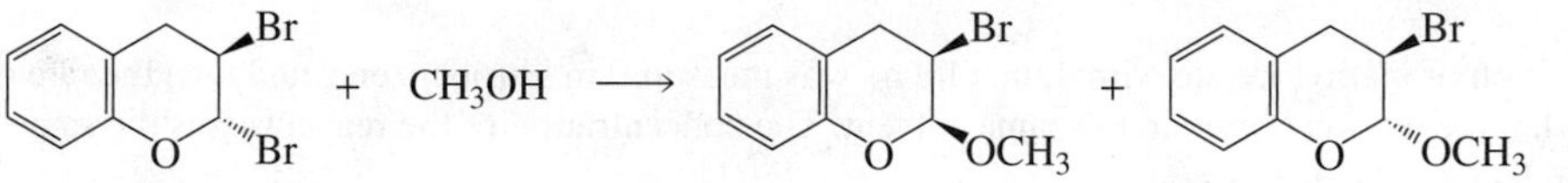

65. Chlordane, like DDT, is an organohalide that was used as an insecticide for crops such as corn and citrus and for lawns. In 1983, it was banned for all uses except against termites, and in 1988 it was banned for use against termites as well. Chlordane can be synthesized from two reactants in one step. One of the reactants is hexachlorocyclopentadiene. What is the other reactant?

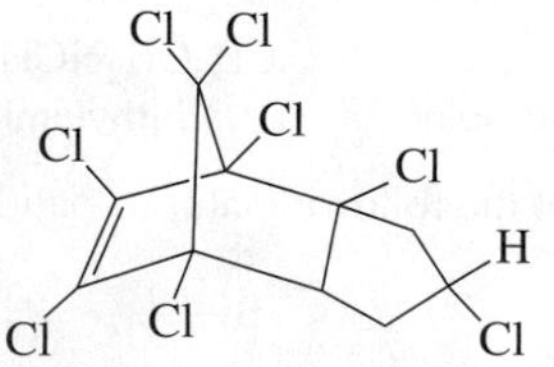

chlordane

66. When equivalent amounts of methyl bromide and sodium iodide are dissolved in methanol, the concentration of iodide ion quickly decreases and then slowly returns to its original concentration. Account for this observation.

67. Explain why the following alkyl halide does not undergo a substitution reaction, regardless of the conditions under which the reaction is run.

68. The reaction of chloromethane with hydroxide ion at 30 °C has a $\Delta G°$ value of −21.7 kcal/mol. What is the equilibrium constant for the reaction?

GLOSSARY

aprotic solvent a solvent that does not have a hydrogen bonded to an oxygen or to a nitrogen.

back-side attack nucleophilic attack on the side of the carbon opposite the side bonded to the leaving group.

base² a heterocyclic compound (a purine or a pyrimidine) in DNA and RNA.

basicity the tendency of a compound to share its electrons with a proton.

bifunctional molecule a molecule with two functional groups.

bimolecular reaction (second-order reaction) a reaction whose rate depends on the concentration of two reactants.

complete racemization the formation of a pair of enantiomers in equal amounts.

dielectric constant a measure of how well a solvent can insulate opposite charges from one another.

elimination reaction a reaction that involves the elimination of atoms (or molecules) from the reactant.

exhaustive methylation reaction of an amine with excess methyl iodide to form a quaternary ammonium iodide.

intermolecular reaction a reaction that takes place between two molecules.

intimate ion pair pair such that the covalent bond that joined the cation and the anion has broken, but the cation and anion are still next to each other.

intramolecular reaction a reaction that takes place within a molecule.

inversion of configuration turning the configuration of a carbon inside out like an umbrella in a windstorm, so that the resulting product has a configuration opposite that of the reactant.

ion–dipole interaction the interaction between an ion and the dipole of a molecule.

kinetics the field of chemistry that deals with the rates of chemical reactions.

leaving group the group that is displaced in a nucleophilic substitution reaction.

nucleophile an electron-rich atom or molecule.

nucleophilicity a measure of how readily an atom or a molecule with a pair of nonbonding electrons attacks an atom.

nucleophilic substitution reaction a reaction in which a nucleophile substitutes for an atom or a group.

partial racemization formation of a pair of enantiomers in unequal amounts.

protic solvent a solvent that has a hydrogen bonded to an oxygen or a nitrogen.

rate constant a measure of how easy or difficult it is to reach the transition state of a reaction (to get over the energy barrier to the reaction).

second-order reaction (bimolecular reaction) a reaction whose rate depends on the concentration of two reactants.

S_N1 reaction a unimolecular nucleophilic substitution reaction.

S_N2 reaction a bimolecular nucleophilic substitution reaction.

solvolysis reaction with the solvent.

steric effects effects due to the fact that groups occupy a certain volume of space.

steric hindrance refers to bulky groups at the site of a reaction that make it difficult for the reactants to approach each other.

α-substitution reaction a reaction that puts a substituent on an α-carbon in place of an α-hydrogen.

PHOTO CREDITS

Photo credits are listed in order of appearance.

Elimination Reactions of Alkyl Halides • Competition Between Substitution and Elimination

Elimination Reactions of Alkyl Halides • Competition Between Substitution and Elimination

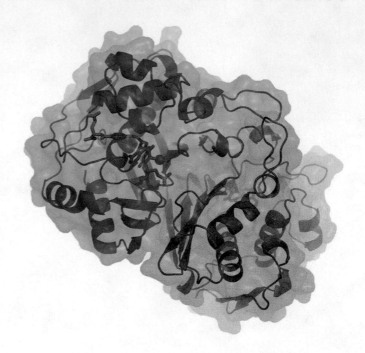

Cancer is characterized by the uncontrolled growth of cells. Because cancer cells cannot multiply if they cannot synthesize DNA, scientists have searched for compounds that interfere with DNA synthesis. 5-Fluoruracil (yellow molecule), one of the oldest drugs used in cancer chemotherapy, is such a compound. It acts by binding to the enzyme (purple and green molecule), that catalyzes the synthesis of thymidine, an essential component of DNA. Once bound to the enzyme, 5-fluorouracil prevents an elimination reaction that is required for the continued synthesis of thymidine.

In addition to undergoing the nucleophilic substitution reactions, alkyl halides also undergo elimination reactions. In an **elimination reaction,** atoms or groups are removed from a reactant.

$$CH_3CH_2CH_2X \; + \; Y^- \quad\begin{cases} \xrightarrow{\text{substitution}} CH_3CH_2CH_2Y \; + \; X^- \\ \\ \xrightarrow{\text{elimination}} CH_3CH{=}CH_2 \; + \; HY \; + \; X^- \end{cases}$$

new double bond

The product of an elimination reaction is an alkene.

Notice that, when an alkyl halide undergoes an elimination reaction, the halogen (X) is removed from one carbon and a hydrogen is removed from an adjacent carbon. A double bond is formed between the two carbons from which the atoms are eliminated. Therefore, *the product of an elimination reaction is an alkene.*

We will start this chapter by looking at the elimination reactions of alkyl halides. Then we will examine the factors that determine whether a given alkyl halide will undergo a substitution reaction, an elimination reaction, or both substitution and elimination reactions.

Naturally Occurring Organohalides That Defend against Predators

For a long time chemists thought that only a few organic compounds containing halogen atoms (organohalides) were found in nature. Now, however, over 5000 naturally occurring organohalides are known. Several marine organisms, including sponges, corals, and algae, synthesize organohalides that they use to deter predators. For example, red algae synthesize a toxic, foul-tasting organohalide that keeps predators from eating them. One predator that is not deterred, however, is a mollusk called a sea hare. After consuming red algae, a sea hare converts the algae's organohalide into a structurally similar compound that the sea hare uses for its own defense. Unlike other mollusks, a sea hare does not have a shell. Its method of defense is to surround itself with a slimy substance that contains the organohalide, thereby protecting itself from carnivorous fish.

Humans also synthesize organohalides to defend against infection. The human immune system has an enzyme that kills invading bacteria—another kind of predator—by halogenating them.

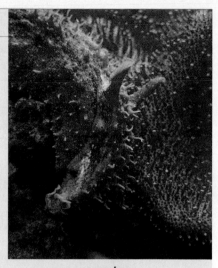

a sea hare

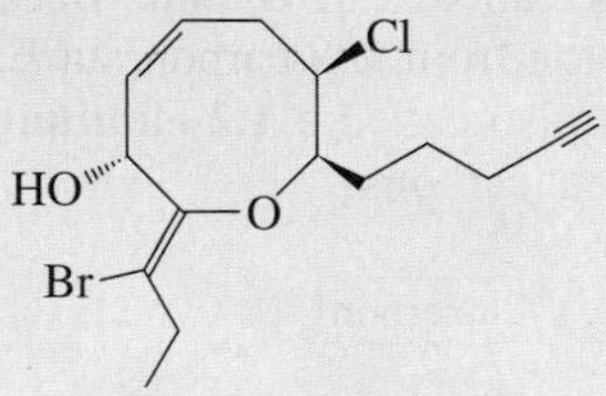

synthesized by red algae 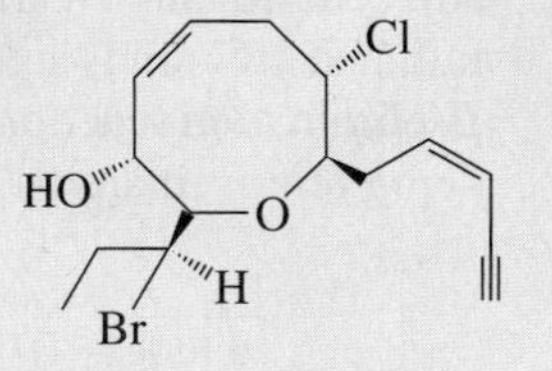   **synthesized by the sea hare**

1 THE E2 REACTION

Just as there are two nucleophilic substitution reactions, S_N1 and S_N2, there are two important elimination reactions, E1 and E2. The following reaction is an example of an **E2 reaction,** where "E" stands for *elimination* and "2" stands for *bimolecular.*

$$CH_3-\underset{\underset{Br}{|}}{\overset{\overset{CH_3}{|}}{C}}-CH_3 + HO^- \longrightarrow CH_2=\underset{}{\overset{\overset{CH_3}{|}}{C}}-CH_3 + H_2O + Br^-$$

2-bromo-2-methylpropane   **2-methylpropene**

The rate of an E2 reaction depends linearly on the concentrations of both the alkyl halide and the base (in this case, hydroxide ion).

$$\text{rate} = k[\text{alkyl halide}][\text{base}]$$

The rate law tells us that both the alkyl halide and the base are involved in the transition state of the rate-determining step, indicating a one-step reaction. The following mechanism—which portrays an E2 reaction as a concerted one-step reaction—agrees with the observed second-order kinetics:

MECHANISM FOR THE E2 REACTION

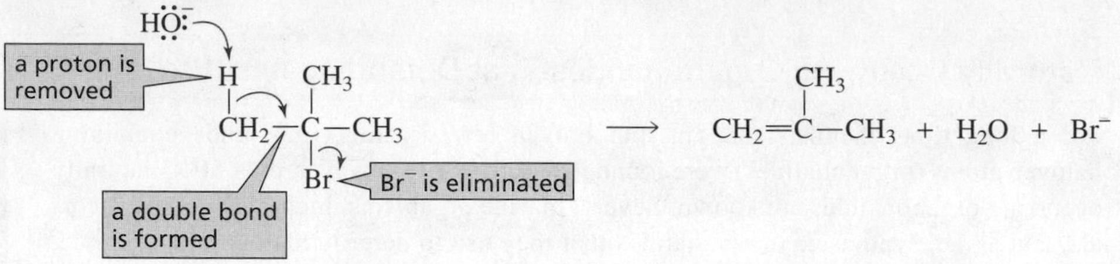

$$
\begin{array}{c}
\text{HO:}^- \\
\boxed{\text{a proton is removed}} \quad \underset{\underset{\displaystyle\boxed{\text{a double bond is formed}}}{\overset{\displaystyle\text{H}}{|}}{\text{CH}_2}\!-\!\underset{\underset{\displaystyle\text{Br}}{\overset{}{|}}}{\overset{\overset{\displaystyle\text{CH}_3}{|}}{\text{C}}}\!-\!\text{CH}_3 \quad \boxed{\text{Br}^- \text{ is eliminated}}
\end{array}
\longrightarrow \quad
\text{CH}_2\!\!=\!\!\underset{}{\overset{\overset{\displaystyle\text{CH}_3}{|}}{\text{C}}}\!-\!\text{CH}_3 \;+\; \text{H}_2\text{O} \;+\; \text{Br}^-
$$

- The base removes a proton from a carbon that is adjacent to the carbon bonded to the halogen. As the proton is removed, the electrons that it shared with carbon move toward the carbon that is bonded to the halogen. As these electrons move toward the carbon, the halogen leaves (because carbon can form no more than four bonds), taking its bonding electrons with it.

When the reaction is over, the electrons that were originally bonded to the hydrogen in the reactant have formed a π bond in the product. The removal of a proton and a halide ion is called **dehydrohalogenation.**

The carbon to which the halogen is attached is called the **α-carbon.** A carbon adjacent to an α-carbon is called a **β-carbon.** Because the elimination reaction is initiated by removing a proton from a β-carbon, an E2 reaction is sometimes called a **β-elimination reaction.** It is also called a **1,2-elimination reaction** because the atoms being removed are on adjacent carbons.

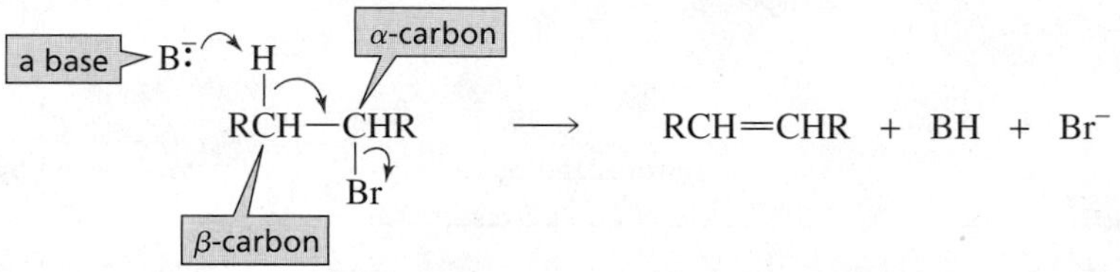

$$
\boxed{\text{a base}}\!\!-\!\!\text{B:}^- \overset{\displaystyle\frown}{\ } \underset{\underset{\displaystyle\boxed{\beta\text{-carbon}}}{}}{\overset{\overset{\displaystyle\text{H}}{|}}{\text{RCH}}}\!-\!\underset{\underset{\displaystyle\text{Br}}{\overset{}{|}}}{\overset{\boxed{\alpha\text{-carbon}}}{\text{CHR}}} \quad \longrightarrow \quad \text{RCH}\!\!=\!\!\text{CHR} \;+\; \text{BH} \;+\; \text{Br}^-
$$

In a series of alkyl halides that have the same alkyl group, alkyl iodides are the most reactive and alkyl fluorides the least reactive in E2 reactions because weaker bases are better leaving groups.

The weaker the base, the better it is as a leaving group.

relative reactivities of alkyl halides in an E2 reaction

$$
\boxed{\text{most reactive}}\!\!- \text{RI} \,>\, \text{RBr} \,>\, \text{RCl} \,>\, \text{RF} \,-\!\!\boxed{\text{least reactive}}
$$

2 AN E2 REACTION IS REGIOSELECTIVE

An alkyl halide such as 2-bromopropane has two β-carbons from which a proton can be removed in an E2 reaction. Because the two β-carbons are identical, the proton can be removed equally easily from either one.

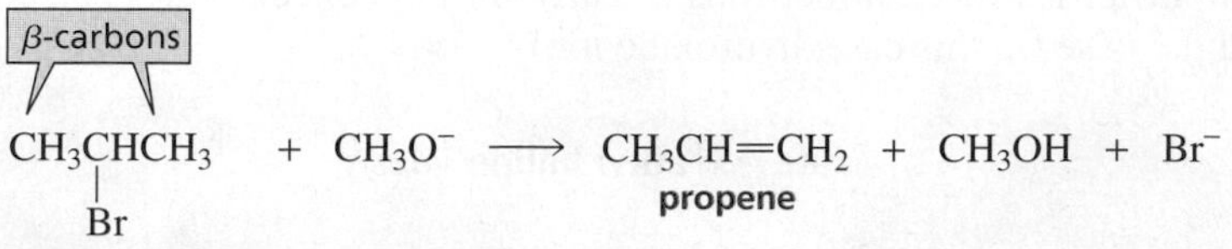

$$
\boxed{\beta\text{-carbons}}
$$

$$
\underset{\underset{\displaystyle\text{Br}}{\overset{}{|}}}{\text{CH}_3\text{CHCH}_3} \;+\; \text{CH}_3\text{O}^- \;\longrightarrow\; \underset{\displaystyle\textbf{propene}}{\text{CH}_3\text{CH}\!\!=\!\!\text{CH}_2} \;+\; \text{CH}_3\text{OH} \;+\; \text{Br}^-
$$

2-bromopropane

In contrast, 2-bromobutane has two structurally different β-carbons from which a proton can be removed. Therefore, when this alkyl halide reacts with a base, two elimination products are formed: 2-butene (80%) and 1-butene (20%). Thus, this E2 reaction is *regioselective* because more of one constitutional isomer is formed than of the other.

β-carbons

$$CH_3CHCH_2CH_3 \;+\; CH_3O^- \;\xrightarrow{\;CH_3OH\;}\; CH_3CH{=}CHCH_3 \;+\; CH_2{=}CHCH_2CH_3 \;+\; CH_3OH \;+\; Br^-$$

	Br		

2-bromobutane   2-butene   1-butene
                                  80%   20%
                        (mixture of *E* and *Z*)

Alkyl Chlorides, Alkyl Bromides, and Alkyl Iodides Preferentially Form the More Stable Product

What factors dictate which of the two alkenes will be formed in greater yield? In other words, what causes the regioselectivity of an E2 reaction? We can answer this question by looking at the reaction coordinate diagram in Figure 1.

In the transition state leading to the alkene, the C—H and C—Br bonds are partially broken and the double bond is partially formed (the partially broken and partially formed

the more stable alkene is the major product because the transition state leading to its formation is more stable

1-butene + CH$_3$OH + Br$^-$
2-butene + CH$_3$OH + Br$^-$

2-bromobutane + CH$_3$O$^-$

Free energy

Progress of the reaction

◀ **Figure 1**

The major product of the E2 reaction of 2-bromobutane and methoxide ion is 2-butene (indicated by the blue line), because the transition state leading to its formation is more stable than the transition state leading to formation of 1-butene (indicated by the red line).

bonds are indicated by dashed lines), giving the transition state an alkene-like structure. Therefore, factors that stabilize the alkene will also stabilize the alkene-like transition state, allowing the more stable alkene to be formed faster.

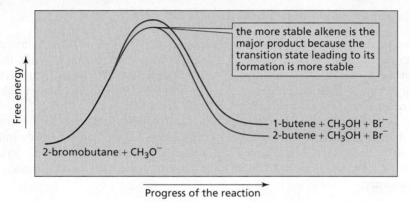

transition state leading to 2-butene

more stable

transition state leading to 1-butene

less stable

Figure 1 shows that the difference in the rate of formation of the two alkenes is not very great. Consequently, both are formed, but the *more stable* alkene is the major product. The stability of an alkene depends on the number of alkyl substituents bonded to its sp^2 carbons: the greater the number of alkyl substituents, the more stable the alkene. Therefore, 2-butene, with two methyl substituents bonded to its sp^2 carbons, is more stable than 1-butene, with one ethyl substituent. Thus, 2-butene is the major product.

The following reaction also forms two elimination products. Because 2-methyl-2-butene is the more substituted alkene (it has a greater number of alkyl substituents bonded to its sp^2 carbons), it is the more stable of the two alkenes and, therefore, is the major product of the elimination reaction.

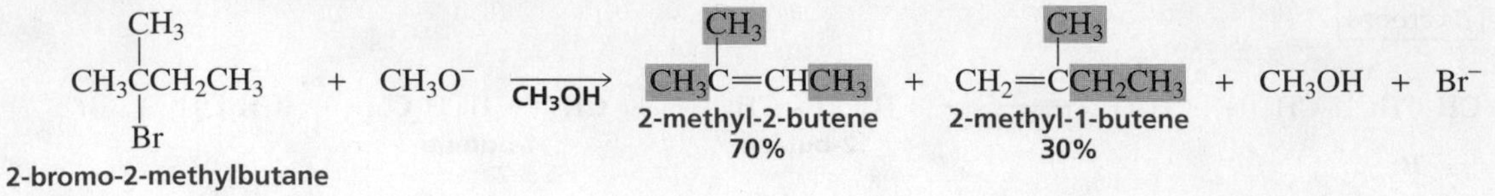

$$\underset{\substack{\text{2-bromo-2-methylbutane}}}{\overset{\substack{CH_3 \\ | \\ CH_3CCH_2CH_3 \\ | \\ Br}}{}} + CH_3O^- \xrightarrow[CH_3OH]{} \underset{\substack{\text{2-methyl-2-butene} \\ \text{70\%}}}{CH_3C{=}CHCH_3} + \underset{\substack{\text{2-methyl-1-butene} \\ \text{30\%}}}{CH_2{=}CCH_2CH_3} + CH_3OH + Br^-$$

Alexander M. Zaitsev, a nineteenth-century Russian chemist, devised a shortcut to predict the more substituted alkene product. He pointed out *that the more substituted alkene is obtained when a hydrogen is removed from the β-carbon that is bonded to the fewest hydrogens.* This is called Zaitsev's rule.

The major product of an E2 reaction is generally the more stable alkene.

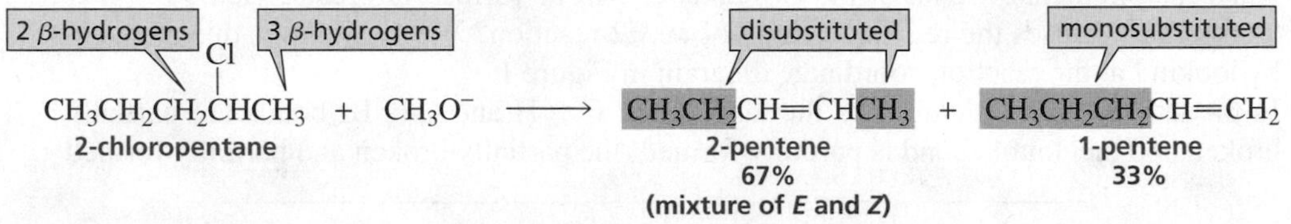

| 2 β-hydrogens | 3 β-hydrogens | | disubstituted | monosubstituted |

$$\underset{\text{2-chloropentane}}{CH_3CH_2CH_2CHCH_3} + CH_3O^- \longrightarrow \underset{\substack{\text{2-pentene} \\ \text{67\%} \\ \text{(mixture of } E \text{ and } Z\text{)}}}{CH_3CH_2CH{=}CHCH_3} + \underset{\substack{\text{1-pentene} \\ \text{33\%}}}{CH_3CH_2CH_2CH{=}CH_2}$$

For example, in the preceding reaction, one β-carbon is bonded to three hydrogens and the other β-carbon is bonded to two hydrogens. According to **Zaitsev's rule,** the more substituted alkene will be the one formed by removing a proton from the β-carbon bonded to two hydrogens. Therefore, 2-pentene (a disubstituted alkene) is the major product and 1-pentene (a monosubstituted alkene) is the minor product.

Because elimination from a tertiary alkyl halide typically leads to a more substituted alkene than does elimination from a secondary alkyl halide, and because elimination from a secondary alkyl halide generally leads to a more substituted alkene than does elimination from a primary alkyl halide, the relative reactivities of alkyl halides in an E2 reaction are:

relative reactivities of alkyl halides in an E2 reaction

tertiary alkyl halide > secondary alkyl halide > primary alkyl halide

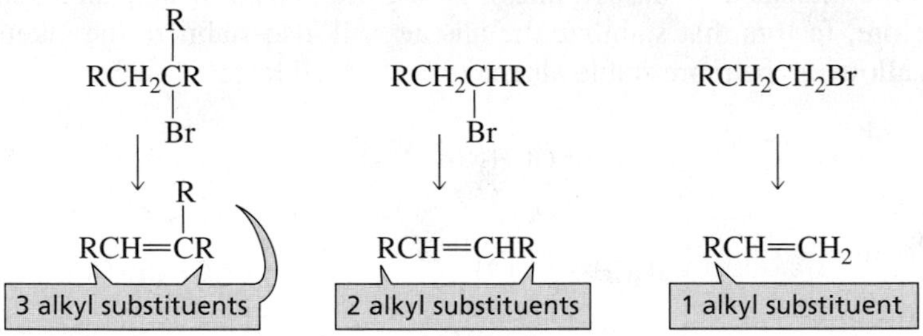

$$\underset{\substack{Br}}{\overset{\substack{R \\ | \\ RCH_2CR \\ |}}{}} \qquad \underset{\substack{| \\ Br}}{RCH_2CHR} \qquad RCH_2CH_2Br$$

$$\downarrow \qquad\qquad \downarrow \qquad\qquad \downarrow$$

$$\underset{\substack{R \\ | \\ RCH{=}CR}}{} \qquad RCH{=}CHR \qquad RCH{=}CH_2$$

| 3 alkyl substituents | 2 alkyl substituents | 1 alkyl substituent |

PROBLEM 1♦

What is the major elimination product obtained from the reaction of each of the following alkyl halides with hydroxide ion?

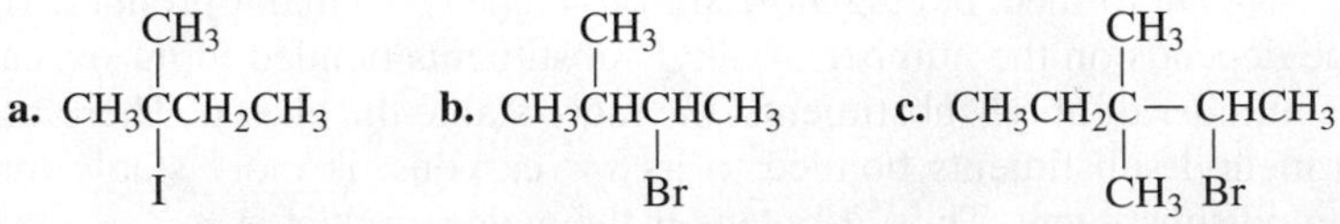

a. $\underset{\substack{| \\ I}}{\overset{\substack{CH_3 \\ | \\ CH_3CCH_2CH_3}}{}}$ **b.** $\underset{\substack{| \\ Br}}{\overset{\substack{CH_3 \\ | \\ CH_3CHCHCH_3}}{}}$ **c.** $\underset{\substack{| \\ CH_3}}{\overset{\substack{CH_3 \\ | \\ CH_3CH_2C}}{}} {-} \underset{\substack{| \\ Br}}{CHCH_3}$

Keep in mind that the major product of an E2 reaction is the *more stable alkene,* and Zaitsev's rule is just a shortcut to determine which of the possible alkene products is the *more substituted alkene.* The more substituted alkene, however, is not always the more stable alkene.

Zaitsev's rule leads to the more substituted alkene.

For example, in each of the following reactions, the major product is the alkene with conjugated double bonds because it is the more stable alkene, even though it is not the more substituted alkene.

$$CH_2=CHCH_2CHCHCH_3 \xrightarrow{CH_3O^-} CH_2=CHCH=CHCHCH_3 + CH_2=CHCH_2CH=CCH_3 + CH_3OH + Cl^-$$

conjugated double bonds

4-chloro-5-methyl-1-hexene

5-methyl-1,3-hexadiene
a conjugated diene
major product

5-methyl-1,4-hexadiene
an isolated diene
minor product

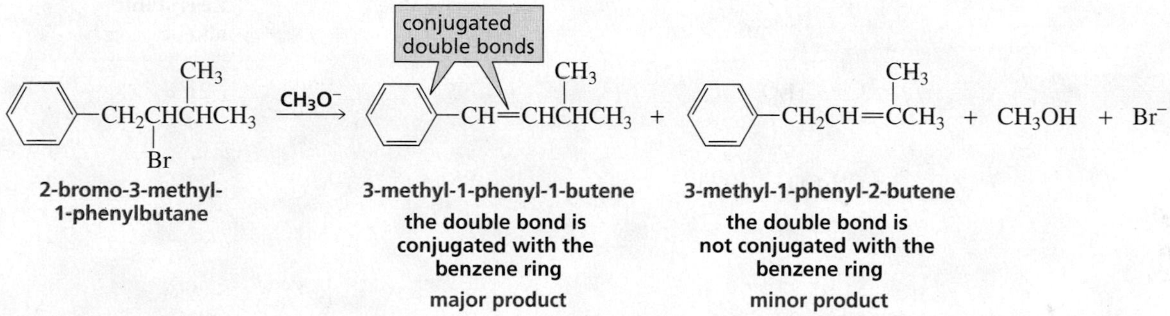

$$\text{Ph}-CH_2CHCHCH_3 \xrightarrow{CH_3O^-} \text{Ph}-CH=CHCHCH_3 + \text{Ph}-CH_2CH=CCH_3 + CH_3OH + Br^-$$

conjugated double bonds

2-bromo-3-methyl-
1-phenylbutane

3-methyl-1-phenyl-1-butene
the double bond is
conjugated with the
benzene ring
major product

3-methyl-1-phenyl-2-butene
the double bond is
not conjugated with the
benzene ring
minor product

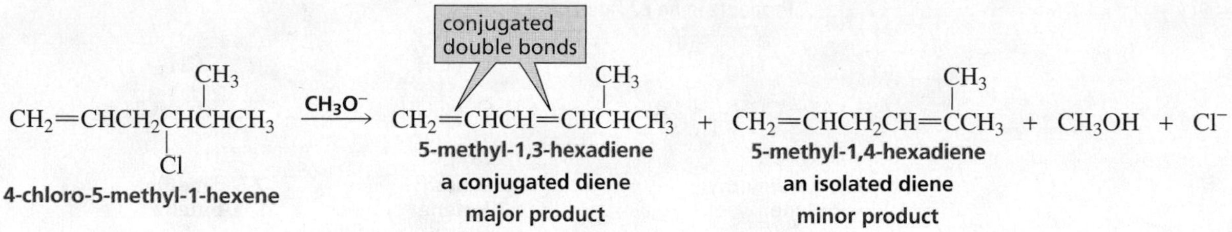

Zaitsev's rule cannot be used to predict the major products of the foregoing reactions because it does not take account of the fact that conjugated double bonds are more stable than isolated double bonds. Therefore, if the alkyl halide has a double bond or a benzene ring, do not use Zaitsev's rule to predict the major product of an elimination reaction.

Compounds with conjugated double bonds are more stable than those with isolated double bonds.

In some E2 reactions, the more stable alkene is not the major product. For example, if the base is bulky and its approach to the hydrogen that would lead to the more stable alkene is sterically hindered, it will preferentially remove the most accessible hydrogen. Therefore, the major product will be the less stable alkene.

less accessible hydrogens   bulky base

more accessible hydrogens

$$CH_3CCH_2CH_3 + CH_3CO^- \xrightarrow{(CH_3)_3COH} CH_3C=CHCH_3 + CH_2=CCH_2CH_3 + CH_3COH + Br^-$$

2-bromo-2-methylbutane   *tert*-butoxide ion

2-methyl-2-butene
28%

2-methyl-1-butene
72%

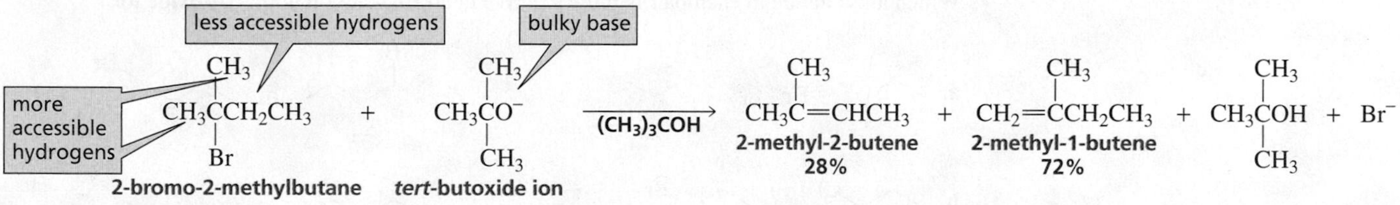

For example, in the preceding reaction, it is easier for the bulky *tert*-butoxide ion to remove one of the more exposed hydrogens from one of the two methyl groups. This leads to formation of the less stable alkene. Because the less stable alkene is more easily formed, it is the major product of the reaction.

If the alkyl halide is not sterically hindered and the base is only moderately hindered, the more stable alkene will be the major product, as expected. In other words, it takes a lot of steric hindrance for the less stable product to be the major product. Thus, the major product of the following reaction is 2-butene.

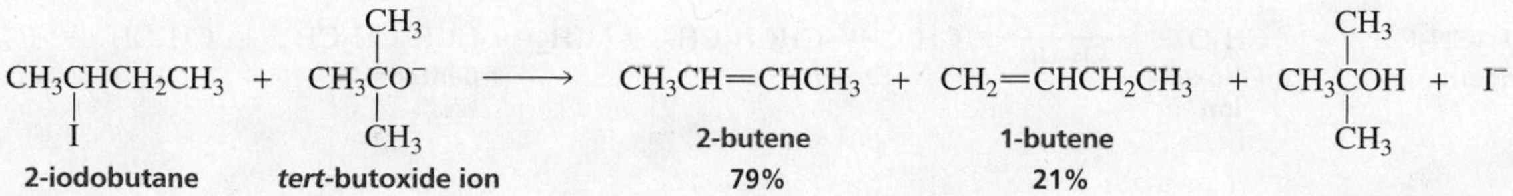

$$CH_3CHCH_2CH_3 + CH_3CO^- \longrightarrow CH_3CH=CHCH_3 + CH_2=CHCH_2CH_3 + CH_3COH + I^-$$

2-iodobutane   *tert*-butoxide ion

2-butene
79%

1-butene
21%

Table 1 shows the amount of each product obtained from the reaction of a sterically hindered alkyl halide with different bases. Notice that the percentage of the less stable (less substituted) alkene increases as the size of the base increases.

Table 1	Effect of the Steric Properties of the Base on the Distribution of Products in an E2 Reaction	

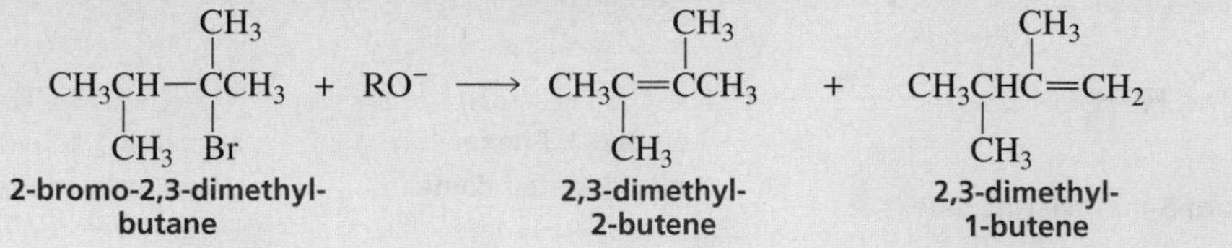

$$CH_3CH-CCH_3 + RO^- \longrightarrow CH_3C=CCH_3 + CH_3CHC=CH_2$$

2-bromo-2,3-dimethyl-butane   2,3-dimethyl-2-butene   2,3-dimethyl-1-butene

Base	More stable alkene	Less stable alkene
$CH_3CH_2O^-$	79%	21%
$CH_3\overset{\underset{\mid}{CH_3}}{\underset{\mid}{C}}O^-$ with CH₃	27%	73%
$CH_3\overset{CH_3}{\underset{CH_2CH_3}{C}}O^-$	19%	81%
$CH_3CH_2\overset{CH_2CH_3}{\underset{CH_2CH_3}{C}}O^-$	8%	92%

PROBLEM 2♦

Which alkyl halide in each pair is more reactive in an E2 reaction with hydroxide ion?

a. 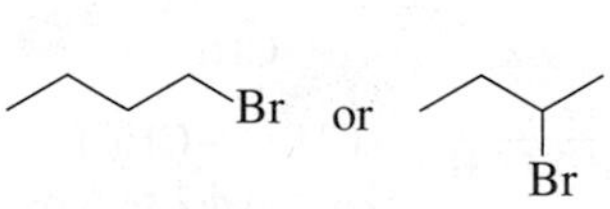 (structure) Br or (structure) Br

b. 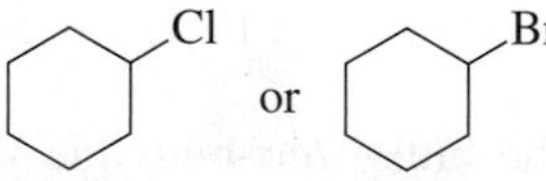 (cyclohexane with Cl) or (cyclohexane with Br)

c. 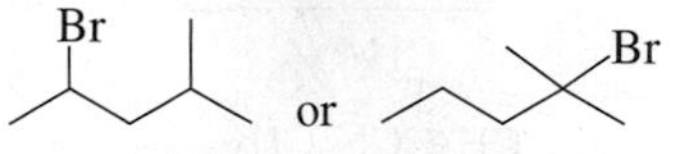 (structure with Br) or (structure with Br)

d. 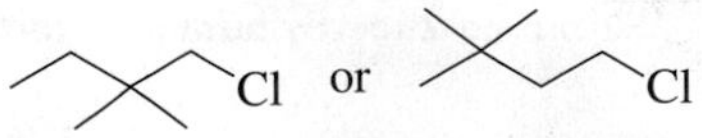 (structure with Cl) or (structure with Cl)

Alkyl Fluorides Preferentially Form the Less Stable Alkene

Although the major product of the E2 reaction of alkyl chlorides, alkyl bromides, and alkyl iodides is normally the more stable alkene, the major product of the E2 reaction of alkyl fluorides is the less stable alkene.

$$CH_3CHCH_2CH_2CH_3 + CH_3O^- \xrightarrow{CH_3OH} CH_3CH=CHCH_2CH_3 + CH_2=CHCH_2CH_2CH_3 + CH_3OH + F^-$$

2-fluoropentane   methoxide ion   2-pentene 30% (mixture of E and Z)   1-pentene 70%

Why do alkyl fluorides form the less stable alkene rather than the more stable alkene? To answer this question, we have to look at the transition states of the reactions.

When a hydrogen and a chlorine, bromine, or iodine are eliminated from an alkyl halide, the halogen starts to leave as soon as the base begins to remove the proton. Consequently, the transition state resembles an alkene (see earlier in this section).

The fluoride ion, however, is the strongest base of the halide ions and therefore the poorest leaving group. So when a base begins to remove a proton from an alkyl fluoride, fluorine does not have as strong a propensity to leave as another halogen would have. As a result, a negative charge develops on the carbon that is losing the proton, causing the transition state to resemble a carbanion rather than an alkene. To determine which of the carbanion-like transition states is more stable, we need to look at the factors that affect carbanion stability.

transition state leading to
1-pentene
more stable

transition state leading to
2-pentene
less stable

We have seen that carbocations, because they are positively charged, are *stabilized* by electron-donating alkyl groups. Thus, tertiary carbocations are the most stable, and methyl cations are the least stable.

relative stabilities of carbocations

tertiary carbocation   secondary carbocation   primary carbocation   methyl cation

Carbanions, on the other hand, are negatively charged, so they are *destabilized* by electron-donating alkyl groups. Therefore, carbanions have the opposite relative stabilities. (Differences in solvation energies also contribute to this trend, since the smaller ions are the most solvated.)

relative stabilities of carbanions

tertiary carbanion   secondary carbanion   primary carbanion   methyl anion

Carbocation stability: 3° > 2° > 1°

Carbanion stability: 1° > 2° > 3°

Looking back at the transition states for the E2 reaction of an alkyl fluoride earlier in this section, we see that the developing negative charge in the transition state leading to 1-pentene is on a primary carbon. Thus, this transition state is more stable than the transition state leading to 2-pentene, in which the developing negative charge is on a secondary carbon. Because the transition state leading to 1-pentene is more stable, 1-pentene is formed more rapidly and is the major product of the E2 reaction.

The data in Table 2 show that as the halide ion increases in basicity (and so decreases in leaving propensity), the yield of the more stable alkene product decreases. However, the more stable alkene remains the major elimination product in all cases, except when the halogen is fluorine.

We can summarize as follows: *the major product of an E2 reaction is the more stable alkene except when the reactants are sterically hindered or the leaving group is poor,* in which case the major product is the less stable alkene. In Section 7, you will see that the more stable alkene is not always the major product in the case of certain cyclic alkyl halides.

Table 2   Products Obtained From the E2 Reaction of CH_3O^- and 2-Halohexanes

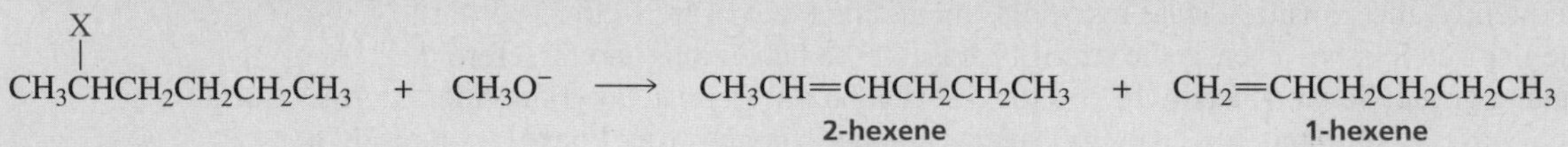

$$\underset{\underset{X}{|}}{CH_3CHCH_2CH_2CH_2CH_3} + CH_3O^- \longrightarrow CH_3CH=CHCH_2CH_2CH_3 + CH_2=CHCH_2CH_2CH_2CH_3$$

2-hexene
(mixture of *E* and *Z*)

1-hexene

Leaving group	Conjugate acid	pK_a	More stable product	Less stable product
X = I	HI	−10	81%	19%
X = Br	HBr	−9	72%	28%
X = Cl	HCl	−7	67%	33%
X = F	HF	3.2	30%	70%

PROBLEM 3♦

What is the major elimination product obtained from an E2 reaction of each of the following alkyl halides with hydroxide ion?

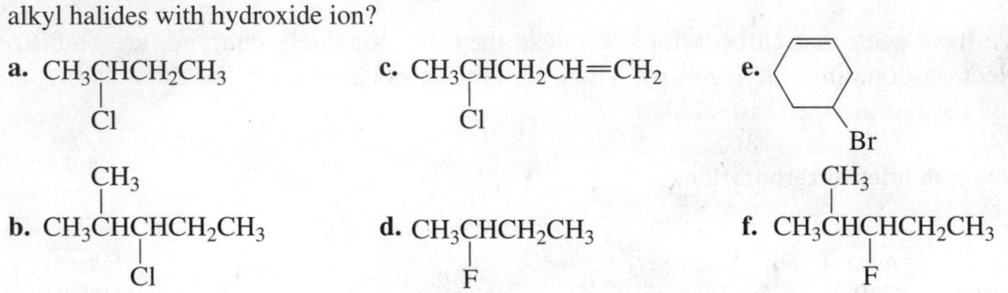

a. $\underset{\underset{Cl}{|}}{CH_3CHCH_2CH_3}$

c. $\underset{\underset{Cl}{|}}{CH_3CHCH_2CH}=CH_2$

e.

b. $\underset{\underset{Cl}{|}}{CH_3\overset{\overset{CH_3}{|}}{C}HCHCH_2CH_3}$

d. $\underset{\underset{F}{|}}{CH_3CHCH_2CH_3}$

f. $CH_3\overset{\overset{CH_3}{|}}{C}H\underset{\underset{F}{|}}{C}HCH_2CH_3$

PROBLEM 4♦

Which alkyl halide in each pair is more reactive in an E2 reaction with hydroxide ion?

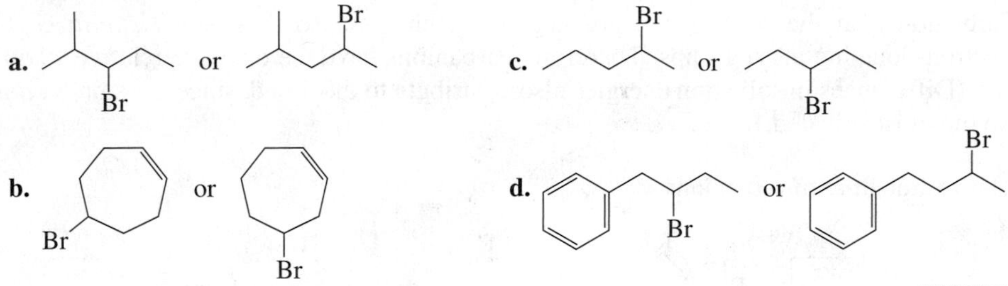

a.

b.

c.

d.

---

3   THE E1 REACTION

The second kind of elimination reaction that alkyl halides can undergo is an **E1 reaction,** where "E" stands for *elimination* and "1" stands for *unimolecular.*

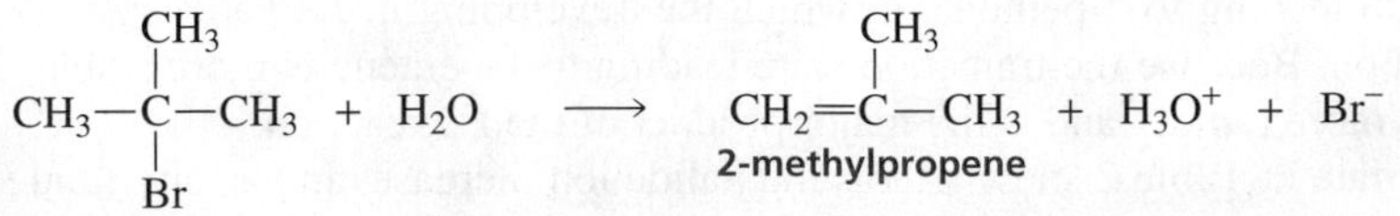

$$CH_3\overset{\overset{CH_3}{|}}{\underset{\underset{Br}{|}}{C}}CH_3 + H_2O \longrightarrow CH_2=\overset{\overset{CH_3}{|}}{C}-CH_3 + H_3O^+ + Br^-$$

2-bromo-2-methylpropane

2-methylpropene

The rate of an E1 reaction depends linearly only on the concentration of the alkyl halide.

$$rate = k[\textbf{alkyl halide}]$$

Therefore, we know that only the alkyl halide takes part in the rate-determining step of the reaction, so an E1 reaction must have at least two steps. The following mechanism agrees with the observed first-order kinetics. Because the first step is the rate-determining step, an increase in the concentration of the base—which participates only in the second step of the reaction—has no effect on the rate of the reaction.

MECHANISM FOR THE E1 REACTION

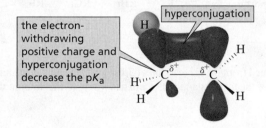

- The alkyl halide dissociates, forming a carbocation.
- The base forms the elimination product by removing a proton from a β-carbon.

The pK_a of a compound such as ethane, which has hydrogens attached only to sp^3 carbons, is >60. How, then, can a weak base such as water remove a proton from an sp^3 carbon in the second step of the preceding reaction?

First of all, the pK_a is greatly reduced by the adjacent positively charged carbon that can accept the electrons left behind when the proton is removed. Second, the β-carbon shares the positive charge as a result of hyperconjugation. Hyperconjugation drains electron density from the C—H bond, thereby weakening it. Note that hyperconjugation (where the σ electrons of a bond attached to a carbon adjacent to a positively charged carbon spread into the empty p orbital) is also responsible for the greater stability of a tertiary carbocation compared with a secondary carbocation.

When more than one alkene can be formed, the E1 reaction, like the E2 reaction, is regioselective. The major product is the *more stable alkene*.

The more stable alkene is the major product because its greater stability causes the transition state leading to its formation to be more stable (Figure 2). Therefore, it is formed more rapidly. Notice that the more stable alkene is formed by removing the hydrogen from the β-carbon bonded to the fewest hydrogens, in accordance with Zaitsev's rule.

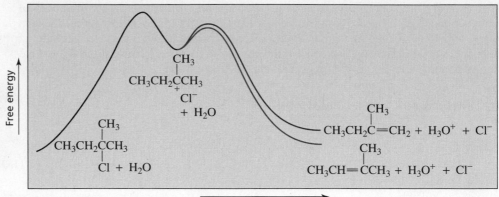

▶ Figure 2

The major product of the E1 reaction is the more stable alkene (green line) because its greater stability causes the transition state leading to its formation to be more stable.

Because the rate-determining step of an E1 reaction is carbocation formation, the rate of an E1 reaction depends both on the ease with which the carbocation is formed *and* on how readily the leaving group leaves. Therefore:

- Tertiary alkyl halides readily undergo E1 reactions because they form relatively stable carbocations, whereas primary and secondary alkyl halides *do not* undergo E1 reactions because their carbocations are less unstable. Primary and secondary alkyl halides can undergo only E2 reactions.

- For a series of alkyl halides with the same alkyl group, alkyl iodides are the most reactive and alkyl fluorides are the least reactive in an E1 reaction, because alkyl iodides have the best leaving group and alkyl fluorides have the worst leaving group (Section 1).

Tertiary alkyl halides undergo E1 and E2 reactions; primary and secondary alkyl halides undergo only E2 reactions.

relative reactivities of alkyl halides in an E1 reaction

most reactive ▷ RI > RBr > RCl > RF ◁ least reactive

PROBLEM 5♦

Four alkenes are formed from the E1 reaction of 3-bromo-2,3-dimethylpentane and methoxide ion. Draw the structures of the alkenes and rank them according to the amount that would be formed.

PROBLEM 6♦

If 2-fluoropentane were to undergo an E1 reaction, would you expect the major product to be the most stable alkene or the least stable alkene? Explain your answer.

PROBLEM 7♦

Which of the following compounds would react faster in an

a. E1 reaction?   **b.** E2 reaction?   **c.** S_N1 reaction?   **d.** S_N2 reaction?

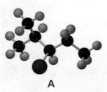

A

B

PROBLEM-SOLVING STRATEGY

Proposing a Mechanism

Propose a mechanism for the following reaction:

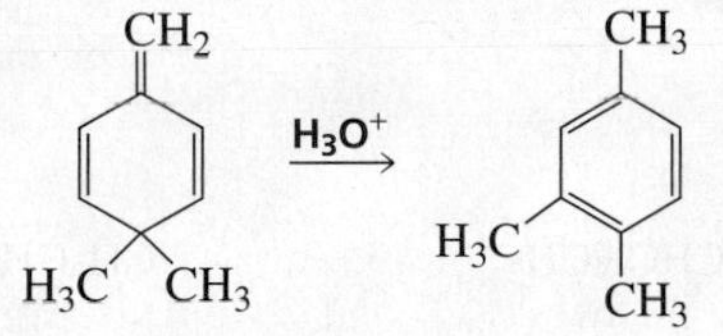

When we see that one of the reactants is an acid, we know we should start by protonating the other reactant. We need to protonate it at the position that allows the most stable carbocation to be formed. Therefore, we protonate the CH_2 group because that forms a tertiary allylic carbocation with a positive charge that is shared by three carbons as a result of electron delocalization. A 1,2-methyl shift results in a carbocation with the desired carbon skeleton. Loss of a proton gives the final product.

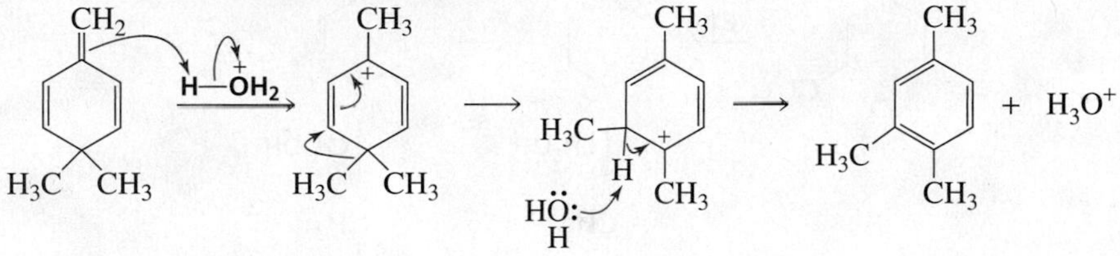

Now use the strategy you have just learned to solve Problem 8.

PROBLEM 8

Propose a mechanism for the following reaction:

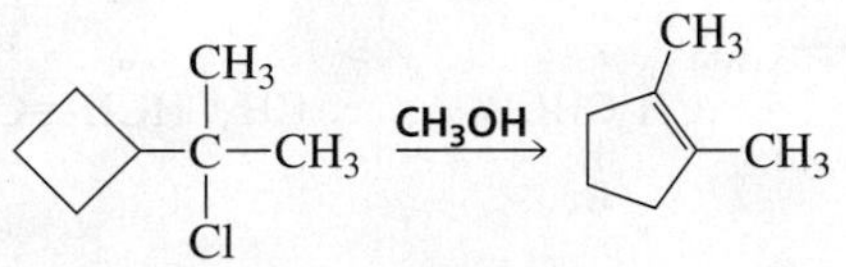

4   BENZYLIC AND ALLYLIC HALIDES

Benzylic and allylic halides readily undergo E2 reactions, because the new double bond in the product is relatively stable, and therefore easily formed, since it is conjugated with a benzene ring or with a double bond.

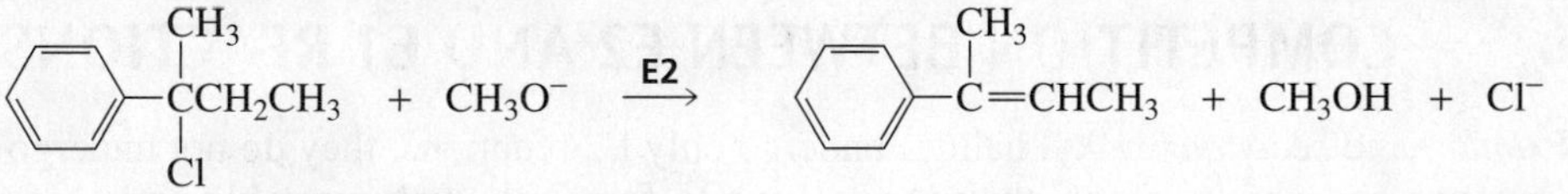

2-chloro-2-phenylbutane

$$CH_3CH=CHCHCH_3 + CH_3O^- \xrightarrow{E2} CH_3CH=CHCH=CH_2 + CH_3OH + Br^-$$
$$\underset{|}{Br}$$

4-bromo-2-pentene

471

Benzylic halides and allylic halides also undergo E1 reactions, because they form relatively stable carbocations.

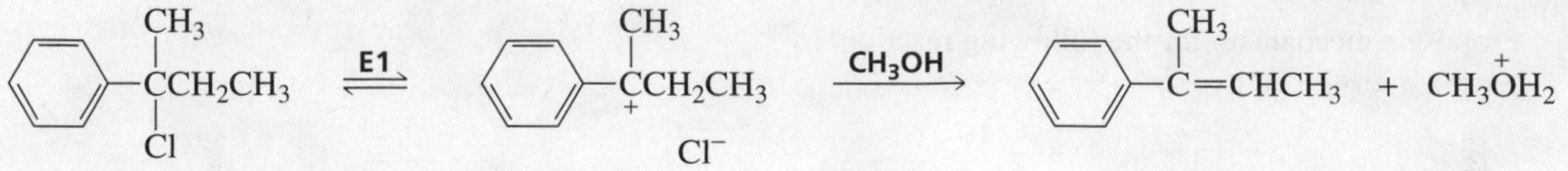

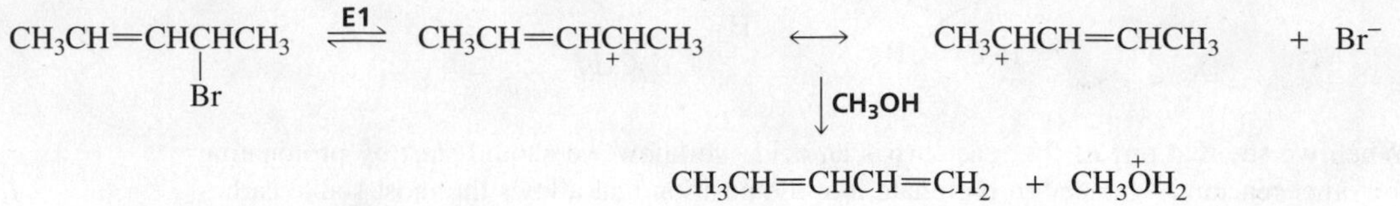

$$CH_3CH=CHCHCH_3 \xrightleftharpoons{E1} CH_3CH=CHCHCH_3 \longleftrightarrow CH_3CHCH=CHCH_3 + Br^-$$

$$\downarrow CH_3OH$$

$$CH_3CH=CHCH=CH_2 + CH_3\overset{+}{O}H_2$$

If the two resonance contributors are not mirror images, as they are in the preceding example, two dienes will be formed.

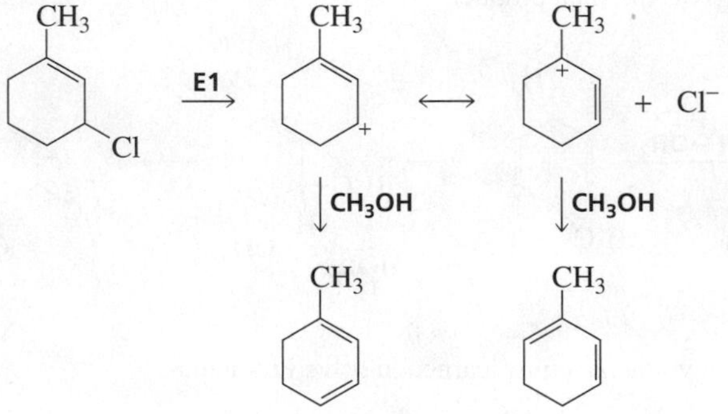

PROBLEM 9♦

What products would be obtained from the E2 reaction of the following alkyl halides?

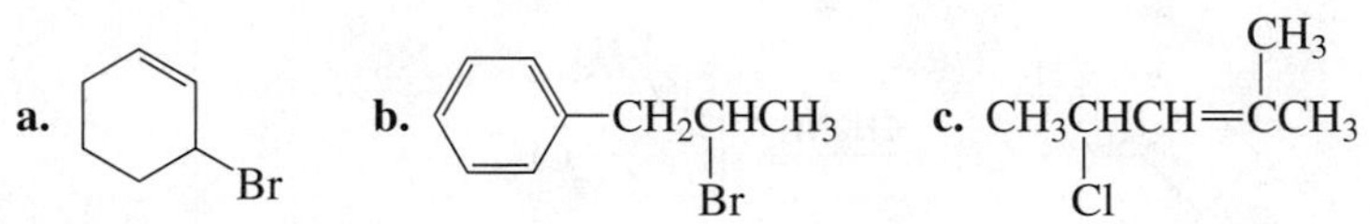

a.   **b.**   **c.** $CH_3CHCH=\overset{\overset{\displaystyle CH_3}{|}}{C}CH_3$
                                                                                     Cl

---

PROBLEM 10♦

What products would be obtained from the E1 reaction of the alkyl halides in Problem 9?

PROBLEM 11♦

Explain why benzyl bromide and allyl bromide cannot undergo either E2 or E1 reactions.

---

5   COMPETITION BETWEEN E2 AND E1 REACTIONS

Primary and *secondary* alkyl halides undergo only E2 reactions; they do not undergo E1 reactions because, to do so, they would have to form relatively unstable carbocations. *Tertiary* alkyl halides and all allylic and benzylic halides (as long as they have a hydrogen bonded to an sp^3 β-carbon) undergo both E2 and E1 reactions (Table 3).

Primary and secondary alkyl halides can undergo only E2 reactions.

Tertiary alkyl halides, allylic halides, and benzylic halides can undergo both E1 and E2 reactions.

Table 3   Summary of the Reactivity of Alkyl Halides in Elimination Reactions

1° and 2° alkyl halides	E2 only
3° alkyl halides and allylic and benzylic halides	E1 and E2

Notice that the same bonds are broken and formed in E2 and E1 reactions. The only difference is the timing—that is, the proton is lost in the first step of an E2 reaction and it is lost in the second step of an E1 reaction.

For a compound that can undergo both E2 and E1 reactions, the E2 reaction is favored by the same factors that favor an S_N2 reaction and the E1 reaction is favored by the same factors that favor an S_N1 reaction. Thus:

> An E2 reaction is favored by a high concentration of a strong base.

> An E1 reaction is favored by a weak base.

Notice that a tertiary alkyl halide and a strong base were chosen to illustrate the E2 reaction in Section 1, whereas a tertiary alkyl halide and a weak base were used to illustrate the E1 reaction in Section 3.

PROBLEM 12

For each of the following reactions, (1) decide whether an E2 or an E1 will occur, and (2) draw the major elimination product:

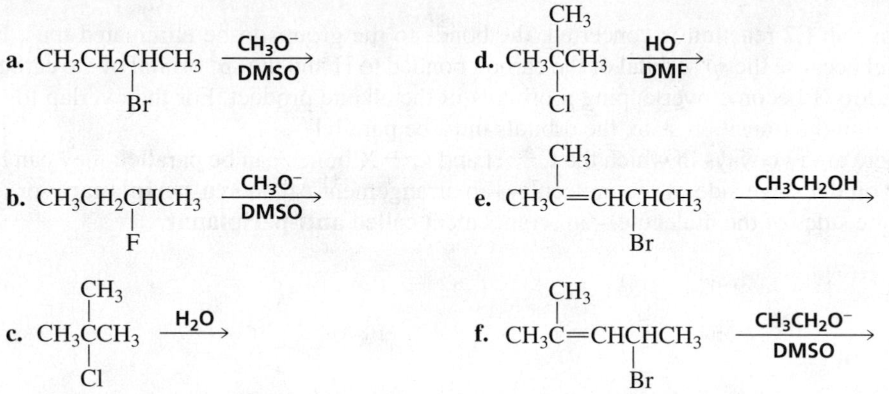

a. $CH_3CH_2CHCH_3$ (with Br) $\xrightarrow[\text{DMSO}]{CH_3O^-}$

d. CH_3CCH_3 (with CH_3 and Cl) $\xrightarrow[\text{DMF}]{HO^-}$

b. $CH_3CH_2CHCH_3$ (with F) $\xrightarrow[\text{DMSO}]{CH_3O^-}$

e. $CH_3C{=}CHCHCH_3$ (with CH_3 and Br) $\xrightarrow{CH_3CH_2OH}$

c. CH_3CCH_3 (with CH_3 and Cl) $\xrightarrow{H_2O}$

f. $CH_3C{=}CHCHCH_3$ (with CH_3 and Br) $\xrightarrow[\text{DMSO}]{CH_3CH_2O^-}$

PROBLEM 13 Solved

The rate law for the reaction of HO^- with *tert*-butyl bromide to form an elimination product in 75% ethanol/25% water at 30 °C is the sum of the rate laws for the E2 and E1 reactions:

$$\text{rate} = 7.1 \times 10^{-5} \left[\textit{tert}\text{-butyl bromide} \right]\left[HO^- \right] + 1.5 \times 10^{-5} \left[\textit{tert}\text{-butyl bromide} \right]$$

What percentage of the reaction takes place by the E2 pathway under the following conditions?

a. $\left[HO^- \right] = 5.0\,M$　　　　　b. $\left[HO^- \right] = 0.0025\,M$

Solution to 13a

$$\frac{E2}{E2 + E1} = \frac{k_2\left[\textit{tert}\text{-butyl bromide}\right]\left[HO^-\right]}{k_2\left[\textit{tert}\text{-butyl bromide}\right]\left[HO^-\right] + k_1\left[\textit{tert}\text{-butyl bromide}\right]}$$

$$\frac{E2}{E2 + E1} = \frac{7.1 \times 10^{-5} \times 5.0}{7.1 \times 10^{-5} \times 5.0 + 1.5 \times 10^{-5}} = \frac{35.5 \times 10^{-5}}{35.5 \times 10^{-5} + 1.5 \times 10^{-5}} = \frac{35.5}{37} = 0.96 = 96\%$$

6　E2 AND E1 REACTIONS ARE STEREOSELECTIVE

We have seen that E2 and E1 reactions are *regioselective,* meaning that more of one constitutional isomer is formed than the other. For example, the following E2 reaction forms more 2-pentene than 1-pentene (Section 2).

$$\underset{\underset{\textbf{2-bromopentane}}{}}{CH_3CH_2CH_2\overset{\overset{\displaystyle Br}{|}}{C}HCH_3} \xrightarrow[\textbf{CH}_3\textbf{CH}_2\textbf{OH}]{\textbf{CH}_3\textbf{CH}_2\textbf{O}^-} \underset{\underset{\underset{\textbf{(mixture of } E \text{ and } Z)}{}}{\underset{\textbf{72\%}}{\textbf{2-pentene}}}}{CH_3CH_2CH{=}CHCH_3} \;+\; \underset{\underset{\textbf{28\%}}{\textbf{1-pentene}}}{CH_3CH_2CH_2CH{=}CH_2}$$

Similarly, the following E1 reaction forms more 2-methyl-2-pentene than 2-methyl-1-pentene (Section 3).

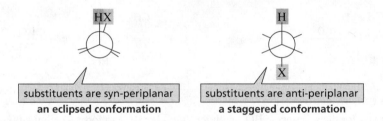

Now we will see that E2 and E1 reactions also are *stereoselective*.

The Stereoisomers Formed in an E2 Reaction

Because an E2 reaction is concerted, the bonds to the groups to be eliminated must be parallel because the sp^3 orbital of the carbon bonded to H and the sp^3 orbital of the carbon bonded to X become overlapping p orbitals in the alkene product. For the overlap to be optimal in the transition state, the orbitals must be parallel.

There are two ways in which the C—H and C—X bonds can be parallel: they can be either on the same side of the molecule—an arrangement called **syn-periplanar**—or on opposite sides of the molecule—an arrangement called **anti-periplanar.**

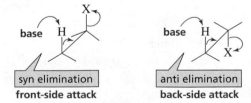

If an elimination reaction removes two substituents from the same side of the molecule, it is called a **syn elimination.** If the substituents are removed from opposite sides of the molecule, the reaction is called an **anti elimination.** Both types of elimination can occur, but anti elimination is highly favored in an E2 reaction. **Sawhorse projections,** which show the C—C bond from an oblique angle, reveal why this is true.

Anti elimination predominates in an E2 reaction.

First, anti elimination requires the molecule to be in a staggered conformation, whereas syn elimination requires it to be in a less stable, eclipsed conformation. Second, in anti elimination, the electrons of the departing hydrogen move to the *back* side of the carbon bonded to X, whereas in syn elimination, the electrons move to the *front* side of the carbon bonded to X. Displacement reactions occur through back-side attack because that's how they achieve the best overlap of the interacting orbitals. Finally, in anti elimination, the electron-rich base is spared the repulsion that it experiences when it is on the same side of the molecule as the electron-rich departing halide ion.

Because of the factors that favor anti elimination, an E2 reaction is *stereoselective*. In other words, more of one stereoisomer is formed than the other. For example, the 2-pentene obtained as the major constitutional isomer from the E2 reaction of 2-bromopentane earlier in this section is actually a pair of stereoisomers, and more (*E*)-2-pentene is formed than (*Z*)-2-pentene.

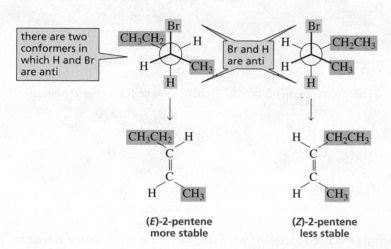

We can make the following general statement about the stereoselectivity of an E2 reaction: if the reactant has *two hydrogens* bonded to the carbon from which a hydrogen is removed, both the *E* and *Z* products will be formed, because the reactant has two conformers in which the groups to be eliminated are anti.

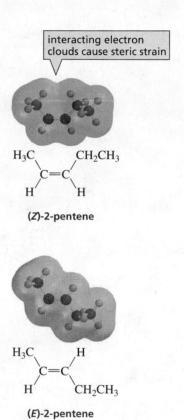

interacting electron clouds cause steric strain

(*Z*)-2-pentene

(*E*)-2-pentene

there are two conformers in which H and Br are anti

Br and H are anti

(*E*)-2-pentene
more stable

(*Z*)-2-pentene
less stable

Of the two stereoisomers, the one with the *largest groups on opposite sides of the double bond* will be formed in greater yield because it is more stable, so it will have the more stable transition state leading to its formation (Figure 3). (The alkene with the largest groups on the *same* side of the double bond is *less stable* because the electron clouds of the large substituents can interfere with each other, causing steric strain).

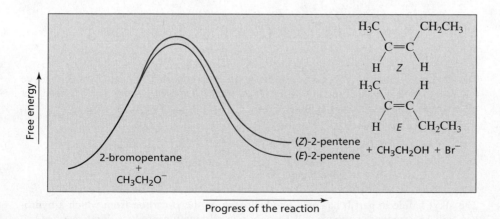

◀ **Figure 3**

The major stereoisomer formed in an E2 reaction is the one with the largest groups on opposite sides of the double bond (here, the *E* isomer). This is because the more stable alkene (indicated by the blue line) has the more stable transition state and therefore is formed more rapidly.

Thus, the following elimination reaction leads predominantly to the *E* isomer because this stereoisomer has the methyl group (the larger group on one sp^2 carbon) opposite the *tert*-butyl group (the larger group on the other sp^2 carbon).

If the β-carbon in an E2 reaction is bonded to two hydrogens, then two alkenes are formed and the major product is the one with the largest substituents on opposite sides of the double bond.

If the β-carbon in an E2 reaction is bonded to only one hydrogen, then one alkene is formed. Its structure depends on the structure of the alkyl halide.

Molecular models can be helpful whenever complex stereochemistry is involved.

$$CH_3CH_2-\underset{\underset{Br}{|}}{\overset{\overset{CH_3}{|}}{C}}-\underset{\underset{CH_3}{|}}{\overset{\overset{CH_3}{|}}{C}}-CH_3 \xrightarrow[CH_3CH_2OH]{CH_3CH_2O^-}$$

3-bromo-2,2,3-trimethyl-pentane

(E)-3,4,4-trimethyl-2-pentene
major product

+

(Z)-3,4,4-trimethyl-2-pentene
minor product

If, however, the β-carbon from which a hydrogen is to be removed is bonded to only *one hydrogen,* then there is only one conformer in which the groups to be eliminated are anti. Therefore, only one alkene product can be formed. The particular stereoisomer formed depends on the configuration of the reactant. For example, anti elimination of HBr from (2S,3S)-2-bromo-3-phenylbutane forms the E isomer.

(2S,3S)-2-bromo-3-phenylbutane

(E)-2-phenyl-2-butene

But anti elimination of HBr from (2S,3R)-2-bromo-3-phenylbutane forms the Z isomer.

(2S,3R)-2-bromo-3-phenylbutane

(Z)-2-phenyl-2-butene

(Notice that the two groups bonded by solid wedges are on the same side of the alkene in the product, as are the two groups bonded by hatched wedges.)

PROBLEM-SOLVING STRATEGY

Determining the Major Product of an E2 Reaction

What is the major product formed when the following compounds undergo an E2 reaction?

a.

b.

The alkyl halide in part **a** has two hydrogens on the β-carbon from which a hydrogen will be removed in the elimination reaction. Therefore, both the E and Z isomers will be formed, but the E isomer will be the major product.

E isomer
major product

Z isomer

The alkyl halide in part **b** has only one hydrogen on the β-carbon from which a hydrogen will be removed, so there will be only one elimination product. First, we need to determine the configuration of the two asymmetric centers. Next, we need to draw the perspective formula showing only the H and Br that will be eliminated. Because we know the configurations of the asymmetric centers, we can add the other groups. When H and Br are eliminated, the two groups bonded to solid wedges will be on the same side of the double bond in the product.

Now use the strategy you have just learned to solve Problem 14.

PROBLEM 14♦

a. What is the major product obtained when each of the following compounds undergoes an E2 reaction? Show the product's configuration.

b. Does the product obtained depend on whether you start with the *R* or *S* enantiomer of the reactant?

1. $CH_3CH_2CHCHCH_3$ **2.** $CH_3CH_2CHCH_2CH=CH_2$ **3.** $CH_3CH_2CHCH_2-$
     Br CH$_3$ Cl   Cl

The Stereoisomers Formed in an E1 Reaction

An E1 reaction, like an E2 reaction, is stereoselective; and again, like an E2 reaction, both the *E* and *Z* products will be formed and the major product will be the one with the *largest groups on opposite sides of the double bond.* Let's see why this is so.

We know that an E1 reaction takes place in two steps. The leaving group leaves in the first step, and a proton is lost from an adjacent carbon in the second step, following Zaitsev's rule for forming the more stable alkene. The carbocation created in the first step is planar, so the electrons from a departing proton can move toward the positively charged carbon from *either side*. Therefore, both syn and anti elimination can occur.

Because both syn and anti elimination can occur in an E1 reaction, both the *E* and *Z* products are formed, *regardless* of whether the β-carbon from which the proton is removed is bonded to one or to two hydrogens. The major product is the one with the largest groups on opposite sides of the double bond, because that is the more stable alkene and, therefore, it is formed more rapidly.

Table 4 it summarizes the stereochemical outcomes of substitution and elimination reactions.

> The major stereoisomer obtained in an E1 reaction is the one with the largest groups on opposite sides of the double bond.

Table 4 Stereochemistry of Substitution and Elimination Reactions

Reaction	Products
S_N2	Only the inverted product is formed.
E2	Both *E* and *Z* stereoisomers are formed (with more of the stereoisomer with the largest groups on opposite sides of the double bond) unless the β-carbon from which the hydrogen is removed is bonded to only one hydrogen, in which case only one stereoisomer is formed. The stereoisomer's configuration depends on the configuration of the reactant.
S_N1	Both stereoisomers (*R* and *S*) are formed (generally with more inverted than retained).
E1	Both *E* and *Z* stereoisomers are formed (with more of the stereoisomer with the largest groups on opposite sides of the double bond).

PROBLEM 15♦ Solved

What is the major product formed when the following compounds undergo an E1 reaction?

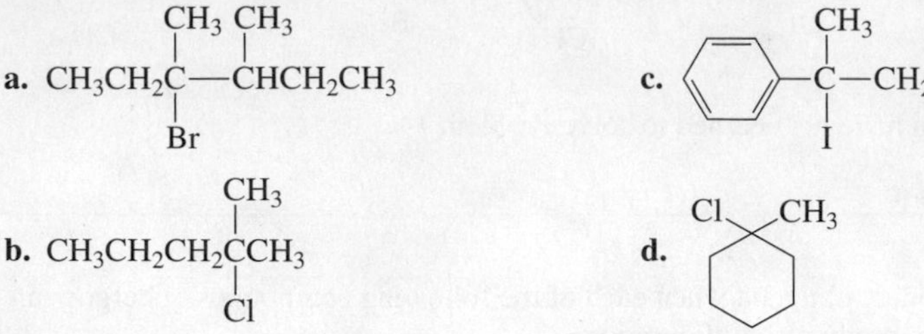

a. CH$_3$CH$_2$C(CH$_3$)—CH(CH$_3$)CH$_2$CH$_3$ with Br

c. phenyl—C(CH$_3$)—CH$_2$CH$_3$ with I

b. CH$_3$CH$_2$CH$_2$C(CH$_3$)CH$_3$ with Cl

d. cyclohexane with Cl and CH$_3$

Solution to 15a First, we need to consider the *regiochemistry* of the reaction: the major product will be 3,4-dimethyl-3-hexene because it is the most stable of the three possible alkene products.

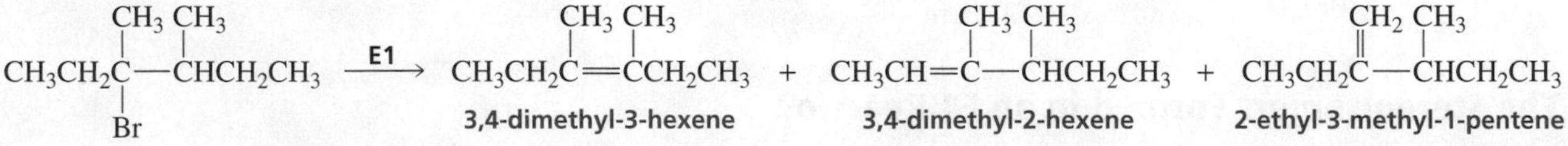

CH$_3$CH$_2$C(CH$_3$)—CH(CH$_3$)CH$_2$CH$_3$ with Br $\xrightarrow{\text{E1}}$ CH$_3$CH$_2$C(CH$_3$)=C(CH$_3$)CH$_2$CH$_3$ + CH$_3$CH=C(CH$_3$)—CH(CH$_3$)CH$_2$CH$_3$ + CH$_3$CH$_2$C(=CH$_2$)—CH(CH$_3$)CH$_2$CH$_3$

3,4-dimethyl-3-hexene **3,4-dimethyl-2-hexene** **2-ethyl-3-methyl-1-pentene**

Next, we need to consider the *stereochemistry* of the reaction: the major product has two stereoisomers and more (*E*)-3,4-dimethyl-3-hexene will be formed because it is more stable than (*Z*)-3,4-dimethyl-3-hexene.

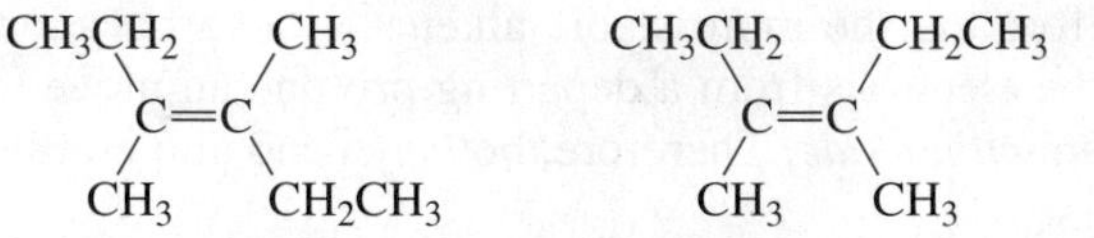

(*E*)-3,4-dimethyl-3-hexene   (*Z*)-3,4-dimethyl-3-hexene

7   ELIMINATION FROM SUBSTITUTED CYCLOHEXANES

Elimination from substituted cyclohexanes follows the same stereochemical rules as elimination from open-chain compounds.

E2 Reactions of Substituted Cyclohexanes

We just saw that to achieve the anti-periplanar geometry favored in an E2 reaction, the two groups being eliminated must be parallel (Section 6). For two groups on a cyclohexane ring to be parallel, they both must be in *axial positions*.

In an E2 reaction of a substituted cyclohexane, the groups being eliminated must both be in axial positions.

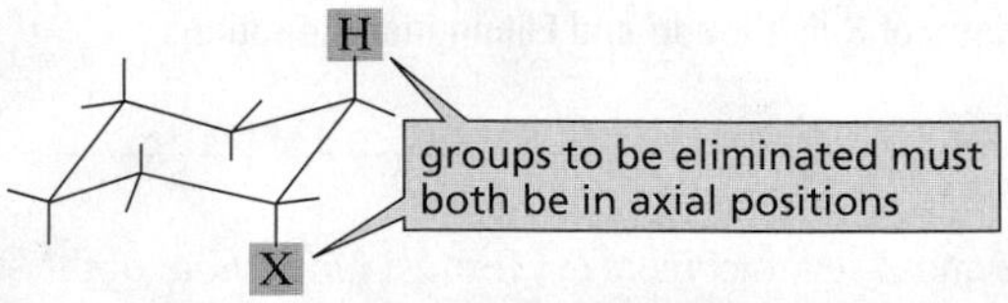

groups to be eliminated must both be in axial positions

The more stable conformer of chlorocyclohexane does not undergo an E2 reaction, because the chloro substituent is in an equatorial position. (The more stable conformer of a monosubstituted cyclohexane is the one in which the substituent is in an equatorial position because there is more room for a substituent in that position.) The less stable conformer, with the chloro substituent in the axial position, readily undergoes an E2 reaction.

more stable   less stable

HO⁻ E2 conditions k' HO⁻ E2 conditions

no reaction $+ \ H_2O \ + \ Cl^-$

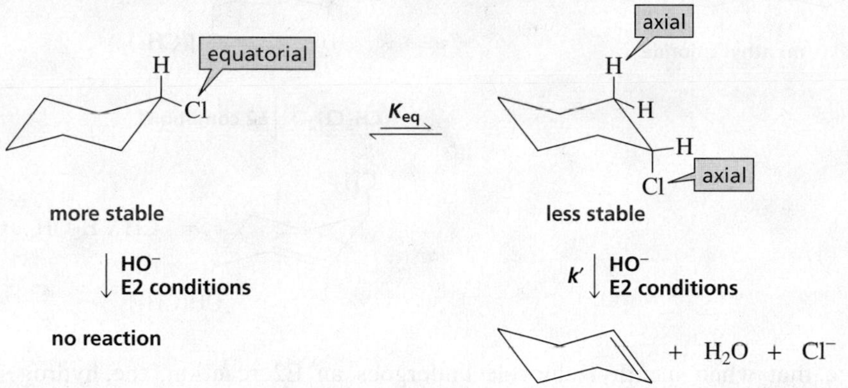

Because only one of the two conformers undergoes an E2 reaction, the rate constant for the elimination reaction is given by $k'K_{eq}$, where k' is a rate constant and $K_{eq} =$ is an equilibrium constant. Therefore, the reaction is faster if K_{eq} is large. Most molecules are in the more stable conformer at any given time. Therefore, K_{eq} will be large if elimination takes place by way of the more stable conformer, and it will be small if elimination takes place by way of the less stable conformer.

$$K_{eq} = \frac{[\textbf{more stable conformer}]}{[\textbf{less stable conformer}]} \qquad K_{eq} = \frac{[\textbf{less stable conformer}]}{[\textbf{more stable conformer}]}$$

for a reaction that takes place
through the more stable conformer

for a reaction that takes place
through the less stable conformer

For example, neomenthyl chloride undergoes an E2 reaction with ethoxide ion about 200 times faster than menthyl chloride does. The conformer of neomenthyl chloride that undergoes elimination is the *more* stable conformer because when the Cl and H are in the required axial positions, the methyl and isopropyl groups are in equatorial positions.

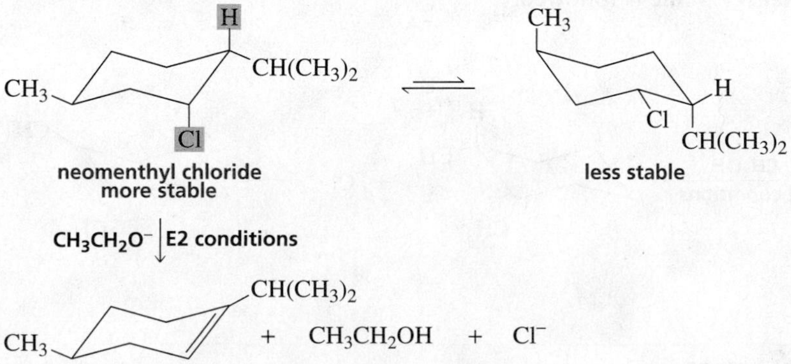

neomenthyl chloride
more stable

less stable

$CH_3CH_2O^-$ E2 conditions

$+ \quad CH_3CH_2OH \quad + \quad Cl^-$

In contrast, the conformer of menthyl chloride that undergoes elimination is the *less* stable conformer because when the Cl and H are in the required axial positions, the methyl and isopropyl groups are also in axial positions.

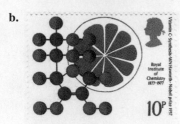

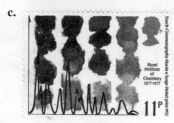

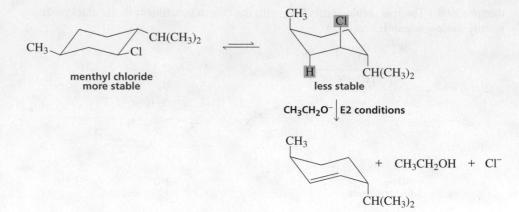

Stamps issued in honor of English Nobel Laureates:

a. Sir Derek Barton for conformational analysis, 1969

b. Sir Walter Haworth for the synthesis of vitamin C, 1937

c. A. J. P. Martin and Richard L. M. Synge for chromatography, 1952

d. William H. Bragg and William L. Bragg for crystallography, 1915 (The Braggs are one of six father–son pairs who both received Nobel Prizes.)

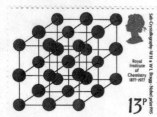

Notice that when menthyl chloride undergoes an E2 reaction, the hydrogen that is eliminated is *not* removed from the β-carbon bonded to the fewest hydrogens. This may seem like a violation of Zaitsev's rule, but this compound has only one β-carbon that is bonded to a hydrogen in an axial position. Therefore, that hydrogen is the one that is removed, even though it is not bonded to the β-carbon with the fewest hydrogens.

PROBLEM 16◆

Why do *cis*-1-bromo-2-ethylcyclohexane and *trans*-1-bromo-2-ethylcyclohexane form different major products when they undergo an E2 reaction?

PROBLEM 17◆

Which isomer reacts more rapidly in an E2 reaction, *cis*-1-bromo-4-*tert*-butylcyclohexane or *trans*-1-bromo-4-*tert*-butylcyclohexane? Explain your answer.

E1 Reactions of Substituted Cyclohexanes

When a substituted cyclohexane undergoes an E1 reaction, the two groups that are eliminated do not both have to be in axial positions because the elimination reaction is not concerted. In the following reaction, a carbocation is formed in the first step. It then loses a proton from the adjacent carbon that is bonded to the fewest hydrogens—in other words, Zaitsev's rule is followed.

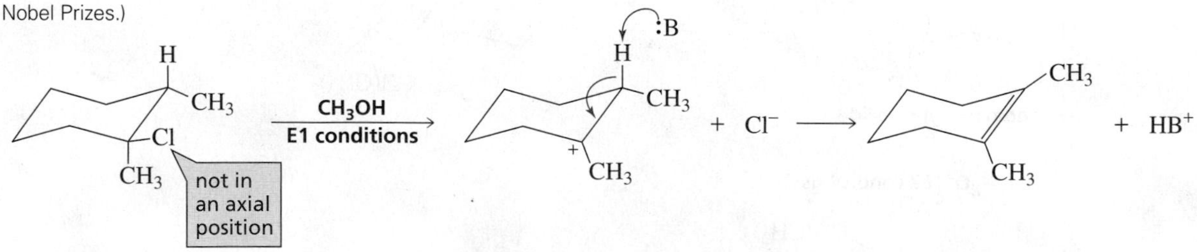

PROBLEM 18

Draw the substitution and elimination products for the following reactions, showing the configuration of each product:

a. *trans*-1-chloro-2-methylcyclohexane + CH_3O^- **c.** 1-chloro-1-methylcyclohexane + CH_3O^-

b. *cis*-1-chloro-2-methylcyclohexane + CH_3O^- **d.** 1-chloro-1-methylcyclohexane + CH_3OH

8 A KINETIC ISOTOPE EFFECT CAN HELP DETERMINE A MECHANISM

A mechanism is a model that accounts for all the experimental evidence that has been accumulated about the reaction. For example, the mechanisms for the S_N1, S_N2, E1, and E2 reactions are based on the rate law for each reaction, the relative reactivities of the reactants, and the structures of the products.

Another piece of experimental evidence that is helpful in determining the mechanism of a reaction is a **deuterium kinetic isotope effect,** which is *the ratio of the rate constant observed for a compound containing hydrogen to the rate constant observed for an identical compound in which one or more of the hydrogens has been replaced by deuterium* (which is an isotope of hydrogen; the nucleus of a hydrogen atom contains only a proton, whereas the nucleus of a deuterium atom has one proton and one neutron.)

$$\text{deuterium kinetic isotope effect} = \frac{k_H}{k_D} = \frac{\text{rate constant for H-containing reactant}}{\text{rate constant for D-containing reactant}}$$

The chemical properties of hydrogen and deuterium are similar, but a C—D bond is about 1.2 kcal/mol (5 kJ/mol) stronger than a C—H bond. As a result, a C—D bond is more difficult to break than a corresponding C—H bond.

When the rate constants (k_H and k_D) for the next two reactions were determined (under identical conditions), k_H was found to be 7.1 times greater than k_D. The deuterium kinetic isotope effect for this reaction, therefore, is 7.1.

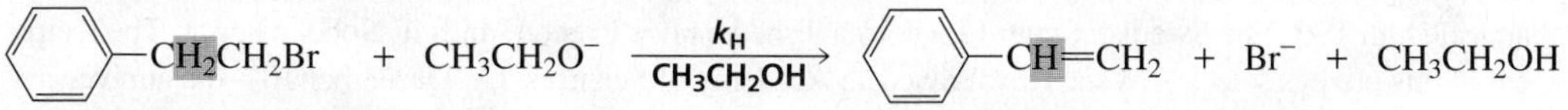

1-bromo-2-phenylethane

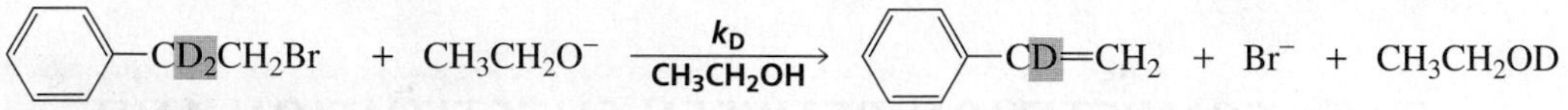

2-bromo-1,1-dideuterio-1-phenylethane

Because the deuterium kinetic isotope effect is greater than 1.0, we know that the change from hydrogen to deuterium affects the rate of the reaction. This tells us that the C—H (or C—D) bond must be broken in the rate-determining step—a fact consistent with the mechanism proposed for an E2 reaction, but not for an E1 reaction.

PROBLEM 19♦

If the two reactions described in this section were E1 reactions, what value would you expect to obtain for the deuterium kinetic isotope effect?

PROBLEM 20♦

List the following compounds in order from most reactive to least reactive in an E2 reaction:

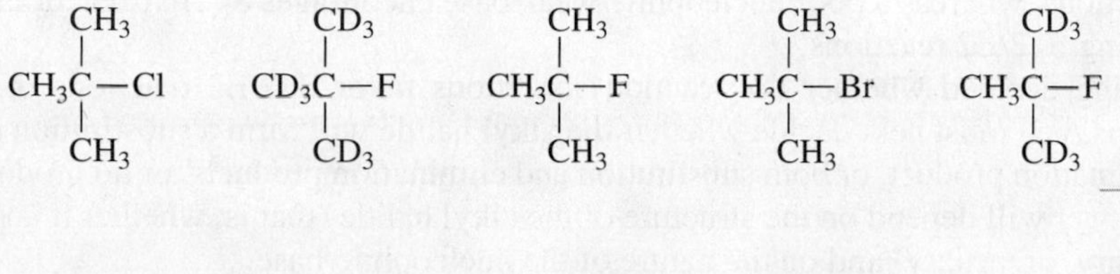

The Nobel Prize

The Nobel Prize is generally considered the highest honor a scientist can receive. These awards were established by **Alfred Bernhard Nobel (1833–1896)** and were first conferred in 1901.

Nobel was born in Stockholm, Sweden. When he was nine, he moved with his parents to St. Petersburg, Russia, where his father worked for the Russian government, manufacturing torpedoes and land and water mines that he had invented. As a young man, Alfred did research on explosives in a factory his father owned near Stockholm. In 1864, an explosion in the factory killed five people, including his younger brother, causing Alfred to look for ways to make explosives easier to handle and transport. After the explosion, the Swedish government would not allow the factory to be rebuilt because so many accidents had occurred there. Nobel, therefore, established an explosives factory in Germany, where, in 1867, he discovered that nitroglycerin mixed with diatomaceous earth can be molded into sticks that cannot be set off without a detonating cap. Thus, Nobel invented dynamite. He also invented blasting gelatin and smokeless powder. Although he was the inventor of the explosives used by the military, he was a strong supporter of peace movements.

Alfred Bernhard Nobel

The 355 patents Nobel held made him a wealthy man. He never married, and when he died, his will stipulated for the bulk of his estate ($9,200,000) to be used to establish prizes to be awarded to those who "have conferred the greatest benefit on mankind." He instructed that the money be invested and the interest earned each year be divided into five equal portions "to be awarded to the persons having made the most important contributions in the fields of chemistry, physics, physiology or medicine, literature, and to the one who had done the most toward fostering fraternity among nations, the abolition of standing armies, and the holding and promotion of peace congresses." Nobel also directed that no consideration be given to the nationality of the prize candidates, that each prize be shared by no more than three persons, and that no prize be awarded posthumously.

Nobel's instructions said that the prizes for chemistry and physics were to be awarded by the Royal Swedish Academy of Sciences, the prizes for physiology or medicine by the Karolinska Institute in Stockholm, the prize for literature by the Swedish Academy, and the prize for peace by a five-person committee appointed by the Norwegian Parliament. The deliberations are secret, and the decisions cannot be appealed. In 1969, the Swedish Central Bank established a prize in economics in Nobel's honor. The recipient of this prize is selected by the Royal Swedish Academy of Sciences. On December 10—the anniversary of Nobel's death—the prizes are awarded in Stockholm, except for the peace prize, which is awarded in Oslo.

The Golden Hall inside the City Hall in Stockholm, where the Nobel prize winners have a celebratory dinner.

9 COMPETITION BETWEEN SUBSTITUTION AND ELIMINATION

We have seen that alkyl halides can undergo four types of reactions: S_N2, S_N1, E2, and E1. As a result, you may feel a bit overwhelmed when you are asked to predict the products of the reaction of a given alkyl halide and a nucleophile/base. Let's pause, therefore, to discuss the reactions of alkyl halides to make it a little easier for you to predict their products. Notice, in the following discussion, that HO^- is called a *nucleophile* in a substitution reaction (because it attacks a carbon) and it is called a *base* in an elimination reaction (because it removes a proton).

To predict the products of the reaction of an alkyl halide with a nucleophile/base, you must first decide whether the reaction conditions favor S_N2/E2 or S_N1/E1 reactions. (Note that the conditions that favor an S_N2 reaction also favor an E2 reaction, and the conditions that favor an S_N1 reaction also favor an E1 reaction.)

That is an easy decision. S_N2/E2 reactions all have good nucleophiles/strong bases (HO^-, CH_3O^-, NH_3), whereas the S_N1/E1 reactions all have poor nucleophiles/weak bases (H_2O, CH_3OH). In other words, a good nucleophile/strong base encourages S_N2/E2 reactions, whereas a poor nucleophile/weak base encourages S_N1/E1 reactions by discouraging S_N2/E2 reactions.

An S_N2/E2 reaction of an alkyl halide is favored by a high concentration of a good nucleophile/strong base.

An S_N1/E1 reaction of an alkyl halide is favored by a poor nucleophile/weak base.

Having decided whether the reaction conditions favor S_N2/E2 reactions or S_N1/E1 reactions, you must next decide whether the alkyl halide will form a substitution product, an elimination product, or both substitution and elimination products, or no product at all. The answer will depend on the structure of the alkyl halide (that is, whether it is primary, secondary, or tertiary) and on the nature of the nucleophile/base.

S$_N$2/E2 Conditions

Let's first consider conditions that lead to S$_N$2/E2 reactions—namely, a high concentration of a good nucleophile/strong base. The following reactions show that hydroxide ion can act as a nucleophile and attack the back side of the α-carbon to form the substitution product, or it can act as a base and remove a hydrogen from a β-carbon to form the elimination product.

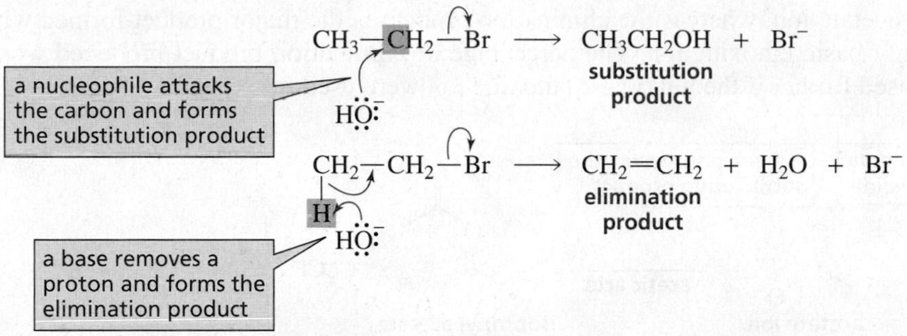

Thus, the two reactions occur at the same time and, therefore, compete with each other. And they both occur for the same reason: the electron-withdrawing halogen gives the carbon to which it is bonded a partial positive charge.

The relative reactivities of alkyl halides in S$_N$2 and E2 reactions are shown here.

In an S$_N$2 reaction: 1° > 2° > 3°   In an E2 reaction: 3° > 2° > 1°

Because a *primary* alkyl halide is the most reactive in an S$_N$2 reaction (the back side of the α-carbon is relatively unhindered) and the least reactive in an E2 reaction (Section 2), a primary alkyl halide forms principally the substitution product in a reaction carried out under conditions that favor S$_N$2/E2 reactions. In other words, substitution wins the competition.

Primary alkyl halides undergo primarily substitution under S$_N$2/E2 conditions.

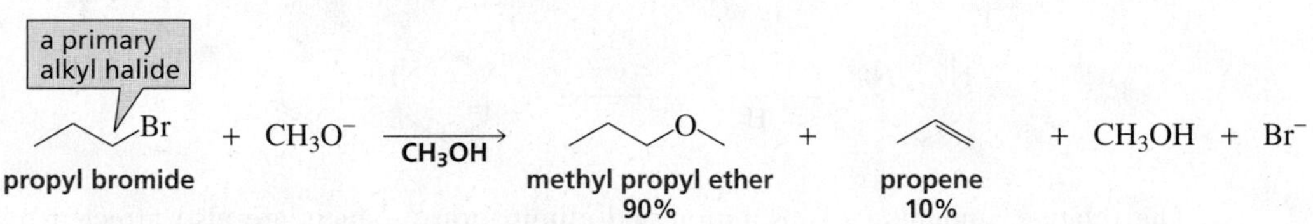

However, if either the primary alkyl halide or the nucleophile/base is sterically hindered, then the nucleophile will have difficulty getting to the back side of the α-carbon but will be able to remove the more accessible proton. As a result, elimination will win the competition.

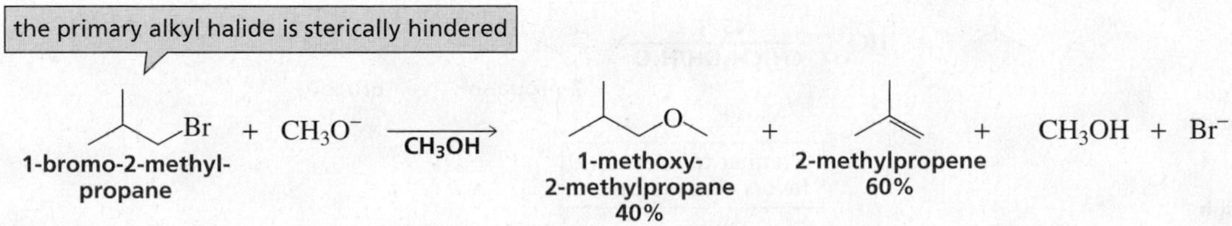

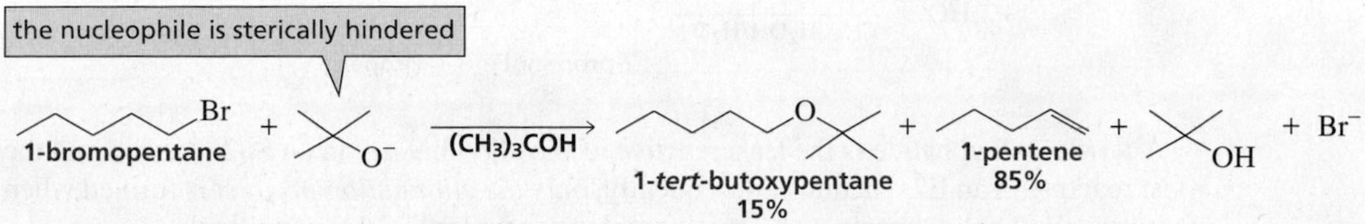

483

A *secondary* alkyl halide, compared with a primary alkyl halide, reacts slower in an S_N2 reaction and faster in an E2 reaction. Thus, a *secondary* alkyl halide forms both substitution and elimination products under S_N2/E2 conditions. The relative amounts of the two products depend on the strength and bulk of the nucleophile/base. *The stronger and bulkier the nucleophile/base, the greater the percentage of the elimination product.*

For example, in the reactions that follow, acetate ion is a weaker base than ethoxide ion because acetic acid is a stronger acid (pK_a = 4.76) than ethanol (pK_a = 15.9). No elimination product is formed from the reaction of 2-chloropropane with the weakly basic acetate ion, whereas the elimination product is the major product formed with the strongly basic ethoxide ion. The percentage of elimination product produced would be increased further if the bulky *tert*-butoxide ion were used instead of ethoxide ion.

a secondary alkyl halide a weak base favors the substitution product

2-chloropropane   acetate ion   → acetic acid → isopropyl acetate 100% + Cl⁻

a strong base favors the elimination product

2-chloropropane + ethoxide ion → CH_3CH_2OH → 2-ethoxypropane 25% + propene 75% + OH + Cl⁻

DBN and DBU are bulky bases commonly used to encourage elimination over substitution. Like other amines, they are relatively strong bases even though they are neutral compounds. These compounds are so bulky that only the elimination reaction occurs with a secondary alkyl halide.

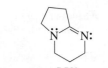

DBN
1,5-diazabicyclo[4.3.0]non-5-ene

DBU
1,8-diazabicyclo[5.4.0]undec-7-ene

100%

The relative amounts of substitution and elimination products are also affected by temperature; higher temperatures favor elimination because of the greater $\Delta S°$ value for the elimination reaction since an elimination reaction forms more product molecules than a substitution reaction: an elimination reaction forms three products (the alkene, the leaving group, and the conjugate acid of the base), whereas a substitution reaction forms two products (the substitution product and the leaving group).

Br + HO⁻ → 45 °C, CH_3CH_2OH/H_2O → 2-propanol 47% + propene 53% + H_2O + Br⁻

Secondary alkyl halides undergo substitution and elimination under S_N2/E2 conditions.

Strong and bulky bases and high temperatures favor elimination over substitution.

a higher temperature favors elimination

Br + HO⁻ → 100 °C, CH_3CH_2OH/H_2O → 2-propanol 29% + propene 71% + H_2O + Br⁻

A *tertiary* alkyl halide is the least reactive of the alkyl halides in an S_N2 reaction and the most reactive in an E2 reaction. Consequently, *only the elimination product* is formed when a tertiary alkyl halide reacts with a nucleophile/base under S_N2/E2 conditions.

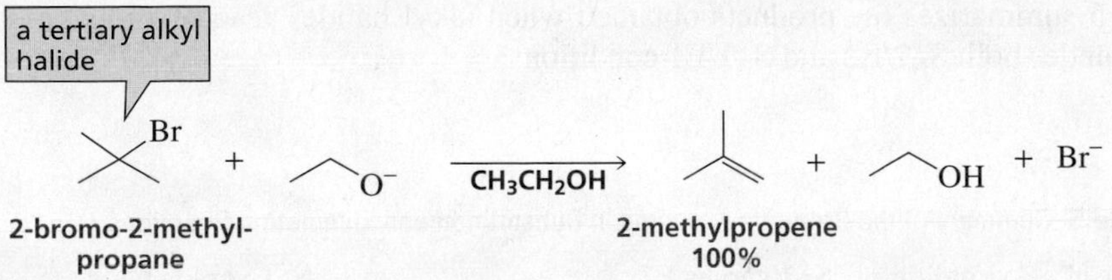

2-bromo-2-methyl-propane **2-methylpropene** **100%**

Tertiary alkyl halides undergo only elimination under $S_N2/E2$ conditions.

PROBLEM 21◆

How would you expect the ratio of substitution product to elimination product formed from the reaction of propyl bromide with CH_3O^- in methanol to change if the nucleophile were changed to CH_3S^-?

PROBLEM 22◆

Explain why only a substitution product and no elimination product is obtained when the following compound reacts with sodium methoxide:

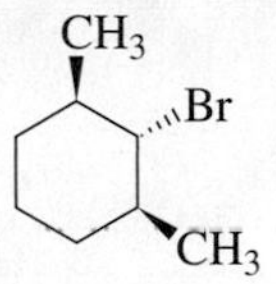

$S_N1/E1$ Conditions

Now let's look at what happens when conditions favor $S_N1/E1$ reactions—that is, a poor nucleophile/weak base. Note that in $S_N1/E1$ reactions, the alkyl halide dissociates to form a carbocation, which can then either combine with the nucleophile to form the substitution product or lose a proton to form the elimination product.

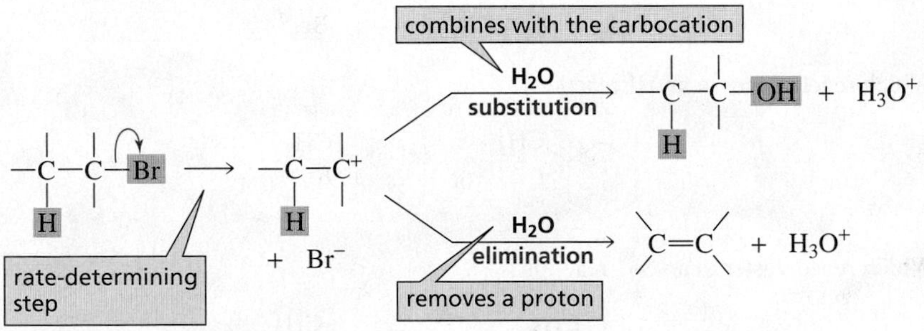

Primary and secondary alkyl halides do not undergo S_N1 and E1 reactions.

S_N1 and E1 reactions both have the same rate-determining step—dissociation of the alkyl halide to form a carbocation. This means that any alkyl halide that reacts under $S_N1/E1$ conditions will form both substitution and elimination products. Remember that primary and secondary alkyl halides do not undergo these reactions because they form relatively unstable carbocations.

It is fortunate that the $S_N1/E1$ reactions of tertiary alkyl halides favor the substitution product, because under $S_N2/E2$ conditions only the elimination product is formed.

81%   **19%**

100%

485

Table 5 summarizes the products obtained when alkyl halides react with nucleophiles/bases under both $S_N2/E2$ and $S_N1/E1$ conditions.

Table 5 Summary of the Products Expected in Substitution and Elimination Reactions		
Class of alkyl halide	**S_N2 versus E2**	**S_N1 versus E1**
Primary alkyl halide	primarily substitution, unless there is steric hindrance in the alkyl halide or nucleophile, in which case elimination is favored	cannot undergo $S_N1/E1$ reactions
Secondary alkyl halide	both substitution and elimination; the stronger and bulkier the base and the higher the temperature, the greater the percentage of elimination	cannot undergo $S_N1/E1$ reactions
Tertiary alkyl halide	only elimination	both substitution and elimination with substitution favored

PROBLEM 23

Why do the $S_N1/E1$ reactions of tertiary alkyl halides favor the substitution product?

PROBLEM 24♦

a. Which reacts faster in an S_N2 reaction?

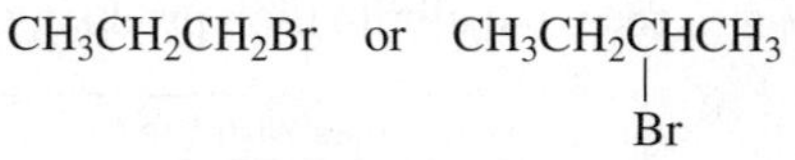

$$CH_3CH_2CH_2Br \quad or \quad CH_3CH_2\underset{\underset{Br}{|}}{C}HCH_3$$

b. Which reacts faster in an E1 reaction?

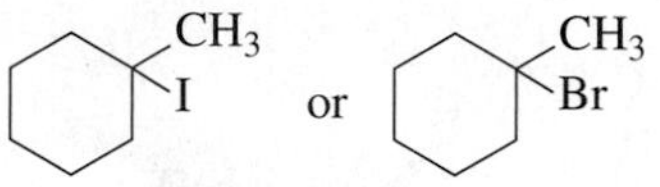

c. Which reacts faster in an S_N1 reaction?

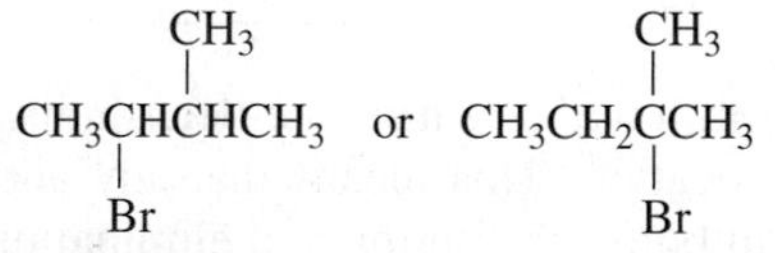

$$CH_3\underset{\underset{Br}{|}}{C}H\overset{\overset{CH_3}{|}}{C}HCH_3 \quad or \quad CH_3CH_2\overset{\overset{CH_3}{|}}{\underset{\underset{Br}{|}}{C}}CH_3$$

d. Which reacts faster in an E2 reaction?

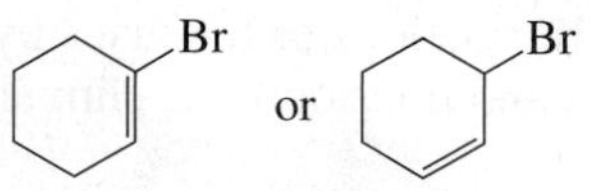

PROBLEM 25♦

Indicate whether the specified alkyl halides will form primarily substitution products, only elimination products, both substitution and elimination products, or no products when they react with the following:

a. methanol under $S_N1/E1$ conditions

b. sodium methoxide under $S_N2/E2$ conditions

    1. 1-bromobutane   **3.** 2-bromobutane

    2. 1-bromo-2-methylpropane   **4.** 2-bromo-2-methylpropane

    a. Explain why 1-bromo-2,2-dimethylpropane has difficulty undergoing both S_N2 and S_N1 reactions.
    b. Can it undergo E2 and E1 reactions?

10   SUBSTITUTION AND ELIMINATION REACTIONS IN SYNTHESIS

When substitution or elimination reactions are used in synthesis, care must be taken to choose reactants and reaction conditions that will maximize the yield of the desired product. Thus, S_N2 and E2 reactions are preferred.

Using Substitution Reactions to Synthesize Compounds

Nucleophilic substitution reactions of alkyl halides can lead to a wide variety of organic compounds. For example, ethers are synthesized by the reaction of an alkyl halide with an alkoxide ion. This reaction, called the Williamson ether synthesis (after Alexander Williamson, who discovered it in 1850) is still considered one of the best ways to synthesize an ether.

Williamson ether synthesis

$$\underset{\text{alkyl halide}}{R-Br} \; + \; \underset{\text{alkoxide ion}}{R-O^-} \longrightarrow \underset{\text{ether}}{R-O-R} + \boxed{Br^-}$$

The alkoxide ion (RO^-) for a **Williamson ether synthesis** can be prepared by using sodium hydride (NaH) to remove a proton from an alcohol.

$$\boxed{ROH} \; + \; NaH \longrightarrow \boxed{RO^-} + Na^+ + H_2$$

The Williamson ether synthesis is a nucleophilic substitution reaction. It requires a high concentration of a good nucleophile, which indicates that it is an S_N2 reaction.

If you want to synthesize an ether such as the one shown next, you have a choice of starting materials: you can use either a propyl halide and butoxide ion or a butyl halide and propoxide ion.

$$\underset{\text{propyl bromide}}{CH_3CH_2CH_2Br} + \underset{\text{butoxide ion}}{CH_3CH_2CH_2CH_2O^-} \longrightarrow CH_3CH_2CH_2OCH_2CH_2CH_2CH_3 + \boxed{Br^-}$$
$$\text{butyl propyl ether}$$

$$\underset{\text{butyl bromide}}{CH_3CH_2CH_2CH_2Br} + \underset{\text{propoxide ion}}{CH_3CH_2CH_2O^-} \longrightarrow CH_3CH_2CH_2CH_2OCH_2CH_2CH_3 + \boxed{Br^-}$$
$$\text{butyl propyl ether}$$

However, if you want to synthesize *tert*-butyl ethyl ether, the starting materials must be an ethyl halide and *tert*-butoxide ion.

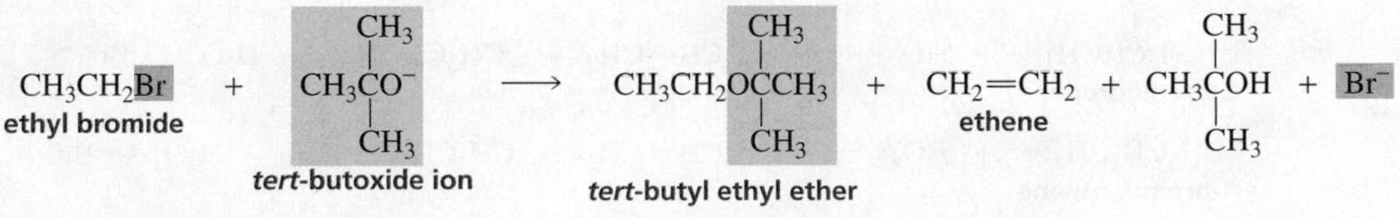

$$\underset{\text{ethyl bromide}}{CH_3CH_2Br} + \underset{\substack{\text{tert-butoxide ion}}}{CH_3\overset{\displaystyle CH_3}{\underset{\displaystyle CH_3}{\overset{|}{\underset{|}{C}}}O^-}} \longrightarrow \underset{\substack{\text{tert-butyl ethyl ether}}}{CH_3CH_2O\overset{\displaystyle CH_3}{\underset{\displaystyle CH_3}{\overset{|}{\underset{|}{C}}}CH_3}} + \underset{\text{ethene}}{CH_2{=}CH_2} + CH_3\overset{\displaystyle CH_3}{\underset{\displaystyle CH_3}{\overset{|}{\underset{|}{C}}}OH} + \boxed{Br^-}$$

If, instead, you used a *tert*-butyl halide and ethoxide ion, you would not obtain any ether because the reaction of a tertiary alkyl halide under S_N2/E2 conditions forms only the elimination product.

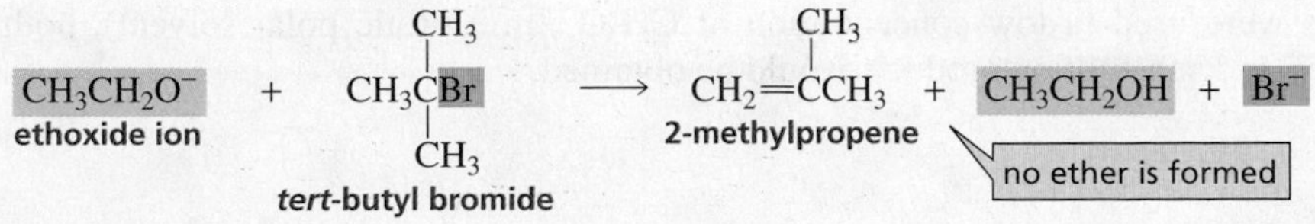

$$\boxed{CH_3CH_2O^-} + \underset{\substack{\text{tert-butyl bromide}}}{CH_3\overset{\displaystyle CH_3}{\underset{\displaystyle CH_3}{\overset{|}{\underset{|}{C}}}Br}} \longrightarrow \underset{\text{2-methylpropene}}{CH_2{=}\overset{\displaystyle CH_3}{\overset{|}{C}}CH_3} + \boxed{CH_3CH_2OH} + \boxed{Br^-}$$

$$\boxed{\text{no ether is formed}}$$

In ether synthesis, the less hindered group should be provided by the alkyl halide.

Consequently, a Williamson ether synthesis should be designed in such a way that the *less hindered alkyl group* is provided by the *alkyl halide* and the *more hindered alkyl group* comes from the *alkoxide ion.*

Alkynes can be synthesized by the reaction of an acetylide anion with an alkyl halide.

$$CH_3CH_2C \equiv C^- \ + \ CH_3CH_2CH_2Br \ \longrightarrow \ CH_3CH_2C \equiv CCH_2CH_2CH_3 \ + \ Br^-$$

Now that you know that this is an S_N2 reaction (the alkyl halide reacts with a high concentration of a good nucleophile), you can understand why you were told that it is best to use methyl halides and primary halides in the reaction. Methyl halides form only substitution products, and primary alkyl halides form mainly the desired substitution product. A tertiary alkyl halide would form only the elimination product, and a secondary alkyl halide would form mainly the elimination product because the acetylide ion is a very strong base.

PROBLEM 27♦

A small amount of another organic product is formed in a Williamson ether synthesis. What is this product when the alkyl halide used in the synthesis of butyl propyl ether is

a. propyl bromide?

b. butyl bromide?

PROBLEM 28

What would be the best way to prepare the following ethers using an alkyl halide and an alcohol?

$$\overset{\displaystyle CH_3}{\underset{\displaystyle |}{}}$$

a. $CH_3CH_2CHOCH_2CH_2CH_3$

$$\overset{\displaystyle CH_3}{\underset{\displaystyle |}{}}$$

b. $CH_3CH_2OCH_2CHCH_2CH_2CH_3$

c. cyclohexyl—OCH_3

d. phenyl—CH_2O—phenyl

Using Elimination Reactions to Synthesize Compounds

If you want to synthesize an alkene, you should choose the most hindered alkyl halide possible in order to maximize the elimination product and minimize the substitution product. For example, 2-bromopropane would be a better starting material than 1-bromopropane for the synthesis of propene because the secondary alkyl halide would give a higher yield of the desired elimination product and a lower yield of the competing substitution product. The percentage of alkene could be further increased by using a sterically hindered base such as *tert*-butoxide ion or DBN instead of hydroxide ion (Section 10).

$$\underset{\text{2-bromopropane}}{\overset{\displaystyle Br}{\underset{\displaystyle |}{CH_3CHCH_3}}} \ + \ HO^- \ \longrightarrow \ \underset{\text{major product}}{CH_3CH=CH_2} \ + \ \underset{\text{minor product}}{\overset{\displaystyle OH}{\underset{\displaystyle |}{CH_3CHCH_3}}} \ + \ H_2O \ + \ Br^-$$

$$\underset{\text{1-bromopropane}}{CH_3CH_2CH_2Br} \ + \ HO^- \ \longrightarrow \ \underset{\text{minor product}}{CH_3CH=CH_2} \ + \ \underset{\text{major product}}{CH_3CH_2CH_2OH} \ + \ H_2O \ + \ Br^-$$

To synthesize 2-methyl-2-butene from 2-bromo-2-methylbutane, you would use $S_N2/E2$ conditions (a high concentration of CH_3O^- in an aprotic polar solvent) because a tertiary alkyl halide forms *only* the elimination product under those conditions. If $S_N1/E1$ conditions were used (a low concentration of CH_3O^- in a protic polar solvent), both elimination and substitution products would be obtained.

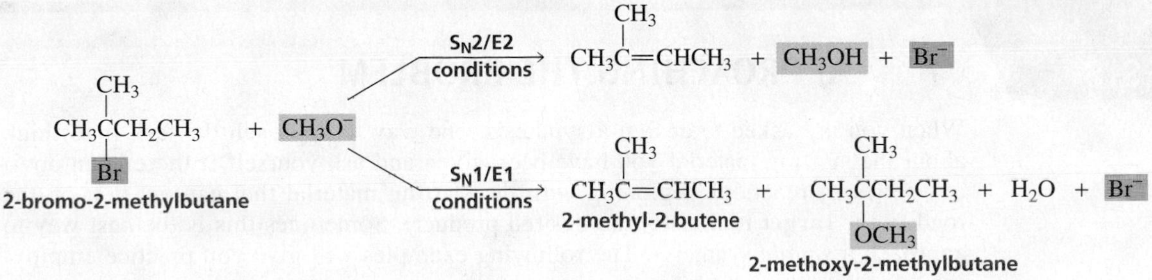

$$CH_3CCH_2CH_3 \;+\; CH_3O^-$$

2-bromo-2-methylbutane

$$\xrightarrow[\text{conditions}]{S_N2/E2} \quad CH_3C{=}CHCH_3 \;+\; CH_3OH \;+\; Br^-$$

$$\xrightarrow[\text{conditions}]{S_N1/E1} \quad CH_3C{=}CHCH_3 \;+\; CH_3CCH_2CH_3 \;+\; H_2O \;+\; Br^-$$

2-methyl-2-butene   2-methoxy-2-methylbutane

PROBLEM 29◆

Identify the three products formed when 2-bromo-2-methylpropane is dissolved in a mixture of 80% ethanol and 20% water.

PROBLEM 30

What products (including stereoisomers, if applicable) would be formed from the reaction of 3-bromo-3-methylpentane with HO^- under $S_N2/E2$ conditions and under $S_N1/E1$ conditions?

If two halogens are on the same carbon (geminal dihalides) or on adjacent carbons (vicinal dihalides), two consecutive E2 reactions will form a triple bond. This is how alkynes are commonly synthesized.

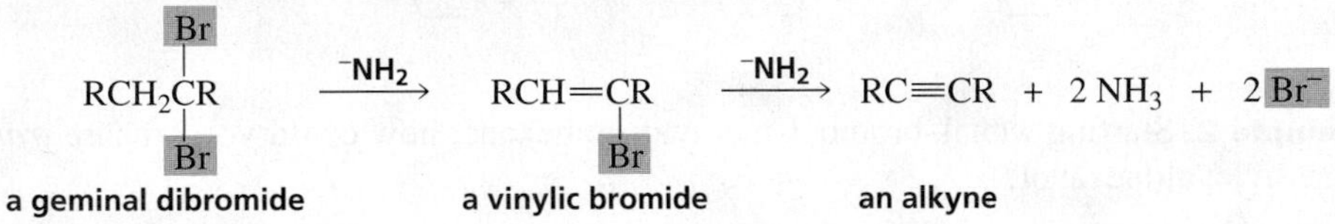

$$RCH_2CR \xrightarrow{^-NH_2} RCH{=}CR \xrightarrow{^-NH_2} RC{\equiv}CR \;+\; 2\,NH_3 \;+\; 2\,Br^-$$

a geminal dibromide   a vinylic bromide   an alkyne

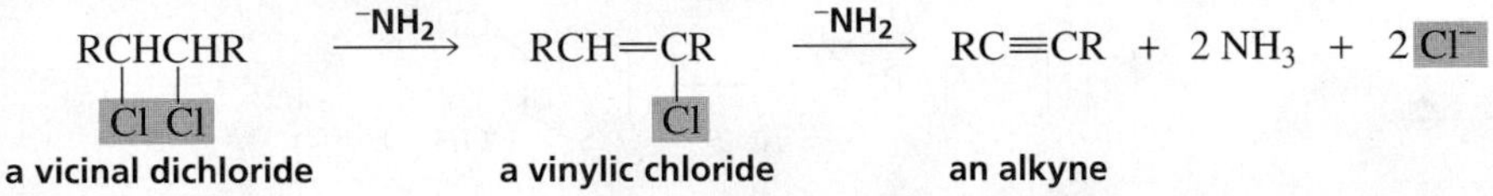

$$RCHCHR \xrightarrow{^-NH_2} RCH{=}CR \xrightarrow{^-NH_2} RC{\equiv}CR \;+\; 2\,NH_3 \;+\; 2\,Cl^-$$

a vicinal dichloride   a vinylic chloride   an alkyne

The vinylic halide intermediates in the preceding reactions are relatively unreactive. Consequently, a very strong base such as $^-NH_2$ is needed for the second elimination reaction. If a weaker base, such as HO^- is used, the reaction will stop at the vinylic halide and no alkyne will be formed.

Because a vicinal dihalide is formed from the reaction of an alkene with Br_2 or Cl_2, you have just learned how to convert a double bond into a triple bond.

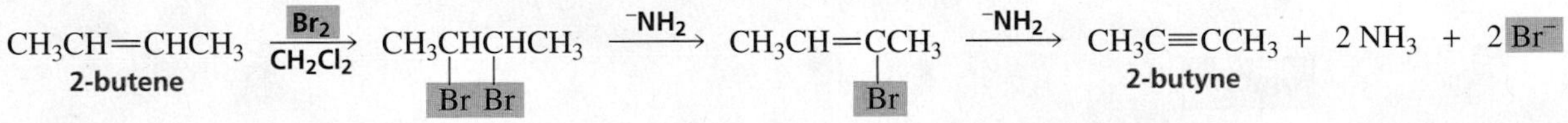

$$CH_3CH{=}CHCH_3 \xrightarrow[\text{CH}_2\text{Cl}_2]{Br_2} CH_3CHCHCH_3 \xrightarrow{^-NH_2} CH_3CH{=}CCH_3 \xrightarrow{^-NH_2} CH_3C{\equiv}CCH_3 \;+\; 2\,NH_3 \;+\; 2\,Br^-$$

2-butene   Br Br   Br   2-butyne

PROBLEM 31

Why is a cumulated diene not formed in the preceding reaction?

PROBLEM 32

What product will be obtained when the following compound undergoes two successive elimination reactions?

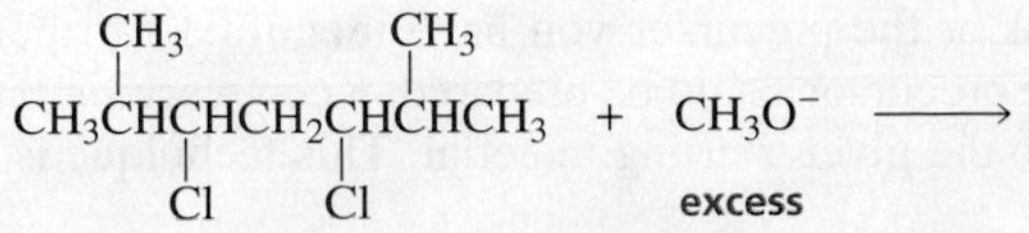

$$CH_3CHCHCH_2CHCHCH_3 \;+\; CH_3O^-\; \longrightarrow$$

    CH₃   CH₃

    Cl   Cl

excess

11   APPROACHING THE PROBLEM

When you are asked to design a synthesis, one way to approach the task is to think about the starting material you have been given and ask yourself if there is an obvious series of reactions beginning with the starting material that can get you on the road to the **target molecule** (the desired product). Sometimes this is the best way to approach a *simple* synthesis. The following examples will give you practice employing this strategy.

Example 1. Using the given starting material, how could you prepare the target molecule?

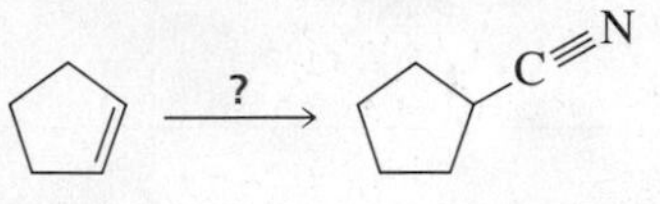

Adding HBr to the alkene would form a compound with a leaving group that can be replaced by a nucleophile. Because ⁻C≡N is a relatively weak base (the pK_a of HC≡N is 9.1), the desired substitution reaction will be favored over the competing elimination reaction.

synthesis

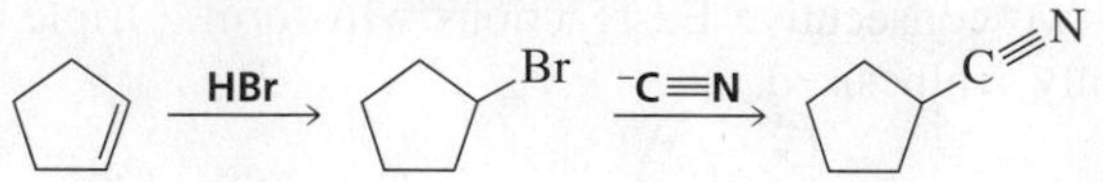

Example 2. Starting with 1-bromo-1-methylcyclohexane, how could you prepare *trans*-2-methylcyclohexanol?

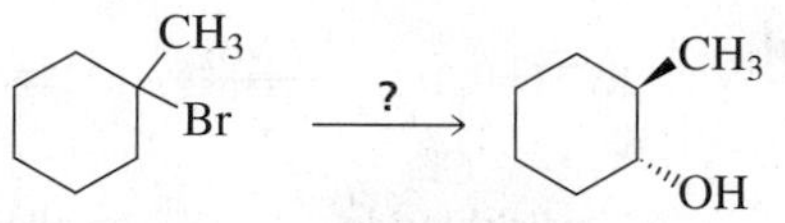

Elimination of HBr from the reactant will form an alkene that can add water *via* an electrophilic addition reaction. The elimination reaction should be carried out under E2 conditions because the tertiary alkyl halide will undergo only elimination, so there will be no competing substitution product. Hydroboration-oxidation will put the OH on the right carbon. Because R_2BH will add preferentially to the less sterically hindered side of the double bond and the overall hydroboration–oxidation reaction results in the syn addition of water, the target molecule (as well as its enantiomer) is obtained.

synthesis

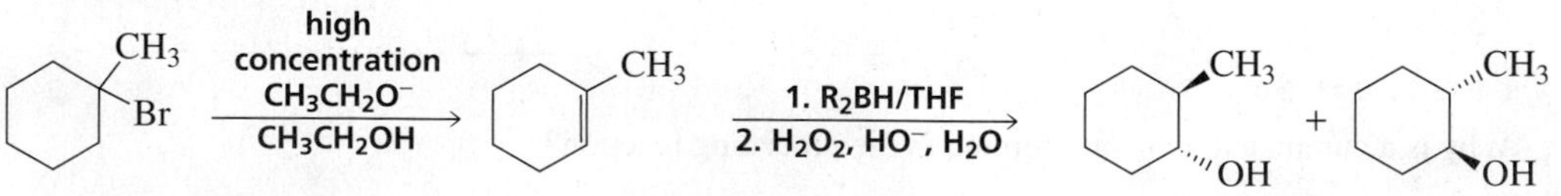

Working backward can be a useful way to design a synthesis, particularly when the starting material does not clearly indicate how to proceed as in Example 3.

Look at the target molecule and ask yourself how it could be prepared. Once you have an answer, look at the precursor you have identified for the target molecule and ask yourself how the precursor could be prepared. Keep working backward one step at a time, until you get to the given starting material. This technique is called *retrosynthetic analysis*.

Example 3. How could you prepare ethyl methyl ketone from 1-bromobutane?

$$CH_3CH_2CH_2CH_2Br \xrightarrow{\text{?}} CH_3CH_2\overset{\displaystyle O}{\overset{\displaystyle \|}{C}}CH_3$$

At this point in your study of organic chemistry, you should be aware of only three ways to synthesize a ketone: (1) the addition of water to an *alkyne*, (2) hydroboration–oxidation of an *alkyne*, and (3) ozonolysis of an *alkene*. Because the target molecule has the same number of carbons as the starting material, we can rule out ozonolysis. Now we know that the precursor molecule must be an alkyne. The alkyne needed to prepare the ketone can be prepared from two successive E2 reactions of a vicinal dihalide, which in turn can be synthesized from an alkene. The desired alkene can be prepared from the given starting material by an elimination reaction, using a bulky base to maximize the elimination product.

retrosynthetic analysis

$$CH_3CH_2\overset{\displaystyle O}{\overset{\displaystyle \|}{C}}CH_3 \implies CH_3CH_2C\equiv CH \implies \underset{\displaystyle Br}{CH_3CH_2CHCH_2Br} \implies CH_3CH_2CH=CH_2 \implies CH_3CH_2CH_2CH_2Br$$

target molecule

open arrow indicates
you are working backwards

Now you can write the reaction sequence in the forward direction, along with the reagents needed to carry out each reaction. Notice that a bulky base is used in the elimination reaction in order to maximize the amount of elimination product.

synthesis

$$CH_3CH_2CH_2CH_2Br \xrightarrow{\text{DBN}} CH_3CH_2CH=CH_2 \xrightarrow[\text{CH}_2\text{Cl}_2]{\text{Br}_2} \underset{\displaystyle Br}{CH_3CH_2CHCH_2Br} \xrightarrow{2\ ^-NH_2} CH_3CH_2C\equiv CH \xrightarrow[\substack{\text{H}_2\text{SO}_4\\\text{HgSO}_4}]{\text{H}_2\text{O}} CH_3CH_2\overset{\displaystyle O}{\overset{\displaystyle \|}{C}}CH_3$$

target molecule

Example 4. How could the following cyclic ether be prepared from the given starting material?

$$BrCH_2CH_2CH_2CH=CH_2 \xrightarrow{\text{?}} \text{(cyclic ether)}\; CH_3$$

To prepare a cyclic ether, the alkyl halide and alcohol must both be part of the same molecule. To determine the precursor to the target molecule, find the new bond that was formed in the target molecule and number the atoms in the ring formed by the new bond.

retrosynthetic analysis

new bond $$\text{(ring)} \implies \underset{\displaystyle 5\,OH}{\overset{1\quad 2\quad 3\quad 4}{BrCH_2CH_2CH_2CHCH_3}} \implies BrCH_2CH_2CH_2CH=CH_2$$

target molecule

new bond

Addition of water to the given starting material will create the required bifunctional compound, which after deprotonation will form the cyclic ether *via* an intramolecular reaction.

synthesis

$$BrCH_2CH_2CH_2CH=CH_2 \xrightarrow[\text{H}_2\text{O}]{\text{H}_2\text{SO}_4} \underset{\displaystyle OH}{BrCH_2CH_2CH_2CHCH_3} \xrightarrow{\text{NaH}} \text{(cyclic ether)}\; CH_3$$

target molecule

PROBLEM 33

How could you have prepared the target molecule in the preceding synthesis using 4-penten-1-ol as the starting material? Which synthesis would give you a higher yield of the target molecule?

PROBLEM 34

For each of the following target molecules, design a multistep synthesis to show how it could be prepared from the given starting material:

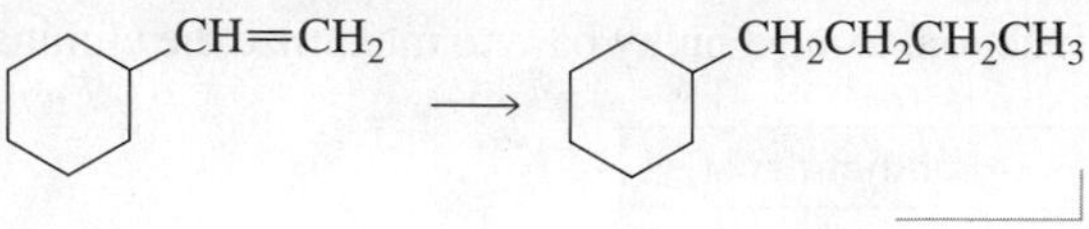

SOME IMPORTANT THINGS TO REMEMBER

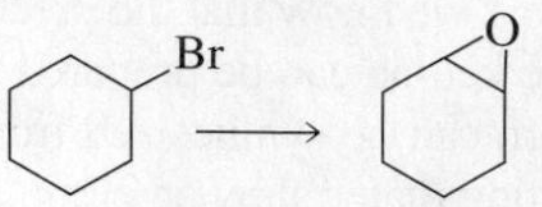

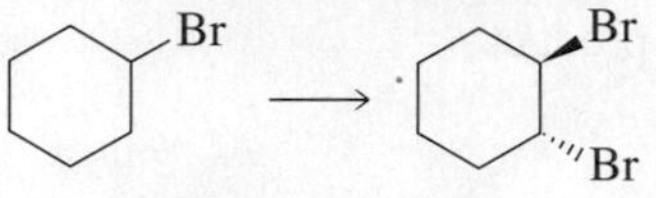

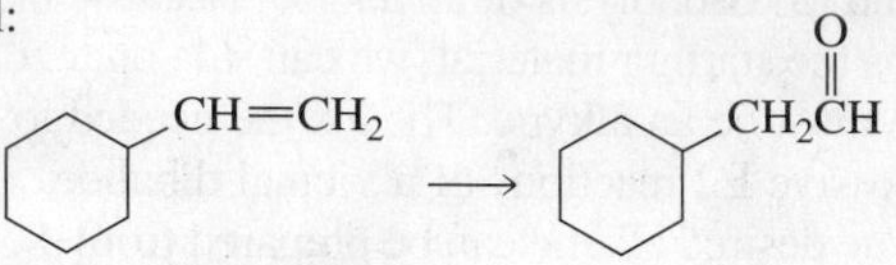

- In addition to undergoing nucleophilic substitution reactions, alkyl halides undergo β-elimination reactions. The product of an **elimination reaction** is an alkene.

- Removal of a proton and a halide ion is called **dehydrohalogenation.**

- There are two important β-**elimination** reactions: E2 and E1.

- An **E2 reaction** is a concerted, one-step reaction in which the proton and the halide ion are removed in the same step.

- An **E1 reaction** is a two-step reaction in which the alkyl halide dissociates, forming a carbocation intermediate. Then, a base removes a proton from a carbon adjacent to the positively charged carbon.

- Primary and secondary alkyl halides undergo only E2 reactions.

- Tertiary alkyl halides, allylic halides, and benzylic halides undergo both E2 and E1 reactions.

- For alkyl halides that can undergo both E2 and E1 reactions, the E2 reaction is favored by the same factors that favor an S_N2 reaction (namely, a high concentration of a strong base) and the E1 reaction is favored by the same factors that favor an S_N1 reaction (namely, a weak base).

- An E2 reaction is regioselective; the major product is the more stable alkene, unless the reactants are sterically hindered or the leaving group is poor.

- The more stable alkene is generally (but not always) the more substituted alkene. The more substituted alkene is predicted by **Zaitsev's rule**: it is the alkene formed when a hydrogen is removed from the β-**carbon** bonded to the fewest hydrogens.

- Alkyl substitution increases the stability of a carbocation and decreases the stability of a carbanion.

- An E2 reaction is also stereoselective: **anti elimination** is favored. If the β-carbon has two hydrogens, then both E and Z products will be formed, but the one with the largest groups on opposite sides of the double bond will be formed in greater yield because it is more stable. If the β-carbon is bonded to only one hydrogen, then only one alkene is formed; its structure depends on the structure of the alkyl halide.

- An E1 reaction is regioselective: the major product is the more stable alkene.

- An E1 reaction is also stereoselective: the major product is the alkene with the largest groups on opposite sides of the double bond.

- The two groups eliminated from a six-membered ring must both be in axial positions in an E2 reaction; elimination is more rapid when H and X are diaxial in the more stable conformer. The two groups do not have to both be in axial positions in an E1 reaction.

- When S_N2/E2 reactions are favored, primary alkyl halides form primarily substitution products unless the nucleophile/base is sterically hindered, in which case elimination products predominate. Secondary alkyl halides form both substitution and elimination products; the percentage of the elimination product increases as the strength and bulk of the base increases and as the temperature increases. Tertiary alkyl halides form only elimination products.

- When $S_N1/E1$ conditions are favored, tertiary alkyl halides, allylic halides, and benzylic halides can form both substitution and elimination products; primary and secondary alkyl halides do not undergo $S_N1/E1$ reactions.

- The **Williamson ether synthesis** prepares ethers from the reaction of an alkyl halide with an alkoxide ion.
- If two halogens are on the same or adjacent carbons, two consecutive E2 dehydrohalogenations form a triple bond.

SUMMARY OF REACTIONS

1. E2 reaction: a one-step mechanism

Relative reactivities of alkyl halides: $3° > 2° > 1°$

Anti elimination only: if the β-carbon from which the hydrogen is removed *is bonded to two hydrogens,* then both *E* and *Z* stereoisomers are formed. The isomer with the largest groups on opposite sides of the double bond is the major product. If the β-carbon from which the hydrogen is removed *is bonded to only one hydrogen,* then only one elimination product is formed. Its configuration depends on the configuration of the reactant.

2. E1 reaction: a two-step mechanism with a carbocation intermediate

Only tertiary alkyl halides, allylic halides, and benzylic halides undergo E1 reactions.

Anti and syn elimination: both *E* and *Z* stereoisomers are formed. The isomer with the largest groups on opposite sides of the double bond is the major product.

Competing S_N2 and E2 Reactions
Primary alkyl halides: primarily substitution. Secondary alkyl halides: substitution and elimination.
Allylic and benzylic halides: substitution and elimination (unless they are tertiary). Tertiary alkyl halides: only elimination.

Competing S_N1 and E1 Reactions
Primary and secondary alkyl halides: neither substitution nor elimination.
Tertiary alkyl halides, allylic halides, and benzylic halides: substitution and elimination.

PROBLEMS

35. Draw the major product obtained when each of the following alkyl halides undergoes an E2 reaction:

a. CH$_3$CHCH$_2$CH$_3$
     |
     Br

c. CH$_3$CHCH$_2$CH$_3$
     |
     Cl

e. CH$_3$CHCH$_2$CH$_2$CH$_3$
     |
     Cl

b.

d.

f.

36. Draw the major product obtained when each alkyl halide in Problem 35 undergoes an E1 reaction.

37. a. Indicate how each of the following factors affects an E1 reaction:
    1. the strength of the base   **2.** the concentration of the base   **3.** the solvent
  b. Indicate how each of the same factors affects an E2 reaction.

38. Which species in each pair is more stable?

a. $CH_3\overset{-}{C}HCH_2CH_3$ or $CH_3CH_2CH_2\overset{-}{C}H_2$

e. $\overset{+}{C}H_2CH_2CH=CH_2$ or $CH_3\overset{+}{C}HCH=CH_2$

b. $CH_3\overset{+}{C}HCH_2CH_3$ or $CH_3CH_2CH_2\overset{+}{C}H_2$

f. $CH_3CH\overset{-}{C}HCH_3$ or $CH_3\overset{-}{C}CH_2CH_3$
   | |
   CH_3 CH_3

c. $\overset{-}{C}H_2CH_2CH=CH_2$ or $CH_3\overset{-}{C}HCH=CH_2$

d. $CH_3CHCH=CH_2$ or $CH_3CH_2C=CH_2$
   | |
   CH_3 CH_3

g. $CH_3CH\overset{+}{C}HCH_3$ or $CH_3\overset{+}{C}CH_2CH_3$
   | |
   CH_3 CH_3

39. A chemist wanted to synthesize the anesthetic 2-ethoxy-2-methylpropane. He used ethoxide ion and 2-chloro-2-methylpropane for his synthesis and ended up with no ether. What was the product of his synthesis? What reagents should he have used?

40. Which reactant in each of the following pairs will undergo an elimination reaction more rapidly? Explain your choice.

a. $(CH_3)_3CCl$ $\xrightarrow[CH_3OH]{CH_3O^-}$ or $(CH_3)_3CBr$ $\xrightarrow[CH_3OH]{CH_3O^-}$

b. 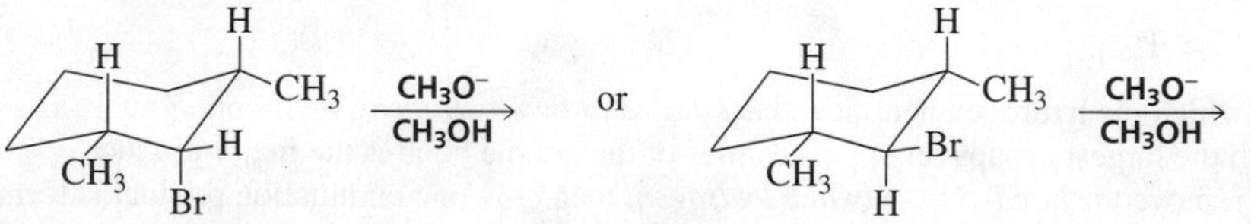 [structure] $\xrightarrow[CH_3OH]{CH_3O^-}$ or   [structure] $\xrightarrow[CH_3OH]{CH_3O^-}$

41. For each of the following reactions, draw the major elimination product; if the product can exist as stereoisomers, indicate which stereoisomer is obtained in greater yield.
a. (*R*)-2-bromohexane + high concentration of CH_3O^-
b. (*R*)-3-bromo-3-methylhexane + CH_3OH
c. *trans*-1-chloro-2-methylcyclohexane + high concentration of CH_3O^-
d. *trans*-1-chloro-3-methylcyclohexane + high concentration of CH_3O^-
e. 3-bromo-3-methylpentane + high concentration of $CH_3CH_2O^-$
f. 3-bromo-3-methylpentane + CH_3CH_2OH

42. a. Which reacts faster in an E2 reaction, 3-bromocyclohexene or bromocyclohexane?
b. Which reacts faster in an E1 reaction?

43. Starting with an alkyl halide, how could the following compounds be prepared?
a. 2-methoxybutane   **b.** 1-methoxybutane   **c.** butylmethylamine

44. Indicate which of the compounds in each pair will give a higher substitution-product to elimination-product ratio when it reacts with isopropyl bromide:
a. ethoxide ion or *tert*-butoxide ion   **b.** ^-OCN or ^-SCN **c.** Cl^- or Br^- **d.** CH_3S^- or CH_3O^-

45. Rank the following compounds in order from most reactive to least reactive in an E2 reaction:

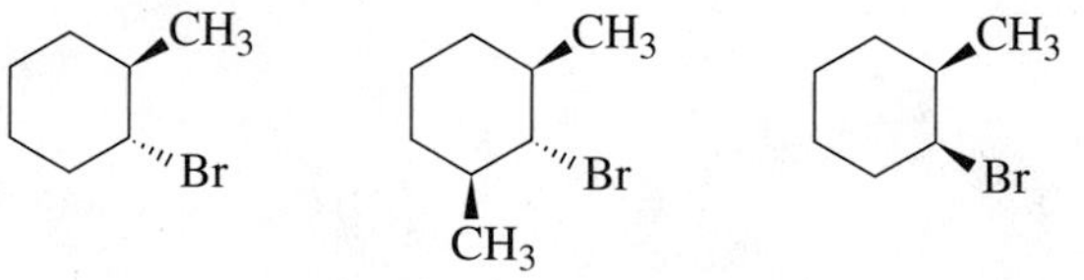

[structures with CH₃ and Br substituents on cyclohexane rings]

46. For each of the following alkyl halides, indicate what stereoisomer would be obtained in greatest yield if it reacts with a high concentration of ethoxide ion.
a. 3-bromo-2,2,3-trimethylpentane   **c.** 3-bromo-2,3-dimethylpentane
b. 4-bromo-2,2,3,3-tetramethylpentane   **d.** 3-bromo-3,4-dimethylhexane

47. Which of following ethers cannot be made by a Williamson ether synthesis?

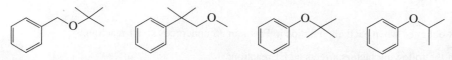

48. When 2-bromo-2,3-dimethylbutane reacts with a base under E2 conditions, two alkenes (2,3-dimethyl-1-butene and 2,3-dimethyl-2-butene) are formed.
   a. Which of the bases (A, B, C, or D) would form the highest percentage of the 1-alkene?
   b. Which would give the highest percentage of the 2-alkene?

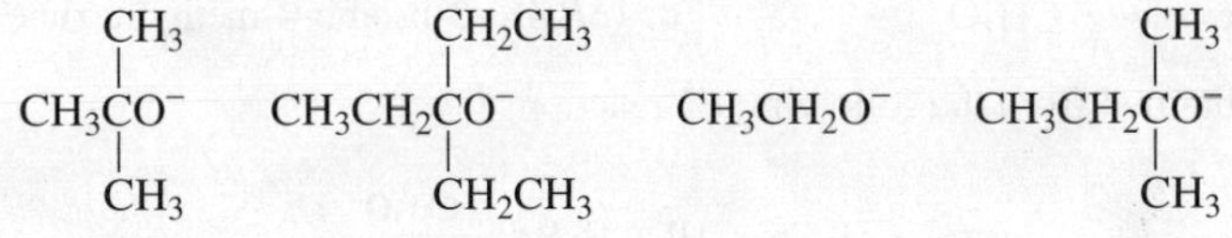

49. a. Draw the structures of the products obtained from the reaction of each enantiomer of *cis*-1-chloro-2-isopropylcyclopentane with a high concentration of sodium methoxide in methanol.
   b. Are all the products optically active?
   c. How would the products differ if the starting material were the trans isomer? Are these products optically active?
   d. Will the cis enantiomers or the trans enantiomers form substitution products more rapidly?
   e. Will the cis enantiomers or the trans enantiomers form elimination products more rapidly?

50. When the following compound undergoes solvolysis in ethanol, three products are obtained. Propose a mechanism to account for the formation of these products.

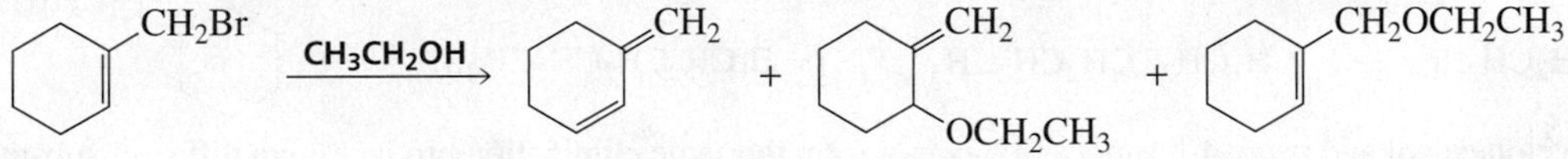

51. *cis*-1-Bromo-4-*tert*-butylcyclohexane and *trans*-1-bromo-4-*tert*-butylcyclohexane both react with sodium ethoxide in ethanol to form 4-*tert*-butylcyclohexene. Explain why the cis isomer reacts much more rapidly than the trans isomer.

52. Draw the elimination products obtained under E2 conditions for each of the following alkyl halides.

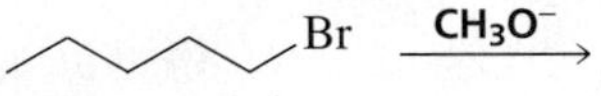

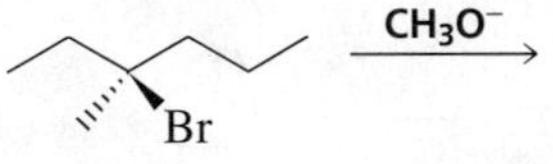

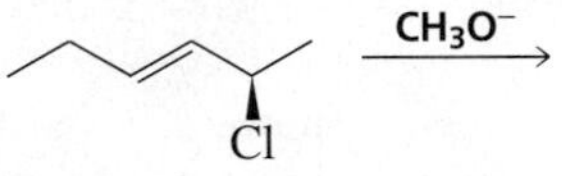

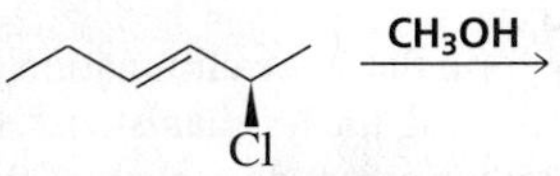

53. Draw the elimination products for each of the following reactions; if the products can exist as stereoisomers, indicate which stereoisomers are obtained.
   a. (2S,3S)-2-chloro-3-methylpentane + high concentration of CH_3O^-
   b. (2S,3R)-2-chloro-3-methylpentane + high concentration of CH_3O^-
   c. (2R,3S)-2-chloro-3-methylpentane + high concentration of CH_3O^-
   d. (2R,3R)-2-chloro-3-methylpentane + high concentration of CH_3O^-
   e. 3-chloro-3-ethyl-2,2-dimethylpentane + high concentration of $CH_3CH_2O^-$

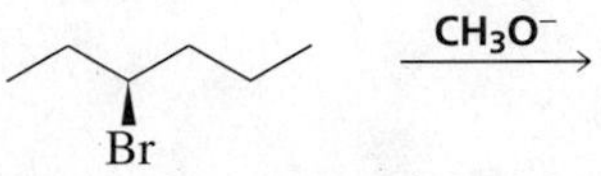

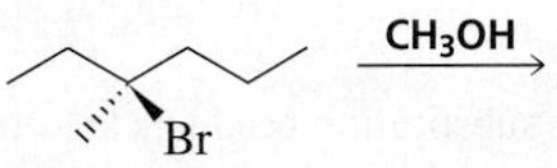

54. Draw the major elimination product that would be obtained from each of the following reactants under E1 and E2 conditions:

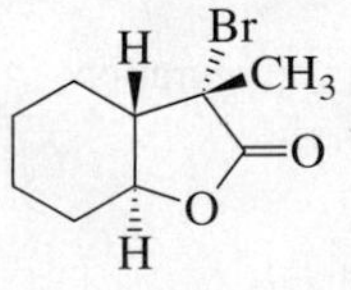

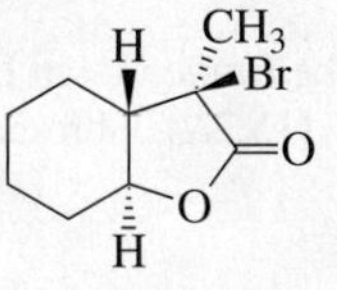

55. Which of the following hexachlorocyclohexanes is the least reactive in an E2 reaction?

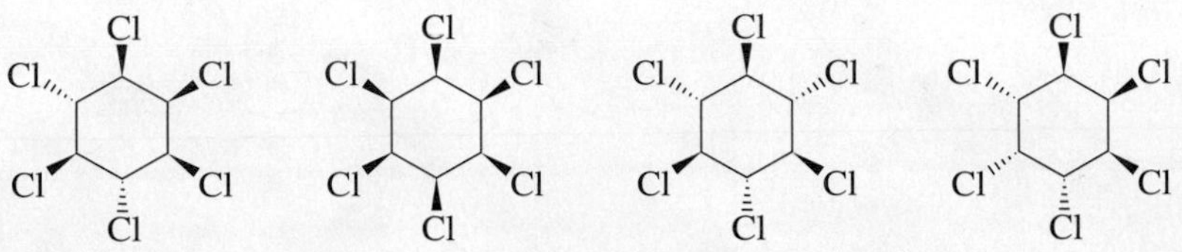

56. Explain why the rate of the reaction of 1-bromo-2-butene with ethanol is increased if silver nitrate is added to the reaction mixture.

57. Draw the products of each of the following reactions carried out under S$_N$2/E2 conditions. If the products can exist as stereoisomers, show which stereoisomers are formed.
    a. (3S,4S)-3-bromo-4-methylhexane + CH$_3$O$^-$
    b. (3R,4R)-3-bromo-4-methylhexane + CH$_3$O$^-$
        c. (3S,4R)-3-bromo-4-methylhexane + CH$_3$O$^-$
        d. (3R,4S)-3-bromo-4-methylhexane + CH$_3$O$^-$

58. Two elimination products are obtained from the following E2 reaction:

$$CH_3CH_2CHDCH_2Br \xrightarrow{CH_3O^-}$$

    a. What are the elimination products?
    b. Which is formed in greater yield? Explain.

59. Draw the structures of the products obtained from the following reaction:

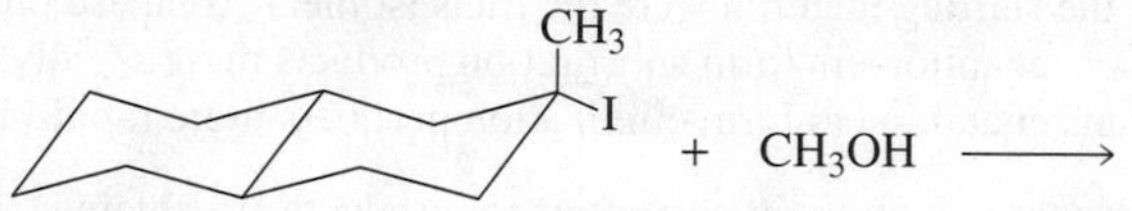

60. How could you prepare the following compounds from the given starting materials?

    a. $CH_3CH_2CH_2CH_2Br \longrightarrow CH_3CH_2\overset{\overset{\textstyle O}{\|}}{C}CH_2CH_2CH_3$
    b. $BrCH_2CH_2CH_2CH_2Br \longrightarrow$

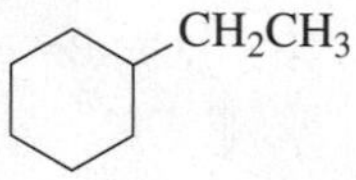

61. *cis*-4-Bromocyclohexanol and *trans*-4-bromocyclohexanol form the same elimination product but a different substitution product when they react with HO$^-$.

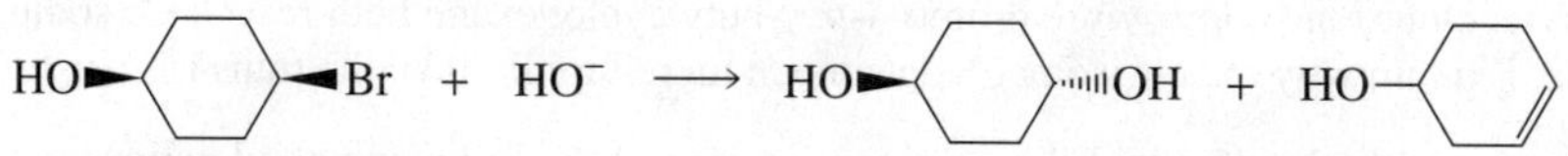

cis-**4-bromocyclohexanol**

trans-**4-bromocyclohexanol**

    a. Why do they form the same elimination product?
    b. Explain, by showing the mechanisms, why different substitution products are obtained.
    c. How many stereoisomers does each of the elimination and substitution reactions form?

62. What products are formed when the following stereoisomer of 2-chloro-1,3-dimethylcyclohexane reacts with methoxide ion in a solvent that encourages S$_N$2/E2 reactions:

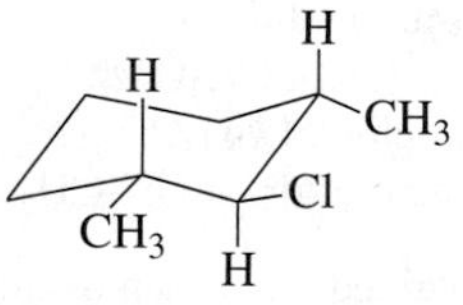

63. For each of the following compounds, draw the product that will be formed in an E2 reaction and indicate its configuration:
    a. (1S,2S)-1-bromo-1,2-diphenylpropane
        b. (1S,2R)-1-bromo-1,2-diphenylpropane

GLOSSARY

anti elimination an elimination reaction in which the two substituents that are eliminated are removed from opposite sides of the molecule.

anti-periplanar parallel substituents on opposite sides of a molecule.

α-carbon a carbon bonded to a leaving group or adjacent to a carbonyl carbon.

β-carbon a carbon bonded an α-carbon.

dehydrohalogenation elimination of a proton and a halide ion.

deuterium kinetic isotope effect ratio of the rate constant obtained for a compound containing hydrogen and the rate constant obtained for an identical compound in which one or more of the hydrogens have been replaced by deuterium.

β-elimination removal of two atoms or groups from adjacent carbons.

elimination reaction a reaction that involves the elimination of atoms (or molecules) from the reactant.

E1 reaction a first-order elimination reaction.

E2 reaction a second-order elimination reaction.

syn elimination an elimination reaction in which the two substituents that are eliminated are removed from the same side of the molecule.

syn-periplanar parallel substituents on the same side of a molecule.

target molecule desired end product of a synthesis.

Williamson ether synthesis formation of an ether from the reaction of an alkoxide ion with an alkyl halide.

Zaitsev's rule the more substituted alkene product is obtained by removing a proton from the β-carbon that is bonded to the fewest hydrogens.

PHOTO CREDITS

Photo credits are listed in order of appearance.

Nikita Tiunov/Shutterstock; Dr. Paula Bruice; (top) Mary Evans Picture Library/Alamy; (bot.) Rolf Adlercreutz/Alamy

Reactions of Alcohols, Ethers, Epoxides, Amines, and Thiols

dried coca leaves

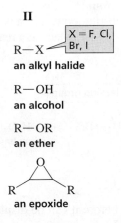

II

X = F, Cl, Br, I

R—X
an alkyl halide

R—OH
an alcohol

R—OR
an ether

R—R (epoxide)
an epoxide

Chemists search the world for plants and berries and search the ocean for flora and fauna that might be used as the source of a lead compound for the development of a new drug. In this chapter, we will see how cocaine, which is obtained from the leaves of Erythroxylon coca—a bush native to the highlands of the South American Andes, was used as the source of a lead compound for the development of some common anesthetics.

Alkyl halides, a family of compounds in Group II, undergo substitution and/or elimination reactions because of their electron-withdrawing halogen atoms. Other families of compounds in Group II also have electron-withdrawing groups, and they too undergo substitution and/or elimination reactions. The relative reactivity of these compounds depends on the electron-withdrawing group—that is, on the leaving group.

The leaving groups of alcohols and ethers (HO^-, RO^-) are much stronger bases than the leaving group of an alkyl halide. Because they are stronger bases, they are poorer leaving groups, and therefore are harder to displace. Consequently, alcohols and ethers are less reactive than alkyl halides in substitution and elimination reactions. We will see that alcohols and ethers have to be "activated" before they can undergo a substitution or an elimination reaction.

pK_a HX = −10 to 3.2

pK_a $R_3\overset{+}{N}H$ = ~10

pK_a H_2O = ~15

pK_a ROH = ~15

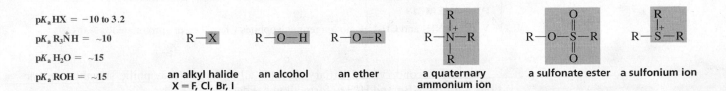

R—X
an alkyl halide
X = F, Cl, Br, I

R—O—H
an alcohol

R—O—R
an ether

R—N—R (with R above and below)
a quaternary ammonium ion

R—O—S—R
a sulfonate ester

R—S—R (with R above)
a sulfonium ion

From Chapter 11 of *Organic Chemistry*, Seventh Edition. Paula Yurkanis Bruice. Copyright © 2014 by Pearson Education, Inc.

Tertiary amines, the leaving groups of **quaternary ammonium ions**, are not only less basic than the leaving groups of alcohols and ethers but they also have a positive charge that enhances their leaving ability. Therefore, quaternary ammonium ions will undergo elimination reactions as long as a strong base is present and the reaction is heated. **Sulfonate esters** and **sulfonium ions** have very good leaving groups, the first because of electron delocalization and the second because of its positive charge. Thus, they undergo substitution and/or elimination reactions with ease.

The stronger the acid, the weaker its conjugate base.

When bases with similar features are compared, it is found that the weaker the base, the more easily it can be displaced.

1 NUCLEOPHILIC SUBSTITUTION REACTIONS OF ALCOHOLS: FORMING ALKYL HALIDES

An **alcohol** has a strongly basic leaving group (HO^-) that cannot be displaced by a nucleophile. Therefore, an alcohol cannot undergo a nucleophilic substitution reaction.

a strongly basic leaving group

$$CH_3-\overset{..}{\underset{..}{O}}H \ + \ Br^- \ \xcancel{\longrightarrow} \ CH_3-Br \ + \ HO^-$$
strong base

However, if the alcohol's OH group is converted into a group that is a weaker base (and therefore a better leaving group), a nucleophilic substitution reaction can occur.

One way to convert an OH group into a weaker base is to protonate it by adding acid to the reaction mixture. Protonation changes the leaving group from HO^- to H_2O, which is a weak enough base to be displaced by a nucleophile. The substitution reaction is slow and requires heat (except in the case of tertiary alcohols) if it is to take place at a reasonable rate.

a weakly basic leaving group

$$CH_3-\overset{..}{\underset{..}{O}}H \ + \ HBr \ \rightleftharpoons \ CH_3-\overset{+}{\underset{..}{O}}{\underset{H}{H}} \ \overset{\Delta}{\longrightarrow} \ CH_3-Br \ + \ H_2O$$
$$Br^-$$
weak base

poor leaving group

good leaving group

Because the OH group of the alcohol has to be protonated before it can be displaced by a nucleophile, only weakly basic nucleophiles (I^-, Br^-, Cl^-) can be used in the substitution reaction. Moderately and strongly basic nucleophiles $(NH_3, RNH_2,$ and $CH_3O^-)$ cannot be used because they too would be protonated in the acidic solution and, once protonated, would no longer be nucleophiles $(^+NH_4, R\overset{+}{N}H_3)$ or would be poor nucleophiles (CH_3OH).

PROBLEM 1♦

Why are NH_3 and CH_3NH_2 no longer nucleophiles when they are protonated?

Primary, secondary, and tertiary alcohols all undergo nucleophilic substitution reactions with HI, HBr, and HCl to form alkyl halides.

$$CH_3CH_2CH_2\boxed{OH} + HI \xrightarrow{\Delta} CH_3CH_2CH_2\boxed{I} + H_2O$$

1-propanol
a primary alcohol

1-iodopropane

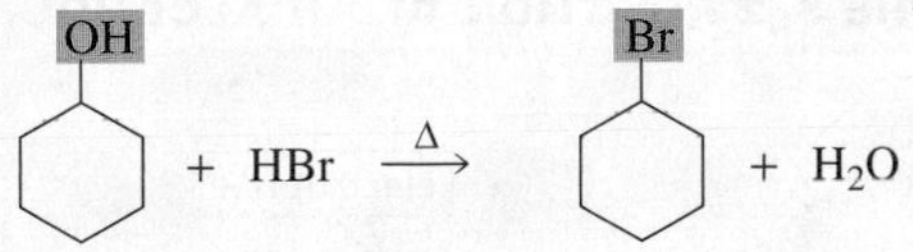

cyclohexanol
a secondary alcohol

bromocyclohexane

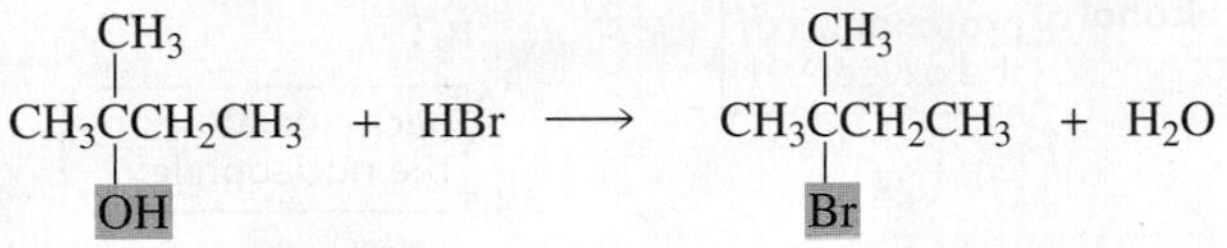

$$\underset{\underset{\boxed{OH}}{|}}{\overset{\overset{CH_3}{|}}{CH_3CCH_2CH_3}} + HBr \longrightarrow \underset{\underset{\boxed{Br}}{|}}{\overset{\overset{CH_3}{|}}{CH_3CCH_2CH_3}} + H_2O$$

2-methyl-2-butanol
a tertiary alcohol

2-bromo-2-methylbutane

The mechanism of the substitution reaction depends on the structure of the alcohol. Secondary and tertiary alcohols undergo S_N1 reactions.

MECHANISM FOR THE S_N1 REACTION OF AN ALCOHOL

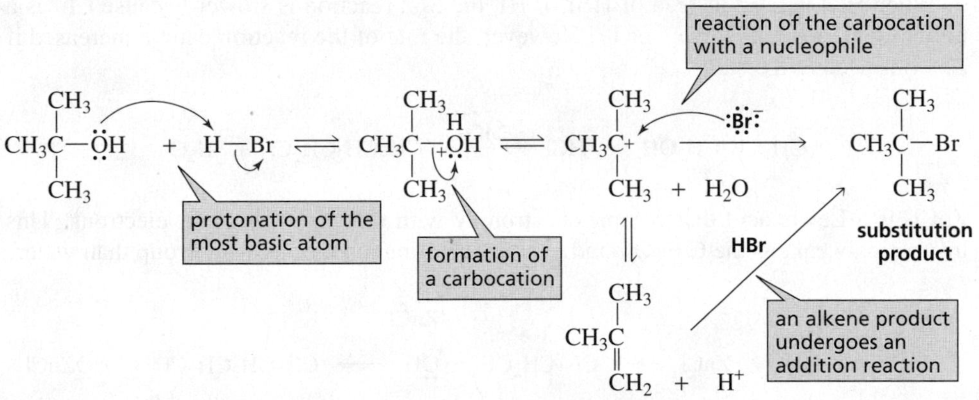

reaction of the carbocation with a nucleophile

protonation of the most basic atom

formation of a carbocation

HBr

substitution product

an alkene product undergoes an addition reaction

elimination product

An acid protonates the most basic atom in a molecule.

- An acid always reacts with an organic molecule in the same way: it protonates the most basic atom in the molecule.
- Weakly basic water is the leaving group that is expelled, forming a carbocation.
- The carbocation, like the carbocation formed when an alkyl halide dissociates in an S_N1 reaction, has two possible fates: it can combine with a nucleophile and form a substitution product, or it can lose a proton and form an elimination product.

Although the reaction can form both a substitution product and an elimination product, little elimination product is actually obtained because the alkene formed in an elimination reaction can undergo a subsequent electrophilic addition reaction with HBr to form more of the substitution product.

Tertiary alcohols undergo substitution reactions with hydrogen halides faster than secondary alcohols do, because tertiary carbocations are more stable and therefore easier to form than secondary carbocations. (Note that alkyl groups stabilize carbocations by hyperconjugation.) As a result, the reaction of a tertiary alcohol with a hydrogen halide proceeds readily at room temperature, whereas the reaction of a secondary alcohol with a hydrogen halide has to be heated to have the reaction occur at a reasonable rate.

Carbocation stability: 3° > 2° > 1°

Primary alcohols cannot undergo S_N1 reactions because primary carbocations are too unstable to be formed, even when the reaction is heated. Therefore, when a primary alcohol reacts with a hydrogen halide, it must do so by an S_N2 reaction.

MECHANISM FOR THE S_N2 REACTION OF AN ALCOHOL

Secondary and tertiary alcohols undergo S_N1 reactions with hydrogen halides.

Primary alcohols undergo S_N2 reactions with hydrogen halides.

A β-carbon is the carbon adjacent to the carbon that is attached to the leaving group.

electrophile

$$CH_3CH_2\overset{..}{\underset{..}{O}}H \quad + \quad H\!-\!Br \quad \rightleftharpoons \quad CH_3CH_2\!-\!\overset{\overset{H}{|}}{\underset{+}{O}}H \quad \longrightarrow \quad CH_3CH_2Br \quad + \quad H_2O$$

ethanol
a primary alcohol

protonation of the oxygen

$:\overset{..}{\underset{..}{Br}}:^-$

back-side attack by the nucleophile

- The acid protonates the most basic atom in the reactant.
- The nucleophile attacks the back side of the carbon and displaces the leaving group.

Only a substitution product is obtained. No elimination product is formed because the halide ion, although a good nucleophile, is a weak base in a reaction mixture that contains alcohol and water (that is, in a polar, protic solvent), and a strong base is required to remove a hydrogen from a β-carbon in an E2 reaction.

When HCl is used instead of HBr or HI, the S_N2 reaction is slower because Cl^- is a poorer nucleophile than Br^- or I^-. However, the rate of the reaction can be increased if $ZnCl_2$ is used as a catalyst.

$$CH_3CH_2CH_2\text{OH} \quad + \quad HCl \quad \xrightarrow[\Delta]{ZnCl_2} \quad CH_3CH_2CH_2Cl \quad + \quad H_2O$$

$ZnCl_2$ is a Lewis acid that complexes strongly with oxygen's lone-pair electrons. This interaction weakens the C—O bond, thereby creating a better leaving group than water.

$$CH_3CH_2CH_2\overset{..}{\underset{..}{O}}H \quad + \quad ZnCl \quad \longrightarrow \quad CH_3CH_2CH_2\!-\!\overset{\overset{ZnCl}{|}}{\underset{+}{O}}H \quad \longrightarrow \quad CH_3CH_2CH_2Cl \quad + \quad HOZnCl$$

$\underset{Cl}{|}$

$:\overset{..}{\underset{..}{Cl}}:^-$

The Lucas Test

Before spectroscopy became available for structural analysis, chemists had to identify compounds by tests that gave visible results. The Lucas test is one such test. It determines whether an alcohol is primary, secondary, or tertiary by taking advantage of the relative rates at which the three classes of alcohols react with $HCl/ZnCl_2$.

To carry out the test, the alcohol is added to a mixture of HCl and $ZnCl_2$ (known as Lucas reagent). Low-molecular-weight alcohols are soluble in Lucas reagent, but the alkyl halide products are not, so they cause the solution to turn cloudy. If the alcohol is tertiary, the solution turns cloudy immediately. If the alcohol is secondary, the solution turns cloudy in approximately one to five minutes. If the alcohol is primary, the solution turns cloudy only if it is heated. Because the test relies on the complete solubility of the alcohol in Lucas reagent, it is limited to alcohols with fewer than six carbons.

PROBLEM 2♦

The observed relative reactivities of primary, secondary, and tertiary alcohols with a hydrogen halide are 3° > 2° > 1°. If secondary alcohols were to undergo an S_N2 reaction rather than an S_N1 reaction with a hydrogen halide, what would be the relative reactivities of the three classes of alcohols?

PROBLEM 3 Solved

Using the pK_a values of the conjugate acids of the leaving groups (the pK_a of HBr is -9, the pK_a of H_2O is 15.7, and the pK_a of H_3O^+ is -1.7), explain the difference in reactivity in substitution reactions between

a. CH_3Br and CH_3OH. **b.** $CH_3\overset{+}{O}H_2$ and CH_3OH.

Solution to 3a The conjugate acid of the leaving group of CH_3Br is HBr; the conjugate acid of the leaving group of CH_3OH is H_2O. Because HBr is a much stronger acid ($pK_a = -9$) than H_2O ($pK_a = 15.7$), Br^- is a much weaker base than HO^-. (Recall that the stronger the acid, the weaker its conjugate base.) Therefore, Br^- is a much better leaving group than HO^-, causing CH_3Br to be much more reactive than CH_3OH in a substitution reaction.

PROBLEM 4 Solved

Show how 1-butanol can be converted into the following compounds:

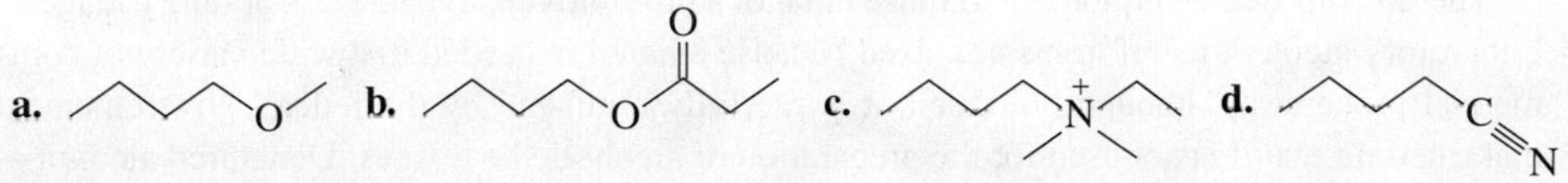

Solution to 4a Because the OH group of 1-butanol is too basic to allow the alcohol to undergo a substitution reaction with CH_3O^-, the alcohol must first be converted into an alkyl halide. The alkyl halide has a leaving group that can be substituted by CH_3O^-, the nucleophile required to obtain the desired product.

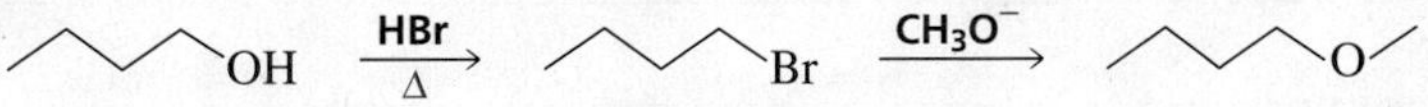

Because the reaction of a secondary alcohol with a hydrogen halide is an S_N1 reaction, a carbocation is formed as an intermediate. Therefore, we must check for the possibility of a carbocation rearrangement when determining the product of the substitution reaction. A carbocation rearrangement will occur if it leads to formation of a more stable carbocation.

For example, the major product of the following reaction is 2-bromo-2-methylbutane, because a 1,2-hydride shift converts the initially formed secondary carbocation into a more stable tertiary carbocation.

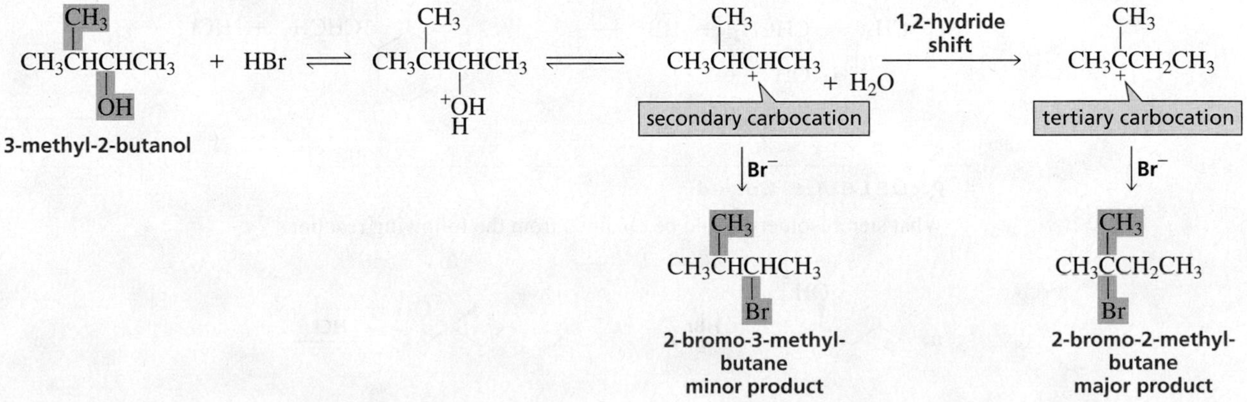

Grain Alcohol and Wood Alcohol

When ethanol is ingested, it acts on the central nervous system. Moderate amounts affect judgment and lower inhibitions. Higher amounts interfere with motor coordination and cause slurred speech and amnesia. Still higher amounts cause nausea and loss of consciousness. Ingesting very large amounts of ethanol interferes with spontaneous respiration and can be fatal.

The ethanol in alcoholic beverages is produced by the fermentation of glucose, generally obtained from grapes or from grains such as corn, rye, and wheat (which is why ethanol is also known as grain alcohol). Grains are cooked in the presence of malt (sprouted barley) to convert much of their starch into glucose. Yeast enzymes are added to convert the glucose into ethanol and carbon dioxide.

$$C_6H_{12}O_6 \xrightarrow{\textbf{yeast enzymes}} 2\ CH_3CH_2OH\ +\ 2\ CO_2$$
$$\text{glucose} \qquad\qquad\qquad \text{ethanol}$$

The kind of beverage produced (white or red wine, beer, scotch, bourbon, champagne) depends on the plant species providing the glucose, whether the CO_2 formed in the fermentation is allowed to escape, whether other substances are added, and how the beverage is purified (by sedimentation, for wines; by distillation, for scotch and bourbon).

The tax imposed on liquor would make ethanol a prohibitively expensive laboratory reagent. Laboratory alcohol, therefore, is not taxed because ethanol is needed in a wide variety of commercial processes. Although not taxed, it is carefully regulated by the federal government to make certain that it is not used for the preparation of alcoholic beverages. Denatured alcohol—ethanol that has been made undrinkable by the addition of a denaturant such as benzene or methanol—is not taxed, but the added impurities make it unfit for many laboratory uses.

Methanol, also known as wood alcohol (because at one time it was obtained by heating wood in the absence of oxygen), is highly toxic. Ingesting even very small amounts can cause blindness, and ingesting as little as an ounce has been fatal.

PROBLEM 5♦

What is the major product of each of the following reactions?

a. $CH_3CH_2CHCH_3$ + HBr $\xrightarrow{\Delta}$
         |
        OH

d. [structure] + HBr $\xrightarrow{\Delta}$

b. [cyclopentane with CH_3 and OH] + HCl $\longrightarrow$

e. [cyclohexane with CH_3 and OH] + HCl $\xrightarrow{\Delta}$

c. $CH_3C\!-\!CHCH_3$ + HBr $\xrightarrow{\Delta}$
with CH_3 above and CH_3, OH below

f. [cyclohexane structure] $CHCH_3$ + HCl $\xrightarrow{\Delta}$

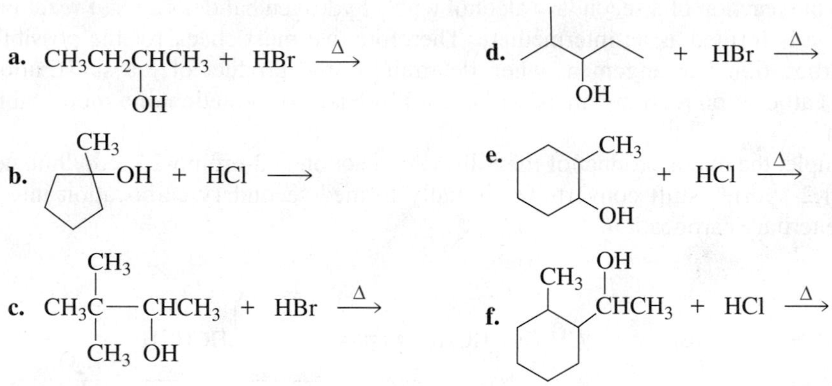

PROBLEM 6 Solved

What stereoisomers would be obtained from the following reactions?

a. [structure with OH] $\xrightarrow[\Delta]{HBr}$

b. [structure with OH] $\xrightarrow[\Delta]{HCl}$

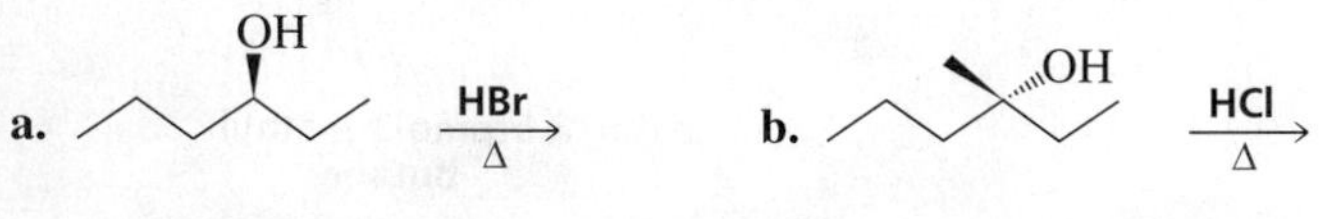

Solution to 6a This must be an S_N1 reaction because the reactant is a secondary alcohol. Therefore, the reaction forms a carbocation intermediate. The bromide ion can attach to the carbocation from either the side from which water left or from the opposite side, so both the R and S stereoisomers are formed.

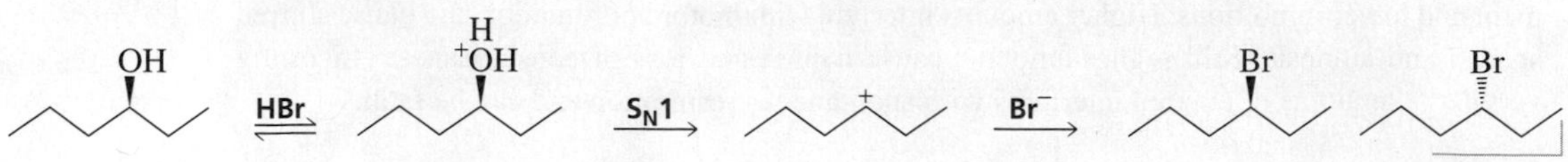

[reaction scheme: OH structure $\xrightarrow{HBr}$ ^+OH structure $\xrightarrow{S_N1}$ carbocation $\xrightarrow{Br^-}$ two Br stereoisomer products]

2 OTHER METHODS USED TO CONVERT ALCOHOLS INTO ALKYL HALIDES

Alcohols are inexpensive and readily available compounds, but they do not undergo nucleophilic substitution because the HO⁻ group is too basic to be displaced by a nucleophile (Section 1). Chemists, therefore, need ways to convert readily available but unreactive alcohols into reactive alkyl halides that can be used as starting materials for the preparation of a wide variety of organic compounds.

$$R-OH \xrightarrow[\Delta]{HX} R-X$$

alcohol   alkyl halide
                X = Cl, Br, I

$$^{-}OCH_3 \longrightarrow R-OCH_3$$

$$^{-}C{\equiv}CCH_3 \longrightarrow R-C{\equiv}CCH_3$$

$$^{-}C{\equiv}N \longrightarrow R-C{\equiv}N$$

We have just seen that an alcohol can be converted into an alkyl halide by treating it with a hydrogen halide. Better yields of the alkyl halide are obtained and carbocation rearrangements can be avoided if a phosphorus trihalide (PCl_3 or PBr_3) or thionyl chloride ($SOCl_2$) is used instead.

$$CH_3CH_2OH + PBr_3 \xrightarrow{\text{pyridine}} CH_3CH_2Br$$

$$CH_3CH_2OH + PCl_3 \xrightarrow{\text{pyridine}} CH_3CH_2Cl$$

$$CH_3CH_2OH + SOCl_2 \xrightarrow{\text{pyridine}} CH_3CH_2Cl$$

These reagents all act in the same way: they convert the alcohol into an intermediate with a leaving group that is readily displaced by a halide ion.

MECHANISM FOR THE CONVERSION OF AN ALCOHOL INTO AN ALKYL BROMIDE (OR ALKYL CHLORIDE) USING PBr₃ (OR PCl₃)

$$CH_3CH_2-\overset{..}{\underset{..}{O}}H + \overset{Br}{\underset{Br}{\overset{|}{P}}}-Br \rightleftharpoons CH_3CH_2-\overset{+..}{O}-PBr_2 \rightleftharpoons CH_3CH_2-\overset{..}{O}-PBr_2 \longrightarrow CH_3CH_2Br$$

phosphorus
tribromide

a bromophosphite group   +   ⁻O—PBr₂

- The first step is an S$_N$2 reaction on phosphorus.

- Pyridine is generally used as a solvent in these reactions because it is a poor nucleophile, but it is sufficiently basic to remove a proton from the intermediate, which prevents the intermediate from reverting to starting materials.

- The bromophosphite group is a weaker base than a halide ion, so it can be easily displaced by a bromide ion.

pyridine

505

MECHANISM FOR THE CONVERSION OF AN ALCOHOL INTO AN ALKYL CHLORIDE USING SOCl$_2$

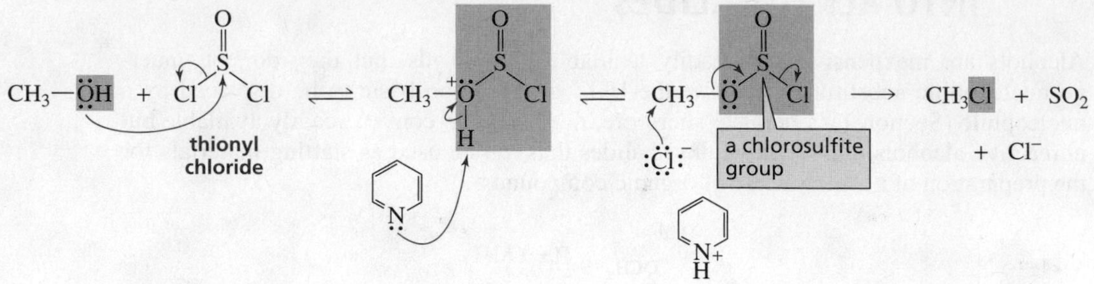

- The first step is an S$_N$2 reaction on sulfur.
- Pyridine removes a proton from the intermediate, which prevents the intermediate from reverting to starting materials.
- The chlorosulfite group is a weaker base than a chloride ion, so it can be easily displaced by a chloride ion.

The foregoing reactions work well for primary and secondary alcohols, but tertiary alcohols give poor yields because the intermediate formed by a tertiary alcohol is sterically hindered to back-side attack by the halide ion.

Table 1 summarizes some of the methods commonly used to convert alcohols into alkyl halides.

Table 1	Commonly Used Methods for Converting Alcohols into Alkyl Halides

$$ROH + HBr \xrightarrow{\Delta} RBr$$

$$ROH + HI \xrightarrow{\Delta} RI$$

$$ROH + HCl \xrightarrow[\Delta]{ZnCl_2} RCl$$

$$ROH + PBr_3 \xrightarrow[pyridine]{} RBr$$

$$ROH + PCl_3 \xrightarrow[pyridine]{} RCl$$

$$ROH + SOCl_2 \xrightarrow[pyridine]{} RCl$$

PROBLEM 7

What stereoisomers would be formed from the following reactions?

a. 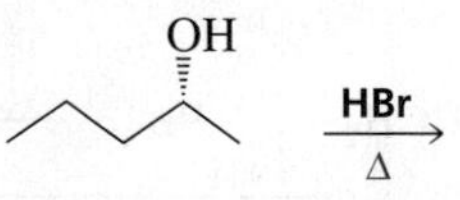 $\xrightarrow[\Delta]{HBr}$

c. 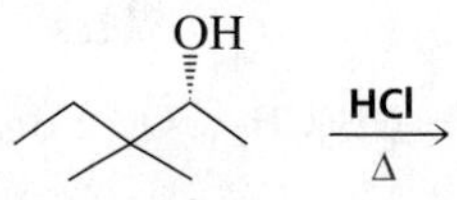 $\xrightarrow[\Delta]{HCl}$

b. 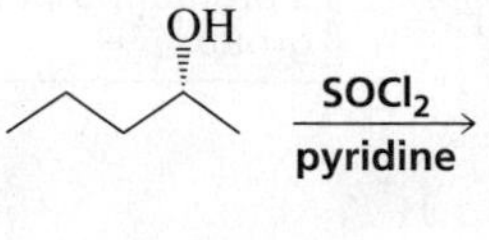 $\xrightarrow[pyridine]{SOCl_2}$

d. 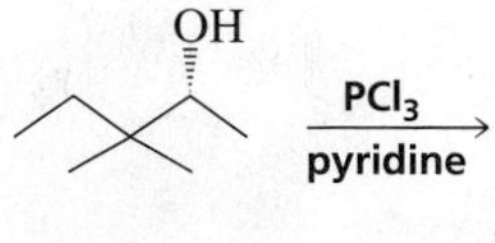 $\xrightarrow[pyridine]{PCl_3}$

---

3   CONVERTING AN ALCOHOL INTO A SULFONATE ESTER

Another way a primary or secondary alcohol can be activated for a subsequent reaction with a nucleophile—instead of converting it into an alkyl halide—is to convert it into a sulfonate ester. A **sulfonate ester** is formed when an alcohol reacts with a sulfonyl chloride. (Notice that sulfur, which is in the third row of the periodic table, has an expanded valence shell—that is, it is surrounded by 12 electrons.)

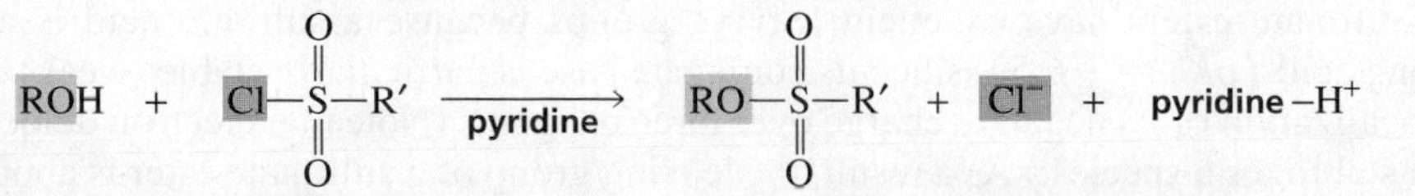

ROH + a sulfonyl chloride → a sulfonate ester + Cl⁻ + pyridine –H⁺

The reaction is a nucleophilic substitution reaction in which the alcohol displaces the chloride ion. Pyridine is the solvent and it is also the base that removes a proton from the intermediate.

MECHANISM FOR THE CONVERSION OF AN ALCOHOL INTO A SULFONATE ESTER

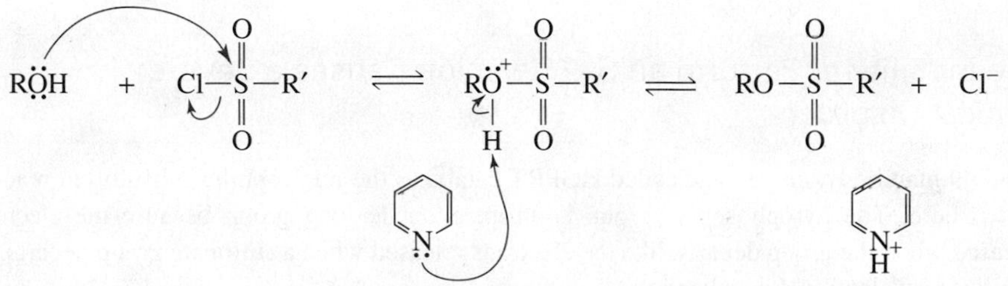

- The first step is an S_N2 reaction on sulfur.
- Pyridine removes a proton from the intermediate, which prevents the intermediate from reverting to starting materials.

Several sulfonyl chlorides are available to activate OH groups. The most common one is *para*-toluenesulfonyl chloride (abbreviated as TsCl). The sulfonate ester formed from the reaction of TsCl and an alcohol is called an **alkyl tosylate** (abbreviated as ROTs).

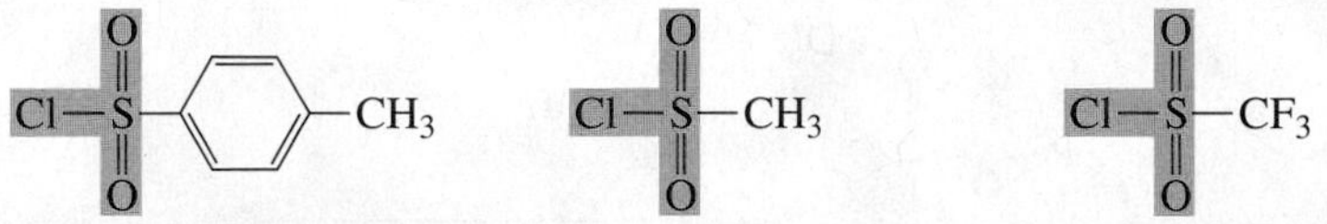

para-toluenesulfonyl chloride

methanesulfonyl chloride

trifluoromethanesulfonyl chloride

Once the alcohol has been activated by being converted into a sulfonate ester, the appropriate nucleophile is added, generally under conditions that favor S_N2 reactions. (Notice that in all cases, the added nucleophile is a *much* better nucleophile than the chloride ion that also is present in the solution, since it was the leaving group in the synthesis of the sulfonate ester.) The reactions take place readily at room temperature because the sulfonate ester has an excellent leaving group. Sulfonate esters react with a wide variety of nucleophiles, so they can be used to synthesize a wide variety of compounds.

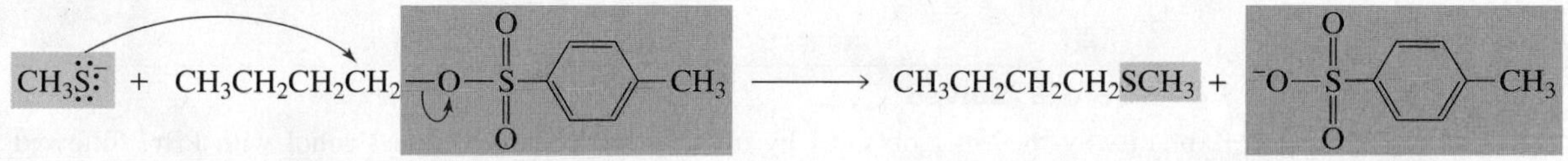

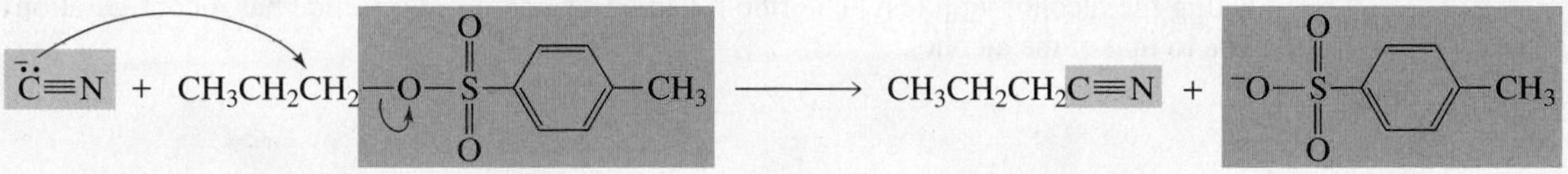

This substitution reaction will not take place if the alcohol that forms the sulfonate ester is tertiary because it would be too sterically hindered to undergo the S_N2 reaction. (Note that tertiary alkyl halides cannot undergo S_N2 reactions.)

Sulfonate esters have excellent leaving groups because a sulfonic acid is a very strong acid ($pK_a = -6.5$), since its conjugate base is particularly stable (weak) due to delocalization of its negative charge over three oxygens. (Note that electron delocalization stabilizes a species.) As a result, the leaving group of a sulfonate ester is about 100 times better as a leaving group than is chloride ion.

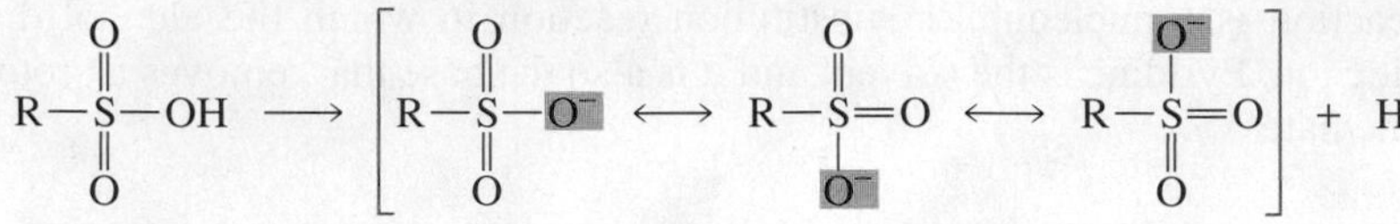

resonance contributors

The Inability to Perform an S_N2 Reaction Causes a Severe Clinical Disorder

In the human body, an enzyme called HGPRT catalyzes the nucleophilic substitution reaction shown here. The pyrophosphate group is an excellent leaving group because the electrons released when the group departs, like the electrons released when a sulfonate group departs, are stabilized by electron delocalization.

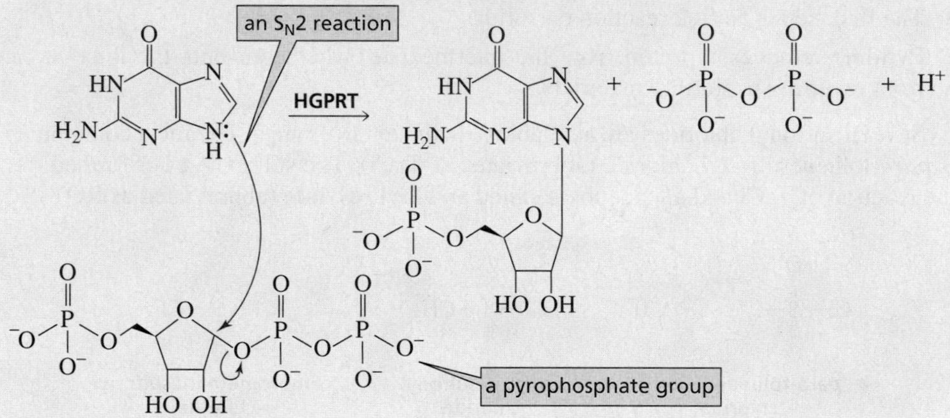

A severe deficiency in HGPRT causes Lesch-Nyhan syndrome. This congenital defect occurs mostly in males and has tragic symptoms—namely, crippling arthritis and severe malfunctions in the nervous system such as mental retardation, highly aggressive and destructive behavior, and self-mutilation. Children with Lesch-Nyhan syndrome have such a compulsive urge to bite their fingers and lips that they have to be restrained. Fortunately, HGPRT deficiencies in fetal cells can be detected by amniocentesis. The condition occurs in 1 in 380,000 live births.

PROBLEM 8 Solved

Explain why the ether obtained by treating an optically active alcohol with PBr$_3$ followed by sodium methoxide has the *same* configuration as the alcohol, whereas the ether obtained by treating the alcohol with tosyl chloride followed by sodium methoxide has a configuration *opposite* to that of the alcohol.

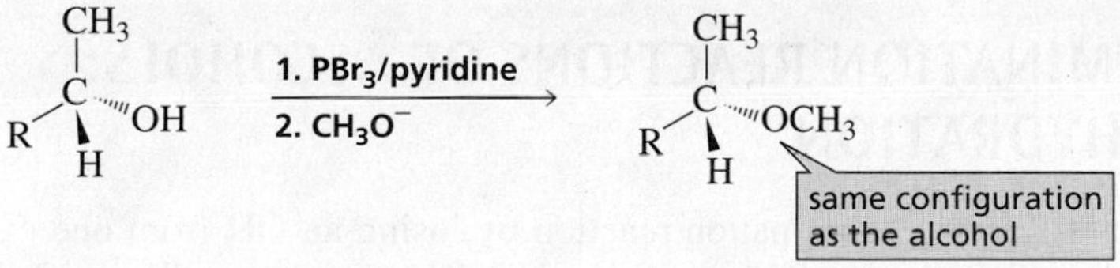

same configuration as the alcohol

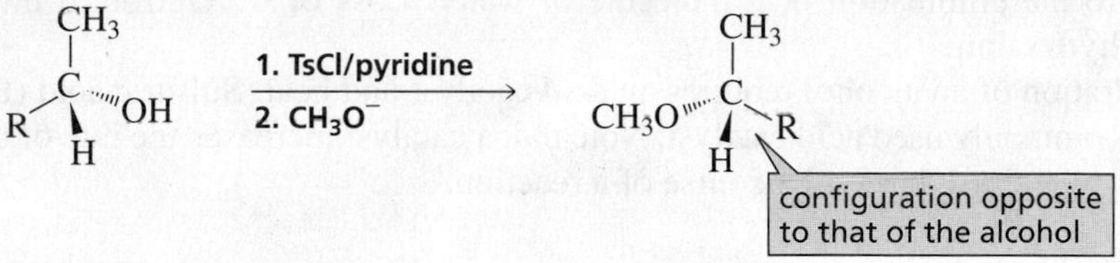

configuration opposite to that of the alcohol

Solution Conversion of the alcohol to the ether by way of an alkyl halide requires two successive S_N2 reactions: (1) attack of Br^- on the bromophosphite and (2) attack of CH_3O^- on the alkyl halide. Each S_N2 reaction takes place with inversion of configuration, so the final product has the same configuration as the starting material.

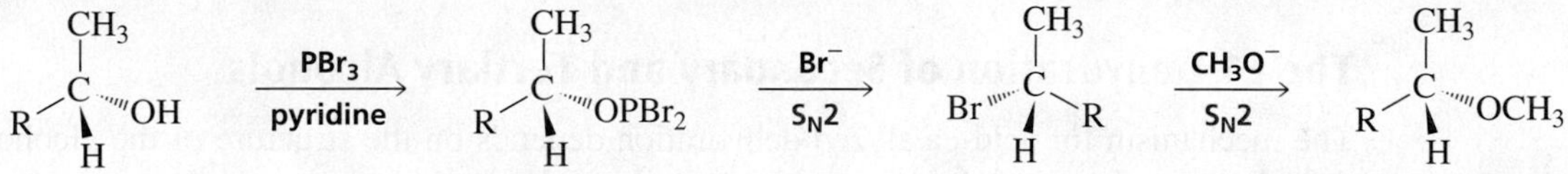

In contrast, conversion of the alcohol to the ether by way of an alkyl tosylate requires only one S_N2 reaction (attack of CH_3O^- on the alkyl tosylate), so the final product and the starting material have opposite configurations.

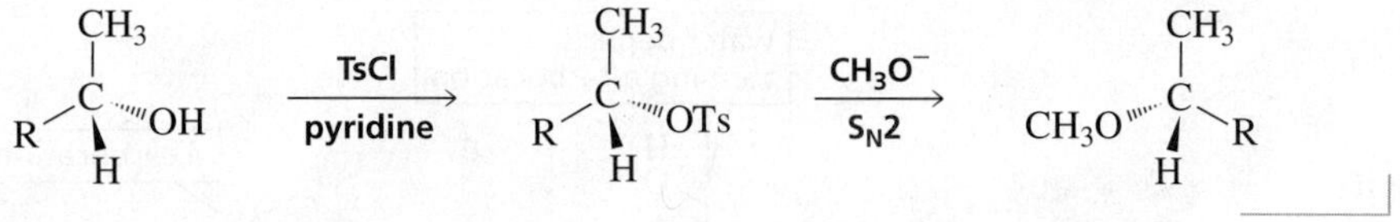

PROBLEM 9

What stereoisomers would be obtained from the following reactions?

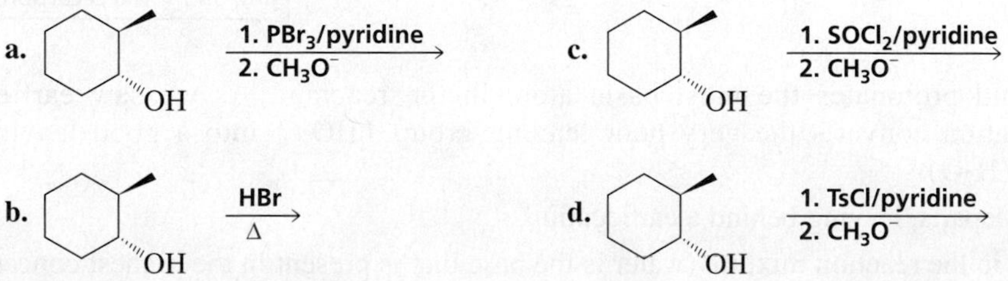

a. 1. PBr_3/pyridine 2. CH_3O^-

c. 1. $SOCl_2$/pyridine 2. CH_3O^-

b. HBr Δ

d. 1. TsCl/pyridine 2. CH_3O^-

PROBLEM 10

Show how 1-propanol can be converted into the following compounds by means of a sulfonate ester:

a. $CH_3CH_2CH_2SCH_2CH_3$

b. $CH_3CH_2CH_2OCH_2CHCH_3$
$\quad\quad\quad\quad\quad\quad\quad\quad\quad | $
$\quad\quad\quad\quad\quad\quad\quad\quad CH_3$

4 ELIMINATION REACTIONS OF ALCOHOLS: DEHYDRATION

An alcohol can undergo an elimination reaction by losing an OH from one carbon and an H from an adjacent carbon. The product of the reaction is an alkene. Overall, this amounts to the elimination of a molecule of water. Loss of water from a molecule is called **dehydration.**

Dehydration of an alcohol requires an acid catalyst and heat. Sulfuric acid (H_2SO_4) is the most commonly used acid catalyst. Note that a catalyst increases the rate of a reaction but is not consumed during the course of a reaction.

acid-catalyzed dehydration

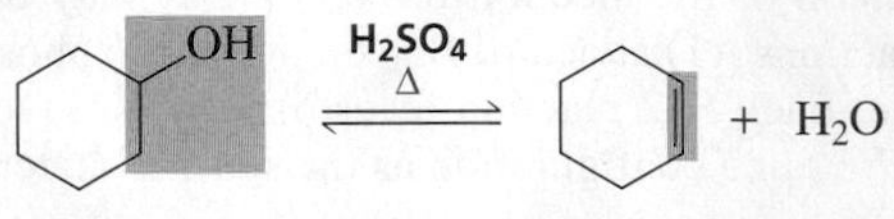

The E1 Dehydration of Secondary and Tertiary Alcohols

The mechanism for acid-catalyzed dehydration depends on the structure of the alcohol; dehydrations of secondary and tertiary alcohols are E1 reactions.

MECHANISM FOR THE E1 DEHYDRATION OF AN ALCOHOL

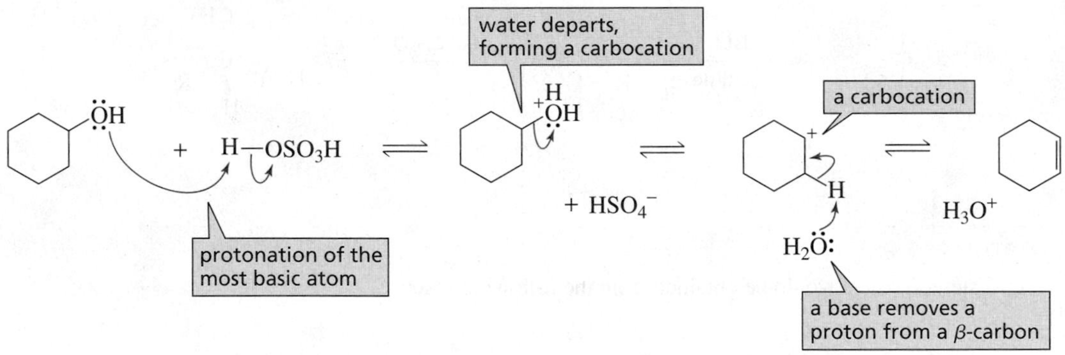

Dehydration of secondary and tertiary alcohols are E1 reactions.

- The acid protonates the most basic atom in the reactant. As we saw earlier, protonation converts the very poor leaving group (HO^-) into a good leaving group (H_2O).
- Water departs, leaving behind a carbocation.
- A base in the reaction mixture (water is the base that is present in the highest concentration) removes a proton from a β-carbon (a carbon adjacent to the positively charged carbon), forming an alkene and regenerating the acid catalyst. Notice that the dehydration reaction is an E1 reaction of a protonated alcohol.

When acid-catalyzed dehydration leads to more than one elimination product, the major product will be the more stable alkene—that is, the one obtained by removing a proton from the β-carbon bonded to the fewest hydrogens. The more stable alkene is the major product because it has the more stable transition state leading to its formation (Figure 1).

Reactions of Alcohols, Ethers, Epoxides, Amines, and Thiols

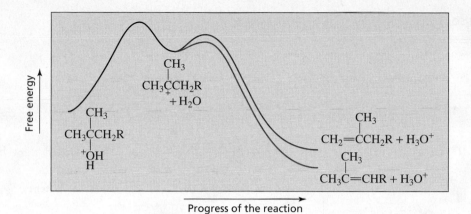

◄ Figure 1
The more stable alkene is the major product obtained from the dehydration of an alcohol because the transition state leading to its formation is more stable (indicated by the green line), allowing it to be formed more rapidly.

Notice that the acid-catalyzed dehydration of an alcohol is the reverse of the acid-catalyzed addition of water to an alkene.

$$\text{RCH}_2\text{CHR} + \text{H}^+ \xrightleftharpoons[\text{hydration}]{\text{dehydration}} \text{RCH}{=}\text{CHR} + \text{H}_2\text{O} + \text{H}^+$$

To prevent the alkene formed in the dehydration reaction from adding water and reforming the alcohol, the alkene is removed by distillation as it is formed, because it has a much lower boiling point than the alcohol. Removing a product displaces the reaction to the right according to Le Châtelier's principle.

Because the rate-determining step in the dehydration of a secondary or a tertiary alcohol is formation of a carbocation intermediate, the rate of dehydration reflects the ease with which the carbocation is formed: tertiary alcohols are the easiest to dehydrate because tertiary carbocations are more stable and are therefore more easily formed than secondary and primary carbocations.

relative ease of dehydration

Be sure to check the structure of the carbocation formed in a dehydration reaction for the possibility of rearrangement. Note that a carbocation will rearrange if rearrangement produces a more stable carbocation. For example, the secondary carbocation formed initially in the following reaction rearranges to a more stable tertiary carbocation:

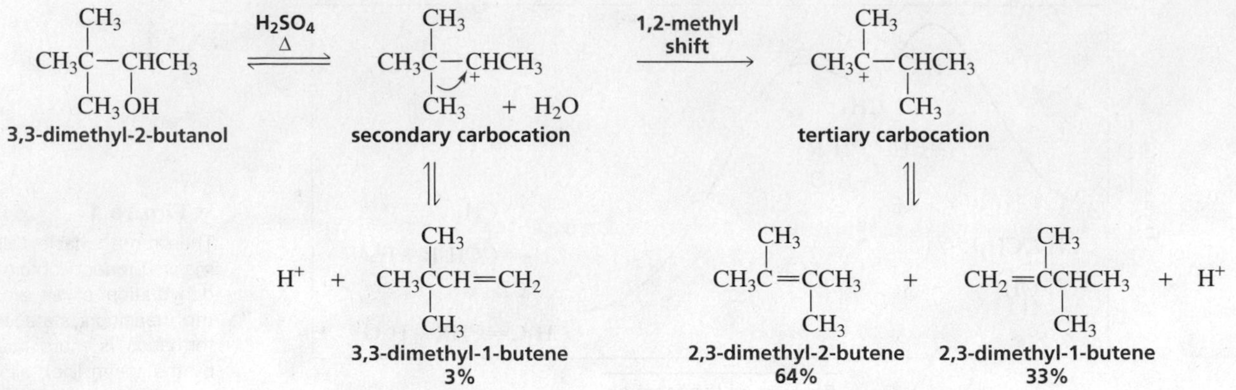

$$CH_3\overset{\overset{\displaystyle CH_3}{|}}{\underset{\underset{\displaystyle CH_3OH}{|}}{C}}-CHCH_3 \quad\xrightleftharpoons[\Delta]{H_2SO_4}\quad CH_3\overset{\overset{\displaystyle CH_3}{|}}{\underset{\underset{\displaystyle CH_3}{|}}{\overset{+}{C}}}-CHCH_3 \;+\; H_2O \quad\xrightarrow{\substack{1,2\text{-methyl}\\ \text{shift}}}\quad CH_3\overset{\overset{\displaystyle CH_3}{|}}{\underset{\underset{\displaystyle CH_3}{|}}{\overset{+}{C}}}-CHCH_3$$

3,3-dimethyl-2-butanol   **secondary carbocation**   **tertiary carbocation**

$$H^+ \;+\; CH_3\overset{\overset{\displaystyle CH_3}{|}}{\underset{\underset{\displaystyle CH_3}{|}}{C}}CH=CH_2 \qquad\qquad CH_3\overset{\overset{\displaystyle CH_3}{|}}{C}=\overset{\underset{\displaystyle CH_3}{|}}{C}CH_3 \;+\; CH_2=\overset{\overset{\displaystyle CH_3}{|}}{\underset{\underset{\displaystyle CH_3}{|}}{C}}CHCH_3 \;+\; H^+$$

3,3-dimethyl-1-butene   **2,3-dimethyl-2-butene**   **2,3-dimethyl-1-butene**
3%   64%   33%

PROBLEM 11◆

Which of the following alcohols would dehydrate the fastest when heated with acid?

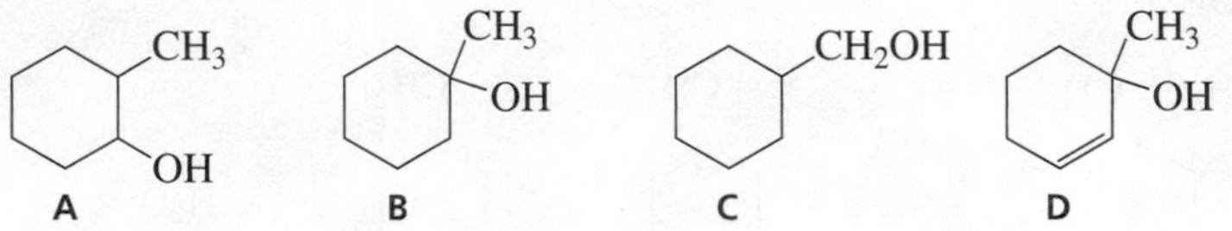

A   B   C   D

The E2 Dehydration of Primary Alcohols

While the dehydration of a secondary or a tertiary alcohol is an E1 reaction, the dehydration of a primary alcohol is an E2 reaction, because primary carbocations are extremely unstable. Any base (B:) in the reaction mixture (ROH, ROR, H_2O, or HSO_4^-) can remove the proton in the elimination reaction. The reaction also forms an ether in a competing S_N2 reaction, since primary alcohols are the ones most likely to form substitution products under $S_N2/E2$ conditions.

MECHANISM FOR THE E2 DEHYDRATION OF A PRIMARY ALCOHOL AND FOR THE COMPETING S_N2 REACTION

Dehydration of a primary alcohol is an E2 reaction.

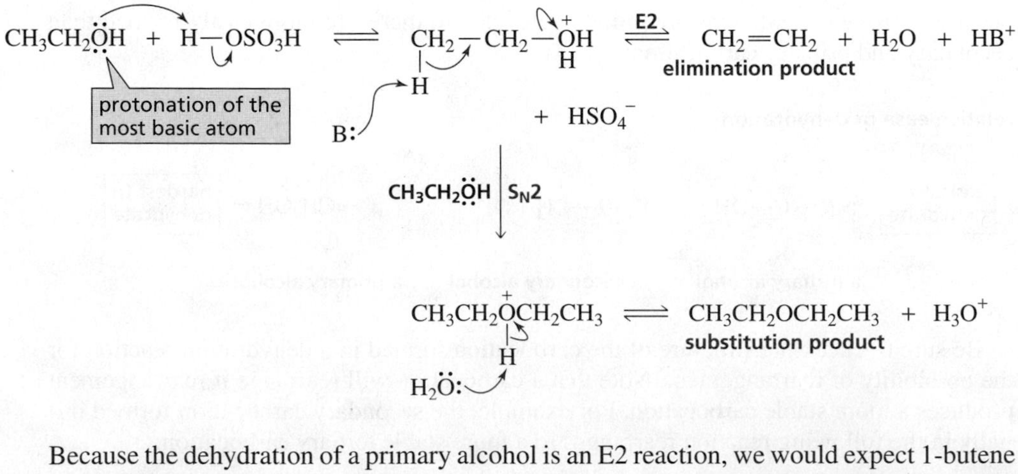

Because the dehydration of a primary alcohol is an E2 reaction, we would expect 1-butene to be the product of the E2 dehydration of 1-butanol. The product, however, is actually 2-butene.

$$CH_3CH_2CH_2CH_2OH \underset{\Delta}{\overset{H_2SO_4}{\rightleftharpoons}} CH_3CH_2CH{=}CH_2 \overset{H^+}{\rightleftharpoons} CH_3CH_2\overset{+}{C}HCH_3 \rightleftharpoons CH_3CH{=}CHCH_3 + H^+$$

1-butanol　　　　　　　　　　**1-butene**　　　　　　　　　　　　　　　　**2-butene**

$$+ \ H_2O$$

2-Butene is the final product because, after 1-butene forms, a proton from the acidic solution adds to the double bond (adding to the sp^2 carbon bonded to the most hydrogens in accordance with the rule that governs electrophilic addition reactions), thereby forming a carbocation. Loss of a proton from the β-carbon bonded to the fewest hydrogens (Zaitsev's rule) forms 2-butene.

> Alcohols undergo $S_N1/E1$ reactions unless they have to form a primary carbocation, in which case they undergo $S_N2/E2$ reactions.

PROBLEM 12

Draw the major product obtained when each of the following alcohols is heated in the presence of H_2SO_4:

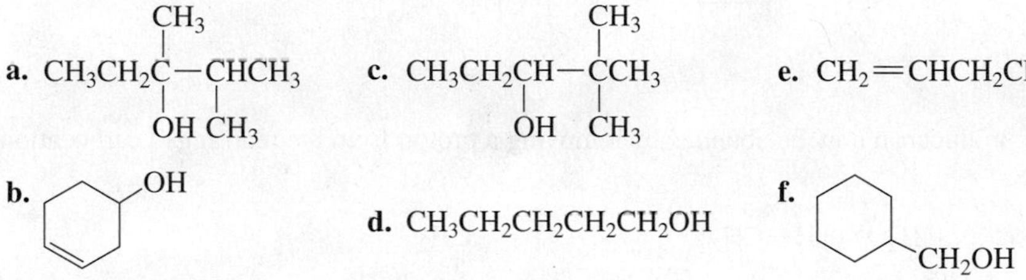

a. $CH_3CH_2\overset{\overset{\textstyle CH_3}{|}}{\underset{\underset{\textstyle OH}{|}}{C}}{-}\overset{\overset{}{}}{\underset{\underset{\textstyle CH_3}{|}}{C}}HCH_3$

c. $CH_3CH_2\overset{}{\underset{\underset{\textstyle OH}{|}}{C}}H{-}\overset{\overset{\textstyle CH_3}{|}}{\underset{\underset{\textstyle CH_3}{|}}{C}}CH_3$

e. $CH_2{=}CHCH_2CH_2OH$

b.

d. $CH_3CH_2CH_2CH_2CH_2OH$

f.

PROBLEM 13

Heating an alcohol with sulfuric acid is a good way to prepare a symmetrical ether such as diethyl ether.

a. Explain why it is not a good way to prepare an unsymmetrical ether such as ethyl propyl ether.
b. How would you synthesize ethyl propyl ether?

PROBLEM-SOLVING STRATEGY

Proposing a Mechanism

Propose a mechanism for the following reaction:

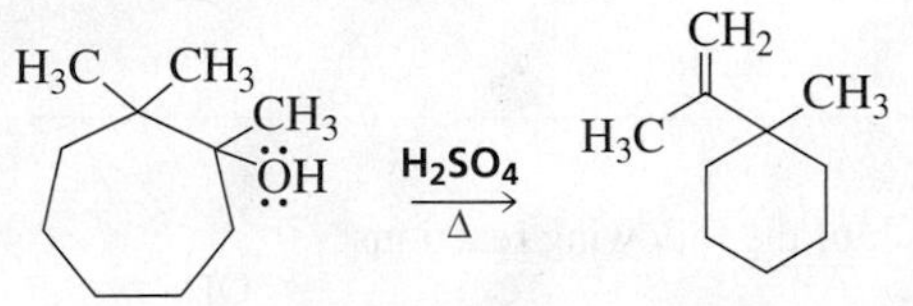

Even the most complicated-looking mechanism can be reasoned out if you proceed one step at a time, always keeping in mind the structure of the final product. Note that when an acid is added to a reactant, it protonates the most basic atom in the reactant. Oxygen is the only basic atom, so that is where protonation occurs. Loss of water forms a tertiary carbocation.

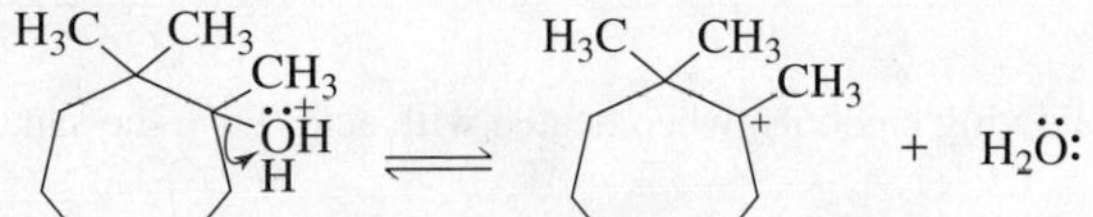

Because the reactant contains a seven-membered ring and the final product has a six-membered ring, a ring-contraction rearrangement must occur to relieve the strain in the six-membered ring. (When doing a *ring-contraction* or a *ring-expansion rearrangement* (Problem 61), you may find it helpful to label the equivalent carbons in the reactant and product, as shown here.) Of the two possible pathways for ring contraction, one leads to a tertiary carbocation while the other leads to a primary carbocation. The correct pathway must be the one that leads to the tertiary carbocation, since that carbocation has the same arrangement of atoms as the product and the primary carbocation would be too unstable to form.

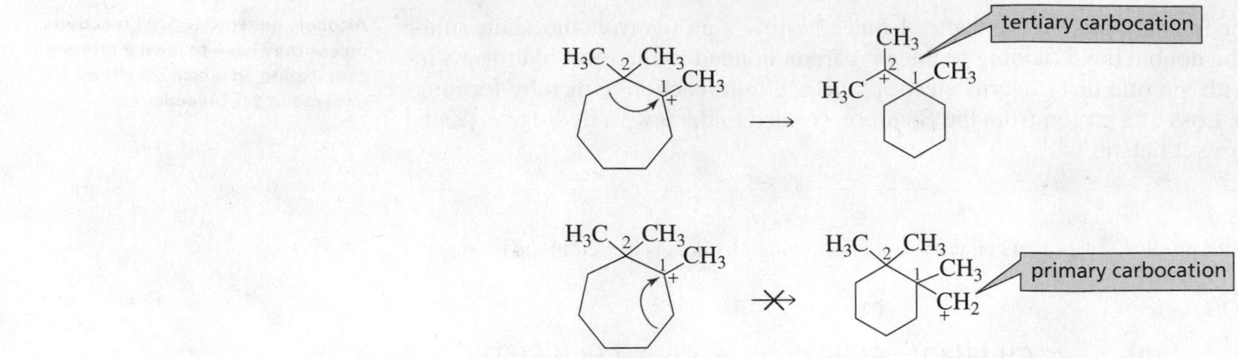

The final product can now be obtained by removing a proton from the rearranged carbocation.

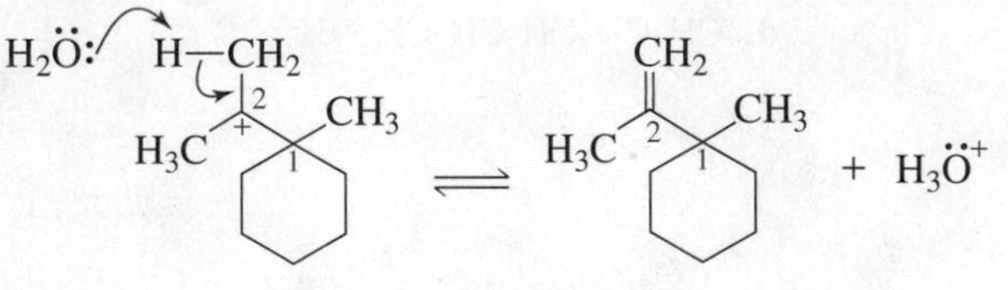

Now use the strategy you have just learned to solve Problem 14.

PROBLEM 14

Propose a mechanism for each of the following reactions:

a.

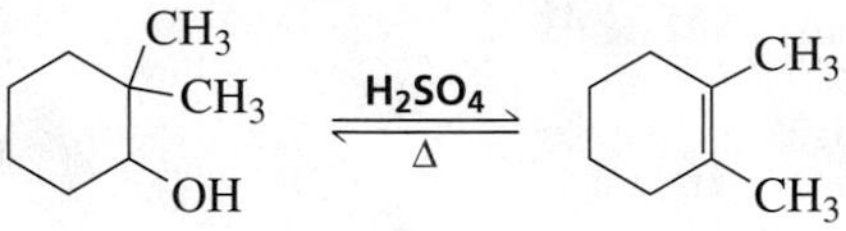

b.

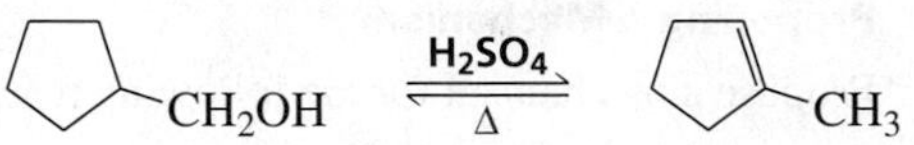

PROBLEM 15

Draw the product of each of the following reactions:

a.

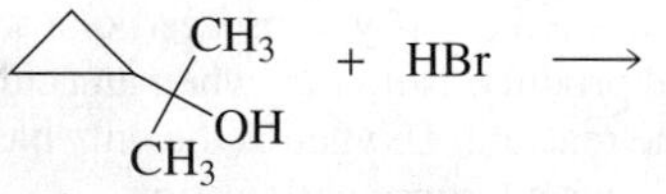

b.

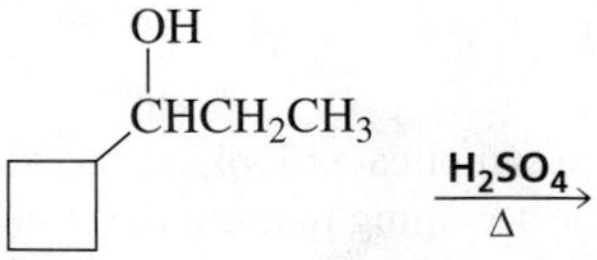

PROBLEM 16

Explain why the following alcohols, when heated with acid, form the same alkene.

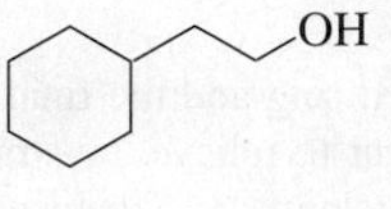

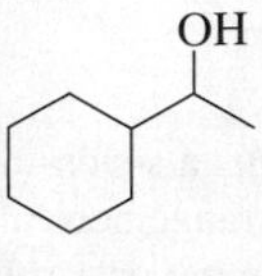

The Stereochemistry of the Dehydration Reaction

The stereochemical outcome of the E1 dehydration of an alcohol is identical to the stereochemical outcome of the E1 dehydrohalogenation of an alkyl halide. That is, both the E and Z stereoisomers are obtained as products, but the major product is the stereoisomer in which the larger group on each of the sp^2 carbons are on opposite sides of the double bond. Because that stereoisomer is more stable, it is formed more rapidly.

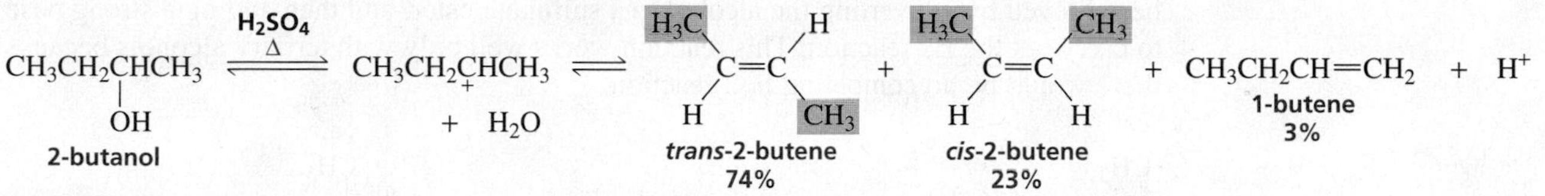

$$CH_3CH_2CHCH_3 \ \overset{H_2SO_4}{\underset{\Delta}{\rightleftharpoons}} \ CH_3CH_2\overset{+}{CHCH_3} \ \rightleftharpoons$$

2-butanol · OH

$+ \ H_2O$

$$\underset{\substack{\textit{trans}\text{-2-butene} \\ 74\%}}{\overset{H_3C \qquad H}{\underset{H \qquad CH_3}{C=C}}} \ + \ \underset{\substack{\textit{cis}\text{-2-butene} \\ 23\%}}{\overset{H_3C \qquad CH_3}{\underset{H \qquad H}{C=C}}} \ + \ \underset{\substack{\text{1-butene} \\ 3\%}}{CH_3CH_2CH=CH_2} \ + \ H^+$$

PROBLEM 17◆

What stereoisomers are formed in the following reactions? Which stereoisomer is the major product?

a. the acid-catalyzed dehydration of 1-pentanol to 2-pentene

b. the acid-catalyzed dehydration of 3,4-dimethyl-3-hexanol to 3,4-dimethyl-3-hexene

---

PROBLEM 18◆

Suppose the following compound is heated in the presence of H_2SO_4.

a. What constitutional isomer is produced in greatest yield?

b. What stereoisomer is produced in greater yield?

Changing an E1 Dehydration into an E2 Dehydration

The relatively harsh conditions (acid and heat) required for alcohol dehydration, and the structural changes resulting from carbocation rearrangements in the E1 reaction may result in low yields of the desired alkene. Dehydration, however, can be carried out under milder conditions that favor E2 reactions by replacing the OH group with a good leaving group.

$$CH_3CH_2CHCH_3 \ \xrightarrow[\text{pyridine, 0 °C}]{POCl_3} \ CH_3CH=CHCH_3$$

· OH

For example, reaction with phosphorus oxychloride ($POCl_3$) converts the OH group of the alcohol into $OPOCl_2$, which is a good leaving group. The reaction conditions (a high concentration of a relatively strong base) favor an E2 reaction and carbocations are not formed in E2 reactions, so carbocation rearrangements do not occur. Pyridine removes a proton in order to prevent the intermediate from reverting to products and is the base employed in the E2 reaction. Pyridine also prevents the buildup of HCl, which would add to the alkene.

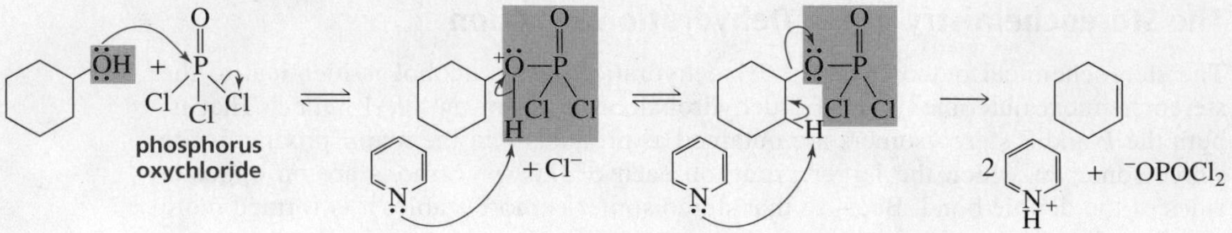

phosphorus
oxychloride

Sulfonate esters also have good leaving groups. So dehydration *via* an E2 reaction can be achieved by converting the alcohol to a sulfonate ester, and then adding a strong base to carry out the E2 reaction. This reaction works well only with tertiary alcohols because there would be no competing S_N2 reaction.

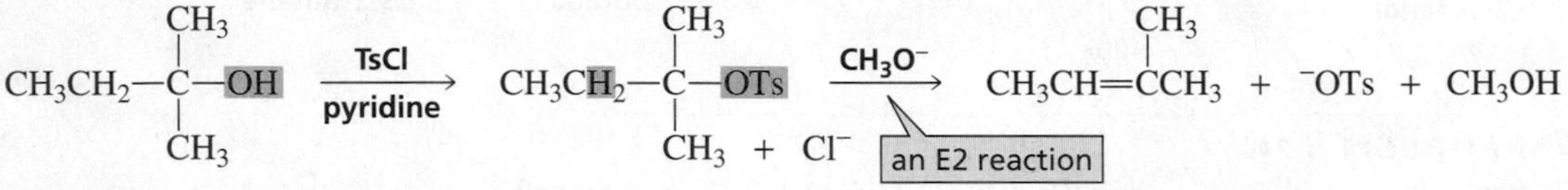

Biological Dehydrations

Dehydration reactions occur in many important biological processes. Instead of being catalyzed by strong acids, which would not be available to a cell, they are catalyzed by enzymes. Enolase, for example, catalyzes the dehydration of α-phosphoglycerate in glycolysis. Glycolysis is a series of reactions that prepare glucose for entry into the citric acid cycle.

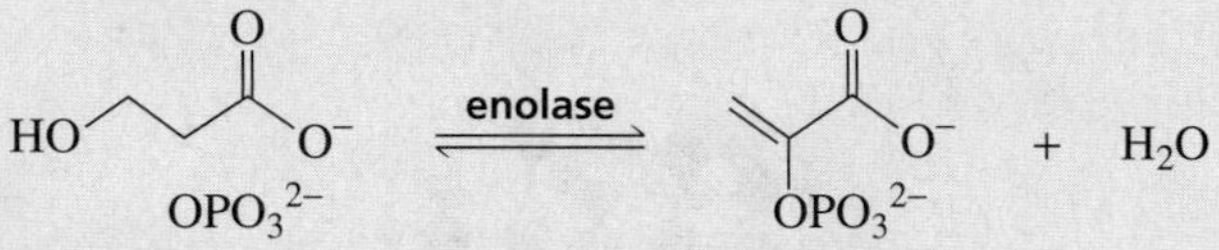

α-phosphoglycerate   **phosphoenolpyruvate**

Fumarase is the enzyme that catalyzes the dehydration of malate in the citric acid cycle. The citric acid cycle is a series of reactions that oxidize compounds derived from carbohydrates, fatty acids, and amino acids.

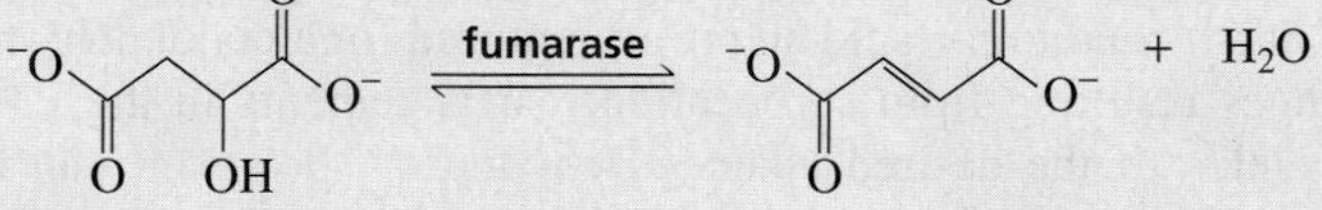

malate   **fumarate**

PROBLEM 19♦

What alcohol would you treat with phosphorus oxychloride and pyridine to form each of the following alkenes?

a. $CH_3CH_2C{=}CH_2$ (with CH_3 substituent)   **b.** 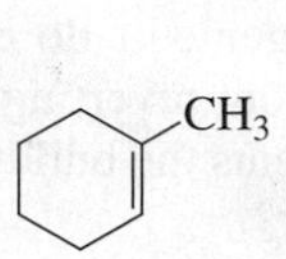   **c.** $CH_3CH{=}CHCH_2CH_3$ **d.**

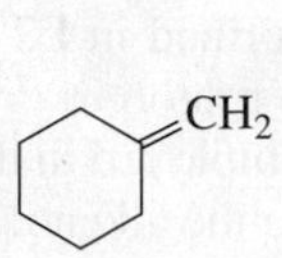

5 OXIDATION OF ALCOHOLS

A variety of compounds are available that oxidize alcohols. For many years, a commonly used reagent was chromic acid (H_2CrO_4), which is formed when sodium dichromate ($Na_2Cr_2O_7$) is dissolved in aqueous acid. Notice that *secondary alcohols* are oxidized to *ketones*.

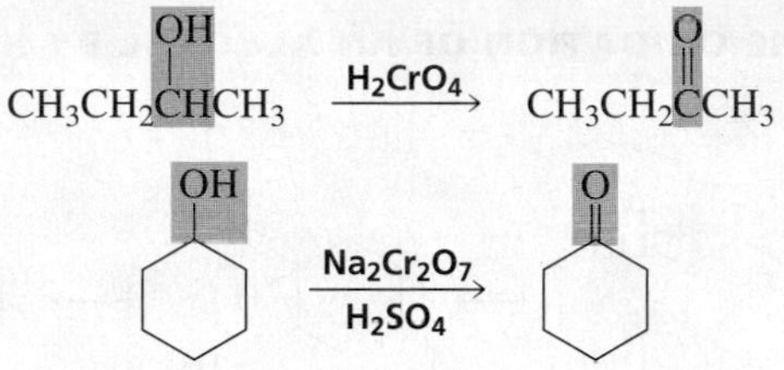

secondary alcohols **ketones**

Primary alcohols are initially oxidized to aldehydes by chromic acid. The reaction, however, does not stop at the aldehyde. Instead, the aldehyde is further oxidized to a carboxylic acid. These reactions are easily recognized as oxidations because the number of C—H bonds in the reactant decreases and the number of C—O bonds increases.

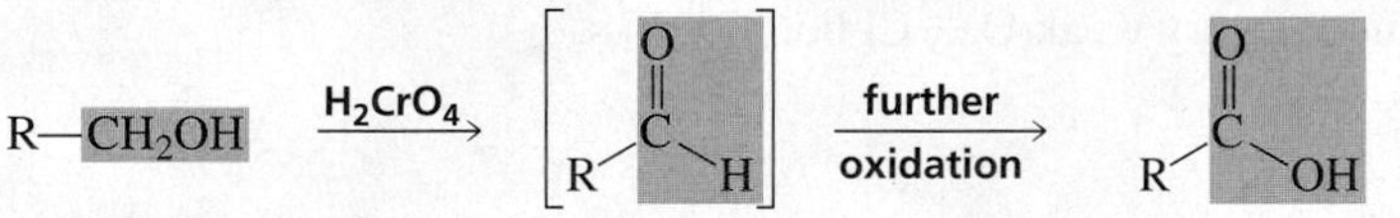

a primary alcohol   **an aldehyde**   **a carboxylic acid**

Pyridine chlorochromate (PCC) is a gentler oxidizing agent. It also oxidizes secondary alcohols to ketones, but it stops the reaction at the aldehyde when it oxidizes primary alcohols. PCC must be used in an anhydrous solvent such as CH_2Cl_2 because if water is present, the aldehyde will be further oxidized to a carboxylic acid.

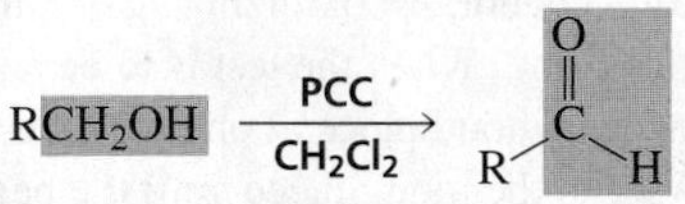

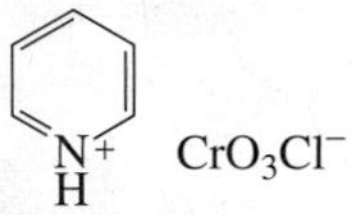

pyridinium chlorochromate PCC

Notice that, in the oxidation of both primary and a secondary alcohols, a hydrogen is removed from the carbon to which the OH is attached. The carbon bearing the OH group in a tertiary alcohol is not bonded to a hydrogen, so its OH group cannot be oxidized to a carbonyl (C=O) group.

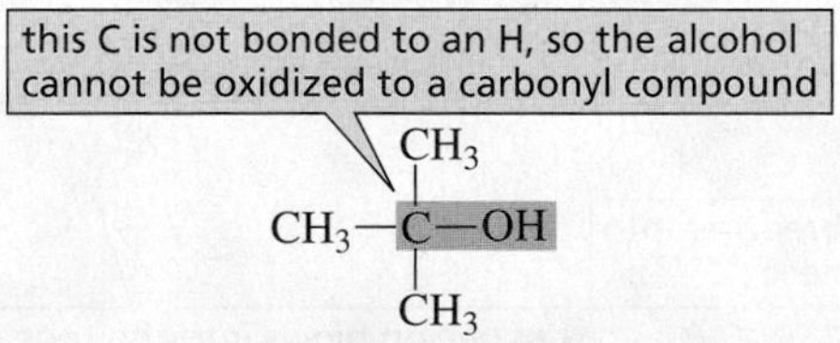

a tertiary alcohol

Because of the toxicity of chromium-based reagents, other reagents for the oxidation of alcohols have been developed. One of the more common is hypochlorous acid (HOCl). HOCl is unstable, so it is generated in situ (in the reaction mixture) by an acid-base reaction between H^+ and $^-$OCl (using CH_3COOH and NaOCl). Secondary alcohols are oxidized to ketones and primary alcohols are oxidized to aldehydes. Also see the Swern oxidation (Problem 70).

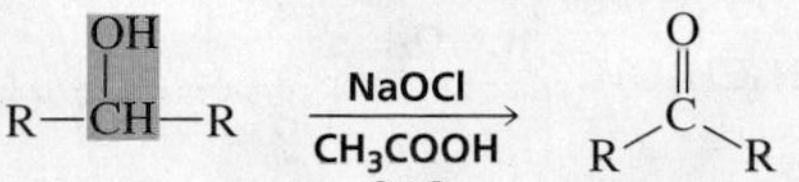

a secondary alcohol   **a ketone**

Primary alcohols are oxidized to aldehydes (or to carboxylic acids).

Secondary alcohols are oxidized to ketones.

$$R-CH_2OH \xrightarrow[\substack{CH_3COOH \\ 0\ °C}]{NaOCl} \underset{\substack{R \quad H \\ \text{an aldehyde}}}{\overset{O}{\parallel}} $$

a primary alcohol

MECHANISM FOR THE OXIDATION OF AN ALCOHOL BY HOCl

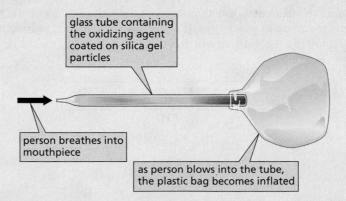

$+ HB^+ + Cl^-$

- The acid protonates the oxygen, the most basic atom in the alcohol.
- Because the reaction is not heated, water does not leave spontaneously but must be kicked out by hypochlorite ion in an S_N2 reaction.
- A base in the reaction mixture removes a proton from the carbon bonded to the O—Cl group and the very weak O—Cl bond breaks.

Blood Alcohol Content

As blood passes through the arteries in our lungs, an equilibrium is established between the alcohol in our blood and the alcohol in our breath. Therefore, if the concentration of one is known, then the concentration of the other can be estimated.

The test that law enforcement agencies use to approximate a person's blood alcohol level is based on the oxidation of breath ethanol. An oxidizing agent impregnated onto an inert material is enclosed within a sealed glass tube. When the test is to be administered, the ends of the tube are broken off and replaced with a mouthpiece at one end and a balloon-type bag at the other. The person being tested blows into the mouthpiece until the bag is filled with air.

glass tube containing the oxidizing agent coated on silica gel particles

person breathes into mouthpiece

as person blows into the tube, the plastic bag becomes inflated

Any breath ethanol is oxidized as it passes through the column. When ethanol is oxidized, the oxidizing agent is reduced to green chromic ion. The greater the concentration of breath alcohol, the farther the green color spreads through the tube.

$$CH_3CH_2OH \xrightarrow{H_2CrO_4} \underset{\substack{CH_3 \qquad OH}}{\overset{O}{\underset{\parallel}{C}}} + \underset{\text{green}}{Cr^{3+}}$$

If the person fails this test—determined by the extent to which the green color spreads through the tube—a more accurate Breathalyzer™ test is administered. The Breathalyzer test also depends on the oxidation of breath ethanol, but it provides more accurate results because it is quantitative. In this test, a known volume of breath is bubbled through a solution of chromic acid, and the concentration of the green chromic ion is measured precisely with a spectrophotometer.

Treating Alcoholism with Antabuse

Disulfiram, most commonly known as Antabuse, is used to treat alcoholism. It causes violently unpleasant effects if ethanol is consumed within two days after taking the drug.

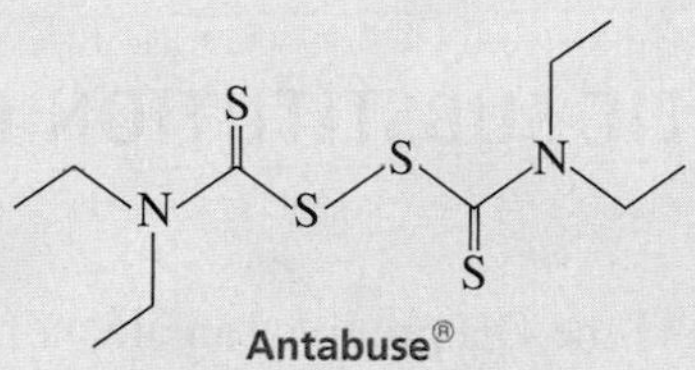

Antabuse®

Antabuse works by inhibiting aldehyde dehydrogenase, the enzyme responsible for oxidizing acetaldehyde (a product of ethanol metabolism) to acetic acid. This causes a buildup of acetaldehyde. It is the acetaldehyde that causes the unpleasant physiological effects of intoxication: intense flushing, nausea, dizziness, sweating, throbbing headaches, decreased blood pressure, and, ultimately, shock. Consequently, Antabuse should be taken only under strict medical supervision.

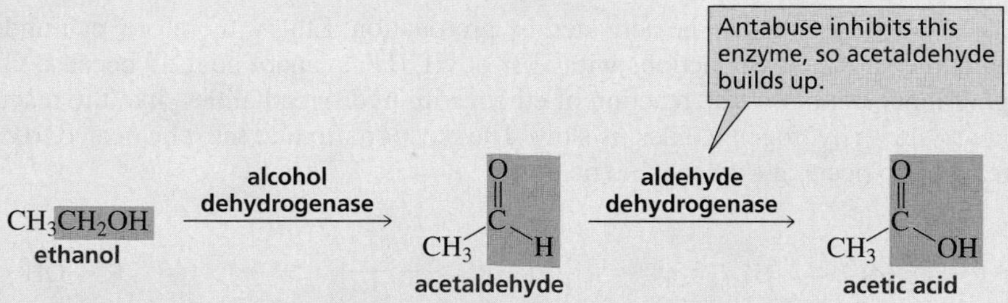

> Antabuse inhibits this enzyme, so acetaldehyde builds up.

In some people, aldehyde dehydrogenase does not function properly even under normal circumstances. Their symptoms in response to ingesting alcohol are nearly the same as those of individuals who are medicated with Antabuse.

Methanol Poisoning

In addition to oxidizing ethanol to acetaldehyde, alcohol dehydrogenase can oxidize methanol to formaldehyde. Formaldehyde is damaging to many tissues and since eye tissue is particularly sensitive, methanol ingestion can cause blindness.

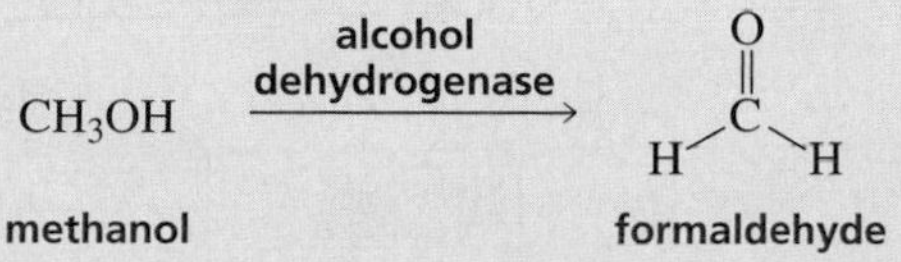

If methanol is ingested, the patient is given ethanol intravenously for several hours. Ethanol competes with methanol for binding at the active site of the enzyme. Binding ethanol minimizes the amount of methanol that can be bound, which minimizes the amount of formaldehyde that can be formed. So ethanol is given to the patient until all the ingested methanol has been excreted in the urine.

PROBLEM 20♦

What product will be obtained from the reaction of each of the following alcohols with HOCl?

a. 3-pentanol   **c.** 2-methyl-2-pentanol   **e.** cyclohexanol

b. 1-pentanol   **d.** 2,4-hexanediol   **f.** 1,4-butanediol

---

6   NUCLEOPHILIC SUBSTITUTION REACTIONS OF ETHERS

The OR group of an **ether** and the OH group of an alcohol have nearly the same basicity, because the conjugate acids of these two groups have similar pK_a values. (The pK_a of CH_3OH is 15.5 and the pK_a of H_2O is 15.7.) Both groups are strong bases, so both are very poor leaving groups. Consequently, ethers, like alcohols, need to be activated before they can undergo a nucleophilic substitution reaction.

$$R-\ddot{O}-H \qquad R-\ddot{O}-R$$

an alcohol   **an ether**

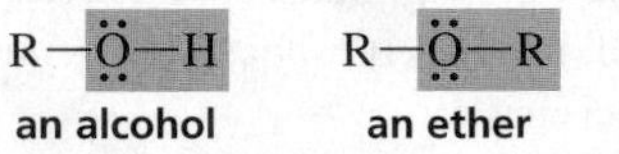

Like alcohols, ethers can be activated by protonation. Ethers, therefore, can undergo nucleophilic substitution reactions with HBr or HI. (HCl cannot be used because Cl^- is too poor a nucleophile.) The reaction of ethers with hydrogen halides, like the reactions of alcohols with hydrogen halides, is slow. The reaction mixture must be heated to cause the reaction to occur at a reasonable rate.

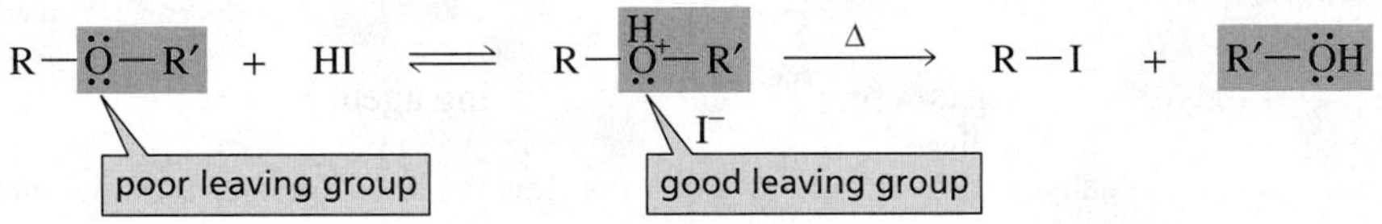

$$R-\ddot{O}-R' + HI \rightleftharpoons R-\overset{H}{\underset{}{\ddot{O}^+}}-R' \xrightarrow{\Delta} R-I + R'-\ddot{O}H$$

poor leaving group I^- **good leaving group**

What happens *after* the ether is protonated depends on the structure of the ether. If departure of ROH creates a relatively stable carbocation (such as a tertiary carbocation), an S_N1 reaction occurs.

MECHANISM FOR ETHER CLEAVAGE: AN S_N1 REACTION

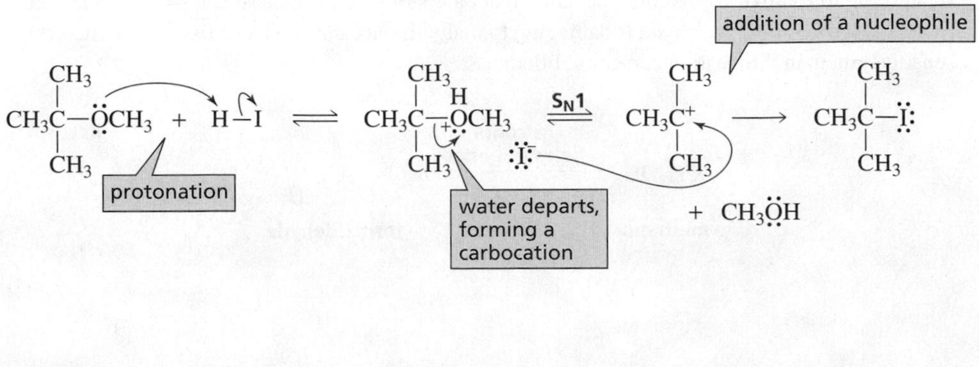

- The acid protonates the oxygen, thereby converting the very basic RO^- leaving group into the less basic ROH leaving group.
- The leaving group departs, forming a carbocation.
- The halide ion combines with the carbocation.

However, if departure of the leaving group would create an unstable carbocation (such as a methyl, vinyl, aryl, or primary carbocation), the leaving group cannot depart. It has to be displaced by the halide ion. In other words, an S_N2 reaction occurs.

MECHANISM FOR ETHER CLEAVAGE: AN S_N2 REACTION

$$CH_3\text{—}\ddot{\underset{..}{O}}\text{—}CH_2CH_2CH_3 + H\text{—}I \rightleftharpoons CH_3\text{—}\overset{H}{\underset{+}{O}}\text{—}CH_2CH_2CH_3 \xrightarrow{S_N2} CH_3\text{—}\ddot{\underset{..}{I}}\text{:} + CH_3CH_2CH_2\text{—}\ddot{\underset{..}{O}}H$$

protonation

:Ï: nucleophile attacks the less sterically hindered carbon

- Protonation converts the very basic RO^- leaving group into the less basic ROH leaving group.
- The halide ion preferentially attacks the less sterically hindered of the two alkyl groups.

Ether cleavage forms only a substitution product because any alkene that would be formed in an elimination reaction would undergo electrophilic addition with HBr or HI to form the same alkyl halide that is obtained from the substitution reaction.

The reagents (such as $SOCl_2$, PCl_3, or TsCl) used to activate alcohols so they can undergo nucleophilic substitution reactions cannot be used to activate ethers. When an alcohol reacts with one of these activating agents, a proton dissociates from the intermediate in the second step of the reaction and a stable product results.

> Ethers are cleaved by an S_N1 reaction unless the instability of the carbocation requires the cleavage to be an S_N2 reaction.

$$\underset{\text{an alcohol}}{\boxed{ROH}} + R'\text{—}\overset{O}{\underset{O}{\overset{\|}{\underset{\|}{S}}}}\text{—}Cl \longrightarrow R'\text{—}\overset{O}{\underset{O}{\overset{\|}{\underset{\|}{S}}}}\text{—}\overset{+}{\boxed{OR}}_{H} + Cl^- \rightleftharpoons \underset{\text{a sulfonate ester}}{R'\text{—}\overset{O}{\underset{O}{\overset{\|}{\underset{\|}{S}}}}\text{—}\boxed{OR}} + H^+$$

However, when an ether reacts with one of these activating agents, the oxygen atom does not have a proton that can dissociate, so a stable product cannot be formed. Instead, the more stable starting materials are reformed.

$$\underset{\text{an ether}}{\boxed{ROR}} + R'\text{—}\overset{O}{\underset{O}{\overset{\|}{\underset{\|}{S}}}}\text{—}Cl \rightleftharpoons R'\text{—}\overset{O}{\underset{O}{\overset{\|}{\underset{\|}{S}}}}\text{—}\overset{+}{\underset{R}{OR}} + \boxed{Cl^-}$$

Because hydrogen halides are the only reagents that react with ethers, ethers are frequently used as solvents. Some common ether solvents are shown in Table 2.

Table 2	Some Ethers Are Used as Solvents

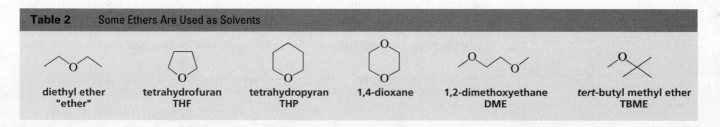

| diethyl ether "ether" | tetrahydrofuran THF | tetrahydropyran THP | 1,4-dioxane | 1,2-dimethoxyethane DME | *tert*-butyl methyl ether TBME |

Anesthetics

Because diethyl ether (commonly known as ether) is a short-lived muscle relaxant, it was at one time widely used as an *inhalation anesthetic*. However, it takes effect slowly and has a slow and unpleasant recovery period, so over time other anesthetics, such as isoflurane, enflurane, and halothane, replaced it. Even so, diethyl ether is still used where trained anesthesiologists are scarce because it is the safest anesthetic for an untrained person to administer. Anesthetics interact with the nonpolar molecules of cell membranes, causing the membranes to swell, which interferes with their permeability.

Sodium pentothal (also called thiopental sodium) is an *intravenous anesthetic*. The onset of anesthesia and the loss of consciousness occur within seconds of its administration. Care must be taken when administering sodium pentothal because the dose for effective anesthesia is 75% of the lethal dose. Because of this high level of toxicity, it cannot be used as the sole anesthetic but, instead, is generally used to induce anesthesia before an inhalation anesthetic is administered. Propofol, in contrast, has all the properties of the "perfect anesthetic": it can be administered as the sole anesthetic by intravenous drip, it has a rapid and pleasant induction period, and it has a wide margin of safety in trained hands. Recovery from the drug is also rapid and pleasant.

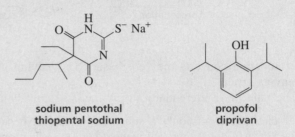

"ether"   isoflurane   enflurane   halothane

sodium pentothal
thiopental sodium

propofol
diprivan

amputation of a leg without anesthetic in 1528

a painting showing the first use of anesthesia (ether) during surgery in 1846 at Massachusetts General Hospital by surgeon John Collins Warren

---

PROBLEM 21 Solved

Explain why methyl propyl ether forms both methyl iodide and propyl iodide when it is heated with excess HI.

Solution The S_N2 reaction of methyl propyl ether with an equivalent amount of HI forms methyl iodide and propyl alcohol because the methyl group is less sterically hindered than the propyl group to attack by the iodide ion. When there is excess HI, the alcohol product of this first reaction can react with HI in another S_N2 reaction. Thus, the products are two alkyl iodides.

$$\sim\!\!\sim\!OCH_3 \xrightarrow[\Delta]{HI} \sim\!\!\sim\!OH \xrightarrow[\Delta]{HI} \sim\!\!\sim\!I + H_2O$$
$$+ \ CH_3I$$

PROBLEM 22♦

Can HF be used to cleave ethers? Explain.

PROBLEM 23 Solved

Draw the major products obtained from heating each of the following ethers with one equivalent of HI:

a. 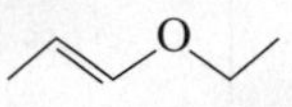   c. 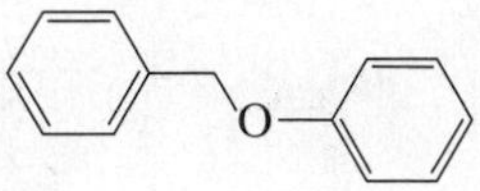   e.

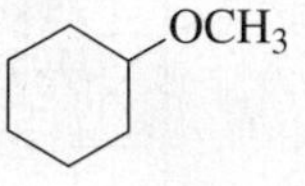

b. 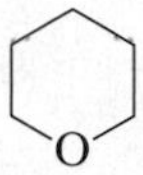   d. 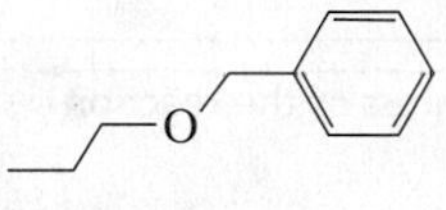   f.

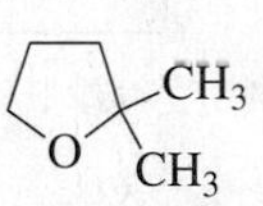

Solution to 23a The reaction takes place by an S_N2 pathway because neither alkyl group will form a relatively stable carbocation (one would be vinylic and the other primary). Iodide ion attacks the carbon of the ethyl group because otherwise it would have to attack a vinylic carbon, and vinylic carbons are not attacked by nucleophiles. Thus, the major products are ethyl iodide and an enol that immediately rearranges to an aldehyde.

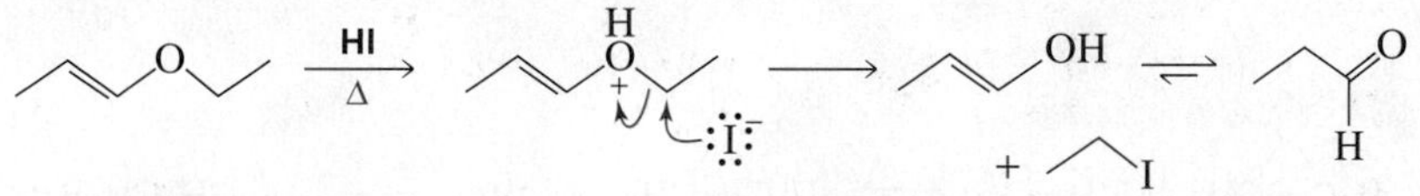

7 NUCLEOPHILIC SUBSTITUTION REACTIONS OF EPOXIDES

An alkene can be converted into an **epoxide,** a three-membered ring ether, by a peroxyacid.

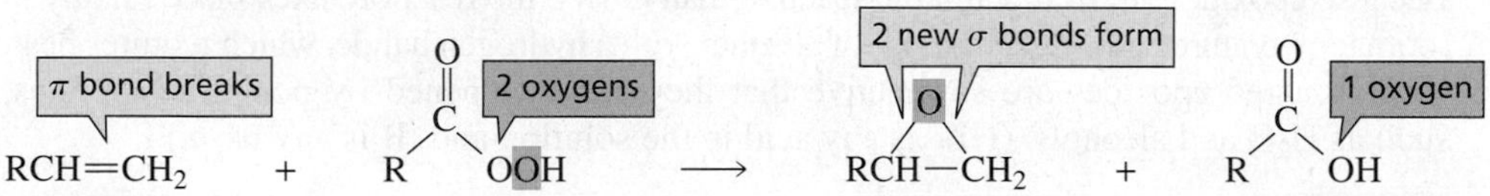

An alkene can also be converted into an epoxide by the addition of HOCl (by using Cl_2 and H_2O) followed by reaction with NaH.

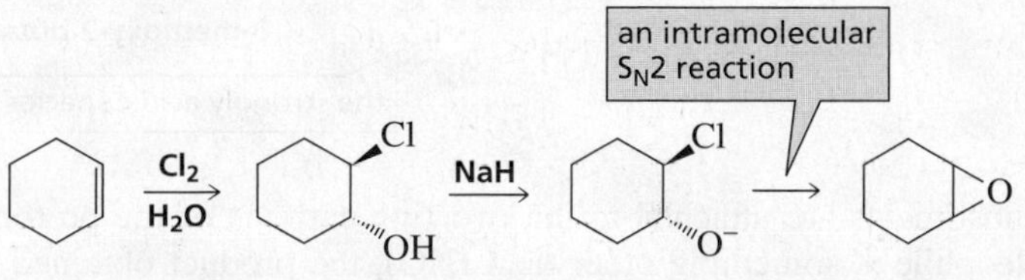

Although an epoxide and an ether have the same leaving group, epoxides are much more reactive than ethers in nucleophilic substitution reactions because the strain in their three-membered ring is relieved when the ring opens (Figure 2). Epoxides, therefore, undergo nucleophilic substitution reactions with a wide variety of nucleophiles.

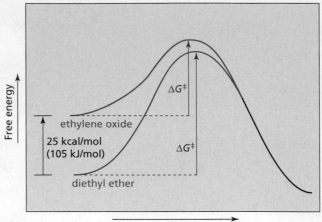

$H_2C - CH_2$
ethylene oxide
(with O bridging the two carbons)

$CH_3CH_2OCH_2CH_3$
diethyl ether

▶ **Figure 2**
The reaction coordinate diagrams for nucleophilic attack of hydroxide ion on ethylene oxide and on diethyl ether. The greater reactivity of the epoxide is a result of the strain in the three-membered ring, which increases the epoxide's free energy.

Nucleophilic Substitution: Acidic Conditions

Epoxides, like other ethers, undergo substitution reactions with hydrogen halides. The mechanism of the reaction depends on whether it is carried out under acidic or neutral/basic conditions. Under acidic conditions, the mechanism shown next is followed.

MECHANISM FOR NUCLEOPHILIC SUBSTITUTION: ACIDIC CONDITIONS

$$H_2C - CH_2 + H - \overset{..}{\underset{..}{Br}}: \rightleftharpoons H_2C - CH_2 + :\overset{..}{\underset{..}{Br}}:^- \longrightarrow H\overset{..}{O}CH_2CH_2\overset{..}{\underset{..}{Br}}:$$

protonation of the epoxide oxygen atom

back-side attack by the nucleophile

- The acid protonates the oxygen of the epoxide.
- The protonated epoxide undergoes back-side attack by the halide ion.

Because epoxides are so much more reactive than ethers, the reaction takes place readily at room temperature, unlike the reaction of an ether with a hydrogen halide, which requires heat.

Protonated epoxides are so reactive that they can be opened by poor nucleophiles, such as H_2O and alcohols. (HB^+ is any acid in the solution and :B is any base.)

$$CH_3CH - CHCH_3 \rightleftharpoons CH_3CH - CHCH_3 \xrightarrow{CH_3\overset{..}{O}H} CH_3CHCHCH_3 \rightleftharpoons CH_3CHCHCH_3 + HB^+$$

(with OH and $:\overset{+}{O}CH_3$ / H groups, and OH, OCH_3 groups)

3-methoxy-2-butanol

the strongly acidic species loses a proton

If different substituents are attached to the two ring carbons of the protonated epoxide (and the nucleophile is something other than H_2O), the product obtained from nucleophilic attack on the 2-position of the oxirane ring will be different than that obtained from nucleophilic attack on the 3-position. The major product is the one resulting from nucleophilic attack on the *more substituted* carbon.

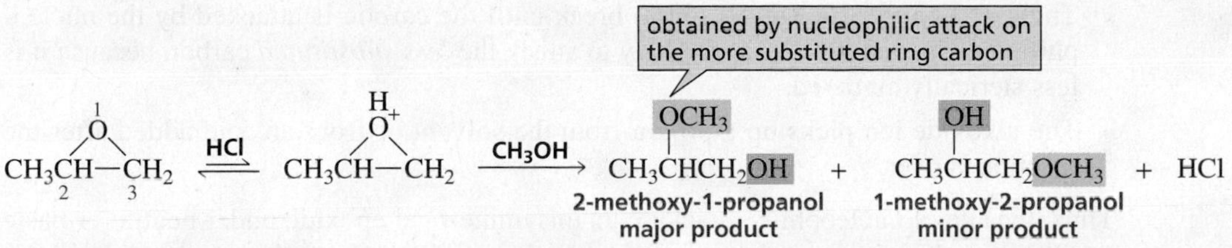

The more substituted carbon is more likely to be attacked because, after the epoxide is protonated, it is so reactive that one of the C—O bonds begins to break even before the nucleophile has an opportunity to attack. As the bond starts to break, a partial positive charge develops on the carbon that is losing its share of oxygen's electrons. Therefore, the protonated epoxide breaks preferentially in the direction that puts the partial positive charge on the more substituted carbon, because a more substituted carbocation is more stable. (Note that tertiary carbocations are more stable than secondary carbocations, which are more stable than primary carbocations.)

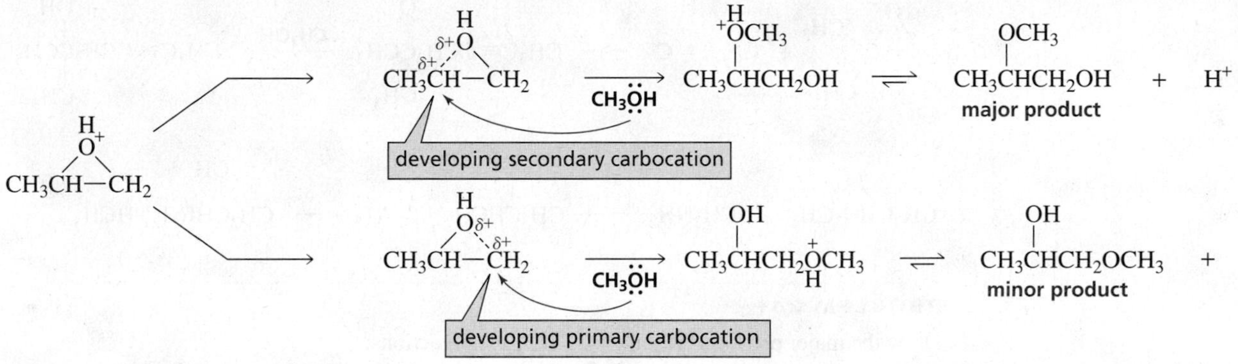

The best way to describe the reaction is to say that it occurs by a pathway that is partially S_N1 and partially S_N2. It is not a pure S_N1 reaction because a carbocation intermediate is not fully formed; it is not a pure S_N2 reaction, either, because the leaving group begins to depart before the compound is attacked by the nucleophile.

Nucleophilic Substitution: Neutral or Basic Conditions

Although an ether must be protonated before it can undergo a nucleophilic substitution reaction (Section 6), the strain in the three-membered ring allows an epoxide to undergo nucleophilic substitution reactions without first being protonated (Figure 2). When a nucleophile attacks an unprotonated epoxide, the reaction is a pure S_N2 reaction.

MECHANISM FOR NUCLEOPHILIC SUBSTITUTION: NEUTRAL OR BASIC CONDITIONS

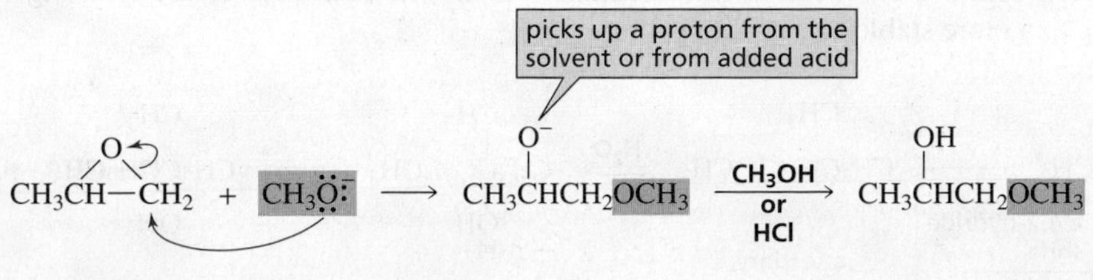

- The C—O bond does not begin to break until the carbon is attacked by the nucleophile. The nucleophile is more likely to attack the *less substituted* carbon because it is less sterically hindered.
- The alkoxide ion picks up a proton from the solvent or from an acid added after the reaction is over.

Thus, the site of nucleophilic attack on an unsymmetrical epoxide under neutral or basic conditions (when the epoxide *is not* protonated) is different from the site of nucleophilic attack under acidic conditions (when the epoxide *is* protonated).

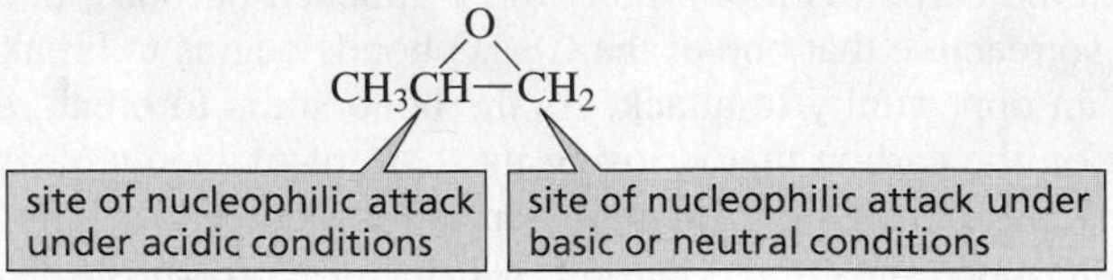

Under acidic conditions, the nucleophile preferentially attacks the more substituted ring carbon.

Under neutral or basic conditions, the nucleophile preferentially attacks the less sterically hindered ring carbon.

site of nucleophilic attack under acidic conditions

site of nucleophilic attack under basic or neutral conditions

Epoxides are useful reagents because they can react with a wide variety of nucleophiles, leading to the formation of a wide variety of products.

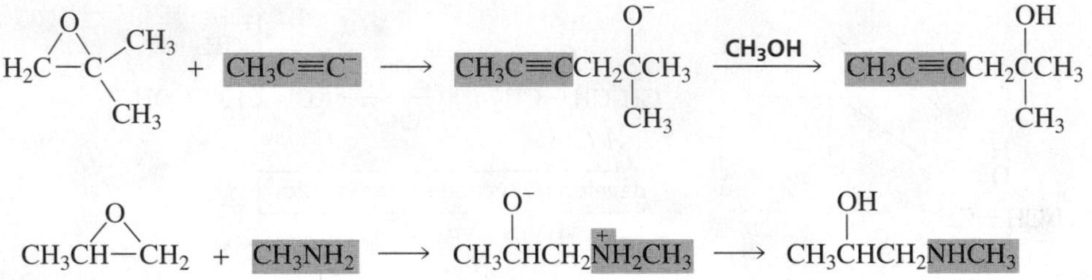

PROBLEM 24◆

Draw the major product of each of the following reactions:

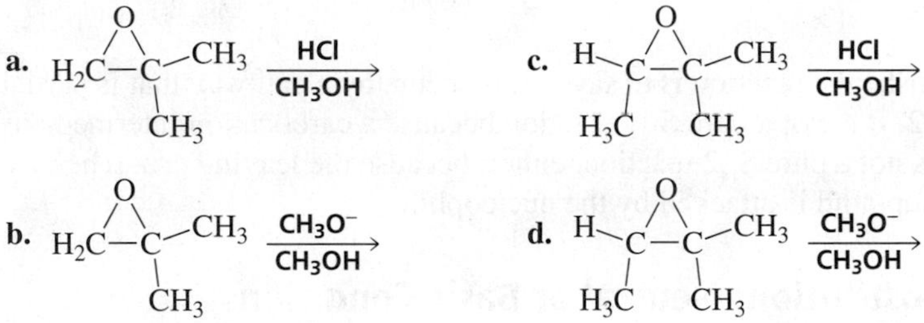

a.

b.

c.

d.

PROBLEM 25◆

Would you expect the reactivity of a five-membered ring ether such as tetrahydrofuran to be more similar to the reactivity of an epoxide or to the reactivity of a noncyclic ether?

Converting an Alkene to an Alcohol Without a Carbocation Rearragement

An *alcohol* can be prepared from the acid-catalyzed addition of water to an *alkene*. The reaction forms a carbocation intermediate, which will rearrange if the rearrangement leads to a more stable carbocation.

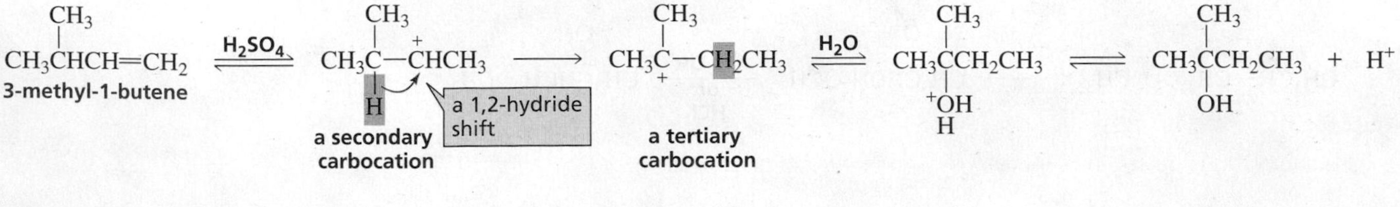

3-methyl-1-butene

a secondary carbocation

a 1,2-hydride shift

a tertiary carbocation

To avoid a carbocation rearrangement, an *alcohol* can be prepared from an *alkene* by first converting the alkene to an epoxide and then treating the epoxide with sodium hydride. Addition of acid protonates the alkoxide ion.

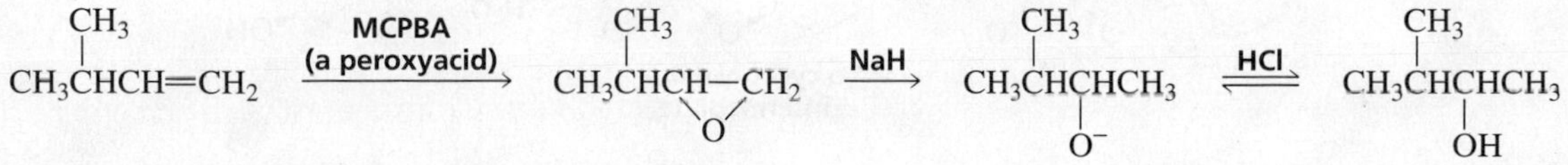

$$CH_3CHCH{=}CH_2 \xrightarrow[\text{(a peroxyacid)}]{\text{MCPBA}} CH_3CHCH{-}CH_2 \xrightarrow{\text{NaH}} CH_3CHCHCH_3 \underset{}{\overset{\text{HCl}}{\rightleftharpoons}} CH_3CHCHCH_3$$

The hydride ion is a nucleophile that attacks the least substituted carbon of the epoxide.

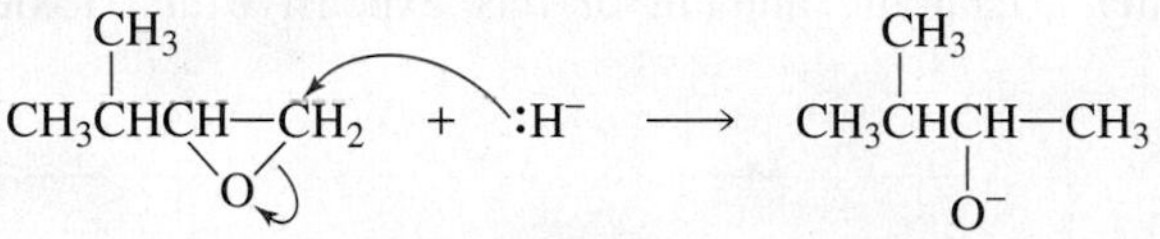

$$CH_3CHCH{-}CH_2 + \ :H^- \longrightarrow CH_3CHCH{-}CH_3$$

PROBLEM 26

How could the following compounds be prepared from 3,3-dimethyl-1-butene?
a. 2,3-dimethyl-2-butanol   **b.** 3,3-dimethyl-2-butanol

Trans and Cis Diols

The reaction of cyclohexene oxide with hydroxide ion leads to a **trans 1,2-diol,** because the S_N2 reaction involves back-side attack. A diol is also called a **glycol.** Because the OH groups are on adjacent carbons, 1,2-diols are also known as **vicinal diols** or **vicinal glycols.** (Note that *vicinal* means that two substituents are on adjacent carbons.)

cyclohexene oxide   trans 1,2-diols

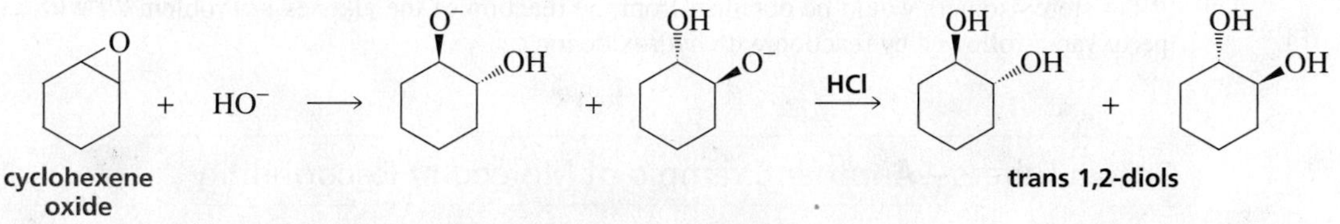

Notice that two stereoisomers are formed since the reaction forms two new asymmetric centers and only anti addition occurs.

PROBLEM 27

What products would be obtained from the reaction of cyclohexene oxide with
a. methoxide ion?   **b.** methylamine?

A **cis 1,2-diol** can be obtained by oxidizing an alkene with osmium tetroxide (OsO_4) followed by hydrolysis with aqueous hydrogen peroxide.

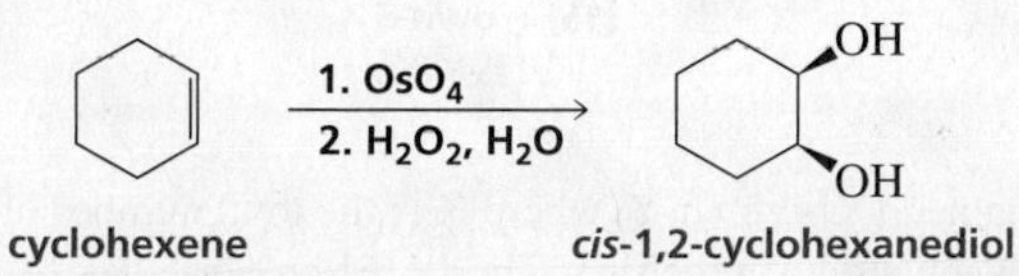

cyclohexene   *cis*-1,2-cyclohexanediol

The 1,2-diol is cis because addition of osmium tetroxide to the alkene is a syn addition—that is, both oxygens are delivered to the same side of the double bond.

Reactions of Alcohols, Ethers, Epoxides, Amines, and Thiols

MECHANISM FOR CIS-GLYCOL FORMATION

a cyclic osmate
intermediate

- Osmium tetroxide forms a cyclic intermediate when it reacts with an alkene.
- The intermediate is hydrolyzed with aqueous hydrogen peroxide. Hydrogen peroxide re-oxidizes the osmium reagent back to osmium tetroxide. (Because osmium tetroxide is recycled, only a catalytic amount of this expensive and toxic oxidizing agent is needed.)

PROBLEM 28

What products would be formed from the reaction of each of the following alkenes with OsO_4 followed by aqueous H_2O_2?

a. $CH_3C=CHCH_2CH_3$
$\quad\quad\ |$
$\quad\quad CH_3$

b. $=CH_2$

PROBLEM 29

What stereoisomers would be obtained from the reaction of each of the following alkenes with OsO_4 followed by aqueous H_2O_2?

a. *trans*-2-butene　**b.** *cis*-2-butene　**c.** *cis*-2-pentene　**d.** *trans*-2-pentene

---

PROBLEM 30

What stereoisomers would be obtained from the reaction of the alkenes in Problem 29 with a peroxyacid followed by reaction with hydroxide ion?

Crown Ethers—Another Example of Molecular Recognition

Crown ethers are cyclic compounds that contain several ether linkages around a central cavity. A crown ether specifically binds certain metal ions or organic molecules, depending on the cavity's size. The crown ether is called the host, and the species it binds is called the guest. Because the ether linkages are chemically inert, the crown ether can bind the guest without reacting with it. The *host–guest complex* is called an **inclusion compound.**

Na⁺
guest
ionic diameter = 1.80 Å

host
[15]-crown-5
cavity diameter = 1.7–2.2 Å

inclusion compound

Crown ethers are named [X]-crown-Y, where X is the total number of atoms in the ring and Y is the number of oxygen atoms in the ring. Thus, [15]-crown-5 has 15 atoms in the ring, five of which are oxygens. [15]-Crown-5 selectively binds Na⁺ because the ether's cavity diameter is 1.7 to 2.2 Å and Na⁺ has an ionic diameter of 1.80 Å. Binding occurs through the interaction of the positively charged ion with the lone-pair electrons of the oxygen atoms that point into the cavity. The ability of a host to bind only certain guests is another example of molecular recognition.

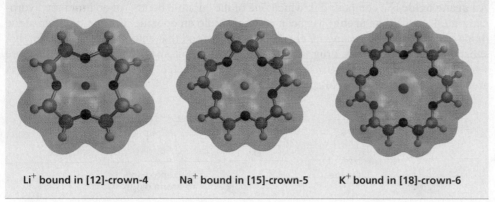

Li$^+$ bound in [12]-crown-4   **Na$^+$ bound in [15]-crown-5**   **K$^+$ bound in [18]-crown-6**

Crown Ethers Can Be Used to Catalyze S$_N$2 Reactions

A problem that often arises in the laboratory is finding a solvent that will dissolve all the reactants needed for a given reaction. For example, if we want cyanide ion to react with 1-bromohexane, we must find a way of mixing sodium cyanide (an ionic compound soluble only in water) with the alkyl halide (an organic compound that is insoluble in water). If we mix an aqueous solution of sodium cyanide with a solution of the alkyl halide in a nonpolar solvent, there will be two distinct phases—an aqueous phase and a nonpolar phase—because the two solutions are immiscible. How, then, can a reaction between sodium cyanide and 1-bromohexane take place?

$$CH_3CH_2CH_2CH_2CH_2CH_2Br \;+\; Na^+\ ^-C\equiv N \;\xrightarrow{?}\; CH_3CH_2CH_2CH_2CH_2CH_2C\equiv N \;+\; Na^+\ Br^-$$
1-bromohexane

The two compounds will be able to react with each other if [15]-crown-5 is added to the reaction mixture.

Na$^+$ will bind in the cavity of [15]-crown-5 and the inclusion compound will be soluble in the nonpolar solvent because the outside of the crown is composed primarily of nonpolar C—H bonds. The inclusion compound must carry a counterion to balance its positive charge. Thus, cyanide ion will also be in the nonpolar solvent, where it will be a powerful nucleophile since it will not be solvated. In this way, nucleophilic substitution reactions with alkyl halides that are soluble only in nonpolar solvents can readily take place.

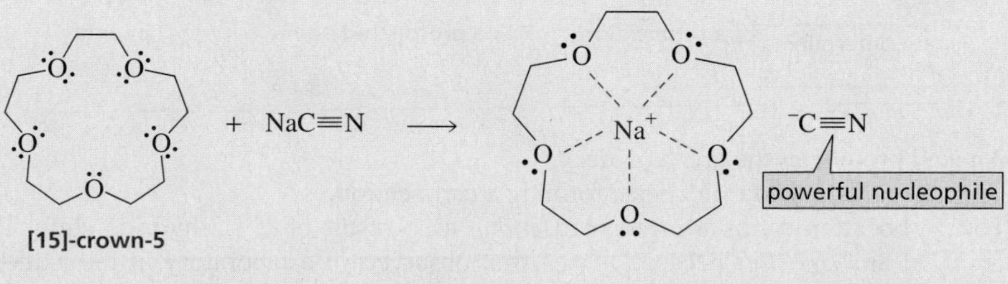

[15]-crown-5   +   NaC≡N ⟶ ... Na$^+$ ... $^-$C≡N

powerful nucleophile

In the next reaction the counter ion is K$^+$, which has a larger ionic diameter than Na$^+$, so a larger crown ether ([18]-crown-6) must be used.

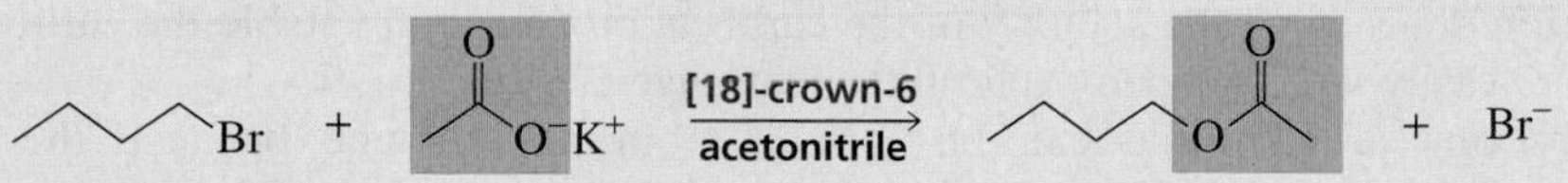

$$\text{Br} \;+\; \underset{O^-K^+}{\overset{O}{\parallel}} \;\xrightarrow[\text{acetonitrile}]{\text{[18]-crown-6}}\; \underset{O}{\overset{O}{\parallel}} \;+\; Br^-$$

8 ARENE OXIDES

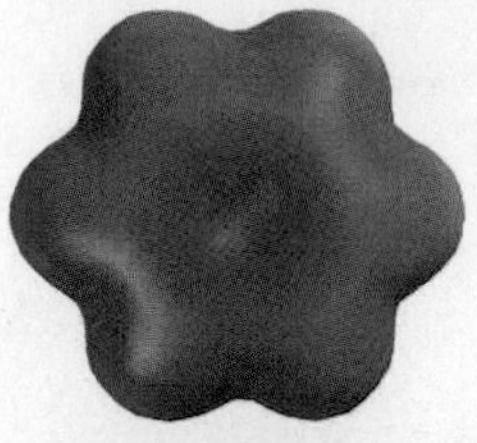

An **arene oxide** is a compound in which one of the "double bonds" of an aromatic hydrocarbon (also called an **arene**) has been converted into an epoxide. Formation of an arene oxide is the first step in changing an aromatic compound that enters the body as a foreign substance (for example, a drug, cigarette smoke, or automobile exhaust) into a more water-soluble compound that can eventually be eliminated. The enzyme that converts arenes into arene oxides is called cytochrome P_{450}.

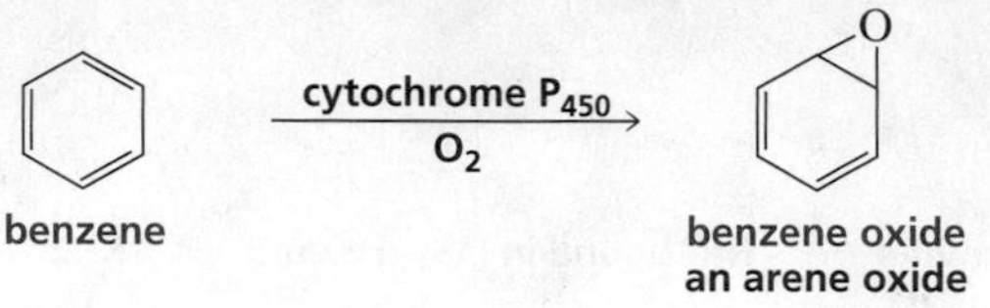

benzene

benzene oxide

An arene oxide can react in two ways. It can react as a typical epoxide, undergoing attack by a nucleophile (Y^-) to form addition products (Section 7). Alternatively, it can rearrange to form a phenol, which other epoxides cannot do.

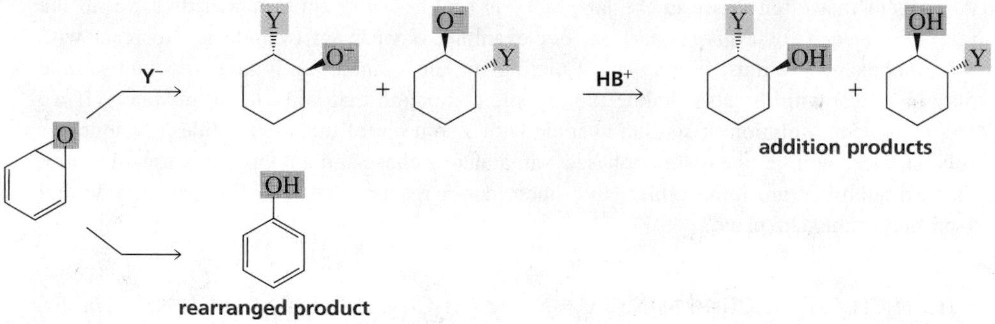

addition products

rearranged product

The mechanism for the rearrangement is shown next.

MECHANISM FOR ARENE OXIDE REARRANGEMENT

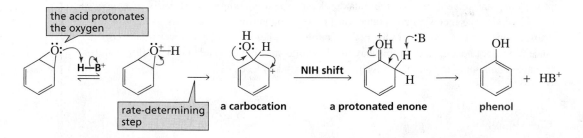

the acid protonates the oxygen

rate-determining step

a carbocation   NIH shift   **a protonated enone**   **phenol**   + HB$^+$

- An acid protonates the arene oxide.
- The three-membered ring opens, forming a carbocation.
- The carbocation forms a protonated *enone* as a result of a 1,2-hydride shift. This is called an *NIH shift* because it was first observed in a laboratory at the National Institutes of Health.
- Removal of a proton from the protonated enone forms phenol.

Because formation of the carbocation is the rate-determining step, the rate of phenol formation depends on the stability of the carbocation. The more stable the carbocation, the more easily the ring opens to form the rearranged product.

Only one arene oxide can be formed from naphthalene because the "double bond" shared by the two rings cannot be epoxidized. Benzene rings are particularly stable, so naphthalene will be epoxidized only at a position that leaves one of the benzene rings intact.

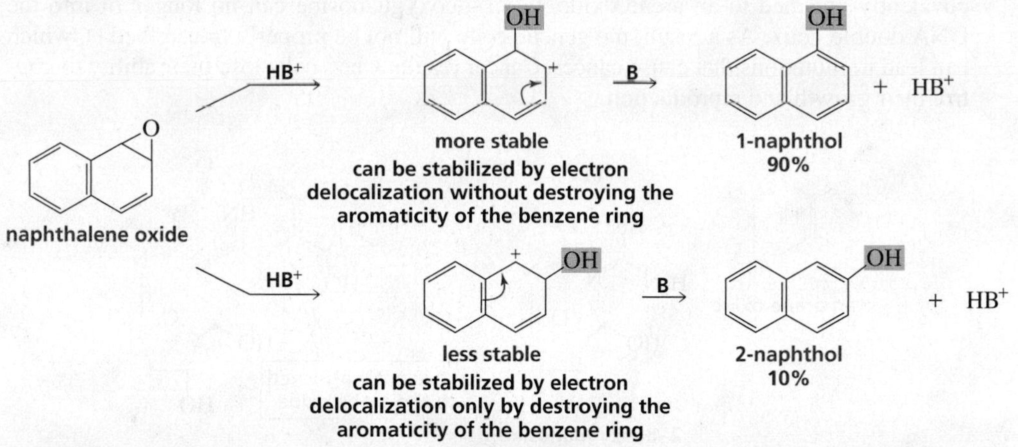

Naphthalene oxide can rearrange to form either 1-naphthol or 2-naphthol. The carbocation leading to 1-naphthol is more stable because its positive charge can be stabilized by electron delocalization without destroying the aromaticity of the benzene ring on the left of the structure. In contrast, the positive charge on the carbocation leading to 2-naphthol can be stabilized by electron delocalization only if the aromaticity of the benzene ring is destroyed. (This can be seen by comparing the predicted stabilities of the resonance contributors of the two carbocations; see Problem 31.) Consequently, rearrangement leads predominantly to 1-naphthol.

PROBLEM 31

Draw all possible resonance contributors for the two carbocations in the preceding reaction. Use the resonance contributors to explain why 1-naphthol is the major product of the reaction.

PROBLEM 32

The existence of the NIH shift was established by determining the major product obtained from rearrangement of the following arene oxide, in which a hydrogen has been replaced by a deuterium.

a. What would be the major product if the NIH shift occurs? (*Hint:* A C—H bond is easier to break than a C—D bond.)

b. What would be the major product if the carbocation forms phenol by losing a proton, rather than by going through the NIH shift?

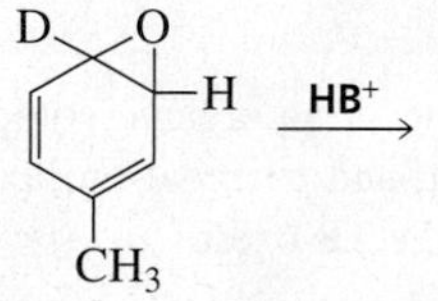

PROBLEM 33

How would the major products obtained from rearrangement of the following arene oxides differ?

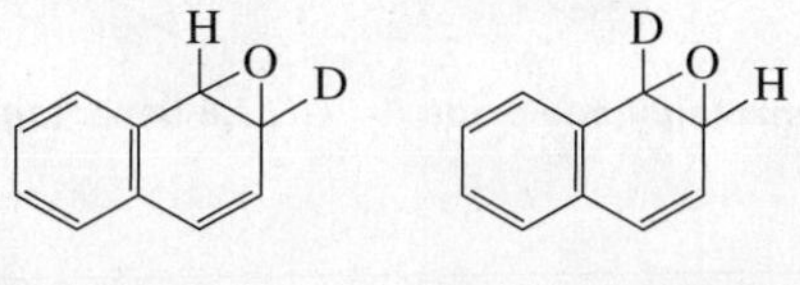

Some aromatic hydrocarbons are carcinogens—that is, compounds that cause cancer. Investigation has revealed, however, that the hydrocarbons themselves are not carcinogenic; the actual carcinogens are the arene oxides into which the hydrocarbons are converted in the body.

A segment of DNA

The more stable the carbocation formed when the arene oxide opens, the less likely it is that the arene oxide is carcinogenic.

How do arene oxides cause cancer? Nucleophiles react with epoxides to form addition products. 2´-Deoxyguanosine, a component of DNA, has a nucleophilic NH_2 group that is known to react with certain arene oxides. Once a molecule of 2´-deoxyguanosine becomes covalently attached to an arene oxide, the 2´-deoxyguanosine can no longer fit into the DNA double helix. As a result, the genetic code will not be properly transcribed (), which can lead to mutations that cause cancer. Cancer results when cells lose their ability to control their growth and reproduction.

an arene oxide

2´-deoxyguanosine

covalently attached to the arene oxide

Not all arene oxides are carcinogenic. Whether a particular arene oxide is carcinogenic depends on the relative rates of its two reaction pathways: rearrangement and reaction with a nucleophile. Arene oxide rearrangement leads to phenols that are not carcinogenic, whereas formation of addition products from nucleophilic attack by DNA can lead to cancer-causing products. Thus, if the rate of arene oxide rearrangement is faster than the rate of nucleophilic attack by DNA, then the arene oxide will be harmless. However, if the rate of nucleophilic attack is faster than the rate of rearrangement, the arene oxide will likely be a carcinogen.

We have seen that the rate-limiting step of arene oxide rearrangement is formation of the carbocation. Thus, the rate of the rearrangement reaction and an arene oxide's cancer-causing potential depend on the stability of the carbocation. If the carbocation is relatively stable, then it will be formed relatively easily, so rearrangement will be fast and the arene oxide will most likely not be carcinogenic. On the other hand, if the carbocation is relatively unstable, then rearrangement will be slow and the arene oxide will more likely exist long enough to be attacked by nucleophiles, and thus be carcinogenic. This means that *the more stable the carbocation formed when the epoxide ring of an arene oxide opens, the less likely it is that the arene oxide is carcinogenic.*

Benzo[a]pyrene and Cancer

Benzo[a]pyrene is one of the most carcinogenic arenes. It is formed whenever an organic compound is not completely burned. For example, benzo[a]pyrene is found in cigarette smoke, automobile exhaust, and charcoal-broiled meat. Several arene oxides can be formed from benzo[a]pyrene. The two most harmful are the 4,5-oxide and the 7,8-oxide.

$$\xrightarrow[\text{O}_2]{\text{cytochrome P}_{450}}$$

benzo[a]pyrene　　　　　**4,5-benzo[a]pyrene oxide**　　+　　**7,8-benzo[a]pyrene oxide**

The 4,5-oxide is harmful because it forms a carbocation that cannot be stabilized by electron delocalization without destroying the aromaticity of an adjacent benzene ring. Thus, the carbocation is relatively unstable, so the epoxide tends not to open until it is attacked by a nucleophile (the carcinogenic pathway). The 7,8-oxide is harmful because it reacts with water (a nucleophile) to form a diol, which then forms a diol epoxide. The diol epoxide does not readily undergo rearrangement (the harmless pathway), because it opens to a carbocation that is destabilized by the electron-withdrawing OH groups. Since carbocation formation is slow, the diol epoxide can exist long enough to be attacked by nucleophiles (the carcinogenic pathway).

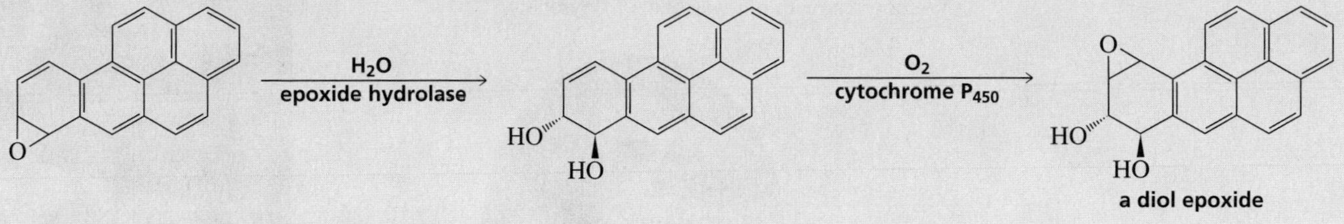

PROBLEM 34 Solved

Which compound in each pair is more likely to be carcinogenic?

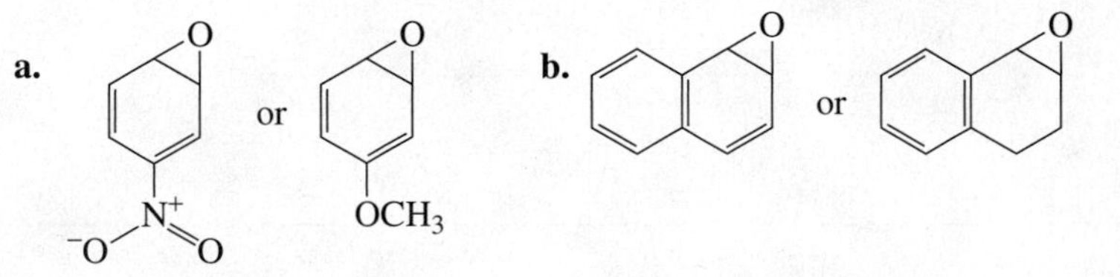

Solution to 34a The nitro-substituted compound is more likely to be carcinogenic. The nitro group destabilizes the carbocation formed when the ring opens by withdrawing electrons from the ring by resonance. In contrast, the methoxy group stabilizes the carbocation by donating electrons to the ring by resonance. Carbocation formation leads to the harmless product, so the nitro-substituted compound with a less stable (less easily formed) carbocation will be less likely to undergo rearrangement to a harmless product. In addition, the electron-withdrawing nitro group increases the arene oxide's susceptibility to nucleophilic attack, which is the cancer-causing pathway.

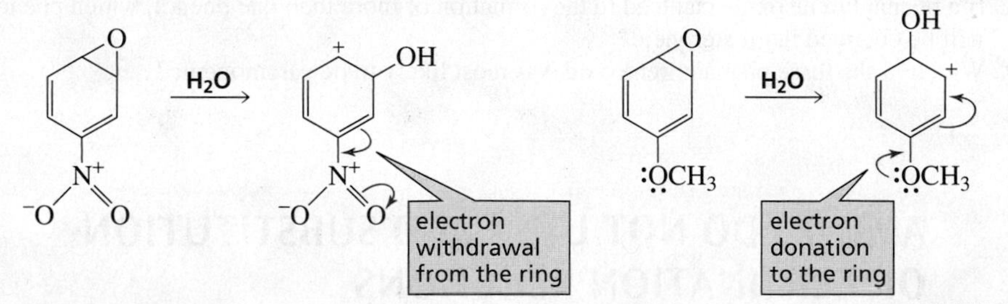

PROBLEM 35

Explain why the two arene oxides in Problem 34a open in opposite directions.

Chimney Sweeps and Cancer

In 1775, British physician Percival Pott became the first to recognize that environmental factors can cause cancer when he observed that chimney sweeps had a higher incidence of scrotum cancer than the male population as a whole. He theorized that something in the chimney soot was causing cancer. We now know that it was benzo[*a*]pyrene.

A Victorian chimney sweep and his assistant—a boy small enough to fit inside narrow passages.

Percival Pott

PROBLEM 36

Three arene oxides can be obtained from phenanthrene.

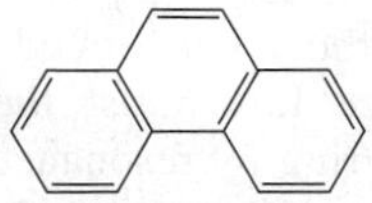

phenanthrene

a. Draw the structures of the three phenanthrene oxides.
b. Draw the structures of the phenols that can be obtained from each phenanthrene oxide.
c. If a phenanthrene oxide can lead to the formation of more than one phenol, which phenol will be obtained in greater yield?
d. Which of the three phenanthrene oxides is most likely to be carcinogenic?

---

9 AMINES DO NOT UNDERGO SUBSTITUTION OR ELIMINATION REACTIONS

Although **amines,** like alkyl halides, alcohols, and ethers, have an electron-withdrawing group bonded to an sp^3 carbon, amines do not undergo substitution and elimination reactions.

An amine's lack of reactivity in substitution and elimination reactions can be understood by comparing the leaving propensity of its electron-withdrawing group with the leaving propensity of the electron-withdrawing groups of the compounds that do undergo substitution and/or elimination reactions.

The relative leaving propensities of the groups can be determined by comparing the pK_a values of their conjugate acids, recalling that the weaker the acid, the stronger its conjugate base and the poorer the base is as a leaving group. The pK_a values of the conjugate acids show that the leaving group of an amine ($^-NH_2$) is such a strong base that amines cannot undergo substitution or elimination reactions. (An alkyl fluoride has the poorest leaving group of the alkyl halides.)

relative reactivities

most reactive	RCH_2F	>	RCH_2OH	>	RCH_2OR	>	RCH_2NH_2	least reactive

HF	H_2O	ROH	NH_3
$pK_a = 3.2$	$pK_a = 15.7$	$pK_a \sim 16$	$pK_a = 36$

The stronger the base, the poorer it is as a leaving group.

Protonating the amino group makes it a better leaving group, but not nearly as good as a protonated alcohol, which is almost 14 pK_a units more acidic than a protonated amine.

$$CH_3CH_2\overset{+}{O}H_2 \quad > \quad CH_3CH_2\overset{+}{N}H_3$$
$$pK_a = -2.4 \qquad\qquad pK_a = 11.2$$

Therefore, unlike protonated alcohols, protonated amines cannot undergo substitution and elimination reactions.

Although they cannot undergo substitution or elimination reactions, amines are extremely important organic compounds. The lone pair on its nitrogen allows an amine to react as both a base and as a nucleophile.

Amines are the most common organic bases. Protonated amines have pK_a values of about 11 and that protonated anilines have pK_a values of about 5. Neutral amines have very high pK_a values. For example, the pK_a of methylamine is 40.

$$CH_3CH_2CH_2\overset{+}{N}H_3 \qquad CH_3\overset{+}{N}H_2 \qquad CH_3CH_2\overset{+}{N}H \qquad \text{(C}_6\text{H}_5)\text{-}\overset{+}{N}H_3 \qquad CH_3\text{-(C}_6\text{H}_4)\text{-}\overset{+}{N}H_3 \qquad CH_3NH_2$$

with the groups $\overset{|}{CH_3}$ and $\overset{CH_2CH_3}{\underset{CH_2CH_3}{|}}$

$pK_a = 10.8$	$pK_a = 10.9$	$pK_a = 11.1$	$pK_a = 4.58$	$pK_a = 5.07$	$pK_a = 40$

Amines react as nucleophiles in a wide variety of reactions. For example, they react as nucleophiles with alkyl halides and epoxides in S_N2 reactions.

an S_N2 reaction

$$CH_3CH_2Br \ + \ CH_3NH_2 \ \longrightarrow \ CH_3CH_2\overset{+}{N}H_2CH_3 \ + \ Br^-$$

$$\overset{O}{\overset{\diagup\diagdown}{CH_3CH-CH_2}} \ + \ CH_3NH_2 \ \longrightarrow \ CH_3\overset{O^-}{\underset{}{CH}}CH_2\overset{+}{N}H_2CH_3 \ \longrightarrow \ CH_3\overset{OH}{\underset{}{CH}}CH_2NHCH_3$$

We will also see that they react as nucleophiles with a wide variety of carbonyl compounds.

Alkaloids

Alkaloids are amines found in the leaves, bark, roots, or seeds of many plants. Examples include caffeine (found in tea leaves, coffee beans, and cola nuts) and nicotine (found in tobacco leaves). Nicotine causes brain cells to release dopamine and endorphins, compounds that makes us feel good, thereby making nicotine addictive. Ephedrine, a bronchodilator, is an alkaloid obtained from *Ephedra sinica,* a plant found in China. Morphine is an alkaloid obtained from opium, a milky fluid exuded by a species of poppy.

coffee beans

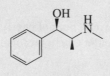
caffeine   nicotine   ephedrine

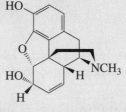
morphine

tobacco leaves

Lead Compounds for the Development of Drugs

Medicinal agents used by humans since ancient times provided the starting point for the development of our current arsenal of drugs. The active ingredients were isolated from the herbs, berries, roots, and bark used by medicine men and women, shamans, and witch doctors. Scientists still search the world for plants, berries, flora, and fauna that might yield new medicinal compounds.

Once a naturally occurring drug is isolated and its structure determined, it can serve as a prototype in a search for other biologically active compounds. The prototype is called a **lead compound** (that is, it plays a leading role in the search). Analogues of the lead compound are synthesized and tested to see if they are more effective or have fewer side effects than the lead compound. An analogue may have a different substituent than the lead compound, a branched chain instead of a straight chain, a different functional group, or some other structural difference. Producing analogues by changing the structure of a lead compound is called **molecular modification.**

In a classic example of molecular modification, a number of synthetic local anesthetics were developed from cocaine, an alkaloid obtained from the leaves of *Erythroxylon coca,* a bush native to the highlands of the South American Andes. Cocaine is a highly effective local anesthetic, but it produces undesirable effects on the central nervous system (CNS), ranging from initial euphoria to severe depression. By dissecting the cocaine molecule step by step—removing the methoxycarbonyl group and cleaving the seven-membered-ring system—scientists identified the portion of the molecule that carries the local anesthetic activity but does not induce the damaging CNS effects. This knowledge provided an improved lead compound.

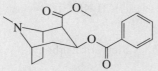
cocaine
lead compound   improved lead compound

Hundreds of related compounds were then synthesized. Successful anesthetics obtained by molecular modification were Benzocaine (a topical anesthetic), Novocain (used by dentists), and Xylocaine (one of the most widely used injectable anesthetics).

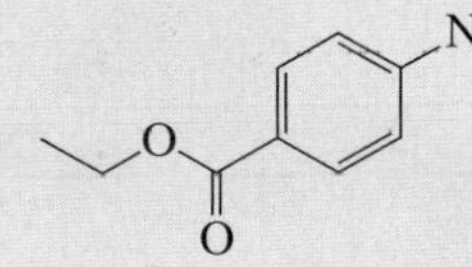

Benzocaine®

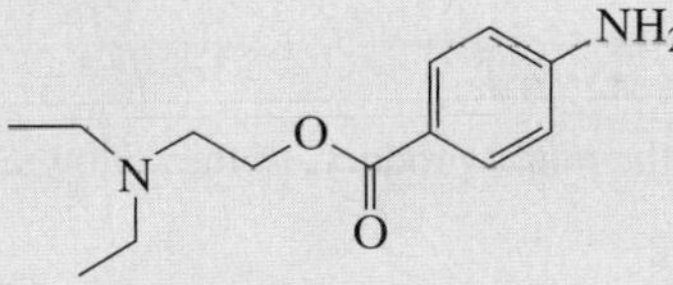

procaine
Novocain®

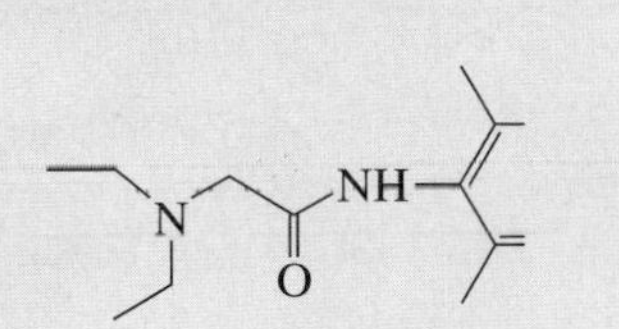

lidocaine
Xylocaine®

PROBLEM 37

Explain why the half-life (the time it takes for one-half of the compound to be metabolized) of Xylocaine is longer than that of Novocaine.

---

10 QUATERNARY AMMONIUM HYDROXIDES UNDERGO ELIMINATION REACTIONS

A **quaternary ammonium ion** can undergo an elimination reaction with a strong base such as hydroxide ion. The reaction is known as a **Hofmann elimination reaction.** The leaving group in a Hofmann elimination reaction is a tertiary amine. Because a tertiary amine is a relatively poor leaving group, the reaction requires heat.

$$CH_3CH_2CH_2-\overset{\overset{CH_3}{|}}{\underset{\underset{CH_3}{|}}{N}CH_3}\ HO^- \xrightarrow{\Delta} CH_3CH{=}CH_2 + :\overset{\overset{CH_3}{|}}{\underset{\underset{CH_3}{|}}{N}CH_3} + H_2O$$

A Hofmann elimination reaction is an E2 reaction, which means the proton and the tertiary amine are removed in the same step. Very little substitution product is formed.

Bitrex®

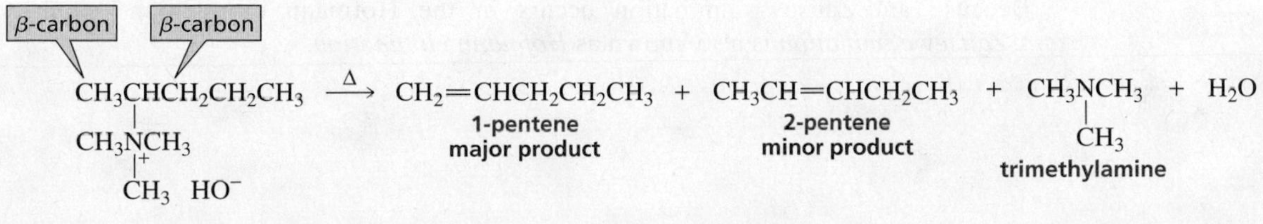

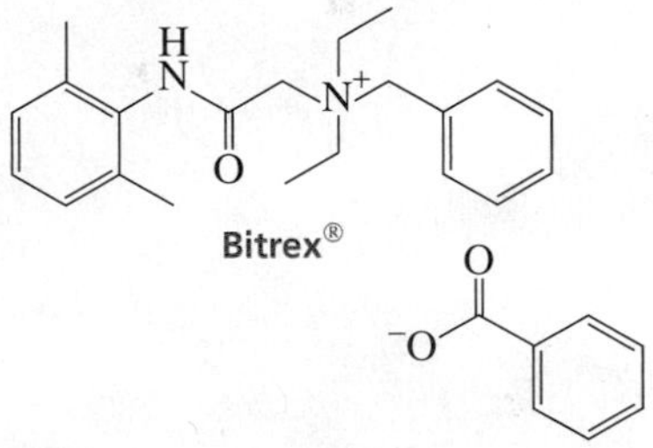

Bitrex, a quaternary ammonium salt, is nontoxic and one of the most bitter-tasting substances known. It is used to encourage deer to look elsewhere for food; it is put on the backs of animals to keep them from biting one another and on children's fingers to persuade them to stop sucking their thumbs or biting their fingernails; and it is added to toxic substances to keep them from being ingested accidentally.

MECHANISM FOR THE HOFMANN ELIMINATION

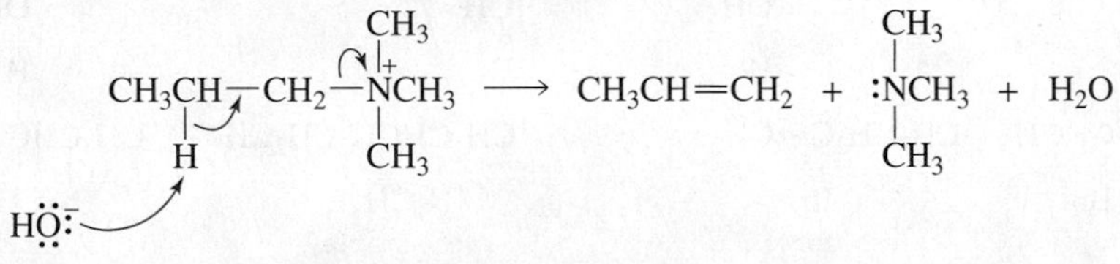

$$CH_3CH-CH_2-\overset{\overset{CH_3}{|}}{\underset{\underset{CH_3}{|}}{\overset{+}{N}}CH_3} \longrightarrow CH_3CH{=}CH_2 + :\overset{\overset{CH_3}{|}}{\underset{\underset{CH_3}{|}}{N}CH_3} + H_2O$$

$$\overset{|}{H}$$

$$HO:^-$$

The tertiary amine is attached to the α-carbon and the proton is removed from the adjacent carbon (the β-carbon). If the quaternary ammonium ion has more than one β-carbon, the major alkene product is the one obtained by removing a proton from the β-carbon bonded to the *most* hydrogens.

For example, in the following reaction, the major alkene product is obtained by removing a proton from the β-carbon bonded to three hydrogens, and the minor alkene product results from removing a proton from the β-carbon bonded to two hydrogens.

$\boxed{\beta\text{-carbon}}$ $\boxed{\beta\text{-carbon}}$

$$CH_3CHCH_2CH_2CH_3 \xrightarrow{\Delta} CH_2{=}CHCH_2CH_2CH_3 + CH_3CH{=}CHCH_2CH_3 + CH_3NCH_3 + H_2O$$
$$\underset{\underset{CH_3\ HO^-}{|}}{\overset{|}{\underset{}{CH_3NCH_3}}}$$

1-pentene
major product   **2-pentene**
minor product

$$\underset{\underset{CH_3}{|}}{\ }$$
trimethylamine

In a Hofmann elimination reaction, the proton is removed from the β-carbon bonded to the most hydrogens.

PROBLEM 38

If a quarternary ammonium ion can undergo an elimination reaction with a strong base, why can't a protonated tertiary amine undergo the same reaction?

PROBLEM 39♦

What are the major products of the following reaction?

$$\text{CH}_3\text{CHCH}_2\overset{\overset{\displaystyle \text{CH}_3}{|}}{\underset{\overset{|}{\text{CH}_3}}{\overset{+}{\text{N}}}}\text{CH}_2\text{CH}_2\text{CH}_3 \xrightarrow[\text{HO}^-]{\Delta}$$

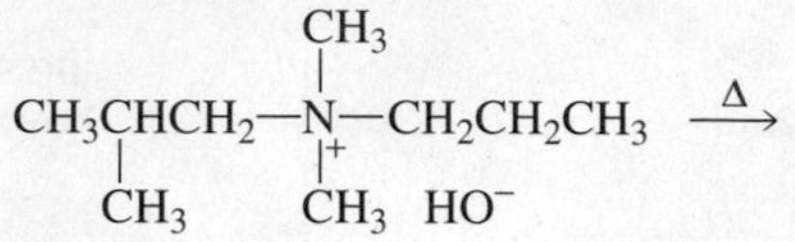

PROBLEM 40♦

What are the minor products of the preceding Hofmann elimination reaction?

PROBLEM 41♦

What is the difference between the reaction that occurs when isopropyltrimethylammonium hydroxide is heated and the reaction that occurs when 2-bromopropane is treated with hydroxide ion?

In an E2 reaction of an alkyl chloride, alkyl bromide, or alkyl iodide, the proton is removed from the β-carbon *bonded to the fewest hydrogens* (known as *Zaitsev's rule*). Now, however, we see that in an E2 reaction of a quaternary ammonium ion, the proton is removed from the β-carbon *bonded to the most hydrogens* (anti-Zaitsev elimination).

Quaternary amines violate Zaitsev's rule for the same reason that alkyl fluorides violate it. Alkyl halides, other than alkyl fluorides, have relatively good leaving groups that immediately start to depart when hydroxide ion starts to remove the proton, forming a transition state with an *alkene-like* structure. The proton is removed *from the β-carbon bonded to the fewest hydrogens* in order to achieve the most stable *alkene-like* transition state.

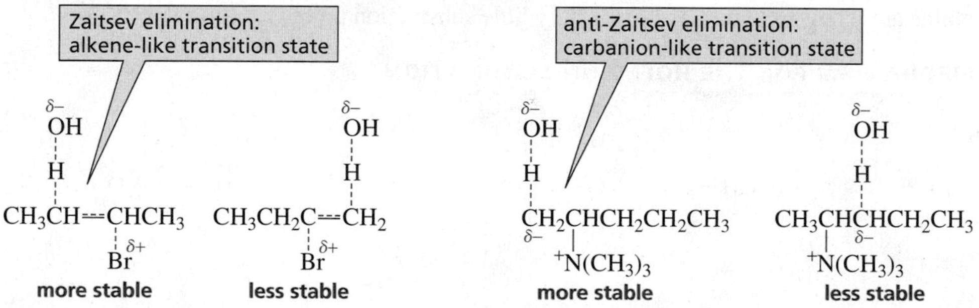

Quaternary ammonium ions and alkyl fluorides have poorer leaving groups that do not start to leave when hydroxide ion starts to remove a proton. Therefore, a partial negative charge builds up on the carbon from which the proton is being removed, giving the transition state a *carbanion-like* structure. The proton is removed *from the β-carbon bonded to the most hydrogens* in order to achieve the more stable *carbanion-like* transition state. (Primary carbanions are more stable than secondary carbanions, which are more stable than tertiary carbanions.) Steric factors also favor anti-Zaitsev elimination.

Because anti-Zaitsev elimination occurs in the Hofmann elimination reaction, *anti-Zaitsev elimination* is also known as *Hofmann elimination*.

For a quaternary ammonium ion to undergo an elimination reaction, the counterion must be hydroxide ion, because a strong base is needed to remove the proton from the β-carbon. Halide ions are weak bases, so quaternary ammonium *halides* cannot undergo Hofmann elimination reactions. However, a quaternary ammonium *halide* can be converted into a quaternary ammonium *hydroxide* by treatment with silver oxide and water. The silver halide precipitates, and the halide ion is replaced by hydroxide ion.

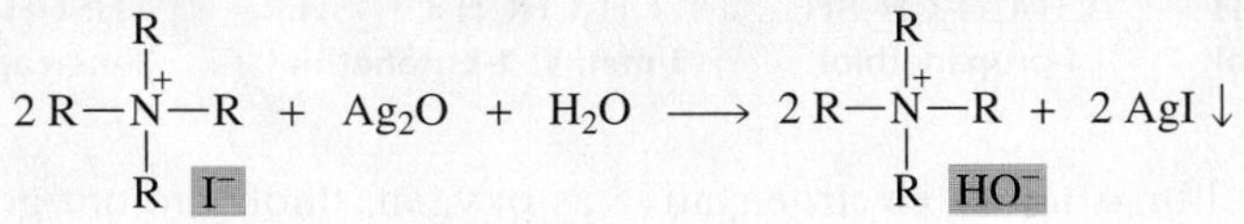

$$2 \, R - \overset{\overset{\displaystyle R}{|+}}{\underset{\underset{\displaystyle R}{|}}{N}} - R \; + \; Ag_2O \; + \; H_2O \longrightarrow 2 \, R - \overset{\overset{\displaystyle R}{|+}}{\underset{\underset{\displaystyle R}{|}}{N}} - R \; + \; 2 \, AgI \downarrow$$

I^- HO^-

PROBLEM 42◆

What is the major product of each of the following reactions?

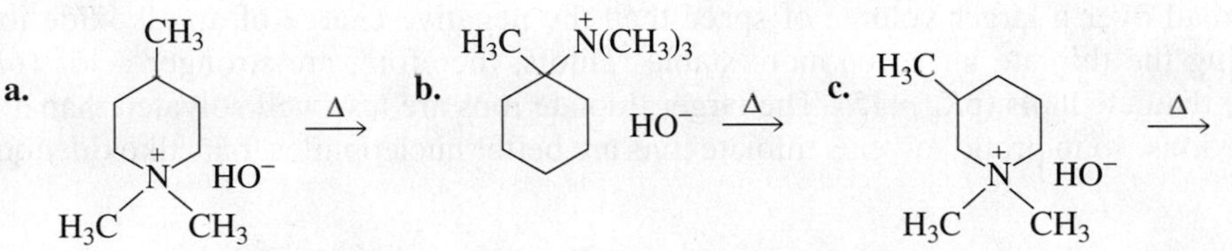

a.   b.   c.

PROBLEM 43 Solved

Describe a synthesis for each of the following compounds, using the given starting material and any necessary reagents:

a. $CH_3CH_2CH_2CH_2NH_2 \longrightarrow CH_3CH_2CH=CH_2$

b. $CH_3CH_2CH_2\underset{\underset{\displaystyle Br}{|}}{C}HCH_3 \longrightarrow CH_3CH_2CH_2CH=CH_2$

c. 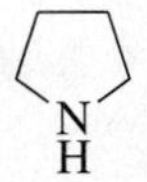 $\longrightarrow CH_2=CH-CH=CH_2$

Solution to 43a Although an amine cannot undergo an elimination reaction, a quaternary ammonium hydroxide can. The amine, therefore, must first be converted into a quaternary ammonium iodide by reacting with excess methyl iodide in a basic solution of potassium carbonate.

Treatment with aqueous silver oxide forms the quaternary ammonium hydroxide. Heat is required for the elimination reaction.

$$CH_3CH_2CH_2CH_2NH_2 \; \xrightarrow[\text{K}_2\text{CO}_3]{\overset{\textstyle \text{CH}_3\text{I}}{\text{excess}}} \; CH_3CH_2CH_2CH_2\overset{+}{N}(CH_3)_3 \; \xrightarrow[\text{H}_2\text{O}]{\text{Ag}_2\text{O}} \; CH_3CH_2CH_2CH_2\overset{+}{N}(CH_3)_3 \; \xrightarrow{\Delta} \; CH_3CH_2CH=CH_2 \; + \; H_2O$$

I^- HO^-

---

11   THIOLS, SULFIDES, AND SULFONIUM SALTS

Thiols are sulfur analogues of alcohols. They used to be called mercaptans because they form strong complexes with heavy metal cations such as arsenic and mercury—that is, they capture mercury.

$$2 \, CH_3CH_2SH \; + \; Hg^{2+} \longrightarrow CH_3CH_2S-Hg-SCH_2CH_3 \; + \; 2 \, H^+$$

a thiol   **mercuric ion**

Thiols are named by adding the suffix *thiol* to the name of the parent hydrocarbon. If there is a second functional group in the molecule that is identified by a suffix, the SH group can be indicated by its substituent name, *mercapto*. Like other substituent names, it is placed before the name of the parent hydrocarbon.

$$CH_3CH_2SH \qquad CH_3CH_2CH_2SH \qquad CH_3\overset{\overset{\displaystyle CH_3}{|}}{C}HCH_2CH_2SH \qquad HSCH_2CH_2OH$$

ethanethiol   1-propanethiol   3-methyl-1-butanethiol   2-mercaptoethanol

Low-molecular-weight thiols are noted for their strong and pungent odors, such as the odors associated with onions, garlic, and skunks. Natural gas is completely odorless and can cause deadly explosions if a leak goes undetected. As a result, a small amount of a thiol is added to natural gas to give it an odor so that gas leaks can be detected.

Because sulfur is not as electronegative as oxygen, thiols are not good at hydrogen bonding. Consequently, they have weaker intermolecular attractions and, therefore, considerably lower boiling points than alcohols. For example, the boiling point of CH_3CH_2SH is 37 °C, whereas the boiling point of CH_3CH_2OH is 78 °C.

Sulfur atoms are larger than oxygen atoms, so the negative charge of the thiolate ion is spread over a larger volume of space than the negative charge of an alkoxide ion, causing the thiolate ion to be more stable. Thiols, therefore, are stronger acids (pK_a ~10) than alcohols (pK_a ~15). The larger thiolate ions are less well solvated than alkoxide ions, so in protic solvents thiolate ions are better nucleophiles than alkoxide ions.

$$CH_3\!-\!\ddot{\underset{..}{S}}{:}^{-} + CH_3CH_2\!-\!Br \xrightarrow{CH_3OH} CH_3\!-\!\ddot{\underset{..}{S}}\!-\!CH_2CH_3 + Br^-$$

The sulfur analogues of ethers are called **sulfides** or **thioethers.** Sulfur is an excellent nucleophile because its electron cloud is polarizable. As a result, sulfides react readily with alkyl halides to form **sulfonium salts,** whereas ethers do not react with alkyl halides because oxygen is not as nucleophilic as sulfur and cannot accommodate a positive charge as well as sulfur can.

$$CH_3\!-\!\ddot{\underset{..}{S}}\!-\!CH_3 + CH_3\!-\!I \longrightarrow CH_3\!-\!\overset{\overset{\displaystyle CH_3}{|}}{\underset{..}{S}}^{+}\!-\!CH_3 \;\; I^-$$

dimethyl sulfide   trimethylsulfonium iodide
a thioether   a sulfonium salt

The positively charged group of a sulfonium ion is an excellent leaving group, so a sulfonium ion readily undergoes nucleophilic substitution reactions. Like other S_N2 reactions, the reaction works best if the group undergoing nucleophilic attack is a methyl group or a primary alkyl group. SAM, a biological methylating agent, is a sulfonium salt.

$$H\ddot{\underset{..}{O}}{:}^{-} + CH_3\!-\!\overset{\overset{\displaystyle CH_3}{|}}{\underset{..}{S}}^{+}\!-\!CH_3 \longrightarrow CH_3\!-\!\ddot{\underset{..}{O}}H + CH_3\!-\!\ddot{\underset{..}{S}}\!-\!CH_3$$

PROBLEM 44

Using an alkyl halide and a thiol as starting materials, how would you prepare the following thioethers?

a. ⌁⌁S⌁⌁   **b.** ⌁S⌁   **c.** $CH_2{=}CHCHSCH_2CH_3$ with $\overset{\overset{\displaystyle CH_3}{|}}{}$ **d.** (phenyl)−S−(benzyl)

Mustard Gas—A Chemical Warfare Agent

Chemical warfare occurred for the first time in 1915, when Germany released chlorine gas against French and British forces in the Battle of Ypres. For the remainder of World War I, both sides used a variety of chemical agents as weapons. One of the more common was mustard gas, a reagent that produces large blisters on exposed skin. Mustard gas is extremely reactive because its highly nucleophilic sulfur atom easily displaces a chloride ion by an intramolecular S_N2 reaction, forming a cyclic sulfonium ion that reacts rapidly with a nucleophile. The sulfonium salt is particularly reactive because of its strained three-membered ring and the excellent (positively charged) leaving group.

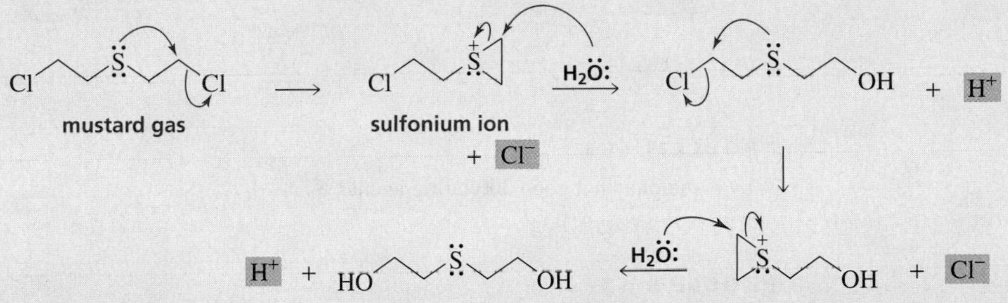

The blistering caused by mustard gas results from the high local concentrations of HCl produced when mustard comes into contact with water—or any other nucleophile—on skin or in lung tissue. Autopsies of soldiers killed by mustard gas in World War I revealed that they had extremely low white blood cell counts and defects in bone marrow development, indicating profound effects on rapidly dividing cells.

Alkylating Agents as Cancer Drugs

Since cancer is characterized by the uncontrolled growth and proliferation of cells, the discovery that mustard gas affected rapidly dividing cells suggested that it might be an effective antitumor agent. Therefore, chemists started looking for less reactive mustards that might be used in chemotherapy—that is, the use of chemicals in the treatment of cancer.

Because mustard gas forms a three-membered ring that can react rapidly with nucleophiles, its clinical reactivity is thought to be due to its ability to alkylate groups on the surface of DNA. Alkylating DNA can destroy it, which means that the rapidly growing cells of cancerous tumors are killed. Unfortunately, compounds used for chemotherapy can also kill normal cells. That is why many side effects, such as nausea and hair loss, are associated with cancer chemotherapy. The challenge for chemists now is to find drugs that will target only cancer cells.

The cancer drugs shown here are all biological alkylating agents—they attach an alkyl group to a nucleophile under physiological conditions.

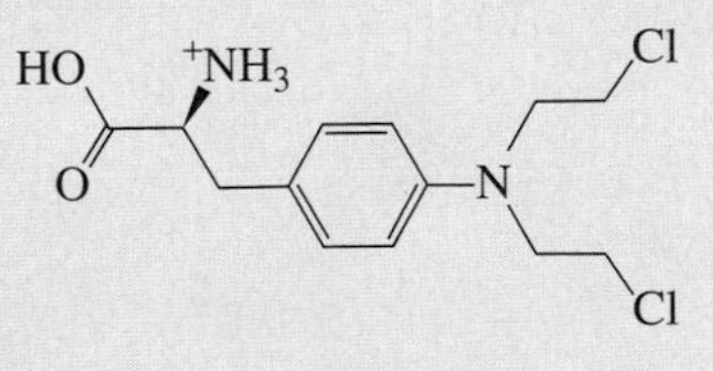

melphalan

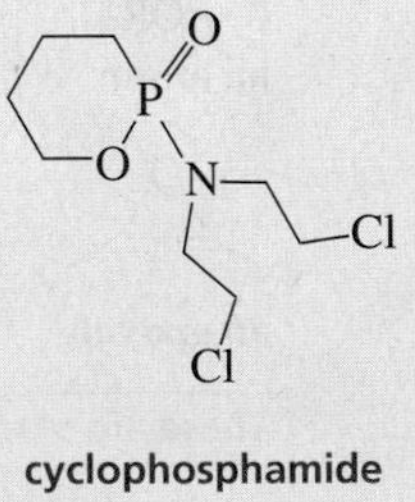

cyclophosphamide

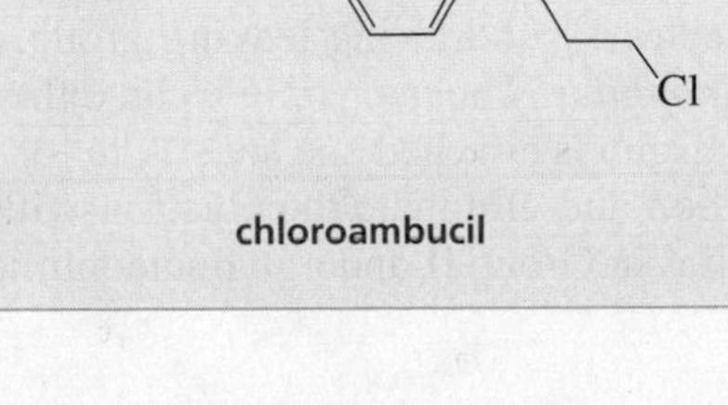

chloroambucil

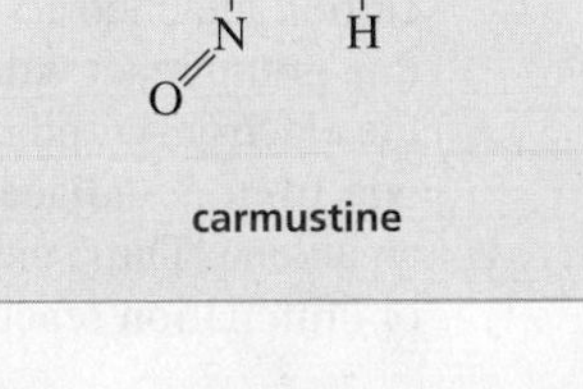

carmustine

PROBLEM 45♦

The following three nitrogen mustards were studied for possible clinical use. One is now used clinically, one was found to be too unreactive, and one was found to be too insoluble in water to be injected intravenously. Which is which? (*Hint:* Draw resonance contributors.)

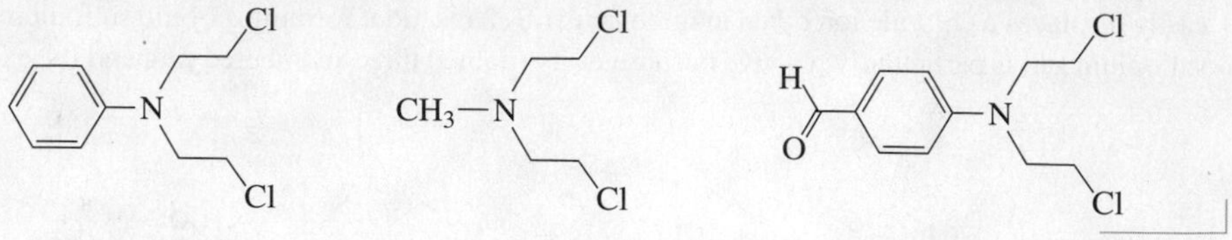

PROBLEM 46♦

Why is melphalan a good alkylating agent?

PROBLEM 47

Mechlorethamine, the drug in Problem 45 that is in clinical use, is highly reactive so can be administered only by physicians who are experienced in its use. Explain why cyclophosphamide, carmustine, and chloroambucil are less reactive alkylating agents.

12 ORGANIZING WHAT WE KNOW ABOUT THE REACTIONS OF ORGANIC COMPOUNDS

Families of organic compounds can be put into one of four groups, and all the members of a group react in similar ways. Let's summarize the families in Group II.

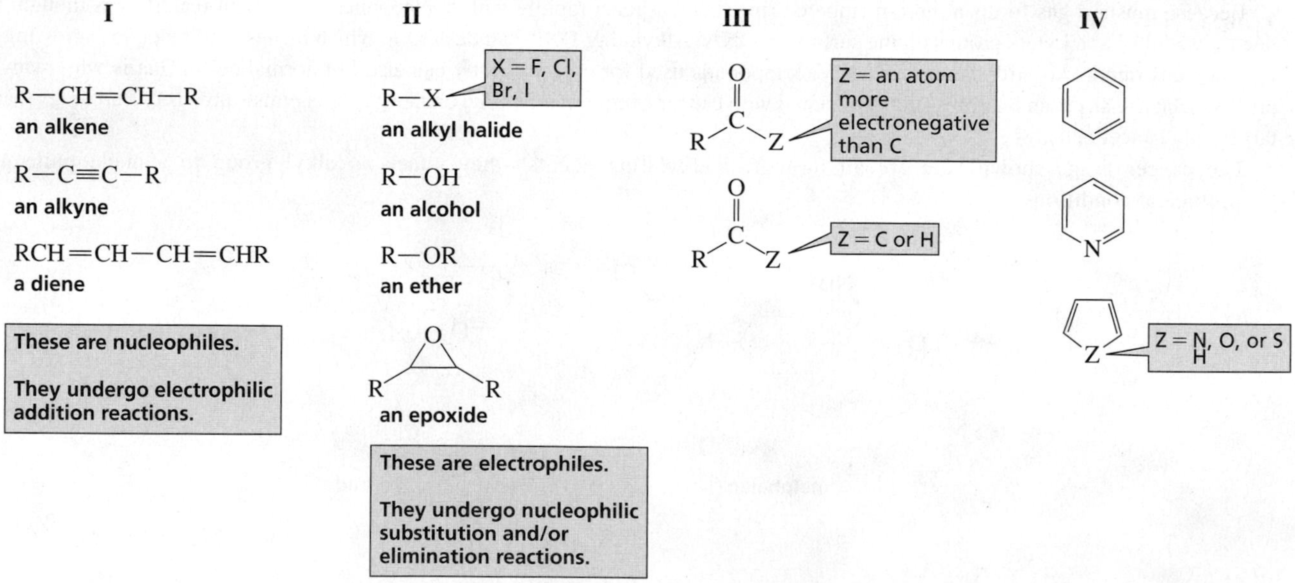

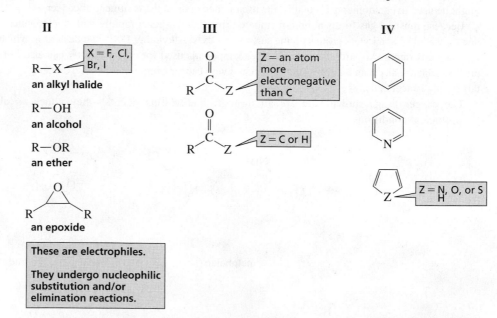

All the families in Group II are *electrophiles,* due to the partial positive charge on the carbon attached to the electron-withdrawing leaving group. As a result, the families in this group react with *nucleophiles.* The nucleophile can either attack the carbon to which the electron-withdrawing group is attached and substitute for it, or it can remove a hydrogen from an adjacent carbon and eliminate the electron-withdrawing group by forming an alkene. Thus, the families in Group II undergo nucleophilic substitution reactions and/or elimination reactions.

- Alkyl halides have excellent leaving groups, so they undergo substitution and/or elimination reactions with ease.

- Alcohols have much poorer leaving groups, so they have to be activated before they can undergo nucleophilic substitution and/or elimination reactions.

- Ethers, like alcohols, have poor leaving groups, but unlike alcohols, they can be activated only by protonation and they undergo only substitution reactions.

- Epoxides are more reactive than acyclic ethers because of the angle strain in the three-membered ring. Thus, they readily undergo substitution reactions whether or not they are activated by protonation.

SOME IMPORTANT THINGS TO REMEMBER

- The leaving groups of **alcohols** and **ethers** are stronger bases than halide ions, so alcohols and ethers have to be "activated" before they can undergo a substitution or an elimination reaction.

- There are several ways to activate an alcohol, but only one way to activate an ether.

- **Epoxides** do not have to be activated, because ring strain increases their reactivity.

- Primary, secondary, and tertiary alcohols undergo nucleophilic substitution reactions with HI, HBr, and HCl to form alkyl halides. These are S_N1 reactions in the case of secondary and tertiary alcohols and S_N2 reactions in the case of primary alcohols.

- An alcohol can also be converted into an alkyl halide by a phosphorus trihalide or thionyl chloride. These reagents convert the alcohol into an intermediate that has a leaving group that is easily displaced by a halide ion.

- Converting an alcohol to a **sulfonate ester** is another way to activate an alcohol for subsequent reaction with a nucleophile.

- Activating an alcohol by converting it to an alkyl halide with a hydrogen halide followed by reaction with a nucleophile forms a substitution product with the same configuration as the alcohol, whereas activating an alcohol by converting it to a sulfonate ester followed by reaction with a nucleophile forms a substitution product with a configuration opposite that of the alcohol.

- An alcohol undergoes **dehydration** (elimination of a water molecule) when it is heated with an acid.

- Dehydration is an E1 reaction in the case of secondary and tertiary alcohols and an E2 reaction in the case of primary alcohols.

- Tertiary alcohols are the easiest to dehydrate, and primary alcohols are the hardest.

- The major product of alcohol dehydration is the more stable alkene.

- If the alkene has stereoisomers, the stereoisomer in which the largest groups are on opposite sides of the double bond predominates.

- Dehydration of secondary and tertiary alcohols can be made to take place by an E2 reaction if a good leaving group is put on the alcohol before the elimination reaction.

- E1 reactions form carbocation intermediates, so carbocation rearrangements can occur.

- Chromic acid oxidizes primary alcohols to carboxylic acids and secondary alcohols to ketones.

- Hypochlorous acid or PCC in CH_2Cl_2 oxidizes primary alcohols to aldehydes and secondary alcohols to ketones.

- Ethers can undergo nucleophilic substitution reactions with HBr or HI and heat; if departure of the leaving group creates a relatively stable carbocation, an S_N1 reaction occurs; otherwise, an S_N2 reaction occurs.

- **Epoxides** undergo nucleophilic substitution reactions. Under acidic conditions, the more substituted ring carbon is attacked; under neutral or basic conditions, the less sterically hindered ring carbon is attacked.

- Aromatic hydrocarbons (**arenes**) are oxidized to **arene oxides** that undergo rearrangement to form phenols, or undergo nucleophilic attack to form addition products.

- The more stable the carbocation formed during rearrangement, the less likely it is that the arene oxide is carcinogenic.

- Amines cannot undergo substitution or elimination reactions because their leaving groups are very strong bases.

- A quaternary ammonium hydroxide can undergo an E2 reaction known as a Hofmann elimination if they are heated. Hydroxide ion removes a proton from the β-carbon bonded to the most hydrogens.

- **Thiols** are sulfur analogues of alcohols. They are stronger acids than alcohols and have lower boiling points.

- Thiolate ions are weaker bases and better nucleophiles than alkoxide ions in protic solvents.

- **Thioethers** react with alkyl halides to form **sulfonium salts,** which have excellent leaving groups, so they undergo substitution reactions with ease.

SUMMARY OF REACTIONS

1. Converting an *alcohol* into an *alkyl halide* (Sections 1 and 2).

$$ROH + HBr \xrightarrow{\Delta} RBr$$

$$ROH + HI \xrightarrow{\Delta} RI$$

$$ROH + HCl \xrightarrow{\Delta} RCl$$

relative rate: tertiary > secondary > primary

$$ROH + PBr_3 \xrightarrow{\text{pyridine}} RBr$$

$$ROH + PCl_3 \xrightarrow{\text{pyridine}} RCl$$

$$ROH + SOCl_2 \xrightarrow{\text{pyridine}} RCl$$

only for primary and secondary alcohols

2. Converting an alcohol into a *sulfonate ester* (Section 3).

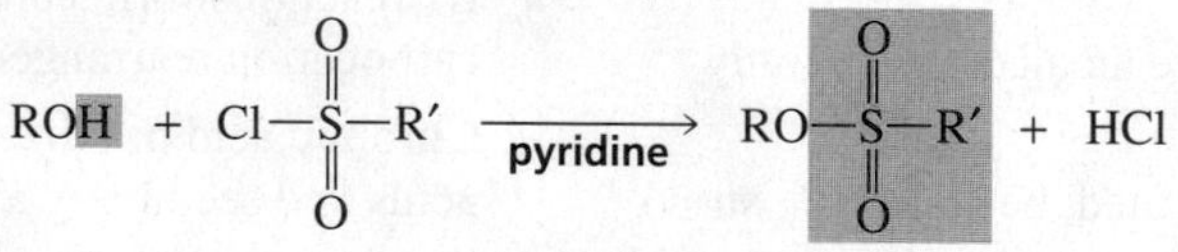

$$ROH + Cl\!-\!\overset{\overset{\displaystyle O}{\|}}{\underset{\underset{\displaystyle O}{\|}}{S}}\!-\!R' \xrightarrow{\text{pyridine}} RO\!-\!\overset{\overset{\displaystyle O}{\|}}{\underset{\underset{\displaystyle O}{\|}}{S}}\!-\!R' + HCl$$

3. Using an S_N2 reaction to convert an *activated alcohol* (an alkyl halide or a sulfonate ester) into a *compound with a new group bonded to the sp³ carbon* (Section 3).

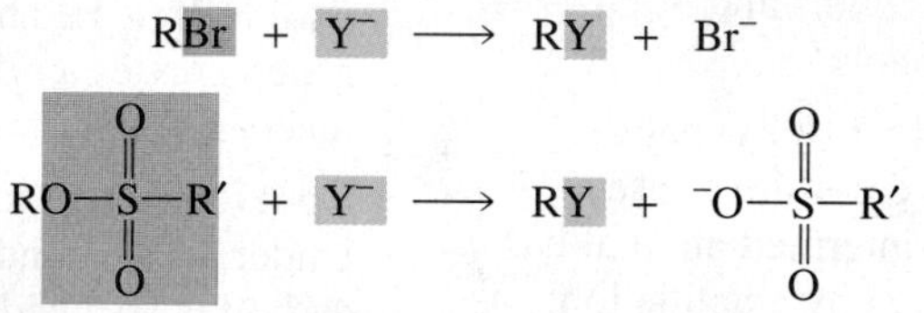

$$RBr + Y^- \longrightarrow RY + Br^-$$

$$RO\!-\!\overset{\overset{\displaystyle O}{\|}}{\underset{\underset{\displaystyle O}{\|}}{S}}\!-\!R' + Y^- \longrightarrow RY + {}^-O\!-\!\overset{\overset{\displaystyle O}{\|}}{\underset{\underset{\displaystyle O}{\|}}{S}}\!-\!R'$$

4. Elimination reactions of alcohols: dehydration (Section 4).

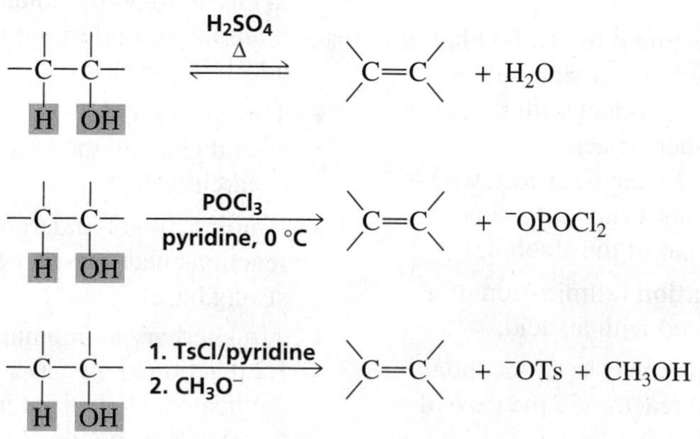

$$-\overset{|}{\underset{|}{C}}\!-\!\overset{|}{\underset{|}{C}}- \xrightarrow[\Delta]{H_2SO_4} \;\; \rightleftharpoons \;\; \overset{}{C}\!=\!\overset{}{C} + H_2O$$
(with H and OH groups)

$$-\overset{|}{\underset{|}{C}}\!-\!\overset{|}{\underset{|}{C}}- \xrightarrow[\text{pyridine, 0 °C}]{POCl_3} \;\; \overset{}{C}\!=\!\overset{}{C} + {}^-OPOCl_2$$
(with H and OH groups)

$$-\overset{|}{\underset{|}{C}}\!-\!\overset{|}{\underset{|}{C}}- \xrightarrow[\text{2. } CH_3O^-]{\text{1. TsCl/pyridine}} \;\; \overset{}{C}\!=\!\overset{}{C} + {}^-OTs + CH_3OH$$
(with H and OH groups)

relative rate: tertiary > secondary > primary

5. Oxidation of alcohols (Section 5).

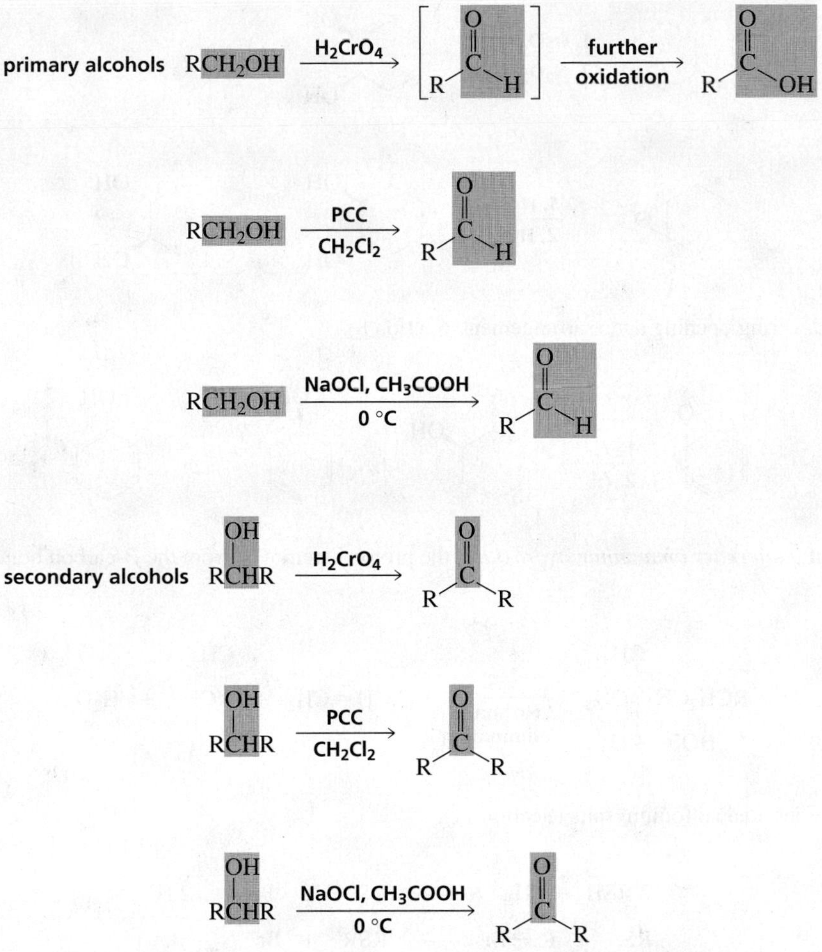

primary alcohols $\quad RCH_2OH \xrightarrow{\textbf{H}_2\textbf{CrO}_4} \left[\begin{matrix} O \\ \parallel \\ R{-}C{-}H \end{matrix} \right] \xrightarrow[\textbf{oxidation}]{\textbf{further}} \begin{matrix} O \\ \parallel \\ R{-}C{-}OH \end{matrix}$

$RCH_2OH \xrightarrow[\textbf{CH}_2\textbf{Cl}_2]{\textbf{PCC}} \begin{matrix} O \\ \parallel \\ R{-}C{-}H \end{matrix}$

$RCH_2OH \xrightarrow[\textbf{0 °C}]{\textbf{NaOCl, CH}_3\textbf{COOH}} \begin{matrix} O \\ \parallel \\ R{-}C{-}H \end{matrix}$

secondary alcohols $\quad \begin{matrix} OH \\ | \\ RCHR \end{matrix} \xrightarrow{\textbf{H}_2\textbf{CrO}_4} \begin{matrix} O \\ \parallel \\ R{-}C{-}R \end{matrix}$

$\begin{matrix} OH \\ | \\ RCHR \end{matrix} \xrightarrow[\textbf{CH}_2\textbf{Cl}_2]{\textbf{PCC}} \begin{matrix} O \\ \parallel \\ R{-}C{-}R \end{matrix}$

$\begin{matrix} OH \\ | \\ RCHR \end{matrix} \xrightarrow[\textbf{0 °C}]{\textbf{NaOCl, CH}_3\textbf{COOH}} \begin{matrix} O \\ \parallel \\ R{-}C{-}R \end{matrix}$

6. Nucleophilic substitution reactions of ethers (Section 6).

$$ROR' + HX \xrightarrow{\Delta} ROH + R'X$$

HX = HBr or HI

7. Nucleophilic substitution reactions of *epoxides* (Section 7).

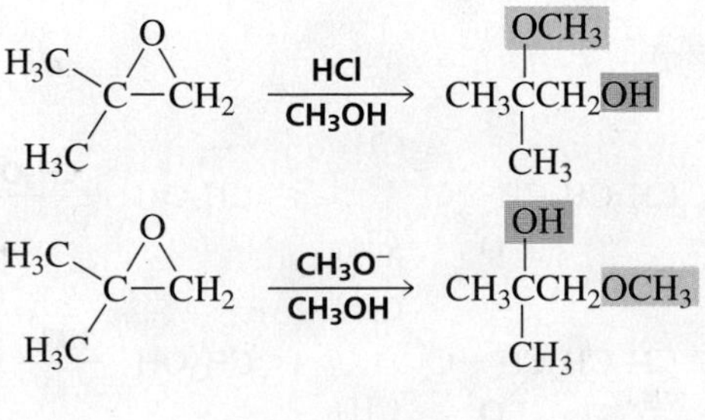

$$H_3C{-}\underset{H_3C}{\overset{O}{\diagup\diagdown}}C{-}CH_2 \xrightarrow[\textbf{CH}_3\textbf{OH}]{\textbf{HCl}} \begin{matrix} OCH_3 \\ | \\ CH_3CCH_2OH \\ | \\ CH_3 \end{matrix}$$

under acidic conditions, the nucleophile attacks the more substituted ring carbon

$$H_3C{-}\underset{H_3C}{\overset{O}{\diagup\diagdown}}C{-}CH_2 \xrightarrow[\textbf{CH}_3\textbf{OH}]{\textbf{CH}_3\textbf{O}^-} \begin{matrix} OH \\ | \\ CH_3CCH_2OCH_3 \\ | \\ CH_3 \end{matrix}$$

under neutral or basic conditions, the nucleophile attacks the less sterically hindered ring carbon

8. Formation of trans and cis 1,2-diols (Section 7).

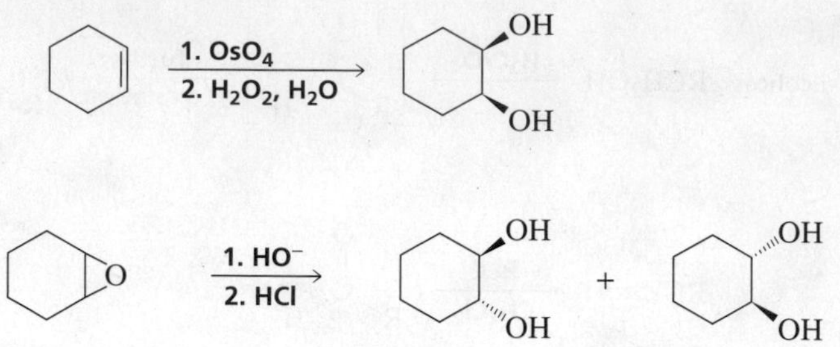

cyclohexene $\xrightarrow[\text{2. } H_2O_2,\ H_2O]{\text{1. } OsO_4}$ cis-1,2-cyclohexanediol (OH, OH)

epoxycyclohexane $\xrightarrow[\text{2. } HCl]{\text{1. } HO^-}$ trans-1,2-cyclohexanediol + trans-1,2-cyclohexanediol

9. Reactions of *arene oxides:* ring opening and rearrangement (Section 8).

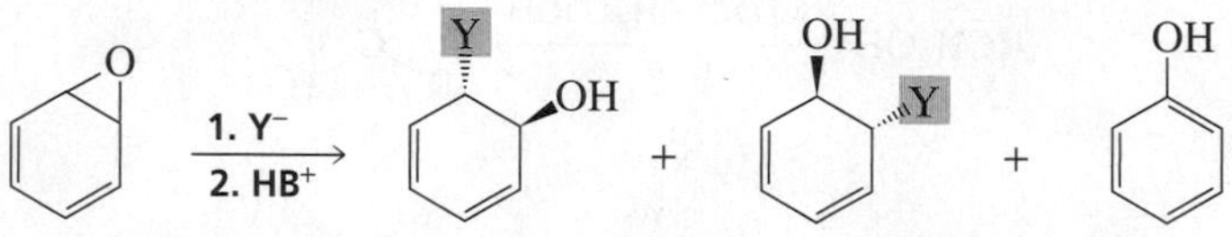

benzene oxide $\xrightarrow[\text{2. } HB^+]{\text{1. } Y^-}$ (Y, OH) + (OH, Y) + phenol (OH)

10. Elimination reactions of *quaternary ammonium hydroxides;* the proton is removed from the β-carbon bonded to the most hydrogens (Section 10).

$$RCH_2CH_2\overset{+}{N}(CH_3)_3 \quad HO^- \xrightarrow[\text{Hofmann elimination}]{\Delta} RCH=CH_2 + N(CH_3)_3 + H_2O$$

11. Reactions of thiols, sulfides, and sulfonium salts (Section 11).

$$2\,RSH + Hg^{2+} \longrightarrow RS-Hg-SR + 2\,H^+$$

$$RS^- + R'-Br \longrightarrow RSR' + Br^-$$

$$R-S-R + R'I \longrightarrow R-\overset{+}{\underset{R'}{S}}-R + I^-$$

$$R-\overset{+}{\underset{R}{S}}-R + Y^- \longrightarrow RY + R-S-R$$

PROBLEMS

48. Draw the product of each of the following reactions:

a. $CH_3CH_2CH_2OH \xrightarrow[\text{2. } CH_3\overset{O}{\overset{\|}{C}}O^-]{\text{1. methanesulfonyl chloride}}$

b. $CH_3CH_2CH_2CH_2OH + PBr_3 \xrightarrow{\text{pyridine}}$

c. $\underset{CH_3}{CH_3CHCH_2CH_2OH} \xrightarrow[\text{2. } \bigcirc-O^-]{\text{1. } p\text{-toluenesulfonyl chloride}}$

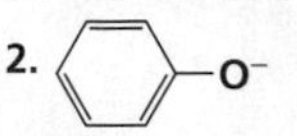

d. $CH_3CH_2CH_2CH_2OH \xrightarrow[\Delta]{H_2SO_4}$

e. $CH_3CH_2CH-\overset{CH_3}{\underset{O}{\overset{|}{C}}}-\overset{}{\underset{CH_3}{}} + CH_3OH \xrightarrow{CH_3O^-}$

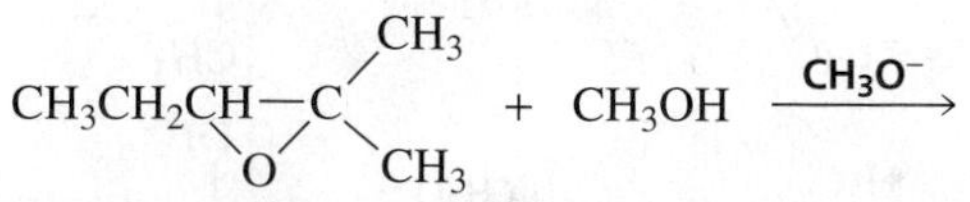

f. $CH_3CH_2CH-\overset{CH_3}{\underset{O}{\overset{|}{C}}}-\overset{}{\underset{CH_3}{}} + CH_3OH \xrightarrow{HCl}$

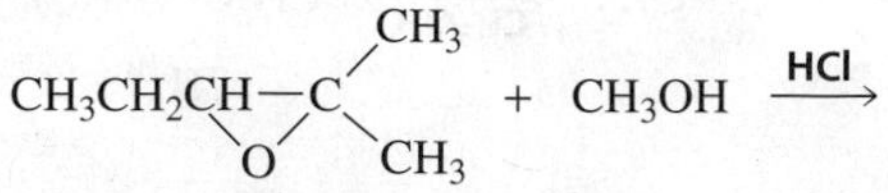

g. $\underset{CH_3}{CH_3CHCH_2CH_2OH} \xrightarrow[\text{pyridine}]{SOCl_2}$

49. Indicate which alcohol in each pair will undergo an elimination reaction more rapidly when heated with H_2SO_4.

a. 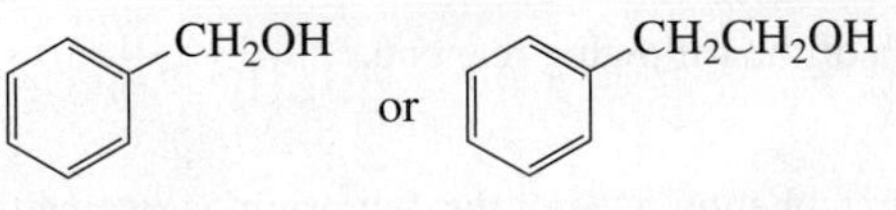 or

b. 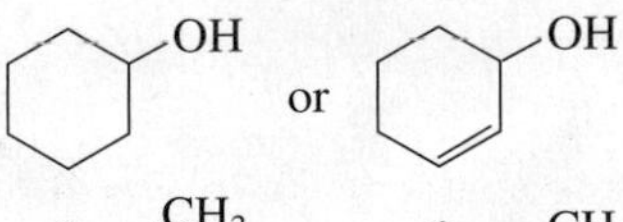 or

c. 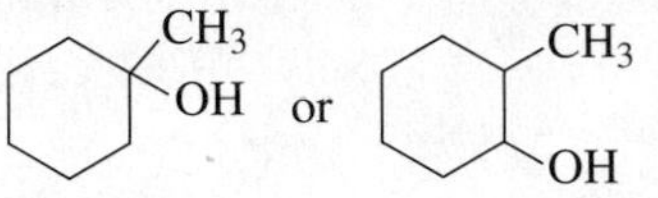 or

d. 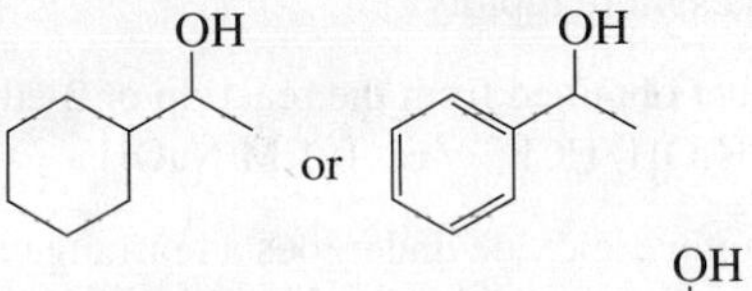 or

e. 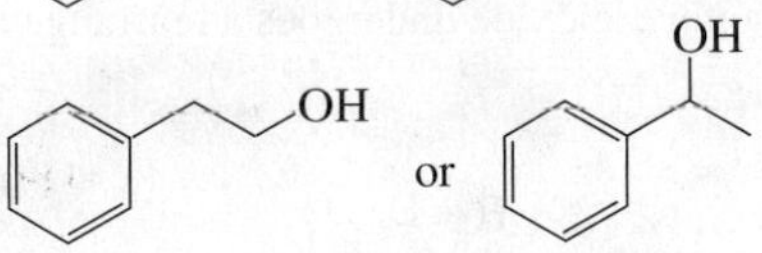 or

50. Identify **A–E.**

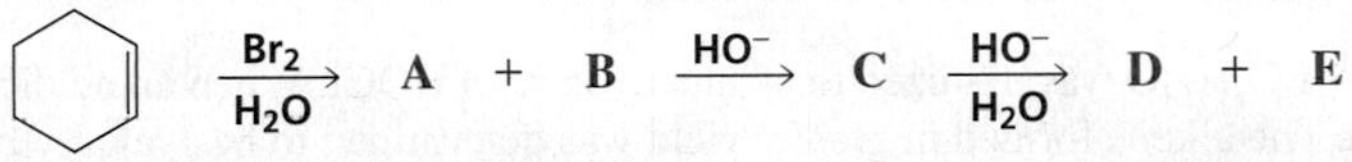

51. Starting with (*R*)-1-deuterio-1-propanol, how could you prepare
  a. (*S*)-1-deuterio-1-propanol? **b.** (*S*)-1-deuterio-1-methoxypropane? **c.** (*R*)-1-deuterio-1-methoxypropane?

52. Write the appropriate reagent over each arrow.

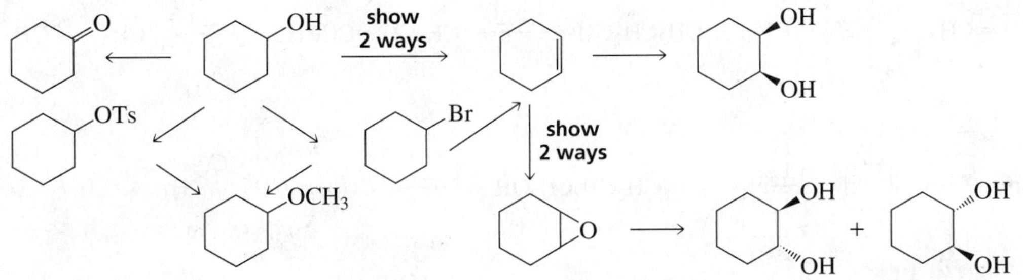

53. What alkenes would you expect to be obtained from the acid-catalyzed dehydration of 1-hexanol?

54. Draw the major product(s) of each of the following reactions:

a. 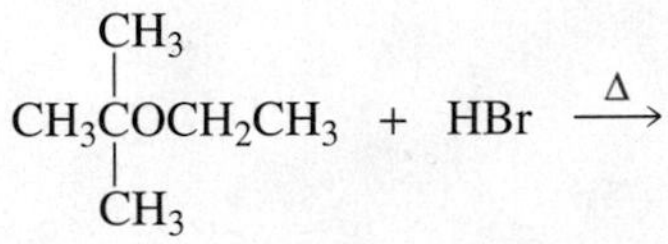 $CH_3\overset{CH_3}{\underset{CH_3}{C}}OCH_2CH_3$ + HBr $\xrightarrow{\Delta}$

b. $CH_3\overset{}{\underset{CH_3}{CH}}CH_2OCH_3$ + HI $\xrightarrow{\Delta}$

c. $\xrightarrow[\Delta]{H_2SO_4}$

d. 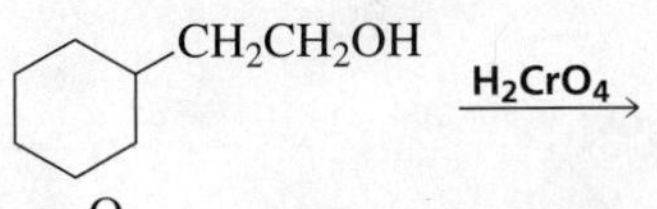 $\xrightarrow{H_2CrO_4}$

e. 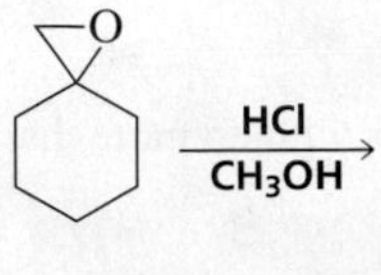 $\xrightarrow[CH_3OH]{HCl}$

f. 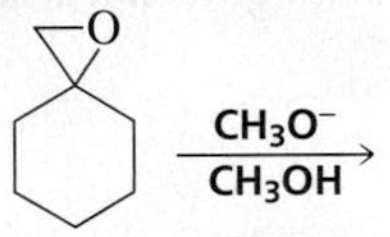 $\xrightarrow[CH_3OH]{CH_3O^-}$

g. 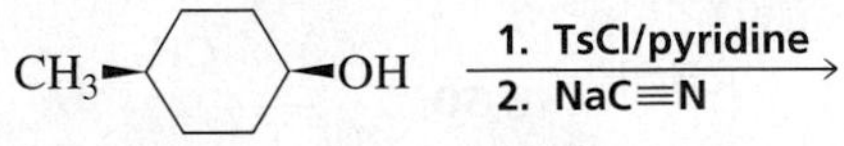 $\xrightarrow[2.\ NaC\equiv N]{1.\ TsCl/pyridine}$

h. $\xrightarrow[\underset{0\ °C}{CH_3COOH}]{NaOCl}$

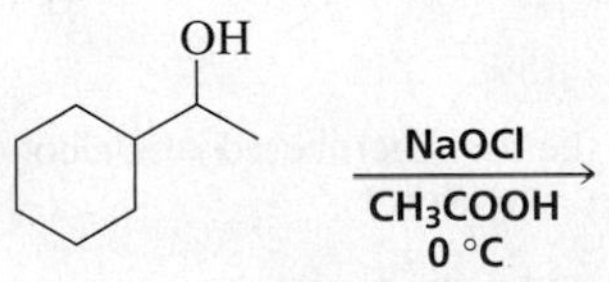

i. 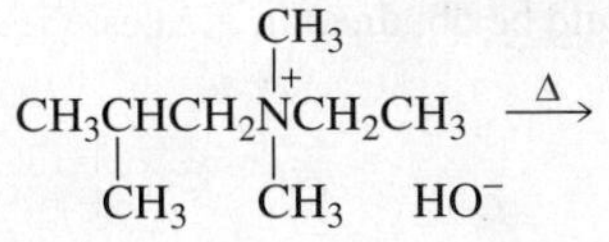 $CH_3\overset{}{\underset{CH_3}{CH}}CH_2\overset{CH_3}{\underset{CH_3}{\overset{+}{N}}}CH_2CH_3 \xrightarrow[HO^-]{\Delta}$

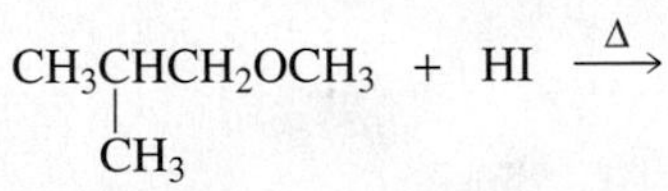

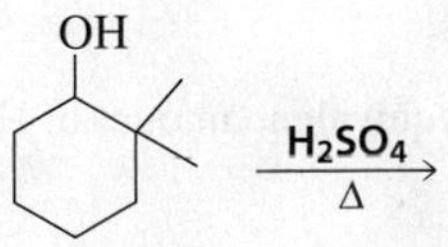

j. 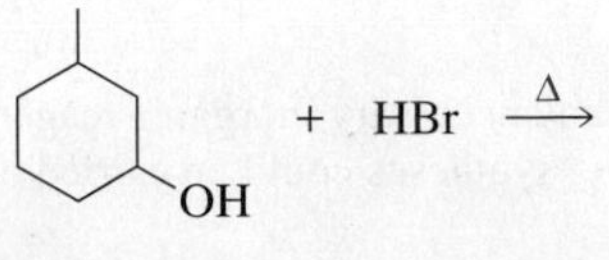 + HBr $\xrightarrow{\Delta}$

55. When heated with H_2SO_4, both 3,3-dimethyl-2-butanol and 2,3-dimethyl-2-butanol are dehydrated to form 2,3-dimethyl-2-butene. Which alcohol dehydrates more rapidly?

56. What is the major product obtained from the reaction of 2-ethyloxirane with each of the following reagents?
   a. 0.1 M HCl   **b.** CH_3OH/HCl **c.** 0.1 M NaOH   **d.** CH_3OH/CH_3O^-

57. When deuterated phenanthrene oxide undergoes a rearrangement in water to form a phenol, 81% of the deuterium is retained in the product.

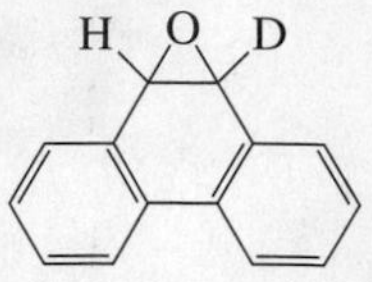

   a. What percentage of the deuterium will be retained if an NIH shift occurs?
   b. What percentage of the deuterium will be retained if an NIH shift does not occur?

58. An unknown alcohol with a molecular formula of $C_7H_{14}O$ was oxidized to an aldehyde with HOCl. When an acidic solution of the alcohol was distilled, two alkenes were formed. The alkene formed in greater yield was determined to be 1-methylcyclohexene. The other alkene formed the original unknown alcohol when treated with BH_3/THF followed by H_2O_2, HO^-, H_2O. Identify the unknown alcohol.

59. Fill in each box with the appropriate reagent:

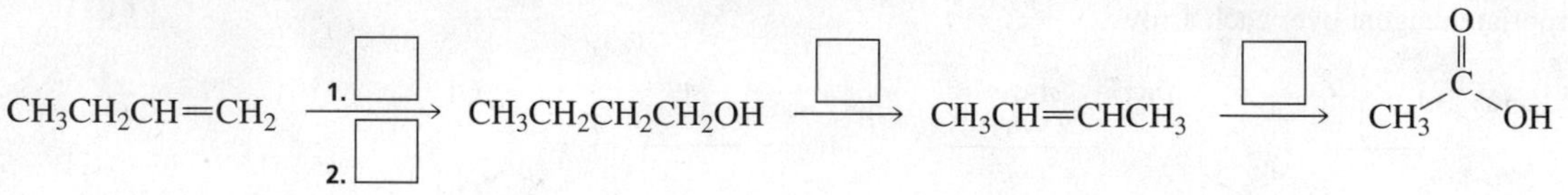

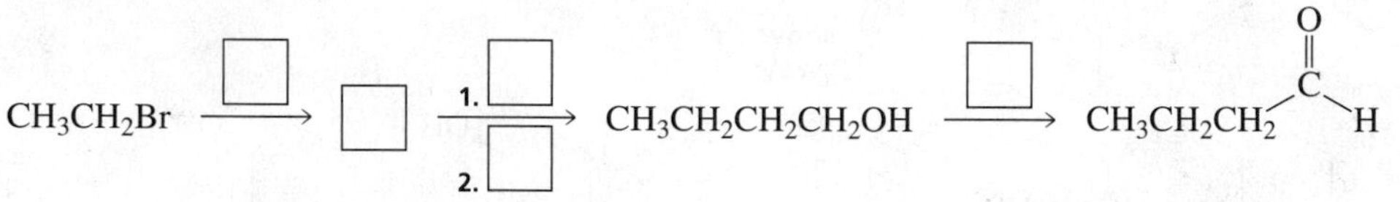

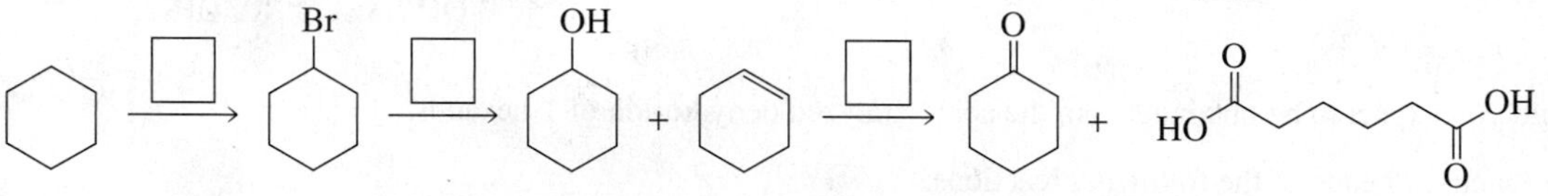

60. Explain why (S)-2-butanol forms a racemic mixture when it is heated in sulfuric acid.

61. Propose a mechanism for the following reaction:

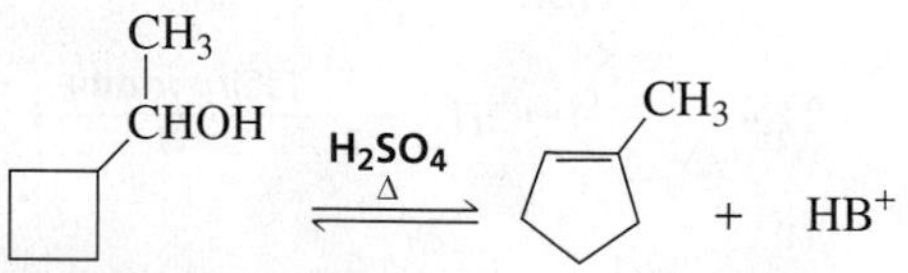

62. What product would be formed if the four-membered ring alcohol in Problem 61 were heated with an equivalent amount of HBr rather than with a catalytic amount of H_2SO_4?

63. Which of the following ethers would be obtained in greatest yield directly from alcohols?

64. Using the given starting material, any necessary inorganic reagents, and any carbon-containing compounds with no more than two carbons, indicate how the following syntheses could be carried out:

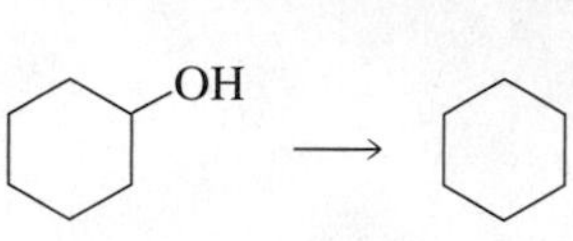

65. When piperidine undergoes the series of reactions shown here, 1,4-pentadiene is obtained as the product. When the four different methyl-substituted piperidines undergo the same series of reactions, each forms a different diene: 1,5-hexadiene, 1,4-pentadiene, 2-methyl-1,4-pentadiene, and 3-methyl-1,4-pentadiene. Which methyl-substituted piperidine forms which diene?

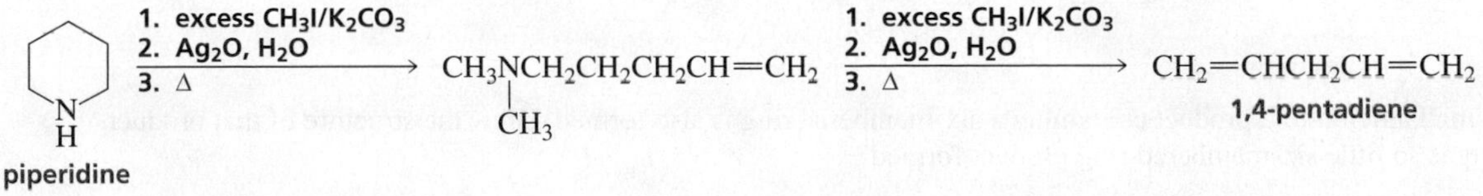

66. When 3-methyl-2-butanol is heated with concentrated HBr, a rearranged product is obtained. When 2-methyl-1-propanol reacts under the same conditions, a rearranged product is not obtained. Explain.

67. Propose a mechanism for each of the following reactions:

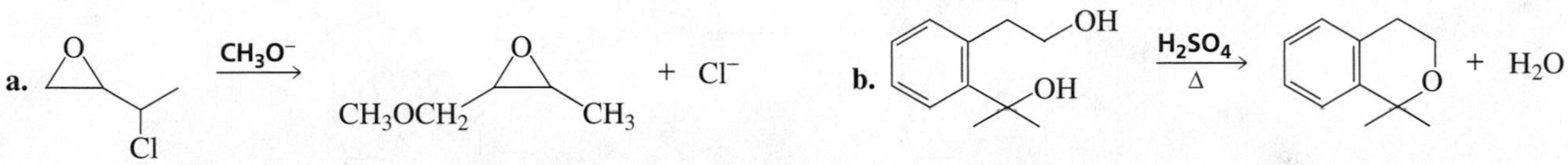

68. How could you synthesize isopropyl propyl ether, using isopropyl alcohol as the only carbon-containing reagent?

69. When the following seven-membered ring alcohol is dehydrated, three alkenes are formed. Propose a mechanism for their formation.

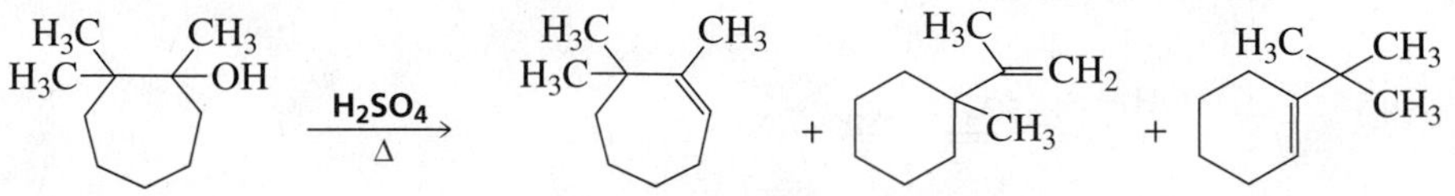

70. Ethylene oxide reacts readily with HO^- because of the strain in the three-membered ring. Explain why cyclopropane, a compound with approximately the same amount of strain, does not react with HO^-.

71. The Swern oxidation oxidizes primary alcohols to aldehydes and secondary alcohols to ketones using dimethyl sulfoxide and oxalyl chloride, followed by reaction with triethylamine. The actual oxidizing agent is the dimethylchlorosulfonium ion, $(CH_3)_2\overset{+}{S}Cl$. compound Propose a mechanism for the oxidation. (*Hint:* the first step is an S_N2 reaction, the last step is an E2 reaction.)

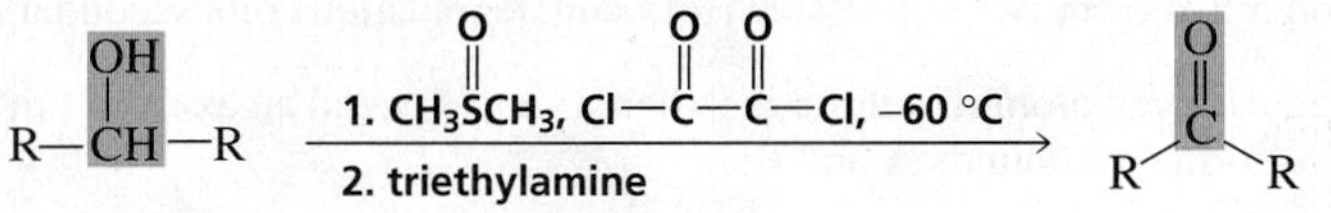

72. Propose a mechanism for each of the following reactions:

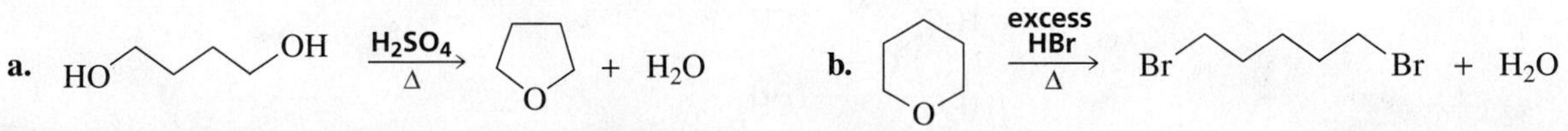

73. Explain why the acid-catalyzed dehydration of an alcohol is a reversible reaction, whereas the base-promoted dehydrohalogenation of an alkyl halide is an irreversible reaction.

74. Triethylene glycol is one of the products obtained from the reaction of ethylene oxide and hydroxide ion. Propose a mechanism for its formation.

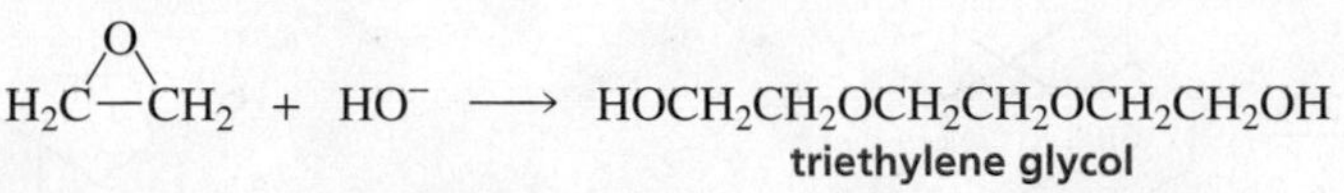

triethylene glycol

75. a. Propose a mechanism for the following reaction:

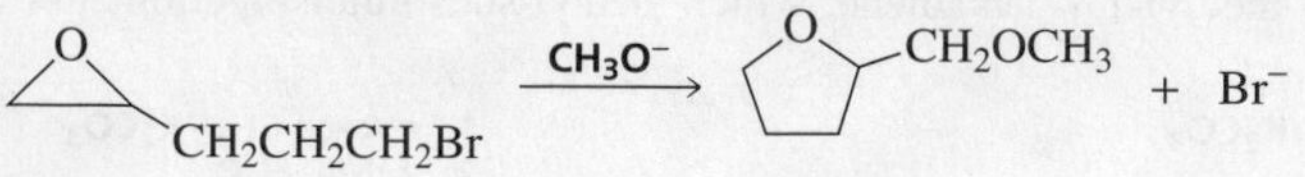

b. A small amount of a product containing a six-membered ring is also formed. Draw the structure of that product.
c. Why is so little six-membered ring product formed?

76. Propose a mechanism for the following reaction:

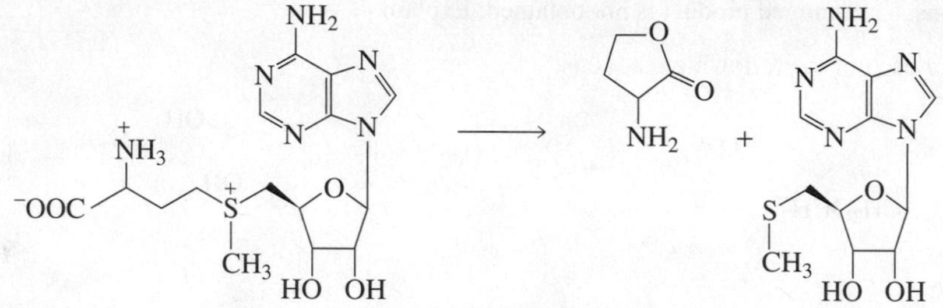

77. When ethyl ether is heated with excess HI for several hours, the only organic product obtained is ethyl iodide. Explain why ethyl alcohol is not obtained as a product.

78. Propose a mechanism for the following reaction:

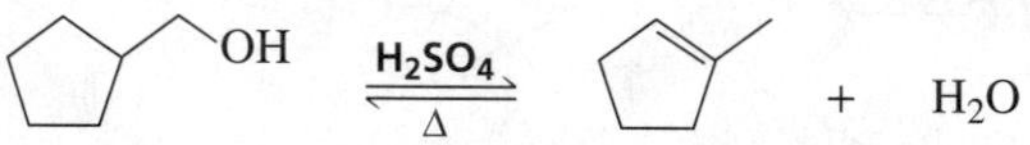

79. Early organic chemists used the Hofmann elimination reaction as the last step of a process known as a Hofmann degradation—a method used to identify amines. In a *Hofmann degradation,* an amine is methylated with excess methyl iodide in a basic solution, treated with silver oxide to convert the quaternary ammonium iodide to a quaternary ammonium hydroxide, and then heated to allow it to undergo a Hofmann elimination. Once the alkene product is identified, working backward gives the structure of the amine. Identify the amine in each of the following cases:
a. 4-Methyl-2-pentene is obtained from the Hofmann degradation of a primary amine.
b. 3-Methyl-1-butene is obtained from the Hofmann degradation of a primary amine.
c. 2-Methyl-1-3-butadiene is obtained from two successive Hofmann degradations of a secondary amine.

80. An ion with a positively charged nitrogen atom in a three-membered ring is called an aziridinium ion. The following aziridinium ion reacts with sodium methoxide to form compounds **A** and **B**:

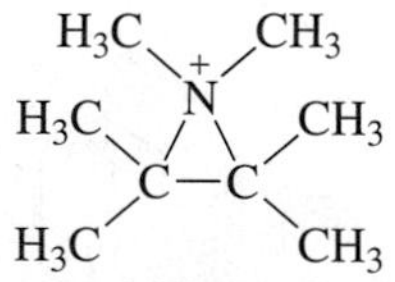

an aziridinium ion

If a small amount of aqueous Br_2 is added to **A**, the reddish color of Br_2 persists, but the color disappears when Br_2 is added to **B**. When the aziridinium ion reacts with methanol, only **A** is formed. Identify **A** and **B**.

81. Propose a mechanism for each of the following reactions:

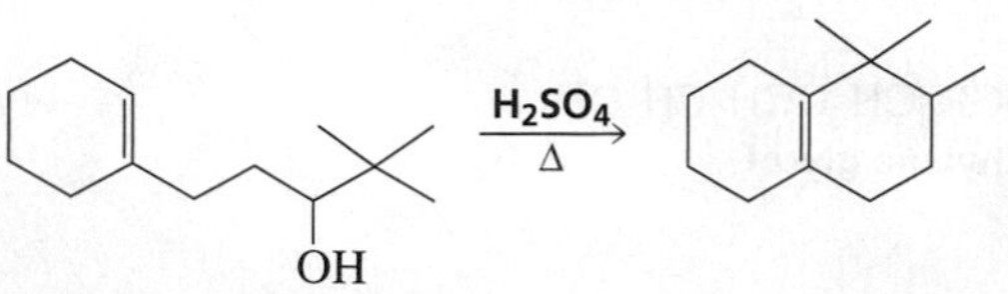

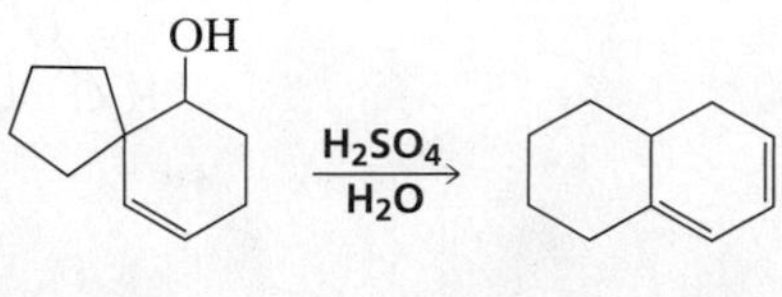

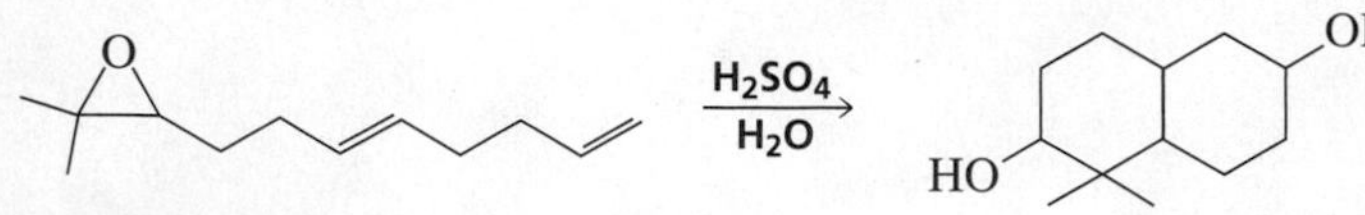

82. Which of the following reactions occurs most rapidly? Why?

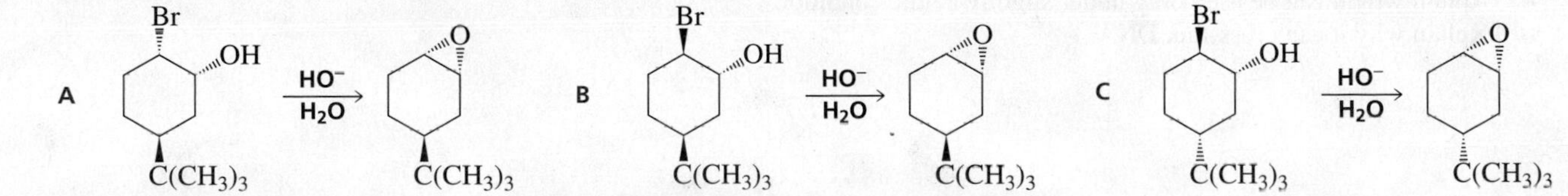

83. The following reaction takes place several times faster than the reaction of 2-chlorobutane with HO^-.

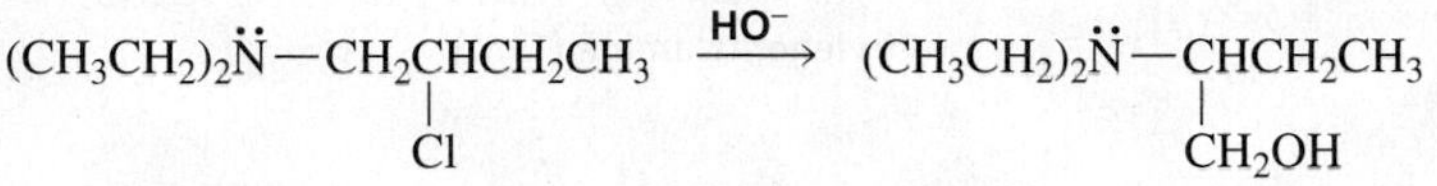

    a. Explain the enhanced reaction rate.
    b. Explain why the OH group in the product is not bonded to the carbon that was bonded to the Cl group in the reactant.

84. A vicinal diol has OH groups on adjacent carbons. The dehydration of a vicinal diol is accompanied by a rearrangement called the pinacol rearrangement. Propose a mechanism for this reaction.

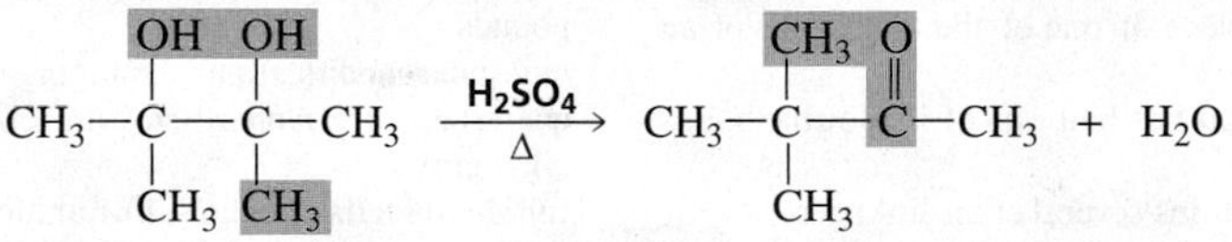

85. Although 2-methyl-1,2-propanediol is an unsymmetrical vicinal diol, only one product is obtained when it is dehydrated in an acidic solution.
    a. What is this product?   **b.** Why is only one product obtained?

86. What product is obtained when the following vicinal diol is heated in an acidic solution?

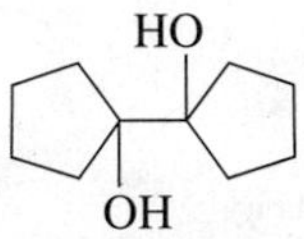

87. Two stereoisomers are obtained from the reaction of cyclopentene oxide and dimethylamine. The *R,R*-isomer is used in the manufacture of eclanamine, an antidepressant. What other isomer is obtained?

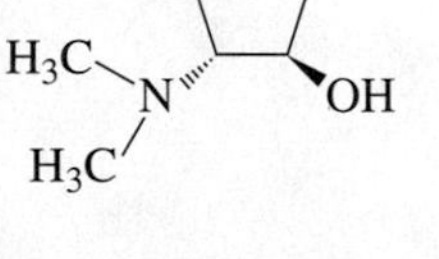

                    R,R-isomer 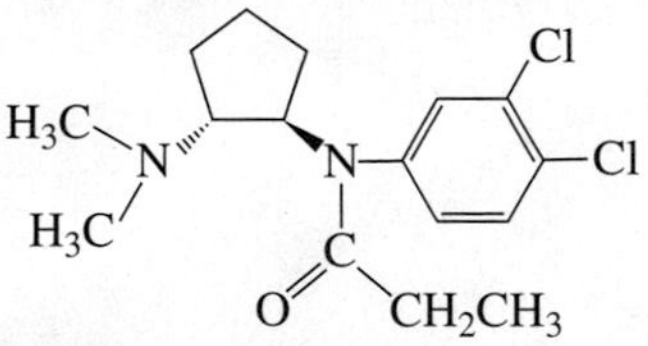   eclanamine

88. Propose a mechanism for each of the following reactions:

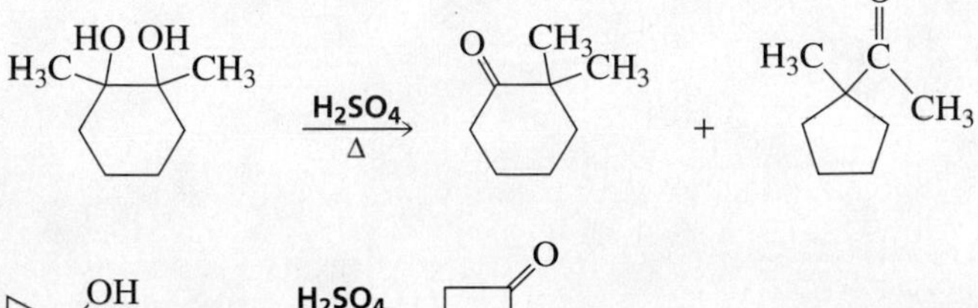

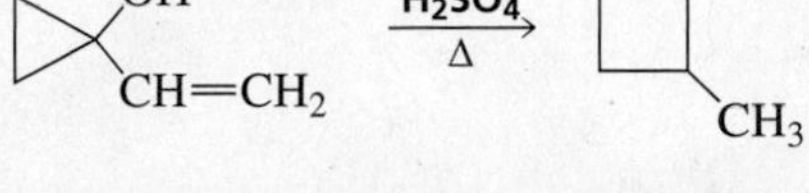

89. Triethylenemelamine (TEM) is an antitumor agent. Its activity is due to its ability to cross-link DNA.
    a. Explain why it can be used only under slightly acidic conditions.
    b. Explain why it can cross-link DNA.

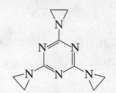

triethylenemelamine (TEM)

GLOSSARY

alcohol a compound with an OH group in place of one of the hydrogens of an alkane; (ROH).

alkaloid a natural product, with one or more nitrogen heteroatoms, found in the leaves, bark, or seeds of plants.

alkyl tosylate an ester of *para*-toluenesulfonic acid.

amine a compound with a nitrogen in place of one of the hydrogens of an alkane; (RNH_2, R_2NH, R_3N)

arene oxide an aromatic compound that has had one of its double bonds converted to an epoxide.

crown ether a cyclic molecule that contains several ether linkages.

dehydration loss of water.

epoxide (oxirane) an ether in which the oxygen is incorporated into a three-membered ring.

ether a compound containing an oxygen bonded to two carbons (ROR).

glycol a compound containing two or more OH groups.

Hofmann elimination reaction elimination of a proton and a tertiary amine from a quaternary ammonium hydroxide.

inclusion compound a compound that specifically binds a metal ion or an organic molecule.

lead compound the prototype in a search for other biologically active compounds.

molecular modification changing the structure of a lead compound.

quaternary ammonium ion an ion containing a nitrogen bonded to four alkyl groups (R_4N^+).

sulfide (thioether) the sulfur analog of an ether (RSR).

sulfonate ester the ester of a sulfonic acid (RSO_2OR).

thioether (sulfide) the sulfur analog of an ether (RSR).

thiol (mercaptan) the sulfur analog of an alcohol (RSH).

vicinal diol (vicinal glycol) a compound with OH groups bonded to adjacent carbons.

PHOTO CREDITS

Photo credits are listed in order of appearance.

Organometallic Compounds

Bombyx mori superimposed on the pheromone binding protein.

In this chapter, you will be asked to design a synthesis for bombykol, the sex pheromone of the silk moth (bombyx mori). Molecules of bombykol diffuse through open pores in the male moth's antenna. When bombykol binds to its receptor, an electrical charge is produced that causes a nerve impulse to be sent to the brain. Bombykol, however, is a nonpolar molecule and has to cross an aqueous solution to get to its receptor. This problem is solved by the pheromone binding protein. The protein binds bombykol in a hydrophobic pocket and then carries it to the receptor. The area around the receptor is relatively acidic, and the decrease in pH causes the pheromone binding protein to unfold and release bombykol to the receptor.

We have seen that the compounds in group II—alkyl halides, alcohols, ethers and epoxides—contain a carbon that is bonded to a *more* electronegative atom. The carbon, therefore, is *electrophilic* and reacts with a nucleophile.

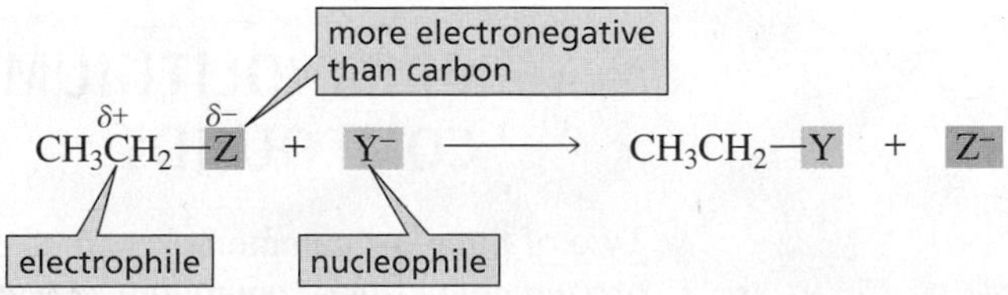

$$\overset{\delta+}{C}H_3CH_2 \overset{\delta-}{-Z} \;+\; Y^- \longrightarrow CH_3CH_2{-}Y \;+\; Z^-$$

more electronegative than carbon

electrophile   nucleophile

But what if you wanted a carbon to react with an electrophile? For that, you would need a compound with a nucleophilic carbon. To be *nucleophilic,* carbon would have to be bonded to a *less* electronegative atom.

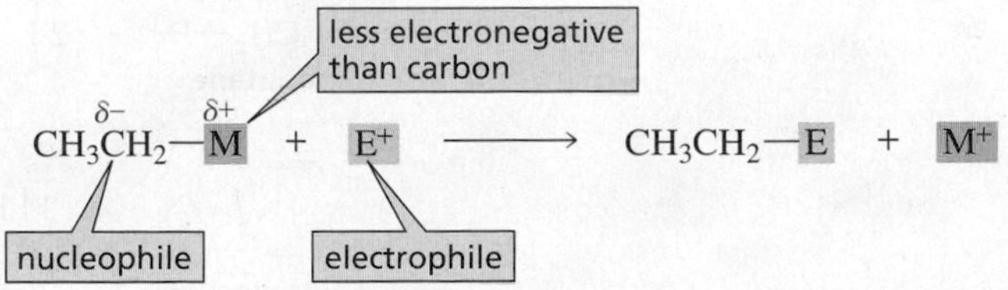

$$\overset{\delta-}{C}H_3CH_2 \overset{\delta+}{-M} \;+\; E^+ \longrightarrow CH_3CH_2{-}E \;+\; M^+$$

less electronegative than carbon

nucleophile   electrophile

From Chapter 12 of *Organic Chemistry*, Seventh Edition. Paula Yurkanis Bruice. Copyright © 2014 by Pearson Education, Inc. All rights reserved.

CH₃Cl

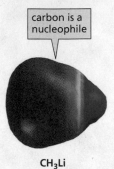

CH₃Li

**A carbon is an electrophile
if it is attached to a
more electronegative atom.**

**A carbon is a nucleophile
if it is attached to a
less electronegative atom.**

Because metals are less electronegative than carbon (Table 1), one way to create a nucleophilic carbon is to attach the carbon to a metal. A compound that contains a carbon–metal bond is called an **organometallic compound.** The electrostatic potential maps show that the carbon attached to the halogen in the alkyl halide is an electrophile (it is blue-green), whereas the carbon attached to the metal (Li) in the organometallic compound is a nucleophile (it is red).

Table 1	The Electronegativities of Some of the Elements[a]								

IA	IIA				IB	IIB	IIIA	IVA	VA	VIA	VIIA
H 2.1											
Li 1.0	Be 1.5						B 2.0	C 2.5	N 3.0	O 3.5	F 4.0
Na 0.9	Mg 1.2						Al 1.5	Si 1.8	P 2.1	S 2.5	Cl 3.0
K 0.8	Ca 1.0	Co 1.7	Ni 1.7	Cu 1.8	Zn 1.7	Ga 1.8	Ge 2.0				Br 2.8
		Rh 2.4	Pd 2.4	Ag 1.8	Cd 1.5		Sn 1.7				I 2.5
					Hg 1.5		Pb 1.6				

[a]From the scale devised by Linus Pauling

A wide variety of metals can be attached to carbon to form organometallic compounds.

$$\overset{\delta^-}{C}-\overset{\delta^+}{Mg} \qquad \overset{\delta^-}{C}-\overset{\delta^+}{Li} \qquad \overset{\delta^-}{C}-\overset{\delta^+}{Cu} \qquad \overset{\delta^-}{C}-\overset{\delta^+}{Cd} \qquad \overset{\delta^-}{C}-\overset{\delta^+}{Pd}$$

$$\overset{\delta^-}{C}-\overset{\delta^+}{Zn} \qquad \overset{\delta^-}{C}-\overset{\delta^+}{Al} \qquad \overset{\delta^-}{C}-\overset{\delta^+}{Pb} \qquad \overset{\delta^-}{C}-\overset{\delta^+}{Hg} \qquad \overset{\delta^-}{C}-\overset{\delta^+}{Ru}$$

The name of an organometallic compound usually begins with the name of the alkyl group, followed by the name of the metal.

$CH_3CH_2CH_2CH_2Li$ CH_3CH_2MgBr $(CH_3CH_2CH_2)_2Cd$ $(CH_3CH_2)_4Pb$
butyllithium   **ethylmagnesium bromide**   **dipropylcadmium**   **tetraethyllead**

1   ORGANOLITHIUM AND ORGANOMAGNESIUM COMPOUNDS

Two of the most common organometallic compounds are *organolithium compounds* and *organomagnesium compounds*. **Organolithium compounds** are prepared by adding lithium to an alkyl halide in a nonpolar solvent such as hexane.

$$CH_3CH_2CH_2CH_2{-}Br + 2\,Li \xrightarrow{\text{hexane}} CH_3CH_2CH_2CH_2{-}Li + LiBr$$
1-bromobutane   **butyllithium** (carbon–metal bond)

$$\text{C}_6\text{H}_5{-}Cl + 2\,Li \xrightarrow{\text{hexane}} \text{C}_6\text{H}_5{-}Li + LiCl$$
chlorobenzene   **phenyllithium**

Organomagnesium compounds (commonly called **Grignard reagents** after their discoverer, Victor Grignard) are prepared by adding an alkyl halide to magnesium shavings being stirred in an ether—usually diethyl ether or tetrahydrofuran (THF)—under anhydrous conditions. The reaction inserts magnesium between the carbon and the halogen.

carbon–metal bond

$$\text{cyclohexyl} \boxed{\text{Br}} + \text{Mg} \xrightarrow{\text{diethyl ether}} \text{cyclohexyl} \boxed{\text{MgBr}}$$

cyclohexyl bromide

cyclohexylmagnesium bromide

$$\text{CH}_2{=}\text{CH}{-}\boxed{\text{Br}} + \text{Mg} \xrightarrow{\text{THF}} \text{CH}_2{=}\text{CH}{-}\boxed{\text{MgBr}}$$

vinyl bromide

vinylmagnesium bromide

The solvent plays a crucial role in the formation of a Grignard reagent. The magnesium atom is surrounded by only four electrons, so it needs four more to form an octet. Solvent molecules provide these electrons. Sharing electrons with a metal is called *coordination*. Coordination of the solvent molecules with the magnesium atom allows the Grignard reagent to dissolve in the solvent, preventing it from coating the magnesium shavings, which would make them unreactive.

$$\begin{array}{c}
\text{CH}_3\text{CH}_2 \quad \text{CH}_2\text{CH}_3 \\
\ddot{\text{O}} \\
\text{CH}_3{-}\text{Mg}{-}\text{Br} \\
\ddot{\text{O}} \\
\text{CH}_3\text{CH}_2 \quad \text{CH}_2\text{CH}_3
\end{array}$$

Alkyl halides, vinylic halides, and aryl halides can all be used to form organolithium and organomagnesium compounds. Alkyl bromides are used most often, because they react more readily than alkyl chlorides and are less expensive than alkyl iodides.

Because carbon is more electronegative than the metal to which it is attached, organolithium and organomagnesium compounds react as if they were carbanions. Thus, they are nucleophiles and bases.

$$\text{CH}_3\text{CH}_2{-}\text{MgBr} \qquad \text{reacts as if it were} \qquad \text{CH}_3\overset{\cdot\cdot}{\overset{-}{\text{C}}}\text{H}_2 \;\; \overset{+}{\text{MgBr}}$$

ethylmagnesium bromide

$$\text{phenyl}{-}\text{Li} \qquad \text{reacts as if it were} \qquad \text{phenyl}{:}^{-} \;\; \text{Li}^{+}$$

phenyllithium

Organomagnesium and organolithium compounds are very strong bases. Therefore, they will react immediately with any acid present in the reaction mixture, even with very weak acids such as water and alcohols. When this happens, the organometallic compound is converted into an alkane. Although this is not how one would normally synthesize an alkane, it is a useful way to prepare deuterated hydrocarbons. Just use D_2O instead of H_2O.

$$\underset{\overset{|}{\text{Br}}}{\text{CH}_3\text{CH}_2\text{CHCH}_3} \xrightarrow[\text{THF}]{\text{Mg}} \underset{\overset{|}{\text{MgBr}}}{\text{CH}_3\text{CH}_2\text{CHCH}_3} \begin{array}{c} \xrightarrow{\text{H}_2\text{O}} \text{CH}_3\text{CH}_2\text{CH}_2\text{CH}_3 \\ \\ \xrightarrow{\text{D}_2\text{O}} \underset{\overset{|}{\text{D}}}{\text{CH}_3\text{CH}_2\text{CHCH}_3} \end{array}$$

This means that Grignard reagents and organolithium compounds cannot be prepared from compounds that contain acidic groups (such as OH, NH_2, NHR, SH, $C{\equiv}CH$, or COOH). Because even trace amounts of moisture can convert an organometallic compound into an alkane, it is important that all reagents are dry when organometallic compounds are being synthesized and when they react with other reagents.

PROBLEM 1♦

Which of the following reactions will favor formation of the products? (For the pK_a values necessary to solve this problem, see the appendix with the table titled "pKa Values". Note that a strong acid reacts to form a weak acid.)

$$CH_3MgBr + H_2O \rightleftharpoons CH_4 + HOMgBr$$

$$CH_3MgBr + CH_3OH \rightleftharpoons CH_4 + CH_3OMgBr$$

$$CH_3MgBr + NH_3 \rightleftharpoons CH_4 + H_2NMgBr$$

$$CH_3MgBr + CH_3NH_2 \rightleftharpoons CH_4 + CH_3NHMgBr$$

$$CH_3MgBr + HC\equiv CH \rightleftharpoons CH_4 + HC\equiv CMgBr$$

2   TRANSMETALLATION

The reactivity of an organometallic compound depends on the polarity of the carbon–metal bond: *the greater the polarity of the bond, the more reactive the compound is as a nucleophile.* The polarity of the bond depends on the difference in electronegativity between the metal and carbon (Table 1).

For example, magnesium has an electronegativity of 1.2, compared with 2.5 for carbon. This large difference in electronegativity makes the carbon–magnesium bond highly polar. Lithium (1.0) is even less electronegative than magnesium. Thus, the carbon–lithium bond is more polar than the carbon–magnesium bond, so an organolithium reagent is a more reactive nucleophile than a Grignard reagent.

> **The more polar the carbon–metal bond, the more reactive the organometallic compound is as a nucleophile.**

An organometallic compound will undergo **transmetallation** (metal exchange) if it is added to a metal halide whose metal is more electronegative than the metal in the organometallic compound. In other words, metal exchange will occur if the alkyl group can be transferred to a metal with an electronegativity closer to that of carbon, thereby forming a less polar carbon–metal bond and, therefore, a less reactive nucleophile.

For example, cadmium is more electronegative (1.5) than magnesium (1.2). Consequently, a carbon–cadmium bond is less polar than a carbon–magnesium bond, so metal exchange occurs.

> **Transmetallation occurs if the alkyl group can be transferred to a more electronegative metal.**

$$2\ CH_3CH_2MgCl + CdCl_2 \longrightarrow (CH_3CH_2)_2Cd + 2\ MgCl_2$$
ethylmagnesium chloride   **diethylcadmium**

PROBLEM 2♦

Which is more reactive, an organolithium compound or an organosodium compound?

PROBLEM 3♦

What organometallic compound will be formed from the reaction of excess methylmagnesium chloride and $GaCl_3$? (*Hint:* See Table 1.)

3   ORGANOCUPRATES

New carbon–carbon bonds can be made using an organometallic reagent that has a *transition metal* as its metal atom. Transition metals are indicated by purple in the periodic table. The reactions are called **coupling reactions** because two CH-containing groups are joined (coupled) together.

> **Coupling reactions join two CH-containing groups.**

The first organometallic compounds used in coupling reactions were **organocuprates** (**R_2CuLi**), also called **Gilman reagents** after their discoverer, Henry Gilman. Organocuprates are less reactive than organolithium reagents or Grignard reagents because a carbon–copper bond is less polar than a carbon–lithium or carbon–magnesium bond—that is, Cu is closer in electronegativity to C than is Li or Mg. Only one of the two alkyl groups in an organocuprate is used as a nucleophile in reactions.

An organocuprate is prepared by the reaction of an organolithium reagent with cuprous iodide in diethyl ether or THF. Notice that because Cu is more electronegative (1.8) than Li (1.0), transmetallation occurs.

$$2\ CH_3Li\ +\ CuI\ \xrightarrow{\text{THF}}\ (CH_3)_2CuLi\ +\ LiI$$

an organolithium
reagent

an organocuprate
a Gilman reagent

When an organocuprate reacts with an alkyl halide (with the exception of alkyl fluorides, which do not undergo this reaction), one of the alkyl groups of the organocuprate replaces the halogen. This means that an alkane can be formed from two alkyl halides—one alkyl halide is used to form the organocuprate, which then reacts with the second alkyl halide in a coupling reaction. The precise mechanism of the substitution reaction is unknown but is thought to involve radicals.

new C–C bond

$$CH_3CH_2CH_2CH_2-Br\ +\ (CH_3CH_2CH_2)_2CuLi\ \xrightarrow{\text{THF}}\ CH_3CH_2CH_2CH_2-CH_2CH_2CH_3\ +\ CH_3CH_2CH_2Cu$$

heptane

$$+\ LiBr$$

an alkyl group of an
organocuprate
replaces a Cl, Br, or I

Organocuprates can react with all alkyl halides except tertiary alkyl halides. Because they can react with vinylic halides and aryl halides (an aryl halide has a hydrogen attached to a benzene ring), they can be used to prepare compounds that cannot be prepared by S_N2 reactions with Grignard reagents or organolithium compounds. (Note that vinylic and aryl halides cannot undergo nucleophilic attack).

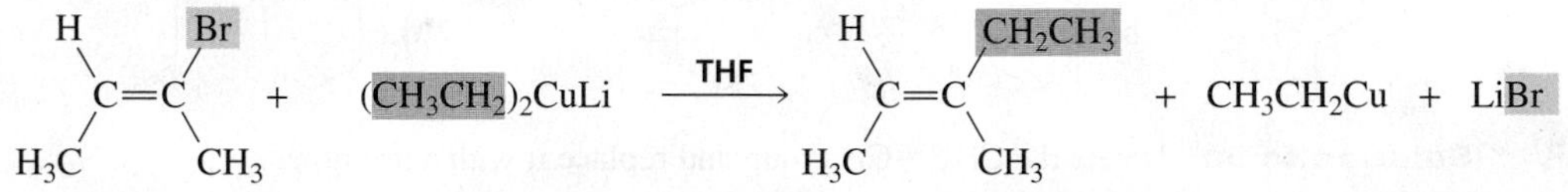

a vinylic halide

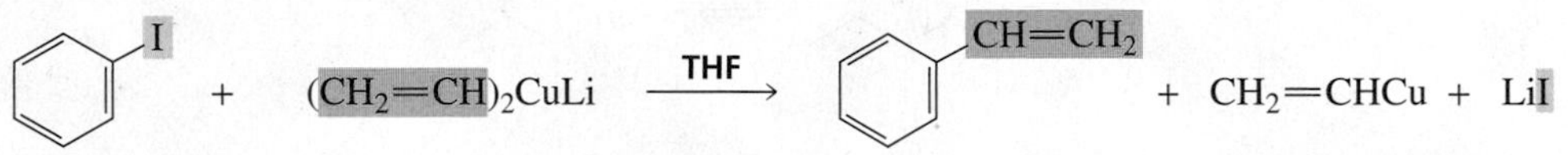

an aryl halide

The substitution reaction is stereospecific. In other words, the configuration of the double bond is retained in the product—that is, the groups bonded to the sp^2 carbons maintain their positions.

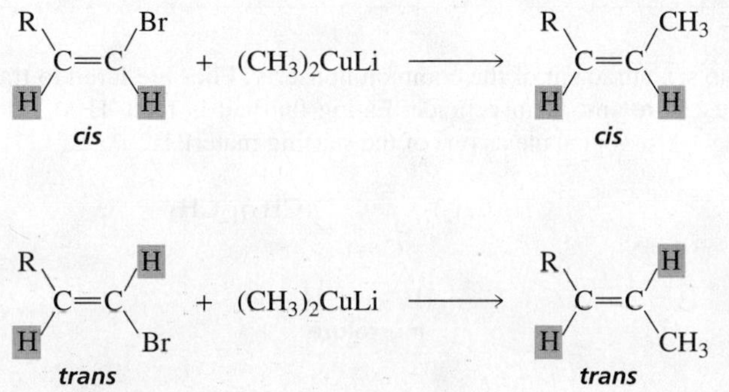

cis

cis

trans

trans

557

Organocuprates can even replace halogens in compounds that contain other functional groups.

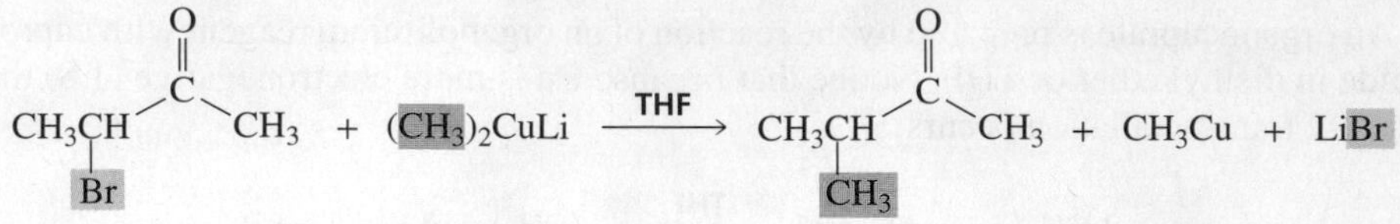

$$CH_3CH(Br)—C(=O)—CH_3 + (CH_3)_2CuLi \xrightarrow{THF} CH_3CH(CH_3)—C(=O)—CH_3 + CH_3Cu + LiBr$$

PROBLEM 4 Solved

Describe two ways to synthesize ethylcyclopentane from alkyl halides.

Solution The ethyl group can be either the organocuprate or the alkyl halide with which the organocuprate reacts. Likewise, the cyclopentyl group can be either the organocuprate or the alkyl halide with which the organocuprate reacts.

$$CH_3CH_2Br \xrightarrow[\text{2. CuI}]{\text{1. Li}} (CH_3CH_2)_2CuLi \longrightarrow$$

$$\text{cyclopentyl-Br} \xrightarrow[\text{2. CuI}]{\text{1. Li}} (\text{cyclopentyl})_2CuLi \xrightarrow{CH_3CH_2Br} \text{ethylcyclopentane}$$

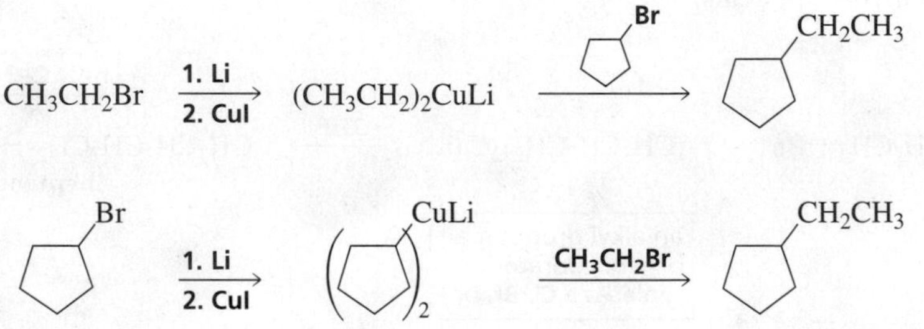

PROBLEM 5 Solved

What alkyl bromide would be required to react with $(CH_2{=}CH)_2CuLi$ in order to form each of the following compounds?

a. 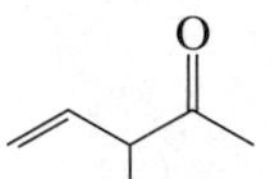   b. 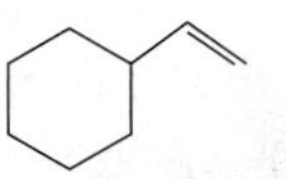   c. 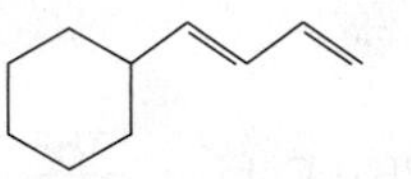   d.

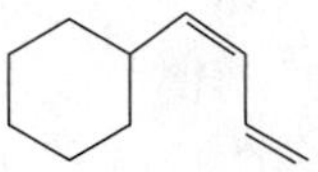

Solution to 5a   Delete the $CH_2{=}CH$ group and replace it with a bromine.

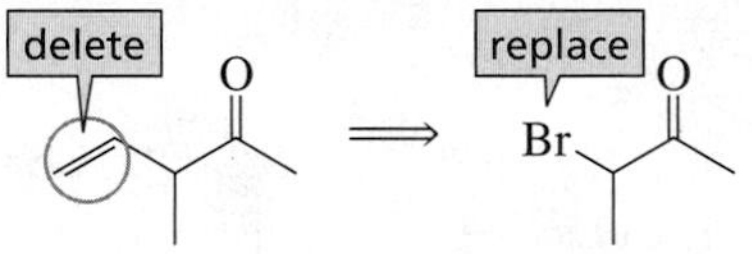

delete   replace

PROBLEM 6♦

Explain why tertiary alkyl halides cannot be used in coupling reactions with Gilman reagents.

PROBLEM 7

Muscalure is the sex attractant of the common housefly. Flies are lured to traps filled with bait that contain muscalure and an insecticide. Eating the bait is fatal. How could you synthesize muscalure using 1-bromopentane as one of the starting materials?

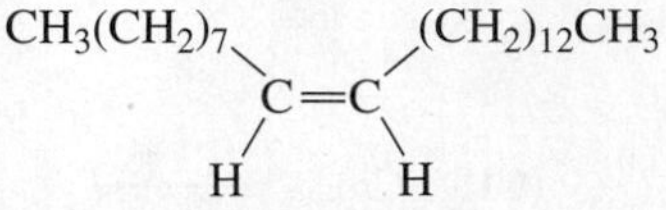

$$CH_3(CH_2)_7 \quad (CH_2)_{12}CH_3$$
$$C{=}C$$
$$H \qquad H$$

muscalure

Since they are nucleophiles, organocuprates react with electrophiles. For example, in the following reactions, the organocuprate reacts with an epoxide in a nucleophilic substitution reaction.

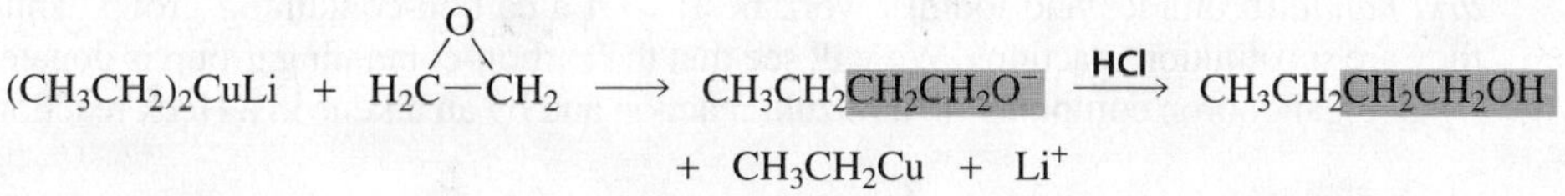

$$(CH_3CH_2)_2CuLi \; + \; H_2\overset{O}{\overset{\triangle}{C}}{-}CH_2 \longrightarrow CH_3CH_2CH_2CH_2O^- \xrightarrow{HCl} CH_3CH_2CH_2CH_2OH$$

$$+ \; CH_3CH_2Cu \; + \; Li^+$$

Notice that when an organocuprate reacts with ethylene oxide, a primary alcohol is formed that contains two more carbons than the organometallic compound.

PROBLEM 8

What are the products of the following reactions?

a. 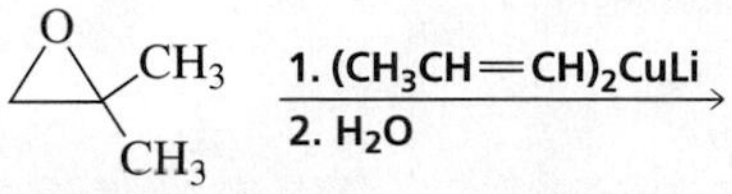 $\xrightarrow[\text{2. H}_2\text{O}]{\text{1. (CH}_3\text{CH}=\text{CH)}_2\text{CuLi}}$

b. 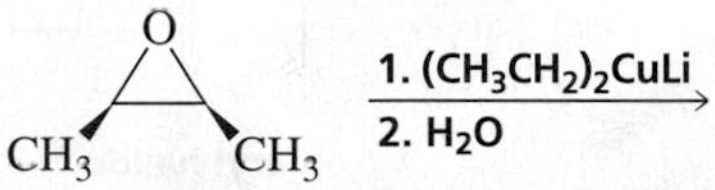 $\xrightarrow[\text{2. H}_2\text{O}]{\text{1. (CH}_3\text{CH}_2)_2\text{CuLi}}$

PROBLEM 9♦

What alcohols would be formed from the reaction of ethylene oxide with the following organocuprates followed by the addition of acid?

a. $(CH_3CH_2CH_2)_2CuLi$

b. 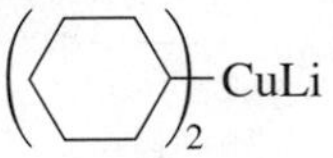 $\left(\langle\bigcirc\rangle\!-\right)_2 CuLi$

c. 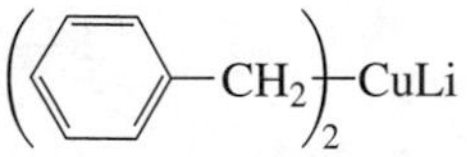 $\left(\langle\bigcirc\rangle\!-CH_2\!-\right)_2 CuLi$

PROBLEM 10

How could the following compounds be prepared, using cyclohexene as a starting material?

a.

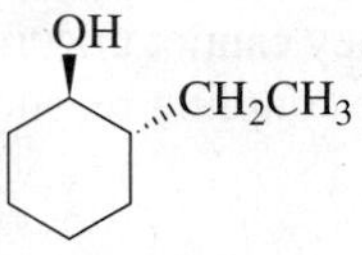

b.

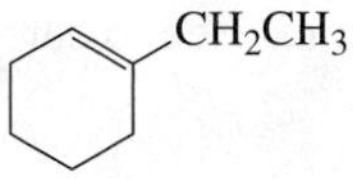

c.

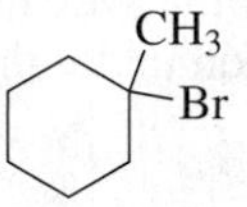

d.

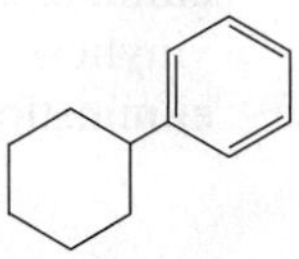

PROBLEM 11 Solved

Using any necessary reagents, show how the following compounds could be prepared using ethylene oxide as one of the reactants:

a. $CH_3CH_2CH_2CH_2OH$
b. $CH_3CH_2CH_2CH_2Br$

c. $CH_3CH_2CH_2CH_2D$
d. $CH_3CH_2CH_2CH_2CH_2CH_2OH$

Solution

a. $CH_3CH_2Br \xrightarrow[\text{2. CuI}]{\text{1. Li}} (CH_3CH_2)_2CuLi \xrightarrow[\text{2. HCl}]{\text{1. } \triangle} CH_3CH_2CH_2CH_2OH$

b. product of a $\xrightarrow{PBr_3} CH_3CH_2CH_2CH_2Br$

c. product of b $\xrightarrow[\text{2. CuI}]{\text{1. Li}} (CH_3CH_2CH_2CH_2)_2CuLi \xrightarrow{D_2O} CH_3CH_2CH_2CH_2D$

d. the same reaction sequence as in part **a,** but with butyl bromide in the first step.

---

4 PALLADIUM-CATALYZED COUPLING REACTIONS

Many new methods of using transition metals to carry out coupling reactions have been developed in recent years. These methods have greatly expanded the synthetic chemist's arsenal of reactions for forming new carbon–carbon bonds. We will look at two reactions that use a palladium catalyst: the *Suzuki reaction* and the *Heck reaction.* The two reactions have several features in common:

- Both reactions form a new C—C bond by replacing a halogen of a *vinylic halide* or an *aryl halide* (bromides and iodides work best) with a carbon-containing group. Thus, they are substitution reactions. We will see that the carbon-containing group is donated by an organoboron compound in a Suzuki reaction and by an alkene in a Heck reaction.

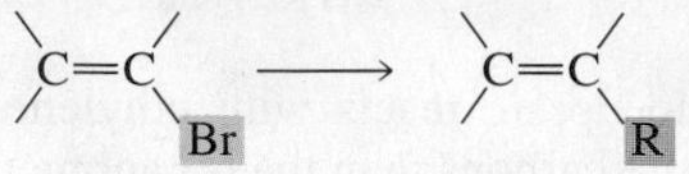

a vinylic halide

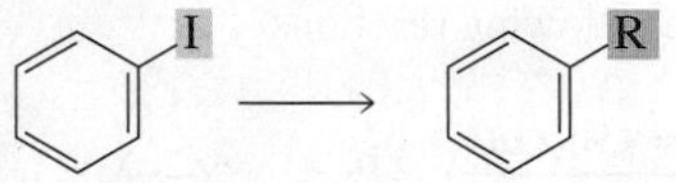

an aryl halide

- The first step in both reactions is insertion of palladium between the carbon and halogen to form an **organopalladium compound.** Notice that the palladium atom is coordinated with ligands, indicated by L. Several different groups can serve as the ligands.

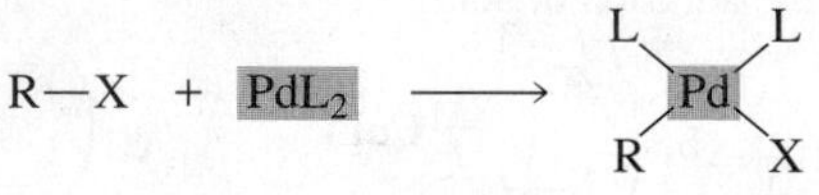

an organopalladium
compound

- If the organopalladium compound has a β-hydrogen, it will rapidly undergo an elimination reaction and the coupling reaction will not occur. This explains why vinylic and aryl halides are the reactants in these reactions—they cannot undergo an elimination reaction under the conditions used to carry out the coupling reactions.

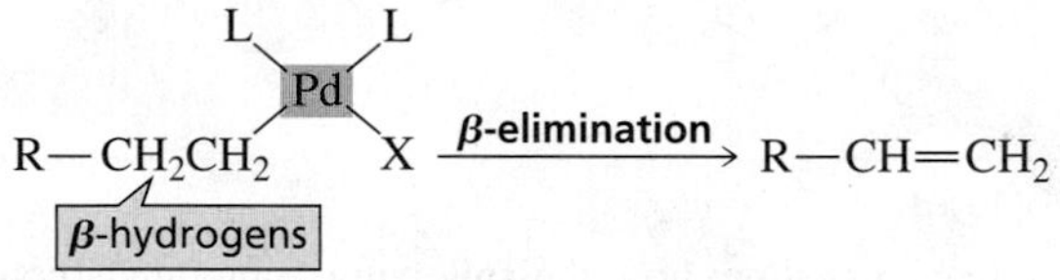

- The reactions can be carried out even if the reactants have other functional groups.
- The reactions are stereospecific: the configuration of the double bond in a vinylic halide is retained in the product.
- The reactions give high yields of products (80–98%).

The Suzuki Reaction

The **Suzuki reaction** couples the R group of a *vinylic or aryl halide* with the R′ group of an **organoboron compound** in a basic solution in the presence of a palladium catalyst (PdL$_2$). The general reaction can be written as follows:

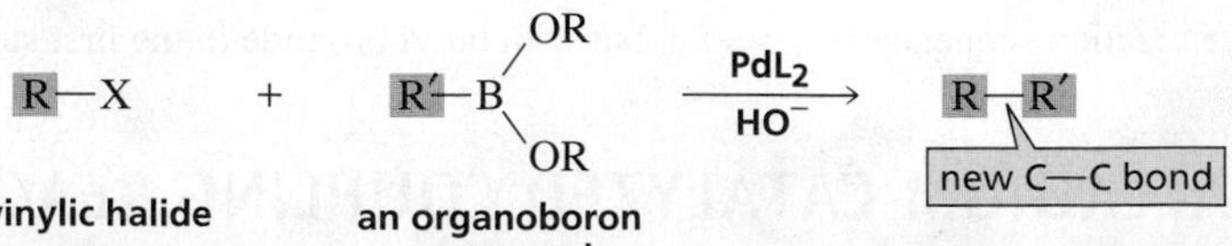

a vinylic halide
or
an aryl halide

an organoboron
compound

The R′ group of the organoboron compound can be either an *alkyl* group, an *alkenyl* group, or an *aryl* group. Notice that the new C—C bond joins the sp^2 carbon that was bonded to the halogen with the carbon that was bonded to boron.

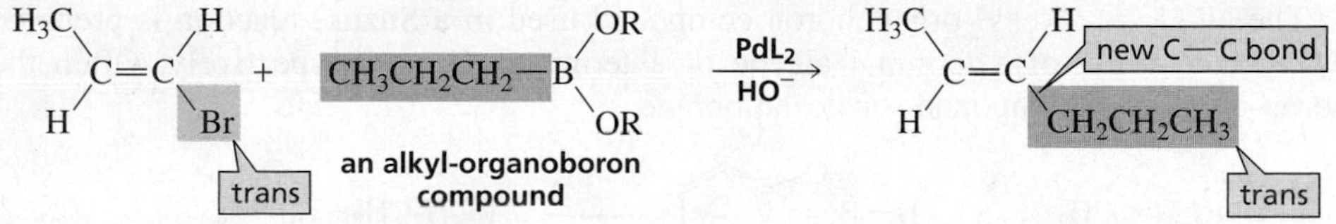

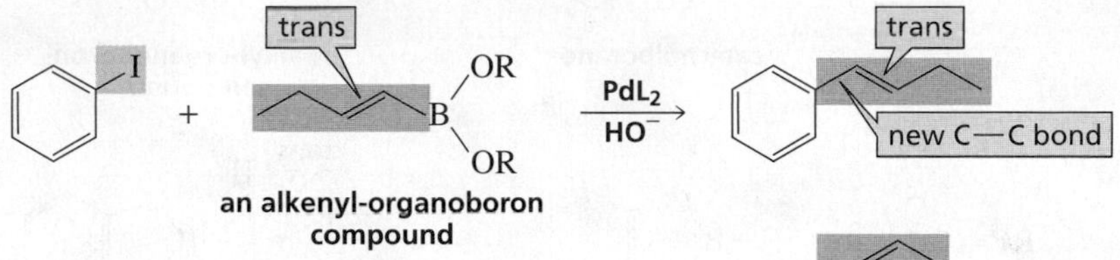

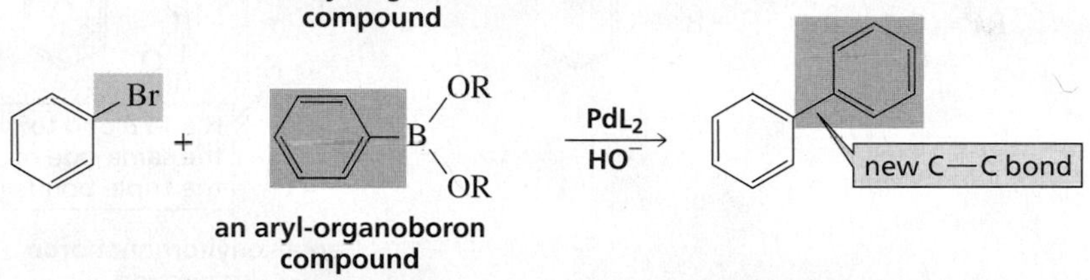

In a Suzuki reaction, the R′ group of an organoboron compound replaces a halogen.

In a Suzuki reaction, the new C—C bond joins the carbon that was bonded to the halogen with the carbon that was bonded to boron.

When the reactant is a vinylic halide, the configuration of its double bond is retained in the product. When an alkenyl-organoboron compound is used, the new double bond in the product will be trans because an alkenyl-organoboron compound always has a *trans configuration*.

The mechanism for the Suzuki reaction is not yet completely understood; the mechanism shown here is based on the experimental evidence collected to date.

MECHANISM FOR THE SUZUKI REACTION

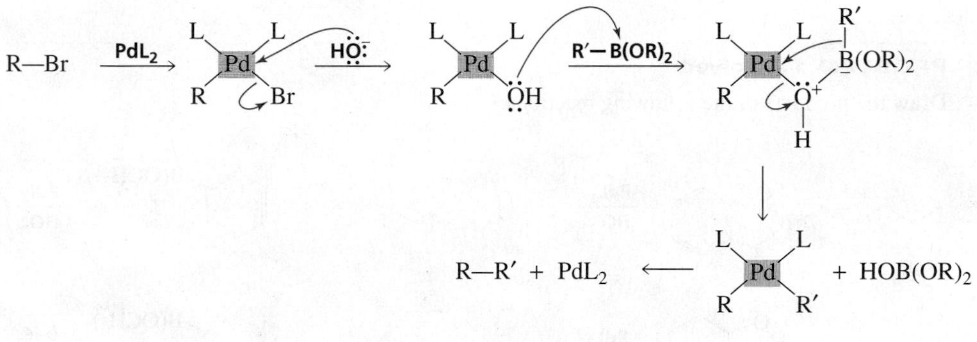

- Palladium is inserted between the carbon and the halogen.

- Hydroxide ion displaces the halide ion.

- In a Lewis acid–Lewis base reaction, a lone pair on oxygen donates a share in an electron pair to boron.

- The R′ group is transferred from boron to palladium, displacing the positively charged leaving group.

- The last step is effectively the reverse of the first step: palladium is eliminated and a new C—C bond is formed. The catalyst can now react with another molecule of alkyl halide.

The insertion of a metal between two atoms is called **oxidative addition**—two new groups are added to the metal. **Reductive elimination** eliminates two groups from the metal. Thus, the first step in a Suzuki reaction is an *oxidative addition,* and the last step is a *reductive elimination*.

Oxidative addition: two groups are added to the metal

Reductive elimination: two groups are eliminated from the metal

The alkyl- or alkenyl-organoboron compound used in a Suzuki reaction is prepared by hydroboration of a terminal alkene or a terminal alkyne, respectively. Often the boron-containing compound is catecholborane.

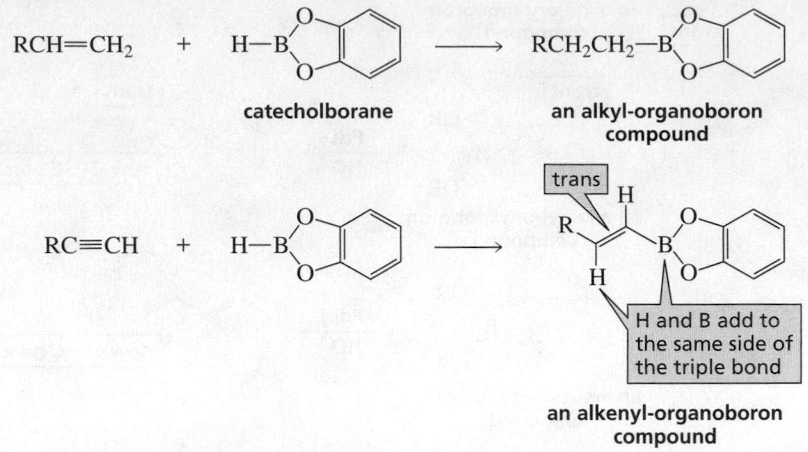

Boron adds to the multiply bonded carbon that is bonded to the most hydrogens and that B and the H add to the same side of the triple bond. Therefore, the boron in an alkenyl-organoboron compound will always be trans to the substituent on the adjacent sp^2 carbon.

An aryl-organoboron compound is prepared from an organolithium compound and trimethylborate.

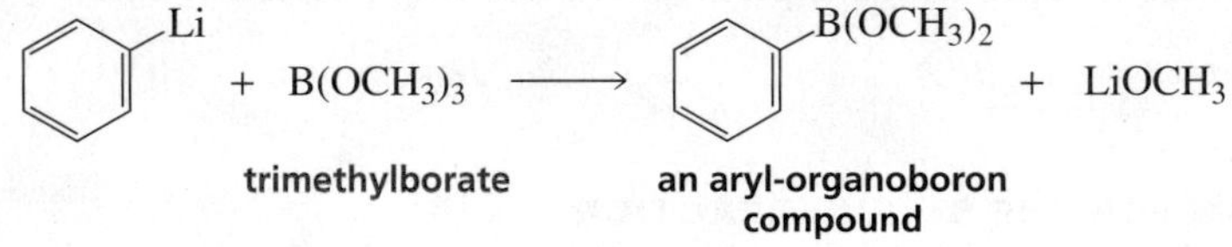

PROBLEM 12 Solved

Draw the products of the following reactions:

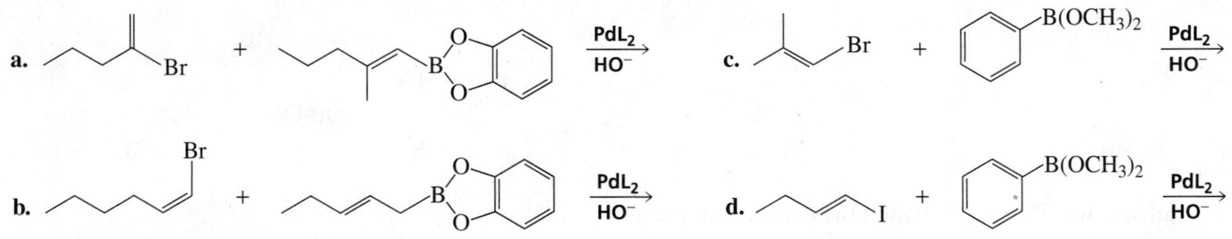

Solution to 12a and 12b Attach the carbon that is bonded to boron to the carbon that is bonded to the halogen, maintaining the configuration of both double bonds.

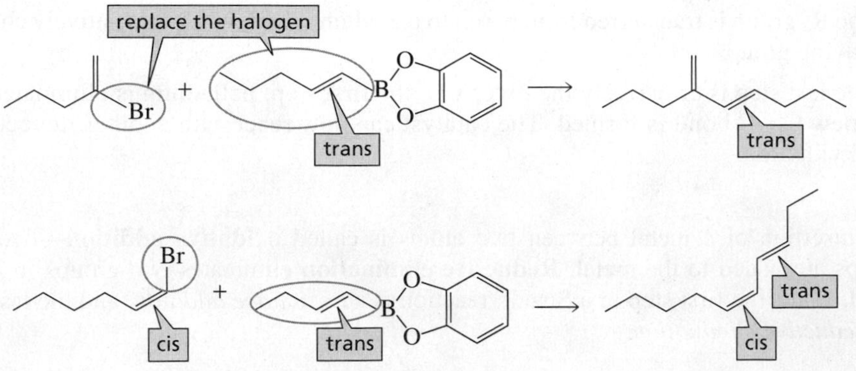

PROBLEM 13♦

What aryl or vinyl halides would you use to synthesize the following compounds, using the alkenyl-organoboron compound shown?

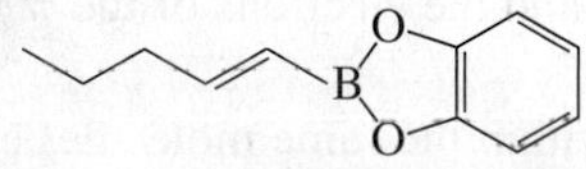

a.

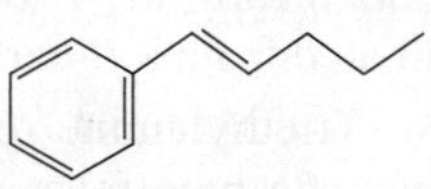

b.

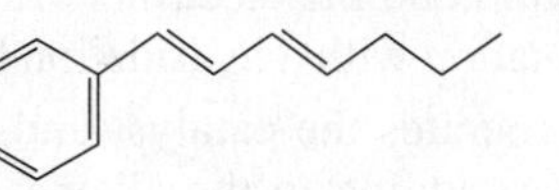

c.

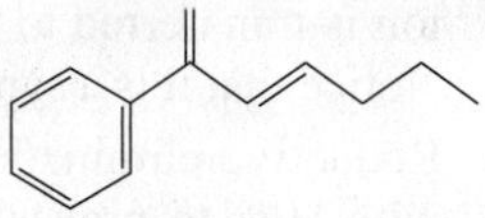

PROBLEM 14♦

What hydrocarbon would you use to prepare the organoboron compound of Problem 13?

The Heck Reaction

The **Heck reaction** couples a *vinylic* or an *aryl halide* with an *alkene* in the presence of a base (such as triethylamine) and a palladium catalyst (PdL_2). Like the Suzuki reaction, the Heck reaction is a substitution reaction: the R group of the halide *replaces a vinylic hydrogen* of an alkene. If there is a substituent attached to the alkene (Z), the R group will be trans to that substituent in the product. Notice that the new C—C bond joins two sp^2 carbons.

H is substituted by R R is trans to Z

R—X + H Z →$\frac{PdL_2}{(CH_3CH_2)_3N}$ R Z

a vinylic halide
or
an aryl halide

new C—C bond

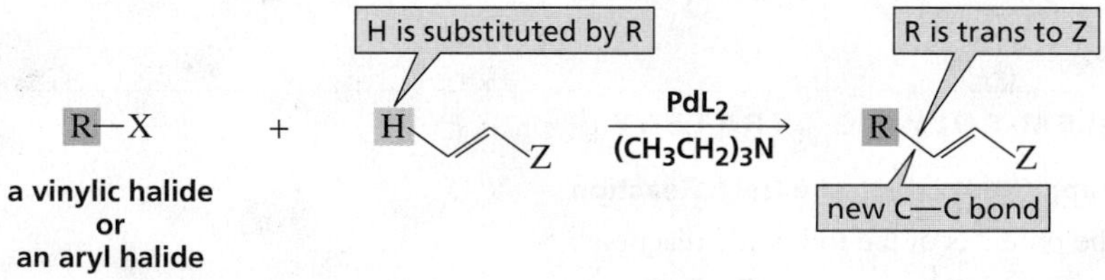

Specific examples of the Heck reaction are shown here:

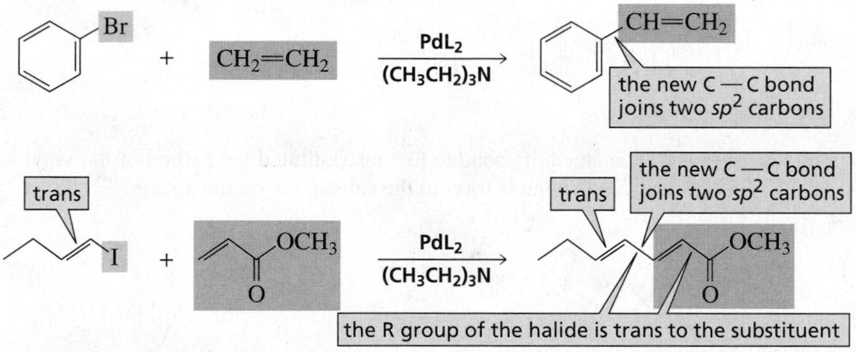

Br + $CH_2\text{=}CH_2$ →$\frac{PdL_2}{(CH_3CH_2)_3N}$ CH=CH₂

the new C—C bond joins two sp^2 carbons

trans I + OCH₃ →$\frac{PdL_2}{(CH_3CH_2)_3N}$ trans OCH₃

O the new C—C bond joins two sp^2 carbons O

the R group of the halide is trans to the substituent

In a Heck reaction, the R group of the halide replaces a vinyl hydrogen of an alkene and the R is trans to the alkene's substituent.

In a Heck reaction, the new C—C bond joins two sp^2 carbons.

When the reactant is a vinylic halide, notice that the configuration of its double bond is retained in the product.

MECHANISM FOR THE HECK REACTION

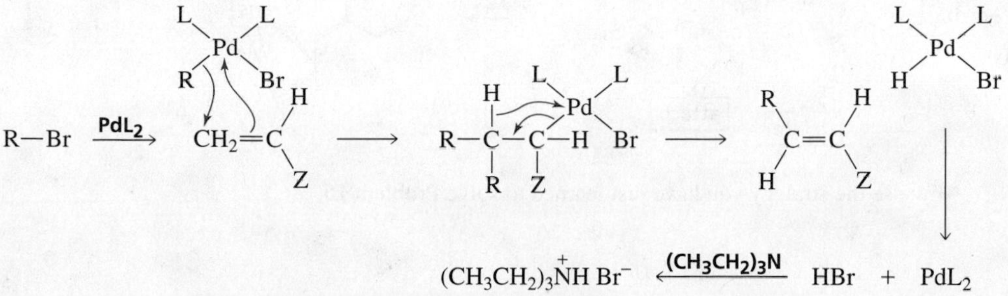

R—Br →$\frac{PdL_2}{}$

$(CH_3CH_2)_3\overset{+}{N}H$ Br⁻ ←$\frac{(CH_3CH_2)_3N}{}$ HBr + PdL_2

- The first step is an oxidative addition: palladium is inserted between the carbon and the halogen.
- Because carbon is more electronegative than palladium, R is a nucleophile. It adds to the sp^2 carbon of the alkene and the electrons of the π bond add to palladium. This is a syn addition reaction.
- Next is an intramolecular (within the same molecule) elimination reaction. A hydride ion is transferred to palladium and the electrons of the Pd—C bond form the π bond. Notice that it is a syn elimination with a hydride ion being eliminated from a β-carbon.
- Reductive elimination regenerates the catalyst and forms HBr. Triethylamine reacts with HBr to prevent it from adding to the alkene. The catalyst can now react with another molecule of alkyl halide.

The nucleophilic R group can add to either sp^2 carbon of the alkene. Therefore, the reaction will lead to a high yield of a single product only in the following situations:

- If the alkene is symmetrical
- If one of the sp^2 carbons is sterically hindered to nucleophilic attack
- If one of the sp^2 carbons is bonded to a group that can withdraw electrons by resonance (such as C=O or C≡N), causing the other sp^2 carbon to have a partial positive charge that will make it more susceptible to nucleophilic attack

---

PROBLEM-SOLVING STRATEGY

Determing the Product of a Heck Reaction

Draw the products of the following reactions:

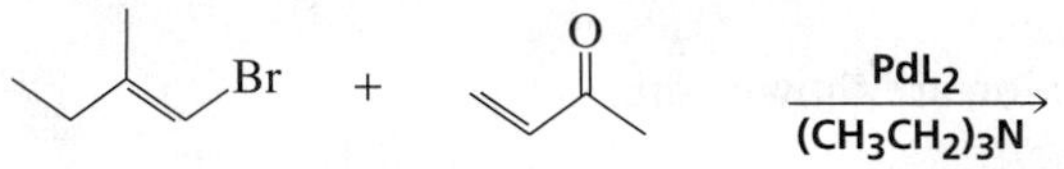

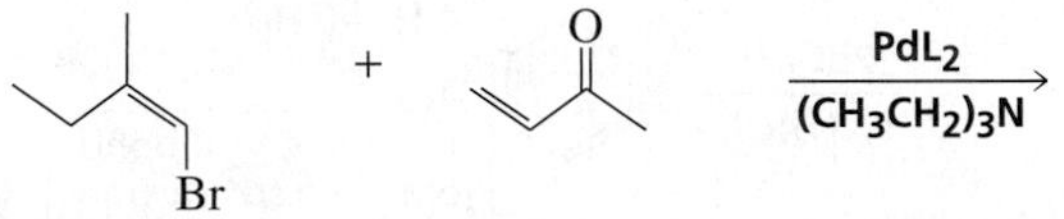

Delete the halogen and then attach its bond to the unsubstituted sp^2 carbon of the vinyl ketone, making sure that the new C—C bond is trans to the substituent on the alkene.

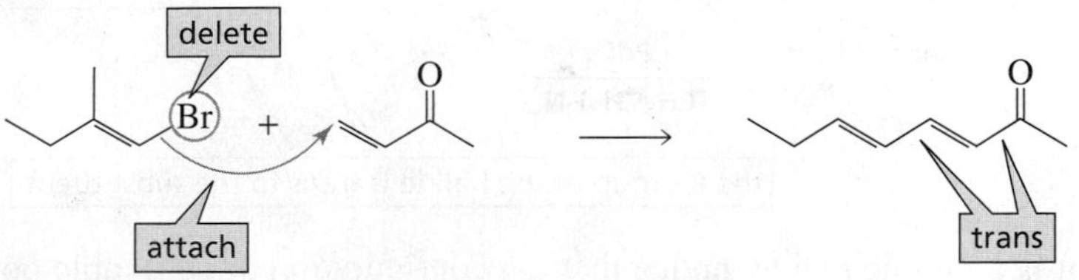

Again, delete the halogen and attach its bond to the unsubstituted sp^2 carbon of the vinyl ketone. Notice that in order to have the new C—C bond trans to the substitutent, the vinyl ketone is redrawn so that the substituent is pointing down.

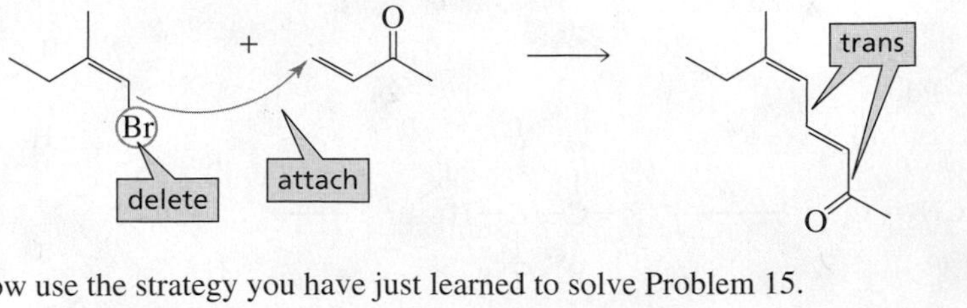

Now use the strategy you have just learned to solve Problem 15.

PROBLEM 15♦

Draw the products of the following reactions:

a. 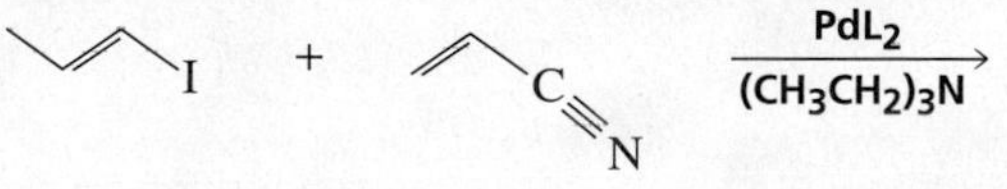 [structure: CH₃CH=CH–I] + [structure: CH₂=CH–C≡N] $\xrightarrow{\text{PdL}_2}$ $\text{(CH}_3\text{CH}_2)_3\text{N}$

b. 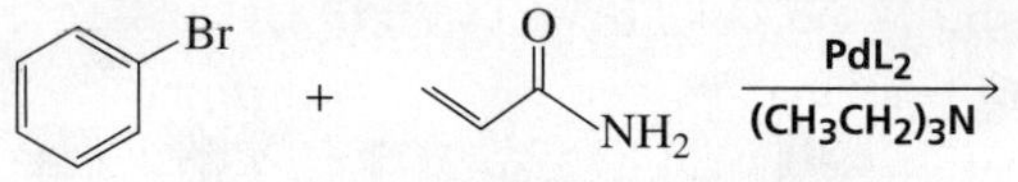 [structure: bromobenzene] + [structure: CH₂=CH–C(=O)–NH₂] $\xrightarrow{\text{PdL}_2}$ $\text{(CH}_3\text{CH}_2)_3\text{N}$

c. [structure: 3-bromoanisole with OCH₃] + [structure: styrene] $\xrightarrow{\text{PdL}_2}$ $\text{(CH}_3\text{CH}_2)_3\text{N}$

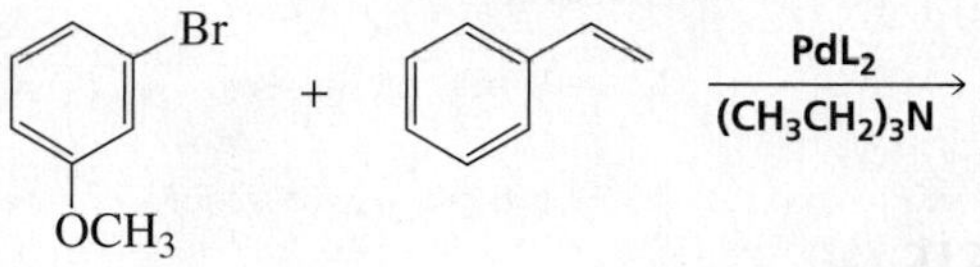

d. 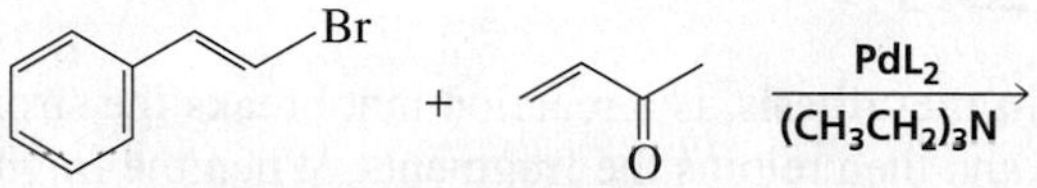 [structure: β-bromostyrene] + [structure: CH₂=CH–C(=O)–CH₃, methyl vinyl ketone] $\xrightarrow{\text{PdL}_2}$ $\text{(CH}_3\text{CH}_2)_3\text{N}$

PROBLEM 16 Solved

What reactants are needed to synthesize each of the following compounds using a Heck reaction?

a. 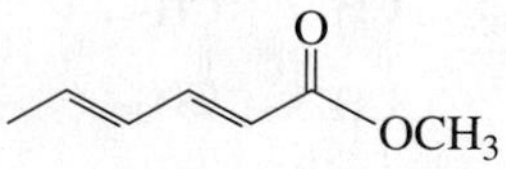 [structure ending in C(=O)OCH₃]

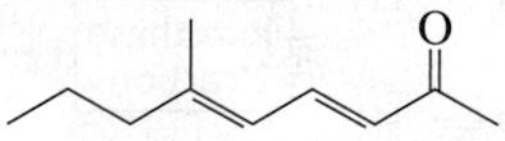

b. 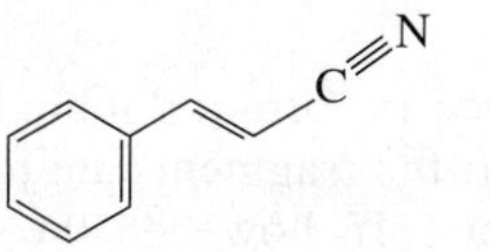 [structure: Ph–CH=CH–CH=CH–C≡N]

c. [structure with branched chain and C=O]

d. 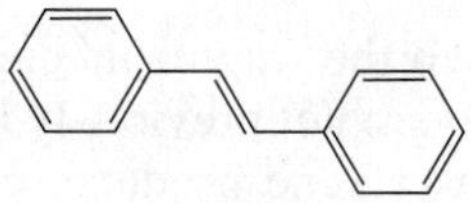 [structure: stilbene-type, Ph–CH=CH–CH₂–Ph]

Solution to 16a Cleave the molecule at the end of the C—C bond that joins two sp^2 carbons and add a halogen to the C—C bond.

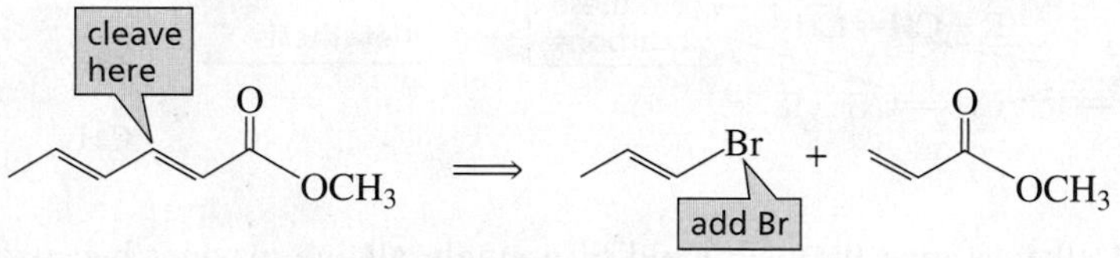

cleave here

[structure with OCH₃] $\implies$ [structure with Br / add Br] + [structure with OCH₃]

Solution to 16b

cleave here

[structure with C≡N] $\implies$ add I [structure with I] + [structure with C≡N]

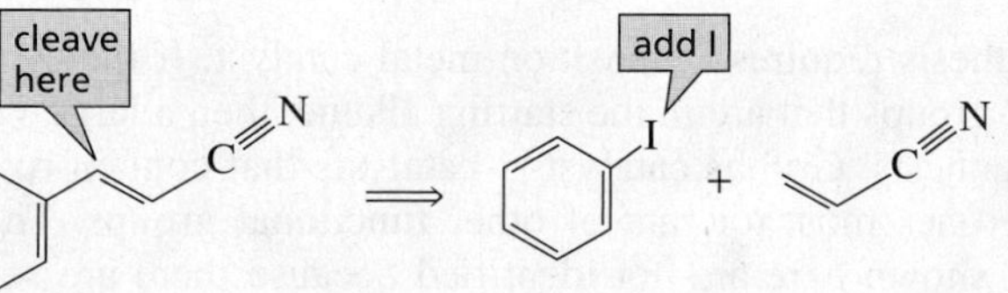

PROBLEM 17

Show how the Suzuki and/or Heck reactions can be used to prepare the following compounds:

a. [structure: Ph–CH=CH–CH₂–CH₃]

b. [structure: Ph–CH₂–CH₂–CH₂–CH₃]

PROBLEM 18♦

Identify two pairs of an alkyl bromide and an alkene that could be used in a Heck reaction to prepare the following compound:

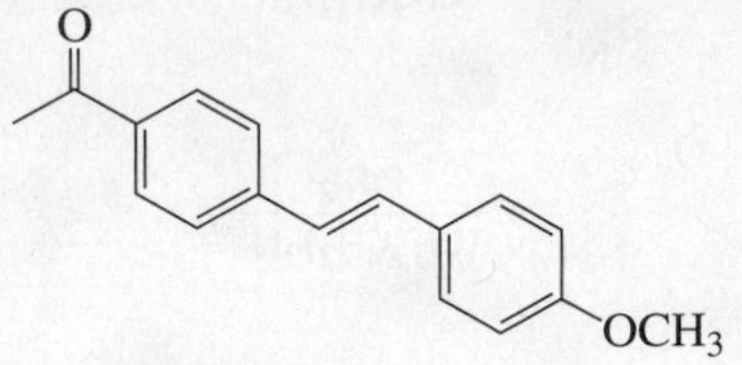

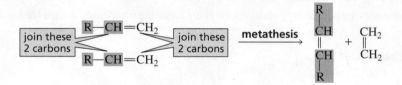

5 ALKENE METATHESIS

Alkene metathesis, also called **olefin metathesis,** is a reaction that breaks the strongest bond in an alkene (the double bond) and then rejoins the fragments. When the fragments are joined, each new double bond is formed between two sp^2 carbons that were not previously bonded. *Metathesis* is a Greek word that means "transposition."

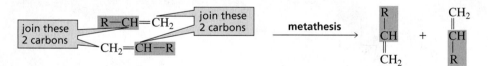

There are two ways in which the fragments can be joined to form a double bond between two sp^2 carbons that were not previously bonded. If the fragments are joined as shown above, then two new alkene products are obtained. If, however, the fragments are joined as shown next, then the starting material is re-formed. All is not lost, however, because the re-formed starting material can undergo another round of metathesis.

Terminal alkenes give the best yield of a single alkene product because one of the new alkene products will be ethene. Ethene is a gas so it can be removed from the reaction mixture as it is formed. This will shift the equilibrium that exists between the two pathways for joining the fragments toward the pathway that forms ethene and the other new alkene.

Alkene metathesis requires a transition metal catalyst. If the catalyst does not affect other functional groups that are in the starting alkene, then a large variety of alkenes can be used for metathesis. Grubbs catalysts—catalysts that contain ruthenium—have been found to be the ones most tolerant of other functional groups. The ligands (L) in the Grubbs catalyst shown here are not identified because there are several generations of Grubbs catalysts, each with different ligands.

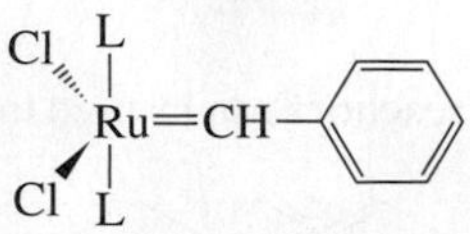

Grubbs catalyst

An example of alkene metathesis is shown next. If *E* and *Z* isomers are possible for the product of metathesis, both will be formed.

$$2 \text{ CH}_3\text{CH}_2\text{CH}=\text{CH}_2 \xrightarrow{\text{Grubbs catalyst}} \text{CH}_3\text{CH}_2\text{CH}=\text{CHCH}_2\text{CH}_3 + \text{CH}_2=\text{CH}_2$$

both *E* and *Z*

Metathesis can also be done using two different alkenes as the starting materials.

$$\text{CH}_3\text{CH}=\text{CHCH}_3 + \text{CH}_3\text{CH}_2\text{CH}=\text{CHCH}_2\text{CH}_3 \longrightarrow 2 \text{ CH}_3\text{CH}=\text{CHCH}_2\text{CH}_3$$

If the reactant is a diene, **ring-closing metathesis** can occur.

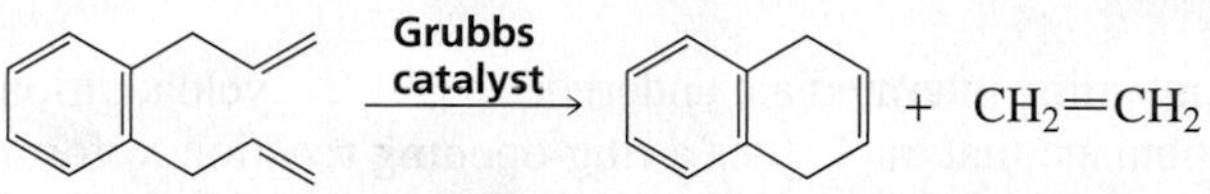

Metathesis occurs in two phases. The first phase creates two intermediates (I and II). Each intermediate has the metal atom of the catalyst in place of a group that was attached by the double bond in the starting alkene.

$$\text{R}-\text{CH}=\text{CH}_2 \xrightarrow{\text{first phase}} \underset{\textbf{I}}{\text{R}-\text{CH}=\text{M}} + \underset{\textbf{II}}{\text{M}=\text{CH}_2}$$

The two-step mechanism for this first phase is shown here.

MECHANISM FOR METATHESIS

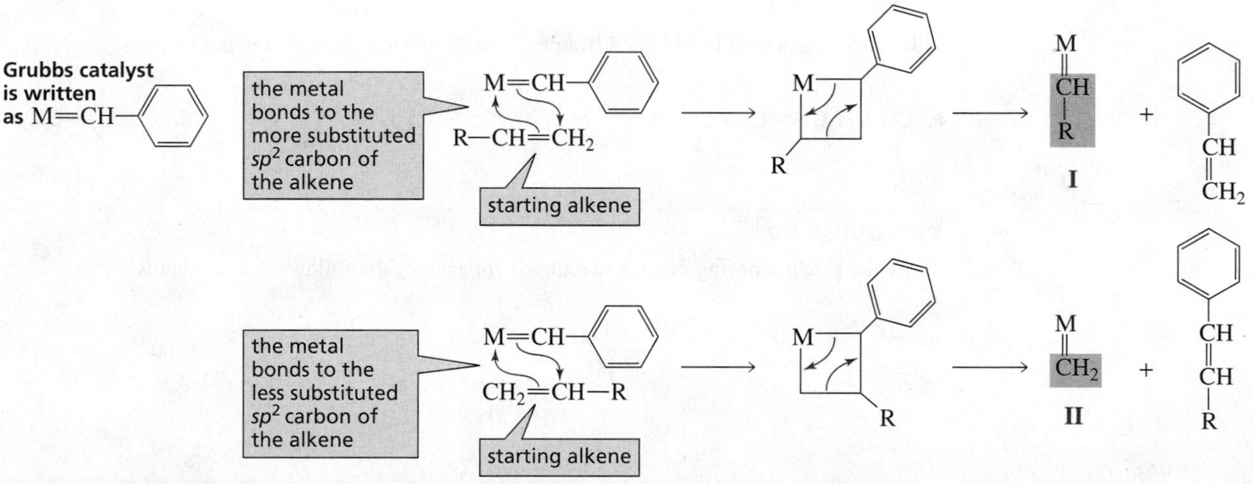

- The Grubbs catalyst and the starting alkene undergo a [2 + 2] cycloaddition reaction. This reaction forms two different metallocyclobutane intermediates because the metal can bond to either sp^2 carbon of the alkene.

- Each of the metallocyclobutane intermediates undergoes a ring-opening reaction. Two metal-containing intermediates are formed (I and II).

In the second phase of metathesis, each of the metal-containing intermediates reacts with the starting alkene to form a new alkene and the other metal-containing intermediate. Notice that the mechanism for each of the two phases is the same.

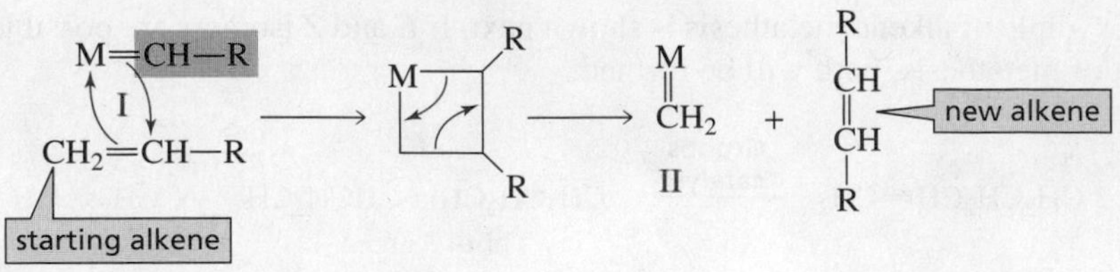

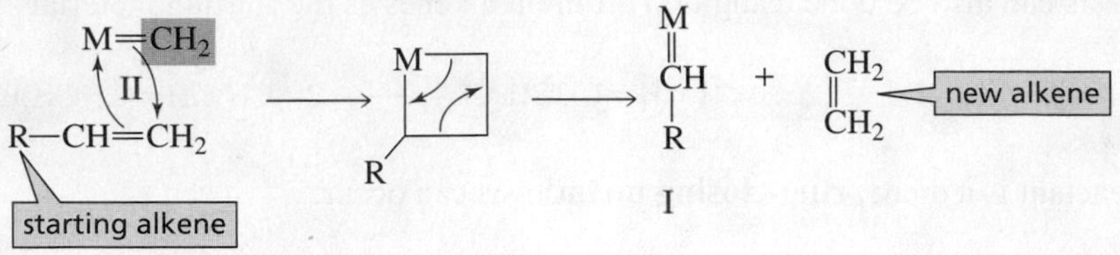

- Each metal-containing intermediate undergoes a [2 + 2] cycloaddition reaction to form a metallocyclobutane that undergoes a ring-opening reaction to form the other metal-containing intermediate and a new alkene. Thus, metal-containing intermediate I reacts to form metal-containing intermediate II, and metal-containing intermediate II reacts to form metal-containing intermediate I.

Because the metal atom of the metal-containing intermediate can bond to either sp^2 carbon of the alkene in the first step of the second phase, two different metallocyclobutanes could be formed. Only if the metal in intermediate I bonds to the *less substituted sp²* carbon of the alkene is a new alkene formed. If the metal bonds to the more substituted sp^2 carbon, then the starting alkene will be re-formed. Similarly, a new alkene is formed only if the metal in intermediate II bonds to the *more substituted sp²* carbon. If the metal bonds to the less substituted sp^2 carbon, then the starting alkene will be re-formed.

Each of the newly formed metal-containing intermediates can now react with another molecule of the alkene starting material.

PROBLEM 19

What products would be obtained from metathesis of each of the following alkenes?

a. $CH_3CH_2CH{=}CH_2$ b. 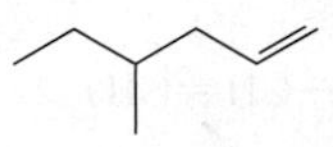   c. $CH_2{=}C\overset{CH_3}{\underset{CH_3}{}}$ d.

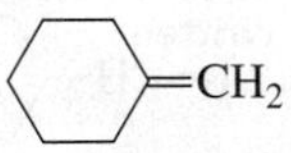

PROBLEM 20

Draw the product of ring-closing metathesis for each of the following compounds:

a. 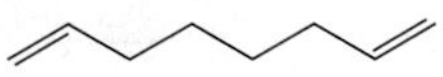   b. 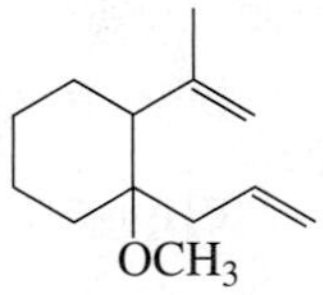   c.

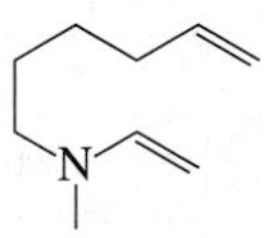

PROBLEM 21 Solved

What alkene will undergo metathesis to form each of the following compounds?

a.    b. 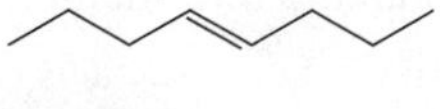   c. 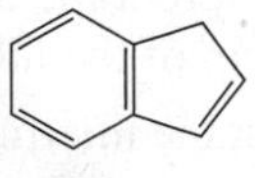   d.

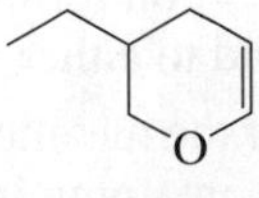

Solution to 21a and 21b Break the double bond that formed during metathesis, and then add $={=}CH_2$ to each end.

a.

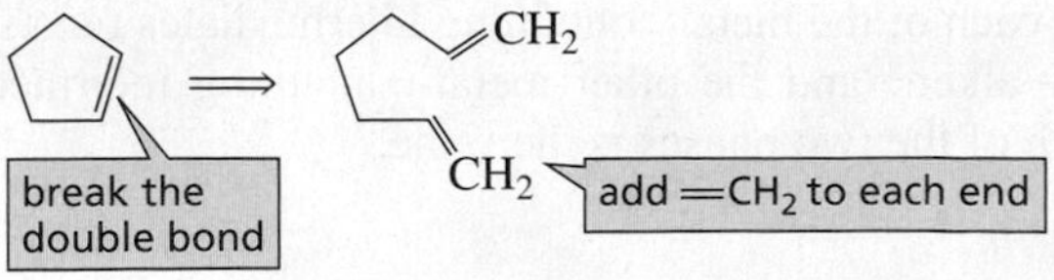

b.

Alkynes can also undergo metathesis. The preferred catalysts for alkyne metathesis are Schrock catalysts—catalysts that contain molybdenum or tungsten as the transition metal.

$$CH_3CH_2C\equiv CH \xrightarrow{\text{Schrock catalyst}} CH_3CH_2C\equiv CCH_2CH_3 + HC\equiv CH$$

PROBLEM 22♦

What products would be obtained from metathesis of the following alkyne?

SOME IMPORTANT THINGS TO REMEMBER

- An **organometallic compound** contains a carbon–metal bond.

- The carbon attached to the halogen in an alkyl halide is an electrophile, whereas the carbon attached to the metal in an organometallic compound is a nucleophile.

- **Organomagnesium compounds (Grignard reagents)** and **organolithium compounds** are the most common organometallic compounds. They cannot be prepared from or react with compounds that contain acidic groups.

- The greater the polarity of the carbon–metal bond, the more reactive the organometallic compound is as a nucleophile.

- **Transmetallation** will occur if an organometallic compound is added to a metal halide whose metal is more electronegative than the metal in the organometallic compound.

- **Coupling reactions** join two carbon-containing groups together.

- An **organocuprate (a Gilman reagent)** substitutes an R group for a Cl, Br, or I of any alkyl halide except a tertiary alkyl halide.

- An organocuprate undergoes a nucleophilic substitution reaction with an epoxide.

- The **Suzuki** and **Heck reactions** are coupling reactions that require a palladium catalyst.

- The first step in both Suzuki and Heck reactions is **oxidative addition**—two groups are added to a metal. The last step in both reactions is a **reductive elimination**—two groups are eliminated from a metal.

- A **Suzuki reaction** couples the R group of a *vinylic or aryl halide* with the R' group of an **organoboron compound.** The new C—C bond joins the carbon that was bonded to the halogen with the carbon that was bonded to boron. When an alkenyl-organoboron compound is used, the new double bond in the product will be trans.

- A **Heck reaction** couples a *vinylic* or an *aryl halide* with an *alkene.* The R group of the vinylic or aryl halide replaces a vinylic hydrogen of an alkene. The new C—C bond joins two sp^2 carbons, and the R group is trans to any substituent on the alkene. If the reactant is a vinylic halide, the configuration of the vinyl group is retained in the product.

- **Alkene metathesis** (or **olefin metathesis**) breaks the double bond of an alkene and then rejoins the fragments. When the fragments are joined, the new double bond is formed between two sp^2 carbons that were not previously bonded. Alkynes also undergo metathesis.

- Terminal alkenes give the best yield of a single alkene product in metathesis because one of the products is ethene, which can be easily removed from the reaction mixture, thus shifting the equilibrium in favor of the other new alkene product.

SUMMARY OF REACTIONS

1. Formation of an *organolithium* or *organomagnesium* compound (Section 1).

$$RBr \xrightarrow[\text{diethyl ether}]{\text{Mg}} RMgBr \qquad\qquad RBr \xrightarrow[\text{hexane}]{\text{Li}} RLi \;+\; Br^-$$

a Grignard　　　　　　　　　　　**an organolithium**
reagent　　　　　　　　　　　　　　**compound**

2. Organocuprates (Section 3)
    a. Coupling of an organocuprate and an alkyl halide

$$2\;R'Li \;+\; CuI \xrightarrow{\text{THF}} R'_2CuLi \;+\; LiI$$

an organocuprate

$$R{-}X \;+\; R_2CuLi \xrightarrow{\text{THF}} R{-}R' \;+\; RCu \;+\; LiX$$

R = primary, secondary, vinylic, aryl　　X = Cl, Br, or I

    b. Reaction of an organocuprate with an epoxide

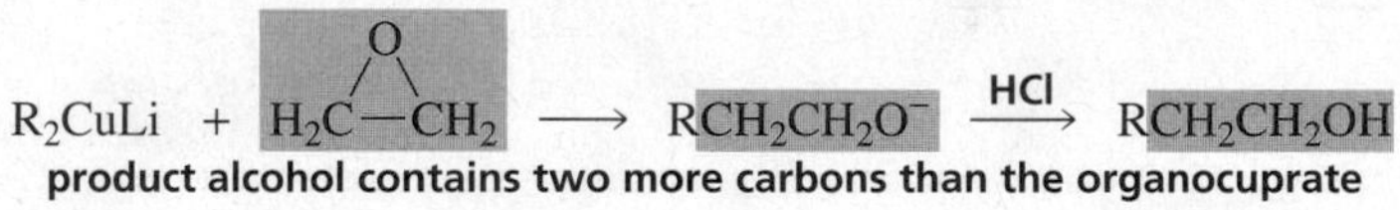

$$R_2CuLi \;+\; \overset{\displaystyle O}{H_2C{-}CH_2} \longrightarrow RCH_2CH_2O^- \xrightarrow{\text{HCl}} RCH_2CH_2OH$$

product alcohol contains two more carbons than the organocuprate

3. Palladium-Catalyzed Coupling Reactions (Section 4)
    a. Coupling of a *vinylic* or an *aryl halide* with an *organoboron compound:* the Suzuki reaction. The mechanism is shown in Section 5.

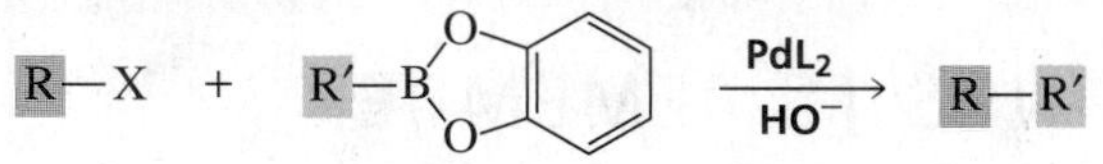

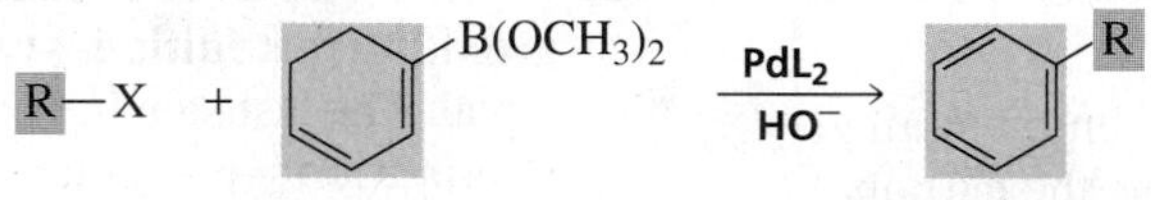

R = vinylic, aryl　　X = Br, I

    b. Coupling of a *vinylic* or an *aryl halide* with an *alkene:* the Heck reaction. The mechanism is shown in Section 5.

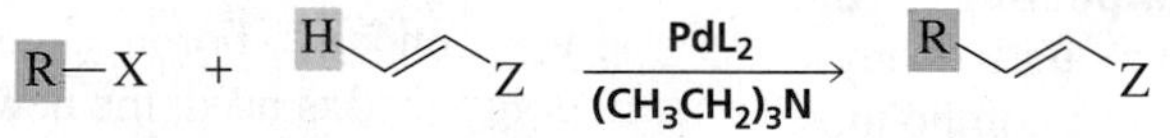

R = vinylic, aryl　　X = Br, I　　　　　　**the product is
a trans alkene**

4. Alkene metathesis (Section 5). The mechanism is shown in Section 6.
    a. The starting material is an alkene.

$$2\;R{-}CH{=}CH_2 \xrightarrow{\substack{\text{Grubbs} \\ \text{catalyst}}} R{-}CH{=}CH{-}R \;+\; CH_2{=}CH_2$$

the reaction works best with a terminal alkene

    b. The starting material is a diene.

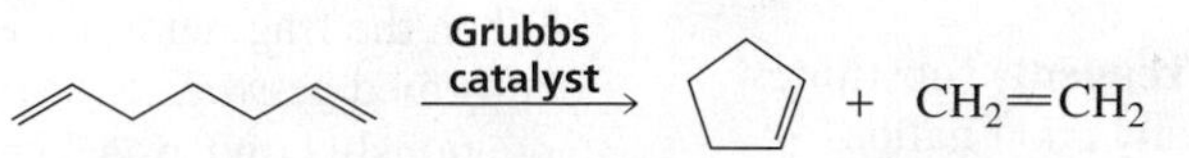

    c. The starting material is an alkyne.

$$RC{\equiv}CH \xrightarrow{\substack{\text{Schrock} \\ \text{catalyst}}} RC{\equiv}CR \;+\; HC{\equiv}CH$$

PROBLEMS

23. Draw the products of each of the following reactions:

a. 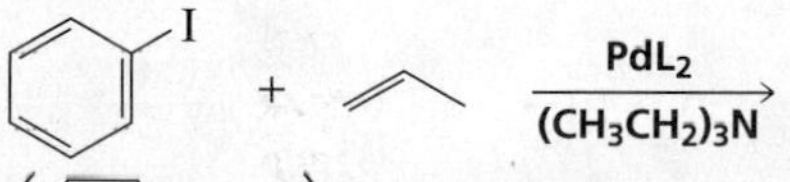 [benzene with I] + [propene] →(PdL₂)/((CH₃CH₂)₃N)

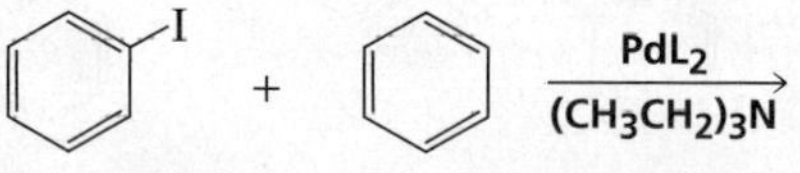

b. 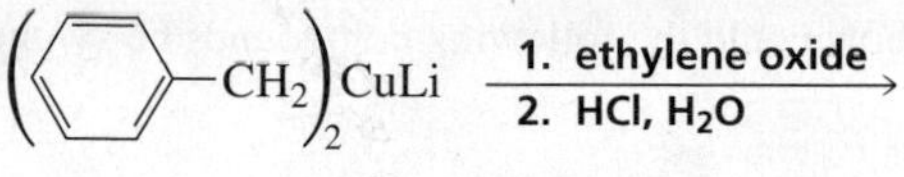 ((benzyl)CH₂)₂CuLi →(1. ethylene oxide)/(2. HCl, H₂O)

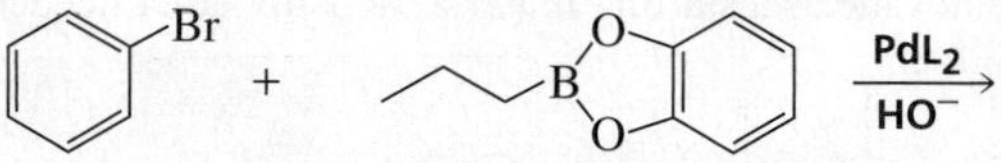

c. 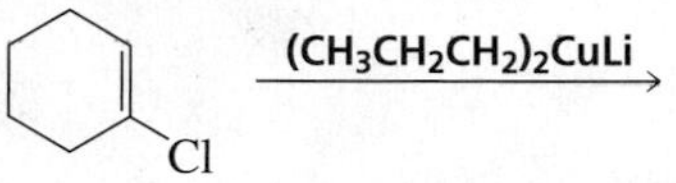 [cyclohexene with Cl] →(CH₃CH₂CH₂)₂CuLi

d. [benzene with I] + [benzene] →(PdL₂)/((CH₃CH₂)₃N)

e. [benzene with Br] + [propyl boronic ester] →(PdL₂)/(HO⁻)

f. 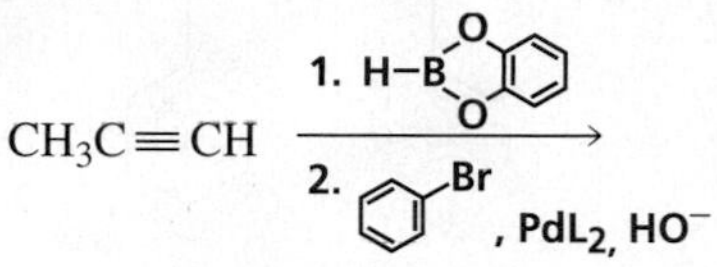 CH₃C≡CH →(1. H–B(catechol))/(2. [benzene-Br], PdL₂, HO⁻)

24. Which of the following alkyl halides could be successfully used to form a Grignard reagent?

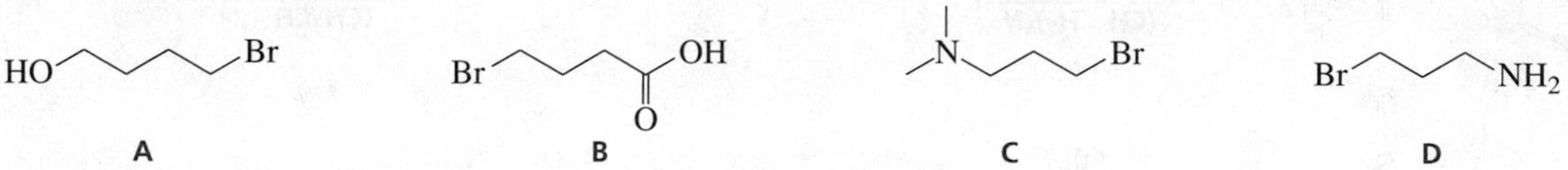

HO⁀⁀⁀Br   Br⁀⁀⁀OH with =O   (CH₃)₂N⁀⁀⁀Br   Br⁀⁀⁀NH₂

     A   **B**   **C**   **D**

25. Identify **A** through **H**.

$$CH_3Br \xrightarrow[2.\ B]{1.\ A} C \xrightarrow[2.\ E]{1.\ D} CH_3CH_2CH_2OH \xrightarrow[\substack{2.\ G \\ 3.\ H}]{1.\ F} CH_3CH_2CH_2OCH_2CH_2OH$$

26. Using the given starting material, any necessary inorganic reagents and catalysts, and any carbon-containing compounds with no more than three carbons, indicate how each of the following compounds can be prepared:

a. 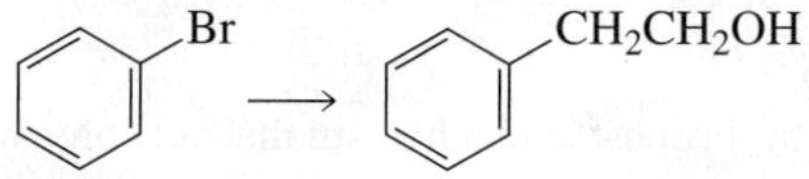 [benzene-Br] → [benzene-CH₂CH₂OH]

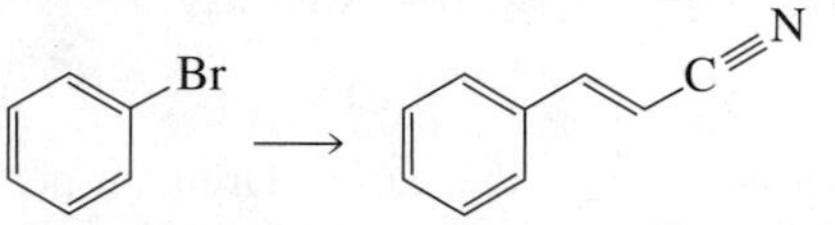

b. CH₃CHCH₂OH → CH₃CHCH₂CH₂CH₂OH
      | |
      CH₃   CH₃

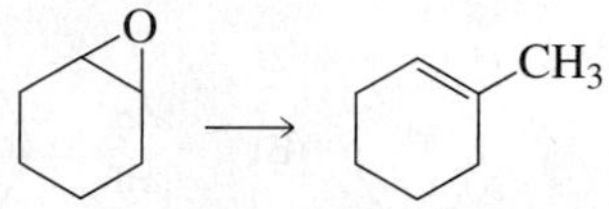

c. CH₃CH₂C≡CH → CH₃CH₂C≡CCH₂CH₂OH

d. [benzene-Br] → [benzene-CH=CH-C≡N]

e. [cyclohexene oxide] → [cyclohexene-CH₃]

f. 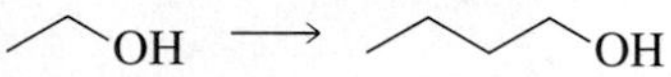 [propenol] → [propanol] OH

27. What alkyl halide would be needed to react with lithium divinylcuprate $\left[(CH_2{=}CH)_2CuLi\right]$ for the synthesis of each of the following compounds?

a. [cyclohexane-CH₂CH=CH₂]   **b.** 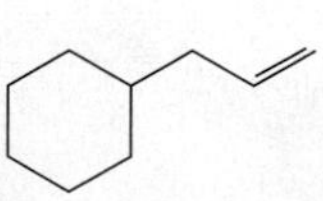 [cyclohexane-CH₂CH₂CH=CH₂]   **c.** 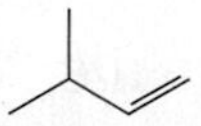 [branched alkene]   **d.** 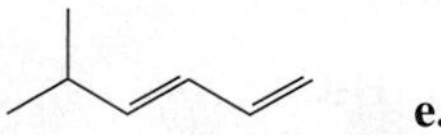 [branched diene]   **e.** 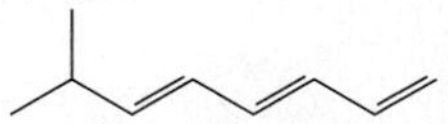 [branched triene]

28. Draw the products of each of the following reactions:

a. 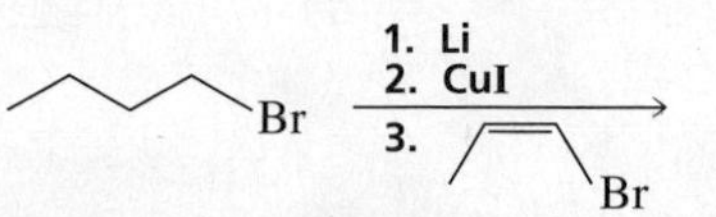 ⁀⁀⁀Br →(1. Li / 2. CuI / 3. ⁀⁀Br)

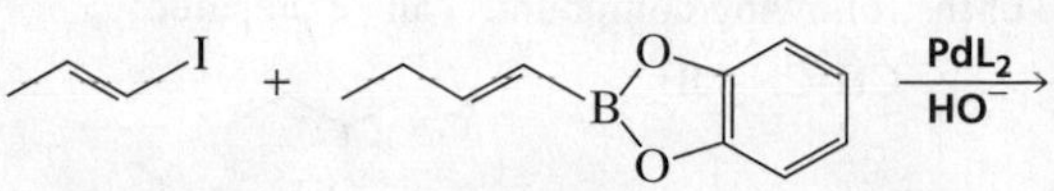

b. 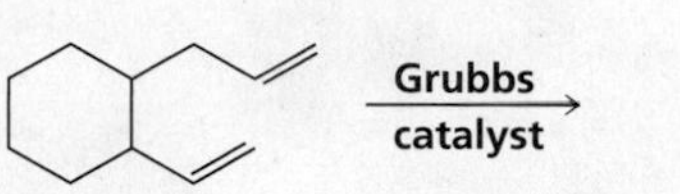 ⁀⁀⁀Br →(1. Li / 2. CuI / 3. [alkene-Br])

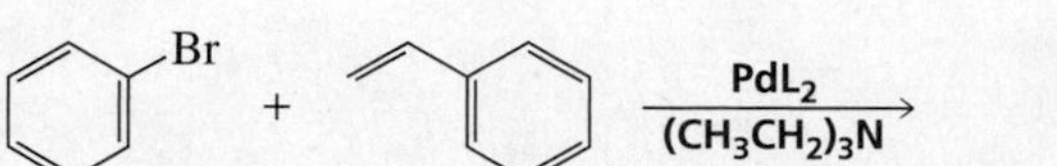

c. →(Grubbs catalyst)

d. [alkene-I] + [butenyl boronic ester] →(PdL₂)/(HO⁻)

e. ⁀⁀I + [butenyl boronic ester] →(PdL₂)/(HO⁻)

f. [benzene-Br with CH₃] + [styrene] →(PdL₂)/((CH₃CH₂)₃N)

29. What vinylic halide would need to be coupled with styrene $(C_6H_5\text{-}CH{=}CH_2)$ for the synthesis of each of the following compounds using a Heck reaction?

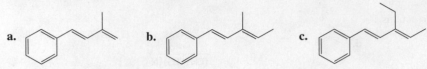

a.　　　　**b.**　　　　**c.**

30. Using ethynylcyclohexane as a starting material and any other needed reagents, how could the following compounds be synthesized?

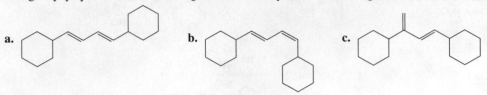

a.　　　　**b.**　　　　**c.**

31. Draw the products of the following reactions:

a. 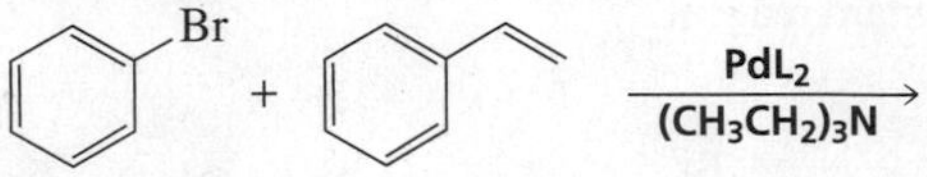 (benzene with Br) + (styrene) $\xrightarrow[\text{(CH}_3\text{CH}_2)_3\text{N}]{\text{PdL}_2}$

c. 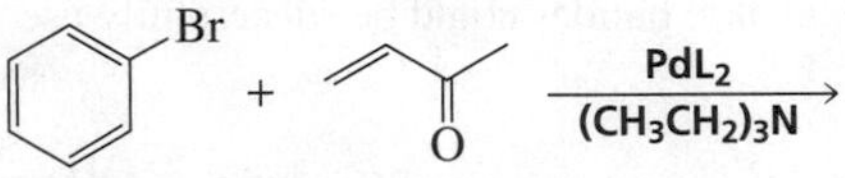 (benzene with Br) + (methyl vinyl ketone) $\xrightarrow[\text{(CH}_3\text{CH}_2)_3\text{N}]{\text{PdL}_2}$

b. 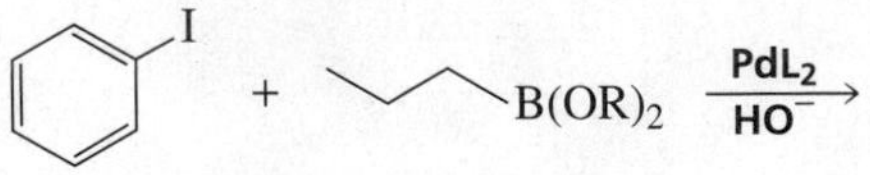 (benzene with I) + $\diagup\!\!\diagdown$ B(OR)$_2$ $\xrightarrow[\text{HO}^-]{\text{PdL}_2}$

d. 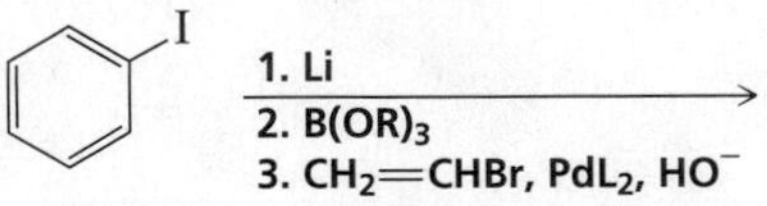 (benzene with I) $\xrightarrow{\begin{array}{l}\text{1. Li}\\\text{2. B(OR)}_3\\\text{3. CH}_2{=}\text{CHBr, PdL}_2\text{, HO}^-\end{array}}$

32. Using the given starting material, any necessary inorganic reagents and catalysts, and any carbon-containing compounds with no more that two carbons, indicate how each of the following compounds can be prepared:

a. $CH_3CH_2CH_2CH_2Br \longrightarrow CH_3CH_2CH_2CH_2CHCH$

b. $CH_3CH{=}CHCH_3 \longrightarrow CH_3CH{=}CCH_2CH_3$ with CH_3 substituent

c. $CH_3CH{=}CHCH_3 \longrightarrow CH_3CH\overset{O}{\overset{\|}{C}}CCH_3$ with CH_3 substituent

33. Dimerization is a side reaction that occurs during the preparation of a Grignard reagent. Propose a mechanism that accounts for the formation of the dimer.

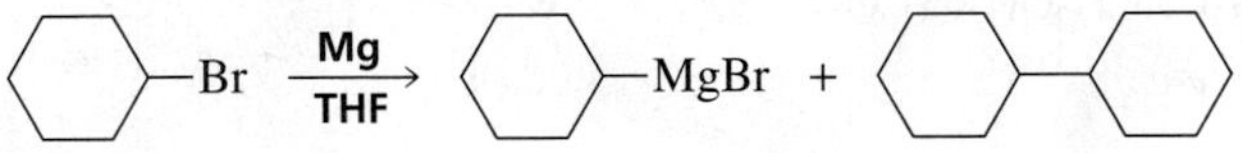

(cyclohexyl)–Br $\xrightarrow[\text{THF}]{\text{Mg}}$ (cyclohexyl)–MgBr + (dicyclohexyl)

a dimer

34. A student added an equivalent of 3,4-epoxy-4-methylcyclohexanol to a solution of methylmagnesium bromide in diethyl ether, and then added dilute hydrochloric acid. He expected that the product would be 1,2-dimethyl-1,4-cyclohexanediol. He did not get any of the expected product. What product did he get?

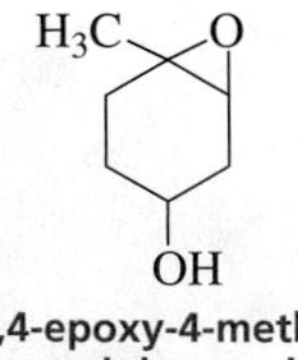

3,4-epoxy-4-methyl-
cyclohexanol

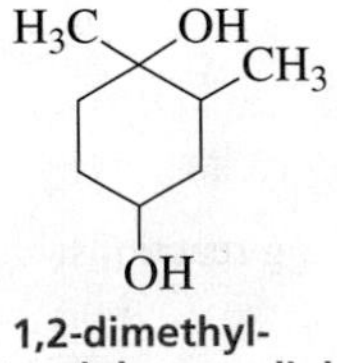

1,2-dimethyl-
1,4-cyclohexanediol

35. Using the given starting material, any necessary inorganic reagents, and any carbon-containing compounds with no more that two carbons, indicate how each of the following compounds can be prepared:

a. 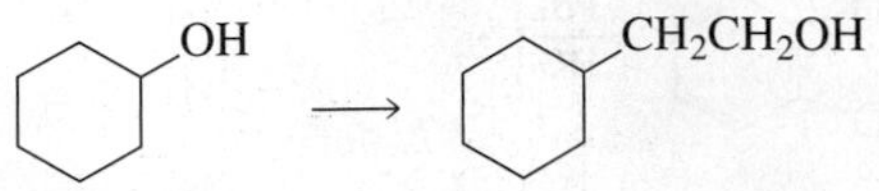 (cyclohexanol with OH) $\longrightarrow$ (cyclohexane with CH$_2$CH$_2$OH)

b.  $\diagup\!\!\diagup\!\!\diagdown$ with OH $\longrightarrow$ $\diagup\!\!\diagup\!\!\diagdown$ with D

36. a. Which of the following compounds cannot be prepared by a Heck reaction?

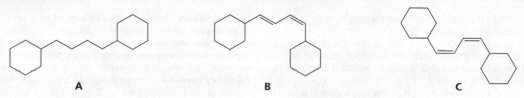

A   B   C

b. For those compounds that can be prepared by a Heck reaction, what starting materials are required?

37. What is the maximum yield of 2-pentene that can be formed in the reaction shown in Section 6?

38. Bombykol is the sex pheromone of the silk moth.

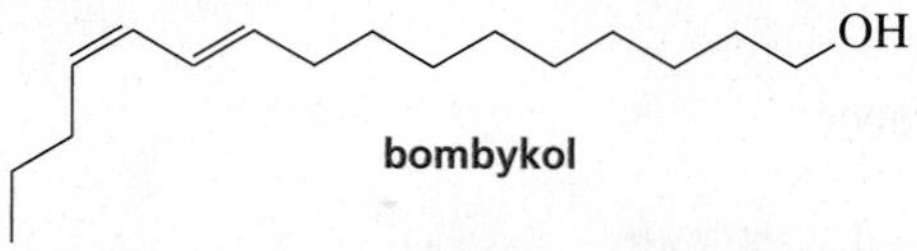

bombykol

Show how bombykol can be synthesized from the following compounds.

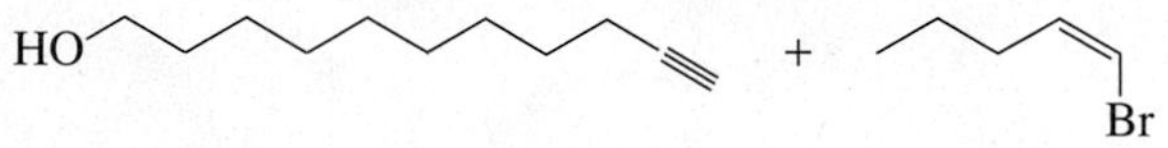

39. a. Metathesis of which of the following sets of alkenes will lead to the highest yield of a single alkene?

 1. 1-butene and 1-pentene **2.** 2-butene and 3-hexene **3.** 2-butene and 1-pentene

b. What kinds of alkenes should be used in metathesis reactions that use two different alkenes as starting materials?

40. A dibromide loses only one bromine when it reacts with sodium hydroxide. The dibromide forms toluene $(C_6H_5\text{-}CH_3)$ when it reacts with magnesium shavings in ether followed by treatment with dilute acid. Give possible structures for the dibromide.

41. What starting material is required in order to synthesize each of the following compounds by ring-closing metathesis?

a. 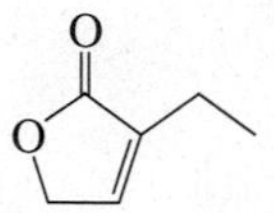   **b.** 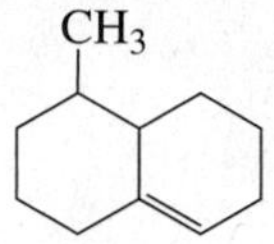   **c.**

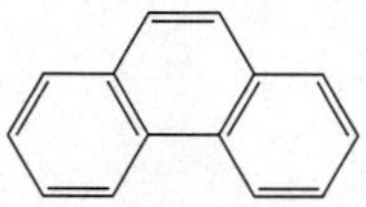

42. What product is obtained from ring-opening metathesis polymerization of each of the following compounds? (*Hint:* In each case, the product is an unsaturated hydrocarbon with a high molecular weight.)

 a.   **b.** 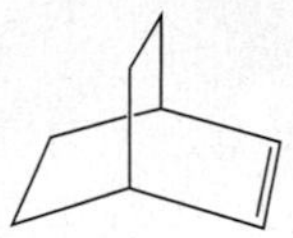   **c.**

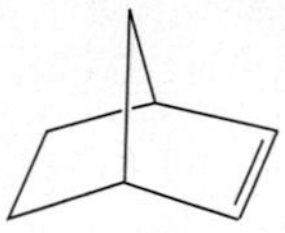

GLOSSARY

alkene metathesis   breaks the double bond of an alkene (or the triple bond of an alkyne) and then rejoins the fragments.

coupling reaction   a reaction that joins two CH-containing groups.

Gilman reagent   an organocuprate, prepared from the reaction of an organolithium reagent with cuprous iodide, used to replace a halogen with an alkyl group.

Grignard reagent   the compound that results when magnesium is inserted between the carbon and halogen of an alkyl halide (RMgBr, RMgCl).

Heck reaction   couples an aryl, benzyl, or vinyl halide or triflate with an alkene in a basic solution in the presence of $Pd(PPh_3)_4$.

olefin metathesis   breaks the double bond of an alkene (or the triple bond of an alkyne) and then rejoins the fragments.

organoboron compound   a compound containing a C — B bond.

organometallic compound   a compound containing a carbon–metal bond.

oxidative addition   the insertion of a metal between two atoms.

reductive elimination   the elimination of two groups attached to a metal.

Suzuki reaction   couples an aryl, a benzyl, or a vinyl halide with an organo-borane in the presence of $Pd(PPh_3)_4$.

transmetallation   metal exchange.

PHOTO CREDITS

Photo credits are listed in order of appearance.

Professor Walter Leal, University of California, Davis

Radicals •
Reactions of Alkanes

From Chapter 13 of *Organic Chemistry*, Seventh Edition. Paula Yurkanis Bruice. Copyright © 2014 by Pearson Education, Inc.
All rights reserved.

Radicals • Reactions of Alkanes

Alkanes are widespread both on Earth and on other planets. The atmospheres of Jupiter, Saturn, Uranus, and Neptune contain large quantities of methane (CH_4), the smallest alkane, which is an odorless and flammable gas. The blue colors of Uranus and Neptune are the result of methane in their atmospheres (see later in this chapter). Alkanes on Earth are found in natural gas and petroleum, which are formed by the decomposition of plant and animal material that have been buried for long periods of time in the Earth's crust, where oxygen is scarce. As a result, natural gas and petroleum are known as fossil fuels.

There are three classes of hydrocarbons: *alkanes,* which contain only carbon–carbon single bonds; *alkenes,* which contain carbon–carbon double bonds; and *alkynes,* which contain carbon–carbon triple bonds. Because **alkanes** do not contain any double or triple bonds, they are called **saturated hydrocarbons,** meaning they are saturated with hydrogen. A few examples of alkanes are

$CH_3CH_2CH_2CH_3$
butane

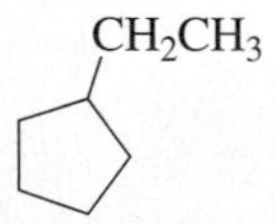

ethylcyclopentane

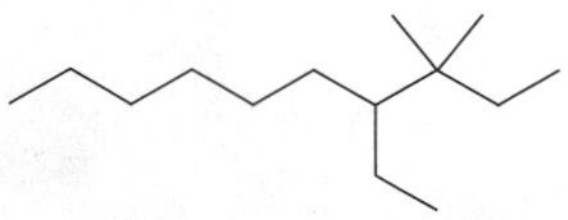

4-ethyl-3,3-dimethyldecane

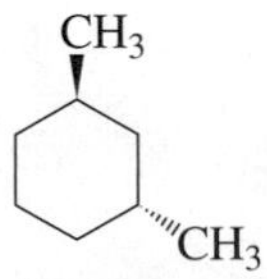

(*R,R*)-1,3-dimethyl-cyclohexane

The world needs a renewable, nonpolluting, and economically affordable source of energy (see later in this chapter)

1   ALKANES ARE UNREACTIVE COMPOUNDS

The carbon–carbon double and triple bonds of *alkenes* and *alkynes* are composed of strong σ bonds and weaker π bonds and that, because of their relatively weak π bonds, alkenes and alkynes undergo electrophilic addition reactions.

Alkanes have only strong σ bonds. In addition, the electrons in the C—C and C—H σ bonds are shared equally by the bonding atoms, so none of the atoms in an alkane has any significant charge. This means that alkanes are neither nucleophiles nor electrophiles, so neither electrophiles nor nucleophiles are attracted to them. Alkanes, therefore, are relatively unreactive compounds. The failure of alkanes to undergo reactions prompted early organic chemists to call them *paraffins,* from the Latin *parum affinis,* which means "little affinity" (for other compounds).

Natural Gas and Petroleum

Natural gas is approximately 75% methane. The remaining 25% is composed of other small alkanes such as ethane, propane, and butane. In the 1950s, natural gas replaced coal as the main energy source for domestic and industrial heating in many parts of the United States.

Petroleum is a complex mixture of alkanes and cycloalkanes that can be separated into fractions by distillation. Natural gas is the fraction that boils off at the lowest temperature (hydrocarbons containing fewer than 5 carbons). The fraction that boils at somewhat higher temperatures (hydrocarbons containing 5 to 11 carbons) is gasoline; the next fraction (9 to 16 carbons) includes kerosene and jet fuel. The fraction with 15 to 25 carbons is used for heating oil and diesel oil, and the highest-boiling fraction is used for lubricants and greases. After distillation, a nonvolatile residue called asphalt or tar is left behind.

The 5- to 11-carbon fraction that is used for gasoline is actually a poor fuel for internal combustion engines. To become a high-performance gasoline, it requires a process known as catalytic cracking. Catalytic cracking converts straight-chain hydrocarbons that are poor fuels into branched-chain compounds that are high-performance fuels. Originally, cracking (also called pyrolysis) required heating the gasoline to very high temperatures in order to obtain hydrocarbons with three to five carbons. Modern cracking methods use catalysts to accomplish the same thing at much lower temperatures.

Fossil Fuels: A Problematic Energy Source

Modern society faces three major problems as a consequence of our dependence on fossil fuels for energy. First, these fuels are a nonrenewable resource and the world's supply is continually decreasing. Second, a group of Middle Eastern and South American countries controls a large portion of the world's supply of petroleum. These countries have formed a cartel called the *Organization of Petroleum Exporting Countries (OPEC)* that controls both the supply and the price of crude oil. Political instability in any OPEC country can seriously affect the world's oil supply.

Third, burning fossil fuels—particularly coal—increases the concentration of CO_2 in the atmosphere; burning coal also increases the concentration of atmospheric SO_2. Scientists have established experimentally that SO_2 causes "acid rain," a threat to plants and, therefore, to our food and oxygen supplies.

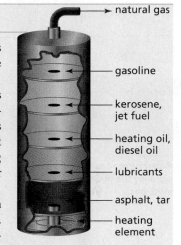

The concentration of atmospheric CO_2 at Mauna Loa, Hawaii, has been periodically measured since 1968. The concentration has increased 25% since the first measurements were taken, causing scientists to predict an increase in the Earth's temperature as a result of the absorption of infrared radiation by CO_2 (the *greenhouse effect*). A steady increase in the temperature of the Earth would have devastating consequences, including the formation of new deserts, massive crop failure, and the melting of glaciers with a concomitant rise in sea level. Clearly, what we need is a renewable, nonpolitical, nonpolluting, and economically affordable source of energy.

2 THE CHLORINATION AND BROMINATION OF ALKANES

Alkanes react with chlorine (Cl_2) or bromine (Br_2) to form alkyl chlorides or alkyl bromides. These **halogenation reactions** take place only at high temperatures or in the presence of light. (Irradiation with light is symbolized by $h\nu$.)

$$CH_4 + Cl_2 \xrightarrow[\substack{or \\ h\nu}]{\Delta} \underset{\text{chloromethane}}{CH_3Cl} + HCl$$

$$CH_3CH_3 + Br_2 \xrightarrow[\substack{or \\ h\nu}]{\Delta} \underset{\text{bromoethane}}{CH_3CH_2Br} + HBr$$

Halogenation and combustion (burning) are the only reactions that alkanes undergo (without the assistance of a metal catalyst). In a **combustion** reaction, alkanes react with oxygen at high temperatures to form carbon dioxide and water.

When a bond breaks so that both of its electrons stay with one of the atoms, the process is called **heterolytic bond cleavage** or **heterolysis.**

> *An arrowhead with two barbs signifies the movement of two electrons.*

| heterolytic bond cleavage | | arrowhead has 2 barbs |

$$A\!-\!B \longrightarrow A^+ + :B^-$$

When a bond breaks so that each of the atoms retains one of the bonding electrons, the process is called **homolytic bond cleavage** or **homolysis.** Homolysis results in the formation of radicals. A **radical** (often called a **free radical**) is a species containing an atom with an unpaired electron. A radical is highly reactive because acquiring an electron will complete its octet.

> *An arrowhead with one barb— sometimes called a fishhook— signifies the movement of one electron.*

| homolytic bond cleavage | | arrowhead has one barb |

$$A\!-\!B \longrightarrow A\cdot + \cdot B$$

radicals

The mechanism for the halogenation of an alkane is well understood. As an example, let's look at the mechanism for the monochlorination of methane. The monochlorination of alkanes other than methane has the same mechanism.

MECHANISM FOR THE MONOCHLORINATION OF METHANE

$$:\!\ddot{C}l\!-\!\ddot{C}l\!: \xrightarrow[\substack{or \\ h\nu}]{\Delta} 2\,:\!\ddot{C}l\cdot \quad \boxed{\text{initiation step}}$$

homolytic cleavage

$$:\!\ddot{C}l\cdot + H\!-\!CH_3 \longrightarrow H\ddot{C}l\!: + \underset{\text{a methyl radical}}{\cdot CH_3}$$

$$\cdot CH_3 + :\!\ddot{C}l\!-\!\ddot{C}l\!: \longrightarrow CH_3Cl + :\!\ddot{C}l\cdot$$

$\Big\}$ $\boxed{\text{propagation steps}}$

$$:\!\ddot{C}l\cdot + :\!\ddot{C}l\cdot \longrightarrow Cl_2$$

$$\cdot CH_3 + \cdot CH_3 \longrightarrow CH_3CH_3$$

$\Big\}$ $\boxed{\text{termination steps}}$

$$:\!\ddot{C}l\cdot + \cdot CH_3 \longrightarrow CH_3Cl$$

> *See the tutorial for this chapter for additional information on drawing curved arrows in radical systems.*

■ Heat or light supplies the energy required to break the Cl—Cl bond homolytically. This is the **initiation step** of the reaction because it creates radicals from a molecule in which all the electrons are paired.

■ The chlorine radical formed in the initiation step removes a hydrogen atom from the alkane (in this case methane), forming HCl and a methyl radical.

■ The methyl radical removes a chlorine atom from Cl_2, forming chloromethane and another chlorine radical, which can then remove a hydrogen atom from another molecule of methane.

Steps 2 and 3 are **propagation steps** because the radical created in the first propagation step reacts in the second propagation step to produce the radical that participates in the first propagation step. A propagation step is one that propagates the chain. Thus, the two propagation steps are repeated over and over. The first propagation step is the rate-determining step of the overall reaction.

■ Any two radicals in the reaction mixture can combine to form a molecule in which all the electrons are paired. The combination of two radicals is called a **termination step** because it helps bring the reaction to an end by decreasing the number of radicals available to propagate the reaction. Any two radicals can combine, so a radical reaction produces a mixture of products.

> Radical chain reactions have initiation, propagation, and termination steps.

Because the reaction has radical intermediates and repeating propagation steps, it is called a **radical chain reaction.** This particular radical chain reaction is called a **radical substitution reaction** because it substitutes a chlorine for one of the hydrogens of the alkane.

In order to maximize the amount of monohalogenated product formed, a radical substitution reaction should be carried out in the presence of excess alkane. Excess alkane in the reaction mixture increases the probability that the halogen radical will collide with a molecule of alkane rather than with a molecule of alkyl halide—even toward the end of the reaction, by which time a considerable amount of alkyl halide will have been formed.

If the halogen radical removes a hydrogen from a molecule of alkyl halide rather than from a molecule of alkane, a dihalogenated product will be obtained.

$$Cl\cdot + CH_3Cl \longrightarrow \cdot CH_2Cl + HCl$$
$$\cdot CH_2Cl + Cl_2 \longrightarrow CH_2Cl_2 + Cl\cdot$$

(alkyl halide → CH_3Cl)

(a dihalogenated compound → CH_2Cl_2)

The bromination of alkanes has the same mechanism as the chlorination of alkanes.

MECHANISM FOR THE MONOBROMINATION OF ETHANE

$$Br{-}Br \xrightarrow[\text{or}]{\Delta \atop h\nu} 2\,Br\cdot \qquad \text{initiation step}$$

$$\dot{Br} + H{-}CH_2CH_3 \longrightarrow HBr + CH_3\dot{C}H_2$$
an ethyl radical

$$CH_3\dot{C}H_2 + Br{-}Br \longrightarrow CH_3CH_2Br + Br\cdot$$

propagation steps

$$Br\cdot + Br\cdot \longrightarrow Br_2$$

$$CH_3\dot{C}H_2 + CH_3\dot{C}H_2 \longrightarrow CH_3CH_2CH_2CH_3$$

$$CH_3\dot{C}H_2 + Br\cdot \longrightarrow CH_3CH_2Br$$

termination steps

Why Radicals No Longer Have to Be Called Free Radicals

At one time an "R" group was called a radical. For example, the OH substituent in CH_3CH_2OH was said to be attached to an ethyl radical. To distinguish this kind of ethyl radical from $CH_3\dot{C}H_2$, which has an unpaired electron and is not attached to a substituent, $CH_3\dot{C}H_2$ was called a "free radical"—it was free from attachment to a substituent. Now that we call "R" a *substituent* or a *group,* instead of a *radical,* we no longer need to call a compound with an unpaired electron a "free radical"; the word *radical* is now unambiguous.

PROBLEM 1

Show the initiation, propagation, and termination steps for the monochlorination of cyclohexane.

PROBLEM 2

Write the mechanism for the formation of tetrachloromethane (CCl_4) from the reaction of methane with $Cl_2 + h\nu$.

3 | RADICAL STABILITY DEPENDS ON THE NUMBER OF ALKYL GROUPS ATTACHED TO THE CARBON WITH THE UNPAIRED ELECTRON

Radicals are classified according to the carbon that bears the unpaired electron. **Primary radicals** have the unshared electron on a primary carbon, **secondary radicals** have the unshared electron on a secondary carbon, and **tertiary radicals** have the unshared electron on a tertiary carbon.

Alkyl groups stabilize radicals the same way they stabilize carbocations—that is, by hyperconjugation. Therefore, the relative stabilities of **primary, secondary,** and **tertiary alkyl radicals** follow the same order as the relative stabilities of primary, secondary, and tertiary carbocations.

relative stabilities of alkyl radicals

Stability of alkyl radicals: 3° > 2° > 1°

tertiary radical   secondary radical   primary radical   methyl radical

The differences in the relative stabilities of the radicals are much smaller than the differences in the relative stabilities of the carbocations because alkyl groups do not stabilize radicals as well as they stabilize carbocations.

The MO diagrams in Figure 1 explain why alkyl groups stabilize carbocations better than they stabilize radicals. Stabilization of a carbocation results from overlap between a filled orbital of a C—H or C—C σ bond and an empty p orbital—a two-electron system (Figure 1a). In contrast, stabilization of a radical results from overlap between a filled orbital of a C—H or C—C σ bond and a p orbital that contains one electron—a three-electron system (Figure 1b).

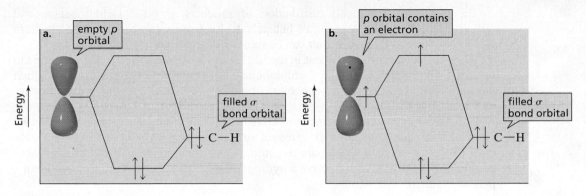

▲ **Figure 1**

MO diagrams showing the stabilization achieved when the electrons of an orbital of a C—H or C—C σ bond overlap with

(a) an empty p orbital

(b) a p orbital that contains one electron

Both electrons are in a bonding MO in the two-electron system, whereas one of the electrons has to go into an antibonding MO in the three-electron system. Overall, the three-electron system is stabilizing because there are more electrons in the bonding MO than in the antibonding MO, but it is not as stabilizing as the two-electron system, which does not have an electron in the antibonding MO. Consequently, an alkyl group stabilizes a carbocation about 5–10 times better than it stabilizes a radical.

PROBLEM 3♦

a. Which of the hydrogens in the following structure is the easiest for a chlorine radical to remove?

b. How many secondary hydrogens does the structure have?

---

4   THE DISTRIBUTION OF PRODUCTS DEPENDS ON PROBABILITY AND REACTIVITY

Two different alkyl halides are obtained from the monochlorination of butane. Substitution of a hydrogen bonded to one of the primary carbons produces 1-chlorobutane, whereas substitution of a hydrogen bonded to one of the secondary carbons forms 2-chlorobutane.

$$CH_3CH_2CH_2CH_3 \;+\; Cl_2 \;\xrightarrow{h\nu}\; CH_3CH_2CH_2CH_2\boxed{Cl} \;+\; CH_3CH_2\overset{\boxed{Cl}}{C}HCH_3 \;+\; HCl$$

butane	1-chlorobutane	2-chlorobutane
	expected = 60%	expected = 40%
	experimental = 29%	experimental = 71%

The expected (statistical) distribution of products is 60% 1-chlorobutane and 40% 2-chlorobutane because six of butane's 10 hydrogens can be substituted to form 1-chlorobutane, whereas only four can be substituted to form 2-chlorobutane.

When the reaction is carried out in the laboratory, however, the products are found to be 29% 1-chlorobutane and 71% 2-chlorobutane. In other words, the product distribution does not depend solely on the *probability* of a chlorine radical colliding with a primary or a secondary hydrogen.

Because more 2-chlorobutane is obtained than expected and *the rate-determining step of the overall reaction is removal of the hydrogen atom,* we can conclude that it is easier to remove a hydrogen atom from a secondary carbon to form a secondary radical than it is to remove a hydrogen atom from a primary carbon to form a primary radical.

We should not be surprised that it is easier to form a secondary radical, because a secondary radical is more stable than a primary radical. The more stable the radical, the more easily it is formed, because the stability of the radical is reflected in the stability of the transition state leading to its formation.

After experimentally determining the amount of each chlorination product obtained from various hydrocarbons, chemists were able to conclude that *at room temperature* it is 5.0 times easier for a chlorine radical to form a tertiary radical than a primary radical, and it is 3.8 times easier to form a secondary radical than a primary radical. The precise ratios differ at different temperatures. (How relative rates are determined experimentally is described in Problem 40.)

relative rates of alkyl radical formation by a chlorine radical at room temperature

tertiary > secondary > primary
5.0   3.8   1.0

increasing rate of formation

Now we see that both *probability* (the number of hydrogens that can be removed that lead to the formation of a particular product) and *reactivity* (the relative rate at which a particular radical is formed) must be taken into account when determining the relative amounts of the products obtained from the radical chlorination of an alkane.

Both probability and reactivity must be taken into account when calculating the relative amounts of products.

relative amount of 1-chlorobutane

number of hydrogens × relative reactivity
$6 \times 1.0 = 6.0$

relative amount of 2-chlorobutane

number of hydrogens × relative reactivity
$4 \times 3.8 = 15$

The anticipated percent yield of each alkyl chloride (as a percentage of all the monochlorinated products) is calculated by dividing the relative amount of the particular product by the sum of the relative amounts of all the alkyl chloride products ($6 + 15 = 21$).

percent yield of 1-chlorobutane   **percent yield of 2-chlorobutane**

$$\frac{6.0}{21} = 29\% \qquad\qquad\qquad \frac{15}{21} = 71\%$$

Because the radical chlorination of an alkane can yield several different monosubstitution products as well as products that contain more than one chlorine atom, it is not the best method to use to synthesize an alkyl halide. The addition of a hydrogen halide to an alkene or the conversion of an alcohol to an alkyl halide are both much better ways to make an alkyl halide.

Radical halogenation of an alkane is still a useful reaction because it is the only way to convert an inert alkane into a reactive compound. And once the halogen is introduced into the alkane, it can be replaced by a variety of other substituents.

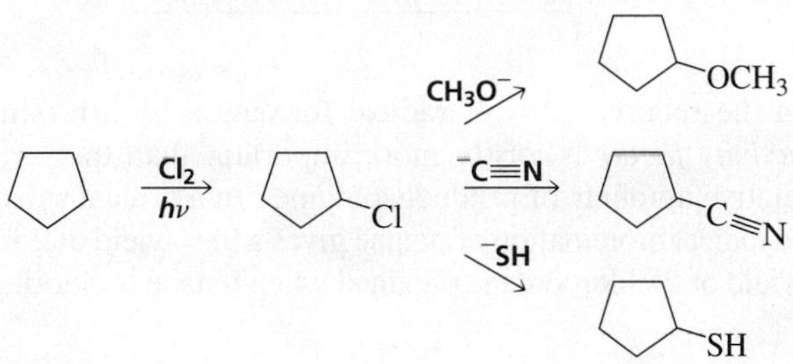

PROBLEM 4♦

How many alkyl chlorides can be obtained from monochlorination of the following alkanes? Disregard stereoisomers.

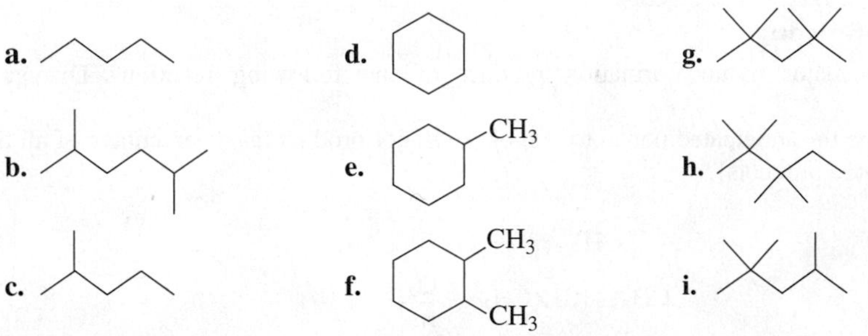

a. **d.** **g.**

b. **e.** **h.**

c. **f.** **i.**

PROBLEM 5 Solved

If cyclopentane reacts with more than one equivalent of Cl₂ at a high temperature, how many dichlorocyclopentanes would you expect to obtain as products?

Solution Seven dichlorocyclopentanes could be obtained as products. Only one isomer is possible for the 1,1-dichloro compound. The 1,2- and 1,3-dichloro compounds have two asymmetric centers. Each has three stereoisomers because the cis isomer is a meso compound and the trans isomer is a pair of enantiomers.

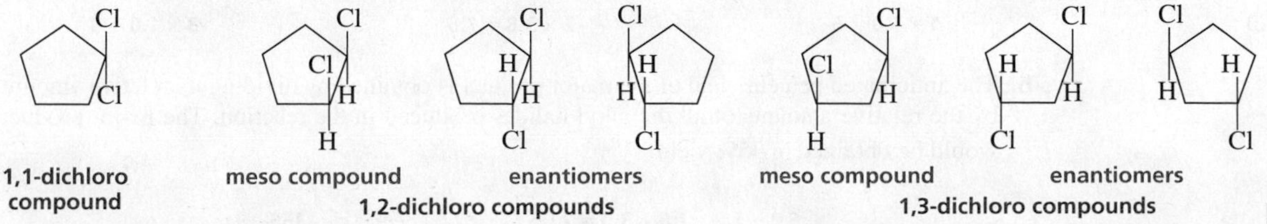

1,1-dichloro compound	meso compound	enantiomers	meso compound	enantiomers
	1,2-dichloro compounds		1,3-dichloro compounds	

5   THE REACTIVITY–SELECTIVITY PRINCIPLE

The relative rates of radical formation by a bromine radical are different from the relative rates of radical formation by a chlorine radical. For example, at 125 °C, a bromine radical forms a tertiary radical 1600 times faster than a primary radical and it forms a secondary radical 82 times faster than a primary radical.

relative rates of alkyl radical formation by a bromine radical at 125 °C

tertiary > secondary > primary
1600   82   1

increasing rate of formation

The differences in the relative rates of radical formation by a bromine radical are so great that the *reactivity factor* is vastly more important than the *probability factor* in determining the relative amounts of products obtained in a radical substitution reaction.

For example, the radical bromination of butane gives a 98% yield of 2-bromobutane, compared with a 71% yield of 2-chlorobutane obtained when butane is chlorinated (Section 4).

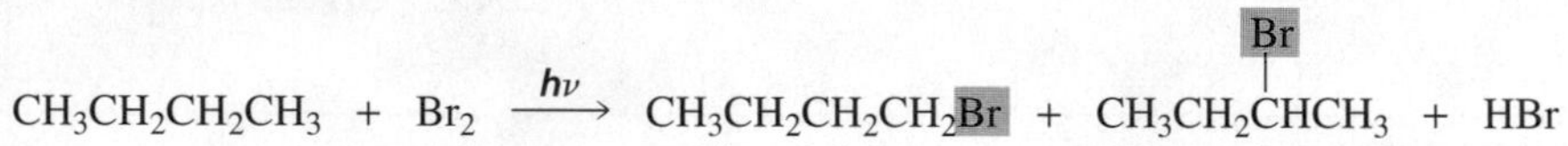

$$CH_3CH_2CH_2CH_3 + Br_2 \xrightarrow{h\nu} CH_3CH_2CH_2CH_2Br + CH_3CH_2\overset{Br}{C}HCH_3 + HBr$$

1-bromobutane 2%   2-bromobutane 98%

PROBLEM 6 Solved

a. What is the major monochlorination product of the following reaction? Disregard stereoisomers.

b. What would be the anticipated percent yield of the major product (as a percentage of all the monochlorinated products)?

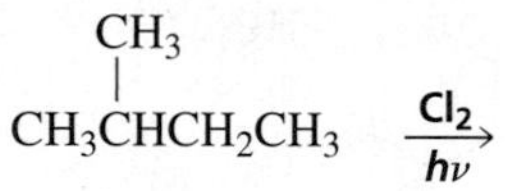

$$CH_3\overset{CH_3}{C}HCH_2CH_3 \xrightarrow[h\nu]{Cl_2}$$

Solution

a. The relative amount of each product is determined by multiplying the number of hydrogens that can be removed that will lead to the formation of that product by the relative rate for formation of the radical. Thus, the major product is 2-chloro-3-methylbutane.

CH₃	CH₃	CH₃	CH₃
CH₂CHCH₂CH₃ Cl	CH₃CCH₂CH₃ Cl	CH₃CHCHCH₃ Cl	CH₃CHCH₂CH₂ Cl
1-chloro-2-methylbutane	2-chloro-2-methylbutane	2-chloro-3-methylbutane	1-chloro-3-methylbutane
6 × 1.0 = 6	1 × 5.0 = 5	2 × 3.8 = 7.6	3 × 1.0 = 3

b. The anticipated percent yield of the major product is obtained by dividing its relative amount by the relative amounts of all the alkyl halides produced in the reaction. The major product would be obtained in 35% yield.

$$6.0 + 5.0 + 7.6 + 3.0 = 21.6 \qquad \frac{7.6}{21.6} = 35\%$$

PROBLEM 7♦

a. What would be the major product of the reaction in Problem 6 if the alkane had reacted with Br_2 instead of with Cl_2? Disregard stereoisomers.

b. What would be the anticipated percent yield of the major product?

Why are the relative rates of radical formation so different when a bromine radical rather than a chlorine radical is used as the hydrogen-removing reagent?

To answer this question, we must compare the $\Delta H°$ values for forming primary, secondary, and tertiary radicals by a chlorine radical and by a bromine radical. If you have the chapter "Alkenes" in your text, use the information in Table 1 to calculate these values.

	$\Delta H°$ (kcal/mol)	$\Delta H°$ (kJ/mol)
Cl· + ⌃⌃ ⟶ ⌃⌃· + HCl	$101 - 103 = -2$	-8
Cl· + ⌃⌃ ⟶ ⌃⌃· + HCl	$99 - 103 = -4$	-17
Cl· + ⋋ ⟶ ⋋· + HCl	$97 - 103 = -6$	-25

$\Delta H°$ is equal to the energy of the bond being broken minus the energy of the bond being formed.

	$\Delta H°$ (kcal/mol)	$\Delta H°$ (kJ/mol)
Br· + ⌃⌃ ⟶ ⌃⌃· + HBr	$101 - 87 = 14$	59
Br· + ⌃⌃ ⟶ ⌃⌃· + HBr	$99 - 87 = 12$	50
Br· + ⋋ ⟶ ⋋· + HBr	$97 - 87 = 10$	42

We must also be aware that bromination is a much slower reaction than chlorination. The activation energy for removing a hydrogen atom by a bromine radical is about 4.5 times greater than that for removing a hydrogen atom by a chlorine radical.

Using the calculated $\Delta H°$ values and the experimental activation energies, we can draw reaction coordinate diagrams for formation of primary, secondary, and tertiary radicals by a chlorine radical (Figure 2a) and by a bromine radical (Figure 2b).

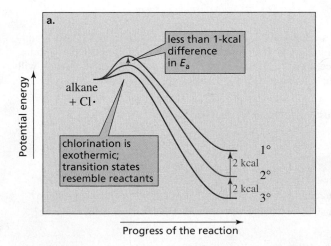

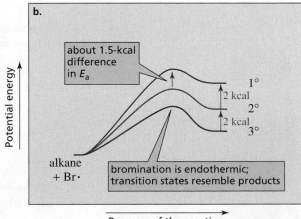

▲ **Figure 2**

(a) Reaction coordinate diagrams for formation of primary, secondary, and tertiary alkyl radicals by a chlorine radical. The transition states have relatively little radical character because they resemble the reactants.

(b) Reaction coordinate diagrams for formation of primary, secondary, and tertiary alkyl radicals by a bromine radical. The transition states have a relatively high degree of radical character because they resemble the products.

Because the reaction of a chlorine radical with an alkane to form a primary, secondary, or tertiary radical is exothermic, the transition states resemble the reactants (see the Hammond postulate, which says that the transition state is more similar in structure to the species to which it is more similar in energy). The reactants all have approximately the same energy, so there is only a small difference in the activation energies for the formation of a primary, secondary, or tertiary radical.

In contrast, the reaction of a bromine radical with an alkane is endothermic, so the transition states resemble the products. Because there are significant differences between the energies of the product radicals—depending on whether they are primary, secondary, or tertiary—there are significant differences between the activation energies.

Therefore, a chlorine radical makes primary, secondary, and tertiary radicals with almost equal ease (Figure 2a), whereas a bromine radical has a clear preference for forming the easiest-to-form tertiary radical (Figure 2b). In other words, because a bromine radical is relatively unreactive, it is highly selective about which radical it forms. In contrast, the much more reactive chlorine radical is considerably less selective.

These observations illustrate the **reactivity–selectivity principle,** which states that *the greater the reactivity of a species, the less selective it will be.*

Because chlorination is relatively nonselective, it is a useful reaction only when there is only one kind of hydrogen in the alkane.

> **A bromine radical is less reactive and more selective than a chlorine radical.**
>
> **The more reactive a species is, the less selective it will be.**

$$\text{cyclohexane} + Cl_2 \xrightarrow{h\nu} \text{chlorocyclohexane} + HCl$$

PROBLEM-SOLVING STRATEGY

Planning the Synthesis of an Alkyl Halide

Would chlorination or bromination of methylcyclohexane produce a greater yield of 1-halo-1-methylcyclohexane?

To solve this kind of problem, first draw the structures of the compounds being discussed.

$$\text{(methylcyclohexane, CH}_3) \xrightarrow[h\nu]{X_2} \text{(1-halo-1-methylcyclohexane, CH}_3, X)$$

1-Halo-1-methylcyclohexane is a tertiary alkyl halide, so the question becomes, "Will bromination or chlorination produce a greater yield of a tertiary alkyl halide?" Because bromination is more selective, it will produce a greater yield of the desired compound. Chlorination will form some of the tertiary alkyl halide, but it will also form significant amounts of primary and secondary alkyl halides.

Now use the strategy you have just learned to solve Problem 8.

PROBLEM 8♦

a. Would chlorination or bromination produce a greater yield of 1-halo-2,3-dimethylbutane?
b. Would chlorination or bromination produce a greater yield of 2-halo-2,3-dimethylbutane?
c. Would chlorination or bromination be a better way to make 1-halo-2,2-dimethylpropane?

By comparing the $\Delta H°$ values for the sum of the two propagating steps in the monohalogenation of methane, we can understand why alkanes undergo chlorination and bromination but not iodination, and why fluorination is too violent a reaction to be useful.

F₂

Cl₂

Br₂

I₂
Halogens

$$F\cdot \ + \ CH_4 \ \longrightarrow \ \cdot CH_3 \ + \ HF \qquad 105 - 136 = -31$$
$$\cdot CH_3 \ + \ F_2 \ \longrightarrow \ CH_3F \ + \ F\cdot \qquad \underline{38 - 115 = -77}$$
$$\Delta H^\circ = -108 \ \text{kcal/mol} \quad (\text{or} -452 \ \text{kJ/mol})$$

$$Cl\cdot \ + \ CH_4 \ \longrightarrow \ \cdot CH_3 \ + \ HCl \qquad 105 - 103 = \ \ 2$$
$$\cdot CH_3 \ + \ Cl_2 \ \longrightarrow \ CH_3Cl \ + \ Cl\cdot \qquad \underline{58 - \ \ 84 = -26}$$
$$\Delta H^\circ = -24 \ \text{kcal/mol} \quad (\text{or} -100 \ \text{kJ/mol})$$

$$Br\cdot \ + \ CH_4 \ \longrightarrow \ \cdot CH_3 \ + \ HBr \qquad 105 - \ \ 87 = \ \ 18$$
$$\cdot CH_3 \ + \ Br_2 \ \longrightarrow \ CH_3Br \ + \ Br\cdot \qquad \underline{46 - \ \ 72 = -26}$$
$$\Delta H^\circ = -8 \ \text{kcal/mol} \quad (\text{or} -23 \ \text{kJ/mol})$$

$$I\cdot \ + \ CH_4 \ \longrightarrow \ \cdot CH_3 \ + \ HI \qquad 105 - \ \ 71 = \ \ 34$$
$$\cdot CH_3 \ + \ I_2 \ \longrightarrow \ CH_3I \ + \ I\cdot \qquad \underline{36 - \ \ 58 = -22}$$
$$\Delta H^\circ = 12 \ \text{kcal/mol} \quad (\text{or} \ 50 \ \text{kJ/mol})$$

The fluorine radical is the most reactive of the halogen radicals—it reacts readily with alkanes ($\Delta H^\circ = -31$ kcal/mol), and the radical it produces reacts violently with F_2 ($\Delta H^\circ = -77$ kcal/mol). In contrast, the iodine radical is the least reactive of the halogen radicals. In fact, it is so unreactive ($\Delta H^\circ = 34$ kcal/mol) that it is unable to remove a hydrogen atom from an alkane. Consequently, it reacts with another iodine radical and reforms I_2.

PROBLEM 9 Solved

How could butanone be prepared from butane?

butane   butanone

Solution We know that the first reaction has to be a radical halogenation because that is the only reaction that an alkane undergoes. Bromination will lead to a greater yield of the desired 2-halo-substituted compound than will chlorination because a bromine radical is more selective than a chlorine radical. A nucleophilic substitution reaction forms the alcohol, which forms the target molecule when it is oxidized.

PROBLEM 10

Show how the following compounds could be prepared from 2-methylpropane:

a. 2-bromo-2-methylpropane   **b.** 2-methyl-2-propanol   **c.** 2-methyl-1-propene

---

6   FORMATION OF EXPLOSIVE PEROXIDES

Ethers are a laboratory hazard because they form explosive peroxides by reacting with O_2 when they are exposed to air. We will see that this reaction is similar to the reaction that causes fats to become rancid (Section 11).

MECHANISM FOR PEROXIDE FORMATION

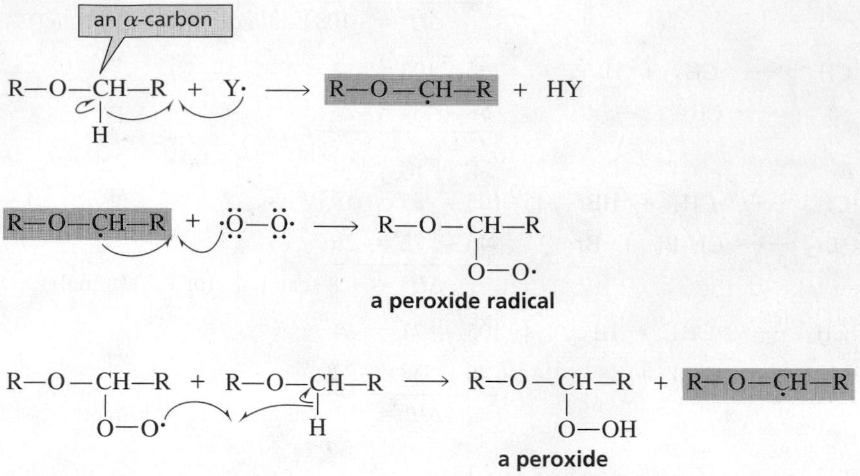

an α-carbon

$$R-O-CH-R \ + \ Y\cdot \ \longrightarrow \ R-O-\dot{C}H-R \ + \ HY$$
$$\overset{|}{H}$$

$$R-O-\dot{C}H-R \ + \ \ddot{\underset{..}{O}}-\ddot{\underset{..}{O}}\cdot \ \longrightarrow \ R-O-CH-R$$
$$\underset{\textbf{a peroxide radical}}{\overset{|}{O-O}\cdot}$$

$$R-O-CH-R \ + \ R-O-\underset{|}{\overset{|}{C}}H-R \ \longrightarrow \ R-O-CH-R \ + \ R-O-\dot{C}H-R$$
$$\underset{O-O\cdot}{|} \qquad \overset{|}{H} \qquad \underset{\textbf{a peroxide}}{\overset{|}{O-OH}}$$

- A chain-initiating radical removes a hydrogen atom from an α-carbon of the ether. (The α-carbon is the carbon attached to the oxygen.) This is an initiation step because it creates the radical that is used in the first propagation step.

- The radical formed in the initiation step reacts with oxygen in a propagation step, forming a peroxide radical.

- In the second propagation step, the peroxide radical removes a hydrogen atom from an α-carbon of another molecule of ether to form a peroxide and regenerate the radical used in the first propagation step.

A **peroxide** is a compound with an O—O bond. Because an O—O bond is easily cleaved homolytically, a peroxide forms radicals that then can create new radicals—it is a **radical initiator.** Thus, the peroxide product of the preceding radical chain reaction can initiate another radical chain reaction—an explosive situation. To prevent the formation of explosive peroxides, ethers contain a stabilizer that traps the chain-initiating radical. Once an ether is purified (in which case it no longer contains the stabilizer), it has to be discarded within 24 hours.

PROBLEM 11♦

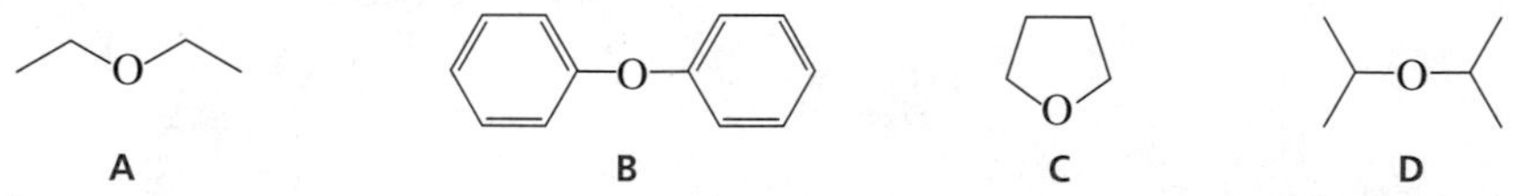

A   B   C   D

a. Which ether is most apt to form a peroxide?   **b.** Which ether is least apt to form a peroxide?

---

7   THE ADDITION OF RADICALS TO AN ALKENE

The addition of HBr to 1-butene forms 2-bromobutane, because the electrophile (H^+) adds to the sp^2 carbon bonded to the most hydrogens. If, however, you want to synthesize 1-bromobutane, then you need to find a way to make bromine an electrophile so it, instead of H^+, will add to the sp^2 carbon bonded to the most hydrogens.

$$CH_3CH_2CH{=}CH_2 \ + \ \boxed{HBr} \ \longrightarrow \ CH_3CH_2CH\overset{\boxed{Br}}{C}H_3$$

1-butene ⟶ **2-bromobutane**

$$CH_3CH_2CH{=}CH_2 \ + \ \boxed{HBr} \ \xrightarrow{\text{peroxide}} \ CH_3CH_2CH_2CH_2Br$$

1-butene ⟶ **1-bromobutane**

The electrophile adds to the sp^2 carbon that is bonded to the most hydrogens.

If a peroxide (ROOR) is added to the reaction mixture, the product of the addition reaction will be the desired 1-bromobutane. Thus, the peroxide changes the mechanism of the reaction in a way that causes a bromine radical to be the electrophile.

The following mechanism for the addition of HBr to an alkene in the presence of a peroxide shows that it is a radical chain reaction with characteristic initiation, propagation, and termination steps:

MECHANISM FOR THE ADDITION OF HBr TO AN ALKENE IN THE PRESENCE OF A PEROXIDE

$$RO{-}OR \ \xrightarrow[\Delta]{\text{light} \atop \text{or}} \ 2 \ RO\cdot$$

a peroxide ⟶ **alkoxy radicals**

$$R{-}\ddot{O}\cdot \ + \ H{-}\ddot{Br}{:} \ \longrightarrow \ R{-}\ddot{O}{-}H \ + \ \cdot\ddot{Br}{:}$$

a bromine radical

initiation steps

$$:\ddot{Br}\cdot \ + \ CH_2{=}CHCH_2CH_3 \ \longrightarrow \ CH_2\dot{C}HCH_2CH_3$$
$$\underset{:\ddot{Br}:}{|}$$

an alkyl radical

$$\underset{\overset{|}{Br}}{CH_2\dot{C}HCH_2CH_3} \ + \ H{-}\ddot{Br}{:} \ \longrightarrow \ \underset{\overset{|}{Br} \ \ \overset{|}{H}}{CH_2{-}CHCH_2CH_3} \ + \ \cdot\ddot{Br}{:}$$

propagation steps

$$:\ddot{Br}\cdot \ + \ \cdot\ddot{Br}{:} \ \longrightarrow \ :\ddot{Br}{-}\ddot{Br}{:}$$

$$BrCH_2\dot{C}HCH_2CH_3 \ + \ :\ddot{Br}\cdot \ \longrightarrow \ \underset{\overset{|}{Br}}{BrCH_2CHCH_2CH_3}$$

termination steps

$$2 \ BrCH_2\dot{C}HCH_2CH_3 \ \longrightarrow \ \underset{\overset{|}{BrCH_2} \ \ \overset{|}{CH_2Br}}{CH_3CH_2CH{-}CHCH_2CH_3}$$

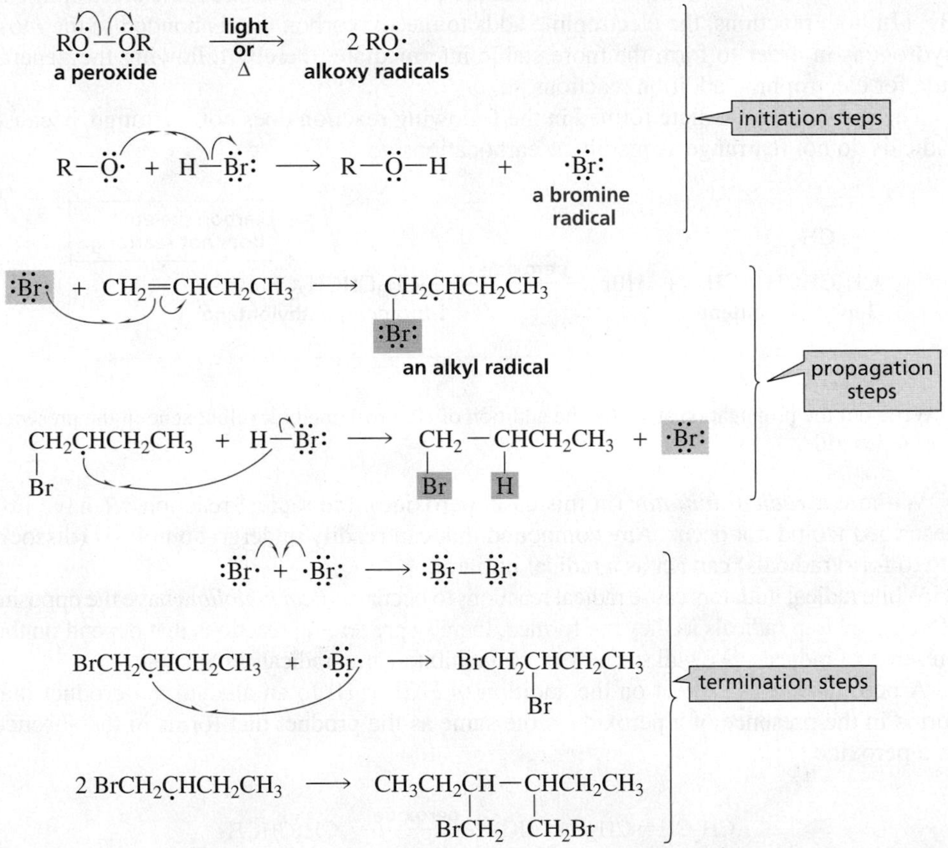

- The weak O—O bond of the peroxide readily breaks homolytically in the presence of light or heat to form alkoxy radicals. This is an initiation step because it creates radicals.

- The alkoxy radical completes its octet by removing a hydrogen atom from a molecule of HBr, thus forming a bromine radical. This too is an initiation step because it creates the radical that is used in the first propagation step.

- The bromine radical now seeks an electron to complete its octet. Because the π bond of an alkene is the weakest bond in the molecule, the bromine radical completes its octet by combining with one of the electrons of the π bond to form a C—Br bond. The second electron of the π bond is the unpaired electron in the resulting alkyl radical.

If the bromine radical adds to the sp^2 carbon of 1-butene that is bonded to the most hydrogens, a secondary alkyl radical is formed. If the bromine radical adds to the other sp^2 carbon, a primary alkyl radical is formed. *The bromine radical, therefore, adds to the* sp^2 *carbon that is bonded to the most hydrogens in order to form the more stable radical.*

- The alkyl radical removes a hydrogen atom from another molecule of HBr to produce a molecule of the alkyl halide product and another bromine radical.

  The preceding two steps are propagation steps. As is characteristic of a pair of propagation steps, a radical (Br·) reacts in the first propagation step to form a radical that reacts in the second propagation step to regenerate the radical (Br·) that is the reactant in the first propagation step.

- The last three steps are termination steps.

Because the first species that adds to the alkene is a radical (Br·), the addition of HBr in the presence of a peroxide is called a **radical addition reaction.**

When HBr reacts with an alkene in the absence of a peroxide, the electrophile—the first species to add to the alkene—is H^+. In the presence of a peroxide, the electrophile is Br·. In both reactions, the electrophile adds to the sp^2 carbon that is bonded to the most hydrogens in order to form the more stable intermediate, thereby following the general rule for electrophilic addition reactions.

The radical intermediate formed in the following reaction does not rearrange, because radicals do not rearrange as readily as carbocations do.

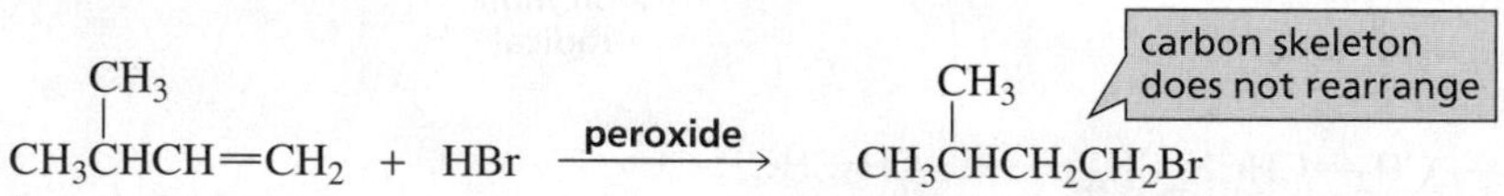

PROBLEM 12

Write out the propagation steps for the addition of HBr to 1-methylcyclohexene in the presence of a peroxide.

Without a *radical initiator* (in this case, peroxide), the radical reaction we have just described would not occur. Any compound that can readily undergo homolysis (dissociate to form radicals) can act as a radical initiator.

While radical initiators cause radical reactions to occur, *radical inhibitors* have the opposite effect; they trap radicals as they are formed, thereby preventing reactions that depend on the presence of radicals. We will see how radical inhibitors trap radicals in Section 11.

A peroxide has no effect on the addition of HCl or HI to an alkene; the product that forms in the presence of a peroxide is the same as the product that forms in the absence of a peroxide.

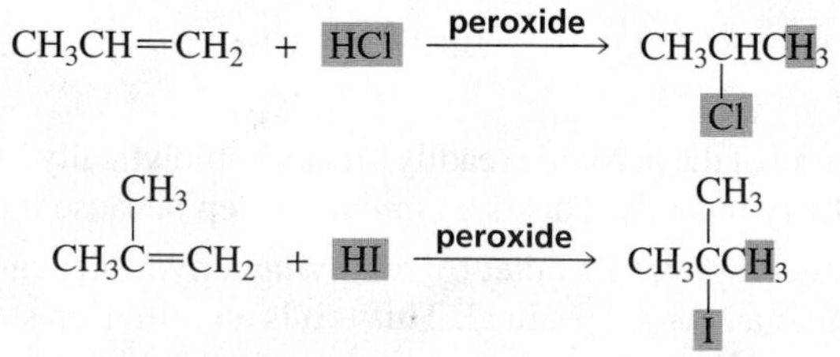

Why is the **peroxide effect** observed for the addition of HBr, but not for the addition of HCl or HI? This question can be answered by calculating the $\Delta H°$ for the two propagation steps in the radical chain reaction. If you have the chapter "Alkenes" in your text, use the information in Table 1 to calculate this.

$Cl\cdot + CH_2{=}CH_2 \longrightarrow ClCH_2\dot{C}H_2$ $\Delta H° = 63 - 85 = -22$ kcal/mol (or -91 kJ/mol)   ◁ exothermic

$ClCH_2\dot{C}H_2 + HCl \longrightarrow ClCH_2CH_3 + Cl\cdot$ $\Delta H° = 103 - 101 = +2$ kcal/mol (or $+8$ kJ/mol)   ◁ endothermic

$Br\cdot + CH_2{=}CH_2 \longrightarrow BrCH_2\dot{C}H_2$ $\Delta H° = 63 - 72 = -9$ kcal/mol (or -38 kJ/mol)

$BrCH_2\dot{C}H_2 + HBr \longrightarrow BrCH_2CH_3 + Br\cdot$ $\Delta H° = 87 - 101 = -14$ kcal/mol (or -59 kJ/mol)   exothermic

$I\cdot + CH_2{=}CH_2 \longrightarrow ICH_2\dot{C}H_2$ $\Delta H° = 63 - 57 = +6$ kcal/mol (or $+25$ kJ/mol)   ◁ endothermic

$ICH_2\dot{C}H_2 + HI \longrightarrow ICH_2CH_3 + I\cdot$ $\Delta H° = 71 - 101 = -30$ kcal/mol (or -126 kJ/mol)   ◁ exothermic

For the radical addition of HCl, the first propagation step is exothermic and the second is endothermic. For the radical addition of HI, the first propagation step is endothermic and the second is exothermic. Only for the radical addition of HBr are both propagation steps exothermic.

In a radical reaction, the steps that propagate the chain reaction compete with the steps that terminate it. Termination steps are always exothermic, because only bond making (and no bond breaking) occurs. Therefore, only when both propagation steps are exothermic can propagation compete successfully with termination. When HCl or HI adds to an alkene in the presence of a peroxide, any chain reaction that is initiated is then terminated rather than propagated because propagation cannot compete successfully with termination. Consequently, the radical chain reaction does not take place, and the only reaction that occurs is ionic addition (H^+ followed by Cl^- or I^-).

PROBLEM 13♦

What will be the major product of the reaction of 2-methyl-2-butene with each of the following reagents?

a. HBr   **b.** HCl   **c.** HBr + peroxide   **d.** HCl + peroxide

---

8   THE STEREOCHEMISTRY OF RADICAL SUBSTITUTION AND RADICAL ADDITION REACTIONS

When a reactant that does not have an asymmetric center undergoes a reaction that forms a product with one asymmetric center, the product will be a racemic mixture. Thus, the following *radical substitution reaction* forms a racemic mixture (that is, an equal amount of each enantiomer).

> When a reactant that does not have an asymmetric center undergoes a reaction that forms a product with one asymmetric center, the product will be a racemic mixture.

an asymmetric center

$$CH_3CH_2CH_2CH_3 + Br_2 \xrightarrow{h\nu} CH_3CH_2\underset{|}{\overset{}{C}}HCH_3 + HBr$$
$$\underset{Br}{}$$

configuration of the products

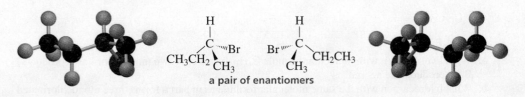

a pair of enantiomers

Similarly, the product of the following *radical addition reaction* is a racemic mixture:

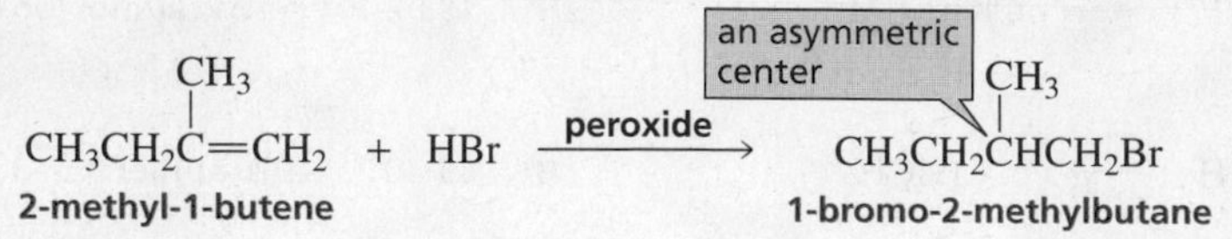

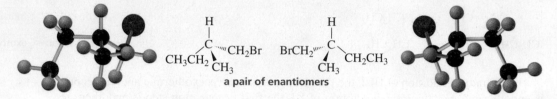

configuration of the products

a pair of enantiomers

Both the radical substitution and radical addition reactions form a racemic mixture because both reactions form a radical intermediate, and the reaction of the intermediate determines the configuration of the products. The radical intermediate in the substitution reaction is formed when the bromine radical removes a hydrogen atom from the reactant; the radical intermediate in the addition reaction is formed when the bromine radical adds to one of the sp^2 carbons of the double bond.

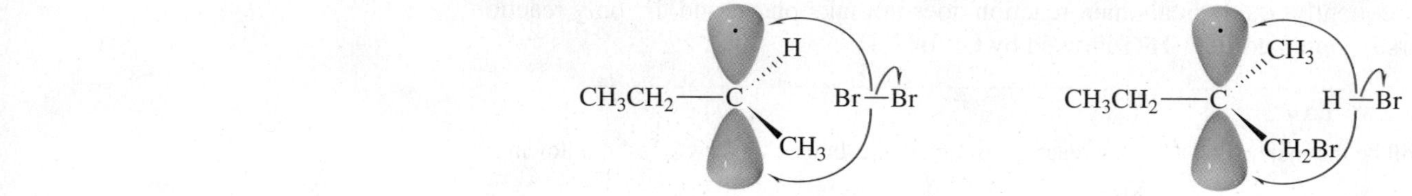

radical intermediate in the substitution reaction

radical intermediate in the addition reaction

The carbon that bears the unpaired electron in the radical intermediate is sp^2 hybridized, so the three atoms to which it is bonded lie in a plane. Therefore, the incoming atom has equal access to both sides of the plane. Consequently, identical amounts of the *R* and *S* enantiomers are formed in both the substitution and addition reactions.

Identical amounts of the *R* and *S* enantiomers are also obtained if a hydrogen bonded to an asymmetric center is substituted by a halogen. Breaking the bond to the asymmetric center destroys the configuration at the asymmetric center and forms a planar radical intermediate. The incoming halogen has equal access to both sides of the plane, so a racemic mixture is formed.

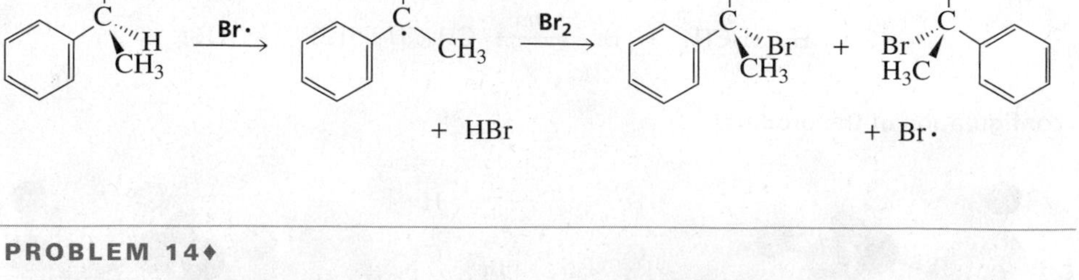

PROBLEM 14◆

a. What hydrocarbon with molecular formula C_4H_{10} forms only two monochlorinated products? Both products are achiral.

b. What hydrocarbon with the same molecular formula as in part **a** forms three monochlorinated products? One is achiral and two are chiral.

PROBLEM 15

Draw the stereoisomers of the major monobromination products obtained from the following reaction.

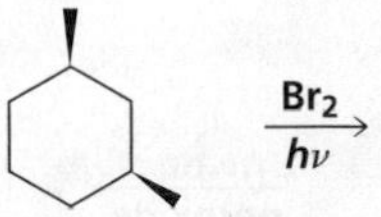

$$\xrightarrow[h\nu]{Br_2}$$

9   RADICAL SUBSTITUTION OF BENZYLIC AND ALLYLIC HYDROGENS

An **allylic radical** has an unpaired electron on an allylic carbon and, like an allylic cation, has two resonance contributors.

$$R\dot{C}H-CH=\!CH_2 \longleftrightarrow RCH=\!CH-\dot{C}H_2$$

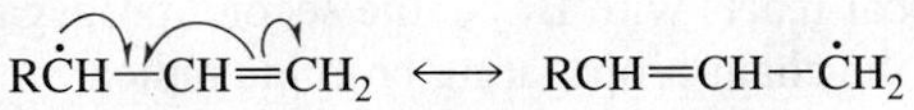

an allylic radical

A **benzylic radical** has an unpaired electron on a benzylic carbon and, like a benzylic cation, has five resonance contributors.

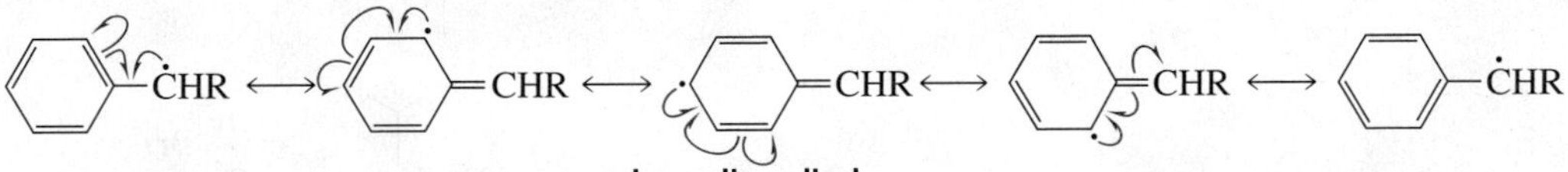

a benzylic radical

Because electron delocalization stabilizes a molecule, allyl and benzyl radicals are both more stable than other primary radicals. They are even more stable than tertiary radicals.

Electron delocalization increases the stability of a molecule.

relative stabilities of radicals

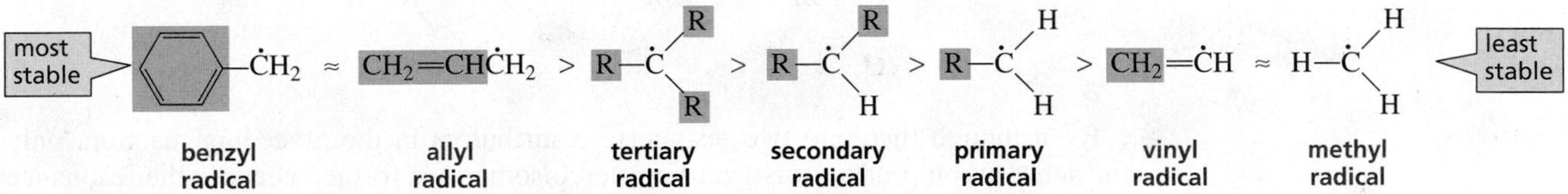

We know that the more stable the radical, the faster it can be formed. This means that a hydrogen bonded to either a benzylic carbon or an allylic carbon will be preferentially substituted in a halogenation reaction. Because bromination is more highly regioselective than chlorination, the percent of substitution at the benzylic or allylic carbon is greater for bromination.

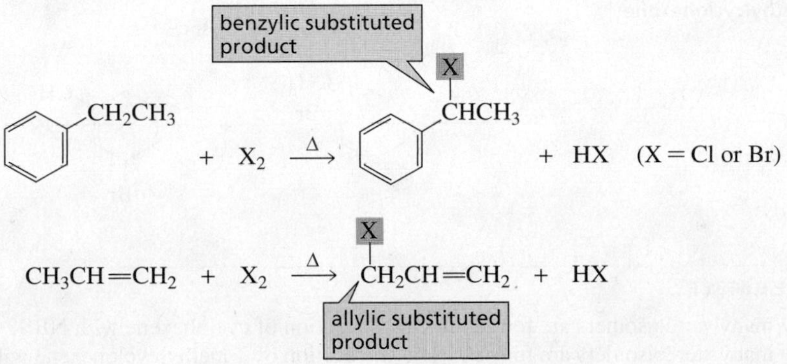

benzylic substituted product

$+ \ X_2 \xrightarrow{\Delta} \quad + \ HX \quad (X = Cl \ or \ Br)$

$$CH_3CH=\!CH_2 \ + \ X_2 \xrightarrow{\Delta} \overset{X}{\underset{}{|}}CH_2CH=\!CH_2 \ + \ HX$$

allylic substituted product

N-Bromosuccinimide (NBS) is frequently used to brominate allylic carbons because it allows a radical substitution reaction to be carried out in the presence of a low concentration of Br_2 and a low concentration of HBr. If a high concentration is present, addition of Br_2 or HBr to the double bond will compete with allylic substitution.

NBS is used to brominate allylic carbons.

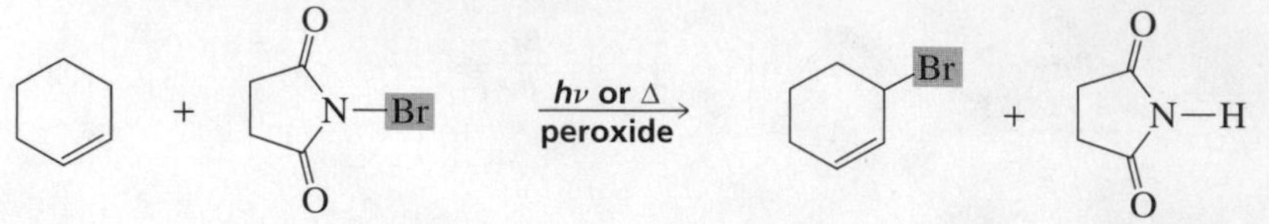

cyclohexene   *N*-bromosuccinimide   3-bromocyclohexene   succinimide
NBS

The bromination reaction begins with homolytic cleavage of the N—Br bond of NBS. This generates the bromine radical needed to initiate the radical reaction. Light or heat and a radical initiator such as a peroxide are used to promote homolytic cleavage.

In the first propagation step, the bromine radical removes an allylic hydrogen to form HBr and an allylic radical. Notice that the allylic radical is stabilized by electron delocalization. The allylic radical reacts with Br_2 in the second propagation step, thus forming the allylic bromide and the chain-propagating bromine radical.

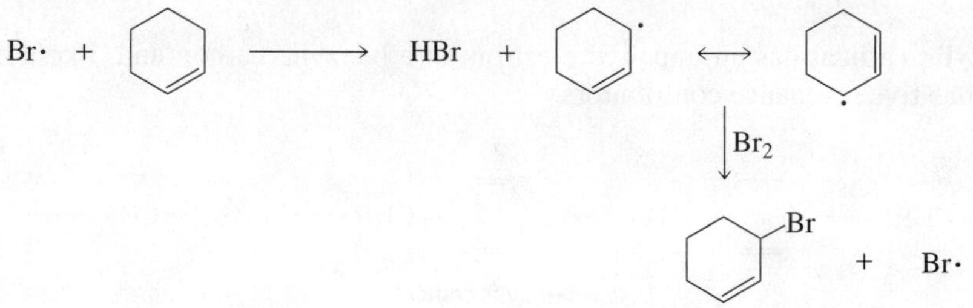

The Br_2 used in the second propagation step is produced in low concentration from a fast ionic reaction between NBS and the HBr that is produced in the first propagation step.

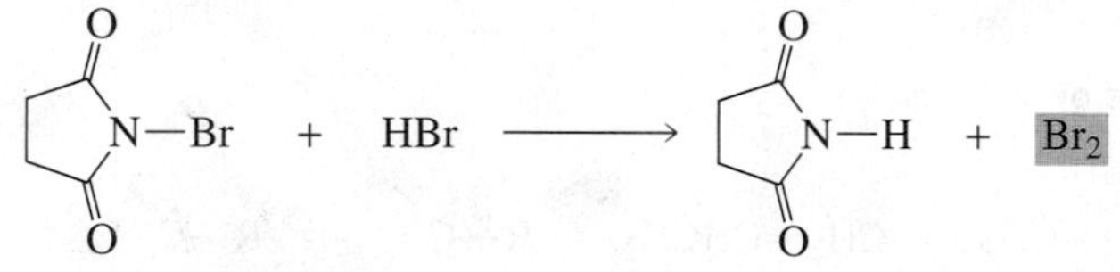

Even though there are two resonance contributors in the preceding reaction, only one substitution product (disregarding stereoisomers) is formed because the resonance contributors are mirror images. However, if the resonance contributors are not mirror images, then two substitution products (disregarding stereoisomers) are formed:

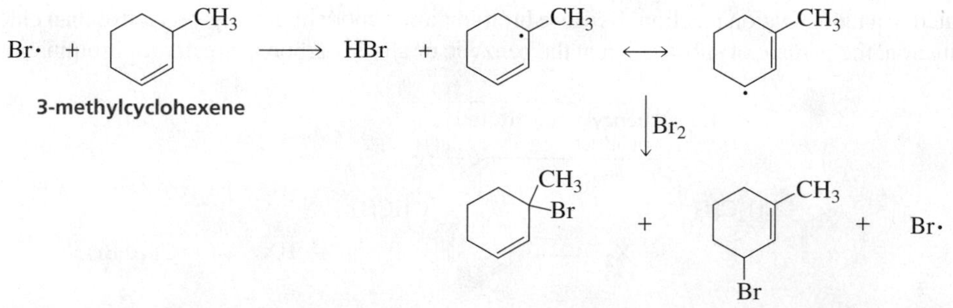

3-methylcyclohexene

PROBLEM 16

a. How many stereoisomers are formed from the reaction of cyclohexene with NBS?
b. How many stereoisomers are formed from the reaction of 3-methylcyclohexene with NBS?

PROBLEM 17

Two products are formed when methylenecyclohexane reacts with NBS. Show how each is formed.

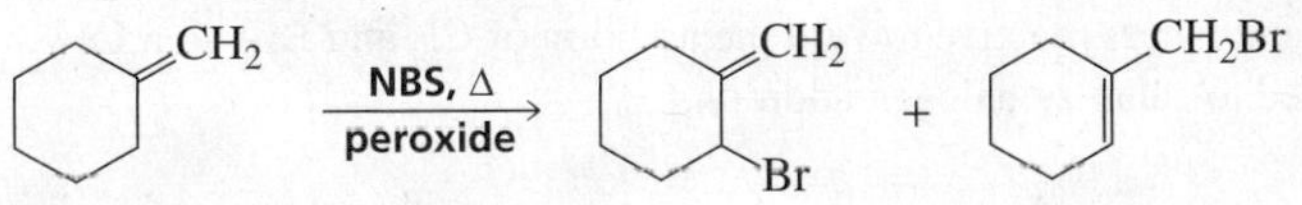

PROBLEM 18 Solved

How many allylic substituted bromoalkenes are formed from the reaction of 2-pentene with NBS? Disregard stereoisomers.

Solution Because of the high selectivity of the bromine radical, it will remove a secondary allylic hydrogen from C-4 of 2-pentene much more easily than it will remove a primary allylic hydrogen from C-1. The resonance contributors of the resulting radical intermediate are mirror images, so only one bromoalkene is formed (ignoring stereoisomers).

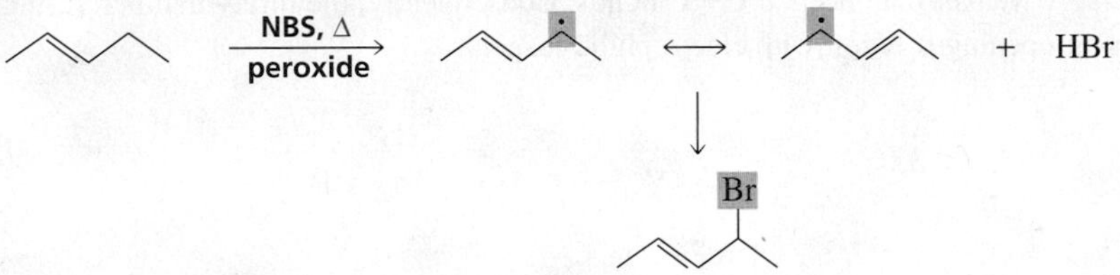

PROBLEM 19

How many allylic substituted bromoalkenes are formed from the reaction in Problem 18 if stereoisomers are included?

PROBLEM 20

Draw the resonance contributors for the following radicals:

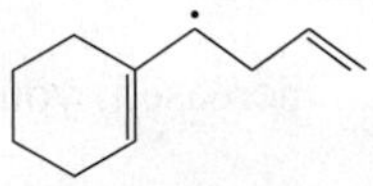

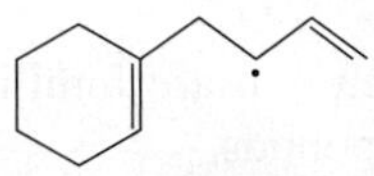

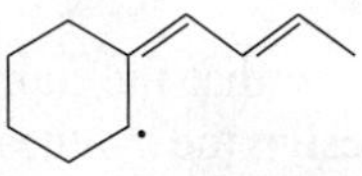

PROBLEM 21

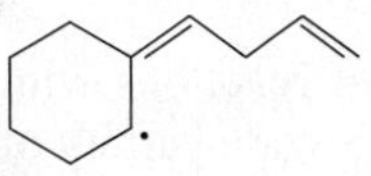

a. Draw the major product(s) of the reaction of 1-methylcyclohexene with the following reagents, disregarding stereoisomers:

   1. NBS/Δ/peroxide   **2.** Br$_2$/CH$_2$Cl$_2$ **3.** HBr   **4.** HBr/peroxide

b. For each reaction, show what stereoisomers are obtained.

Why does the bromine radical generated from NBS remove an allylic hydrogen, whereas the bromine radical generated from HBr + peroxide adds to the double bond? The bromine radical can in fact do both. When NBS is used, however, there is little HBr present to complete the addition reaction after the bromine radical has added to the double bond. Because addition of a bromine radical to a double bond is reversible, the reactant is reformed and allylic substitution becomes the major reaction pathway.

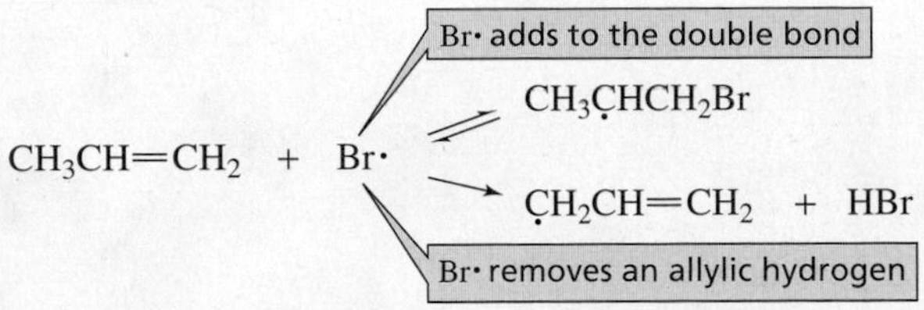

Cyclopropane

Although it is an alkane, cyclopropane undergoes electrophilic addition reactions as if it were an alkene. Cyclopropane is more reactive than propene toward the addition of acids such as HBr and HCl, but is less reactive toward the addition of Cl_2 and Br_2, so a Lewis acid ($FeCl_3$ or $FeBr_3$) is needed to catalyze halogen addition.

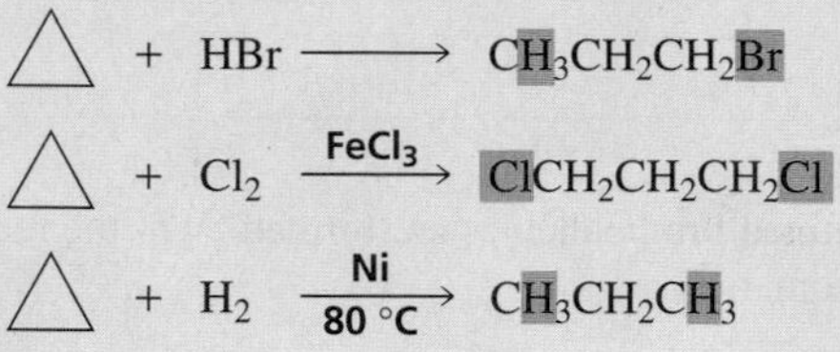

$$\triangle + HBr \longrightarrow CH_3CH_2CH_2Br$$

$$\triangle + Cl_2 \xrightarrow{FeCl_3} ClCH_2CH_2CH_2Cl$$

$$\triangle + H_2 \xrightarrow[80\,°C]{Ni} CH_3CH_2CH_3$$

It is the strain in the small ring that makes it possible for cyclopropane to undergo electrophilic addition reactions. Because of the 60° bond angles in its three-membered ring, the compound's sp^3 orbitals cannot overlap head-on. Thus, the C—C bonds in cyclopropane are considerably weaker than normal C—C bonds. Consequently, the three-membered ring undergoes a ring-opening reaction with electrophilic reagents.

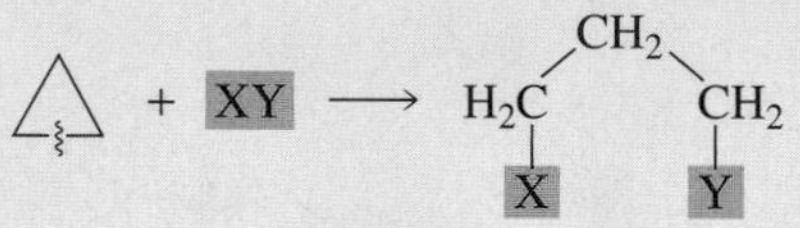

$$\triangle + XY \longrightarrow H_2C{-}CH_2{-}CH_2$$
  X   Y

DESIGNING A SYNTHESIS

10 PRACTICE WITH MULTISTEP SYNTHESIS

Now that the number of reactions with which you are familiar has increased, you can design the synthesis of a wide variety of compounds.

Example 1. Starting with the ether shown here, how could you prepare the aldehyde?

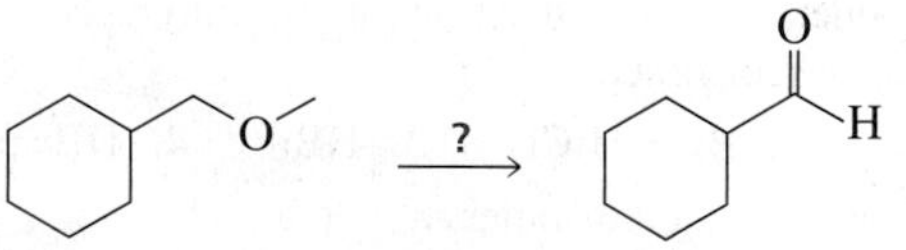

Heating the ether with one equivalent of HI forms a primary alcohol that, when oxidized, forms the desired aldehyde.

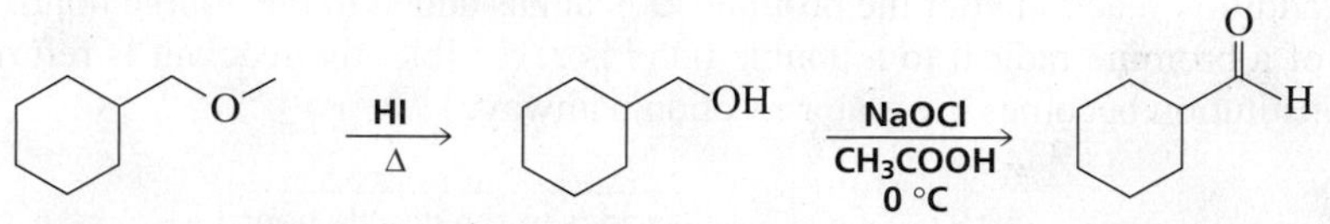

Example 2. Suggest a way to prepare 1,3-cyclohexadiene from cyclohexane.

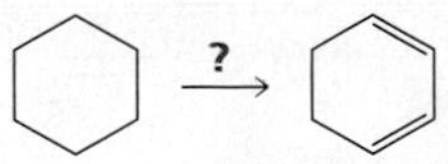

Deciding what the first reaction should be is easy, because the only reaction that an alkane can undergo is a radical substitution reaction with Cl_2 or Br_2. Next, an E2 reaction, using a high concentration of a strong and bulky base and carried out at a relatively high temperature to encourage elimination over substitution, will form cyclohexene. Radical bromination of cyclohexene forms an allylic bromide, which will form the desired target molecule by undergoing another E2 reaction.

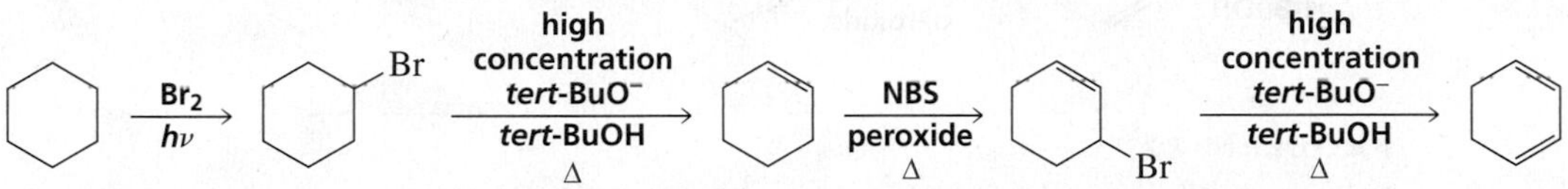

Example 3. Starting with methylcyclohexane, how could the following vicinal *trans*-dihalide be prepared?

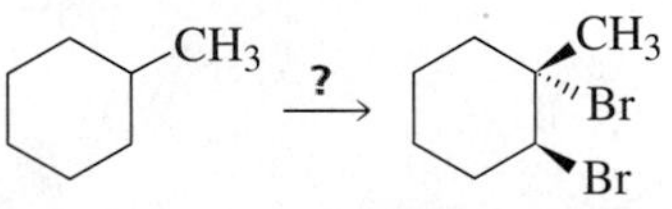

Again, since the starting material is an alkane, the first reaction must be a radical substitution. Bromination leads to selective substitution of the tertiary hydrogen. Under E2 conditions, tertiary alkyl halides undergo only elimination, so there will be no competing substitution product formed in the next reaction. A relatively unhindered base should be used to favor removal of a proton from the secondary carbon over removal of a proton from the methyl group. The final step is addition of Br_2; only anti addition occurs in this reaction, so the target molecule (along with its enantiomer) is obtained.

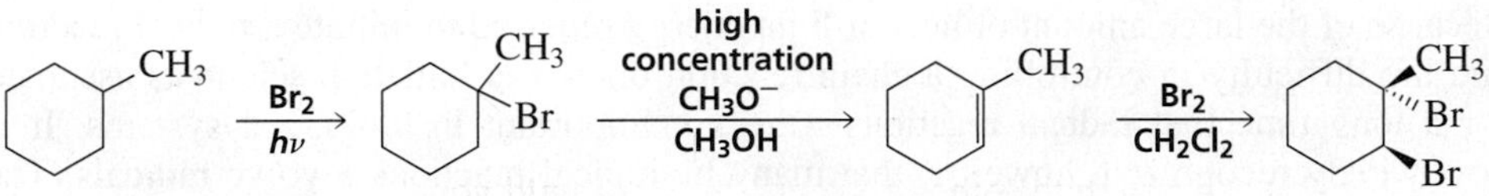

Example 4. Design a synthesis for the target molecule from the indicated starting material.

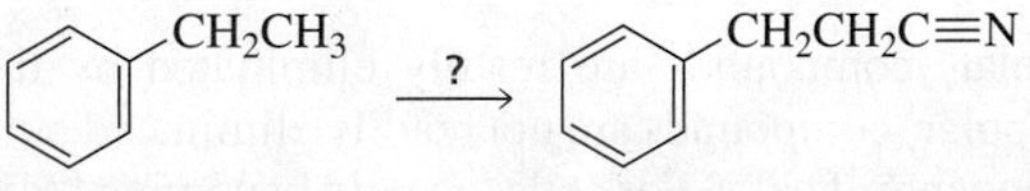

It is not immediately obvious how to carry out this synthesis, so let's use retrosynthetic analysis to find a way. The only method you know for introducing a $C\equiv N$ group into a molecule is nucleophilic substitution. The alkyl halide for that substitution reaction can be obtained from the addition of HBr to an alkene in the presence of a peroxide. The alkene for that addition reaction can be obtained from an elimination reaction using an alkyl halide obtained by benzylic substitution.

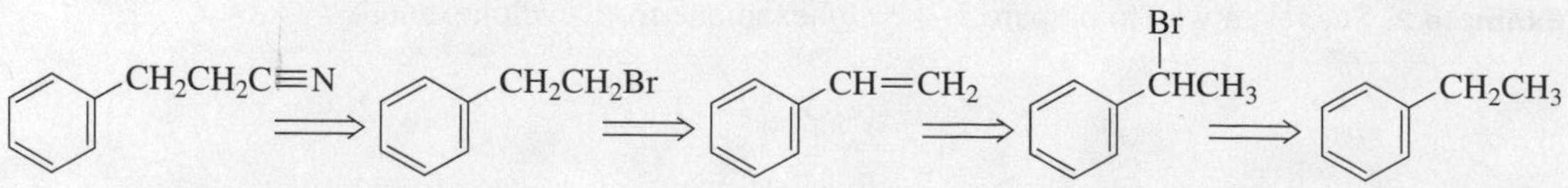

The reaction sequence can now be written in the forward direction along with the reagents required to carry out each step. Notice that a bulky base is used to encourage elimination over substitution.

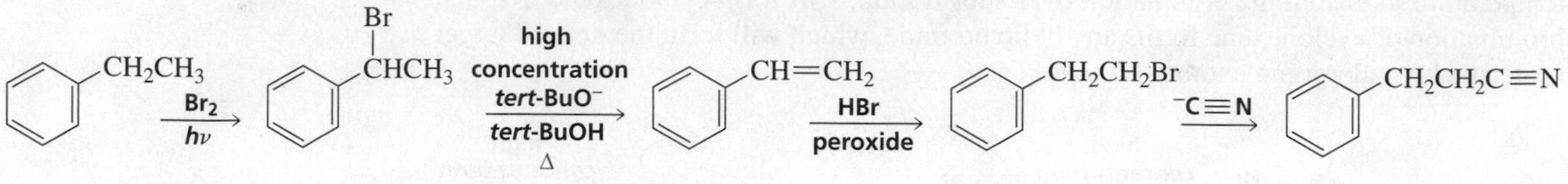

PROBLEM 22

Design a multistep synthesis to show how each of the following compounds could be prepared from the given starting material:

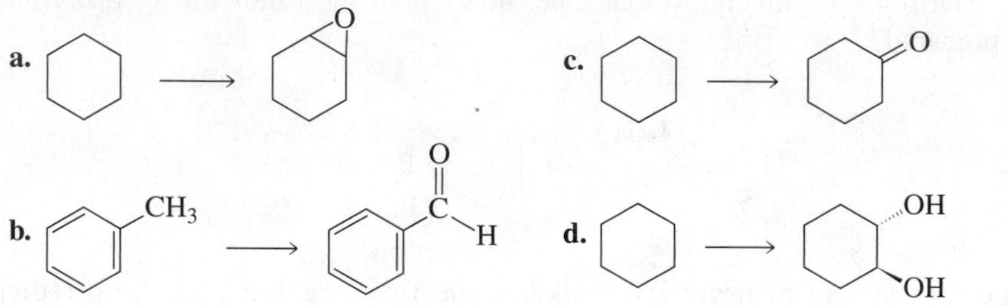

11 RADICAL REACTIONS OCCUR IN BIOLOGICAL SYSTEMS

Because of the large amount of heat or light energy required to initiate a radical reaction and the difficulty in controlling a chain reaction once it is initiated, scientists assumed for a long time that radical reactions were not important in biological systems. It is now widely recognized, however, that many biological reactions involve radicals. The radicals in these reactions, instead of being generated by heat or light, are formed by the interaction of organic molecules with metal ions. The radical reactions take place at the active sites of enzymes. Containing the chain reaction at a specific site allows the reaction to be controlled.

Water-soluble (polar) compounds are readily eliminated by the body. In contrast, water-insoluble (nonpolar) compounds are not readily eliminated but, instead, accumulate in the nonpolar components of cells. For cells to avoid becoming "toxic dumps," nonpolar compounds that are ingested (such as drugs, foods, and environmental pollutants) must be converted into polar compounds that can be excreted.

A radical reaction carried out in the liver converts nonpolar hydrocarbons into less toxic polar alcohols by substituting an H in the hydrocarbon with an OH. The reaction is catalyzed by an iron-containing enzyme called cytochrome P_{450}.

A radical intermediate is created when Fe^V=O removes a hydrogen atom from the alkane. Then Fe^{IV}—OH dissociates homolytically into Fe^{III} and ȮH, and the ȮH immediately combines with the radical intermediate to form the alcohol.

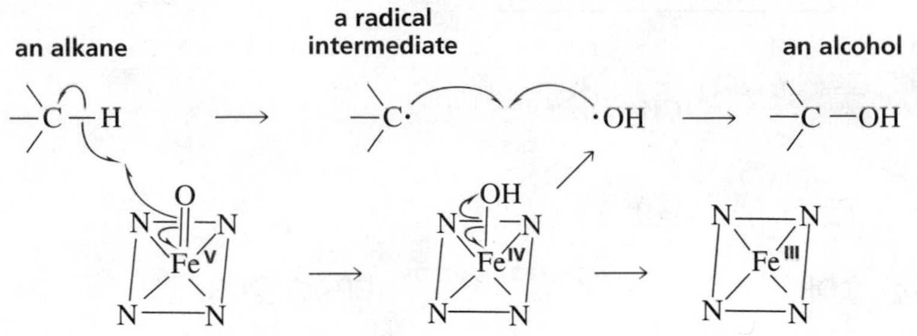

This reaction can also have the opposite toxicological effect. For example, studies found that when animals inhale dichloromethane (CH_2Cl_2), it becomes a carcinogen as a result of an H being substituted by an OH.

Decaffeinated Coffee and the Cancer Scare

Animal studies revealing that dichloromethane becomes a carcinogen when inhaled immediately led to a study of thousands of workers who inhaled dichloromethane daily. However, no increased risk of cancer was found in this group. (This shows that the results of studies done on humans do not always agree with the results of those done on laboratory animals.)

Because dichloromethane was the solvent used to extract caffeine from coffee beans in the manufacture of decaffeinated coffee, a study was done to see what happened to animals that drank dichloromethane. When dichloromethane was added to the drinking water given to laboratory rats and mice, researchers found no toxic effects, even in rats that had consumed an amount of dichloromethane equivalent to the amount that would be ingested by drinking 120,000 cups of decaffeinated coffee per day and in mice that had consumed an amount equivalent to drinking 4.4 million cups of decaffeinated coffee per day.

However, because of the initial concern, researchers sought alternative methods for extracting caffeine from coffee beans. Extraction by CO_2 at supercritical temperatures and pressures was found to be a better method because it extracts caffeine without simultaneously extracting some of the flavor compounds, as dichloromethane does. This was one of the first green (environmentally benign) commercial chemical processes to be developed. After the caffeine has been removed, the CO_2 can be recycled, whereas dichloromethane is not a substance that should be released into the environment.

Fats and oils are easily oxidized by O_2 by means of a radical chain reaction to form compounds that have strong odors. These compounds are responsible for the unpleasant taste and smell associated with sour milk and rancid butter.

Fats and oils have double bonds (—CH=CH—CH$_2$—CH=CH—) that are separated by two single bonds. The mechanism for their oxidation by O_2 is shown next. Notice the similarity of this mechanism to that shown for peroxide formation in Section 6.

MECHANISM FOR THE OXIDATION OF FATS AND OILS BY OXYGEN

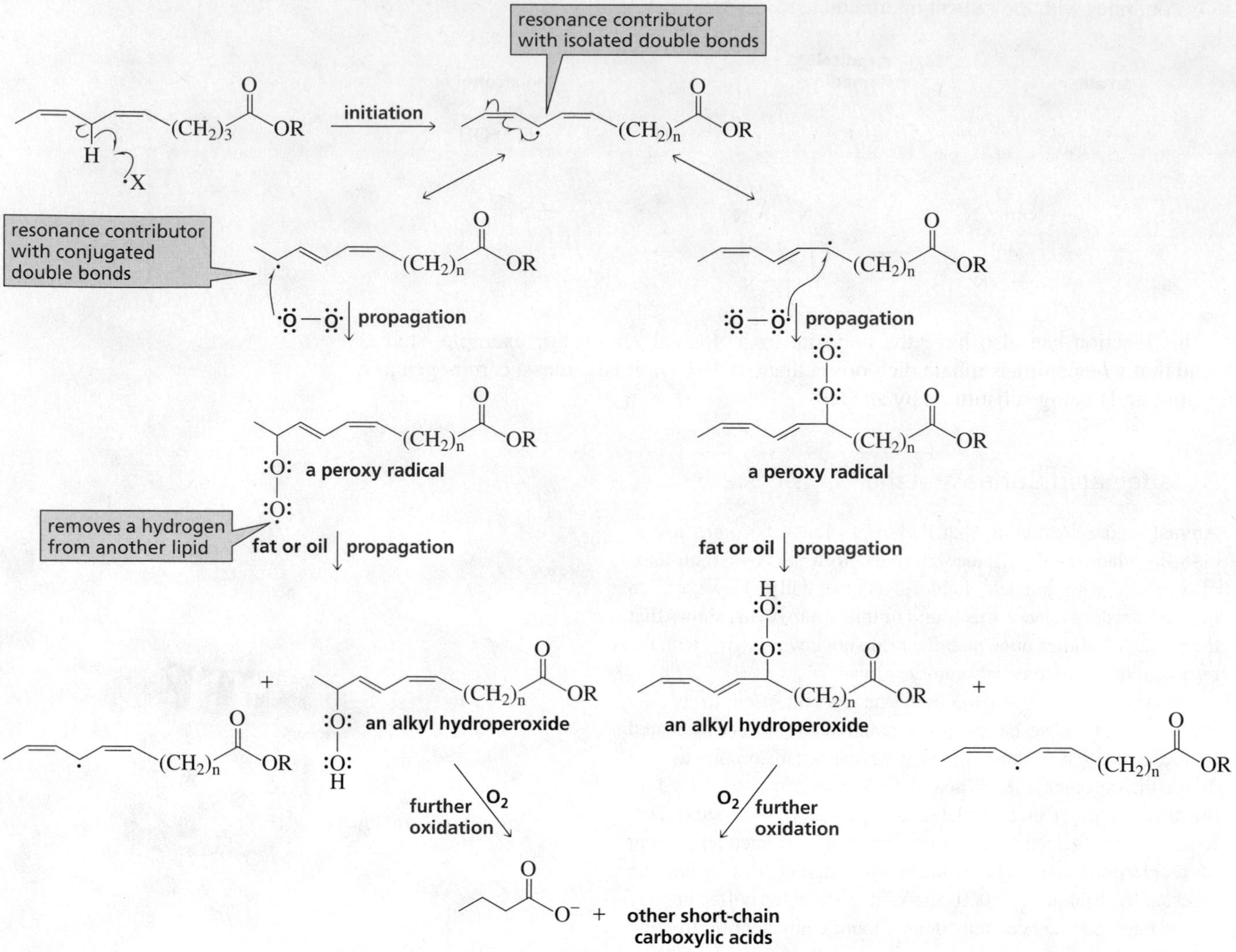

- In the initiation step, a radical removes a hydrogen atom from a methylene group that is flanked by two double bonds. This hydrogen is the one most easily removed because the resulting radical is relatively stable, since the unpaired electron is shared by three carbons.

- In the first propagation step, the radical created in the initiation step reacts with O_2, forming a peroxy radical.

- In the second propagation step, the peroxy radical removes a hydrogen atom from another molecule of fat or oil. The two propagation steps are repeated over and over.

- The alkyl hydroperoxide undergoes further oxidation to form butyric acid and other short-chain carboxylic acids.

The molecules that form cell membranes can undergo this same radical reaction, which leads to their degradation. Radical reactions in cells have been implicated in the aging process.

Clearly, unwanted radicals in cells must be destroyed before they damage the cells. Radical reactions can be prevented by **radical inhibitors,** compounds that destroy reactive radicals by converting them into unreactive radicals or into compounds with only paired electrons. Radical inhibitors are *antioxidants*—that is, they prevent oxidation reactions such as the one just shown.

Hydroquinone is an example of a radical inhibitor. When hydroquinone traps a radical, it forms semiquinone, a radical that is stabilized by electron delocalization, so it is less reactive than other radicals. Moreover, semiquinone can trap another radical and form quinone, a compound whose electrons are all paired. Hydroquinones are found in the cells of all aerobic organisms.

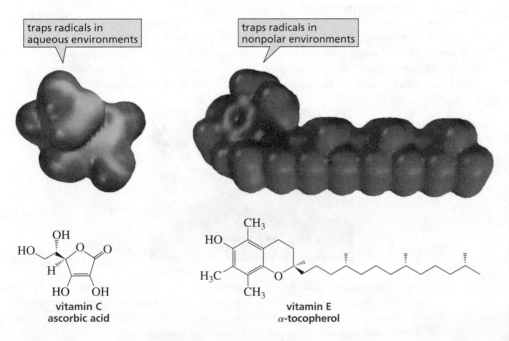

Two other examples of radical inhibitors in living systems are vitamins C and E. Like hydroquinone, they form relatively stable (unreactive) radicals.

Vitamin C (also called ascorbic acid) is a water-soluble compound that traps radicals formed in the interior of cells and in blood plasma (both of which have aqueous environments).

Vitamin E (also called α-tocopherol) is a fat-soluble compound that traps radicals formed in cell membranes, which are nonpolar. Vitamin E is the primary antioxidant for fat tissue in humans and is, therefore, important in preventing the development of atherosclerosis.

Why one vitamin functions in aqueous environments and the other in nonaqueous environments is apparent from their structures and their electrostatic potential maps; both show that vitamin C is a relatively polar compound and vitamin E is a nonpolar compound.

Because radicals are implicated in the aging process, many products are available that contain antioxidants.

Nuts are a natural source of vitamin E.

601

Food Preservatives

Radical inhibitors found in food are known as *preservatives*. They preserve food by preventing radical chain reactions. Vitamin E is a naturally occurring preservative found in such things as vegetable oil, sunflower seeds, and spinach. BHA and BHT are synthetic preservatives that are added to many packaged foods. Notice that, like hydroquinone, vitamin E and all the synthetic preservatives are phenols.

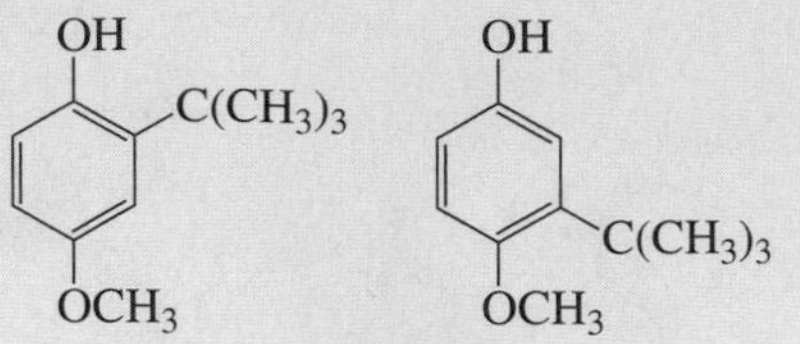

butylated hydroxyanisole
BHA

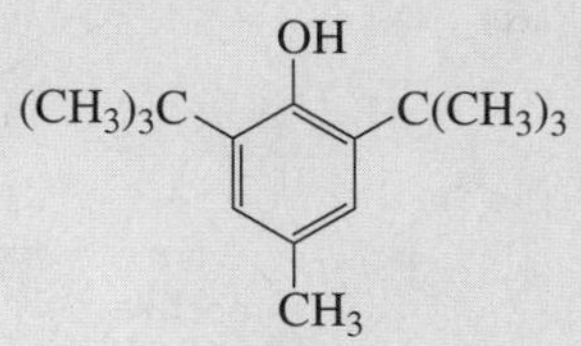

butylated hydroxytoluene
BHT

food preservatives

Is Chocolate a Health Food?

We have long been told that our diets should include lots of fruits and vegetables because they are good sources of antioxidants. Antioxidants protect against cardiovascular disease, cancer, and cataracts, and they are thought to slow the effects of aging. Chocolate is made up of hundreds of organic compounds, including high levels of antioxidants called catechins. (Catechins are also phenols.)

On a weight basis, the concentration of antioxidants in chocolate is higher than in red wine or green tea, and 20 times higher than in tomatoes. Another piece of good news for chocolate lovers is that stearic acid, the main fatty acid in chocolate, does not appear to raise blood cholesterol levels the way other saturated fatty acids do. Dark chocolate contains more than twice the level of antioxidants found in milk chocolate. Unfortunately, white chocolate contains no antioxidants.

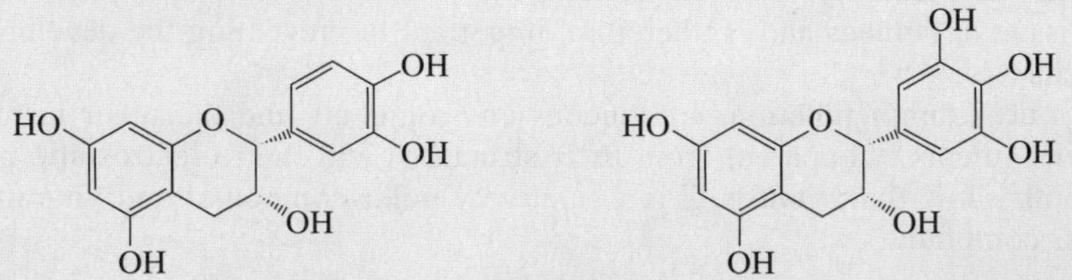

catechins

PROBLEM 23

How many atoms share the unpaired electron in semiquinone?

PROBLEM 24

Using resonance structures, explain why a catechin is an antioxidant.

12 RADICALS AND STRATOSPHERIC OZONE

Ozone (O_3), a major constituent of smog, is a health hazard at ground level—it inflames the airways, worsens lung ailments, and increases the risk of death from heart or lung disease. In the stratosphere, however, a layer of ozone shields the Earth from harmful solar radiation, with the greatest concentrations lying between 12 and 15 miles above the Earth's surface.

In the stratosphere, ozone acts as a filter for biologically harmful ultraviolet light that otherwise would reach the surface of the Earth. Among other effects, short-wavelength ultraviolet light can damage DNA in skin cells, causing mutations that trigger skin cancer. We owe our very existence to this protective ozone layer. According to current theories of evolution, life could not have developed on land without it. Instead, most if not all living things would have had to remain in the ocean, where water screens out the harmful ultraviolet light.

The ozone layer is thinnest at the equator and densest toward the poles. Since about 1985, scientists have noted a precipitous drop in stratospheric ozone over Antarctica. This area of ozone depletion, dubbed the "ozone hole," is unprecedented in the history of ozone observations. Scientists subsequently noted a similar decrease in ozone over Arctic regions; then, in 1988, they detected a depletion of ozone over the United States for the first time. Three years later, scientists determined that the rate of ozone depletion was two to three times faster than originally anticipated.

Strong circumstantial evidence implicated synthetic chlorofluorocarbons (CFCs)— alkanes in which all the hydrogens have been replaced by fluorine and chlorine—as a major cause of ozone depletion. These gases, known commercially as Freon, had been used extensively as cooling fluids in refrigerators and air conditioners. They were also once widely used as propellants in aerosol spray cans (deodorant, hair spray, and so on) because of their odorless, nontoxic, and nonflammable properties and because, being chemically inert, they do not react with the contents of the can. Now such use is banned and propane and butane are used as propellants instead.

The global agreement to phase out CFCs and other ozone-depleting agents seems to be working. The ozone layer is no longer depleting and it is hoped that it will regain its density by 2048.

Ozone is formed from the interaction of molecular oxygen with very short wavelength ultraviolet light.

Polar stratospheric clouds increase the rate of ozone destruction. These clouds form over Antarctica during the cold winter months. Ozone depletion in the Arctic is less severe because the temperature generally does not get low enough for stratospheric clouds to form there.

$$O_2 \xrightarrow{h\nu} O + O$$

$$O + O_2 \longrightarrow \boxed{O_3}$$
$$\textbf{ozone}$$

Chlorofluorocarbons are stable until they reach the stratosphere. There they encounter wavelengths of ultraviolet light that cause the C—Cl bond to break homolytically, generating chlorine radicals.

$$
\begin{array}{c}
\overset{\displaystyle Cl}{\underset{\displaystyle F}{F-C-Cl}} \xrightarrow{h\nu} \overset{\displaystyle Cl}{\underset{\displaystyle F}{F-C\cdot}} + Cl\cdot
\end{array}
$$

These chlorine radicals are the ozone-removing agents. They react with ozone to form chlorine monoxide radicals and molecular oxygen. The chlorine monoxide radical then reacts with more ozone to form chlorine dioxide, which dissociates to regenerate a chlorine radical. These three steps—two of which each destroy an ozone molecule—are the propagating steps that are repeated over and over. It has been calculated that a single chlorine atom destroys 100,000 ozone molecules!

$$Cl\cdot \ + \ O_3 \longrightarrow ClO\cdot \ + \ O_2$$
$$ClO\cdot \ + \ O_3 \longrightarrow \cdot ClO_2 \ + \ O_2$$
$$\cdot ClO_2 \longrightarrow Cl\cdot \ + \ O_2$$

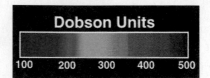

The growth of the Antarctic ozone hole, located mostly over the continent of Antarctica, since 1979. The images were made from data supplied by total ozone-mapping spectrometers (TOMSs). The color scale depicts the total ozone values in Dobson units, with the lowest ozone densities represented by dark blue. (The ozone over a given area is compressed to 0 °C and 1 atm pressure, and the thickness of the slab is measured. 1 Dobson unit = 0.01 mm thickness.)

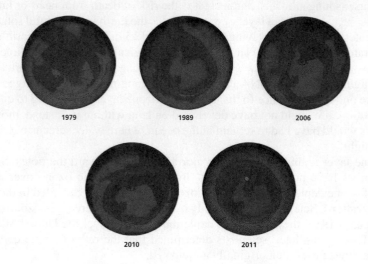

1979   1989   2006

2010   2011

Artificial Blood

Clinical trials are underway to test the use of perfluorocarbons—alkanes in which all the hydrogens have been replaced by fluorines—as compounds to replace blood volume and mimic hemoglobin's ability to carry oxygen to cells and transport carbon dioxide to the lungs.

These compounds are not a true blood substitute, because blood performs many functions that artificial blood cannot. For example, white blood cells fight against infection and platelets are involved in blood clotting. However, artificial blood has several advantages in trauma situations until an actual transfusion can be done: it is safe from disease, it can be administered to any blood type, its availability does not depend on blood donors, and it can be stored longer than whole blood, which is good for only about 40 days.

SOME IMPORTANT THINGS TO REMEMBER

- **Alkanes** are called **saturated hydrocarbons**. Because they do not contain any carbon–carbon double or triple bonds, they are saturated with hydrogen.

- Alkanes are unreactive compounds because they have only strong σ bonds and atoms with no partial charges.

- In **heterolytic bond cleavage,** a bond breaks so that one of the atoms retains both of the bonding electrons; in **homolytic bond cleavage,** a bond breaks so that each of the atoms retains one of the bonding electrons.

- Alkanes undergo **radical substitution reactions** with chlorine (Cl_2) or bromine (Br_2) at high temperatures or in the presence of light to form alkyl chlorides or alkyl bromides. This substitution reaction is a **radical chain reaction** with **initiation, propagation,** and **termination steps.**

- The rate-determining step of a radical substitution reaction is removal of a hydrogen atom to form an alkyl radical.

- The relative rates of radical formation are benzylic $\sim$ allyl $> 3° > 2° > 1° >$ vinyl $\sim$ methyl.

- Calculation of the relative amounts of products obtained from the radical halogenation of an alkane must take into account both probability and the relative rate at which a particular radical is formed.

- The **reactivity–selectivity principle** states that the more reactive a species is, the less selective it will be.

- A bromine radical is *less reactive* than a chlorine radical, so a bromine radical is *more selective* about which hydrogen atom it removes.

- Ethers form explosive peroxides when they are exposed to air.

- A **peroxide** is a **radical initiator** because it creates radicals.

- A **radical inhibitor** (an **antioxidant**) destroys reactive radicals by creating radicals that are less reactive or by creating compounds that have only paired electrons.

- **Radical addition reactions** are chain reactions with **initiation, propagation,** and **termination steps.**

- A peroxide reverses the order of addition of H and Br to an alkene because it causes $Br\cdot$, instead of H^+, to be the electrophile. The **peroxide effect** is observed only for the addition of HBr.

- If a reactant that does not have an asymmetric center undergoes a radical substitution or a radical addition reaction that forms a product with an asymmetric center, then a racemic mixture will be obtained.

- A racemic mixture is also obtained if a hydrogen bonded to an asymmetric center is substituted by a halogen.

- *N*-Bromosuccinimide (NBS) is used to brominate allylic carbons.

- Some biological reactions involve radicals formed by the interaction of organic molecules with metal ions. The reactions take place at the active sites of enzymes.

- Fats, oils, and membranes are oxidized by O_2 in a radical chain reaction.

- The interaction of CFCs with UV light generates chlorine radicals, which are ozone-removing agents.

SUMMARY OF REACTIONS

1. *Alkanes* undergo radical substitution reactions with Cl_2 or Br_2 in the presence of heat or light (Sections 2–5).

$$CH_3CH_3 \ + \ Cl_2 \ \xrightarrow{\Delta \text{ or } h\nu} \ CH_3CH_2Cl \ + \ HCl$$
excess

$$CH_3CH_3 \ + \ Br_2 \ \xrightarrow{\Delta \text{ or } h\nu} \ CH_3CH_2Br \ + \ HBr$$
excess
bromination is more selective than chlorination

2. A radical initiator removes a hydrogen atom from an α-carbon of an ether to form a peroxide (Section 6).

$$R-O-\underset{\underset{H}{|}}{C}H-R \ + \ O_2 \ \longrightarrow \ R-O-\underset{\underset{O-O-H}{|}}{C}H-R$$

3. *Alkenes* undergo radical addition of hydrogen bromide in the presence of a peroxide (Br· is the electrophile; Section 7).

$$RCH{=}CH_2 \ + \ \boxed{HBr} \ \xrightarrow{\text{peroxide}} \ RCH_2CH_2Br$$

4. *Alkyl-substituted benzenes* undergo radical substitution at the benzylic position (Section 9).

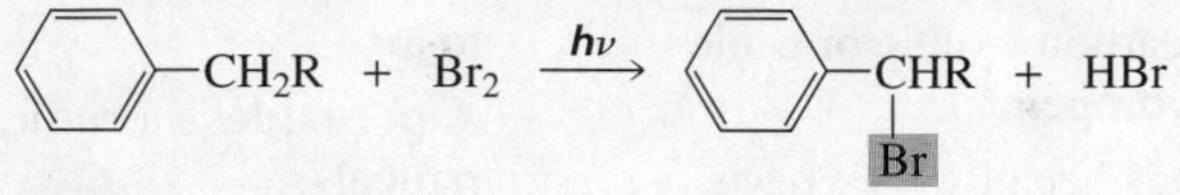

5. *Alkenes* undergo radical substitution at allylic carbons. NBS is used for bromination at allylic carbons (Section 9).

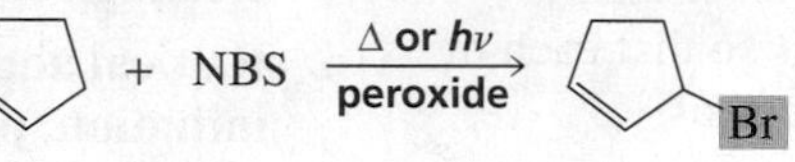

$$RCH_2CH{=}CH_2 \ + \ NBS \ \xrightarrow[\text{peroxide}]{\Delta \text{ or } h\nu} \ RCHCH{=}CH_2 \ + \ RCH{=}CHCH_2 \ + \ HBr$$
$$\quad\quad\quad\quad\quad\quad\quad\quad\quad\quad\quad\quad\quad \boxed{Br} \quad\quad\quad\quad\quad\quad\quad \boxed{Br}$$

PROBLEMS

25. Draw the product(s) of each of the following reactions, disregarding stereoisomers:

a. $CH_2{=}CHCH_2CH_2CH_3 \ + \ NBS \ \xrightarrow[\text{peroxide}]{\Delta}$

b.
$$\begin{array}{c} CH_3 \\ | \\ CH_3C{=}CHCH_3 \end{array} \ + \ NBS \ \xrightarrow[\text{peroxide}]{\Delta}$$

c. 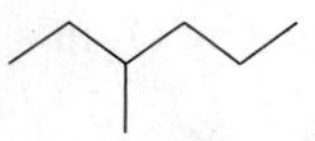 $\ + \ Br_2 \ \xrightarrow{h\nu}$

d. 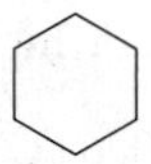 $\ + \ Cl_2 \ \xrightarrow{h\nu}$

e.  $\ + \ Cl_2 \ \xrightarrow{CH_2Cl_2}$

f. 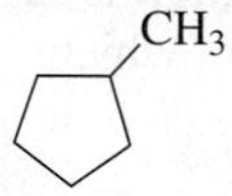 $\ + \ Cl_2 \ \xrightarrow{h\nu}$

26. a. What alkane, with molecular formula C_5H_{12}, forms only one monochlorinated product when it is heated with Cl_2?

b. What alkane, with molecular formula C_7H_{16}, forms seven monochlorinated products (disregarding stereoisomers) when heated with Cl_2?

27. What is the major product that would be obtained from treating an excess of each of the following compounds with Cl_2 in the presence of light at room temperature? Disregard stereoisomers.

a.

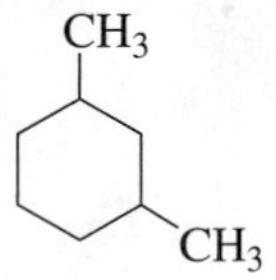

b.

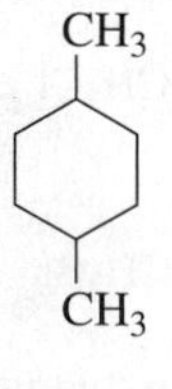

c.

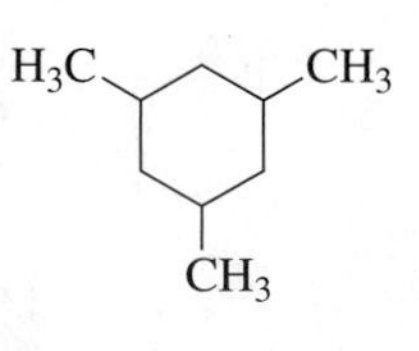

28. What would the answers be to Problem 27 if the same compounds were treated with Br_2 at 125 °C?

29. Draw the major product of each of the following reactions, disregarding stereoisomers:

a.  [cyclopentane] + NBS $\xrightarrow[\text{peroxide}]{\Delta}$

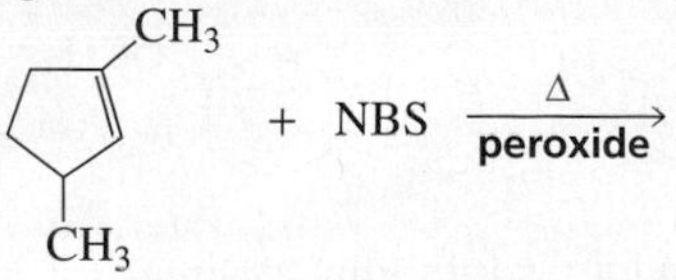

b.  [cyclopentadiene] + NBS $\xrightarrow[\text{peroxide}]{\Delta}$

c. 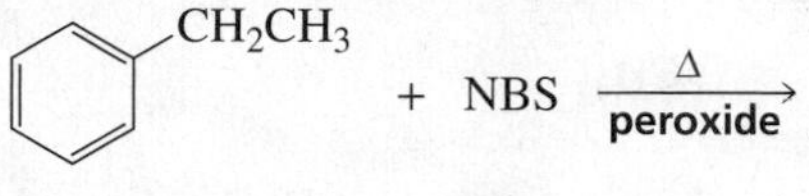 [ethylbenzene, CH₂CH₃] + NBS $\xrightarrow[\text{peroxide}]{\Delta}$

d. [1,5-dimethylcyclopentene with CH₃ groups] + NBS $\xrightarrow[\text{peroxide}]{\Delta}$

e. [3-methylcyclohexene with CH₃] + NBS $\xrightarrow[\text{peroxide}]{\Delta}$

f. 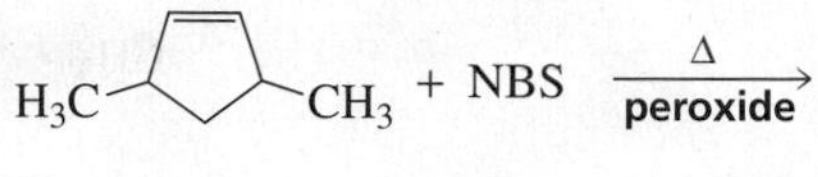 [H₃C—cyclopentene—CH₃] + NBS $\xrightarrow[\text{peroxide}]{\Delta}$

30. When 2-methylpropane is monochlorinated in the presence of light at room temperature, 36% of the product is 2-chloro-2-methylpropane and 64% is 1-chloro-2-methylpropane. From these data, calculate how much easier it is to remove a hydrogen atom from a tertiary carbon than from a primary carbon under these conditions.

31. Iodine (I_2) does not react with ethane, even though I_2 is more easily cleaved homolytically than the other halogens. Explain.

32. Propose a mechanism to account for the products formed in the following reaction:

[octahydronaphthalene] + NBS $\xrightarrow[\text{peroxide}]{\Delta}$ [product with Br] + [product with Br]

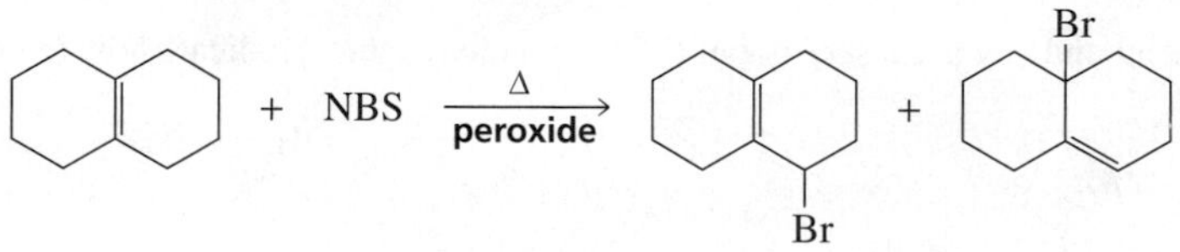

33. The deuterium kinetic isotope effect for the halogenation of an alkane is defined in the following equation, where $X\cdot = Cl\cdot$ or $Br\cdot$

$$\frac{\textbf{deuterium kinetic}}{\textbf{isotope effect}} = \frac{\textbf{rate of homolytic cleavage of a C—H bond by X·}}{\textbf{rate of homolytic cleavage of a C—D bond by X·}}$$

Predict whether chlorination or bromination would have a greater deuterium kinetic isotope effect.

34. a. How many monochlorination products could be obtained from the radical chlorination of methylcyclohexane? Disregard stereoisomers.
b. Which product would be obtained in greatest yield? Explain.
c. How many monochlorination products would be obtained if all stereoisomers are included?

35. Draw the alkyl halide that would be obtained in greatest yield. Include stereoisomers.

a. $CH_3\overset{\overset{\displaystyle CH_3}{|}}{C}HCH_3$ + Cl_2 $\xrightarrow{h\nu}$
excess

b. $CH_3\overset{\overset{\displaystyle CH_3}{|}}{C}HCH_3$ + Br_2 $\xrightarrow{h\nu}$
excess

c. 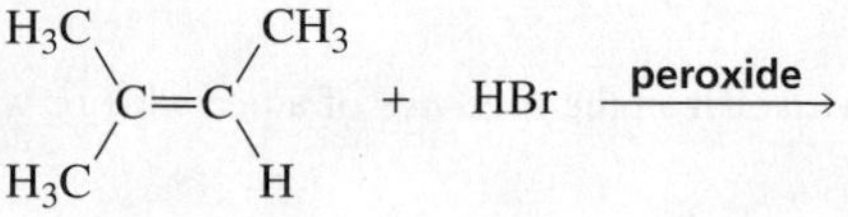 $\underset{H_3C}{\overset{H_3C}{>}}C=C\underset{H}{\overset{CH_3}{<}}$ + HBr $\xrightarrow{\text{peroxide}}$

d. $CH_2{=}CHCH_2CH_2\overset{\overset{\displaystyle CH_3}{|}}{C}{=}CH_2$ $\xrightarrow[\text{peroxide}]{\text{HBr}}$

e. 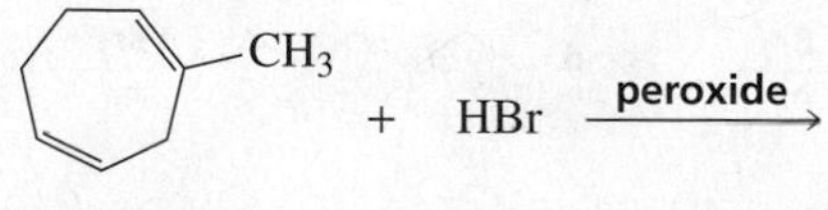 [cyclooctene with CH₃] + HBr $\xrightarrow{\text{peroxide}}$

f. 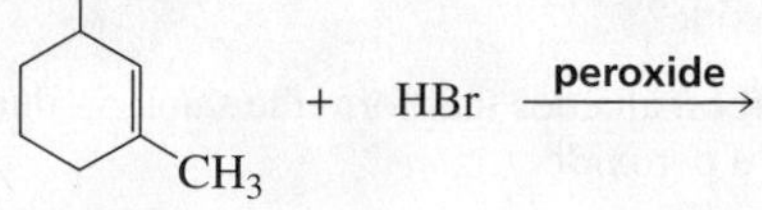 [cyclohexene with CH=CH₂ and CH₃] + HBr $\xrightarrow{\text{peroxide}}$

36. Starting with cyclohexane, how could the following compounds be prepared?

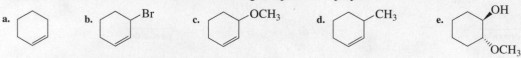

a. b. c. d. e.

37. a. Propose a mechanism for the following reaction:

$$CH_3CH_3 \ + \ CH_3-\overset{\overset{\displaystyle CH_3}{|}}{\underset{\underset{\displaystyle CH_3}{|}}{C}}-OCl \ \xrightarrow{\Delta} \ CH_3CH_2Cl \ + \ CH_3-\overset{\overset{\displaystyle CH_3}{|}}{\underset{\underset{\displaystyle CH_3}{|}}{C}}-OH$$

b. Given that the $\Delta H°$ value for the reaction is -42 kcal/mol and the bond dissociation energies for the C—H, C—Cl, and O—H bonds are 101, 85, and 105 kcal/mol respectively, calculate the bond dissociation energy of the O—Cl bond.

38. What stereoisomers would be obtained from the following reaction?

$$\underset{H_3C}{\overset{CH_3CH_2}{>}}C=C\underset{CH_3}{\overset{CH_2CH_3}{<}} \ + \ HBr \ \xrightarrow{\text{peroxide}}$$

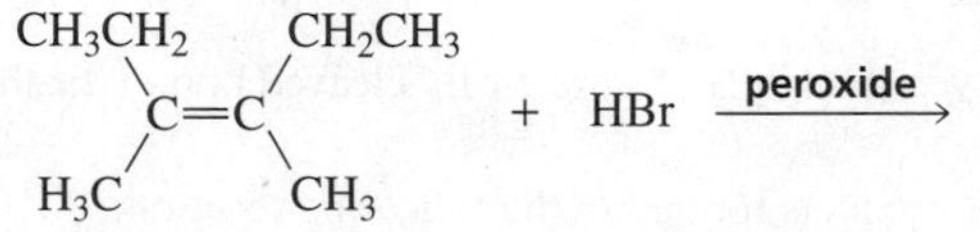

39. Using the given starting material and any necessary organic or inorganic reagents, indicate how the desired compounds could be synthesized:

a. 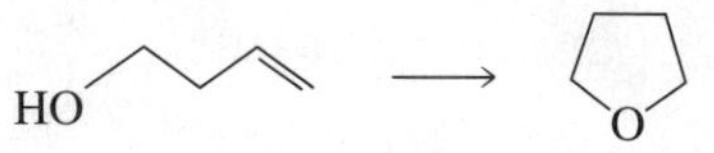 (structure) $\longrightarrow$ (structure)

c.  (structure) $\longrightarrow$ (structure)

b.  (structure) $\longrightarrow$ (structure)

d.  (structure) $\longrightarrow$ (structure)

40. A chemist wanted to determine experimentally the relative ease of removing a hydrogen atom from a tertiary, a secondary, and a primary carbon by a chlorine radical. He allowed 2-methylbutane to undergo chlorination at 300 °C and obtained as products 36% 1-chloro-2-methylbutane, 18% 2-chloro-2-methylbutane, 28% 2-chloro-3-methylbutane, and 18% 1-chloro-3-methylbutane. What values did he obtain for the relative ease of removing a hydrogen atom from tertiary, secondary, and primary hydrogen carbons by a chlorine radical under the conditions of his experiment?

41. At 600 °C, the ratio of the relative rates of formation of a tertiary, a secondary, and a primary radical by a chlorine radical is 2.6 : 2.1 : 1. Explain the change in the degree of regioselectivity compared to what was found in Problem 40.

42. Draw the products of the following reactions, including all stereoisomers:

a. 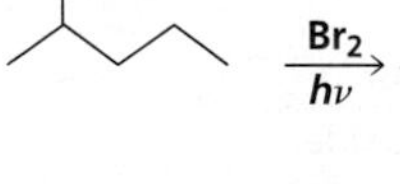 (structure) $\xrightarrow[hv]{Br_2}$

c. 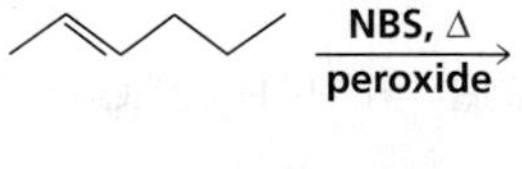 (structure) $\xrightarrow[\text{peroxide}]{\text{NBS, }\Delta}$

e. 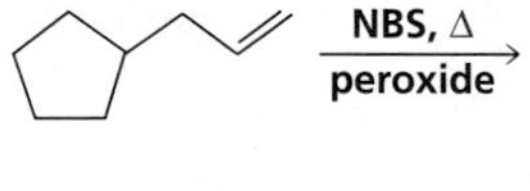 (structure) $\xrightarrow[\text{peroxide}]{\text{NBS, }\Delta}$

b.  (structure) $\xrightarrow[hv]{Br_2}$

d. 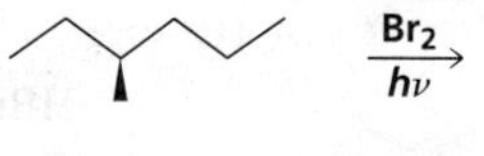 (structure) $\xrightarrow[hv]{Br_2}$

f. 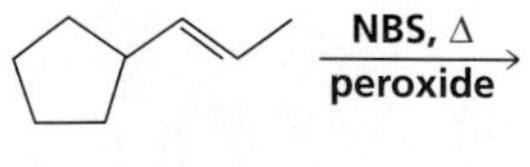 (structure) $\xrightarrow[\text{peroxide}]{\text{NBS, }\Delta}$

43. a. What five-carbon alkene will form the same product whether it reacts with HBr in the *presence* of a peroxide or with HBr in the *absence* of a peroxide?

b. Give three six-carbon alkenes that form the same product, whether they react with HBr in the *presence* of a peroxide or with HBr in the *absence* of a peroxide.

44 a. Calculate the $\Delta H°$ value for the following reaction:

$$CH_4 \;+\; Cl_2 \;\xrightarrow{h\nu}\; CH_3Cl \;+\; HCl$$

b. Calculate the sum of the $\Delta H°$ values for the following two propagation steps:

$$CH_3\!-\!H \;+\; \cdot Cl \;\longrightarrow\; \cdot CH_3 \;+\; H\!-\!Cl$$
$$\cdot CH_3 \;+\; Cl\!-\!Cl \;\longrightarrow\; CH_3\!-\!Cl \;+\; \cdot Cl$$

c. Why do both calculations give you the same value of $\Delta H°$?

45. A possible alternative mechanism to that shown in Problem 44 for the monochlorination of methane would involve the following propagation steps:

$$CH_3\!-\!H \;+\; \cdot Cl \;\longrightarrow\; CH_3\!-\!Cl \;+\; \cdot H$$
$$\cdot H \;+\; Cl\!-\!Cl \;\longrightarrow\; H\!-\!Cl \;+\; \cdot Cl$$

How do you know that the reaction does not take place by this mechanism?

46. Propose a mechanism for the following reaction:

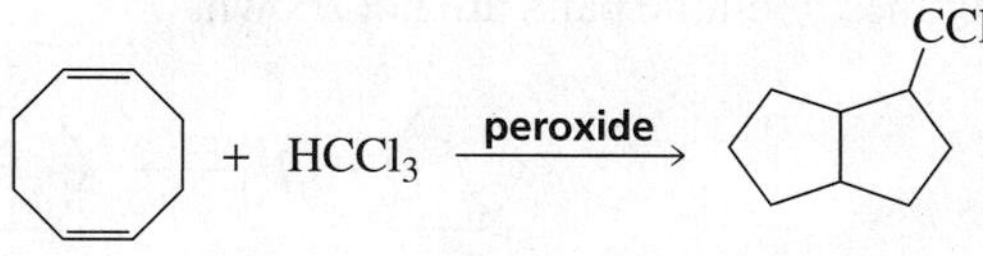

47. Explain why the rate of bromination of methane decreases if HBr is added to the reaction mixture.

48. Enediynes are natural products with potent antitumor properties because they are able to cleave DNA. Their cytotoxic properties are due to the enediyne undergoing a cyclization to form a highly reactive diradical intermediate. The intermediate abstracts hydrogen atoms from the backbone of DNA, which triggers its damage. Draw the structure of the diradical intermediate.

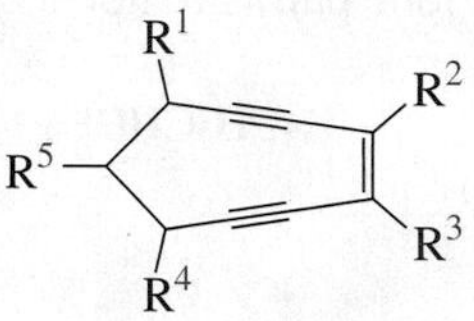

an enediyne

DRAWING CURVED ARROWS IN RADICAL SYSTEMS

We have seen that chemists use curved arrows

1. to show how electrons move when going from a reactant to a product.
2. to show how electrons move when going from one resonance contributor to the next.

The movement of two electrons is indicated by an arrowhead with two barbs. Electron movement in radical systems involves the movement of only one electron. The movement of a single electron is indicated by an arrowhead with one barb.

DRAWING CURVED ARROWS IN RADICAL REACTIONS

When a bond breaks in a way that allows each of the bonded atoms to retain one of the bonding electrons, an arrowhead with one barb is used to represent the movement of each of the single electrons.

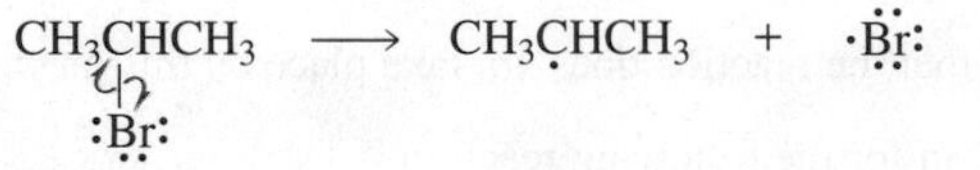

$$CH_3CHCH_3 \longrightarrow CH_3\overset{\cdot}{C}HCH_3 + \cdot\overset{..}{\underset{..}{Br}}:$$
$$\overset{|}{:\overset{..}{Br}}:$$

Sometimes, the lone pairs are not shown.

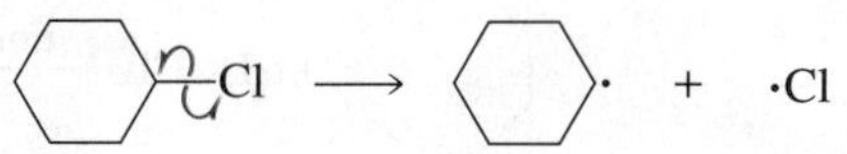

When a bond is formed using one electron from one atom and one electron from another, an arrowhead with one barb is used to represent the movement of each of the single electrons.

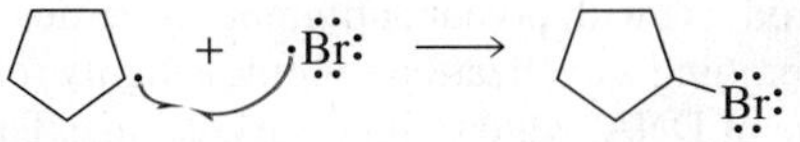

Sometimes, the lone pairs are not shown.

$$CH_3CH_2\overset{\cdot}{C}HCH_3 + \cdot Cl \longrightarrow CH_3CH_2\overset{|}{C}HCH_3$$
$$\overset{|}{Cl}$$

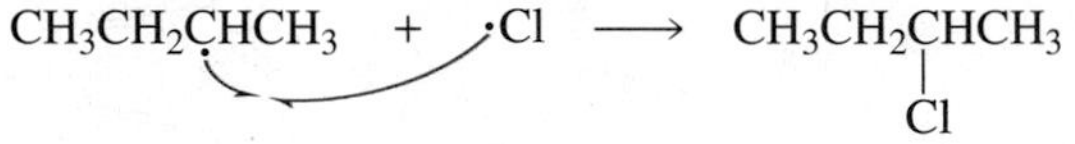

PROBLEM 1 Draw curved arrows to show the movement of the electrons as the bond breaks.

a. $CH_3CH_2\!-\!\overset{..}{\underset{..}{C}l}: \longrightarrow CH_3\overset{\cdot}{C}H_2 + \cdot\overset{..}{\underset{..}{C}l}:$ **c.** $CH_3\overset{..}{\underset{..}{O}}\!-\!\overset{..}{\underset{..}{O}}CH_3 \longrightarrow CH_3\overset{..}{\underset{..}{O}}\cdot + \cdot\overset{..}{\underset{..}{O}}CH_3$

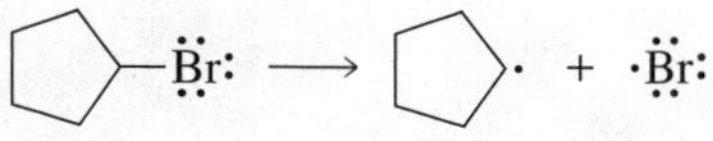

b. $-\overset{..}{\underset{..}{Br}}: \longrightarrow$ $\cdot + \cdot\overset{..}{\underset{..}{Br}}:$

PROBLEM 2 Draw curved arrows to show the movement of the electrons as the bond forms.

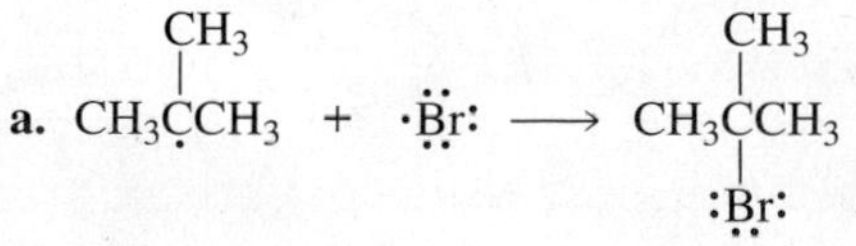

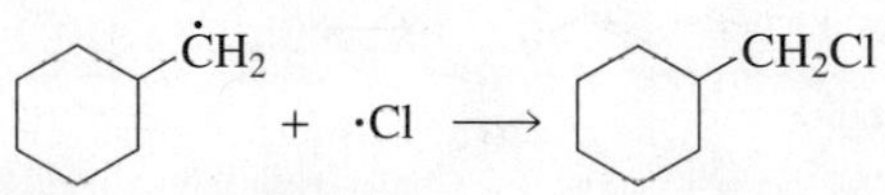

a. $CH_3\overset{CH_3}{\underset{}{C}}CH_3$ + ·B̤r: ⟶ $CH_3\overset{CH_3}{\underset{:\overset{..}{B}r:}{C}}CH_3$ c. + ·Cl ⟶

b. ·Br + ·Br ⟶ Br_2

Often one bond breaks and another bond forms in the same step.

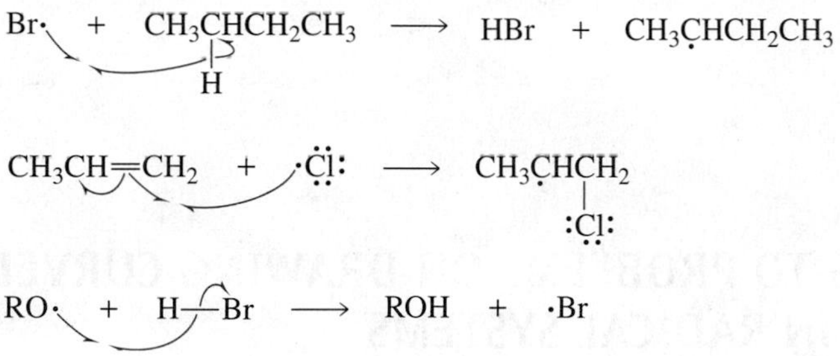

$$Br· + CH_3\overset{}{\underset{H}{C}}HCH_2CH_3 \longrightarrow HBr + CH_3\dot{C}HCH_2CH_3$$

$$CH_3CH{=}CH_2 + ·\overset{..}{C}l: \longrightarrow CH_3\dot{C}HCH_2\underset{:\overset{..}{C}l:}{}$$

$$RO· + H{-}Br \longrightarrow ROH + ·Br$$

PROBLEM 3 Draw curved arrows to show the movement of the electrons as one (or more) bond breaks and another bond forms.

a. + ·B̤r: ⟶

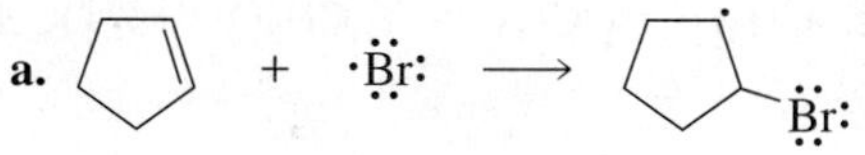

b. $CH_3\dot{C}HCH_2CH_3$ + H—Cl ⟶ $CH_3CH_2CH_2CH_3$ + ·Cl

c. $CH_3\dot{C}H_2$ + Br—Br ⟶ CH_3CH_2Br + ·Br

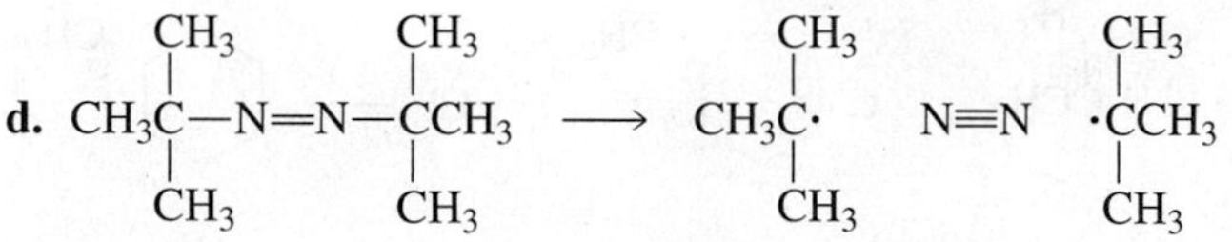

d. $CH_3\overset{CH_3}{\underset{CH_3}{C}}{-}N{=}N{-}\overset{CH_3}{\underset{CH_3}{C}}CH_3$ ⟶ $CH_3\overset{CH_3}{\underset{CH_3}{C}}·$ N≡N $·\overset{CH_3}{\underset{CH_3}{C}}CH_3$

DRAWING CURVED ARROWS IN CONTRIBUTING RESONANCE STRUCTURES THAT ARE RADICALS

The arrows that represent electron movement in resonance contributors of radicals have only one barb on the arrowhead because the arrow represents the movement of only one electron.

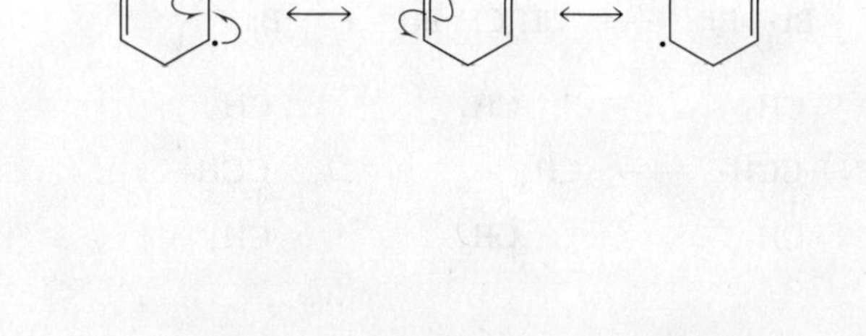

$$CH_3CH{=}CH{-}\dot{C}HCH_3 \longleftrightarrow CH_3\dot{C}H{-}CH{=}CHCH_3$$

PROBLEM 4 Draw curved arrows to show the movement of the electrons as one resonance contributor is converted to the next.

a.  ⟷

b. 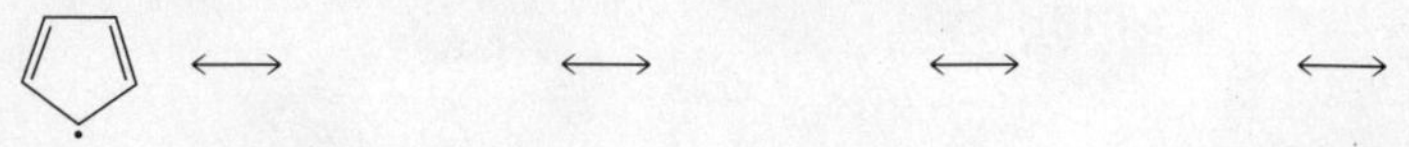 ⟷   ⟷   ⟷   ⟷

c. ⟷   ⟷   ⟷

d.  ⟷   ⟷   ⟷   ⟷

ANSWERS TO PROBLEMS ON DRAWING CURVED ARROWS IN RADICAL SYSTEMS

PROBLEM 1

a. 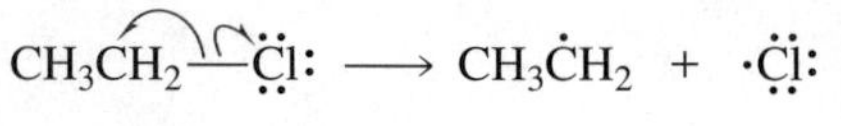 CH_3CH_2—$\ddot{C}l:$ ⟶ $CH_3\dot{C}H_2$ + $\cdot\ddot{C}l:$ c. $CH_3\ddot{O}$—$\ddot{O}CH_3$ ⟶ $CH_3\ddot{O}\cdot$ + $\cdot\ddot{O}CH_3$

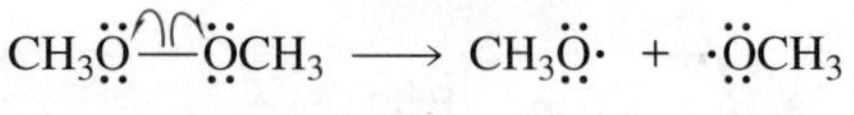

b. 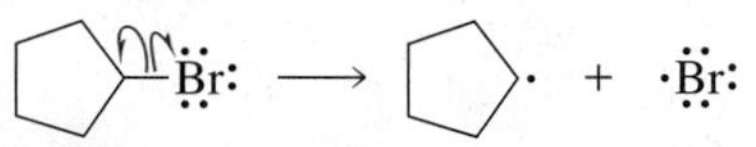 $\ddot{B}r:$ ⟶ $\cdot$ + $\cdot\ddot{B}r:$

PROBLEM 2

a. 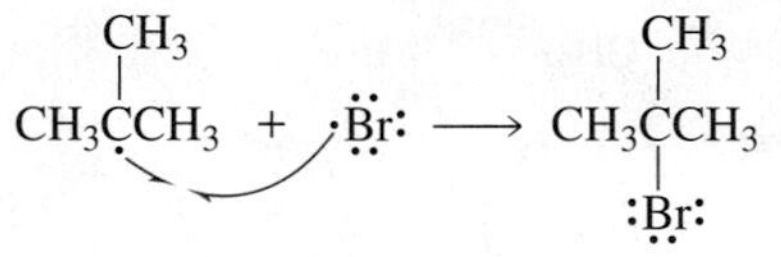   c.

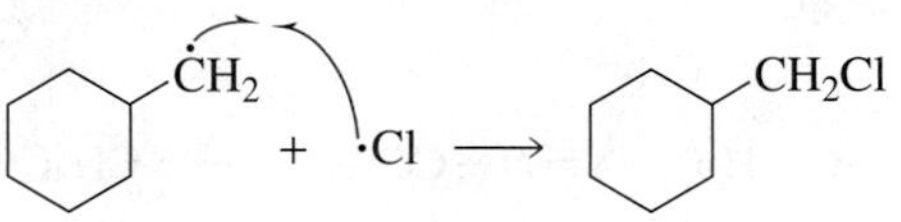

b. 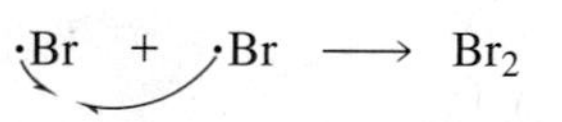 $\cdot Br$ + $\cdot Br$ ⟶ Br_2

PROBLEM 3

a. 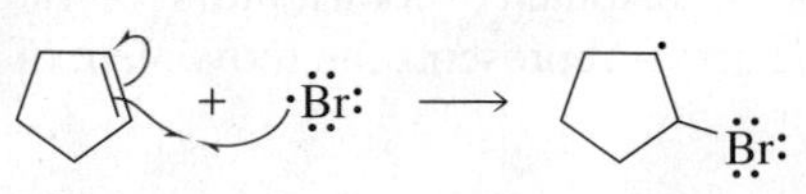 + $\cdot\ddot{B}r:$ ⟶ $\ddot{B}r:$

b. $CH_3\dot{C}HCH_2CH_3$ + H—Cl ⟶ $CH_3CH_2CH_2CH_3$ + $\cdot Cl$

c. $CH_3\dot{C}H_2$ + Br—Br ⟶ CH_3CH_2Br + $\cdot Br$

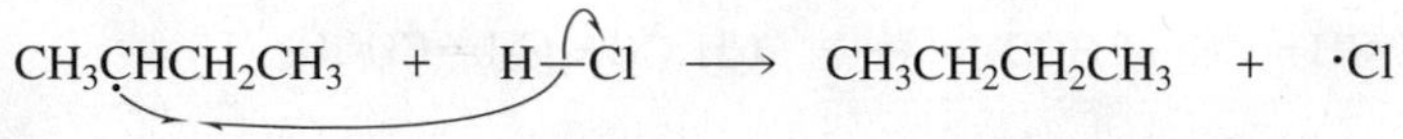

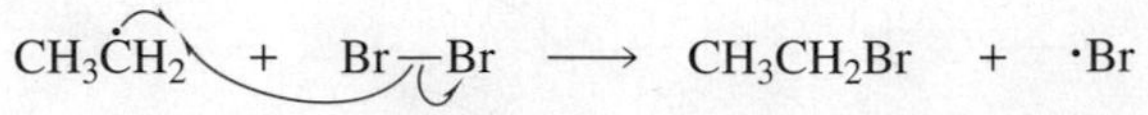

d.

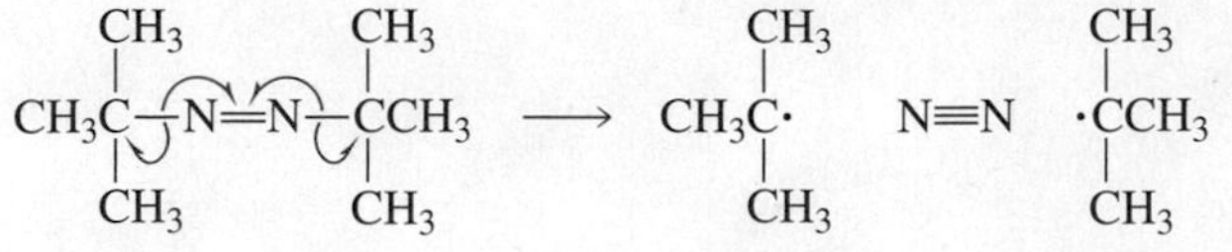

PROBLEM 4

a.

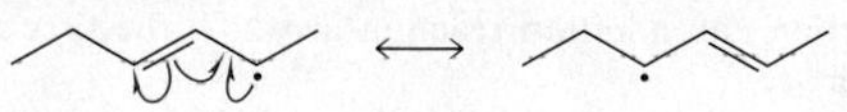

b.

c.

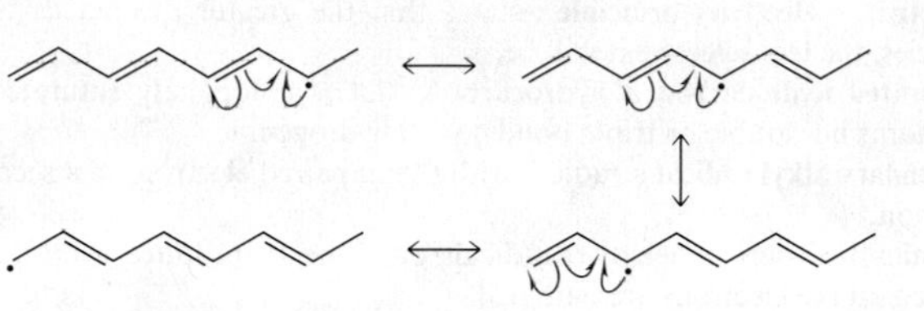

d.

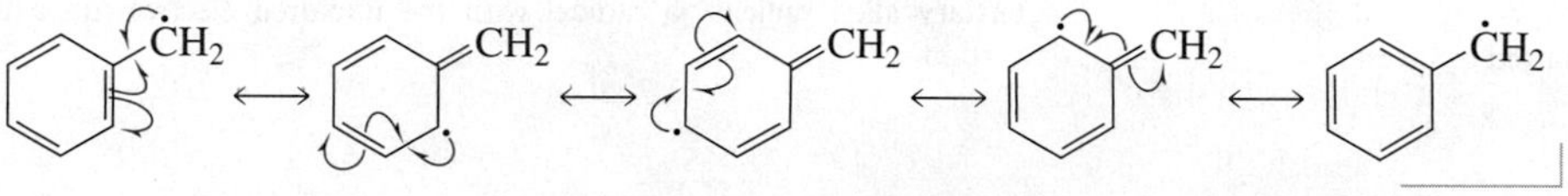

GLOSSARY

alkane a hydrocarbon that contains only single bonds.

halogenation reaction with halogen (Br2 , Cl2 , I2).

heat of combustion the amount of heat given off when a carbon-containing compound reacts completely with O_2 to form CO_2 and H_2O.

heterolytic bond cleavage (heterolysis) breaking a bond with the result that both bonding electrons stay with one of the atoms.

homolytic bond cleavage (homolysis) breaking a bond with the result that each of the atoms gets one of the bonding electrons.

initiation step the step in which radicals are created, or the step in which the radical needed for the first propagation step is created.

primary alkyl radical a radical with the unpaired electron on a primary carbon.

propagation step in the first of a pair of propagation steps, a radical (or an electrophile or a nucleophile) reacts to produce another radical (or an electrophile or a nucleophile) that reacts in the second to produce the radical (or the electrophile or the nucleophile) that was the reactant in the first propagation step.

radical an atom or a molecule with an unpaired electron.

radical addition reaction an addition reaction in which the first species that adds is a radical.

radical chain reaction a reaction in which radicals are formed and react in repeating propagating steps.

radical inhibitor a compound that traps radicals.

radical initiator a compound that creates radicals.

radical substitution reaction a substitution reaction that has a radical intermediate.

reactivity–selectivity principle states that the greater the reactivity of a species, the less selective it will be.

saturated hydrocarbon a hydrocarbon that is completely saturated (i.e., contains no double or triple bonds) with hydrogen.

secondary alkyl radical a radical with the unpaired electron on a secondary carbon.

termination step when two radicals combine to produce a molecule in which all the electrons are paired.

tertiary alkyl radical a radical with the unpaired electron on a tertiary carbon.

PHOTO CREDITS

Photo credits are listed in order of appearance.

minogindmitriy/Fotolia; Colin Anderson/Getty Images; Maslov Dmitry/Shutterstock; (top) PRNewsFoto/POM Wonderful/Associated Press; (bot.) Larina Natalia/Shutterstock; chepatchet/iStockphoto; Philippe Plailly/SPL/Photo Researchers, Inc.; NASA

Mass Spectrometry, Infrared Spectroscopy, and Ultraviolet/Visible Spectroscopy

The red, purple, and blue colors of many flowers, fruits, vegetables are due to a class of compounds called anthocyanins (see later in this chapter).

Determining the structures of organic compounds is an important part of organic chemistry. Whenever a chemist synthesizes a compound, its structure must be confirmed. For example, a ketone is formed when an alkyne undergoes the acid-catalyzed addition of water, but how was it determined that the product of that reaction is actually a ketone? Scientists search the world for new compounds with physiological activity. If a promising compound is found, its structure needs to be determined. Without knowing its structure, chemists cannot design ways to synthesize the compound, nor can they undertake studies to provide insights into its physiological behavior.

Before the structure of a compound can be determined, the compound must be isolated. For example, the product of a reaction carried out in the laboratory must first be isolated from the solvent used to run the reaction, from any unreacted starting materials, and from any side products that might have formed. A compound found in nature must be isolated from the organism that manufactures it.

Isolating products and figuring out their structures used to be daunting tasks. The only tools chemists had for isolating products were distillation (for liquids) and sublimation or fractional recrystallization (for solids). Today, a variety of chromatographic techniques allow compounds to be isolated with relative ease. You will learn about these techniques in your laboratory course.

At one time, determining the structure of an organic compound required finding out its molecular formula by elemental analysis, determining the compound's physical properties (its melting point, boiling point, and so on), and conducting simple chemical tests that indicate the presence (or absence) of certain functional groups.

Unfortunately, these simple procedures were inadequate for characterizing molecules with complex structures, and because a relatively large sample of the compound was needed in order to perform all the tests, they were impractical for the analysis of compounds that were difficult to obtain in large amounts.

Today, a number of different instrumental techniques are used to identify organic compounds. These techniques can be performed quickly on small amounts of a compound and can provide much more information about the compound's structure than simple chemical tests can give.

- **Mass spectrometry** allows us to determine the *molecular mass* and the *molecular formula* of a compound, as well as some of its *structural features.*
- **Infrared (IR) spectroscopy** tells us the *kinds of functional groups* a compound has.
- **Ultraviolet/Visible (UV/Vis) spectroscopy** provides information about organic compounds with conjugated double bonds.
- **Nuclear magnetic resonance (NMR) spectroscopy** provides information about the carbon–hydrogen framework of an organic compound.

Sometimes more than one technique is required to deduce the structure of a compound. You will encounter problems that require you to use two or three techniques at the same time. In this chapter, we will look at three instrumental techniques: mass spectrometry, infrared spectroscopy, and ultraviolet/visible spectroscopy.

We will be referring to different classes of organic compounds as we discuss various instrumental techniques; they are listed in Table 1.

Table 1	Classes of Organic Compounds		
Alkane	$-\overset{\mid}{\underset{\mid}{C}}-$ contains only C—C and C—H bonds	Aldehyde	$\overset{O}{\underset{R}{\parallel}}\diagdown H$
Alkene	$\diagup C = C \diagdown$	Ketone	$\overset{O}{\underset{R}{\parallel}}\diagdown R$
Alkyne	$-C \equiv C-$	Carboxylic acid	$\overset{O}{\underset{R}{\parallel}}\diagdown OH$
Nitrile	$-C \equiv N$	Ester	$\overset{O}{\underset{R}{\parallel}}\diagdown OR$
Alkyl halide	R—X X = F, Cl, Br, or I	Amides {	$\overset{O}{\underset{R}{\parallel}}\diagdown NH_2$
Ether	R—O—R		$\overset{O}{\underset{R}{\parallel}}\diagdown NHR$
Alcohol	R—OH		$\overset{O}{\underset{R}{\parallel}}\diagdown NR_2$
Benzene	(benzene ring)	Amine (primary)	R—NH_2
Phenol	(benzene ring)—OH	Amine (secondary)	$R—\overset{R}{\underset{}{N}}H$
Aniline	(benzene ring)—NH_2	Amine (tertiary)	$R—\overset{R}{\underset{}{N}}—R$

1 MASS SPECTROMETRY

At one time, the molecular weight of a compound was determined by its vapor density or freezing-point depression, and molecular formulas were determined by elemental analysis, a technique for measuring the relative proportions of the elements in the compound. These were long and tedious procedures that required a relatively large amount of a very pure sample of the compound. Today, molecular weights and molecular formulas can be rapidly determined from a very small sample by mass spectrometry.

In mass spectrometry, a small amount of a compound is introduced into an instrument called a mass spectrometer, where it is vaporized and then ionized (an electron is removed from each molecule). The sample can be ionized in several ways. Electron ionization (EI), the most common method, bombards the vaporized molecules with a beam of high-energy electrons. The energy of the beam can be varied but is typically about 70 electron volts. When the electron beam hits a molecule, it knocks out an electron, producing a **molecular ion.** A molecular ion is a **radical cation,** a species with an unpaired electron *and* a positive charge.

$$M \xrightarrow[\text{beam}]{\text{electron}} M^{+\cdot} + e^-$$

molecule   **molecular ion**   **electron**
a radical cation

Electron bombardment injects so much kinetic energy into the molecular ions that most of them break apart (fragment) into cations, radicals, neutral molecules, and other radical cations. Not surprisingly, the bonds most likely to break are the weakest ones and those that result in formation of the most stable products.

All the *positively charged fragments* of the molecule are drawn between two negatively charged plates, which accelerate the fragments into an analyzer tube (Figure 1). Neutral fragments are not attracted to the negatively charged plates and therefore are not accelerated. They are eventually pumped out of the spectrometer.

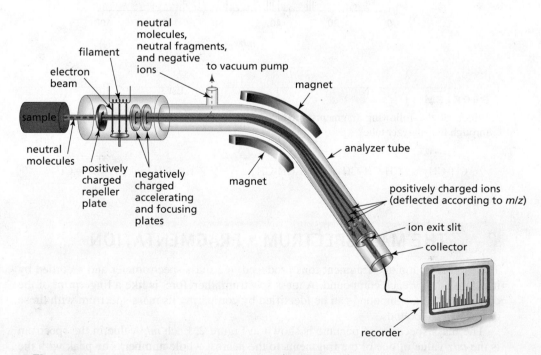

▲ Figure 1

Schematic diagram of an electron ionization mass spectrometer. A beam of high-energy electrons causes molecules to ionize and fragment. Positively charged fragments pass through the analyzer tube. Changing the magnetic field strength makes it possible to separate fragments of varying mass-to-charge ratios.

The analyzer tube is surrounded by a magnet whose magnetic field deflects the positively charged fragments in a curved path. At a given magnetic field strength, the degree to which the path is curved depends on the mass-to-charge ratio (m/z) of the fragment: the path of a fragment with a smaller m/z value will bend more than that of a heavier fragment. In this way, the particles with the same m/z values can be separated from all the others. If a fragment's path matches the curvature of the analyzer tube, the fragment will pass through the tube and out the ion exit slit.

A collector records the relative number of fragments with a particular m/z value passing through the slit. The more stable the fragment, the more likely it is to arrive at the collector without breaking down further. The strength of the magnetic field is gradually increased, so fragments with progressively larger m/z values are guided through the tube and out the exit slit.

The mass spectrometer records a **mass spectrum**—a graph of the relative abundance of each fragment plotted against its m/z value (Figure 2). Because the charge (z) on essentially all the fragments that reach the collector plate is $+1$, m/z is the mass (m) of the fragment. *Remember that only positively charged species reach the collector.*

A mass spectrum records only positively charged fragments.

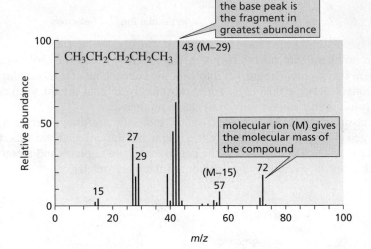

▶ **Figure 2**

The mass spectrum of pentane. The base peak represents the fragment that appears in greatest abundance. The m/z value of the molecular ion (M) gives the molecular mass of the compound.

PROBLEM 1♦

Which of the following fragments produced in a mass spectrometer would be accelerated through the analyzer tube?

$$CH_3\dot{C}H_2 \qquad CH_3CH_2\overset{+}{C}H_2 \qquad [CH_3CH_2CH_3]^{\overset{+}{\cdot}} \qquad \dot{C}H_2CH{=}CH_2 \qquad \overset{+}{C}H_2CH{=}CH_2$$

---

2   THE MASS SPECTRUM • FRAGMENTATION

The molecular ion and fragment ions produced in a mass spectrometer and recorded by it are unique for each compound. A mass spectrum, therefore, is like a fingerprint of the compound, so a compound can be identified by comparing its mass spectrum with those of known compounds.

The mass spectrum of pentane is shown in Figure 2. Each m/z value in the spectrum is the m/z value of one of the fragments to the nearest whole number. The peak with the highest m/z value in the spectrum—in this case, at $m/z = 72$—is the molecular ion (M), the fragment that results when an electron is knocked out of a molecule. (The extremely tiny peak at $m/z = 73$ will be explained later.) *The m/z value of the molecular ion gives the molecular mass of the compound.*

Because it is not known what bond loses the electron, the molecular ion is written in brackets and the positive charge and unpaired electron are assigned to the entire structure.

$$CH_3CH_2CH_2CH_2CH_3 \xrightarrow{\text{electron beam}} [CH_3CH_2CH_2CH_2CH_3]^{+\cdot} + e^-$$

molecular ion
$m/z = 72$

The *m/z* value of the molecular ion gives the molecular mass of the compound.

Peaks with smaller *m/z* values—called **fragment ion peaks**—represent positively charged fragments of the molecular ion. The **base peak** is the tallest peak, because it has the greatest relative abundance. It has the greatest relative abundance because it is generally the most stable positively charged fragment. The base peak is assigned a relative abundance of 100%, and the relative abundance of each of the other peaks is shown as a percentage of the base peak.

A mass spectrum gives us structural information about the compound because the *m/z* values and relative abundances of the fragments depend on the strength of the molecular ion's bonds and the stability of the fragments. *Weak bonds break in preference to strong bonds, and bonds that break to form more stable fragments break in preference to those that form less stable fragments.*

For example, all the C—C bonds in the molecular ion formed from pentane have about the same strength. However, the C-2–C-3 bond is the one most likely to break because it leads to formation of a *primary* carbocation and a *primary* radical, which together are more stable than the *primary* carbocation and *methyl* radical (or *primary* radical and *methyl* cation) obtained from C-1–C-2 fragmentation.

Ions formed by C-2–C-3 fragmentation have *m/z* values of 43 and 29, whereas ions formed by C-1–C-2 fragmentation have *m/z* values 57 and 15. The base peak of 43 in the mass spectrum of pentane indicates the greater likelihood of C-2–C-3 fragmentation.

The way a molecular ion fragments depends on the strength of its bonds and the stability of the fragments.

A method for identifying fragment ions makes use of the difference between the *m/z* value of a given fragment ion and that of the molecular ion. For example, the fragment ion with *m/z* = 43 in the mass spectrum of pentane is 29 units smaller than the molecular ion (73 – 43 = 29). An ethyl radical (CH_3CH_2) has a *m/z* value of 29 (because the mass numbers of C and H are 12 and 1, respectively). Thus, the peak at 43 can be attributed to loss of an ethyl radical from the molecular ion. Similarly, the fragment ion with *m/z* = 57 can be attributed to loss of a methyl radical from the molecular ion (73 – 57 = 15). Peaks at *m/z* = 15 and *m/z* = 29 are readily recognizable as being due to methyl and ethyl cations, respectively. Appendix: "Spectroscopy Tables," contains a table of common fragment ions and a table of common fragments lost.

Peaks are commonly observed at *m/z* values two units below the *m/z* values of the carbocations, because a carbocation can lose two hydrogen atoms.

$$\overset{+}{CH_3CH_2CH_2} \longrightarrow \overset{+}{CH_2CH}=CH_2 + 2\,H\cdot$$

$m/z = 43 \qquad\qquad m/z = 41$

2-Methylbutane has the same molecular formula as pentane. Thus, it too has a molecular ion with an *m/z* value of 72 (Figure 3). Its mass spectrum is similar to that of pentane, with one notable exception: the peak at $m/z = 57$ indicating loss of a methyl radical, is much more intense than the same peak in pentane.

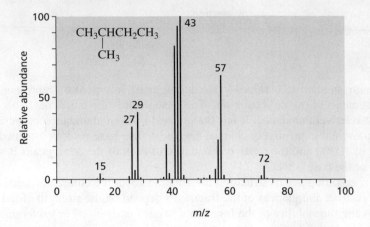

▲ **Figure 3**

The mass spectrum of 2-methylbutane. The peak at $m/z = 57$ corresponds to the loss of a methyl group and the formation of a relatively stable secondary carbocation.

2-Methylbutane is more likely than pentane to lose a methyl radical because, when it does, a *secondary* carbocation is formed. In contrast, when pentane loses a methyl radical, a less stable *primary* carbocation is formed.

$$\begin{bmatrix} CH_3 \\ | \\ CH_3CHCH_2CH_3 \end{bmatrix}^{\ddagger} \longrightarrow CH_3\overset{+}{C}HCH_2CH_3 + \cdot CH_3$$

molecular ion
$m/z = 72$ $m/z = 57$

PROBLEM 2

What would distinguish the mass spectrum of 2,2-dimethylpropane from the mass spectra of pentane and 2-methylbutane?

PROBLEM 3◆

What is the likeliest m/z value for the base peak in the mass spectrum of 3-methylpentane?

---

3   USING THE m/z VALUE OF THE MOLECULAR ION TO CALCULATE THE MOLECULAR FORMULA

The **rule of 13** allows possible molecular formulas to be determined from the m/z value of the molecular ion. Remember that the m/z value of the molecular ion gives the molecular mass of the compound.

First the **base value** must be determined. To do this, divide the m/z value of the molecular ion by 13. The answer gives the number of carbons in the compound. For example, if the m/z value is 142, dividing 142 by 13 gives 10 as the number of carbons (with 12 left over). The number of Hs is determined by adding the number left over to the number of carbons ($10 + 12 = 22$). Thus, the base value is $C_{10}H_{22}$.

If the compound has one oxygen, then one O (16 amu) must be added to the base value and one C and four Hs (16 amu) must be subtracted from it. Thus, the molecular formula is $C_9H_{18}O$. If the compound has two oxygens, the process must be repeated, in which case the molecular formula is $C_8H_{14}O_2$. (Notice that in order to maintain the m/z value, you need to subtract the same number of atomic mass units that you add.)

PROBLEM 4 Solved

Draw possible structures for an ester that has a molecular ion with an m/z value of 74.

Solution Dividing 74 by 13 gives 5 with 9 left over. Thus, the base value is C_5H_{14}. We know that an ester has two oxygens. For each oxygen, add one O and subtract one C and four Hs. This gives a molecular formula of $C_3H_6O_2$. Possible structures are

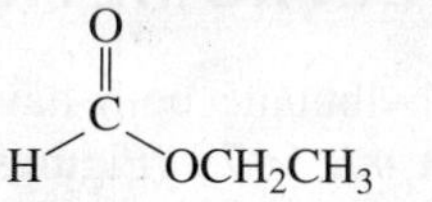

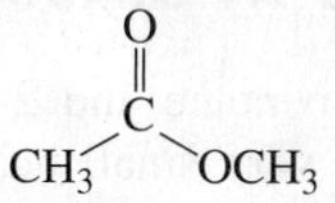

$$\underset{\text{H}}{\overset{\overset{\displaystyle O}{\parallel}}{\underset{}{C}}}\text{OCH}_2\text{CH}_3 \qquad \underset{\text{CH}_3}{\overset{\overset{\displaystyle O}{\parallel}}{\underset{}{C}}}\text{OCH}_3$$

PROBLEM 5♦

Determine the molecular formula for each of the following:

a. a compound that contains only C and H and has a molecular ion with an m/z value of 72
b. a compound that contains C, H, and one O and has a molecular ion with an m/z value of 100
c. a compound that contains C, H, and two Os and has a molecular ion with an m/z value of 102
d. an amide that has a molecular ion with an m/z value of 115

PROBLEM 6

If a compound has a molecular ion with an odd-numbered mass, then the compound contains an odd number of nitrogen atoms. This is known as the **nitrogen rule.**

a. Calculate the m/z value for the molecular ion of each of the following compounds:

1. 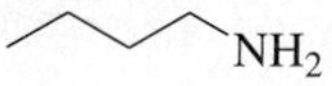 $\diagup\diagdown\diagup$NH$_2$ **2.** H$_2$N$\diagup\diagdown\diagup$NH$_2$

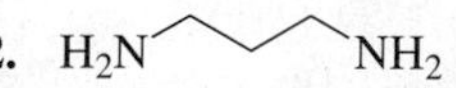

b. Explain why the nitrogen rule holds.
c. State the rule in terms of a molecular ion with an even-numbered mass.

PROBLEM 7♦

a. Suggest possible molecular formulas for a compound that has a molecular ion with an m/z value of 86.
b. Can one of the possible molecular formulas contain a nitrogen atom?

PROBLEM-SOLVING STRATEGY

Using Mass Spectra to Determine Structures

The mass spectra of two very stable cycloalkanes both show a molecular ion peak at $m/z = 98$. One spectrum shows a base peak at $m/z = 69$ whereas the other shows a base peak at $m/z = 83$. Identify the cycloalkanes.

First, let's determine the molecular formula of the compounds from the m/z value of their molecular ions. Dividing 98 by 13 results in 7 with 7 left over. Thus, each of their molecular formulas is C_7H_{14}.

Now let's see what fragment is lost to give the base peak. A base peak of 69 means the loss of an ethyl radical $(98 - 69 = 29)$, whereas a base peak of 83 means the loss of a methyl radical $(98 - 83 = 15)$.

Because the two cycloalkanes are known to be very stable, we can assume they do not have three- or four-membered rings. A seven-carbon cycloalkane with a base peak signifying the loss of an ethyl radical must be ethylcyclopentane. A seven-carbon cycloalkane with a base peak signifying the loss of a methyl radical must be methylcyclohexane.

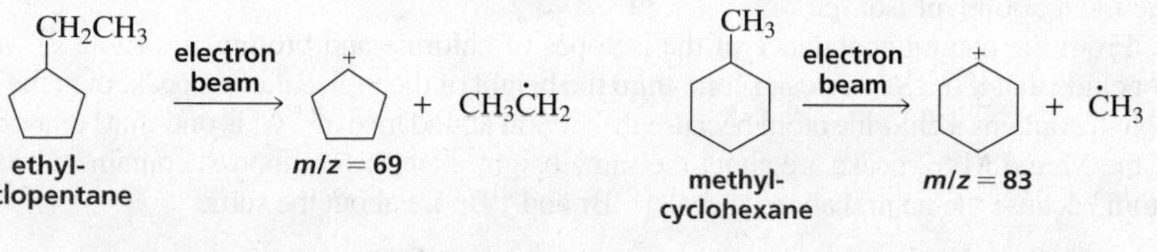

Now use the strategy you have just learned to solve Problem 8.

PROBLEM 8♦

Identify the hydrocarbon that has a molecular ion with an *m/z* value of 128, a base peak with an *m/z* value of 43, and significant peaks with *m/z* values of 57, 71, and 85.

4 ISOTOPES IN MASS SPECTROMETRY

The molecular ions of pentane and 2-methylbutane both have *m/z* values of 72, but each spectrum shows a very small peak at $m/z = 73$ (Figures 2 and 3). This peak is called the M+1 peak because the ion responsible for it is one unit heavier than the molecular ion. The M+1 peak owes its presence to the fact that there are two naturally occurring isotopes of carbon: ^{12}C (98.89% of naturally occurring carbon) and ^{13}C (1.11%). Because mass spectrometry records individual molecules, any molecule containing a ^{13}C will appear at M+1.

The isotopic distributions of several elements commonly found in organic compounds are listed in Table 2.

Table 2 The Natural Abundance of Isotopes Commonly Found in Organic Compounds

Element	Natural abundance			
Carbon	^{12}C	^{13}C		
	98.89%	1.11%		
Hydrogen	^{1}H	^{2}H		
	99.99%	0.01%		
Nitrogen	^{14}N	^{15}N		
	99.64%	0.36%		
Oxygen	^{16}O	^{17}O	^{18}O	
	99.76%	0.04%	0.20%	
Sulfur	^{32}S	^{33}S	^{34}S	^{36}S
	95.0%	0.76%	4.22%	0.02%
Fluorine	^{19}F			
	100%			
Chlorine	^{35}Cl		^{37}Cl	
	75.77%		24.23%	
Bromine	^{79}Br		^{81}Br	
	50.69%		49.31%	
Iodine	^{127}I			
	100%			

Mass spectra can also show M+2 peaks due to ^{18}O or from having two heavy isotopes in the same molecule (such as ^{13}C and ^{2}H, or two ^{13}Cs). These situations are unusual, though, so the M+2 peaks tend to be very small. The presence of a large M+2 peak is evidence of a compound containing either chlorine or bromine, because each of these elements has a high percentage of a naturally occurring isotope that is two units heavier than the most abundant isotope.

From the natural abundance of the isotopes of chlorine and bromine in Table 2, we can conclude that if the M+2 peak is one-third the height of the molecular ion peak, then the compound contains a chlorine atom because the natural abundance of ^{37}Cl is one-third that of ^{35}Cl. If the M and M+2 peaks are about the same height, then the compound contains a bromine atom because the natural abundances of ^{79}Br and ^{81}Br are about the same.

In calculating the *m/z* values of molecular ions and fragments, the *atomic mass* of a single isotope of the atom must be used (for example, Cl = 35 or 37) because mass spectrometry records the *m/z* value of an *individual* fragment. The *atomic weights* in the periodic table (Cl = 35.453) cannot be used, because they are the *weighted averages* of all the naturally occurring isotopes for that element.

PROBLEM 9♦

Predict the relative intensities of the molecular ion peak, the M+2 peak, and the M+4 peak for a compound that contains two bromine atoms.

---

5   HIGH-RESOLUTION MASS SPECTROMETRY CAN REVEAL MOLECULAR FORMULAS

All the mass spectra shown in this text were produced with a low-resolution mass spectrometer. Such spectrometers give the *m/z* value of a fragment to the nearest whole number. High-resolution mass spectrometers can determine the *exact mass* of a fragment to a precision of 0.0001 amu, making it possible to distinguish between compounds that have the same molecular mass to the nearest whole number. For example, the following listing shows six compounds that have a molecular mass of 122 amu, but each of them has a different exact molecular mass.

Some Compounds with a Molecular Mass of 122 amu and Their Exact Molecular Masses and Molecular Formulas

Exact molecular mass (amu)	122.1096	122.0845	122.0732	122.0368	122.0579	122.0225
Molecular formula	C_9H_{14}	$C_7H_{10}N_2$	$C_8H_{10}O$	$C_7H_6O_2$	$C_4H_{10}O_4$	$C_4H_{10}S_2$

The exact masses of some common isotopes are listed in Table 3. There are computer programs that can determine the molecular formula of a compound from the compound's exact molecular mass.

Table 3   The Exact Masses of Some Common Isotopes

Isotope	Mass	Isotope	Mass
^1H	1.007825 amu	^{32}S	31.9721 amu
^{12}C	12.00000 amu	^{35}Cl	34.9689 amu
^{14}N	14.0031 amu	^{79}Br	78.9183 amu
^{16}O	15.9949 amu		

PROBLEM 10♦

Which molecular formula has an exact molecular mass of 86.1096 amu: C_6H_{14}, $C_4H_{10}N_2$, or $C_4H_6O_2$?

PROBLEM 11

a. Can a low-resolution mass spectrometer distinguish between $C_2H_5^+$ and CHO^+?
b. Can a high-resolution mass spectrometer distinguish between them?

6 THE FRAGMENTATION PATTERNS OF FUNCTIONAL GROUPS

Each functional group has characteristic fragmentation patterns that can help identify a compound. The patterns began to be recognized after the mass spectra of many compounds containing a particular functional group had been studied. We will look at the fragmentation patterns of alkyl halides, ethers, alcohols, and ketones as examples.

Alkyl Halides

Let's look first at the mass spectrum of 1-bromopropane (Figure 4). The relative heights of the M and M+2 peaks are about equal, which we have seen is characteristic of a compound containing a bromine atom. Electron bombardment is most likely to dislodge a lone-pair electron if the molecule has any, because a molecule does not hold onto its lone-pair electrons as tightly as it holds onto its bonding electrons. Thus, the molecular ion is created when electron bombardment dislodges one of bromine's lone-pair electrons.

$$CH_3CH_2CH_2 \overset{79}{-} \ddot{\overset{..}{Br}}: \; + \; CH_3CH_2CH_2 \overset{81}{-} \ddot{\overset{..}{Br}}: \; \xrightarrow{-e^-} \; CH_3CH_2CH_2 \overset{79}{\underset{\curvearrowleft}{-}} \overset{+}{\ddot{Br}}: \; + \; CH_3CH_2CH_2 \overset{81}{\underset{\curvearrowleft}{-}} \overset{+}{\ddot{Br}}: \; \longrightarrow \; CH_3CH_2\overset{+}{C}H_2 \; + \; {}^{79}\ddot{\overset{..}{Br}}: \; + \; {}^{81}\ddot{\overset{..}{Br}}:$$

1-bromopropane $m/z = 122$ $m/z = 124$ $m/z = 43$

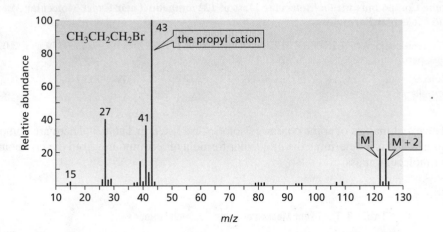

▲ **Figure 4**

The mass spectrum of 1-bromopropane. The M and M+2 peaks are about equal because ^{79}Br and ^{81}Br have almost equal abundances.

The weakest bond in the molecular ion is the one most apt to break. In this case, the C—Br bond is the weakest. When the C—Br bond breaks, it breaks heterolytically, with both electrons going to the more electronegative of the atoms that were joined by the bond, forming a propyl cation and a bromine atom. As a result, the base peak in the mass spectrum of 1-bromopropane is at $m/z = 43$ (M − 79 = 43 or [M+2] − 81 = 43).

The mass spectrum of 2-chloropropane is shown in Figure 5. The M+2 peak is one-third the height of the molecular ion peak, which is consistent with a compound that contains a chlorine atom (Section 4).

The C—Cl (85 kcal/mol) and C—C (89 kcal/mol) bonds have similar strengths, so both bonds can break. The C—Cl bond breaks heterolytically, giving a base peak at $m/z = 43$. The C—C bond breaks homolytically; the peaks at $m/z = 63$ and

The way a molecular ion fragments depends on the strength of its bonds and the stability of the fragments.

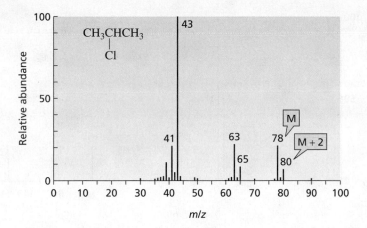

◀ **Figure 5**

The mass spectrum of 2-chloropropane. The M+2 peak is one-third the size of the M peak because ^{37}Cl has one-third the natural abundance of ^{35}Cl.

$m/z = 65$ have a 3:1 ratio, indicating that these fragments contain a chlorine atom. They result from *homolytic cleavage* (called **α-cleavage**) of a group bonded to the α-carbon. (The α-carbon is the carbon bonded to the chlorine).

$$2\ CH_3\overset{\underset{|}{CH_3}}{CH} \ +\ \ ^{35}\!\overset{\cdot\cdot}{\underset{\cdot\cdot}{Cl}}\!: \ + \ ^{37}\!\overset{\cdot\cdot}{\underset{\cdot\cdot}{Cl}}\!:$$
$$m/z = \mathbf{43}$$

↑ heterolytic cleavage

the α-carbon

$$CH_3\overset{\underset{|}{CH_3}}{CH}\!-\!^{35}\!\overset{\cdot\cdot}{\underset{\cdot\cdot}{Cl}}\!: \ + \ CH_3\overset{\underset{|}{CH_3}}{CH}\!-\!^{37}\!\overset{\cdot\cdot}{\underset{\cdot\cdot}{Cl}}\!: \ \xrightarrow{-e^-} \ CH_3\overset{\underset{|}{CH_3}}{CH}\!\overset{+}{-}\!^{35}\!\overset{\cdot\cdot}{\underset{\cdot\cdot}{Cl}}\!: \ + \ CH_3\overset{\underset{|}{CH_3}}{CH}\!\overset{+}{-}\!^{37}\!\overset{\cdot\cdot}{\underset{\cdot\cdot}{Cl}}\!:$$

2-chloropropane $m/z = \mathbf{78}$ $m/z = \mathbf{80}$

↓ homolytic cleavage (α-cleavage)

$$CH_3CH\!=\!\!^{35}\!\overset{+}{\underset{\cdot\cdot}{Cl}}\!: \ + \ CH_3CH\!=\!\!^{37}\!\overset{+}{\underset{\cdot\cdot}{Cl}}\!: \ + \ 2\ ^{\cdot}CH_3$$
$$m/z = \mathbf{63} \qquad\qquad m/z = \mathbf{65}$$

α-Cleavage occurs in alkyl chlorides because the C—C and C—Cl bonds have similar strengths.

α-Cleavage is less likely to occur in alkyl bromides because the C—C bond is much stronger than the C—Br bond.

α-Cleavage occurs because it leads to a cation that is relatively stable since its positive charge is shared by two atoms:

product of α-cleavage   **2 atoms share the positive charge**

$$CH_3\overset{\overset{CH_3}{\curvearrowleft}}{CH}\!-\!\overset{+}{\underset{\cdot\cdot}{Cl}}\!: \ \xrightarrow{\text{α-cleavage}} \ CH_3\overset{\cdot}{CH}\!-\!\overset{+}{\underset{\cdot\cdot}{Cl}}\!: \ \longleftrightarrow \ CH_3CH\!=\!\overset{+}{\underset{\cdot\cdot}{Cl}}\!: \ \longleftrightarrow \ CH_3\overset{+}{CH}\!-\!\overset{\cdot\cdot}{\underset{\cdot\cdot}{Cl}}$$

An arrowhead with one barb represents the movement of one electron.

α-Cleavage is much less likely to occur in *alkyl bromides* because the C—C (89 kcal/mol) bond is much stronger than the C—Br bond (71 kcal/mol).

PROBLEM 12

Sketch the mass spectrum expected for 1-chloropropane.

Ethers

The mass spectrum of 2-isopropoxybutane is shown in Figure 6.

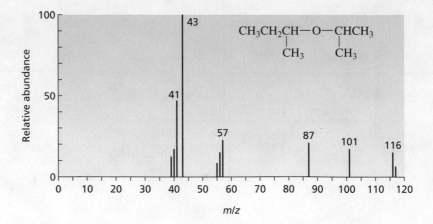

▶ Figure 6

The mass spectrum of 2-isopropoxybutane.

The fragmentation pattern of an ether is similar to that of an alkyl halide.

1. Electron bombardment dislodges one of the lone-pair electrons from oxygen.

2. Fragmentation of the resulting molecular ion occurs mainly in two ways:

   a. A C—O bond is cleaved *heterolytically*, with the electrons going to the more electronegative oxygen atom.

$$CH_3CH_2\overset{\overset{\displaystyle CH_3}{|}}{CH}-\overset{..}{\overset{..}{O}}-\overset{\overset{\displaystyle CH_3}{|}}{CH}CH_3 \quad \xrightarrow{-e^-} \quad CH_3CH_2\overset{\overset{\displaystyle CH_3}{|}}{CH}-\overset{+}{\overset{..}{O}}-\overset{\overset{\displaystyle CH_3}{|}}{CH}CH_3$$

2-isopropoxybutane   **m/z = 116**

$$CH_3CH_2\overset{+}{\overset{\overset{\displaystyle CH_3}{|}}{CH}} \;+\; :\overset{..}{\overset{}{O}}-\overset{\overset{\displaystyle CH_3}{|}}{CH}CH_3$$

m/z = 57

$$CH_3CH_2\overset{\overset{\displaystyle CH_3}{|}}{CH}-\overset{..}{\overset{..}{O}}: \;+\; \overset{+}{\overset{\overset{\displaystyle CH_3}{|}}{CH}}CH_3$$

m/z = 43

   b. A C—C bond is cleaved *homolytically* at an α-carbon because this leads to a relatively stable cation in which the positive charge is shared by two atoms (a carbon and an oxygen). The alkyl group most easily cleaved is the one that forms the most stable radical. Thus, the peak at *m/z* = 87 is more abundant than the one at *m/z* = 101 because a primary radical is more stable than a methyl radical, even though the compound has three methyl groups bonded to α-carbons that can be cleaved to produce the peak at *m/z* = 101

$$CH_3CH_2\!\!-\!\!\overset{\overset{\displaystyle CH_3}{|}}{CH}\!-\!\overset{+}{\overset{..}{O}}\!-\!\overset{\overset{\displaystyle CH_3}{|}}{CH}CH_3 \quad \xrightarrow{\alpha\text{-cleavage}} \quad \overset{\overset{\displaystyle CH_3}{|}}{CH}=\overset{+}{\overset{..}{O}}-\overset{\overset{\displaystyle CH_3}{|}}{CH}CH_3 \;+\; CH_3\overset{.}{C}H_2$$

m/z = 87

$$CH_3CH_2\overset{\overset{\overset{\displaystyle CH_3}{|}}{}}{CH}\!\!-\!\!\overset{+}{\overset{..}{O}}\!-\!\overset{\overset{\displaystyle CH_3}{|}}{CH}CH_3 \quad \xrightarrow{\alpha\text{-cleavage}} \quad CH_3CH_2CH=\overset{+}{\overset{..}{O}}-\overset{\overset{\displaystyle CH_3}{|}}{C}CH_3 \;+\; \overset{.}{C}H_3$$

m/z = 101

$$CH_3CH_2\overset{\overset{\displaystyle CH_3}{|}}{CH}\!-\!\overset{+}{\overset{..}{O}}\!\overset{\overset{\displaystyle CH_3}{|}}{CH}CH_3 \quad \xrightarrow{\alpha\text{-cleavage}} \quad CH_3CH_2\overset{\overset{\displaystyle CH_3}{|}}{CH}\!-\!\overset{+}{\overset{..}{O}}=CHCH_3 \;+\; \overset{.}{C}H_3$$

α-carbon   **m/z = 101**

PROBLEM 13◆

The mass spectra of 1-methoxybutane, 2-methoxybutane, and 2-methoxy-2-methylpropane are shown in Figure 7. Match each compound with its spectrum.

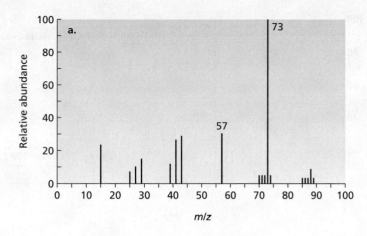

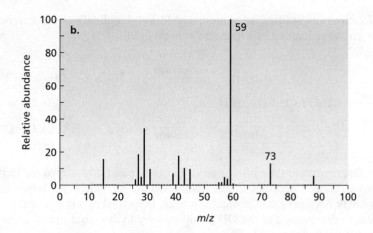

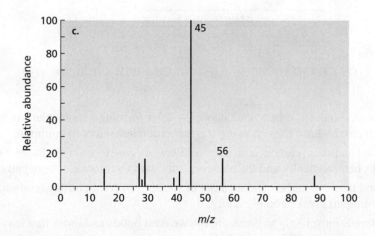

◀ **Figure 7**

The mass spectra for Problem 13.

Alcohols

As with alkyl halides and ethers, the molecular ion of an alcohol is created by knocking out a lone-pair electron. The molecular ions obtained from alcohols fragment so readily that few of them survive to reach the collector. As a result, the mass spectrum of an alcohol shows a small molecular ion peak, if any. Notice the small molecular ion peak at $m/z = 102$ in the mass spectrum of 2-hexanol (Figure 8).

The molecular ion of an alcohol, like those of alkyl halides and ethers, undergoes α-cleavage. Consequently, the mass spectrum of 2-hexanol shows a base peak at

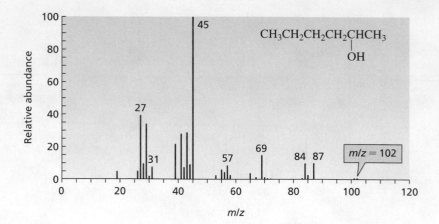

▶ **Figure 8**

The mass spectrum of 2-hexanol. The molecular ion peak at $m/z = 102$ is small because alcohols fragment so easily.

$m/z = 45$ (α-cleavage leading to a more stable butyl radical) and a smaller peak at $m/z = 87$ (α-cleavage leading to a less stable methyl radical).

In all the fragmentations we have seen so far, only one bond is broken. In the case of alcohols, however, an important fragmentation occurs in which two bonds are broken. Two bonds break because a stable water molecule is formed as a result of the fragmentation. The water is composed of the OH group and a γ-hydrogen. Loss of water results in a fragmentation peak at $m/z = M{-}18$.

Notice that alkyl halides, ethers, and alcohols—after forming a molecular ion by losing a lone-pair electron—have the following fragmentation behaviors in common:

1. A bond between carbon and a *more electronegative* atom (halogen or oxygen) breaks heterolytically and the electrons stay with the more electronegative atom.

2. A bond between carbon and an atom of *similar electronegativity* (carbon, chlorine, or hydrogen) breaks homolytically.

3. The bonds most likely to break are the weakest bonds and those that lead to formation of the most stable products. (Look for a fragmentation that results in a cation with a positive charge shared by two atoms.)

PROBLEM 14♦

Primary alcohols have a strong peak at $m/z = 31$. What fragment is responsible for this peak?

Ketones

A ketone generally has an intense molecular ion peak that is formed by knocking out a lone-pair electron. The molecular ion fragments homolytically at the C—C bond adjacent to the C—O bond because this cleavage results in formation of a cation with a positive charge shared by two atoms. The alkyl group leading to the more stable radical is the one more easily cleaved. Thus, the peak at $m/z = 43$ will be more abundant than the one at $m/z = 71$.

$$CH_3CH_2CH_2\overset{\overset{\displaystyle \ddot{O}:}{\|}}{C}CH_3 \xrightarrow{-e^-} CH_3CH_2CH_2\overset{\overset{\displaystyle \ddot{O}\overset{+}{\cdot}}{\|}}{C}CH_3$$

2-pentanone $m/z = 86$

$$\longrightarrow CH_3CH_2\dot{C}H_2 \;+\; CH_3C\!\!\equiv\!\!\overset{+}{O}:$$
$$m/z = 43$$

$$\longrightarrow CH_3CH_2CH_2C\!\!\equiv\!\!\overset{+}{O}: \;+\; \dot{C}H_3$$
$$m/z = 71$$

If one of the alkyl groups attached to a carbonyl carbon has a γ-hydrogen, then a cleavage which goes through a favorable six-membered-ring transition state may occur. This fragmentation—known as a **McLafferty rearrangement**—also breaks two bonds. The bond between the α-carbon and the β-carbon breaks homolytically and a hydrogen atom from the γ-carbon migrates to the oxygen. Again, fragmentation has occurred in a way that produces a cation with a positive charge shared by two atoms.

$$\xrightarrow[\text{rearrangement}]{\text{McLafferty}} H_2C\!\!=\!\!CH_2 \;+\; \dot{C}H_2\!-\!\overset{\overset{\displaystyle H\diagdown \overset{+}{O}:}{\|}}{C}\!-\!CH_3$$

$m/z = 86$ $m/z = 58$

PROBLEM 15♦

Identify the ketones responsible for the mass spectra in Figure 9.

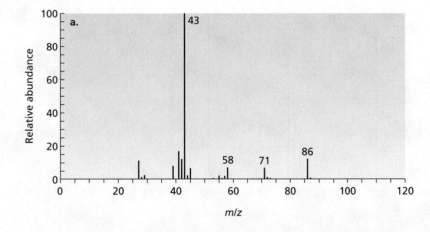

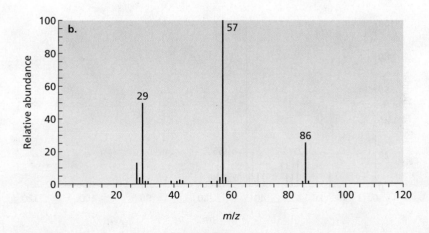

The mass spectra for Problem 15.

PROBLEM 16

How could their mass spectra be used to distinguish between the following compounds?

$$CH_3CH_2\overset{\overset{\displaystyle O}{\|}}{C}CH_2CH_3 \qquad CH_3\overset{\overset{\displaystyle O}{\|}}{C}CH_2CH_2CH_3 \qquad CH_3\overset{\overset{\displaystyle O}{\|}}{C}\underset{\underset{\displaystyle CH_3}{|}}{CHCH_3}$$

PROBLEM 17

Using curved arrows, show the principal fragments that would be observed in the mass spectrum of each of the following compounds:

a. $CH_3CH_2CH_2CH_2CH_2OH$

c. $CH_3CH_2O\underset{\underset{\displaystyle CH_3}{|}}{\overset{\overset{\displaystyle CH_2CH_3}{|}}{C}}CH_2CH_2CH_3$

e. $CH_3CH_2\underset{\underset{\displaystyle CH_3}{|}}{CHCl}$

b. $CH_3CH_2\underset{\underset{\displaystyle OH}{|}}{CH}CH_2CH_2CH_2CH_3$

d. $CH_3\overset{\overset{\displaystyle O}{\|}}{C}CH_2CH_2CH_2CH_3$

f. $CH_3\overset{\overset{\displaystyle CH_3}{|}}{\underset{\underset{\displaystyle CH_3}{|}}{C}}Br$

PROBLEM 18◆

The reaction of (Z)-2-pentene with water and a trace of H_2SO_4 forms two products. Identify the products from their mass spectra that are shown in Figure 10.

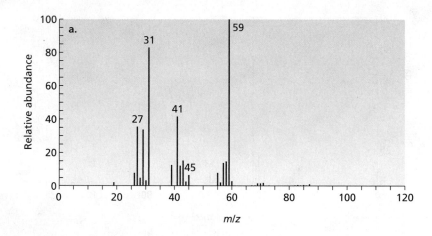

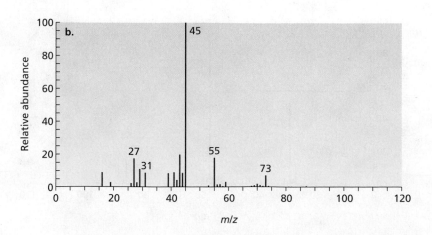

▶ **Figure 10**

The mass spectra for Problem 18.

7   OTHER IONIZATION METHODS

Methods other than electron bombardment (EI–MS) can be used to obtain mass spectral data. In chemical ionization–mass spectrometry (CI–MS), the sample is sprayed with a pre-ionized gas such as methane or ammonia that causes the sample to ionize by electron transfer or proton transfer from the gas to the sample. Because the molecular ions produced by this technique are less apt to undergo fragmentation, the ability to obtain the molecular mass (and therefore the molecular formula) of the sample is enhanced.

Both EI–MS and CI–MS require the sample to be vaporized before it is ionized. Therefore, these methods can be used only for samples with relatively low molecular weights. Mass spectra of large molecules, even large biological molecules such as enzymes or oligonucleotides, can be obtained by desorption ionization (DI). In this technique, the sample is dissolved on a matrix compound, which is then ionized by one of several methods. The matrix compound transfers energy to the sample, causing some of the sample molecules to ionize and be ejected from the matrix. The ejected ions are accelerated in the mass spectrometer and recorded to give a mass spectrum as in EI–MS (Section 1). Ionization methods include bombarding the matrix compound with a high-energy beam of ions (called secondary ion MS [SIMS]), photons (called matrix-assisted laser desorption ionization [MALDI]), or neutral atoms (called fast-atom bombardment [FAB]).

8   GAS CHROMATOGRAPHY–MASS SPECTROMETRY

Mixtures of compounds are often analyzed using gas chromatography and mass spectrometry (GC–MS) at the same time. The sample is injected into a gas chromatograph and the various components of the mixture travel through the column at different rates, based on their boiling points. The lowest boiling component of the mixture exits first. As each compound exits, it enters the mass spectrometer where it is ionized, forming a molecular ion and fragments of the molecular ion. The mass spectrometer records a mass spectrum for each of the components of the mixture. GC–MS is widely used to analyze forensic samples.

Mass Spectrometry in Forensics

Forensic science is the application of science for the purpose of answering questions related to a civil or criminal case. Mass spectrometry is an important tool of the forensic scientist. It is used to analyze body fluids for the presence and levels of drugs and other toxic substances. It can also identify the presence of drugs in hair, which increases the window of detection from hours and days (after which body fluids are no longer useful) to months and even years. It was employed for the first time at an athletic event in 1955 to detect drugs in athletes at a cycling competition in France. (Twenty percent of those tests were positive.) Mass spectrometry is also used to identify residues of arson fires and explosives from post-explosion residues, and to analyze paints, adhesives, and fibers.

9   SPECTROSCOPY AND THE ELECTROMAGNETIC SPECTRUM

Spectroscopy is the study of the interaction of matter and electromagnetic radiation. A continuum of different types of **electromagnetic radiation**—each associated with a particular energy range—makes up the electromagnetic spectrum (Figure 11). Visible light is the electromagnetic radiation we are most familiar with, but it represents only a fraction of the full electromagnetic spectrum. X-rays, microwaves, and radio waves are other familiar types of electromagnetic radiation.

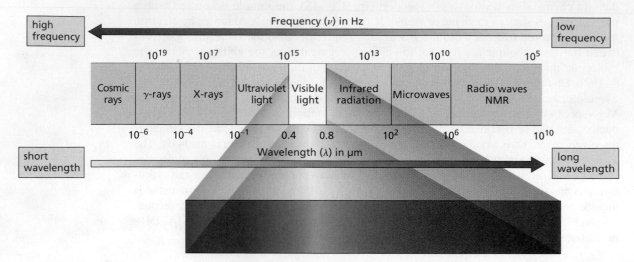

▲ **Figure 11**

The electromagnetic spectrum. Electromagnetic radiation with the highest energy (the highest frequency and the shortest wavelength) is located on the left, whereas electromagnetic radiation with the lowest energy (the lowest frequency and the longest wavelength) is located on the right.

The various kinds of electromagnetic radiation can be characterized briefly as follows:

- *Cosmic rays* are discharged by the sun; they have the highest energy of the various kinds of electromagnetic radiation.

- *γ-Rays* (gamma rays) are emitted by the nuclei of certain radioactive elements. Because of their high energy, they can severely damage biological organisms.

- *X-Rays,* somewhat lower in energy than γ-rays, are less harmful, except in high doses. Low-dose X-rays are used to examine the internal structure of organisms. The denser the tissue, the more it blocks X-rays.

- *Ultraviolet (UV) light,* a component of sunlight, causes sunburns, and repeated exposure to it can cause skin cancer by damaging DNA molecules in skin cells.

- *Visible light* is the electromagnetic radiation we see.

- We feel *infrared radiation* as heat.

- We cook with *microwaves* and use them in radar.

- *Radio waves* have the lowest energy of the various kinds of electromagnetic radiation. We use them for radio and television communication, digital imaging, remote control devices, and wireless linkages for computers. Radio waves are also used in NMR spectroscopy and in magnetic resonance imaging (MRI).

Each spectroscopic technique discussed in this text employs a different kind of electromagnetic radiation. In this chapter, we will look at infrared (IR) spectroscopy and ultraviolet/visible (UV/Vis) spectroscopy.

Because electromagnetic radiation has wave-like properties, it can be characterized, as a wave can, by either its frequency (ν) or its wavelength (λ). **Frequency** is defined as the number of wave crests that pass by a given point in one second; frequency has units of hertz (Hz). **Wavelength** is the distance from any point on one wave to the corresponding point on the next wave; wavelength is generally measured in nanometers (nm; one nm = 10^{-9} meter).

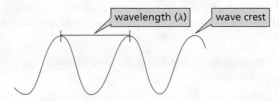

The relationship between the energy (E) and the frequency (ν) or wavelength (λ) of the electromagnetic radiation is described by the equation

$$E = h\nu = \frac{hc}{\lambda}$$

where h is *Planck's constant,* a proportionality constant named after the German physicist who discovered the relationship, and c is the speed of light ($c = 3.0 \times 10^{10}$ cm/s). The equation shows that short wavelengths have high energies and high frequencies, and long wavelengths have low energies and low frequencies.

Another way to describe the *frequency* of the electromagnetic radiation—and the one most often used in infrared spectroscopy—is **wavenumber** ($\tilde{\nu}$), which is the number of waves in 1 cm. Wavenumbers, therefore, have units of reciprocal centimeters (cm^{-1}). The relationship between wavenumber (in cm^{-1}) and wavelength (in micrometers) is given by the equation

$$\tilde{\nu}\ (\mathbf{cm^{-1}}) = \frac{10^4}{\lambda(\boldsymbol{\mu m})} \qquad \text{because 1 } \mu m = 10^{-6} \text{ meters} = 10^{-4} \text{ cm}$$

So *high frequencies, large wavenumbers,* and *short wavelengths* are associated with *high energies.*

PROBLEM 19◆

One of the following depicts the waves associated with infrared radiation, and one depicts the waves associated with visible light. Which is which?

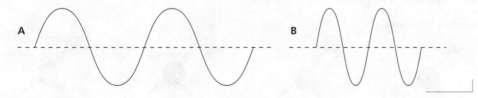

High frequencies, large wavenumbers, and short wavelengths are associated with high energies.

PROBLEM 20◆

a. Which is higher in energy, electromagnetic radiation with wavenumber 100 cm^{-1} or with wavenumber 2000 cm^{-1}?

b. Which is higher in energy, electromagnetic radiation with wavelength 9 μm or with wavelength 8 μm?

c. Which is higher in energy, electromagnetic radiation with wavenumber 2000 cm^{-1} or with wavelength 2 μm?

PROBLEM 21◆

a. What is the wavenumber of radiation that has a wavelength of 4 μm?

b. What is the wavelength of radiation that has a wavenumber of 200 cm^{-1}?

10 INFRARED SPECTROSCOPY

The length reported for a bond between two atoms is an average length, because in reality a bond behaves as if it were a vibrating spring. A bond vibrates with both stretching and bending motions.

A *stretch* is a vibration occurring along the line of the bond; a stretching vibration changes the bond length.

A *bend* is a vibration that does *not* occur along the line of the bond; a bending vibration changes the bond angle.

A diatomic molecule such as H—Cl can undergo only a **stretching vibration** because it has no bond angles.

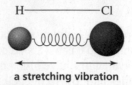

a stretching vibration

The vibrations of a molecule containing three or more atoms are more complex because they include stretches and bends (Figure 12). The stretching and bending vibrations, moreover, can be symmetric or asymmetric, and the bending vibrations can be either in-plane or out-of-plane. **Bending vibrations** are often referred to by the descriptive terms *rock, scissor, wag*, and *twist*.

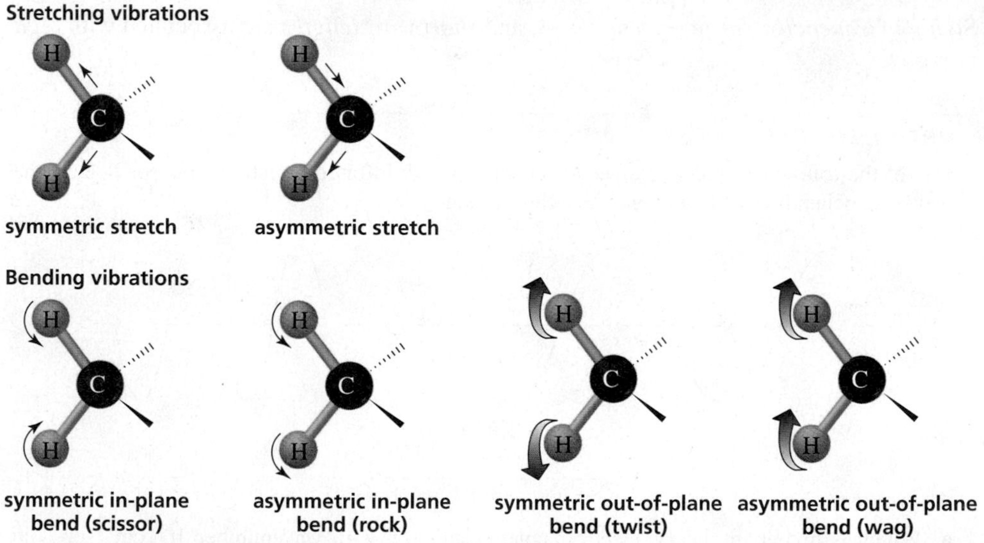

Stretching vibrations

symmetric stretch   asymmetric stretch

Bending vibrations

symmetric in-plane
bend (scissor)

asymmetric in-plane
bend (rock)

symmetric out-of-plane
bend (twist)

asymmetric out-of-plane
bend (wag)

▶ **Figure 12**

Stretching and bending vibrations of bonds in organic molecules containing three or more atoms.

Infrared radiation has just the right range of frequencies—4000 to 600 cm^{-1}—to correspond to the frequencies of the stretching and bending vibrations of the bonds in organic molecules. It is just below (to the right of) the "red region" of visible light, meaning that infrared light has lower energy than visible light. (*Infra* is Latin for "below.")

Each stretching and bending vibration of a given bond occurs with a characteristic frequency. When a molecule is bombarded with radiation of a frequency that exactly matches the frequency of the vibration of one of its bonds, the molecule absorbs energy. This allows the bond to stretch and bend a bit more. By experimentally determining the wavenumbers of the energy absorbed by a particular compound, we can ascertain what kinds of bonds it has. For example, the stretching vibration of a $C{=}O$ bond absorbs energy with wavenumber ~1700 cm^{-1}, whereas the stretching vibration of an $O{-}H$ bond absorbs energy with wavenumber ~3400 cm^{-1} (Figure 13).

$$C{=}O \qquad\qquad O{-}H$$
$$\tilde{\nu} = \textbf{~1700 cm}^{-1} \qquad \tilde{\nu} = \textbf{~3400 cm}^{-1}$$

The Infrared Spectrum

An **infrared spectrum,** obtained by passing infrared radiation through a sample of a compound, is a plot of the percent transmission of radiation versus the wavenumber (or wavelength) of the radiation transmitted (Figure 13). The instrument used to obtain an infrared spectrum is called an IR spectrophotometer. At 100% transmission all the energy of the radiation (of a particular wavenumber) passes through the molecule. Lower values of percent transmission mean that some of the energy is being absorbed by the compound. Each downward spike in an IR spectrum represents absorption of energy. The spikes are called **absorption bands.** Most chemists report the location of absorption bands using wavenumbers.

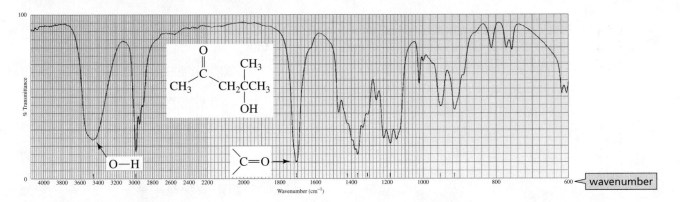

▲ **Figure 13**

An infrared spectrum shows the percent transmission of radiation versus the wavenumber of the radiation. The (C═O) stretch absorbs at 1705 cm^{-1} and the (O─H) stretch absorbs at 3450 cm^{-1}.

A newer type of IR spectrophotometer, called a Fourier transform IR (FT-IR) spectrophotometer, has several advantages over conventional IR spectrophotometers. Its sensitivity is better because, instead of scanning through the frequencies sequentially, it measures all frequencies simultaneously. This allows time for multiple measurements, which are averaged. The information is then digitized and Fourier transformed by a computer to produce an FT-IR spectrum. A conventional IR spectrophotometer can take 2 to 10 minutes to scan through all the frequencies, whereas FT-IR spectra can be taken in 1 to 2 seconds. The infrared spectra shown in this text are FT-IR spectra.

The Functional Group and Fingerprint Regions

An infrared spectrum can be divided into two areas. The area on the left (4000−1400 cm^{-1}) is where most of the functional groups show absorption bands. This is called the **functional group region.**

The area on the right (1400−600 cm^{-1}) is called the **fingerprint region** because it is characteristic of the compound as a whole, just as a fingerprint is characteristic of an individual. Even if two different molecules have the same functional groups, their IR spectra will not be identical because the functional groups are not in exactly the same environment in both compounds. This difference will be reflected in the patterns of the absorption bands in the fingerprint region.

For example, 2-pentanol and 3-pentanol have the same functional groups, so they show similar absorption bands in the functional group region. Their fingerprint regions are different, however, because the compounds are different (Figure 14). Thus, a compound can be positively identified by comparing its fingerprint region with the fingerprint region of the spectrum of a known sample of the compound.

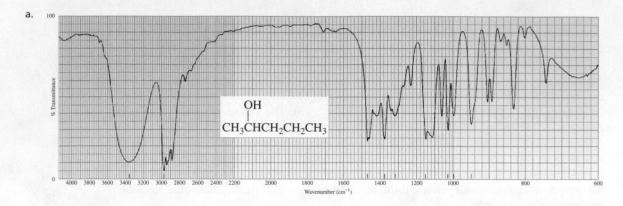

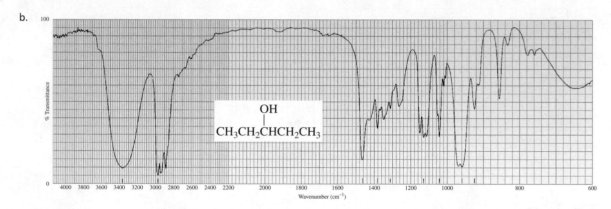

▲ **Figure 14**

The IR spectra of (a) 2-pentanol and (b) 3-pentanol. The functional group regions (4000–1400 cm^{-1}) are very similar because the two compounds have the same functional groups, but the fingerprint region (1400–600 cm^{-1}) is unique for each compound.

11   CHARACTERISTIC INFRARED ABSORPTION BANDS

IR spectra can be quite complex because the stretching and bending vibrations of each bond in a molecule can produce an absorption band. Organic chemists, however, do not try to identify all the absorption bands in an IR spectrum. They tend to focus on the functional groups. In this chapter, we will look at several characteristic absorption bands so you will be able to tell something about the structure of a compound that gives a particular IR spectrum.

You can find an extensive table of characteristic functional group frequencies in the Study Area of MasteringChemistry. When identifying an unknown compound, chemists often use IR spectroscopy in conjunction with information obtained from other spectroscopic techniques. Some of the problems in this chapter provide practice in using information from two or more instrumental methods to identify compounds.

More energy is required to stretch a bond than to bend it, so absorption bands for stretching vibrations are found in the functional group region (4000−1400 cm^{-1}), whereas those for bending vibrations are typically found in the fingerprint region (1400−600 cm^{-1}). As a result, stretching vibrations are the ones most often used to determine what kinds of bonds a molecule has. The *frequencies of the stretching vibrations* associated with different types of bonds are listed in Table 4 and are discussed in Sections 12 and 13.

It takes more energy to stretch a bond than to bend it, so stretching vibrations are found at higher wavenumbers than bending vibrations.

Table 4	Frequencies of Important IR Stretching Vibrations	
Type of bond	**Wavenumber (cm^{-1})**	**Intensity**
C≡N	2260–2220	medium
C≡C	2260–2100	medium to weak
C=C	1680–1600	medium
C=N	1650–1550	medium
(benzene ring)	~1600 and ~1500–1430	strong to weak
C=O	1780–1650	strong
C—O	1250–1050	strong
C—N	1230–1020	medium
O—H (alcohol)	3650–3200	strong, broad
O—H (carboxylic acid)	3300–2500	strong, very broad
N—H	3500–3300	medium, broad
C—H	3300–2700	medium

12 THE INTENSITY OF ABSORPTION BANDS

When a bond stretches, the increasing distance between the atoms increases its dipole moment. The intensity of the absorption band depends on the size of this change in dipole moment: *the greater the change in dipole moment, the more intense the absorption.*

For example, absorption bands for the stretching vibrations of C=O and C=C bonds appear at similar frequencies, but they are easily distinguished: the one for C=O is much more intense because it is associated with a much greater change in dipole moment since the bond is more polar. (Compare the C=O absorption bands in Figures 15–18 with the C=C absorption band in Figure 22.)

The stretching vibration of an O—H bond is associated with a greater change in dipole moment than that of an N—H bond, because the O—H bond is more polar. Consequently, an O—H bond shows more intense absorption than an N—H bond. Similarly, an N—H bond shows more intense absorption than a C—H bond, because the N—H bond is more polar.

Note that the dipole moment of a bond is equal to the magnitude of the charge on one of the bonded atoms multiplied by the distance between the two bonded atoms.

The greater the change in the dipole moment, the more intense the absorption.

relative bond polarities
relative intensities of IR absorption

most polar → O—H > N—H > C—H ← least polar
most intense → → ← least intense

The more polar the bond, the more intense the absorption.

The intensity of an absorption band also depends on the number of bonds responsible for the absorption. For example, the absorption band for a C—H stretch will be more intense for a compound such as octyl iodide, which has 17 C—H bonds, than for methyl iodide, which has only three C—H bonds.

The concentration of the sample used to obtain an IR spectrum also affects the intensity of the absorption bands. Concentrated samples have greater numbers of absorbing molecules and, therefore, more intense absorption bands. In the chemical literature, you will find intensities referred to as strong (s), medium (m), weak (w), broad, and sharp.

13 / THE POSITION OF ABSORPTION BANDS

The frequency of a stretching vibration—the amount of energy required to stretch a bond—depends on the *strength* of the bond and the *masses* of the bonded atoms. The stronger the bond, the greater the energy required to stretch it. The frequency of the stretching vibration is also inversely related to the mass of the atoms joined by the bond; thus, heavier atoms vibrate at lower frequencies.

Hooke's Law

The approximate wavenumber of an absorption band can be calculated from the following equation derived from **Hooke's law,** which describes the motion of a vibrating spring:

$$\tilde{\nu} = \frac{1}{2\pi c}\left[\frac{f(m_1 + m_2)}{m_1 m_2}\right]^{1/2}$$

where $\tilde{\nu}$ is the wavenumber of an absorption band, c is the speed of light, f is the force constant of the bond (a measure of the strength of the bond), and m_1 and m_2 are the masses of the atoms (in grams) joined by the bond. This equation shows that *stronger bonds* and *lighter atoms* give rise to higher frequencies.

Lighter atoms show absorption bands at larger wavenumbers.

C—H
~3000 cm^{-1}
C—D
~2200 cm^{-1}
C—O
~1100 cm^{-1}
C—Cl
~700 cm^{-1}

The Originator of Hooke's Law

Robert Hooke (1635–1703) was born on the Isle of Wight off the southern coast of England. A brilliant scientist, he contributed to almost every scientific field. He was the first to suggest that light had wave-like properties. He discovered that Gamma Arietis is a double star, and he discovered Jupiter's Great Red Spot. In a lecture published posthumously, he suggested that earthquakes are caused by the cooling and contracting of the Earth. He examined cork under a microscope and coined the term *cell* to describe what he saw. He wrote about evolutionary development based on his studies of microscopic fossils, and he produced some highly regarded studies of insects. Hooke also invented the balance spring for watches and the universal joint currently used in automobiles.

Robert Hooke's drawing of a "blue fly" appeared in *Micrographia,* the first book on microscopy, published by Hooke in 1665.

The Effect of Bond Order

Bond order—whether a bond is a single bond, a double bond, or a triple bond—affects bond strength. Therefore, bond order affects the position of absorption bands.

A C≡C bond is stronger than a C=C bond, so a C≡C bond stretches at a higher frequency (~2100 cm^{-1}) than does a C=C bond (~1650 cm^{-1}); C—C bonds show stretching vibrations in the region from 1300 to 800 cm^{-1}, but since they are weak and very common, these vibrations are of little value in identifying compounds.

Similarly, a C=O bond stretches at a higher frequency (~1700 cm^{-1}) than does a C—O bond (~1100 cm^{-1}), and a C≡N bond stretches at a higher frequency (~2200 cm^{-1}) than does a C=N bond (~1600 cm^{-1}), which in turn stretches at a higher frequency than does a C—N bond (~1100 cm^{-1}) (Table 4).

Stronger bonds show absorption bands at larger wavenumbers.

C≡N
~2200 cm^{-1}
C=N
~1600 cm^{-1}
C—N
~1100 cm^{-1}

PROBLEM 22♦

a. Which will occur at a larger wavenumber:
   1. a $C \equiv C$ stretch or a $C = C$ stretch?   3. a $C - N$ stretch or a $C = N$ stretch?
   2. a $C - H$ stretch or a $C - H$ bend?   4. a $C = O$ stretch or a $C - O$ stretch?

b. Assuming that the force constants are the same, which will occur at a larger wavenumber:
   1. a $C - O$ stretch or a $C - Cl$ stretch?   2. a $C - O$ stretch or a $C - C$ stretch?

14   THE POSITION AND SHAPE OF AN ABSORPTION BAND IS AFFECTED BY ELECTRON DELOCALIZATION, ELECTRON DONATION AND WITHDRAWAL, AND HYDROGEN BONDING

Table 4 shows a range of wavenumbers for the frequency of the stretching vibration for each functional group because the exact position and shape of a group's absorption band depends on other structural features of the molecule, such as electron delocalization, the electronic effect of neighboring substituents, and hydrogen bonding. In fact, important details about the structure of a compound can be revealed by the exact positions and shape of its absorption bands.

For example, the IR spectrum in Figure 15 shows that the carbonyl group ($C = O$) of 2-pentanone absorbs at 1720 cm^{-1}, whereas the IR spectrum in Figure 16 shows that the carbonyl group of 2-cyclohexenone absorbs at a lower frequency (1680 cm^{-1}). 2-Cyclohexenone's carbonyl group absorbs at a lower frequency because it has more single-bond character due to electron delocalization. A single bond is weaker than a double bond, so a carbonyl group with significant single-bond character will stretch at a lower frequency than will one with little or no single-bond character.

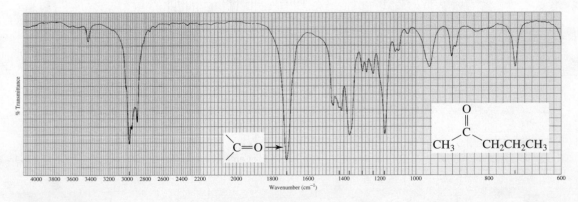

▲ **Figure 15**

The intense absorption band at ~1720 cm^{-1} indicates a $C = O$ bond.

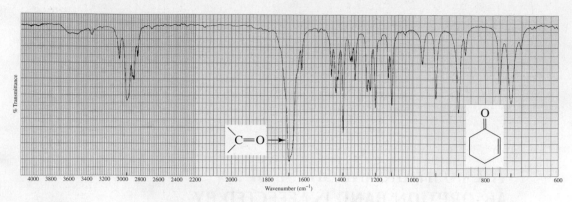

▲ **Figure 16**

Electron delocalization gives the carbonyl group less double-bond character, so it absorbs at a lower frequency ($\sim$1680 cm^{-1}) than does a carbonyl group with localized electrons ($\sim$1720 cm^{-1}).

Putting an atom other than carbon next to the carbonyl group also causes the position of the carbonyl absorption band to shift. Whether it shifts to a lower or to a higher frequency depends on whether the predominant effect of the atom is to donate electrons by resonance or to withdraw electrons inductively.

The predominant effect of the nitrogen of an amide is electron donation by resonance. In contrast, oxygen is less able than nitrogen to accommodate a positive charge because of oxygen's greater electronegativity, so the predominant effect of the oxygen of an ester is inductive electron withdrawal. As a result, the carbonyl group of an ester has less single-bond character, so it requires more energy to stretch (1740 cm^{-1} in Figure 17) than the carbonyl group of an amide (1660 cm^{-1} in Figure 18).

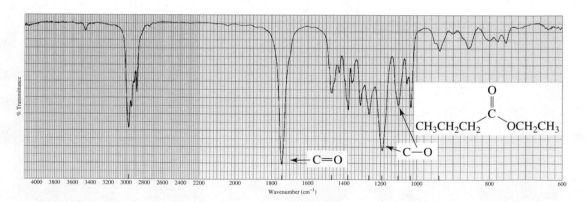

▲ **Figure 17**

The electron-withdrawing oxygen atom makes the carbonyl group of an ester harder to stretch ($\sim$1740 cm^{-1}) than the carbonyl group of a ketone ($\sim$1720 cm^{-1}).

When we compare the frequency for the stretching vibration of the carbonyl group of an ester (1740 cm^{-1} in Figure 17) with that of the carbonyl group of a ketone (1720 cm^{-1} in Figure 15), we can see how important inductive electron withdrawal is to the position of the stretching vibration of the carbonyl group.

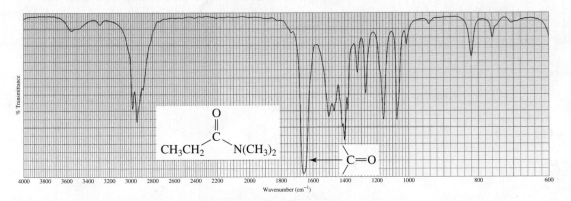

▲ **Figure 18**

The carbonyl group of an amide has less double-bond character than the carbonyl group of a ketone, so the carbonyl group of an amide stretches more easily (~1660 cm^{-1}) than the carbonyl group of a ketone (~1720 cm^{-1}).

A C—O bond shows a stretching vibration between 1250 and 1050 cm^{-1}. If the C—O bond is in an alcohol or an ether, the stretch will lie toward the lower end of the range (Figure 19). If, however, the C—O bond is in a carboxylic acid, the stretch will lie at the higher end of the range (Figure 20).

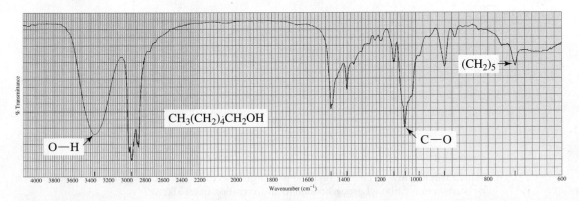

The position of the C—O absorption varies because the C—O bond in an *alcohol* or in an *ether* is a pure single bond, whereas the C—O bond in a *carboxylic acid* has partial double-bond character due to electron delocalization. *Esters* show C—O stretches at both ends of the range because esters have two C—O single bonds: one that is a pure single bond and one that has partial double-bond character (Figure 17) .

▲ **Figure 19**

The IR spectrum of 1-hexanol. The C—O absorption lies at the lower end of the range of C—O stretches (1250–1050 cm^{-1}) because the C—O bond in an alcohol is a pure single bond.

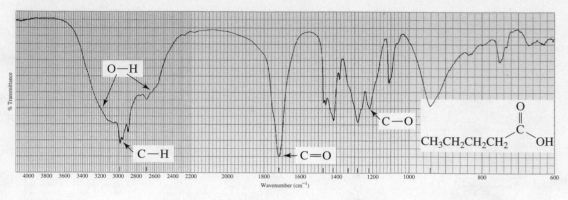

▲ **Figure 20**

The IR spectrum of pentanoic acid. The C—O absorption lies at the higher end of the range of C—O stretches (1250–1050 cm⁻¹) because the C—O bond in a carboxylic acid has some double-bond character.

PROBLEM-SOLVING STRATEGY

Differences in IR Spectra

Which will occur at a larger wavenumber, the C—N stretch of an amine or the C—N stretch of an amide?

To answer this question, we need to determine what effect electron delocalization has on the C—N bond in amines and amides. When we do that we see that the C—N bond of the amine is a pure single bond, whereas electron delocalization causes the C—N bond of the amide to have partial double-bond character. The C—N stretch of an amide, therefore, will occur at a larger wavenumber.

$$R - \ddot{N}H_2$$

no electron delocalization   **electron delocalization causes the C—N bond to have partial double-bond character**

Now use the strategy you have just learned to solve Problem 23.

PROBLEM 23♦

Which will occur at a larger wavenumber:

a. the C—O stretch of phenol or the C—O stretch of cyclohexanol?
b. the C=O stretch of a ketone or the C=O stretch of an amide?
c. the C—N stretch of cyclohexylamine or the C—N stretch of aniline?

PROBLEM 24♦

Which would show an absorption band at a larger wavenumber: a carbonyl group bonded to an sp^3 carbon or a carbonyl group bonded to an sp^2 carbon of an alkene?

PROBLEM 25♦

Why is the C—O absorption band of 1-hexanol at a smaller wavenumber (1060 cm⁻¹) than the C—O absorption band of pentanoic acid (1220 cm⁻¹)?

PROBLEM 26♦

List the following compounds in order from highest wavenumber to lowest wavenumber for their C=O absorption bands:

a. [structures] b. [structures]

O—H and N—H Absorption Bands

Because O—H bonds are polar, they show intense absorption bands that can be quite broad (Figures 19 and 20). Both the position and the shape of an O—H absorption band depend on hydrogen bonding. It is easier for an O—H bond to stretch if it is hydrogen bonded, because the hydrogen is attracted to the oxygen of a neighboring molecule. Hydrogen-bonded OH groups also have broader absorption bands because hydrogen bonds vary in strength.

Carboxylic acids exist as hydrogen-bonded dimers. The additional hydrogen bonding of carboxylic acids compared with the hydrogen bonding of alcohols causes the O—H stretch of a carboxylic acid to occur at a lower frequency and to be broader ($3300–2500 \text{ cm}^{-1}$) than the O—H stretch of an alcohol ($3550–3200 \text{ cm}^{-1}$).

hydrogen-bonded dimer

$$R—C \quad \quad C—R$$

$3300–2500 \text{ cm}^{-1}$

hydrogen bond

H

$R—O—H\text{------}O—R$
concentrated solution
$3550–3200 \text{ cm}^{-1}$

$R—O—H$
dilute solution
$3650–3590 \text{ cm}^{-1}$

Because the position and breadth of an O—H stretch depend on hydrogen bonding, they also depend on the concentration of the solution. The more concentrated the solution, the more likely it is for the OH-containing molecules to form intermolecular hydrogen bonds. Therefore, the O—H stretch in a concentrated (hydrogen-bonded) solution of an alcohol occurs at $3550–3200 \text{ cm}^{-1}$ whereas the O—H stretch in a dilute solution (with little or no hydrogen bonding) occurs at $3650–3590 \text{ cm}^{-1}$.

N—H bonds are less polar and form weaker hydrogen bonds than O—H bonds, so the absorption band for an N—H stretch is less intense and narrower than that for an O—H stretch (Figure 25).

The position, intensity, and shape of an absorption band are helpful in identifying functional groups.

PROBLEM 27♦

Which will show an O—H stretch at a larger wavenumber: ethanol dissolved in carbon disulfide or an undiluted sample of ethanol?

C—H Absorption Bands

Important information about the identity of a compound is provided by the stretching and bending vibrations of its C—H bonds.

Stretching Vibrations

The strength of a C—H bond depends on the hybridization of the carbon—namely, the greater the s character of the carbon, the stronger is the bond that it forms.

A C—H bond, therefore, is stronger when the carbon is sp hybridized than when it is sp^2 hybridized, which in turn is stronger than when the carbon is sp^3 hybridized. Because more energy is needed to stretch a stronger bond, the absorption band for a C—H stretch is at ~3300 cm^{-1} for an *sp* carbon, at ~3100 cm^{-1} for an sp^2 carbon, and at ~2900 cm^{-1} for an sp^3 carbon (Table 5).

Table 5	IR Absorptions of Carbon–Hydrogen Bonds
Carbon–Hydrogen Stretching Vibrations	**Wavenumber (cm^{-1})**
C≡C—H	⊠3300
C=C—H	3100–3020
C—C—H	2960–2850
R—C(=O)—H	⊠2820 and ⊠2720
Carbon–Hydrogen Bending Vibrations	**Wavenumber (cm^{-1})**
CH$_3$— —CH$_2$— —CH—	1450–1420
CH$_3$—	1385–1365
trans	980–960
cis	730–675
trisubstituted	840–800
terminal alkene (disubstituted)	890
terminal alkene (monosubstituted)	990 and 910

A useful step in the analysis of a spectrum is to look at the absorption bands in the vicinity of 3000 cm^{-1}. The only absorption band in the vicinity of 3000 cm^{-1} in Figure 21 is slightly to the right of that value. This tells us that the compound has

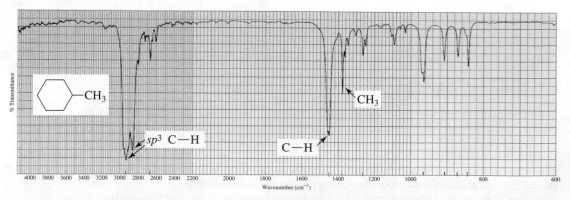

▲ **Figure 21**

The IR spectrum of methylcyclohexane. The absorptions at 2940 and 2860 cm^{-1} indicate that methylcyclohexane has hydrogens bonded to sp^3 carbons.

hydrogens bonded to sp^3 carbons, but none bonded to sp^2 or to sp carbons. Both Figure 22 and 23 show absorption bands slightly to the left and slightly to the right of 3000 cm^{-1}, indicating that the compounds that produced those spectra contain hydrogens bonded to both sp^2 and sp^3 carbons.

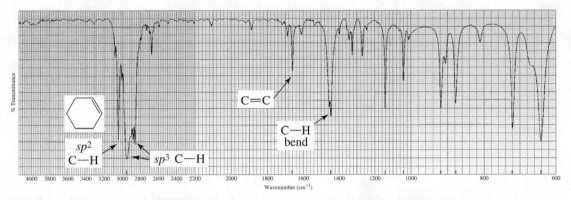

▲ **Figure 22**

The IR spectrum of cyclohexene. The absorptions at 3040, 2950, and 2860 cm^{-1} indicate that cyclohexene has hydrogens bonded to both sp^2 and sp^3 carbons.

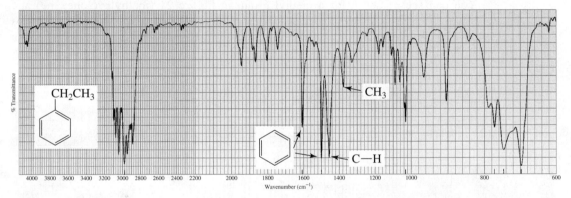

▲ **Figure 23**

The IR spectrum of ethylbenzene. The absorptions in the 3100–2880 cm^{-1} region indicate that ethylbenzene has hydrogens bonded to both sp^2 and sp^3 carbons. The two sharp absorptions at 1610 and 1500 cm^{-1} indicate that the sp^2 carbons are due to a benzene ring.

Once we know that a compound has hydrogens bonded to sp^2 carbons, we need to determine whether those carbons are the sp^2 carbons of an alkene or of a benzene ring. A benzene ring is indicated by two sharp absorption bands, one at ~1600 cm^{-1} and one at 1500−1430 cm^{-1}, whereas an alkene is indicated by a band only at ~1600 cm^{-1} (Table 4). The compound whose spectrum is shown in Figure 22 is, therefore, an alkene, whereas the one whose spectrum is shown in Figure 23 has a benzene ring. (If you also have an NMR spectrum of the compound, the presence of a benzene ring is very easy to detect)

Be aware that N—H bending vibrations also occur at 1600 cm^{-1}, so absorption at that wavelength does not always indicate a C═C bond. However, absorption bands resulting from N—H bends tend to be broader (due to hydrogen bonding) and more intense (due to being more polar) than those caused by C═C stretches, and they will be accompanied by N—H stretches at 3500−3300 cm^{-1} (Figure 25).

The stretch of the C—H bond of an aldehyde group shows two absorption bands—one at ~2820 cm^{-1} and the other at ~2720 cm^{-1} (Figure 24). This makes aldehydes relatively easy to identify because essentially no other absorption occurs at these wavenumbers.

Bending Vibrations

If a compound has sp^3 carbons, a look at 1400 cm^{-1} will tell you whether the compound has a methyl group. All hydrogens bonded to sp^3 carbons show a C—H bending vibration slightly to the *left* of 1400 cm^{-1}. Methyl groups show an additional C—H bending vibration

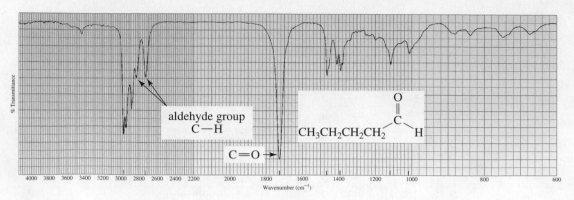

▲ **Figure 24**

The absorptions at ~2820 cm⁻¹ and ~2720 cm⁻¹ readily identify an aldehyde group. Note also the intense absorption band at ~1730 cm⁻¹ indicating a C═O bond.

slightly to the *right* of 1400 cm⁻¹. So, if a compound has a methyl group, absorption bands will appear *both* to the left and to the right of 1400 cm⁻¹; otherwise, only the band to the left of 1400 cm⁻¹ will be present.

You can see evidence for a methyl group in Figure 21 (methylcyclohexane) and in Figure 23 (ethylbenzene), but not in Figure 22 (cyclohexene). Two methyl groups attached to the same carbon can sometimes be detected by a split in the methyl peak at ~1380 cm⁻¹ (Figure 25).

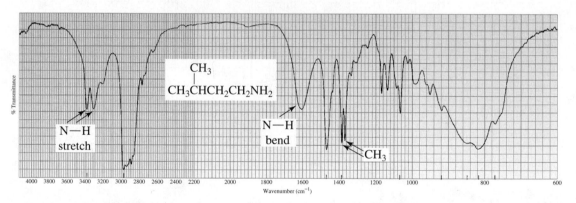

▲ **Figure 25**

The IR spectrum of isopentylamine. The two-peak absorption band at ~1380 cm⁻¹ indicates the presence of an isopropyl group. The N—H bend around 1600 cm⁻¹ is broad due to intermolecular hydrogen bonding.

The C—H bending vibrations for hydrogens bonded to sp^2 carbons give rise to absorption bands in the 1000−600 cm⁻¹ region. As Table 5 shows, the frequency of the C—H bending vibration of an alkene depends on the number of alkyl groups attached to the sp^2 carbons and on the configuration of the alkene. It is important to realize that these absorption bands can be shifted out of the characteristic regions if strongly electron-withdrawing or electron-donating substituents are close to the double bond. Acyclic compounds with more than four adjacent methylene (CH₂) groups show a characteristic absorption band at 720 cm⁻¹ that results from in-phase rocking of the methylene groups (Figure 19).

15 THE ABSENCE OF ABSORPTION BANDS

The absence of an absorption band can be as useful as the presence of one in identifying a compound by IR spectroscopy.

For example, the spectrum in Figure 26 shows a strong absorption at ~1100 cm⁻¹, indicating the presence of a C—O bond. Clearly, the compound is not an alcohol because there is no absorption above 3100 cm^{-1}. Nor is it a carbonyl compound because there is no absorption at $\sim 1700 \text{ cm}^{-1}$. The compound has no $C \equiv C$, $C = C$, $C \equiv N$, $C = N$, or $C - N$ bonds. We may deduce, then, that the compound is an ether. Its $C - H$ absorption bands show that it has hydrogens only on sp^3 carbons (2950 cm^{-1}) and that it has a methyl group (1385 cm^{-1}). We also know that the compound has fewer than four adjacent methylene groups, because there is no absorption at $\sim 720 \text{ cm}^{-1}$. The compound is in fact diethyl ether.

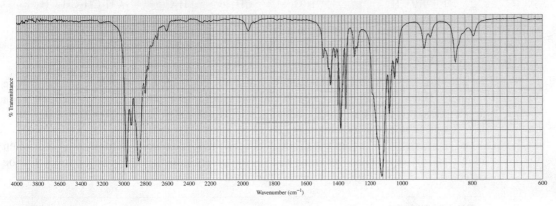

▲ **Figure 26**
The IR spectrum of diethyl ether.

PROBLEM 28♦

How do we know that the absorption band at $\sim 1100 \text{ cm}^{-1}$ in Figure 26 is due to a $C - O$ bond and not to a $C - N$ bond?

PROBLEM 29♦

a. An oxygen-containing compound shows an absorption band at $\sim 1700 \text{ cm}^{-1}$ and no absorption bands at $\sim 3300 \text{ cm}^{-1}$, $\sim 2700 \text{ cm}^{-1}$, or $\sim 1100 \text{ cm}^{-1}$. What class of compound is it?

b. A nitrogen-containing compound shows no absorption band at $\sim 3400 \text{ cm}^{-1}$ and no absorption bands between $\sim 1700 \text{ cm}^{-1}$ and $\sim 1600 \text{ cm}^{-1}$. What class of compound is it?

PROBLEM 30

How could IR spectroscopy be used to distinguish between the following compounds?

a. a ketone and an aldehyde   **d.** *cis*-2-hexene and *trans*-2-hexene
b. a cyclic ketone and an open-chain ketone   **e.** cyclohexene and cyclohexane
c. benzene and cyclohexene   **f.** a primary amine and a tertiary amine

---

PROBLEM 31

For each of the following pairs of compounds, name one absorption band that could be used to distinguish between them.

a. $CH_3CH_2CH_2CH_3$ and $CH_3CH_2OCH_3$ **d.** $CH_3CH_2C \equiv CCH_3$ and $CH_3CH_2C \equiv CH$

b. [structure: propanoate ester with OCH₃] and [structure: carboxylic acid with OH]   **e.** [structure: acid with OH] and [structure with OH]

c. [cyclohexane ring] and [methylcyclohexane ring with CH₃]   **f.** [benzene ring] and [toluene ring with CH₃]

---

16   SOME VIBRATIONS ARE INFRARED INACTIVE

In order for the vibration of a bond to absorb IR radiation, the dipole moment of the bond must change when it vibrates. Therefore, not all bond vibrations give rise to an absorption band.

For example, the C=C bond in 1-butene has a dipole moment because the molecule is not symmetrical. So when the bond stretches, the dipole moment changes and an absorption band is observed.

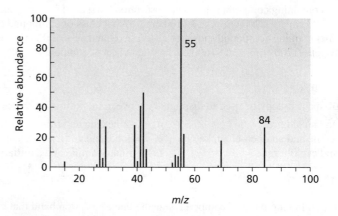

In contrast, 2,3-dimethyl-2-butene is symmetrical, so its C=C bond has no dipole moment. When the bond stretches, it still has no dipole moment. Since stretching is not accompanied by a change in dipole moment, no absorption band is observed. The vibration is said to be *infrared inactive*.

2,3-Dimethyl-2-heptene experiences a very small change in dipole moment when its C=C bond stretches, so only an extremely weak absorption band (if any) will be detected for the stretching vibration of the bond.

PROBLEM 32♦

Which of the following compounds has a vibration that is infrared inactive?
1-butyne, 2-butyne, H_2, H_2O, Cl_2, and ethene

PROBLEM 33♦

Identify the compound that gives the mass spectrum and infrared spectrum shown in Figure 27 and 28.

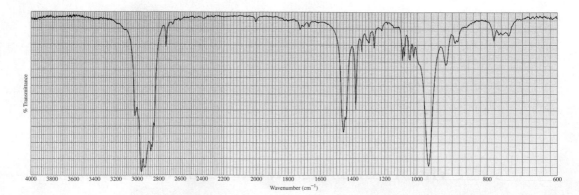

▶ **Figure 27**

The mass spectrum for Problem 33.

▶ **Figure 28**

The IR spectrum for Problem 33.

17   HOW TO INTERPRET AN INFRARED SPECTRUM

We will now look at some IR spectra and see what we can deduce about the structures of the compounds that give rise to the spectra. We might not be able to identify the compound precisely, but when we are told what it is, its structure should fit our observations.

Compound 1. The absorptions in the 3000 cm^{-1} region in Figure 29 indicate that hydrogens are attached both to sp^2 carbons (3075 cm^{-1}) and to sp^3 carbons (2950 cm^{-1}). Now we need to determine whether the sp^2 carbons belong to an alkene or to a benzene ring. The absorption at 1650 cm^{-1}, the absence of absorption at 1500−1430 cm^{-1}, and the absorption at ~890 cm^{-1} (Table 5) suggest that the compound is a terminal alkene with two alkyl substituents at the 2-position. The absence of absorption at ~720 cm^{-1} indicates that the compound has fewer than four adjacent methylene groups. We are not surprised to find that the compound is 2-methyl-1-pentene.

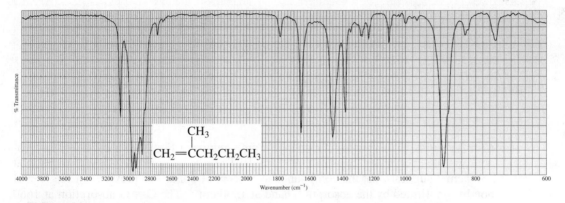

▲ **Figure 29**

The IR spectrum of Compound 1.

Compound 2. The absorptions in the 3000 cm^{-1} region in Figure 30 indicate that hydrogens are attached to sp^2 carbons (3050 cm^{-1}) but not to sp^3 carbons. The sharp absorptions at 1600 cm^{-1} and 1460 cm^{-1} indicate that the compound has a benzene ring. The absorptions at 2810 cm^{-1} and 2730 cm^{-1} show that the compound is an aldehyde. The characteristically strong absorption band for the carbonyl group (C=O) is lower (~1700 cm^{-1}) than normal (1720 cm^{-1}), so the carbonyl group has partial single-bond character. Thus, it must be attached directly to the benzene ring, so electron delocalization from the ring can reduce the double bond character of the carbonyl group. The compound is benzaldehyde.

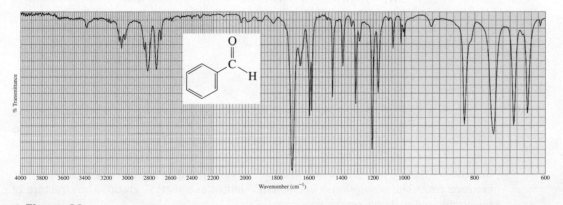

▲ **Figure 30**

The IR spectrum of Compound 2.

Compound 3. The absorptions in the 3000 cm^{-1} region in Figure 31 indicate that hydrogens are attached to sp^3 carbons (2950 cm^{-1}) but not to sp^2 carbons. The shape of the strong absorption band at 3300 cm^{-1} is characteristic of an O—H group of an alcohol. The absorption at 2100 cm^{-1} indicates that the compound has a triple bond. The sharp absorption band at 3300 cm^{-1} indicates that the compound has a hydrogen attached to an sp carbon, so we know it is a terminal alkyne. The structure of the compound is shown on the spectrum.

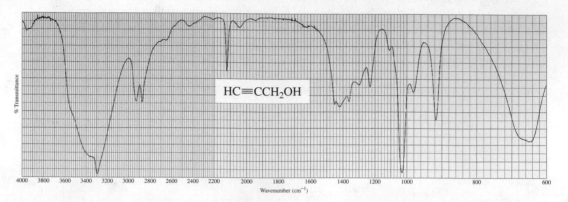

▲ **Figure 31**

The IR spectrum of Compound 3.

Compound 4. The absorptions in the 3000 cm^{-1} region in Figure 32 indicate that hydrogens are attached to sp^3 carbons (2950 cm^{-1}). The relatively strong absorption band at 3300 cm^{-1} suggests that the compound has an N—H bond. The presence of the N—H bond is confirmed by the absorption band at 1560 cm^{-1}. The C=O absorption at 1660 cm^{-1} indicates that the compound is an amide. The structure of the compound is shown on the spectrum.

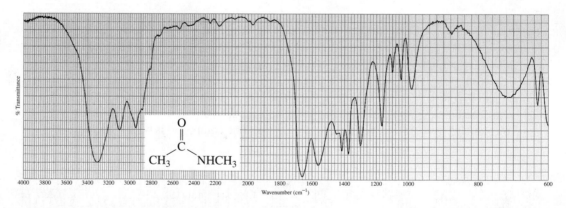

▲ **Figure 32**

The IR spectrum of Compound 4.

Compound 5. The absorptions in the 3000 cm^{-1} region in Figure 33 indicate that the compound has hydrogens attached to sp^2 carbons (>3000 cm^{-1}) and to sp^3 carbons (<3000 cm^{-1}). The absorptions at 1605 cm^{-1} and 1500 cm^{-1} indicate that the compound contains a benzene ring. The absorption at 1720 cm^{-1} for the carbonyl group indicates that the compound is a ketone and that the carbonyl group is not directly attached to the benzene ring. The absorption at ~1380 cm^{-1} indicates a methyl group. The structure of the compound is shown on the spectrum.

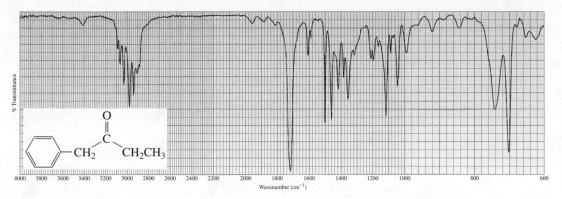

▲ **Figure 33**

The IR spectrum of Compound 5.

PROBLEM 34♦

A compound with molecular formula C_4H_6O gives the infrared spectrum shown in Figure 34. Identify the compound.

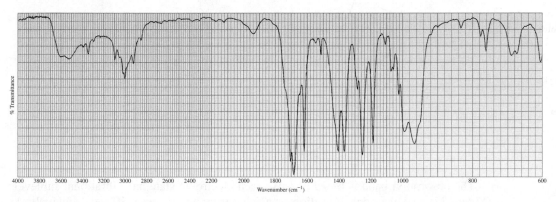

▲ **Figure 34**

The IR spectrum for Problem 34.

18   ULTRAVIOLET AND VISIBLE SPECTROSCOPY

Ultraviolet and visible (UV/Vis) spectroscopy provides information about compounds that have conjugated double bonds. Ultraviolet light and visible light have just the right energy to cause an electronic transition in a molecule—that is, to promote an electron from one molecular orbital to another of higher energy.

Depending on the energy needed for the electronic transition, a molecule will absorb either ultraviolet or visible light. If it absorbs **ultraviolet light,** a UV spectrum is obtained; if it absorbs **visible light,** a visible spectrum is obtained. *Ultraviolet light has* wavelengths ranging from 180 to 400 nm (nanometers); *visible light has* wavelengths ranging from 400 to 780 nm.

Wavelength (λ) is inversely related to the energy of the radiation, so the shorter the wavelength, the greater the energy of the radiation. Ultraviolet light, therefore, has greater energy than visible light.

In the ground-state electronic configuration of a molecule, all the electrons are in the lowest-energy molecular orbitals. When a molecule absorbs light with sufficient energy to promote an electron to a higher-energy molecular orbital—that is, when it undergoes an **electronic transition**—the molecule is then in an excited state. Ultraviolet and visible light has just enough energy to promote an electron from

> The shorter the wavelength, the greater the energy of the radiation.

$$E = \frac{hc}{\lambda}$$

a π bonding MO to a π^* antibonding MO (Figure 35). Therefore, *only compounds with π bonds can produce UV/Vis spectra.*

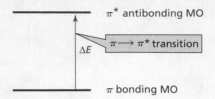

▶ **Figure 35**

Ultraviolet and visible light promote an electron from a π bonding MO to a π* antibonding MO.

The UV spectrum of acetone is shown in Figure 36. The λ_{max} (stated as "lambda max") is the wavelength at which the absorption band for the $\pi \longrightarrow \pi^*$ transition (stated as "π to π^*") has its maximum absorbance. For acetone, $\lambda_{max} = 195\,nm$.

Only compounds with π electrons can produce ultraviolet and visible spectra.

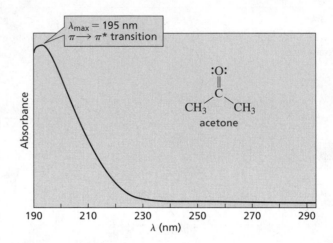

▶ **Figure 36**

The UV spectrum of acetone.

The absorption band is broad because each electronic state has vibrational sublevels (Figure 37), and electronic transitions can occur from and to these different sublevels. As a result, the electronic transitions span a range of wavelengths.

A **chromophore** is the part of a molecule that absorbs UV or visible light. The carbonyl group is the chromophore of acetone. The following four compounds all have the same chromophore, so they all have approximately the same λ_{max}.

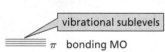

▲ **Figure 37**

UV/Vis absorption bands are broad because each electronic state has vibrational sublevels.

Ultraviolet Light and Sunscreens

Exposure to ultraviolet (UV) light stimulates specialized cells in the skin to produce a black pigment known as melanin, which causes the skin to look tan. Melanin absorbs UV light, so it protects our bodies from the harmful effects of the sun. If more UV light reaches the skin than melanin can absorb, the light will "burn" the skin and cause photochemical reactions that can result in skin cancer.

UV-A is the lowest-energy UV light (315 to 400 nm). It is the light that causes skin to wrinkle. Much of the more dangerous, higher-energy light—namely, UV-B (290 to 315 nm) and UV-C (180 to 290 nm)—is filtered out by the ozone layer in the stratosphere, which is why the thinning of the ozone layer became such an important issue.

Applying a sunscreen can protect skin from UV light. The amount of protection from UV-B light (the light that causes skin to burn) is indicated by the sunscreen's SPF (sun protection factor); the higher the SPF, the greater the protection. Some sunscreens contain an inorganic component, such as zinc oxide, which reflects the light as it reaches the skin. Others contain a compound that absorbs UV light.

para-Aminobenzoic acid (PABA) was the first commercially available UV-absorbing sunscreen. It absorbs UV-B light, but is not very soluble in oily skin lotions. Thus, the next generation of sunscreens contained Padimate O, a less polar compound. Subsequent research showed that sunscreens need to absorb both UV-B and UV-A light in order to give adequate protection against skin cancer. Now the FDA requires that sunscreens, such as Give-Tan F, protect against both UV-A and UV-B light.

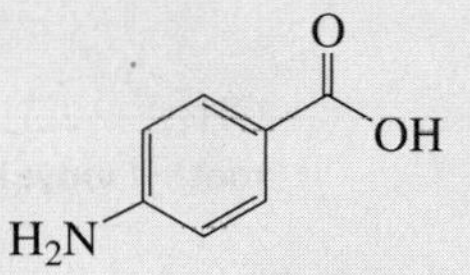

para-aminobenzoic acid
PABA

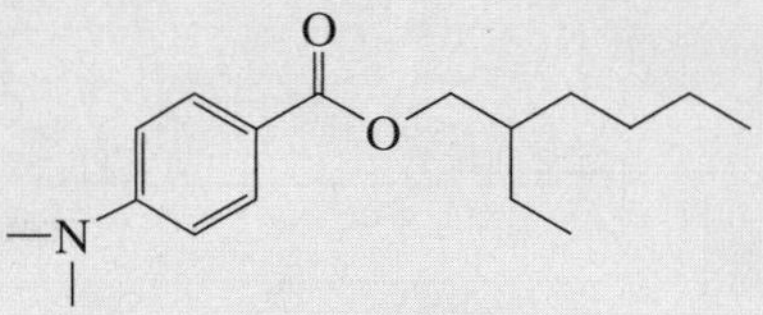

2-ethylhexyl 4-(dimethylamino)benzoate
Padimate O

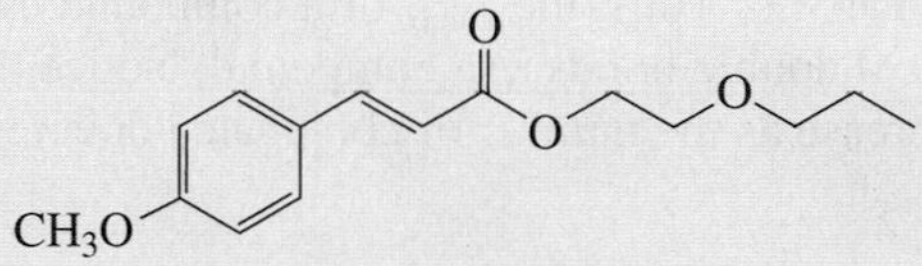

2-ethoxyethyl (*E*)-3-(4-methoxyphenyl)-2-propenoate
Give-Tan F

19   THE BEER–LAMBERT LAW

Wilhelm Beer and Johann Lambert independently proposed that the absorbance of a sample at a given wavelength depends on the amount of absorbing species that the light encounters as it passes through a solution of the sample. In other words, absorbance depends on both the concentration of the sample and the length of the light path through the sample. The relationship between absorbance, concentration, length of the light path, and molar absorptivity, known as the **Beer–Lambert law,** is given by

$$A = c\, l\, \varepsilon, \text{ where}$$

A = absorbance of the sample

c = concentration of the sample, in moles/liter

l = length of the light path through the sample, in centimeters

ε = molar absorptivity ($M^{-1}cm^{-1}$)

The **molar absorptivity** (ε) is a constant that is characteristic of the compound at a particular wavelength. It is the absorbance that would be observed for a 1.00 M solution in a cell with a 1.00-cm path length. (The abbreviation ε comes from the fact that molar absorptivity was formerly called the extinction coefficient.)

For example, the molar absorptivity of acetone dissolved in hexane is $9000\,M^{-1}\,cm^{-1}$ at 195 nm. The solvent in which the sample is dissolved is reported because molar absorptivity is not exactly the same in all solvents. Therefore, the UV spectrum of acetone in hexane would be reported as $\lambda_{max} = 195$ nm ($\varepsilon_{max} = 9000$, hexane).

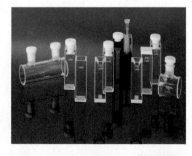

The solution whose UV or visible spectrum is to be taken is put into a cell, such as one of those shown here. Most cells have 1-cm path lengths. Either glass or quartz cells can be used for visible spectra, but quartz cells (made of high-purity fused silica) must be used for UV spectra because glass absorbs UV light.

PROBLEM 35♦

A solution of a compound in ethanol shows an absorbance of 0.52 at 236 nm in a cell with a 1-cm light path. Its molar absorptivity in ethanol at that wavelength is $12,600\,M^{-1}\,cm^{-1}$. What is the concentration of the compound?

PROBLEM 36♦

A 4.0×10^{-5} M solution of a compound in hexane shows an absorbance of 0.40 at 252 nm in a cell with a 1-cm light path. What is the molar absorptivity of the compound in hexane at 252 nm?

20   THE EFFECT OF CONJUGATION ON λ_{max}

The wavelength at which the $\pi \rightarrow \pi^*$ transition occurs increases as the number of conjugated double bonds in the compound increases. For example, the λ_{max} for methyl vinyl ketone is at a longer wavelength (219 nm) than the λ_{max} for acetone (195 nm), because methyl vinyl ketone has two conjugated double bonds, whereas acetone has only one double bond.

:O:
‖
C
CH₃   CH₃
acetone
λ_{max} = **195 nm**

:O:
‖
C
CH₃   CH=CH₂
methyl vinyl ketone
λ_{max} = **219 nm**

The λ_{max} values of the $\pi \rightarrow \pi^*$ transition for several conjugated polyenes are listed in Table 6. Thus, the λ_{max} of a compound can be used to estimate the number of conjugated double bonds in a compound. Notice that both the λ_{max} and the molar absorptivity increase as the number of conjugated double bonds increases.

The λ_{max} increases as the number of conjugated double bonds increases.

Table 6	Values of λ_{max} and ε for Ethylene and Conjugated Polyenes	
Compound	λ_{max} **(nm)**	ε **($M^{-1}\ cm^{-1}$)**
$H_2C{=}CH_2$	165	15,000
∿	217	21,000
∿	256	50,000
∿	290	85,000
∿	334	125,000
∿	364	138,000

Conjugation raises the energy of the **HOMO (highest occupied molecular orbital)** and lowers the energy of the **LUMO (lowest unoccupied molecular orbital)**, so less energy is required for an electronic transition in a conjugated system than in a nonconjugated system (Figure 38). Therefore, the more conjugated double bonds in a compound, the less energy required for the electronic transition so the longer the wavelength at which it occurs.

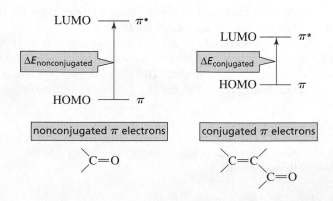

▶ Figure 38

Conjugation raises the energy of the HOMO and lowers the energy of the LUMO.

If a compound has enough conjugated double bonds, it will absorb visible light (light with wavelengths > 400 nm) and the compound will be colored. For example, β-carotene, a precursor of vitamin A with a λ_{max} = 455 nm, is an orange substance found in carrots, apricots, and the feathers of flamingos. Lycopene with a $\lambda_{max} = 474$ nm—found in tomatoes, watermelon, and pink grapefruit—is red.

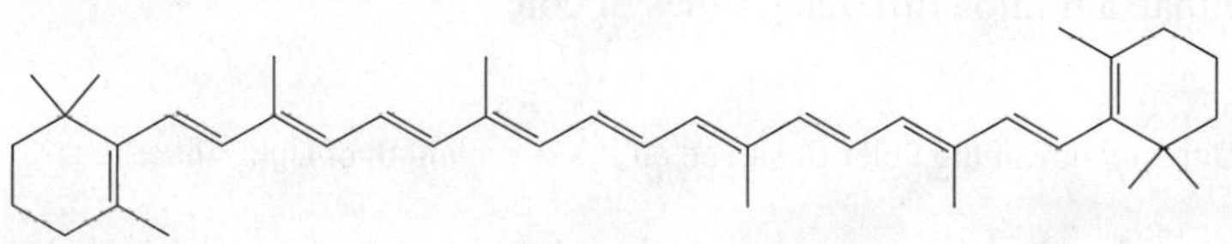

β-carotene
$\lambda_{max} = 455$ nm

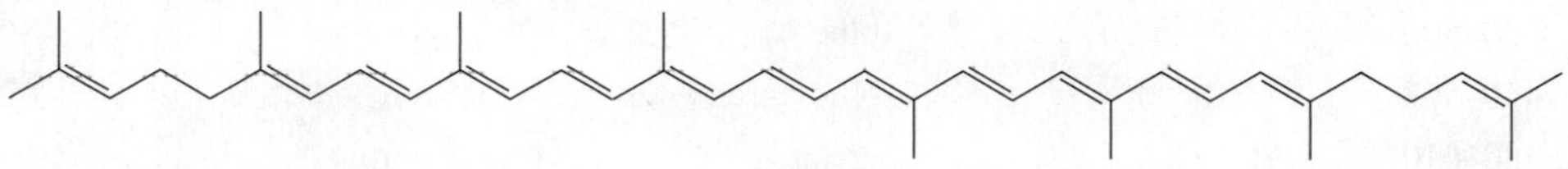

lycopene
$\lambda_{max} = 474$ nm

An **auxochrome** is a substituent that, when attached to a chromophore, alters both the λ_{max} and the intensity of the absorption, usually increasing both. For example, OH and NH_2 groups are auxochromes. The lone-pair electrons on oxygen and nitrogen in the compounds shown here are available to interact with the π electron cloud of the benzene ring; such an interaction increases λ_{max}.

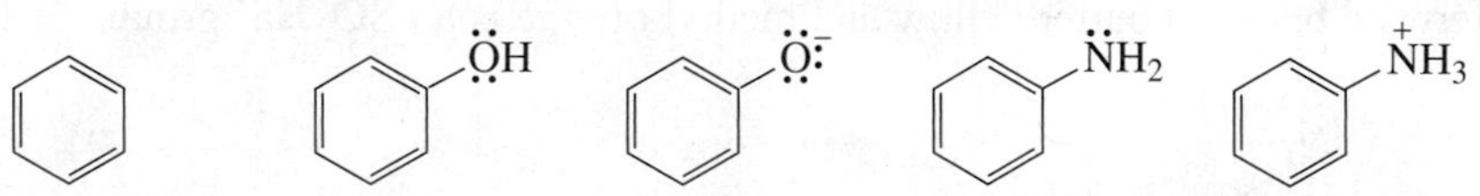

benzene	phenol	phenolate ion	aniline	anilinium ion
$\lambda_{max} = $ 255 nm	270 nm	287 nm	280 nm	254 nm

Removing a proton from phenol increases the λ_{max} because the phenolate ion has an additional lone pair. Protonating aniline decreases the λ_{max} because the lone pair is no longer available to interact with the π cloud of the benzene ring. Because the anilinium ion does not have an auxochrome, its λ_{max} is similar to that of benzene.

PROBLEM 37♦

Predict the λ_{max} of the following compound:

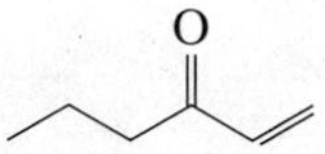

PROBLEM 38♦

Rank each set of compounds in order of decreasing λ_{max}:

a.

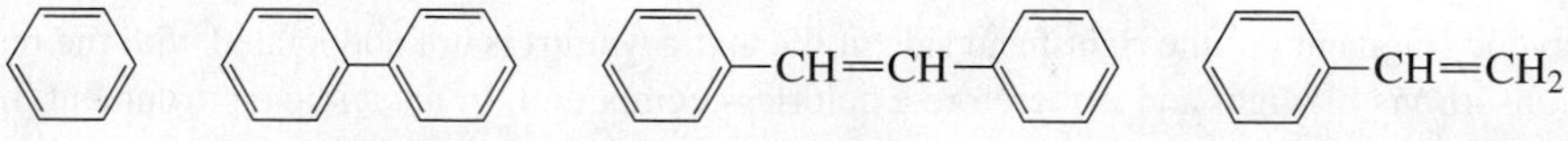

b.

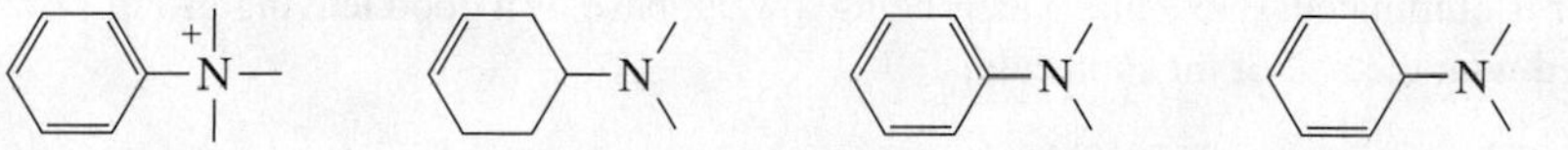

---

21 THE VISIBLE SPECTRUM AND COLOR

White light is a mixture of all visible wavelengths. If any of these wavelengths are removed from white light, the eye registers the remaining light as colored. Therefore, any compound that absorbs visible light appears colored. The perceived color depends on the precise wavelengths reaching the eye. The wavelengths that the compound does *not* absorb are reflected back to the viewer, producing the color the viewer sees.

The relationship between the wavelengths of the light that a substance absorbs and the substance's observed color is shown in Table 7. Notice that two absorption bands

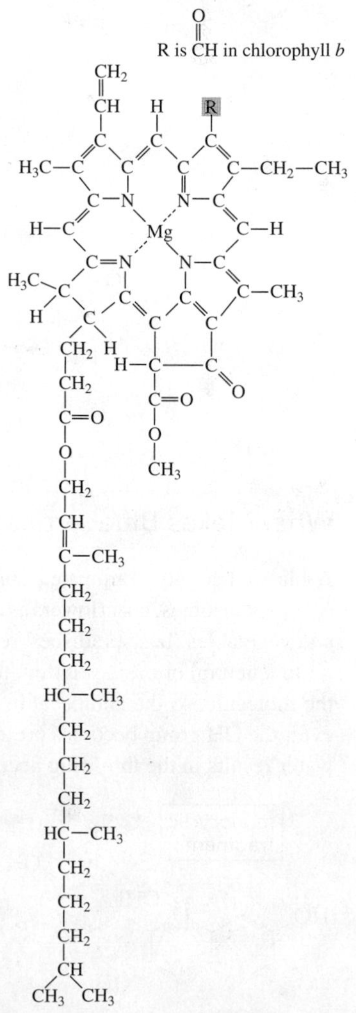

R is CH_3 in chlorophyll *a*

R is $\overset{O}{\overset{\|}{CH}}$ in chlorophyll *b*

Chlorophyll *a* and *b* are highly conjugated compounds that absorb visible light, causing green light to be reflected from the surface tissues of plants.

are necessary to produce green. Most colored compounds have fairly broad absorption bands, but vivid colors have narrow absorption bands. The human eye is able to distinguish more than a million different shades of color!

Table 7 Dependence of the Color Observed on the Wavelength of Light Absorbed

Wavelengths absorbed (nm)	Color absorbed	Color observed
380–460	blue-violet	yellow
380–500	blue	orange
440–560	blue-green	red
480–610	green	purple
540–650	orange	blue
380–420 and 610–700	purple	green

Azobenzenes (benzene rings connected by an $N=N$ bond) have an extended conjugated system that causes them to absorb visible light. The two shown here are used commercially as dyes. Changing the number of conjugated double bonds and the substituents attached to them creates a large number of different colors. Notice that the only difference between butter yellow and methyl orange is an $SO_3^-Na^+$ group.

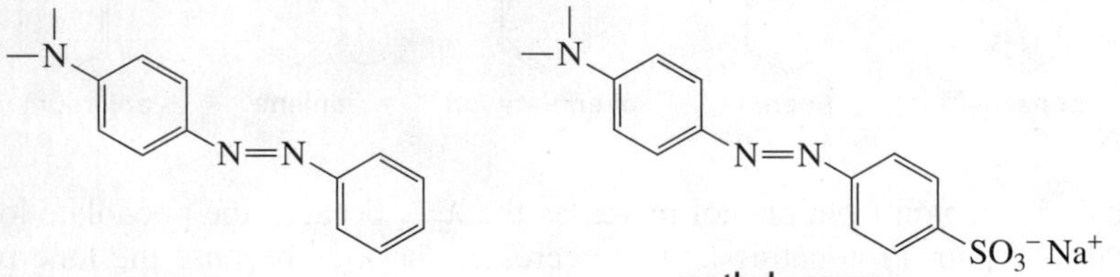

butter yellow
an azobenzene

methyl orange
an azobenzene

When margarine was first produced, it was colored with butter yellow to make it look more like butter. (White margarine would not be very appetizing.) This dye was abandoned after it was found to be carcinogenic. β-Carotene (see earlier in this section) is now used to color margarine. Methyl orange is a commonly used acid–base indicator (see Problem 71).

What Makes Blueberries Blue and Strawberries Red?

A class of highly conjugated compounds called *anthocyanins* is responsible for the red, purple, and blue colors of many flowers (poppies, peonies, cornflowers), fruits (cranberries, rhubarb, strawberries, blueberries, the red skin of apples, the purple skin of grapes), and vegetables (beets, radishes, red cabbage).

In a neutral or basic solution, the monocyclic fragment (on the right-hand side of the anthocyanin) is not conjugated with the rest of the molecule, so the anthocyanin does not absorb visible light and is therefore a colorless compound. In an acidic environment, however, the OH group becomes protonated and water is eliminated. (Note that water, being a weak base, is a good leaving group.) Loss of water results in the third ring becoming conjugated with the rest of the molecule.

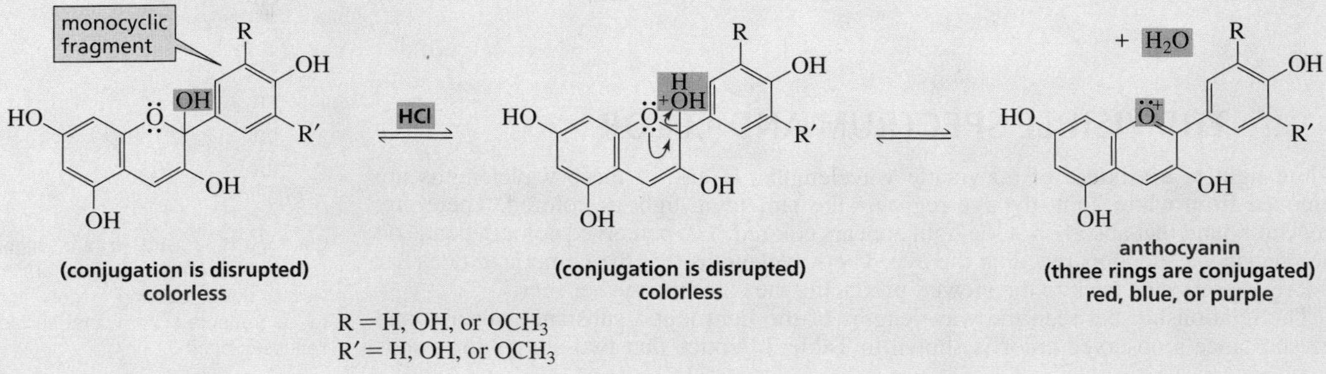

(conjugation is disrupted)
colorless

(conjugation is disrupted)
colorless

anthocyanin
(three rings are conjugated)
red, blue, or purple

R = H, OH, or OCH$_3$
R' = H, OH, or OCH$_3$

As a result of the increase in conjugation, the anthocyanin absorbs visible light with wavelengths between 480 and 550 nm. The exact wavelength of light absorbed depends on the substituents (R and R′) on the anthocyanin. Thus, the flower, fruit, or vegetable appears red, purple, or blue, depending on what R and R′ are. You can see this color change if you alter the pH of cranberry juice so that it is no longer acidic.

PROBLEM 39♦

a. At pH = 7, one of the ions shown here is purple and the other is blue. Which is which?

b. What would be the difference in the colors of the compounds at pH = 3?

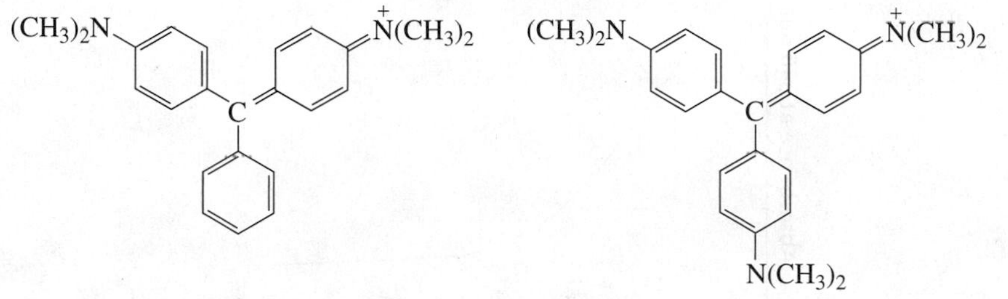

PROBLEM 40

Predict from Table 7 the two colors that when mixed together produce green.

22　SOME USES OF UV/VIS SPECTROSCOPY

UV/Vis spectroscopy is not nearly as useful as other instrumental techniques for determining the structures of organic compounds. However, UV/Vis spectroscopy has many other important uses.

　UV/Vis spectroscopy is often used to measure reaction rates. The rate of any reaction can be measured, as long as one of the reactants or one of the products absorbs UV or visible light at a wavelength at which the other reactants and products have little or no absorbance.

　For example, the anion of nitroethane has a λ_{max} at 240 nm, but neither the other product (H_2O) nor the reactants show any significant absorbance at this wavelength. In order to measure the rate at which hydroxide ion removes a proton from nitroethane to form the nitroethane anion, the UV spectrophotometer is adjusted to measure absorbance at 240 nm as a function of time (Figure 39) instead of absorbance as a function of wavelength (Figure 36).

Lycopene, β-carotene, and anthocyanins are found in the leaves of trees, but their characteristic colors are usually obscured by the green color of chlorophyll. Chlorophyll is an unstable molecule, so plants must continually synthesize it. Its synthesis requires sunlight and warm temperatures. As the weather becomes colder in the fall, plants can no longer replace chlorophyll as it degrades, so the other colors become apparent.

$$CH_3CH_2NO_2 \;+\; HO^- \;\rightleftharpoons\; CH_3\bar{C}HNO_2 \;+\; H_2O$$

　　　nitroethane　　　　　　　　　　nitroethane anion

　　　　　　　　　　　　　　　　　$\lambda_{max} = 240$ nm

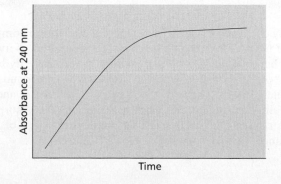

◄ **Figure 39**

The rate at which a proton is removed from nitroethane is determined by monitoring the increase in absorbance at 240 nm.

The enzyme lactate dehydrogenase catalyzes the reduction of pyruvate to lactate by NADH. NADH is the only species in the reaction mixture that absorbs light at 340 nm, so the rate of the reaction can be determined by monitoring the decrease in absorbance at 340 nm (Figure 40).

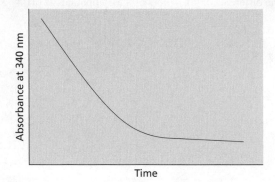

▶ Figure 40

The rate of reduction of pyruvate by NADH is measured by monitoring the decrease in absorbance at 340 nm.

PROBLEM 41♦

Describe a way to determine the rate of the alcohol dehydrogenase catalyzed oxidation of ethanol by NAD^+.

The Henderson–Hasselbalch equation states that the pK_a of a compound is the pH at which half the compound exists in its acidic form and half exists in its basic form.

The pK_a of a compound can be determined by UV/Vis spectroscopy if either the acidic form or the basic form of the compound absorbs UV or visible light. For example, the phenolate ion has a λ_{max} at 287 nm. If the absorbance at 287 nm is monitored as a function of pH, the pK_a of phenol can be ascertained by determining the pH at which exactly one-half the increase in absorbance has occurred (Figure 41). At this pH, half of the phenol has been converted into phenolate ion, so this pH is equal to the pK_a of the compound.

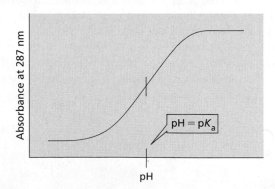

▶ Figure 41

The absorbance of an aqueous solution of phenol as a function of pH.

UV spectroscopy can also be used to estimate the nucleotide composition of DNA. The two strands of DNA are held together by hydrogen bonds between the bases in one strand and the bases in the other: each guanine forms three hydrogen bonds with a cytosine (a G–C pair); each adenine forms two hydrogen bonds with a thymine (an A–T pair). When DNA is heated, the strands break apart and the absorbance increases because single-stranded DNA has a greater molar absorptivity at 260 nm than does double-stranded DNA. The melting temperature (T_m) of DNA is the midpoint of an absorbance-versus-temperature curve (Figure 42).

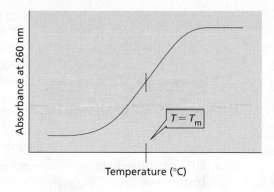

◀ **Figure 42**

The absorbance of a solution of DNA as a function of temperature.

The T_m increases with increasing numbers of G–C pairs, because they are held together by three hydrogen bonds, whereas A–T pairs are held together by only two hydrogen bonds. Therefore, T_m can be used to estimate the number of G–C pairs. These are just a few examples of the many uses of UV/Vis spectroscopy.

PROBLEM 42◆

The absorbance of a solution of a weak acid was measured under the same conditions at a series of pH values. Its conjugate base is the only species in the solution that absorbs UV light at the wavelength used. Estimate the pK_a of the acid from the data obtained.

pH	1.0	2.0	3.0	4.0	5.0	6.0	7.0	8.0	9.0	10.0
Absorbance	0	0	0.10	0.50	0.80	1.10	1.50	1.60	1.60	1.60

SOME IMPORTANT THINGS TO REMEMBER

- **Mass spectrometry** allows us to determine the *molecular mass* and the *molecular formula* of a compound and some of its structural features.

- The **molecular ion** (a **radical cation**), which is formed by removing an electron from a molecule, can break apart. The bonds most likely to break are the weakest ones and those that result in the formation of the most stable products.

- A **mass spectrum** is a graph of the relative abundance of each positively charged fragment plotted against its *m/z* value. The *m/z* value of the molecular ion (M) gives the molecular mass of the compound.

- Peaks with smaller *m/z* values—**fragment ion peaks**—represent positively charged fragments of the molecular ion. The **base peak** is the peak with the greatest abundance. It the most stable fragment.

- The **rule of 13** allows possible molecular formulas to be determined from the *m/z* value of the molecular ion.

- The **nitrogen rule** states that if a compound has a molecular ion with an odd-numbered mass, the compound contains an odd number of nitrogen atoms.

- High-resolution mass spectrometers determine the exact molecular mass, which allows a compound's molecular formula to be determined.

- The M + 1 peak occurs because of the naturally occurring ^{13}C isotope of carbon.

- If the M + 2 peak is one-third the height of the M peak, the compound contains a chlorine atom; if the M and M + 2 peaks are about the same height, the compound contains a bromine atom.

- Electron bombardment is most likely to dislodge a lone-pair electron. A bond between carbon and a more electronegative atom breaks *heterolytically,* with the electrons going to the more electronegative atom, whereas a bond between carbon and an atom of similar electronegativity breaks *homolytically.*

- *α*-**Cleavage** occurs because the species it forms has a positive charge that is shared by two atoms.

- **Spectroscopy** is the study of the interaction of matter and **electromagnetic radiation.**

- High-energy radiation is associated with *high frequencies, large wavenumbers,* and *short wavelengths.*

- **Infrared (IR) spectroscopy** identifies the kinds of functional groups in a compound. To absorb IR radiation, the dipole moment of the bond must change when the vibration occurs.

- It takes more energy to stretch a bond than to bend it.

- Stronger bonds show **absorption bands** at larger wavenumbers.

- The position of an absorption band depends on bond order, hybridization, inductive electron donation and withdrawal, electron delocalization, and hydrogen bonding.

- The intensity of an absorption band depends on the size of the change in dipole moment (more polar bonds show more intense absorptions), the number of bonds giving rise to the absorption, and the concentration.

- The shape of an absorption band depends on hydrogen bonding. Hydrogen bonds vary in strength so hydrogen-bonded groups show broader absorption bands.

- The frequency of an absorption band is inversely related to the *mass* of the atoms that form the bond, so heavier atoms vibrate at lower frequencies.

- **Ultraviolet** and **visible (UV/Vis) spectroscopy** provide information about compounds with conjugated double bonds—the more conjugated double bonds in a compound, the longer the λ_{max} at which absorption occurs.

- UV light has greater energy than visible light—the shorter the wavelength, the greater the energy.

- UV and visible light cause $\pi \rightarrow \pi^*$ **electronic transitions.**

- A **chromophore** is the part of a molecule that absorbs UV or visible light.

- The **Beer–Lambert law** is the relationship between absorbance, concentration, length of the light path, and molar absorptivity: $A = c l \varepsilon$.

PROBLEMS

43. In the mass spectrum of the following compounds, which would be more intense—the peak at $m/z = 57$ or the peak at $m/z = 71$?

   a. 3-methylpentane   **b.** 2-methylpentane

44. List three factors that influence the intensity of an IR absorption band.

45. For each of the following pairs of compounds, identify one IR absorption band that could be used to distinguish between them:

a. 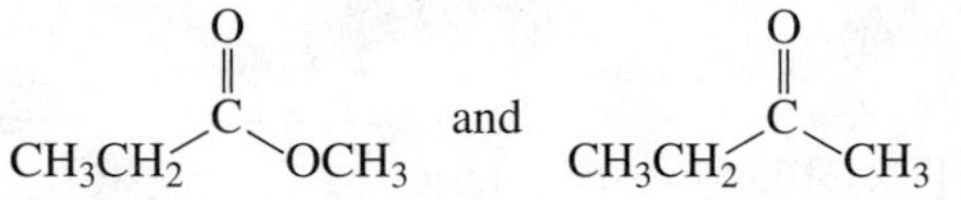 CH_3CH_2—C(=O)—OCH_3 and CH_3CH_2—C(=O)—CH_3

b. 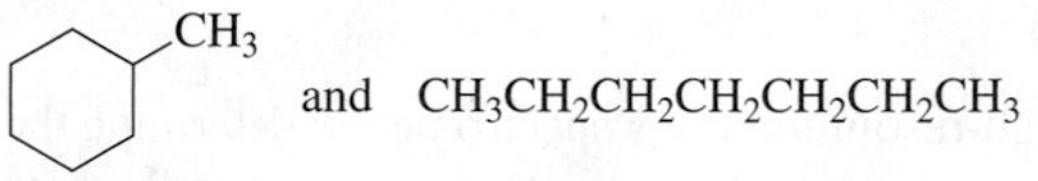 (cyclohexane with CH_3) and $CH_3CH_2CH_2CH_2CH_2CH_2CH_3$

c. $CH_3CH_2CH_2OH$ and $CH_3CH_2OCH_3$

d. 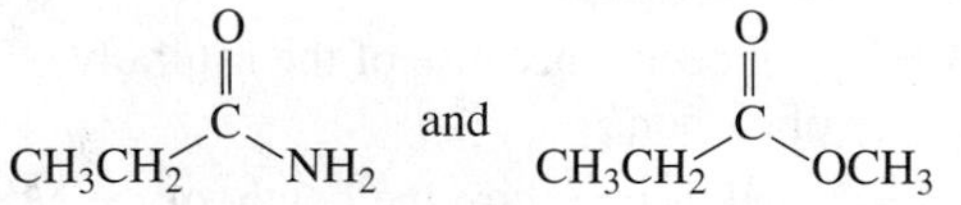 CH_3CH_2—C(=O)—NH_2 and CH_3CH_2—C(=O)—OCH_3

e. 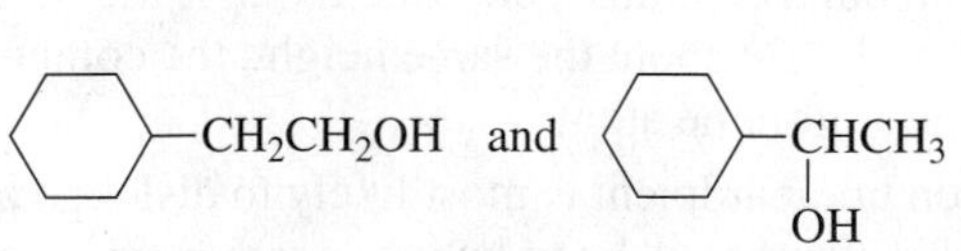 (cyclohexyl)—CH_2CH_2OH and (cyclohexyl)—CH(OH)CH_3

f. *cis*-2-butene and *trans*-2-butene

g. 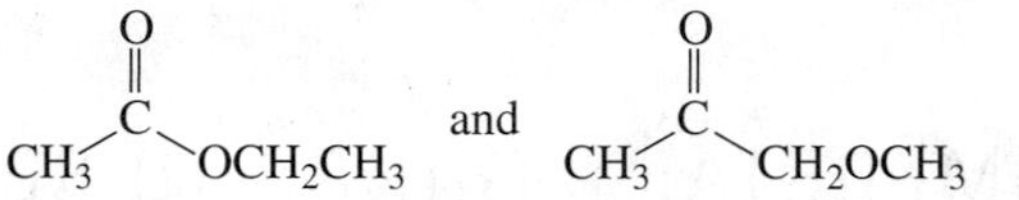 CH_3—C(=O)—OCH_2CH_3 and CH_3—C(=O)—CH_2OCH_3

h. 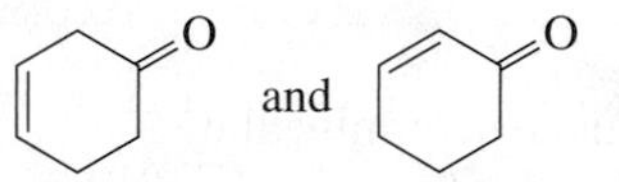 (cyclohexanone) and (cyclohexenone)

i. CH_3CH_2CH—$CHCH_3$ and CH_3CH_2C—CCH_3

j. 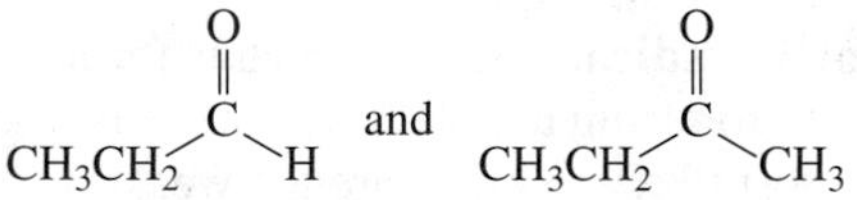 CH_3CH_2—C(=O)—H and CH_3CH_2—C(=O)—CH_3

k. 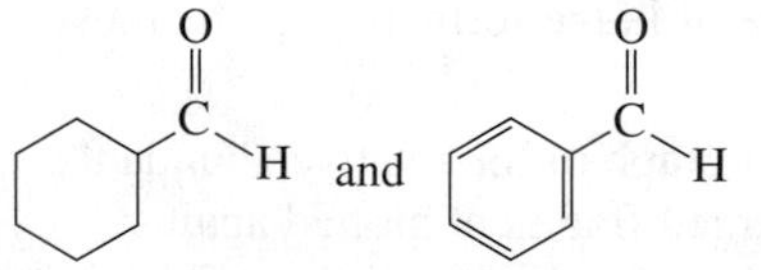 (cyclohexyl)—C(=O)—H and (phenyl)—C(=O)—H

l. CH_3CH_2CH=CH_2 and CH_3CH_2CH=$C(CH_3)CH_3$

46. Draw the structure of an unsaturated hydrocarbon that has a molecular ion with an m/z value of 128.

47. a. How could you use IR spectroscopy to determine whether the following reaction had occurred?

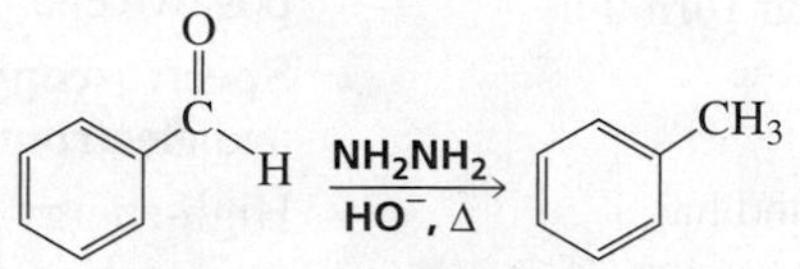

(phenyl)—C(=O)—H $\xrightarrow[\text{HO}^-, \Delta]{NH_2NH_2}$ (phenyl)—CH_3

   b. After purifying the product, how could you determine whether all the NH_2NH_2 had been removed?

48. Assuming that the force constant is approximately the same for C—C, C—N, and C—O bonds, predict the relative positions of their stretching vibrations in an IR spectrum.

49. In the following boxes, list the types of bonds and the approximate wavenumber at which each type of bond is expected to show an IR absorption:

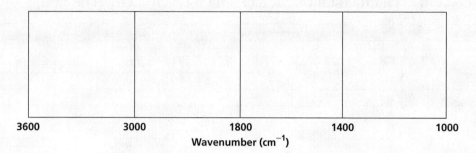

| 3600 | 3000 | 1800 | 1400 | 1000 |

Wavenumber (cm^{-1})

50. A mass spectrum shows significant peaks at $m/z = 87$, 115, 140, and 143. Which of the following compounds is responsible for that mass spectrum?

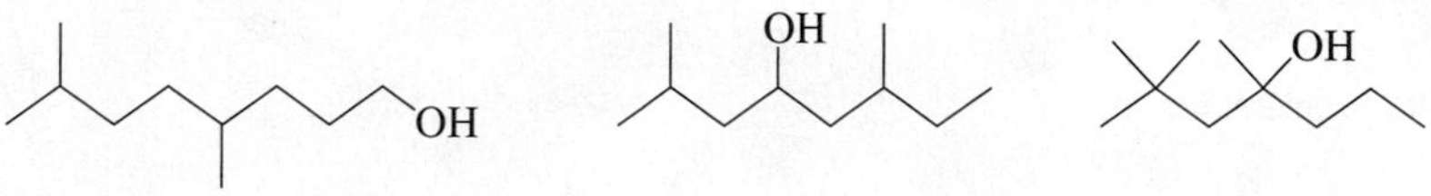

51. How could IR spectroscopy distinguish between 1,5-hexadiene and 2,4-hexadiene?

52. A compound gives a mass spectrum with essentially only three peaks at $m/z = 77$ (40%), 112 (100%), 114 (33%). Identify the compound.

53. What hydrocarbons that contain a six-membered ring will have a molecular ion peak at $m/z = 112$?

54. How could you use UV spectroscopy to distinguish between the compounds in each of the following pairs?

a. 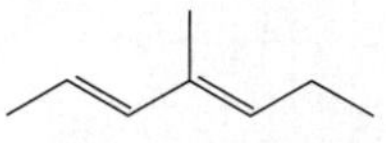 [structure] and [structure]

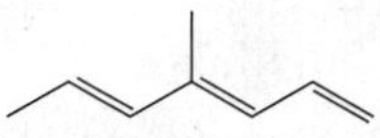

b. 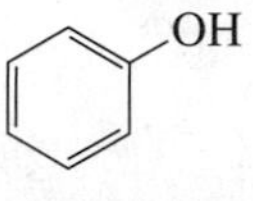 [structure with OH] and [structure with OCH$_3$]

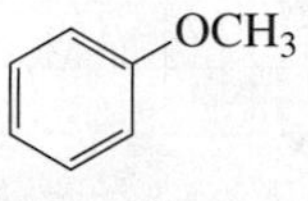

c. 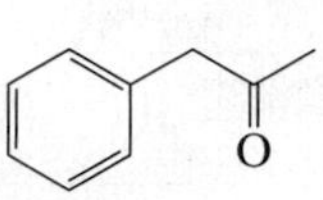 [structure] and [structure]

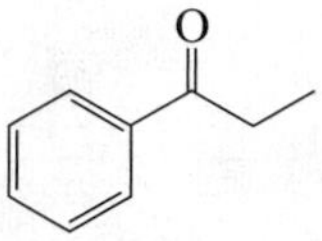

55. List the following compounds in order from highest wavenumber to lowest wavenumber for their C=O absorption bands:

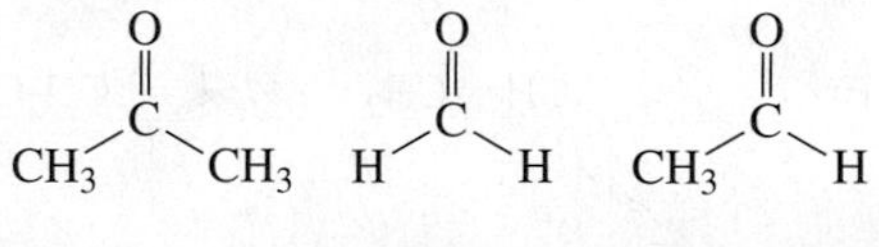

56. List the following compounds in order from highest wavenumber to lowest wavenumber for their C—O absorption bands:

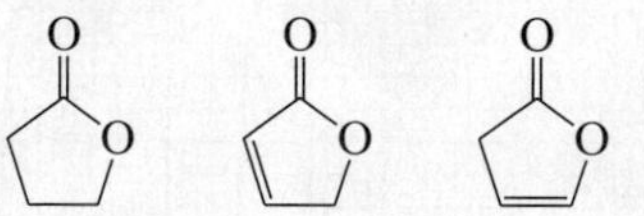

57. What peaks in their mass spectra could be used to distinguish between 4-methyl-2-pentanone and 2-methyl-3-pentanone?

661

58. Each of the IR spectra presented in Figures 43, 44, and 45 is accompanied by a set of four compounds. In each case, indicate which of the four compounds is responsible for the spectrum.

a. CH₃CH₂CH₂C≡CCH₃   CH₃CH₂CH₂CH₂OH   CH₃CH₂CH₂CH₂C≡CH   CH₃CH₂CH₂COH

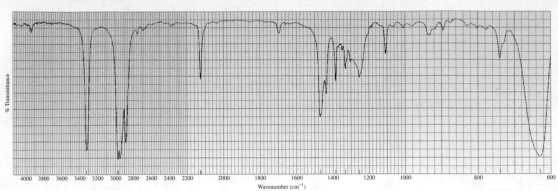

▲ **Figure 43**

The IR spectrum for Problem 58a.

b.

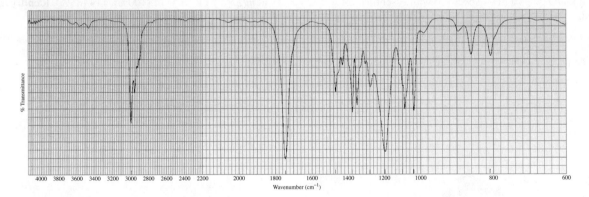

▲ **Figure 44**

The IR spectrum for Problem 58b.

c. C(CH₃)₃   CH₂CH₂Br   CH=CH₂   CH₂OH

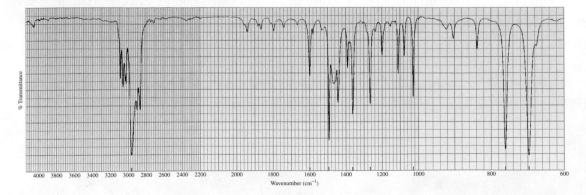

▲ **Figure 45**

The IR spectrum for Problem 58c.

59. Five compounds are shown for each of the IR spectra in Figures 46, 47, and 48. Indicate which of the five compounds is responsible for each spectrum.

a.

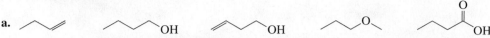

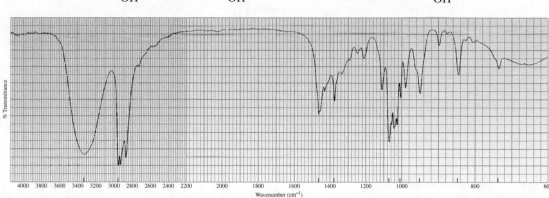

▲ **Figure 46**

The IR spectrum for Problem 59a.

b.

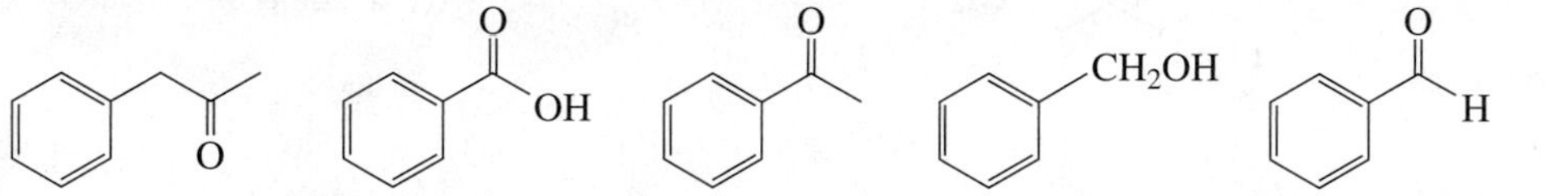

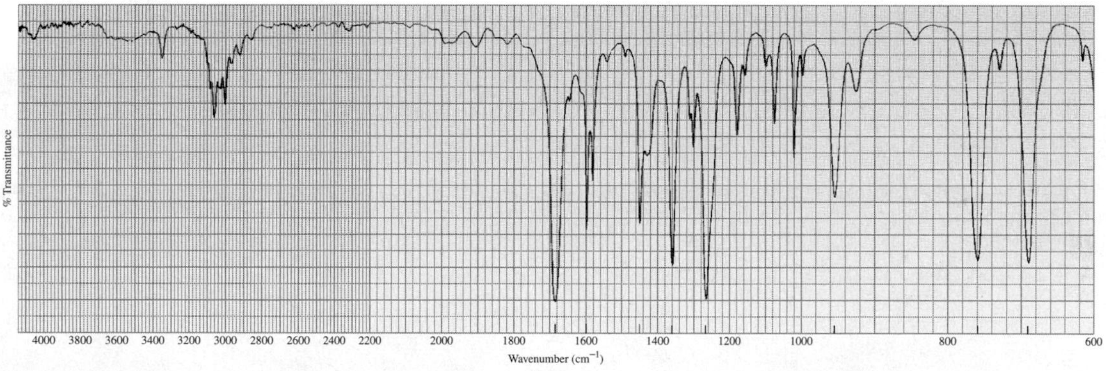

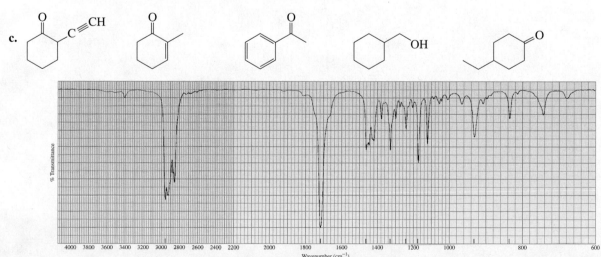

▲ **Figure 47**

The IR spectrum for Problem 59b.

c.

▲ **Figure 48**

The IR spectrum for Problem 59c.

60. A compound is known to be one of those shown here. What absorption bands in its IR spectrum would allow you to identify this compound?

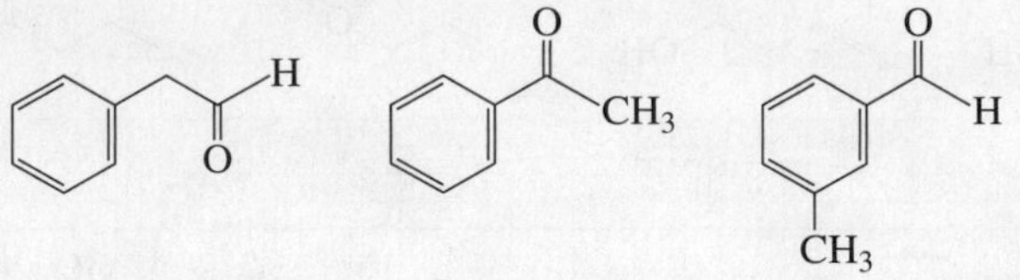

A   B   C

61. How could IR spectroscopy distinguish between 1-hexyne, 2-hexyne, and 3-hexyne?

62. Draw the structure of a carboxylic acid that has a molecular ion with an m/z value of 114.

63. A solution of ethanol has been contaminated with benzene—a technique employed to make ethanol unfit to drink. Benzene has a molar absorptivity of 230 M^{-1} cm^{-1} at 260 nm in ethanol, and ethanol shows no absorbance at 260 nm. How could the concentration of benzene in the solution be determined?

64. Give approximate wavenumbers for the major characteristic IR absorption bands that would be given by each of the following compounds:

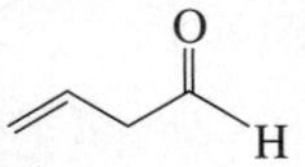

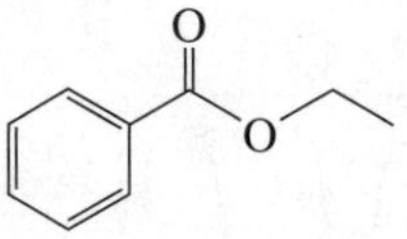

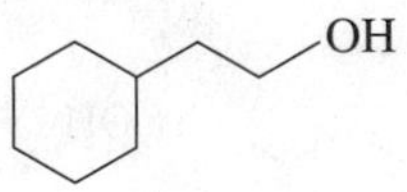

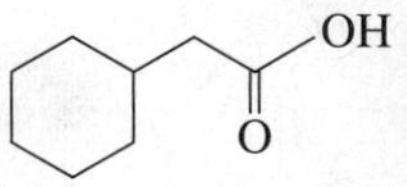

65. Given that the force constants are similar for C—H and C—C bonds, explain why the stretching vibration of a C—H bond occurs at a larger wavenumber.

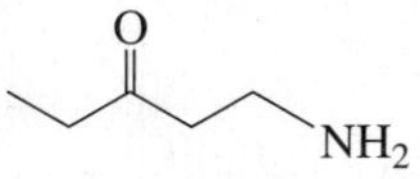

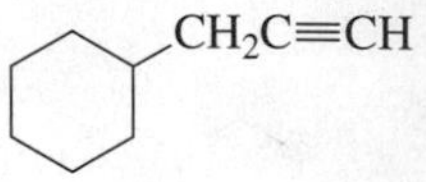

66. Some credit card sales slips have a top sheet of "carbonless paper" that transfers an imprint of a signature to a sheet lying underneath (the customer receipt). The paper contains tiny capsules filled with the following colorless compound. When you press on the paper, the capsules burst, and the colorless compound comes into contact with the acid-treated bottom sheet, forming a highly colored compound. What is the structure of the colored compound?

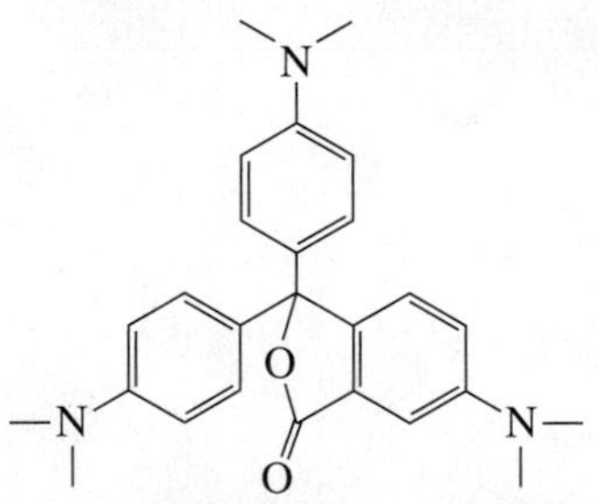

67. Each of the IR spectra shown in Figure 49 is the spectrum of one of the following compounds. Identify the compound that produced each spectrum.

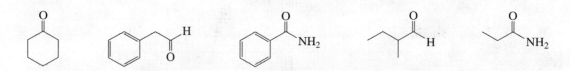

a.

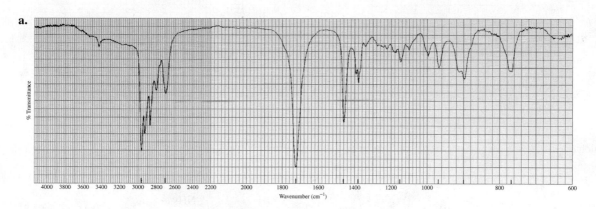

b.

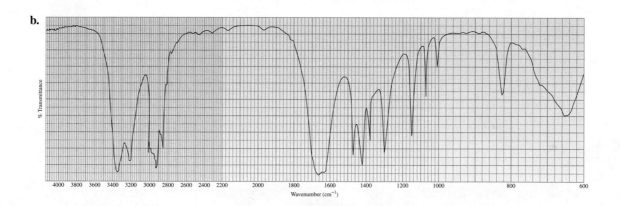

c.

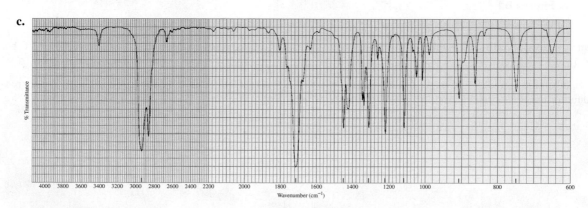

▲ **Figure 49**

The IR spectra for Problem 67.

68. The IR spectrum of a compound with molecular formula C_5H_8O is shown in Figure 50. Identify the compound.

▲ **Figure 50**

The IR spectrum for Problem 68.

69. Which one of the following five compounds produced the IR spectrum shown in Figure 51?

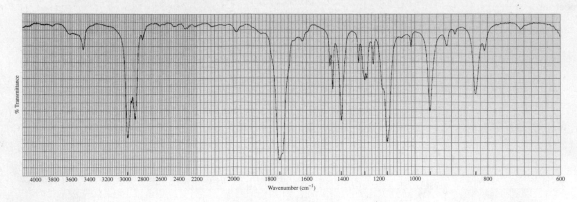

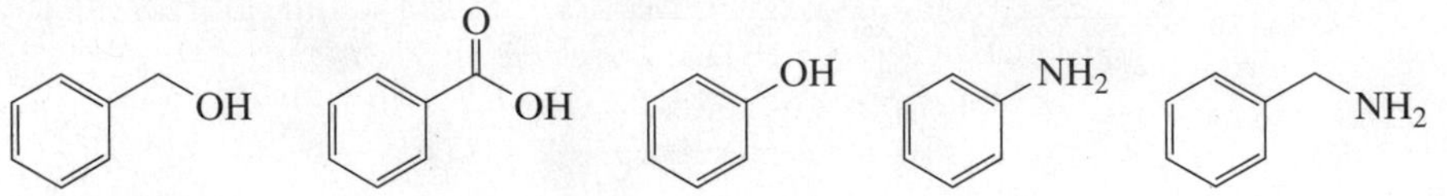

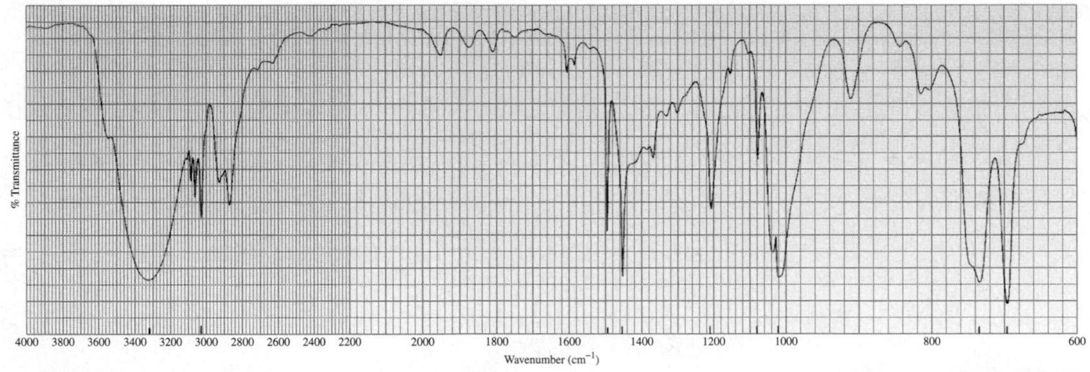

▲ **Figure 51**

The IR spectrum for Problem 69.

70. Calculate the approximate wavenumber at which a $C\!=\!C$ stretch will occur, given that the force constant for the $C\!=\!C$ bond is 10×10^5 g s^{-2}.

71. Phenolphthalein is an acid–base indicator. In solutions of pH < 8.5, it is colorless; in solutions of pH > 8.5, it is deep red-purple. Account for the change in color.

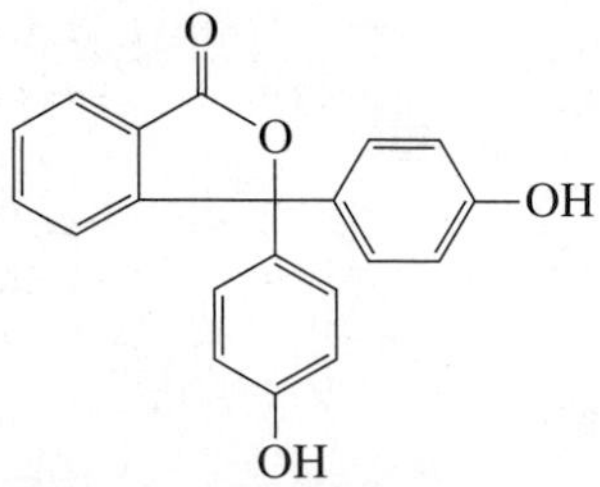

phenolphthalein

72. Which one of the following five compounds produced the IR spectrum shown in Figure 52?

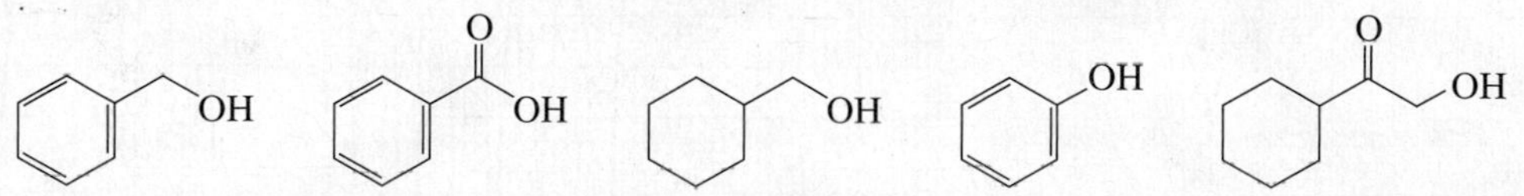

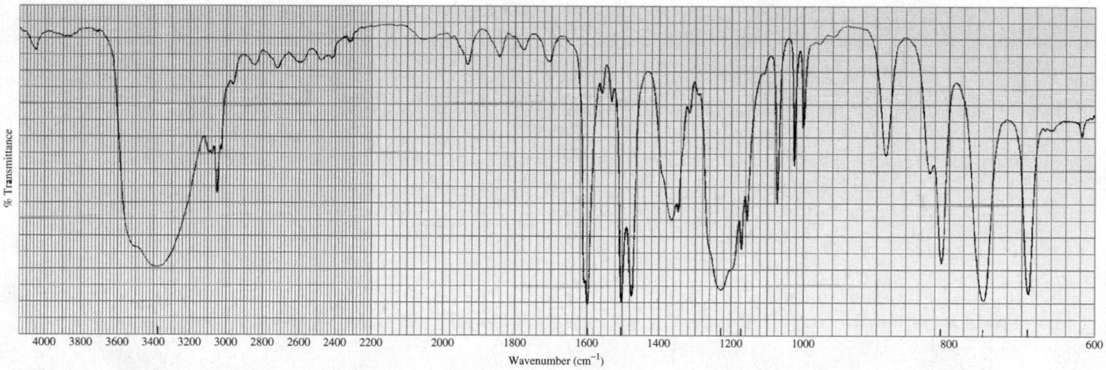

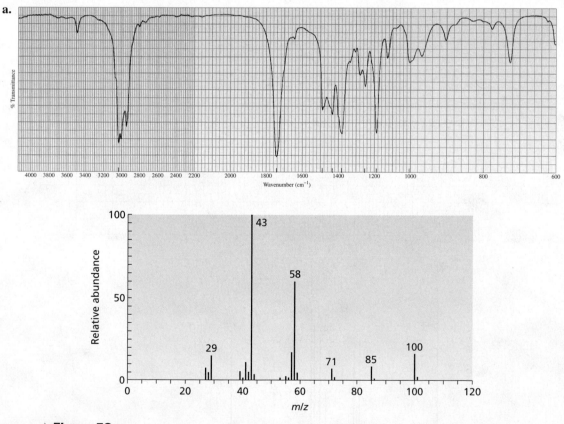

▲ **Figure 52**

The IR spectrum for Problem 72.

73. The IR and mass spectra for three different compounds are shown in Figures 53–55. Identify each compound.

a.

▲ **Figure 53**

The IR and mass spectra for Problem 73a.

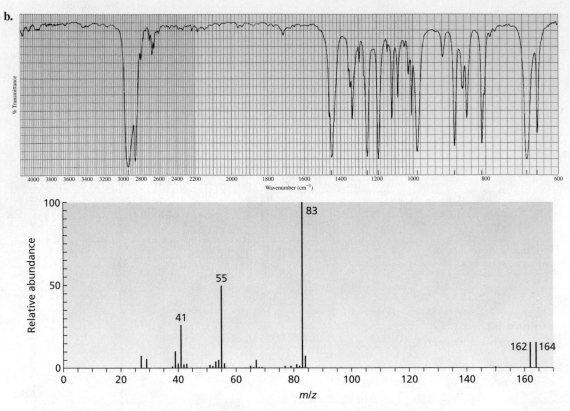

▲ Figure 54

The IR and mass spectra for Problem 73b.

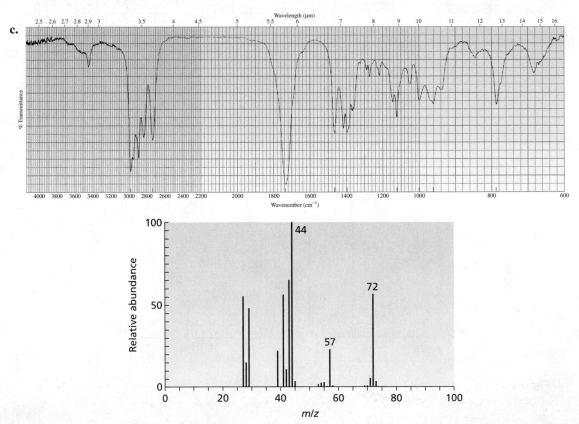

▲ Figure 55

The IR and mass spectra for Problem 73c.

GLOSSARY

absorption band a peak in a spectrum that occurs as a result of the absorption of energy.

auxochrome a substituent that when attached to a chromophore, alters the λ_{max} and intensity of absorption of UV/Vis radiation.

base peak the peak with the greatest abundance in a mass spectrum.

Beer–Lambert law relationship among the absorbance of UV/Vis light, the concentration of the sample, the length of the light path, and the molar absorptivity ($A = cl\varepsilon$).

bending vibration a vibration that does not occur along the line of the bond. It results in changing bond angles.

chromophore the part of a molecule responsible for a UV or visible spectrum.

α-cleavage homolytic cleavage of an alpha substituent.

electromagnetic radiation radiant energy that displays wave properties.

electronic transition promotion of an electron from its HOMO to its LUMO.

fingerprint region the right-hand third of an IR spectrum where the absorption bands are characteristic of the compound as a whole.

frequency the velocity of a wave divided by its wavelength (in units of cycles/s).

functional group region the left-hand two-thirds of an IR spectrum where most functional groups show absorption bands.

highest occupied molecular orbital (HOMO) the molecular orbital of highest energy that contains an electron.

Hooke's law an equation that describes the motion of a vibrating spring.

infrared radiation electromagnetic radiation familiar to us as heat.

infrared spectroscopy uses infrared energy to provide a knowledge of the functional groups in a compound.

infrared (IR) spectrum a plot of percent transmission versus wave number (or wavelength) of infrared radiation.

lowest unoccupied molecular orbital (LUMO) the molecular orbital of lowest energy that does not contain an electron.

mass spectrometry provides a knowledge of the molecular weight, molecular formula, and certain structural features of a compound.

mass spectrum a plot of the relative abundance of the positively charged fragments produced in a mass spectrometer versus their m/z values.

McLafferty rearrangement rearrangement of the molecular ion of a ketone. The bond between the α- and β-carbons breaks, and a γ-hydrogen migrates to the oxygen.

molar absorptivity the absorbance obtained from a 1.00 M solution in a cell with a 1.00-cm light path.

molecular ion (parent ion) peak in the mass spectrum with the greatest m/z.

NMR spectroscopy the absorption of electromagnetic radiation to determine the structural features of an organic compound. In the case of NMR spectroscopy, it determines the carbon–hydrogen framework.

radical cation a species with a positive charge and an unpaired electron.

spectroscopy study of the interaction of matter and electromagnetic radiation.

stretching vibration a vibration occurring along the line of a bond.

ultraviolet light electromagnetic radiation with wavelengths ranging from 180 to 400 nm.

UV/Vis spectroscopy the absorption of electromagnetic radiation in the ultraviolet and visible regions of the spectrum; used to determine information about conjugated systems.

visible light electromagnetic radiation with wavelengths ranging from 400 to 780 nm.

wavelength distance from any point on one wave to the corresponding point on the next wave (usually in units of μm or nm).

wavenumber the number of waves in 1 cm.

PHOTO CREDITS

Credits are listed in order of appearance.

NMR Spectroscopy

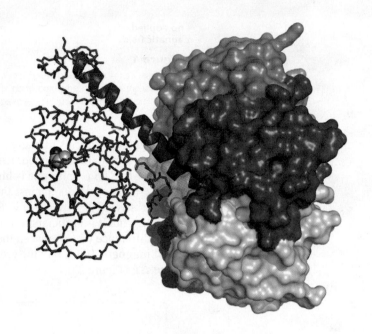

You should already be aware of three instrumental techniques that are used to determine the structures of organic compounds: mass spectrometry, IR spectroscopy, and ultraviolet/visible spectroscopy. Now we will look at a fourth technique: *nuclear magnetic resonance (NMR) spectroscopy*. **NMR spectroscopy** helps to identify the carbon–hydrogen framework of an organic compound.

The advantage of NMR spectroscopy over the other instrumental techniques is that it not only allows us to identify the functionality at a specific carbon, it also enables us to connect neighboring carbons. In many cases, NMR spectroscopy can be used to determine a molecule's entire structure.

1 AN INTRODUCTION TO NMR SPECTROSCOPY

NMR spectroscopy was developed by physicists in the late 1940s to study the properties of atomic nuclei. In 1951, chemists realized that NMR spectroscopy could also be used to study the structures of organic compounds.

Nuclei that have an odd number of protons or an odd number of neutrons (or both) have a property called spin that allows them (^1H, ^{13}C, ^{15}N, ^{19}F, and ^{31}P) to be studied by NMR. Nuclei such as ^{12}C and ^{16}O do not have spin and therefore cannot be studied by NMR. Because hydrogen nuclei (protons) were the first nuclei studied by NMR, the acronym *NMR* is generally assumed to mean ^{1}**H NMR (proton magnetic resonance).**

As a result of its charge, a nucleus with spin has a magnetic moment and generates a magnetic field similar to the magnetic field generated by a small bar magnet. In the absence of an applied magnetic field, the magnetic moments of the nuclei are randomly oriented. However, when placed between the poles of a strong magnet, the magnetic moments of the nuclei align either *with* or *against* the applied magnetic field (Figure 1).

There are additional spectroscopy problems in the *Study Guide and Solutions Manual.*

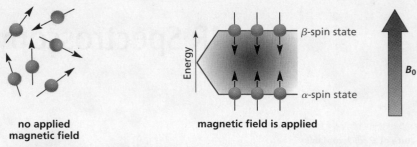

▲ Figure 1

In the absence of an applied magnetic field, the magnetic moments of the nuclei are randomly oriented. In the presence of an applied magnetic field, the magnetic moments of the nuclei line up with (the α-spin state) or against (the β-spin state) the applied magnetic field.

Nuclei with magnetic moments that align with the field are in the lower-energy **α-spin state,** whereas those with magnetic moments that align against the field are in the higher-energy **β-spin state.** The β-spin state is higher in energy because more energy is needed to align the magnetic moments against the field than with it. As a result, more nuclei are in the α-spin state. The difference in the populations is very small (about 20 out of 1 million protons), but it is sufficient to form the basis of NMR spectroscopy.

The energy difference (ΔE) between the α- and β-spin states depends on the strength of the **applied magnetic field** (B_0): the greater the strength of the applied magnetic field, the greater the ΔE (Figure 2).

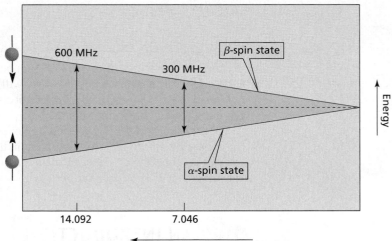

▶ Figure 2

The difference in energy between the α- and β-spin states increases as the strength of the applied magnetic field increases.

When a sample is subjected to a pulse of radiation whose energy corresponds to the difference in energy (ΔE) between the α- and β-spin states, nuclei in the α-spin state are promoted to the β-spin state. This transition is called "flipping" the spin.

With currently available magnets, the energy difference between the α- and β-spin states is small, so only a small amount of energy is needed to flip the spin. The radiation used to supply this energy is in the radio frequency (rf) region of the electromagnetic spectrum and is called **rf radiation**. When the nuclei absorb rf radiation and flip their spins, they generate signals whose frequency depends on the difference in energy (ΔE) between the α- and β-spin states. The NMR spectrometer detects these signals and plots their frequency versus their intensity; this plot is an NMR spectrum.

The nuclei are said to be *in resonance* with the rf radiation, hence the term **nuclear magnetic resonance.** In this context, "resonance" refers to the nuclei flipping back and forth between the α- and β-spin states in response to the rf radiation; it has nothing to do with the "resonance" associated with electron delocalization.

The following equation shows that the energy difference between the spin states (ΔE) depends on the operating frequency of the spectrometer (ν), which depends, in turn, on the strength of the magnetic field (B_0), measured in tesla (T), and the *gyromagnetic ratio* (γ); h is Planck's constant.

$$\Delta E = h\nu = h\frac{\gamma}{2\pi}B_0$$

The **gyromagnetic ratio** is a constant that depends on the particular kind of nucleus. In the case of a proton, $\gamma = 2.675 \times 10^8\,\text{T}^{-1}\text{s}^{-1}$; in the case of a ^{13}C nucleus, it is $6.688 \times 10^7\,\text{T}^{-1}\text{s}^{-1}$. Canceling Planck's constant on both sides of the equation gives

$$\nu = \frac{\gamma}{2\pi}B_0$$

The following calculation shows that if an ^{1}H NMR spectrometer is equipped with a magnet that generates a magnetic field of 7.046 T, then the spectrometer will require an operating frequency of 300 MHz (megahertz):

$$\begin{aligned}
v &= \frac{\gamma}{2\pi}B_0 \\
&= \frac{2.675 \times 10^8}{2(3.1416)}\,\text{T}^{-1}\text{s}^{-1} \times 7.046\,\text{T} \\
&= 300 \times 10^6\,\text{Hz} = 300\,\text{MHz}
\end{aligned}$$

> Earth's magnetic field is 5×10^{-5} T, measured at the equator.
> Its maximum surface magnetic field is 7×10^{-5} T, measured at the south magnetic pole.

> The magnetic field is proportional to the operating frequency of the spectrometer.

The equation shows that the strength of the *magnetic field (B_0) is proportional to the operating frequency* (MHz) of the spectrometer. Therefore, if the spectrometer has a more powerful magnet, then it must have a higher **operating frequency.** For example, a magnetic field of 14.092 T requires an operating frequency of 600 MHz.

Today's **NMR spectrometers** operate at frequencies between 300 and 1000 MHz. The resolution of the NMR spectrum increases as the operating frequency of the instrument— and the strength of the magnet—increases.

Because each kind of nucleus has its own gyromagnetic ratio, different frequencies are required to bring different kinds of nuclei into resonance. For example, an NMR spectrometer that requires a frequency of 300 MHz to flip the spin of an ^{1}H nucleus requires a frequency of 75 MHz to flip the spin of a ^{13}C nucleus. NMR spectrometers are equipped with radiation sources that can be tuned to different frequencies so that they can be used to obtain NMR spectra of different kinds of nuclei (^{1}H, ^{13}C, ^{15}N, ^{19}F, ^{31}P).

Nikola Tesla (1856–1943)

The tesla, used to measure the strength of a magnetic field, was named in honor of Nikola Tesla. Tesla was born in Croatia, emigrated to the United States in 1884, and became a citizen in 1891. He was a proponent of the use of alternating current to distribute electricity and bitterly fought Thomas Edison, who promoted direct current. Tesla was granted a patent for developing the radio in 1900, but Guglielmo Marconi was also given a patent for its development in 1904. Not until 1943—a few months after his death—was Tesla's patent upheld by the U.S. Supreme Court.

Tesla held over 800 patents and is given credit for developing neon and fluorescent lighting, the electron microscope, the refrigerator motor, and the Tesla coil (a type of transformer for changing the voltage of alternating current). Perhaps his most important contribution was polyphase

Nikola Tesla in his laboratory

electric power, which became the prototype for all large power systems. He made most of his equipment himself, including insulators, a technology that was kept classified until recently because the same technology was being used for part of the U.S. Strategic Defense Initiative. Telsa frequently staged flamboyant high-voltage demonstrations, which may explain why he did not receive proper recognition for his work.

PROBLEM 1♦

What frequency (in MHz) is required to cause a proton to flip its spin when it is exposed to a magnetic field of 1 T?

PROBLEM 2♦

a. Calculate the magnetic field (in tesla) required to flip an ^{1}H nucleus in an NMR spectrometer that operates at 360 MHz.

b. What strength magnetic field is required when a 500-MHz instrument is used for ^{1}H NMR?

---

2   FOURIER TRANSFORM NMR

To obtain an **NMR spectrum,** a small amount of a compound is dissolved in about 0.5 mL of solvent. This solution is put into a long, thin glass tube, which is then placed within a powerful magnetic field (Figure 3). Spinning the sample tube about its long axis averages the position of the molecules in the magnetic field, which increases the resolution of the spectrum.

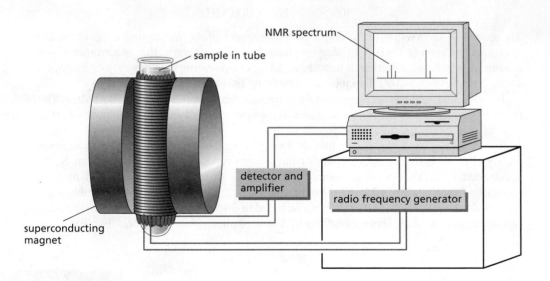

▶ **Figure 3**

Schematic diagram of an NMR spectrometer

In modern instruments called *pulsed Fourier transform (FT) spectrometers,* the magnetic field is held constant and an rf pulse of short duration excites all the protons simultaneously. The rf pulse covers a range of frequencies, so each nucleus can absorb the frequency it requires to come into resonance (flip its spin) and produce a signal—called a free induction decay (FID)—at a frequency corresponding to ΔE. The intensity of the FID signal decays as the nuclei lose the energy they gained from the rf pulse.

A computer measures the change in intensity over time and converts it into intensity-versus-frequency data, in a mathematical operation known as a *Fourier transform,* to produce a spectrum called a **Fourier transform NMR (FT–NMR) spectrum.** An FT–NMR spectrum can be recorded in about 2 seconds—and large numbers of FIDs can be averaged in a few minutes—using less than 5 mg of compound. The NMR spectra in this text are FT–NMR spectra that were taken on a spectrometer with an operating frequency of 300 MHz.

3   SHIELDING CAUSES DIFFERENT HYDROGENS TO SHOW SIGNALS AT DIFFERENT FREQUENCIES

We have seen that the frequency of an NMR signal depends on the strength of the magnetic field experienced by the nucleus (Figure 2). Therefore, if all the hydrogens in a compound were to experience the same magnetic field, they would all give signals of the same frequency. If this were the case, all NMR spectra would consist of one signal, which would tell us nothing about the structure of the compound, except that it contains hydrogens.

A nucleus, however, is embedded in a cloud of electrons that partly *shields* it from the applied magnetic field. Fortunately for chemists, the **shielding** varies for different hydrogens in a molecule. In other words, all the hydrogens do not experience the same magnetic field.

What causes shielding? In a magnetic field, the electrons circulate about the nuclei and induce a local magnetic field that acts in opposition to the applied magnetic field and, therefore, subtracts from it. As a result, the **effective magnetic field**—the amount of magnetic field that the nuclei actually "sense" through the surrounding electrons—is somewhat smaller than the applied magnetic field:

$$B_{effective} = B_{applied} - B_{local}$$

This means that the greater the electron density of the environment in which the proton[*] is located, the more the proton is shielded from the applied magnetic field and the greater is B_{local}. This type of shielding is called **diamagnetic shielding.**

Thus, protons in electron-rich environments sense a *smaller effective magnetic field.* Therefore, they require a *lower frequency* to come into resonance—that is, flip their spin—because ΔE is smaller (Figure 2). Protons in electron-poor environments sense a *larger effective magnetic field* and so require a *higher frequency* to come into resonance because ΔE is larger.

An NMR spectrum exhibits a signal for each proton in a different environment. Protons in electron-rich environments are more shielded and appear at lower frequencies (on the right-hand side of the spectrum; Figure 4). Protons in electron-poor environments are less shielded and appear at higher frequencies (on the left-hand side of the spectrum). Notice that high frequency in an NMR spectrum is on the left-hand side, just as it is in IR and UV/Vis spectra.

> The electron density of the environment in which the proton is located shields the proton from the applied magnetic field.

> The larger the magnetic field sensed by the proton, the higher the frequency of the signal.

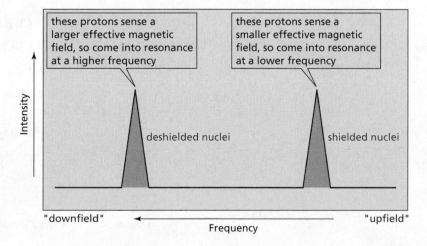

> deshielded = less shielded

◀ **Figure 4**

Shielded protons come into resonance at lower frequencies than deshielded nuclei.

---

[*]In discussions of NMR spectroscopy, the terms *proton* and *hydrogen* are both used to describe a covalently bonded hydrogen.

The terms *upfield* and *downfield,* which came into use when continuous-wave (CW) spectrometers were used (before the advent of Fourier transform spectrometers), are so entrenched in the vocabulary of NMR that you should know what they mean. **Upfield** means farther to the right-hand side of the spectrum, and **downfield** means farther to the left-hand side of the spectrum.

4 THE NUMBER OF SIGNALS IN AN ^{1}H NMR SPECTRUM

Protons in the same environment are called **chemically equivalent protons.** For example, 1-bromopropane has three different sets of chemically equivalent protons: the three methyl protons are chemically equivalent because of rotation about the C—C bond; the two methylene (CH_2) protons on the middle carbon are chemically equivalent; and the two methylene protons on the carbon bonded to the bromine make up the third set of chemically equivalent protons.

Each set of chemically equivalent protons produces an NMR signal.

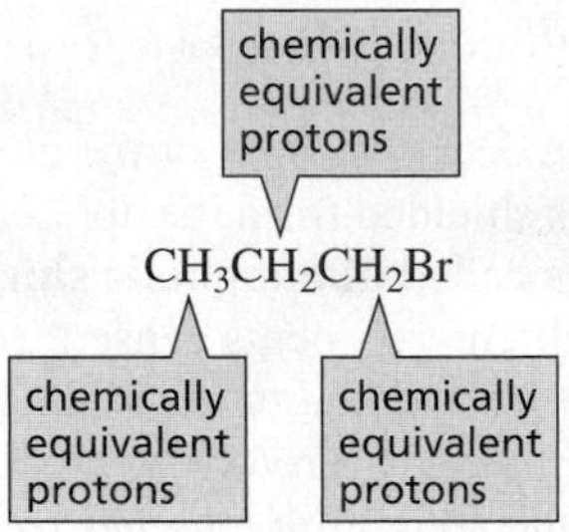

chemically equivalent protons

$CH_3CH_2CH_2Br$

chemically equivalent protons

chemically equivalent protons

Each set of chemically equivalent protons in a compound produces a separate signal in its ^{1}H NMR spectrum. Thus, 1-bromopropane has three signals in its ^{1}H NMR spectrum because it has three sets of chemically equivalent protons. (Sometimes the signals are not sufficiently separated and overlap each other. When this happens, one sees fewer signals than anticipated.)

2-Bromopropane has two sets of chemically equivalent protons, so it has two signals in its ^{1}H NMR spectrum. The six methyl protons are equivalent so they produce only one signal, and the hydrogen bonded to the middle carbon gives the second signal.

You can tell how many sets of chemically equivalent protons a compound has from the number of signals in its ^{1}H NMR spectrum.

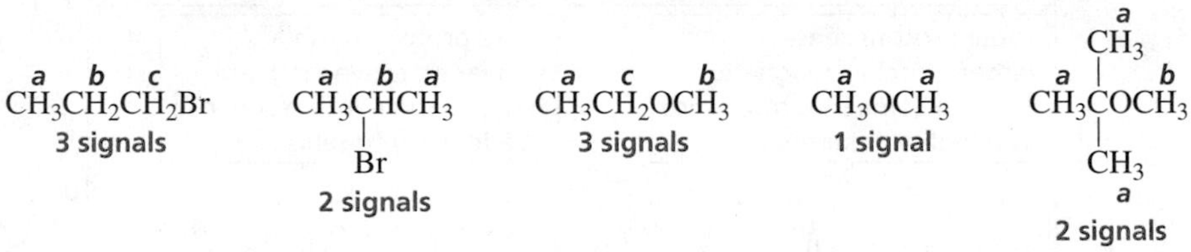

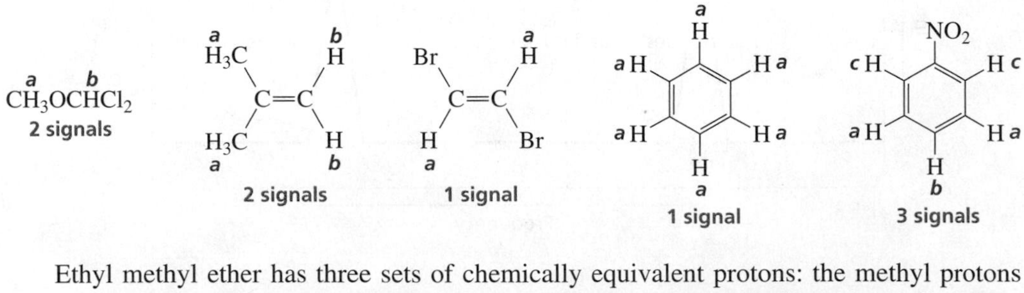

Ethyl methyl ether has three sets of chemically equivalent protons: the methyl protons on the carbon adjacent to the oxygen, the methylene protons on the carbon adjacent to the oxygen, and the methyl protons on the carbon that is one carbon removed from the oxygen. The chemically equivalent protons in the compounds shown here are designated by the same letter.

If bonds are prevented from freely rotating, as in a compound with a double bond or in a cyclic compound, two protons on the same carbon may not be equivalent. For example, the H_a and H_b protons of bromoethene are not equivalent because they are not in the same environment: H_a is trans to Br, whereas H_b is cis to Br. Thus, the ^1H NMR spectrum of bromoethene has three signals.

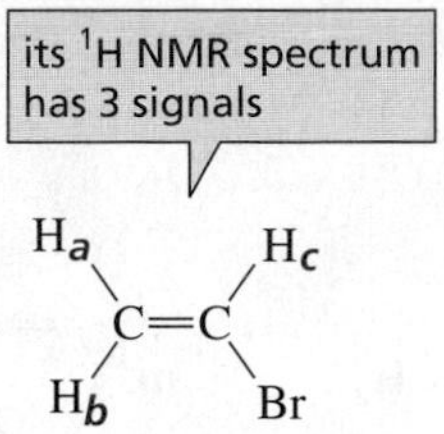

bromoethene
H_a and H_b are not equivalent

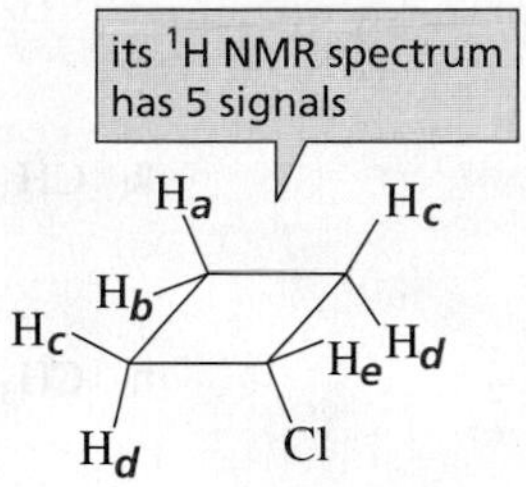

chlorocyclobutane
H_a and H_b are not equivalent
H_c and H_d are not equivalent

The ^1H NMR spectrum of chlorocyclobutane has five signals. The H_a and H_b protons are not equivalent: H_a is trans to Cl, whereas H_b is cis to Cl. Similarly, the H_c and H_d protons are not equivalent.

PROBLEM-SOLVING STRATEGY

Determining the Number of Signals in an ^1H NMR Spectrum

How many signals would you expect to see in the ^1H NMR spectrum of ethylbenzene?

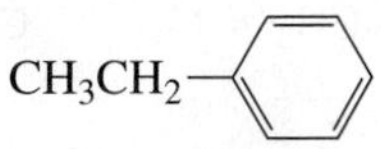

To determine the number of signals you would expect to see in the spectrum, replace each hydrogen in turn by another atom (here we use Br) and name the resulting compound. The number of different names corresponds to the number of signals in the ^1H NMR spectrum. We get five different names for the bromosubstituted ethylbenzenes, so we expect to see five signals in the ^1H NMR spectrum of ethylbenzene.

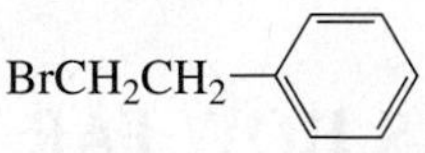

1-bromo-2-phenylethane 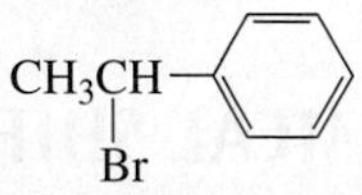   **1-bromo-1-phenylethane** 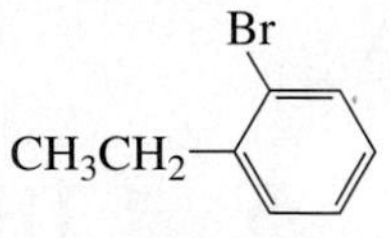   **1-bromo-2-ethylbenzene**

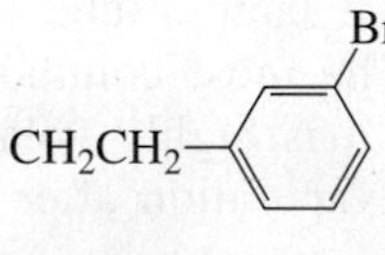

1-bromo-3-ethylbenzene 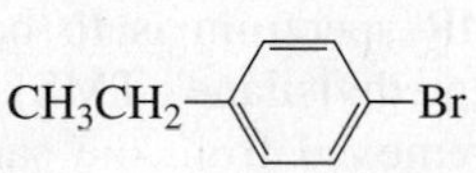   **1-bromo-4-ethylbenzene** 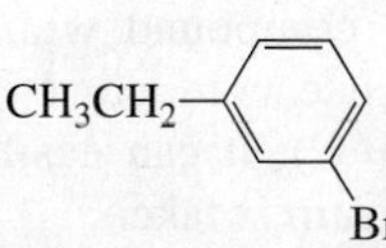   **1-bromo-3-ethylbenzene** 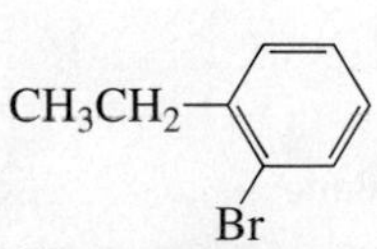   **1-bromo-2-ethylbenzene**

Now use the strategy you have just learned to solve Problems 3 and 4.

PROBLEM 3

How many signals would you expect to see in the ^1H NMR spectrum of each of the five compounds with molecular formula C_6H_{14}?

PROBLEM 4♦

How many signals would you expect to see in the ^{1}H NMR spectrum of each of the following compounds?

a. $CH_3CH_2CH_2CH_3$

f. $CH_3CH_2CH_2 \overset{\overset{\displaystyle O}{\|}}{C} CH_3$

k. $CH_3\!-\!\!\langle\rangle\!\!-\!CH_3$

b. $BrCH_2CH_2Br$

g. $CH_3CH_2\underset{\underset{\displaystyle Cl}{|}}{CH}CH_2CH_3$

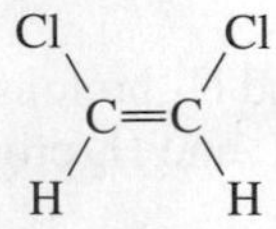

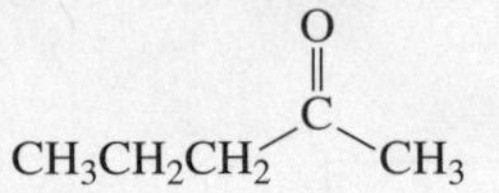

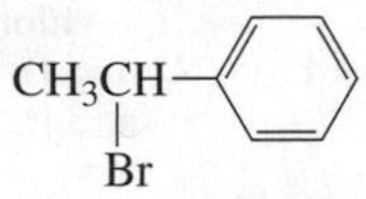

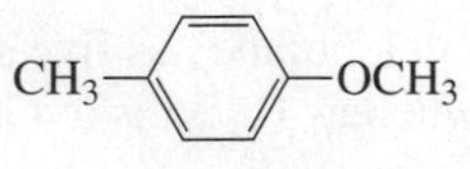

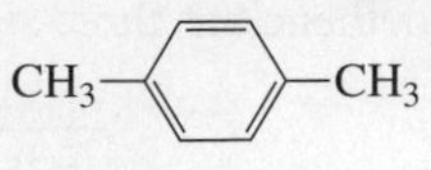

l. (benzene with two Br substituents, meta)

c. $CH_2\!=\!CCl_2$

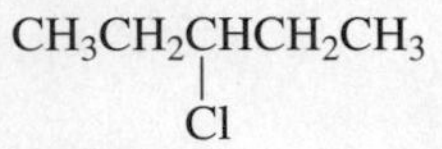

h. 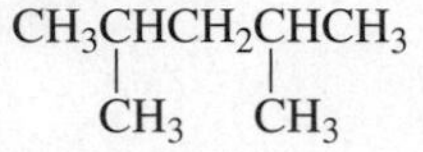 $CH_3\underset{\underset{\displaystyle CH_3}{|}}{CH}CH_2\underset{\underset{\displaystyle CH_3}{|}}{CH}CH_3$

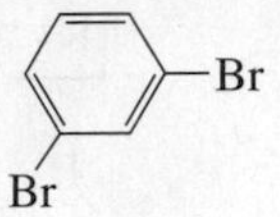

m. 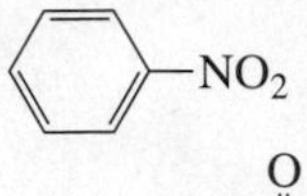 (benzene with NO_2)

d. (cyclohexene)

i. $CH_3\underset{\underset{\displaystyle Br}{|}}{CH}\!-\!\langle\rangle$

n. 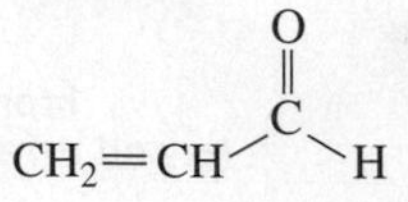 $CH_2\!=\!CH\!-\!\overset{\overset{\displaystyle O}{\|}}{C}\!-\!H$

e. $\underset{\overset{\displaystyle |}{H}}{\overset{\overset{\displaystyle Cl}{|}}{C}}\!=\!\underset{\overset{\displaystyle |}{H}}{\overset{\overset{\displaystyle Cl}{|}}{C}}$

j. $CH_3\!-\!\!\langle\rangle\!\!-\!OCH_3$

o. $\underset{\overset{\displaystyle |}{H}}{\overset{\overset{\displaystyle Cl}{|}}{C}}\!=\!\underset{\overset{\displaystyle |}{H}}{\overset{\overset{\displaystyle CH_3}{|}}{C}}$

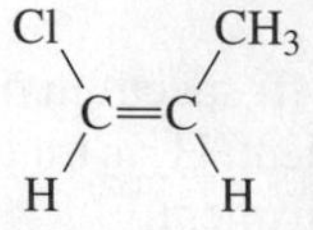

PROBLEM 5

How could you distinguish the ^{1}H NMR spectra of the following compounds?

$CH_3OCH_2OCH_3$ CH_3OCH_3 $CH_3OCH_2\underset{\underset{\displaystyle CH_3}{\overset{\overset{\displaystyle CH_3}{|}}{|}}}{C}CH_2OCH_3$

        A   **B**   **C**

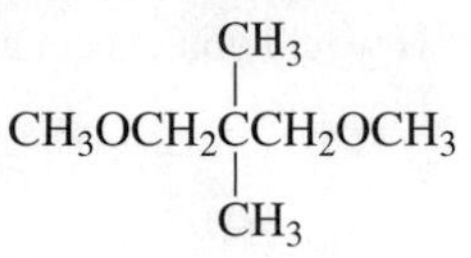

PROBLEM 6

Draw an isomer of dichlorocyclopropane that gives an ^{1}H NMR spectrum

a. with one signal.   **b.** with two signals.   **c.** with three signals.

---

5   THE CHEMICAL SHIFT TELLS HOW FAR THE SIGNAL IS FROM THE REFERENCE SIGNAL

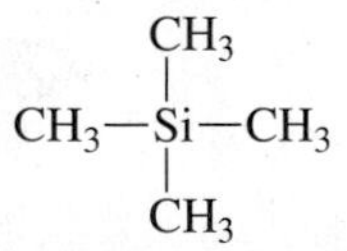

$$CH_3\!-\!\underset{\underset{\displaystyle CH_3}{|}}{\overset{\overset{\displaystyle CH_3}{|}}{Si}}\!-\!CH_3$$

**tetramethylsilane
TMS**

A small amount of an inert **reference compound** is added to the sample tube containing the compound whose NMR spectrum is to be taken. The most commonly used reference compound is tetramethylsilane (TMS). Because it is highly volatile (bp = 26.5 °C), it can easily be removed from the sample by evaporation after the NMR spectrum is taken.

The methyl protons of TMS are in a more electron-rich environment than are most protons in organic molecules because silicon is less electronegative than carbon (their electronegativities are 1.8 and 2.5, respectively). Consequently, the signal for the methyl protons of TMS is at a lower frequency than most other signals (that is, the TMS signal appears to the right of the other signals).

The position at which a signal occurs in an NMR spectrum is called the *chemical shift*. The **chemical shift** is a measure of how far the signal is from the signal for the reference compound. The most common scale for chemical shifts is the δ (delta) scale. The TMS signal defines the zero position on the δ scale (Figure 5).

The chemical shift is determined by measuring the distance from the TMS peak in hertz and dividing by the operating frequency of the instrument in megahertz. Because the units are Hz/MHz, a chemical shift has units of parts per million (ppm) of the operating frequency:

$$\delta = \text{chemical shift (ppm)} = \frac{\textbf{distance downfield from TMS (Hz)}}{\textbf{operating frequency of the spectrometer (MHz)}}$$

The ^{1}H NMR spectrum in Figure 5 shows that the chemical shift (δ) of the methyl protons is at 1.05 ppm and the chemical shift of the methylene protons, which are deshielded by the electron-withdrawing bromine, is at 3.28 ppm. *Notice that low-frequency (shielded) signals have small δ (ppm) values, whereas high-frequency (deshielded) signals have large δ values.*

Most proton chemical shifts are between 0 and 12 ppm.

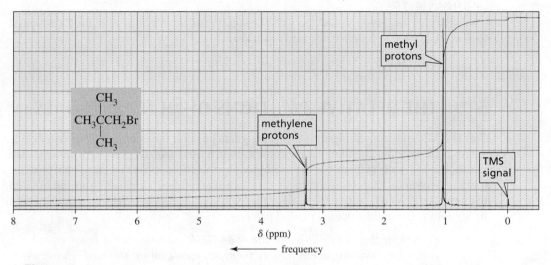

▲ **Figure 5**

The ^{1}H NMR spectrum of 1-bromo-2,2-dimethylpropane. The TMS signal is a reference signal from which chemical shifts are measured; it defines the zero position on the scale.

The advantage of the δ scale is that the chemical shift is *independent of the operating frequency of the NMR spectrometer.* Thus, the chemical shift of the methyl protons of 1-bromo-2,2-dimethylpropane is at 1.05 ppm in both a 300-MHz and a 500-MHz instrument. If the chemical shift were reported in hertz instead, it would be at 315 Hz in a 300-MHz instrument and at 525 Hz in a 500-MHz instrument (315/300 = 1.05, 525/500 = 1.05).

The following diagram will help you keep track of the terms associated with NMR spectroscopy:

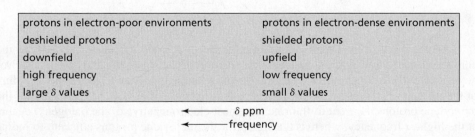

protons in electron-poor environments	protons in electron-dense environments
deshielded protons	shielded protons
downfield	upfield
high frequency	low frequency
large δ values	small δ values

The greater the value of the chemical shift (δ), the higher the frequency.

PROBLEM 7♦

How many hertz downfield from the TMS signal would be the signal occurring at 2.0 ppm

a. in a 300-MHz spectrometer?   **b.** in a 500-MHz spectrometer?

The chemical shift (δ) is independent of the operating frequency of the spectrometer.

PROBLEM 8♦

A signal is seen at 600 Hz downfield from the TMS signal in an NMR spectrometer with a 300-MHz operating frequency.

a. What is the chemical shift of the signal?

b. What would its chemical shift be in an instrument operating at 500 MHz?

c. How many hertz downfield from TMS would the signal be in a 500-MHz spectrometer?

PROBLEM 9♦

a. If two signals differ by 1.5 ppm in a 300-MHz spectrometer, by how much do they differ in a 500-MHz spectrometer?

b. If two signals differ by 90 Hz in a 300-MHz spectrometer, by how much do they differ in a 500-MHz spectrometer?

PROBLEM 10♦

Where would you expect to find the ^1H NMR signal of $(CH_3)_2Mg$ relative to the TMS signal? (*Hint:* Magnesium is less electronegative than silicon.)

6 THE RELATIVE POSITIONS OF ^1H NMR SIGNALS

The ^1H NMR spectrum in Figure 5 has two signals because the compound has two different kinds of protons. The methylene protons are in a *less electron-rich environment* than are the methyl protons because the methylene protons are closer to the electron-withdrawing bromine. Therefore, the methylene protons are *less shielded* from the applied magnetic field. As a result, the signal for these protons occurs at a higher frequency than the signal for the more shielded methyl protons.

> *Remember that the right-hand side of an NMR spectrum is the low-frequency side, where protons in electron-rich environments (more shielded) show a signal. The left-hand side is the high-frequency side, where protons in electron-poor environments (less shielded) show a signal (Figure 4).*

Protons in electron-poor environments show signals at high frequencies.

We expect the ^1H NMR spectrum of 1-nitropropane to have three signals because the compound has three different kinds of protons. The closer the protons are to the electron-withdrawing nitro group, the less they are shielded from the applied magnetic field, so the higher the frequency at which their signal will appear. Thus, the protons closest to the nitro group show a signal at the highest frequency (4.37 ppm), and the ones farthest from the nitro group show a signal at the lowest frequency (1.04 ppm).

Electron withdrawal causes NMR signals to appear at higher frequencies (at larger δ values).

$$\boxed{1.04 \text{ ppm}} \quad \boxed{2.07 \text{ ppm}}$$
$$\boxed{4.37 \text{ ppm}}$$
$$CH_3CH_2CH_2NO_2$$

Compare the chemical shifts of the methylene protons immediately adjacent to the halogen in the following alkyl halides. The position of the signal depends on the electronegativity of the halogen—as the electronegativity of the halogen increases, the shielding of the protons decreases, so the frequency of the signal increases. Thus, the signal for the methylene protons adjacent to fluorine (the most electronegative of the halogens) occurs at the highest frequency, whereas the signal for the methylene protons adjacent to iodine (the least electronegative of the halogens) occurs at the lowest frequency.

$CH_3CH_2CH_2CH_2CH_2F$ $CH_3CH_2CH_2CH_2CH_2Cl$ $CH_3CH_2CH_2CH_2CH_2Br$ $CH_3CH_2CH_2CH_2CH_2I$

$\boxed{4.50 \text{ ppm}}$ $\boxed{3.50 \text{ ppm}}$ $\boxed{3.40 \text{ ppm}}$ $\boxed{3.20 \text{ ppm}}$

PROBLEM 11♦

a. Which set of protons in each of the following compounds is the least shielded?

b. Which set of protons in each compound is the most shielded?

1. $CH_3CH_2CH_2Cl$

2. 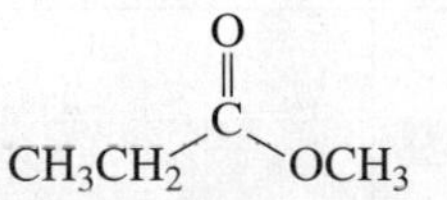 $CH_3CH_2\overset{\displaystyle O}{\overset{\|}{C}}OCH_3$

3. $CH_3\underset{Br}{\underset{|}{C}}H\underset{Br}{\underset{|}{C}}HBr$

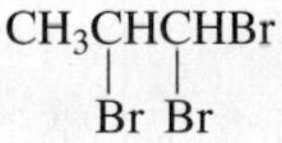

7 THE CHARACTERISTIC VALUES OF CHEMICAL SHIFTS

Approximate values of chemical shifts for different kinds of protons are listed in Table 1. (A more extensive compilation can be found in Appendix: "Spectroscopy Tables.")

Table 1	Approximate Values of Chemical Shifts for ^1H NMR[a]		
Type of proton	**Approximate chemical shift (ppm)**	**Type of proton**	**Approximate chemical shift (ppm)**
$-CH_3$	0.85	$I-C-H$	2.5–4
$-CH_2-$	1.20	$Br-C-H$	2.5–4
$-CH-$	1.55	$Cl-C-H$	3–4
$-C=C-CH_3$	1.7	$F-C-H$	4–4.5
$-\overset{\displaystyle O}{\overset{\|}{C}}-CH_3$	2.1	$R-NH_2$	Variable, 1.5–4
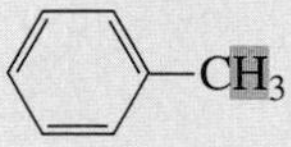 $\langle\!\!\langle\bigcirc\rangle\!\!\rangle-CH_3$	2.3	$R-OH$	Variable, 2–5
		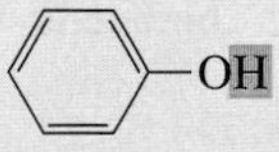 $\langle\!\!\langle\bigcirc\rangle\!\!\rangle-OH$	Variable, 4–7
$-C\equiv C-H$	2.4	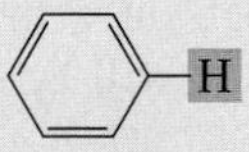 $\langle\!\!\langle\bigcirc\rangle\!\!\rangle-H$	6.5–8
$R-O-CH_3$	3.3	$-\overset{\displaystyle O}{\overset{\|}{C}}-H$	9.0–10
$R-\underset{R}{\underset{\|}{C}}=CH_2$	4.7	$-\overset{\displaystyle O}{\overset{\|}{C}}-OH$	Variable, 10–12
$R-\underset{R}{\underset{\|}{C}}=\underset{R}{\underset{\|}{C}}-H$	5.3	$-\overset{\displaystyle O}{\overset{\|}{C}}-NH_2$	Variable, 5–8

[a]The values are approximate because they are affected by neighboring substituents.

An ^{1}H NMR spectrum can be divided into seven regions, one of which is empty. If you can remember the kinds of protons that appear in each region, you will be able to tell what kinds of protons a molecule has from a quick look at its NMR spectrum.

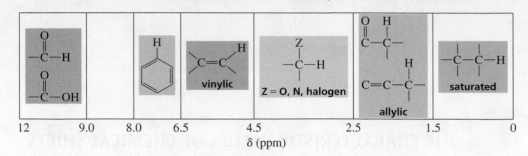

δ (ppm)

—CH—
methine

—CH$_2$—
methylene

—CH$_3$
methyl

Carbon is more electronegative than hydrogen. Therefore, the chemical shift of a **methine proton** (a hydrogen bonded to an sp^3 carbon that is attached to *three* carbons) is more deshielded and so shows a chemical shift at a higher frequency than the chemical shift of **methylene protons** (hydrogens bonded to an sp^3 carbon that is attached to *two* carbons) in a similar environment. Likewise, the chemical shift of methylene protons is at a higher frequency than the chemical shift of **methyl protons** (hydrogens bonded to an sp^3 carbon that is attached to *one* carbon) in a similar environment (Table 1).

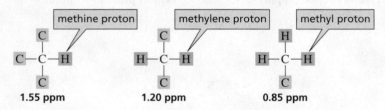

For example, the ^{1}H NMR spectrum of butanone shows three signals. The signal at the lowest frequency is the signal for the **a** protons; these protons are farthest from the electron-withdrawing carbonyl group. The **b** and **c** protons are the same distance from the carbonyl group, but the signal for the **c** protons is at a higher frequency than the signal for the **b** protons because methylene protons appear at a higher frequency than do methyl protons in a similar environment.

In a similar environment, the signal for a methine proton occurs at a higher frequency than the signal for methylene protons, which occurs in turn at a higher frequency than the signal for methyl protons.

butanone

2-methoxypropane

(In correlating an NMR spectrum with a structure, the set of protons responsible for the signal at the lowest frequency will be labeled **a,** the next set will be labeled **b,** the next set **c,** and so on.)

The signal for the **a** protons of 2-methoxypropane is the one at the lowest frequency because these protons are farthest from the electron-withdrawing oxygen. The **b** and **c** protons are the same distance from the oxygen, but the signal for the **c** protons appears at a higher frequency because, in a similar environment, a methine proton appears at a higher frequency than do methyl protons.

PROBLEM 12♦

Which of the underlined protons (or sets of protons) has the greater chemical shift (that is, the higher frequency signal)?

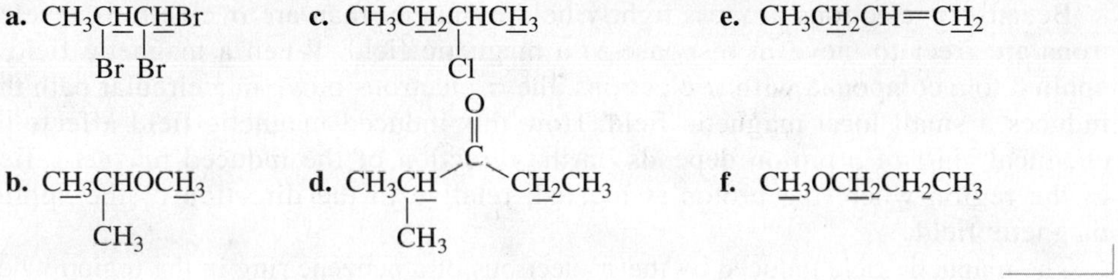

a. CH$_3$C̲HC̲HBr
      | |
      Br Br

c. CH$_3$CH$_2$C̲HCH$_3$
             |
             Cl

e. CH$_3$C̲H$_2$CH=CH̲$_2$

b. CH$_3$C̲HOCH̲$_3$
      |
      CH$_3$

d.
$$CH_3\overset{|}{\underset{CH_3}{C}}H\overset{O}{\overset{||}{C}}CH_2CH_3$$

f. C̲H$_3$OCH$_2$CH$_2$CH$_3$

PROBLEM 13♦

Which of the underlined protons (or sets of protons) has the greater chemical shift (that is, the higher frequency signal)?

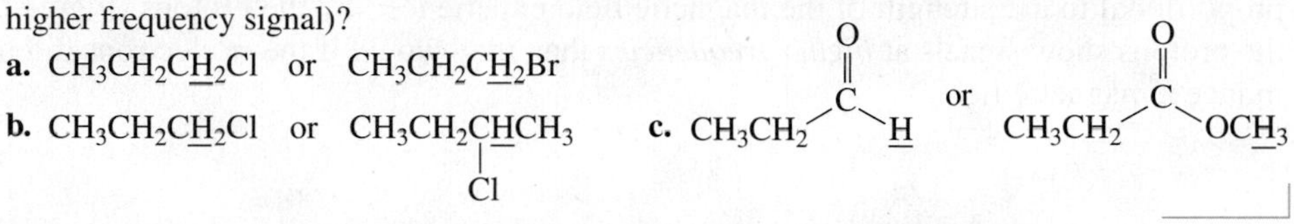

a. CH$_3$CH$_2$C̲H$_2$Cl or CH$_3$CH$_2$C̲H$_2$Br

b. CH$_3$CH$_2$C̲H$_2$Cl or CH$_3$CH$_2$C̲HCH$_3$
                                |
                                Cl

c.
$$CH_3CH_2\overset{O}{\overset{||}{C}}H \quad or \quad CH_3CH_2\overset{O}{\overset{||}{C}}OC̲H_3$$

PROBLEM 14

Without referring to Table 1, label the proton or set of protons in each compound that gives the signal at the lowest frequency **a,** at the next lowest **b,** and so on.

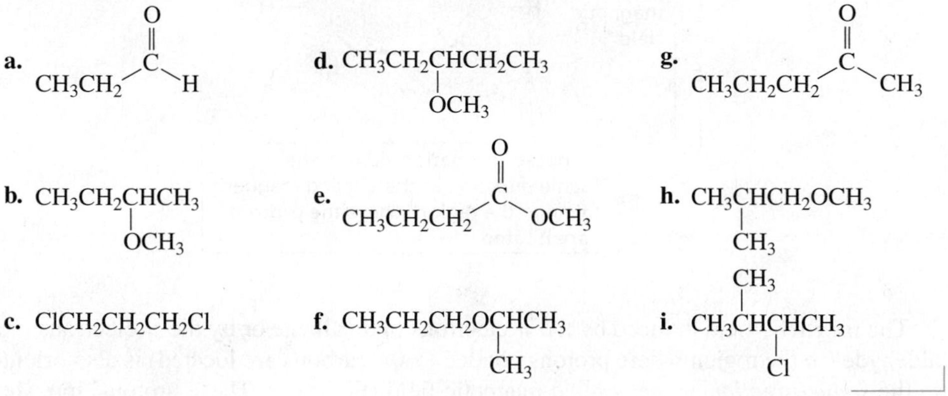

a.
$$CH_3CH_2\overset{O}{\overset{||}{C}}H$$

d. CH$_3$CH$_2$CHCH$_2$CH$_3$
               |
               OCH$_3$

g.
$$CH_3CH_2CH_2\overset{O}{\overset{||}{C}}CH_3$$

b. CH$_3$CH$_2$CHCH$_3$
          |
          OCH$_3$

e.
$$CH_3CH_2CH_2\overset{O}{\overset{||}{C}}OCH_3$$

h. CH$_3$CHCH$_2$OCH$_3$
       |
       CH$_3$

c. ClCH$_2$CH$_2$CH$_2$Cl

f. CH$_3$CH$_2$CH$_2$OCHCH$_3$
                   |
                   CH$_3$

i. CH$_3$CHCHCH$_3$
         CH$_3$
         |
  CH$_3$CHCHCH$_3$
           |
           Cl

8 DIAMAGNETIC ANISOTROPY

The chemical shifts of hydrogens bonded to sp^2 carbons are at higher frequencies than one would predict from the electronegativities of the sp^2 carbons. For example, a hydrogen bonded to a benzene ring appears at 6.5 to 8.0 ppm, a hydrogen bonded to the terminal sp^2 carbon of an alkene appears at 4.7 to 5.3 ppm, and a hydrogen bonded to a carbonyl carbon appears at 9.0 to 10.0 ppm (Table 1).

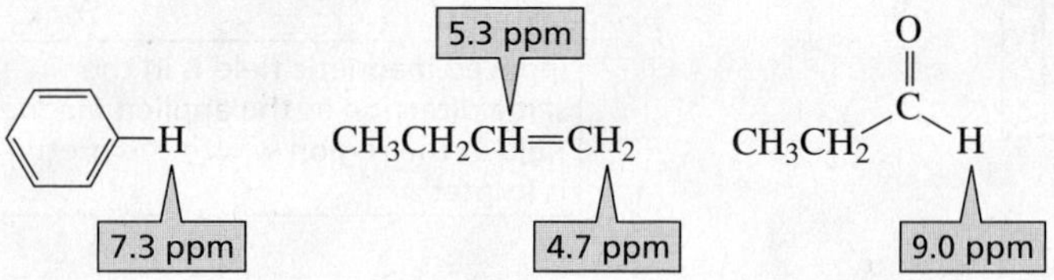

5.3 ppm

$$\text{⬡—H} \qquad CH_3CH_2CH=CH_2 \qquad CH_3CH_2\overset{O}{\overset{||}{C}}H$$

7.3 ppm   4.7 ppm   9.0 ppm

The unusual chemical shifts associated with hydrogens bonded to carbons that form π bonds are due to **diamagnetic anisotropy.** This term describes an environment in which different magnetic fields are found at different points in space. (*Anisotropic* is Greek for "different in different directions.")

Because π electrons are less tightly held by nuclei than are σ electrons, π electrons are freer to move in response to a magnetic field. When a magnetic field is applied to a compound with π electrons, the π electrons move in a circular path that induces a small local magnetic field. How this induced magnetic field affects the chemical shift of a proton depends on the direction of the induced magnetic field in the region where the proton is located, relative to the direction of the applied magnetic field.

The magnetic field induced by the π electrons of a benzene ring in the region where benzene's protons are located is oriented in the same direction as the applied magnetic field (Figure 6). As a result, the protons sense a larger effective magnetic field—namely, the sum of the strengths of the applied field and the induced field. Because frequency is proportional to the strength of the magnetic field experienced by the protons (Figure 2), the protons show signals at *higher frequencies* than they would if the π electrons did not induce a magnetic field.

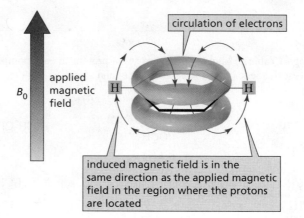

▶ **Figure 6**

The magnetic field induced by the π electrons of a benzene ring in the vicinity of the protons attached to the sp^2 carbons has the same direction as the applied magnetic field. As a result, these protons sense a larger effective magnetic field, so their signals appear at higher frequencies.

The magnetic field induced by the π electrons of an alkene or by the π electrons of an aldehyde (in the region where protons bonded to sp^2 carbons are located) is also oriented in the *same direction* as the applied magnetic field (Figure 7). These protons, too, show signals at higher than expected frequencies.

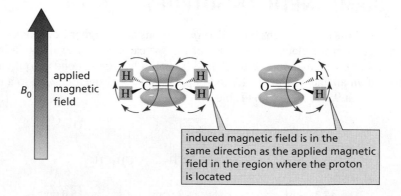

▶ **Figure 7**

The magnetic fields induced by the π electrons of an alkene and by the π electrons of a carbonyl group in the vicinity of the vinylic and aldehydic protons have the same direction as the applied magnetic field. Because a larger effective magnetic field is sensed by the protons, their signals appear at higher frequencies.

In contrast, the chemical shift of a hydrogen bonded to an sp carbon is at a lower frequency (~1.9 ppm) than one would predict from the electronegativity of the sp carbon. This is because the direction of the magnetic field induced by the alkyne's cylinder of π electrons, in the region where the proton is located, is *opposite* to the direction of the applied magnetic field (Figure 8). Thus, the proton senses a smaller effective magnetic field and therefore shows a signal at a *lower frequency* than it would if the π electrons did not induce a magnetic field.

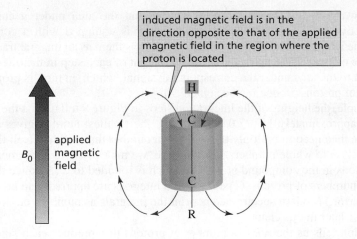

induced magnetic field is in the direction opposite to that of the applied magnetic field in the region where the proton is located

B_0 applied magnetic field

◄ **Figure 8**

The magnetic field induced by the π electrons of an alkyne in the vicinity of the proton bonded to the *sp* carbon is in a direction opposite to that of the applied magnetic field. Because a smaller effective magnetic field is sensed by the proton, its signal appears at a lower frequency.

PROBLEM 15♦

[18]-Annulene shows two signals in its ^{1}H NMR spectrum: one at 9.25 ppm and the other to the right of the TMS signal at -2.88 ppm. What hydrogens are responsible for each of the signals? (*Hint:* Look at the direction of the induced magnetic field outside and inside the benzene ring in Figure 6.)

[18]-annulene

9 THE INTEGRATION OF NMR SIGNALS REVEALS THE RELATIVE NUMBER OF PROTONS CAUSING EACH SIGNAL

The two signals in the ^{1}H NMR spectrum in Figure 9 are not the same size because *the area under each signal is proportional to the number of protons producing the signal.* The area under the signal occurring at the lower frequency is larger because the signal is produced by *nine* methyl protons, whereas the smaller, higher-frequency signal is produced by *two* methylene protons.

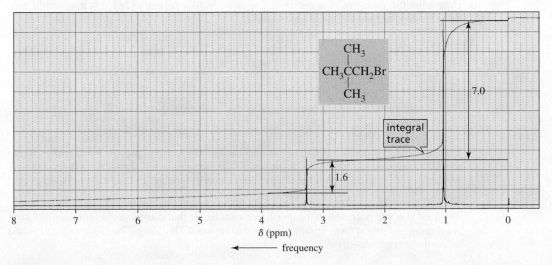

integral trace

7.0

1.6

δ (ppm)

◄ frequency

▲ **Figure 9**

Analysis of the integration line in the ^{1}H NMR spectrum of 1-bromo-2,2-dimethylpropane. The peak at 3.3 ppm has a smaller integral trace than the peak at 1.0 ppm because the peak at 3.3 ppm is produced by two methylene protons, whereas the peak at 1.0 ppm is produced by nine methyl protons.

You may remember from a calculus course that the area under a curve can be determined by an integral. An NMR spectrometer is equipped with a computer that calculates the integrals electronically and then displays them as an integral trace superimposed on the original spectrum (Figure 9). The height of each step in the integral trace is proportional to the area under the corresponding signal, which, in turn, is proportional to the number of protons producing the signal.

For example, the heights of the integration steps in Figure 9 tell us that the ratio of the integrals is approximately 1.6 : 7.0. Dividing by the smallest number gives a new ratio (1 : 4.4). We then need to multiply this ratio by a number that will make all the numbers in the ratio close to whole numbers—in this case, we multiply by 2. This means that the ratio of protons in the compound is 2 : 8.8, which is rounded to 2 : 9, since there can be only whole numbers of protons. (The measured integrals are approximate because of experimental error.) Modern spectrometers print the integrals as numbers on the spectrum; see Figure 11 later in this chapter.

Integration tells us the *relative* number of protons that produce each signal, not the *absolute* number. In other words, integration could not distinguish between the following two compounds because both would show an integral ratio of 1 : 3.

$$CH_3-CH-Cl \qquad\qquad CH_3-\overset{\overset{\displaystyle CH_3}{|}}{\underset{\underset{\displaystyle Cl}{|}}{C}}-CH_2Cl$$

$$\underset{\displaystyle Cl}{|}$$

1,1-dichloroethane **1,2-dichloro-2-methylpropane**
ratio of protons = 1 : 3 **ratio of protons = 2 : 6 = 1 : 3**

PROBLEM 16◆

How would integration distinguish the ^1H NMR spectra of the following compounds?

$$CH_3-\overset{\overset{\displaystyle CH_3}{|}}{\underset{\underset{\displaystyle CH_3}{|}}{C}}-CH_2Br \qquad CH_3-\overset{\overset{\displaystyle CH_3}{|}}{\underset{\underset{\displaystyle Br}{|}}{C}}-CH_2Br \qquad CH_3-\overset{\overset{\displaystyle CH_2Br}{|}}{\underset{\underset{\displaystyle CH_2Br}{|}}{C}}-CH_2Br$$

PROBLEM 17 Solved

a. Calculate the ratios of the different kinds of protons in a compound with an integral ratio of 6 : 4 : 18.4 (going from left to right across the spectrum).

b. Determine the structure of a compound with molecular formula $C_7H_{14}O_2$ that would give these relative integrals in the observed order.

Solution

a. Divide each number in the ratio by the smallest number:

$$\frac{6}{4} = 1.5 \qquad \frac{4}{4} = 1 \qquad \frac{18.4}{4} = 4.6$$

Multiply the resulting numbers by a number that will make them close to whole numbers:

$$1.5 \times 2 = 3 \qquad 1 \times 2 = 2 \qquad 4.6 \times 2 = 9$$

The ratio 3 : 2 : 9 gives the relative numbers of the different kinds of protons. Because the sum of the relative number of protons (14) is the same as the actual number of protons in the compound, we know that the actual ratio is the relative ratio and not some multiple of the relative ratio.

b. The "3" suggests a methyl group, the "2" a methylene group, and the "9" a *tert*-butyl group. The methyl group is closest to the group in the molecule that causes deshielding, and the *tert*-butyl group is farthest away from the group that causes deshielding. The following compound meets these requirements:

$$CH_3\overset{\overset{\displaystyle CH_3}{|}}{\underset{\underset{\displaystyle CH_3}{|}}{C}}CH_2 \overset{\overset{\displaystyle O}{\|}}{C}{\diagdown}_{OCH_3}$$

PROBLEM 18♦

Which of the following compounds is responsible for the ^{1}H NMR spectrum shown in Figure 10?

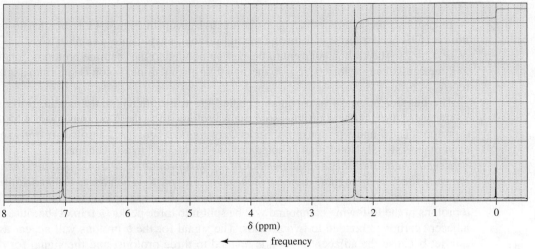

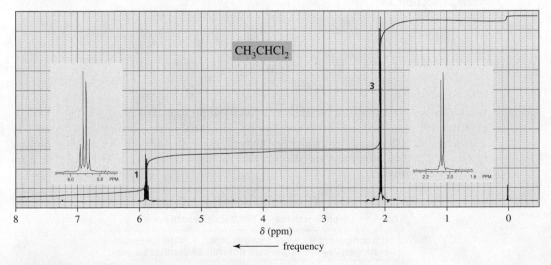

▲ **Figure 10**

The ^{1}H NMR spectrum for Problem 18.

10 THE SPLITTING OF SIGNALS IS DESCRIBED BY THE $N + 1$ RULE

Notice that the shapes of the signals in the ^{1}H NMR spectrum in Figure 11 are different from the shapes of the signals in the ^{1}H NMR spectrum in Figure 9. Both signals in Figure 9 are **singlets,** meaning each is composed of a single peak. In contrast, the signal for the methyl protons in Figure 11 (the lower-frequency signal) is split into two peaks (a **doublet**), and the signal for the methine proton is split into four peaks (a **quartet**). (Magnifications of the frequency axis for the doublet and quartet are shown as insets in Figure 11; integration numbers are shown in green.)

▲ **Figure 11**

The ^{1}H NMR spectrum of 1,1-dichloroethane. The higher-frequency signal (due to CHCl$_2$) is an example of a quartet; the lower-frequency signal (due to CH_3) is a doublet.

Splitting is caused by protons bonded to *adjacent* carbons. The splitting of a signal is described by the **N + 1 rule,** where *N* is the number of equivalent protons bonded to *adjacent* carbons that are not equivalent to the proton producing the signal. Both signals in Figure 9 are singlets; the three methyl groups give an unsplit signal because they are attached to a carbon that is not bonded to a hydrogen; the methylene group also gives an unsplit signal because it too is attached to a carbon that is not bonded to a hydrogen ($N = 0$, so $N + 1 = 1$).

In contrast, the carbon adjacent to the methyl group in Figure 11 is bonded to one proton ($CHCl_2$), so the signal for the methyl protons is split into a doublet ($N = 1$, so $N + 1 = 2$). The carbon adjacent to the carbon bonded to the methine proton is bonded to three equivalent protons (CH_3), so the signal for the methine proton is split into a quartet ($N = 3$, so $N + 1 = 4$).

The number of peaks in a signal is called the **multiplicity** of the signal. Splitting is always mutual: if the *a* protons split the *b* protons, then the *b* protons must split the *a* protons. The *a* and *b* protons, in this case, are *coupled protons*. **Coupled protons** split each other's signal. Notice that coupled protons are bonded to adjacent carbons.

Keep in mind that it is not the number of protons producing a signal that determines the multiplicity of the signal; rather, it is the number of protons bonded to the immediately adjacent carbons that determines the multiplicity. For example, the signal for the *a* protons in the following compound will be split into three peaks (a **triplet**) because the adjacent carbon is bonded to two protons. The signal for the *b* protons will appear as a quartet because the adjacent carbon is bonded to three protons, and the signal for the *c* protons will be a singlet.

The signal for the *a* protons in the following compound will be a triplet, the signal for the *c* protons will also be a triplet, and the signal for the *d* proton will be a singlet since there are no hydrogens bonded to the adjacent carbon. Because the *a* and *c* protons are not equivalent, the $N + 1$ rule has to be applied separately to each set to determine the splitting of the *b* protons. Thus, the signal for the *b* protons will be split into a quartet by the *a* protons, and each of the four peaks of the quartet will be split into a triplet by the *c* protons: $(N_a+1)(N_b+1) = (4)(3) = 12$. As a result, the signal for the *b* protons is a **multiplet** (a signal that is more complex than a triplet, quartet, etc.). The actual number of peaks seen in the multiplet depends on how many of them overlap (Section 14).

A signal for a proton is never split by *equivalent* protons. For example, the ^1H NMR spectrum of bromomethane shows one singlet. The three methyl protons are chemically equivalent, and chemically equivalent protons do not split each other's signal. The four protons in 1,2-dichloroethane are also chemically equivalent, so its ^1H NMR spectrum also shows one singlet.

CH_3Br **bromomethane** $ClCH_2CH_2Cl$ **1,2-dichloroethane**

each compound shows one singlet in its ^1H NMR spectrum because equivalent protons do not split each other's signals

An ^1H NMR signal is split into $N + 1$ peaks, where *N* is the number of equivalent protons bonded to adjacent carbons.

Coupled protons split each other's signal.

Coupled protons are bonded to adjacent carbons.

Equivalent protons do not split each other's signal.

PROBLEM 19

One of the spectra in Figure 12 is produced by 1-chloropropane and the other by 1-iodopropane. Which is which?

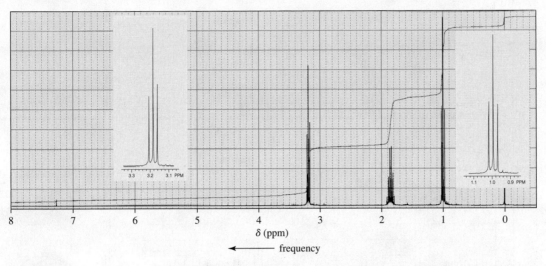

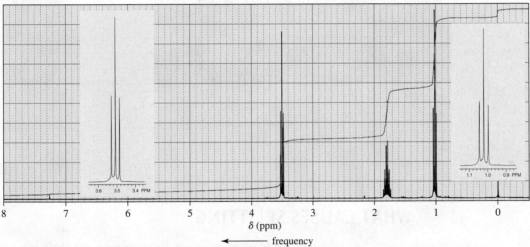

▲ **Figure 12**

The ^{1}H NMR spectra for Problem 19.

PROBLEM 20

Explain how the following compounds, each with the same molecular formula, could be distinguished by their ^{1}H NMR spectra.

$$ClCH_2-CH_2-CHCl_2 \qquad CH_3-CH-CHCl_2 \qquad CH_3-CH_2-CCl_3$$
$$\underset{\textbf{A}}{} \qquad\qquad \underset{\underset{\textbf{B}}{\overset{|}{Cl}}}{} \qquad\qquad \underset{\textbf{C}}{}$$

PROBLEM 21♦

The ^{1}H NMR spectra of two carboxylic acids with molecular formula $C_3H_5O_2Cl$ are shown in Figure 13. Identify the carboxylic acids. (The "offset" notation means that the farthest-left signal has been moved to the right by the indicated amount in order to fit on the spectrum; thus, the signal at 9.8 ppm offset by 2.4 ppm has an actual chemical shift of 9.8 + 2.4 = 12.2 ppm.)

a.

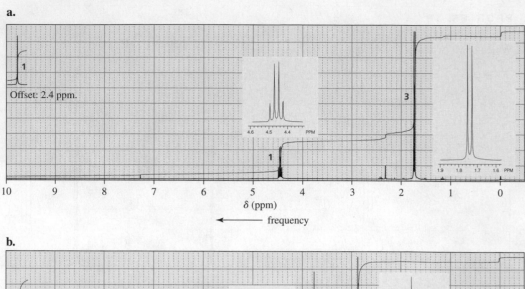

b.

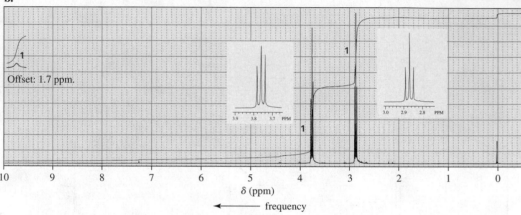

▲ **Figure 13**

The ¹H NMR spectra for Problem 21.

11 / WHAT CAUSES SPLITTING?

Splitting occurs when different kinds of protons are close enough for their magnetic fields to influence one another—a situation called **spin–spin coupling.**

For example, the frequency at which the methyl protons of 1,1-dichloroethane show a signal is influenced by the magnetic field of the methine proton (Figure 14). If the magnetic field of the methine proton aligns *with* that of the applied magnetic field (indicated in Figure 14 by the up arrow), then it will add to the applied magnetic field, which will

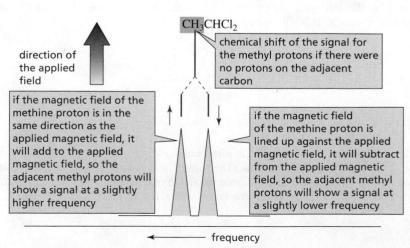

▶ **Figure 14**

The signal for the methyl protons of 1,1-dichloroethane is split into a doublet by the methine proton.

cause the methyl protons to show a signal at a slightly higher frequency. (Recall that the frequency is proportional to the strength of the magnetic field.) On the other hand, if the magnetic field of the methine proton aligns *against* the applied magnetic field (indicated in Figure 14 by the down arrow), then it will subtract from the applied magnetic field and the methyl protons will show a signal at a lower frequency.

Therefore, the signal for the methyl protons is split into two peaks, one at a higher frequency and one at a lower frequency. Because the α- and β-spin states have almost the same population, about half the methine protons are lined up with the applied magnetic field and about half are lined up against it. As a result, the two peaks of the *doublet* have approximately the same height and area.

Similarly, the frequency at which the methine proton shows a signal is influenced by the magnetic fields of the three protons bonded to the adjacent carbon. The magnetic fields of all three methyl protons can align with the applied magnetic field, two can align with the field and one against it, one can align with it and two against it, or all three can align against it (Figure 15). Because the magnetic field that the methine proton senses is affected in four different ways, its signal is a *quartet* (Figure 16).

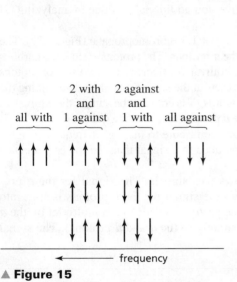

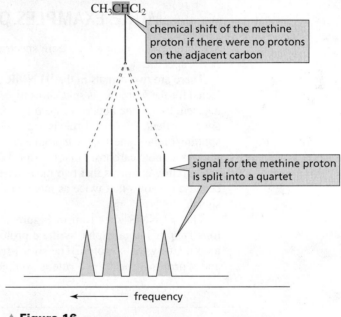

CH$_3$CHCl$_2$

chemical shift of the methine proton if there were no protons on the adjacent carbon

signal for the methine proton is split into a quartet

▲ **Figure 15**

The different ways in which the magnetic fields of three protons can be aligned.

▲ **Figure 16**

The signal for the methine proton of 1,1-dichloroethane is split into a quartet by the methyl protons.

Why does the signal for the methine proton in Figure 16 have peaks of different intensities—that is, why are the inner peaks more intense than the outer peaks?

The relative intensities of the peaks in a signal reflect the number of ways the neighboring protons can be aligned relative to the applied magnetic field (Figure 15). There is only one way to align the magnetic fields of three protons so that they are all lined up with the applied magnetic field and only one way to align them so that they are all lined up against the applied field. However, there are three ways to align the magnetic fields of three protons so that two are lined up with the applied field and one is lined up against it, and there are three ways to align them so that one is lined up with the applied field and two are lined up against it. Therefore, a quartet has relative peak intensities of $1:3:3:1$.

Normally, *nonequivalent* protons split each other's signal only if they are on the same or *adjacent* carbons. Splitting is a "through-bond" effect, not a "through-space" effect, and it is rarely observed if the protons are separated by more than three σ bonds. However, if they are separated by more than three bonds and one of the bonds is a double or a triple bond, then splitting is sometimes observed. This phenomenon is called **long-range coupling.**

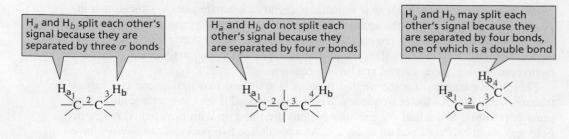

PROBLEM 22

Draw a diagram like the one shown in Figure 15 to predict

a. the relative intensities of the peaks in a triplet.
b. the relative intensities of the peaks in a quintet.

12 MORE EXAMPLES OF ¹H NMR SPECTRA

We will now look at a few more spectra to give you additional practice in analyzing ¹H NMR spectra.

There are two signals in the ¹H NMR spectrum of 1,3-dibromopropane (Figure 17). The signal for the **b** protons is split into a triplet by the **a** protons. The protons on the two carbons adjacent to the one bonded to the **a** protons are equivalent. Because the two sets of protons are equivalent, the $N + 1$ rule is applied to both sets at the same time when determining the splitting of the signal for the **a** protons. In other words, N is equal to the sum of the equivalent protons on both carbons. Thus, the signal for the **a** protons is split into a quintet $(4 + 1 = 5)$. Integration confirms that two methylene groups contribute to the higher frequency signal because it shows that twice as many protons produce that signal than the lower frequency signal.

The ¹H NMR spectrum in Figure 18 shows five signals. The signal for the **a** protons is split into a triplet by the **c** protons and the signal for the **b** protons is split into a doublet by the **e** proton. The signal for the **c** protons is split into a multiplet by the **a** and **d** protons (the $N + 1$ rule is applied separately to the **a** and **d** protons). The signal

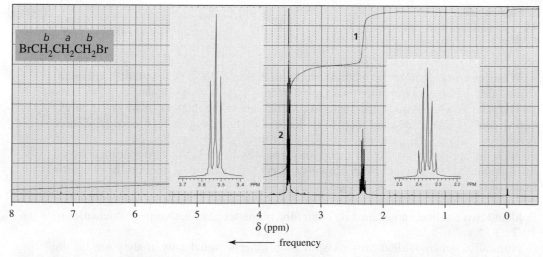

▲ **Figure 17**

The ¹H NMR spectrum of 1,3-dibromopropane. The quintet corresponds to H$_a$ and the triplet corresponds to H$_b$.

for the **d** protons is split into a triplet by the **c** protons; and the signal for the **e** proton is split into a septet by the **b** protons (the $N + 1$ rule is applied to both sets of **b** protons at the time).

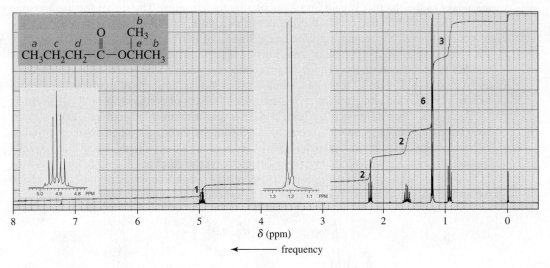

▲ **Figure 18**
The ^{1}H NMR spectrum of isopropyl butanoate.

PROBLEM 23

Indicate the number of signals and the multiplicity of each signal in the ^{1}H NMR spectrum of each of the following compounds:

a. $CH_3CH_2CH_2CH_2CH_2CH_3$ **b.** $ICH_2CH_2CH_2Br$ **c.** $ClCH_2CH_2CH_2Cl$ **d.** $ICH_2CH_2CHBr_2$

The ^{1}H NMR spectrum of 3-bromo-1-propene shows four signals (Figure 19). The signal for the **a** protons is split into a doublet by the **d** proton. Although the **b** and **c** protons are bonded to the same carbon, they are not equivalent (one is cis to the bromomethyl group, and the other is trans to the bromomethyl group), so each produces a separate signal. The signal for the **d** proton is a multiplet because it is split separately by the **a, b,**

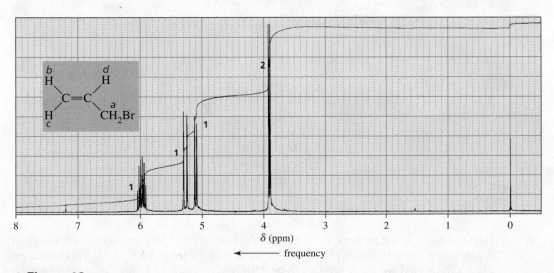

▲ **Figure 19**
The ^{1}H NMR spectrum of 3-bromo-1-propene.

and **c** protons. Notice that the signals for the three vinylic protons are at relatively high frequencies because of diamagnetic anisotropy (Section 8).

Because the **b** and **c** protons are not equivalent, they split one another's signal. This means that the signal for the **b** proton is split into a doublet by the **d** proton and that each of the peaks in the doublet is split into a doublet by the **c** proton. The signal for the **b** proton should therefore be what is called a **doublet of doublets,** and so should the signal for the **c** proton. However, the mutual splitting of the signals of two nonidentical protons bonded to the same carbon—called **geminal coupling**—is often too small to be observed if they are bonded to an sp^2 carbon (see Table 2 later in this chapter). Therefore, the signals for the **b** and **c** protons in Figure 19 appear as doublets rather than as doublets of doublets. (If the signals were expanded along the frequency axis, the doublets of doublets would be observed.)

There is a clear difference between a quartet and a doublet of doublets, even though both have four peaks. A quartet results from splitting by *three equivalent* adjacent protons; it therefore has relative peak intensities of 1 : 3 : 3 : 1, and the individual peaks are equally spaced (Figures 15 and 16). A doublet of doublets, on the other hand, results from splitting by *two nonequivalent* adjacent protons; it has relative peak intensities of 1 : 1 : 1 : 1, and the individual peaks are not necessarily equally spaced (see Figure 26 later in this chapter).

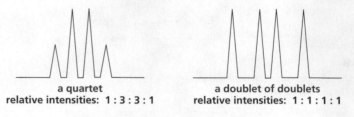

a quartet
relative intensities: **1 : 3 : 3 : 1**

a doublet of doublets
relative intensities: **1 : 1 : 1 : 1**

Ethylbenzene has five sets of chemically equivalent protons (Figure 20). We see the expected triplet for the **a** protons and the quartet for the **b** protons. (This is the characteristic pattern for an ethyl group.) We expect the signal for the **c** protons to be a doublet and the signal for the **e** proton to be a triplet.

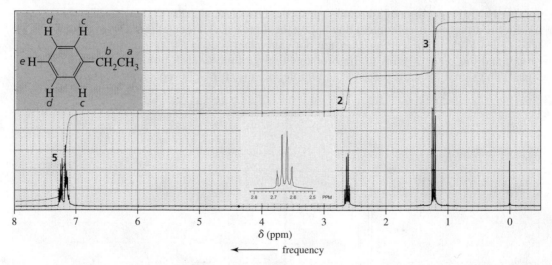

▲ **Figure 20**

The ^{1}H NMR spectrum of ethylbenzene. The signals for the **c, d,** and **e** protons overlap.

Because the **c** and **e** protons are not equivalent, they must be considered separately in determining the splitting of the signal for the **d** protons $(N_c + 1)(N_e + 1)$. Therefore, we expect the signal for the **d** protons to be split into a doublet by the **c** protons and each peak of the doublet to be split into another doublet by the *e* proton, forming a doublet of doublets. However, we do not see three distinct signals for the *c, d,* and *e* protons in Figure 20. Instead, we see overlapping signals. Apparently, the electronic effect (that is, the electron-donating or electron-withdrawing ability) of an ethyl substituent is not sufficiently different from that of a hydrogen to cause a difference in the environments of the *c, d,* and *e* protons that is large enough to allow them to appear as separate signals.

Unlike the benzene protons of ethylbenzene (*c, d,* and *e*), the benzene protons of nitrobenzene (*a, b,* and *c*) show three distinct signals (Figure 21), and the multiplicity of each signal is what we would have predicted for the signals for the benzene ring protons in ethylbenzene (**c** is a doublet, **b** is a triplet, and **a** is a doublet of doublets with some overlapping peaks). In contrast to the ethyl substituent, the nitro substituent is sufficiently electron withdrawing to cause the *a, b,* and *c* protons to be in sufficiently different environments for their signals not to overlap.

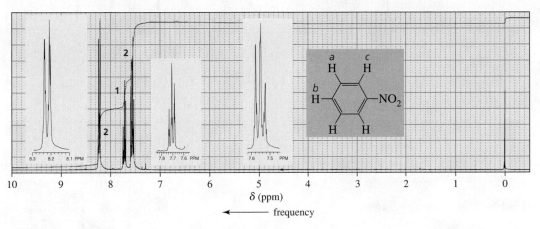

▲ **Figure 21**

The ^{1}H NMR spectrum of nitrobenzene. The signals for the *a, b,* and *c* protons do not overlap and their multiplicities are what the $N + 1$ rule predicts.

Notice that the signals for the benzene ring protons in Figures 20 and 21 occur in the 7.0 to 8.5 ppm region. Other kinds of protons usually do not resonate in this region, so signals in this region of an ^{1}H NMR spectrum indicate that the compound probably contains a benzene ring.

PROBLEM 24

How could their ^{1}H NMR spectra distinguish the following compounds?

A   **B**   **C**

PROBLEM 25

How would the ^{1}H NMR spectra for the four compounds with molecular formula $C_3H_6Br_2$ differ?

PROBLEM 26♦

Identify each compound from its molecular formula and its ^1H NMR spectrum:

a. C_9H_{12}

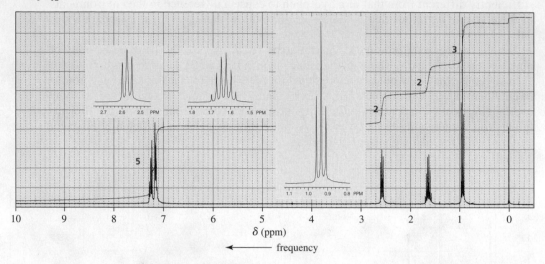

b. $C_5H_{10}O$

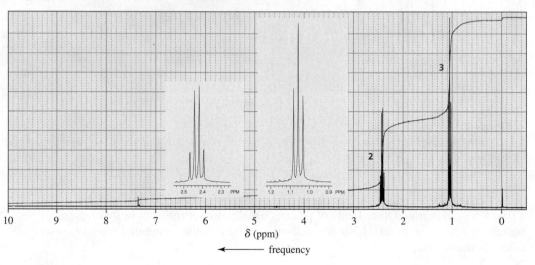

c. $C_9H_{10}O_2$

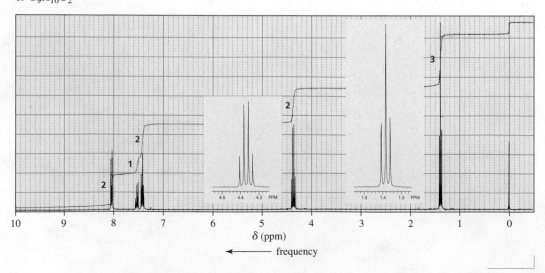

PROBLEM 27

Predict the splitting patterns for the signals given by compounds **a–m** in Problem 4.

PROBLEM 28

Identify the following compounds. (Relative integrals are given from left to right across the spectrum.)

a. The ^{1}H NMR spectrum of a compound with molecular formula $C_4H_{10}O_2$ has two singlets with an area ratio of 2 : 3.

b. The ^{1}H NMR spectrum of a compound with molecular formula $C_6H_{10}O_2$ has two singlets with an area ratio of 2 : 3.

c. The ^{1}H NMR spectrum of a compound with molecular formula $C_8H_6O_2$ has two singlets with an area ratio of 1 : 2.

PROBLEM 29

Describe the ^{1}H NMR spectrum you would expect for each of the following compounds, indicating the relative positions of the signals:

a. $BrCH_2CH_2CH_2CH_2Br$

b. $CH_3OCH_2CH_2CH_2Br$

c. O=⟨ ⟩=O

d. $CH_3\overset{\overset{\displaystyle CH_3}{|}}{\underset{\underset{\displaystyle Br}{|}}{C}}CH_2CH_3$

e. $CH_3\overset{\overset{\displaystyle O}{||}}{C}CH_2\overset{\overset{\displaystyle O}{||}}{C}OCH_3$

f. $\overset{\displaystyle H \quad\quad H}{\underset{\displaystyle H \quad\quad Cl}{C=C}}$

g. $CH_3CH_2OCH_2CH_3$

h. $CH_3CH_2OCH_2Cl$

i. $CH_3\overset{\overset{\displaystyle }{|}}{\underset{\underset{\displaystyle Cl}{|}}{CH}}CHCl_2$

j. ⊏O⊐ (oxetane)

k. $CH_3\overset{\underset{\displaystyle CH_3}{|}}{CH}\overset{\overset{\displaystyle O}{||}}{C}H$

l. $CH_3OCH_2CH_2CH_2OCH_3$

m. $\overset{\displaystyle Cl \quad\quad Cl}{\underset{\displaystyle H \quad\quad Cl}{C=C}}$

n. $\overset{\displaystyle Cl \quad\quad H}{\underset{\displaystyle H \quad\quad Cl}{C=C}}$

o. ⬠ (cyclopentane)

p. ⬠—Cl (chlorocyclopentane)

---

13   COUPLING CONSTANTS IDENTIFY COUPLED PROTONS

The distance, in hertz, between two adjacent peaks of a split NMR signal is called the **coupling constant** (denoted by **J**). The coupling constant for the **a** protons being split by the **b** proton is denoted by J_{ab}. The signals of coupled protons (protons that split each other's signal) have the same coupling constant; in other words, $J_{ab} = J_{ba}$ (Figure 22). Coupling constants are useful in analyzing complex NMR spectra because protons on adjacent carbons can be identified by their identical coupling constants.

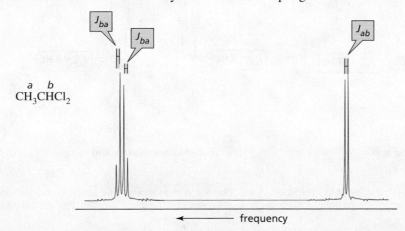

$$\overset{a\ \ b}{CH_3CHCl_2}$$

← frequency

◀ **Figure 22**

The **a** and **b** protons of 1,1-dichloroethane are coupled protons—that is, they split each others signals, so their signals have the same coupling constant, $J_{ab} = J_{ba}$.

J has the same value regardless of the operating frequency of the spectrometer. The magnitude of a coupling constant is a measure of how strongly the nuclear spins of the coupled protons influence each other. It depends, therefore, on the number of bonds and the type of bonds that connect the coupled protons, as well as on the geometric relationship of the protons. Characteristic coupling constants are shown in Table 2; they range from 0 to 15 Hz.

Table 2	Approximate Values of Coupling Constants		
Types of protons	**Approximate value of J_{ab} (Hz)**	**Types of protons**	**Approximate value of J_{ab} (Hz)**
H$_a$ —C— H$_b$	12	H$_a$ C=C H$_b$	15 (trans)
C=C H$_a$ H$_b$	2	H$_a$ H$_b$ C=C	10 (cis)
H$_a$ H$_b$ —C—C—	7	H$_a$ C=C C H$_b$	1 (long-range coupling)
H$_a$ H$_b$ —C—C—C—	0		

The coupling constant for two nonequivalent hydrogens bonded to the *same* sp^3 carbon is large (12 Hz). In contrast, the coupling constant for two nonequivalent hydrogens bonded to the *same* sp^2 carbon is often too small to see (2 Hz) (see Figure 19), but it is large if the nonequivalent hydrogens are bonded to *adjacent* sp^2 carbons (10–15 Hz). Apparently, the interaction between the hydrogens is affected by the intervening π electrons. We have seen that π electrons also allow long-range coupling—that is, coupling through four or more bonds (Section 11).

Coupling constants can be used to distinguish between the ^{1}H NMR spectra of cis and trans alkenes. The coupling constant of trans vinylic protons is significantly greater than that of cis vinylic protons, because the coupling constant depends on the dihedral angle between the two C—H bonds in the H—C=C—H unit (Figure 23). The coupling

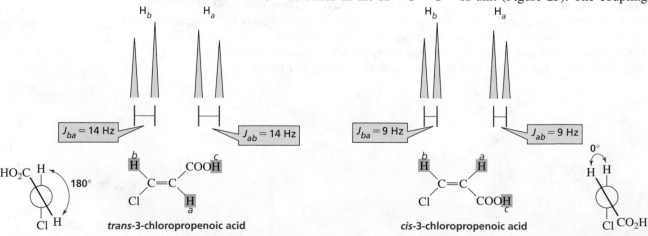

▲ Figure 23

The doublets observed for the *a* and *b* protons in the ^{1}H NMR spectra of a trans alkene and a cis alkene. The coupling constant for trans protons (14 Hz) is greater than the coupling constant for cis protons (9 Hz) because it depends on the dihedral angle (180° for trans protons; 0° for cis protons).

constant is greatest when the angle between the two C—H bonds is 180° (trans) and smallest when the angle is 0° (cis).

The coupling constant for trans vinylic protons is greater than that for cis vinylic protons.

The peaks in the doublets in Figure 23 are not the same height because of **leaning**—an arrow drawn from the top of the smaller peak of the doublet to the top of the larger peak will point at the signal of the proton responsible for the splitting. Thus, the figure shows that H_a is split by H_b and H_b is split by H_a.

Let's now summarize the kind of information that can be obtained from an ^1H NMR spectrum:

1. The number of signals indicates the number of different kinds of protons in the compound.

2. The position of a signal indicates the kind of proton(s) that produce the signal (methyl, methylene, methine, allylic, vinylic, benzene, and so on) and the kinds of neighboring substituents.

3. The integration of the signal tells the relative number of protons that produce the signal.

4. The multiplicity of the signal $(N + 1)$ tells the number of protons (N) bonded to adjacent carbons.

5. The coupling constants identify coupled protons.

PROBLEM 30♦

Why is there no coupling between the **a** and **c** protons or between the **b** and **c** protons in the cis and trans alkenes shown in Figure 23?

PROBLEM-SOLVING STRATEGY

Using IR and ^1H NMR Spectra to Deduce a Chemical Structure

Identify the compound with molecular formula $C_9H_{10}O$ that gives the IR and ^1H NMR spectra in Figure 24.

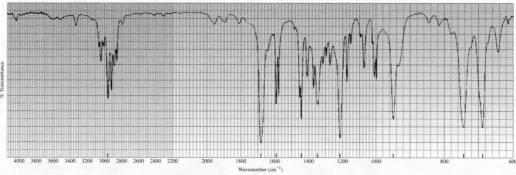

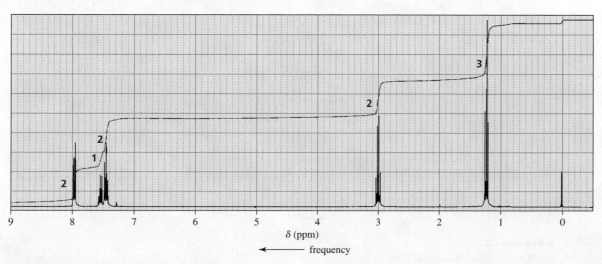

▲ **Figure 24**

The IR and ^1H NMR spectra for this problem-solving strategy.

One way to approach this kind of problem is to identify whatever structural features you can from the ^{1}H NMR spectrum and then use the information from the molecular formula and the IR spectrum to expand on that knowledge.

The signals in the 7.4 to 8.0 ppm region of the NMR spectrum indicate a benzene ring; since the signals integrate to 5H, we know it is a monosubstituted benzene ring. The triplet at ~1.2 ppm and the quartet at ~3.0 ppm indicate an ethyl group that is attached to an electron-withdrawing group.

From the molecular formula and the IR spectrum, we learn that the compound is a ketone: it has a carbonyl group at ~1680 cm^{-1}, only one oxygen, and no absorption bands at ~2820 and ~2720 cm^{-1} that would indicate an aldehyde. The carbonyl group absorption band is at a lower frequency than is typical, which suggests that it has partial single-bond character as a result of electron delocalization, indicating that it is attached to an sp^2 carbon. Now we can conclude that the compound is the ketone shown here. The integration ratio (5 : 2 : 3) confirms this answer.

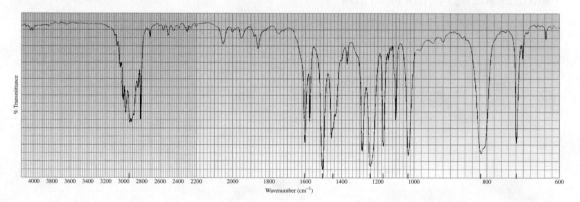

Now use the strategy you have just learned to solve Problem 31.

PROBLEM 31♦

Identify the compound with molecular formula $C_8H_{10}O$ that gives the IR and ^{1}H NMR spectra shown in Figure 25.

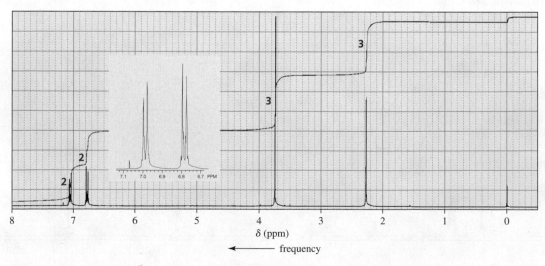

▲ **Figure 25**

The IR and ^{1}H NMR spectra for Problem 31.

14 SPLITTING DIAGRAMS EXPLAIN THE MULTIPLICITY OF A SIGNAL

The splitting pattern obtained when a signal is split by more than one set of protons can best be understood by using a splitting diagram. In a **splitting diagram** (also called a **splitting tree**), the NMR peaks are shown as vertical lines, and the effect of each of the splittings is shown one at a time.

For example, the splitting diagram in Figure 26 depicts the splitting of the signal for the **c** proton of 1,1,2-trichloro-3-methylbutane into a doublet of doublets by the **b** and **d** protons. Notice that we start the diagram with the largest (J_{cb}) of the two J values.

$$
\underset{a}{CH_3}\overset{b}{\underset{a}{\underset{CH_3}{|}}}\overset{\text{Cl Cl}}{\underset{c\ \ d}{CHCHCHCl}}
$$

1,1,2-trichloro-3-methylbutane

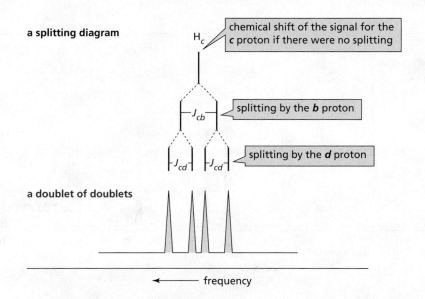

a splitting diagram

chemical shift of the signal for the **c** proton if there were no splitting

splitting by the **b** proton

splitting by the **d** proton

a doublet of doublets

← frequency

◀ **Figure 26**

A splitting diagram for a doublet of doublets. The diagram begins with J_{cb} because $J_{cb} > J_{cd}$.

The splitting diagram in Figure 27 depicts the splitting of the signal for the **b** protons of 1-bromopropane. It is split into a quartet by the **a** protons, and each of the resulting four peaks is split into a triplet by the **c** protons. How many of the 12 peaks are actually seen in the spectrum depends on how many overlap one another, which in turn depends on the relative magnitudes of the two coupling constants, J_{ba} and J_{bc}.

$$
\underset{a}{CH_3}\underset{b}{CH_2}\underset{c}{CH_2}Br
$$

1-bromopropane

For example, Figure 27 shows that there are 12 peaks when J_{ba} is much greater than J_{bc}, 9 peaks when $J_{ba} = 2J_{bc}$, and only 6 peaks when $J_{ba} = J_{bc}$. When peaks overlap, their intensities add together.

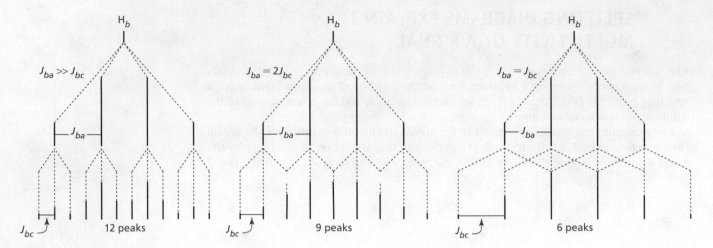

▲ Figure 27

A splitting diagram for a quartet of triplets. The number of peaks actually observed when a signal is split by two sets of protons depends on the relative magnitudes of the two coupling constants.

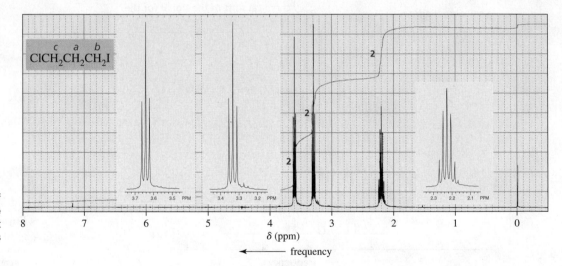

▲ Figure 28

The ^{1}H NMR spectrum of 1-chloro-3-iodopropane. The signal at 2.2 ppm is a quintet instead of a triplet of triplets because $J_{ab} = J_{ac}$.

We expect the signal for the **a** protons of 1-chloro-3-iodopropane to be a triplet of triplets (nine peaks) because the signal would be split into a triplet by the **b** protons and each of the three resulting peaks would be split into a triplet by the **c** protons. The signal, however, in Figure 28 is actually a quintet.

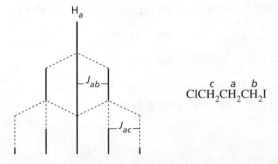

Finding that the signal for the **a** protons is a quintet indicates that J_{ab} and J_{ac} have about the same value. The splitting diagram shows that a quintet results if $J_{ab} = J_{ac}$.

We can conclude that *when two different sets of protons split a signal, the multiplicity of the signal can be determined by using*

- the *N + 1 rule simultaneously for both sets, if the two sets are equivalent and, therefore, have the same coupling constants.*
- the *N + 1 rule simultaneously for both sets, if the two sets are nonequivalent and have similar coupling constants.*
- the *N + 1 rule separately for each set of protons—that is,* $(N_a+1)(N_b+1)$, *if the two sets are nonequivalent and have different coupling constants.*

PROBLEM 32

Draw a splitting diagram for H_b, where

a. $J_{ba} = 12$ Hz and $J_{bc} = 6$ Hz   **b.** $J_{ba} = 12$ Hz and $J_{bc} = 12$ Hz

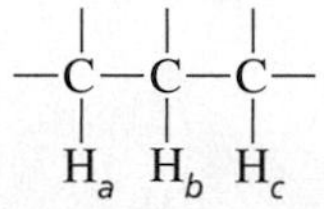

$$-\overset{|}{C}-\overset{|}{C}-\overset{|}{C}-$$
$$\;\;H_a\;\;H_b\;\;H_c$$

---

15   DIASTEREOTOPIC HYDROGENS ARE NOT CHEMICALLY EQUIVALENT

If a carbon is bonded to two hydrogens *and* to two nonidentical groups, then the two hydrogens are called **enantiotopic hydrogens.** For example, H_a and H_b in the CH_2 group of ethanol are enantiotopic hydrogens because the other two groups bonded to the carbon (CH_3 and OH) are not identical. They are called enantiotopic hydrogens, because replacing each of them in turn with deuterium (or another group) creates a pair of enantiomers.

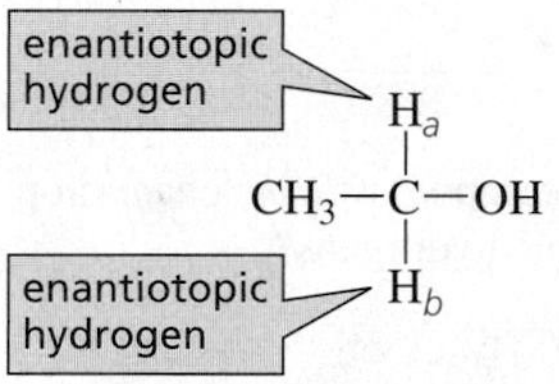

Enantiotopic hydrogens show one NMR signal because they are chemically equivalent.

Enantiotopic hydrogens are chemically equivalent, so they show only one NMR signal.

The carbon to which the enantiotopic hydrogens are attached is called a **prochiral carbon.** If H_a is replaced by a deuterium, the asymmetric center will have the *R* configuration. Thus, H_a is called the **pro-*R*-hydrogen.** Similarly, H_b is called the **pro-*S*-hydrogen** because replacing it with a deuterium would generate an asymmetric center with the *S* configuration.

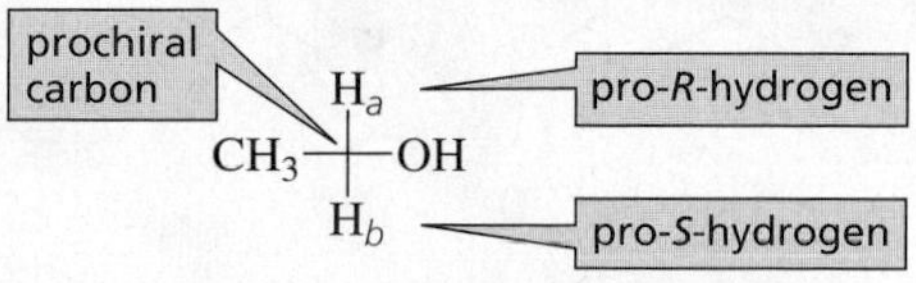

If a carbon is bonded to two hydrogens in a compound that has an asymmetric center, then the two hydrogens are called **diastereotopic hydrogens,** because replacing each of them in turn with deuterium (or another group) creates a pair of diastereomers.

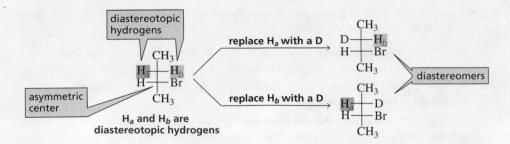

We know that diasterotopic hydrogens are not chemically equivalent because they do not react at the same rate with achiral reagents. For example, removal of H_b and Br to form (E)-2-butene occurs more rapidly than removal of H_a and Br to form (Z)-2-butene because (E)-2-butene is more stable than (Z)-2-butene. Because diastereotopic hydrogens are not equivalent, the $N + 1$ rule has to be applied to them separately.

Diasterotopic hydrogens show two NMR signals because they are chemically nonequivalent.

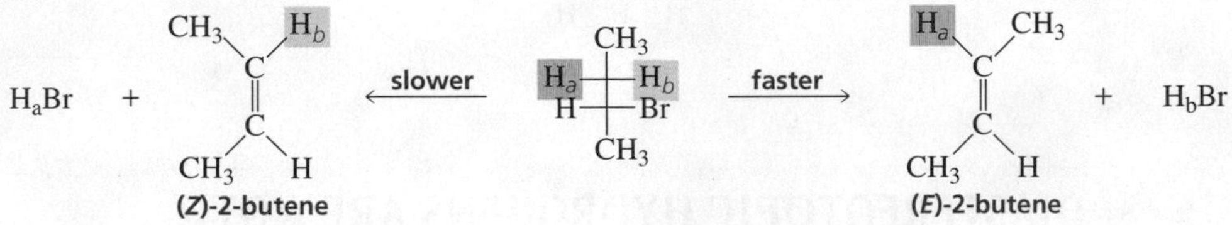

Diastereotopic hydrogens are not chemically equivalent, so they show two NMR signals.

Typically, the chemical shifts of diastereotopic hydrogens are similar and, like other nonequivalent hydrogens, can even be the same by chance. The further the diastereotopic hydrogens are from the asymmetric center, the more similar their chemical shifts are expected to be.

PROBLEM 33♦

a. Which of the following compounds have enantiotopic hydrogens?
b. Which have diastereotopic hydrogens?

$$CH_3CH_2\overset{\overset{\displaystyle H_a}{|}}{\underset{\underset{\displaystyle H_b}{|}}{C}}CH_3$$

A

B

C

D

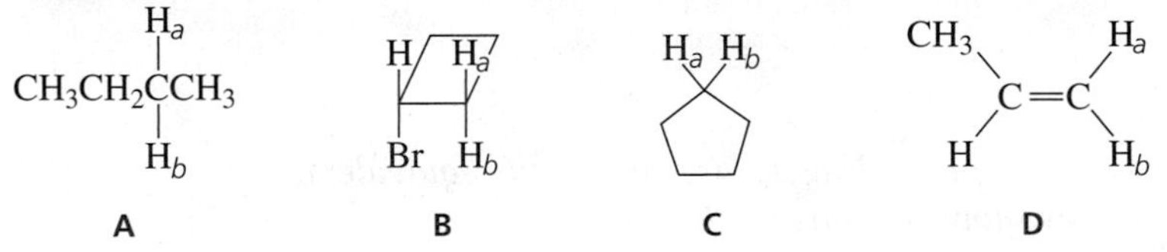

PROBLEM 34 Solved

How many signals would you expect to see for the indicated hydrogens in the following compound's ^1H NMR spectrum?

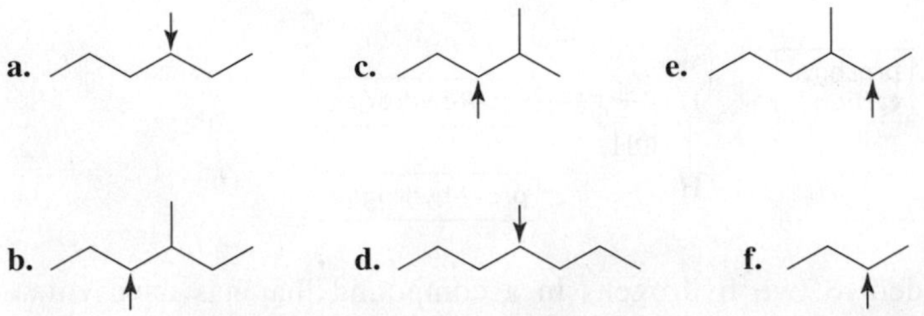

a.

c.

e.

b.

d.

f.

Solution We know that, unless the two hydrogens of a CH_2 group are diastereotopic, they are chemically equivalent and will give one signal. If the two hydrogens are diastereotopic, then they are not chemically equivalent and will give two signals.

  a. These hydrogens are enantiotopic, so they will give one signal.
  b. These hydrogens are diastereotopic, so they will give two signals.
  c. These hydrogens are enantiotopic, so they will give one signal.
  d. These hydrogens are neither enantiotopic nor diastereotopic, so they will give one signal.
  e. These hydrogens are diastereotopic, so they will give two signals.
  f. These hydrogens are enantiotopic, so they will give one signal.

PROBLEM 35 Solved

We expect the signal for the methyl protons adjacent to the diastereotopic hydrogens of 2-bromobutane to be a doublet of doublets as a result of splitting by the nonequivalent diastereotopic hydrogens. The signal, however, is a triplet. Use a splitting diagram to explain why it is a triplet rather than a doublet of doublets.

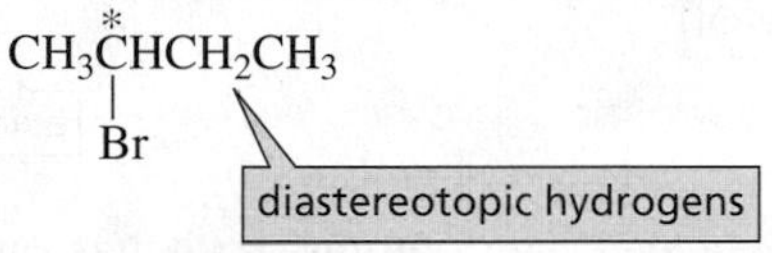

$$\overset{*}{CH_3CH}CH_2CH_3$$
$$|$$
$$Br$$

diastereotopic hydrogens

Solution The observation of a triplet means that the $N + 1$ rule did not have to be applied to the diastereotopic hydrogens separately, but could have been applied to the two protons as a set ($N = 2$, so $N + 1 = 3$). This indicates that the coupling constant for splitting of the methyl signal by one of the diastereotopic hydrogens is similar to the coupling constant for splitting by the other diastereotopic hydrogen—that is, $J_{ab} \approx J_{ac}$.

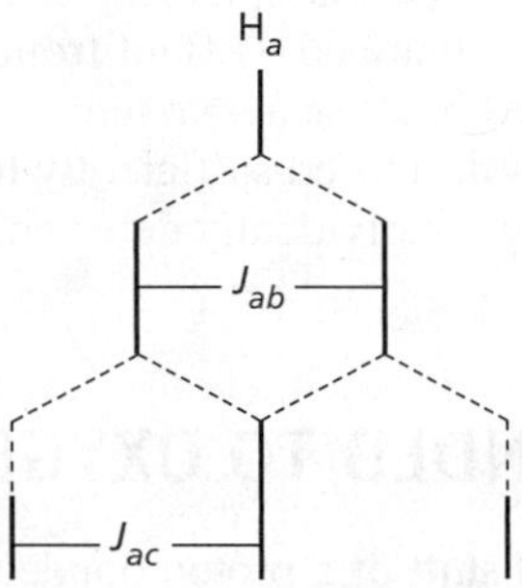

---

16   THE TIME DEPENDENCE OF NMR SPECTROSCOPY

The three methyl hydrogens of ethyl bromide produce just one signal in the ^1H NMR spectrum because they are chemically equivalent due to rotation about the C—C bond. At any given instant, however, the three hydrogens can be in quite different environments. For example, one can be anti to the bromine, one can be gauche to the bromine, and one can be eclipsed with the bromine.

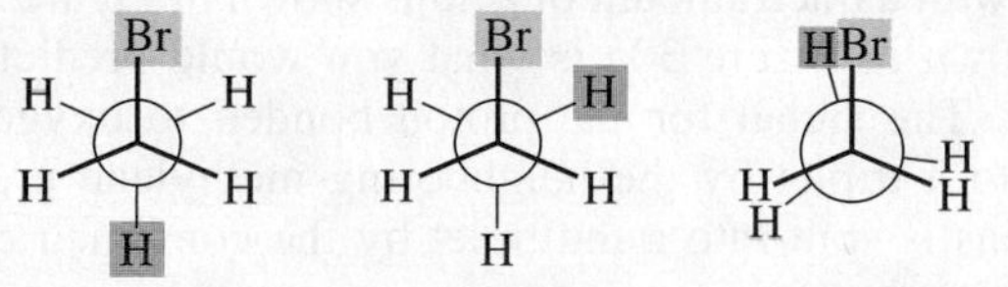

anti   gauche   eclipsed

The average time to take an NMR spectrum is about one second. Any process (in this case rotation about a C—C bond) that happens faster than once every second will cause the signals to average. Therefore, we see only one signal in the ^1H NMR spectrum for the three methyl hydrogens of ethyl bromide. The signal represents an average of their environments.

Similarly, the ^{1}H NMR spectrum of cyclohexane shows only one signal, even though cyclohexane has both axial and equatorial hydrogens. There is only one signal because the chair conformers of cyclohexane undergo ring flip too rapidly at room temperature for the NMR spectrometer to detect them individually. Because axial hydrogens in one chair conformer are equatorial hydrogens in the other chair conformer, all the hydrogens in cyclohexane have the same average environment on the NMR time scale, so the NMR spectrum shows one signal.

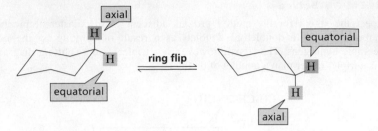

Cyclohexane-d_{11} has 11 deuterium atoms, so it has only one hydrogen. Several ^{1}H NMR spectra of cyclohexane-d_{11} taken at various temperatures are shown in Figure 29. Cyclohexane with only one hydrogen was used for this experiment in order to prevent splitting, which would have complicated the spectrum. Deuterium signals are not detectable in ^{1}H NMR (Section 18).

At 30 °C, the ^{1}H NMR spectrum of cyclohexane-d_{11} shows one sharp signal, which is an average for the axial hydrogen of one chair and the equatorial hydrogen of the other chair. As the temperature decreases, the signal becomes broader and eventually separates into two signals, which are equidistant from the original signal. At −89 °C, two sharp singlets are observed because at that temperature the rate of ring flip, which is temperature dependent, has decreased sufficiently to allow the two kinds of hydrogens (axial and equatorial) to be individually detected on the NMR time scale.

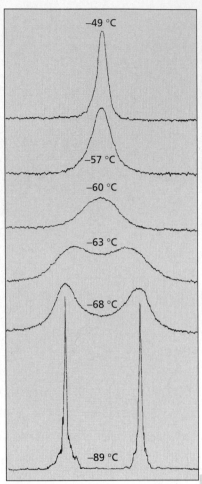

▲ **Figure 29**

A series of ^{1}H NMR spectra of cyclohexane-d_{11} obtained as the temperature is lowered from −49 °C to −89 °C.

17   PROTONS BONDED TO OXYGEN AND NITROGEN

Table 1 shows that the chemical shift of a proton bonded to an oxygen or a nitrogen occurs within a range of chemical shifts. For example, the chemical shift of the OH proton of an alcohol ranges from 2 to 5 ppm. The chemical shift is variable because it depends on the extent of hydrogen bonding that the proton experiences. The chemical shift increases as the extent of hydrogen bonding increases, because hydrogen bonding decreases the electron density around the proton.

PROBLEM 36

Explain why the chemical shift of the OH proton of a carboxylic acid is at a higher frequency than the chemical shift of an OH proton of an alcohol.

The ^{1}H NMR spectrum of pure dry ethanol is shown in Figure 30a, and the ^{1}H NMR spectrum of ethanol with a trace amount of acid is shown in Figure 30b.

The spectrum shown in Figure 30a is what you would predict based on what you have learned so far. The signal for the proton bonded to oxygen is farthest downfield and is split into a triplet by the neighboring methylene protons; the signal for the methylene protons is split into a multiplet by the combined effects of the methyl protons and the OH proton.

However, the spectrum shown in Figure 30b is the type of spectrum most often obtained for alcohols. The signal for the proton bonded to oxygen is not split, and this proton does not split the signal of the adjacent methylene protons. Therefore, the signal for the OH proton is a singlet, and the signal for the methylene protons is a quartet because it is split only by the three methyl protons.

The two spectra differ because protons bonded to oxygen undergo **proton exchange,** which means that they are transferred from one molecule to another. Whether the OH

a.

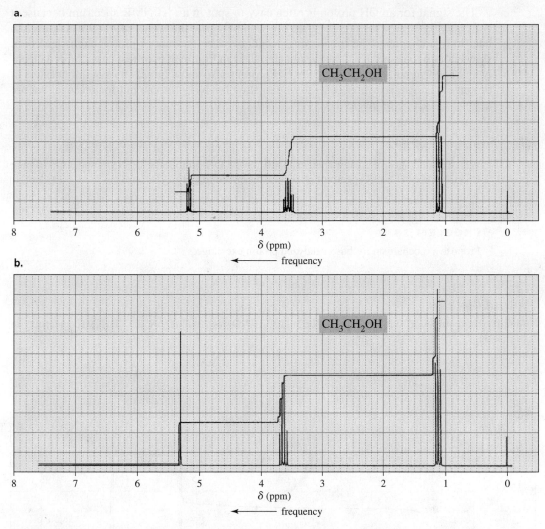

b.

CH₃CH₂OH

▲ Figure 30

(a) The ¹H NMR spectrum of pure ethanol.

(b) The ¹H NMR spectrum of ethanol containing a trace amount of acid.

proton and the methylene protons split each other's signals depends on how long a particular proton stays attached to the oxygen.

In a sample of pure alcohol, the rate of proton exchange is very slow. This causes the spectrum to look no different from one that would be obtained if proton exchange did not occur.

Acids and bases catalyze proton exchange, so if the alcohol is contaminated with just a trace amount of acid or base, proton exchange becomes rapid. When proton exchange is rapid, the spectrum records only an average of all possible environments. Therefore, a rapidly exchanging proton is recorded as a singlet. The effect of a rapidly exchanging proton on adjacent protons is also averaged. Thus, not only is its signal not split by adjacent protons, a rapidly exchanging proton does not cause splitting.

MECHANISM FOR ACID-CATALYZED PROTON EXCHANGE

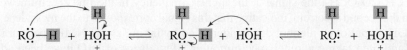

Unless the sample is pure, the hydrogen of an OH group is not split by its neighbors and does not split its neighbors.

- A proton is transferred to the alcohol.
- A different proton is transferred from the alcohol.

The signal for an OH proton is often easy to spot in an ^{1}H NMR spectrum because it is frequently somewhat broader than other signals (see the signal at δ 4.9 in Figure 32b later in this chapter). The broadening occurs because the rate of proton exchange is not slow enough to result in a cleanly split signal, as in Figure 30a, or fast enough for a cleanly averaged signal, as in Figure 30b. NH protons also show broad signals, not because of proton exchange, which is generally quite slow for NH protons, but for reasons that are beyond the scope of this text.

PROBLEM 37◆

Which would show a greater chemical shift for the OH proton, the ^{1}H NMR spectrum of pure ethanol or the ^{1}H NMR spectrum of ethanol dissolved in CH_2Cl_2?

PROBLEM 38

Propose a mechanism for base-catalyzed proton exchange.

PROBLEM 39◆

Identify the compound with molecular formula C_3H_7NO responsible for the ^{1}H NMR spectrum in Figure 31.

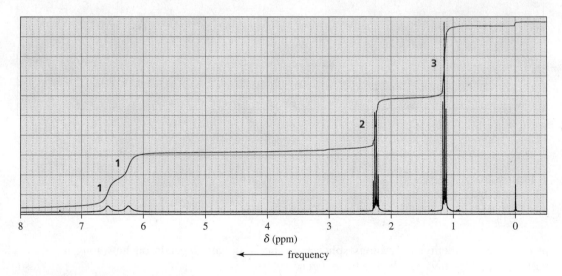

▲ **Figure 31**
The ^{1}H NMR spectrum for Problem 39.

18 THE USE OF DEUTERIUM IN ^{1}H NMR SPECTROSCOPY

Deuterium signals are not seen in an ^{1}H NMR spectrum. Therefore, substituting a deuterium for a hydrogen is a technique used to identify signals and to simplify ^{1}H NMR spectra. For example, if the ^{1}H NMR spectrum of $CH_3CH_2OCH_3$ were compared with the ^{1}H NMR spectrum of $CH_3CD_2OCH_3$, the signal at the highest frequency in the first spectrum would be absent in the second spectrum, indicating that this signal corresponds to the methylene group.

The OH signal of an alcohol can be identified by taking an ^{1}H NMR spectrum of the alcohol and then taking another spectrum after a few drops of D_2O have been added to the sample. The OH signal will be the one that becomes less intense (or disappears) in the second spectrum because of the proton exchange process just discussed. This technique can be used with any proton that undergoes exchange.

R—O—H + D—O—D $\longrightarrow$ R—O—D + D—O—H

| seen in ^{1}H NMR | | not seen in ^{1}H NMR |

The sample used to obtain an ^{1}H NMR spectrum is made by dissolving the compound in an appropriate solvent. Solvents with protons cannot be used because the signals for solvent protons would be very intense, since there is much more solvent than compound in a solution. Instead, deuterated solvents such as $CDCl_3$ (rather than $CHCl_3$) and D_2O (rather than H_2O) are commonly used.

19 THE RESOLUTION OF ^{1}H NMR SPECTRA

An ^{1}H NMR spectrum taken on a 60-MHz NMR spectrometer is shown in Figure 32a (60-MHz spectrophotometers are no longer being manufactured.); an ^{1}H NMR spectrum of the same compound taken on a 300-MHz instrument is shown in Figure 32b. Why is the resolution of the second spectrum so much better?

To produce separate signals with "clean" splitting patterns, the difference in the chemical shifts (Δv in Hz) of two coupled protons must be *at least 10 times* the value of the coupling constant (J).

a.

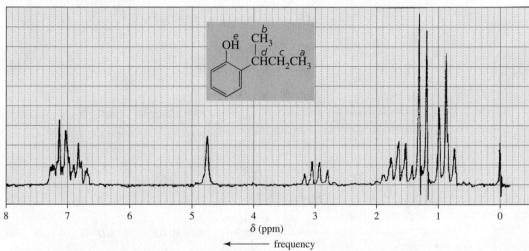

b.

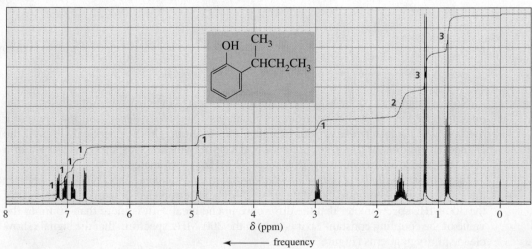

▲ **Figure 32**

(a) A 60-MHz ^{1}H NMR spectrum of 2-*sec*-butylphenol.

(b) A 300-MHz ^{1}H NMR spectrum of 2-*sec*-butylphenol.

Figure 33 shows that as $\Delta v/J$ decreases, the two signals produced by the H_a and H_b protons in an ethyl group appear closer to each other and the outer peaks of the signals become less intense while the inner peaks become more intense. The quartet and triplet of the ethyl group are clearly observed only when $\Delta v/J$ is greater than 10.

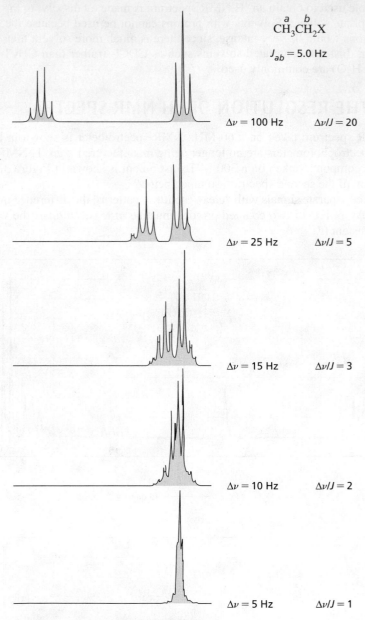

$$\overset{a}{C}H_3\overset{b}{C}H_2X$$

$J_{ab} = 5.0$ Hz

$\Delta v = 100$ Hz $\Delta v/J = 20$

$\Delta v = 25$ Hz $\Delta v/J = 5$

$\Delta v = 15$ Hz $\Delta v/J = 3$

$\Delta v = 10$ Hz $\Delta v/J = 2$

$\Delta v = 5$ Hz $\Delta v/J = 1$

▶ **Figure 33**

The splitting pattern of an ethyl group as a function of $\Delta v/J$.

Now we can understand why the resolution of an ^{1}H NMR spectrum increases as the operating frequency of the spectrometer increases. *Δv values depend on the operating frequency of the spectrometer.* For example, the difference in the chemical shifts of the H_a and H_c protons of 2-*sec*-butylphenol is 0.8 ppm, which corresponds to 240 Hz in a 300-MHz spectrometer but only 48 Hz in a 60-MHz spectrometer.

In contrast, *J values are independent of the operating frequency,* so J_{ac} is 7 Hz, whether the spectrum is taken on a 300-MHz or a 60-MHz spectrometer. It is only in the case of the 300-MHz spectrometer that the difference in chemical shift is more than 10 times the value of the coupling constant, so it is only in the 300-MHz spectrum that the signals show clean splitting patterns (Figure 32b).

in a 300-MHz spectrometer

$$\frac{\Delta v}{J} = \frac{240}{7} = 34$$

in a 60-MHz spectrometer

$$\frac{\Delta v}{J} = \frac{48}{7} = 6.9$$

20 ^{13}C NMR SPECTROSCOPY

The number of signals in a ^{13}C **NMR** spectrum tells how many different kinds of carbons a compound has—just as the number of signals in an ^{1}H NMR spectrum tells how many different kinds of hydrogens a compound has. The principles behind ^{1}H NMR and ^{13}C NMR spectroscopy are essentially the same.

The use of ^{13}C NMR spectroscopy as a routine analytical procedure was not possible until computers were available that could carry out a Fourier transform. ^{13}C NMR requires Fourier transform techniques because the signals obtained from a single scan are too weak to be distinguished from background electronic noise. However, ^{13}C FT–NMR scans can be repeated rapidly, so a large number of scans can be recorded and added together. When hundreds of scans are combined, ^{13}C signals stand out because electronic noise is random, so its sum is close to zero.

The individual ^{13}C signals are weak for two reasons. First, the isotope of carbon (^{13}C) that produces ^{13}C NMR signals constitutes only 1.11% of the carbon in nature. (The most abundant isotope of carbon, ^{12}C, has no nuclear spin and therefore cannot produce an NMR signal.) The low abundance of ^{13}C would cause the intensities of the signals in ^{13}C NMR to be weaker than those in ^{1}H NMR by a factor of approximately 100. Second, the gyromagnetic ratio (γ) of ^{1}H is about four times that of ^{13}C. The intensity of a signal is proportional to γ^3, so the overall intensity of a ^{1}H signal is about 6400 times ($100 \times 4 \times 4 \times 4$) stronger than the intensity of an ^{13}C signal.

One advantage to ^{13}C NMR spectroscopy is that the chemical shifts of carbon atoms range over about 220 ppm (Table 3), compared with about 12 ppm for hydrogens (Table 1). This means that signals for carbons in different environments are more easily distinguished. For example, the data in Table 3 show that aldehyde (190 to 200 ppm) and ketone (205 to 220 ppm) carbonyl groups can be distinguished from each other and from other carbonyl groups.

Table 3	Approximate Values of Chemical Shifts for ^{13}C NMR		
Type of carbon	**Approximate chemical shift (ppm)**	**Type of carbon**	**Approximate chemical shift (ppm)**
$(CH_3)_4Si$	0	C—I	−20–10
R—CH$_3$	0–35	C—Br	10–40
		C—Cl	25–50
R—CH$_2$—R	15–55	C—N	40–60
		C—O	50–90
R—CH—R (R)	25–55	R, N—C=O	165–175
R—C—R (R, R)	30–40	R, RO—C=O	165–175
C≡C	70–90	R, HO—C=O	175–185
C≡N	110–120		
C=C	80–145	R, H—C=O	190–200
C=N	150–170		
(benzene ring) C	110–170	R, R—C=O	205–220

The reference compound used in ^{13}C NMR is TMS, the same reference compound used in ^{1}H NMR. You will find it helpful when analyzing a ^{13}C NMR spectrum to divide it into five regions and remember the kind of carbons that show signals in each.

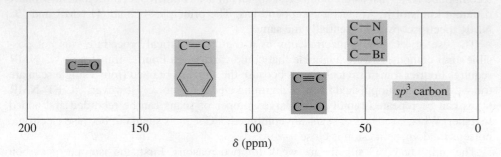

A disadvantage of ^{13}C NMR spectroscopy is that, unless special techniques are used, the area under a ^{13}C NMR signal is *not* proportional to the number of carbons that produce the signal. Thus, the number of carbons that produce a particular ^{13}C NMR signal cannot routinely be determined by integration.

The ^{13}C NMR spectrum of 2-butanol shows four signals (Figure 34), so we know that it has carbons in four different environments. The relative positions of the signals depend on the same factors that determine the relative positions of the proton signals in an ^{1}H NMR spectrum—namely, carbons in electron-rich environments produce low-frequency signals, whereas carbons close to electron-withdrawing groups produce high-frequency signals. This means that the signals for the carbons of 2-butanol are in the same relative order as the signals of the protons bonded to those carbons in its ^{1}H NMR spectrum.

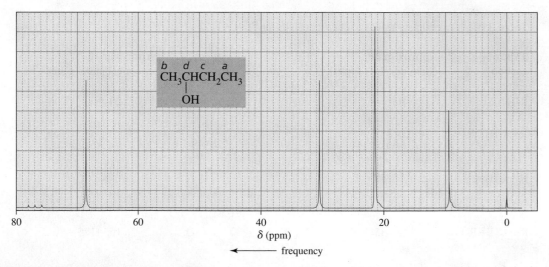

▲ **Figure 34**

The ^{13}C NMR spectrum of 2-butanol.

Thus, the carbon of the methyl group farthest away from the electron-withdrawing OH group gives the lowest-frequency signal. The other methyl carbon comes next in order of increasing frequency, followed by the methylene carbon; the carbon attached to the OH group gives the highest-frequency signal.

The signals in ^{13}C NMR are not normally split by neighboring carbons because there is little likelihood of an adjacent carbon being a ^{13}C; the probability of two ^{13}C carbons being next to each other is 1.11% × 1.11% (or about 1 in 10,000). ^{12}C does not have a magnetic moment, so it cannot split the signal of an adjacent ^{13}C.

The signals in a ^{13}C NMR spectrum can be split by hydrogens, but such splitting is not usually observed because the spectra are recorded using spin-decoupling, which obliterates the carbon–proton interactions. Thus, all the signals are singlets in an ordinary ^{13}C NMR spectrum (Figure 34).

However, if the spectrometer is run in a *proton-coupled* mode, then each signal will be split by the *hydrogens* bonded to the carbon that produces the signal. The multiplicity of the signal is determined by the $N + 1$ rule.

The coupling between ^{13}C and its attached hydrogens is called **heteronuclear coupling** because the coupling is between different types of nuclei. (The coupling between nuclei of the same type, such as the coupling between adjacent protons in ^1H NMR, is called **homonuclear coupling.**)

The **proton-coupled ^{13}C NMR spectrum** of 2-butanol is shown in Figure 35. The signals for the methyl carbons are each split into a quartet because each methyl carbon is bonded to three hydrogens ($3 + 1 = 4$). The signal for the methylene carbon is split into a triplet because the carbon is bonded to two hydrogens ($2 + 1 = 3$), and the signal for the carbon bonded to the OH group is split into a doublet because the carbon is bonded to one hydrogen ($1 + 1 = 2$). (The signal at 78 ppm is produced by the solvent, $CDCl_3$.)

> If the spectrometer is run in a proton-coupled mode, splitting by the directly attached protons is observed in a ^{13}C NMR spectrum.

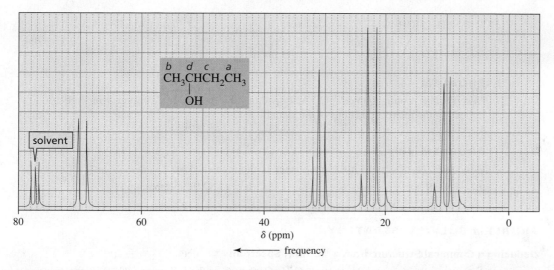

▲ **Figure 35**

The proton-coupled ^{13}C NMR spectrum of 2-butanol. Each signal is split by the hydrogens bonded to the carbon that produces the signal, according to the $N + 1$ rule.

In the following ^{13}C NMR spectrum, the three methyl groups at one end of the molecule are equivalent, so they give one signal (Figure 36). Because the intensity of a

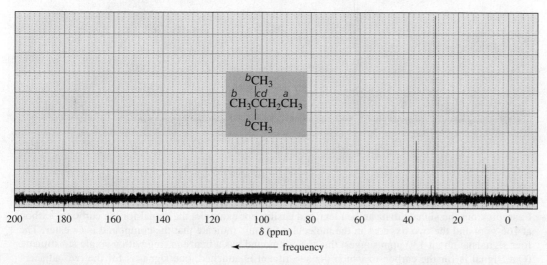

▲ **Figure 36**

The ^{13}C NMR spectrum of 2,2-dimethylbutane.

signal is somewhat related to the number of carbons producing it (and somewhat related to the number of hydrogens on the carbons), the signal for these three methyl groups is the most intense signal in the spectrum. The tiny signal at ~31 ppm is for the quaternary carbon; carbons that are not attached to hydrogens give very small signals.

PROBLEM 40

Answer the following questions for each of the following compounds:

a. How many signals are in its ^{13}C NMR spectrum?
b. Which signal is at the lowest frequency?

1. $CH_3CH_2CH_2Br$

2. $(CH_3)_2C=CH_2$

3. CH_3CHCH_3
   |
   Br

4.

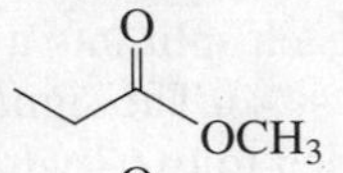

5.

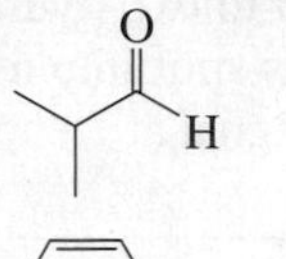

6.

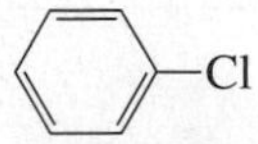

7.

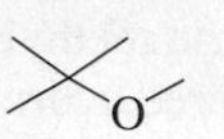

8.

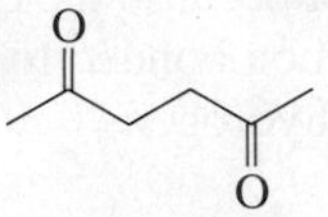

9. $CH_2=CHBr$

PROBLEM 41

Describe the proton-coupled ^{13}C NMR spectra for compounds 1, 3, and 5 in Problem 40, indicating the relative positions of the signals.

PROBLEM 42

How can 1,2-, 1,3-, and 1,4-dinitrobenzene be distinguished by

a. ^1H NMR spectroscopy?   **b.** ^{13}C NMR spectroscopy?

PROBLEM-SOLVING STRATEGY

Deducing a Chemical Structure from a ^{13}C NMR Spectrum

Identify the compound with molecular formula $C_9H_{10}O_2$ that gives the following ^{13}C NMR spectrum:

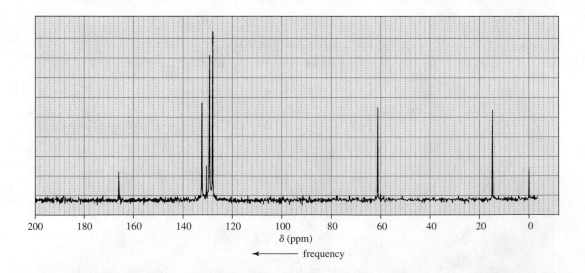

First, pick out the signals that can be identified easily. For example, the signal for the carbonyl carbon at 166 ppm and the two oxygens in the molecular formula indicate that the compound is an ester. The four signals at about 130 ppm suggest that the compound has a benzene ring with a single substituent. (One signal is for the carbon to which the substituent is attached, one signal is for the two adjacent carbons, and so on.) Subtracting those fragments (C_6H_5 and CO_2) from the molecular formula of the compound leaves C_2H_5, the molecular formula of an ethyl substituent. Therefore, we know that the compound is one of the following two compounds.

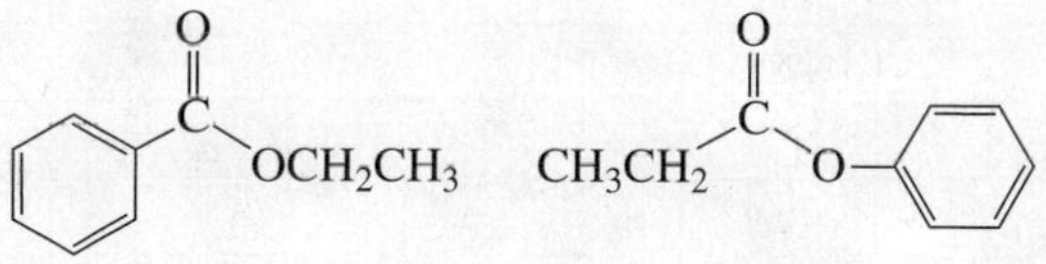

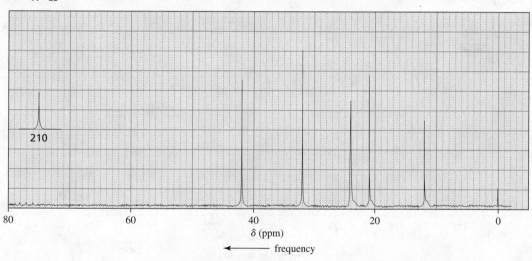

Since the signal for the methylene group is at ~60 ppm, it must be adjacent to an oxygen. Thus, the compound is the one on the left.

Now use the strategy you have just learned to solve Problem 43.

PROBLEM 43♦

Identify each compound in Figure 37 from its molecular formula and its ^{13}C NMR spectrum.

a. $C_{11}H_{22}O$

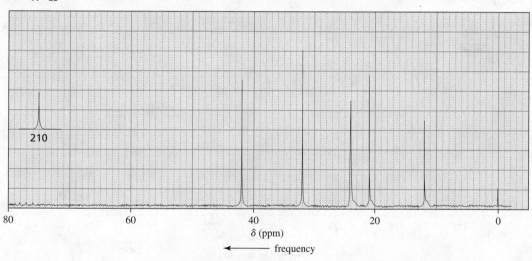

b. C_8H_9Br

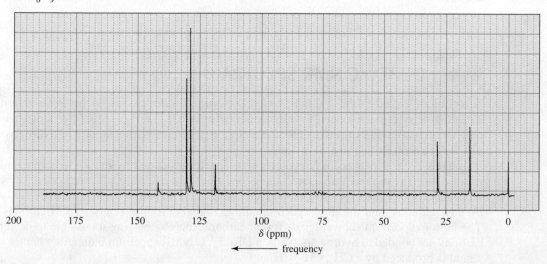

c. $C_6H_{10}O$

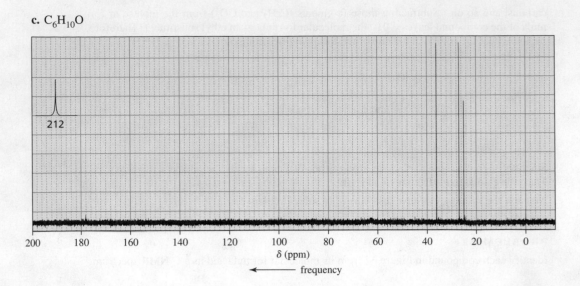

d. C_6H_{12}

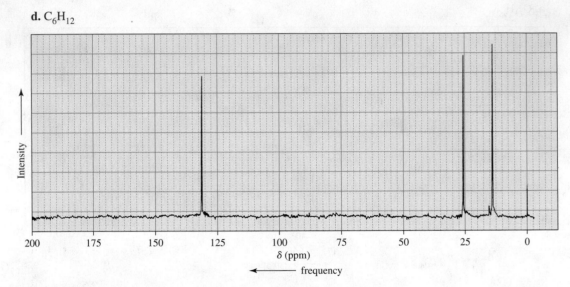

▲ **Figure 37**

The ^{13}C NMR spectra for Problem 43.

21   DEPT ^{13}C NMR SPECTRA

A technique called DEPT ^{13}C NMR (for Distortionless Enhancement by Polarization Transfer) has been developed to distinguish between CH_3, CH_2, and CH groups. It is now much more widely used than proton coupling to determine the number of hydrogens attached to each carbon in a compound.

A **DEPT ^{13}C NMR** recording shows four spectra produced by the same compound (Figure 38). The top spectrum is run under conditions that allow signals produced only by CH_3 carbons. The next spectrum is run under conditions that allow signals produced only by CH_2 carbons, and the third spectrum shows signals produced only by CH carbons. The bottom spectrum shows signals for all carbons, thereby making it possible to detect carbons *not* bonded to hydrogens. Thus, a DEPT ^{13}C NMR spectrum indicates whether a signal is produced by a CH_3, CH_2, CH, or C.

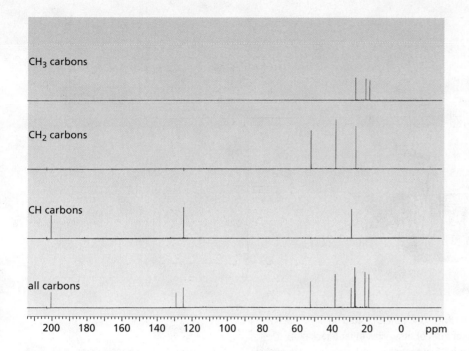

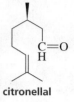

citronellal

◀ **Figure 38**
The DEPT ^{13}C NMR spectrum of citronellal.

22 TWO-DIMENSIONAL NMR SPECTROSCOPY

Complex molecules such as proteins and nucleic acids are difficult to analyze by NMR because the signals in their spectra overlap. Such compounds can be analyzed by two-dimensional NMR spectroscopy (2-D NMR), a technique that allows the structures of complex molecules in solution to be determined. This is a particularly important technique for studying biological molecules whose properties depend on how they fold in water.

More recently, 3-D and 4-D NMR spectroscopy have been developed and can be used to determine the structures of highly complex molecules. A thorough discussion of 2-D NMR is beyond the scope of this text, but the chapter would not be complete without a brief introduction to this increasingly important spectroscopic technique.

The ^{1}H NMR and ^{13}C NMR spectra discussed in the preceding sections have *one frequency* axis and one intensity axis; **2-D NMR** spectra have *two frequency* axes and one intensity axis. The most common 2-D spectra involve ^{1}H–^{1}H shift correlations, which identify coupled protons (that is, protons that split each other's signal). This is called ^{1}H–^{1}H shift-COrrelated SpectroscopY, known by the acronym COSY.

A portion of the **COSY spectrum** of ethyl vinyl ether is shown in Figure 39a; it looks like a mountain range because intensity is the third axis. These "mountain-like" spectra, called *stack plots,* are not the spectra actually used to identify a compound. Instead, the compound is identified using a contour plot (Figure 39b) that represents each mountain in Figure 39a by a large dot (as if its top had been cut off). The two mountains shown in Figure 39a correspond to the dots labeled Y and Z in Figure 39b.

In Figure 39b, the usual one-dimensional ^{1}H NMR spectrum is plotted on both the *x*-and *y*-axes. To analyze the spectrum, a diagonal line is drawn through the dots that bisect the spectrum. Dots that are *not* on the diagonal (X, Y, Z) are called *cross peaks;* they reveal pairs of protons that are coupled.

For example, if we start at the cross peak labeled X and draw a straight line, parallel to the *y*-axis, from X to the diagonal, we hit the dot on the diagonal at ~1.1 ppm, the signal produced by the H$_a$ protons. If we then go back to X and draw a straight line, parallel to the *x*-axis, from X to the diagonal, we hit the dot on the diagonal at ~3.8 ppm, the signal produced by the H$_b$ protons. This means that the H$_a$ and H$_b$ protons are coupled.

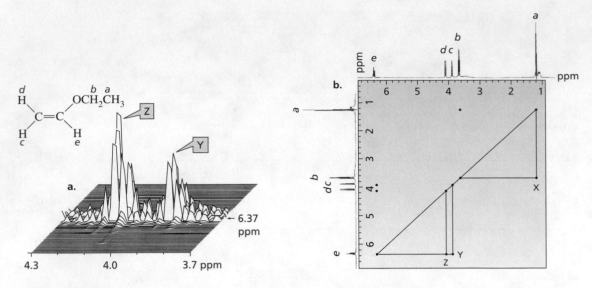

▲ Figure 39

In a COSY spectrum, an ^1H NMR spectrum is plotted on both the *x*- and *y*-axes.

(a) A portion of a COSY spectrum of ethyl vinyl ether (stack plot).

(b) A COSY spectrum of ethyl vinyl ether (contour plot). Cross peaks Y and Z represent the two mountains in (a).

If we go next to the cross peak labeled Y and draw another two perpendicular lines back to the diagonal, we see that the H_c and H_e protons are coupled. Similarly, the cross peak labeled Z shows that the H_d and H_e protons are coupled.

Although we have been using cross peaks below the diagonal as examples, the cross peaks above the diagonal give the same information because a COSY spectrum is symmetrical with respect to the diagonal. The absence of a cross peak due to coupling of H_c and H_d is consistent with expectations—the two protons bonded to an sp^2 carbon of ethyl vinyl ether are also not coupled in the one-dimensional ^1H NMR spectrum shown in Figure 19. The power of a COSY experiment is that it reveals coupled protons without having to do coupling constant analysis.

The COSY spectrum of 1-nitropropane is shown in Figure 40. Cross peak X shows that the H_a and H_b protons are coupled, and cross peak Y shows that the H_b and H_c protons are coupled. Notice that the two triangles in the figure have a common vertex, since the H_b protons are coupled to both the H_a and H_c protons.

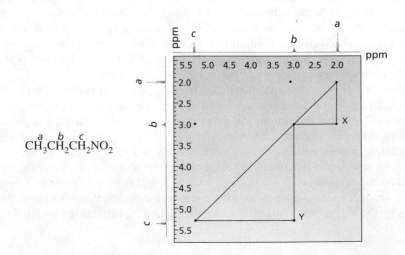

$$\overset{a}{CH_3}\overset{b}{CH_2}\overset{c}{CH_2}NO_2$$

▶ Figure 40

The COSY spectrum of 1-nitropropane. A COSY spectrum identifies coupled protons, such as H_a/H_b and H_b/H_c in 1-nitropropane.

PROBLEM 44

Identify pairs of coupled protons in the compound whose COSY spectrum is shown in Figure 41.

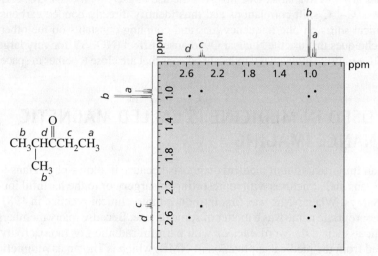

▲ **Figure 41**

The COSY spectrum for Problem 44.

HETCOR spectra (from HETeronuclear CORrelation) are 2-D NMR spectra that show ^{13}C–^1H shift correlations and thus reveal coupling between protons and the carbon to which they are attached.

In a HETCOR spectrum, a compound's ^{13}C NMR spectrum is shown on the *x*-axis, and its ^1H NMR spectrum is shown on the *y*-axis. The cross peaks in a HETCOR spectrum identify which hydrogens are attached to which carbons.

For example, cross peak W in Figure 42 indicates that the hydrogens that produce a signal at ~0.9 ppm in the ^1H NMR spectrum are bonded to the carbon that produces a signal at ~8 ppm in the ^{13}C NMR spectrum. Similarly, cross peak Y shows that the hydrogens that produce a signal at ~2.5 ppm are bonded to the carbon that produces a signal at ~34 ppm.

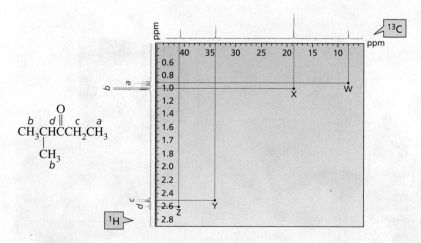

▲ **Figure 42**

The HETCOR spectrum for the compound shown. A HETCOR spectrum reveals coupling between protons and the carbons to which they are attached.

PROBLEM 45♦

What does cross peak X in Figure 42 tell you?

Clearly, 2-D NMR techniques are not necessary for interpreting the NMR spectra of simple compounds such as those shown here. However, in the case of many complicated molecules, signals cannot be assigned without the aid of 2-D NMR. A number of different 2-D NMR techniques are now available. One involves the use of 2-D ^{13}C INADEQUATE[*] spectra, which show ^{13}C–^{13}C shift correlations and thus identify directly bonded carbons. Another plots chemical shifts on one frequency axis and coupling constants on the other. Finally, there are techniques that use the Nuclear Overhauser Effect (NOESY for very large molecules, ROESY for midsize molecules)[†] to locate protons that are close together in space.

23 | NMR USED IN MEDICINE IS CALLED MAGNETIC RESONANCE IMAGING

NMR has become an important tool in medical diagnosis because it allows physicians to examine internal organs and structures without resorting to surgery or to the harmful ionizing radiation of X-rays. When NMR was first introduced into clinical practice in 1981, the selection of an appropriate name was a matter of some debate. Because many members of the general public associate the word *nuclear* with harmful radiation or radioactivity, the "N" was dropped from the medical application of NMR, which is known as **magnetic resonance imaging (MRI).** The spectrometer is called an **MRI scanner.**

An MRI scanner consists of a magnet large enough to surround a person, along with an apparatus for exciting the nuclei, modifying the magnetic field, and receiving signals. (By comparison, the NMR spectrometer used by chemists is only large enough to accommodate a 5-mm glass tube.) Different tissues yield different signals, which are separated into components by Fourier transform analysis (Section 2). Each component can be attributed to a specific location within the part of the body being scanned, so that a set of images through the scanned volume is generated. MRI can produce an image showing any cross section of the body, regardless of the person's position within the machine, which allows optimal visualization of the anatomical feature of interest.

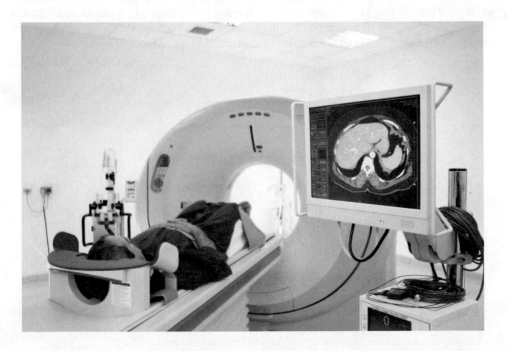

---

* INADEQUATE is an acronym for Incredible Natural Abundance Double Quantum Transfer Experiment.
† NOESY and ROESY are the acronyms for Nuclear Overhauser Effect SpectroscopY and Rotation-frame Overhauser Effect SpectroscopY, respectively.

Most of the signals in an MRI scan originate from the hydrogens of water molecules because tissues contain far more of these hydrogens than they do hydrogens of organic compounds. The difference in the way water is bound in different tissues is what produces much of the variation in signal between different organs, as well as the variation between healthy and diseased tissue. MRI scans, therefore, can provide much more information than images obtained by other means.

For example, MRI can provide detailed images of blood vessels. Flowing fluids, such as blood, respond differently to excitation in an MRI scanner than do stationary tissues, and proper processing will result in the display of only the moving fluids. The quality of these images has become high enough that it can often eliminate the need for more invasive diagnostic techniques.

The versatility of MRI has been enhanced by using gadolinium as a contrast agent. Gadolinium modifies the magnetic field in its immediate vicinity, altering the signal from nearby hydrogens. The distribution of gadolinium, which is infused into a patient's veins, may be affected by certain disease processes such as cancer and inflammation. Any abnormal patterns of distribution are revealed in the MRI images.

A brain tumor and a brain abscess may have very similar appearances in an MRI (Figure 43a). Suppressing the signal from water makes it possible to detect signals from specific compounds such as choline and acetate. A tumor will produce an elevated choline signal, whereas an abscess is more likely to produce an elevated acetate signal (Figure 43b).

a.
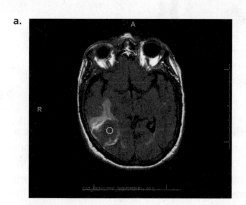

b.
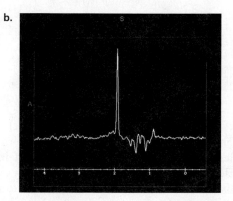

▲ **Figure 43**

(a) The white circle indicates a brain lesion that could be caused by either a tumor (elevated choline) or an abscess (elevated acetate).

(b) The major peak in the spectrum corresponds to acetate, supporting the diagnosis of an abscess.

24 X-RAY CRYSTALLOGRAPHY

X-ray crystallography is a technique used to determine the arrangement of atoms within a crystal. With this technique, the structure of any material that can form a crystal can be determined. Dorothy Crowfoot Hodgkin was a pioneer of X-ray crystallography, using it to determine the structures of cholesterol (1937), penicillin (1945), vitamin B_{12} (1954), and insulin (1969). In 1953, Rosalind Franklin's X-ray images allowed James Watson and Francis Crick to accurately describe the structure of DNA.

A crystal is a solid composed of atoms and molecules in a regular three-dimensional array, called a unit cell, that is repeated indefinitely throughout the crystal. Often the most difficult aspect of X-ray crystallography is obtaining the crystal, which should be at least 0.5 mm long in all three dimensions in order to provide good structural data. Fortunately,

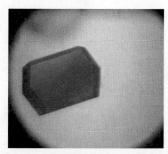

A crystal of lysozyme from hen egg white. It appears blue because it was photographed under polarized light.

crystallographers can get a head start by looking up over 14,000 crystallization conditions on the *Biological Macromolecule Crystallization Database* (xpdb.nist.gov:8060/BMCD4).

In order to resolve atomic distances, the wavelength of the light used and the distance between the atoms to be resolved must be about the same order of magnitude. The length of a C—C bond is 1.54 Å. Visible light, therefore, with wavelengths of 4000–8000 Å cannot be used. The X-rays produced by bombarding a Cu-anode with electrons in a vacuum tube have a wavelength of 1.542 Å, which make them ideal for resolving most atomic distances. For high-resolution structures, crystallographers use more intense sources of X-rays. The high intensity coupled with the superior dispersion of the X-rays in a synchrotron allows experiments to be carried out at less than 1 Å resolution.

Once a suitable crystal has been obtained, it is bombarded with X-rays while being gradually rotated. X-rays, however, cannot be focused by conventional lenses to form an image of a molecule. This is where X-ray diffraction comes in. While most of the X-rays pass through the crystal, some are scattered by the electron clouds of the atoms and land on a detector, creating a diffraction pattern of regularly spaced dots (Figure 44).

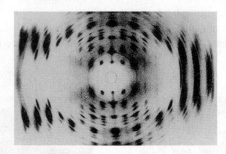

▲ **Figure 44**
an X-ray diffraction pattern

By means of complex mathematical methods, now facilitated by computers, a three-dimensional model of the electron density within the crystal is produced from the two-dimensional **X-ray diffraction** images taken at different angles of rotation. Electrons are found around atoms, so any two regions of electron density within bonding distance of each other can be assumed to represent atoms bonded to one another. The greater the density of the electron cloud around the atom, the more precisely the location of the atom can be assigned. Therefore, the position of electron-deficient hydrogen atoms cannot be assigned unambiguously. However, hydrogen atoms now can be located in the final structure by computational methods.

Structural Databases

There are a variety of databases that provide structural information to chemists. Two important examples are the Chemical Abstracts Services Database (CAS) and the Research Collaboratory for Structural Bioinformatics Protein Data Bank (RCSB PDB).

CAS (http://www.cas.org/) is a subscription-based service with over 100 million organic, inorganic, protein, and nucleic acid structures. It can be accessed through SciFinder, which allows searches via a structural drawing program as well as searches by subject, author, or structure name.

The RCSB Protein Data Bank (www.pdb.org) is the repository of more than 57,000 three-dimensional structures that have been determined by X-ray crystallography (shown on the opening page of this chapter) or multidimensional NMR (Section 22). The structures are publicly available for download in a PDB format. The structures can be visualized and manipulated with free programs such as Visual Molecular Dynamics (http://www.ks.uiuc.edu/Research/vmd/) or Chimera (http://www.cgl.ucsf.edu/chimera).

Dorothy Crowfoot Hodgkin (1910–1994) *was born in Egypt to English parents. She received an undergraduate degree from Somerville College at Oxford University and earned a Ph.D. from Cambridge University. She performed the first three-dimensional calculations in crystallography, and she was the first to use computers to determine the structures of compounds. She received the 1964 Nobel Prize in Chemistry for her work on vitamin B_{12}. Hodgkin was a professor of chemistry at Somerville, where one of her research students was Margaret Thatcher, the future prime minister of England. Hodgkin was also a founding member of Pugwash, an organization whose purpose was to further communication between scientists on both sides of the Iron Curtain.*

SOME IMPORTANT THINGS TO REMEMBER

- **NMR spectroscopy** identifies the carbon–hydrogen framework of an organic compound.

- Each set of chemically equivalent protons produces a signal, so the number of signals in an ^{1}H NMR spectrum indicates the number of different kinds of protons in a compound.

- The **chemical shift** (δ), which is independent of the operating frequency of the spectrometer, is a measure of how far the signal is from the reference TMS signal.

- The larger the magnetic field sensed by the proton, the higher the frequency of its signal.

- The electron density of the environment in which the proton is located **shields** the proton from the applied magnetic field. Therefore, a proton in an electron-dense environment shows a signal at a lower frequency than a proton near electron-withdrawing groups.

- Low-frequency signals have small δ (ppm) values; high-frequency signals have large δ values. Thus, the position of a signal indicates the kind of proton(s) responsible for the signal and the kinds of neighboring substituents.

- In a similar environment, the chemical shift of a methine proton is at a higher frequency than that of methylene protons, which is at a higher frequency than that of methyl protons.

- **Integration** tells us the relative number of protons that produce each signal.

- **Diamagnetic anisotropy** causes unusual chemical shifts for hydrogens bonded to carbons that form π bonds.

- The **multiplicity** of a signal indicates the number of protons bonded to adjacent carbons. Multiplicity is described by the $N+1$ **rule,** where N is the number of equivalent protons bonded to an adjacent carbon.

- Coupled protons split each other's signal.

- A **splitting diagram** describes the splitting pattern obtained when a signal is split by more than one set of protons.

- The **coupling constant (J),** which is independent of the operating frequency of the spectrometer, is the distance between two adjacent peaks of a split NMR signal. Coupled protons have the same coupling constant.

- The coupling constant for trans alkene protons is greater than that for cis alkene protons.

- When two different sets of protons split a signal, the multiplicity of the signal is determined by using the $N+1$ rule separately $(N_a+1)(N_b+1)$ for each set of protons when the coupling constants for the two sets are different. When the coupling constants are similar, the $N+1$ rule is applied to both sets at the same time.

- The chemical shift of a proton bonded to an O or to an N depends on the extent to which the proton is hydrogen bonded.

- In the presence of trace amounts of acid or base, protons bonded to oxygen undergo **proton exchange.** In that case, the signal for a proton bonded to an O is not split and does not split the signal of adjacent protons.

- The number of signals in a ^{13}C NMR spectrum corresponds to the number of different kinds of carbons in the compound. Carbons in electron-rich environments produce low-frequency signals, whereas carbons close to electron-withdrawing groups produce high-frequency signals.

- ^{13}C NMR signals are not split by attached protons unless the spectrometer is run in a proton-coupled mode.

- A DEPT ^{13}C NMR spectrum tells whether a signal is produced by CH_3, CH_2, CH, or C.

- NMR (known in medical applications as **MRI**) is an important tool in medical diagnosis because it allows internal structures to be examined without surgery or harmful X-rays.

- **2-D NMR** and **X-ray crystallography** are techniques that can be used to determine the structures of large molecules.

PROBLEMS

46. How many signals are produced by each of the following compounds in its

   a. ^{1}H NMR spectrum?   **b.** ^{13}C NMR spectrum?

47. Draw a splitting diagram for the H$_b$ proton and give its multiplicity if

   a. $J_{ba} = J_{bc}$ **b.** $J_{ba} = 2J_{bc}$

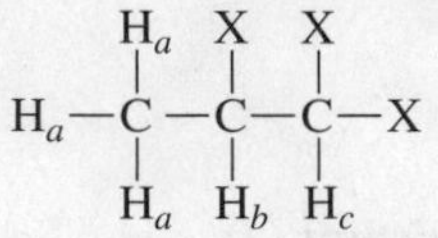

$$H_a{-}\underset{\underset{H_a}{|}}{\overset{\overset{H_a}{|}}{C}}{-}\underset{\underset{H_b}{|}}{\overset{\overset{X}{|}}{C}}{-}\underset{\underset{H_c}{|}}{\overset{\overset{X}{|}}{C}}{-}X$$

48. Label each set of chemically equivalent protons, using **a** for the set that will be at the lowest frequency (farthest upfield) in the ^{1}H NMR spectrum, **b** for the next lowest, and so on. Indicate the multiplicity of each signal.

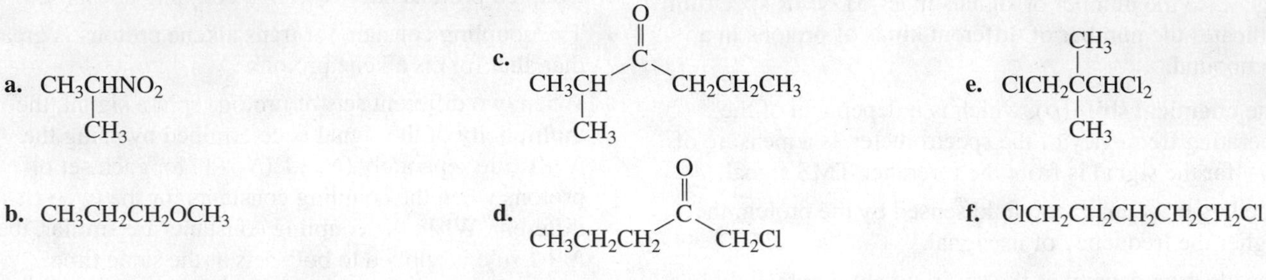

a. CH_3CHNO_2
         |
        CH_3

b. $CH_3CH_2CH_2OCH_3$

c. $CH_3\underset{\underset{CH_3}{|}}{C}H{-}\overset{\overset{O}{\|}}{C}{-}CH_2CH_2CH_3$

d. $CH_3CH_2CH_2{-}\overset{\overset{O}{\|}}{C}{-}CH_2Cl$

e. $ClCH_2\underset{\underset{CH_3}{|}}{\overset{\overset{CH_3}{|}}{C}}CHCl_2$

f. $ClCH_2CH_2CH_2CH_2CH_2Cl$

49. Match each of the ^{1}H NMR spectra with one of the following compounds:

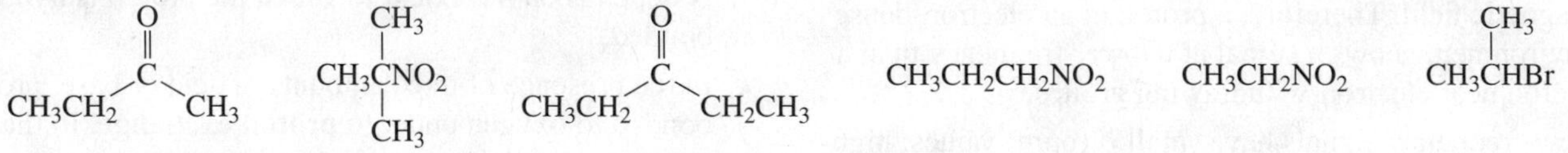

$CH_3CH_2{-}\overset{\overset{O}{\|}}{C}{-}CH_3$ $CH_3\underset{\underset{CH_3}{|}}{\overset{\overset{CH_3}{|}}{C}}NO_2$ $CH_3CH_2{-}\overset{\overset{O}{\|}}{C}{-}CH_2CH_3$ $CH_3CH_2CH_2NO_2$ $CH_3CH_2NO_2$ $CH_3\underset{}{\overset{\overset{CH_3}{|}}{C}}HBr$

a.

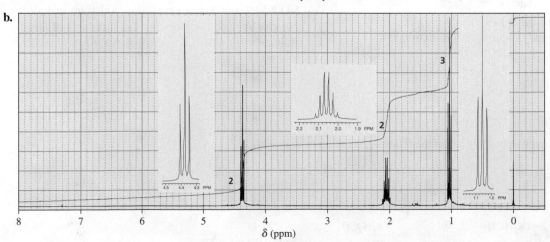

b.

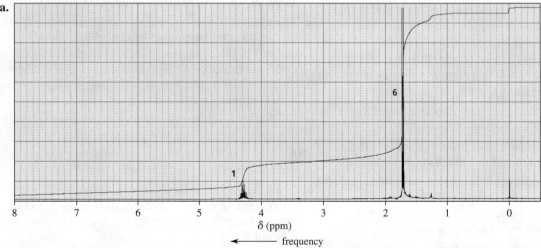

c.

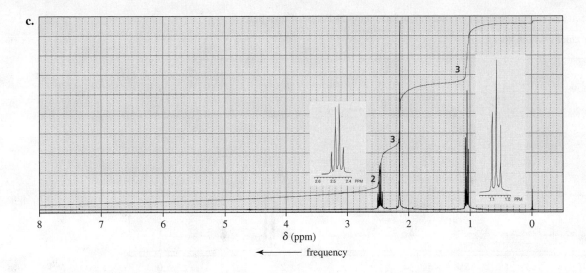

50. Determine the ratios of the chemically nonequivalent protons in a compound if the steps of the integration curves measure 40.5, 27, 13, and 118 mm, from left to right across the spectrum. Draw the structure of a compound whose ^1H NMR spectrum would show these integrals in the observed order.

51. How could ^1H NMR distinguish between the compounds in each of the following pairs?

a. 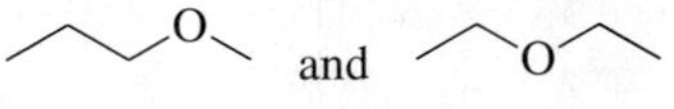 (CH₃CH₂—O—CH₂CH₃ structure) and (CH₃CH₂—O—CH₂CH₃ structure)

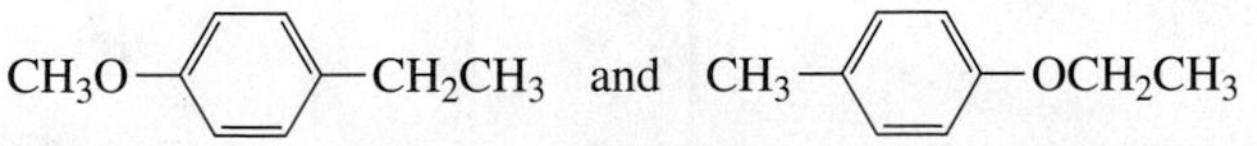

f. 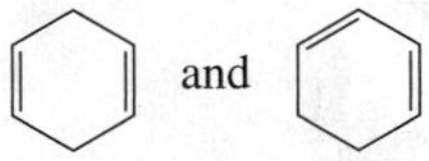 (cyclohexene) and (1,3-cyclohexadiene)

b. Br—CH₂CH₂CH₂—Br and Br—CH₂CH₂CH₂—NO₂

g. 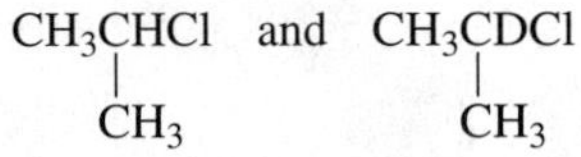 CH₃CHCl and CH₃CDCl
         | |
         CH₃   CH₃

c. CH₃CH—CHCH₃ and CH₃CCH₂CH₃
         | | |
        CH₃ CH₃   CH₃
                            CH₃

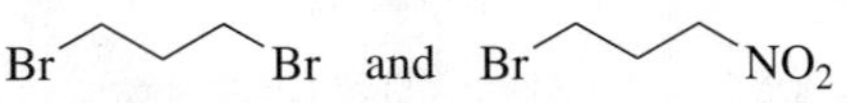

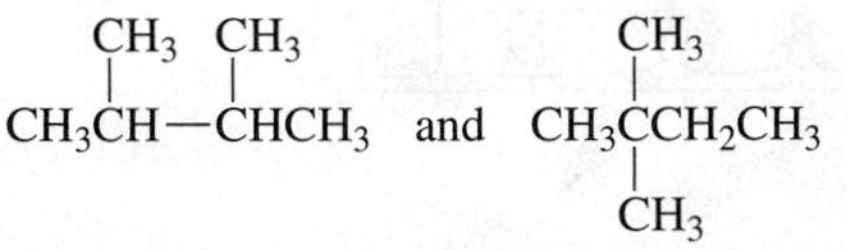

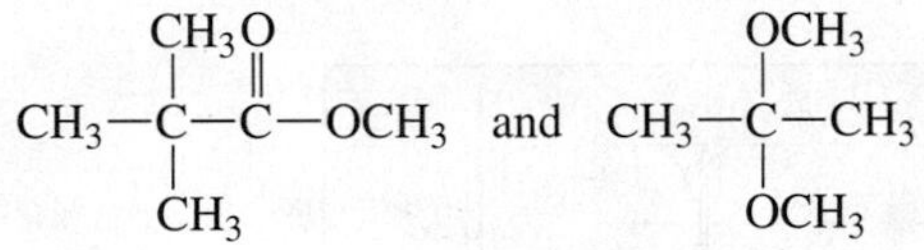

h. 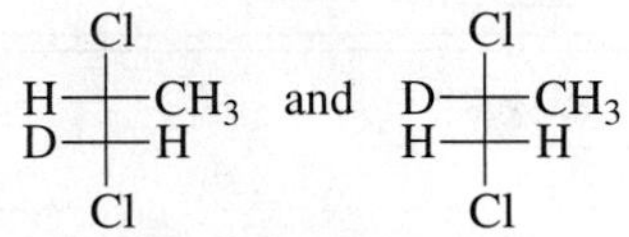   Cl   Cl
         | |
       H——CH₃ and   D——CH₃
       D——H   H——H
         | |
         Cl   Cl

d.
       CH₃ O
        | ‖
   CH₃—C—C—OCH₃   and
        |
       CH₃

       OCH₃
        |
   CH₃—C—CH₃
        |
       OCH₃

i. 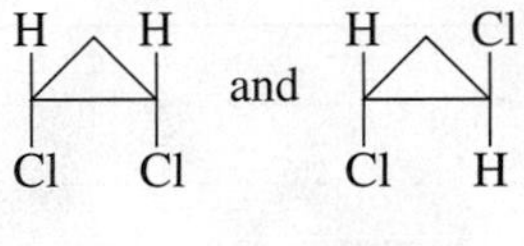 (cyclopropane with H H / Cl Cl) and (cyclopropane with H Cl / Cl H)

e. CH₃O—⟨benzene⟩—CH₂CH₃ and CH₃—⟨benzene⟩—OCH₂CH₃

j. 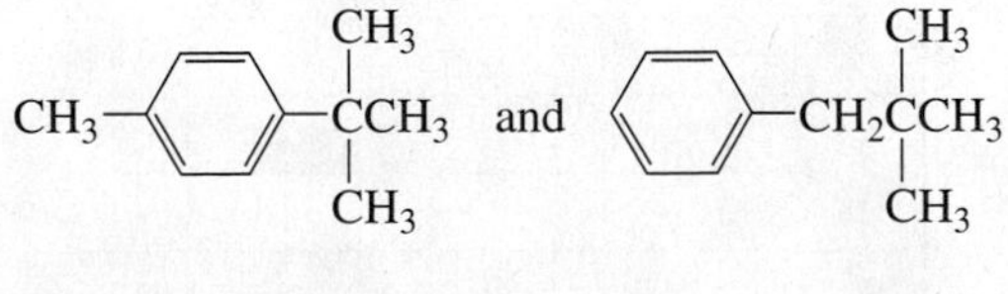
   CH₃—⟨benzene⟩—C(CH₃)₃   and   ⟨benzene⟩—CH₂CCH₃
                                              CH₃

52. Answer the following questions:

  a. What is the relationship between chemical shift in ppm and operating frequency?
  b. What is the relationship between chemical shift in hertz and operating frequency?
  c. What is the relationship between coupling constant in hertz and operating frequency?
  d. How does the operating frequency in NMR spectroscopy compare with the operating frequency in IR and UV/Vis spectroscopy?

53. The ^{1}H NMR spectra of three isomers with molecular formula C_4H_9Br are shown here.

Which isomer produces which spectrum?

a.

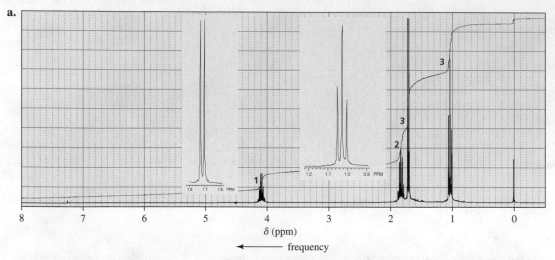

b.

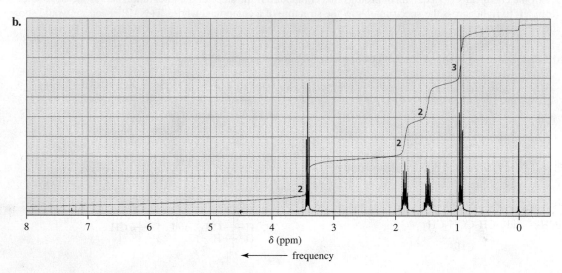

c.

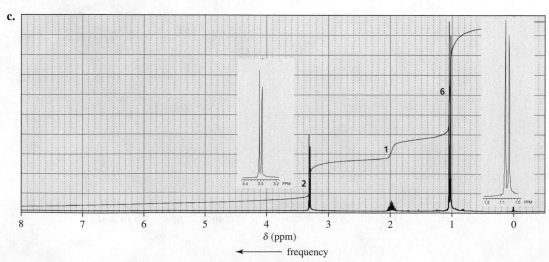

54. How many signals are produced by each of the following compounds in its

    a. ^{1}H NMR spectrum?   **b.** ^{13}C NMR spectrum?

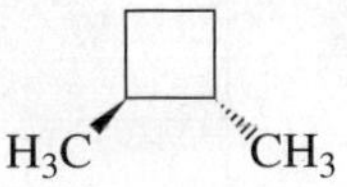

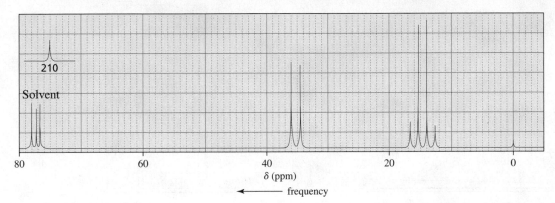

55. Identify each of the following compounds from the ^{1}H NMR data and molecular formula. The number of hydrogens responsible for each signal is shown in parentheses.

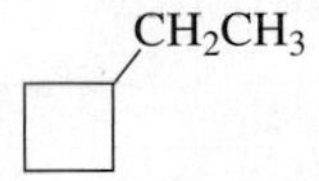

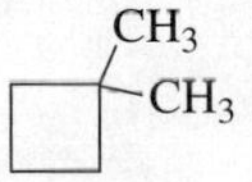

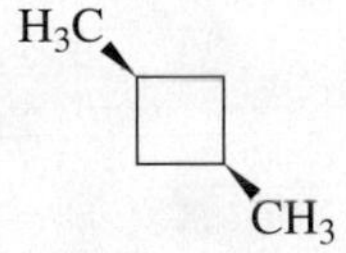

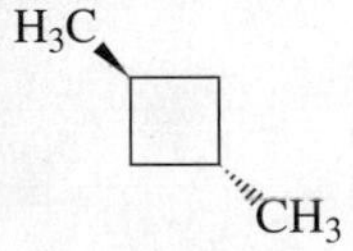

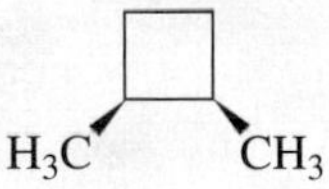

    a. $C_4H_8Br_2$ 1.97 ppm (6) singlet
                 3.89 ppm (2) singlet

    b. C_8H_9Br 2.01 ppm (3) doublet
                 5.14 ppm (1) quartet
                 7.35 ppm (5) broad singlet

    c. $C_5H_{10}O_2$ 1.15 ppm (3) triplet
                    1.25 ppm (3) triplet
                    2.33 ppm (2) quartet
                    4.13 ppm (2) quartet

56. Identify the compound with molecular formula $C_7H_{14}O$ that gives the following proton-coupled ^{13}C NMR spectrum:

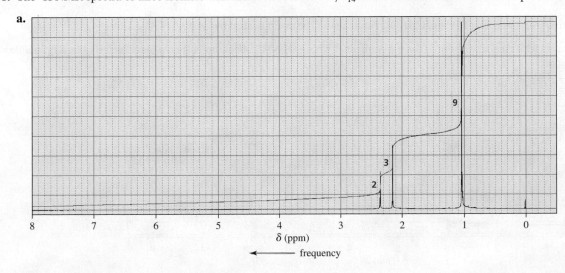

δ (ppm)

← frequency

57. Compound **A**, with molecular formula C_4H_9Cl, shows two signals in its ^{13}C NMR spectrum. Compound **B**, an isomer of compound **A**, shows four signals, and in the proton-coupled mode, the signal farthest downfield is a doublet. Identify compounds **A** and **B**.

58. The ^{1}H NMR spectra of three isomers with molecular formula $C_7H_{14}O$ are shown here. Which isomer produces which spectrum?

    a.

δ (ppm)

← frequency

b.

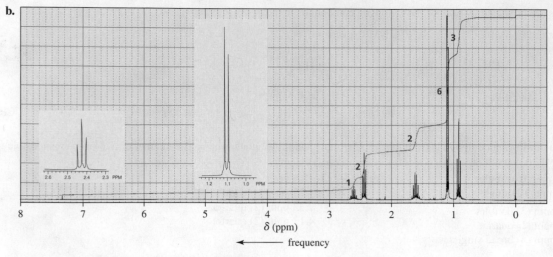

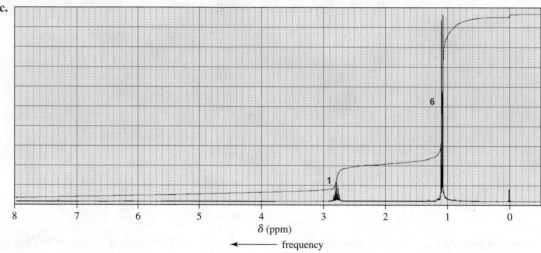

c.

59. Would it be better to use ^{1}H NMR or ^{13}C NMR spectroscopy to distinguish among 1-butene, *cis*-2-butene, and 2-methylpropene? Explain your answer.

60. There are four esters with molecular formula $C_4H_8O_2$. How can they be distinguished by ^{1}H NMR?

61. An alkyl halide reacts with an alkoxide ion to form a compound whose ^{1}H NMR spectrum is shown here. Identify the alkyl halide and the alkoxide ion.

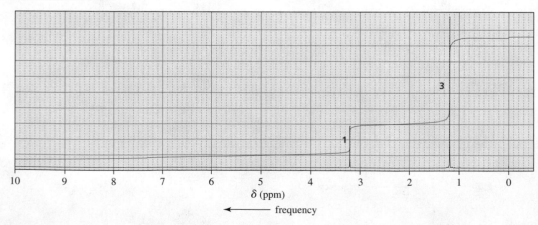

62. Determine the structure of each of the following unknown compounds based on its molecular formula and its IR and ^{1}H NMR spectra.

a. $C_5H_{12}O$

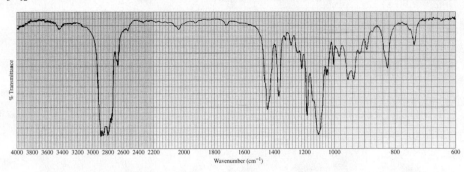

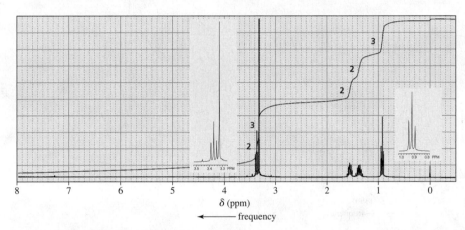

b. $C_6H_{12}O_2$

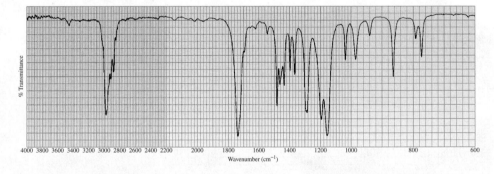

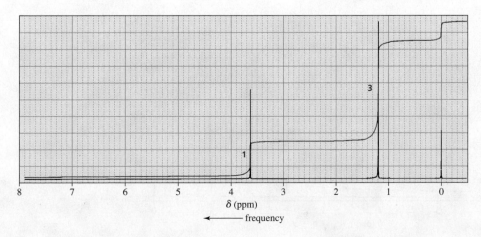

c. $C_4H_7ClO_2$

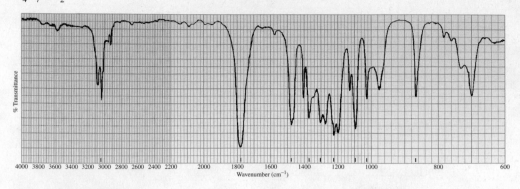

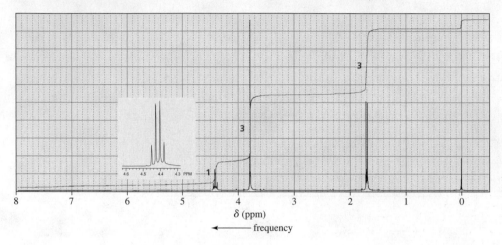

δ (ppm)

← frequency

d. $C_4H_8O_2$

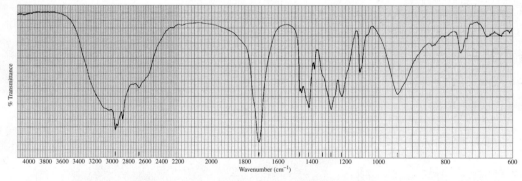

Wavenumber (cm^{-1})

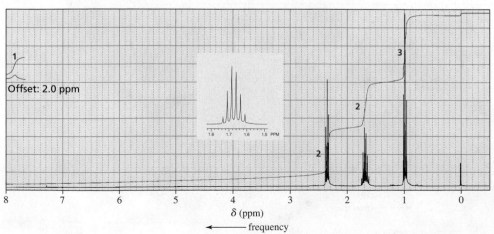

Offset: 2.0 ppm

δ (ppm)

← frequency

63. Determine the structure of each of the following compounds based on its molecular formula and its ^{13}C NMR spectrum:

a. $C_4H_{10}O$

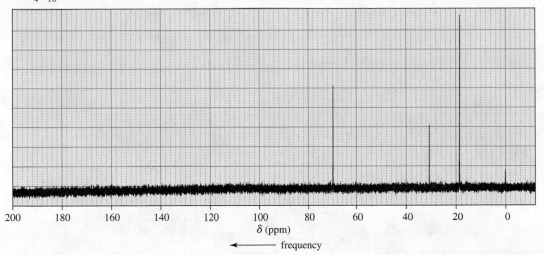

b. $C_6H_{12}O$

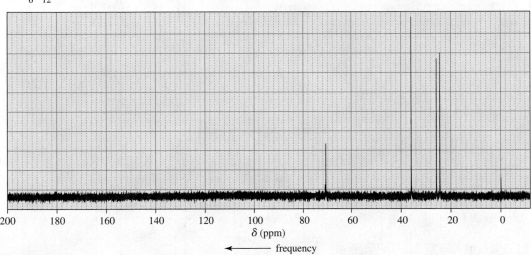

64. The ^1H NMR spectrum of 2-propen-1-ol is shown here. Indicate the protons in the molecule that give rise to each of the signals in the spectrum.

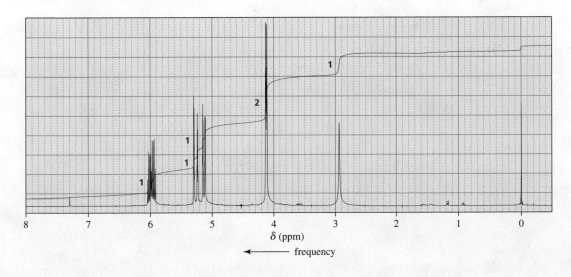

731

65. How can the signals in the 6.5 to 8.1 ppm region of their ^{1}H NMR spectra distinguish the following compounds?

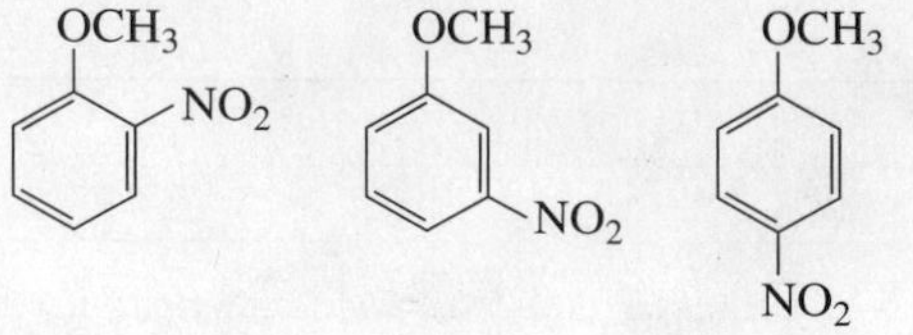

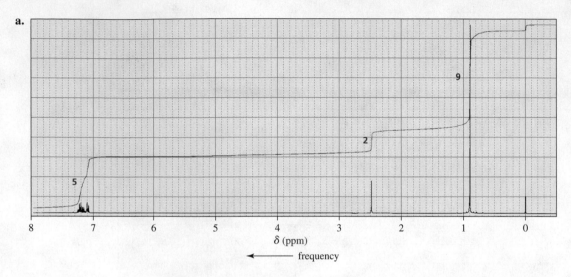

66. The ^{1}H NMR spectra of two compounds, each with molecular formula $C_{11}H_{16}$, are shown here. Identify the compounds.

a.

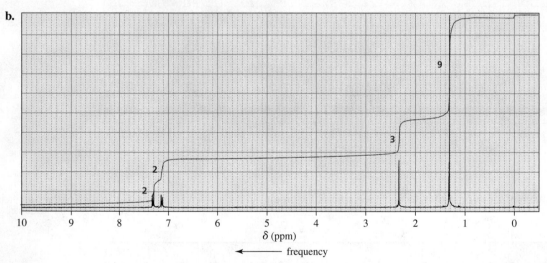

b.

67. Draw a splitting diagram for the H_b proton if $J_{bc} = 10$ and $J_{ba} = 5$.

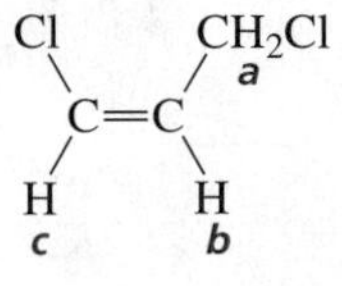

68. Sketch the following spectra that would be obtained for 2-chloroethanol:

    a. The ^{1}H NMR spectrum for an anhydrous sample of the alcohol.
    b. The ^{1}H NMR spectrum for a sample of the alcohol that contains a trace amount of acid.
    c. The ^{13}C NMR spectrum.
    d. The proton-coupled ^{13}C NMR spectrum.
    e. The four parts of a DEPT ^{13}C NMR spectrum.

69. How can ^{1}H NMR be used to prove that the addition of HBr to propene follows the rule that says that the electrophile adds to the sp^2 carbon bonded to the greater number of hydrogens?

70. Identify each of the following compounds from its molecular formula and its ^{1}H NMR spectrum:

    a. C_8H_8

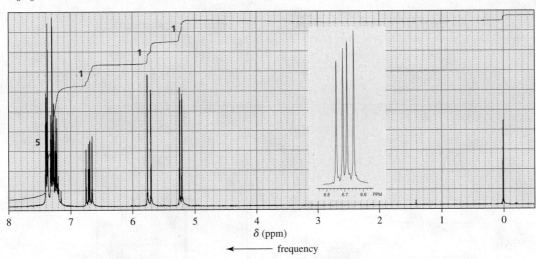

    b. $C_6H_{12}O$

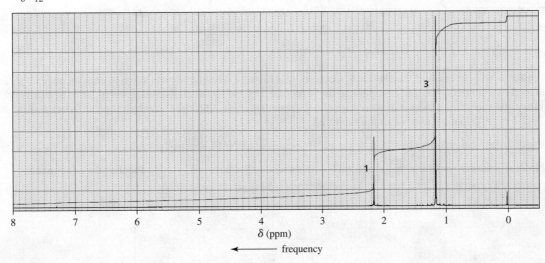

c. $C_9H_{18}O$

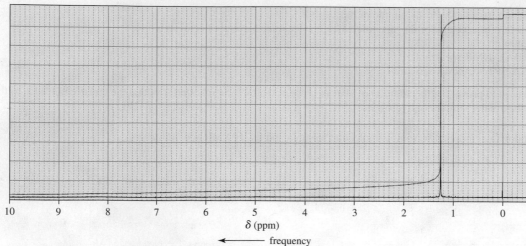

d. C_4H_8O

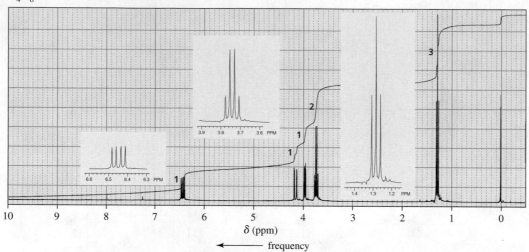

71. Dr. N. M. Arr was called in to help analyze the ^1H NMR spectrum of a mixture of compounds known to contain only C, H, and Br. The mixture showed two singlets—one at 1.8 ppm and the other at 2.7 ppm—with relative integrals of 1 : 6, respectively. Dr. Arr determined that the spectrum was that of a mixture of bromomethane and 2-bromo-2-methylpropane. What was the ratio of bromomethane to 2-bromo-2-methylpropane in the mixture?

72. Calculate the amount of energy (in calories) required to flip an ^1H nucleus in an NMR spectrometer that operates at 300 MHz.

73. The following ^1H NMR spectra are for four compounds, each with molecular formula of $C_6H_{12}O_2$. Identify the compounds.

a.

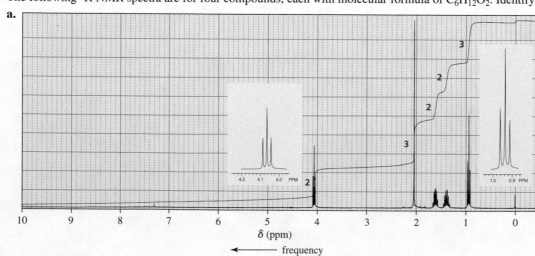

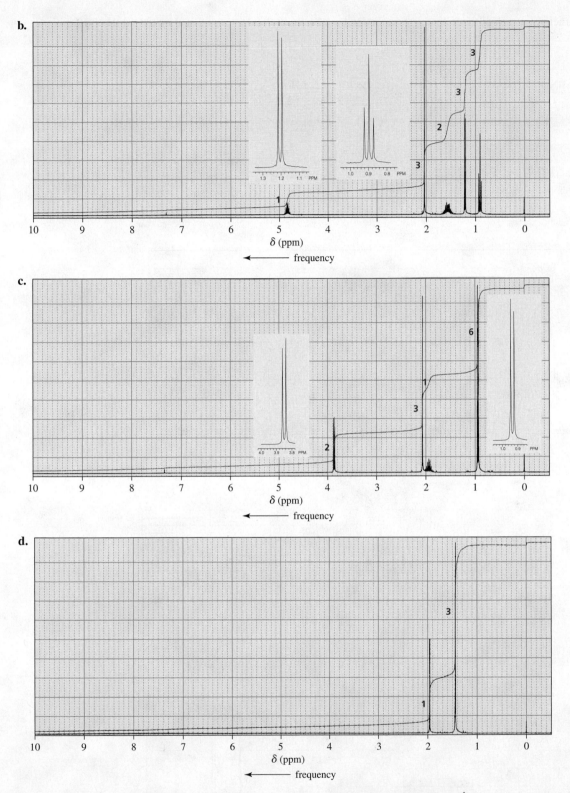

b.

c.

d.

74. When compound **A** ($C_5H_{12}O$) is treated with HBr, it forms compound **B** ($C_5H_{11}Br$). The ^{1}H NMR spectrum of compound **A** has one singlet (1), two doublets (3, 6), and two multiplets (both 1). (The relative areas of the signals are indicated in parentheses.) The ^{1}H NMR spectrum of compound **B** has a singlet (6), a triplet (3), and a quartet (2). Identify compounds **A** and **B**.

75. Determine the structure of each of the following compounds, based on its molecular formula and its IR and ^1H NMR spectra:

a. $C_6H_{12}O$

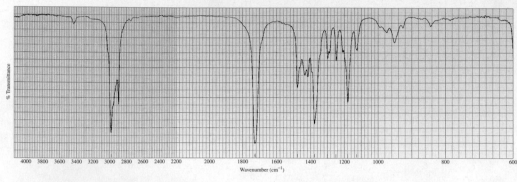

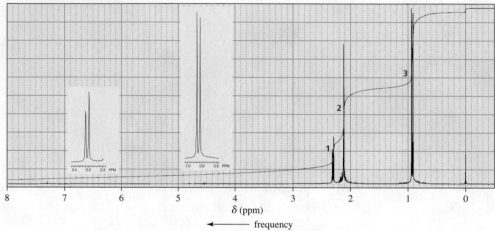

b. $C_6H_{14}O$

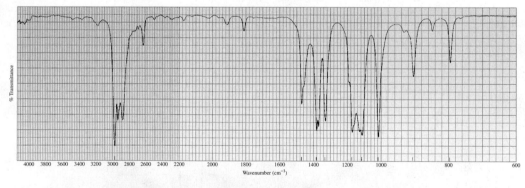

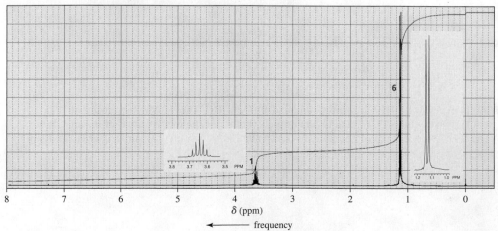

c. $C_{10}H_{13}NO_3$

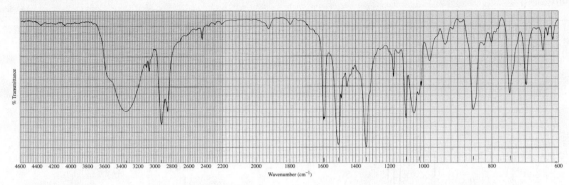

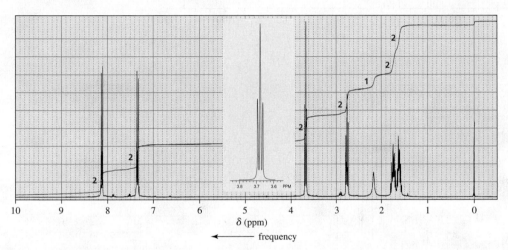

d. $C_{11}H_{14}O_2$

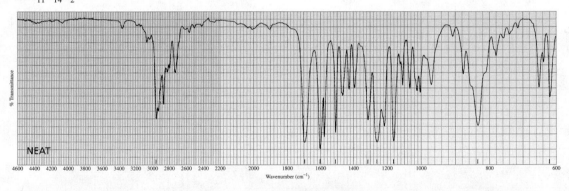

NEAT

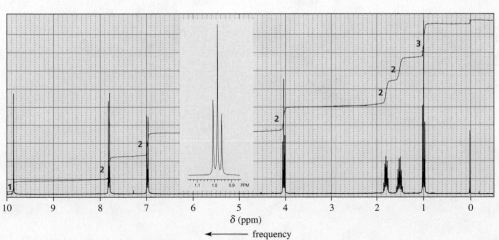

76. Determine the structure of each of the following compounds, based on the compound's mass spectrum, IR spectrum, and ^{1}H NMR spectrum:

a.

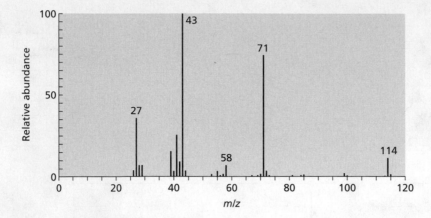

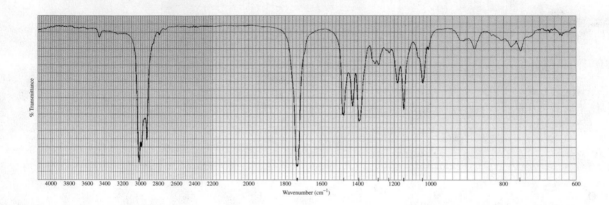

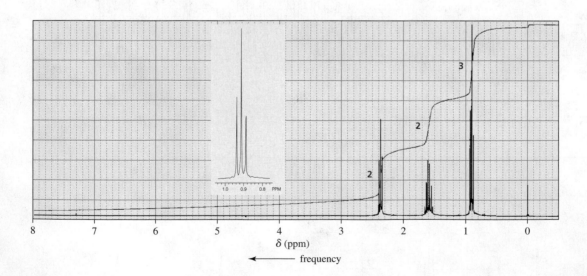

b.

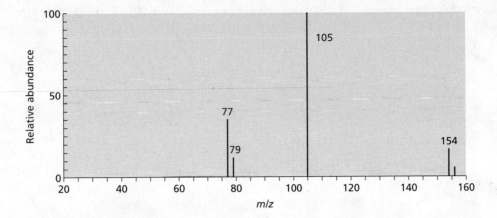

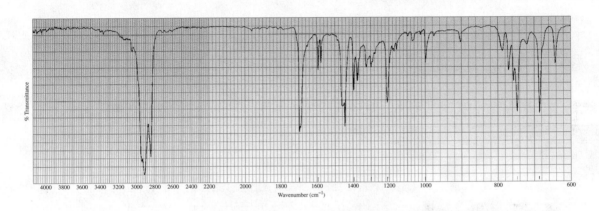

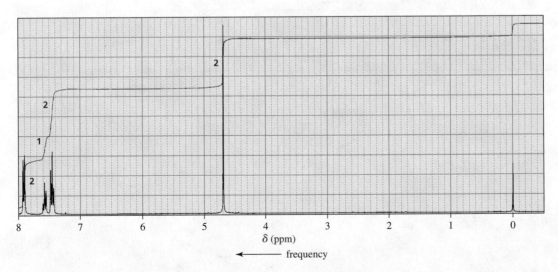

77. Identify the compound with molecular formula $C_6H_{10}O$ that is responsible for the following DEPT ^{13}C NMR spectrum:

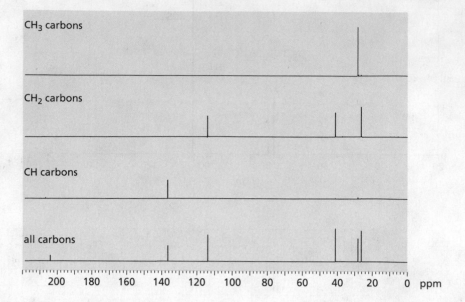

78. Identify the compound with molecular formula C_6H_{14} that is responsible for the following ^1H NMR spectrum:

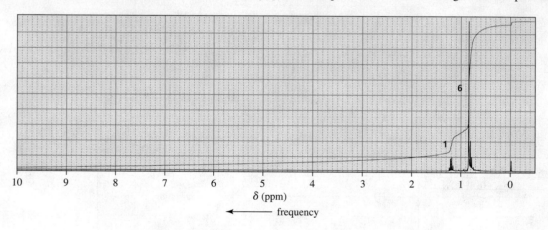

GLOSSARY

applied magnetic field the externally applied magnetic field.

chemically equivalent protons protons with the same connectivity relationship to the rest of the molecule.

chemical shift the location of a signal in an NMR spectrum. It is measured downfield from a reference compound (most often, TMS).

COSY spectrum a 2-D NMR spectrum that shows coupling between sets of protons.

coupled protons protons that split each other. Coupled protons have the same coupling constant.

coupling constant the distance (in hertz) between two adjacent peaks of a split NMR signal.

DEPT ^{13}C NMR spectrum a series of four spectra that distinguishes among —CH$_3$, —CH$_2$, and —CH groups.

diastereotopic hydrogens two hydrogens bonded to a carbon that when replaced in turn with a deuterium, result in a pair of diastereomers.

doublet an NMR signal split into two peaks.

doublet of doublets an NMR signal split into four peaks of approximately equal height. Caused by splitting a signal into a doublet by one hydrogen and into another doublet by another (nonequivalent) hydrogen.

effective magnetic field the magnetic field that a proton "senses" through the surrounding cloud of electrons.

enantiotopic hydrogens two hydrogens bonded to a carbon that is bonded to two other groups that are nonidentical.

Fourier transform NMR a technique in which all the nuclei are excited simultaneously by an rf pulse, their relaxation is monitored, and the data are mathematically converted to a spectrum.

geminal coupling the mutual splitting of two nonidentical protons bonded to the same carbon.

HETCOR spectrum a 2-D NMR spectrum that shows coupling between protons and the carbons to which they are attached.

leaning when a line drawn over the outside peaks of an NMR signal points in the direction of the signal given by the protons that cause the splitting.

long-range coupling splitting of a proton by a proton more than three σ bonds away.

magnetic anisotropy the term used to describe the greater freedom of a π electron cloud to move in response to a magnetic field as a consequence of the greater polarizability of π electrons compared with σ electrons.

magnetic resonance imaging (MRI) NMR used in medicine. The difference in the way water is bound in different tissues produces a variation in signal between organs as well as between healthy and diseased tissue.

MRI scanner an NMR spectrometer used in medicine for whole-body NMR.

multiplet an NMR signal split into more than seven peaks.

multiplicity the number of peaks in an NMR signal.

$N + 1$ rule an ^{1}H NMR signal for a hydrogen with N equivalent hydrogens bonded to an adjacent carbon is split into $N + 1$ peaks. A ^{13}C NMR signal for a carbon bonded to N hydrogens is split into $N + 1$ peaks.

NMR spectroscopy the absorption of electromagnetic radiation to determine the structural features of an organic compound. In the case of NMR spectroscopy, it determines the carbon–hydrogen framework.

operating frequency the frequency at which an NMR spectrometer operates.

pro-R hydrogen replacing this hydrogen with deuterium creates an asymmetric center with the R configuration.

pro-S hydrogen replacing this hydrogen with deuterium creates an asymmetric center with the S configuration.

proton-decoupled ^{13}C NMR spectrum a ^{13}C NMR spectrum in which all the signals appear as singlets because there is no coupling between the nucleus and its bonded hydrogens.

quartet an NMR signal split into four peaks.

reference compound a compound added to a sample whose NMR spectrum is to be taken. The positions of the signals in the NMR spectrum are measured from the position of the signal given by the reference compound.

rf radiation radiation in the radiofrequency region of the electromagnetic spectrum.

shielding phenomenon caused by electron donation to the environment of a proton. The electrons shield the proton from the full effect of the applied magnetic field. The more a proton is shielded, the farther to the right its signal appears in an NMR spectrum.

singlet an unsplit NMR signal.

spin–spin coupling the splitting of a signal in an NMR spectrum described by the $N + 1$ rule.

α-spin state nuclei in this spin state have their magnetic moments oriented in the same direction as the applied magnetic field.

β-spin state nuclei in this spin state have their magnetic moments oriented opposite the direction of the applied magnetic field.

splitting diagram a diagram that describes the splitting of a set of protons.

triplet an NMR signal split into three peaks.

X-ray crystallography a technique used to determine the arrangement of atoms within a crystal.

X-ray diffraction a technique used to obtain images used to determine the electron density within a crystal.

PHOTO CREDITS

Photo credits are listed in order of appearance.

Reactions of Carboxylic Acids and Carboxylic Acid Derivatives

Reactions of Carboxylic Acids and Carboxylic Acid Derivatives

Some of the things you will learn in this chapter are the purpose of the large deposit of fat in a whale's head, how aspirin decreases inflammation and fever, why Dalmatians are the only dogs that excrete uric acid, how bacteria become resistant to penicillin, and why young people sleep better than adults.

Families of organic compounds can be placed in one of four groups, and all the families in a group react in similar ways. This chapter begins our discussion of the familes of compounds in Group III—compounds that contain a carbonyl group.

The **carbonyl group** (a carbon doubly bonded to an oxygen) is probably the most important functional group. Compounds containing carbonyl groups—called **carbonyl** ("car-bo-neel") **compounds**—are abundant in nature, and many play important roles in biological processes. Vitamins, amino acids, proteins, hormones, drugs, and flavorings are just a few of the carbonyl compounds that affect us daily. An **acyl group** consists of a carbonyl group attached to an alkyl group (R) or to an aromatic group (Ar), such as benzene.

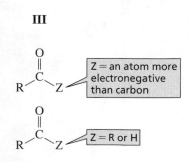

III

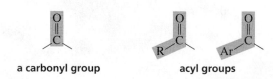

a carbonyl group   acyl groups

The group (or atom) attached to the acyl group strongly affects the reactivity of the carbonyl compound. In fact, carbonyl compounds can be divided into two classes determined by that group. The first class are those in which the acyl group is attached to a group (or atom) that *can* be replaced by another group. Carboxylic acids, acyl halides, esters, and amides belong to this class. All of these compounds contain a group (OH, Cl, OR, NH_2, NHR, NR_2) that can be replaced by a nucleophile.

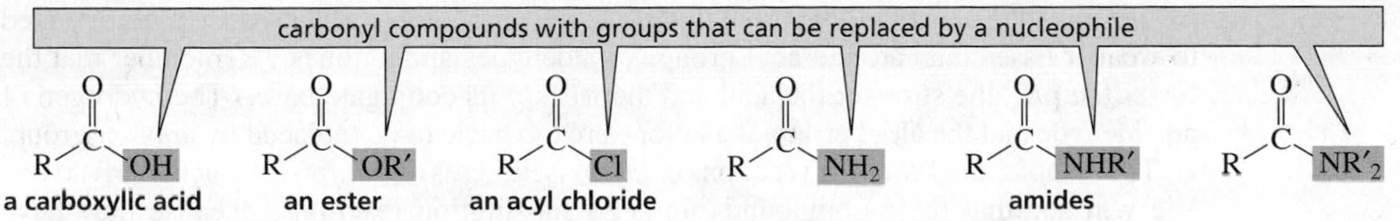

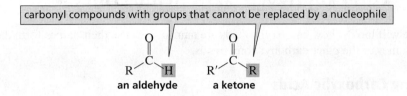

carbonyl compounds with groups that can be replaced by a nucleophile

a carboxylic acid	an ester	an acyl chloride	amides		
R—C(=O)—OH	R—C(=O)—OR′	R—C(=O)—Cl	R—C(=O)—NH₂	R—C(=O)—NHR′	R—C(=O)—NR′₂

Esters, acyl chlorides, and amides are called **carboxylic acid derivatives** because they differ from a carboxylic acid only in the nature of the group or atom that has replaced the OH group of the carboxylic acid.

The second class of carbonyl compounds are those in which the acyl group is attached to a group that *cannot* be readily replaced by another group. Aldehydes and ketones belong to this class. The H bonded to the acyl group of an aldehyde and the R group bonded to the acyl group of a ketone cannot be readily replaced by a nucleophile.

carbonyl compounds with groups that cannot be replaced by a nucleophile

an aldehyde	a ketone
R—C(=O)—H	R′—C(=O)—R

When comparing bases of the same type, weak bases are good leaving groups and strong bases are poor leaving groups. The pK_a values of the conjugate acids of the leaving groups of various carbonyl compounds are listed in Table 1.

Table 1 The pK_a Values of the Conjugate Acids of the Leaving Groups of Carbonyl Compounds

Carbonyl compound	Leaving group	Conjugate acid of the leaving group	pK_a
Carboxylic Acids and Carboxylic Acid Derivatives			
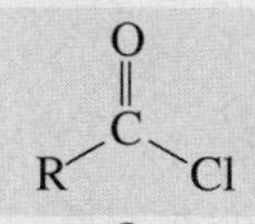 R—C(=O)—Cl	Cl⁻	HCl	-7
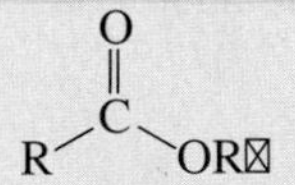 R—C(=O)—OR☒	⁻OR☒	R☒OH	~15–16
R—C(=O)—OH	⁻OH	H₂O	15.7
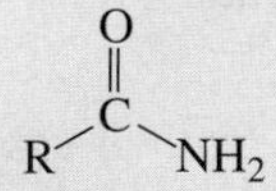 R—C(=O)—NH₂	⁻NH₂	NH₃	36*
Aldehydes and Ketones			
R—C(=O)—H	H⁻	H₂	35
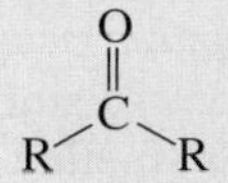 R—C(=O)—R	R⁻	RH	> 60

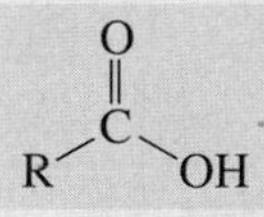

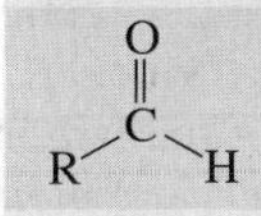

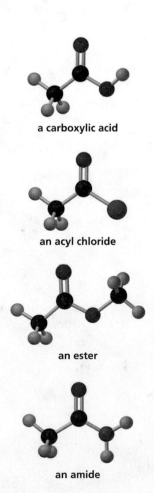

a carboxylic acid

an acyl chloride

an ester

an amide

---

*An amide can undergo substitution reactions only when its leaving group is converted to NH_3, giving its conjugate acid ($^+NH_4$) a pK_a value of 9.4.

Notice that the acyl groups of carboxylic acids and carboxylic acid derivatives are attached to weaker bases than are the acyl groups of aldehydes and ketones. (Remember that the lower the pK_a, the stronger the acid and the weaker its conjugate base.) The hydrogen of an aldehyde and the alkyl group of a ketone are too basic to be replaced by another group.

This chapter discusses the reactions of carboxylic acids and carboxylic acid derivatives. We will see that these compounds undergo substitution reactions, because they have an acyl group attached to a group that can be replaced by a nucleophile. The reactions of aldehydes and ketones are not discussed in this chapter, but note that these compounds *do not* undergo substitution reactions because their acyl group is attached to a group that *cannot* be replaced by a nucleophile.

1 THE NOMENCLATURE OF CARBOXYLIC ACIDS AND CARBOXYLIC ACID DERIVATIVES

First we will look at how carboxylic acids are named, because their names form the basis of the names of the other carbonyl compounds.

Naming Carboxylic Acids

The functional group of a carboxylic acid is called a **carboxyl group.**

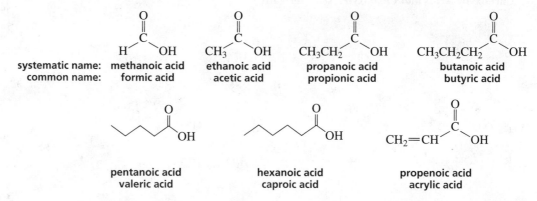

In systematic (IUPAC) nomenclature, a **carboxylic acid** is named by replacing the terminal "e" of the alkane name with "oic acid." For example, the one-carbon alkane is methan*e*, so the one-carbon carboxylic acid is methan*oic acid.*

systematic name:	methanoic acid	ethanoic acid	propanoic acid	butanoic acid
common name:	formic acid	acetic acid	propionic acid	butyric acid

pentanoic acid	hexanoic acid	propenoic acid
valeric acid	caproic acid	acrylic acid

Carboxylic acids containing six or fewer carbons are frequently called by their common names. These names were chosen by early chemists to describe some feature of the compound, usually its origin. For example, formic acid is found in ants, bees, and other stinging insects; its name comes from *formica,* which is Latin for "ant." Acetic acid—contained in vinegar—got its name from *acetum,* the Latin word for "vinegar." Propionic acid is the smallest acid that shows some of the characteristics of the larger fatty acids (Section 4); its name comes from the Greek words *pro* ("the first") and *pion* ("fat"). Butyric acid is found in rancid butter; the Latin word for "butter" is *butyrum.* Valeric acid got its name from *valerian,* an herb that has been used as a sedative since Greco/Roman times. Caproic acid is found in goat's milk. If you have ever smelled a goat, then you know what caproic acid smells like. *Caper* is the Latin word for "goat."

In systematic nomenclature, the position of a substituent is designated by a number. The carbonyl carbon is always the C-1 carbon. In common nomenclature, the position of a substituent is designated by a lowercase Greek letter, and the carbonyl carbon is not given a designation. Thus, the carbon adjacent to the carbonyl carbon is the α-carbon, the carbon adjacent to the α-carbon is the β-carbon, and so on.

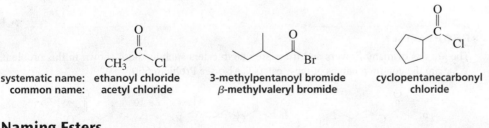

$$CH_3CH_2CH_2CH_2CH_2\overset{\displaystyle O}{\underset{\text{6 \quad 5 \quad 4 \quad 3 \quad 2}}{C}}OH \qquad CH_3CH_2CH_2CH_2CH_2\overset{\displaystyle O}{\underset{\varepsilon \quad \delta \quad \gamma \quad \beta \quad \alpha}{C}}OH$$

| systematic nomenclature | common nomenclature |

α = alpha
β = beta
γ = gamma
δ = delta
ε = epsilon

Take a careful look at the following examples to make sure that you understand the difference between systematic (IUPAC) and common nomenclature:

systematic name: 2-methoxybutanoic acid	3-bromopentanoic acid	4-chlorohexanoic acid
common name: α-methoxybutyric acid	β-bromovaleric acid	γ-chlorocaproic acid

Carboxylic acids in which a carboxyl group is attached to a ring are named by adding "carboxylic acid" to the name of the cyclic compound.

cyclohexanecarboxylic acid	benzenecarboxylic acid benzoic acid	1,2-benzenedicarboxylic acid

Naming Acyl Chlorides

Acyl chlorides have a Cl in place of the OH group of a carboxylic acid. Acyl chlorides are named by replacing "ic acid" of the acid name with "yl chloride." For cyclic acids that end with "carboxylic acid," "carboxylic acid" is replaced with "carbonyl chloride." (Acyl bromides exist too, but are less common than acyl chlorides.)

systematic name: ethanoyl chloride	3-methylpentanoyl bromide	cyclopentanecarbonyl
common name: acetyl chloride	β-methylvaleryl bromide	chloride

Naming Esters

An **ester** has an OR group in place of the OH group of a carboxylic acid. In naming an ester, the name of the group (R′) attached to the **carboxyl oxygen** is stated first, followed by the name of the acid, with "ic acid" replaced by "ate." (The prime on R′ indicates that the alkyl group it designates does not have to be the same as the alkyl group designated by R.) Recall the difference between a phenyl group and a benzyl group.

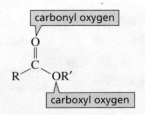

carbonyl oxygen

carboxyl oxygen

The double-bonded oxygen is the carbonyl oxygen; the single-bonded oxygen is the carboxyl oxygen.

systematic name: ethyl ethanoate	phenyl propanoate	methyl 3-bromobutanoate	ethyl cyclohexanecarboxylate
common name: ethyl acetate	phenyl propionate	methyl β-bromobutyrate	

Salts of carboxylic acids are named in the same way. That is, the cation is named first, followed by the name of the acid, again with "ic acid" replaced by "ate."

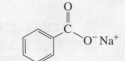

systematic name:	sodium methanoate	potassium ethanoate	sodium benzenecarboxylate
common name:	sodium formate	potassium acetate	sodium benzoate

Frequently, the name of the cation is omitted.

acetate   pyruvate   (S)-(+)-lactate

Cyclic esters are called **lactones.** In systematic nomenclature, they are named as "2-oxacycloalkanones" ("oxa" designates the oxygen atom.) For their common names, the length of the carbon chain is indicated by the common name of the carboxylic acid, and a Greek letter specifies the carbon to which the oxygen is attached. Thus, six-membered ring lactones are δ-lactones (the carboxyl oxygen is on the δ-carbon), five-membered ring lactones are γ-lactones, and four-membered ring lactones are β-lactones.

2-oxacyclopentanone	2-oxacyclohexanone	3-methyl-2-oxacyclohexanone	3-ethyl-2-oxacyclopentanone
γ-butyrolactone	δ-valerolactone	δ-caprolactone	γ-caprolactone
a γ-lactone	a δ-lactone	a δ-lactone	a γ-lactone

PROBLEM 1♦

The aromas of many flowers and fruits are due to esters such as those shown in this problem. What are the common names of these esters? (Also see Problem 66.)

a.   b.   c.

jasmine   banana   apple

α-Hydroxycarboxylic acids are found in skin products that claim to reduce wrinkles by penetrating the top layer of the skin, causing it to flake off.

PROBLEM 2

The word "lactone" has its origin in lactic acid, a three-carbon carboxylic acid with an OH group on the α-carbon. Ironically, lactic acid (for its structure, see the structure of lactate near the top of this page) cannot form a lactone. Why not?

Naming Amides

An **amide** has an NH₂, NHR, or NR₂ group in place of the OH group of a carboxylic acid. Amides are named by replacing "oic acid," "ic acid," or "ylic acid" of the acid name with "amide."

systematic name:	ethanamide	4-chlorobutanamide	benzenecarboxamide
common name:	acetamide	γ-chlorobutyramide	benzamide

If a substituent is bonded to the nitrogen, the name of the substituent is stated first (if there is more than one substituent bonded to the nitrogen, they are stated alphabetically), followed by the name of the amide. The name of each substituent is preceded by an *N* to indicate that the substituent is bonded to a nitrogen.

N-cyclohexylpropanamide   **N-ethyl-N-methylpentanamide**   **N,N-diethylbutanamide**

Cyclic amides are called **lactams.** Their nomenclature is similar to that of lactones. In systematic nomenclature, they are named as "2-azacycloalkanones" ("aza" designates the nitrogen atom). For their common names, the length of the carbon chain is indicated by the common name of the carboxylic acid, and a Greek letter specifies the carbon to which the nitrogen is attached.

2-azacyclohexanone	2-azacyclopentanone	2-azacyclobutanone
δ-valerolactam	γ-butyrolactam	β-propiolactam
a δ-lactam	a γ-lactam	a β-lactam

Nature's Sleeping Pill

Melatonin, a naturally occurring amide, is a hormone synthesized by the pineal gland from the amino acid tryptophan. An amino acid is an α-aminocarboxylic acid. Melatonin regulates the dark–light clock in our brains that governs such things as the sleep–wake cycle, body temperature, and hormone production.

Melatonin levels increase from evening to night and then decrease as morning approaches. People with high levels of melatonin sleep longer and more soundly than those with low levels. The concentration of the hormone in our bodies varies with age—6-year-olds have more than five times the concentration that 80-year-olds have—which is one of the reasons young people have less trouble sleeping than older people. Melatonin supplements are used to treat insomnia, jet lag, and seasonal affective disorder.

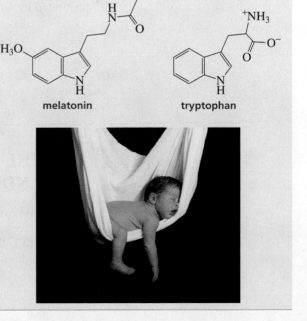

melatonin   tryptophan

PROBLEM 3♦

Name the following compounds:

a.

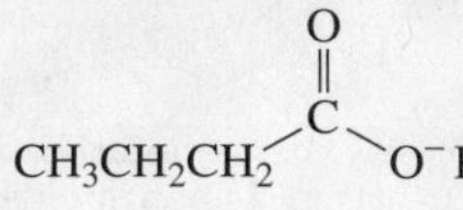

$$CH_3CH_2CH_2 - \overset{\displaystyle O}{\overset{\displaystyle \|}{C}} - O^- \; K^+$$

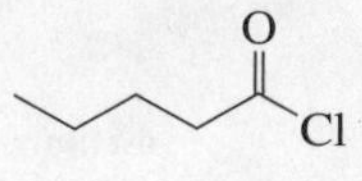

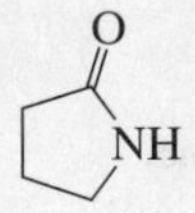

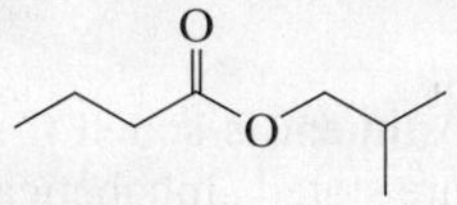

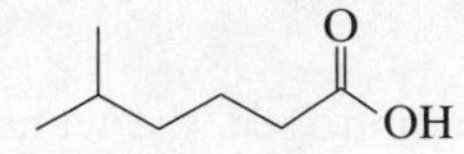

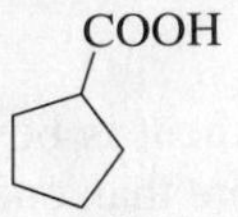

b.

c.

d.

e.

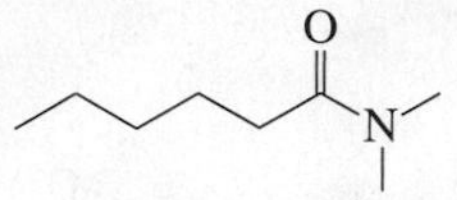

f.

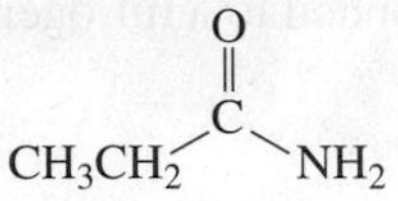

$$CH_3CH_2 - \overset{\displaystyle O}{\overset{\displaystyle \|}{C}} - NH_2$$

g.

h. COOH

i. 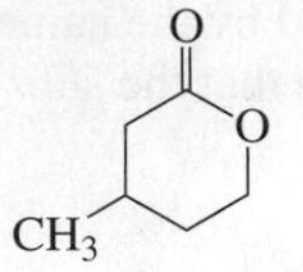 CH₃

PROBLEM 4

Draw the structure of each of the following:

a. phenyl acetate

b. γ-caprolactam

c. N-benzylethanamide

d. γ-methylcaproic acid

e. ethyl 2-chloropentanoate

f. β-bromobutyramide

g. cyclohexanecarbonyl chloride

h. α-chlorovaleric acid

Derivatives of Carbonic Acid

Carbonic acid—a compound with two OH groups bonded to a carbonyl carbon—is unstable, readily breaking down to CO_2 and H_2O. The reaction is reversible, so carbonic acid is formed when CO_2 is bubbled into water.

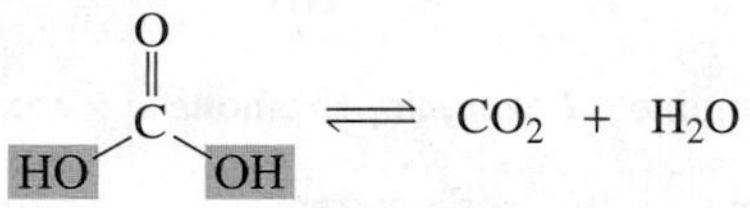

$$\underset{\textbf{carbonic acid}}{HO - \overset{\displaystyle O}{\overset{\displaystyle \|}{C}} - OH} \;\; \rightleftharpoons \;\; CO_2 \;+\; H_2O$$

The OH groups of carbonic acid, just like the OH group of a carboxylic acid can be substituted by other groups.

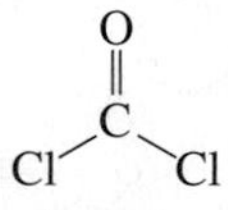

$Cl-\overset{O}{\overset{\|}{C}}-Cl$	$CH_3O-\overset{O}{\overset{\|}{C}}-OCH_3$	$H_2N-\overset{O}{\overset{\|}{C}}-NH_2$	$H_2N-\overset{O}{\overset{\|}{C}}-OH$	$H_2N-\overset{O}{\overset{\|}{C}}-OCH_3$
phosgene	**dimethyl carbonate**	**urea**	**carbamic acid**	**methyl carbamate**

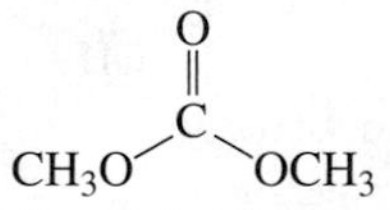

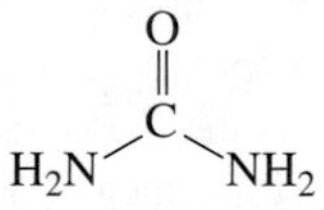

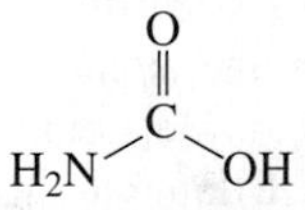

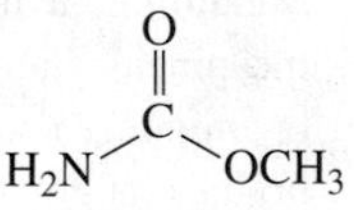

2 THE STRUCTURES OF CARBOXYLIC ACIDS AND CARBOXYLIC ACID DERIVATIVES

The **carbonyl carbon** in carboxylic acids and carboxylic acid derivatives is sp^2 hybridized. It uses its three sp^2 orbitals to form σ bonds to the carbonyl oxygen, the α-carbon, and a substituent (Y). The three atoms attached to the carbonyl carbon are in the same plane, and the bond angles are each approximately 120°.

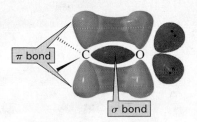

▲ Figure 1

Bonding in a carbonyl group. The π bond is formed by the side-to-side overlap of a p orbital of carbon with a p orbital of oxygen.

The **carbonyl oxygen** is also sp^2 hybridized. One of its sp^2 orbitals forms a σ bond with the carbonyl carbon, and each of the other two sp^2 orbitals contains a lone pair. The remaining p orbital of the carbonyl oxygen overlaps the remaining p orbital of the carbonyl carbon to form a π bond (Figure 1).

Esters, carboxylic acids, and amides each have two resonance contributors. The resonance contributor on the right makes an insignificant contribution to an acyl chloride (Section 6), so it is not shown here.

The resonance contributor on the right makes a greater contribution to the hybrid in the amide than in the ester or the carboxylic acid, because the amide's resonance contributor is more stable. It is more stable because nitrogen is less electronegative than oxygen, so nitrogen can better accommodate a positive charge.

PROBLEM 5♦

Which is a correct statement?

A The delocalization energy of an ester is about 18 kcal/mol, and the delocalization energy of an amide is about 10 kcal/mol.

B The delocalization energy of an ester is about 10 kcal/mol, and the delocalization energy of an amide is about 18 kcal/mol.

PROBLEM 6♦

Which is longer, the carbon–oxygen single bond in a carboxylic acid or the carbon–oxygen bond in an alcohol? Why?

PROBLEM 7♦

There are three carbon–oxygen bonds in methyl acetate.

a. What are their relative lengths?
b. What are the relative infrared (IR) stretching frequencies of these bonds?

PROBLEM 8♦

Match the compound to the appropriate carbonyl IR absorption band:

acyl chloride 1800 cm^{-1}
ester 1640 cm^{-1}
amide 1730 cm^{-1}

3 THE PHYSICAL PROPERTIES OF CARBONYL COMPOUNDS

Carboxylic acids have pK_a values of approximately 5. Carbonyl compounds have the following relative boiling points:

relative boiling points

amide > carboxylic acid > nitrile ≫ ester ~ acyl chloride ~ ketone ~ aldehyde

The boiling points of an ester, acyl chloride, ketone, and aldehyde of comparable molecular weight are similar and are *lower* than the boiling point of an alcohol of similar molecular weight because only the alcohol molecules can form hydrogen bonds with each other. The boiling points of these four carbonyl compounds are *higher* than the boiling point of the same-sized ether because of the dipole–dipole interactions between the polar carbonyl groups.

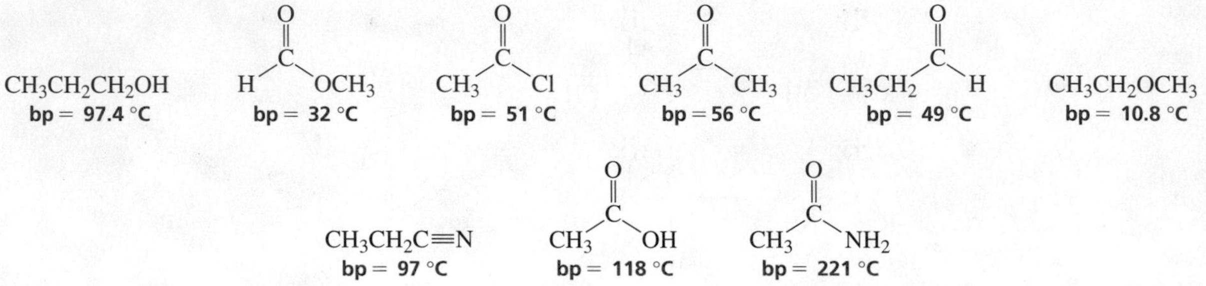

$CH_3CH_2CH_2OH$
bp = 97.4 °C

H–C(=O)–OCH₃
bp = 32 °C

CH₃–C(=O)–Cl
bp = 51 °C

CH₃–C(=O)–CH₃
bp = 56 °C

CH₃CH₂–C(=O)–H
bp = 49 °C

$CH_3CH_2OCH_3$
bp = 10.8 °C

$CH_3CH_2C{\equiv}N$
bp = 97 °C

CH₃–C(=O)–OH
bp = 118 °C

CH₃–C(=O)–NH₂
bp = 221 °C

The strong dipole–dipole interactions of a nitrile give it a boiling point similar to that of an alcohol. Carboxylic acids have relatively high boiling points because each molecule has two groups that can form hydrogen bonds. Amides have the highest boiling points because they have strong dipole–dipole interactions, since the resonance contributor with separated charges contributes significantly to the overall structure of the compound (Section 2). In addition, if the nitrogen of an amide is bonded to a hydrogen, hydrogen bonds can form between the molecules.

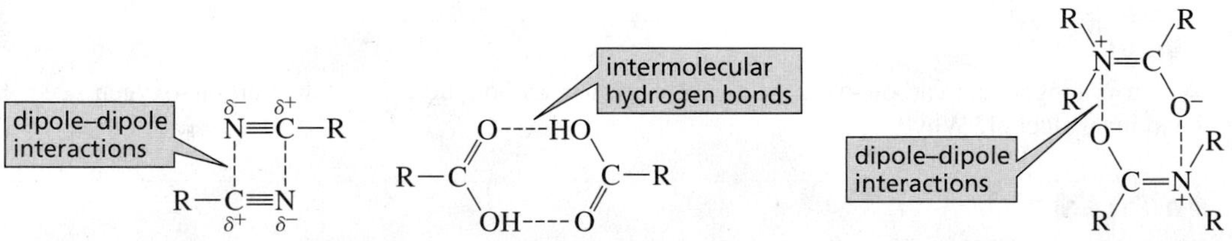

Carboxylic acid derivatives are soluble in solvents such as ethers, chloroalkanes, and aromatic hydrocarbons. Like alcohols and ethers, carbonyl compounds with fewer than four carbons are soluble in water. Tables of physical properties can be found in the Study Area of MasteringChemistryAppendix: "Physical Properties of Organic Compounds."

Esters, *N,N*-disubstituted amides, and nitriles are often used as solvents because they are polar but do not have reactive OH or NH_2 groups. Note that dimethylformamide (DMF) is a common aprotic polar solvent.

4 | FATTY ACIDS ARE LONG-CHAIN CARBOXYLIC ACIDS

Fatty acids are carboxylic acids with long hydrocarbon chains that are found in nature (Table 2). They are unbranched and contain an even number of carbons because they are synthesized from acetate, a compound with two carbons.

Table 2	Common Naturally Occurring Fatty Acids			
Number of carbons	**Common name**	**Systematic name**	**Structure**	**Melting point (°C)**
Saturated				
12	lauric acid	dodecanoic acid	 COOH	44
14	myristic acid	tetradecanoic acid	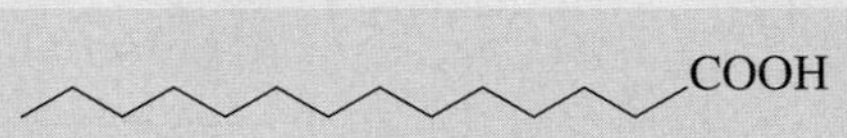 COOH	58
16	palmitic acid	hexadecanoic acid	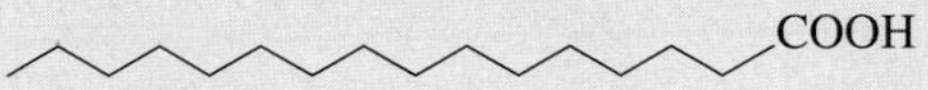 COOH	63
18	stearic acid	octadecanoic acid	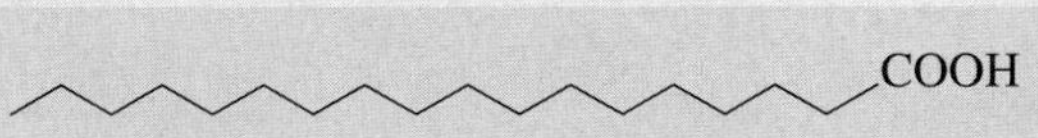 COOH	69
20	arachidic acid	eicosanoic acid	 COOH	77
Unsaturated				
16	palmitoleic acid	(9Z)-hexadecenoic acid	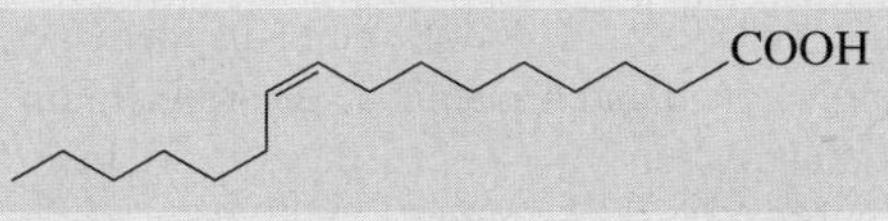 COOH	0
18	oleic acid	(9Z)-octadecenoic acid	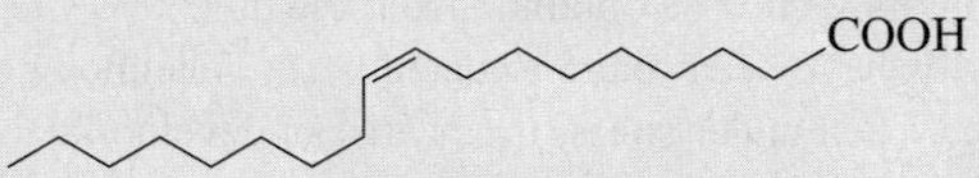 COOH	13
18	linoleic acid	(9Z,12Z)-octadecadienoic acid	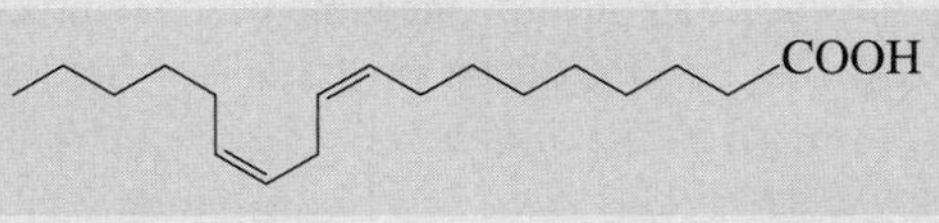 COOH	–5
18	linolenic acid	(9Z,12Z,15Z)-octadecatrienoic acid	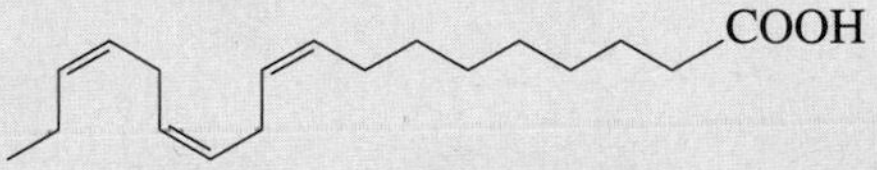 COOH	–11
20	arachidonic acid	(5Z,8Z,11Z,14Z)-eicosatetraenoic acid	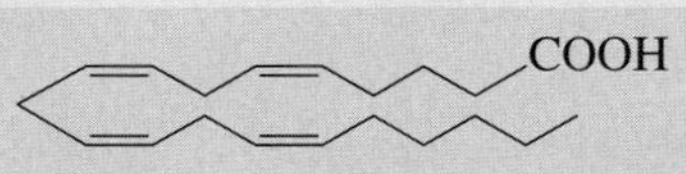 COOH	–50
20	EPA	(5Z,8Z,11Z,14Z,17Z)-eicosapentaenoic acid	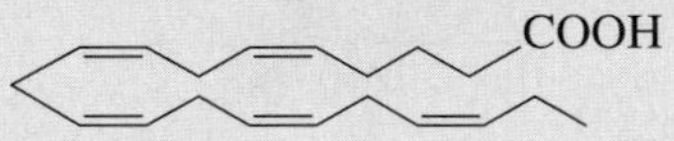 COOH	–50

Fatty acids can be saturated with hydrogen (and therefore have no carbon–carbon double bonds) or unsaturated (and have carbon–carbon double bonds). Fatty acids with more than one double bond are called **polyunsaturated fatty acids.**

The melting points of saturated fatty acids increase with increasing molecular weight because of increased van der Waals interactions between the molecules. The melting points of unsaturated fatty acids with the same number of double bonds also increase with increasing molecular weight (Table 2).

The double bonds in naturally occurring unsaturated fatty acids have the cis configuration and are always separated by one CH_2 group. The cis double bond produces a bend in the molecule, which prevents unsaturated fatty acids from packing together as tightly as saturated fatty acids. As a result, unsaturated fatty acids have fewer intermolecular interactions and therefore have lower melting points than saturated fatty acids with comparable molecular weights (Table 2).

Unsaturated fatty acids have lower melting points than saturated fatty acids.

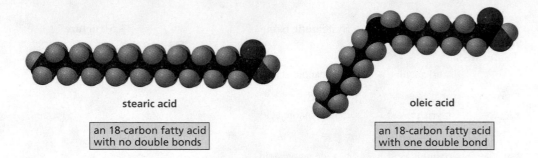

stearic acid

an 18-carbon fatty acid
with no double bonds

oleic acid

an 18-carbon fatty acid
with one double bond

Omega Fatty Acids

Omega indicates the position of the first double bond in an unsaturated fatty acid, counting from the methyl end. For example, linoleic acid is an omega-6 fatty acid because its first double bond is after the sixth carbon, and linolenic acid is an omega-3 fatty acid because its first double bond is after the third carbon. Mammals lack the enzyme that introduces a double bond beyond C-9, counting from the carbonyl carbon. Linoleic acid and linolenic acids are therefore *essential fatty acids* for mammals: mammals cannot synthesize them, but since they are needed for normal body function, they must be obtained from the diet.

Omega-3 fatty acids have been found to decrease the likelihood of sudden death due to a heart attack. When under stress, the heart can develop fatal disturbances in its rhythm. Omega-3 fatty acids are incorporated into cell membranes in the heart and apparently have a stabilizing effect on heart rhythm. These fatty acids are found in fatty fish such as herring, mackerel, and salmon.

Linoleic and linolenic acids are essential fatty acids for mammals.

omega-6
fatty acid linoleic acid

omega-3
fatty acid

linolenic acid

COOH

COOH

PROBLEM 9

Explain the difference in the melting points of the following fatty acids:

a. palmitic acid and stearic acid

b. palmitic acid and palmitoleic acid

PROBLEM 10

What products are formed when arachidonic acid reacts with excess ozone followed by treatment with dimethyl sulfide?

---

5 HOW CARBOXYLIC ACIDS AND CARBOXYLIC ACID DERIVATIVES REACT

The reactivity of carbonyl compounds is due to the polarity of the carbonyl group, which results from oxygen being more electronegative than carbon. The carbonyl carbon is therefore electron deficient (an electrophile), so it reacts with nucleophiles.

When a nucleophile adds to the carbonyl carbon of a carboxylic acid derivative, the weakest bond in the molecule—the π bond—breaks, and an intermediate is formed. It is called a **tetrahedral intermediate** because the sp^2 carbon in the reactant has become an sp^3 carbon in the intermediate.

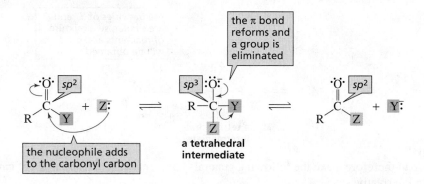

A compound that has an sp^3 carbon bonded to an oxygen atom generally will be unstable if the sp^3 carbon is bonded to another electronegative atom.

The tetrahedral compound is an intermediate rather than a final product because it is not stable. Generally, *a compound that has an* sp^3 *carbon bonded to an oxygen atom will be unstable if the* sp^3 *carbon is bonded to another electronegative atom.* The tetrahedral intermediate, therefore, is unstable because Y and Z are both electronegative atoms. A lone pair on the oxygen re-forms the π bond, and either Y^- or Z^- is eliminated along with its bonding electrons. (Here we show Y^- being eliminated.)

Whether Y^- or Z^- is eliminated from the tetrahedral intermediate depends on their relative basicities. The weaker base is eliminated preferentially—remember, when comparing bases of the same type, *the weaker base is a better leaving group.* Because a weak base does not share its electrons as well as a strong base does, a weaker base forms a weaker bond—one that is easier to break.

If Z^- is a much weaker base than Y^-, then Z^- will be eliminated.

The weaker the base, the better it is as a leaving group.

In this case, no new product is formed. The nucleophile adds to the carbonyl carbon, but the tetrahedral intermediate eliminates the nucleophile and re-forms the reactants.

On the other hand, if Y^- is a much weaker base than Z^-, then Y^- will be eliminated and a new product will be formed.

Y⁻ is a weaker base than Z⁻, so Y⁻ is eliminated and the products are formed

$$R-C-Y + Z: \rightleftharpoons R-C-Y \rightleftharpoons R-C-Z + Y:$$

a tetrahedral
intermediate

This reaction is a **nucleophilic acyl substitution reaction** because a nucleophile (Z^-) has replaced the substituent (Y^-) that was attached to the acyl group in the reactant. It is also called an **acyl transfer reaction** because an acyl group has been transferred from one group to another. Most chemists, however, prefer to call it a **nucleophilic addition–elimination reaction** to emphasize the two-step nature of the reaction: a nucleophile adds to the carbonyl carbon in the first step, and a group is eliminated in the second step.

If the basicities of Y^- and Z^- are similar, some molecules of the tetrahedral intermediate will eliminate Y^- and others will eliminate Z^-. When the reaction is over, both the reactant and the product will be present.

the basicities of Y^- and Z^- are similar, so a mixture of reactants and products will be obtained

a tetrahedral
intermediate

We can therefore make the following general statement about the reactions of carboxylic acid derivatives:

> *A carboxylic acid derivative will undergo a nucleophilic addition–elimination reaction, provided that the newly added group in the tetrahedral intermediate is not a much weaker base than the group attached to the acyl group in the reactant.*

Let's compare this two-step addition–elimination reaction with a one-step S_N2 reaction. When a nucleophile attacks a carbon, the weakest bond in the molecule breaks. The weakest bond in an S_N2 reaction is the bond to the leaving group, so this is the bond that breaks in the first and only step of the reaction. In contrast, the weakest bond in an addition–elimination reaction is the π bond, so this bond breaks first and the leaving group is eliminated in a subsequent step.

$$CH_3CH_2-Y + Z: \longrightarrow CH_3CH_2-Z + Y:$$

an S_N2 reaction

Let's now look at a molecular orbital description of how carbonyl compounds react. It's important to note that because oxygen is more electronegative than carbon, the $2p$ orbital of oxygen contributes more to the π bonding molecular orbital (it is closer to it in energy) and the $2p$ orbital of carbon contributes more to the π^* antibonding molecular orbital. As a result, the π^* antibonding orbital is largest at the carbon atom, so that is where the nucleophile's nonbonding orbital, in which the lone pair resides, overlaps. This allows the greatest amount of orbital overlap, and greater overlap means greater stability. When two orbitals overlap, the result is a molecular orbital—in this case, a σ molecular orbital—that is more stable than either of the overlapping orbitals (Figure 2).

A carboxylic acid derivative will undergo a nucleophilic addition–elimination reaction if the newly added group in the tetrahedral intermediate is not a much weaker base than the group attached to the acyl group in the reactant.

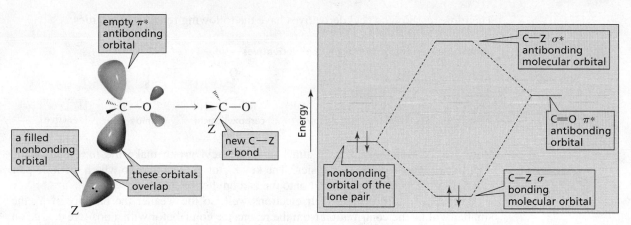

▲ Figure 2

The filled nonbonding orbital containing the nucleophile's lone pair overlaps the empty π^* antibonding orbital of the carbonyl group, forming the new σ bond in the tetrahedral intermediate.

PROBLEM-SOLVING STRATEGY

Using Basicity to Predict the Outcome of a Nucleophilic Addition–Elimination Reaction

What is the product of the reaction of acetyl chloride with CH_3O^-? The pK_a of HCl is -7; the pK_a of CH_3OH is 15.5.

To identify the product of the reaction, we need to compare the basicities of the two groups in the tetrahedral intermediate so that we can determine which one will be eliminated. Because HCl is a stronger acid than CH_3OH, Cl^- is a weaker base than CH_3O^-. Therefore, Cl^- will be eliminated from the tetrahedral intermediate and methyl acetate will be the product of the reaction.

acetyl chloride   **methyl acetate**

Now use the strategy you have just learned to solve Problem 11.

PROBLEM 11♦

a. What is the product of the reaction of acetyl chloride with HO^-? The pK_a of HCl is -7; the pK_a of H_2O is 15.7.

b. What is the product of the reaction of acetamide with HO^-? The pK_a of NH_3 is 36; the pK_a of H_2O is 15.7.

---

6   THE RELATIVE REACTIVITIES OF CARBOXYLIC ACIDS AND CARBOXYLIC ACID DERIVATIVES

We have just seen that there are two steps in a nucleophilic addition–elimination reaction: *formation* of a tetrahedral intermediate and *collapse* of the tetrahedral intermediate. The weaker the base attached to the acyl group (Table 1), the easier it is for *both steps* of the reaction to take place.

relative basicities of the leaving groups

$$\text{weakest base} \quad Cl^- \ < \ ^-OR \ \approx \ ^-OH \ < \ ^-NH_2 \quad \text{strongest base}$$

Therefore, carboxylic acid derivatives have the following relative reactivities:

relative reactivities of carboxylic acid derivatives

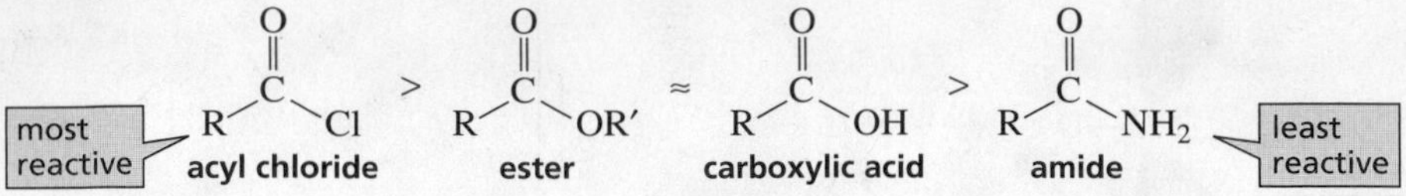

How does having a weak base attached to the acyl group make the *first* step of the addition–elimination reaction easier? The key factor is the extent to which the lone-pair electrons on Y can be delocalized onto the carbonyl oxygen.

Weak bases do not share their electrons well, so the weaker the basicity of Y, the smaller will be the contribution from the resonance contributor with a positive charge on Y. In addition, when Y = Cl, delocalization of chlorine's lone pair is minimal due to the poor orbital overlap between the large $3p$ orbital on chlorine and the smaller $2p$ orbital on carbon. The less the contribution from the resonance contributor with the positive charge on Y, the more electrophilic the carbonyl carbon. Thus, weak bases cause the carbonyl carbon to be more electrophilic and, therefore, more reactive toward nucleophiles.

relative reactivity: acyl chloride > ester ~ carboxylic acid > amide

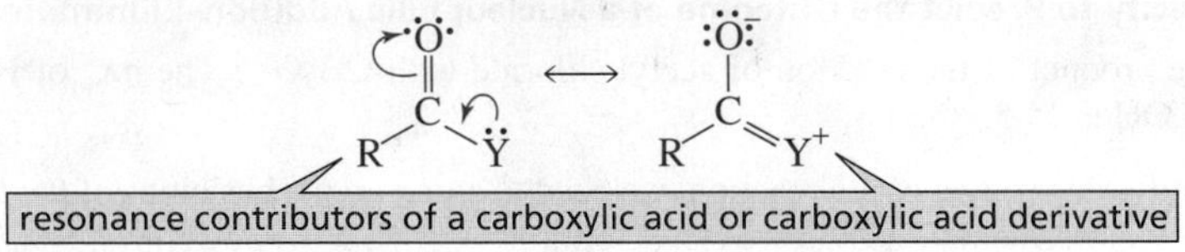

resonance contributors of a carboxylic acid or carboxylic acid derivative

---

PROBLEM 12◆

a. Which compound will have the stretching vibration for its carbonyl group at the highest frequency: acetyl chloride, methyl acetate, or acetamide?

b. Which one will have the stretching vibration for its carbonyl group at the lowest frequency?

A weak base attached to the acyl group also makes the *second* step of the addition–elimination reaction easier, because weak bases are easier to eliminate when the tetrahedral intermediate collapses.

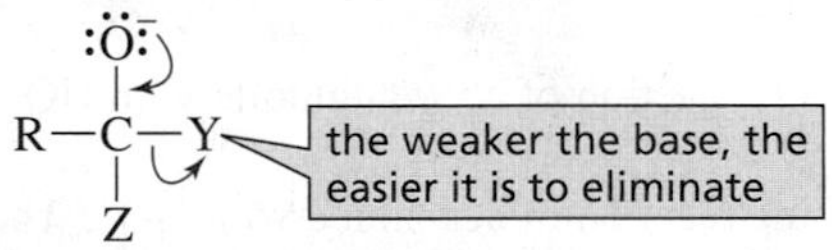

the weaker the base, the easier it is to eliminate

In Section 5 we saw that in a nucleophilic addition–elimination reaction, the nucleophile that adds to the carbonyl carbon must be a stronger base than the substituent that is attached to the acyl group. This means that *a carboxylic acid derivative can be converted into a less reactive carboxylic acid derivative in a nucleophilic addition–elimination reaction, but not into one that is more reactive.* For example, an acyl chloride can be converted into an ester because an alkoxide ion is a stronger base than a chloride ion.

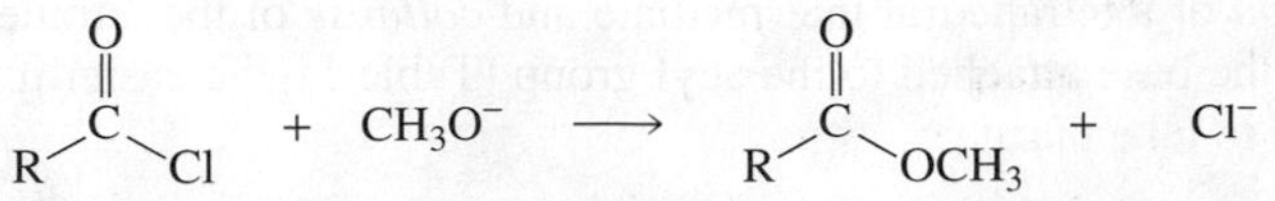

An ester, however, cannot be converted into an acyl chloride because a chloride ion is a weaker base than an alkoxide ion.

$$\underset{\substack{R \quad OCH_3 \\ \\ }}{\overset{\displaystyle O}{\underset{\displaystyle \|}{C}}} + Cl^- \longrightarrow \text{ no reaction}$$

Reaction coordinate diagrams for nucleophilic addition–elimination reactions with nucleophiles of varying basicity are shown in Figure 3 (where TI is the tetrahedral intermediate).

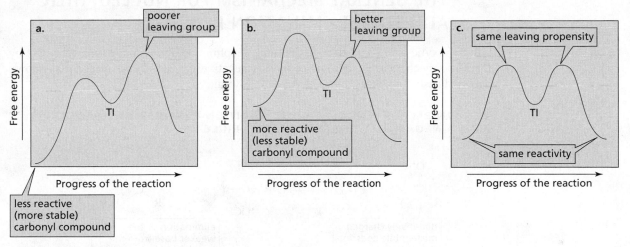

▲ **Figure 3**

(a) The nucleophile is a weaker base than the group attached to the acyl group in the reactant.

(b) The nucleophile is a stronger base than the group attached to the acyl group in the reactant.

(c) The nucleophile and the group attached to the acyl group in the reactant have similar basicities.

1. To synthesize a more reactive compound from a less reactive compound, the new group in the tetrahedral intermediate will have to be a weaker base than the group attached to the acyl group in the reactant. The lower energy pathway will be for the tetrahedral intermediate (TI) to eliminate the newly added group and re-form the reactants, so no reaction takes place (Figure 3a).

2. To synthesize a less reactive compound from a more reactive compound, the new group in the tetrahedral intermediate will have to be a stronger base than the group attached to the acyl group in the reactant. The lower energy pathway will be for the tetrahedral intermediate (TI) to eliminate the group attached to the acyl group in the reactant and form a substitution product (Figure 3b).

3. If the reactant and product have similar reactivities, then both groups in the tetrahedral intermediate will have similar basicities. In this case, the tetrahedral intermediate can eliminate either group with similar ease, so a mixture of the reactant and the substitution product will result (Figure 3c).

PROBLEM 13♦

Using the pK_a values listed in Table 1, predict the products of the following reactions:

a. $\underset{\substack{CH_3 \quad OCH_3}}{\overset{\displaystyle O}{\underset{\displaystyle \|}{C}}} + NaCl \longrightarrow$ c. $\underset{\substack{CH_3 \quad NH_2}}{\overset{\displaystyle O}{\underset{\displaystyle \|}{C}}} + NaCl \longrightarrow$

b. $\underset{\substack{CH_3 \quad Cl}}{\overset{\displaystyle O}{\underset{\displaystyle \|}{C}}} + NaOH \longrightarrow$ d. $\underset{\substack{CH_3 \quad NH_2}}{\overset{\displaystyle O}{\underset{\displaystyle \|}{C}}} + NaOH \longrightarrow$

PROBLEM 14♦

Is the following statement true or false?

If the newly added group in the tetrahedral intermediate is a stronger base than the group attached to the acyl group in the reactant, then formation of the tetrahedral intermediate is the rate-limiting step of a nucleophilic addition–elimination reaction.

7 THE GENERAL MECHANISM FOR NUCLEOPHILIC ADDITION–ELIMINATION REACTIONS

All carboxylic acid derivatives undergo nucleophilic addition–elimination reactions by the same mechanism. If the nucleophile is negatively charged, the mechanism shown here and described later in this chapter is followed:

MECHANISM FOR A NUCLEOPHILIC ADDITION–ELIMINATION REACTION WITH A NEGATIVELY CHARGED NUCLEOPHILE

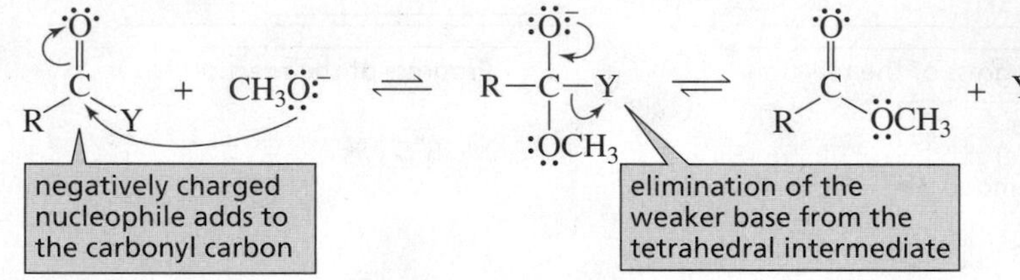

- The nucleophile adds to the carbonyl carbon, forming a tetrahedral intermediate.
- The weaker of the two bases is eliminated—either the group that was attached to the acyl group in the reactant or the newly added group—and the π bond re-forms.

If the nucleophile is not charged, then the mechanism has an additional step.

MECHANISM FOR A NUCLEOPHILIC ADDITION–ELIMINATION REACTION WITH A NEUTRAL NUCLEOPHILE

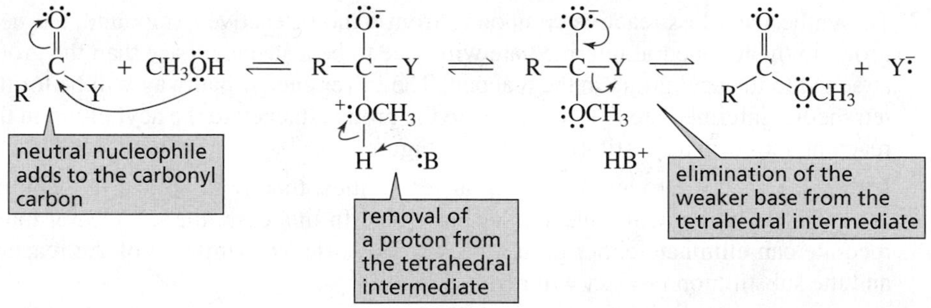

:B represents any species in the solution that is capable of removing a proton, and HB^+ represents any species in the solution that is capable of donating a proton.

- The nucleophile adds to the carbonyl carbon, forming a tetrahedral intermediate.
- A proton is removed from the tetrahedral intermediate, resulting in a tetrahedral intermediate like the one formed by a negatively charged nucleophile. (Proton transfers to and from oxygen are extremely fast steps.)
- The weaker of the two bases is eliminated—either the newly added group after it has lost a proton or the group that was attached to the acyl group in the reactant—and the π bond re-forms.

The remaining sections of this chapter show specific examples of these general principles. Keep in mind that *all the nucleophilic addition–elimination reactions follow the same mechanism.* Therefore, you can always determine the outcome of the reactions of carboxylic acids and carboxylic acid derivatives presented in this chapter by examining the tetrahedral intermediate and remembering that the weaker base is preferentially eliminated (Section 5).

PROBLEM 15♦

What will be the product of a nucleophilic addition–elimination reaction—a new carboxylic acid derivative, a mixture of two carboxylic acid derivatives, or no reaction—if the new group in the tetrahedral intermediate is the following?

a. a stronger base than the substituent that was attached to the acyl group
b. a weaker base than the substituent that was attached to the acyl group
c. similar in basicity to the substituent that was attached to the acyl group

---

8 THE REACTIONS OF ACYL CHLORIDES

Acyl chlorides react with alcohols to form esters, with water to form carboxylic acids, and with amines to form amides because, in each case, the incoming nucleophile is a stronger base than the departing halide ion (Table 1).

acetyl chloride

$$\underset{R}{\overset{O}{\underset{|}{\overset{\|}{C}}}}\text{—Cl} \;+\; CH_3OH \;\longrightarrow\; \underset{R}{\overset{O}{\underset{|}{\overset{\|}{C}}}}\text{—OCH}_3 \;+\; HCl$$

$$\underset{R}{\overset{O}{\underset{|}{\overset{\|}{C}}}}\text{—Cl} \;+\; H_2O \;\longrightarrow\; \underset{R}{\overset{O}{\underset{|}{\overset{\|}{C}}}}\text{—OH} \;+\; HCl$$

$$\underset{R}{\overset{O}{\underset{|}{\overset{\|}{C}}}}\text{—Cl} \;+\; 2\,CH_3NH_2 \;\longrightarrow\; \underset{R}{\overset{O}{\underset{|}{\overset{\|}{C}}}}\text{—NHCH}_3 \;+\; CH_3\overset{+}{N}H_3\,Cl^-$$

All the reactions follow the general mechanism described earlier in this chapter.

MECHANISM FOR THE REACTION OF AN ACYL CHLORIDE WITH AN ALCOHOL

the weaker base is eliminated

formation of a tetrahedral intermediate

a proton is removed

The weaker base is eliminated from the tetrahedral intermediate.

- The nucleophilic alcohol adds to the carbonyl carbon of the acyl chloride, forming a tetrahedral intermediate.

- Because the protonated ether group is a strong acid, the tetrahedral intermediate loses a proton. (Proton transfers to and from oxygen are diffusion controlled, so they occur very rapidly.)

- The chloride ion is eliminated from the deprotonated tetrahedral intermediate because chloride ion is a weaker base than the alkoxide ion.

Notice that the reaction of an acyl chloride with an amine (shown earlier) or with ammonia (shown next) to form an amide is carried out with twice as much amine or ammonia as acyl chloride because the HCl formed as a product of the reaction will protonate any amine or ammonia that has yet to react. Once protonated, it is no longer a nucleophile, so it cannot react with the acyl chloride. Using twice as much amine or ammonia as acyl chloride guarantees that there will be enough unprotonated amine to react with all the acyl chloride.

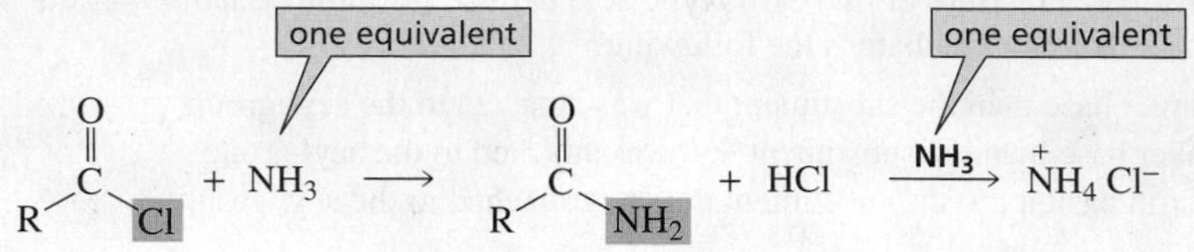

PROBLEM 16

Starting with acetyl chloride, what neutral nucleophile would you use to make each of the following compounds?

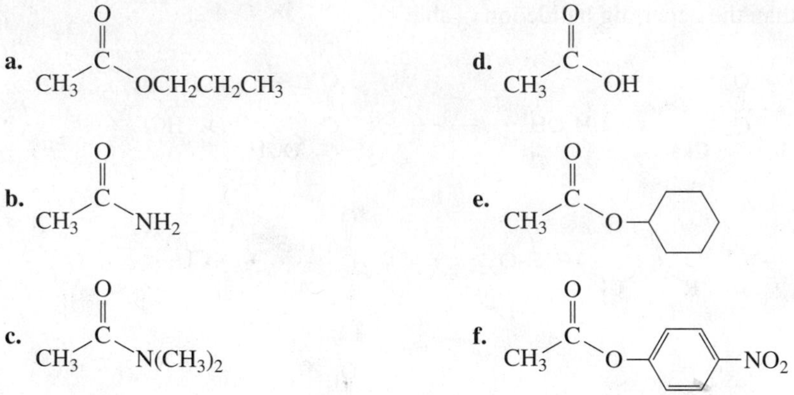

a.

$$CH_3\!-\!\overset{\overset{\displaystyle O}{\|}}{C}\!-\!OCH_2CH_2CH_3$$

b.

$$CH_3\!-\!\overset{\overset{\displaystyle O}{\|}}{C}\!-\!NH_2$$

c.

$$CH_3\!-\!\overset{\overset{\displaystyle O}{\|}}{C}\!-\!N(CH_3)_2$$

d.

$$CH_3\!-\!\overset{\overset{\displaystyle O}{\|}}{C}\!-\!OH$$

e.

$$CH_3\!-\!\overset{\overset{\displaystyle O}{\|}}{C}\!-\!O\!-\!\text{cyclohexyl}$$

f.

$$CH_3\!-\!\overset{\overset{\displaystyle O}{\|}}{C}\!-\!O\!-\!\text{phenyl}\!-\!NO_2$$

PROBLEM 17 Solved

a. What two amides are obtained from the reaction of acetyl chloride with an equivalent of ethylamine and an equivalent of propylamine?

b. Why is only one amide obtained from the reaction of acetyl chloride with an equivalent of ethylamine and an equivalent of triethylamine?

Solution to 17a Either of the amines can react with acetyl chloride, so both *N*-ethylacetamide and *N*-propylacetamide are formed.

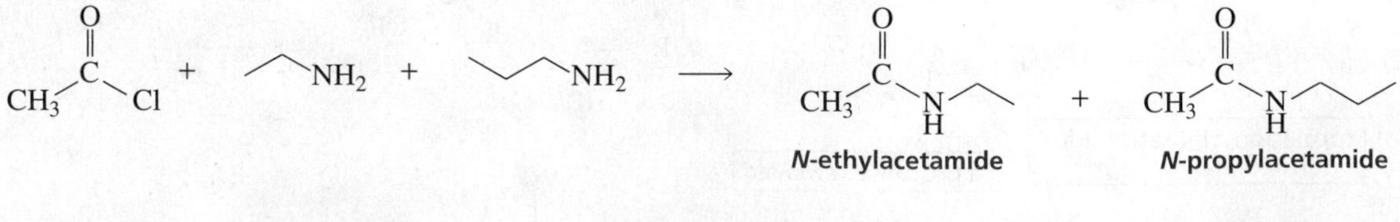

N-ethylacetamide   **N-propylacetamide**

Solution to 17b Two compounds are formed initially. However, the compound formed by triethylamine is very reactive because it has a positively charged nitrogen, which is an excellent leaving group. Therefore, the compound will react immediately with any unreacted ethylamine, so *N*-ethylacetamide is the only amide product of the reaction.

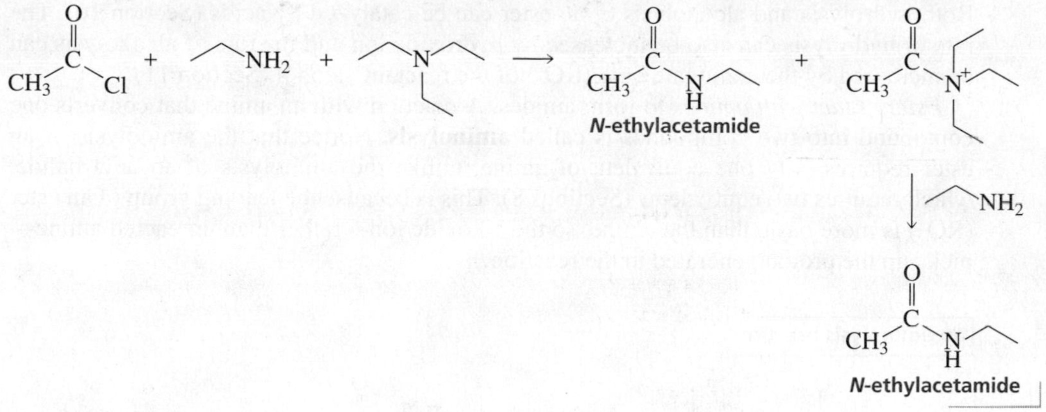

PROBLEM 18

Write the mechanism for each of the following reactions:

a. the reaction of acetyl chloride with water to form acetic acid

b. the reaction of benzoyl chloride with excess methylamine to form *N*-methylbenzamide

9 THE REACTIONS OF ESTERS

Esters do not react with chloride ion because it is a much weaker base than the RO^- group of the ester, so Cl^- (not RO^-) would be the base eliminated from the tetrahedral intermediate (Table 1).

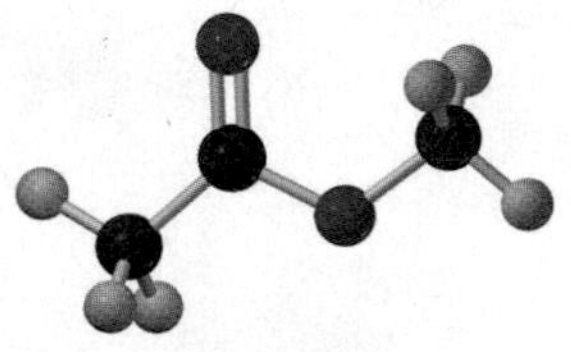

methyl acetate

An ester reacts with water to form a carboxylic acid and an alcohol. This is an example of a hydrolysis reaction. A **hydrolysis reaction** is a reaction with water that converts one compound into two compounds (*lysis* is Greek for "breaking down").

a hydrolysis reaction

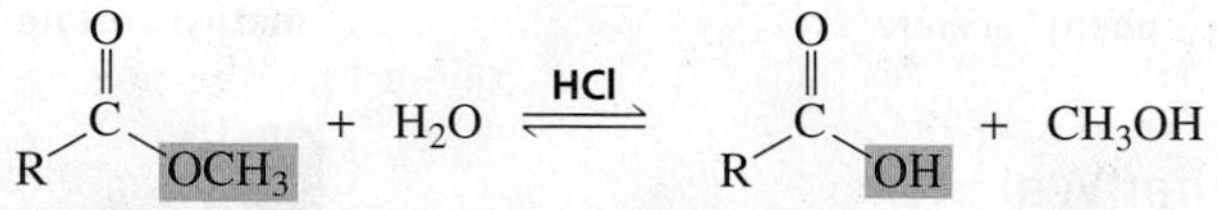

An ester reacts with an alcohol to form a new ester and a new alcohol. This is an example of an **alcoholysis reaction**—a reaction with an alcohol that converts one compound into two compounds. This particular alcoholysis reaction is also called a **transesterification reaction** because one ester is converted to another ester.

a transesterification reaction

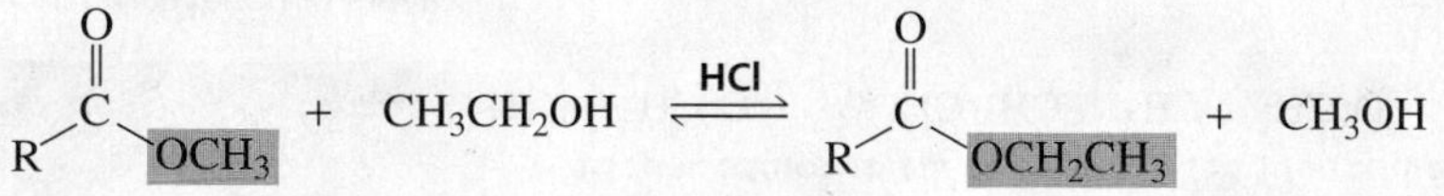

Both the hydrolysis and the alcoholysis of an ester are very slow reactions because water and alcohols are poor nucleophiles and the RO^- group of an ester is a poor leaving group. Therefore, these reactions are always catalyzed when carried out in the laboratory.

Both hydrolysis and alcoholysis of an ester can be catalyzed by acids (Section 10). The rate of hydrolysis can also be increased by hydroxide ion and the rate of alcoholysis can be increased by the conjugate base (RO⁻) of the reactant alcohol (Section 11).

Esters react with amines to form amides. A reaction with an amine that converts one compound into two compounds is called **aminolysis.** Notice that the aminolysis of an ester requires only one equivalent of amine, unlike the aminolysis of an acyl halide, which requires two equivalents (Sections 8). This is because the leaving group of an ester (RO⁻) is more basic than the amine, so the alkoxide ion—rather than unreacted amine—picks up the proton generated in the reaction.

an aminolysis reaction

$$\underset{R}{\overset{O}{\underset{\|}{C}}}OCH_2CH_3 \; + \; CH_3NH_2 \;\; \xrightarrow{\Delta} \;\; \underset{R}{\overset{O}{\underset{\|}{C}}}NHCH_3 \; + \; CH_3CH_2OH$$

The reaction of an ester with an amine is not as slow as the reaction of an ester with water or an alcohol because an amine is a better nucleophile. This is fortunate because the reaction cannot be catalyzed by an acid. The acid would protonate the amine, and a protonated amine is not a nucleophile. The rate of the reaction, however, can be increased by heat.

Phenol is a stronger acid than alcohol.

$$\text{C}_6\text{H}_5-\text{OH} \qquad\qquad CH_3OH$$

$$\textbf{p}\textit{K}_\textbf{a} = \textbf{10.0} \qquad\qquad \textbf{p}\textit{K}_\textbf{a} = \textbf{15.5}$$

Therefore, a phenolate ion (ArO⁻) is a weaker base than an alkoxide ion (RO⁻), so a phenyl ester is more reactive than an alkyl ester.

$$\underset{\textbf{phenyl acetate}}{CH_3\overset{O}{\overset{\|}{C}}O-\text{C}_6\text{H}_5} \qquad \text{is more reactive than} \qquad \underset{\textbf{methyl acetate}}{CH_3\overset{O}{\overset{\|}{C}}OCH_3}$$

Waxes Are Esters That Have High-Molecular Weights

Waxes are esters formed from long-chain carboxylic acids and long-chain alcohols. For example, beeswax, the structural material of beehives, has a 26-carbon carboxylic acid component and a 30-carbon alcohol component. The word *wax* comes from the Old English *weax,* meaning "material of the honeycomb." Carnauba wax is a particularly hard wax because of its relatively high molecular weight; it has a 32-carbon carboxylic acid component and a 34-carbon alcohol component. Carnauba wax is widely used as a car wax and in floor polishes.

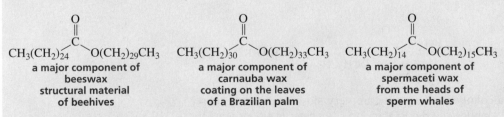

$$\underset{\substack{\textbf{a major component of}\\ \textbf{beeswax}\\ \textbf{structural material}\\ \textbf{of beehives}}}{CH_3(CH_2)_{24}\overset{O}{\overset{\|}{C}}O(CH_2)_{29}CH_3} \qquad \underset{\substack{\textbf{a major component of}\\ \textbf{carnauba wax}\\ \textbf{coating on the leaves}\\ \textbf{of a Brazilian palm}}}{CH_3(CH_2)_{30}\overset{O}{\overset{\|}{C}}O(CH_2)_{33}CH_3} \qquad \underset{\substack{\textbf{a major component of}\\ \textbf{spermaceti wax}\\ \textbf{from the heads of}\\ \textbf{sperm whales}}}{CH_3(CH_2)_{14}\overset{O}{\overset{\|}{C}}O(CH_2)_{15}CH_3}$$

Waxes are common in the biological world. The feathers of birds are coated with wax to make them water repellent. Some vertebrates secrete wax in order to keep their fur lubricated and water repellent. Insects secrete a waterproof, waxy layer on the outside of their exoskeletons. Wax is also found on the surfaces of certain leaves and fruits, where it serves as a protectant against parasites and minimizes the evaporation of water.

layers of honeycomb in a beehive

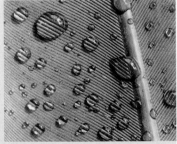

raindrops on a feather

PROBLEM 19

PROBLEM 19

We have seen that it is necessary to use excess amine in the reaction of an acyl chloride with an amine. Explain why it is not necessary to use excess alcohol in the reaction of an acyl chloride with an alcohol.

PROBLEM 20

Write a mechanism for each of the following reactions:

a. the noncatalyzed hydrolysis of methyl propionate.

b. the aminolysis of phenyl formate, using methylamine.

PROBLEM 21♦

a. State three factors cause the uncatalyzed hydrolysis of an ester to be a slow reaction.

b. Which is faster, the hydrolysis of an ester or the aminolysis of the same ester? Explain your answer.

PROBLEM 22 Solved

List the following esters in order from most reactive to least reactive toward hydrolysis:

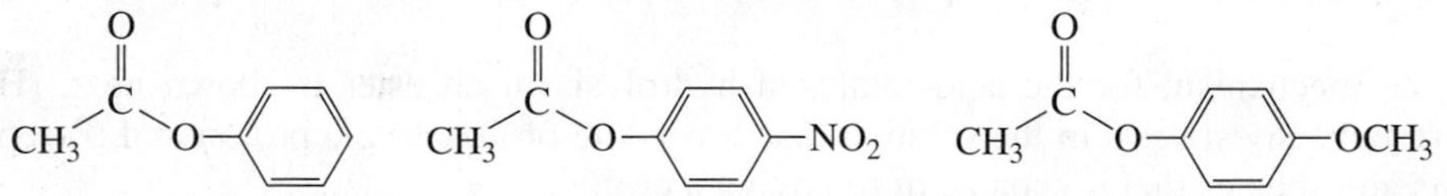

Solution We know that the reactivity of a carboxylic acid derivative depends on the basicity of the group attached to the acyl group—the weaker the base, the easier it is for *both steps* of the reaction to take place (Section 6). So now we need to compare the basicities of the three phenolate ions.

The nitro-substituted phenolate ion is the weakest base because the nitro group withdraws electrons inductively and by resonance, which decreases the concentration of negative charge on the oxygen. The methoxy-substituted phenolate ion is the strongest base because the methoxy group donates electrons by resonance more than it withdraws electrons inductively, so the concentration of negative charge on the oxygen is increased. Therefore, the three esters have the following relative reactivity toward hydrolysis.

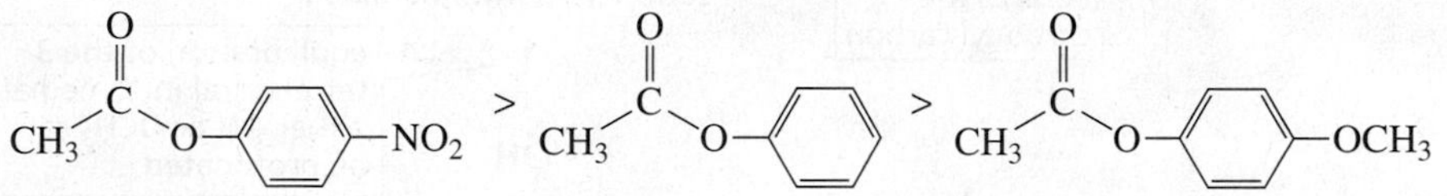

10 ACID-CATALYZED ESTER HYDROLYSIS AND TRANSESTERIFICATION

We have seen that esters hydrolyze slowly because water is a poor nucleophile and esters have relatively basic leaving groups. The rate of hydrolysis can be increased by either acid or hydroxide ion. When you examine the mechanisms for these reactions, notice the following features that hold for all organic reactions:

All organic intermediates and products in acidic solutions are positively charged or neutral; *negatively charged organic intermediates and products are not formed in acidic solutions.*

All organic intermediates and products in basic solutions are negatively charged or neutral; *positively charged organic intermediates and products are not formed in basic solutions.*

Hydrolysis of An Ester with a Primary or Secondary Alkyl Group

When an acid is added to a reaction, the first thing that happens is the acid protonates the atom in the reactant that has the greatest electron density. Therefore, when an acid is added to an ester, the acid protonates the carbonyl oxygen.

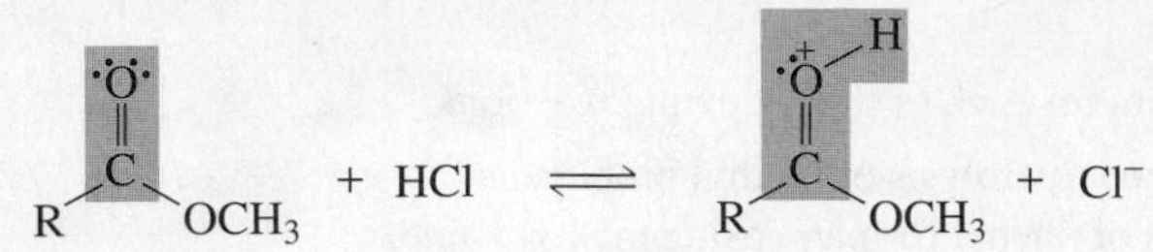

The resonance contributors of the ester show why the carbonyl oxygen is the atom with the greatest electron density.

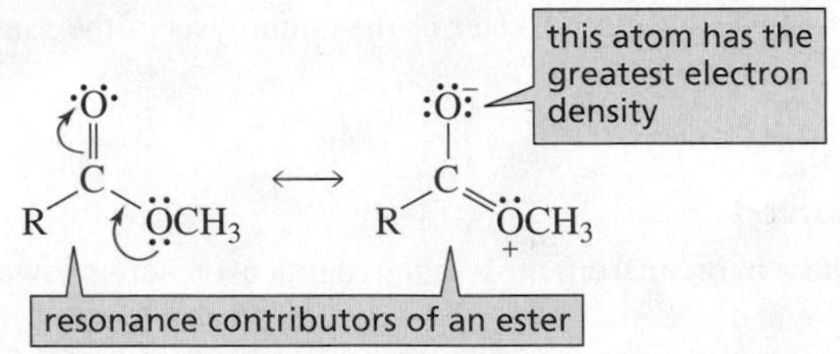

The mechanism for the acid-catalyzed hydrolysis of an ester is shown next. (HB⁺ represents any species in the solution that is capable of donating a proton and :B represents any species that is capable of removing a proton.)

MECHANISM FOR ACID-CATALYZED ESTER HYDROLYSIS

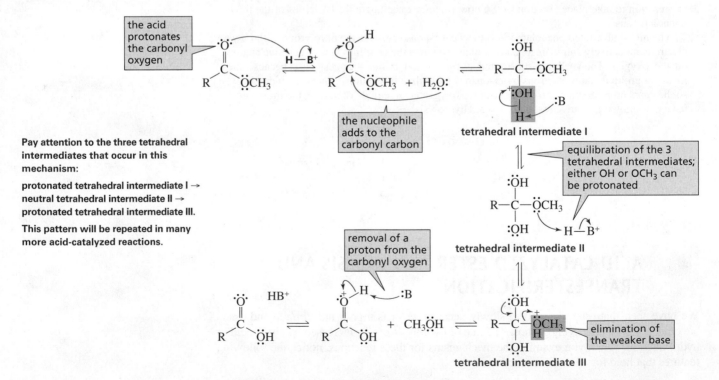

Pay attention to the three tetrahedral intermediates that occur in this mechanism:

protonated tetrahedral intermediate I →
neutral tetrahedral intermediate II →
protonated tetrahedral intermediate III.

This pattern will be repeated in many more acid-catalyzed reactions.

- The acid protonates the carbonyl oxygen.
- The nucleophile (H_2O) adds to the carbonyl carbon of the protonated carbonyl group, forming a protonated tetrahedral intermediate.
- The protonated tetrahedral intermediate (I) is in equilibrium with its nonprotonated form (II).

- The nonprotonated tetrahedral intermediate can be re-protonated on OH, which re-forms tetrahedral intermediate I, or protonated on OCH_3, which forms tetrahedral intermediate III. (The relative amounts of the three tetrahedral intermediates depend on the pH of the solution and the pK_a values of the protonated intermediates.)

- When tetrahedral intermediate I collapses, it eliminates H_2O in preference to CH_3O^- (because H_2O is a weaker base), and re-forms the ester. When tetrahedral intermediate III collapses, it eliminates CH_3OH rather than HO^- (because CH_3OH is a weaker base) and forms the carboxylic acid. Because H_2O and CH_3OH have approximately the same basicity, it will be as likely for tetrahedral intermediate I to collapse to re-form the ester as it will for tetrahedral intermediate III to collapse to form the carboxylic acid. (Tetrahedral intermediate II is much less likely to collapse because both HO^- and CH_3O^- are strong bases and, therefore, poor leaving groups.)

- Removal of a proton from the protonated carboxylic acid forms the carboxylic acid and re-forms the acid catalyst.

Because tetrahedral intermediates I and III are equally likely to collapse, both ester and carboxylic acid will be present when the reaction has reached equilibrium. Excess water can be used to force the equilibrium to the right (Le Châtelier's principle). Or, if the boiling point of the product alcohol is significantly lower than the boiling points of the other components of the reaction, the reaction can be driven to the right by distilling off the alcohol as it is formed.

In Section 14, we will see that the mechanism for the acid-catalyzed reaction of a carboxylic acid and an alcohol to form an ester and water is the exact reverse of the mechanism for the acid-catalyzed hydrolysis of an ester to form a carboxylic acid and an alcohol.

PROBLEM 23♦

What products would be formed from the acid-catalyzed hydrolysis of the following esters?

PROBLEM 24

Using the mechanism for the acid-catalyzed hydrolysis of an ester as your guide, write the mechanism—showing all the curved arrows—for the acid-catalyzed reaction of acetic acid and methanol to form methyl acetate. Use HB^+ and $:B$ to represent the proton-donating and proton-removing species, respectively.

Now let's see how the acid catalyst increases the rate of ester hydrolysis. For a catalyst to increase the rate of a reaction, it must increase the rate of the slow step of the reaction because changing the rate of a fast step will not affect the rate of the overall reaction. Four of the six steps in the mechanism for acid-catalyzed ester hydrolysis are proton transfer steps. Proton transfer to or from an electronegative atom such as oxygen or nitrogen is always a fast step. The other two steps in the mechanism—namely, formation of the tetrahedral intermediate and collapse of the tetrahedral intermediate—are relatively slow. The acid increases the rates of both these steps.

The acid increases *the rate of formation of the tetrahedral intermediate* by protonating the carbonyl oxygen. Protonated carbonyl groups are more susceptible than nonprotonated carbonyl groups to nucleophilic addition, because a positively charged oxygen is more electron withdrawing than an uncharged oxygen. Increased electron withdrawal by the positively charged oxygen makes the carbonyl carbon more electron deficient, which increases its reactivity toward nucleophiles.

protonation of the carbonyl oxygen increases the susceptibility of the carbonyl carbon to nucleophilic addition

An acid catalyst increases the reactivity of a carbonyl group.

The acid increases *the rate of collapse of the tetrahedral intermediate* by decreasing the basicity of the leaving group, which makes it easier to eliminate: in the acid-catalyzed hydrolysis of an ester, the leaving group is CH_3OH, which is a weaker base than CH_3O^-, the leaving group in the uncatalyzed reaction.

An acid catalyst increases the leaving propensity of a group.

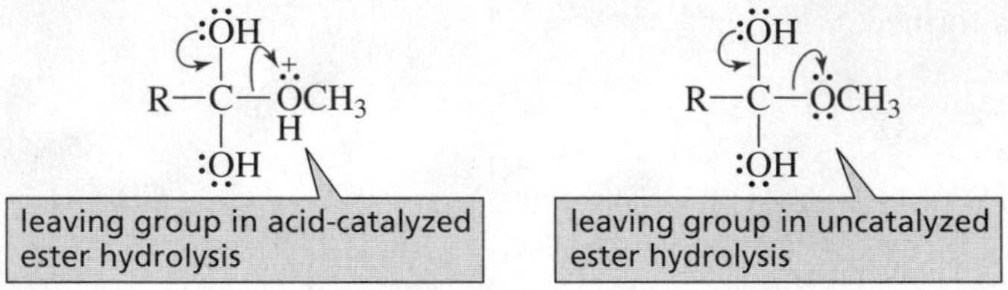

PROBLEM 25♦

In the mechanism for the acid-catalyzed hydrolysis of an ester,

a. what species could be represented by HB^+?
b. what species could be represented by $:B$?
c. what species is HB^+ most likely to be in the hydrolysis reaction?
d. what species is HB^+ most likely to be in the reverse reaction?

PROBLEM 26 Solved

How could the target molecule (butanone) be prepared from butane?

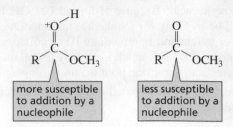

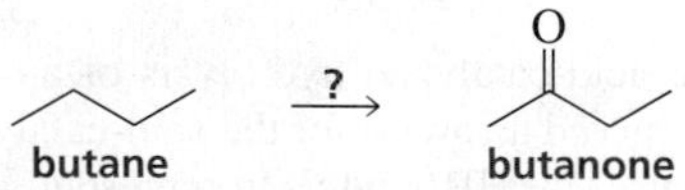

butane   butanone

Solution We know that the first reaction has to be a radical halogenation because that is the only reaction that an alkane undergoes. Bromination will lead to a greater yield of the desired 2-halo-substituted compound than chlorination will because a bromine radical is more selective than a chlorine radical. To maximize the yield of the desired substitution product over the elimination product, a weak base (acetate ion) is used for the substitution reaction. The ester is hydrolyzed to an alcohol that forms the target molecule when it is oxidized.

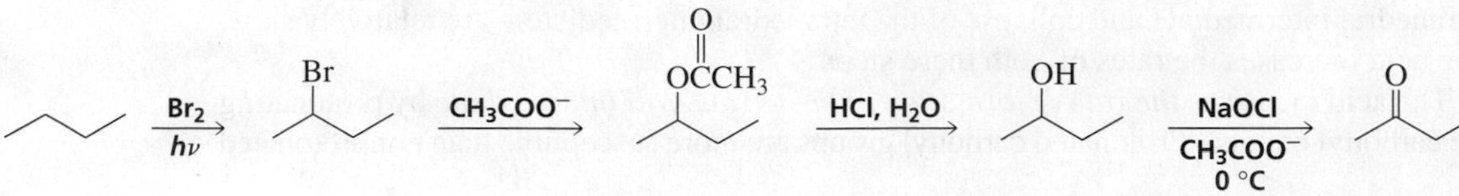

Hydrolysis of An Ester with a Tertiary Alkyl Group

The hydrolysis of an ester with a *tertiary alkyl group* forms the same products as the hydrolysis of an ester with *a primary or secondary alkyl group*—namely, a carboxylic acid and an alcohol—but does so by a completely different mechanism. The hydrolysis of an ester with a tertiary alkyl group is an S_N1 reaction rather than a nucleophilic addition–elimination reaction, because the carboxylic acid leaves behind a relatively stable tertiary carbocation.

MECHANISM FOR THE HYDROLYSIS OF AN ESTER WITH A TERTIARY ALKYL GROUP

departure of the leaving group to form a tertiary carbocation

reaction of the carbocation with a nucleophile

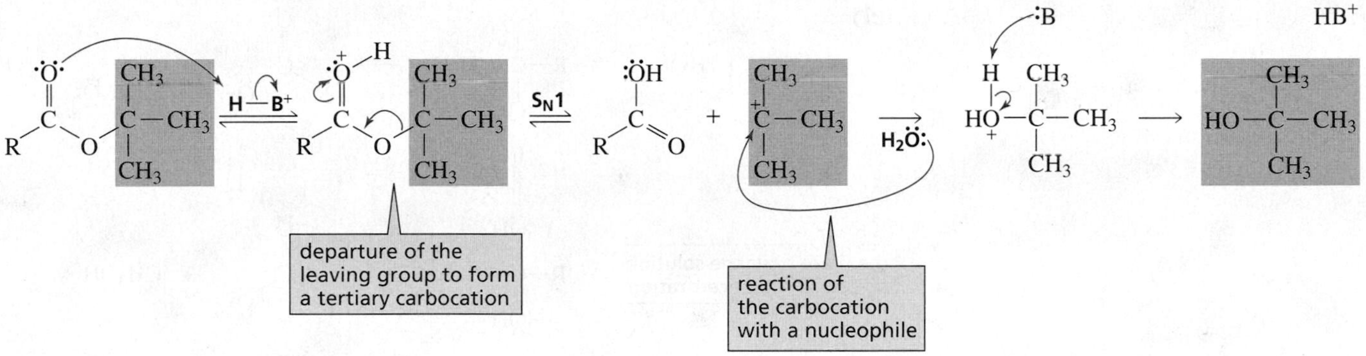

- An acid protonates the carbonyl oxygen.
- The leaving group departs, forming a tertiary carbocation.
- The nucleophile (H_2O) reacts with the carbocation.
- A base removes a proton from the strongly acidic protonated alcohol.

Transesterification

Transesterification—the reaction of an ester with an alcohol—is also catalyzed by acid. The mechanism for acid-catalyzed transesterification is identical to the mechanism for acid-catalyzed ester hydrolysis, except that the nucleophile is ROH rather than H_2O. As in ester hydrolysis, the leaving groups in the tetrahedral intermediate have approximately the same basicity. Consequently, an excess of the reactant alcohol must be used to produce a good yield of the desired product.

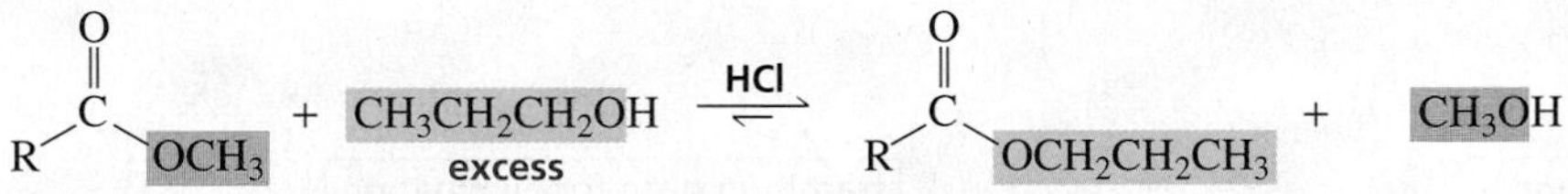

PROBLEM 27♦

What products would be obtained from the following reactions?

a. ethyl benzoate + excess isopropanol + HCl

b. phenyl acetate + excess ethanol + HCl

PROBLEM 28

Write the mechanism for the acid-catalyzed transesterification of ethyl acetate with methanol.

PROBLEM 29

Write the mechanism for the acid-catalyzed transesterification of *tert*-butyl acetate with methanol.

11 HYDROXIDE-ION-PROMOTED ESTER HYDROLYSIS

The rate of hydrolysis of an ester can be increased by hydroxide ion. Like an acid catalyst, hydroxide ion increases the rates of the two slow steps of the reaction—namely, formation of the tetrahedral intermediate and collapse of the tetrahedral intermediate.

MECHANISM FOR THE HYDROXIDE-ION-PROMOTED HYDROLYSIS OF AN ESTER

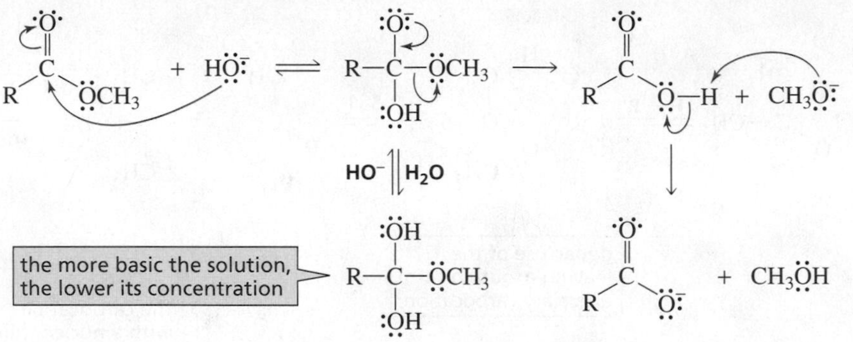

- Hydroxide ion adds to the carbonyl carbon of the ester.
- The two potential leaving groups in the tetrahedral intermediate (HO^- and CH_3O^-) have the same leaving propensity. Elimination of HO^- re-forms the ester, whereas elimination of CH_3O^- forms a carboxylic acid
- The final products are not the carboxylic acid and methoxide ion because if only one base is protonated, it will be the stronger base. Therefore, the final products are the carboxylate ion and methanol because CH_3O^- is more basic than $RCOO^-$.

Hydroxide ion is a better nucleophile than water.

Hydroxide ion increases the rate of formation of the tetrahedral intermediate because HO^- is a better nucleophile than H_2O. Hydroxide ion increases the rate of collapse of the tetrahedral intermediate because the transition state for expulsion of CH_3O^- by a negatively charged oxygen is more stable than the transition state for expulsion of CH_3O^- by a neutral oxygen since, in the former, the oxygen does not develop a partial positive charge.

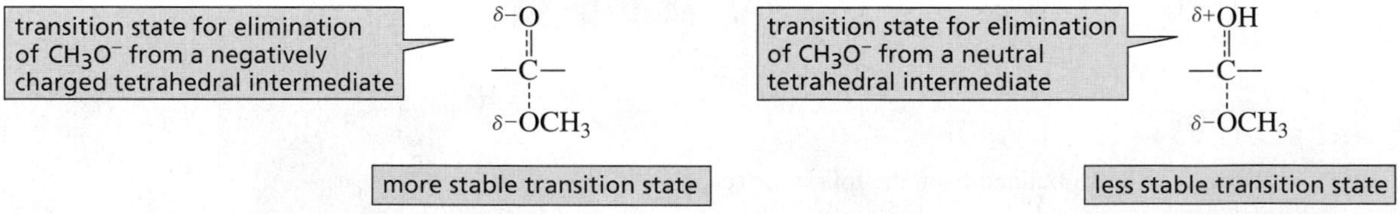

In addition, a smaller fraction of the negatively charged tetrahedral intermediate becomes protonated in a basic solution. (The relative amounts of the neutral and negatively charged tetrahedral intermediates depend on the pH of the solution and the pK_a value of the neutral tetrahedral intermediate).

Because carboxylate ions are negatively charged, they do not react with nucleophiles. Therefore, the hydroxide-ion-promoted hydrolysis of an ester, unlike the acid-catalyzed hydrolysis of an ester, is *not* a reversible reaction.

The hydrolysis of an ester in the presence of hydroxide ion is called a *hydroxide-ion-promoted reaction* rather than a base-catalyzed reaction because hydroxide ion increases the rate of the first step of the reaction by being a better nucleophile than water—not by being a stronger base than water—and because hydroxide ion is consumed in the overall reaction. To be a catalyst, a species must not be changed by or consumed in the reaction. Therefore, hydroxide ion is actually a reagent rather than a catalyst, so it is more accurate to call the reaction a hydroxide-ion-*promoted* reaction than a hydroxide-ion-*catalyzed* reaction.

Hydroxide ion promotes only hydrolysis reactions; it does not promote transesterification reactions or aminolysis reactions. Hydroxide ion cannot promote reactions of carboxylic acid derivatives with alcohols or with amines because one function of hydroxide ion is to provide a good nucleophile for the first step of the reaction. When the nucleophile is supposed to be an alcohol or an amine, nucleophilic addition by hydroxide ion would form a different product from the one that would be formed by the alcohol or amine. Hydroxide can be used to promote a hydrolysis reaction because the same product is formed, whether the nucleophile that adds to the carbonyl carbon is H_2O or HO^-.

Reactions in which the nucleophile is an alcohol can be promoted by the conjugate base of the alcohol. The function of the alkoxide ion is to provide a good nucleophile for the reaction, so only reactions in which the nucleophile is an alcohol can be promoted by the conjugate base of the alcohol.

Aspirin, NSAIDs, and COX-2 Inhibitors

Salicylic acid, found in willow bark and myrtle leaves, is perhaps the oldest known drug. As early as the fifth century B.C., Hippocrates wrote about the curative powers of willow bark. In 1897, Felix Hoffmann, a scientist working at Bayer and Co. in Germany, found that acylating salicylic acid produced a more potent drug to control fever and pain. They called it *aspirin*; "a" for acetyl, "spir" for the spiraea flower that also contains salicylic acid, and "in" was a common ending for drugs at that time. It soon became the world's best-selling drug. However, its mode of action was not discovered until 1971, when it was found that the anti-inflammatory and fever-reducing activity of aspirin was due to a transesterification reaction that blocks the synthesis of prostaglandins.

Prostaglandins have several different physiological functions. One is to stimulate inflammation and another to induce fever. The enzyme prostaglandin synthase catalyzes the conversion of arachidonic acid into PGH_2, a precursor of all prostaglandins and the related thromboxanes.

$$\text{arachidonic acid} \xrightarrow{\textbf{prostaglandin synthase}} \text{PGH}_2 \begin{cases} \nearrow \text{prostaglandins} \\ \searrow \text{thromboxanes} \end{cases}$$

Prostaglandin synthase is composed of two enzymes. One of them—cyclooxygenase—has a CH_2OH group at its active site that is necessary for enzymatic activity. When the CH_2OH group reacts with aspirin in a transesterification reaction, the enzyme is inactivated. This prevents prostaglandins from being synthesized, so inflammation is suppressed and fever is reduced. Notice that the carboxyl group of aspirin is a basic catalyst. It removes a proton from the CH_2OH group, which makes it a better nucleophile. This is why aspirin is maximally active in its basic form. (The red arrows show the formation of the tetrahedral intermediate; the blue arrows show its collapse.)

Because aspirin inhibits the formation of PGH_2, it also inhibits the synthesis of thromboxanes, compounds involved in blood clotting. Presumably, this is why low levels of aspirin have been reported to reduce the incidence of strokes and heart attacks that result from the formation of blood clots. Because of aspirin's activity as an anticoagulant, doctors caution patients not to take aspirin for several days before surgery.

Other NSAIDs (nonsteroidal anti-inflammatory drugs), such as ibuprofen (the active ingredient in Advil, Motrin, and Nuprin) and naproxen (the active ingredient in Aleve), also inhibit the synthesis of prostaglandins.

There are two forms of prostaglandin synthase: one carries out the normal production of prostaglandin, and the other synthesizes additional prostaglandin in response to inflammation. NSAIDs inhibit the synthesis of all prostaglandins. One prostaglandin regulates the production of acid in the stomach, so when prostaglandin synthesis stops, the acidity of the stomach can rise above normal levels. Celebrex, a relatively new drug, inhibits only the prostaglandin synthase that produces prostaglandin in response to inflammation. Thus, inflammatory conditions now can be treated without some of the harmful side effects.

Celebrex®

PROBLEM 30♦

a. What species other than an acid can be used to increase the rate of the transesterification reaction that converts methyl acetate to propyl acetate?

b. Explain why the rate of aminolysis of an ester cannot be increased by H^+, HO^-, or RO^-.

12 HOW THE MECHANISM FOR NUCLEOPHILIC ADDITION–ELIMINATION WAS CONFIRMED

We have seen that nucleophilic addition–elimination reactions take place by a mechanism in which a tetrahedral intermediate is formed and subsequently collapses. The tetrahedral intermediate, however, is too unstable to be isolated. How, then, do we know that it is formed? How do we know that the reaction doesn't take place by a one-step direct-displacement mechanism (similar to the mechanism for an S_N2 reaction) in which the incoming nucleophile attacks the carbonyl carbon and displaces the leaving group—a mechanism that does not break the π bond and so does not form a tetrahedral intermediate?

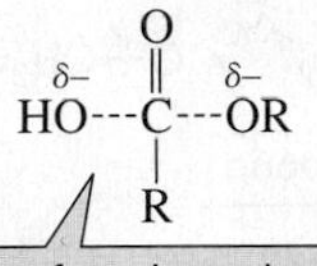

transition state for a hypothetical one-step direct-displacement mechanism

To answer this question, Myron Bender investigated the hydroxide-ion-promoted hydrolysis of ethyl benzoate after first labeling the carbonyl oxygen of ethyl benzoate with ^{18}O, an isotope of ^{16}O. When he isolated ethyl benzoate from the reaction mixture, he found that some of the ester was no longer labeled. If the reaction had taken place by a one-step direct-displacement mechanism, all the isolated ester would have remained labeled because the carbonyl group would not have participated in the reaction. On the other hand, if the mechanism involved a tetrahedral intermediate, some of the isolated ester would no longer be labeled because some of the label would have been transferred to the hydroxide ion. Bender's experiment, therefore, provided evidence for the reversible formation of a tetrahedral intermediate.

unlabeled ester

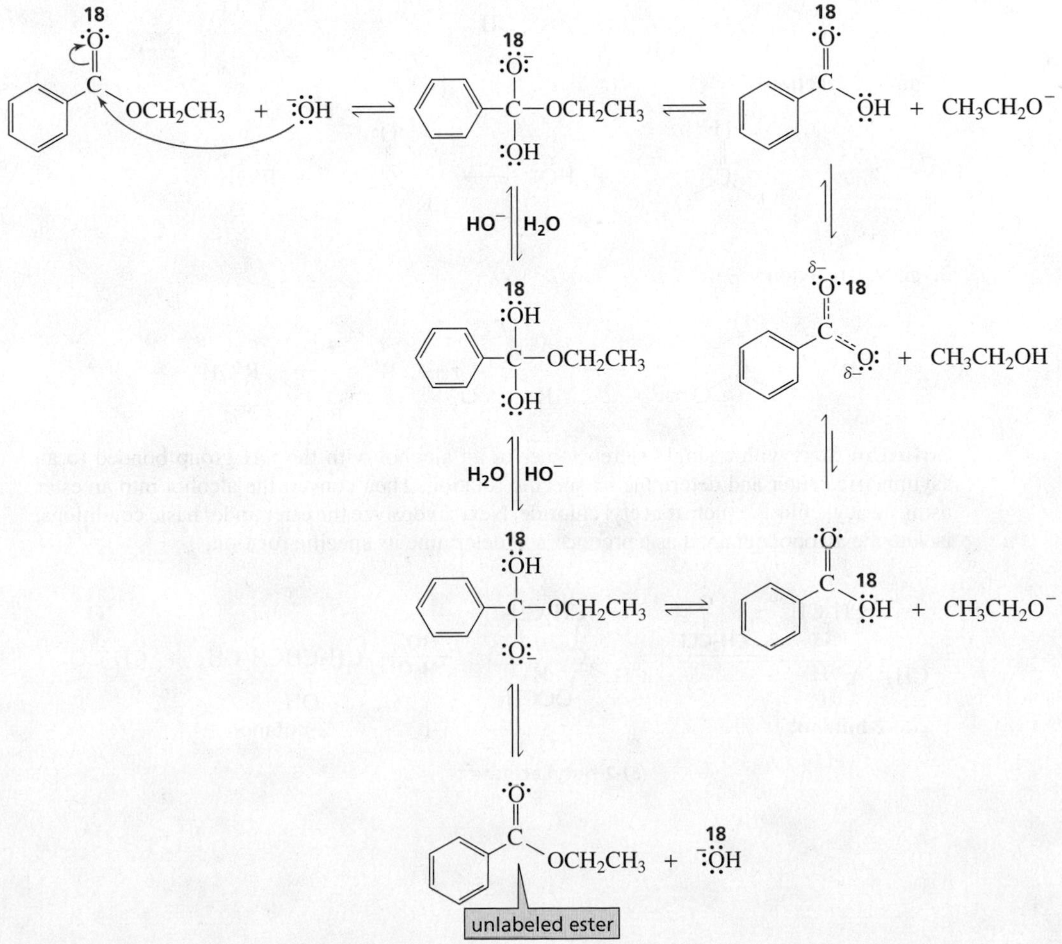

PROBLEM 31♦

When butanoic acid and ^{18}O-labeled methanol react under acidic conditions, what compounds will be labeled when the reaction has reached equilibrium?

PROBLEM 32♦

D. N. Kursanov, a Russian chemist, proved that the bond that is broken in the hydroxide-ion-promoted hydrolysis of an ester is the acyl C—O bond, rather than the alkyl C—O bond, by studying the hydrolysis of the following ester under basic conditions:

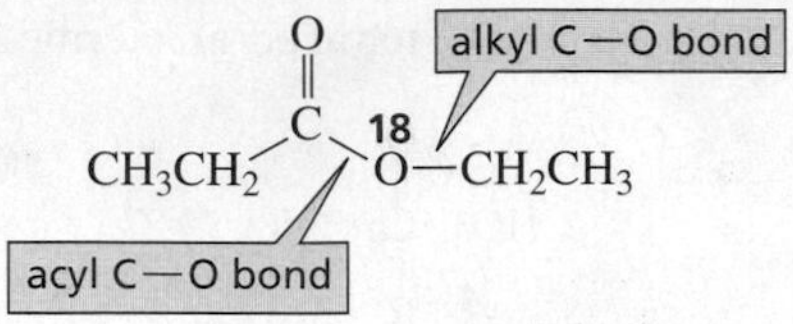

a. What products contained the ^{18}O label?
b. What product would have contained the ^{18}O label if the alkyl C—O bond had broken?

PROBLEM 33 Solved

Early chemists could envision three possible mechanisms for hydroxide-ion-promoted ester hydrolysis. Devise an experiment that would show which of the three is the actual mechanism.

1. a nucleophilic addition–elimination reaction

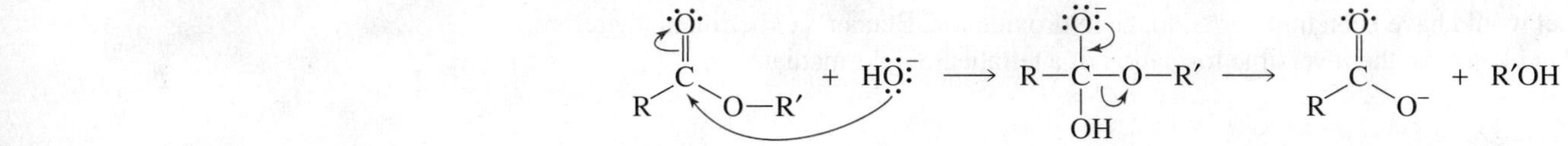

2. an S_N2 reaction

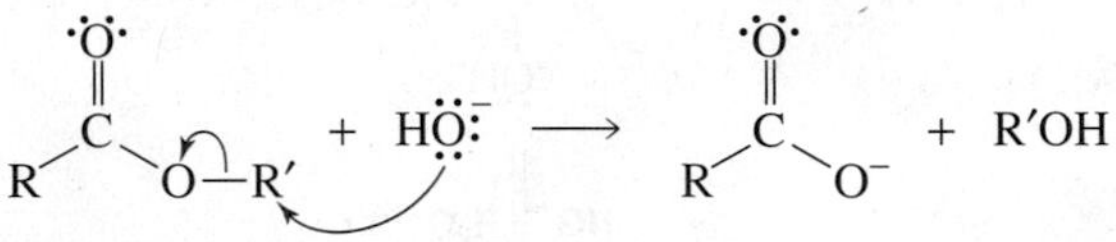

3. an S_N1 reaction

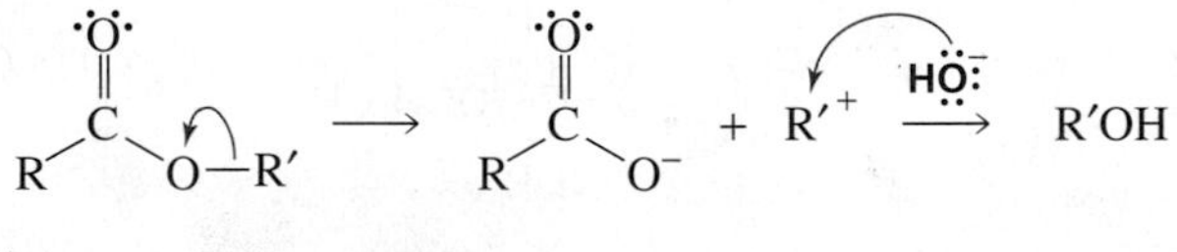

Solution Start with a single stereoisomer of an alcohol with the OH group bonded to an asymmetric center and determine its specific rotation. Then convert the alcohol into an ester using an acyl chloride such as acetyl chloride. Next, hydrolyze the ester under basic conditions, isolate the alcohol obtained as a product, and determine its specific rotation.

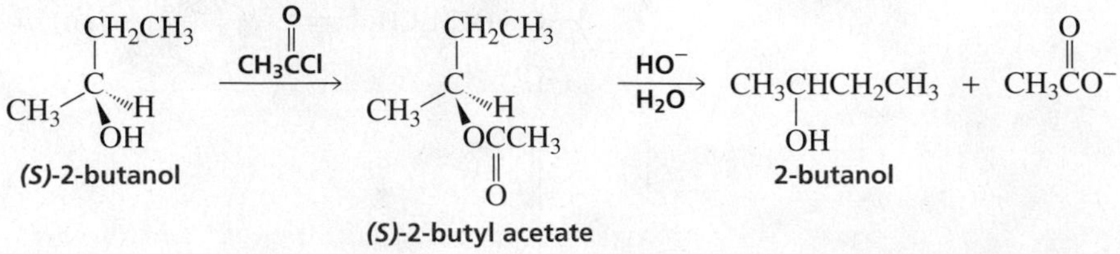

If the reaction is a nucleophilic addition–elimination reaction, the product alcohol will have the same specific rotation as the reactant alcohol because no bonds to the asymmetric center are broken during formation or hydrolysis of the ester.

If the reaction is an S_N2 reaction, the product alcohol and the reactant alcohol will have opposite specific rotations because the mechanism requires back-side attack of hydroxide ion on the asymmetric center.

If the reaction is an S_N1 reaction, the product alcohol will have a small (or zero) specific rotation because the mechanism requires carbocation formation, which leads to racemization of the alcohol.

13 FATS AND OILS ARE TRIGLYCERIDES

Triglycerides (also called **triacylglycerols**) are compounds in which each of the three OH groups of glycerol has formed an ester with a fatty acid. If the three fatty acid components of a triglyceride are the same, the compound is called a **simple triglyceride. Mixed triglycerides** contain two or three different fatty acid components and are more common than simple triglycerides.

Triglycerides that are solids or semisolids at room temperature are called **fats.** Most fats are obtained from animals and are composed largely of triglycerides with fatty acid components that are either saturated or have only one double bond. The saturated fatty acid tails pack closely together, giving these triglycerides relatively high melting points (Table 2). Therefore, they are solids at room temperature.

a fat   an oil

Liquid triglycerides are called **oils.** Oils typically come from plant products such as corn, soybeans, olives, and peanuts. They are composed primarily of triglycerides with unsaturated fatty acids and therefore cannot pack tightly together. Consequently, they have relatively low melting points and so are liquids at room temperature. All triglyceride molecules from a single source are not necessarily identical; most substances such as lard and olive oil, for example, are mixtures of several different mixed triglycerides.

Some or all of the double bonds of polyunsaturated oils can be reduced by catalytic hydrogenation. Margarine and shortening are prepared by hydrogenating vegetable oils, such as soybean oil or safflower oil, until they have the desired consistency. The hydrogenation reaction must be carefully controlled, however, because reducing all the carbon–carbon double bonds would produce a hard fat with the consistency of beef tallow. Trans fats can be formed during hydrogenation.

Vegetable oils have become popular in food preparation because some studies have linked the consumption of saturated fats with heart disease. However, recent studies have shown that *un*saturated fats may also be implicated in heart disease. One unsaturated fatty acid—a 20-carbon fatty acid with five double bonds, known as EPA and found in high concentrations in fish oils—is thought to lower the chance of developing certain forms of heart disease. This puffin's diet is high in fish oil.

Fats, oils, waxes, and fatty acids are all lipids. **Lipids** are naturally occurring organic compounds that are soluble in nonpolar solvents. Their solubility in nonpolar solvents results from their significant hydrocarbon component. The word *lipid* comes from the Greek *lipos,* which means "fat."

Whales and Echolocation

Whales have enormous heads, accounting for 33% of their total weight. They have large deposits of fat in their heads and lower jaws. This fat is very different from both the whale's normal body fat and its dietary fat. Because major anatomical modifications were necessary to accommodate this fat, it must have some important function for the animal.

It is now believed that the fat is used for echolocation—emitting sounds in pulses to gain information by analyzing the returning echoes. The fat in the whale's head focuses the emitted sound waves in a directional beam, and the echoes are received by the fat organ in the lower jaw. This organ transmits the sound to the brain for processing and interpretation, providing the whale with information about the depth of the water, changes in the sea floor, and the location of the coastline. The fat deposits in the whale's head and jaw therefore give the animal a unique acoustic sensory system and allow it to compete successfully for survival with the shark, which also has a well-developed sense of sound direction.

Soaps and Micelles

When the ester groups of a fat or an oil are hydrolyzed in a basic solution, glycerol and fatty acids are formed. Because the solution is basic, the fatty acids are in their basic forms—namely, RCO_2^-.

$$
\begin{array}{ccccc}
\text{a fat or an oil} & + \, H_2O & \xrightarrow{\text{NaOH}} & \text{glycerol} & + \text{sodium salts of fatty acids} \\
& & & & \text{soap}
\end{array}
$$

The sodium or potassium salts of fatty acids are what we know as **soap.** Consequently, the hydrolysis of an ester in a basic solution is called **saponification** (the Latin word for "soap" is *sapo*). After hydrolysis, sodium chloride is added to precipitate the soap, which is dried and pressed into bars. Perfume can be added for scented soaps, dyes can be added for colored soaps, sand can be added for scouring soaps, and air can be blown into the soap to make it float in water. Three of the most common soaps are:

sodium stearate

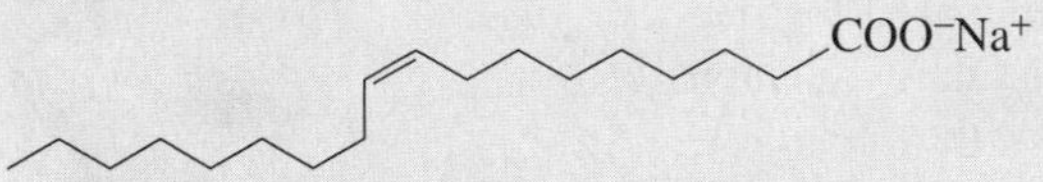

COO^-Na^+

sodium oleate

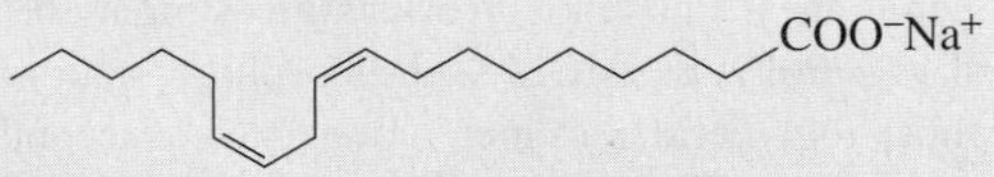

COO^-Na^+

sodium linoleate

COO^-Na^+

Long-chain carboxylate ions do not exist as individual ions in aqueous solution. Instead, they arrange themselves in spherical clusters called **micelles.** Each micelle contains 50 to 100 long-chain carboxylate ions and resembles a large ball. The polar heads of the carboxylate ions, each accompanied by a counterion, are on the outside of the ball because of their attraction for water, whereas the nonpolar tails are buried in the interior of the ball to minimize their contact with water. The hydrophobic interactions between the nonpolar tails increase the stability of the micelle.

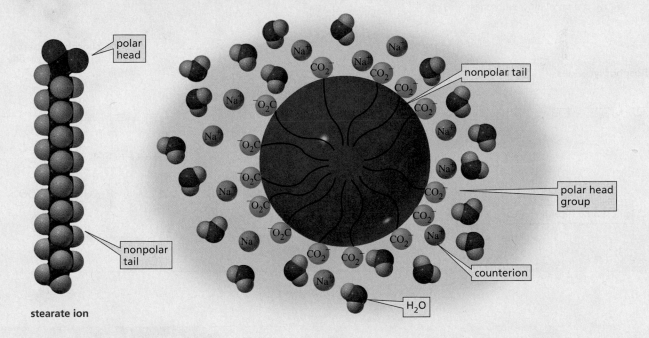

polar head

nonpolar tail

nonpolar tail

polar head group

stearate ion

counterion

H_2O

Water by itself is not a very effective cleaner because dirt is carried by nonpolar oil molecules. Soap has cleansing ability because the nonpolar oil molecules dissolve in the nonpolar interior of the micelles and are washed away with the micelle during rinsing.

Because the surface of the micelle is charged, the individual micelles repel each other instead of clustering together to form larger aggregates. However, in "hard" water—water containing high concentrations of calcium and magnesium ions—micelles do form aggregates, which we know as "bathtub ring" or "soap scum."

Phosphoglycerides Are Components of Membranes

For organisms to operate properly, some of their parts must be separated from other parts. On a cellular level, for example, the outside of the cell must be separated from the inside. "Greasy" lipid **membranes** serve as the barrier. In addition to isolating the cell's contents, membranes allow the selective transport of ions and organic molecules into and out of the cell.

Phosphoglycerides (also called **phosphoacylglycerols**) are the major components of cell membranes. Phosphoglycerides are similar to triglycerides except that a terminal OH group of glycerol is esterified with phosphoric acid rather than with a fatty acid. The most common phosphoglycerides in membranes have a second phosphate ester linkage—thus, they are phosphodiesters. Phosphoglycerides form **membranes** by arranging themselves in a **lipid bilayer**.

The alcohols most commonly used to form the second ester group are ethanolamine, choline, and serine. Phosphatidylethanolamines are also called *cephalins,* and phosphatidylcholines are called *lecithins.* Lecithins are added to foods such as mayonnaise to prevent the aqueous and fat components from separating.

The fluidity of a membrane is controlled by the fatty acid components of the phosphoglycerides. Saturated fatty acids decrease membrane fluidity because their hydrocarbon chains pack closely together. Unsaturated fatty acids increase fluidity because they pack less closely together. Cholesterol also decreases fluidity. Only animal membranes contain cholesterol, so they are more rigid than plant membranes.

phosphatidylserine
a phosphoglyceride

phosphoglycerides

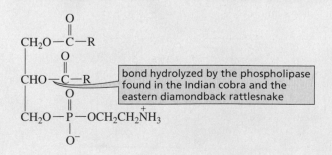

a phosphatidylethanolamine
a cephalin

a phosphatidylcholine
a lecithin

a phosphatidylserine

Snake Venom

The venom of some poisonous snakes contains a phospholipase, an enzyme that hydrolyzes an ester group of a phosphoglyceride. For example, both the eastern diamondback rattlesnake and the Indian cobra contain a phospholipase that hydrolyzes an ester bond of cephalins, which causes the membranes of red blood cells to rupture.

an eastern diamondback rattlesnake

bond hydrolyzed by the phospholipase found in the Indian cobra and the eastern diamondback rattlesnake

PROBLEM 34♦

An oil obtained from coconuts is unusual in that all three fatty acid components are identical. The molecular formula of the oil is $C_{45}H_{86}O_6$. What is the molecular formula of the carboxylate ion obtained when the oil is saponified?

PROBLEM 35♦

Which has a higher melting point, glyceryl tripalmitoleate or glyceryl tripalmitate? (The structures of the fatty acid constituents of these compounds can be found in Table 2.)

PROBLEM 36

Draw the structure of an optically inactive fat that, when hydrolyzed under acidic conditions, forms glycerol, one equivalent of lauric acid, and two equivalents of stearic acid.

PROBLEM 37

Draw the structure of an optically active fat that, when hydrolyzed under acidic conditions, forms the same products as the fat in Problem 36.

14   REACTIONS OF CARBOXYLIC ACIDS

Carboxylic acids can undergo nucleophilic addition–elimination reactions only when they are in their acidic forms. The basic form of a carboxylic acid is not reactive because its negative charge makes it resistant to approach by a nucleophile. Therefore, carboxylate ions are even less reactive than amides in nucleophilic addition–elimination reactions.

acetic acid

relative reactivities toward nucleophilic addition–elimination

most reactive — structure > structure > structure — least reactive

Carboxylic acids have approximately the same reactivity as esters, because the HO^- leaving group of a carboxylic acid has about the same basicity as the RO^- leaving group of an ester.

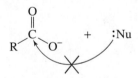

Carboxylic acids, therefore, react with alcohols to form esters. The reaction must be carried out in an acidic solution, not only to catalyze the reaction but also to keep the carboxylic acid in its acidic form so that the nucleophile will react with it. Because the tetrahedral intermediate formed in this reaction has two potential leaving groups with approximately the same basicity, the reaction must be carried out with excess alcohol to drive it toward products.

Emil Fischer was the first to discover that an ester could be prepared by treating a carboxylic acid with excess alcohol in the presence of an acid catalyst, so the reaction is called a **Fischer esterification.** Its mechanism is the exact reverse of the mechanism for the acid-catalyzed hydrolysis of an ester shown earlier in this chapter. Also see Problem 24.

Carboxylic acids do not undergo nucleophilic addition–elimination reactions with amines. A carboxylic acid is an acid and an amine is a base, so the carboxylic acid immediately loses a proton to the amine when the two compounds are mixed. The resulting ammonium carboxylate salt is the final product of the reaction; the carboxylate ion is not reactive and the protonated amine is not a nucleophile.

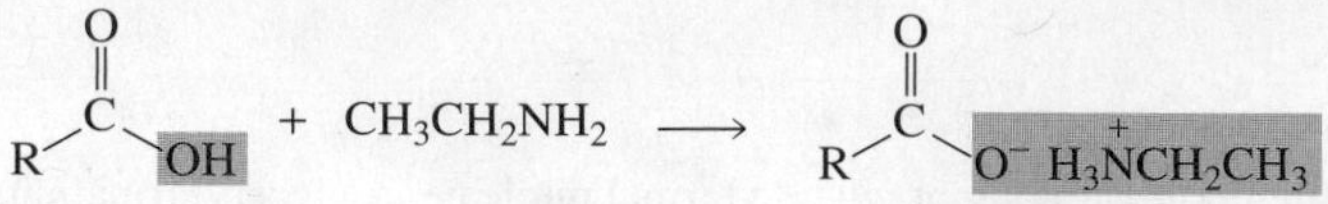

$$\underset{R}{\overset{O}{\underset{}{\|}}} \hspace{-0.3cm}\overset{}{\underset{\text{OH}}{C}} \;+\; CH_3CH_2NH_2 \;\longrightarrow\; \underset{R}{\overset{O}{\underset{}{\|}}} \hspace{-0.3cm}\overset{}{\underset{O^-\; H_3\overset{+}{N}CH_2CH_3}{C}}$$

**an ammonium
carboxylate salt**

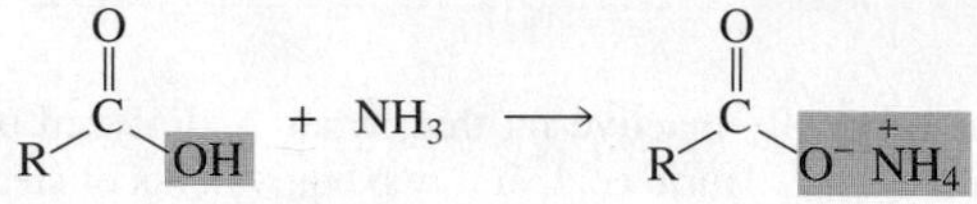

$$\underset{R}{\overset{O}{\underset{}{\|}}} \hspace{-0.3cm}\overset{}{\underset{\text{OH}}{C}} \;+\; NH_3 \;\longrightarrow\; \underset{R}{\overset{O}{\underset{}{\|}}} \hspace{-0.3cm}\overset{}{\underset{O^-\; \overset{+}{N}H_4}{C}}$$

An ammonium carboxylate salt can lose water to form an amide if it is heated.

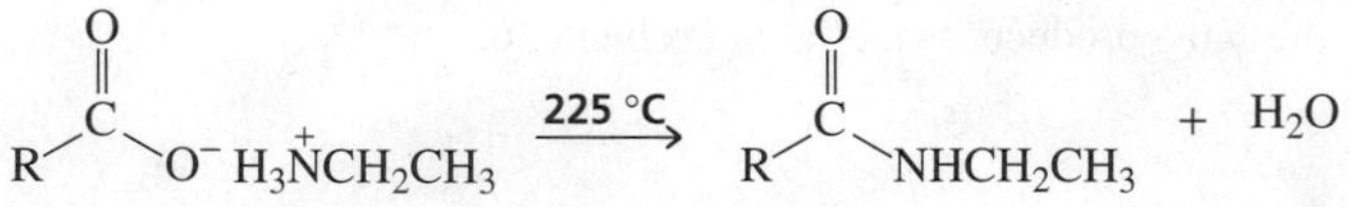

$$\underset{R}{\overset{O}{\underset{}{\|}}} \hspace{-0.3cm}\overset{}{\underset{O^-\; H_3\overset{+}{N}CH_2CH_3}{C}} \xrightarrow{\textbf{225 °C}} \underset{R}{\overset{O}{\underset{}{\|}}} \hspace{-0.3cm}\overset{}{\underset{NHCH_2CH_3}{C}} \;+\; H_2O$$

PROBLEM 38♦

Show how each of the following esters could be prepared using a carboxylic acid as one of the starting materials:

a. methyl butyrate (odor of apples)   **b.** octyl acetate (odor of oranges)

PROBLEM-SOLVING STRATEGY

Proposing a Mechanism

Propose a mechanism for the following reaction:

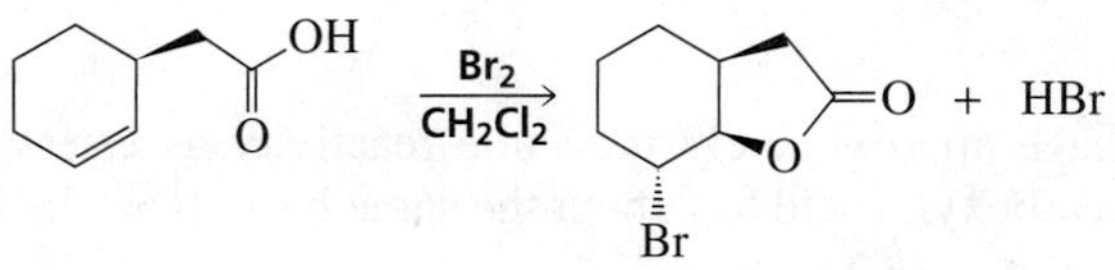

When you are asked to propose a mechanism, look carefully at the reactants in order to determine the first step. One of the reactants has two functional groups: a carboxyl group and a carbon–carbon double bond. The other reactant, Br_2, does not react with carboxylic acids, but does react with alkenes. Approach to one side of the double bond is sterically hindered by the carboxyl group, so Br_2 will add to the other side of the double bond, forming a cyclic bromonium ion.

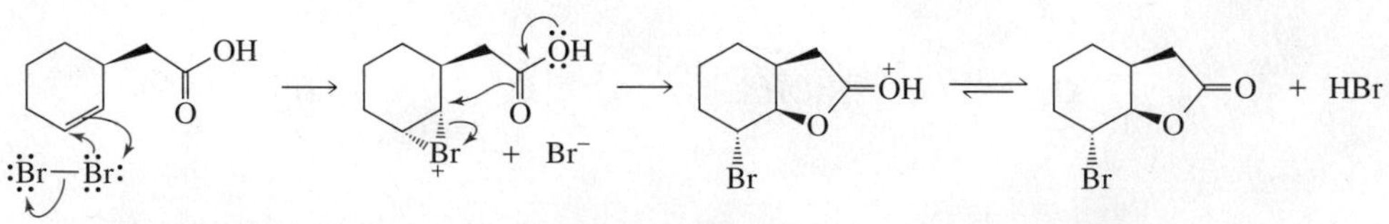

We know that in the second step of this addition reaction, a nucleophile will attack the bromonium ion. Of the two nucleophiles present, the carbonyl oxygen is better positioned than the bromide ion to attack the back side of the bromonium ion, resulting in a compound with the observed configuration. Loss of a proton forms the final product of the reaction.

Now use the strategy you have just learned to solve Problem 39.

PROBLEM 39

Propose a mechanism for the following reaction. (*Hint:* Number the carbons to help you see where they end up in the product.)

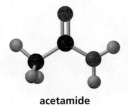

$$CH_2{=}CHCH_2CH_2CH{=}CCH_3 \quad + \quad \underset{CH_3}{\overset{O}{\underset{\quad}{\overset{\parallel}{C}}}}OH \quad \xrightarrow{\ H_2SO_4\ } \quad$$

---

15 REACTIONS OF AMIDES

Amides are very unreactive compounds, which is comforting, since the proteins that impart strength to biological structures and catalyze the reactions that take place in cells are composed of amino acids linked together by amide bonds. Amides do not react with halide ions, alcohols, or water because, in each case, the incoming nucleophile is a weaker base than the leaving group of the amide (Table 1).

$$\underset{R}{\overset{O}{\underset{\quad}{\overset{\parallel}{C}}}}NHCH_2CH_2CH_3 \quad + \quad Cl^- \quad \longrightarrow \quad \text{no reaction}$$

$$\underset{R}{\overset{O}{\underset{\quad}{\overset{\parallel}{C}}}}NHCH_3 \quad + \quad CH_3OH \quad \longrightarrow \quad \text{no reaction}$$

$$\underset{R}{\overset{O}{\underset{\quad}{\overset{\parallel}{C}}}}NHCH_2CH_3 \quad + \quad H_2O \quad \longrightarrow \quad \text{no reaction}$$

We will see, however, that amides do react with water and alcohols under acidic conditions (Section 16).

Molecular orbital theory can explain why amides are unreactive. In Section 2, we saw that amides have an important resonance contributor in which nitrogen shares its lone pair with the carbonyl carbon. The orbital that contains the lone pair overlaps the empty π^* antibonding orbital of the carbonyl carbon. This overlap lowers the energy of the lone pair—so that it is neither basic nor nucleophilic—and raises the energy of the π^* antibonding orbital of the carbonyl group, making it less reactive to nucleophiles (Figure 4).

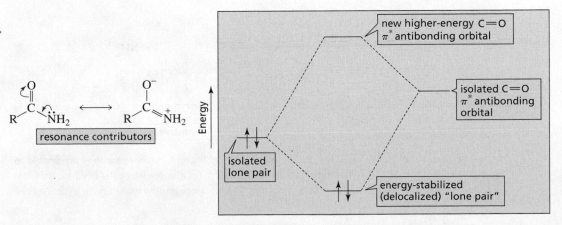

◀ **Figure 4**

The filled nonbonding orbital containing nitrogen's lone pair overlaps the empty π^* antibonding orbital of the carbonyl carbon. This stabilizes the lone pair and raises the energy of the π^* orbital of the carbonyl carbon.

781

The Discovery of Penicillin

Sir Alexander Fleming was a professor of bacteriology at the University of London. The story is told that one day Fleming was about to throw away a culture of staphylococcal bacteria that had been contaminated by a rare strain of the mold *Penicillium notatum.* He noticed that the bacteria had disappeared wherever there was a particle of mold. This suggested to him that the mold must have produced an antibacterial substance. Ten years later, in 1938, Howard Florey and Ernest Chain isolated the active substance—penicillin G—but the delay allowed the sulfa drugs to be the first antibiotics. After penicillin G was found to cure bacterial infections in mice, it was used successfully in 1941 on nine cases of human bacterial infections. By 1943, it was being produced for the military and was first used for war casualties in Sicily and Tunisia. The drug became available to the civilian population in 1944. The pressure of the war made the determination of penicillin G's structure a priority because once its structure was determined, large quantities of the drug could conceivably be synthesized.

Fleming, Florey, and Chain shared the 1945 Nobel Prize in Physiology or Medicine. Chain also discovered penicillinase, the enzyme that destroys penicillin (see the box "Penicillin and Drug Resistance"). Although Fleming is generally given credit for the discovery of penicillin, there is clear evidence that the germicidal activity of the mold was recognized in the nineteenth century by Lord Joseph Lister (1827–1912), the English physician renowned for the introduction of aseptic surgery in 1865. Unfortunately, it took several years for the surgical profession to follow his example.

Penicillin and Drug Resistance

Penicillin G

the reactive part of a penicillin is its β-lactam ring

The antibiotic activity of penicillin results from its ability to acylate (put an acyl group on) a CH_2OH group of an enzyme that has a role in the synthesis of bacterial cell walls. Acylation occurs by a nucleophilic addition–elimination reaction: the CH_2OH group adds to the carbonyl carbon of the β-lactam, forming a tetrahedral intermediate (red arrows); when the π bond re-forms, the strain in the four-membered ring increases the leaving propensity of the amino group (blue arrows).

Acylation inactivates the enzyme, and actively growing bacteria die because they are unable to synthesize functional cell walls. Penicillin has no effect on mammalian cells because they are not enclosed by cell walls. Penicillins are stored at cold temperatures to minimize hydrolysis of the β-lactam.

Bacteria that are resistant to penicillin secrete penicillinase, an enzyme that catalyzes hydrolysis of the β-lactam. The ring-opened product has no antibacterial activity.

penicillin + H_2O → [penicillinase] → penicillinoic acid

no antibacterial activity

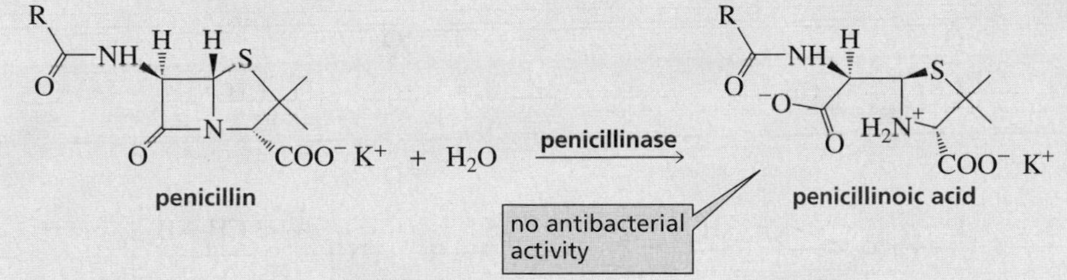

Penicillins in Clinical Use

More than 10 different penicillins are currently in clinical use. They differ only in the group (R) attached to the carbonyl group. The variable groups (R) of these penicillins are shown here. In addition to their structural differences, the penicillins differ in the organisms against which they are most effective. They also differ in their susceptibility to penicillinase. For example, methicillin, a synthetic penicillin, is effective against bacteria that are resistant to penicillin G, a naturally occurring penicillin. Almost 19% of humans are allergic to penicillin G.

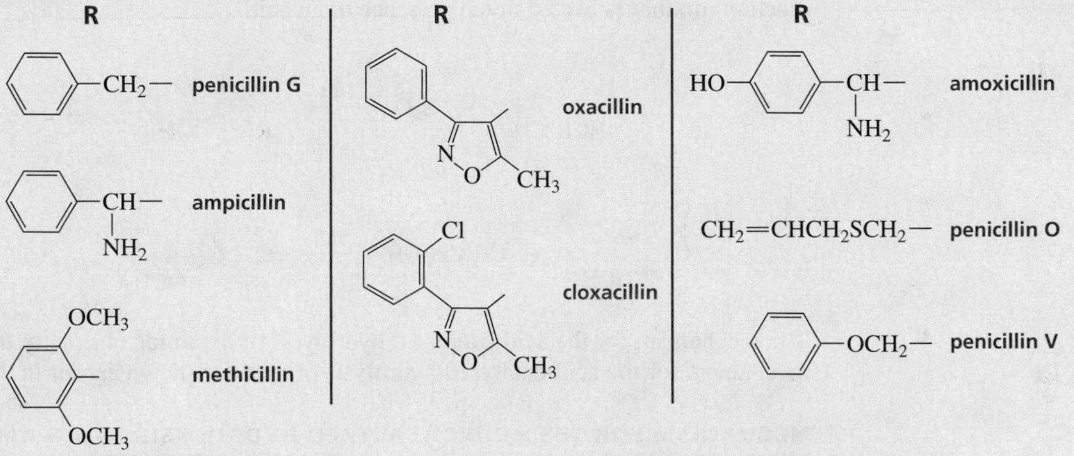

penicillin G

ampicillin

methicillin

oxacillin

cloxacillin

amoxicillin

penicillin O $CH_2=CHCH_2SCH_2-$

penicillin V

A Semisynthetic Penicillin

Penicillin V is a semisynthetic penicillin in clinical use. It is not a naturally occurring penicillin, but it is also not a true synthetic penicillin because chemists do not synthesize it. The *Penicillium* mold synthesizes it after being fed 2-phenoxyethanol, the compound needed for the R group.

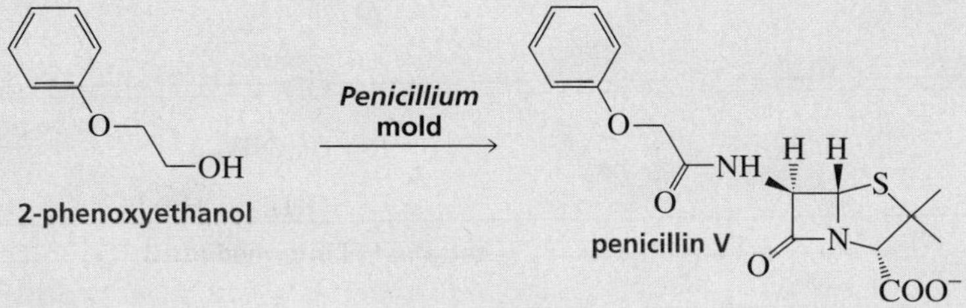

2-phenoxyethanol → [*Penicillium* mold] → penicillin V

PROBLEM 40♦

What acyl chloride and what amine would be required to synthesize the following amides?

a. *N*-ethylbutanamide
b. *N,N*-dimethylbenzamide

PROBLEM 41♦

Which of the following reactions would lead to the formation of an amide?

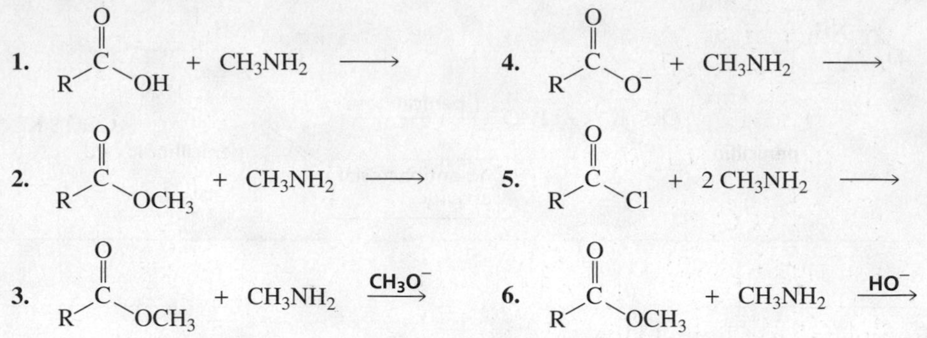

1. $\underset{R}{\overset{O}{\underset{}{\parallel}}}\!\!C\!\!-\!\!OH$ + CH$_3$NH$_2$ ⟶

2. $\underset{R}{\overset{O}{\parallel}}C\!\!-\!\!OCH_3$ + CH$_3$NH$_2$ ⟶

3. $\underset{R}{\overset{O}{\parallel}}C\!\!-\!\!OCH_3$ + CH$_3$NH$_2$ $\xrightarrow{CH_3O^-}$

4. $\underset{R}{\overset{O}{\parallel}}C\!\!-\!\!O^-$ + CH$_3$NH$_2$ ⟶

5. $\underset{R}{\overset{O}{\parallel}}C\!\!-\!\!Cl$ + 2 CH$_3$NH$_2$ ⟶

6. $\underset{R}{\overset{O}{\parallel}}C\!\!-\!\!OCH_3$ + CH$_3$NH$_2$ $\xrightarrow{HO^-}$

16 ACID-CATALYZED AMIDE HYDROLYSIS AND ALCOHOLYSIS

Amides react with water to form carboxylic acids and with alcohols to form esters, if the reaction mixture is heated in the presence of an acid.

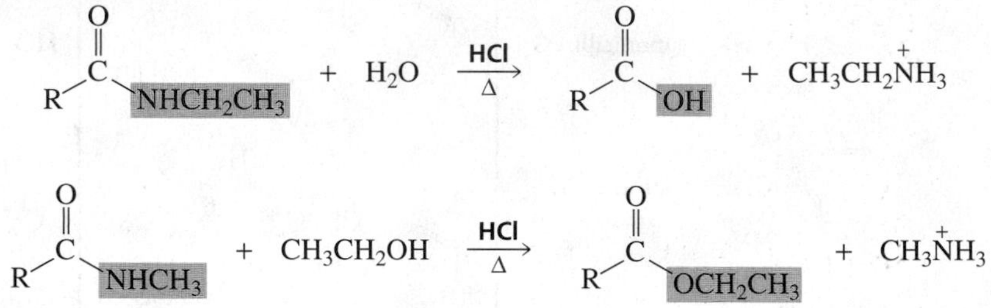

$$\underset{R}{\overset{O}{\parallel}}C\!\!-\!\!NHCH_2CH_3 + H_2O \xrightarrow[\Delta]{HCl} \underset{R}{\overset{O}{\parallel}}C\!\!-\!\!OH + CH_3CH_2\overset{+}{N}H_3$$

$$\underset{R}{\overset{O}{\parallel}}C\!\!-\!\!NHCH_3 + CH_3CH_2OH \xrightarrow[\Delta]{HCl} \underset{R}{\overset{O}{\parallel}}C\!\!-\!\!OCH_2CH_3 + CH_3\overset{+}{N}H_3$$

The mechanism for the acid-catalyzed hydrolysis of an amide is exactly the same as the mechanism for the acid-catalyzed hydrolysis of an ester shown earlier in this chapter.

MECHANISM FOR THE ACID-CATALYZED HYDROLYSIS OF AN AMIDE

Notice again the pattern of the three tetrahedral intermediates:

protonated tetrahedral intermediate →
neutral tetrahedral intermediate →
protonated tetrahedral intermediate.

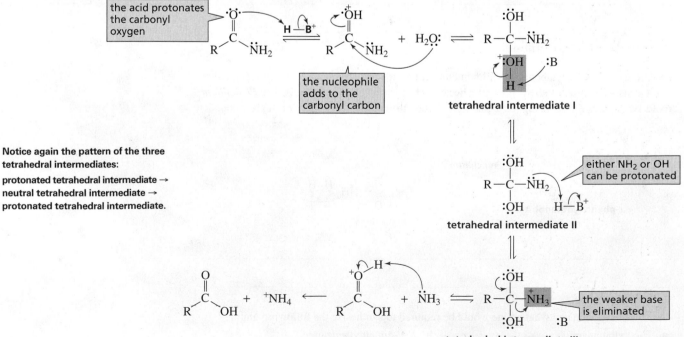

- The acid protonates the carbonyl oxygen, which increases the susceptibility of the carbonyl carbon to nucleophilic addition.

- Addition of the nucleophile (H_2O) to the carbonyl carbon leads to tetrahedral intermediate I, which is in equilibrium with its nonprotonated form, tetrahedral intermediate II.

- Re-protonation can occur either on oxygen to re-form tetrahedral intermediate I or on nitrogen to form tetrahedral intermediate III. Protonation on nitrogen is favored because the NH_2 group is a stronger base than the OH group.

- Of the two possible leaving groups in tetrahedral intermediate III (HO^- and NH_3), NH_3 is the weaker base, so it is the one eliminated.

- Because the reaction is carried out in an acidic solution, NH_3 will be protonated after it is eliminated from the tetrahedral intermediate. This prevents the reverse reaction from occurring since $^+NH_4$ is not a nucleophile.

Let's take a minute to see why an amide cannot be hydrolyzed without a catalyst. In an uncatalyzed reaction, the amide would not be protonated. Therefore, water, a very poor nucleophile, would have to add to a neutral amide that is much less susceptible to nucleophilic addition than a protonated amide would be. More importantly in the uncatalyzed reaction, the NH_2 group of the tetrahedral intermediate would not be protonated. Therefore, HO^- would be eliminated from the tetrahedral intermediate (because HO^- is a weaker base than $^-NH_2$), which would re-form the amide.

| the leaving group in acid-catalyzed amide hydrolysis | the leaving group in (unsuccessively) uncatalyzed amide hydrolysis |

When an amide reacts with an alcohol in the presence of acid to form an ester, it follows the same mechanism as it does when it reacts with water to form a carboxylic acid.

Dalmatians: Do Not Fool with Mother Nature

When amino acids are metabolized, the excess nitrogen is concentrated into uric acid, a compound with five amide bonds. A series of enzyme-catalyzed hydrolysis reactions degrade uric acid—one amide bond at a time—all the way to ammonium ion. The extent to which uric acid is degraded depends on the species. Primates, birds, reptiles, and insects excrete excess nitrogen as uric acid. Other mammals excrete excess nitrogen as allantoin. Excess nitrogen in aquatic animals is excreted as allantoic acid, urea, or as ammonium salts.

Dalmatians, unlike other dogs, excrete high levels of uric acid. This is because breed-ers of Dalmatians have selected dogs that have no white hairs in their black spots, and the gene that causes the white hairs is linked to the gene that causes uric acid to be hydrolyzed to allantoin. Dalmatians, therefore, are susceptible to gout, a painful buildup of uric acid in joints.

PROBLEM 42

Write the mechanism for the acid-catalyzed reaction of an amide with an alcohol to form an ester.

PROBLEM 43◆

List the following amides in order from greatest reactivity to least reactivity toward acid-catalyzed hydrolysis:

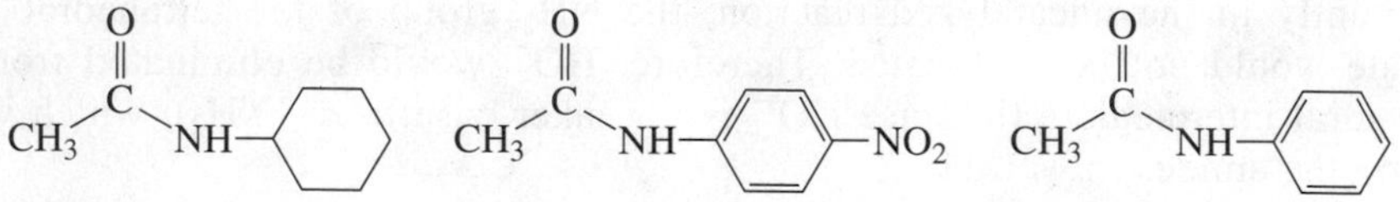

17   HYDROXIDE-ION-PROMOTED HYDROLYSIS OF AMIDES

Amides can also be hydrolyzed when heated under strongly basic conditions.

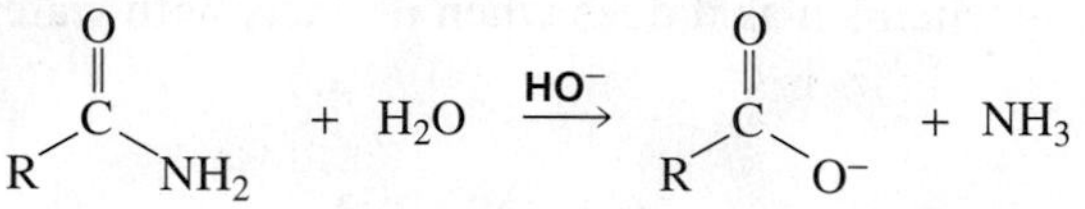

MECHANISM FOR THE HYDROXIDE-ION-PROMOTED HYDROLYISIS OF AN AMIDE

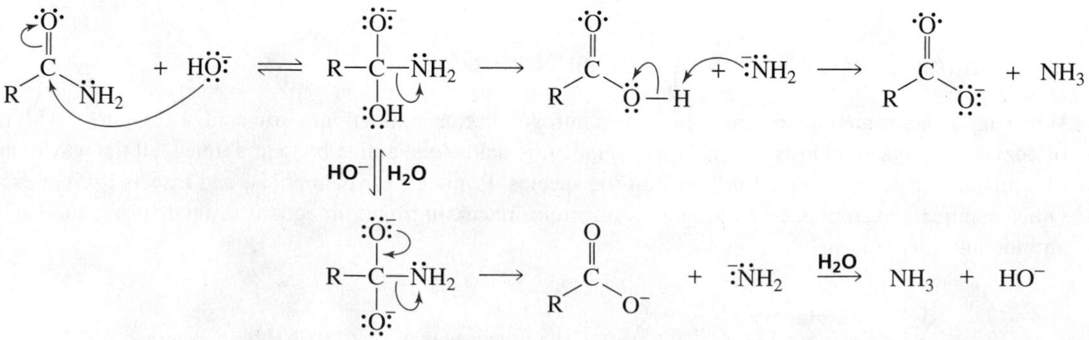

- Hydroxide ion is the nucleophile instead of water. Because it is a better nucleophile than water, it forms the tetrahedral intermediate more rapidly.

- Of the two potential leaving groups in the tetrahedral intermediate, $^-$OH is the weaker base and therefore the one more likely to be eliminated, thereby reforming the amide.

- Occasionally, an $^-$NH$_2$ is eliminated. When this happens, the carboxylic acid that is formed immediately loses a proton. Since this step is irreversible (the negatively charged carboxylate ion is not reactive), it disturbs the equilibrium and drives the reaction toward products.

- In strongly basic solutions, the reaction is second order in hydroxide ion. That is, two equivalents of hydroxide ion participate in the reaction. The second equivalent of hydroxide ion removes a proton from the initially formed tetrahedral intermediate.

- Now the possible leaving groups are $^-$NH$_2$ and O^{2-}. Because $^-$NH$_2$ is the weaker base, it is eliminated and the carboxylate ion is formed.

Notice that one equivalent of hydroxide ion is not a catalyst (it is not regenerated) but the second equivalent is regenerated, so it is a catalyst.

18 THE HYDROLYSIS OF AN IMIDE: A WAY TO SYNTHESIZE PRIMARY AMINES

Although primary amines can be prepared by S$_N$2 reactions with alkyl halides, the yields are poor because it is difficult to stop the reaction after one alkyl group has been placed on the nitrogen. A much better way to prepare a primary amine from an alkyl halide is by means of a **Gabriel synthesis.** The Gabriel synthesis involves the hydrolysis of an imide. An **imide** is a compound with two acyl groups bonded to a nitrogen.

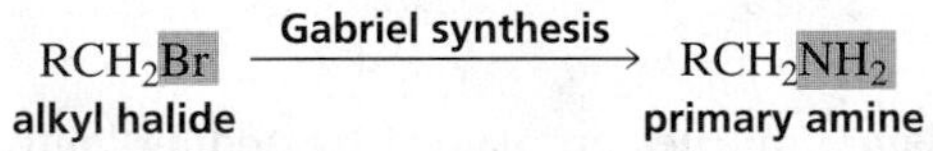

The steps involved in the synthesis are shown next. Notice that the alkyl group of the alkyl halide used in the second step of the reaction is identical to the alkyl group of the desired primary amine.

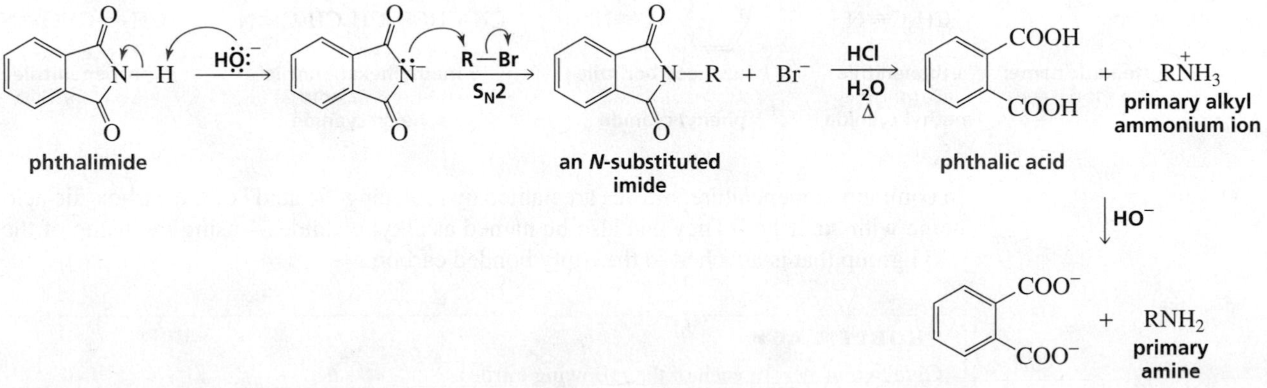

- A base removes a proton from the nitrogen of phthalimide.

- The resulting nucleophile reacts with an alkyl halide. Because this is an S$_N$2 reaction, it works best with primary alkyl halides.

- Hydrolysis of the two amide bonds of the *N*-substituted imide is catalyzed by acid. Because the solution is acidic, the final products are a primary alkyl ammonium ion and phthalic acid.

- Reaction of the alkyl ammonium ion with base forms the primary amine.

Only one alkyl group can be placed on the nitrogen because there is only one hydrogen bonded to the nitrogen of phthalimide. This means that the Gabriel synthesis can be used only for the preparation of primary amines.

PROBLEM 44♦

What alkyl bromide would you use in a Gabriel synthesis to prepare each of the following amines?

a. pentylamine   **b.** isohexylamine   **c.** benzylamine   **d.** cyclohexylamine

---

PROBLEM 45

Primary amines can also be prepared by the reaction of an alkyl halide with azide ion, followed by catalytic hydrogenation. What advantage do this method and the Gabriel synthesis have over the synthesis of a primary amine using an alkyl halide and ammonia?

$$CH_3CH_2CH_2Br \xrightarrow{\ ^-N_3\ } CH_3CH_2CH_2N{=}\overset{+}{N}{=}\overset{-}{N} \xrightarrow[Pd/C]{H_2} CH_3CH_2CH_2NH_2 + N_2$$

---

19 NITRILES

acetonitrile

Nitriles are compounds that contain a cyano (C≡N) group. They are considered to be carboxylic acid derivatives because, like all the other carboxylic acid derivatives, they can be hydrolyzed to carboxylic acids.

Naming Nitriles

In systematic nomenclature, nitriles are named by adding "nitrile" to the name of the parent alkane. Notice in the following examples that the triple-bonded carbon of the nitrile group is included in the number of carbons in the longest continuous chain.

	$CH_3C{\equiv}N$	phenyl—C≡N	$CH_3CHCH_2CH_2CH_2C{\equiv}N$ (CH$_3$)	$CH_2{=}CHC{\equiv}N$
systematic name:	ethanenitrile	benzenecarbonitrile	5-methylhexanenitrile	propenenitrile
common name:	acetonitrile	benzonitrile	δ-methylcapronitrile	acrylonitrile
	methyl cyanide	phenyl cyanide	isohexyl cyanide	

In common nomenclature, nitriles are named by replacing "ic acid" of the carboxylic acid name with "onitrile." They can also be named as alkyl cyanides—using the name of the alkyl group that is attached to the triply bonded carbon.

PROBLEM 46♦

Give two names for each of the following nitriles:
a. $CH_3CH_2CH_2C{\equiv}N$ **b.** $CH_3CHCH_2CH_2C{\equiv}N$ with CH_3

Reactions of Nitriles

Nitriles are even harder to hydrolyze than amides, but they slowly hydrolyze to carboxylic acids when heated with water and an acid.

$$RC{\equiv}N \xrightarrow[\Delta]{HCl,\ H_2O} \underset{R}{\overset{O}{\underset{\ }{C}}}{-}OH$$

MECHANISM FOR THE ACID-CATALYZED HYDROLYSIS OF A NITRILE

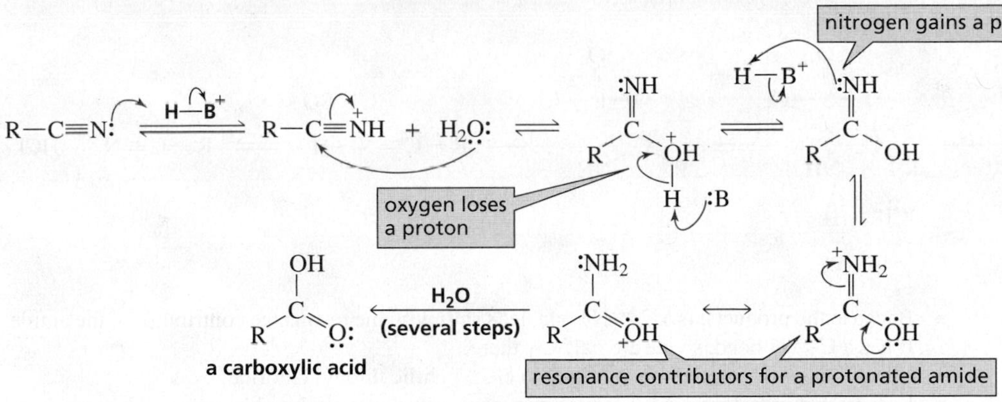

Notice again the pattern of the three intermediates:

protonated intermediate →
neutral intermediate →
protonated intermediate.

- The acid protonates the nitrogen of the cyano group, which makes the carbon of the cyano group more susceptible to the addition of water. (The addition of water to a protonated cyano group is analogous to the addition of water to a protonated carbonyl group.)

- A base removes a proton from oxygen, forming a neutral species that can be reprotonated on oxygen or protonated on nitrogen. Protonation on nitrogen forms a protonated amide, whose two resonance contributors are shown. (Notice that one of the resonance contributors is a protonated amide.)

- The protonated amide is immediately hydrolyzed to a carboxylic acid—because an amide is easier to hydrolyze than a nitrile—by means of the acid-catalyzed mechanism shown earlier in this chapter.

Nitriles can be prepared from an S_N2 reaction of alkyl halide with cyanide ion. Because a nitrile can be hydrolyzed to a carboxylic acid, you now know how to convert an alkyl halide into a carboxylic acid. Notice that the carboxylic acid has one more carbon than the alkyl halide.

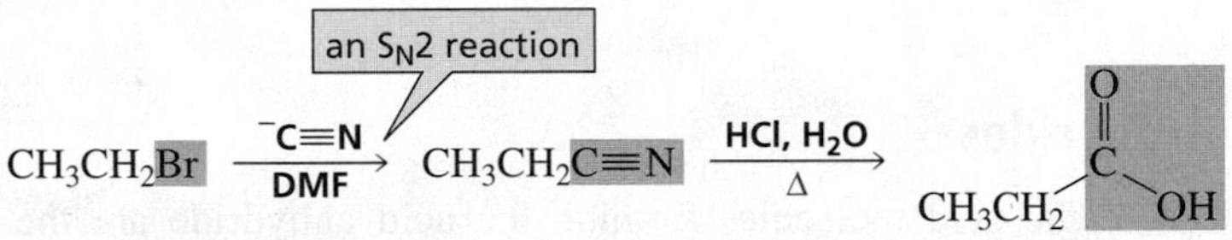

Catalytic hydrogenation of a nitrile is another way to make a primary amine. Raney nickel is the preferred metal catalyst for this reduction.

$$RC{\equiv}N \xrightarrow[\text{Raney nickel}]{H_2} RCH_2NH_2$$

PROBLEM 47♦

Which alkyl halides form the carboxylic acids listed here after reaction with sodium cyanide followed by heating the product in an acidic aqueous solution?

a. butyric acid   **b.** isovaleric acid   **c.** cyclohexanecarboxylic acid

PROBLEM 48 Solved

An amide with an NH_2 group can be dehydrated to a nitrile with thionyl chloride ($SOCl_2$). Propose a mechanism for this reaction.

Solution

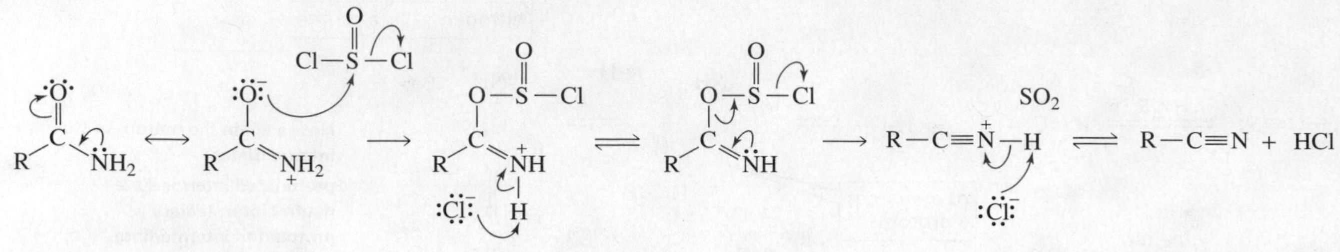

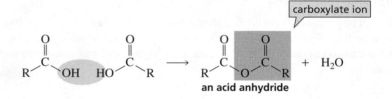

- Because the product has a C≡N bond, let's start with the resonance contributor of the amide. It has a C=N bond, so we are halfway there.
- The nucleophilic oxygen reacts with the electrophilic thionyl chloride.
- Loss of a proton forms a compound that can form the triple bond because of its very good leaving group.
- Loss of another proton forms the nitrile.

20 ACID ANHYDRIDES

Loss of water from two molecules of a carboxylic acid results in an **acid anhydride.** "Anhydride" means "without water." An anhydride is a *carboxylic acid derivative*, because the OH of a carboxylic acid has been replaced by a carboxylate ion.

acetic anhydride

Naming Anhydrides

If the two carboxylic acid molecules forming the acid anhydride are the same, then the anhydride is a **symmetrical anhydride.** If they are different, then it is a **mixed anhydride.** Symmetrical anhydrides are named by replacing "acid" in the acid name with "anhydride." Mixed anhydrides are named by stating the names of both acids in alphabetical order, followed by "anhydride."

systematic name:	ethanoic anhydride
common name:	acetic anhydride
	a symmetrical anhydride

ethanoic methanoic anhydride
acetic formic anhydride
a mixed anhydride

Reactions of Anhydrides

The leaving group of an anhydride is a carboxylate ion (its conjugate acid has a pK_a of ~5), which means that an anhydride is less reactive than an acyl chloride but more reactive than an ester or a carboxylic acid.

relative reactivities of carboxylic acid derivatives

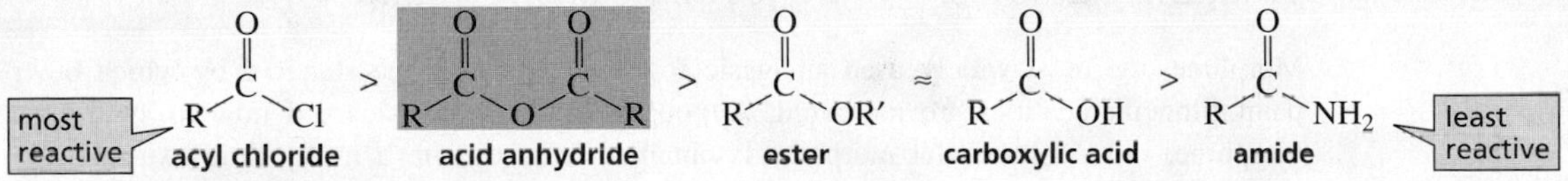

Therefore, an acid anhydride reacts with an alcohol to form an ester and a carboxylic acid, with water to form two equivalents of a carboxylic acid, and with an amine to form an amide and a carboxylate ion. In each case, the incoming nucleophile—after it loses a proton—is a stronger base than the departing carboxylate ion. (Recall that a carboxylic acid derivative can be converted to one that is less reactive but not to one that is more reactive.)

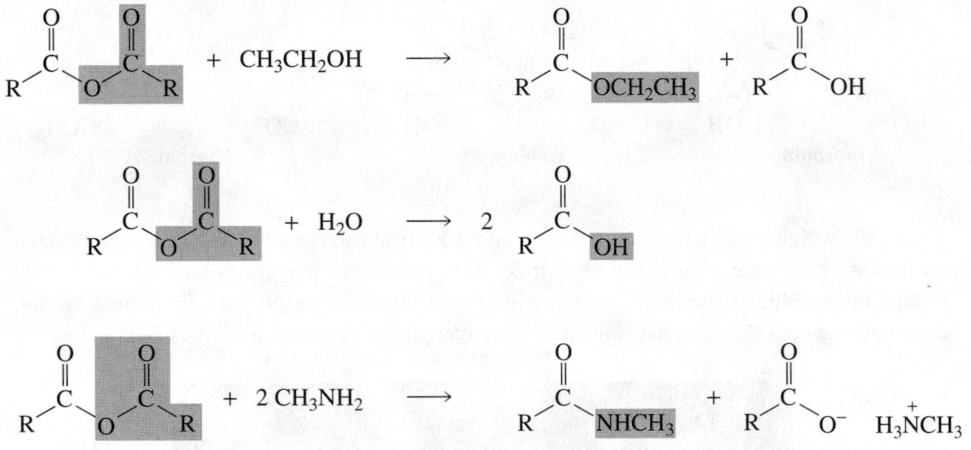

In the reaction of an amine with an anhydride, two equivalents of amine must be used so that sufficient amine will be present to react with both the carbonyl compound and the proton produced in the reaction (Section 8).

The reactions of acid anhydrides follow the general mechanism described in Section 7. For example, compare the mechanism for the reaction of an acid anhydride with an alcohol to the mechanism for the reaction of an acyl chloride with an alcohol earlier in this chapter.

MECHANISM FOR THE REACTION OF AN ACID ANHYDRIDE WITH AN ALCOHOL

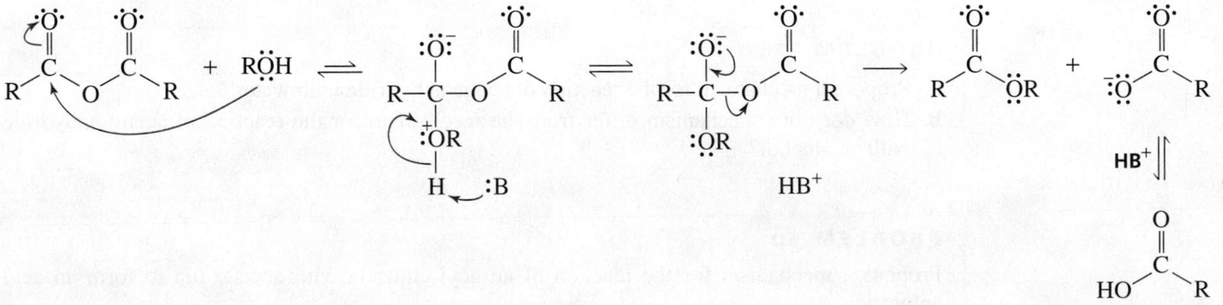

- The nucleophile adds to the carbonyl carbon, forming a tetrahedral intermediate.
- A proton is removed from the tetrahedral intermediate.
- The carboxylate ion, the weaker of the two bases in the tetrahedral intermediate, is eliminated.

What Drug-Enforcement Dogs Are Really Detecting

Morphine, the most widely used analgesic for severe pain, is the standard by which other painkilling medications are measured. Although scientists have learned how to synthesize morphine, most commercial morphine is obtained from opium, a milky fluid exuded by a species of poppy. Morphine occurs in opium at concentrations as high as 10%. Opium was used for its analgesic properties as early as 4000 B.C. In Roman times both opium use and opium addiction were widespread. Methylating one of the OH groups of morphine produces codeine, which has one-tenth the analgesic activity of morphine. Codeine profoundly inhibits the cough reflex.

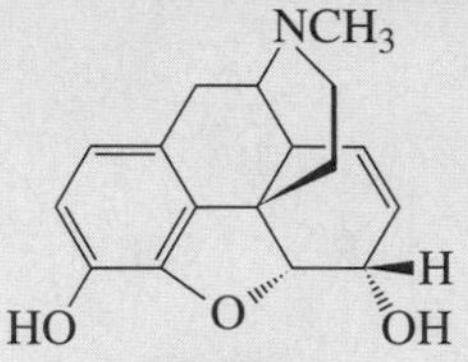

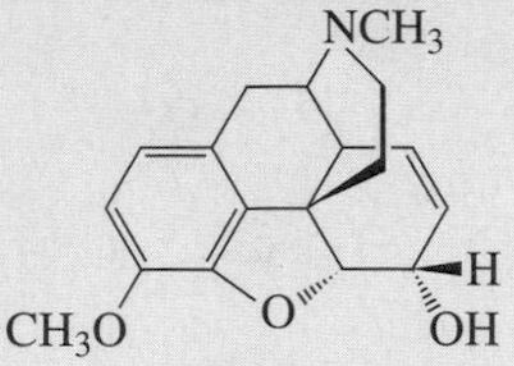

Heroin, which is much more potent (and more widely abused) than morphine, is synthesized by treating morphine with acetic anhydride. This puts an acetyl group on each of the OH groups of morphine. Therefore, acetic acid is also formed as a product. To detect heroin, drug-enforcement agencies use dogs trained to recognize the pungent odor of acetic acid.

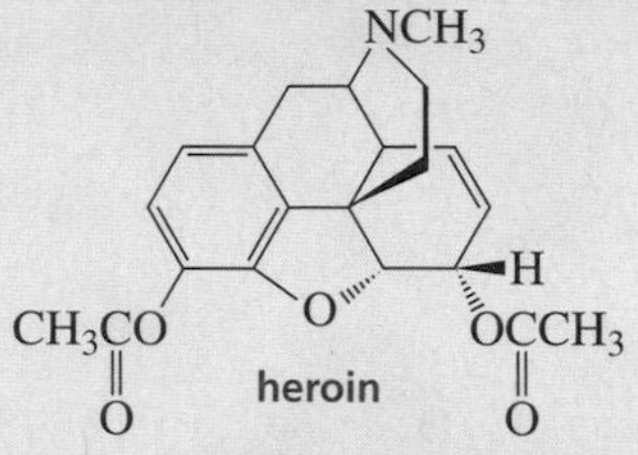

PROBLEM 49

a. Propose a mechanism for the reaction of acetic anhydride with water.

b. How does this mechanism differ from the mechanism for the reaction of acetic anhydride with an alcohol?

PROBLEM 50

Propose a mechanism for the reaction of an acyl chloride with acetate ion to form an acid anhydride.

PROBLEM 51♦

We have seen that acid anhydrides react with alcohols, water, and amines. In which of these reactions can the tetrahedral intermediate eliminate the carboxylate ion even if it does not lose a proton before the elimination step? Explain.

21 / DICARBOXYLIC ACIDS

The structures of some common dicarboxylic acids and their pK_a values are listed in Table 3.

Table 3 Structures, Names, and pK_a Values of Some Simple Dicarboxylic Acids

Dicarboxylic acid	Common name	pK_{a1}	pK_{a2}
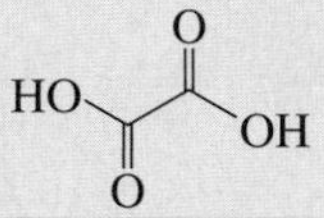	Oxalic acid	1.27	4.27
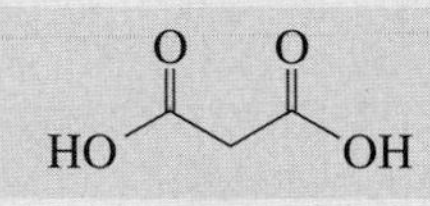	Malonic acid	2.86	5.70
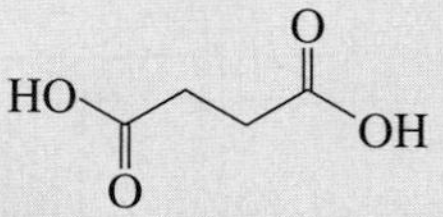	Succinic acid	4.21	5.64
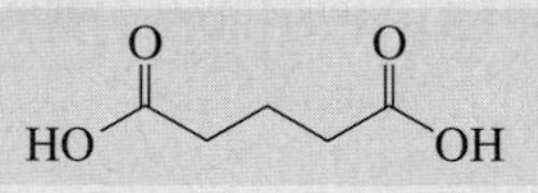	Glutaric acid	4.34	5.27
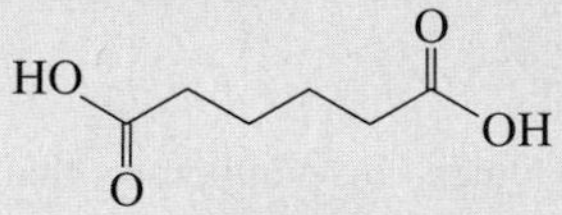	Adipic acid	4.41	5.28
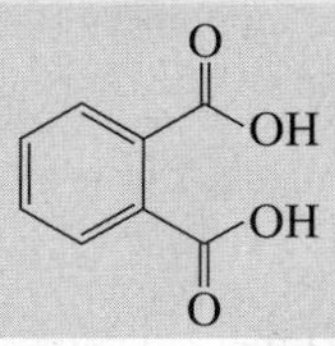	Phthalic acid	2.95	5.41

Although the two carboxyl groups of a dicarboxylic acid are identical, the two pK_a values are different because the protons are lost one at a time and therefore leave from different species—namely, the first proton is lost from a neutral molecule, then the second proton is lost from a negatively charged ion.

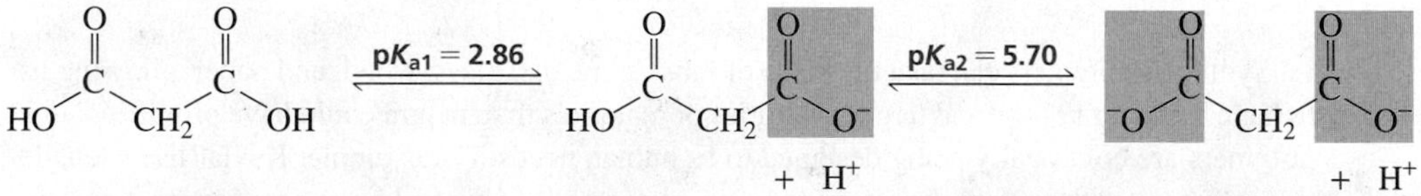

A COOH group withdraws electrons (compared to an H) and therefore increases the acidity of the first COOH group. The pK_a values of the dicarboxylic acids show that the acid-strengthening effect of the COOH group decreases as the separation between the two carboxyl groups increases.

Dicarboxylic acids readily lose water when heated, if they can form a cyclic anhydride with a five- or a six-membered ring.

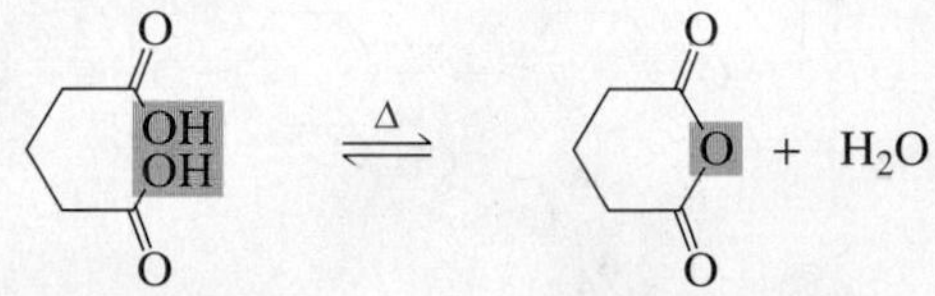

glutaric acid   glutaric anhydride

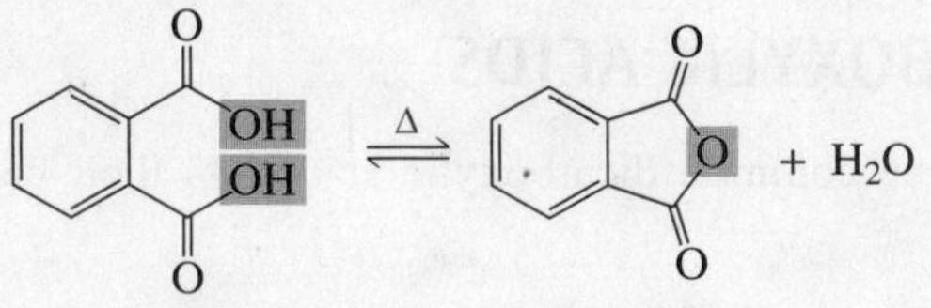

phthalic acid   phthalic anhydride

Cyclic anhydrides are more easily prepared if the dicarboxylic acid is heated in the presence of acetyl chloride or acetic anhydride.

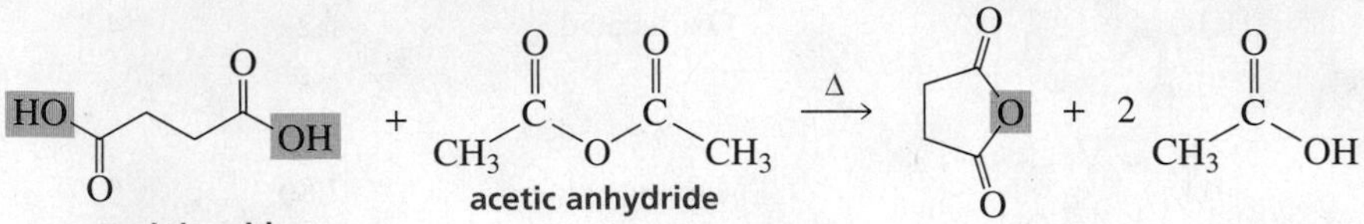

succinic acid

acetic anhydride

succinic anhydride

PROBLEM 52

a. Propose a mechanism for the formation of succinic anhydride in the presence of acetic anhydride.
b. How does acetic anhydride make it easier to form the anhydride?

Synthetic Polymers

Synthetic polymers play important roles in our daily lives. Polymers are compounds that are made by linking together many small molecules called monomers. The monomers of many synthetic polymers are held together by ester and amide bonds. For example, Dacron is a polyester and nylon is a polyamide.

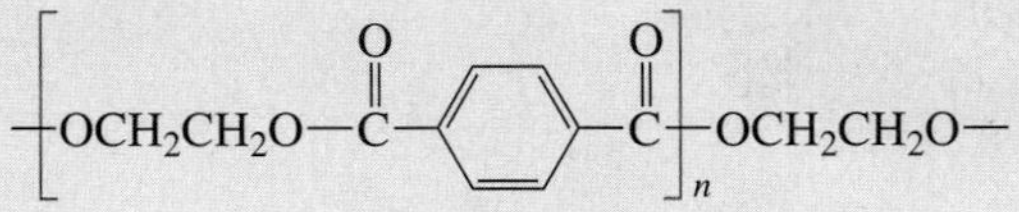

Dacron®

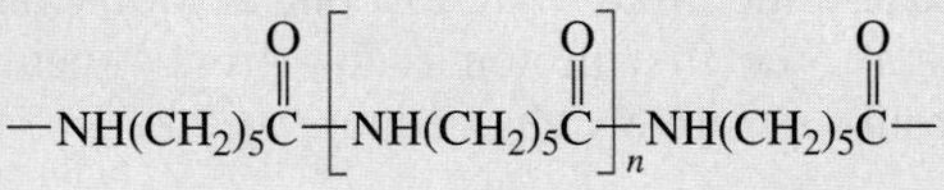

nylon 6

Synthetic polymers can take the place of fabrics, metals, glass, wood, and paper, allowing us to have a greater variety and larger quantities of materials than nature could have provided. New polymers are continually being designed to fit human needs. For example, Kevlar has a tensile strength greater than steel. It is used for high-performance skis and bulletproof vests. Lexan is a strong and transparent polymer used for such things as traffic light lenses and compact disks.

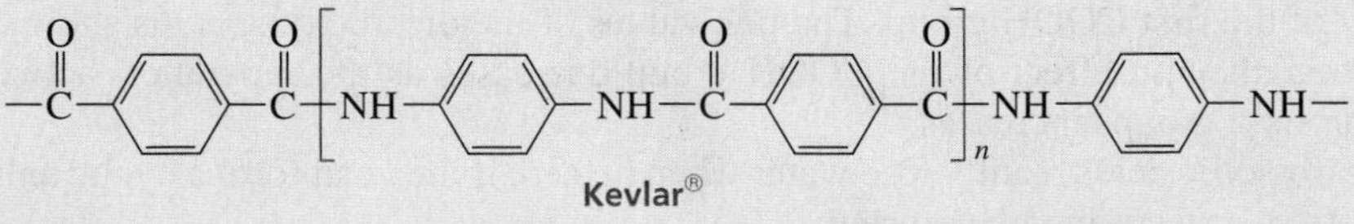

Kevlar®

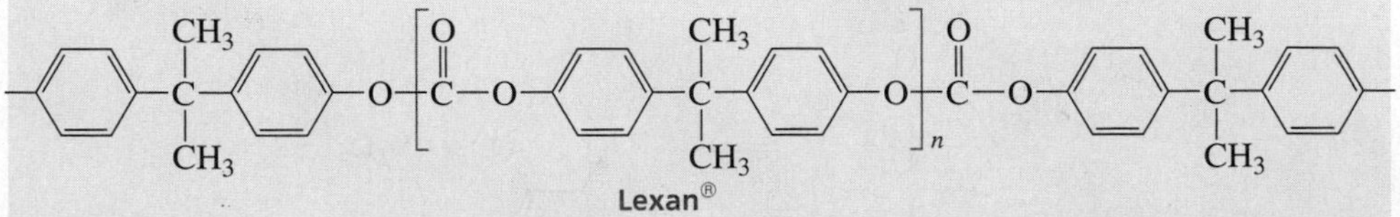

Lexan®

Dissolving Sutures

Dissolving sutures, such as dexon and poly(dioxanone) (PDS), are synthetic polymers that are now routinely used in surgery. The many ester groups they contain are slowly hydrolyzed to small molecules that are then metabolized to compounds easily excreted by the body. Patients no longer have to undergo a second medical procedure to remove the sutures that was required when traditional suture materials were used.

Depending on their structures, these synthetic sutures lose 50% of their strength after two to three weeks and are completely absorbed within three to six months.

$$\left[OCH_2C-OCH_2C \right]_n \qquad \left[OCH_2CH_2OCH_2C-OCH_2CH_2OCH_2C \right]_n$$

Dexon® **PDS®**

PROBLEM 53♦

One of the two polymers used for sutures shown in the box "Synthetic Polymers" loses 50% of its strength in two weeks, and the other loses that much strength in three weeks. Which suture material lasts longer?

---

22 HOW CHEMISTS ACTIVATE CARBOXYLIC ACIDS

Of the various classes of carbonyl compounds discussed in this chapter—acyl halides, acid anhydrides, esters, carboxylic acids, and amides—carboxylic acids are the most commonly available, both in the laboratory and in cells. Therefore, carboxylic acids are the reagents most likely to be available when a chemist or a cell needs to synthesize a carboxylic acid derivative. However, we have seen that carboxylic acids are relatively unreactive toward nucleophilic addition–elimination reactions because the OH group of a carboxylic acid is a strong base and therefore a poor leaving group. And at physiological pH (pH = 7.4), a carboxylic acid is even more resistant to nucleophilic addition–elimination reactions because it exists predominantly in its unreactive, negatively charged basic form. Therefore, both organic chemists and cells need a way to activate carboxylic acids so that they can readily undergo nucleophilic addition–elimination reactions. First we will look at how chemists activate carboxylic acids, and then we will see how cells do it.

One way organic chemists activate carboxylic acids is by converting them into acyl chlorides, the most reactive of the carboxylic acid derivatives. A carboxylic acid can be converted into an acyl chloride by being heated either with thionyl chloride (SOCl₂) or with phosphorus trichloride (PCl₃).

$$\underset{R}{\overset{O}{\underset{\|}{C}}}\!\!-\!\!O^- + SOCl_2 \xrightarrow{\Delta} \underset{R}{\overset{O}{\underset{\|}{C}}}\!\!-\!\!Cl + SO_2 + Cl^-$$

thionyl chloride

$$\underset{R}{\overset{O}{\underset{\|}{C}}}\!\!-\!\!O^- + PCl_3 \xrightarrow{\Delta} \underset{R}{\overset{O}{\underset{\|}{C}}}\!\!-\!\!Cl + {}^-OPCl_2$$

phosphorus trichloride

These reagents convert the leaving group of a carboxylic acid into a better leaving group than the chloride ion.

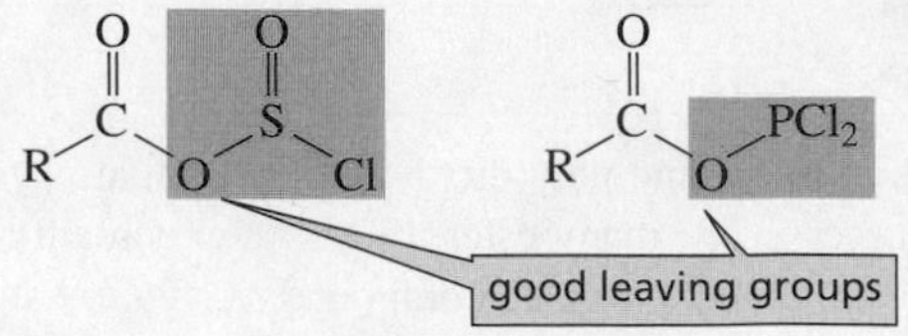

good leaving groups

As a result, when the chloride ion subsequently adds to the carbonyl carbon and forms a tetrahedral intermediate, the chloride ion is *not* the group that is eliminated.

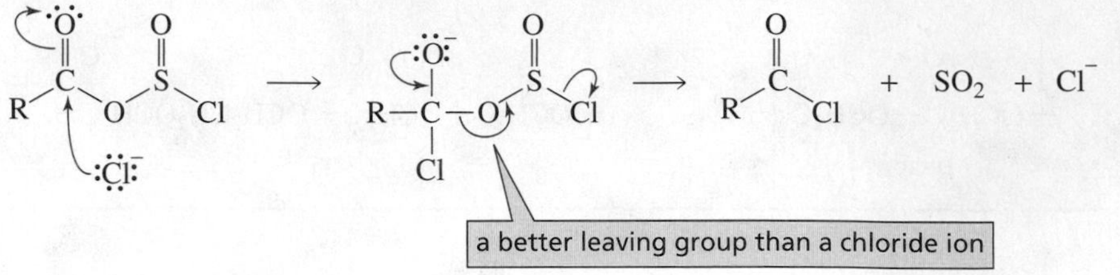

a better leaving group than a chloride ion

Notice that these reagents are the same reagents that cause the OH group of an alcohol to be replaced by a chlorine.

Once the acyl halide has been prepared, a wide variety of carboxylic acid derivatives can be synthesized by adding the appropriate nucleophile.

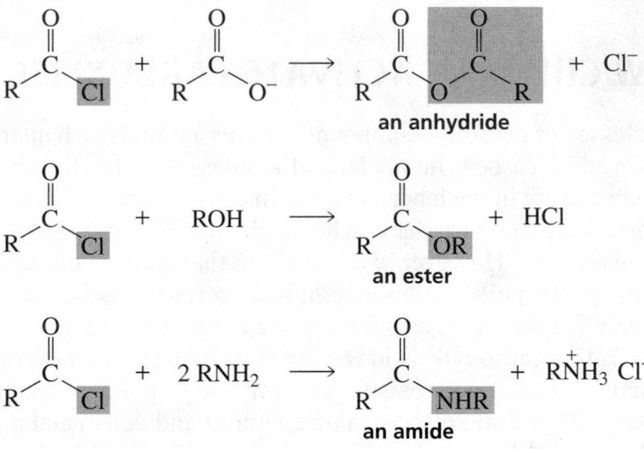

an anhydride

an ester

an amide

Carboxylic acids can also be activated for nucleophilic addition–elimination reactions by being converted into acid anhydrides by a dehydrating agent, such as P_2O_5.

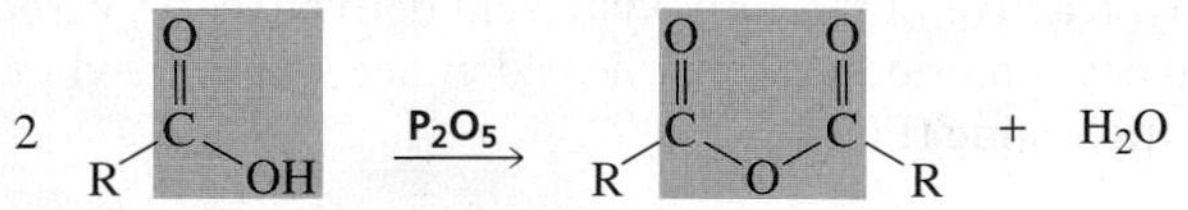

Carboxylic acids and carboxylic acid derivatives can also be prepared by methods other than nucleophilic addition–elimination reactions. A summary of the methods used to synthesize these compounds is provided in the Study Area of MasteringChemistry Appendix: "Summary of Methods Used to Synthesize a Particular Functional Group."

PROBLEM 54♦

How would you synthesize the following compounds starting with a carboxylic acid?

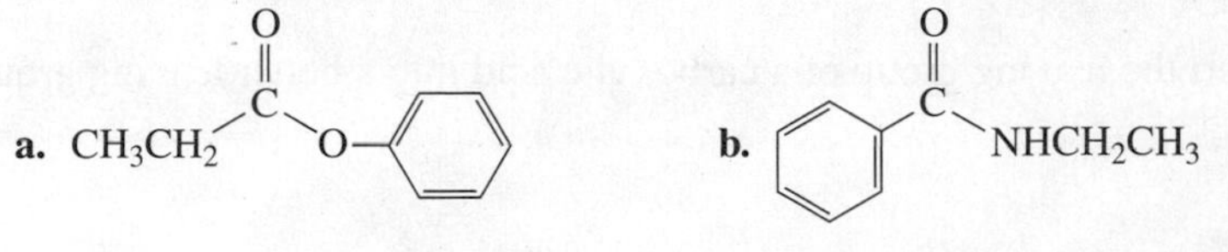

a. CH_3CH_2

b.

23 HOW CELLS ACTIVATE CARBOXYLIC ACIDS

The synthesis of compounds by a living organism is called **biosynthesis.** Acyl chlorides and acid anhydrides are too reactive to be used as reagents in cells. Cells live in a predominantly aqueous environment, and acyl halides and acid anhydrides are rapidly hydrolyzed in water. So cells must activate carboxylic acids in a different way.

When phosphoric acid is heated with P_2O_5, it loses water, forming a phosphoanhydride called pyrophosphoric acid. Its name comes from *pyr*, the Greek word for "fire," since pyrophosphoric acid is prepared by "fire"—that is, by heating. Triphosphoric acid and higher polyphosphoric acids are also formed.

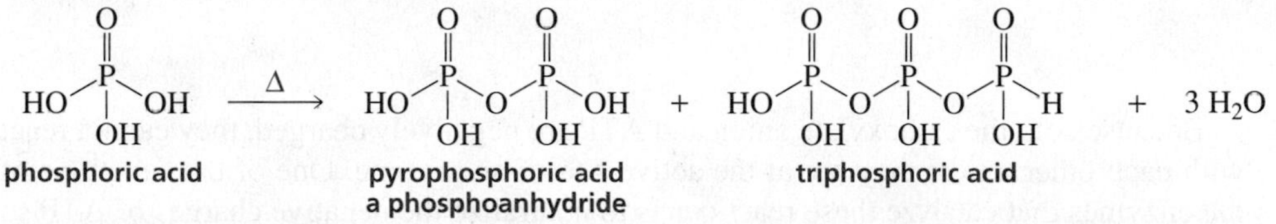

phosphoric acid   pyrophosphoric acid   triphosphoric acid
                         a phosphoanhydride

One way cells can activate a carboxylic acid is to use adenosine triphosphate (ATP) to convert the carboxylic acid into an **acyl phosphate** or an **acyl adenylate**—carbonyl compounds with good leaving groups. ATP is an ester of triphosphoric acid. Its structure is shown here, both in its entirety and with "Ad" in place of the adenosyl group.

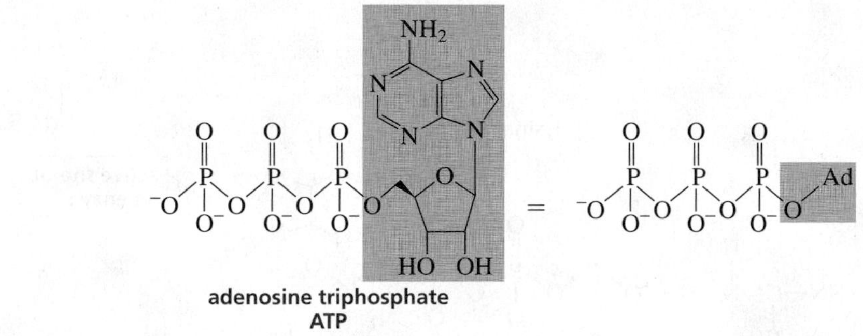

adenosine triphosphate
ATP

Acyl phosphates and acyl adenylates are mixed anhydrides of a carboxylic acid and phosphoric acid.

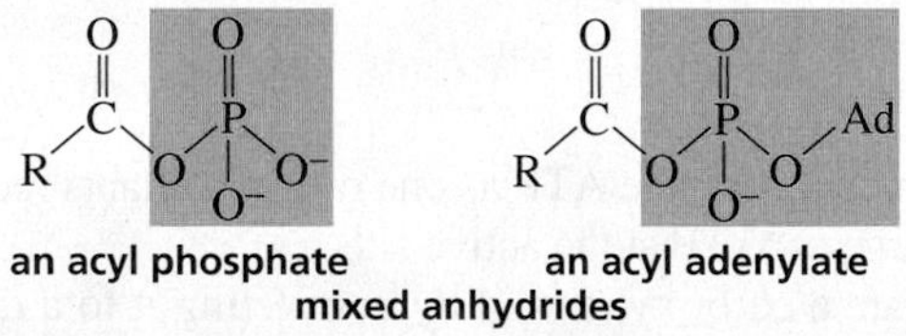

an acyl phosphate   an acyl adenylate
            mixed anhydrides

An acyl phosphate is formed by nucleophilic attack of a carboxylate ion on the γ-phosphorus (the phosphorus farthest away from the adenosyl group) of ATP. Attack of the nucleophile breaks the **phosphoanhydride bond** (rather than the π bond), so an intermediate is not formed. Essentially, it is an S_N2 reaction with an adenosine diphosphate leaving group.

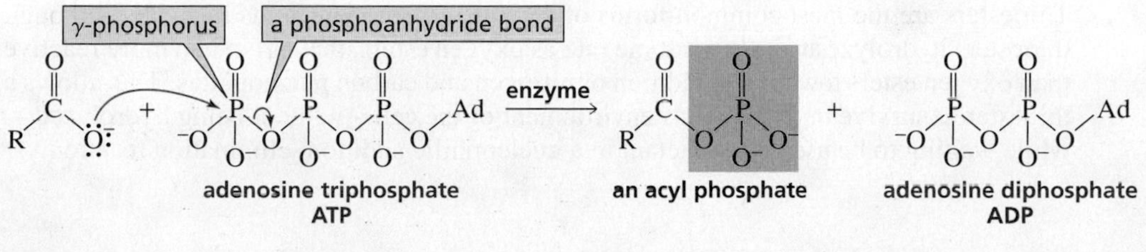

adenosine triphosphate   an acyl phosphate   adenosine diphosphate
        ATP   ADP

An acyl adenylate is formed by nucleophilic attack of a carboxylate ion on the α-phosphorus of ATP. The enzyme that catalyzes the reaction determines which phosphorus is attacked by the nucleophile, and therefore, whether an acyl phosphate or an acyl adenylate is formed. The reactions that require acyl phosphate intermediates differ from those that require acyl adenylate intermediates.

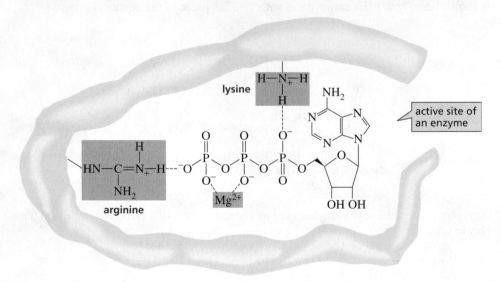

Because both the carboxylate anion and ATP are negatively charged, they cannot react with each other unless they are at the active site of an enzyme. One of the functions of the enzymes that catalyze these reactions is to neutralize the negative charges of ATP so it can react with a nucleophile (Figure 5). Another function of the enzyme is to exclude water from the active site where the reaction takes place. Otherwise, hydrolysis of the mixed anhydride formed by the reaction of a carboxylate ion and ATP would compete with the desired nucleophilic substitution reaction.

▶ Figure 5

The interactions between ATP, Mg^{2+}, and positively charged groups at the active site of an enzyme.

Enzyme-catalyzed reactions that have ATP as one of the reactants require Mg^{2+}, which helps reduce the negative charge on ATP at the active site.

Cells can also activate a carboxylic acid by converting it to a thioester. A **thioester** is an ester with a sulfur in place of the alkoxy oxygen.

Thioesters are the most common forms of activated carboxylic acids in a cell. Although thioesters hydrolyze at about the same rate as oxygen esters, they are much more reactive than oxygen esters toward the addition of nitrogen and carbon nucleophiles. This allows a thioester to survive in the aqueous environment of the cell—without being hydrolyzed—while waiting to be used as a reactant in a nucleophilic addition–elimination reaction.

The carbonyl carbon of a thioester is more susceptible to nucleophilic addition than is the carbonyl carbon of an oxygen ester, because electron delocalization onto the carbonyl oxygen, which reduces the carbonyl group's reactivity, is weaker when Y is S than when Y is O. Electron delocalization is weaker because less overlap occurs between the $3p$ orbital of sulfur and the $2p$ orbital of carbon than between the $2p$ orbital of oxygen and the $2p$ orbital of carbon. In addition, the tetrahedral intermediate formed from a thioester undergoes elimination more rapidly than the tetrahedral intermediate formed from an oxygen ester because a thiolate ion is a weaker base and is therefore easier to eliminate than an alkoxide ion.

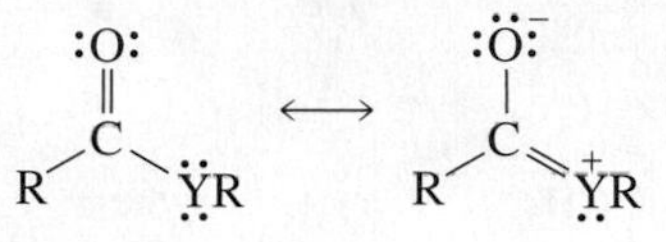

$$CH_3CH_2SH \qquad CH_3CH_2OH$$
$$pK_a = 10.5 \qquad pK_a = 15.9$$

The thiol used in biological systems for the formation of thioesters is coenzyme A. The compound is written "CoASH" to emphasize that the thiol group is the reactive part of the molecule. CoASH is composed of a decarboxylated cysteine (an amino acid), pantothenate (a vitamin), and phosphorylated adenosine diphosphate.

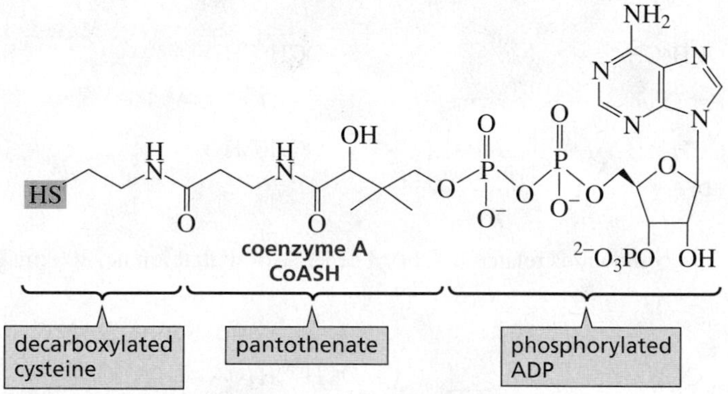

When a cell converts a carboxylic acid into a thioester, it first converts the carboxylic acid into an acyl adenylate. The acyl adenylate then reacts with CoASH to form the thioester. The most common thioester in cells is acetyl-CoA.

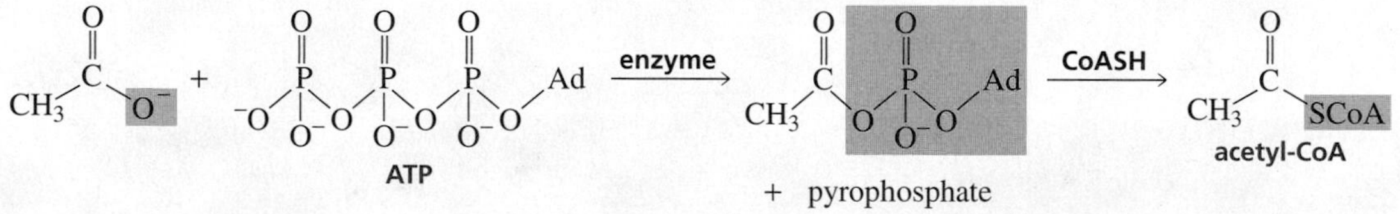

Acetylcholine (an ester) is an example of a compound that cells synthesize using acetyl-CoA. Acetylcholine is a *neurotransmitter*—that is, it transmits nerve impulses across the synapses (spaces) between nerve cells.

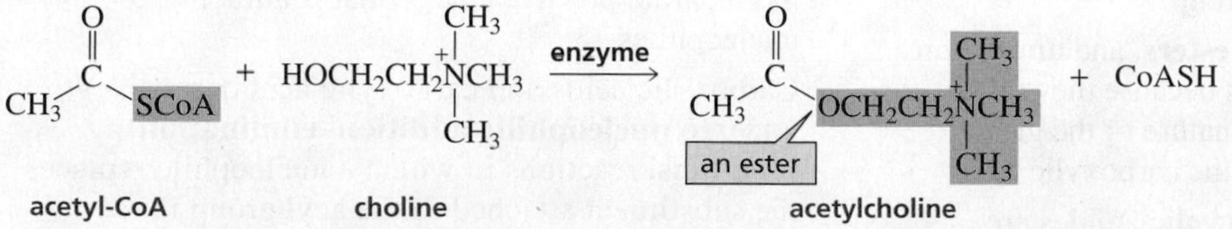

Nerve Impulses, Paralysis, and Insecticides

After an impulse is transmitted between two nerve cells, acetylcholine must be hydrolyzed immediately to enable the recipient cell to receive another impulse. Acetylcholinesterase, the enzyme that catalyzes this hydrolysis, has a CH_2OH group that is necessary for its catalytic activity. The CH_2OH group participates in a transesterification reaction with acetylcholine, which releases choline. Hydrolysis of the ester group attached to the enzyme restores its active form.

Diisopropyl fluorophosphate (DFP), a military nerve gas used during World War II, inactivates acetylcholinesterase by reacting with its CH_2OH group. When the enzyme is inactivated, nerve impulses cannot be transmitted properly and paralysis occurs. DFP is extremely toxic. Its LD_{50} (the lethal dose for 50% of the test animals) is only 0.5 mg/kg of body weight.

Malathion and parathion, widely used as insecticides, are compounds related to DFP. The LD_{50} of malathion is 2800 mg/kg. Parathion is much more toxic, with an LD_{50} of 2 mg/kg.

malathion   parathion

SOME IMPORTANT THINGS TO REMEMBER

- A **carbonyl group** is a carbon double bonded to an oxygen; an **acyl group** is a carbonyl group attached to an alkyl (R) or an aromatic (Ar) group.

- **Acyl chlorides, acid anhydrides, esters,** and **amides** are called **carboxylic acid derivatives** because they differ from a carboxylic acid only in the nature of the group that has replaced the OH group of the carboxylic acid.

- Cyclic esters are called **lactones;** cyclic amides are called **lactams.**

- The reactivity of carbonyl compounds resides in the polarity of the carbonyl group; the carbonyl carbon has a partial positive charge that is attractive to nucleophiles.

- Carboxylic acids and carboxylic acid derivatives undergo **nucleophilic addition–elimination reactions,** reactions in which a nucleophile replaces the substituent attached to the acyl group in the reactant.

- A carboxylic acid or carboxylic acid derivative will undergo a nucleophilic addition–elimination reaction provided that the newly added group in the tetrahedral intermediate is not a much weaker base than the group attached to the acyl group in the reactant.

- Generally, a compound with an sp^3 carbon bonded to an oxygen is unstable if the sp^3 carbon is bonded to another electronegative atom.

- The weaker the base attached to the acyl group, the more easily both steps of the nucleophilic addition–elimination reaction can take place.

- The relative reactivities toward nucleophilic addition–elimination are acyl chlorides > acid anhydrides > esters ~ carboxylic acids > amides > carboxylate ions.

- **Hydrolysis, alcoholysis,** and **aminolysis** are reactions in which water, alcohols, and amines, respectively, convert one compound into two compounds.

- A **transesterification reaction** converts one ester to another ester.

- Treating a carboxylic acid with excess alcohol and an acid catalyst is called a **Fischer esterification.**

- The hydrolysis of an ester with a tertiary alkyl group occurs via an S_N1 reaction.

- The rate of hydrolysis can be increased by acid or by HO^-; the rate of alcoholysis can be increased by acid or by RO^-.

- An acid increases the rate of formation of the tetrahedral intermediate by protonating the carbonyl oxygen, which increases the electrophilicity of the carbonyl carbon.

- An acid decreases the basicity of the leaving group by protonating it, which makes it easier to eliminate.

- Hydroxide (or alkoxide) ion increases the rate of formation of the tetrahedral intermediate—it is a better nucleophile than water (or an alcohol)—and increases the rate of collapse of the tetrahedral intermediate by stabilizing the transition state.

- Hydroxide ion promotes only hydrolysis reactions; alkoxide ion promotes only alcoholysis reactions.

- In an acid-catalyzed reaction, all organic reactants, intermediates, and products are positively charged or neutral; in hydroxide-ion- or alkoxide-ion-promoted reactions, all organic reactants, intermediates, and products are negatively charged or neutral.

- **Fats** and **oils** are triesters of glycerol.

- Amides are unreactive compounds but do react with water and alcohols if the reaction mixture is heated in an acidic solution. Amides are also hydrolyzed in strongly basic solutions.

- Nitriles are harder to hydrolyze than amides.

- The **Gabriel synthesis,** which converts an alkyl halide into a primary amine, involves the hydrolysis of an **imide.**

- Organic chemists activate carboxylic acids by converting them into acyl chlorides or acid anhydrides.

- Cells activate carboxylic acids by converting them into **acyl phosphates, acyl adenylates,** or **thioesters.**

SUMMARY OF REACTIONS

The general mechanisms for nucleophilic addition–elimination reactions are shown in Section 7.

1. Reactions of acyl chlorides (Section 8).

2. Reactions of esters (Sections 9–12).

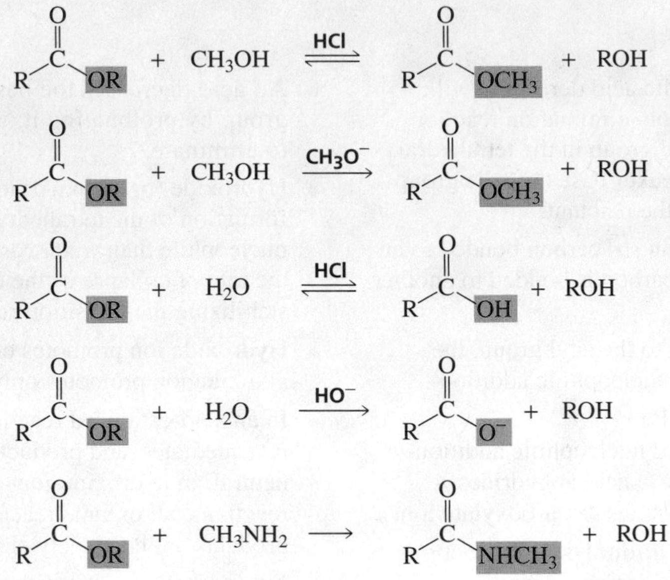

$$R-\overset{\overset{O}{\|}}{C}-OR + CH_3OH \; \underset{}{\overset{HCl}{\rightleftharpoons}} \; R-\overset{\overset{O}{\|}}{C}-OCH_3 + ROH$$

$$R-\overset{\overset{O}{\|}}{C}-OR + CH_3OH \; \overset{CH_3O^-}{\longrightarrow} \; R-\overset{\overset{O}{\|}}{C}-OCH_3 + ROH$$

$$R-\overset{\overset{O}{\|}}{C}-OR + H_2O \; \underset{}{\overset{HCl}{\rightleftharpoons}} \; R-\overset{\overset{O}{\|}}{C}-OH + ROH$$

$$R-\overset{\overset{O}{\|}}{C}-OR + H_2O \; \overset{HO^-}{\longrightarrow} \; R-\overset{\overset{O}{\|}}{C}-O^- + ROH$$

$$R-\overset{\overset{O}{\|}}{C}-OR + CH_3NH_2 \; \longrightarrow \; R-\overset{\overset{O}{\|}}{C}-NHCH_3 + ROH$$

3. Reactions of carboxylic acids (Section 14).

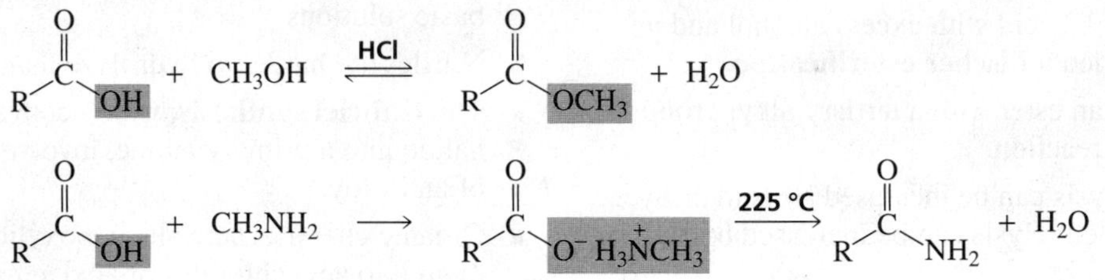

$$R-\overset{\overset{O}{\|}}{C}-OH + CH_3OH \; \underset{}{\overset{HCl}{\rightleftharpoons}} \; R-\overset{\overset{O}{\|}}{C}-OCH_3 + H_2O$$

$$R-\overset{\overset{O}{\|}}{C}-OH + CH_3NH_2 \; \longrightarrow \; R-\overset{\overset{O}{\|}}{C}-O^- \overset{+}{H_3}NCH_3 \; \overset{225\,°C}{\longrightarrow} \; R-\overset{\overset{O}{\|}}{C}-NH_2 + H_2O$$

4. Reactions of amides (Sections 15–17).

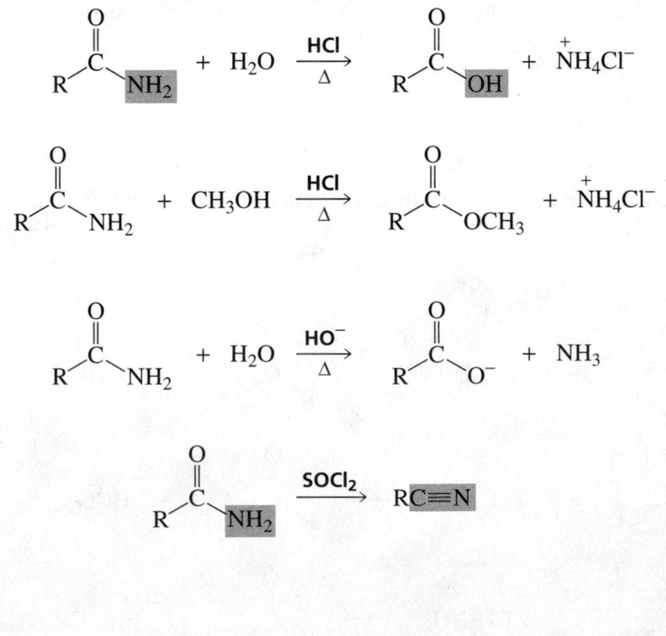

$$R-\overset{\overset{O}{\|}}{C}-NH_2 + H_2O \; \overset{HCl}{\underset{\Delta}{\longrightarrow}} \; R-\overset{\overset{O}{\|}}{C}-OH + \overset{+}{N}H_4Cl^-$$

$$R-\overset{\overset{O}{\|}}{C}-NH_2 + CH_3OH \; \overset{HCl}{\underset{\Delta}{\longrightarrow}} \; R-\overset{\overset{O}{\|}}{C}-OCH_3 + \overset{+}{N}H_4Cl^-$$

$$R-\overset{\overset{O}{\|}}{C}-NH_2 + H_2O \; \overset{HO^-}{\underset{\Delta}{\longrightarrow}} \; R-\overset{\overset{O}{\|}}{C}-O^- + NH_3$$

$$R-\overset{\overset{O}{\|}}{C}-NH_2 \; \overset{SOCl_2}{\longrightarrow} \; RC\equiv N$$

5. Gabriel synthesis of a primary amine (Section 18)

$$\text{phthalimide} \xrightarrow[\substack{\text{3. HCl, H}_2\text{O, }\Delta \\ \text{4. HO}^-}]{\substack{\text{1. HO}^- \\ \text{2. RCH}_2\text{Br}}} \text{RCH}_2\overset{+}{\text{NH}_2}$$

6. Hydrolysis of nitriles (Section 19).

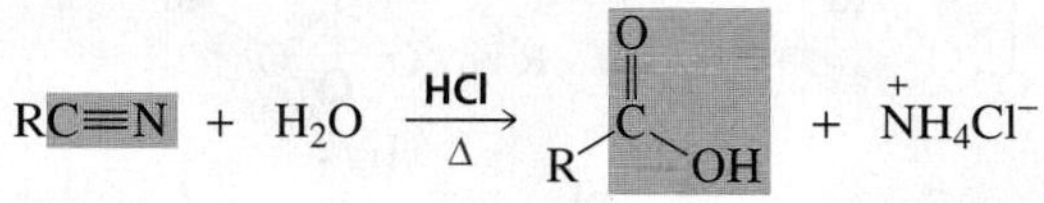

7. Reactions of acid anhydrides (Section 20).

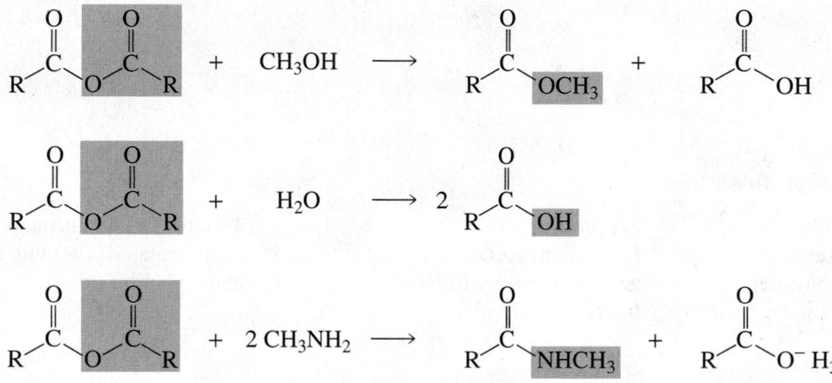

8. Reactions of dicarboxylic acids (Section 21)

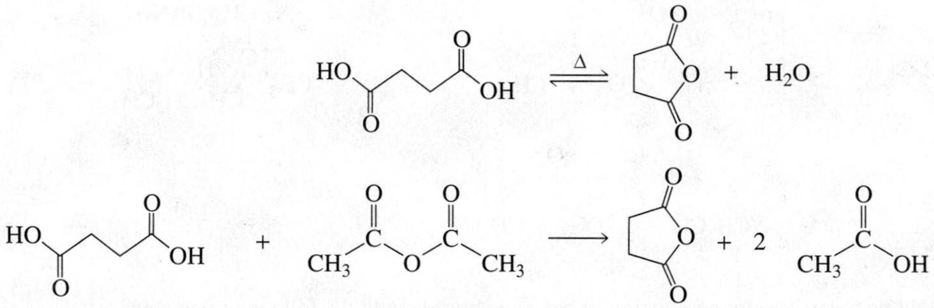

9. Activation of carboxylic acids by chemists (Section 22).

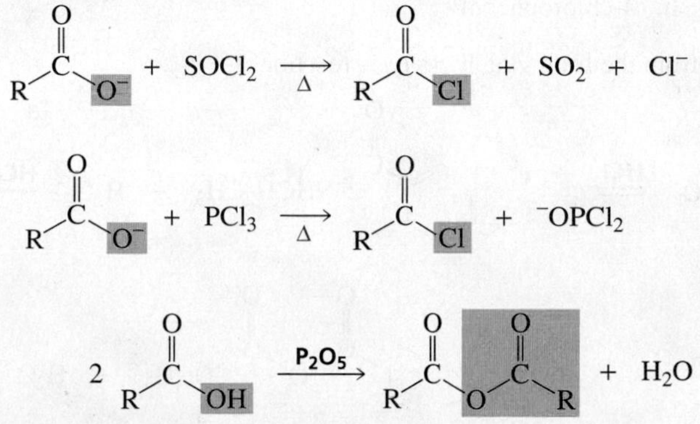

10. Activation of carboxylic acids by cells (Section 23).

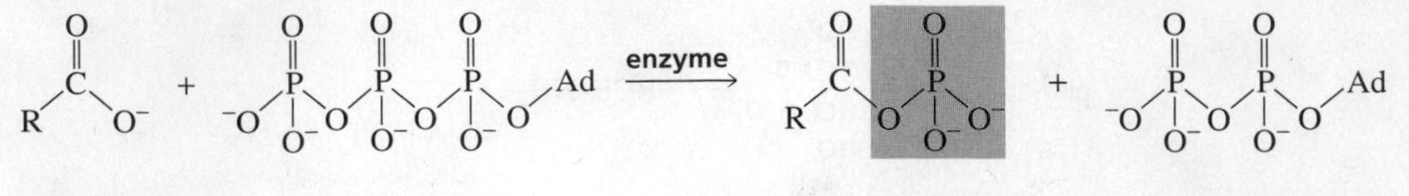

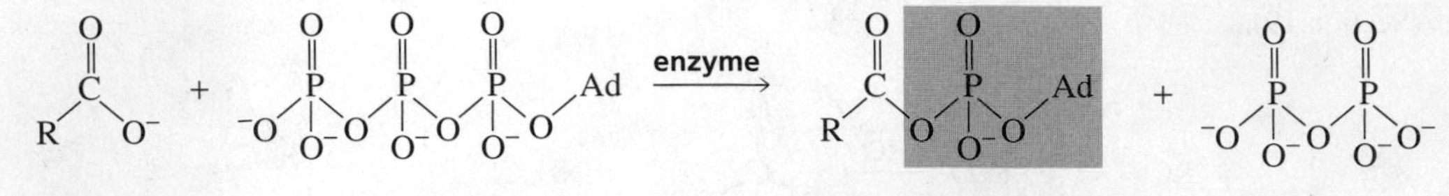

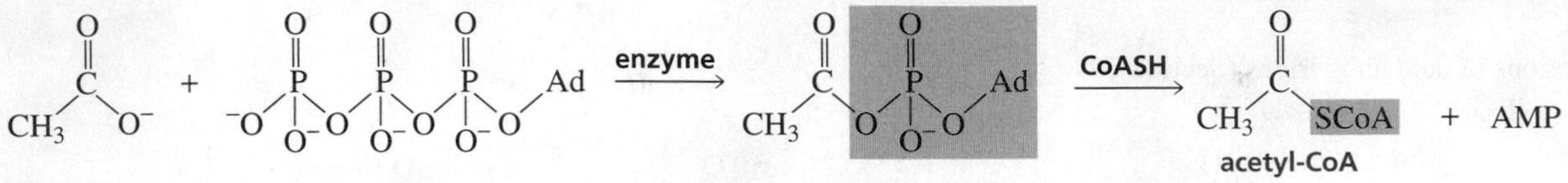

PROBLEMS

55. Write a structure for each of the following:

a. *N,N*-dimethylhexanamide
b. 3,3-dimethylhexanamide
c. cyclohexanecarbonyl chloride
d. propanenitrile

e. propionamide
f. sodium acetate
g. benzoic anhydride
h. β-valerolactone

i. 3-methylbutanenitrile
j. cycloheptanecarboxylic acid
k. benzoyl chloride

56. Name the following compounds:

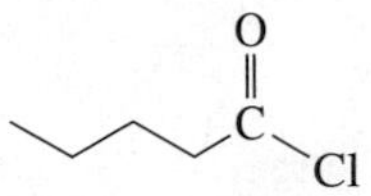

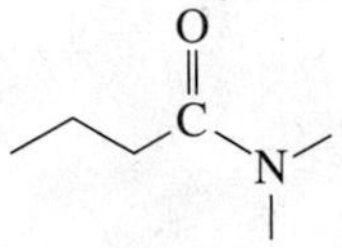

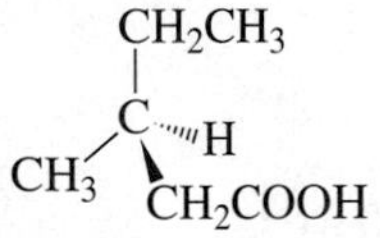

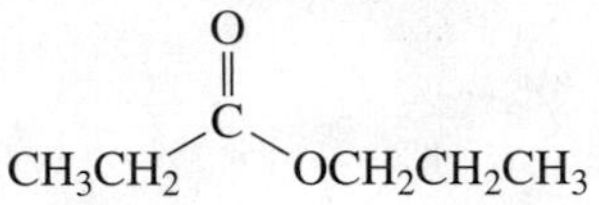

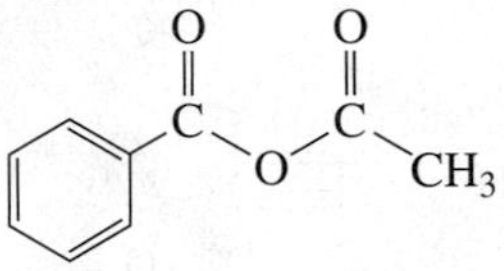

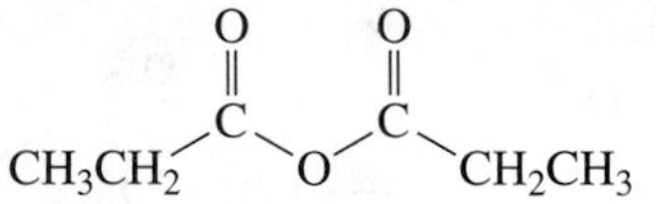

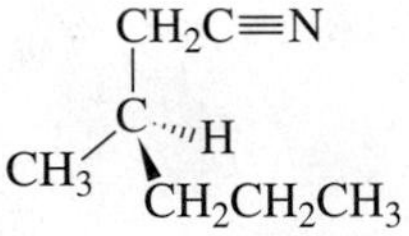

57. What products would be formed from the reaction of benzoyl chloride with the following reagents?

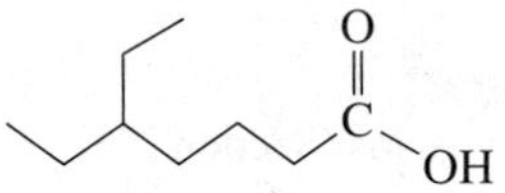

a. sodium acetate
b. water
c. excess dimethylamine
d. aqueous HCl

e. aqueous NaOH
f. cyclohexanol
g. excess benzylamine
h. 4-chlorophenol

i. isopropyl alcohol
j. excess aniline
k. potassium formate

58. What products would be obtained from the following hydrolysis reactions?

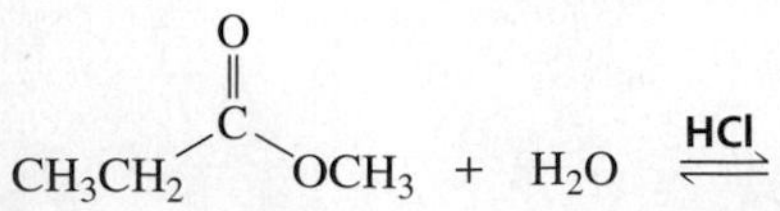

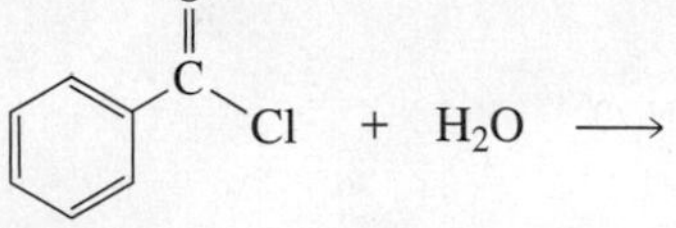

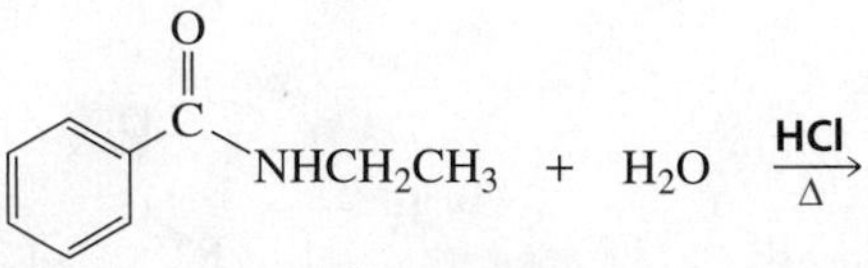

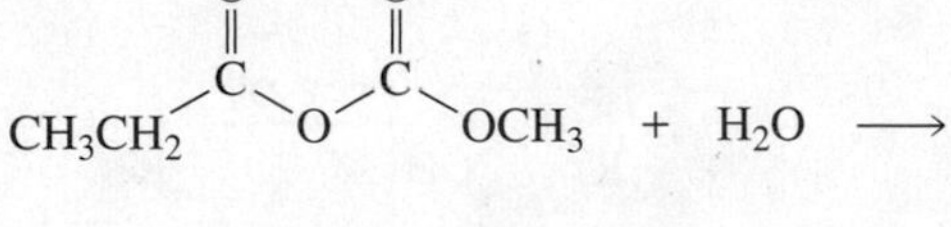

59. a. List the following esters in order of decreasing reactivity in the first slow step of a nucleophilic addition-elimination reaction (formation of the tetrahedral intermediate):

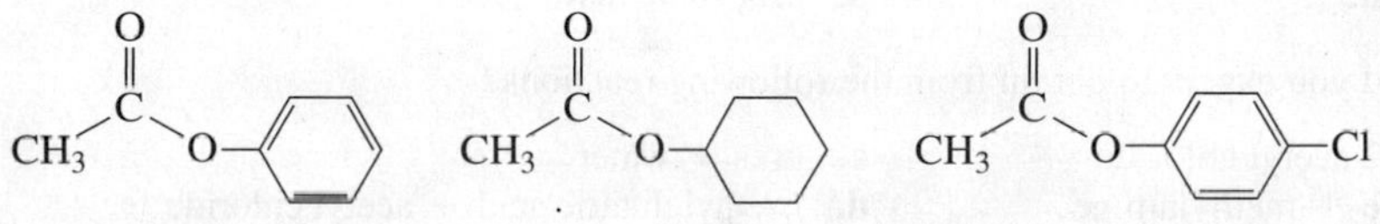

b. List the same esters in order of decreasing reactivity in the second slow step of a nucleophilic addition–elimination reaction (collapse of the tetrahedral intermediate).

60. Because bromocyclohexane is a secondary alkyl halide, both cyclohexanol and cyclohexene are formed when the alkyl halide reacts with hydroxide ion. Suggest a method to synthesize cyclohexanol from bromocyclohexane that would form little or no cyclohexene.

61. a. Which compound would you expect to have a higher dipole moment, methyl acetate or butanone?

methyl acetate   butanone

b. Which would you expect to have a higher boiling point?

62. How could you use ^{1}H NMR spectroscopy to distinguish among the following esters?

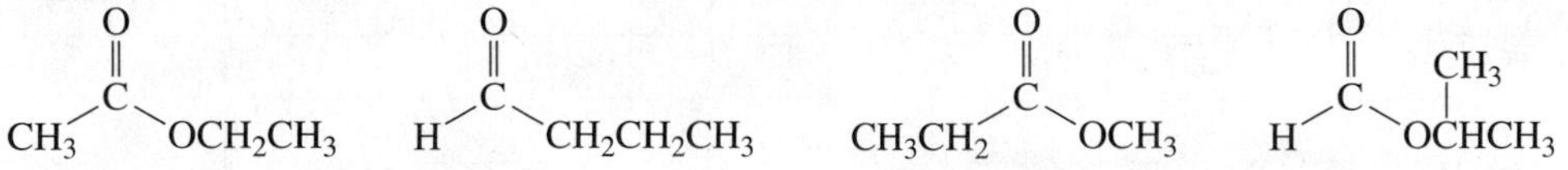

63. If propionyl chloride is added to one equivalent of methylamine, only a 50% yield of *N*-methylpropanamide is obtained. If, however, the acyl chloride is added to two equivalents of methylamine, the yield of *N*-methylpropanamide is almost 100%. Explain these observations.

64. a. When a carboxylic acid is dissolved in isotopically labeled water ($H_2{}^{18}O$) and an acid catalyst is added, the label is incorporated into both oxygens of the acid. Propose a mechanism to account for this.

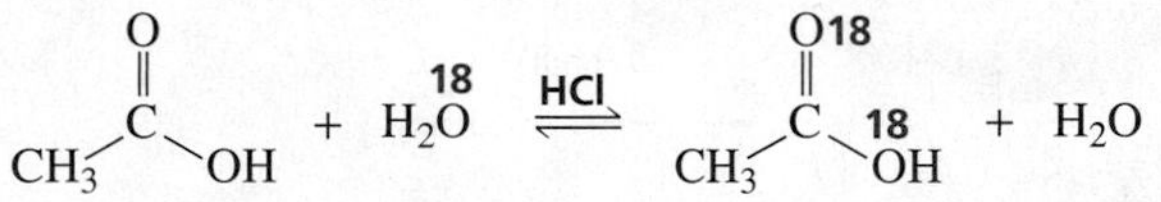

b. If a carboxylic acid is dissolved in isotopically labeled methanol ($CH_3{}^{18}OH$) and an acid catalyst is added, where will the label reside in the product?

c. If an ester is dissolved in isotopically labeled water ($H_2{}^{18}O$) and an acid catalyst is added, where will the label reside in the product?

65. List the following compounds in order of decreasing frequency of the carbon–oxygen double-bond stretch:

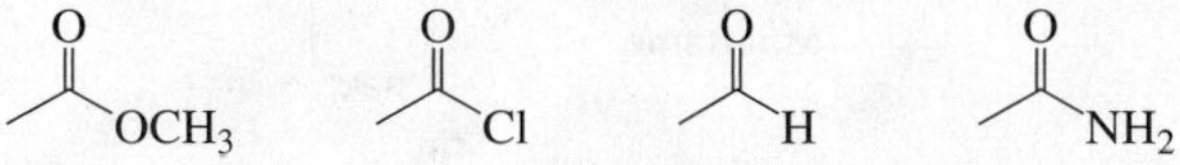

66. Using an alcohol for one method and an alkyl halide for the other, show two ways to make each of the following esters:

a. propyl acetate (odor of pears)
b. isopentyl acetate (odor of bananas)

c. ethyl butyrate (odor of pineapple)
d. methyl phenylethanoate (odor of honey)

67. What reagents would you use to convert methyl propanoate into the following compounds?

a. isopropyl propanoate
b. sodium propanoate
c. *N*-ethylpropanamide
d. propanoic acid

68. What products would you expect to obtain from the following reactions?

a. malonic acid + 2 acetyl chloride
b. methyl carbamate + methylamine
c. urea + water
d. β-ethylglutaric acid + acetyl chloride + Δ

69. A compound with molecular formula $C_5H_{10}O_2$ gives the following IR spectrum. When it undergoes acid-catalyzed hydrolysis, the compound with the ^1H NMR spectrum shown here is formed. Identify the compounds.

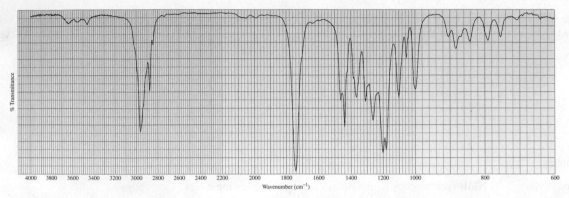

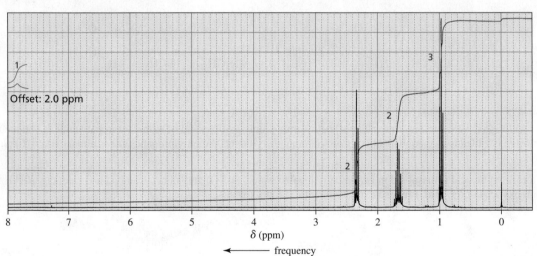

70. Aspartame, the sweetener used in the commercial products NutraSweet and Equal, is 200 times sweeter than sucrose. What products would be obtained if aspartame were hydrolyzed completely in an aqueous solution of HCl?

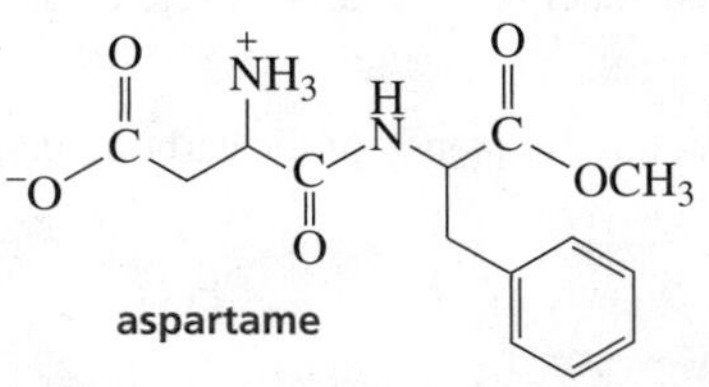

aspartame

71. a. Which of the following reactions will not give the carbonyl product shown?

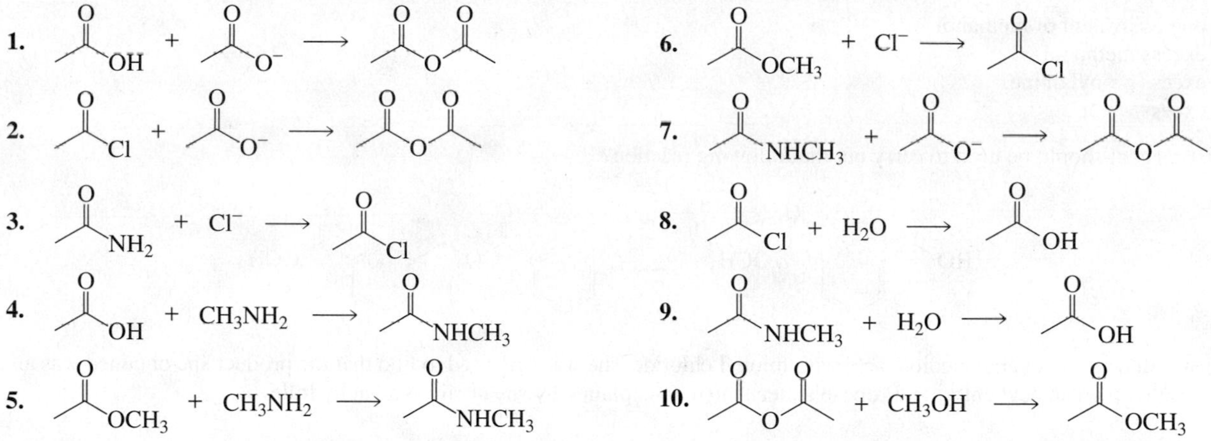

1. CH_3COOH + $CH_3CO\text{-}O^-$ $\longrightarrow$ acetic anhydride

2. CH_3COCl + $CH_3CO\text{-}O^-$ $\longrightarrow$ acetic anhydride

3. CH_3CONH_2 + Cl^- $\longrightarrow$ CH_3COCl

4. CH_3COOH + CH_3NH_2 $\longrightarrow$ $CH_3CONHCH_3$

5. CH_3COOCH_3 + CH_3NH_2 $\longrightarrow$ $CH_3CONHCH_3$

6. CH_3COOCH_3 + Cl^- $\longrightarrow$ CH_3COCl

7. $CH_3CONHCH_3$ + $CH_3CO\text{-}O^-$ $\longrightarrow$ acetic anhydride

8. CH_3COCl + H_2O $\longrightarrow$ CH_3COOH

9. $CH_3CONHCH_3$ + H_2O $\longrightarrow$ CH_3COOH

10. acetic anhydride + CH_3OH $\longrightarrow$ CH_3COOCH_3

b. Which of the reactions that do not occur can be made to occur if an acid catalyst is added to the reaction mixture?

72. 1,4-Diazabicyclo[2.2.2]octane (abbreviated DABCO) is a tertiary amine that catalyzes transesterification reactions. Propose a mechanism to show how it does this.

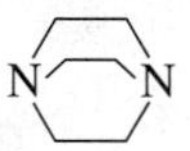

1,4-diazabicyclo[2.2.2]octane
DABCO

73. Two products, **A** and **B**, are obtained from the reaction of 1-bromobutane with NH_3. Compound **A** reacts with acetyl chloride to form **C**, and compound **B** reacts with acetyl chloride to form **D.** The IR spectra of **C** and **D** are shown. Identify **A, B, C,** and **D.**

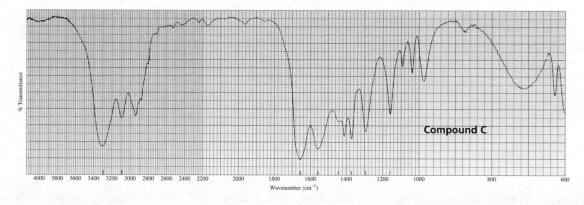

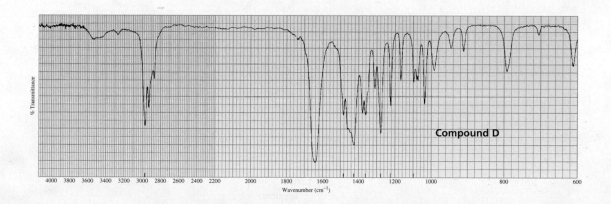

74. Phosgene (COCl$_2$) was used as a poison gas in World War I. Give the product that would be formed from the reaction of phosgene with each of the following reagents:

   a. one equivalent of methanol
   b. excess methanol
   c. excess propylamine
   d. excess water

75. What reagent should be used to carry out the following reaction?

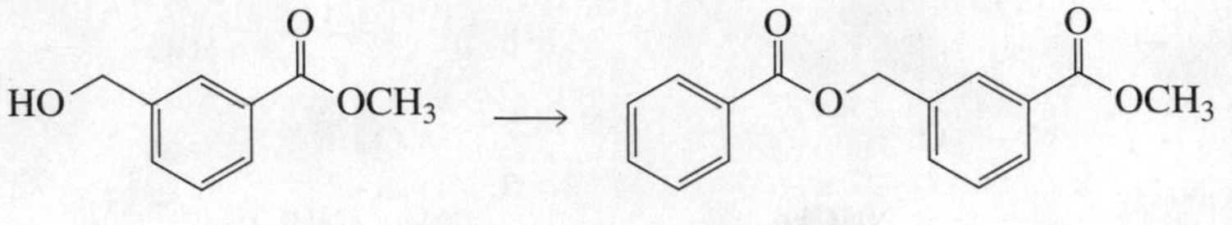

76. When a student treated butanedioic acid with thionyl chloride, she was surprised to find that the product she obtained was an anhydride rather than an acyl chloride. Propose a mechanism to explain why she obtained an anhydride.

77. What are the products of the following reactions?

   a. 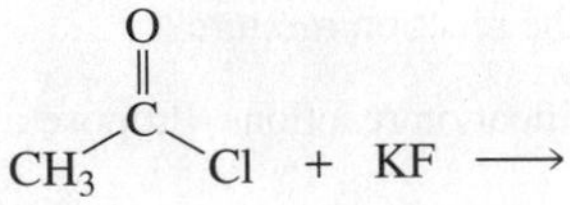 CH_3—C(=O)—Cl + KF $\longrightarrow$

   b. 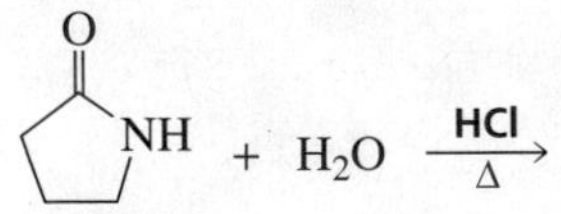 (lactam)—NH + H$_2$O $\xrightarrow{\text{HCl}}{\Delta}$

   c. 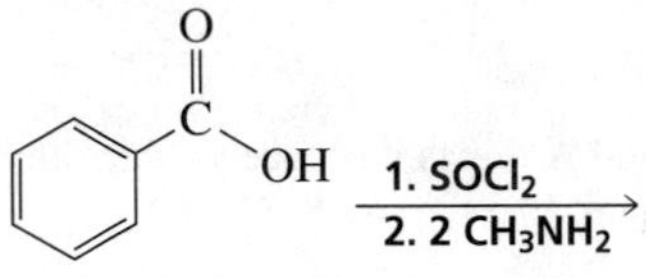 C$_6$H$_5$—C(=O)—OH $\xrightarrow[\text{2. 2 CH}_3\text{NH}_2]{\text{1. SOCl}_2}$

   d. 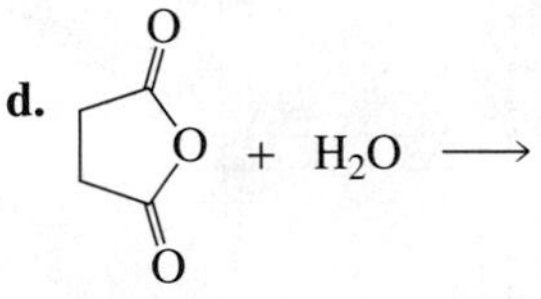 (succinic anhydride) + H$_2$O $\longrightarrow$

   e. Cl—C(=O)—Cl + (catechol: benzene with two OH) $\longrightarrow$

   f. 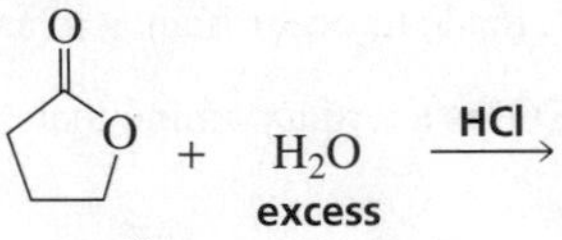 (γ-butyrolactone) + H$_2$O $\xrightarrow{\text{HCl}}$ excess

   g. CH_3—C(=O)—O—CH$_3$...—C(=O)... + CH$_3$OH $\xrightarrow{\text{CH}_3\text{O}^-}$ excess

   h. (benzene with CH$_2$COH and COH groups) + (CH$_3$C)$_2$O $\xrightarrow{\Delta}$

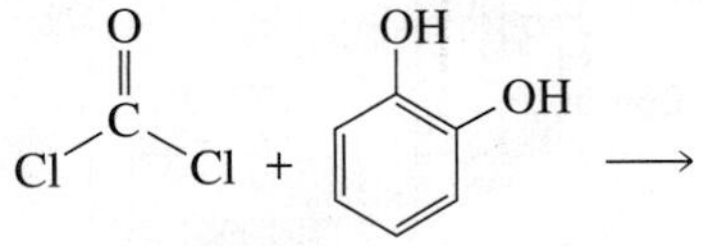

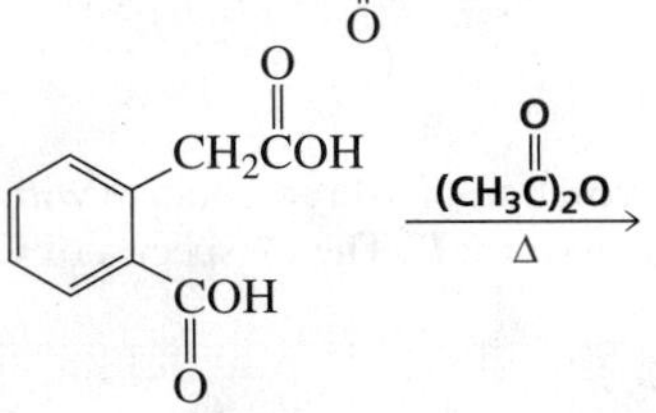

   i. 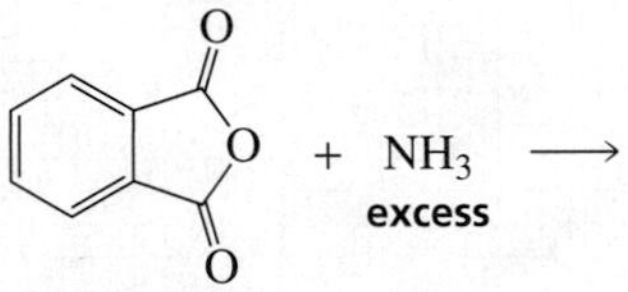 (phthalic anhydride) + NH$_3$ $\longrightarrow$ excess

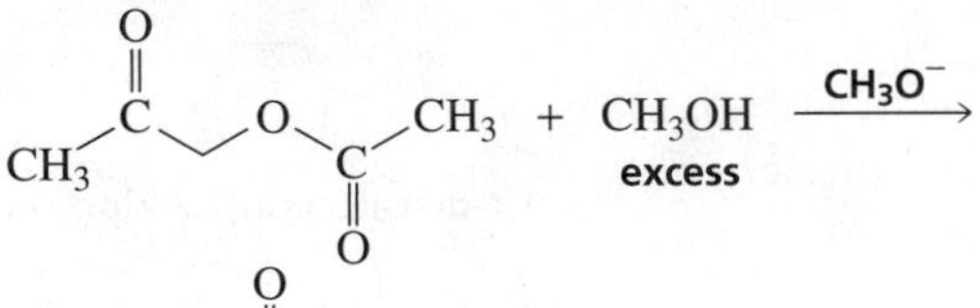

   j. 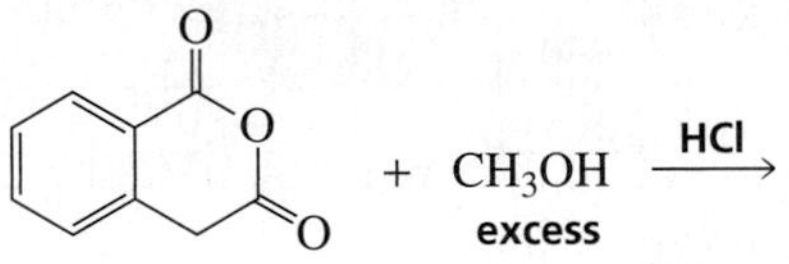 (isochromanone) + CH$_3$OH $\xrightarrow{\text{HCl}}$ excess

78. When treated with an equivalent of methanol, compound **A,** with molecular formula C$_4$H$_6$Cl$_2$O, forms the compound whose ^{1}H NMR spectrum is shown here. Identify compound **A.**

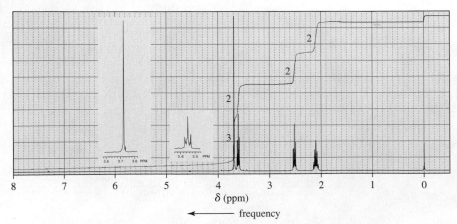

79. a. Identify the two products obtained from the following reaction:

$$CH_3-\overset{O}{\underset{}{C}}-O-\overset{O}{\underset{}{C}}-CH_3 + CH_3CHCH_2CH_2OH \longrightarrow$$
$$\text{excess} \qquad\qquad \underset{NH_2}{}$$

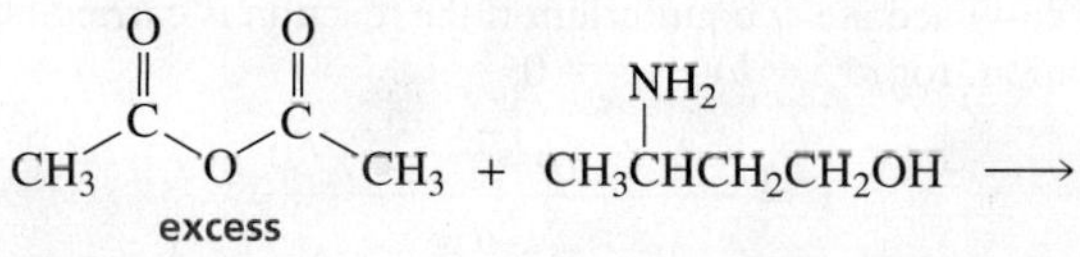

b. A student carried out the preceding reaction, but stopped it before it was half over, whereupon he isolated the major product. He was surprised to find that the product he isolated was neither of the products obtained when the reaction was allowed to go to completion. What product did he isolate?

80. An aqueous solution of a primary or secondary amine reacts with an acyl chloride to form an amide as the major product. However, if the amine is tertiary, an amide is not formed. What product *is* formed? Explain.

81. Identify the major and minor products of the following reaction:

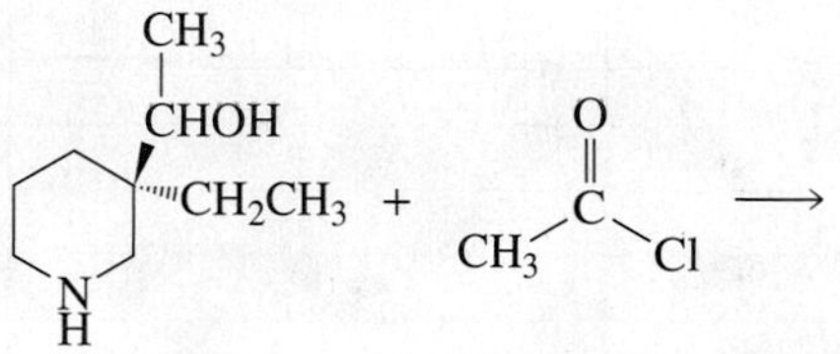

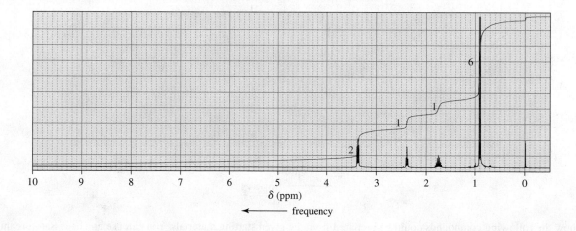

82. a. A student did not obtain any ester when he added 2,4,6-trimethylbenzoic acid to an acidic solution of methanol. Why? (*Hint:* Build models.)
b. Would he have encountered the same problem if he had tried to synthesize the methyl ester of 4-methylbenzoic acid in the same way?

83. When a compound with molecular formula $C_{11}H_{14}O_2$ undergoes acid-catalyzed hydrolysis, one of the products that is isolated gives the following ^1H NMR spectrum. Identify the compound.

84. Cardura, a drug used to treat hypertension, is synthesized as shown here.

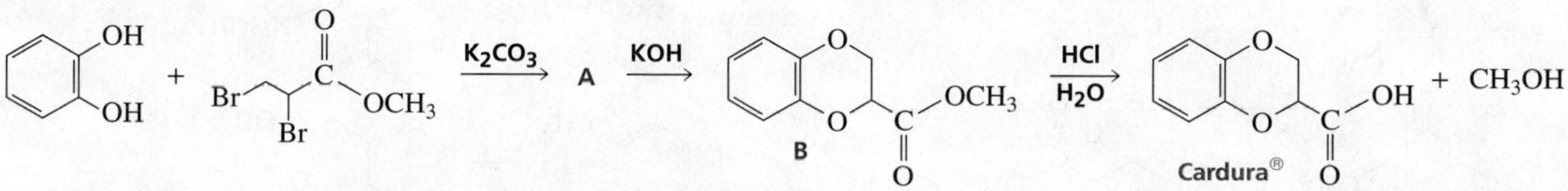

a. Identify the intermediate (**A**) and show the mechanism for its formation.
b. Show the mechanism for conversion of **A** to **B**. Which would be formed more rapidly, **A** or **B**?

85. a. If the equilibrium constant for the reaction of acetic acid and ethanol to form ethyl acetate is 4.02, what will be the concentration of ethyl acetate at equilibrium if the reaction is carried out with equal amounts of acetic acid and ethanol?

b. What will be the concentration of ethyl acetate at equilibrium if the reaction is carried out with 10 times more ethanol than acetic acid? *Hint:* Use the quadratic equation: for $ax^2 + bx + c = 0$,

$$x = \frac{-b \pm (b^2 - 4ac)^{1/2}}{2a}$$

c. What will be the concentration of ethyl acetate at equilibrium if the reaction is carried out with 100 times more ethanol than acetic acid?

86. The ^{1}H NMR spectra for two esters with molecular formula $C_8H_8O_2$ are shown next. If each of the esters is added to an aqueous solution with a pH of 10, which of the esters will be hydrolyzed more rapidly?

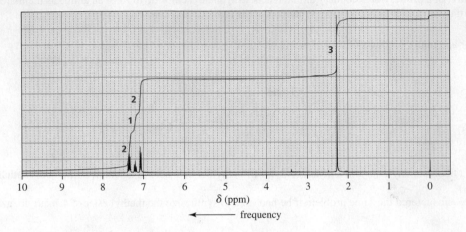

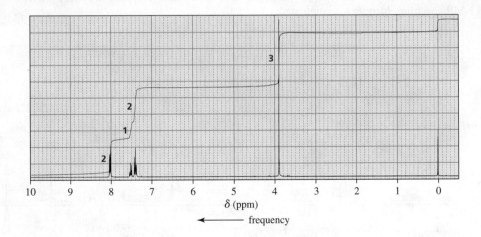

87. Show how the following compounds could be prepared from the given starting materials. You can use any necessary organic or inorganic reagents.

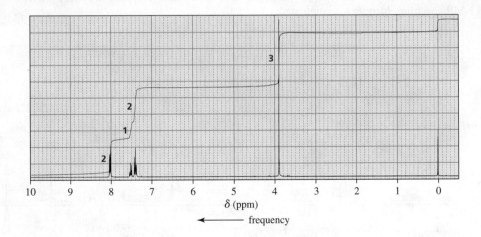

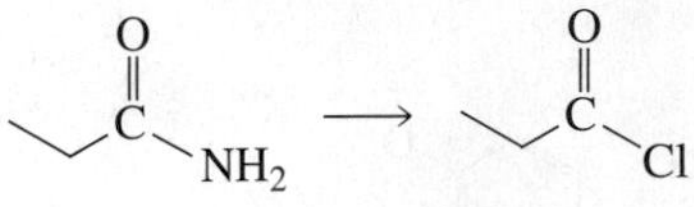

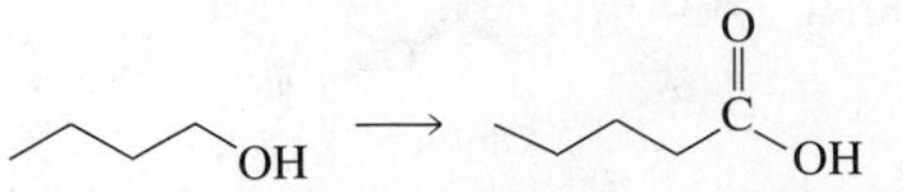

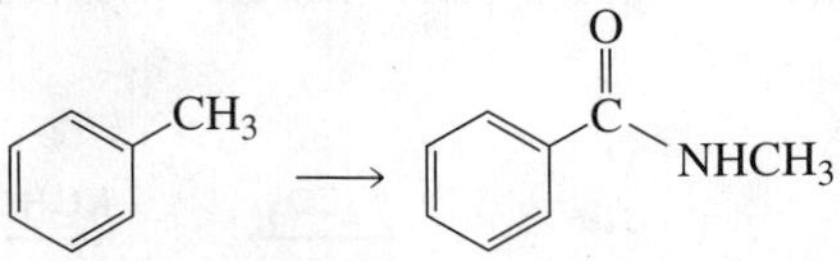

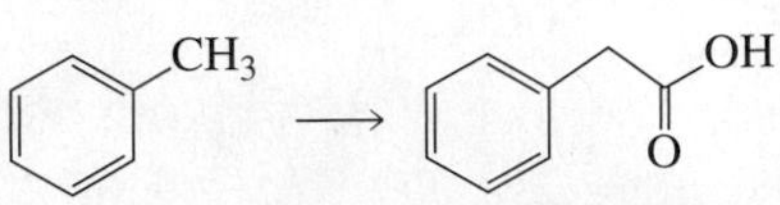

88. Is the acid-catalyzed hydrolysis of acetamide a reversible or an irreversible reaction? Explain.

89. What product would you expect to obtain from each of the following reactions?

a. 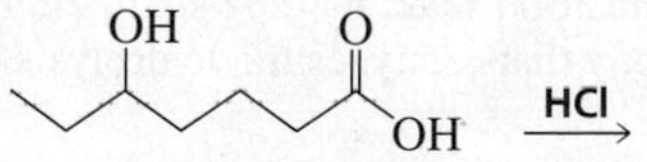 $\xrightarrow{\text{HCl}}$

b. 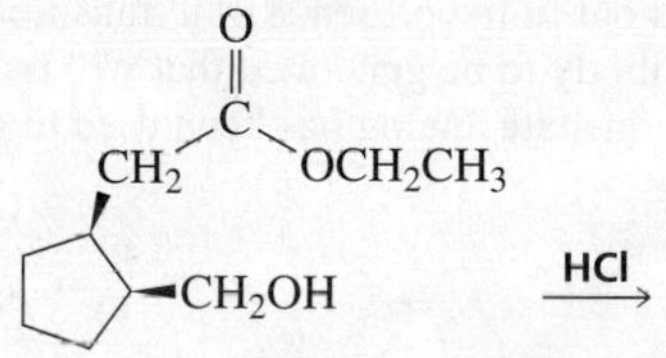 $\xrightarrow{\text{HCl}}$

90. The reaction of a nitrile with an alcohol in the presence of a strong acid forms an *N*-substituted amide. This reaction, known as the *Ritter reaction,* does not work with primary alcohols.

$$RC\equiv N \ + \ R'OH \ \xrightarrow{\text{HCl}} \ \underset{R}{\overset{O}{\underset{\quad}{\|}}}C \underset{NHR'}{}$$

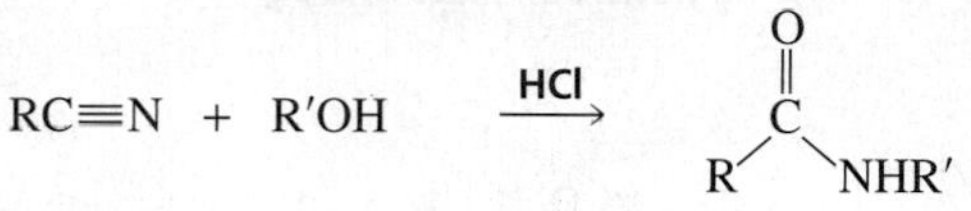

the Ritter reaction

   a. Propose a mechanism for the Ritter reaction.
   b. Why does the Ritter reaction not work with primary alcohols?
   c. How does the Ritter reaction differ from the acid-catalyzed hydrolysis of a nitrile to form an amide?

91. The intermediate shown here is formed during the hydroxide-ion-promoted hydrolysis of the ester group. Propose a mechanism for the reaction.

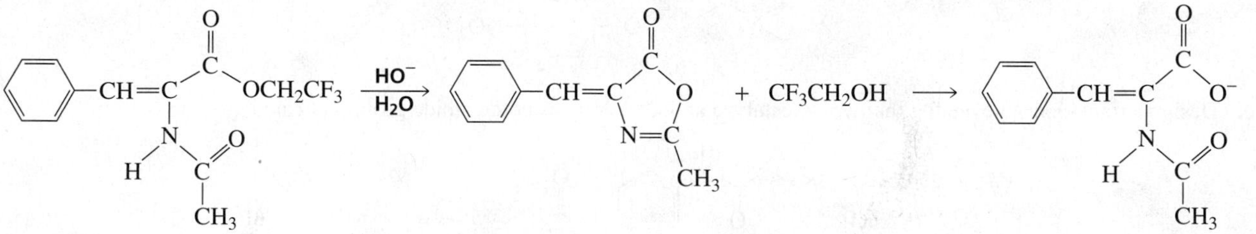

92. The following compound has been found to be an inhibitor of penicillinase. The enzyme can be reactivated by hydroxylamine (NH_2OH). Propose a mechanism to account for the inhibition and for the reactivation.

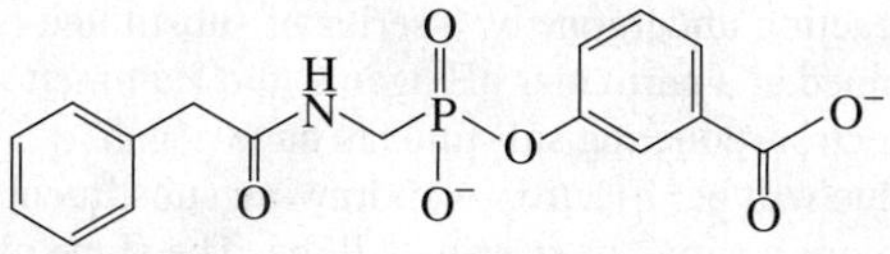

93. Propose a mechanism that will account for the formation of the product.

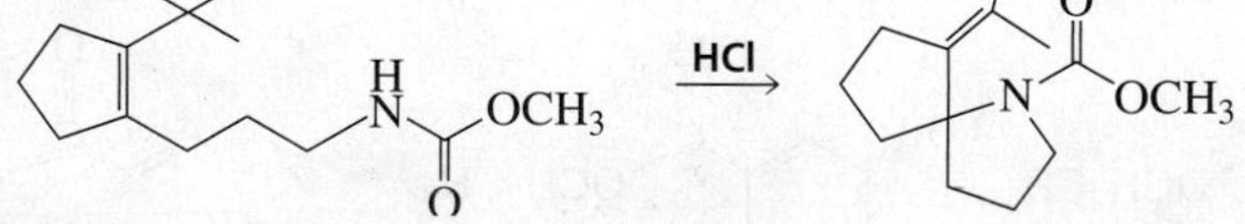

94. Catalytic antibodies catalyze a reaction by forcing the conformation of the substrate in the direction of the transition state. The synthesis of the antibody is carried out in the presence of a transition state analog—a stable molecule that structurally resembles the transition state. This causes an antibody to be generated that will recognize and bind to the transition state, thereby stabilizing it. For example, the following transition state analog has been used to generate a catalytic antibody that catalyzes the hydrolysis of the structurally similar ester:

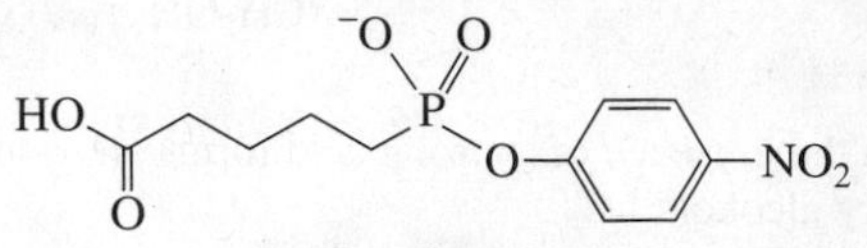

transition state analog

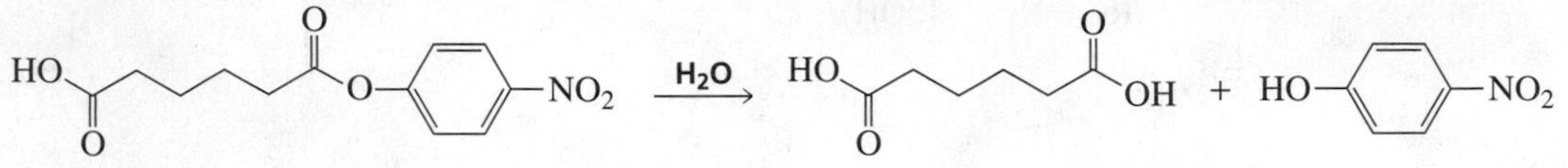

a. Draw the transition state for the hydrolysis reaction.
b. The following transition state analog is used to generate a catalytic antibody for the catalysis of ester hydrolysis. Draw the structure of an ester whose rate of hydrolysis would be increased by this catalytic antibody.

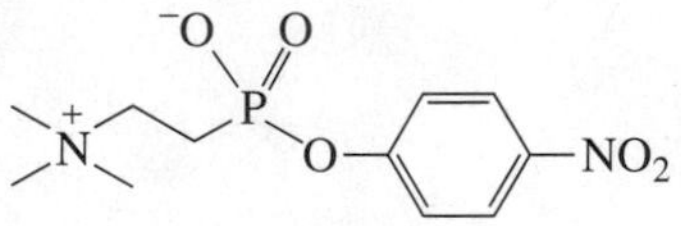

c. Design a transition state analog that would catalyze amide hydrolysis at the amide group indicated.

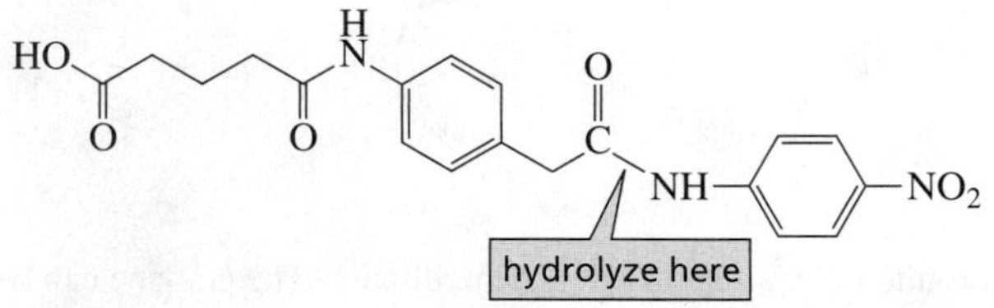

hydrolyze here

95. Information about the mechanism of the reaction undergone by a series of substituted benzenes can be obtained by plotting the logarithm of the observed rate constant determined at a particular pH against the Hammett substituent constant (σ) for the particular substituent. The σ value for hydrogen is 0. Electron-donating substituents have negative σ values; the more strongly electron donating the substituent, the more negative its σ value will be. Electron-withdrawing substituents have positive σ values; the more strongly electron withdrawing the substituent, the more positive its σ value will be. The slope of a plot of the logarithm of the rate constant versus σ is called the ρ (rho) value. The ρ value for the hydroxide-ion-promoted hydrolysis of a series of meta- and para-substituted ethyl benzoates is +2.46; the ρ value for amide formation for the reaction of a series of meta- and para-substituted anilines with benzoyl chloride is −2.78.

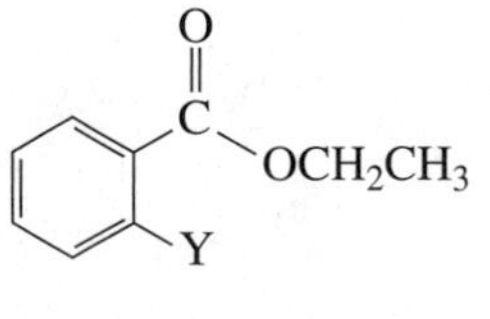

| ortho-substituted | meta-substituted | para-substituted |

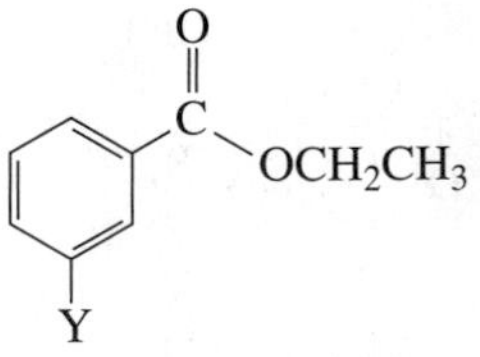

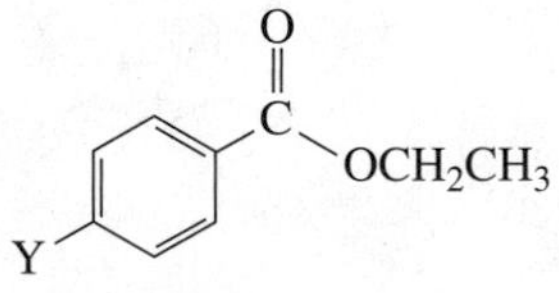

a. Why does one set of experiments give a positive ρ value, whereas the other set of experiments gives a negative ρ value?
b. Why were ortho-substituted compounds not included in the experiment?
c. What would you predict the sign of the ρ value to be for the ionization of a series of meta- and para-substituted benzoic acids?

GLOSSARY

acid anhydride

acyl adenylate   a carboxylic acid derivative with AMP as the leaving group.

acyl group   a carbonyl group bonded to an alkyl group or to an aryl group.

acyl chloride

acyl phosphate   a carboxylic acid derivative with a phosphate leaving group.

alcoholysis   reaction with an alcohol.

amide

aminolysis   reaction with an amine.

biosynthesis   synthesis in a biological system.

carbonyl carbon   the carbon of a carbonyl group.

carbonyl compound   a compound that contains a carbonyl group.

carbonyl group   a carbon doubly bonded to an oxygen.

carbonyl oxygen   the oxygen of a carbonyl group.

carboxyl group   COOH

carboxylic acid

carboxylic acid derivative   a compound that is hydrolyzed to a carboxylic acid.

carboxyl oxygen   the single-bonded oxygen of a carboxylic acid or an ester.

ester

fat   a triester of glycerol that exists as a solid at room temperature.

fatty acid   a long-chain carboxylic acid.

Fischer esterification reaction   the reaction of a carboxylic acid with alcohol in the presence of an acid catalyst to form an ester.

Gabriel synthesis   conversion of an alkyl halide into a primary amine, using phthalimide as a starting material.

hydrolysis   reaction with water.

lactam   a cyclic amide.

lactone   a cyclic ester.

lipid   a water-insoluble compound found in a living system.

lipid bilayer   two layers of phosphoacylglycerols arranged so that their polar heads are on the outside and their nonpolar fatty acid chains are on the inside.

membrane   the material that surrounds a cell in order to isolate its contents.

micelle   a spherical aggregation of molecules, each with a long hydrophobic tail and a polar head, arranged so that the polar head points to the outside of the sphere.

mixed anhydride   an anhydride formed from two different acids.

nitrile   a compound that contains a carbon–nitrogen triple bond ($RC \equiv N$).

nucleophilic acyl substitution reaction   a reaction in which a group bonded to an acyl or aryl group is substituted by another group.

nucleophilic addition–elimination reaction   a nucleophilic addition reaction that is followed by an elimination reaction. Imine formation is an example: an amine adds to the carbonyl carbon, and water is eliminated.

oil   a triester of glycerol that exists as a liquid at room temperature.

phosphoacylglycerol (phosphoglyceride)   a compound formed when two OH groups of glycerol form esters with fatty acids and the terminal OH group forms a phosphate ester.

phosphoanhydride bond   the bond holding two phosphoric acid molecules together.

polyunsaturated fatty acid   a fatty acid with more than one double bond.

saponification   hydrolysis of an ester (such as a fat) under basic conditions.

soap   a sodium or potassium salt of a fatty acid.

symmetrical anhydride   an acid anhydride with identical R groups:

tetrahedral intermediate   the intermediate formed in a nucleophilic acyl substitution reaction.

thioester   the sulfur analog of an ester:

transesterification reaction   the reaction of an ester with an alcohol to form a different ester.

triacylglycerol   the compound formed when the three OH groups of glycerol are esterified with fatty acids.

wax   an ester formed from a long-chain carboxylic acid and a long-chain alcohol.

PHOTO CREDITS

Photo credits are listed in order of appearance.

Robert Plotz/Fotolia; Pearson Science/Eric Schrader; Kelly Richardson/Shutterstock; Dr. Paula Bruice Andre Nantel/Shutterstock; Arvedphoto/iStockphoto; AtWaG/iStockphoto age fotostock/SuperStock; Uwe Bumann/Photos.com; Daily Mail/Rex/Alamy

Reactions of Aldehydes and Ketones • More Reactions of Carboxylic Acid Derivatives • Reactions of α,β-Unsaturated Carbonyl Compounds

a yew tree forest

From Chapter 17 of *Organic Chemistry*, Seventh Edition. Paula Yurkanis Bruice. Copyright © 2014 by Pearson Education, Inc. All rights reserved.

III

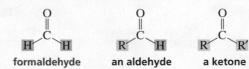

Taxol, a compound extracted from the bark of yew trees, was found to be an effective drug against several kinds of cancer. However, removing the bark kills the tree, the trees grow very slowly, and the bark of one tree provides only a small amount of the drug. Yew tree forests, moreover, are the home of the spotted owl—an endangered species. When chemists determined the structure of the drug they were disappointed to find that it would be a very difficult compound to synthesize. Nevertheless, the current supply of taxol is sufficient to meet medical demands. In this chapter, you will see how this was accomplished.

This chapter discusses the families of compounds in group III. Here we will look at the reactions of aldehydes and ketones—that is, carbonyl compounds that do not have a group that can be substituted by another group, and we will compare their reactions with those of some carboxylic acid derivatives.

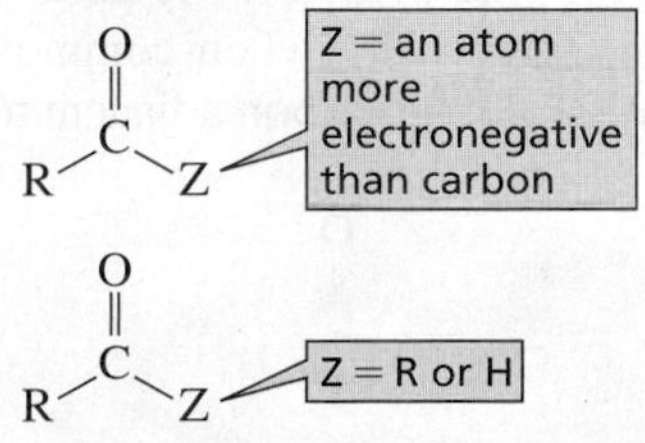

The carbonyl carbon of the simplest aldehyde, formaldehyde, is bonded to two hydrogens. The carbonyl carbon of all other **aldehydes** is bonded to a hydrogen and to an alkyl (or aryl) group (R). The carbonyl carbon of a **ketone** is bonded to two R groups.

815

Many compounds found in nature have aldehyde or ketone functional groups. Aldehydes have pungent odors, whereas ketones tend to smell sweet. Vanillin and cinnamaldehyde are examples of naturally occurring aldehydes. A whiff of vanilla extract will allow you to appreciate the pungent odor of vanillin. The ketones camphor and carvone are responsible for the characteristic sweet odors of the leaves of camphor trees, spearmint leaves, and caraway seeds.

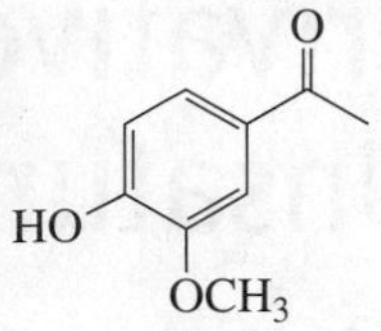

vanillin
vanilla flavoring

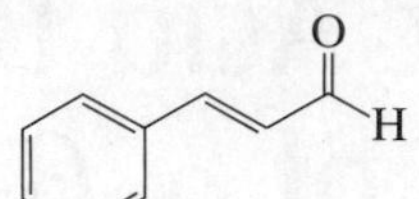

cinnamaldehyde
cinnamon flavoring

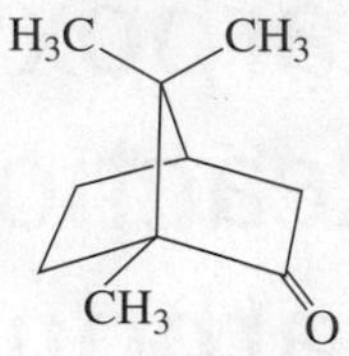

camphor

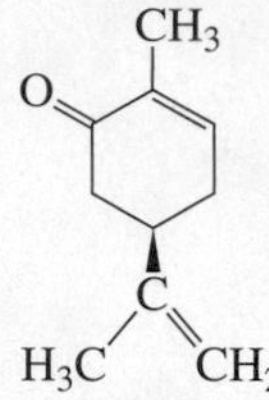

(R)-(−)-carvone
spearmint oil

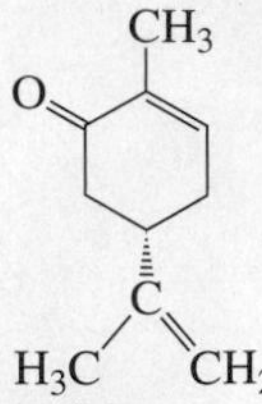

(S)-(+)-carvone
caraway seed oil

Progesterone and testosterone are two biologically important ketones that illustrate how a small difference in structure can be responsible for a large difference in biological activity. Both are sex hormones, but progesterone is synthesized primarily in the ovaries, whereas testosterone is synthesized primarily in the testes.

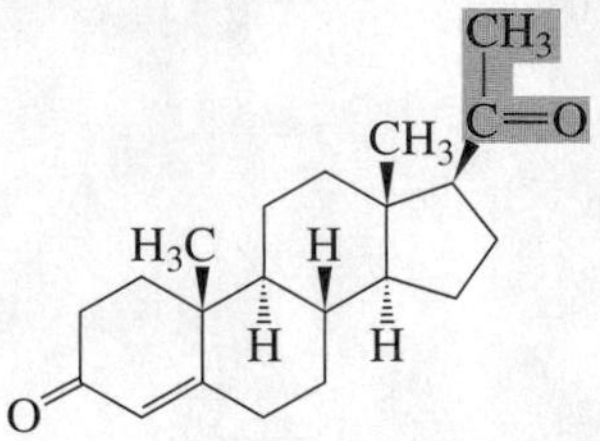

progesterone

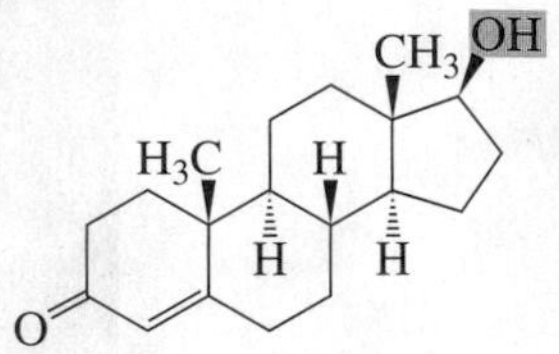

testosterone

The physical properties of aldehydes and ketones are discussed in Appendix: "Physical Properties of Organic Compounds," and the methods used to prepare aldehydes and ketones are summarized in Appendix: "Summary of Methods Used to Synthesize a Particular Functional Group."

1 THE NOMENCLATURE OF ALDEHYDES AND KETONES

Naming Aldehydes

The systematic (IUPAC) name of an aldehyde is obtained by replacing the final "e" on the name of the parent hydrocarbon with "al." For example, a one-carbon aldehyde is called methan*al,* and a two-carbon aldehyde is called ethan*al.* The position of the carbonyl carbon does not have to be designated because it is always at the end of the parent hydrocarbon (or else the compound would not be an aldehyde), so it always has the 1-position.

The common name of an aldehyde is the same as the common name of the corresponding carboxylic acid, except that "aldehyde" is substituted for "oic acid" (or "ic acid"). The position of a substituent is designated by a lowercase Greek letter when common names are used. The carbonyl carbon is not given a designation, so the carbon adjacent to the carbonyl carbon is the α-carbon.

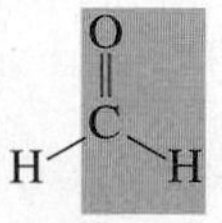

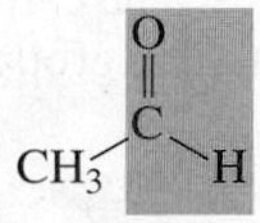

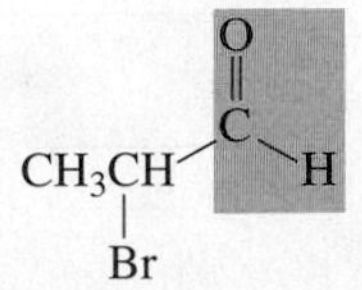

systematic name:
common name:

methanal
formaldehyde

ethanal
acetaldehyde

2-bromopropanal
α-bromopropionaldehyde

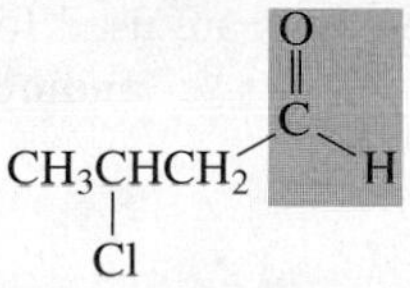

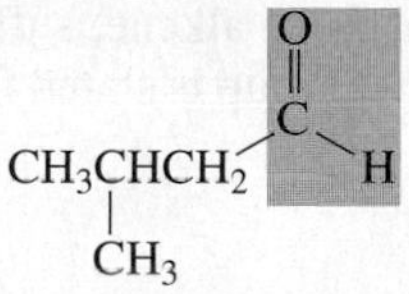

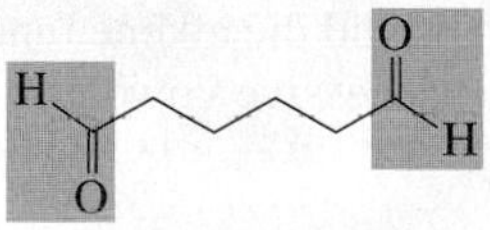

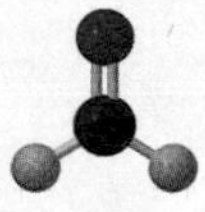

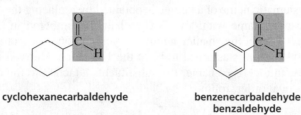

systematic name:	3-chlorobutanal	3-methylbutanal
common name:	β-chlorobutyraldehyde	isovaleraldehyde

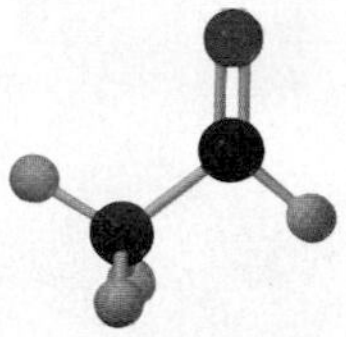

hexanedial

formaldehyde

acetaldehyde

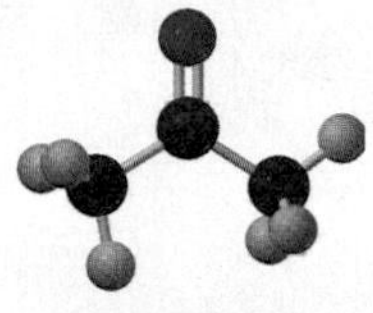

Notice that the terminal "e" of the parent hydrocarbon is not removed in hexanedial. (The "e" is removed only to avoid two successive vowels.)

If the aldehyde group is attached to a ring, then the aldehyde is named by adding "carbaldehyde" to the name of the cyclic compound.

systematic name:	cyclohexanecarbaldehyde	benzenecarbaldehyde
common name:		benzaldehyde

acetone

A carbonyl group has a higher nomenclature priority than an alcohol or an amino-group. However, all carbonyl compounds do not have the same priority. Nomenclature priorities of the various functional groups, including carbonyl groups, are listed in Table 1.

Table 1   Functional Group Nomenclature

Class	Suffix name	Prefix name
Carboxylic acid	-oic acid	Carboxy
Ester	-oate	Alkoxycarbonyl
Amide	-amide	Amido
Nitrile	-nitrile	Cyano
Aldehyde	-al	Oxo (=O)
Aldehyde	-al	Formyl (CH=O)
Ketone	-one	Oxo (=O)
Alcohol	-ol	Hydroxy
Amine	-amine	Amino
Alkene	-ene	Alkenyl
Alkyne	-yne	Alkynyl
Alkane	-ane	Alkyl
Ether	—	Alkoxy
Alkyl halide	—	Halo

increasing priority (↑)

If a compound has two functional groups, the one with the lower priority is indicated by a prefix and the one with the higher priority by a suffix (unless one of the functional groups is an alkene).

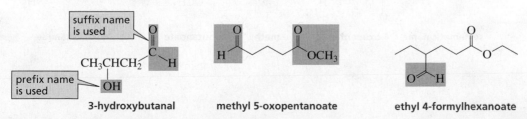

3-hydroxybutanal	methyl 5-oxopentanoate	ethyl 4-formylhexanoate

If one of the functional groups is an alkene, suffix endings are used for both functional groups and the alkene functional group is stated first, with its "e" ending omitted to avoid two successive vowels.

$$CH_3CH=CHCH_2 \quad \overset{\displaystyle O}{\underset{\displaystyle H}{\|}}C$$

3-pentenal

Naming Ketones

The systematic name of a ketone is obtained by replacing the "e" on the end of the parent hydrocarbon name with "one." The chain is numbered in the direction that gives the carbonyl carbon the smaller number. Cyclic ketones do not need a number because the carbonyl carbon is assumed to be at the 1-position. Derived names can also be used for ketones. In a derived name, the substituents attached to the carbonyl group are cited in alphabetical order, followed by "ketone."

Aldehydes and ketones are named using a functional group suffix.

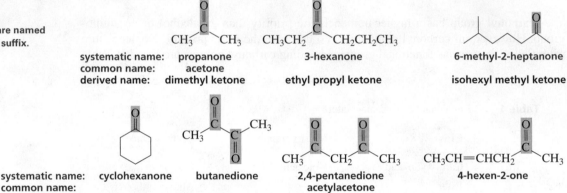

systematic name:	propanone	3-hexanone	6-methyl-2-heptanone
common name:	acetone		
derived name:	dimethyl ketone	ethyl propyl ketone	isohexyl methyl ketone

systematic name:	cyclohexanone	butanedione	2,4-pentanedione	4-hexen-2-one
common name:			acetylacetone	

Only a few ketones have common names. The smallest ketone, propanone, is usually referred to by its common name, acetone. Acetone is a widely used laboratory solvent. Common names are also used for some phenyl-substituted ketones; the number of carbons (other than those of the phenyl group) is indicated by the common name of the corresponding carboxylic acid, substituting "ophenone" for "ic acid."

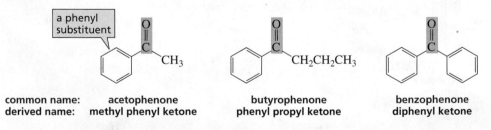

a phenyl substituent

common name:	acetophenone	butyrophenone	benzophenone
derived name:	methyl phenyl ketone	phenyl propyl ketone	diphenyl ketone

If a ketone has a second functional group of higher naming priority, the ketone oxygen is indicated by the prefix "oxo."

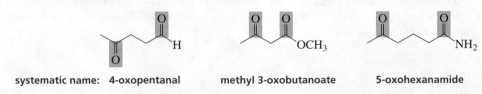

systematic name:	4-oxopentanal	methyl 3-oxobutanoate	5-oxohexanamide

Butanedione: An Unpleasant Compound

Fresh perspiration is odorless. The smells we associate with perspiration result from a chain of events initiated by bacteria that are always present on our skin. These bacteria produce lactic acid, which creates an acidic environment that allows other bacteria to break down the components of perspiration, producing compounds with the unappealing odors we associate with armpits and sweaty feet. One such compound is butanedione.

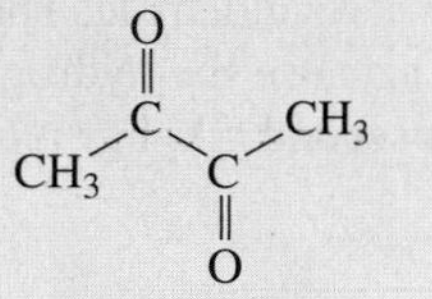

butanedione

PROBLEM 1♦

Why are numbers not used to designate the position of the functional group in propanone and butanedione?

PROBLEM 2♦

Give two names for each of the following compounds:

a.

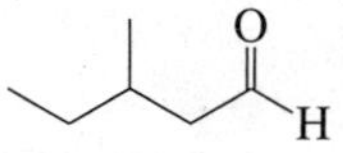

b.

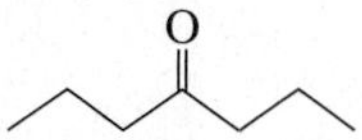

c.

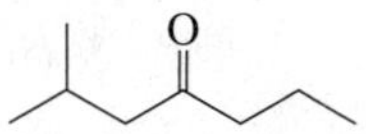

d.

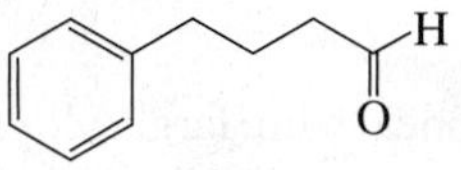

e.

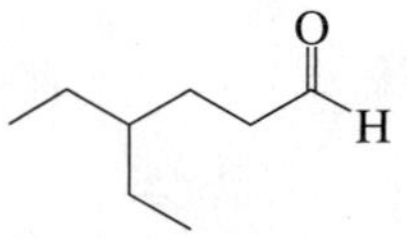

f.

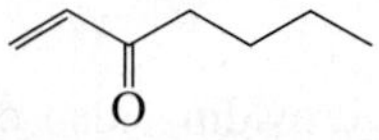

PROBLEM 3

Name the following compounds:

a.

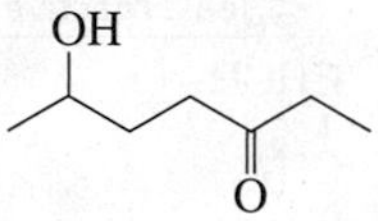

b.

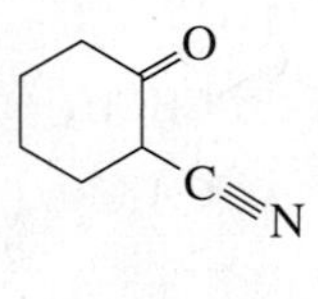

c.

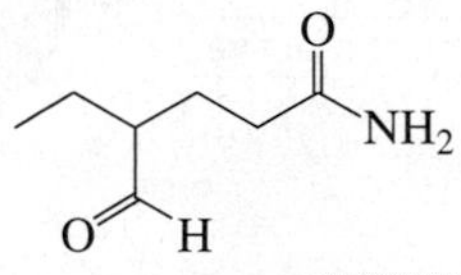

2 THE RELATIVE REACTIVITIES OF CARBONYL COMPOUNDS

The carbonyl group is polar because oxygen is more electronegative than carbon, so oxygen has a greater share of the double bond's electrons. As a result, the carbonyl carbon is electron deficient (it is an electrophile), and it reacts with nucleophiles. The electron deficiency of the carbonyl carbon is indicated by the blue region in the electrostatic potential maps.

formaldehyde

acetaldehyde

acetone

An aldehyde has a greater partial positive charge on its carbonyl carbon than a ketone does because a hydrogen is more electron withdrawing than an alkyl group. An aldehyde, therefore, is more reactive than a ketone toward nucleophilic addition. Steric factors also contribute to the greater reactivity of an aldehyde. The carbonyl carbon of an aldehyde is more accessible to a nucleophile because the hydrogen attached to the carbonyl carbon of an aldehyde is smaller than the second alkyl group attached to the carbonyl carbon of a ketone.

relative reactivities

most reactive

formaldehyde > an aldehyde > a ketone   least reactive

Steric factors are also important in the transition state, which is tetrahedral (so it has bond angles of 109.5°). This causes the alkyl groups to be closer to one another than they are in the carbonyl compound, where the bond angles are 120°. As a result of the greater steric crowding in their transition states, ketones have less stable transition states than aldehydes have. In summary, alkyl groups stabilize the carbonyl compound and destabilize the transition state. Both factors increase $\Delta G^{\ddagger}$, causing ketones to be less reactive than aldehydes.

120°   109.5°

Aldehydes are more reactive than ketones.

Steric crowding also causes ketones with large alkyl groups bonded to the carbonyl carbon to be less reactive than those with small alkyl groups.

relative reactivities

most reactive >   >   least reactive

PROBLEM 4♦

Which ketone in each pair is more reactive?

a. 2-heptanone or 4-heptanone
b. bromomethyl phenyl ketone or chloromethyl phenyl ketone

Aldehydes and ketones are less reactive than acyl chlorides and acid anhydrides but they are more reactive than esters, carboxylic acids, and amides.

Aldehydes and ketones are *less* reactive than acyl halides and anhydrides, but they are *more* reactive than esters, carboxylic acids, and amides.

relative reactivities of carbonyl compounds

acyl halide > acid anhydride > aldehyde > ketone > ester ~ carboxylic acid > amide > carboxylate ion

most reactive   least reactive

820

3   HOW ALDEHYDES AND KETONES REACT

The carbonyl group of a carboxylic acid or a carboxylic acid derivative is attached to a group that can be replaced by another group. As a result, these compounds undergo **nucleophilic addition–elimination reactions**—the nucleophile adds to the carbonyl carbon and a group is eliminated from the tetrahedral intermediate. Overall, it is a substitution reaction in which Z^- substitutes for Y^-.

Carboxylic acid derivatives undergo nucleophilic addition–elimination reactions with nucleophiles.

In contrast, the carbonyl group of an aldehyde or a ketone is attached to a group that is too strong a base (H^- or R^-) to be eliminated under normal conditions, so it cannot be replaced by another group. Aldehydes and ketones, therefore, do *not* form substitution products when they react with nucleophiles.

The addition of a nucleophile to the carbonyl carbon of an aldehyde or a ketone forms a tetrahedral compound. If the nucleophile is a strong base, such as R^- or H^-, then the tetrahedral compound will be stable because it will not have a group that can be eliminated. Thus, the reaction will be an *irreversible* **nucleophilic addition reaction.** (Note that a tetrahedral compound is unstable only if the sp^3 carbon is attached to an oxygen *and* to another electronegative atom).

Aldehydes and ketones undergo irreversible nucleophilic addition reactions with nucleophiles that are strong bases.

If the attacking atom of the nucleophile is an oxygen or a nitrogen and there is enough acid in the solution to protonate the OH group of the tetrahedral compound, then water can be eliminated from the addition product by a lone pair on the oxygen or nitrogen. The reaction is reversible because there are two electronegative groups attached to the tetrahedral intermediate, either one of which can be protonated and therefore eliminated. We will see that the fate of the dehydrated product depends on the identity of Z.

Aldehydes and ketones undergo nucleophilic addition–elimination reactions with nucleophiles that have a lone pair on the attacking atom.

This is also called a **nucleophilic addition–elimination reaction.** However, unlike carboxylic acid derivatives that undergo nucleophilic addition–eliminations that eliminate a group attached to the acyl group, aldehydes and ketones undergo nucleophilic addition–elimination reactions that eliminate water.

4 THE REACTIONS OF CARBONYL COMPOUNDS WITH GRIGNARD REAGENTS

Addition of a Grignard reagent to a carbonyl compound is a versatile reaction that forms a new C—C bond. This reaction can produce compounds with a variety of structures because both the structure of the carbonyl compound and the structure of the Grignard reagent can be varied.

A Grignard reagent is prepared by adding an alkyl halide to magnesium shavings in diethyl ether under anhydrous conditions. We also saw that a Grignard reagent reacts as if it were a carbanion; therefore, it is a strongly basic nucleophile.

$$CH_3CH_2Br \xrightarrow[Et_2O]{Mg} CH_3CH_2MgBr$$

$$CH_3CH_2MgBr \quad \text{reacts as if it were} \quad CH_3\overset{..}{\overset{-}{C}}H_2 \quad \overset{+}{Mg}Br$$

Consequently, aldehydes and ketones undergo nucleophilic addition reactions with Grignard reagents (Section 3).

The Reactions of Aldehydes and Ketones with Grignard Reagents

The reaction of an aldehyde or a ketone with a Grignard reagent is a nucleophilic addition reaction—the nucleophile adds to the carbonyl carbon. The tetrahedral alkoxide ion is stable because it does not have a group that can be eliminated. (A tetrahedral compound is unstable only if the sp^3 carbon is attached to an oxygen *and* to another electronegative atom.)

MECHANISM FOR THE REACTION OF AN ALDEHYDE OR A KETONE WITH A GRIGNARD REAGENT

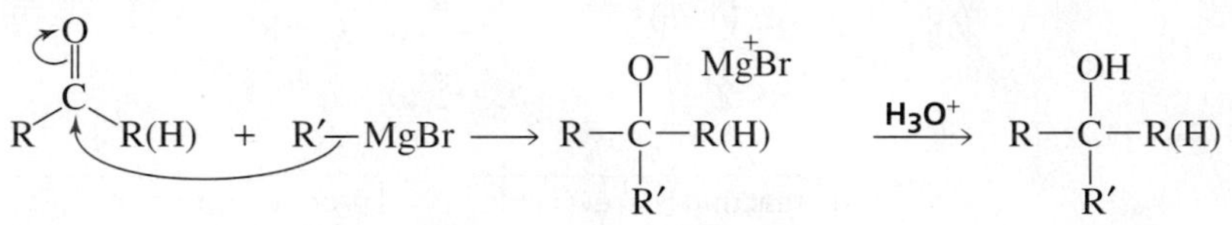

- Nucleophilic addition of the Grignard reagent to the carbonyl carbon forms an alkoxide ion that is complexed with the magnesium ion.
- Addition of dilute acid breaks up the complex.

When a Grignard reagent reacts with formaldehyde, the product of the nucleophilic addition reaction is a *primary alcohol*.

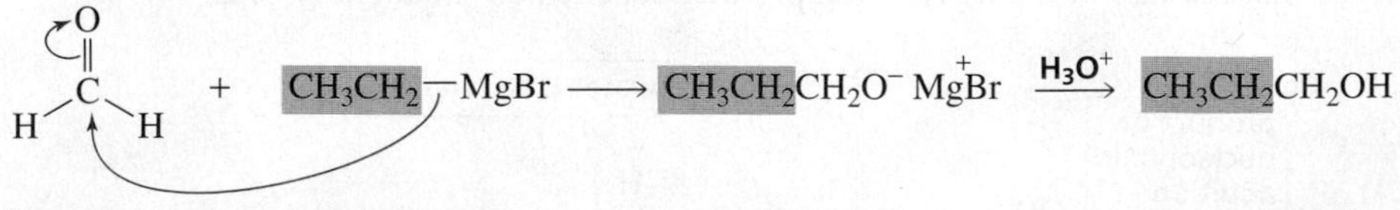

| formaldehyde | ethylmagnesium bromide | an alkoxide ion | 1-propanol a primary alcohol |

When a Grignard reagent reacts with an aldehyde other than formaldehyde, the product of the nucleophilic addition reaction is a *secondary alcohol*.

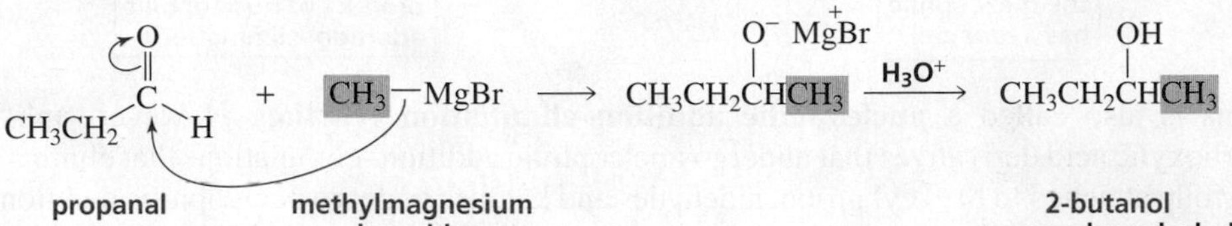

| propanal | methylmagnesium bromide | | 2-butanol a secondary alcohol |

When a Grignard reagent reacts with a ketone, the product of the nucleophilic addition reaction is a *tertiary alcohol*.

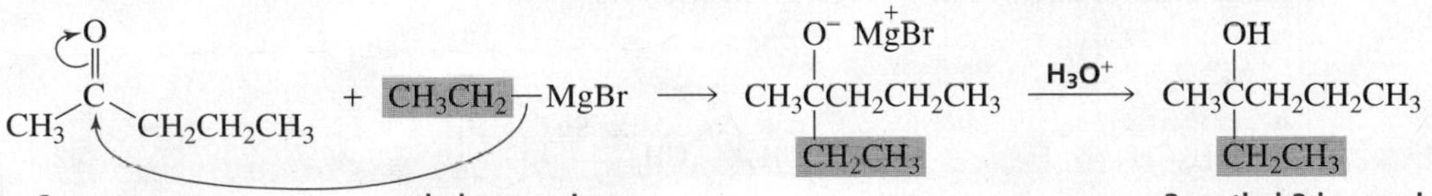

2-pentanone   ethylmagnesium bromide   3-methyl-3-hexanol a tertiary alcohol

A Grignard reagent can also react with carbon dioxide. The product of the reaction is a carboxylic acid that has one more carbon than the Grignard reagent.

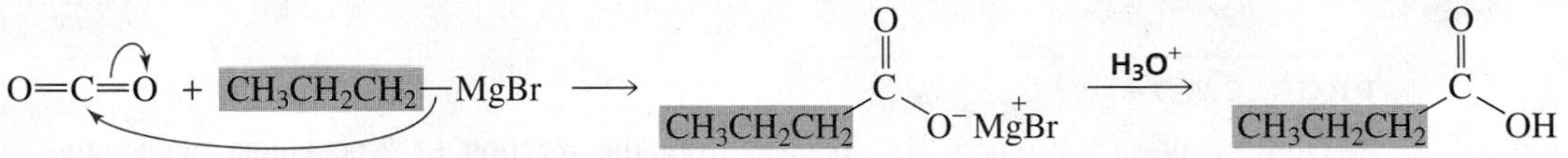

carbon dioxide   propylmagnesium bromide   butanoic acid

In the following reactions, the reagents above and below the reaction arrows are numbered in order of use, indicating that the acid is not added until after the Grignard reagent has reacted with the carbonyl compound.

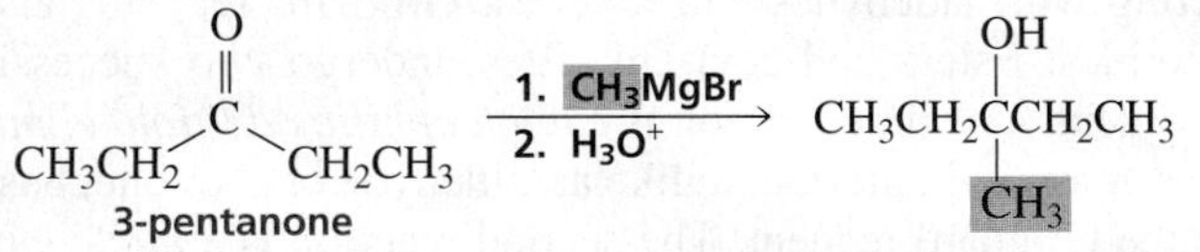

3-pentanone   3-methyl-3-pentanol

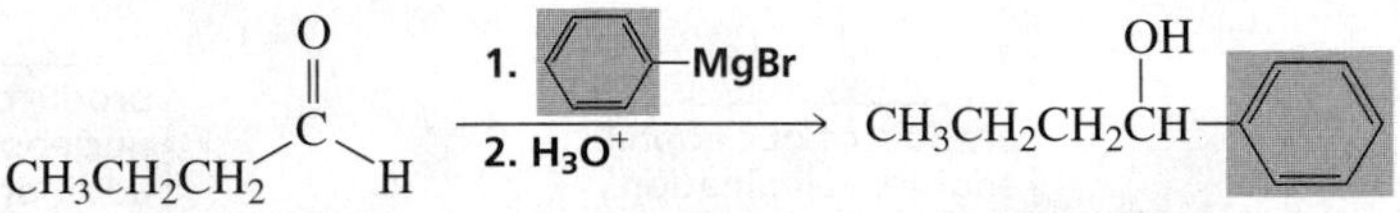

butanal   1-phenyl-1-butanol

If the reaction with the carbonyl compound forms a product with an asymmetric center, such as the preceding reaction that forms 1-phenyl-1-butanol, the product will be a racemic mixture. (When a reactant without an asymmetric center undergoes a reaction that forms an asymmetric center, then the product will be a racemic mixture.)

Enzyme-Catalyzed Carbonyl Additions

If the reaction with a carbonyl compound that forms a product with an asymmetric center is catalyzed by an enzyme, then a racemic mixture is not formed. Instead, only one of the enantiomers is formed, because the enzyme can block one face of the carbonyl compound so that it cannot be attacked, or the enzyme can position the nucleophile so it can attack the carbonyl group from only one side.

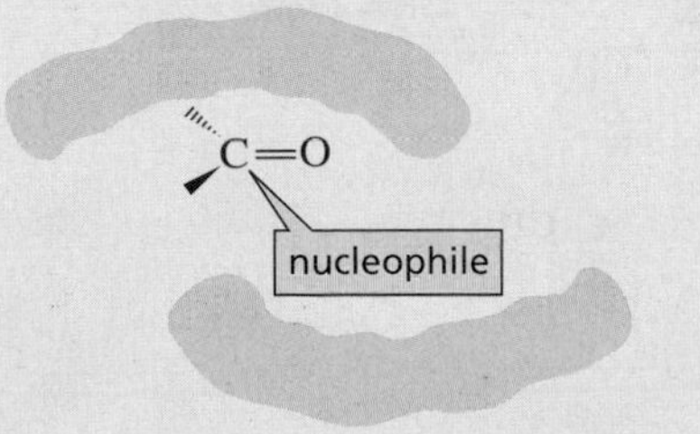

nucleophile

The Reactions of Esters and Acyl Chlorides with Grignard Reagents

In addition to reacting with aldehydes and ketones, Grignard reagents also react with esters and acyl chlorides. Esters and acyl chlorides undergo two successive reactions with the Grignard reagent. The first reaction is a *nucleophilic addition–elimination reaction* because an ester or an acyl chloride, unlike an aldehyde or a ketone, has a group that can be replaced by the Grignard reagent. The second reaction is a *nucleophilic addition reaction.*

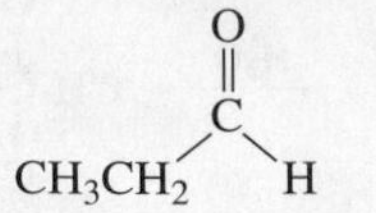

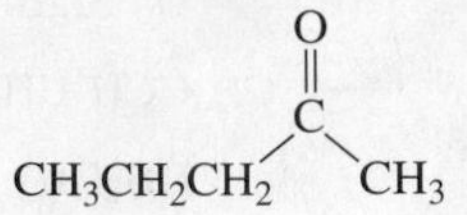

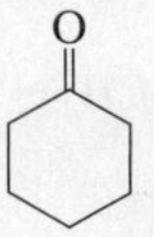

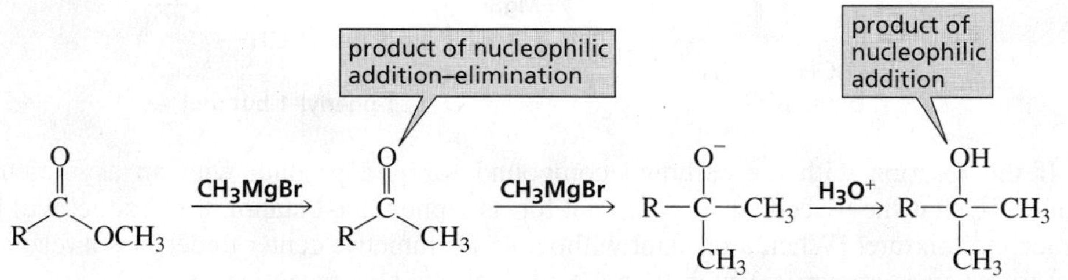

product of nucleophilic addition–elimination

product of nucleophilic addition

$$R-\overset{\overset{\displaystyle O}{\|}}{C}-OCH_3 \xrightarrow{CH_3MgBr} R-\overset{\overset{\displaystyle O}{\|}}{C}-CH_3 \xrightarrow{CH_3MgBr} R-\overset{\overset{\displaystyle O^-}{|}}{\underset{\underset{\displaystyle CH_3}{|}}{C}}-CH_3 \xrightarrow{H_3O^+} R-\overset{\overset{\displaystyle OH}{|}}{\underset{\underset{\displaystyle CH_3}{|}}{C}}-CH_3$$

an ester   a ketone   a tertiary alcohol

The product of the reaction of an ester with a Grignard reagent is a tertiary alcohol. Because the tertiary alcohol is formed from two successive reactions with the Grignard reagent, the alcohol has at least two identical alkyl groups bonded to the tertiary carbon.

MECHANISM FOR THE REACTION OF AN ESTER WITH A GRIGNARD REAGENT

a second addition of the Grignard reagent occurs

$$R-\overset{\overset{\displaystyle \ddot{O}:}{\|}}{C}-OCH_3 + CH_3-MgBr \longrightarrow R-\overset{\overset{\displaystyle :\ddot{O}:^-\overset{+}{M}gBr}{|}}{\underset{\underset{\displaystyle CH_3}{|}}{C}}-OCH_3 \longrightarrow R-\overset{\overset{\displaystyle \ddot{O}:}{\|}}{C}-CH_3 \xrightarrow{CH_3-MgBr} R-\overset{\overset{\displaystyle :\ddot{O}:^-\overset{+}{M}gBr}{|}}{\underset{\underset{\displaystyle CH_3}{|}}{C}}-CH_3 \xrightarrow{H_3O^+} R-\overset{\overset{\displaystyle :\ddot{O}H}{|}}{\underset{\underset{\displaystyle CH_3}{|}}{C}}-CH_3$$

an ester   a ketone   a tertiary alcohol

a group is eliminated from the tetrahedral intermediate

$+ CH_3O^-$

$+ CH_3OH$

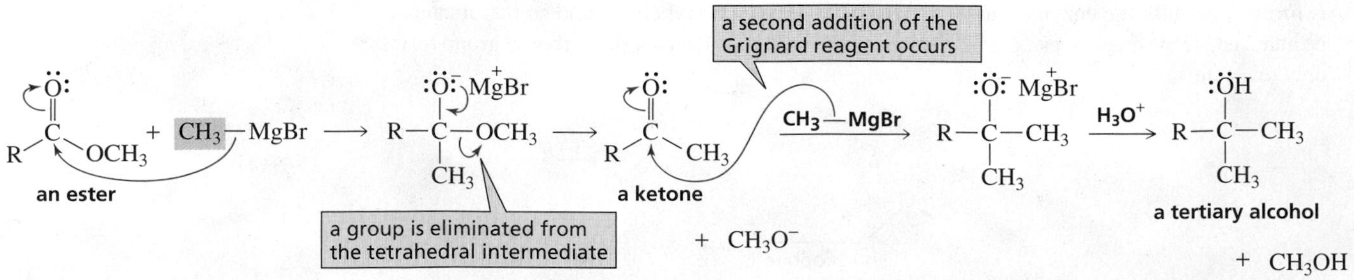

- Nucleophilic addition of the Grignard reagent to the carbonyl carbon forms a tetrahedral intermediate that is unstable because it has a group that can be eliminated.

- The tetrahedral intermediate eliminates methoxide ion, forming a ketone.

- The ketone reacts with a second molecule of the Grignard reagent, forming an alkoxide ion that forms a tertiary alcohol after protonation.

Tertiary alcohols are also formed from the reaction of two equivalents of a Grignard reagent with an acyl chloride. The mechanism for the reaction of a Grignard reagent with an acyl chloride is the same as the mechanism for the reaction a Grignard reagent with an ester.

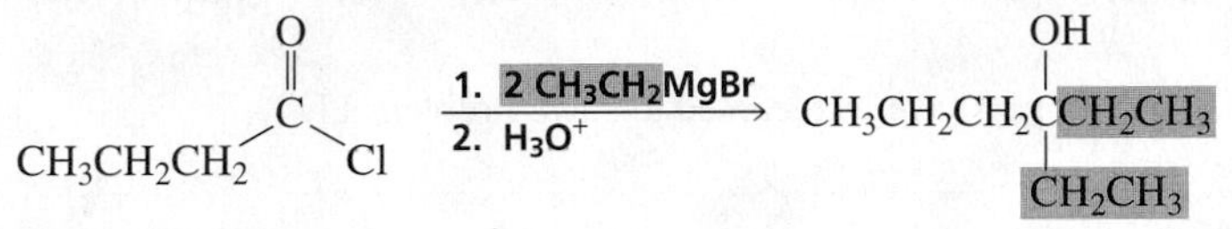

butyryl chloride

3-ethyl-3-hexanol

PROBLEM 8 Solved

a. Which of the following tertiary alcohols cannot be prepared by the reaction of an ester with excess Grignard reagent?

1. $CH_3CCH_2CH_3$ with OH and CH_3

2. CH_3CCH_3 with OH and CH_3

3. $CH_3CH_2CCH_2CH_2CH_3$ with OH and CH_3

4. $CH_3CH_2CCH_3CH_3$ with OH and CH_3

5. $CH_3CCH_2CH_2CH_2CH_3$ with OH and CH_2CH_3

6. (diphenyl) C with OH and CH_3

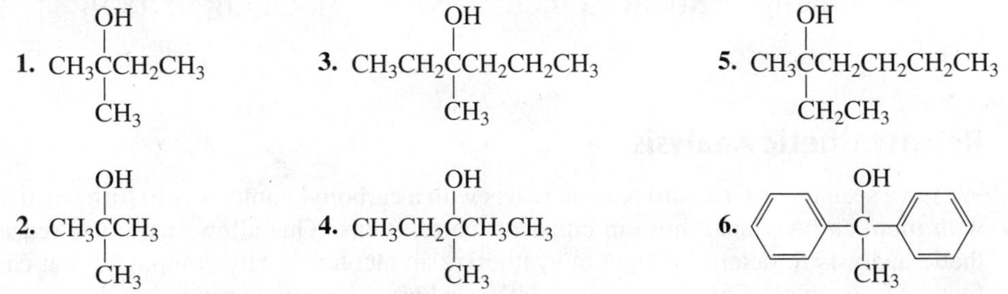

b. For those alcohols that can be prepared by the reaction of an ester with excess Grignard reagent, what ester and what Grignard reagent should be used?

Solution to 8a A tertiary alcohol prepared by the reaction of an ester with two equivalents of a Grignard reagent must have at least two identical substituents bonded to the carbon to which the OH is attached because two of the three substituents come from the Grignard reagent. Alcohols 3 and 5 do not have two identical substituents, so they cannot be prepared in this way.

Solution to 8b (1) An ester of propanoic acid and excess methylmagnesium bromide.

PROBLEM 9♦

Which of the following secondary alcohols can be prepared by the reaction of methyl formate with excess Grignard reagent?

$CH_3CH_2CHCH_3$ CH_3CHCH_3 $CH_3CHCH_2CH_2CH_3$ $CH_3CH_2CHCH_2CH_3$

OH   OH   OH   OH

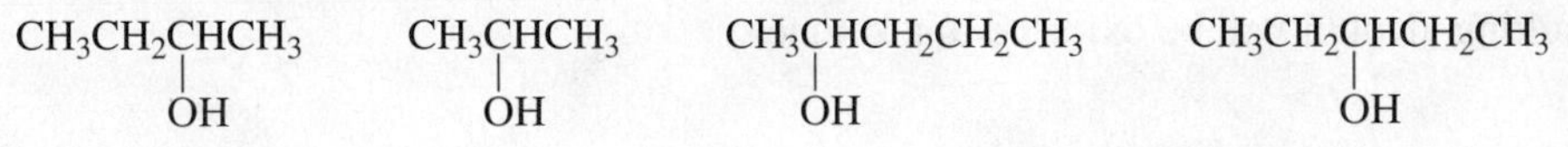

A   B   C   D

PROBLEM 10

Write the mechanism for the reaction of acetyl chloride with ethylmagnesium bromide.

PROBLEM-SOLVING STRATEGY

Predicting Reactions with Grignard Reagents

Why doesn't a Grignard reagent add to the carbonyl carbon of a carboxylic acid?

We know that Grignard reagents add to carbonyl carbons, so if we find that a Grignard reagent does not add to the carbonyl carbon, we can conclude that it must react more rapidly with another part of the molecule. A carboxylic acid has an acidic proton that reacts rapidly with the Grignard reagent, converting it to an alkane.

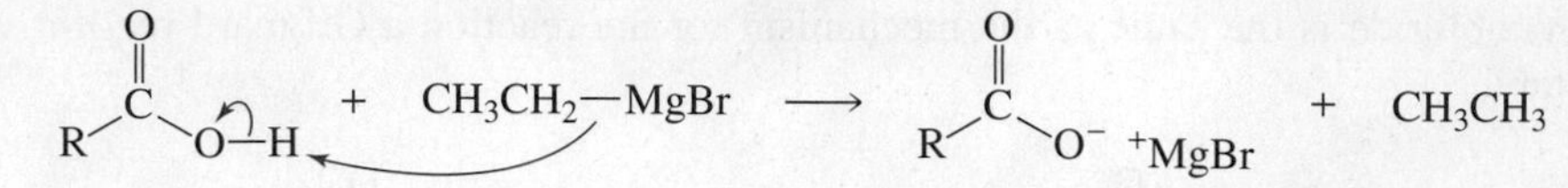

$$\underset{R}{\overset{\displaystyle O}{\overset{\|}{C}}}\text{—O—H} \quad + \quad CH_3CH_2\text{—MgBr} \quad \longrightarrow \quad \underset{R}{\overset{\displaystyle O}{\overset{\|}{C}}}\text{—O}^- \ ^+MgBr \quad + \quad CH_3CH_3$$

Now use the strategy you have just learned to solve Problem 11.

PROBLEM 11♦

Which of the following compounds will not undergo a nucleophilic addition reaction with one equivalent of a Grignard reagent?

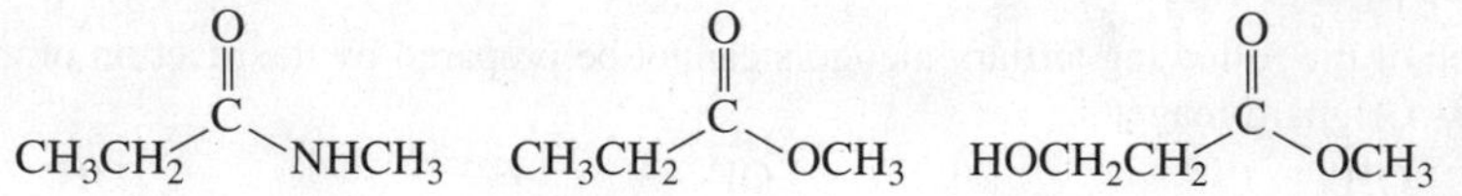

$$\underset{\textbf{A}}{CH_3CH_2\overset{\displaystyle O}{\overset{\|}{C}}\text{NHCH}_3} \qquad \underset{\textbf{B}}{CH_3CH_2\overset{\displaystyle O}{\overset{\|}{C}}\text{OCH}_3} \qquad \underset{\textbf{C}}{HOCH_2CH_2\overset{\displaystyle O}{\overset{\|}{C}}\text{OCH}_3}$$

Retrosynthetic Analysis

We have seen that a Grignard reagent reacts with a carbonyl compound to form an alcohol with more carbons than those in either of the reactants. This allows us to use retrosynthetic analysis to determine how to synthesize an alcohol or any compound that can be formed from an alcohol. For example, let's see how 3-hexanone can be synthesized from 1-propanol and no other carbon-containing reagents.

3-Hexanone can be synthesized from 3-hexanol, which can be synthesized from a three-carbon aldehyde and a three-carbon Grignard reagent. Oxidation of 1-propanol forms the three-carbon aldehyde, and conversion of 1-propanol to a propyl halide allows the three-carbon Grignard reagent to be synthesized.

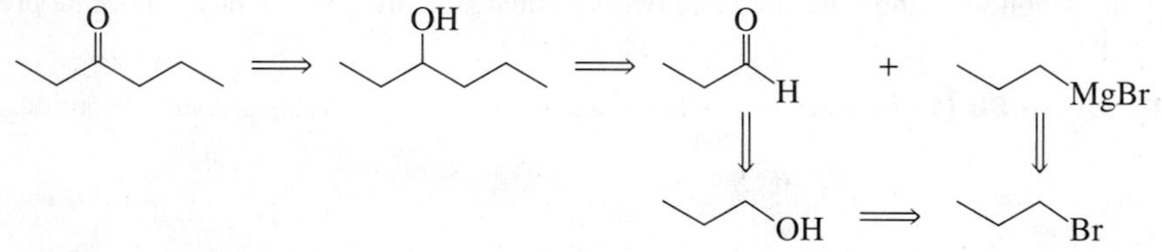

Now we can write the reaction in the forward direction, showing the reagents needed for each step of the synthesis. Note that acetic acid will protonate the alkoxide ion and the resulting alcohol will be oxidized to a ketone.

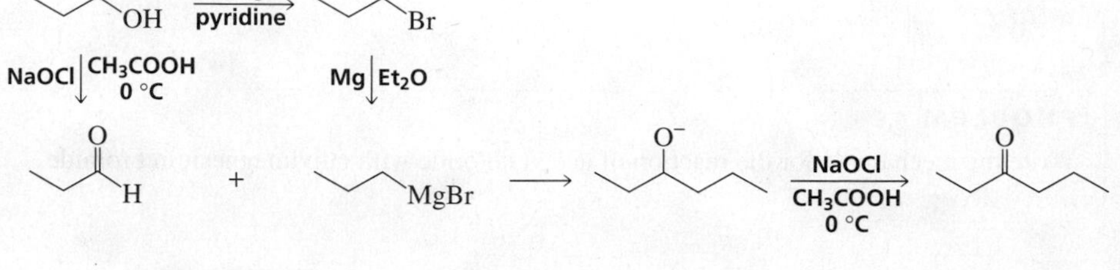

PROBLEM 12

Show how the following compounds could be synthesized from cyclohexanol.

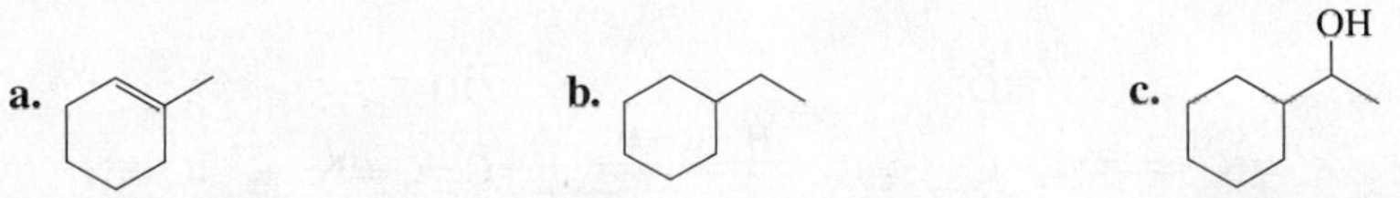

a.   b.   c.

5   THE REACTIONS OF CARBONYL COMPOUNDS WITH ACETYLIDE IONS

Note that a terminal alkyne can be converted into an acetylide ion by a strong base.

$$CH_3C\equiv CH \xrightarrow{\text{NaNH}_2} CH_3C\equiv C^{:-}$$

An acetylide ion is another example of a carbon nucleophile that reacts with an aldehyde or a ketone to form a nucleophilic addition product. When the reaction is over, a weak acid (one that will not react with the triple bond, such as the pyridinium ion shown here) is added to the reaction mixture to protonate the alkoxide ion.

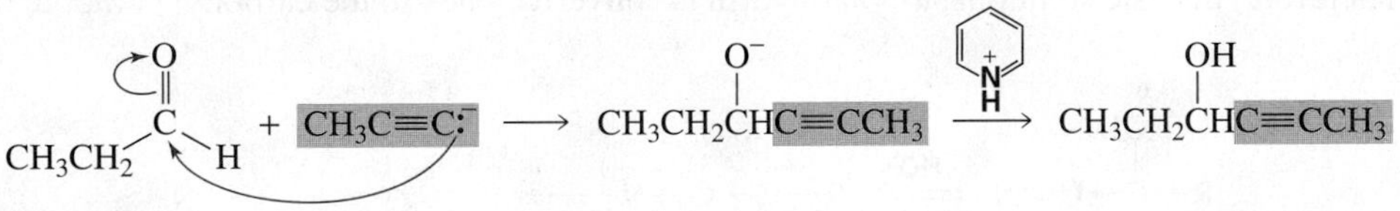

PROBLEM 13

Show how the following compounds can be prepared, using ethyne as one of the starting materials. Explain why ethyne should be alkylated before, rather than after, nucleophilic addition.

a. 1-pentyn-3-ol   b. 1-phenyl-2-butyn-1-ol   c. 2-methyl-3-hexyn-2-ol

PROBLEM 14♦

What is the product of the reaction of an ester with excess acetylide ion followed by the addition of pyridinium chloride?

6   THE REACTIONS OF ALDEHYDES AND KETONES WITH CYANIDE ION

Cyanide ion is another carbon nucleophile that can add to an aldehyde or a ketone. The product of the reaction is a **cyanohydrin.** Unlike the addition reactions of other carbon nucleophiles, the addition of cyanide ion has to be carried out under acidic conditions. Excess cyanide ion is used to ensure that some cyanide ion is not protonated by the acid and, therefore, is available to act as a nucleophile.

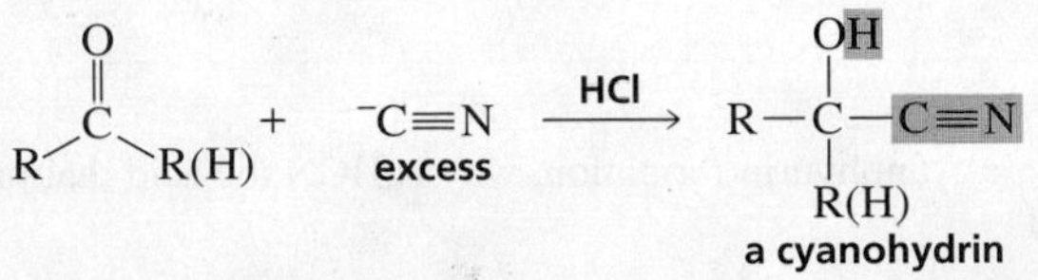

MECHANISM FOR THE REACTION OF AN ALDEHYDE OR KETONE WITH HYDROGEN CYANIDE

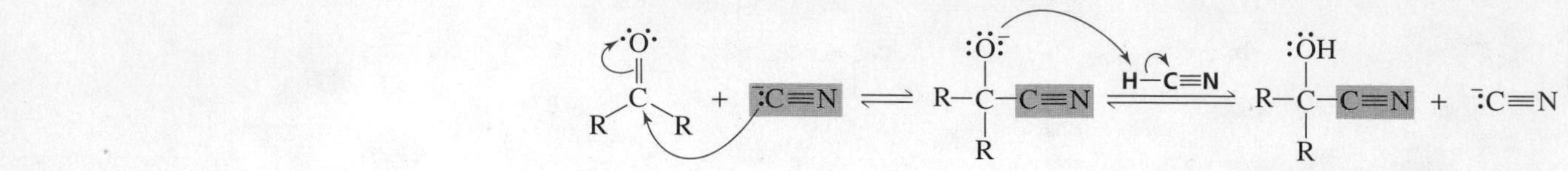

acetone cyanohydrin

- Cyanide ion adds to the carbonyl carbon.
- The alkoxide ion is protonated by an undissociated molecule of hydrogen cyanide.

Compared with Grignard reagents and acetylide ions, cyanide ion is a relatively weak base. The pK_a of CH_3CH_3 is > 60 and the pK_a of $RC\equiv CH$ is 25, but the pK_a of $HC\equiv N$ is 9.14. (The stronger the acid, the weaker the conjugate base.) Therefore, the cyano group, unlike the R^- or $RC\equiv C^-$ groups, can be eliminated from the addition product.

Cyanohydrins, however, are stable. The OH group will not eliminate the cyano group because the transition state for the elimination reaction would have a partial *positive* charge on the oxygen, thereby making it relatively unstable.

If, however, the OH group loses its proton, then the cyano group will be eliminated because the oxygen atom would then have a partial *negative* charge in the transition state. Therefore, in basic solutions, a cyanohydrin is converted back to the carbonyl compound.

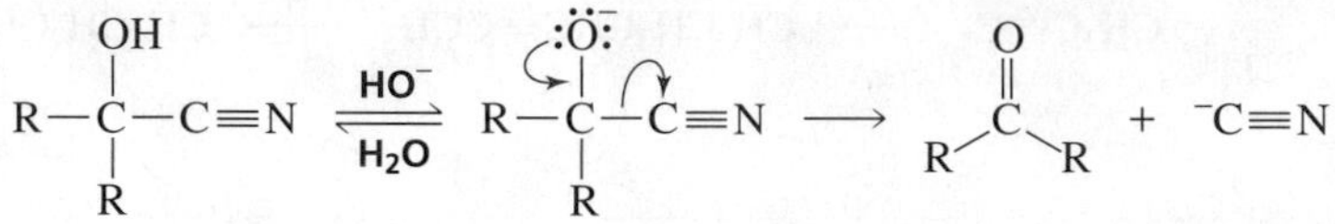

Cyanide ion does not react with esters because the cyanide ion is a weaker base than an alkoxide ion, so the cyanide ion would be eliminated from the tetrahedral intermediate. (The incoming nucleophile cannot be a much weaker base than the substituent attached to the acyl group).

The addition of hydrogen cyanide to aldehydes and ketones is a synthetically useful reaction because of the subsequent reactions that can be carried out on the cyanohydrin. For example, the acid-catalyzed hydrolysis of a cyanohydrin forms an α-hydroxycarboxylic acid.

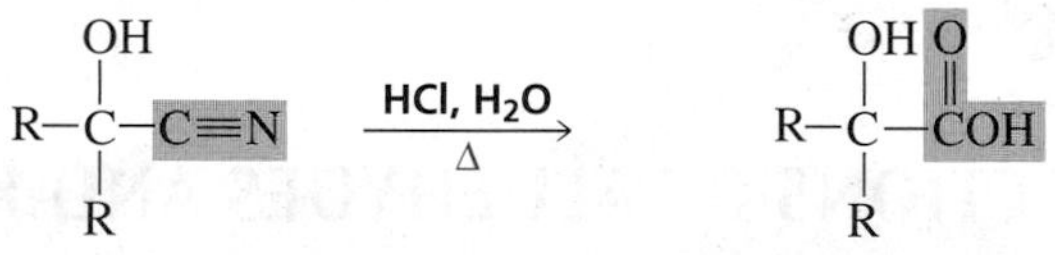

a cyanohydrin   an α-hydroxycarboxylic acid

The catalytic addition of two moles of hydrogen to the triple bond of a cyanohydrin produces a primary amine with an OH group on the β-carbon.

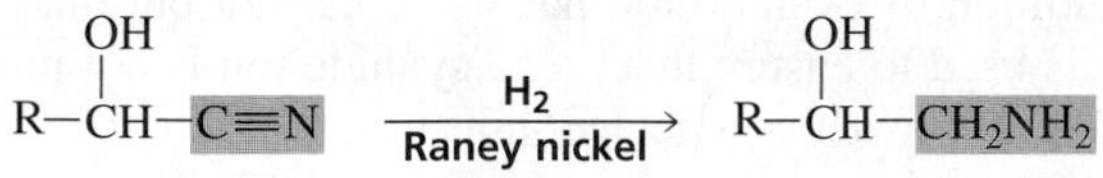

PROBLEM 15◆

In the mechanism for cyanohydrin formation, why is HCN the acid that protonates the alkoxide ion, instead of HCl?

PROBLEM 16◆

Can a cyanohydrin be prepared by treating a ketone with sodium cyanide?

PROBLEM 17♦

Explain why aldehydes and ketones react with a weak acid such as hydrogen cyanide but do not react with strong acids such as HCl or H_2SO_4 (other than being protonated by them).

PROBLEM 18 Solved

How can the following compounds be prepared from a carbonyl compound that has one less carbon than the desired product?

a. $HOCH_2CH_2NH_2$

b. 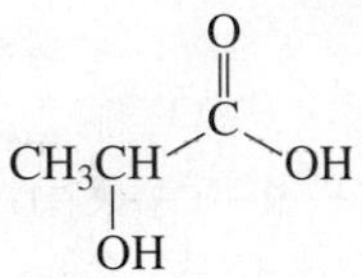 $CH_3CH\overset{\displaystyle O}{\underset{\displaystyle OH}{|}}\overset{||}{C}-OH$ with OH below

Solution to 18a The starting material for the synthesis of this two-carbon compound must be formaldehyde. Addition of hydrogen cyanide followed by addition of H_2 to the triple bond of the cyanohydrin forms the target molecule.

$$H-\overset{O}{\overset{||}{C}}-H \xrightarrow[\text{HCl}]{\text{NaC}\equiv\text{N}} HOCH_2C\equiv N \xrightarrow[\text{Raney nickel}]{H_2} HOCH_2CH_2NH_2$$

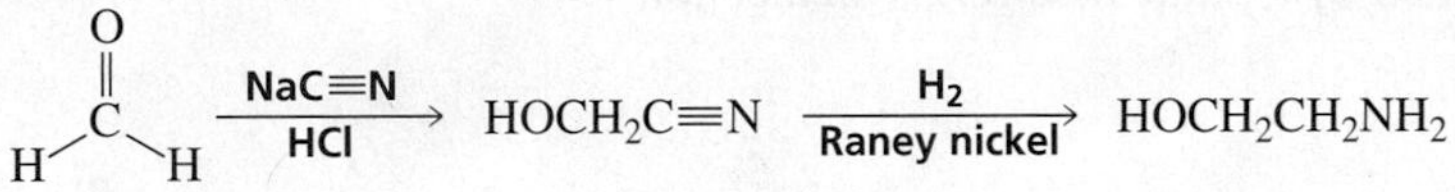

Solution to 18b The addition of cyanide ion adds one carbon to the reactant, so the starting material for the synthesis of this three-carbon α-hydroxycarboxylic acid must be acetaldehyde. Addition of hydrogen cyanide, followed by hydrolysis of the resulting cyanohydrin, forms the target molecule.

$$CH_3-\overset{O}{\overset{||}{C}}-H \xrightarrow[\text{HCl}]{\text{NaC}\equiv\text{N}} CH_3\underset{\underset{\displaystyle OH}{|}}{CH}C\equiv N \xrightarrow[\Delta]{\text{HCl, H}_2\text{O}} CH_3\underset{\underset{\displaystyle OH}{|}}{CH}\overset{O}{\overset{||}{C}}-OH$$

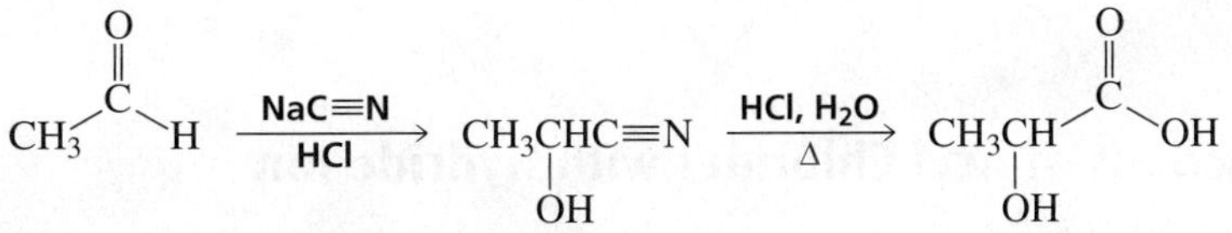

PROBLEM 19

Show two ways to convert an alkyl halide into a carboxylic acid that has one more carbon than the alkyl halide.

7 THE REACTIONS OF CARBONYL COMPOUNDS WITH HYDRIDE ION

The Reactions of Aldehydes and Ketones with Hydride Ion

A hydride ion is another strongly basic nucleophile that reacts with an aldehyde or a ketone to form a nucleophilic addition product. Usually, sodium borohydride ($NaBH_4$) is used for the source of the hydride ion.

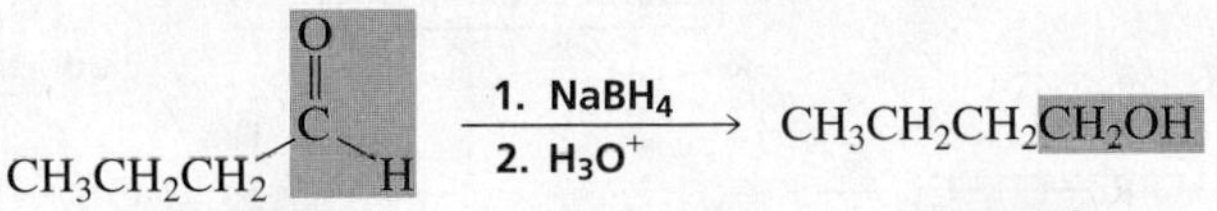

$$CH_3CH_2CH_2-\overset{O}{\overset{||}{C}}-H \xrightarrow[\text{2. H}_3\text{O}^+]{\text{1. NaBH}_4} CH_3CH_2CH_2CH_2OH$$

butanal
an aldehyde

1-butanol
a primary alcohol

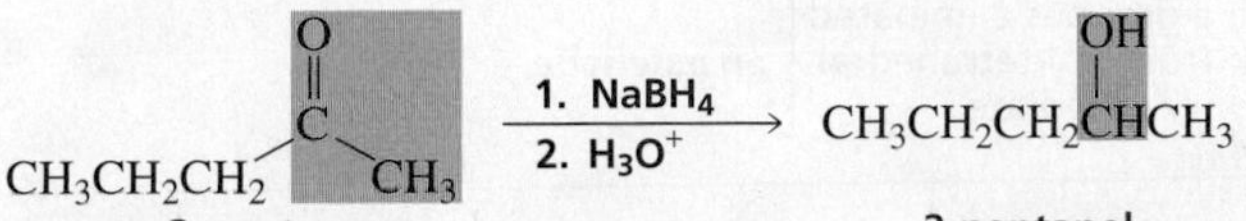

$$CH_3CH_2CH_2-\overset{O}{\overset{||}{C}}-CH_3 \xrightarrow[\text{2. H}_3\text{O}^+]{\text{1. NaBH}_4} CH_3CH_2CH_2\underset{\underset{\displaystyle }{}}{\overset{OH}{\overset{|}{C}}}HCH_3$$

2-pentanone
a ketone

2-pentanol
a secondary alcohol

The addition of hydrogen to a compound is a **reduction reaction**. Aldehydes are reduced to primary alcohols, and ketones are reduced to secondary alcohols. Notice that the acid is not added to the reaction mixture until after the hydride ion has reacted with the carbonyl compound.

MECHANISM FOR THE REACTION OF AN ALDEHYDE OR A KETONE WITH HYDRIDE ION

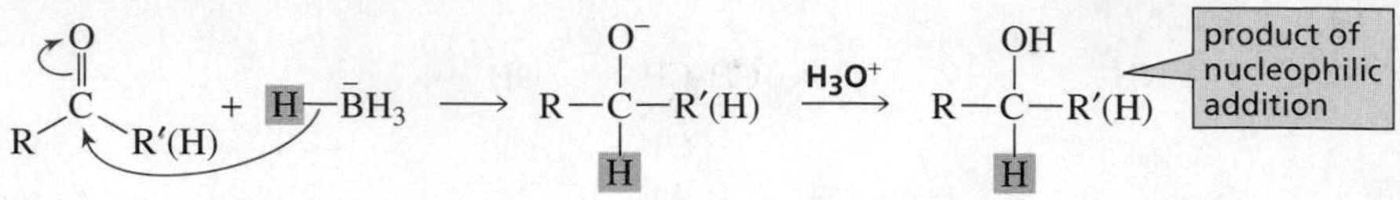

- Addition of a hydride ion to the carbonyl carbon of an aldehyde or a ketone forms an alkoxide ion.
- Protonation by a dilute acid forms an alcohol.

PROBLEM 20♦

What alcohols are obtained from the reduction of the following compounds with sodium borohydride?

a. 2-methylpropanal

b. cyclohexanone

c. 4-*tert*-butylcyclohexanone

d. acetophenone

The Reactions of Acyl Chlorides with Hydride Ion

Because an acyl chloride has a group that can be replaced by another group, it undergoes two successive reactions with hydride ion, just as it undergoes two successive reactions with a Grignard reagent (Section 4). Therefore, the reaction of an acyl chloride with sodium borohydride forms a primary alcohol with the same number of carbons as the acyl chloride.

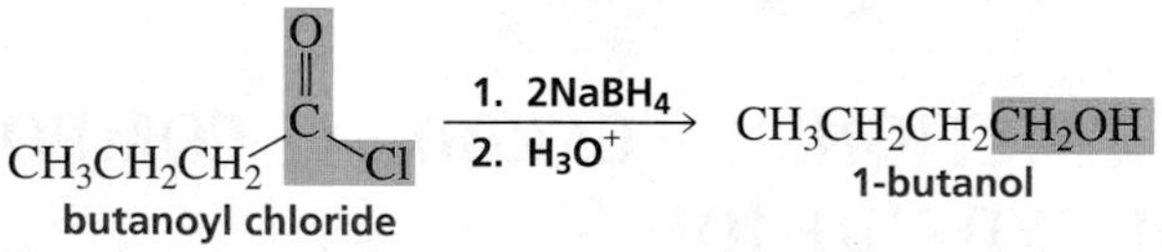

MECHANISM FOR THE REACTION OF AN ACYL CHLORIDE WITH HYDRIDE ION

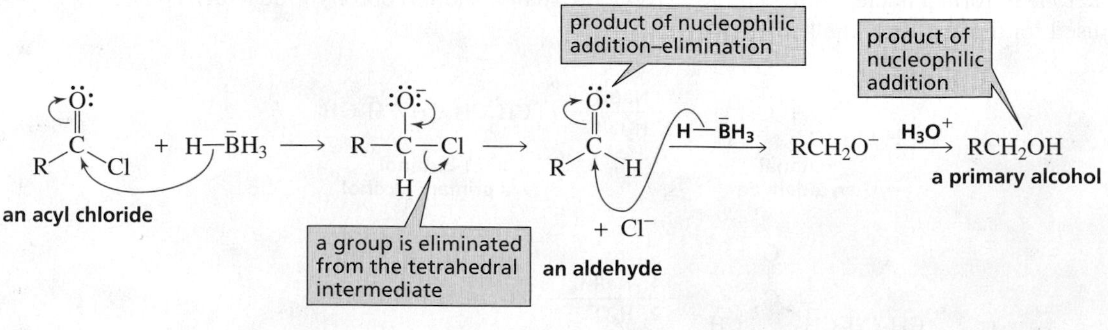

- The acyl chloride undergoes a nucleophilic addition–elimination reaction because it has a group (Cl^-) that can be replaced by hydride ion. The product of this reaction is an aldehyde.

- The aldehyde undergoes a nucleophilic addition reaction with a second equivalent of hydride ion, forming an alkoxide ion.

- Protonation of the alkoxide ion forms a primary alcohol.

Replacing some of the hydrogens of $LiAlH_4$ with alkoxy (OR) groups decreases the reactivity of the metal hydride. For example, lithium tri-*tert*-butoxyaluminum hydride reduces an acyl chloride only as far as the aldehyde, whereas $NaBH_4$ reduces the acyl chloride all the way to an alcohol.

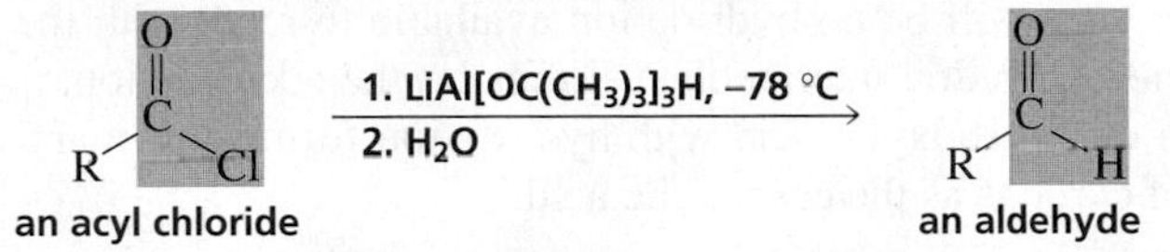

The Reactions of Esters and Carboxylic Acids with Hydride Ion

Sodium borohydride ($NaBH_4$) is not a sufficiently strong hydride donor to react with carbonyl compounds that are less reactive than aldehydes and ketones. Therefore, esters, carboxylic acids, and amides must be reduced with lithium aluminum hydride ($LiAlH_4$), a more reactive hydride donor. Lithium aluminum hydride is not as safe or as easy to use as sodium borohydride. It reacts violently with protic solvents, so it must be used in a dry, aprotic solvent, and it is never used if $NaBH_4$ can be used instead.

The reaction of an ester with $LiAlH_4$ produces two alcohols, one corresponding to the acyl portion of the ester and one corresponding to the alkyl portion.

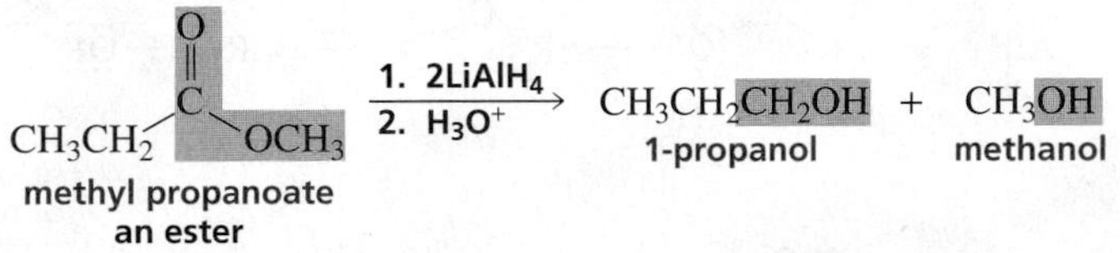

MECHANISM FOR THE REACTION OF AN ESTER WITH HYDRIDE ION

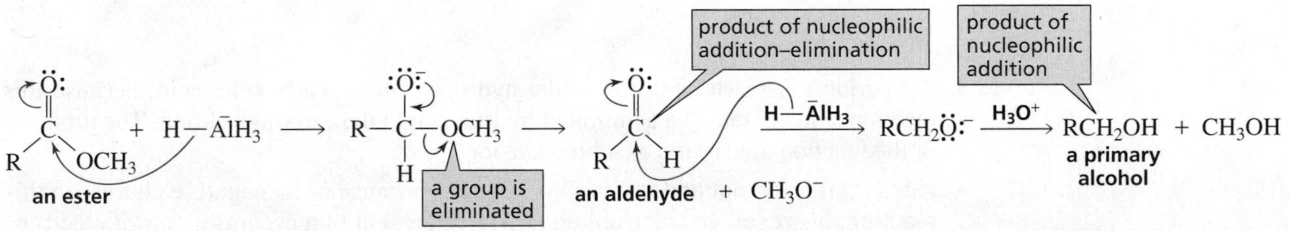

- The ester undergoes a nucleophilic addition–elimination reaction because an ester has a group (CH_3O^-) that can be replaced by hydride ion. The product of this reaction is an aldehyde.

- The aldehyde undergoes a nucleophilic addition reaction with a second equivalent of hydride ion, forming an alkoxide ion.

- Protonation of the two alkoxide ions forms two alcohols.

The reaction of an ester with hydride ion cannot be stopped at the aldehyde because an aldehyde is more reactive than an ester toward nucleophilic addition (Section 2).

Acyl chlorides and esters undergo two successive reactions with hydride ion and with Grignard reagents.

831

However, if diisobutylaluminum hydride (DIBALH) is used as' the hydride donor at a low temperature, then the reaction can be stopped at the aldehyde. DIBALH, therefore, makes it possible to convert esters into aldehydes.

CH₃CHCH₂—Al—CH₂CHCH₃
  | | |
 CH₃   H   CH₃
diisobutylaluminum hydride DIBALH

$$CH_3CH_2CH_2-\overset{\overset{O}{\|}}{C}-OCH_3 \xrightarrow[\text{2. H}_2\text{O}]{\text{1. [(CH}_3)_2\text{CHCH}_2]_2\text{AlH, }-78\,°\text{C}} CH_3CH_2CH_2-\overset{\overset{O}{\|}}{C}-H + CH_3OH$$

methyl butanoate   **butanal**   **methanol**

The reaction is carried out at −78 °C (the temperature of a dry ice/acetone bath). At this temperature, the initially formed tetrahedral intermediate is stable, so it does not eliminate the alkoxide ion. If all of the unreacted hydride donor is removed before the solution warms up, then there will be no hydride ion available to react with the aldehyde that is formed when the tetrahedral intermediate eliminates the alkoxide ion.

The reaction of a carboxylic acid with hydride ion forms a primary alcohol with the same number of carbons as the carboxylic acid.

The reaction of a carboxylic acid with LiAlH₄ forms a primary alcohol.

$$CH_3CH_2-\overset{\overset{O}{\|}}{C}-OH \xrightarrow[\text{2. H}_3\text{O}^+]{\text{1. LiAlH}_4} CH_3CH_2CH_2OH$$

propanoic acid   **1-propanol**

MECHANISM FOR THE REACTION OF A CARBOXYLIC ACID WITH HYDRIDE ION

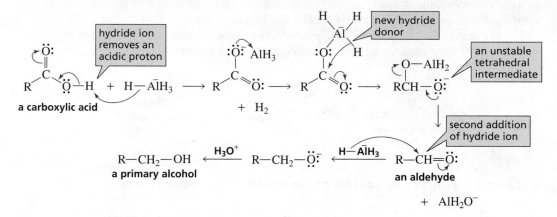

- A hydride ion reacts with the acidic hydrogen of the carboxylic acid, because this reaction is faster than the addition of hydride ion to the carbonyl carbon. The products of the reaction are H₂ and a carboxylate ion.

- Nucleophiles do not add to a carboxylate ion because of its negative charge. In this reaction, however, an electrophile (AlH₃) is present that accepts a pair of electrons from the carboxylate ion, neutralizing it and forming a new hydride donor.

- Addition of hydride ion followed by elimination from the unstable tetrahedral compound forms an aldehyde.

- Addition of hydride ion to the aldehyde forms the primary alcohol.

PROBLEM 21◆

What products would be obtained from the reaction of the following compounds with LiAlH₄ followed by treatment with dilute acid?

a. ethyl butanoate   **c.** methyl benzoate
b. benzoic acid   **d.** pentanoic acid

The Reactions of Amides with Hydride Ion

Amides also undergo two successive additions of hydride ion when they react with $LiAlH_4$. Overall, the reaction converts a carbonyl group into a methylene (CH_2) group, so the product of the reaction is an amine. Primary, secondary, or tertiary amines can be formed, depending on the number of substituents bonded to the nitrogen of the amide. (Notice that H_2O rather than H_3O^+ is used in the second step of the reaction. If H_3O^+ were used, the product would be an ammonium ion rather than an amine.)

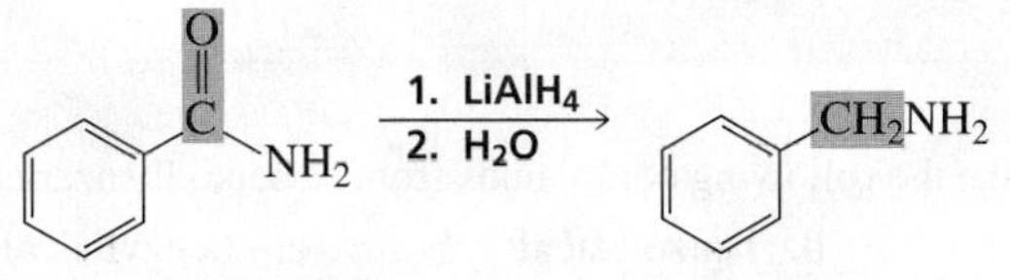

benzamide → benzylamine
a primary amine

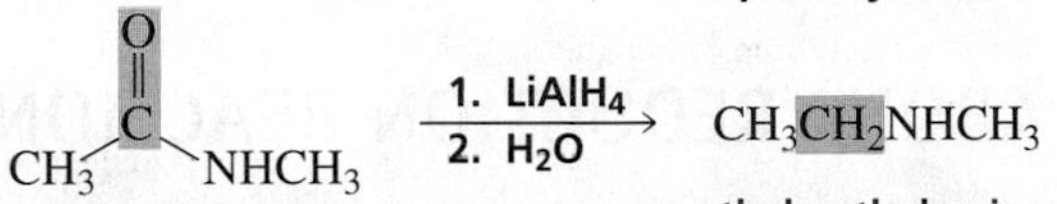

1. LiAlH₄
2. H₂O

N-methylacetamide → ethylmethylamine
a secondary amine

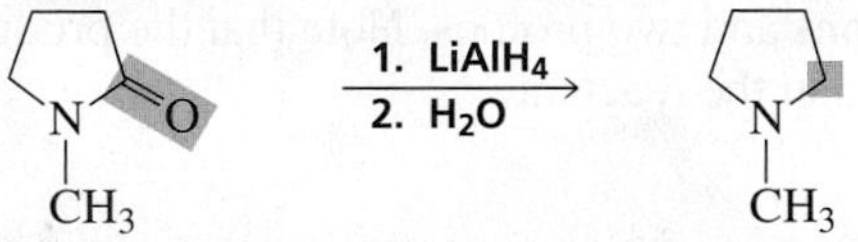

1. LiAlH₄
2. H₂O

N-methyl-γ-butyrolactam → *N*-methylpyrrolidine
a tertiary amine

1. LiAlH₄
2. H₂O

The mechanism for the reaction shows why the product of the reaction is an amine. Take a minute to note the similarities between this mechanism and the mechanism for the reaction of hydride ion with a carboxylic acid.

MECHANISM FOR THE REACTION OF AN *N*-SUBSTITUTED AMIDE WITH HYDRIDE ION

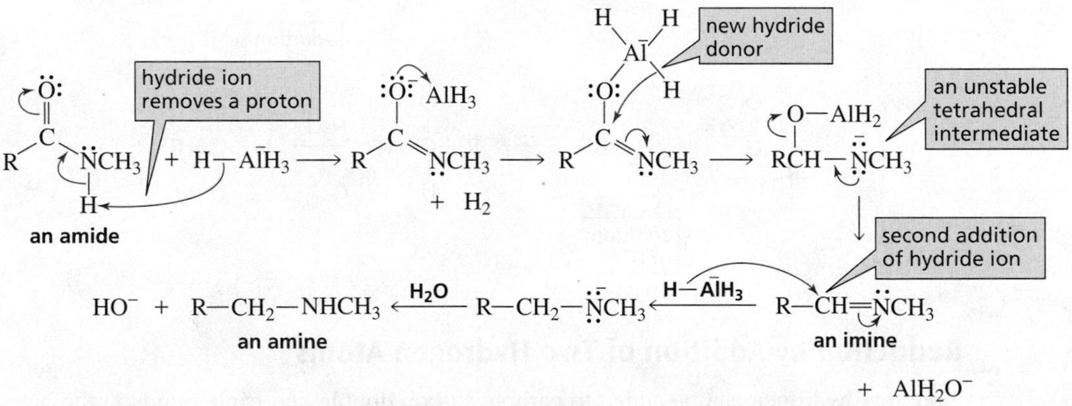

- A hydride ion removes the acidic hydrogen from the nitrogen of the amide, and the electrons left behind are delocalized onto oxygen.
- An electrophile (AlH_3) accepts a pair of electrons from the anion and forms a new hydride donor.
- Addition of hydride ion, elimination from the unstable tetrahedral intermediate, and a second addition of hydride ion followed by protonation forms the amine.

The mechanisms for the reaction of $LiAlH_4$ with unsubstituted and *N,N*-disubstituted amides are somewhat different, but have the same result—the conversion of a carbonyl group into a methylene group.

Biological reactions also need reagents that can deliver hydride ions to carbonyl groups. Cells use NADH and NADPH as hydride donors. (Sodium borohydride and lithium aluminum hydride are too reactive to be used in biological reactions.)

PROBLEM 22♦

What amides would you react with $LiAlH_4$ to form the following amines?

a. benzylmethylamine   **c.** diethylamine
b. ethylamine   **d.** triethylamine

PROBLEM 23

How would you make the following compounds from *N*-benzylbenzamide?

a. dibenzylamine   **b.** benzoic acid   **c.** benzyl alcohol

---

8   MORE ABOUT REDUCTION REACTIONS

An organic compound is reduced when hydrogen (H_2) is added to it. A molecule of H_2 can be thought of as being composed of (1) a hydride ion and a proton, (2) two hydrogen atoms, or (3) two electrons and two protons. Note that the product of a reduction reaction has more C—H bonds than the reactant.

components of H:H

$$H:^- \quad H^+ \qquad\qquad H\cdot \quad \cdot H \qquad\qquad \cdot^- \quad H^+ \quad \cdot^- \quad H^+$$

a hydride ion and a proton   two hydrogen atoms   two electrons and two protons

Reduction by Addition of a Hydride Ion and a Proton

When aldehydes and ketones are reduced to alcohols by $NaBH_4$, the reduction occurs as a result of the addition of a hydride ion followed by the addition of a proton.

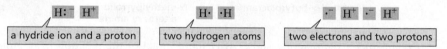

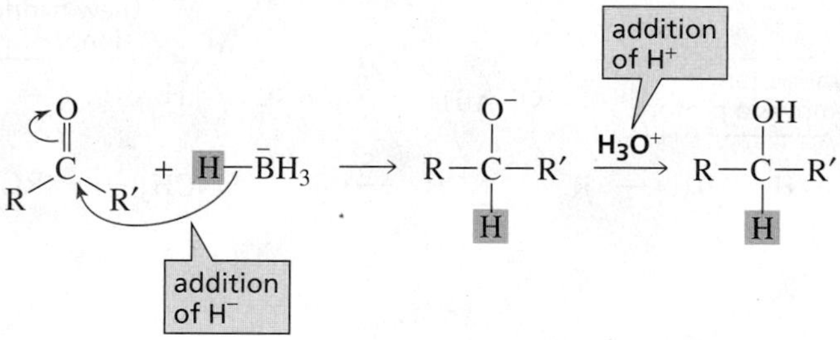

Reduction by Addition of Two Hydrogen Atoms

Note that hydrogen can be added to carbon–carbon double and triple bonds in the presence of a metal catalyst. In these reactions, called **catalytic hydrogenations,** the H—H bond breaks homolytically, so reduction results from the addition of two hydrogen atoms to the reactant.

$$CH_3CH_2CH{=}CH_2 + H_2 \xrightarrow{\text{Pd/C}} CH_3CH_2CH_2CH_3$$

1-butene   butane

$$CH_3CH_2CH_2C{\equiv}CH + 2 H_2 \xrightarrow{\text{Pd/C}} CH_3CH_2CH_2CH_2CH_3$$

1-pentyne   pentane

Catalytic hydrogenation can also be used to reduce carbon–nitrogen double and triple bonds. The reaction products are amines (Section 10).

$$CH_3CH_2CH{=}NCH_3 \;+\; H_2 \xrightarrow{\text{Pd/C}} CH_3CH_2CH_2NHCH_3$$
methylpropylamine

$$CH_3CH_2CH_2C{\equiv}N \;+\; 2\,H_2 \xrightarrow{\text{Raney nickel}} CH_3CH_2CH_2CH_2NH_2$$
butylamine

The $C{=}O$ group of ketones and aldehydes can also be reduced by catalytic hydrogenation. A palladium catalyst is not very effective at reducing these carbonyl groups. However, they are reduced easily over a nickel catalyst. (Raney nickel is finely dispersed nickel with adsorbed hydrogen, so an external source of H_2 is not needed.) Aldehydes are reduced to primary alcohols, and ketones are reduced to secondary alcohols.

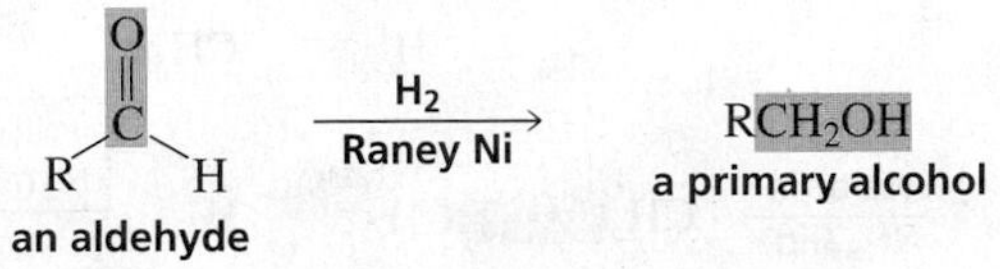

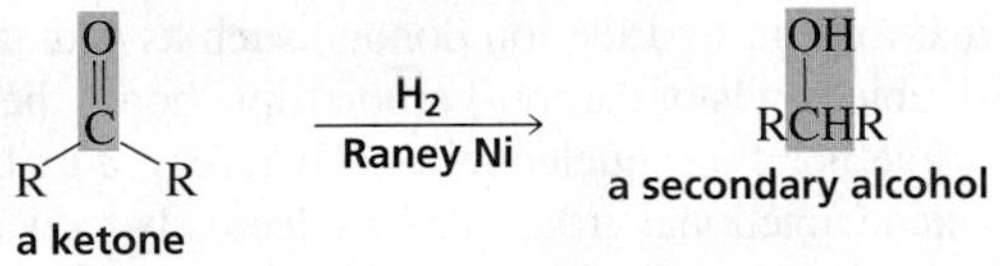

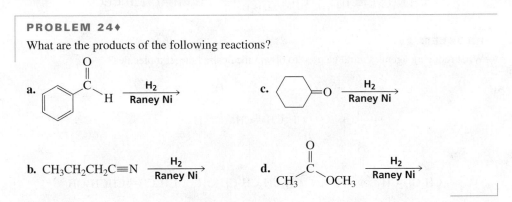

The $C{=}O$ group of carboxylic acids, esters, and amides are less reactive than the $C{=}O$ group of aldehydes and ketones and are, therefore, harder to reduce (Section 7). They cannot be reduced by catalytic hydrogenation (except under extreme conditions).

PROBLEM 24♦

What are the products of the following reactions?

a. [benzaldehyde structure] $\xrightarrow[\text{Raney Ni}]{H_2}$

c. [cyclohexanone] $=O$ $\xrightarrow[\text{Raney Ni}]{H_2}$

b. $CH_3CH_2CH_2C{\equiv}N \xrightarrow[\text{Raney Ni}]{H_2}$

d. [methyl acetate structure] $CH_3\!-\!\overset{O}{\underset{}{C}}\!-\!OCH_3$ $\xrightarrow[\text{Raney Ni}]{H_2}$

Reduction by Addition of an Electron, a Proton, an Electron, and a Proton

An alkyne can be reduced to a trans alkene using sodium in liquid ammonia. In this reaction, called a **dissolving metal reduction,** sodium donates an electron to the alkyne and ammonia donates a proton. This sequence is then repeated, so the overall reaction adds an electron, a proton, an electron, and a proton to the alkyne.

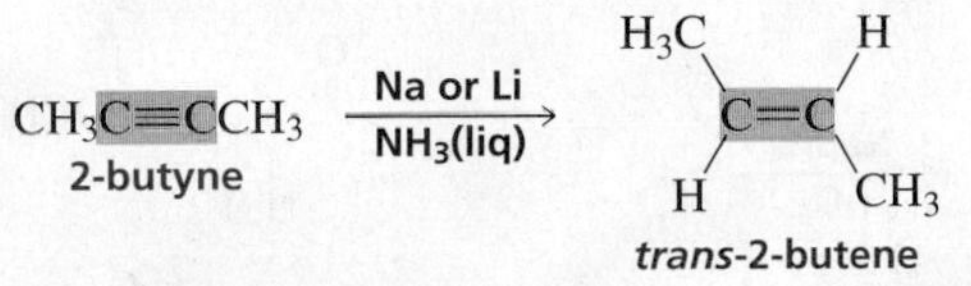

$$CH_3C{\equiv}CCH_3 \xrightarrow[\text{NH}_3\text{(liq)}]{\text{Na or Li}}$$
2-butyne

[trans-2-butene structure]

trans-2-butene

9 CHEMOSELECTIVE REACTIONS

A **chemoselective reaction** is a reaction in which a reagent reacts with one functional group in preference to another. For example, because sodium borohydride cannot reduce an ester, an amide, or a carboxylic acid, it can be used to selectively reduce an aldehyde or a ketone in a compound that also contains a less reactive carbonyl group. Water, not aqueous acid, is used in the second step of the reaction to avoid hydrolyzing the ester.

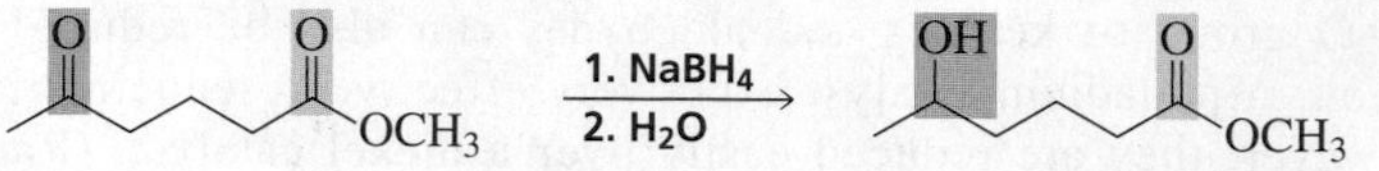

Sodium (or lithium) in liquid ammonia can reduce a carbon–carbon triple bond, but *not* a carbon–carbon double bond. This reagent is therefore useful for reducing a triple bond in a compound that also contains a double bond.

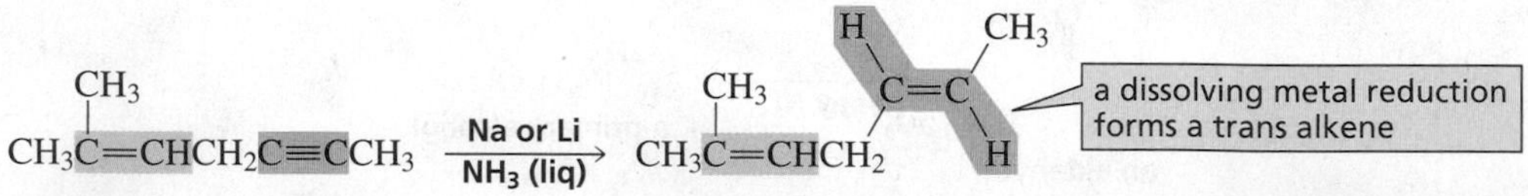

a dissolving metal reduction forms a trans alkene

Reducing reagents that that are hydride ion donors, such as sodium borohydride, cannot reduce carbon–carbon double bonds or carbon–carbon triple bonds, because both the hydride ion and the double or triple bond are nucleophiles. Therefore, a carbonyl group in a compound that also has an alkene functional group can be selectively reduced. Water, not aqueous acid, is used in the second step of the reaction to avoid addition of the acid to the double bond.

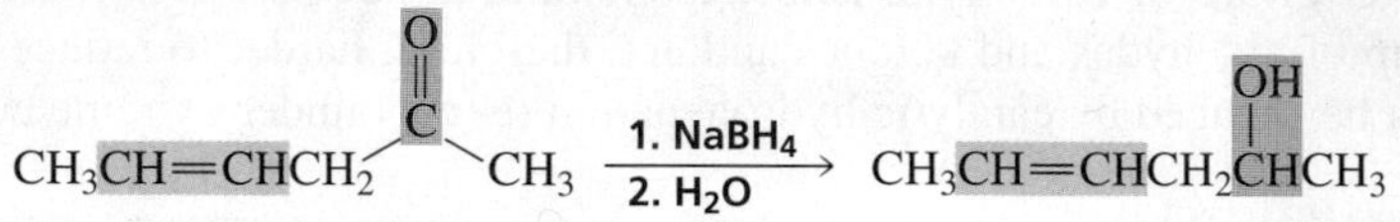

PROBLEM 25

What reducing agents should be used to obtain the desired target molecules?

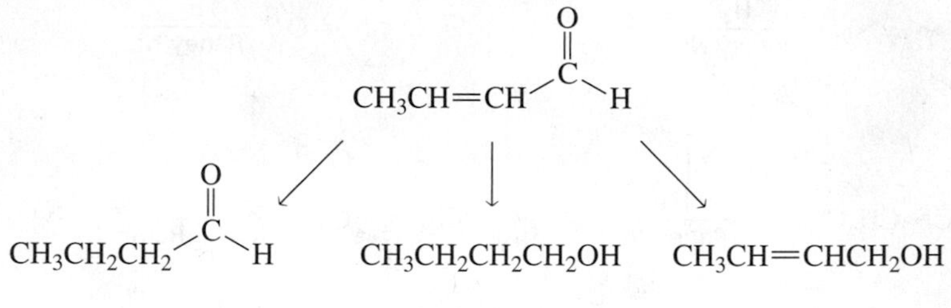

PROBLEM 26◆

Explain why carbon–nitrogen double bonds and carbon–nitrogen triple bonds can be reduced by hydride donors such as $NaBH_4$, while carbon–carbon double bonds and carbon–carbon triple bonds cannot.

PROBLEM 27

What are the products of the following reactions?

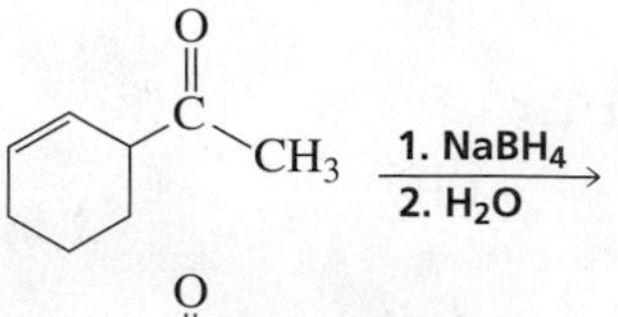

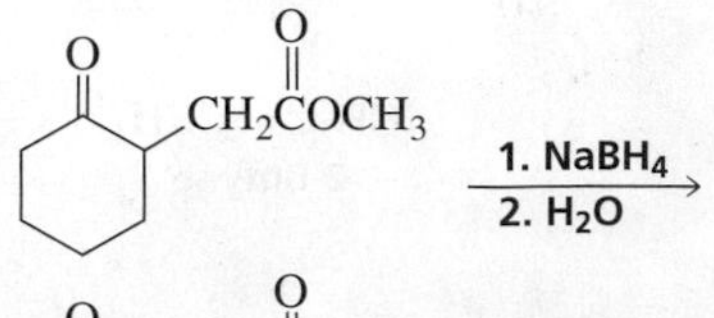

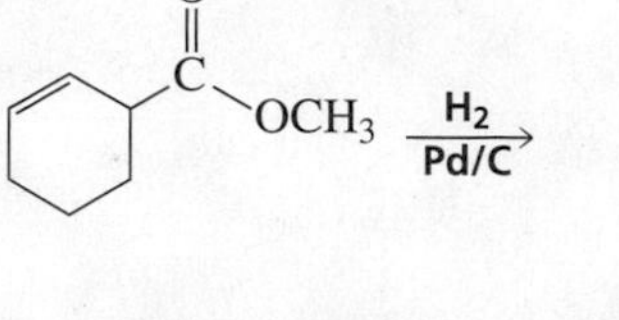

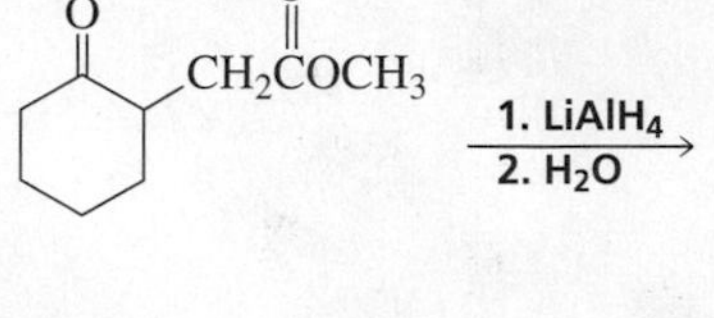

10 THE REACTIONS OF ALDEHYDES AND KETONES WITH AMINES

Aldehydes and Ketones Form Imines with Primary Amines

An aldehyde or a ketone reacts with a *primary* amine to form an imine (sometimes called a **Schiff base**). An **imine** is a compound with a carbon–nitrogen double bond. The reaction requires a trace amount of acid. Notice that imine formation replaces a C=O with a C=NR.

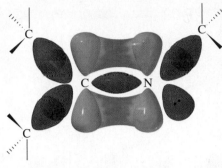

▲ **Figure 1**
Bonding in an imine. The π bond is formed by side-to-side overlap of a p orbital of carbon with a p orbital of nitrogen; it is perpendicular to the orange orbitals.

A C=N group (Figure 1) is similar to a C=O group. The imine nitrogen is sp^2 hybridized. One of its sp^2 orbitals forms a σ bond with the imine carbon, one forms a σ bond with a substituent, and the third sp^2 orbital contains a lone pair. The p orbital of nitrogen and the p orbital of carbon overlap to form a π bond.

The mechanism for imine formation is shown next. We will see that the pH of the reaction mixture must be carefully controlled. (Note that HB$^+$ represents any species in the solution that is capable of donating a proton, and :B represents any species in the solution that is capable of removing a proton.)

MECHANISM FOR IMINE FORMATION

Notice the pattern of three tetrahedral intermediates that occurs in this mechanism:

protonated tetrahedral intermediate →
neutral tetrahedral intermediate →
protonated tetrahedral intermediate.

- The amine adds to the carbonyl carbon.
- Protonation of the alkoxide ion and deprotonation of the ammonium ion form a neutral tetrahedral intermediate.
- The neutral tetrahedral intermediate, called a *carbinolamine,* is in equilibrium with two protonated forms because either its oxygen (forward step) or its nitrogen (reverse step) can be protonated.
- Because the nucleophile has a lone pair, water is eliminated from the oxygen-protonated intermediate, thereby forming a protonated imine.
- A base removes a proton from the nitrogen to form the imine.

Aldehydes and ketones react with primary amines to form imines.

Unlike the stable tetrahedral compounds that are formed when a Grignard reagent or a hydride ion adds to an aldehyde or a ketone, the tetrahedral compound formed when an amine adds to an aldehyde or a ketone is unstable because it contains a group that can be protonated and thereby become a weak enough base to be eliminated by the lone pair on the other electronegative atom.

OH	OH	H⁺OH	H⁺OH
R—C—R	R—C—R	R—C—R	R—C—R
H	CH₃	:NHCH₃	:ÖCH₃

stable tetrahedral compounds   **unstable tetrahedral compounds**

Imine formation is reversible because there are two protonated tetrahedral intermediates that can eliminate a group. The equilibrium favors the nitrogen-protonated tetrahedral intermediate because nitrogen is more basic than oxygen. However, the equilibrium can be forced toward the oxygen-protonated tetrahedral intermediate and therefore toward the imine by removing water as it is formed.

Overall, the addition of an amine to an aldehyde or a ketone is a *nucleophilic addition–elimination reaction:* nucleophilic addition of an amine to form a tetrahedral intermediate, followed by elimination of water.

Controlling the pH

The pH at which imine formation is carried out must be carefully controlled. There must be sufficient acid present to protonate the tetrahedral intermediate so that H_2O rather than the much more basic HO^- can be the leaving group. If too much acid is present, however, it will protonate all of the reactant amine. Protonated amines are not nucleophiles, so they cannot react with carbonyl groups. Therefore, there is not enough acid present to protonate the carbonyl group in the first step of the reaction (see Problem 30).

A plot of the observed rate constant for the reaction of acetone with hydroxylamine to form an imine as a function of the pH of the reaction mixture is shown in Figure 2.

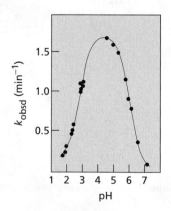

► **Figure 2**

A pH-rate profile for the reaction of acetone with hydroxylamine. It shows the dependence of the reaction rate on the pH of the reaction mixture.

at about pH 4.5, which is 1.5 pH units below the pK_a of protonated hydroxylamine. This type of plot is called a **pH-rate profile**. For this reaction, the maximum rate occurs at about pH 4.5, which is 1.5 pH units below the pK_a of protonated hydroxylamine ($pK_a = 6.0$). As the acidity increases below pH 4.5, the rate of the reaction decreases because more and more of the amine becomes protonated. As a result, less and less of the amine is present in the nucleophilic nonprotonated form. As the acidity decreases above pH 4.5, the rate decreases because less and less of the tetrahedral intermediate is present in the reactive protonated form.

An imine can be hydrolyzed back to the carbonyl compound and the amine in an acidic solution. This reaction is irreversible because the amine is protonated in the acidic solution, so it is unable to react with the carbonyl compound to re-form the imine.

$$\begin{array}{c} R \\ \diagdown \\ C = NCH_2CH_3 \\ \diagup \\ R \end{array} + H_2O \xrightarrow{\text{HCl}} \begin{array}{c} R \\ \diagdown \\ C = O \\ \diagup \\ R \end{array} + CH_3CH_2\overset{+}{N}H_3$$

An imine undergoes acid-catalyzed hydrolysis to form a carbonyl compound and a primary amine.

Imine formation and hydrolysis are important reactions in biological systems. For example, all the reactions that require vitamin B_6 involve imine formation, and imine hydrolysis is the reason that DNA contains T nucleotides instead of U nucleotides.

PROBLEM 28

A ketone can be prepared from the reaction of a nitrile with a Grignard reagent. Describe the intermediate formed in this reaction, and show how it can be converted to a ketone.

PROBLEM 29♦

Why is the pK_a value of protonated hydroxylamine (6.0) so much lower than the pK_a value of a protonated primary amine such as protonated methylamine (10.7)?

PROBLEM 30♦

At what pH should imine formation be carried out if the amine's protonated form has a pK_a value of 10.0?

PROBLEM 31♦

The pK_a of protonated acetone is about -7.5, and the pK_a of protonated hydroxylamine is 6.0.

a. In a reaction with hydroxylamine at pH 4.5 (Figure 2), what fraction of acetone will be present in its acidic, protonated form?
b. In a reaction with hydroxylamine at pH 1.5, what fraction of acetone will be present in its acidic, protonated form?
c. In a reaction with acetone at pH 1.5 (Figure 2), what fraction of hydroxylamine will be present in its reactive basic form?

Formation of Imine Derivatives

Compounds such as hydroxylamine and hydrazine are similar to primary amines in that they have an NH_2 group. Thus, like primary amines, they react with aldehydes and ketones to form imines—called *imine derivatives* because the substituent attached to the imine nitrogen is not an R group. The imine derivative obtained from the reaction with hydroxylamine is called an **oxime,** and the imine derivative obtained from the reaction with hydrazine is called a **hydrazone.**

$$\begin{array}{c} R \\ \diagdown \\ C = O \\ \diagup \\ R \end{array} + \underset{\text{hydroxylamine}}{H_2NOH} \underset{\text{acid}}{\overset{\text{trace}}{\rightleftharpoons}} \begin{array}{c} R \\ \diagdown \\ C = NOH \\ \diagup \\ R \end{array} + H_2O$$
$$\text{an oxime}$$

$$\begin{array}{c} R \\ \diagdown \\ C = O \\ \diagup \\ R \end{array} + \underset{\text{hydrazine}}{H_2NNH_2} \underset{\text{acid}}{\overset{\text{trace}}{\rightleftharpoons}} \begin{array}{c} R \\ \diagdown \\ C = NNH_2 \\ \diagup \\ R \end{array} + H_2O$$
$$\text{hydrazone}$$

PROBLEM 32

Imines can exist as stereoisomers. The isomers are named by the *E,Z* system of nomenclature. The lone pair has the lowest priority.

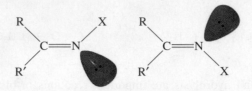

Draw the structure of each of the following compounds:

a. the (*E*)-hydrazone of benzaldehyde   **b.** the (*Z*)-oxime of propiophenone

Aldehydes and Ketones Form Enamines with Secondary Amines

An aldehyde or a ketone reacts with a *secondary amine* to form an enamine (pronounced "ENE-amine"). An **enamine** is an α,β-unsaturated tertiary amine—that is, a tertiary amine with a double bond in the α,β-position relative to the nitrogen. Notice that the double bond is in the part of the molecule that is provided by the aldehyde or ketone, not in the part that is provided by the secondary amine. The name comes from joining "ene" and "amine," with the second "e" in "ene" omitted in order to avoid two successive vowels. Like imine formation, the reaction requires a trace amount of an acid catalyst.

Aldehydes and ketones react with secondary amines to form enamines.

Notice that the mechanism for enamine formation is exactly the same as that for imine formation, except for the last step.

MECHANISM FOR ENAMINE FORMATION

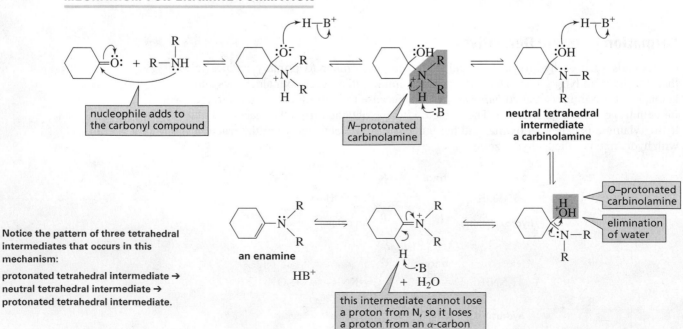

Notice the pattern of three tetrahedral intermediates that occurs in this mechanism:

**protonated tetrahedral intermediate →
neutral tetrahedral intermediate →
protonated tetrahedral intermediate.**

- The amine adds to the carbonyl carbon.
- Protonation of the alkoxide ion and deprotonation of the ammonium ion forms a neutral tetrahedral intermediate (a carbinolamine).
- The neutral tetrahedral intermediate is in equilibrium with two protonated forms because either its oxygen or its nitrogen can be protonated.
- Because the nucleophile has a lone pair, water is eliminated from the oxygen-protonated intermediate, thereby forming a compound with a positively charged nitrogen.
- When a primary amine reacts with an aldehyde or a ketone, a proton is removed from the positively charged nitrogen in the last step of the mechanism, forming a neutral imine. When the amine is secondary, however, the positively charged nitrogen is not bonded to a hydrogen. In this case, a stable neutral molecule can be obtained only by removing a proton from the α-carbon of the compound provided by the carbonyl compound. An enamine is the result.

As with imine formation, water must be removed as it is formed in order to force the equilibrium toward the enamine.

In an aqueous acidic solution, an enamine is hydrolyzed back to the carbonyl compound and secondary amine. This reaction is similar to the acid-catalyzed hydrolysis of an imine back to the carbonyl compound and primary amine (see earlier in this section).

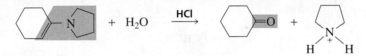

An enamine undergoes acid-catalyzed hydrolysis to form a carbonyl compound and a secondary amine.

PROBLEM 33

a. Write the mechanism for the following reactions:
   1. the acid-catalyzed hydrolysis of an imine to a carbonyl compound and a primary amine
   2. the acid-catalyzed hydrolysis of an enamine to a carbonyl compound and a secondary amine
b. How do the two mechanisms differ?

PROBLEM 34

What are the products of the following reactions? (A trace amount of acid is present in each case.)

a. cyclopentanone + ethylamine   **c.** acetophenone + hexylamine
b. cyclopentanone + diethylamine   **d.** acetophenone + cyclohexylamine

---

Reductive Amination

The imine formed from the reaction of an aldehyde or a ketone with ammonia is relatively unstable because it does not have a substituent other than a hydrogen attached to the nitrogen. Nevertheless, such an imine is a useful intermediate.

For example, if the reaction with ammonia is carried out in the presence of a reducing agent such as H_2 and a metal catalyst, then the double bond will be reduced as it is formed, forming a primary amine. The reaction of an aldehyde or a ketone with excess ammonia in the presence of a reducing agent is called **reductive amination.**

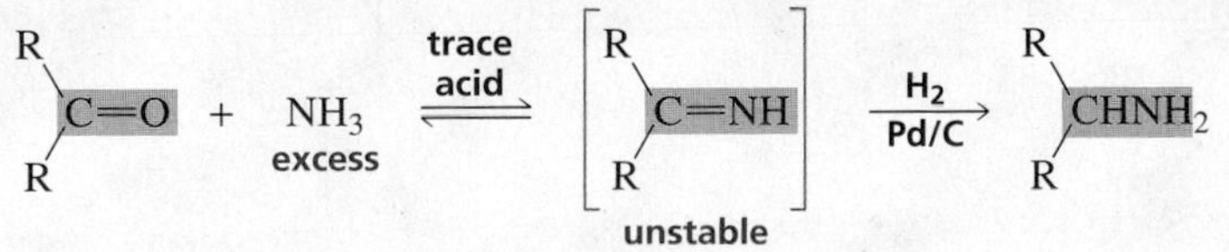

$$\begin{matrix} R \\ \diagdown \\ C=O \\ \diagup \\ R \end{matrix} + \underset{\text{excess}}{NH_3} \xrightleftharpoons{\substack{\text{trace} \\ \text{acid}}} \underset{\text{unstable}}{\left[\begin{matrix} R \\ \diagdown \\ C=NH \\ \diagup \\ R \end{matrix} \right]} \xrightarrow{\substack{H_2 \\ Pd/C}} \begin{matrix} R \\ \diagdown \\ CHNH_2 \\ \diagup \\ R \end{matrix}$$

The double bond of an imine or enamine is reduced more rapidly than a $C=O$ bond, so reduction of the carbonyl group does not compete with reduction of the imine in these reactions.

Secondary and tertiary amines can be prepared from imines and enamines by reducing the imine or enamine. Sodium cyanoborohydride ($NaBH_3CN$) is a commonly used reducing agent for these reactions because it can be handled easily and it is stable even in acidic solutions. (Note that $NaBH_3CN$ differs from $NaBH_4$ in having a $C\equiv N$ group in place of one of the hydrogens.)

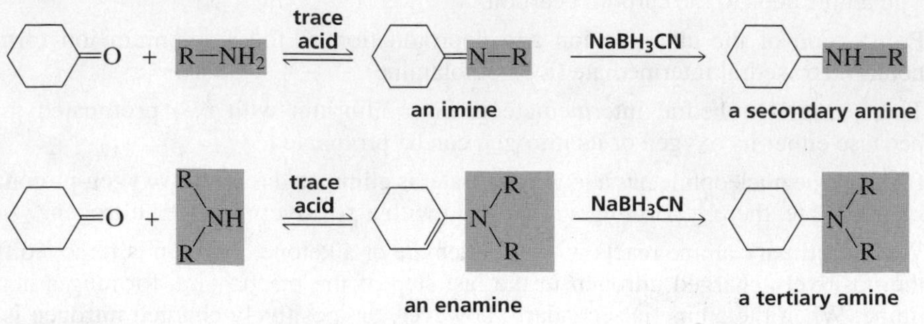

PROBLEM 35♦

Excess ammonia must be used when a primary amine is synthesized by reductive amination. What product will be obtained if the reaction is carried out with excess carbonyl compound?

PROBLEM 36

The compounds commonly known as "amino acids" are actually α-aminocarboxylic acids. What carbonyl compounds should be used to synthesize the two amino acids shown here?

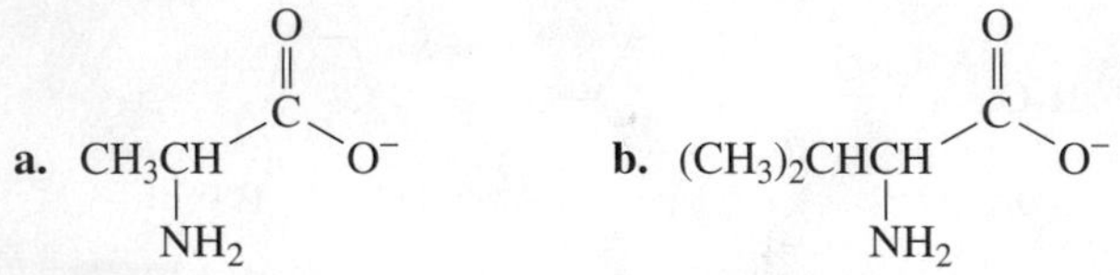

Serendipity in Drug Development

Many drugs have been discovered accidentally. Librium, a tranquilizer, is one example of such a drug. Leo Sternbach, a research chemist at Hoffmann-LaRoche, synthesized a series of quinazoline 3-oxides, but none of them showed any pharmacological activity. One of the compounds was not submitted for testing because it was not the quinazoline 3-oxide Sternbach had set out to synthesize. Two years after the project was abandoned, a laboratory worker came across this compound while cleaning up the lab, and Sternbach decided that he might as well submit it for testing before it was thrown away. The compound was found to have tranquilizing properties and, when its structure was investigated, was discovered to be a benzodiazepine 4-oxide.

Methylamine, instead of displacing the chloro substituent in an S_N2 reaction to form a quinazoline 3-oxide, had added to the imine group of the six-membered ring. This caused the ring to open and then reclose to form a benzodiazepine. The compound was given the brand name Librium when it was put into clinical use in 1960.

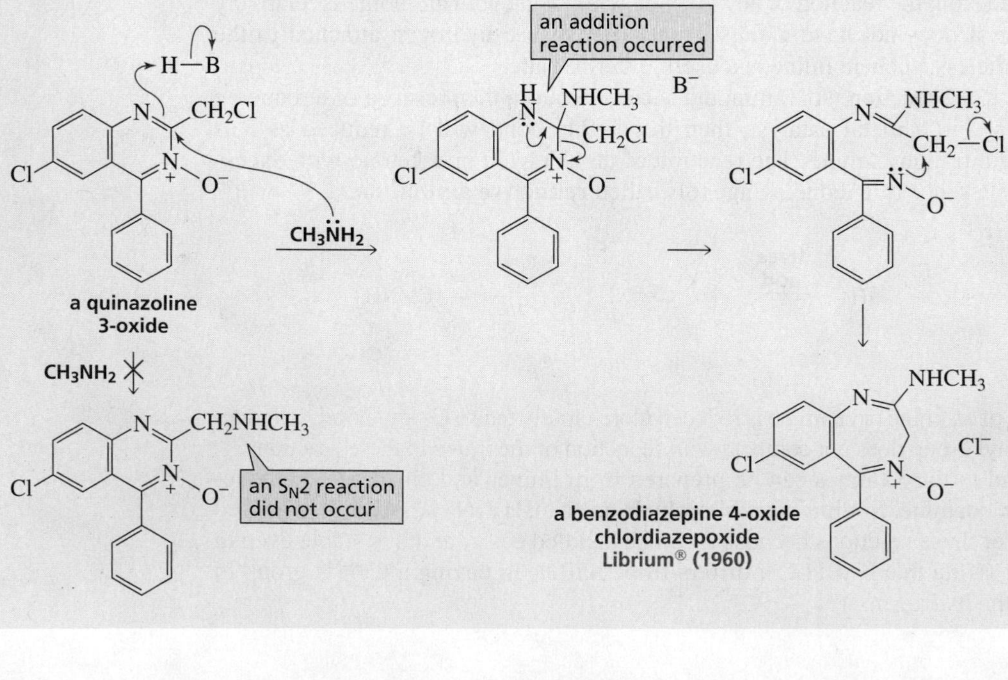

Librium was structurally modified in an attempt to find other tranquilizers. One successful modification produced Valium, a tranquilizer almost 10 times more potent than Librium. Currently, there are 8 benzodiazepines in clinical use as tranquilizers in the United States and some 15 others abroad. Rohypnol is one of the so-called date-rape drugs.

diazepam
Valium® (1963)

flunitrazepam
Rohypnol® (1963)

alprazolam
Xanax® (1970)

flurazepam
Dalmane® (1970)

clonazepam
Klonopin® (1975)

lorazepam
Ativan® (1977)

Viagra is a recent example of a drug that was discovered accidentally. Viagra was in clinical trials as a drug for heart ailments. The clinical trials were canceled when Viagra was found to be ineffective as a heart drug. However, those enrolled in the trials refused to return the remaining tablets. The pharmaceutical company then realized that the drug had other marketable effects.

11 THE REACTIONS OF ALDEHYDES AND KETONES WITH WATER

The addition of water to an aldehyde or a ketone forms a *hydrate*. A **hydrate** is a molecule with two OH groups bonded to the same carbon. Hydrates are also called **gem-diols** (*gem* comes from *geminus*, Latin for "twin").

$$\underset{\substack{\text{an aldehyde or} \\ \text{a ketone}}}{R-\overset{\displaystyle O}{\overset{\|}{C}}-R\,(H)} \;+\; H_2O \;\underset{\text{HCl}}{\rightleftharpoons}\; \underset{\substack{\text{a } gem\text{-diol} \\ \text{a hydrate}}}{R-\overset{\displaystyle OH}{\underset{\displaystyle OH}{\overset{|}{\underset{|}{C}}}}-R\,(H)}$$

Water is a poor nucleophile and therefore adds relatively slowly to a carbonyl group. The rate of the reaction can be increased by an acid catalyst (Figure 3). Keep in mind that a catalyst affects the *rate* at which an aldehyde or a ketone is converted to a hydrate; it has no effect on the *amount* of aldehyde or ketone converted to hydrate.

MECHANISM FOR ACID-CATALYZED HYDRATE FORMATION

O
‖
CH₃—C—H

+OH
‖
CH₃—C—H

▲ Figure 3

The electrostatic potential maps show that the carbonyl carbon of the protonated aldehyde is more electrophilic (the blue is more intense) than the carbonyl carbon of the unprotonated aldehyde.

- The acid protonates the carbonyl oxygen, which makes the carbonyl carbon more susceptible to nucleophilic attack (Figure 3).
- Water adds to the carbonyl carbon.
- Removal of a proton from the protonated tetrahedral intermediate forms the hydrate.

PROBLEM 37

Hydration of an aldehyde can also be catalyzed by hydroxide ion. Propose a mechanism for the reaction.

The extent to which an aldehyde or a ketone is hydrated in an aqueous solution depends on the substituents attached to the carbonyl group. For example, only 0.2% of acetone is hydrated at equilibrium, but 99.9% of formaldehyde is hydrated.

			K_{eq}
acetone 99.8%	+ H₂O ⇌	CH₃—C(OH)(OH)—CH₃ 0.2%	2×10^{-3}
acetaldehyde 42%	+ H₂O ⇌	CH₃—C(OH)(H)—H 58%	1.4
formaldehyde 0.1%	+ H₂O ⇌	H—C(OH)(H)—H 99.9%	2.3×10^{3}

Why is there such a difference in the extent of hydration? The equilibrium constant for a reaction depends on the relative stabilities of the reactants and products. The equilibrium constant for hydrate formation, which is a measure of the extent of hydration, depends, therefore, on the relative stabilities of the carbonyl compound and the hydrate.

$$K_{eq} = \frac{[\text{products}]}{[\text{reactants}]} = \frac{[\text{hydrate}]}{[\text{carbonyl compound}][\text{H}_2\text{O}]}$$

We have seen that electron-donating alkyl groups make a carbonyl compound *more stable* (less reactive) (Section 2).

In contrast, alkyl groups make the hydrate *less stable* because of steric interactions between the alkyl groups when the bond angles change from 120° to 109.5° (Section 2).

$$
\underset{\text{least stable}}{
\begin{array}{c}
\text{OH} \\
| \\
\text{CH}_3-\text{C}-\text{CH}_3 \\
| \\
\text{OH}
\end{array}
}
<
\begin{array}{c}
\text{OH} \\
| \\
\text{CH}_3-\text{C}-\text{H} \\
| \\
\text{OH}
\end{array}
<
\begin{array}{c}
\text{OH} \\
| \\
\text{H}-\text{C}-\text{H} \\
| \\
\text{OH} \; 109.5°
\end{array}
$$

Alkyl groups, therefore, shift the equilibrium to the left (toward reactants) because they stabilize the carbonyl compound and destabilize the hydrate, which makes K_{eq} smaller. As a result, less acetone than formaldehyde is hydrated at equilibrium. The relative stability of hydrates of aldehydes is why "generally" was used in the following statement: "*generally, a compound that has an* sp³ *carbon bonded to an oxygen atom will be unstable if the* sp³ *carbon is bonded to another electronegative atom.*"

In summary, the percentage of hydrate present in solution at equilibrium depends on both electronic and steric effects. Electron-donating substituents and bulky substituents (such as the methyl groups of acetone) *decrease* the percentage of hydrate present at equilibrium, whereas electron-withdrawing substituents and small substituents (the hydrogens of formaldehyde) *increase* the percent of hydration present at equilibrium.

If the amount of hydrate formed from the reaction of water with a ketone is too small to detect, how do we know that the reaction has even occurred? We can prove that it occurs by adding the ketone to ¹⁸O-labeled water and isolating the ketone after equilibrium has been established. Finding that ¹⁸O has been incorporated into the ketone indicates that hydration has occurred.

Notice that the pattern of three tetrahedral intermediates occurs again in this mechanism—namely,

protonated tetrahedral intermediate → neutral tetrahedral intermediate → protonated tetrahedral intermediate.

Preserving Biological Specimens

A 37% solution of formaldehyde in water, known as *formalin,* was commonly used to preserve biological specimens. Formaldehyde is an eye and skin irritant, however, so formalin has been replaced in most biology laboratories by other preservatives. One frequently used preparation is a solution of 2–5% phenol in ethanol with added antimicrobial agents.

PROBLEM 38♦

Which ketone forms the most hydrate in an aqueous solution?

A

B

O_2N ... NO_2

PROBLEM 39

When trichloroacetaldehyde is dissolved in water, almost all of it is converted to the hydrate. Chloral hydrate, the product of the reaction, is a sedative that can be lethal. A cocktail laced with it is known—in detective novels, at least—as a "Mickey Finn." Explain why an aqueous solution of trichloroacetaldehyde is almost all hydrate.

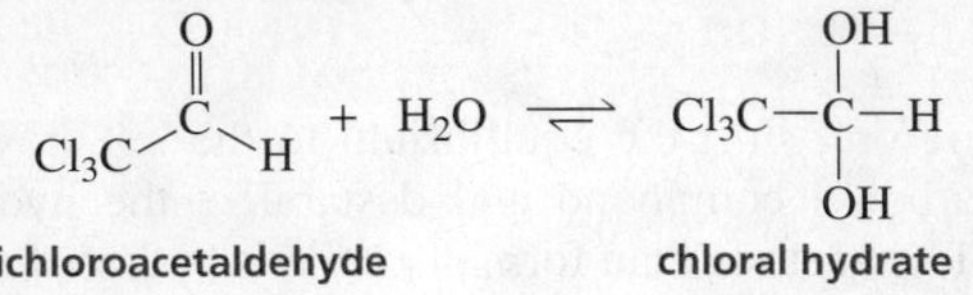

trichloroacetaldehyde   chloral hydrate

12 THE REACTIONS OF ALDEHYDES AND KETONES WITH ALCOHOLS

The product formed when one equivalent of an alcohol adds to an *aldehyde* or a *ketone* is called a **hemiacetal.** The product formed when a second equivalent of alcohol is added is called an **acetal** (ass-ett-AL). Like water, an alcohol is a poor nucleophile, so an acid catalyst is required for the reaction to take place at a reasonable rate. (Occasionally you see the terms *hemiketal* and *ketal* used instead of *hemiacetal* and *acetal* for the products of the reaction of an alcohol with a ketone.)

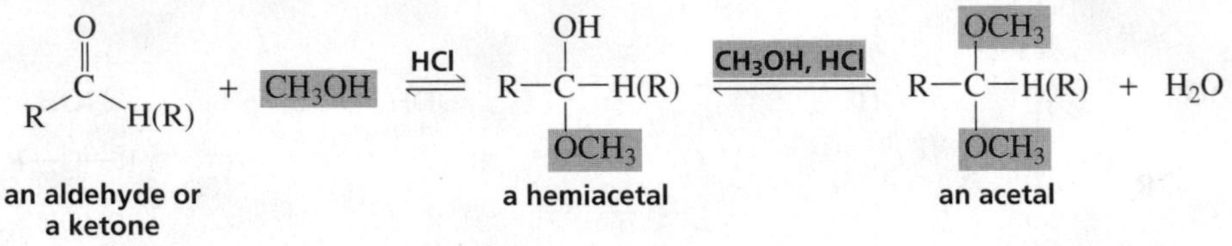

an aldehyde or   a hemiacetal   an acetal
a ketone

Notice that the pattern of three tetrahedral intermediates occurs again in this mechanism—namely,

protonated tetrahedral intermediate →
neutral tetrahedral intermediate →
protonated tetrahedral intermediate.

Hemi is the Greek word for "half." When one equivalent of alcohol has added to an aldehyde or a ketone, the compound is halfway to the final acetal, which contains groups from two equivalents of alcohol.

MECHANISM FOR ACID-CATALYZED ACETAL FORMATION

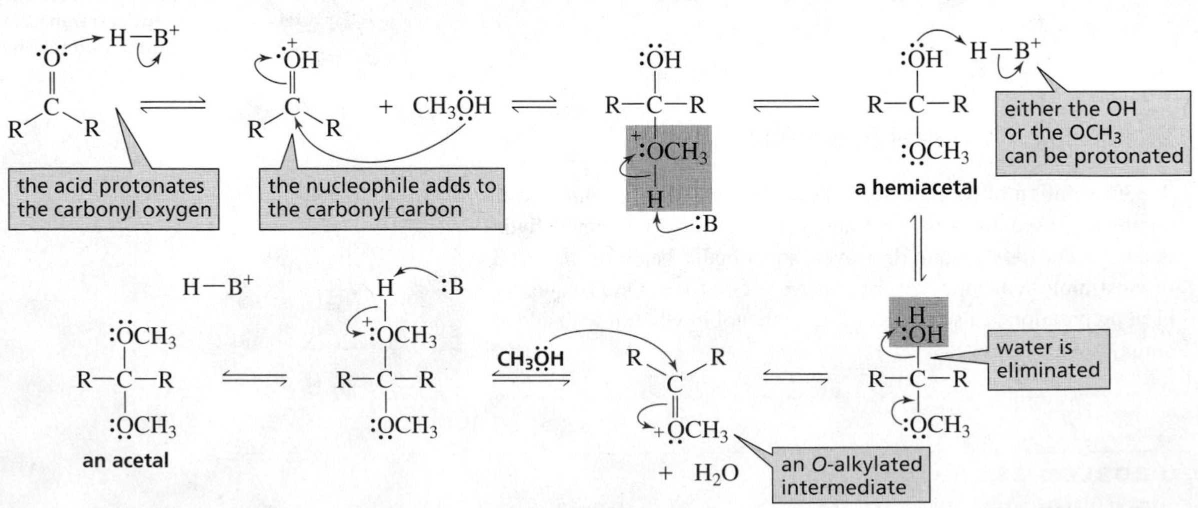

- The acid protonates the carbonyl oxygen, making the carbonyl carbon more susceptible to nucleophilic addition (Figure 3).
- The alcohol adds to the carbonyl carbon.

- Loss of a proton from the protonated tetrahedral intermediate forms the neutral tetrahedral intermediate (the hemiacetal).

- The hemiacetal is in equilibrium with its protonated form. The two oxygen atoms of the hemiacetal are equally basic, so either one can be protonated.

- Because the nucleophile has a lone pair, water is eliminated from the second protonated intermediate, thereby forming an intermediate that is very reactive because of its positively charged oxygen.

- Nucleophilic addition to this intermediate by a second molecule of alcohol, followed by loss of a proton, forms the acetal.

Although the sp^3 carbon of an acetal is bonded to two oxygens, which suggests that it is not stable, the acetal can be isolated if the water that is eliminated is removed from the reaction mixture. If water is not available, the only compound the acetal can form is the O-alkylated intermediate, which is even less stable than the acetal.

The acetal can be hydrolyzed back to the aldehyde or ketone in an acidic aqueous solution.

$$\begin{array}{c} OCH_3 \\ | \\ R-C-R \\ | \\ OCH_3 \end{array} + \underset{\text{excess}}{H_2O} \;\overset{HCl}{\rightleftharpoons}\; \underset{R \quad R}{\overset{O}{\overset{\|}{C}}} + 2\,CH_3OH$$

Notice that the mechanisms for imine, enamine, hydrate, and acetal formation are similar. The nucleophile in each reaction has a lone pair on its attacking atom. After the nucleophile has added to the carbonyl carbon, water is eliminated from a protonated tetrahedral intermediate, forming a positively charged species. In imine and hydrate formation, a neutral product is achieved by loss of a proton from a nitrogen and an oxygen, respectively. (In hydrate formation, the neutral product is the original aldehyde or ketone.) In enamine formation, a neutral product is achieved by the loss of a proton from an α-carbon. In acetal formation, a neutral compound is achieved by the addition of a second equivalent of alcohol.

Carbohydrates

Individual sugar units in a carbohydrate are held together by acetal groups. For example, the reaction of the aldehyde group and an alcohol group of D-glucose forms a cyclic compound that is a hemiacetal. Molecules of the cyclic compound are then hooked together by the reaction of the hemiacetal group of one molecule with an OH group of another, resulting in the formation of an acetal. Hundreds of cyclic glucose molecules hooked together by acetal groups is a major component of both starch and cellulose.

PROBLEM 40♦

Which of the following are

a. hemiacetals?   **b.** acetals?   **c.** hydrates?

1. 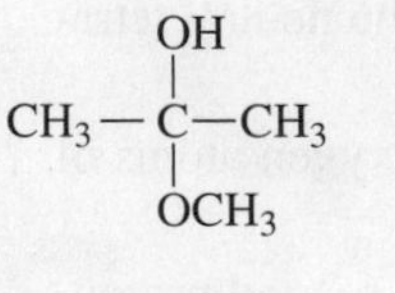 OH
    |
CH_3-C-CH_3
    |
  OCH_3

3. OCH_3
    |
CH_3-C-H
    |
  OCH_3

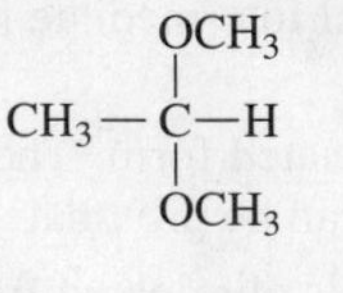

5. 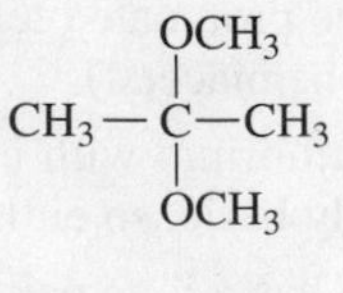 OCH_3
    |
CH_3-C-CH_3
    |
  OCH_3

7. 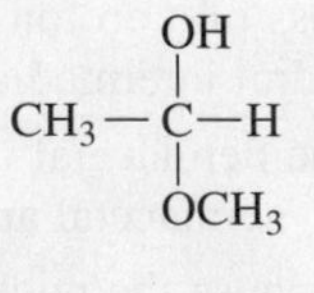 OH
    |
CH_3-C-H
    |
  OCH_3

2. 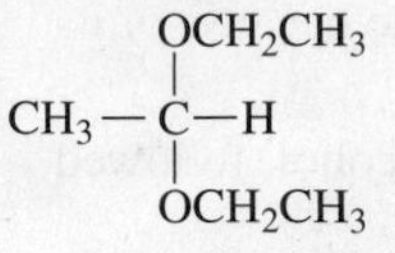 OCH_2CH_3
    |
CH_3-C-H
    |
  OCH_2CH_3

4. 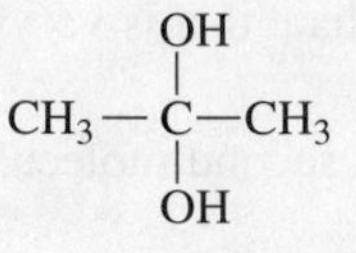 OH
    |
CH_3-C-CH_3
    |
  OH

6. 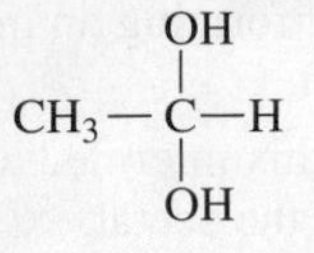 OH
    |
CH_3-C-H
    |
  OH

8. 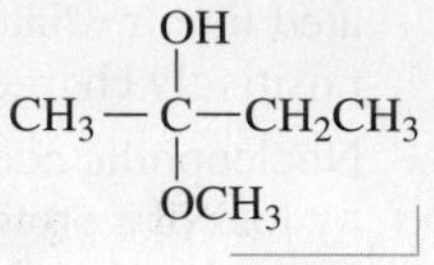 OH
    |
$CH_3-C-CH_2CH_3$
    |
  OCH_3

PROBLEM-SOLVING STRATEGY

Analyzing the Behavior of Acetals

Why are acetals hydrolyzed back to the aldehyde or ketone in acidic aqueous solutions, but are stable in basic aqueous solutions?

The best way to approach this question is to write out the mechanism that describes the situation to which the question is referring. When the mechanism is written, the answer may become apparent.

In an acidic solution, the acid protonates an oxygen of the acetal. This creates a weak base (CH_3OH) that can be eliminated. When the group is eliminated, water can attack the reactive intermediate, and you are on your way back to the ketone (or aldehyde).

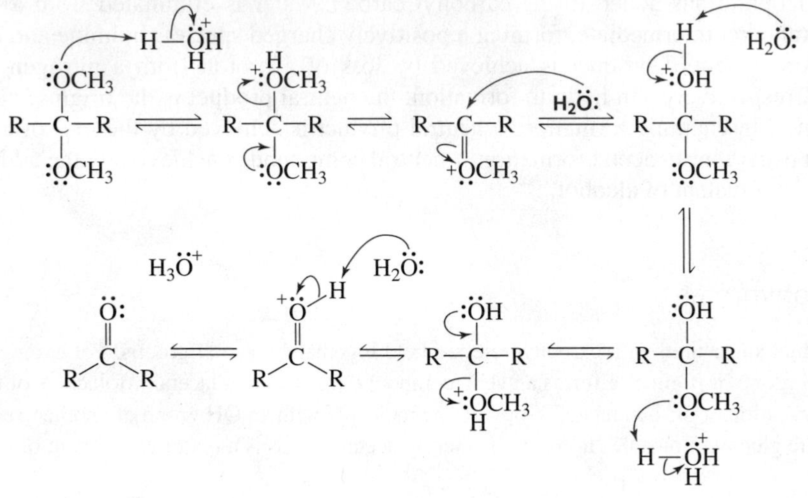

In a basic solution, the CH_3O group of the acetal cannot be protonated. Therefore, a very basic CH_3O^- group has to be eliminated to re-form the ketone (or aldehyde), and the transition state of the elimination reaction would be very unstable because it would have a partial positive charge.

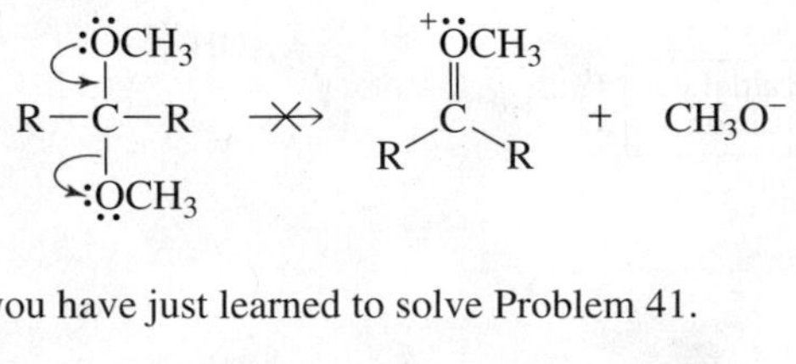

Now use the strategy you have just learned to solve Problem 41.

PROBLEM 41

a. Would you expect hemiacetals to be stable in basic solutions? Explain your answer.

b. Acetal formation must be catalyzed by an acid. Explain why it cannot be catalyzed by CH_3O^-.

c. Can the rate of hydrate formation be increased by hydroxide ion as well as by acid? Explain.

PROBLEM 42

Explain why an acetal can be isolated but most hydrates cannot be isolated.

13   PROTECTING GROUPS

A ketone (or an aldehyde) reacts with a 1,2-diol to form a five-membered ring acetal and it reacts with a 1,3-diol to form six-membered ring acetal. Five- and six-membered rings are formed relatively easily. The mechanism is the same as that shown in Section 12 for acetal formation, except that instead of reacting with two separate molecules of alcohol, the carbonyl compound reacts with the two alcohol groups of a single molecule of the diol.

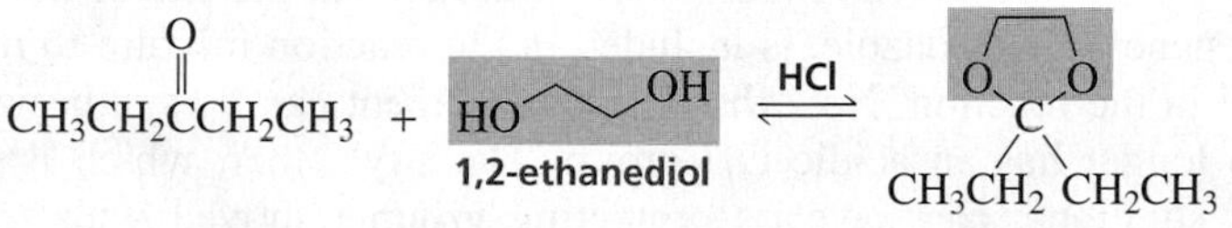

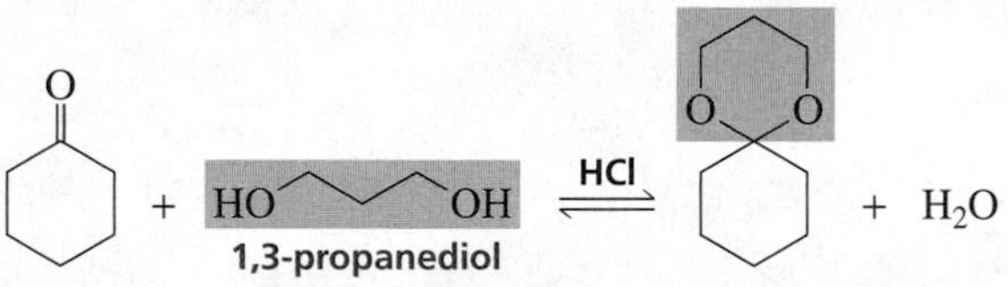

If a compound has two functional groups that will react with a given reagent, but you want only one of them to react, then you must protect the other functional group from the reagent. A group that protects a functional group from a synthetic operation that the functional group would not otherwise survive is called a **protecting group.**

If you have ever painted a room with a spray gun, you may have taped over the things you did not want to paint, such as baseboards and window frames. In a similar way, 1,2-diols and 1,3-diols are used to protect ("tape over") the carbonyl group of aldehydes and ketones.

For example, suppose you want to convert the keto ester shown next into a hydroxy-ketone. Both functional groups of the keto ester will be reduced by LiAlH$_4$, and the one that you don't want to react—the keto group—is the more reactive of the two groups.

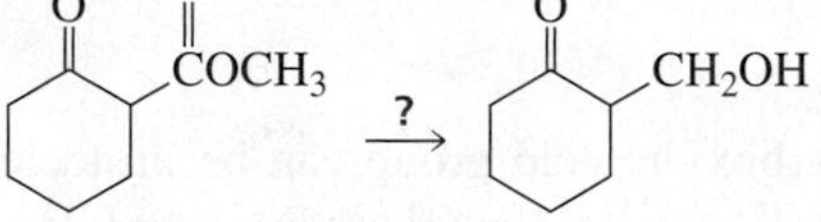

However, if the keto group is first converted to an acetal, only the ester group will react with LiAlH$_4$. The protecting group can then be removed (called "deprotection") by acid-catalyzed hydrolysis after the ester has been reduced. It is critical that the conditions used to remove a protecting group do not affect other groups in the molecule. Acetals are good protecting groups because they are similar to ethers, and like ethers do not react with bases, reducing agents, or oxidizing agents.

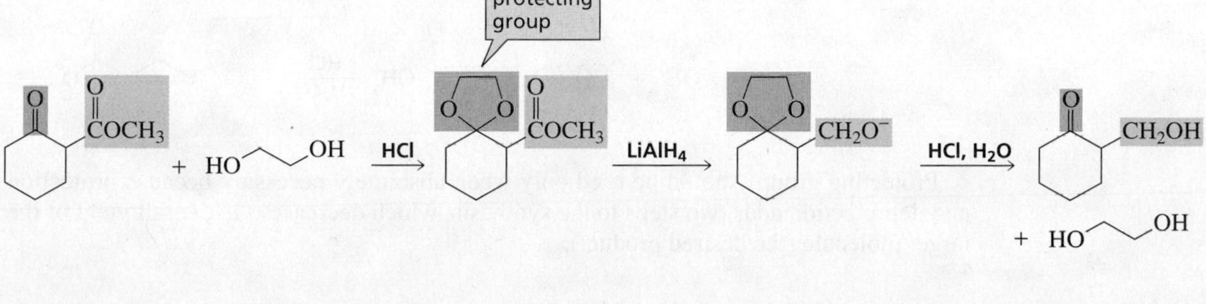

PROBLEM 43◆

a. What would have been the product of the preceding reaction with $LiAlH_4$ if the keto group had not been protected?

b. What reagent could you use to reduce only the keto group?

PROBLEM 44

Explain why acetals do not react with nucleophiles.

One of the best ways to protect an OH group of an alcohol is to convert it to a silyl ether. *tert*-Butyldimethylsilyl chloride is a commonly used reagent for this conversion. The TBDMS ether is formed by an S_N2 reaction. Although a tertiary alkyl halide does not undergo an S_N2 reaction, the tertiary silyl compound does because Si—C bonds are longer than C—C bonds, which reduces steric hindrance at the site of nucleophilic attack. An amine, generally imidazole, is included in the reaction mixture to react with the HCl generated in the reaction. Now the Grignard reagent can be synthesized since the compound no longer has an acidic OH group. The silyl ether, which is stable in neutral and basic solutions, can have its protecting group removed with tetrabutyl-ammonium fluoride.

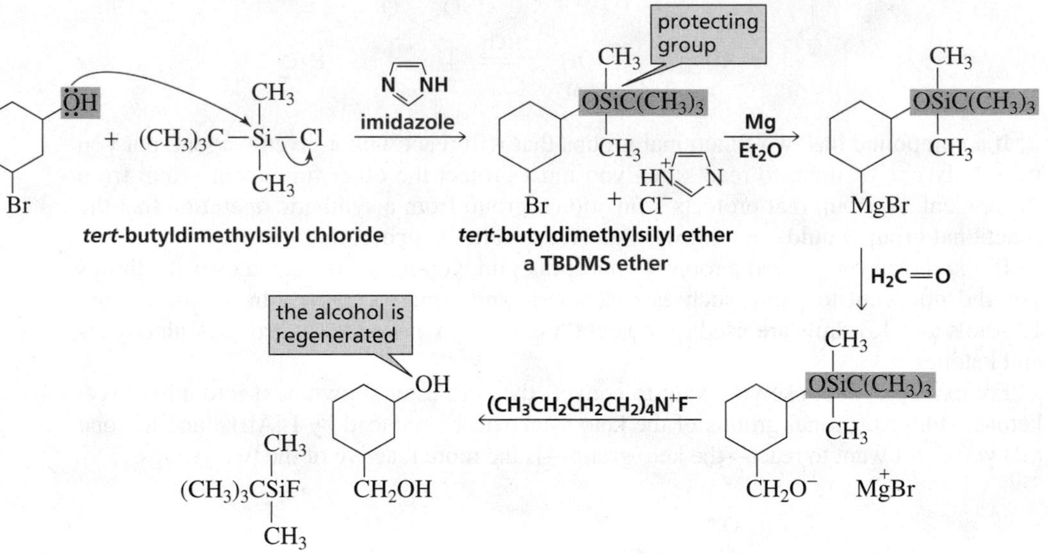

tert-butyldimethylsilyl chloride

tert-butyldimethylsilyl ether
a TBDMS ether

the alcohol is regenerated

The OH group of a carboxylic acid group can be protected by converting the carboxylic acid into an ester. The ester, then, has only one OH group that will react with thionyl chloride.

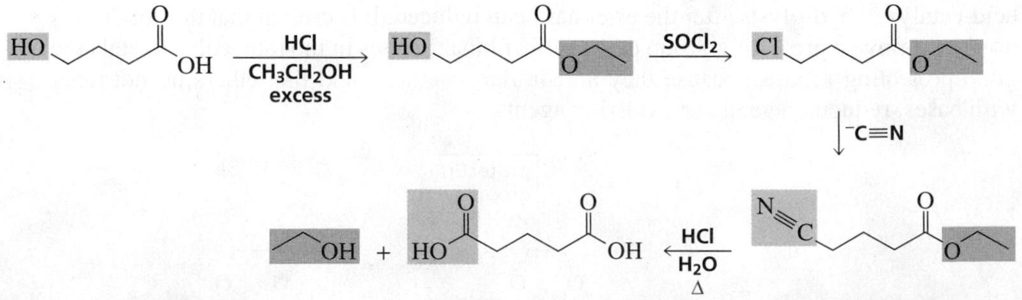

Protecting groups should be used only when absolutely necessary because protection and deprotection adds two steps to the synthesis, which decreases the overall yield of the target molecule (the desired product).

PROBLEM 45

What products would be formed from the preceding reaction if the carboxylic acid group were not protected?

PROBLEM 46♦

a. In a six-step synthesis, what is the yield of the target molecule if each of the reactions employed gives an 80% yield?

b. What would the yield be if two more steps (each with an 80% yield) were added to the synthesis?

PROBLEM 47

Show how each of the following compounds could be prepared from the given starting material. In each case, you will need to use a protecting group.

a. HO⌒⌒Br ⟶ HO⌒⌒⌒OH

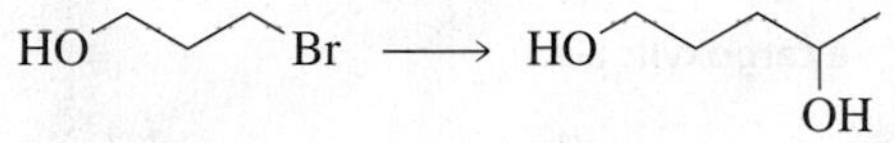

b. HO⌒⌒⌒(C=O) ⟶ HO⌒⌒⌒⌒OH

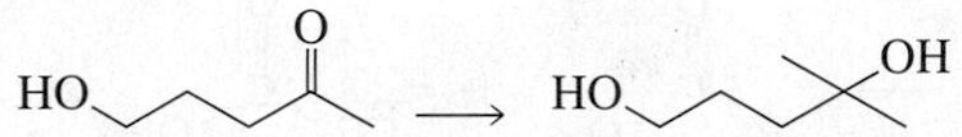

c.

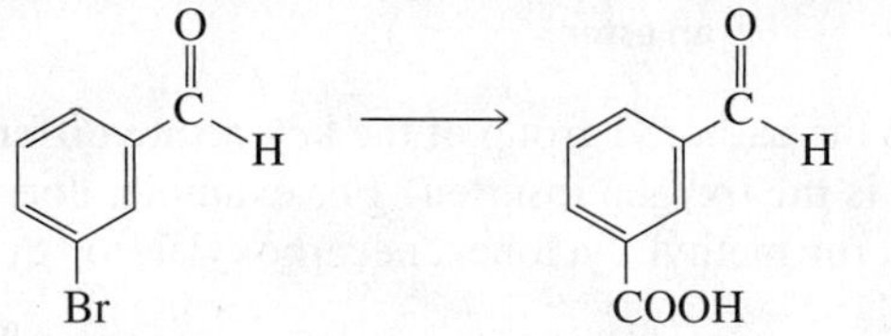

14 THE ADDITION OF SULFUR NUCLEOPHILES

Aldehydes and ketones react with thiols (the sulfur analogues of alcohols) to form thioacetals. The mechanism for the addition of a thiol is the same as the mechanism for the addition of an alcohol (Section 12).

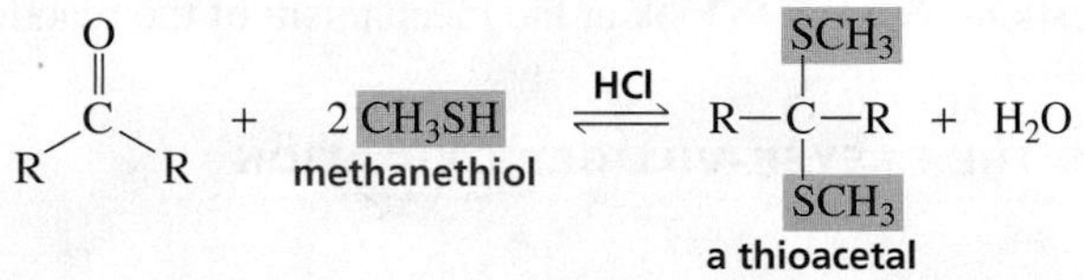

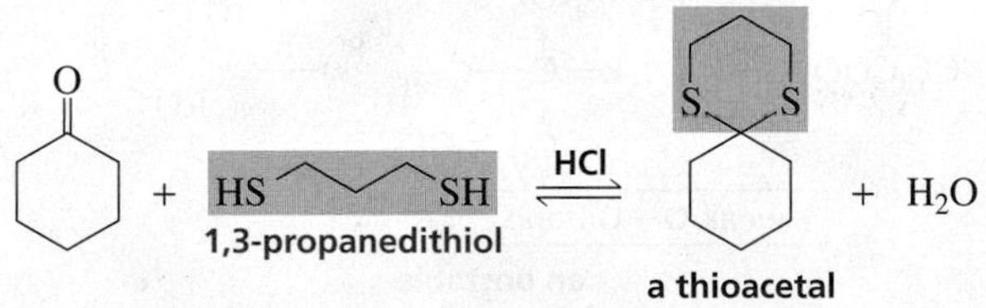

$$R-\underset{R}{\overset{O}{C}} + 2\,CH_3SH \xrightarrow{HCl} R-\underset{SCH_3}{\overset{SCH_3}{C}}-R + H_2O$$

methanethiol   a thioacetal

1,3-propanedithiol   a thioacetal

Thioacetal formation is useful in organic synthesis because a thioacetal is desulfurized when it reacts with H_2 and Raney nickel. **Desulfurization** replaces the C—S bonds with C—H bonds.

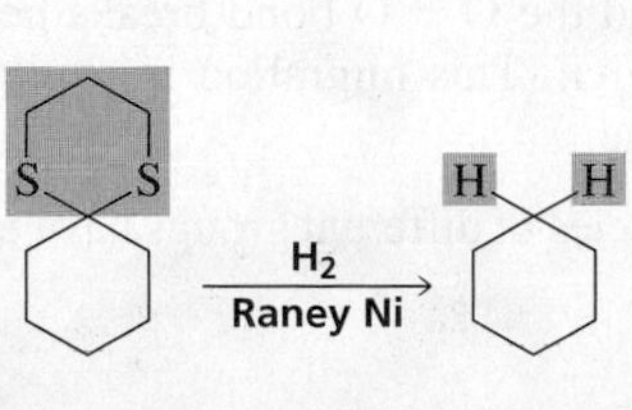

Thus, thioacetal formation followed by desulfurization provides a way to convert a carbonyl group into a methylene group.

15 THE REACTIONS OF ALDEHYDES AND KETONES WITH A PEROXYACID

Aldehydes *and* ketones react with the conjugate base of a peroxyacid to form carboxylic acids and esters, respectively. A **peroxyacid** contains one more oxygen than a carboxylic acid, and it is this oxygen that is inserted between the carbonyl carbon and the H of an aldehyde or the R of a ketone. The reaction is called a **Baeyer–Villiger oxidation.** It is an oxidation reaction because the number of C—O bonds increases. A particularly good reagent for a Baeyer–Villiger oxidation is peroxytrifluoroacetate ion.

Aldehydes are oxidized to carboxylic acids by a peroxyacid.

Ketones are oxidized to esters by a peroxyacid.

an aldehyde + CF₃C OO⁻ ⟶ a carboxylic acid + CF₃C O⁻

a ketone + CF₃C OO⁻ ⟶ an ester + CF₃C O⁻

If the two alkyl substituents attached to the carbonyl group of the ketone are different, then on what side of the carbonyl carbon is the oxygen inserted? For example, does the oxidation of cyclohexyl methyl ketone form methyl cyclohexanecarboxylate or cyclohexyl acetate?

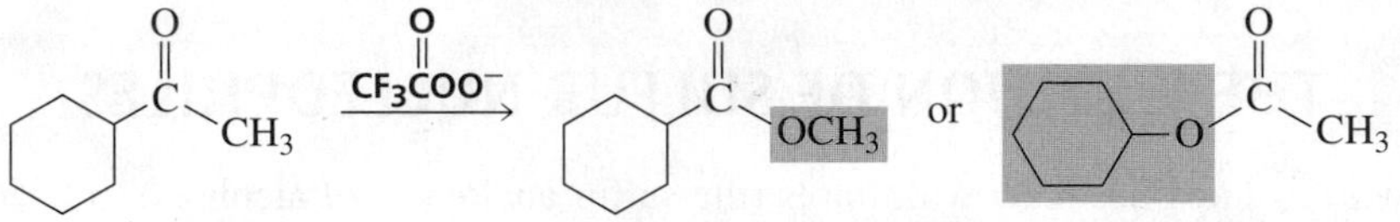

cyclohexyl methyl ketone   →(CF₃COO⁻)   methyl cyclohexanecarboxylate   or   cyclohexyl acetate

To answer this question, we need to look at the mechanism of the reaction.

MECHANISM FOR THE BAEYER–VILLIGER OXIDATION

a weak O—O bond

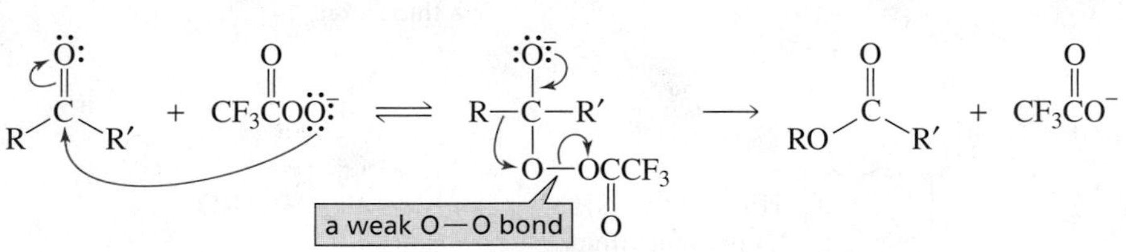

an unstable intermediate

- The nucleophilic oxygen of the peroxyacid adds to the carbonyl carbon and forms an unstable tetrahedral intermediate with a weak O—O bond.
- As the π bond re-forms and the O—O bond breaks heterolytically, one of the alkyl groups migrates to an oxygen. This migration is similar to the 1,2-shifts that occur when carbocations rearrange.

Studies of the migration tendencies of different groups have established the following order:

relative migration tendencies

most likely to migrate → H > *tert*-alkyl > *sec*-alkyl ~ phenyl > primary alkyl > methyl ← least likely to migrate

852

Therefore, the product of the Baeyer–Villiger oxidation of cyclohexyl methyl ketone will be cyclohexyl acetate because a secondary alkyl group (the cyclohexyl group) is more likely to migrate than a methyl group. Aldehydes are always oxidized to carboxylic acids, since H has the greatest tendency to migrate.

PROBLEM 48

What is the product of each of the following reactions?

a.

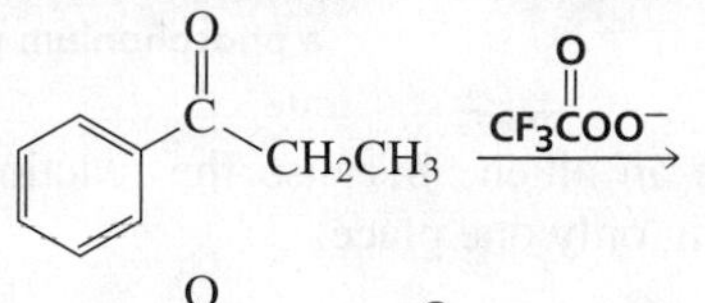

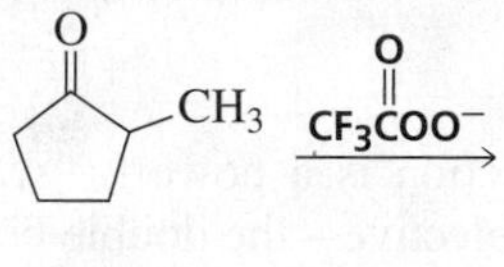

b.

c.

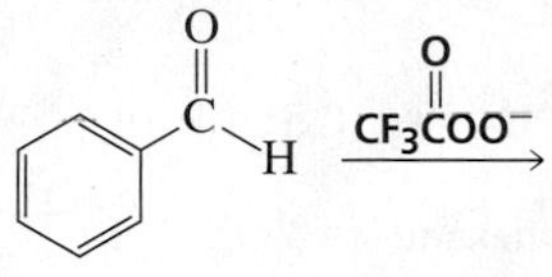

d.

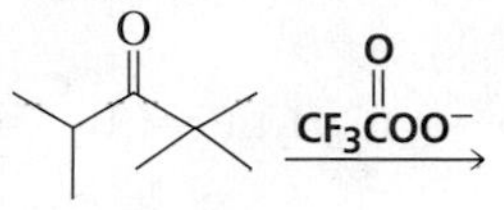

16 THE WITTIG REACTION FORMS AN ALKENE

An aldehyde or a ketone reacts with a *phosphonium ylide* ("ILL-id") to form an alkene. This reaction, called a **Wittig reaction,** interchanges the double-bonded oxygen of the carbonyl compound with the double-bonded carbon group of the phosphonium ylide.

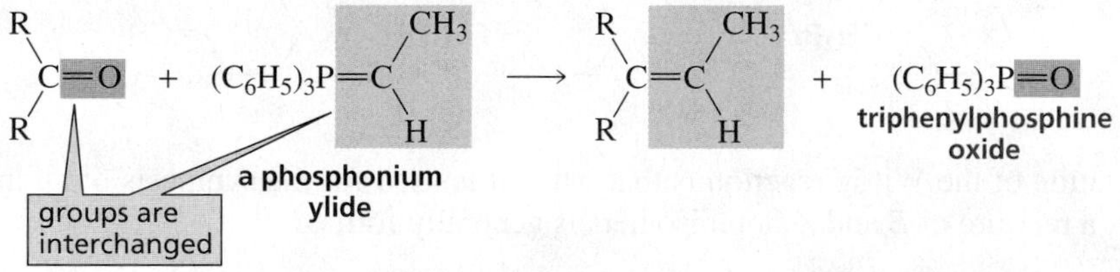

An **ylide** is a compound with opposite charges on adjacent covalently bonded atoms that have complete octets. The ylide can be written in the double-bonded form because phosphorus can have more than eight valence electrons.

$$(C_6H_5)_3\overset{+}{P}-\overset{..}{\overset{-}{C}}HR \longleftrightarrow (C_6H_5)_3P{=}CHR$$

a phosphonium ylide

The Wittig reaction is a concerted [2 + 2] cycloaddition reaction. It is called a [2 + 2] cycloaddition reaction because, of the four electrons involved in the cyclic transition state, two come from the carbonyl group and two come from the ylide.

MECHANISM FOR THE WITTIG REACTION

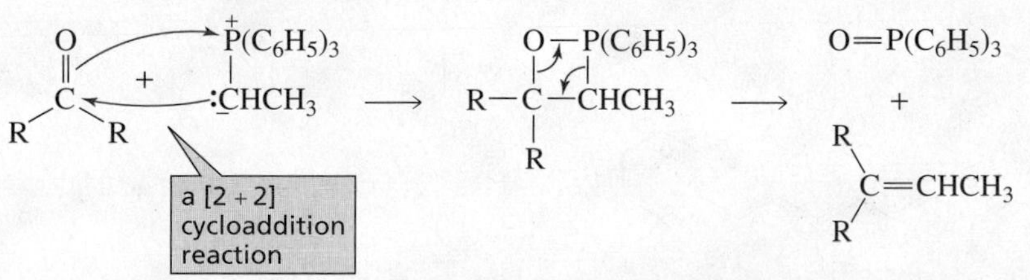

- The nucleophilic carbon of the ylide adds to the carbonyl carbon while the carbonyl oxygen adds to the electrophilic phosphorus.
- Elimination of triphenylphosphine oxide forms the alkene product.

The phosphonium ylide needed for a particular synthesis is obtained by an S_N2 reaction between triphenylphosphine and an alkyl halide with the appropriate number of carbons. A proton on the carbon adjacent to the positively charged phosphorus atom is sufficiently acidic ($pK_a = 35$) to be removed by a strong base such as butyllithium.

$$(C_6H_5)_3P: \quad + \quad CH_3CH_2{-}Br \quad \xrightarrow{\ S_N2\ } \quad (C_6H_5)_3\overset{+}{P}{-}CH_2CH_3 \quad \xrightarrow{\ CH_3CH_2CH_2\overset{..}{C}H_2\ \overset{+}{Li}\ } \quad (C_6H_5)_3\overset{+}{P}{-}\overset{..}{\overset{-}{C}}HCH_3$$
$$Br^-$$
triphenylphosphine $\qquad\qquad\qquad\qquad\qquad\qquad\qquad\qquad\qquad\qquad$ a phosphonium ylide

The Wittig reaction is a powerful way to make an alkene because the reaction is completely regioselective—the double bond will be in only one place.

$$\bigcirc{=}O \quad + \quad (C_6H_5)_3P{=}CH_2 \quad \longrightarrow \quad \bigcirc{=}CH_2 \quad + \quad (C_6H_5)_3P{=}O$$
methylenecyclohexane

The Wittig reaction is the best way to make a terminal alkene, such as the one just shown, because other methods would form a terminal alkene only as a minor product.

$$\bigcirc\!\!\!\!\begin{smallmatrix}CH_3\\Br\end{smallmatrix} \quad \xrightarrow{\ HO^-\ } \quad \bigcirc{-}CH_3 \quad + \quad \bigcirc{=}CH_2$$
$\qquad\qquad\qquad\qquad\qquad\qquad\qquad\qquad\qquad$ minor

$$\bigcirc{-}CH_2Br \quad \xrightarrow{\ HO^-\ } \quad \bigcirc{-}CH_2OH \quad + \quad \bigcirc{=}CH_2$$
$\qquad\qquad\qquad\qquad\qquad\qquad\qquad\qquad\qquad$ minor

A limitation of the Wittig reaction is that when it is used for the synthesis of an internal alkene, a mixture of E and Z stereoisomers is generally formed.

β-Carotene

β-Carotene is found in orange and yellow-orange fruits and vegetables such as apricots, mangoes, carrots, and sweet potatoes. It is also responsible for the characteristic color of flamingos. β-Carotene is used in the food industry to color margarine. The synthesis of β-carotene from vitamin A for use in foods is an important application of the Wittig reaction in industry.

vitamin A aldehyde

β-carotene

Many people take β-carotene as a dietary supplement because there is some evidence associating high levels of β-carotene with a low incidence of cancer. More recent evidence, however, suggests that β-carotene taken in pill form does not have the cancer-preventing effects of β-carotene obtained from the diet.

Retrosynthetic Analysis

When synthesizing an alkene using a Wittig reaction, the first thing you must do is decide which part of the alkene should come from the carbonyl compound and which part should come from the ylide. If both sets of carbonyl compound and ylide are available, the better choice is the set that requires the less sterically hindered alkyl halide for the synthesis of the ylide via an S_N2 reaction. (The more sterically hindered the alkyl halide, the less reactive it is in an S_N2 reaction.)

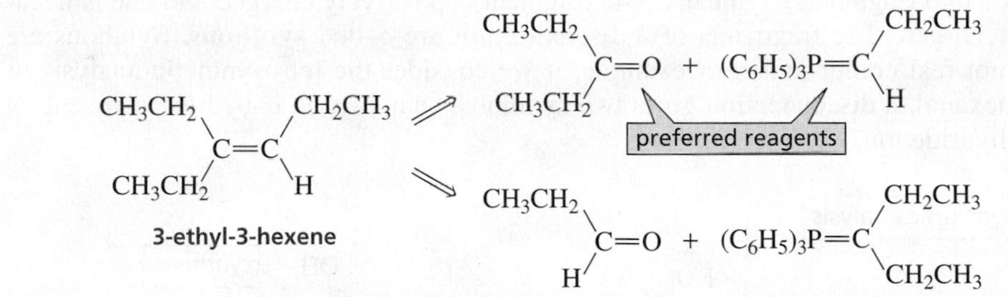

For the synthesis of 3-ethyl-3-hexene, for example, it is better to use a three-carbon alkyl halide for the ylide and the five-carbon carbonyl compound than the five-carbon alkyl halide for the ylide and the three-carbon carbonyl compound, because it is easier to form an ylide from a primary alkyl halide (1-bromopropane) than from a secondary alkyl halide (3-bromopentane).

PROBLEM 49 Solved

a. What two sets of reagents (each consisting of a carbonyl compound and phosphonium ylide) can be used for the synthesis of each of the following alkenes?

1. $CH_3CH_2CH_2CH=CCH_3$
　　　　　　　　　　$|$
　　　　　　　　　　CH_3

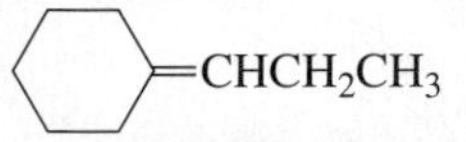

3. $(C_6H_5)_2C=CHCH_3$

2. $=CHCH_2CH_3$

4. 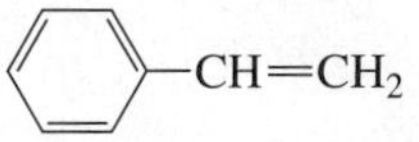 $-CH=CH_2$

b. What alkyl halide is required to prepare each of the phosphonium ylides in part **a**?

c. What would be the best set of reagents to use for each of the syntheses?

Solution to 49a (1) The atoms on either side of the double bond can come from the carbonyl compound, so two pairs of compounds could be used.

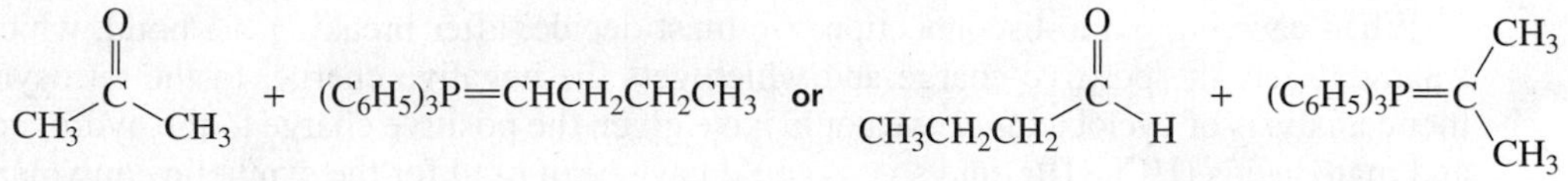

Solution to 49b (1) The alkyl halide required to make the phosphonium ylide would be 1-bromobutane for the first pair of reagents or 2-bromopropane for the second pair.

$$CH_3CH_2CH_2CH_2Br \quad \textbf{or} \quad CH_3CHCH_3$$
$$| \qquad$$
$$Br$$

Solution to 49c (1) The primary alkyl halide would be more reactive in the S_N2 reaction required to make the ylide, so the best method would be to use the first set of reagents (acetone and the ylide obtained from 1-bromobutane).

17　DISCONNECTIONS, SYNTHONS, AND SYNTHETIC EQUIVALENTS

The route to the synthesis of a complicated molecule from simple starting materials is not always obvious. It is often easier to work backward from the desired product—a process called *retrosynthetic analysis*. In a retrosynthetic analysis, the chemist dissects a molecule into smaller and smaller pieces to arrive at read-ily available starting materials.

retrosynthetic analysis

$$\text{target molecule} \implies Y \implies X \implies W \implies \text{starting materials}$$

A useful step in a retrosynthetic analysis is a **disconnection**—breaking a bond to produce two fragments. Typically, one fragment is positively charged and one is nega-tively charged. The fragments of a disconnection are called **synthons.** Synthons are often not real compounds. For example, if we consider the retrosynthetic analysis of cyclohexanol, a disconnection gives two synthons—namely, an α-hydroxycarbocation and a hydride ion.

retrosynthetic analysis

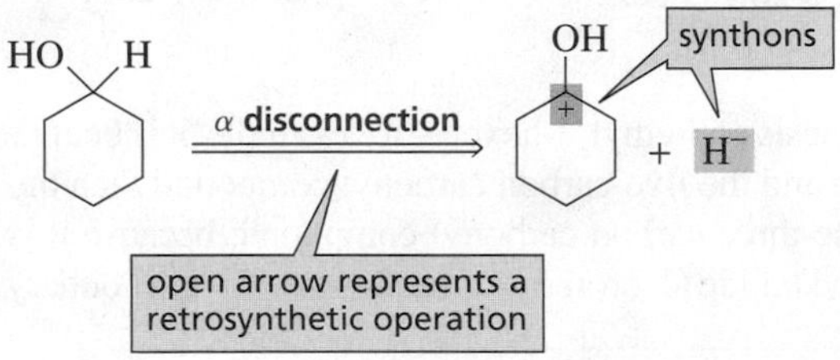

A **synthetic equivalent** is the reagent that is actually used as the source of a synthon. In the synthesis of cyclohexanol, cyclohexanone is the synthetic equivalent for the α-hydroxycarbocation, and sodium borohydride is the synthetic equivalent for hydride ion. Thus, cyclohexanol, the target molecule, can be prepared by treating cyclohexanone with sodium borohydride.

synthesis

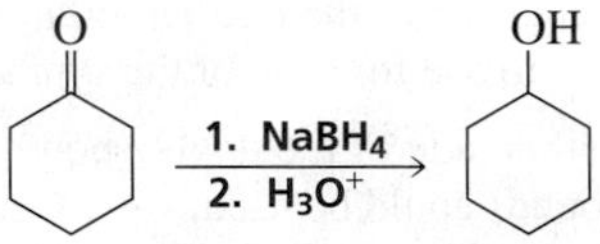

When carrying out a disconnection, we must decide, after breaking the bond, which fragment gets the positive charge and which gets the negative charge. In the retrosyn-thetic analysis of cyclohexanol, we could have given the positive charge to the hydrogen, and many acids (HCl, HBr, and so on) could have been used for the synthetic equivalent for H^+. However, we would have been at a loss to find a synthetic equivalent for an α-hydroxycarbanion. Therefore, when we carried out the disconnection, we assigned the positive charge to the carbon and the negative charge to the hydrogen.

Cyclohexanol can also be disconnected by breaking the C—O bond instead of the C—H bond, forming a carbocation and hydroxide ion.

retrosynthetic analysis

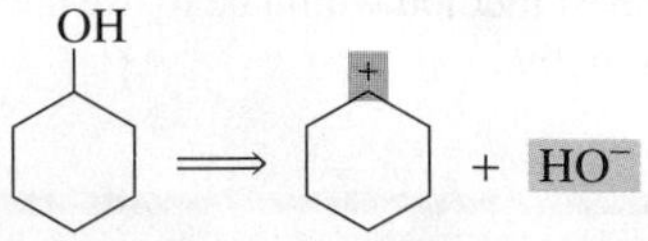

The problem then becomes choosing a synthetic equivalent for the carbocation. A syn-thetic equivalent for a positively charged synthon needs an electron-withdrawing group at just the right place. Cyclohexyl bromide, with an electron-withdrawing bromine, is a synthetic equivalent for the cyclohexyl carbocation. Cyclohexanol, therefore, can be prepared by treating cyclohexyl bromide with hydroxide ion. This method, however, is not as satisfactory as the first synthesis we proposed—reduction of cyclohexanone—because some of the alkyl halide is converted to an alkene, so the overall yield of the target molecule is lower.

synthesis

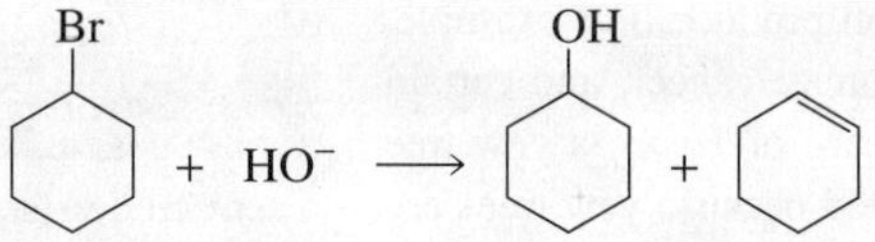

Retrosynthetic analysis shows that 1-methylcyclohexanol can be formed from the reaction of cyclohexanone, the synthetic equivalent for the α-hydroxycarbocation, with methylmagnesium bromide, the synthetic equivalent for the methyl anion.

retrosynthetic analysis

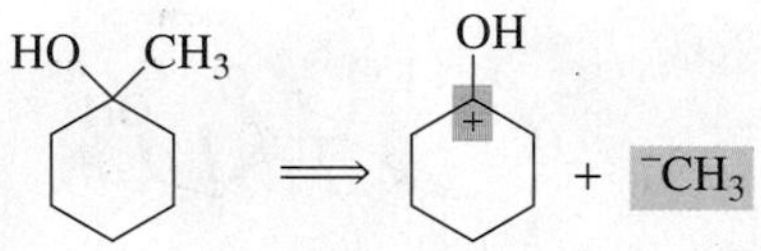

synthesis

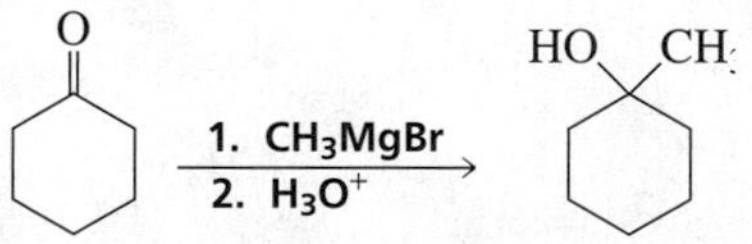

Other disconnections of 1-methylcyclohexanol are possible because any bond to carbon can serve as a disconnection site. For example, one of the ring C—C bonds could be broken. However, these are not useful disconnections because the synthetic equivalents of the synthons they produce are not easily prepared. A disconnection must lead to readily obtainable starting materials.

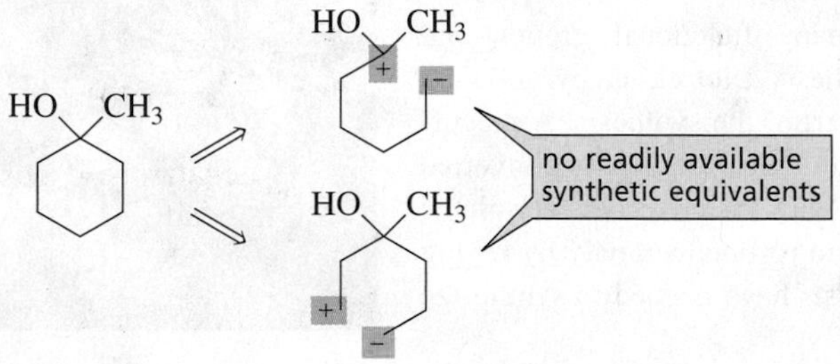

no readily available synthetic equivalents

For additional practice using restrosynthetic analysis, see the tutorial.

PROBLEM 50

Using bromocyclohexane as a starting material, how could you synthesize the following compounds?

a. 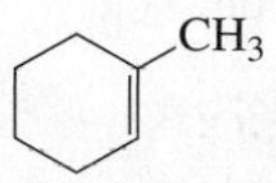   b. 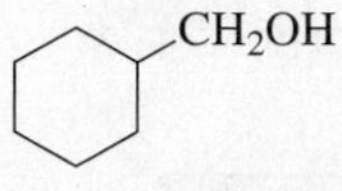   c. 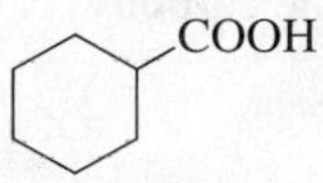   d. 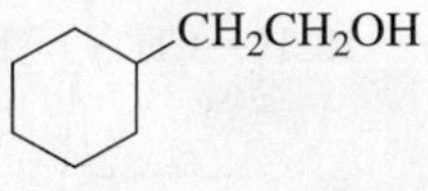   e. 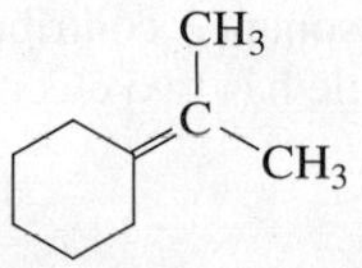   f.

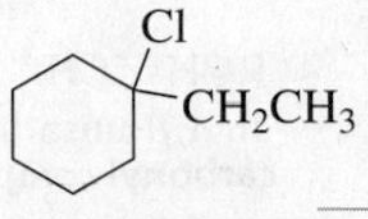

Synthesizing Organic Compounds

Organic chemists synthesize compounds for many reasons: to study their properties, to answer a variety of chemical questions, or to take advantage of one or more useful properties. One reason chemists synthesize a natural product—that is, a compound synthesized in nature—is to provide us with a larger supply of the compound than nature can produce. For example, Taxol—a compound that has successfully treated ovarian cancer, breast cancer, and certain forms of lung cancer by inhibiting mitosis—is extracted from the bark of *Taxus,* a yew tree found in the Pacific Northwest. The supply of natural Taxol is limited because yew trees are uncommon, they grow very slowly, and stripping the bark kills the tree. Moreover, the bark of a 40-foot tree, which may have taken 200 years to grow, provides only 0.5 g of the drug. In addition, *Taxus* forests serve as habitats for the spotted owl, an endangered species, so harvesting the trees would accelerate the owl's demise. Once chemists determined the structure of Taxol, efforts were undertaken to synthesize it in order to make it more widely available as an anticancer drug. Several syntheses have been successful.

a spotted owl (*strix occidentalis*) taking off from a falconer's glove

yew tree bark

Taxol®

Once a compound has been synthesized, chemists can study its properties to learn how it works. Then they may be able to design and synthesize analogues to obtain safer or more potent drugs.

Semisynthetic Drugs

Taxol is difficult to synthesize because of its many functional groups and 11 asymmetric centers. Chemists have made the synthesis a lot easier by allowing the common English yew shrub to carry out the first part of the synthesis. A precursor of the drug is extracted from the shrub's needles, and the precursor is converted to Taxol in a four-step procedure in the laboratory. Thus, the precursor is isolated from a renewable resource, whereas the drug itself could be obtained only by killing a slow-growing tree. This is an example of how chemists have learned to synthesize compounds jointly with nature.

a yew shrub

18 NUCLEOPHILIC ADDITION TO α,β-UNSATURATED ALDEHYDES AND KETONES

The resonance contributors for an α,β-unsaturated carbonyl compound show that the molecule has two electrophilic sites: the carbonyl carbon and the β-carbon.

an α,β-unsaturated carbonyl compound

electrophilic site   electrophilic site

This means that a nucleophile can add either to the carbonyl carbon or to the β-carbon.

Nucleophilic addition to the carbonyl carbon is called **direct addition** or 1,2-addition.

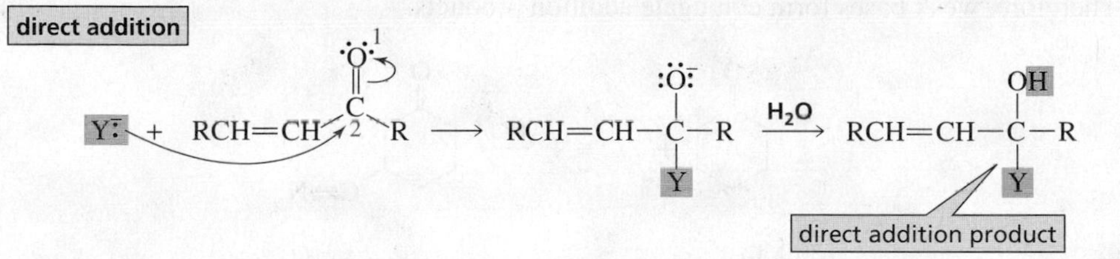

direct addition

direct addition product

Nucleophilic addition to the β-carbon is called **conjugate addition** or 1,4-addition, because it occurs at the 1- and 4-positions. The product of 1,4-addition is an enol, which tautomerizes to a ketone or to an aldehyde. Thus, the overall reaction is addition to the carbon–carbon double bond, with the nucleophile adding to the β-carbon and a proton from the reaction mixture adding to the α-carbon.

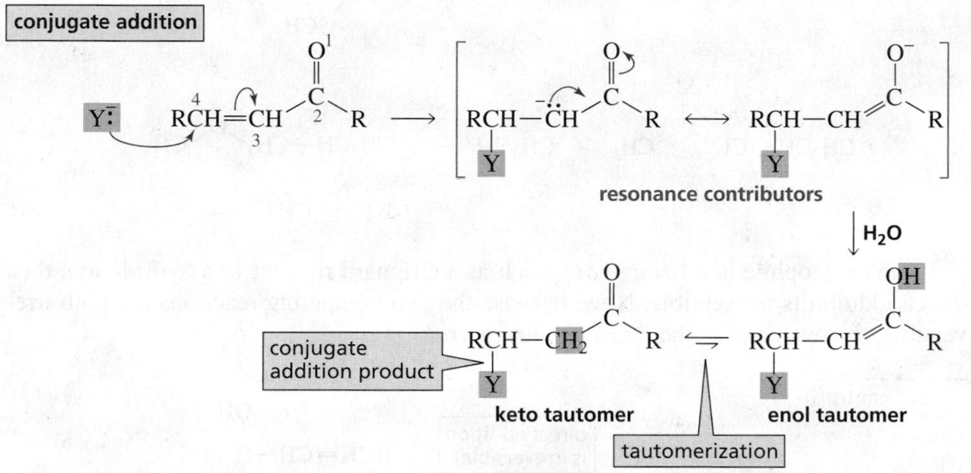

conjugate addition

resonance contributors

conjugate addition product

keto tautomer   **enol tautomer**

tautomerization

Whether the product obtained from nucleophilic addition to an α,β-unsaturated aldehyde or ketone is the direct addition product or the conjugate addition product depends on the nature of the nucleophile and the structure of the carbonyl compound.

If two competing reactions are both irreversible, the reaction will be under kinetic control, and if one or both of the reactions is reversible, the reaction will be under thermodynamic control. Addition to the β-carbon (conjugate addition) is generally irreversible. Direct addition can be reversible or irreversible.

If the nucleophile is a *weak base,* such as a halide ion, a cyanide ion, a thiol, an alcohol, or an amine, then direct addition is *reversible,* because a weak base is a good leaving group. Therefore, the reaction will be under *thermodynamic control.*

thermodynamic control

direct addition is reversible

direct addition

weakly basic nucleophile

conjugate addition

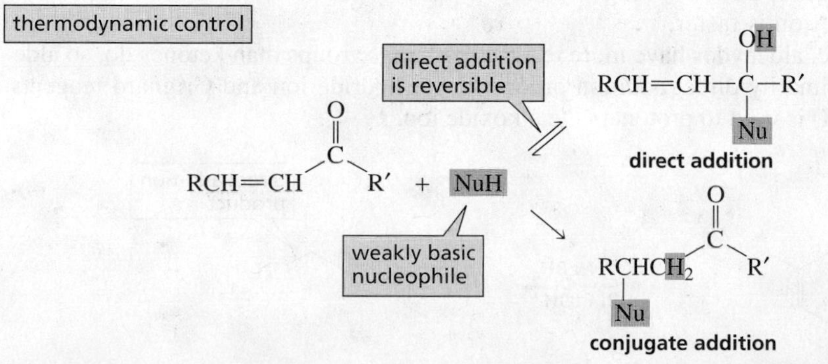

The reaction that prevails when the reaction is under thermodynamic control is the one that forms the more stable product. The conjugate addition product is always the more stable product because it retains the very stable carbonyl group. Therefore, weak bases form conjugate addition products.

Nucleophiles that are weak bases form conjugate addition products.

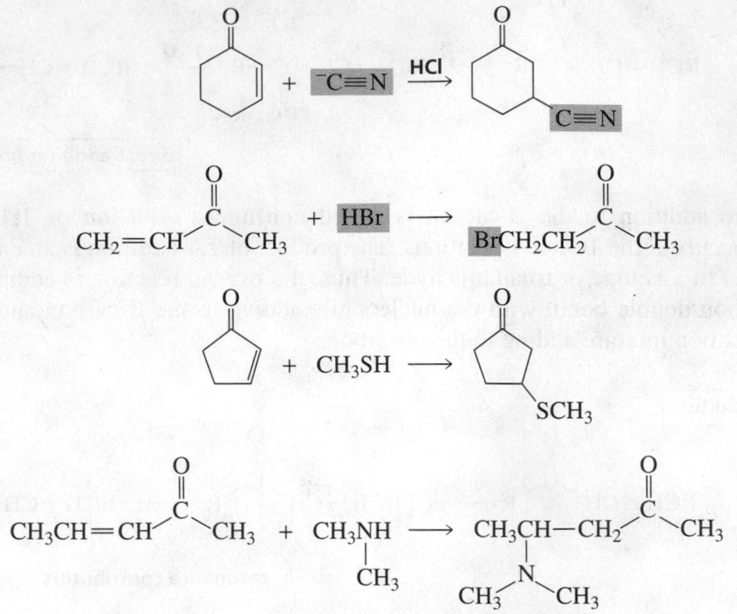

If the nucleophile is a *strong base,* such as a Grignard reagent or a hydride ion, then direct addition is irreversible. Now, because the two competing reactions are both irreversible, the reaction will be under *kinetic control.*

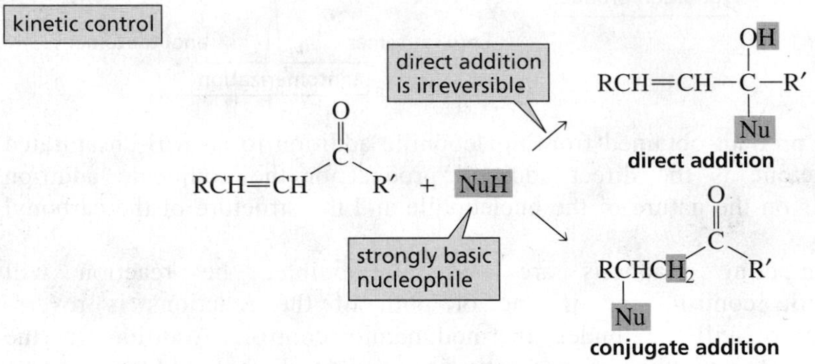

The reaction that prevails when the reaction is under kinetic control is the one that is faster. Therefore, the product depends on the reactivity of the carbonyl group. Compounds with reactive carbonyl groups form primarily direct addition products because for *those* compounds, direct addition is faster. Compounds with less reactive carbonyl groups form conjugate addition products because for *those* compounds, conjugate addition is faster.

For example, aldehydes have more reactive carbonyl groups than ketones do, so aldehydes form primarily direct addition products with hydride ion and Grignard reagents. Ethanol (EtOH) is used to protonate the alkoxide ion.

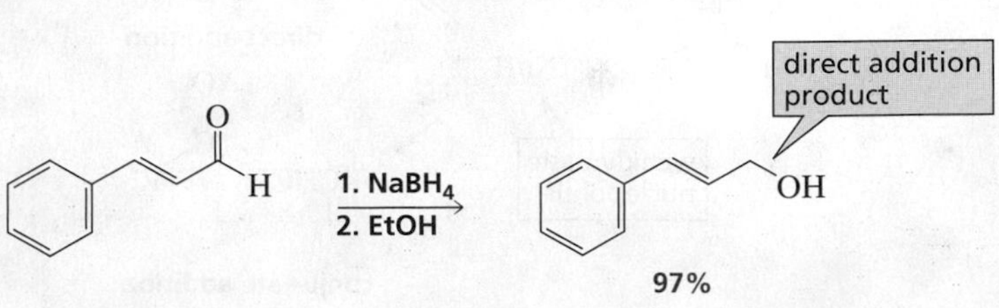

Compared with aldehydes, ketones form less of the direct addition product and more of the conjugate addition product, because ketones are more sterically hindered than aldehydes.

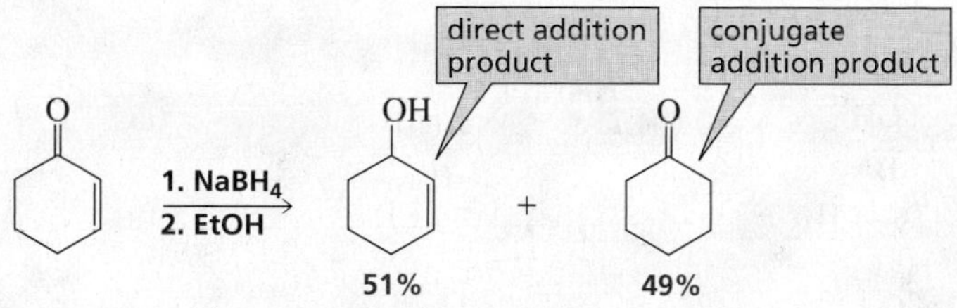

direct addition product

conjugate addition product

51%   +   49%

Nucleophiles that are strong bases form direct addition products with reactive carbonyl groups and form conjugate addition products with less reactive carbonyl groups.

Like hydride ions, Grignard reagents are strong bases and therefore add irreversibly to carbonyl groups. Thus, the reaction is under kinetic control. If the carbonyl compound is reactive, reaction with the Grignard reagent will form the direct addition product.

direct addition product

$$CH_2=CH-\overset{\overset{\displaystyle OH}{|}}{C}HCH_3$$

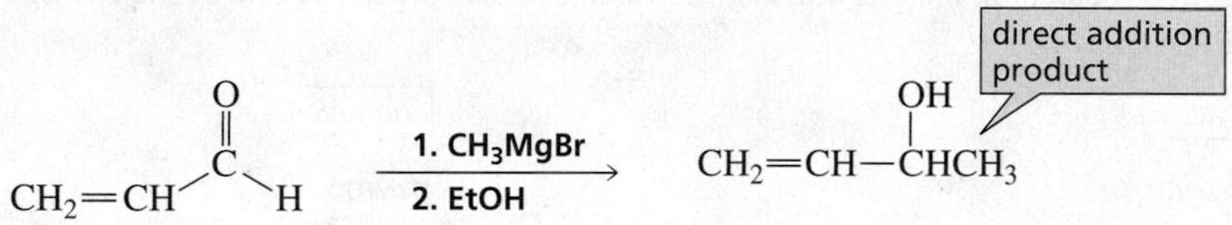

If, however, the rate of direct addition is slowed down by steric hindrance, reaction with the Grignard reagent will form the conjugate addition product because conjugate addition then becomes the faster reaction.

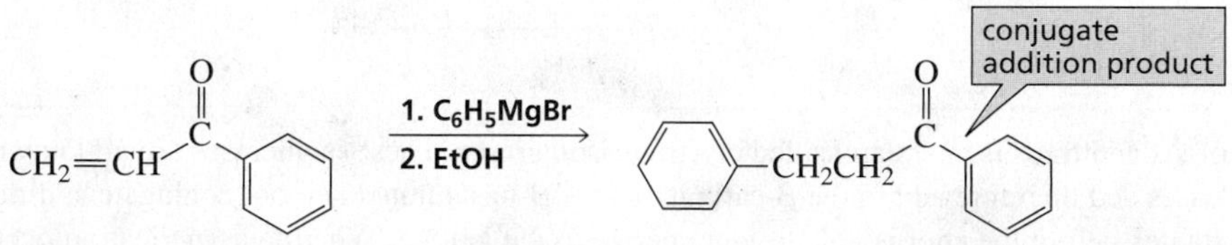

conjugate addition product

Only conjugate addition occurs when organocuprates (Gilman reagents) react with α,β-unsaturated aldehydes and ketones. Therefore, Grignard reagents should be used when you want to add an alkyl group to the carbonyl carbon, whereas Gilman reagents should be used when you want to add an alkyl group to the β-carbon.

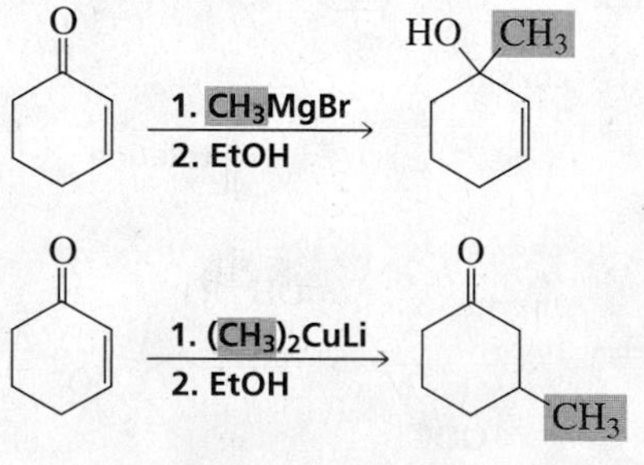

Electrophiles and nucleophiles can be classified as either *hard* or *soft*. Hard electrophiles and nucleophiles are more polarized than soft ones. Hard nucleophiles prefer to react with hard electrophiles, and soft nucleophiles prefer to react with soft electrophiles. Therefore, a Grignard reagent with a highly polarized C—Mg bond prefers to react with the harder C=O bond, whereas a Gilman reagent with a much less polarized C—Cu bond prefers to react with the softer C=C bond.

Cancer Chemotherapy

Two compounds—vernolepin and helenalin—owe their effectiveness as anticancer drugs to conjugate addition reactions.

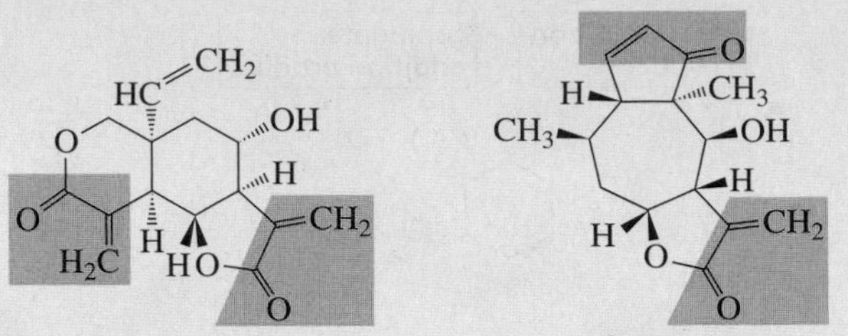

vernolepin   helenalin

Cancer cells are cells that have lost the ability to control their growth, so they proliferate rapidly. DNA polymerase is an enzyme that a cell needs in order to make a copy of its DNA for a new cell. DNA polymerase has an SH group at its active site, and each of these drugs has two α,β-unsaturated carbonyl groups. When an SH group of DNA polymerase adds irreversibly to the β-carbon of one of the α,β-unsaturated carbonyl groups of vernolepin or helenalin, the enzyme is inactivated because the active site of the enzyme is now blocked by the drug, so the enzyme cannot bind its substrate.

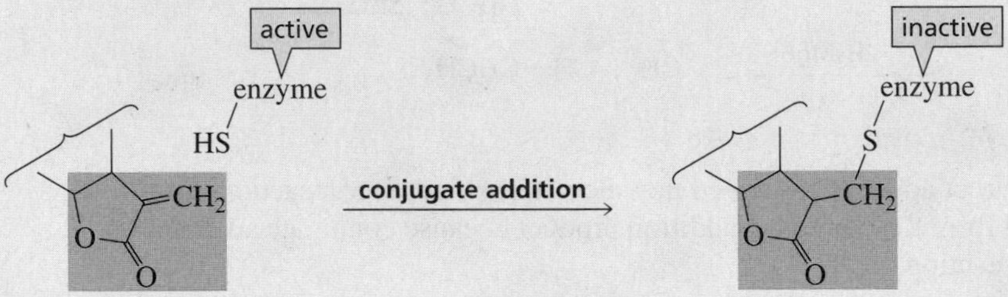

Enzyme-Catalyzed Cis–Trans Interconversion

Enzymes that catalyze the interconversion of cis and trans isomers are called cis–trans isomerases. These isomerases are all known to contain thiol (SH) groups. Thiols are weak bases and therefore add to the β-carbon of an α,β-unsaturated ketone (conjugate addition), forming a carbon–carbon single bond that rotates before the enol is able to tautomerize to the ketone. When tautomerization occurs, the absence of a proton at the active site of the enzyme in the vicinity of the α-carbon prevents the addition of a proton to the α-carbon. Therefore, the thiol is eliminated, leaving the compound as it was originally except for the configuration of the double bond.

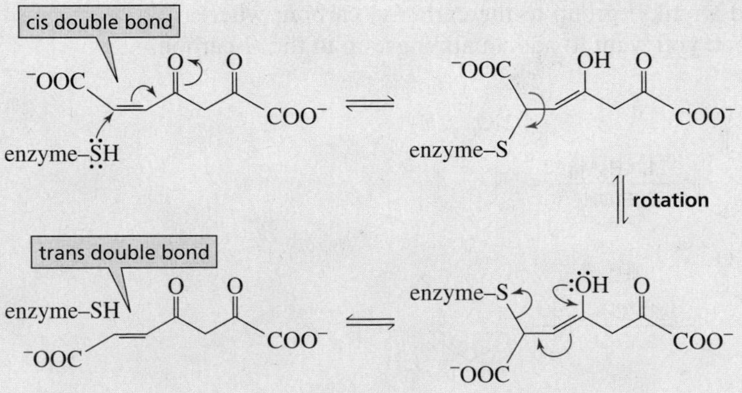

PROBLEM 51

What is the major product of each of the following reactions?

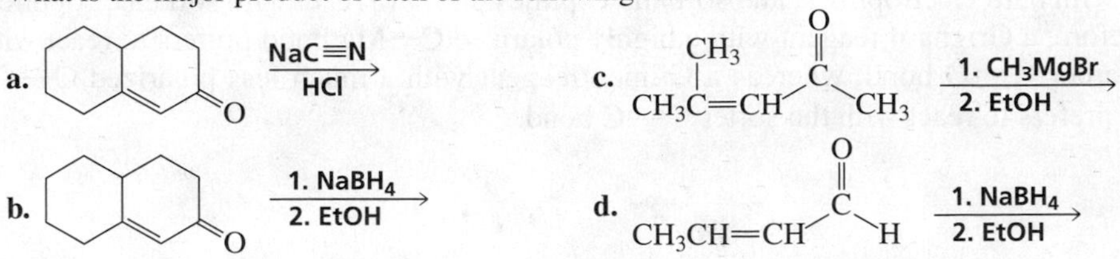

a. $\xrightarrow[\text{HCl}]{\text{NaC}\equiv\text{N}}$

b. $\xrightarrow[\text{2. EtOH}]{\text{1. NaBH}_4}$

c. $\xrightarrow[\text{2. EtOH}]{\text{1. CH}_3\text{MgBr}}$

d. $\text{CH}_3\text{CH}=\text{CH}-\overset{\overset{\displaystyle O}{\|}}{\text{C}}-\text{H} \xrightarrow[\text{2. EtOH}]{\text{1. NaBH}_4}$

19 NUCLEOPHILIC ADDITION TO α,β-UNSATURATED CARBOXYLIC ACID DERIVATIVES

α,β-Unsaturated carboxylic acid derivatives, like α,β-unsaturated aldehydes and ketones, have two electrophilic sites for nucleophilic addition. They can undergo *conjugate addition* or *nucleophilic addition–elimination*. Notice that α,β-unsaturated carboxylic acid derivatives undergo *nucleophilic addition–elimination* rather than *direct nucleophilic addition* because they have a group that can be replaced by a nucleophile. In other words, direct nucleophilic addition becomes nucleophilic addition–elimination if the carbonyl group is attached to a group that can be replaced by a nucleophile (Section 3).

Nucleophiles react with α,β-unsaturated carboxylic acid derivatives *with reactive carbonyl groups,* such as acyl chlorides, at the carbonyl group, forming nucleophilic addition–elimination products. Conjugate addition products are formed from the reaction of nucleophiles *with less reactive carbonyl groups,* such as esters and amides.

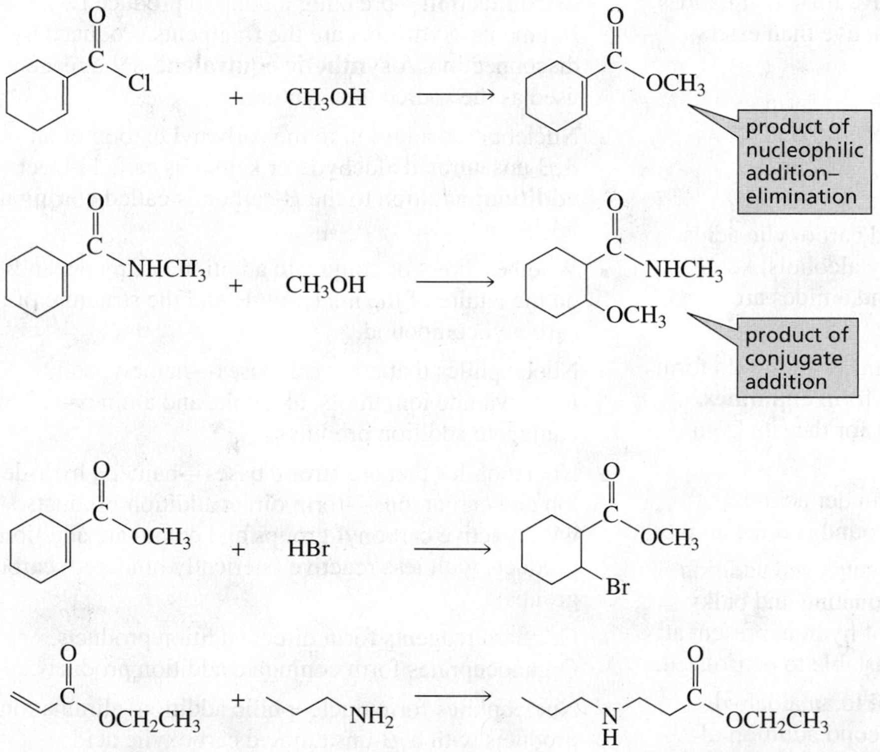

PROBLEM 52

What is the major product of each of the following reactions?

a. 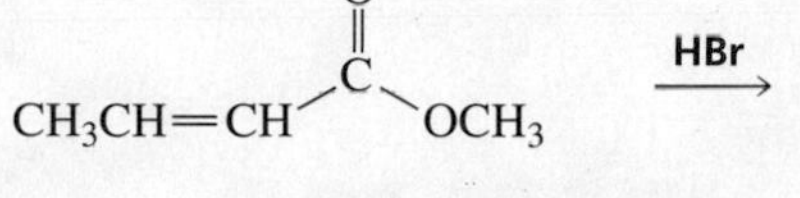 $CH_3CH{=}CH{-}\overset{\displaystyle O}{\overset{\|}{C}}{-}OCH_3 \xrightarrow{HBr}$

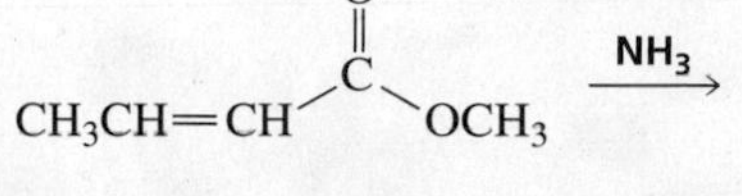

b. 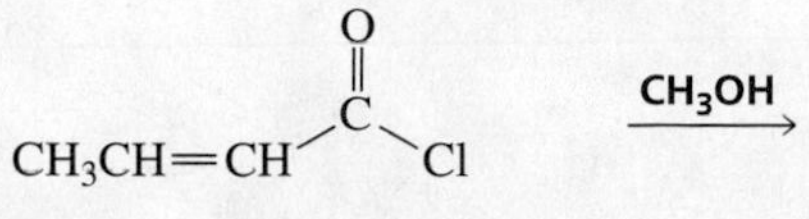 $CH_3CH{=}CH{-}\overset{\displaystyle O}{\overset{\|}{C}}{-}Cl \xrightarrow{CH_3OH}$

c. $CH_3CH{=}CH{-}\overset{\displaystyle O}{\overset{\|}{C}}{-}OCH_3 \xrightarrow{NH_3}$

d. 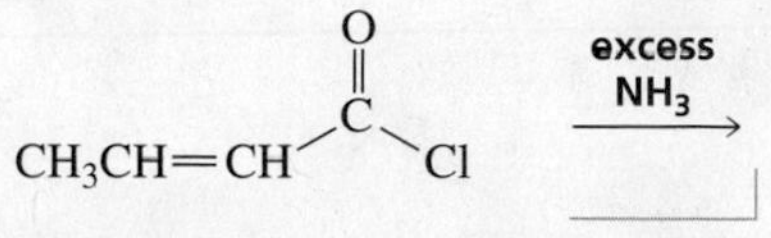 $CH_3CH{=}CH{-}\overset{\displaystyle O}{\overset{\|}{C}}{-}Cl \xrightarrow{\text{excess } NH_3}$

SOME IMPORTANT THINGS TO REMEMBER

- **Aldehydes** and **ketones** have an acyl group attached to an H and an R, respectively.

- Aldehydes and ketones undergo **nucleophilic addition reactions** with strongly basic (R⁻ and H⁻) nucleophiles, and undergo **nucleophilic addition–elimination reactions** with O and N nucleophiles.

- Acyl chlorides and esters undergo a nucleophilic addition–elimination reaction with strongly basic nucleophiles (R⁻ and H⁻) to form a ketone or an aldehyde, which then undergoes a nucleophilic addition reaction with a second equivalent of the nucleophile.

- Electronic and steric factors cause an aldehyde to be more reactive than a ketone toward nucleophilic addition.

- Aldehydes and ketones are less reactive than acyl halides and acid anhydrides and are more reactive than esters, carboxylic acids, and amides.

- Grignard reagents react with aldehydes to form secondary alcohols, with ketones, esters, and acyl halides to form tertiary alcohols, and with carbon dioxide to form carboxylic acids.

- Aldehydes, acyl chlorides, esters, and carboxylic acids are reduced by hydride ion to primary alcohols; ketones are reduced to secondary alcohols; and amides are reduced to amines.

- Aldehydes and ketones react with primary amines to form **imines** and with secondary amines to form **enamines.** The mechanisms are the same, except for the site from which a proton is lost in the last step.

- Imines and enamines are hydrolyzed under acidic conditions back to the carbonyl compound and the amine.

- Aldehydes and ketones undergo acid-catalyzed addition of water to form hydrates. Electron-donating and bulky substituents decrease the percentage of hydrate present at equilibrium. Most hydrates are too unstable to be isolated.

- Acid-catalyzed addition of an alcohol to an aldehyde or a ketone forms a **hemiacetal;** a second addition of alcohol forms an **acetal.** Acetal formation is reversible.

- The carbonyl group of an aldehyde or a ketone can be protected by being converted to an acetal; the OH group

of an alcohol can be protected by being converted to a TBDMS ether; and the OH group of a carboxylic acid can be protected by being converted to an ester.

- Aldehydes and ketones react with thiols to form thioacetals; desulfurization replaces the C—S bonds with C—H bonds.

- Aldehydes and ketones are oxidized by the conjugate base of a peroxyacid to carboxylic acids and esters, respectively.

- An aldehyde or a ketone reacts with a phosphonium ylide in a **Wittig reaction** to form an alkene. A Wittig reaction is a concerted [2 + 2] cycloaddition reaction.

- A useful step in a retrosynthetic analysis is a **disconnection**—breaking a bond to produce two fragments. **Synthons** are the fragments produced by a disconnection. A **synthetic equivalent** is the reagent used as the source of a synthon.

- Nucleophilic addition to the carbonyl carbon of an α,β-unsaturated aldehyde or ketone is called **direct addition;** addition to the β-carbon is called **conjugate addition.**

- Whether direct or conjugate addition occurs depends on the nature of the nucleophile and the structure of the carbonyl compound.

- Nucleophiles that are weak bases—namely, halide ions, cyanide ion, thiols, alcohols, and amines—form conjugate addition products.

- Nucleophiles that are strong bases—namely, hydride ion and carbanions—form direct addition products with reactive carbonyl groups and conjugate addition products with less reactive (sterically hindered) carbonyl groups.

- Grignard reagents form direct addition products. Organocuprates form conjugate addition products.

- Nucleophiles form nucleophilic addition–elimination products with α,β-unsaturated carboxylic acid derivatives that have reactive carbonyl groups and form conjugate addition products with compounds that have less reactive carbonyl groups.

SUMMARY OF REACTIONS

1. Reactions of *carbonyl compounds* with Grignard reagents (Section 4)
   a. Reaction of *formaldehyde* with a Grignard reagent forms a primary alcohol.

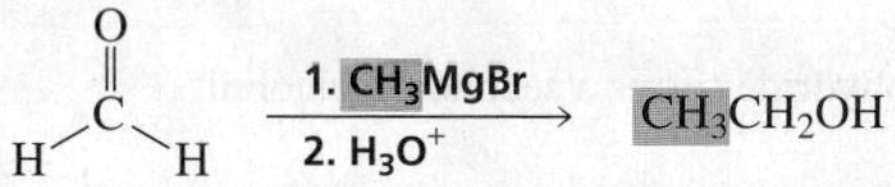

   b. Reaction of an *aldehyde* (other than formaldehyde) with a Grignard reagent forms a secondary alcohol.

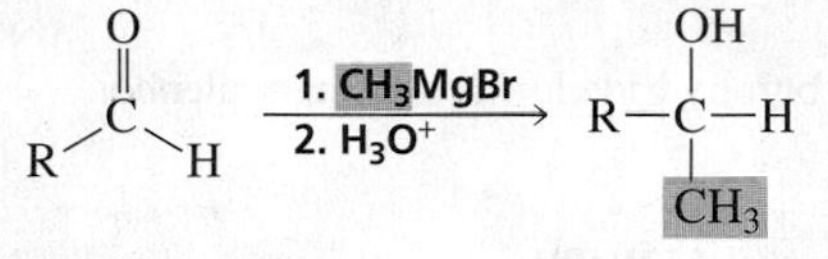

   c. Reaction of a *ketone* with a Grignard reagent forms a tertiary alcohol.

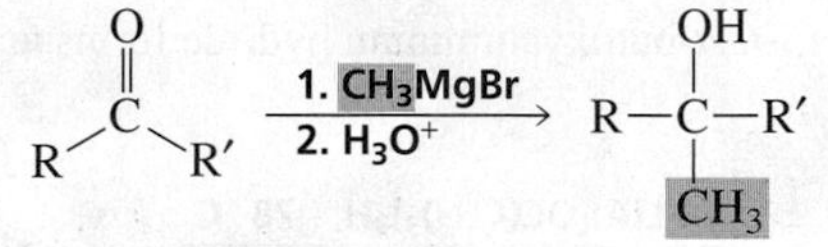

   d. Reaction of CO_2 with a Grignard reagent forms a carboxylic acid.

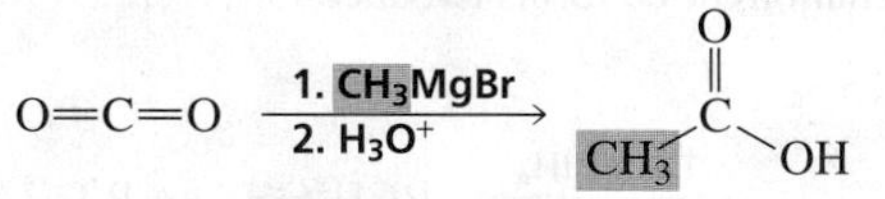

   e. Reaction of an *ester* with excess Grignard reagent forms a tertiary alcohol with two identical subunits.

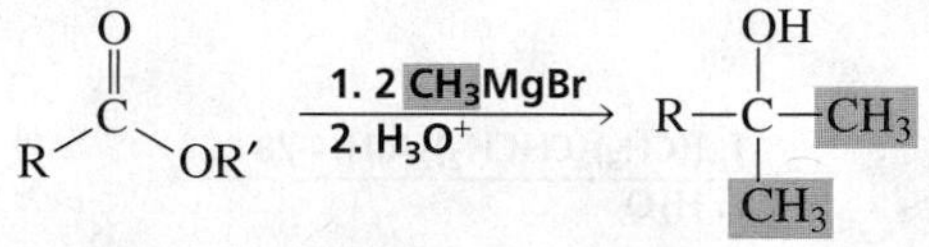

   f. Reaction of an *acyl chloride* with excess Grignard reagent forms a tertiary alcohol with two identical substituents.

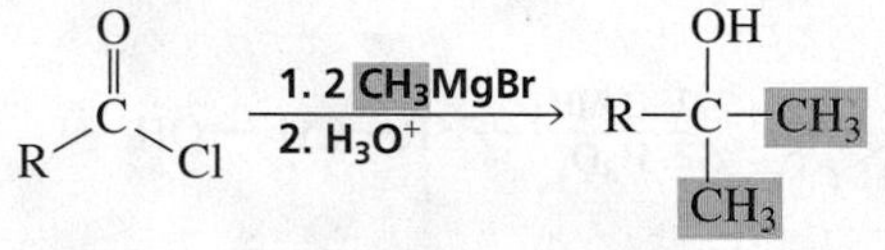

2. Reaction of *carbonyl compounds* with acetylide ions (Section 5).

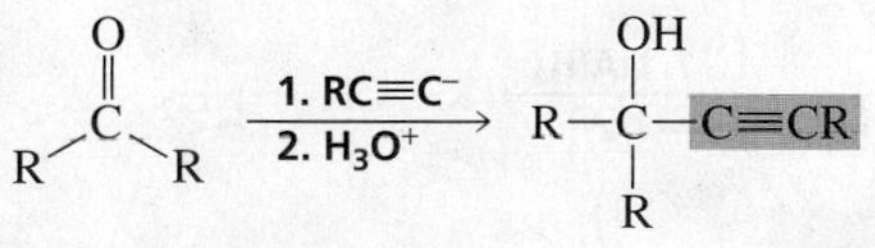

3. Reaction of an *aldehyde* or a *ketone* with cyanide ion under acidic conditions forms a cyanohydrin (Section 6).

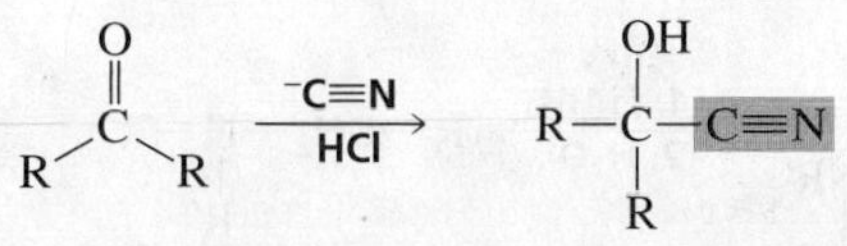

4. Reactions of *carbonyl compounds* with hydride ion donors (Section 7)

    a. Reaction of an *aldehyde* with sodium borohydride forms a primary alcohol.

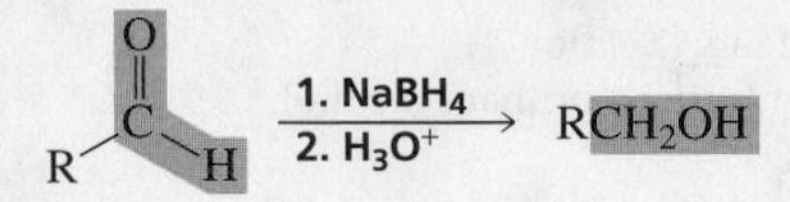

    b. Reaction of a *ketone* with sodium borohydride forms a secondary alcohol.

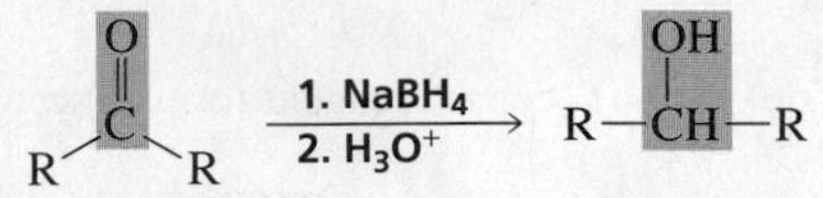

    c. Reaction of an *acyl chloride* with sodium borohydride forms a primary alcohol.

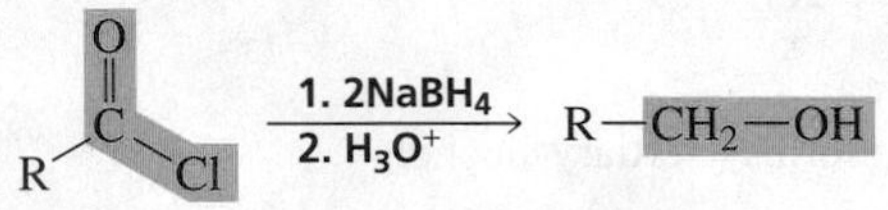

    d. Reaction of an *acyl chloride* with lithium tri-*tert*-butoxyaluminum hydride forms an aldehyde.

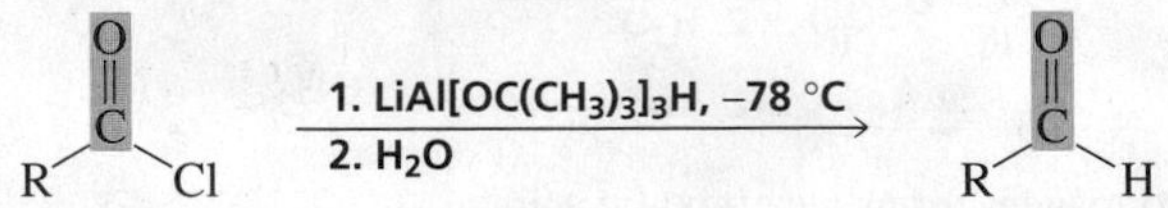

    e. Reaction of an *ester* with lithium aluminum hydride forms two alcohols.

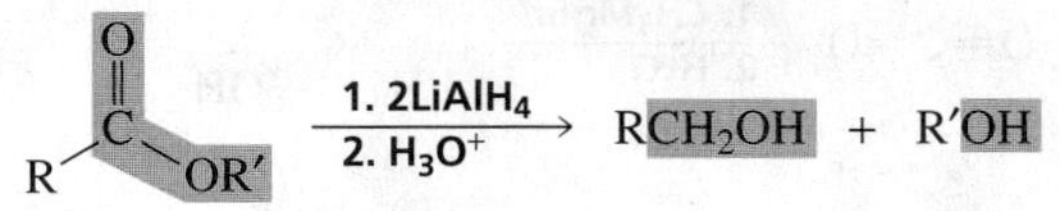

    f. Reaction of an *ester* with diisobutylaluminum hydride (DIBALH) forms an aldehyde.

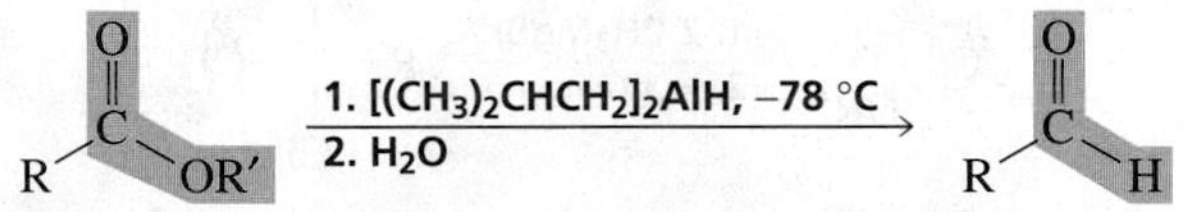

    g. Reaction of a *carboxylic acid* with lithium aluminum hydride forms a primary alcohol.

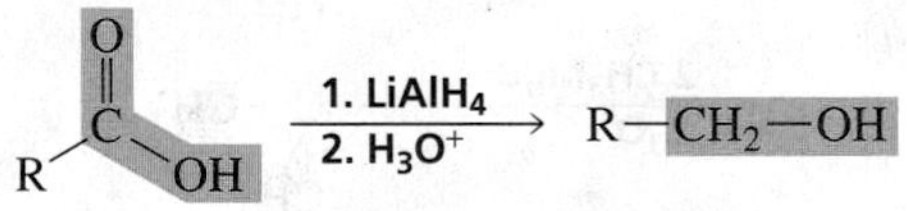

    h. Reaction of an *amide* with lithium aluminum hydride forms an amine.

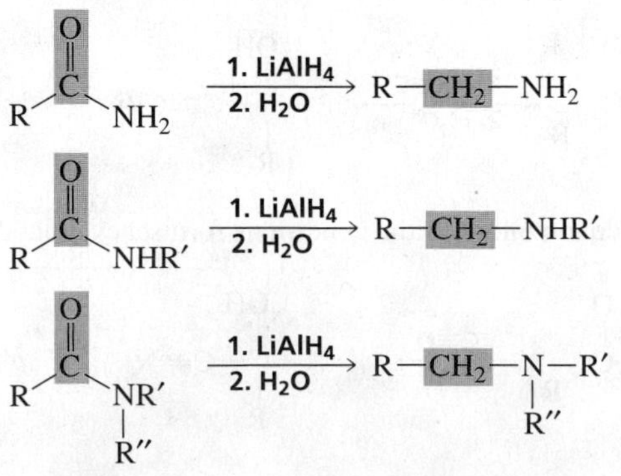

5. Reactions of *aldehydes* and *ketones* with amines and amine derivatives (Section 10)

    a. Reaction with a *primary amine* forms an imine.

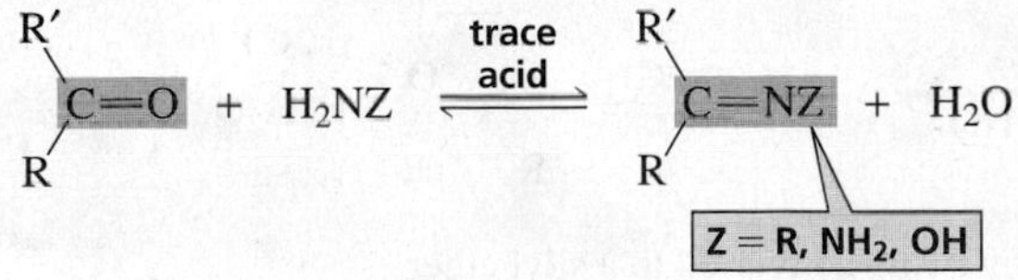

    b. Reaction with a *secondary amine* forms an enamine.

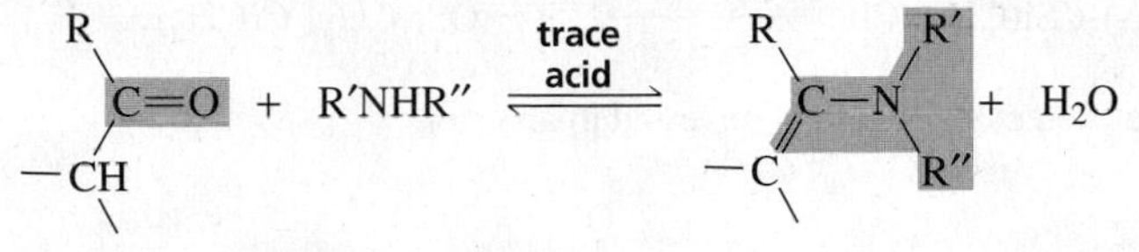

6. *Reductive amination:* the imines and enamines formed from the reaction of aldehydes and ketones with ammonia and primary and secondary amines are reduced to primary, secondary, and tertiary amines (Section 10).

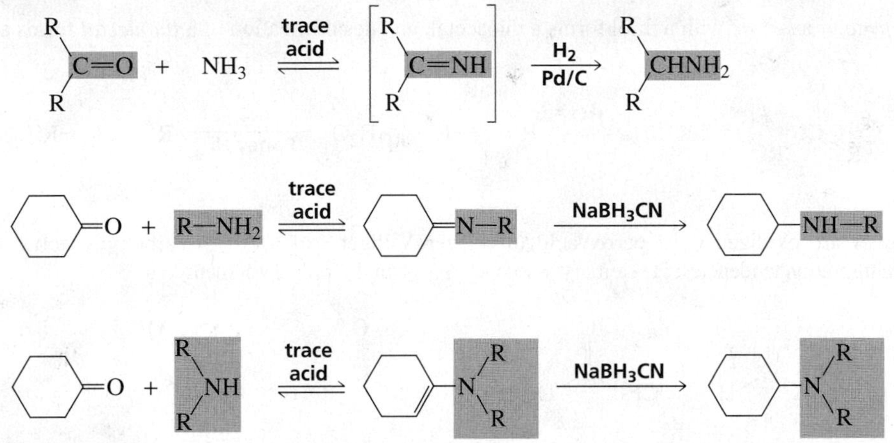

7. Reaction of an *aldehyde* or a *ketone* with water forms a hydrate (Section 11).

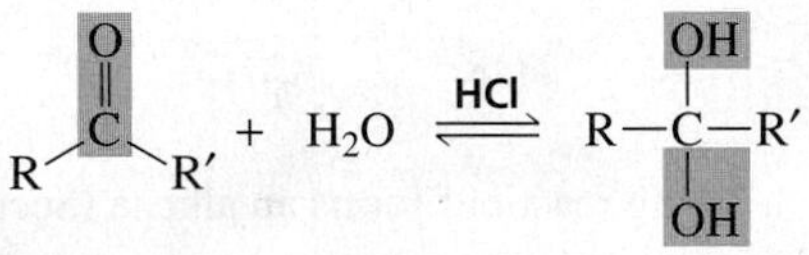

8. Reaction of an *aldehyde* or a *ketone* with excess alcohol forms first a hemiacetal and then an acetal (Section 12).

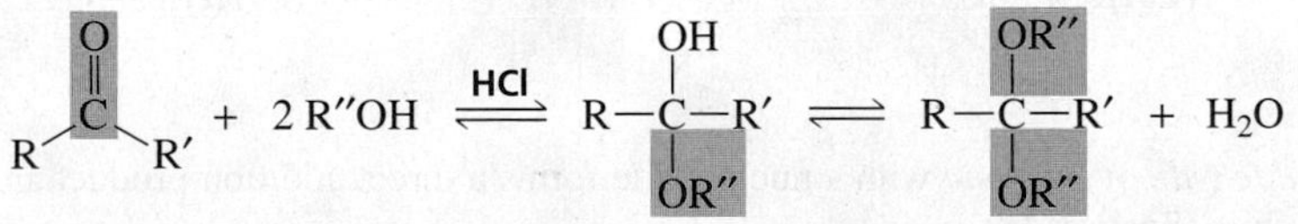

9. Protecting groups (Section 13)

a. *Aldehydes* and *ketones* can be protected by being converted to acetals.

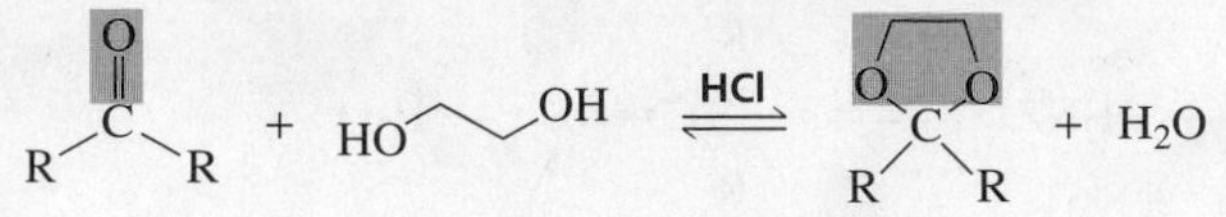

b. The OH group of an *alcohol* can be protected by being converted to a silyl ether.

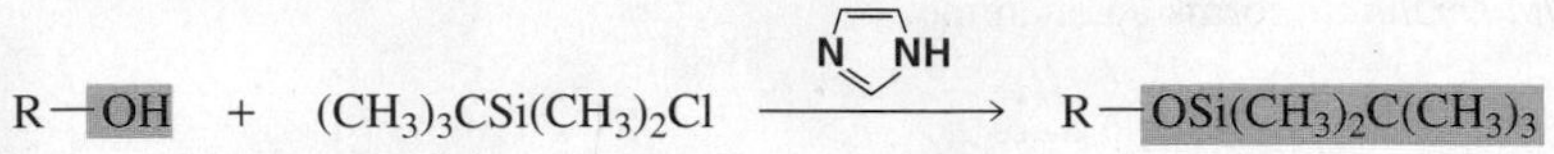

$$R-OH \ + \ (CH_3)_3CSi(CH_3)_2Cl \ \xrightarrow{\overset{NNH}{}} \ R-OSi(CH_3)_2C(CH_3)_3$$

c. The OH group of a *carboxylic acid* can be protected by being converted to an ester.

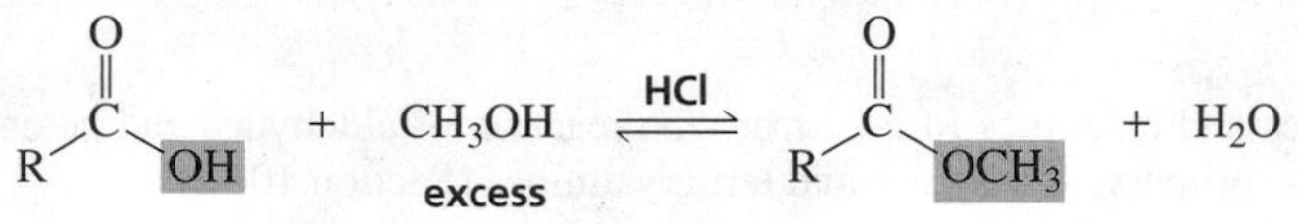

10. Reaction of an *aldehyde* or a *ketone* with a thiol forms a thioacetal, and desulfurization of a *thioacetal* forms an alkane (Section 14).

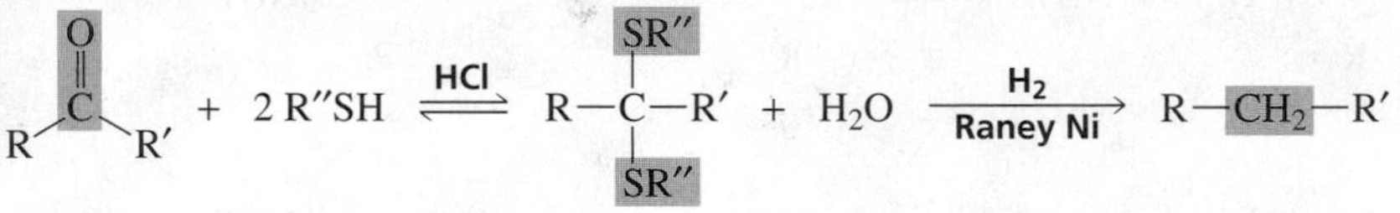

11. *Aldehydes* and *ketones* are oxidized by a peroxyacid (a Baeyer–Villiger oxidation) to carboxylic acids and esters, respectively (Section 15). Relative migration tendencies: H > tertiary > secondary ~ phenyl > primary > methyl

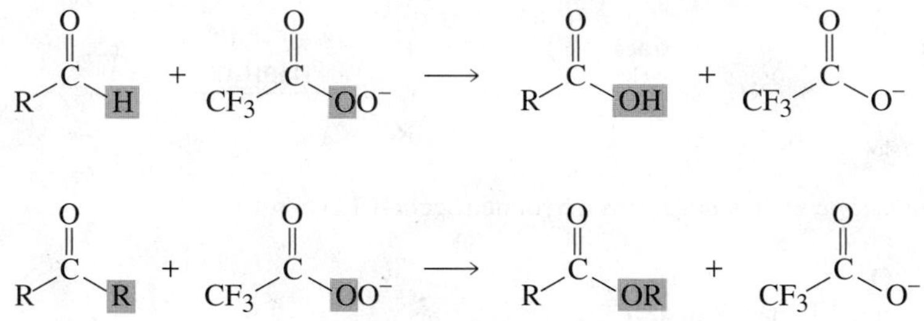

12. Reaction of an *aldehyde* or a *ketone* with a phosphonium ylide (a Wittig reaction) forms an alkene (Section 16).

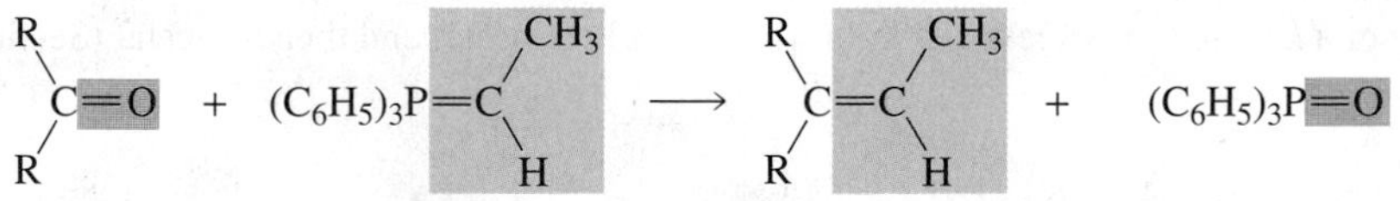

13. Reaction of an α,β-*unsaturated aldehyde* or a *ketone* with a nucleophile forms a direct addition product and/or a conjugate addition product, depending on the nucleophile (Section 18).

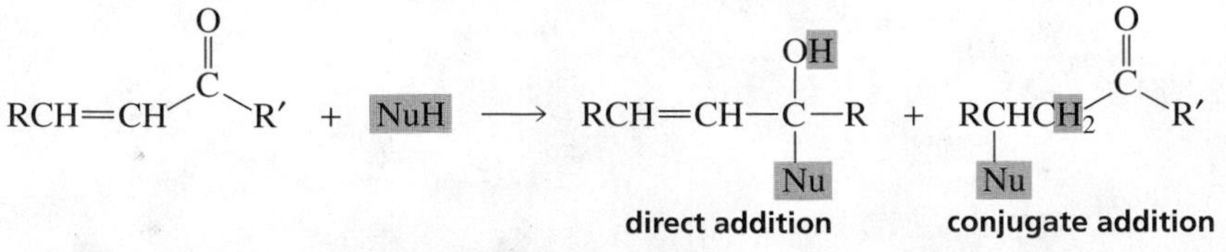

Nucleophiles that are weak bases ($^-C\equiv N$, RSH, RNH_2, Br^-) form conjugate addition products. Nucleophiles that are strong bases (RLi, RMgBr, H^-) form direct addition products with reactive carbonyl groups and conjugate addition products with less reactive carbonyl groups. Organocuprates (R_2CuLi) form conjugate addition products.

14. Reaction of an α,β-*unsaturated carboxylic acid derivative* with a nucleophile forms a nucleophilic addition–elimination product with a reactive carbonyl group and a conjugate addition product with a less reactive carbonyl group (Section 19).

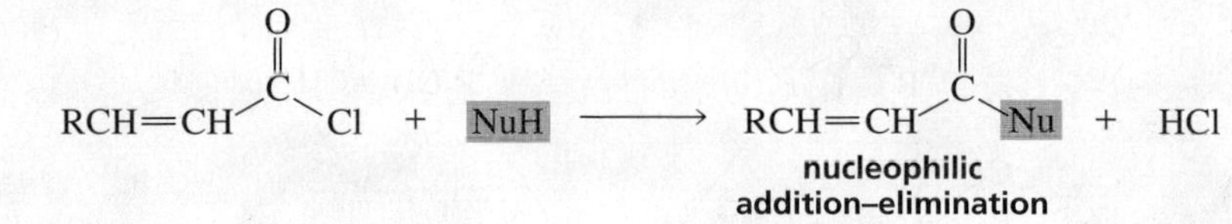

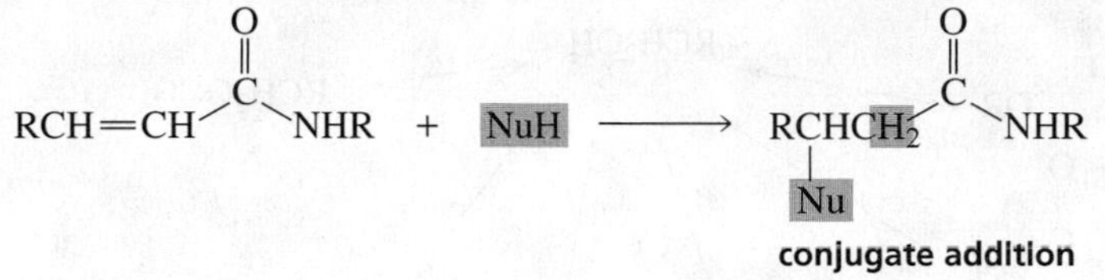

$$RCH{=}CH{-}\overset{\overset{\displaystyle O}{\|}}{C}{-}Cl \ + \ \boxed{NuH} \ \longrightarrow \ RCH{=}CH{-}\overset{\overset{\displaystyle O}{\|}}{C}{-}\boxed{Nu} \ + \ HCl$$

**nucleophilic
addition–elimination**

$$RCH{=}CH{-}\overset{\overset{\displaystyle O}{\|}}{C}{-}NHR \ + \ \boxed{NuH} \ \longrightarrow \ RCHCH_2{-}\overset{\overset{\displaystyle O}{\|}}{C}{-}NHR$$

$$\boxed{Nu}$$

conjugate addition

PROBLEMS

53. Draw the structure for each of the following:

a. isobutyraldehyde

b. 4-hexenal

c. diisopentyl ketone

d. 3-methylcyclohexanone

e. 2,4-pentanedione

f. 4-bromo-3-heptanone

g. γ-bromocaproaldehyde

h. 2-ethylcyclopentanecarbaldehyde

i. 4-methyl-5-oxohexanal

54. What are the products of the following reactions?

a. 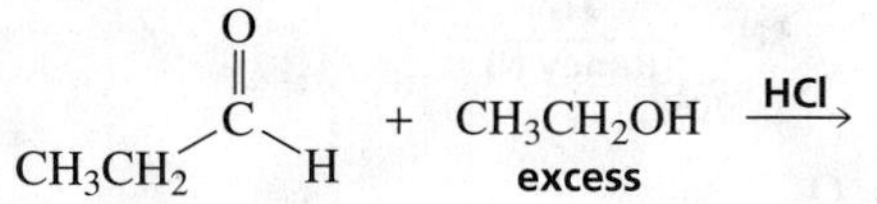 $CH_3CH_2\overset{\overset{\displaystyle O}{\|}}{C}H \ + \ CH_3CH_2OH \ \overset{HCl}{\longrightarrow} \ $ **excess**

b. (phenyl)$\overset{\overset{\displaystyle O}{\|}}{C}CH_2CH_3 \ + \ NH_2NH_2 \ \overset{\text{trace acid}}{\longrightarrow}$

c. $CH_3CH_2\overset{\overset{\displaystyle O}{\|}}{C}CH_3 \ \overset{\text{1. NaBH}_4}{\underset{\text{2. H}_3O^+}{\longrightarrow}}$

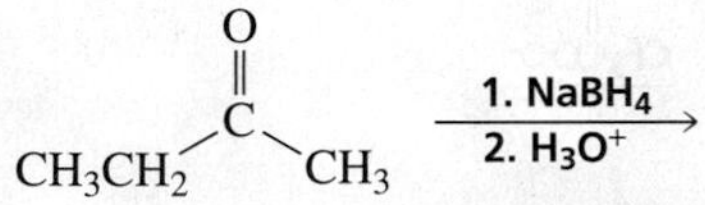

d. 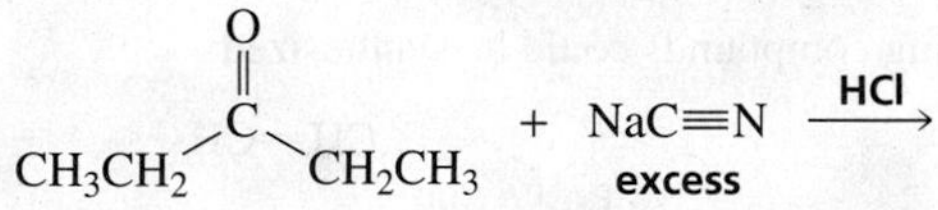 $CH_3CH_2\overset{\overset{\displaystyle O}{\|}}{C}CH_2CH_3 \ + \ NaC{\equiv}N \ \overset{HCl}{\longrightarrow}$ **excess**

e. 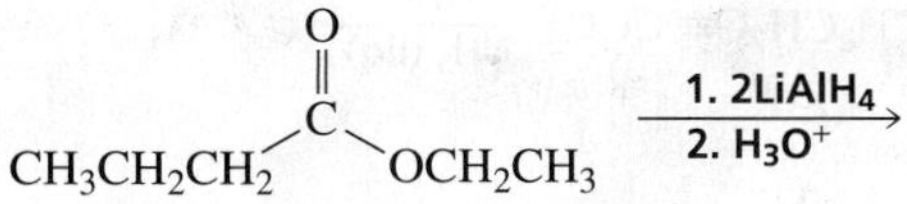 $CH_3CH_2CH_2\overset{\overset{\displaystyle O}{\|}}{C}OCH_2CH_3 \ \overset{\text{1. 2LiAlH}_4}{\underset{\text{2. H}_3O^+}{\longrightarrow}}$

f. 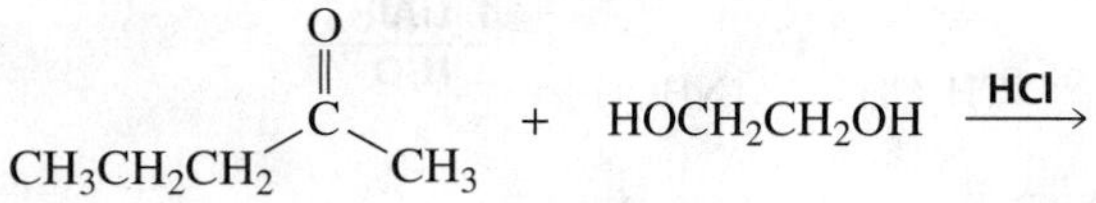 $CH_3CH_2CH_2\overset{\overset{\displaystyle O}{\|}}{C}CH_3 \ + \ HOCH_2CH_2OH \ \overset{HCl}{\longrightarrow}$

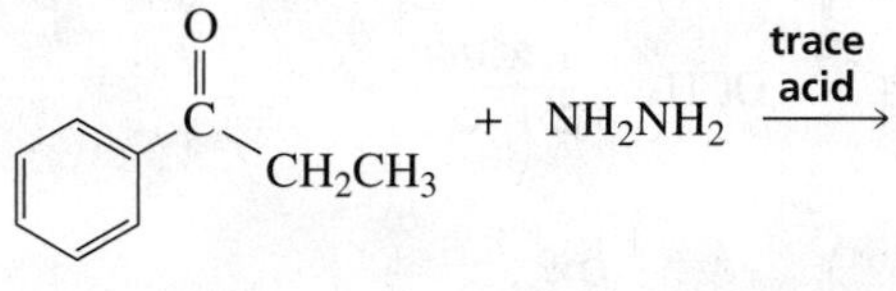

g. 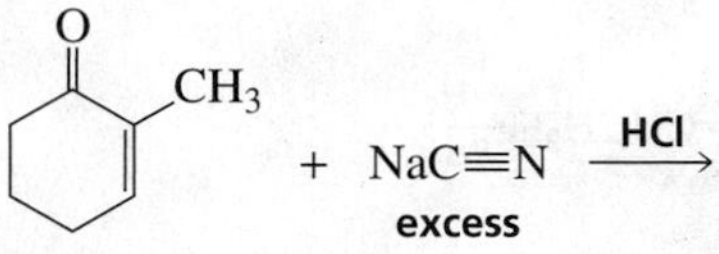 (2-methylcyclohex-2-enone) $\ + \ NaC{\equiv}N \ \overset{HCl}{\longrightarrow}$ **excess**

55. List the following compounds in order from most reactive to least reactive toward nucleophilic addition:

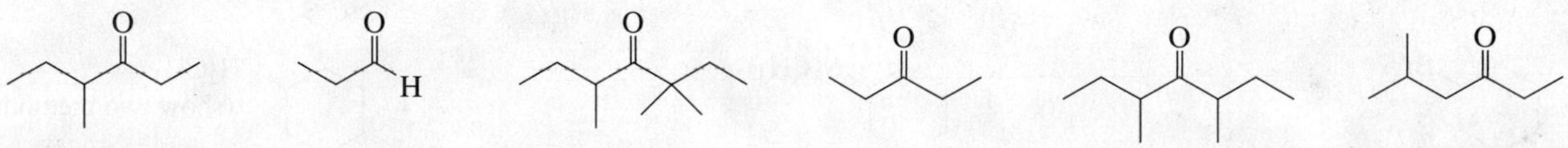

56. a. Show the reagents required to form the primary alcohol in each of the following reactions:

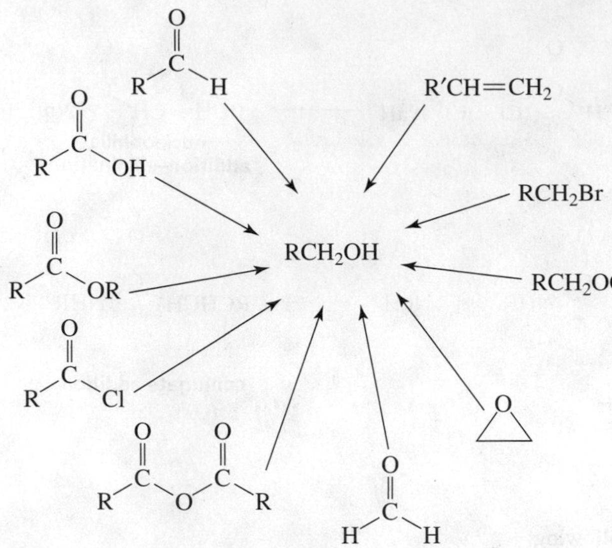

b. Which of the reactions cannot be used for the synthesis of isobutyl alcohol?

57. Draw the products of the following reactions. Indicate whether each reaction is an oxidation or a reduction.

a. $CH_3CH_2CH_2C{\equiv}CCH_3$ $\xrightarrow[\text{NH}_3 \text{ (liq)}]{\text{Na}}$

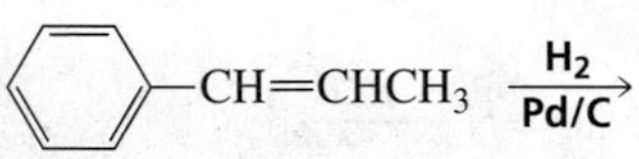

d. 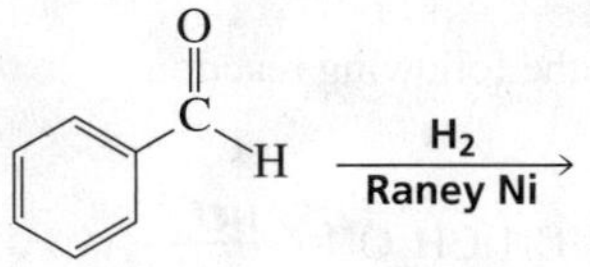 $\xrightarrow[\text{Raney Ni}]{\text{H}_2}$

b. $\xrightarrow[\text{2. H}_2\text{O}]{\text{1. LiAlH}_4}$

e. 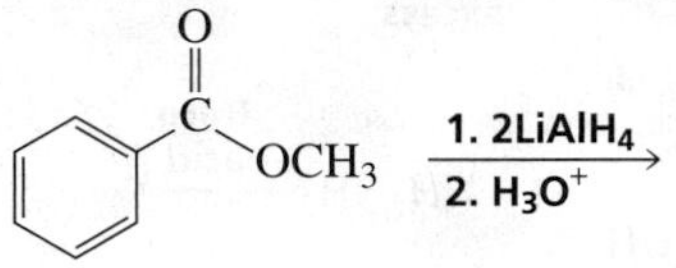 $\xrightarrow[\text{2. H}_3\text{O}^+]{\text{1. 2LiAlH}_4}$

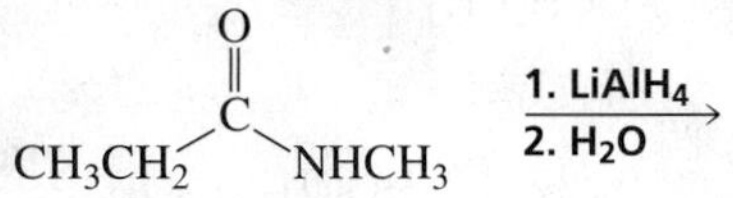

c. $\xrightarrow[\text{Pd/C}]{\text{H}_2}$

f. 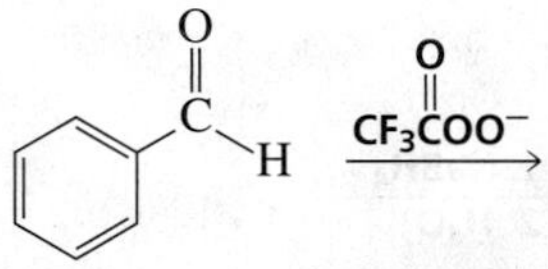 $\xrightarrow{\text{CF}_3\text{COO}^-}$

58. Using cyclohexanone as the starting material, describe how each of the following compounds could be synthesized:

a.

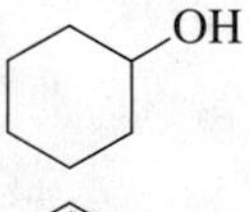

b.

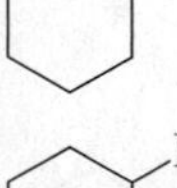

c.

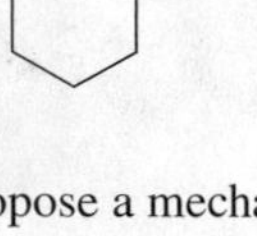

d.

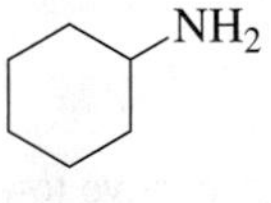

e.

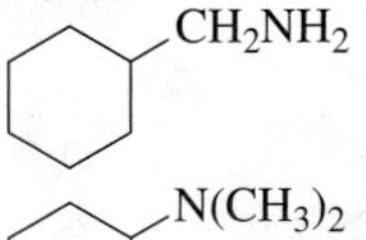

f.

g.

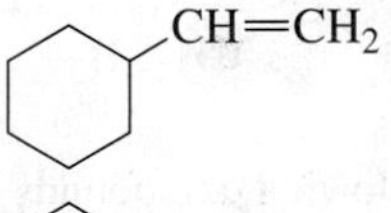

h.  (show two methods)

i. 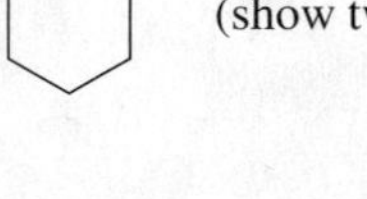 (show two methods)

59. Propose a mechanism for the following reaction:

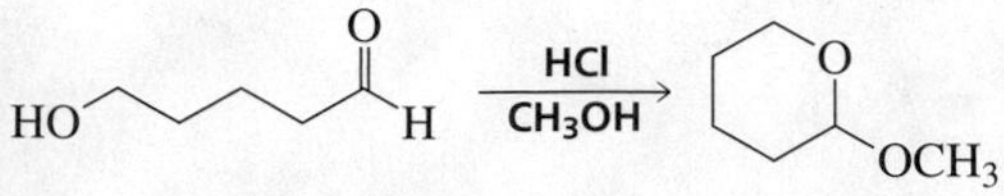

60. Show how each of the following compounds could be prepared, using the given starting material:

a. 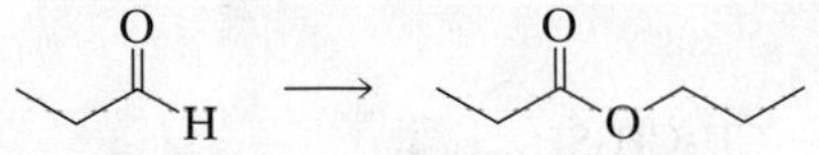

b. 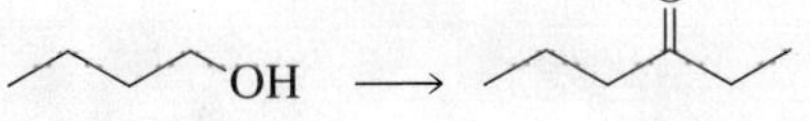

c. 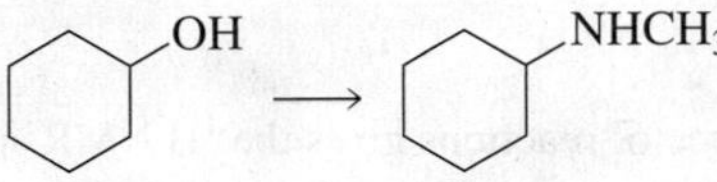

d. 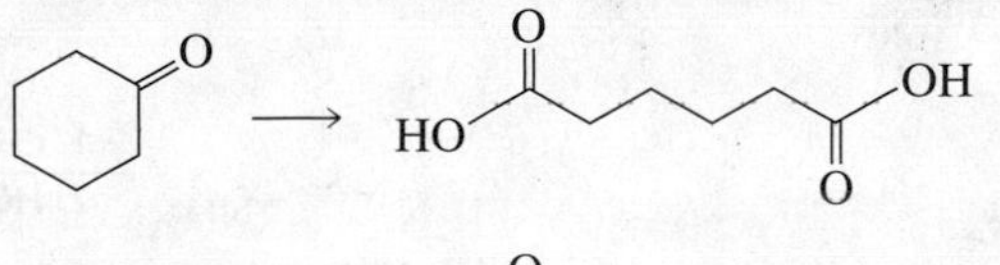

e. 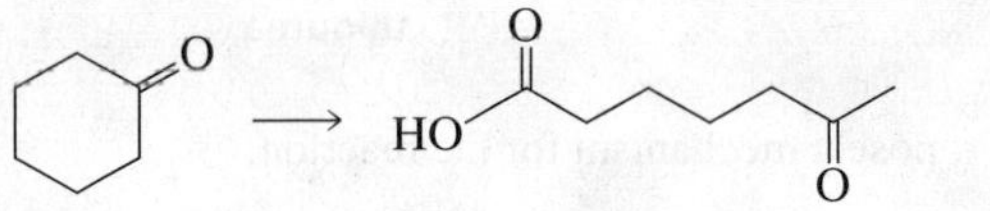

61. Fill in the boxes with the appropriate reagents:

a. CH_3OH —[ ]→ CH_3Br —[ ] / [ ]→ [ ] —1. [ ] 2. [ ]→ CH_3CH_2OH

b. CH_4 —[ ]→ CH_3Br —[ ] / [ ]→ [ ] —1. [ ] 2. [ ]→ $CH_3CH_2CH_2OH$

62. What are the products of the following reactions?

a. 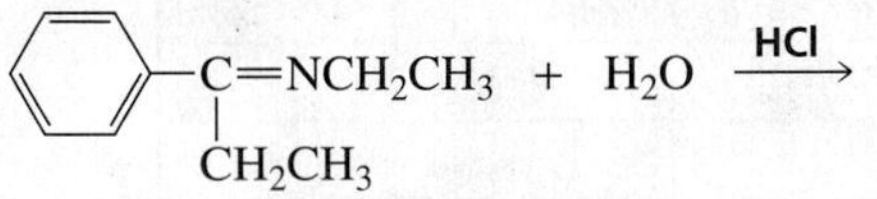

b. $CH_3CH_2C(=O)CH_3$ —1. CH_3CH_2MgBr 2. H_3O^+→

c. 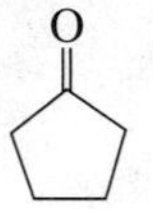 + $(C_6H_5)_3P{=}CHCH_3$ →

d. $CH_3CH_2C(=O)OCH_3$ —1. CH_3CH_2MgBr excess 2. H_3O^+→

e. 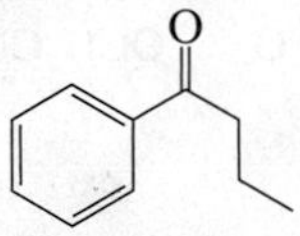 + CH_3OH —HCl→

f. 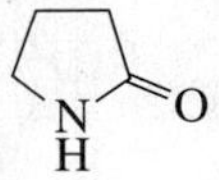 —1. $LiAlH_4$ 2. H_2O→

g. 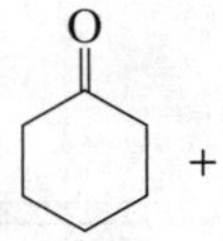 + $CH_3CH_2NH_2$ —trace acid→

h. 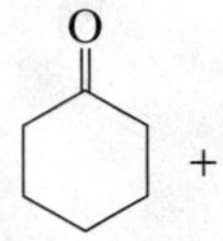 + $(CH_3CH_2)_2NH$ —trace acid→

i. 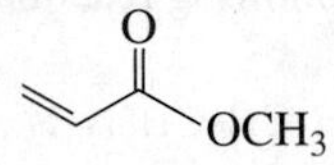 + CH_3NH_2 →

j. 2 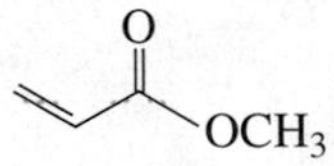 + CH_3NH_2 →

63. Identify **A** through **O**:

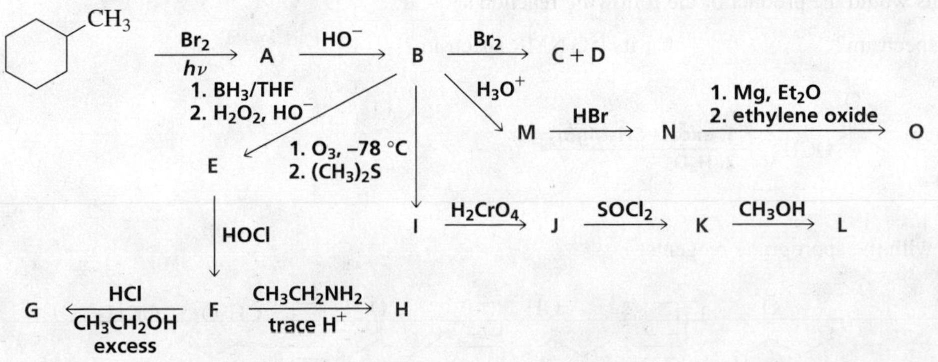

64. Thiols can be prepared from the reaction of thiourea with an alkyl halide, followed by hydroxide-ion-promoted hydrolysis.

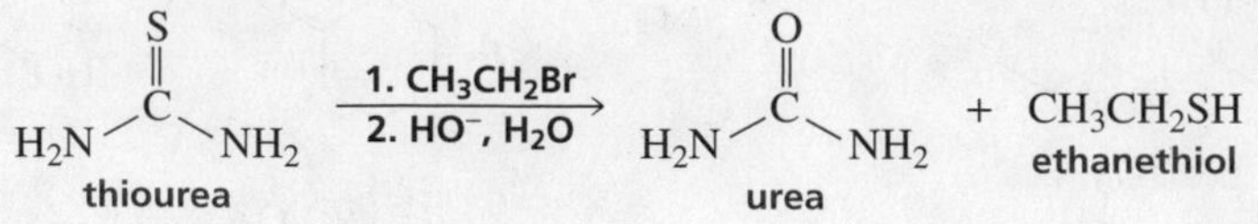

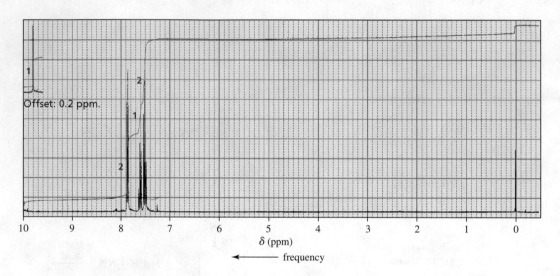

a. Propose a mechanism for the reaction.

b. What thiol would be formed if the alkyl halide employed were pentyl bromide?

65. The only organic compound obtained when compound **Z** undergoes the following sequence of reactions gives the ^{1}H NMR spectrum shown. Identify compound **Z**.

Compound Z $\xrightarrow{\substack{\textbf{1. phenylmagnesium bromide} \\ \textbf{2. H}_3\textbf{O}^+}}$ $\xrightarrow{\substack{\textbf{NaOCl} \\ \textbf{CH}_3\textbf{COOH} \\ \textbf{0 °C}}}$

Offset: 0.2 ppm.

δ (ppm)

← frequency

66. Propose a mechanism for each of the following reactions:

a. $\xrightarrow{\substack{\textbf{HCl} \\ \textbf{H}_2\textbf{O}}}$ $CH_2CH_2CH_2\overset{+}{N}H_3$

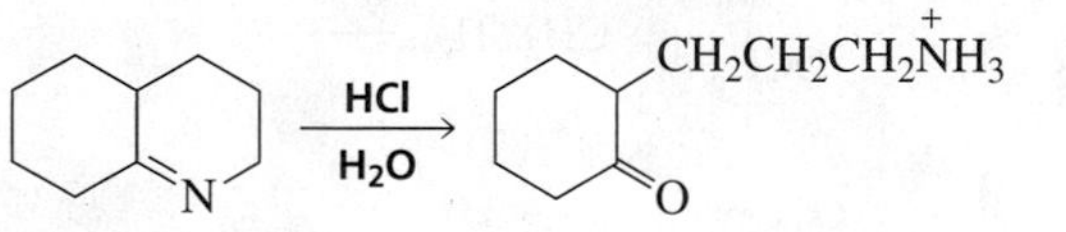

c. $+ CH_3CH_2OH \xrightarrow{\textbf{HCl}}$ OCH$_2$CH$_3$

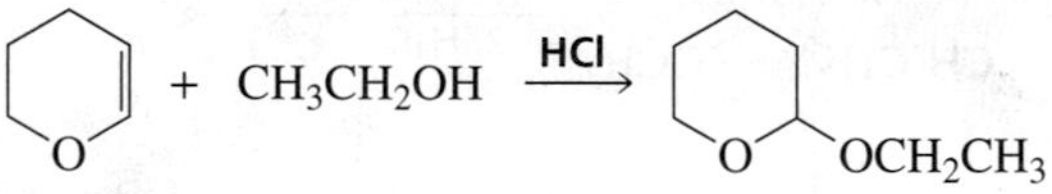

b. OCH$_3$ $\xrightarrow{\substack{\textbf{HCl} \\ \textbf{H}_2\textbf{O}}}$

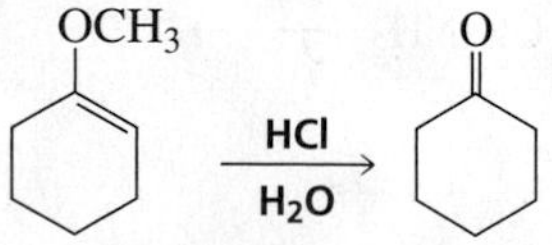

67. How many signals would the product of the following reaction show in

a. its ^{1}H NMR spectrum?   **b.** its ^{13}C NMR spectrum?

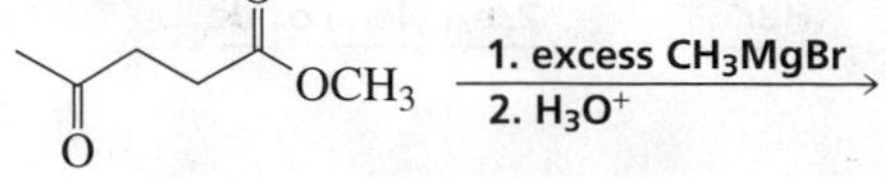

OCH$_3$ $\xrightarrow{\substack{\textbf{1. excess CH}_3\textbf{MgBr} \\ \textbf{2. H}_3\textbf{O}^+}}$

68. Fill in the boxes with the appropriate reagents:

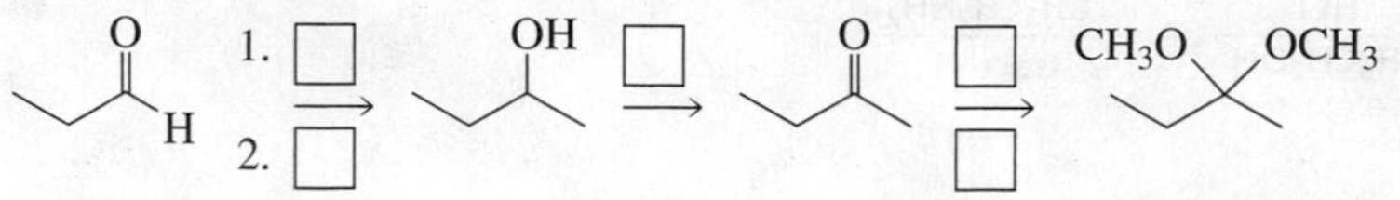

H $\overset{1.\ \square}{\underset{2.\ \square}{\Longrightarrow}}$ OH $\square \rightarrow$ O $\square$ $\overset{\square}{\underset{\square}{\Longrightarrow}}$ CH$_3$O OCH$_3$

69. How could you convert *N*-methylbenzamide into the following compounds?

a. *N*-methylbenzylamine **b.** benzoic acid **c.** methyl benzoate **d.** benzyl alcohol

70. What are the products of the following reactions? Show all stereoisomers that are formed.

a.
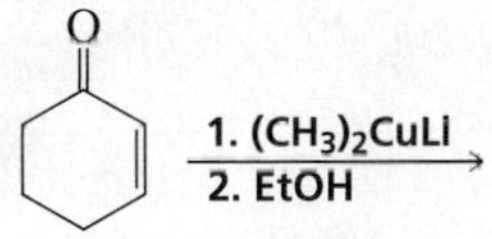
O
1. $(CH_3)_2CuLi$
2. EtOH

c.
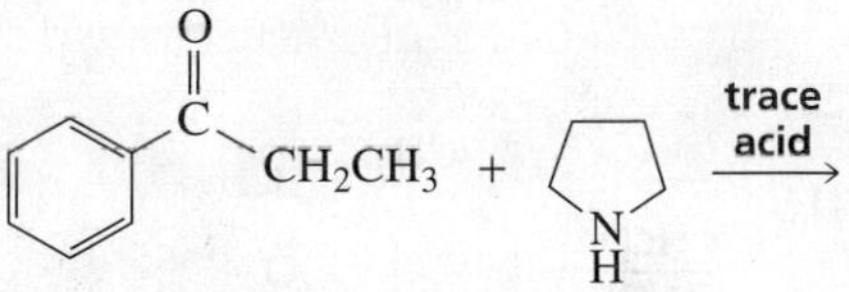
O
C
CH_2CH_3 +
N
H
trace
acid

b.
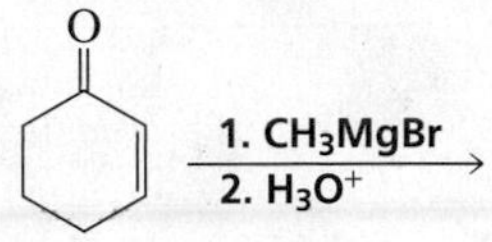
O
1. CH_3MgBr
2. H_3O^+

d.
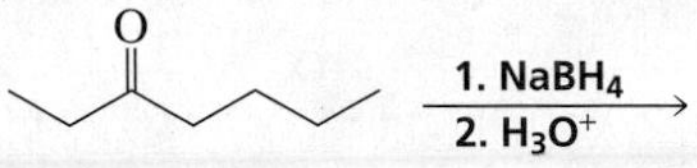
O
1. $NaBH_4$
2. H_3O^+

71. List three different sets of reagents (each set consisting of a carbonyl compound and a Grignard reagent) that could be used to prepare each of the following tertiary alcohols:

a.
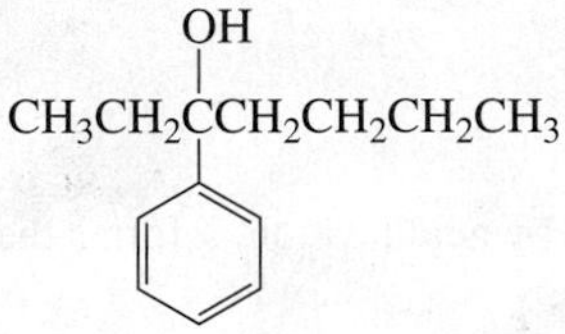
OH
$CH_3CH_2CCH_2CH_2CH_2CH_3$

b.
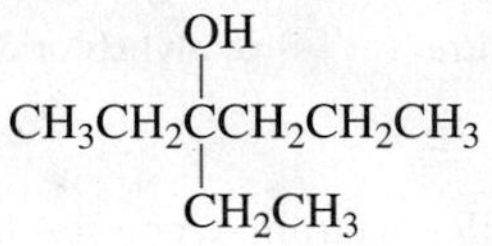
OH
$CH_3CH_2CCH_2CH_2CH_3$
CH_2CH_3

72. What product is formed when 3-methyl-2-cyclohexenone reacts with each of the following reagents?

a. CH_3MgBr followed by H_3O^+ **c.** HBr

b. $(CH_3CH_2)_2CuLi$ followed by H_3O^+ **d.** CH_3CH_2SH

73. Norlutin and Enovid are ketones that suppress ovulation, so they have been used clinically as contraceptives. For which of these compounds would you expect the infrared carbonyl absorption (C=O stretch) to be at a higher frequency? Explain.

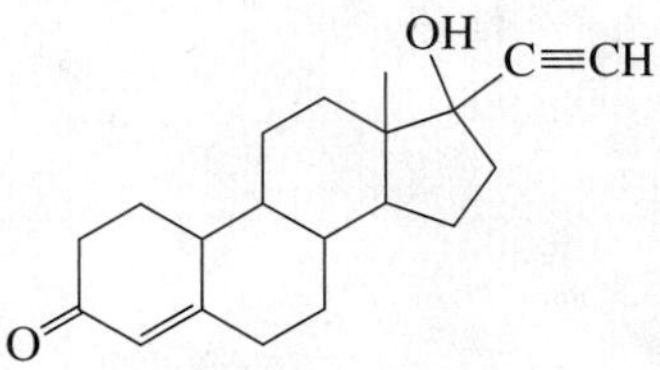
OH
C≡CH
O

Norlutin®

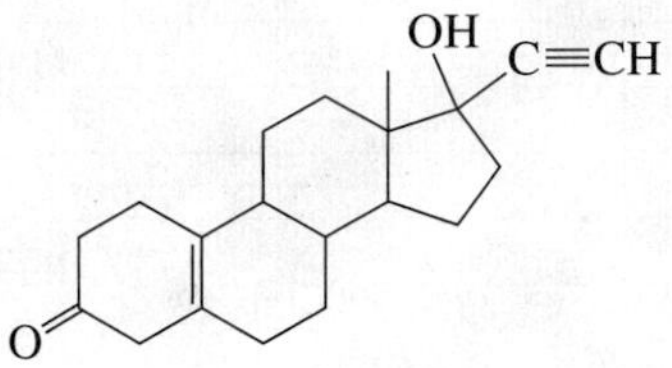
OH
C≡CH
O

Enovid®

74. What is the product of each of the following reactions?

a.
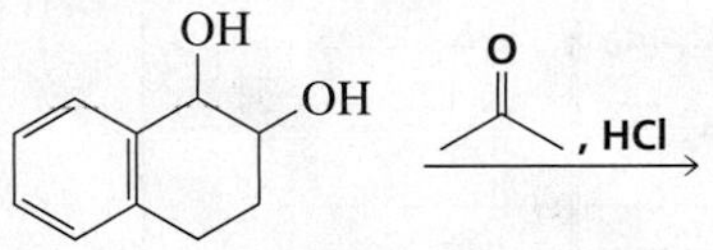
OH
OH
O
, HCl

b.
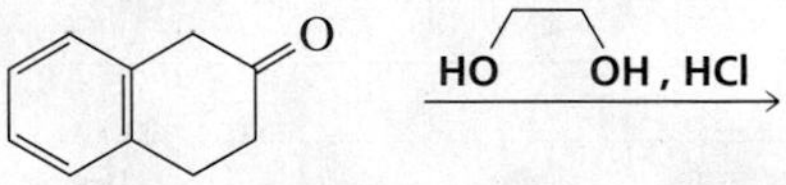
O
HO OH, HCl

75. What is the product of each of the following reactions?

a.
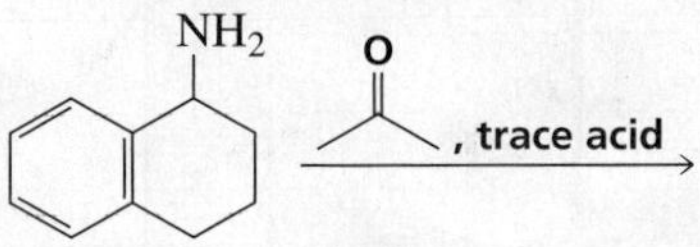
NH_2
O
, trace acid

c.
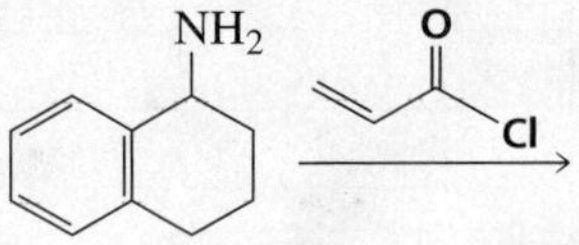
NH_2
O
Cl

b.
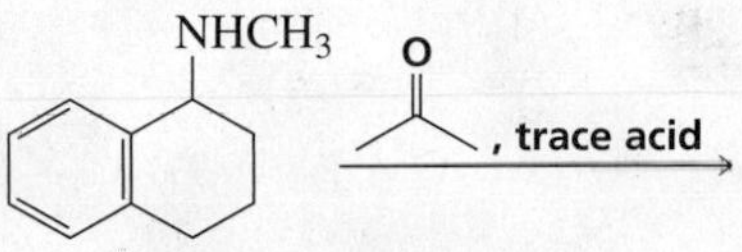
$NHCH_3$
O
, trace acid

d.
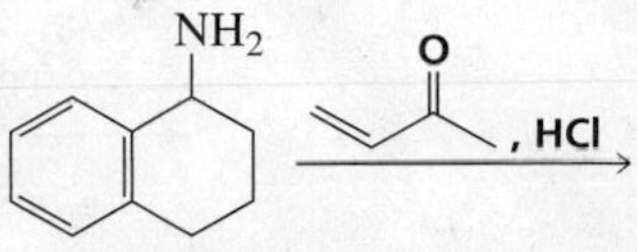
NH_2
O
, HCl

76. Propose a reasonable mechanism for each of the following reactions:

a.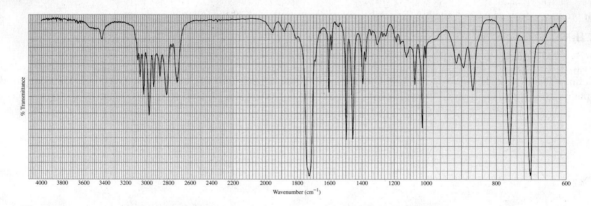

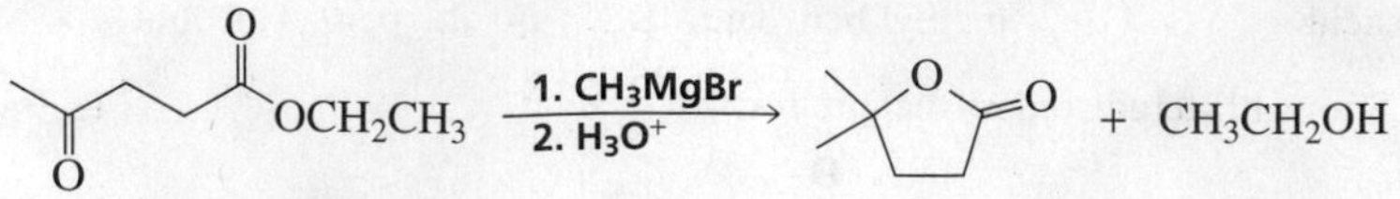

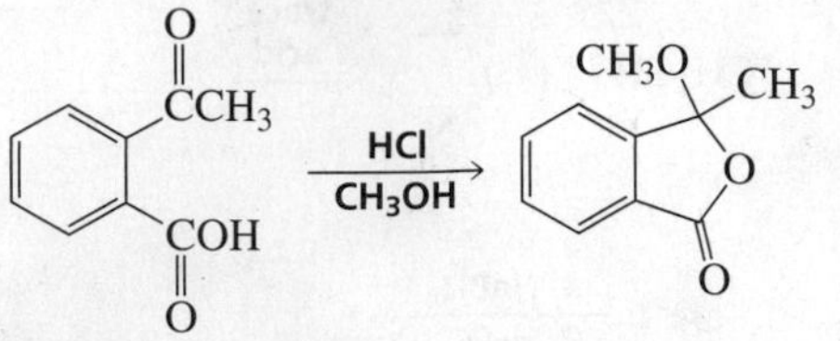

77. Propose a mechanism to explain how dimethyl sulfoxide and oxalyl chloride react to form the dimethylchlorosulfonium ion used as the oxidizing agent in the Swern oxidation.

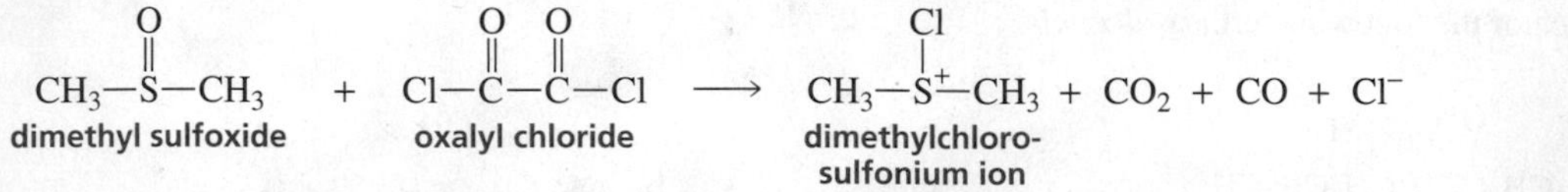

$$CH_3-\underset{\underset{O}{\|}}{S}-CH_3 \quad + \quad Cl-\underset{\underset{O}{\|}}{C}-\underset{\underset{O}{\|}}{C}-Cl \quad \longrightarrow \quad CH_3-\underset{\overset{+}{S}}{\underset{|}{\overset{Cl}{}}}-CH_3 + CO_2 + CO + Cl^-$$

dimethyl sulfoxide · · · · · · · · oxalyl chloride · · · · · · · · dimethylchlorosulfonium ion

78. A compound gives the following IR spectrum. Upon reaction with sodium borohydride followed by acidification, it forms the product with the ^1H NMR spectrum shown here. Identify the starting material and the product.

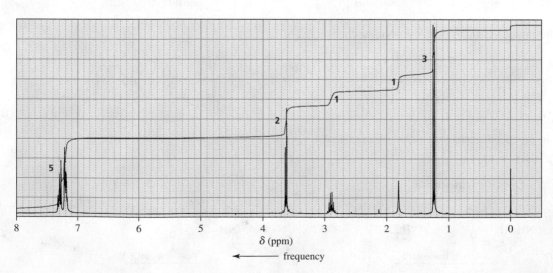

79. a. Propose a mechanism for the following reaction:

$$CH_2=CHCHC\equiv N \quad \xrightarrow[\text{H}_2\text{O}]{\text{HO}^-} \quad N\equiv CCH_2CH_2CH$$

(with OH on the third carbon of the starting material and O double-bonded at the end of the product)

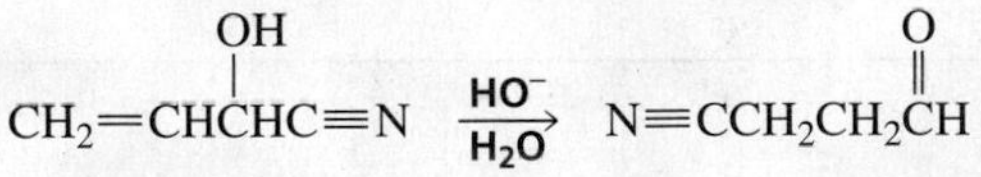

b. What is the product of the following reaction?

$$CH_3\overset{\overset{\displaystyle OH}{|}}{\underset{\underset{\displaystyle C\equiv N}{|}}{C}}CH=CH_2 \quad \xrightarrow[\text{H}_2\text{O}]{\text{HO}^-}$$

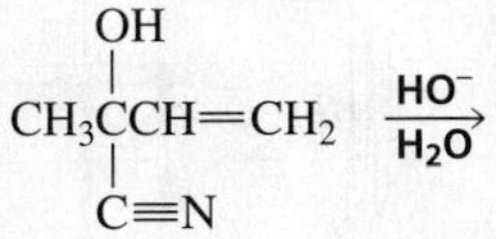

80. Unlike a phosphonium ylide, which reacts with an aldehyde or a ketone to form an alkene, a sulfonium ylide reacts with an aldehyde or a ketone to form an epoxide. Explain why one ylide forms an alkene, whereas the other forms an epoxide.

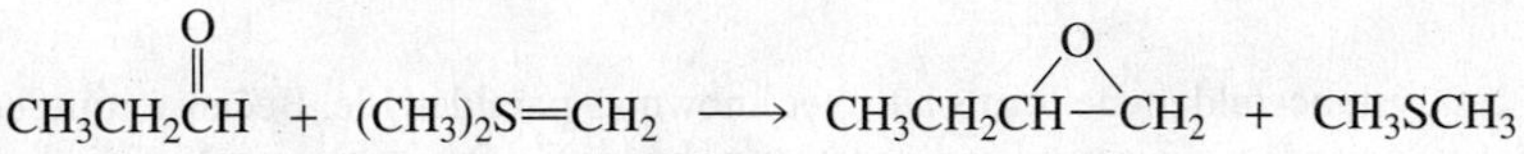

$$CH_3CH_2\overset{\overset{\displaystyle O}{\|}}{C}H \; + \; (CH_3)_2S=CH_2 \; \longrightarrow \; CH_3CH_2CH\overset{O}{\diagdown}CH_2 \; + \; CH_3SCH_3$$

81. Indicate how the following compounds could be prepared from the given starting materials:

a.

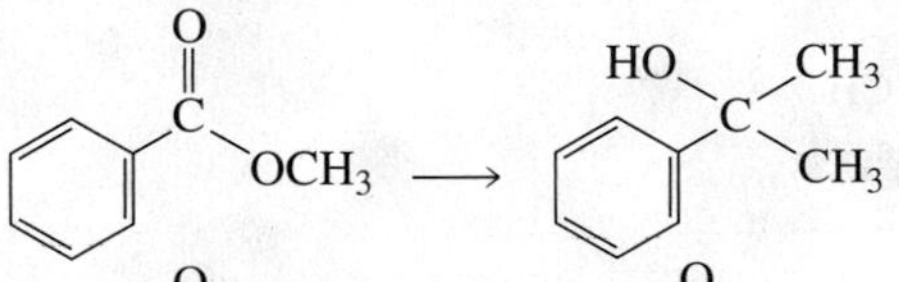

b.

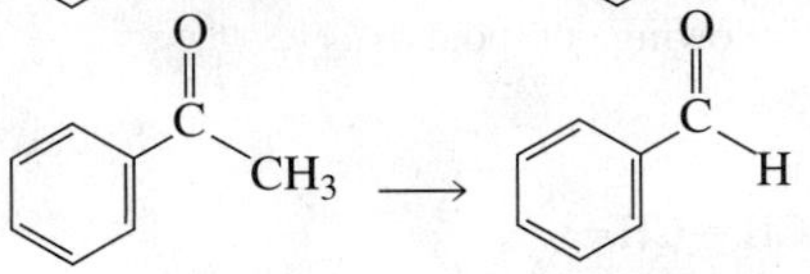

c.

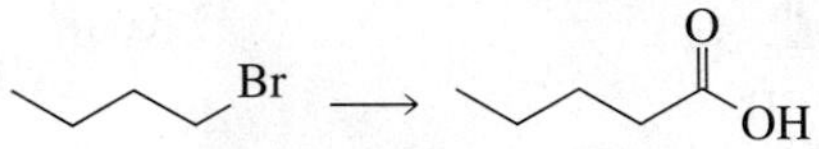

d.

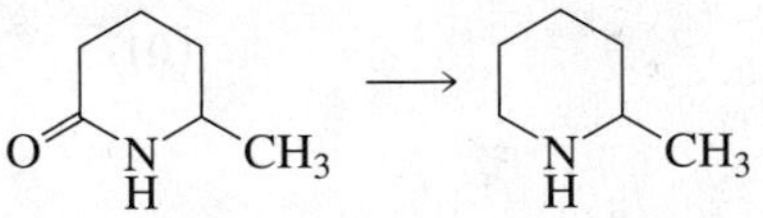

e.

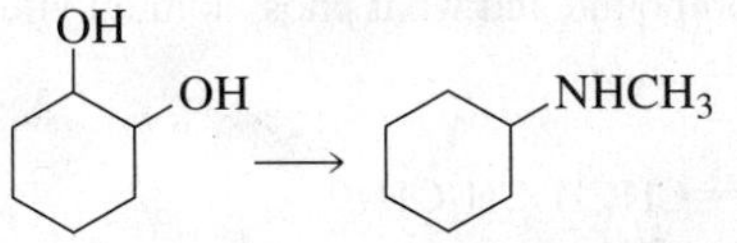

82. Propose a reasonable mechanism for each of the following reactions:

a.

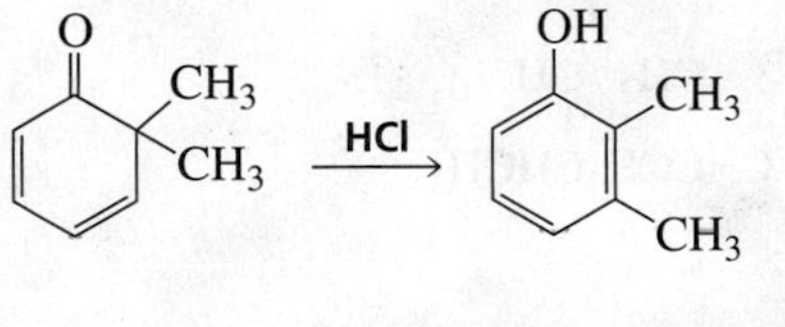

b.

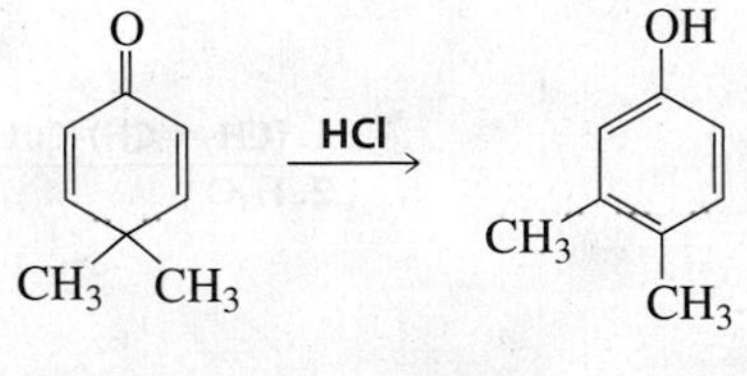

83. a. In an aqueous solution, D-glucose exists in equilibrium with two six-membered ring compounds. Draw the structures of these compounds.

b. Which of the six-membered ring compounds will be present in greater amount?

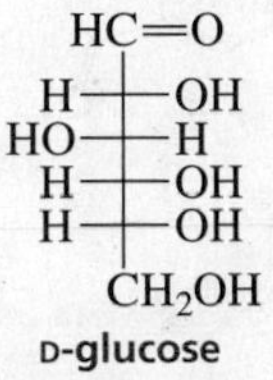

$$\begin{array}{c}
HC=O \\
H-\!\!\!-OH \\
HO-\!\!\!-H \\
H-\!\!\!-OH \\
H-\!\!\!-OH \\
CH_2OH
\end{array}$$

D-glucose

84. Shown here is the ^{1}H NMR spectrum of the alkyl bromide used to make the phosphonium ylide that reacts with a ketone in a Wittig reaction to form a compound with molecular formula $C_{11}H_{14}$. What product is obtained from the Wittig reaction?

3.6 3.5 3.4 PPM
3.2 3.1 3.0 PPM
10 9 8 7 6 5 4 3 2 1 0
δ (ppm)
frequency

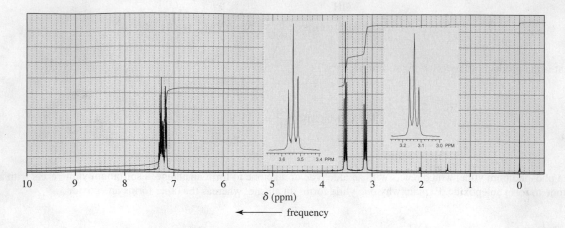

85. In the presence of an acid catalyst, acetaldehyde forms a trimer known as paraldehyde. Because it induces sleep when it is administered to animals in large doses, paraldehyde is used as a sedative or hypnotic. Propose a mechanism for the formation of paraldehyde.

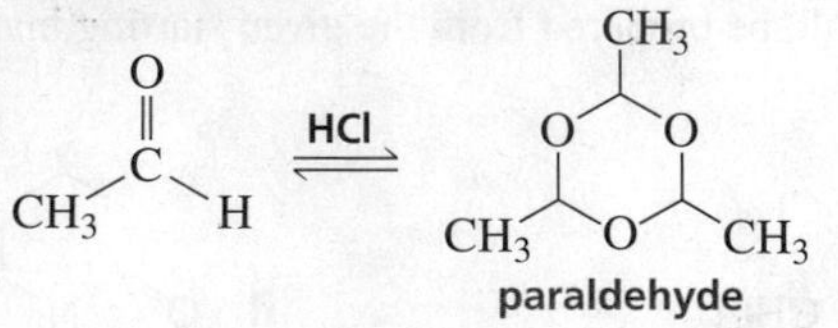

HCl
paraldehyde

86. What carbonyl compound and what phosphonium ylide are needed to synthesize the following compounds?

a.
CH=CHCH2CH2CH3

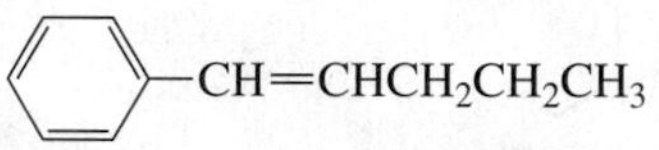

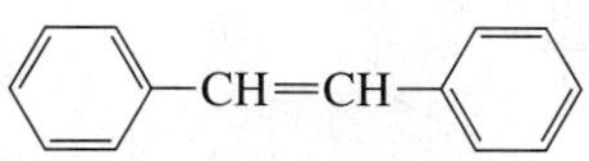

c.
CH=CH

b.
CHCH2CH3

d.
=CH2

87. Identify compounds **A** and **B**:

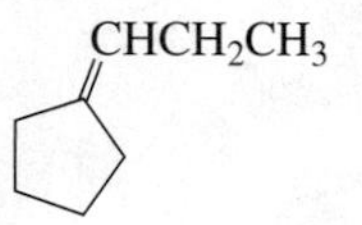

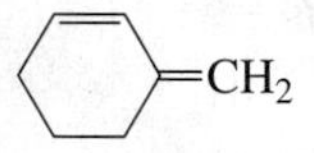

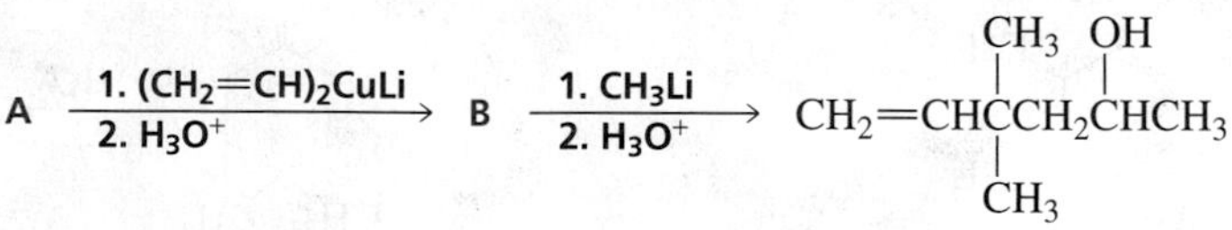

A
1. (CH2=CH)2CuLi
2. H3O+
B
1. CH3Li
2. H3O+
CH3 OH
CH2=CHCCH2CHCH3
CH3

88. Propose a reasonable mechanism for each of the following reactions:

a.
CH3O
HCl
Δ
CH3O

b.
NH
OH
+
trace
acid
N
OH

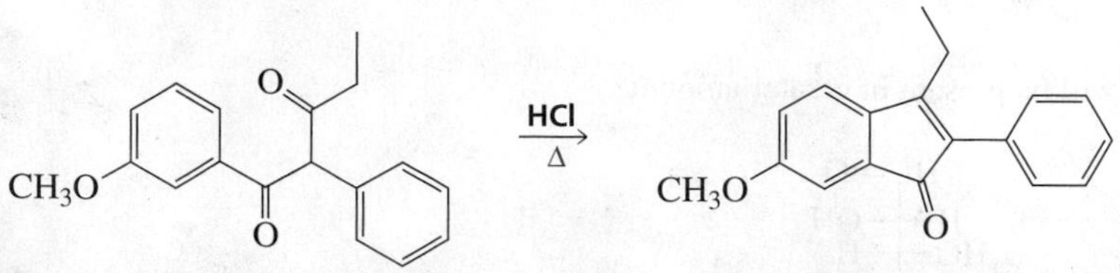

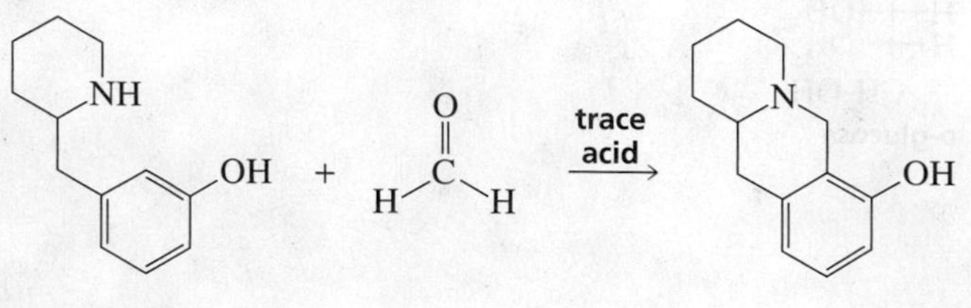

89. A compound reacts with methylmagnesium bromide followed by acidification to form the product with the following ^{1}H NMR spectrum. Identify the compound.

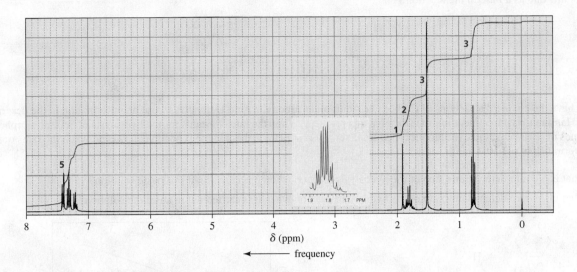

δ (ppm)

← frequency

90. Show how each of the following compounds can be prepared from the given starting material. In each case, you will need to use a protecting group.

a. CH₃CHCH₂COCH₃ → CH₃CHCH₂CCH₃
    OH   OH OH

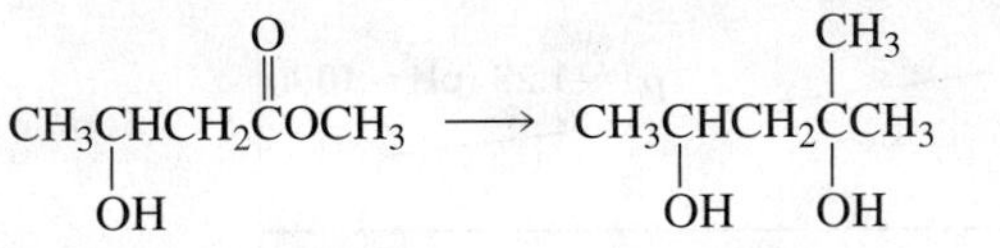

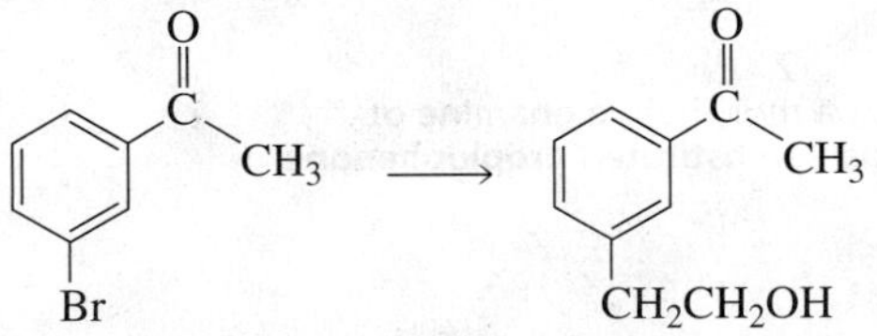

b.

c.

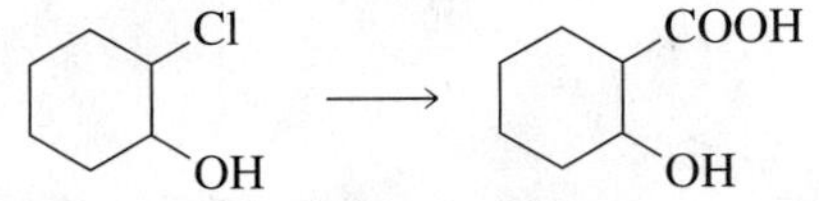

91. When a cyclic ketone reacts with diazomethane, the next larger cyclic ketone is formed. This is called a ring-expansion reaction. Draw a mechanism for the following ring-expansion reaction.

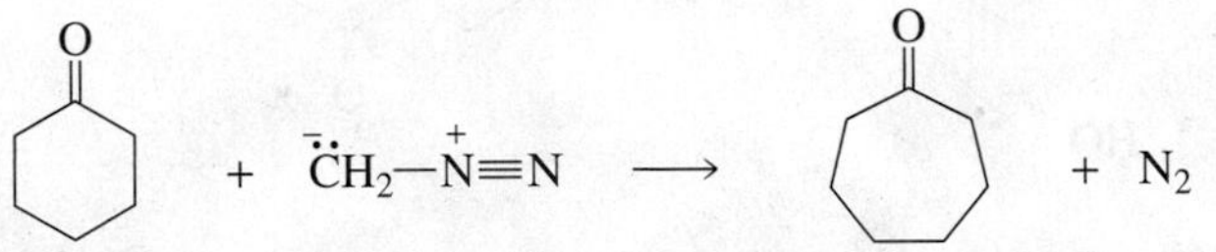

cyclohexanone   **diazomethane**   **cycloheptanone**   + N₂

92. Describe how 1-ethylcyclohexanol can be prepared from cyclohexane. You can use any inorganic reagents, any solvents, and any organic reagents as long as they contain no more than two carbons.

93. The pK_a values of the carboxylic acid groups of oxaloacetic acid are 2.22 and 3.98.

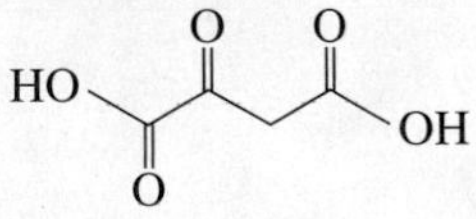

oxaloacetic acid

a. Which carboxyl group is more acidic?

b. The amount of hydrate present in an aqueous solution of oxaloacetic acid depends on the pH of the solution: 95% at pH 0, 81% at pH 1.3, 35% at pH 3.1, 13% at pH 4.7, 6% at pH 6.7, and 6% at pH 12.7. Explain this pH dependence.

94. The Baylis–Hillman reaction is a DABCO (1,4-diazabicyclo[2.2.2]octane) catalyzed reaction of an α,β-unsaturated carbonyl compound with an aldehyde to form an allylic alcohol. Propose a mechanism for the reaction. (*Hint:* DABCO serves as both a nucleophile and as a base in the reaction.)

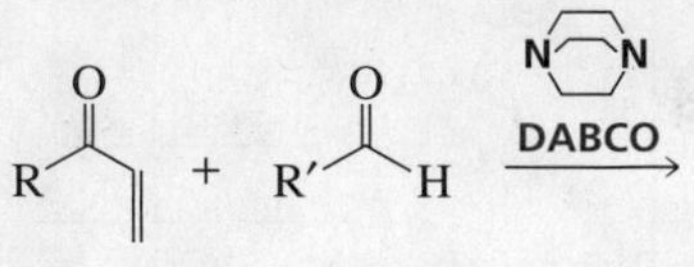

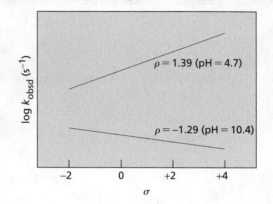

95. If you have the chapter "Reactions of Carboxylic Acids and Carboxylic Acid Derivatives," use the information in the description of the Hammett σ, ρ treatment in order to solve this problem. When the rate constants for the hydrolysis of several morpholine enamines of para-substituted propiophenones are determined at pH 4.7, the ρ value is positive; however, when the rates of hydrolysis are determined at pH 10.4, the ρ value is negative.

a. What is the rate-determining step of the hydrolysis reaction when it is carried out in a basic solution?

b. What is the rate-determining step of the hydrolysis reaction when it is carried out in an acidic solution?

a morpholine enamine of a para-substituted propiophenone

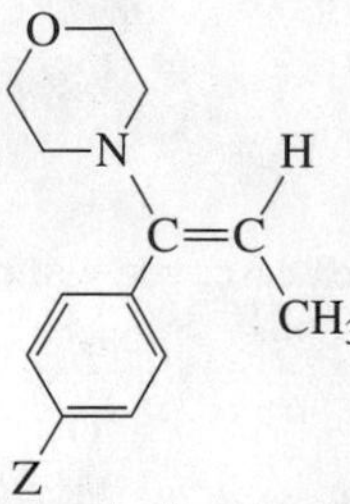

96. Propose a mechanism for each of the following reactions:

a.

b.

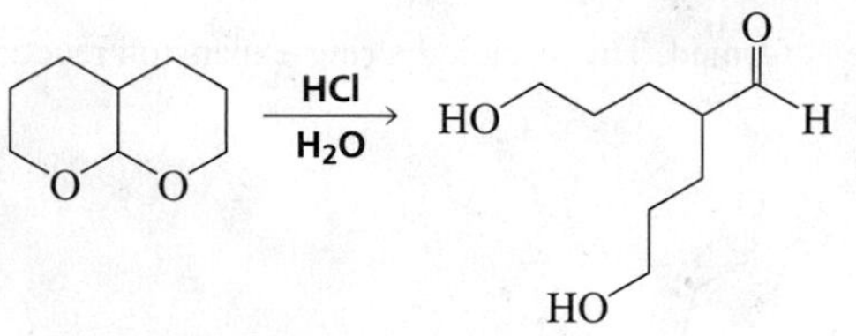

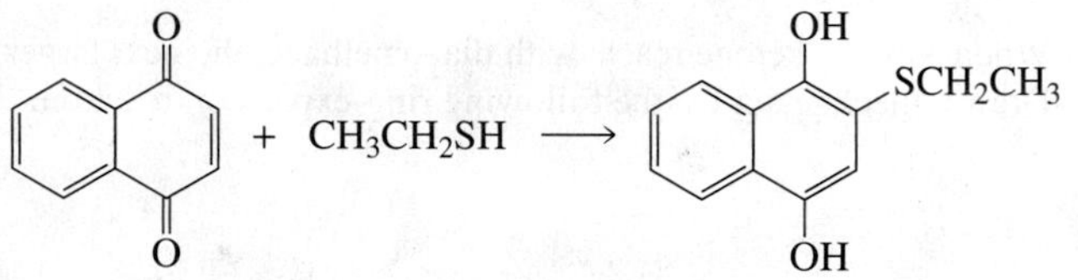

GLOSSARY

acetal

$$R{-}\overset{\displaystyle OR}{\underset{\displaystyle OR}{\vert\;\overset{\vert}{C}\;\vert}}{-}H \quad \text{or} \quad R{-}\overset{\displaystyle OR}{\underset{\displaystyle OR}{\vert\;\overset{\vert}{C}\;\vert}}{-}R$$

aldehyde

$$\overset{\displaystyle O}{\underset{\displaystyle}{R{-}\overset{\Vert}{C}{-}H}}$$

Baeyer–Villiger oxidation oxidation of aldehydes or ketones with H_2O_2 to form carboxylic acids or esters, respectively.

catalytic hydrogenation the addition of hydrogen to a double or a triple bond with the aid of a metal catalyst.

conjugate addition 1,4-addition to an α,β,-unsaturated carbonyl compound.

cyanohydrin

$$R{-}\overset{\displaystyle OH}{\underset{\displaystyle C\equiv N}{\vert\;\overset{\vert}{C}\;\vert}}{-}R(H)$$

direct addition 1,2-addition.

disconnection breaking a bond to carbon to give a simpler species.

dissolving-metal reduction a reduction brought about by the use of sodium or lithium metal dissolved in liquid ammonia.

enamine an α,β-unsaturated tertiary amine.

gem-diol (hydrate) a compound with two OH groups on the same carbon.

hemiacetal

$$R{-}\overset{\displaystyle OH}{\underset{\displaystyle OR}{\vert\;\overset{\vert}{C}\;\vert}}{-}H \quad \text{or} \quad R{-}\overset{\displaystyle OH}{\underset{\displaystyle OR}{\vert\;\overset{\vert}{C}\;\vert}}{-}R$$

hydrate (gem-diol)

$$R{-}\overset{\displaystyle OH}{\underset{\displaystyle OH}{\vert\;\overset{\vert}{C}\;\vert}}{-}R(H)$$

hydrazone $R_2C{=}NNH_2$

imine $R_2C{=}NR$

ketone

$$\overset{\displaystyle O}{\underset{\displaystyle}{R{-}\overset{\Vert}{C}{-}R}}$$

nucleophilic addition–elimination–nucleophilic addition reaction a nucleophilic addition reaction that is followed by an elimination reaction that is followed by a nucleophilic addition reaction. Acetal formation is an example: an alcohol adds to the carbonyl carbon, water is eliminated, and a second molecule of alcohol adds to the dehydrated product.

nucleophilic addition–elimination reaction a nucleophilic addition reaction that is followed by an elimination reaction. Imine formation is an example: an amine adds to the carbonyl carbon, and water is eliminated.

nucleophilic addition reaction a reaction that involves the addition of a nucleophile to a reagent.

oxime $R_2C{=}NOH$

peroxyacid a carboxylic acid with an OOH group instead of an OH group.

pH-rate profile a plot of the observed rate of a reaction as a function of the pH of the reaction mixture.

protecting group a reagent that protects a functional group from a synthetic operation that it would otherwise not survive.

reduction reaction a reaction in which the number of C—H bonds increases or the number of C—O, C—N, or C—X (X = a halogen) decreases.

reductive amination the reaction of an aldehyde or a ketone with ammonia or with a primary amine in the presence of a reducing agent (H_2/Raney Ni).

Schiff base $R_2C{=}NR$

synthetic equivalent the reagent actually used as the source of a synthon.

synthon a fragment of a disconnection.

Wittig reaction the reaction of an aldehyde or a ketone with a phosphonium ylide, resulting in the formation of an alkene.

ylide a compound with opposite charges on adjacent covalently bonded atoms with complete octets.

PHOTO CREDITS

Photo credits are listed in order of appearance.

Reactions at the α-Carbon of Carbonyl Compounds

a paper mulberry tree

Fifteen aromatase inhibitors, compounds used in the treatment of breast cancer, have been isolated from the leaves of the paper mulberry tree (see later in this chapter).

The site of reactivity of carbonyl compounds is the partially positively charged carbonyl carbon to which a nucleophile adds.

Many carbonyl compounds have a second site of reactivity—namely, a hydrogen bonded to *a carbon that is adjacent to the carbonyl carbon*. This hydrogen is sufficiently acidic to be removed by a strong base. The carbon adjacent to a carbonyl carbon is called an **α-carbon** so a hydrogen bonded to an α-carbon is called an **α-hydrogen.**

In Section 1, you will find out why a hydrogen bonded to an α-carbon is more acidic than hydrogens bonded to other sp^3 carbons, and you will look at some reactions that result from this acidity. Later in the chapter, you will see that a hydrogen is not the only substituent that can be removed from an α-carbon: a carboxyl group bonded to an α-carbon can be removed as CO_2. At the end of the chapter, you will be introduced to some important biological reactions that rely on the ability to remove protons and carboxyl groups from α-carbons.

1 THE ACIDITY OF AN α-HYDROGEN

Hydrogen and carbon have similar electronegativities, which means that the two atoms share the electrons that bond them together almost equally. Consequently, a hydrogen bonded to a carbon is usually not acidic. This is particularly true for hydrogens bonded to sp^3 carbons because these carbons are the most similar to hydrogen in electronegativity. For example, the pK_a of ethane is greater than 60.

$$CH_3CH_3 \quad pK_a > 60$$

A hydrogen bonded to an sp^3 carbon that is adjacent to a carbonyl carbon, however, is much more acidic than hydrogens bonded to other sp^3 carbons. For example, the pK_a value for dissociation of a proton from the α-carbon of an aldehyde or a ketone ranges from 16 to 20, and the pK_a value for dissociation of a proton attached to the α-carbon of an ester is about 25 (Table 1). A compound that contains a relatively acidic hydrogen bonded to an sp^3 carbon is called a **carbon acid.**

The α-hydrogen of a ketone or an aldehyde is more acidic than the α-hydrogen of an ester.

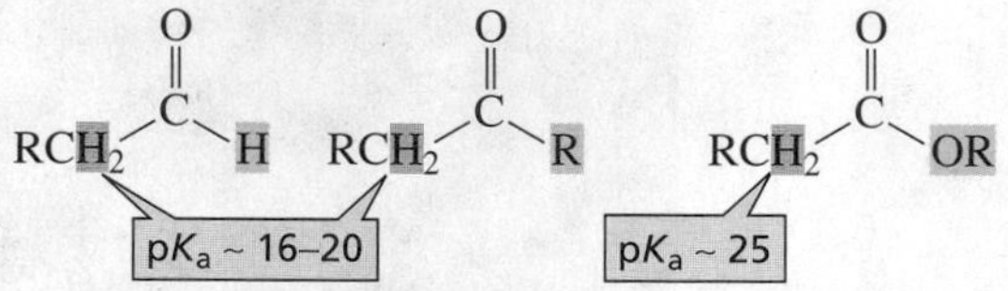

Table 1 The pK_a Values of Some Carbon Acids

	pK_a		pK_a
$H{-}CH_2C(=O)N(CH_3)_2$	30	$N{\equiv}CCH(H)C{\equiv}N$	11.8
$H{-}CH_2C(=O)OCH_2CH_3$	25	$CH_3C(=O)CH(H)C(=O)OCH_2CH_3$	10.7
$H{-}CH_2C{\equiv}N$	25	$C_6H_5C(=O)CH(H)C(=O)CH_3$	9.4
$H{-}CH_2C(=O)CH_3$	20	$CH_3C(=O)CH(H)C(=O)CH_3$	8.9
$H{-}CH_2C(=O)H$	17	$CH_3CH(H)NO_2$	8.6
$CH_3CH_2OC(=O)CH(H)C(=O)OCH_2CH_3$	13.3	$CH_3C(=O)CH(H)C(=O)H$	5.9
		$O_2NCH(H)NO_2$	3.6

A hydrogen bonded to an α-carbon is more acidic than hydrogens bonded to other sp^3 carbons because the base formed when a proton is removed from an α-carbon is relatively stable. And the more stable the base, the stronger is its conjugate acid.

Why is the base formed by removing a proton from an α-carbon more stable than bases formed by removing a proton from other sp^3 carbons? When a proton is removed from ethane, the electrons left behind reside solely on a carbon. This localized carbanion is unstable because carbon is not very electronegative. As a result, the pK_a of its conjugate acid is very high.

$$CH_3CH_3 \; \rightleftharpoons \; CH_3\overset{..}{C}H_2 \; + \; H^+$$

localized electrons

In contrast, when a proton is removed from an α-carbon, two factors combine to increase the stability of the base that is formed. First, the electrons left behind when the proton is removed are delocalized, and electron delocalization increases stability. More importantly, the electrons are delocalized onto an oxygen, an atom that is better able to accommodate them because it is more electronegative than carbon.

electrons are better accommodated on O than on C

delocalized electrons

resonance contributors

Now we can understand why aldehydes and ketones ($pK_a = 16\text{–}20$) are more acidic than esters ($pK_a = 25$). The electrons left behind when a proton is removed from the α-carbon of an ester are not as readily delocalized onto the carbonyl oxygen (indicated by the red arrows) as they would be in an aldehyde or a ketone. This is because the oxygen of the OR group of the ester also has a lone pair that can be delocalized onto the carbonyl oxygen (indicated by the green arrows). Thus, the two pairs of electrons compete for delocalization onto the same oxygen.

delocalization of a lone pair on oxygen

delocalization of the α-carbon's negative charge

resonance contributors

Nitroalkanes, nitriles, and N,N-disubstituted amides also have a relatively acidic α-hydrogen (Table 1) because in each case the electrons left behind when the proton is removed can be delocalized onto an atom that is more electronegative than carbon.

$CH_3\overset{..}{C}H_2NO_2$
nitroethane
$pK_a = 8.6$

$CH_3\overset{..}{C}H_2C\equiv N$
propanenitrile
$pK_a = 26$

$$\underset{\substack{\text{N,N-dimethylacetamide} \\ pK_a = 30}}{\overset{\displaystyle O}{\underset{\displaystyle}{CH_3-\overset{\|}{C}-N(CH_3)_2}}}$$

PROBLEM 1♦

Explain why the pK_a of a hydrogen bonded to the sp^3 carbon of propene is greater ($pK_a = 42$) than that of any of the carbon acids listed in Table 1 but is less than the pK_a of an alkane ($pK_a > 60$).

If the α-carbon is *between* two carbonyl groups, the acidity of its α-hydrogen is even greater (Table 1). For example, the pK_a value for dissociation of a proton from the α-carbon of 2,4-pentanedione, a compound with an α-carbon between two ketone carbonyl groups, is 8.9. And the pK_a value for dissociation of a proton from the α-carbon of ethyl 3-oxobutyrate, which is between a ketone carbonyl group and an ester carbonyl group, is 10.7. Ethyl 3-oxobutyrate is classified as a **β-keto ester** because the ester has a carbonyl group at the β-position; 2,4-pentanedione is a **β-diketone.**

$pK_a = 8.9$

2,4-pentanedione
acetylacetone
a β-diketone

$pK_a = 10.7$

ethyl 3-oxobutyrate
ethyl acetoacetate
a β-keto ester

The acidity of α-hydrogens bonded to carbons flanked by two carbonyl groups increases because the electrons left behind when the proton is removed can be delocalized onto either of *two* oxygens. β-Diketones have lower pK_a values than β-keto esters because, as we have seen, electrons are more readily delocalized onto the carbonyl oxygen of a ketone than they are onto the carbonyl oxygen of an ester.

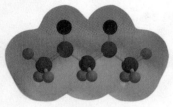

2,4-pentanedione

resonance contributors for the 2,4-pentanedione anion

PROBLEM 2♦

Give an example for each of the following:

a. a β-keto nitrile   **b.** a β-diester   **c.** a β-keto aldehyde

---

PROBLEM-SOLVING STRATEGY

The Acid–Base Behavior of a Carbonyl Compound

Explain why a base cannot remove a proton from the α-carbon of a carboxylic acid.

If a base cannot remove a proton from the α-carbon of a carboxylic acid, then the base must react with another portion of the molecule more rapidly. Because the proton on the carboxyl group is more acidic ($pK_a \sim 5$) than the proton on the α-carbon, we can conclude that the base removes a proton from the carboxyl group rather than from the α-carbon.

Now use the strategy you have just learned to solve Problem 3.

PROBLEM 3♦

Explain why a base can remove a proton from the α-carbon of *N,N*-dimethylethanamide but not from the α-carbon of either *N*-methylethanamide or ethanamide.

N,N-dimethylethanamide   **N-methylethanamide**   **ethanamide**

PROBLEM 4♦

Explain why the α-hydrogen of an *N,N*-disubstituted amide is less acidic ($pK_a = 30$) than the α-hydrogen of an ester ($pK_a = 25$).

PROBLEM 5♦

List the compounds in each of the following groups in order from strongest acid to weakest acid:

a. $CH_2{=}CH_2$ CH_3CH_3 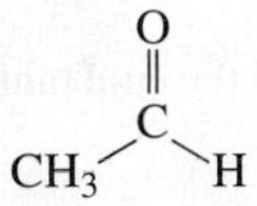 $HC{\equiv}CH$

b.

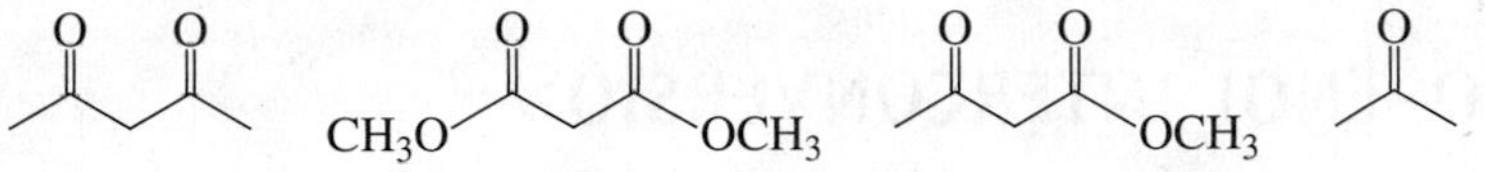

c.

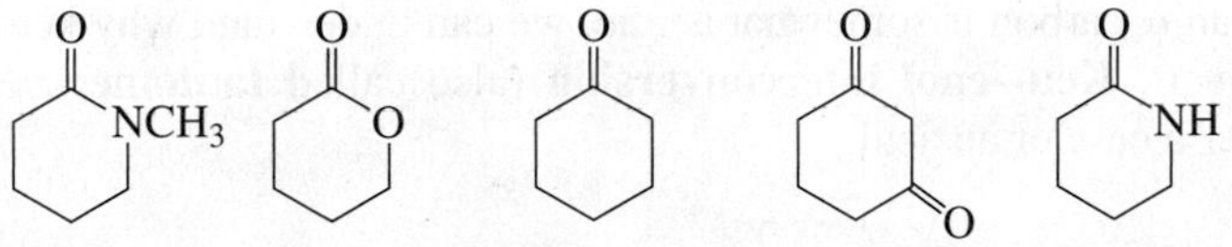

---

2   KETO–ENOL TAUTOMERS

A ketone exists in equilibrium with its enol tautomer. **Tautomers** are isomers that are in rapid equilibrium. Keto–enol tautomers differ in the location of a double bond and a hydrogen.

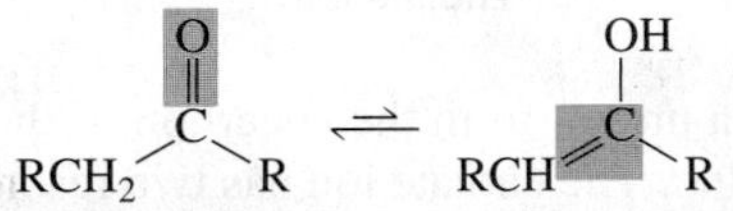

keto tautomer   enol tautomer

For most ketones, the **enol tautomer** is much less stable than the **keto tautomer**. For example, an aqueous solution of acetone exists as an equilibrium mixture of more than 99.9% keto tautomer and less than 0.1% enol tautomer.

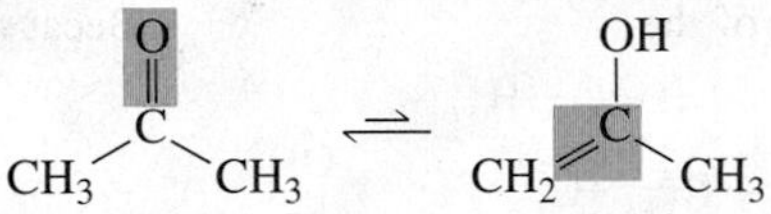

> 99.9%   < 0.1%
keto tautomer   enol tautomer

The fraction of the enol tautomer in an aqueous solution is considerably greater for a β-diketone because the enol tautomer is stabilized both by intramolecular hydrogen bonding and by conjugation of the carbon–carbon double bond with the second carbonyl group.

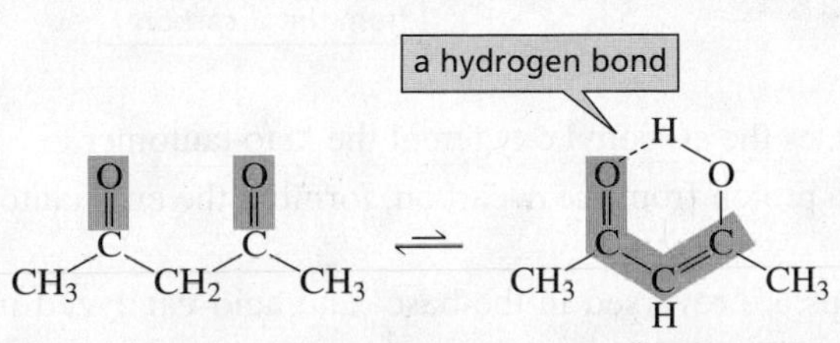

85%   15%
keto tautomer   enol tautomer

Phenol is unusual in that its enol tautomer is *more* stable than its keto tautomer because the enol tautomer is aromatic, but the keto tautomer is not.

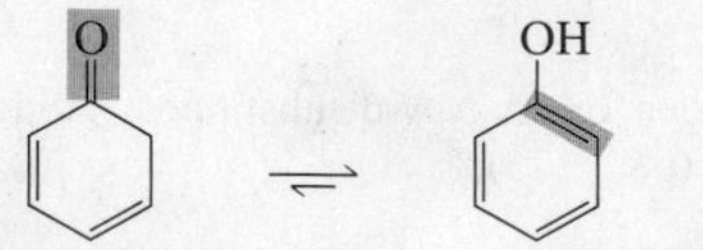

keto tautomer
not aromatic

enol tautomer
aromatic

PROBLEM 6

Explain why 92% of 2,4-pentanedione exists as the enol tautomer in hexane but only 15% of this compound exists as the enol tautomer in water.

---

3   KETO–ENOL INTERCONVERSION

Now that we know that an α-carbon is somewhat acidic, we can understand why keto and enol tautomers interconvert. **Keto–enol interconversion** (also called **tautomerization**) can be catalyzed by either a base or an acid.

MECHANISM FOR BASE-CATALYZED KETO–ENOL INTERCONVERSION

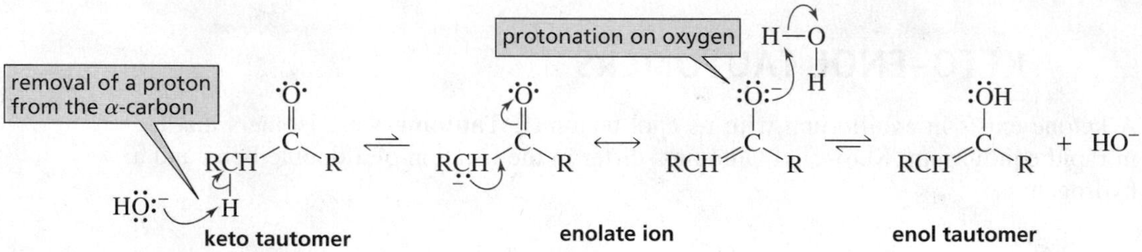

- Hydroxide ion removes a proton from the α-carbon of the keto tautomer, forming an anion called an **enolate ion.** The enolate ion has two resonance contributors.
- Protonating the oxygen forms the enol tautomer, whereas protonating the α-carbon re-forms the keto tautomer.

MECHANISM FOR ACID-CATALYZED KETO–ENOL INTERCONVERSION

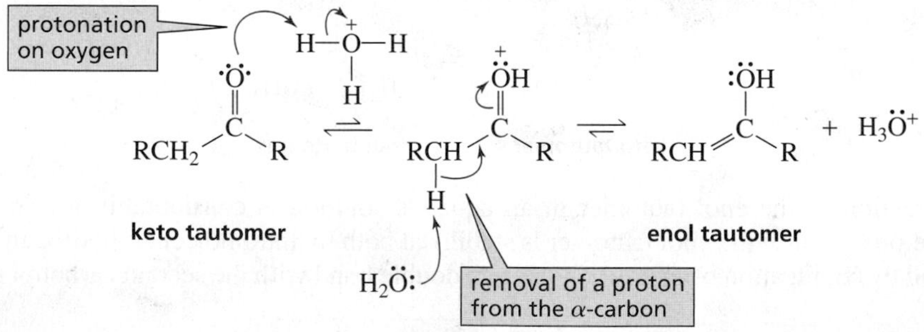

- The acid protonates the carbonyl oxygen of the keto tautomer.
- Water removes a proton from the α-carbon, forming the enol tautomer.

Notice that the steps are reversed in the base- and acid-catalyzed interconversions. In the base-catalyzed reaction, the base removes a proton from an α-carbon in the first step and the oxygen is protonated in the second step. In the acid-catalyzed reaction, the oxygen is protonated in the first step and the proton is removed from the α-carbon in the second step. Notice also that the catalyst is regenerated in both the acid- and base-catalyzed reactions.

PROBLEM 7♦

Draw the enol tautomers for each of the following compounds. For compounds that have more than one enol tautomer, indicate which one is more stable.

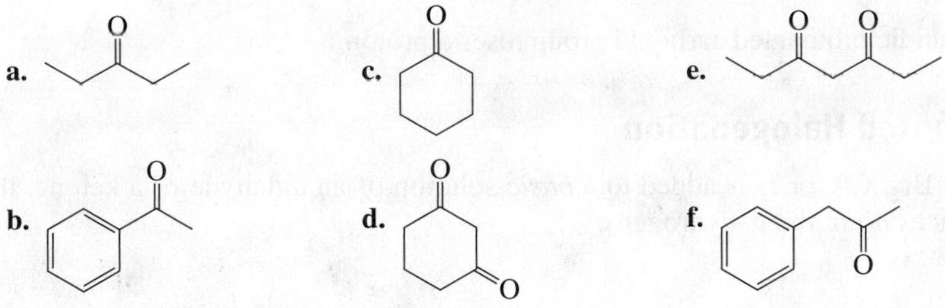

a.   c.   e.

b.   d.   f.

PROBLEM 8

When a dilute solution of acetaldehyde in D_2O containing NaOD is shaken, explain why the methyl hydrogens are exchanged with deuterium but the hydrogen attached to the carbonyl carbon is not.

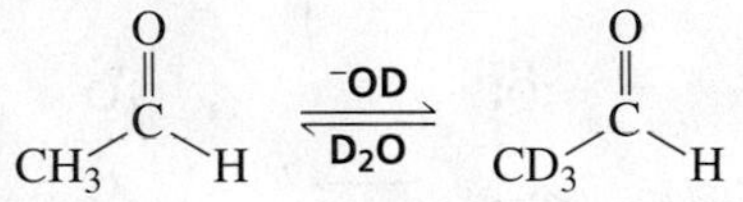

4   HALOGENATION OF THE α-CARBON OF ALDEHYDES AND KETONES

When Br_2, Cl_2, or I_2 is added to a solution of an aldehyde or a ketone, a halogen replaces *one or more* of the α-hydrogens of the carbonyl compound. The reaction can be catalyzed by either an acid or a base. This is an α-**substitution reaction** because one electrophile (Br^+) is substituted for another (H^+) on the α-carbon.

Acid-Catalyzed Halogenation

In the acid-catalyzed reaction, the halogen replaces *one* of the α-hydrogens:

Under acidic conditions, *one* α-hydrogen is substituted for a halogen.

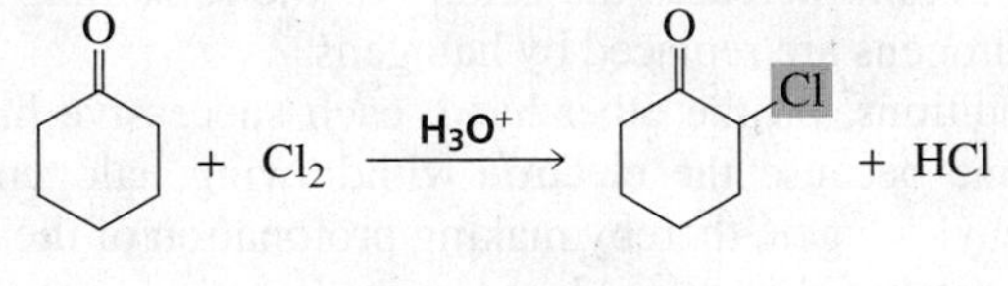

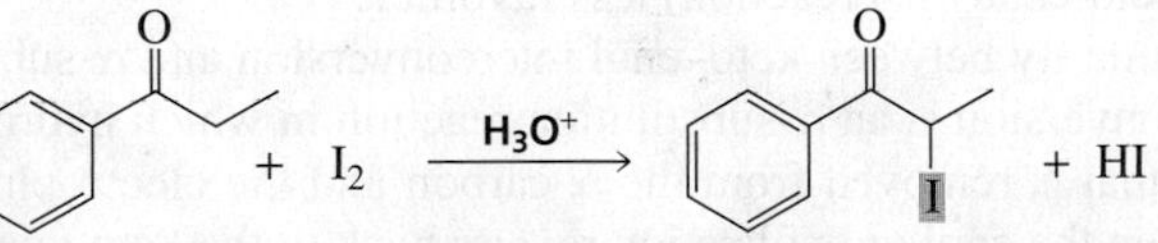

MECHANISM FOR ACID-CATALYZED HALOGENATION

887

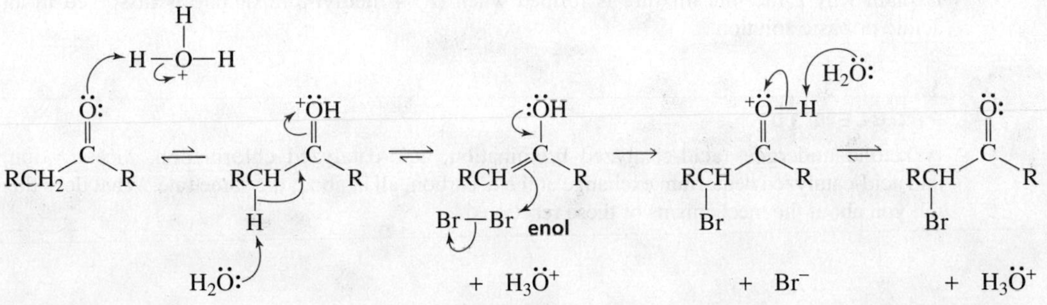

- The carbonyl oxygen is protonated.
- Water removes a proton from the α-carbon, forming an enol.
- The enol reacts with an electrophilic halogen; the other halogen atom keeps the bonding electrons.
- The very acidic protonated carbonyl group loses a proton.

Base-Promoted Halogenation

When excess Br_2, Cl_2, or I_2 is added to a *basic* solution of an aldehyde or a ketone, the halogen replaces *all* of the α-hydrogens.

Under basic conditions, *all* of the α-hydrogens are substituted for halogens.

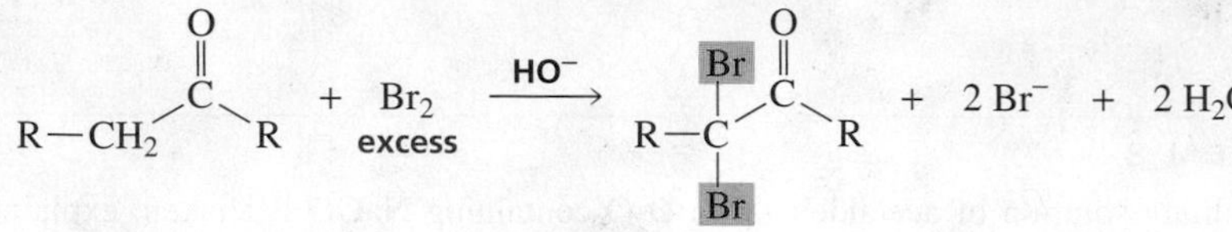

MECHANISM FOR BASE-PROMOTED HALOGENATION

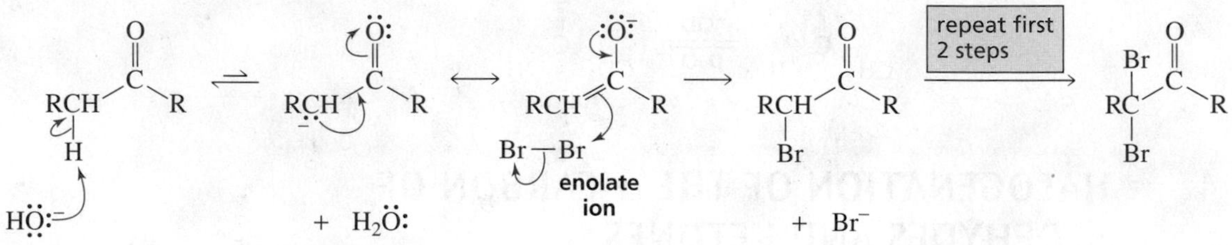

- Hydroxide ion removes a proton from the α-carbon, forming an enolate ion.
- The enolate ion reacts with the electrophilic halogen; the other halogen atom keeps the bonding electrons. Notice that hydroxide ion is not regenerated, so it is promoting the reaction, not catalyzing it.

These two steps are repeated until all of the α-hydrogens are replaced by a halogen. Each successive halogenation is *more rapid* than the previous one because the electron-withdrawing halogen atom increases the acidity of the remaining α-hydrogens. This is why *all* of the α-hydrogens are replaced by halogens.

Under acidic conditions, on the other hand, each successive halogenation is *slower* than the previous one because the electron-withdrawing halogen atom decreases the basicity of the carbonyl oxygen, thereby making protonation of the carbonyl oxygen (the first step in the acid-catalyzed reaction) less favorable.

Notice the similarity between keto–enol interconversion and α-substitution. Actually, keto–enol interconversion is an α-substitution reaction in which hydrogen serves as both the electrophile that is removed from the α-carbon and the electrophile that is added to the α-carbon when the enol or enolate ion reverts back to the keto tautomer.

PROBLEM 9

Explain why a racemic mixture is formed when (R)-4-methyl-3-hexanone is dissolved in an acidic or basic solution.

PROBLEM 10♦

A ketone undergoes acid-catalyzed bromination, acid-catalyzed chlorination, racemization, and acid-catalyzed deuterium exchange at the α-carbon, all at about the same rate. What does this tell you about the mechanisms of these reactions?

5 HALOGENATION OF THE α-CARBON OF CARBOXYLIC ACIDS: THE HELL–VOLHARD–ZELINSKI REACTION

Carboxylic acids cannot undergo substitution reactions at the α-carbon, because a base will remove a proton from the OH group instead of from the α-carbon, since the OH group is more acidic. If, however, a carboxylic acid is treated with PBr_3 and Br_2, the α-carbon can be brominated. This halogenation reaction is called the **Hell–Volhard–Zelinski reaction** or, more simply, the **HVZ reaction.** (Red phosphorus can be used in place of PBr_3 since P and excess Br_2 react to form PBr_3.)

the HVZ reaction

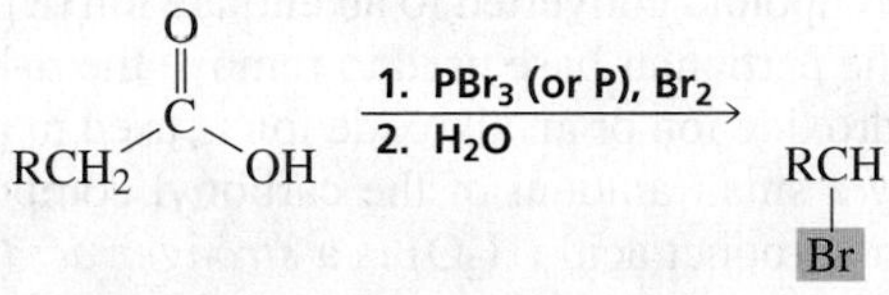

You will see, when you examine the reaction, that α-substitution occurs because an acyl bromide, rather than a carboxylic acid, is the compound that undergoes α-substitution.

STEPS IN THE HELL–VOLHARD–ZELINSKI REACTION

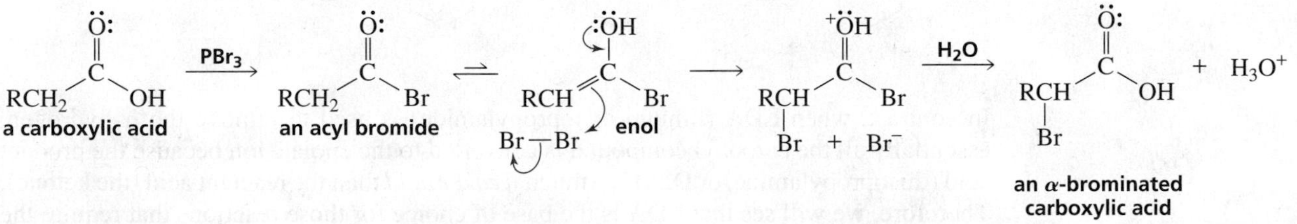

- PBr_3 converts the carboxylic acid into an acyl bromide, which is in equilibrium with its enol.

- Bromination of the enol forms a protonated α-brominated acyl bromide, which is hydrolyzed to a α-brominated carboxylic acid.

The bromine attached to the α-carbon of *aldehydes and ketones* can be replaced only by weakly basic nucleophiles such as a carboxylate ion. Strongly basic nucleophiles form enolate ions, which then undergo other reactions.

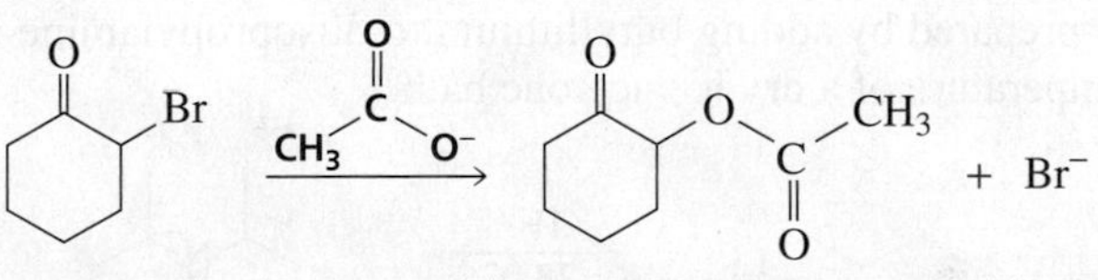

However, the bromine on the α-carbon of *carboxylate ions* can be replaced by basic nucleophiles, because carboxylate ions do not form enolate ions. Forming an enolate ion would require putting a second negative charge on the compound.

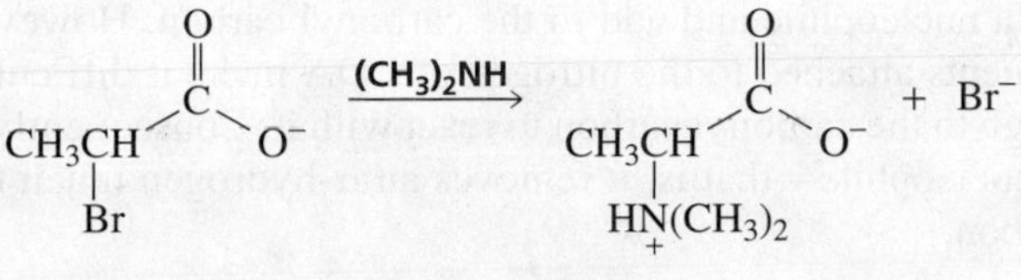

PROBLEM 11

Show how the following compounds could be prepared from the given starting material:

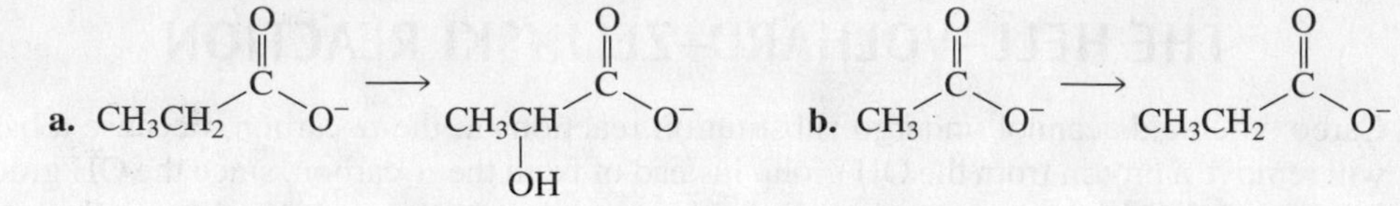

a. CH₃CH₂—C(=O)—O⁻ ⟶ CH₃CH(OH)—C(=O)—O⁻

b. CH₃—C(=O)—O⁻ ⟶ CH₃CH₂—C(=O)—O⁻

6 FORMING AN ENOLATE ION

The amount of carbonyl compound converted to an enolate ion depends on the pK_a of the carbonyl compound and the particular base used to remove the α-hydrogen.

For example, when hydroxide ion or an alkoxide ion is used to remove an α-hydrogen from cyclohexanone, only a small amount of the carbonyl compound is converted into the enolate ion because the product acid (H_2O) is a *stronger acid* than the reactant acid (the ketone). (The equilibrium of an acid–base reaction favors dissociation of the strong acid and formation of the weak acid.)

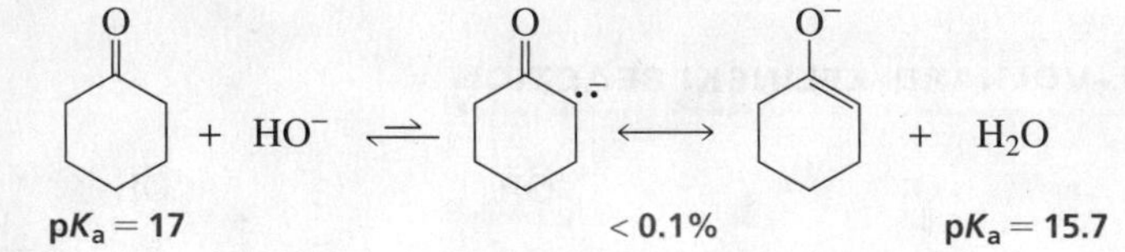

pK_a = 17   < 0.1%   pK_a = 15.7

In contrast, when LDA (lithium diisopropylamide) is used to remove the α-hydrogen, essentially all the carbonyl compound is converted to the enolate ion because the product acid (diisopropylamine, or DIA) is a much *weaker acid* than the reactant acid (the ketone). Therefore, we will see that LDA is the base of choice for those reactions that require the carbonyl compound to be completely converted to an enolate ion before it reacts with an electrophile (Section 7).

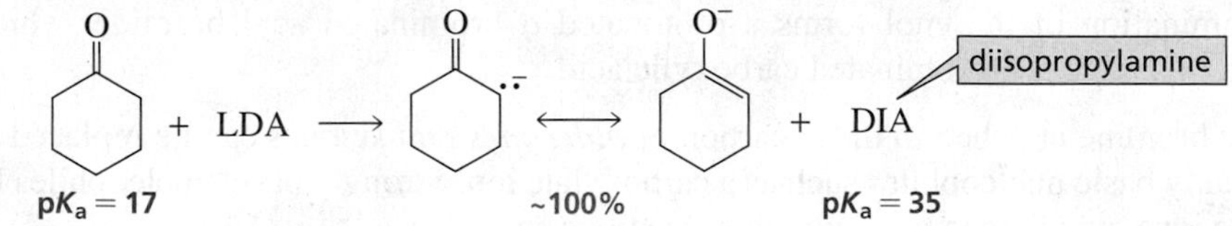

pK_a = 17   ~100%   pK_a = 35   diisopropylamine

LDA is easily prepared by adding butyllithium to diisopropylamine in THF at −78 °C (that is, at the temperature of a dry ice/acetone bath.)

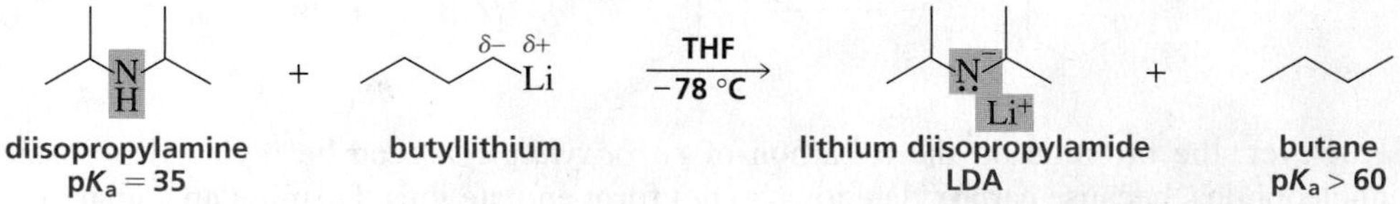

diisopropylamine   butyllithium   THF   lithium diisopropylamide   butane
pK_a = 35   −78 °C   LDA   pK_a > 60

Using a nitrogen base to form an enolate ion can be a problem because a nitrogen base can also react as a nucleophile and add to the carbonyl carbon. However, the two bulky isopropyl substituents attached to the nitrogen of LDA make it difficult for the nitrogen to get close enough to the carbonyl carbon to react with it. Consequently, LDA is a strong base but a poor nucleophile—that is, it removes an α-hydrogen much faster than it adds to a carbonyl carbon.

PROBLEM 12♦

What compound is formed when a dilute solution of cyclohexanone is shaken with NaOD in D_2O for several hours?

7 ALKYLATING THE α-CARBON OF CARBONYL COMPOUNDS

Putting an alkyl group on the α-carbon of a carbonyl compound is an important reaction because it gives us another way to form a carbon–carbon bond. Alkylation is carried out by first removing a proton from the α-carbon with a strong base, such as LDA, and then adding the appropriate alkyl halide. Because the alkylation is an S_N2 reaction, it works best with methyl halides and primary alkyl halides.

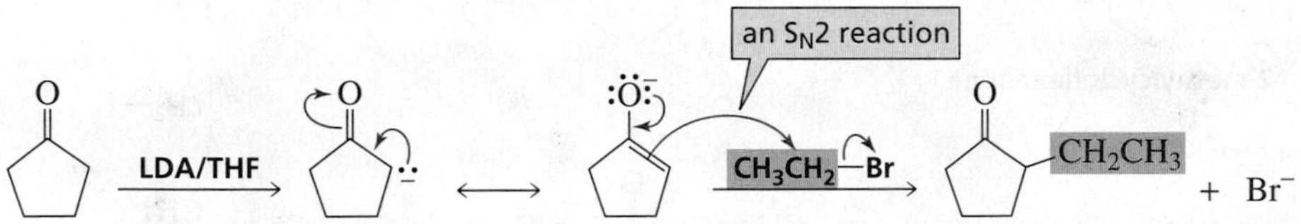

Although an enolate ion has two resonance contributors, for the sake of simplicity, only the resonance contributor with the negative charge on carbon will be shown in many of the reactions in this chapter. For example, compare the following presentation with the preceding one.

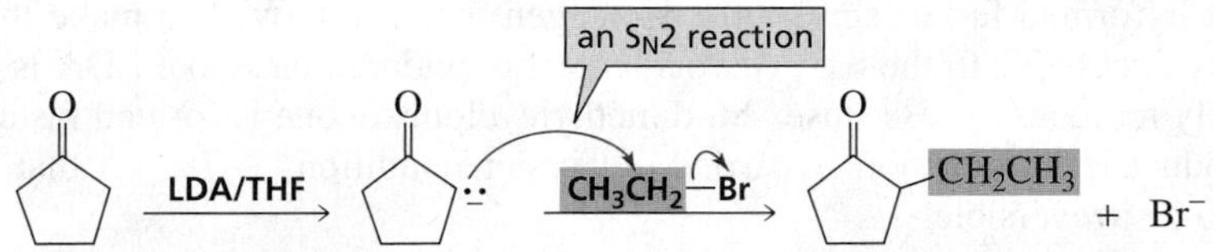

It is important that a strong base such as LDA is used to form the enolate ion. If a weaker base such as hydroxide ion or an alkoxide ion is used, very little of the desired monoalkylated product will be obtained. We have seen that these weaker bases form only a small amount of enolate ion at equilibrium (Section 6). Therefore, after monoalkylation has occurred, there is still a lot of base present in solution, which can form an enolate ion from the monoalkyated ketone as well as from the unalkylated ketone. As a result, di-, tri-, and even tetra-alkylated products may be produced.

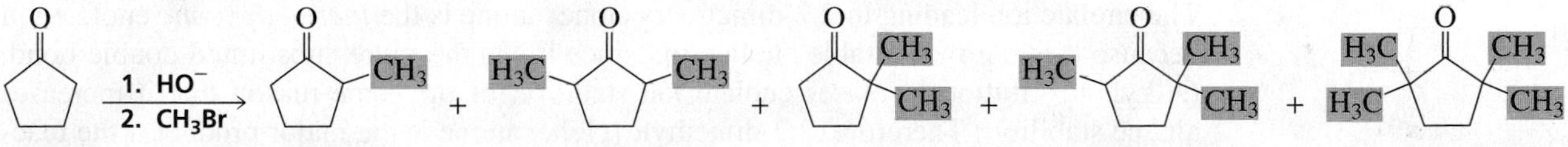

Esters and nitriles can also be alkylated on the α-carbon.

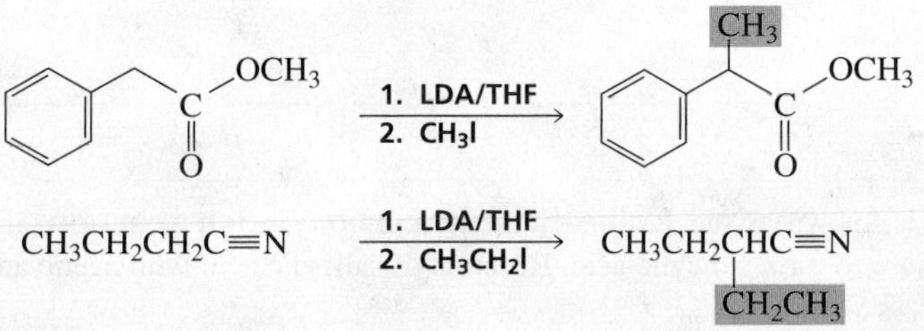

Ketones, aldehydes, esters, and nitriles can be alkylated on the α-carbon.

Explain why alkylation of an α-carbon works best if the alkyl halide used in the reaction is a primary alkyl halide, and why alkylation does not work at all if a tertiary alkyl halide is used.

Alkylating Unsymmetrical Ketones

If the ketone is unsymmetrical and has hydrogens on both α-carbons, two monoalkylated products can be obtained because either α-carbon can be alkylated. For example, methylation of 2-methylcyclohexanone with one equivalent of methyl iodide forms both 2,6-dimethylcyclohexanone and 2,2-dimethylcyclohexanone. The relative amounts of the two products depend on the reaction conditions.

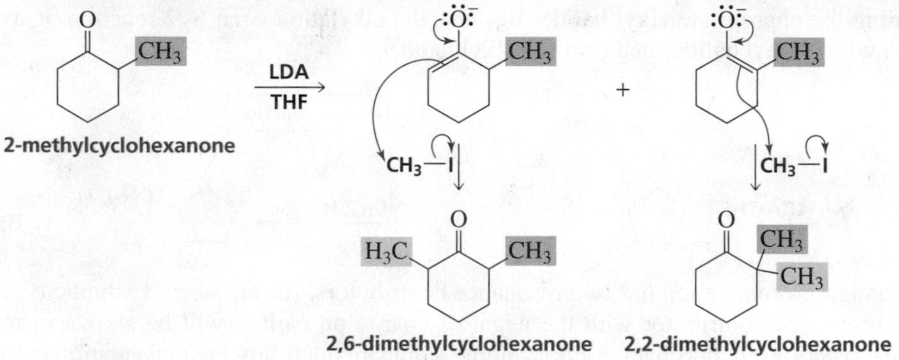

The enolate ion leading to 2,6-dimethylcyclohexanone is the *kinetic* enolate ion because it is formed faster, since the α-hydrogen that is removed to make this enolate ion is more accessible to the base (particularly if a hindered base like LDA is used) and it is slightly more acidic. Because, 2,6-dimethylcyclohexanone is formed faster, it is the major product if the reaction is carried out under conditions (-78 °C) that cause the reaction to be irreversible.

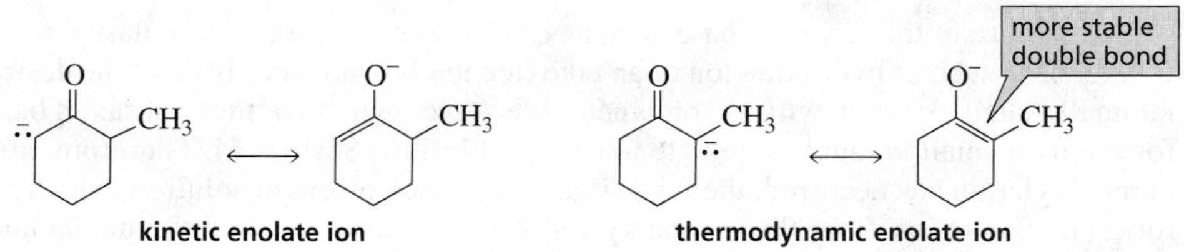

The enolate ion leading to 2,2-dimethylcyclohexanone is the *thermodynamic* enolate ion because it is the more stable enolate ion since it has the more substituted double bond. (Alkyl substitution increases enolate ion stability for the same reason that it increases alkene stability.) Therefore, 2,2-dimethylcyclohexanone is the major product if the reaction is carried out under conditions that cause enolate ion formation to be reversible.

The Synthesis of Aspirin

The first step in the industrial synthesis of aspirin is known as the **Kolbe–Schmitt carboxylation reaction.** In this step, the phenolate ion reacts with carbon dioxide under pressure to form salicylic acid. Reaction of salicylic acid with acetic anhydride forms acetylsalicylic acid (aspirin).

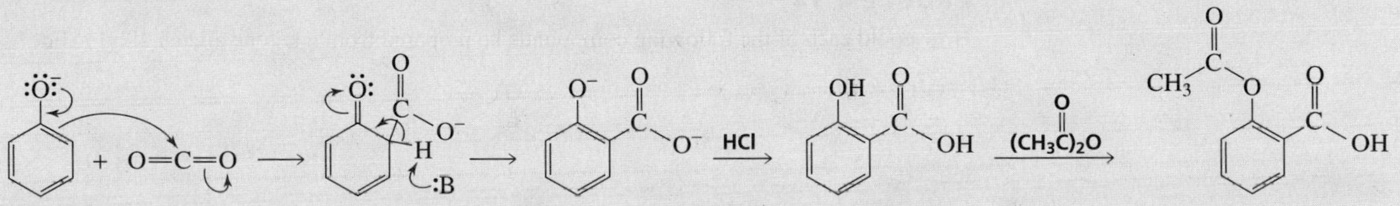

salicylic acid
o-hydroxybenzoic acid

acetylsalicylic acid
aspirin

During World War I, the Bayer Company bought as much phenol as it could on the international market, knowing that eventually all of it could be used to manufacture aspirin. This left little phenol available for other countries to purchase for the synthesis of 2,4,6-trinitrophenol (picric acid), a common explosive at that time.

PROBLEM-SOLVING STRATEGY

Alkylating a Carbonyl Compound

How could 4-methyl-3-hexanone be prepared from a ketone containing no more than six carbons?

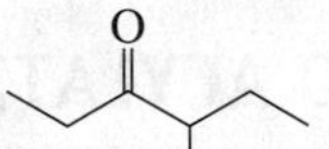

4-methyl-3-hexanone

Either of two sets of ketone and alkyl halide could be used for the synthesis: one set is 3-hexanone and a methyl halide; the other is 3-pentanone and an ethyl halide.

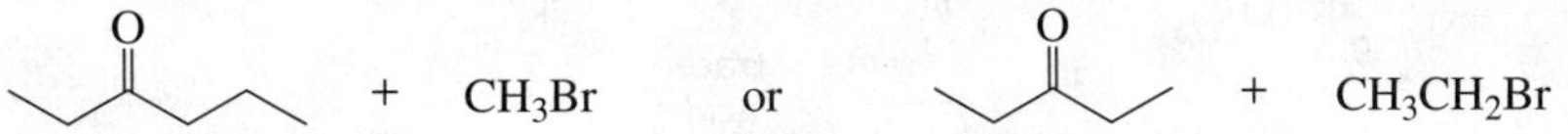

3-hexanone 3-pentanone

3-Pentanone and an ethyl halide are the preferred starting materials because 3-pentanone is symmetrical, so only one α-substituted ketone will be formed.

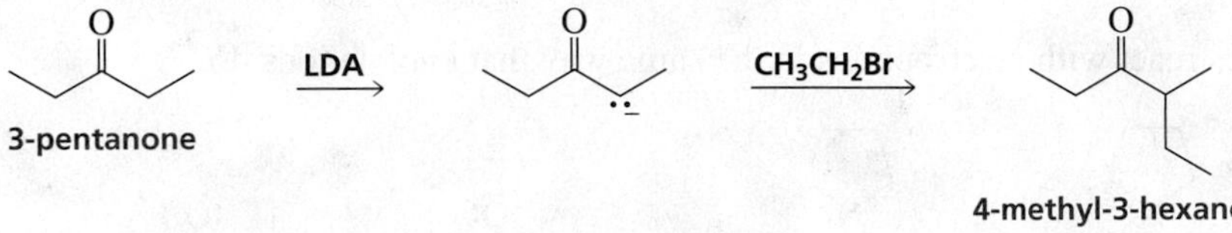

3-pentanone

4-methyl-3-hexanone

In contrast, 3-hexanone can form two different enolate ions, so two α-substituted ketones can be formed, decreasing the yield of he target molecule.

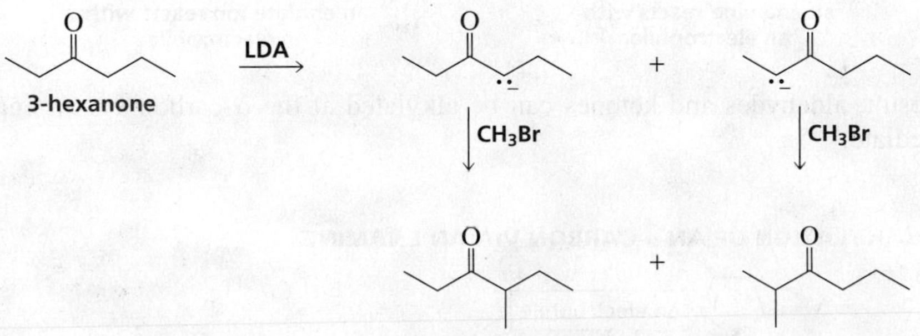

3-hexanone

4-methyl-3-hexanone 2-methyl-3-hexanone

Now use the strategy you have just learned to solve Problem 14.

PROBLEM 14

How could each of the following compounds be prepared from a ketone and an alkyl halide?

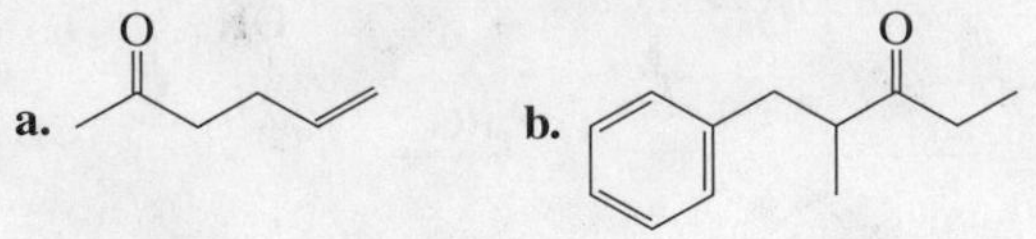

a.

b.

PROBLEM 15♦

How many stereoisomers are obtained from each of the syntheses described in Problem 14?

PROBLEM 16

How could each of the following compounds be prepared from cyclohexanone?

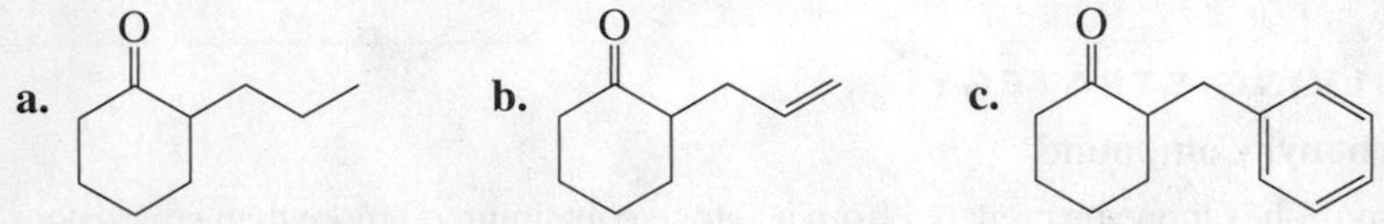

a.

b.

c.

8 ALKYLATING AND ACYLATING THE α-CARBON USING AN ENAMINE INTERMEDIATE

An enamine is formed when an aldehyde or a ketone reacts with a secondary amine.

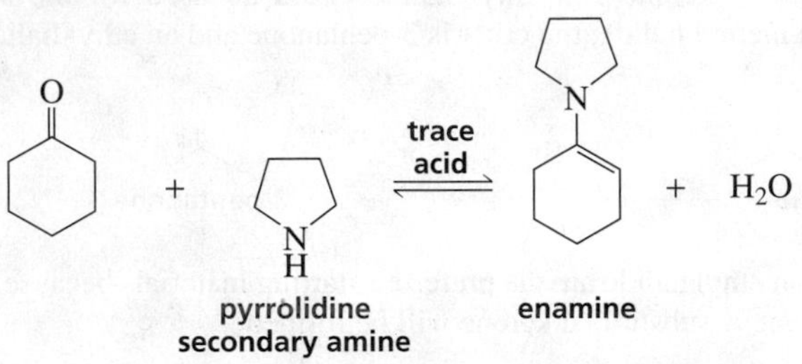

pyrrolidine
secondary amine   enamine

Enamines react with electrophiles in the same way that enolate ions do.

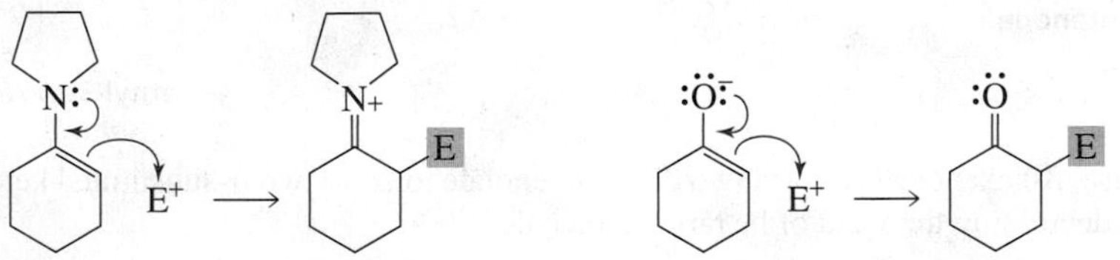

an enamine reacts with
an electrophile

an enolate ion reacts with
an electrophile

As a result, aldehydes and ketones can be alkylated at the α-carbon via an enamine intermediate:

STEPS IN THE ALKYLATION OF AN α-CARBON VIA AN ENAMINE

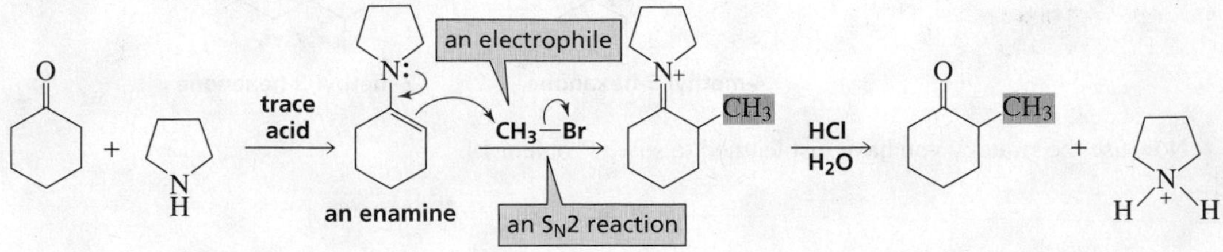

an enamine

an electrophile

an S$_N$2 reaction

- The carbonyl compound is converted to an enamine (by treating it with a secondary amine in the presence of a trace amount of acid).

- The enamine reacts with the alkyl halide in an S_N2 reaction.

- The imine is hydrolyzed to an α-alkylated ketone and the secondary amine that was used to form the enamine.

Because the alkylation step is an S_N2 reaction, only primary alkyl halides or methyl halides should be used. The main advantage to using an enamine intermediate to alkylate an aldehyde or a ketone is that it forms a monoalkylated product without having to use a strong base (LDA).

In addition to using an enamine to alkylate the α-carbon of an aldehyde or ketone, it can also be used to acylate the α-carbon.

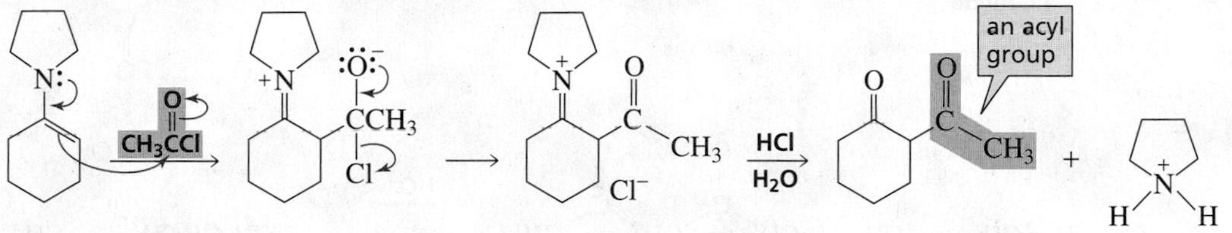

Enamines can also be used to attach the α-carbon of an aldehyde or ketone to the β-carbon of an α, β-unsaturated carbonyl compound via conjugate addition.

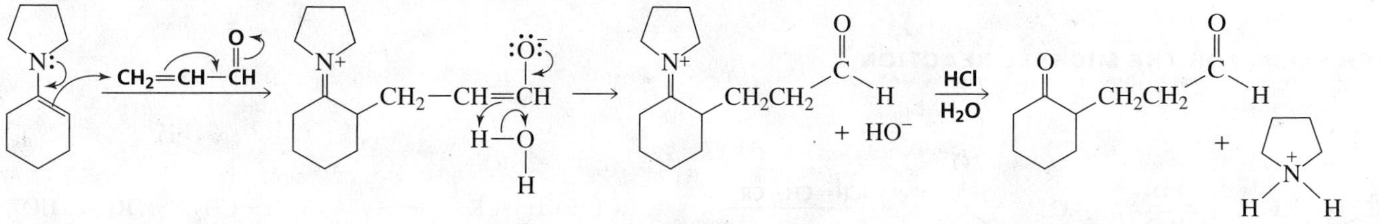

PROBLEM 17

Describe how the following compounds could be prepared from cyclohexanone using an enamine intermediate:

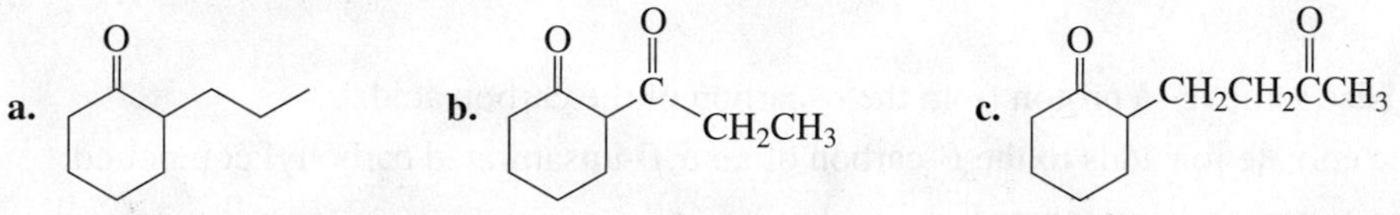

9   ALKYLATING THE β-CARBON: THE MICHAEL REACTION

Nucleophiles react with α,β-unsaturated aldehydes and ketones, forming either direct (nucleophilic) addition products or conjugate addition products.

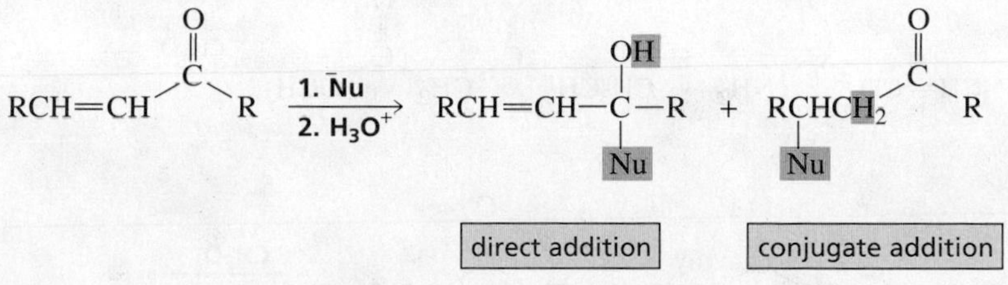

When the nucleophile in this reaction is an enolate ion, the addition reaction is called a **Michael reaction.** The enolate ions that work best in Michael reactions are those formed from carbon acids that are flanked by two electron-withdrawing groups—that is, enolate ions of β-diketones, β-diesters, β-keto esters, and β-keto nitriles.

Because these enolate ions are relatively weak bases, conjugate addition occurs—that is, addition to the β-carbon of α,β-unsaturated aldehydes and ketones. These enolate ions also add to the β-carbon of α,β-unsaturated esters and amides because of the low reactivity of the carbonyl group. Notice that a *Michael reaction forms a 1,5-dicarbonyl compound.*

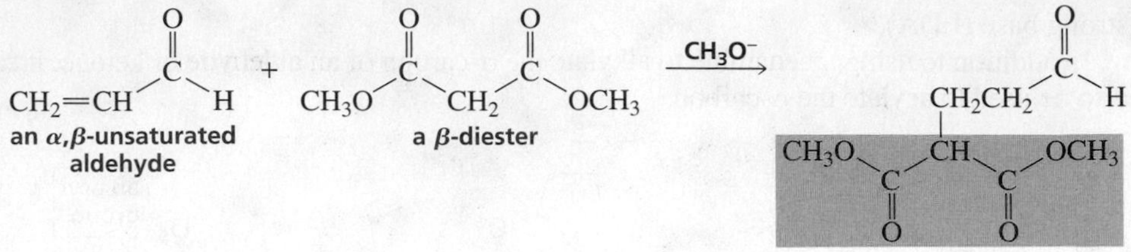

A Michael reaction forms a 1,5-dicarbonyl compound.

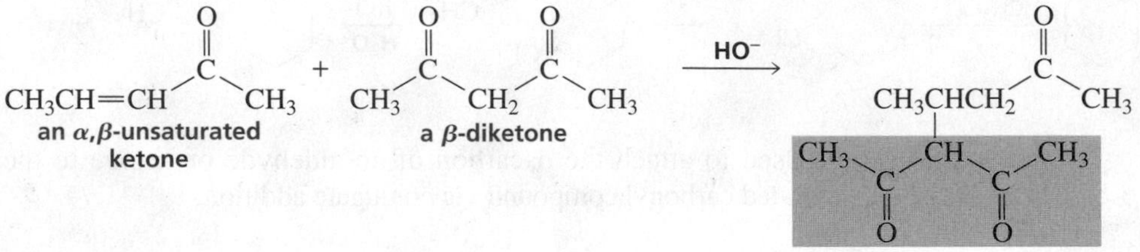

MECHANISM FOR THE MICHAEL REACTION

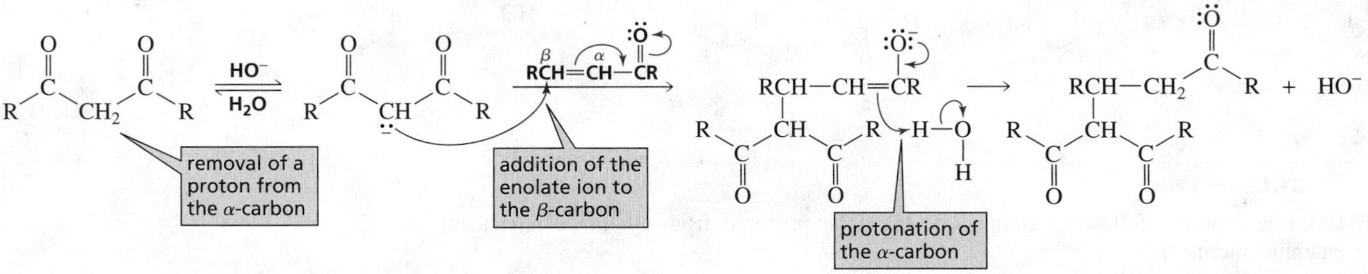

- A base removes a proton from the α-carbon of the carbon acid.
- The enolate ion adds to the β-carbon of an α,β-unsaturated carbonyl compound.
- The α-carbon is protonated.

Notice that if either of the reactants in a Michael reaction has an ester group, then the base used to remove the α-proton must be the same as the leaving group of the ester (Section 13).

PROBLEM 18

Draw the products of the following reactions:

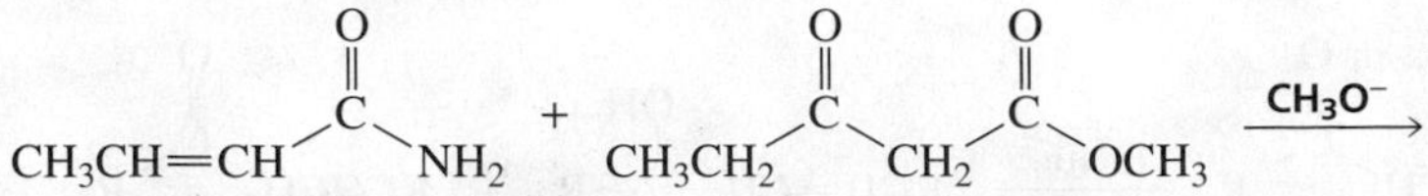

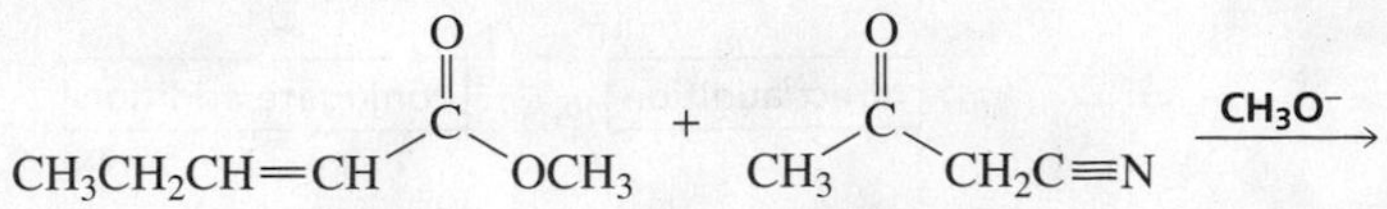

What reagents should be used to prepare the following compounds?

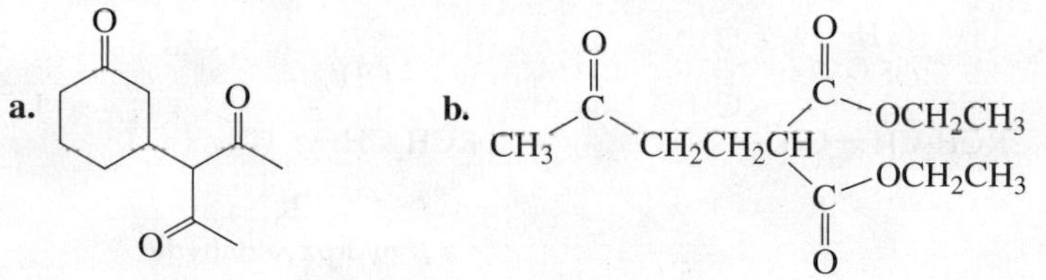

a.

b.

10 AN ALDOL ADDITION FORMS β-HYDROXYALDEHYDES OR β-HYDROXYKETONES

The carbonyl carbon of an aldehyde or a ketone is an electrophile. We have just seen that a proton can be removed from the α-carbon of an aldehyde or a ketone, converting the α-carbon into a nucleophile. An **aldol addition** is a reaction in which *both* of these reactivities are observed. That is, one molecule of a carbonyl compound—after a proton is removed from an α-carbon—reacts as a *nucleophile* and adds to the *electrophilic* carbonyl carbon of a second molecule of the carbonyl compound.

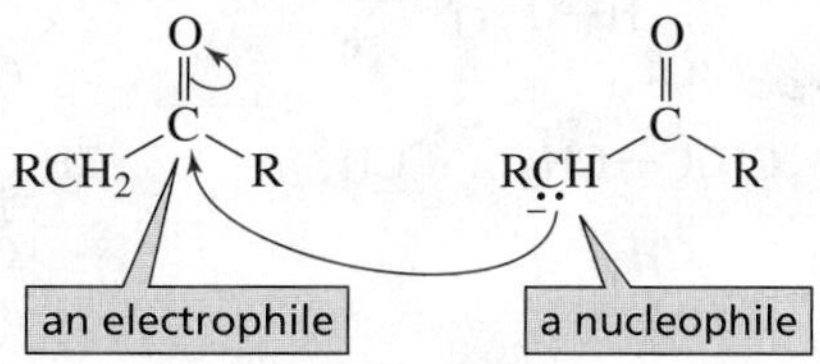

An Aldol Addition

An aldol addition is a reaction between two molecules of an *aldehyde* or two molecules of a *ketone*. When the reactant is an aldehyde, the product is a **β-hydroxyaldehyde,** which is why the reaction is called an aldol addition ("ald" for *aldehyde*, "ol" for *alcohol*). When the reactant is a ketone, the product is a **β-hydroxyketone.** Notice that the reaction forms a new C—C bond that connects the α-carbon of one molecule and the carbon that was originally the carbonyl carbon of the other molecule.

aldol additions

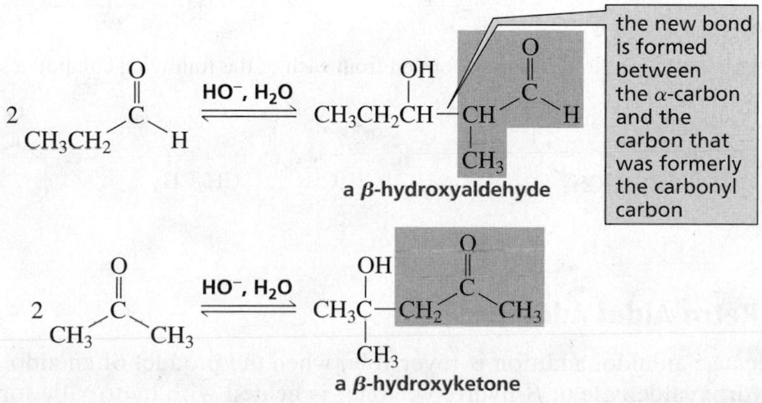

a β-hydroxyaldehyde

a β-hydroxyketone

Because the addition reaction is reversible, good yields of the addition product are obtained only if it is removed from the solution as it is formed.

MECHANISM FOR THE ALDOL ADDITION

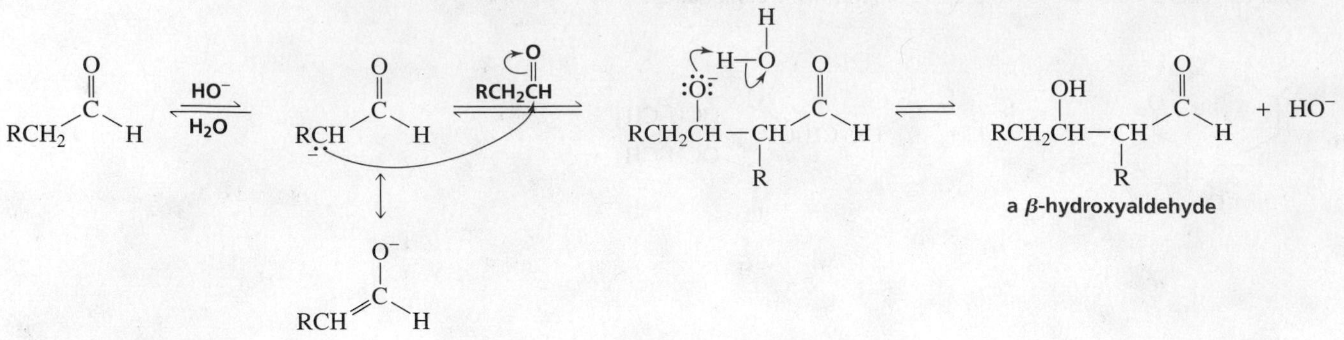

- A base removes a proton from the α-carbon, creating an enolate ion.
- The enolate ion adds to the carbonyl carbon of a second molecule of the carbonyl compound.
- The negatively charged oxygen is protonated.

Ketones are less susceptible than aldehydes to attack by nucleophiles, so aldol additions occur more slowly with ketones.

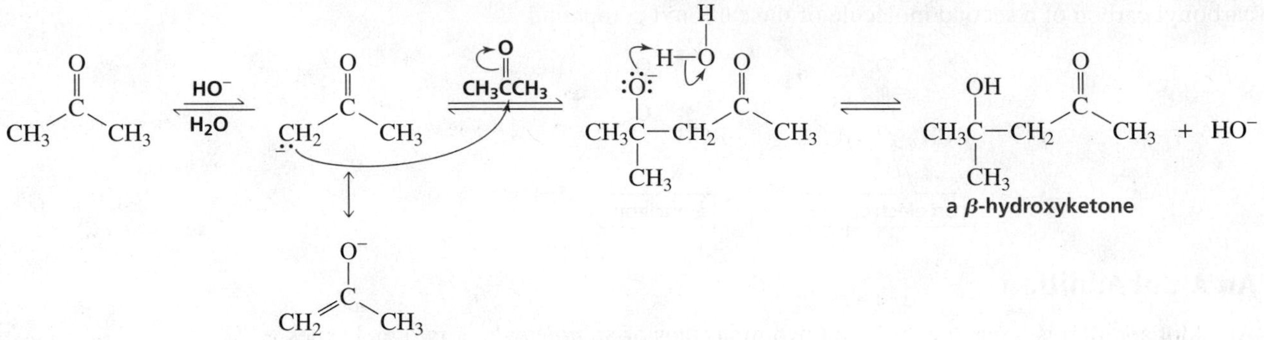

An aldol addition forms a β-hydroxyaldehyde or a β-hydroxyketone.

Notice that the aldol addition is a nucleophilic addition reaction. It is just like the nucleophilic addition reactions that aldehydes and ketones undergo with other carbon nucleophiles. Because an aldol addition occurs between two molecules of the same carbonyl compound, the product has twice as many carbons as the reacting aldehyde or ketone.

PROBLEM 20

What aldol addition product is formed from each of the following compounds?

a. 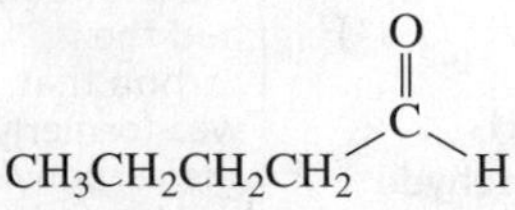 $CH_3CH_2CH_2CH_2$ — C(=O) — H

b. 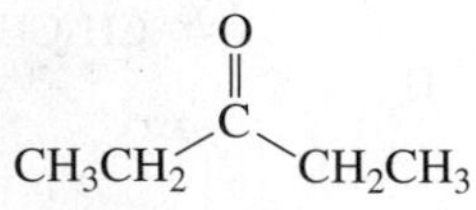 CH_3CH_2 — C(=O) — CH_2CH_3

c.

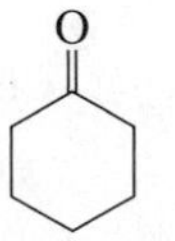

A Retro-Aldol Addition

Because an aldol addition is reversible, when the product of an aldol addition (the β-hydroxyaldehyde or β-hydroxyketone) is heated with hydroxide ion and water, the aldehyde or ketone that formed the aldol addition product can be regenerated. In Section 21 we will see that a retro-aldol addition is an important reaction in glycolysis.

keto-enol
interconversion

PROBLEM 21◆

What aldehyde or ketone would be obtained when each of the following compounds is heated in a basic aqueous solution?

a. 2-ethyl-3-hydroxyhexanal

b. 4-hydroxy-4-methyl-2-pentanone

c. 2,4-dicyclohexyl-3-hydroxybutanal

d. 5-ethyl-5-hydroxy-4-methyl-3-heptanone

---

11   THE DEHYDRATION OF ALDOL ADDITION PRODUCTS FORMS α,β-UNSATURATED ALDEHYDES AND KETONES

Alcohols dehydrate when they are heated with acid. The β-hydroxyaldehyde and β-hydroxyketone products of aldol addition reactions are easier to dehydrate than many other alcohols because the double bond formed when the compound is dehydrated is conjugated with a carbonyl group. Conjugation increases the stability of the product and therefore makes it easier to form.

When the product of an aldol addition is dehydrated, the overall reaction is called an **aldol condensation.** A **condensation reaction** is a reaction that combines two molecules by forming a new C—C bond while removing a small molecule (usually water or an alcohol). Notice that an aldol condensation forms an α,β-**unsaturated** aldehyde or an α,β-**unsaturated ketone,** called enones ("ene" for the double bond and "one" for the carbonyl group).

conjugated double bonds

a β-hydroxyaldehyde   a α,β-unsaturated aldehyde
an enone

An aldol addition product loses water to form an aldol condensation product.

Unlike alcohols that can be dehydrated only under acidic conditions, β-hydroxyaldehydes and β-hydroxyketones can also be dehydrated under basic conditions. Thus, heating the product of an aldol addition in either acid or base leads to dehydration.

a β-hydroxyketone   an α,β-unsaturated ketone
an enone

An aldol condensation forms an α,β-unsaturated aldehyde or an α,β-unsaturated ketone.

E1 reactions are two-step elimination reactions that form a carbocation intermediate and E2 reactions are concerted elimination reactions. The preceding base-catalyzed dehydration represents the third kind of elimination reaction—namely, an **E1cB** (elimination unimolecular conjugate base) **reaction,** a two-step elimination reaction that forms a carbanion intermediate. E1cB reactions occur only if the carbanion can be stabilized by electron delocalization.

MECHANISM FOR THE E1cB REACTION

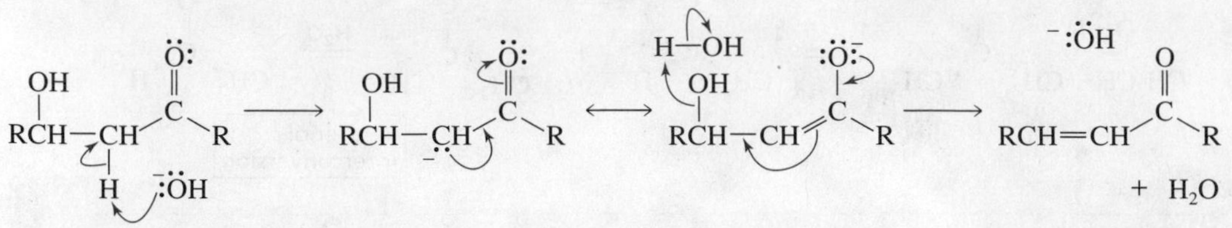

- Hydroxide ion removes a proton from the α-carbon, thereby forming an enolate ion.
- The enolate ion eliminates the OH group, which picks up a proton as it leaves, thereby making it a better leaving group.

Sometimes dehydration occurs under the conditions in which the aldol addition is carried out, without requiring additional heat. For example, in the following reaction, the aldol addition product loses water as soon as it is formed because the new double bond is conjugated not only with the carbonyl group but also with the benzene ring. (Note that the more stable the alkene, the easier it is formed.)

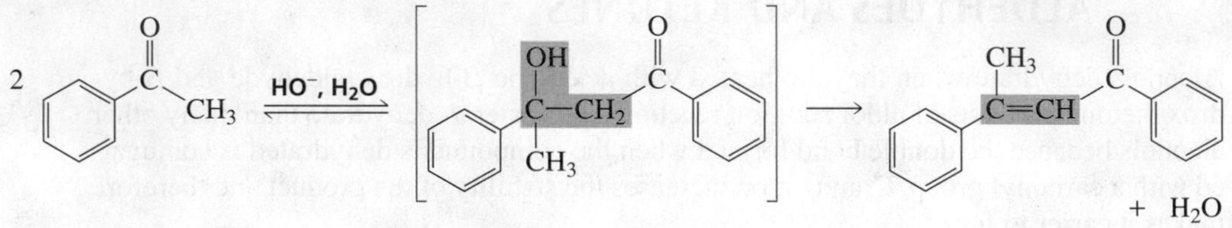

PROBLEM 22♦

What product is obtained from the aldol condensation of cyclohexanone?

PROBLEM 23 Solved

How could you prepare the following compounds using a starting material containing no more than three carbons?

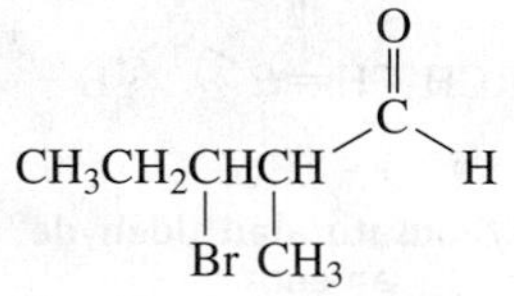

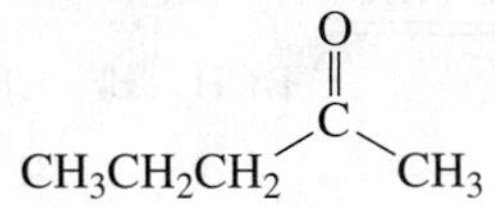

Solution to 23a A compound with the correct six-carbon skeleton can be obtained if a three-carbon aldehyde undergoes an aldol addition. Dehydration of the addition product forms an α,β-unsaturated aldehyde. Conjugate addition of HBr forms the target molecule.

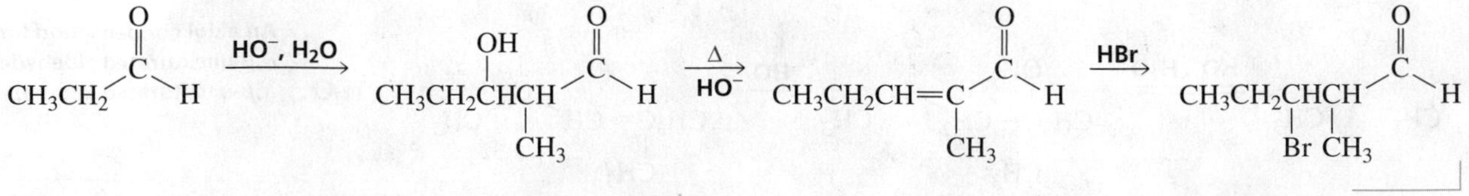

---

12   A CROSSED ALDOL ADDITION

If two different carbonyl compounds are used in an aldol addition—known as a **crossed aldol addition**—four products can be formed because reaction with hydroxide ion can form two different enolate ions (A^- and B^-) and each enolate ion can react with either of the two carbonyl compounds (A or B). A reaction that forms four products clearly is not a synthetically useful reaction.

$$\text{CH}_3\text{CH}_2\overset{\text{O}^-}{\text{CH}}-\overset{\text{CH}_3}{\text{CH}}-\overset{\text{O}}{\underset{}{\text{C}}}\text{H} \underset{\text{HO}^-}{\overset{\text{H}_2\text{O}}{\rightleftharpoons}} \text{CH}_3\text{CH}_2\overset{\text{OH}}{\text{CH}}-\overset{\text{CH}_3}{\text{CH}}-\overset{\text{O}}{\underset{}{\text{C}}}\text{H}$$

$$\overset{\text{O}}{\underset{}{\text{CH}_3\overset{-}{\text{CH}}}}\text{C}\text{H} \quad \text{A} \nearrow$$

$$\underset{\text{H}_2\text{O}}{\overset{\text{HO}^-}{\rightleftharpoons}} \quad \text{A}^- \nearrow \quad \searrow \text{B}$$

$$\text{CH}_3\overset{\text{O}^-}{\text{CH}}-\overset{\text{CH}_3}{\text{CH}}-\overset{\text{O}}{\underset{}{\text{C}}}\text{H} \underset{\text{HO}^-}{\overset{\text{H}_2\text{O}}{\rightleftharpoons}} \text{CH}_3\overset{\text{OH}}{\text{CH}}-\overset{\text{CH}_3}{\text{CH}}-\overset{\text{O}}{\underset{}{\text{C}}}\text{H}$$

$$\overset{\text{O}}{\underset{}{\text{CH}_3\text{CH}_2}}\text{C}\text{H} \quad + \quad \overset{\text{O}}{\underset{}{\text{CH}_3}}\text{C}\text{H}$$
$$\text{A} \qquad\qquad \text{B}$$

$$\underset{\text{H}_2\text{O}}{\overset{\text{HO}^-}{\rightleftharpoons}}$$

$$\text{CH}_3\text{CH}_2\overset{\text{O}^-}{\text{CH}}-\text{CH}_2-\overset{\text{O}}{\underset{}{\text{C}}}\text{H} \underset{\text{HO}^-}{\overset{\text{H}_2\text{O}}{\rightleftharpoons}} \text{CH}_3\text{CH}_2\overset{\text{OH}}{\text{CH}}-\text{CH}_2-\overset{\text{O}}{\underset{}{\text{C}}}\text{H}$$

$$\overset{\text{O}}{\underset{}{\overset{-}{\text{CH}_2}}}\text{C}\text{H} \quad \text{A} \nearrow$$

$$\text{B}^- \searrow \quad \searrow \text{B}$$

$$\text{CH}_3\overset{\text{O}^-}{\text{CH}}-\text{CH}_2-\overset{\text{O}}{\underset{}{\text{C}}}\text{H} \underset{\text{HO}^-}{\overset{\text{H}_2\text{O}}{\rightleftharpoons}} \text{CH}_3\overset{\text{OH}}{\text{CH}}-\text{CH}_2-\overset{\text{O}}{\underset{}{\text{C}}}\text{H}$$

Primarily one product can be obtained from a crossed aldol addition if one of the aldehydes does not have any α-hydrogens and, therefore, cannot form an enolate ion. That cuts the possible products from four to two. Then, if the aldehyde with α-hydrogens is added slowly to a solution of the aldehyde without α-hydrogens and hydroxide ion, the chance that the aldehyde with α-hydrogens, after forming an enolate ion, will then react with another molecule of its parent carbonyl compound will be minimized, so the possible products are cut to essentially one.

$$\overset{\text{O}}{\underset{}{\text{C}_6\text{H}_5}}\text{C}\text{H} \quad + \quad \text{CH}_3\overset{\text{O}}{\underset{}{\text{C}}}\text{C(CH}_3)_3 \underset{\text{H}_2\text{O/EtOH}}{\overset{\text{HO}^-}{\longrightarrow}} \text{C}_6\text{H}_5\text{CH}=\text{CH}\overset{\text{O}}{\underset{}{\text{C}}}\text{C(CH}_3)_3 \quad + \quad \text{H}_2\text{O}$$

add slowly

90%

> If one of the carbonyl compounds does not have any α-hydrogens, then the compound with α-hydrogens is added slowly to a solution of the compound without α-hydrogens and a base.

This crossed aldol condensation is sufficiently important to be given its own name— the **Claisen-Schmidt condensation.**

If both aldehydes have α-hydrogens, primarily one aldol addition product can be formed if LDA is used to remove the α-hydrogen that creates the enolate ion. Because LDA is a strong base, all of the carbonyl compound will be converted into an enolate ion, so none of that carbonyl compound will be left for the enolate ion to react with in an aldol addition. Therefore, an aldol addition will not be able to occur until the second carbonyl compound is added to the reaction mixture. If the second carbonyl compound is added slowly, the chance that it will form an enolate ion and then react with another molecule of its parent carbonyl compound will be minimized.

$$\overset{\text{O}}{\bigcirc} \xrightarrow{\text{LDA}} \overset{\text{O}}{\bigcirc}{}^- \xrightarrow[\text{2. H}_2\text{O}]{\overset{\text{1. CH}_3\text{CH}_2\text{CH}_2\overset{\text{O}}{\text{CH}}}{\text{add slowly}}} \overset{\text{O}}{\bigcirc}\overset{\text{OH}}{\underset{}{\text{CHCH}_2\text{CH}_2\text{CH}_3}}$$

> If both reactants undergoing a condensation reaction have α-hydrogens, then LDA is used to form the enolate ion and the other carbonyl carbon is added slowly.

Retrosynthetic Analysis

To determine the starting materials needed for the synthesis of a compound formed by an *aldol addition*, first locate the new carbon–carbon bond that was formed between the α-carbon of one molecule of carbonyl compound and the carbonyl carbon of the other molecule. Then draw the two carbonyl compounds that were the reactants in the aldol addition.

the bond formed by the aldol addition

$$CH_3CH_2CH(OH)-CH_2-CHO \implies CH_3CH_2-CHO + CH_3-CHO$$

To synthesize the target molecule, add LDA to the carbonyl compound whose α-carbon will be the nucleophile. Then add the other carbonyl compound slowly to the enolate ion.

$$CH_3-CHO \xrightarrow[\text{THF}]{\text{LDA}} {}^-CH_2-CHO \xrightarrow[\text{2. H}_2\text{O}]{\substack{\text{1. CH}_3\text{CH}_2\text{CHO} \\ \text{add slowly}}} CH_3CH_2CH(OH)-CH_2-CHO$$

To determine the starting materials needed for the synthesis of a compound formed by an *aldol condensation*, first convert the α,β-unsaturated carbonyl compound to a β-hydroxycarbonyl compound and then proceed as just described.

$$CH_3CH_2=C(CH_2CH_3)-CHO \implies CH_3CH_2-C(CH_2CH_3)(OH)-CHO \implies CH_3-CHO + CH_3CH_2CH_2-CHO$$

a β-hydroxyaldehyde

Breast Cancer and Aromatase Inhibitors

Current statistics show that one in eight women will develop breast cancer. Men also get breast cancer but are 100 times less likely to do so than women. There are several different types of tumors that cause breast cancer, some of which are estrogen dependent. An estrogen-dependent tumor has receptors that bind estrogen. Without estrogen, the tumor cannot grow.

the A ring $\xrightarrow{\text{aromatase}}$ estrone $\xrightarrow{\text{reduction}}$ estradiol

an aromatic ring

The A ring of estrogen hormones (estrone and estradiol) is an aromatic phenol. One of the last steps in the biosynthesis of estrogen hormones from cholesterol is catalyzed by an enzyme called aromatase. Aromatase catalyzes the reaction that causes the A ring to become aromatic. Therefore, one approach to the treatment of breast cancer is to administer drugs that will inhibit aromatase. If aromatase is inhibited, the estrogen hormones cannot be synthesized but the biosynthesis of other important hormones from cholesterol will not be affected. There are several aromatase inhibitors on the market, and scientists continue to search for more potent ones. Fifteen different aromatase inhibitors have been isolated from the leaves of the paper mulberry tree (*Broussonetia papyifera*), one of which is morachalcone A.

morachalcone A

PROBLEM 24

Describe how the following compounds could be prepared using an aldol addition in the first step of the synthesis:

a.

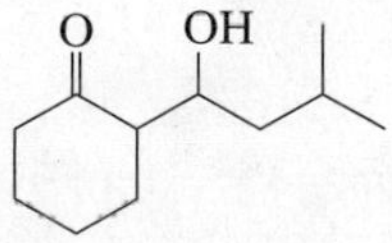

c.

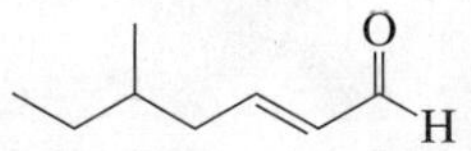

b.

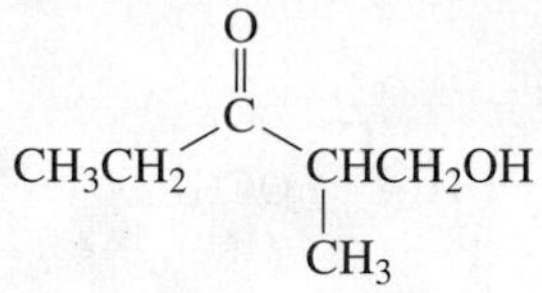

CH₃CH₂ C CHCH₂OH
         CH₃

d.

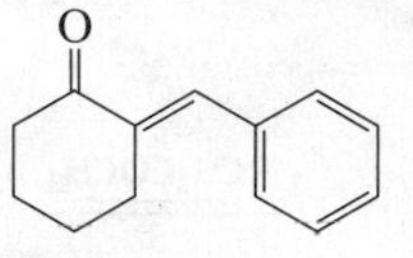

PROBLEM 25

What two carbonyl compounds are required for the synthesis of morachalcone A, via a Claisen–Schmidt condensation?

PROBLEM 26

Propose a mechanism for the following reaction:

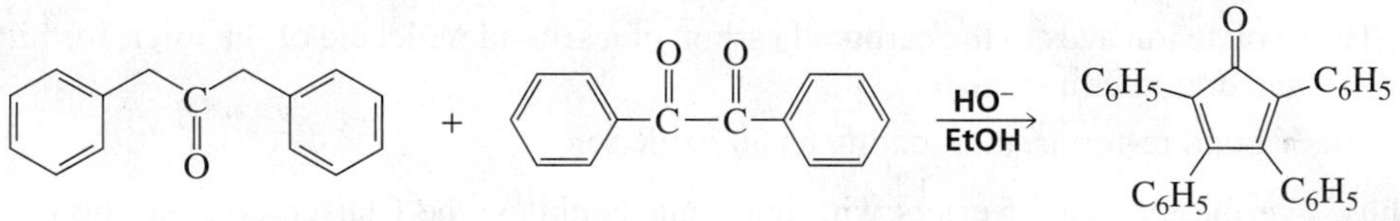

$\xrightarrow[\text{EtOH}]{\text{HO}^-}$

C₆H₅ C₆H₅

C₆H₅   C₆H₅

13 A CLAISEN CONDENSATION FORMS A β-KETO ESTER

When two molecules of an *ester* undergo a condensation reaction, the reaction is called a **Claisen condensation.** The product of a Claisen condensation is a **β-keto ester.** In a Claisen condensation, as in an aldol addition, one molecule of carbonyl compound is the nucleophile and a second molecule is the electrophile. And, as in an aldol addition, the new C—C bond connects the α-carbon of one molecule to the carbon that was formerly the carbonyl carbon of the other molecule.

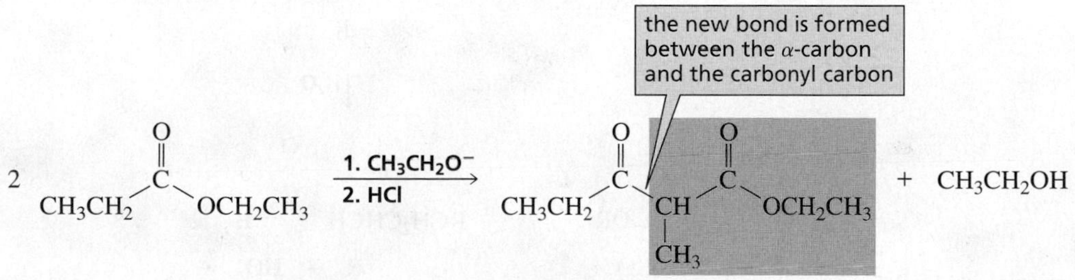

the new bond is formed between the α-carbon and the carbonyl carbon

2

$\xrightarrow[\text{2. HCl}]{\text{1. CH}_3\text{CH}_2\text{O}^-}$

+ CH₃CH₂OH

a β-keto ester

The base used to remove the proton from the α-carbon in a Claisen condensation should be the same as the leaving group of the ester. This is necessary because the base, in addition to being able to remove a proton from an α-carbon, can react as a nucleophile and add to the carbonyl group of the ester. If the nucleophile is identical to the OR group of the ester, then nucleophilic addition to the carbonyl group will not change the reactant.

$$CH_3-\overset{\overset{\displaystyle :\ddot{O}:}{\|}}{C}-\ddot{O}CH_3 \quad + \quad CH_3\ddot{O}:^- \quad \rightleftharpoons \quad CH_3-\overset{\overset{\displaystyle :\ddot{O}:^-}{|}}{\underset{\underset{\displaystyle :\ddot{O}CH_3}{|}}{C}}-\ddot{O}CH_3$$

MECHANISM FOR THE CLAISEN CONDENSATION

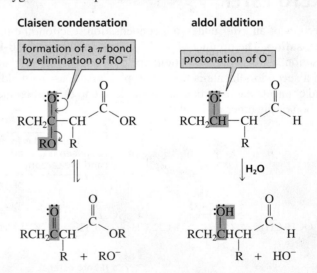

A Claisen condensation forms a β-keto ester.

- A base removes a proton from the α-carbon, creating an enolate ion. The base employed corresponds to the leaving group of the ester.

- The enolate ion adds to the carbonyl carbon of a second molecule of the ester, forming a tetrahedral intermediate.

- The π bond re-forms, eliminating an alkoxide ion.

Thus, like the reaction of esters with other nucleophiles, the Claisen condensation is a nucleophilic addition–elimination reaction.

Notice that, after nucleophilic addition, the Claisen condensation and the aldol addition reactions differ. In the Claisen condensation, the negatively charged oxygen re-forms the carbon–oxygen π bond and eliminates the ^-OR group. In the aldol addition, the negatively charged oxygen obtains a proton from the solvent.

Claisen condensation   **aldol addition**

formation of a π bond by elimination of RO^- protonation of O^-

The last step of the Claisen condensation is different than the last step of the aldol addition because esters are different than aldehydes or ketones. The carbon bonded to the negatively charged oxygen in an ester is also bonded to a group that can be eliminated, whereas the carbon bonded to the negatively charged oxygen in an aldehyde or a ketone is not bonded to such a group. Thus, the Claisen condensation is a nucleophilic addition–elimination reaction, whereas the aldol addition is a nucleophilic addition reaction.

The Claisen condensation is reversible and favors the reactant since it is more stable than the β-keto ester. The condensation reaction can be driven to completion, however, if a proton is removed from the β-keto ester (Le Châtelier's principle). A proton is easily removed because the central α-carbon of the β-keto ester is flanked by two carbonyl groups, making its α-hydrogen much more acidic than the α-hydrogen of the ester.

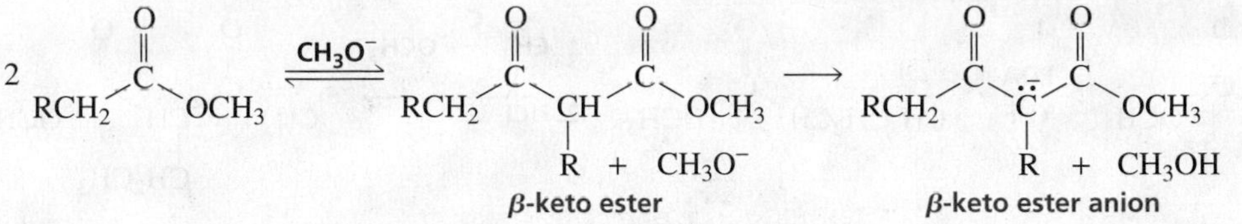

$$2 \; RCH_2\overset{O}{\underset{}{\overset{\|}{C}}}OCH_3 \quad \underset{}{\overset{CH_3O^-}{\rightleftharpoons}} \quad RCH_2\overset{O}{\overset{\|}{C}}\underset{R \;+\; CH_3O^-}{CH}\overset{O}{\overset{\|}{C}}OCH_3 \quad \longrightarrow \quad RCH_2\overset{O}{\overset{\|}{C}}\underset{R \;+\; CH_3OH}{\overset{..}{C}}\overset{O}{\overset{\|}{C}}OCH_3$$

β-keto ester   **β-keto ester anion**

Consequently, a successful Claisen condensation requires an ester with two α-hydrogens and an equivalent amount of base rather than a catalytic amount of base. When the reaction is over, addition of acid to the reaction mixture reprotonates the β-keto ester anion and protonates the alkoxide ion so the reverse reaction cannot occur.

PROBLEM 27◆

Draw the products of the following reactions:

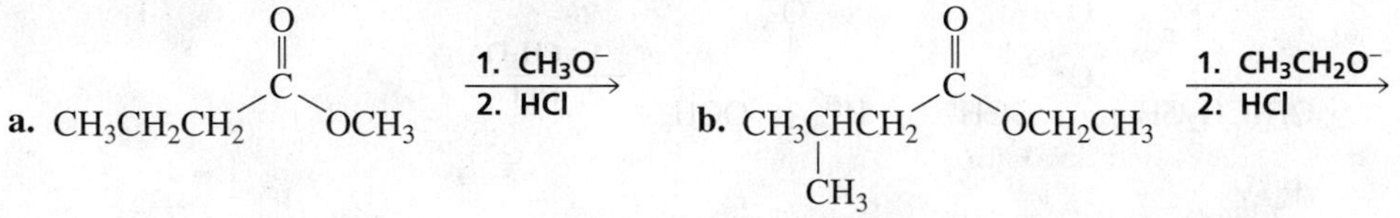

a. $CH_3CH_2CH_2\overset{O}{\overset{\|}{C}}OCH_3$ $\xrightarrow{\substack{1.\; CH_3O^- \\ 2.\; HCl}}$

b. $CH_3\underset{CH_3}{\overset{|}{CH}}CH_2\overset{O}{\overset{\|}{C}}OCH_2CH_3$ $\xrightarrow{\substack{1.\; CH_3CH_2O^- \\ 2.\; HCl}}$

PROBLEM 28◆

Which of the following esters cannot undergo a Claisen condensation?

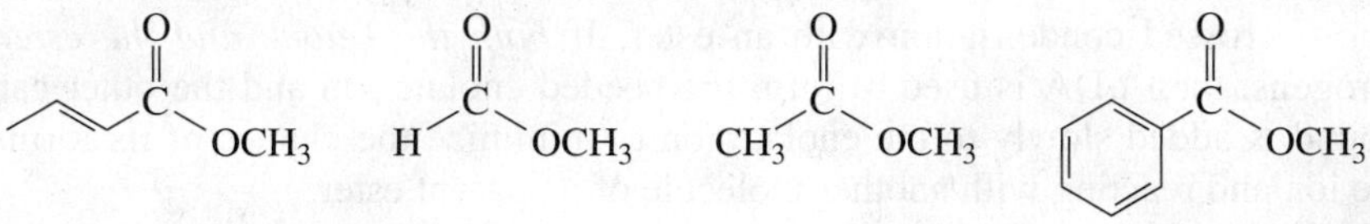

A: $CH=CH-\overset{O}{\overset{\|}{C}}-OCH_3$
B: $H-\overset{O}{\overset{\|}{C}}-OCH_3$
C: $CH_3-\overset{O}{\overset{\|}{C}}-OCH_3$
D: $C_6H_5-\overset{O}{\overset{\|}{C}}-OCH_3$

   A   B   C   D

A Crossed Claisen Condensation

A **crossed Claisen condensation** is a condensation reaction between two different esters. Like a crossed aldol addition, a crossed Claisen condensation is a useful reaction only if it is carried out under conditions that foster the formation of primarily one product. Otherwise, the reaction will form a mixture of products that are difficult to separate.

Primarily one product will be formed from a crossed Claisen condensation if one of the esters has no α-hydrogens (and therefore cannot form an enolate ion) and the ester with α-hydrogens is added slowly to a solution of the ester without α-hydrogens and the alkoxide ion.

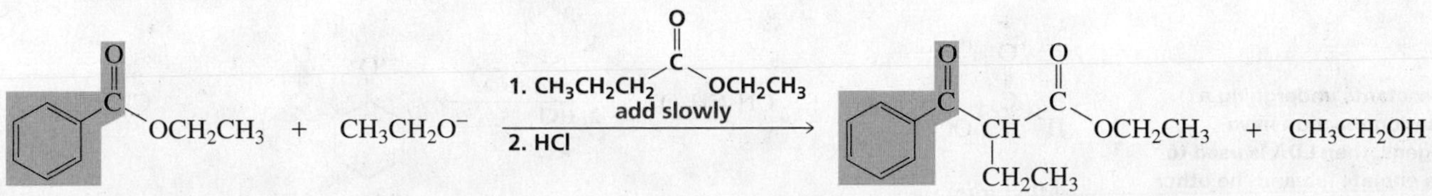

$$C_6H_5\overset{O}{\overset{\|}{C}}OCH_2CH_3 \;+\; CH_3CH_2O^- \quad \xrightarrow[2.\; HCl]{\substack{1.\; CH_3CH_2CH_2\overset{O}{\overset{\|}{C}}OCH_2CH_3 \\ add\; slowly}} \quad C_6H_5\overset{O}{\overset{\|}{C}}\underset{CH_2CH_3}{\overset{|}{CH}}\overset{O}{\overset{\|}{C}}OCH_2CH_3 \;+\; CH_3CH_2OH$$

If both esters have α-hydrogens, primarily one product can be formed if LDA is used to remove the α-hydrogen to form the ester enolate ion. The other ester is then added slowly to minimize the chance that it will form an enolate ion and react with another molecule of its parent ester.

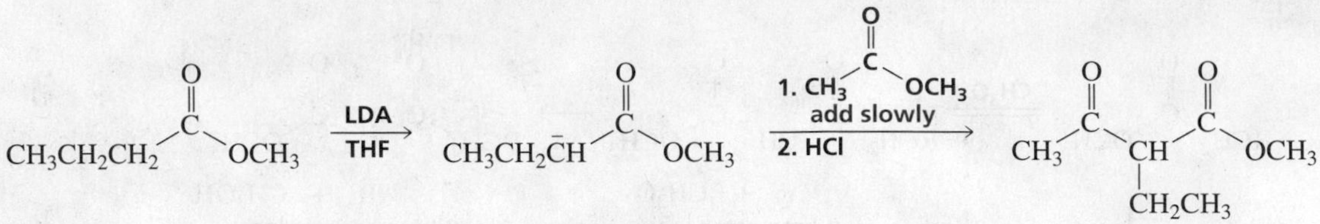

PROBLEM 29

Draw the product of each of the following reactions:

a.

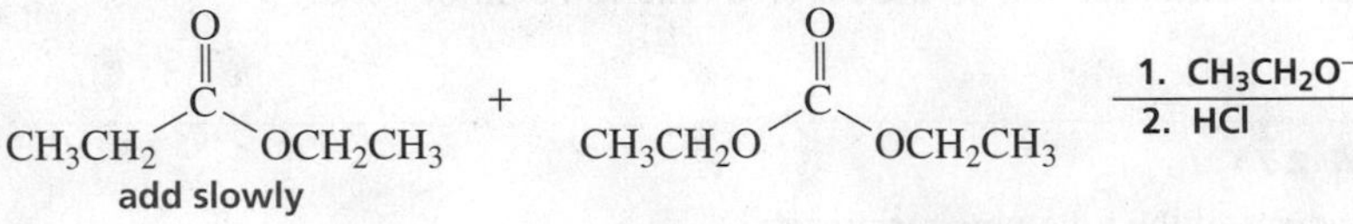

b.

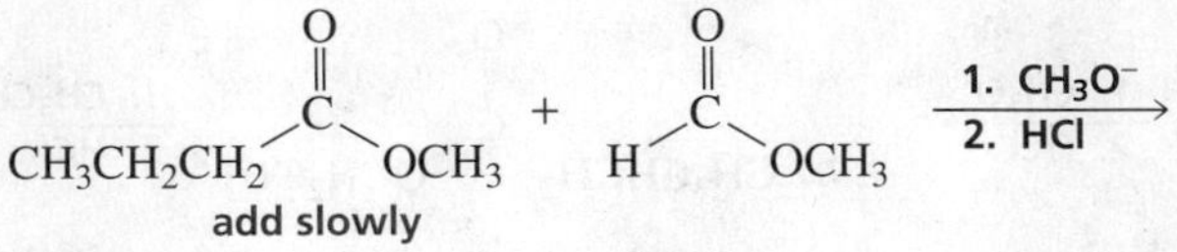

---

14   OTHER CROSSED CONDENSATIONS

In addition to crossed aldol additions and crossed Claisen condensations, a ketone can undergo a crossed condensation with an ester. If *both the ketone and the ester have* α-hydrogens, then LDA is used to form the needed enolate ion and the other carbonyl compound is added slowly to the enolate ion to minimize the chance of its forming an enolate ion and reacting with another molecule of its parent ester.

A *β-diketone* is formed when a ketone condenses with an *ester*.

If one of the carbonyl compounds does not have any α-hydrogens, then the compound with α-hydrogens is added slowly to a solution of the compound without α-hydrogens and a base.

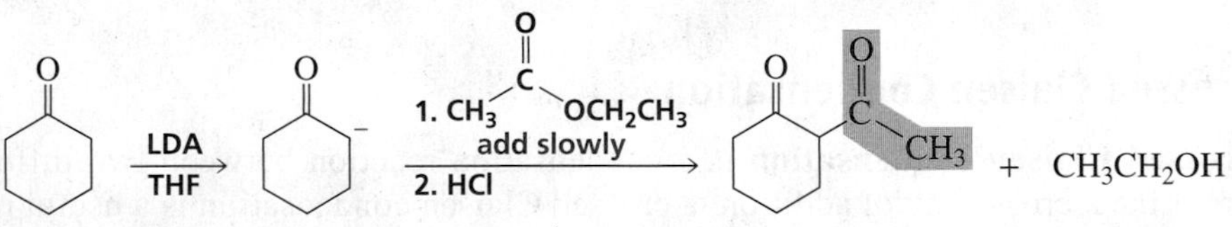

If *one of the carbonyl compounds does not have* any α-hydrogens, then the compound with α-hydrogens is added slowly to a solution of the compound without α-hydrogens and a base. A *β-keto aldehyde* is formed when a ketone condenses with a *formate ester*.

A condensation between a ketone and an ester forms a 1,3-dicarbonyl compound.

If both reactants undergoing a condensation reaction have α-hydrogens, then LDA is used to form the enolate ion and the other carbonyl carbon is added slowly.

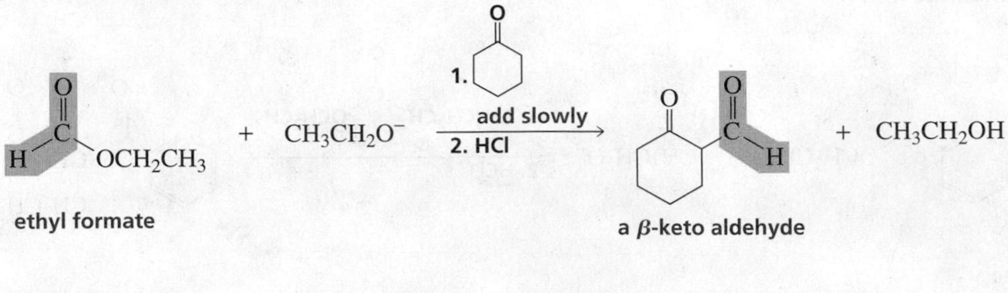

ethyl formate

a β-keto aldehyde

A β-keto ester is formed when a ketone condenses with *diethyl carbonate*.

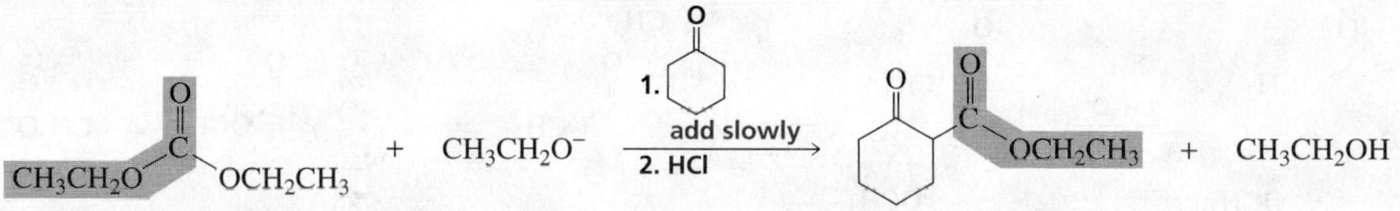

diethyl carbonate   a β-keto ester

PROBLEM 30

Show how each of the following compounds could be prepared from methyl phenyl ketone:

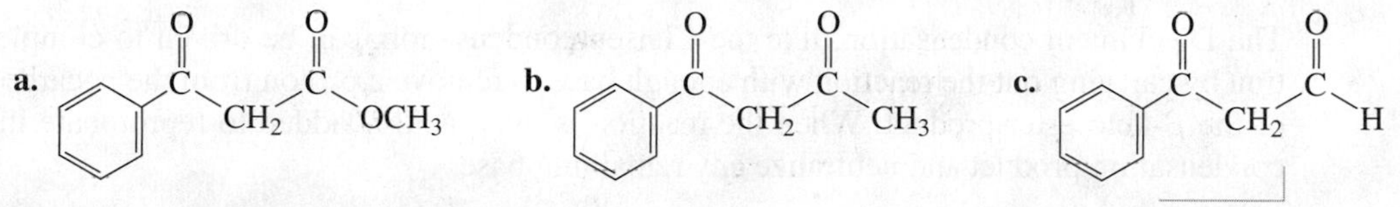

a.   b.   c.

---

15   INTRAMOLECULAR CONDENSATIONS AND INTRAMOLECULAR ALDOL ADDITIONS

If a compound has two functional groups that can react with each other, an intramolecular reaction readily occurs if the reaction leads to the formation of a five-or a six-membered rings. Consequently, a compound with two ester, aldehyde, or ketone groups can undergo an intramolecular reaction if a product with a five- or six-membered ring can be formed.

Intramolecular Claisen Condensations

The addition of base to a 1,6-diester causes the diester to undergo an intramolecular Claisen condensation, thereby forming a five-membered ring β-keto ester. An intramolecular Claisen condensation is called a **Dieckmann condensation.**

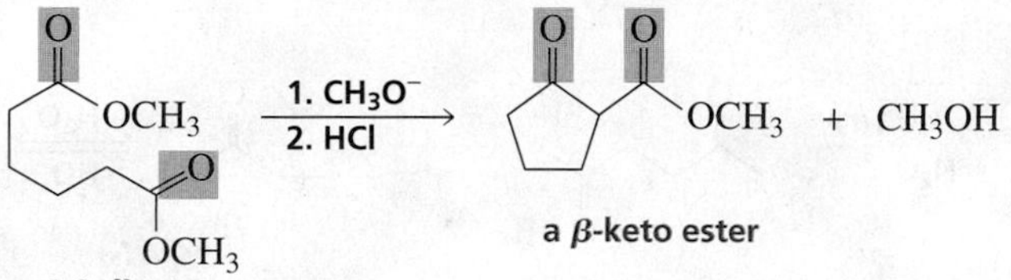

a 1,6-diester   a β-keto ester

A six-membered ring β-keto ester is formed from the Dieckmann condensation of a 1,7-diester.

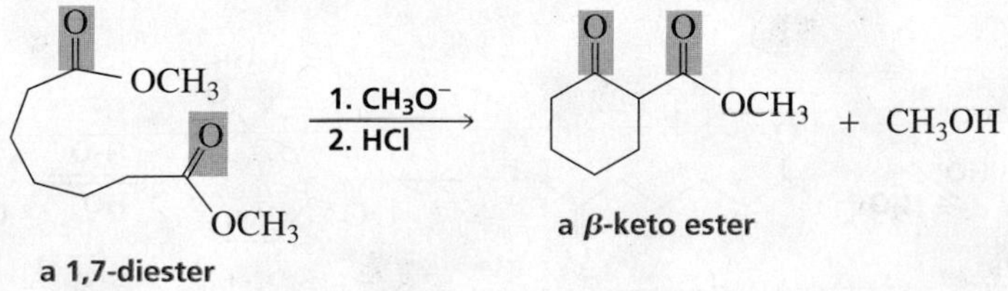

a 1,7-diester   a β-keto ester

    The mechanism for the Dieckmann condensation is the same as the mechanism for the Claisen condensation. The only difference in the two reactions is that the enolate ion and the carbonyl group undergoing nucleophilic addition–elimination are in different molecules in the Claisen condensation and are in the same molecule in the Dieckmann condensation.

MECHANISM FOR THE DIECKMANN CONDENSATION

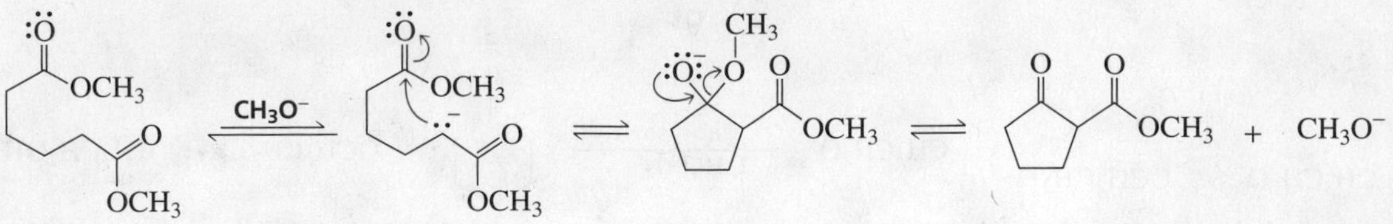

- A base removes a proton from an α-carbon, creating an enolate ion.
- The enolate ion adds to a carbonyl carbon.
- The π bond re-forms, eliminating an alkoxide ion.

The Dieckmann condensation, like the Claisen condensation, can be driven to completion by carrying out the reaction with enough base to remove a proton from the α-carbon of the β-keto ester product. When the reaction is over, acid is added to reprotonate the condensation product and neutralize any remaining base.

PROBLEM 31

Write the mechanism for the base-catalyzed formation of a cyclic β-keto ester from a 1,7-diester.

Intramolecular Aldol Additions

Because a 1,4-diketone has two different sets of α-hydrogens, two different β-hydroxyketones can potentially form, one with a five-membered ring and one with a three-membered ring. The greater stability of the five-membered ring causes it to be formed preferentially. In fact, the five-membered ring product is the only product formed from the intramolecular aldol addition of a 1,4-diketone.

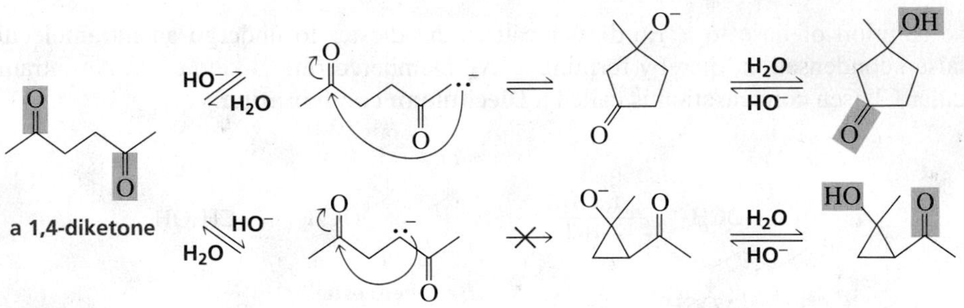

The intramolecular aldol addition of a 1,6-diketone can potentially lead to either a seven- or a five-membered ring product. Again, the more stable product—the one with the five-membered ring—is the only product of the reaction.

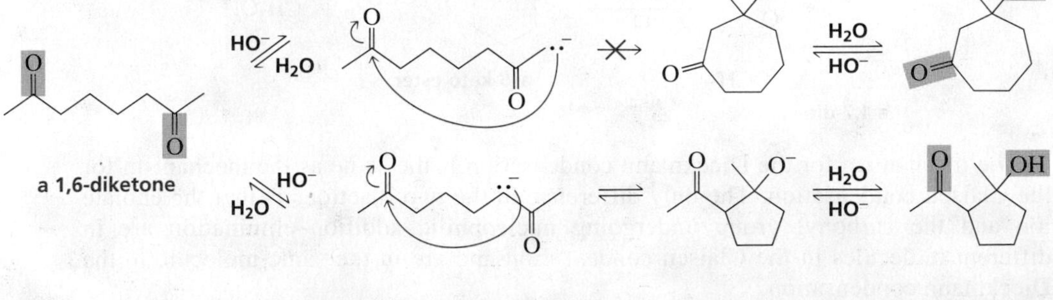

1,5-Diketones and 1,7-diketones undergo intramolecular aldol additions to form six-membered ring β-hydroxyketones.

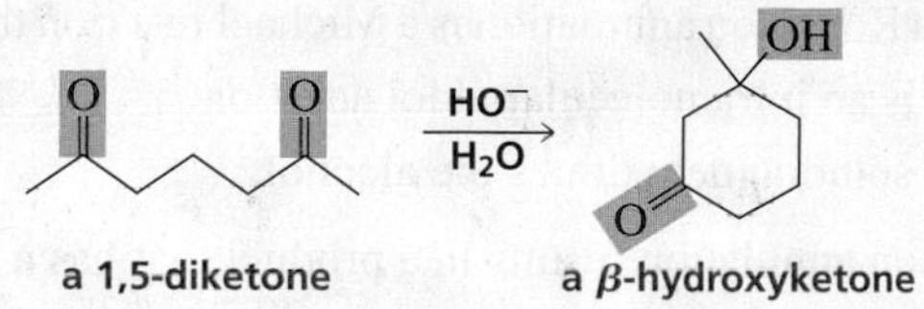

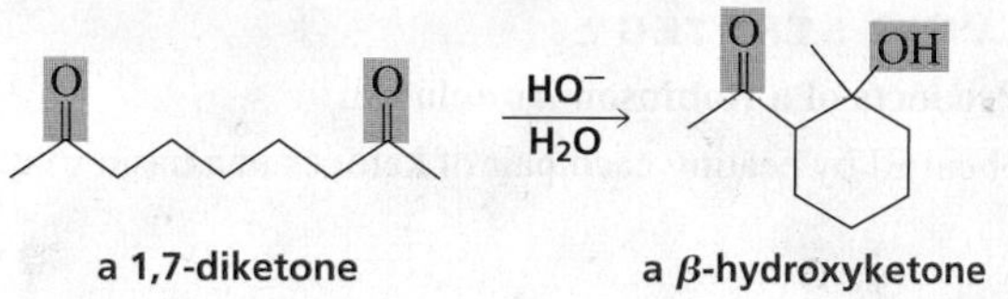

a 1,5-diketone   a β-hydroxyketone

a 1,7-diketone   a β-hydroxyketone

PROBLEM 32♦

If the preference for formation of a six-membered ring were not so great, what other cyclic product would be formed from the intramolecular aldol addition of

a. 2,6-heptanedione?   **b.** 2,8-nonanedione?

PROBLEM 33

Can 2,4-pentanedione undergo an intramolecular aldol addition? If so, why? If not, why not?

PROBLEM 34♦

Draw the product of the reaction of each of the following compounds with a base:

a.

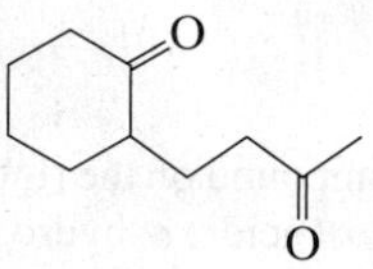

c.

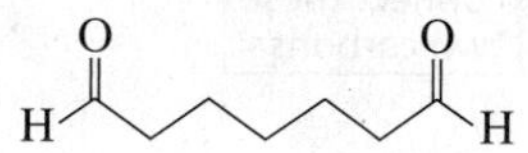

b.

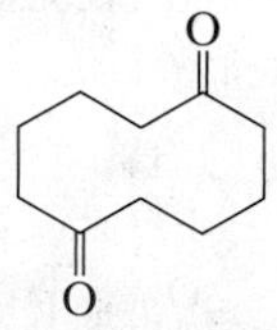

d.

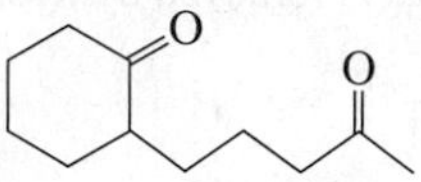

---

16   THE ROBINSON ANNULATION

We have seen that Michael reactions and aldol additions both form new carbon–carbon bonds. The **Robinson annulation** is a reaction that puts these two carbon–carbon bond-forming reactions together in order to form an α,β-unsaturated cyclic ketone. "Annulation" comes from *annulus,* Latin for "ring," so an **annulation reaction** is a ring-forming reaction. The Robinson annulation makes it possible to synthesize many complicated organic molecules.

STEPS IN THE ROBINSON ANNULATION

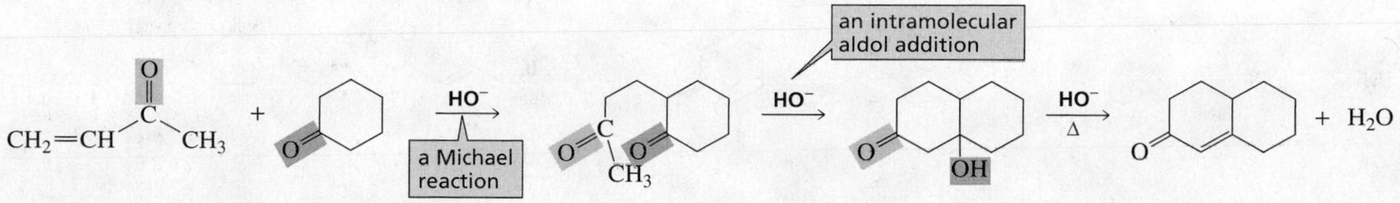

- The first stage of a Robinson annulation is a Michael reaction that forms a 1,5-diketone.
- The second stage is an intramolecular aldol addition.
- Heating the basic solution dehydrates the alcohol.

Notice that a Robinson annulation results in a product that has a 2-cyclohexenone ring.

PROBLEM-SOLVING STRATEGY

Determining the Products of a Robinson Annulation

Draw the product obtained by heating each pair of ketones in a basic solution.

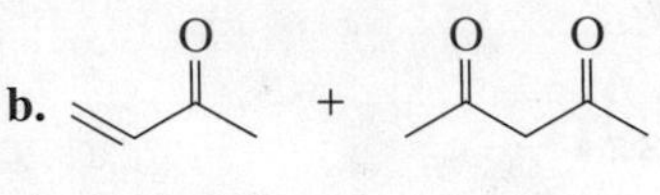

a. 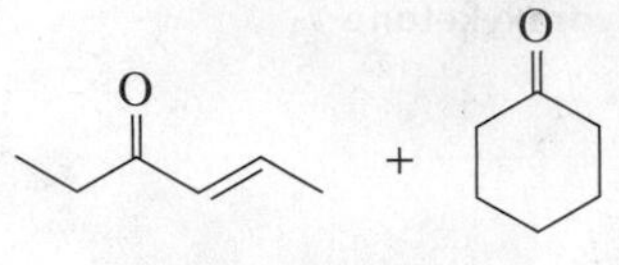 First align the α,β-unsaturated carbonyl compound, so the carbonyl oxygen points at seven o'clock and the double bond is at the top of the structure. Draw the second carbonyl compound to the right of the first with its carbonyl oxygen also pointing at seven o'clock. Connect the two pairs of carbons and draw the double bond that forms as a result of dehydration.

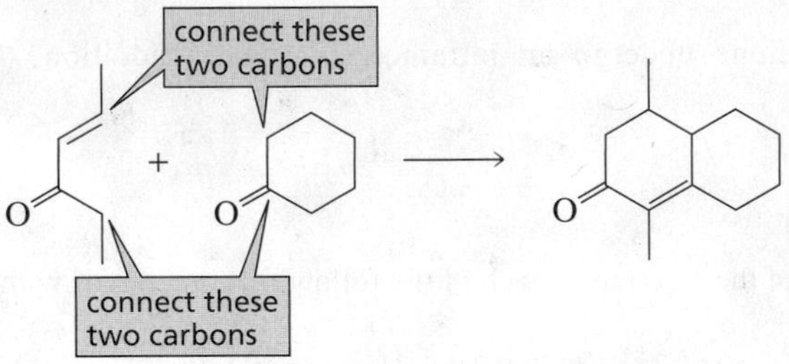

b. Align the carbonyl compounds as described in part **a**. The carbonyl compound on the right has two different α-carbons, so make certain that the α-carbon with the most acidic α-hydrogen is the one that points at the β-carbon of the α,β-unsaturated carbonyl compound, because that will be the α-hydrogen that is removed by hydroxide ion.

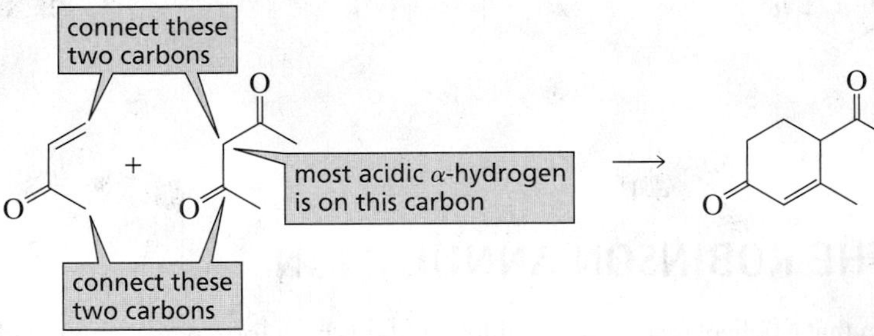

Now use the strategy you have just learned to solve Problem 36.

PROBLEM 35

Draw the product obtained by heating each pair of ketones in a basic solution.

A 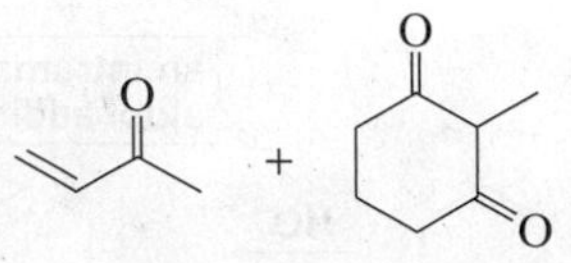 Robinson annulation forms an α,β-unsaturated cyclic ketone.

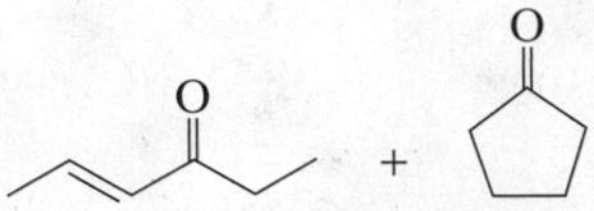

Retrosynthetic Analysis

To determine the starting materials needed for the synthesis of a 2-cyclohexenone by a Robinson annulation, first draw a line that bisects both the double bond and the σ bond between the β- and γ-carbons on the other side of the molecule. The β-carbon comes from the α,β-unsaturated carbonyl compound; the γ-carbon is the α-carbon of the enolate ion of the second carbonyl compound that will add to the β-carbon in the Michael reaction. The double bond was formed by the reaction of the enolate ion of the α,β-unsaturated carbonyl compound and the carbonyl carbon of the second carbonyl compound.

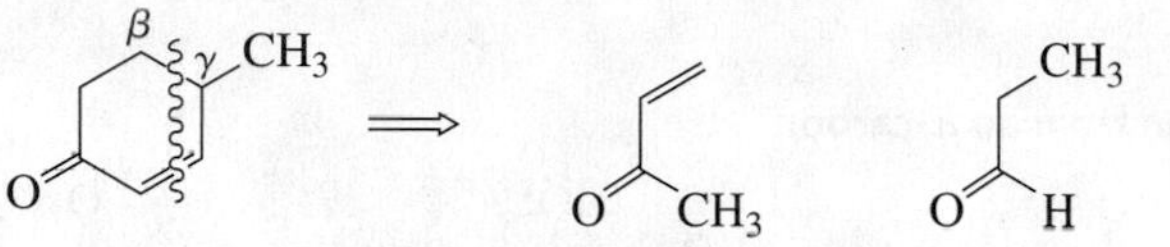

α,β-unsaturated
carbonyl compound

By cutting through the following compound in the same way, we can determine the required reactants for its synthesis:

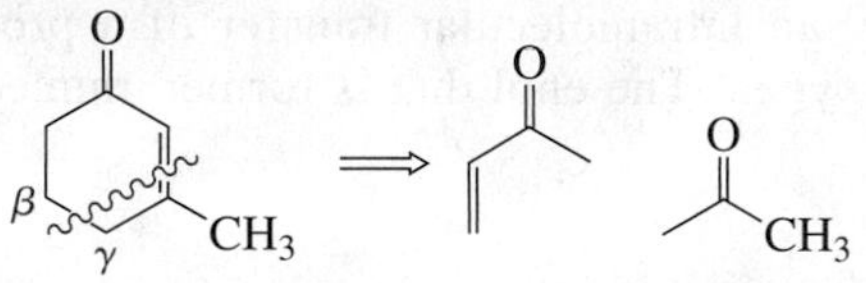

PROBLEM 36

What two carbonyl compounds are needed to synthesize each of the following compounds, using a Robinson annulation?

a. 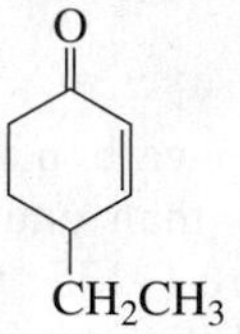   b. 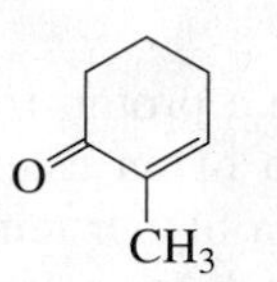   c. 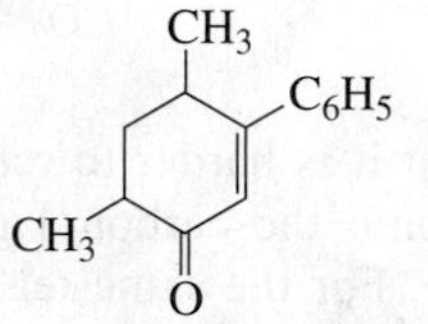   d.

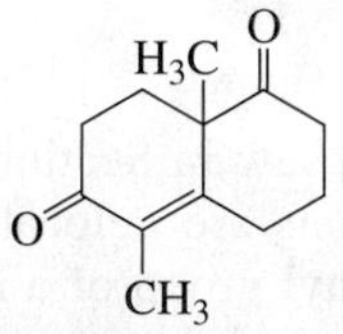

---

17   CARBOXYLIC ACIDS WITH A CARBONYL GROUP AT THE 3-POSITION CAN BE DECARBOXYLATED

Carboxylate ions do not lose CO_2, for the same reason that alkanes such as ethane do not lose a proton—namely, the leaving group would be a carbanion. Carbanions are very strong bases, which makes them very poor leaving groups.

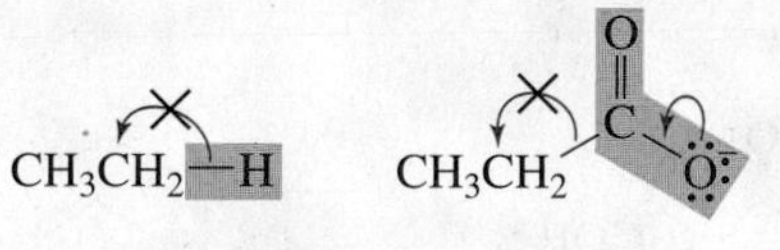

If, however, the CO_2 group is attached to a carbon that is adjacent to a carbonyl carbon, the CO_2 group can be removed, because the electrons left behind can be delocalized onto the carbonyl oxygen. Consequently, carboxylate ions with a carbonyl group at the 3-position (3-oxocarboxylate ions) lose CO_2 with gentle heating (~50 °C). Loss of CO_2 from a molecule is called **decarboxylation.**

removing CO_2 from an α-carbon

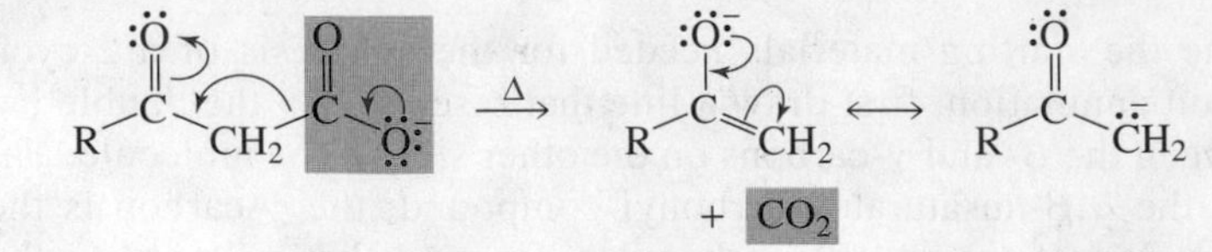

Notice the similarity between removing CO_2 from a 3-oxocarboxylate ion and removing a proton from an α-carbon. In both reactions, a substituent—CO_2 in one case, H^+ in the other—is removed from an α-carbon, and its bonding electrons are delocalized onto an oxygen.

removing a proton from an α-carbon

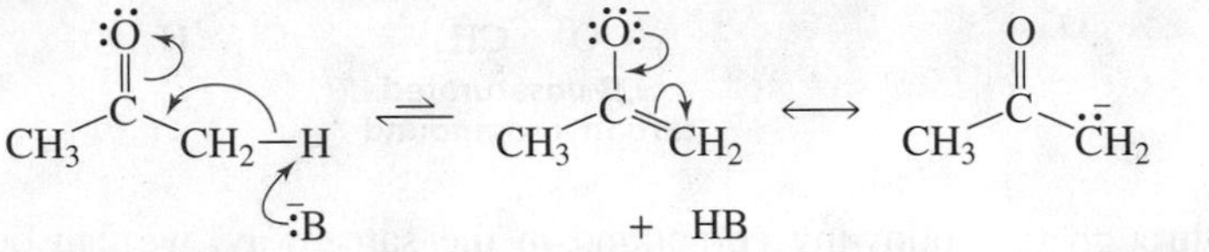

Decarboxylation is even easier (~30 °C) under acidic conditions because the reaction is catalyzed by an intramolecular transfer of a proton from the carboxyl group to the carbonyl oxygen. The enol that is formed immediately tautomerizes to a ketone.

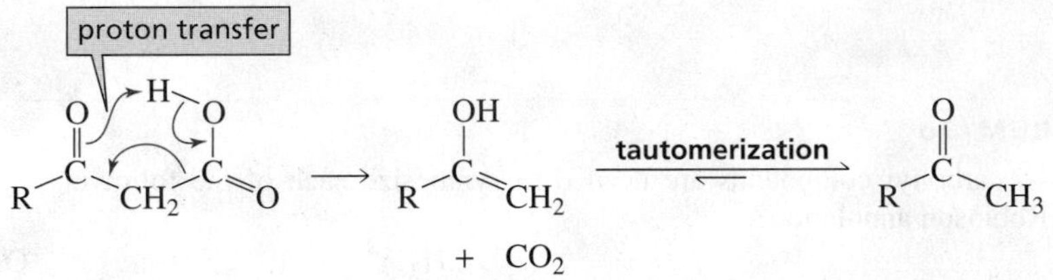

We saw in Section 1 that it is harder to remove a proton from an α-carbon if the electrons are delocalized onto the carbonyl group of an ester rather than onto the carbonyl group of a ketone. For the same reason, a higher temperature (~135 °C) is required to decarboxylate a β-dicarboxylic acid such as malonic acid than to decarboxylate a β-keto acid.

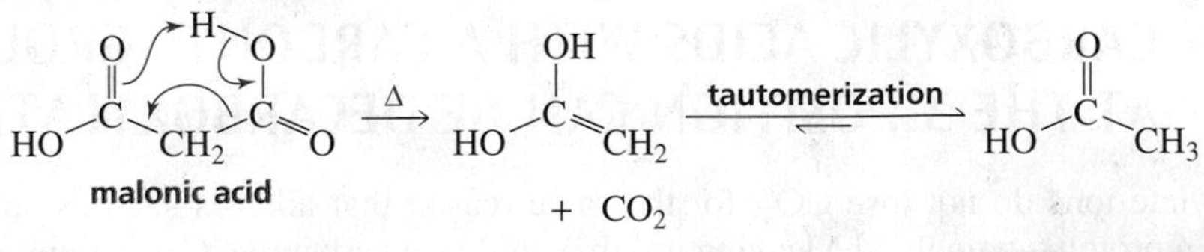

In summary, carboxylic acids with a carbonyl group at the 3-position lose CO_2 when they are heated.

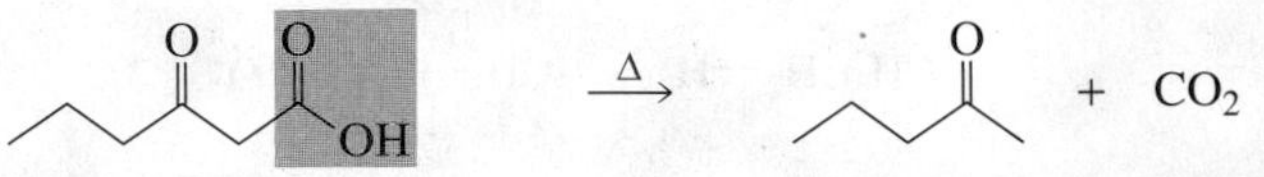

3-oxohexanoic acid   **2-pentanone** $+ CO_2$

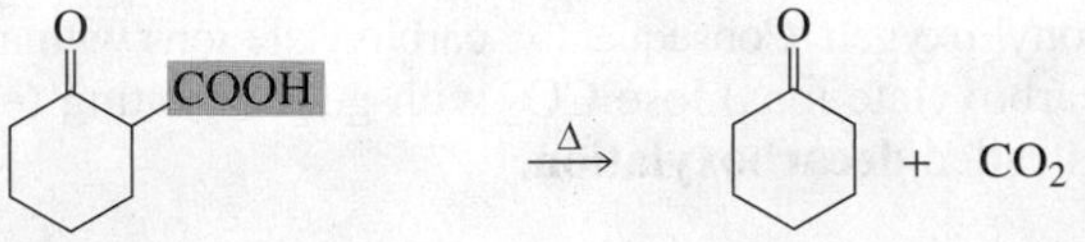

2-oxocyclohexanecarboxylic acid   **cyclohexanone** $+ CO_2$

PROBLEM 37♦

Which of the following compounds would be expected to decarboxylate when heated?

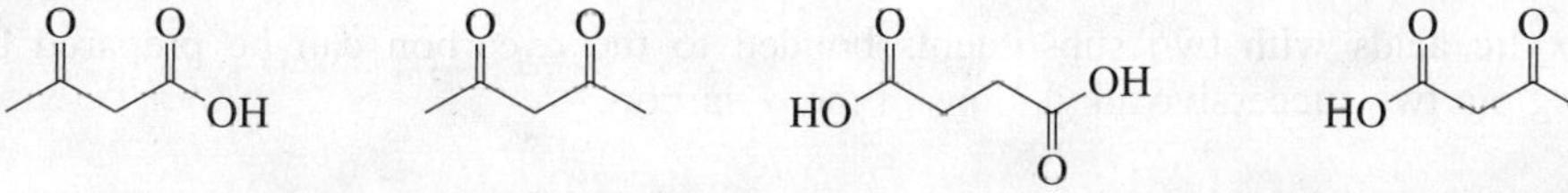

A   B   C   D

---

18   THE MALONIC ESTER SYNTHESIS: A WAY TO SYNTHESIZE A CARBOXYLIC ACID

A combination of two of the reactions discussed in this chapter—namely, alkylation of an α-carbon and decarboxylation of a 3-oxocarboxylic acid—can be used to prepare carboxylic acids of any desired chain length. The procedure is called the **malonic ester synthesis** because the starting material for the synthesis of the carboxylic acid is the diester of malonic acid.

The carbonyl carbon and the α-carbon of the carboxylic acid being synthesized come from malonic ester, and the rest of the carboxylic acid comes from the alkyl halide used in the second step of the synthesis. Thus, a malonic ester synthesis forms a carboxylic acid with two more carbons than the alkyl halide.

malonic ester synthesis

diethyl malonate
malonic ester

from malonic ester

from the alkyl halide

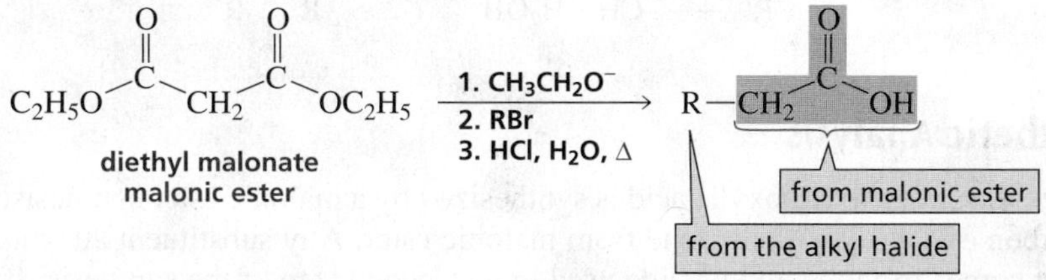

STEPS IN THE MALONIC ESTER SYNTHESIS

alkylation of the α-carbon

removal of a proton from the α-carbon

an α-substituted malonic ester

HCl, H$_2$O | Δ ← hydrolysis

decarboxylation

an α-substituted malonic acid

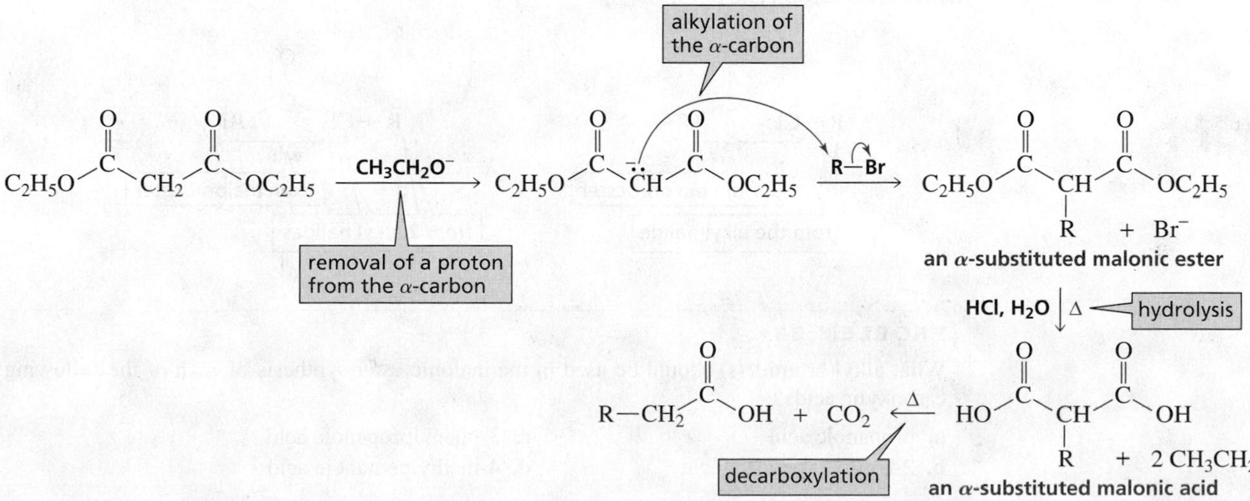

- A proton is easily removed from the α-carbon because it is flanked by two carbonyl groups.
- The resulting α-carbanion reacts with an alkyl halide, forming an α-substituted malonic ester. Because alkylation is an S$_N$2 reaction, it works best with primary alkyl halides and methyl halides.

- Heating the α-substituted malonic ester in an acidic aqueous solution hydrolyzes both ester groups to carboxylic acid groups, forming an α-substituted malonic acid.
- Further heating decarboxylates the 3-oxocarboxylic acid.

Carboxylic acids with two substituents bonded to the α-carbon can be prepared by carrying out two successive alkylations of the α-carbon.

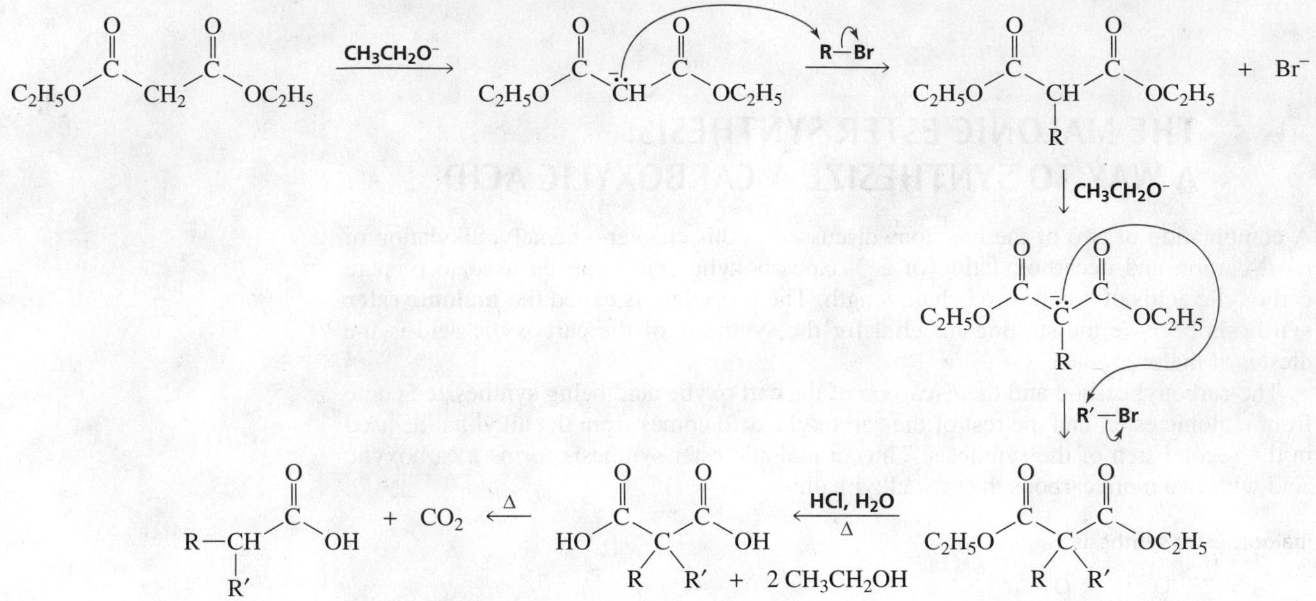

Retrosynthetic Analysis

We have seen that when a carboxylic acid is synthesized by a malonic ester synthesis, the carbonyl carbon and the α-carbon come from malonic ester. Any substituent attached to the α-carbon comes from the alkyl halide used in the second step of the synthesis. If the α-carbon has two substituents, then two successive alkylations of the α-carbon will form the desired carboxylic acid.

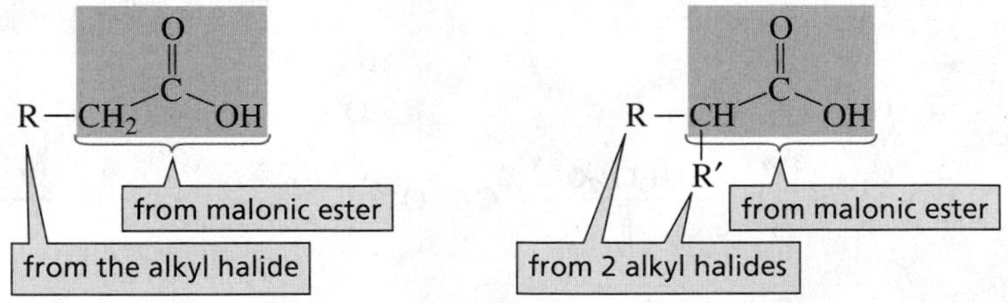

PROBLEM 38♦

What alkyl bromide(s) should be used in the malonic ester synthesis of each of the following carboxylic acids?

a. propanoic acid

b. 2-methylpropanoic acid

c. 3-phenylpropanoic acid

d. 4-methylpentanoic acid

PROBLEM 39

Explain why the following carboxylic acids cannot be prepared by a malonic ester synthesis:

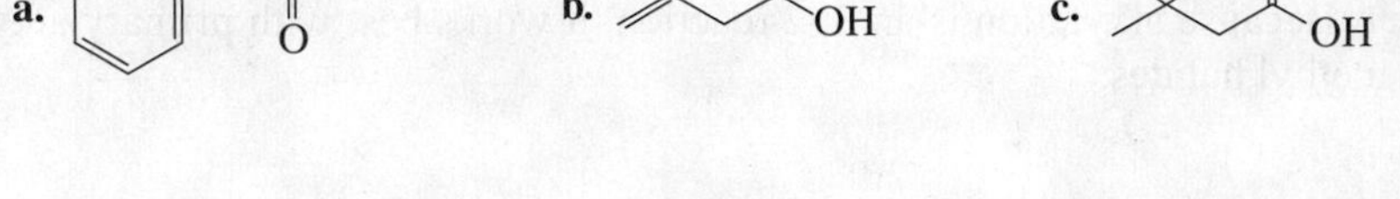

19 THE ACETOACETIC ESTER SYNTHESIS: A WAY TO SYNTHESIZE A METHYL KETONE

The only difference between the acetoacetic ester synthesis and the malonic ester synthesis is the use of acetoacetic ester rather than malonic ester as the starting material. The difference in starting material causes the product of the **acetoacetic ester synthesis** to be a *methyl ketone* rather than a *carboxylic acid*. The carbonyl group of the methyl ketone and the carbons on either side of it come from acetoacetic ester; the rest of the ketone comes from the alkyl halide used in the second step of the synthesis.

acetoacetic ester synthesis

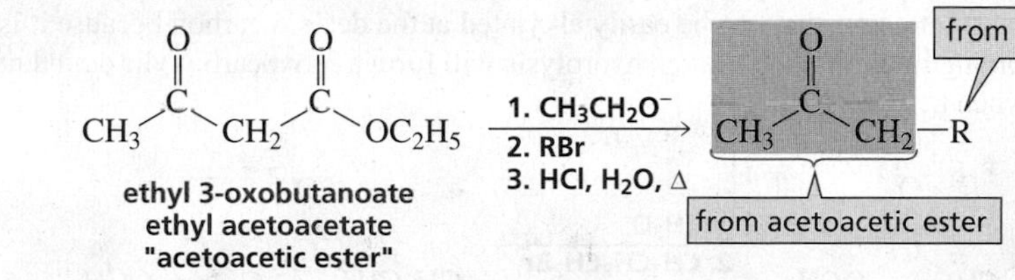

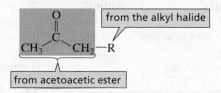

ethyl 3-oxobutanoate
ethyl acetoacetate
"acetoacetic ester"

from the alkyl halide

from acetoacetic ester

An acetoacetic ester synthesis forms a methyl ketone with three more carbons than the alkyl halide.

The steps in the acetoacetic ester synthesis are the same as those in the malonic ester synthesis.

STEPS IN THE ACETOACETIC ESTER SYNTHESIS

alkylation of the α-carbon

removal of a proton from the α-carbon

hydrolysis

decarboxylation

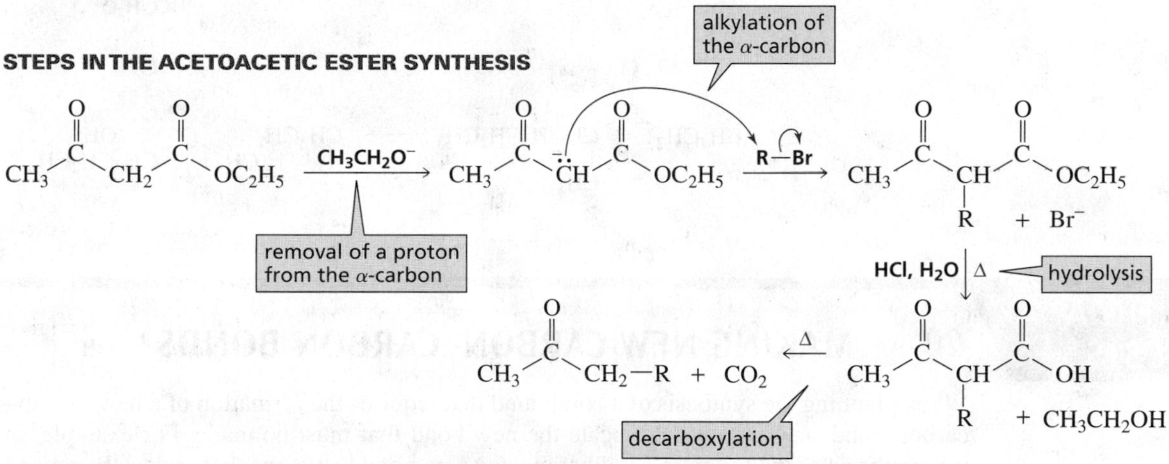

- A proton is removed from the α-carbon that is flanked by two carbonyl groups.
- The α-carbanion undergoes an S_N2 reaction with the alkyl halide.
- Hydrolysis and decarboxylation forms the methyl ketone.

Retrosynthetic Analysis

When a methyl ketone is synthesized by an acetoacetic ester synthesis, the carbonyl carbon and the carbons on either side of it come from malonic ester. Any substituent attached to the α-carbon comes from the alkyl halide used in the second step of the synthesis.

from the alkyl halide

from acetoacetic ester

PROBLEM 40♦

What alkyl bromide should be used in the acetoacetic ester synthesis of each of the following methyl ketones?

a. 2-pentanone   **b.** 2-octanone   **c.** 4-phenyl-2-butanone

PROBLEM 41 Solved

Starting with methyl propanoate, how could you prepare 4-methyl-3-heptanone?

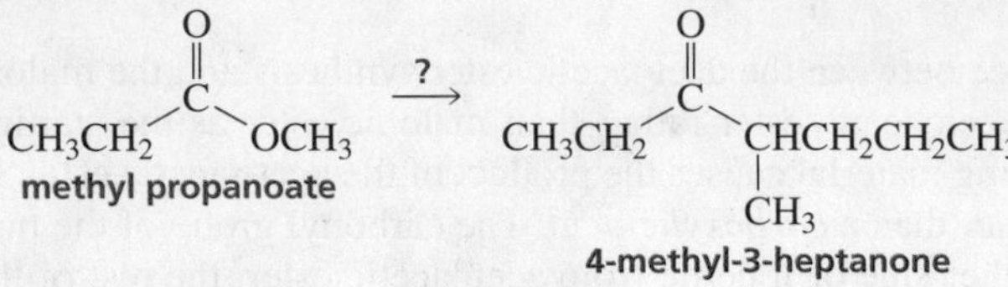

$$\text{CH}_3\text{CH}_2\overset{\displaystyle O}{\underset{}{\text{C}}}\text{OCH}_3 \quad \xrightarrow{?} \quad \text{CH}_3\text{CH}_2\overset{\displaystyle O}{\underset{}{\text{C}}}\underset{\underset{\displaystyle \text{CH}_3}{|}}{\text{CH}}\text{CH}_2\text{CH}_2\text{CH}_3$$

methyl propanoate   **4-methyl-3-heptanone**

Because the starting material is an ester and the target molecule has more carbons than the starting material, a Claisen condensation appears to be a good way to start this synthesis. The Claisen condensation forms a β-keto ester that can be easily alkylated at the desired carbon because it is flanked by two carbonyl groups. Acid-catalyzed hydrolysis will form a 3-oxocarboxylic acid that will decarboxylate when heated.

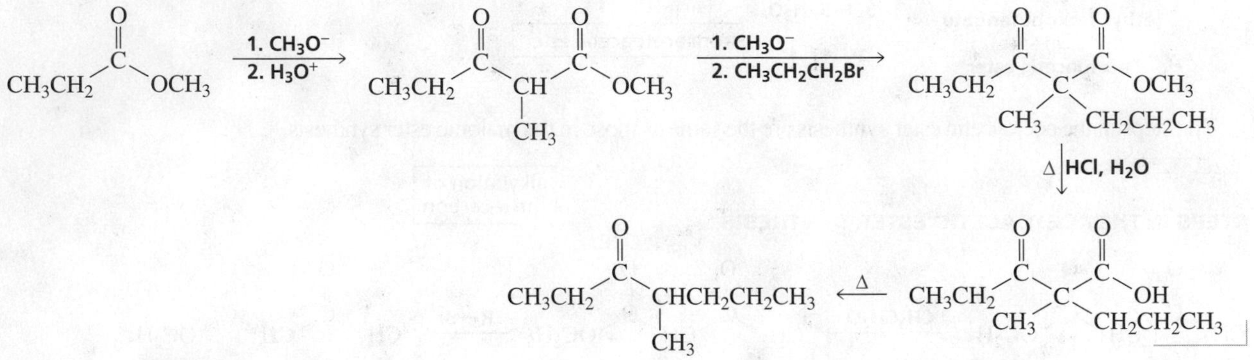

$$\text{CH}_3\text{CH}_2\overset{\displaystyle O}{\underset{}{\text{C}}}\text{OCH}_3 \quad \xrightarrow[\text{2. H}_3\text{O}^+]{\text{1. CH}_3\text{O}^-} \quad \text{CH}_3\text{CH}_2\overset{\displaystyle O}{\underset{}{\text{C}}}\underset{\underset{\displaystyle \text{CH}_3}{|}}{\text{CH}}\overset{\displaystyle O}{\underset{}{\text{C}}}\text{OCH}_3 \quad \xrightarrow[\text{2. CH}_3\text{CH}_2\text{CH}_2\text{Br}]{\text{1. CH}_3\text{O}^-} \quad \text{CH}_3\text{CH}_2\overset{\displaystyle O}{\underset{}{\text{C}}}\underset{\underset{\displaystyle \text{CH}_3 \quad \text{CH}_2\text{CH}_2\text{CH}_3}{|}}{\text{C}}\overset{\displaystyle O}{\underset{}{\text{C}}}\text{OCH}_3$$

$$\Big\downarrow \Delta \mid \text{HCl, H}_2\text{O}$$

$$\text{CH}_3\text{CH}_2\overset{\displaystyle O}{\underset{}{\text{C}}}\underset{\underset{\displaystyle \text{CH}_3}{|}}{\text{CH}}\text{CH}_2\text{CH}_2\text{CH}_3 \quad \xleftarrow{\Delta} \quad \text{CH}_3\text{CH}_2\overset{\displaystyle O}{\underset{}{\text{C}}}\underset{\underset{\displaystyle \text{CH}_3 \quad \text{CH}_2\text{CH}_2\text{CH}_3}{|}}{\text{C}}\overset{\displaystyle O}{\underset{}{\text{C}}}\text{OH}$$

DESIGNING A SYNTHESIS

20   MAKING NEW CARBON–CARBON BONDS

When planning the synthesis of a compound that requires the formation of a new carbon–carbon bond, we first need to locate the new bond that must be made. For example, in the synthesis of the following β-diketone, the new bond is the one that makes the second five-membered ring:

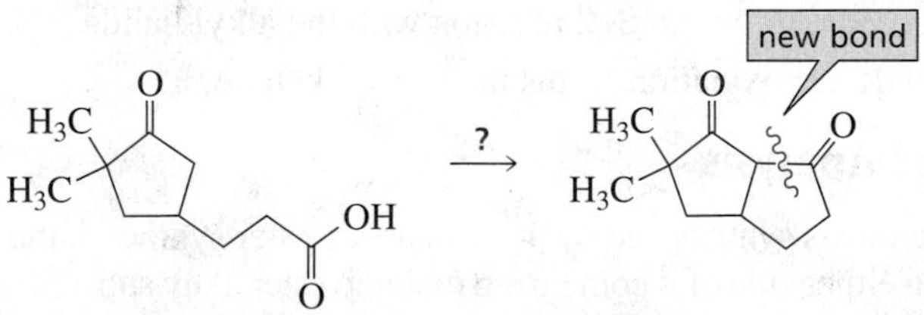

Next, we need to determine which of the atoms that form the bond should be the nucleophile and which should be the electrophile. In this case, it is easy to choose between the two possibilities because we know that a carbonyl carbon is an electrophile.

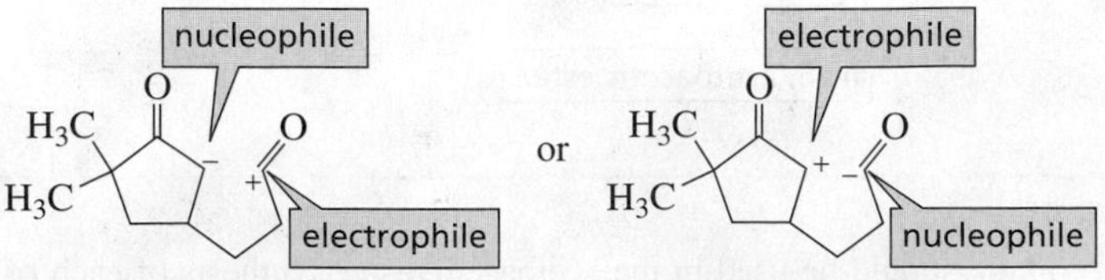

Now we need to determine what compound would give the desired electrophilic and nucleophilic sites. If we know what the starting material is, we can use it as a clue to arrive at the desired compound. For example, an ester carbonyl group would be a good electrophile for this synthesis because it has a group that would be eliminated. Moreover, the α-hydrogens of the ketone are more acidic than the α-hydrogens of the ester, so the desired nucleophile would be easy to obtain. Thus, preparation of the ester from the starting material, followed by an intramolecular condensation, forms the target molecule.

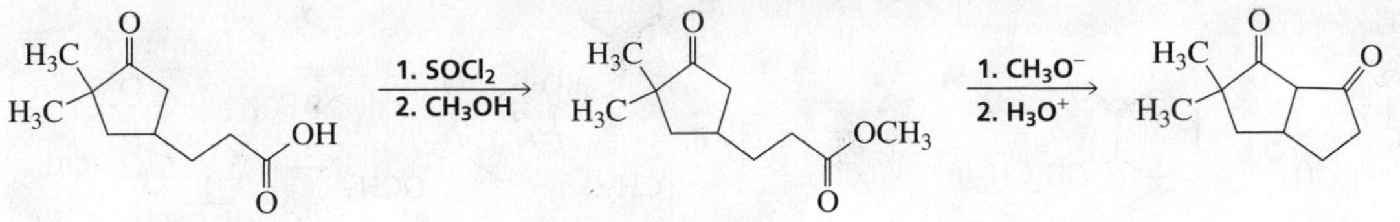

In the next synthesis, two new carbon–carbon bonds must be formed.

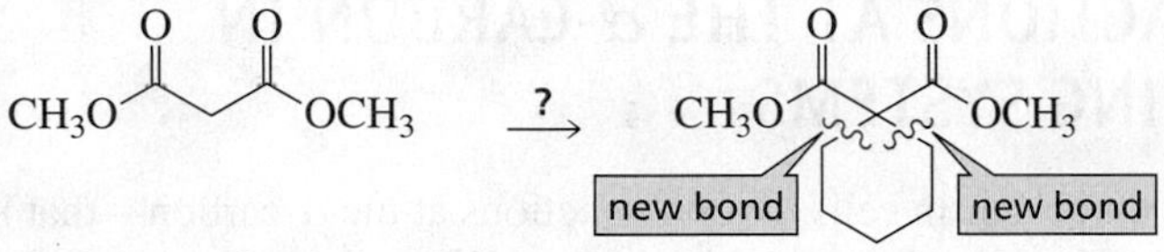

After we have identified the electrophilic and nucleophilic sites, we can see that two successive alkylations of a diester of malonic acid, using 1,5-dibromopentane for the alkyl halide, will produce the target molecule.

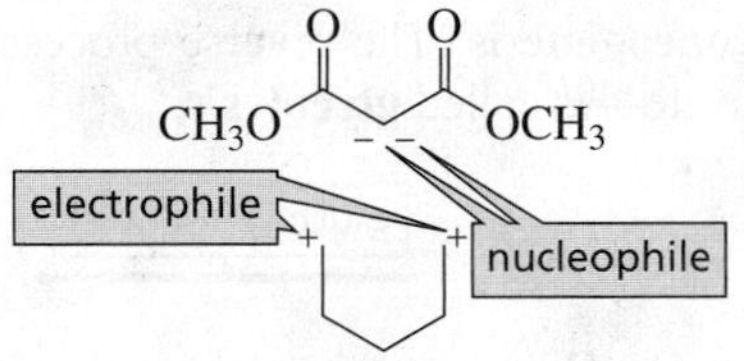

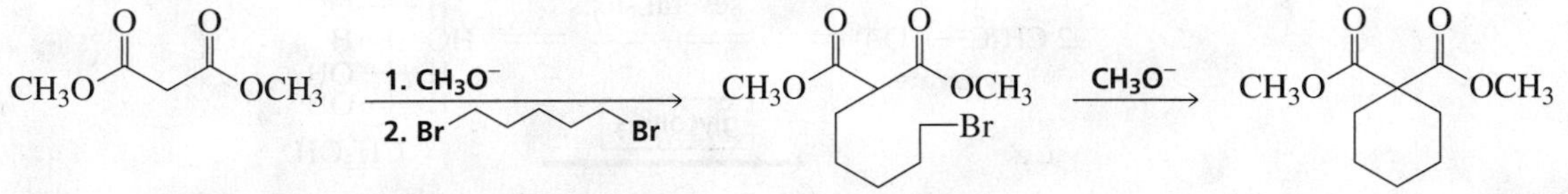

In planning the following synthesis, the diester given as the starting material suggests that a Dieckmann condensation should be used to obtain the cyclic compound:

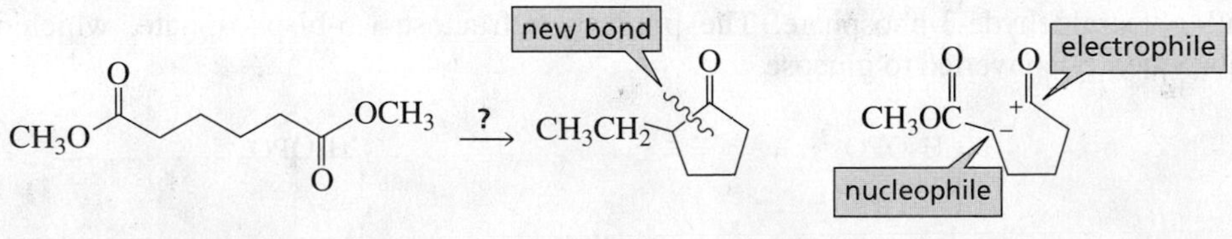

The Dieckmann condensation is followed by alkylation of the α-carbon of the cyclopentanone ring. Hydrolysis of the β-keto ester and decarboxylation forms the target molecule.

917

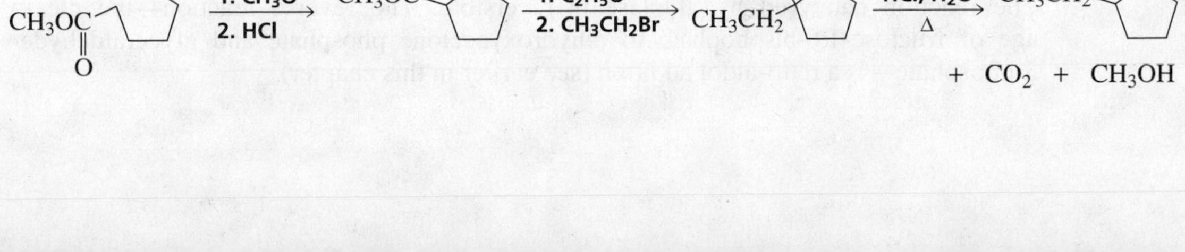

PROBLEM 42

Design a synthesis for each of the following compounds using the given starting material:

a.

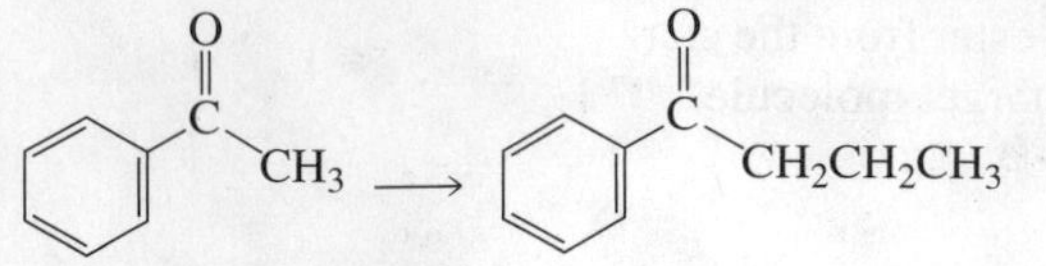

c.

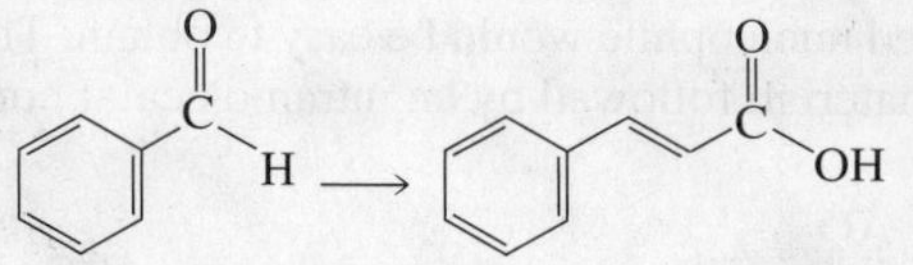

b.

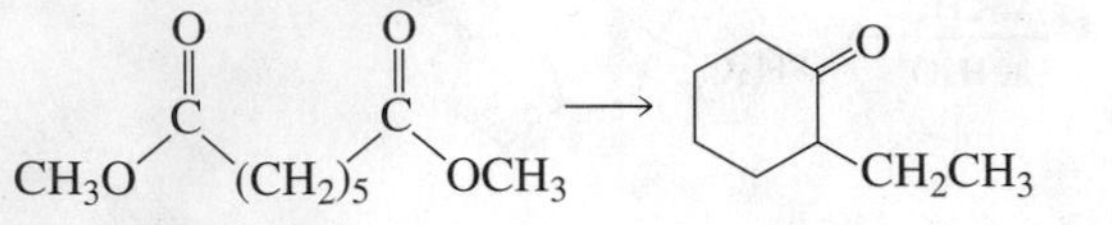

d.

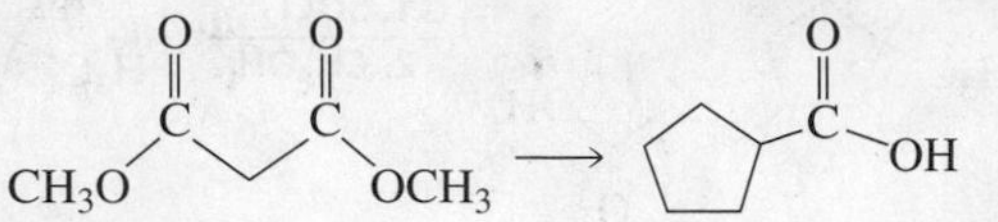

REACTIONS AT THE α-CARBON IN LIVING SYSTEMS

Many reactions that occur in cells involve reactions at the α-carbon—that is, the kinds of reactions you have been studying in this chapter. We will now look at a few examples.

A Biological Aldol Addition

Glucose, the most abundant sugar found in nature, is synthesized in cells from two molecules of pyruvate. The series of reactions that convert two molecules of pyruvate into glucose is called **gluconeogenesis**. The reverse process—the breakdown of glucose into two molecules of pyruvate—is called **glycolysis**.

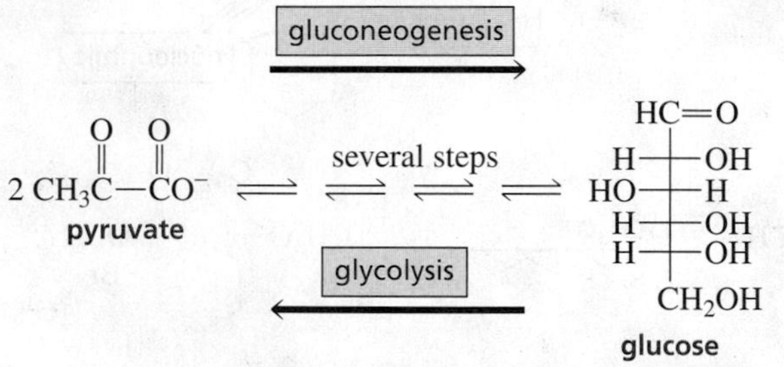

Because glucose has twice as many carbons as pyruvate, you should not be surprised to learn that one of the steps in the biosynthesis of glucose is an aldol addition. An enzyme called aldolase catalyzes an aldol addition between dihydroxyacetone phosphate and glyceraldehyde-3-phosphate. The product is fructose-1,6-bisphosphate, which is subsequently converted to glucose.

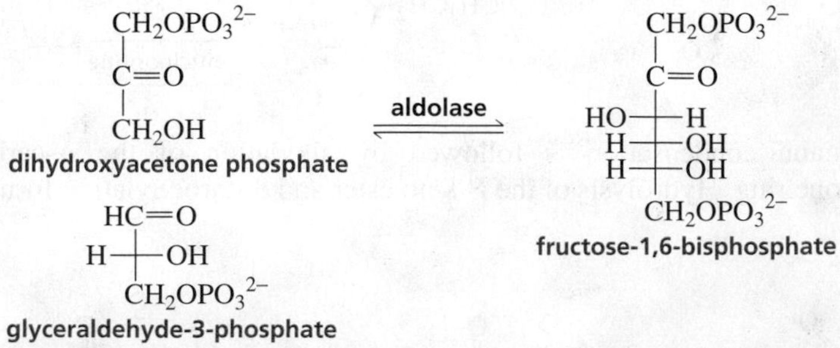

The reaction catalyzed by aldolase is reversible. The reverse reaction—the cleavage of fructose-1,6-bisphophate to dihydroxyacetone phosphate and glyceraldehyde-3-phosphate—is a retro-aldol addition (see earlier in this chapter).

PROBLEM 43

Propose a mechanism for the formation of fructose-1,6-bisphosphate from dihydroxyacetone phosphate and glyceraldehyde-3-phosphate, using hydroxide ion as the catalyst.

A Biological Aldol Condensation

Collagen is the most abundant protein in mammals, amounting to about one-fourth of the total protein. It is the major fibrous component of bone, teeth, skin, cartilage, and tendons. Individual collagen molecules—called tropocollagen—can be isolated only from tissues of young animals. As animals age, the individual collagen molecules become cross-linked, which is why meat from older animals is tougher than meat from younger ones. Collagen cross-linking is an example of an aldol condensation.

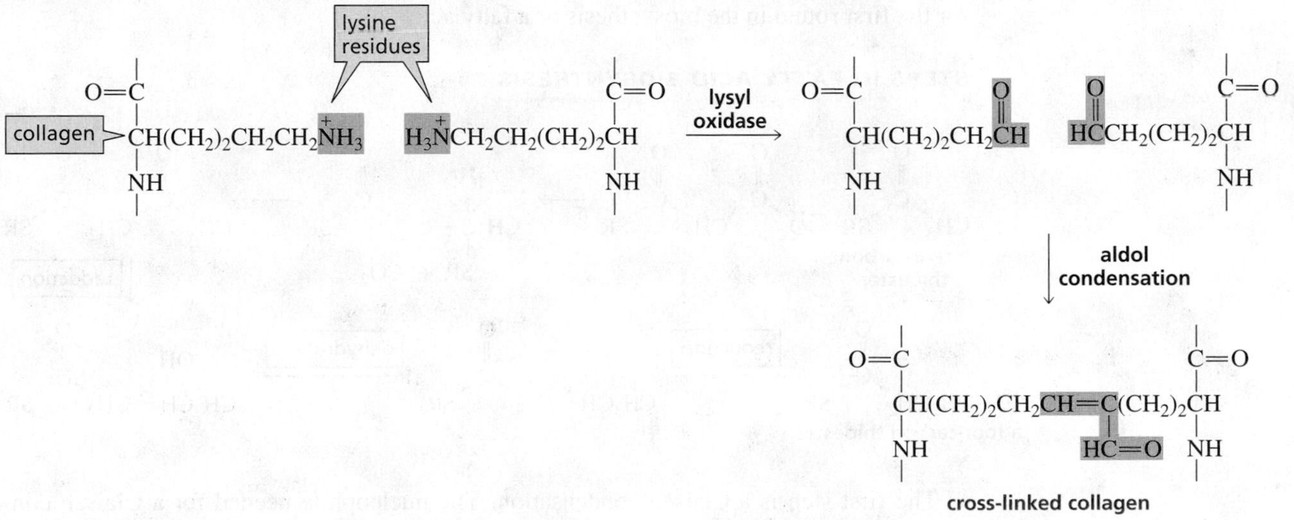

Before collagen molecules can cross-link, the primary amino groups of the lysine residues of collagen must be converted to aldehyde groups. (Lysine is an amino acid.) The enzyme that catalyzes this reaction is called lysyl oxidase. An aldol condensation between two aldehyde groups results in a cross-linked protein.

A Biological Claisen Condensation

Fatty acids are long-chain carboxylic acids. Naturally occurring fatty acids are unbranched and contain an even number of carbons because they are synthesized from acetate, a compound with two carbons.

Carboxylic acids can be activated in cells by being converted to thioesters of coenzyme A.

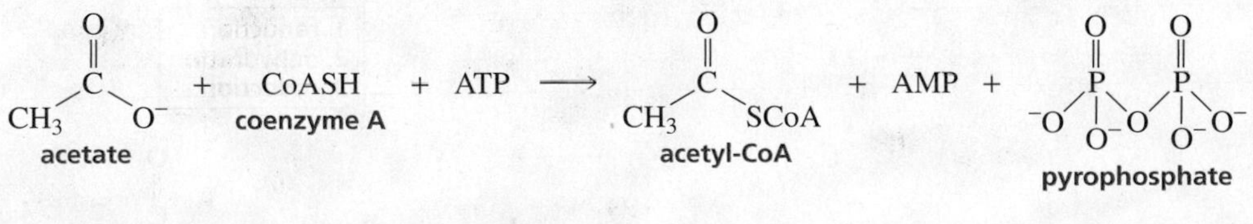

One of the necessary reactants for fatty acid synthesis is malonyl-CoA, which is obtained by carboxylation of acetyl-CoA.

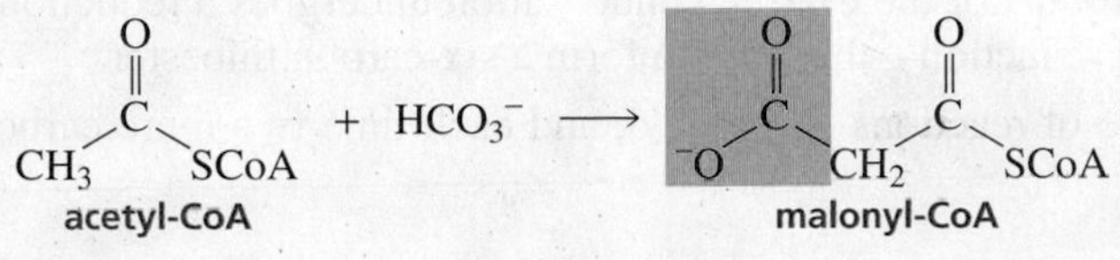

Before fatty acid synthesis can occur, the acyl groups of acetyl-CoA and malonyl-CoA are transferred to other thiols by means of a transesterification reaction.

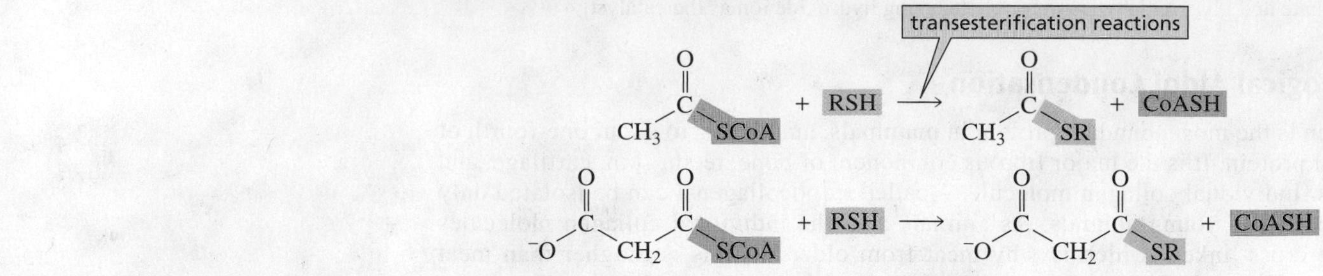

A molecule of acetyl thioester and a molecule of malonyl thioester are the reactants for the first round in the biosynthesis of a fatty acid.

STEPS IN FATTY ACID BIOSYNTHESIS

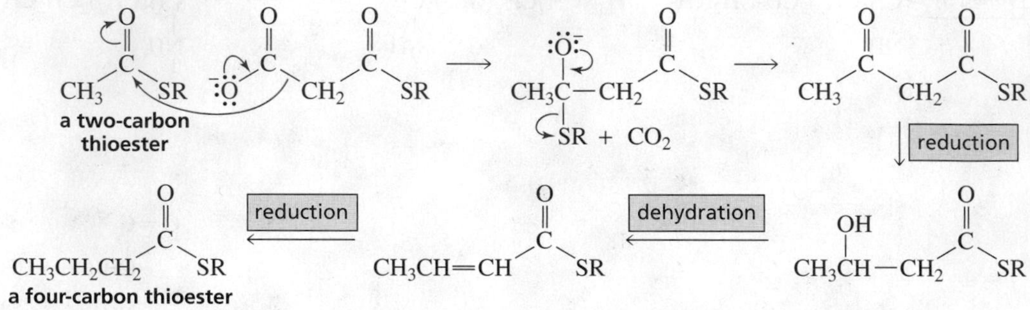

- The first step is a Claisen condensation. The nucleophile needed for a Claisen condensation is obtained by removing CO_2—rather than a proton—from the α-carbon of malonyl thioester. (Recall that 3-oxocarboxylic acids are easily decarboxylated; Section 17.) Loss of CO_2 also drives the condensation reaction to completion.

- The product of the condensation reaction undergoes a reduction, a dehydration, and a second reduction to form a four-carbon thioester. (Recall that a ketone is easier to reduced than an ester.) Each reaction is catalyzed by a different enzyme.

The four-carbon thioester and another molecule of malonyl thioester are the reactants for the second round.

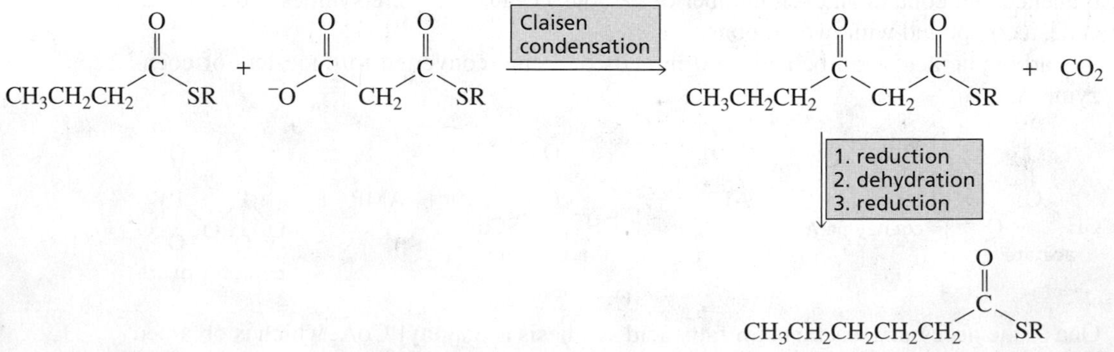

- Again, the product of the Claisen condensation undergoes a reduction, a dehydration, and a second reduction—this time to form a six-carbon thioester.

- The sequence of reactions is repeated, and each time two more carbons are added to the chain.

This sequence of reactions explains why naturally occurring fatty acids are unbranched and contain an even number of carbons. Once a thioester with the appropriate number of carbons is obtained, it undergoes a transesterification reaction with glycerol-3-phosphate in order to form fats, oils, and phospholipids.

PROBLEM 44♦

Palmitic acid is a straight-chain saturated 16-carbon fatty acid. How many moles of malonyl-CoA are required for the synthesis of one mole of palmitic acid?

PROBLEM 45♦

a. If the biosynthesis of palmitic acid were carried out with CD_3COSR and nondeuterated malonyl thioester, how many deuterium atoms would be incorporated into palmitic acid?

b. If the biosynthesis of palmitic acid were carried out with $^-OOCCD_2COSR$ and nondeuterated acetyl thioester, how many deuterium atoms would be incorporated into palmitic acid?

A Biological Decarboxylation

An example of a decarboxylation that occurs in cells is the decarboxylation of acetoacetate.

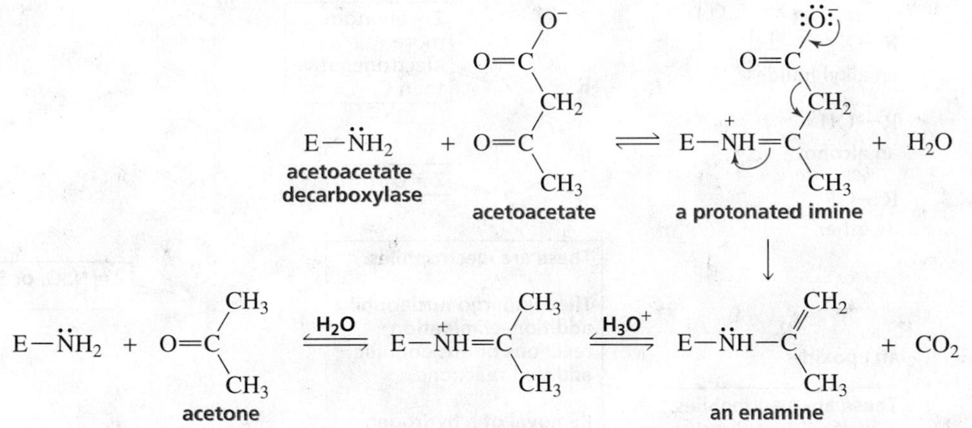

- Acetoacetate decarboxylase, the enzyme that catalyzes the reaction, forms an imine with acetoacetate.

- Under physiological conditions, the imine is protonated and the positively charged nitrogen readily accepts the pair of electrons left behind when the substrate loses CO_2.

- Decarboxylation forms an enamine.

- Protonation of the enamine followed by hydrolysis produces the decarboxylated product (acetone) and regenerates the enzyme.

In ketosis, a pathological condition that can occur in people with diabetes, the body produces more acetoacetate than can be metabolized. The excess acetoacetate is decarboxylated, so ketosis can be recognized by the smell of acetone on a person's breath.

PROBLEM 46

When the enzymatic decarboxylation of acetoacetate is carried out in $H_2{}^{18}O$, all the acetone that is formed contains ^{18}O. What does this tell you about the mechanism of the reaction?

---

22 ORGANIZING WHAT WE KNOW ABOUT THE REACTIONS OF ORGANIC COMPOUNDS

The families of organic compounds can be put into one of four groups, and all the members of a group react in similar ways. Let's go over the families in Group III (families of compounds that contain a carbonyl group).

Both families in Group III have carbonyl groups, and since the carbonyl carbon is an *electrophile*, both families in this group react with *nucleophiles*.

- The first family (carboxylic acids and carboxylic acid derivatives) has a group attached to the carbonyl carbon that can be replaced by another group. Therefore, this family undergoes nucleophilic addition–elimination reactions.

- The second family (aldehyde and ketones) does not have a group attached to the carbonyl carbon that can be replaced by another group. Therefore, this family undergoes nucleophilic addition reactions with strongly basic nucleophiles such as R^- or H^-. If the attacking atom of the nucleophile is an oxygen or a nitrogen and there is enough acid in the solution to protonate the OH group of the tetrahedral compound formed by the nucleophilic addition reaction, then water is eliminated from the addition product.

- Aldehydes, ketones, esters, and *N,N*-disubstituted amides have a hydrogen on an α-carbon that can be removed by a strong base. Removal of a hydrogen from an α-carbon creates an enolate ion that can react with electrophiles.

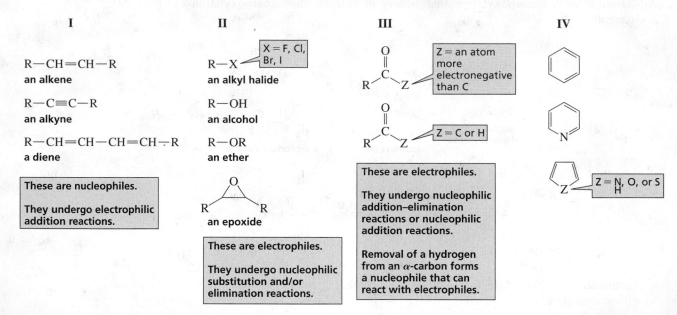

SOME IMPORTANT THINGS TO REMEMBER

- A hydrogen bonded to an α-carbon of an aldehyde, ketone, ester, or *N,N*-disubstituted amide is sufficiently acidic to be removed by a strong base.

- A **carbon acid** is a compound with a relatively acidic hydrogen bonded to an sp^3 carbon.

- Aldehydes and ketones (pK_a ~16 to 20) are more acidic than esters (pK_a ~25). **β-Diketones** (pK_a ~ 9) and **β-keto esters** (pK_a ~11) are even more acidic.

- **Keto–enol interconversion** can be catalyzed by acids or by bases. Generally, the **keto tautomer** is more stable.

- Aldehydes and ketones react with Br_2, Cl_2, or I_2: under acidic conditions, a halogen replaces one α-hydrogen; under basic conditions, halogens replace all of the α-hydrogens.

- The **HVZ reaction** brominates the α-carbon of a carboxylic acid.

- LDA (a strong, bulky base but a poor nucleophile) is used to form an enolate ion in reactions that require the carbonyl compound to be completely converted to the enolate ion before it reacts with an electrophile.

- If the electrophile is an alkyl halide, the enolate ion is alkylated. The less substituted α-carbon is alkylated when the reaction is under kinetic control, whereas the more substituted α-carbon is alkylated when the reaction is under thermodynamic control.

- Aldehydes and ketones can be alkylated or acylated on the α-carbon via an enamine intermediate.

- Enolate ions of β-diketones, β-diesters, β-keto esters, and β-keto nitriles undergo **Michael reactions** with α,β-unsaturated carbonyl compounds. Michael reactions form 1,5-dicarbonyl compounds.

- In an **aldol addition,** the enolate ion of an aldehyde or a ketone reacts with the carbonyl carbon of a second molecule of aldehyde or ketone, forming a β-hydroxyaldehyde or a β-hydroxyketone. The new C—C bond forms between the α-carbon of one molecule and the carbon that formerly was the carbonyl carbon of the other molecule.

- The product of an aldol addition can be dehydrated under acidic or basic conditions to form an **aldol condensation** product. Base-catalyzed dehydration is an **E1cB reaction.**

- In a **Claisen condensation,** the enolate ion of an ester reacts with a second molecule of ester, eliminating an ⁻OR group to form a β-keto ester. A **Dieckmann condensation** is an intramolecular Claisen condensation.

- A **Robinson annulation** is a ring-forming reaction in which a Michael reaction and an intramolecular aldol condensation occur sequentially.

- Carboxylic acids with a carbonyl group at the 3-position **decarboxylate** when they are heated.

- Carboxylic acids can be prepared by a **malonic ester synthesis;** the resulting carboxylic acid has two more carbons than the alkyl halide.

- Methyl ketones can be prepared by an **acetoacetic ester synthesis;** the resulting methyl ester has three more carbons (the carbonyl group and the carbon on either side of it) than the alkyl halide.

SUMMARY OF REACTIONS

1. Keto–enol interconversion (Section 3).

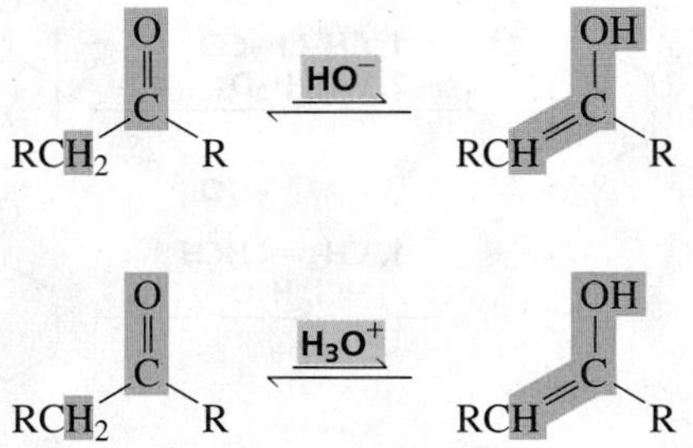

2. Halogenation of the α-carbon of aldehydes and ketones (Section 4).

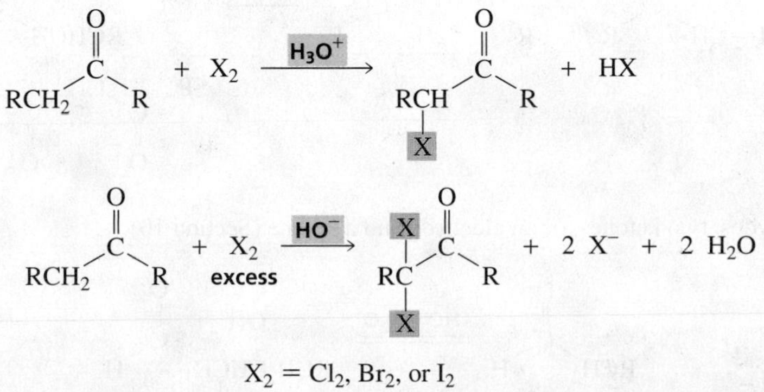

$$X_2 = Cl_2, \ Br_2, \ or \ I_2$$

3. Halogenation of the α-carbon of carboxylic acids: the Hell–Volhard–Zelinski reaction (Section 5).

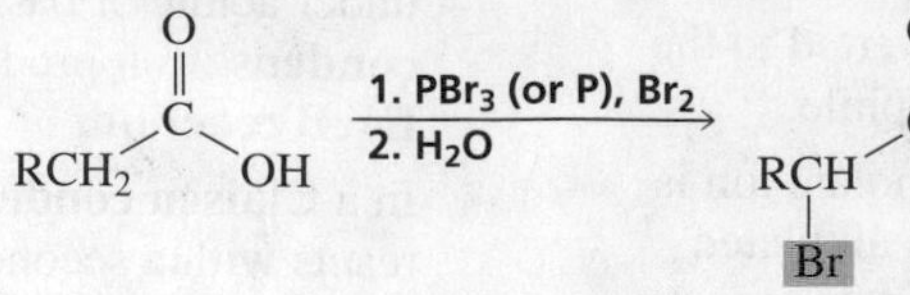

4. Alkylating the α-carbon of carbonyl compounds (Section 7).

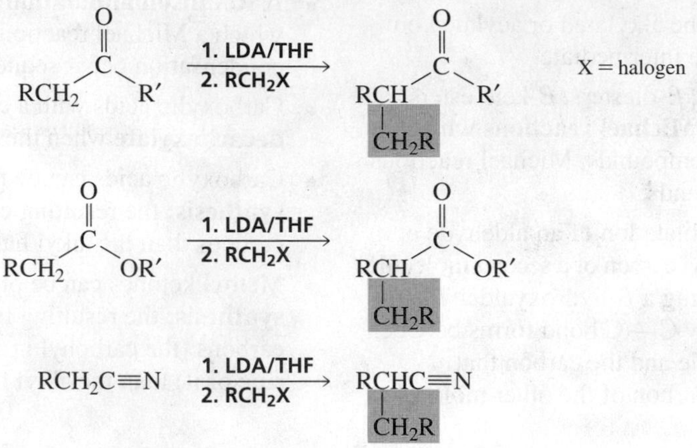

X = halogen

5. Alkylating or acylating the α-carbon of aldehydes and ketones by means of an enamine intermediate (Section 8).

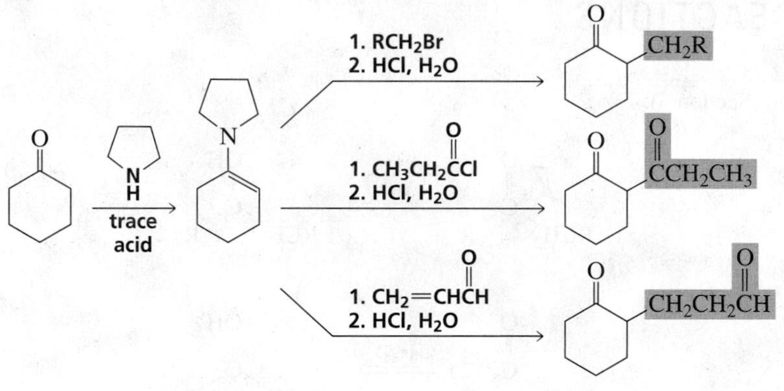

6. The Michael reaction: addition of an enolate ion to an α,β-unsaturated carbonyl compound (Section 9).

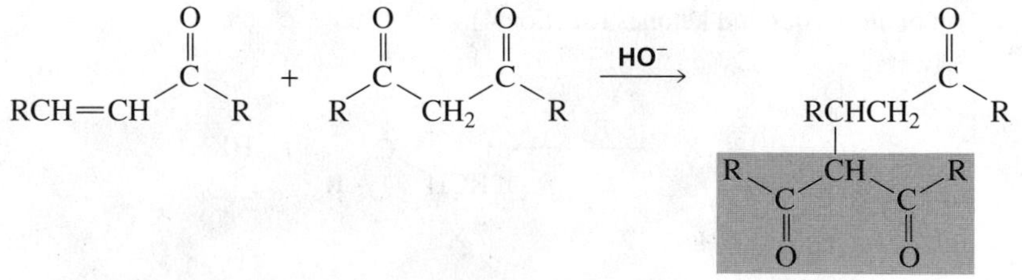

7. Aldol addition of two aldehydes, two ketones, or an aldehyde and a ketone (Section 10).

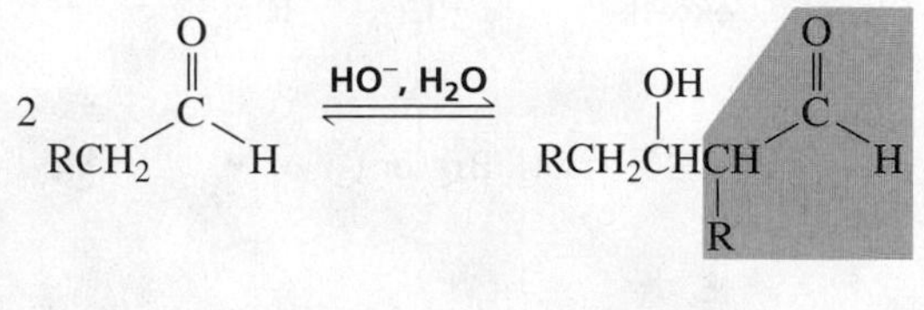

8. Aldol condensation is an aldol addition followed by acid-catalyzed or base-catalyzed dehydration (Section 11).

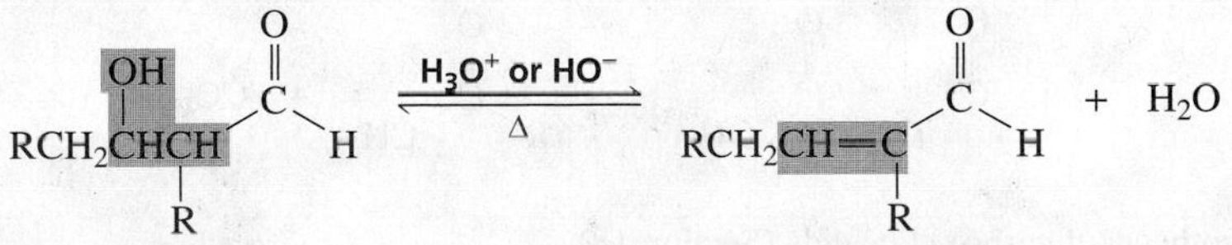

9. Claisen condensation of two esters (Section 13).

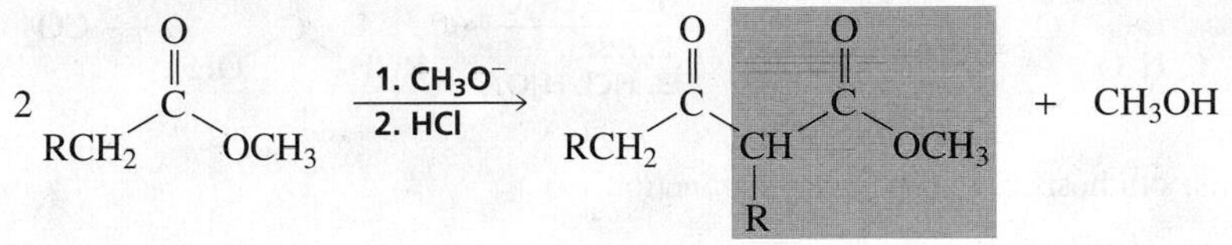

10. Crossed addition and condensation reactions:
   a. when one of the carbonyl compounds does not have any α-hydrogens (Sections 12 and 14).
   b. when both carbonyl compounds have α-hydrogens (Sections 12 and 14).

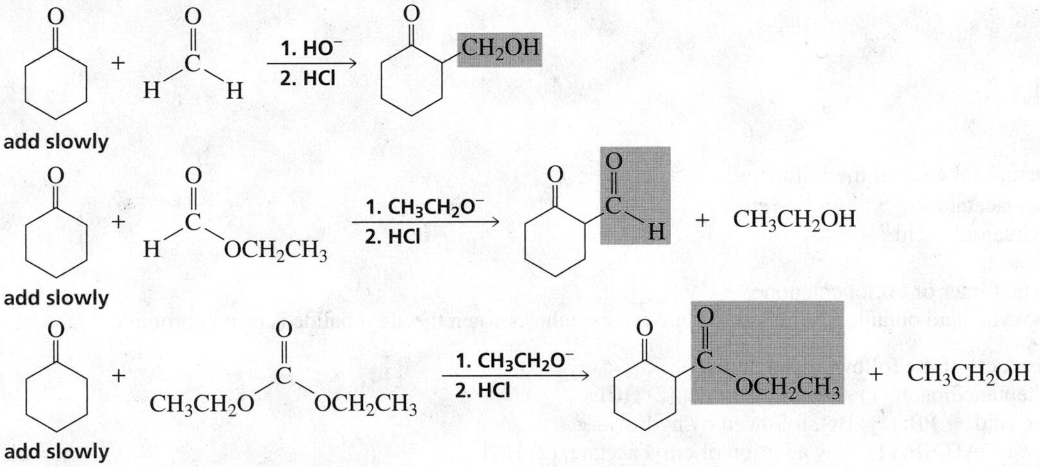

11. Robinson annulation (Section 16).

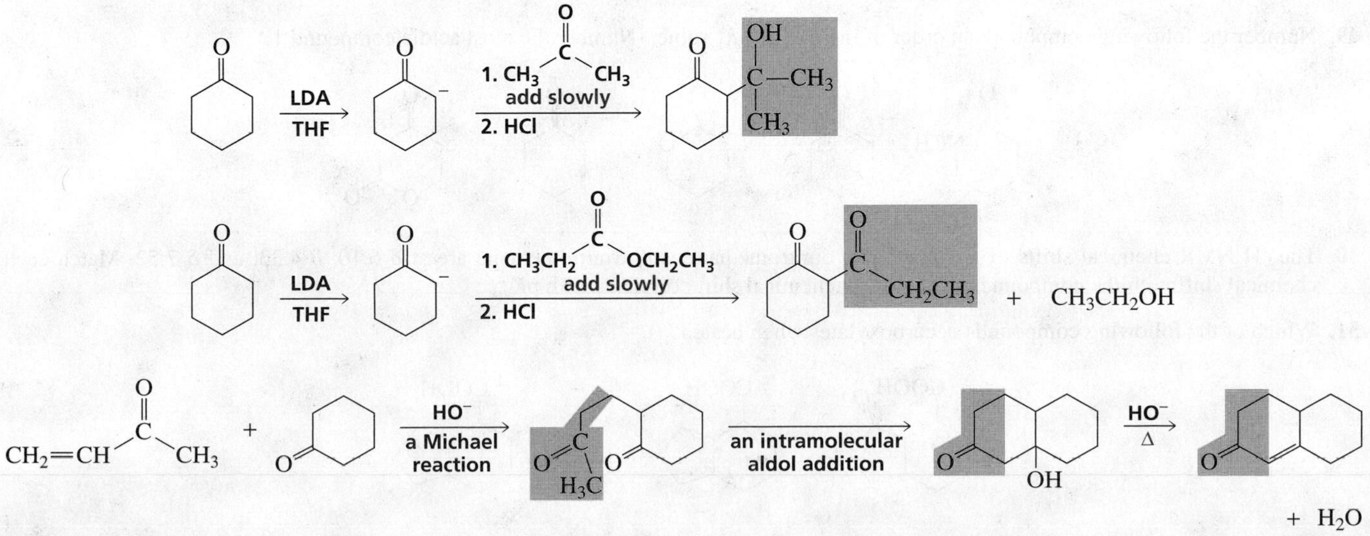

12. Decarboxylation of 3-oxocarboxylic acids (Section 17).

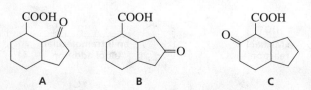

13. Malonic ester synthesis: synthesis of carboxylic acids (Section 18).

14. Acetoacetic ester synthesis: synthesis of methyl ketones (Section 19).

PROBLEMS

47. Draw a structure for each of the following:
  a. ethyl acetoacetate
  b. α-methylmalonic acid
  c. a β-keto ester
  d. the enol tautomer of cyclopentanone
  e. the carboxylic acid obtained from the malonic ester synthesis when the alkyl halide is propyl bromide

48. Draw the products of the following reactions:
  a. diethyl heptanedioate: (1) sodium ethoxide; (2) HCl
  b. pentanoic acid + PBr$_3$ + Br$_2$, followed by hydrolysis
  c. acetone + LDA/THF: (1) slow addition of ethyl acetate; (2) HCl
  d. diethyl 2-ethylhexanedioate: (1) sodium ethoxide; (2) HCl
  e. diethyl malonate: (1) sodium ethoxide; (2) isobutyl bromide; (3) HCl, H$_2$O + Δ
  f. acetophenone + LDA/THF: (1) slow addition of diethyl carbonate; (2) HCl

49. Number the following compounds in order of increasing pK_a value. (Number the most acidic compound 1.)

50. The ^{1}H NMR chemical shifts of nitromethane, dinitromethane, and trinitromethane are at δ 6.10, δ 4.33, and δ 7.52. Match each chemical shift with the compound. Explain how chemical shift correlates with pK_a.

51. Which of the following compounds decarboxylates when heated?

A   B   C

52. Draw the products of the following reactions:
  a. 1,3-cyclohexanedione + LDA/THF, followed by allyl bromide
  b. γ-butyrolactone + LDA/THF, followed by methyl iodide
  c. 2,7-octanedione + sodium hydroxide
  d. diethyl 1,2-benzenedicarboxylate + sodium ethoxide: (1) slow addition of ethyl acetate; (2) HCl

53. A racemic mixture of 2-methyl-1-phenyl-1-butanone is formed when (R)-2-methyl-1-phenyl-1-butanone is dissolved in an acidic or basic aqueous solution. Give an example of another ketone that would undergo acid- or base-catalyzed racemization.

54. Draw the products of the following reactions:

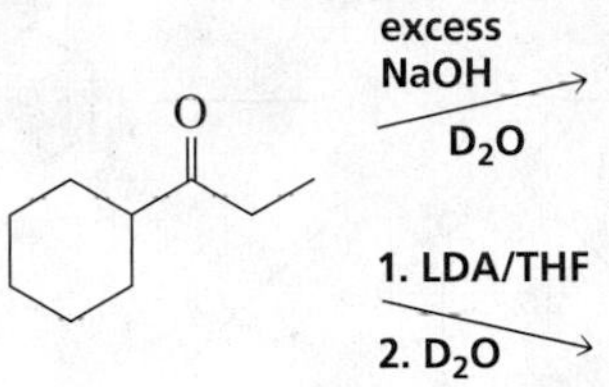

55. What is the product of the following reaction?

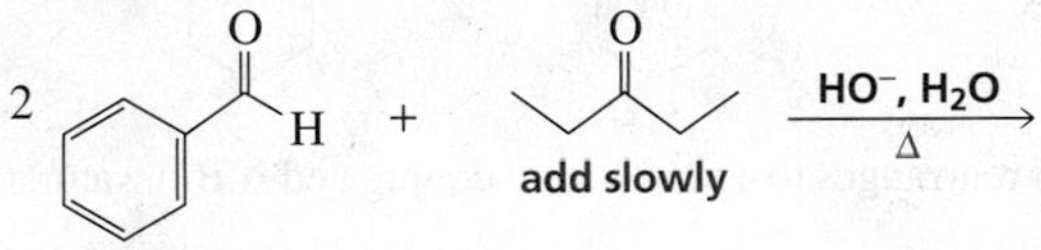

56. An aldol addition can be catalyzed by acids as well as by bases. Propose a mechanism for the acid-catalyzed aldol addition of propanal.

57. In the presence of excess base and excess halogen, a methyl ketone is converted into a carboxylate ion. The reaction is known as the haloform reaction because one of the products is haloform—either chloroform, bromoform, or iodoform. Before spectroscopy became a routine analytical tool, the haloform reaction served as a test for methyl ketones: the formation of iodoform, a bright yellow compound, signaled that a methyl ketone was present. Why do only methyl ketones form a haloform?

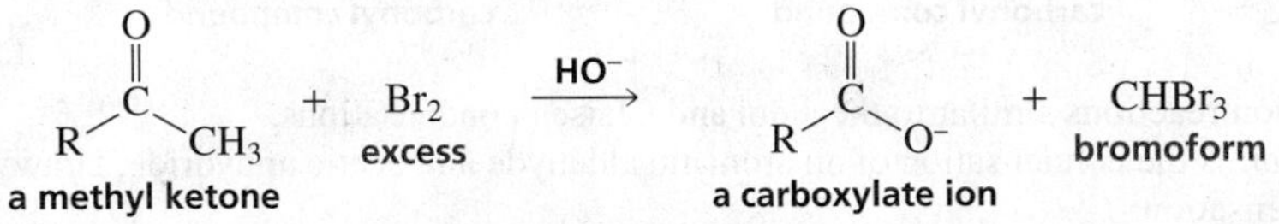

58. Identify **A–L**. (*Hint:* **A** shows three singlets in its ^1H NMR spectrum with integral ratios 3 : 2 : 3 and gives a positive iodoform test; see Problem 57.)

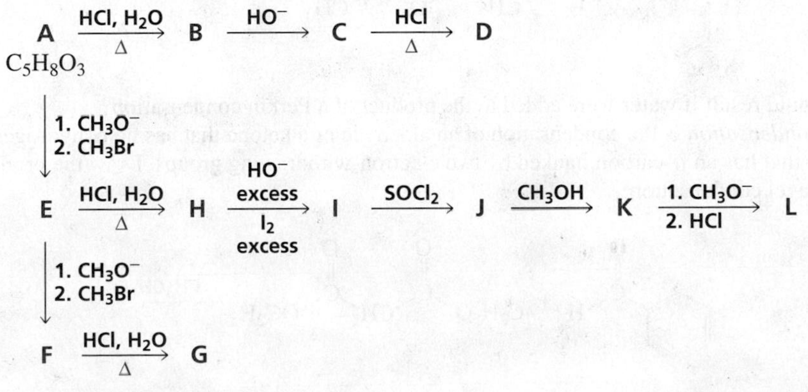

59. Using cyclopentanone as the reactant, show the product of
  a. acid-catalyzed keto–enol interconversion. **b.** an aldol addition. **c.** an aldol condensation.

60. Show how the following compounds could be prepared from cyclohexanone:

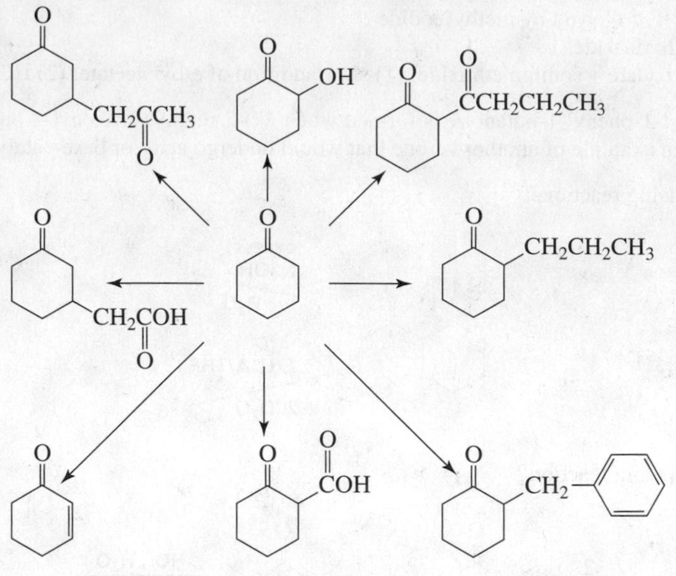

61. A β,γ-unsaturated carbonyl compound rearranges to a more stable conjugated α,β-unsaturated compound in the presence of either acid or base.

    a. Propose a mechanism for the base-catalyzed rearrangement.   **b.** Propose a mechanism for the acid-catalyzed rearrangement.

a β,γ-unsaturated
carbonyl compound

an α,β-unsaturated
carbonyl compound

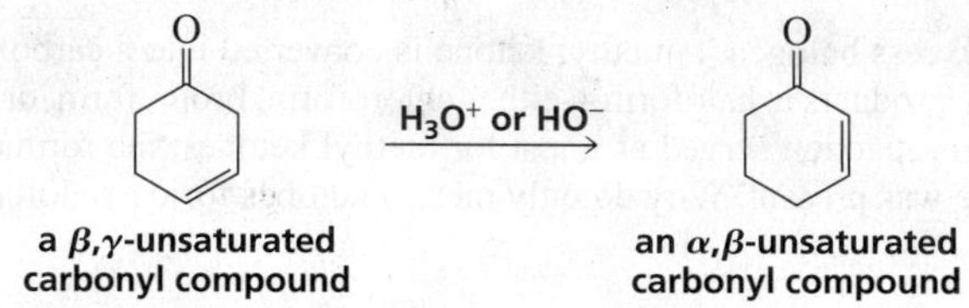

62. There are other condensation reactions similar to the aldol and Claisen condensations:

    a. The *Perkin condensation* is the condensation of an aromatic aldehyde and acetic anhydride. Draw the product obtained from the following Perkin condensation:

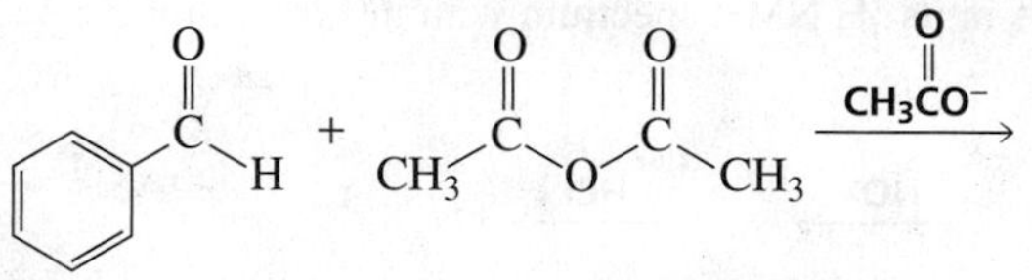

    b. What compound would result if water were added to the product of a Perkin condensation?

    c. The *Knoevenagel condensation* is the condensation of an aldehyde or a ketone that has no α-hydrogens and a compound such as diethyl malonate that has an α-carbon flanked by two electron-withdrawing groups. Draw the product obtained from the following Knoevenagel condensation:

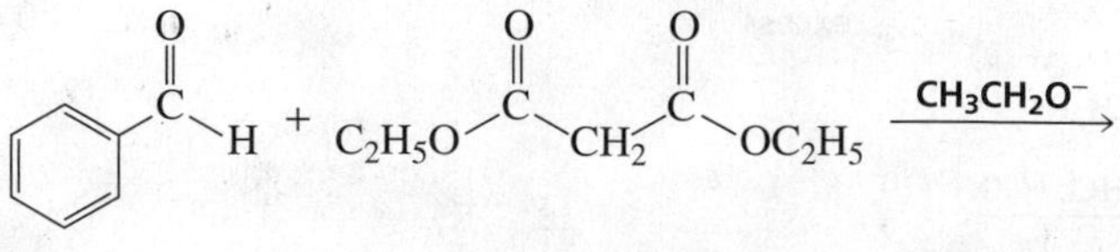

    d. What product would be obtained if the product of a Knoevenagel condensation were heated in an aqueous acidic solution?

63. The *Reformatsky reaction* is an addition reaction in which an organozinc reagent is used instead of a Grignard reagent to add to the carbonyl group of an aldehyde or a ketone. Because the organozinc reagent is less reactive than a Grignard reagent, a nucleophilic addition to the ester group does not occur. The organozinc reagent is prepared by treating an α-bromo ester with zinc.

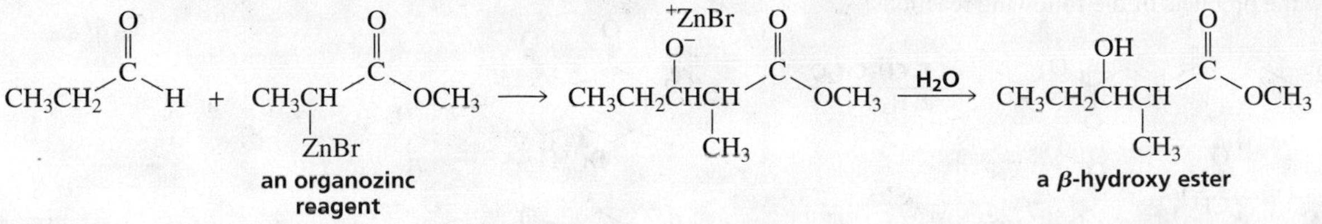

$$CH_3CH_2\overset{\displaystyle O}{\underset{}{C}}H \ + \ CH_3\underset{\underset{ZnBr}{|}}{CH}\overset{\displaystyle O}{\underset{}{C}}OCH_3 \longrightarrow CH_3CH_2\underset{\underset{CH_3}{|}}{CH}\overset{\overset{+ZnBr}{|}}{\underset{}{O^-}}CH\overset{\displaystyle O}{\underset{}{C}}OCH_3 \xrightarrow{H_2O} CH_3CH_2\underset{\underset{CH_3}{|}}{CH}\overset{\overset{OH}{|}}{CH}\overset{\displaystyle O}{\underset{}{C}}OCH_3$$

an organozinc
reagent

a β-hydroxy ester

Describe how each of the following compounds could be prepared, using a Reformatsky reaction:

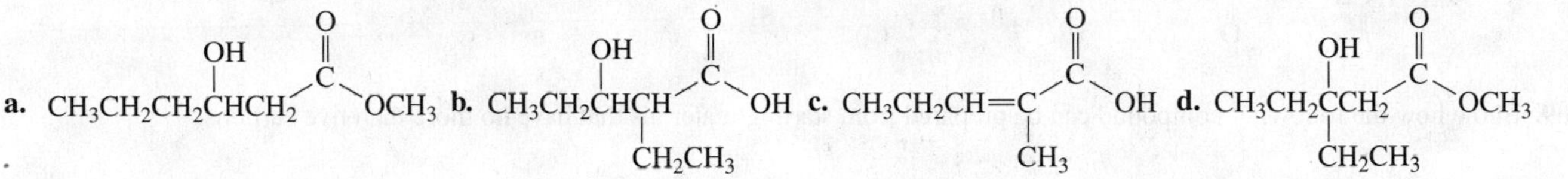

a. $CH_3CH_2CH_2\underset{\underset{}{|OH}}{CH}CH_2\overset{\displaystyle O}{\underset{}{C}}OCH_3$
b. $CH_3CH_2\underset{\underset{CH_2CH_3}{|}}{\overset{\overset{OH}{|}}{CH}CH}\overset{\displaystyle O}{\underset{}{C}}OH$
c. $CH_3CH_2CH{=}\underset{\underset{CH_3}{|}}{C}\overset{\displaystyle O}{\underset{}{C}}OH$
d. $CH_3CH_2\underset{\underset{CH_2CH_3}{|}}{\overset{\overset{OH}{|}}{C}CH_2}\overset{\displaystyle O}{\underset{}{C}}OCH_3$

64. The ketone whose ^{1}H NMR spectrum is shown here was obtained as the product of an acetoacetic ester synthesis. What alkyl halide was used in the synthesis?

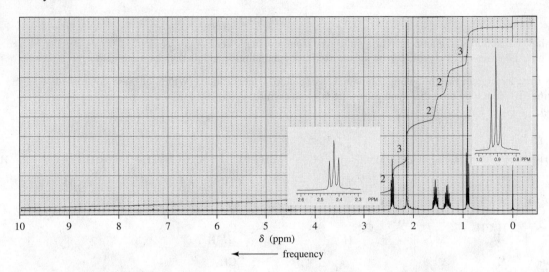

65. Indicate how the following compounds could be synthesized from cyclohexanone and any other necessary reagents:

a. (cyclohexanone with $CH_2CH_2CH_2CH_3$ substituent)

b. (cyclohexanone with $\overset{O}{C}CH_2CH_2CH_3$ substituent) **(two ways)**

c. (cyclohexanone with $\overset{O}{C}OCH_2CH_3$ substituent)

d. (cyclohexanone with $\overset{O}{C}H$ substituent)

e. (decalone structure)

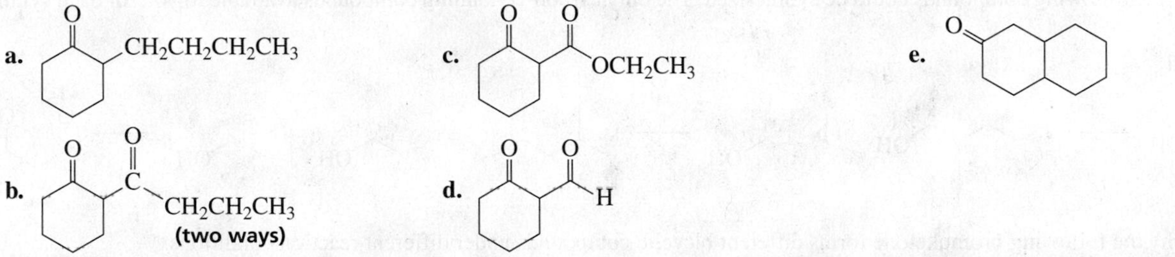

66. Compound **A** with molecular formula C_6H_{10} has two peaks in its ^{1}H NMR spectrum, both of which are singlets (with ratio 9 : 1). Compound **A** reacts with an acidic aqueous solution containing mercuric sulfate to form compound **B**, which gives a positive iodoform test (Problem 57) and has an ^{1}H NMR spectrum that shows two singlets (with ratio 3 : 1). Identify **A** and **B**.

67. Indicate how each of the following compounds could be synthesized from the given starting material and any other necessary reagents:

a. $CH_3\overset{\displaystyle O}{\underset{}{C}}CH_3 \longrightarrow CH_3\overset{\displaystyle O}{\underset{}{C}}CH_2\overset{\displaystyle O}{\underset{}{C}}CH_3$

b. $CH_3\overset{\displaystyle O}{\underset{}{C}}(CH_2)_3\overset{\displaystyle O}{\underset{}{C}}OCH_3 \longrightarrow$ (cyclohexane-1,3-dione with two CH_3 groups)

c. $CH_3\overset{\displaystyle O}{\underset{}{C}}CH_2\overset{\displaystyle O}{\underset{}{C}}OCH_2CH_3 \longrightarrow CH_3\overset{\displaystyle O}{\underset{}{C}}CH_2{-}$(cyclopentyl)

d. $CH_3CH_2O\overset{\displaystyle O}{\underset{}{C}}(CH_2)_4\overset{\displaystyle O}{\underset{}{C}}OCH_2CH_3 \longrightarrow$ (cyclopentanone with $(CH_2)_2CH_3$ substituent)

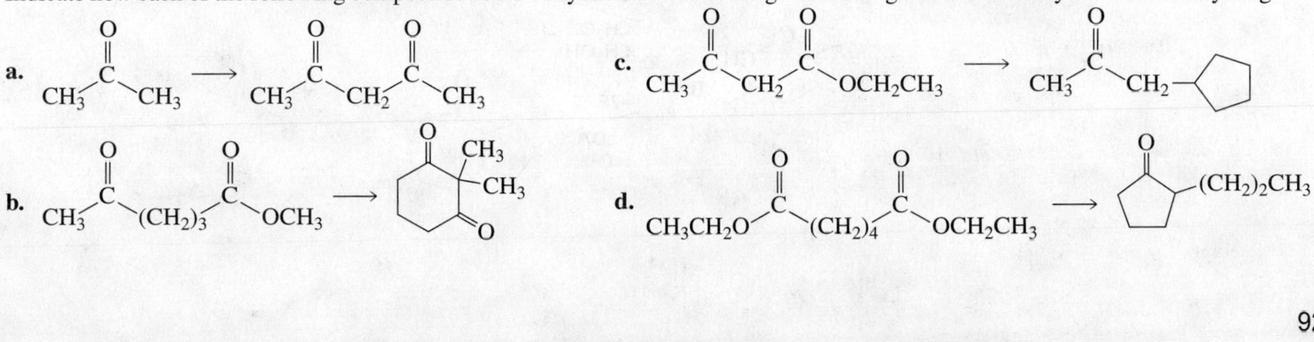

68. Draw the products of the following reactions:

a. 2 $\xrightarrow[\text{2. H}_3\text{O}^+]{\text{1. CH}_3\text{CH}_2\text{O}^-}$

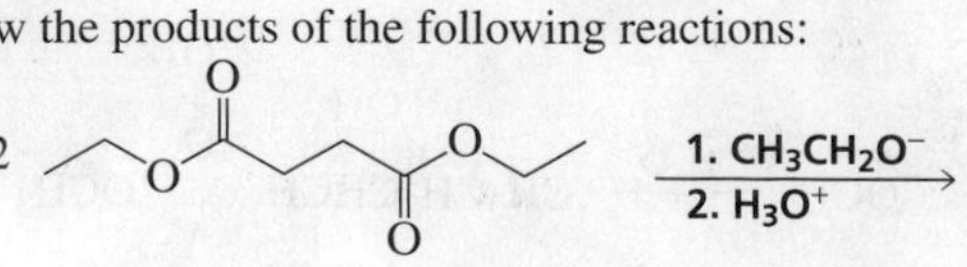

c. 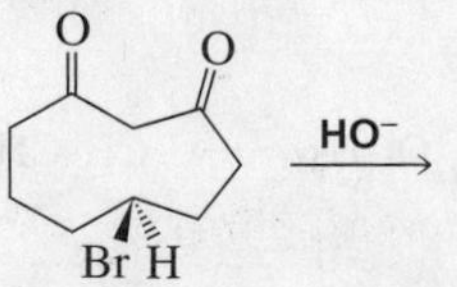 $\xrightarrow{\text{HO}^-}$

b. 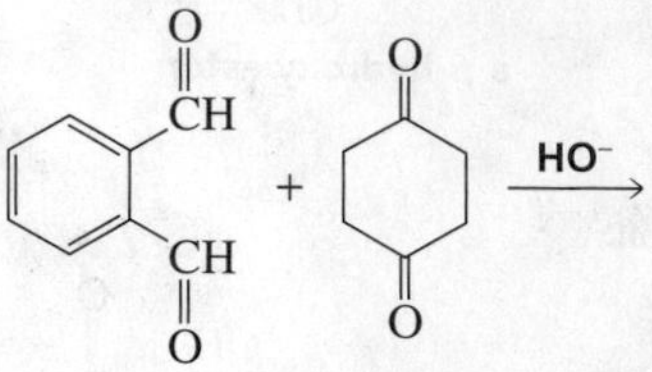 $\xrightarrow{\text{HO}^-}$

d.

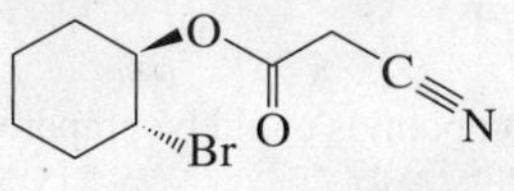

69. Show how the following compound can be prepared from starting materials that have no more than five carbons:

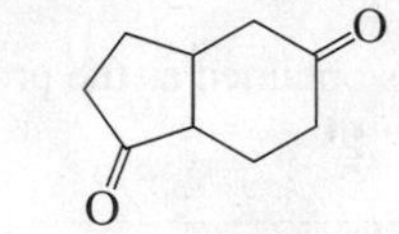

70. a. Show how the amino acid alanine can be synthesized from propanoic acid.

b. Show how the amino acid glycine can be synthesized from phthalimide and diethyl 2-bromomalonate.

71. A student tried to prepare the following compounds using aldol condensations. Which of these compounds was she successful in synthesizing? Explain why the other syntheses were not successful.

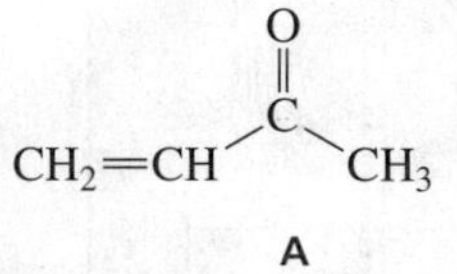

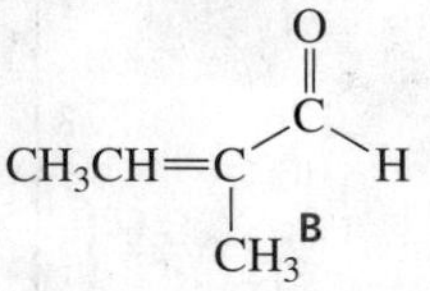

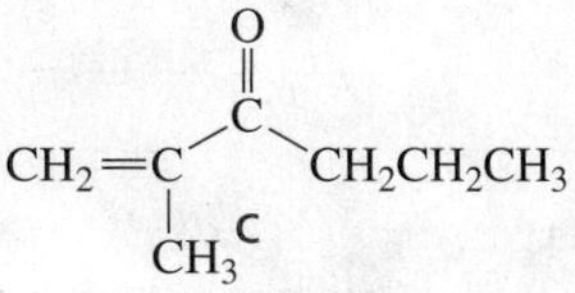

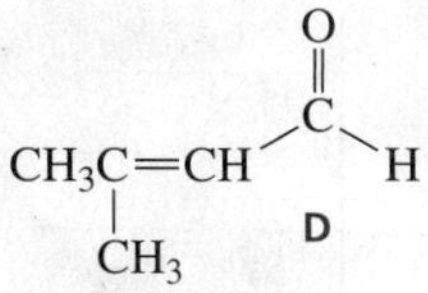

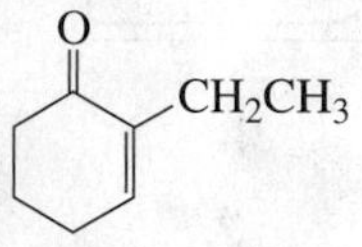

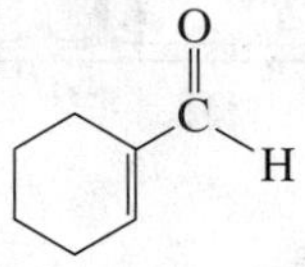

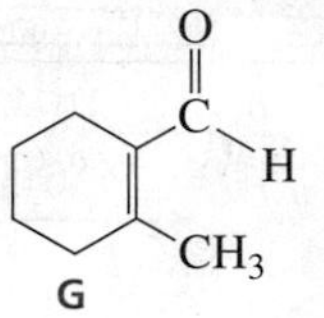

72. Show how the following compounds could be synthesized. The only carbon-containing compounds available to you for each synthesis are shown.

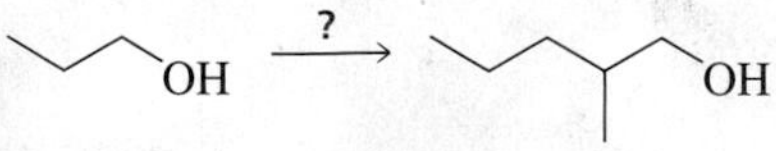

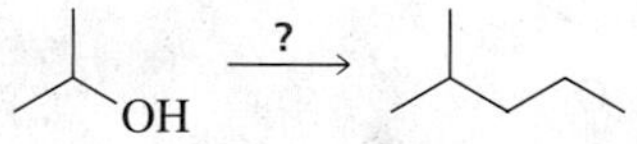

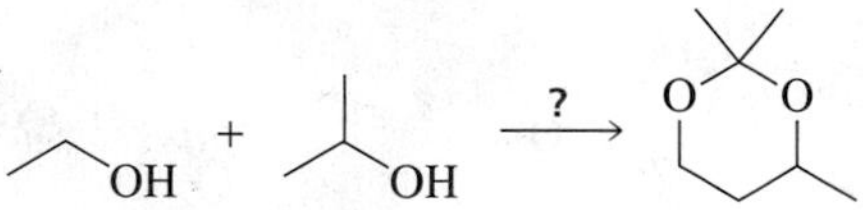

73. Explain why the following bromoketone forms different bicyclic compounds under different reaction conditions:

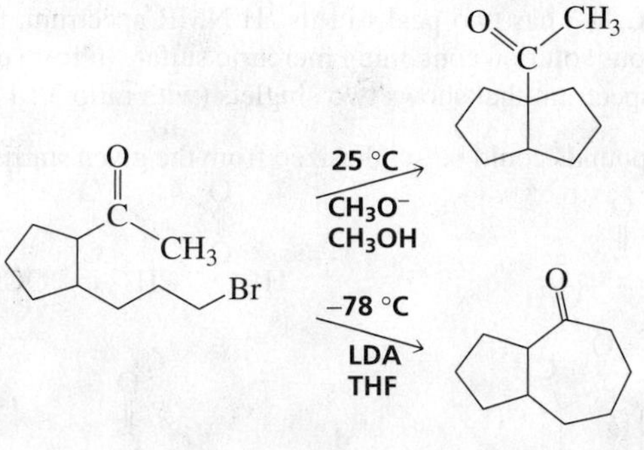

74. A *Mannich reaction* puts a R$_2$NCH$_2$— group on the α-carbon of a carbon acid. Propose a mechanism for the reaction.

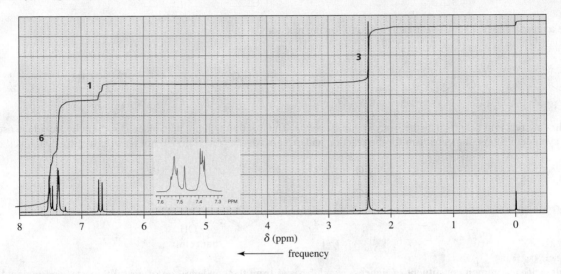

75. What carbonyl compounds are required to prepare a compound with molecular formula $C_{10}H_{10}O$ whose ^1H NMR spectrum is shown?

76. Ninhydrin reacts with an amino acid to form a purple-colored compound. Propose a mechanism to account for the formation of the colored compound.

77. A carboxylic acid is formed when an α-haloketone reacts with hydroxide ion. This reaction is called a *Favorskii reaction*. Propose a mechanism for the following Favorskii reaction. (*Hint:* In the first step, HO$^-$ removes a proton from the α-carbon that is not bonded to Br; a three-membered ring is formed in the second step; and HO$^-$ is a nucleophile in the third step.)

78. An α,β-unsaturated carbonyl compound can be prepared by a reaction known as a *selenenylation–oxidation reaction*. A selenoxide is formed as an intermediate. Propose a mechanism for the reaction.

79. a. What carboxylic acid would be formed if the malonic ester synthesis were carried out with one equivalent of malonic ester, one equivalent of 1,5-dibromopentane, and two equivalents of base?

b. What carboxylic acid would be formed if the malonic ester synthesis were carried out with two equivalents of malonic ester, one equivalent of 1,5-dibromopentane, and two equivalents of base?

80. A *Cannizzaro reaction* is the reaction of an aldehyde that has no α-hydrogens with concentrated aqueous sodium hydroxide. In this reaction, half the aldehyde is converted to a carboxylic acid and the other half is converted to an alcohol. Propose a reasonable mechanism for the following Cannizzaro reaction:

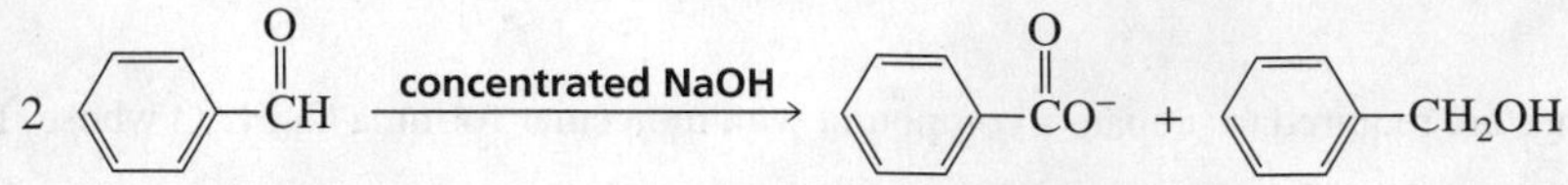

81. Propose a reasonable mechanism for each of the following reactions:

a.

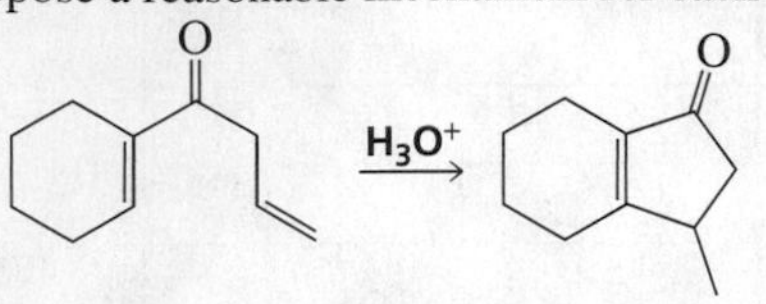

b.

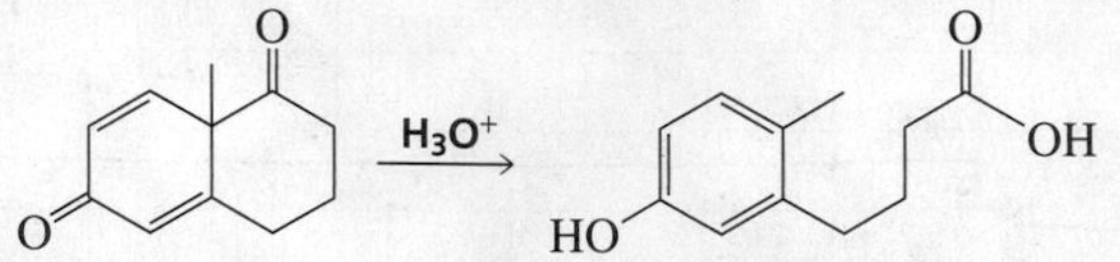

82. The following reaction is known as the *benzoin condensation*. The reaction will not take place if sodium hydroxide is used instead of sodium cyanide. Propose a mechanism for the reaction and explain why the reaction will not occur if hydroxide ion is the base.

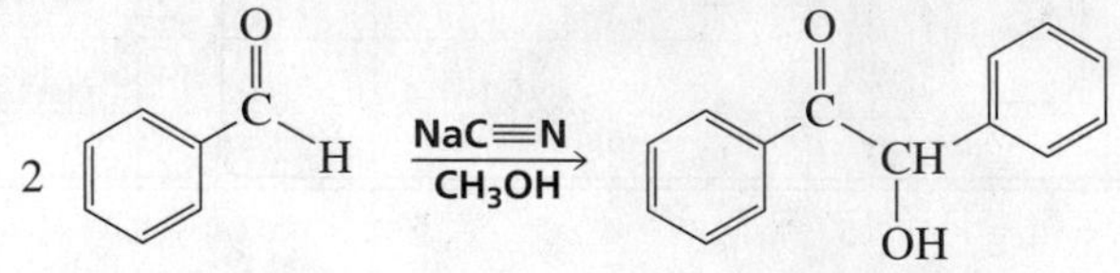

benzoin

83. Orsellinic acid, a common constituent of lichens, is synthesized from the condensation of acetyl thioester and malonyl thioester. If a lichen were grown on a medium containing acetate that was radioactively labeled with ^{14}C at the carbonyl carbon, which carbons would be labeled in orsellinic acid?

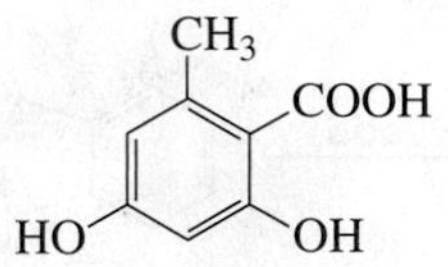

orsellinic acid

84. Propose a mechanism for the following reaction. (*Hint:* The intermediate has a cumulated double bond.)

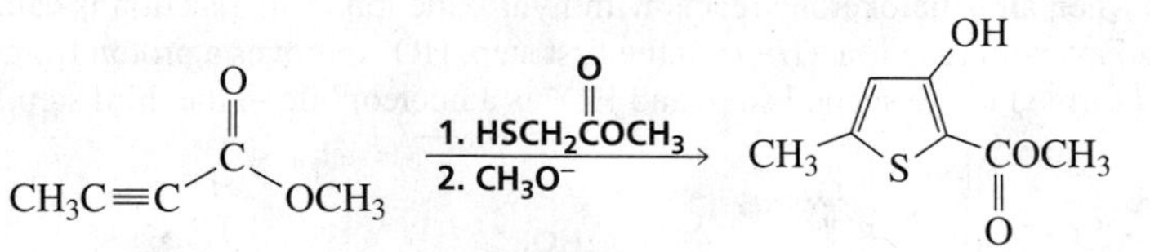

85. A compound known as *Hagemann's ester* can be prepared by treating a mixture of formaldehyde and ethyl acetoacetate first with base and then with acid and heat. Write the structure for the product of each of the steps.

a. The first step is an aldol-like condensation.

b. The second step is a Michael addition.

c. The third step is an intramolecular aldol addition.

d. The final transformation includes a dehydration and a hydrolysis followed by a decarboxylation.

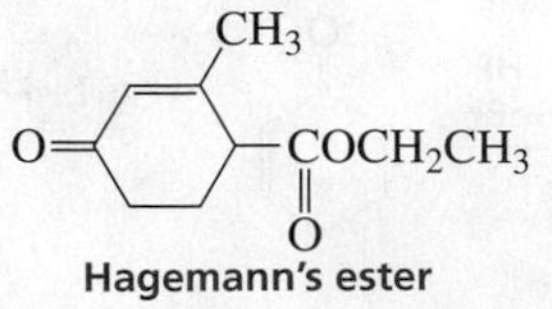

Hagemann's ester

86. Amobarbital is a sedative marketed under the trade name Amytal. Propose a synthesis of amobarbital, using diethyl malonate and urea as two of the starting materials.

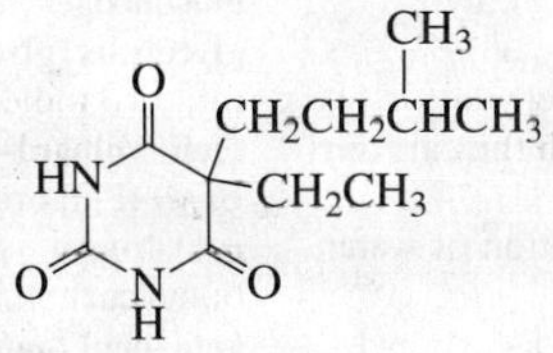

Amytal®

87. Propose a reasonable mechanism for the following reaction:

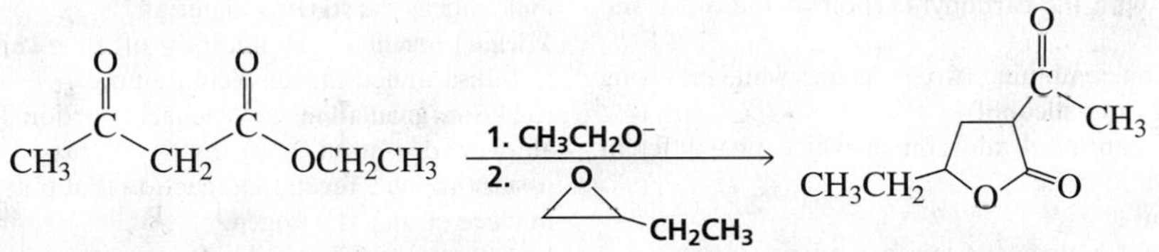

GLOSSARY

acetoacetic ester synthesis synthesis of a methyl ketone, using ethyl acetoacetate as the starting material.

aldol addition a reaction between two molecules of an aldehyde (or two molecules of a ketone) that connects the α-carbon of one with the carbonyl carbon of the other.

aldol condensation an aldol addition followed by the elimination of water.

annulation reaction a ring-forming reaction.

α-carbon a carbon bonded to a leaving group or adjacent to a carbonyl carbon.

carbon acid a compound containing a carbon that is bonded to a relatively acidic hydrogen.

Claisen condensation a reaction between two molecules of an ester that connects the α-carbon of one with the carbonyl carbon of the other and eliminates an alkoxide ion.

condensation reaction a reaction combining two molecules while removing a small molecule (usually water or an alcohol).

crossed (mixed) aldol addition an aldol addition in which two different carbonyl compounds are used.

decarboxylation loss of carbon dioxide.

Dieckmann condensation an intramolecular Claisen condensation.

β-diketone a ketone with a second carbonyl group at the b -position.

gluconeogenesis the synthesis of D-glucose from pyruvate.

glycolysis (glycolytic cycle) the series of reactions that converts D-glucose into two molecules of pyruvate.

Hell–Volhard–Zelinski (HVZ) reaction heating a carboxylic acid with $Br_2 + P$ in order to convert it into an α-bromocarboxylic acid.

α-hydrogen usually, a hydrogen bonded to the carbon adjacent to a carbonyl carbon.

keto–enol tautomers a ketone and its isomeric α,β-unsaturated alcohol.

β-keto ester an ester with a second carbonyl group at the β-position.

Kolbe–Schmitt carboxylation reaction a reaction that uses CO_2 to carboxylate phenol.

malonic ester synthesis the synthesis of a carboxylic acid, using diethyl malonate as the starting material.

Michael reaction the addition of an α-carbanion to the β-carbon of an α,β-unsaturated carbonyl compound.

Robinson annulation a Michael reaction followed by an intramolecular aldol condensation.

α-substitution reaction a reaction that puts a substituent on an a -carbon in place of an a -hydrogen.

tautomers rapidly equilibrating isomers that differ in the location of their bonding electrons.

PHOTO CREDITS

Photo credits are listed in order of appearance.

Reactions of Benzene and Substituted Benzenes

gas lamps along a street

The compound we know as benzene was first isolated in 1825 by Michael Faraday. He extracted it from the liquid residue obtained after heating whale oil under pressure to produce the gas then being used in gas lamps. In 1834, Eilhard Mitscherlich correctly determined the molecular formula (C_6H_6) of Faraday's compound and named it benzin because of its relationship to benzoic acid, a known substituted form of the compound. Later its name was changed to benzene.

The families of organic compounds can be placed in one of four groups, and that all the families in a group react in similar ways. This chapter begins our discussion of aromatic compounds—the families of compounds in Group IV.

Benzene is an aromatic compound that can be represented by two resonance contributors. In order to be aromatic, a compound must be cyclic and planar and have an uninterrupted cloud of π electrons (called a π cloud) above and below the plane of the molecule, *and* the π cloud must contain an odd number of pairs of π electrons. In addition, aromatic compounds are unusually stable compounds.

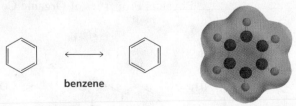

benzene

electrostatic potential
map for benzene

From Chapter 19 of *Organic Chemistry*, Seventh Edition. Paula Yurkanis Bruice. Copyright © 2014 by Pearson Education, Inc.

Many substituted benzenes are found in nature. A few that have physiological activity are shown here.

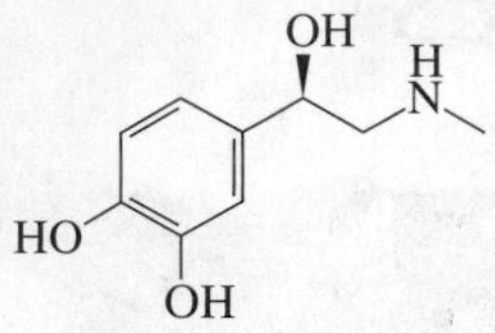

adrenaline
epinephrine
a hormone released
by the body in
response to stress

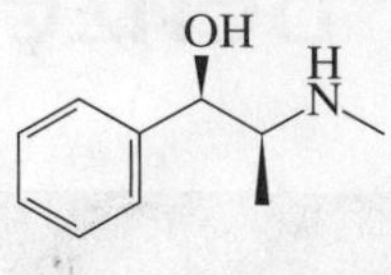

ephedrine
a bronchodilator

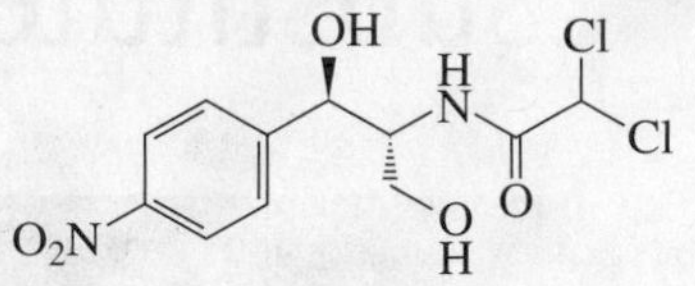

chloramphenicol
an antibiotic that is particularly
effective against typhoid fever

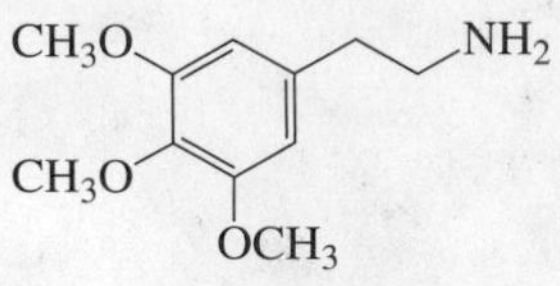

mescaline
active agent of
the peyote cactus

Many other physiologically active substituted benzenes are not found in nature, but exist because chemists have synthesized them. In fact, more than two-thirds of the top 400 brand name and generic drugs contain a benzene ring. Three of the most commonly prescribed drugs are shown here.

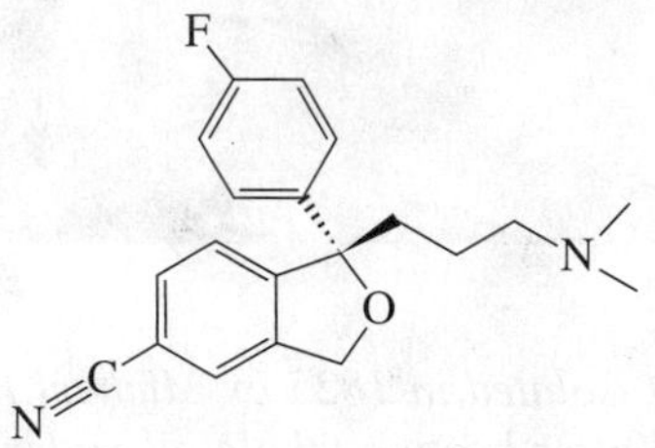

Lexapro®
an antidepressant

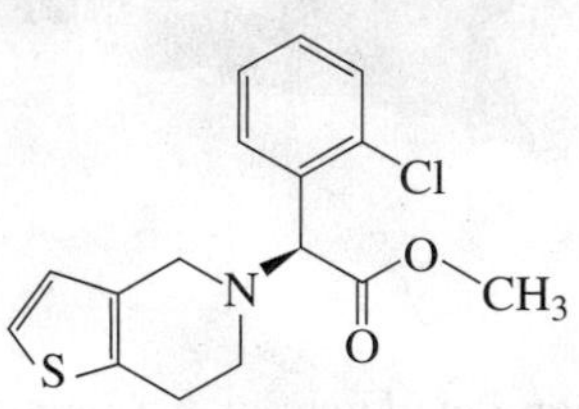

Plavix®
an inhibitor of platelet aggregation

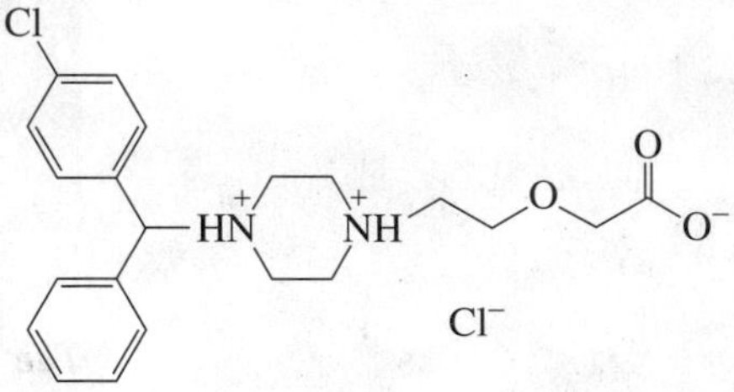

Zyrtec®
an antihistamine

When naturally occurring compounds are found to have desirable physiological activities, chemists attempt to synthesize structurally similar compounds with the hope of developing them into useful products. For example, chemists have synthesized compounds with structures similar to that of adrenaline, producing amphetamine, a central nervous system stimulant, and the closely related methamphetamine (methylated amphetamine); both are used clinically as appetite suppressants. Methamphetamine, known as "speed" because of its rapid and intense psychological effects, is also made and sold illegally. The compounds shown here represent just a few of the many substituted benzenes that have been synthesized for commercial use. The physical properties of several substituted benzenes are given in Appendix: "Physical Properties of Organic Compounds."

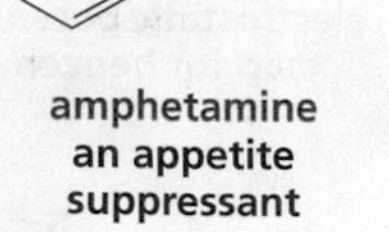

amphetamine
an appetite
suppressant

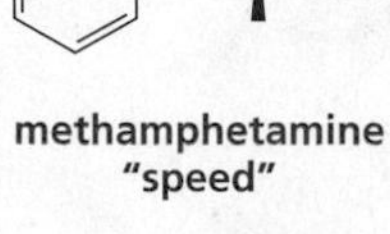

methamphetamine
"speed"

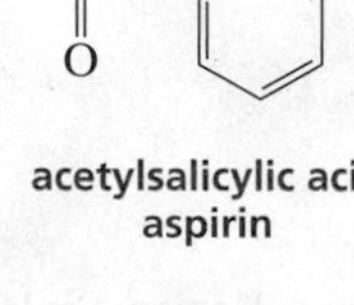

acetylsalicylic acid
aspirin

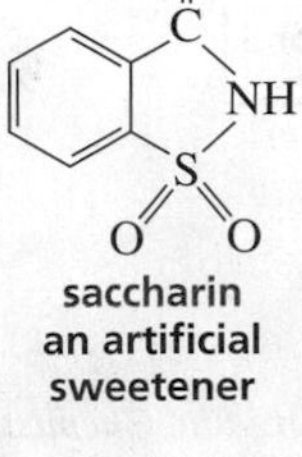

saccharin
an artificial
sweetener

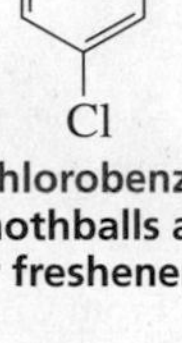

p-dichlorobenzene
in mothballs and
air fresheners

Measuring Toxicity

Agent Orange, a defoliant widely used in the Vietnam War, is a mixture of two synthetic substituted benzenes: 2,4-D and 2,4,5-T. Dioxin (TCDD), a contaminant formed during the manufacture of Agent Orange, has been implicated as the causative agent of the various symptoms suffered by those exposed to the defoliant during the war.

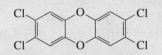

2,4-dichlorophenoxyacetic acid
2,4-D

2,4,5-trichlorophenoxyacetic acid
2,4,5-T

2,3,7,8-tetrachlorodibenzo[b,e][1,4]dioxin
TCDD

The toxicity of a compound is indicated by its LD_{50} value—the dosage found to kill 50% of the test animals exposed to it. Dioxin, with an LD_{50} value of 0.0006 mg/kg for guinea pigs, is extremely toxic. For comparison, the LD_{50} values of some well-known but far less toxic poisons are 0.96 mg/kg for strychnine and 15 mg/kg for both arsenic trioxide and sodium cyanide. One of the most toxic agents known is the toxin that causes botulism; it has an LD_{50} value of about 1×10^{-7} mg/kg.

1 THE NOMENCLATURE OF MONOSUBSTITUTED BENZENES

Some monosubstituted benzenes are named simply by adding the name of the substituent to "benzene."

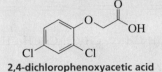

bromobenzene   chlorobenzene   nitrobenzene
used as a solvent
in shoe polish   ethylbenzene

Some monosubstituted benzenes have names that incorporate the substituent. Unfortunately, these names have to be memorized.

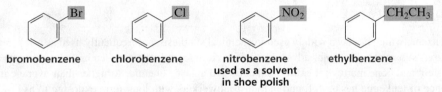

toluene   phenol   aniline   benzenesulfonic acid

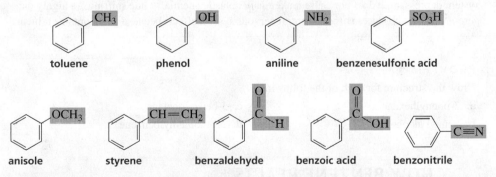

anisole   styrene   benzaldehyde   benzoic acid   benzonitrile

When a benzene ring is a substituent, it is called a **phenyl group.** A benzene ring with a methylene group is called a **benzyl group**.

a phenyl group   a benzyl group

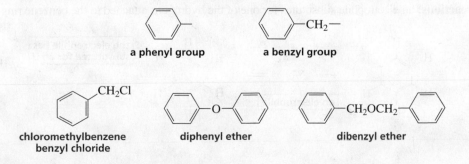

chloromethylbenzene
benzyl chloride   diphenyl ether   dibenzyl ether

With the exception of toluene, benzene rings with an alkyl substituent are named as alkyl-substituted benzenes (when the alkyl substituent has a name), or as phenyl-substituted alkanes (when the name of the alkyl substituent can be used). For example, *sec*-pentyl cannot be used as a substituent name because both compounds that appear on the right would be called *sec*-pentylbenzene, and a name must specify only one compound.

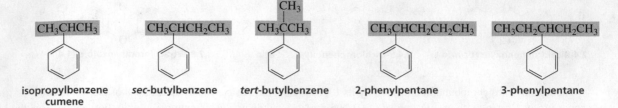

| isopropylbenzene cumene | *sec*-butylbenzene | *tert*-butylbenzene | 2-phenylpentane | 3-phenylpentane |

Aryl (Ar) is the general term for either a phenyl group or a substituted phenyl group, just as alkyl (R) is the general term for a group derived from an alkane. Thus, ArOH could be used to designate any of the following phenols:

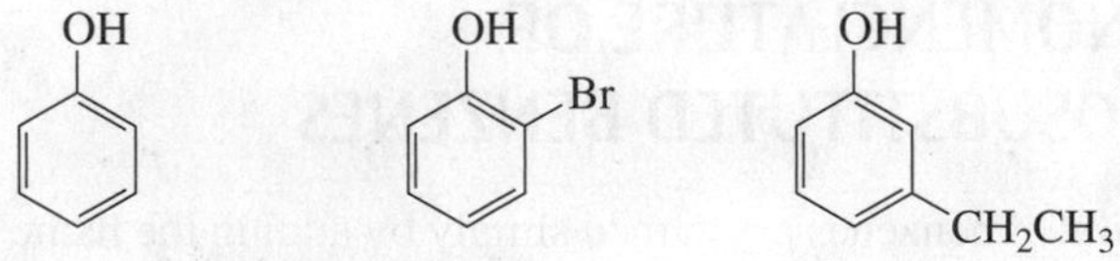

The Toxicity of Benzene

Benzene, which has been widely used in chemical synthesis and frequently used as a solvent, is a toxic substance. The major adverse effects of chronic exposure are seen in the central nervous system and bone marrow; it causes leukemia and aplastic anemia. A higher-than-average incidence of leukemia has been found in industrial workers with long-term exposure to as little as 1 ppm benzene in the atmosphere.

Toluene has replaced benzene as a solvent because, although it too is a central nervous system depressant, it does not cause leukemia or aplastic anemia. "Glue sniffing," a highly dangerous activity, produces narcotic central nervous system effects because glue contains toluene.

PROBLEM 1

Draw the structure for each of the following:

a. 2-phenylhexane   **c.** 3-benzylpentane

b. benzyl alcohol   **d.** bromomethylbenzene

2   HOW BENZENE REACTS

Aromatic compounds such as benzene undergo **electrophilic aromatic substitution reactions:** an electrophile substitutes for one of the hydrogens attached to the benzene ring.

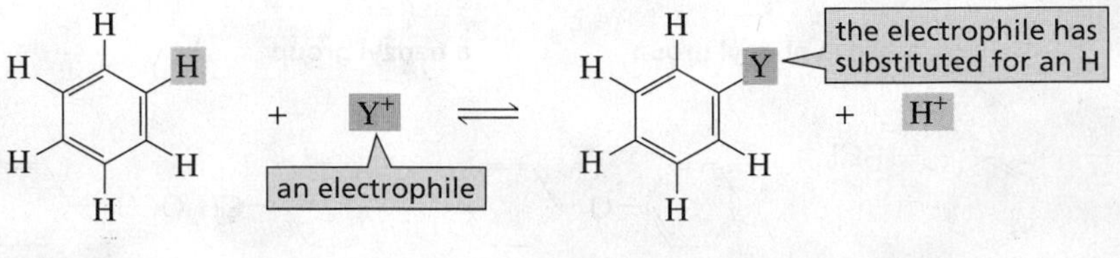

Now let's look at why this substitution reaction occurs. The cloud of π electrons above and below the plane of its ring makes benzene a nucleophile, so it reacts with an electrophile (Y^+). When an electrophile attaches itself to a benzene ring, a carbocation intermediate is formed.

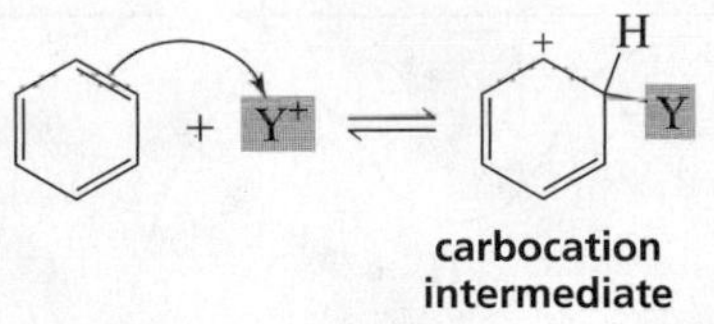

carbocation intermediate

This description is like the first step in an *electrophilic addition reaction* of an alkene: the nucleophilic alkene reacts with an electrophile and forms a carbocation intermediate. In the second step of the reaction, the carbocation reacts with a nucleophile (Z^-) to form an addition product.

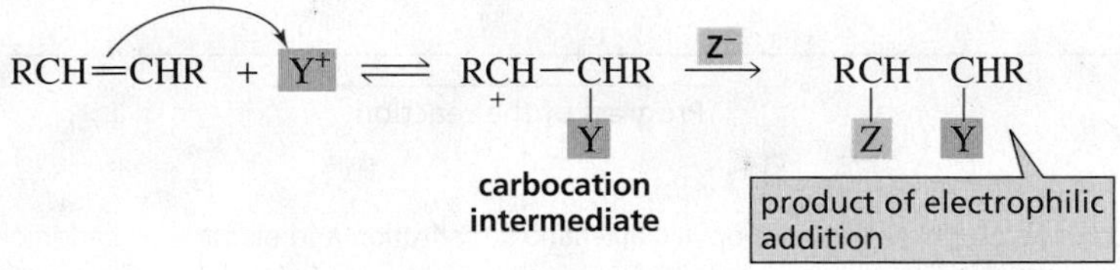

carbocation intermediate

product of electrophilic addition

If the carbocation intermediate that is formed from the reaction of benzene with an electrophile were to react similarly with a nucleophile (depicted as path *a* in Figure 1), then the *addition product* would not be aromatic. But, if the carbocation instead were to lose a proton from the site of electrophilic addition and form a *substitution product* (depicted as path *b* in Figure 1), then the aromaticity of the benzene ring would be restored.

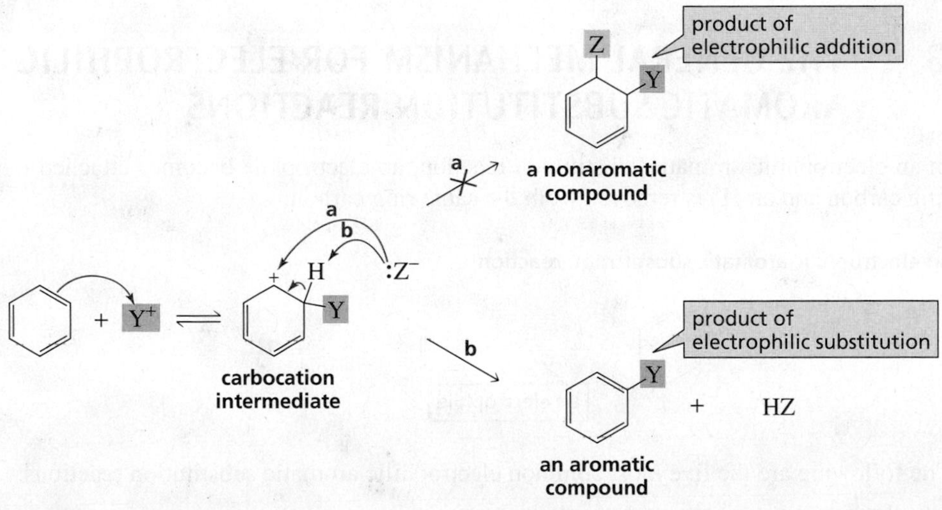

product of electrophilic addition

a nonaromatic compound

carbocation intermediate

product of electrophilic substitution

an aromatic compound

▲ **Figure 1**

Reaction of benzene with an electrophile. Because of the greater stability of the aromatic product, the reaction procedes by an electrophilic substitution reaction (path *b*) rather than by an electrophilic addition reaction (path *a*).

Because the aromatic substitution product is much more stable than the nonaromatic addition product (Figure 2), benzene undergoes *electrophilic substitution reactions* that preserve aromaticity, rather than *electrophilic addition reactions*—the reactions characteristic of alkenes—that would destroy aromaticity. The substitution reaction is more accurately called an **electrophilic aromatic substitution reaction,** since the electrophile substitutes for a hydrogen of an aromatic compound.

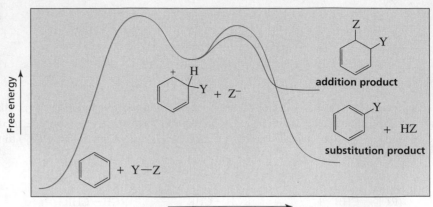

▲ **Figure 2**

Reaction coordinate diagrams for electrophilic aromatic substitution and electrophilic addition.

PROBLEM 2

If electrophilic addition to benzene is overall an endergonic reaction, how can electrophilic addition to an alkene be overall an exergonic reaction?

---

3 THE GENERAL MECHANISM FOR ELECTROPHILIC AROMATIC SUBSTITUTION REACTIONS

In an electrophilic aromatic substitution reaction, an electrophile becomes attached to a ring carbon and an H^+ is removed from the same ring carbon.

an electrophilic aromatic substitution reaction

The following are the five most common electrophilic aromatic substitution reactions:

1. **Halogenation:** A bromine (Br), a chlorine (Cl), or an iodine (I) substitutes for a hydrogen.

2. **Nitration:** A nitro group (NO_2) substitutes for a hydrogen.

3. **Sulfonation:** A sulfonic acid group (SO_3H) substitutes for a hydrogen.

4. **Friedel–Crafts acylation:** An acyl group ($RC{=}O$) substitutes for a hydrogen.

5. **Friedel–Crafts alkylation:** An alkyl group (R) substitutes for a hydrogen.

All five of these reactions take place by the same two-step mechanism.

the proton is removed from the carbon that has formed the bond with the electrophile

a base in the reaction mixture

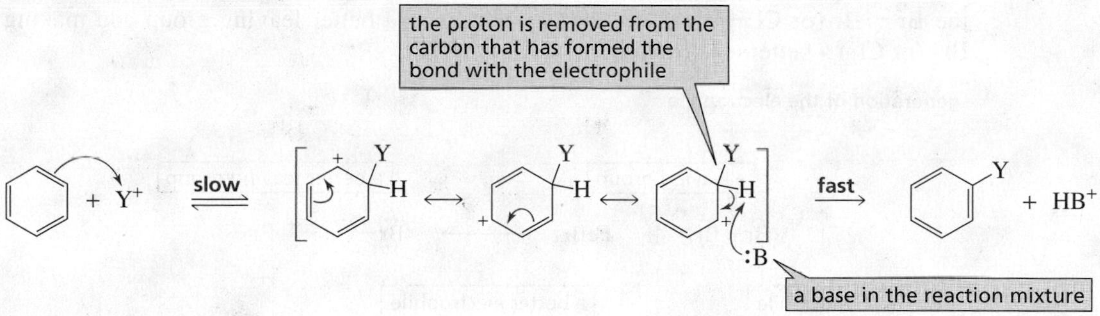

- The electrophile (Y^+) adds to the nucleophilic benzene ring, thereby forming a carbocation intermediate. The structure of the carbocation intermediate can be approximated by three resonance contributors.

- A base in the reaction mixture (:B) removes a proton from the carbocation intermediate, and the electrons that held the proton move into the ring to reestablish its aromaticity. Notice that *the proton is always removed from the carbon that has formed the bond with the electrophile*.

The first step is relatively slow and endergonic because an aromatic compound is being converted into a much less stable nonaromatic intermediate (Figure 2). The second step is fast and strongly exergonic because this step restores the stability-enhancing aromaticity.

We will look at each of these five electrophilic aromatic substitution reactions individually. As you study them, notice that they differ only in how the electrophile (Y^+) that is needed to start the reaction is generated. Once the electrophile is formed, all five reactions follow the same two-step mechanism for electrophilic aromatic substitution that was just shown.

4   THE HALOGENATION OF BENZENE

The bromination or chlorination of benzene requires a Lewis acid catalyst such as ferric bromide or ferric chloride. A *Lewis acid* is a compound that accepts a share in an electron pair.

bromination

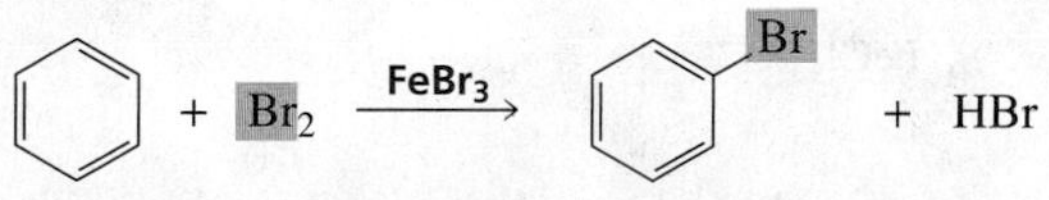

bromobenzene

chlorination

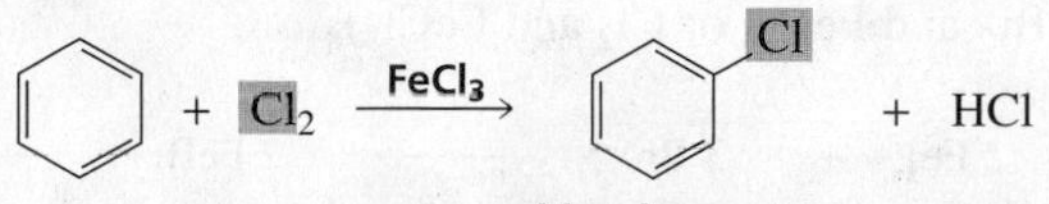

chlorobenzene

Why does the reaction of benzene with Br_2 or Cl_2 require a catalyst when the reaction of an alkene with these reagents does not require a catalyst? Benzene's aromaticity makes it much more stable and therefore much less reactive than an alkene. Therefore, benzene requires a better electrophile. Donating a lone pair to the Lewis acid weakens the Br—Br (or Cl—Cl) bond, thereby providing a better leaving group and making Br_2 (or Cl_2) a better electrophile.

generation of the electrophile

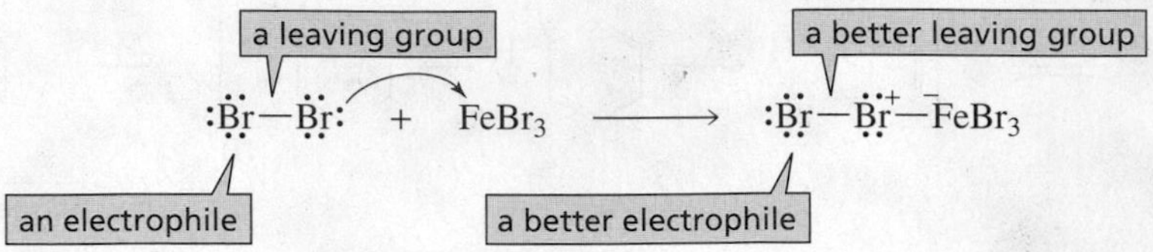

| a leaving group | | a better leaving group |

$$:\ddot{Br}-\ddot{Br}: \ + \ FeBr_3 \ \longrightarrow \ :\ddot{Br}-\overset{+}{\ddot{Br}}-\overline{F}eBr_3$$

| an electrophile | | a better electrophile |

MECHANISM FOR BROMINATION

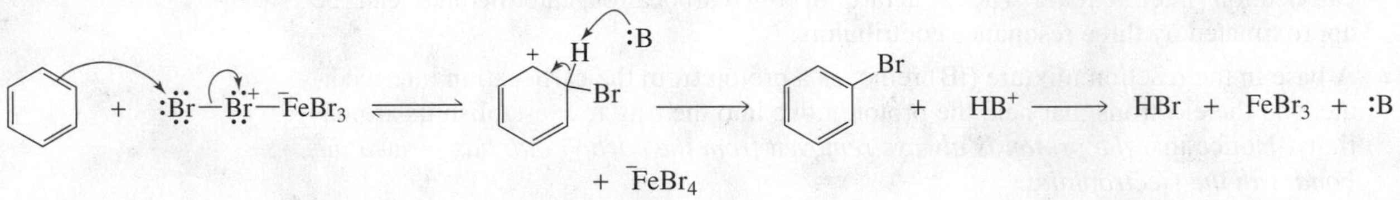

- The electrophile adds to the benzene ring.
- A base (:B) in the reaction mixture (such as $\overline{F}eBr_4$ or the solvent) removes the proton from the carbon that formed the bond with the electrophile. Notice that the catalyst is regenerated.

For the sake of clarity, only one of the three resonance contributors of the carbocation intermediate is shown in this and subsequent mechanisms for electrophilic aromatic substitution reactions. Bear in mind, however, that each carbocation intermediate actually has the three resonance contributors shown in Section 3.

The mechanism for chlorination of benzene is the same as that for bromination.

MECHANISM FOR CHLORINATION

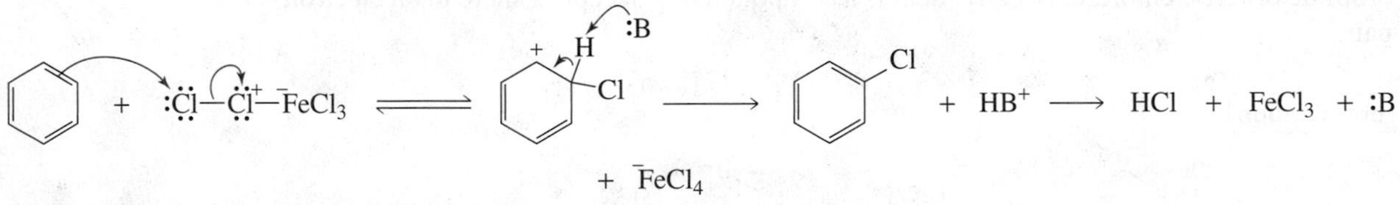

The ferric bromide and ferric chloride catalysts react readily with moisture in the air during handling, which inactivates them. To avoid this problem, they are generated in situ (in the reaction mixture) from iron filings and bromine (or chlorine). As a result, the halogen component of the Lewis acid will be the same as the halogen used in the substitution reaction (Br_2 and $FeBr_3$ or Cl_2 and $FeCl_3$).

$$2 \ Fe \ + \ 3 \ Br_2 \ \longrightarrow \ 2 \ FeBr_3$$

$$2 \ Fe \ + \ 3 \ Cl_2 \ \longrightarrow \ 2 \ FeCl_3$$

PROBLEM 3

Why does hydration inactivate $FeBr_3$?

Iodobenzene can be prepared using I_2 and an oxidizing agent under acidic conditions. Hydrogen peroxide is commonly used as the oxidizing agent.

iodination

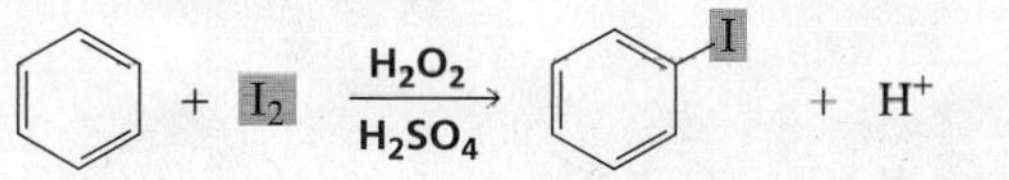

The oxidizing agent converts I_2 into the electrophilic iodonium ion (I^+).

generation of the electrophile

$$I_2 \xrightarrow{\text{oxidizing agent}} 2\,I^+$$

Once the electrophile is formed, iodination of benzene occurs by the same mechanism as bromination and chlorination.

MECHANISM FOR IODINATION

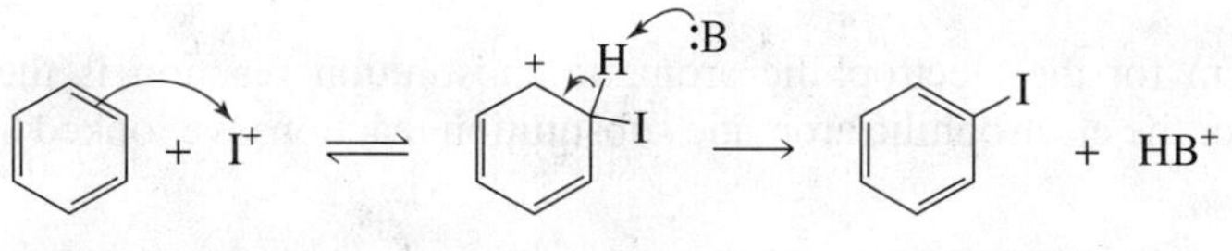

- The electrophile adds to the benzene ring.
- A base (:B) in the reaction mixture removes the proton from the carbon that formed the bond with the electrophile.

Thyroxine

Thyroxine, a hormone produced by the thyroid gland, increases the rate at which fats, carbohydrates, and proteins are metabolized. Humans obtain thyroxine from tyrosine (an amino acid) and iodine. The thyroid gland is the only part of the body that uses iodine, which we acquire primarily from seafood or iodized salt.

An enzyme called iodoperoxidase converts the I^- that we ingest to I^+, the electrophile needed to place an iodo substituent on a benzene ring. A deficiency in iodine is the number one cause of preventable intellectual disability in children.

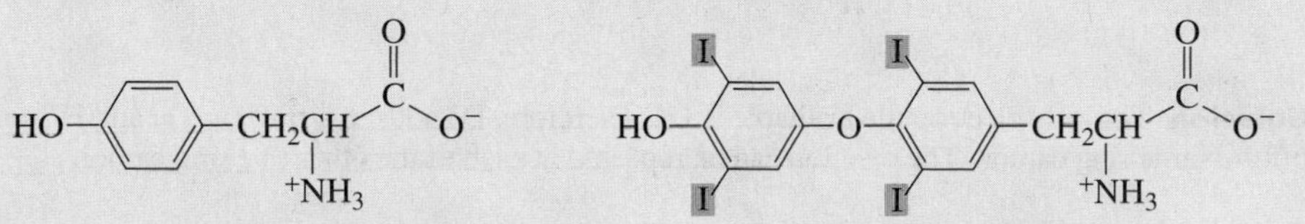

tyrosine   thyroxine

Chronically low levels of thyroxine cause enlargement of the thyroid gland as it tries in vain to make more thyroxine, a condition known as goiter. Low thyroxine levels can be corrected by taking thyroxine orally. Synthroid, the most popular brand of thyroxine, is currently one of the most-prescribed drugs in the United States.

5   THE NITRATION OF BENZENE

The nitration of benzene with nitric acid requires sulfuric acid as a catalyst.

nitration

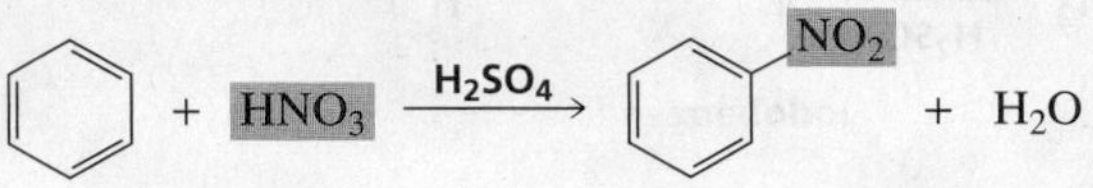

nitric acid

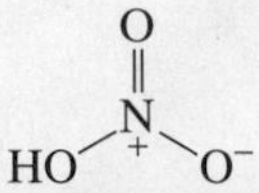

nitronium ion

$$\text{benzene} + \boxed{\text{HNO}_3} \xrightarrow{\text{H}_2\text{SO}_4} \text{nitrobenzene} \; (\text{NO}_2) + \text{H}_2\text{O}$$

To generate the necessary electrophile, sulfuric acid protonates nitric acid. Protonated nitric acid then loses water to form a nitronium ion, the electrophile required for nitration.

generation of the electrophile

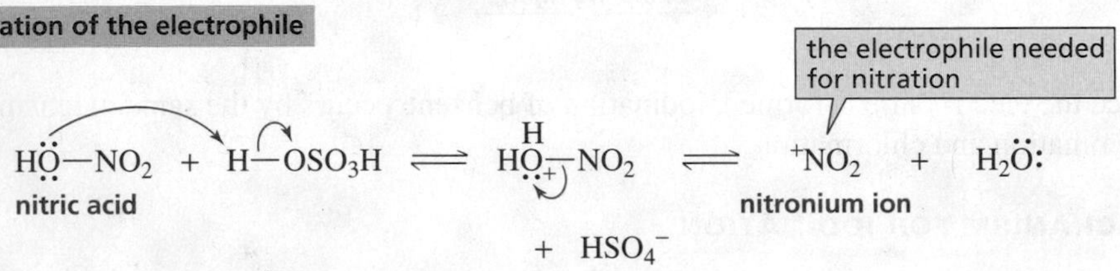

the electrophile needed for nitration

$$\text{HO–NO}_2 + \text{H–OSO}_3\text{H} \rightleftharpoons \text{HO}\overset{\text{H}}{+}\text{NO}_2 \rightleftharpoons {}^{+}\text{NO}_2 + \text{H}_2\text{O:}$$

nitric acid   nitronium ion

$$+ \; \text{HSO}_4^{-}$$

The mechanism for the electrophilic aromatic substitution reaction is the same as the mechanisms for the electrophilic aromatic substitution reactions we looked at in Section 4.

MECHANISM FOR NITRATION

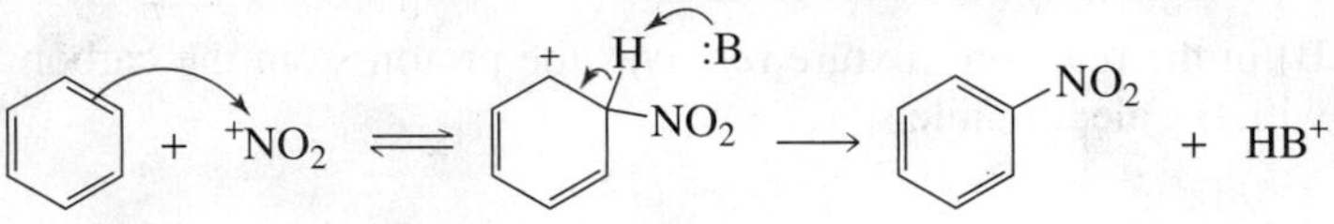

$$\text{benzene} + {}^{+}\text{NO}_2 \rightleftharpoons \overset{+}{\underset{\text{NO}_2}{\text{(ring with H, :B)}}} \longrightarrow \text{(ring)–NO}_2 + \text{HB}^{+}$$

- The electrophile adds to the benzene ring.
- A base (:B) in the reaction mixture (for example, H$_2$O, HSO$_4^{-}$, or the solvent) removes the proton from the carbon that formed the bond with the electrophile.

PROBLEM 4 Solved

Propose a mechanism for the following reaction:

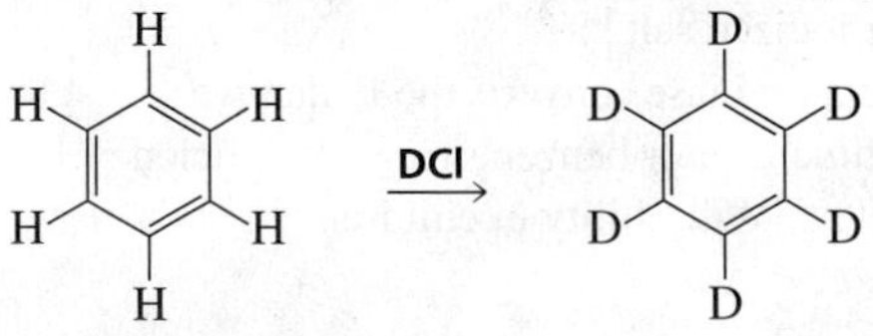

(benzene with H's) $\xrightarrow{\text{DCl}}$ (benzene with D's)

Solution The only electrophile available is D^{+}. Therefore, D^{+} adds to a ring carbon and H^{+} comes off the same ring carbon. The reaction can be repeated at each of the other five ring carbons.

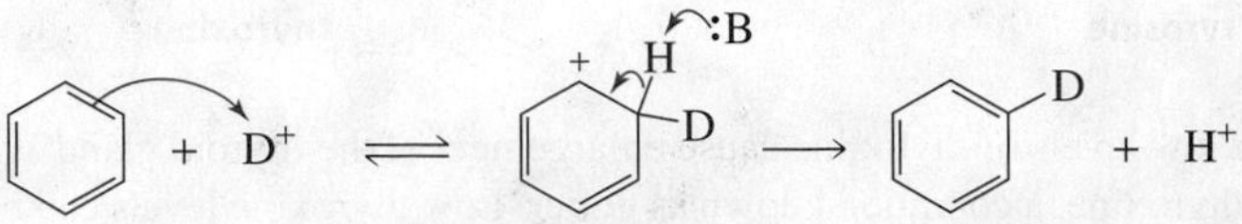

$$\text{benzene} + \text{D}^{+} \rightleftharpoons \overset{+}{\underset{\text{D}}{\text{(ring with H, :B)}}} \longrightarrow \text{(ring)–D} + \text{H}^{+}$$

6   THE SULFONATION OF BENZENE

Concentrated sulfuric acid or fuming sulfuric acid (a solution of SO_3 in sulfuric acid) is used to sulfonate aromatic rings.

sulfonation

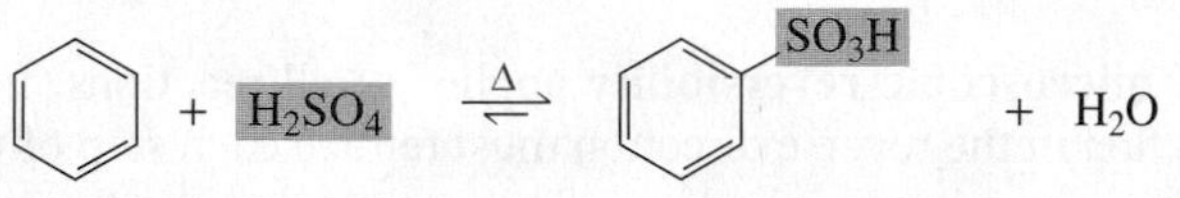

benzenesulfonic acid

Take a minute to note the similarities in the mechanisms for forming the $^+SO_3H$ electrophile for sulfonation and the $^+NO_2$ electrophile for nitration. A substantial amount of electrophilic sulfur trioxide (SO_3) is generated when concentrated sulfuric acid is heated, due to the $^+SO_3H$ electrophile losing a proton.

generation of the electrophile

the electrophile needed for sulfonation

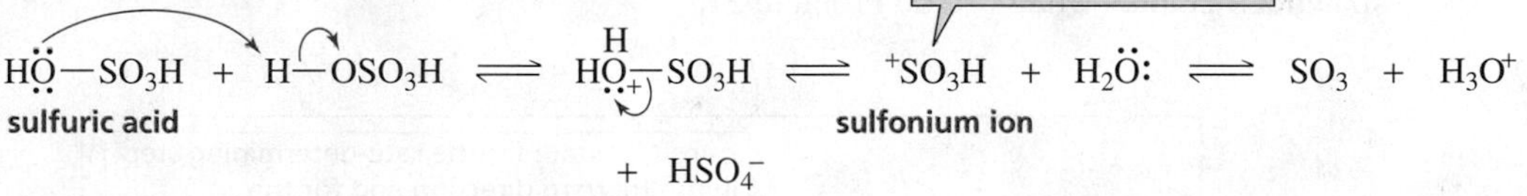

sulfuric acid   sulfonium ion

MECHANISM FOR SULFONATION

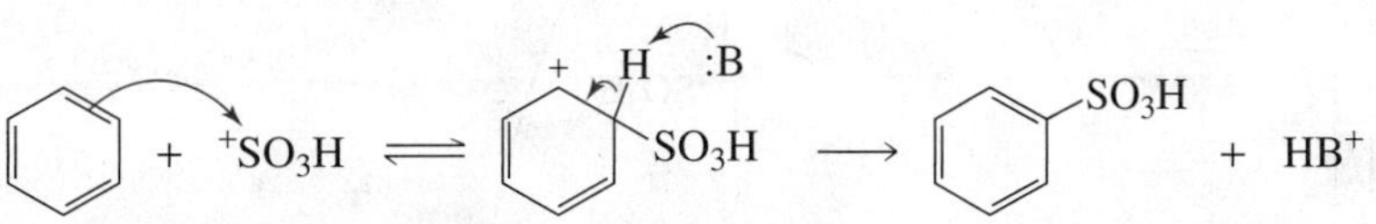

- The electrophile attaches to the ring.
- A base (:B) in the reaction mixture (for example, H_2O, HSO_4^-, or the solvent) removes the proton from the carbon that formed the bond with the electrophile.

A sulfonic acid is a strong acid because its conjugate base is particularly stable (weak) due to delocalization of its negative charge over three oxygens.

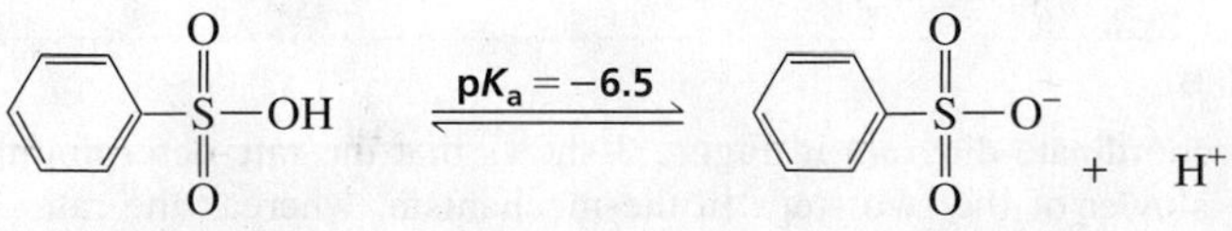

benzenesulfonic acid   benzenesulfonate ion

Sulfonation is reversible. If benzenesulfonic acid is heated in dilute acid, an H^+ adds to the ring and the sulfonic acid group comes off the ring.

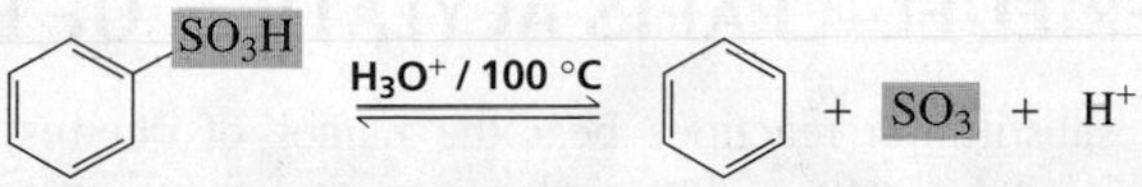

MECHANISM FOR DESULFONATION

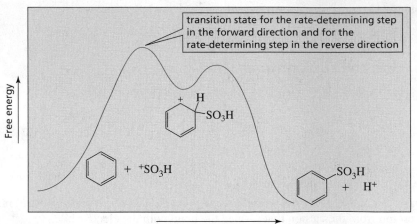

The **principle of microscopic reversibility** applies to all reactions. It states that the mechanism of a reaction in the reverse direction must retrace each step of the mechanism in the forward direction in microscopic detail. This means that the forward and reverse reactions must have the same intermediates, and that the "hill" with the highest point on the reaction coordinate represents the rate-determining step in both the forward and reverse directions.

For example, sulfonation is described by the reaction coordinate diagram in Figure 3, going from left to right. Therefore, desulfonation is described by the same reaction coordinate diagram going from right to left. In sulfonation, the rate-determining step is addition of the $^+SO_3H$ ion to benzene. In desulfonation, the rate-limiting step is loss of the $^+SO_3H$ ion from the carbocation intermediate. You will see how desulfonation can be useful in synthesizing compounds when you do Problem 31.

transition state for the rate-determining step in the forward direction and for the rate-determining step in the reverse direction

Free energy

Progress of the reaction

▲ **Figure 3**

Reaction coordinate diagram for the sulfonation of benzene (left to right) and for the desulfonation of benzenesulfonic acid (right to left).

PROBLEM 5

The reaction coordinate diagram in Figure 3 shows that the rate-determining step for sulfonation is the slower of the two steps in the mechanism, whereas the rate-determining step for desulfonation is the faster of the two steps. Explain how the faster step can be the rate-determining step.

---

7   THE FRIEDEL–CRAFTS ACYLATION OF BENZENE

an acyl group   **an alkyl group**

Two electrophilic substitution reactions bear the names of chemists Charles Friedel and James Crafts. *Friedel–Crafts acylation* places an acyl group on a benzene ring, and *Friedel–Crafts alkylation* places an alkyl group on a benzene ring.

An acyl chloride or an acid anhydride is used as the source of the acyl group needed for a Friedel–Crafts acylation.

Friedel–Crafts acylation

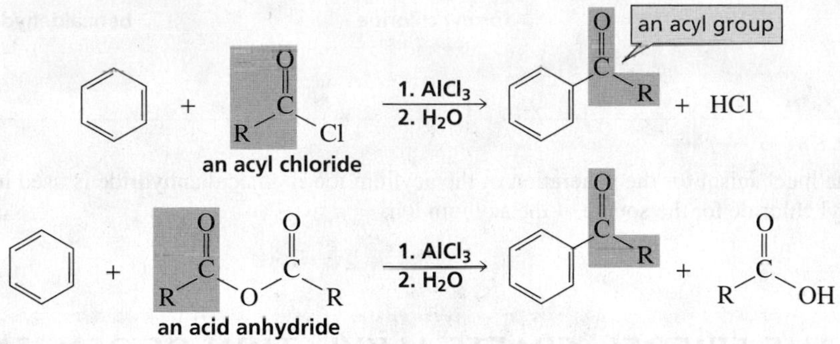

The electrophile (an acylium ion) required for the reaction is formed by the reaction of an acyl chloride (or an acid anhydride) with $AlCl_3$, a Lewis acid. Oxygen and carbon share the positive charge in the acylium ion.

generation of the electrophile

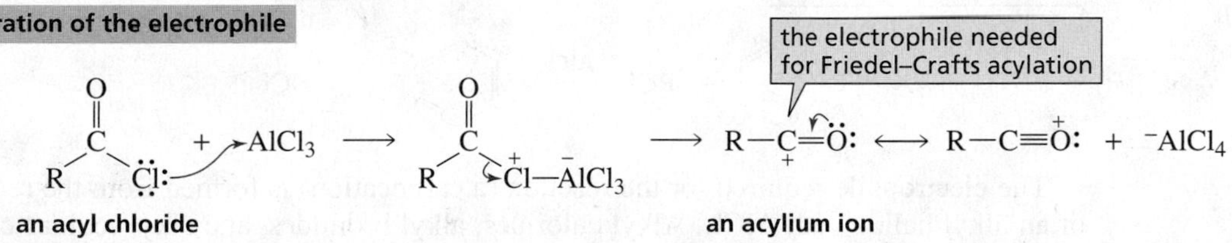

MECHANISM FOR FRIEDEL–CRAFTS ACYLATION

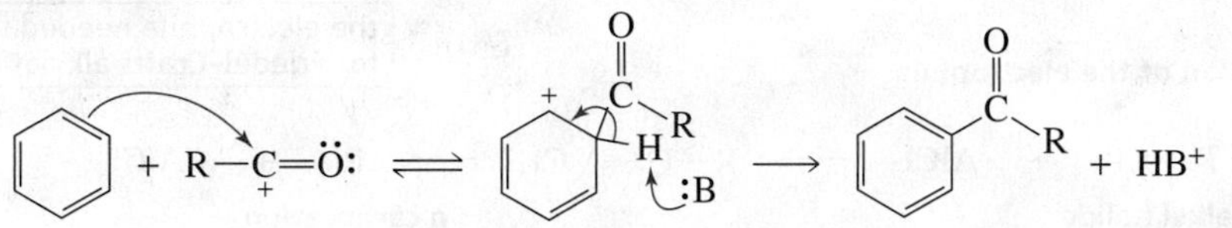

- The electrophile (in this case, an acylium ion) adds to the benzene ring.
- A base (:B) in the reaction mixture removes the proton from the carbon that formed the bond with the electrophile.

Because the product of a Friedel–Crafts acylation contains a carbonyl group that can complex with $AlCl_3$, Friedel–Crafts acylations must be carried out with more than one equivalent of $AlCl_3$. When the reaction is over, water is added to the reaction mixture to liberate the product from the complex.

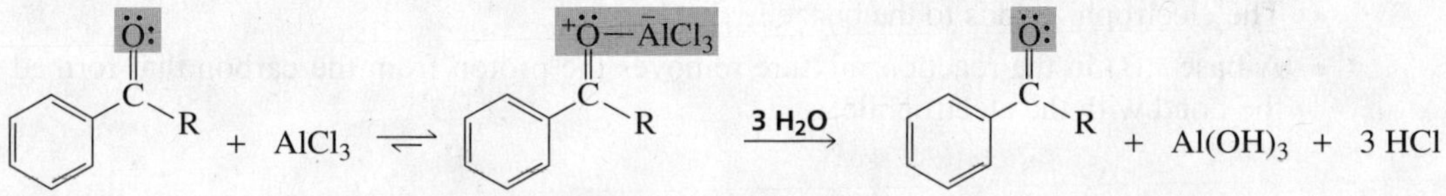

The synthesis of benzaldehyde from benzene poses a problem because formyl chloride, the acyl halide required for the reaction, is unstable and must be generated in situ. The **Gatterman–Koch reaction** uses a high-pressure mixture of carbon monoxide and HCl to generate formyl chloride and an aluminum chloride–cuprous chloride catalyst for the acylation.

947

$$CO \quad + \quad HCl \quad \underset{\text{pressure}}{\overset{\text{high}}{\rightleftharpoons}} \quad \left[\begin{array}{c} O \\ \parallel \\ H \end{array} \begin{array}{c} \\ C \\ \end{array} Cl \right] \quad \xrightarrow{\text{AlCl}_3/\text{CuCl}} \quad$$

formyl chloride
unstable

benzaldehyde

PROBLEM 6

Show the mechanism for the generation of the acylium ion if an acid anhydride is used instead of an acyl chloride for the source of the acylium ion.

---

8   THE FRIEDEL–CRAFTS ALKYLATION OF BENZENE

Friedel–Crafts alkylation substitutes an alkyl group for a hydrogen of benzene.

Friedel–Crafts alkylation

an alkyl group

$$+ \quad \text{RCl} \quad \xrightarrow{\text{AlCl}_3} \quad \text{R} \quad + \quad \text{HCl}$$

The electrophile required for the reaction (a carbocation) is formed from the reaction of an alkyl halide with $AlCl_3$. Alkyl chlorides, alkyl bromides, and alkyl iodides can all be used. Vinylic halides and aryl halides cannot be used because their carbocations are too unstable to be formed.

generation of the electrophile

the electrophile needed for Friedel–Crafts alkylation

$$R-\ddot{\underset{..}{Cl}}: \; + \; AlCl_3 \quad \longrightarrow \quad R-\overset{+}{Cl}-\overset{-}{AlCl_3} \quad \longrightarrow \quad R^+ \quad + \quad {}^-AlCl_4$$

an alkyl halide

a carbocation

MECHANISM FOR FRIEDEL–CRAFTS ALKYLATION

$$+ \quad R^+ \quad \longrightarrow \quad \overset{+}{\underset{R}{\bigcirc}}\overset{H \;\; :B}{} \quad \longrightarrow \quad \overset{R}{\bigcirc} \quad + \quad HB^+$$

- The electrophile adds to the benzene ring.
- A base (:B) in the reaction mixture removes the proton from the carbon that formed the bond with the electrophile.

In Section 14, we will see that an alkyl-substituted benzene is more reactive than benzene. Therefore, a large excess of benzene is used in Friedel–Crafts alkylations to prevent the alkyl-substituted product from being alkylated instead of benzene. When benzene is in excess, the electrophile is more likely to encounter a molecule of benzene than a molecule of the more reactive alkyl-substituted benzene.

A carbocation will rearrange if rearrangement leads to a more stable carbocation. When the carbocation employed in a Friedel–Crafts alkylation reaction rearranges, the major product will be the product with the rearranged alkyl group on the benzene ring. For example, when benzene reacts with 1-chlorobutane, 60%–80% of the product (the actual percentage depends on the reaction conditions) has the rearranged alkyl substituent. (See the box on "Incipient Primary Carbocations" later in this section).

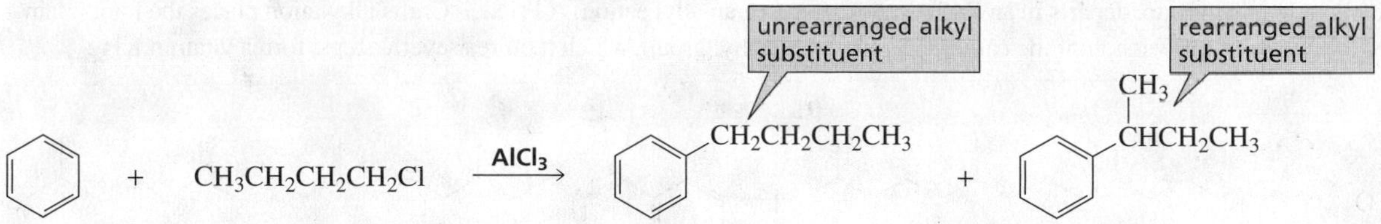

excess **1-chlorobutane** **1-phenylbutane** **2-phenylbutane 60 - 80%**

rearrangement of the carbocation

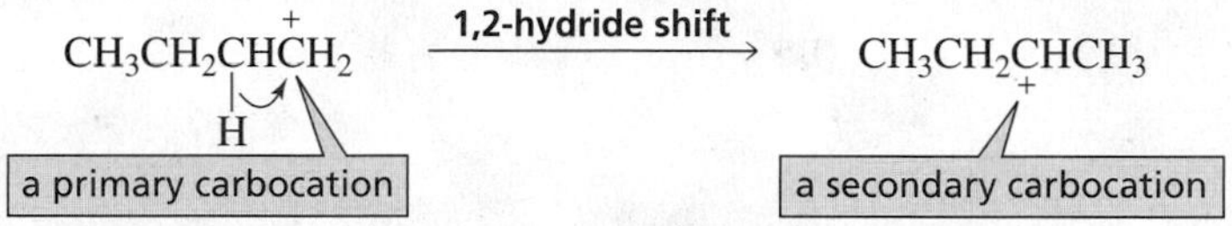

When benzene reacts with 1-chloro-2,2-dimethylpropane, the initially formed primary carbocation rearranges to a tertiary carbocation; in this case, 100% of the product (under all reaction conditions) has the rearranged alkyl substituent.

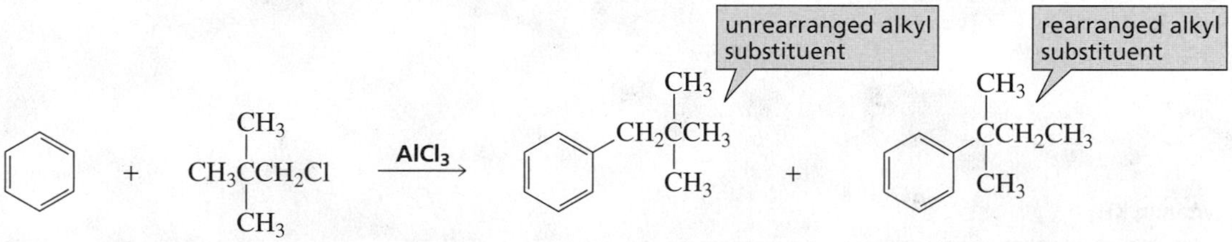

excess **1-chloro-2,2-dimethylpropane** **2,2-dimethyl-1-phenylpropane 0%** **2-methyl-2-phenylbutane 100%**

rearrangement of the carbocation

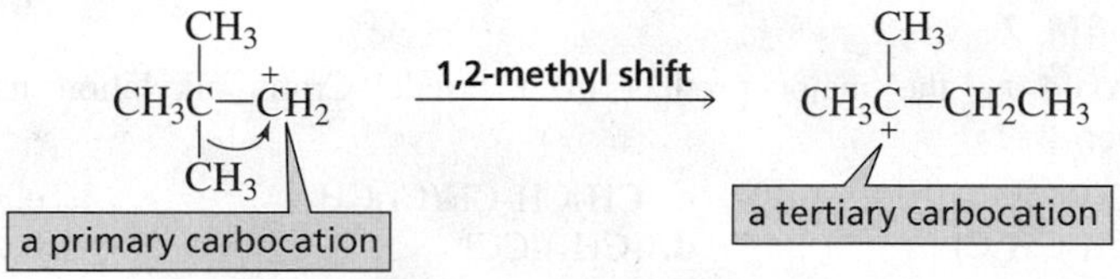

Incipient Primary Carbocations

For simplicity, the two reactions shown earlier that involve carbocation rearrangements were written showing the formation of a primary carbocation. However, primary carbocations are too unstable to be formed. Not surprisingly, a true primary carbocation is never formed in a Friedel–Crafts alkylation. Instead, the carbocation remains complexed with the Lewis acid—a so-called "incipient" carbocation. A carbocation rearrangement occurs because the incipient carbocation has sufficient carbocation character to permit the rearrangement.

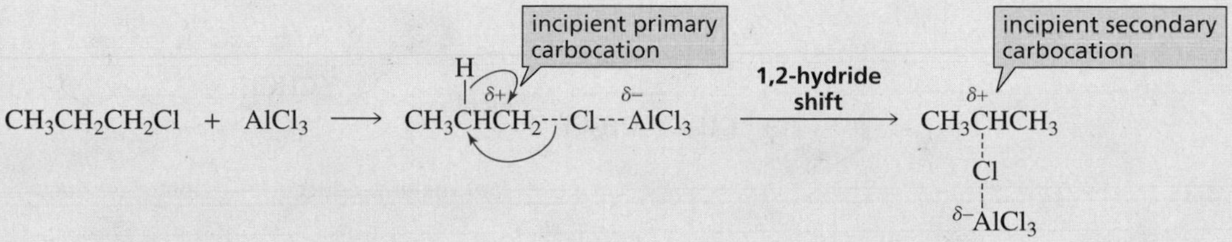

A Biological Friedel–Crafts Alkylation

A Friedel–Crafts alkylation is one of the steps in the biosynthesis of vitamin KH_2, the coenzyme required to form blood clots. The pyrophosphate group, an excellent leaving group, departs in an S_N1 reaction, forming an allyl cation. A Friedel–Crafts alkylation places the long-chain alkyl group on the benzene ring. Conversion of the carboxyl group to a methyl group, which requires several steps, forms vitamin KH_2.

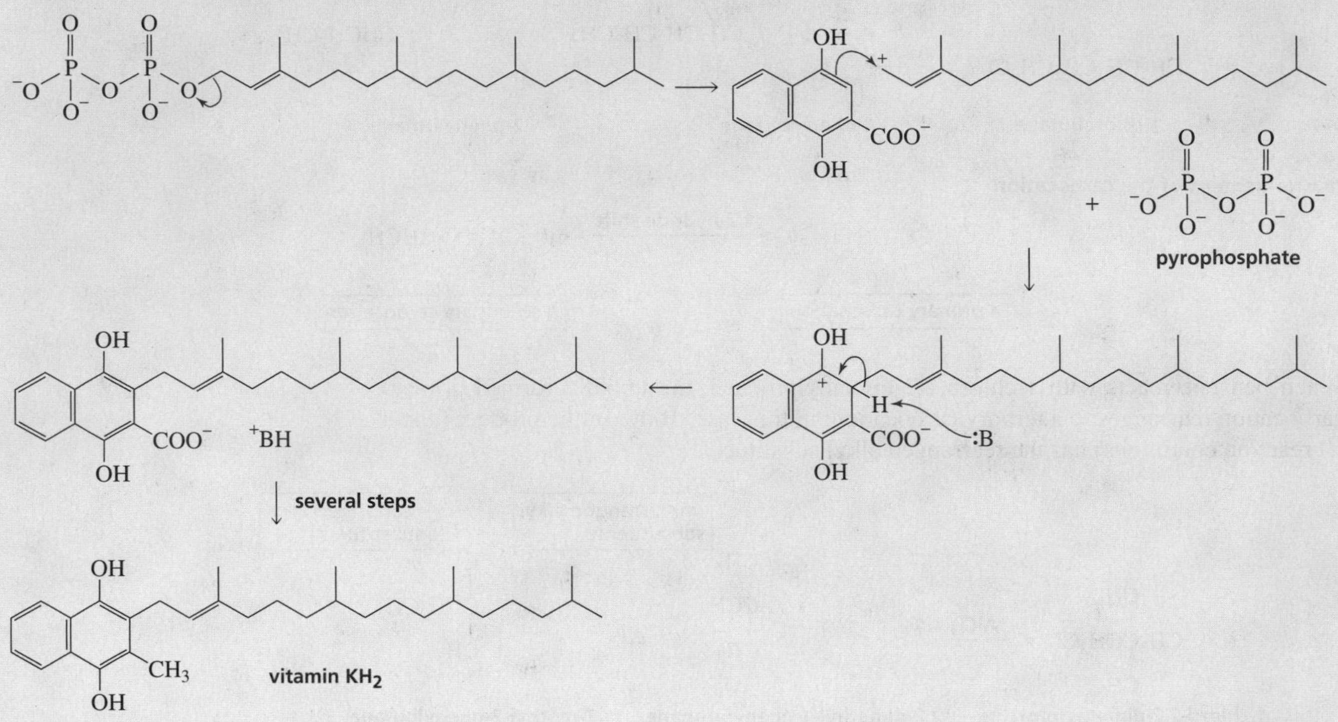

pyrophosphate

several steps

vitamin KH₂

PROBLEM 7

What would be the major product of a Friedel–Crafts alkylation using the following alkyl chlorides?

a. CH_3CH_2Cl **c.** $CH_3CH_2CH(Cl)CH_3$ **e.** $(CH_3)_2CHCH_2Cl$

b. $CH_3CH_2CH_2Cl$ **d.** $(CH_3)_3CCl$ **f.** $CH_2{=}CHCH_2Cl$

---

9   THE ALKYLATION OF BENZENE BY ACYLATION–REDUCTION

A Friedel–Crafts alkylation cannot produce a good yield of an alkylbenzene containing a straight-chain alkyl group because the incipient primary carbocation will rearrange to a more stable carbocation.

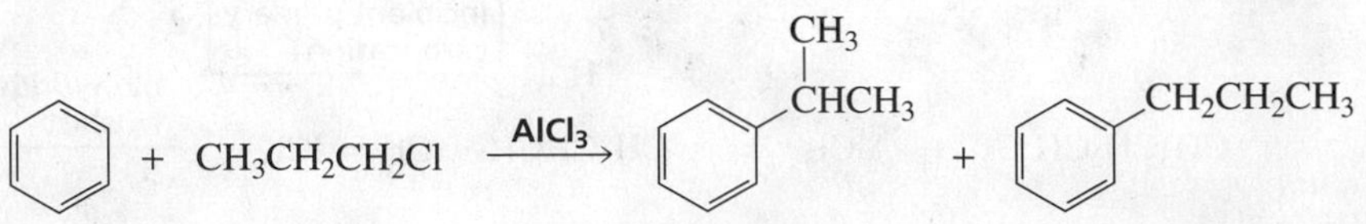

major product   minor product

Acylium ions, however, do not rearrange. Consequently, a straight-chain alkyl group can be placed on a benzene ring by means of a Friedel–Crafts acylation, followed by reduction of the carbonyl group to a methylene group. Conversion of a carbonyl group to a methylene group is a reduction reaction because the two C—O bonds are replaced by two C—H bonds. Only a ketone carbonyl group that is adjacent to a benzene ring can be reduced to a methylene group by catalytic hydrogenation (H_2 + Pd/C).

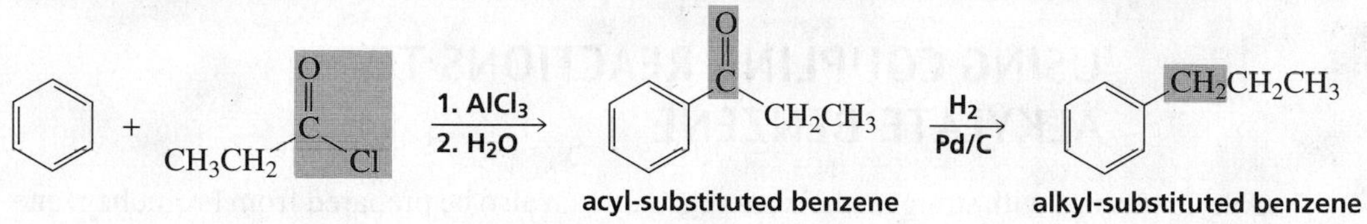

Besides avoiding carbocation rearrangements, another advantage to preparing alkyl-substituted benzenes by acylation–reduction rather than by direct alkylation is that a large excess of benzene does not have to be used (Section 8). Unlike alkyl-substituted benzenes, which are more reactive than benzene, acyl-substituted benzenes are less reactive than benzene, so they will not undergo a second Friedel–Crafts reaction (Section 14).

Several other methods are available for reducing the carbonyl group of a ketone to a methylene group. These methods reduce all ketone carbonyl groups, not just those adjacent to benzene rings. Two of the most effective are the **Clemmensen reduction,** which uses an acidic solution of zinc dissolved in mercury as the reducing reagent, and the **Wolff–Kishner reduction,** which uses hydrazine (H_2NNH_2) under basic conditions.

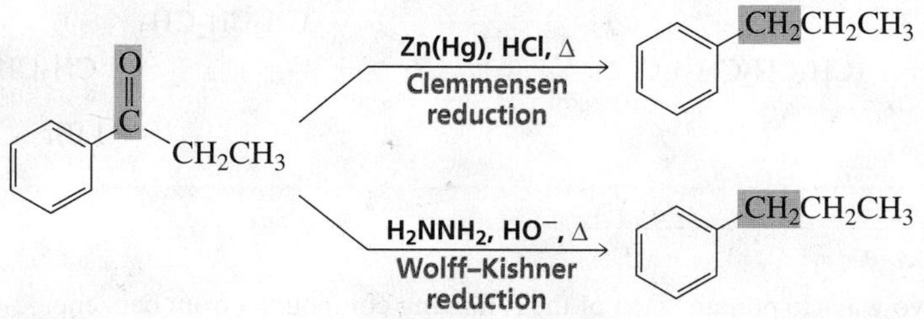

Hydrazine reacts with a ketone to form a hydrazone. Hydroxide ion and heat differentiate the Wolff–Kishner reduction from ordinary hydrazone formation.

MECHANISM FOR THE WOLFF–KISHNER REDUCTION

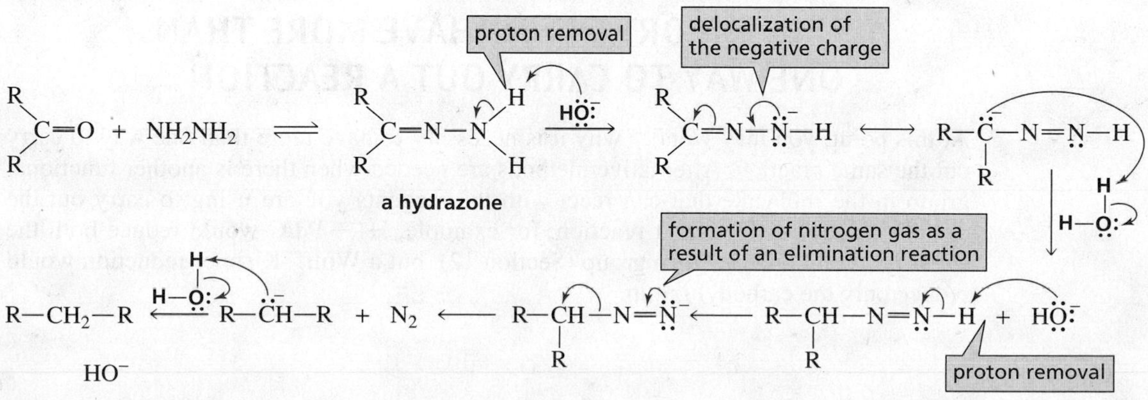

■ The ketone reacts with hydrazine to form a hydrazone.

- Hydroxide ion removes a proton from the NH_2 group of the hydrozone. The reaction requires heat because this proton is only weakly acidic.
- The negative charge can be delocalized onto carbon, which removes a proton from water.
- The last two steps are repeated to form the deoxygenated product and nitrogen gas.

10 USING COUPLING REACTIONS TO ALKYLATE BENZENE

Alkylbenzenes with straight-chain alkyl groups can also be prepared from bromobenzene or chlorobenzene, using a Suzuki reaction or an organocuprate.

the Suzuki reaction

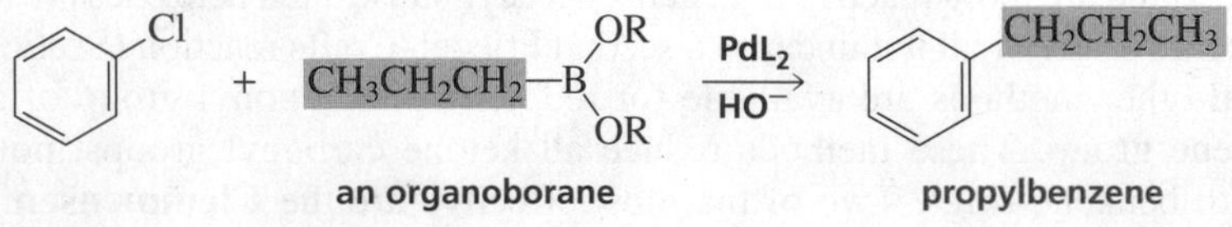

an organoborane — propylbenzene

reaction with an organocuprate

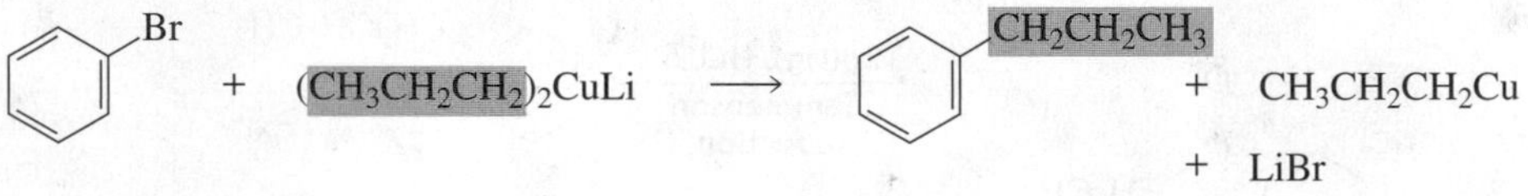

PROBLEM 8

Describe two ways to prepare each of the following compounds from benzene.

a. 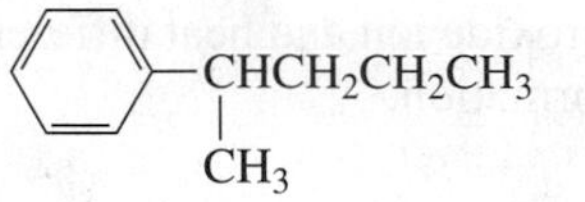 [structure: phenyl-CHCH$_2$CH$_2$CH$_3$ with CH$_3$ branch]

b. 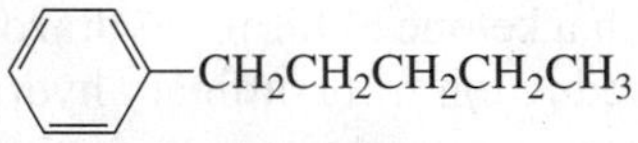 [structure: phenyl-CH$_2$CH$_2$CH$_2$CH$_2$CH$_3$]

---

11 IT IS IMPORTANT TO HAVE MORE THAN ONE WAY TO CARRY OUT A REACTION

At this point, you may wonder why it is necessary to have more than one way to carry out the same reaction. Alternative methods are needed when there is another functional group in the molecule that can react with the reagents you are using to carry out the desired reaction. In the next reaction, for example, $H_2 + Pd/C$ would reduce both the carbonyl group and the nitro group (Section 12), but a Wolff–Kishner reduction would reduce only the carbonyl group.

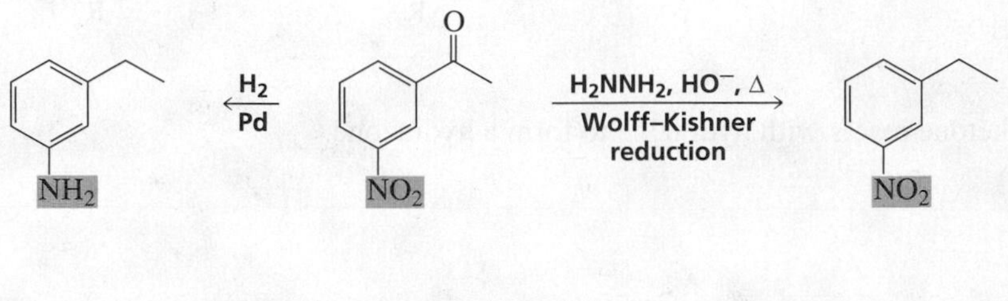

12 HOW SOME SUBSTITUENTS ON A BENZENE RING CAN BE CHEMICALLY CHANGED

Benzene rings with substituents other than those placed on a benzene ring by the five reactions listed in Section 3 can be prepared by first synthesizing one of those substituted benzenes and then chemically changing the substituent. Several of these reactions may be familiar to you.

Reactions of Alkyl Substituents

A bromine will selectively substitute for a benzylic hydrogen in a radical substitution reaction.

propylbenzene 1-bromo-1-phenylpropane

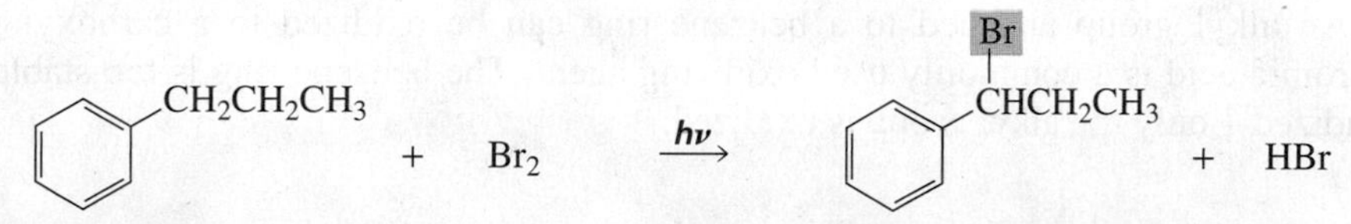

Once a halogen has been placed in the benzylic position, it can be replaced by a nucleophile via an S_N2 or S_N1 reaction. A wide variety of substituted benzenes can be prepared this way.

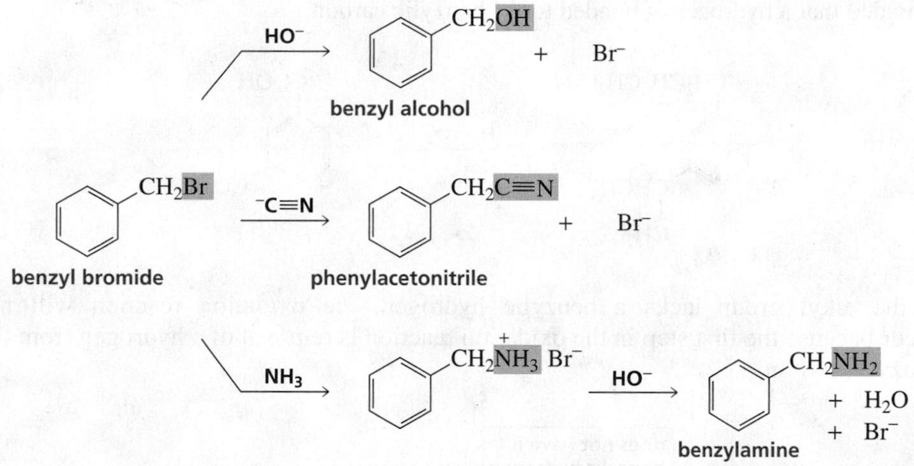

Alkyl halides can also undergo E2 and E1 reactions. Notice that a bulky base (*tert*-BuO$^-$) is used to encourage elimination over substitution.

1-bromo-1-phenylethane styrene

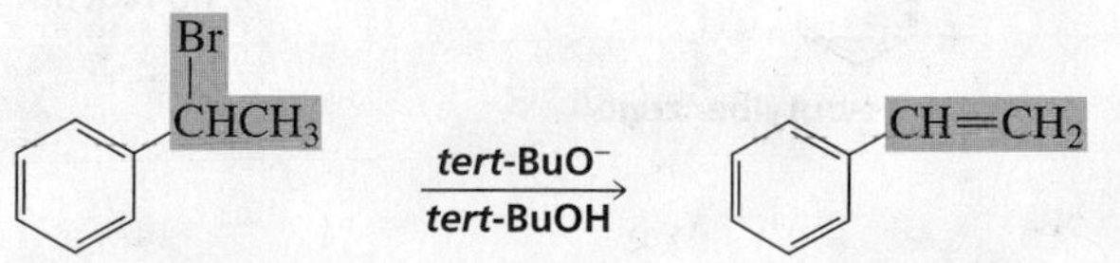

Substituents with double and triple bonds can be reduced by catalytic hydrogenation.

styrene ethylbenzene

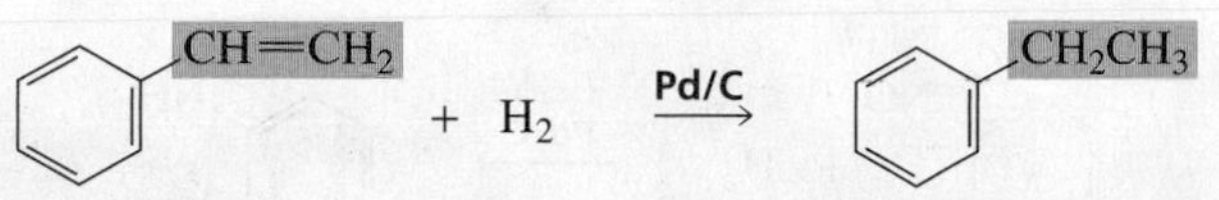

benzonitrile + 2 H$_2$ $\xrightarrow{\text{Raney Ni}}$ **benzylamine**

Because benzene is an unusually stable compound, it can be reduced only at high temperature and pressure, so only the substituents are reduced in the preceding reactions.

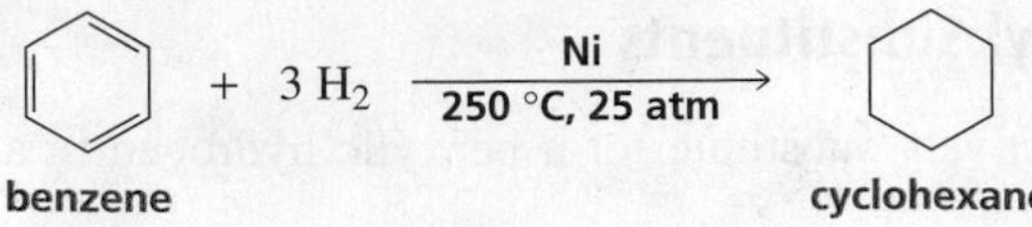

benzene + 3 H$_2$ $\xrightarrow[\text{250 °C, 25 atm}]{\text{Ni}}$ **cyclohexane**

An alkyl group attached to a benzene ring can be oxidized to a carboxyl group. Chromic acid is a commonly used oxidizing agent. The benzene ring is too stable to be oxidized—only the alkyl group is oxidized.

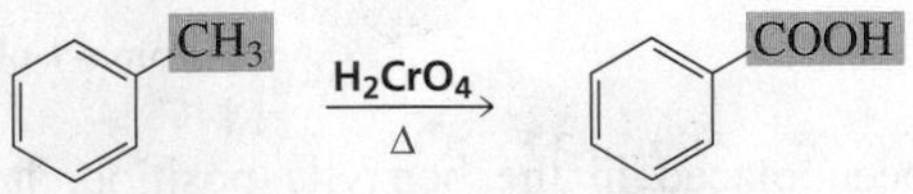

toluene $\xrightarrow[\Delta]{\text{H}_2\text{CrO}_4}$ **benzoic acid**

Regardless of the length of the alkyl substituent, it will be oxidized to a COOH group, provided that a hydrogen is bonded to the benzylic carbon.

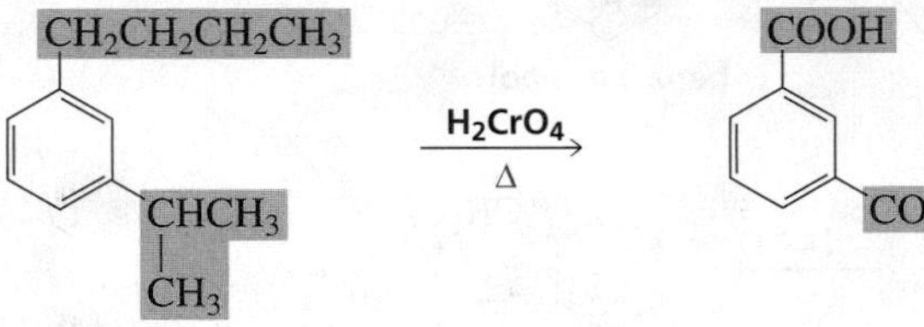

$\xrightarrow[\Delta]{\text{H}_2\text{CrO}_4}$

If the alkyl group lacks a benzylic hydrogen, the oxidation reaction will not occur because the first step in the oxidation reaction is removal of a hydrogen from the benzylic carbon.

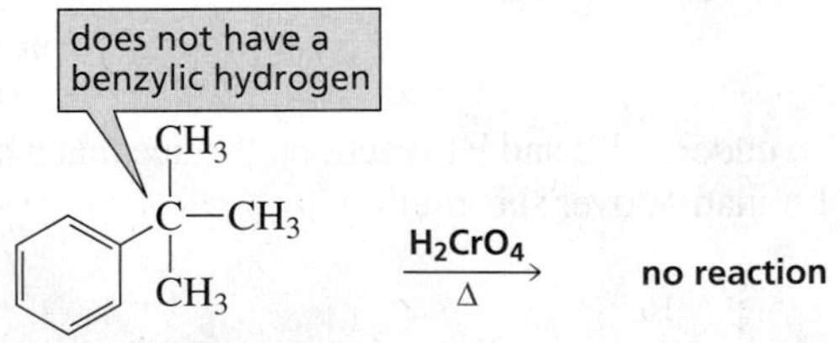

does not have a benzylic hydrogen

tert-butylbenzene $\xrightarrow[\Delta]{\text{H}_2\text{CrO}_4}$ **no reaction**

Reducing a Nitro Substituent

A nitro substituent can be reduced to an amino substituent. Catalytic hydrogenation is commonly used to carry out this reaction.

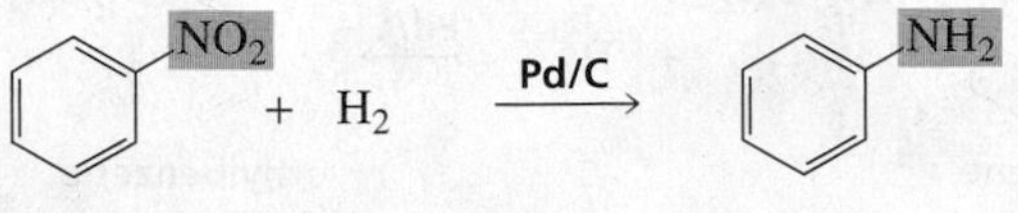

+ H$_2$ $\xrightarrow{\text{Pd/C}}$

PROBLEM 9♦

What are the products of the following reactions?

a.

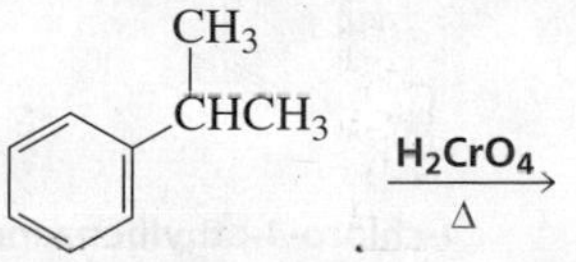

CH₃
|
CHCH₃

$\xrightarrow{\text{H}_2\text{CrO}_4}$
Δ

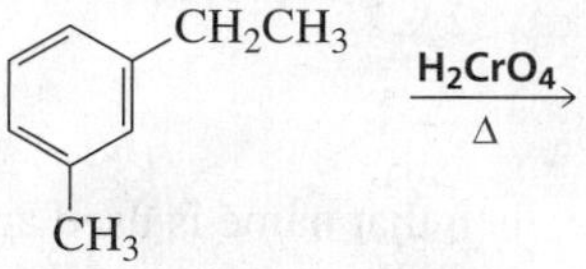

c.

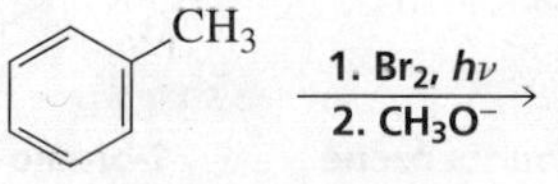

CH₃

1. Br₂, hν
$\xrightarrow{\hspace{1.5cm}}$
2. CH₃O⁻

b.

CH₂CH₃

$\xrightarrow{\text{H}_2\text{CrO}_4}$
Δ

CH₃

d.

CH₃

1. Br₂, hν
2. ⁻C≡N
$\xrightarrow{\hspace{1.5cm}}$
3. H₂, Raney Ni

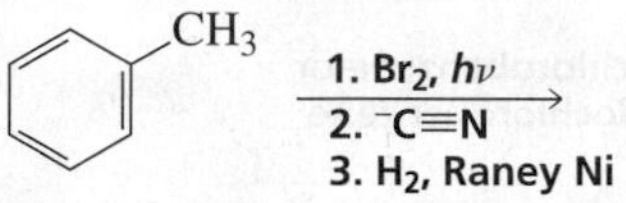

PROBLEM 10 Solved

Show how the following compounds could be prepared from benzene:

a. benzaldehyde   c. 1-bromo-2-phenylethane   e. aniline

b. styrene   d. 2-phenylethanol   f. benzoic acid

Solution to 10a Benzaldehyde can be prepared by the Gatterman–Koch reaction (see earlier in this chapter) or by the following sequence of reactions:

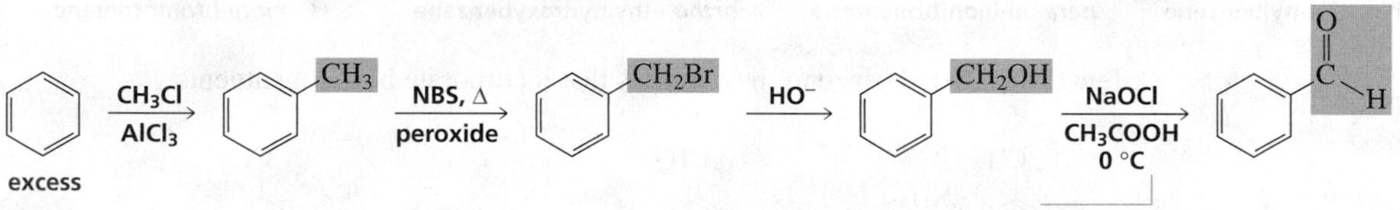

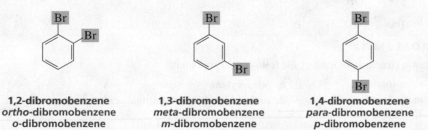

13   THE NOMENCLATURE OF DISUBSTITUTED AND POLYSUBSTITUTED BENZENES

Section 1 described how monosubstituted benzenes are named. Now we will look at how benzene rings that have more than one substituent are named.

Naming Disubstituted Benzenes

The relative positions of two substituents on a benzene ring can be indicated either by numbers or by the prefixes *ortho, meta,* and *para.* Adjacent substituents are called *ortho,* substituents separated by one carbon are called *meta,* and substituents located opposite one another are called *para.* Often, the abbreviations for these prefixes (*o, m, p*) are used in compounds' names.

1,2-dibromobenzene
ortho-dibromobenzene
o-dibromobenzene

1,3-dibromobenzene
meta-dibromobenzene
m-dibromobenzene

1,4-dibromobenzene
para-dibromobenzene
p-dibromobenzene

If the two substituents are different, they are listed in alphabetical order, each preceded by it assigned number. The first mentioned substituent is given the 1-position, and the ring is numbered in the direction that gives the second substituent the lowest possible number.

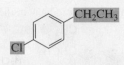

1-chloro-3-iodobenzene
meta-chloroiodobenzene
not
1-iodo-3-chlorobenzene or
meta-**iodochlorobenzene**

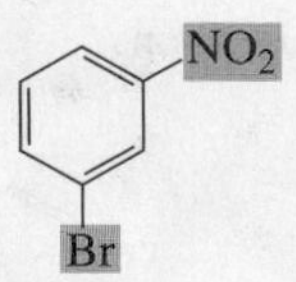

1-bromo-3-nitrobenzene
meta-bromonitrobenzene

1-chloro-4-ethylbenzene
para-chloroethylbenzene

If one of the substituents can be incorporated into a name, then that name is used and the incorporated substituent is given the 1-position. However, methylbenzene, not toluene, is used to name a compound that has a second substituent on the ring.

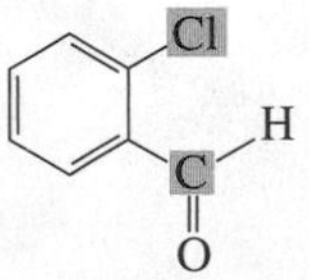

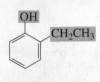

2-chlorobenzaldehyde
ortho-chlorobenzaldehyde
not
ortho-**chloroformylbenzene**

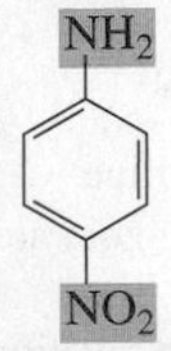

4-nitroaniline
para-nitroaniline
not
para-**aminonitrobenzene**

2-ethylphenol
ortho-ethylphenol
not
ortho-**ethylhydroxybenzene**

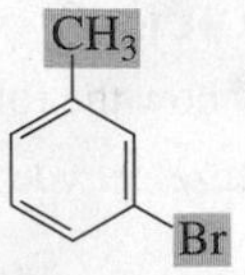

3-bromo-1-methylbenzene
meta-bromomethylbenzene
not
meta-**bromotoluene**

A few disubstituted benzenes have names that incorporate both substituents.

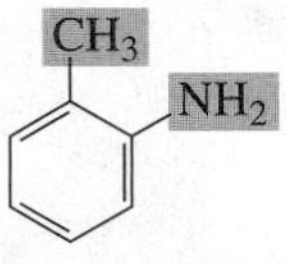

ortho-toluidine

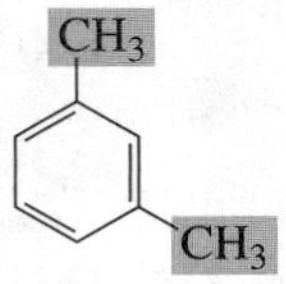

meta-xylene

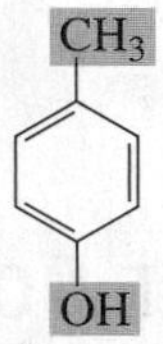

para-cresol
used as a wood preservative
until prohibited for
environmental reasons

PROBLEM 11♦

Name the following compounds:

a. 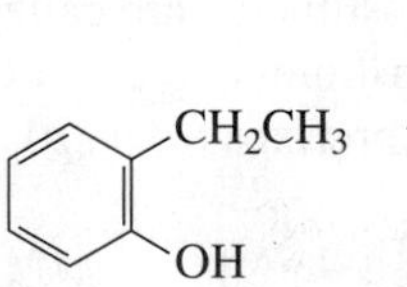 ![structure] CH₂CH₃ / OH   **b.** 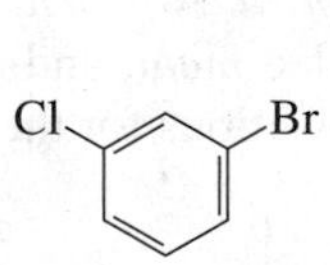 Cl ... Br   **c.** 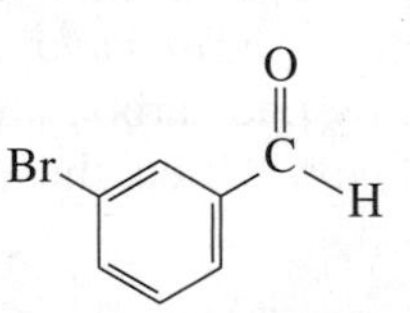 Br ... C=O H   **d.** 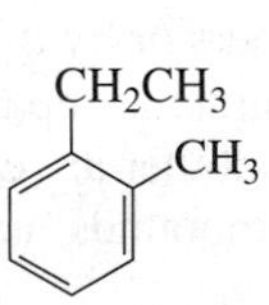 CH₂CH₃ / CH₃

PROBLEM 12

Draw a structure for each of the following compounds:

a. *para*-toluidine

b. *meta*-cresol

c. *para*-xylene

d. *ortho*-chlorobenzenesulfonic acid

Naming Polysubstituted Benzenes

If the benzene ring has more than two substituents, the substituents are numbered in the direction that results in the lowest possible numbers in the name of the compound. The substituents are listed in alphabetical order, each preceded by its assigned number.

2-bromo-4-chloro-1-nitrobenzene

4-bromo-1-chloro-2-nitrobenzene

1-bromo-4-chloro-2-nitrobenzene

As with disubstituted benzenes, if one of the substituents can be incorporated into a name, that name is used and the incorporated substituent is given the 1-position. The ring is then numbered in the direction that results in the lowest possible numbers in the name of the compound.

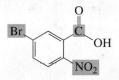

5-bromo-2-nitrobenzoic acid

3-bromo-4-chlorophenol

2-ethyl-4-iodoaniline

PROBLEM 13

Draw the structure for each of the following:

a. *m*-chloromethylbenzene
b. *p*-bromophenol
c. *o*-nitroaniline

d. *m*-chlorobenzonitrile
e. 2-bromo-4-iodophenol
f. *m*-dichlorobenzene

g. 2,5-dinitrobenzaldehyde
h. 4-bromo-3-chloroaniline
i. *o*-xylene

PROBLEM 14♦

Correct the following incorrect names:

a. 2,4,6-tribromobenzene
b. 3-hydroxynitrobenzene

c. *para*-methylbromobenzene
d. 1,6-dichlorobenzene

---

14   THE EFFECT OF SUBSTITUENTS ON REACTIVITY

Like benzene, substituted benzenes undergo the five electrophilic aromatic substitution reactions listed in Section 13: halogenation, nitration, sulfonation, and Friedel–Crafts acylation and aklyation.

Now we need to find out whether a substituted benzene is more or less reactive than benzene itself. The answer depends on the substituent. Some substituents make the ring more reactive toward electrophilic aromatic substitution than benzene, and some make it less reactive.

The slow step of an electrophilic aromatic substitution reaction is the addition of an electrophile to the nucleophilic aromatic ring to form a carbocation intermediate (see Figure 2). A substituent that makes benzene a better nucleophile will make it more attractive to electrophiles; a substituent that makes benzene a poorer nucleophile will make it less attractive to electrophiles.

In addition, because the transition state for the slow step is closer in energy to the carbocation intermediate than to benzene (Figure 2), the transition state resembles the carbocation intermediate. Therefore, the transition state has a partial positive charge.

As a result, substituents that donate electrons to the benzene ring increase benzene's nucleophilicity and stabilize the partially positively charged transition state, thereby increasing the rate of electrophilic aromatic substitution; these are called **activating substituents.** In contrast, substituents that withdraw electrons from the benzene ring decrease benzene's nucleophilicity and destabilize the transition state, thereby decreasing the rate of electrophilic aromatic substitution; these are called **deactivating substituents.**

relative rates of electrophilic aromatic substitution

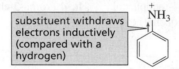

Z donates electrons to the benzene ring

Y withdraws electrons from the benzene ring

Let's now review the ways a substituent can donate or withdraw electrons.

Inductive Electron Withdrawal

If a substituent that is bonded to a benzene ring is *more electron withdrawing than a hydrogen,* then it will draw the σ electrons away from the benzene ring more strongly than a hydrogen will. Withdrawal of electrons through a σ bond is called **inductive electron withdrawal.** The $^+NH_3$ group is an example of a substituent that withdraws electrons inductively because it is more electronegative than a hydrogen.

substituent withdraws electrons inductively (compared with a hydrogen)

$\overset{+}{N}H_3$

Electron Donation by Hyperconjugation

Alkyl substituents (such as CH_3) stabilize carbocations by hyperconjugation—that is, by donating electrons to an empty *p* orbital.

Electron Donation by Resonance

If a substituent has a lone pair on the atom directly attached to the benzene ring, then the lone pair can be delocalized into the ring. These substituents are said to **donate electrons by resonance.** Substituents such as NH_2, OH, OR, and Cl donate electrons by resonance. These substituents also withdraw electrons inductively because the atom attached to the benzene ring is more electronegative than a hydrogen.

donation of electrons by resonance into a benzene ring

anisole

Electron Withdrawal by Resonance

If a substituent is attached to the benzene ring by an atom that is doubly or triply bonded to a more electronegative atom, then the electrons of the ring can be delocalized onto the substituent; these substituents are said to **withdraw electrons by resonance.** Substituents such as $C=O$, $C\equiv N$, SO_3H, and NO_2 withdraw electrons by resonance. These substituents also withdraw electrons inductively because the atom attached to the benzene ring has a full or partial positive charge and is therefore more electronegative than a hydrogen.

withdrawal of electrons by resonance from a benzene ring

nitrobenzene

PROBLEM 15♦

For each of the following substituents, indicate whether it withdraws electrons inductively, donates electrons by hyperconjugation, withdraws electrons by resonance, or donates electrons by resonance. (Effects should be compared with that of a hydrogen; note that many substituents can be characterized in more than one way.)

    a. Br   **b.** CH_2CH_3 **c.** $\overset{\displaystyle O}{\overset{\|}{C}}CH_3$ **d.** $NHCH_3$ **e.** OCH_3 **f.** $^+N(CH_3)_3$

Relative Reactivity of Substituted Benzenes

The substituents in Table 1 are listed according to how they affect the reactivity of a benzene ring toward electrophilic aromatic substitution compared with benzene—where the substituent is a hydrogen. *The activating substituents make the benzene ring more reactive toward electrophilic aromatic substitution, and the deactivating substituents make the benzene ring less reactive.* Note that activating substituents donate electrons to the ring and deactivating substituents withdraw electrons from the ring.

All the *strongly activating substituents* donate electrons to the ring by resonance, because they have a lone pair on the atom attached to the ring. Additionally, they all withdraw electrons from the ring inductively, because the atom attached to the ring is more electronegative than hydrogen. The fact that these substituents have been found experimentally to make the benzene ring more reactive indicates that their electron donation to the ring by resonance is more significant than their inductive electron withdrawal from the ring.

strongly activating substituents

959

Electron-donating substituents increase the reactivity of the benzene ring toward electrophilic aromatic substitution.

Electron-withdrawing substituents decrease the reactivity of the benzene ring toward electrophilic aromatic substitution.

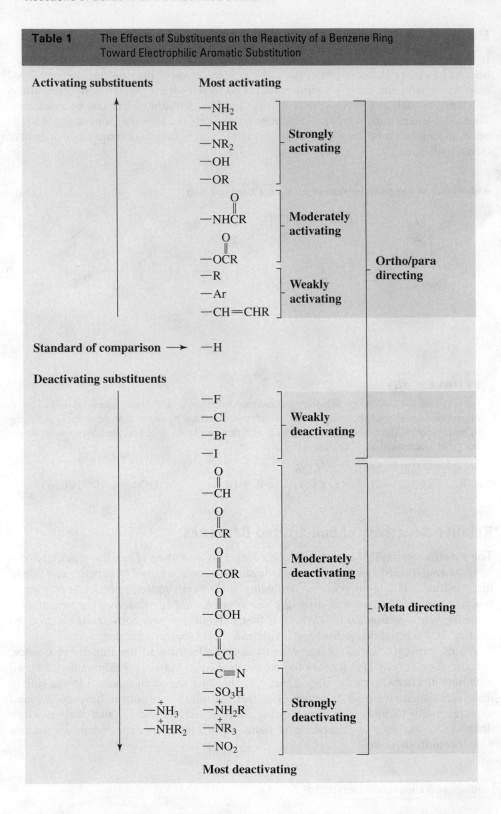

Table 1 The Effects of Substituents on the Reactivity of a Benzene Ring Toward Electrophilic Aromatic Substitution

The *moderately activating substituents* also both donate electrons to the ring by resonance and withdraw electrons from the ring inductively. Because they are only moderately activating, we know that they donate electrons to the ring by resonance less effectively than do the strongly activating substituents.

moderately activating substituents

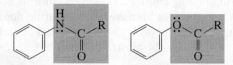

These substituents are less effective at donating electrons to the ring by resonance because, unlike the strongly activating substituents that donate electrons only to the ring, the moderately activating substituents donate electrons by resonance in two competing directions: to the ring and away from the ring. The fact that these substituents increase the reactivity of the benzene ring indicates that, despite their diminished resonance electron donation to the ring, overall they donate electrons by resonance more strongly than they withdraw electrons inductively.

Alkyl, aryl, and CH=CHR groups are *weakly activating substituents*. An alkyl substituent donates electrons to the ring by hyperconjugation (see Figure 5). Aryl and CH=CHR groups donate electrons to the ring by resonance and also withdraw electrons from the ring by resonance. The fact that they are weak activators indicates that they are slightly more electron donating than they are electron withdrawing.

weakly activating substituents

The halogens are weakly deactivating substituents. Like all the strongly and moderately activating substituents, the halogens donate electrons to the ring by resonance and withdraw electrons from the ring inductively. Because halogen substituents have been found experimentally to make benzene less reactive, they must withdraw electrons inductively more strongly than they donate electrons by resonance.

weakly deactivating substituents

Let's look at why the halogens are *weakly* deactivating, whereas OH and OCH_3 are *strongly* activating. The electronegativities of chlorine and oxygen are similar, so they have similar inductive electron-withdrawing abilities. However, chlorine does not donate electrons by resonance as well as oxygen does because when chlorine donates electrons by resonance, it uses a $3p$ orbital to form a π bond with the $2p$ orbital of carbon. A $3p$–$2p$ orbital overlap is less effective than the $2p$–$2p$ orbital overlap that forms the π bond between oxygen and carbon.

Fluorine, like oxygen, uses a $2p$ orbital, so fluorine, like oxygen, donates electrons by resonance better than chlorine does. However, this is outweighed by fluorine's greater electronegativity, which causes it to strongly withdraw electrons inductively. Bromine and iodine are less effective than chlorine at withdrawing electrons inductively, but they are also less effective at donating electrons by resonance because they use $4p$ and $5p$ orbitals, respectively. Thus, all the halogens withdraw electrons inductively more strongly than they donate electrons by resonance (see Problem 17).

The moderately deactivating substituents all have a carbonyl group directly attached to the benzene ring. A carbonyl group withdraws electrons from a benzene ring both inductively and by resonance.

The strongly deactivating substituents are powerful electron withdrawers. Except for the ammonium ions ($^+NH_3$, $^+NH_2R$, $^+NHR_2$, and $^+NR_3$), these substituents withdraw electrons both inductively and by resonance. The ammonium ions have no resonance effect, but the positive charge on the nitrogen atom causes them to strongly withdraw electrons inductively.

Take a minute to compare the electrostatic potential maps for anisole, benzene, and nitrobenzene. Notice that an electron-donating substituent (OCH_3) makes the ring more red (more negative), whereas an electron-withdrawing substituent (NO_2) makes the ring less red (less negative).

PROBLEM 16◆

List the compounds in each set from most reactive to least reactive toward electrophilic aromatic substitution:

a. benzene, phenol, toluene, nitrobenzene, bromobenzene
b. dichloromethylbenzene, difluoromethylbenzene, toluene, chloromethylbenzene

PROBLEM 17 Solved

Explain why the halo-substituted benzenes have the relative reactivities shown in Table 1.

Solution Table 1 shows that fluorine is the least deactivating of the halogen substituents and iodine is the most deactivating. We know that fluorine is the most electronegative of the halogens, which means that it is best at withdrawing electrons inductively. Fluorine is also best at donating electrons by resonance because its $2p$ orbital—compared with the $3p$ orbital of chlorine, the $4p$ orbital of bromine, or the $5p$ orbital of iodine—can better overlap the $2p$ orbital of carbon when it forms a π bond. So fluorine is best both at donating electrons by resonance and at withdrawing electrons inductively. Since fluorine is the weakest deactivator of the halogens, electron donation by resonance must be the more important factor in determining the relative reactivities of halo-substituted benzenes.

15 THE EFFECT OF SUBSTITUENTS ON ORIENTATION

When a substituted benzene undergoes an electrophilic aromatic substitution reaction, where does the new substituent attach itself? In other words, is the product of the reaction the ortho isomer, the meta isomer, or the para isomer?

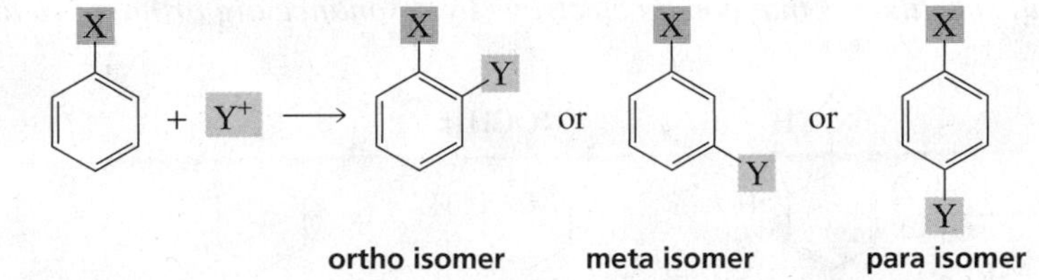

ortho isomer   **meta isomer**   **para isomer**

The substituent already attached to the benzene ring determines the location of the new substituent. The attached substituent will have one of two effects: it will direct an incoming substituent to the ortho *and* para positions, or it will direct an incoming substituent to the meta position.

All activating substituents and the weakly deactivating halogens are **ortho–para directors,** and all substituents that are more deactivating than the halogens are **meta directors.** Thus, the substituents can be divided into three groups:

1. All *activating substituents* direct an incoming electrophile to the ortho and para positions.

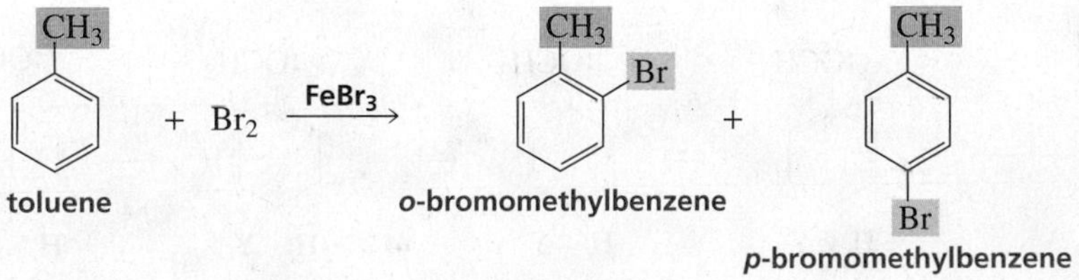

toluene   o-bromomethylbenzene   p-bromomethylbenzene

All activating substituents are ortho–para directors.

2. The *weakly deactivating* halogens also direct an incoming electrophile to the ortho and para positions.

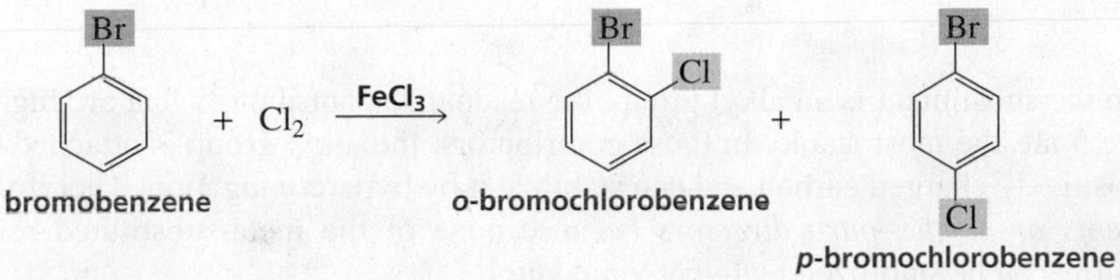

bromobenzene   o-bromochlorobenzene   p-bromochlorobenzene

The weakly deactivating halogens are ortho–para directors.

3. All *moderately and strongly deactivating* substituents direct an incoming electrophile to the meta position.

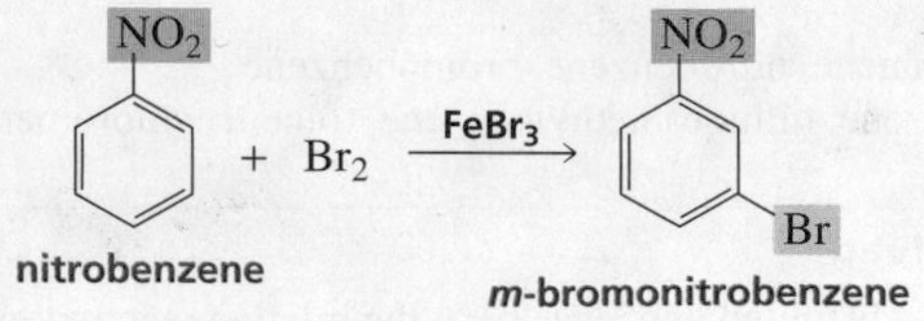

nitrobenzene

m-bromonitrobenzene

All deactivating substituents (except the halogens) are meta directors.

To understand why a substituent directs an incoming electrophile to a particular position, we must look at the stability of the carbocation intermediate. Since the transition state of the rate-limiting step resembles the carbocation intermediate, anything that stabilizes the carbocation intermediate will stabilize the transition state for its formation (Figure 3).

When a substituted benzene undergoes an electrophilic aromatic substitution reaction, three different carbocation intermediates can be formed: an *ortho*-substituted carbocation, a *meta*-substituted carbocation, and a *para*-substituted carbocation. The relative stabilities of the three carbocations enable us to determine the preferred pathway of the reaction because the more stable the carbocation, the more stable the transition state for its formation will be, and therefore the more rapidly it will be formed.

When the substituent is one that can donate electrons by *resonance*, the carbocations formed by putting the incoming electrophile on the ortho and para positions have a fourth resonance contributor (Figure 4). This is an especially stable resonance contributor because it is the only one whose atoms (except for hydrogen) all have complete octets. It is obtained only by directing an incoming substituent to the ortho and para positions. Therefore, *all substituents that donate electrons by resonance are ortho–para directors.*

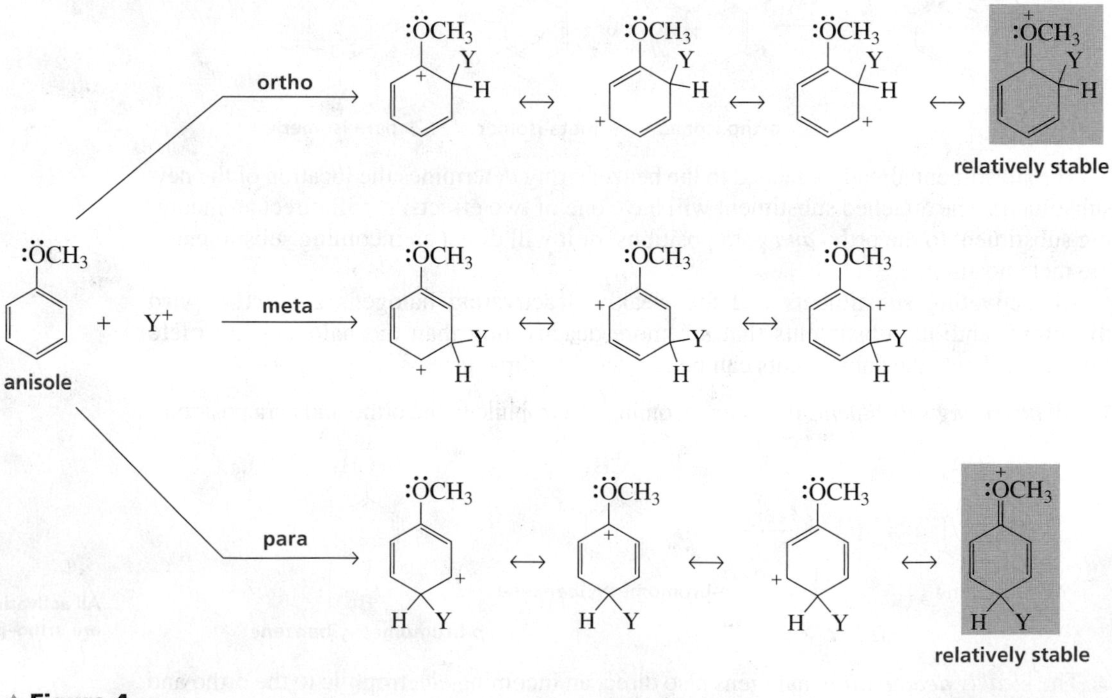

▲ **Figure 4**

The structures of the carbocation intermediates formed from the reaction of an electrophile with anisole at the ortho, meta, and para positions.

When the substituent is an alkyl group, the resonance contributors that are highlighted in Figure 5 are the most stable. In those contributors, the alkyl group is attached directly to the positively charged carbon and can stabilize it by hyperconjugation. Therefore, *alkyl substituents are ortho–para directors* because none of the meta-substituted resonance contributors can be stabilized by hyperconjugation.

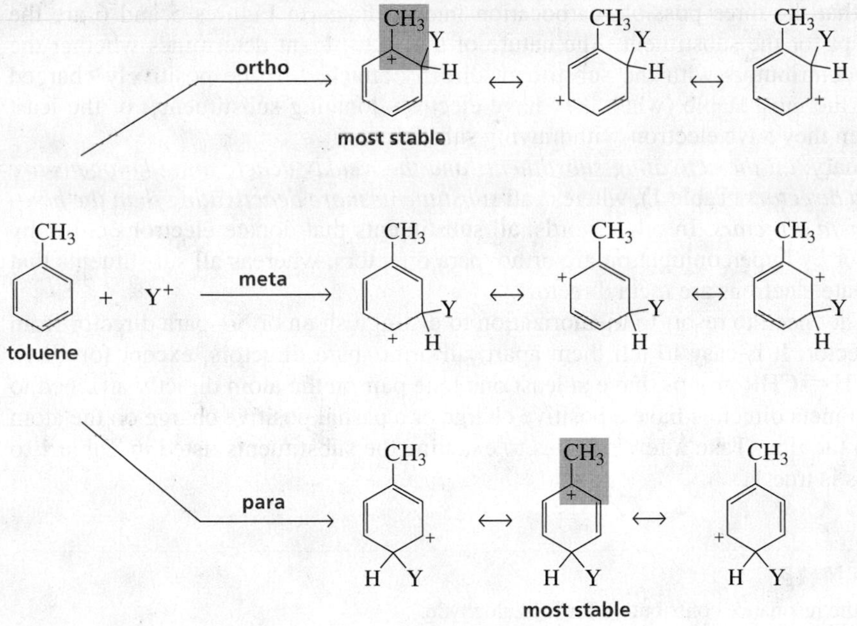

▲ **Figure 5**

The structures of the carbocation intermediates formed from the reaction of an electrophile with toluene at the ortho, meta, and para positions.

Substituents with a positive charge or a partial positive charge on the atom attached to the benzene ring will withdraw electrons inductively from the benzene ring, and most will withdraw electrons by resonance as well. For all of these substituents, the resonance contributors highlighted in Figure 6 are the least stable because they have a positive charge on each of two adjacent atoms, so the most stable carbocation is formed when the incoming electrophile is directed to the meta position. Thus, *all substituents that withdraw electrons (except for the halogens, which are ortho–para directors because they donate electrons by resonance) are meta directors.*

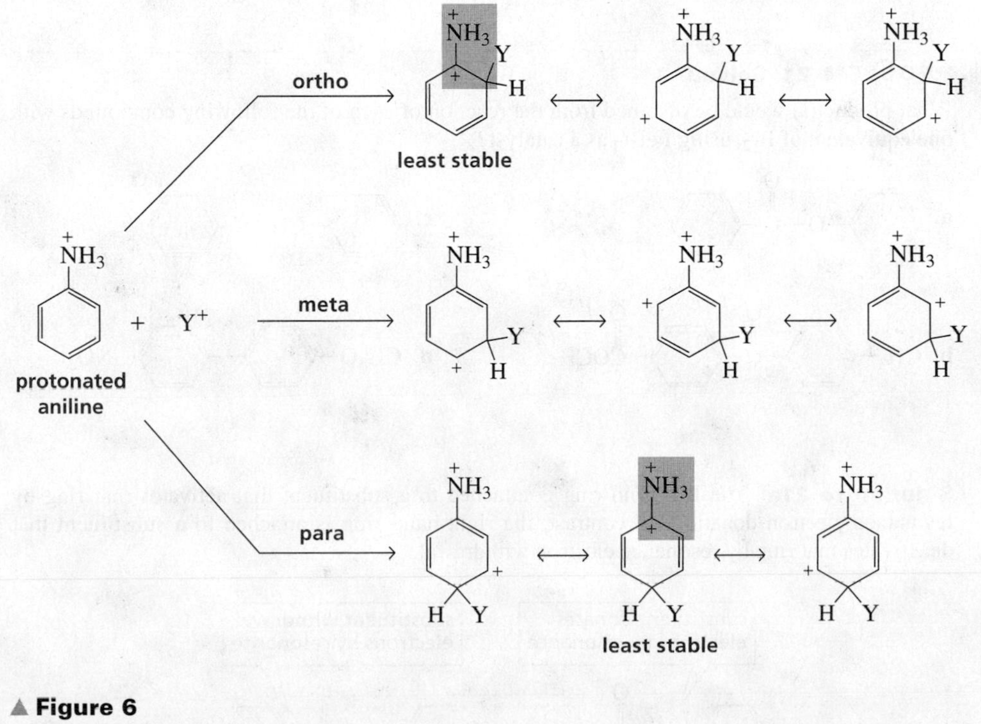

▲ **Figure 6**

The structures of the carbocation intermediates formed from the reaction of an electrophile with protonated aniline at the ortho, meta, and para positions.

Notice that the three possible carbocation intermediates in Figures 5 and 6 are the same, except for the substituent. The nature of that substituent determines whether the resonance contributors with the substituent directly attached to the positively charged carbon are the most stable (when they have electron-donating substituents) or the least stable (when they have electron-withdrawing substituents).

In summary, *all the activating substituents and the weakly deactivating halogens are ortho–para directors* (Table 1), whereas all *substituents more deactivating than the halogens are meta directors.* In other words, all substituents that donate electrons either by resonance or by hyperconjugation are ortho–para directors, whereas all substituents that cannot donate electrons are meta directors.

You do not need to resort to memorization to distinguish an ortho–para director from a meta director. It is easy to tell them apart: all ortho–para directors, except for alkyl, aryl, and CH=CHR groups, have at least one lone pair on the atom directly attached to the ring; all meta directors have a positive charge or a partial positive charge on the atom attached to the ring. Take a few minutes to examine the substituents listed in Table 1 to see that this is true.

> **All substituents that donate electrons either by resonance or by hyperconjugation are ortho–para directors.**

> **All substituents that cannot donate electrons are meta directors.**

PROBLEM 18

a. Draw the resonance contributors for benzaldehyde.
b. Draw the resonance contributors for chlorobenzene.

PROBLEM 19♦

What product(s) would result from nitration of each of the following?

a. propylbenzene   **c.** benzaldehyde   **e.** benzenesulfonic acid
b. bromobenzene   **d.** benzonitrile   **f.** cyclohexylbenzene

---

PROBLEM 20♦

Are the following substituents ortho–para directors or meta directors?

a. CH=CHC≡N **b.** NO_2 **c.** CH_2OH **d.** $COOH$ **e.** CF_3 **f.** N=O

PROBLEM 21 Solved

What product(s) would be obtained from the reaction of each of the following compounds with one equivalent of Br_2, using $FeBr_3$ as a catalyst?

Solution to 21a The left-hand ring is attached to a substituent that activates that ring by resonance electron donation. In contrast, the right-hand ring is attached to a substituent that deactivates that ring by resonance electron withdrawal.

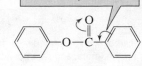

substituent donates electrons by resonance

substituent withdraws electrons by resonance

Thus, the left-hand ring is more reactive toward electrophilic aromatic substitution. The activating substituent will direct the bromine to the positions that are ortho and para to it.

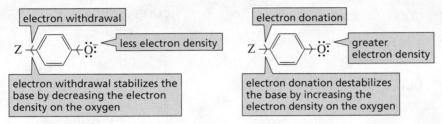

16   THE EFFECT OF SUBSTITUENTS ON pK_a

If a substituent can withdraw electrons from or donate electrons to a benzene ring, then the pK_a values of substituted phenols, benzoic acids, and protonated anilines will change to reflect this withdrawal or donation.

Electron-withdrawing groups stabilize a base and therefore increase the strength of its conjugate acid; electron-donating groups destabilize a base, which decreases the strength of its conjugate acid. The stronger the acid, the more stable (weaker) its conjugate base.

electron withdrawal	electron donation
Z — ⬡ — Ö⁻ ← less electron density	Z — ⬡ — Ö⁻ ← greater electron density
electron withdrawal stabilizes the base by decreasing the electron density on the oxygen	electron donation destabilizes the base by increasing the electron density on the oxygen

As an example, look at the pK_a values of phenol and substituted phenols. The pK_a of phenol is 9.95. The pK_a of *para*-nitrophenol is lower (7.14) because the nitro substituent withdraws electrons from the ring, whereas the pK_a of *para*-methoxyphenol is higher (10.20) because the methoxy substituent donates electrons to the ring.

OH	OH	OH	OH	OH	OH
OCH$_3$	CH$_3$		Cl	HC=O	NO$_2$
pK_a = 10.20	pK_a = 10.19	pK_a = 9.95	pK_a = 9.38	pK_a = 7.66	pK_a = 7.14
		phenol			

Take a minute to compare the effect a substituent has on the reactivity of a benzene ring toward electrophilic aromatic substitution with the effect the same substituent has on the pK_a of phenol. Notice that the more strongly deactivating the substituent (for example, NO$_2$), the lower the pK_a of the phenol, and the more strongly activating the substituent (for example, OCH$_3$), the higher the pK_a of the phenol.

> *Electron withdrawal decreases reactivity toward electrophilic aromatic substitution and increases acidity, whereas electron donation increases reactivity toward electrophilic aromatic substitution and decreases acidity.*

A similar substituent effect on pK_a is observed for substituted benzoic acids and substituted protonated anilines—that is, electron-withdrawing substituents increase acidity, whereas electron-donating substituents decrease acidity.

COOH	COOH	COOH	COOH	COOH	COOH
OCH$_3$	CH$_3$		Br	CH$_3$C=O	NO$_2$
pK_a = 4.47	pK_a = 4.34	pK_a = 4.20	pK_a = 4.00	pK_a = 3.70	pK_a = 3.44

The more deactivating (electron withdrawing) the substituent, the more it increases the acidity of a COOH, an OH, or an ⁺NH$_3$ group attached to a benzene ring.

967

The more activating (electron donating) the substituent, the more it decreases the acidity of a COOH, an OH, or an $^+NH_3$ group attached to a benzene ring.

NH_3 OCH$_3$ $pK_a = 5.29$

NH_3 CH$_3$ $pK_a = 5.07$

NH_3 $pK_a = 4.58$

NH_3 Br $pK_a = 3.91$

NH_3 HC=O $pK_a = 1.76$

NH_3 NO$_2$ $pK_a = 0.98$

PROBLEM 22◆

Which species in each of the following pairs is more acidic?

a. CH_3COOH or $ClCH_2COOH$

b. O_2NCH_2COOH or $O_2NCH_2CH_2COOH$

c. CH_3CH_2COOH or $H_3\overset{+}{N}CH_2COOH$

d.

COOH or COOH
OCH$_3$

e. $HOOCCH_2COOH$ or $^-OOCCH_2COOH$

f. $HCOOH$ or CH_3COOH

g. FCH_2COOH or $ClCH_2COOH$

h.

COOH or COOH
F Cl

PROBLEM-SOLVING STRATEGY

Explaining the Effect of Substituents on pK_a

The *para*-nitroanilinium ion is 3.60 pK_a units more acidic than the anilinium ion (pK_a = 0.98 versus 4.58), but *para*-nitrobenzoic acid is only 0.76 pK_a unit more acidic than benzoic acid (pK_a = 3.44 versus 4.20). Explain why the nitro substituent causes a large change in pK_a in one case and a small change in the other.

Do not expect to be able to solve this kind of problem simply by reading it. First, you need to note that the acidity of a compound depends on the stability of its conjugate base. Next, draw structures of the conjugate bases in question in order to compare their stabilities.

When a proton is lost from the *para*-nitroanilinium ion, the electrons it leaves behind are shared by five atoms. (Draw resonance contributors if you want to see which atoms share the electrons.) In contrast, when a proton is lost from *para*-nitrobenzoic acid, the electrons it leaves behind are shared by two atoms. In other words, loss of a proton leads to greater electron delocalization in one base than in the other. Electron delocalization stabilizes a compound, so the difference in electron delocalization explains why a nitro substituent has a greater effect on the acidity of an anilinium ion than on the acidity of benzoic acid.

Now use the strategy you have just learned to solve Problem 23.

PROBLEM 23

Explain why the pK_a of *p*-nitrophenol is 7.14, whereas the pK_a of *m*-nitrophenol is 8.39.

PROBLEM 24◆

a. Which is the strongest acid, benzoic acid, *o*-fluorobenzoic acid, or *o*-chlorobenzoic acid? (*Hint:* see Problem 17.)

b. Which of these compounds is the weakest acid?

17 THE ORTHO–PARA RATIO

When a benzene ring with an ortho–para-directing substituent undergoes an electrophilic aromatic substitution reaction, what percentage of the product is the ortho isomer and what percentage is the para isomer?

Solely on the basis of probability, more of the ortho product is expected because two ortho positions are available to the incoming electrophile but only one para position is available. The ortho positions, however, are sterically hindered by the substituent on the ring, whereas the para position is not. Consequently, the para isomer will be formed preferentially if either the substituent on the ring or the incoming electrophile is large. The following nitration reactions illustrate the decrease in the ortho–para ratio as the size of the substituent on the ring increases:

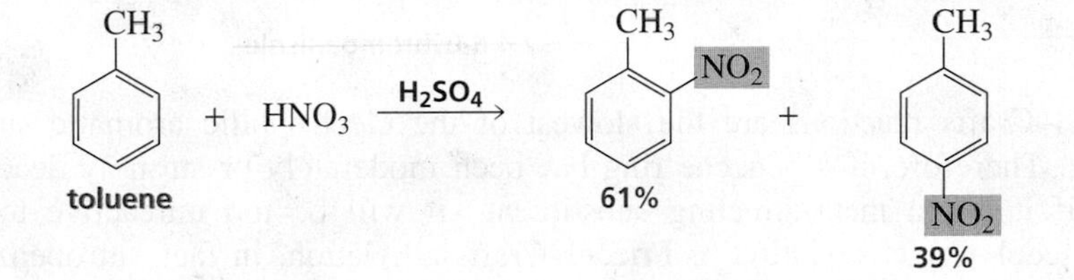

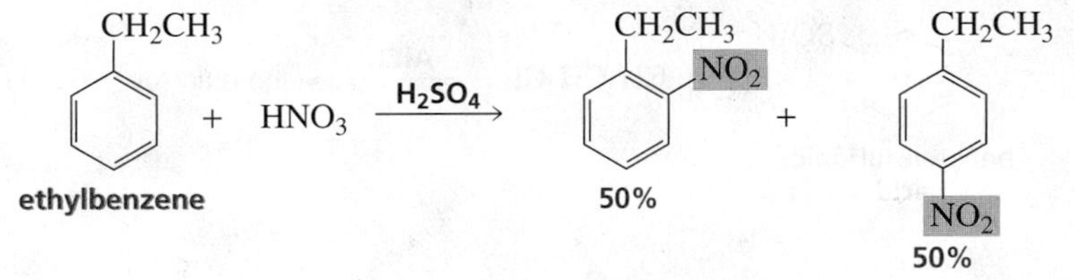

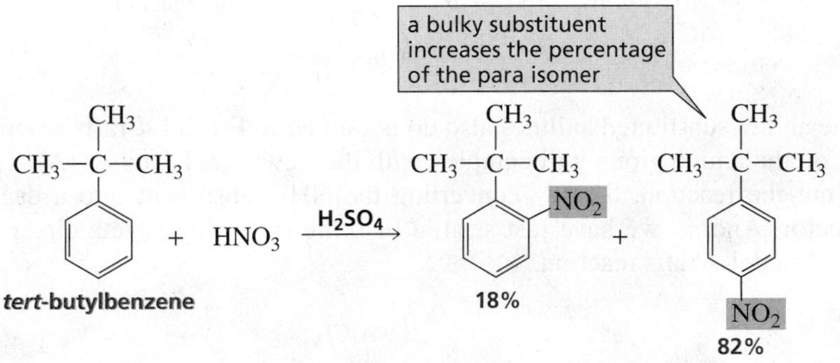

Fortunately, the physical properties of the ortho- and para-substituted isomers differ enough to allow them to be easily separated. Consequently, electrophilic aromatic substitution reactions that lead to both ortho and para isomers are useful in syntheses because the desired product can be easily separated from the reaction mixture.

18 ADDITIONAL CONSIDERATIONS REGARDING SUBSTITUENT EFFECTS

It is important to know whether a substituent is activating or deactivating in order to determine the conditions needed to carry out a reaction. Halogenation is the fastest of the electrophilic aromatic substitution reactions, so halogenation should be carried out without the Lewis acid catalyst (FeBr$_3$ or FeCl$_3$) if the ring has a strongly activating substituent.

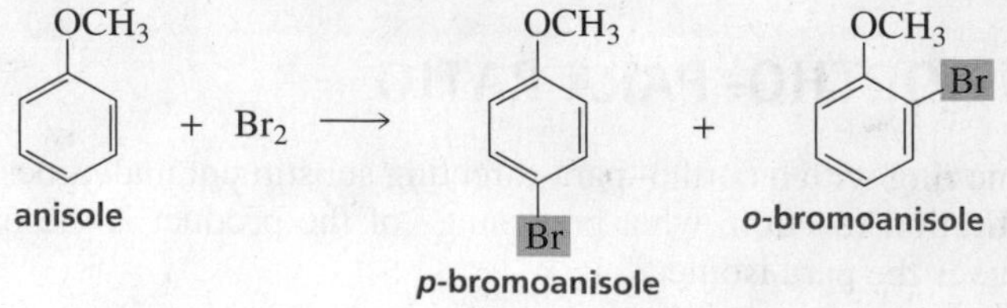

If the Lewis acid catalyst and excess bromine are used, the tribromo compound is obtained.

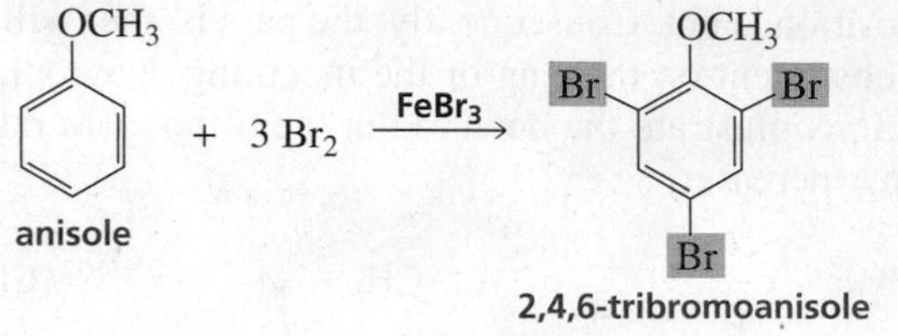

Friedel–Crafts reactions are the slowest of the electrophilic aromatic substitution reactions. Therefore, if a benzene ring has been moderately or strongly deactivated—that is, if it has a meta-directing substituent—it will be too unreactive to undergo either Friedel–Crafts acylation or Friedel–Crafts alkylation. In fact, nitrobenzene is so unreactive that it is often used as a solvent for Friedel–Crafts reactions.

A benzene ring with a meta director cannot undergo a Friedel–Crafts reaction.

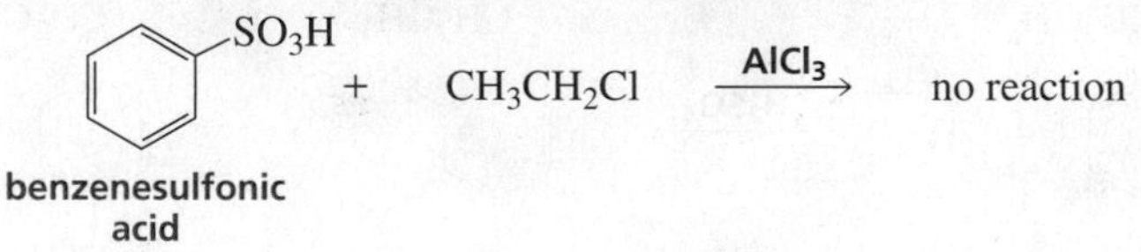

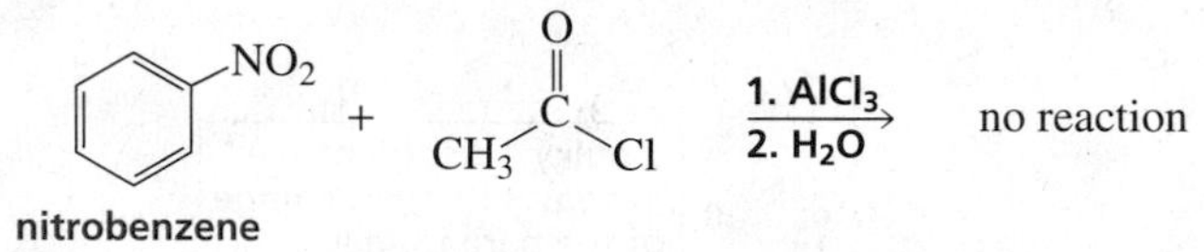

Aniline and *N*-substituted anilines also do not undergo Friedel–Crafts reactions. The lone pair on the amino group will complex with the Lewis acid catalyst ($AlCl_3$) needed to carry out the reaction, thereby converting the NH_2 substituent into a deactivating meta director. And, as we have just seen, a benzene ring with a meta director cannot undergo a Friedel–Crafts reaction.

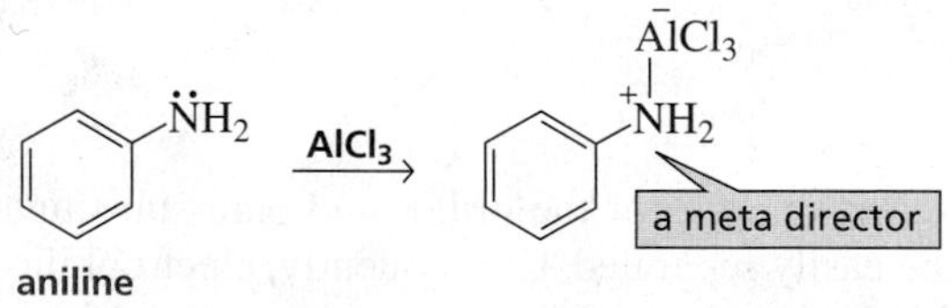

Phenol and anisole can undergo Friedel–Crafts reactions—at the ortho and para positions—because oxygen, being a weaker base than nitrogen, does not complex with the Lewis acid. (A Lewis acid complexes with a carbonyl oxygen because a carbonyl oxygen has a partial negative charge, unlike an oxygen directly attached to a benzene ring, which has a partial positive charge.)

Aniline also cannot be nitrated, because nitric acid is an oxidizing agent and an NH_2 group is easily oxidized. (Nitric acid and aniline can be an explosive combination.) However, if the amino group is protected, then the ring can be nitrated. The acetyl group can subsequently be removed by acid-catalyzed hydrolysis.

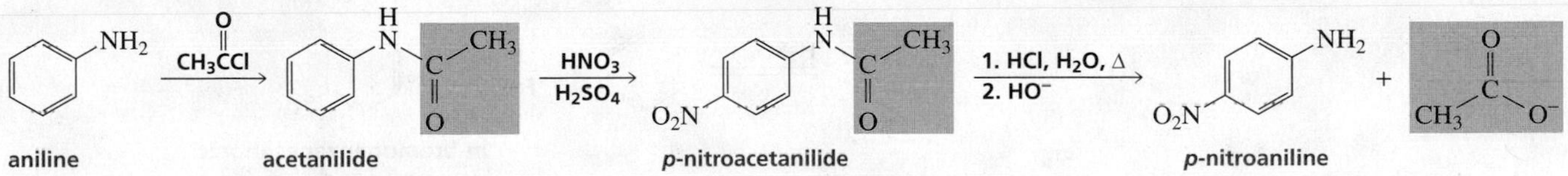

aniline   acetanilide   *p*-nitroacetanilide   *p*-nitroaniline

---

PROBLEM 25

Show how the following compounds could be synthesized from benzene:

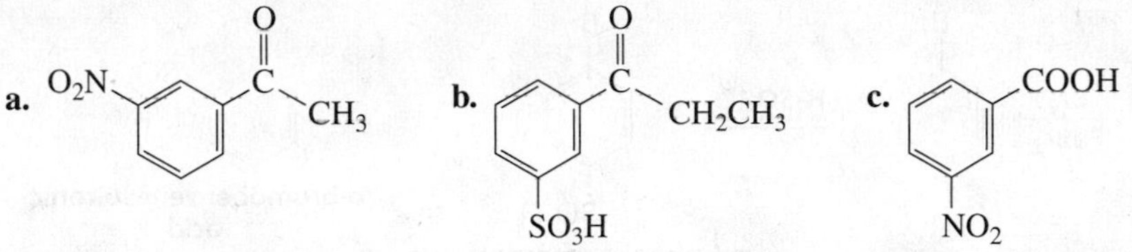

a.   b.   c.

PROBLEM 26♦

Give the products, if any, of each of the following reactions:

a. benzonitrile + methyl chloride + AlCl₃   c. benzoic acid + CH₃CH₂Cl + AlCl₃

b. aniline + Br₂   d. benzene + 2 CH₃Cl + AlCl₃

---

19   SYNTHESIS OF MONOSUBSTITUTED AND DISUBSTITUTED BENZENES

As the number of reactions you know increases, so do the possibilities you are able to choose from when you design a synthesis. For example, you can now design two very different routes for the synthesis of 2-phenylethanol from benzene.

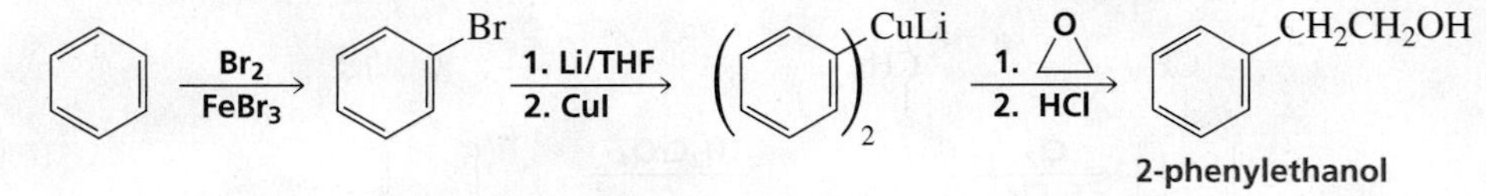

2-phenylethanol

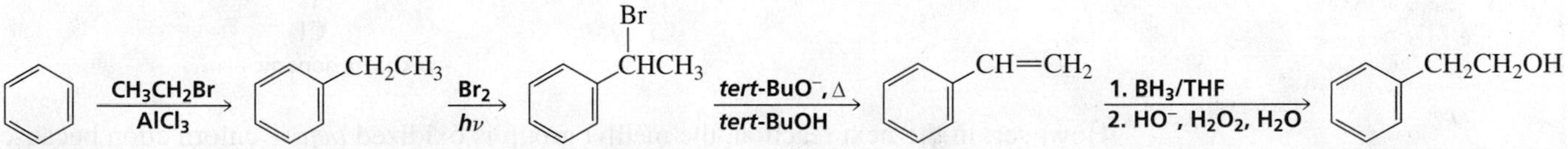

excess   2-phenylethanol

The preferred route depends on such factors as convenience, expense, and the expected yield of the target molecule (the desired product). For example, the first route shown for the synthesis of 2-phenylethanol is the better procedure, because the second route has more steps, requires excess benzene to prevent polyalkylation, and uses a radical reaction that can produce unwanted side products. Moreover, the yield of the elimination reaction is not high (because some substitution product is formed as well).

Designing the synthesis of a disubstituted benzene requires careful consideration of the order in which the substituents are to be placed on the ring. For example, if you want to synthesize *m*-bromobenzenesulfonic acid, the sulfonic acid group has to be placed on the ring first, because that group will direct the bromo substituent to the desired meta position.

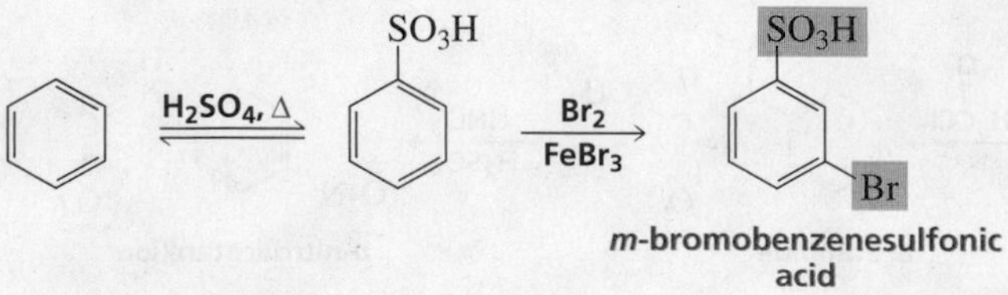

m-bromobenzenesulfonic acid

However, if the desired product is *p*-bromobenzenesulfonic acid, then the order of the two reactions must be reversed because only the bromo substituent is an ortho–para director.

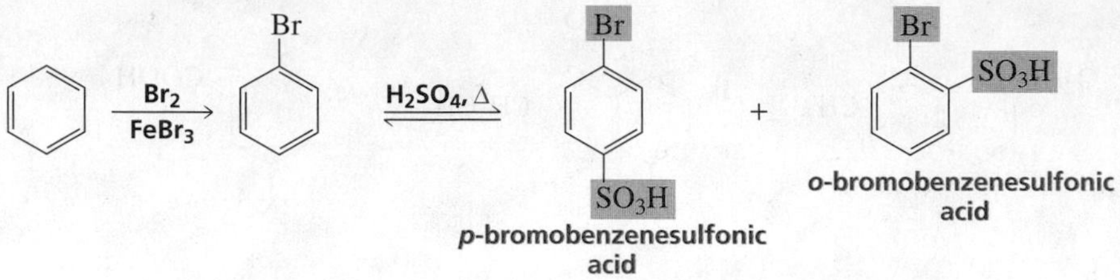

p-bromobenzenesulfonic acid

o-bromobenzenesulfonic acid

Both substituents of *m*-nitroacetophenone are meta directors. However, the Friedel–Crafts acylation reaction must be carried out first because nitrobenzene cannot undergo a Friedel–Crafts reaction (Section 18).

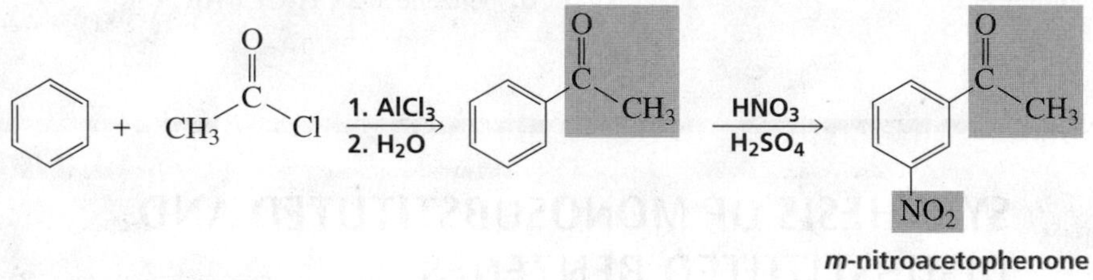

m-nitroacetophenone

Another question that needs to be considered is, at what point in a reaction sequence should a substituent be chemically modified? For example, in the following reaction, the methyl group is oxidized *after* it directs the chloro substituent to the para position. (*o*-Chlorobenzoic acid is also formed in this reaction.)

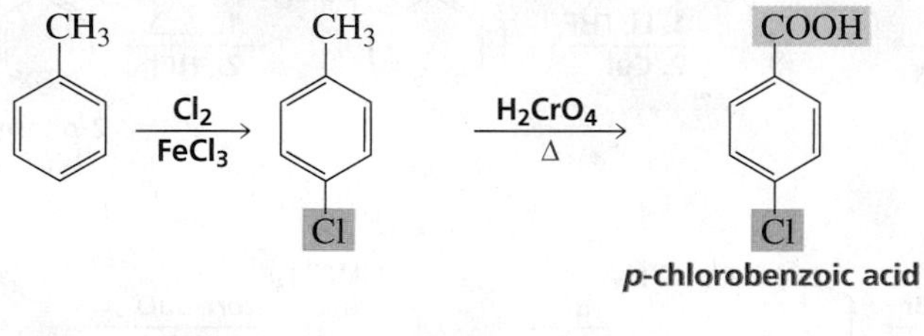

p-chlorobenzoic acid

However, in the next reaction, the methyl group is oxidized *before* chlorination because a meta director is needed to obtain the desired product.

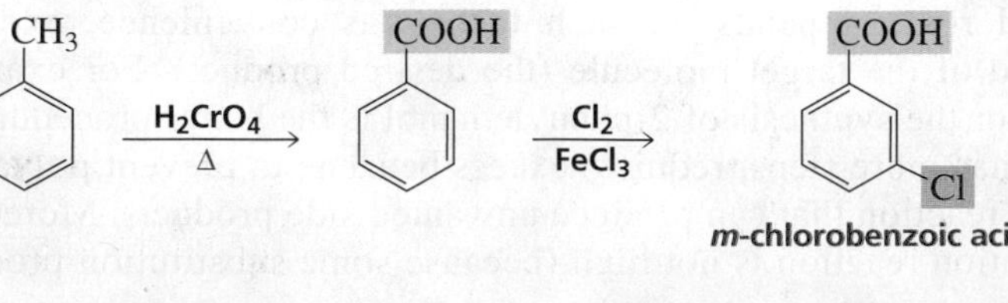

m-chlorobenzoic acid

Although chemists often have several ways to carry out a reaction, there may be factors that require the use of one particular method. For example, we have seen that there is more than one way to get a straight-chain alkyl group onto a benzene ring (Sections 9 and 10). In the next synthesis, a Friedel–Crafts alkylation/reduction can be used to put the first alkyl group on the ring but not the second one. A meta-directing group is required to get the second alkyl substituent onto the ring, and a Friedel–Crafts reaction cannot be done on a ring with a meta director. Instead, the target molecule has to be prepared using a coupling reaction.

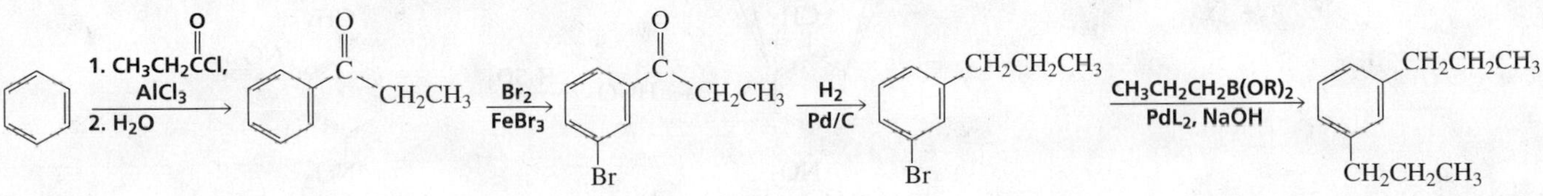

m-dipropylbenzene

PROBLEM 27♦

a. Does a coupling reaction have to be used to synthesize *p*-dipropylbenzene?
b. Can a coupling reaction be used to synthesize *p*-dipropylbenzene?

In the next synthesis, the type of reaction employed, the order in which the substituents are put on the benzene ring, and the point at which a substituent is chemically modified must all be considered. The straight-chain propyl substituent must be put on the ring by a Friedel–Crafts acylation rather than by an alkylation reaction, in order to avoid the carbocation rearrangement that would occur with the alkylation reaction. The Friedel–Crafts acylation must be carried out before sulfonation because acylation cannot be carried out on a ring with a meta-directing sulfonic acid substituent. Finally, the sulfonic acid group must be put on the ring after the carbonyl group is reduced to a methylene group, so that the sulfonic acid group will be directed primarily to the para position by the alkyl group.

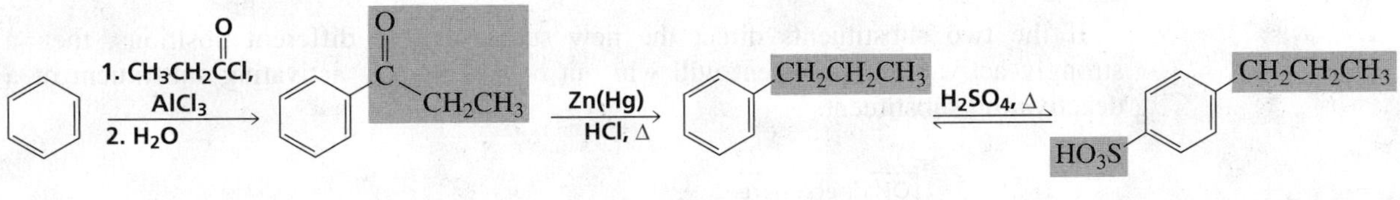

p-propylbenzenesulfonic acid

Alternatively, the propylbenzene intermediate could be prepared using bromobenzene and a coupling reaction.

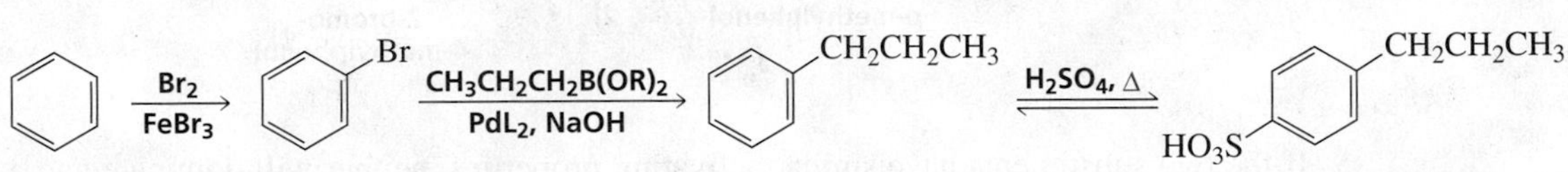

p-propylbenzenesulfonic acid

PROBLEM 28

Show how each of the following compounds can be synthesized from benzene:

a. *p*-chloroaniline
b. *m*-chloroaniline
c. *m*-xylene
d. 2-phenylpropene
e. *m*-nitrobenzoic acid
f. *p*-nitrobenzoic acid
g. *m*-bromopropylbenzene
h. *o*-bromopropylbenzene
i. 1-phenyl-2-propanol

20 THE SYNTHESIS OF TRISUBSTITUTED BENZENES

When a disubstituted benzene undergoes an electrophilic aromatic substitution reaction, the directing effects of both substituents have to be considered. If both substituents direct the incoming substituent to the same position, the product of the reaction is easily predicted.

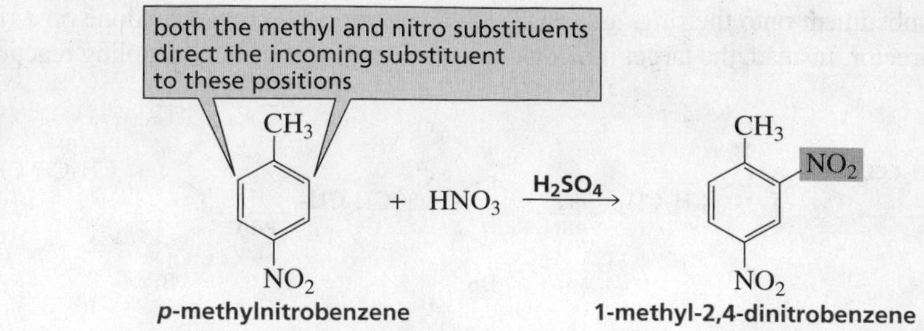

both the methyl and nitro substituents
direct the incoming substituent
to these positions

p-methylnitrobenzene   + HNO₃ $\xrightarrow{\text{H}_2\text{SO}_4}$ 1-methyl-2,4-dinitrobenzene

Notice that three positions are activated in the next reaction, but the new substituent ends up primarily on only two of the three. Steric hindrance makes the position between the substituents less accessible, so only a very small amount of the third product will be formed.

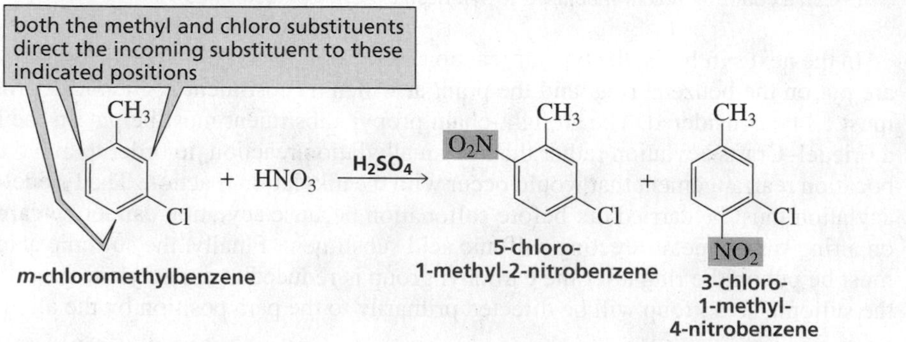

both the methyl and chloro substituents
direct the incoming substituent to these
indicated positions

m-chloromethylbenzene   + HNO₃ $\xrightarrow{\text{H}_2\text{SO}_4}$ 5-chloro-1-methyl-2-nitrobenzene   +   3-chloro-1-methyl-4-nitrobenzene

If the two substituents direct the new substituent to different positions, then a strongly activating substituent will win out over a weakly activating substituent or a deactivating substituent.

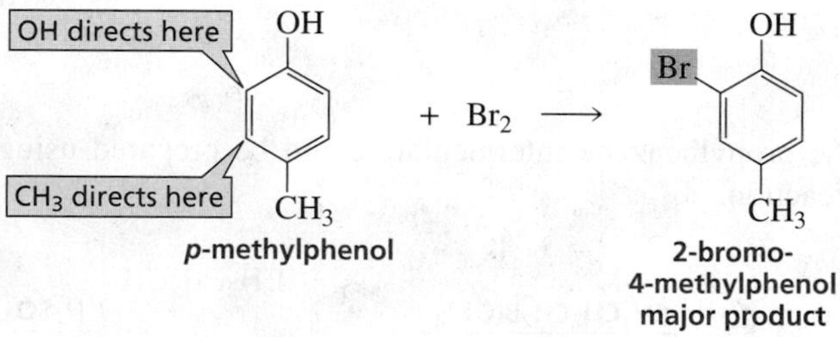

OH directs here

CH₃ directs here

p-methylphenol   + Br₂ $\longrightarrow$ 2-bromo-4-methylphenol
major product

If the two substituents have similar activating properties, neither will dominate and a mixture of products will be obtained.

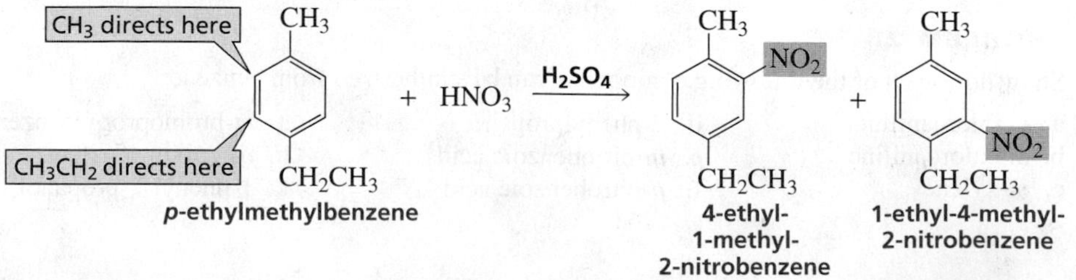

CH₃ directs here

CH₃CH₂ directs here

p-ethylmethylbenzene   + HNO₃ $\xrightarrow{\text{H}_2\text{SO}_4}$ 4-ethyl-1-methyl-2-nitrobenzene   +   1-ethyl-4-methyl-2-nitrobenzene

---

PROBLEM 29♦

What is the major product(s) of each of the following reactions?

a. bromination of *p*-methylbenzoic acid

b. chlorination of *o*-benzenedicarboxylic acid

c. bromination of *p*-chlorobenzoic acid

d. nitration of *p*-fluoroanisole

e. nitration of *p*-methoxybenzaldehyde

f. nitration of *p*-*tert*-butylmethylbenzene

PROBLEM 30◆

A student had prepared three ethyl-substituted benzaldehydes, but neglected to label them. The student at the next bench said they could be identified by brominating a sample of each and determining how many bromo-substituted products were formed. Is the student's advice sound?

PROBLEM 31 Solved

When phenol is treated with Br_2, a mixture of *ortho*-bromophenol, *para*-bromophenol, as well as dibromo- and tribromophenols is obtained. Design a synthesis that would convert phenol primarily to *ortho*-bromophenol.

Solution In the first step, the bulky sulfonic acid group will add preferentially to the para position. Both the OH and SO_3H groups will direct bromine to the position ortho to the OH group. Heating in dilute acid removes the sulfonic acid group (Section 6).

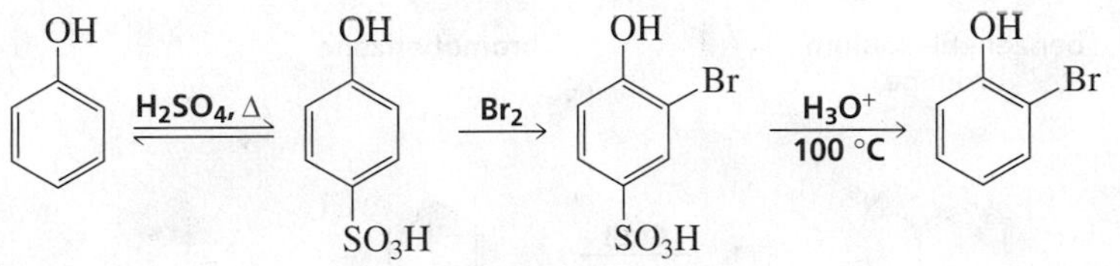

Use of a sulfonic acid group to block the para position is a common strategy for synthesizing high yields of ortho-substituted compounds.

21 THE SYNTHESIS OF SUBSTITUTED BENZENES USING ARENEDIAZONIUM SALTS

So far, we have learned how to place a limited number of different substituents on a benzene ring—the substituents listed in Section 3 and those that can be obtained from these substituents by chemical conversion (Section 12). The list of substituents that can be placed on a benzene ring can be expanded using **arenediazonium salts.**

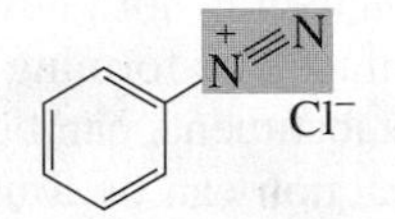

an arenediazonium salt

Displacing the leaving group of a diazonium ion by a nucleophile occurs readily because it results in formation of nitrogen gas (indicated by an upward pointing arrow), which is very stable. Some displacements involve phenyl cations, whereas others involve radicals—the actual mechanism depends on the particular nucleophile.

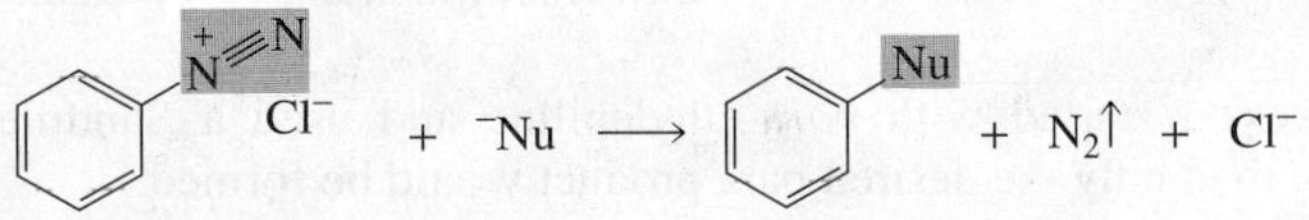

benzenediazonium chloride

Aniline can be converted into an arenediazonium salt by treatment with nitrous acid (HNO_2). Because nitrous acid is unstable, it must be formed in situ, using an aqueous solution of sodium nitrite and an acid; indeed, N_2 is such a good leaving group that the diazonium salt is synthesized at 0 °C and used immediately without isolation. The mechanism for conversion of a primary amino group (NH_2) to a diazonium group ($^+N{\equiv}N$) is shown in Section 23.

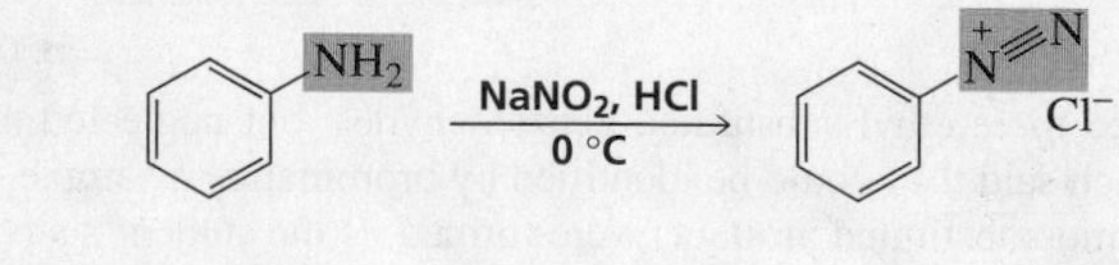

The nucleophiles $^-C\equiv N$, Cl^-, and Br^- will replace the diazonium group if the appropriate copper(I) salt is added to the solution containing the arenediazonium salt. The reaction of an arenediazonium salt with a copper(I) salt is known as a **Sandmeyer reaction.**

Sandmeyer reactions

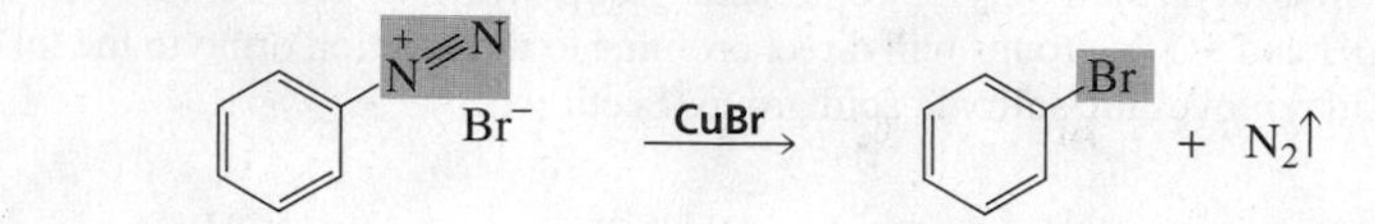

benzenediazonium bromide $\xrightarrow{\text{CuBr}}$ bromobenzene $+ N_2\uparrow$

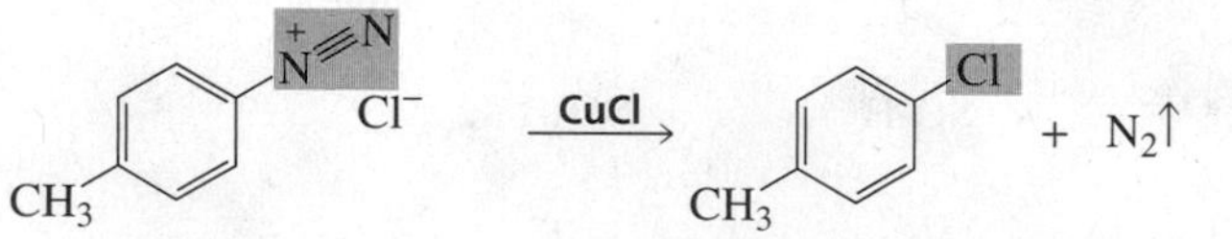

p-toluenediazonium chloride $\xrightarrow{\text{CuCl}}$ p-chloromethylbenzene $+ N_2\uparrow$

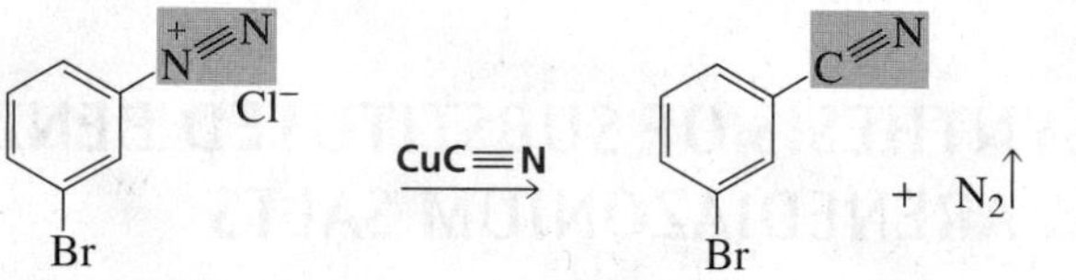

m-bromobenzenediazonium chloride $\xrightarrow{\text{CuC}\equiv\text{N}}$ m-bromobenzonitrile $+ N_2\uparrow$

KCl and KBr cannot be used in place of CuCl and CuBr in Sandmeyer reactions. The cuprous salts are required, which indicates that the copper(I) ion has a role in the reaction. Although the precise mechanism is not known, it is thought that the copper(I) ion donates an electron to the diazonium salt, forming an aryl radical and nitrogen gas.

Although chloro and bromo substituents can be placed directly on a benzene ring by halogenation, the Sandmeyer reaction can be a useful alternative. For example, if you wanted to make *para*-chloroethylbenzene, the chlorination of ethylbenzene leads to a mixture of ortho and para isomers.

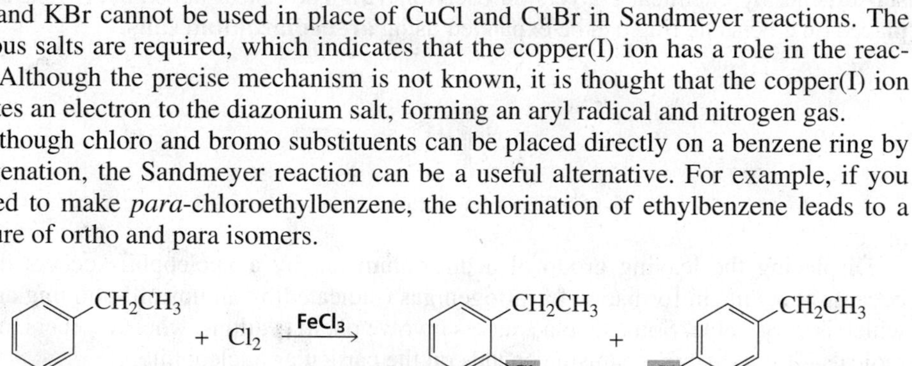

ethylbenzene $+ Cl_2 \xrightarrow{\text{FeCl}_3}$ o-chloroethylbenzene $+$ p-chloroethylbenzene

If, however, you started with *para*-ethylaniline and used a Sandmeyer reaction for chlorination, then only the desired para product would be formed.

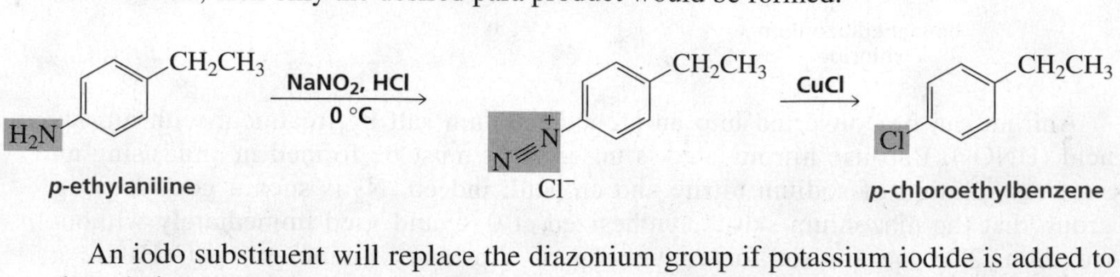

p-ethylaniline $\xrightarrow[\text{0 °C}]{\text{NaNO}_2,\text{ HCl}}$ (diazonium salt) $\xrightarrow{\text{CuCl}}$ p-chloroethylbenzene

An iodo substituent will replace the diazonium group if potassium iodide is added to the solution containing the diazonium ion.

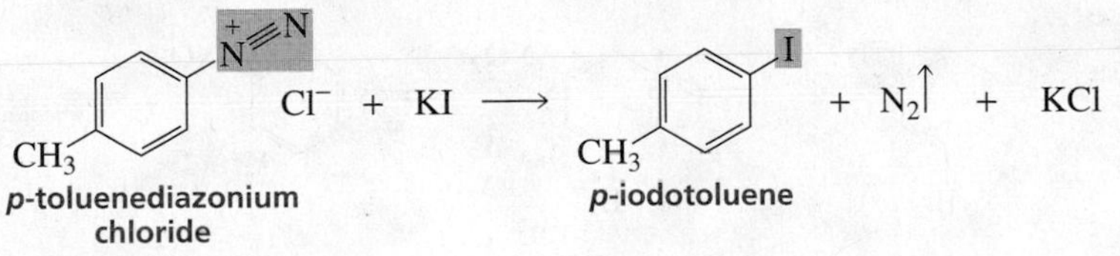

Fluoro substitution occurs if the arenediazonium salt is heated with fluoroboric acid (HBF_4). This is known as the **Schiemann reaction.**

Schiemann reaction

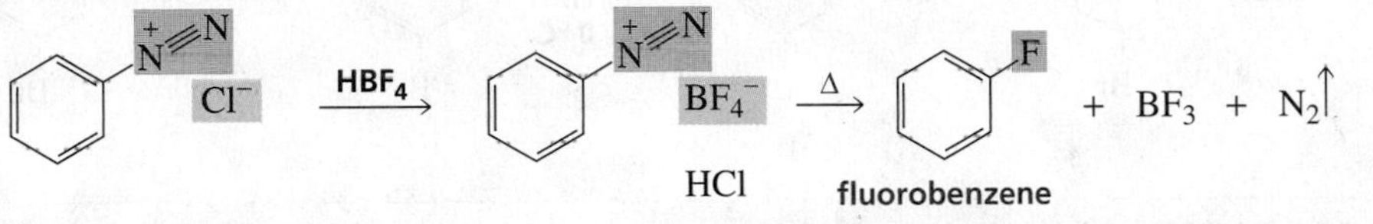

If the acidic aqueous solution in which the diazonium salt has been synthesized is allowed to warm up, then an OH group will replace the diazonium group. (H_2O is the nucleophile.)

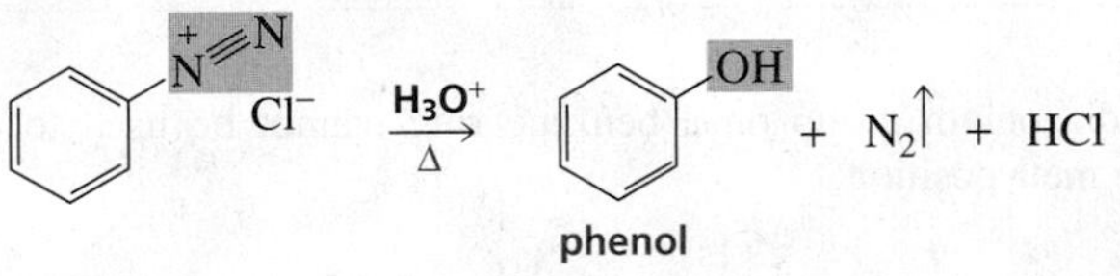

A better yield of phenol is obtained if aqueous copper(I) oxide and copper(II) nitrate are added to the cold solution.

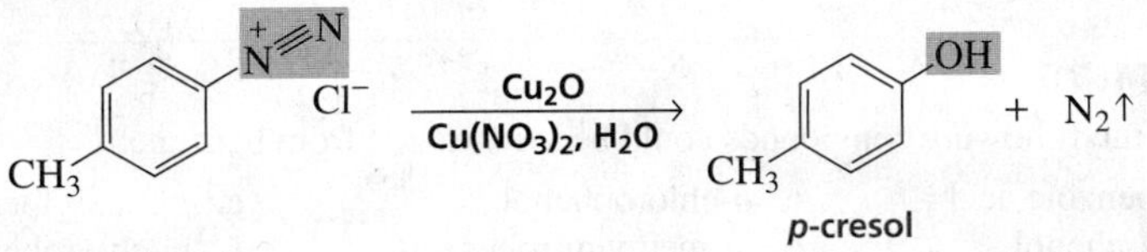

A hydrogen will replace a diazonium group if the arenediazonium salt is treated with hypophosphorous acid (H_3PO_2). This is a useful reaction if an amino group or a nitro group is needed for directing purposes and subsequently must be removed. For example, it is difficult to envision how 1,3,5-tribromobenzene could be synthesized without such a reaction.

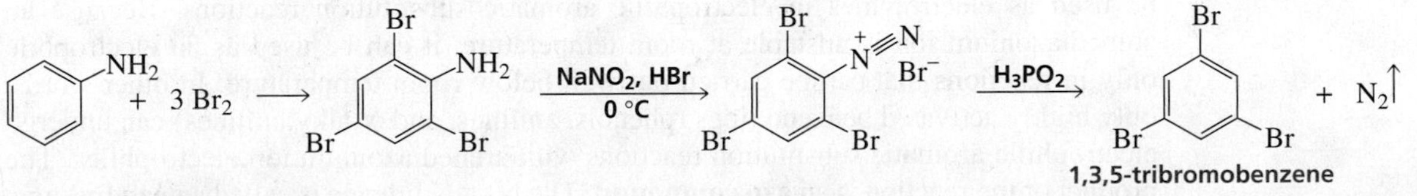

Reactions involving arenediazonium ions must be carried out at $0\,°C$ because they are unstable at higher temperatures.

Retrosynthetic Analysis

We have seen that when designing a complicated synthesis, it is often easier to work backwards. For example, when planning a synthesis for *meta*-dibromobenzene, we realize that, because a bromo substituent is an ortho–para director, halogenation cannot be used to put both bromo substituents on the ring. Knowing that a bromo substituent can be put on a benzene ring with a Sandmeyer reaction and that the bromo substituent in a Sandmeyer reaction replaces what originally was a meta-directing nitro substituent, we have a route to the synthesis of the target molecule.

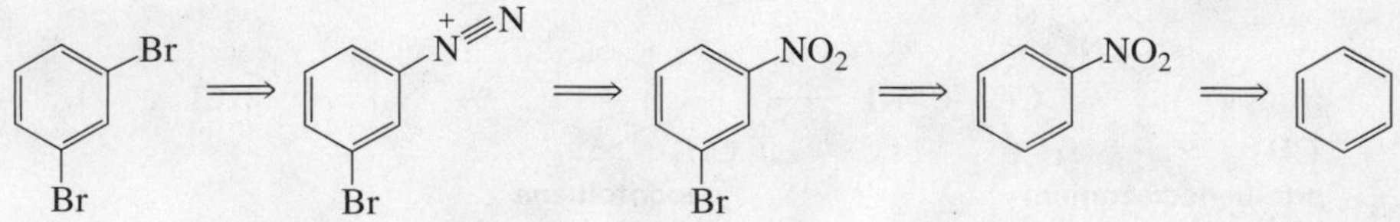

Now we can write the reaction in the forward direction, including the reagents necessary to carry out each step.

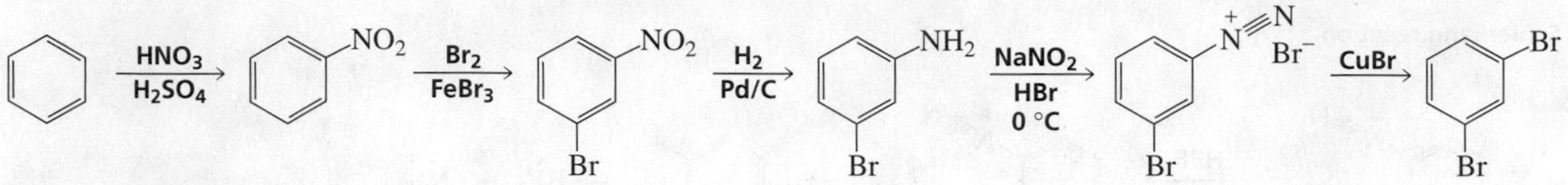

PROBLEM 32♦

Why isn't FeBr$_3$ used as a catalyst in the first step of the synthesis of 1,3,5-tribromobenzene?

PROBLEM 33

Explain why a diazonium group on a benzene ring cannot be used to direct an incoming substituent to the meta position.

PROBLEM 34

Write the sequence of steps required for the conversion of benzene into benzenediazonium chloride.

PROBLEM 35

Show how the following compounds could be synthesized from benzene:

a. *m*-nitrobenzoic acid   **c.** *o*-chlorophenol   **e.** *p*-methylbenzonitrile

b. *m*-bromophenol   **d.** *m*-methylnitrobenzene   **f.** *m*-chlorobenzaldehyde

---

22   THE ARENEDIAZONIUM ION AS AN ELECTROPHILE

In addition to being used to synthesize substituted benzenes, arenediazonium ions can be used as electrophiles in electrophilic aromatic substitution reactions. Because an arenediazonium ion is unstable at room temperature, it can be used as an electrophile only in reactions that can be carried out well below room temperature. In other words, only highly activated benzene rings (phenols, anilines, and *N*-alkylanilines) can undergo electrophilic aromatic substitution reactions with arenediazonium ion electrophiles. The product of the reaction is an **azo compound.** The N=N linkage is called an **azo linkage.**

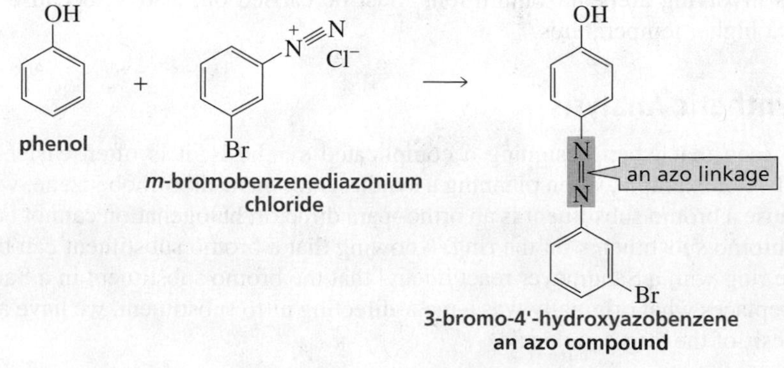

Because the electrophile is so large, substitution takes place preferentially at the less sterically hindered para position. However, if the para position is blocked, then substitution will occur at an ortho position.

The mechanism for electrophilic aromatic substitution with an arenediazonium ion electrophile is the same as the mechanism for electrophilic aromatic substitution with any other electrophile.

MECHANISM FOR ELECTROPHILIC AROMATIC SUBSTITUTION USING AN ARENEDIAZONIUM ION ELECTROPHILE

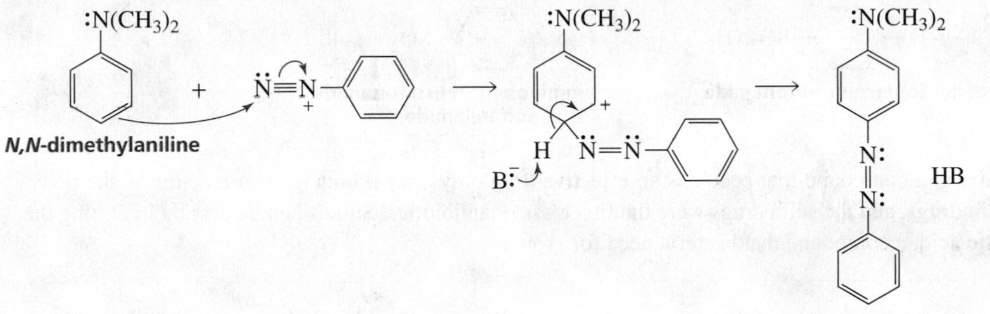

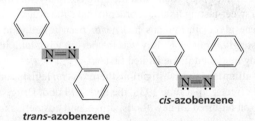

N,N-dimethylaniline

p-N,N-dimethylaminoazobenzene

- The electrophile adds to the benzene ring.
- A base in the solution removes a proton from the carbon that formed the bond with the electrophile.

Azo compounds, like alkenes, can exist in cis and trans forms. The trans isomer is more stable because the cis isomer has steric strain.

trans-azobenzene

cis-azobenzene

Azobenzenes are colored compounds because of their extended conjugation. They are used commercially as dyes.

Discovery of the First Antibiotic

The observation that azo dyes effectively dyed wool fibers (animal protein) suggested that such dyes might bind to bacterial proteins, too, and perhaps harm the bacteria in the process. Well over 10,000 dyes were screened *in vitro* ("in glass"), but none showed any antibiotic activity. At that point, the investigators decided to screen the dyes *in vivo* ("in an organism").

Now the luck of the investigators improved. *In vivo* studies were done in mice that had been infected with a bacterial culture. Several dyes turned out to counteract Gram-positive infections. The least toxic of these, Prontosil (a bright red dye), became the first drug to treat bacterial infections.

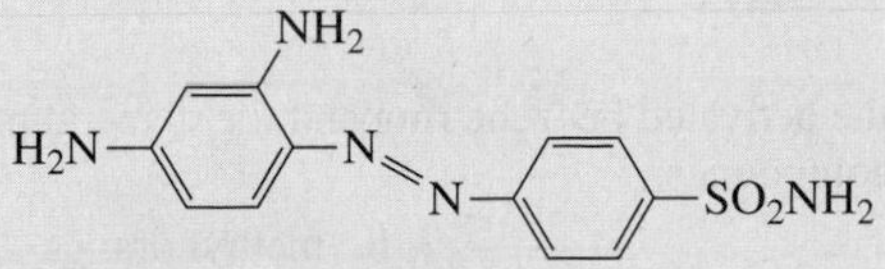

Prontosil®

The fact that Prontosil was inactive *in vitro* but active *in vivo* should have been recognized as a sign that the dye was converted to an active compound by the mouse, but this did not occur to the investigators, who were content to have found a useful antibiotic.

When scientists at the Pasteur Institute later investigated Prontosil, they noted that mice given the drug did not excrete a red compound. Urine analysis showed that the mice excreted *para*-acetamidobenzenesulfonamide, a colorless compound, instead. Chemists knew that anilines are acetylated *in vivo,* so they prepared the nonacetylated compound (sulfanilamide). When sulfanilamide was tested in mice infected with streptococcus, all the mice were cured, whereas untreated control mice died.

$$CH_3-\overset{\overset{\displaystyle O}{\|}}{C}-NH-\!\!\!\!\bigcirc\!\!\!\!-SO_2NH_2 \qquad H_2N-\!\!\!\!\bigcirc\!\!\!\!-SO_2NH_2$$

| *para*-acetamidobenzenesulfonamide | *para*-aminobenzenesulfonamide sulfanilamide |

Protonsil is an example of a **prodrug,** a compound that becomes an effective drug only after it undergoes a reaction in the body. Sulfanilamide was the first of the sulfa drugs, and the sulfa drugs were the first class of antibiotics. Sulfanilamide acts by inhibiting the bacterial enzyme that synthesizes folic acid, a compound that bacteria need for growth.

Drug Safety

In October 1937, patients who had obtained sulfanilamide from a company in Tennessee were experiencing excruciating abdominal pains before slipping into fatal comas. The FDA asked Eugene Geiling, a pharmacologist at the University of Chicago, and his graduate student Frances Kelsey to investigate. They found that the drug company was dissolving sulfanilamide in diethylene glycol, a sweet-tasting liquid, in order to make the drug easy to swallow. However, the safety of diethylene glycol in humans had never been tested, and it turned out to be a deadly poison. Interestingly, Frances Kelsey was the person who, later in her career, prevented thalidomide from being marketed in the United States.

At the time of the sulfanilamide investigation, there was no legislation to prevent the sale of medicines with lethal effects, so in June 1938, the Federal Food, Drug, and Cosmetic Act was enacted. This legislation required all new foods, drugs, and cosmetics to be thoroughly tested for effectiveness and safety before being marketed. The laws are amended from time to time to reflect changing circumstances.

PROBLEM 36
What product is formed from the reaction of *p*-methylphenol with benzenediazonium chloride?

PROBLEM 37
In the mechanism for electrophilic aromatic substitution with a diazonium ion as the electrophile, why does nucleophilic attack occur on the terminal nitrogen of the diazonium ion rather than on the nitrogen that has the formal positive charge?

PROBLEM 38
Draw the structure of the activated benzene ring and the diazonium ion used in the synthesis of each of the following compounds.

a. butter yellow

b. methyl orange

23 THE MECHANISM FOR THE REACTION OF AMINES WITH NITROUS ACID

Conversion of an NH_2 group to a diazonium group requires a *nitrosonium ion*. A nitrosonium ion is formed when water is eliminated from protonated nitrous acid, similar to the way the nitronium ion is generated from nitric acid (Section 5) and the sulfonium ion is generated from sulfuric acid (Section 6).

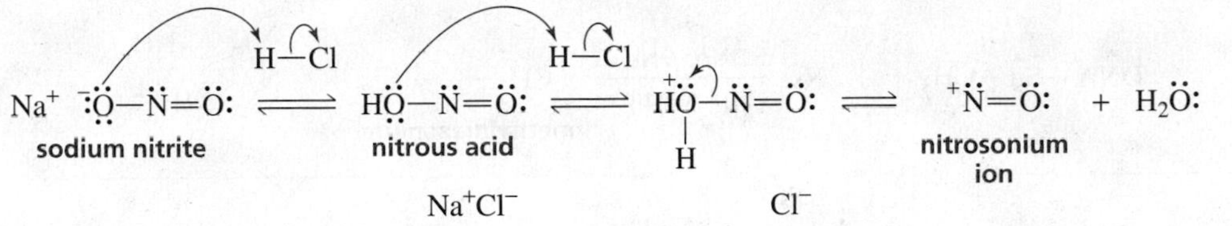

MECHANISM FOR FORMATION OF A DIAZONIUM ION FROM ANILINE

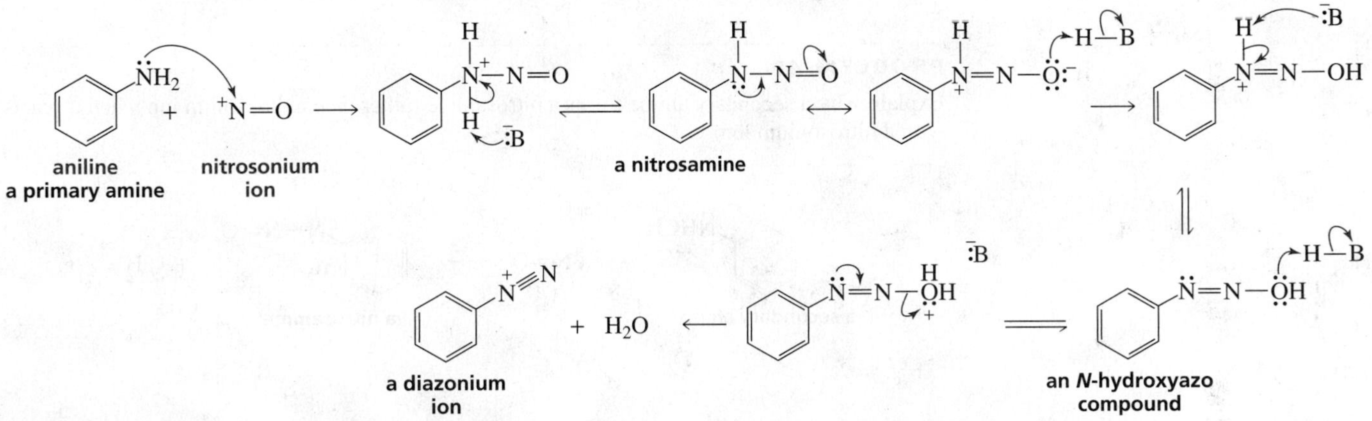

- Aniline shares an electron pair with the nitrosonium ion.
- Removal of a proton from nitrogen forms a **nitrosamine.**
- Delocalization of nitrogen's lone pair and protonation on oxygen form a protonated *N*-hydroxyazo compound.
- The protonated *N*-hydroxyazo compound is in equilibrium with its nonprotonated form.
- The *N*-hydroxyazo compound can be reprotonated on nitrogen (the reverse reaction) or protonated on oxygen (the forward reaction).
- Elimination of water forms the diazonium ion.

Notice the pattern of the three intermediates:

protonated intermediate →
neutral intermediate →
protonated intermediate.

A New Cancer-Fighting Drug

Temozolomide, the drug used to treat the late Senator Ted Kennedy's brain tumor, is a relatively new member in the arsenal of cancer-fighting drugs. The drug can be taken orally. While in the circulatory system, it is converted to its active form by reacting with water and losing CO_2. Once inside the cell, the drug eliminates a methyldiazonium ion, a very reactive methylating agent. Methylating DNA triggers the death of cancer cells. Temozolomide is another example of a prodrug.

The reaction scheme shows the conversion of temozolomide to methyldiazonium:

temozolomide → (H₂O) → activated form ⇌ intermediate

$$\text{DNA}-\overset{+}{\text{Nu}}-CH_3 + N_2 \xleftarrow[\text{S}_N2 \text{ reaction}]{\text{DNA}-\overset{..}{\text{Nu}}} CH_3-\overset{+}{N}\equiv N + \text{(imidazole product)}$$

methyldiazonium

PROBLEM 39♦

Which amide bond is hydrolyzed in the first step of the conversion of temozolomide to methyldiazonium?

PROBLEM 40

Explain why a secondary amine forms a nitrosamine rather than a diazonium ion when it reacts with a nitrosonium ion.

a secondary amine + $^+N{=}O$ → a nitrosamine + H^+

Nitrosamines and Cancer

A 1962 outbreak of food poisoning in sheep in Norway was traced to their ingestion of nitrite-treated fish meal. This incident immediately raised concerns about human consumption of nitrite-treated foods, because sodium nitrite, a commonly used food preservative, can react with naturally occurring secondary amines that are present in food, to produce nitrosamines, which are known to be carcinogenic. Smoked fish, cured meats, and beer all contain nitrosamines. Nitrosamines are also found in cheese because some cheeses are preserved with sodium nitrite and cheese is rich in secondary amines. When consumer groups in the United States asked the Food and Drug Administration to ban the use of sodium nitrite as a preservative, the request was vigorously opposed by the meat-packing industry.

Despite extensive investigations, it has not yet been determined whether the small amounts of nitrosamines present in our food pose a hazard to our health. Until this question is answered, it will be hard to avoid sodium nitrite in our diet. Meanwhile, it is worrisome to note that Japan has both one of the highest gastric cancer rates and the highest average ingestion of sodium nitrite. Some good news, however, is that the concentration of nitrosamines present in bacon has been considerably reduced in recent years by the addition of ascorbic acid—a nitrosamine inhibitor—to the curing mixture. Also, improvements in the malting process have reduced the level of nitrosamines in beer. Dietary sodium nitrite does have a redeeming feature: there is some evidence that it protects against botulism, a type of severe food poisoning.

PROBLEM 41

Diazomethane can be used to convert a carboxylic acid into a methyl ester. Propose a mechanism for this reaction.

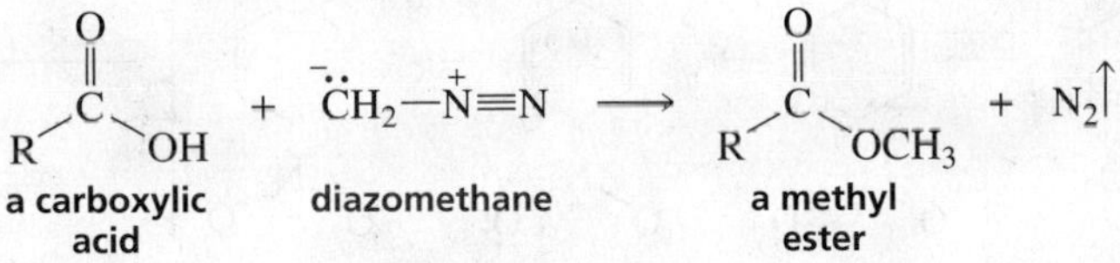

| a carboxylic acid | diazomethane | a methyl ester |

---

24 NUCLEOPHILIC AROMATIC SUBSTITUTION: AN ADDITION–ELIMINATION REACTION

Aryl halides do not react with nucleophiles because the π electron cloud repels the approach of a nucleophile.

a nucleophile is repelled by the π electron cloud

nucleophile

If, however, the aryl halide has one or more substituents that strongly withdraw electrons from the ring by resonance, a **nucleophilic aromatic substitution** reaction can take place. The electron-withdrawing groups must be positioned ortho or para to the halogen. The greater the number of electron-withdrawing substituents ortho and para to the halogen, the more easily the nucleophilic aromatic substitution reaction occurs. Notice the conditions required for each reaction.

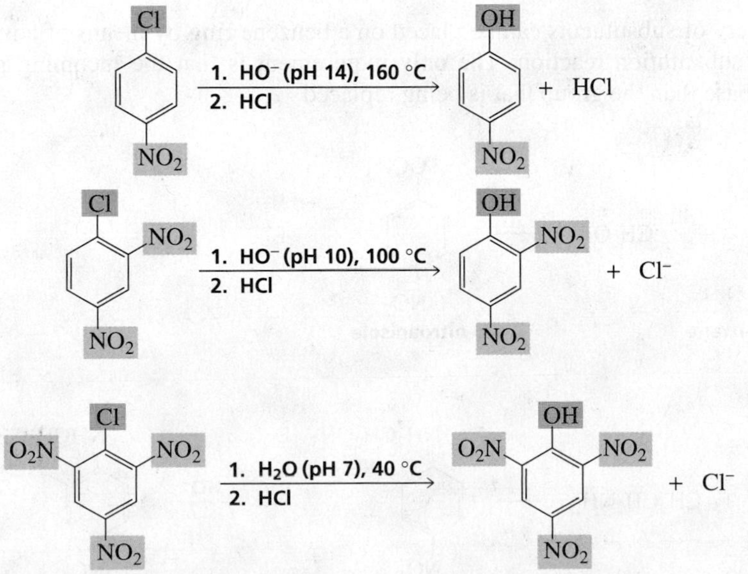

Nucleophilic aromatic substitution takes place by a two-step mechanism. The reaction is called an **S$_N$Ar reaction** (substitution nucleophilic aromatic).

GENERAL MECHANISM FOR NUCLEOPHILIC AROMATIC SUBSTITUTION

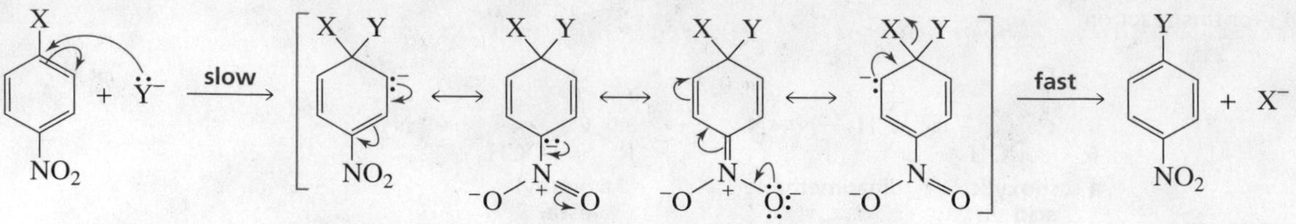

a Meisenheimer complex

- The nucleophile attacks the carbon bearing the leaving group from a trajectory that is nearly perpendicular to the aromatic ring. (Groups cannot be displaced from sp^2 carbon atoms by back-side attack.) Nucleophilic attack forms a resonance-stabilized carbanion intermediate called a *Meisenheimer complex*.
- The leaving group is eliminated, reestablishing the aromaticity of the ring.

In a nucleophilic aromatic substitution reaction, the incoming nucleophile must be a stronger base than the substituent that is being replaced, because the weaker of the two bases will be the one eliminated from the intermediate.

The electron-withdrawing substituent must be ortho or para to the site of nucleophilic attack because the electrons of the attacking nucleophile can be delocalized onto the substituent only if the substituent is in one of those positions.

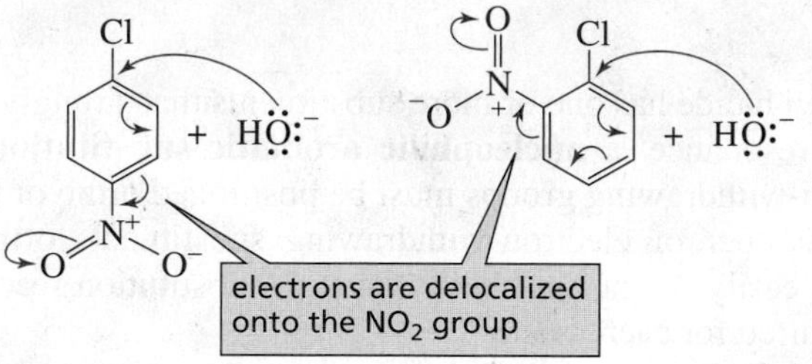

electrons are delocalized onto the NO_2 group

A variety of substituents can be placed on a benzene ring by means of a nucleophilic aromatic substitution reaction. The only requirement is that the incoming group be a stronger base than the group that is being replaced.

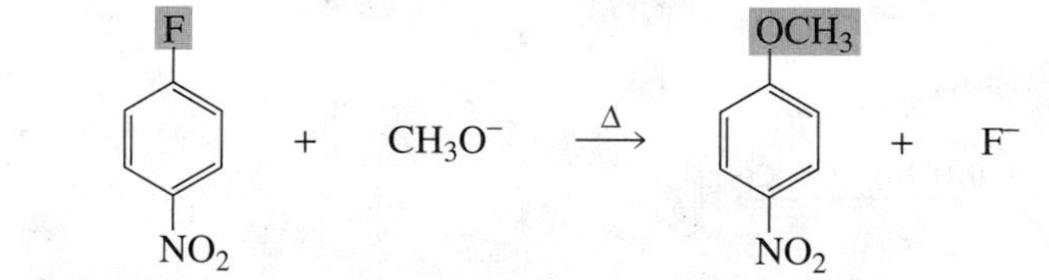

p-fluoronitrobenzene · · · *p*-nitroanisole

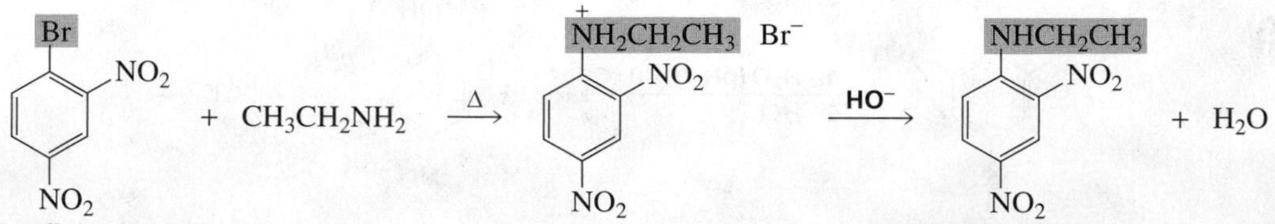

1-bromo-2,4-dinitrobenzene · · · *N*-ethyl-2,4-dinitro-aniline

Notice that the strongly electron-withdrawing nitro substituent that *activates* the benzene ring toward *nucleophilic aromatic substitution, deactivates* the ring toward *electrophilic aromatic substitution* (Table 1). In other words, making the ring less electron rich makes it more reactive toward a nucleophile but less reactive toward an electrophile.

An electron-withdrawing substituent increases the reactivity of the benzene ring toward nucleophilic substitution and decreases the reactivity of the benzene ring toward electrophilic substitution.

PROBLEM 42

Draw resonance contributors for the carbanion that would be formed if *meta*-chloronitrobenzene were to react with hydroxide ion. Why doesn't the reaction occur?

PROBLEM 43◆

a. List the following compounds in order from greatest tendency to least tendency to undergo nucleophilic aromatic substitution:

chlorobenzene   1-chloro-2,4-dinitrobenzene   *p*-chloronitrobenzene

b. List the same compounds in order from greatest tendency to least tendency to undergo electrophilic aromatic substitution.

PROBLEM 44

Show how each of the following compounds could be synthesized from benzene:

a. *o*-nitrophenol   **b.** *p*-nitroaniline   **c.** *p*-bromoanisole   **d.** anisole

---

25   THE SYNTHESIS OF CYCLIC COMPOUNDS

DESIGNING A SYNTHESIS

Most of the reactions covered in this text are intermolecular reactions—that is, the two reacting groups are in different molecules. *Cyclic compounds are formed from intramolecular reactions*—reactions in which the two reacting groups are in the same molecule. Intramolecular reactions are particularly favored if the reaction forms a compound with a five- or a six-membered ring.

In designing the synthesis of a cyclic compound, we must determine the kinds of reactive groups that are necessary for a successful synthesis. For example, we know that a ketone is formed from the Friedel–Crafts acylation of a benzene ring with an acyl chloride (Section 7). Therefore, a cyclic ketone will result if a Lewis acid (AlCl$_3$) is added to a compound that contains both a benzene ring and an acyl chloride. The size of the ring is determined by the number of carbons between the two groups.

An ester is formed from the reaction of a carboxylic acid with an alcohol. Therefore, a lactone (cyclic ester) can be prepared from a reactant that has a carboxylic acid group and an alcohol group in the same molecule separated by the appropriate number of carbons.

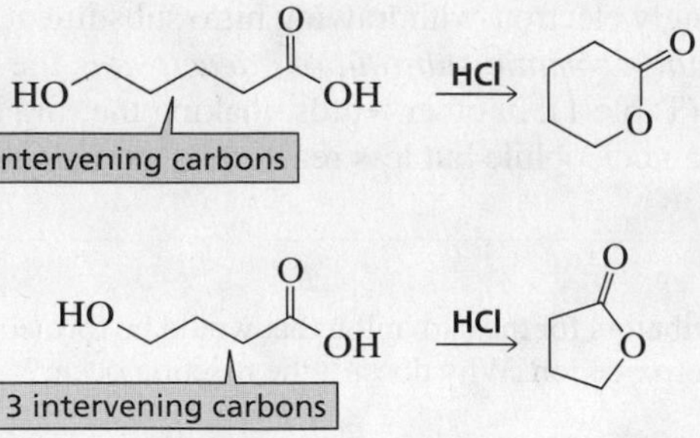

$$HO\underbrace{}_{\text{4 intervening carbons}}\overset{O}{\underset{}{\|}}C{-}OH \xrightarrow{\text{HCl}}$$

$$HO\underbrace{}_{\text{3 intervening carbons}}\overset{O}{\underset{}{\|}}C{-}OH \xrightarrow{\text{HCl}}$$

A cyclic ether can be prepared by an intramolecular Williamson ether synthesis.

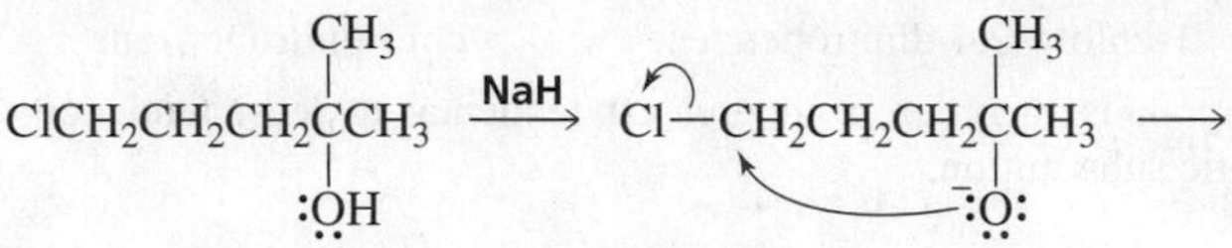

A cyclic ether can also be prepared by an intramolecular electrophilic addition reaction.

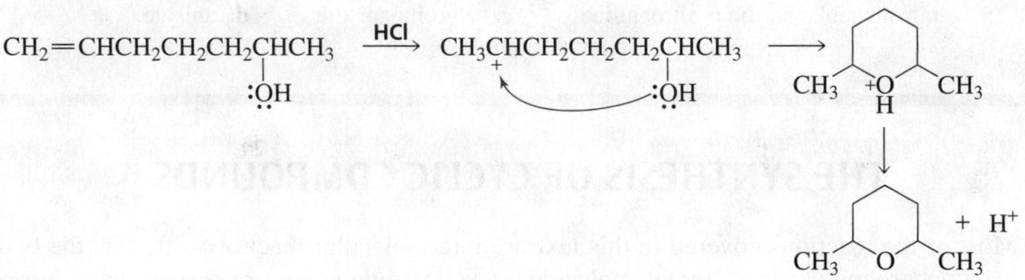

The product obtained from an intramolecular reaction can undergo further reactions, making it possible to synthesize many different compounds. For example, the cyclic alkyl bromide formed as the product of the next reaction could undergo an elimination reaction, or it could undergo substitution with a wide variety of nucleophiles, or it could be converted to a Grignard reagent that could react with many different electrophiles.

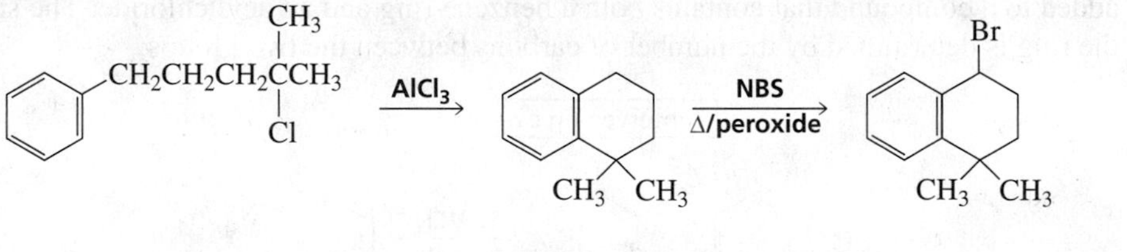

PROBLEM 45

Design a synthesis for each of the following, using an intramolecular reaction:

a.

b.

c.

d.

e.

f.

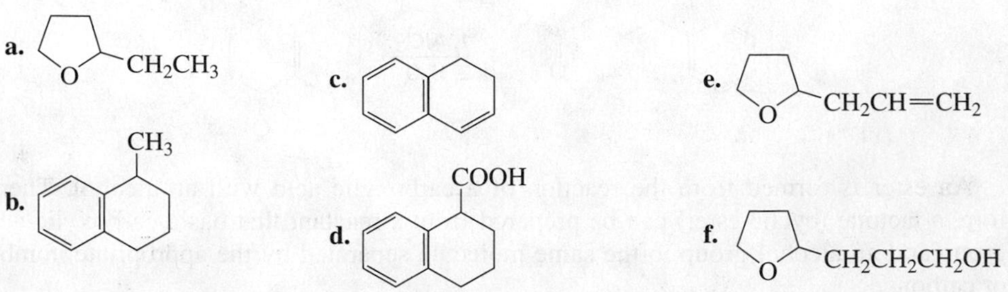

SOME IMPORTANT THINGS TO REMEMBER

- Some monosubstituted benzenes are named as substituted benzenes (for example, bromobenzene, nitrobenzene); others have names that incorporate the substituent (for example, toluene, phenol, aniline, anisole).

- Benzene is aromatic so it undergoes **electrophilic aromatic substitution reactions.**

- The most common electrophilic aromatic substitution reactions are halogenation, nitration, sulfonation, and Friedel–Crafts acylation and alkylation.

- Once the electrophile is generated, all electrophilic aromatic substitution reactions take place by the same two-step mechanism: (1) benzene forms a bond with the electrophile, forming a carbocation intermediate; and (2) a base removes a proton from the carbon that formed the bond with the electrophile.

- According to the **principle of microscopic reversibility,** the mechanism of a reaction in the reverse direction must retrace each step of the mechanism in the forward direction in microscopic detail.

- If the carbocation formed from the alkyl halide used in a **Friedel–Crafts alkylation** reaction can rearrange, the major product will be the product with the rearranged alkyl group.

- A straight-chain alkyl group can be placed on a benzene ring via a **Friedel–Crafts acylation** reaction, followed by reduction of the carbonyl group by catalytic hydrogenation, a **Clemmensen reduction,** or a **Wolff–Kishner reduction.**

- Alkylbenzenes with straight-chain alkyl groups can also be prepared from bromobenzene by a coupling reaction.

- The relative positions of two substituents on a benzene ring are indicated in the compound's name either by numbers or by the prefixes *ortho, meta,* and *para.*

- The nature of the substituent affects the reactivity of the benzene ring: the rate of electrophilic aromatic substitution is increased by electron-donating substituents and decreased by electron-withdrawing substituents.

- Substituents can withdraw electrons **inductively,** donate electrons by **hyperconjugation,** and withdraw and donate electrons **by resonance.**

- Electron-withdrawing substituents increase the acidity (decrease the pK_a values) of substituted phenols,

benzoic acids, and protonated anilines, whereas electron-donating substituents decrease their acidity (increase the pK_a values).

- The nature of the substituent affects the placement of an incoming substituent. All activating substituents and the weakly deactivating halogens are **ortho–para directors;** all substituents more deactivating than the halogens are **meta directors.**

- Ortho–para directors, with the exception of alkyl, aryl, and CH=CHR, have a lone pair on the atom attached to the ring; meta directors have a positive or partial positive charge on the atom attached to the ring.

- Ortho–para-directing substituents form the para isomer preferentially if either the substituent or the incoming electrophile is large.

- When planning the synthesis of a disubstituted benzene, the order in which the substituents are placed on the ring and the point in a reaction sequence at which a substituent is chemically modified are important considerations.

- When a disubstituted benzene undergoes an electrophilic aromatic substitution reaction, the directing effect of both substituents has to be considered.

- Benzene rings with meta-directing substituents cannot undergo Friedel–Crafts reactions.

- The kinds of substituents that can be placed on a benzene ring are expanded by reactions of arene diazonium salts and nucleophilic aromatic substitution reactions.

- Aniline reacts with nitrous acid to form an **arenediazonium salt;** a diazonium group can be displaced by a nucleophile.

- Arenediazonium ions can be used as electrophiles with highly activated benzene rings to form **azo compounds** that can exist in cis and trans forms.

- An aryl halide with one or more strongly electron-withdrawing groups ortho or para to the leaving group can undergo a **nucleophilic aromatic substitution** (S_NAr) reaction. The incoming nucleophile must be a stronger base than the halide ion that is replaced.

- A substituent that deactivates a benzene ring toward electrophilic substitution activates it toward nucleophilic substitution and vice versa.

SUMMARY OF REACTIONS

1. Electrophilic aromatic substitution reactions
   a. Halogenation (Section 4). The mechanisms are shown earlier in the chapter.

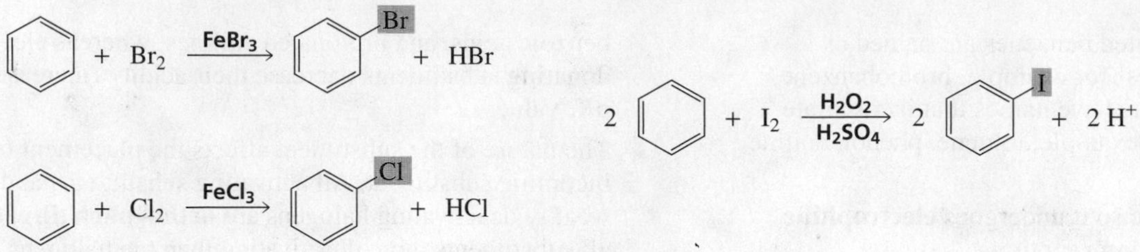

   b. Nitration, sulfonation, and desulfonation (Sections 5 and 6).

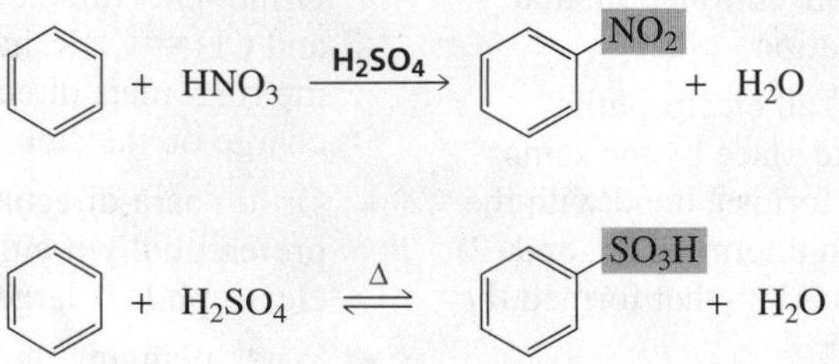

   c. Friedel–Crafts acylation and alkylation (Sections 7 and 8).

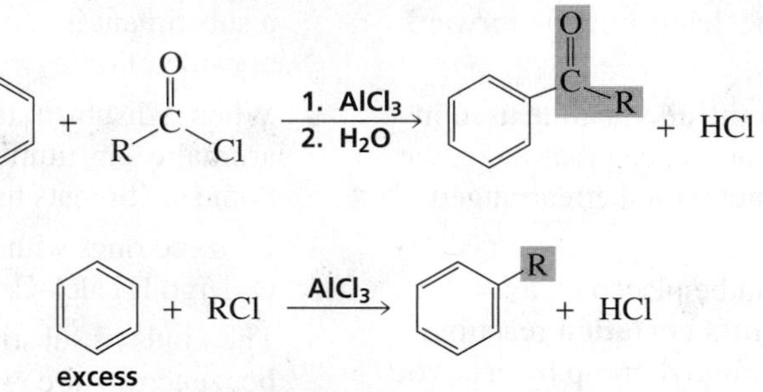

   d. Formation of benzaldehyde by a Gatterman–Koch reaction (Section 8).

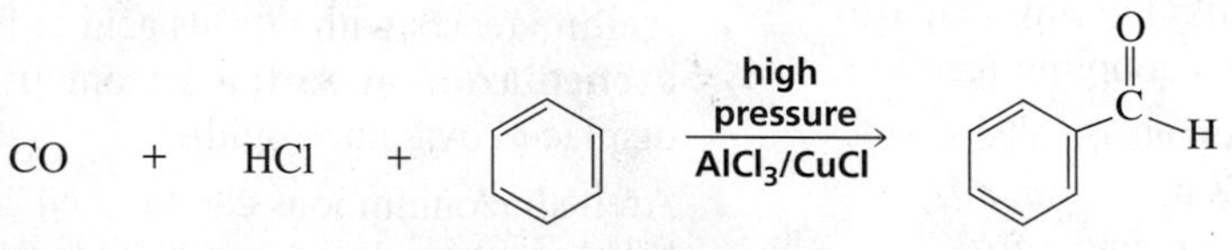

2. Reduction of a carbonyl group to a methylene group (Section 9).

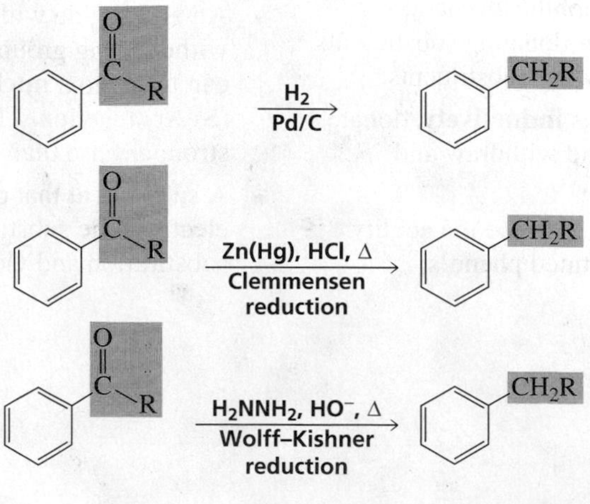

3. Alkyation via coupling reactions (Section 10)
   a. Alkylation by a Suzuki reaction

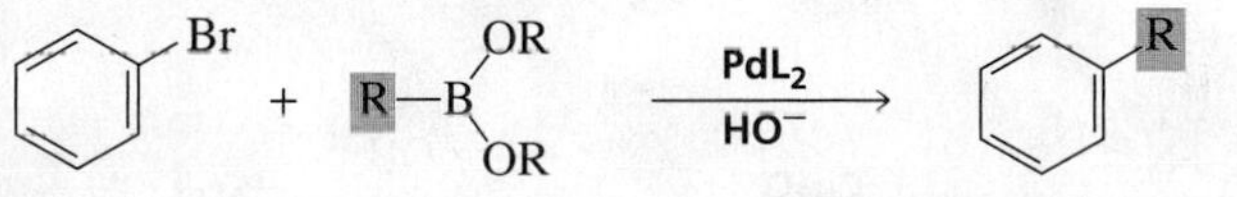

   b. Alkylation with an organocuprate

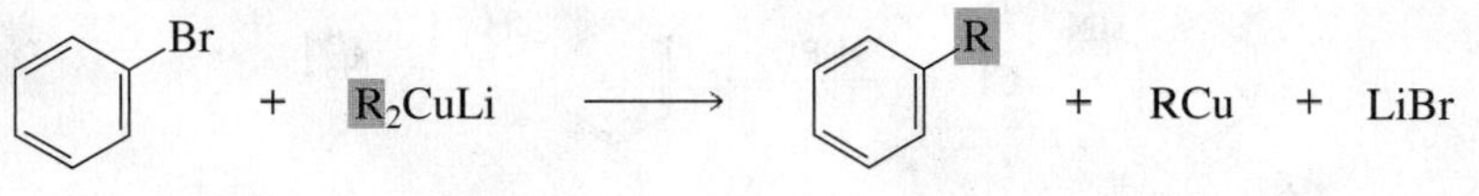

4. Reactions of a substituent on a benzene ring (Section 12)

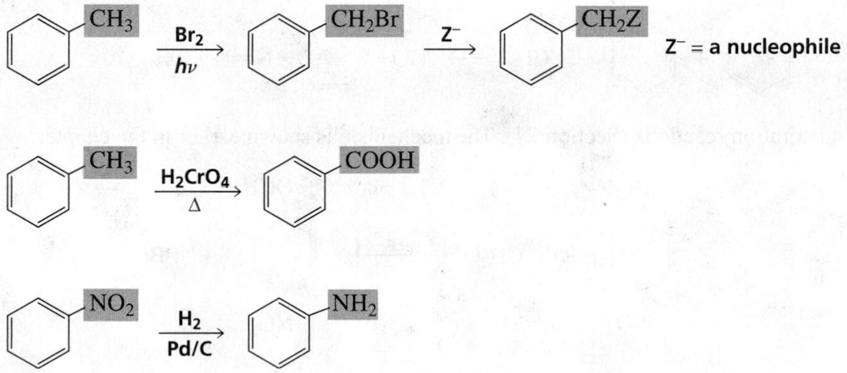

Z^- = a nucleophile

5. Reaction of aniline with nitrous acid (Section 23). The mechanism is shown earlier in the chapter.

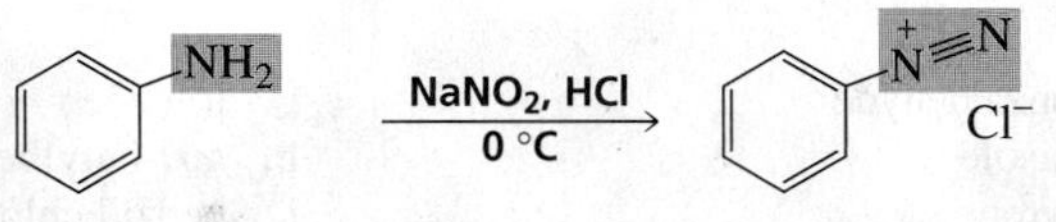

6. Replacement of a diazonium group (Section 21)

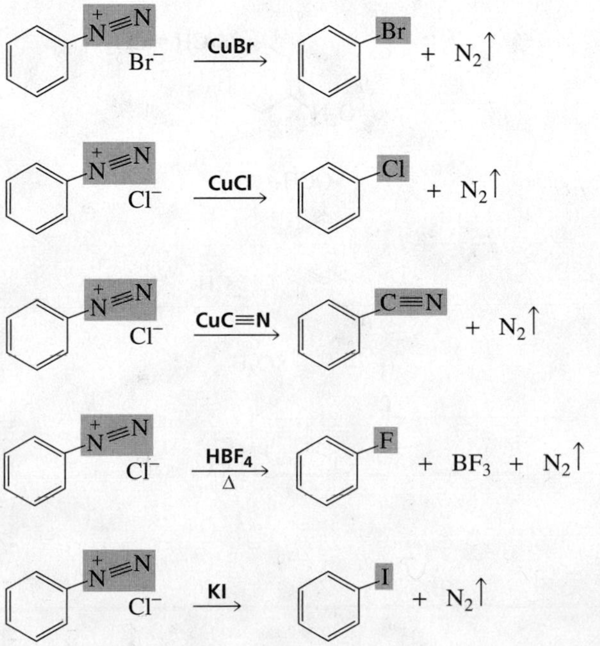

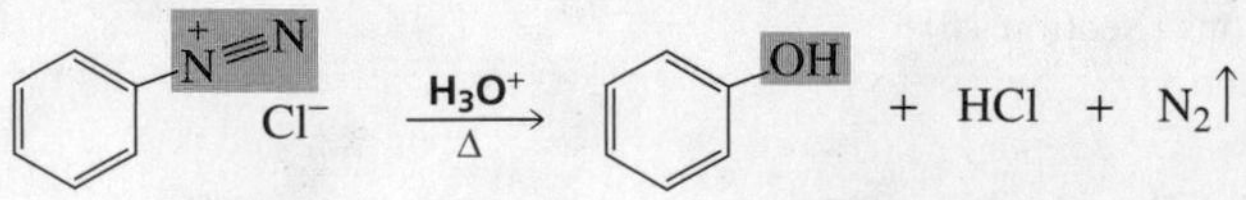

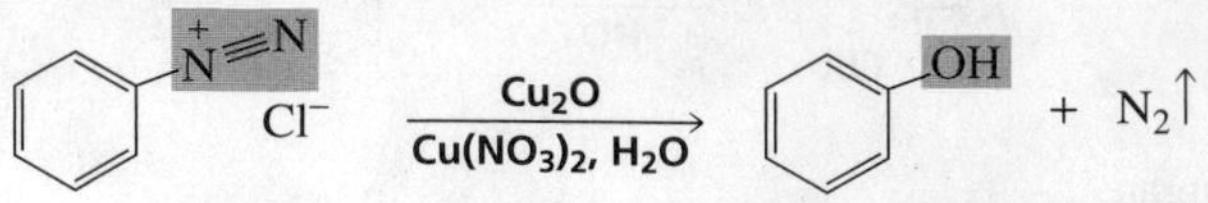

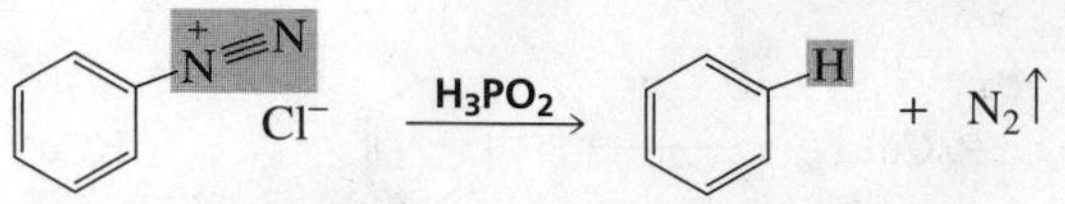

7. Formation of an azo compound (Section 22). The mechanism is shown earlier in the chapter.

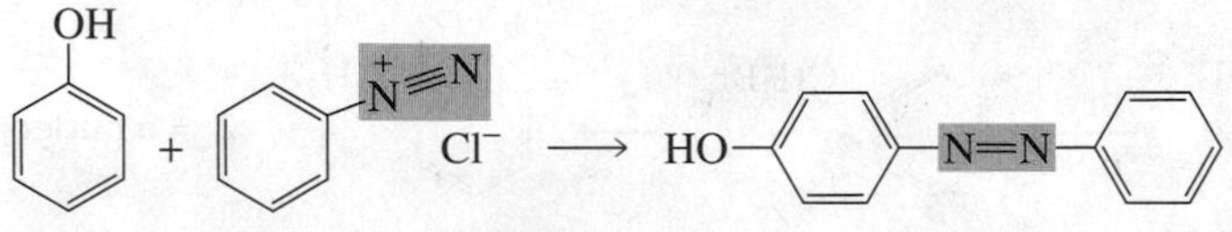

8. Nucleophilic aromatic substitution reactions (Section 24). The mechanism is shown earlier in the chapter.

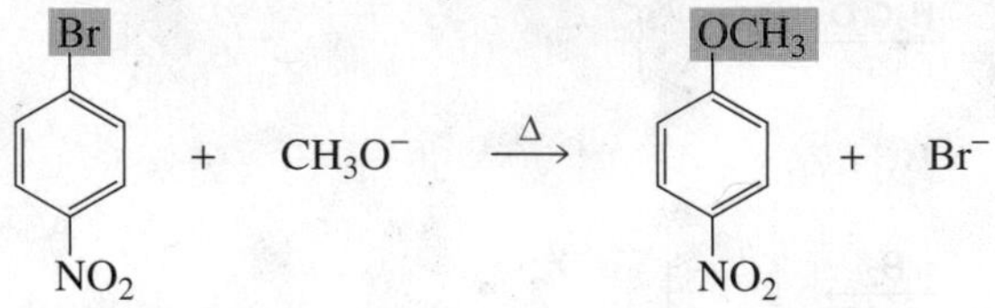

PROBLEMS

46. Draw the structure for each of the following:

a. phenol	**d.** benzaldehyde	**g.** toluene
b. benzyl phenyl ether	**e.** anisole	**h.** *tert*-butylbenzene
c. benzonitrile	**f.** styrene	**i.** benzyl chloride

47. Name the following:

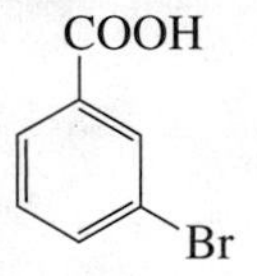

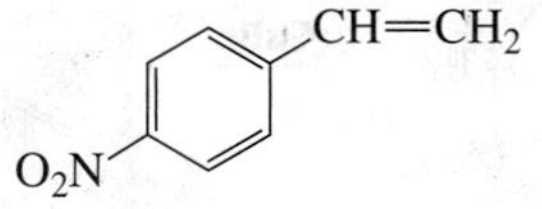

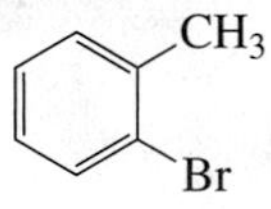

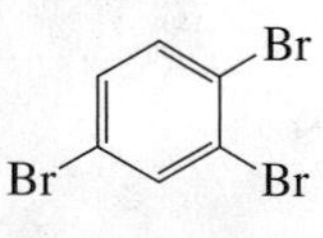

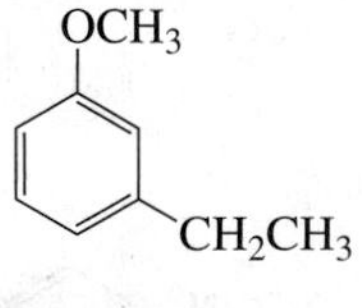

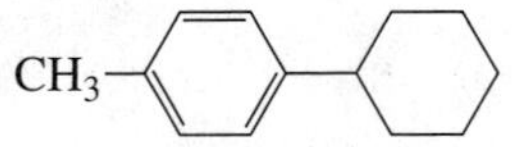

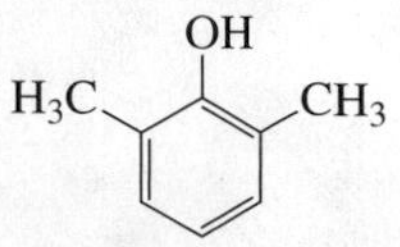

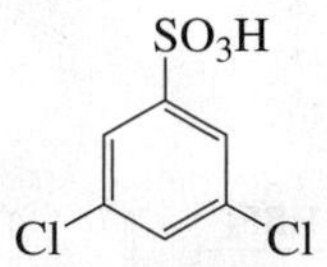

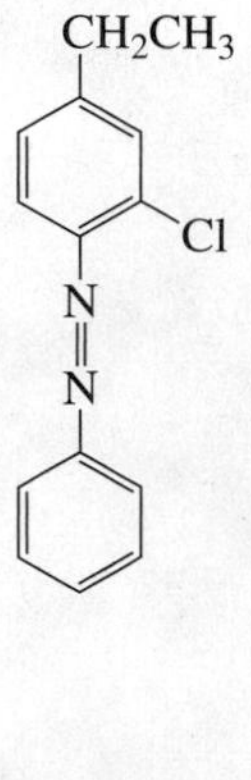

48. Provide the necessary reagents next to the arrows.

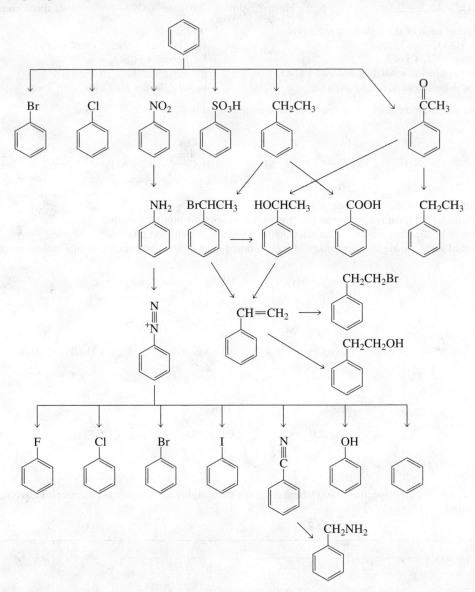

49. Draw the structure for each of the following:
  a. *m*-ethylphenol
  b. *p*-nitrobenzenesulfonic acid
  c. (*E*)-2-phenyl-2-pentene
  d. *o*-bromoaniline
  e. 4-bromo-1-chloro-2-methylbenzene
  f. *m*-chlorostyrene
  g. *o*-nitroanisole
  h. 2,4-dichloromethylbenzene
  i. *m*-chlorobenzoic acid

50. For each of the statements in Column I, choose a substituent from Column II that fits the description for the compound on the right:

Column I	Column II
a. Z donates electrons by hyperconjugation and does not donate or withdraw electrons by resonance.	OH
b. Z withdraws electrons inductively and withdraws electrons by resonance.	Br
c. Z deactivates the ring and directs ortho–para.	$^{+}NH_3$
d. Z withdraws electrons inductively, donates electrons by resonance, and activates the ring.	CH_2CH_3
e. Z withdraws electrons inductively and does not donate or withdraw electrons by resonance.	NO_2

51. What product would be obtained from the reaction of excess benzene with each of the following reagents?
   a. isobutyl chloride + AlCl₃   **b.** 1-chloro-2,2-dimethylpropane + AlCl₃   **c.** dichloromethane + AlCl₃

52. Draw the product(s) of each of the following reactions:
   a. benzoic acid + HNO₃/H₂SO₄   **e.** cyclohexyl phenyl ether + Br₂
   b. isopropylbenzene + Cl₂ + FeCl₃   **f.** phenol + H₂SO₄ + Δ
   c. *p*-xylene + acetyl chloride + AlCl₃ followed by H₂O   **g.** ethylbenzene + Br₂/FeBr₃
   d. *o*-methylaniline + benzenediazonium chloride   **h.** *m*-xylene + Na₂Cr₂O₇ + HCl + Δ

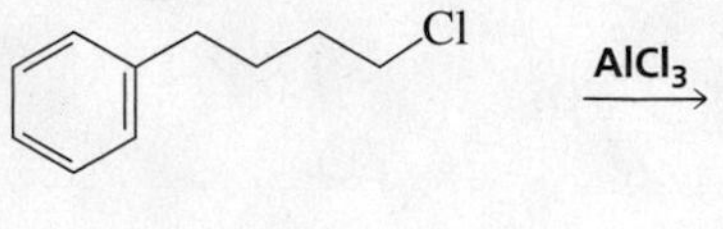

53. Rank the following substituted anilines in order from most basic to least basic:

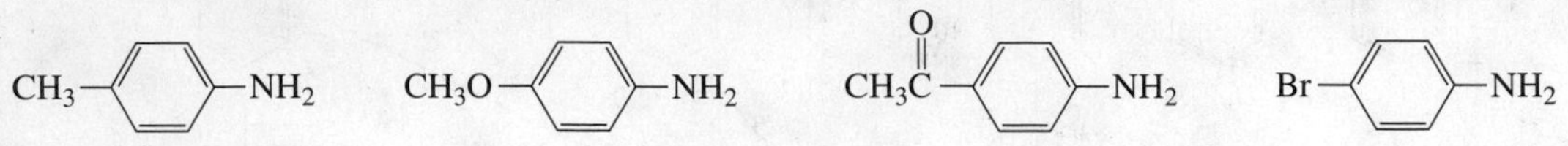

$$CH_3-\!\!\!\bigcirc\!\!\!-NH_2 \qquad CH_3O-\!\!\!\bigcirc\!\!\!-NH_2 \qquad CH_3\overset{O}{\overset{\|}{C}}-\!\!\!\bigcirc\!\!\!-NH_2 \qquad Br-\!\!\!\bigcirc\!\!\!-NH_2$$

54. For each horizontal row of substituted benzenes, indicate
   a. the one that would be the most reactive in an electrophilic aromatic substitution reaction.
   b. the one that would be the least reactive in an electrophilic aromatic substitution reaction.
   c. the one that would yield the highest percentage of meta product in an electrophilic aromatic substitution reaction.

$$\bigcirc\!\!-CH_3 \qquad \bigcirc\!\!-CHF_2 \qquad \bigcirc\!\!-CF_3$$

$$\bigcirc\!\!-\overset{+}{N}(CH_3)_3 \qquad \bigcirc\!\!-CH_2\overset{+}{N}(CH_3)_3 \qquad \bigcirc\!\!-CH_2CH_2\overset{+}{N}(CH_3)_3$$

$$\bigcirc\!\!-OCH_2CH_3 \qquad \bigcirc\!\!-CH_2OCH_3 \qquad \bigcirc\!\!-\overset{O}{\overset{\|}{C}}OCH_3$$

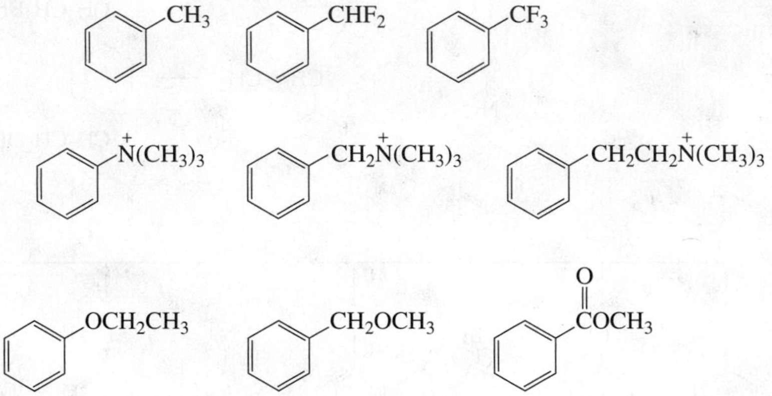

55. The compound with the ¹H NMR spectrum shown here is known to be highly reactive toward electrophilic aromatic substitution. Identify the compound.

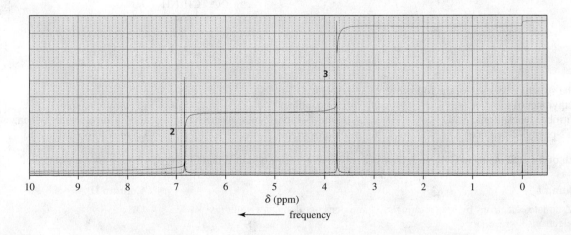

56. Draw the product of each of the following reactions:

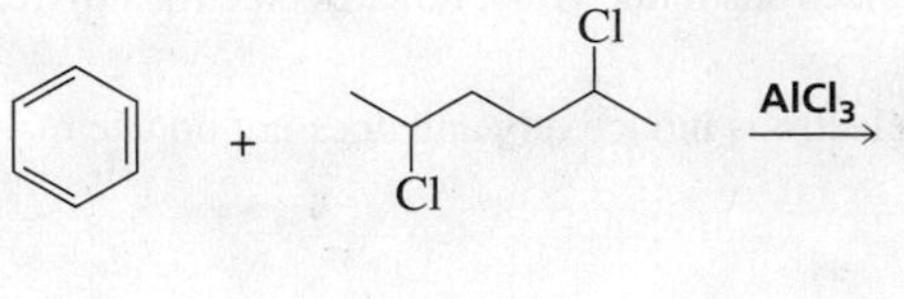

57. Show how the following compounds could be synthesized from benzene:
  a. *m*-chlorobenzenesulfonic acid
  b. *m*-chloroethylbenzene
  c. *m*-bromobenzonitrile
  d. 1-phenylpentane
  e. *m*-bromobenzoic acid
  f. *m*-hydroxybenzoic acid
  g. *p*-cresol
  h. benzyl alcohol
  i. benzylamine

58. How many electrophilic aromatic substitution products are obtained from the chlorination of
  a. *o*-xylene?
  b. *p*-xylene?
  c. *m*-xylene?

59. Arrange the following groups of compounds in order from most reactive to least reactive toward electrophilic aromatic substitution:
  a. benzene, ethylbenzene, chlorobenzene, nitrobenzene, anisole
  b. 1-chloro-2,4-dinitrobenzene, 2,4-dinitrophenol, 1-methyl-2,4-dinitrobenzene
  c. toluene, *p*-cresol, benzene, *p*-xylene
  d. benzene, benzoic acid, phenol, propylbenzene
  e. *p*-methylnitrobenzene, 2-chloro-1-methyl-4-nitrobenzene, 1-methyl-2,4-dinitrobenzene, *p*-chloromethylbenzene
  f. bromobenzene, chlorobenzene, fluorobenzene, iodobenzene

60. What are the products of the following reactions?

a. + HNO$_3$ $\xrightarrow[\Delta]{H_2SO_4}$

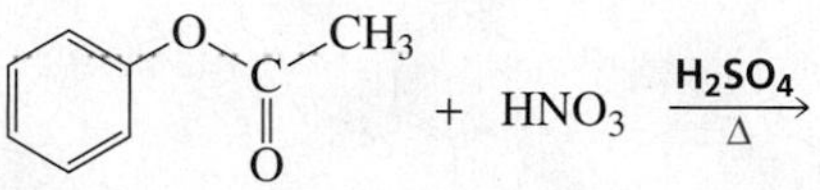

b. $\xrightarrow{\text{1. Mg/Et}_2\text{O} \quad \text{2. D}_2\text{O}}$

c. + $\xrightarrow{\text{1. AlCl}_3 \quad \text{2. H}_2\text{O}}$

d. + Br$_2$ $\longrightarrow$

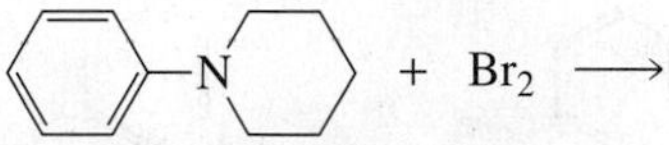

e. $\xrightarrow{\begin{array}{l}\text{1. Br}_2,\, h\nu \\ \text{2. Mg/Et}_2\text{O} \\ \text{3. ethylene oxide} \\ \text{4. HCl}\end{array}}$

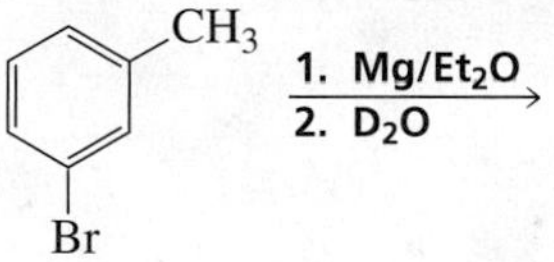

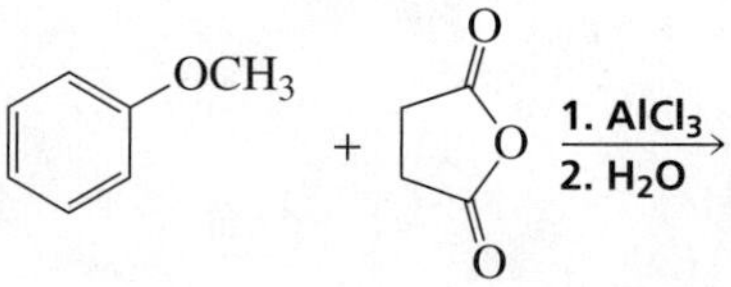

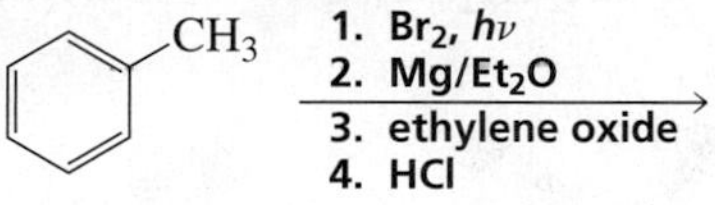

f. + Cl$_2$ $\xrightarrow{\text{FeCl}_3}$

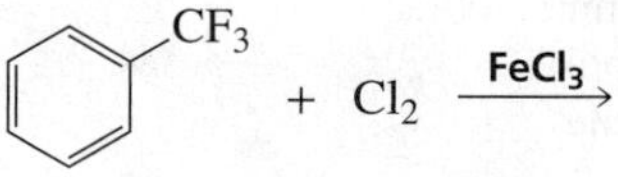

61. Describe two ways to prepare anisole from benzene.

62. For each of the following compounds, indicate the ring carbon that would be nitrated if the compound is treated with HNO$_3$/H$_2$SO$_4$:

a. 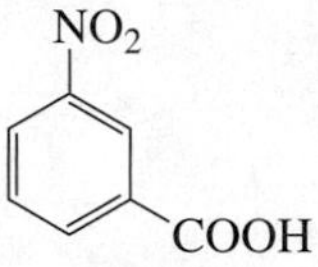
b. 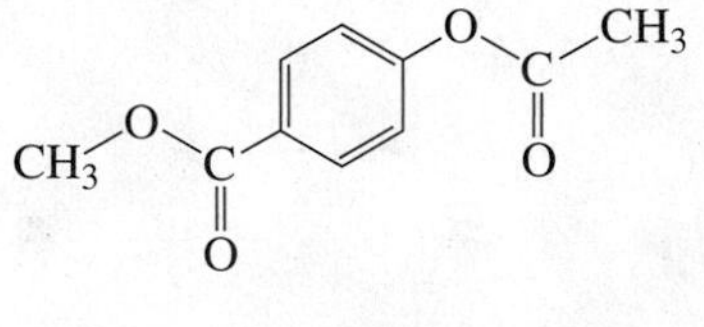
c. 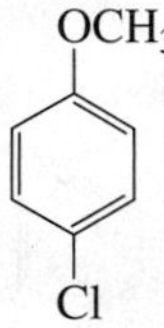
d. 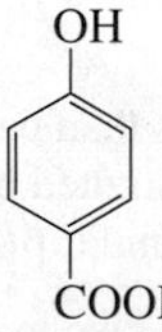
e. 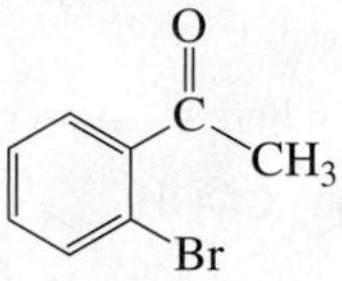
f. 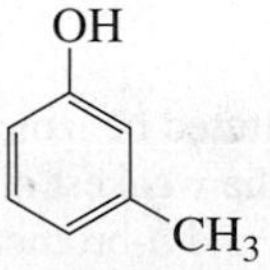
g. 
h.

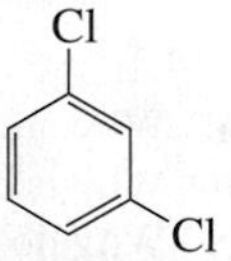

63. Show two ways that the following compound could be synthesized:

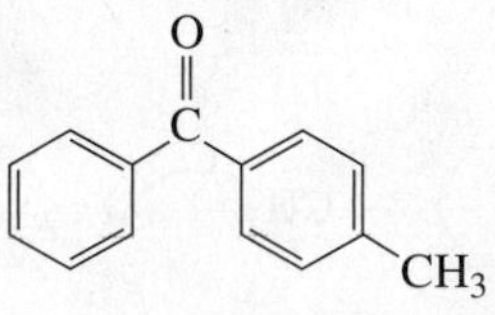

64. Why is anisole nitrated more rapidly than thioanisole under the same conditions?

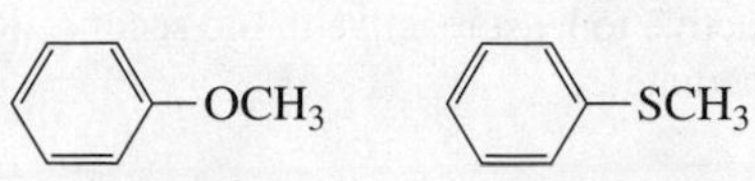

anisole   thioanisole

993

65. If anisole is allowed to sit in D_2O that contains a small amount of D_2SO_4, what products will be formed?

66. Which of the following compounds will react with HBr more rapidly?

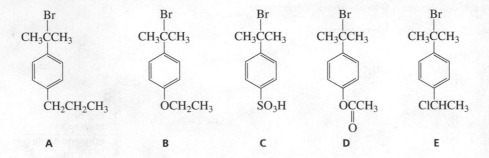

67. An aromatic hydrocarbon with a molecular formula of $C_{13}H_{20}$ has an ^1H NMR spectrum with a signal at ~7 ppm that integrates to 5H. It also has two singlets; one of the singlets has 1.5 times the area of the second. What is the structure of the aromatic hydrocarbon?

68. The following tertiary alkyl bromides undergo an S_N1 reaction in aqueous acetone to form the corresponding tertiary alcohols. List the alkyl bromides in order from most reactive to least reactive.

69. Show how the following compounds could be synthesized from benzene:
  a. *N,N,N*-trimethylanilinium iodide
  b. 2-methyl-4-nitrophenol
  c. *p*-benzylchlorobenzene
  d. benzyl methyl ether
  e. *p*-nitroaniline
  f. *m*-bromoiodobenzene
  g. *p*-dideuteriobenzene
  h. *p*-nitro-*N*-methylaniline
  i. 1-bromo-3-nitrobenzene

70. Use the four compounds shown here to answer the following questions:

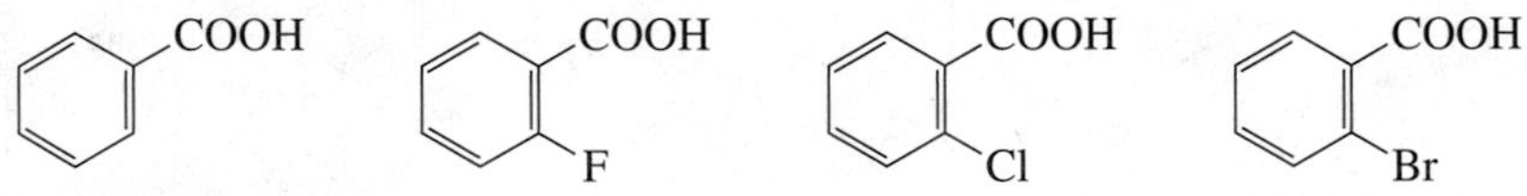

$pK_a = 4.2$ $pK_a = 3.3$ $pK_a = 2.9$ $pK_a = 2.8$

  a. Why are the *ortho*-halo-substituted benzoic acids stronger acids than benzoic acid?
  b. Why is *o*-fluorobenzoic acid the weakest of the *ortho*-halo-substituted benzoic acids?
  c. Why do *o*-chlorobenzoic acid and *o*-bromobenzoic acid have similar pK_a values?

71. a. List the following esters in order from most reactive to least reactive in the first slow step of a nucleophilic addition–elimination reaction (formation of the tetrahedral intermediate):

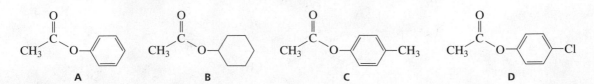

  b. List the same esters in order from most reactive to least reactive in the second slow step of a nucleophilic addition–elimination reaction (collapse of the tetrahedral intermediate).

72. A mixture of 0.10 mol benzene and 0.10 mol *p*-xylene was allowed to react with 0.10 mol nitronium ion until all the nitronium ion was gone. Two products were obtained: 0.002 mol of one and 0.098 mol of the other.
    a. What was the major product?
    b. Why was more of one product obtained than of the other?

73. What are the products of the following reactions?

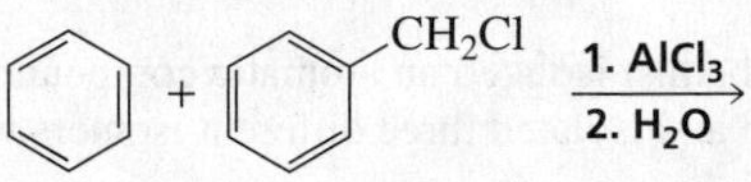

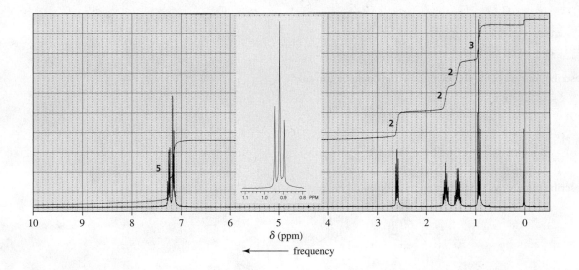

74. Benzene underwent a Friedel–Crafts acylation followed by a Clemmensen reduction. The product gave the following ¹H NMR spectrum. What acyl chloride was used in the Friedel–Crafts acylation?

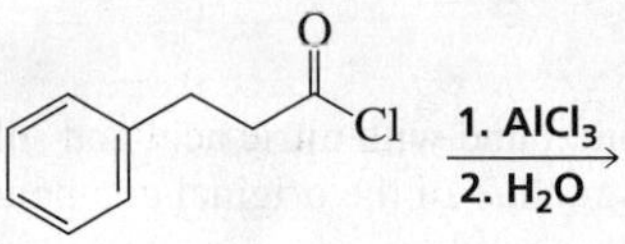

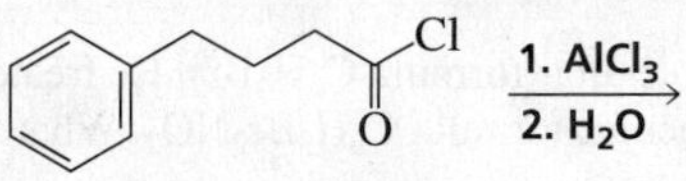

75. Would *m*-xylene or *p*-xylene react more rapidly with $Cl_2 + FeCl_3$? Explain your answer.

76. What products would be obtained from the reaction of the following compounds with $H_2CrO_4 + \Delta$?

  a. 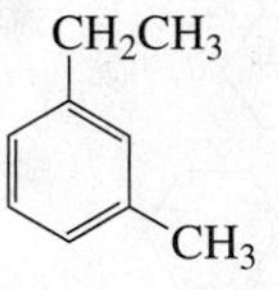 3-ethyl-methylbenzene
  b. 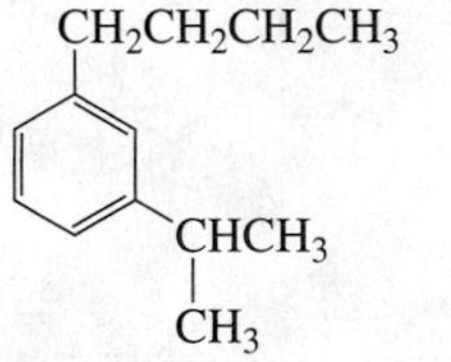 1-butyl-3-isopropylbenzene
  c. 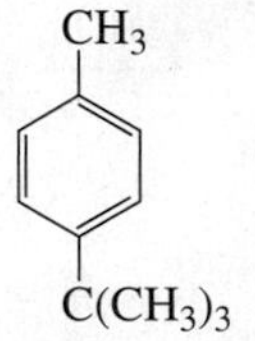 4-tert-butyltoluene

77. Which set of underlined hydrogens will have its ¹H NMR signal at a higher frequency?

    a. $CH_3CH_2C\underline{H}_3$ or $CH_3OCH_2C\underline{H}_3$ **b.** $CH_3CH{=}C\underline{H}_2$ or $CH_3OCH{=}C\underline{H}_2$

78. Friedel–Crafts alkylations can be carried out with carbocations formed from reactions other than the reaction of an alkyl halide with $AlCl_3$. Propose a mechanism for the following reaction:

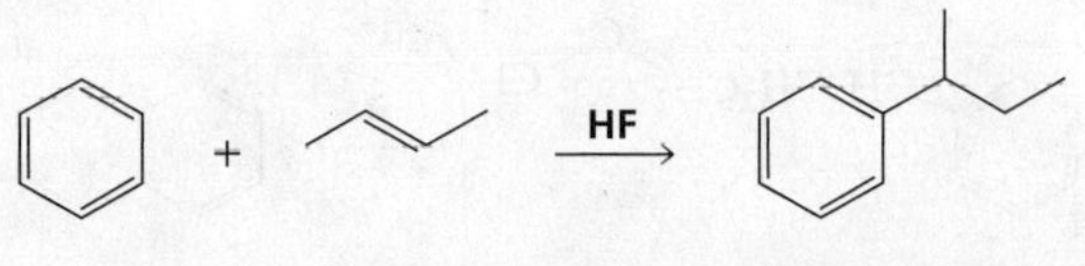

79. Show how the following compounds could be prepared from the given starting materials. You can use any necessary organic or inorganic reagents.

a.

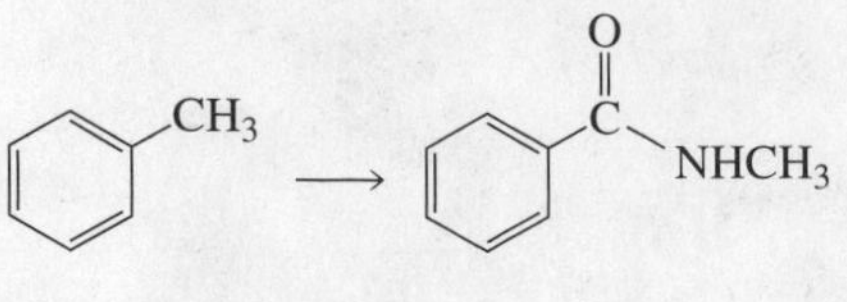

b.

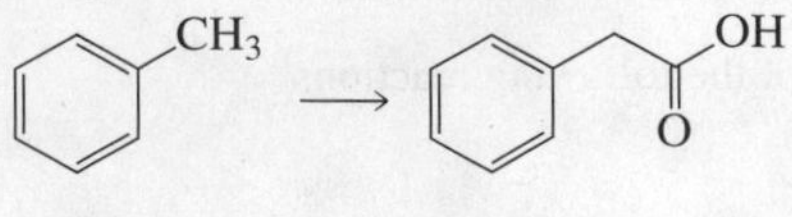

80. A chemist isolated an aromatic compound with molecular formula $C_6H_4Br_2$. He treated this compound with nitric acid and sulfuric acid and isolated three different isomers with molecular formula $C_6H_3Br_2NO_2$. What was the structure of the original compound?

81. List the following compounds in order from largest K_{eq} to smallest K_{eq} for hydrate formation:

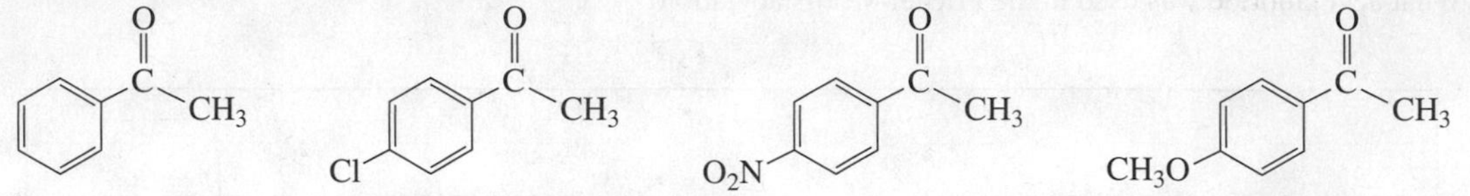

82. a. Describe four ways the following reaction could be carried out.

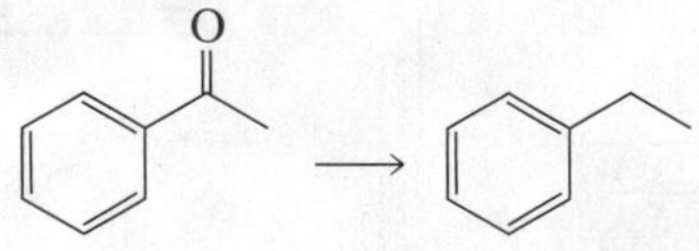

b. Describe three ways the following reaction could be carried out.

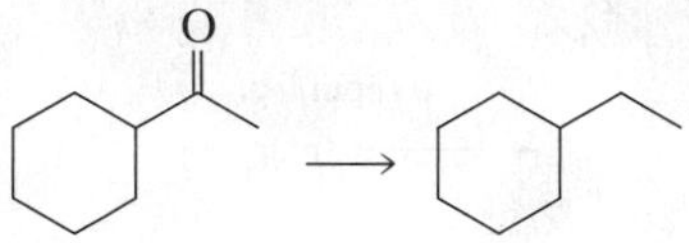

83. Propose a mechanism for each of the following reactions:

a.

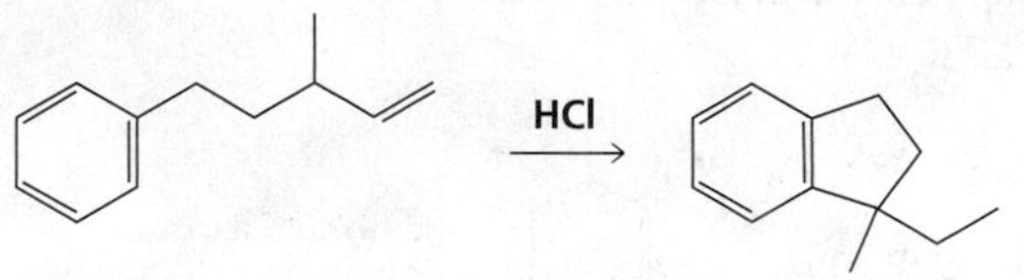

b.

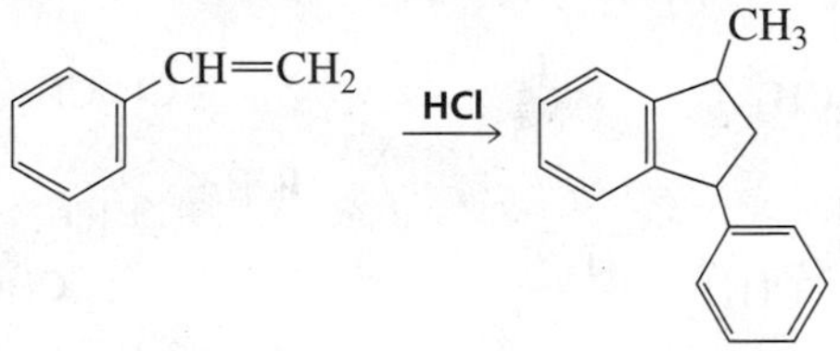

84. How could you prepare the following compounds with benzene as one of the starting materials?

a.

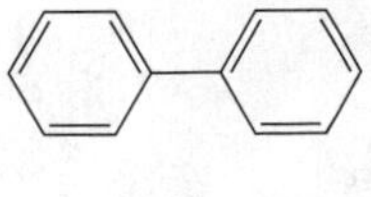

b.

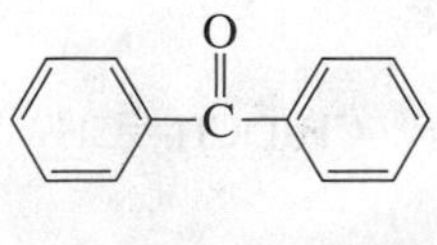

85. Describe how naphthalene could be prepared from the given starting material.

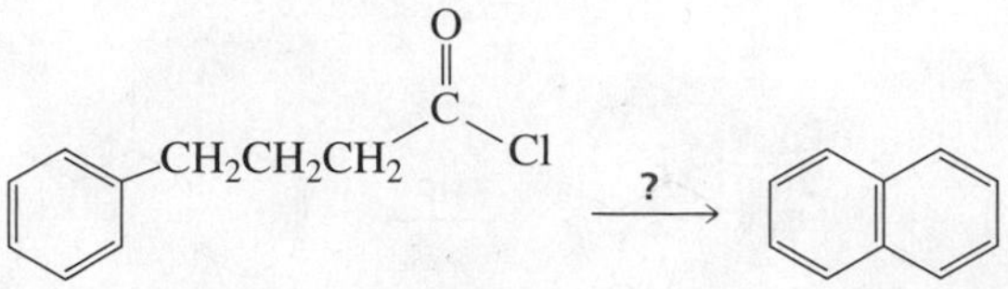

86. Identify **A–J**:

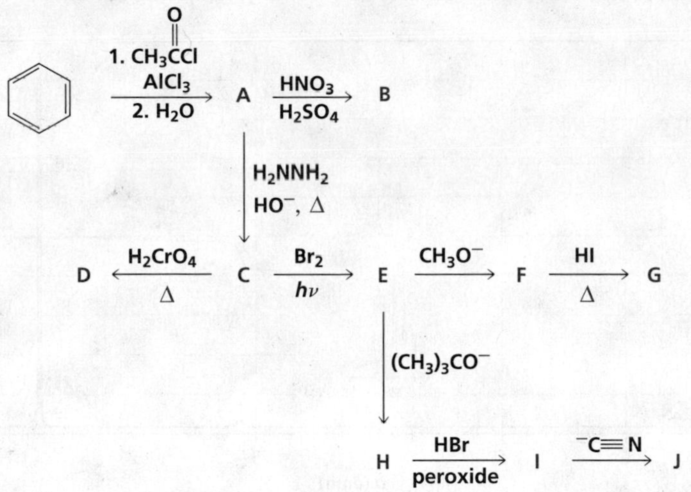

87. Using resonance contributors for the intermediate carbocation, explain why a phenyl group is an ortho–para director.

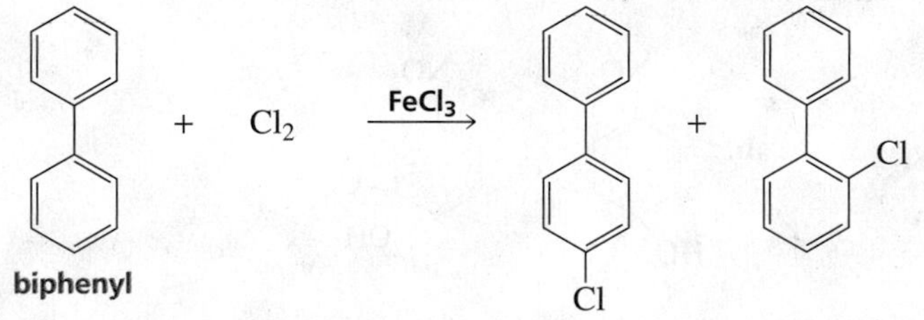

biphenyl

88. The pK_a values of a few ortho-, meta-, and para-substituted benzoic acids are shown here:

pK_a = 2.94   pK_a = 3.83   pK_a = 3.99   pK_a = 2.17   pK_a = 3.49   pK_a = 3.44

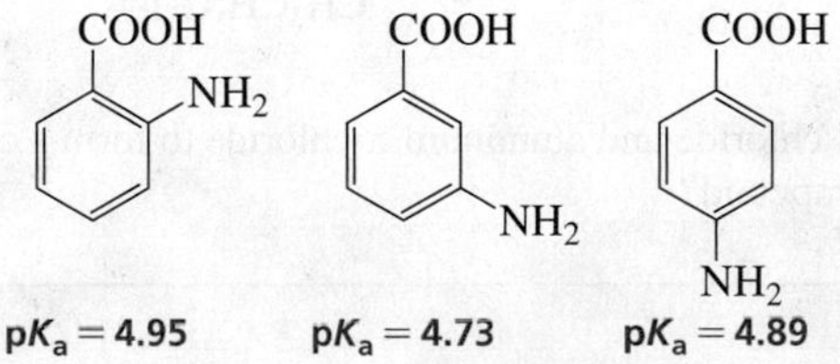

pK_a = 4.95   pK_a = 4.73   pK_a = 4.89

The relative pK_a values depend on the substituent. For chloro-substituted benzoic acids, the ortho isomer is the most acidic and the para isomer is the least acidic; for nitro-substituted benzoic acids, the ortho isomer is the most acidic and the meta isomer is the least acidic; and for amino-substituted benzoic acids, the meta isomer is the most acidic and the ortho isomer is the least acidic. Explain these relative acidities.

a. Cl: ortho > meta > para   **b.** NO$_2$: ortho > para > meta   **c.** NH$_2$: meta > para > ortho

89. When heated with chromic acid, compound **A** forms benzoic acid. Identify compound **A** from its ^{1}H NMR spectrum.

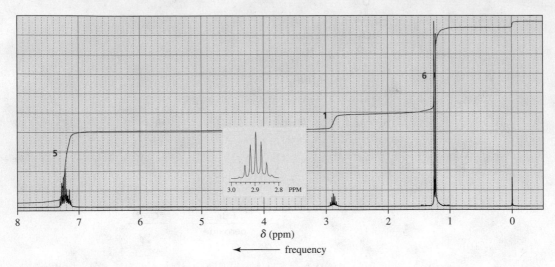

90. Describe two synthetic routes for the preparation of *p*-methoxyaniline from benzene.

91. Which is a more stable intermediate in each pair?

a. OH   CH$_3$ b. NO$_2$ NO$_2$
              or   or
    H NO$_2$ H NO$_2$ HO Cl   Cl
                                              OH

92. What reagents would be required to carry out the following transformations?

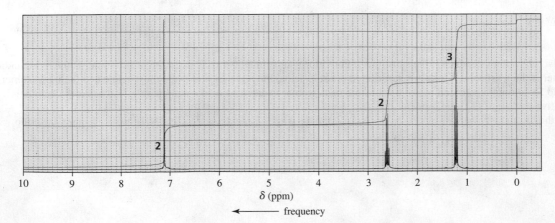

93. Show how the following compounds could be prepared from benzene:

a. CH$_3$ ... CH$_3$C=CH$_2$ b. OCH$_3$, Br, NO$_2$ c. CH$_3$-C(=O)..SO$_3$H, CH$_2$CH(CH$_3$)$_2$ d. C(CH$_3$)$_3$, CH$_3$CH$_2$, Br

94. An unknown compound reacts with ethyl chloride and aluminum trichloride to form a compound that has the following ^{1}H NMR spectrum. What is the structure of the compound?

95. How could you distinguish the following compounds, using
   a. their infrared spectra?   **b.** their ^{1}H NMR spectra?

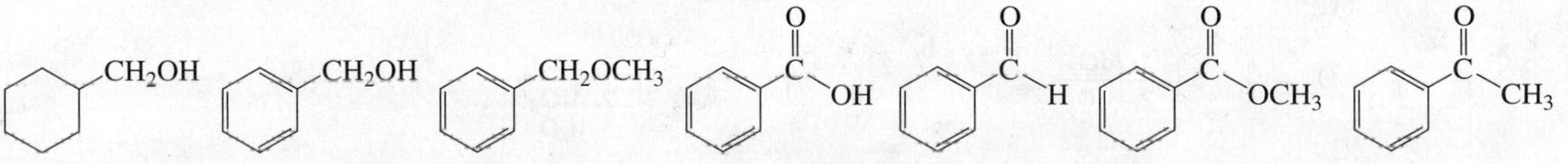

96. *p*-Fluoronitrobenzene is more reactive toward hydroxide ion than is *p*-chloronitrobenzene. What does this tell you about the rate-determining step for nucleophilic aromatic substitution?

97. **a.** Explain why the following reaction leads to the products shown:

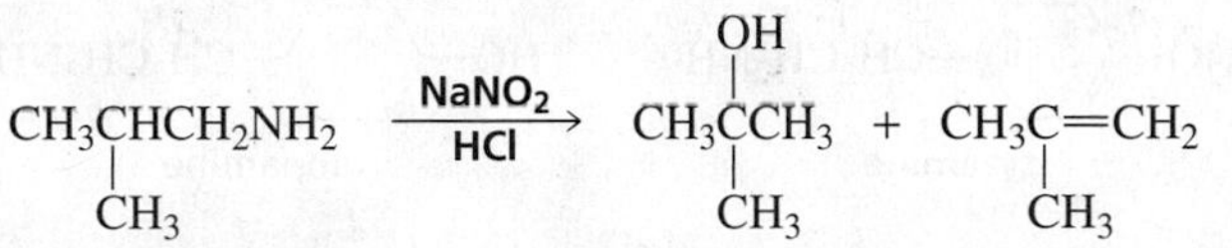

   b. What product would be obtained from the following reaction?

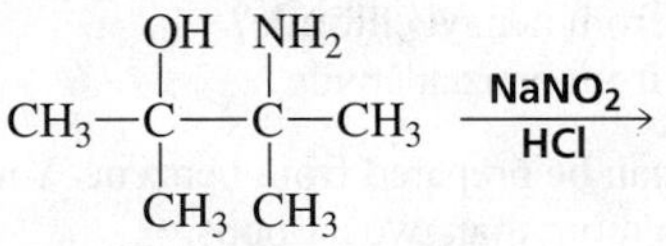

98. Describe how mescaline could be synthesized from benzene.

99. Propose a mechanism for the following reaction that explains why the configuration of the asymmetric center in the reactant is retained in the product:

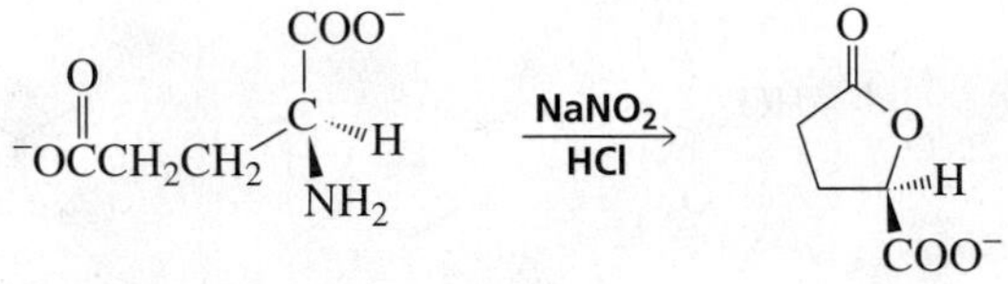

100. Explain why hydroxide ion catalyzes the reaction of piperidine with 2,4-dinitroanisole but has no effect on the reaction of piperidine with 1-chloro-2,4-dinitrobenzene.

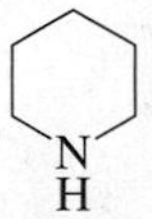

piperidine

101. Propose a reasonable mechanism for each of the following reactions:

a.

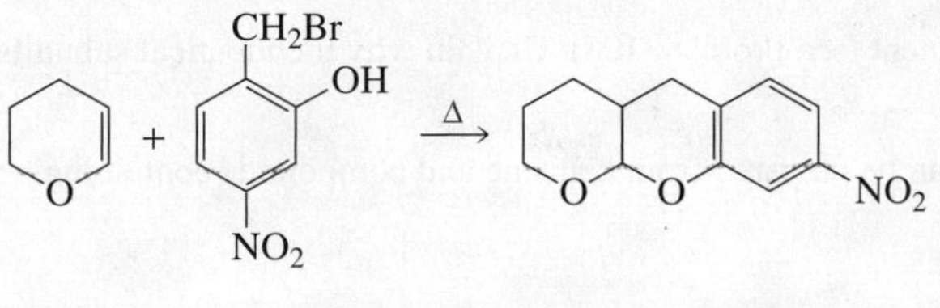

b.

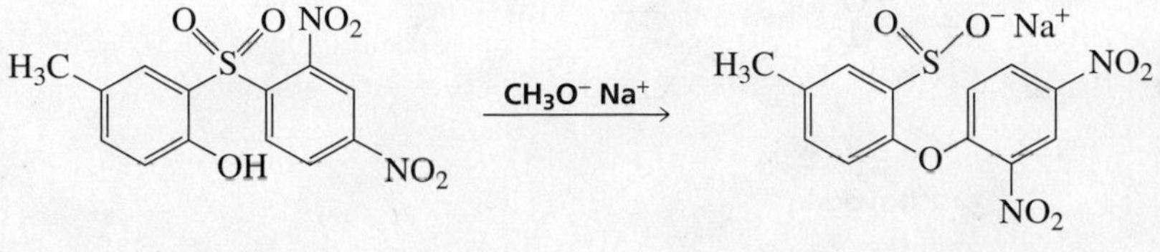

999

102. What are the products of the following reactions?

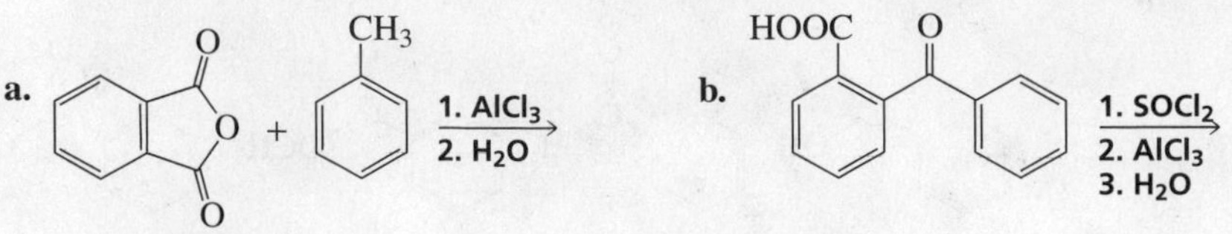

a.

1. AlCl₃
2. H₂O

b.

1. SOCl₂
2. AlCl₃
3. H₂O

103. Tyramine is an alkaloid found in mistletoe and ripe cheese. Dopamine is a neurotransmitter involved in the regulation of the central nervous system.

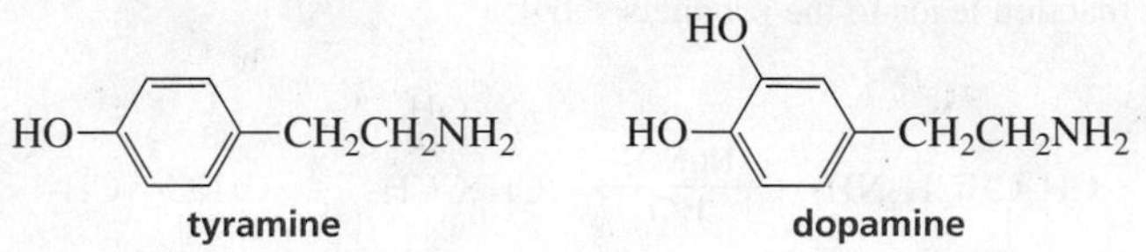

tyramine   dopamine

a. How can tyramine be prepared from β-phenylethylamine?
b. How can dopamine be prepared from tyramine?
c. Give two ways to prepare β-phenylethylamine from β-phenylethyl chloride.
d. How can β-phenylethylamine be prepared from benzyl chloride?
e. How can β-phenylethylamine be prepared from benzaldehyde?

104. Describe how 3-methyl-1-phenyl-3-pentanol can be prepared from benzene. You can use any inorganic reagents and solvents, and any organic reagents provided they contain no more than two carbons.

105. **a.** How could aspirin be synthesized from benzene?
b. Ibuprofen is the active ingredient in pain relievers such as Advil, Motrin, and Nuprin. How could ibuprofen be synthesized from benzene?
c. Acetaminophen is the active ingredient in Tylenol. How could acetominophen be synthesized from benzene?

aspirin   ibuprofen   acetaminophen

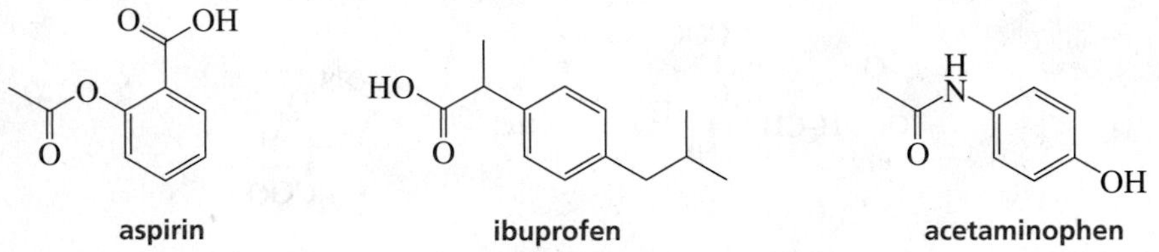

106. **a.** Ketoprofen, like ibuprofen, is an anti-inflammatory analgesic. How could ketoprofen be synthesized from the given starting material?

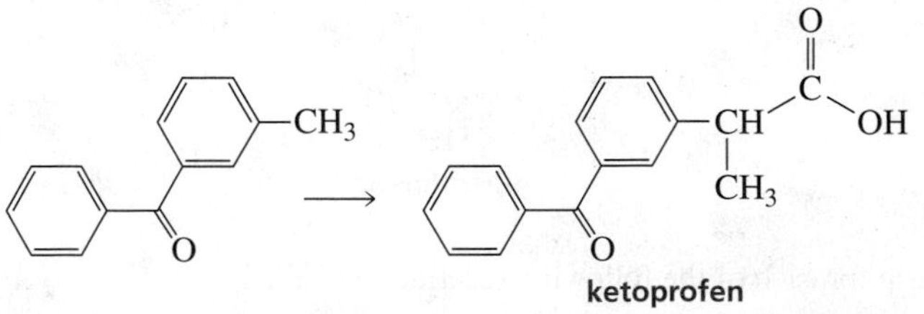

ketoprofen

b. Ketoprofen and ibuprofen both have a propanoic acid substituent (see Problem 105). Explain why the identical subunits are synthesized in different ways.

107. Show how Novocain, a painkiller used frequently by dentists, can be prepared from benzene and compounds containing no more than four carbons.

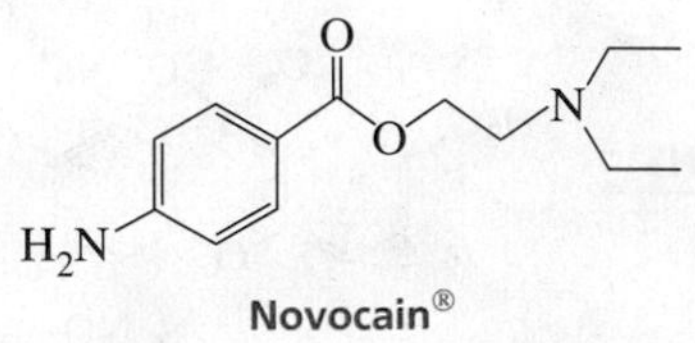

Novocain®

108. Show how lidocaine, one of the most widely used injectable anesthetics, can be prepared from benzene and compounds containing no more than four carbons.

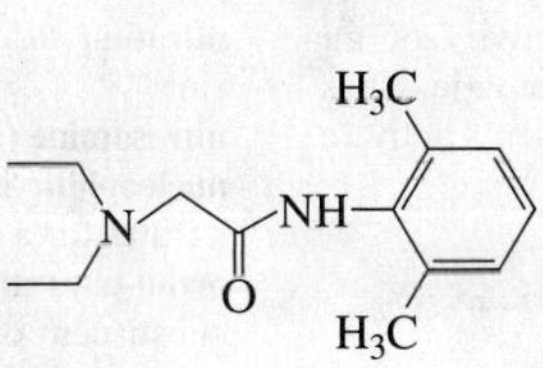

lidocaine

109. Saccharin, an artificial sweetener, is about 300 times sweeter than sucrose. Describe how saccharin could be prepared from benzene.

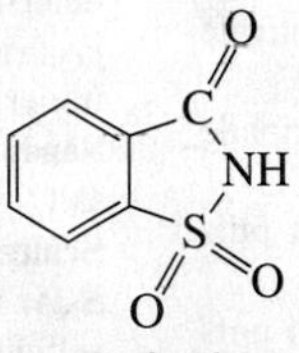

saccharin

GLOSSARY

activating substituent   a substituent that increases the reactivity of an aromatic ring. Electron-donating substituents activate aromatic rings toward electrophilic attack, and electron-withdrawing substituents activate aromatic rings toward nucleophilic attack.

azo linkage   an —N=N— bond.

benzyl group   —CH_2—

Clemmensen reduction   a reaction that reduces the carbonyl group of a ketone to a methylene group using Zn(Hg)/ HCl.

deactivating substituent   a substituent that decreases the reactivity of an aromatic ring. Electron-withdrawing substituents deactivate aromatic rings toward electrophilic attack, and electron-donating substituents deactivate aromatic rings toward nucleophilic attack.

electrophilic aromatic substitution   a reaction in which an electrophile substitutes for a hydrogen of an aromatic ring.

Friedel–Crafts acylation   an electrophilic substitution reaction that puts an acyl group on a benzene ring.

Friedel–Crafts alkylation   an electrophilic substitution reaction that puts an alkyl group on a benzene ring.

halogenation   reaction with halogen (Br_2, Cl_2, I_2).

inductive electron withdrawal   withdrawal of electrons through a σ bond.

meta-directing substituent   a substituent that directs an incoming substituent meta to an existing substituent.

nitration   substitution of a nitro group (NO_2) for a hydrogen of a benzene ring.

nitrosamine (N-nitroso compound) $R_2 NN{=}O$

nucleophilic aromatic substitution   a reaction in which a nucleophile substitutes for a substituent of an aromatic ring.

ortho-para-directing substituent   a substituent that directs an incoming substituent ortho and para to an existing substituent.

phenyl group   — C_6H_5—

principle of microscopic reversibility   states that the mechanism for a reaction in the forward direction has the same intermediates and the same rate-determining step as the mechanism for the reaction in the reverse direction.

prodrug   a compound that does not become an effective drug until it undergoes a reaction in the body

Sandmeyer reaction   the reaction of an aryl diazonium salt with a cuprous salt.

Schiemann reaction   the reaction of an arenediazonium salt with HBF_4.

S_NAr reaction   a nucleophilic aromatic substitution reaction.

sulfonation   substitution of a hydrogen of a benzene ring by a sulfonic acid group (SO_3H).

Wolff–Kishner reduction   a reaction that reduces the carbonyl group of a ketone to a methylene group with the use of NH_2NH_2/HO^-.

PHOTO CREDITS

Photo credits are listed in order of appearance.

Chee-Onn Leong/Shutterstock; pablo h. caridad/Fotolia; pressmaster/Fotolia; Baronb/Fotolia

TUTORIAL

Enhanced by MasteringChemistry®

SYNTHESIS AND RETROSYNTHETIC ANALYSIS

Organic synthesis is the preparation of organic com pounds from other organic compounds. The word *synthesis* comes from the Greek word *synthesis,* which means "a putting together." There are many aspects of organic synthesis the synthesis of a lot of organic compounds. In this tutorial, we will examine some of the strategies chemists use when designing a synthesis.

The most important factor in designing a synthesis is to have a good command of organic reactions. The more reactions you know, the better your chances of coming up with a useful synthesis. The guiding factor in planning a synthesis is to keep it as simple as possible. The simpler the synthetic plan, the greater the chance it will be successful.

CHANGING THE FUNCTIONAL GROUP

The first thing to do when designing a synthesis is to compare both the carbon skeleton and the positions of the functional group in the reactant and the product. If they both are the same, then all you need to determine is how to convert the functional group in the reactant to the functional group in the product..

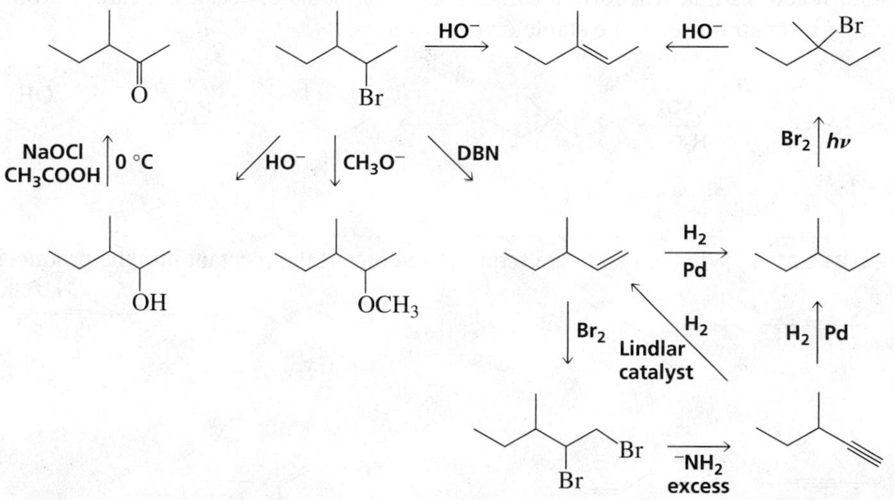

Notice that HO^- is used as the base to form a double bond between C-2 and C-3, whereas a bulky base (DBN) is needed to put the double bond between C-1 and C-2. Only two S_N2 reactions are shown here (using CH_3O^- and HO^-), but many other products could be synthesized just by changing the nucleophile. Notice, too, how single, double, and triple bonds can be interconverted.

FUNCTIONALIZING A CARBON

A carbon can be functionalized by a radical reaction.

CHANGING THE POSITION OF THE FUNCTIONAL GROUP

If the carbon skeleton has not changed but the position of the functional group has, then you need to consider reactions that will change the position of the functional group. In the electrophilic addition reactions shown here, •Br and R_2BH (which is replaced by OH) are electrophiles, so they add to the sp^2 carbon bonded to the most hydrogens.

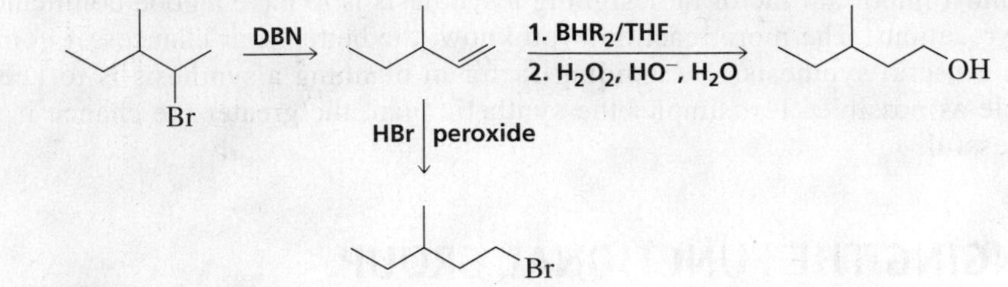

CHANGING THE CARBON SKELETON

If the carbon skeleton has changed but the number of carbons has not, then you need to consider reactions that will form a carbocation intermediate, because a carbocation will rearrange if it can form a more stable carbocation.

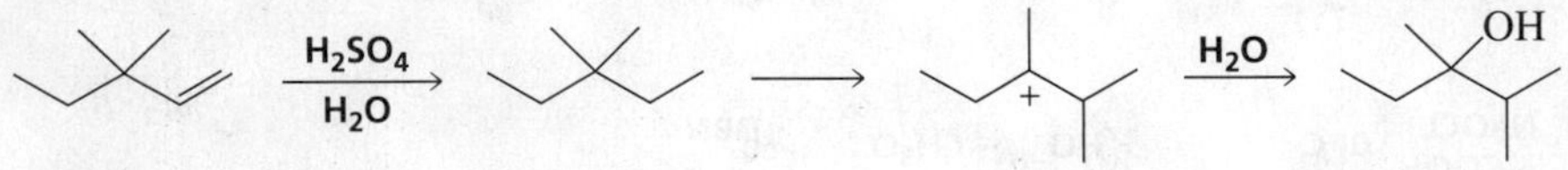

PROBLEM 1 What reagents are required to convert the reactant into the product?

a.

e.

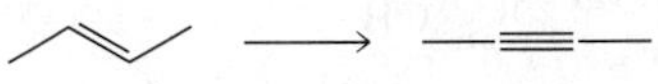

b.

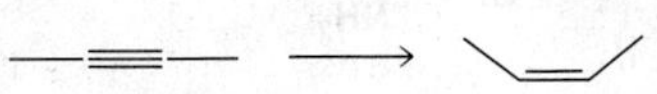

f.

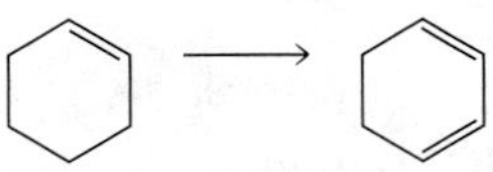

c.

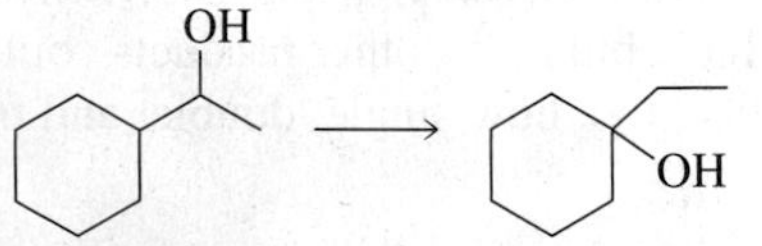

g.

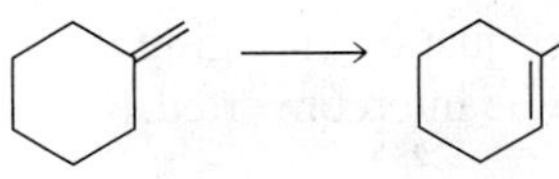

d.

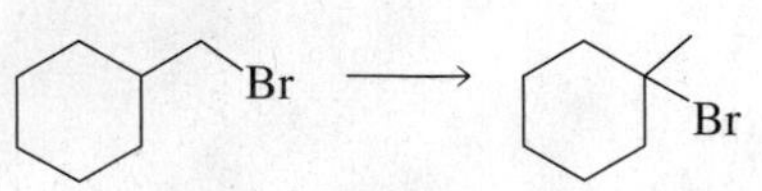

h.

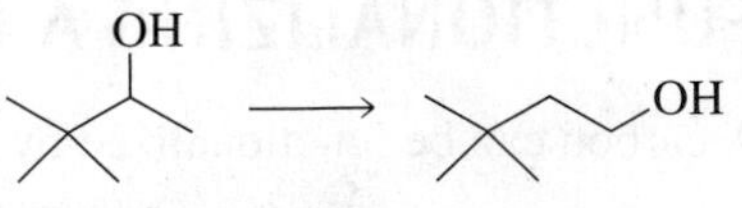

ADDING ONE CARBON TO THE CARBON SKELETON

There are several ways to create a product with one more carbon than the reactant. The method you choose will depend on the functional group that you want to end up with in the product.

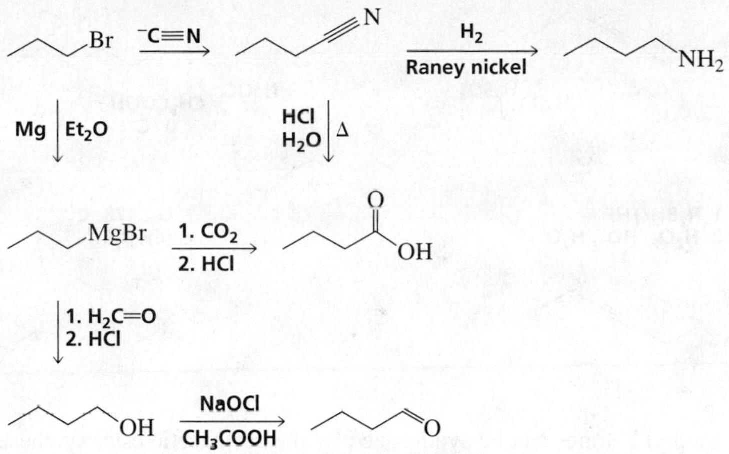

ADDING MORE THAN ONE CARBON TO THE CARBON SKELETON

There are many different ways to increase the carbon skeleton by more than one carbon. Acetylide ions, epoxides, Grignard reagents, aldol additions, Wittig reactions, and coupling reactions are just some of the methods that can be used. Common methods used to form new C—C bonds are summarized in Appendix: "Summary of Methods Employed to Form Carbon–Carbon Bonds."

PROBLEM 2 Starting with bromocyclohexane, how can each of the following compounds be prepared?

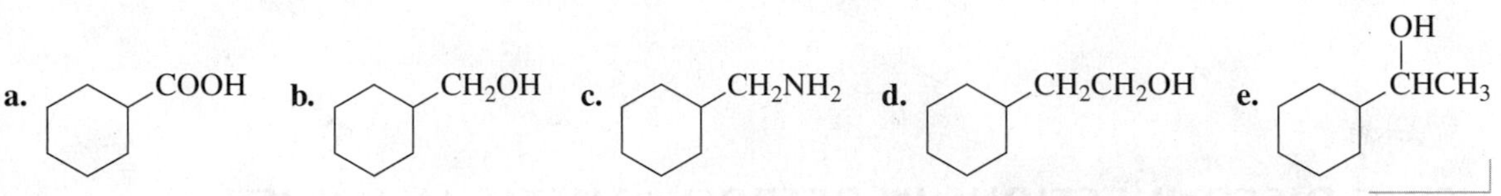

PROBLEM 3 Describe how the following compounds could be prepared from compounds containing no more than six carbons. (You can also use triphenylphosphine.)

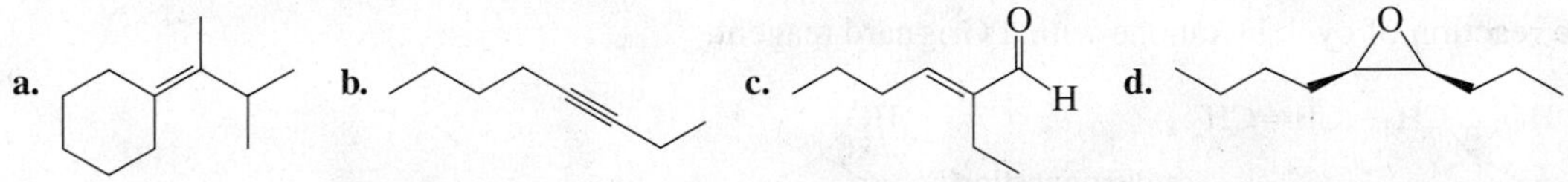

USING RETROSYNTHETIC ANALYSIS TO CREATE A FUNCTIONAL GROUP

When you know what functional group you want to create, you can try to note the various ways it can be synthesized. For example, a ketone can be synthesized by the acid-catalyzed addition of water to an alkyne, hydroboration–oxidation of an alkyne, oxidation of a secondary alcohol, and ozonolysis of an alkene. Notice that ozonolysis *decreases* the number of carbons in a molecule.

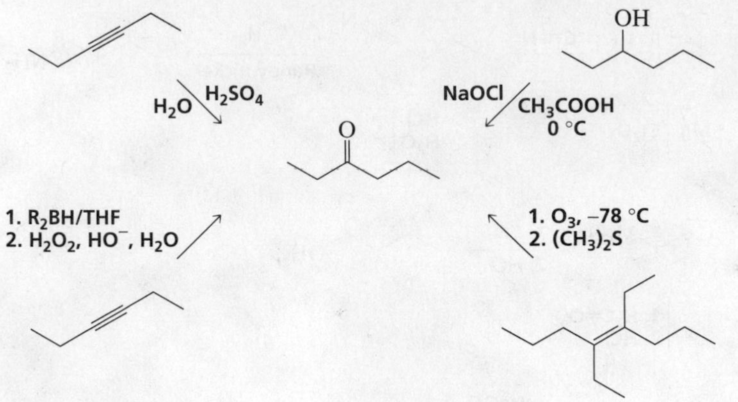

In addition, methyl ketones can be synthesized by the acetoacetic ester synthesis, aromatic ketones can be synthesized by a Friedel–Crafts acylation, and a cyclic ketone, when treated with diazomethane, forms the next-size-larger cyclic ketone. Making note of all the methods that can be used to synthesize a particular functional group might be challenging. Therefore, they are listed for you in Appendix: "Summary of Methods Used to Synthesize a Particular Functional Group."

PROBLEM 4 How many ways can you synthesize

a. an ether?　　**b.** an aldehyde?　　**c.** an alkene?　　**d.** an amine?

PROBLEM 5 Describe three ways to synthesize the following compound:

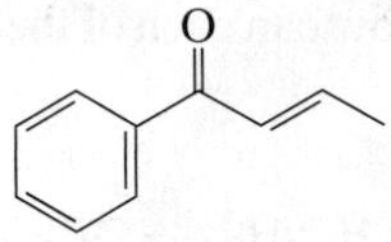

USING DISCONNECTIONS IN RETROSYNTHETIC ANALYSIS

A disconnection can be a useful step in a retrosynthetic analysis. A disconnection involves breaking a bond to give two fragments and then adding a positive charge to the end of one fragment and a negative charge to the end of the other. If the following compound is disconnected at bond **a,** then we can see that the target molecule can be prepared from the reaction of cyclohexanone with a Grignard reagent.

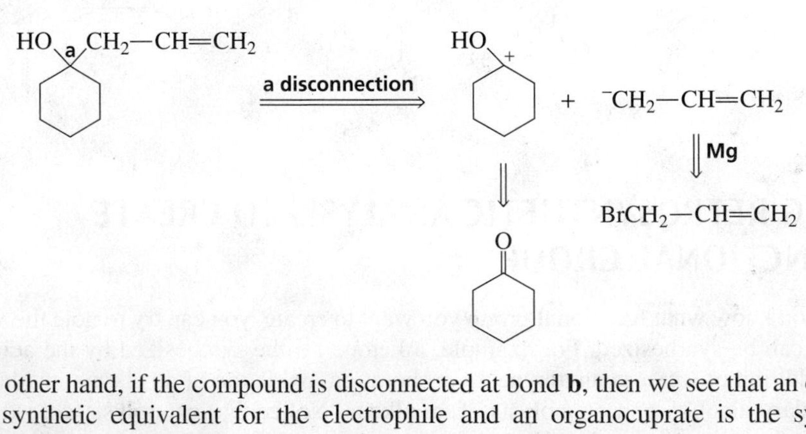

On the other hand, if the compound is disconnected at bond **b**, then we see that an epoxide is the synthetic equivalent for the electrophile and an organocuprate is the synthetic equivalent for the nucleophile.

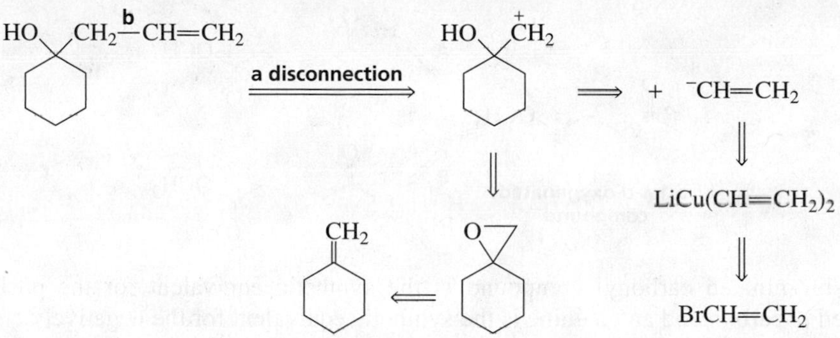

PROBLEM 6 Do a retrosynthetic analysis on each of the following compounds, ending with the given starting materials:

a.

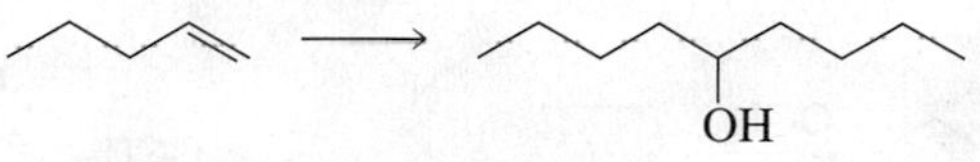

b.

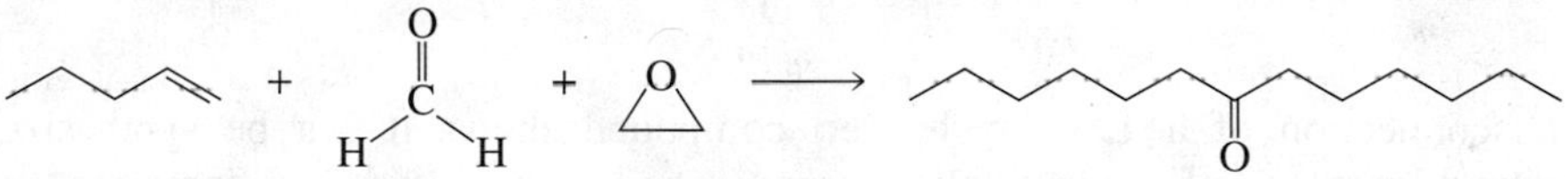

USING THE RELATIVE POSITIONS OF TWO FUNCTIONAL GROUPS TO DESIGN A SYNTHESIS

If a compound has two functional groups, then the relative positions of the two groups can provide a valuable hint as to how to approach the synthesis. For example, the following synthesis forms a **1,2-dioxygenated** compound.

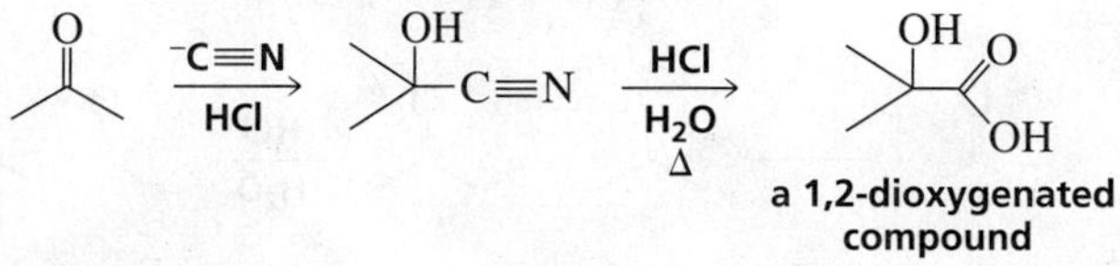

a 1,2-dioxygenated compound

Retrosynthetic analysis shows that a **1,3-dioxygenated** compound can be formed by an aldol addition.

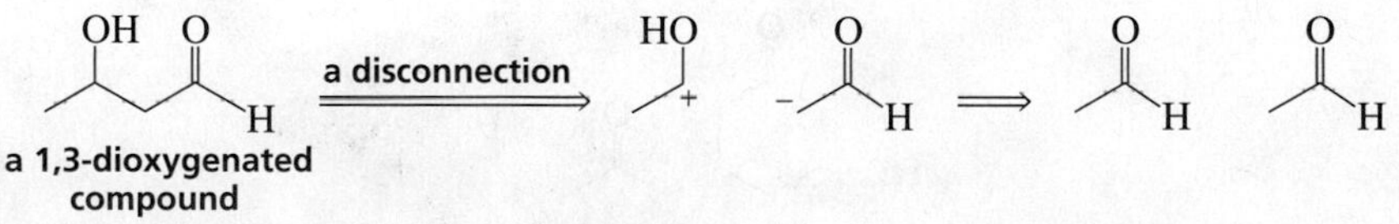

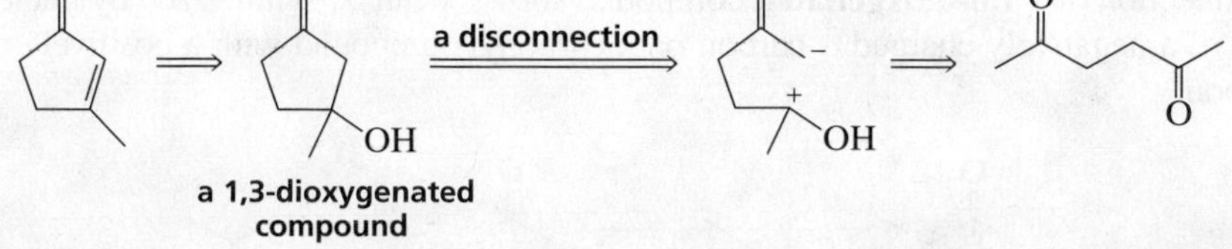

a 1,3-dioxygenated compound

a 1,3-dioxygenated compound

Disconnection of a **1,4-dioxgenated** compound shows it can be synthesized by nucleophilic attack of a negatively charged α-carbon (an enolate ion) on a compound with a positively charged α-carbon.

1007

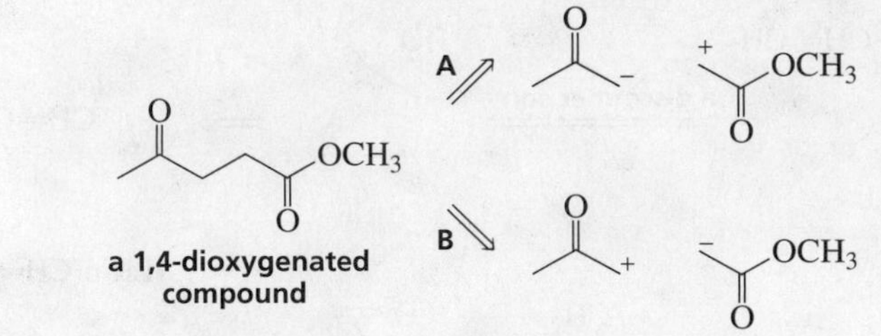

a 1,4-dioxygenated compound

An α-brominated carbonyl compound is the synthetic equivalent for the positively charged α-carbon and an enamine is the synthetic equivalent for the negatively charged α-carbon. An enamine avoids the requirement for a strong base that could remove a proton from the brominated carbonyl compound rather than from the nonbrominated carbonyl carbon. Because esters cannot form enamines, path A is the preferred disconnection. Fortunately, the iminium ion hydrolyzes more readily than the ester.

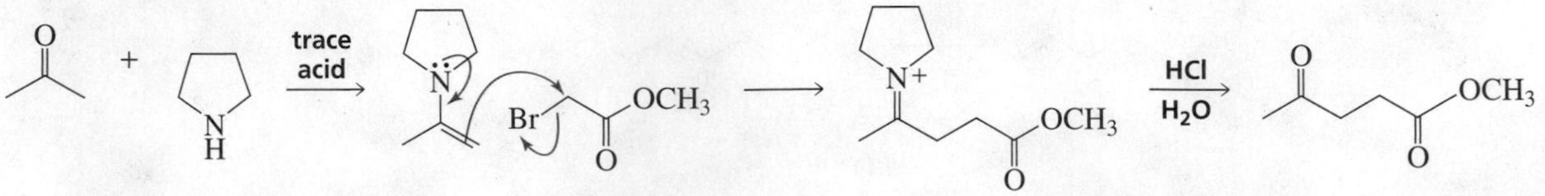

Disconnection of a **1,5-dioxgenated** compound shows it can be synthesized by nucleophilic attack of a negatively charged α-carbon on a carbonyl compound with a positively charged β-carbon.

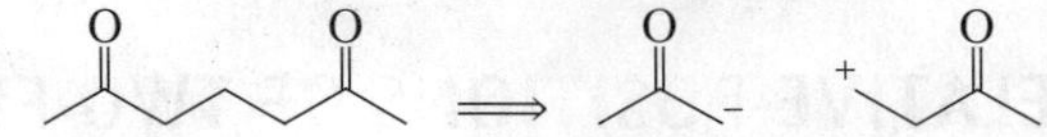

a 1,5-dioxygenated compound

An α,β-unsaturated carbonyl compound is the synthetic equivalent of a compound with a positively charged β-carbon. And we have seen that an enamine is the synthetic equivalent for the negatively charged α-carbon.

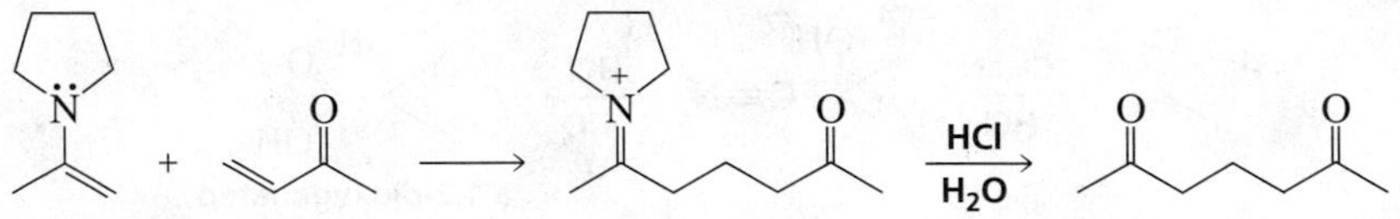

PROBLEM 7 Describe how the following compound could be synthesized from compounds containing no more than six carbons.

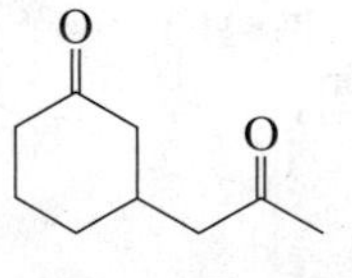

Disconnection of a **1,6-dioxgenated** compound shows it can be synthesized by nucleophilic attack of a negatively charged α-carbon on a carbonyl compound with a positively charged γ-carbon.

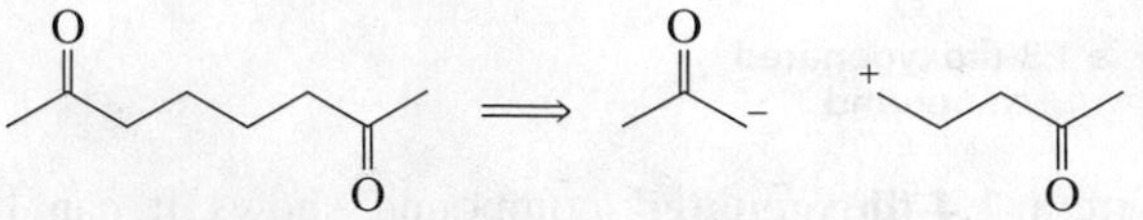

a 1,6-dioxygenated compound

There is no synthetic equivalent for a carbonyl compound with a positively charged γ-carbon, so we have to consider another route. Recognizing that a 1,6-dioxygenated compound can be prepared by oxidative cleavage of a cyclohexene provides an easy route to the target molecule, because a compound with a six-membered ring can be readily prepared by a Diels–Alder reaction.

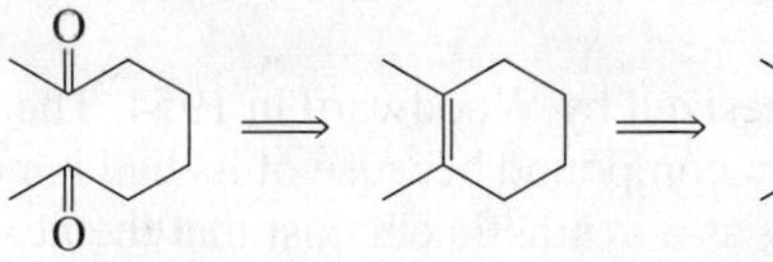

PROBLEM 8 Do a retrosynthetic analysis on each of the following compounds, ending with available starting materials.

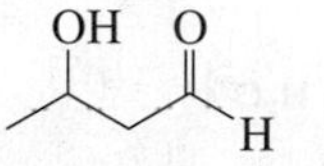

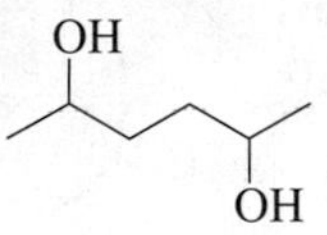

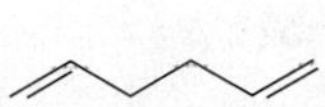

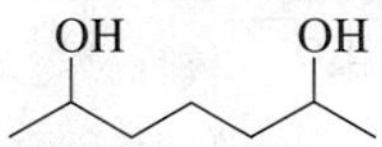

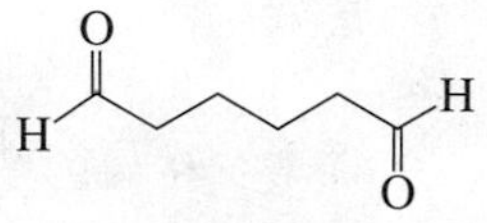

PROBLEM 9 Use retrosynthetic analysis to plan a synthesis of the following target molecules. The only carbon-containing compounds available for the syntheses are cyclohexanol, ethanol, and carbon dioxide.

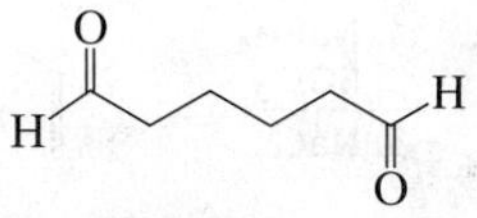

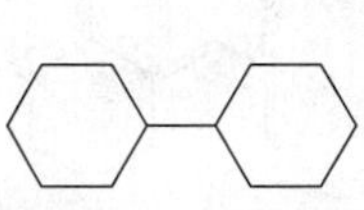

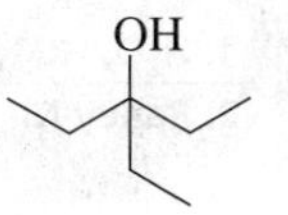

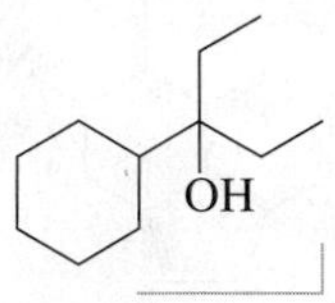

PROBLEM 10 How could you synthesize the following compounds from starting materials containing no more than four carbons?

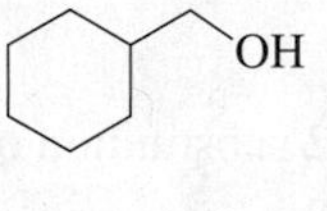

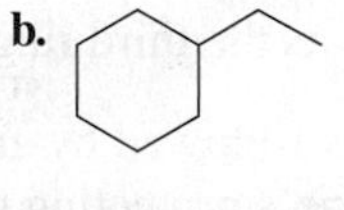

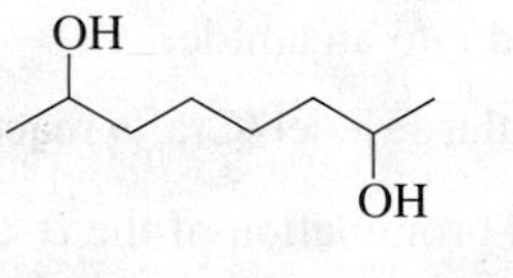

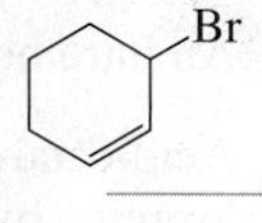

PROBLEM 11 Show how the following compounds could be synthesized from the given starting materials:

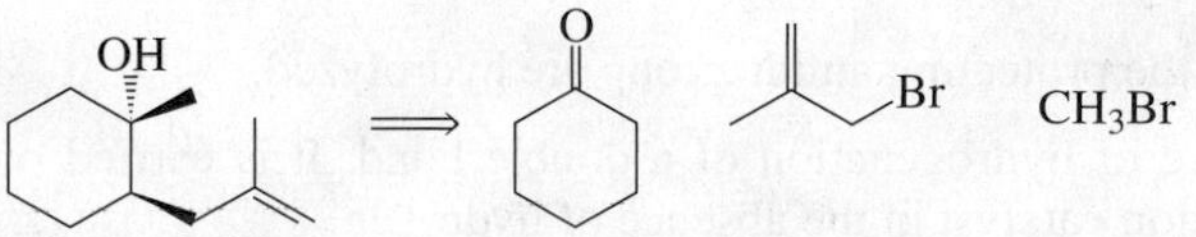

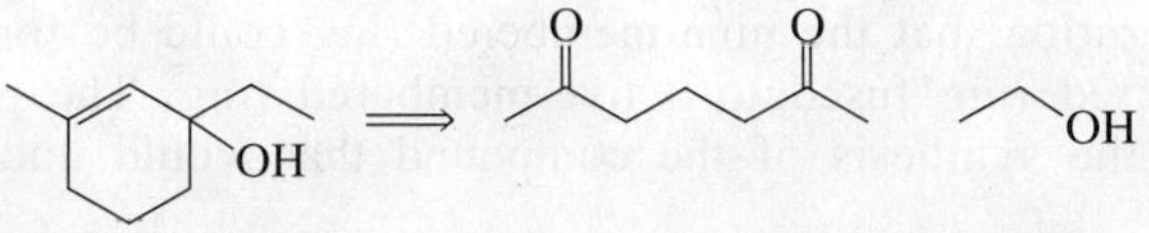

EXAMPLES OF MULTISTEP ORGANIC SYNTHESIS

To give you an idea of the kind of thinking required for the synthesis of a complicated molecule, we will look at the synthesis of lysergic acid, which was done by R. B. Woodward, and the synthesis of caryophyllene, which was done by E. J. Corey. Woodward (in 1965) and Corey (in 1990) both received the Nobel Prize for their contributions to synthetic organic chemistry.

Lysergic acid was first synthesized by Woodward in 1954. The diethylamide of lysergic acid (LSD) is a better known compound because of its hallucinogenic properties. It is a tribute to Woodward's ability as a synthetic chemist that the next synthesis of lysergic acid was not accomplished until 1969. Lysergic acid has an indole ring. Because indole is unstable under acidic conditions, Woodward designed a synthesis that did not form the indole portion until the final step.

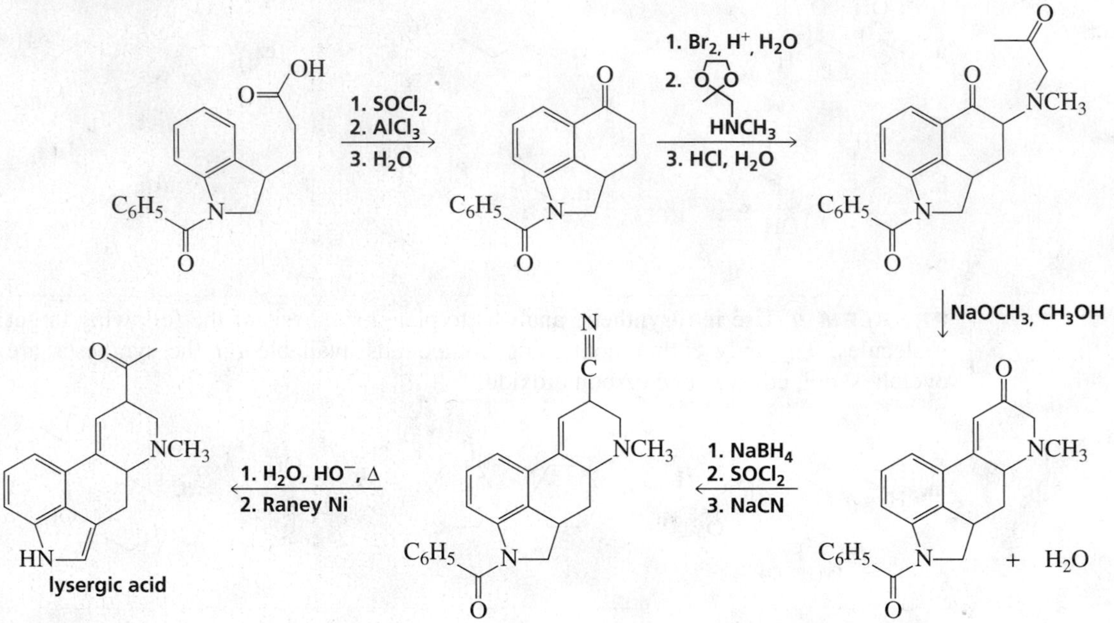

- The starting material is a dihydroindole carboxylic acid with its nitrogen protected by being converted into an amide.

- An intramolecular Friedel–Crafts reaction forms the third ring.

- Acid-catalyzed bromination of the α-carbon is followed by an S_N2 substitution of the bromine by an amine and removal of the ketone's protecting group.

- An aldol condensation forms the fourth ring.

- The ketone carbonyl group is reduced to an OH group, which is activated by thionyl chloride and replaced by a cyano group.

- Both the cyano group and the protecting amide group are hydrolyzed.

- The last step is the reverse of hydrogenation of a double bond. It is carried out by using a typical hydrogenation catalyst in the absence of hydrogen.

Caryophyllene is an oil found in cloves. The most important strategic element in its synthesis was the realization that the nine-membered ring could be formed by fragmenting a six-membered ring fused to a five-membered ring. The problem then became designing the synthesis of the compound that would undergo fragmentation.

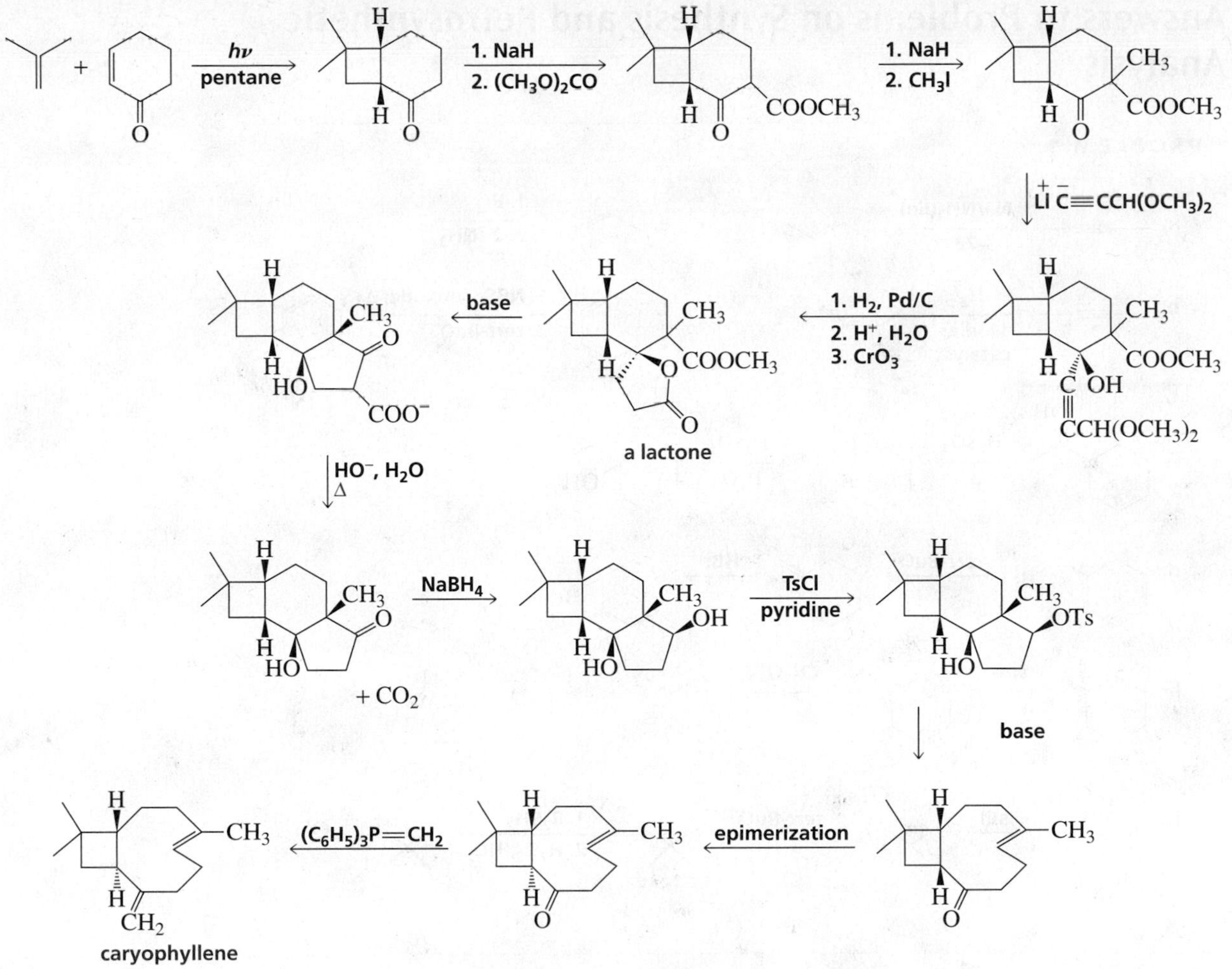

caryophyllene

- The first step is a photochemical [2+2] cycloaddition reaction.

- One α-hydrogen of the resulting ketone is substituted with a methoxycarbonyl group, and the second α-hydrogen on the same carbon is substituted with a methyl group. (Today we would use LDA, not NaH.)

- An acetylide ion is used to add a three-carbon fragment to the carbonyl group of the ketone.

- The triple bond is reduced, the acetal is hydrolyzed, and the resulting aldehyde is oxidized to a carboxylic acid that forms a lactone.

- A Claisen condensation and hydrolysis of the lactone forms the 3-oxocarboxylic acid that is decarboxylated to give the cyclic ketone.

- Reduction of the ketone forms an alcohol.

- Because the secondary alcohol is more reactive than the sterically hindered tertiary alcohol, the desired tosylate is formed.

- The stage is now set for the desired fragmentation reaction. This reaction involves removing a proton from the OH group, which allows the oxygen to form a carbonyl group and eliminate the tosyl group.

- Epimerization of the asymmetric center adjacent to the carbonyl group occurs once the ketone has been formed.

- A Wittig reaction forms the exocyclic double bond.

Answers to Problems on Synthesis and Retrosynthetic Analysis

PROBLEM 1

a. 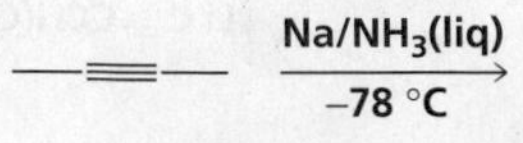 $\quad$ (HC≡C−CH₃) $\xrightarrow[\text{−78 °C}]{\text{Na/NH}_3\text{(liq)}}$ (trans-alkene)

e. 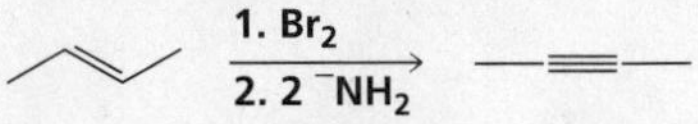 $\quad$ (alkene) $\xrightarrow[\text{2. 2 }^-\text{NH}_2]{\text{1. Br}_2}$ (alkyne)

b. $\quad$ (alkyne) $\xrightarrow[\substack{\text{Lindlar} \\ \text{catalyst}}]{\text{H}_2}$ (cis-alkene)

f. 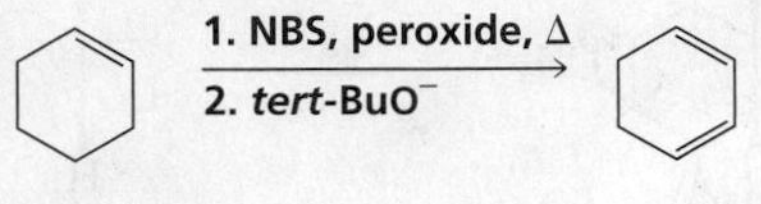 $\quad$ (cyclohexene) $\xrightarrow[\text{2. }\textit{tert}\text{-BuO}^-]{\text{1. NBS, peroxide, }\Delta}$ (1,3-cyclohexadiene)

c. 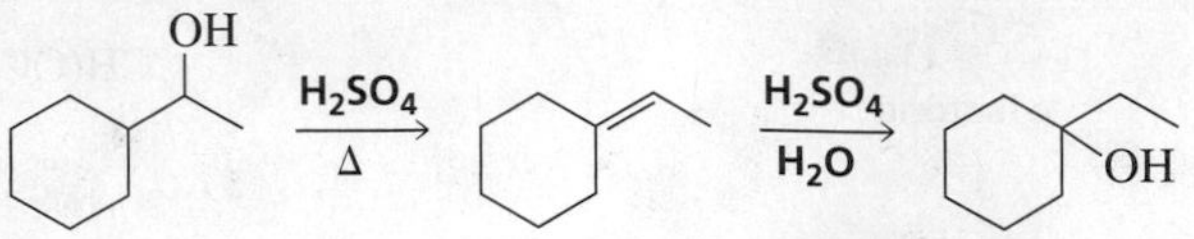 $\quad$ (1-cyclohexylethanol, OH) $\xrightarrow[\Delta]{\text{H}_2\text{SO}_4}$ (ethylidenecyclohexane) $\xrightarrow[\text{H}_2\text{O}]{\text{H}_2\text{SO}_4}$ (1-ethylcyclohexanol, OH)

d. 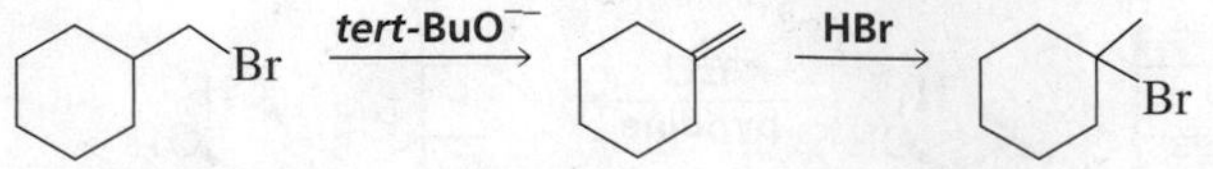 $\quad$ (cyclohexylmethyl bromide, Br) $\xrightarrow{\textit{tert}\text{-BuO}^-}$ (methylenecyclohexane) $\xrightarrow{\text{HBr}}$ (1-methyl-1-bromocyclohexane, Br)

g. 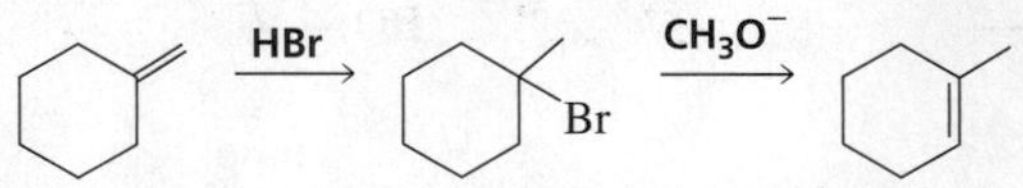 $\quad$ (methylenecyclohexane) $\xrightarrow{\text{HBr}}$ (1-methyl-1-bromocyclohexane, Br) $\xrightarrow{\text{CH}_3\text{O}^-}$ (1-methylcyclohexene)

h. $\quad$ (2,3-dimethyl-2-butanol, OH) $\xrightarrow{\text{TsCl}}$ (OTs) $\xrightarrow{\textit{tert}\text{-BuO}^-}$ (alkene) $\xrightarrow[\text{2. H}_2\text{O}_2\text{, HO}]{\text{1. R}_2\text{BH}_2}$ (alcohol, OH)

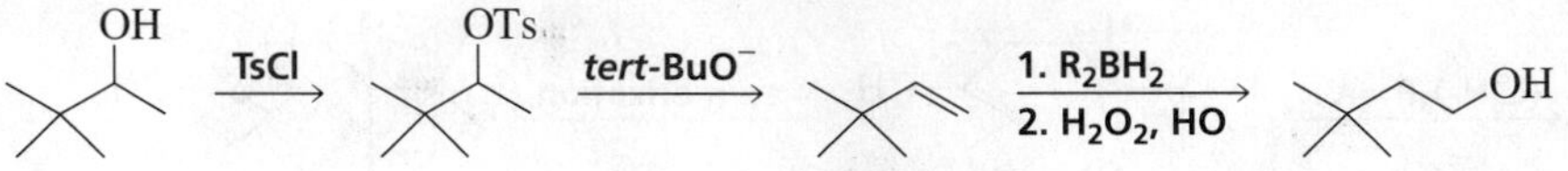

PROBLEM 2

a. 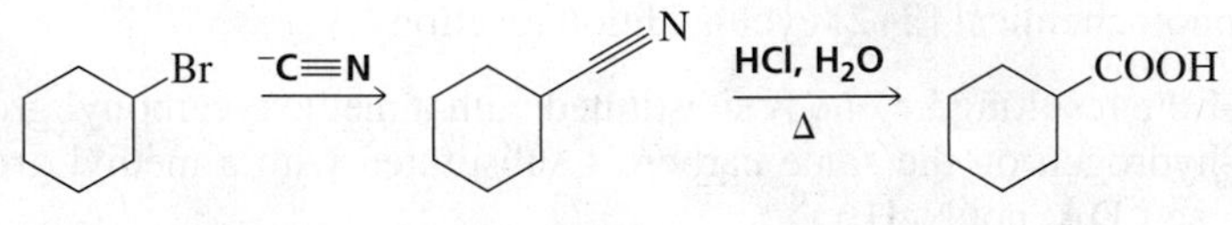 $\quad$ (bromocyclohexane, Br) $\xrightarrow{^-\text{C≡N}}$ (cyclohexyl nitrile, ≡N) $\xrightarrow[\Delta]{\text{HCl, H}_2\text{O}}$ (cyclohexanecarboxylic acid, COOH)

b. 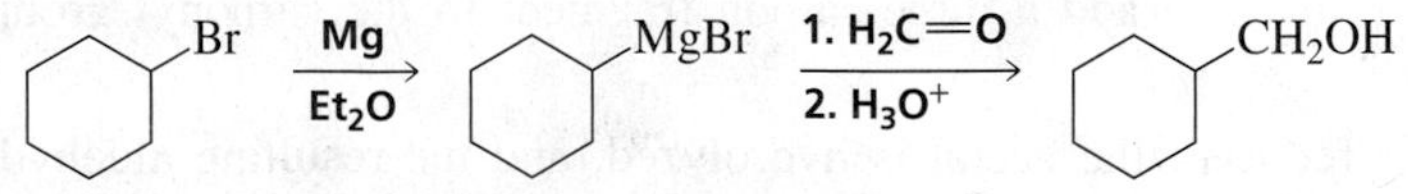 $\quad$ (bromocyclohexane, Br) $\xrightarrow[\text{Et}_2\text{O}]{\text{Mg}}$ (MgBr) $\xrightarrow[\text{2. H}_3\text{O}^+]{\text{1. H}_2\text{C=O}}$ (CH₂OH)

c. $\quad$ (bromocyclohexane, Br) $\xrightarrow{^-\text{C≡N}}$ (cyclohexyl nitrile, ≡N) $\xrightarrow[\substack{\text{Raney} \\ \text{Nickel}}]{\text{H}_2}$ (CH₂NH₂)

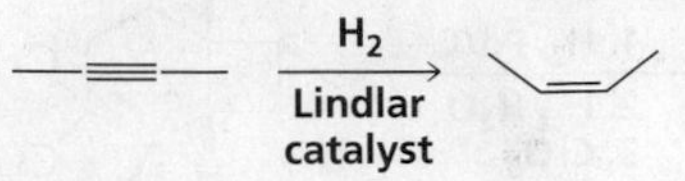

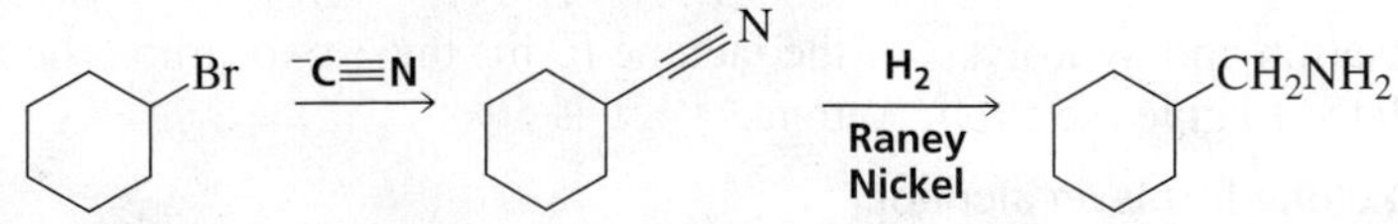

d. 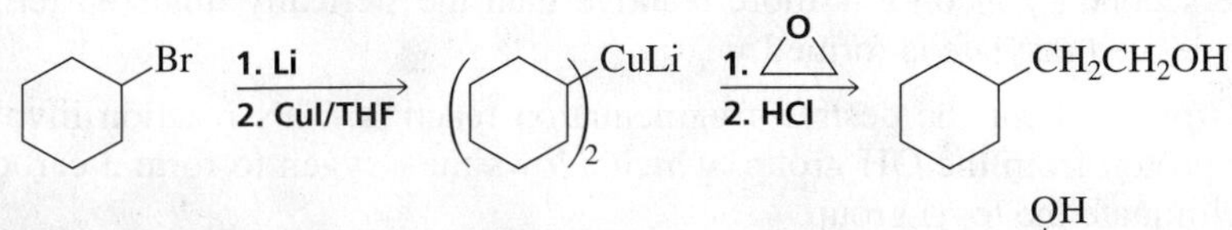 $\quad$ (bromocyclohexane, Br) $\xrightarrow[\text{2. CuI/THF}]{\text{1. Li}}$ (dicyclohexylcuprate, CuLi)₂ $\xrightarrow[\text{2. HCl}]{\text{1. }\overset{\text{O}}{\triangle}}$ (CH₂CH₂OH)

e. 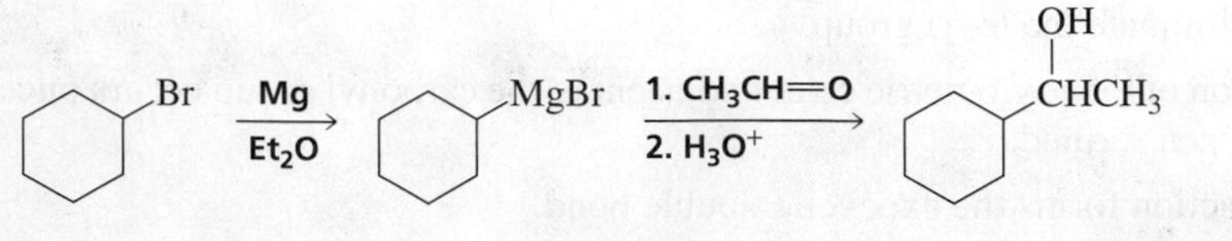 $\quad$ (bromocyclohexane, Br) $\xrightarrow[\text{Et}_2\text{O}]{\text{Mg}}$ (MgBr) $\xrightarrow[\text{2. H}_3\text{O}^+]{\text{1. CH}_3\text{CH=O}}$ (CHCH₃ with OH)

PROBLEM 3

a. An alkene can be made in one step by a Wittig reaction.

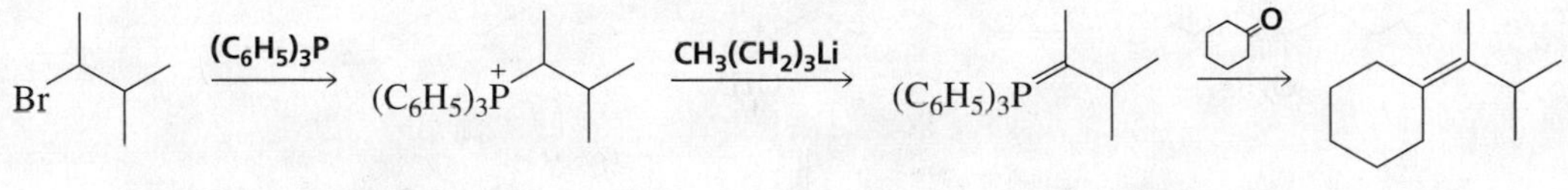

b. The alkyne can prepared from 1-hexyne and ethyl bromide as shown here or it can be prepared from 1-butyne and butyl bromide.

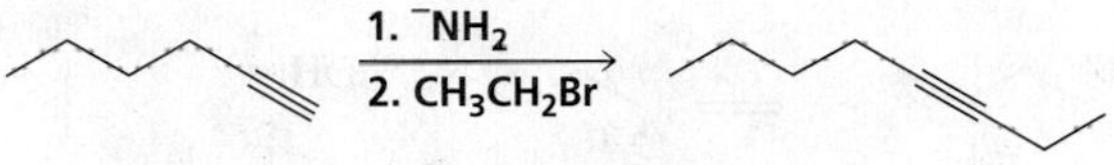

c. The desired aldehyde has eight carbons and a structure that suggests it can be prepared by an aldol condensation using an aldehyde with four carbons.

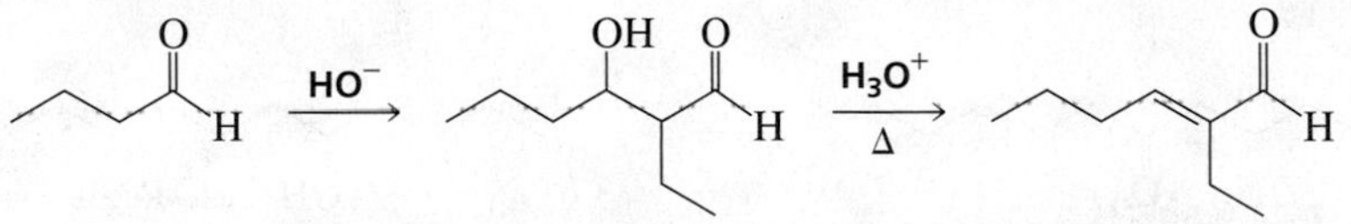

d.

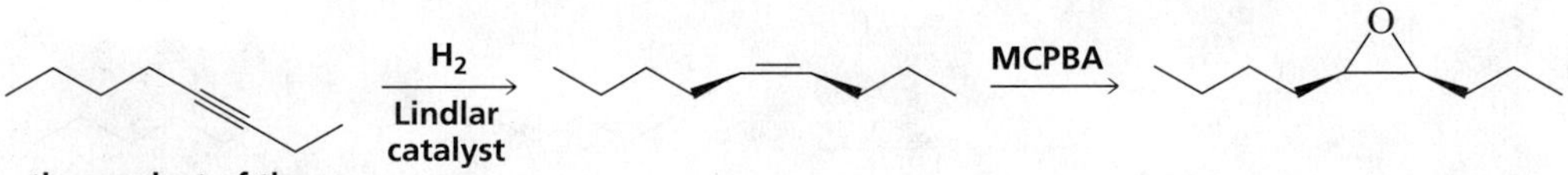

the product of the
synthesis in part b

PROBLEM 4 See Appendix: "Summary of Methods Used to Synthesize a Particular Functional Group."

PROBLEM 5

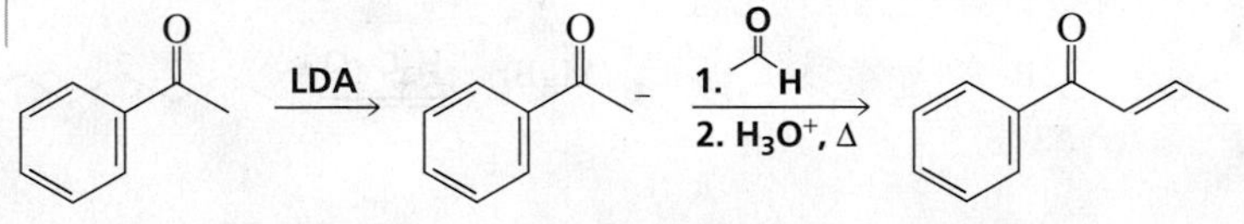

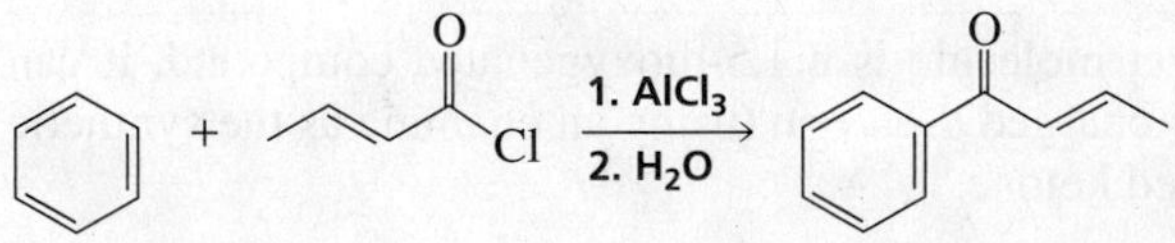

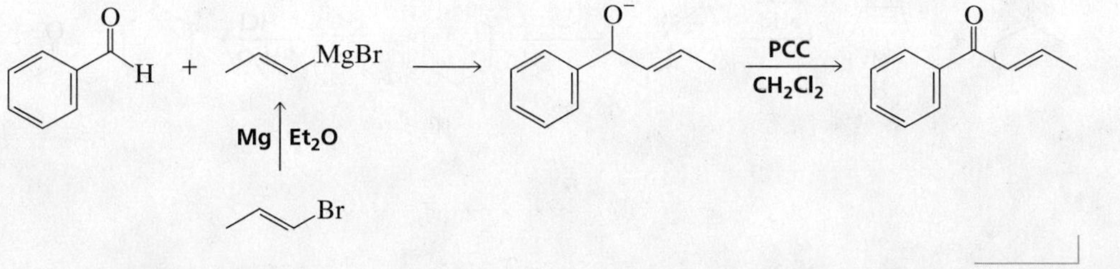

PROBLEM 6

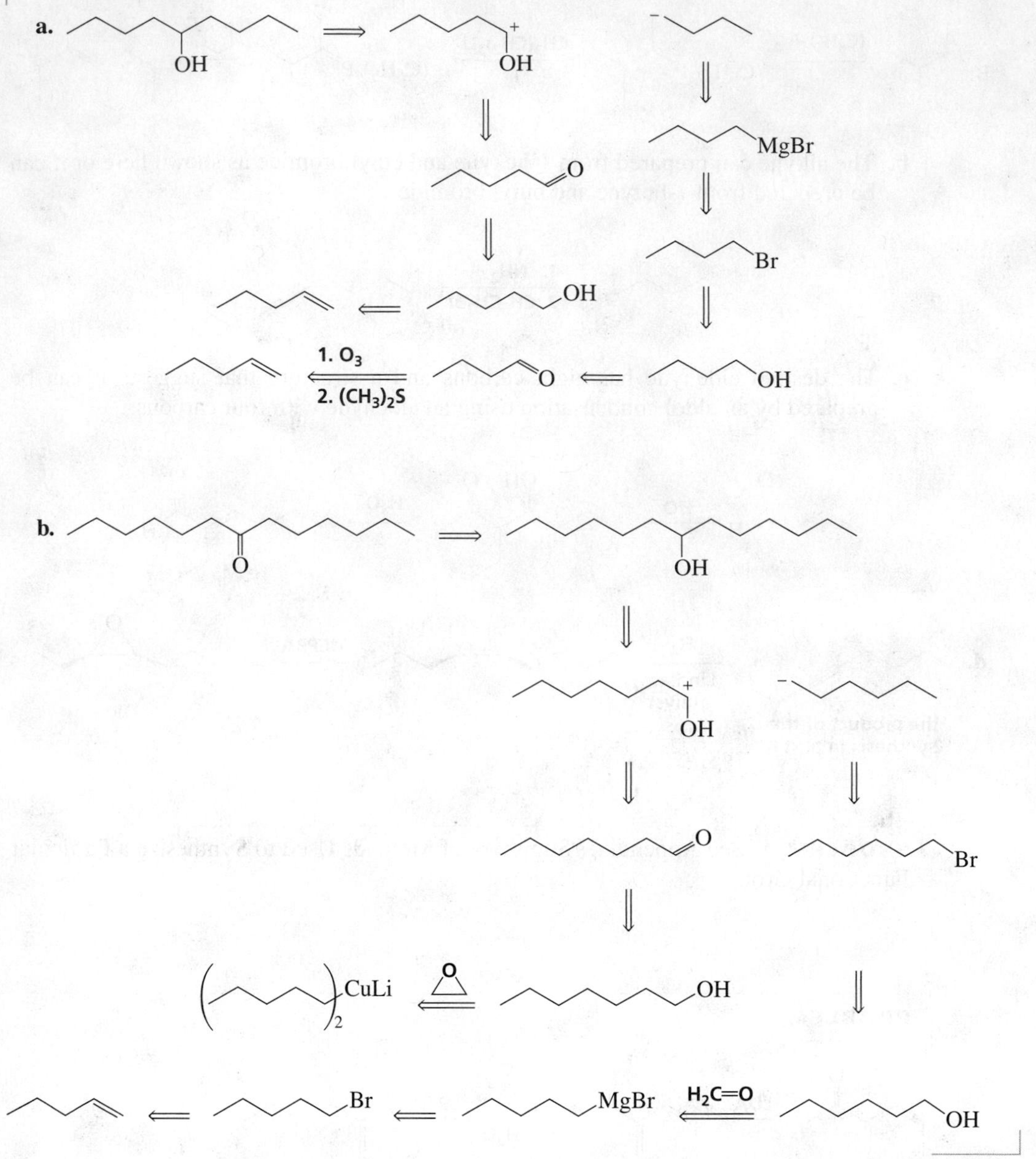

a.

b.

PROBLEM 7 Because the target molecule is a 1,5-dioxygenated compound, it can be synthesized from a negatively charged α-carbon (using an enamine as the synthetic equivalent) and an α,β-unsaturated ketone.

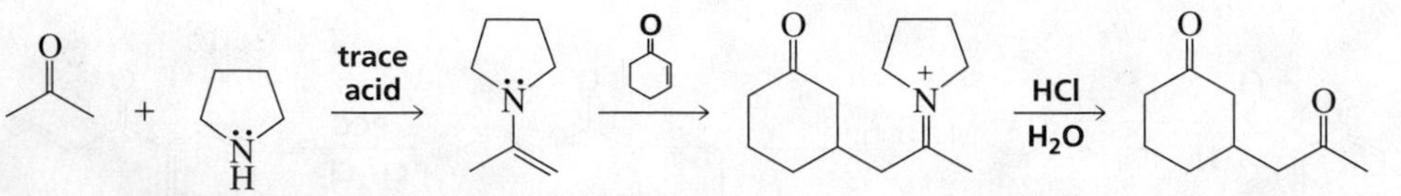

PROBLEM 8

a. 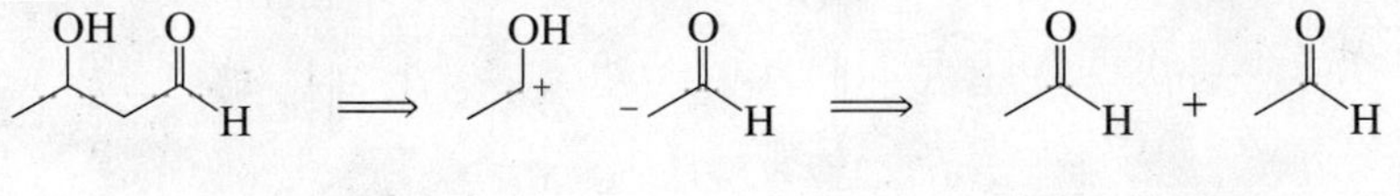

b. 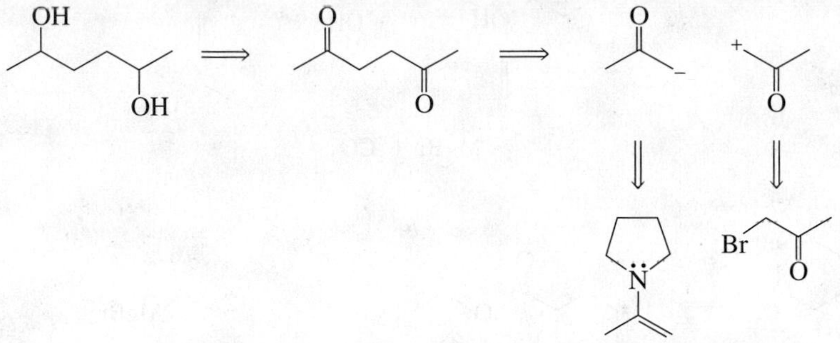

c. 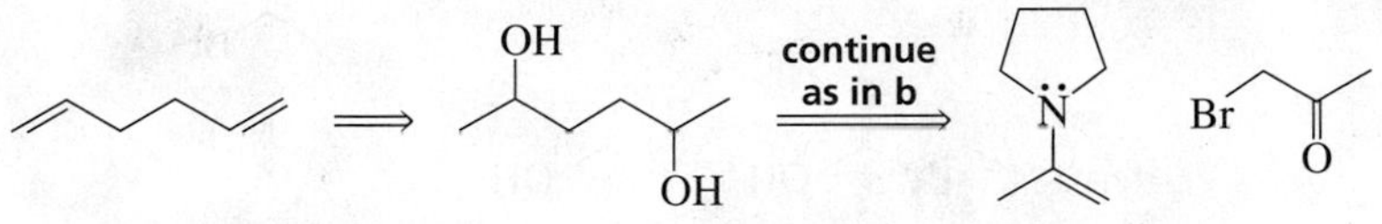

d. 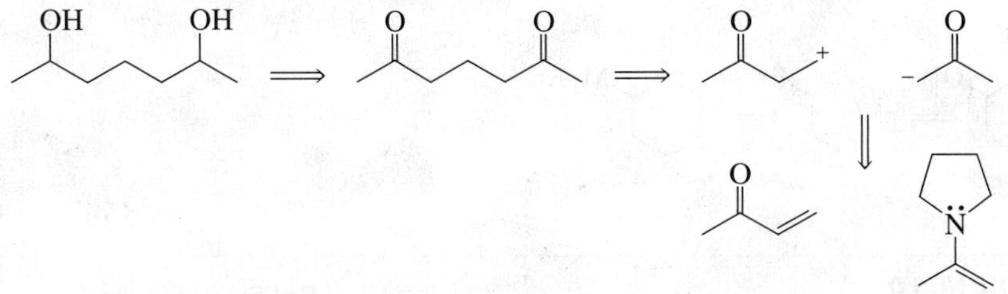

e. 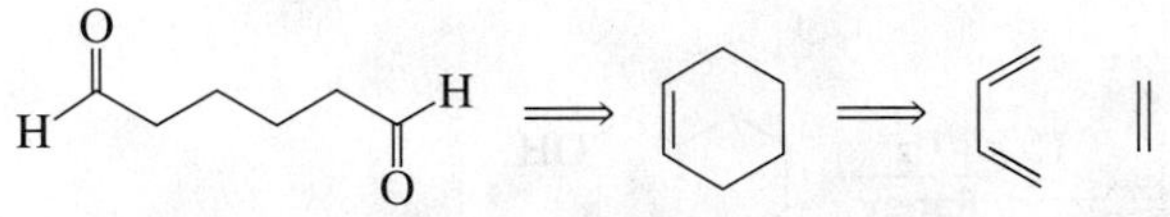

PROBLEM 9

a. 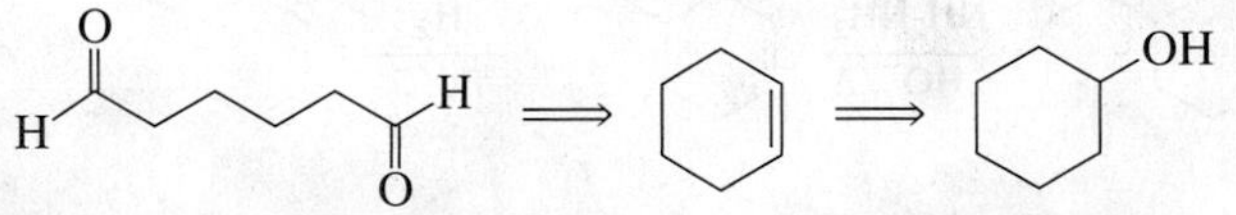

b. 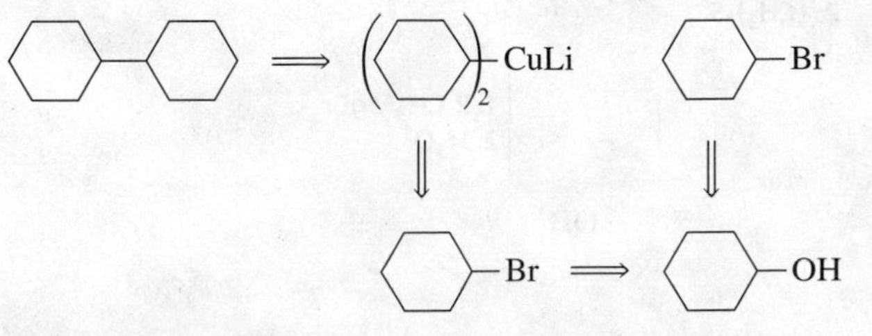

c.

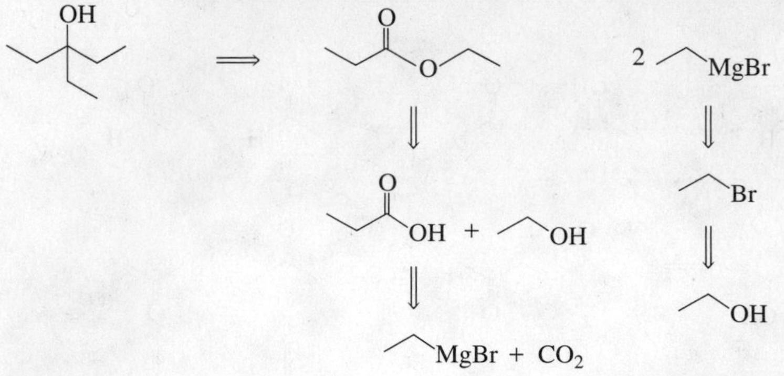

OH

⟹

O

2   MgBr

⟱

O

OH   +   OH

⟱

O

Br

⟱

OH

MgBr   +   CO$_2$

d.

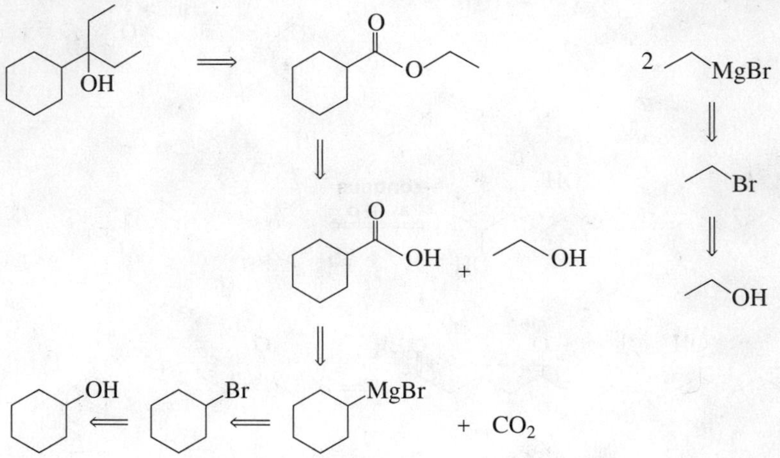

OH

⟹

O

O

2   MgBr

⟱

O

Br

⟱

O

OH   +   OH

⟱

OH

OH   ⟸   Br   ⟸   MgBr   +   CO$_2$

PROBLEM 10

a.

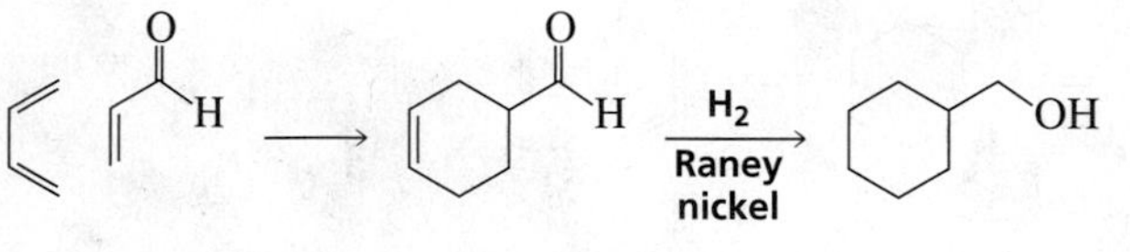

O
H
→

O
H

$\xrightarrow[\text{nickel}]{\text{Raney}}^{\text{H}_2}$

OH

b.

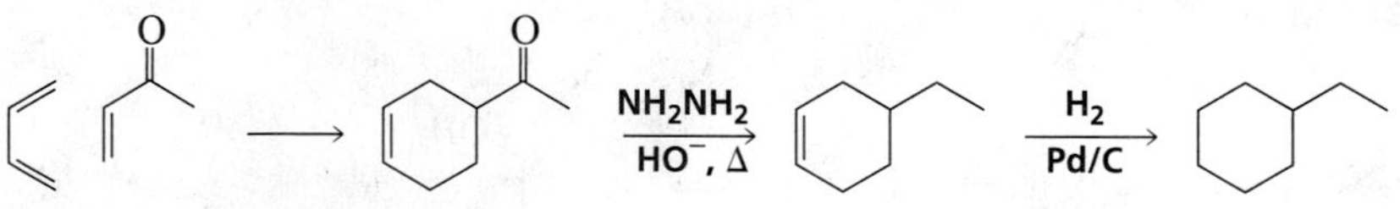

O
→

O

$\xrightarrow[\text{HO}^-, \Delta]{\text{NH}_2\text{NH}_2}$

$\xrightarrow{\text{H}_2}$
Pd/C

c.

$\xrightarrow{\Delta}$

$\xrightarrow[\text{2. (CH}_3\text{)}_2\text{S}]{\text{1. O}_3, -78\ ^\circ\text{C}}$

O
H   H
O

1. 2 CH$_3$MgBr
2. H$_3$O$^+$
↓

OH

OH

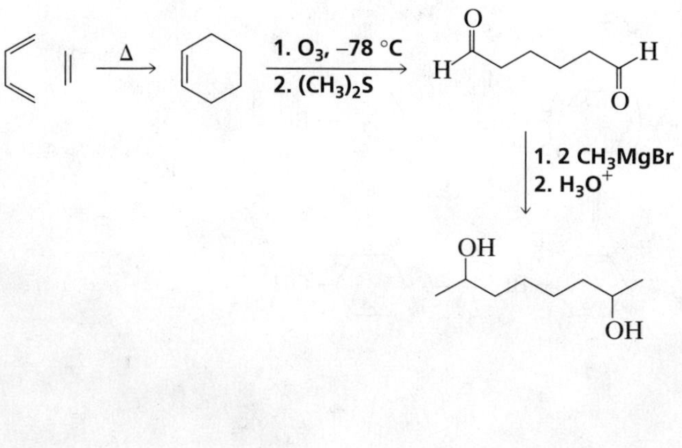

d.

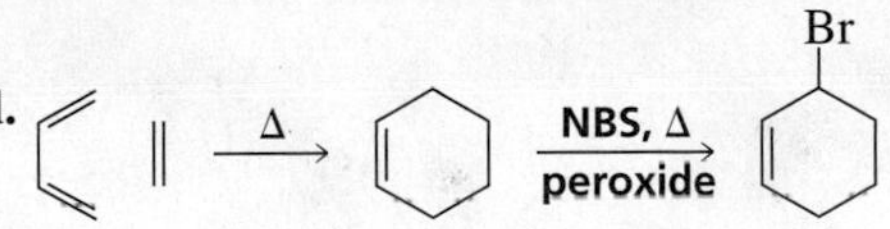

PROBLEM 11

a.

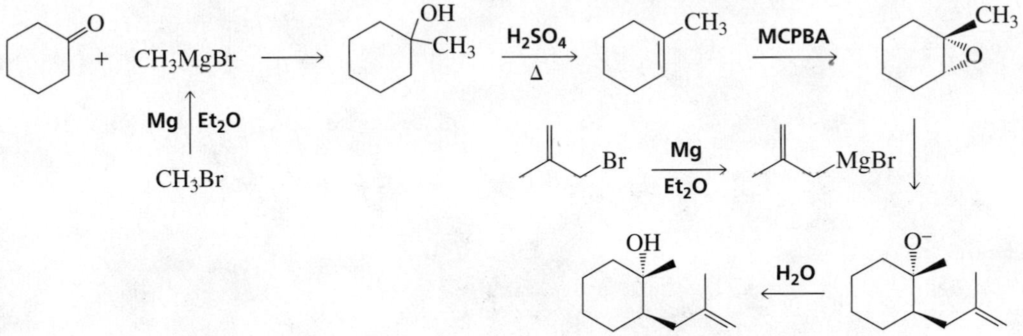

b.

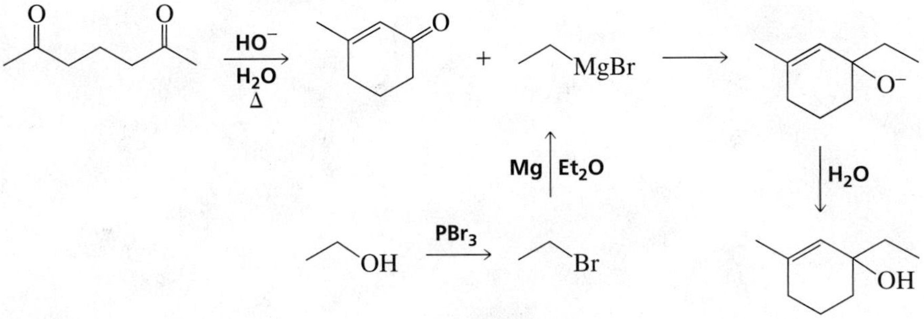

MasteringChemistry®

for Organic Chemistry

MasteringChemistry tutorials guide you through the toughest topics in chemistry with self-paced tutorials that provide individualized coaching. These assignable, in-depth tutorials are designed to coach you with hints and feedback specific to your individual misconceptions. For additional practice on Synthesis and Retrosynthetic Analysis, go to MasteringChemistry where the following tutorials are available:

- Synthesis and Retrosynthetic Analysis: Functional Groups
- Synthesis and Retrosynthetic Analysis: Carbon Chain
- Synthesis and Retrosynthetic Analysis: Retrosynthesis of 2-Pentanone Using Reactions of Carbonyl Compounds

More About Amines • Reactions of Heterocyclic Compounds

Auletta

In 1994, a team from UC Santa Cruz reported the structures of milnamides A and B, compounds the team had obtained from the sponge Auletta. Subsequently, other researchers in various countries isolated these compounds from three other coral reef sponges. Collaborative research between scientists at the University of British Columbia and Wyeth Pharmaceuticals lead to an explanation of why breaking the saturated heterocyclic ring of milnamide A resulted in a compound (milnamide B) with greater potency against solid tumor cancer cells. Using milnamide B as a lead compound, chemists at Eisai Inc. designed and synthesized E7974. This compound shows broad activity against a variety of human tumors and exhibits better efficacy than the classic anticancer drug, 5-fluorouracil. It is now being evaluated as an anticancer drug in Phase I clinical trials. Many believe this sponge-inspired compound may someday be a blockbuster drug.

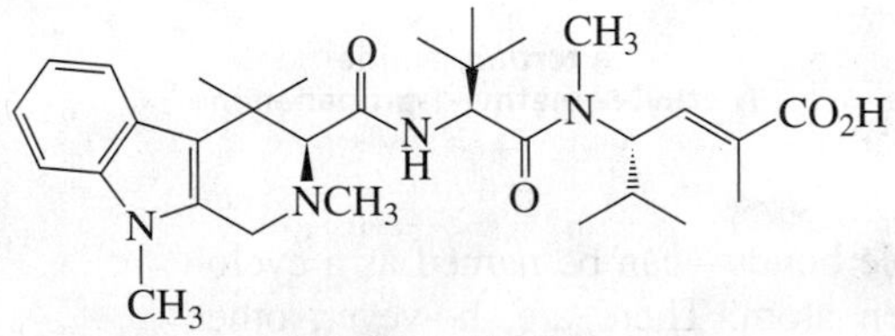

milnamide A

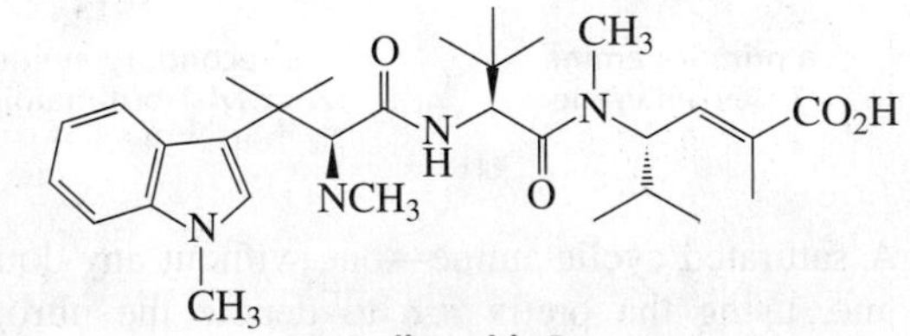

milnamide B

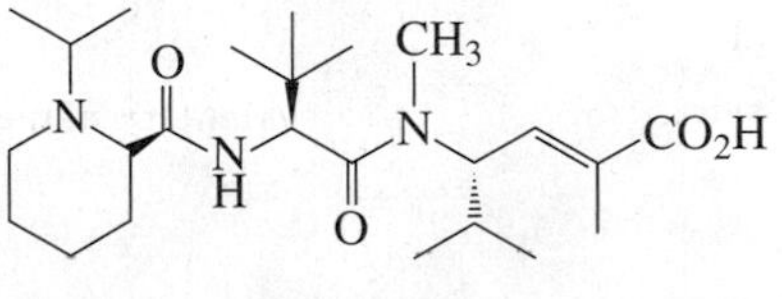

E7974

From Chapter 20 of *Organic Chemistry*, Seventh Edition. Paula Yurkanis Bruice. Copyright © 2014 by Pearson Education, Inc. All rights reserved.

Amines, compounds in which one or more of the hydrogens of ammonia (NH_3) have been replaced by an alkyl group, are among some of the most abundant compounds in the biological world. The biological importance of these compounds is shown in the structures and properties of amino acids and proteins; in how enzymes catalyze reactions; in the ways in which coenzymes—compounds derived from vitamins—help enzymes catalyze reactions; and in nucleic acids (DNA and RNA).

Amines are exceedingly important compounds to organic chemists. For example, the nitrogen in amines is sp^3 hybridized with the lone pair residing in an sp^3 orbital, and amines invert rapidly at room temperature through a transition state in which the sp^3 nitrogen becomes an sp^2 nitrogen. The physical properties of amines—their hydrogen-bonding properties, boiling points, and solubilities are also very important—as is knowing how amines are named. Most importantly, the lone-pair electrons of the nitrogen atom cause amines to react as bases (that is, to share their lone pair with a proton) and as nucleophiles (that is, to share their lone pair with an atom other than a proton).

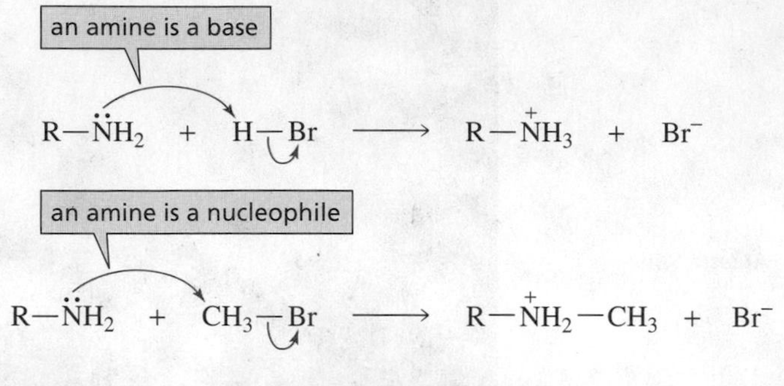

an amine is a base

$$R—\overset{..}{N}H_2 \; + \; H—Br \longrightarrow R—\overset{+}{N}H_3 \; + \; Br^-$$

an amine is a nucleophile

$$R—\overset{..}{N}H_2 \; + \; CH_3—Br \longrightarrow R—\overset{+}{N}H_2—CH_3 \; + \; Br^-$$

In this chapter, we will visit some of these topics and look at other aspects of amines and their chemistry.

Some amines are **heterocyclic compounds** (or **heterocycles**)—cyclic compounds in which one or more of the atoms of the ring are **heteroatoms**. A variety of atoms, such as N, O, S, Se, P, Si, B, and As, can be incorporated into ring structures.

Heterocycles are an extraordinarily important class of compounds, making up more than half of all known organic compounds. Most drugs, most vitamins, and many natural products are heterocycles. In this chapter, we will consider the most prevalent heterocyclic compounds—the ones containing the heteroatom N, O, or S.

1   MORE ABOUT AMINE NOMENCLATURE

Amines are classified as primary, secondary, or tertiary, depending on whether one, two, or three hydrogens of ammonia have been replaced by an alkyl group. In addition, amines have both common and systematic names. Common names are obtained by citing the names of the alkyl substituents attached to the nitrogen (in alphabetical order) followed by "amine." Systematic names employ "amine" as a functional group suffix.

a primary amine	**a secondary amine**	**a tertiary amine**
systematic name: 1-pentanamine	*N*-ethyl-1-butanamine	*N*-ethyl-*N*-methyl-1-propanamine
common name: pentylamine	butylethylamine	ethylmethylpropylamine

A saturated cyclic amine—one without any double bonds—can be named as a cyclo-alkane, using the prefix *aza* to denote the nitrogen atom. There are, however, other acceptable names. Some of the more commonly used names are shown here. Notice that heterocyclic rings are numbered to give the heteroatom the lowest possible number.

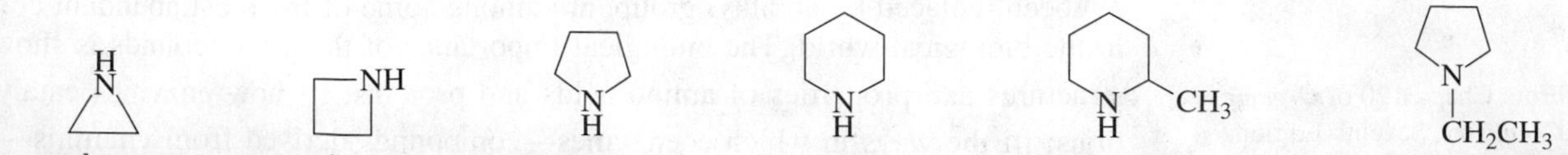

azacyclopropane **aziridine**	**azacyclobutane** **azetidine**	**azacyclopentane** **pyrrolidine**	**azacyclohexane** **piperidine**	**2-methylazacyclohexane** **2-methylpiperidine**	**N-ethylazacyclopentane** **N-ethylpyrrolidine**

Saturated heterocycles with oxygen and sulfur heteroatoms are named similarly. The prefix for oxygen is *oxa* and the prefix for sulfur is *thia*.

oxacyclopropane
oxirane
ethylene oxide

thiacyclopropane
thiirane

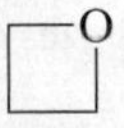

oxacyclobutane
oxetane

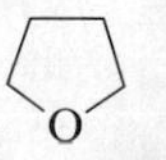

oxacyclopentane

tetrahydrofuran

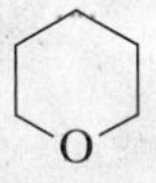

oxacyclohexane

tetrahydropyran

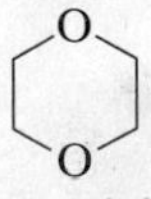

1,4-dioxacyclohexane

1,4-dioxane

PROBLEM 1◆

Name the following compounds:

a.

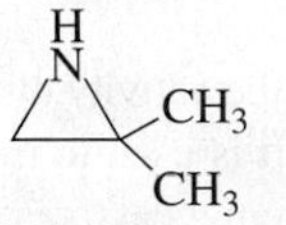

b.

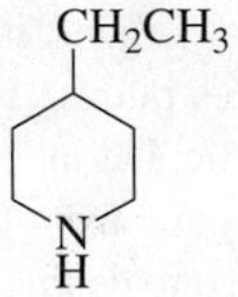

c.

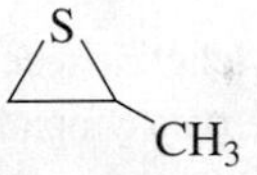

d.

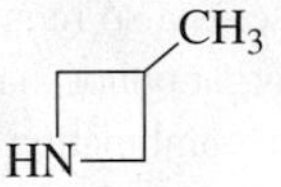

e.

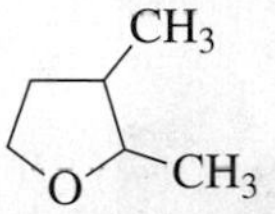

f.

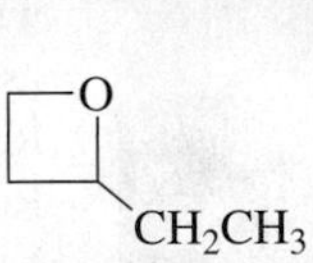

---

2 MORE ABOUT THE ACID–BASE PROPERTIES OF AMINES

Amines are the most common organic bases. Ammonium ions have pK_a values of about 10 and anilinium ions have pK_a values of about 5. The greater acidity of anilinium ions compared with ammonium ions is due to the greater stability of the conjugate bases of the anilinium ions as a result of electron delocalization. Amines have very high pK_a values. For example, the pK_a of methylamine is 40.

$CH_3CH_2CH_2\overset{+}{N}H_3$

an ammonium ion
$pK_a = 10.8$

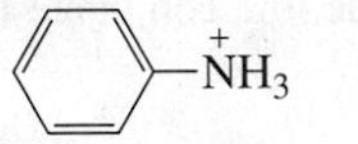

$\overset{+}{N}H_3$

an anilinium ion
$pK_a = 4.58$

CH_3NH_2

an amine
$pK_a = 40$

Saturated amine heterocycles containing five or more atoms have physical and chemical properties like those of acyclic amines. For example, pyrrolidine, piperidine, and morpholine are typical secondary amines, and *N*-methylpyrrolidine and quinuclidine are typical tertiary amines. The conjugate acids of these amines have the pK_a values expected for ammonium ions.

the ammonium ions of:

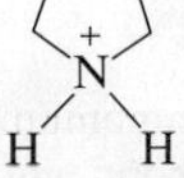

pyrrolidine
$pK_a = 11.27$

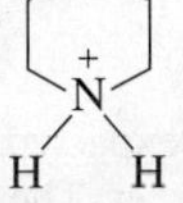

piperidine
$pK_a = 11.12$

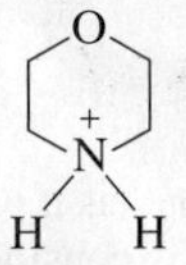

morpholine
$pK_a = 9.28$

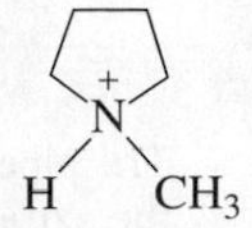

N-methylpyrrolidine
$pK_a = 10.32$

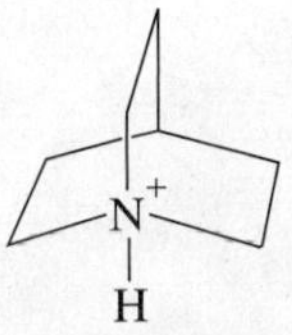

quinuclidine
$pK_a = 11.38$

Atropine

Atropa belladonna

Atropine is a naturally occurring heterocyclic compound found in Jimson weed and nightshade (*Atropa belladonna*). The *R* isomer is the one nature synthesizes, but it racemizes during isolation. Atropine has a variety of medicinal usages, thus landing it on the list of drugs the World Health Organization deems needed for basic health care.

Atropine blocks acetylcholine receptor sites, so it lowers the activity of all muscles regulated by the parasympathetic nervous system. It is used to treat low heart rates, spasms from the gastrointestinal tract, and tremors associated with Parkinson's disease. It also reduces the secretions of many organs, it is used as an antidote to organophosphate poisoning, and it dilates pupils. The Romans used atropine in combination with opium as an anesthetic. During the Renaissance, women enhanced their appearance by using the juice from the berries of nightshade to dilate their pupils. Cleopatra was reported to do this as well. Note that *belladonna* is Italian for "beautiful woman."

PROBLEM 2♦

Why is the conjugate acid of morpholine more acidic than the conjugate acid of piperidine?

PROBLEM 3♦

a. Draw the structure of 3-quinuclidinone.
b. What is the approximate pK_a of its conjugate acid?
c. Which has a lower pK_a value, the conjugate acid of 3-bromoquinuclidine or the conjugate acid of 3-chloroquinuclidine?

PROBLEM 4 Solved

Explain the difference in the pK_a values of the piperidinium and aziridinium ions:

piperidinium ion
$pK_a = 11.1$

aziridinium ion
$pK_a = 8.0$

The piperidinium ion has a pK_a value expected for an ammonium ion, whereas the pK_a value of the aziridinium ion is considerably lower. The internal bond angle of the three-membered ring is smaller than usual, which causes the external bond angles to be larger than usual. Because of the larger bond angles, the orbital used by nitrogen to overlap the orbital of hydrogen has more *s* character than that of a typical sp^3 nitrogen. This makes nitrogen in the aziridinium ion more electronegative, which lowers the pK_a.

3   AMINES REACT AS BASES AND AS NUCLEOPHILES

The leaving group of an amine ($^-NH_2$) is such a strong base that amines cannot undergo the substitution and elimination reactions that alkyl halides, alcohols, and ethers do. The relative reactivities of these compounds—each with an electron-withdraw ing group bonded to an sp^3 carbon—can be appreciated by comparing the pK_a values of the conjugate acids of their leaving groups, keeping in mind that the weaker the acid, the stronger its conjugate base. And, when comparing bases of the same type, the stronger the base, the poorer it is as a leaving group.

relative reactivities

$$\boxed{\text{most reactive}} \quad RCH_2F \; > \; RCH_2OH \; \sim \; RCH_2OCH_3 \; > \; RCH_2NH_2 \quad \boxed{\text{least reactive}}$$

$$\boxed{\begin{array}{l}\text{strongest acid,}\\ \text{has the weakest}\\ \text{conjugate base}\end{array}} \quad \begin{array}{c} HF \\ pK_a = 3.2 \end{array} \quad \begin{array}{c} H_2O \\ pK_a = 15.7 \end{array} \quad \begin{array}{c} RCH_2OH \\ pK_a = 15.5 \end{array} \quad \begin{array}{c} NH_3 \\ pK_a = 36 \end{array} \quad \boxed{\begin{array}{l}\text{weakest acid,}\\ \text{has the strongest}\\ \text{conjugate base}\end{array}}$$

Amines react as bases in proton transfer reactions and in elimination reactions.

The lone pair on the nitrogen of an amine makes it nucleophilic as well as basic. Amines react as nucleophiles in a number of different reactions; for example, in nucleophilic substitution reactions that *alkylate* the amine

$$CH_3CH_2Br \; + \; \underset{\textbf{methylamine}}{CH_3NH_2} \; \longrightarrow \; \underset{Br^-}{CH_3CH_2\overset{+}{N}H_2CH_3} \; \rightleftharpoons \; \underset{\textbf{ethylmethylamine}}{CH_3CH_2NHCH_3} \; + \; HBr$$

and in nucleophilic addition–elimination reactions that *acylate* the amine.

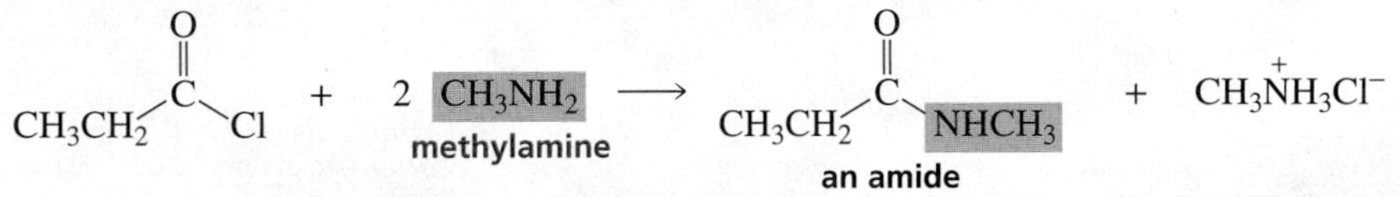

Aldehydes and ketones react with primary amines to form imines and with secondary amines to form enamines

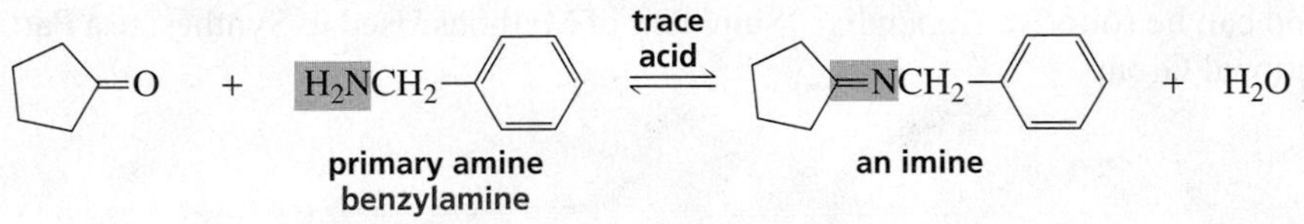

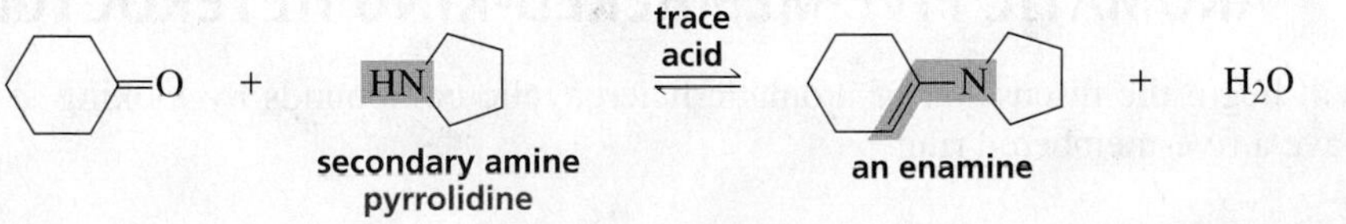

and amines are nucleophiles in conjugate addition reactions.

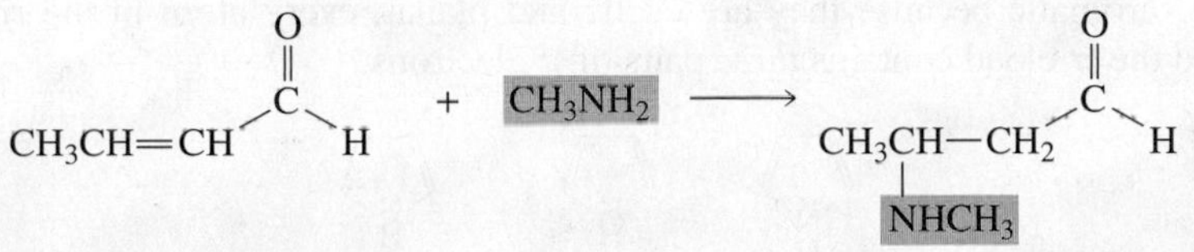

Primary arylamines react with nitrous acid to form stable arenediazonium salts. Arenediazonium salts are useful to synthetic chemists because the diazonium group can be replaced by a variety of nucleophiles.

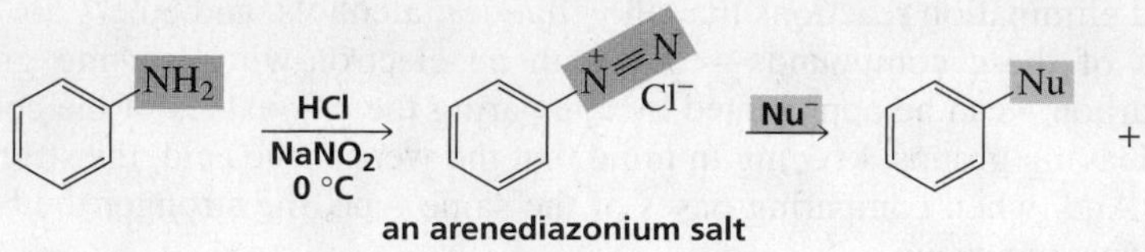

$$\text{ArNH}_2 \xrightarrow[\substack{\text{NaNO}_2 \\ 0\,°C}]{\text{HCl}} \text{Ar—N}{\overset{+}{\equiv}}\text{N} \ \text{Cl}^- \xrightarrow{\text{Nu}^-} \text{Ar—Nu} + \text{N}_2 + \text{Cl}^-$$

an arenediazonium salt

PROBLEM 5

Draw the product of each of the following reactions:

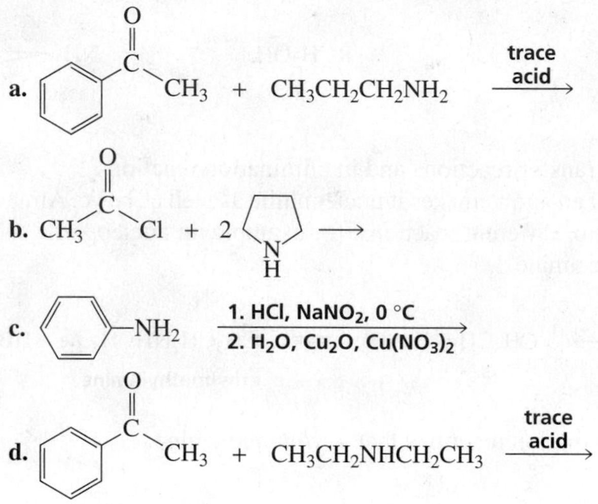

a. PhC(=O)CH$_3$ + CH$_3$CH$_2$CH$_2$NH$_2$ $\xrightarrow{\text{trace acid}}$

b. CH$_3$C(=O)Cl + 2 pyrrolidine(N-H) $\longrightarrow$

c. Ph—NH$_2$ $\xrightarrow[\text{2. H}_2\text{O, Cu}_2\text{O, Cu(NO}_3)_2]{\text{1. HCl, NaNO}_2,\ 0\,°C}$

d. PhC(=O)CH$_3$ + CH$_3$CH$_2$NHCH$_2$CH$_3$ $\xrightarrow{\text{trace acid}}$

---

4 THE SYNTHESIS OF AMINES

There are many different ways to synthesize amines and the mechanisms of these reactions. These reactions are summarized and the references to where they are discussed can be found in Appendix: "Summary of Methods Used to Synthesize a Particular Functional Group."

---

5 AROMATIC FIVE-MEMBERED-RING HETEROCYCLES

We will begin the discussion of aromatic heterocyclic compounds by looking at those that have a five-membered ring.

Pyrrole, Furan, and Thiophene

Pyrrole, furan, and **thiophene** are heterocycles with five-membered rings. These compounds are aromatic because they are cyclic and planar, every atom in the ring has a *p* orbital, and the π cloud contains *three* pairs of π electrons.

pyrrole   **furan**   **thiophene**

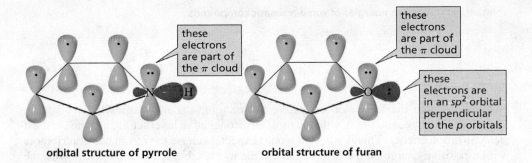

orbital structure of pyrrole   **orbital structure of furan**

Pyrrole is an extremely weak base because the electrons shown as a lone pair in the structure are part of the π cloud. That is, the nitrogen donates the lone-pair into the five-membered ring (as shown by the resonance contributors). Therefore, protonating pyrrole destroys its aromaticity. As a result, the conjugate acid of pyrrole is a very strong acid ($pK_a = -3.8$).

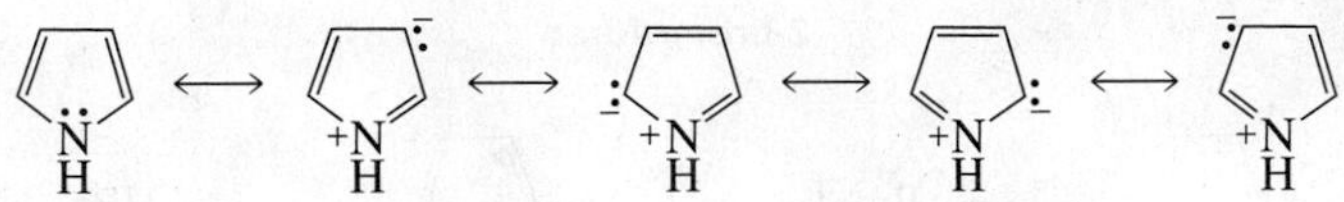

resonance contributors of pyrrole

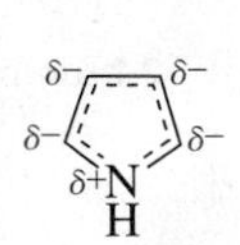

resonance hybrid

PROBLEM 6

Draw arrows to show the movement of electrons in going from one resonance contributor to the next in pyrrole.

Pyrrole has a dipole moment of 1.80 D. The saturated amine with a five-membered ring—pyrrolidine—has a slightly smaller dipole moment of 1.57 D, but as we see from the electrostatic potential maps, the two dipole moments are in opposite directions. (The red area is in the ring in pyrrole and on the nitrogen in pyrrolidine.) The dipole moment in pyrrolidine is due to inductive electron withdrawal by the nitrogen. Apparently, the ability of pyrrole's nitrogen to donate electrons into the ring by resonance more than makes up for its inductive electron withdrawal.

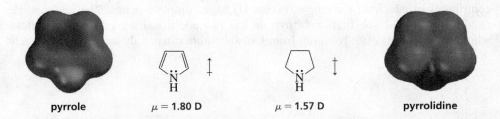

pyrrole $\mu = 1.80$ D $\mu = 1.57$ D   **pyrrolidine**

A compound's delocalization energy increases as the resonance contributors become more stable and more nearly equivalent. The delocalization energies of pyrrole, furan, and thiophene are not as great as the delocalization energies of benzene or the cyclopentadienyl anion, each a compound for which the resonance contributors are all equivalent.

relative delocalization energies of some aromatic compounds

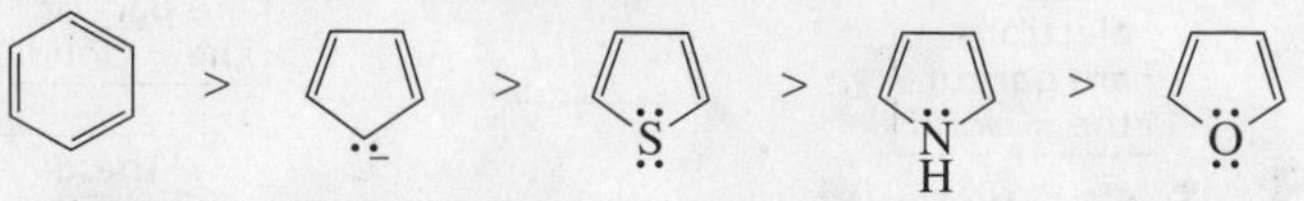

Thiophene, with the least electronegative heteroatom, has the greatest delocalization energy of the three, and furan, with the most electronegative heteroatom, has the smallest delocalization energy. This is what we would expect, because the resonance contributors with a positive charge on the heteroatom are the most stable for the compound with the least electronegative heteroatom and the least stable for the compound with the most electronegative heteroatom.

Because pyrrole, furan, and thiophene are aromatic, they undergo electrophilic aromatic substitution reactions.

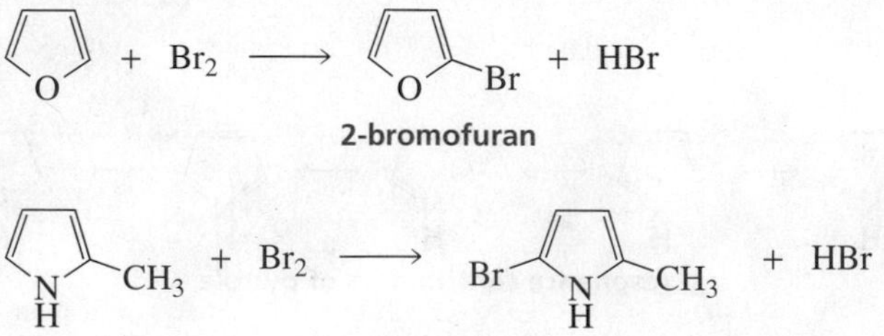

2-bromofuran

2-bromo-5-methylpyrrole

MECHANISM FOR ELECTROPHILIC AROMATIC SUBSTITUTION

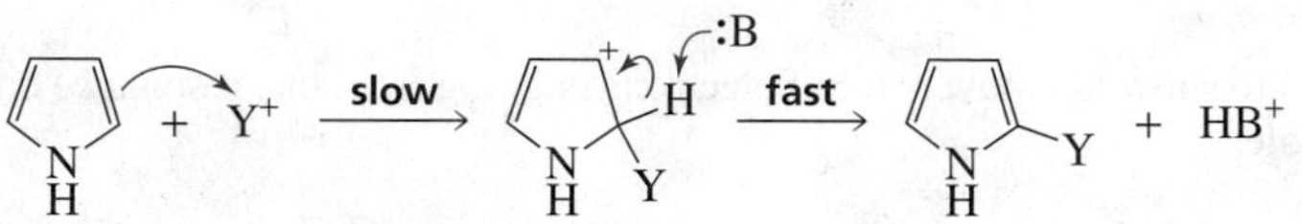

- The electrophile adds to the pyrrole ring.
- A base (:B) in the reaction mixture removes the proton from the carbon that formed the bond with the electrophile.

Substitution occurs preferentially at C-2 because the intermediate obtained by adding a substituent to this position is more stable than the intermediate obtained by adding a substituent to C-3 (Figure 1). Both intermediates have a relatively stable resonance contributor in which all the atoms (except H) have complete octets. The intermediate resulting from C-2 substitution of pyrrole has *two* additional resonance contributors, whereas the intermediate resulting from C-3 substitution has only *one* additional resonance contributor.

Pyrrole, furan, and thiophene undergo electrophilic aromatic substitution, preferentially at C-2.

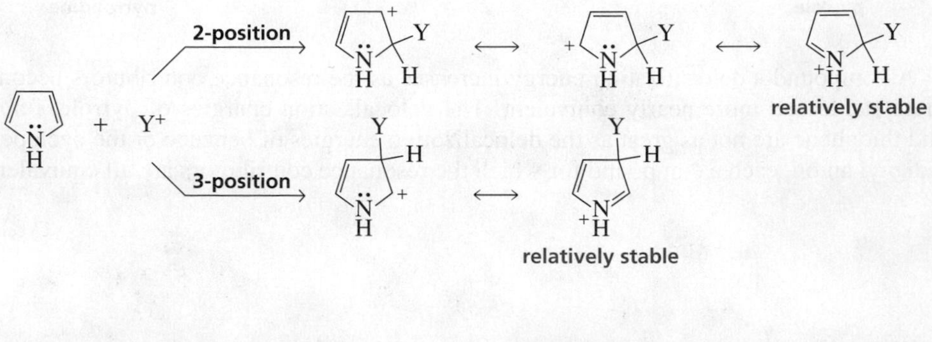

▶ **Figure 1**

Structures of the intermediates that would be formed from the reaction of an electrophile with pyrrole at C-2 and C-3.

If both positions adjacent to the heteroatom are occupied, electrophilic substitution will take place at C-3.

$$CH_3 \overset{S}{\diagup} CH_3 + Br_2 \longrightarrow CH_3 \overset{Br}{\underset{S}{\diagup}} CH_3 + HBr$$

3-bromo-2,5-dimethylthiophene

Pyrrole, furan, and thiophene are all more reactive than benzene toward electrophilic aromatic substitution because they are better able to stabilize the positive charge on the carbocation intermediate, since the lone pair on the heteroatom can donate electrons into the ring by resonance (Figure 1).

Furan is not as reactive as pyrrole in electrophilic aromatic substitution reactions. The oxygen of furan is more electronegative than the nitrogen of pyrrole, so oxygen is not as effective as nitrogen in stabilizing the carbocation. Thiophene is less reactive than furan because the $3p$ orbital of sulfur overlaps less effectively than the $2p$ orbital of nitrogen or oxygen with the $2p$ orbital of carbon. The electrostatic potential maps in the margin illustrate the different electron densities of the three rings.

relative reactivity toward electrophilic aromatic substitution

$$\underset{\substack{N \\ H \\ \text{pyrrole}}}{\diagup} > \underset{\substack{\ddot{O} \\ \text{furan}}}{\diagup} > \underset{\substack{\ddot{S} \\ \text{thiophene}}}{\diagup} > \underset{\text{benzene}}{\bigcirc}$$

Pyrrole, furan, and thiophene are more reactive than benzene toward electrophilic aromatic substitution.

The relative reactivities of pyrrole, furan, and thiophene are reflected in the Lewis acid required to catalyze a Friedel–Crafts acylation. Benzene requires $AlCl_3$, a relatively strong Lewis acid. Thiophene is more reactive than benzene, so it can undergo a Friedel–Crafts reaction using $SnCl_4$, a weaker Lewis acid. An even weaker Lewis acid, BF_3, can be used when the reactant is furan. Pyrrole is so reactive that an anhydride is used instead of a more reactive acyl chloride, and no catalyst is necessary.

$$\bigcirc + \underset{CH_3}{\overset{O}{\overset{\|}{C}}} Cl \xrightarrow[\text{2. } H_2O]{\text{1. } AlCl_3} \bigcirc \overset{O}{\overset{\|}{C}} CH_3 + HCl$$

phenylethanone

$$\overset{S}{\diagup} + \underset{CH_3}{\overset{O}{\overset{\|}{C}}} Cl \xrightarrow[\text{2. } H_2O]{\text{1. } SnCl_4} \overset{S}{\diagup} \overset{CH_3}{\underset{O}{\overset{\|}{C}}} + HCl$$

2-acetylthiophene

$$\overset{O}{\diagup} + \underset{CH_3}{\overset{O}{\overset{\|}{C}}} Cl \xrightarrow[\text{2. } H_2O]{\text{1. } BF_3/THF} \overset{O}{\diagup} \overset{CH_3}{\underset{O}{\overset{\|}{C}}} + HCl$$

2-acetylfuran

$$\underset{\substack{N \\ H}}{\diagup} + \underset{CH_3}{\overset{O}{\overset{\|}{C}}} O \overset{O}{\overset{\|}{C}} CH_3 \longrightarrow \underset{\substack{N \\ H}}{\diagup} \overset{CH_3}{\underset{O}{\overset{\|}{C}}} + CH_3 \overset{O}{\overset{\|}{C}} OH$$

2-acetylpyrrole

pyrrole

furan

thiophene

1027

The resonance hybrid of pyrrole (see earlier in this chapter) shows that there is a partial positive charge on the nitrogen and a partial negative charge on each of the carbons. As a result, pyrrole is protonated on C-2 rather than on nitrogen. This should be expected because a proton is an electrophile, and we have just seen that electrophiles attach to the C-2 position of pyrrole.

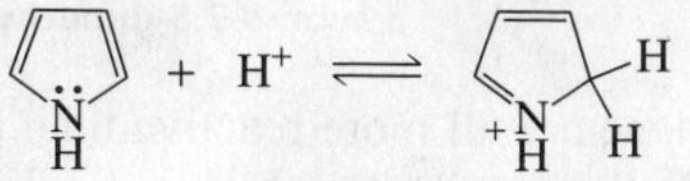

$$pK_a = -3.8$$

Pyrrole is unstable in strongly acidic solutions because, once protonated, it can readily polymerize.

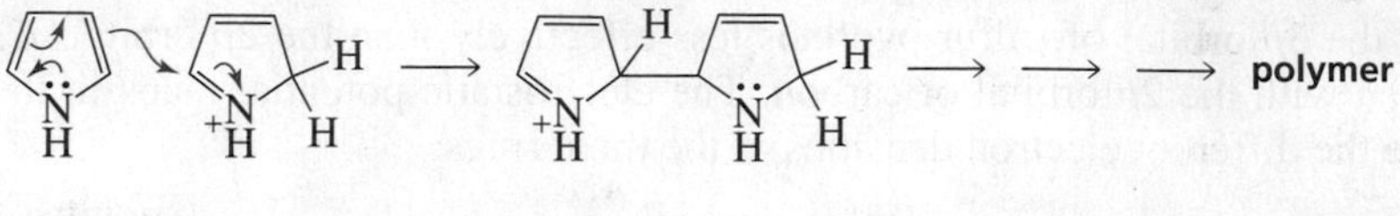

The sp^2 nitrogen in pyrrole is more electronegative than the sp^3 nitrogen in a saturated amine. As a result, pyrrole ($pK_a \sim 17$) is more acidic than the analogous saturated amine ($pK_a \sim 36$). The partial positive charge on the nitrogen atom (apparent in the structure of the resonance hybrid) also contributes significantly to pyrrole's increased acidity.

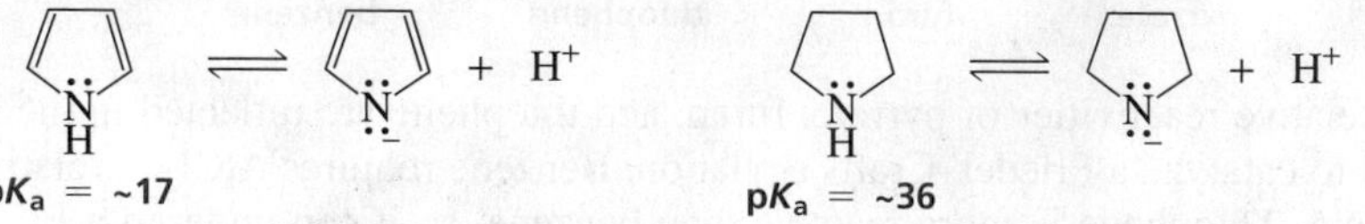

$$pK_a = {\sim}17 \qquad\qquad pK_a = {\sim}36$$

The pK_a values of several nitrogen-containing heterocycles are listed in Table 1.

Table 1   The pK_a Values of Several Nitrogen-Containing Heterocycles

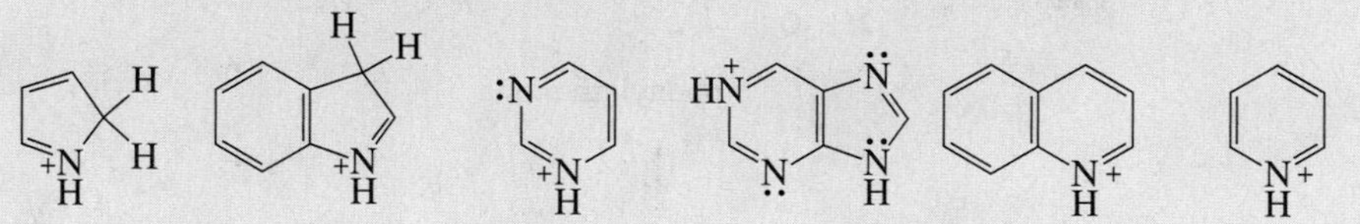

$pK_a = -3.8$	$pK_a = -2.4$	$pK_a = 1.0$	$pK_a = 2.5$	$pK_a = 4.85$	$pK_a = 5.16$
$pK_a = 6.8$	$pK_a = 8.0$	$pK_a = 11.1$	$pK_a = 14.4$	$pK_a \approx 17$	$pK_a \approx 36$

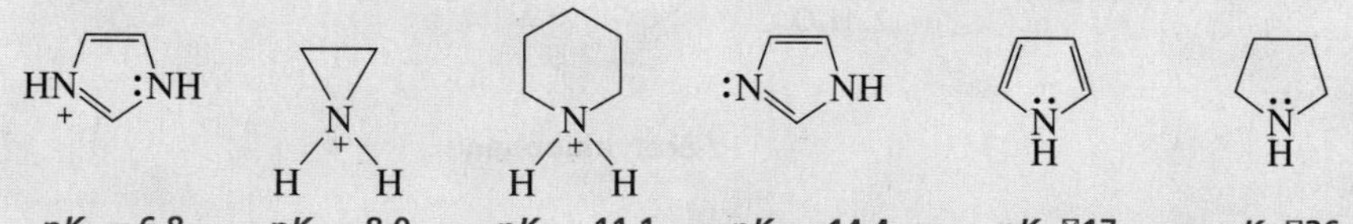

PROBLEM 7

When pyrrole is added to a dilute solution of D_2SO_4 in D_2O, 2-deuteriopyrrole is formed. Propose a mechanism to account for the formation of this compound.

PROBLEM-SOLVING STRATEGY

Determining Relative Basicity

Rank the following compounds in order from least basic to most basic.

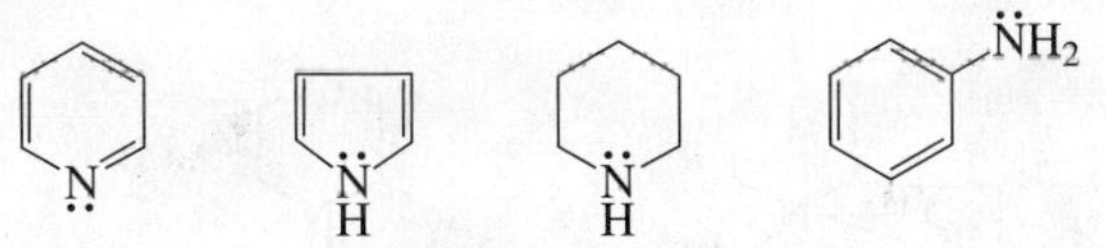

First, we need to see whether any of the compounds will lose their aromaticity if they are protonated. Such compounds will be very difficult to protonate, so they will be very weak bases. Pyrrole is one such compound. Next, we need to see whether any of the compounds has a lone pair that can be delocalized. A delocalized lone pair will be harder to protonate than one that is localized. Aniline is such a compound. Finally, we need to look at the hybridization of the nitrogens. The nitrogen of the unsaturated six-membered ring is sp^2 hybridized, whereas the nitrogen of the saturated six-membered ring is sp^3 hybridized. sp^2 nitrogens are more electronegative and therefore harder to protonate than sp^3 nitrogens. Thus, the order of basicity is

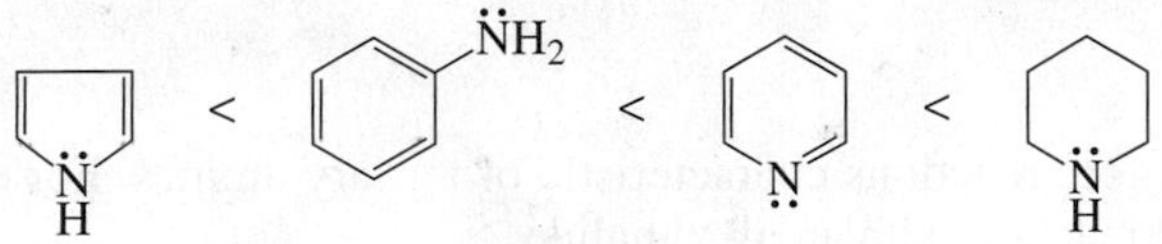

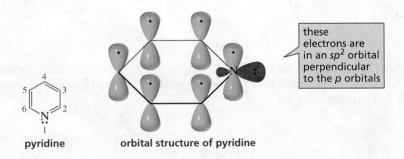

Now use the strategy you have just learned to solve Problem 8.

PROBLEM 8♦

Explain why cyclopentadiene ($pK_a = 15$) is more acidic than pyrrole ($pK_a \sim 17$), even though nitrogen is more electronegative than carbon.

6 AROMATIC SIX-MEMBERED-RING HETEROCYCLES

Now we will look at heterocycles that have a six-membered aromatic ring.

Pyridine

Pyridine is an aromatic compound with a nitrogen in place of one of the carbons in a benzene ring.

pyridine   orbital structure of pyridine

these electrons are in an sp^2 orbital perpendicular to the p orbitals

The pyridinium ion is a stronger acid than a typical ammonium ion because its proton is attached to an sp^2 nitrogen, which is more electronegative than an sp^3 nitrogen.

pyridinium ion
pK_a = 5.16

pyridine

piperidinium ion
pK_a = 11.12

piperidine

benzene

pyridine

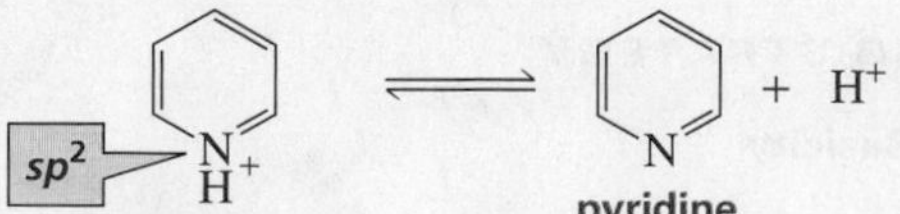

The dipole moment of pyridine is 1.57 D. As the electrostatic potential map indicates, the electron-withdrawing nitrogen is the negative end of the dipole.

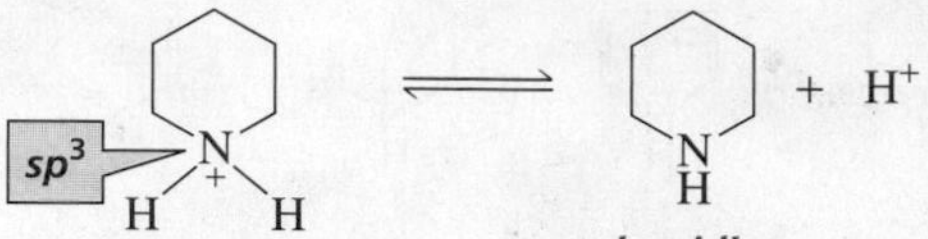

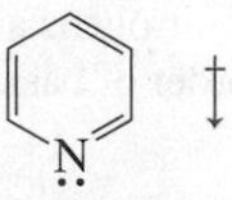

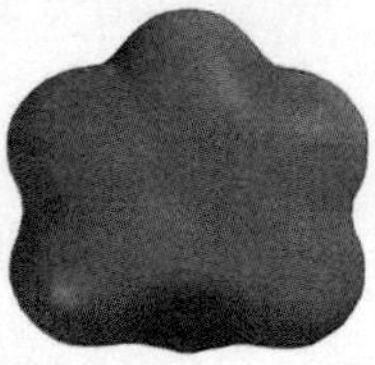

μ = 1.57 D

Pyridine undergoes reactions characteristic of tertiary amines. For example, pyridine undergoes an S_N2 reaction with an alkyl halide.

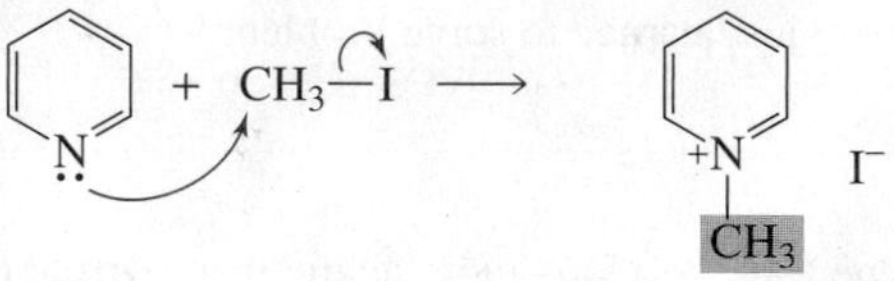

N-methylpyridinium iodide

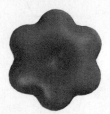

PROBLEM 9 Solved

Will an amide be the final product obtained from the reaction of an acyl chloride with pyridine in an aqueous solution? Explain your answer.

Solution The positively charged nitrogen in the initially formed carbonyl compound causes pyridine to be an excellent leaving group. Therefore, the compound hydrolyzes rapidly. As a result, the final product of the reaction is a carboxylic acid. (If the final pH of the solution is greater than the pK_a of the carboxylic acid, the carboxylic acid will be predominantly in its basic form.)

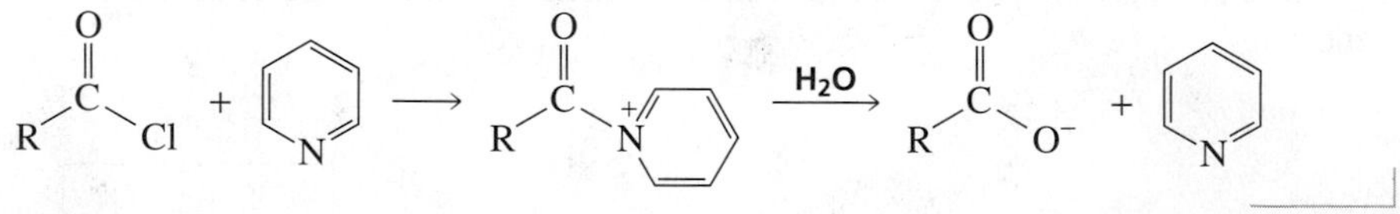

Pyridine is aromatic. Like benzene, it has two uncharged resonance contributors. Because of the electron-withdrawing nitrogen, pyridine has three charged resonance contributors that benzene does not have.

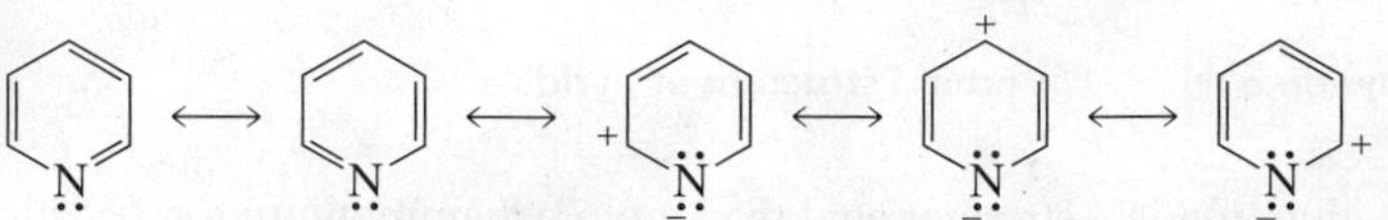

resonance contributors of pyridine

Because it is aromatic, pyridine (like benzene) undergoes electrophilic aromatic substitution reactions.

MECHANISM FOR ELECTROPHILIC AROMATIC SUBSTITUTION

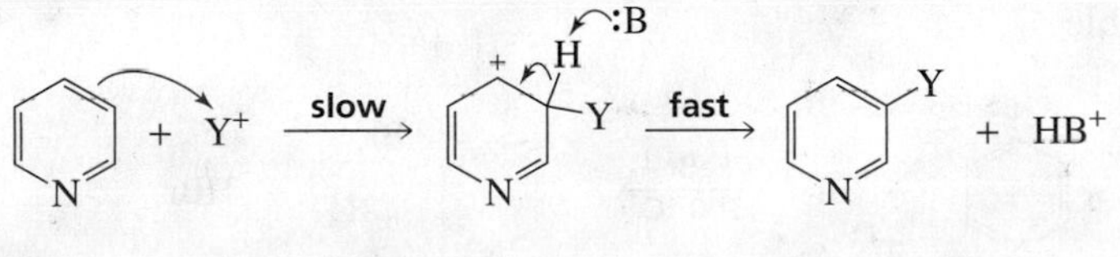

- The electrophile adds to the pyridine ring.
- A base (:B) in the reaction mixture removes the proton from the carbon that formed the bond with the electrophile.

Electrophilic aromatic substitution of pyridine takes place at C-3 because the most stable intermediate is obtained by placing an electrophilic substituent at that position (Figure 2). When the substituent is placed at C-2 or C-4, one of the resulting resonance contributors is particularly unstable because its nitrogen atom has an incomplete octet *and* a positive charge.

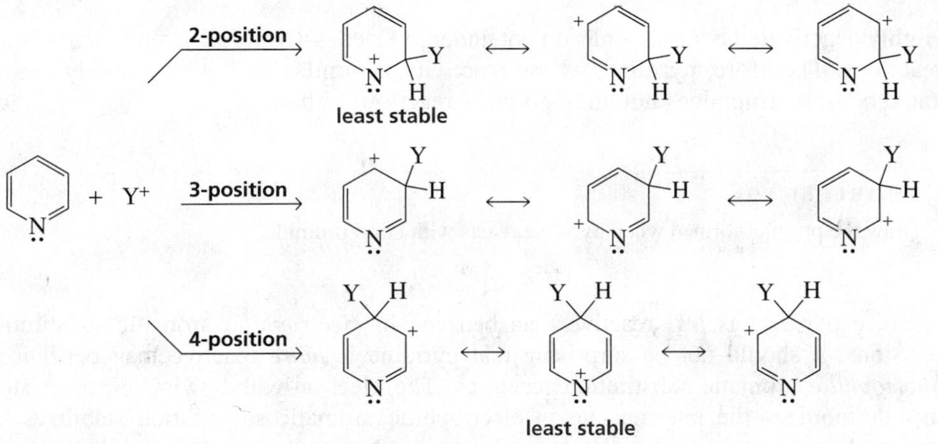

▲ **Figure 2**
Structures of the intermediates that would be formed from the reaction of an electrophile with pyridine.

The electron-withdrawing nitrogen atom makes the intermediate obtained from electrophilic aromatic substitution of pyridine less stable than the carbocation intermediate obtained from electrophilic aromatic substitution of benzene. Pyridine, therefore, is less reactive than benzene. Indeed, it is even less reactive than nitrobenzene. (Note that an electron-withdrawing nitro group strongly deactivates a benzene ring toward electrophilic aromatic substitution.)

relative reactivity toward electrophilic aromatic substitution

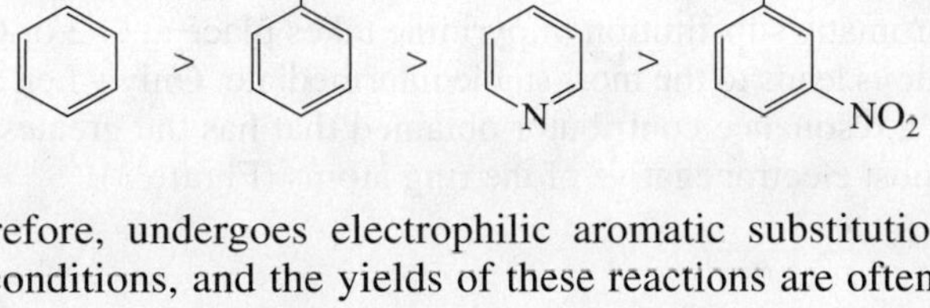

Pyridine, therefore, undergoes electrophilic aromatic substitution reactions only under vigorous conditions, and the yields of these reactions are often quite low. If the nitrogen becomes protonated under the reaction conditions, the reactivity decreases further because a positively charged nitrogen would make the carbocation intermediate even less stable.

Pyridine undergoes electrophilic aromatic substitution at C-3.

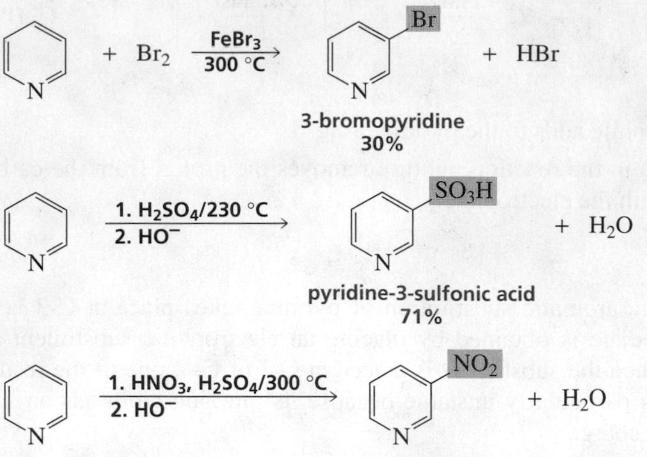

3-bromopyridine
30%

pyridine-3-sulfonic acid
71%

3-nitropyridine
22%

Highly deactivated benzene rings do not undergo Friedel–Crafts alkylation or acylation reactions. Therefore, pyridine, whose reactivity is similar to that of a highly deactivated benzene ring, does not undergo these reactions either.

PROBLEM 10♦

Draw the product formed when pyridine reacts with ethyl bromide.

Since pyridine is *less* reactive than benzene in *electrophilic* aromatic substitution reactions, it should not be surprising that pyridine is *more* reactive than benzene in *nucleophilic* aromatic substitution reactions. The electron-withdrawing nitrogen atom that destabilizes the intermediate in electrophilic aromatic substitution stabilizes the intermediate in nucleophilic aromatic substitution. Notice that, in nucleophilic aromatic substitution reactions, the ring has a leaving group that can be replaced by a nucleophile.

Pyridine is *less* reactive than benzene toward electrophilic aromatic substitution and *more* reactive than benzene toward nucleophilic aromatic substitution.

MECHANISM FOR NUCLEOPHILIC AROMATIC SUBSTITUTION

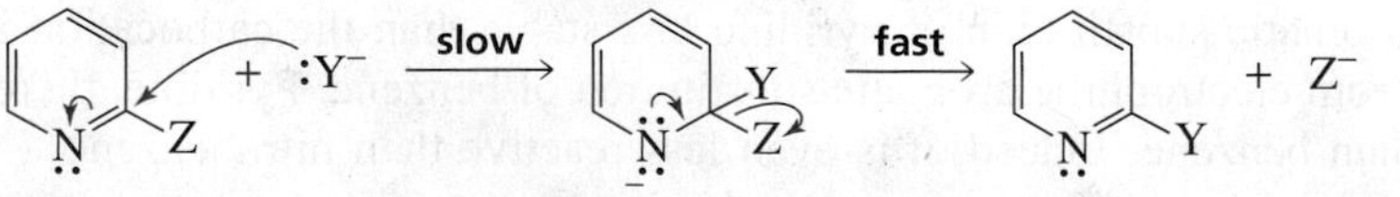

- The nucleophile adds to the ring carbon attached to the leaving group.
- The leaving group is eliminated.

Nucleophilic aromatic substitution of pyridine takes place at C-2 or C-4 because addition to these positions leads to the most stable intermediate. Only when addition occurs to these positions is a resonance contributor obtained that has the greatest electron density on nitrogen, the most electronegative of the ring atoms (Figure 3).

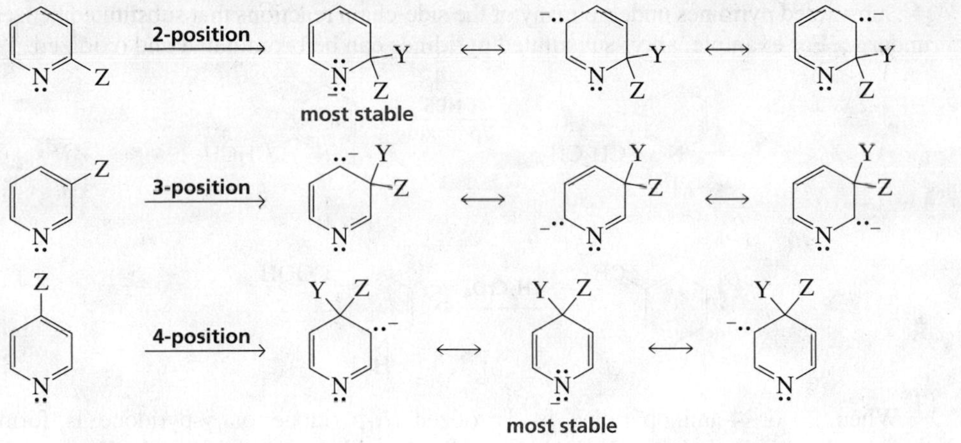

◀ **Figure 3**
Structures of the intermediates that would be formed from the reaction of a nucleophile with a substituted pyridine.

If the leaving groups at C-2 and C-4 are different, then the incoming nucleophile will preferentially substitute for the weaker base (the better leaving group).

Pyridine undergoes nucleophilic aromatic substitution at C-2 or C-4.

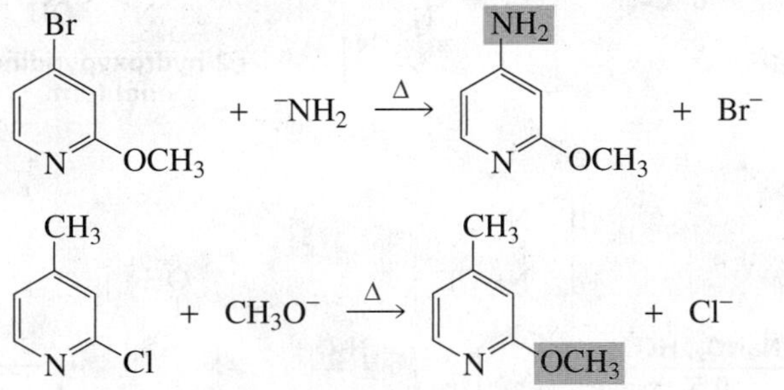

PROBLEM 11

Compare the mechanisms of the following reactions:

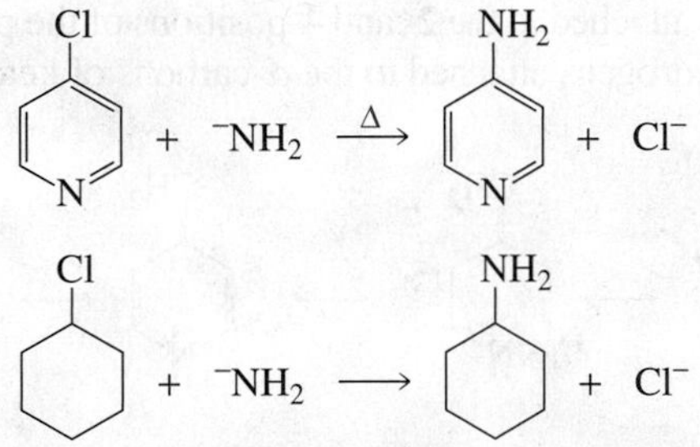

PROBLEM 12

a. Propose a mechanism for the following reaction:

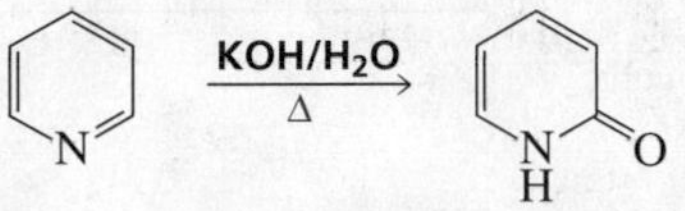

b. What other product is formed in this reaction?

Substituted pyridines undergo many of the side-chain reactions that substituted benzenes undergo. For example, alkyl-substituted pyridines can be brominated and oxidized.

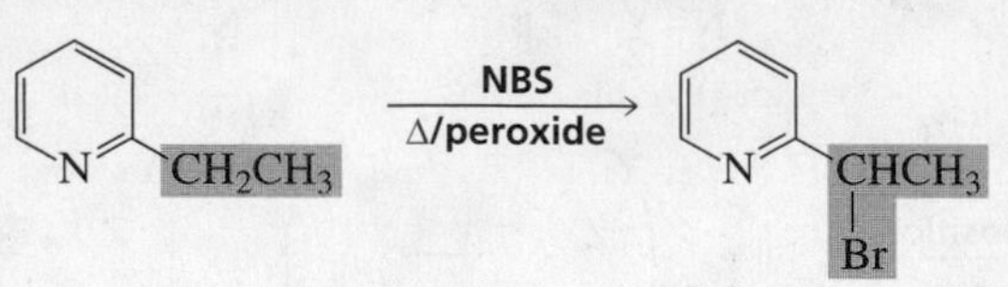

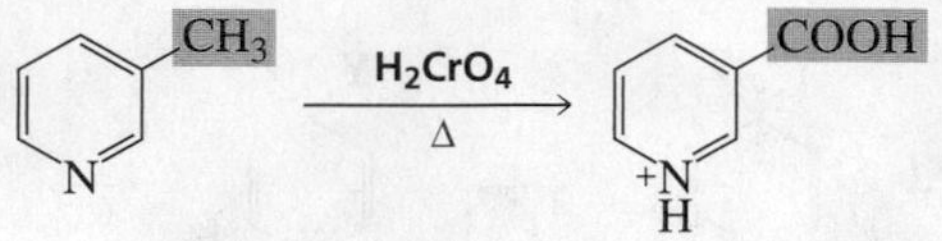

When 2- or 4-aminopyridine is diazotized, α-pyridone or γ-pyridone is formed. Apparently, the diazonium salt reacts immediately with water to form a hydroxypyridine. The product of the reaction is a pyridone because the keto form of a hydroxypyridine is more stable than the enol form.

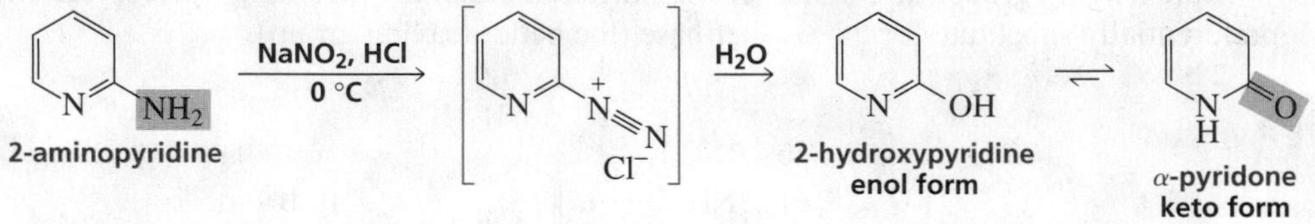

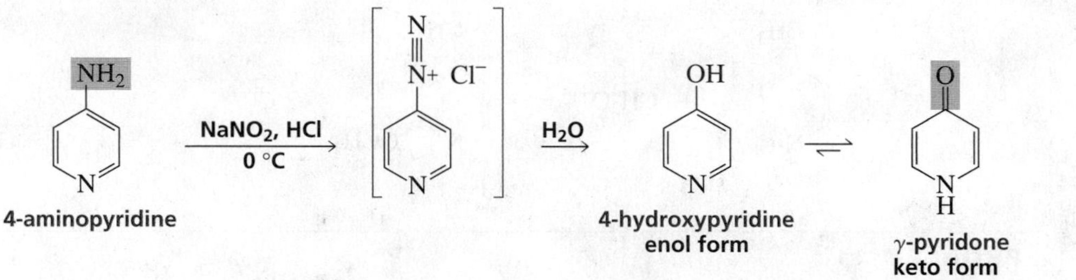

The electron-withdrawing nitrogen and the ability to delocalize the negative charge causes the hydrogens on carbons attached to the 2- and 4-positions of the pyridine ring to have about the same acidity as the hydrogens attached to the α-carbons of ketones.

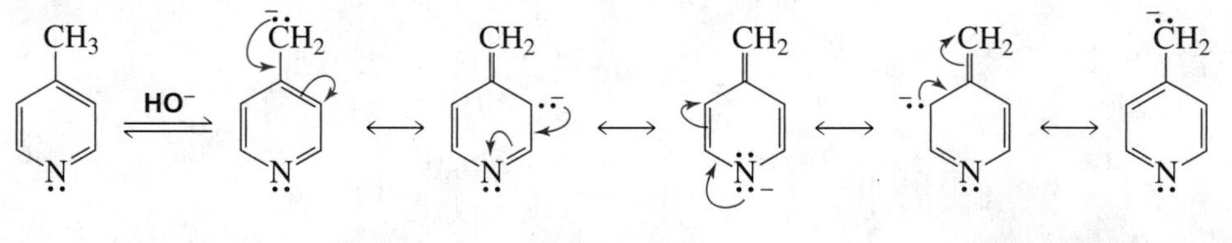

Consequently, the hydrogens can be removed by base, and the resulting carbanions can react as nucleophiles.

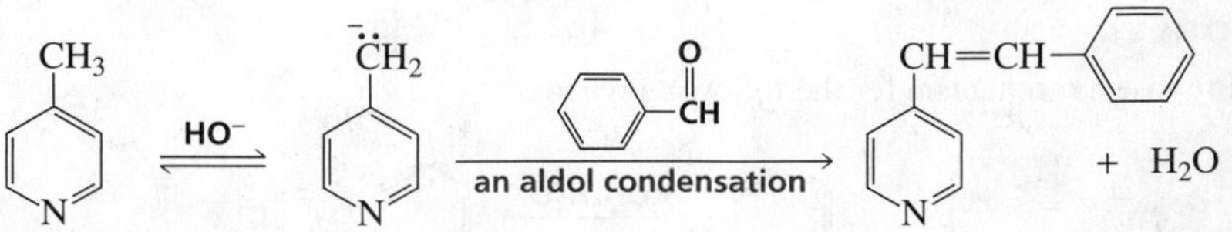

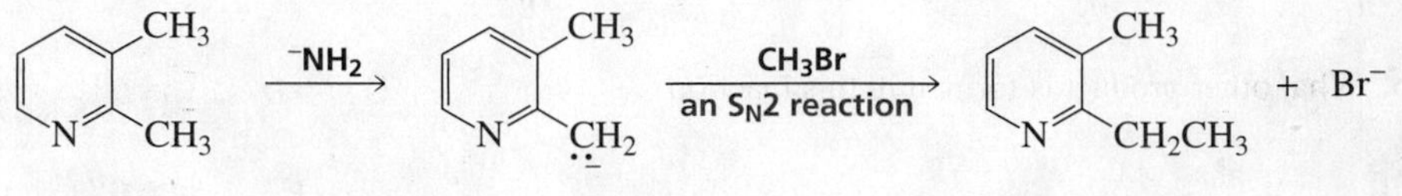

PROBLEM 13♦

Rank the following compounds in order from easiest to hardest at removing a proton from a methyl group:

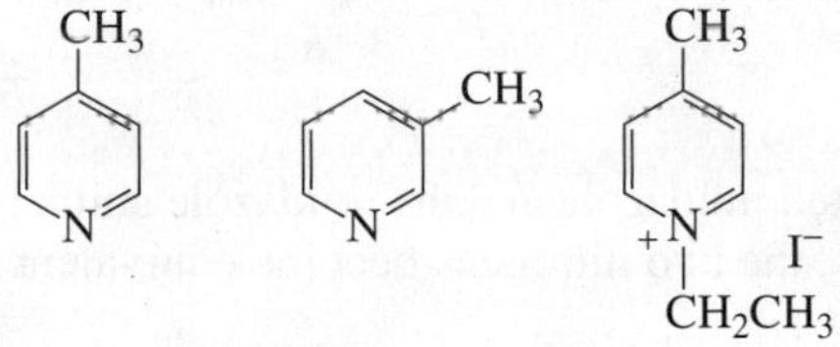

7 SOME AMINE HETEROCYCLES HAVE IMPORTANT ROLES IN NATURE

Proteins are naturally occurring polymers of α-amino acids. Three of the 20 most common amino acids in proteins contain heterocyclic rings. Proline has a pyrrolidine ring; typtophan has an indole ring; and histidine has an imidazole ring.

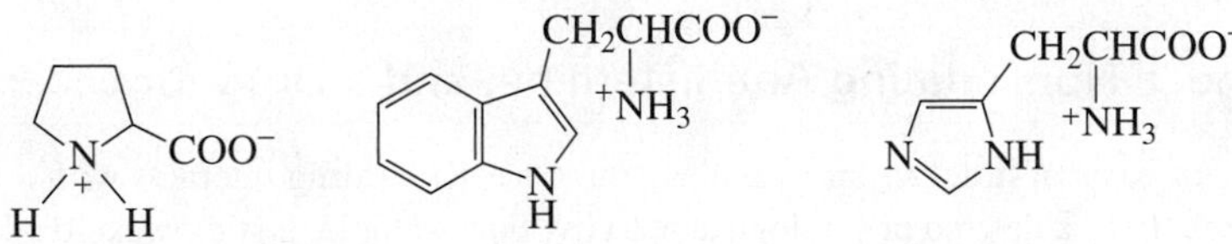

proline   **tryptophan**   **histidine**

Imidazole

Imidazole, the heterocyclic ring of histidine, is an aromatic compound because it is cyclic and planar, every atom in the ring has a p orbital, and the π cloud contains *three* pairs of π electrons. The electrons drawn as lone-pair electrons on N-1 are part of the π cloud because they are in a p orbital, whereas the lone-pair electrons on N-3 are not part of the π cloud because they are in an sp^2 orbital that is perpendicular to the p orbitals.

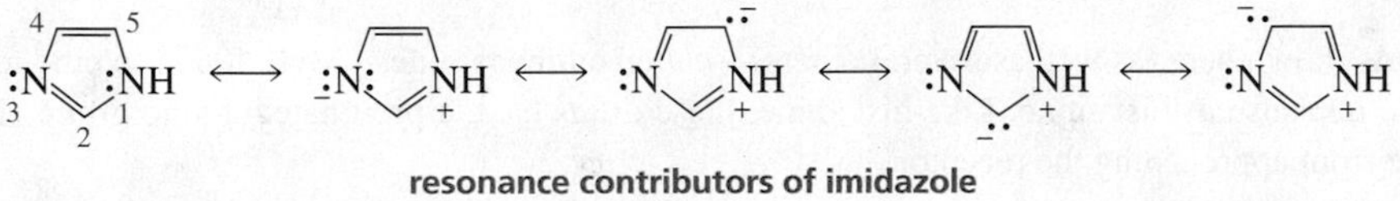

resonance contributors of imidazole

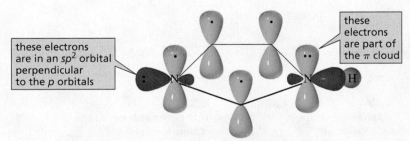

these electrons are in an sp^2 orbital perpendicular to the p orbitals

these electrons are part of the π cloud

orbital structure of imidazole

Because the lone-pair electrons in the sp^2 orbital are not part of the π cloud, imidazole is protonated in acidic solutions. The conjugate acid of imidazole has a $pK_a = 6.8$. Thus, imidazole exists in both the protonated and unprotonated forms at physiological pH (7.4).

This is one of the reasons that histidine, the imidazole-containing amino acid, is a catalytic component of many enzymes.

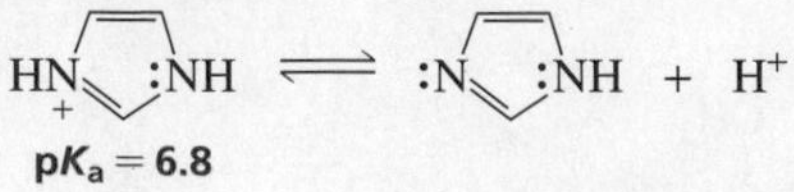

$$HN \overset{+}{\underset{}{}} :NH \rightleftharpoons :N :NH + H^+$$

$$pK_a = 6.8$$

Notice that both protonated imidazole and the imidazole anion have two equivalent resonance contributors. Thus, the two nitrogens become equivalent when imidazole is either protonated or deprotonated.

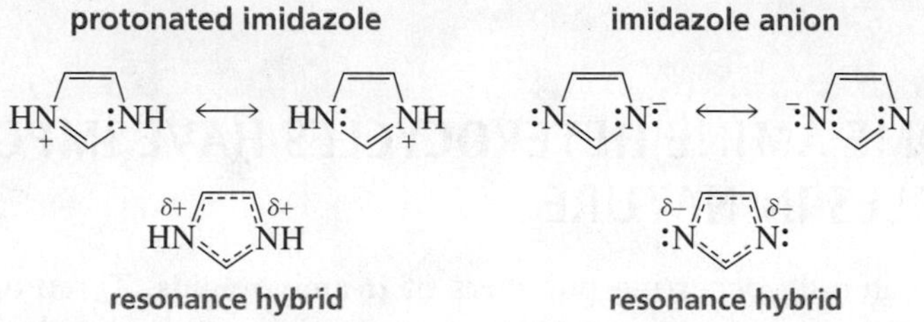

protonated imidazole

imidazole anion

resonance hybrid

resonance hybrid

Searching for Drugs: An Antihistamine, a Nonsedating Antihistamine, and a Drug for Ulcers

When scientists know something about the molecular basis of drug action—such as how a particular drug interacts with a receptor—they can design and synthesize compounds that might have a desired physiological activity. For example, histidine can be decarboxylated in an enzyme-catalyzed reaction to form histamine. When the body produces excess histamine, it causes the symptoms associated with the common cold and allergies. Knowing that the amine is positively charged at physiological pH gave scientists one clue as to how it might interact with its receptor.

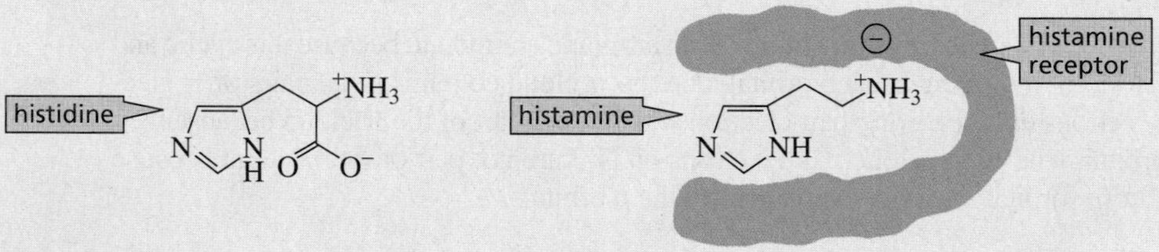

After an extensive search, the compounds shown here (as well as several others)—called antihistamines—were found to bind to the histamine receptor but not trigger the same response as histamine. Like histamine, these drugs have a protonated amino group. They also have bulky groups that keep histamine from approaching the receptor.

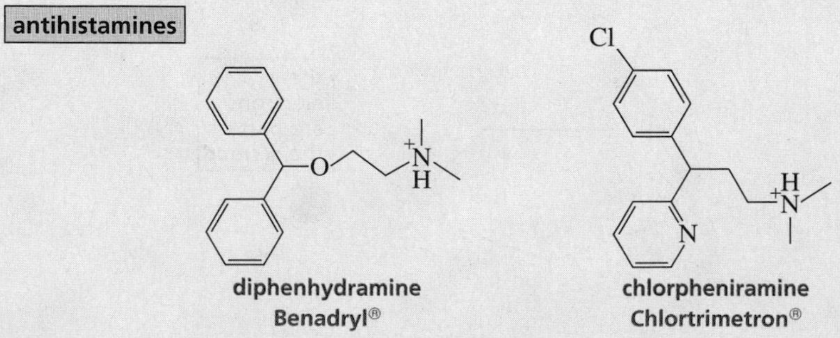

antihistamines

diphenhydramine
Benadryl®

chlorpheniramine
Chlortrimetron®

These compounds, however, are able to cross the blood–brain barrier and bind to receptors in the central nervous system. Binding to these receptors causes drowsiness, the well-known side effect associated with these drugs. So now the search was on for compounds that would bind to the histamine receptor but not cross the blood–brain barrier. Because the blood–brain barrier is nonpolar, this was achieved by putting polar groups on the compounds. Allegra, Claritin, and Zyrtec are nonsedating antihistamines. Zyrtec became available as an over-the-counter drug in 2007.

nonsedating antihistamines

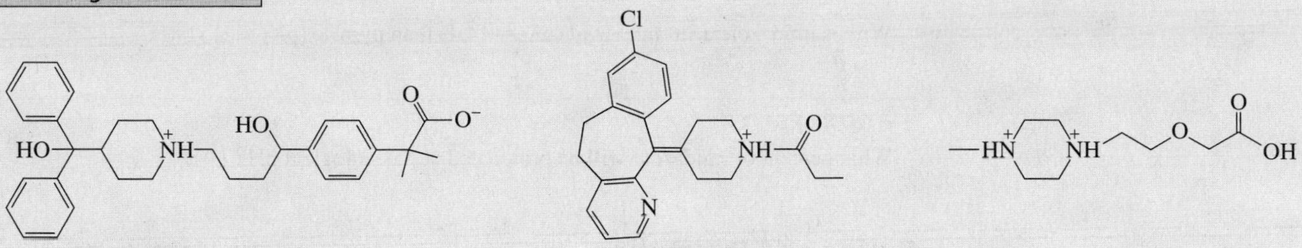

fexofenadine Allegra®	**loratadine** Claritin®	**cetirizine** Zyrtec®

In addition to causing allergic responses, excess histamine production by the body also causes the hypersecretion of HCl by the cells of the stomach lining, which leads to the development of ulcers. The antihistamines that prevent allergic responses have no effect on HCl production. This suggested that a second kind of histamine receptor—called a histamine H_2-receptor—causes the secretion of HCl.

Because 4-methylhistamine was found to cause weak inhibition of HCl secretion, it was used as a lead compound. More than 500 molecular modifications were performed over a 10-year period before a compound was found that would bind to the histamine H_2-receptor.

Tagamet, introduced in 1976, was the first drug for the treatment of peptic ulcers. Previously, the only treatment was extensive bed rest, a bland diet, and antacids. Zantac followed in 1981, and by 1988 it became the world's best-selling prescription drug. Protonix became available in 1994.

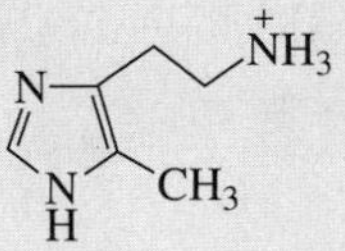

4-methylhistamine

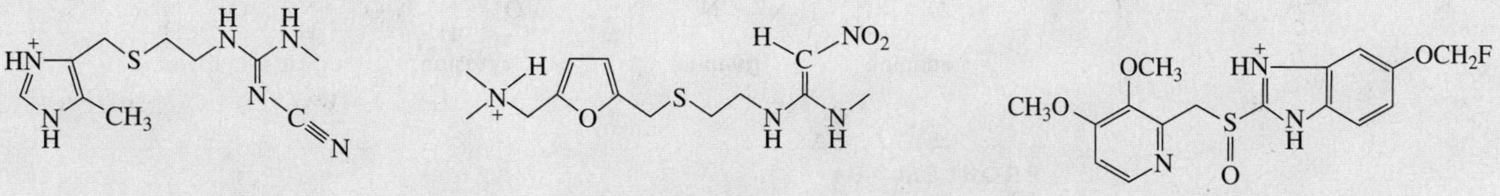

cimetidine Tagamet®	**ranitidine** Zantac®	**pantropazole** Protonix®

PROBLEM 14♦

What is the major product of the following reaction?

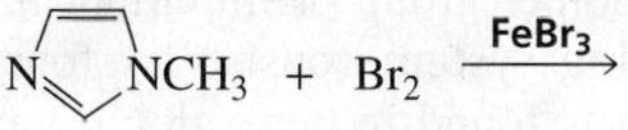

PROBLEM 15♦

List imidazole, pyrrole, and benzene in order from most reactive to least reactive toward electrophilic aromatic substitution.

PROBLEM 16♦

Imidazole boils at 257 °C, whereas *N*-methylimidazole boils at 199 °C. Explain this difference in boiling points.

PROBLEM 17

Why is imidazole a stronger acid (pK_a = 14.4) than pyrrole (pK_a ~ 17)?

PROBLEM 18

What percent of imidazole will be protonated at physiological pH (7.4)?

Purine and Pyrimidine

Nucleic acids (DNA and RNA) contain substituted **purines** and substituted **pyrimidines**; DNA contains adenine, guanine, cytosine, and thymine (abbreviated A, G, C, and T); RNA contains adenine, guanine, cytosine, and uracil (A, G, C, and U). Why DNA contains T instead of U is beyond the scope of this chapter. Unsubstituted purine and pyrimidine are not found in nature. Guanine (a hydroxypurine) and cytosine, uracil, and thymine (hydroxypyrimidines) are more stable in the keto form than in the enol form, so the keto forms are shown here. We will see that the preference for the keto form is crucial for proper base pairing in DNA.

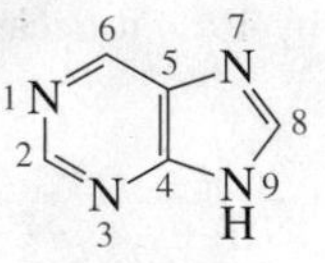

purine

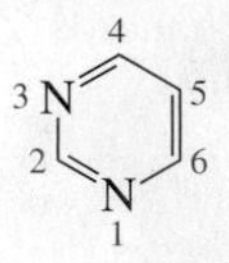

pyrimidine

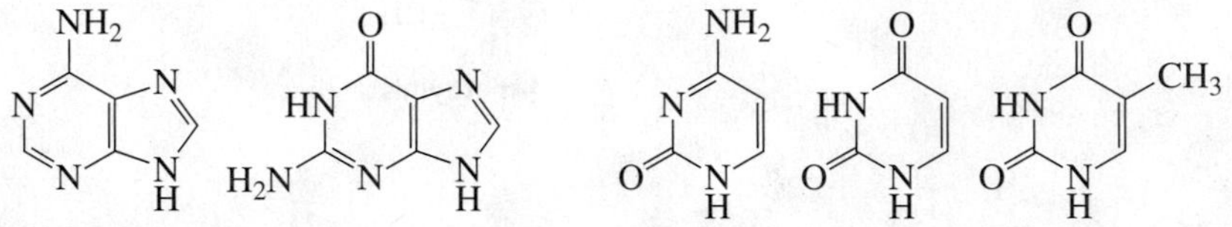

adenine   guanine   cytosine   uracil   thymine

PROBLEM 19♦

Draw guanine and cytosine in the enol form.

PROBLEM 20♦

Why is protonated pyrimidine (pK_a = 1.0) more acidic than protonated pyridine (pK_a = 5.2)?

Porphyrin

Substituted *porphyrins* are another group of important naturally occurring heterocyclic compounds. A **porphyrin ring system** consists of four pyrrole rings joined by one-carbon bridges. Heme, which is found in hemoglobin and myoglobin, contains an iron ion (Fe^{II}) ligated by the four nitrogens of a porphyrin ring system. **Ligation** is the sharing of lone-pair electrons with a metal ion.

a porphyrin ring system

heme

Hemoglobin has four polypeptide chains and four heme groups; myoglobin has one polypeptide chain and one heme group. Hemoglobin is responsible for transporting oxygen to cells and carbon dioxide away from cells, whereas myoglobin is responsible for storing oxygen in cells. The iron atoms in hemoglobin and myoglobin, in addition to being ligated to the four nitrogens of the porphyrin ring, are also ligated to a histidine of the protein component (globin), and the sixth ligand is oxygen or carbon dioxide.

Carbon monoxide is about the same size and shape as O_2, but CO binds more tightly than O_2 to iron. In addition, once a hemoglobin molecule has bound two CO molecules, it can no longer achieve the configuration necessary to bind O_2. Consequently, breathing carbon monoxide can be fatal because it interferes with the transport of oxygen through the bloodstream.

The extensive conjugated system of heme gives blood its characteristic red color. Its high molar absorptivity (about 160,000 $M^{-1}cm^{-1}$) allows concentrations as low as 1×10^{-8} M to be detected by UV spectroscopy.

The ring system in chlorophyll a, the substance responsible for the green color of plants, is similar to porphyrin but it contains a cyclopentanone ring and one of its pyrrole rings is partially reduced. The metal ion in chlorophyll a is magnesium (Mg^{II}).

Vitamin B_{12} also has a ring system similar to porphyrin, but in this case, the metal ion is cobalt (Co^{III}).

Porphyrin, Bilirubin, and Jaundice

The average human body breaks down about 6 g of hemoglobin each day. The protein portion (globin) and the iron are reutilized, but the porphyrin ring is cleaved between the A and B rings to form a linear tetrapyrrole called biliverdin (a green compound). The bridge between the C and D ring is reduced, forming bilirubin (a red-orange compound). You can witness heme degradation by observing the changing colors of a bruise.

Enzymes in the large intestine convert bilirubin to urobilinogen. Some urobilinogen is transported to the kidney, where it is converted to urobilin (a yellow compound), which is excreted. This is the compound that gives urine its characteristic color. If more bilirubin is formed than can be metabolized and excreted by the liver, it accumulates in the blood. When its concentration there reaches a certain level, it diffuses into the tissues, giving them a yellow appearance. This condition is known as jaundice.

8 ORGANIZING WHAT WE KNOW ABOUT THE REACTIONS OF ORGANIC COMPOUNDS

The families of organic compounds can be put into one of four groups, and all the members of a group react in similar ways. Now that we have discussed the families in Group IV, let's go over how these compounds react.

I

R—CH=CH—R
an alkene

R—C≡C—R
an alkyne

R—CH=CH—CH=CH—R
a diene

These are nucleophiles.

They undergo electrophilic addition reactions.

II

R—X $X = F, Cl, Br, I$
an alkyl halide

R—OH
an alcohol

R—OR
an ether

R—R   (epoxide)
an epoxide

These are electrophiles.

They undergo nucleophilic substitution and/or elimination reactions.

III

$\underset{R}{\overset{O}{\|}}C{-}Z$ $Z =$ an atom more electronegative than C
carboxylic acid and carboxylic acid derivatives

$\underset{R}{\overset{O}{\|}}C{-}Z$ $Z = C$ or H
aldehydes and ketones

These are electrophiles.

They undergo nucleophilic addition–elimination reactions or nucleophilic addition reactions.

Removal of a hydrogen from an α-carbon forms a nucleophile that can react with electrophiles.

IV

benzene

pyridine

$Z = N, O,$ or S (H)
pyrrole, furan, thiophene

These are nucleophiles.

They undergo electrophilic and/or nucleophilic aromatic substitution reactions.

All the compounds in Group IV are aromatic. In order to preserve the aromaticity of the rings, these compounds undergo electrophilic aromatic substitution reactions and/or nucleophilic aromatic substitution reactions.

- Benzene and substituted benzenes are nucleophiles, so they react with electrophiles in electrophilic aromatic substitution reactions. If the electron density of the benzene ring is reduced by strongly electron-withdrawing groups, a halo-substituted benzene can undergo a nucleophilic aromatic substitution reaction.

- Pyrrole, furan, and thiophene are much *more* nucleophilic than benzene, so they are more reactive in electrophilic aromatic substitution reactions.

- Pyridine is much *less* nucleophilic than benzene, so it undergoes electrophilic aromatic substitution reactions only under rigorous conditions. However, halo-substituted benzenes readily undergo nucleophilic aromatic substitution reactions.

SOME IMPORTANT THINGS TO REMEMBER

- Amines are classified as primary, secondary, or tertiary, depending on whether one, two, or three hydrogens of ammonia have been replaced by alkyl groups.

- Some amines are **heterocyclic compounds**—cyclic compounds in which one or more of the atoms of the ring is an atom other than carbon.

- Heterocyclic rings are numbered so that the **heteroatom** has the lowest possible number.

- The lone pair on the nitrogen causes amines to be both bases and nucleophiles.

- Amines react as bases in acid–base reactions and in elimination reactions.

- Amines react as nucleophiles in nucleophilic substitution reactions, in nucleophilic addition–elimination reactions, and in conjugate addition reactions.

- Saturated heterocycles have physical and chemical properties like those of acyclic compounds that contain the same heteroatom.

- **Pyrrole, furan,** and **thiophene** are heterocyclic aromatic compounds that undergo electrophilic aromatic substitution reactions preferentially at C-2. They are more reactive than benzene toward electrophilic aromatic substitution.

- Protonating pyrrole destroys its aromaticity. Pyrrole polymerizes in strongly acidic solutions.

- Replacing one of benzene's carbons with a nitrogen forms **pyridine,** a heterocyclic aromatic compound that undergoes electrophilic aromatic substitution reactions at C-3 and nucleophilic aromatic substitution reactions at C-2 or C-4. Pyridine is *less* reactive than benzene toward electrophilic aromatic substitution and *more* reactive than benzene toward nucleophilic aromatic substitution.

- Three amino acids have heterocyclic rings: histidine has an **imidazole** ring, proline has a **pyrrolidine** ring, and tryptophan has an **indole** ring.

- Nucleic acids (DNA and RNA) contain substituted **purines** and substituted **pyrimidines.** Hydroxypurines and hydroxypyrimidines are more stable in the keto form.

- A **porphyrin ring system** consists of four pyrrole rings joined by one-carbon bridges; in hemoglobin and myoglobin, the four nitrogen atoms are ligated to Fe^{II}. The metal ion in chlorophyll a is magnesium, and the metal ion in vitamin B_{12} is cobalt.

SUMMARY OF REACTIONS

1. Reactions of amines as nucleophiles (Section 3)
   a. In alkylation reactions
   b. In reactions with a carbonyl group that has a leaving group
   c. In reactions with an aldehyde or ketone to form an imine or an enamine
   d. In conjugate addition reactions
   e. In reactions with nitrous acid: primary arylamines form arenediazonium salts

2. Electrophilic aromatic substitution reactions
   a. Pyrrole, furan, and thiophene (Section 5):

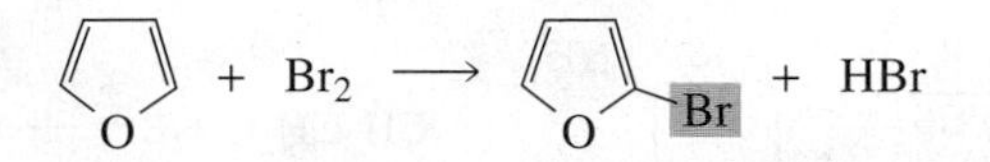

   b. Pyridine (Section 6): the mechanism is shown earlier in the chapter.

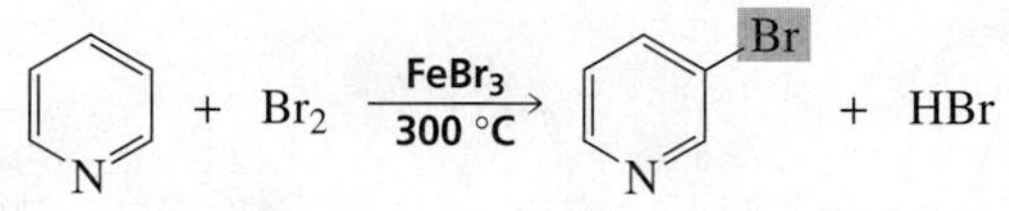

3. Nucleophilic aromatic substitution reactions of pyridine (Section 6). The mechanism is shown earlier in the chapter.

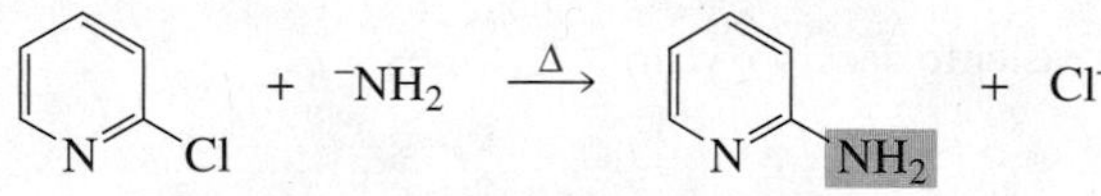

4. Reactions of substituents on pyridine (Section 6)

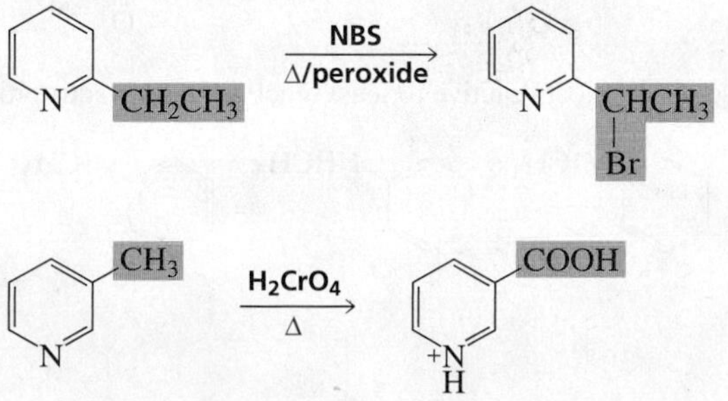

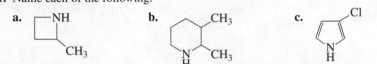

PROBLEMS

21. Name each of the following:

a. [structure: azetidine with NH and CH₃]

b. [structure: piperidine ring with CH₃ groups]

c. [structure: pyrrole with Cl]

d. [structure: piperidine with CH₃CH₂ and CH₃]

22. What are the products of the following reactions?

a. [pyrrolidine] + [benzoyl chloride] ⟶

b. [2-bromopyridine] + HO⁻ $\xrightarrow{\Delta}$

c. CH₃CH₂CH₂CH₂Br $\xrightarrow[\text{2. H}_2\text{, Raney Ni}]{\text{1. }^-\text{C}\equiv\text{N}}$

d. [furan with CH₃] + [acetyl chloride] $\xrightarrow[\text{2. H}_2\text{O}]{\text{1. BF}_3\text{/THF}}$

e. [2-chloropyridine] $\xrightarrow{\text{C}_6\text{H}_5\text{Li}}$

f. [cyclohexylamine, NH₂] + CH₃CH₂CH₂Br ⟶

g. [pyrrole] + C₆H₅N⁺≡N ⟶

h. [2-methylpyridine] $\xrightarrow[\text{2. CH}_3\text{CH}_2\text{CH}_2\text{Br}]{\text{1. }^-\text{NH}_2}$

i. [2-bromothiophene] $\xrightarrow[\substack{\text{2. CO}_2 \\ \text{3. HCl}}]{\text{1. Mg/Et}_2\text{O}}$

23. List the following compounds in order from strongest acid to weakest acid:

[seven structures: protonated pyridine, protonated pyrrole, piperidinium, protonated imidazole, imidazole, pyrrole, protonated piperidine]

24. Which of the following compounds is easier to decarboxylate?

[pyridine-2-carboxylic acid, COH with O]   or   [pyridine-2-acetic acid, CH₂COH with O]

25. Rank the following compounds in order from most reactive to least reactive in an electrophilic aromatic substitution reaction:

[benzene-OCH₃]   [benzene-NHCH₃]   [benzene-SCH₃]

26. One of the following compounds undergoes electrophilic aromatic substitution predominantly at C-3, and one undergoes electrophilic aromatic substitution predominantly at C-4. Which is which?

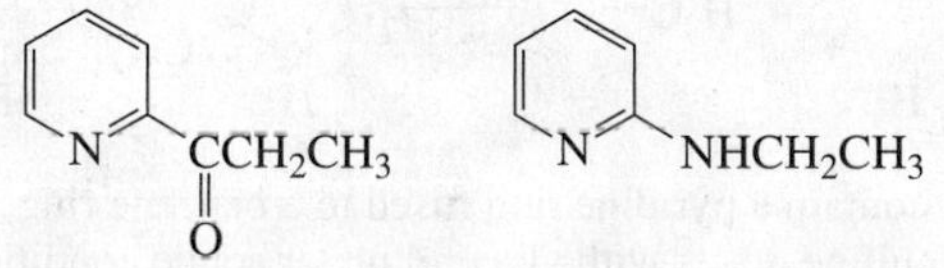

27. Benzene undergoes electrophilic aromatic substitution reactions with aziridines in the presence of a Lewis acid such as $AlCl_3$.

a. What are the major and minor products of the following reaction?

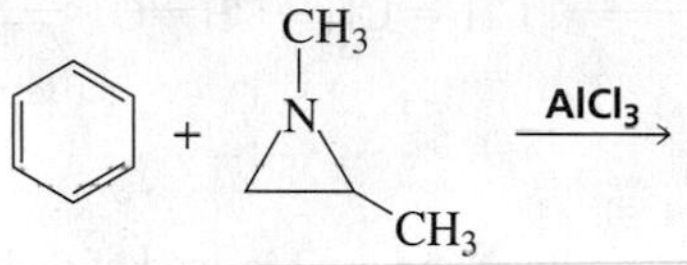

b. Would you expect epoxides to undergo similar reactions?

28. The dipole moments of furan and tetrahydrofuran are in the same direction. One compound has a dipole moment of 0.70 D, and the other has a dipole moment of 1.73 D. Which is which?

29. A tertiary amine reacts with hydrogen peroxide to form a tertiary amine oxide.

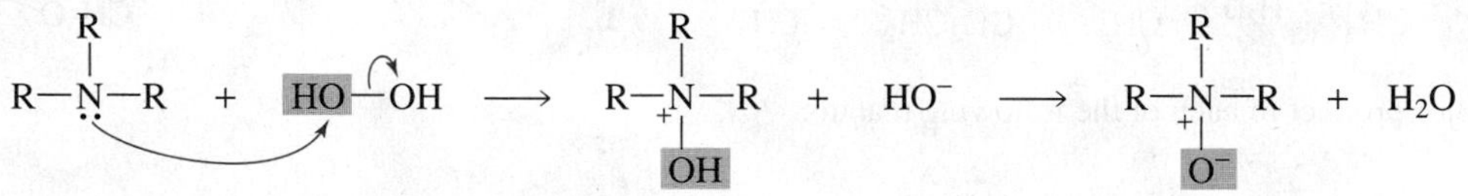

a tertiary amine **a tertiary amine oxide**

Tertiary amine oxides undergo a reaction similar to the Hofmann elimination reaction, called a **Cope elimination.** In this reaction, a tertiary amine oxide, rather than a quaternary ammonium ion, undergoes elimination. A strong base is not needed for a Cope elimination because the amine oxide acts as its own base.

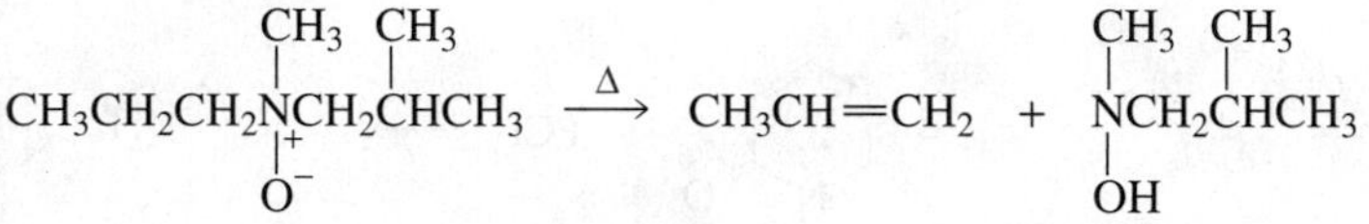

Does the Cope elimination have an alkene-like transition state or a carbanion-like transition state?

30. What products would be obtained by treating the following tertiary amines with hydrogen peroxide followed by heat?

a. 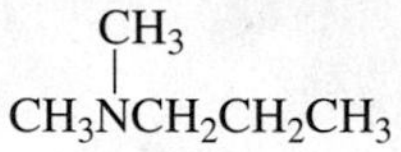

b. 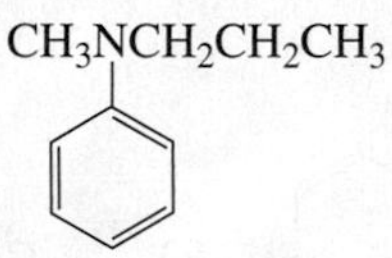

c. 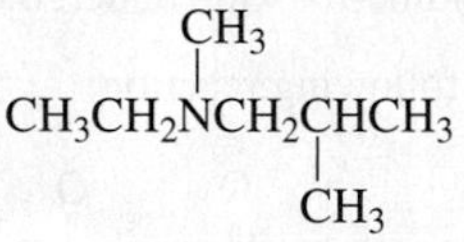

d. 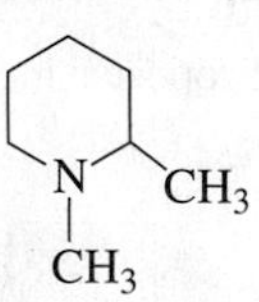

31. The chemical shifts of the C-2 hydrogen in the spectra of pyrrole, pyridine, and pyrrolidine are 2.82 ppm, 6.42 ppm, and 8.50 ppm. Match each heterocycle with its chemical shift.

32. Explain why protonating aniline has a dramatic effect on the compound's UV spectrum, whereas protonating pyridine has only a small effect on that compound's UV spectrum.

33. Explain why pyrrole ($pK_a \sim 17$) is a much stronger acid than ammonia ($pK_a = 36$).

pK_a ~17 **$pK_a = 36$**

34. Propose a mechanism for the following reaction:

$$2 \text{ pyrrole} + H_2C=O \xrightarrow{HCl} \text{product}$$

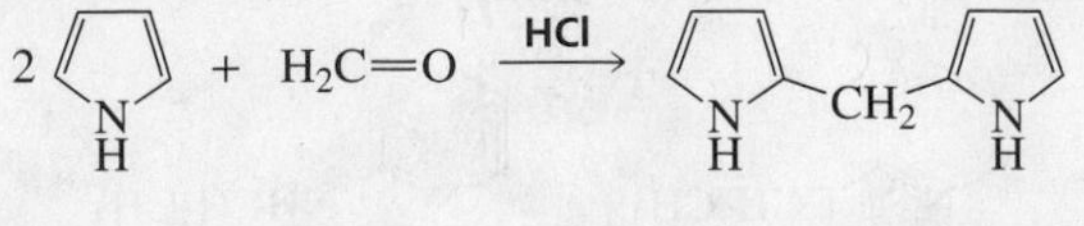

35. Quinolines, heterocyclic compounds that contain a pyridine ring fused to a benzene ring, are commonly synthesized by a method known as the Skraup synthesis, in which aniline reacts with glycerol under acidic conditions. Nitrobenzene is added to the reaction mixture to serve as an oxidizing agent. The first step in the synthesis is the dehydration of glycerol to propenal.

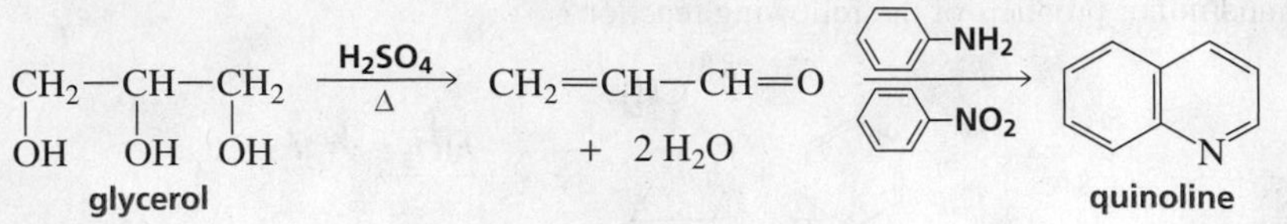

$$\underset{\substack{|\quad|\quad| \\ \text{OH} \ \ \text{OH} \ \ \text{OH} \\ \textbf{glycerol}}}{CH_2-CH-CH_2} \xrightarrow[\Delta]{H_2SO_4} \ CH_2=CH-CH=O \ + \ 2\,H_2O \xrightarrow{} \underset{\textbf{quinoline}}{\text{quinoline}}$$

a. What product would be obtained if *para*-ethylaniline were used instead of aniline?
b. What product would be obtained if 3-hexen-2-one were used instead of glycerol?
c. What starting materials are needed for the synthesis of 2,7-diethyl-3-methylquinoline?

36. Propose a mechanism for each of the following reactions:

a. 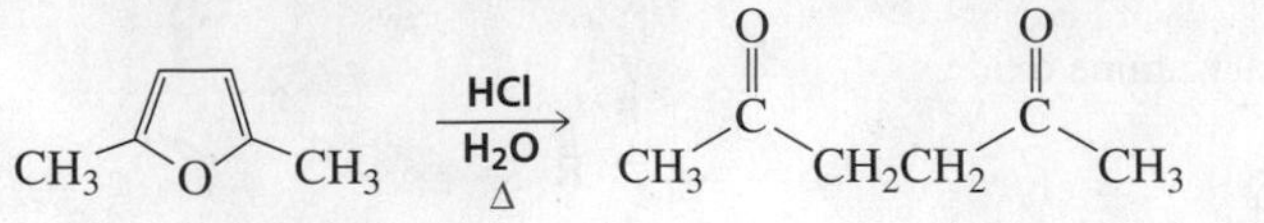 $\xrightarrow[\Delta]{\substack{HCl \\ H_2O}}$

b. 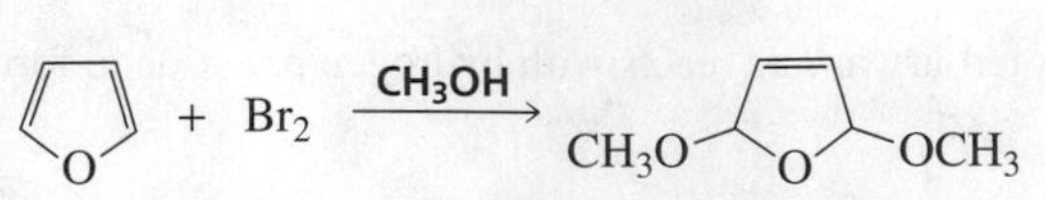 $+ \ Br_2 \xrightarrow{CH_3OH}$

37. What is the major product of each of the following reactions?

a. 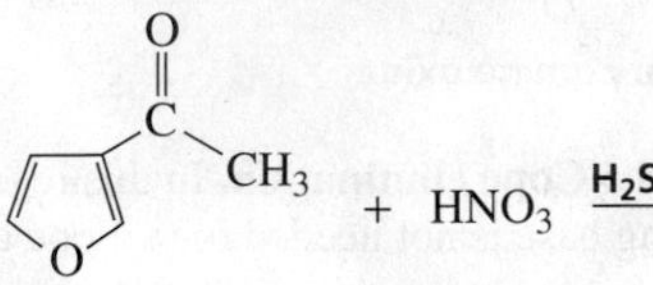 $+ \ HNO_3 \xrightarrow{H_2SO_4}$

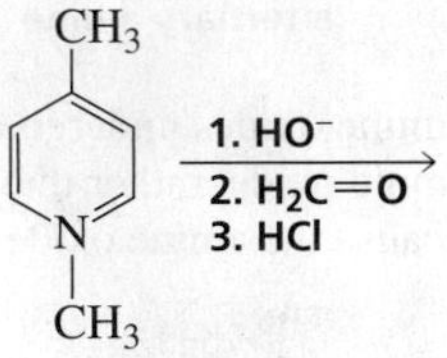

b. 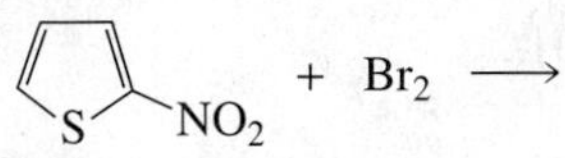 $+ \ Br_2 \longrightarrow$

c. $+ \ CH_3I \longrightarrow$

d. 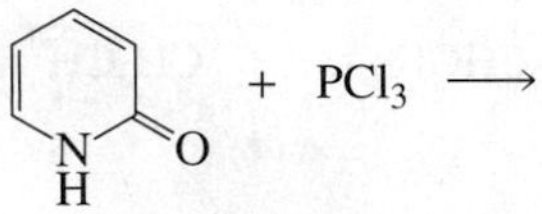 $+ \ PCl_3 \longrightarrow$

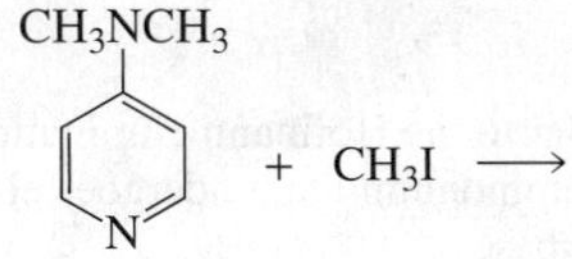

e. $\xrightarrow{\substack{1.\ HO^- \\ 2.\ H_2C=O \\ 3.\ HCl}}$

f. 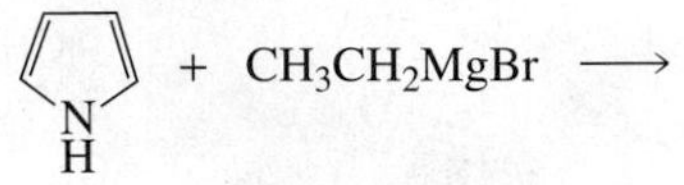 $+ \ CH_3CH_2MgBr \longrightarrow$

38. a. Draw resonance contributors to show why pyridine-*N*-oxide is more reactive than pyridine toward electrophilic aromatic substitution.
b. At what position does pyridine-*N*-oxide undergo electrophilic aromatic substitution?

39. Propose a mechanism for the following reaction:

$$+ \ CH_3\underset{O}{\overset{O}{C}}O\underset{O}{\overset{O}{C}}CH_3 \longrightarrow + \ CH_3\underset{O}{\overset{O}{C}}OH$$

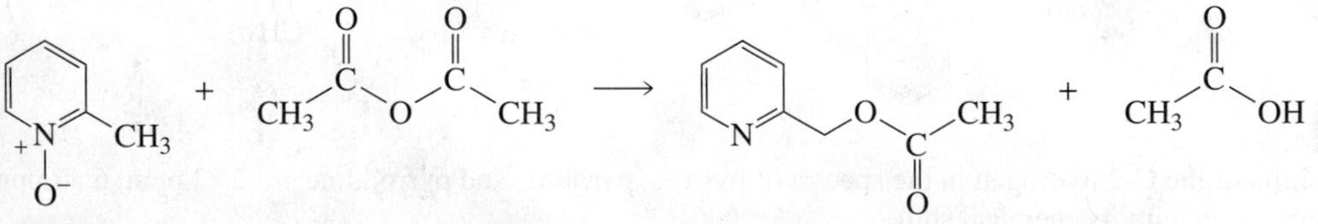

40. Pyrrole reacts with excess *para*-(*N*,*N*-dimethylamino)benzaldehyde to form a highly colored compound. Draw the structure of the colored compound.

41. 2-Phenylindole is prepared from the reaction of acetophenone and phenylhydrazine, a method known as the Fischer indole synthesis. Propose a mechanism for this reaction. (*Hint:* the reactive intermediate is the enamine tautomer of the phenylhydrazone.)

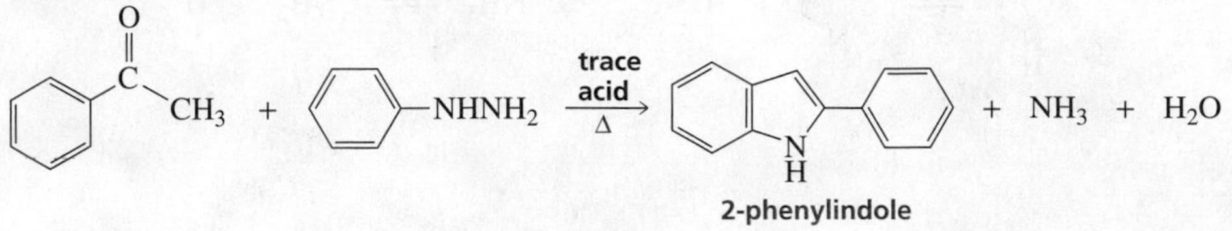

$$\underset{O}{\overset{O}{C}}CH_3 \ + \ \text{—NHNH}_2 \xrightarrow[\Delta]{\text{trace acid}} \underset{\textbf{2-phenylindole}}{\text{2-phenylindole}} \ + \ NH_3 \ + \ H_2O$$

42. What starting materials are required to synthesize the following compounds, using the Fischer indole synthesis? (*Hint:* see Problem 41.)

a.

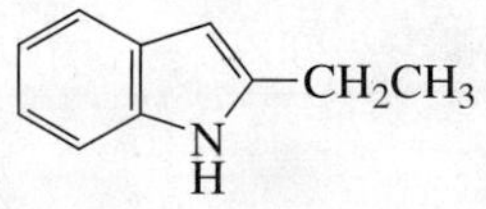

b.

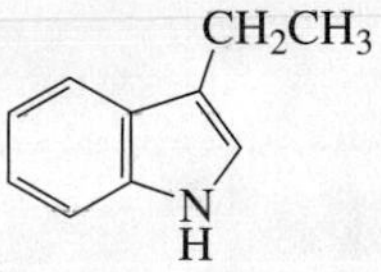

c.

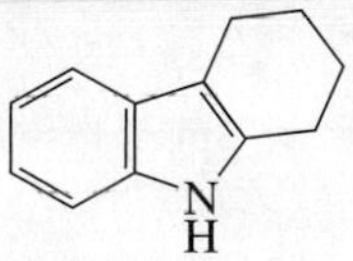

43. Organic chemists work with tetraphenylporphyrins rather than with porphyrins because tetraphenylporphyrins are much more resistant to air oxidation. Tetraphenylporphyrin can be prepared by the reaction of benzaldehyde with pyrrole. Propose a mechanism for the formation of the ring system shown here:

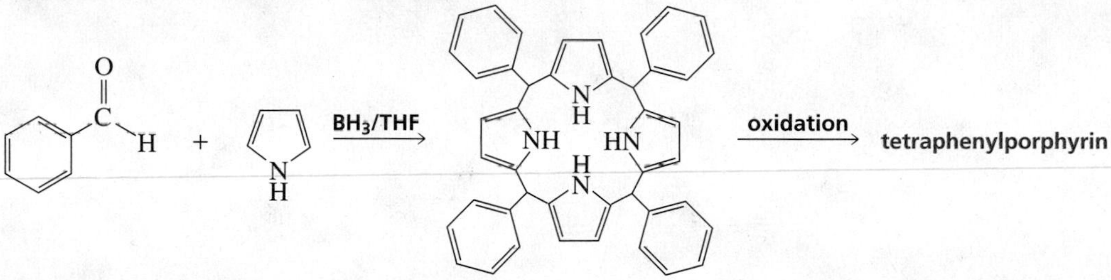

GLOSSARY

amine   a compound with a nitrogen in place of one of the hydrogens of an alkane; (RNH$_2$, R$_2$NH, R$_3$N)

Cope elimination reaction   elimination of a proton and a hydroxyl amine from an amine oxide.

ligation   sharing of nonbonding electrons with a metal ion.

heteroatom   an atom other than carbon or hydrogen.

heterocyclic compound (heterocycle)   a cyclic compound in which one or more of the atoms of the ring are heteroatoms.

porphyrin ring system   consists of four pyrrole rings joined by one-carbon bridges.

PHOTO CREDITS

Photo credits are listed in order of appearance.

Paula Bruice; kanusommer/Shutterstock

The Organic Chemistry of Carbohydrates

a field of sugar cane

BIOORGANIC COMPOUNDS ARE ORGANIC COMPOUNDS FOUND IN LIVING SYSTEMS. The first group of bioorganic compounds we will look at are carbohydrates—the most abundant class of compounds in the biological world, making up more than 50% of the dry weight of the Earth's biomass. Carbohydrates are important constituents of all living organisms and have a variety of different functions. Some are important structural components of cells; others act as recognition sites on cell surfaces. For example, the first event in our lives was a sperm recognizing a carbohydrate on the outer surface of an egg. Other carbohydrates serve as a major source of metabolic energy. The leaves, fruits, seeds, stems, and roots of plants, for instance, contain carbohydrates that plants use for their own metabolic needs and these also serve the metabolic needs of the animals that eat them.

The structures of **bioorganic compounds** can be quite complex, yet their reactivity is governed by the same principles that govern the reactivity of comparatively simple organic molecules. In other words, the organic reactions that chemists carry out in the laboratory are in many ways just like those performed by nature inside a cell. Thus, bioorganic reactions can be thought of as organic reactions that take place in tiny flasks called cells.

Although most bioorganic compounds have more complicated structures than those of the organic compounds you are now used to seeing, do not let the structures fool you into thinking that their chemistry must be equally complicated. One reason the structures of bioorganic compounds are more complicated is that they must be able to recognize each other. Much of their structure is for that very purpose, a function called **molecular recognition.**

Early chemists noted that carbohydrates have molecular formulas that make them appear to be hydrates of carbon, $C_n(H_2O)_n$—hence the name. Structural studies later revealed that these compounds are not hydrates because they do not contain intact water molecules. Nevertheless, the term *carbohydrate* persisted.

Carbohydrates are polyhydroxy aldehydes such as glucose, polyhydroxy ketones such as fructose, and compounds such as sucrose formed by linking polyhydroxy aldehydes or polyhydroxy ketones together (Section 15). The chemical structures of carbohydrates are

From Chapter 21 of *Organic Chemistry*, Seventh Edition. Paula Yurkanis Bruice. Copyright © 2014 by Pearson Education, Inc. All rights reserved.

ᴅ-glucose

ᴅ-fructose

commonly represented by Fischer projections. Notice that both glucose and fructose have the molecular formula $C_6H_{12}O_6$, consistent with the general formula $C_6(H_2O)_6$ that made early chemists think that these compounds were hydrates of carbon. Notice, too, that the structures of glucose and fructose differ only at the top two carbons.

Fischer projection
D-glucose
a polyhydroxy aldehyde

Fischer projection
D-fructose
a polyhydroxy ketone

All the horizontal bonds in Fischer projections point toward the viewer.

The most abundant carbohydrate in nature is glucose. Animals obtain glucose from food that contains glucose, such as plants. Plants produce glucose by *photosynthesis.* During photosynthesis, plants take up water through their roots and use carbon dioxide from the air to synthesize glucose and oxygen. Because photosynthesis is the reverse of the process used by organisms to obtain energy—specifically, the oxidation of glucose to carbon dioxide and water—plants require energy to carry out photosynthesis. They obtain that energy from sunlight, which is captured by chlorophyll molecules in green plants. Photosynthesis uses the CO_2 that animals exhale as waste and generates the O_2 that animals inhale to sustain life. Nearly all the oxygen in the atmosphere has been released by photosynthetic processes.

$$C_6H_{12}O_6 \ + \ 6\,O_2 \ \underset{\text{photosynthesis}}{\overset{\text{oxidation}}{\rightleftharpoons}} \ 6\,CO_2 \ + \ 6\,H_2O$$
glucose

1 CLASSIFICATION OF CARBOHYDRATES

The terms *carbohydrate, saccharide,* and *sugar* are used interchangeably. *Saccharide* comes from the word for sugar in several early languages (*sarkara* in Sanskrit, *sakcharon* in Greek, and *saccharum* in Latin).

Simple carbohydrates are **monosaccharides** (single sugars); **complex carbohydrates** contain two or more monosaccharides linked together. **Disaccharides** have two mono-saccharides linked together, **oligosaccharides** have 3 to 10 (*oligos* is Greek for "few"), and **polysaccharides** have 10 or more. Disaccharides, oligosaccharides, and polysaccharides can be broken down to monosaccharides by hydrolysis.

a monosaccharide subunit

$$-M-M-M-M-M-M-M-M-M- \quad \overset{\text{hydrolysis}}{\longrightarrow} \quad x\,M$$
polysaccharide $\qquad\qquad\qquad\qquad\qquad\qquad$ monosaccharide

A *monosaccharide* can be a polyhydroxy aldehyde such as glucose or a polyhydroxy ketone such as fructose. Polyhydroxy aldehydes are called **aldoses** ("ald" is for aldehyde; "ose" is the suffix for a sugar); polyhydroxy ketones are called **ketoses.**

Monosaccharides are also classified according to the number of carbons they contain: those with three carbons are **trioses,** those with four carbons are **tetroses,** those with five carbons are **pentoses,** and those with six and seven carbons are **hexoses** and **heptoses.** Therefore, a six-carbon polyhydroxy aldehyde such as glucose is an aldohexose, whereas a six-carbon polyhydroxy ketone such as fructose is a ketohexose.

PROBLEM 1♦

Classify the following monosaccharides:

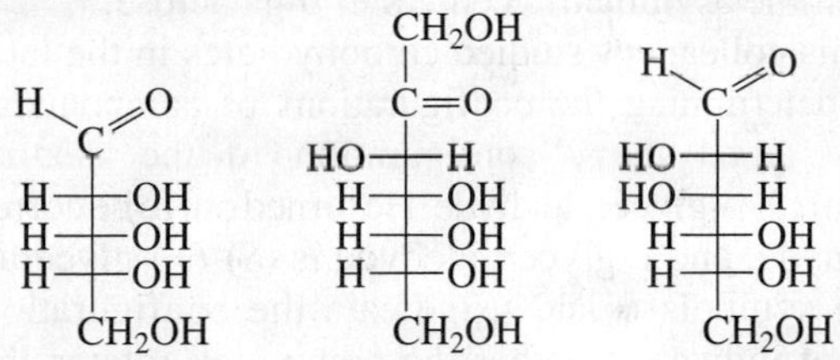

$$
\begin{array}{ccc}
\text{H}\diagdown\!\!\!\!\!\!\!\!\diagup\text{O} & \text{CH}_2\text{OH} & \text{H}\diagdown\!\!\!\!\!\!\!\!\diagup\text{O} \\
\text{C} & \text{C}\!=\!\text{O} & \text{C} \\
\text{H}-\!\!\!-\text{OH} & \text{HO}-\!\!\!-\text{H} & \text{HO}-\!\!\!-\text{H} \\
\text{H}-\!\!\!-\text{OH} & \text{H}-\!\!\!-\text{OH} & \text{HO}-\!\!\!-\text{H} \\
\text{H}-\!\!\!-\text{OH} & \text{H}-\!\!\!-\text{OH} & \text{H}-\!\!\!-\text{OH} \\
\text{CH}_2\text{OH} & \text{H}-\!\!\!-\text{OH} & \text{H}-\!\!\!-\text{OH} \\
 & \text{CH}_2\text{OH} & \text{CH}_2\text{OH} \\
\text{D-ribose} & \text{D-sedoheptulose} & \text{D-mannose}
\end{array}
$$

2   THE D AND L NOTATION

The smallest aldose, and the only one whose name does not end in "ose," is glyceralde-hyde, an aldotriose.

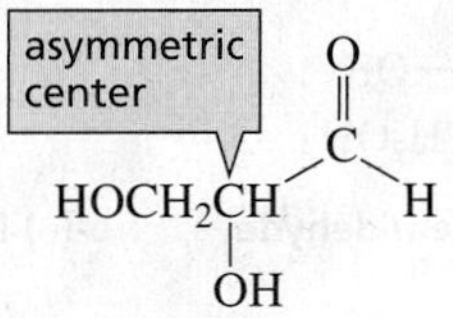

asymmetric center

$$\text{HOCH}_2\text{CH} \overset{\text{O}}{\underset{\text{OH}}{\diagup}} \text{C} \diagdown \text{H}$$

glyceraldehyde

A carbon to which four different groups are attached is an asymmetric center.

Because glyceraldehyde has an asymmetric center, it can exist as a pair of enantio-mers. The isomer on the left has the *R* configuration because an arrow drawn from the highest priority substituent (OH) to the next highest priority substituent (HC=O) is clockwise and the lowest priority group is on a hatched wedge. The *R* and *S* enantiomers drawn as Fischer projections are shown on the right.

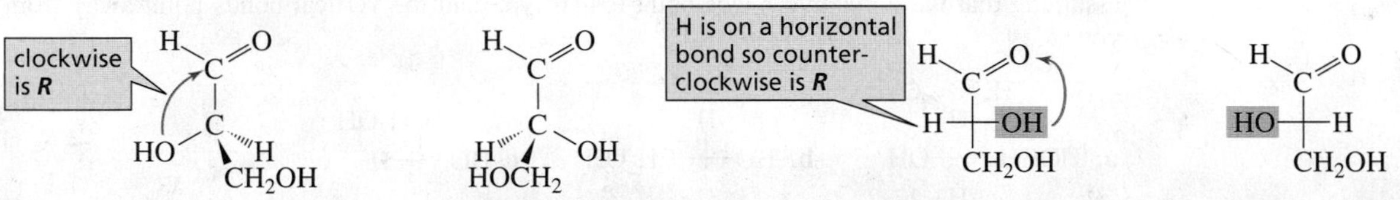

clockwise is *R*

H is on a horizontal bond so counter-clockwise is *R*

(R)-(+)-glyceraldehyde   **(S)-(−)-glyceraldehyde**   **(R)-(+)-glyceraldehyde**   **(S)-(−)-glyceraldehyde**

perspective formulas   **Fischer projections**

The notations D and L are used to describe the configurations of carbohydrates. In a Fischer projection of a monosaccharide, the carbonyl group is always placed on top (in the case of aldoses) or as close to the top as possible (in the case of ketoses). Examine the Fischer projection of galactose shown below and note that the compound has four asymmetric centers (C-2, C-3, C-4, and C-5). *If the OH group attached to the bottommost asymmetric center (the carbon second from the bottom) is on the right, then the compound is a D-sugar. If that same OH group is on the left, then the compound is an L-sugar.* Almost all sugars found in nature are D-sugars. The mirror image of a D-sugar is an L-sugar.

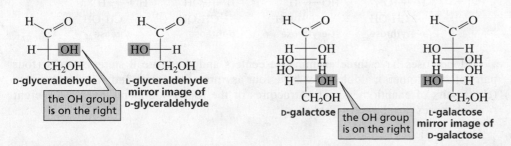

D-glyceraldehyde   L-glyceraldehyde mirror image of D-glyceraldehyde

the OH group is on the right

D-galactose   the OH group is on the right   L-galactose mirror image of D-galactose

The common name of the monosaccharide, together with the D or L designation, completely defines its structure, because the configurations of all the asymmetric centers are implicit in the common name. Thus, the structure of L-galactose is obtained by changing the configuration of *all* the asymmetric centers in D-galactose.

Emil Fischer and his colleagues studied carbohydrates in the late nineteenth century, when techniques for determining the configurations of compounds were not available. Fischer arbitrarily assigned the *R*-configuration to the dextrorotatory isomer of glyceraldehyde that we call D-glyceraldehyde. He turned out to be correct: D-glyceraldehyde is (*R*)-(+)-glyceraldehyde, and L-glyceraldehyde is (*S*)-(−)-glyceraldehyde.

Like *R* and *S*, the symbols D and L indicate the configuration of an asymmetric center, but they do not indicate whether the compound rotates the plane of polarization of polarized light to the right (+) or to the left (−). For example, D-glyceraldehyde is dextrorotatory, whereas D-lactic acid is levorotatory. In other words, optical rotation, like melting points or boiling points, is a physical property of a compound, whereas "*R*, *S*, D, and L" are conventions humans use to indicate the configuration about an asymmetric center.

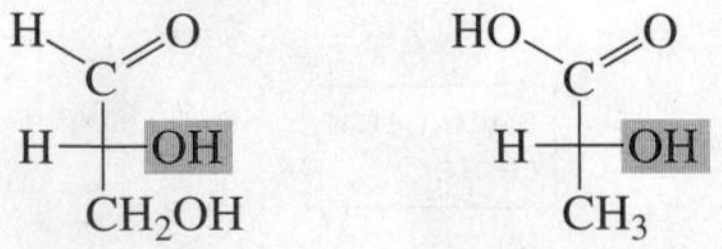

D-(+)-glyceraldehyde   D-(−)-lactic acid

PROBLEM 2

Draw Fischer projections of L-glucose and L-fructose.

PROBLEM 3♦

Indicate whether each of the following structures is D-glyceraldehyde or L-glyceraldehyde, assuming that the horizontal bonds point toward you and the vertical bonds point away from you:

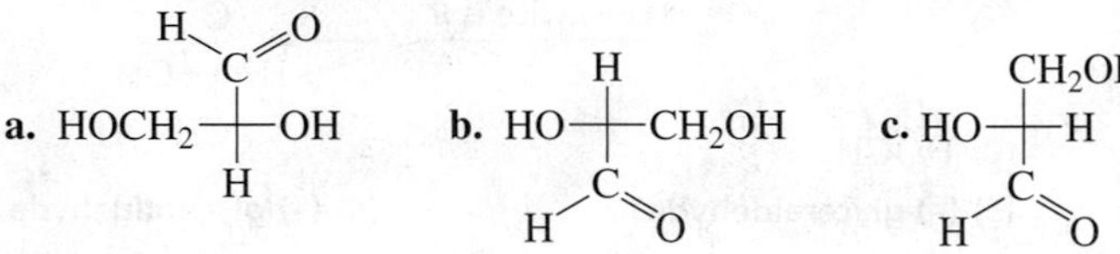

---

3   **THE CONFIGURATIONS OF ALDOSES**

Aldotetroses have two asymmetric centers and therefore four stereoisomers. Two of the stereoisomers are D-sugars and two are L-sugars. The names of the aldotetroses—erythrose and threose—were used to name the erythro and threo pairs of enantiomers.

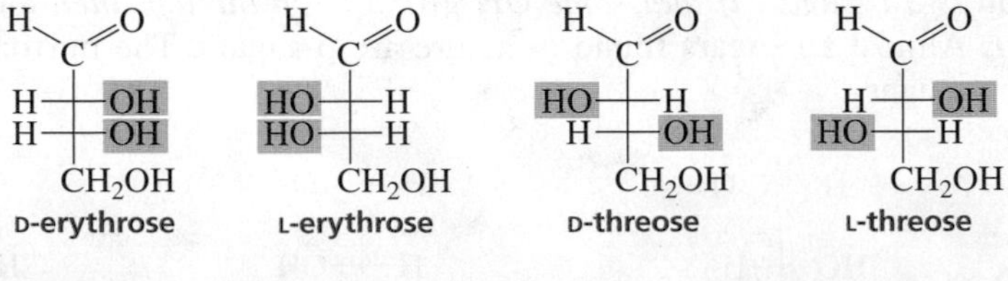

D-erythrose   L-erythrose   D-threose   L-threose

Aldopentoses have three asymmetric centers and therefore 8 stereoisomers (four pairs of enantiomers); aldohexoses have four asymmetric centers and 16 stereoisomers (eight pairs of enantiomers). The structures of the four D-aldopentoses and the eight D-aldohexoses are shown in Table 1.

Table 1   Configurations of the D-Aldoses

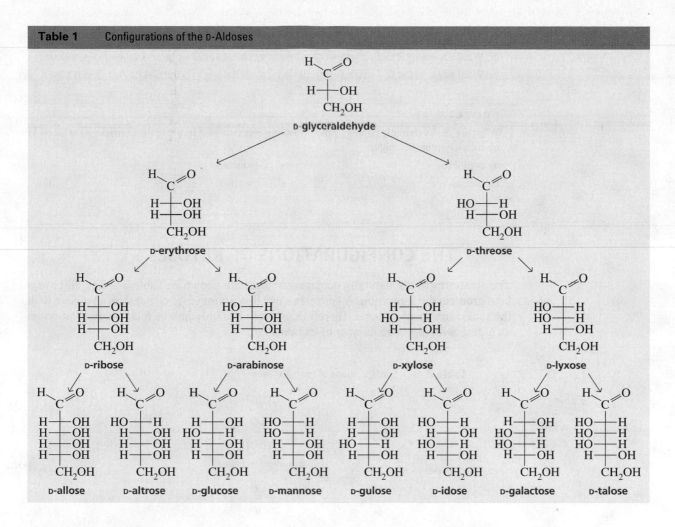

Diastereomers that differ in configuration at only one asymmetric center are called **epimers.** For example, D-ribose and D-arabinose are C-2 epimers because they differ in configuration only at C-2; D-idose and D-talose are C-3 epimers. An epimer is a particular kind of diastereomer. (Diastereomers are stereoisomers that are not enantiomers.)

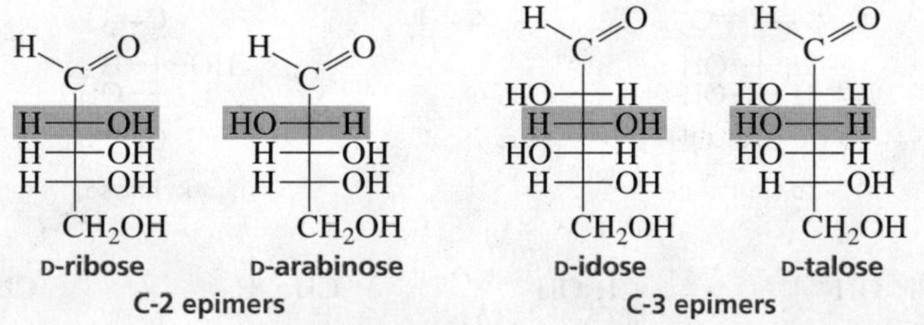

D-Glucose, D-mannose, and D-galactose are the most common aldohexoses in living systems. An easy way to learn their structures is to memorize the structure of D-glucose and then note that D-mannose is the C-2 epimer of D-glucose and D-galactose is the C-4 epimer of D-glucose.

D-Mannose is the C-2 epimer of D-glucose.

D-Galactose is the C-4 epimer of D-glucose.

Diastereomers are stereoisomers that are not enantiomers.

PROBLEM 4♦

a. Are D-erythrose and L-erythrose enantiomers or diastereomers?

b. Are L-erythrose and L-threose enantiomers or diastereomers?

PROBLEM 5♦

a. What sugar is the C-3 epimer of D-xylose?
b. What sugar is the C-5 epimer of D-allose?
c. What sugar is the C-4 epimer of L-gulose?
d. What sugar is the C-4 epimer of D-lyxose?

PROBLEM 6♦

What are the systematic names of the following compounds? Indicate the configuration (*R* or *S*) of each asymmetric center.

a. D-glucose
b. D-mannose
c. D-galactose
d. L-glucose

4 THE CONFIGURATIONS OF KETOSES

The structures of the naturally occurring ketoses are shown in Table 2—they all have a keto group in the 2-position. A ketose has one less asymmetric center than an aldose with the same number of carbons. Therefore, a ketose has only half as many stereoisomers as an aldose with the same number of carbons.

Table 2 Configurations of the D-Ketoses

PROBLEM 7♦

What sugar is the C-3 epimer of D-fructose?

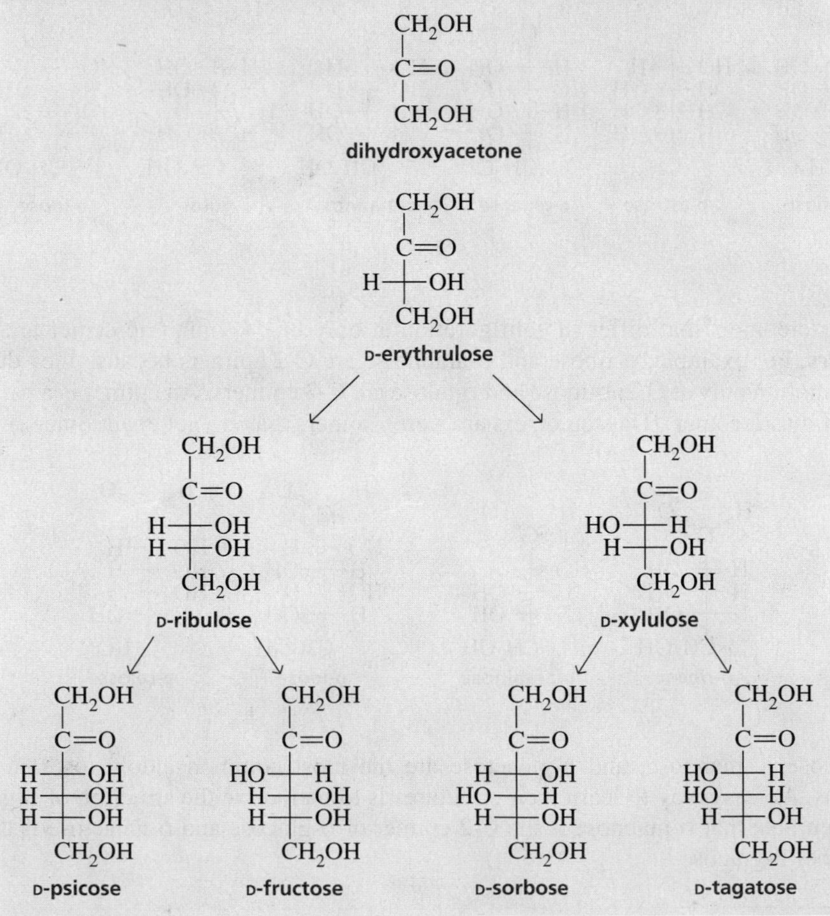

PROBLEM 8♦

How many stereoisomers are possible for

a. a ketoheptose? **b.** an aldoheptose? **c.** a ketotriose?

5 THE REACTIONS OF MONOSACCHARIDES IN BASIC SOLUTIONS

In a basic solution, a monosaccharide is converted to a mixture of polyhydroxy aldehydes and polyhydroxy ketones. Let's look at what happens to D-glucose in a basic solution, beginning with its conversion to its C-2 epimer.

MECHANISM FOR THE BASE-CATALYZED EPIMERIZATION OF A MONOSACCHARIDE

D-glucose an enolate ion D-mannose

- The base removes a proton from an α-carbon, forming an enolate ion. Notice that C-2 in the enolate ion is no longer an asymmetric center.
- When C-2 is reprotonated, the proton can come from the top or the bottom of the planar sp^2 carbon, forming both D-glucose and D-mannose (C-2 epimers).

Because the reaction forms a pair of epimers, it is called epimerization. **Epimerization** changes the configuration of a carbon by removing a proton from it and then reprotonating it.

In addition to forming its C-2 epimer in a basic solution, D-glucose also undergoes an **enediol rearrangement,** which forms D-fructose and other ketohexoses.

MECHANISM FOR THE BASE-CATALYZED ENEDIOL REARRANGEMENT OF A MONOSACCHARIDE

D-glucose an enolate ion an enediol an enolate ion D-fructose

- The base removes a proton from an α-carbon, forming an enolate ion.
- The enolate ion is protonated to form an enediol.
- The enediol has two OH groups that can form a carbonyl group. Tautomerization of the OH at C-1 (as in base-catalyzed epimerization shown earlier) re-forms D-glucose or forms D-mannose; tautomerization of the OH group at C-2 forms D-fructose.

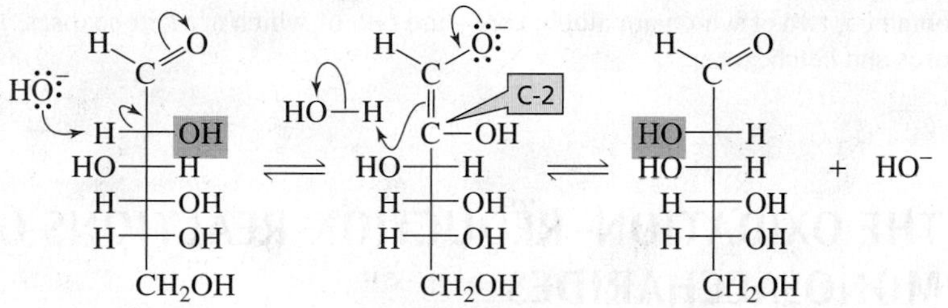

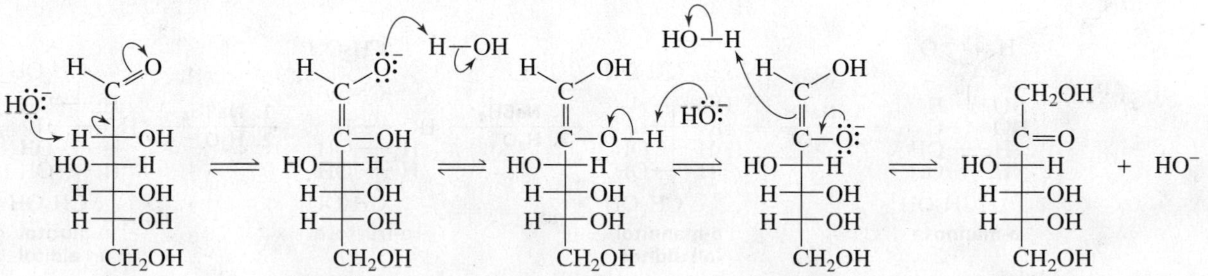

In a basic solution an aldose forms a C-2 epimer and one or more ketoses.

Another enediol rearrangement, initiated by a base removing a proton from C-3 of D-fructose, forms an enediol that can tautomerize to give a ketose with the carbonyl group at C-2 or C-3. Thus, the carbonyl group can be moved up and down the chain.

PROBLEM 9

Show how an enediol rearrangement can move the carbonyl carbon of fructose from C-2 to C-3.

PROBLEM 10

Write the mechanism for the base-catalyzed conversion of D-fructose into D-glucose and D-mannose.

PROBLEM 11♦

When D-tagatose is added to a basic aqueous solution, an equilibrium mixture of monosaccharides is obtained, two of which are aldohexoses and two of which are ketohexoses. Identify the aldohexoses and ketohexoses.

---

6 THE OXIDATION–REDUCTION REACTIONS OF MONOSACCHARIDES

Because they contain *alcohol* functional groups and *aldehyde* or *ketone* functional groups, the reactions of monosaccharides are an extension of what you may already know about the reactions of alcohols, aldehydes, and ketones. For example, an aldehyde group in a monosaccharide can be oxidized or reduced and can react with nucleophiles to form imines, hemiacetals, and acetals.

Reduction

The carbonyl group in aldoses and ketoses can be reduced by $NaBH_4$. The product of the reduction is a polyalcohol, known as an **alditol.** Reduction of an aldose forms one alditol. For example, the reduction of D-mannose forms D-mannitol, the alditol found in mushrooms, olives, and onions.

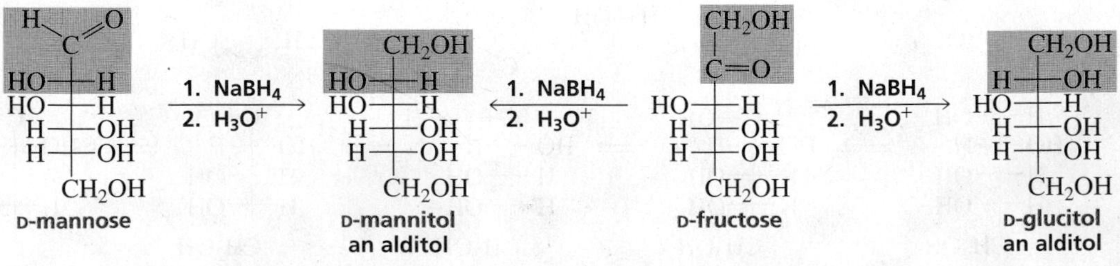

The reduction of a ketose forms two alditols because the reaction creates a new asymmetric center in the product. For example, the reduction of D-fructose forms D-mannitol and D-glucitol, the C-2 epimer of D-mannitol. D-Glucitol—also called sorbitol—is about 60% as sweet as sucrose. It is found in plums, pears, cherries, and berries. It is also used as a sugar substitute in the manufacture of candy; we will see why later in this chapter.

PROBLEM 12♦

What products are obtained from the reduction of

a. D-idose?   b. D-sorbose?

PROBLEM 13♦

a. What other monosaccharide is reduced only to the alditol obtained from the reduction of

1. D-talose?   2. D-glucose?   3. D-galactose?

b. What monosaccharide is reduced to two alditols, one of which is the alditol obtained from the reduction of

1. D-talose?   2. D-allose?

Oxidation

Aldoses can be distinguished from ketoses by observing what happens to the color of an aqueous solution of Br_2 when it is added to the sugar. Br_2 is a mild oxidizing agent and easily oxidizes the aldehyde group, but it cannot oxidize ketones or alcohols. Consequently, if a small amount of an aqueous solution of Br_2 is added to an unknown monosaccharide, the reddish-brown color of Br_2 will disappear if the monosaccharide is an aldose because Br_2 will be reduced to Br^-, which is colorless. If the red color persists, indicating no reaction with Br_2, then the monosaccharide is a ketose. The product of the oxidation reaction is an **aldonic acid.**

D-glucose + Br_2 (red) $\xrightarrow{H_2O}$ D-gluconic acid (an aldonic acid) + 2 Br^- (colorless)

Both aldoses and ketoses are oxidized to aldonic acids by Tollens reagent (Ag^+, NH_3, HO^-), so Tollens reagent cannot be used to distinguish them. Tollens reagent only oxidizes aldehydes, but since the oxidation reaction is carried out in a basic solution, a ketose is converted into an aldose by an enediol rearrangement (Section 5), and the aldose is then oxidized by Tollens reagent.

a ketose $\xrightleftharpoons{HO^-}$ an aldose $\xrightarrow[HO^-]{Ag^+,\ NH_3}$ a carboxylate ion

Dilute nitric acid (HNO_3) is a stronger oxidizing agent than those discussed earlier. It oxidizes aldehydes *and* primary alcohols, but it does not oxidize secondary alcohols. The product obtained when both the aldehyde and the primary alcohol groups of an aldose are oxidized is called an **aldaric acid.**

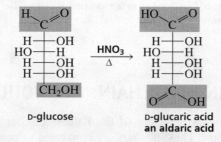

D-glucose $\xrightarrow[\Delta]{HNO_3}$ D-glucaric acid (an aldaric acid)

In an ald*onic* acid, *one* end is oxidized.

In an ald*aric* acid, both ends *are* oxidized.

7 LENGTHENING THE CHAIN: THE KILIANI–FISCHER SYNTHESIS

The carbon chain of an aldose can be increased by one carbon by a modified **Kiliani–Fischer synthesis.** Thus, tetroses can be converted into pentoses, and pentoses can be converted into hexoses.

STEPS IN THE MODIFIED KILIANI–FISCHER SYNTHESIS

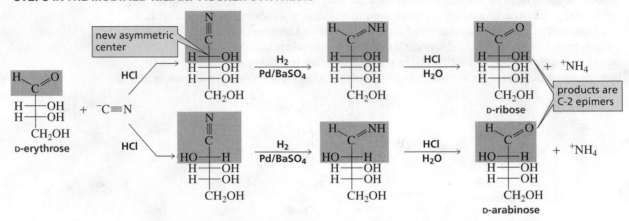

- In the first step, hydrogen cyanide adds to the carbonyl group. This reaction converts the carbonyl carbon in the starting material to an asymmetric center. Consequently, two products are formed that differ only in their configuration at C-2. The configurations of the other asymmetric centers do not change because no bond to any of the asymmetric centers is broken during the course of the reaction.

- The C≡N bond is reduced to an imine, using a partially deactivated palladium catalyst so that the imine is not further reduced to an amine.

- The two imines are hydrolyzed to two aldoses.

Notice that the modified Kiliani–Fischer synthesis leads to a pair of C-2 epimers. The two epimers are not obtained in equal amounts because the first step of the reaction produces a pair of diastereomers, and diastereomers are generally formed in unequal amounts.

The Kiliani–Fischer synthesis leads to a pair of C-2 epimers.

8 SHORTENING THE CHAIN: THE WOHL DEGRADATION

The **Wohl degradation**—the opposite of the Kiliani–Fischer synthesis—shortens an aldose chain by one carbon. Thus, hexoses are converted to pentoses, and pentoses are converted to tetroses.

STEPS IN THE WOHL DEGRADATION

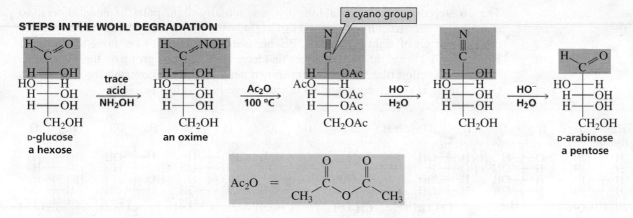

- In the first step, the aldehyde reacts with hydroxylamine to form an oxime.
- Heating with acetic anhydride dehydrates the oxime, forming a nitrile; all the OH groups are converted to esters as a result of reacting with acetic anhydride.
- In a basic aqueous solution, all the ester groups are hydrolyzed and the cyano group is eliminated.

Measuring the Blood Glucose Levels in Diabetes

Glucose in the bloodstream reacts with an NH_2 group of hemoglobin to form an imine that subsequently undergoes an irreversible rearrangement to a more stable α-aminoketone known as hemoglobin-A_{1c}.

Insulin is the hormone that regulates the level of glucose—and thus the amount of hemoglobin-A_{1c}—in the blood. Diabetes is a condition in which the body does not produce sufficient insulin, or in which the insulin it produces does not function properly. Because people with untreated diabetes have increased blood glucose levels, they also have a higher concentration of hemoglobin-A_{1c} than people without diabetes. Thus, measuring the hemoglobin-A_{1c} level is a way to determine whether the blood glucose level of a diabetic patient is being controlled.

Cataracts, a common complication in diabetes, are caused by the reaction of glucose with the NH_2 group of proteins in the lens of the eye. Some think the arterial rigidity common in old age may be attributable to a similar reaction of glucose with the NH_2 group of proteins.

PROBLEM 16♦

What two monosaccharides can be degraded to

a. D-ribose? **b.** D-arabinose? **c.** L-ribose?

9 THE STEREOCHEMISTRY OF GLUCOSE: THE FISCHER PROOF

Emil Fischer's determination of the stereochemistry of glucose, done in 1891, is a example of brilliant reasoning. He chose (+)-glucose for his study because it is the most common monosaccharide found in nature.

Fischer knew that (+)-glucose is an aldohexose, but 16 different structures can be written for an aldohexose. Which of them represents the structure of (+)-glucose?

The 16 stereoisomers of an aldohexose are actually eight pairs of enantiomers, so if we know the structures of one set of eight, we automatically know the structures of the other set of eight. Therefore, Fischer needed to consider only one set of eight. He considered the eight stereoisomers that have their C-5 OH group on the right in the Fischer projection (the stereoisomers shown here, which we now call the D-aldoses). One of these is (+)-glucose, and its mirror image is (−)-glucose.

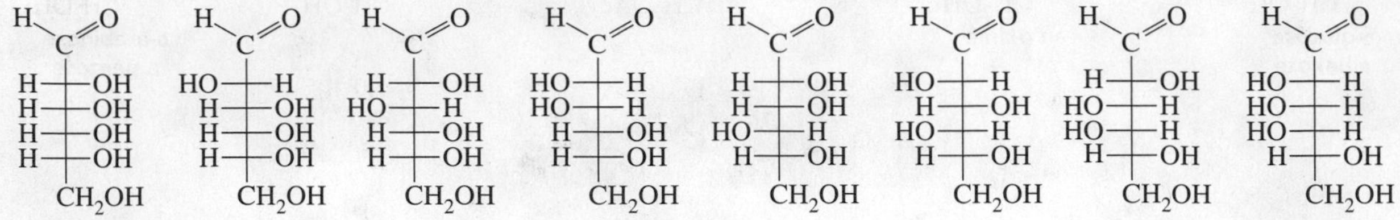

Emil Fischer (1852–1919) *was born in a village near Cologne, Germany. He became a chemist against the wishes of his father, a successful merchant, who wanted him to enter the family business. Fischer was a professor of chemistry at the Universities of Erlangen, Würzburg, and Berlin and in 1902 he received the Nobel Prize in Chemistry for his work on sugars. Fischer organized German chemical production during World War I. He lost two of his three sons in that war.*

Fischer used the following information to determine glucose's stereochemistry—that is, to determine the configuration of each of its asymmetric centers.

1. When the Kiliani–Fischer synthesis is performed on the sugar known as (−)-arabinose, the two sugars known as (+)-glucose and (+)-mannose are obtained. This means that (+)-glucose and (+)-mannose are C-2 epimers. Consequently, (+)-glucose and (+)-mannose have to be one of the following pairs: sugars 1 and 2, 3 and 4, 5 and 6, or 7 and 8.

2. (+)-Glucose and (+)-mannose are both oxidized by nitric acid to optically active aldaric acids. The aldaric acids of sugars 1 and 7 would not be optically active because each has a plane of symmetry. (A compound containing a plane of symmetry is achiral.) Excluding sugars 1 and 7 means that (+)-glucose and (+)-mannose must be sugars 3 and 4 or 5 and 6.

3. Because (+)-glucose and (+)-mannose are the products obtained when the Kiliani–Fischer synthesis is carried out on (−)-arabinose, Fischer knew that if (−)-arabinose has the structure shown below on the left, then (+)-glucose and (+)-mannose are sugars 3 and 4. On the other hand, if (−)-arabinose has the structure shown on the right, then (+)-glucose and (+)-mannose are sugars 5 and 6:

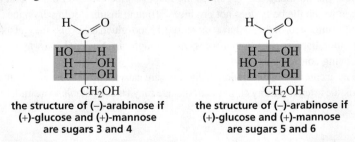

the structure of (−)-arabinose if (+)-glucose and (+)-mannose are sugars 3 and 4

the structure of (−)-arabinose if (+)-glucose and (+)-mannose are sugars 5 and 6

When (−)-arabinose is oxidized with nitric acid, it forms an optically active aldaric acid. This means that the aldaric acid does *not* have a plane of symmetry. Therefore, (−)-arabinose must have the structure shown on the left because the aldaric acid of the sugar on the right would have a plane of symmetry. Thus, (+)-glucose and (+)-mannose are represented by sugars 3 and 4.

4. Now the only question remaining is whether (+)-glucose is sugar 3 or sugar 4. To answer this, Fischer had to develop a chemical method for interchanging the aldehyde and primary alcohol groups of an aldohexose. When he chemically interchanged those groups on the sugar known as (+)-glucose, he obtained an aldohexose that was different from (+)-glucose, but when he interchanged those groups on (+)-mannose, he still had (+)-mannose. Therefore, he was able to conclude that (+)-glucose is sugar 3 because interchanging its aldehyde and primary alcohol groups leads to a different sugar (L-gulose).

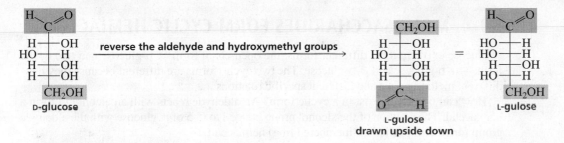

If (+)-glucose is sugar 3, then (+)-mannose must be sugar 4. As predicted, when the aldehyde and hydroxymethyl groups of sugar 4 are interchanged, the same sugar is obtained.

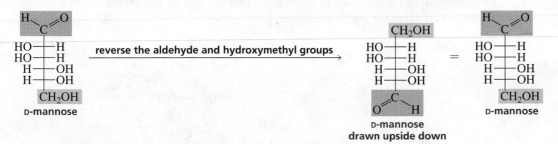

Using similar reasoning, Fischer went on to determine the stereochemistry of the other aldohexoses. He received the Nobel Prize in Chemistry in 1902 for this achievement. His original guess that (+)-glucose is a D-sugar was shown to be correct in 1951, using X-ray crystallography and a new technique known as anomalous dispersion, so all of his structures are correct. If he had been wrong and (+)-glucose had been an L-sugar, his contribution to the stereochemistry of aldoses would still have had the same importance, but all his stereochemical assignments would have had to be reversed.

Glucose/Dextrose

André Dumas first used the term *glucose* in 1838 to refer to the sweet compound that comes from honey and grapes. Later, Kekulé decided that it should be called dextrose because it was dextrorotatory. When Fischer studied the sugar, he called it glucose, and chemists have called it glucose ever since, although dextrose is often found on food labels.

PROBLEM 17 Solved

Aldohexoses **A** and **B** are formed from aldopentose **C** via a Kilani–Fischer synthesis. Nitric acid oxidizes **A** to an optically active aldaric acid, **B** to an optically inactive aldaric acid, and **C** to an optically active aldaric acid. Wohl degradation of **C** forms **D**, which is oxidized by nitric acid to an optically active aldaric acid. Wohl degradation of **D** forms (+)-glyceraldehyde. Identify **A, B, C,** and **D.**

Solution This is the kind of problem that should be solved by working backward. The bottommost asymmetric center in **D** must have the OH group on the right because **D** is degraded to (+)-glyceraldehyde. Since **D** is oxidized to an optically active aldaric acid, **D** must be D-threose. The two bottommost asymmetric centers in **C** and **D** have the same configuration because **C** is degraded to **D**. Since **C** is oxidized to an optically active aldaric acid, **C** must be D-lyxose. Compounds **A** and **B**, therefore, must be D-galactose and D-talose. Because **A** is oxidized to an optically active aldaric acid, it must be D-talose and **B** must be D-galactose.

PROBLEM 18♦

Identify **A, B, C,** and **D** in the preceding problem if **D** is oxidized to an optically *inactive* aldaric acid; if **A, B,** and **C** are oxidized to optically active aldaric acids; and if interchanging the aldehyde and alcohol groups of **A** leads to a different sugar.

10   MONOSACCHARIDES FORM CYCLIC HEMIACETALS

D-Glucose exists in three different forms: the open-chain form of D-glucose and two cyclic forms—α-D-glucose and β-D-glucose. The two cyclic forms are different because they have different melting points and different specific rotations.

How can D-glucose exist in a cyclic form? An aldehyde reacts with an alcohol to form a hemiacetal. The reaction of the alcohol group bonded to C-5 of D-glucose with the aldehyde group forms two cyclic (six-membered ring) hemiacetals.

To see that the OH group on C-5 is in the proper position to attack the aldehyde group, we need to convert the Fischer projection of D-glucose to a flat ring structure. To do this, draw the primary alcohol group *up* from the back left-hand corner. Groups on the *right* in a Fischer projection are *down* in the cyclic structure, and groups on the *left* in a Fischer projection are *up* in the cyclic structure.

Groups on the *right* in a Fischer projection are *down* in a Haworth projection.

Groups on the *left* in a Fischer projection are *up* in a Haworth projection.

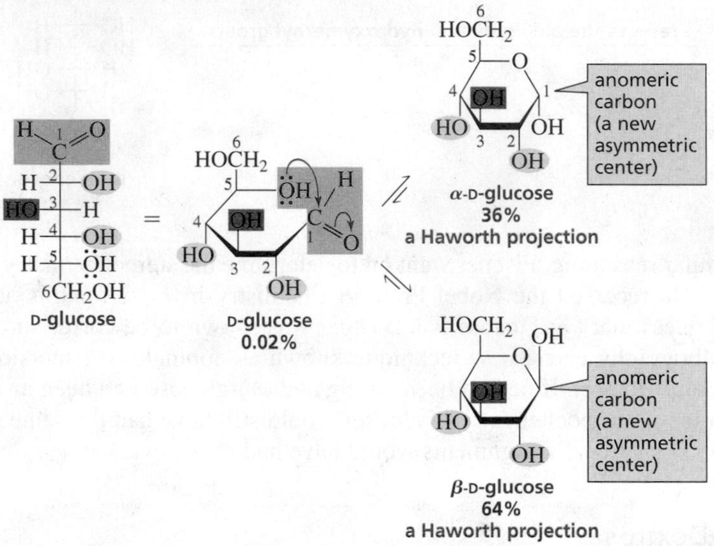

The cyclic hemiacetals shown here are drawn as Haworth projections. In a **Haworth projection,** the six-membered ring is represented as flat and is viewed edge-on. The ring oxygen is always placed in the back right-hand corner of the ring, with C-1 on the right-hand side, and the primary alcohol group attached to C-5 is drawn *up* from the back left-hand corner.

There are two different cyclic hemiacetals because the carbonyl carbon of the open-chain aldehyde becomes a new asymmetric center in the cyclic hemiacetal. If the OH group bonded to the new asymmetric center is down (trans to the primary alcohol group at C-5), then the hemiacetal is α-D-glucose; if the OH group is up (cis to the primary alcohol group at C-5), then the hemiacetal is β-D-glucose. The mechanism for cyclic hemiacetal formation is the same as the mechanism for hemiacetal formation between individual aldehyde and alcohol molecules.

α-D-Glucose and β-D-glucose are anomers. **Anomers** are two sugars that differ in configuration only at the carbon that was the carbonyl carbon in the open-chain form. This carbon is called the **anomeric carbon.** The prefixes α- and β- denote the configuration about the anomeric carbon. Because anomers, like epimers, differ in configuration at only one carbon, they too are a particular kind of diastereomer. Notice that the anomeric carbon is the only carbon in the molecule that is bonded to two oxygens.

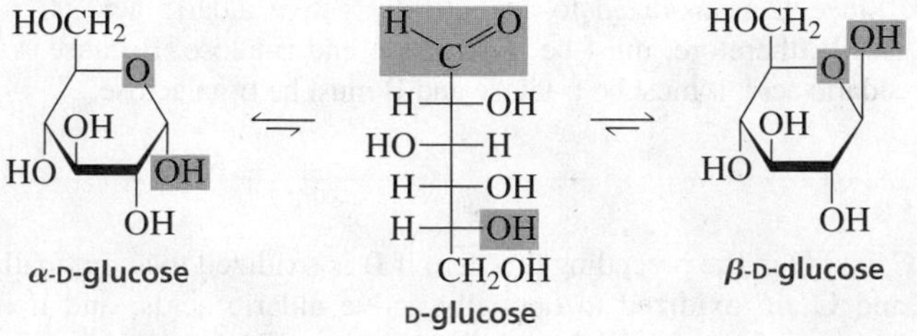

In an aqueous solution, the open-chain form of D-glucose is in equilibrium with the two cyclic hemiacetals. Because formation of the cyclic hemiacetals proceeds nearly to completion (unlike formation of acyclic hemiacetals), very little glucose is in the open-chain form (about 0.02%). Even so, the sugar still undergoes the reactions of oxidation, reduction, imine formation, etc. because the reagents react with the small amount of open-chain aldehyde that is present. As the open-chain compound reacts, the equilibrium shifts to produce more open-chain aldehyde, which can then undergo reaction. Eventually, all the glucose molecules react by way of the open-chain form.

When crystals of pure α-D-glucose are dissolved in water, the specific rotation gradually changes from +112.2 to +52.7. When crystals of pure β-D-glucose are dissolved in water, the specific rotation gradually changes from +18.7 to +52.7. This change in rotation occurs because in water, the hemiacetal opens to form the aldehyde, and both α-D-glucose and β-D-glucose are formed when the aldehyde recyclizes. Eventually, the three forms of glucose reach equilibrium concentrations. At equilibrium, there is almost twice as much β-D-glucose (64%) as α-D-glucose (36%). The specific rotation of the equilibrium mixture is +52.7. This is why the same specific rotation results whether the crystals originally dissolved in water are α-D-glucose or β-D-glucose or any mixture of the two. A slow change in optical rotation to an equilibrium value is called **mutarotation.**

If an aldose can form a five- or a six-membered-ring, it will exist predominantly as a cyclic hemiacetal in solution. Whether a five- or a six-membered-ring is formed depends on their relative stabilities. D-Ribose is an example of an aldose that forms five-membered-ring hemiacetals: α-D-ribose and β-D-ribose. The Haworth projection of a five-membered-ring sugar is viewed edge-on, with the ring oxygen pointing away from the viewer. The anomeric carbon is again on the right-hand side of the molecule, and the primary alcohol group is drawn up from the back left-hand corner. Again, notice that the anomeric carbon is the only carbon in the molecule that is bonded to two oxygens.

D-ribose = D-ribose ⇌ α-D-ribose + β-D-ribose

Haworth projections

Six-membered-ring sugars are called **pyranoses,** and five-membered-ring sugars are called **furanoses.** These names come from *pyran* and *furan,* the names of the cyclic ethers shown in the margin. Consequently, α-D-glucose is also called α-D-glucopyranose, and α-D-ribose is also called α-D-ribofuranose. The prefix "α" indicates the configuration about the anomeric carbon, and *pyranose* or *furanose* indicates the size of the ring.

pyran

furan

α-D-glucose
α-D-glucopyranose

α-D-ribose
α-D-ribofuranose

Ketoses also exist in solution predominantly in cyclic forms. For example, D-fructose forms a five-membered-ring hemiketal because its C-5 OH group reacts with its ketone carbonyl group. If the OH group bonded to the new asymmetric center is trans to the primary alcohol group, then the compound is α-D-fructofuranose; if it is cis to the primary alcohol group, the compound is β-D-fructofuranose. Notice that the anomeric carbon is C-2 in ketoses, not C-1 as in aldoses.

α-D-fructofuranose β-D-fructofuranose α-D-fructopyranose β-D-fructopyranose

D-Fructose can also form a six-membered ring by using its C-6 OH group. The pyranose form predominates in the monosaccharide, whereas the furanose form predominates when the sugar is part of a disaccharide. (See the structure of sucrose later in this chapter.)

Haworth projections are useful because they show clearly whether the OH groups on the ring are cis or trans to each other. Five-membered rings are nearly planar, so furanoses are represented fairly accurately by Haworth projections. Haworth projections, however, are structurally misleading for pyranoses because a six-membered ring is not flat—it exists preferentially in a chair conformation.

PROBLEM 19 Solved

4-Hydroxy- and 5-hydroxyaldehydes exist primarily as cyclic hemiacetals. Draw the structure of the cyclic hemiacetal formed by each of the following:

a. 4-hydroxybutanal **b.** 4-hydroxypentanal **c.** 5-hydroxypentanal **d.** 4-hydroxyheptanal

Solution to 19a Draw the reactant with its alcohol and carbonyl groups on the same side of the molecule, then look to see what size ring will form. Two cyclic products are obtained because the carbonyl carbon of the reactant has been converted into a new asymmetric center in the product.

PROBLEM 20

Draw the following sugars using Haworth projections:

a. β-D-galactopyranose   **b.** α-D-tagatopyranose   **c.** α-L-glucopyranose

---

PROBLEM 21

D-Glucose most often exists as a pyranose, but it can also exist as a furanose. Draw the Haworth projection of α-D-glucofuranose.

PROBLEM 22♦

Draw the anomers of D-erythrofuranose.

---

11 GLUCOSE IS THE MOST STABLE ALDOHEXOSE

Drawing D-glucose in its chair conformation shows why it is the most common aldohexose in nature. To convert the Haworth projection of D-glucose into a chair conformer, start by drawing the chair so that the backrest is on the left and the footrest is on the right. Then place the ring oxygen at the back right-hand corner and the primary alcohol group in the equatorial position. The primary alcohol group is the largest of all the substituents, and large substituents are more stable in the equatorial position because there is less steric strain in that position.

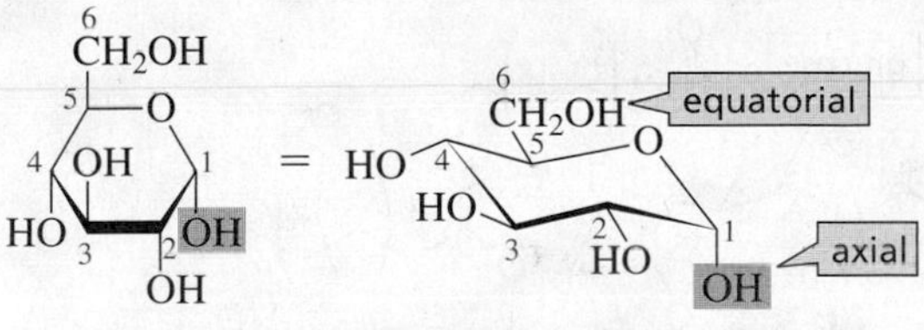

α-D-glucose
chair conformer

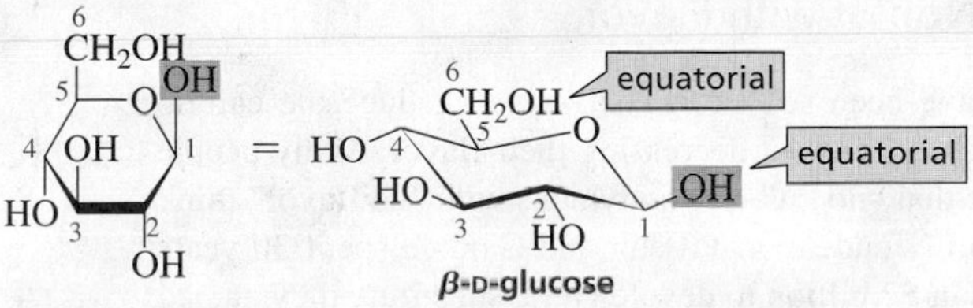

β-D-glucose
chair conformer

Because the OH group bonded to C-4 is trans to the primary alcohol group (this is easily seen in the Haworth projection), the C-4 OH group is also in the equatorial position. (1,2-Di-equatorial substituents are trans to one another.) The C-3 OH group is trans to the C-4 OH group, so the C-3 OH group is also in the equatorial position. As you move around the ring, you will find that all the OH substituents in β-D-glucose are in equatorial positions. The axial positions are all occupied by hydrogens, which require little space and therefore experience little steric strain. No other aldohexose exists in such a strain-free conformation. This means that β-D-glucose is the most stable of all the aldohexoses, so we should not be surprised that it is the most prevalent aldohexose in nature.

The OH group bonded to the anomeric carbon is in the equatorial position in β-D-glucose, whereas it is in the axial position in α-D-glucose. Therefore, β-D-glucose is more stable than α-D-glucose, so β-D-glucose predominates at equilibrium in an aqueous solution.

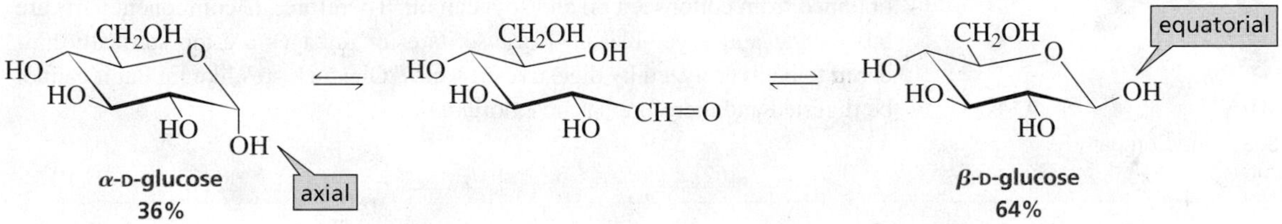

α-D-glucose
36%

β-D-glucose
64%

If you remember that all the OH groups in β-D-glucose are in equatorial positions, you will find it easy to draw the chair conformer of any other pyranose. For example, if you want to draw α-D-galactose, you would put all the OH groups in equatorial positions except the OH group at C-4 (because galactose is a C-4 epimer of glucose) and the OH group at C-1 (because you want the α-anomer), which both go in axial positions.

The α-position is
down in a Haworth projection and
axial in a chair conformation.

The β-position is
up in a Haworth projection and
equatorial in a chair conformation.

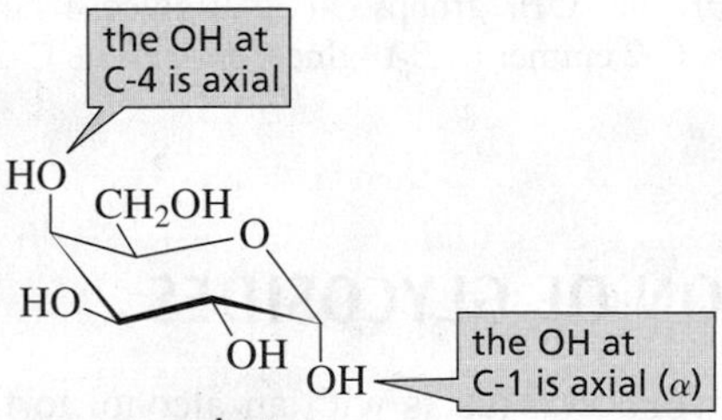

α-D-galactose

the OH at
C-4 is axial

the OH at
C-1 is axial (α)

To draw an L-pyranose, draw the D-pyranose first, and then draw its mirror image. For example, to draw β-L-gulose, first draw β-D-gulose, then draw its mirror image. (Gulose differs from glucose at C-3 and C-4, so the OH groups at these positions go in axial positions.)

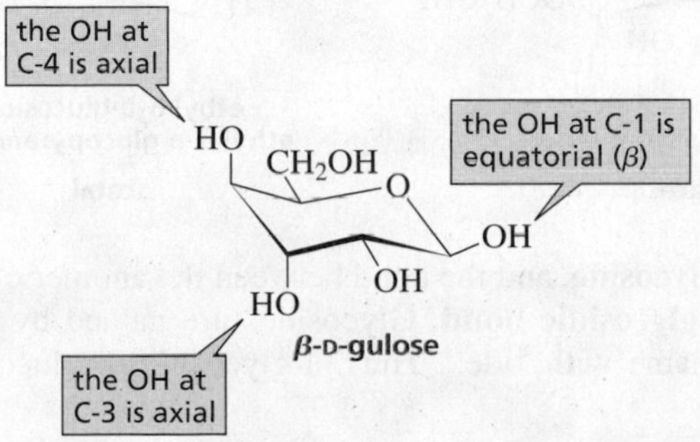

the OH at
C-4 is axial

the OH at C-1 is
equatorial (β)

β-D-gulose

the OH at
C-3 is axial

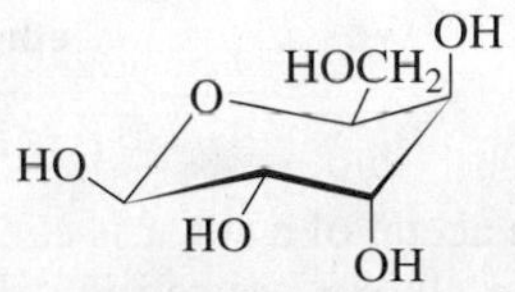

β-L-gulose

Olestra: Nonfat with Flavor

Chemists have been searching for ways to reduce the caloric content of foods without decreasing their flavor. Many people who believe that "no fat" is synonymous with "no flavor" think this is a worthy endeavor. Procter and Gamble spent 30 years and more than $2 billion to develop a fat substitute they named Olestra (also called Olean). After reviewing the results of more than 150 studies, in 1996 the U. S. Food and Drug Administration approved the limited use of Olestra in snack foods.

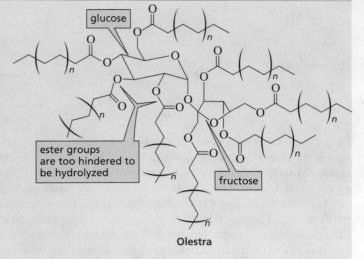

Olestra

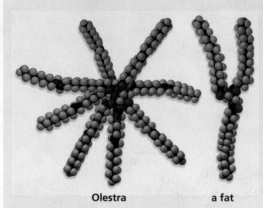

Olestra   **a fat**

courtesy of Procter & Gamble Company

Olestra is a semisynthetic compound because it does not exist in nature, but its components do. Developing a compound that can be made from units that are a normal part of our diet decreases the likelihood that the new compound will be toxic. Olestra is made by esterifying all the OH groups of sucrose with fatty acids obtained from cottonseed oil and soybean oil. Therefore, its component parts are table sugar and vegetable oil. Because its ester linkages are too sterically hindered to be hydrolyzed by digestive enzymes, Olestra tastes like fat but it cannot be digested and therefore has no caloric value.

PROBLEM 23 ♦ Solved

Which OH groups are in the axial position in

a. β-D-mannopyranose?   **b.** β-D-idopyranose?   **c.** α-D-allopyranose?

Solution to 23a All the OH groups in β-D-glucose are in equatorial positions. Because β-D-mannose is a C-2 epimer of β-D-glucose, only the C-2 OH group of β-D-mannose is in the axial position.

---

12 FORMATION OF GLYCOSIDES

In the same way that a hemiacetal reacts with an alcohol to form an acetal, the cyclic hemiacetal formed by a monosaccharide can react with an alcohol to form two acetals.

β-D-glucose
β-D-glucopyranose

ethyl β-D-glucoside
ethyl β-D-glucopyranoside
acetal

ethyl α-D-glucoside
ethyl α-D-glucopyranoside
acetal

a glycosidic bond

The acetal of a sugar is called a **glycoside,** and the bond between the anomeric carbon and the alkoxy oxygen is called a **glycosidic bond.** Glycosides are named by replacing the "e" ending of the sugar's name with "ide." Thus, a glycoside of glucose is a glucoside, a glycoside of galactose is a galactoside, and so on. If the pyranose or furanose name is used, the acetal is called a **pyranoside** or a **furanoside.**

Notice that the reaction of a single anomer with an alcohol leads to the formation of both the α- and β-glycosides. The mechanism of the reaction shows why both glycosides are formed.

MECHANISM FOR GLYCOSIDE FORMATION

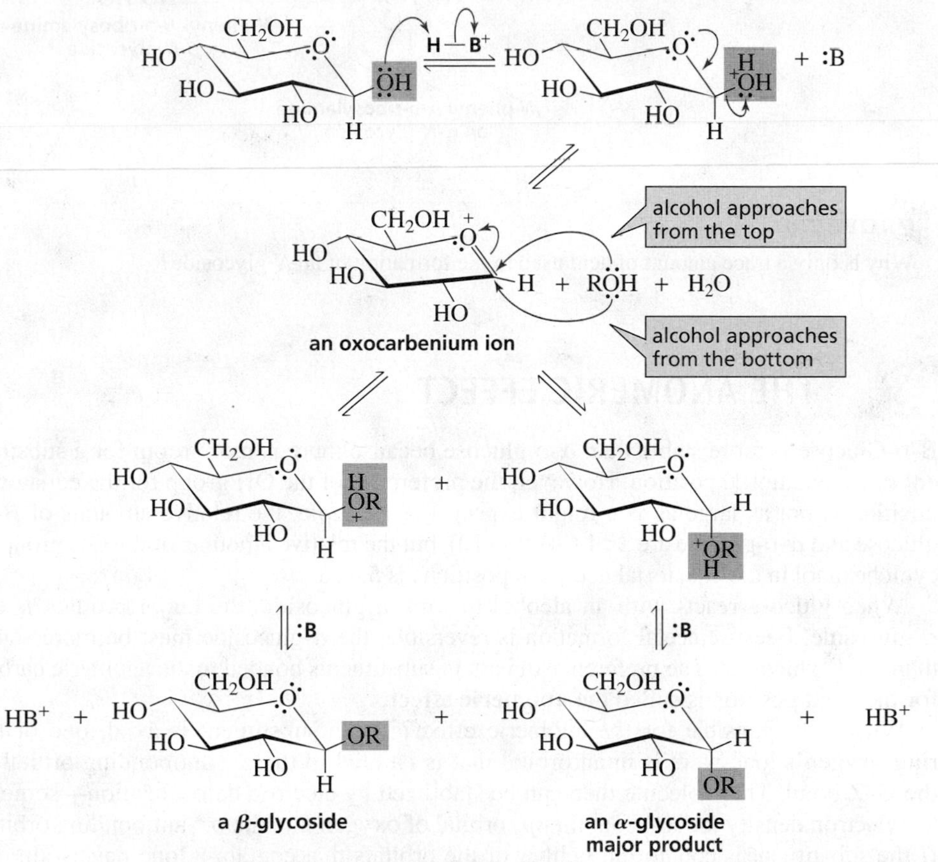

an oxocarbenium ion

alcohol approaches from the top

alcohol approaches from the bottom

a β-glycoside

an α-glycoside
major product

- The acid protonates the OH group bonded to the anomeric carbon.

- A lone pair on the ring oxygen helps eliminate a molecule of water. The anomeric carbon in the resulting oxocarbenium ion is sp^2 hybridized, so that part of the molecule is planar. (An **oxocarbenium** ion has a positive charge that is shared by a carbon and an oxygen.)

- When the alcohol approaches from the top of the plane, the β-glycoside is formed; when it approaches from the bottom of the plane, the α-glycoside is formed.

The mechanism is the same as that for acetal formation. Surprisingly, D-glucose forms more α-glycoside than β-glycoside. The reason for this is explained in the next section.

PROBLEM 24♦

Draw the products formed when β-D-galactose reacts with ethanol and HCl.

The reaction of a monosaccharide with an amine in the presence of a trace amount of acid is similar to the reaction of a monosaccharide with an alcohol. The product of the reaction is an *N*-glycoside. An **N-glycoside** has a nitrogen in place of the oxygen at the glycosidic linkage. The subunits of DNA and RNA are β-N-glycosides.

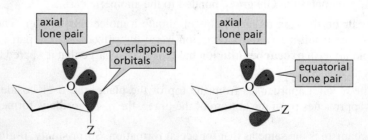

N-phenyl α-D-ribosylamine
an α-N-glycoside

N-phenyl β-D-ribosylamine
a β-N-glycoside

PROBLEM 25♦

Why is only a trace amount of acid used in the formation of an N-glycoside?

13   THE ANOMERIC EFFECT

β-D-Glucose is more stable than α-D-glucose because there is more room for a substituent in the equatorial position. However, the preference of the OH group for the equatorial position is not as large as you might expect. For example, the relative amounts of β-D-glucose and α-D-glucose are 2 : 1 (Section 10), but the relative amounts of the OH group of cyclohexanol in the equatorial and axial positions is 5.4 : 1.

When glucose reacts with an alcohol to form a glucoside, the major product is the α-glucoside. Because acetal formation is reversible, the α-glucoside must be more stable than the β-glucoside. The preference of certain substituents bonded to the anomeric carbon for the axial position is called the **anomeric effect.**

What is responsible for the anomeric effect? If the substituent is axial, one of the ring oxygen's lone pairs is in an orbital that is parallel to the σ^* antibonding orbital of the C–Z bond. The molecule then can be stabilized by electron delocalization—some of the electron density moves from the sp^3 orbital of oxygen into the σ^* antibonding orbital. If the substituent is equatorial, neither of the orbitals that contain a lone pair is aligned correctly for overlap.

14   REDUCING AND NONREDUCING SUGARS

Because glycosides are acetals, they are not in equilibrium with the open-chain aldehyde (or ketone) in aqueous solutions. Without being in equilibrium with a compound that has a carbonyl group, they cannot be oxidized by Ag^+. Glycosides, therefore, are nonreducing sugars—they cannot reduce Ag^+.

Hemiacetals, on the other hand, are in equilibrium with the open-chain sugars in aqueous solution, so they can reduce Ag^+. In summary, as long as a sugar has an aldehyde, a ketone, or a hemiacetal group, it can reduce Ag^+ and therefore is classified as a **reducing sugar.** An acetal is a **nonreducing sugar.**

A sugar with an aldehyde, a ketone, or a hemiacetal group is a reducing sugar. An acetal is a nonreducing sugar.

PROBLEM 26 ♦ Solved

Name the following compounds and indicate whether each is a reducing sugar or a nonreducing sugar:

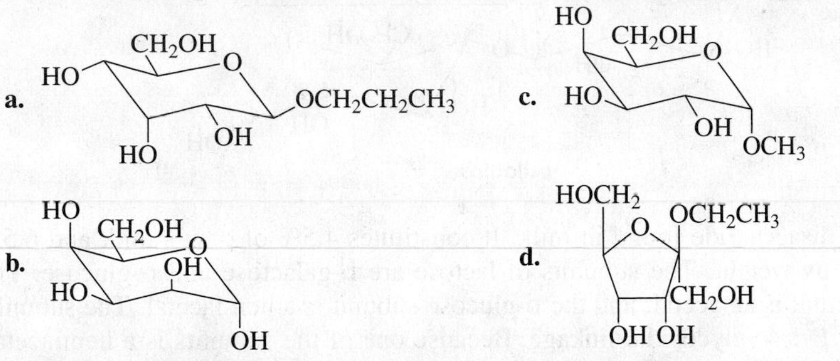

a. [chemical structure: pyranose ring with CH₂OH, HO, HO, OH substituents and OCH₂CH₂CH₃ at anomeric carbon]

b. [chemical structure: pyranose ring with HO, CH₂OH, OH, HO, OH substituents]

c. [chemical structure: pyranose ring with HO, CH₂OH, HO, OH substituents and OCH₃ at anomeric carbon]

d. [chemical structure: furanose ring with HOCH₂, OCH₂CH₃, CH₂OH, OH OH substituents]

Solution to 26a The only OH group in an axial position in part **a** is the one at C-3. Therefore, this sugar is the C-3 epimer of D-glucose, which is D-allose. The substituent at the anomeric carbon is in the β-position. Thus, the sugar's name is propyl β-D-alloside or propyl β-D-allopyranoside. Because the sugar is an acetal, it is a nonreducing sugar.

15   DISACCHARIDES

If the hemiacetal group of a monosaccharide forms an acetal by reacting with an alcohol group of another monosaccharide, the glycoside that is formed is a disaccharide. **Disaccharides** are compounds that consist of two monosaccharide subunits hooked together by a glycosidic linkage. For example, maltose, a disaccharide obtained from the hydrolysis of starch, contains two D-glucose subunits connected by a glycosidic linkage. This particular linkage is called an $\boldsymbol{\alpha}$**-1,4$'$-glycosidic linkage** because the linkage is between C-1 of one sugar subunit and C-4 of the other, and the oxygen bonded to the anomeric carbon in the glycosidic linkage is in the α-position. The prime superscript indicates that C-4 is not in the same ring as C-1.

Remember that the α-position is axial and the β-position is equatorial when a sugar is shown in a chair conformation.

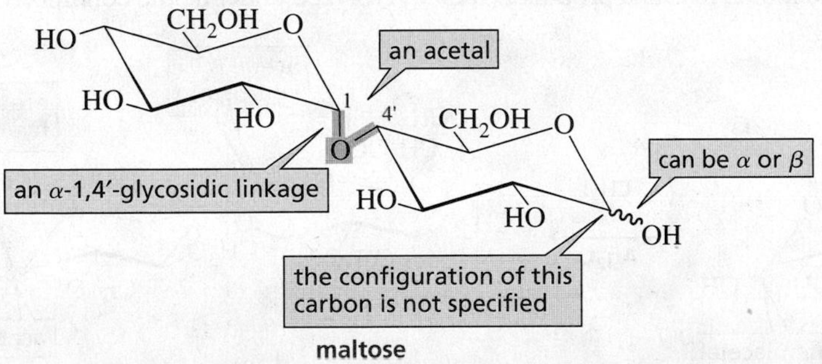

[chemical structure of maltose with labels: "an acetal", "an α-1,4$'$-glycosidic linkage", "the configuration of this carbon is not specified", "can be α or β"]

maltose

Notice that the structure of maltose does not specify the configuration of the anomeric carbon that is not an acetal (the anomeric carbon of the subunit on the right marked with a wavy line) because maltose can exist in both the α and β forms. In α-maltose, the OH group bonded to this anomeric carbon is in the axial position. In β-maltose, the OH group is in the equatorial position. Because maltose can exist in both α and β forms, mutarotation occurs when crystals of one form are dissolved in a solvent. Maltose is a reducing sugar because the right-hand subunit is a hemiacetal and therefore is in equilibrium with the open-chain aldehyde that is easily oxidized.

Cellobiose, a disaccharide obtained from the hydrolysis of cellulose, also contains two D-glucose subunits. Cellobiose is different from maltose, however, because the two glucose subunits are hooked together by a $\boldsymbol{\beta}$**-1,4$'$-glycosidic linkage.** Thus, the only difference in the structures of maltose and cellobiose is the configuration of the glycosidic linkage. Like maltose, cellobiose exists in both α and β forms because the OH group bonded to the anomeric carbon not involved in acetal formation can be in either the axial position (in α-cellobiose) or the equatorial position (in β-cellobiose). Cellobiose is a reducing sugar because the subunit on the right is a hemiacetal.

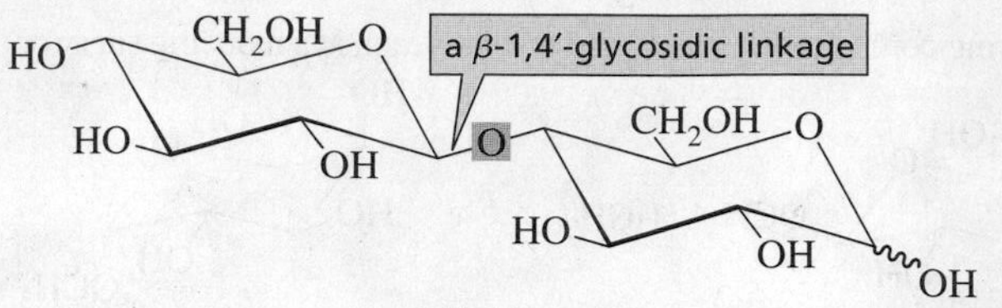

cellobiose

Lactose is a disaccharide found in milk. It constitutes 4.5% of cow's milk and 6.5% of human milk by weight. The subunits of lactose are D-galactose and D-glucose. The D-galactose subunit is an acetal, and the D-glucose subunit is a hemiacetal. The subunits are joined by a β-1,4'-glycosidic linkage. Because one of the subunits is a hemiacetal, lactose is a reducing sugar and undergoes mutarotation.

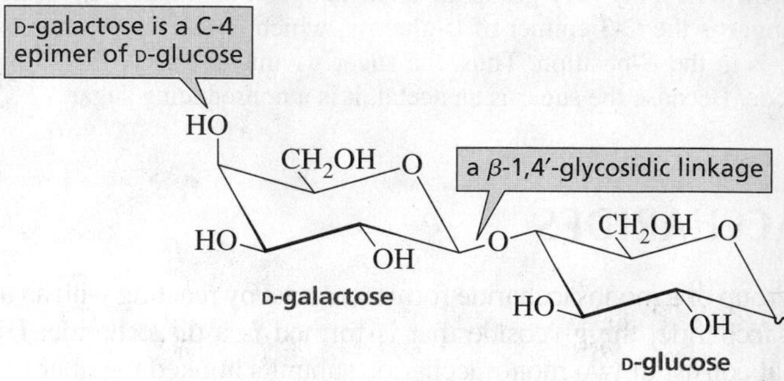

lactose

A simple experiment can prove that the hemiacetal group in lactose belongs to the glucose residue and not to the galactose residue. The disaccharide is treated with excess methyl iodide in the presence of Ag_2O, reagents that methylate all the OH groups *via* S_N2 reactions. Because the OH group is a relatively poor nucleophile, silver oxide is used to increase the leaving propensity of the iodide ion. The product is then hydrolyzed under acidic conditions.

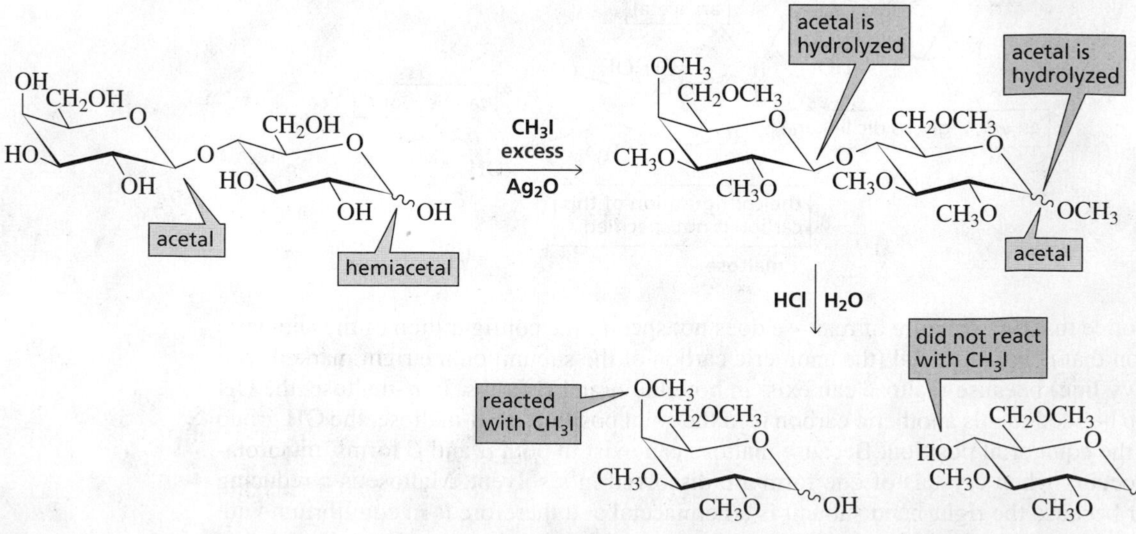

2,3,4,6-tetra-*O*-methylgalactose   **2,3,6-tri-*O*-methylglucose**

This treatment hydrolyzes the two acetal groups, but the ethers, formed by methylating the OH groups, are untouched. Identification of the products shows that the glucose residue contained the hemiacetal group in the disaccharide because its C-4 OH group was not able to react with methyl iodide (since it was used to form the acetal with galactose). The C-4 OH of galactose, on the other hand, was able to react with methyl iodide.

Lactose Intolerance

Lactase is an enzyme that specifically breaks the β-1,4'-glycosidic linkage of lactose. Cats and dogs lose their intestinal lactase when they become adults; they are then no longer able to digest lactose. Consequently, when they are fed milk or milk products, the undegraded lactose causes digestive problems such as bloating, abdominal pain, and diarrhea. These problems occur because only monosaccharides can pass into the bloodstream, so lactose, a disaccharide, has to pass undigested into the large intestine.

When humans have stomach flu or other intestinal disturbances, they can temporarily lose their lactase, thereby becoming lactose intolerant. About 75% of adults lose their lactase permanently as they mature—explaining why "lactose-free" products are so common. Those intolerant to lactose can take lactase in pill form before eating products that contain lactose.

Lactose intolerance is most common in people whose ancestors came from nondairy-producing countries. For example, only 3% of Danes are lactose intolerant, compared with 90% of all Chinese and Japanese and 97% of Thais. This is why you are not likely to find dairy items on Chinese menus.

Galactosemia

After lactose is degraded into glucose and galactose, the galactose must be converted into glucose before it can be used by cells. Individuals who do not have the enzyme that converts galactose into glucose have the genetic disease known as galactosemia. Without this enzyme, galactose accumulates in the bloodstream, a condition that can cause mental retardation and even death in infants. Galactosemia is treated by excluding galactose from the diet.

The most common disaccharide, sucrose, is the substance we know as table sugar. Obtained from sugar beets and sugar cane, sucrose consists of a D-glucose subunit and a D-fructose subunit linked by a glycosidic bond between C-1 of glucose (in the α-position) and C-2 of fructose (in the β-position). About 90 million tons of sucrose are produced commercially throughout the world each year.

Unlike some other disaccharides, sucrose is not a reducing sugar and does not exhibit mutarotation because its glycosidic bond is between the anomeric carbon of glucose and the anomeric carbon of fructose. Sucrose, therefore, does not have a hemiacetal group, so it is not in equilibrium with the readily oxidized open-chain aldehyde or ketone form in aqueous solution.

α linkage at glucose

β linkage at fructose

sucrose

a mixture of sucrose, glucose, and fructose

Sucrose has a specific rotation of +66.5. When it is hydrolyzed, the resulting 1 : 1 mixture of glucose and fructose has a specific rotation of −22.0. Because the sign of the rotation changes when sucrose is hydrolyzed, a 1 : 1 mixture of glucose and fructose is called *invert sugar*. The enzyme that catalyzes the hydrolysis of sucrose is called *invertase*. Honeybees have invertase, so the honey they produce is a mixture of sucrose, glucose, and fructose. Because fructose is sweeter than sucrose, invert sugar is also sweeter than sucrose.

Some "lite" foods contain fructose instead of sucrose, which means that they achieve the same level of sweetness with a lower sugar (lower calorie) content.

PROBLEM 27♦

What is the specific rotation of an equilibrium mixture of fructose? (*Hint:* The specific rotation of an equilibrium mixture of glucose is +52.7.)

---

16   POLYSACCHARIDES

Polysaccharides contain as few as 10 or as many as several thousand monosaccharide units joined together by glycosidic linkages. The most common polysaccharides are starch and cellulose.

Starch is the major component of flour, potatoes, rice, beans, corn, and peas. It is a mixture of two different polysaccharides: amylose (~20%) and amylopectin (~80%). Amylose is composed of unbranched chains of D-glucose units joined by α-1,4′-glycosidic linkages.

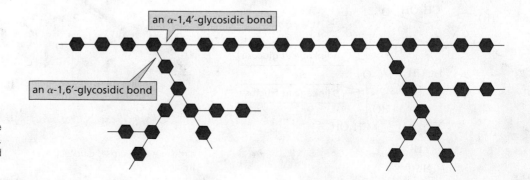

Amylopectin is a branched polysaccharide. Like amylose, it is composed of chains of D-glucose units joined by α-1,4′-glycosidic linkages. Unlike amylose, however, amylopectin also contains **α-1,6′-glycosidic linkages.** These linkages create branches in the polysaccharide (Figure 1). Amylopectin can contain up to 10^6 glucose units, making it one of the largest molecules found in nature.

▶ **Figure 1**

Branching in amylopectin. The hexagons represent glucose units. They are joined by α-1,4′- and α-1,6′-glyclosidic bonds.

Cells oxidize D-glucose in the first of a series of processes that provide them with energy. When animals have more D-glucose than they need for energy, they convert the excess D-glucose into a polymer called glycogen. Glycogen has a structure similar to that of amylopectin, but glycogen has more branches (Figure 2). The branch points in glycogen occur about every 10 residues, whereas those in amylopectin occur about every 25 residues. The high degree of branching in glycogen has important physiological consequences. When an animal needs energy, many individual glucose units can be simultaneously removed from the ends of many branches. Plants convert excess D-glucose into starch.

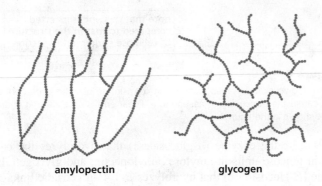

amylopectin   **glycogen**

◄ **Figure 2**

A comparison of the branching in amylopectin and glycogen.

Why the Dentist Is Right

Bacteria found in the mouth have an enzyme that converts sucrose into a polysaccharide called dextran. Dextran is made up of glucose units joined mainly through α-1,3'- and α-1,6'-glycosidic linkages. About 10% of dental plaque is composed of dextran, and bacteria hidden in the plaque attack tooth enamel. This is the chemical basis for your dentist's warning not to eat candy. This is also why sorbitol and mannitol are the saccharides added to "sugarless" gum—they cannot be converted to dextran.

Cellulose is the major structural component of plants. Cotton, for example, is composed of about 90% cellulose, and wood is about 50% cellulose. Like amylose, cellulose is composed of unbranched chains of D-glucose units. Unlike amylose, however, the glucose units in cellulose are joined by β-1,4'-glycosidic linkages rather than by α-1,4'-glycosidic linkages.

three subunits of cellulose

The different glycosidic linkages in starch and cellulose give these compounds very different physical properties. The α-linkages in starch cause amylose to form a helix that promotes hydrogen bonding of its OH groups to water molecules (Figure 3). As a result, starch is soluble in water.

On the other hand, the β-linkages in cellulose promote the formation of intramolecular hydrogen bonds. Consequently, these molecules form linear arrays (Figure 4), held together by hydrogen bonds between adjacent chains. These large aggregates cause cellulose to be insoluble in water. The strength of these bundles of polymer chains makes cellulose an effective structural material. Processed cellulose is also used for the production of paper and cellophane.

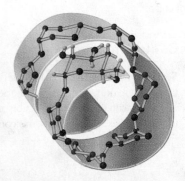

▲ **Figure 3**

The α-1,4'-glycosidic linkages in amylose cause it to form a left-handed helix. Many of its OH groups form hydrogen bonds with water molecules.

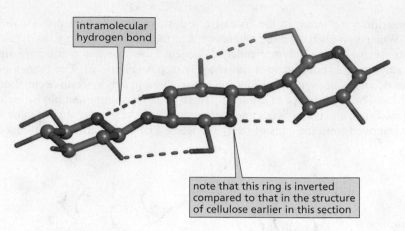

intramolecular hydrogen bond

note that this ring is inverted compared to that in the structure of cellulose earlier in this section

strands of cellulose in a plant cell wall

▲ **Figure 4**

The β-1,4'-glycosidic linkages in cellulose form intramolecular hydrogen bonds, which cause the molecules to assemble in linear arrays. (Notice that hydrogens are not shown in the figure.)

All mammals have the enzyme (α-glucosidase) that hydrolyzes the α-1,4'-glycosidic linkages that join glucose units in amylose, amylopectin, and glycogen, but they do *not* have the enzyme (β-glucosidase) that hydrolyzes β-1,4'-glycosidic linkages. As a result, mammals *cannot* obtain the glucose they need by eating cellulose. However, bacteria that possess β-glucosidase inhabit the digestive tracts of grazing animals, so cows can eat grass and horses can eat hay to meet their nutritional requirements for glucose. Termites also harbor bacteria that break down the cellulose in the wood they eat.

Chitin (KY-tin) is a polysaccharide that is structurally similar to cellulose. It is the major structural component of the shells of crustaceans (such as lobsters, crabs, and shrimp) and the exoskeletons of insects and other arthropods, and it is also the structural material of fungi. Like cellulose, chitin has β-1,4'-glycosidic linkages. Unlike cellulose, chitin has an *N*-acetylamino group instead of an OH group at the C-2 position. The β-1,4'-glycosidic linkages give chitin its structural rigidity.

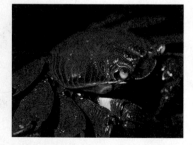

The shell of this bright orange crab from Australia is composed largely of chitin.

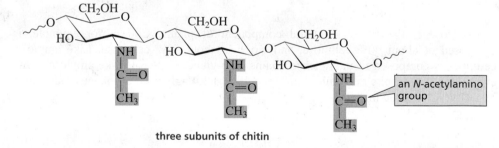

an *N*-acetylamino group

three subunits of chitin

Controlling Fleas

Several different drugs have been developed to help pet owners control fleas. One of these drugs is lufenuron, the active ingredient in Program. Lufenuron interferes with the flea's production of chitin. The consequences are fatal for the flea because its exoskeleton is composed primarily of chitin.

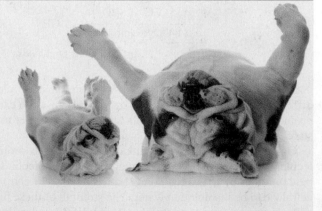

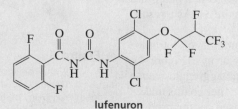

lufenuron

PROBLEM 28♦

What is the main structural difference between

a. amylose and cellulose?   **c.** amylopectin and glycogen?

b. amylose and amylopectin?   **d.** cellulose and chitin?

17 SOME NATURALLY OCCURRING COMPOUNDS DERIVED FROM CARBOHYDRATES

Deoxy sugars are sugars in which one of the OH groups is replaced by a hydrogen (*deoxy* means "without oxygen"). 2-Deoxyribose is a deoxy sugar that is missing the oxygen at the C-2 position. D-Ribose is the sugar component of ribonucleic acid (RNA), whereas 2-deoxyribose is the sugar component of deoxyribonucleic acid (DNA).

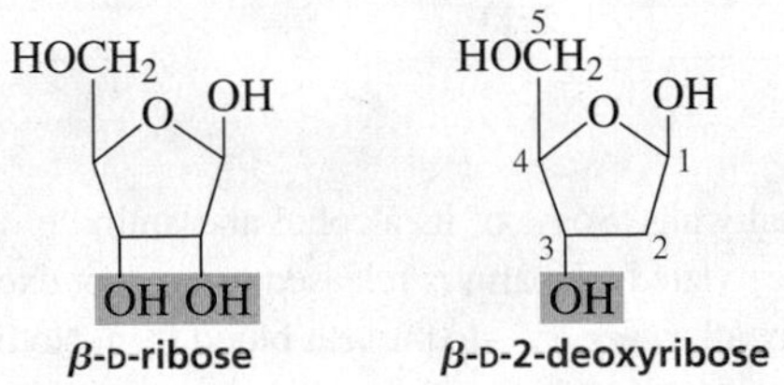

β-D-ribose β-D-2-deoxyribose

In **amino sugars,** one of the OH groups is replaced by an amino group. *N*-Acetyl-glucosamine—the subunit of chitin and one of the subunits of bacterial cell walls—is an example of an amino sugar.

Some important antibiotics contain amino sugars. For example, two of the three subunits of the antibiotic gentamicin are deoxyamino sugars. Notice that the middle subunit is missing the ring oxygen, so it is not a sugar.

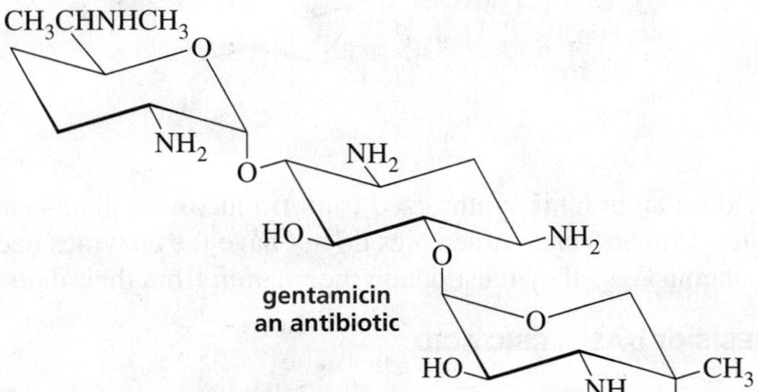

gentamicin an antibiotic

Gentamicin is one of several aminoglycoside antibiotics; streptomycin and neomycin are others. The antibiotics work by binding to a bacterial ribosome on which protein synthesis takes place. As a result, the bacteria are not able to synthesize proteins.

Bacterial Resistance

A bacterial strain typically takes 15 to 20 years to become resistant to an antibiotic. For example, penicillin became widely available in 1944 and by 1952, 60% of all *Staphylococcus aureus* infections were penicillin resistant. As a result of the widespread use of the aminoglycoside antibiotics, some bacteria developed enzymes that can acetylate or phosphorylate the OH and NH_2 groups of the antibiotic. When this happens, the antibiotic can no longer bind to the bacterial ribosome, so it has no effect on the bacteria.

Until relatively recently, the last discovery of a new class of antibiotics was in the 1970s, so **drug resistance** became an increasingly important problem in medicinal chemistry. Vancomycin had been the antibiotic of last resort because there were no reported cases of vancomycin-resistant bacteria until 1989, when more and more bacteria became resistant.

The approval of Zyvox by the FDA in April 2000 was met with great relief by the medical community because it was the first in a new family of antibiotics—namely, the oxazolidinones. In clinical trials, Zyvox was found to cure 75% of the patients infected with bacteria that had become resistant to all other antibiotics. Another new class of antibiotic became available in 2005 when the FDA approved Cubicin, the first of the cyclic lipopeptide antibiotics.

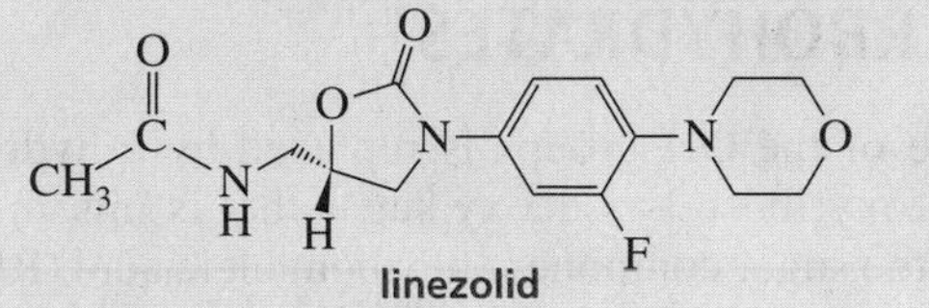

linezolid
Zyvox®

Heparin—A Natural Anticoagulant

Heparin is a polysaccharide found principally in cells that line arterial walls. Some of its alcohol and amino groups are sulfonated, some of its alcohol groups are oxidized, and some of its amino groups are acetylated. Heparin is released to prevent excessive blood clot formation when an injury occurs. Heparin is widely used clinically—particularly after surgery—to prevent blood from clotting.

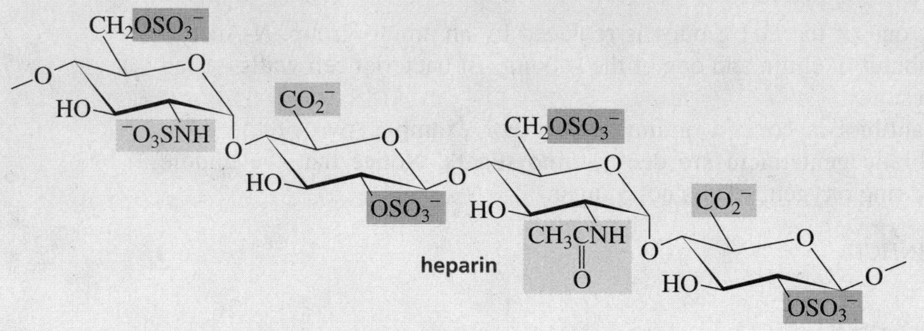

heparin

L-Ascorbic acid (vitamin C) is synthesized from D-glucose in plants and in the livers of most vertebrates. Primates and guinea pigs do not have the enzymes necessary for the biosynthesis of vitamin C, so they must obtain the vitamin from their diets.

STEPS IN THE SYNTHESIS OF L-ASCORBIC ACID

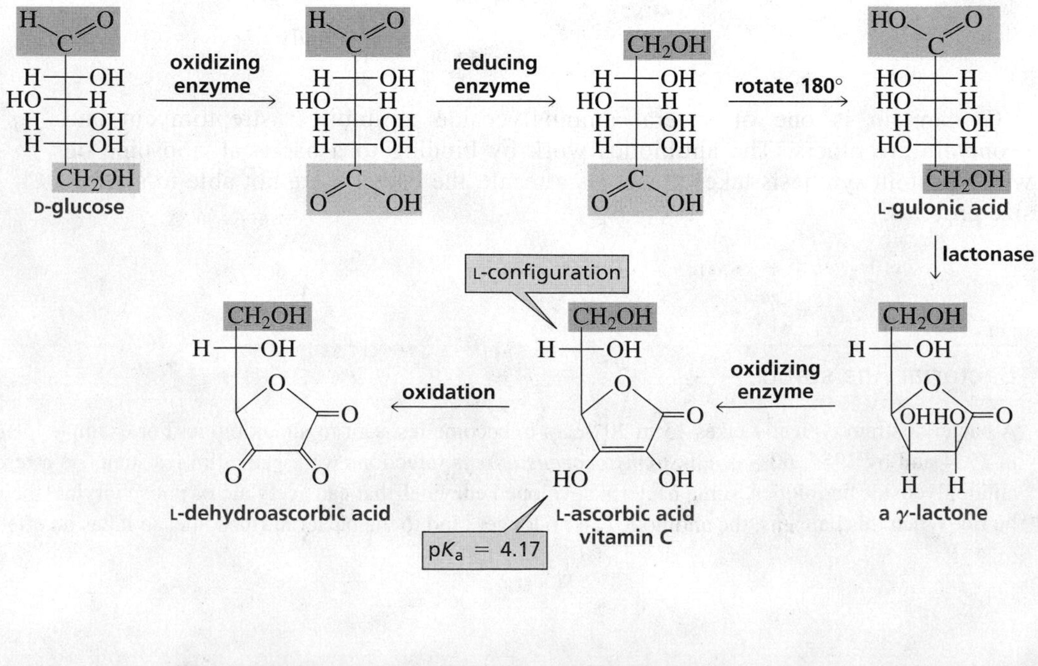

D-glucose

L-gulonic acid

L-configuration

L-dehydroascorbic acid

L-ascorbic acid
vitamin C

$pK_a = 4.17$

a γ-lactone

- The biosynthesis of vitamin C involves the conversion of D-glucose to L-gulonic acid, reminiscent of the last step in the Fischer proof.
- L-Gulonic acid is converted into a γ-lactone by the enzyme lactonase.
- The lactone is oxidized to L-ascorbic acid. The L-designation of ascorbic acid refers to the configuration at C-5, which was C-2 in D-glucose and C-5 in L-gulonic acid.

Although L-ascorbic acid does not have a carboxylic acid group, it is an acidic compound because the pK_a of the C-3 OH group is 4.17. L-Ascorbic acid is readily oxidized to L-dehydroascorbic acid, which is also physiologically active. If the lactone ring is opened by hydrolysis, all vitamin C activity is lost. Therefore, not much intact vitamin C survives in food that has been thoroughly cooked. And if food is cooked in water and then drained, the water-soluble vitamin is thrown out with the water!

Vitamin C

Vitamin C is an antioxidant because it traps radicals formed in aqueous environments, preventing harmful oxidation reactions the radicals would cause. Not all the physiological functions of vitamin C are known. However, we do know it is required for collagen fibers to form properly. Collagen is the structural protein of skin, tendons, connective tissue, and bone.

Vitamin C is abundant in citrus fruits and tomatoes. When the vitamin is not present in the diet, lesions appear on the skin, severe bleeding occurs about the gums, in the joints, and under the skin, and any wound heals slowly. The condition, known as *scurvy,* was the first disease to be treated by adjusting the diet. British sailors who shipped

an English sailor circa 1829

out to sea after the late 1700s were required to eat limes to prevent scurvy (which is how they came to be called "limeys"). Not until 200 years later did it become known that the substance preventing scurvy was vitamin C. *Scorbutus* is Latin for "scurvy"; *ascorbic,* therefore, means "no scurvy."

PROBLEM 29♦

Explain why the C-3 OH group of vitamin C is more acidic than the C-2 OH group.

18 CARBOHYDRATES ON CELL SURFACES

Many cells have short oligosaccharide chains on their surface that enable the cells to recognize and interact with other cells and with invading viruses and bacteria. These oligosaccharides are linked to the surface of the cell by the reaction of an OH or an NH_2 group of a cell-membrane protein with the anomeric carbon of a cyclic sugar. Proteins attached to oligosaccharides are called **glycoproteins.** The percentage of carbohydrate in glycoproteins is variable; some glycoproteins contain as little as 1% carbohydrate by weight, whereas others contain as much as 80%.

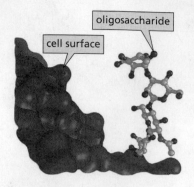

Carbohydrates on the surfaces of cells provide a way for cells to recognize one another, serving as points of attachment for other cells, viruses, and toxins. Therefore, surface carbohydrates have been found to play a role in activities as diverse as infection, prevention of infection, fertilization, inflammatory diseases such as rheumatoid arthritis and septic shock, and blood clotting. The fact that several known antibiotics contain amino sugars (Section 17) suggests that the antibiotics function by recognizing target cells. Carbohydrate interactions also are involved in the regulation of cell growth, so changes in membrane glycoproteins are thought to be correlated with malignant transformations.

Differences in blood type (A, B, or O) are actually differences in the sugars bound to the surfaces of red blood cells. Each type of blood is associated with a different carbohydrate structure (Figure 5). Type AB blood is a mixture of type A blood and type B blood.

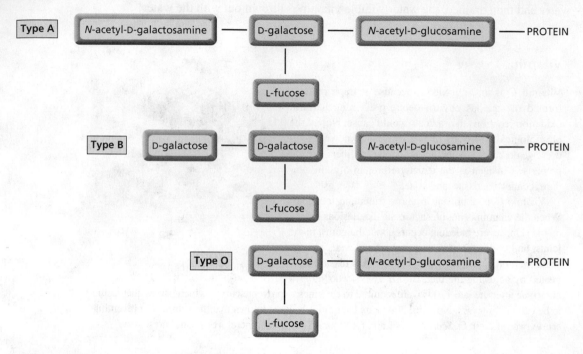

▲ **Figure 5**

Blood type is determined by the sugars on the surfaces of red blood cells. Fucose is 6-deoxygalactose.

Antibodies are proteins that are synthesized by the body in response to foreign substances called *antigens*. Interaction with the antibody causes the antigen to either precipitate or be flagged for destruction by immune system cells. This is why blood cannot be transferred from one person to another unless the blood types of the donor and acceptor are compatible. Otherwise the donated blood will be considered a foreign substance and will provoke an immune response.

Looking at Figure 5, we can see why the immune system of people with type A blood recognizes type B blood as foreign and vice versa. The immune system of people with type A, B, or AB blood does not recognize type O blood as foreign because the carbohydrate in type O blood is also a component of types A, B, and AB. Thus, anyone can accept type O blood, so people with that blood type are called universal donors. People with type AB blood can accept types AB, A, B, and O blood, so they are referred to as universal acceptors.

PROBLEM 30

Refer to Figure 5 to answer the following questions:

a. People with type O blood can donate blood to anyone, but they cannot receive blood from everyone. From whom can they *not* receive blood?

b. People with type AB blood can receive blood from anyone, but they cannot give blood to everyone. To whom can they *not* give blood?

19 ARTIFICIAL SWEETENERS

For a molecule to taste sweet, it must bind to a receptor on a taste bud cell on the tongue. The binding of this molecule causes a nerve impulse to pass from the taste bud to the brain, where the molecule is interpreted as being sweet. Sugars differ in their degree of "sweetness." Compared with the sweetness of glucose, which is assigned a relative value of 1.00, the sweetness of sucrose is 1.45, and that of fructose, the sweetest of all sugars, is 1.65.

Developers of artificial sweeteners must evaluate potential products in terms of several factors—such as toxicity, stability, and cost—in addition to taste. Saccharin (Sweet'N Low), the first synthetic sweetener, was discovered accidentally by Ira Remsen in 1879. One evening he noticed that the dinner rolls initially tasted sweet and then bitter. Because his wife did not notice that the rolls had an unusual taste, Remsen tasted his fingers and found they had the same odd taste. The next day he tasted the chemicals he had been working with the day before and found one that had an extremely sweet taste. (As strange as it may seem today, at one time it was common for chemists to taste compounds in order to characterize them.) He called this compound saccharin; it was eventually found to be about 300 times sweeter than glucose. Notice that, in spite of its name, saccharin is *not* a saccharide.

saccharin   dulcin   acesulfame potassium

aspartame   sucralose

sodium cyclamate

Because it has little caloric value, saccharin became an important substitute for sucrose when it became commercially available in 1885. The chief nutritional problem in the West was—and still is—the overconsumption of sugar and its consequences: obesity, heart disease, and dental decay. Saccharin is also a boon to people with diabetes, who must limit their consumption of sucrose and glucose. Although the toxicity of saccharin had not been studied carefully when it was first marketed (our current concern with toxicity is a fairly recent development), extensive studies since then have shown saccharin to be harmless. In 1912, saccharin was temporarily banned in the United States, not because of any concern about its toxicity, but because of a concern that people would miss out on the nutritional benefits of sugar.

Dulcin was the second synthetic sweetener to be discovered (in 1884). Even though it did not have the bitter, metallic aftertaste associated with saccharin, it never achieved much popularity. Dulcin was taken off the market in 1951 in response to concerns about its toxicity.

Sodium cyclamate became a widely used nonnutritive sweetener in the 1950s, but was banned in the United States some 20 years later in response to two studies that appeared to show that large amounts of sodium cyclamate cause liver cancer in mice.

Aspartame (NutraSweet, Equal), about 200 times sweeter than sucrose, was approved by the U.S. Food and Drug Administration (FDA) in 1981. Because aspartame contains a phenylalanine subunit, it should not be used by people with the genetic disease known as phenylketonuria (PKU).

Acesulfame potassium (Sweet and Safe, Sunette, Sweet One) was approved in 1988. Also called acesulfame-K, it too is about 200 times sweeter than glucose. It has less aftertaste than saccharine and is more stable than aspartame at high temperatures.

Sucralose (Splenda), 600 times sweeter than glucose, is the most recently approved (1991) synthetic sweetener. It maintains its sweetness in foods stored for long periods and at temperatures used in baking. Sucralose is made from sucrose by selectively replacing three of sucrose's OH groups with chlorines. During chlorination, the 4-position of the glucose ring becomes inverted, so sucralose is a galactopyranoside, not a glucopyranoside. Sucralose is the only artificial sweetener that has a carbohydrate-like structure. However, because of the chlorine atoms, the body does not recognize it as a carbohydrate, so it is eliminated from the body instead of being metabolized.

The fact that these synthetic sweeteners have such different structures shows that the sensation of sweetness is not induced by a single molecular shape.

Acceptable Daily Intake

The FDA has established an acceptable daily intake (ADI) value for many of the food ingredients it clears for use. The ADI is the amount of the substance a person can consume safely, each day of his or her life. For example, the ADI for acesulfame-K is 15 mg/kg/day. This means that each day a 132-lb person can consume the amount of acesulfame-K that would be found in two gallons of an artificially sweetened beverage. The ADI for sucralose is also 15 mg/kg/day.

SOME IMPORTANT THINGS TO REMEMBER

- **Bioorganic compounds**—organic compounds found in living systems—obey the same chemical principles that smaller organic molecules do.

- Much of the structure of bioorganic compounds exists for the purpose of **molecular recognition.**

- **Carbohydrates** are polyhydroxy aldehydes (**aldoses**) and polyhydroxy ketones (**ketoses**), or compounds formed by linking up aldoses and ketoses.

- The notations D and L describe the configuration of the bottommost asymmetric center of a **monosaccharide** in a Fischer projection. Most naturally occurring sugars are D-sugars.

- Naturally occurring ketoses have the ketone group in the 2-position.

- **Epimers** differ in configuration at only one asymmetric center: D-mannose is the C-2 epimer of D-glucose and D-galactose is the C-4 epimer of D-glucose.

- In a basic solution, a monosaccharide is converted to a mixture of polyhydroxy aldehydes and polyhydroxy ketones.

- Reduction of an aldose forms one **alditol;** reduction of a ketose forms two alditols.

- Br_2 oxidizes aldoses but not ketoses; Tollens reagent oxidizes both.

- Aldoses are oxidized to **aldonic acids** or to **aldaric acids.**

- The **Kiliani–Fischer synthesis** increases the carbon chain of an aldose by one carbon; it forms C-2 epimers.

- The **Wohl degradation** decreases the carbon chain by one carbon.

- The OH groups of monosaccharides react with methyl iodide/silver oxide to form ethers.

- The aldehyde or keto group of a **monosaccharide** reacts with one of its OH groups to form cyclic hemiacetals: glucose forms α-D-glucose and β-D-glucose. More β-D-glucose is present than α-D-glucose in an aqueous solution at equilibrium.

- α-D-Glucose and β-D-glucose are **anomers**—they differ in configuration only at the **anomeric carbon,** which is the carbon that was the carbonyl carbon in the open-chain form.

- A slow change in optical rotation to an equilibrium value is called **mutarotation.**

- The α-position is axial when a sugar is shown in a chair conformation and down when the sugar is shown in a

- Haworth projection; the β-position is equatorial when a sugar is shown in a chair conformation and up when the sugar is shown in a Haworth projection.

- Six-membered-ring sugars are **pyranoses;** five-membered-ring sugars are **furanoses.**

- The most abundant monosaccharide in nature is D-glucose. All the OH groups in β-D-glucose are in equatorial positions.

- A cyclic hemiacetal can react with an alcohol to form an acetal, called a **glycoside.** If the name "pyranose" or "furanose" is used, the acetal is called a **pyranoside** or a **furanoside,** respectively.

- The bond between the anomeric carbon and the alkoxy oxygen is called a **glycosidic bond.**

- The preference for the axial position by certain substituents bonded to the anomeric carbon is called the **anomeric effect.**

- If a sugar has an aldehyde, ketone, hemiacetal, or hemiketal group, it is a **reducing sugar.** Acetals are not reducing sugars.

- **Disaccharides** consist of two monosaccharides hooked together by a glycosidic linkage. Maltose has an **α-1,4$'$-glycosidic linkage** between two glucose subunits; cellobiose has a **β-1,4$'$-glycosidic linkage** between two glucose subunits.

- The most common disaccharide is sucrose; it has a D-glucose subunit and a D-fructose subunit linked by their anomeric carbons.

- **Polysaccharides** contain as few as 10 or as many as several thousand monosaccharides joined together by glycosidic linkages.

- Starch is composed of amylose and amylopectin. Amylose has unbranched chains of D-glucose units joined by α-1,4$'$-glycosidic linkages.

- Amylopectin also has chains of D-glucose units joined by α-1,4$'$-glycosidic linkages, but it also has α-1,6$'$-glycosidic linkages that create branches. Glycogen is similar to amylopectin but has more branches.

- Cellulose has unbranched chains of D-glucose units joined by β-1,4$'$-glycosidic linkages.

- The α-linkages cause amylose to form a helix and be water soluble; the β-linkages allow the molecules of cellulose to form linear arrays and be water insoluble.

- The surfaces of many cells contain short oligosaccharides that allow the cells to interact with each other. The oligosaccharides are attached to the cell surface by protein groups.

- Proteins bonded to oligosaccharides are called **glycoproteins.**

SUMMARY OF REACTIONS

1. Epimerization (Section 5).

2. Enediol rearrangement (Section 5).

3. Reduction (Section 6)

4. Oxidation (Section 6)

a.

$$\underset{\text{CH}_2\text{OH}}{\overset{\overset{\text{H}}{\underset{|}{\text{C}}=\text{O}}}{(\text{CHOH})_n}} \xrightarrow[\text{HO}^-]{\text{Ag}^+,\ \text{NH}_3} \underset{\text{CH}_2\text{OH}}{\overset{\overset{^-\text{O}}{\underset{|}{\text{C}}=\text{O}}}{(\text{CHOH})_n}} + \text{Ag}$$

c.

$$\underset{\text{CH}_2\text{OH}}{\overset{\overset{\text{H}}{\underset{|}{\text{C}}=\text{O}}}{(\text{CHOH})_n}} \xrightarrow[\text{H}_2\text{O}]{\text{Br}_2} \underset{\text{CH}_2\text{OH}}{\overset{\overset{\text{HO}}{\underset{|}{\text{C}}=\text{O}}}{(\text{CHOH})_n}} + 2\ \text{Br}^-$$

b.

$$\underset{\text{CH}_2\text{OH}}{\overset{\overset{\text{CH}_2\text{OH}}{\underset{|}{\text{C}}=\text{O}}}{(\text{CHOH})_n}} \xrightarrow[\text{HO}^-]{\text{Ag}^+,\ \text{NH}_3} \underset{\text{CH}_2\text{OH}}{\overset{\overset{^-\text{O}}{\underset{|}{\text{C}}=\text{O}}}{\overset{\text{CHOH}}{(\text{CHOH})_n}}} + \text{Ag}$$

d.

$$\underset{\text{CH}_2\text{OH}}{\overset{\overset{\text{H}}{\underset{|}{\text{C}}=\text{O}}}{(\text{CHOH})_n}} \xrightarrow[\Delta]{\text{HNO}_3} \underset{\overset{|}{\underset{\text{O}}{\text{C}}}\overset{\diagdown}{\text{OH}}}{\overset{\overset{\text{HO}}{\underset{|}{\text{C}}=\text{O}}}{(\text{CHOH})_n}}$$

5. Chain elongation: the Kiliani–Fischer synthesis (Section 7)

$$\underset{\text{CH}_2\text{OH}}{\overset{\overset{\text{H}}{\underset{|}{\text{C}}=\text{O}}}{(\text{CHOH})_n}} \xrightarrow[\text{3. HCl/H}_2\text{O}]{\overset{\text{1. NaC}\equiv\text{N/HCl}}{\text{2. H}_2,\ \text{Pd/BaSO}_4}} \underset{\text{CH}_2\text{OH}}{\overset{\overset{\text{H}}{\underset{|}{\text{C}}=\text{O}}}{(\text{CHOH})_{n+1}}}$$

6. Chain shortening: the Wohl degradation (Section 8)

$$\underset{\text{CH}_2\text{OH}}{\overset{\overset{\text{H}}{\underset{|}{\text{C}}=\text{O}}}{(\text{CHOH})_n}} \xrightarrow[\substack{\text{1. NH}_2\text{OH/trace acid} \\ \text{2. Ac}_2\text{O, 100 °C} \\ \text{3. HO}^-,\ \text{H}_2\text{O}}]{} \underset{\text{CH}_2\text{OH}}{\overset{\overset{\text{H}}{\underset{|}{\text{C}}=\text{O}}}{(\text{CHOH})_{n-1}}}$$

7. Hemiacetal formation (Section 11)

8. Glycoside formation (Section 12).

PROBLEMS

31. What product or products are obtained when D-galactose reacts with each of the following substances?

a. nitric acid + Δ
b. Ag^+, NH_3, HO^-
c. NaBH_4, followed by H_3O^+

d. excess CH_3I + Ag_2O
e. Br_2 in water
f. ethanol + HCl

g. 1. hydroxylamine/trace acid
   2. acetic anhydride/Δ
   3. $\text{HO}^-/\text{H}_2\text{O}$

32. Name the epimers of D-glucose.

33. Identify the sugar in each description.

    a. An aldopentose that is not D-arabinose forms D-arabinitol when it is reduced with NaBH₄.
    b. A sugar that is not D-altrose forms D-altraric acid when it is oxidized with nitric acid.
    c. A ketose that, when reduced with NaBH₄, forms D-altritol and D-allitol.

34. D-Xylose and D-lyxose are formed when D-threose undergoes a Kiliani–Fischer synthesis. D-Xylose is oxidized to an optically inactive aldaric acid, whereas D-lyxose forms an optically active aldaric acid. What are the structures of D-xylose and D-lyxose?

35. Answer the following questions about the eight aldopentoses.

    a. Which are enantiomers?
    b. Which are C-2 epimers?
    c. Which form an optically active compound when oxidized with nitric acid?

36. What is the configuration of each of the asymmetric centers in the Fischer projection of

    a. D-glucose?   **b.** D-galactose?   **c.** D-ribose?   **d.** D-xylose?   **e.** D-sorbose?

37. The reaction of D-ribose with one equivalent of methanol plus HCl forms four products. Draw the products.

38. Name the following compounds:

39. A student isolated a monosaccharide and determined that it had a molecular weight of 150. Much to his surprise, he found that it was not optically active. What is the structure of the monosaccharide?

40. Propose a mechanism for the formation of D-allose from D-glucose in a basic solution.

41. Treatment with sodium borohydride converts aldose **A** into an optically inactive alditol. Wohl degradation of **A** forms **B**, whose alditol is optically inactive. Wohl degradation of **B** forms D-glyceraldehyde. Identify **A** and **B**.

42. A hexose was obtained after (+)-glyceraldehyde underwent three successive Kiliani–Fischer syntheses. Identify the hexose from the following experimental information: oxidation with nitric acid forms an optically active aldaric acid; a Wohl degradation followed by oxidation with nitric acid forms an optically inactive aldaric acid; and a second Wohl degradation forms erythrose.

43. Draw the mechanism for the interconversion of α-D-glucose and β-D-glucose in dilute HCl.

44. An unknown β-D-aldohexose has only one axial substituent. A Wohl degradation forms a compound which, when treated with sodium borohydride, forms an optically active alditol. This information allows you to arrive at two possible structures for the β-D-aldohexose. What experiment can you carry out to distinguish between the two possibilities?

45. The ¹H NMR spectrum of D-glucose in D₂O exhibits two high-frequency doublets. What is responsible for these doublets?

46. D-Glucuronic acid is found widely in plants and animals. One of its functions is to detoxify poisonous HO-containing compounds by reacting with them in the liver to form glucuronides. Glucuronides are water soluble and therefore readily excreted. After ingestion of a poison such as turpentine or phenol, the glucuronides of these compounds are found in the urine. Draw the structure of the glucuronide formed by the reaction of β-D-glucuronic acid and phenol.

β-D-glucuronic acid

47. Determine the structure of D-galactose, using arguments similar to those used by Fischer to prove the structure of D-glucose.

48. A D-aldopentose is oxidized by nitric acid to an optically active aldaric acid. A Wohl degradation of the aldopentose leads to a monosaccharide that is oxidized by nitric acid to an optically inactive aldaric acid. Identify the D-aldopentose.

49. Draw the mechanism for the formation of β-maltose from α-D-galactose and β-D-glucose in dilute HCl.

50. Hyaluronic acid, a component of connective tissue, is the fluid that lubricates the joints. It is a polymer of alternating *N*-acetyl-D-glucosamine and D-glucuronic acid subunits joined by β-1,3′-glycosidic linkages. Draw a short segment of hyaluronic acid.

51. In order to synthesize D-galactose, a student went to the stockroom to get some D-lyxose to use as a starting material. She found that the labels had fallen off the bottles containing D-lyxose and D-xylose. How could she determine which bottle contains D-lyxose?

52. The aldonic acid of D-glucose forms a five-membered ring lactone. Draw its structure.

53. The aldaric acid of D-glucose forms two five-membered ring lactones. Draw their structures.

54. Draw the mechanism for the acid-catalyzed hydrolysis of β-maltose.

55. How many aldaric acids are obtained from the 16 aldohexoses?

56. A hexose is obtained when the residue of a shrub *Sterculia setigeria* undergoes acid-catalyzed hydrolysis. Identify the hexose from the following experimental information: it undergoes mutarotation; it does not react with Br_2; and D-galactonic acid and D-talonic acid are formed when it reacts with Tollens reagent.

57. When D-fructose is dissolved in D_2O and the solution is made basic, the D-fructose recovered from the solution has an average of 1.7 deuterium atoms attached to the C-1 carbon per molecule. Show the mechanism that accounts for the incorporation of these deuterium atoms into D-fructose.

58. Draw each of the following compounds:

   a. β-D-talopyranose   **c.** α-D-tagatopyranose   **e.** β-L-talopyranose

   b. α-D-idopyranose   **d.** β-D-psicofuranose   **f.** α-L-tagatopyranose

59. Calculate the percentages of α-D-glucose and β-D-glucose present at equilibrium from the specific rotations of α-D-glucose, β-D-glucose, and the equilibrium mixture. Compare your values with those given in Section 10. (*Hint:* The specific rotation of the mixture equals the specific rotation of α-D-glucose times the fraction of glucose present in the α-form plus the specific rotation of β-D-glucose times the fraction of glucose present in the β-form.)

60. The specific rotation of α-D-galactose is 150.7° and that of β-D-galactose is 52.8°. When an aqueous mixture that was initially 70% α-D-galactose and 30% β-D-galactose reaches equilibrium, the specific rotation is 80.2°. What is the concentration of the anomers at equilibrium?

61. An unknown disaccharide gives a positive Tollens test. A glycosidase hydrolyzes it to D-galactose and D-mannose. When the disaccharide is treated with methyl iodide and Ag_2O and then hydrolyzed with dilute HCl, the products are 2,3,4,6-tetra-*O*-methylgalactose and 2,3,4-tri-*O*-methylmannose. Propose a structure for the disaccharide.

62. Predict whether D-altrose exists preferentially as a pyranose or a furanose. (*Hint:* In the most stable arrangement for a five-membered ring, all the adjacent substituents are trans.)

63. Trehalose, $C_{12}H_{22}O_{11}$, is a nonreducing sugar that is only 45% as sweet as sugar. When hydrolyzed by aqueous acid or the enzyme maltase, it forms only D-glucose. When it is treated with excess methyl iodide in the presence of Ag_2O and then hydrolyzed with water under acidic conditions, only 2,3,4,6-tetra-*O*-methyl-D-glucose is formed. Draw the structure of trehalose.

64. Propose a mechanism for the rearrangement that converts an α-hydroxyimine into an α-aminoketone in the presence of a trace amount of acid.

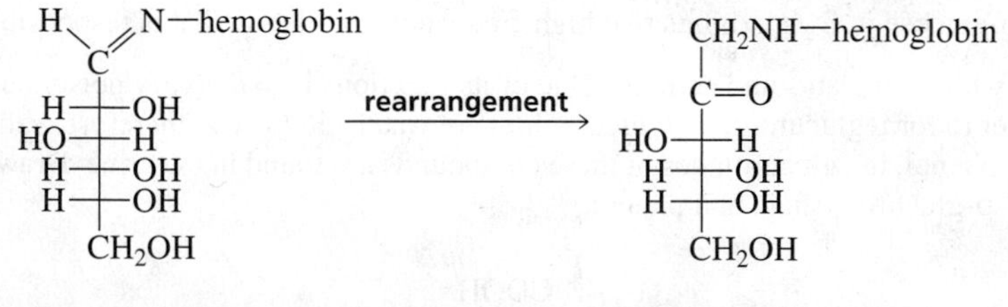

65. All the glucose units in dextran have six-membered rings. When a sample of dextran is treated with methyl iodide and Ag_2O and the product is hydrolyzed under acidic conditions, the final products are 2,3,4,6-tetra-*O*-methyl-D-glucose, 2,4,6-tri-*O*-methyl-D-glucose, 2,3,4-tri-*O*-methyl-D-glucose, and 2,4-di-*O*-methyl-D-glucose. Draw a short segment of dextran.

66. When a pyranose is in the chair conformation in which the CH_2OH group and the C-1 OH group are both in axial positions, the two groups can react to form an acetal. This is called the anhydro form of the sugar (it has "lost water"). The anhydro form of D-idose is shown here. Explain why about 80% of D-idose exists in the anhydro form in an aqueous solution at 100 °C, but only about 0.1% of D-glucose exists in the anhydro form under the same conditions.

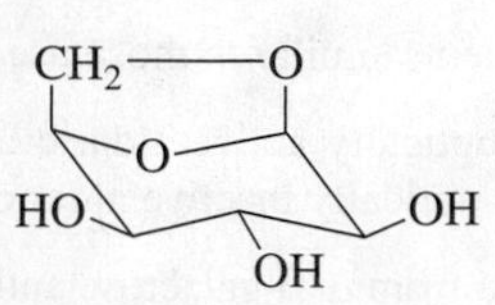

anhydro form of D-idose

GLOSSARY

aldaric acid   a dicarboxylic acid with an OH group bonded to each carbon. Obtained by oxidizing the aldehyde and primary alcohol groups of an aldose.

alditol   a compound with an OH group bonded to each carbon. Obtained by reducing an aldose or a ketose.

aldonic acid   a carboxylic acid with an OH group bonded to each carbon. Obtained by oxidizing the aldehyde group of an aldose.

aldose   a polyhydroxyaldehyde.

amino sugar   a sugar in which one of the OH groups is replaced by an NH_2 group.

anomeric carbon   the carbon in a cyclic sugar that is the carbonyl carbon in the open-chain form.

anomeric effect   the preference for the axial position shown by certain substituents bonded to the anomeric carbon of a six-membered-ring sugar.

anomers   two cyclic sugars that differ in configuration only at the carbon that is the carbonyl carbon in the open-chain form.

bioorganic compound   an organic compound found in biological systems.

carbohydrate   a sugar or a saccharide. Naturally occurring carbohydrates have the D configuration.

complex carbohydrate   a carbohydrate containing two or more sugar molecules linked together.

Curtius rearrangement   conversion of an acyl chloride into a primary amine with the use of azide ion ($^-N_3$).

deoxy sugar   a sugar in which one of the OH groups has been replaced by an H.

disaccharide   a compound containing two sugar molecules linked together.

drug resistance   biological resistance to a particular drug.

enediol rearrangement   interconversion of an aldose and one or more ketoses.

epimerization   changing the configuration of an asymmetric center by removing a proton from it and then reprotonating the molecule at the same site.

epimers   monosaccharides that differ in configuration at only one carbon.

furanose   a five-membered-ring sugar.

furanoside   a five-membered-ring glycoside.

glycoprotein   a protein that is covalently bonded to a polysaccharide.

glycoside   the acetal of a sugar.

N-glycoside   a glycoside with a nitrogen instead of an oxygen at the glycosidic linkage.

glycosidic bond   the bond between the anomeric carbon and the alcohol in a glycoside.

α-1,4′-glycosidic linkage   a linkage between the C-1 oxygen of one sugar and the C-4 of a second sugar with the oxygen atom of the glycosidic linkage in the axial position.

β-1,4′-glycosidic linkage   a linkage between the C-1 oxygen of one sugar and the C-4 of a second sugar with the oxygen atom of the glycosidic linkage in the equatorial position.

Haworth projection   a way to show the structure of a sugar; the five- and six-membered rings are represented as being flat.

heptose   a monosaccharide with seven carbons.

hexose   a monosaccharide with six carbons.

ketose   a polyhydroxyketone.

Kiliani–Fischer synthesis   a method used to increase the number of carbons in an aldose by one, resulting in the formation of a pair of C-2 epimers.

molecular recognition   the recognition of one molecule by another as a result of specific interactions; for example, the specificity of an enzyme for its substrate.

monosaccharide (simple carbohydrate)   a single sugar molecule.

mutarotation   a slow change in optical rotation to an equilibrium value.

nonreducing sugar   a sugar that cannot be oxidized by reagents such as Ag^+ and Cu^+. Nonreducing sugars are not in equilibrium with the open-chain aldose or ketose.

oligosaccharide   3 to 10 sugar molecules linked by glycosidic bonds.

pentose   a monosaccharide with five carbons.

polysaccharide   a compound containing more than 10 sugar molecules linked together.

pyranose   a six-membered-ring sugar.

pyranoside   a six-membered-ring glycoside.

reducing sugar   a sugar that can be oxidized by reagents such as Ag^+ or Br_2. Reducing sugars are in equilibrium with the open-chain aldose or ketose.

simple carbohydrate (monosaccharide)   a single sugar molecule.

tetrose   a monosaccharide with four carbons.

triose   a monosaccharide with three carbons.

Wohl degradation   a method used to shorten an aldose by one carbon.

PHOTO CREDITS

Credits are listed in order of appearance.

The Organic Chemistry of Amino Acids, Peptides, and Proteins

Cobwebs, silk, muscles, and wool are all proteins. In this chapter you will find out why muscles and wool can be stretched but cobwebs and silk cannot. You also will see how a reduction reaction followed by an oxidation reaction can make hair (another protein) either straight or curly.

The three kinds of polymers prevalent in nature are polysaccharides, proteins, and nucleic acids. In this chapter we will turn our attention to proteins and the structurally similar, but shorter, peptides.

Peptides and **proteins** are polymers of amino acids. The amino acids are linked together by amide bonds. An **amino acid** is a carboxylic acid with a protonated amino group on the α-carbon.

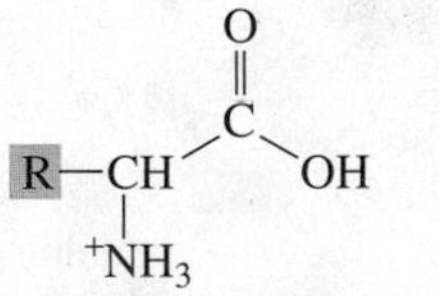

a protonated
α-aminocarboxylic acid
an amino acid

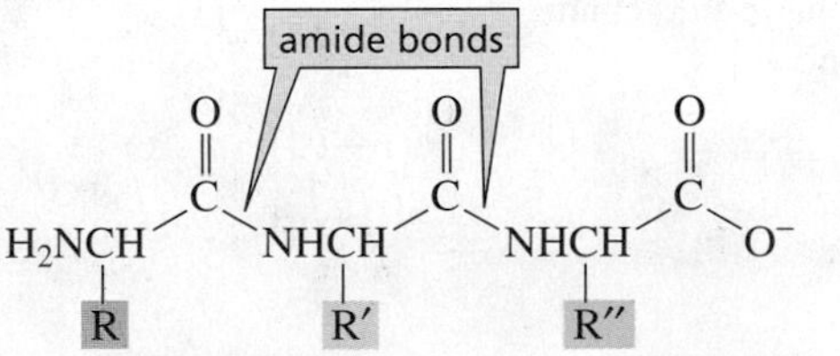

amide bonds

amino acids are linked together by amide bonds

a tripeptide

Amino acid polymers can be composed of any number of amino acids. A **dipeptide** contains two amino acids linked together, a **tripeptide** contains three, an **oligopeptide** contains 4 to 10, and a **polypeptide** contains many. Proteins are naturally occurring polypeptides made up of 40 to 4000 amino acids. Proteins serve many functions in biological systems (Table 1).

Table 1   Examples of the Diverse Functions of Proteins in Biological Systems

Structural proteins	These proteins impart strength to biological structures or protect organisms from their environment. For example, collagen is the major component of bones, muscles, and tendons; keratin is the major component of hair, hooves, feathers, fur, and the outer layer of skin.
Protective proteins	Snake venoms and plant toxins are proteins that protect their owners from predators. Blood-clotting proteins protect the vascular system when it is injured. Antibodies and peptide antibiotics protect us from disease.
Enzymes	Enzymes are proteins that catalyze the reactions that occur in cells.
Hormones	Some hormones, compounds that regulate the reactions that occur in living systems, are proteins.
Proteins with physiological functions	These proteins include those that transport and store oxygen in the body, store oxygen in the muscles, and contract muscles.

Proteins can be classified as either fibrous or globular. **Fibrous proteins** contain long chains of polypeptides arranged in threadlike bundles; these proteins are insoluble in water. **Globular proteins** tend to have roughly spherical shapes and most are soluble in water. All structural proteins are fibrous proteins; most enzymes are globular proteins.

1   THE NOMENCLATURE OF AMINO ACIDS

The structures of the 20 most common naturally occurring amino acids and the frequency with which each occurs in proteins are shown in Table 2. Other amino acids occur in nature but only infrequently. Notice that the amino acids differ only in the substituent (R) that is attached to the α-carbon. The wide variation in these substituents (called **side chains**) is what gives proteins their great structural diversity and, as a consequence, their great functional diversity. Notice too that all amino acids except proline contain a primary amino group. Proline contains a secondary amino group incorporated into a five-membered ring.

Table 2   The Most Common Naturally Occurring Amino Acids Shown in the Form That Predominates as Physiological pH (7.4)

	Formula	Name	Abbreviations		Average relative abundance in proteins
Aliphatic side-chain amino acids		Glycine	Gly	G	7.5%
		Alanine	Ala	A	9.0%
		Valine*	Val	V	6.9%

Formula	Name	Abbreviations		Average relative abundance in proteins

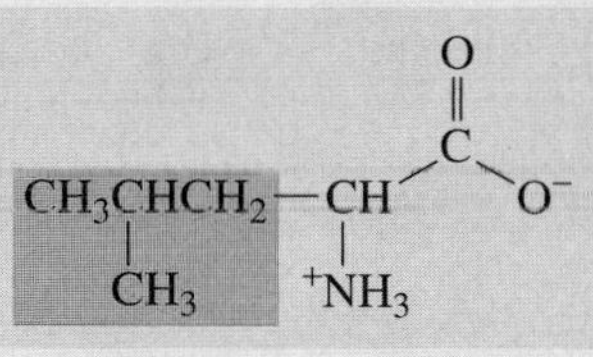

Leucine[*] — Leu — L — 7.5%

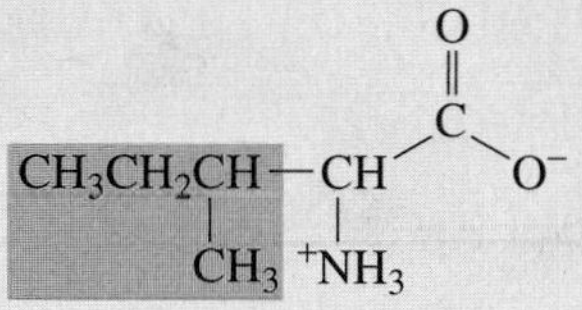

$CH_3CHCH_2-CH-C(=O)O^-$, with CH_3 and $^+NH_3$

Isoleucine[*] — Ile — I — 4.6%

$CH_3CH_2CH-CH-C(=O)O^-$, with CH_3 and $^+NH_3$

Hydroxy-containing amino acids

Serine — Ser — S — 7.1%

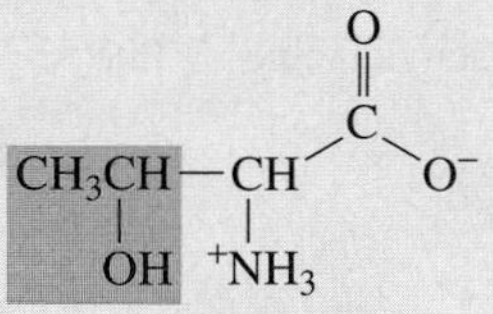

$HOCH_2-CH-C(=O)O^-$, with $^+NH_3$

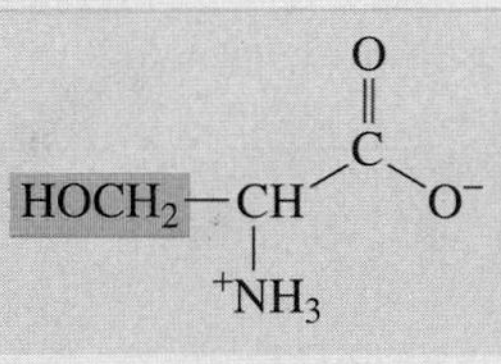

Threonine[*] — Thr — T — 6.0%

$CH_3CH-CH-C(=O)O^-$, with OH and $^+NH_3$

Sulfur-containing amino acids

Cysteine — Cys — C — 2.8%

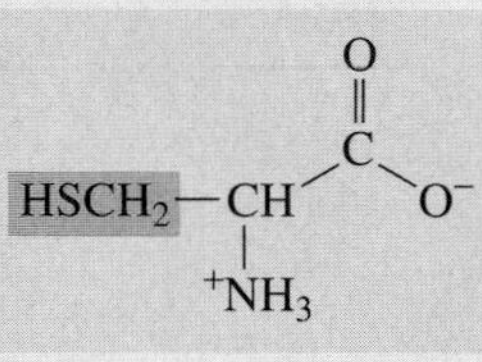

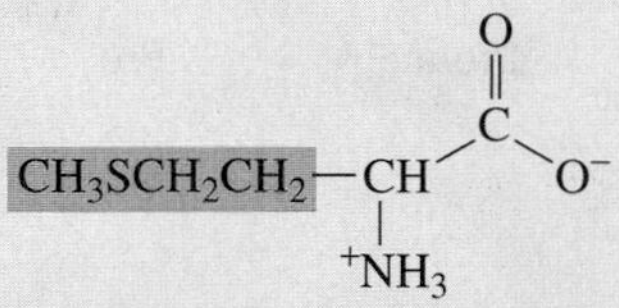

$HSCH_2-CH-C(=O)O^-$, with $^+NH_3$

Methionine[*] — Met — M — 1.7%

$CH_3SCH_2CH_2-CH-C(=O)O^-$, with $^+NH_3$

Acidic amino acids

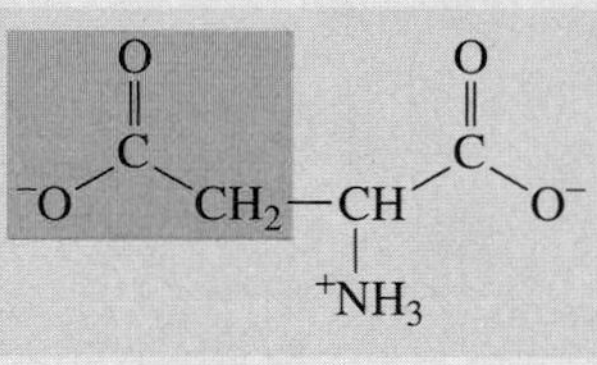

Aspartate (aspartic acid) — Asp — D — 5.5%

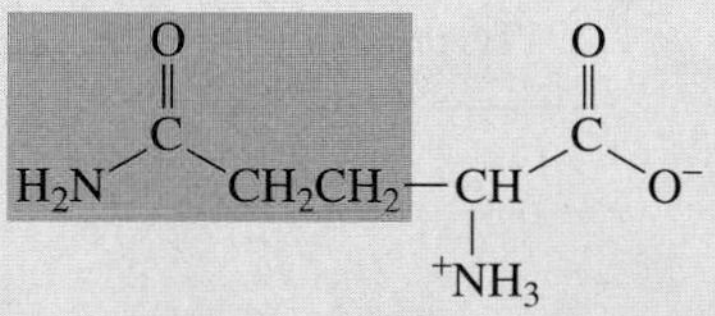

$^-O-C(=O)-CH_2-CH-C(=O)O^-$, with $^+NH_3$

Glutamate (glutamic acid) — Glu — E — 6.2%

$H_2N-C(=O)-CH_2CH_2-CH-C(=O)O^-$, with $^+NH_3$

Amides of acidic amino acids

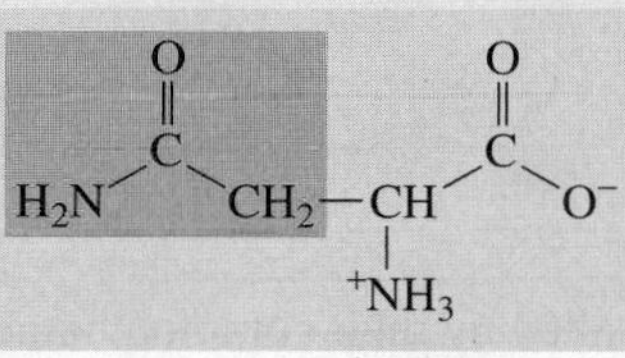

Asparagine — Asn — N — 4.4%

$H_2N-C(=O)-CH_2-CH-C(=O)O^-$, with $^+NH_3$

(Continued)

Formula	Name	Abbreviations		Average relative abundance in proteins
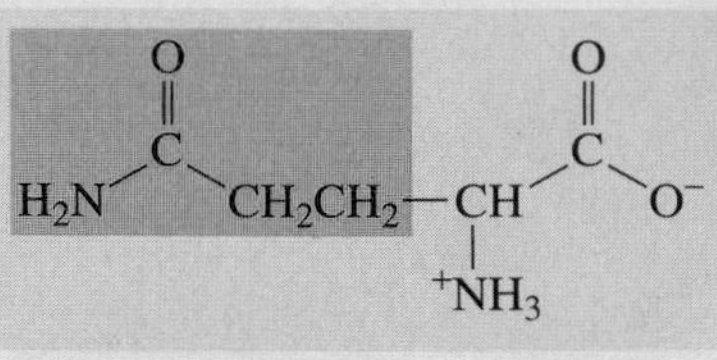 Glutamine structure: $H_2N-C(=O)-CH_2CH_2-CH(^+NH_3)-C(=O)-O^-$	Glutamine	Gln	Q	3.9%
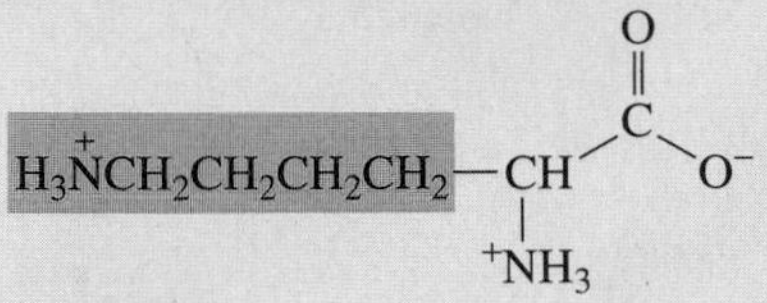 Lysine structure: $H_3^+NCH_2CH_2CH_2CH_2-CH(^+NH_3)-C(=O)-O^-$	Lysine*	Lys	K	7.0%
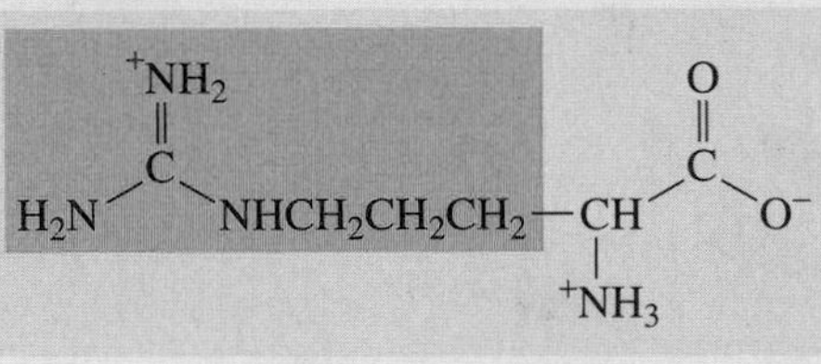 Arginine structure: $H_2N-C(=^+NH_2)-NHCH_2CH_2CH_2-CH(^+NH_3)-C(=O)-O^-$	Arginine*	Arg	R	4.7%
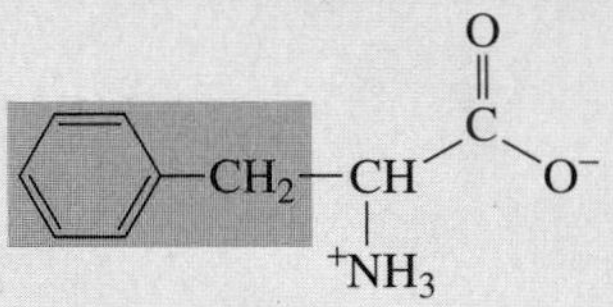 Phenylalanine structure: phenyl–$CH_2-CH(^+NH_3)-C(=O)-O^-$	Phenylalanine*	Phe	F	3.5%
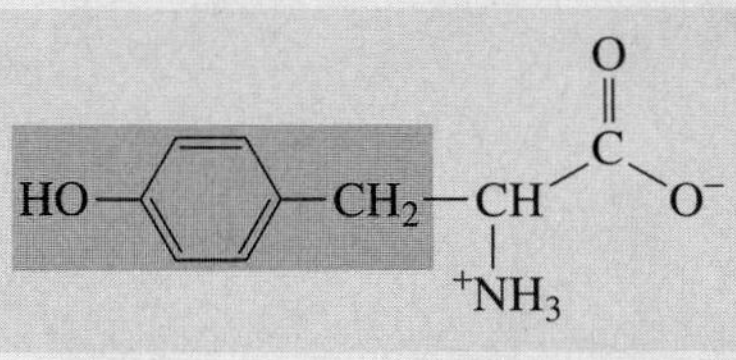 Tyrosine structure: HO–phenyl–$CH_2-CH(^+NH_3)-C(=O)-O^-$	Tyrosine	Tyr	Y	3.5%
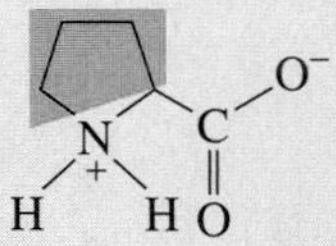 Proline structure (pyrrolidine ring with $C(=O)O^-$)	Proline	Pro	P	4.6%
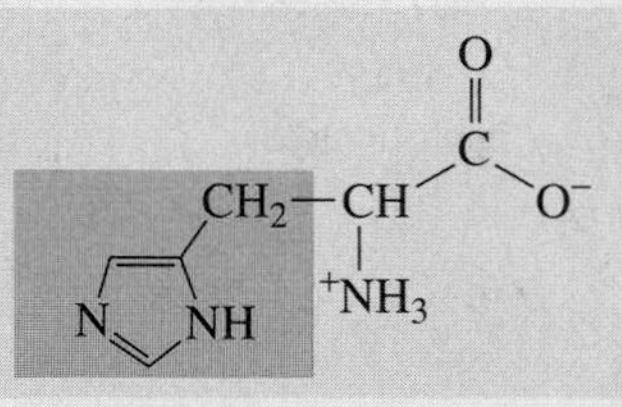 Histidine structure: imidazole–$CH_2-CH(^+NH_3)-C(=O)-O^-$	Histidine*	His	H	2.1%
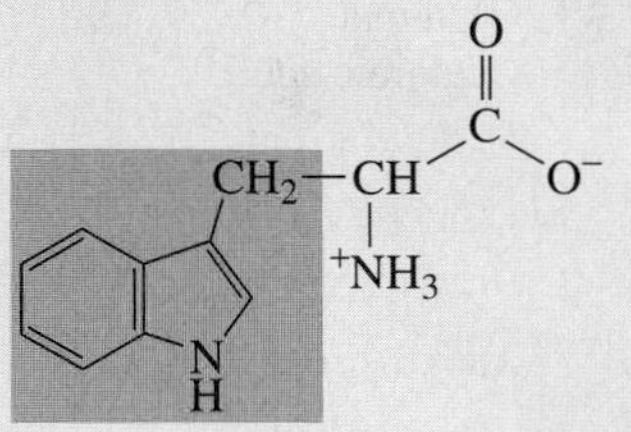 Tryptophan structure: indole–$CH_2-CH(^+NH_3)-C(=O)-O^-$	Tryptophan*	Trp	W	1.1%

Basic amino acids

Benzene-containing amino acids

Heterocyclic amino acids

*Essential amino acid

The amino acids are always called by their common names. Often, the name tells you something about the amino acid. For example, glycine got its name from its sweet taste (*glykos* is Greek for "sweet"), and valine, like valeric acid, has five carbons. Asparagine was first found in asparagus, and tyrosine was isolated from cheese (*tyros* is Greek for "cheese").

Dividing the amino acids into classes, as in Table 2, makes them easier to learn. The aliphatic side-chain amino acids include glycine, the amino acid in which R = H, and four amino acids with alkyl side chains. Alanine is the amino acid with a methyl side chain, and valine has an isopropyl side chain. Notice that, in spite of its name, isoleucine has a *sec*-butyl substituent, not an isobutyl substituent. Leucine is the amino acid that has an isobutyl substituent. Each of the amino acids has both a three-letter abbreviation (in most cases, the first three letters of the name) and a single-letter abbreviation.

Two amino acid side chains—serine and threonine—contain alcohol groups. Serine is an HO-substituted alanine, and threonine has a branched ethanol substituent. There are also two sulfur-containing amino acids: cysteine is an HS-substituted alanine, and methionine has a 2-(methylthio)ethyl substituent.

There are two acidic amino acids (amino acids with two carboxylic acid groups): aspartate and glutamate. Aspartate is a carboxy-substituted alanine, and glutamate has one more methylene group than aspartate. (If their carboxyl groups are protonated, they are called aspartic acid and glutamic acid) Two amino acids—asparagine and glutamine—are amides of the acidic amino acids; asparagine is the amide of aspartate, and glutamine is the amide of glutamate. Notice that the obvious one-letter abbreviations cannot be used for these four amino acids because A and G are used for alanine and glycine. Instead, aspartate and glutamate are abbreviated D and E, and asparagine and glutamine are abbreviated N and Q.

There are two basic amino acids (amino acids with two basic nitrogen-containing groups): lysine and arginine. Lysine has an ε-amino group, and arginine has a δ-guanidino group. At physiological pH, these groups are protonated. Use the ε and δ to remind you how many methylene groups each amino acid has.

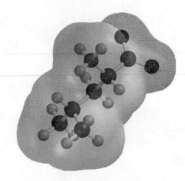

glycine

leucine

aspartate

lysine

Two amino acids—phenylalanine and tyrosine—contain benzene rings. As its name indicates, phenylalanine is phenyl-substituted alanine. Tyrosine is phenylalanine with a *para*-hydroxy substituent.

Proline, histidine, and tryptophan are heterocyclic amino acids. Proline, with its nitrogen incorporated into a five-membered ring, is the only amino acid that contains a secondary amino group. Histidine is an imidazole-substituted alanine. Imidazole is an aromatic compound because it is cyclic and planar, each of its ring atoms has a *p* orbital, and it has three pairs of delocalized π electrons. The pK_a of a protonated imidazole ring is 6.0, so the ring can exist in both the acidic form and the basic form at physiological pH (7.4).

Tryptophan is an indole-substituted alanine. Like imidazole, indole is an aromatic compound. Because the lone pair on the nitrogen of indole is needed for the compound's aromaticity, indole is a very weak base. (The pK_a of protonated indole is −2.4.) Therefore, the ring nitrogen in tryptophan is never protonated under physiological conditions.

indole

The 10 amino acids denoted in Table 2 by asterisks (*) are **essential amino acids.** Humans must obtain them from their diet because they either cannot synthesize them at all or they cannot synthesize them in adequate amounts. For example, humans must have a dietary source of phenylalanine because they cannot synthesize benzene rings. However, they do not need tyrosine in their diet because they can synthesize it from phenylalanine. Although humans can synthesize arginine, it is needed for growth in greater amounts than can be synthesized. So arginine is considered an essential amino acid for children but not for adults.

Proteins and Nutrition

Proteins are an important component of our diets. Dietary protein is hydrolyzed in the body to individual amino acids. Some of these amino acids are used to synthesize proteins needed by the body, some are broken down (metabolized) to supply energy to the body, and some are used as starting materials for the synthesis of nonprotein compounds that the body needs, such as thyroxine, adrenaline, and melanin.

Complete proteins (meat, fish, eggs, and milk) contain all 10 essential amino acids. Incomplete proteins contain too little of one or more essential amino acids to support human growth. For example, beans and peas are deficient in methionine, corn is deficient in lysine and tryptophan, and rice is deficient in lysine and threonine. Vegetarians, therefore, must have a diet that includes proteins from different sources.

PROBLEM 1

a. Explain why, when the imidazole ring of histidine is protonated, the double-bonded nitrogen is the nitrogen that accepts the proton.

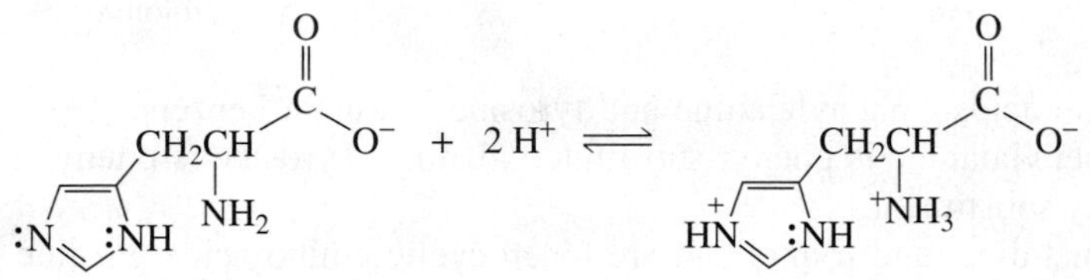

b. Explain why, when the guanidino group of arginine is protonated, the double-bonded nitrogen is the nitrogen that accepts the proton.

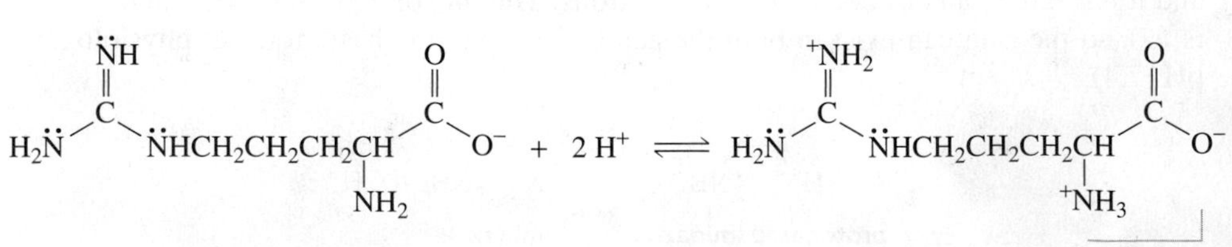

2 THE CONFIGURATION OF AMINO ACIDS

The α-carbon of all the naturally occurring amino acids (except glycine) is an asymmetric center. Therefore, 19 of the 20 amino acids listed in Table 2 can exist as enantiomers. The D and L notation used for monosaccharides is also used for amino acids.

An amino acid drawn in a Fischer projection with the carboxyl group at the top and the R group at the bottom of the vertical axis is a **D-amino acid** if the amino group is on the right and an **L-amino acid** if the amino group is on the left. Unlike monosaccharides, where the D isomer is the one found in nature, most amino acids found in nature have the L configuration. To date, D-amino acids have been found only in a few peptide antibiotics and in some small peptides attached to the cell walls of bacteria.

Naturally occurring monosaccharides have the D configuration.

Naturally occurring amino acids have the L configuration.

D-glyceraldehyde

L-glyceraldehyde

D-amino acid

L-amino acid

L-alanine
an amino acid

Why D-sugars and L-amino acids? Although it made no difference which isomer nature "selected" to be synthesized, it was important that only one was selected. For example, proteins that contain both D- and L-amino acids do not fold properly, and without proper folding there can be no catalysis (Section 15). It was also important that the same isomer be synthesized by all organisms. For example, since mammals ended up having L-amino acids, L-amino acids must be the isomers synthesized by the organisms that mammals depend on for food.

Amino Acids and Disease

The Chamorro people of Guam have a high incidence of a syndrome that resembles amyotrophic lateral sclerosis (ALS, or Lou Gehrig's disease) with elements of Parkinson's disease and dementia. This syndrome developed during World War II when, as a result of food shortages, the tribe ate large quantities of *Cycas circinalis* seeds. These seeds contain β-methylamino-L-alanine, an amino acid that binds to cell receptors that bind L-glutamate. When monkeys are given β-methylamino-L-alanine, they develop some of the features of this syndrome. There is hope that, by studying the mechanism of action of β-methylamino-L-alanine, we may gain an understanding of how ALS and Parkinson's disease arise.

L-alanine

β-methylamino-L-alanine

L-glutamate

PROBLEM 2♦

a. Which isomer—(R)-alanine or (S)-alanine—is D-alanine?

b. Which isomer—(R)-aspartate or (S)-aspartate—is D-aspartate?

c. Can a general statement be made relating R and S to D and L?

A Peptide Antibiotic

Gramicidin S, an antibiotic produced by a strain of bacteria, is a cyclic decapeptide. Notice that one of its 10 amino acids is ornithine, an amino acid not listed in Table 2 because it occurs rarely in nature. Ornithine resembles lysine but has one less methylene group in its side chain. Notice that gramicidin S contains two D-amino acids.

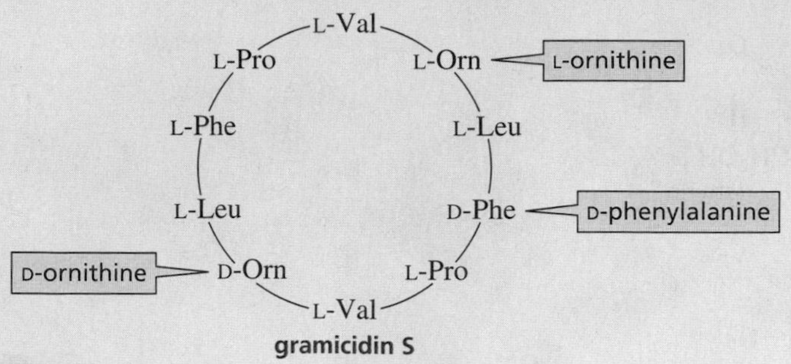

gramicidin S

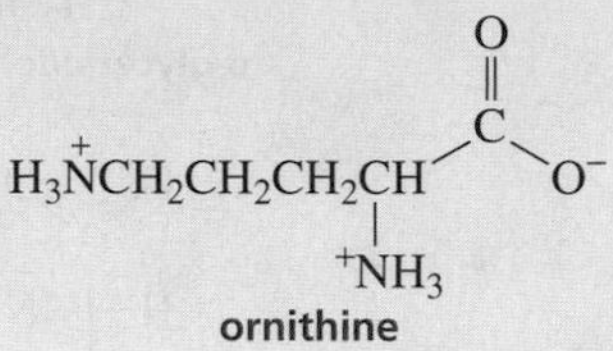

$$H_3\overset{+}{N}CH_2CH_2CH_2\underset{\underset{+NH_3}{|}}{CH}\overset{\overset{O}{\parallel}}{C}\;O^-$$

ornithine

PROBLEM 3 Solved

Threonine has two asymmetric centers and therefore has four stereoisomers. Naturally occurring L-threonine is (2S,3R)-threonine. Which of the following stereoisomers is L-threonine?

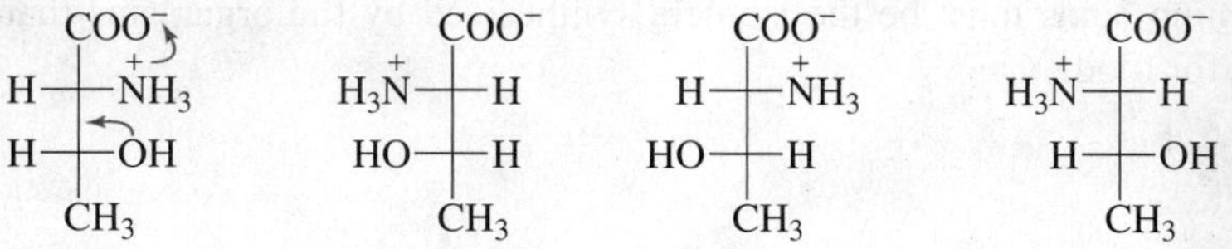

Solution Stereoisomer **A** has the R configuration at both C-2 and C-3 because in both cases the arrow drawn from the highest to the next-highest-priority substituent is counterclockwise. In both cases, counterclockwise signifies R because the lowest priority substituent (H) is on a horizontal bond. Therefore, the configuration of (2S,3R)-threonine is the opposite of that in stereoisomer **A** at C-2 and the same as that in stereoisomer **A** at C-3. Thus, L-threonine is stereoisomer **D**. Notice that the $^+NH_3$ group is on the left, just as we would expect for the Fischer projection of a naturally occurring L-amino acid.

PROBLEM 4◆

Do any other amino acids in Table 2 have more than one asymmetric center?

---

3   THE ACID–BASE PROPERTIES OF AMINO ACIDS

Every amino acid has a carboxyl group and an amino group, and each group can exist in an acidic form or a basic form, depending on the pH of the solution in which the amino acid is dissolved.

Compounds exist primarily in their acidic forms (that is, with their protons attached) in solutions that are more acidic than their pK_a values, and primarily in their basic forms (that is, without their protons) in solutions that are more basic than their pK_a values.

The acidic form (with the proton) predominates if the pH of the solution is less than the pK_a of the ionizable group, and the basic form (without the proton) predominates if the pH of the solution is greater than the pK_a of the ionizable group.

The Henderson–Hasselbalch equation states that half the group is in its acidic form and half is in its basic form at pH = pK_a.

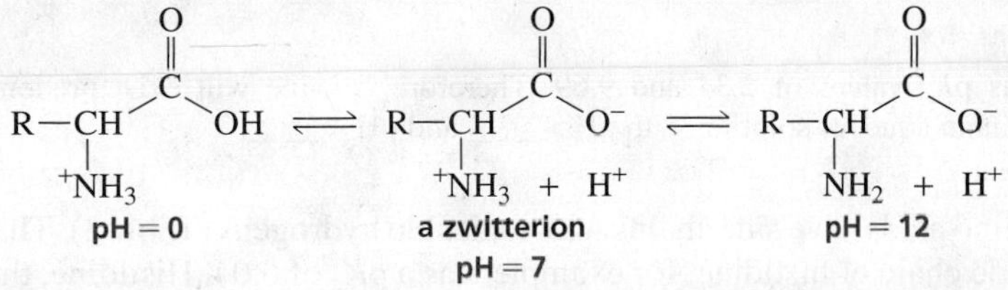

The carboxyl groups of the amino acids have pK_a values of approximately 2, and the protonated amino groups have pK_a values near 9 (Table 3). Both groups, therefore, will be in their acidic forms in a very acidic solution (pH ~ 0). At pH = 7, the pH of the solution is greater than the pK_a of the carboxyl group, but less than the pK_a of the protonated amino group, so the carboxyl group will be in its basic form and the amino group will be in its acidic form. In a strongly basic solution (pH ~ 12), both groups will be in their basic forms.

Table 3 The pK_a Values of Amino Acids

Amino acid	pK_a α-COOH	pK_a α-$\overset{+}{N}H_3$	pK_a Side chain
Alanine	2.34	9.69	—
Arginine	2.17	9.04	12.48
Asparagine	2.02	8.84	—
Aspartic acid	2.09	9.82	3.86
Cysteine	1.92	10.46	8.35
Glutamic acid	2.19	9.67	4.25
Glutamine	2.17	9.13	—
Glycine	2.34	9.60	—
Histidine	1.82	9.17	6.04
Isoleucine	2.36	9.68	—
Leucine	2.36	9.60	—
Lysine	2.18	8.95	10.79
Methionine	2.28	9.21	—
Phenylalanine	2.16	9.18	—
Proline	1.99	10.60	—
Serine	2.21	9.15	—
Threonine	2.63	9.10	—
Tryptophan	2.38	9.39	—
Tyrosine	2.20	9.11	10.07
Valine	2.32	9.62	—

Notice that an amino acid can never exist as an uncharged compound, regardless of the pH of the solution. To be uncharged, an amino acid would have to lose a proton from an $^{+}NH_3$ group with a pK_a of about 9 before it would lose a proton from a COOH group with a pK_a of about 2. This is impossible because a weak acid (pK_a = 9) cannot lose a proton easier than a strong acid (pK_a = 2) can. Therefore, at physiological pH (7.4), an amino acid such as alanine exists as a dipolar ion, called a zwitterion. A **zwitterion** is a compound that has a negative charge on one atom and a positive charge on a nonadjacent atom. (The name comes from *zwitter,* German for "hermaphrodite" or "hybrid.")

PROBLEM 5◆

Alanine has pK_a values of 2.34 and 9.69. Therefore, alanine will exist predominately as a zwitterion in an aqueous solution with pH > ____ and pH < ____.

A few amino acids have side chains with ionizable hydrogens (Table 3). The protonated imidazole side chain of histidine, for example, has a pK_a of 6.04. Histidine, therefore, can exist in four different forms, and the form that predominates depends on the pH of the solution.

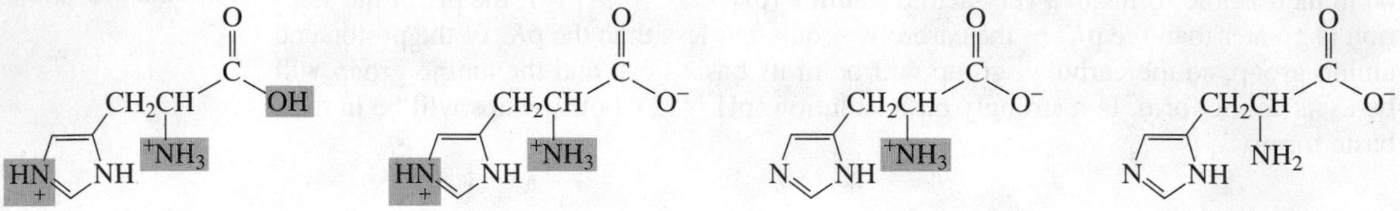

histidine

PROBLEM 6◆

Why are the carboxylic acid groups of the amino acids so much more acidic (pK_a ~ 2) than a carboxylic acid such as acetic acid (pK_a = 4.76)?

PROBLEM 7 Solved

Draw the predominant form for each of the following amino acids at physiological pH (7.4):

a. aspartate   **c.** glutamine   **e.** arginine
b. histidine   **d.** lysine   **f.** tyrosine

Solution to 7a Both carboxyl groups are in their basic forms because the pH of the solution is greater than their pK_a values. The protonated amino group is in its acidic form because the pH of the solution is less than its pK_a value.

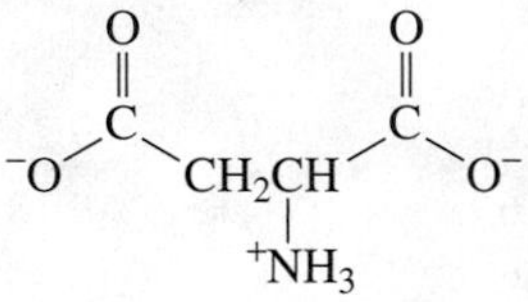

PROBLEM 8◆

Draw the predominant form for glutamate in a solution with the following pH:

a. 0   **b.** 3   **c.** 6   **d.** 11

PROBLEM 9

a. Why is the pK_a of the glutamate side chain greater than the pK_a of the aspartate side chain?
b. Why is the pK_a of the arginine side chain greater than the pK_a of the lysine side chain?

4 THE ISOELECTRIC POINT

The **isoelectric point** (pI) of an amino acid is the pH at which it has no net charge. In other words, it is the pH at which the amount of positive charge on an amino acid exactly balances the amount of negative charge:

pI = pH at which there is no net charge

The pI of an amino acid that does *not have an ionizable side chain*—such as alanine—is midway between its two pK_a values. This is because half of the molecules have a negatively charged carboxyl group and half have an uncharged carboxyl group at pH = 2.34, whereas half of the molecules have a positively charged amino group and half have an uncharged amino group at pH = 9.69. As the pH increases from 2.34, the carboxyl group of more molecules becomes negatively charged; as the pH decreases from 9.69, the amino group of more molecules becomes positively charged. Therefore, the number of negatively charged groups equals the number of positively charged groups at the intersection (average) of the two pK_a values.

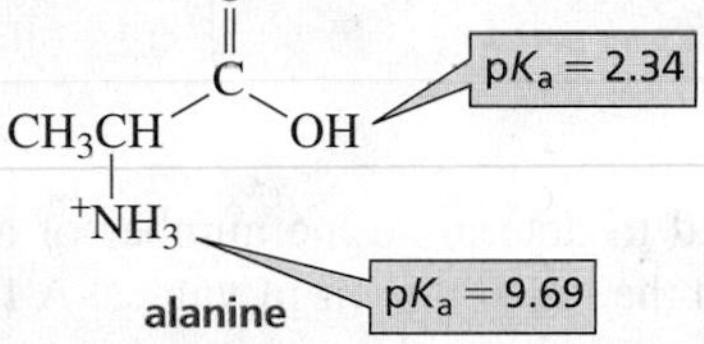

$$pI = \frac{2.34 + 9.69}{2} = \frac{12.03}{2} = 6.02$$

The pI of most amino acids (see Problem 13) that *have an ionizable side chain* is the average of the pK_a values of the similarly ionizing groups (either positively charged groups ionizing to uncharged groups or uncharged groups ionizing to negatively charged groups). For example, the pI of lysine is the average of the pK_a values of the two groups that are positively charged in their acidic form and uncharged in their basic form. The pI of glutamic acid, on the other hand, is the average of the pK_a values of the two groups that are uncharged in their acidic form and negatively charged in their basic form.

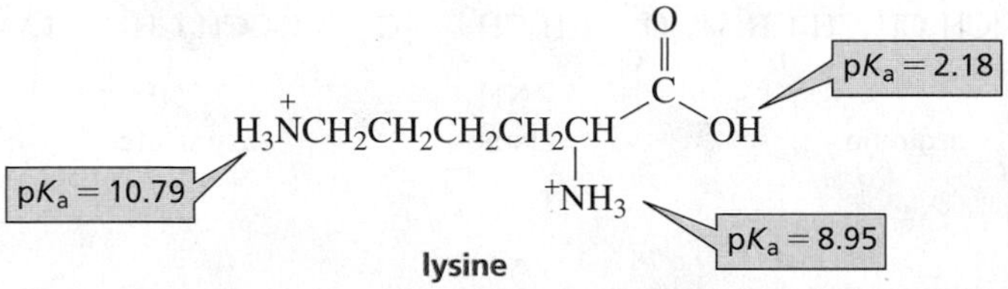

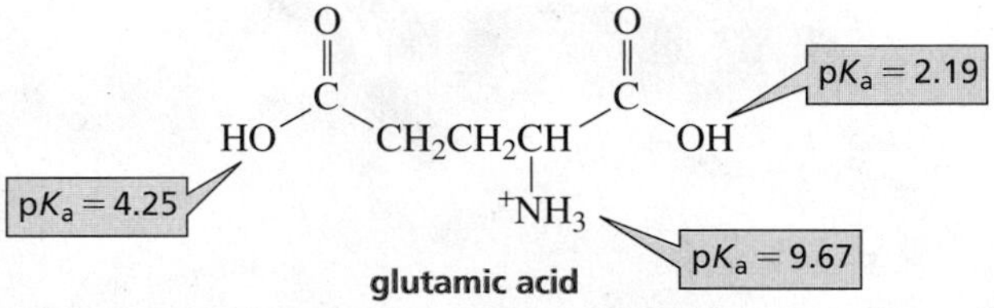

$$pI = \frac{8.95 + 10.79}{2} = \frac{19.74}{2} = 9.87$$

$$pI = \frac{2.19 + 4.25}{2} = \frac{6.44}{2} = 3.22$$

PROBLEM 10

Explain why the pI of lysine is the average of the pK_a values of its two protonated amino groups.

PROBLEM 11♦

Calculate the pI of each of the following amino acids:

a. asparagine   **b.** arginine   **c.** serine   **d.** aspartate

---

PROBLEM 12♦

a. Which amino acid has the lowest pI value?
b. Which amino acid has the highest pI value?
c. Which amino acid has the greatest amount of negative charge at pH = 6.20?
d. Which amino acid has a greater negative charge at pH = 6.20, glycine or methionine?

5 SEPARATING AMINO ACIDS

A mixture of amino acids can be separated by several different techniques, such as electrophoresis, paper/thin-layer chromatography, and ion-exchange chromatography.

Electrophoresis

An amino acid will be positively charged if the pH of the solution is less than the pI of the amino acid and it will be negatively charged if the pH of the solution is greater than the pI of the amino acid.

Electrophoresis can be used to determine the number of amino acids in a mixture. It separates the amino acids on the basis of their pI values. A few drops of a solution of an amino acid mixture are applied to the middle of a piece of filter paper or to a gel. When the paper or the gel is placed in a buffered solution between two electrodes and an electric field is applied (Figure 1), an amino acid with a pI greater than the pH of the solution will have an overall *positive charge* and will migrate toward the *cathode (the negative electrode)*.

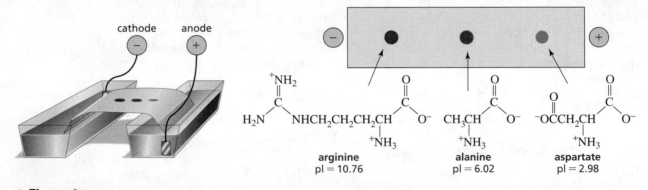

▲ Figure 1

Arginine, alanine, and aspartate are separated by electrophoresis at pH = 5.

The farther the amino acid's pI is from the pH of the buffer, the more positive the amino acid will be and the farther it will migrate toward the cathode in a given amount of time. An amino acid with a pI less than the pH of the buffer will have an overall *negative charge* and will migrate toward the *anode (the positive electrode)*. If two molecules have the same charge, the larger one will move more slowly during electrophoresis because the same charge has to move a greater mass.

Considering that amino acids are colorless, how can we detect them after they have been separated? After the amino acids have been separated by electrophoresis, the filter paper is painted with a solution of ninhydrin and dried in a warm oven. Most amino acids form a purple product when heated with ninhydrin. The number of amino acids in the mixture is determined by the number of colored spots on the filter paper. The individual amino acids can be identified by their location on the paper compared with a standard.

The mechanism for formation of the colored product is shown here, omitting the mechanisms for dehydration, imine formation, and imine hydrolysis.

The Organic Chemistry of Amino Acids, Peptides, and Proteins

STEPS IN THE REACTION OF AN AMINO ACID WITH NINHYDRIN TO FORM A COLORED PRODUCT

ninhydrin + H₂O

an amino acid

+ H₂O

+ H₂O

+ HO⁻

+ CO₂

purple-colored product

- Loss of water from the hydrate forms a ketone that reacts with the amino acid to form an imine.
- Decarboxylation occurs because the electrons left behind can be delocalized onto a carbonyl oxygen.
- Tautomerization followed by hydrolysis of an imine forms the deaminated amino acid and a ninhydrin-amine.
- Reaction of this amine with another molecule of ninhydrin forms an imine. Loss of a proton forms a highly conjugated (colored) product.

PROBLEM 14◆

What aldehyde is formed when valine is treated with ninhydrin?

Painting a paper with a solution of ninhydrin allows latent fingerprints (as a consequence of amino acids left behind by the fingers) to be developed.

Paper/Thin-Layer Chromatography

Paper chromatography once played an important role in biochemical analysis because it provided a method for separating amino acids using very simple equipment. Although more modern techniques are generally employed today, we will describe the principles behind paper chromatography because many of the same principles are employed in modern separation techniques.

Paper chromatography separates amino acids on the basis of polarity. A few drops of a solution of an amino acid mixture are applied to the bottom of a strip of filter paper. The edge of the paper is then placed in a solvent. The solvent moves up the paper by capillary action, carrying the amino acids with it. Depending on their polarities, the amino acids have different affinities for the mobile (solvent) and stationary (paper) phases and therefore some travel up the paper farther than others.

When a solvent less polar than the paper is employed, the more polar the amino acid, the more strongly it is adsorbed onto the relatively polar paper. The less polar amino acids travel farther up the paper, since they have a greater affinity for the less polar mobile phase. Therefore, when the paper is developed with ninhydrin, the colored spot closest to the origin is the most polar amino acid and the spot farthest away from the origin is the least polar amino acid (Figure 2).

Less polar amino acids travel more rapidly if the solvent is less polar than the paper.

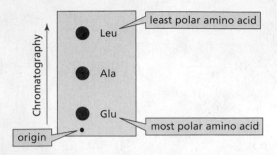

▶ **Figure 2**

Separation of glutamate, alanine, and leucine by paper chromatography.

The most polar amino acids are those with charged side chains, the next most polar are those with side chains that can form hydrogen bonds, and the least polar are those with hydrocarbon side chains. For amino acids with hydrocarbon side chains, the polarity of the amino acid decreases as the size of the alkyl group increases. In other words, leucine $[R = -CH_2CH(CH_3)_2]$ is less polar than valine $[R = -CH(CH_3)_2]$.

Paper chromatography has largely been replaced by **thin-layer chromatography** (TLC), which differs from paper chromatography in that TLC uses a plate with a coating of solid material instead of filter paper. How the amino acids separate depend on the solid material and the solvent chosen for the mobile phase.

Chromatography separates amino acids based on their polarity and electrophoresis separates them based on their charge. The two techniques can be applied on the same piece of filter paper—a technique called **fingerprinting**—to give a two-dimensional separation (that is, the amino acids are separated according to both their polarity and their charge. (See Problems 54 and 67.)

PROBLEM 15♦

A mixture of seven amino acids (glycine, glutamate, leucine, lysine, alanine, isoleucine, and aspartate) is separated by chromatography. Explain why only six spots show up when the chromatographic plate is coated with ninhydrin and heated.

Ion-Exchange Chromatography

A technique called **ion-exchange chromatography** is able to both separate and identify amino acids and to determine the relative amount of each amino acid in a mixture. This technique employs a column packed with an insoluble resin. A solution of a mixture of amino acids is loaded onto the top of the column, and a series of buffer solutions of increasing pH are poured through the column. The amino acids separate because they flow through the column at different rates.

The resin is a chemically inert material with charged groups. The structure of a commonly used resin is shown in Figure 3. If a mixture of lysine and glutamate in a solution of pH = 6 were to be loaded onto the column, glutamate would travel down the column rapidly because its negatively charged side chain would be repelled by the negatively charged sulfonic acid groups of the resin. The positively charged side chain of lysine, on the other hand, would cause that amino acid to be retained on the column. This kind of resin is called a **cation-exchange resin** because it exchanges the Na⁺ counterions of the SO_3^- groups for the positively charged species that is traveling through the column. In addition, the relatively nonpolar nature of the column causes it to retain nonpolar amino acids longer than polar amino acids.

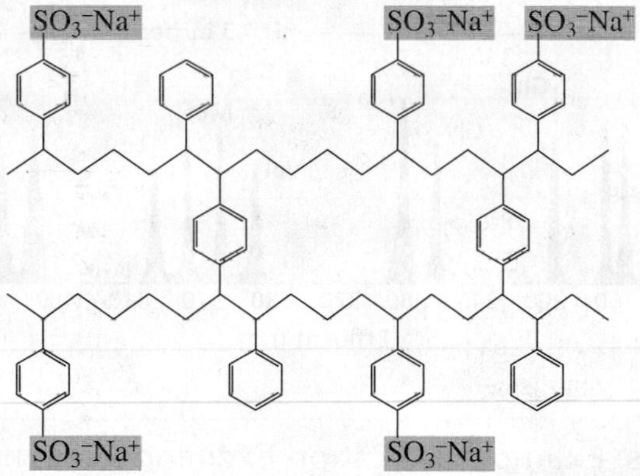

◀ **Figure 3**

A section of a cation-exchange resin. This particular resin is called Dowex 50.

Cations bind most strongly to cation-exchange resins.

Anions bind most strongly to anion-exchange resins.

Resins with positively charged groups are called **anion-exchange resins** because they impede the flow of anions by exchanging their negatively charged counterions for negatively charged species traveling through the column. A common anion-exchange resin, Dowex 1, has —$CH_2N^+(CH_3)_3Cl^-$ groups in place of the —$SO_3^-Na^+$ groups in Figure 3.

An **amino acid analyzer** is an instrument that automates ion-exchange chromatography. When a solution of amino acids passes through the column of an amino acid analyzer containing a cation-exchange resin, the amino acids move through the column at different rates, depending on their overall charge. The solution that flows out of the column (the effluent) is collected in fractions. These are collected often enough that a different amino acid ends up in each one (Figure 4).

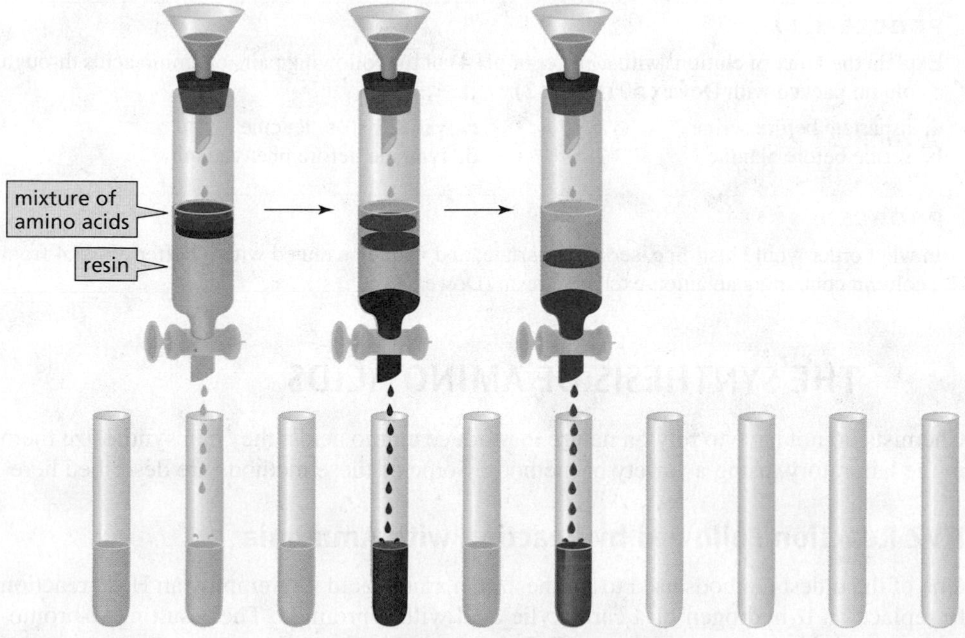

mixture of amino acids

resin

fractions are sequentially collected

◀ **Figure 4**

Separation of amino acids by ion-exchange chromatography.

If ninhydrin is added to each of the fractions, the concentration of amino acid in each fraction can be determined by the amount of absorbance at 570 nm because the colored compound formed by the reaction of an amino acid with ninhydrin has a λ_{max} of 570 nm. This information combined with each fraction's rate of passage through the column allows the identity and relative amount of each amino acid in the mixture to be determined (Figure 5).

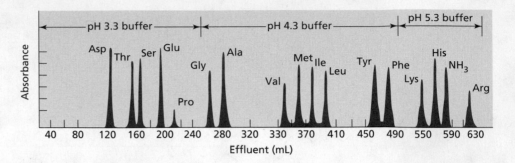

▶ **Figure 5**

A typical chromatogram obtained from the separation of a mixture of amino acids using an automated amino acid analyzer.

Water Softeners: Examples of Cation-Exchange Chromatography

Water-softening systems contain a column packed with a cation-exchange resin that has been flushed with concentrated sodium chloride. When "hard water" (water with high levels of Ca^{2+} and Mg^{2+}) passes through the column, the resin binds Ca^{2+} and Mg^{2+} more tightly than it binds Na^+. Thus, the water-softening system removes Ca^{2+} and Mg^{2+} from the water and replaces them with Na^+. The resin must be recharged from time to time by being flushed with concentrated sodium chloride, thereby replacing the bound Ca^{2+} and Mg^{2+} with Na^+.

PROBLEM 16

Why are buffer solutions of increasingly higher pH used to elute the column that generates the chromatogram shown in Figure 5? (*Elute* means wash out with a solvent.)

PROBLEM 17

Explain the order of elution (with a buffer of pH 4) of the following pairs of amino acids through a column packed with Dowex 50 (Figure 3):

a. aspartate before serine

b. serine before alanine

c. valine before leucine

d. tyrosine before phenylalanine

PROBLEM 18◆

In what order would histidine, serine, aspartate, and valine be eluted with a buffer of pH 4 from a column containing an anion-exchange resin (Dowex 1)?

6 THE SYNTHESIS OF AMINO ACIDS

Chemists do not have to rely on nature to produce amino acids; they can synthesize them in the laboratory, using a variety of methods. Some of these methods are described here.

HVZ Reaction Followed by Reaction with Ammonia

One of the oldest methods used to synthesize an amino acid is to employ an HVZ reaction to replace an α-hydrogen of a carboxylic acid with a bromine. The resulting α-bromocarboxylic acid can then undergo an S_N2 reaction with ammonia to form the amino acid.

PROBLEM 19♦

Why is excess ammonia used in the preceding reaction?

Reductive Amination

Amino acids can also be synthesized by reductive amination of an α-keto acid.

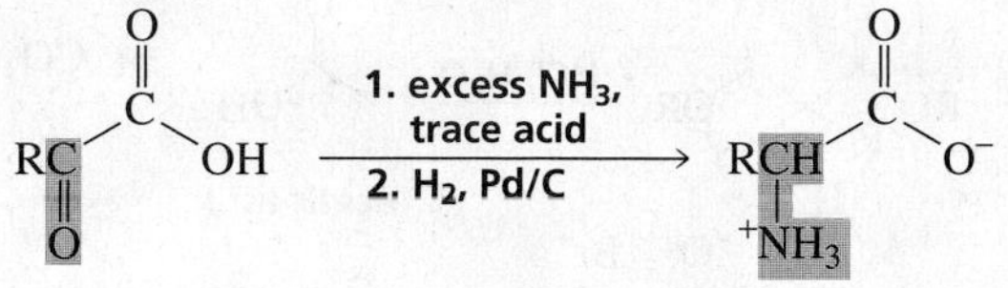

PROBLEM 20♦

Cells can also convert α-keto acids into amino acids, but because the reagents organic chemists use for this reaction are not available in cells, they carry out this reaction by a different mechanism.

a. What amino acid is obtained from the reductive amination of each of the following metabolic intermediates in a cell?

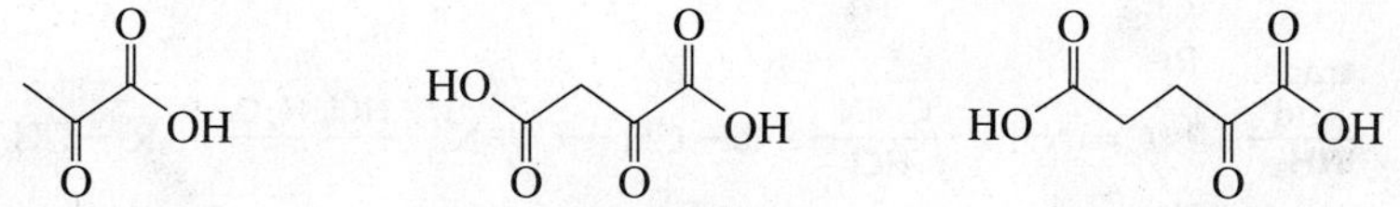

pyruvic acid   oxaloacetic acid α-ketoglutaric acid

b. What amino acids are obtained from the same metabolic intermediates when the amino acids are synthesized in the laboratory?

N-Phthalimidomalonic Ester Synthesis

Amino acids can be synthesized with much better yields than those obtained by the previous two methods via an *N*-phthalimidomalonic ester synthesis, a method that combines the malonic ester synthesis and the Gabriel synthesis.

THE STEPS IN THE *N*-PHTHALIMIDOMALONIC ESTER SYNTHESIS

α-bromomalonic ester   potassium phthalimide   *N*-phthalimidomalonic ester

phthalic acid   an amino acid

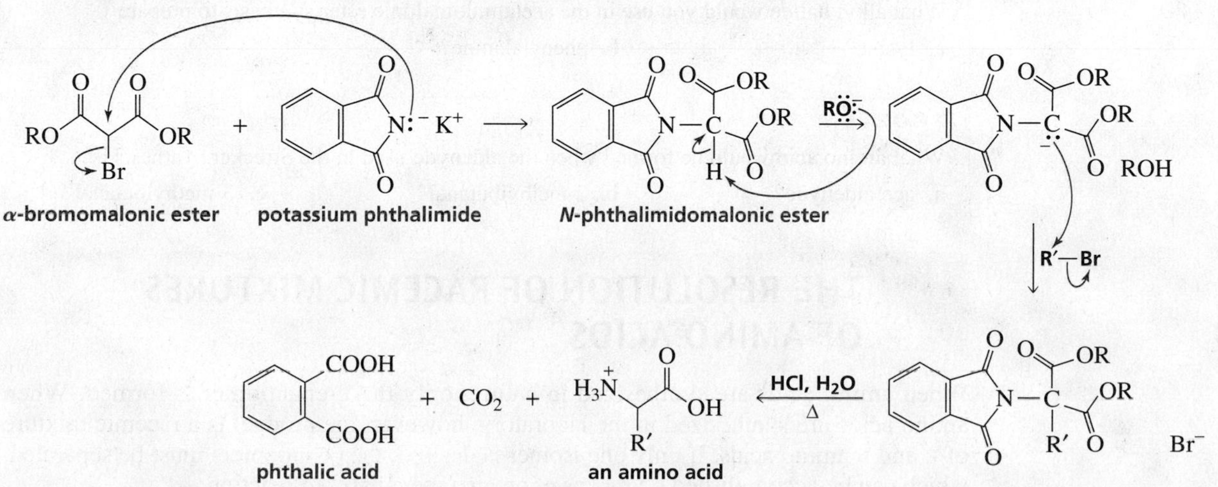

- α-Bromomalonic ester and potassium phthalimide undergo an S_N2 reaction.
- A proton is easily removed from the α-carbon of *N*-phthalimidomalonic ester because it is flanked by two carbonyl groups.

- The resulting carbanion undergoes an S_N2 reaction with an alkyl halide.
- Heating in an acidic aqueous solution hydrolyzes the two ester groups and the two amide bonds and decarboxylates the 3-oxocarboxylic acid.

A variation of the *N*-phthalimidomalonic ester synthesis uses acetamidomalonic ester in place of *N*-phthalimidomalonic ester.

acetamidomalonic ester

acetic acid

an amino acid

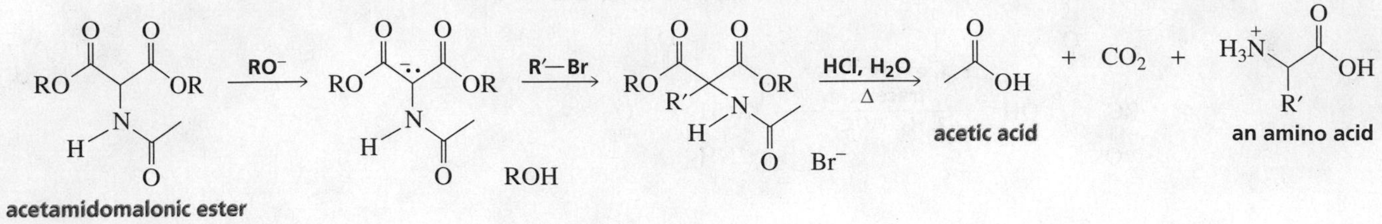

Strecker Synthesis

In the Strecker synthesis, an aldehyde reacts with ammonia to form an imine. An addition reaction with cyanide ion forms an intermediate, which, when hydrolyzed, forms the amino acid. Compare this reaction with the Kiliani–Fischer synthesis of aldoses.

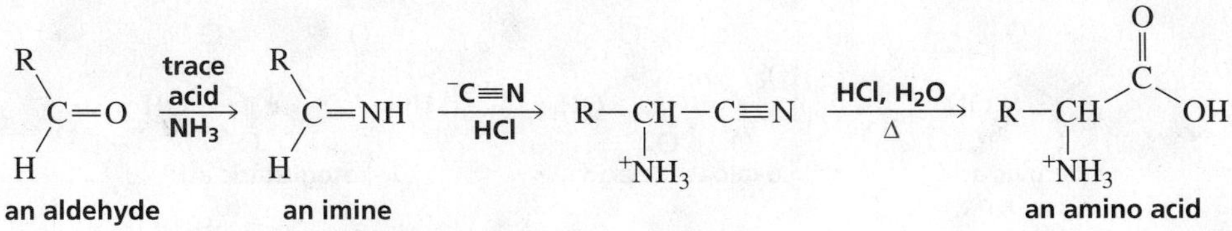

an aldehyde   an imine   an amino acid

PROBLEM 21♦

What amino acid would be formed using the *N*-phthalimidomalonic ester synthesis when the following alkyl halides are used in the third step?

a. CH$_3$CHCH$_2$Br
         |
         CH$_3$

b. CH$_3$SCH$_2$CH$_2$Br

PROBLEM 22♦

What alkyl halide would you use in the acetamidomalonic ester synthesis to prepare

a. lysine?

b. phenylalanine?

PROBLEM 23♦

What amino acid would be formed when the aldehyde used in the Strecker synthesis is

a. acetaldehyde?

b. 2-methylbutanal?

c. 3-methylbutanal?

---

7   THE RESOLUTION OF RACEMIC MIXTURES OF AMINO ACIDS

When amino acids are synthesized in nature, only the L-enantiomer is formed. When amino acids are synthesized in the laboratory, however, the product is a racemic mixture of D and L amino acids. If only one isomer is desired, the enantiomers must be separated, which can be accomplished by means of an enzyme-catalyzed reaction.

Because an enzyme is chiral, it will react at a different rate with each of the enantiomers. For example, pig kidney aminoacylase is an enzyme that catalyzes the hydrolysis of *N*-acetyl-L-amino acids, but not *N*-acetyl-D-amino acids.

Therefore, if the racemic mixture of amino acids is converted to a pair of *N*-acetylamino acids and the *N*-acetylated mixture is hydrolyzed with pig kidney aminoacylase, then the products will be the L-amino acid and *N*-acetyl-D-amino acid, which are easily separated.

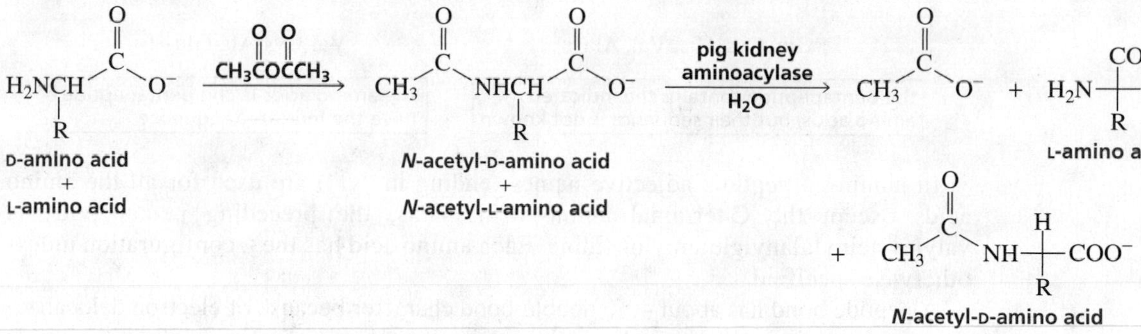

Because the resolution (separation) of the enantiomers depends on the difference in the rates of reaction of the enzyme with the two *N*-acetylated compounds, this technique is known as a **kinetic resolution.** A racemic mixture of amino acids can also be separated by the enzyme D-amino acid oxidase.

PROBLEM 24

Pig liver esterase is an enzyme that catalyzes the hydrolysis of esters. It hydrolyzes esters of L-amino acids more rapidly than esters of D-amino acids. How can this enzyme be used to separate a racemic mixture of amino acids?

---

8   PEPTIDE BONDS AND DISULFIDE BONDS

Peptide bonds and disulfide bonds are the only covalent bonds that join amino acids together in a peptide or a protein.

Peptide Bonds

The amide bonds that link amino acids are called **peptide bonds.** By convention, peptides and proteins are written with the free amino group (of the **N-terminal amino acid**) on the left and the free carboxyl group (of the **C-terminal amino acid**) on the right.

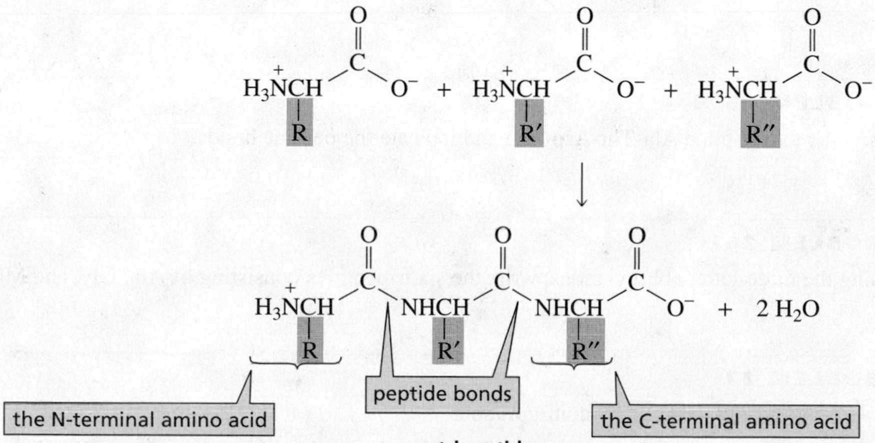

a tripeptide

When the identities of the amino acids in a peptide are known but their sequence is not, the amino acids are written separated by commas. When their sequence is known, the amino acids are written connected by hyphens. For example, in the pentapeptide represented on the right, valine is the N-terminal amino acid and histidine is the C-terminal amino acid. The amino acids are numbered starting with the N-terminal end. Alanine is therefore referred to as Ala 3 because it is the third amino acid from the N-terminal end.

Glu, Cys, His, Val, Ala

> the pentapeptide contains the indicated amino acids, but their sequence is not known

Val-Cys-Ala-Glu-His

> the amino acids in the pentapeptide have the indicated sequence

In naming a peptide, adjective names (ending in "yl") are used for all the amino acids except the C-terminal amino acid. Thus, the preceding pentapeptide is valylcysteinylalanylglutamylhistidine. Each amino acid has the L configuration unless otherwise specified.

A peptide bond has about 40% double-bond character because of electron delocalization. Steric strain in the cis configuration causes the trans configuration about the amide linkage to be more stable, so the α-carbons of adjacent amino acids are trans to each other.

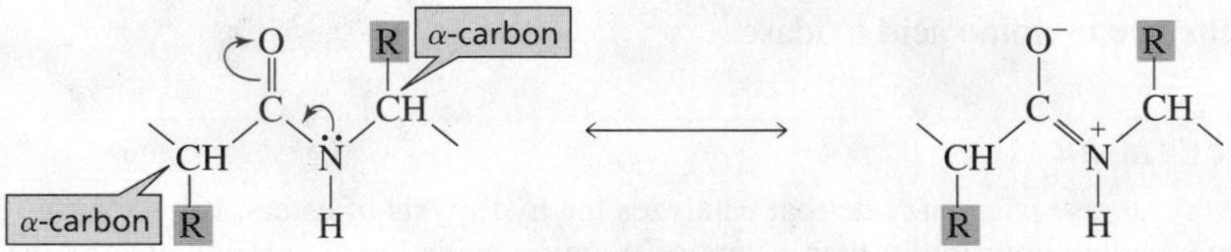

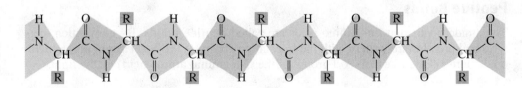

resonance contributors

The partial double-bond character prevents free rotation about the peptide bond, so the carbon and nitrogen atoms of the peptide bond and the two atoms to which each is attached are held rigidly in a plane (Figure 6). This regional planarity affects the way a chain of amino acids can fold; this has important implications for the three-dimensional shapes of peptides and proteins (Section 14).

▶ **Figure 6**

A segment of a polypeptide chain. Colored squares indicate the plane defined by each peptide bond. Notice that the R groups bonded to the α-carbons are on alternate sides of the peptide backbone.

PROBLEM 25

Draw the tetrapeptide Ala-Thr-Asp-Asn and indicate the peptide bonds.

PROBLEM 26♦

Using the three-letter abbreviations, write the six tripeptides consisting of Ala, Gly, and Met.

PROBLEM 27

Draw a peptide bond in the cis configuration.

PROBLEM 28♦

Which bonds in the backbone of a peptide can rotate freely?

Disulfide Bonds

When thiols are oxidized under mild conditions, they form a **disulfide**—a compound with an S—S bond. (Like C—H bonds, the number of S—H bonds decreases in an oxidation reaction and increases in a reduction reaction.)

$$2 \text{ R}-\text{SH} \xrightarrow{\text{mild oxidation}} \text{RS}-\text{SR}$$

$$\text{a thiol} \qquad\qquad\qquad \text{a disulfide}$$

An oxidizing agent commonly used for this reaction is Br_2 (or I_2) in a basic solution.

MECHANISM FOR THE OXIDATION OF A THIOL TO A DISULFIDE

$$R-\overset{..}{\underset{..}{S}}H \underset{H_2O}{\overset{HO^-}{\rightleftharpoons}} R-\overset{..}{\underset{..}{S}}{:}^- \xrightarrow{Br-Br} R-S-Br \xrightarrow{R-\overset{..}{S}{:}^-} R-\overset{..}{\underset{..}{S}}-\overset{..}{\underset{..}{S}}-R + Br^-$$
$$+ Br^-$$

- A thiolate ion attacks the electrophilic bromine of Br_2.
- A second thiolate ion attacks the sulfur and eliminates Br^-.

Because thiols can be oxidized to disulfides, disulfides can be reduced to thiols.

$$\text{RS}-\text{SR} \xrightarrow{\text{reduction}} 2 \text{ R}-\text{SH}$$

$$\text{a disulfide} \qquad\qquad \text{a thiol}$$

Disulfides are reduced to thiols.

The amino acid cysteine contains a thiol group, so two cysteine molecules can be oxidized to a disulfide. The disulfide is called cystine.

$$2 \; \underset{\overset{|}{\underset{+NH_3}{}}}{HSCH_2CH}\overset{\overset{O}{\parallel}}{C}-O^- \xrightarrow{\text{mild oxidation}} {}^-O-\overset{\overset{O}{\parallel}}{C}-\underset{\overset{|}{\underset{+NH_3}{}}}{CHCH_2S}-SCH_2\underset{\overset{|}{\underset{+NH_3}{}}}{CH}-\overset{\overset{O}{\parallel}}{C}-O^-$$

cysteine $\qquad\qquad\qquad\qquad\qquad\qquad$ cystine

Thiols are oxidized to disulfides.

Two cysteines in a protein can be oxidized to a disulfide, creating a bond known as a **disulfide bridge.** Disulfide bridges are the only covalent bonds that are found between nonadjacent amino acids in peptides and proteins. They contribute to the overall shape of a protein by linking cysteines found in different parts of the peptide backbone, as shown in Figure 7.

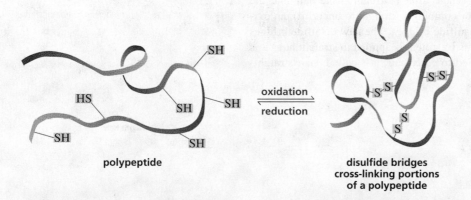

polypeptide

oxidation ⇌ reduction

disulfide bridges
cross-linking portions
of a polypeptide

◀ **Figure 7**

Disulfide bridges cross-linking portions of a polypeptide.

The hormone insulin, synthesized in the pancreas by cells known as the islets of Langerhans, maintains the proper level of glucose in the blood. Insulin is a polypeptide with two peptide chains; one contains 21 amino acids and the other 30 amino acids. The two chains are connected to each other by two **interchain disulfide bridges** (between two chains). Insulin also has an **intrachain disulfide bridge** (within a chain).

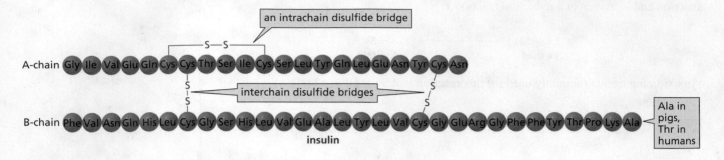

A-chain: Gly Ile Val Glu Gln Cys Cys Thr Ser Ile Cys Ser Leu Tyr Gln Leu Glu Asn Tyr Cys Asn

an intrachain disulfide bridge

interchain disulfide bridges

B-chain: Phe Val Asn Gln His Leu Cys Gly Ser His Leu Val Glu Ala Leu Tyr Leu Val Cys Gly Glu Arg Gly Phe Phe Tyr Thr Pro Lys Ala

Ala in pigs, Thr in humans

insulin

Diabetes

Diabetes is the third leading cause of death (heart disease and cancer are the first and second) in the United States. It is caused either by insufficient secretion of insulin (type 1 diabetes) or its inability to stimulate its target cells (type 2 diabetes). Injections of insulin can control some of the symptoms associated with diabetes.

Until genetic engineering techniques became available (Section 11), pigs were the primary source of insulin for people with diabetes. The insulin was effective, but there were concerns about whether enough could be obtained over the long term for the growing population of diabetics. In addition, the C-terminal amino acid of the B-chain is alanine in pig insulin and threonine in human insulin, which caused some people to have allergic reactions. Now, however, mass quantities of synthetic insulin, chemically identical to human insulin, are produced from genetically modified host cells.

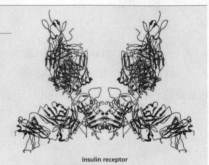

insulin receptor

Insulin binds to the insulin receptor on the surface of cells, telling the cell to transport glucose from the blood into the cell.

Hair: Straight or Curly?

Hair is made up of a protein called keratin that contains an unusually large number of cysteines (about 8% of the amino acids compared to an average of 2.8% for other proteins). These cysteines furnish keratin with many disulfide bridges that preserve its three-dimensional structure.

People can alter the structure of their hair (if they think it is either too straight or too curly) by changing the location of these disulfide bridges. This can be accomplished by first applying a reducing agent to the hair to reduce all the disulfide bridges in the protein strands. Then, after rearranging the hair into the desired shape (using curlers to curl it or combing it straight to uncurl it), an oxidizing agent is applied to form new disulfide bridges. The new disulfide bridges hold the hair in its new shape. When this treatment is applied to straight hair, it is called a "permanent." When it is applied to curly hair, it is called "hair straightening."

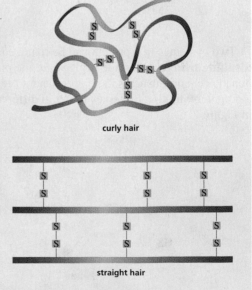

curly hair

straight hair

9   SOME INTERESTING PEPTIDES

Several peptide hormones, including β-endorphin, leucine enkephalin, and methionine enkephalin are synthesized by the body to control pain. β-Endorphin has a chain of 31 amino acids, whereas the two enkephalins are pentapeptides. The five amino acids at the N-terminal end of β-endorphin are the same as those in methionine enkephalin. These peptides control the body's sensitivity to pain by binding to receptors in certain brain cells. Part of the their three-dimensional structures must be similar to that of morphine because they bind to the same receptors. The phenomenon known as "runner's high" that kicks in after vigorous exercise and the relief of pain through acupuncture are thought to be due to the release of these peptides.

Tyr-Gly-Gly-Phe-Leu   Tyr-Gly-Gly-Phe-Met
leucine enkephalin   **methionine enkephalin**

The nonapeptides bradykinin, vasopressin, and oxytocin are also peptide hormones. Bradykinin inhibits the inflammation of tissues. Vasopressin controls blood pressure by regulating the contraction of smooth muscle; it is also an antidiuretic. Oxytocin induces labor in pregnant women by stimulating the uterine muscle to contract, and it also stimulates milk production in nursing mothers. Vasopressin and oxytocin, like β-endorphin and the enkephalins, also act on the brain. Vasopressin is a "fight-or-flight" hormone, whereas oxytocin has the opposite effect: it calms the body and promotes social bonding. In spite of their very different physiological effects, vasopressin and oxytocin differ only by two amino acids.

bradykinin   Arg-Pro-Pro-Gly-Phe-Ser-Pro-Phe-Arg

vasopressin   Cys-Tyr-Phe-Gln-Asn-Cys-Pro-Arg-Gly-NH$_2$
                  S————————S

oxytocin   Cys-Tyr-Ile-Gln-Asn-Cys-Pro-Leu-Gly-NH$_2$
                  S————————S

Vasopressin and oxytocin both have an intrachain disulfide bridge, and their C-terminal amino acids contain amide rather than carboxyl groups. Notice that the C-terminal amide group is indicated by writing "NH$_2$" after the name of the C-terminal amino acid.

The synthetic sweetener aspartame (or Equal or NutraSweet) is about 200 times sweeter than sucrose. Aspartame is the methyl ester of a dipeptide of L-aspartate and L-phenylalanine. The ethyl ester of the same dipeptide is not sweet. If a D-amino acid is substituted for either of the L-amino acids of aspartame, the resulting dipeptide is bitter rather than sweet.

aspartame
NutraSweet®

PROBLEM 29♦

What is the configuration about each of the asymmetric centers in aspartame?

PROBLEM 30

Glutathione is a tripeptide whose function is to destroy harmful oxidizing agents in the body. Oxidizing agents are thought to be responsible for some of the effects of aging and to play a causative role in cancer. Glutathione removes oxidizing agents by reducing them. In the process, glutathione is oxidized, resulting in the formation of a disulfide bond between two glutathione molecules. An enzyme subsequently reduces the disulfide bond, returning glutathione to its original condition so it can react with another oxidizing agent.

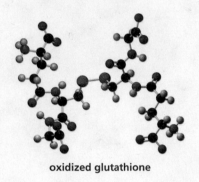

oxidized glutathione

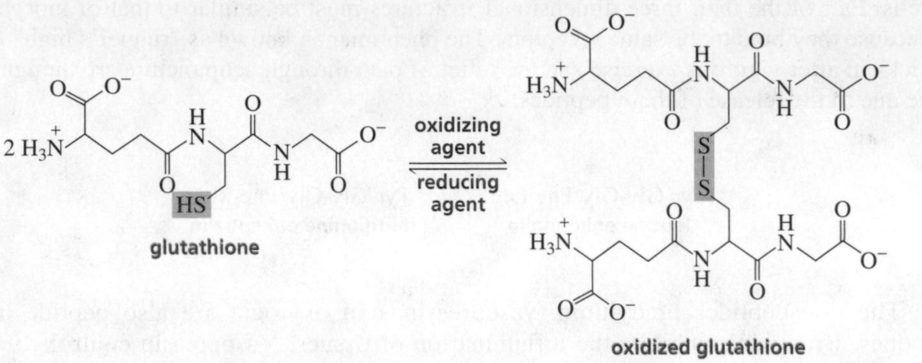

a. What amino acids make up glutathione?
b. What is unusual about glutathione's structure? (If you cannot answer this question, draw the structure you would expect for the tripeptide, and compare your structure with the actual structure.)

10   THE STRATEGY OF PEPTIDE BOND SYNTHESIS: N-PROTECTION AND C-ACTIVATION

One difficulty in synthesizing a polypeptide is that the amino acids have two functional groups, enabling them to combine in different ways. Suppose, for example, that you wanted to make the dipeptide Gly-Ala. That dipeptide is only one of four possible dipeptides that could be formed by heating a mixture of alanine and glycine.

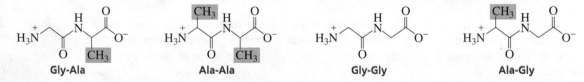

| Gly-Ala | Ala-Ala | Gly-Gly | Ala-Gly |

If the amino group of the amino acid that is to be on the N-terminal end (in this case, Gly) is protected, then it will not be available to form a peptide bond. If its carboxyl group is activated before the second amino acid is added, then the amino group of the added amino acid (in this case, Ala) will react with the activated carboxyl group of glycine in preference to reacting with a nonactivated carboxyl group of another alanine.

The N-terminal amino acid must have its amino group protected and its carboxyl group activated.

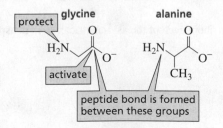

The reagent most often used to protect the amino group of an amino acid is di-*tert*-butyl dicarbonate. Notice that the amino group rather than the carboxylate group of the amino acid reacts with di-*tert*-butyl dicarbonate because the amino group is a better nucleophile. When glycine reacts with *di-tert*-butyl dicarbonate in an addition-elimination reaction, the anhydride bond breaks, forming CO_2 and *tert*-butyl alcohol.

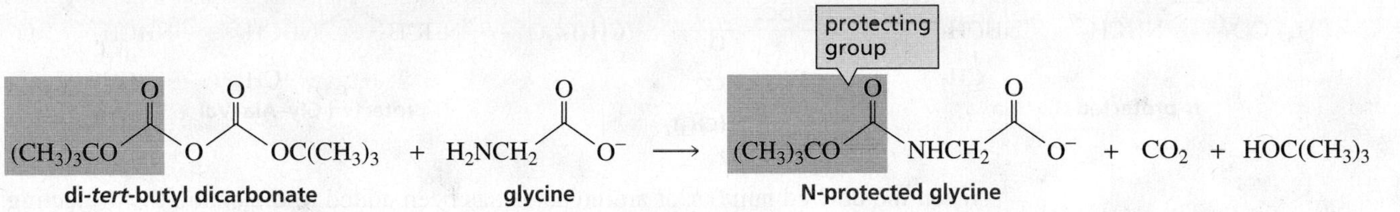

The reagent most often used to activate the carboxyl group is dicyclohexylcarbodiimide (DCC). DCC activates a carboxyl group by putting a good leaving group on the carbonyl carbon.

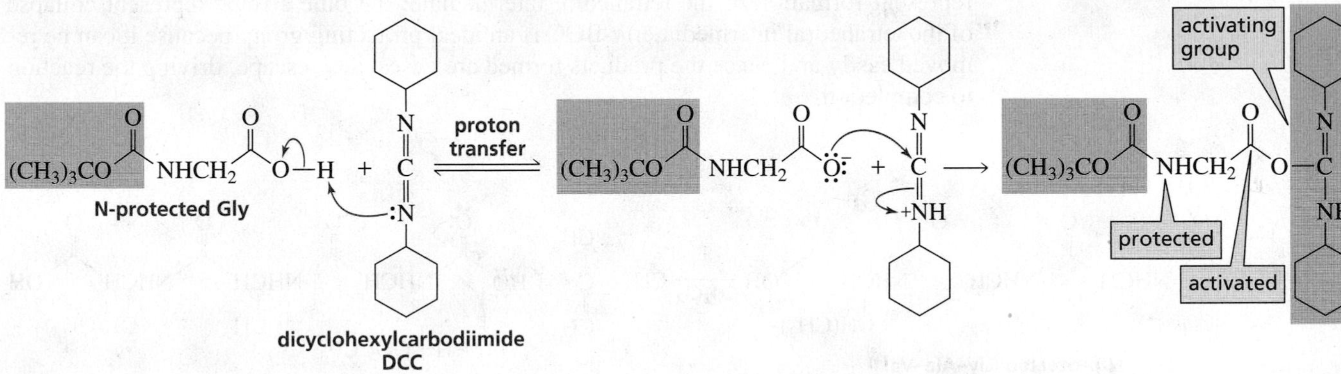

After the amino acid's N-terminal group is protected and its C-terminal group is activated, the second amino acid is added. The unprotected amino group of the second amino acid adds to the activated carboxyl group, forming a tetrahedral intermediate. The C—O bond of the tetrahedral intermediate is easily broken because the bonding electrons are delocalized. Breaking this bond forms dicyclohexylurea, a stable diamide.

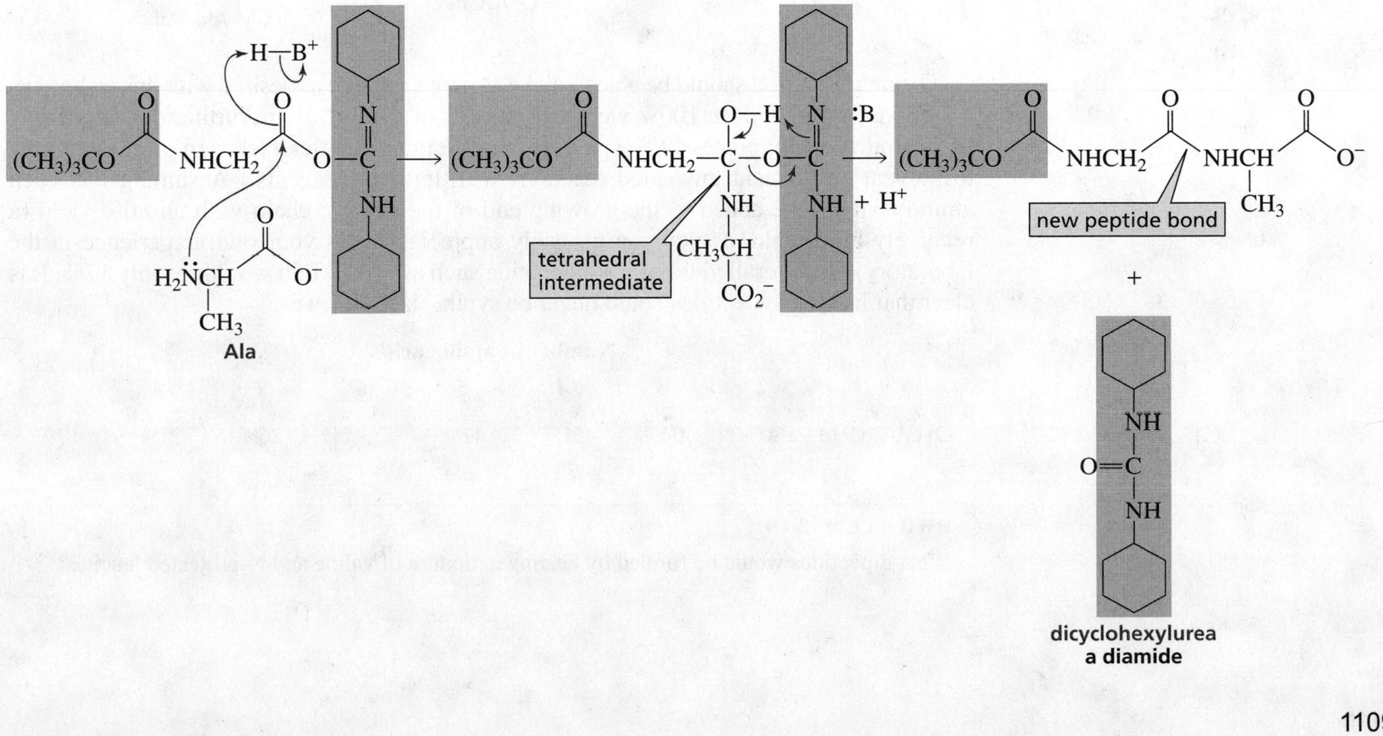

Amino acids can be added to the growing C-terminal end by repeating the same two steps: activating the carboxyl group of the C-terminal amino acid of the peptide by treating it with DCC and then adding a new amino acid.

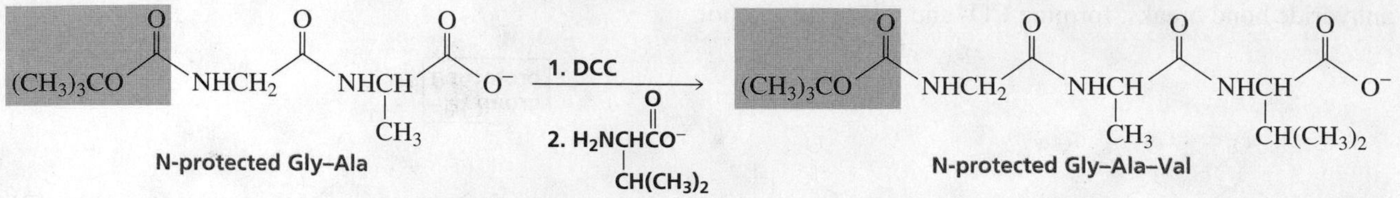

When the desired number of amino acids has been added to the chain, the protecting group, known by the acronym t-Boc ($tert$-butyloxycarbonyl; pronounced "tee-bok"), on the N-terminal amino acid is removed with trifluoroacetic acid in dichloromethane, a reagent that will not break any other covalent bonds. The protecting group is removed by an elimination reaction, forming isobutylene and carbon dioxide. (The red arrows represent formation of the tetrahedral intermediate; the blue arrows represent collapse of the tetrahedral intermediate.) t-BOC is an ideal protecting group because it can be removed easily and, since the products formed are gases, they escape, driving the reaction to completion.

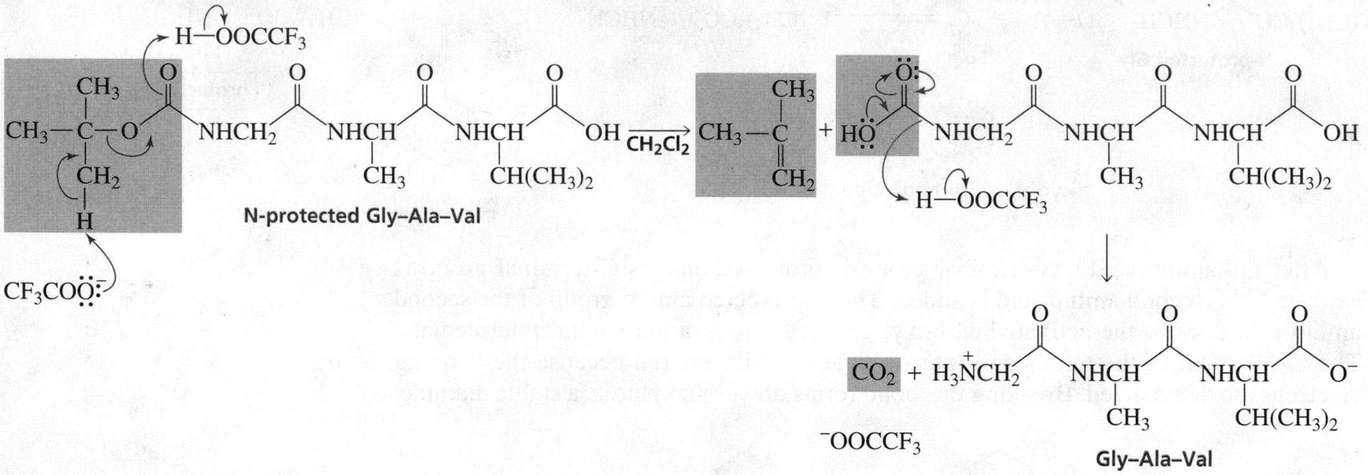

Theoretically, you should be able to make as long a peptide as desired with this technique. Reactions never produce 100% yields, however, and the yields are further decreased during the purification process. (The peptide must be purified after each step of the synthesis to prevent subsequent unwanted reactions with leftover reagents.) Assuming that each amino acid can be added to the growing end of the peptide chain with an 80% yield (a relatively high yield, as you can probably appreciate from your own experience in the laboratory), the overall yield of a nonapeptide such as bradykinin would be only 17%. It is clear that large polypeptides could never be synthesized this way.

	Number of amino acids							
	2	3	4	5	6	7	8	9
Overall yield	80%	64%	51%	41%	33%	26%	21%	17%

PROBLEM 31♦

What dipeptides would be formed by heating a mixture of valine and N-protected leucine?

PROBLEM 32

Suppose you are trying to synthesize the dipeptide Val-Ser. Compare the product that would be obtained if thionyl chloride were used to activate the carboxyl group of N-protected valine with the product that would be obtained if it were activated with DCC.

PROBLEM 33

Show the steps in the synthesis of the tetrapeptide Leu-Phe-Ala-Val.

PROBLEM 34♦

a. Calculate the overall yield of bradykinin when the yield for the addition of each amino acid to the chain is 70%.

b. What would be the overall yield of a peptide containing 15 amino acids if the yield for the incorporation of each is 80%?

11   AUTOMATED PEPTIDE SYNTHESIS

In addition to producing low overall yields, the method of peptide synthesis described in Section 10 is extremely time consuming because the product must be purified at each step of the synthesis. In 1969, Bruce Merrifield described a method that revolutionized the synthesis of peptides because it provided a much faster way to produce peptides in much higher yields. Furthermore, because it is automated, the synthesis requires fewer hours of direct attention. With this technique, bradykinin was synthesized in 27 hours with an overall yield of 85%. Subsequent refinements in the technique now allow a reasonable yield of a peptide containing 100 amino acids to be synthesized in four days.

In the Merrifield method, the synthesis is done on a solid support in a column. The solid support is similar to the one used in ion-exchange chromatography (Section 5), except that the benzene rings have chloromethyl substituents instead of sulfonic acid substituents.

Before the C-terminal amino acid is added to the solid support, its amino group is protected with *t*-BOC to prevent the amino group from reacting with the solid support. The C-terminal amino acid is attached to the solid support by means of an S_N2 reaction—its carboxyl group attacks a benzyl carbon of the resin, displacing a chloride ion.

After the C-terminal amino acid is attached to the resin, the *t*-BOC protecting group is removed (Section 10). The next amino acid, with its amino group protected with *t*-BOC and its carboxyl group activated with DCC, is added to the column, and then its protecting group is removed. Then the next N-terminal-protected and C-activated amino acid is added to the column. Thus, the protein is synthesized from the C-terminal end to the N-terminal end. (Proteins are synthesized in nature from the N-terminal end to the C-terminal end.) Because the process uses a solid support and is automated, Merrifield's method of protein synthesis is called **automated solid-phase peptide synthesis**.

THE STEPS IN THE MERRIFIELD AUTOMATED SOLID-PHASE SYNTHESIS OF A TRIPEPTIDE

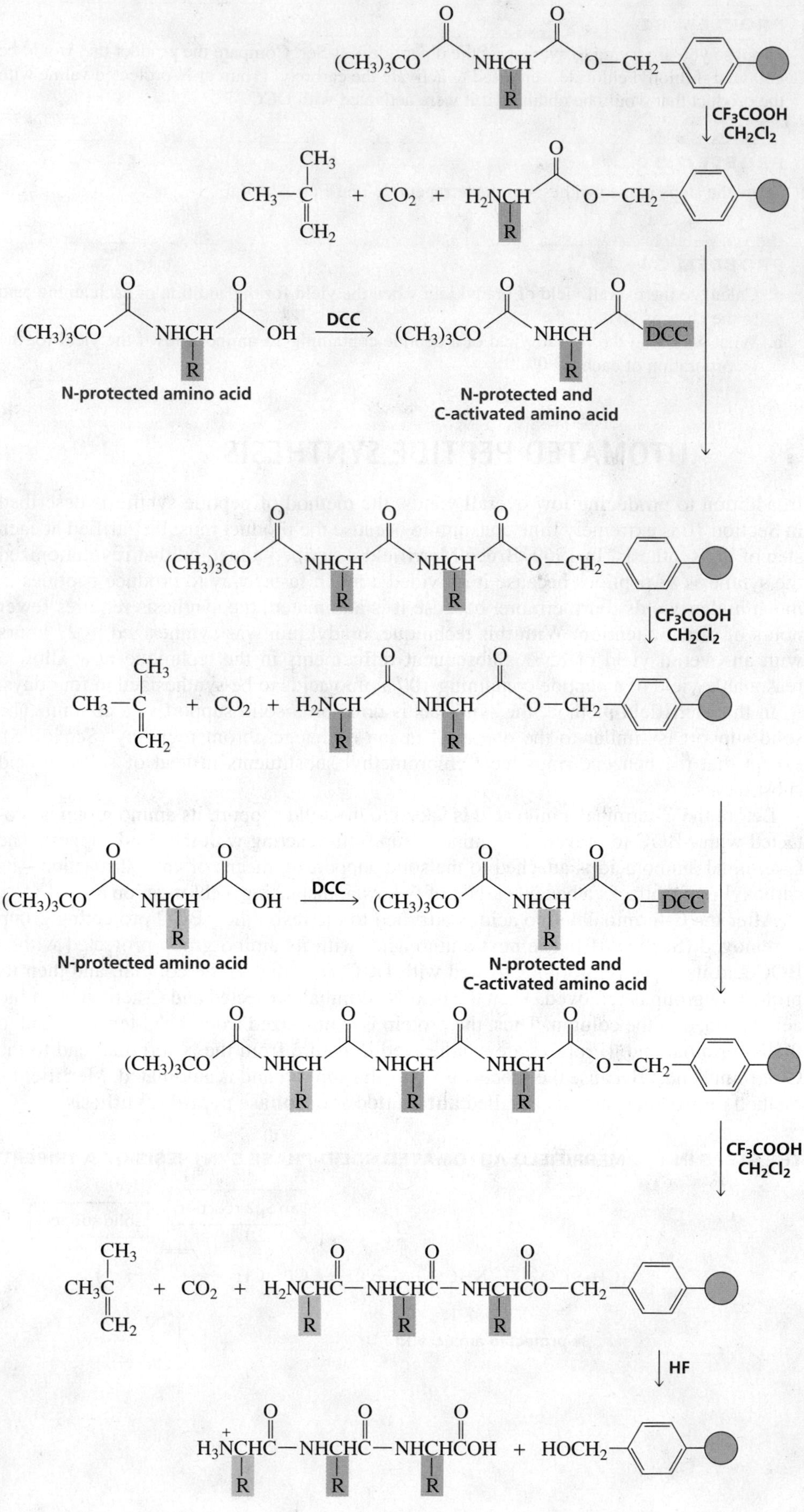
CF3COOH
CH2Cl2
DCC
N-protected amino acid
N-protected and
C-activated amino acid
CF3COOH
CH2Cl2
DCC
N-protected amino acid
N-protected and
C-activated amino acid
CF3COOH
CH2Cl2
HF

A huge advantage of the Merrifield method of peptide synthesis is that the growing peptide can be purified by washing the column with an appropriate solvent after each step of the procedure. The impurities are washed out of the column because they are not attached to the solid support. Because the peptide is covalently attached to the resin, none of it is lost in the purification step, leading to high yields of purified product.

After the required amino acids have been added one by one, the peptide can be removed from the resin by treatment with HF under mild conditions that do not break any peptide bonds.

Over time, Merrifield's technique has been improved so that peptides can be made more rapidly and more efficiently. However, it still cannot begin to compare with nature. A bacterial cell is able to synthesize a protein containing thousands of amino acids in seconds and can simultaneously synthesize thousands of different proteins with no mistakes.

Since the early 1980s, it has been possible to synthesize proteins by **genetic engineering techniques**. Strands of DNA, introduced into host cells, cause cells to produce large amounts of a desired protein. Genetic engineering techniques have also been useful in synthesizing proteins that differ in one or a few amino acids from a natural protein. Such synthetic proteins have been used, for example, to learn how a change in a single amino acid affects the properties of a protein.

PROBLEM 35

Show the steps in the synthesis of the tetrapeptide in Problem 33, using Merrifield's method.

---

12 AN INTRODUCTION TO PROTEIN STRUCTURE

Proteins are described by four levels of structure, called primary, secondary, tertiary, and quaternary.

- The **primary structure** of a protein is the sequence of the amino acids in the chain and the location of all the disulfide bridges.

- **Secondary structures** are regular conformations assumed by segments of the protein's backbone when it folds.

- The **tertiary structure** is the three-dimensional structure of the entire protein.

- If a protein has more than one polypeptide chain, it also has a quaternary structure. The **quaternary structure** is the way the individual polypeptide chains are arranged with respect to one another.

Primary Structure and Taxonomic Relationship

When scientists examine the primary structures of proteins that carry out the same function in different organisms, they can correlate the number of amino acid differences in the proteins to the closeness of the taxonomic relationship between the species. For example, cytochrome *c*, a protein that transfers electrons in biological oxidations, has about 100 amino acids. Yeast cytochrome *c* differs by 48 amino acids from horse cytochrome *c*, whereas duck cytochrome *c* differs by only two amino acids from chicken cytochrome *c*. Ducks and chickens, therefore, have a much closer taxonomic relationship than horses and yeast. Likewise, the cytochrome *c* in chickens and turkeys have identical primary structures. Human cytochrome *c* and chimpanzee cytochrome *c* are also identical and differ by one amino acid from the cytochrome *c* of the rhesus monkey.

13 HOW TO DETERMINE THE PRIMARY STRUCTURE OF A POLYPEPTIDE OR A PROTEIN

The first step in determining the sequence of amino acids in a polypeptide (or a protein) is to reduce the disulfide bridges in order to obtain a single polypeptide chain. A commonly used reducing agent is 2-mercaptoethanol. Notice that when it reduces a disulfide bridge, 2-mercaptoethanol is oxidized to a disulfide (Section 8). Reaction of the protein thiol groups with iodoacetic acid prevents the disulfide bridges from re-forming as a result of oxidation.

reducing disulfide bridges

PROBLEM 36

Write the mechanism for the reaction of a cysteine side chain with iodoacetic acid.

Now we need to determine the number and kinds of amino acids in the polypeptide chain. To do this, a sample of the polypeptide is dissolved in 6 M HCl and heated at 100 °C for 24 hours. This treatment hydrolyzes all the amide bonds in the polypeptide, including the amide bonds in the side chains of asparagine and glutamine.

$$\text{protein} \xrightarrow[\substack{100\ °C \\ 24\ h}]{6\ M\ HCl} \text{amino acids}$$

The mixture of amino acids is then passed through an amino acid analyzer to identify the amino acids and determine how many of each kind are in the polypeptide (Section 5).

Because all the asparagines and glutamines have been hydrolyzed to aspartates and glutamates, the number of aspartates or glutamates in the amino acid mixture tells us the number of aspartates plus asparagines or glutamates plus glutamines in the original protein. A separate test must be conducted to distinguish between aspartate and asparagine or between glutamate and glutamine in the original polypeptide.

The strongly acidic conditions used for hydrolysis destroy all the tryptophans because the indole ring is unstable in acid. However, the tryptophan content can be determined by hydroxide-ion-promoted hydrolysis of the peptide. This is not a general method for peptide bond hydrolysis because the strongly basic conditions destroy several amino acids.

Determining the N-Terminal Amino Acid

One of the most widely used methods to identify the N-terminal amino acid of a polypeptide is to treat the polypeptide with phenyl isothiocyanate (PITC), more commonly known as **Edman's reagent.** Edman's reagent reacts with the N-terminal amino group, and the resulting thiazolinone derivative is cleaved from the polypeptide under mildly acidic conditions, leaving behind a polypeptide with one less amino acid.

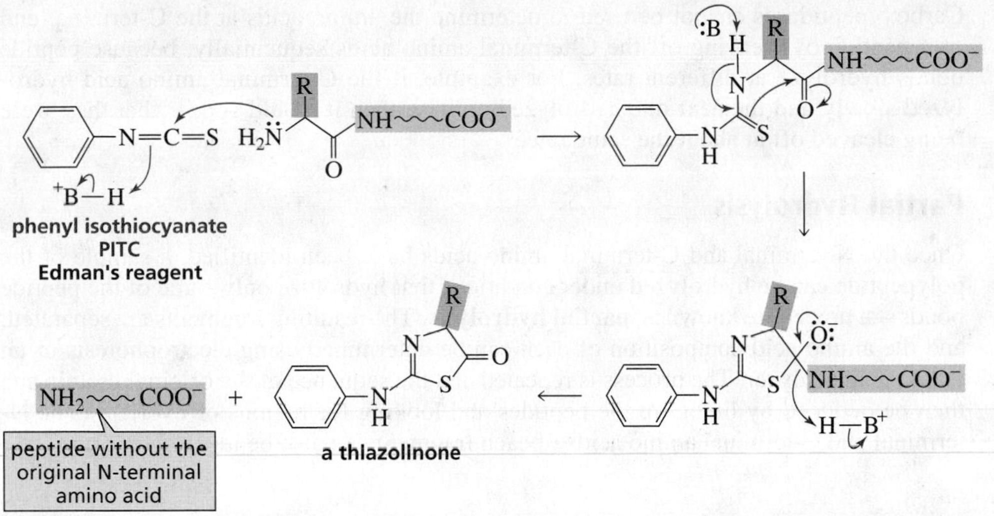

The thiazolinone rearranges in dilute acid to a more stable phenylthiohydantoin (PTH) (see Problem 72).

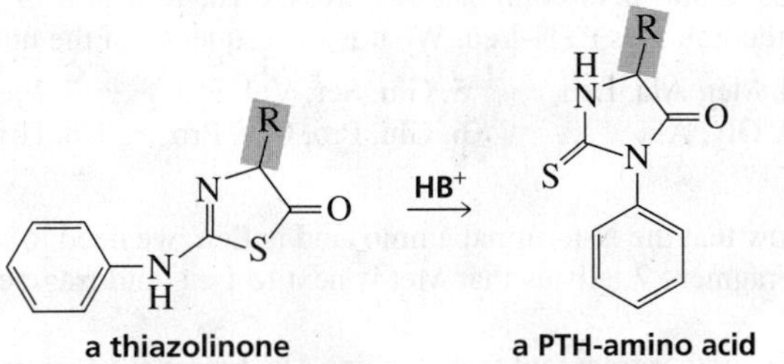

Because each amino acid has a different side chain (R), each amino acid forms a different PTH–amino acid. The particular PTH–amino acid can be identified by chromatography using known standards.

An automated instrument known as a *sequencer* allows about 50 successive Edman degradations of a polypeptide to be performed (100 with more advanced instruments). The entire primary structure cannot be determined in this way, however, because side products accumulate that interfere with the results.

PROBLEM 37♦

In determining the primary structure of insulin, what would lead you to conclude that insulin had more than one polypeptide chain?

Determining the C-Terminal Amino Acid

The C-terminal amino acid of a polypeptide can be identified using a carboxypeptidase, an enzyme that catalyzes the hydrolysis of the C-terminal peptide bond, thereby cleaving off the C-terminal amino acid. Carboxypeptidase A cleaves off the C-terminal amino acid, as long as it is *not* arginine or lysine. On the other hand, carboxypeptidase B cleaves off the C-terminal amino acid, *only* if it is arginine or lysine. Carboxypeptidases are **exopeptidases,** enzymes that catalyze the hydrolysis of a peptide bond at the end of a peptide chain.

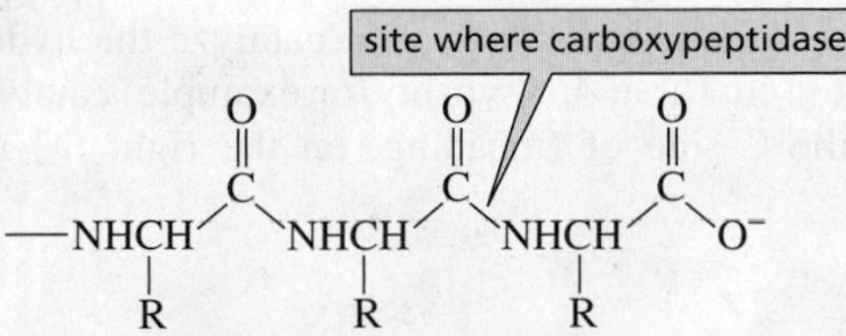

Carboxypeptidases cannot be used to determine the amino acids at the C-terminal end of a peptide by cleaving off the C-terminal amino acids sequentially, because peptide bonds hydrolyze at different rates. For example, if the C-terminal amino acid hydrolyzed slowly and the next one hydrolyzed rapidly, then it would appear that they were being cleaved off at about the same rate.

Partial Hydrolysis

Once the N-terminal and C-terminal amino acids have been identified, a sample of the polypeptide can be hydrolyzed under conditions that hydrolyze only some of the peptide bonds—a procedure known as **partial hydrolysis.** The resulting fragments are separated, and the amino acid composition of each can be determined using electrophoresis or an amino acid analyzer. The process is repeated and the sequence of the original protein can then be deduced by lining up the peptides and looking for regions of overlap. (The N-terminal and C-terminal amino acids of each fragment can also be identified, if needed.)

PROBLEM-SOLVING STRATEGY

Sequencing an Oligopeptide

When a nonapeptide undergoes partial hydrolysis, it forms dipeptides, a tripeptide and two tetrapeptides whose amino acid compositions are shown. Reaction of the intact nonapeptide with Edman's reagent releases PTH-Leu. What is the sequence of the nonapeptide?

1. Pro, Ser   **3.** Met, Ala, Leu   **5.** Glu, Ser, Val, Pro   **7.** Met, Leu
2. Gly, Glu   **4.** Gly, Ala   **6.** Glu, Pro, Gly, Pro   **8.** His, Val

- Because we know that the N-terminal amino acid is Leu, we need to look for a fragment that contains Leu. Fragment **7** tells us that Met is next to Leu, and fragment **3** tells us that Ala is next to Met.
- Now we look for another fragment that contains Ala. Fragment **4** contains Ala and tells us that Gly is next to Ala.
- From fragment **2**, we know that Glu comes next; Glu is in both fragments **5** and **6**.
- Fragment **5** has three amino acids we have yet to place in the growing peptide (Ser, Val, Pro), but fragment **6** has only one, so we know from fragment **6** that Pro is the next amino acid.
- Fragment **1** indicates that the next amino acid is Ser, so we can now use fragment **5**. Fragment **5** indicates that the next amino acid is Val, and fragment **8** tells us that His is the last (C-terminal) amino acid.
- Thus, the amino acid sequence of the nonapeptide is Leu-Met-Ala-Gly-Glu-Pro-Ser-Val-His

Now use the strategy you have just learned to solve Problem 38.

PROBLEM 38♦

A decapeptide undergoes partial hydrolysis to give peptides whose amino acid compositions are shown. Reaction of the intact decapeptide with Edman's reagent releases PTH-Gly. What is the sequence of the decapeptide?

1. Ala, Trp   **3.** Pro, Val   **5.** Trp, Ala, Arg   **7.** Glu, Ala, Leu
2. Val, Pro, Asp   **4.** Ala, Glu   **6.** Arg, Gly   **8.** Met, Pro, Leu, Glu

---

Hydrolysis Using Endopeptidases

A polypeptide can also be partially hydrolyzed using **endopeptidases**—enzymes that catalyze the hydrolysis of a peptide bond that is *not* at the end of a peptide chain. Trypsin, chymotrypsin, and elastase are endopeptidases that catalyze the hydrolysis of only the specific peptide bonds listed in Table 4. Trypsin, for example, catalyzes the hydrolysis of the peptide bond on the C-side of (meaning, on the right of) positively charged side chains (arginine or lysine). These enzymes belong to the group of enzymes known as **digestive enzymes.**

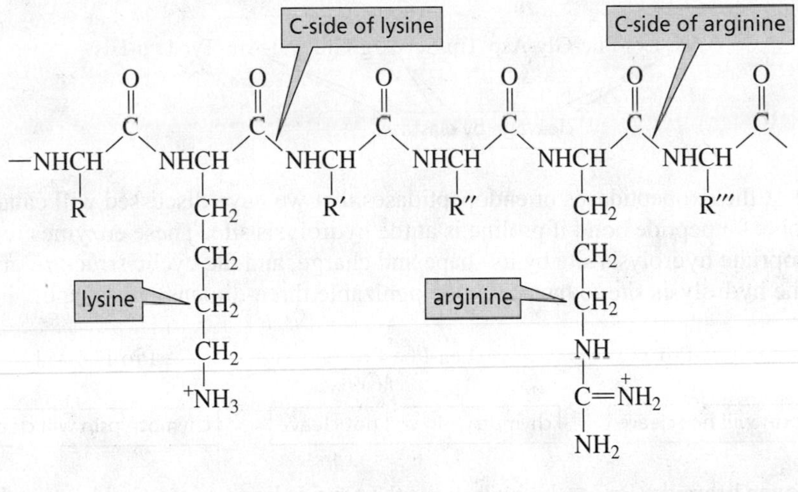

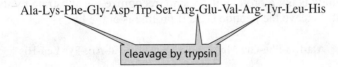

trypsin (see the legend to Figure 10)

Thus, trypsin will catalyze the hydrolysis of three peptide bonds in the following polypeptide, creating a hexapeptide, a dipeptide, and two tripeptides.

Ala-Lys-Phe-Gly-Asp-Trp-Ser-Arg-Glu-Val-Arg-Tyr-Leu-His

cleavage by trypsin

Table 4	Specificity of Peptide or Protein Cleavage
Reagent	**Specificity**
Chemical reagents	
Edman's reagent	removes the N-terminal amino acid
Cyanogen bromide	hydrolyzes on the C-side of Met
Exopeptidases*	
Carboxypeptidase A	removes the C-terminal amino acid (not if it is Arg or Lys)
Carboxypeptidase B	removes the C-terminal amino acid (only if it is Arg or Lys)
Endopeptidases*	
Trypsin	hydrolyzes on the C-side of Arg and Lys
Chymotrypsin	hydrolyzes on the C-side of amino acids that contain aromatic six-membered rings (Phe, Tyr, Trp)
Elastase	hydrolyzes on the C-side of small amino acids (Gly, Ala, Ser, and Val)
Thermolysin	hydrolyzes on the C-side of Ile, Met, Phe, Trp, Tyr, and Val

*Cleavage will not occur if Pro is at the hydrolysis site.

Chymotrypsin catalyzes the hydrolysis of the peptide bond on the C-side of amino acids that contain aromatic six-membered rings (Phe, Tyr, Trp).

Ala-Lys-Phe-Gly-Asp-Trp-Ser-Arg-Glu-Val-Arg-Tyr-Leu-His

cleavage by chymotrypsin

Elastase catalyzes the hydrolysis of peptide bonds on the C-side of the two smallest amino acids (Gly, Ala, Ser, and Val). Chymotrypsin and elastase are much less specific than trypsin.

None of the exopeptidases or endopeptidases that we have discussed will catalyze the hydrolysis of a peptide bond if proline is at the hydrolysis site. These enzymes recognize the appropriate hydrolysis site by its shape and charge, and the cyclic structure of proline causes the hydrolysis site to have an unrecognizable three-dimensional shape.

Ala-Lys-Pro	Leu-Phe-Pro	Pro-Phe-Val
trypsin will not cleave	chymotrypsin will not cleave	chymotrypsin will cleave

Cyanogen bromide (BrC≡N) hydrolyzes the peptide bond on the C-side of methionine. Cyanogen bromide is more specific than the endopeptidases about which peptide bonds it cleaves, so it provides more reliable information about the primary structure. Because cyanogen bromide is not a protein, it does not recognize the substrate by its shape. As a result, it will still cleave the peptide bond if proline is at the hydrolysis site.

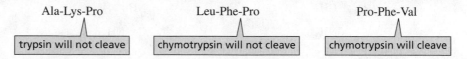

MECHANISM FOR THE CLEAVAGE OF A PEPTIDE BOND BY CYANOGEN BROMIDE

- The nucleophilic sulfur of methionine attacks the carbon of cyanogen bromide and displaces a bromide ion.
- Nucleophilic attack by oxygen on the methylene group, resulting in departure of the weakly basic leaving group, forms a five-membered ring.
- Acid-catalyzed hydrolysis of the imine cleaves the protein.
- Further hydrolysis causes the lactone (a cyclic ester) to open to a carboxyl group and an alcohol group.

The last step in determining the primary structure of a protein is to figure out the location of any disulfide bonds. This is done by hydrolyzing a sample of the protein that has intact disulfide bonds. From a determination of the amino acids in the cysteine-containing fragments, the locations of the disulfide bonds in the protein can be established (Problem 61).

PROBLEM 39

Why doesn't cyanogen bromide cleave on the C-side of cysteine?

PROBLEM 40♦

Indicate the peptides that would result from cleavage by the indicated reagent:

a. His-Lys-Leu-Val-Glu-Pro-Arg-Ala-Gly-Ala by trypsin
b. Leu-Gly-Ser-Met-Phe-Pro-Tyr-Gly-Val by chymotrypsin

PROBLEM 41 Solved

Determine the amino acid sequence of a polypeptide from the following data:

Acid-catalyzed hydrolysis gives Ala, Arg, His, 2 Lys, Leu, 2 Met, Pro, 2 Ser, Thr, and Val.

Carboxypeptidase A releases Val.

Edman's reagent releases PTH-Leu.

Treatment with cyanogen bromide gives three peptides with the following amino acid compositions:

1. His, Lys, Met, Pro, Ser   **2.** Thr, Val   **3.** Ala, Arg, Leu, Lys, Met, Ser

Trypsin-catalyzed hydrolysis gives three peptides and a single amino acid:

1. Arg, Leu, Ser   **3.** Lys
2. Met, Pro, Ser, Thr, Val   **4.** Ala, His, Lys, Met

Solution   Acid-catalyzed hydrolysis shows that the polypeptide has 13 amino acids. The N-terminal amino acid is Leu (revealed by Edman's reagent), and the C-terminal amino acid is Val (revealed by carboxypeptidase A).

Leu __ __ __ __ __ __ __ __ __ __ __ Val

- Because cyanogen bromide cleaves on the C-side of Met, any peptide containing Met must have Met as its C-terminal amino acid. Thus, the peptide that does not contain Met must be the C-terminal peptide, so we know that the twelfth amino acid is Thr. We know that peptide **3** is the N-terminal peptide because it contains Leu. Because peptide **3** is a hexapeptide, we know that the sixth amino acid in the polypeptide is Met. We also know that the eleventh amino acid is Met because cyanogen bromide cleavage gave the dipeptide Thr, Val.

Ala, Arg, Lys, Ser   His, Lys, Pro, Ser

Leu __ __ __ __ Met __ __ __ __ Met Thr Val

- Because trypsin cleaves on the C-side of Arg and Lys, any peptide containing Arg or Lys must have that amino acid as its C-terminal amino acid. Therefore, Arg is the C-terminal amino acid of peptide **1**, so we now know that the first three amino acids are Leu-Ser-Arg. We also know that the next two are Lys-Ala because if they were Ala-Lys, then trypsin cleavage would give an Ala, Lys dipeptide. The trypsin data also identify the positions of His and Lys.

Pro, Ser

Leu Ser Arg Lys Ala Met His Lys __ __ Met Thr Val

- Finally, because trypsin successfully cleaves on the C-side of Lys, Pro cannot be adjacent to Lys. Thus, the amino acid sequence of the polypeptide is

Leu Ser Arg Lys Ala Met His Lys Ser Pro Met Thr Val

PROBLEM 42♦

Determine the primary structure of an octapeptide from the following data:

Acid-catalyzed hydrolysis gives 2 Arg, Leu, Lys, Met, Phe, Ser, and Tyr.

Carboxypeptidase A releases Ser.

Edman's reagent releases Leu.

Treatment with cyanogen bromide forms two peptides with the following amino acid compositions:

1. Arg, Phe, Ser   **2.** Arg, Leu, Lys, Met, Tyr

Trypsin-catalyzed hydrolysis forms the following two amino acids and two peptides:

1. Arg   **2.** Ser   **3.** Arg, Met, Phe   **4.** Leu, Lys, Tyr

---

PROBLEM 43

Three peptides were obtained from a trypsin digestion of two different polypeptides. In each case, indicate the possible sequences from the given data and tell what further experiment should be carried out in order to determine the primary structure of the polypeptide.

a. 1. Val-Gly-Asp-Lys   **2.** Leu-Glu-Pro-Ala-Arg   **3.** Ala-Leu-Gly-Asp

b. 1. Val-Leu-Gly-Glu   **2.** Ala-Glu-Pro-Arg   **3.** Ala-Met-Gly-Lys

---

14   SECONDARY STRUCTURE

Secondary structure describes the repetitive conformations assumed by segments of the backbone chain of a peptide or protein. In other words, the secondary structure describes how segments of the backbone fold. Three factors determine the secondary structure of a segment of protein:

- The regional planarity about each peptide bond (as a result of the partial double-bond character of the amide bond), which limits the possible conformations of the peptide chain (Section 8)

- Minimizing energy by maximizing the number of peptide groups that engage in hydrogen bonding (that is, that form a hydrogen bond between the carbonyl oxygen of one amino acid and the amide hydrogen of another).

- The need for adequate separation between neighboring R groups to avoid steric strain and repulsion of like charges

α-Helix

One type of secondary structure is an **α-helix.** In an α-helix, the backbone of the polypeptide coils around the long axis of the protein molecule. The substituents on the α-carbons of the amino acids protrude outward from the helix, thereby minimizing steric strain (Figure 8a). The helix is stabilized by hydrogen bonds—each hydrogen attached to an amide nitrogen is hydrogen bonded to a carbonyl oxygen of an amino acid four amino acids away (Figure 8b).

Because the amino acids have the L-configuration, the α-helix is a right-handed helix—that is, it rotates in a clockwise direction as it spirals down (Figure 8c).

Each turn of the helix contains 3.6 amino acids, and the repeat distance of the helix is 5.4 Å.

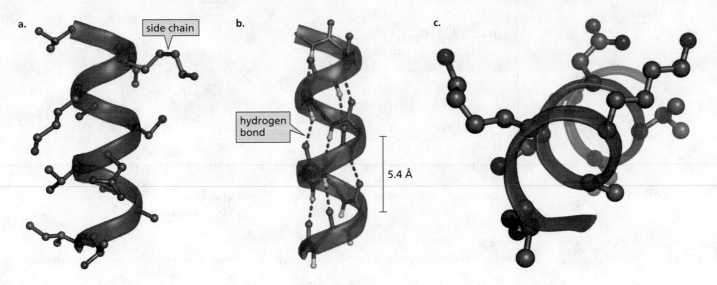

▲ Figure 8
(a) A segment of a protein in an α-helix.
(b) The helix is stabilized by hydrogen bonding between peptide groups.
(c) Looking at the longitudinal axis of an α-helix.

Not all amino acids are able to fit into an α-helix. Proline, for example, causes a distortion in a helix because the bond between the proline nitrogen and the α-carbon cannot rotate to let proline fit into a helix properly. Similarly, two adjacent amino acids that have more than one substituent on a β-carbon (valine, isoleucine, or threonine) cannot fit into a helix because of steric crowding between the R groups. Finally, two adjacent amino acids with like-charged substituents cannot fit into a helix because of electrostatic repulsion between the R groups. The percentage of amino acids coiled into an α-helix varies from protein to protein, but on average about 25% of the amino acids in a globular protein are in α-helices.

Right-Handed and Left-Handed Helices

The α-helix, composed of a chain of L-amino acids, is a right-handed helix. When scientists synthesized a chain of D-amino acids, they found that it folded into a left-handed helix that was the mirror image of the right-handed α-helix. When they synthesized a peptidase that contained only D-amino acids, they found that the enzyme was just as catalytically active as a naturally occurring peptidase with L-amino acids. However, the peptidase that contained D-amino acids would cleave peptide bonds only in polypeptide chains composed of D-amino acids.

β-Pleated Sheet

The second type of secondary structure is a **β-pleated sheet.** In a β-pleated sheet, the polypeptide backbone is extended in a zigzag structure resembling a series of pleats. The hydrogen bonding in a β-pleated sheet occurs between neighboring peptide chains, and these chains can run in the same direction or in opposite directions—called a **parallel β-pleated sheet** and an **antiparallel β-pleated sheet** (Figure 9). A β-pleated sheet is almost fully extended; the average two amino acid repeat distance is 7.0 Å.

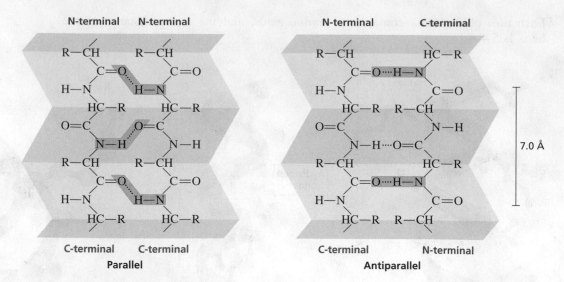

▲ Figure 9

Segments of a parallel β-pleated sheet and an antiparallel β-pleated sheet drawn to illustrate their pleated character.

Because the substituents (R) on the α-carbons of the amino acids on adjacent chains are close to each other, the substituents must be small if the chains are to nestle closely enough together to maximize hydrogen-bonding interactions. Silk, for example, contains a large proportion of relatively small amino acids (glycine and alanine) and therefore has large segments of β-pleated sheet. The number of side-by-side strands in a β-pleated sheet ranges from 2 to 15 in a globular protein. The average strand in a β-pleated sheet section of a globular protein contains six amino acids.

Wool and the fibrous protein of muscle have secondary structures that are almost all α-helices. Consequently, these proteins can be stretched. In contrast, proteins with secondary structures that are predominantly β-pleated sheets, such as silk and spider webs, cannot be stretched because a β-pleated sheet is almost fully extended already.

Coil Conformation

Generally, less than half of the protein's backbone is arranged in a defined secondary structure—an α-helix or a β-pleated sheet (Figure 10). Most of the rest of the protein, though highly ordered, is nonrepetitive and therefore difficult to describe. Many of these ordered polypeptide fragments are said to be in **coil** or **loop conformations.**

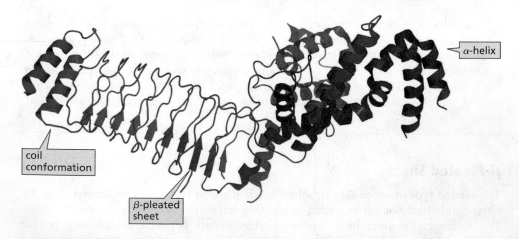

▲ Figure 10

The backbone structure of an enzyme called ligase: a β-pleated sheet is indicated by a flat arrow pointing in the N → C direction, an α-helix by a flat helical ribbon, and a coil or loop conformation by a thin tube.

PROBLEM 44♦

How long is an α-helix that contains 74 amino acids? How long is a fully extended peptide chain that contains the same number of amino acids? (The distance between consecutive amino acids in a fully extended chain is 3.5 Å; the repeat distance of an α-helix is 5.4 Å.)

β-Peptides: An Attempt to Improve on Nature

β-Peptides are polymers of β-amino acids, so they have backbones that are one carbon longer than the peptides nature synthesizes using α-amino acids. Therefore, each β-amino acid has two carbons to which side chains can be attached.

Like α-polypeptides, β-polypeptides fold into relatively stable helical and pleated sheet conformations, so scientists are trying to find out whether biological activity might be possible with such peptides. Recently, a β-peptide with biological activity has been synthesized that mimics the activity of the hormone somatostatin. There is hope that β-polypeptides will provide a source of new drugs and catalysts. Surprisingly, the peptide bonds in β-polypeptides are resistant to the enzymes that catalyze the hydrolysis of peptide bonds in α-polypeptides. This resistance to hydrolysis suggests that a β-polypeptide drug would have a longer duration of action in the bloodstream.

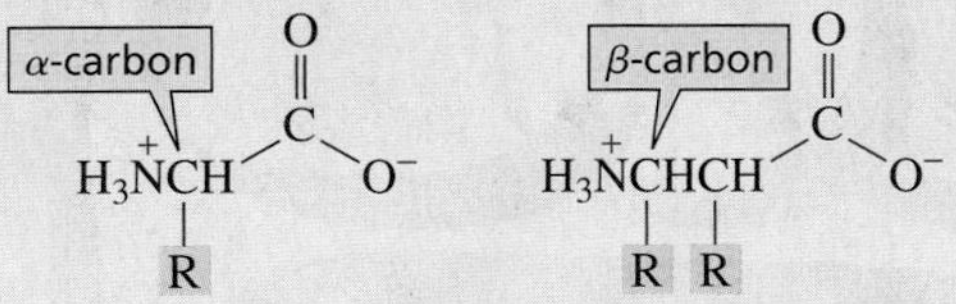

15   TERTIARY STRUCTURE

The *tertiary structure* of a protein is the three-dimensional arrangement of all the atoms in the protein (Figure 11). Proteins fold spontaneously in solution to maximize their stability. Every time there is a stabilizing interaction between two atoms, free energy is released. The more free energy released (the more negative the $\Delta G°$), the more stable the protein. Consequently, a protein tends to fold in a way that maximizes the number of stabilizing interactions.

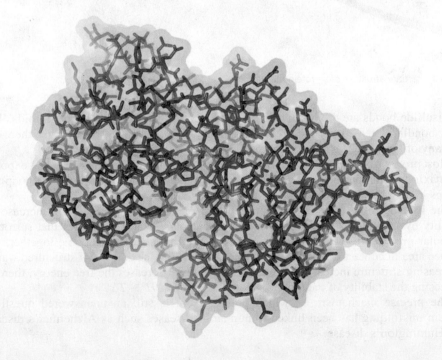

◄ **Figure 11**

The three-dimensional structure of thermolysin (an endopeptidase).

The stabilizing interactions in a protein include disulfide bonds, hydrogen bonds, electrostatic attractions (attractions between opposite charges), and hydrophobic (van der Waals) interactions. Stabilizing interactions can occur between peptide groups (atoms in the backbone of the protein), between side-chain groups (α-substituents), and between peptide and side-chain groups (Figure 12). Because the side-chain groups help determine how a protein folds, the tertiary structure of a protein is determined by its primary structure.

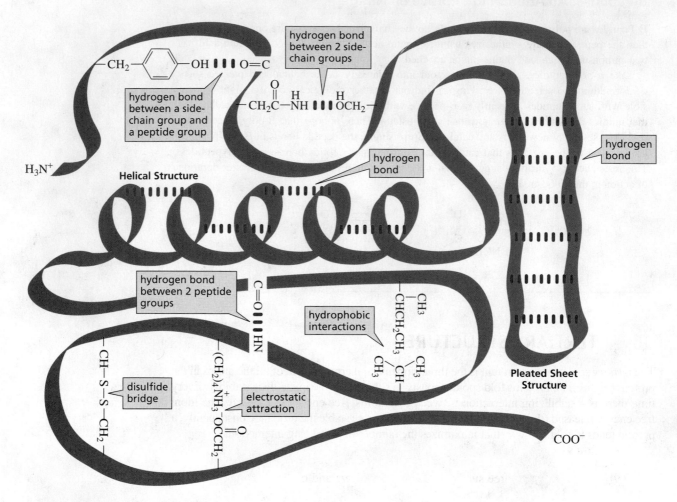

▲ Figure 12

Stabilizing interactions responsible for the tertiary structure of a protein.

Disulfide bonds are the only covalent bonds that can form when a protein folds. The other bonding interactions that occur in folding are much weaker, but because there are so many of them, they are important in determining how a protein folds.

Most proteins exist in aqueous environments, so they tend to fold in a way that exposes the maximum number of polar groups to the surrounding water and buries the nonpolar groups in the protein's interior, away from water.

The **hydrophobic interactions** between nonpolar groups in the protein increase its stability by increasing the entropy of water molecules. Water molecules that surround nonpolar groups are highly structured. When two nonpolar groups come together, the surface area in contact with water diminishes, decreasing the amount of structured water. Decreasing structure increases entropy, which in turn decreases the free energy, thereby increasing the stability of the protein. (Note that $\Delta G° = \Delta H° - T\Delta S°$.)

The precise mechanism by which proteins fold is still an unanswered question. Protein misfolding has been linked to numerous diseases such as Alzheimer's disease and Huntington's disease.

PROBLEM 45

How would a protein that resides in the nonpolar interior of a membrane fold compared with the water-soluble protein just discussed?

Diseases Caused by a Misfolded Protein

Bovine spongiform encephalopathy (BSE), commonly known as mad cow disease, is unlike most other diseases because it is not caused by a microorganism. Instead, it caused by a misfolded protein in the brain called a prion. It is not yet known what causes the prion to become misfolded. The misfolded prion causes cells to deteriorate until the brain has a sponge-like appearance. The deterioration causes the loss of mental function, which makes cows with the disease act strangely (hence the name *mad cow disease*). It is not curable and it is fatal, but it is not contagious. It takes several years from the time of first exposure until the first signs of the disease appear, but then it progresses quickly.

There are other diseases caused by misfolded prions that have similar symptoms. Kuru—transmitted through cannibalism—has been found to occur in the Fore people of Papua New Guinea (*kuru* means "trembling"). Scrapie affects sheep and goats. This disease got its name from the tendency of sheep to scrape off their wool on fences as they lean on them in an attempt to stay upright. It is thought that mad cow disease, first reported in the U.K. in 1985, was caused by cows eating bone meal made from scrapie-infected sheep.

The human form of the disease is called Creutzfeldt–Jakob disease (CJD), which is rare and apparently arises spontaneously. The average age of onset of CJD is 64. In 1994, however, several cases of the disease appeared in young adults in the U.K. To date, 200 cases have been reported. This new variant Creutzfeldt–Jakob disease (vCJD) is caused by ingesting meat products of an animal infected with the disease.

16   QUATERNARY STRUCTURE

Some proteins have more than one polypeptide chain. The individual chains are called **subunits.** A protein with a single subunit is called a *monomer;* one with two subunits is called a *dimer;* one with three subunits is called a *trimer;* and one with four subunits is called a *tetramer.* The quaternary structure of a protein describes the way the subunits are arranged with respect to each other.

The subunits can be the same or different. Hemoglobin, for example, is a tetramer; it has two different kinds of subunits and each hemoglobin molecule has two of each kind (Figure 13).

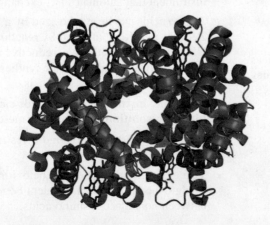

◀ **Figure 13**

The quaternary structure of hemoglobin. The two α-subunits are green and the two β-subunits are purple. The porphyrin rings are blue.

The subunits of a protein are held together by the same kinds of interactions that hold the individual protein chains in a particular three-dimensional conformation—namely, hydrophobic interactions, hydrogen bonding, and electrostatic attractions.

PROBLEM 46♦

a. Which would have the greatest percentage of polar amino acids, a spherical protein, a cigar-shaped protein, or a subunit of a hexamer?

b. Which would have the smallest percentage of polar amino acids?

---

17   PROTEIN DENATURATION

Destroying the highly organized tertiary structure of a protein is called **denaturation.** Anything that breaks the bonds maintaining the three-dimensional shape of the protein will cause the protein to denature (unfold). Because these bonds are weak, proteins are easily denatured. The following are some of the ways that proteins can be denatured:

- Changing the pH denatures proteins because it changes the charges on many of the side chains. This disrupts electrostatic attractions and hydrogen bonds.

- Certain reagents such as urea and guanidine hydrochloride denature proteins by forming hydrogen bonds to protein groups that are stronger than the hydrogen bonds formed between the groups.

- Organic solvents or detergents such as sodium dodecyl sulfate denature proteins by associating with the nonpolar groups of the protein, thereby disrupting the normal hydrophobic interactions.

- Proteins can also be denatured by heat or by agitation. Both increase molecular motion, which can disrupt the attractive forces. A well-known example is the change that occurs to the white of an egg when it is heated or whipped.

SOME IMPORTANT THINGS TO REMEMBER

- **Peptides** and **proteins** are polymers of **amino acids** linked together by **peptide** (amide) bonds.

- The **amino acids** differ only in the substituent attached to the α-carbon.

- Almost all amino acids found in nature have the L-configuration.

- The carboxyl groups of the amino acids have pK_a values of ~2, and the protonated amino groups have pK_a values of ~9. At physiological pH (7.4), an amino acid exists as a **zwitterion.**

- The **isoelectric point** (pI) of an amino acid is the pH at which the amino acid has no net charge.

- A mixture of amino acids can be separated based on their pI values by **electrophoresis** or based on their polarities by **paper chromatography** or **thin-layer chromatography.**

- Separation can also be done by **ion-exchange chromatography** employing an **anion-** or a **cation-exchange resin.** An **amino acid analyzer** is an instrument that automates ion-exchange chromatography.

- Amino acids can be synthesized by a Hell–Volhard–Zelinski reaction (followed by reaction with excess NH_3), a Strecker synthesis, reductive amination, a N-phthalimidomalonic ester synthesis, or an acetamidomalonic ester synthesis.

- A racemic mixture of amino acids can be separated by a **kinetic resolution** using an enzyme-catalyzed reaction.

- Rotation about a peptide bond is restricted because of its partial double-bond character.

- Two cysteine side chains can be oxidized to a **disulfide bridge,** the only kind of covalent bond that is found between nonadjacent amino acids.

- By convention, peptides and proteins are written with the free amino group (of the **N-terminal amino acid**) on the left and the free carboxyl group (of the **C-terminal amino acid**) on the right.

- To synthesize a peptide bond, the amino group of the N-terminal amino acid must be protected (by *t*-BOC) and its carboxyl group activated (with DCC). A second amino acid is added to form a dipeptide. Amino acids can be added to the growing C-terminal end by repeating the same two steps: activating the carboxyl group of the C-terminal amino acid with DCC and adding a new amino acid.

- **Automated solid-phase peptide synthesis** allows peptides to be made more rapidly and in higher yields.

- The **primary structure** of a protein is the sequence of its amino acids and the location of all its disulfide bridges.

- The N-terminal amino acid can be determined with **Edman's reagent.** The C-terminal amino acid can be identified with a carboxypeptidase.

- An **exopeptidase** catalyzes the hydrolysis of a peptide bond at the end of a peptide chain. An **endopeptidase** catalyzes the hydrolysis of a peptide bond that is not at the end of a peptide chain.

- The **secondary structure** of a protein describes how local segments of the protein's backbone fold. An **α-helix** and a **β-pleated sheet** are two types of secondary structure.

- A protein folds so as to maximize the number of stabilizing interactions—namely, disulfide bonds, hydrogen bonds, **electrostatic attractions,** and **hydrophobic interactions.**

- The **tertiary structure** of a protein is the three-dimensional arrangement of all the atoms in the protein.

- The **quaternary structure** of a protein describes the way the peptide chains (subunits) of a protein with more than one peptide chain are arranged with respect to each other.

PROBLEMS

47. Explain why amino acids, unlike most amines and carboxylic acids, are insoluble in diethyl ether.

48. Alanine has pK_a values of 2.34 and 9.69. At what pH will alanine exist in the indicated form?

a.
H_2N ... O^- (50%) $H_3\overset{+}{N}$ ... O^- (50%)
b. $H_3\overset{+}{N}$ ... O^- (100%)
c. $H_3\overset{+}{N}$ ... O^- (50%) $H_3\overset{+}{N}$ ... OH (50%)

49. Show the peptides that would result from cleavage by the indicated reagent:
a. Val-Arg-Gly-Met-Arg-Ala-Ser by carboxypeptidase A
b. Ser-Phe-Lys-Met-Pro-Ser-Ala-Asp by cyanogen bromide
c. Arg-Ser-Pro-Lys-Lys-Ser-Glu-Gly by trypsin

50. Which would have a higher percentage of negative charge at physiological pH (7.4), leucine with pI = 5.98 or asparagine with pI = 5.43?

51. Aspartame has a pI of 5.9. Draw its prevailing form at physiological pH (7.4).

52. Draw the form of aspartate that predominates at the following pH values:
a. pH = 1.0   **b.** pH = 2.6   **c.** pH = 6.0   **d.** pH = 11.0

53. A professor was preparing a manuscript for publication in which she reported that the pI of the tripeptide Lys-Lys-Lys was 10.6. One of her students pointed out that there must be an error in her calculations because the pK_a of the ε-amino group of lysine is 10.8 and the pI of the tripeptide has to be greater than any of its individual pK_a values. Was the student correct?

54. A mixture of amino acids that do not separate sufficiently when a single technique is used can often be separated by two-dimensional chromatography. In this technique, the mixture of amino acids is applied to a piece of filter paper and separated by chromatographic techniques. The paper is then rotated 90°, and the amino acids are further separated by electrophoresis, producing a type of chromatogram called a *fingerprint*. Identify the spots in the fingerprint obtained from a mixture of Ser, Glu, Leu, His, Met, and Thr.

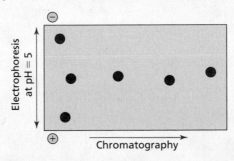

55. Determine the amino acid sequence of a polypeptide from the following data:

Complete hydrolysis of the peptide yields Arg, 2 Gly, Ile, 3 Leu, 2 Lys, 2 Met, 2 Phe, Pro, Ser, 2 Tyr, and Val.

Treatment with Edman's reagent releases PTH-Gly.

Carboxypeptidase A releases Phe.

Treatment with cyanogen bromide yields the following three peptides:
1. Gly-Leu-Tyr-Phe-Lys-Ser-Met
2. Gly-Leu-Tyr-Lys-Val-Ile-Arg-Met
3. Leu-Pro-Phe

Treatment with trypsin yields the following four peptides:
1. Gly-Leu-Tyr-Phe-Lys
2. Ser-Met-Gly-Leu-Tyr-Lys
3. Val-Ile-Arg
4. Met-Leu-Pro-Phe

56. Explain the difference in the pK_a values of the carboxyl groups of alanine, serine, and cysteine.

57. Which would be a more effective buffer at physiological pH, a solution of 0.1 M glycylglycylglycylglycine or a solution of 0.2 M glycine?

58. Identify the location and type of charge on the hexapeptide Lys-Ser-Asp-Cys-His-Tyr at each of the following pH values:
a. pH = 1   **b.** pH = 5   **c.** pH = 7   **d.** pH = 12

59. Draw the product obtained when a lysine side chain in a polypeptide reacts with maleic anhydride.

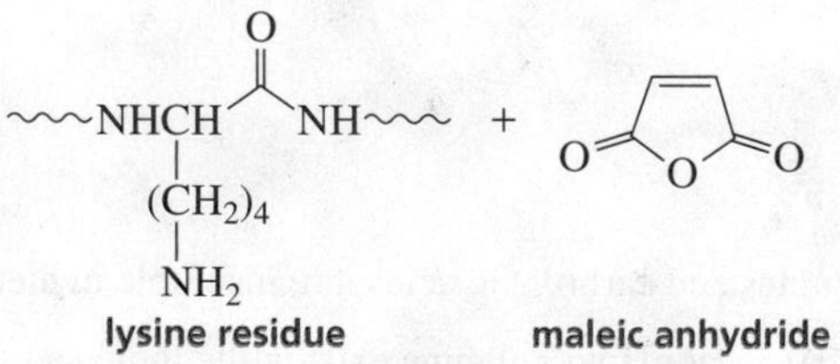

60. After the polypeptide shown here was treated with maleic anhydride, it was hydrolyzed by trypsin. (After a polypeptide is treated with maleic anhydride, trypsin will cleave the polypeptide only on the C-side of arginine.)

Gly-Ala-Asp-Ala-Leu-Pro-Gly-Ile-Leu-Val-Arg-Asp-Val-Gly-Lys-Val-Glu-Val-Phe-Glu-Ala-Gly-
Arg-Ala-Glu-Phe-Lys-Glu-Pro-Arg-Leu-Val-Met-Lys-Val-Glu-Gly-Arg-Pro-Val-Gly-Ala-Gly-Leu-Trp

a. After a polypeptide is treated with maleic anhydride, why does trypsin no longer cleave it on the C-side of lysine?
b. How many fragments are obtained from the polypeptide?
c. In what order would the fragments be eluted from an anion-exchange column using a buffer of pH = 5?

61. Treatment of a polypeptide with 2-mercaptoethanol yields two polypeptides with the following primary structures:

Val-Met-Tyr-Ala-Cys-Ser-Phe-Ala-Glu-Ser
Ser-Cys-Phe-Lys-Cys-Trp-Lys-Tyr-Cys-Phe-Arg-Cys-Ser

Treatment of the original intact polypeptide with chymotrypsin yields the following peptides:
1. Ala, Glu, Ser
2. 2 Phe, 2 Cys, Ser
3. Tyr, Val, Met
4. Arg, Ser, Cys
5. Ser, Phe, 2 Cys, Lys, Ala, Trp
6. Tyr, Lys

Determine the positions of the disulfide bridges in the original polypeptide.

62. Show how aspartame can be synthesized using DCC.

63. α-Amino acids can be prepared by treating an aldehyde with ammonia/trace acid, followed by hydrogen cyanide, followed by acid-catalyzed hydrolysis.
a. Draw the structures of the two intermediates formed in this reaction.
b. What amino acid is formed when the aldehyde that is used is 3-methylbutanal?
c. What aldehyde would be needed to prepare isoleucine?

64. Reaction of a polypeptide with carboxypeptidase A releases Met. The polypeptide undergoes partial hydrolysis to give the following peptides. What is the sequence of the polypeptide?

1. Ser, Lys, Trp	**4.** Leu, Glu, Ser	**7.** Glu, His	**10.** Glu, His, Val
2. Gly, His, Ala	**5.** Met, Ala, Gly	**8.** Leu, Lys, Trp	**11.** Trp, Leu, Glu
3. Glu, Val, Ser	**6.** Ser, Lys, Val	**9.** Lys, Ser	**12.** Ala, Met

65. a. How many different octapeptides can be made from the 20 naturally occurring amino acids?
   b. How many different proteins containing 100 amino acids can be made from the 20 naturally occurring amino acids?

66. Glycine has pK_a values of 2.3 and 9.6. Would you expect the pK_a values of glycylglycine to be higher or lower than these values?

67. A mixture of 15 amino acids gave the fingerprint shown here (see Problem 54). Identify the spots. (*Hint 1:* Pro reacts with ninhydrin to produce a yellow color; Phe and Tyr produce a green color. *Hint 2:* Count the number of spots before you start.)

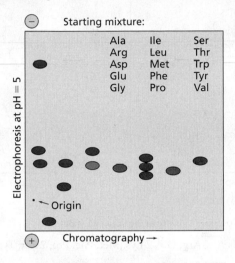

68. Write the mechanism for the reaction of an amino acid with di-*tert*-butyl dicarbonate.

69. Dithiothreitol reacts with disulfide bridges in the same way that 2-mercaptoethanol does. With dithiothreitol, however, the equilibrium lies much more to the right. Explain.

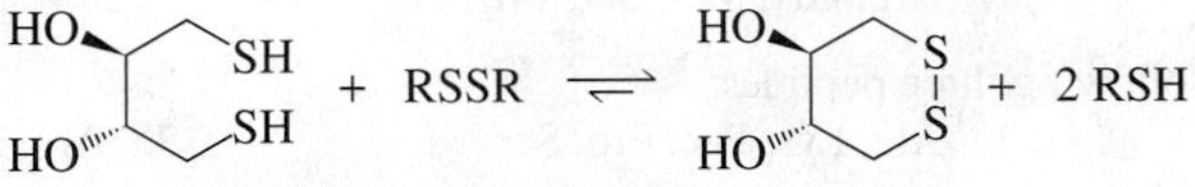

70. Show how valine can be prepared by
   a. a Hell–Volhard–Zelinski reaction.
   b. a Strecker synthesis.
   c. a reductive amination.
   d. a *N*-phthalimidomalonic ester synthesis.
   e. an acetamidomalonic ester synthesis.

71. A chemist wanted to test his hypothesis that the disulfide bridges that form in many proteins do so after the minimum energy conformation of the protein has been achieved. He treated a sample of an enzyme that contained four disulfide bridges, with 2-mercaptoethanol and then added urea to denature the enzyme. He slowly removed these reagents so that the enzyme could re-fold and re-form the disulfide bridges. The enzyme he recovered had 80% of its original activity. What would be the percent activity in the recovered enzyme if disulfide bridge formation were entirely random rather than determined by the tertiary structure? Does this experiment support his hypothesis?

72. Propose a mechanism for the rearrangement of the thiazoline obtained from the reaction of Edman's reagent with a peptide to a PTH-amino acid (Section 13). (*Hint:* Thioesters are susceptible to hydrolysis.)

73. A normal polypeptide and a mutant of the polypeptide were hydrolyzed by an endopeptidase under the same conditions. The normal and mutant polypeptide differ by one amino acid. The fingerprints of the peptides obtained from the two polypeptides are shown here. What kind of amino acid substitution occurred as a result of the mutation? (That is, is the substituted amino acid more or less polar than the original amino acid? Is its pI lower or higher?) (*Hint*: Photocopy the fingerprints, cut them out, and overlay them.)

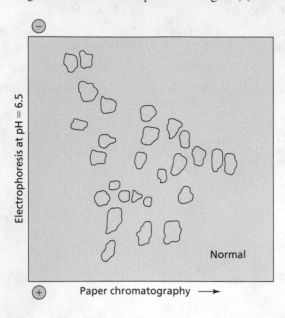
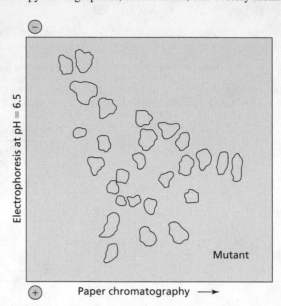

74. Determine the amino acid sequence of a polypeptide from the following data:

Complete hydrolysis of the peptide yields Ala, Arg, Gly, 2 Lys, Met, Phe, Pro, 2 Ser, Tyr, and Val.

Treatment with Edman's reagent releases PTH-Val.

Carboxypeptidase A releases Ala.

Treatment with cyanogen bromide yields the following two peptides:
1. Ala, 2 Lys, Phe, Pro, Ser, Tyr   **2.** Arg, Gly, Met, Ser, Val

Treatment with trypsin yields the following three peptides:
1. Gly, Lys, Met, Tyr   **2.** Ala, Lys, Phe, Pro, Ser   **3.** Arg, Ser, Val

Treatment with chymotrypsin yields the following three peptides:
1. 2 Lys, Phe, Pro   **2.** Arg, Gly, Met, Ser, Tyr, Val   **3.** Ala, Ser

GLOSSARY

amino acid   an α-aminocarboxylic acid. Naturally occurring amino acids have the *l* configuration.

amino acid analyzer   an instrument that automates the ion-exchange separation of amino acids.

anion-exchange resin   a positively charged resin used in ion-exchange chromatography.

automated solid-phase peptide synthesis   an automated technique that synthesizes a peptide while its C-terminal amino acid is attached to a solid support.

cation-exchange resin   a negatively charged resin used in ion-exchange chromatography.

coil conformation (loop conformation)   that part of a protein that is highly ordered, but not in an α-helix or a β-pleated sheet.

C-terminal amino acid   the terminal amino acid of a peptide (or protein) that has a free carboxyl group.

denaturation   destruction of the highly organized tertiary structure of a protein.

dipeptide   two amino acids linked by an amide bond.

disulfide bridge   a disulfide (—S—S—) bond in a peptide or protein.

Edman's reagent   phenyl isothiocyanate. A reagent used to determine the N-terminal amino acid of a polypeptide.

electrophoresis   a technique that separates amino acids on the basis of their pI values.

endopeptidase   an enzyme that hydrolyzes a peptide bond that is not at the end of a peptide chain.

essential amino acid   an amino acid that humans must obtain from their diet because they cannot synthesize it at all or cannot synthesize it in adequate amounts.

exopeptidase   an enzyme that hydrolyzes a peptide bond at the end of a peptide chain.

fibrous protein   a water-insoluble protein in which the polypeptide chains are arranged in bundles.

globular protein   a water-soluble protein that tends to have a roughly spherical shape.

α-helix   the backbone of a polypeptide coiled in a right-handed spiral with hydrogen bonding occurring within the helix.

hydrophobic interactions   interactions between nonpolar groups. These interactions increase stability by decreasing the amount of structured water (increasing entropy).

interchain disulfide bridge   a disulfide bridge between two cysteine residues in different peptide chains.

intrachain disulfide bridge a disulfide bridge between two cysteine residues in the same peptide chain.

ion-exchange chromatography a technique that uses a column packed with an insoluble resin to separate compounds on the basis of their charges and polarities.

isoelectric point (pI) the pH at which there is no net charge on an amino acid.

kinetic resolution separation of enantiomers on the basis of the difference in their rate of reaction with an enzyme.

N-terminal amino acid the terminal amino acid of a peptide (or protein) that has a free amino group.

oligopeptide 3 to 10 amino acids linked by amide bonds.

partial hydrolysis a technique that hydrolyzes only some of the peptide bonds in a polypeptide.

peptide polymer of amino acids linked together by amide bonds. A peptide contains fewer amino acid residues than a protein does.

peptide bond the amide bond that links the amino acids in a peptide or protein.

β-pleated sheet the backbone of a polypeptide that is extended in a zigzag structure with hydrogen bonding between neighboring chains.

polypeptide many amino acids linked by amide bonds.

primary structure (of a protein) the sequence of amino acids in a protein.

protein a polymer containing 40 to 4000 amino acids linked by amide bonds.

quaternary structure a description of the way the individual polypeptide chains of a protein are arranged with respect to each other.

secondary structure a description of the conformation of the backbone of a protein.

subunit an individual chain of an oligomer.

tertiary structure a description of the three-dimensional arrangement of all the atoms in a protein.

thin-layer chromatography a technique that separates compounds on the basis of their polarity.

tripeptide three amino acids linked by amide bonds.

zwitterion a compound with a negative charge and a positive charge on nonadjacent atoms.

PHOTO CREDITS

Photo credits are listed in order of appearance.

Fancy/Alamy; Kateryna Larina/Shutterstock; by-studio/Fotolia

Catalysis in Organic Reactions and in Enzymatic Reactions

lysozyme

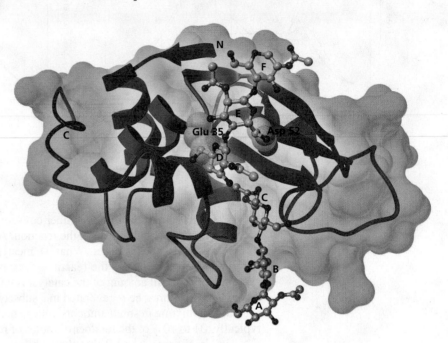

Lysozyme is an enzyme that catalyzes the hydrolysis of bacterial cell walls. C indicates its C-terminal end, N indicates its N-terminal end, and the cell wall is indicated by A–F. Lysozyme's catalytic groups (Glu 35 and Asp 52) are also shown. (See Section 10.)

A **catalyst** is a substance that increases the rate of a chemical reaction without itself being consumed or changed in the reaction. In this chapter we will look at the types of catalysts used in organic reactions and the ways in which they provide an energetically more favorable pathway for a reaction. We will then see how the same catalysts are used in reactions that take place in living systems—that is, in enzyme-catalyzed reactions. We will also see why enzymes are extraordinarily good catalysts—they can increase the rate of an intermolecular reaction by as much as 10^{16}. In contrast, rate enhancements achieved by nonbiological catalysts in intermolecular reactions are seldom greater than 10,000-fold.

The rate of a chemical reaction depends on the energy barrier of the rate-determining step that must be overcome in the process of converting reactants into products. The height of the "energy hill" is indicated by the free energy of activation ($\Delta G^{\ddagger}$). A catalyst increases the rate of a chemical reaction by providing a pathway with a lower $\Delta G^{\ddagger}$. A catalyst can decrease $\Delta G^{\ddagger}$ in one of three ways:

1. The catalyzed and uncatalyzed reactions can have different, but similar, mechanisms, with the catalyst providing a way to make the *reactant more reactive (less stable)* (Figure 1a).

2. The catalyzed and uncatalyzed reactions can have different, but similar, mechanisms, with the catalyst providing a way to make *the transition state more stable* (Figure 1b).

3. The catalyst can completely *change the mechanism* of the reaction, providing an alternative pathway with a smaller $\Delta G^{\ddagger}$ than that for the uncatalyzed reaction (Figure 2).

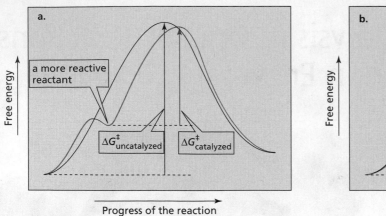
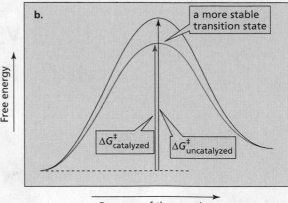

▲ **Figure 1**

Reaction coordinate diagrams for an uncatalyzed reaction (black) and for a catalyzed reaction (green).

(a) The catalyst converts the reactant to a more reactive species.

(b) The catalyst stabilizes the transition state.

When we say that a catalyst is neither consumed nor changed by a reaction, we do not mean that it does not participate in the reaction. A catalyst *must* participate in the reaction if it is going to make it go faster. What we mean is that a catalyst has the same form after the reaction as it had before the reaction. Because the catalyst is not used up during the reaction, only a small amount of the catalyst is needed. (If a catalyst is used up in one step of the reaction, it must be regenerated in a subsequent step.) Therefore, a catalyst is added to a reaction mixture in small amounts, much smaller than the number of moles of reactant (typically .01 to 10% of the number of moles of reactant). We call this a *catalytic* amount.

Notice in Figures 1 and 2 that the stability of the original reactants and final products is the same in both the catalyzed and corresponding uncatalyzed reactions. In other words, the catalyst does not change the equilibrium constant of the reaction. Because the catalyst does not change the equilibrium constant, it does not change the *amount* of product formed when the reaction has reached equilibrium. It changes only the *rate* at which the product is formed.

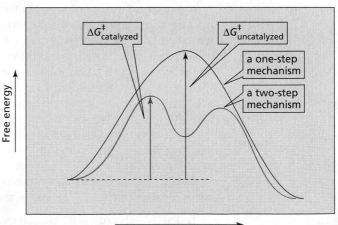

▶ **Figure 2**

Reaction coordinate diagrams for an uncatalyzed reaction (black) and for a catalyzed reaction (green). The catalyzed reaction takes place by an alternative and energetically more favorable pathway.

PROBLEM 1♦

Which of the following parameters would be different for a reaction carried out in the presence of a catalyst, compared with the same reaction carried out in the absence of a catalyst?

$$\Delta G°, \Delta H^{\ddagger}, E_a, \Delta S^{\ddagger}, \Delta H°, K_{eq}, \Delta G^{\ddagger}, \Delta S°, k_{rate}$$

1   CATALYSIS IN ORGANIC REACTIONS

There are several ways a catalyst can provide a more favorable pathway for an organic reaction:

- It can increase the reactivity of an electrophile so that it is more susceptible to reaction with a nucleophile.

- It can increase the reactivity of a nucleophile.

- It can increase the leaving propensity of a group by converting it into a weaker base.

- It can increase the stability of a transition state.

Now we will look at some of the most common catalysts—namely, acid catalysts, base catalysts, nucleophilic catalysts, and metal-ion catalysts—and the ways in which they catalyze the reactions of organic compounds.

2   ACID CATALYSIS

An **acid catalyst** increases the rate of a reaction by donating a proton to a reactant. There are many examples of acid catalysis. For example, an acid provides the electrophile needed for the addition of water or an alcohol to an alkene. Also, an alcohol cannot undergo substitution and elimination reactions unless an acid is present to protonate the OH group, which increases its leaving propensity by making it a weaker base.

To examine some of the important ways an acid can catalyze a reaction, let's look at the mechanism for the acid-catalyzed hydrolysis of an ester. The reaction has two slow steps: formation of the tetrahedral intermediate and collapse of the tetrahedral intermediate. Donation of a proton to, and removal of a proton from, an electronegative atom such as oxygen are always fast steps.

MECHANISM FOR ACID-CATALYZED ESTER HYDROLYSIS

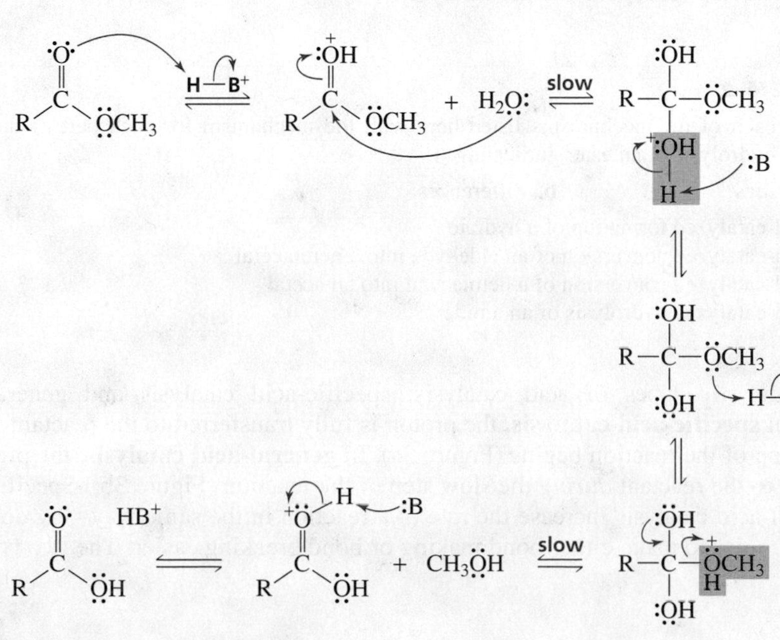

In an acid-catalyzed reaction, a proton is donated to the reactant.

A catalyst must increase the rate of a slow step because increasing the rate of a fast step will not increase the rate of the overall reaction. The acid increases the rates of both slow steps of the hydrolysis reaction. It increases the rate of formation of the tetrahedral intermediate by protonating the carbonyl oxygen, thereby making it more susceptible to nucleophilic addition than an unprotonated carbonyl group would be. Increasing the reactivity of the carbonyl group by protonating it is an example of providing a way to convert the reactant into a more reactive species (Figure 1a).

> A catalyst must increase the rate of a slow step. Increasing the rate of a fast step will not increase the rate of the overall reaction.

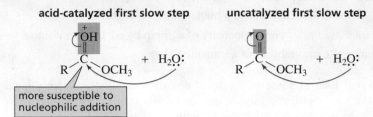

The acid increases the rate of the second slow step by decreasing the basicity—and thereby increasing the leaving propensity—of the group that is eliminated when the tetrahedral intermediate collapses. In the presence of an acid, methanol is eliminated; in the absence of an acid, methoxide ion is eliminated. Methanol is a much weaker base than methoxide ion, so methanol is much more easily eliminated.

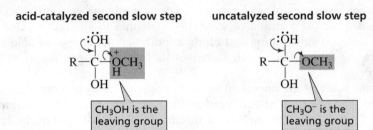

The mechanism for the acid-catalyzed hydrolysis of an ester shows that the reaction can be divided into two distinct parts: formation of a tetrahedral intermediate and collapse of a tetrahedral intermediate. There are three steps in each part. Notice that in each part, the first step is a fast protonation step, the second step is a slow acid-catalyzed step that involves either breaking a π bond or forming a π bond, and the last step is a fast deprotonation step (to regenerate the catalyst).

PROBLEM 2

Compare each of the mechanisms listed here with the mechanism for each part of the acid-catalyzed hydrolysis of an ester, indicating

a. similarities.   **b.** differences.

    1. acid-catalyzed formation of a hydrate
    2. acid-catalyzed conversion of an aldehyde into a hemiacetal
    3. acid-catalyzed conversion of a hemiacetal into an acetal
    4. acid-catalyzed hydrolysis of an amide

There are two types of acid catalysis: specific-acid catalysis and general-acid catalysis. In **specific-acid catalysis,** the proton is fully transferred to the reactant *before* the slow step of the reaction begins (Figure 3a). In **general-acid catalysis,** the proton is transferred to the reactant *during* the slow step of the reaction (Figure 3b). Specific-acid and general-acid catalysis increase the rate of a reaction in the same way—by donating a proton in order to make either bond making or bond breaking easier. The two types of acid catalysis differ only in the extent to which the proton is transferred in the transition state of the slow step of the reaction.

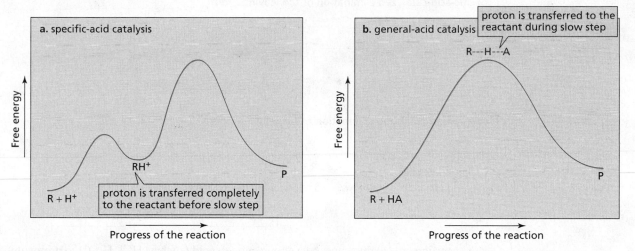

▲ Figure 3

(a) Reaction coordinate diagram for a specific-acid-catalyzed reaction. The proton is transferred completely to the reactant before the slow step of the reaction begins (R = reactant; P = product).

(b) Reaction coordinate diagram for a general-acid-catalyzed reaction. The proton is transferred to the reactant during the slow step of the reaction.

In the examples that follow, notice the difference in the extent to which the proton has been transferred when the nucleophile adds to the reactant.

- In the specific-acid-catalyzed addition of water to a carbonyl group, the nucleophile adds to a fully protonated carbonyl group. In the general-acid-catalyzed addition of water to a carbonyl group, the carbonyl group becomes protonated as the nucleophile adds to it.

specific-acid-catalyzed addition of water

general-acid-catalyzed addition of water

The proton is donated to the reactant *before* the slow step in a specific-acid-catalyzed reaction, and *during* the slow step in a general-acid-catalyzed reaction.

- In the specific-acid-catalyzed collapse of a tetrahedral intermediate, a fully protonated leaving group is eliminated, whereas in the general-acid-catalyzed collapse of a tetrahedral intermediate, the leaving group picks up a proton as the group is eliminated.

specific-acid-catalyzed elimination of the leaving group

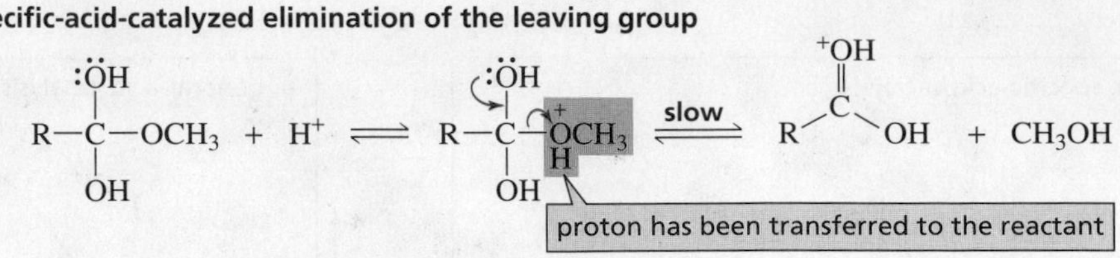

proton has been transferred to the reactant

general-acid-catalyzed elimination of the leaving group

proton is being transferred to the reactant

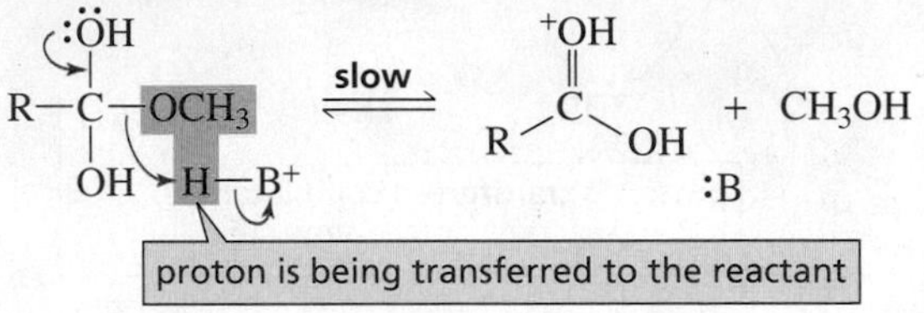

A specific-acid catalyst must be a strong enough acid (such as HCl, H_3O^+) to protonate the reactant fully before the slow step begins. A general-acid catalyst can be a weaker acid because it has only a partially transferred proton in the transition state of the slow step.

All the mechanisms are written as specific-acid-catalyzed reactions. Weaker acids could have been used to catalyze many of the reactions, in which case the protonation step and the subsequent slow step would have been shown as a single step. A list of acids and their pK_a values is given in the Appendix containing the table "pK_a Values."

PROBLEM 3◆

Are the slow steps for the acid-catalyzed hydrolysis of an ester earlier in this chapter general-acid catalyzed or specific-acid catalyzed?

PROBLEM 4

a. Draw the mechanism for the following reaction if it involves specific-acid catalysis.
b. Draw the mechanism if it involves general-acid catalysis.

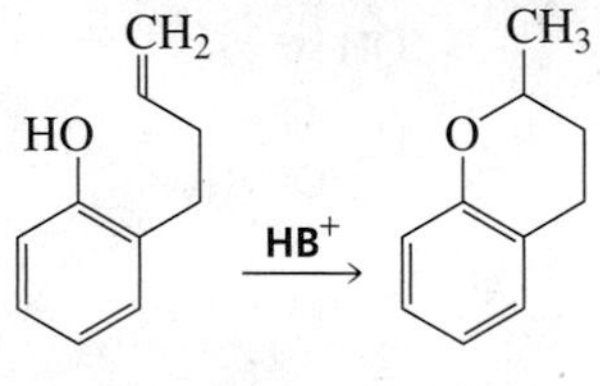

PROBLEM 5 Solved

An alcohol will not react with aziridine unless an acid is present. Why is the acid necessary?

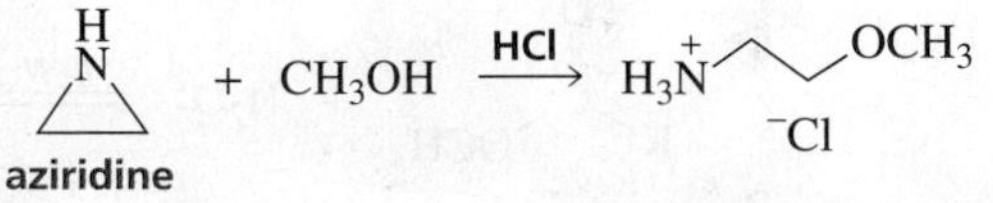

Solution Although relief of ring strain is sufficient by itself to cause an epoxide to undergo a ring-opening reaction, it is not sufficient to cause an aziridine to undergo a ring-opening reaction. A negatively charged nitrogen is a stronger base, and therefore a poorer leaving group, than a negatively charged oxygen. An acid, therefore, is needed to protonate the ring nitrogen to make it a better leaving group.

3 BASE CATALYSIS

There are several base-catalyzed reactions, such as the interconversion of keto and enol tautomers, the Claisen condensation, and the enediol rearrangement. A **base catalyst** increases the rate of a reaction by removing a proton from the reactant. For example, the dehydration of a hydrate in the presence of hydroxide ion is a base-catalyzed reaction. Hydroxide ion (the base) increases the rate of the reaction by removing a proton from the neutral hydrate.

specific-base-catalyzed dehydration

a hydrate

In a base-catalyzed reaction, a proton is removed from the reactant.

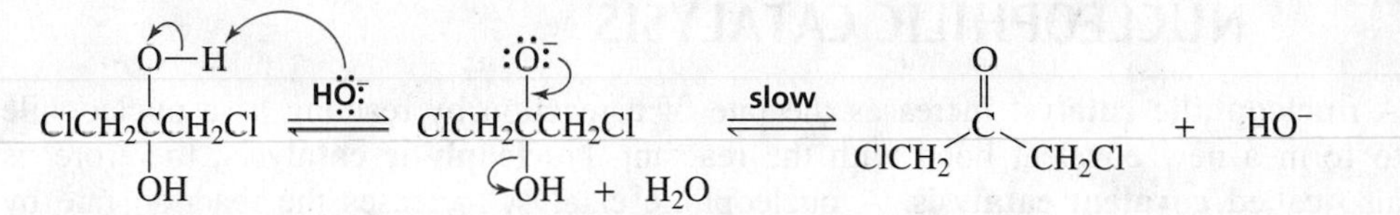

Removing a proton from the hydrate increases the rate of dehydration by providing a pathway with a more sta4ble transition state. The transition state for the elimination of HO⁻ from a negatively charged tetrahedral intermediate is more stable because a positive charge does not develop on the electronegative oxygen atom, as it does in the transition state for the elimination of HO⁻ from a neutral tetrahedral intermediate.

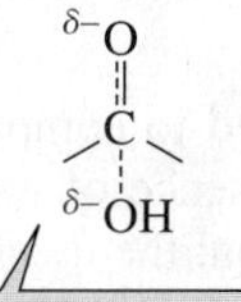

transition state for elimination of HO⁻ from a negatively charged tetrahedral intermediate

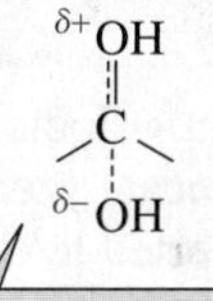

transition state for elimination of HO⁻ from a neutral tetrahedral intermediate

The foregoing base-catalyzed dehydration of a hydrate is an example of specific-base catalysis. In **specific-base catalysis,** the proton is completely removed from the reactant *before* the slow step of the reaction begins. In **general-base catalysis,** on the other hand, the proton is removed from the reactant *during* the slow step of the reaction. Compare the extent of proton transfer in the slow step of the preceding specific-base-catalyzed dehydration with the extent of proton transfer in the slow step of the following general-base-catalyzed dehydration:

general-base-catalyzed dehydration

a hydrate

The proton is removed from the reactant *before* the slow step in a specific-base-catalyzed reaction and *during* the slow step in a general-base-catalyzed reaction.

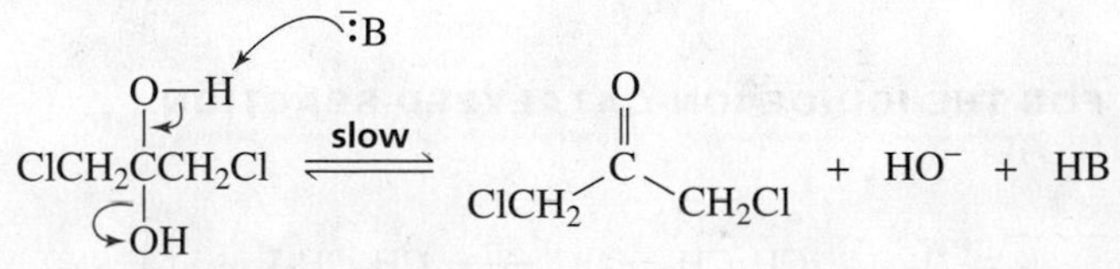

In specific-base catalysis, the base has to be strong enough to remove a proton from the reactant completely before the slow step begins. In general-base catalysis, the base can be weaker because the proton is only partially transferred to the base in the transition state of the slow step.

We will see that enzymes catalyze reactions using general-acid and general-base catalysis because at physiological pH (7.4), the concentration of $H^+(\sim 1 \times 10^{-7} M)$ is too small for specific-acid catalysis and the concentration of HO⁻ is too small for specific-base catalysis.

PROBLEM 6

a. Draw the mechanism for the following reaction if it involves specific-base catalysis.
b. Draw the mechanism if it involves general-base catalysis.

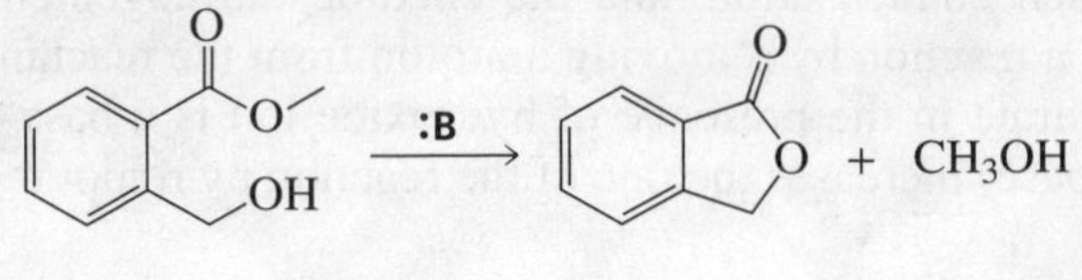

4 NUCLEOPHILIC CATALYSIS

A **nucleophilic catalyst** increases the rate of a reaction by reacting as a nucleophile to form a new covalent bond with the reactant. **Nucleophilic catalysis,** therefore, is also called **covalent catalysis.** A nucleophilic catalyst increases the reaction rate by completely changing the mechanism of the reaction.

In the following reaction, iodide ion increases the rate of conversion of ethyl chloride to ethyl alcohol by acting as a nucleophilic catalyst:

A nucleophilic catalyst forms a covalent bond with the reactant.

a nucleophilic catalyst

$$CH_3CH_2Cl + HO^- \xrightarrow[H_2O]{I^-} CH_3CH_2OH + Cl^-$$

To understand how iodide ion catalyzes this reaction, we need to compare the mechanisms for the uncatalyzed and catalyzed reactions. In the absence of iodide ion, ethyl chloride is converted to ethyl alcohol in a one-step S_N2 reaction: the nucleophile is HO^- and the leaving group is Cl^-.

MECHANISM FOR THE UNCATALYZED REACTION

$$HO^- + CH_3CH_2{-}Cl \longrightarrow CH_3CH_2OH + Cl^-$$

If iodide ion is present in the reaction mixture, then the reaction takes place by two successive S_N2 reactions.

MECHANISM FOR THE IODIDE-ION-CATALYZED REACTION

I⁻ is a better nucleophile than HO⁻

$$:I^- + CH_3CH_2{-}Cl \longrightarrow CH_3CH_2I + Cl^-$$

I⁻ is a better leaving group than Cl⁻

$$HO^- + CH_3CH_2{-}I \longrightarrow CH_3CH_2OH + I^-$$

- The first S_N2 reaction in the catalyzed reaction is faster than the uncatalyzed S_N2 reaction because in a protic solvent, iodide ion is a better nucleophile than hydroxide ion, the nucleophile in the uncatalyzed reaction.

- The second S_N2 reaction in the catalyzed reaction is also faster than the uncatalyzed S_N2 reaction because iodide ion is a weaker base and therefore a better leaving group than chloride ion, the leaving group in the uncatalyzed reaction.

Thus, iodide ion increases the rate of formation of ethanol by changing a relatively slow S_N2 reaction to one that involves two relatively fast S_N2 reactions (Figure 2).

Iodide ion is a nucleophilic catalyst because it reacts as a nucleophile, forming a covalent bond with the reactant. The iodide ion consumed in the first reaction is regenerated in the second, so it comes out of the reaction unchanged.

Another reaction in which a nucleophilic catalyst provides a more favorable pathway by changing the mechanism of the reaction is the imidazole-catalyzed hydrolysis of an ester.

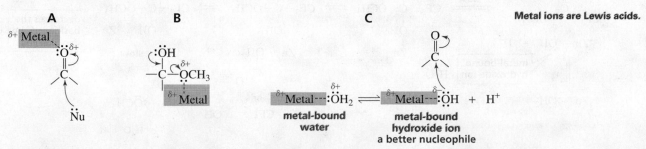

Imidazole is a better nucleophile than water, so imidazole reacts faster with the ester than water would. The acyl imidazole that is formed is particularly reactive because the positively charged nitrogen makes imidazole a very good leaving group. Therefore, it is hydrolyzed much more rapidly than the ester would have been. Because formation of the acyl imidazole and its subsequent hydrolysis are both faster than ester hydrolysis, imidazole increases the rate of hydrolysis of the ester.

5 METAL-ION CATALYSIS

Metal ions exert their catalytic effect by complexing with atoms that have lone-pair electrons. **Metal ions,** therefore, are Lewis acids. A *metal ion* can increase the rate of a reaction in the following ways:

- It can make a reaction center more susceptible to receiving electrons (that is, more electrophilic) as in A.

- It can make a leaving group a weaker base, and therefore a better leaving group, as in B.

- It can increase the rate of a hydrolysis reaction by increasing the nucleophilicity of water, as in C.

Metal ions are Lewis acids.

In A and B, the metal ion exerts the same kind of catalytic effect as a proton does. However, a metal ion can be a much more effective catalyst than a proton, because metal ions can have a charge greater than +1, and a high concentration of a metal ion can be made available at neutral pH.

In C, metal-ion complexation increases water's nucleophilicity by converting it to metal-bound hydroxide ion. That is, the metal-ion increases water's tendency to lose a proton as shown by the pK_a values listed in Table 1. (The pK_a of water is 15.7.) Metal-bound hydroxide ion, while not as good a nucleophile as hydroxide ion, is a better nucleophile than water. Metal ions are important catalysts in living systems because hydroxide ion itself would not be available at physiological pH (7.4).

Table 1	The pK_a of Metal-Bound Water		
M^{2+}	**pK_a**	**M^{2+}**	**pK_a**
Ca^{2+}	12.7	Co^{2+}	8.9
Cd^{2+}	11.6	Zn^{2+}	8.7
Mg^{2+}	11.4	Fe^{2+}	7.2
Mn^{2+}	10.6	Cu^{2+}	6.8
Ni^{2+}	9.4	Be^{2+}	5.7

Now we will look at some examples of metal-ion-catalyzed organic reactions. The decarboxylation of dimethyloxaloacetate can be catalyzed by either Cu^{2+} or Al^{3+}.

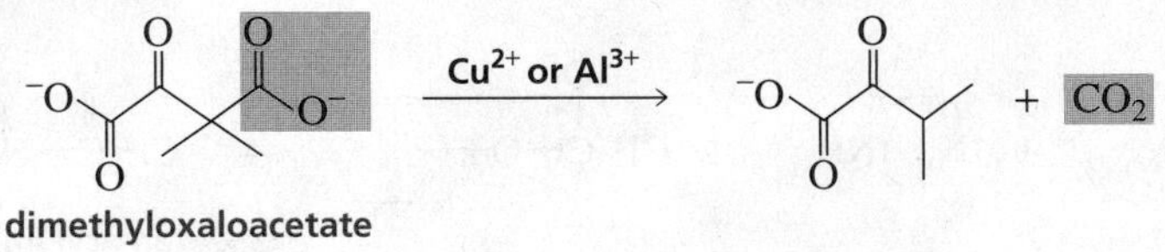

dimethyloxaloacetate

In this reaction, the metal ion complexes with two oxygen atoms of the reactant. Complexation increases the rate of decarboxylation by making the carbonyl oxygen more susceptible to receiving the electrons left behind when CO_2 is eliminated.

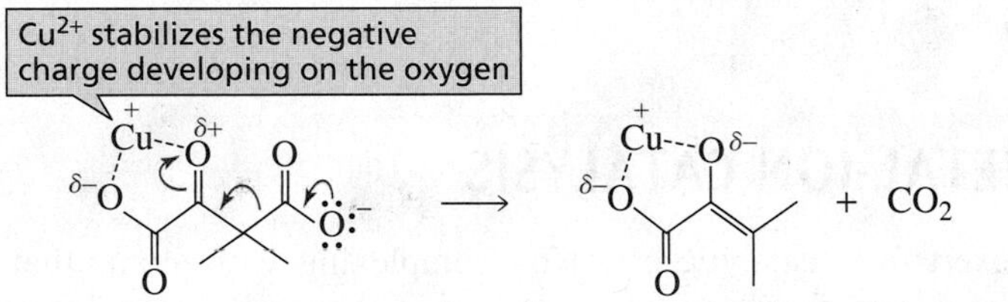

Cu^{2+} stabilizes the negative charge developing on the oxygen

The hydrolysis of the ester shown next has two slow steps. Zn^{2+} increases the rate of the first slow step by providing metal-bound hydroxide ion, a better nucleophile than water. Zn^{2+} increases the rate of the second slow step by decreasing the basicity of the group that is eliminated from the tetrahedral intermediate.

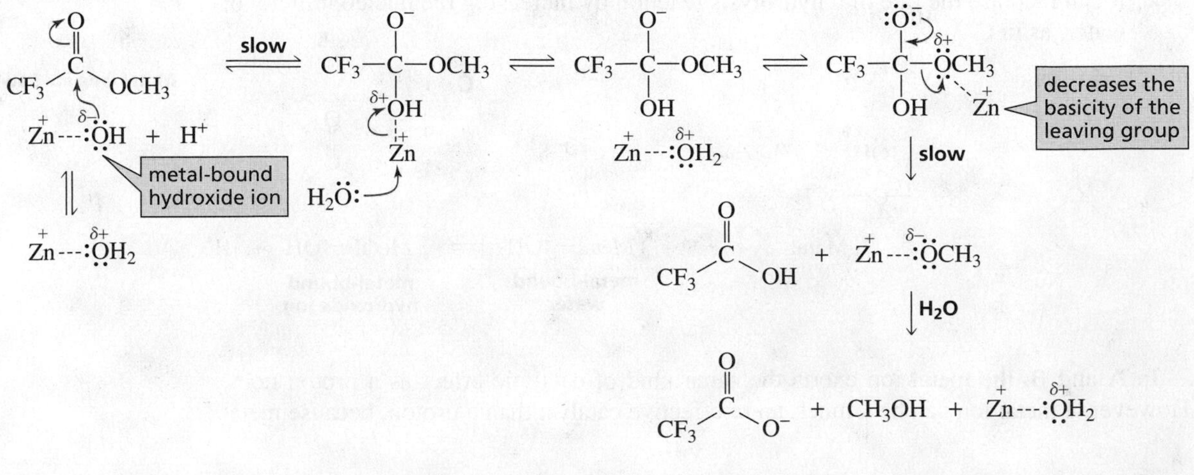

metal-bound hydroxide ion

decreases the basicity of the leaving group

Although metal ions increase the rate of decarboxylation of dimethyloxaloacetate, they have no effect on the rate of decarboxylation of either the monoethyl ester of dimethyloxaloacetate or acetoacetate. Explain why this is so.

dimethyloxaloacetate   **monoethyl ester of**   **acetoacetate**
dimethyloxaloacetate

Propose a mechanism for the Co^{2+}-catalyzed hydrolysis of glycinamide.

6   INTRAMOLECULAR REACTIONS

The rate of a chemical reaction is determined by the number of molecular collisions with sufficient energy *and* with the proper orientation in a given period of time:

$$\text{rate of reaction} = \frac{\text{number of collisions}}{\text{unit of time}} \times \frac{\text{fraction with}}{\text{sufficient energy}} \times \frac{\text{fraction with}}{\text{proper orientation}}$$

Because a catalyst decreases the energy barrier of a reaction, it increases the fraction of collisions that occur with sufficient energy to overcome the barrier.

The rate of a reaction can also be increased by increasing the frequency of the collisions, which can be achieved by increasing the concentration of the reactants. In addition, an *intramolecular reaction* that forms a five- or a six-membered ring occurs more readily than the analogous *intermolecular reaction*. This is because an intramolecular reaction has the advantage of the reacting groups being tied together in the same molecule, which gives them a better chance of finding each other than if they were in two different molecules in a solution of the same concentration. As a result, the frequency of the collisions increases.

If, in addition to being in the same molecule, the reacting groups are arranged in a way that increases the probability that they will collide with each other in the proper orientation, then the rate of the reaction will be further increased. The relative rates shown in Table 2 demonstrate the enormous increase in the rate of a reaction when the reacting groups are properly oriented.

Rate constants for a series of reactions are generally compared in terms of relative rates because relative rates allow us to see immediately how much faster one reaction is than another. **Relative rates** are obtained by dividing the rate constant for each of the reactions by the rate constant for the slowest reaction in the series. The slowest reaction in Table 2 is an *intermolecular reaction;* all the others are *intramolecular reactions.*

Because an intramolecular reaction is a first-order reaction (with units of time^{-1}) and an intermolecular reaction is a second-order reaction (with units of M^{-1} time^{-1}), the relative rates in Table 2 have units of molarity (M).

$$\text{relative rate} = \frac{\text{first-order rate constant}}{\text{second-order rate constant}} \times \frac{\text{time}^{-1}}{\text{time}^{-1}M^{-1}} = M$$

Relative rates are also called *effective molarities.* **Effective molarity** is the concentration of the reactant that would be required in an *intermolecular* reaction for it to have the same rate as the *intramolecular* reaction. In other words, the effective

Table 2 Relative Rates of an Intermolecular Reaction and Five Intramolecular Reactions

Reaction	Relative rate
A	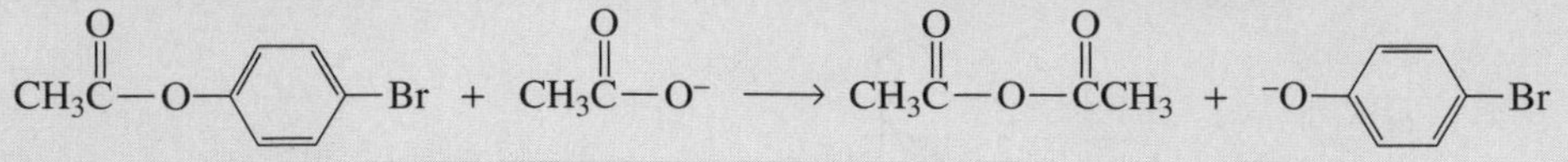 1.0
B	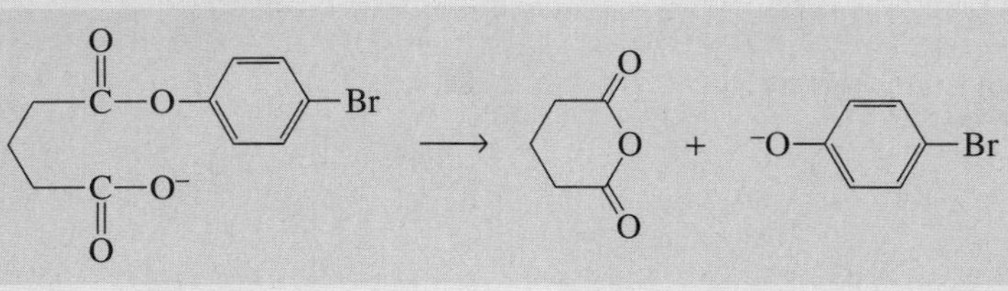 1×10^3 M
C	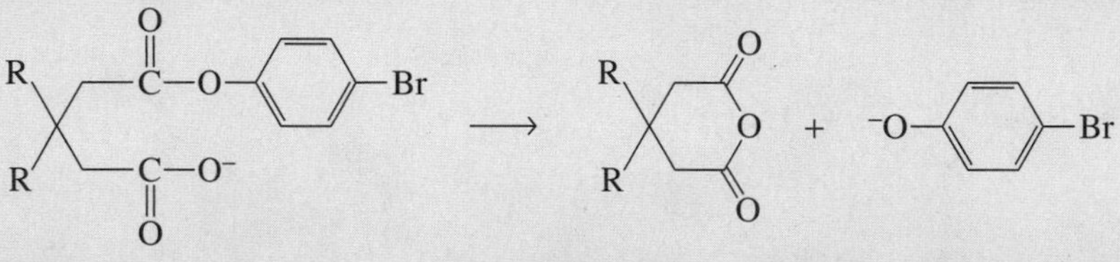 2.3×10^4 M R = CH_3 1.3×10^6 M R = $(CH_3)_2CH$
D	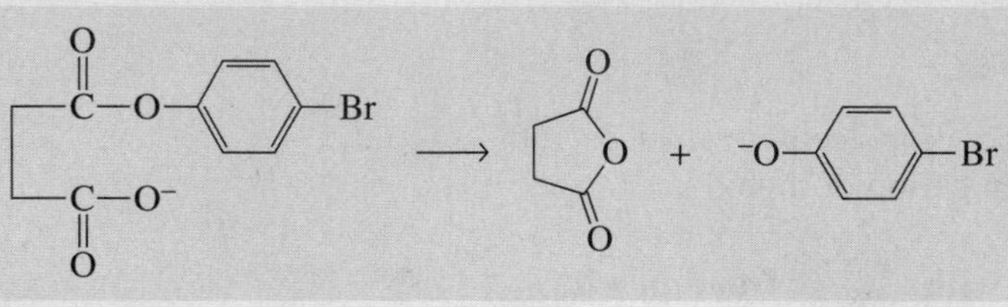 2.2×10^5 M
E	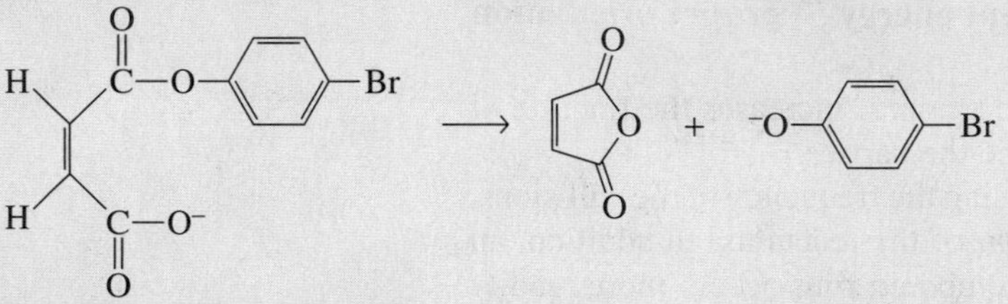 1×10^7 M
F	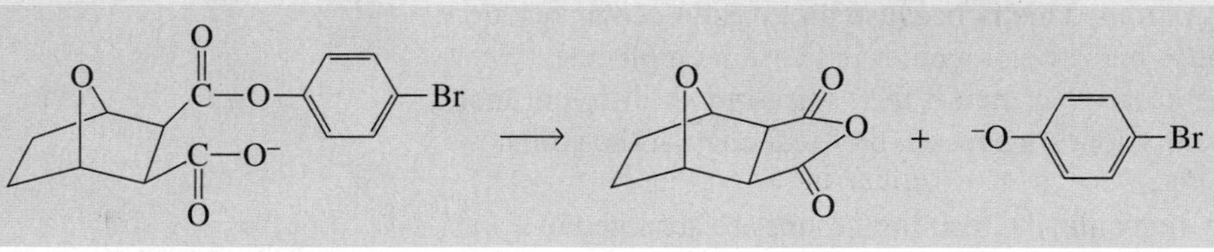 5×10^7 M

molarity is the advantage given to a reaction by having the reacting groups in the same molecule. In some cases, juxtaposing the reacting groups provides such an enormous increase in rate that the effective molarity is greater than the concentration of the reactant in its solid state!

Reaction **A**, the first reaction in Table 2, is an intermolecular reaction between an ester and a carboxylate ion. The second reaction, **B**, has the same two reacting groups in a single molecule. The rate of the intramolecular reaction is 1000 times faster than the rate of the intermolecular reaction.

The reactant in **B** has four C—C bonds that are free to rotate, whereas the reactant in **D** has only three such bonds. Conformers in which the large groups are rotated away from each other are more stable. However, when these groups are pointed away from each other, they are in an unfavorable conformation for reaction. Because the reactant in **D** has fewer bonds that are free to rotate, the groups are more apt to be in a conformation that is favorable for a reaction. Therefore, reaction **D** is faster than reaction **B**.

the reacting groups are in a favorable conformation for a reaction

four carbon–carbon bonds can rotate

three carbon–carbon bonds can rotate

Reaction **C** is faster than reaction **B** because the alkyl substituents of the reactant in **C** decrease the available space for the reactive groups to rotate away from each other. Thus, there is a greater probability that the molecule will be in a conformation that has the reacting groups positioned for ring closure. This is called the *gem-dialkyl effect* because the two alkyl substituents are bonded to the same (geminal) carbon. Comparing the rate when the substituents are methyl groups with the rate when the substituents are isopropyl groups, we see that the rate is further increased when the size of the alkyl groups is increased.

The increased rate of reaction of **E** is due to the double bond that prevents the reacting groups from rotating away from each other. The bicyclic compound in **F** reacts even faster because the reacting groups are locked in the proper orientation for reaction.

PROBLEM 9♦

The relative rate of reaction of the cis alkene (**E**) is given in Table 2. What would you expect the relative rate of reaction of the trans isomer to be?

7   INTRAMOLECULAR CATALYSIS

Just as having two reacting groups in the same molecule increases the rate of a reaction compared with having the groups in separate molecules, having a *reacting group* and a *catalyst* in the same molecule increases the rate of a reaction compared with having them in separate molecules. When a catalyst is part of the reacting molecule, the catalysis is called **intramolecular catalysis.** Intramolecular general-acid or general-base catalysis, intramolecular nucleophilic catalysis, and intramolecular metal-ion catalysis are all possible.

When chlorocyclohexane reacts with an aqueous solution of ethanol, an alcohol and an ether are formed. Two products are formed because there are two nucleophiles (H_2O and CH_3CH_2OH) in the solution.

A 2-thio-substituted chlorocyclohexane undergoes the same reaction. However, the rate of the reaction depends on whether the thio substituent is cis or trans to the chloro substituent. If it is trans, then the 2-thio-substituted compound reacts about 70,000 times faster than the unsubstituted compound. But if it is cis, the 2-thio-substituted compound reacts at about the same rate as the unsubstituted compound.

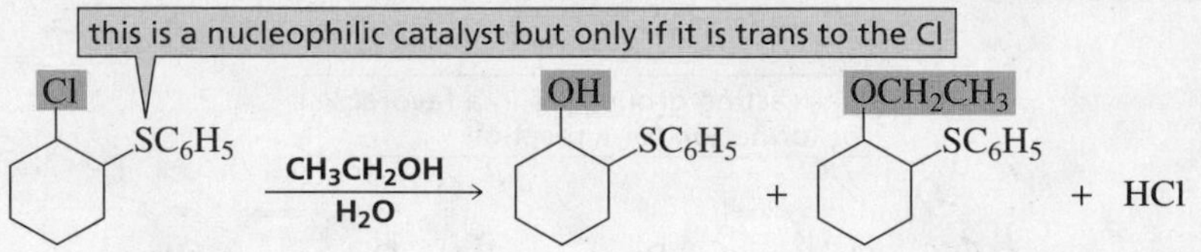

What accounts for the much faster reaction of the trans-substituted compound? When the thio substituent is trans to the chlorine, the substituent can be an intramolecular nucleophilic catalyst—it displaces the chloro substituent by attacking the back side of the carbon to which the chloro substituent is attached (an S_N2 reaction). Back-side attack requires both substituents to be in axial positions, and only the trans isomer can have both of its substituents in axial positions. Subsequent attack by water or ethanol on the sulfonium ion is rapid because breaking the three-membered ring releases strain and the positively charged sulfur is an excellent leaving group.

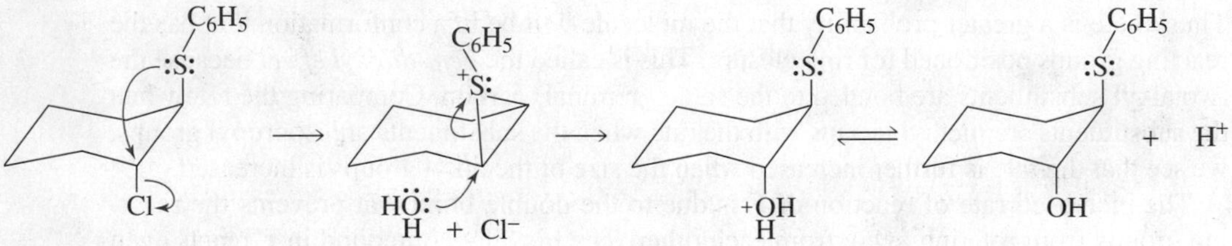

PROBLEM 10♦

Show all the products, including their configurations, that would be obtained from the reaction illustrated in the preceding diagram.

The rate of hydrolysis of phenyl acetate is increased about 150-fold at neutral pH by putting a carboxylate ion in the ortho position. The *ortho*-carboxyl-substituted ester is commonly known as aspirin. In the following reactions, each reactant and product is shown in the form that predominates at physiological pH (7.4).

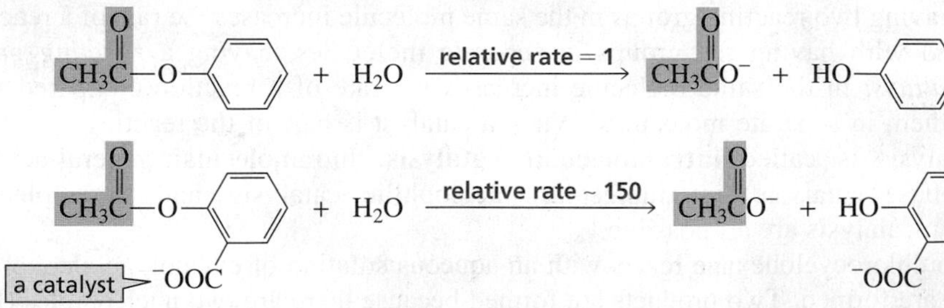

The *ortho*-carboxylate group is an intramolecular general-base catalyst that increases the nucleophilicity of water, thereby increasing the rate of formation of the tetrahedral intermediate.

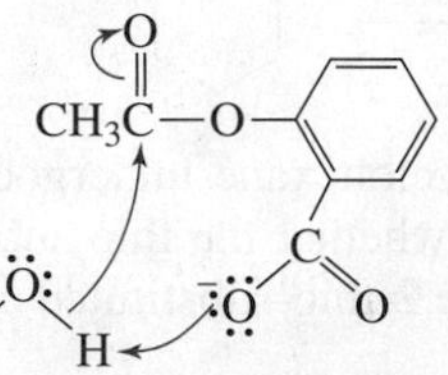

If there are nitro groups on the benzene ring, the *ortho*-carboxyl substituent acts as an intramolecular *nucleophilic catalyst* instead of an intramolecular *general-base catalyst*. In this case, the carboxyl group increases the rate of hydrolysis by converting the ester to an anhydride, which is more rapidly hydrolyzed than an ester.

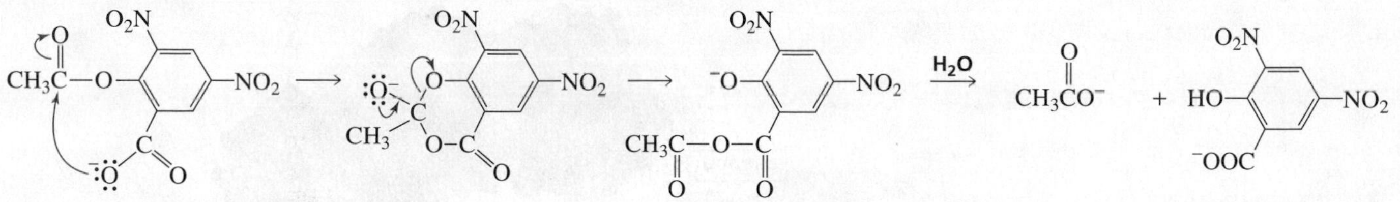

an ester **an anhydride**

PROBLEM 11 Solved

Why is the *ortho*-carboxyl substituent a general-base catalyst in one reaction and a nucleophilic catalyst in another?

Solution Because of its location, the *ortho*-carboxyl substituent will form a tetrahedral intermediate. If the tetrahedral intermediate's carboxyl group is a better leaving group than its phenoxy group, then the carboxyl group will be eliminated preferentially (blue arrow). This will re-form the starting material (path **A**), which then will be hydrolyzed by a general-base-catalyzed mechanism (seen earlier in this chapter). However, if the phenoxy group is a better leaving group than the carboxyl group, the phenoxy group will be eliminated (green arrow), thereby forming an anhydride (path **B**), and the reaction will have occurred via nucleophilic catalysis.

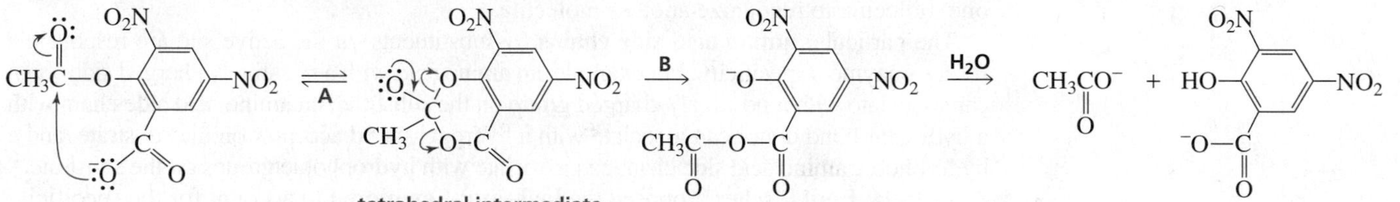

tetrahedral intermediate

PROBLEM 12◆

Why do the nitro groups change the relative leaving tendencies of the carboxyl and phenoxy groups in the tetrahedral intermediate in Problem 11?

PROBLEM 13

Whether the *ortho*-carboxyl substituent acts as an intramolecular general-base catalyst or as an intramolecular nucleophilic catalyst can be determined by carrying out the hydrolysis of aspirin with ^{18}O-labeled water and determining whether ^{18}O is incorporated into *ortho*-carboxyl-substituted phenol. Explain the results that would be obtained with the two types of catalysis.

---

8 CATALYSIS IN BIOLOGICAL REACTIONS

Essentially all organic reactions that occur in cells require a catalyst. Most biological catalysts are **enzymes,** which are globular proteins. Each biological reaction is catalyzed by a different enzyme.

Binding the Substrate

The reactant of an enzyme-catalyzed reaction is called a **substrate.**

$$\text{substrate} \xrightarrow{\text{enzyme}} \text{product}$$

The enzyme binds its substrate in its **active site,** a pocket in the cleft of the enzyme. All the bond-making and bond-breaking steps that convert the substrate to the product occur while the substrate is bound to the active site.

enzyme

the substrate is bound at the active site

Enzymes differ from nonbiological catalysts in that they are specific for the substrate whose reaction they catalyze. Enzymes, however, have different degrees of specificity. Some enzymes are specific for a single compound. For example, glucose-6-phosphate isomerase catalyzes the isomerization of glucose-6-phosphate only. On the other hand, some enzymes catalyze the reactions of several compounds with similar structures. For instance, hexokinase catalyzes the phosphorylation of any D-hexose. The specificity of an enzyme for its substrate is another example of **molecular recognition**—the ability of one molecule to recognize another molecule.

The particular **amino acid side chains** (α-substituents) at the active site are responsible for the enzyme's specificity. For example, an amino acid with a negatively charged side chain can associate with a positively charged group on the substrate, an amino acid side chain with a hydrogen-bond donor can associate with a hydrogen-bond acceptor on the substrate, and a hydrophobic amino acid side chain can associate with hydrophobic groups on the substrate.

In 1894, Emil Fischer proposed the **lock-and-key model** to account for the specificity of an enzyme for its substrate. This model related the specificity of an enzyme for its substrate to the specificity of a lock for a correctly shaped key.

lock-and-key model   **induced-fit model**

In 1958, Daniel Koshland proposed the **induced-fit model** of substrate binding. In this model, the shape of the active site does not become completely complementary to the shape of the substrate until the enzyme has bound the substrate. The energy released as a result of binding the substrate can be used to induce a change in the conformation of the enzyme, leading to more precise binding between the substrate and the active site. An example of induced fit is shown in Figure 4.

Catalyzing the Reaction

There is no single explanation for the remarkable catalytic ability of enzymes. Each enzyme is unique in the factors it employs to catalyze a reaction. Some of the factors most enzymes have in common are:

- Reacting groups are brought together at the active site in the proper orientation for reaction. This is analogous to the way proper positioning of reacting groups increases the rate of an intramolecular reaction (Section 6).

- Some of the amino acid side chains of the enzyme serve as acid, base, and nucleophilic catalysts, and many enzymes also have metal ions at their active site that act

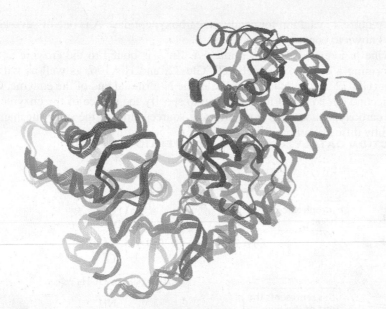

◀ **Figure 4**
The structure of hexokinase before binding its substrate is shown in red. The structure of hexokinase after binding its substrate is shown in green.

as catalysts. These species are positioned relative to the substrate precisely where they are needed for catalysis. This factor is analogous to the way intramolecular catalysis by acids, bases, nucleophiles, and metal ions enhances reaction rates (Section 7).

- Amino acid side chains can stabilize transition states and intermediates—by van der Waals interactions, electrostatic interactions, and hydrogen bonding—which makes them easier to form (Figure 1b).

Now we will look at the mechanisms of five enzyme-catalyzed reactions. Notice that the modes of catalysis used by enzymes are the same as the modes of catalysis used in organic reactions. Thus, if you refer back to sections referenced throughout this chapter, you will be able to see that much of the organic chemistry you have learned also applies to the reactions of compounds found in the biological world. The remarkable catalytic ability of enzymes stems in part from their ability to use several modes of catalysis in the same reaction.

9 THE MECHANISMS FOR TWO ENZYME-CATALYZED REACTIONS THAT ARE REMINISCENT OF ACID-CATALYZED AMIDE HYDROLYSIS

The names of most enzymes end in "ase," and the enzyme's name tells you something about the reaction it catalyzes. For example, carboxypeptidase A catalyzes the hydrolysis of the C-terminal (carboxy-terminal) peptide bond in polypeptides, releasing the terminal amino acid.

$$—NHCH(R)—C(=O)—NHCH(R')—C(=O)—NHCH(R'')—C(=O)—O^- + H_2O$$

carboxypeptidase A

$$—NHCH(R)—C(=O)—NHCH(R')—C(=O)—O^- + H_3\overset{+}{N}CH(R'')—C(=O)—O^-$$

Carboxypeptidase A is a *metalloenzyme,* which is an enzyme that contains a tightly bound metal ion. The metal ion in carboxypeptidase A is Zn^{2+}. About one-third of all enzymes require a metal ion for catalysis; carboxypeptidase A is one of several hundred enzymes known to contain zinc.

In bovine pancreatic carboxypeptidase A, Zn^{2+} is bound to the enzyme at its active site by forming a complex with His 69, Glu 72, and His 196, as well as with a water molecule. (Glu 72 means that, starting from the N-terminal end of the enzyme, Glu is the seventy-second amino acid.) It is common to specify the source of the enzyme because, although carboxypeptidase As from different sources employ the same mechanism, they have slightly different primary structures.

PROPOSED MECHANISM FOR THE REACTION CATALYZED BY CARBOXYPEPTIDASE A

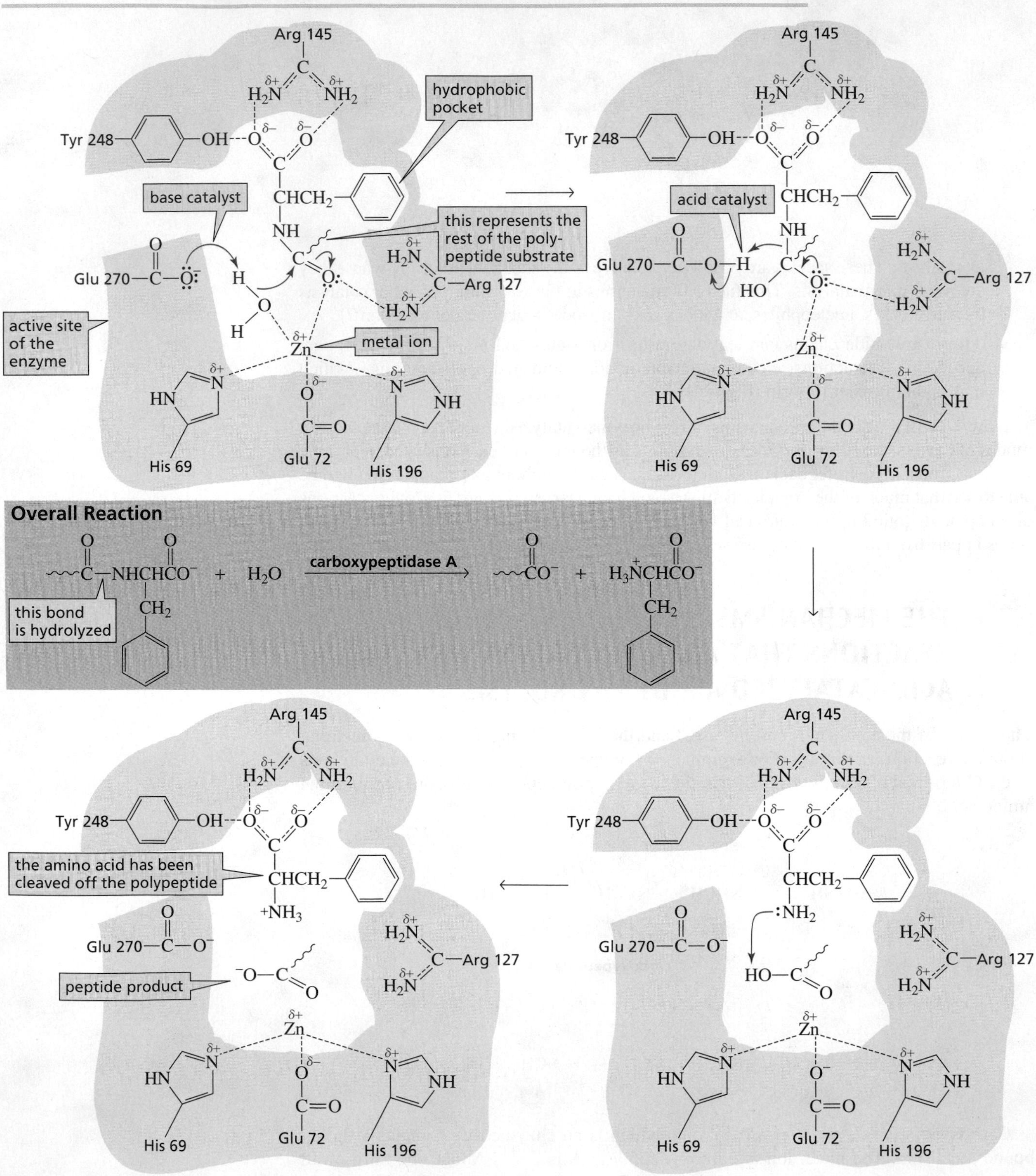

Overall Reaction

Several amino acid side chains at the active site of carboxypeptidase A participate in binding the substrate in the optimal position for reaction. Arg 145 forms two hydrogen bonds and Tyr 248 forms one hydrogen bond with the C-terminal carboxyl group of the substrate. (In this example, the C-terminal amino acid is phenylalanine.) The side chain of the C-terminal amino acid is positioned in a hydrophobic pocket, which is why carboxypeptidase A is not active if the C-terminal amino acid is arginine or lysine. Apparently, the long, positively charged side chains of these amino acids cannot fit into the nonpolar pocket.

- When the substrate binds to the active site, Zn^{2+} partially complexes with the oxygen of the carbonyl group of the amide that will be hydrolyzed. Thus, Zn^{2+} polarizes the carbon–oxygen double bond, making the carbonyl carbon more susceptible to nucleophilic addition and stabilizing the negative charge that develops on the oxygen atom in the transition state that leads to the tetrahedral intermediate. Arg 127 also increases the carbonyl group's electrophilicity and stabilizes the developing negative charge on the oxygen atom. In addition, Zn^{2+} complexes with water, making it a better nucleophile. Glu 270 is a general base-catalyst, further increasing water's nucleophilicity.

- In the next step, Glu 270 is a general-acid catalyst, increasing the leaving tendency of the amino group. When the reaction is over, the amino acid (phenylalanine in this example) and the peptide with one less amino acid dissociate from the enzyme, and another molecule of substrate binds to the active site.

The unfavorable electrostatic interaction between the negatively charged carboxyl group of the peptide product and the negatively charged Glu 270 side chain may facilitate the release of the product from the enzyme.

Notice that the protons are being donated and removed during (rather than before) the other bond-making and bond-breaking processes in these enzyme-catalyzed reactions, so the catalysis that occurs is general-acid and general-base catalysis (Sections 2 and 3). At physiological pH (7.4), the concentration of H^+ or HO^- ($\sim 1 \times 10^{-7}$ M) is too small for specific-acid or specific-base catalysis.

PROBLEM 14 Solved

Which of the following amino acid side chains can aid the departure of a leaving group?

Solution Side chains **1** and **2** do not have an acidic proton, so they cannot aid the departure of a leaving group by protonating it. Side chains **3** and **4** each have an acidic proton, so they can aid the departure of a leaving group.

PROBLEM 15♦

Which of the following amino acid side chains can help remove a proton from the α-carbon of an aldehyde?

PROBLEM 16◆

Which of the following C-terminal peptide bonds would be more readily cleaved by carboxy-peptidase A? Explain your choice.

Ser-Ala-Leu   or   Ser-Ala-Asp

PROBLEM 17

Carboxypeptidase A has esterase activity as well as peptidase activity, so it can hydrolyze ester bonds as well as peptide bonds. When carboxypeptidase A hydrolyzes ester bonds, Glu 270 acts as a nucleophilic catalyst instead of a general-base catalyst. Propose a mechanism for the carboxypeptidase A–catalyzed hydrolysis of an ester bond.

Trypsin, chymotrypsin, and elastase are members of a large group of *endopeptidases* known collectively as serine proteases. (An endopeptidase cleaves a peptide bond that is not at the end of a peptide chain. They are called *proteases* because they catalyze the hydrolysis of protein peptide bonds. They are called *serine proteases* because each one has a serine side chain at the active site that participates in the catalysis.)

The various serine proteases have similar primary structures, suggesting that they are evolutionarily related. Although they all have the same three catalytic side chains at the active site (that is, Asp, His, and Ser), they have one important difference—namely, the composition of the pocket at the active site that binds the side chain of the amino acid in the peptide bond that undergoes hydrolysis (Figure 5). This pocket is what gives the serine proteases their different specificities.

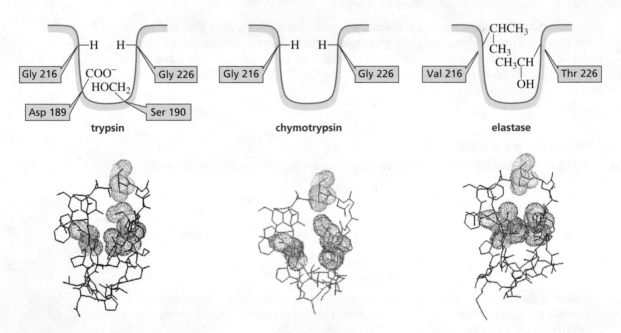

▲ **Figure 5**

The binding pockets in trypsin, chymotrypsin, and elastase. The negatively charged aspartate is shown in red, and the relatively nonpolar amino acids are shown in green. The structures of the binding pockets explain why trypsin binds long, positively charged amino acids; chymotrypsin binds flat, nonpolar amino acids; and elastase binds only small amino acids.

The pocket in trypsin is narrow and has a serine and a negatively charged aspartate carboxyl group at its bottom. The shape and charge of the binding pocket cause it to bind long, positively charged amino acid side chains (Lys and Arg). This is why trypsin hydrolyzes only peptide bonds on the C-side of arginine and lysine. The pocket in chymotrypsin is narrow and is lined with nonpolar amino acids, so chymotrypsin cleaves on the C-side of amino acids with flat, nonpolar side chains (Phe, Tyr, and Trp).

In elastase, two glycines on the sides of the pocket in trypsin and in chymotrypsin are replaced by relatively bulky valine and threonine. Consequently, only small amino acids can fit into the pocket. Elastase, therefore, hydrolyzes peptide bonds on the C-side of small amino acids (Gly, Ala, Ser, and Val).

The proposed mechanism for bovine chymotrypsin-catalyzed hydrolysis of a peptide bond is shown here. The other serine proteases follow the same mechanism.

PROPOSED MECHANISM FOR THE REACTION CATALYZED BY SERINE PROTEASES

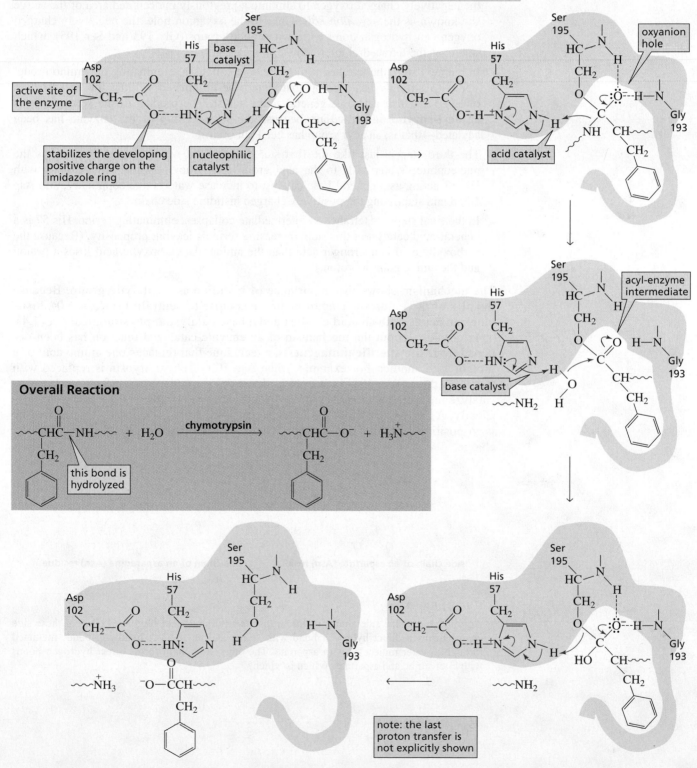

note: the last proton transfer is not explicitly shown

1153

- As a consequence of binding the flat, nonpolar side chain in the hydrophobic pocket, the amide linkage to be hydrolyzed is positioned very close to Ser 195. His 57 is a general-base catalyst, increasing the nucleophilicity of serine, which adds to the carbonyl group. This step is helped by Asp 102, which does not remove a proton from imidazole but remains as a carboxylate ion, using its negative charge to stabilize the developing positive charge on His 57 and to position the five-membered ring so that its basic N atom is close to the OH group of serine. The stabilization of a charge by an opposite charge is called **electrostatic catalysis.** Formation of the tetrahedral intermediate causes a slight change in the conformation of the protein. This allows the negatively charged oxygen to slip into a previously unoccupied area of the active site known as the *oxyanion hole.* Once in the oxyanion hole, the negatively charged oxygen can hydrogen bond with two peptide groups (Gly 193 and Ser 195), which stabilizes the tetrahedral intermediate.

- In the next step, the tetrahedral intermediate collapses, eliminating the amino group. This is a strongly basic group that cannot be eliminated without the participation of His 57, which acts as a general-acid catalyst. The product of the second step is an **acyl-enzyme intermediate** because the serine group of the enzyme has been acylated—that is, an acyl group has been put on it.

- The third step is just like the first step, except that water instead of serine is the nucleophile. Water adds to the acyl group of the acyl-enzyme intermediate, with His 57 acting as a general-base catalyst to increase water's nucleophilicity, and Asp 102 again stabilizing the positively charged histidine side chain.

- In the final step, the tetrahedral intermediate collapses, eliminating serine. His 57 is a general-acid catalyst in this step, increasing serine's leaving propensity. (Because the carboxylic acid is a stronger acid than the amine, the carboxylic acid loses a proton and the amine gains a proton.)

The mechanism shows the importance of histidine as a catalytic group. Because the pKa of the imidazole ring of histidine is close to neutrality ($pK_a = 6.0$), histidine can act as both an acid catalyst and a base catalyst at physiological pH (7.4).

Information about the mechanism of an enzyme-catalyzed reaction has been obtained from **site-specific mutagenesis,** a technique that replaces one amino acid of a protein with another. For example, when Asp 102 of chymotrypsin is replaced with Asn 102, the enzyme's ability to bind the substrate is unchanged, but its ability to catalyze the reaction decreases to less than 0.05% of the value for the native enzyme. Clearly, Asp 102 must be involved in the catalytic process. We just saw that its role is to position histidine and use its negative charge to stabilize histidine's positive charge.

side chain of an aspartate (Asp) residue

side chain of an asparagine (Asn) residue

PROBLEM 18♦

Arginine and lysine side chains fit into trypsin's binding pocket (Figure 5). One of these side chains forms a direct hydrogen bond with serine and an indirect hydrogen bond (mediated through a water molecule) with aspartate. The other side chain forms direct hydrogen bonds with both serine and aspartate. Which is which?

Explain why serine proteases do not catalyze hydrolysis if the amino acid at the hydrolysis site is a D-amino acid. Trypsin, for example, cleaves on the C-side of L-Arg and L-Lys, but not on the C-side of D-Arg and D-Lys.

10   THE MECHANISM FOR AN ENZYME-CATALYZED REACTION THAT INVOLVES TWO SEQUENTIAL S_N2 REACTIONS

Lysozyme is an enzyme that destroys bacterial cell walls. These cell walls are composed of alternating *N*-acetylmuramic acid (NAM) and *N*-acetylglucosamine (NAG) units linked by β-1,4'-glycosidic linkages. Lysozyme destroys the cell wall by catalyzing the hydrolysis of the NAM–NAG bond.

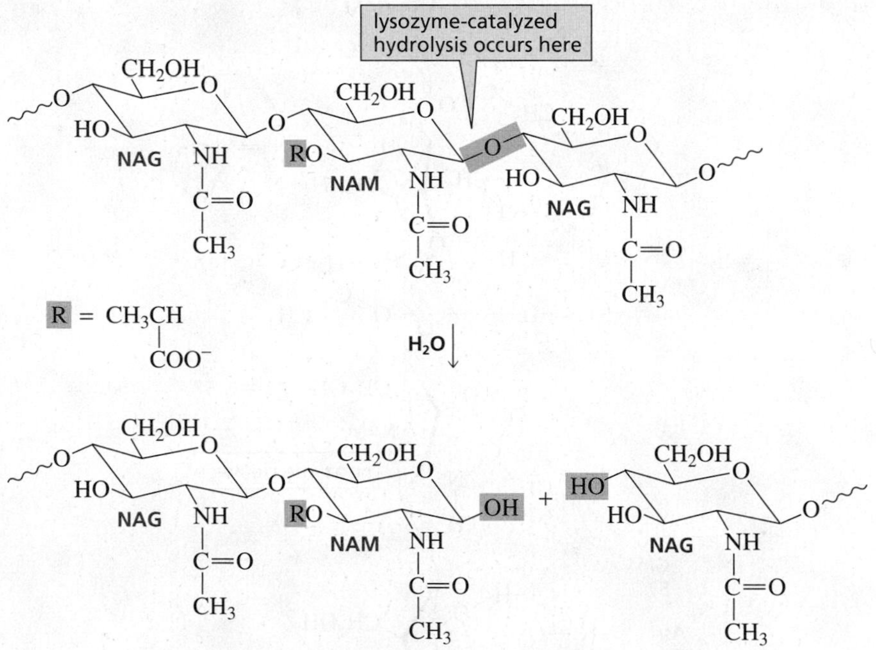

The active site of hen egg-white lysozyme binds six sugar residues of the substrate. They are labeled A, B, C, D, E, and F in Figure 6. The many amino acid side chains involved in binding the substrate in the correct position in the active site are also shown in Figure 6. The carboxylic acid substituent of the RO group of NAM cannot fit into the binding site for C or E. This means that NAM units must bind at the sites for B, D, and F. Hydrolysis occurs between D and E.

Lysozyme has two catalytic groups at the active site: Glu 35 and Asp 52. The discovery that the enzyme-catalyzed reaction takes place with retention of configuration at the anomeric carbon indicates that it cannot be a one-step S_N2 reaction. (An S_N2 reaction takes place with inversion of configuration.) Therefore, the reaction must involve either two sequential S_N2 reactions or an S_N1 reaction with the enzyme blocking one face of an oxocarbenium ion intermediate to nucleophilic attack. Although lysozyme was the first enzyme to have its mechanism studied—and it has been studied extensively for over 40 years—only recently have data been obtained that support the mechanism involving two sequential S_N2 reactions.

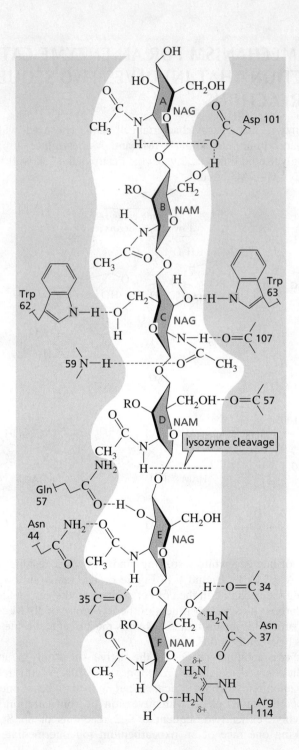

▶ **Figure 6**

The amino acids at the active site of lysozyme that are involved in binding the substrate.

PROPOSED MECHANISM FOR THE REACTION CATALYZED BY LYSOZYME

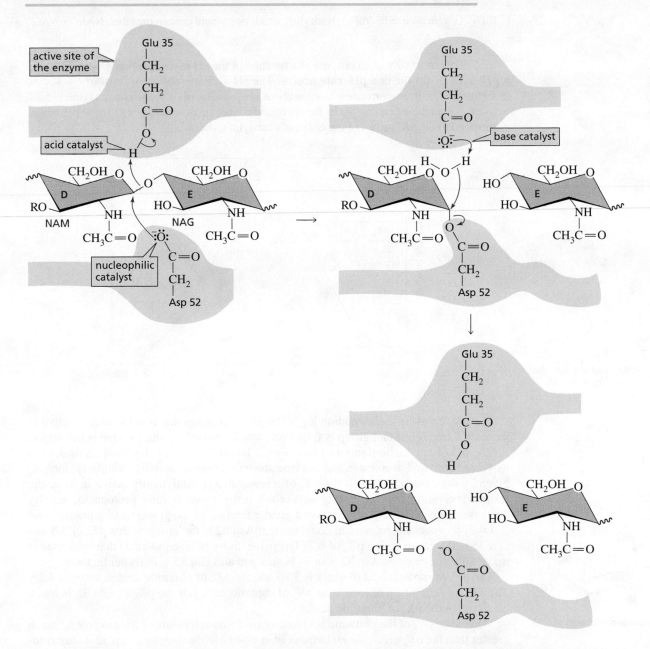

- In the first S_N2 reaction, Asp 52 is a nucleophilic catalyst that attacks the anomeric carbon (C-1) of the NAM residue, displacing the leaving group. Glu 35 is a general-acid catalyst, protonating the leaving group and thereby making it a weaker base and a better leaving group.

  Site-specific mutagenesis studies show that when Glu 35 is replaced by Asp, the enzyme has only weak activity. Apparently, Asp does not lie at the optimal distance from, and angle to, the oxygen atom to easily protonate it. When Glu 35 is replaced by Ala, an amino acid that cannot act as an acid catalyst, the activity of the enzyme is completely lost.

- In the second S_N2 reaction, Glu 35 is a general-base catalyst to increase water's nucleophilicity.

PROBLEM 20♦

If $H_2^{18}O$ were used to hydrolyze lysozyme, which ring would contain the label, NAM or NAG?

A plot of the activity of an enzyme as a function of the pH of the reaction mixture is called a **pH–activity profile** or a **pH–rate profile**. The pH–activity profile for lysozyme is shown in Figure 7. The maximum rate occurs at about pH 5.3. The pH at which the enzyme is 50% active is 3.8 on the ascending leg of the curve and 6.7 on the descending leg. These pH values correspond to the pK_a values of the enzyme's catalytic groups.

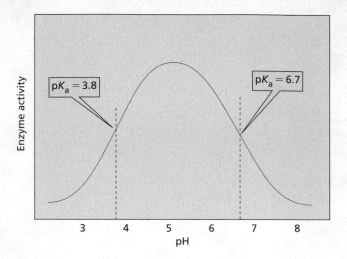

▶ **Figure 7**

Lysozyme's activity depends on the pH of the reaction mixture.

The pK_a given by the ascending leg is the pK_a of a group that is catalytically active in its basic form. When that group is fully protonated (~pH = 2), the enzyme is not active. As the pH of the reaction mixture increases, a larger fraction of the group is present in its basic form, and as a result, the enzyme shows increasing activity. Similarly, the pK_a given by the descending leg is the pK_a of a group that is catalytically active in its acidic form. Maximum catalytic activity occurs when the group is fully protonated; activity decreases with increasing pH because a greater fraction of the group lacks a proton.

From the mechanism, we can conclude that Asp 52 is the group with a pK_a of 3.8 and Glu 35 is the group with a pK_a of 6.7. The pH–activity profile indicates that lysozyme is maximally active when Asp 52 is in its basic form and Glu 35 is in its acidic form.

The pK_a of aspartic acid in water is 3.86 and the pK_a of glutamic acid in water is 4.25. The pK_a of Asp 52 agrees with the pK_a of aspartic acid, but the pK_a of Glu 35 is much greater than the pK_a of glutamic acid.

Why is the pK_a of the glutamic acid side chain at the active site of the enzyme so much greater than the pK_a given for glutamic acid in water? In the enzyme, Asp 52 is surrounded by polar groups, which means that its pK_a should be close to the pK_a determined in water, a polar solvent. Glu 35, however, is situated in a predominantly nonpolar pocket, so its pK_a should be greater than the pK_a determined in water. The pK_a of a carboxylic acid is greater in a nonpolar solvent because there is less tendency to form charged species in nonpolar solvents.

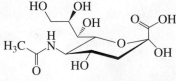

N-acetylneuramic acid

Tamiflu®

How Tamiflu Works

Tamiflu is one of the few antiviral drugs currently available. It is used for the prevention and treatment of influenza A and B. Before a virus particle can be released from its host cell, an enzyme called neuraminidase must cleave off a sugar residue (*N*-acetylneuramic acid) from a glycoprotein on the surface of the cell. Because *N*-acetylneuramic acid and Tamiflu have similar shapes, the enzyme cannot distinguish between them. Therefore, it can bind either one at its active site. When the enzyme binds Tamiflu, it cannot bind *N*-acetylneuramic acid. Thus, the virus particles are prevented from being released from their host cells, in which case they cannot infect new cells. Early treatment with Tamiflu is important because it will be less effective if a lot of cells have already been infected. In the past 10 years, 500 million people have been treated with Tamiflu.

Part of the catalytic efficiency of lysozyme results from its ability to provide different solvent environments at the active site. This allows one catalytic group to exist in its acidic form at the same surrounding pH at which a second catalytic group exists in its basic form. This property is unique to enzymes; chemists cannot provide different solvent environments for different parts of nonenzymatic systems.

PROBLEM 21♦

When apples that have been cut are exposed to oxygen, an enzyme-catalyzed reaction causes them to turn brown. Explain why coating them with lemon juice (pH ∼ 3.5) as soon as they are cut prevents the color change.

---

11 THE MECHANISM FOR AN ENZYME-CATALYZED REACTION THAT IS REMINISCENT OF THE BASE-CATALYZED ENEDIOL REARRANGEMENT

Glycolysis is the name given to the series of enzyme-catalyzed reactions responsible for converting glucose into two molecules of pyruvate. The second reaction in glycolysis is an isomerization reaction that converts glucose-6-phosphate to fructose-6-phosphate. Glucose is an aldohexose, whereas fructose is a ketohexose, so the enzyme that catalyzes this reaction—glucose-6-phosphate isomerase—converts an aldose to a ketose. Notice that the mechanism of this reaction is exactly the same as the mechanism for the enediol rearrangement.

PROPOSED MECHANISM FOR THE REACTION CATALYZED BY GLUCOSE-6-PHOSPHATE ISOMERASE

acid catalyst

active site of the enzyme

base catalyst

base catalyst

acid catalyst

Overall Reaction

glucose-6-phosphate isomerase

glucose-6-phosphate

fructose-6-phosphate

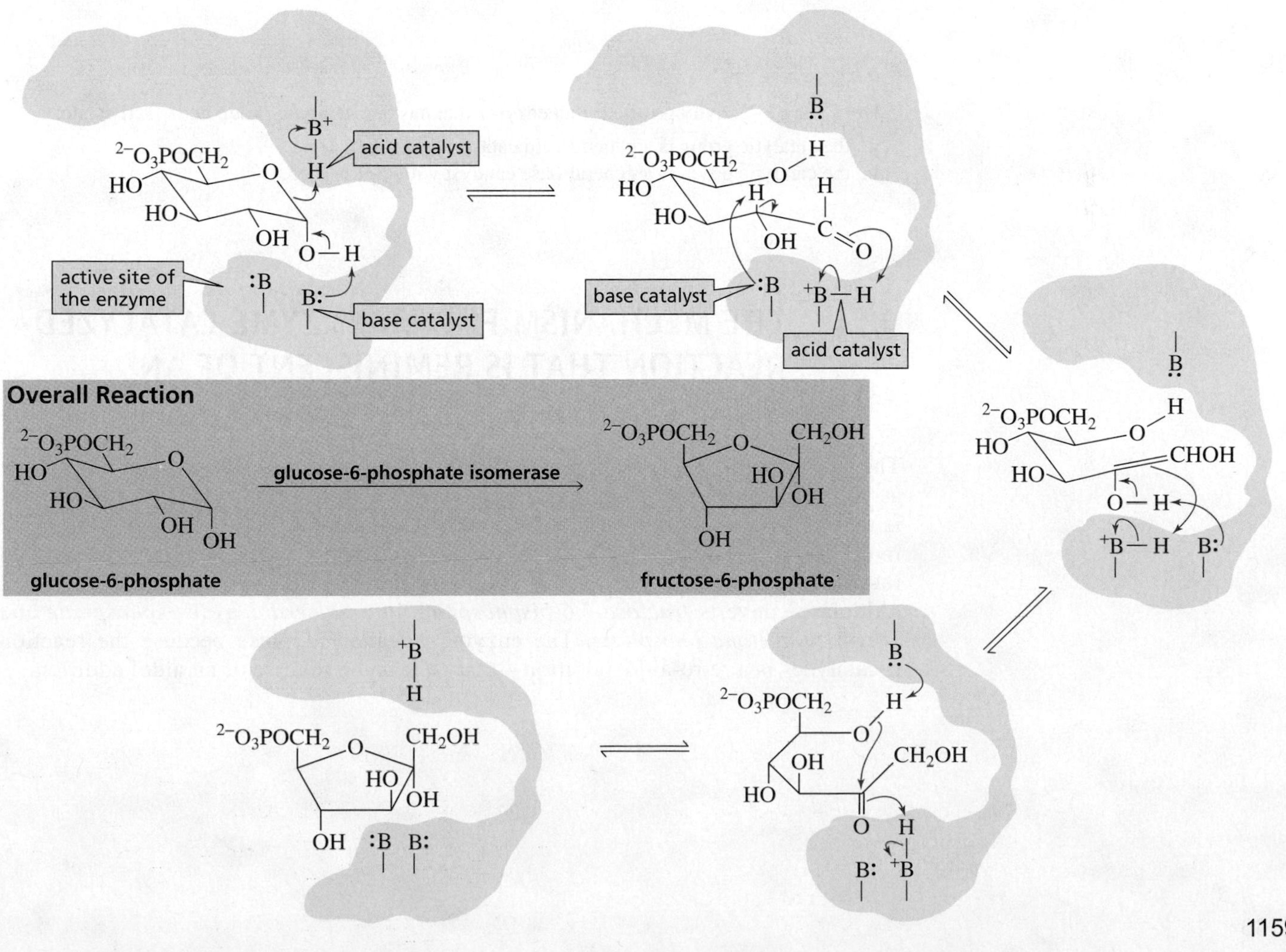

- The first step is a ring-opening reaction. A general-base catalyst (thought to be a histidine side chain) removes a proton from the OH group, and a general-acid catalyst (a protonated lysine side chain) aids the departure of the leaving group by protonating it, thereby making it a weaker base and therefore a better leaving group.

- In the second step of the reaction, a general-base catalyst (a glutamate side chain) removes a proton from the α-carbon of the aldehyde and a general-acid catalyst protonates the oxygen, forming an enediol. Note that the α-hydrogen of an aldehyde is relatively acidic.

- In the next step, the enediol is converted to a ketone.

- In the final step, the conjugate base of the acid catalyst employed in the first step and the conjugate acid of the base catalyst employed in the first step catalyze ring closure.

PROBLEM 22

When glucose undergoes base-catalyzed isomerization in the absence of the enzyme, mannose is one of the products that are formed. Why is mannose not formed in the enzyme-catalyzed reaction?

PROBLEM 23

The pH–activity profile for glucose-6-phosphate isomerase indicates the participation of a group with a $pK_a = 6.7$ as a basic catalyst and a group with a $pK_a = 9.3$ as an acid catalyst. Draw the pH–activity profile and identify the amino acids that participate in the catalysis.

PROBLEM 24

Draw the pH–activity profile for an enzyme that has one catalytic group at the active site:

a. the catalytic group is a general-acid catalyst with a $pK_a = 5.6$.
b. the catalytic group is a general-base catalyst with a $pK_a = 7.2$.

---

12 THE MECHANISM FOR AN ENZYME CATALYZED-REACTION THAT IS REMINISCENT OF AN ALDOL ADDITION

The substrate for the first enzyme-catalyzed reaction in the series of reactions known as glycolysis is D-glucose (a six-carbon compound). The final product of glycolysis is two molecules of pyruvate (a three-carbon compound). Therefore, at some point in the series of enzyme-catalyzed reactions, a six-carbon compound must be cleaved into two three-carbon compounds. The enzyme *aldolase* catalyzes this cleavage. (Aldolase converts *fructose-1,6-bisphosphate* into *glyceraldehyde-3-phosphate* and *dihydroxyacetone phosphate*. The enzyme is called aldolase because the reaction it catalyzes is a retro-aldol addition—that is, it is the reverse of an aldol addition.

PROPOSED MECHANISM FOR THE REACTION CATALYZED BY ALDOLASE

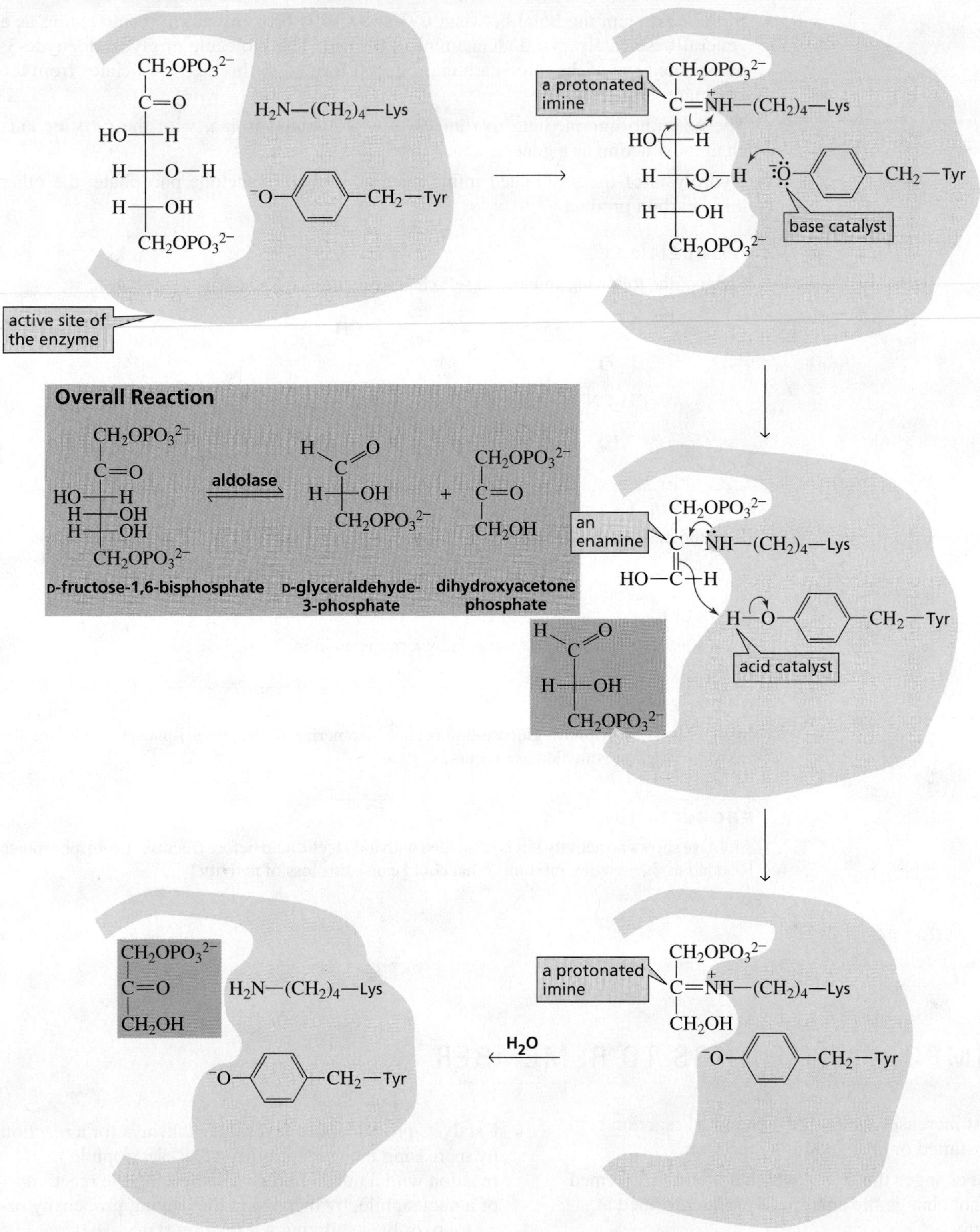

active site of the enzyme

a protonated imine

base catalyst

Overall Reaction

D-fructose-1,6-bisphosphate ⇌ D-glyceraldehyde-3-phosphate + dihydroxyacetone phosphate

aldolase

an enamine

acid catalyst

a protonated imine

H₂O

- In the first step, fructose-1,6-bisphosphate forms a protonated imine with a lysine side chain at the active site of the enzyme.

- In the next step, the bond between C-3 and C-4 is broken, with tyrosine acting as a general-base catalyst, and an enamine is formed. The molecule of glyceraldehyde-3-phosphate (one of the three-carbon products) formed in this step dissociates from the enzyme.

- The enamine intermediate rearranges to a protonated imine, with the tyrosine side chain now acting as a general-acid catalyst.

- Hydrolysis of the protonated imine releases dihydroxyacetone phosphate, the other three-carbon product.

PROBLEM 25◆

Which of the following amino acid side chains can form an imine with a substrate?

$$-CH_2CNH_2 \qquad -(CH_2)_4NH_2 \qquad \qquad -CH_2OH$$

      1   **2**   **3**   **4**

PROBLEM 26

Draw the mechanism for the hydroxide ion–catalyzed cleavage of fructose-1,6-bisphosphate.

PROBLEM 27

What advantage does the enzyme gain by forming an imine?

PROBLEM 28

In glycolysis, why must glucose-6-phosphate isomerize to fructose-6-phosphate before the cleavage reaction with aldolase occurs?

PROBLEM 29◆

Aldolase shows no activity if it is incubated with iodoacetic acid before fructose-1,6-bisphosphate is added to the reaction mixture. What could cause this loss of activity?

SOME IMPORTANT THINGS TO REMEMBER

- A **catalyst** increases the rate of a chemical reaction but is not consumed or changed in the reaction.

- A catalyst changes the rate at which a product is formed but does not change the amount of product formed at equilibrium.

- A catalyst must increase the rate of a slow step. It does this by providing a pathway with a smaller $\Delta G^{\ddagger}$, either by converting the reactant into a more reactive species, by making the transition state more stable, or by completely changing the mechanism of the reaction.

- Catalysts provide more favorable pathways for a reaction by increasing the susceptibility of an electrophile to reaction with a nucleophile, by increasing the reactivity of a nucleophile, by increasing the leaving propensity of a group, or by stabilizing a transition state.

- An acid catalyst increases the rate of a reaction by donating a proton to a reactant. In **specific-acid catalysis,** the proton is fully transferred to the reactant before the slow step of the reaction; in **general-acid catalysis,** the proton is transferred during the slow step.

- A **base catalyst** increases the rate of a reaction by removing a proton from a reactant. In **specific-base catalysis,** the proton is completely removed from the reactant before the slow step of the reaction; in **general-base catalysis,** the proton is removed during the slow step.

- A **nucleophilic catalyst** increases the rate of a reaction by acting as a nucleophile: it forms an intermediate by forming a covalent bond with a reactant.

- Stabilization of a charge by an opposite charge is called **electrostatic catalysis.**

- A **metal ion** can increase the rate of a reaction by making a reaction center more receptive to electrons, by making a leaving group a weaker base, or by increasing the nucleophilicity of water.

- The rate of a chemical reaction is determined by the number of collisions between two molecules or between two groups in the same molecule, with sufficient energy and with the proper orientation in a given period of time.

- An **intramolecular reaction** that forms a five- or a six-membered ring occurs more readily than the analogous intermolecular reaction because both the frequency of the collisions and the probability that collisions will occur in the proper orientation increases.

- **Effective molarity** is the concentration of the reactant that would be required to give the corresponding intermolecular reaction the same rate as the intramolecular reaction.

- When a catalyst is part of the reacting molecule, the catalysis is called **intramolecular catalysis.**

- Most biological catalysts are **enzymes.** The reactant of an enzyme-catalyzed reaction is called a **substrate.** The enzyme specifically binds the substrate at the **active site** of the enzyme; all the bond-making and bond-breaking steps of the reaction occur while it is at the active site.

- The specificity of an enzyme for its substrate is an example of **molecular recognition.**

- **Induced fit** is the change in conformation of the enzyme that occurs when it binds the substrate.

- Two important factors contribute to the remarkable catalytic ability of enzymes: 1. the reacting groups are brought together at the active site in the proper orientation for reaction, and 2. the amino acid side chains (and a metal ion in the case of some enzymes) are well positioned relative to the substrate, where they are needed for catalysis.

- A **pH–activity profile** is a plot of the activity of an enzyme as a function of the pH of the reaction mixture.

PROBLEMS

30. Which of the following two compounds would eliminate HBr more rapidly in a basic solution?

31. Which compound would form an anhydride more rapidly?

32. Which compound has the greatest rate of hydrolysis at pH = 3.5: benzamide, *o*-carboxybenzamide, *o*-formylbenzamide, or *o*-hydroxybenzamide?

33. Indicate the type of catalysis that is occurring in the slow step in each of the following reaction sequences:

a.

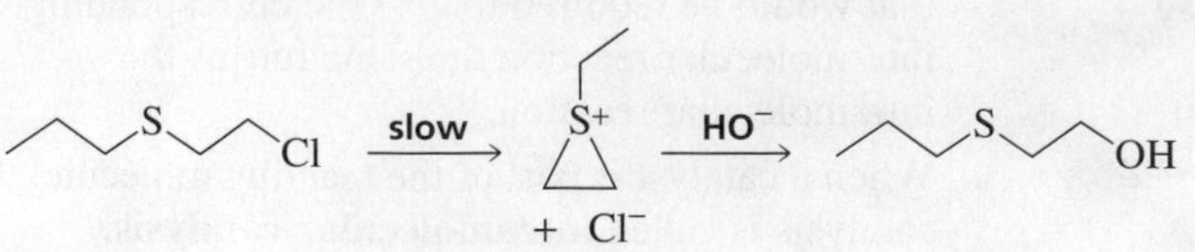

b.

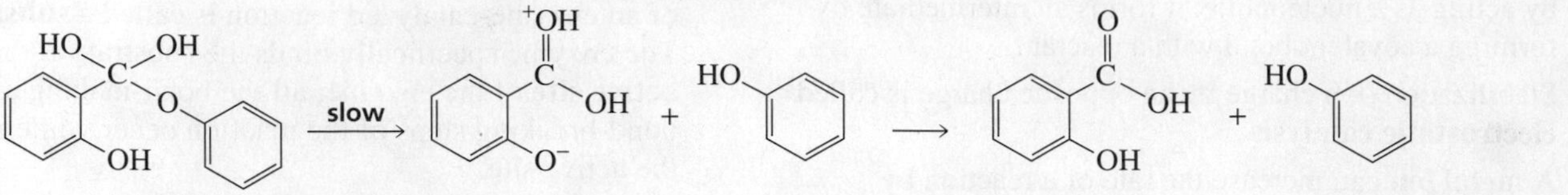

34. The deuterium kinetic isotope effect (k_{H_2O}/k_{D_2O}) for the hydrolysis of aspirin is 2.2. What does this tell you about the kind of catalysis exerted by the *ortho*-carboxyl substituent? (*Hint:* It is easier to break an O—H bond than an O—D bond.)

35. The rate constant for the uncatalyzed reaction of two molecules of glycine ethyl ester to form glycylglycine ethyl ester is 0.6 $M^{-1}s^{-1}$. In the presence of Co^{2+}, the rate constant is 1.5×10^6 $M^{-1}s^{-1}$. What rate enhancement does the catalyst provide?

36. A Co^{2+} complex catalyzes the hydrolysis of the lactam shown here. Propose a mechanism for the metal-ion catalyzed reaction.

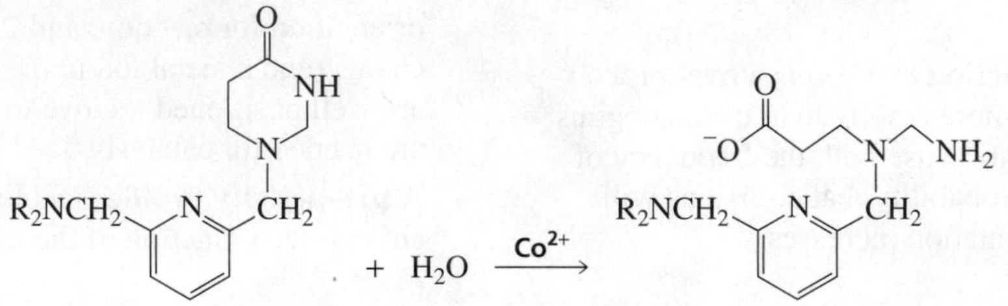

37. There are two kinds of aldolases. Class I aldolases are found in animals and plants, whereas class II aldolases are found in fungi, algae, and some bacteria. Only class I aldolases form an imine. Class II aldolases have a metal ion (Zn^{2+}) at the active site. The mechanism for catalysis by class I aldolases was described in Section 12. Propose a mechanism for catalysis by class II aldolases.

38. Propose a mechanism for the following reaction. (*Hint:* The rate of the reaction is much slower if the nitrogen atom is replaced by CH.)

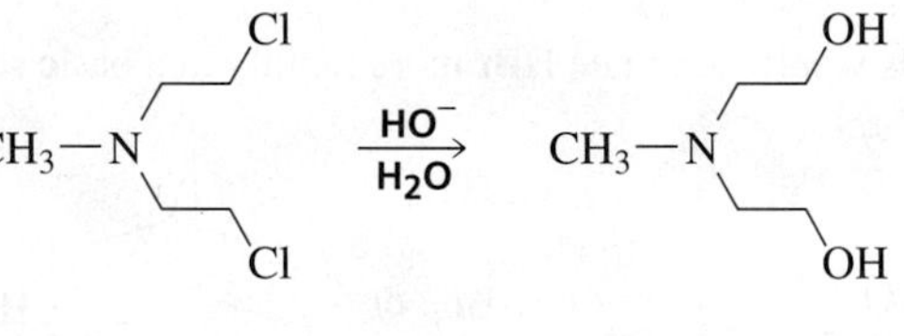

39. The hydrolysis of the ester shown here is catalyzed by morpholine, a secondary amine. Propose a mechanism for the reaction. (*Hint:* The pK_a of the conjugate acid of morpholine is 9.3, so morpholine is too weak a base to function as a base catalyst.)

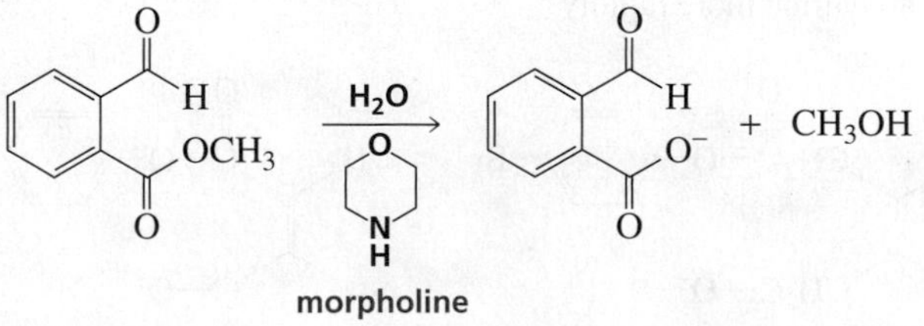

morpholine

40. Carbonic anhydrase is an enzyme that catalyzes the conversion of carbon dioxide into bicarbonate ion. It is a metalloenzyme, with Zn^{2+} coordinated at the active site by three histidine side chains. Propose a mechanism for the reaction.

$$CO_2 + H_2O \xrightarrow{\text{carbonic anhydrase}} HCO_3^- + H^+$$

41. At pH = 12, the rate of hydrolysis of ester **A** is greater than the rate of hydrolysis of ester **B**. At pH = 8, the relative rates reverse (that is, ester **B** hydrolyzes faster than ester **A**). Explain these observations.

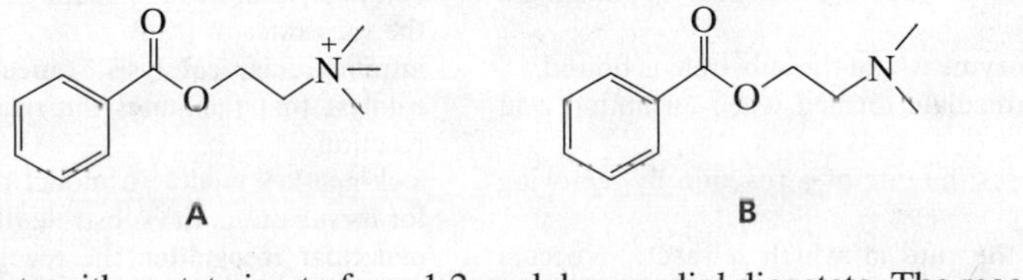

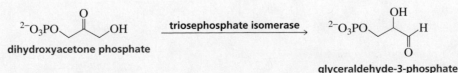

A   **B**

42. 2-Acetoxycyclohexyl tosylate reacts with acetate ion to form 1,2-cyclohexanediol diacetate. The reaction is stereospecific—that is, the stereoisomers obtained as products depend on the stereoisomer used as a reactant. Note that because 2-acetoxycyclohexyl tosylate has two asymmetric centers, it has four stereoisomers—two are cis and two are trans. Explain the following observations:
  a. Both cis reactants form an optically active trans product, but each cis reactant forms a different trans product.
  b. Both trans reactants form the same racemic mixture.
  c. A trans reactant is more reactive than a cis reactant.

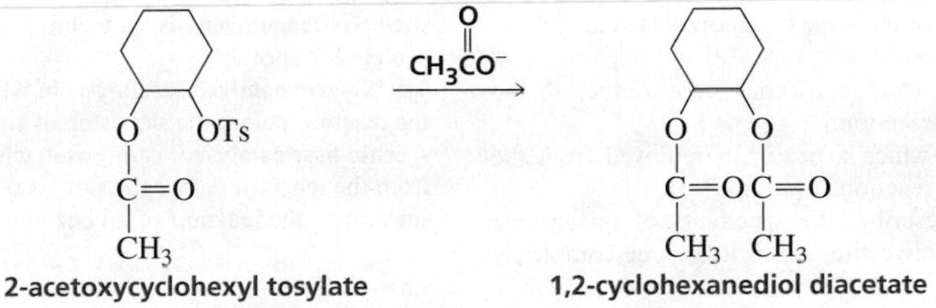

2-acetoxycyclohexyl tosylate   **1,2-cyclohexanediol diacetate**

43. Proof that an imine was formed between aldolase and its substrate was obtained by using D-fructose-1,6-bisphosphate labeled at the C-2 position with ^{14}C as the substrate. $NaBH_4$ was added to the reaction mixture. A radioactive product was isolated from the reaction mixture and hydrolyzed in an acidic solution. Draw the structure of the radioactive product obtained from the acidic solution. (*Hint:* $NaBH_4$ reduces an imine linkage.)

44. 3-Amino-2-oxindole catalyzes the decarboxylation of α-keto acids.
  a. Propose a mechanism for the catalyzed reaction.
  b. Would 3-aminoindole be equally effective as a catalyst?

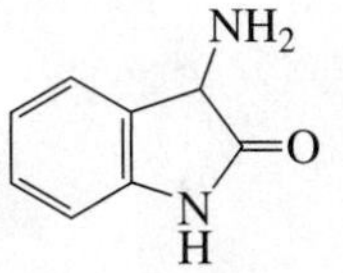

3-amino-2-oxindole

45. a. Explain why the alkyl halide shown here reacts much more rapidly with guanine than does a primary alkyl halide (such as pentyl chloride).

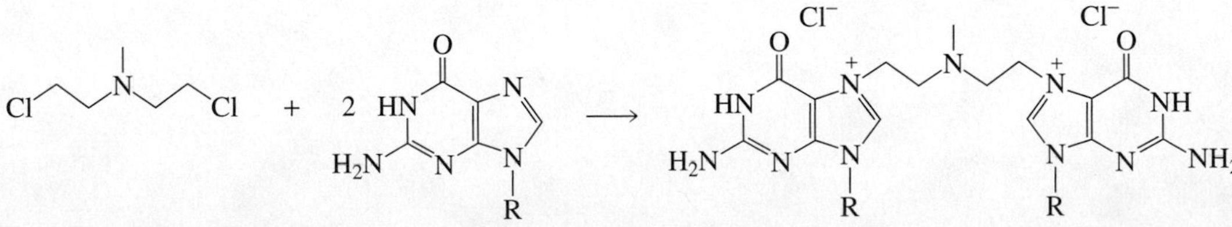

  b. The alkyl halide can react with two guanines, each in a different DNA chain, thereby cross-linking the chains. Propose a mechanism for the cross-linking reaction.

46. Triosephosphate isomerase (TIM) catalyzes the conversion of dihydroxyacetone phosphate to glyceraldehyde-3-phosphate. The enzyme's catalytic groups are Glu 165 and His 95. In the first step of the reaction, these catalytic groups function as a general-base and a general-acid catalyst, respectively. Propose a mechanism for the reaction.

dihydroxyacetone phosphate   triosephosphate isomerase   **glyceraldehyde-3-phosphate**

GLOSSARY

acid catalyst   a catalyst that increases the rate of a reaction by donating a proton.

active site   a pocket or cleft in an enzyme where the substrate is bound.

acyl–enzyme intermediate   an intermediate formed when an amino acid residue of an enzyme is acetylated.

base catalyst   a catalyst that increases the rate of a reaction by removing a proton.

catalyst   a species that increases the rate at which a reaction occurs without being consumed in the reaction. Because it does not change the equilibrium constant of the reaction, it does not change the amount of product that is formed.

covalent catalysis (nucleophilic catalysis)   catalysis that occurs as a result of a nucleophile forming a covalent bond with one of the reactants.

effective molarity   the concentration of the reagent that would be required in an intermolecular reaction for it to have the same rate as an intramolecular reaction.

electrostatic catalysis   stabilization of a charge by an opposite charge.

enzyme   a protein that is a catalyst.

general-acid catalysis   catalysis in which a proton is transferred to the reactant during the slow step of the reaction.

general-base catalysis   catalysis in which a proton is removed from the reactant during the slow step of the reaction.

induced-fit model   a model that describes the specificity of an enzyme for its substrate: the shape of the active site does not become completely complementary to the shape of the substrate until after the enzyme binds the substrate.

intramolecular catalysis (anchimeric assistance)   catalysis in which the catalyst that facilitates the reaction is part of the molecule undergoing reaction.

lock-and-key model   a model that describes the specificity of an enzyme for its substrate: the substrate fits the enzyme as a key fits a lock.

molecular recognition   the recognition of one molecule by another as a result of specific interactions; for example, the specificity of an enzyme for its substrate.

nucleophilic catalyst   a catalyst that increases the rate of a reaction by acting as a nucleophile.

pH-activity profile   a plot of the activity of an enzyme as a function of the pH of the reaction mixture.

relative rate   obtained by dividing the actual rate constant by the rate constant of the slowest reaction in the group being compared.

site-specific mutagenesis   a technique that substitutes one amino acid of a protein for another.

specific-acid catalysis   catalysis in which the proton is fully transferred to the reactant before the slow step of the reaction takes place.

specific-base catalysis   catalysis in which the proton is completely removed from the reactant before the slow step of the reaction takes place.

substrate   the reactant of an enzyme-catalyzed reaction.

The Organic Chemistry of the Coenzymes, Compounds Derived from Vitamins

From Chapter 24 of *Organic Chemistry*, Seventh Edition. Paula Yurkanis Bruice. Copyright © 2014 by Pearson Education, Inc.
All rights reserved.

The Organic Chemistry of the Coenzymes, Compounds Derived from Vitamins

limes: vitamins A, C, K, and folate

green beans: vitamins A, B$_1$, B$_6$, C, K, folate, and riboflavin

apples: vitamins A, C, K, and folate

cucumbers: vitamins A, C, K, and pantothenate

asparagus: vitamins A, B$_1$, B$_6$, C, E, K, folate, niacin, and riboflavin

*A **vitamin** is a substance needed in a small amount for normal body function that the body cannot synthesize. In this chapter, we will look at several organic reactions occurring in cells that require the participation of a vitamin.*

Many enzymes cannot catalyze a reaction without the help of a **cofactor.** Some cofactors are *metal ions*, while others are *organic molecules*.

A metal ion cofactor acts as a Lewis acid in a variety of ways to help an enzyme catalyze a reaction. It can help bind the substrate to the active site of the enzyme; it can complex with the substrate to increase its reactivity; it can coordinate with groups on the enzyme, causing them to align in a way that facilitates the reaction; and it can increase the nucleophilicity of water at the active site. Zn^{2+} plays an important role in the hydrolysis reaction catalyzed by carboxypeptidase A.

PROBLEM 1♦

How does the metal ion in carboxypeptidase A increase the enzyme's catalytic activity?

Cofactors that are organic molecules are called **coenzymes.** Coenzymes are derived from organic compounds commonly known as *vitamins*. The first such compound recognized to be essential in the diet was an amine (vitamin B$_1$), which led scientists to conclude incorrectly that all such compounds were amines. As a result, they were originally called vitamines ("amines required for life"). The *e* was later dropped from the name. Table 1 lists the vitamins and their chemically active coenzyme forms.

The acids, bases, and nucleophiles that catalyze organic reactions in the laboratory are similar to the acidic, basic, and nucleophilic side chains that enzymes use to catalyze organic reactions that occur in cells. In this chapter, we will see that coenzymes play a variety of chemical roles that the amino acid side chains of enzymes cannot play. Some coenzymes function as oxidizing and reducing agents, some allow electrons to be delocalized, some activate groups for further reaction, and some provide good nucleophiles or strong bases needed for a particular reaction.

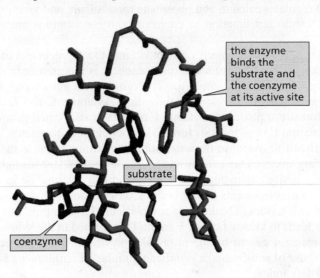

Because it would be highly inefficient for the body to use a compound only once and then discard it, coenzymes are recycled. Therefore, we will see that any coenzyme that is changed during the course of a reaction is subsequently converted back to its original form.

Table 1 The Vitamins, the Coenzymes for Which They Are Precursors, and the Chemical Functions of the Coenzymes

Vitamin	Coenzyme	Reaction catalyzed	Human deficiency disease
Water-Soluble Vitamins			
Niacin (vitamin B_3)	NAD^+, $NADP^+$	Oxidation	Pellagra
	NADH, NADPH	Reduction	
Riboflavin (vitamin B_2)	FAD	Oxidation	Skin inflammation
	$FADH_2$	Reduction	
Thiamine (vitamin B_1)	Thiamine pyrophosphate (TPP)	Acyl group transfer	Beriberi
Lipoic acid (lipoate)	Lipoate	Oxidation	—
	Dihydrolipoate	Reduction	
Pantothenic acid (pantothenate)	Coenzyme A (CoASH)	Acyl group transfer	—
Biotin (vitamin H)	Biotin	Carboxylation	—
Pyridoxine (vitamin B_6)	Pyridoxal phosphate (PLP)	Decarboxylation	Anemia
		Transamination	
		Racemization	
		C_α—C_β bond cleavage	
		α,β-Elimination	
		β-Substitution	
Vitamin B_{12}	Coenzyme B_{12}	Isomerization	Pernicious anemia
Folic acid (folate)	Tetrahydrofolate (THF)	One-carbon transfer	Megaloblastic anemia
Ascorbic acid (vitamin C)	—	—	Scurvy
Water-Insoluble Vitamins			
Vitamin A	—	—	—
Vitamin D	—	—	Rickets
Vitamin E	—	—	—
Vitamin K	Vitamin KH_2	Carboxylation	—

Early nutritional studies divided vitamins into two classes: water-soluble vitamins and water-insoluble vitamins (Table 1). Vitamin K is the only *water-insoluble vitamin* currently known to be a precursor of a coenzyme. Vitamin A is required for proper vision, vitamin D regulates calcium and phosphate metabolism, and vitamin E is an antioxidant. Because they do not function as coenzymes, these vitamins are not discussed in this chapter.

All the *water-soluble vitamins* except vitamin C are precursors of coenzymes. In spite of its name, vitamin C is not a vitamin because it is required in fairly high amounts and most mammals are able to synthesize it. Primates and guinea pigs cannot synthesize it, however, so it must be included in their diets. Vitamins C and E are radical inhibitors and therefore are antioxidants: vitamin C traps radicals formed in aqueous environments, whereas vitamin E traps radicals formed in nonpolar environments.

It is difficult to overdose on water-soluble vitamins because the body can generally eliminate any excess. One can, however, overdose on water-insoluble vitamins because they are *not* easily eliminated by the body and can accumulate in cell membranes and other nonpolar components of the body. Excess vitamin D, for example, causes calcification of soft tissues. The kidneys are particularly susceptible to calcification, which eventually leads to kidney failure. Vitamin D is formed in the skin as a result of a photochemical reaction caused by the ultraviolet rays from the sun. Because of the current, widespread use of sunscreens, a significant number of children are being found to have a vitamin D deficiency.

Vitamin B₁

Christiaan Eijkman (1858–1930) was a member of a medical team that was sent to the East Indies to study an outbreak of beriberi in 1886. At that time, all diseases were thought to be caused by microorganisms. When the microorganism that caused beriberi could not be found, the team left the East Indies but Eijkman stayed behind to become the director of a new bacteriological laboratory. There, in 1896, Eijkman accidentally discovered the cause of the disease when he noticed that laboratory chickens had developed the symptoms characteristic of beriberi. He found that the symptoms had appeared when a cook started feeding the chickens polished (white) rice meant for hospital patients. The symptoms disappeared when a new cook resumed giving brown rice to the chickens. Later it was recognized that thiamine (vitamin B₁) is present in rice husks, which are removed when the rice is polished.

1   NIACIN: THE VITAMIN NEEDED FOR MANY REDOX REACTIONS

Any enzyme that catalyzes an oxidation or a reduction reaction requires a coenzyme because none of the amino acid side chains are oxidizing or reducing agents. The coenzyme serves as the oxidizing or reducing agent. The enzyme's role is to hold the substrate and coenzyme together so that the oxidation or reduction reaction can take place (see the model earlier in this chapter).

The Pyridine Nucleotide Coenzymes

The coenzyme most commonly used by enzymes to catalyze an oxidation reaction is **nicotinamide adenine dinucleotide (NAD⁺)**. The coenzyme most commonly used to catalyze a reduction reaction is reduced **nicotinamide adenine dinucleotide phosphate (NADPH).**

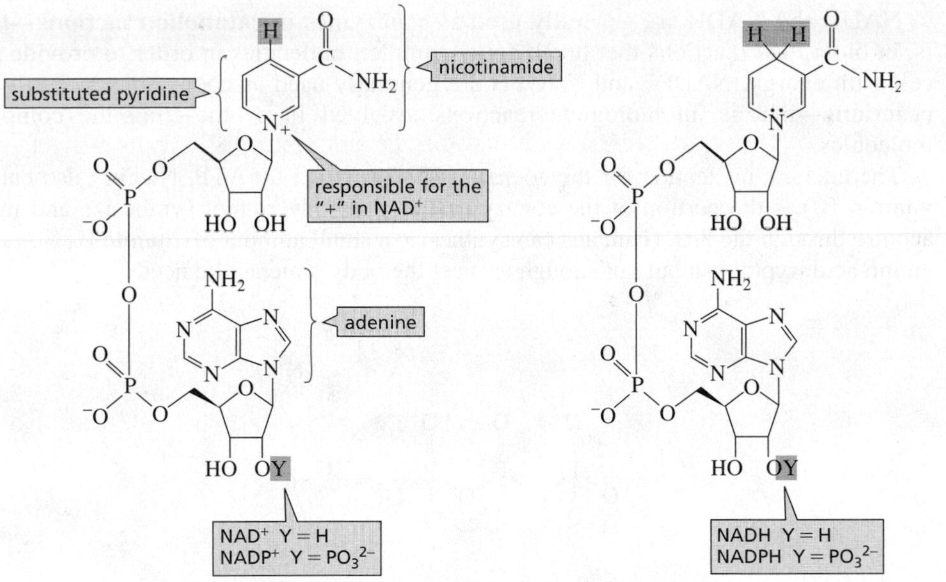

NAD⁺ is an oxidizing agent.

NADPH is a reducing agent.

When NAD^+ oxidizes a substrate, the coenzyme is reduced to NADH. When NADPH reduces a substrate, it is oxidized to $NADP^+$. Enzymes that catalyze oxidation reactions bind NAD^+ more tightly than they bind NADH. Once the oxidation reaction is completed, the relatively loosely bound NADH dissociates from the enzyme. Similarly, enzymes that catalyze reduction reactions bind NADPH more tightly than they bind $NADP^+$. Once the reduction reaction is complete, the relatively loosely bound $NADP^+$ dissociates from the enzyme. Because each of the coenzymes has a pyridine ring, they are known as the **pyridine nucleotide coenzymes.** We will see why NAD^+ is the most common oxidizing agent, whereas NADPH is the most common reducing agent later in this section.

$$\text{substrate}_{\text{reduced}} + \boxed{NAD^+} \xrightleftharpoons{\text{enzyme}} \text{substrate}_{\text{oxidized}} + \boxed{NADH} + H^+$$

$$\text{substrate}_{\text{oxidized}} + \boxed{NADPH} + H^+ \xrightleftharpoons{\text{enzyme}} \text{substrate}_{\text{reduced}} + \boxed{NADP^+}$$

NAD^+ is composed of two nucleotides linked together through their phosphate groups. A **nucleotide** consists of a heterocyclic compound attached, in a β-linkage, to C-1 of a phosphorylated ribose. A **heterocyclic compound** has one or more ring atoms that are not carbons.

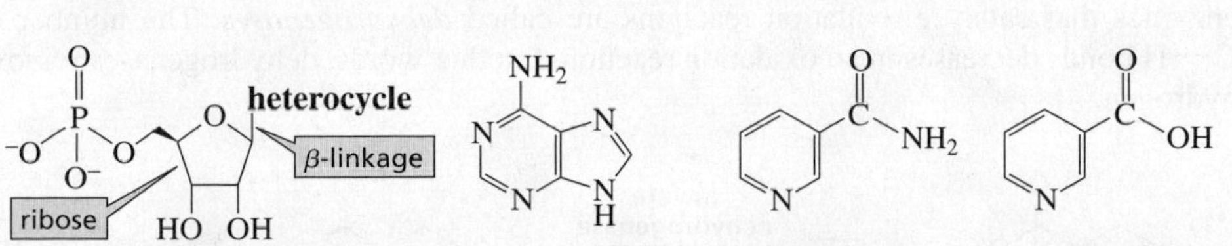

| a nucleotide | adenine | niacinamide
nicotinamide | niacin
nicotinic acid
vitamin B₃ |

The heterocyclic component of one of the nucleotides of NAD$^+$ is nicotinamide, and the heterocyclic component of the other is adenine. This accounts for the coenzyme's name (**n**icotinamide **a**denine **d**inucleotide). The positive charge in the NAD$^+$ abbreviation indicates the positively charged nitrogen of the pyridine ring. NADP$^+$ differs from NAD$^+$ only in having a phosphate group bonded to the 2'-OH group of the ribose of the adenine nucleotide—hence the addition of "P" to the name.

NAD$^+$ and NADH are generally used as coenzymes in **catabolic reactions**—that is, in biological reactions that break down complex molecules in order to provide the cell with energy. NADP$^+$ and NADPH are generally used as coenzymes in **anabolic reactions**—that is, in biological reactions involved in synthesizing the complex molecules.

The adenine nucleotide for the coenzymes is provided by ATP. Niacin (also called vitamin B$_3$) is the portion of the coenzyme that the body cannot synthesize and must acquire through the diet. (Humans can synthesize a small amount of vitamin B$_3$ from the amino acid tryptophan but not enough to meet the body's metabolic needs.)

**adenosine triphosphate
ATP**

Niacin Deficiency

A deficiency in niacin causes pellagra, a disease that begins with dermatitis and ultimately causes insanity and death. More than 120,000 cases of pellagra were reported in the United States in 1927, mainly among poor people with unvaried diets. A factor known to be present in preparations of vitamin B$_3$ prevented pellagra, but it was not until 1937 that the factor was identified as nicotinic acid. Mild deficiencies slow down metabolism, which is a potential contributing factor in obesity.

When bread companies started adding nicotinic acid to their bread, they insisted that its name be changed to niacin because nicotinic acid sounded too much like nicotine and they did not want their vitamin-enriched bread to be associated with a harmful substance.

Malate dehydrogenase is the enzyme that catalyzes the oxidation of the *secondary alcohol group* of malate to a *ketone*. (This is one of the reactions in the catabolic pathway known as the citric acid cycle.) The oxidizing agent in this reaction is NAD$^+$. Most enzymes that catalyze oxidation reactions are called *dehydrogenases*. The number of C—H bonds decreases in an oxidation reaction. In other words, dehydrogenases remove hydrogen.

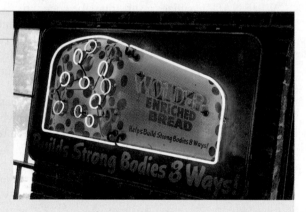

malate + NAD$^+$ $\xrightarrow{\text{malate dehydrogenase}}$ oxaloacetate + NADH + H$^+$

the alcohol is oxidized to a ketone

β-Aspartate-semialdehyde is reduced to homoserine in an anabolic pathway. NADPH is the reducing agent.

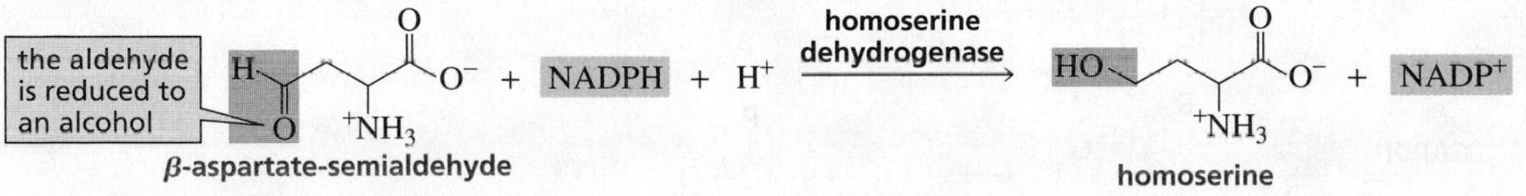

The differentiation between the coenzymes used in catabolism and those used in anabolism results from the strong specificity that each of the enzymes that catalyze these oxidation–reduction reactions exhibits for a particular coenzyme. For example, an enzyme that catalyzes an oxidation reaction can readily tell the difference between NAD^+ and $NADP^+$; if the enzyme is in a catabolic pathway, it will bind NAD^+ but not $NADP^+$.

The relative concentrations of the coenzymes in a cell also encourage the appropriate coenzyme to be bound. For example, catabolic reactions are predominately oxidation reactions, and anabolic reactions are predominately reduction reactions. The cell maintains its $[NAD^+]/[NADH]$ ratio near 1000 and its $[NADP^+]/[NADPH]$ ratio at about 0.01. Thus, NAD^+ is the oxidizing coenzyme and NADPH is the reducing coenzyme most available in a cell.

How Does NAD^+ Oxidize a Substrate?

All the chemistry of the pyridine nucleotide coenzymes takes place at the 4-position of the pyridine ring. The purpose of the rest of the molecule is to recognize the coenzyme and bind it to the proper site on the enzyme.

A substrate that is being *oxidized* donates a hydride ion (H^-) to the 4-position of the pyridine ring. In the following reaction, for example, the secondary alcohol is oxidized to a ketone. A basic amino acid side chain of the enzyme can help the reaction by removing a proton from the oxygen atom of the substrate. (In the mechanisms shown in this chapter, HB and :B^- represent amino acid side chains at the active site of the enzyme that can donate a proton or remove a proton, respectively.)

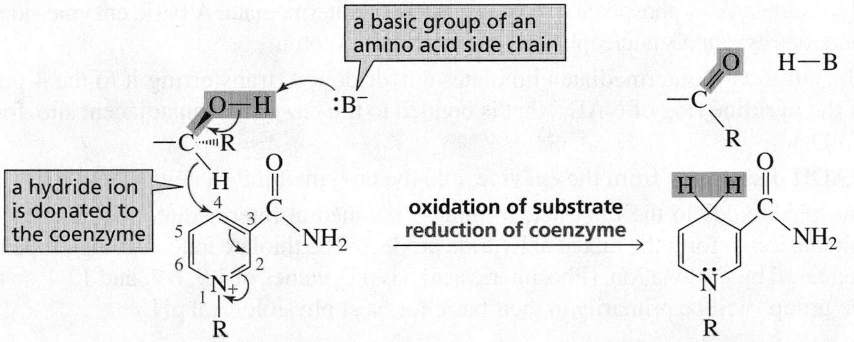

Glyceraldehyde-3-phosphate dehydrogenase (GAPDH) is another enzyme that uses NAD^+ to oxidize a substrate. The enzyme catalyzes the oxidation of the aldehyde group of glyceraldehyde-3-phosphate to a mixed anhydride of a carboxylic acid and phosphoric acid.

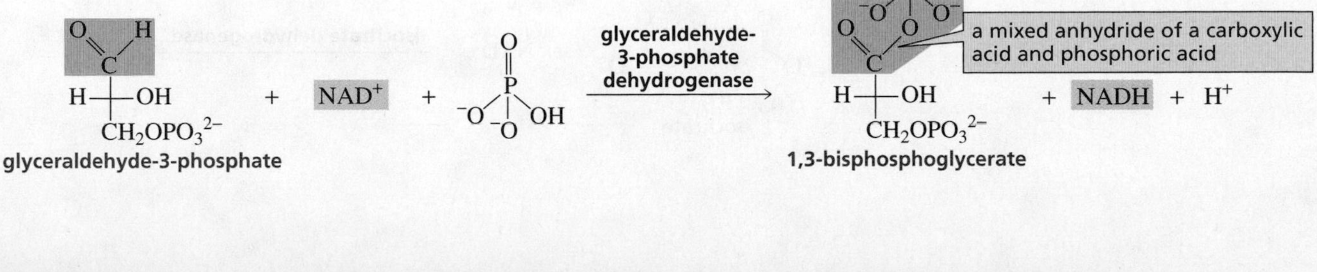

MECHANISM FOR THE REACTION CATALYZED BY GLYCERALDEHYDE-3-PHOSPHATE DEHYDROGENASE (GAPDH)

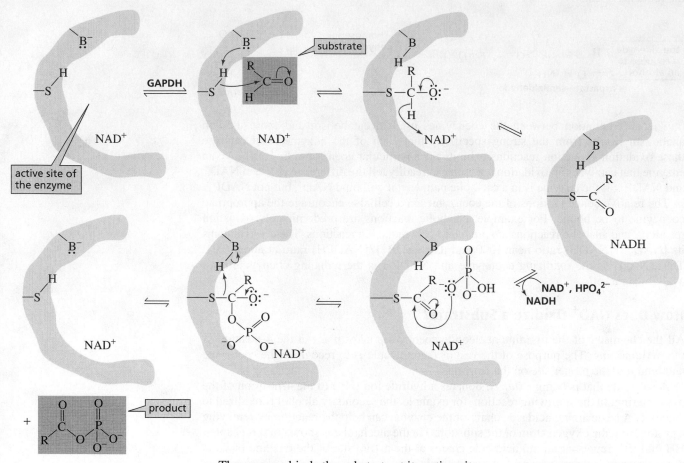

- The enzyme binds the substrate at its active site.
- An SH group (a nucleophile) of a cysteine side chain adds to the carbonyl carbon of glyceraldehyde-3-phosphate to form a tetrahedral intermediate. A basic enzyme side chain increases cysteine's nucleophilicity by removing a proton.
- The tetrahedral intermediate eliminates a hydride ion, transferring it to the 4-position of the pyridine ring of NAD^+ that is bonded to the enzyme at an adjacent site, forming NADH.
- NADH dissociates from the enzyme, and the enzyme binds a new NAD^+.
- Phosphate adds to the thioester, forming a tetrahedral intermediate that eliminates the thiolate ion to form the mixed anhydride product. The thiolate ion's leaving propensity is increased by protonation. (Phosphoric acid has pK_a values of 1.9, 6.7, and 12.4, so two of the groups will be primarily in their basic forms at physiological pH.)

Notice that at the end of the reaction the enzyme is exactly as it was at the beginning of the reaction, so another molecule of glyceraldehyde-3-phosphate can be converted to 1,3-bisphosphoglycerate.

PROBLEM 2♦

What is the product of the following reaction?

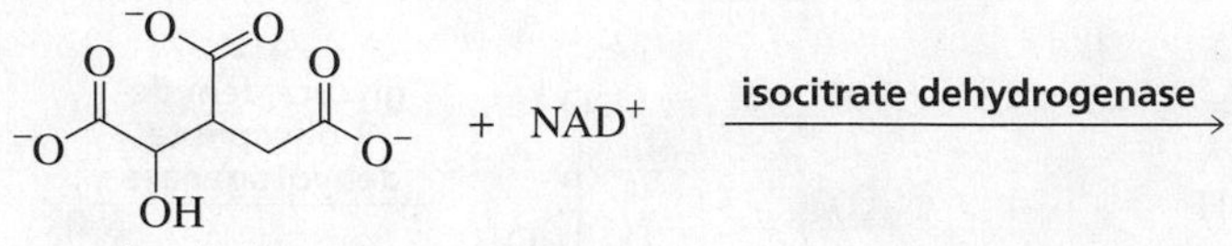

isocitrate $+ NAD^+$ $\xrightarrow{\text{isocitrate dehydrogenase}}$

How Does NADPH Reduce a Substrate?

The mechanism for reduction by NADPH is the reverse of the mechanism for oxidation by NAD^+. When a substrate is being *reduced,* the dihydropyridine ring donates a hydride ion from its 4-position to the substrate. An acidic amino acid side chain of the enzyme aids the reaction by protonating the substrate.

Because NADPH reduces compounds by donating a hydride ion, it can be considered the biological equivalent of $NaBH_4$ or $LiAlH_4$, the hydride donors that are used as reducing reagents in nonbiological reactions.

Why are the structures of biological oxidizing and reducing reagents so much more complicated than the structures of the reagents used to carry out the same reactions in the laboratory? NADPH is certainly more complicated than $NaBH_4$, although both reagents reduce compounds by donating a hydride ion. Much of the structural complexity of a coenzyme is for **molecular recognition**—to allow the coenzyme to be recognized and bound by the enzyme. As you study the coenzymes in this chapter, do not let the complexity of their structures intimidate you. Notice that only a small part of the coenzyme is actually involved in the chemical reaction.

Another reason for the difference in complexity is that reagents used in cells must be highly selective and therefore less reactive than a laboratory reagent. For example, a biological reducing agent cannot reduce just any reducible compound with which it comes into contact. Biological reactions have to be much more carefully controlled than that. Because the coenzymes are relatively unreactive compared with nonbiological reagents, the reaction between the substrate and the coenzyme does not occur at all or takes place very slowly without the enzyme. For example, NADPH will not reduce an aldehyde or a ketone unless an enzyme is present. $NaBH_4$ and $LiAlH_4$ are more reactive hydride donors—in fact, much too reactive to even exist in the aqueous environment of the cell. Similarly, NAD^+ is a much less reactive oxidizing agent than the typical oxidizing agent used in the laboratory; for example, NAD^+ will oxidize an alcohol only in the presence of an enzyme.

Because a biological reducing agent must be recycled (rather than being discarded in its oxidized form, as is the fate of a reducing agent used in a laboratory), the equilibrium constant for its oxidized and reduced forms is generally close to unity. Therefore, biological redox reactions are not highly exergonic. Rather, they are equilibrium reactions driven in the appropriate direction by the removal of reaction products as a result of their participation in subsequent reactions.

Unlike a laboratory reagent, an enzyme that catalyzes an oxidation reaction can distinguish between the two hydrogens on the carbon from which it catalyzes removal of a hydride ion. For example, alcohol dehydrogenase removes only the pro-*R* hydrogen (H_a) of ethanol. It is called the pro-*R* hydrogen because if it were replaced by deuterium, the asymmetric center would have the *R* configuration; H_b is the pro-*S* hydrogen.

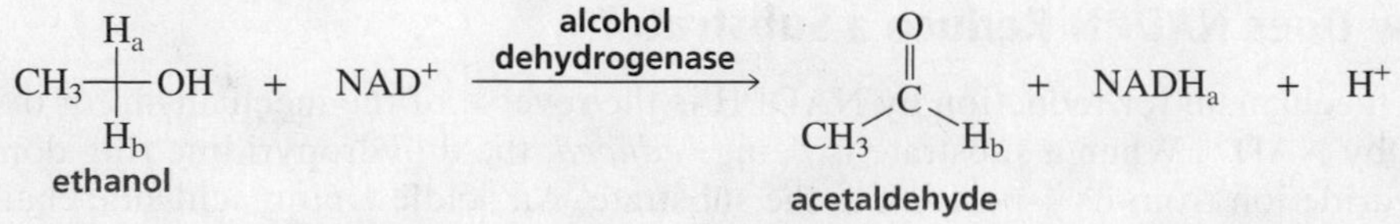

$$\underset{\substack{\text{ethanol}}}{\text{CH}_3\!\!-\!\!\overset{\displaystyle \text{H}_a}{\underset{\displaystyle \text{H}_b}{\text{C}}}\!\!-\!\!\text{OH}} + \text{NAD}^+ \xrightarrow{\substack{\text{alcohol} \\ \text{dehydrogenase}}} \underset{\substack{\text{acetaldehyde}}}{\overset{\displaystyle \text{O}}{\underset{}{\text{CH}_3\!\!-\!\!\overset{\|}{\text{C}}\!\!-\!\!\text{H}_b}}} + \text{NADH}_a + \text{H}^+$$

Similarly, an enzyme that catalyzes a reduction reaction can distinguish between the two hydrogens at the 4-position of the nicotinamide ring of NADPH. An enzyme has a specific binding site for the coenzyme and, when it binds the coenzyme, it blocks one of its sides. If the enzyme blocks the B-side of NADPH, the substrate will bind to the A-side and the H_a hydride ion will be transferred to the substrate. If the enzyme blocks the A-side of the coenzyme, then the substrate will have to bind to the B-side, and the H_b hydride ion will be transferred.

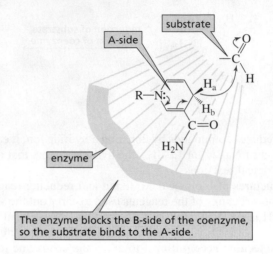

The enzyme blocks the B-side of the coenzyme, so the substrate binds to the A-side.

PROBLEM 3♦

What is the product of the following reaction?

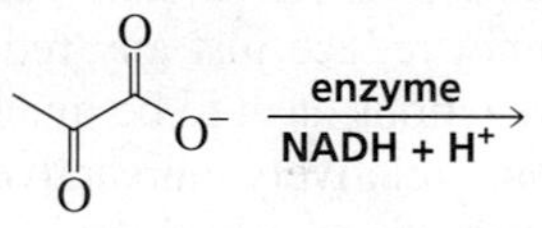

$$\underset{}{\overset{\displaystyle \text{O}}{\underset{\displaystyle \text{O}}{\text{C}}}\!\!-\!\!\text{O}^-} \xrightarrow[\text{NADH + H}^+]{\text{enzyme}}$$

---

2   RIBOFLAVIN: ANOTHER VITAMIN USED IN REDOX REACTIONS

Flavin adenine dinucleotide (FAD), like NAD$^+$, is a coenzyme used to oxidize a substrate. As its name indicates, FAD is a dinucleotide in which one of the heterocyclic components is flavin and the other is adenine. Notice that instead of ribose, FAD has a reduced ribose (a ribitol group). Flavin plus ribitol is the vitamin known as *riboflavin* or vitamin B$_2$. Flavin is a bright yellow compound; *flavus* is Latin for "yellow." A vitamin B$_2$ deficiency causes inflammation of the skin.

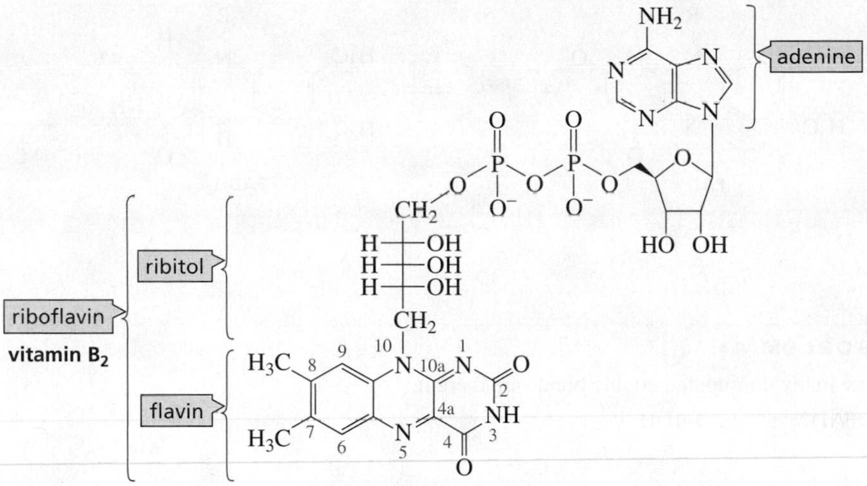

FAD

A *flavoprotein* is an enzyme that contains FAD. In most flavoproteins, FAD is bound quite tightly. Tight binding allows the enzyme to control the oxidation potential of the coenzyme. (The more positive the oxidation potential, the stronger is the oxidizing agent.) Consequently, some flavoproteins are stronger oxidizing agents than others.

How can we tell which enzymes use FAD and which use NAD^+ as the oxidizing coenzyme? A rough guideline is that NAD^+ is the coenzyme used in enzyme-catalyzed oxidation reactions involving carbonyl compounds (alcohols being oxidized to ketones, aldehydes, or carboxylic acids), whereas FAD is the coenzyme used in other types of oxidations. For example, in the following reactions, FAD oxidizes a dithiol to a disulfide, a saturated alkyl group to an alkene, and an amine to an imine. (This is only an approximate guideline, however, because FAD is used in some oxidations that involve carbonyl compounds, and NAD^+ is used in some oxidations that do not involve carbonyl compounds.)

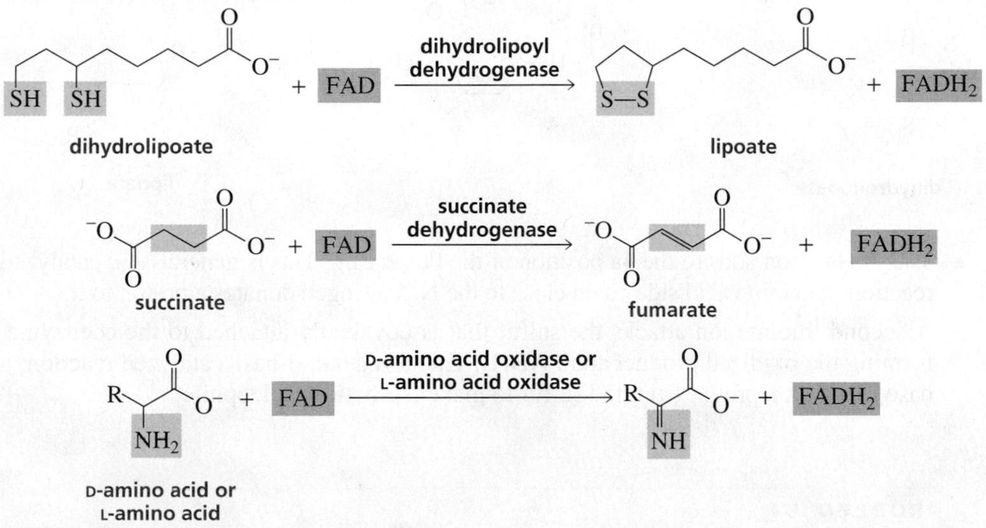

When FAD oxidizes a substrate, the coenzyme is reduced to $FADH_2$. $FADH_2$, like NADPH, is a reducing agent. All the oxidation–reduction chemistry takes place on the flavin ring. Reduction of the flavin ring disrupts the conjugated system, so the reduced coenzymes are less colored than their oxidized forms.

FAD is an oxidizing agent.

FADH₂ is a reducing agent.

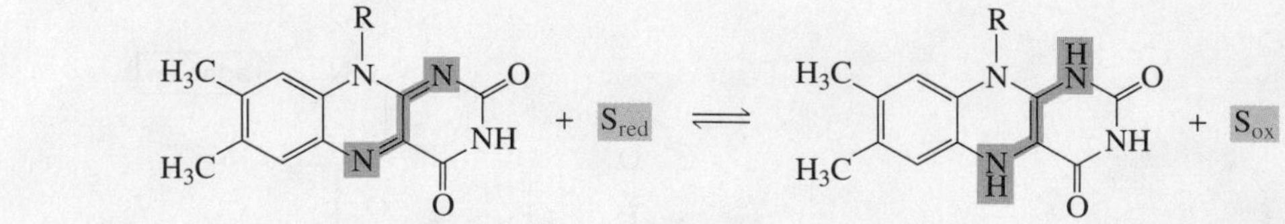

FAD + S_{red} ⇌ FADH₂ + S_{ox}

PROBLEM 4◆

How many conjugated double bonds are there in

a. FAD?   **b.** FADH₂?

The mechanism proposed for the FAD-catalyzed oxidation of dihydrolipoate to lipoate is shown here.

MECHANISM PROPOSED FOR THE REACTION CATALYZED BY DIHYDROLIPOYL DEHYDROGENASE

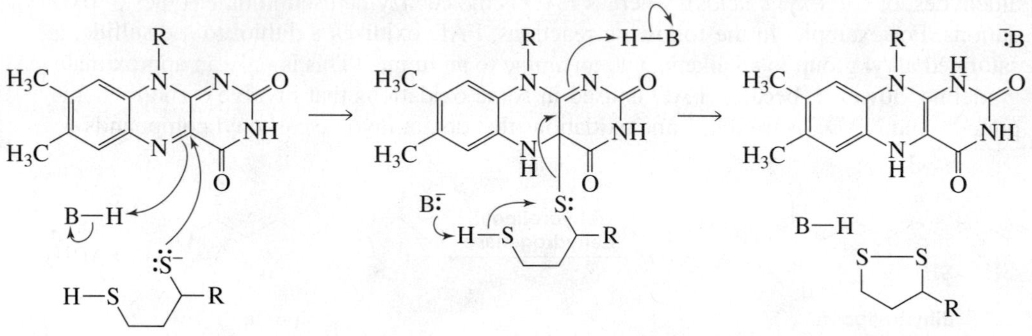

dihydrolipoate   lipoate

- The thiolate ion adds to the 4a position of the flavin ring. This is general-acid catalyzed reaction: an amino acid side chain close to the N-5 nitrogen donates a proton to it.

- A second thiolate ion attacks the sulfur that is covalently attached to the coenzyme, forming the oxidized product and FADH₂. This is a general-base catalyzed reaction: a base removes a proton from the sulfur to make it a better nucleophile.

PROBLEM 5◆

Instead of adding to the 4a position and protonating N-5, the thiolate ion could have added to the 10a position and protonated N-1. Why is addition to the 4a position favored? (*Hint:* Which nitrogen is a stronger base?)

The mechanism proposed for the FAD-catalyzed oxidation of succinate to fumarate is similar to the mechanism you have just seen for the FAD-catalyzed oxidation of dihydrolipoate to lipoate.

MECHANISM PROPOSED FOR THE REACTION CATALYZED BY SUCCINATE DEHYDROGENASE

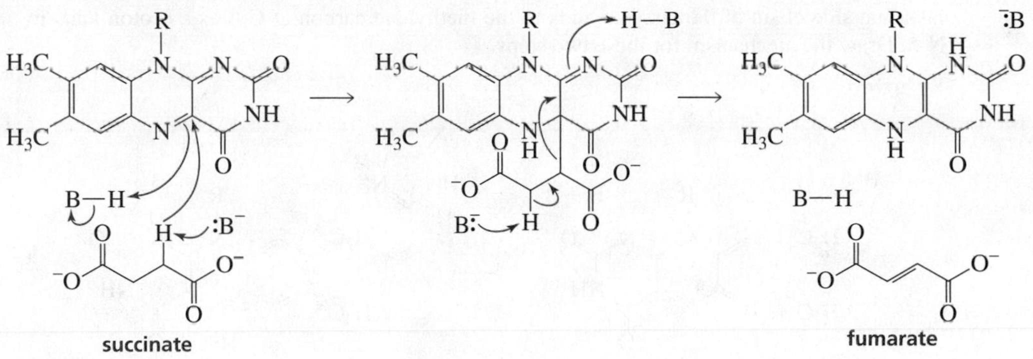

succinate   fumarate

- The A base removes a proton from the α-carbon, creating a nucleophile that adds to the 4a position of the flavin ring. A proton is donated simultaneously to the N-5 nitrogen.

- A base removes a proton from the other α-carbon, forming the oxidized product and $FADH_2$.

The mechanism proposed for the FAD-catalyzed oxidation of an amino acid to an imino acid is quite different from the two preceding FAD-catalyzed reactions. Notice that it is a concerted reaction—that is, all bond breaking and bond making occur in the same step.

MECHANISM PROPOSED FOR THE REACTION CATALYZED BY D- OR L-AMINO ACID OXIDASE

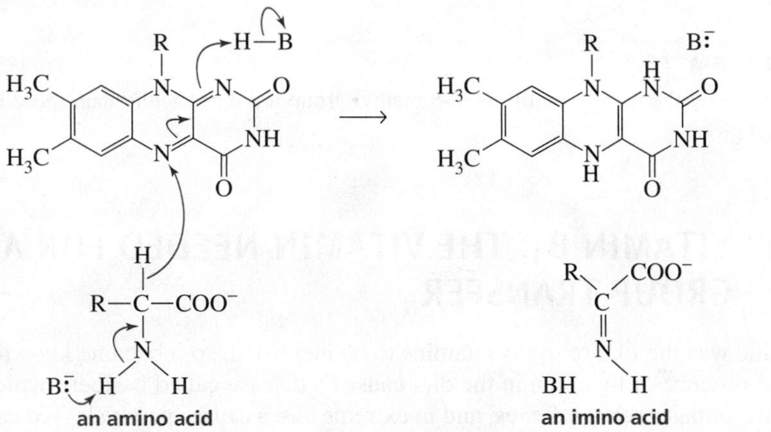

an amino acid   an imino acid

- As a basic amino acid side chain removes a proton from the nitrogen of the amino acid, the hydride ion adds to the N-5 position of the flavin ring, and an acidic amino acid side chain protonates N-1.

These mechanisms show that FAD is a more versatile coenzyme than NAD^+. Unlike NAD^+, which always uses the same mechanism, flavin coenzymes use many different mechanisms to carry out oxidation reactions.

Cells contain very low concentrations of FAD and much higher concentrations of NAD^+. Therefore, NAD^+ is only loosely bound to its enzyme and, after being reduced to NADH, it dissociates from the enzyme. In contrast, FAD is tightly bound to its enzyme. If it remains bound after being reduced to $FADH_2$, it has to be reoxidized to FAD before the enzyme can begin another round of catalysis. The oxidizing agent used for this reaction is NAD^+ or O_2. Therefore, an enzyme that uses an oxidizing coenzyme other than NAD^+ may still require NAD^+ to oxidize the reduced coenzyme.

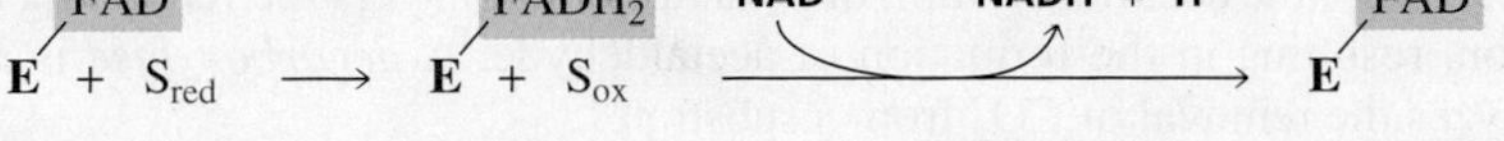

PROBLEM 6 Solved

In succinate dehydrogenase, FAD is covalently bound to its enzyme as a result of a base removing a proton from the C-8 methyl group and an acid donating a proton to N-1. Then a histidine side chain of the enzyme adds to the methylene carbon at C-8 as a proton adds to N-5. Draw the mechanism for these two steps.

Solution

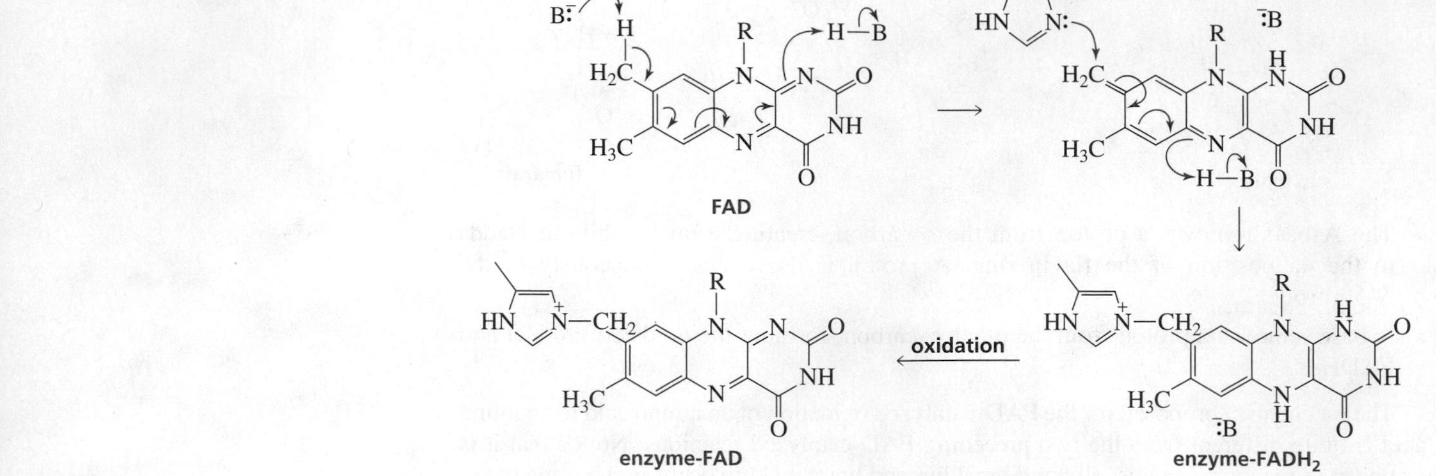

FAD

enzyme-FAD   oxidation   **enzyme-FADH$_2$**

Notice that FAD is reduced to FADH$_2$ during the process of being attached to the enzyme. It is subsequently oxidized back to FAD. Once the coenzyme is covalently attached to the enzyme, it does not come off.

PROBLEM 7

Explain why the hydrogens of the C-8 methyl group are more acidic than those of the C-7 methyl group.

3 VITAMIN B$_1$: THE VITAMIN NEEDED FOR ACYL GROUP TRANSFER

Thiamine was the first of the B vitamins to be identified, so it became known as vitamin B$_1$. The absence of thiamine in the diet causes a disease called beriberi, which damages the heart, impairs nerve reflexes, and in extreme cases causes paralysis (see earlier in this chapter).

Vitamin B$_1$ is used to form the coenzyme **thiamine pyrophosphate (TPP)**. TPP is the coenzyme required by enzymes that catalyze the transfer of an acyl group from one group to another.

**thiamine
vitamin B$_1$**   **thiamine pyrophosphate
TPP**

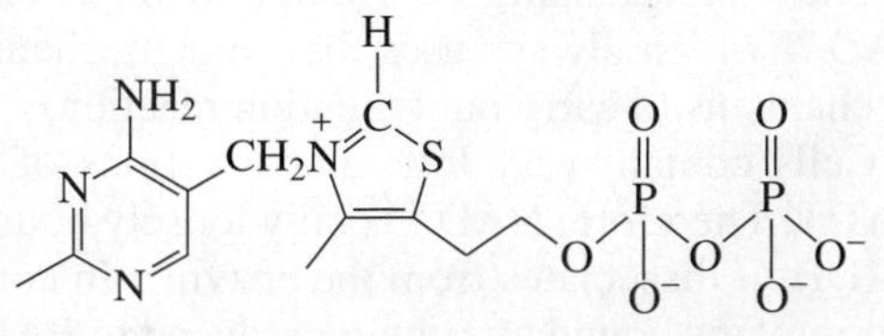

Pyruvate decarboxylase is an enzyme that requires thiamine pyrophosphate. This enzyme catalyzes the decarboxylation of pyruvate and transfers the remaining acyl group to a proton, resulting in the formation of acetaldehyde. A *decarboxylase* is an enzyme that catalyzes the removal of CO_2 from a substrate.

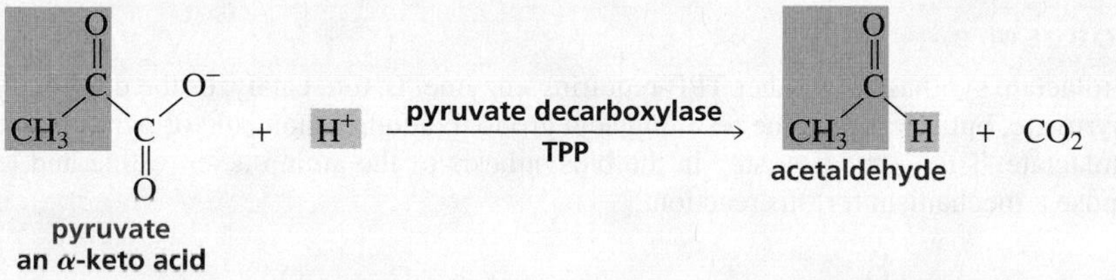

You may be wondering how an α-keto acid such as pyruvate can be decarboxylated, since the electrons left behind when CO_2 is removed cannot be delocalized onto the carbonyl oxygen. We will see that the thiazolium ring of the coenzyme provides the site to which the electrons can be delocalized.

The hydrogen bonded to the imine carbon of TPP is relatively acidic ($pK_a = 12.7$) compared to a hydrogen attached to other sp^2 carbons, because the ylide formed when the proton is removed is stabilized by the adjacent positively charged nitrogen. The TPP ylide is a good nucleophile.

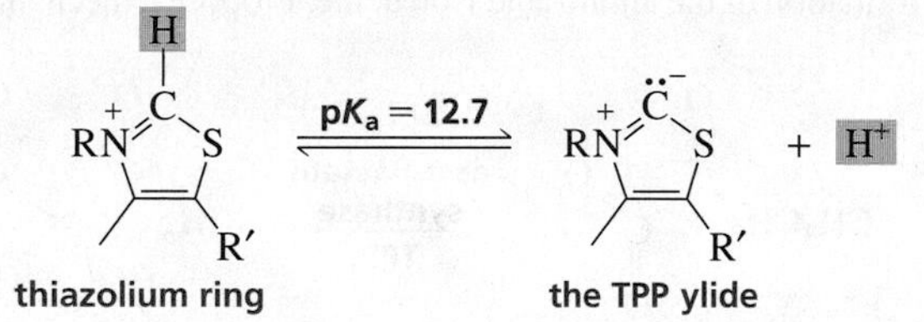

> Thiamine pyrophosphate (TPP) is required by enzymes that catalyze the transfer of an acyl group from one group to another.

MECHANISM FOR THE REACTION CATALYZED BY PYRUVATE DECARBOXYLASE

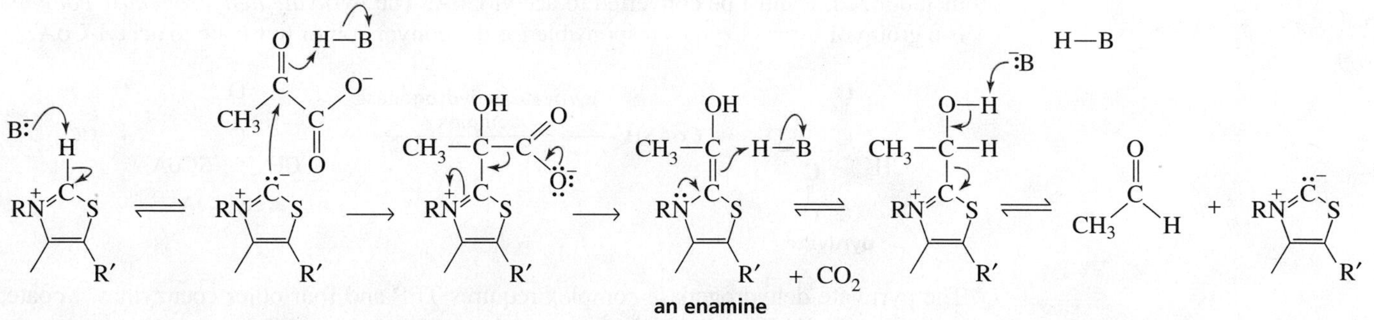

- After the proton is removed, the TPP ylide adds to the carbonyl carbon of the α-keto acid. An acid side chain of the enzyme increases the electrophilicity of the carbonyl carbon.

- The tetrahedral intermediate thus formed can easily undergo decarboxylation because the electrons left behind when CO_2 is removed can be delocalized onto the positively charged nitrogen. The decarboxylated product is an enamine.

- Protonation of the enamine on carbon and a subsequent elimination reaction forms acetaldehyde and regenerates the TPP ylide.

A site to which electrons can be delocalized is called an **electron sink.** The positively charged nitrogen of TPP is a more effective electron sink than the β-keto group of a β-keto acid, a class of compounds that we have seen are fairly easily decarboxylated.

PROBLEM 8

Draw structures that show the similarity between the decarboxylation of the pyruvate–TPP intermediate and the decarboxylation of a β-keto acid.

PROBLEM 9

Acetolactate synthase is another TPP-requiring enzyme. It, too, catalyzes the decarboxylation of pyruvate, but it transfers the resulting acyl group to another molecule of pyruvate, forming acetolactate. This is the first step in the biosynthesis of the amino acids valine and leucine. Propose a mechanism for this reaction.

pyruvate — acetolactate

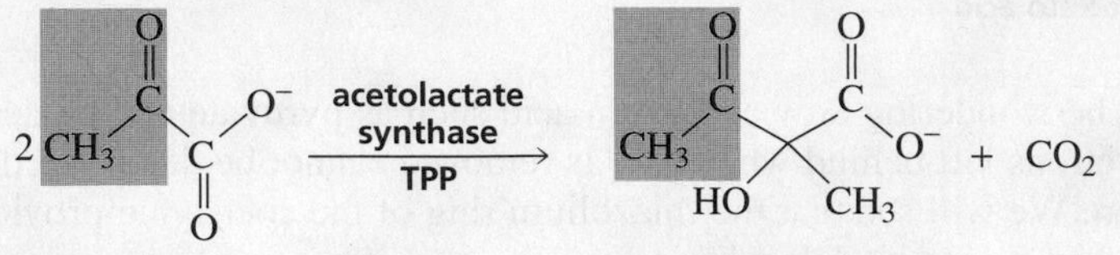

PROBLEM 10

Acetolactate synthase can also transfer the acyl group from pyruvate to α-ketobutyrate. This is the first step in the formation of the amino acid isoleucine. Propose a mechanism for this reaction.

α-ketobutyrate — α-aceto-α-hydroxybutyrate

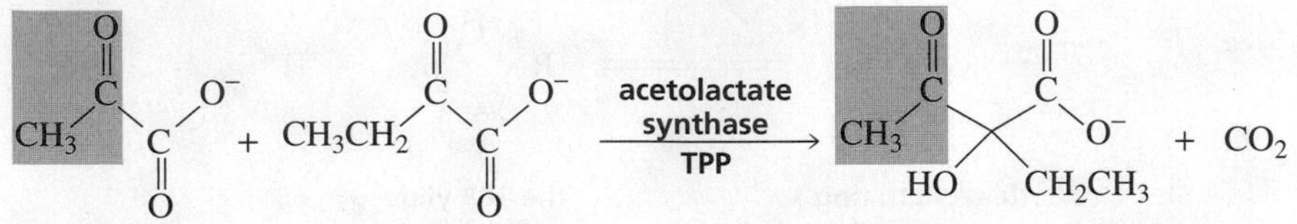

The final product of carbohydrate metabolism is pyruvate. For pyruvate to be further metabolized, it must be converted to acetyl-CoA. The *pyruvate dehydrogenase complex* is a group of three enzymes responsible for the conversion of pyruvate to acetyl-CoA.

pyruvate — acetyl-CoA

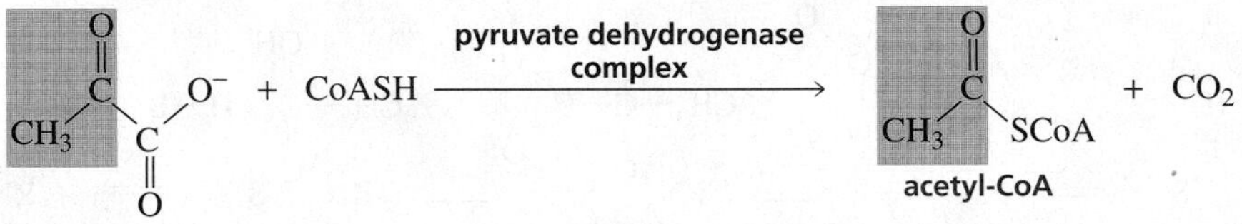

The pyruvate dehydrogenase complex requires TPP and four other coenzymes: lipoate, coenzyme A, FAD, and NAD^+.

MECHANISM FOR THE REACTION CATALYZED BY THE PYRUVATE DEHYDROGENASE COMPLEX

enamine

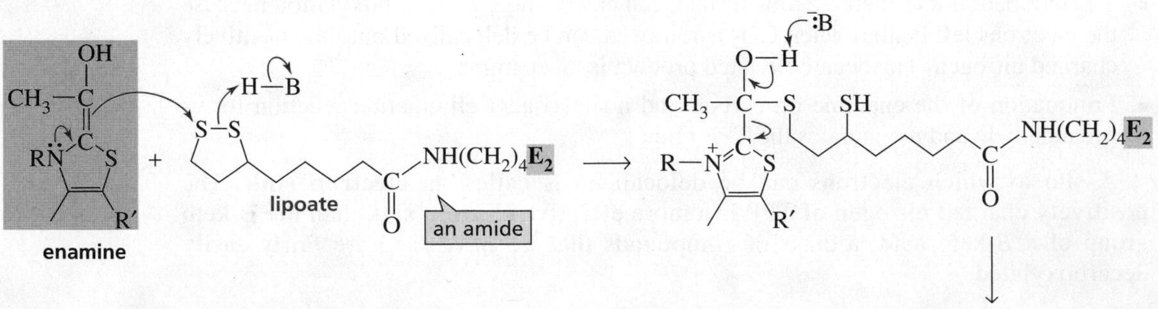

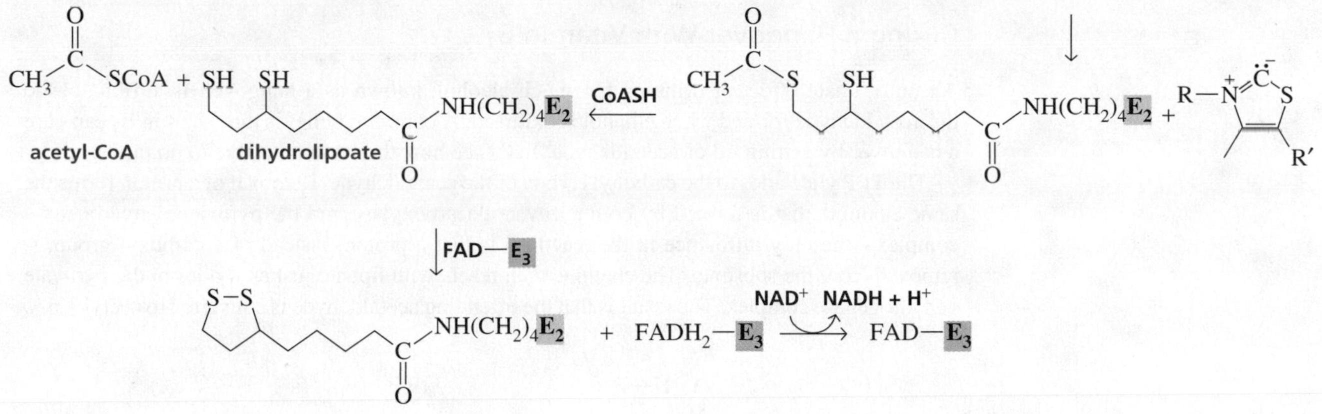

- The first enzyme in the complex catalyzes the reaction of the TPP ylide with pyruvate to form an enamine identical to the one formed by pyruvate decarboxylase and by the enzyme in Problems 9 and 10.

- The second enzyme of the complex (E_2) uses a lysine side chain to form an amide with its coenzyme (**lipoate**). The disulfide bond of lipoate is broken when it undergoes nucleophilic attack by the enamine.

- The TPP ylide is eliminated from the tetrahedral intermediate.

- **Coenzyme A (CoASH)** reacts with the thioester in a transthioesterification reaction (one thioester is converted to another), substituting coenzyme A for dihydrolipoate. At this point, the final reaction product (acetyl-CoA) has been formed.

- Before another catalytic cycle can occur, dihydrolipoate must be oxidized back to lipoate. This is done by the third enzyme (E_3), an FAD-requiring enzyme. We saw the mechanism for this reaction in Section 2. When dihydrolipoate is oxidized by FAD, the coenzyme is reduced to $FADH_2$.

- NAD^+ oxidizes $FADH_2$ back to FAD.

The vitamin needed to make coenzyme A is pantothenate (see Table 1). In coenzyme A, pantothenate is attached to a decarboxylated cysteine and a phosphorylated ADP. CoASH is used in biological systems to activate carboxylic acids by converting them to thioesters.

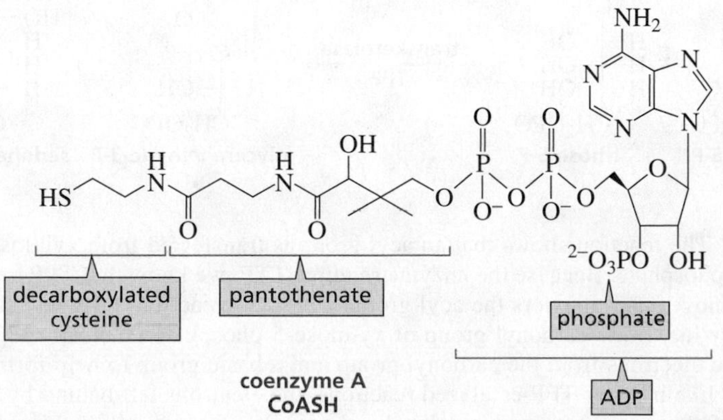

Curing a Hangover With Vitamin B$_1$

An unfortunate effect of drinking too much alcohol, known as a hangover, is attributable to the acetaldehyde formed when ethanol is oxidized. A common belief is that vitamin B$_1$ can cure a hangover by getting rid of acetaldehyde. Let's see how the vitamin is able to do this.

The TPP ylide adds to the carbonyl carbon of the acetaldehyde. Removal of a proton forms the same enamine that is formed by both pyruvate decarboxylase and the pyruvate dehydrogenase complex—the only difference in the reactions is that a proton, instead of a carboxyl group, is removed from the substrate. The enamine then reacts with lipoate just as it does in the pyruvate dehydrogenase complex. The result is that the offending acetaldehyde is converted to acetyl-CoA.

an enamine

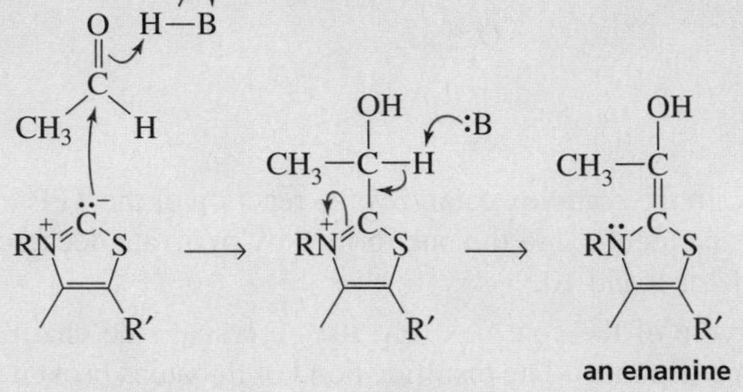

There is a limit to the amount of acetaldehyde that can be converted to acetyl-CoA in a given amount of time, so the vitamin can cure only hangovers that result from moderate drinking.

PROBLEM 11 Solved

TPP is a coenzyme for transketolase, the enzyme that catalyzes the conversion of a ketopentose (xylulose-5-phosphate) and an aldopentose (ribose-5-phosphate) into an aldotriose (glyceraldehyde-3-phosphate) and a ketoheptose (sedoheptulose-7-phosphate). Notice that the total number of carbons in the reactants and products is the same ($5 + 5 = 3 + 7$). Propose a mechanism for this reaction.

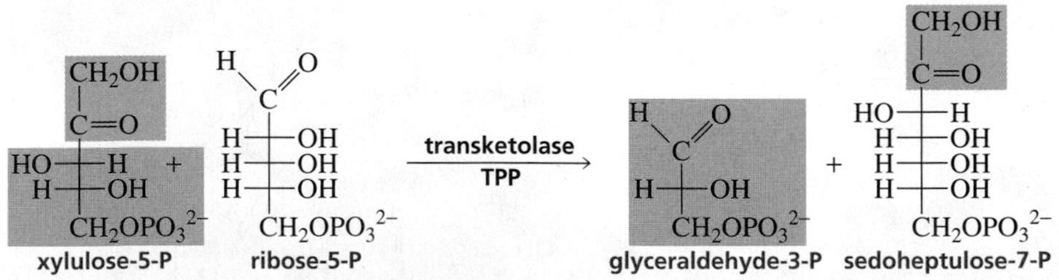

xylulose-5-P   ribose-5-P   glyceraldehyde-3-P   sedoheptulose-7-P

Solution The reaction shows that an acyl group is transferred from xylulose-5-phosphate to ribose-5-phosphate. Because the enzyme requires TTP, we know that TPP must be the species that removes and transfers the acyl group. Thus, the reaction must start by the addition of the TPP ylide to the carbonyl group of xylulose-5-phosphate. We can add an acid group to accept the electrons from the carbonyl group and a basic group to help form the enamine. Notice that, like in other TPP-catalyzed reactions, the electrons left behind by the group that is removed from the acyl group are delocalized onto the nitrogen of the thiazolium ring. The enamine then transfers the acyl group to the carbonyl carbon of ribose-5-phosphate. Again, an acid group accepts the electrons from the carbonyl group, and a basic group helps eliminate the TPP ylide.

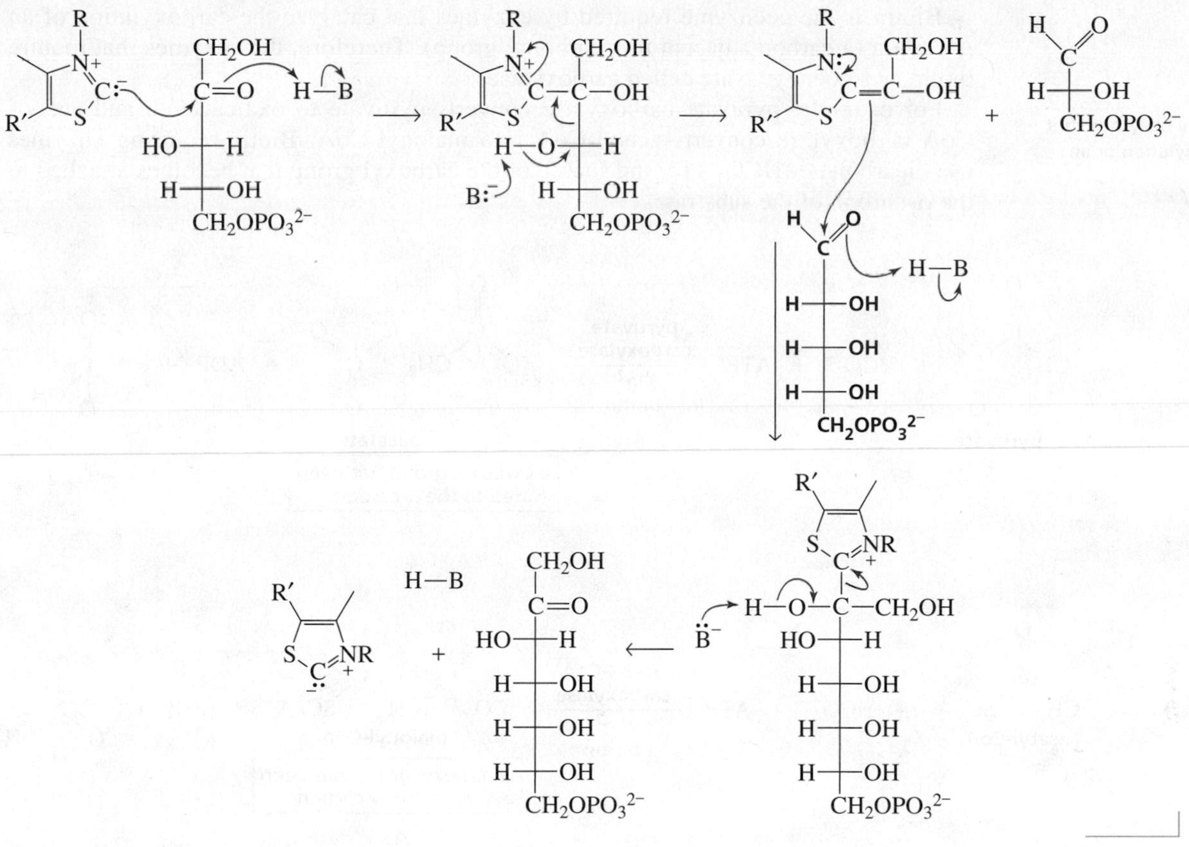

Notice the similar function of TPP in all TPP-requiring enzymes. In each reaction, the TPP ylide adds to a carbonyl carbon of the substrate and allows a bond to that carbon to be broken because the electrons left behind can be delocalized into the thiazolium ring. The acyl group is then transferred—to a proton in the case of pyruvate decarboxylase, to coenzyme A (via lipoate) in the pyruvate dehydrogenase system, and to a carbonyl group in Problems 9, 10, and 11.

4   VITAMIN H: THE VITAMIN NEEDED FOR CARBOXYLATION OF AN α-CARBON

Biotin (vitamin H) is an unusual vitamin because it is synthesized by bacteria that live in the intestine. Consequently, biotin does not have to be included in our diet, and deficiencies are rare. Biotin deficiencies, however, can be found in people who maintain a diet high in raw eggs. Egg whites contain a protein (called avidin) that binds biotin tightly and thereby prevents it from acting as a coenzyme. When eggs are cooked, avidin is denatured, and the denatured protein does not bind biotin. Biotin, like lipoic acid, is attached to its enzyme (E) by forming an amide with the amino group of a lysine side chain.

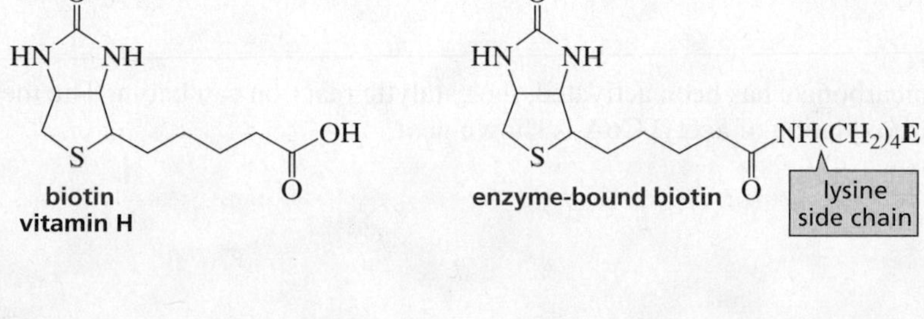

biotin
vitamin H

enzyme-bound biotin   lysine side chain

1185

Biotin is the coenzyme required by enzymes that catalyze the carboxylation of an α-carbon (a carbon adjacent to a carbonyl group). Therefore, the enzymes that require biotin as a coenzyme are called carboxylases.

For example, pyruvate carboxylase converts pyruvate to oxaloacetate and acetyl-CoA carboxylase converts acetyl-CoA into malonyl-CoA. Biotin-requiring enzymes use bicarbonate (HCO_3^-) for the source of the carboxyl group that becomes attached to the α-carbon of the substrate.

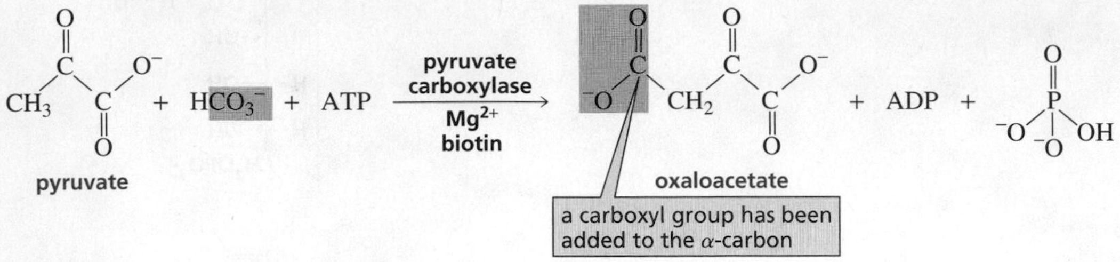

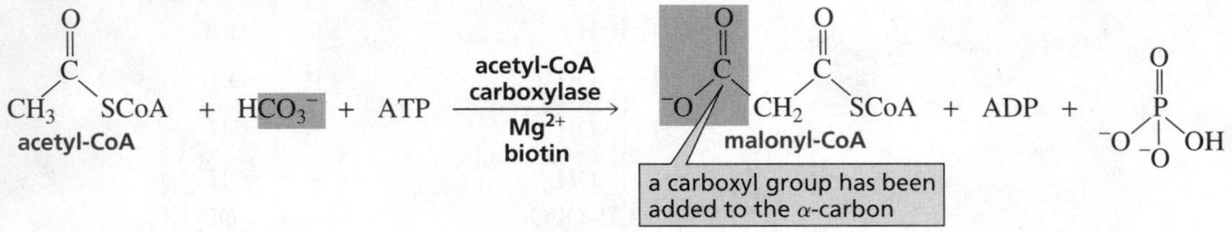

a carboxyl group has been added to the α-carbon

a carboxyl group has been added to the α-carbon

In addition to bicarbonate, biotin-requiring enzymes also require ATP and Mg^{2+}. The function of Mg^{2+} is to decrease the overall negative charge on ATP by complexing with two of its negatively charged oxygens. Unless its negative charge is reduced, ATP cannot be approached by a nucleophile.

The function of ATP is to increase the reactivity of bicarbonate by converting it to "activated bicarbonate"—a compound with a good leaving group. To form "activated bicarbonate," bicarbonate attacks the γ-phosphorous of ATP and expels ADP. Notice that "activated bicarbonate" is a mixed anhydride of carbonic acid and phosphoric acid.

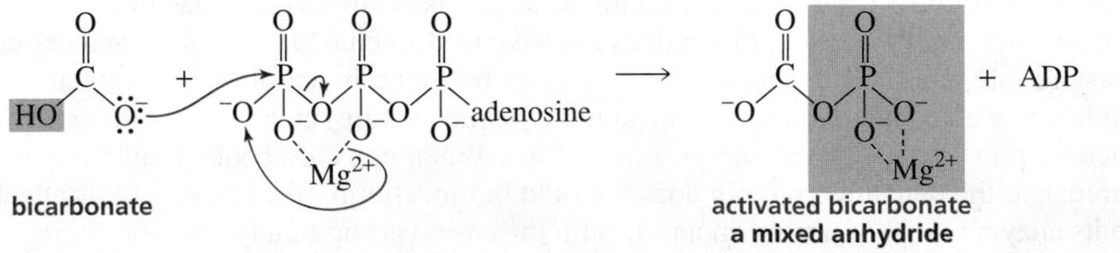

Once bicarbonate has been activated, the catalytic reaction can begin. The mechanism for the carboxylation of acetyl CoA is shown next.

MECHANISM FOR THE REACTION CATALYZED BY ACETYL-COA CARBOXYLASE

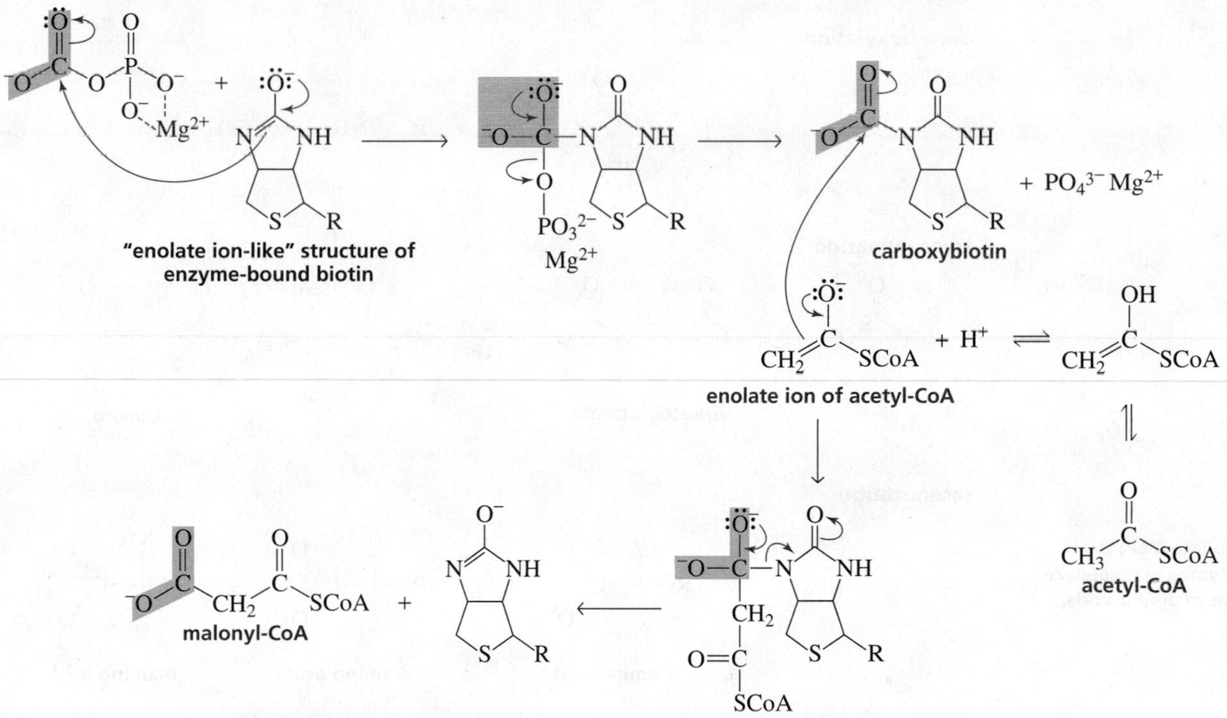

"enolate ion-like" structure of enzyme-bound biotin

carboxybiotin

$+ PO_4^{3-} Mg^{2+}$

enolate ion of acetyl-CoA

$+ H^+ \rightleftharpoons$

acetyl-CoA

malonyl-CoA

- Biotin reacts with activated bicarbonate in a nucleophilic addition–elimination reaction to form carboxybiotin. Because the nitrogen of an amide is not nucleophilic, the active form of biotin has an enolate ion–like structure.

- The substrate (in this case, the enolate ion of acetyl-CoA) reacts with carboxybiotin in a second nucleophilic addition–elimination reaction that transfers the carboxyl group from carboxybiotin to the substrate.

All biotin-requiring enzymes follow the same three steps: activation of bicarbonate by ATP, reaction of activated bicarbonate with biotin to form carboxybiotin, and transfer of the carboxyl group from carboxybiotin to the substrate.

5 VITAMIN B$_6$: THE VITAMIN NEEDED FOR AMINO ACID TRANSFORMATIONS

The coenzyme **pyridoxal phosphate (PLP)** is derived from vitamin B$_6$, which is also known as pyridoxine. (Pyridoxal's "al" suffix indicates the coenzyme is an aldehyde.) A deficiency in vitamin B$_6$ causes anemia; severe deficiencies can cause seizures and death.

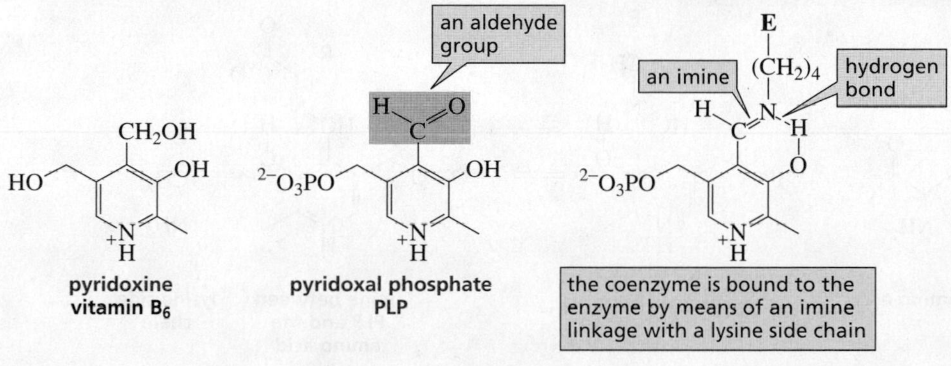

an aldehyde group

an imine

hydrogen bond

E (CH$_2$)$_4$

pyridoxine vitamin B$_6$

pyridoxal phosphate PLP

the coenzyme is bound to the enzyme by means of an imine linkage with a lysine side chain

PLP is required by enzymes that catalyze certain reactions of amino acids, such as decarboxylation, transamination, racemization, C_α—C_β bond cleavage, and α,β-elimination.

decarboxylation

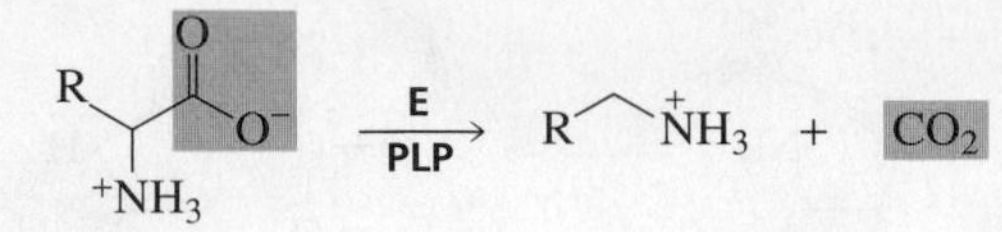

transamination

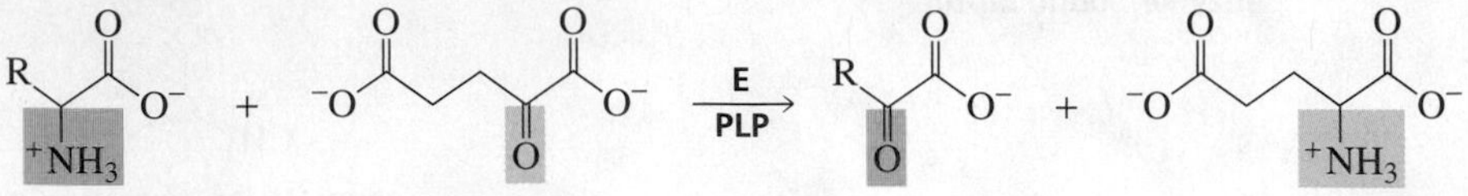

α-ketoglutarate

glutamate

racemization

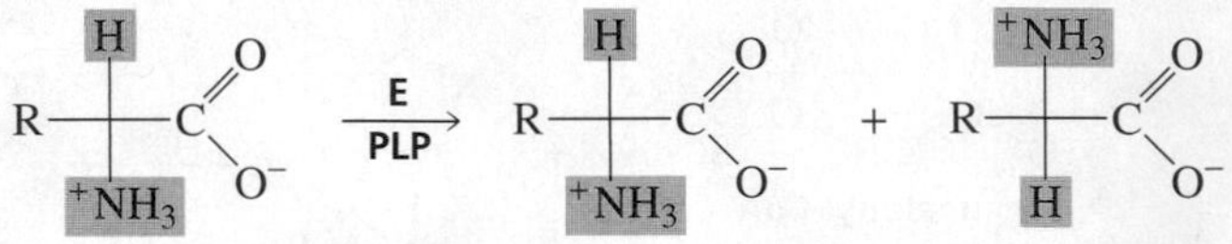

L-amino acid

L-amino acid

D-amino acid

C_α—C_β bond cleavage

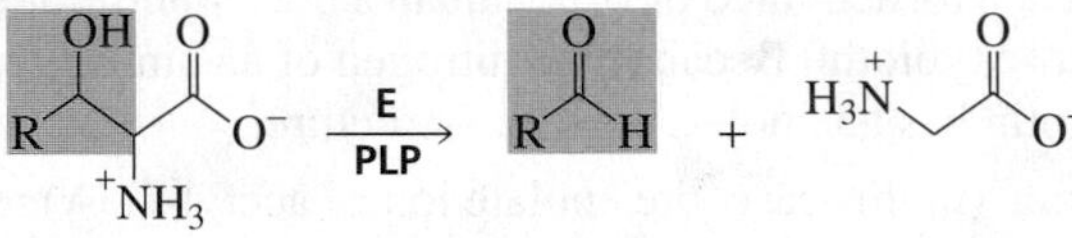

α,β-elimination

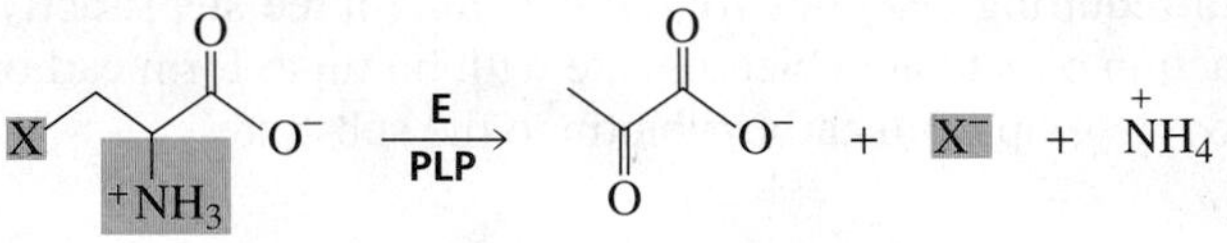

> Pyridoxal phosphate (PLP) is required by enzymes that catalyze certain reactions of amino acids.

PLP becomes attached to its enzyme by forming an imine with the amino group of a lysine side chain. The first step in all reactions catalyzed by PLP-requiring enzymes is a **transimination** reaction—a reaction that converts one imine into another imine. Thus, the amino acid substrate reacts with the imine formed by PLP and the lysine side chain, and forms an imine with PLP and releases the lysine side chain.

a transimination reaction

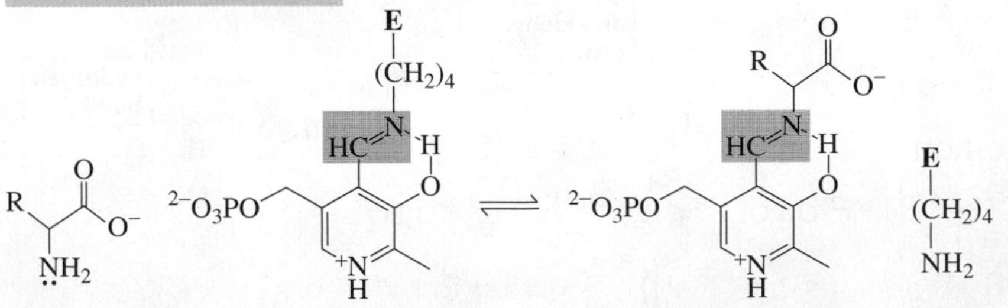

amino acid

imine between
PLP and the lysine
side chain

imine between
PLP and the
amino acid

lysine side
chain

Once the amino acid has formed an imine with PLP, the next step is to break a bond to the α-carbon of the amino acid. Decarboxylation breaks the bond joining the carboxyl group to the α-carbon; transamination, racemization, and α,β-elimination break the bond joining the hydrogen to the α-carbon; and C_α—C_β bond cleavage breaks the bond joining the R group to the α-carbon.

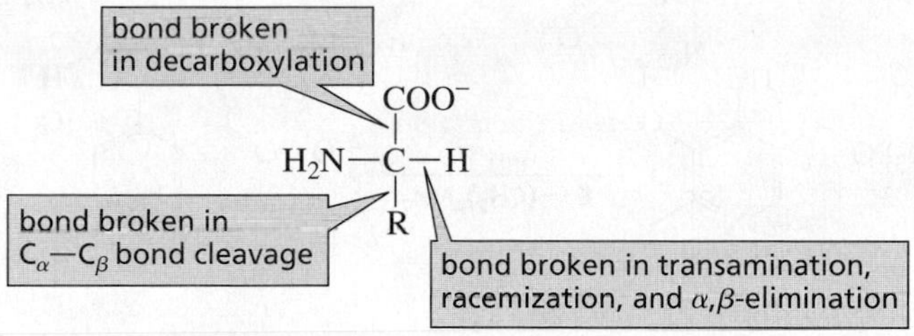

A bond to the α-carbon can be broken because the electrons left behind when the bond breaks can be delocalized onto the positively charged protonated nitrogen of the pyridinium ring. Thus, the protonated nitrogen is an electron sink. The coenzyme loses much of its activity if the OH group is removed from the pyridine ring. Apparently, the hydrogen bond formed by the OH group helps weaken the bond to the α-carbon.

Decarboxylation

All enzymes that catalyze the decarboxylation of an amino acid do so by the mechanism shown here.

MECHANISM FOR THE PLP-CATALYZED DECARBOXYLATION OF AN AMINO ACID

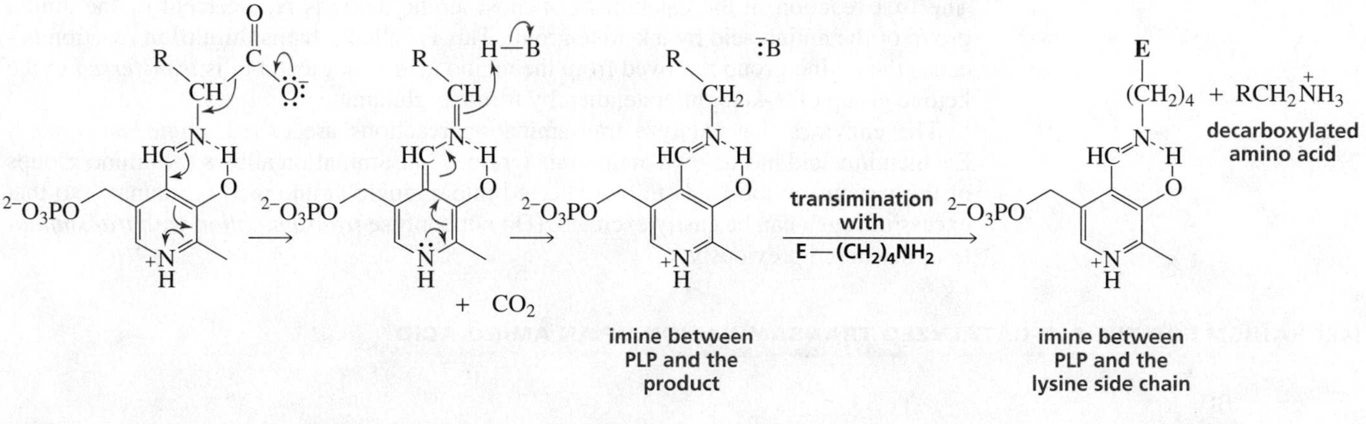

- The carboxyl group is removed from the α-carbon in the first step; the electrons left behind are delocalized onto the positively charged nitrogen.
- The aromaticity of the pyridinium ring is reestablished by electron rearrangement and protonation of what was the α-carbon of the amino acid.
- The last step in all PLP-requiring enzymes is another transimination reaction. The imine formed by PLP and the product of the enzyme-catalyzed reaction reacts with the lysine side chain of the enzyme, forming an imine with PLP and releasing the product.

Racemization

The mechanism for the PLP-catalyzed racemization of an L-amino acid is shown next. Notice that the mechanism for the interconversion of the enantiomers is the same as the mechanism for decarboxylation except for the group removed from the α-carbon of the amino acid in the first step.

MECHANISM FOR THE PLP-CATALYZED RACEMIZATION OF AN L-AMINO ACID

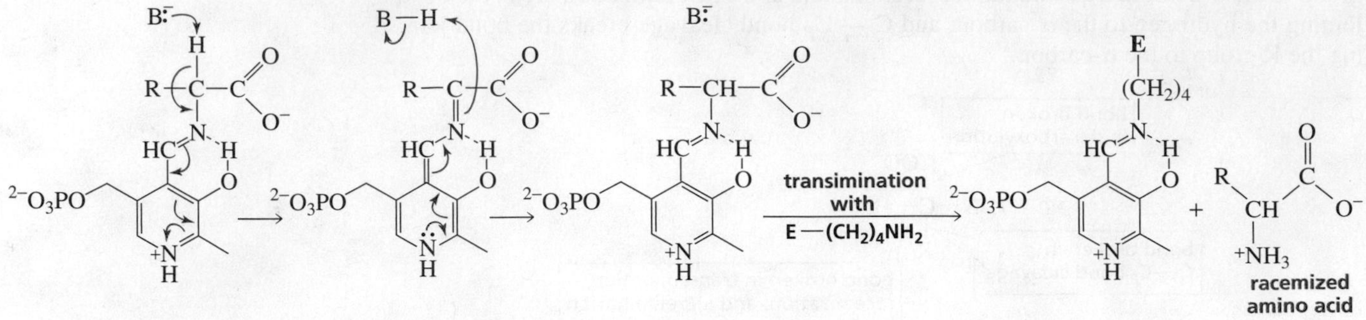

- A proton is removed from the α-carbon, and the electrons left behind are delocalized onto the positively charged nitrogen.

- The aromaticity of the pyridinium ring is reestablished by electron rearrangement and protonation of what was the α-carbon of the amino acid.

- A transimination reaction with a lysine side chain releases the product of the reaction (the racemized amino acid) and regenerates the imine between the enzyme and PLP.

In the second step of the reaction, the proton can be donated to the sp^2 carbon from either side of the plane defined by the double bond. Consequently, both D- and L-amino acids are formed. In other words, the L-amino acid is racemized.

Transamination

The first reaction in the catabolism of most amino acids is replacement of the amino group of the amino acid by a ketone group. This is called a **transamination** reaction because the amino group removed from the amino acid is not lost, but is *transferred* to the ketone group of α-ketoglutarate, thereby forming glutamate.

The enzymes that catalyze transamination reactions are called *aminotransferases*. Each amino acid has its own aminotransferase. Transamination allows the amino groups of the various amino acids to be collected into a single amino acid (glutamate) so that excess nitrogen can be easily excreted. (Do not confuse *transamination* with *transimination*, discussed previously.)

MECHANISM FOR THE PLP-CATALYZED TRANSAMINATION OF AN AMINO ACID

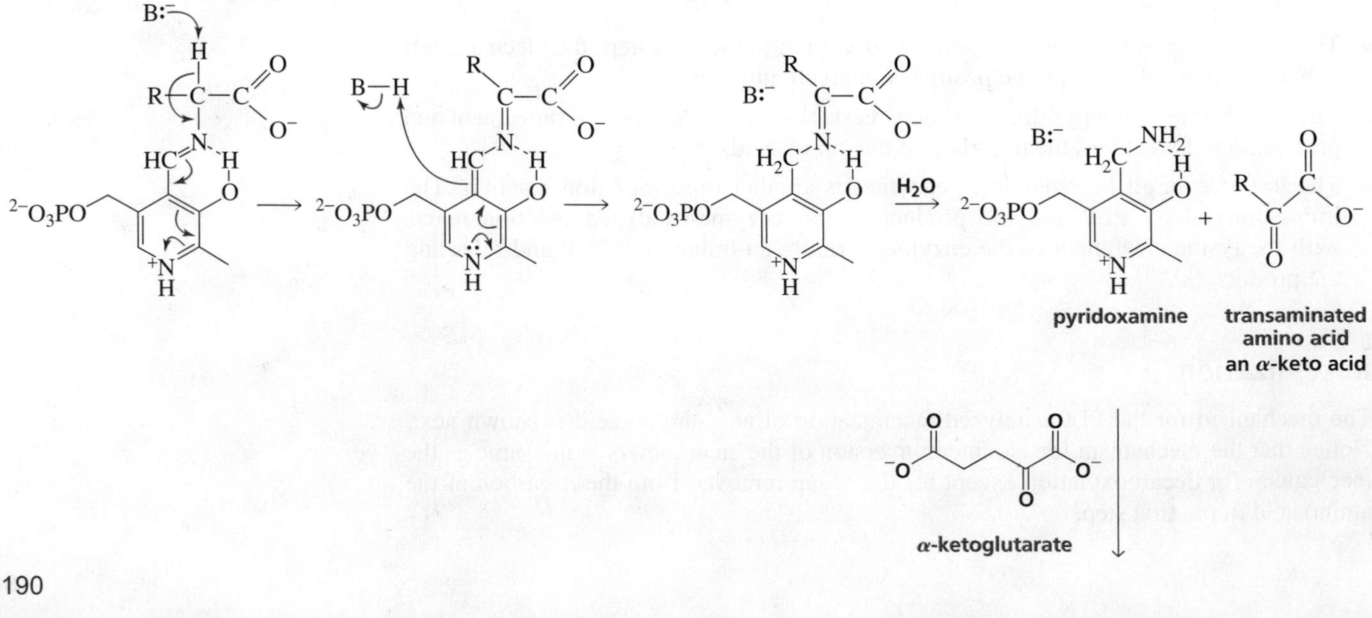

glutamate

- In the first step, a proton is removed from the α-carbon and the electrons left behind are delocalized onto the positively charged nitrogen.
- The aromaticity of the pyridinium ring is reestablished by electron rearrangement and protonation of the carbon attached to the pyridine ring.
- Hydrolysis of the imine forms the transaminated amino acid (an α-keto acid) and pyridoxamine.

At this point, the amino group has been removed from the amino acid, but the amino group of pyridoxamine has to be converted to a carbonyl group that can form an imine with the lysine side chain of the enzyme before another round of catalysis can occur.

- Pyridoxamine forms an imine with α-ketoglutarate, the second substrate of the reaction.
- A proton is removed from the carbon attached to the pyridine ring and the electrons left behind are delocalized onto the positively charged nitrogen.
- The aromaticity of the pyridinium ring is reestablished by electron rearrangement and protonation of the α-carbon.
- A transimination reaction with a lysine side chain releases the product of the reaction (glutamate) and regenerates the imine between PLP and the lysine side chain of the enzyme.

Notice that the two proton transfer steps are reversed in the two phases of the reaction. Transfer of the amino group of the amino acid to PLP requires removal of the proton from the α-carbon (of the amino acid) and donation of a proton to the carbon bonded to the pyridine ring. The steps are reversed in the transfer of the amino group of pyridoxamine to α-ketoglutarate: a proton is removed from the carbon bonded to the pyridinium ring, and a proton is donated to the α-carbon (of α-ketoglutarate).

Compare the second step in a PLP-catalyzed transamination with the second step in a PLP-catalyzed decarboxylation or racemization. In an enzyme that catalyzes transamination, an acidic group at the active site is in position to donate a proton to the carbon attached to the pyridinium ring. The enzyme that catalyzes decarboxylation or racemization does not have this acidic group, so the substrate is reprotonated at the α-carbon. In other words, the *coenzyme* carries out the chemical reaction, but the *enzyme* determines the course of the reaction.

Assessing the Damage After a Heart Attack

After a heart attack, aminotransferases and other enzymes leak from the damaged cells of the heart into the bloodstream. The severity of damage done to the heart can be determined from the concentrations of alanine aminotransferase and aspartate aminotransferase in the blood.

PROBLEM 12♦

α-Keto acids other than α-ketoglutarate can accept the amino group from pyridoxamine in enzyme-catalyzed transamination reactions. What amino acids are formed when the following α-keto acids accept the amino group?

a. 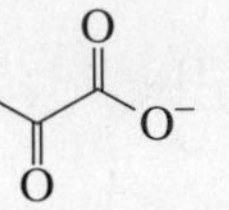 pyruvate

b. 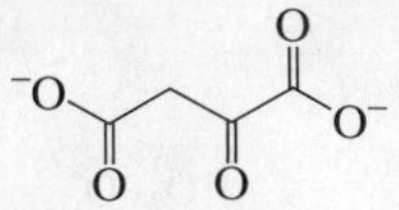 oxaloacetate

PROBLEM 13

The PLP-requiring enzyme that catalyzes the C_α—C_β bond cleavage reaction that converts serine to glycine (see earlier in this section) removes the substituent bonded to the α-carbon in the first step of the reaction. Starting with PLP bound to serine in an imine linkage, propose a mechanism for this reaction. (*Hint:* The first step involves general-base catalyzed removal of the proton from serine's OH group.)

PROBLEM 14

Propose a mechanism for the PLP-catalyzed α,β-elimination reaction shown earlier in this section.

PROBLEM 15♦

Which compound is more easily decarboxylated?

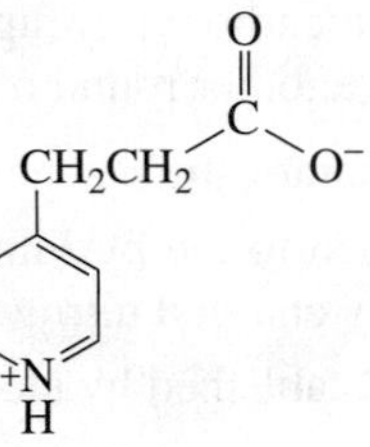

or

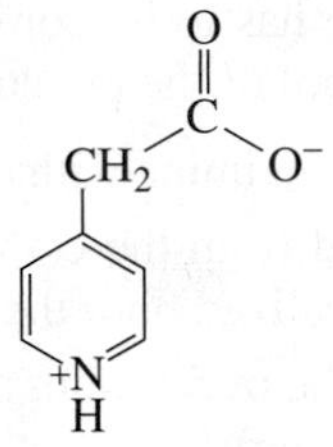

PROBLEM 16♦

Explain why the ability of PLP to catalyze an amino acid transformation is greatly reduced if a PLP-requiring enzymatic reaction is carried out at a pH at which the pyridine nitrogen is not protonated.

PROBLEM 17♦

Explain why the ability of PLP to catalyze an amino acid transformation is greatly reduced if the OH substituent of pyridoxal phosphate is replaced by OCH_3.

6 VITAMIN B$_{12}$: THE VITAMIN NEEDED FOR CERTAIN ISOMERIZATIONS

Enzymes that catalyze certain isomerization reactions require **coenzyme B$_{12}$,** a coenzyme derived from vitamin B$_{12}$. The structure of vitamin B$_{12}$ was determined by Dorothy Crowfoot Hodgkin using X-ray crystallography. The vitamin has a cyano group (or HO^- or H_2O) coordinated with cobalt. This group is replaced by a 5'-deoxyadenosyl group in the coenzyme.

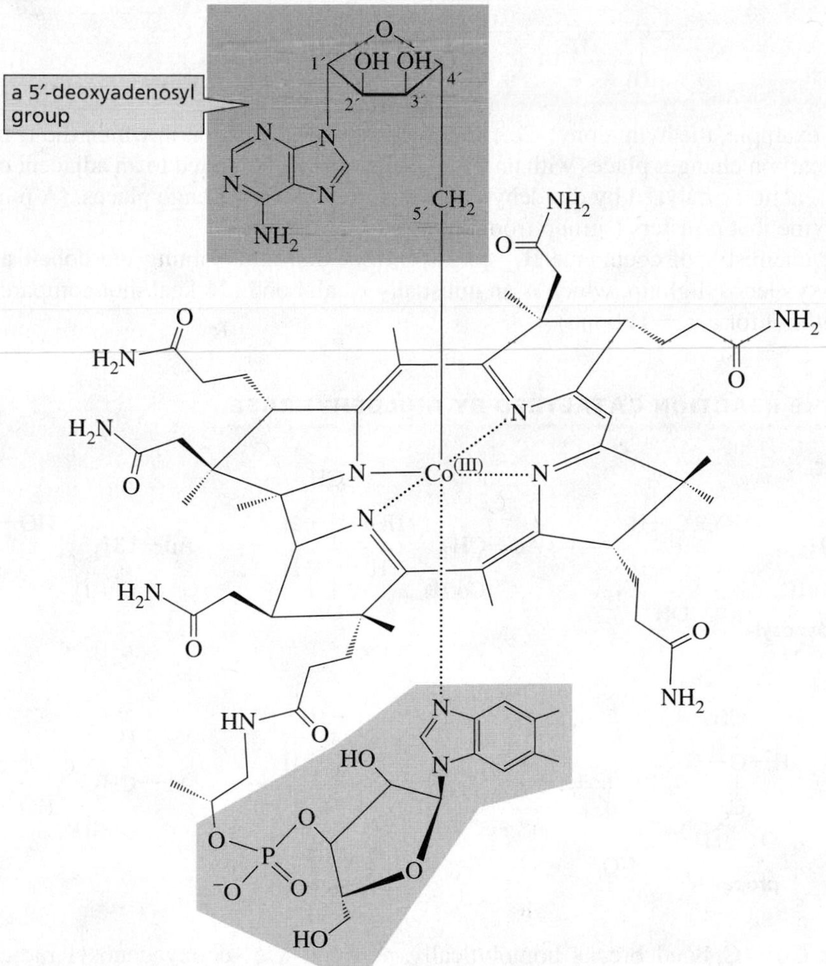

coenzyme B$_{12}$

Animals and plants cannot synthesize vitamin B$_{12}$. In fact, only a few species of bacteria can synthesize it. Humans must obtain all their vitamin B$_{12}$ from their diet, particularly from meat. A deficiency causes pernicious anemia. Because vitamin B$_{12}$ is needed in only very small amounts, deficiencies caused by the consumption of insufficient amounts of the vitamin are rare but have been found in vegetarians who eat no animal products. Most deficiencies are caused by the inability of the intestines to absorb the vitamin.

The following are examples of enzyme-catalyzed reactions that require coenzyme B$_{12}$.

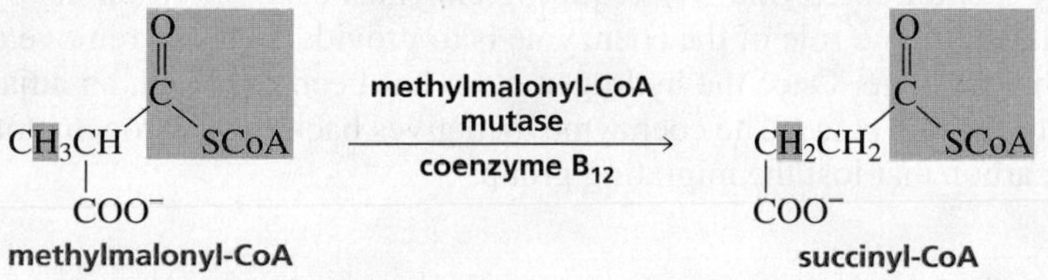

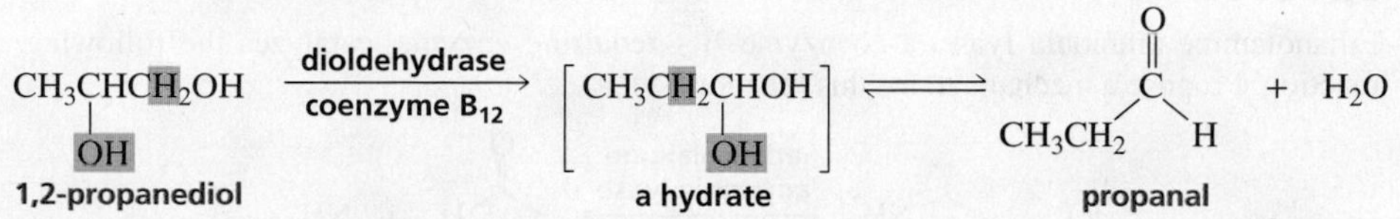

In each of these coenzyme B_{12}–requiring reactions, a hydrogen bonded to one carbon exchanges places with a group (Y) bonded to an adjacent carbon.

Coenzyme B_{12} is required by enzymes that catalyze the exchange of a hydrogen bonded to one carbon with a group bonded to an adjacent carbon.

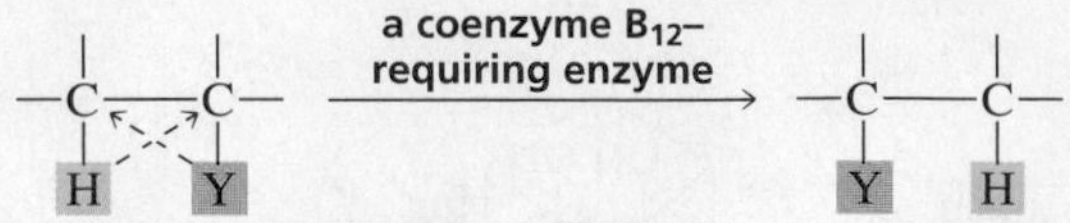

For example, methylmalonyl-CoA mutase catalyzes a reaction in which the H bonded to one carbon changes places with an C(=O)SCoA group bonded to an adjacent carbon. In the reaction catalyzed by dioldehydrase, an H and an OH change places. (A mutase is an enzyme that transfers a group from one position to another.)

The chemistry of coenzyme B_{12} takes place at the bond joining the cobalt and the 5′-deoxyadenosyl group, which is an unusually weak bond (26 kcal/mol compared with 99 kcal/mol for a C—H bond).

MECHANISM FOR THE REACTION CATALYZED BY DIOLDEHYDRASE

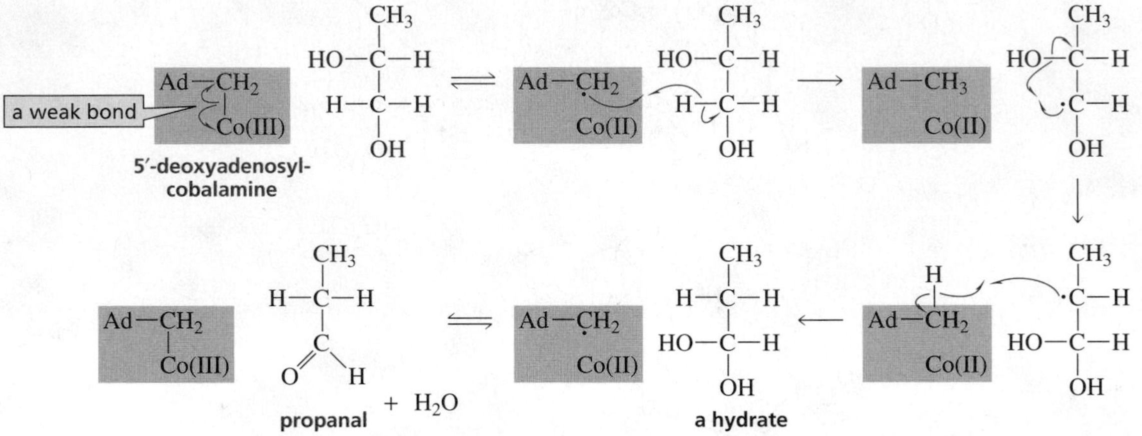

- The Co—C bond breaks homolytically, forming a 5′-deoxyadenosyl radical and reducing Co(III) to Co(II).
- The 5′-deoxyadenosyl radical removes the hydrogen atom (in this case, from the C-1 carbon) that will change place with another group, thereby becoming 5′-deoxy-adenosine.
- A hydroxyl radical (•OH) migrates from C-2 to C-1, creating a radical at C-2.
- The radical at C-2 removes a hydrogen atom from 5′-deoxyadenosine, forming the rearranged product and regenerating the 5′-deoxyadenosyl radical.
- The 5′-deoxyadenosyl radical recombines with Co(II) to regenerate the coenzyme. The enzyme–coenzyme complex is then ready for another catalytic cycle.

It is likely that all coenzyme B_{12}–requiring enzymes catalyze reactions by the same general mechanism. The role of the coenzyme is to provide a way to remove a hydrogen atom from the substrate. Once the hydrogen atom has been removed, an adjacent group can migrate to take its place. The coenzyme then gives back the hydrogen atom, delivering it to the carbon that lost the migrating group.

PROBLEM 18

Ethanolamine ammonia lyase, a coenzyme B_{12}–requiring enzyme, catalyzes the following reaction. Propose a mechanism for this reaction.

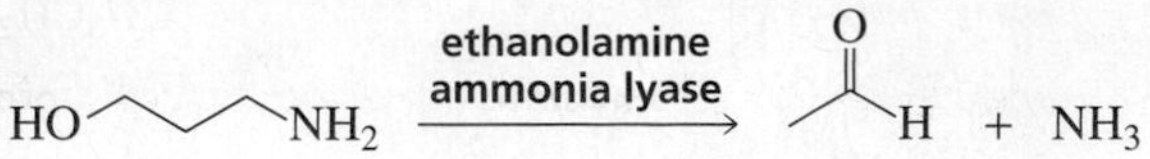

A fatty acid with an even number of carbons is metabolized to acetyl-CoA, which can then enter the citric acid cycle to be further metabolized. A fatty acid with an odd number of carbons is metabolized to acetyl-CoA and one equivalent of propionyl-CoA. Propionyl-CoA cannot enter the citric acid cycle. Two coenzyme-requiring enzymes are needed to convert it into succinyl-CoA, a compound that can enter the citric acid cycle. Write the two enzyme-catalyzed reactions and include the names of the required coenzymes.

7 FOLIC ACID: THE VITAMIN NEEDED FOR ONE-CARBON TRANSFER

Tetrahydrofolate (THF) is the coenzyme used by enzymes that catalyze reactions that transfer a group containing a single carbon to their substrates. The one-carbon group can be a methyl group (CH_3), a methylene group (CH_2), or a formyl group ($HC\!=\!O$). Tetrahydrofolate is produced by the reduction of two double bonds of folic acid (folate), its precursor vitamin. Bacteria can synthesize folate, but mammals cannot.

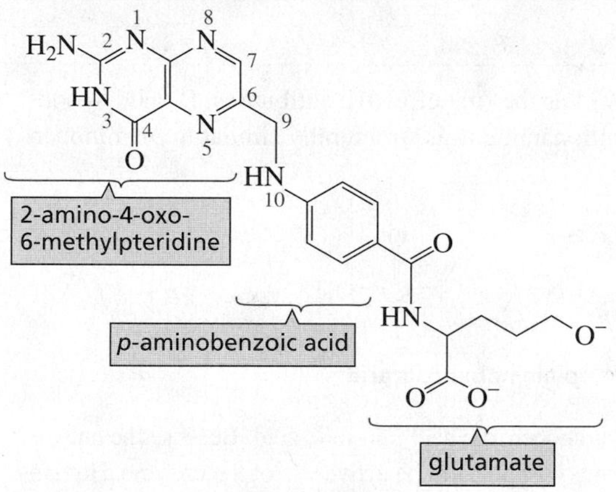

folic acid (folate)

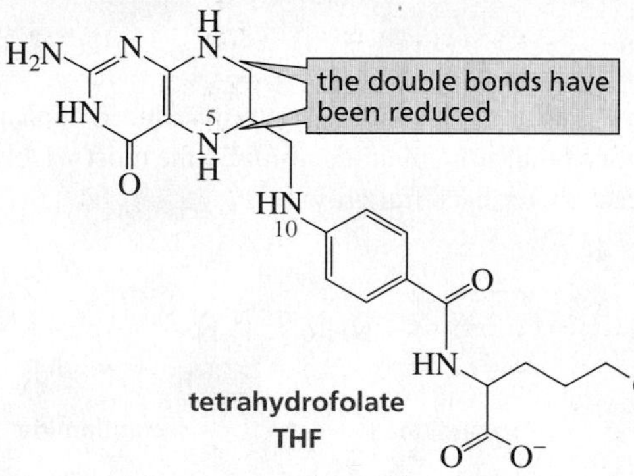

tetrahydrofolate
THF

There are six THF-coenzymes. N^5-Methyl-THF transfers a methyl group, N^5,N^{10}-methylene-THF transfers a methylene group, and N^5,N^{10}-methenyl-THF transfers a formyl group. The other three coenzymes also transfer a formyl group.

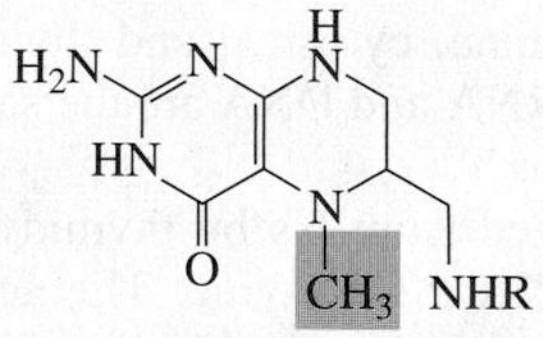

N^5-methyl-THF

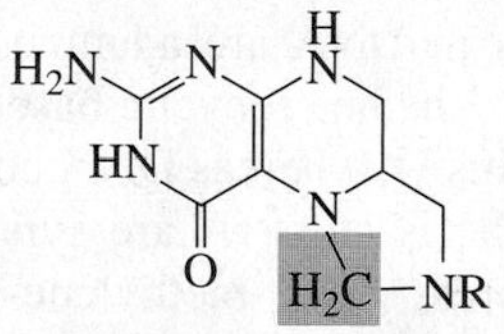

N^5,N^{10}-methylene-THF

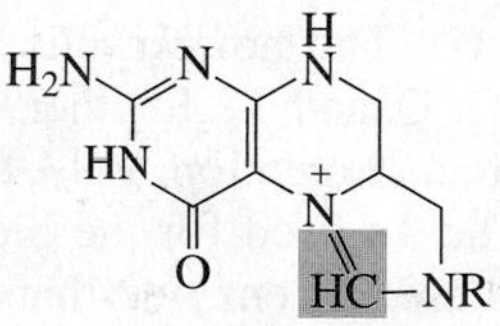

N^5,N^{10}-methenyl-THF

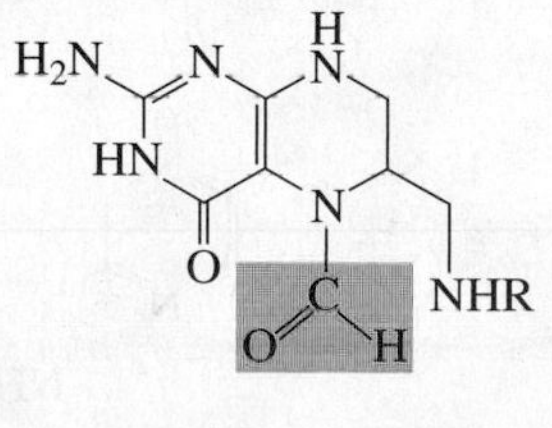

N^5-formyl-THF

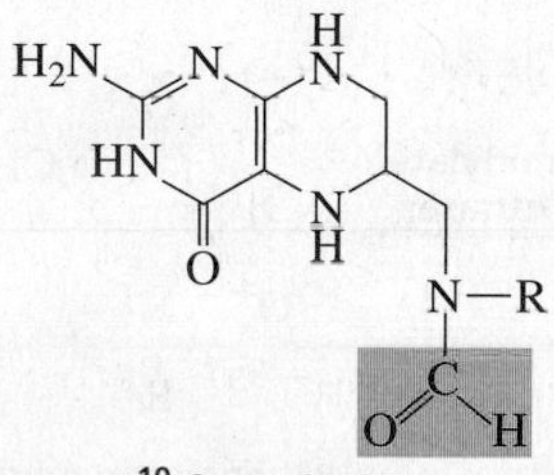

N^{10}-formyl-THF

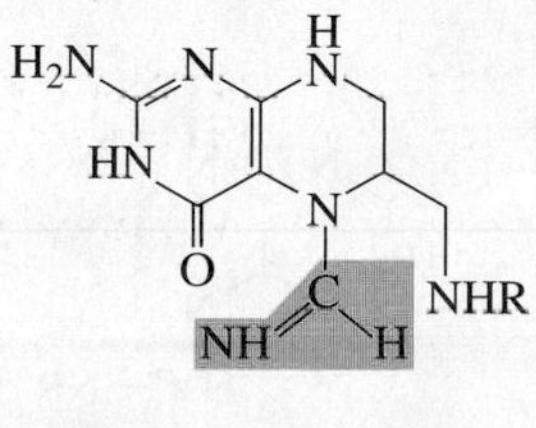

N^5-formimino-THF

A tetrahydrofolate (THF) coenzyme is required by enzymes that catalyze the transfer of a group containing one carbon to their substrates.

Glycinamide ribonucleotide (GAR) transformylase is an enzyme that requires a THF-coenzyme. The formyl group that is given to the substrate eventually ends up being the C-8 carbon of the purine nucleotides, two of the four heterocyclic bases found in DNA and RNA.

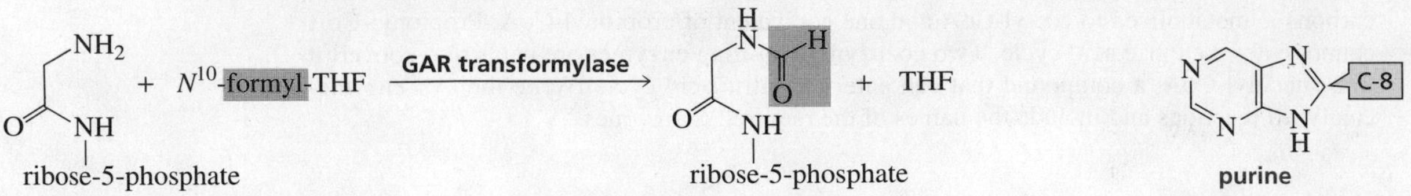

NH_2 ... + N^{10}-formyl-THF → **GAR transformylase** → ... + THF

ribose-5-phosphate ... ribose-5-phosphate ... purine [C-8]

Homocysteine methyl transferase, an enzyme required for the synthesis of methionine (an amino acid), also requires a THF-coenzyme.

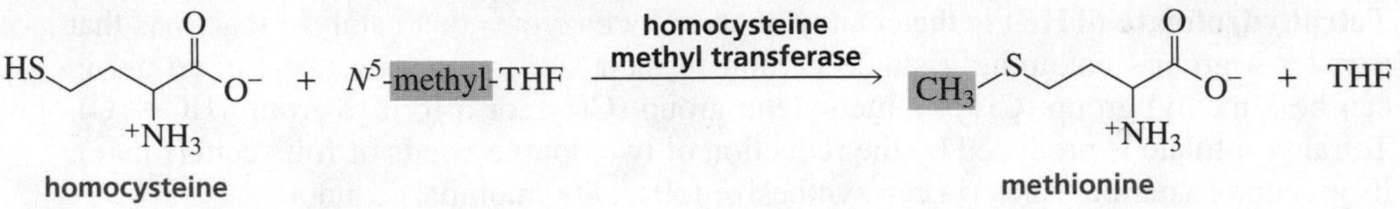

HS ... + N^5-methyl-THF → **homocysteine methyl transferase** → CH_3—S ... + THF

homocysteine ... methionine

The First Antibiotics

Sulfonamides—commonly known as sulfa drugs—were introduced clinically in 1934 as the first effective antibiotics. Donald Woods, a British bacteriologist, noticed that sulfanilamide, initially the most widely used sulfonamide, was structurally similar to *p*-aminobenzoic acid, a compound necessary for bacterial growth.

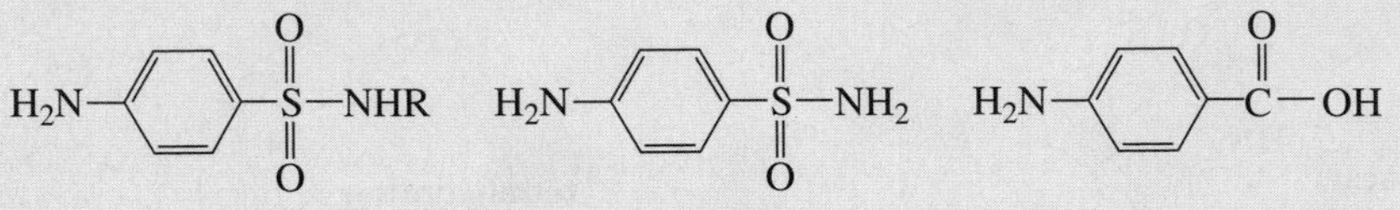

H_2N—⟨⟩—S—NHR H_2N—⟨⟩—S—NH_2 H_2N—⟨⟩—C—OH

a sulfonamide   **sulfanilamide**   ***p*-aminobenzoic acid**

This suggested that sulfanilamide acts by inhibiting the enzyme that incorporates *p*-aminobenzoic acid into folic acid. Because the enzyme cannot tell the difference between sulfanilamide and *p*-aminobenzoic acid, both compounds compete for the active site of the enzyme. Humans are not adversely affected by the drug because they do not synthesize folate; instead, they get all their folate from their diet.

Thymidylate Synthase: The Enzyme That Converts U to T

The heterocyclic bases in RNA are adenine, guanine, cytosine, and uracil (A, G, C, and U). The heterocyclic bases in DNA are adenine, guanine, cytosine, and thymine (A, G, C, and T). In other words, the heterocyclic bases in RNA and DNA are the same, with one exception: RNA contains U, whereas DNA contains T.

The Ts used for the biosynthesis of DNA are synthesized from Us by thymidylate synthase, an enzyme that requires N^5,N^{10}-methylene-THF as a coenzyme. The actual substrate is dUMP (2'-deoxyuridine 5'-monophosphate) and the product is dTMP (2'-deoxythymidine 5'-monophosphate).

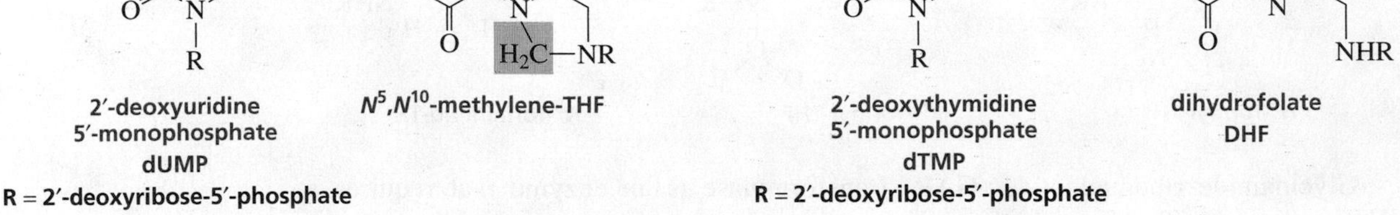

... + ... → **thymidylate synthase** → ... + ...

2'-deoxyuridine 5'-monophosphate dUMP ... N^5,N^{10}-**methylene-THF** ... **2'-deoxythymidine 5'-monophosphate dTMP** ... **dihydrofolate DHF**

R = 2'-deoxyribose-5'-phosphate ... **R = 2'-deoxyribose-5'-phosphate**

Even though the only structural difference between T and U is a *methyl* group, T is synthesized by first transferring a *methylene* group to a U. The methylene group is then reduced to a methyl group. The mechanism of the reaction is shown here.

MECHANISM FOR THE REACTION CATALYZED BY THYMIDYLATE SYNTHASE

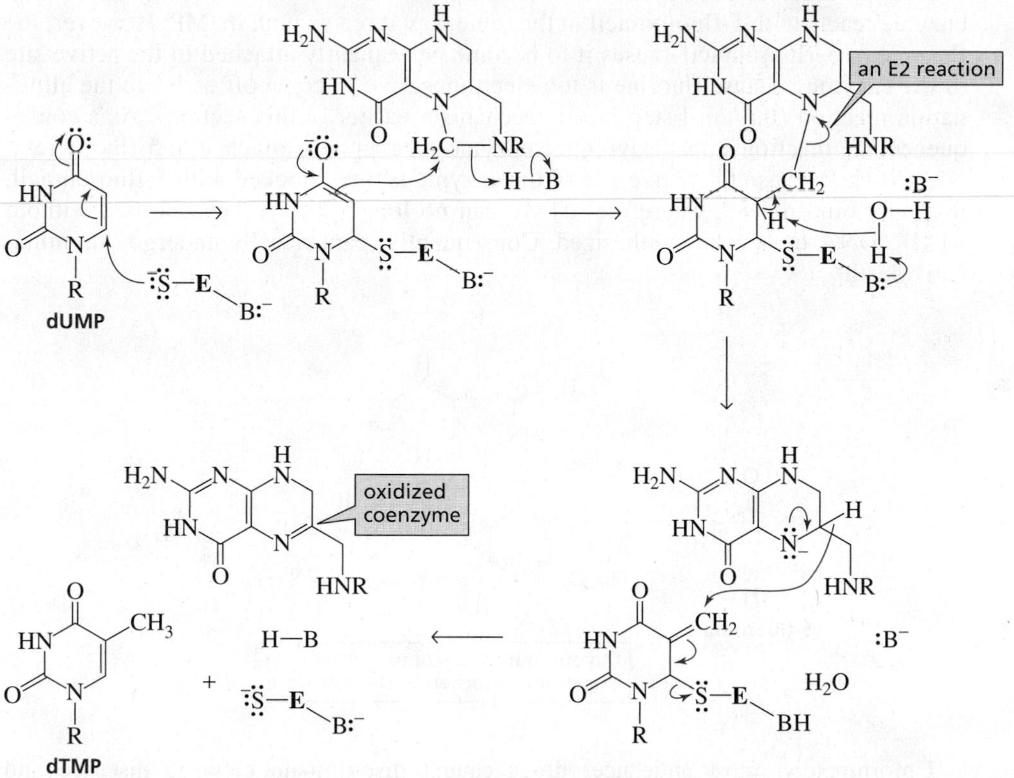

- A nucleophilic cysteine at the active site of the enzyme adds to the β-carbon of dUMP. (This is an example of conjugate addition.)

- Nucleophilic attack by the enolate ion of dUMP on the methylene group of N^5,N^{10}-methylene-THF forms a covalent bond between dUMP and the coenzyme. This is an S_N2 reaction. The leaving group has to be protonated to improve its leaving propensity.

- A base removes a proton from the α-carbon and the coenzyme is eliminated in an E2 reaction. The base is thought to be a water molecule whose basicity is increased by the OH group of a tyrosine side chain of the enzyme (here written as :B$^-$).

- Transfer of a hydride ion from the coenzyme to the methylene group, followed by elimination of the enzyme, forms dTMP and dihydrofolate (DHF).

Notice that the coenzyme that transfers the methylene group to the substrate is also the reagent that subsequently reduces the methylene group to a methyl group. Because the coenzyme is the reducing agent, it is simultaneously oxidized to dihydrofolate. (Oxidation decreases the number of C—H bonds.) Dihydrofolate must then be reduced back to methylene-THF by the enzyme dihydrofolate reductase so that tetrahydrofolate continues to be available as a coenzyme.

$$\text{dihydrofolate} + \text{NADPH} + \text{H}^+ \xrightarrow[\text{reductase}]{\text{dihydrofolate}} \text{tetrahydrofolate} + \text{NADP}^+$$

Cancer Chemotherapy

Cancer is the uncontrolled growth and proliferation of cells. Because cells cannot multiply if they cannot synthesize DNA, scientists have long searched for compounds that would inhibit thymidylate synthase or dihydrofolate reductase. If a cell cannot make Ts, it cannot synthesize DNA. Inhibiting dihydrofolate reductase also prevents the synthesis of Ts because cells have a limited amount of tetrahydrofolate. If they cannot convert dihydrofolate back to tetrahydrofolate, they cannot continue to synthesize Ts.

5-Fluorouracil is a common anticancer drug that inhibits thymidylate synthase. The enzyme reacts with 5-fluorouracil in the same way it reacts with dUMP. However, the fluorine of 5-fluorouracil causes it to become permanently attached to the active site of the enzyme because fluorine is too electronegative to come off as F^+ in the elimination reaction (the third step in the mechanism earlier in this section). As a consequence, the reaction stops, leaving the enzyme permanently attached to 5-fluorouracil (Figure 1). Because the active site of the enzyme is now blocked with 5-fluorouracil, it cannot bind dUMP. Therefore, dTMP can no longer be synthesized and, without dTMP, DNA cannot be synthesized. Consequently, cancer cells undergo "thymineless" death.

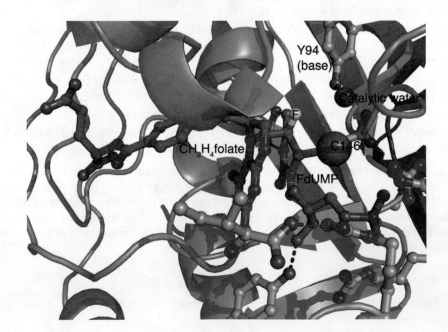

5-fluorouracil
5-FU

the enzyme has become irreversibly attached to the substrate

Unfortunately, most anticancer drugs cannot discriminate between diseased and normal cells, so most cancer chemotherapy is accompanied by debilitating side effects. However, because cancer cells divide more rapidly than normal cells, they are harder hit by cancer-fighting chemotherapeutic agents than normal cells are.

▶ **Figure 1**

The tetrahydrofolate coenzyme and 5-fluoro-dUMP that is covalently bonded to cysteine 146 of the enzyme are shown in yellow. The fluorine (turquoise), tyrosine 94, and the catalytic water molecule are also visible. The side chains that hold the coenzyme in the proper position at the active site are pink.

5-Fluorouracil is a **mechanism-based inhibitor:** it inactivates the enzyme by taking part in the normal catalytic mechanism. It is also called a **suicide inhibitor** because the enzyme effectively "commits suicide" by reacting with the inhibitor. The therapeutic use of 5-fluorouracil illustrates the importance of knowing the mechanism for enzyme-catalyzed reactions. If scientists know the mechanism of a reaction, they may be able to design an inhibitor to turn the reaction off at a certain step.

Competitive Inhibitors

Aminopterin and methotrexate are anticancer drugs that are inhibitors of dihydrofolate reductase. Because their structures are similar to that of dihydrofolate, they compete with it for binding to the active site of the enzyme. Because they bind 1000 times more tightly to the enzyme than dihydrofolate does, they win the competition and therefore inhibit the enzyme's activity. These two compounds are examples of **competitive inhibitors.**

aminopterin R = H
methotrexate R = CH₃

trimethoprim

Because aminopterin and methotrexate inhibit the synthesis of THF, they interfere with the synthesis of any compound that requires a THF-coenzyme in one of the steps of its synthesis. Thus, not only do they prevent the synthesis of thymidine, but they also inhibit the synthesis of adenine and guanine—other heterocyclic compounds needed for the synthesis of DNA—because their synthesis also requires a THF-coenzyme. One clinical technique used to fight cancer calls for the patient to be given a lethal dose of methotrexate. Then, after cancer cells have died, the patient is "saved" by being given N^5-formyl-THF.

Trimethoprim is used as an antibiotic because it binds to bacterial dihydrofolate reductase much more tightly than to mammalian dihydrofolate reductase.

Cancer Drugs and Side Effects

Scientists are searching for drugs that can discriminate between diseased and normal cells, so cancer chemotherapy will not be accompanied by debilitating side effects. A new drug now in clinical trials, is able to deliver a very toxic agent only to cancer cells.

Herceptin has been used since 1998 to treat certain kinds of breast cancers. Recently, scientists have been able to attach it to another anticancer drug that is so toxic that it cannot be used directly. Once Herceptin has bound to a breast cancer cell, it releases the poisonous agent to kill the cell. The survival time for women with advanced breast cancer enrolled in the clinical trials for this combined drug has been found to be almost a year longer, with much fewer side effects, than for women treated with Herceptin and other cancer drugs.

PROBLEM 20♦

How do the structures of tetrahydrofolate and aminopterin differ?

PROBLEM 21

What is the source of the methyl group in thymidine?

8 | VITAMIN K: THE VITAMIN NEEDED FOR CARBOXYLATION OF GLUTAMATE

Vitamin K is required for proper blood clotting. The letter K comes from *koagulation*, which is German for "clotting." Vitamin K is found in the leaves of green plants. Deficiencies are rare because the vitamin is also synthesized by intestinal bacteria. **Vitamin KH₂** (the hydroquinone of vitamin K) is the coenzyme form of the vitamin.

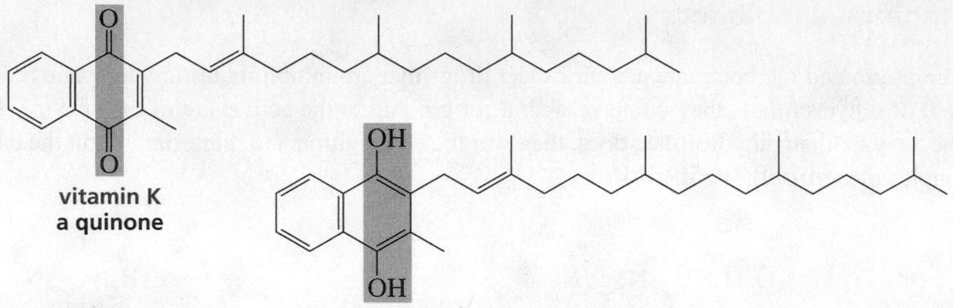

**vitamin K
a quinone**

**vitamin KH₂
a hydroquinone**

A series of reactions involving six proteins causes blood to clot. The process requires these proteins to bind Ca^{2+}. γ-Carboxyglutamates bind Ca^{2+} much more effectively than glutamates do.

Vitamin KH₂ is the coenzyme for the enzyme that converts a glutamate to a γ-carboxyglutamate. The enzyme uses CO_2 for the carboxyl group that it puts on the glutamate side chain. All the proteins responsible for blood clotting have several glutamates near their N-terminal ends. For example, prothrombin, a blood-clotting protein, has glutamates at positions 7, 8, 15, 17, 20, 21, 26, 27, 30, and 33.

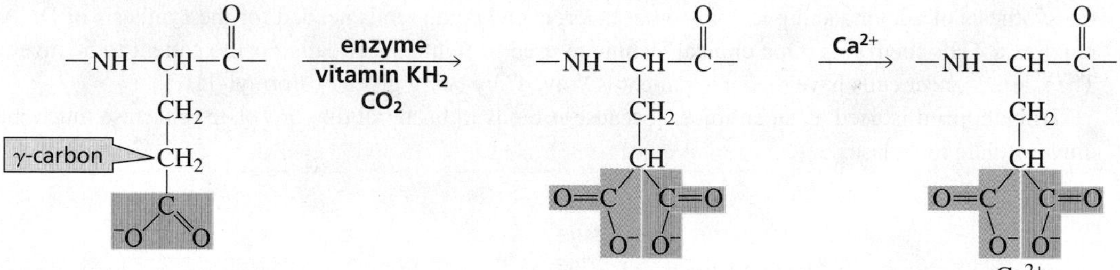

glutamate side chain

γ-carboxyglutamate side chain

calcium complex

Vitamin KH₂ is required by the enzyme that catalyzes the carboxylation of the γ-carbon of a glutamate side chain.

The mechanism for the vitamin KH₂–catalyzed carboxylation of glutamate had puzzled chemists because the γ-proton that must be removed from glutamate before it reacts with CO_2 is not very acidic. The mechanism, therefore, must involve the creation of a strong base that can remove this proton. A mechanism that has been proposed is shown here.

MECHANISM FOR THE VITAMIN KH₂–DEPENDENT CARBOXYLATION OF GLUTAMATE

a dioxetane

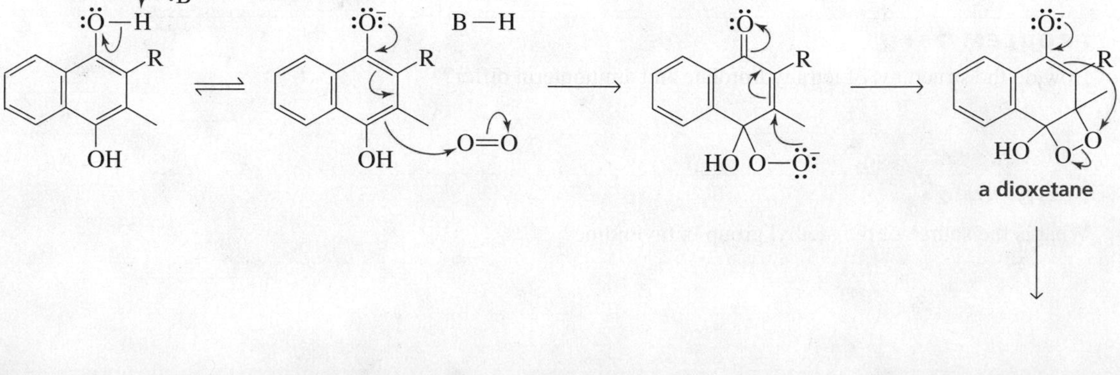

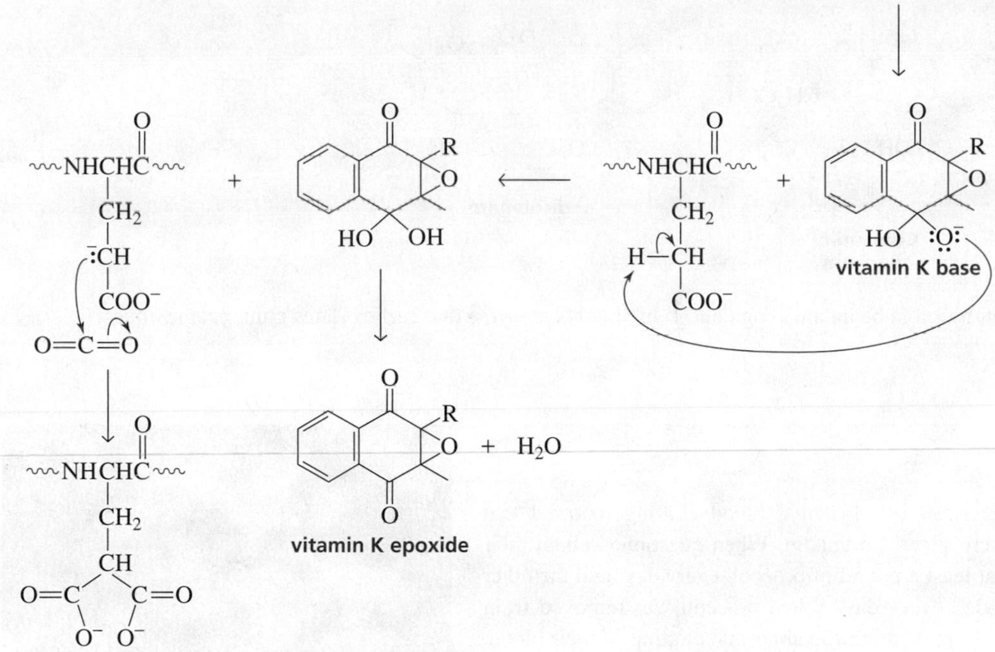

γ-carboxyglutamate

- The vitamin loses a proton from an OH group.
- The base that is formed attacks molecular oxygen.
- A conjugate addition reaction forms a dioxetane, which is a heterocyclic compound with a four-membered ring comprising two carbons and two oxygens.
- The dioxetane rearranges to an epoxide that has a strongly basic alkoxide ion called a vitamin K base.
- The vitamin K base is a strong enough base to remove a proton from the γ-carbon of glutamate.
- The glutamate carbanion adds to CO_2 to form γ-carboxyglutamate, and the protonated vitamin K base (a hydrate) loses water, forming vitamin K epoxide.

Vitamin K epoxide must now be reduced back to vitamin KH_2 so another round of catalysis can occur. The reducing agent for this enzyme-catalyzed reaction is the coenzyme, dihydrolipoate. Dihydrolipoate first reduces the epoxide to vitamin K, which is then further reduced by another dihydrolipoate to vitamin KH_2.

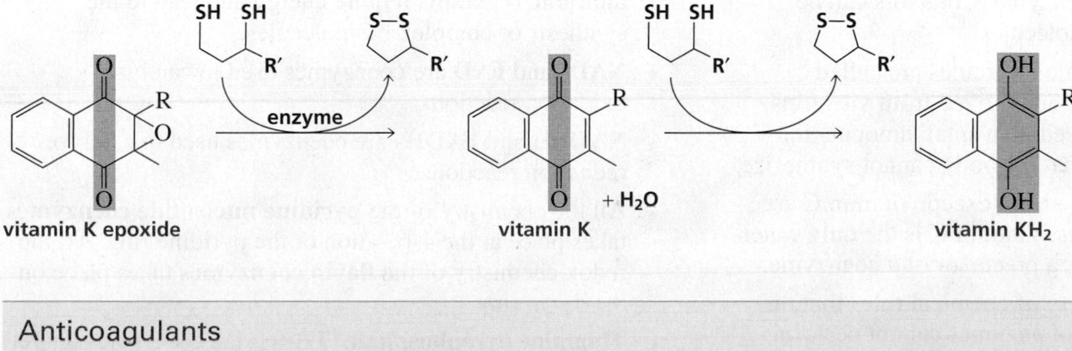

Anticoagulants

Warfarin and dicoumarol are used clinically as anticoagulants. These compounds prevent blood clotting because they inhibit the enzyme that converts vitamin K epoxide to vitamin KH_2 by binding to the enzyme's active site. Because the enzyme cannot tell the difference between these two compounds and vitamin K epoxide, the compounds are *competitive inhibitors*. Warfarin is also commonly used as a rat poison, causing death by internal bleeding.

Coumadin®
warfarin

dicoumarol

Vitamin E, too, has recently been found to be an anticoagulant. It inhibits the enzyme that carboxylates glutamate residues.

Too Much Broccoli

Two women with diseases characterized by abnormal blood clotting reported that they did not improve when they were given Coumadin. When questioned about their diets, one woman said that she ate at least a pound of broccoli every day, and the other ate broccoli soup and a broccoli salad every day. When broccoli was removed from their diets, Coumadin was effective in preventing the abnormal clotting of their blood. Because broccoli is high in vitamin K, these patients had been getting enough dietary vitamin K to compete successfully with the drug for the enzyme's active site, thereby making the drug ineffective.

PROBLEM 22

Thiols such as ethanethiol and propanethiol can be used to reduce vitamin K epoxide to vitamin KH$_2$, but they react much more slowly than dihydrolipoate. Explain why this is so.

SOME IMPORTANT THINGS TO REMEMBER

- **Cofactors** assist enzymes in catalyzing a variety of reactions that cannot be catalyzed solely by the amino acid side chains of the enzyme. Cofactors can be metal ions or organic molecules.

- Cofactors that are organic molecules are called **coenzymes;** coenzymes are derived from **vitamins,** which are substances needed in small amounts for normal body function that the body cannot synthesize.

- All the water-soluble vitamins except vitamin C are precursors of coenzymes. Vitamin K is the only water-insoluble vitamin that is a precursor of a coenzyme.

- Coenzymes play a variety of chemical roles that the amino acid side chains of enzymes cannot perform. Some function as oxidizing and reducing agents; some allow electrons to be delocalized; some activate groups for further reaction; and some provide good nucleophiles or strong bases needed for a reaction.

- **Catabolic reactions** break down complex biomolecules to provide energy and simple molecules, whereas **anabolic reactions** require energy and lead to the synthesis of complex biomolecules.

- **NAD$^+$** and **FAD** are coenzymes used to catalyze oxidation reactions.

- **NADPH** and **FADH$_2$** are coenzymes used to catalyze reduction reactions.

- All the chemistry of the **pyridine nucleotide coenzymes** takes place at the 4-position of the pyridine ring. All the redox chemistry of the **flavin coenzymes** takes place on the flavin ring.

- **Thiamine pyrophosphate (TPP)** is the coenzyme required by enzymes that catalyze the transfer of an acyl group.

- **Biotin** is the coenzyme required by enzymes that catalyze the carboxylation of a carbon adjacent to a carbonyl group.

- **Pyridoxal phosphate (PLP)** is the coenzyme required by enzymes that catalyze certain reactions of amino acids, such as decarboxylation, transamination, racemization, C_α—C_β bond cleavage, and α,β-elimination.

- In a **transimination reaction,** one imine is converted into another imine; in a **transamination reaction,** the amino group is removed from a substrate and transferred to another molecule, leaving a keto group in its place.

- In a **coenzyme B$_{12}$**–requiring enzymatic reaction, a group bonded to one carbon changes places with a hydrogen bonded to an adjacent carbon.

- **Tetrahydrofolate (THF)** is the coenzyme used by enzymes catalyzing reactions that transfer a group containing a single carbon—methyl, methylene, or formyl—to their substrates.

- **Vitamin KH$_2$** is the coenzyme for the enzyme that catalyzes the carboxylation of the γ-carbon of glutamate side chains, a reaction required for blood clotting.

- A **suicide inhibitor** inactivates an enzyme by taking part in the catalytic mechanism.

- A **competitive inhibitor** competes with the substrate for binding at the active site of the enzyme.

PROBLEMS

23. Answer the following:
   a. What coenzyme transfers an acyl group from one substrate to another?
   b. What is the function of FAD in the pyruvate dehydrogenase complex?
   c. What is the function of NAD$^+$ in the pyruvate dehydrogenase complex?
   d. What reaction necessary for proper blood clotting is catalyzed by vitamin KH$_2$?
   e. What coenzymes are used for decarboxylation reactions?
   f. What kinds of substrates do the decarboxylating coenzymes work on?
   g. What coenzymes are used for carboxylation reactions?
   h. What kinds of substrates do the carboxylating coenzymes work on?

24. Name the coenzymes that
   a. allow electrons to be delocalized.   **c.** provide a strong base.
   b. are oxidizing agents.   **d.** donate one-carbon groups.

25. For each of the following reactions, name both the enzyme that catalyzes the reaction and the required coenzyme:

a.

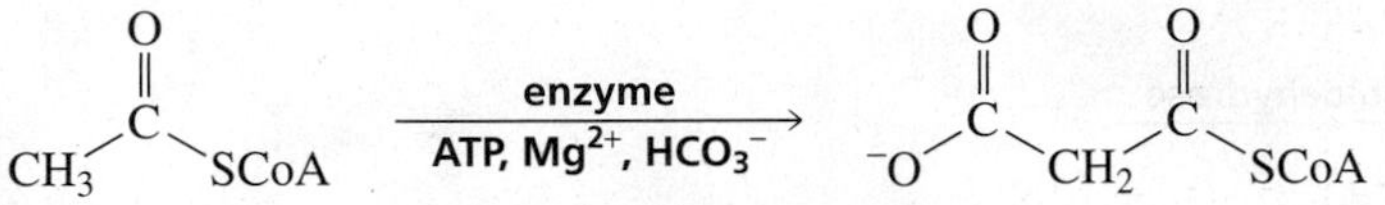

b.

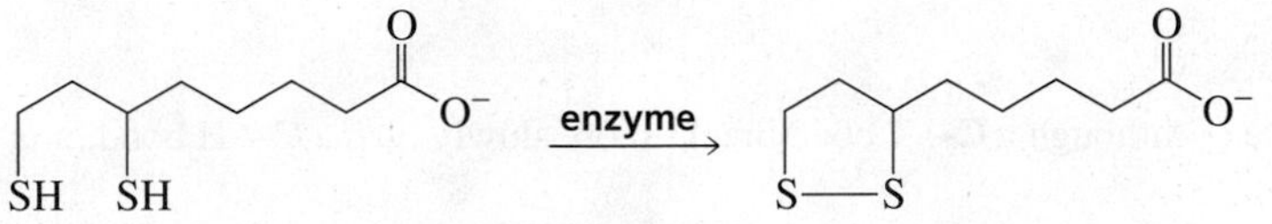

c.

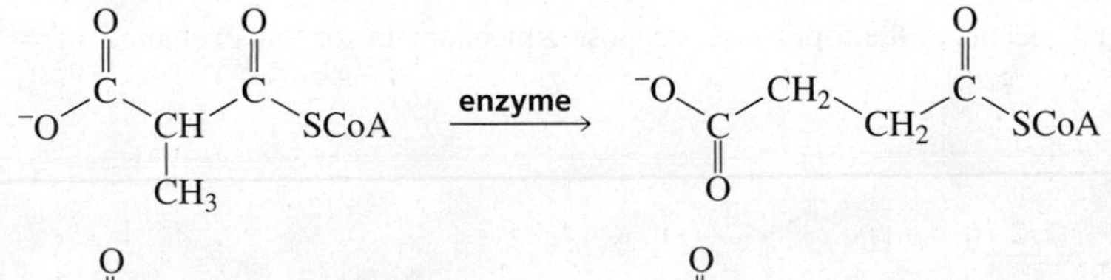

d.

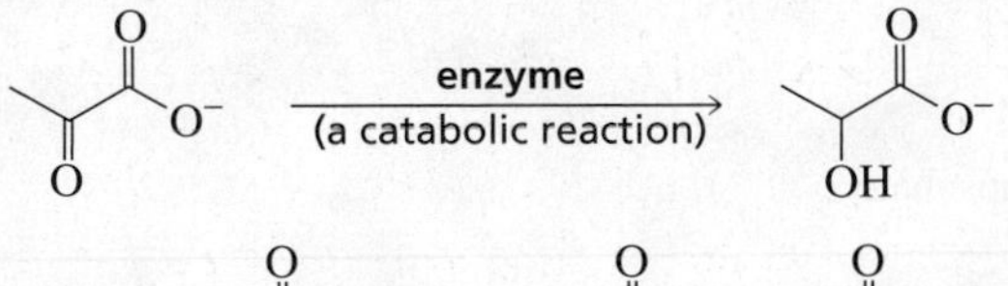

e.

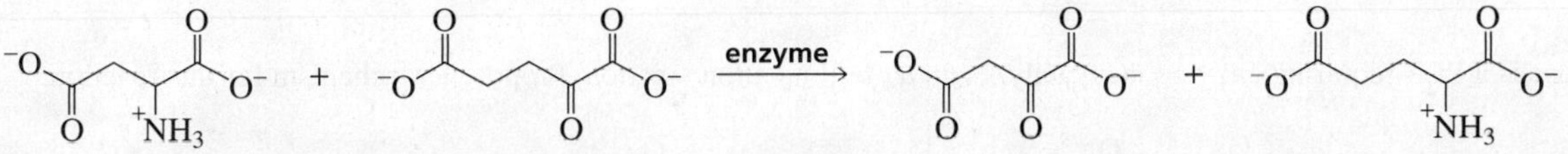

f.

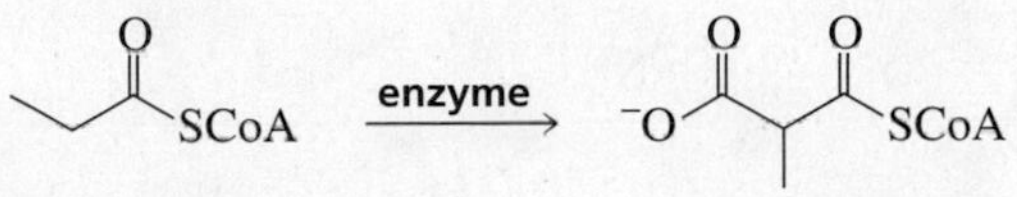

26. *S*-Adenosylmethionine (SAM) is formed from the reaction of methionine with ATP. The other product of the reaction is triphosphate. Propose a mechanism for this reaction.

27. Five coenzymes are required by α-ketoglutarate dehydrogenase, the enzyme in the citric acid cycle that converts α-ketoglutarate to succinyl-CoA.

a. Identify the coenzymes. **b.** Propose a mechanism for the reaction.

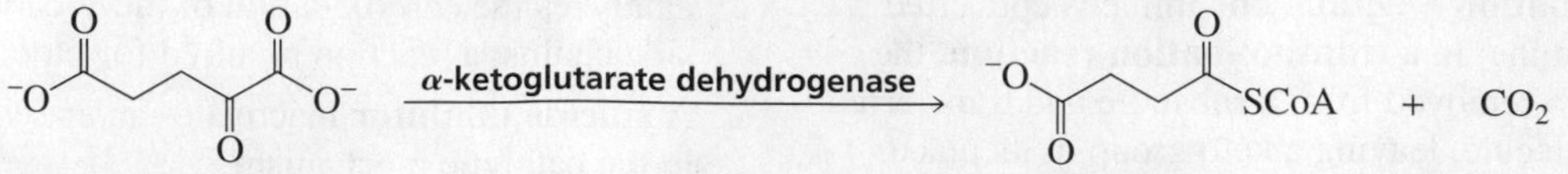

α-ketoglutarate — succinyl-CoA

28. What acyl groups have we seen transferred by reactions that require thiamine pyrophosphate as a coenzyme? (*Hint:* See Problems 9, 10, 27, and 30.)

29. Propose a mechanism for the following reaction:

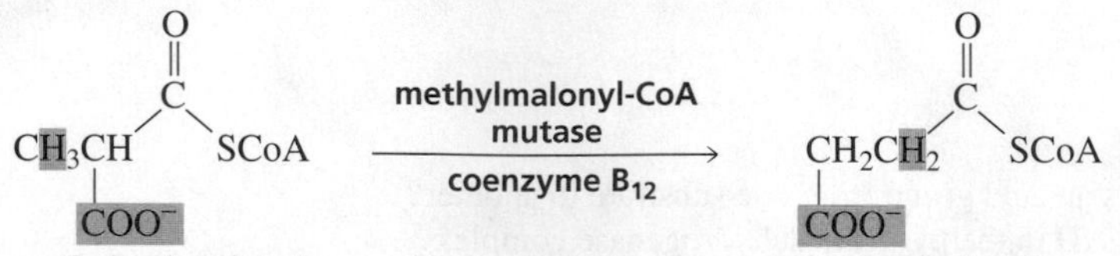

methylmalonyl-CoA — succinyl-CoA

30. When transaminated, the three branched-chain amino acids (valine, leucine, and isoleucine) form compounds that have the characteristic odor of maple syrup. An enzyme known as branched-chain α-keto acid dehydrogenase converts these compounds into CoA esters. People who do not have this enzyme have the genetic disease known as maple syrup urine disease, so called because their urine smells like maple syrup.

a. Draw the compounds that smell like maple syrup.
b. Draw the CoA esters.
c. Branched-chain α-keto acid dehydrogenase has five coenzymes. Identify them.
d. Suggest a way to treat maple syrup urine disease.

31. Draw the products of the following reaction, where T is tritium:

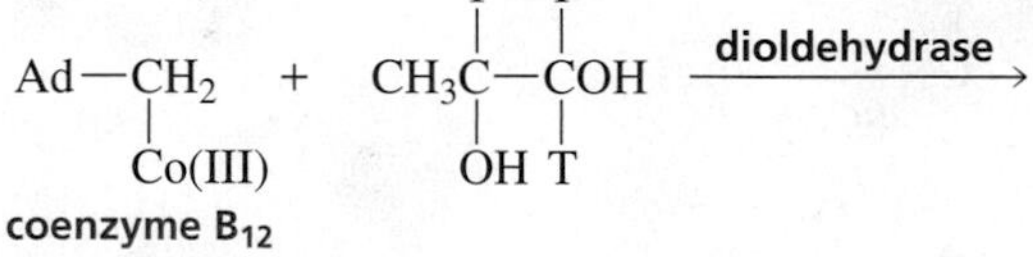

coenzyme B_{12}

(*Hint:* Tritium is ^{3}H, a hydrogen atom with two neutrons. Although a C—T bond breaks more slowly than a C—H bond, it is still the first bond in the substrate that breaks.)

32. When UMP is dissolved in T_2O, exchange of T for H occurs at the 5-position. Propose a mechanism for this exchange. (T = ^{3}H; see Problem 31).

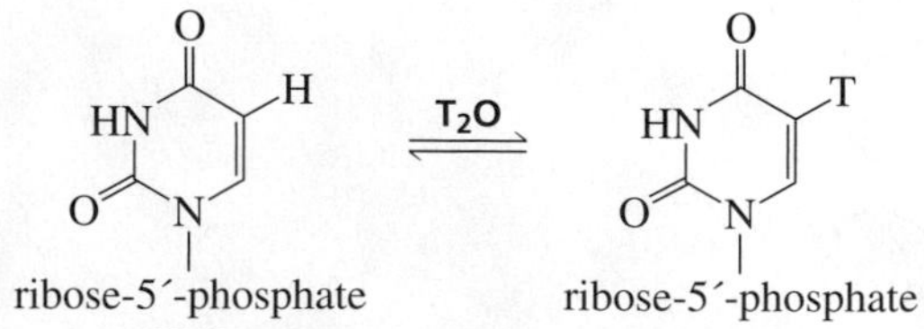

UMP

33. Dehydratase is a PLP-requiring enzyme that catalyzes an α,β-elimination reaction. Propose a mechanism for this reaction.

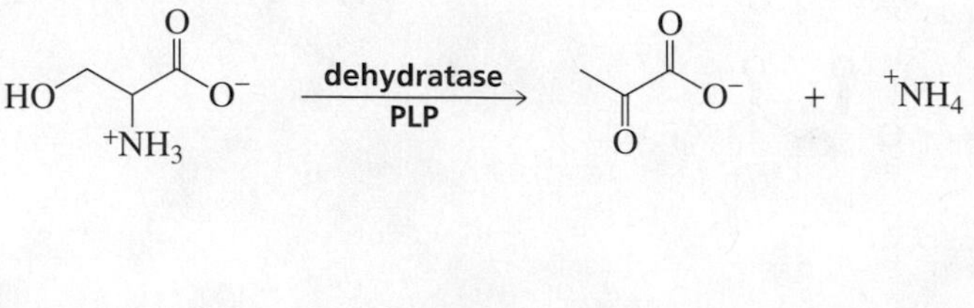

34. In addition to the reactions mentioned in Section 5, PLP can catalyze β-substitution reactions. Propose a mechanism for the following PLP-catalyzed β-substitution reaction:

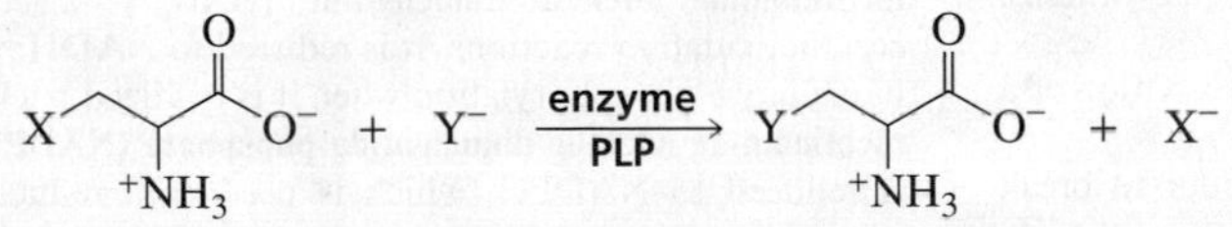

35. PLP can catalyze both α,β-elimination reactions (Problem 33) and β,γ-elimination reactions. Propose a mechanism for the following PLP-catalyzed β,γ-elimination:

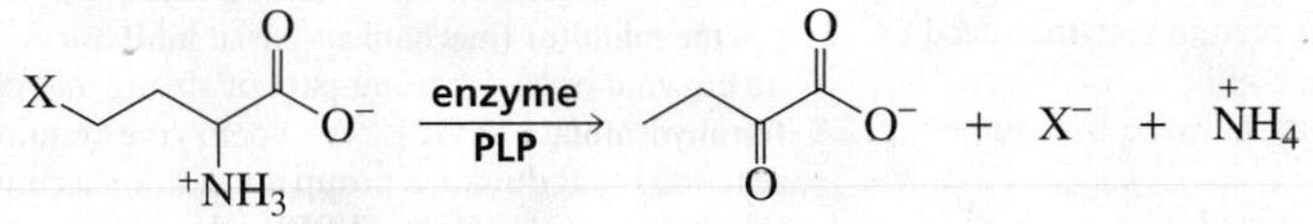

36. The glycine cleavage system is a group of four enzymes that together catalyze the following reaction:

$$\text{glycine} \;+\; \text{THF} \;\xrightarrow{\text{glycine cleavage system}}\; N^5,N^{10}\text{-methylene-THF} \;+\; CO_2$$

Use the following information to determine the sequence of reactions carried out by the glycine cleavage system:
a. The first enzyme is a PLP-requiring decarboxylase.
b. The second enzyme is aminomethyltransferase. This enzyme has a lipoate coenzyme.
c. The third enzyme synthesizes N^5,N^{10}-methylene-THF and also forms $^+NH_4$.
d. The fourth enzyme is an FAD-requiring enzyme.
e. The cleavage system also requires NAD^+.

37. Nonenzyme-bound FAD is a stronger oxidizing agent than NAD^+. How, then, can NAD^+ oxidize the reduced flavoenzyme in the pyruvate dehydrogenase complex?

38. $FADH_2$ reduces α,β-unsaturated thioesters to saturated thioesters. The reaction is thought to take place by a mechanism that involves radicals. Propose a mechanism for this reaction.

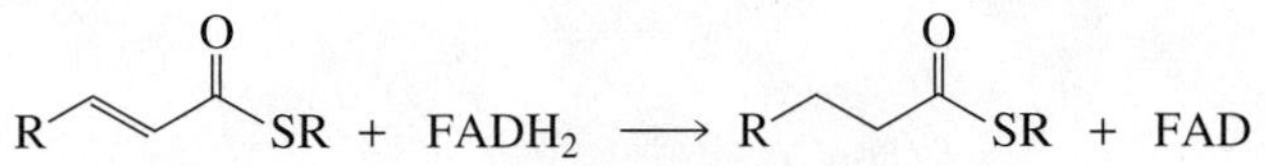

GLOSSARY

anabolism   reactions that living organisms carry out in order to synthesize complex molecules from simple precursor molecules.

biotin   the coenzyme required by enzymes that catalyze carboxylation of a carbon adjacent to an ester or a keto group.

catabolism   reactions that living organisms carry out in order to break down complex molecules into simple molecules and energy.

coenzyme   a cofactor that is an organic molecule.

coenzyme A   a thiol used by biological organisms to form thioesters.

coenzyme B$_{12}$ the coenzyme required by enzymes that catalyze certain rearrangement reactions.

cofactor   an organic molecule or a metal ion that certain enzymes need to catalyze a reaction.

competitive inhibitor   a compound that inhibits an enzyme by competing with the substrate for binding at the active site.

electron sink   site to which electrons can be delocalized.

flavin adenine dinucleotide (FAD)   a coenzyme required in certain oxidation reactions. It is reduced to FADH$_2$, which forms 1.5 ATPs in oxidative phosphorylation when it is oxidized back to FAD.

heterocyclic compound (heterocycle)   a cyclic compound in which one or more of the atoms of the ring are heteroatoms.

lipoate   a coenzyme required in certain oxidation reactions.

mechanism-based inhibitor (suicide inhibitor)   a compound that inactivates an enzyme by undergoing part of its normal catalytic mechanism.

molecular recognition   the recognition of one molecule by another as a result of specific interactions; for example, the specificity of an enzyme for its substrate.

nicotinamide adenine dinucleotide (NAD$^+$)   a coenzyme required in certain oxidation reactions. It is reduced to NADH, which forms 2.5 ATPs in oxidative phosphorylation when it is oxidized back to NAD$^+$.

nicotinamide adenine dinucleotide phosphate (NADP$^+$)   a coenzyme that is reduced to NADPH, which is used as a reducing agent in anabolic reactions.

nucleotide   a heterocycle attached in the b -position to a phosphorylated ribose or deoxyribose.

pyridoxal phosphate   the coenzyme required by enzymes that catalyze certain transformations of amino acids.

suicide inhibitor (mechanism-based inhibitor)   a compound that inactivates an enzyme by undergoing part of its normal catalytic mechanism.

tetrahydrofolate (THF)   the coenzyme required by enzymes that catalyze reactions that donate a group containing a single carbon to their substrates.

thiamine pyrophosphate (TPP)   the coenzyme required by enzymes, which catalyze a reaction that transfers a two-carbon fragment to a substrate.

transamination   a reaction in which an amino group is transferred from one compound to another.

transimination   the reaction of a primary amine with an imine to form a new imine and a primary amine derived from the original imine.

vitamin KH$_2$ the coenzyme required by the enzyme that catalyzes the carboxylation of glutamate side chains.

PHOTO CREDITS

Photo credits are listed in order of appearance.

The Organic Chemistry of the Metabolic Pathways • Terpene Biosynthesis

From Chapter 25 of *Organic Chemistry*, Seventh Edition. Paula Yurkanis Bruice. Copyright © 2014 by Pearson Education, Inc. All rights reserved.

The Organic Chemistry of the Metabolic Pathways • Terpene Biosynthesis

You are what you eat.

The reactions that living organisms carry out to obtain the energy they need and to synthesize the compounds they require are collectively known as **metabolism.** *Metabolism can be divided into two parts: catabolism and anabolism. Catabolic reactions convert complex nutrient molecules into simple molecules that can be used for synthesis. Anabolic reactions synthesize complex biomolecules from simpler precursor molecules. Catabolism comes from the Greek word* katabol, *which means "throwing down."*

A catabolic pathway is a series of sequential reactions that converts a complex molecule into simple molecules. Catabolic pathways involve *oxidation reactions* and produce energy. An anabolic pathway is a series of sequential reactions that converts simple molecules into a complex molecule. Anabolic pathways involve *reduction reactions* and require energy.

> ***catabolism:*** *complex molecule* → *simple molecules + energy*
>
> ***anabolism:*** *simple molecules + energy* → *complex molecule*

It is important to remember that almost every reaction that occurs in a living system is catalyzed by an enzyme. The enzyme holds the reactants and any necessary coenzymes in place, orienting the reacting functional groups and the amino acid side chain catalysts in a way that allows the enzyme-catalyzed reaction to take place.

Most of the reactions described in this chapter might be familiar to you. Many of the organic reactions done by cells are the same as the organic reactions done by chemists.

Differences in Metabolism

Humans do not necessarily metabolize compounds in the same way as other species do. This becomes a significant problem when drugs are tested on animals. For example, chocolate is metabolized to different compounds in humans and in dogs. The metabolites produced in humans are nontoxic, whereas those produced in dogs can be highly toxic. Differences in metabolism have been found even within the same species. For example, isoniazid—an antituberculosis drug—is metabolized by Eskimos much faster than by Egyptians. Current research is showing that men and women metabolize certain drugs differently. For example, kappa opioids—a class of painkillers—have been found to be about twice as effective in women as they are in men.

1 ATP IS USED FOR PHOSPHORYL TRANSFER REACTIONS

All cells require energy to live and reproduce. They get the energy they need from nutrients that they convert into a chemically useful form. The most important repository of chemical energy is **adenosine 5′-triphosphate (ATP).** The importance of ATP to biological reactions is reflected in its turnover rate in humans: each day, a person uses an amount of ATP equivalent to his or her body weight.

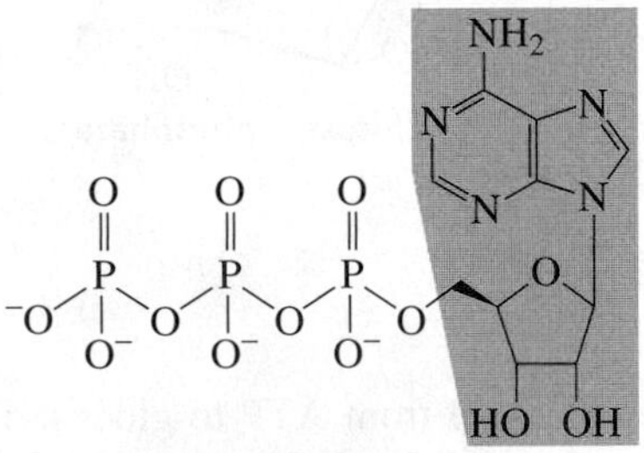

adenosine triphosphate
ATP

Phosphoric acid can be dehydrated to pyrophosphoric acid and triphosphoric acid, which are commonly known as **phosphoanhydrides**.

phosphoric acid **pyrophosphoric acid** **triphosphoric acid**
phosphoanhydrides

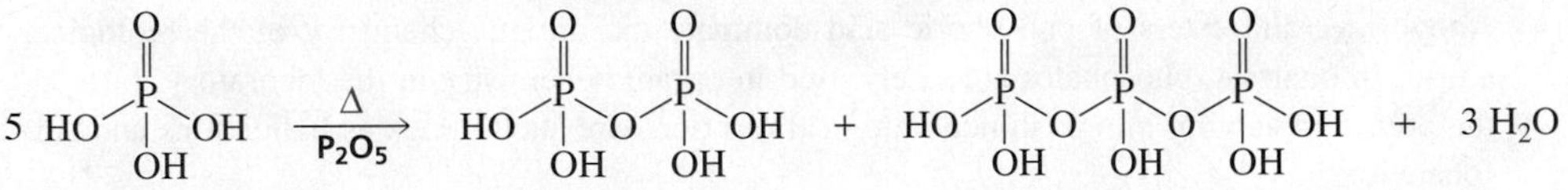

Esters of phosphoric acid, pyrophosphoric acid, and triphosphoric acid are important biological molecules. Each OH group of the acid can be esterified: monoesters and diesters are the most common biological molecules. ATP is a monoester of triphosphoric acid. The components of cell membranes are monoesters or diesters of phosphoric acid, and many of the coenzymes are monoesters or diesters of either phosphoric acid or pyrophosphoric acid. In addition, DNA and RNA are diesters of phosphoric acid.

phosphoric acid **a phosphomonoester** **a phosphodiester** **a phosphotriester**

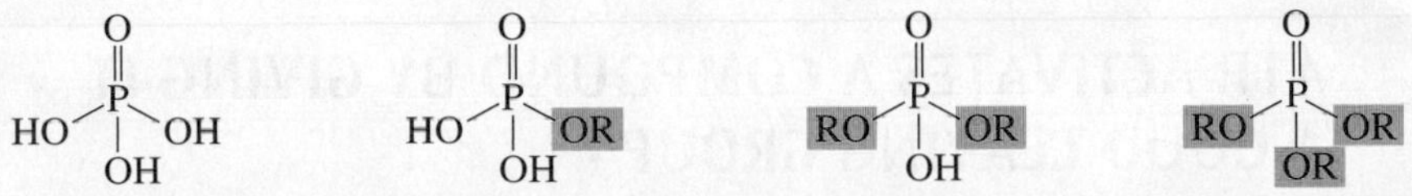

Without ATP, many important biological reactions could not occur. For example, the reaction of glucose with hydrogen phosphate does not form glucose-6-phosphate because the 6-OH group of glucose would have to displace a very basic HO⁻ group from hydrogen phosphate.

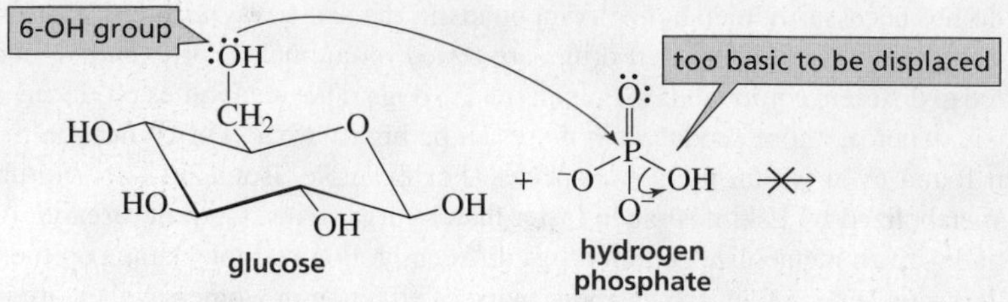

Glucose-6-phosphate, however, can be formed from the reaction of glucose with ATP, because the 6-OH group of glucose can attack the terminal phosphate of ATP, breaking a **phosphoanhydride bond.** The phosphoanhydride bond is a weaker bond than the $P=O$ π bond, so the reaction is a simple S_N2 reaction, rather than an addition–elimination reaction on phosphorus. We will see why ADP is a good leaving group in Section 4.

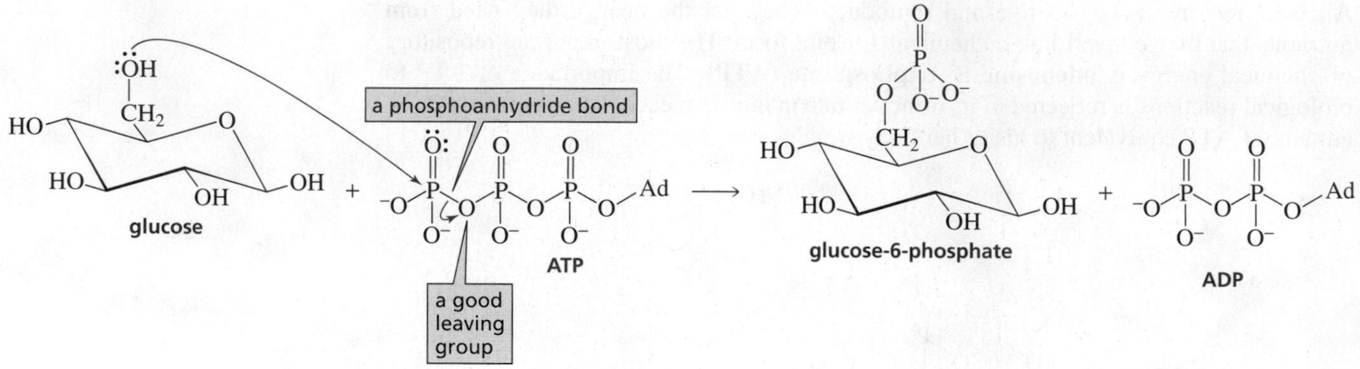

The transfer of a phosphate group from ATP to glucose is one of many **phosphoryl transfer reactions** that occur in cells. In all these reactions, an electrophilic phosphoryl group is transferred to a nucleophile as a result of breaking a phosphoanhydride bond.

Why Did Nature Choose Phosphates?

Anhydrides and esters of phosphoric acid dominate the organic chemistry of the biological world. In contrast, phosphates are rarely used in organic chemistry in the laboratory. Instead, the preferred leaving groups in nonbiological reactions are such things as halide ions and sulfonate ions.

Why did nature choose phosphates? There are several reasons. To keep a molecule from leaking through a membrane, it should be charged; to prevent reactive nucleophiles from approaching a molecule, it should be negatively charged; and to link the bases in RNA and DNA, the linking molecule needs two functional groups. Phosphoric acid, with its three OH groups, fits all these requirements; it can use two of its OH groups to link the bases, and the third OH group is negatively charged at physiological pH. In addition, reactions of nucleophiles with phosphoanhydrides can be irreversible, which is an important attribute in many biological reactions.

---

2 | ATP ACTIVATES A COMPOUND BY GIVING IT A GOOD LEAVING GROUP

Phosphoryl transfer reactions can be used to activate a compound for a reaction. For example, a carboxylate ion does not react with a nucleophile because a carboxylate ion is negatively charged and its leaving group is a very strong base.

$$R-C(=O)-\ddot{O}:^- \quad + \quad ROH \quad \longrightarrow \quad \text{no reaction}$$

However, ATP can activate a carboxylate ion in one of two ways. The carboxylate ion can attack the γ-phosphorus (the terminal phosphorus) of ATP, forming an **acyl phosphate.** This is just like the reaction of glucose-6-phosphate with ATP that we looked at.

nucleophilic attack on the γ-phosphorus

a phosphoanhydride bond

ATP → an acyl phosphate a mixed anhydride + ADP

The carboxylate ion can also attack the α-phosphorus of ATP, forming an **acyl adenylate.**

nucleophilic attack on the α-phosphorus

ATP → an acyl adenylate a mixed anhydride + pyrophosphate

Each of the preceding reactions is an S_N2 reaction that forms a mixed anhydride of a carboxylic acid and a phosphoric acid. Thus a leaving group has been put on the carboxylate ion that can easily be displaced by a nucleophile in an enzyme-catalyzed reaction.

Nucleophiles other than carboxylate ions are also activated by ATP in cells. Whether a nucleophile attacks the γ-phosphorus or the α-phosphorus depends on the enzyme catalyzing the reaction. Notice that when a nucleophile attacks the γ-phosphorus, the side product is ADP, but when it attacks the α-phosphorus, the side product is pyrophosphate. When pyrophosphate is formed, it is subsequently hydrolyzed to two equivalents of hydrogen phosphate. Removing a reaction product from the reaction mixture drives the reaction to the right, which ensures its irreversibility (refer to Le Châtelier's principle).

pyrophosphate $+ H_2O \longrightarrow 2$ phosphate

Therefore, in an enzyme-catalyzed reaction in which complete irreversibility is essential, the nucleophile will attack the α-phosphorus of ATP. For example, both the reaction that links nucleotide subunits to form DNA and RNA and the reaction that binds an amino acid to a tRNA (the first step in translating RNA into a protein) involve nucleophilic attack on the α-phosphorus of ATP. If these reactions were reversible, the genetic information in DNA would not be preserved and proteins would be synthesized that would not have the correct sequence of amino acids.

A carboxylate ion can also be activated by being converted into a thioester. ATP is needed for this reaction too. The carboxylate ion reacts with ATP to form an acyl adenylate, giving the carboxylate ion a good leaving group, so it can then react with the thiol.

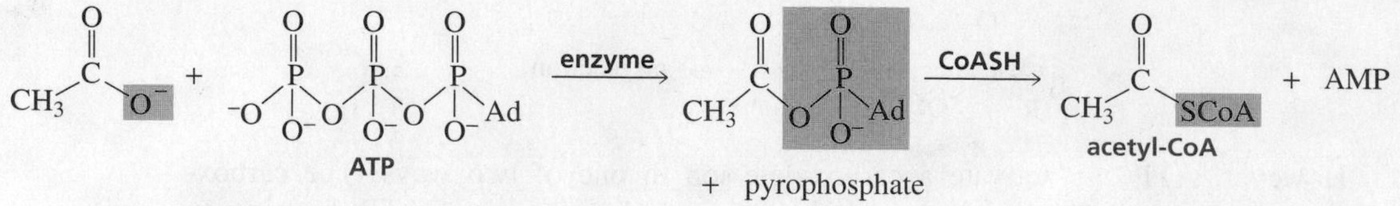

The phosphoryl transfer reactions discussed here demonstrate the actual chemical function of ATP:

> *ATP provides a reaction pathway involving a good leaving group for a reaction that cannot occur (or would occur very slowly) because of a poor leaving group.*

3 WHY ATP IS KINETICALLY STABLE IN A CELL

ATP reacts readily in enzyme-catalyzed reactions, but it reacts quite slowly in the absence of an enzyme. For example, a carboxylic acid anhydride hydrolyzes in a matter of minutes, but ATP (a phosphoric acid anhydride) takes several weeks to hydrolyze. The low rate of ATP hydrolysis is important because it allows ATP to exist in the cell until it is needed for an enzyme-catalyzed reaction.

The negative charges on ATP are what make it relatively unreactive. These negative charges repel the approach of a nucleophile. When ATP is bound at an active site of an enzyme, it complexes with magnesium (Mg^{2+}), which decreases ATP's overall negative charge. (This is why ATP-requiring enzymes also require Mg^{2+}.) Interactions with positively charged groups, such as arginine or lysine side chains, at the active site of the enzyme further reduce ATP's negative charge. Thus, when bound at the active site of an enzyme, ATP can be readily approached by nucleophiles.

For the same reason, acyl phosphates and acyl adenylates are unreactive unless they are at the active site of an enzyme. It is important to note that the leaving groups of acyl phosphates and acyl adenylates are weaker bases than their pK_a values would indicate, because they are coordinated with magnesium ion. Metal coordination decreases their basicity, which increases their leaving propensity.

4 THE "HIGH-ENERGY" CHARACTER OF PHOSPHOANHYDRIDE BONDS

Because the reaction of a nucleophile (such as ROH) with a phosphoanhydride bond is a highly exergonic reaction, phosphoanhydride bonds are called **high-energy bonds.** The term *high-energy* in this context means that a lot of energy is released when a nucleophile reacts with ATP. Do not confuse it with *bond energy,* the term chemists use to describe how difficult it is to break a bond.

Why is the reaction of a nucleophile with a phosphoanhydride bond so exergonic? In other words, why is the $\Delta G^{\circ\prime}$ value large and negative? A large negative $\Delta G^{\circ\prime}$ means that the products of the reaction are much more stable than the reactants. Let's look at the reactants and products of the following reaction to see why this is so.

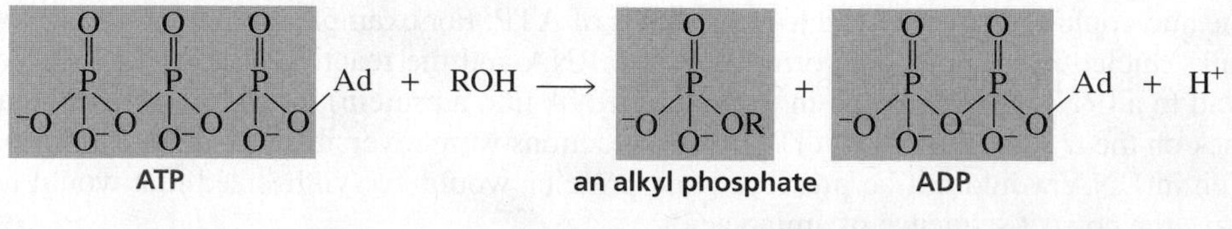

Three factors contribute to the greater stability of ADP and the alkyl phosphate compared to ATP:

1. **Greater electrostatic repulsion in ATP.** At physiological pH (7.4), ATP has 3.3 negative charges, ADP has 2.8 negative charges, and the alkyl phosphate has 1.8 negative charges (see Problem 1). Because of ATP's greater negative charge, the electrostatic repulsion is greater in ATP than in either of the products. Electrostatic repulsions destabilize a molecule.

2. **Greater solvation stabilization in the products.** Negatively charged ions are stabilized in an aqueous solution by solvation. Because the reactant has 3.3 negative charges, whereas the sum of the negative charges on the products is 4.6 (2.8 + 1.8), there is greater solvation stabilization in the products than in the reactant.

3. **Greater electron delocalization in the products.** A lone pair on the oxygen joining two phosphorus atoms is not effectively delocalized because delocalization would put a partial positive charge on an oxygen. When the phosphoanhydride bond breaks, one additional lone pair can be effectively delocalized. Electron delocalization stabilizes a molecule.

four lone pairs can be effectively delocalized

five lone pairs can be effectively delocalized

Similar factors explain the large negative $\Delta G^{\circ\prime}$ when a nucleophile reacts with ATP to form a substituted AMP and pyrophosphate, and when pyrophosphate is hydrolyzed to two equivalents of phosphate.

PROBLEM 1 Solved

ATP has pK_a values of 0.9, 1.5, 2.3, and 7.7; ADP has pK_a values of 0.9, 2.8, and 6.8; and phosphoric acid has pK_a values of 1.9, 6.7, and 12.4.

Do the calculation showing that at pH 7.4

a. the charge on ATP is −3.3.

b. the charge on ADP is −2.8.

c. the charge on the alkyl phosphate is −1.8.

Solution to 1a Because pH 7.4 is much greater than the pK_a values of the first three ionizations of ATP, these three groups will be entirely in their basic forms at that pH, giving ATP three negative charges. So to answer the question, we need to calculate what fraction of the group with a pK_a value of 7.7 will be in its basic form at pH 7.4.

$$\frac{\text{concentration of the basic form}}{\text{total concentration}} = \frac{[A^-]}{[A^-] + [HA]}$$

$[A^-] = $ concentration of the basic form

$[HA] = $ concentration of the acidic form

Because this equation has two unknowns, one of the unknowns must be expressed in terms of the other unknown. Using the definition of the acid dissociation constant (K_a), we can define [HA] in terms of $[A^-]$, K_a, and $[H^+]$ and now we have only one unknown.

$$K_a = \frac{[A^-][H^+]}{[HA]}$$

$$[HA] = \frac{[A^-][H^+]}{K_a}$$

$$\frac{[A^-]}{[A^-] + [HA]} = \frac{[A^-]}{[A^-] + \dfrac{[A^-][H^+]}{K_a}} = \frac{K_a}{K_a + [H^+]}$$

Now we can calculate the fraction of the group with a pK_a value of 7.7 that will be in the basic form. (Notice that K_a is calculated from pK_a and that $[H^+]$ is calculated from pH.)

$$\frac{K_a}{K_a + [H^+]} = \frac{2.0 \times 10^{-8}}{2.0 \times 10^{-8} + 4.0 \times 10^{-8}} = 0.3$$

Thus, the total negative charge on ATP = 3.0 + 0.3 = 3.3.

5 THE FOUR STAGES OF CATABOLISM

The reactants required for all life processes ultimately come from our diet. In that sense, we really are what we eat. Mammalian nutrition requires fats, carbohydrates, and proteins in addition to vitamins and minerals.

Catabolism can be divided into four stages (Figure 1). The *first stage of catabolism* is called digestion. In this stage, the fats, carbohydrates, and proteins we consume are hydrolyzed to fatty acids, monosaccharides, and amino acids, respectively. These reactions occur in the mouth, stomach, and small intestine.

In the first stage of catabolism, fats, carbohydrates, and proteins are hydrolyzed to fatty acids, monosaccharides, and amino acids.

In the second stage of catabolism, the products obtained from the first stage are converted to compounds that can enter the citric acid cycle.

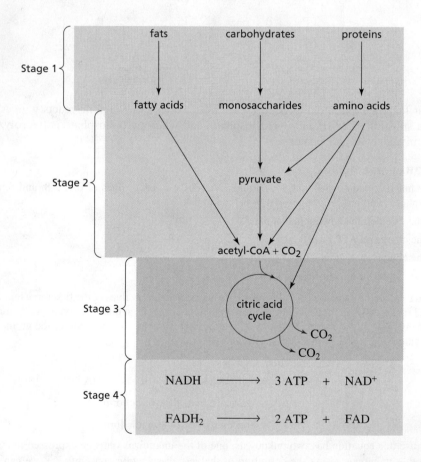

▶ **Figure 1**

The four stages of catabolism: 1. digestion; 2. conversion of the products of the first stage to compounds that can enter the citric acid cycle; 3. the citric acid cycle; and 4. oxidative phosphorylation.

In the *second stage of catabolism,* the products obtained from the first stage—fatty acids, monosaccharides, and amino acids—are converted to compounds that can enter the citric acid cycle. The only such compounds are (1) those that take part in the cycle itself (that is, citric acid cycle intermediates), (2) *acetyl-CoA,* and (3) *pyruvate* (because it can be converted to acetyl-CoA).

Neither pyruvate nor acetyl-CoA can enter the citric acid cycle directly because they are not citric acid cycle intermediates. Pyruvate is first converted to acetyl-CoA. We will see that acetyl-CoA then enters the cycle by being converted to citrate, one of the citric acid cycle intermediates.

The *third stage of catabolism* is the citric acid cycle. In this cycle, the acetyl group of each molecule of acetyl-CoA is converted to two molecules of CO_2.

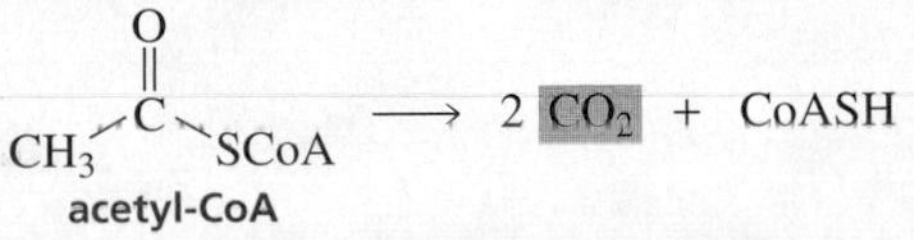

$$CH_3-\overset{\overset{O}{\|}}{C}-SCoA \longrightarrow 2\ CO_2\ +\ CoASH$$

acetyl-CoA

Cells get the energy they need by using nutrient molecules to make ATP. Only a small amount of ATP is formed in the first three stages of catabolism—most ATP is formed in the fourth stage. (You will be able to see this when you finish the chapter and can compare the answers to Problems 52, 54, and 55.)

Many catabolic reactions are oxidation reactions. In the *fourth stage of catabolism*, every molecule of NADH formed in one of the earlier stages of catabolism (formed when NAD^+ is used to carry out an oxidation reaction) is converted to 2.5 molecules of ATP in a process called *oxidative phosphorylation*. Oxidative phosphorylation also converts every molecule of $FADH_2$ formed in the earlier stages of catabolism (when FAD is used to carry out an oxidation reaction) to 1.5 molecules of ATP. Thus, most of the energy (ATP) that is provided by fats, carbohydrates, and proteins is obtained in the fourth stage of catabolism.

6 THE CATABOLISM OF FATS

We have seen that in the first two stages of catabolism, fats, carbohydrates, and proteins are converted to compounds that can enter the citric acid cycle (Section 5). Here we will look at the reactions that allow fats to enter the citric acid cycle.

In the first stage of fat catabolism, the fat's three ester groups are hydrolyzed by an enzyme-catalyzed reaction to glycerol and three fatty acid molecules.

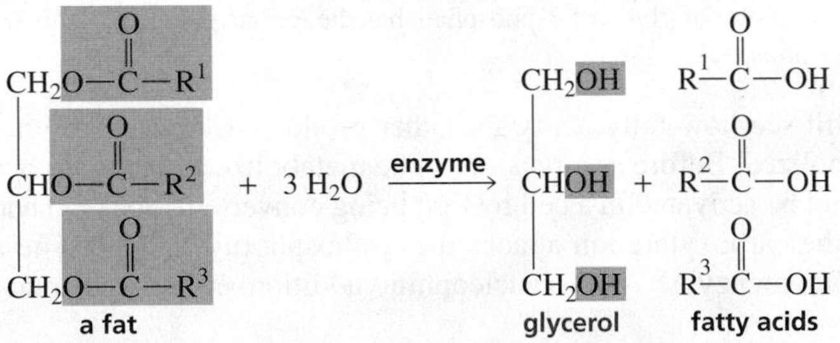

a fat   **glycerol**   **fatty acids**

The following reaction sequence shows what happens to glycerol, one of the products of the preceding reaction, in the second stage of catabolism.

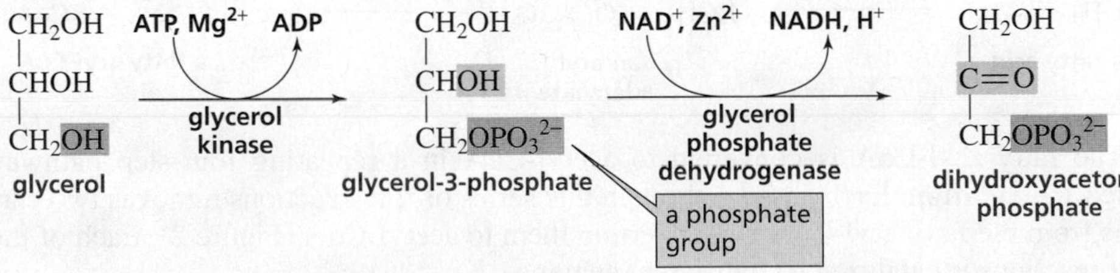

- Glycerol reacts with ATP to form glycerol-3-phosphate in the same way that glucose reacts with ATP to form glucose-6-phosphate (Section 1). The enzyme that catalyzes this reaction is called glycerol kinase. A *kinase* is an enzyme that puts a phosphate group on its substrate; thus, glycerol kinase puts a phosphate group on glycerol. Notice that this ATP-requiring enzyme also requires Mg^{2+}.

- The secondary alcohol group of glycerol-3-phosphate is then oxidized by NAD^+ to a ketone. The enzyme that catalyzes this reaction is called glycerol phosphate dehydrogenase. A *dehydrogenase* is an enzyme that oxidizes its substrate. We have seen that when a substrate is oxidized by NAD^+, the substrate donates a hydride ion to the 4-position of the pyridine ring. Zn^{2+} is a cofactor for the reaction; it increases the acidity of the secondary alcohol's proton by coordinating to the oxygen.

oxidation of substrate
reduction of the coenzyme

The product of the reaction sequence, dihydroxyacetone phosphate, is one of the intermediates in the glycolytic pathway, so it can enter that pathway directly and be broken down further (Section 7).

Notice that when biochemical reactions are written, the only structures typically shown are those of the primary reactant and primary product. (See the last reaction earlier in this chapter.) The names of other reactants and products are abbreviated and placed on a curved arrow that intersects the reaction arrow.

PROBLEM 2

Show the mechanism for the reaction of glycerol with ATP to form glycerol-3-phosphate.

PROBLEM 3

The asymmetric center of glycerol-3-phosphate has the R configuration. Draw the structure of (R)-glycerol-3-phosphate.

Now we will see how fatty acids, the other products formed from the hydrolysis of fats, are metabolized. Before a fatty acid can be metabolized, it must be activated. A carboxylic acid can be activated in a cell by first being converted to an acyl adenylate, which occurs when the carboxylate ion attacks the α-phosphorus of ATP. The acyl adenylate then reacts with coenzyme A in a nucleophilic addition–elimination reaction to form a thioester.

The fatty acyl-CoA is converted to acetyl-CoA in a repeating four-step pathway called **β-oxidation.** Each passage through this series of four reactions removes two carbons from the fatty acyl-CoA by converting them to acetyl-CoA (Figure 2). Each of the four reactions is catalyzed by a different enzyme.

1. The first reaction is an oxidation reaction that removes hydrogens from the α- and β-carbons, forming an α,β-unsaturated fatty acyl-CoA. The oxidizing agent is FAD. The enzyme that catalyzes this reaction has been found to be deficient in 10% of babies that experience sudden infant death syndrome (SIDS), a condition where an apparently healthy baby dies, generally while sleeping. Glucose is the cell's primary fuel immediately after a meal and then the cell switches to a combination of glucose and fatty acids. The inability to oxidize fats may be what causes the infant's distress.

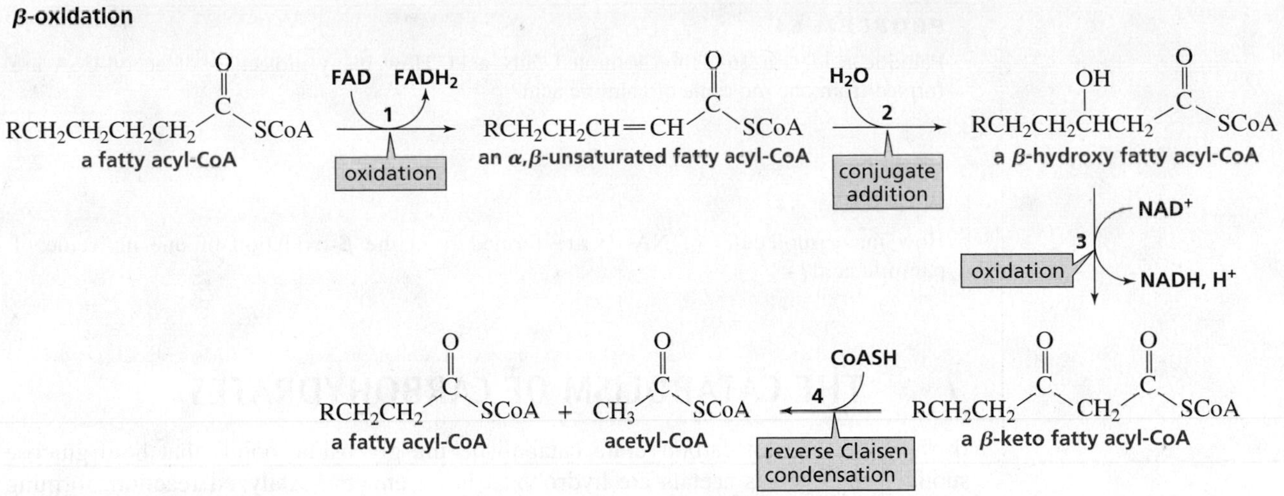

β-oxidation

▲ Figure 2

In β-oxidation a series of four enzyme-catalyzed reactions is repeated until the entire fatty acyl-CoA molecule has been converted to acetyl-CoA molecules. The enzymes that catalyze the reactions are 1. acyl-CoA dehydrogenase; 2. enoyl-CoA hydratase; 3. 3-L-hydroxyacyl-CoA dehydrogenase; and 4. β-ketoacyl-CoA thiolase

2. The second reaction, whose mechanism is shown next, is the conjugate addition of water to the α,β-unsaturated fatty acyl-CoA. A glutamate side chain of the enzyme removes a proton from water, making it a better nucleophile; the enolate ion is protonated by glutamic acid.

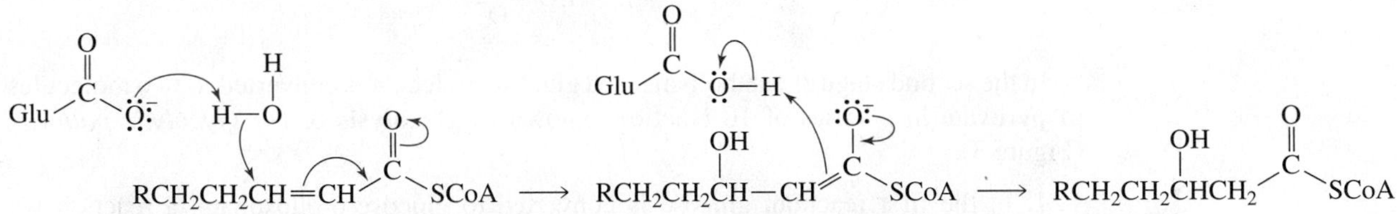

3. The third reaction is another oxidation reaction: NAD$^+$ oxidizes the secondary alcohol to a ketone. The mechanism for all NAD$^+$ oxidations involves donation of a hydride ion from the substrate to the 4-positon of the pyridine ring of NAD$^+$.

4. The fourth reaction is the reverse of a Claisen condensation, followed by conversion of the enolate ion to the keto tautomer. The mechanism for this reaction is shown here. The final product is acetyl-CoA and a fatty acyl-CoA *with two fewer carbons* than the starting fatty acyl-CoA.

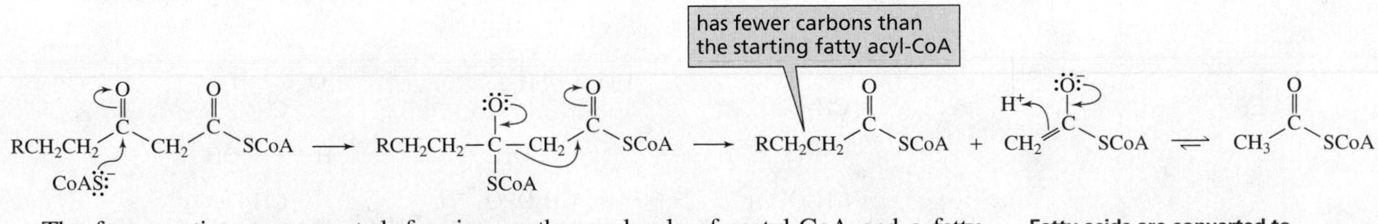

The four reactions are repeated, forming another molecule of acetyl-CoA and a fatty acyl-CoA that is now four carbons shorter than it was originally. Each time the series of four reactions is repeated, two more carbons are removed (as acetyl-CoA) from the fatty acyl-CoA. The series of reactions is repeated until the entire fatty acid has been converted into acetyl-CoA molecules. In Section 10 we will see how acetyl-CoA enters the citric acid cycle.

Fatty acids are converted to molecules of acetyl-CoA.

PROBLEM 4♦

Why does the OH group add to the β-carbon rather than to the α-carbon in the second reaction in the β-oxidation of fats?

Palmitic acid is a 16-carbon saturated fatty acid. How many molecules of acetyl-CoA are formed from one molecule of palmitic acid?

How many molecules of NADH are formed from the β-oxidation of one molecule of palmitic acid?

7   THE CATABOLISM OF CARBOHYDRATES

In the first stage of carbohydrate catabolism, the glycosidic bonds that hold glucose subunits together as acetals are hydrolyzed in an enzyme-catalyzed reaction, forming individual glucose molecules.

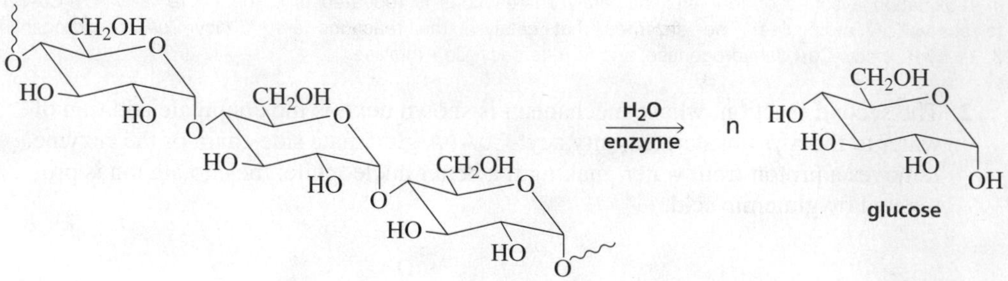

In the second stage of catabolism, each glucose molecule is converted to two molecules of pyruvate in a series of 10 reactions known as **glycolysis** or the *glycolytic pathway* (Figure 3).

1. In the first reaction, glucose is converted to glucose-6-phosphate, a reaction we looked at.

2. Glucose-6-phosphate then isomerizes to fructose-6-phosphate.

3. In the third reaction, ATP puts a second phosphate group on fructose-6-phosphate, forming fructose-1,6-bisphosphate. The mechanism of this reaction is the same as the one that converts glucose to glucose-6-phosphate.

4. The fourth reaction is the reverse of an aldol addition.

5. Dihydroxyacetone phosphate, produced in the fourth reaction, forms an enediol that then forms glyceraldehyde-3-phosphate (if the OH group at C-1 is the one that ketonizes) or reforms dihydroxyacetone phosphate (if the OH group at C-2 is the one that ketonizes).

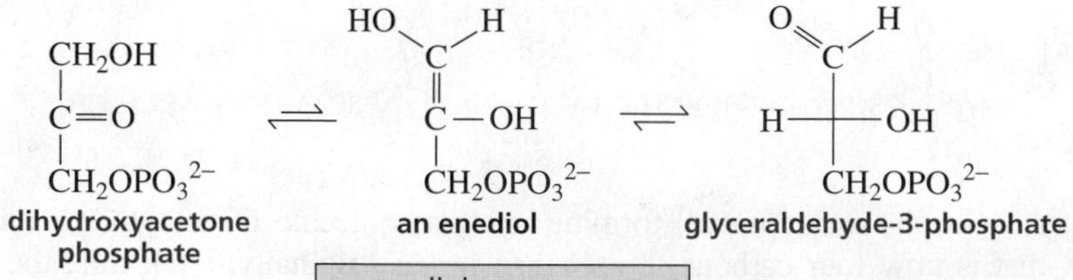

The mechanism for this reaction shows that a glutamate side chain of the enzyme removes a proton from the α-carbon, and a protonated histidine side chain donates a proton to the carbonyl oxygen. In the second step, the histidine removes a proton from the C-1 OH group, and a glutamic acid protonates C-2. Compare this mechanism with the one for the enediol rearrangement shown later in this section.

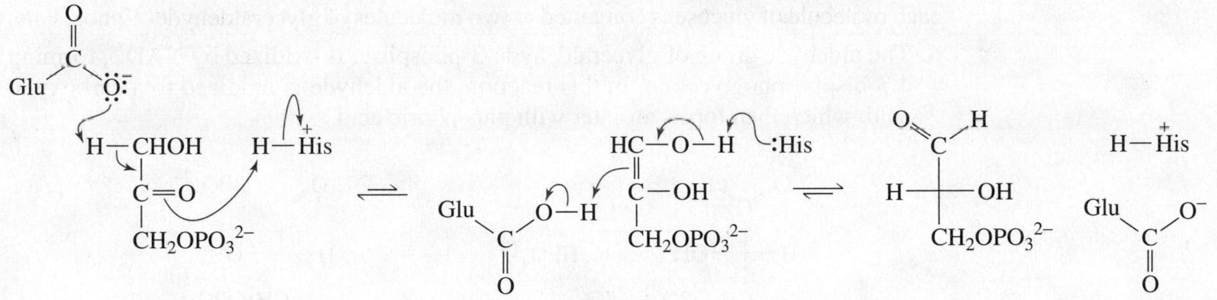

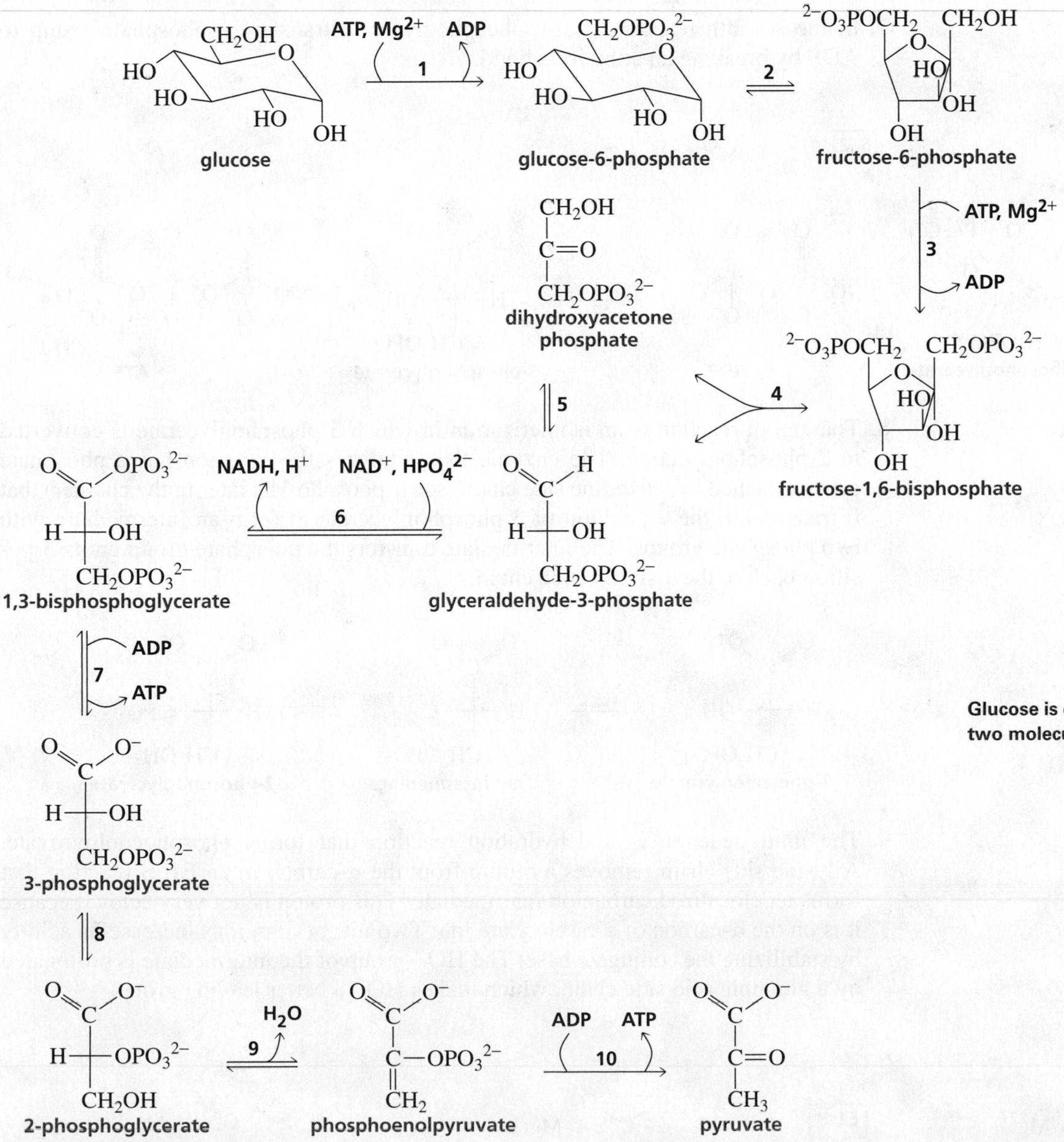

Glucose is converted to two molecules of pyruvate.

▲ Figure 3
Glycolysis, the series of enzyme-catalyzed reactions responsible for converting 1 mol of glucose to 2 mol of pyruvate. The enzymes that catalyze the reactions are 1. hexokinase; 2. phosphoglucose isomerase; 3. phosphofructokinase; 4. aldolase; 5. triosephosphate isomerase; 6. glyceraldehyde-3-phosphate dehydrogenase; 7. phosphoglycerate kinase; 8. phosphoglycerate mutase; 9. enolase; and 10. pyruvate kinase.

Because each molecule of glucose is converted to a molecule of glyceraldehyde-3-phosphate *and* a molecule of dihydroxyacetone phosphate, and each molecule of dihydroxyacetone phosphate is converted to glyceraldehyde-3-phosphate, then overall each molecule of glucose is converted to two molecules of glyceraldehyde-3-phosphate.

6. The aldehyde group of glyceraldehyde-3-phosphate is oxidized by NAD$^+$, forming 1,3-bisphosphoglycerate. In this reaction, the aldehyde is oxidized to a carboxylic acid, which then forms an ester with phosphoric acid.

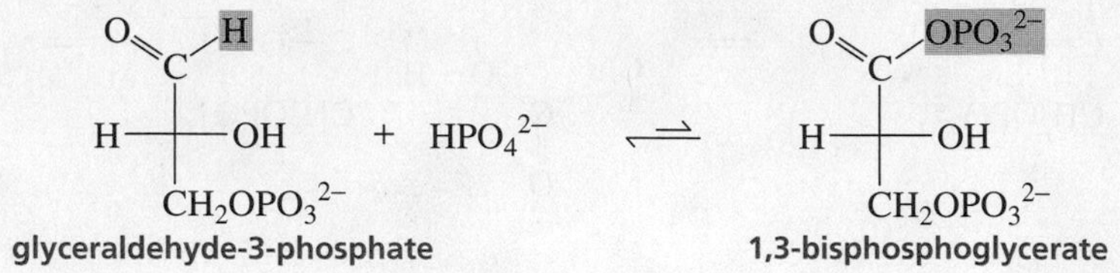

glyceraldehyde-3-phosphate   **1,3-bisphosphoglycerate**

7. In the seventh reaction, 1,3-bisphosphoglycerate transfers a phosphate group to ADP by breaking an anhydride bond.

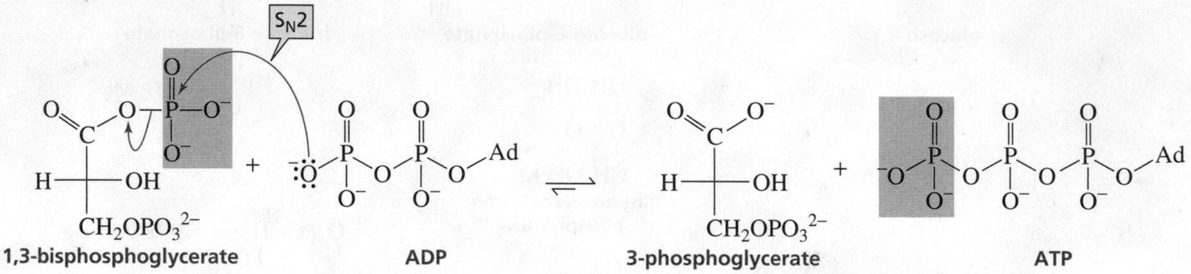

1,3-bisphosphoglycerate   **ADP**   **3-phosphoglycerate**   **ATP**

8. The eighth reaction is an isomerization in which 3-phosphoglycerate is converted to 2-phosphoglycerate. The enzyme that catalyzes this reaction has a phosphate group attached to a histidine side chain (see 3-phospho-His later in the chapter) that it transfers to the 2-position of 3-phosphoglycerate to form an intermediate with two phosphate groups. The intermediate transfers the phosphate group on its 3-position back to the histidine side chain.

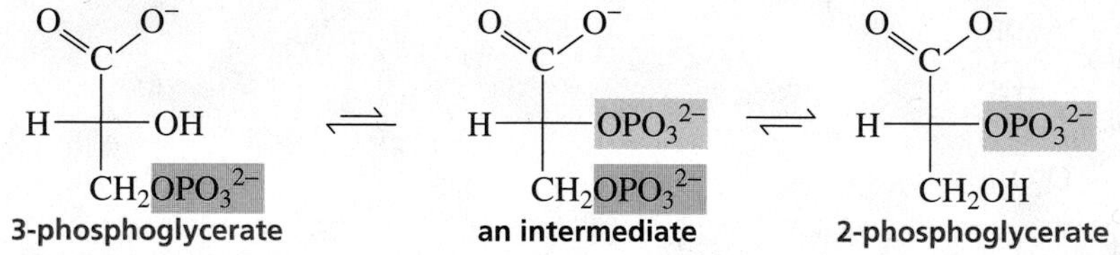

3-phosphoglycerate   **an intermediate**   **2-phosphoglycerate**

9. The ninth reaction is a dehydration reaction that forms phosphoenolpyruvate. A lysine side chain removes a proton from the α-carbon in an E1cB reaction that forms a delocalized carbanion intermediate. This proton is not very acidic because it is on the α-carbon of a carboxylate ion. Two magnesium ions increase its acidity by stabilizing the conjugate base. The HO$^-$ group of the intermediate is protonated by a glutamic acid side chain, which makes HO$^-$ a better leaving group.

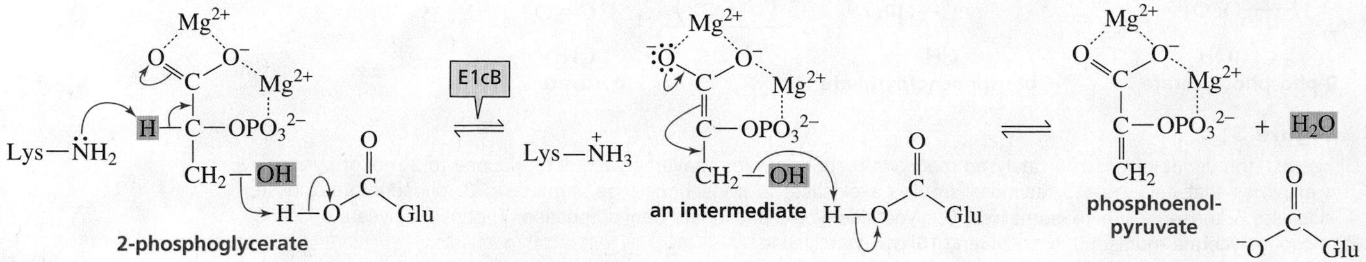

2-phosphoglycerate   **an intermediate**   **phosphoenol-pyruvate**

10. In the last reaction of the glycolytic pathway, phosphoenolpyruvate transfers its phosphate group to ADP, forming pyruvate and ATP.

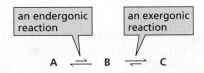

Phosphorylating glucose in the first reaction of glycolysis and phosphorylating fructose-6-phospate in the third reaction do not make glucose or fructose-6-phosphate any more reactive. The purpose of phosphorylation is to put a group on each of those compounds that allows enzymes both to recognize them (and recognize the subsequent intermediates formed in glycolysis), so they can be bound at their active site. The two molecules of ATP that are used to put these "handles" on the sugar molecules are re-formed in the last step of glycolysis—namely, in the conversion of two molecules of phosphoenolpyruvate to two molecules of pyruvate.

Glycolysis is exergonic overall, but all the reactions in the pathway are not themselves exergonic. For example, the conversion of glyceraldehyde-3-phosphate to 1,3-bisphosphoglycerate (the sixth reaction) is an endergonic reaction. However, the subsequent reaction (the conversion of 1,3-bisphosphoglycerate to 3-phosphoglycerate) is highly exergonic. Therefore, as the second reaction converts B to C, the first reaction will replenish the equilibrium concentration of B. In doing so, the exergonic reaction *drives* the reaction that precedes it. The two reactions (an endergonic reaction followed by an exergonic reaction) are called **coupled reactions.** Coupled reactions are the thermodynamic basis for how metabolic pathways operate since they are composed of both endergonic and exergonic reactions.

an endergonic reaction   an exergonic reaction

$$A \rightleftharpoons B \rightleftharpoons C$$

PROBLEM 7

Draw the mechanism for the third reaction in glycolysis—the reaction of fructose-6-phosphate with ATP to form fructose-1,6-bisphosphate.

PROBLEM 8♦

a. Which steps in glycolysis consume ATP?   **b.** Which steps in glycolysis produce ATP?

PROBLEM 9

The oxidation of glyceraldehyde-3-phosphate to 1,3-bisphosphoglycerate is an endergonic reaction, but the flow through this point in glycolysis proceeds smoothly. How is the unfavorable equilibrium constant overcome?

PROBLEM-SOLVING STRATEGY

Calculating Production of ATP

How many molecules of ATP are obtained from each molecule of glucose that is metabolized to pyruvate?

First we need to count the number of ATPs used to convert glucose to pyruvate. We see that two are used: one to form glucose-6-phosphate and the other to form fructose-1,6-bisphosphate.

Next, we need to know many ATPs are formed. Each glyceraldehyde-3-phosphate that is metabolized to pyruvate forms two ATPs. Because each molecule of glucose forms two molecules of glyceraldehyde-3-phosphate, four molecules of ATP are formed from each molecule of glucose. Subtracting the two molecules used, we find that each molecule of glucose that is metabolized to pyruvate forms two molecules of ATP.

Now continue on to Problem 10.

PROBLEM 10♦

How many molecules of NAD^+ are required to convert one molecule of glucose to pyruvate?

8 THE FATE OF PYRUVATE

We have just seen that NAD^+ is used as an oxidizing agent in glycolysis. If glycolysis is to continue, the NADH that is produced must to be oxidized back to NAD^+. Otherwise, no NAD^+ will be available as an oxidizing agent.

Under normal (aerobic) conditions, when oxygen is present, oxygen oxidizes NADH back to NAD^+ (this happens in *the fourth stage of catabolism*), and pyruvate is converted to acetyl-CoA, which then enters the citric acid cycle. This occurs via a series of reactions catalyzed by a complex of three enzymes and five coenzymes, known collectively as the pyruvate dehydrogenase complex. The overall result of this series of reactions is to transfer the acetyl group of pyruvate to coenzyme A.

$$CH_3-\underset{\underset{O}{\|}}{C}-\underset{\|}{\overset{O}{\|}}C-O^- + CoASH \xrightarrow{\text{pyruvate dehydrogenase complex}} CH_3-\underset{\underset{O}{\|}}{C}-\overset{O}{\|}C-SCoA + CO_2$$

pyruvate   acetyl-CoA

When oxygen is in short supply, such as when intense activity depletes all the oxygen in muscle cells, pyruvate (the product of glycolysis) oxidizes NADH back to NAD^+. In the process, pyruvate is reduced to lactate (lactic acid). The need to replenish oxygen is why people breathe hard during exercise.

$$CH_3-\underset{\underset{O}{\|}}{C}-\overset{O}{\|}C-O^- \underset{\text{lactate dehydrogenase}}{\overset{\text{NADH, } H^+ \quad NAD^+}{\rightleftharpoons}} CH_3-\underset{\underset{OH}{|}}{CH}-\overset{O}{\|}C-O^-$$

pyruvate   lactate

Although pyruvate is reduced to lactate under anaerobic (oxygen-free) conditions in animals, it has a different fate in yeast—namely, it is decarboxylated to acetaldehyde by pyruvate decarboxylase (an enzyme that is not present in animals).

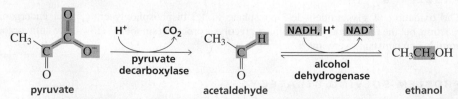

$$CH_3-\underset{\underset{O}{\|}}{C}-\overset{O}{\|}C-O^- \underset{\substack{\text{pyruvate} \\ \text{decarboxylase}}}{\overset{H^+ \quad CO_2}{\longrightarrow}} CH_3-\overset{O}{\underset{\|}{C}}-H \underset{\substack{\text{alcohol} \\ \text{dehydrogenase}}}{\overset{\text{NADH, } H^+ \quad NAD^+}{\rightleftharpoons}} CH_3CH_2OH$$

pyruvate   acetaldehyde   ethanol

In this case, acetaldehyde is the compound that oxidizes NADH back to NAD^+ and in the process is reduced to ethanol. This reaction has been used by humankind for thousands of years to produce wine, beer, and other fermented drinks. (Notice that enzyme names can refer either to the forward or to the reverse reaction. For example, pyruvate decarboxylase refers to the forward reaction, whereas alcohol dehydrogenase refers to the reverse reaction.)

PROBLEM 11♦

Suggest a name for alcohol dehydrogenase that would refer to the forward reaction, that is to the conversion of acetaldehyde to ethanol.

PROBLEM 12♦

What functional group of pyruvate is reduced when pyruvate is converted to lactate?

PROBLEM 13♦

What coenzyme is required to convert pyruvate to acetaldehyde?

PROBLEM 14

Propose a mechanism for the reduction of acetaldehyde by NADH to ethanol.

9　THE CATABOLISM OF PROTEINS

In the first stage of protein catabolism, proteins are hydrolyzed in an enzyme-catalyzed reaction to amino acids.

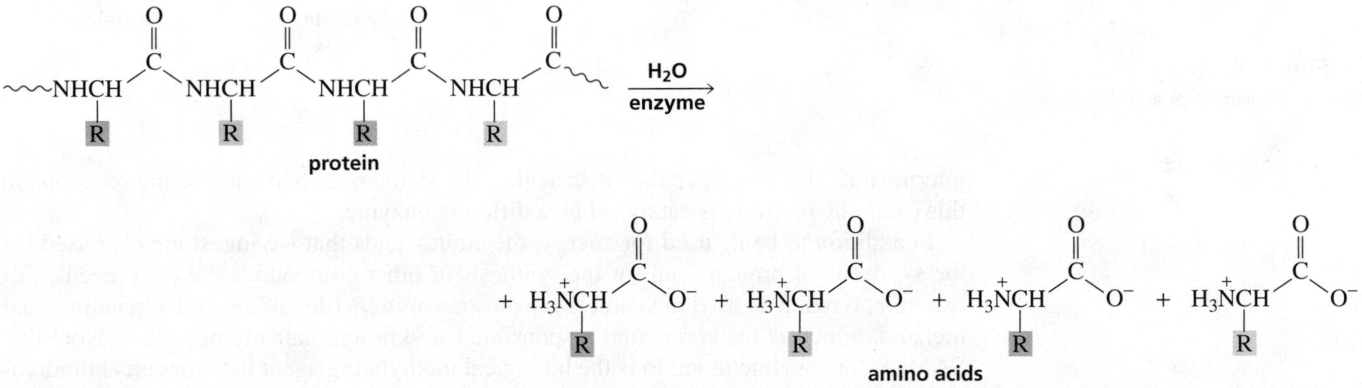

In the second stage of catabolism, the amino acids are converted to acetyl-CoA, pyruvate, and/or citric acid cycle intermediates, depending on the amino acid. The products of the second stage of catabolism then enter the citric acid cycle—the third stage of catabolism—and are further metabolized.

We will use the catabolism of phenylalanine as an example of how an amino acid is metabolized (Figure 4). Phenylalanine is one of the essential amino acids, so it must be included in our diet. The enzyme phenylalanine hydroxylase converts phenylalanine to tyrosine. Thus, tyrosine is not an essential amino acid, unless the diet lacks phenylalanine.

The first reaction in the catabolism of most amino acids is transamination, a reaction that requires the coenzyme pyridoxal pyrophosphate. A transamination reaction replaces the amino group of the amino acid with a ketone group. *para*-Hydroxyphenylpyruvate, the product of the transamination of tyrosine, is converted by a series of reactions to fumarate and acetyl-CoA.

Fumarate is a citric acid cycle intermediate, so it can enter the citric acid cycle directly. We will see in Section 10 that acetyl-CoA is the only non-citric acid cycle

Amino acids are converted to acetyl-CoA, pyruvate, and/or citric acid cycle intermediates.

1223

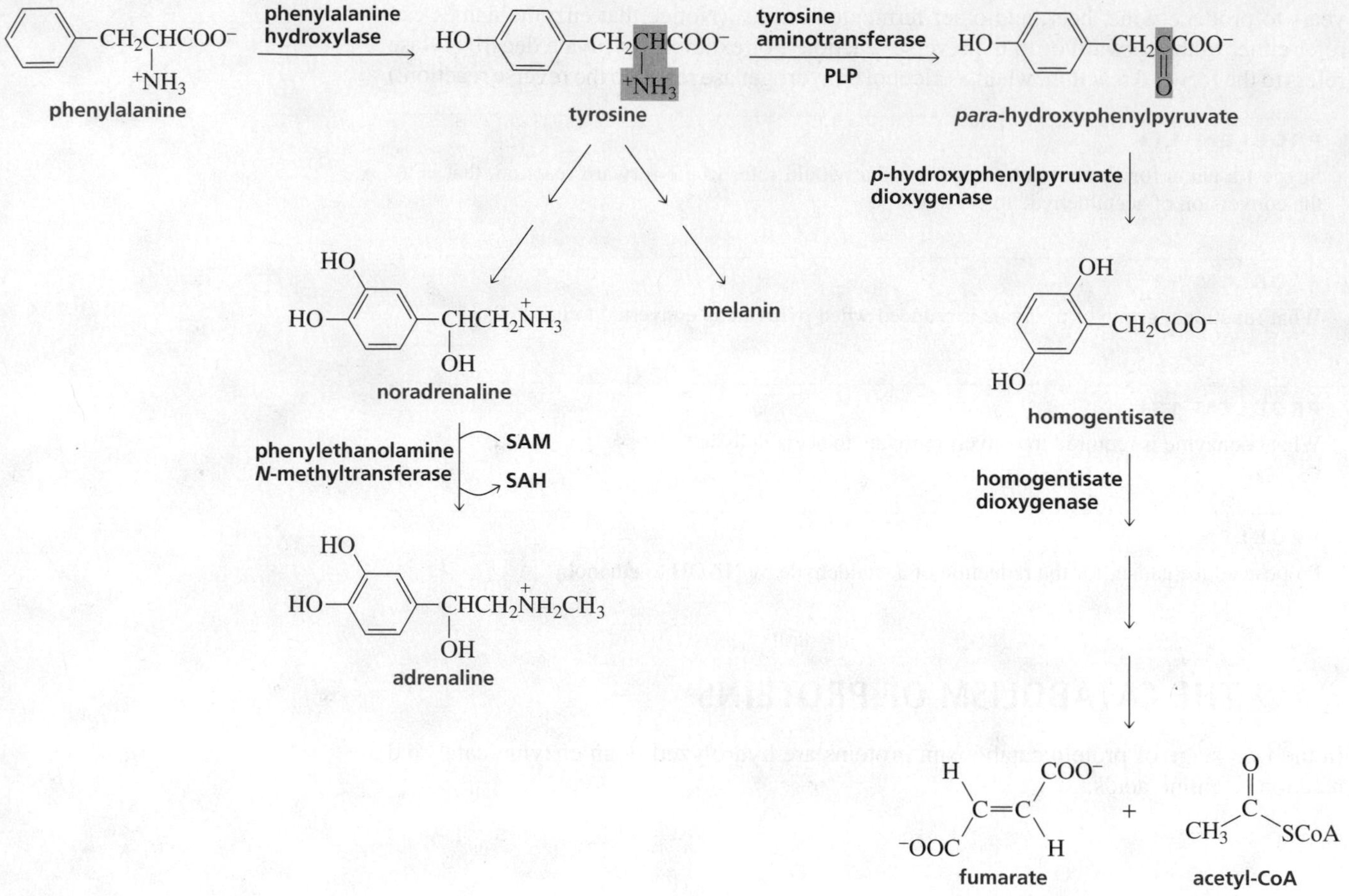

▲ Figure 4

The catabolism of phenylalanine.

intermediate that can enter the citric acid cycle. Remember that each of the reactions in this catabolic pathway is catalyzed by a different enzyme.

In addition to being used for energy, the amino acids that we ingest are also used for the synthesis of proteins and for the synthesis of other compounds the body needs. For example, tyrosine is used to synthesize neurotransmitters (dopamine and adrenaline) and melanin, which is the compound responsible for skin and hair pigmentation. Note that SAM (S-adenosylmethionine) is the biological methylating agent that converts noradrenaline to adrenaline.

Phenylketonuria (PKU): An Inborn Error of Metabolism

About one in every 20,000 babies is born without phenylalanine hydroxylase, the enzyme that converts phenylalanine to tyrosine. This genetic disease is called phenylketonuria (PKU). Without phenylalanine hydroxylase, the level of phenylalanine builds up; when it reaches a high concentration, it is transaminated to phenylpyruvate, a compound that interferes with normal brain development. The high level of phenylpyruvate found in urine gives the disease its name.

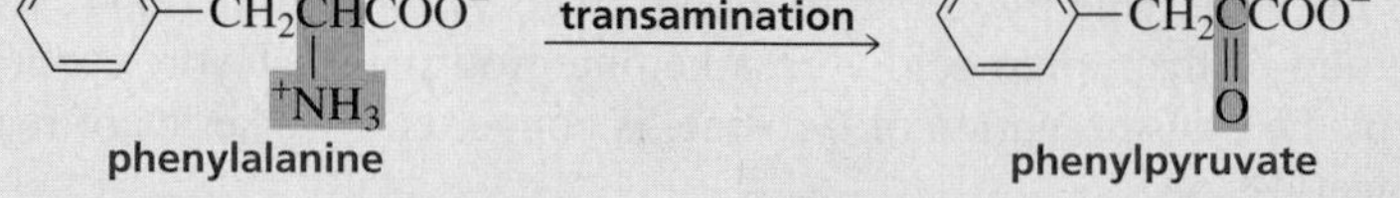

Within 24 hours after birth, all babies born in the United States are tested for high serum phenylalanine levels, which indicate a buildup of phenylalanine caused by an absence of phenylalanine hydroxylase. Babies with high levels are immediately put on a diet low in phenylalanine and high in tyrosine. As long as the phenylalanine level is kept under careful control for the first 5 to 10 years of life, the child will experience no adverse effects. You may have noticed the warning on containers of foods that contain NutraSweet, announcing that it contains phenylalanine. (Note that this sweetener is a methyl ester of a dipeptide of L-aspartate and L-phenylalanine).

If phenylalanine in the diet is not controlled, then the baby will be severely mentally retarded by the time he or she is a few months old. Untreated children have paler skin and fairer hair than other members of their family because, without tyrosine, they cannot synthesize melanin, a skin and hair pigment. Half of untreated phenylketonurics die by age 20. When a woman with PKU becomes pregnant, she must return to the low-phenylalanine diet she had as a child because a high level of phenylalanine can cause abnormal development of the fetus.

Alcaptonuria

Another genetic disease that results from a deficiency of an enzyme in the pathway for phenylalanine degradation is alcaptonuria, which is caused by lack of homogentisate dioxygenase. The only ill effect of this enzyme deficiency is black urine. The urine of those afflicted with alcaptonuria turns black because the homogentisate they excrete immediately oxidizes in the air, forming a black compound.

PROBLEM 15♦

What coenzyme is required for transamination?

PROBLEM 16♦

What compound is formed when alanine is transaminated?

10   THE CITRIC ACID CYCLE

The **citric acid cycle** (third stage of catabolism) is a series of eight reactions in which the acetyl group of each molecule of acetyl-CoA—formed by the catabolism of fats, carbohydrates, and amino acids—is converted to two molecules of CO_2 (Figure 5).

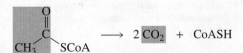

$$\longrightarrow \quad 2\ CO_2 \quad + \quad CoASH$$

The series of reactions is called a *cycle* because, unlike the reactions of other metabolic pathways, they comprise a closed loop in which the product of the eighth reaction (oxaloacetate) is the reactant for the first reaction.

The acetyl group of each molecule of acetyl-CoA that enters the citric acid cycle is converted to two molecules of CO_2.

1. In the first reaction of the citric acid cycle, acetyl-CoA reacts with oxaloacetate to form citrate. The mechanism for the reaction shows that an aspartate side chain of the enzyme removes a proton from the α-carbon of acetyl-CoA, creating an enolate ion. This enolate ion adds to the keto carbonyl carbon of oxaloacetate and the carbonyl oxygen picks up a proton from a histidine side chain. This is similar to an aldol addition where the α-carbanion (enolate ion) of one molecule is the nucleophile and the carbonyl carbon of another is the electrophile. The intermediate (a thioester) that results is hydrolyzed to citrate in a nucleophilic addition–elimination reaction.

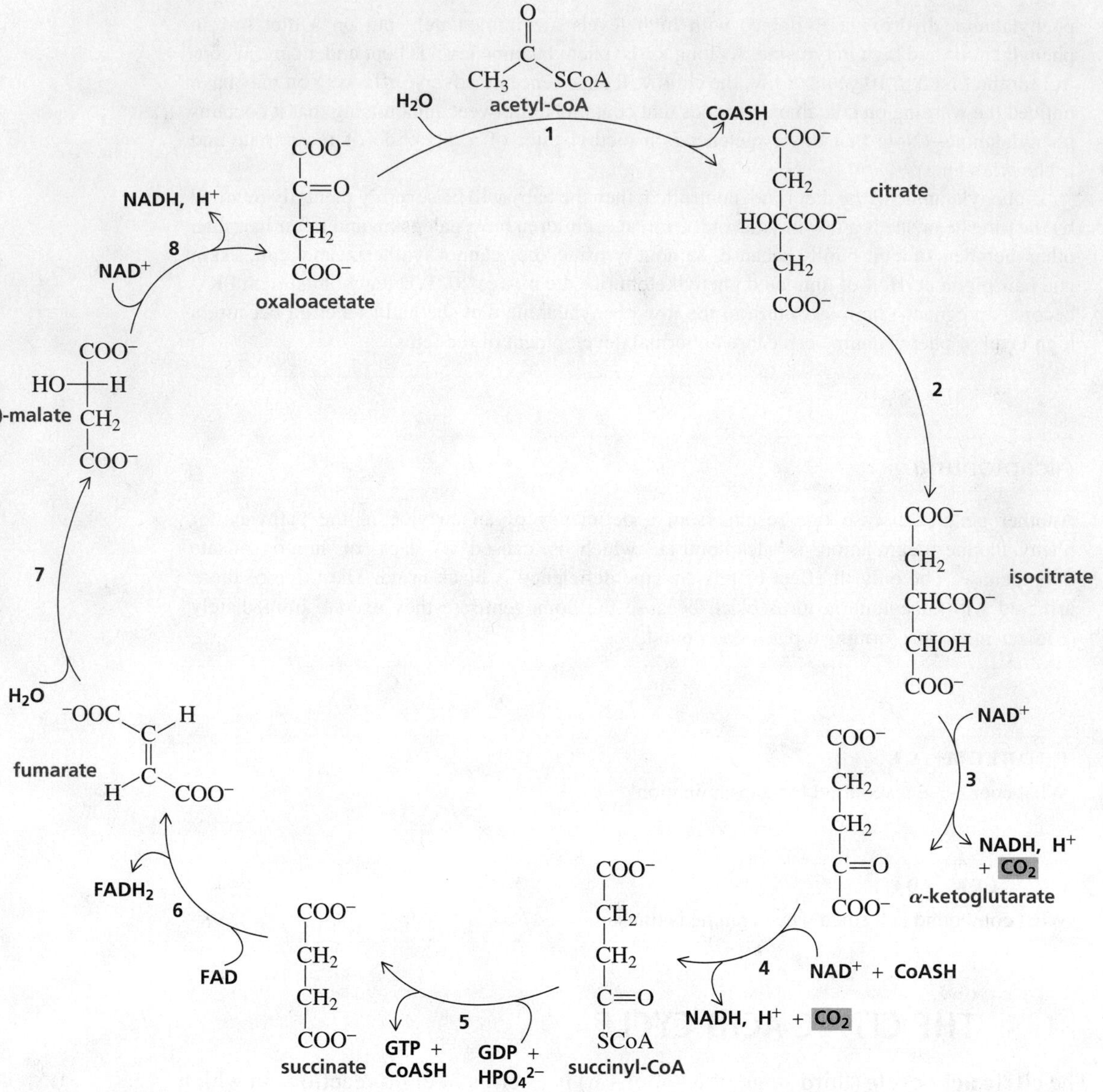

▲ Figure 5

The citric acid cycle is the series of enzyme-catalyzed reactions responsible for the oxidation of the acetyl group of acetyl-CoA to two molecules of CO_2. The enzymes that catalyze the reactions are 1. citrate synthase; 2. aconitase; 3. isocitrate dehydrogenase; 4. α-ketoglutarate dehydrogenase; 5. succinyl-CoA synthetase; 6. succinate dehydrogenase; 7. fumarase; and 8. malate dehydrogenase.

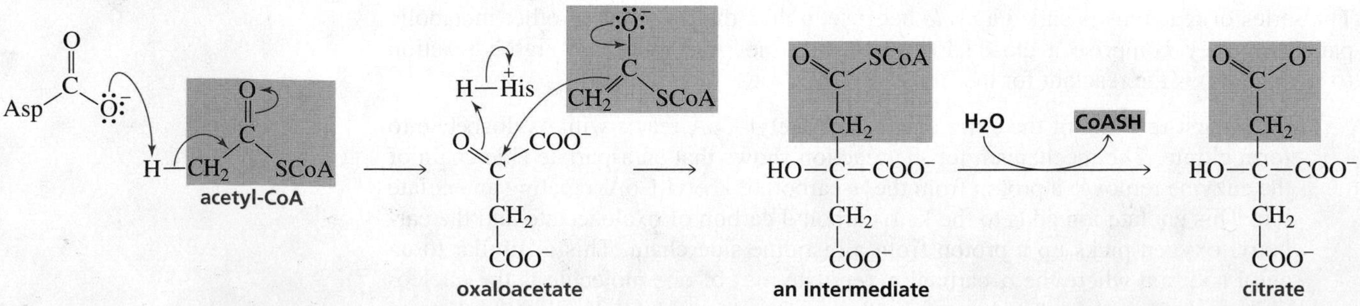

2. In the second reaction, citrate is converted to isocitrate, its isomer. The reaction takes place in two steps: water is removed in the first step and then re-added in the second step. The first step is an E2 dehydration in which a serine side chain removes a proton, and the OH leaving group is protonated by a histidine side chain to make it a weaker base (H_2O) and therefore a better leaving group. In the second step, conjugate addition of water to the intermediate forms isocitrate.

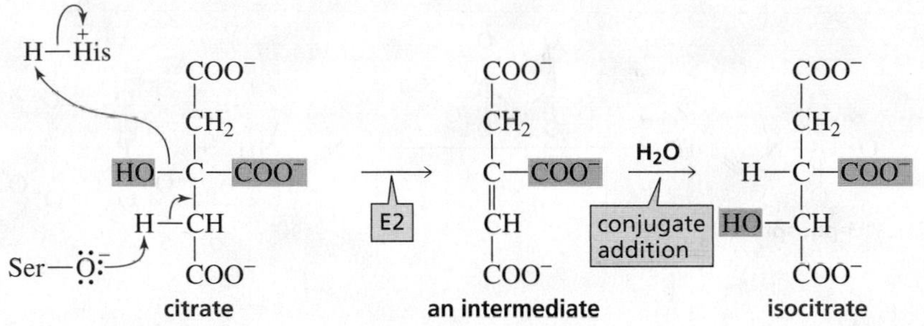

3. The third reaction is the one that releases the first molecule of CO_2. It also has two steps. In the first, the secondary alcohol group of isocitrate is oxidized to a ketone by NAD^+. In the second, the ketone loses CO_2, with Mg^{2+} acting as a catalyst. A CO_2 group bonded to a carbon adjacent to a carbonyl carbon can be removed because the electrons left behind can be delocalized onto the carbonyl oxygen. The enolate ion tautomerizes to a ketone.

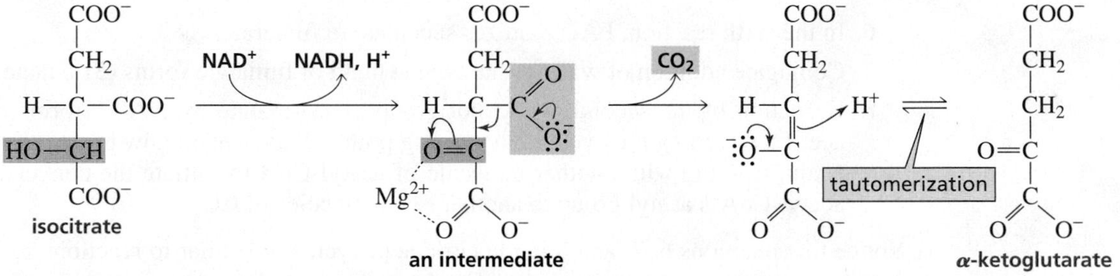

4. The fourth reaction is the one that releases the second molecule of CO_2. The reaction requires a group of enzymes and the same five coenzymes required by the pyruvate dehydrogenase complex that forms acetyl-CoA. Like the reaction catalyzed by the pyruvate dehydrogenase complex, the overall result of this reaction is the transfer of an acyl group to CoASH. Thus, the product of the reaction is succinyl-CoA.

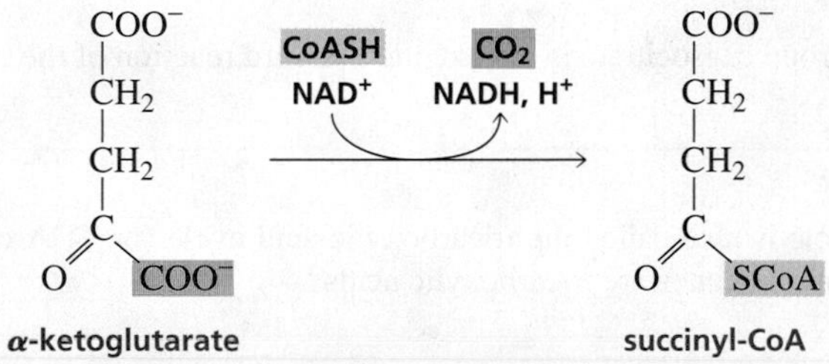

5. The fifth reaction takes place in two steps. First, hydrogen phosphate reacts with succinyl-CoA in a nucleophilic addition–elimination reaction to form an intermediate, which then transfers its phosphate group to GDP.

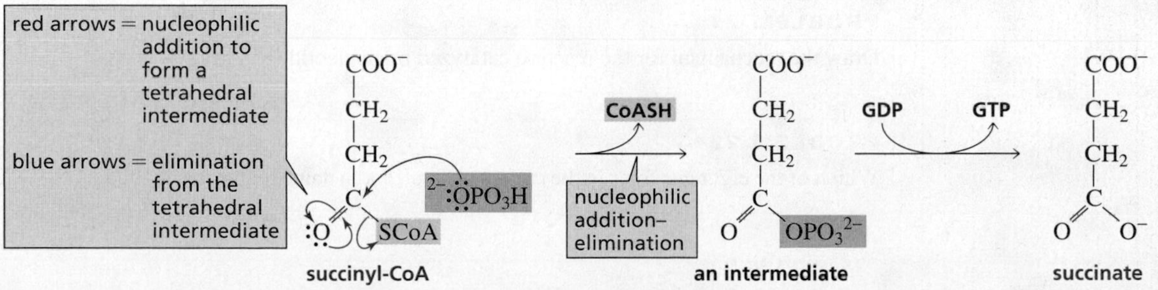

The intermediate does not transfer its phosphate group directly to GDP. Instead, it transfers the phosphate group to a histidine side chain of the enzyme, forming 3-phospho-His, which then transfers the phosphate group to GDP.

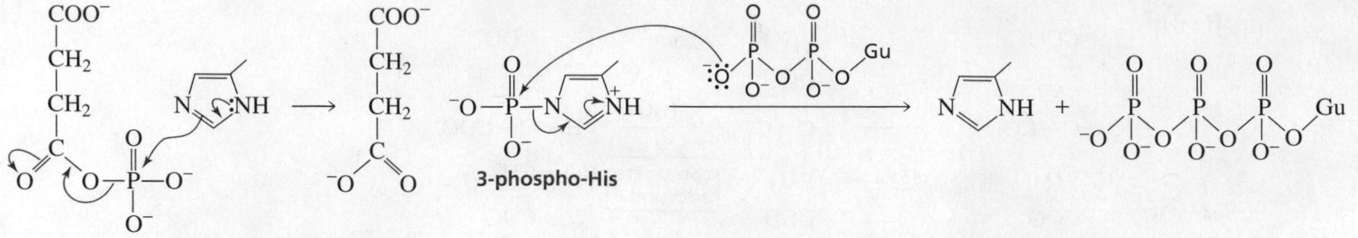

3-phospho-His

Once formed, GTP transfers a phosphate group to ADP to form ATP. The rapid interconversion of GTP and ATP is catalyzed by an enzyme called nucleotide diphosphate kinase.

$$GTP \ + \ ADP \ \rightleftharpoons \ GDP \ + \ ATP$$

At this point, the citric acid cycle has accomplished the required transformation—that is, acetyl-CoA has been converted to CoASH and two molecules of CO_2. What remains to be done is to convert succinate to oxaloacetate, so oxaloacetate can begin the cycle again by reacting with another molecule of acetyl-CoA.

6. In the sixth reaction, FAD oxidizes succinate to fumarate.

7. Conjugate addition of water to the double bond of fumarate forms (S)-malate.

8. Oxidation of the secondary alcohol group of (S)-malate by NAD^+ forms oxaloacetate, returning the cycle to its starting point. Oxaloacetate now begins the cycle again, reacting with another molecule of acetyl-CoA to initiate the conversion of acetyl-CoA's acetyl group to another two molecules of CO_2.

Notice that reactions 6, 7, and 8 in the citric acid cycle are similar to reactions 1, 2, and 3 in the β-oxidation of fatty acids (Section 6).

PROBLEM 17

Acid-catalyzed dehydration reactions are normally E1 reactions. Why is the acid-catalyzed dehydration in the second reaction of the citric acid cycle an E2 reaction?

PROBLEM 18♦

What functional group of isocitrate is oxidized in the third reaction of the citric acid cycle?

PROBLEM 19♦

The citric acid cycle is also called the tricarboxylic acid cycle (or TCA cycle). Which of the citric acid cycle intermediates are tricarboxylic acids?

PROBLEM 20♦

What acyl group is transferred by thiamine pyrophosphate in the fourth reaction of the citric acid cycle?

PROBLEM 21

Draw the mechanism for the reaction catalyzed by nucleotide diphosphate kinase.

PROBLEM 22♦

Which of the eight enzymes in the citric acid cycle has a name that refers to the reverse reaction?

---

11   OXIDATIVE PHOSPHORYLATION

The NADH and FADH$_2$ molecules formed in the second and third stages of catabolism undergo **oxidative phosphorylation**—the fourth stage of catabolism—which oxidizes them back to NAD$^+$ and FAD so that they can participate in more oxidation reactions.

The electrons lost when NAD$^+$ and FAD are oxidized are transferred to a system of linked electron acceptors. One of the first electron acceptors is coenzyme Q$_{10}$, which is a quinone. When a quinone gains electrons (is reduced), it forms a hydroquinone. When hydroquinone passes electrons to the next electron acceptor, it is oxidized back to quinone. The last electron acceptor is O$_2$. When O$_2$ accepts electrons, it is reduced to water. This chain of oxidation–reduction reactions supplies the energy that is used to convert ADP to ATP.

coenzyme Q$_{10}$

For each NADH that undergoes oxidative phosphorylation, 2.5 molecules of ATP are formed, and for each FADH$_2$ that undergoes oxidative phosphorylation, 1.5 molecules of ATP are formed.

$$NADH \longrightarrow NAD^+ + 2.5\ ATP$$

$$FADH_2 \longrightarrow FAD + 1.5\ ATP$$

> In oxidative phosphorylation, each molecule of NADH is converted to 2.5 molecules of ATP and each molecule of FADH$_2$ is converted to 1.5 molecules of ATP.

Each round of the citric acid cycle forms 3 molecules of NADH, 1 molecule of FADH$_2$, and 1 molecule of ATP. Therefore, for every molecule of acetyl-CoA that enters the citric acid cycle, 7.5 molecules of ATP are formed from NADH, 1.5 molecules of ATP are formed from FADH$_2$, and 1 molecule of ATP is formed in the cycle, for a total of 10 ATP.

$$3\ NADH + FADH_2 \longrightarrow 3\ NAD^+ + FAD + 9\ ATP \xrightarrow[\quad\quad]{GTP,\ ADP\quad GDP} 10\ ATP$$

Basal Metabolic Rate

Your basal metabolic rate (BMR) is the number of calories you would burn if you stayed in bed all day. A BMR is affected by gender, age, and genetics: it is greater for men than for women, it is greater for young people than for old people, and some people are born with a faster metabolic rate than others. The BMR is also affected by the percentage of body fat: the higher the percentage, the lower the BMR. For humans, the average BMR is about 1600 kcal/day.

In addition to consuming sufficient calories to sustain your basal metabolism, you must also consume calories for the energy needed to carry out physical activities. The more active you are, the more calories you must consume in order to maintain your current weight. People who consume more calories than required by their BMR plus their level of physical activity will gain weight; if they consume fewer calories, they will lose weight.

PROBLEM 23♦

How many molecules of ATP are obtained from the conversion of one molecule of glycerol to pyruvate

a. not including the fourth stage of catabolism?
b. including the fourth stage of catabolism?

12   ANABOLISM

Anabolism is the reverse of catabolism. In anabolism, acetyl-CoA, pyruvate, citric acid cycle intermediates, and intermediates formed in glycolysis are the starting materials for the synthesis of fatty acids, carbohydrates, and proteins.

For example, cells use acetyl-CoA to synthesize fatty acyl-CoAs. Once the fatty acyl-CoAs are synthesized, they can form fats or oils by esterifying glycerol-3-phosphate, which is obtained by reducing dihydroxyacetone phosphate, an intermediate formed in glycolysis.

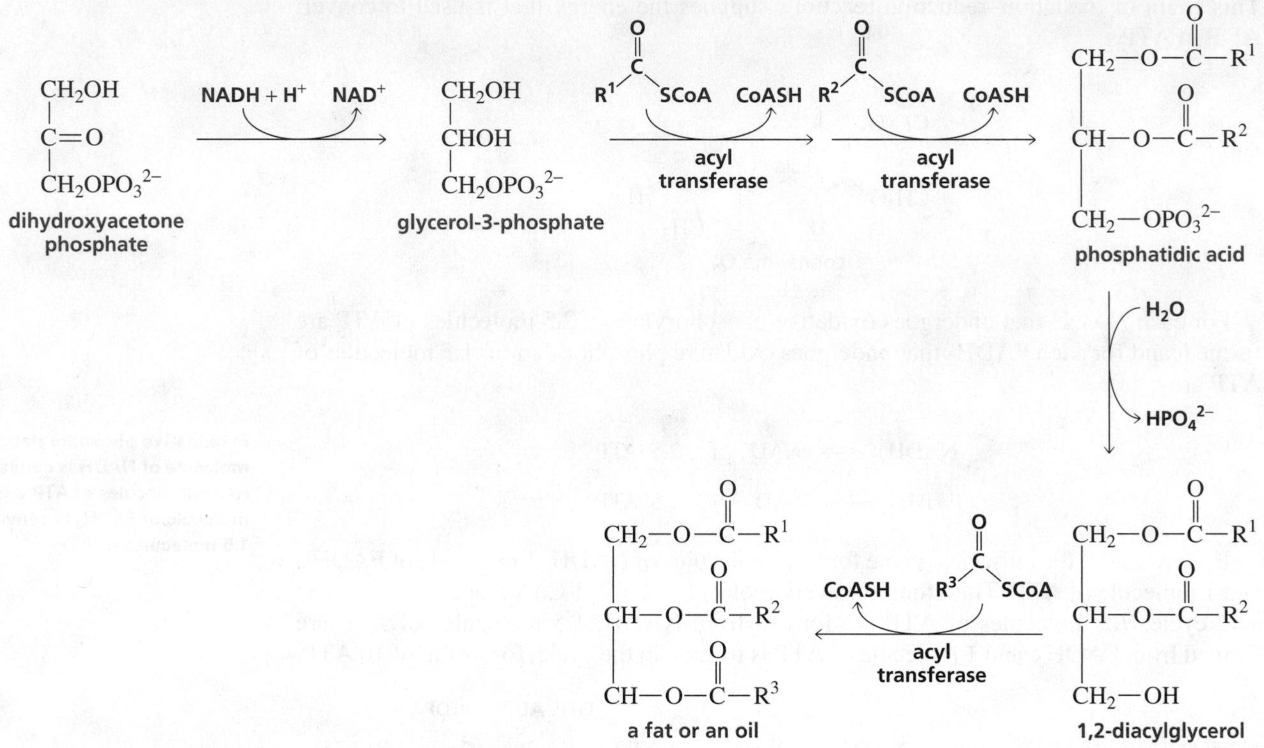

PROBLEM 24♦

a. What is the name of the enzyme that converts glycerol to glycerol-3-phosphate?

b. What is the name of the enzyme that converts phosphatidic acid to 1,2-diacylglycerol?

13   GLUCONEOGENESIS

Gluconeogenesis—the synthesis of glucose from pyruvate—is an anabolic pathway. Glucose is the primary fuel for the body. But in times of prolonged exercise or fasting, the body runs out of glucose and has to use fat for its fuel. The brain, however, cannot metabolize fat, so it has to have a continuous supply of glucose. Therefore, the body needs to have a way to synthesize glucose when a sufficient supply is not available.

As you can see from Figure 6, many of the reactions involved in the synthesis of glucose are carried out by the same enzymes that catalyze the breakdown of glucose to pyruvate in glycolysis—they are just operating in reverse. However, all the reactions in gluconeogenesis cannot be just the reverse of those operating in glycolysis. Some of the enzymes in each pathway catalyze essentially irreversible reactions, and detours have to be made around these reactions when going in the other direction. By the use of different enzymes for the forward and reverse irreversible reactions, both pathways become thermodynamically favorable.

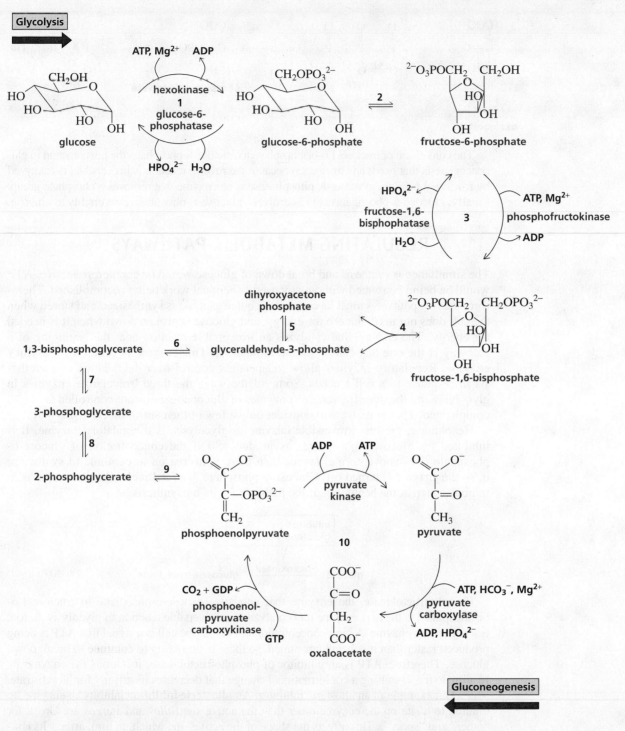

▲ Figure 6

Glycolysis (the conversion of glucose to pyruvate) and gluconeogenesis (the biosynthesis of glucose from pyruvate).

Reactions 1, 3, and 10 in glycolysis are irreversible (Figure 6). Therefore, a different enzyme is needed to catalyze the reverse of these reactions in gluconeogenesis. The reverse of the last irreversible reaction in glycolysis is actually two successive enzyme-catalyzed reactions. First pyruvate is converted to oxaloacetate by pyruvate carboxylase, a biotin-dependent enzyme. Oxaloacetate is then converted to phosphoenol-pyruvate. In this reaction, the 3-oxocarboxylic acid is decarboxylated and the oxygen of the enolate ion attacks the γ-phosphorus of GTP.

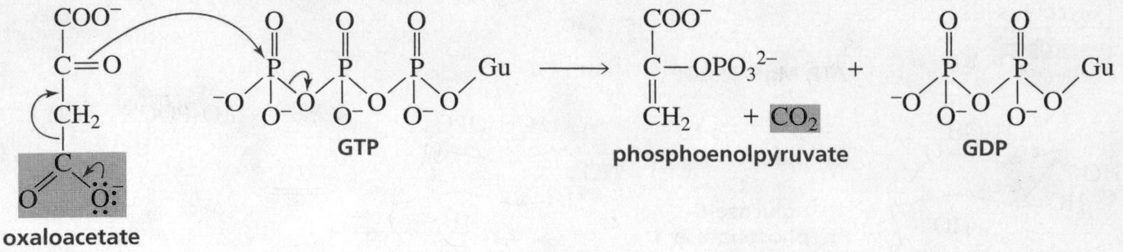

The conversion of fructose-1,6-bisphosphate to fructose-6-phosphate, the next reaction in gluconeogenesis that needs an enzyme (3) because the reverse reaction is irreversible, is catalyzed by fructose-1,6-bisphosphatase. A **phosphatase** is an enzyme that removes a phosphate group. Finally, glucose-6-phosphatase (1) hydrolyzes glucose-6-phosphate irreversibly to glucose.

14 REGULATING METABOLIC PATHWAYS

The simultaneous synthesis and breakdown of glucose would be counterproductive. ATP would be being consumed without any useful chemical work being accomplished. Therefore, the two pathways must be controlled so that glucose is synthesized and stored when the cell does not need glucose for energy, and glucose is broken down when it is needed for energy. The enzyme that catalyzes an irreversible reaction near the beginning of a pathway is the one that can be turned on and off. This enzyme is called a **regulatory enzyme.** Regulatory enzymes allow independent control over degradation and synthesis in response to a cell's needs. Some of the ways the three irreversible enzymes in glycolysis and the three irreversible enzymes in gluconeogenesis are controlled are quite complicated. Therefore, we will consider only a few of the control mechanisms here.

Hexokinase, the first irreversible enzyme in glycolysis, is a regulatory enzyme. It is inhibited by glucose-6-phosphate, its product. So if the concentration of glucose-6-phosphate rises above normal levels, then there is no reason to continue to synthesize it, so the enzyme is turned off. Glucose-6-phosphate is a **feedback inhibitor**—that is, it inhibits a step at the beginning of the pathway for its biosynthesis.

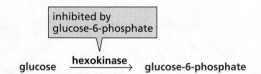

Phosphofructokinase, the enzyme that converts fructose-6-phosphate to fructose-1,6-bisphosphate, is the next enzyme that catalyzes an irreversible reaction in glycolysis. It, too, is a regulatory enzyme. A high concentration of ATP in the cell is a signal that ATP is being produced faster than it is being consumed, so there is no reason to continue to break down glucose. Therefore, ATP is an inhibitor of phosphofructokinase. It inhibits the enzyme by binding to it and causing a conformational change that decreases its affinity for its substrate. ATP is an example of an allosteric inhibitor. An **allosteric inhibitor** inhibits an enzyme by binding to a site on the enzyme other than the active site (*allos* and *stereos* are Greek for "other" and "space"). This affects the shape of the active site, which, in turn, affects its ability to catalyze a reaction. On the other hand, when the concentration of ADP and AMP in the cell are high, this is a signal that ATP is being consumed faster than it is being produced. Therefore, ADP and AMP are **allosteric activators** of phosphofructokinase. They bind to the enzyme and reverse the inhibition that was brought on by binding ATP.

Citrate is also an allosteric inhibitor of phosphofructokinase. A high concentration of citrate (a citric acid cycle intermediate) in a cell signals that the cell is currently meeting its energy needs by the oxidation of fats and proteins, so the oxidation of carbohydrates can be stopped temporarily.

Pyruvate carboxylase, the first irreversible enzyme in gluconeogenesis, also is a regulatory enzyme. Pyruvate can be converted to oxaloacetate (by pyruvate carboxylase) or it can be converted to acetyl-CoA (by the pyruvate dehydrogenase complex), which then enters the citric acid cycle. Acetyl-CoA is an allosteric activator of pyruvate carboxylase and a feedback inhibitor of the pyruvate dehydrogenase complex. A high concentration of acetyl-CoA signals that the oxidation of glucose for energy is not necessary, so pyruvate is converted to glucose rather than prepared to enter the citric acid cycle.

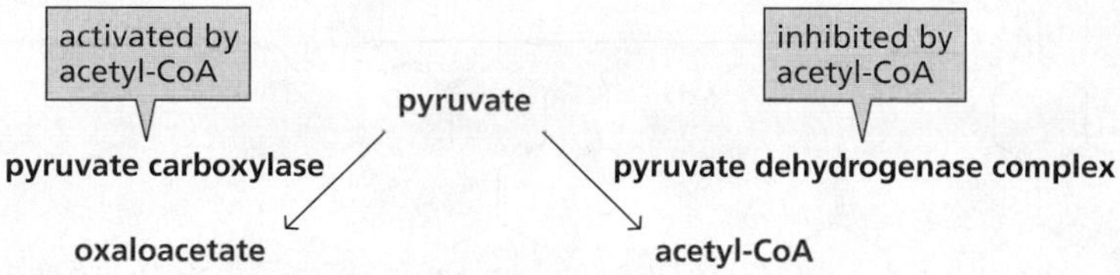

15   AMINO ACID BIOSYNTHESIS

The only amino acids synthesized by the body are the 10 nonessential amino acids—the others must be obtained from food. All the nonessential amino acids are biosynthesized from one of four metabolic intermediates: pyruvate, oxaloacetate, α-ketoglutarate, and 3-phosphoglycerate. Each amino acid has its own pathway for its biosynthesis.

For example, glutamate is biosynthesized by a transamination reaction using an amino acid as the nitrogen donor and α-ketoglutarate as the nitrogen acceptor. Alanine and aspartate also are biosynthesized by a transamination reaction using an amino acid as the nitrogen donor.

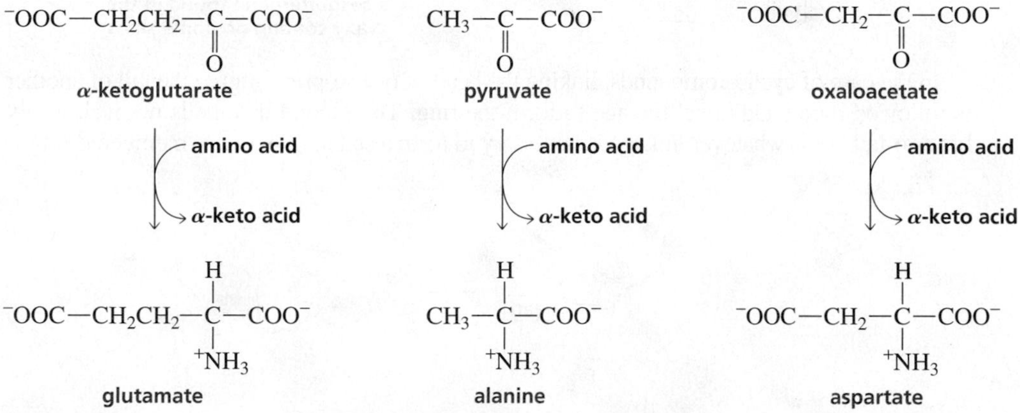

Serine is biosynthesized by oxidizing 3-phosphoglycerate (an intermediate in glycolysis), transaminating the product with glutamate and then hydrolyzing off the phosphate group.

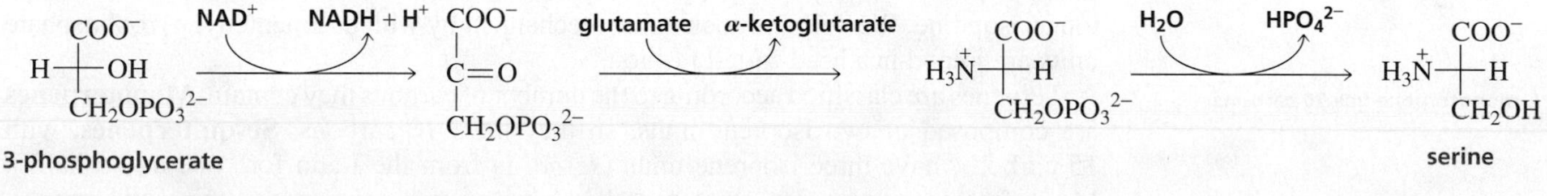

PROBLEM 25

Glutamine is biosynthesized from glutamate in two steps using ATP and ammonia. Propose a mechanism for this biosynthesis.

---

16   TERPENES CONTAIN CARBON ATOMS IN MULTIPLES OF FIVE

Terpenes are a diverse class of compounds that contain 10, 15, 20, 25, 30, or 40 carbons. More than 20,000 terpenes are known. Many are found in oils extracted from fragrant plants. Terpenes can be hydrocarbons, or they can contain oxygen and be alcohols, ketones, or aldehydes. Oxygen-containing terpenes are sometimes called **terpenoids.** Terpenes and terpenoids have been used as spices, perfumes, and medicines for thousands of years.

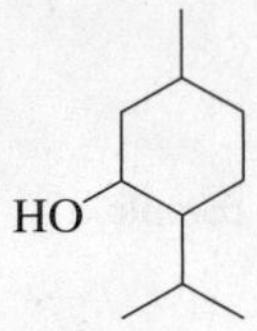

menthol	geraniol	zingiberene	β-selinene
peppermint oil	geranium oil	oil of ginger	oil of celery

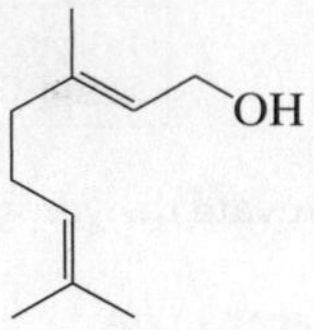

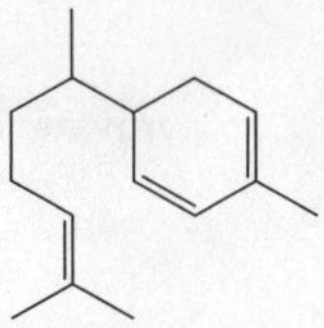

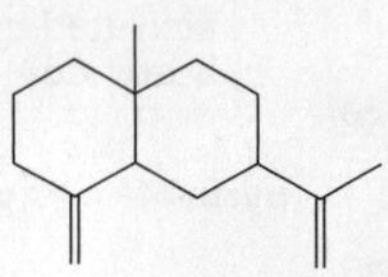

The structures of terpenes are consistent with how they are made: by joining together five-carbon isoprene units, usually in a head-to-tail fashion. (The branched end of isoprene is called the head, and the unbranched end is called the tail.) Isoprene is the common name for 2-methyl-1,3-butadiene, a compound with five carbons.

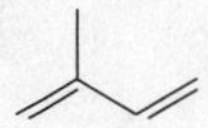

**2-methyl-1,3-butadiene
isoprene**

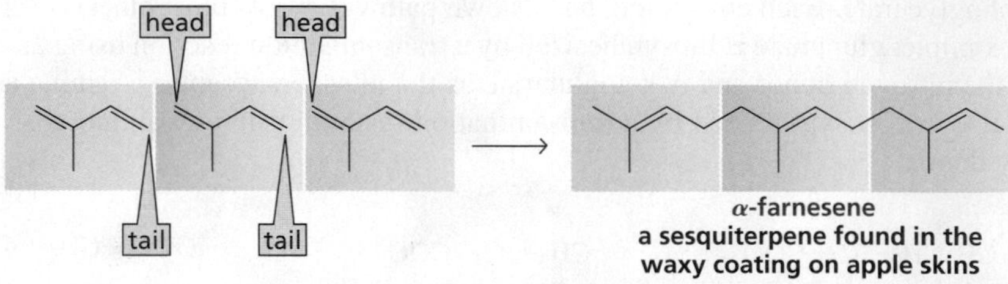

α-farnesene
a sesquiterpene found in the
waxy coating on apple skins

In the case of cyclic compounds, linking the head of one isoprene unit to the tail of another is followed by an additional linkage to form the ring. The second linkage is not necessarily head-to-tail but is whatever linkage is necessary to form a stable five- or six-membered ring.

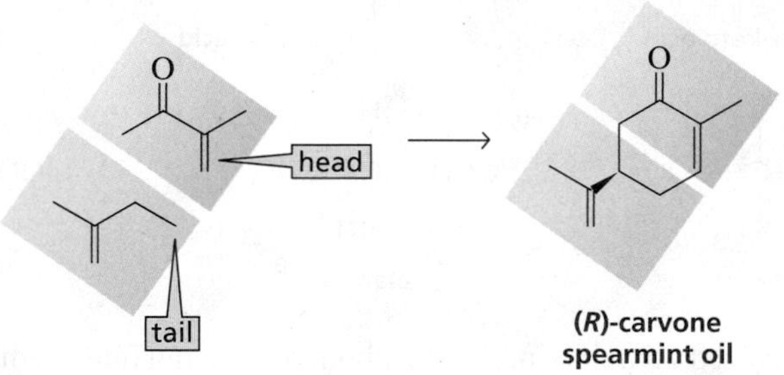

(R)-carvone
spearmint oil
a monoterpene

In Section 17, we will see that the compound actually used in the biosynthesis of terpenes is not isoprene, but isopentenyl pyrophosphate, a compound with the same carbon skeleton as isoprene. We will also look at the mechanism by which isopentenyl pyrophosphate units are joined in a head-to-tail fashion.

A monoterpene has 10 carbons.

Terpenes are classified according to the number of carbons they contain. **Monoterpenes** are composed of two isoprene units, so they have 10 carbons. **Sesquiterpenes,** with 15 carbons, have three isoprene units (*sesqui* is from the Latin for "one and a half"). Many fragrances and flavorings found in plants are monoterpenes or sesquiterpenes. These compounds are known as *essential oils.*

Triterpenes (30 carbons) and **tetraterpenes** (40 carbons) have important biological roles. For example, **squalene,** a triterpene, is the precursor of cholesterol, which is the precursor of all the other steroid hormones. Lycopene and carotene, compounds responsible for the red and orange colors of many fruits and vegetables, are tetraterpenes.

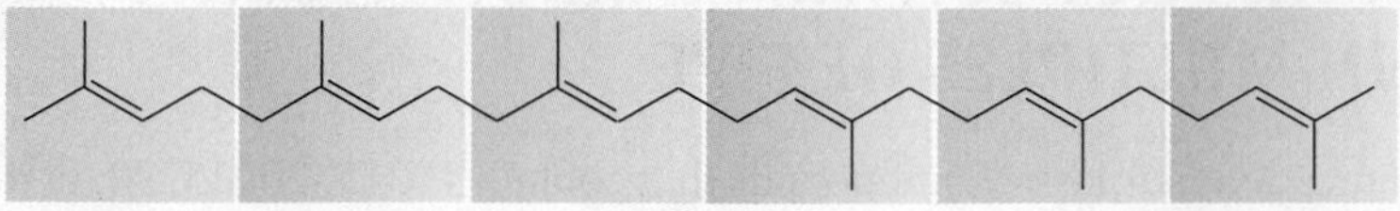

squalene

PROBLEM 26◆

One of the linkages in squalene is tail-to-tail, not head-to-tail. What does this suggest about how squalene is synthesized in nature? (*Hint:* Locate the position of the tail-to-tail linkage.)

PROBLEM 27

Mark off the isoprene units in lycopene and β-carotene. Can you detect a similarity in the way in which squalene, lycopene, and β-carotene are biosynthesized?

17   HOW TERPENES ARE BIOSYNTHESIZED

The five-carbon compound used for the biosynthesis of terpenes is 3-methyl-3-butenyl pyrophosphate, more commonly known as isopentenyl pyrophosphate. Each step in its biosynthesis is catalyzed by a different enzyme.

STEPS IN THE BIOSYNTHESIS OF ISOPENTENYL PYROPHOSPHATE

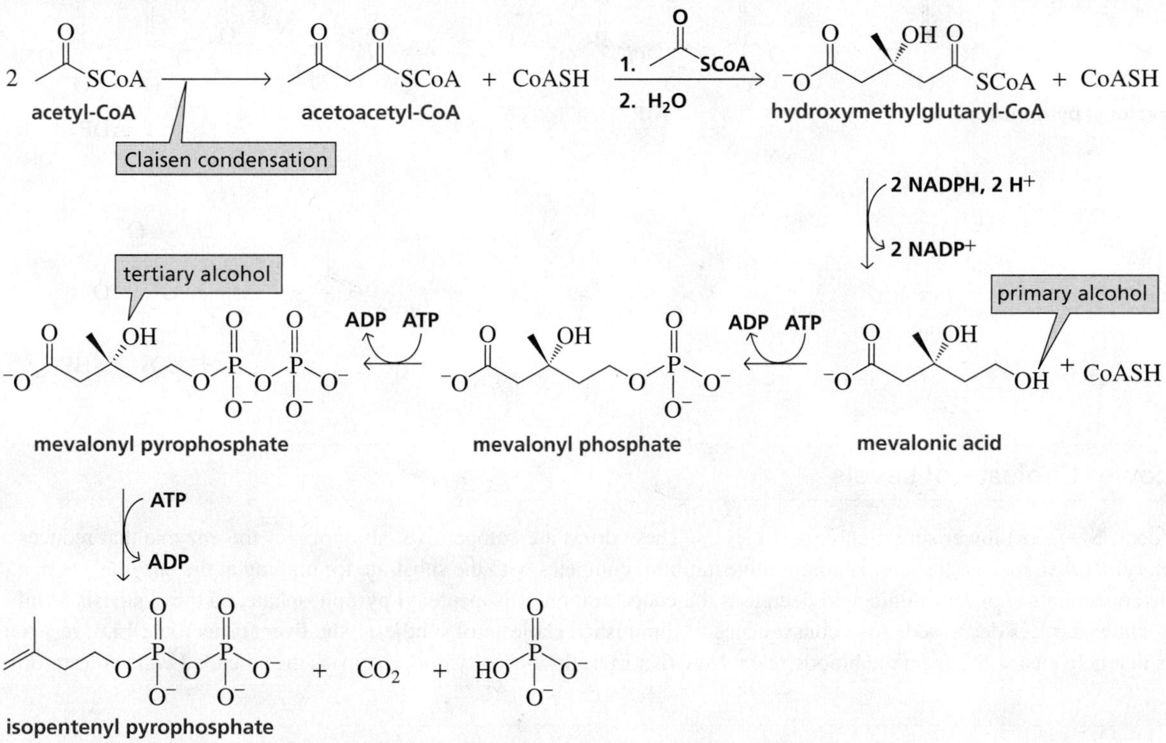

- The first step a Claisen condensation.
- The second step is an aldol addition with a third molecule of acetyl-CoA, which is followed by hydrolysis of one of the thioester groups (see Problem 28).
- The thioester is reduced by NADPH to form mevalonic acid (see Problem 29).
- A pyrophosphate group is added by means of two successive phosphorylations of the primary alcohol with ATP (see Problem 30).
- The tertiary alcohol is phosphorylated with ATP. Subsequent decarboxylation and loss of the phosphate group forms isopentenyl pyrophosphate (see Problem 31).

PROBLEM 28

Propose mechanisms for the Claisen condensation and aldol addition that comprise the first two steps of the biosynthesis of isopentenyl pyrophosphate.

PROBLEM 29♦

Why are two equivalents of NADPH required to oxidize hydroxymethylglutaryl-CoA to mevalonic acid?

PROBLEM 30

Propose a mechanism for the conversion of mevalonic acid to mevalonyl pyrophosphate.

PROBLEM 31 Solved

Propose a mechanism for the last step in the biosynthesis of isopentenyl pyrophosphate, showing why ATP is required.

Solution In order to form the double bond in isopentenyl pyrophosphate, elimination of CO_2 needs to be accompanied by elimination of hydroxide ion. But hydroxide ion, a strong base, is a poor leaving group. Therefore, ATP is used to convert the OH group into a phosphate group, which is easily eliminated because it is a weak base.

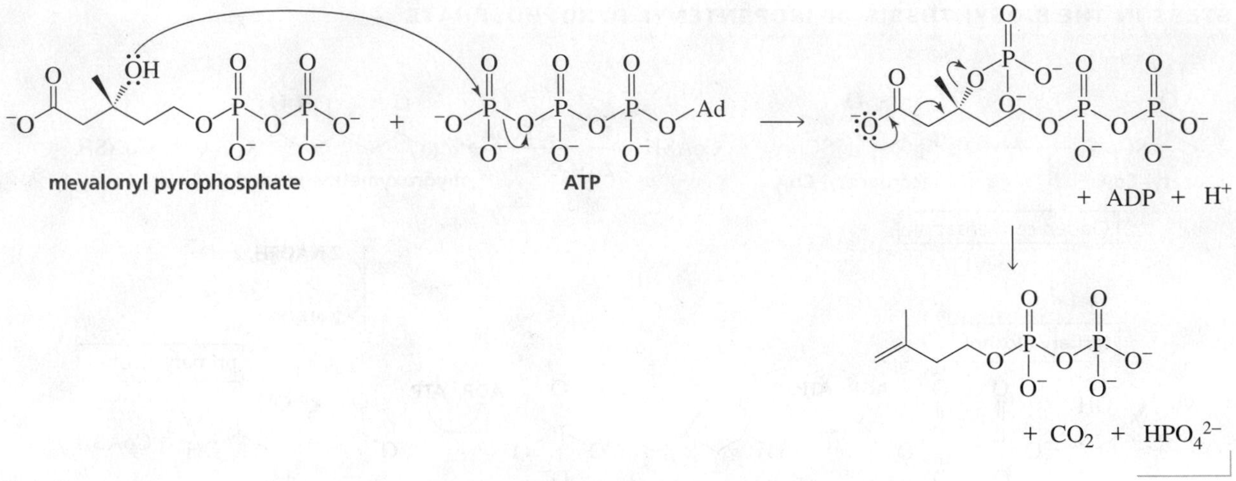

mevalonyl pyrophosphate   ATP

$+ \ ADP \ + \ H^+$

$+ \ CO_2 \ + \ HPO_4^{2-}$

How Statins Lower Cholesterol Levels

Statins (Lipitor, Zocor, Mevacor) lower serum cholesterol levels. These drugs are competitive inhibitors for the enzyme that reduces hydroxymethylglutaryl-CoA to mevalonic acid. A competitive inhibitor competes with the substrate for binding at the enzyme's active site. Decreasing the concentration of mevalonic acid decreases the concentration of isopentenyl pyrophosphate, so the synthesis of all terpenes, including cholesterol, is decreased. As a consequence of diminished cholesterol synthesis, the liver forms more LDL receptors—the receptors that help clear LDL from the bloodstream. Note that LDL (low-density lipoprotein) is the so-called *bad* cholesterol.

Both **isopentenyl pyrophosphate** and **dimethylallyl pyrophosphate** are needed for the biosynthesis of terpenes. Therefore, some isopentenyl pyrophosphate is converted to dimethylallyl pyrophosphate in a two-step enzyme-catalyzed reaction.

MECHANISM FOR THE CONVERSION OF ISOPENTYL PYROPHOSPHATE TO DIMETHYLALLYL PYROPHOSPHATE

isopentenyl pyrophosphate   dimethylallyl pyrophosphate

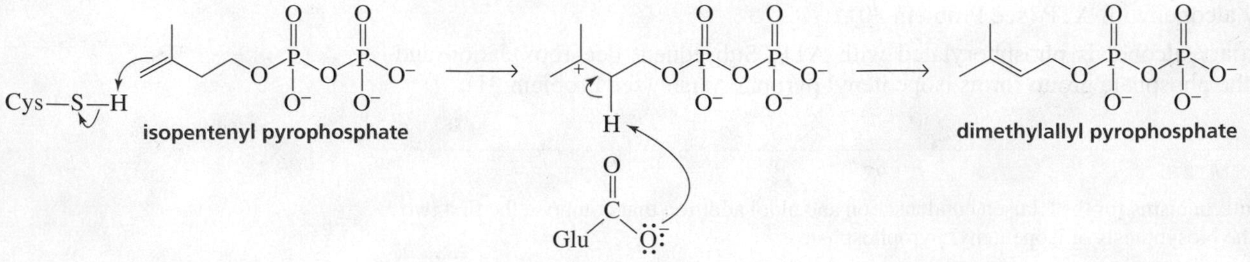

- A cysteine side chain is in the proper position at the enzyme's active site to donate a proton to the sp^2 carbon of the alkene that is bonded to the most hydrogens.

- A glutamate side chain removes a proton from the β-carbon of the carbocation intermediate that is bonded to the fewest hydrogens (in accordance with Zaitsev's rule).

> **Adding a proton and then removing a proton converts isopentenyl pyrophosphate to dimethylallyl pyrophosphate.**

The enzyme-catalyzed reaction of dimethylallyl pyrophosphate with isopentenyl pyrophosphate forms geranyl pyrophosphate, a 10-carbon compound.

MECHANISM FOR THE BIOSYNTHESIS OF TERPENES

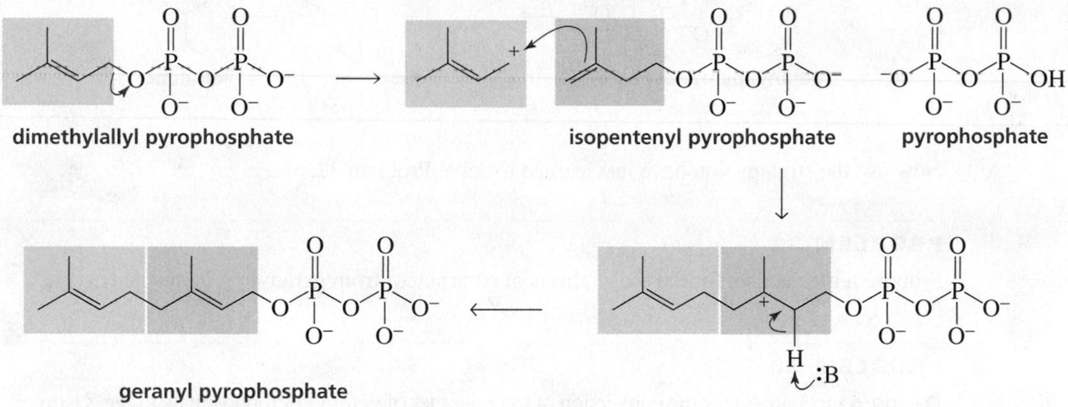

dimethylallyl pyrophosphate isopentenyl pyrophosphate pyrophosphate

geranyl pyrophosphate

- Experimental evidence suggests that this is an S_N1 reaction (see Problem 34). Thus, the leaving group departs, forming an allylic cation.

- Isopentenyl pyrophosphate is the nucleophile that adds to the allylic cation.

- A base removes a proton, forming geranyl pyrophosphate.

The scheme shown here shows how some of the many monoterpenes can be synthesized from geranyl pyrophosphate:

geranyl pyrophosphate $\xrightarrow{\text{H}_2\text{O}}$ geraniol
in rose and geranium oils $\xrightarrow{\text{reduction}}$ citronellol
in rose and geranium oils $\xrightarrow{\text{oxidation}}$ citronellal
in lemon oil

$\downarrow \text{H}_3\text{O}^+$

$\xrightarrow{\text{H}_2\text{O}}$ α-terpineol
in juniper oil $\xrightarrow{\text{H}_3\text{O}^+}$ terpin hydrate
a common constituent
of cough medicine

limonene
in orange and
lemon oils $\xrightarrow{\text{oxidation}}$ $\xrightarrow{\text{reduction}}$ menthol
in peppermint oil

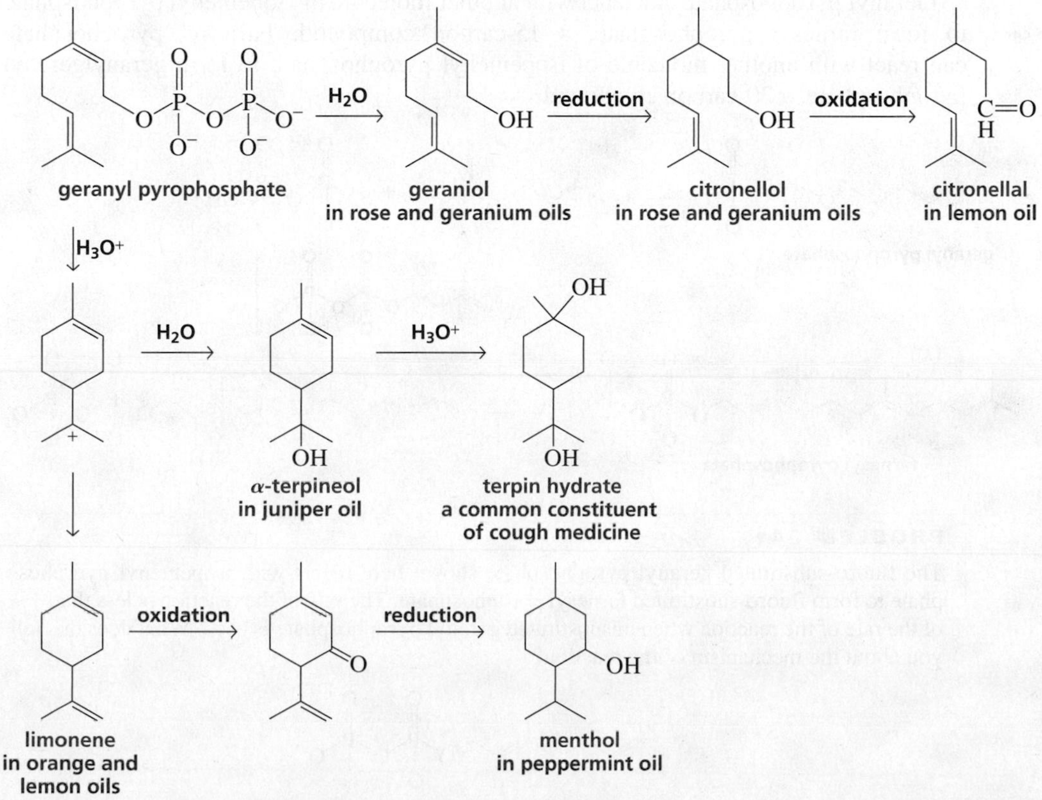

PROBLEM-SOLVING STRATEGY

Proposing a Mechanism for Biosynthesis

Propose a mechanism for the biosynthesis of limonene from geranyl pyrophosphate.

Assuming that geranyl pyrophosphate reacts like dimethylallyl pyrophosphate, the leaving group departs in an S_N1 reaction. The electrons of the π bond attack the allylic cation, forming the six-membered ring and a new carbocation. A base removes a proton to form the required double bond.

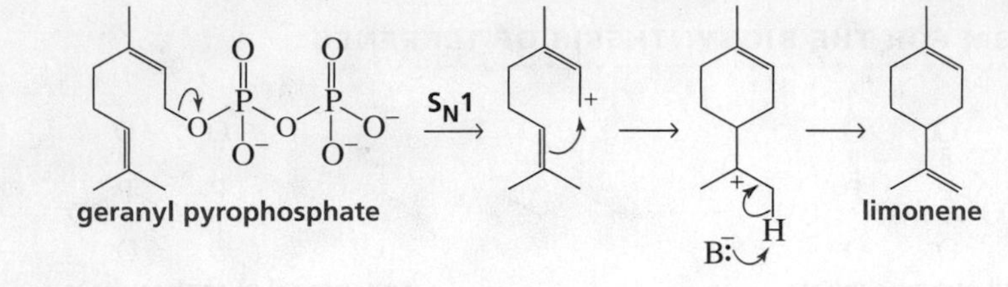

Now use the strategy you have just learned to solve Problem 32.

PROBLEM 32

Propose a mechanism for the biosynthesis of α-terpineol from geranyl pyrophosphate.

PROBLEM 33

Propose a mechanism for the conversion of the *E* isomer of geranyl pyrophosphate to the *Z* isomer.

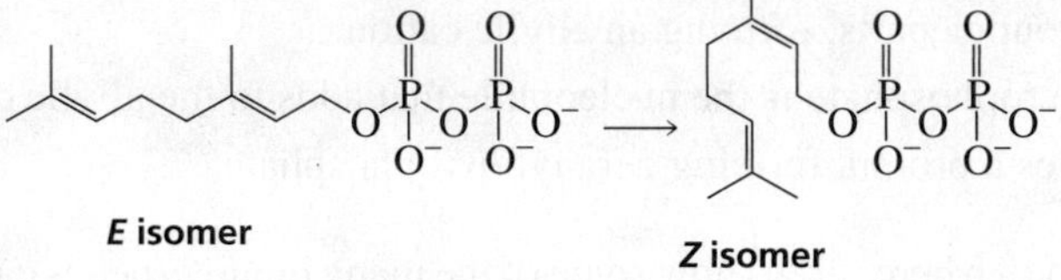

E isomer   **Z isomer**

Geranyl pyrophosphate can react with another molecule of isopentenyl pyrophosphate to form farnesyl pyrophosphate, a 15-carbon compound. Farnesyl pyrophosphate can react with another molecule of isopentenyl pyrophosphate to form geranylgeranyl pyrophosphate, a 20-carbon compound.

dimethyallyl = 5 carbons

isopentenyl = 5 carbons

geranyl = 10 carbons

farnesyl = 15 carbons

geranylgeranyl = 20 carbons

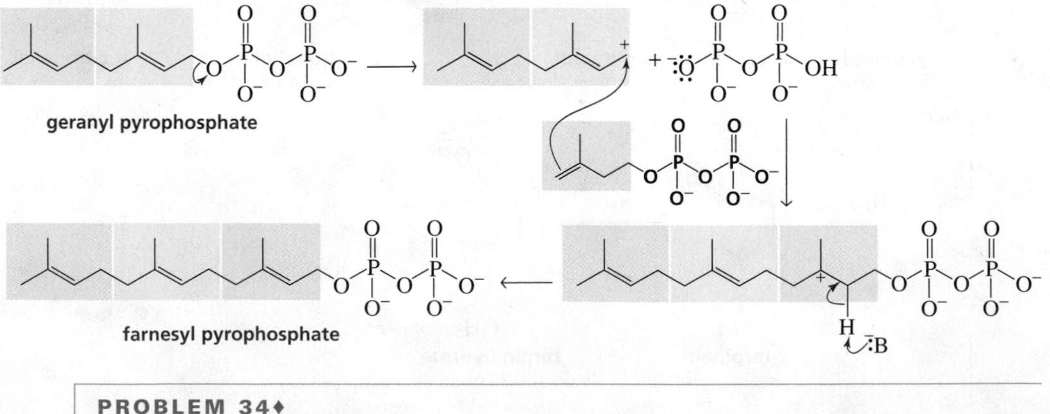

geranyl pyrophosphate

farnesyl pyrophosphate

PROBLEM 34◆

The fluoro-substituted geranyl pyrophosphate shown here reacts with isopentenyl pyrophosphate to form fluoro-substituted farnesyl pyrophosphate. The rate of the reaction is less than 1% of the rate of the reaction when unsubstituted geranyl pyrophosphate is used. What does this tell you about the mechanism of the reaction?

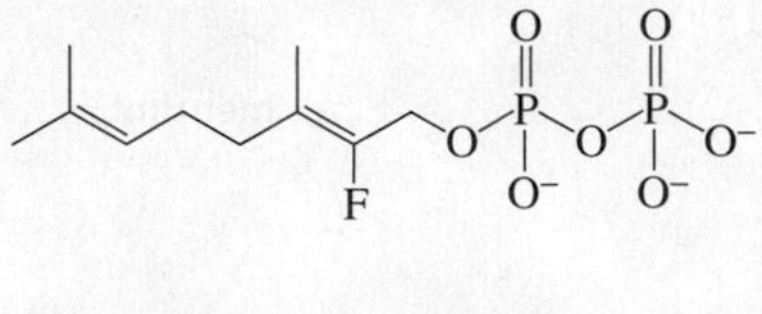

Two molecules of farnesyl pyrophosphate form squalene, a 30-carbon compound. The reaction is catalyzed by the enzyme squalene synthase, which joins the two molecules in a tail-to-tail linkage. As we noted earlier, squalene is the precursor of cholesterol, and cholesterol is the precursor of all the steroid hormones.

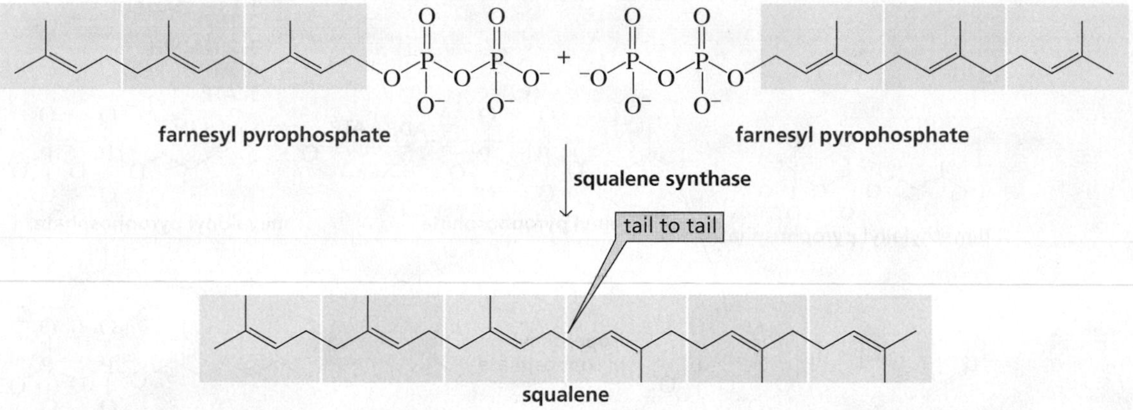

farnesyl pyrophosphate + farnesyl pyrophosphate

squalene synthase

tail to tail

squalene

Protein Prenylation

Farnesyl and geranylgeranyl groups are put on proteins to allow them to become anchored to membranes. Because "isopentenyl" is also called "prenyl," attaching these polymers of isopentenyl units to proteins is known as **protein prenylation.** The most common prenylation site on a protein is the tetramer CaaX, where "C" is a cysteine, "a" is an aliphatic amino acid, and "X" can be one of several amino acids. A farnesyl group is put on the protein when X is Gln, Met, or Ser, whereas a geranylgeranyl group is put on when X is Leu.

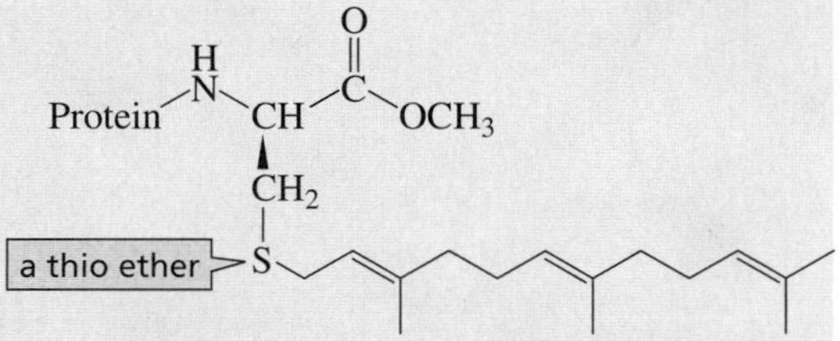

a thio ether

S-farnesyl cysteine methyl ester

The cysteine side chain of CaaX is the nucleophile that reacts with farnesyl pyrophosphate or geranylgeranyl pyrophosphate, forming a thioether and eliminating pyrophosphate. Once the protein has been prenylated, the C—aaX amide bond is hydrolyzed, causing cysteine to become the C-terminal amino acid. Cysteine's carboxyl group is then esterified with a methyl group.

PROBLEM 35

Farnesyl pyrophosphate forms the sesquiterpene shown here. Propose a mechanism for this reaction.

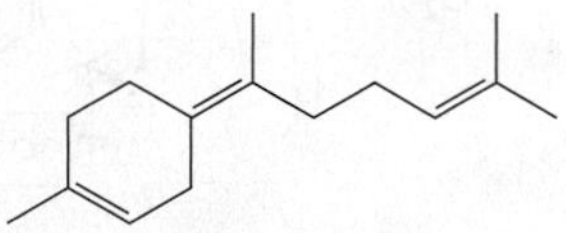

PROBLEM 36 Solved

If squalene were synthesized in a medium containing acetate with a ^{14}C-labeled carbonyl carbon, which carbons in squalene would be labeled?

Solution Acetate reacts with ATP to form acetyl adenylate, which then reacts with CoASH to form acetyl-CoA. Two equivalents of acetyl-CoA form acetoacetyl-CoA as a result of a Claisen condensation. Examining each step of the mechanism for the biosynthesis of isopentenyl pyrophosphate from acetyl-CoA allows you to determine the locations of the radioactively labeled carbons in isopentenyl pyrophosphate. Similarly, the locations of the radioactively labeled carbons in geranyl pyrophosphate can be determined from the mechanism for its biosynthesis from isopentenyl pyrophosphate, and the locations of the radioactively labeled carbons in farnesyl pyrophosphate can be determined from the mechanism for its biosynthesis from geranyl pyrophosphate. Knowing that squalene is obtained from a tail-to-tail linkage of two farnesyl pyrophosphates tells you which carbons in squalene will be labeled.

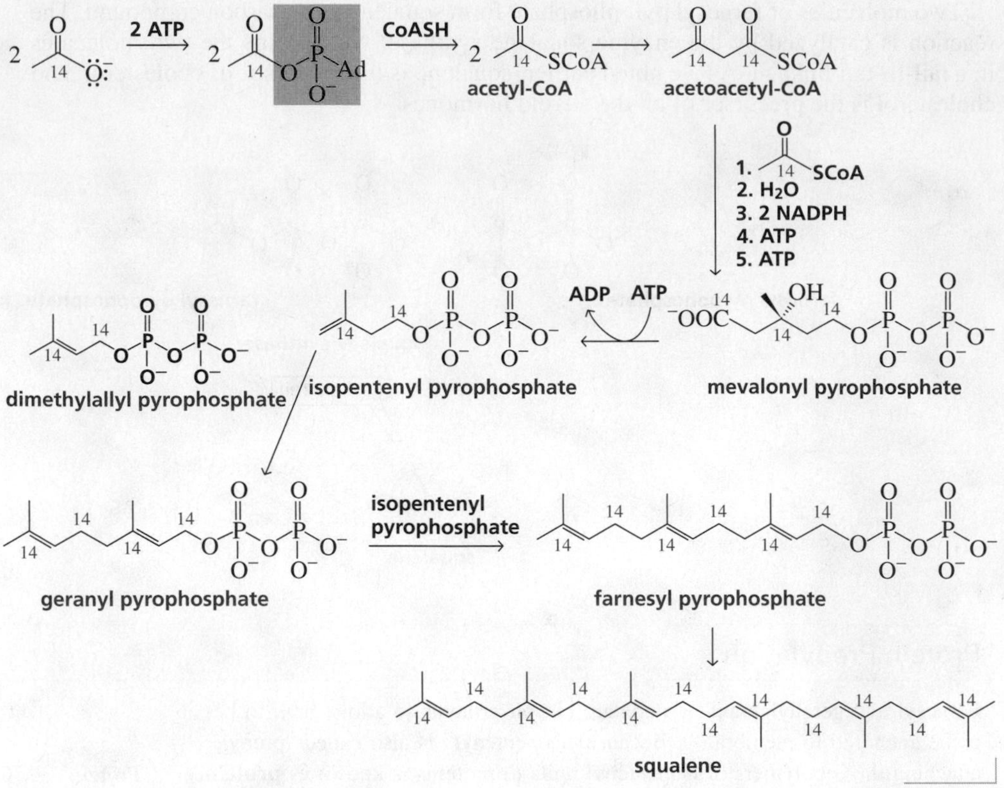

18   HOW NATURE SYNTHESIZES CHOLESTEROL

How is cholesterol, the precursor of all the steroid hormones, biosynthesized? The starting material for the biosynthesis is the triterpene squalene, which must first be converted to lanosterol, which is converted to cholesterol in a series of 19 steps.

STEPS IN THE BIOSYNTHESIS OF LANOSTEROL AND CHOLESTEROL

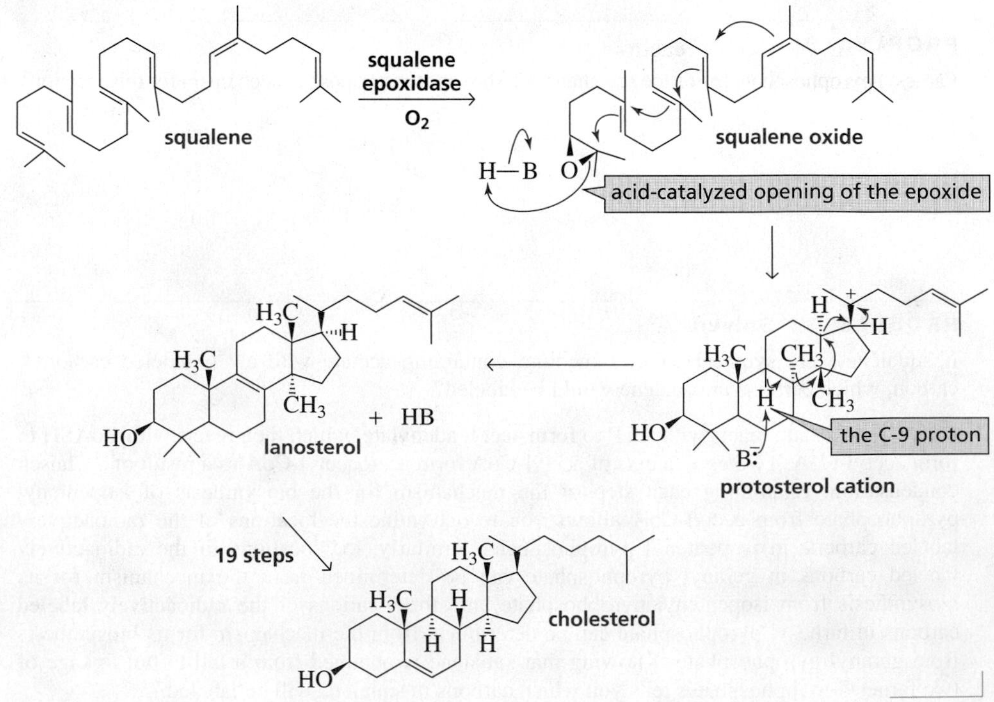

- The first step is epoxidation of the 2,3-double bond of squalene.

- Acid-catalyzed opening of the epoxide initiates a series of cyclizations resulting in the protosterol cation.

- Elimination of a C-9 proton from the cation initiates a series of 1,2-hydride and 1,2-methyl shifts, resulting in lanosterol.

Converting lanosterol to cholesterol requires removing three methyl groups from lanosterol, reducing two double bonds, and creating a new double bond. Removing methyl groups from carbon atoms is not easy, and many different enzymes are required to carry out the 19 steps. So why does nature bother? Why not just use lanosterol instead of cholesterol? Konrad Bloch answered that question when he found that membranes containing lanosterol instead of cholesterol are much more permeable. Small molecules are able to pass easily through lanosterol-containing membranes. As each methyl group is removed from lanosterol, the membrane becomes less and less permeable.

PROBLEM 37♦

Draw the individual 1,2-hydride and 1,2-methyl shifts responsible for conversion of the protosterol cation to lanosterol. How many hydride shifts are involved? How many methyl shifts?

SOME IMPORTANT THINGS TO REMEMBER

- **Metabolism** is the set of reactions living organisms carry out to obtain energy and to synthesize the compounds they need. Metabolism can be divided into catabolism and anabolism.

- A **catabolic pathway** is a series of reactions that breaks down a compound to provide energy and simpler compounds.

- An **anabolic pathway** is a series of reactions that leads to the synthesis of a compound from simpler compounds.

- ATP is a cell's most important source of chemical energy; ATP provides a reaction pathway involving a good leaving group for a reaction that would not otherwise occur because of a poor leaving group. This occurs by way of a phosphoryl transfer reaction.

- A **phosphoryl transfer reaction** involves the transfer of a phosphoryl group of ATP to a nucleophile as a result of breaking a **phosphoanhydride bond.**

- A phosphoryl transfer reaction forms one of two intermediates—an **acyl (or alkyl) phosphate** or an **acyl (or alkyl) adenylate.**

- The reaction of a nucleophile with a phosphoanhydride bond is highly exergonic because of electrostatic repulsion, solvation, and electron delocalization.

- **Catabolism** can be divided into four stages. In the *first stage,* fats, carbohydrates, and proteins are hydrolyzed to fatty acids, monosaccharides, and amino acids.

- In the *second stage,* the products obtained from the first stage are converted to compounds that can enter the citric acid cycle. In order to enter the citric acid cycle, a compound must be either a citric acid cycle

intermediate, acetyl-CoA, or pyruvate (because it can be converted to acetyl-CoA).

- In the *second stage,* a fatty acyl-CoA is converted to acetyl-CoA in a pathway called **β-oxidation.** The series of four reactions is repeated until the entire fatty acid has been converted to acetyl-CoA molecules.

- In the *second stage,* glucose is converted to two molecules of pyruvate in a series of 10 reactions known as **glycolysis.**

- Under aerobic conditions, pyruvate is converted to acetyl-CoA, which then enters the citric acid cycle.

- In the second stage, amino acids are metabolized to pyruvate, acetyl-CoA, and/or citric acid cycle intermediates, depending on the amino acid.

- The **citric acid cycle** is *the third stage* of catabolism. It is a series of eight reactions that converts the acetyl group of each molecule of acetyl-CoA that enters the cycle to two molecules of CO_2.

- In *the fourth stage* of catabolism, called **oxidative phosphorylation,** each molecule of NADH and $FADH_2$ formed in oxidation reactions in the second and third stages of catabolism is converted into 2.5 molecules of ATP and 1.5 molecules of ATP, respectively.

- **Anabolism** is the reverse of catabolism. In anabolism, acetyl-CoA, pyruvate, glycolytic intermediates, and citric acid cycle intermediates are the starting materials for the synthesis of fatty acids, carbohydrates, and proteins.

- A **kinase** is an enzyme that puts a phosphate group on its substrate.

- A **phosphatase** is an enzyme that takes a phosphate group off its substrate

- Many of the reactions involved in the synthesis of glucose from pyruvate—**gluconeogenesis**—are carried out by the same enzymes that catalyze the reactions in glycolysis—they are just operating in reverse.

- Some of the enzymes near the beginning of each pathway catalyze essentially irreversible reactions, and those reactions have to be detoured around when going in the other direction.

- The enzyme that catalyzes an irreversible reaction near the beginning of the pathway is a **regulatory enzyme**—it can be activated and inhibited.

- A feedback inhibitor inhibits a step at the beginning of the pathway for its biosynthesis.

- An **allosteric inhibitor** or **activator** inhibits or activates an enzyme by binding to a site on the enzyme other than the active site, which affects the function of the active site.

- All the nonessential amino acids are biosynthesized from one of four metabolic intermediates: pyruvate, oxaloacetate, α-ketoglutarate, or 3-phosphoglycerate.

- **Terpenes** are made by joining five-carbon units, usually in a head-to-tail fashion.

- **Monoterpenes**—terpenes with two isoprene units—have 10 carbons, **sesquiterpenes** have 15, **triterpenes** have 30, and **tetraterpenes** have 40.

- Isopentenyl pyrophosphate is the five-carbon compound used for the biosynthesis of terpenes.

- The reaction of **dimethylallyl pyrophosphate** (formed from isopentenyl pyrophosphate) with **isopentenyl pyrophosphate** forms geranyl pyrophosphate, a 10-carbon compound.

- Geranyl pyrophosphate can react with another molecule of isopentenyl pyrophosphate to form farnesyl pyrophosphate, a 15-carbon compound.

- Farnesyl pyrophosphate can react with another molecule of isopentenyl pyrophosphate to form geranylgeranyl pyrophosphate, a 20-carbon compound.

- Two molecules of farnesyl pyrophosphate form **squalene,** a 30-carbon compound.

- Squalene is the precursor of **lanosterol,** which is the precursor of cholesterol.

- **Cholesterol** is the precursor of all the steroid hormones.

PROBLEMS

38. Indicate whether an anabolic pathway or a catabolic pathway does the following:

    a. produces energy in the form of ATP   b. involves primarily oxidation reactions

39. Galactose can enter the glycolytic cycle but it must first react with ATP to form galactose-1-phosphate. Propose a mechanism for this reaction.

galactose   ATP   galactose-1-phosphate   ADP

40. When pyruvate is reduced by NADH to lactate, which hydrogen in lactate comes from NADH?

41. Which of the ten reactions in glycolysis are
    a. phsophorylations?   b. isomerizations?   c. reductions?   d. dehydrations?

42. What reactions in the citric acid cycle form a product with a new asymmetric center?

43. Acyl-CoA synthase is the enzyme that activates a fatty acid by converting it to a fatty acyl-CoA (Section 6) in a series of two reactions. In the first reaction, the fatty acid reacts with ATP, and one of the products formed is ADP. The other product reacts in a second reaction with CoASH to form the fatty acyl-CoA. Propose a mechanism for each of the reactions.

44. In some brain cancers, isocitrate dehydrogenase, instead of catalyzing the oxidation of the secondary alcohol of isocitrate, catalyzes the reduction of α-ketoglutarate. Draw the product of the reaction.

45. If the phosphorus atom in 3-phosphoglycerate is radioactively labeled, where will the label be when the reaction that forms 2-phosphoglycerate is over?

46. What carbon atoms of glucose end up as a carboxyl group in pyruvate?

D-glucose

47. What carbon atoms of glucose end up in ethanol under anaerobic conditions in yeast?

48. How would blood glucose levels be affected before and after a 24-hour fast if there is a deficiency of fructose-1,6-bisphosphatase?

49. Explain why the conversion of pyruvate to lactate is a reversible reaction but the conversion of pyruvate to acetaldehyde is not reversible.

50. How many molecules of acetyl-CoA are obtained from the β-oxidation of one molecule of a 16-carbon saturated fatty acyl-CoA?

51. How many molecules of CO_2 are obtained from the complete metabolism of one molecule of a 16-carbon saturated fatty acyl-CoA?

52. How many molecules of ATP are obtained from the β-oxidation of one molecule of a 16-carbon saturated fatty acyl-CoA?

53. How many molecules of NADH and $FADH_2$ are obtained from the β-oxidation of one molecule of a 16-carbon saturated fatty acyl-CoA?

54. How many molecules of ATP are obtained from the NADH and $FADH_2$ formed in the β-oxidation of one molecule of a 16-carbon saturated fatty acyl-CoA?

55. How many molecules of ATP are obtained from the complete (including the fourth stage of catabolism) metabolism of one molecule of a 16-carbon saturated fatty acyl-CoA?

56. How many molecules of ATP are obtained from complete (including the fourth stage of catabolism) metabolism of one molecule of glucose?

57. What are four possible fates of pyruvate in a mammalian cell?

58. Most fatty acids have an even number of carbons and therefore are completely metabolized to acetyl-CoA. A fatty acid with an odd number of carbons is metabolized to acetyl-CoA and one equivalent of propionyl-CoA. The following two reactions convert propionyl-CoA into succinyl-CoA, a citric acid cycle intermediate, so it can be further metabolized. Each of the reactions requires a coenzyme. Identify the coenzyme for each step. From what vitamins are the coenzymes derived?

propionyl-CoA   methylmalonyl-CoA   succinyl-CoA

59. If glucose is labeled with ^{14}C in the indicated position, where will the label be in pyruvate?

  a. glucose-1-^{14}C **c.** glucose-3-^{14}C **e.** glucose-5-^{14}C
  b. glucose-2-^{14}C **d.** glucose-4-^{14}C **f.** glucose-6-^{14}C

60. Write the reactions for the synthesis of citrate from two equivalents of pyruvate. What enzymes are required for the reactions?

61. Under conditions of starvation, acetyl-CoA, instead of being degraded in the citric acid cycle, is converted to acetone and 3-hydroxybutyrate, which are compounds called ketone bodies that the brain can use as a temporary fuel. Propose a mechanism for their formation.

acetone   3-hydroxybutyrate

62. Shortly after adding ^{14}C-labeled glyceraldehyde-3-phosphate to a yeast extract, fructose-1,6-bisphosphate labeled at C-3 and C-4 can be isolated. Where was the ^{14}C-label in glyceraldehyde-3-phosphate? How did fructose-1,6-bisphosphate get the second label?

63. UDP-galactose-4-epimerase converts UDP-galactose to UDP-glucose. The reaction requires NAD^+ as a coenzyme.

  a. Propose a mechanism for the reaction.   **b.** Why is the enzyme called an epimerase?

UDP-galactose   UDP-glucose

64. A student is trying to determine the mechanism for a reaction that uses ATP to activate a carboxylate ion, which then reacts with a thiol. If the carboxylate ion attacks the γ-phosphorus of ATP, the reaction products are the thioester, ADP, and phosphate. However, whether it attacks the α-phosphorus or the β-phosphorus of ATP cannot be determined from the reaction products because the thioester, AMP, and pyrophosphate would be the products in both reactions. The mechanisms can be distinguished by a labeling experiment in which the enzyme, the carboxylate ion, ATP, and radioactively labeled pyrophosphate are incubated, and ATP is isolated. If the isolated ATP is radioactive, attack occurred on the α-phosphorus. If it is not radioactive, then attack occurred on the β-phosphorus. Explain these conclusions.

65. What would be the results of the experiment in Problem 64 if radioactive AMP were added to the incubation mixture instead of radioactive pyrophosphate?

66. If the following sesquiterpene were synthesized in a medium containing acetate with a ^{14}C-labeled carbonyl carbon, which carbons would be labeled?

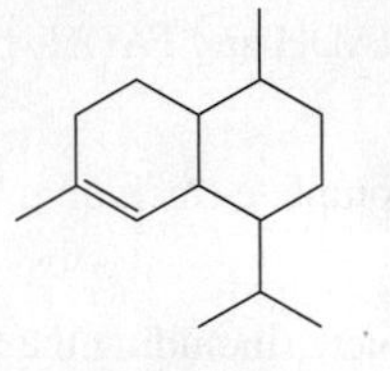

67. Propose a mechanism for the biosynthesis of α-pinene from geranyl pyrophosphate.

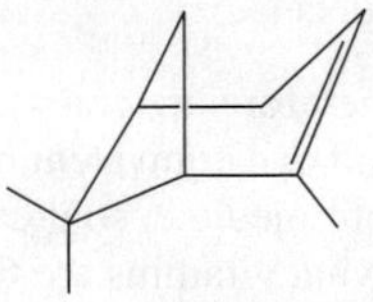

α-pinene

68. Eudesmol is a sesquiterpene found in eucalyptus. Propose a mechanism for its biosynthesis.

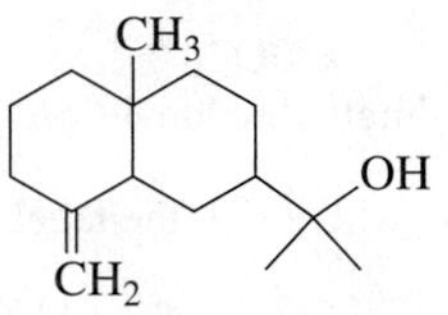

eudesmol

GLOSSARY

allosteric activator a compound that activates an enzyme when it binds to a site on the enzyme (other than the active site).

citric acid cycle (Krebs cycle) a series of reactions that converts the acetyl group of acetyl-CoA into two molecules of CO_2.

feedback inhibitor a compound that inhibits a step at the beginning of the pathway for its biosynthesis.

high-energy bond a bond that releases a great deal of energy when it is broken.

oxidative phosphorylation a series of reactions that converts a molecule of NADH and a molecule of $FADH_2$ into 2.5 and 1.5 molecules of ATP, respectively.

phosphatase an enzyme that removes a phosphate group from its substrate.

phosphoryl transfer reaction the transfer of a phosphate group from one compound to another.

protein prenylation attaching isopentenyl units to proteins.

regulatory enzyme an enzyme that can be turned on and off.

sesquiterpene a terpene that contains 15 carbons.

squalene a triterpene that is a precursor of steroid molecules.

terpene a lipid, isolated from a plant, that contains carbon atoms in multiples of five.

terpenoid a terpene that contains oxygen.

tetraterpene a terpene that contains 40 carbons.

triterpene a terpene that contains 30 carbons.

PHOTO CREDITS

Credits are listed in order of appearance.

Erich Lessing/Art Resource; Anthony Stanley/Actionplus/Newscom; Tsurukame Design/Shutterstock

The Chemistry of the Nucleic Acids

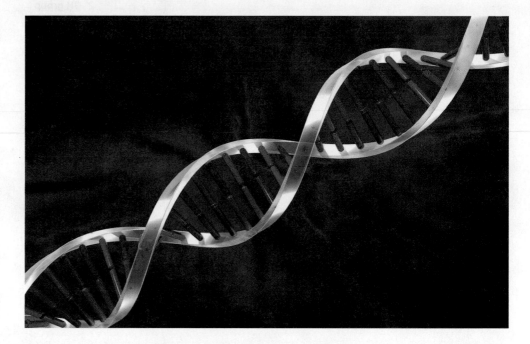

a double helix

There are three major kinds of biopolymers: polysaccharides, proteins, and nucleic acids. In this chapter we will look at nucleic acids. There are two types of nucleic acids: deoxyribonucleic acid (DNA) and ribonucleic acid (RNA). DNA encodes an organism's entire hereditary information and controls the growth and division of cells. In all organisms (except certain viruses), the genetic information stored in DNA is transcribed into RNA. This information can then be translated for the synthesis of all the proteins needed for cellular structure and function.

DNA was first isolated in 1869 from the nuclei of white blood cells. Because it was found in the nucleus and was acidic, it was called *nucleic acid*. Eventually, scientists found that the nuclei of all cells contain DNA, but not until it was shown in 1944 that DNA could be transferred from one species to another, along with inheritable traits, did they realize that DNA is the carrier of genetic information. In 1953, James Watson and Francis Crick described the three-dimensional structure of DNA—the famed double helix.

1 NUCLEOSIDES AND NUCLEOTIDES

Nucleic acids are chains of five-membered-ring sugars linked by phosphate groups. Notice that the linkages are **phosphodiesters** (Figure 1). In RNA, the five-membered-ring sugar is D-ribose. In DNA it is 2′-deoxy-D-ribose (D-ribose without an OH group in the 2′-position).

The anomeric carbon of each sugar is bonded to a nitrogen of a heterocyclic compound in a β-glycosidic linkage. (A β-linkage is one in which the substituents at C-1 and C-4 are on the same side of the furanose ring.) Because the heterocyclic compounds are amines, they are commonly referred to as **bases.**

From Chapter 26 of *Organic Chemistry*, Seventh Edition. Paula Yurkanis Bruice. Copyright © 2014 by Pearson Education, Inc. All rights reserved.

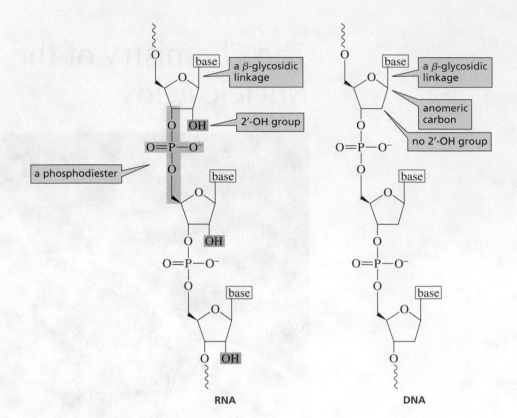

▶ Figure 1

Nucleic acids consist of a chain of five-membered-ring sugars linked by phosphate groups. Each sugar (D-ribose in RNA, 2′-deoxy-D-ribose in DNA) is bonded to a heterocyclic amine (a base) in a β-glycosidic linkage.

The vast differences in heredity between different species and between different members of the same species are determined by the sequence of the bases in DNA. Surprisingly, there are only four bases in DNA: two are substituted purines (adenine and guanine), and two are substituted pyrimidines (cytosine and thymine).

purine   pyrimidine

adenine   guanine   cytosine   uracil   thymine

RNA also contains only four bases. Three (adenine, guanine, and cytosine) are the same as those in DNA, but the fourth base in RNA is uracil instead of thymine. Notice that thymine and uracil differ only by a methyl group. (Thymine is 5-methyluracil.) The reason DNA contains thymine instead of uracil is explained in Section 10.

The anomeric carbon of the furanose ring is bonded to purines at N-9 and to pyrimidines at N-1. A compound containing a base bonded to D-ribose or to 2′-deoxy-D-ribose is called a **nucleoside.** The ring positions of the sugar component of a nucleoside are indicated by primed numbers to distinguish them from the ring positions of the base. This is why the sugar component of DNA is referred to as 2′-deoxy-D-ribose. The nucleosides of RNA— where the sugar is D-ribose— are more precisely called ribonucleosides, whereas the nucleosides of DNA—where the sugar is 2′-deoxy-D-ribose—are called deoxyribonucleosides.

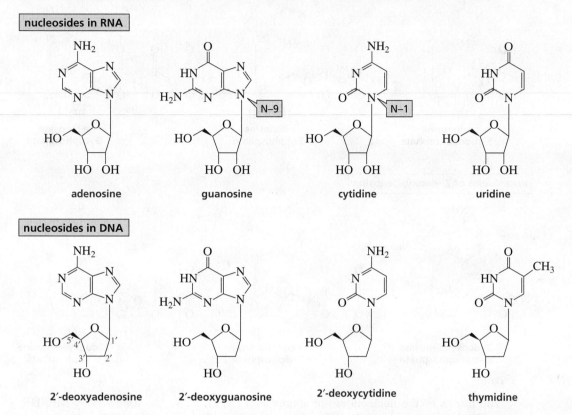

nucleosides in RNA

adenosine   guanosine   cytidine   uridine

nucleosides in DNA

2′-deoxyadenosine   2′-deoxyguanosine   2′-deoxycytidine   thymidine

Notice the difference in the base names and their corresponding nucleoside names in Table 1. For example, adenine is the base, whereas adenosine is the nucleoside; similarly, cytosine is the base, whereas cytidine is the nucleoside, and so forth. Because uracil is found only in RNA, it is shown attached to D-ribose but not to 2′-deoxy-D-ribose; because thymine is found only in DNA, it is shown attached to 2′-deoxy-D-ribose but not to D-ribose.

Table 1 The names of the Bases, the Nucleosides, and the Nucleotides

Base	Ribonucleoside	Deoxyribonucleoside	Ribonucleotide	Deoxyribonucleotide
Adenine	Adenosine	2′-Deoxyadenosine	Adenosine 5′-phosphate	2′-Deoxyadenosine 5′-phosphate
Guanine	Guanosine	2′-Deoxyguanosine	Guanosine 5′-phosphate	2′-Deoxyguanosine 5′-phosphate
Cytosine	Cytidine	2′-Deoxycytidine	Cytidine 5′-phosphate	2′-Deoxycytidine 5′-phosphate
Thymine	—	Thymidine	—	Thymidine 5′-phosphate
Uracil	Uridine	—	Uridine 5′-phosphate	

A **nucleotide** is a nucleoside with an OH group of the sugar bonded in an ester linkage to phosphoric acid. The nucleotides of RNA are more precisely called **ribonucleotides,** and those of DNA are called **deoxyribonucleotides.**

Nucleoside = base + sugar

Nucleotide = base + sugar + phosphate

Because phosphoric acid can form an anhydride, nucleotides can exist as monophosphates, diphosphates, and triphosphates. They are named by adding *monophosphate* or *diphosphate* or *triphosphate* to the name of the nucleoside.

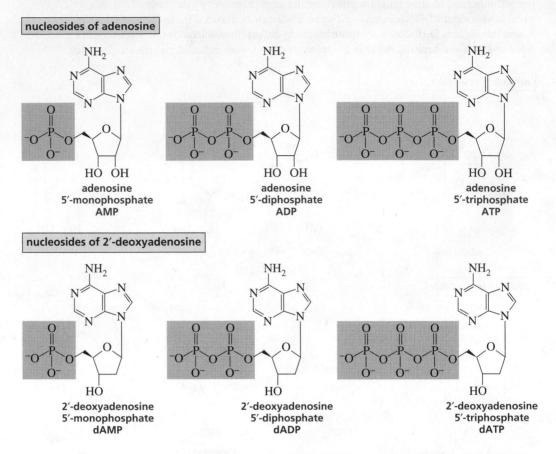

The names of the nucleotides are abbreviated (A, G, C, T, U—followed by MP, DP, or TP, depending on whether it is a monophosphate, diphosphate, or triphosphate—with a d in front if it contains 2′-deoxy-D-ribose instead of D-ribose).

The Structure of DNA: Watson, Crick, Franklin, and Wilkins

James D. Watson was born in Chicago in 1928. He graduated from the University of Chicago at the age of 19 and received a Ph.D. three years later from Indiana University. In 1951, as a postdoctoral fellow at Cambridge University, Watson worked on determining the three-dimensional structure of DNA.

Francis H. C. Crick (1916–2004) was born in Northampton, England. Originally trained as a physicist, Crick did research on radar during World War II. After the war, deciding that the most interesting problem in science was the physical basis of life, he entered Cambridge University to study the structure of biological molecules by X-ray analysis. He was a graduate student when he carried out his portion of the work that led to the proposal of the double helical structure of DNA. He received a Ph.D. in chemistry in 1953.

Rosalind Franklin (1920–1958) was born in London. She graduated from Cambridge University and studied X-ray diffraction techniques in Paris. In 1951 she returned to England and accepted a position to develop an X-ray diffraction unit in the biophysics department at King's College. Her X-ray studies showed that DNA was a helix with the sugars and phosphate groups on the outside of the molecule. Tragically, Franklin never protected herself from her X-ray source and died without knowing the role her work had played in determining the structure of DNA, and without being recognized for her contribution.

Francis Crick (*left*) and James Watson (*right*)

Rosalind Franklin

Watson and Crick shared the 1962 Nobel Prize in Medicine or Physiology with Maurice Wilkins for determining the double helical structure of DNA. Wilkins (1916–2004), who contributed X-ray studies that confirmed the double helical structure, was born in New Zealand to Irish immigrants and moved to England six years later with his parents. He received a Ph.D. from Birmingham University. During World War II he joined other British scientists who were working with American scientists on the development of the atomic bomb. He returned to England in 1945 and, having lost interest in physics, turned his attention to biology.

PROBLEM 1

In acidic solutions, nucleosides are hydrolyzed to a sugar and a heterocyclic base. Propose a mechanism for this reaction.

PROBLEM 2

Draw the structure for each of the following:

a. dCDP   c. dUMP   e. guanosine triphosphate
b. dTTP   d. UDP   f. adenosine monophosphate

---

2 OTHER IMPORTANT NUCLEOTIDES

ATP is the most important source of chemical energy. ATP, however, is not the only biologically important nucleotide. Guanosine $5'$-triphosphate (GTP) is used in place of ATP in some phosphoryl transfer reactions. Also, dinucleotides are used as oxidizing agents (NAD^+ and FAD) and reducing agents (NADPH, and $FADH_2$).

Another important nucleotide is adenosine $3',5'$-monophosphate, commonly known as cyclic AMP. Cyclic AMP is called a "second messenger" because it serves as a link between several hormones (the first messengers) and enzymes that regulate cellular function. Secretion of certain hormones, such as adrenaline, activates adenylate cyclase, the enzyme responsible for the synthesis of cyclic AMP from ATP. Cyclic AMP then activates an enzyme, generally by phosphorylating it. Cyclic nucleotides are very important in regulating cellular reactions.

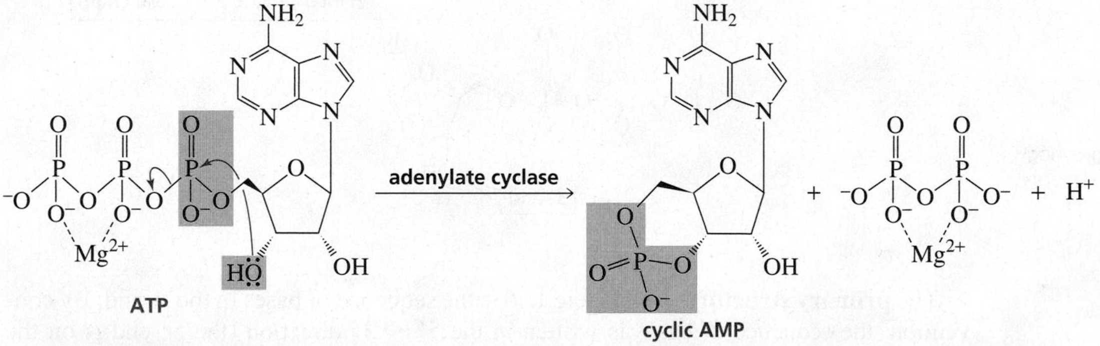

PROBLEM 3

What products would be obtained from the hydrolysis of cyclic AMP?

---

3 NUCLEIC ACIDS ARE COMPOSED OF NUCLEOTIDE SUBUNITS

Nucleic acids are composed of long strands of nucleotide subunits (Figure 1). A **dinucleotide** contains two nucleotide subunits, an **oligonucleotide** contains 3 to 10 subunits, and a **polynucleotide** contains many subunits. DNA and RNA are polynucleotides.

Nucleotide triphosphates are the starting materials for the biosynthesis of nucleic acids. DNA is synthesized by enzymes called *DNA polymerases;* RNA is synthesized by enzymes called *RNA polymerases*. The nucleotides are linked as a result of nucleophilic attack by a 3'-OH group of one nucleotide triphosphate on the α-phosphorus of another nucleotide triphosphate, breaking a phosphoanhydride bond and eliminating pyrophosphate (Figure 2). Thus, the phosphodiester joins the 3'-OH group of one nucleotide and the 5'-OH group of the next nucleotide, and the growing polymer is synthesized in the 5' → 3' direction. In other words, new nucleotides are added to the 3'-end. Pyrophosphate is subsequently hydrolyzed, which makes the reaction irreversible. Irreversibility is important if the genetic information in DNA is to be preserved. RNA strands are biosynthesized in the same way, using ribonucleotides instead of 2'-deoxyribonucleotides.

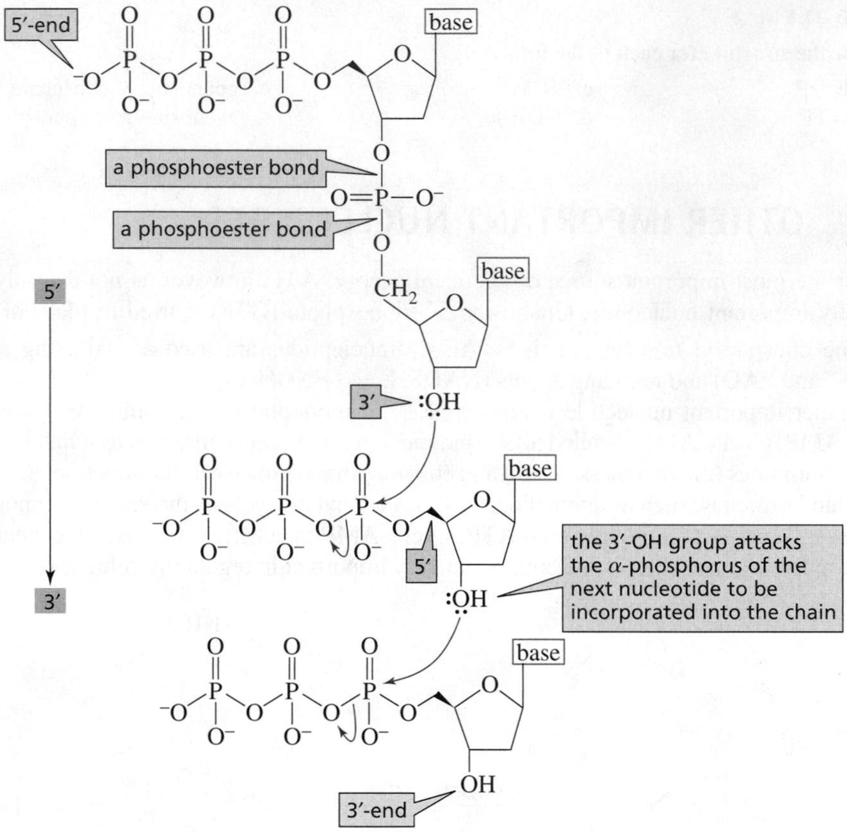

▶ **Figure 2**

Addition of nucleotides to a growing strand of DNA. Biosynthesis occurs in the 5' → 3' direction.

The **primary structure** of a nucleic acid is the sequence of bases in the strand. By convention, the sequence of bases is written in the 5' → 3' direction (the 5'-end is on the left). Note that the nucleotide at the 5'-end of the strand has an unlinked 5'-triphosphate group, and the nucleotide at the 3'-end has an unlinked 3'-hydroxyl group.

ATGAGCCATGTAGCCTAATCGGC

5'-end   3'-end

DNA is synthesized in the 5' → 3' direction.

Watson and Crick concluded that DNA consists of two strands of nucleotides, with the sugar–phosphate backbone on the outside and the bases on the inside. The strands are antiparallel (they run in opposite directions) and are held together by hydrogen bonds between the bases on one strand and the bases on the other strand (Figure 3).

Experiments carried out by Erwin Chargaff were critical to Watson and Crick's proposal for the structure of DNA. These experiments showed that the number of adenines in DNA equals the number of thymines, and the number of guanines equals the number of cytosines. Chargaff also noted that the number of adenines and thymines relative to the number of guanines and cytosines is characteristic of a given species but varies from species to species. In human DNA, for example, 60.4% of the bases are adenines and thymines, whereas 74.2% of the bases are adenines and thymines in the DNA of the bacterium *Sarcina lutea.*

Chargaff's data showing that [adenine] = [thymine] and [guanine] = [cytosine] could be explained if adenine (A) always paired with thymine (T), and guanine (G) always paired with cytosine (C). This means the two strands are *complementary:* where there is an A in one strand, there is a T in the opposing strand; and where there is a G in one strand, there is a C in the other strand (Figure 3). Thus, if you know the sequence of bases in one strand, you can figure out the sequence of bases in the other strand.

Why does A pair with T? Why does G pair with C? First of all, the width of the double-stranded molecule is relatively constant, so a purine must pair with a pyrimidine. If the larger purines paired, the strands would bulge; if the smaller pyrimidines paired, the strands would have to pull in to bring the two pyrimidines close enough to form hydrogen bonds. But what causes A to pair with T rather than with C (the other pyrimidine)?

The base pairing is dictated by hydrogen bonding. Learning that the bases exist in the keto form and not the enol form allowed Watson to explain the pairing.* Adenine forms two hydrogen bonds with thymine but would form only one hydrogen bond with cytosine. Guanine forms three hydrogen bonds with cytosine but would form only one hydrogen bond with thymine (Figure 4).

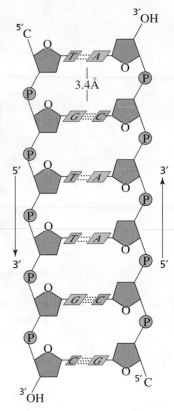

▲ **Figure 3**

The sugar–phosphate backbone of DNA is on the outside and the bases are on the inside. As pair with Ts and Gs pair with Cs. The two strands are antiparallel—that is, they run in opposite directions.

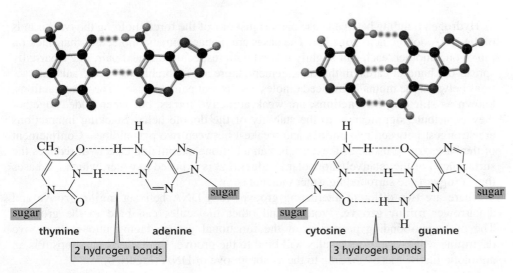

▲ **Figure 4**

Base pairing in DNA: adenine and thymine form two hydrogen bonds; cytosine and guanine form three hydrogen bonds.

The two antiparallel DNA strands are not linear but are twisted into a helix around a common axis (Figure 5a). The base pairs are planar and parallel to each other on the inside of the helix (Figure 5c). The secondary structure is therefore known as a **double helix.** The double helix resembles a circular staircase: the base pairs are the rungs, and the sugar–phosphate backbones are the handrails (see earlier in this section). The OH group of the phosphodiester linkages has a pK_a of about 2, so it is in its basic form (negatively charged) at physiological pH (Figure 2). The negatively charged backbone repels nucleophiles, thereby preventing cleavage of the phosphodiester bonds.

*Watson was having difficulty understanding the base pairing in DNA because he thought the bases existed in the enol form (see Problem 5). When Jerry Donohue, an American crystallographer, informed him that the bases more likely existed in the keto form, Chargaff's data could easily be explained by hydrogen bonding between adenine and thymine and between guanine and cytosine.

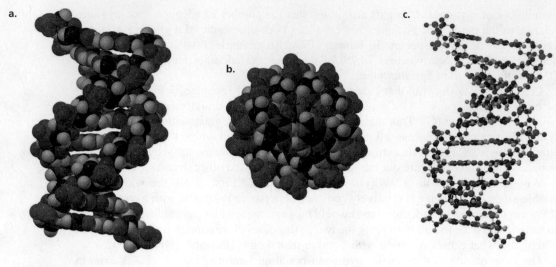

▲ **Figure 5**
a. The DNA double helix.
b. The view looking down the long axis of the helix.
c. The bases are planar and parallel on the inside of the helix.

Hydrogen bonding between base pairs is just one of the forces holding the two strands of the DNA double helix together. The bases are planar aromatic molecules that stack on top of one another, each pair slightly rotated with respect to the next pair, like a partially spread-out hand of cards. In this arrangement, there are favorable van der Waals interactions between the mutually induced dipoles of adjacent pairs of bases. These interactions, known as **stacking interactions,** are weak attractive forces, but when added together they contribute significantly to the stability of the double helix. Stacking interactions are strongest between two purines and weakest between two pyrimidines. Confinement of the bases to the inside of the helix has an additional stabilizing effect—it reduces the surface area of the relatively nonpolar residues that is exposed to water, which increases the entropy of the surrounding water molecules.

There are two different alternating grooves in a DNA helix; a **major groove** and a narrower **minor groove.** Proteins and other molecules can bind to the grooves. The hydrogen-bonding properties of the functional groups facing into each groove determine what kind of molecules will bind to the groove. For example, netropsin is an antibiotic that works by binding to the minor groove of DNA (Figure 6).

▲ **Figure 6**

The antibiotic netropsin bound in the minor groove of DNA.

PROBLEM 4

Indicate whether each functional group of the five heterocyclic bases in nucleic acids can function as a hydrogen bond acceptor (A), a hydrogen bond donor (D), or both (D/A).

PROBLEM 5

Using the D, A, and D/A designations in Problem 4, indicate how base pairing would be affected if the bases existed in the enol form.

PROBLEM 6♦

If one of the strands of DNA has the following sequence of bases running in the $5' \rightarrow 3'$ direction,

$$5'—G—G—A—C—A—A—T—C—T—G—C—3'$$

a. what is the sequence of bases in the complementary strand?
b. what base is closest to the 5′-end in the complementary strand?

4 WHY DNA DOES NOT HAVE A 2′-OH GROUP

Unlike DNA, RNA is not stable because the 2′-OH group of ribose acts as a nucleophilic catalyst for the cleavage of RNA (Figure 7). This explains why the 2′-OH group is absent in DNA. DNA must remain intact throughout the life span of a cell in order to preserve the genetic information. Easy cleavage of DNA would have disastrous consequences for the cell and for life itself. RNA, in contrast, is synthesized as it is needed and is degraded once it has served its purpose.

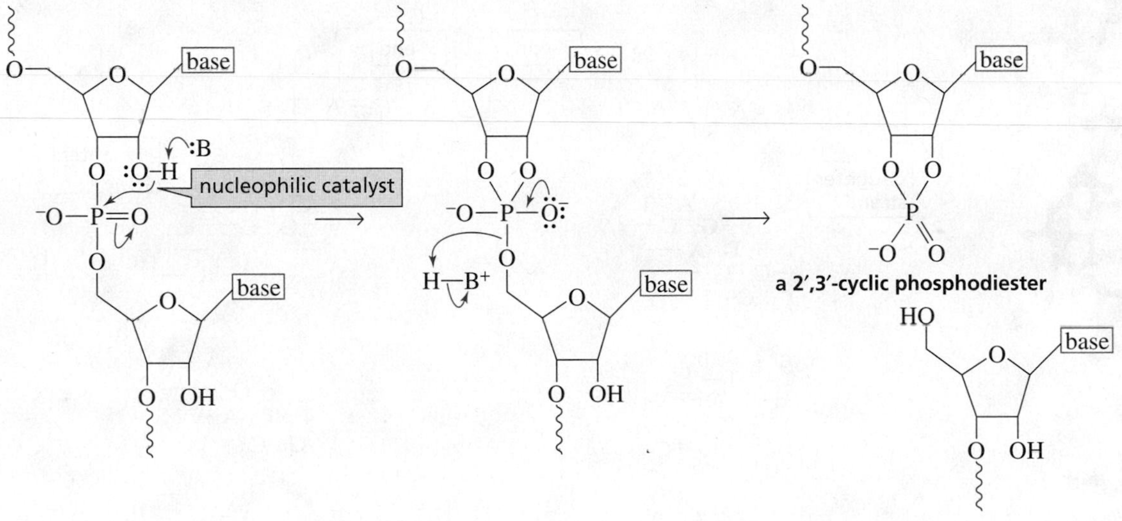

▲ **Figure 7**
Catalysis of RNA cleavage by the 2′-OH group. RNA undergoes cleavage 3 billion times faster than DNA.

PROBLEM 7

The 2′,3′-cyclic phosphodiester that is formed (Figure 7) when RNA is cleaved reacts with water, forming a mixture of nucleotide 2′- and 3′-phosphates. Propose a mechanism for this reaction.

5 THE BIOSYNTHESIS OF DNA IS CALLED REPLICATION

The genetic information of a human cell is contained in 23 pairs of chromosomes. Each chromosome is composed thousands of **genes** (segments of DNA). The total DNA from a human cell—the **human genome**—contains 3.1 billion base pairs.

Part of the excitement created by Watson and Crick's proposed structure for DNA was due to the fact that the structure immediately suggested how DNA is able to pass on genetic information to succeeding generations. Because the two strands are complementary, both carry the same genetic information. Thus, when organisms reproduce, DNA molecules can be copied using the same base-pairing principle that is fundamental to their structure—that is, each strand can serve as the template for the synthesis of a complementary new strand (Figure 8). The new (daughter) DNA molecules are identical to the original (parent) molecule, so they contain all the original genetic information. The synthesis of identical copies of DNA is called **replication.**

All the reactions involved in nucleic acid synthesis are catalyzed by enzymes. The synthesis of DNA takes place in a region of the molecule where the strands have started to separate. Because a nucleic acid can be synthesized only in the 5′ → 3′ direction, only the daughter

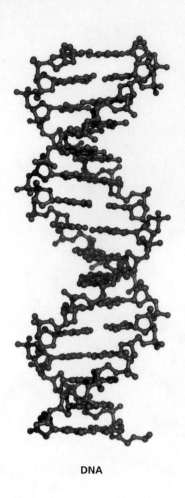

DNA

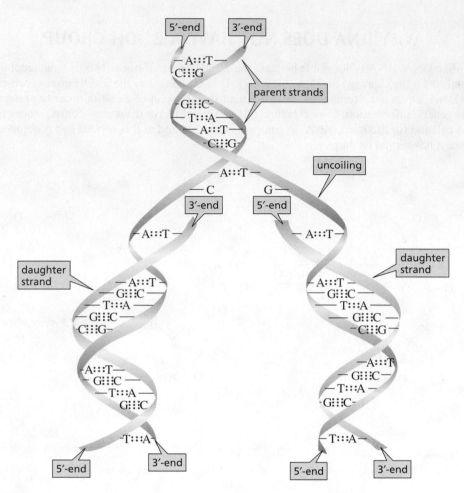

▲ **Figure 8**

Replication of DNA. The green daughter strand on the left is synthesized continuously in the 5′ → 3′ direction; the green daughter strand on the right is synthesized discontinuously in the 5′ → 3′ direction.

strand on the left in Figure 8 is synthesized continuously in a single piece (because it is synthesized in the 5′ → 3′ direction). The other daughter strand needs to grow in a 3′ → 5′ direction, so it is synthesized discontinuously in small pieces. Each piece is synthesized in the 5′ → 3′ direction, and the fragments are joined together by an enzyme called DNA ligase. Each of the two new molecules of DNA—called daughter molecules—contains one of the original parent strands (blue strand in Figure 8) plus a newly synthesized strand (green strand). This process is called **semiconservative replication.**

PROBLEM 8

Using a dark line for the original parental DNA and a wavy line for DNA synthesized from parental DNA, show what the population of DNA molecules would look like in the fourth generation. (Parental DNA is the first generation.)

---

6   DNA AND HEREDITY

If DNA contains hereditary information, there must be a method to decode that information. The decoding occurs in two steps.

1. The sequence of bases in DNA provides a blueprint for the synthesis of RNA; the synthesis of RNA from a DNA blueprint is called **transcription** (Section 7).

2. The sequence of bases in RNA determines the sequence of amino acids in a protein; the synthesis of a protein from an RNA blueprint is called **translation** (Section 9).

Transcription: DNA → RNA

Translation: mRNA → protein

Don't confuse transcription and translation: these words are used just as they are used in English. Transcription (DNA to RNA) is copying *within the same language*—in this case the language of nucleotides. Translation (RNA to protein) is *changing to another language*—the language of amino acids. First we will look at transcription.

Natural Products That Modify DNA

More than three-quarters of clinically approved anticancer drugs are natural products—compounds derived from plants, marine organisms, or microbes—that interact with DNA. Because cancer is characterized by the uncontrolled growth and proliferation of cells, compounds that interfere with the replication or transcription of DNA stop the growth of cancer cells. These drugs can interact with DNA by binding between the base pairs (called intercalation) or by binding to either its major or minor groove. The three anticancer drugs discussed here were isolated from *Streptomyces* bacteria found in soil.

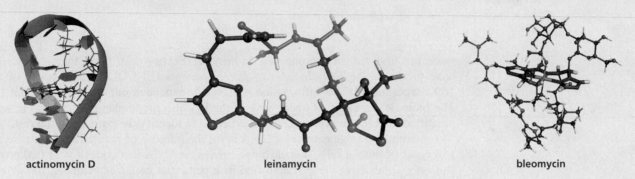

| actinomycin D | leinamycin | bleomycin |

Because intercalating compounds become sandwiched between the stacked bases in DNA, they are planar and often aromatic. Their binding to DNA is stabilized by stacking interactions with neighboring base pairs. Actinomycin D is an example of an intercalator. When this drug binds to DNA, it distorts the double helix, inhibiting both the replication and transcription of DNA. Actinomycin D has been used to treat a variety of cancers.

Drugs that bind to the major and minor grooves of DNA do so by a combination of hydrogen bonding, van der Waals interactions, and electrostatic attractions—the same forces proteins use to bind their substrates. Leinamycin is an example of an anticancer drug that binds to the major groove. Once leinamycin is bound, it alkylates the N-7 position of a purine ring.

Bleomycin binds to the minor groove of DNA. Once in the minor groove, it uses a bound iron atom to remove a hydrogen atom from DNA, the first step in cleaving DNA. This drug has been approved for the treatment of Hodgkin lymphoma.

7 THE BIOSYNTHESIS OF RNA IS CALLED TRANSCRIPTION

Transcription starts when DNA unwinds at a particular site—called a *promoter site*—to form two single strands. One of the strands is called the **sense strand.** The complementary strand is called the **template strand.** In order for RNA to be synthesized in the $5' \rightarrow 3'$ direction, the template strand is read in the $3' \rightarrow 5'$ direction (Figure 9). The bases in the template strand specify the bases that need to be incorporated into RNA, following the same base-pairing principle used in the replication of DNA. For example, each guanine in the template strand specifies the incorporation of a cytosine into RNA, and each adenine in the template strand specifies the incorporation of a uracil into RNA. (Note that in RNA, uracil is used instead of thymine.) Because both RNA and the sense strand of DNA are complementary to the template strand, RNA and the sense strand of DNA have the same base sequence, except that RNA has a uracil wherever the sense strand has a thymine. Just as there are promoter sites in DNA that signal where to start RNA synthesis, there are sites signaling that no more bases should be added to the growing RNA chain.

Until recently, it was thought that only about 2% of the DNA in our cells was used to make proteins and the rest had no informational content. However, in the decade

RNA is synthesized in the $5' \rightarrow 3'$ direction.

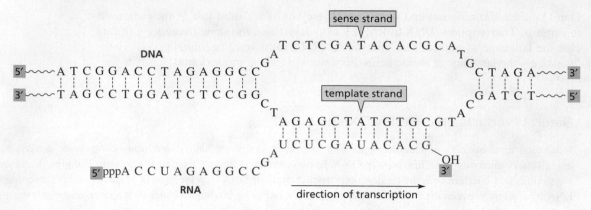

▲ Figure 9

Transcription: using DNA as a blueprint for the synthesis of RNA.

since the first human genome was sequenced (Section 12), more than 400 scientists have produced the *Encyclopedia of DNA Elements* (ENCODE), as a result of about 1600 experiments. This information has greatly expanded our knowledge about DNA. The biological purpose of about 80% of the DNA in the human genome had now been identified, and future experiments are expected to identify the purpose of the rest.

Apparently, a large amount of DNA is for the purpose of regulation. There are about 150 types of human cells, and each one carries the DNA that codes for 21,000 proteins. But only a subset of these is activated in a particular cell. For example, the gene that makes hair is not activated in a cell that makes insulin and vice versa.

It is now known that there are about 30,000 additional genes that make RNA that is not subsequently translated to make proteins. Instead, the RNA is used for regulation. In other words, these RNA strands appear to be the switches that turn genes on and off. The enormous number of switches has surprised scientists, and more have yet to be identified. Now the problem is to find out how these switches work.

There Are More Than Four Bases in DNA

For some time it has been known that a fifth base is found in DNA—namely, 5-methylcytosine. This base silences genes so they are no longer transcribed. Recently, an enzyme was discovered that converts 5-methylcytosine to 5-hydroxymethylcytosine. It appears that 5-hydroxymethylcytosine also plays a role in turning genes on and off.

5-methylcytosine **5-hydroxymethylcytosine** **5-formylcysteine**

Now yet another base has been discovered. Again it is a 5-subsituted cysteine—in this case, it is 5-formylcysteine. 5-Formylcysteine was found in embryonic stem cell DNA. Its function is yet to be determined but it is expected that it plays a role in transforming a fertilized egg into an embryonic stem cell.

PROBLEM 9♦

Why do both thymine and uracil specify the incorporation of adenine?

8 | THE RNAs USED FOR PROTEIN BIOSYNTHESIS

RNA molecules are much shorter than DNA molecules and are generally single-stranded. Although DNA molecules have billions of base pairs, RNA molecules rarely have more than 10,000 nucleotides. There are several kinds of RNA. The RNAs used for protein biosynthesis are:

- **messenger RNA (mRNA),** whose sequence of bases determines the sequence of amino acids in a protein
- **ribosomal RNA (rRNA),** a structural component of ribosomes, which are the particles on which the biosynthesis of proteins takes place
- **transfer RNA (tRNA),** the carrier of amino acids for protein synthesis

tRNA molecules are much smaller than mRNA or rRNA molecules. A tRNA contains only 70 to 90 nucleotides. The single strand of tRNA is folded into a characteristic cloverleaf structure, with three loops and a little bulge next to the right-hand loop (Figure 10a). There are at least four regions with complementary base pairing. The three bases at the bottom of the loop directly opposite the 5'- and 3'-ends are called an **anticodon.** All tRNAs have a CCA sequence at the 3'-end (Figures 10a and 10b).

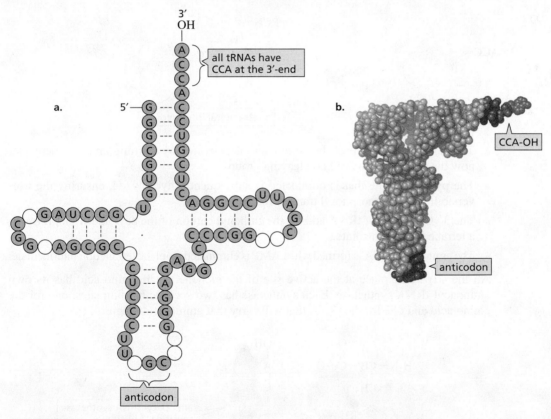

▲ **Figure 10**

a. A transfer RNA. Compared with other RNAs, tRNA contains a high percentage of unusual bases (shown as empty circles). These bases result from enzymatic modification of the four normal bases.

b. A transfer RNA: the anticodon is green; the CCA at the 3'-end is red.

Each tRNA can carry an amino acid bound as an ester to its terminal 3'-OH group. The amino acid will be inserted into a protein during protein biosynthesis. Each tRNA can carry only one particular amino acid. A tRNA that carries alanine is designated as tRNA^Ala.

The attachment of an amino acid to a tRNA is catalyzed by an enzyme called aminoacyl-tRNA synthetase. The mechanism for the reaction is shown here.

MECHANISM FOR THE ATTACHMENT OF AN AMINO ACID TO A tRNA

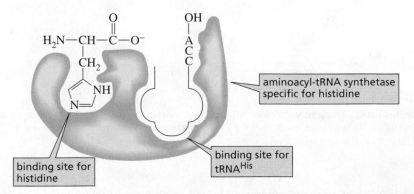

- The carboxylate group of the amino acid is activated by forming an acyl adenylate; now the amino acid has a good leaving group.
- The pyrophosphate that is eliminated is subsequently hydrolyzed, ensuring the irreversibility of the phosphoryl transfer reaction.
- The 3′-OH group of tRNA adds to the carbonyl carbon of the acyl adenylate, forming a tetrahedral intermediate.
- The aminoacyl tRNA is formed when AMP is eliminated from the tetrahedral intermediate.

All the steps take place at the active site of the enzyme. Each amino acid has its own aminoacyl-tRNA synthetase. Each synthetase has two specific binding sites, one for the amino acid and one for the tRNA that will carry that amino acid (Figure 11).

▶ **Figure 11**
An aminoacyl-tRNA synthetase has a binding site for the amino acid and a binding site for the tRNA that will carry that amino acid. In this example, histidine is the amino acid, and tRNA^{His} is the tRNA.

It is critical that the correct amino acid be attached to the tRNA. Otherwise, the protein will not be synthesized correctly. Fortunately, the synthetases correct their own mistakes. For example, valine and threonine are approximately the same size, but threonine has an

OH group in place of a CH_3 group of valine. Both amino acids, therefore, can bind at the amino acid–binding site of the aminoacyl-tRNA synthetase for valine, and both can then be activated by reacting with ATP to form an acyl adenylate. The aminoacyl-tRNA synthetase for valine has two adjacent catalytic sites. In addition to the site for attaching the acyl adenylate to tRNA, it has a site for hydrolyzing the acyl adenylate.

The site for attaching the acyl adenylate to tRNA is hydrophobic, so the valine acyl adenylate binds preferentially to that site. The site for hydrolyzing the acyl adenylate is polar, so the threonine acyl adenylate binds to that site. Thus, if threonine is activated by the aminoacyl-tRNA synthetase for valine, it will be hydrolyzed rather than transferred to the tRNA.

9 THE BIOSYNTHESIS OF PROTEINS IS CALLED TRANSLATION

A protein is biosynthesized from its N-terminal end to its C-terminal end by a process that reads the bases along the mRNA strand in the $5' \rightarrow 3'$ direction. The amino acid that is to be incorporated into a protein is specified by a three-base sequence called a **codon.** The bases are read consecutively and are never skipped. The three-base sequences and the amino acid that each sequence codes for are known as the **genetic code** (Table 2). A codon is written with the $5'$-nucleotide on the left. For example, the codon UCA on mRNA codes for the amino acid serine, whereas CAG codes for glutamine.

Table 2 The Genetic Code

5'-Position	Middle position				3'-Position
	U	C	A	G	
U	Phe	Ser	Tyr	Cys	U
	Phe	Ser	Tyr	Cys	C
	Leu	Ser	Stop	Stop	A
	Leu	Ser	Stop	Trp	G
C	Leu	Pro	His	Arg	U
	Leu	Pro	His	Arg	C
	Leu	Pro	Gln	Arg	A
	Leu	Pro	Gln	Arg	G
A	Ile	Thr	Asn	Ser	U
	Ile	Thr	Asn	Ser	C
	Ile	Thr	Lys	Arg	A
	Met	Thr	Lys	Arg	G
G	Val	Ala	Asp	Gly	U
	Val	Ala	Asp	Gly	C
	Val	Ala	Glu	Gly	A
	Val	Ala	Glu	Gly	G

Because there are four bases and the codons are triplets, 4^3 (or 64) different codons are possible. This is many more than are needed to specify the 20 different amino acids, so all the amino acids—except methionine and tryptophan—have more than one codon.

It is not surprising, therefore, that methionine and tryptophan are the least abundant amino acids in proteins. Actually, 61 of the codons specify amino acids, and three codons are stop codons. **Stop codons** tell the cell to "stop protein synthesis here."

How the information in mRNA is translated into a polypeptide is shown in Figure 12. In this figure, serine, specified by the codon AGC, was the last amino acid incorporated into the growing polypeptide chain.

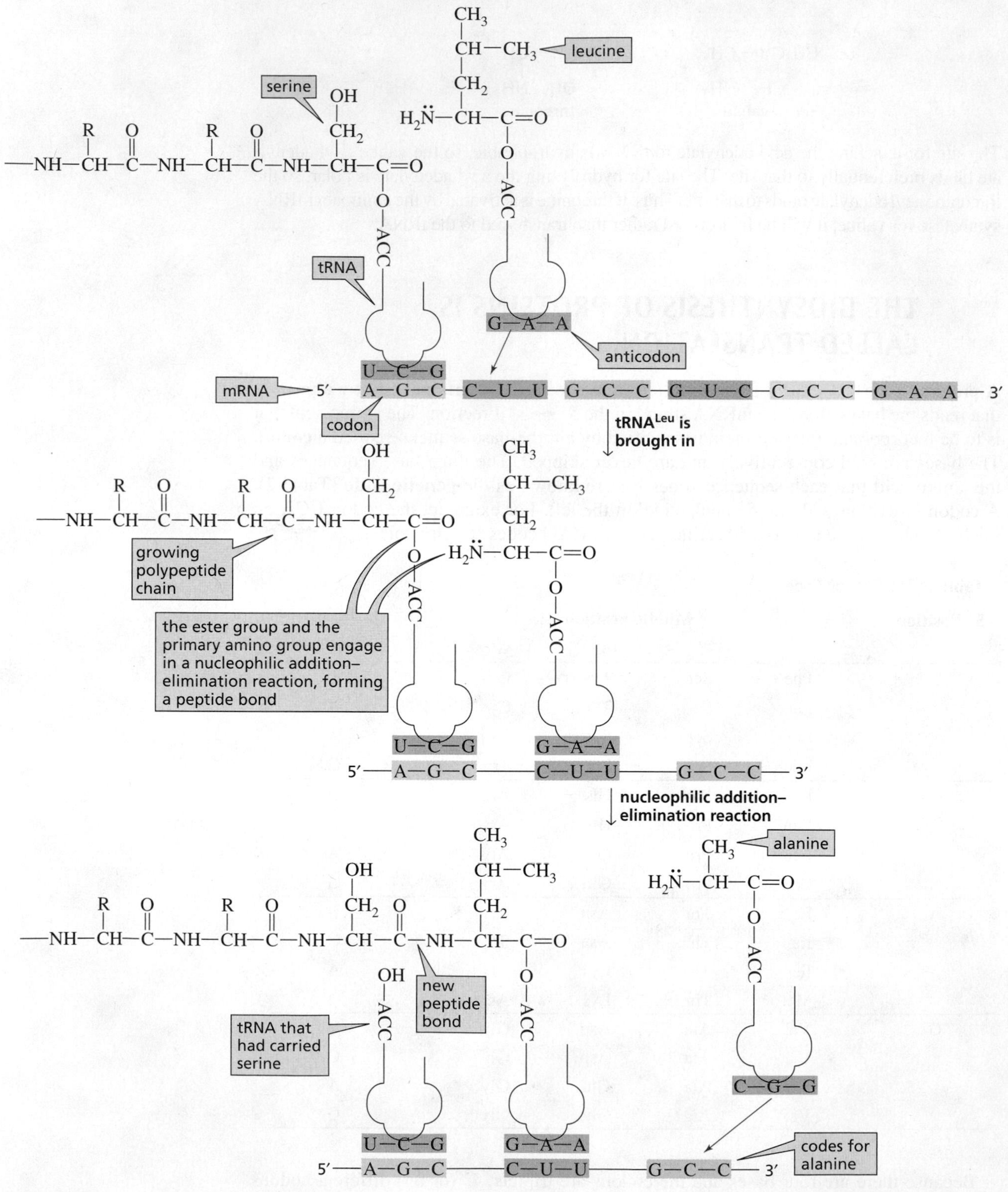

▲ **Figure 12**

Translation: the sequence of bases in mRNA determines the sequence of amino acids in a protein.

- Serine was specified by the AGC codon in mRNA because the anticodon of the tRNA that carries serine is GCU (3′-UCG-5′). (Note that a base sequence is read in the 5′ → 3′ direction, so the sequence of bases in an anticodon must be read from right to left.)

- The next codon, CUU, signals for a tRNA with an anticodon of AAG (3′-GAA-5′). That particular tRNA carries leucine. The amino group of leucine reacts in an enzyme-catalyzed nucleophilic addition–elimination reaction with the ester on the adjacent serine-carrying tRNA, displacing the tRNA that brought in serine.

- The next codon (GCC) specifies a tRNA carrying alanine. The amino group of alanine reacts in an enzyme-catalyzed nucleophilic addition–elimination reaction with the ester group on the adjacent leucine-carrying tRNA, displacing the tRNA that brought in leucine.

Subsequent amino acids are brought in one at a time in the same way, with the codon in mRNA specifying the amino acid to be incorporated by complementary base pairing with the anticodon of the tRNA that carries that amino acid.

A protein is biosynthesized in the N-terminal → C-terminal direction

PROBLEM 10♦

If methionine is the first amino acid incorporated into an oligopeptide, what oligopeptide is coded for by the following stretch of mRNA?

5′—G—C—A—U—G—G—A—C—C—C—C—G—U—U—A—U—U—A—A—A—C—A—C—3′

PROBLEM 11♦

Four Cs occur in a row in the segment of mRNA shown in Problem 10. What oligopeptide would be formed from the mRNA if one of the four Cs were cut out of the strand?

PROBLEM 12

UAA is a stop codon. Why does the UAA sequence in the segment of mRNA in Problem 10 not cause protein synthesis to stop?

Protein synthesis takes place on a ribosome, which is composed of ribosomal RNA(rRNA) and protein (Figure 13). The ribosome has three binding sites for RNA molecules. It binds the mRNA whose base sequence is to be read, the tRNA that carries the growing peptide chain, and the tRNA that carries the next amino acid to be incorporated into the protein.

1. Transcription of DNA occurs in the nucleus of the cell. The initial RNA transcript is the precursor of all RNA.

2. The initially formed RNA often must be chemically modified before it acquires biological activity. Modification can entail removing nucleotide segments, adding nucleotides to the 5′- or 3′-ends, or chemically altering certain nucleotides.

3. Proteins are added to rRNA to form ribosomes. tRNA, mRNA, and ribosomes leave the nucleus.

4. Each tRNA binds the appropriate amino acid.

5. tRNA, mRNA, and a ribosome work together to translate the information in mRNA into a protein.

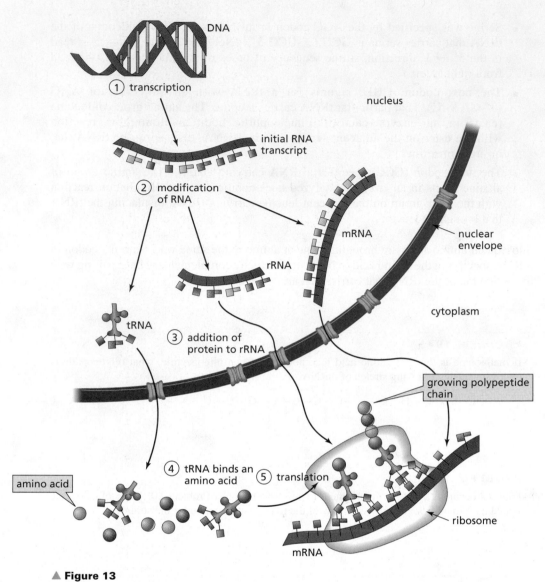

▲ **Figure 13**

Transcription and translation.

Sickle Cell Anemia

Sickle cell anemia is an example of the damage that can be caused by a change in a single base of DNA. It is a hereditary disease caused when a GAG triplet becomes a GTG triplet in the sense strand of a section of DNA that codes for the β-subunit of hemoglobin. As a consequence, the mRNA codon becomes GUG—which signals for incorporation of valine—rather than GAG, which would have

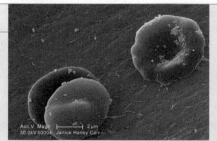

normal red blood cells

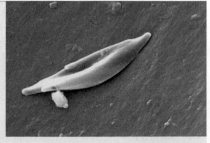

sickle red blood cell

signaled for incorporation of glutamate. The change from a polar glutamate to a nonpolar valine is sufficient to change the shape of the deoxyhemoglobin molecule. The change in shape stiffens the cells, making it difficult for them to squeeze through capillaries. Blockage of capillaries causes severe pain and can be fatal.

Antibiotics That Act by Inhibiting Translation

Puromycin is a naturally occurring antibiotic, one of several that acts by inhibiting translation. It does so by mimicking the 3′-CCA-aminoacyl portion of a tRNA, fooling the enzyme into transferring the growing peptide chain to the NH_2 group of puromycin rather than to the NH_2 group of the incoming 3′-CCA-aminoacyl tRNA. As a result, protein synthesis stops. Because puromycin blocks protein synthesis in eukaryotes as well as in prokaryotes, it is poisonous to humans and therefore is not a clinically useful antibiotic. To be clinically useful, an antibiotic must affect protein synthesis only in prokaryotic cells.

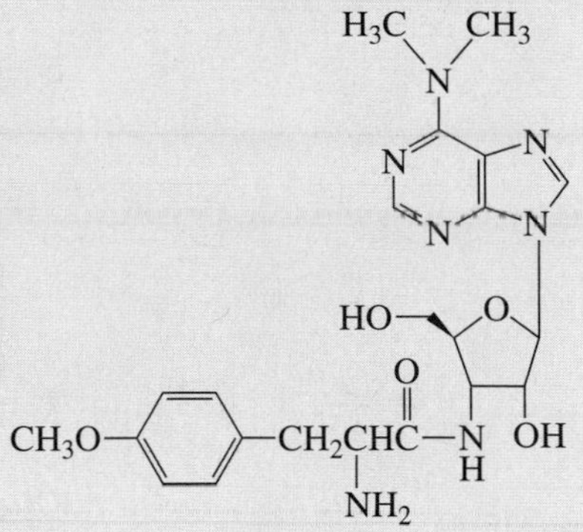

Clinically useful antibiotics	Mode of action
Tetracycline	Prevents the aminoacyl-tRNA from binding to the ribosome
Erythromycin	Prevents the incorporation of new amino acids into the protein
Streptomycin	Inhibits the initiation of protein synthesis
Chloramphenicol	Prevents the new peptide bond from being formed

puromycin

PROBLEM 13♦
A change in which base of a codon would be least likely to cause a mutation?

PROBLEM 14♦
Write the sequences of bases in the sense strand of DNA that resulted in the mRNA in Problem 10.

PROBLEM 15
List the possible codons on mRNA that specify each amino acid in Problem 10 and the anticodon on the tRNA that carries that amino acid.

10 WHY DNA CONTAINS THYMINE INSTEAD OF URACIL

dTMP is formed by methylating dUMP, with coenzyme N^5,N^{10}-methylenetetrahydrofolate supplying the methyl group.

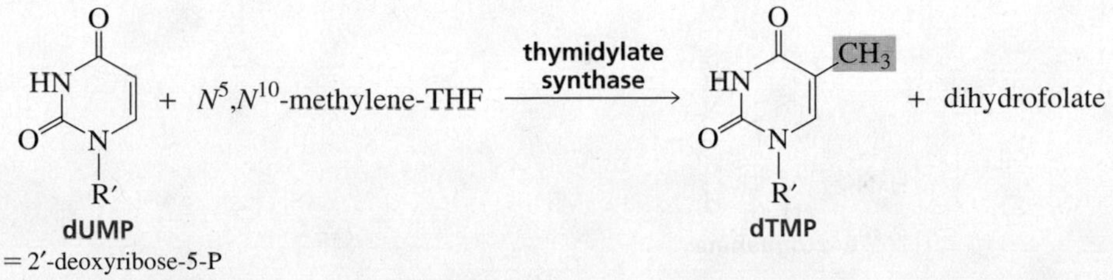

dUMP $+ N^5,N^{10}$-methylene-THF $\xrightarrow{\text{thymidylate synthase}}$ dTMP $+$ dihydrofolate

R′ = 2′-deoxyribose-5-P

Because the incorporation of the methyl group into uracil oxidizes tetrahydrofolate to dihydrofolate, dihydrofolate must be reduced back to tetrahydrofolate to prepare the coenzyme for another catalytic reaction. The reducing agent is NADPH.

$$\text{dihydrofolate} + \text{NADPH} + \text{H}^+ \xrightarrow{\text{dihydrofolate reductase}} \text{tetrahydrofolate} + \text{NADP}^+$$

The NADP$^+$ formed in this reaction has to be reduced back to NADPH by NADH. Every NADH formed in a cell can result in the formation of 2.5 ATPs. Therefore, reducing dihydrofolate comes at the expense of ATP. This means that the synthesis of thymine is energetically expensive, so there must be a good reason for DNA to contain thymine instead of uracil.

The presence of thymine instead of uracil in DNA prevents potentially lethal mutations. Cytosine can tautomerize to form an imine, which can be hydrolyzed to uracil. The overall reaction is called a **deamination** because it removes an amino group.

cytosine
amino tautomer   imino tautomer   uracil

If a C in DNA is deaminated to a U, the U will specify incorporation of an A into the daughter strand during replication instead of the G that would have been specified by C, and all the progeny of the daughter strand would have the same mutated chromosome. Fortunately, there is an enzyme that recognizes a U in DNA as a "mistake" and replaces it with a C before an incorrect base can be inserted into the daughter strand. The enzyme cuts out the U and replaces it with a C. If Us were normally found in DNA, the enzyme would not be able to distinguish between a normal U and a U formed by deamination of a cytosine. Having Ts in place of Us in DNA allows the Us that are found in DNA to be recognized as mistakes.

Unlike DNA, which replicates itself, any mistake in RNA does not survive for long because RNA is continually degraded and then resynthesized from the DNA template. Therefore, changing a C to a U in RNA could lead to some copies of a defective protein, but most would not be defective. Thus, it is not worth incurring the loss of ATP to incorporate Ts into RNA.

Antibiotics Act by a Common Mechanism

Recently, it has been found that three different classes of antibiotics (a β-lactam, a quinolone, and an aminoglycoside) all kill bacteria in the same way. The antibiotics trigger the production of hydroxide radicals. The hydroxide radicals oxidize guanines to 8-oxoguanines. The cell is able to recognize 8-oxoguanines as mistakes and replace them with guanines. However, if there are too many 8-oxoguanines in DNA, the cell's repair mechanism becomes overwhelmed. Then, instead of cutting out the 8-oxoguanines, it breaks the DNA strand, which leads to cell death. This finding suggests that new antibiotics might be found that specifically target guanine oxidation.

8-oxoguanine

PROBLEM 16♦

Adenine can be deaminated to hypoxanthine, and guanine can be deaminated to xanthine. Draw structures for hypoxanthine and xanthine.

PROBLEM 17

Explain why thymine cannot be deaminated.

11 ANTIVIRAL DRUGS

Relatively few clinically useful drugs have been developed for viral infections. The slow progress of this endeavor is due to the nature of viruses and the way they replicate. Viruses are smaller than bacteria and consist of nucleic acid—either DNA or RNA—surrounded by a coat of protein. Some viruses penetrate the host cell; others merely inject their nucleic acid into the cell. In either case, the viral nucleic acid is transcribed by the host and is integrated into the host genome.

Most **antiviral drugs** are analogues of nucleosides, interfering with the virus's nucleic acid synthesis. In this way, they prevent the virus from replicating. For example, acyclovir, the drug used against herpes viruses, has a three-dimensional shape similar to guanine. Therefore, acyclovir can fool the virus into incorporating the drug instead of guanine into its DNA. Once this happens, the DNA strand can no longer grow because acyclovir lacks a ribose with a 3′-OH group. The terminated DNA remains bound to DNA polymerase, which irreversibly inactivates the enzyme (Section 3).

| acyclovir Aclovir® used against herpes simplex infections | cytarabine Cytosar® used against acute myelocytic leukemia | ribavirin Viramid® used to treat hepatitis C | idoxuridine Herplex® approved for topical ophthalmic use |

Cytarabine, used for acute myelocytic leukemia, competes with cytosine for incorporation into viral DNA. Cytarabine contains an arabinose rather than a ribose. Because the 2′-OH group is in the β-position (note that the 2′-OH group of a natural ribonucleoside is in the α-position), the bases in cytarabine-modified DNA are not able to stack properly (Section 3).

A step in the metabolic pathway responsible for the synthesis of guanosine triphosphate (GTP) converts inosine monophosphate (IMP) into xanthosine monophosphate (XMP). Ribavirin is a competitive inhibitor of the enzyme that catalyzes this step. Thus, ribavirin interferes with the synthesis of GTP and, therefore, with the synthesis of nucleic acids. It is used to treat hepatitis C.

Idoxuridine is approved (in the United States) only for the topical treatment of ocular infections, although it is used for herpes infections in other countries. Idoxuridine has an iodo group in place of the methyl group of thymine and is incorporated into DNA in place of thymine. Chain elongation can continue because idoxuridine has a 3′-OH group, but the resulting DNA is more easily broken and is also not transcribed properly. (Also see AZT later in this chapter.)

Influenza Pandemics

Every year we face an outbreak of influenza (the flu). Most of the time it is a virus that is already present in the population and therefore can be controlled by flu shots. But every once in awhile, a new influenza virus appears, which can cause a worldwide pandemic because it is not affected by any immunity a person may have to older strains of influenza and can therefore spread rapidly and infect a large number of people. And almost no effective antiviral drugs are available for the flu.

The Russian flu of 1889–1890 was the first of the flu pandemics. It killed about 1 million people. The Spanish flu that broke out in 1918–1919 killed over 50 million people worldwide. The Asian flu of 1956–1958 killed about 2 million people before a vaccine was developed in 1957 to contain it. The Hong Kong flu of 1968–1969—so called because it affected 15% of the population of Hong Kong—had a much lower death rate—only about 750,000 people died—because people who had had the Asian flu had some immunity. Because this was the last worldwide pandemic, public health officials worry that another may occur soon.

Recent flu outbreaks that have been causes for concern are the avian flu (bird flu), discovered in 1997, and the swine flu, discovered in 2009. The avian flu was linked to chickens but it was subsequently transmitted to hundreds of people, 60% of whom died. The swine flu is a respiratory disease of pigs but it has been known to affect people. There are concerns that either of these flus could become a worldwide pandemic.

The carbohydrates attached to the surface of the viral protein account for the biggest difference in virus strains. The symptoms caused by viruses that bind primarily to sugars in the nose and throat are not as severe as those caused by viruses that bind to sugars deep in the lungs.

12 HOW THE BASE SEQUENCE OF DNA IS DETERMINED

In June 2000, two teams of scientists (one from a private biotechnology company and one from the publicly funded Human Genome Project) announced that they had completed the first draft of the sequence of the 3.1 billion base pairs in human DNA. This was an enormous accomplishment.

Clearly, DNA molecules are too large to sequence as a unit. Therefore, DNA is first cleaved at specific base sequences, and the resulting DNA fragments are then sequenced individually.

restriction enzyme	recognition sequence
*Alu*I	AG CT TC GA
*Fnu*DI	GG CC CC GG
*Pst*I	CTGCA G G ACGTC

The enzymes that cleave DNA at specific base sequences are called **restriction endonucleases,** and the DNA fragments they produce are called **restriction fragments.** Several hundred restriction endonucleases are now known; a few examples, the base sequence that each recognizes, and the point of cleavage in that base sequence are shown in the margin.

The base sequences that most restriction endonucleases recognize are *palindromes.* A palindrome is a word or a group of words that reads the same forward and backward. "Toot" and "race car" are examples of palindromes, as is "Was it a car or a cat I saw?" A restriction endonuclease recognizes a piece of DNA in which *the template strand is a palindrome of the sense strand.* In other words, the sequence of bases in the template strand (reading from right to left) is identical to the sequence of bases in the sense strand (reading from left to right).

PROBLEM 18♦

Which of the following base sequences would most likely be recognized by a restriction endonuclease?

a. ACGCGT   c. ACGGCA   e. ACATCGT

b. ACGGGT   d. ACACGT   f. CCAACC

Frederick Sanger (who received the 1953 Nobel Prize in Chemistry for being the first to determine the primary sequence of a protein) and Walter Gilbert shared the 1980 Nobel Prize in Chemistry for their work on DNA sequencing. Over the years, their procedure has been replaced by other methods. A currently used technique is an automated procedure called **pyrosequencing.** In this method, a small piece of DNA primer is added to the restriction fragment whose sequence is to be determined. Nucleotides are then added to the primer by base pairing with the restriction fragment. This method—known as *sequencing-by-synthesis*—detects the identity of each base that adds to the primer.

Pyrosequencing requires DNA polymerase—the enzyme that adds nucleotides to a strand of DNA—and two additional enzymes that cause light to be emitted when pyrophosphate is detected.

DNA to be sequenced

$$3'—\text{AGGCTCCAGTGATCCG} —5'$$

DNA polymerase
2 additional enzymes
3'-protected
2'-deoxyribonucleotide
5'-triphosphate

P—TC

a restriction fragment:
primer hybrid

primer

Pyrosequencing also requires the four 2'-deoxyribonucleotide 5'-triphosphates, each with a protected 3'-OH group.

base

a protecting group

a 3'-protected 2'-deoxyribonucleotide triphosphate

The restriction fragment:primer hybrid is attached to a solid support, similar to the solid support used in the automated synthesis of polypeptides. The steps involved in pyrosequencing are:

- The enzymes and one of the four 3'-protected 2'-deoxyribonucleotide 5'-triphosphates (for example, 3'-protected dATP) are added to the column.
- The reagents are washed from the solid support.
- The process is repeated with a different 3'-protected 2'-deoxynucleotide 5'-triphosphate (for example, 3'-protected dGTP).
- The process is repeated with 3'-protected dCTP, and then repeated again with 3'-protected dTTP.
- The sequencer keeps track of which of the four nucleotides caused light to be observed—in other words, which nucleotide released pyrophosphate as a result of being added to the primer.
- The protecting group on the 3'-OH is removed.

The steps are repeated to determine the identity of the next nucleotide that adds to the primer. Pyrosequencing can determine the base sequence of a restriction fragment with as many as 500 nucleotides.

Once the sequence of bases in a restriction fragment is determined, the results can be checked by using the same process to obtain the base sequence of the fragment's complementary strand. The base sequence of the original piece of DNA can be determined by repeating the entire procedure with a different restriction endonuclease and noting overlapping fragments.

The X PRIZE

The successful sequencing of the first human genome, with 3.1 million base pairs, cost ~ $2.7 billion and took about 10 years to complete. Using the next-generation sequencing technology (pyrosequencing), a human genome was sequenced in two months at a cost of ~ $1 million.

To stimulate the development of new technology that will increase the speed and decrease the cost of sequencing human genomes, the X PRIZE (worth $10 million) has been created. It will be given to the first team that can build a device that can sequence the whole genome of 100 human subjects in 30 days or less with the following criteria:

- It must have an accuracy of no more than one error per million base pairs sequenced.
- It must cost no more than $1,000 per genome.

The genomes of 100 people who have reached their 100th birthday will be used. These people are expected to carry rare genes that protect them against diseases associated with old age.

Since the X PRIZE was first announced, sequencing technologies, both in terms of time and cost, have continued to advance. When a human genome can be rapidly sequenced at a reasonable cost, the era of personalized medicine can begin. We will then understand what makes people more susceptible to certain diseases and why drugs work differently on different people. Eventually, people will be able to be given drugs that fit their genetic profile.

13 THE POLYMERASE CHAIN REACTION (PCR)

The PCR (**polymerase chain reaction**) technique, developed in 1983, allows scientists to amplify DNA—that is, to make billions of copies—in a very short time. With PCR, sufficient DNA for analysis can be obtained from a single hair follicle or sperm.

PCR is done by adding the following to a solution containing the segment of DNA to be amplified (the target DNA):

- a large excess of primers (short pieces of DNA) that base pair (anneal) with the nucleotide sequence at the two ends of the piece of DNA to be amplified
- the four deoxyribonucleotide triphosphates (dATP, dGTP, dCTP, dTTP)
- a heat-stable DNA polymerase

The following three steps are then carried out (Figure 14):

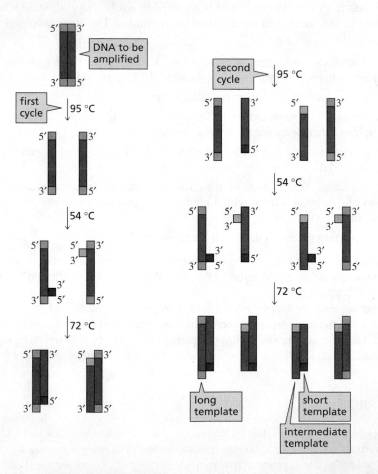

▶ **Figure 14**

Two cycles of the polymerase chain reaction.

- *Strand separation:* The solution is heated to 95 °C, which causes double-stranded DNA to separate into two single strands.
- *Base pairing of the primers:* The solution is cooled to 54 °C, a temperature at which the primers (red and yellow boxes in Figure 14) can pair with the bases at the 3′-end of the target (green) DNA.

- *DNA synthesis:* The solution is heated to 72 °C, a temperature at which DNA polymerase catalyzes the addition of nucleotides to the primer. Notice that because the primers attach to the 3′-end of the target DNA, copies of the target DNA are synthesized in the required 5′ → 3′ direction.

The solution is then heated to 95 °C to begin a second cycle. The second cycle produces four copies of double-stranded DNA. The third cycle therefore starts with 8 single strands of DNA and produces 16 single strands. It takes about an hour to complete enough cycles to amplify DNA a billion-fold. The amplification is carried out in a thermocycler—a devise that automatically changes the temperature so that each of the three steps can be carried out.

PCR has a wide range of clinical uses. It can be used to detect mutations that lead to cancer, to diagnose genetic diseases, to reveal the presence of HIV that would be missed by an antibody assay, to monitor cancer chemotherapy, and to rapidly identify an infectious disease.

The DNA from a 40-million-year-old leaf preserved in amber was amplified by PCR and then sequenced.

DNA Fingerprinting

Forensic chemists use PCR to compare DNA samples collected at the scene of a crime with the DNA of the suspected perpetrator. The base sequences of segments of noncoding DNA vary from individual to individual. Forensic laboratories have identified 13 of these segments that are the most accurate for identification purposes. If the sequences of bases from two DNA samples are the same, the chance is about 80 billion to 1 that they are from the same individual. DNA fingerprinting is also used to establish paternity, which accounts for about 100,000 DNA profiles a year.

14   GENETIC ENGINEERING

Recombinant DNA molecules are DNA molecules (natural or synthetic) that have been attached to a small piece of carrier DNA in a compatible host cell and allowed to replicate millions of times. Recombinant DNA technology—also known as **genetic engineering**—has many practical applications. For example, replicating the DNA that codes for human insulin makes it possible to synthesize large amounts of the protein.

Agriculture is benefiting from genetic engineering as crops are being produced with new genes that increase their resistance to drought and insects. For example, genetically engineered cotton crops are resistant to the cotton bollworm, and genetically engineered corn is resistant to the corn rootworm. Genetically modified organisms (GMOs) have been responsible for a nearly 50% reduction in agricultural chemical sales in the United States. Recently, corn has been genetically modified to boost ethanol production, and apples have been genetically modified to prevent them from turning brown when they are cut.

Resisting Herbicides

Glyphosate, the active ingredient in a well-known herbicide called Roundup, kills weeds by inhibiting an enzyme that plants need to synthesize phenylalanine and tryptophan, amino acids they require for growth. Corn and cotton have been genetically engineered to tolerate the herbicide. Then, when fields are sprayed with glyphosate, the weeds are killed but not the crops.

These crops have been given a gene that produces an enzyme that uses acetyl-CoA to acetylate glyphosate in a nucleophilic addition–elimination reaction. Unlike glyphosphate, *N*-acetylglyphosphate does not inhibit the enzyme that synthesizes phenylalanine and tryptophan.

Corn genetically engineered to resist the herbicide glyphosate by acetylating it.

glyphosate
an herbicide

+

enzyme

N-acetylglyphosate
harmless to plants

+ CoASH

SOME IMPORTANT THINGS TO REMEMBER

- **Deoxyribonucleic acid** (**DNA**) encodes an organism's hereditary information and controls the growth and division of cells.

- A **nucleoside** contains a base bonded to D-ribose or to 2′-deoxy-D-ribose. A **nucleotide** is a nucleoside with an OH group of the sugar bonded to phosphoric acid by an ester linkage.

- **Nucleic acids** are composed of long strands of nucleotide subunits linked by phosphodiester bonds. These linkages join the 3′-OH group of one nucleotide to the 5′-OH group of the next nucleotide.

- DNA contains 2′-deoxy-D-ribose, whereas RNA contains D-ribose. The difference in the sugars causes DNA to be stable and RNA to be easily cleaved.

- The **primary structure** of a nucleic acid is the sequence of bases in its strand. DNA contains **A, G, C,** and **T;** RNA contains **A, G, C,** and **U.**

- The presence of **T** instead of **U** in DNA prevents mutations caused by tautomerization and imine hydrolysis of C to form U.

- DNA is double-stranded. The strands run in opposite directions and are twisted into a double helix, giving DNA a major groove and a minor groove.

- The bases are confined to the inside of the helix, and the sugar and phosphate groups are on the outside. The strands are held together by hydrogen bonds between bases of opposing strands as well as by **stacking interactions.**

- The two strands—one is called a **sense strand** and the other a **template strand**—are complementary: **A** pairs with **T,** and **G** pairs with **C.**

- DNA is synthesized in the 5′ → 3′ direction by a process called **semiconservative replication.**

- The sequence of bases in DNA provides the blueprint for the synthesis (**transcription**) of RNA. RNA is synthesized in the 5′ → 3′ direction by reading the bases along the DNA template strand in the 3′ → 5′ direction.

- The RNAs used in protein biosynthesis are: messenger RNA, ribosomal RNA, and transfer RNA.

- Each three-base sequence of mRNA—a **codon**—specifies the particular amino acid to be incorporated into a protein. The codons and the amino acids they specify are known as the **genetic code.**

- Protein synthesis (**translation**) proceeds from the N-terminal end to the C-terminal end by reading the bases along the mRNA strand in the 5′ → 3′ direction.

- A tRNA carries the amino acid bound as an ester to its terminal 3′-position.

- Cytosines can be deaminated to uracils. A **deamination** is a reaction that removes an amino group.

- **Restriction endonucleases** cleave DNA at specific palindromes, forming **restriction fragments.**

- **Pyrosequencing** is a method used to determine the sequence of bases in the restriction fragments.

- The **polymerase chain reaction** (**PCR**) amplifies segments of DNA.

- A large amount of a particular protein can be synthesized by **genetic engineering.**

- The human genome contains 3.1 billion base pairs.

PROBLEMS

19. Name the following compounds:

a.

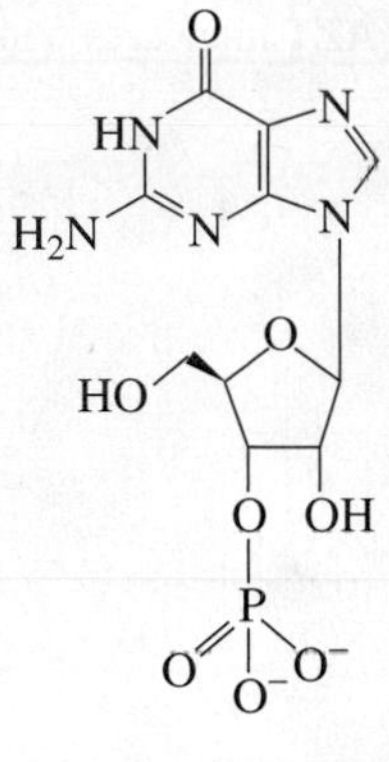

b.

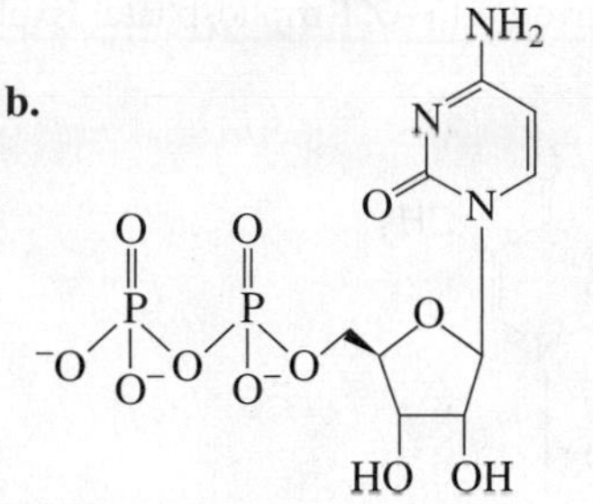

c.

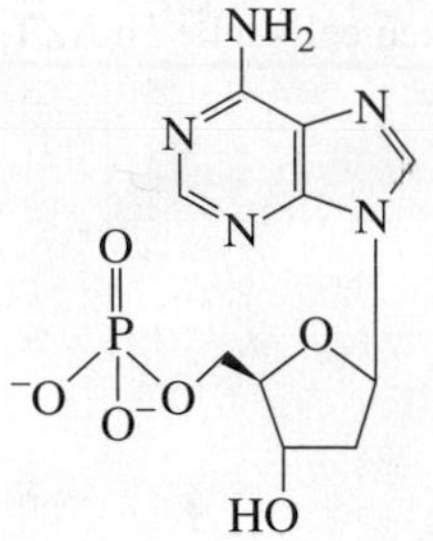

d.

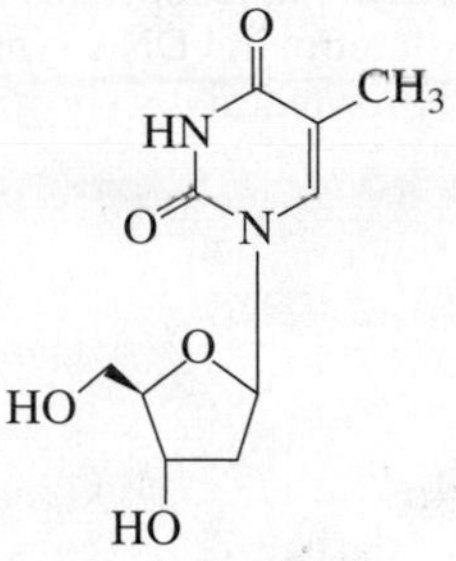

20. What nonapeptide is coded for by the following fragment of mRNA?

$$5'—AAA—GUU—GGC—UAC—CCC—GGA—AUG—GUG—GUC—3'$$

21. What is the sequence of bases in the template strand of DNA that codes for the mRNA in Problem 20?

22. What is the sequence of bases in the sense strand of DNA that codes for the mRNA in Problem 20?

23. What would be the C-terminal amino acid if the codon at the 3'-end of the mRNA in Problem 20 underwent the following mutations?
   a. The first base is changed to A.
   b. The second base is changed to A.
   c. The third base is changed to A.
   d. The third base is changed to G.

24. What would be the base sequence of the segment of DNA responsible for the biosynthesis of the following hexapeptide?

Gly-Ser-Arg-Val-His-Glu

25. Propose a mechanism for the following reaction:

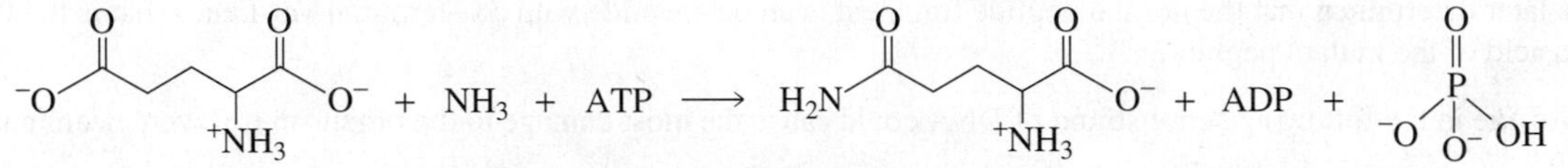

26. A segment of DNA has 18 base pairs. It has 7 cytosines in the segment.
   a. How many uracils are in the segment?
   b. How many guanines are in the segment?

27. Match the codon with the anticodon:

Codon	Anticodon
AAA	ACC
GCA	CCU
CUU	UUU
AGG	AGG
CCU	UGA
GGU	AAG
UCA	GUC
GAC	UGC

28. If your book contains the chapter "The Organic Chemistry of Amino Acids, Peptides, and Proteins," use the single-letter abbreviations for the amino acids in Table 2, and write the sequence of amino acids in a tetrapeptide represented by the first four different letters in your first name. Do not use any letter twice. (Because not all letters are assigned to amino acids, you might have to use one or two letters in your last name.) Write one of the sequences of bases in mRNA that would result in the synthesis of that tetrapeptide. Write the sequence of bases in the sense strand of DNA that would result in formation of that fragment of mRNA.

1273

29. Which of the following pairs of dinucleotides are present in equal amounts in DNA?

   a. CC and GG   **c.** CA and TG   **e.** GT and CA

   b. CG and GT   **d.** CG and AT   **f.** TA and AT

30. Human immunodeficiency virus (HIV) is the retrovirus that causes AIDS. AZT was one of the first drugs designed to interfere with retroviral DNA synthesis. When cells take up AZT, they convert it to AZT-triphosphate. Explain how AZT interferes with DNA synthesis.

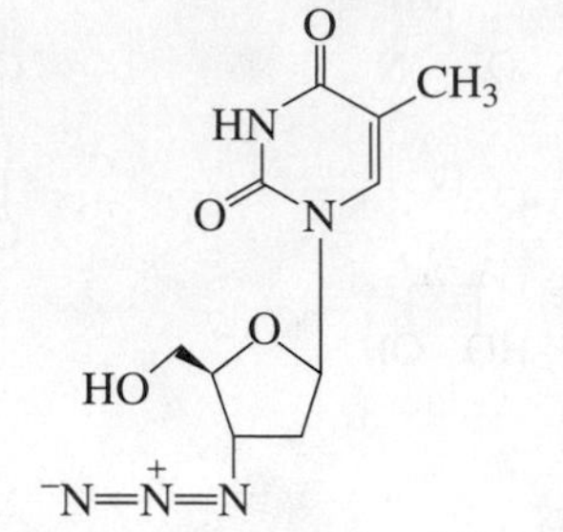

3′-azido-2′-deoxythymidine
AZT

31. If an mRNA contained only U and G in random sequence, what amino acids would be present in the protein when the mRNA is translated?

32. Why is the codon a triplet rather than a doublet or a quartet?

33. RNAase, the enzyme that catalyzes the hydrolysis of RNA, has two catalytically active histidine residues at its active site. One of the histidine residues is catalytically active in its acidic form, and the other is catalytically active in its basic form. Propose a mechanism for RNAase.

34. The amino acid sequences of peptide fragments obtained from a normal protein were compared with those obtained from the same protein synthesized by a defective gene. They were found to differ in only one peptide fragment. Their amino acid sequences are shown here:

Normal: Gln-Tyr-Gly-Thr-Arg-Tyr-Val
Mutant: Gln-Ser-Glu-Pro-Gly-Thr

   a. What is the defect in DNA?

   b. It was later determined that the normal peptide fragment is an octapeptide with a C-terminal Val-Leu. What is the C-terminal amino acid of the mutant peptide?

35. Which cytosine in the following sense strand of DNA could cause the most damage to the organism if it were deaminated?

A—T—G—T—C—G—C—T—A—A—T—C

36. 5-Bromouracil, a highly mutagenic compound (that is, a compound that causes changes in DNA), is used in cancer chemotherapy. When administered to a patient, it is converted to the triphosphate and incorporated into DNA in place of thymine, which it resembles sterically. Why does it cause mutations? (*Hint:* The bromo substituent increases the stability of the enol tautomer.)

37. Sodium nitrite, a common food preservative, is capable of causing mutations in an acidic environment by converting cytosines to uracils. Explain how this occurs.

38. Why does DNA not unravel completely before replication begins?

39. *Staphylococcus* nuclease is an enzyme that catalyzes the hydrolysis of DNA.
The reaction is catalyzed by Ca^{2+}, Glu 43, and Arg 87. Propose a mechanism for this reaction. Note that the nucleotides in DNA have phosphodiester linkages.

40. The first amino acid incorporated into a polypeptide chain during its biosynthesis in prokaryotes is *N*-formylmethionine. Explain the purpose of the formyl group. (*Hint:* The ribosome has a binding site for the growing peptide chain and a binding site for the incoming amino acid.)

GLOSSARY

anticodon   the three bases at the bottom of the middle loop in tRNA.

antiviral drug   a drug that interferes with DNA or RNA synthesis in order to prevent a virus from replicating.

codon   a sequence of three bases in mRNA that specifies the amino acid to be incorporated into a protein.

deoxyribonucleotide   a nucleotide in which the sugar component is D-2′ deoxyribose.

dinucleotide   two nucleotides linked by phosphodiester bonds.

gene   a segment of DNA.

genetic code   the amino acid specified by each three-base sequence of mRNA.

genetic engineering   recombinant DNA technology.

α-helix   the backbone of a polypeptide coiled in a right-handed spiral with hydrogen bonding occurring within the helix.

human genome   the total DNA of a human cell.

major groove   the wider and deeper of the two alternating grooves in DNA.

minor groove   the narrower and more shallow of the two alternating grooves in DNA.

nucleic acid   the two kinds of nucleic acid are DNA and RNA.

nucleoside   a heterocyclic base (a purine or a pyrimidine) bonded to the anomeric carbon of a sugar (D-ribose or D-2′-deoxyribose).

nucleotide   a heterocycle attached in the β-position to a phosphorylated ribose or deoxyribose.

oligonucleotide   3 to 10 nucleotides linked by phosphodiester bonds.

polymerase chain reaction (PCR)   a method that amplifies segments of DNA.

polynucleotide   many nucleotides linked by phosphodiester bonds.

primary structure (of a nucleic acid)   the sequence of bases in a nucleic acid.

primary structure (of a protein)   the sequence of amino acids in a protein.

pyrosequencing   a technique used to determine the sequence of bases in a polynucleotide by detecting the identity of each base that adds to a primer.

recombinant DNA   DNA that has been incorporated into a host cell.

replication   the synthesis of identical copies of DNA.

restriction endonuclease   an enzyme that cleaves DNA at a specific base sequence.

restriction fragment   a fragment that is formed when DNA is cleaved by a restriction endonuclease.

ribonucleotide   a nucleotide in which the sugar component is D-ribose.

semiconservative replication   the mode of replication that results in a daughter molecule of DNA having one of the original DNA strands plus a newly synthesized strand.

sense strand (informational strand)   the strand in DNA that is not read during transcription; it has the same sequence of bases as the synthesized mRNA strand (with a U, T difference).

stacking interactions   van der Waals interactions between the mutually induced dipoles of adjacent pairs of bases in DNA.

stop codon   a codon at which protein synthesis is stopped.

template strand (antisense strand)   the strand in DNA that is read during transcription.

transcription   the synthesis of mRNA from a DNA blueprint.

translation   the synthesis of a protein from an mRNA blueprint.

PHOTO CREDITS

Photo credits are listed in order of appearance.

Paul Fleet/Shutterstock; (rt.) National Library of Medicine; (left) A. Barrington Brown/Science Source/Photo Researchers, Inc.; CDC; (top) Professor George O. Poinar/Oregon State University; (bot.) Pioneer Hi-Bred, International, Inc.

Synthetic Polymers

From Chapter 27 of *Organic Chemistry*, Seventh Edition. Paula Yurkanis Bruice. Copyright © 2014 by Pearson Education, Inc.

Synthetic Polymers

boots made of synthetic rubber

P*robably no group of synthetic compounds is more important to modern life than synthetic polymers. Unlike small organic molecules, which are of interest because of their chemical properties, these giant molecules—with molecular weights ranging from thousands to millions—are interesting primarily because of their physical properties that make them useful in everyday life. Some synthetic polymers resemble natural substances, but most are quite different from materials found in nature. Such diverse products as photographic film, compact discs, rugs, food wrap, artificial joints, Super Glue, toys, plastic bottles, weather stripping, automobile body parts, and shoe soles are made of synthetic polymers.*

A **polymer** is a large molecule made by linking together repeating units of small molecules called **monomers.** The process of linking them together is called **polymerization.**

$$n \, \text{M} \xrightarrow{\text{polymerization}} \text{—M—M—M—M—M—M—M—M—M—}$$

monomer   polymer

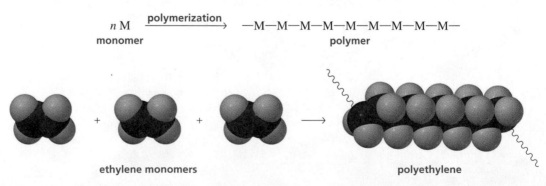

ethylene monomers   polyethylene

Polymers can be divided into two broad groups: **synthetic polymers** and **biopolymers.** Synthetic polymers are synthesized by scientists, whereas biopolymers are synthesized by cells. Examples of biopolymers are DNA—the storage molecule for genetic information; RNA and proteins—the molecules that facilitate biochemical transformations; and polysaccharides—compounds that store energy and also function as structural materials. In this chapter, we will explore synthetic polymers.

Humans first relied on *biopolymers* for clothing, wrapping themselves in animal skins and furs. Later, they learned to spin natural fibers into thread and to weave the thread into cloth. Today, much of our clothing is made of *synthetic polymers* (such as nylon, polyester, and polyacrylonitrile). It has been estimated that if synthetic polymers were not available, all the arable land in the United States would have to be used for the production of cotton and wool for clothing

A **plastic** is a polymer capable of being molded. The first commercial plastic was celluloid. Invented in 1856, it was a mixture of nitrocellulose and camphor. Celluloid was used in the manufacture of billiard balls and piano keys, replacing scarce ivory and providing a reprieve for many elephants. Celluloid was also used for motion picture film until it was replaced by cellulose acetate, a more stable polymer.

The first synthetic fiber was rayon. In 1865, the French silk industry was threatened by an epidemic that killed many silkworms. Louis Pasteur determined the source of the disease, but it was his assistant, Louis Chardonnet, who realized that a substitute for silk would be a desirable commercial item. Chardonnet accidentally discovered the starting material for a synthetic fiber when, while wiping up some spilled nitrocellulose from a table, he noticed long silk-like strands adhering to both the cloth and the table. "Chardonnet silk" was introduced at the Paris Exposition in 1891. It was called rayon because it was so shiny that it appeared to give off rays of light. The "rayon" used today does not contain any nitro groups.

The first synthetic rubber was synthesized by German chemists in 1917 in response to a severe shortage of raw materials as a result of blockading during World War I.

Hermann Staudinger was the first to recognize that the various polymers being produced were not disorderly conglomerates of monomers, but instead were made up of chains of monomers held together by covalent bonds. Today, the synthesis of polymers has grown from a process carried out with little chemical understanding to a sophisticated science in which molecules are engineered to predetermined specifications in order to produce new materials tailored to fit human needs. Examples include Lycra, a fabric with elastic properties, and Dyneema, the strongest fiber commercially available.

Polymer chemistry is part of the larger discipline of **materials science,** which involves the creation of new materials with improved properties to add to the metals, glass, fabrics, and woods we currently have. Polymer chemistry has evolved into a trillion-dollar industry. More than 2.5×10^{13} kilograms of synthetic polymers are produced in the United States each year, and approximately 30,000 polymer patents are currently in force. We can expect scientists to develop many more new materials in the years to come.

1 THERE ARE TWO MAJOR CLASSES OF SYNTHETIC POLYMERS

Synthetic polymers can be divided into two major classes. **Chain-growth polymers,** also known as **addition polymers,** are made by **chain reactions**—the addition of monomers to the end of a growing chain. The end of a growing chain is reactive because it is a radical, a cation, or an anion. Polystyrene—used for hot drink cups, egg cartons, and insulation, among other things—is an example of a chain-growth polymer. Polystyrene is pumped full of air to produce the material used as insulation in house construction.

styrene

polystyrene
a chain-growth polymer

Chain-growth polymers are also called addition polymers.

Chain-growth polymers are made by chain reactions.

1279

Step-growth polymers, also called **condensation polymers,** are made by linking monomers as a result of removing (in most cases) a small molecule, generally water or an alcohol. The monomers have reactive functional groups at each end. Unlike chain-growth polymerization, which requires the individual monomers to add to the end of a growing chain, step-growth polymerization allows any two reactive monomers or oligomers to be linked. Dacron is an example of a step-growth polymer. It was one of the first synthetic polymers to have a medical application—namely, for arterial grafts.

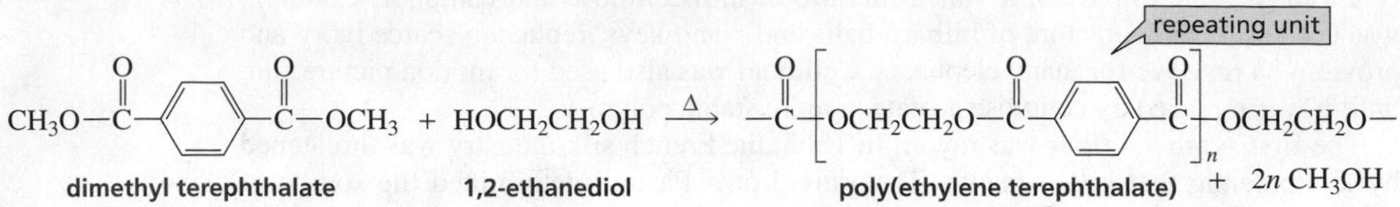

dimethyl terephthalate   1,2-ethanediol   poly(ethylene terephthalate)
Dacron®
a step-growth polymer

$+$ $2n$ CH$_3$OH

2   CHAIN-GROWTH POLYMERS

The monomers used most commonly in chain-growth polymerization are ethylene (ethene) and substituted ethylenes (CH$_2$=CHR). Polymers formed from ethylene or substituted ethylenes are called **vinyl polymers**. Some of the many vinyl polymers synthesized by chain-growth polymerization are listed in Table 1.

Table 1 Some Important Chain-Growth Polymers and Their Uses

Monomer	Repeating unit	Polymer name	Uses
CH$_2$=CH$_2$	—CH$_2$—CH$_2$—	polyethylene	toys, water bottles, grocery bags
CH$_2$=CH—Cl	—CH$_2$—CH—Cl	poly(vinyl chloride)	shampoo bottles, pipe, siding, flooring, clear food packaging
CH$_2$=CH—CH$_3$	—CH$_2$—CH—CH$_3$	polypropylene	molded caps, margarine tubs, indoor/outdoor carpeting, upholstery
CH$_2$=CH 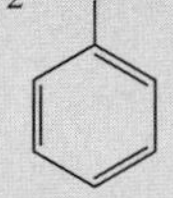 (phenyl)	—CH$_2$—CH— 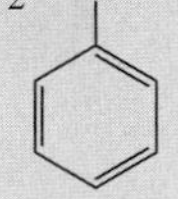 (phenyl)	polystyrene	compact disc jackets, egg cartons, hot drink cups, insulation
CF$_2$=CF$_2$	—CF$_2$—CF$_2$—	poly(tetrafluoroethylene) Teflon	nonstick surfaces, liners, cable insulation
CH$_2$=CH—C≡N	—CH$_2$—CH—C≡N	poly(acrylonitrile) Orlon, Acrilan	rugs, blankets, yarn, apparel, simulated fur
CH$_2$=C—CH$_3$, COCH$_3$, O	—CH$_2$—C—CH$_3$, COCH$_3$, O	poly(methyl methacrylate) Plexiglas, Lucite	lighting fixtures, signs, solar panels, skylights
CH$_2$=CH—OCCH$_3$, O	—CH$_2$—CH—OCCH$_3$, O	poly(vinyl acetate)	latex paints, adhesives

Chain-growth polymerization proceeds by one of three mechanisms: **radical polymerization, cationic polymerization,** or **anionic polymerization.** Each mechanism has three distinct phases: an *initiation step* that starts the polymerization, *propagation steps* that allow the chain to grow, and *termination steps* that stop the growth of the chain. The mechanism for a given chain-growth reaction depends on the structure of the monomer *and* on the initiator used to activate the monomer.

Radical Polymerization

Radical polymerization consists of chain-initiating, chain-propagating, and chain-terminating steps.

For chain-growth polymerization to occur by a radical mechanism, a radical initiator must be added to the monomer to generate the radical that will start the growth of the polymer chain.

MECHANISM FOR RADICAL POLYMERIZATION

initiation steps

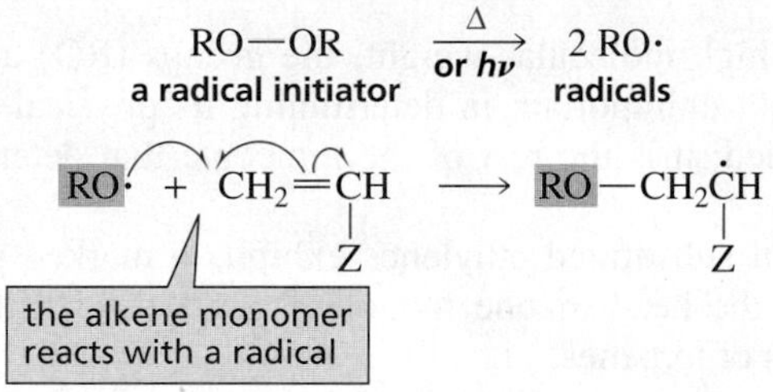

propagation steps

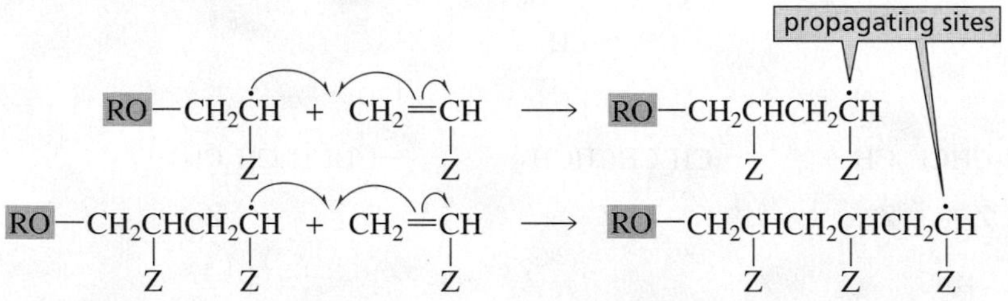

- The initiator breaks homolytically into radicals, and each of those radicals can react with a monomer.

- In the first propagation step, the monomer radical reacts with another monomer, converting it into a radical.

- This radical reacts with another monomer, adding a new subunit to the chain. The unpaired electron is now at the end of the unit most recently added to the chain. This is called the **propagating site.**

The propagation steps are repeated over and over. Hundreds or even thousands of alkene monomers can add one at a time to the growing chain. Eventually, the chain reaction stops because the propagating sites are destroyed in a termination step. Propagating sites are destroyed when

- two chains combine at their propagating sites.

- two chains undergo *disproportionation,* with one chain being oxidized to an alkene and the other being reduced to an alkane as a result of hydrogen atom being transferred from one chain to another.

termination steps

chain combination

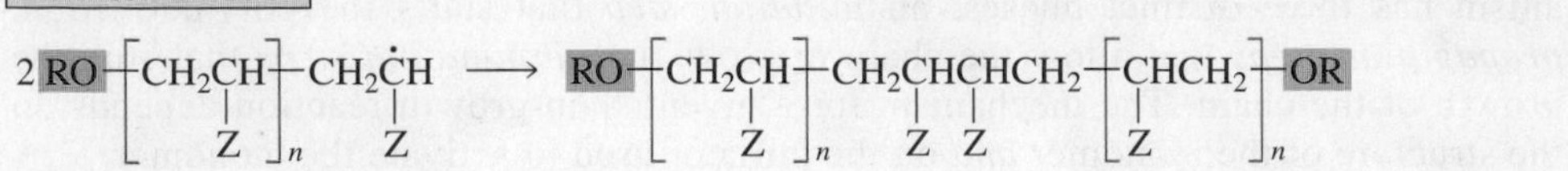

$$2 \; RO \left[CH_2CH \right]_n CH_2\dot{C}H \longrightarrow RO \left[CH_2CH \right]_n CH_2CHCHCH_2 \left[CHCH_2 \right]_n OR$$

disproportionation

$$2 \; RO \left[CH_2CH \right]_n CH_2\dot{C}H \longrightarrow RO \left[CH_2CH \right]_n CH=CH + RO \left[CH_2CH \right]_n CH_2CH_2$$

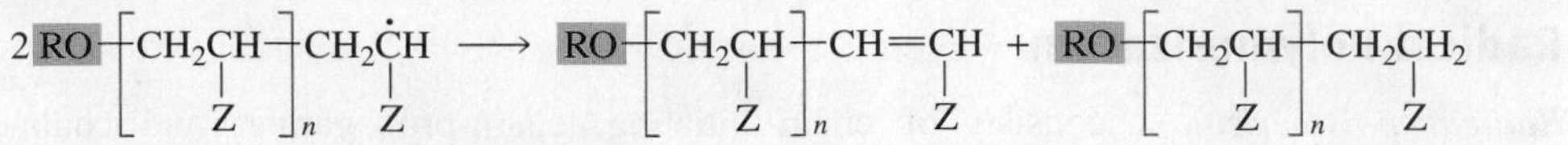

The molecular weight of the polymer can be controlled by a process known as **chain transfer.** In chain transfer, the growing chain reacts with a molecule XY in a manner that allows X· to terminate the chain, leaving behind Y· to initiate a new chain. Molecule XY can be a solvent, a radical initiator, or any molecule with a bond that can readily be cleaved homolytically.

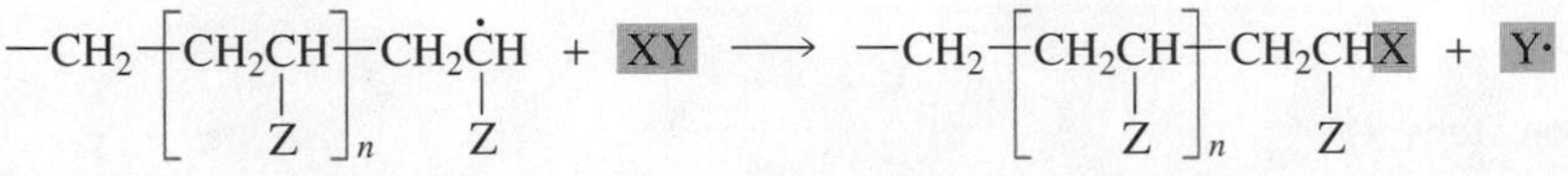

$$-CH_2 \left[CH_2CH \right]_n CH_2\dot{C}H + XY \longrightarrow -CH_2 \left[CH_2CH \right]_n CH_2CHX + Y\cdot$$

As long as the polymer has a high molecular weight, the groups (RO) at the ends of the polymer chains are relatively unimportant in determining its physical properties and are generally not even specified; it is the rest of the molecule that determines the properties of the polymer.

Chain-growth polymerization of substituted ethylenes exhibits a marked preference for **head-to-tail addition,** where the head of one monomer is attached to the tail of another, similar to the biosynthesis of terpenes.

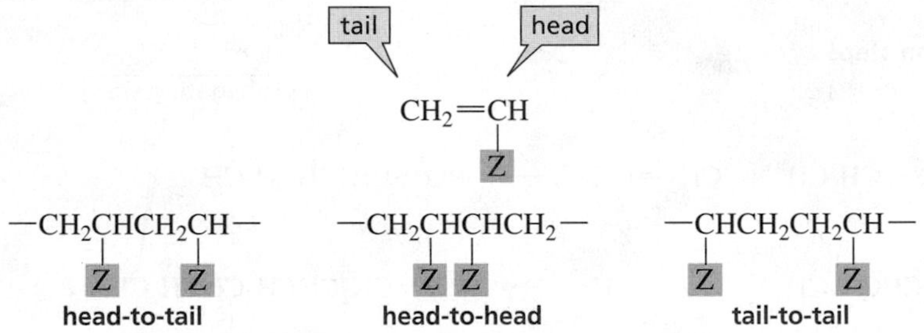

tail   head

$$CH_2=CH$$
$$Z$$

$$-CH_2CHCH_2CH-$$
$$Z \quad\quad Z$$
head-to-tail

$$-CH_2CHCHCH_2-$$
$$Z \; Z$$
head-to-head

$$-CHCH_2CH_2CH-$$
$$Z \quad\quad Z$$
tail-to-tail

Two factors favor head-to-tail addition. First, there is less steric hindrance at the unsubstituted sp^2 carbon of the alkene; therefore, the propagating site attacks it preferentially. (Notice that head-to-tail addition of a substituted ethylene results in a polymer in which every other carbon bears a substituent.)

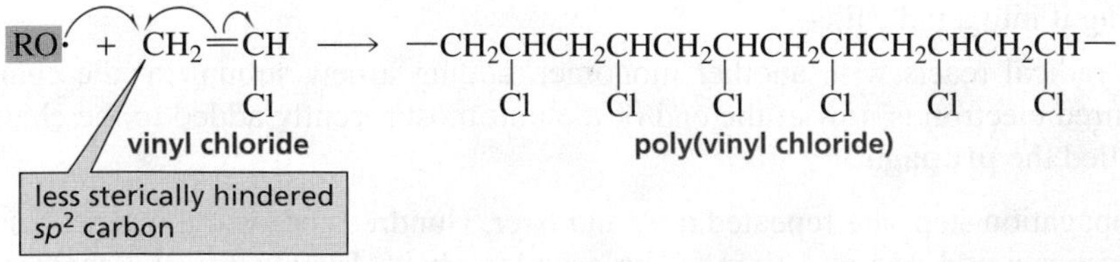

$$RO\cdot + CH_2=CH \longrightarrow -CH_2CHCH_2CHCH_2CHCH_2CHCH_2CHCH_2CH-$$
$$Cl$$
vinyl chloride
$$Cl \quad Cl \quad Cl \quad Cl \quad Cl \quad Cl$$
poly(vinyl chloride)

less sterically hindered sp^2 carbon

Second, radicals formed by addition to the unsubstituted sp^2 carbon can be stabilized by the substituent attached to the other sp^2 carbon. For example, when Z is a phenyl substituent, the benzene ring stabilizes the radical by electron delocalization.

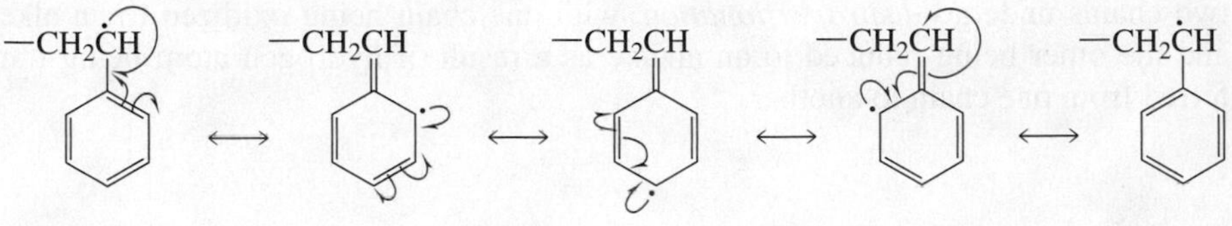

In cases where Z is small (so that steric considerations are less important) and is less able to stabilize the growing end of the chain by electron delocalization, some head-to-head addition and some tail-to-tail addition occurs. This has been observed primarily when Z is fluorine. Abnormal addition, however, has never been found to constitute more than 10% of the overall chain.

Monomers that most readily undergo chain-growth polymerization by a radical mechanism are those in which the substituent Z is a group able to stabilize the growing radical species by electron delocalization (top row in Table 2) or is an electron-withdrawing group (bottom row in Table 2). Radical polymerization of methyl methacrylate forms a clear plastic known as Plexiglas. The largest window in the world, made of a single piece of Plexiglas (54 ft long, 18 ft high, 13 in. thick), houses the sharks and barracudas in the Monterey Bay Aquarium.

Plexiglas windows

Table 2	Examples of Alkenes That Undergo Radical Polymerization	
$CH_2=CH$ (phenyl) styrene	$CH_2=CH$ — $C \equiv N$ acrylonitrile	$CH_2=CH$ — $CH=CH_2$ 1,3-butadiene
$CH_2=CH$ — Cl vinyl chloride	$CH_2=CH$ — $OCCH_3$ (C=O) vinyl acetate	$CH_2=CCH_3$ — $COCH_3$ (C=O) methyl methacrylate

The initiator for radical polymerization can be any compound with a weak bond that readily undergoes homolytic cleavage by heat or ultraviolet light to form radicals sufficiently energetic to convert an alkene into a radical. Several radical initiators are shown in Table 3. In all but one, the weak bond is an oxygen–oxygen bond.

Table 3	Some Radical Initiators

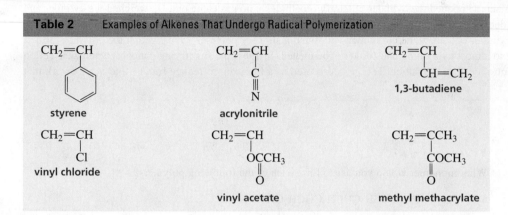

Two factors enter into the choice of radical initiator for a particular chain-growth polymerization. The first is the solubility of the initiator. For example, potassium persulfate is often used if the initiator needs to be soluble in water, whereas an initiator with several carbons is chosen if the initiator needs to be soluble in a nonpolar solvent. The second factor is the temperature at which the polymerization reaction is to be carried out. For example, a *tert*-butoxy radical is relatively stable, so an initiator that forms a *tert*-butoxy radical is used for polymerizations carried out at relatively high temperatures.

Teflon: An Accidental Discovery

Teflon is a polymer of tetrafluoroethylene (Table 1). In 1938, a scientist needed some tetrafluoroethylene for the synthesis of what he hoped would be a new refrigerant. When he opened the cylinder of tetrafluoroethylene, no gas came out. He weighed the cylinder and found that it weighed more than an identical empty cylinder. In fact, it weighed the same as what a cylinder full of tetrafluoroethylene would weigh. Wondering what the cylinder contained, he cut it open and found a slippery polymer. Investigating the polymer further, he found that it was chemically inert to almost everything and could not be melted. In 1961, the first frying pan with a nonstick Teflon coating—"The Happy Pan"— was introduced to the public. Teflon is also used as a lubricant to reduce friction and in pipework that carries corrosive chemicals.

PROBLEM 1♦

What monomer would you use to form each of the following polymers?

a. $-CH_2CHCH_2CHCH_2CHCH_2CHCH_2CH-$
   with Cl substituents

b. polymer with CH_3 groups, $C=O$, and OCH_3 groups

c. $-CF_2CF_2CF_2CF_2CF_2CF_2CF_2CF_2CF_2CF_2-$

PROBLEM 2♦

Which polymer would be more apt to contain abnormal head-to-head linkages: poly(vinyl chloride) or polystyrene?

PROBLEM 3

Draw a segment of polystyrene that contains two head-to-head, two tail-to-tail, and two head-to-tail linkages.

PROBLEM 4

Show the mechanism for the formation of a segment of poly(vinyl chloride) that contains three units of vinyl chloride and is initiated by hydrogen peroxide.

Branching of the Polymer Chain

If the propagating site removes a hydrogen atom from a chain, a branch can grow off the chain at that point. The propagating site can remove a hydrogen atom from a different polymer chain or from the same polymer chain.

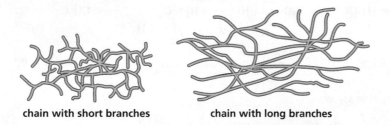

a hydrogen atom is removed from a different chain

$$-CH_2CH_2CH_2\dot{C}H_2 \quad + \quad -CH_2CH_2\overset{\overset{\displaystyle H}{|}}{C}HCH_2CH_2CH_2-$$

$$-CH_2CH_2CH_2CH_2 \quad + \quad -CH_2CH_2\dot{C}HCH_2CH_2CH_2- \quad \xrightarrow{CH_2=CH_2} \quad -CH_2CH_2CHCH_2CH_2CH_2-$$

a hydrogen atom is removed from the same chain

Removing a hydrogen atom from a carbon near the end of a chain leads to short branches, whereas removing a hydrogen atom from a carbon near the middle of a chain results in long branches. Short branches are more likely to be formed than long ones because the ends of the chain are more accessible.

chain with short branches   chain with long branches

Branching greatly affects the physical properties of the polymer. Unbranched chains can pack together more closely than branched chains can. Consequently, linear polyethylene (known as high-density polyethylene) is a relatively hard plastic, used for the production of such things as artificial hip joints; whereas branched polyethylene (low-density polyethylene) is a much more flexible polymer, used for trash bags and dry-cleaning bags.

Branched polymers are more flexible.

Recycling Symbols

When plastics are recycled, the various types must be separated from one another. To aid in the separation, many states require manufacturers to place a recycling symbol on their products to indicate the type of plastic it is. You are probably familiar with these symbols, which are often embossed on the bottom of plastic containers. The symbols consist of three arrows around one of seven numbers; an abbreviation below the symbol indicates the type of polymer from which the container is made. The lower the number in the middle of the symbol, the greater is the ease with which the material can be recycled: 1 (PET) stands for poly(ethylene terephthalate), 2 (HDPE) for high-density polyethylene, 3 (V) for poly(vinyl chloride), 4 (LDPE) for low-density polyethylene, 5 (PP) for polypropylene, 6 (PS) for polystyrene, and 7 for all other plastics.

recycling symbols

PROBLEM 5♦

Polyethylene can be used for the production of beach chairs as well as beach balls. Which of these items is made from more highly branched polyethylene?

PROBLEM 6

Draw a short segment of branched polystyrene.

Cationic Polymerization

In cationic polymerization, the initiator is an electrophile (generally a proton) that adds to the monomer, causing it to become a carbocation. The initiator cannot be an acid such as HCl because Cl^- will be able to react with the carbocation. Thus, the initiator most often used in cationic polymerization is a Lewis acid, such as BF_3, together with a proton-donating Lewis base, such as water. Notice that the reaction follows the rule that governs electrophilic addition reactions—that is, the electrophile (the initiator) adds to the sp^2 carbon bonded to the most hydrogens.

MECHANISM FOR CATIONIC POLYMERIZATION

initiation steps

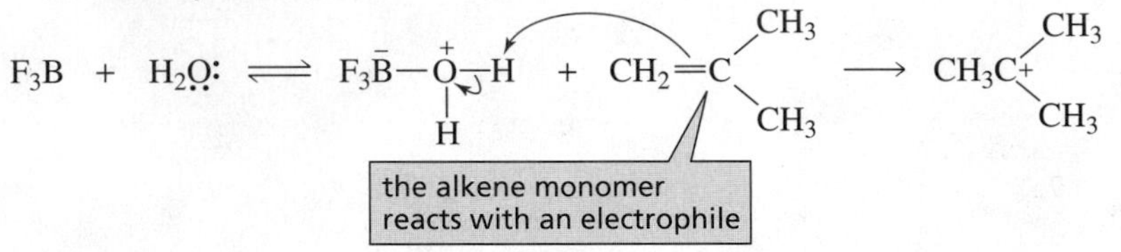

the alkene monomer reacts with an electrophile

propagating steps

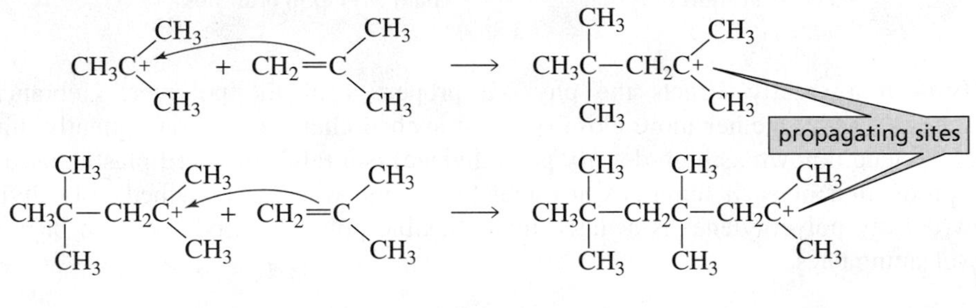

propagating sites

- The cation formed in the initiation step reacts with a second monomer, forming a new cation that reacts in turn with a third monomer. As each subsequent monomer adds to the chain, the new positively charged propagating site is at the end of the most recently added unit.

Cationic polymerization can be terminated by

- loss of a proton;
- addition of a nucleophile to the propagating site;
- a chain-transfer reaction with the solvent (XY).

termination steps

loss of a proton

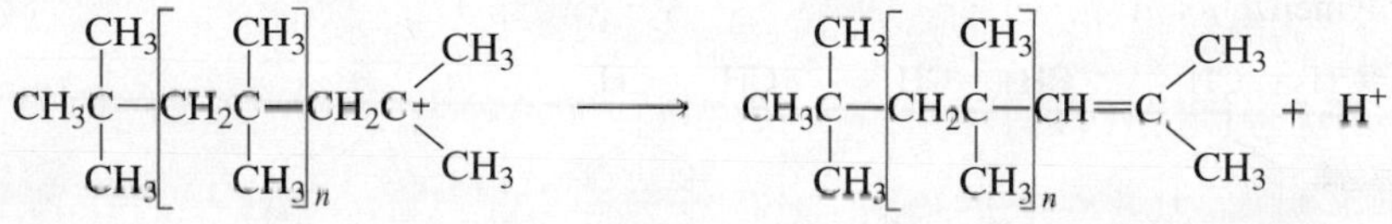

$$CH_3C\!-\!\left[CH_2C\!-\!\right]_n\!CH_2C^+ \longrightarrow CH_3C\!-\!\left[CH_2C\!-\!\right]_n\!CH\!=\!C + H^+$$

(with CH_3 substituents as shown)

reaction with a nucleophile

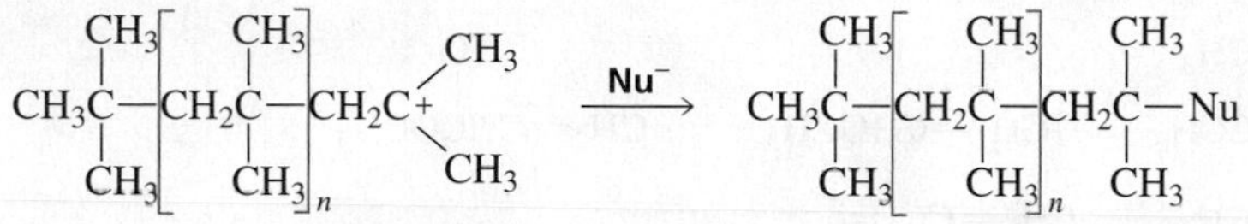

$$CH_3C\!-\!\left[CH_2C\!-\!\right]_n\!CH_2C^+ \xrightarrow{Nu^-} CH_3C\!-\!\left[CH_2C\!-\!\right]_n\!CH_2C\!-\!Nu$$

chain-transfer reaction with the solvent

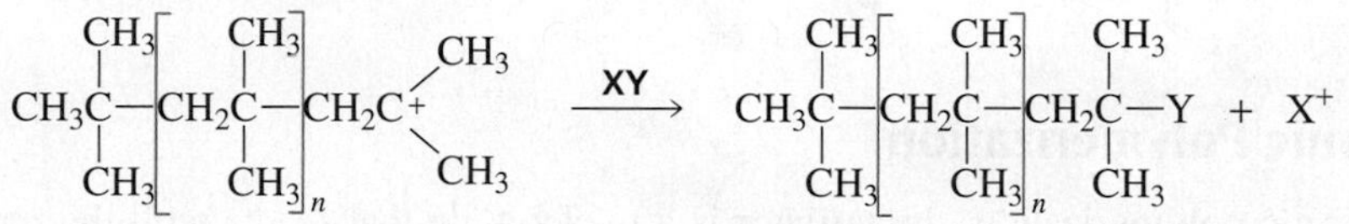

$$CH_3C\!-\!\left[CH_2C\!-\!\right]_n\!CH_2C^+ \xrightarrow{XY} CH_3C\!-\!\left[CH_2C\!-\!\right]_n\!CH_2C\!-\!Y + X^+$$

The carbocation intermediates formed during cationic polymerization, like any other carbocations, can undergo rearrangement by either a 1,2-hydride shift or a 1,2-methyl shift if rearrangement leads to a more stable carbocation. For example, the polymer formed from the cationic polymerization of 3-methyl-1-butene contains both unrearranged and rearranged units.

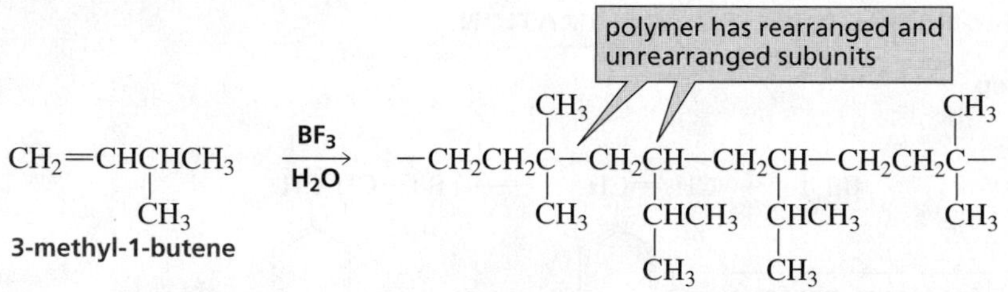

polymer has rearranged and unrearranged subunits

$$CH_2\!=\!CHCHCH_3 \xrightarrow[\text{H}_2\text{O}]{\textbf{BF}_3} -CH_2CH_2C\!-\!CH_2CH\!-\!CH_2CH\!-\!CH_2CH_2C-$$

3-methyl-1-butene

The unrearranged propagating site is a secondary carbocation, whereas the rearranged propagating site, obtained by a 1,2-hydride shift, is a more stable tertiary carbocation. The extent of rearrangement depends on the reaction temperature.

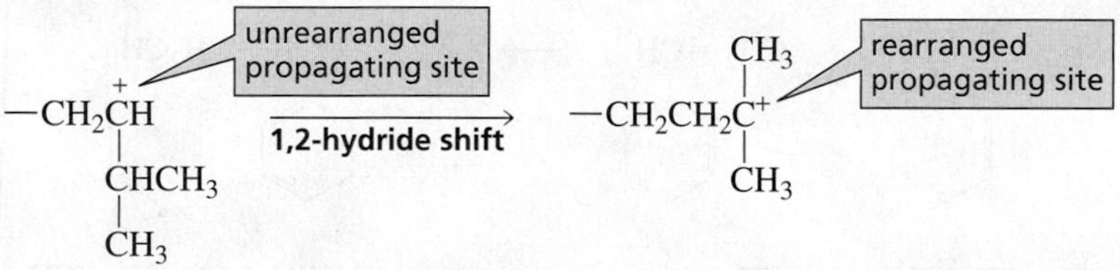

unrearranged propagating site

$$-CH_2\overset{+}{CH} \xrightarrow{\textbf{1,2-hydride shift}} -CH_2CH_2\overset{+}{C}$$

rearranged propagating site

Monomers that are best able to undergo polymerization by a cationic mechanism are those with substituents that can stabilize the positive charge at the propagating site either by hyperconjugation (the first compound in Table 4) or by donating electrons by resonance (the other two compounds in the table).

Table 4	Examples of Alkenes That Undergo Cationic Polymerization	
$CH_2\!=\!CCH_3$ $\vert$ CH_3 **isobutylene**	$CH_2\!=\!CH$ $\vert$ OCH_3 **methyl vinyl ether**	$CH_2\!=\!CH$ **styrene**

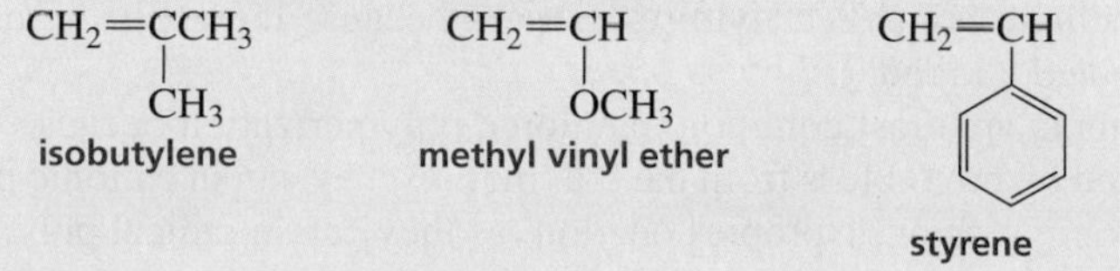

PROBLEM 7♦

List the following groups of monomers in order from most able to least able to undergo cationic polymerization:

a. CH$_2$=CH
(p-NO$_2$-phenyl)   CH$_2$=CH (p-CH$_3$-phenyl)   CH$_2$=CH (p-OCH$_3$-phenyl)

        NO$_2$ CH$_3$ OCH$_3$

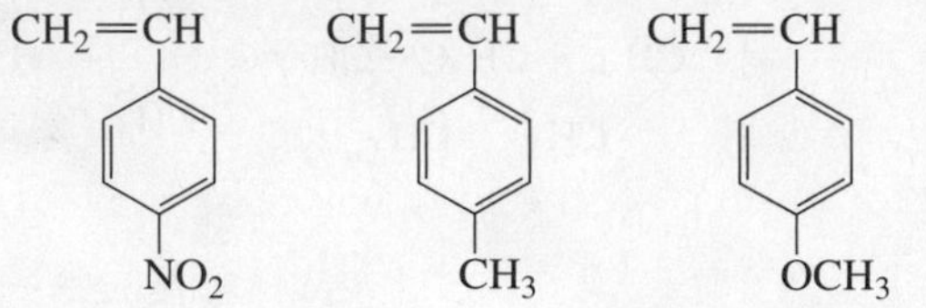

b. CH$_2$=CCH$_3$ (with CH$_3$)   CH$_2$=CHOCH$_3$ CH$_2$=CHCOCH$_3$ (with O double bond)

c. CH$_2$=CH (phenyl)   CH$_2$=CCH$_3$ (phenyl)

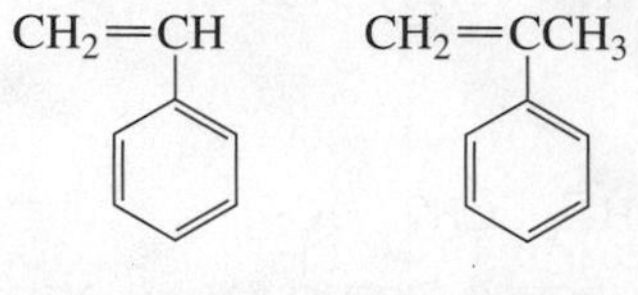

Anionic Polymerization

In anionic polymerization, the initiator is a nucleophile that reacts with the monomer to form a propagating site that is an anion. Nucleophilic attack on an alkene does not occur readily because alkenes are themselves electron rich. Therefore, the initiator must be a very good nucleophile, such as sodium amide or butyllithium, and the alkene must contain a substituent that can withdraw electrons by resonance, which will decrease the electron density of the double bond.

MECHANISM FOR ANIONIC POLYMERIZATION

initiation step

the alkene monomer reacts with a nucleophile

propagation steps

propagating sites

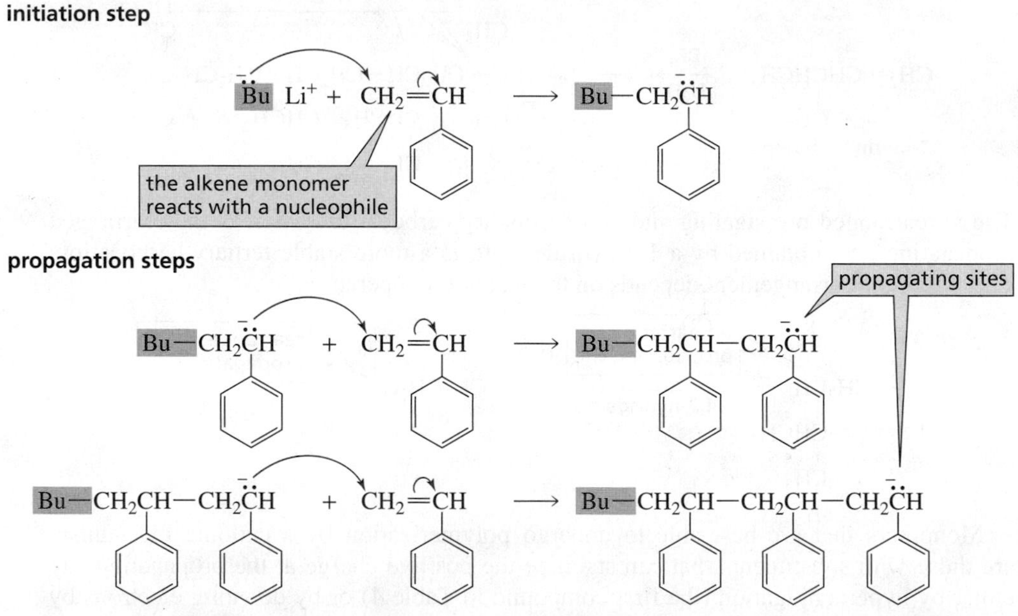

The chain can be terminated by reaction with an impurity in the reaction mixture. If all impurities are rigorously excluded, chain propagation will continue until all the monomer has been consumed. At this point, the propagating site will still be active, so the polymerization reaction will continue if more monomer is added to the system. Such nonterminated chains are called **living polymers** because the chains remain active until they are terminated ("killed").

Living polymers are most common in anionic polymerization because the chains cannot be terminated by proton loss from the polymer (as they can in cationic polymerization) or by chain combination or disproportionation (as they can in radical polymerization).

Alkenes that undergo polymerization by an anionic mechanism are those that can stabilize the negatively charged propagating site by resonance electron withdrawal (Table 5).

Table 5	Examples of Alkenes That Undergo Anionic Polymerization

$$CH_2{=}CH \qquad\qquad CH_2{=}CCH_3 \qquad\qquad CH_2{=}CH$$
$$\overset{\displaystyle |}{\underset{\displaystyle O}{CH}} \qquad\qquad \overset{\displaystyle |}{\underset{\displaystyle O}{COCH_3}}$$

acrolein   methyl methacrylate   styrene

Super Glue is a polymer of methyl α-cyanoacrylate. Because the monomer has two electron-withdrawing groups, it requires only a moderately good nucleophile to initiate anionic polymerization, such as surface-absorbed water. You may well have experienced this reaction if you have ever spilled a drop of Super Glue on your fingers. A nucleophilic group on the surface of the skin initiates the polymerization reaction, with the result that two fingers can become firmly glued together. The ability to form covalent bonds with groups on the surfaces of the objects to be glued together is what gives Super Glue its amazing strength. Polymers similar to Super Glue—namely, butyl, isobutyl, or octyl esters rather than methyl esters—are used by surgeons to close wounds.

methyl α-cyanoacrylate   Super Glue®

PROBLEM 8♦

List the following groups of monomers in order from most able to least able to undergo anionic polymerization:

a.

$$CH_2{=}CH \qquad CH_2{=}CH \qquad CH_2{=}CH$$

(with para-substituted benzene rings bearing) NO_2 CH_3 OCH_3

b. $CH_2{=}CHCH_3 \qquad CH_2{=}CHCl \qquad CH_2{=}CHC{\equiv}N$

What Determines the Mechanism?

The substituent attached to the alkene determines the best mechanism for chain-growth polymerization. Alkenes with substituents that can stabilize radicals readily undergo radical polymerization, alkenes with electron-donating substituents that can stabilize cations undergo cationic polymerization, and alkenes with electron-withdrawing substituents that can stabilize anions undergo anionic polymerizations. Cationic polymerization is the least common of the three chain-growth mechanisms.

Some alkenes undergo polymerization by more than one mechanism. For example, styrene can undergo polymerization by radical, cationic, and anionic mechanisms because the phenyl group can stabilize benzylic radicals, benzylic cations, and benzylic anions. The mechanism followed for its polymerization depends on the nature of the initiator chosen to start the reaction.

PROBLEM 9♦

Why does methyl methacrylate not undergo cationic polymerization?

Ring-Opening Polymerizations

Although ethylene and substituted ethylenes are the monomers most commonly used for the synthesis of chain-growth polymers, other compounds can polymerize as well. For example, epoxides can undergo chain-growth polymerization.

If the initiator is a nucleophile, polymerization occurs by an anionic mechanism. Note that the nucleophile attacks the less sterically hindered carbon of the epoxide.

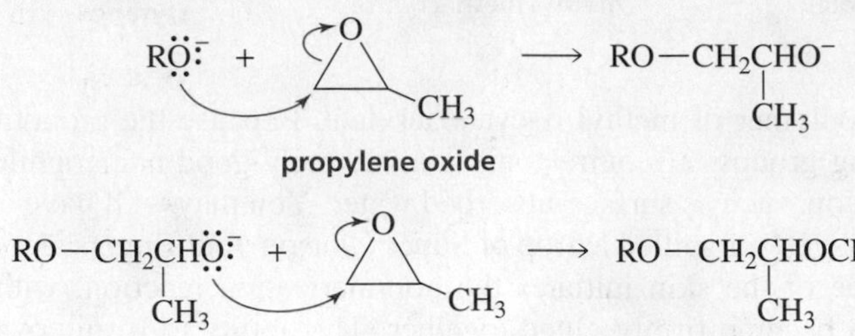

propylene oxide

If the initiator is a Lewis acid or a proton-donating acid, epoxides are polymerized by a cationic mechanism. Polymerization reactions that involve ring-opening reactions are called **ring-opening polymerizations.** Notice that under acidic conditions the nucleophile attacks the more substituted carbon of the epoxide.

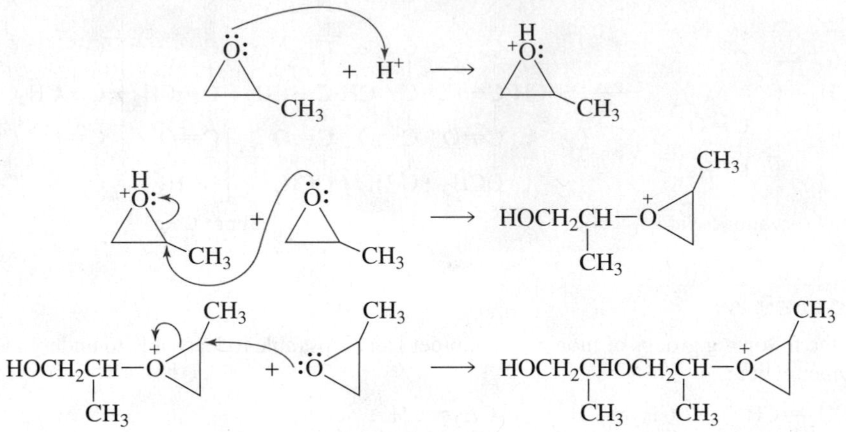

PROBLEM 10

Explain why, when propylene oxide undergoes anionic polymerization, nucleophilic attack occurs at the less substituted carbon of the epoxide, but when it undergoes cationic polymerization, nucleophilic attack occurs at the more substituted carbon.

PROBLEM 11

Describe the polymerization of 2,2-dimethyloxirane by

a. an anionic mechanism.   **b.** a cationic mechanism.

PROBLEM 12♦

Which monomer and which type of initiator would you use to synthesize each of the following polymers?

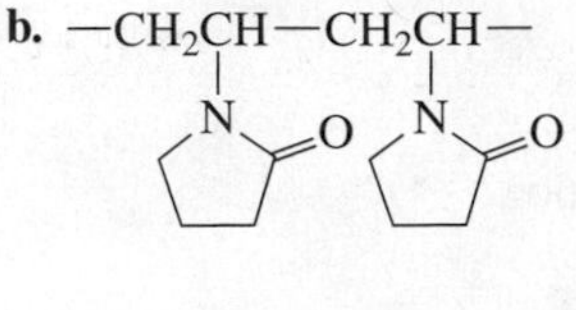

PROBLEM 13◆

Draw a short segment of the polymer formed from cationic polymerization of 3,3-dimethyloxa-cyclobutane.

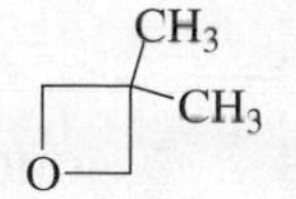

3,3-dimethyloxacyclobutane

3   STEREOCHEMISTRY OF POLYMERIZATION • ZIEGLER–NATTA CATALYSTS

Polymers formed from monosubstituted ethylenes can exist in three configurations: isotactic, syndiotactic, and atactic. An **isotactic polymer** has all of its substituents on the same side of the fully extended carbon chain. (*Iso* and *taxis* are Greek for "the same" and "order," respectively.) In a **syndiotactic polymer** (*syndio* means "alternating"), the substituents regularly alternate on both sides of the carbon chain. The substituents in an **atactic polymer** are randomly oriented.

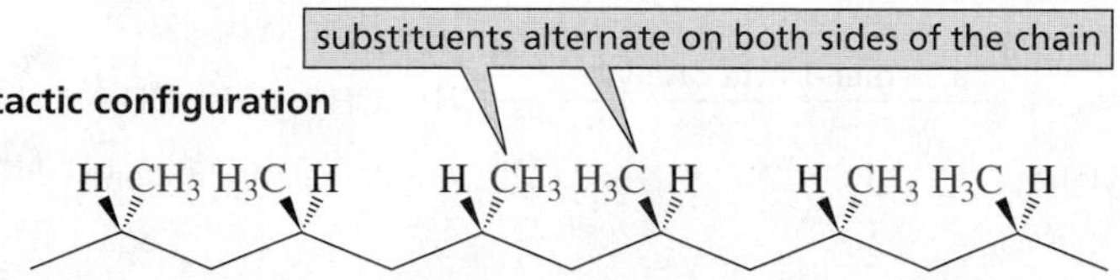

The configuration of the polymer affects its physical properties. Polymers in the isotactic or syndiotactic configuration are more likely to be crystalline solids (Section 8) because positioning the substituents in a regular order allows for a more regular packing arrangement. Polymers in the atactic configuration are more disordered and cannot pack together as well, so these polymers are less rigid and therefore softer.

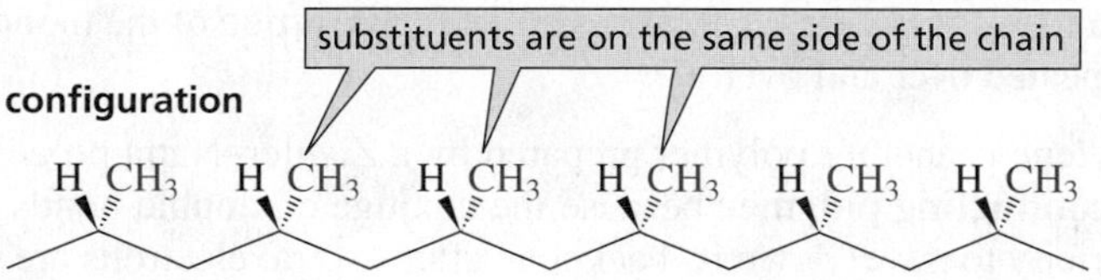

The configuration of the polymer depends on the mechanism by which polymerization occurs. In general, radical polymerization leads primarily to branched polymers in the atactic configuration. Anionic polymerization can produce polymers with the most stereoregularity. The percentage of chains in the isotactic or syndiotactic configuration increases as the polymerization temperature decreases and the solvent polarity decreases.

In 1953, Karl Ziegler and Giulio Natta found that the structure of a polymer could be controlled if the growing end of the chain and the incoming monomer were coordinated with an aluminum–titanium initiator. These initiators are now called **Ziegler–Natta catalysts.**

Long, unbranched polymers with either the isotactic or the syndiotactic configuration can be prepared using Ziegler–Natta catalysts. Whether the chain is isotactic or syndiotactic depends on the particular catalyst employed. These catalysts revolutionized the field of polymer chemistry because they allow the synthesis of stronger and stiffer polymers that have greater resistance to cracking and heat. High-density polyethylene is prepared using a Ziegler–Natta process.

A proposed mechanism for Ziegler–Natta-catalyzed polymerization of a substituted ethylene is shown here.

MECHANISM FOR ZIEGLER–NATTA–CATALYZED POLYMERIZATION OF A SUBSTITUTED ETHYLENE

- The monomer forms a complex with titanium at an open coordination site—that is, a site available to accept electrons.
- The coordinated alkene is inserted between the titanium and the growing polymer, thereby lengthening the polymer chain.
- Because a new coordination site opens up during insertion of the monomer, the process can be repeated over and over.

Polyacetylene is another polymer prepared by a Ziegler–Natta process. It can be converted to a **conducting polymer** because the conjugated double bonds in polyacetylene enable electricity to travel down its backbone after several electrons are removed from or added to the backbone.

$$HC\equiv CH \xrightarrow{\text{a Ziegler–Natta catalyst}} -CH=CH-\!\!\left[CH=CH\right]_n\!\!-CH=CH-$$

acetylene   polyacetylene

4 POLYMERIZATION OF DIENES • THE MANUFACTURE OF RUBBER

When the bark of a rubber tree is cut, a sticky white liquid oozes out. This same liquid is found inside the stalks of dandelions and milkweed. In fact, more than 400 plants produce this substance. The sticky material is *latex,* a suspension of rubber particles in water. Its biological function is to protect the plant after an injury by covering the wound like a bandage.

Natural rubber is a polymer of 2-methyl-1,3-butadiene, also called isoprene. On average, a molecule of rubber contains 5000 isoprene units. As in the case of other naturally occurring terpenes, the five-carbon compound actually used in the biosynthesis of rubber is isopentenylpyrophosphate.

All the double bonds in natural rubber have the Z configuration. Rubber is a waterproof material because its tangled hydrocarbon chains have no affinity for water. Charles Macintosh was the first to use rubber as a waterproof coating for raincoats.

isoprene units   (Z)-poly(2-methyl-1,3-butadiene) natural rubber

Gutta-percha (from the Malaysian words *getah,* meaning "gum," and *percha,* meaning "tree") is a naturally occurring isomer of rubber in which all the double bonds have the

latex being collected from a rubber tree

E configuration. Like rubber, gutta-percha is exuded by certain trees, but it is much less common. It is also harder and more brittle than rubber. Gutta-percha is the filling material that dentists use in root canals. At one time, it was used for the casing of golf balls, but it became brittle in cold weather and tended to split on impact.

PROBLEM 14

Draw a short segment of gutta-percha.

By mimicking nature, scientists have learned to make synthetic rubbers with properties tailored to meet human needs. These materials have some of the properties of natural rubber, including being waterproof and elastic, but they have some improved properties as well; they are tougher, more flexible, and more durable than natural rubber.

Synthetic rubbers have been made by radical polymerization of dienes other than isoprene in an emulsion of water and a detergent. The detergent forms micelles, and the monomers dissolve in the micelles, where polymerization takes place. Therefore, the monomers can find each other easily and termination reactions are less likely.

One synthetic rubber, in which all the double bonds are cis, is formed by polymerizing 1,3-butadiene. Two reasons cause 1,4-polymerization to predominate over 1,2-polymerization. First, there is less steric hindrance at the 4-position than at the 2-position. Second, the product of 1,4-polymerization is more stable since it has a substituent attached to each of the sp^2 carbons, whereas the product of 1,2-polymerization has a substituent attached to only one of the sp^2 carbons.

1,3-butadiene monomers → **(Z)-poly(1,3-butadiene)**
a synthetic rubber

Neoprene is a synthetic rubber made by polymerizing 2-chloro-1,3-butadiene. It is used to make wet suits, shoe soles, tires, hoses, and coated fabrics.

$$CH_2{=}\overset{\displaystyle Cl}{C}CH{=}CH_2 \longrightarrow$$

2-chloro-1,3-butadiene
chloroprene

neoprene

A problem common to both natural and most synthetic rubbers is that the polymers are soft and sticky. However, they can be hardened by *vulcanization*. Charles Goodyear discovered this process while looking for ways to improve the properties of rubber. He accidentally spilled a mixture of rubber and sulfur on a hot stove. To his surprise, the mixture became hard but flexible. He called the heating of rubber with sulfur vulcanization, after Vulcan, the Roman god of fire.

Heating rubber with sulfur causes **cross-linking** of the separate polymer chains through disulfide bonds (Figure 1). Thus, the vulcanized

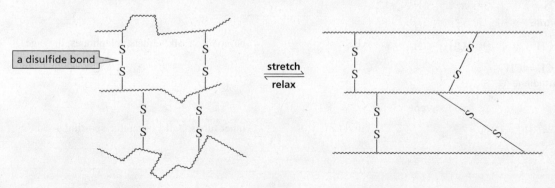

a disulfide bond

stretch ⇌ relax

◀ **Figure 1**

The rigidity of rubber is increased by cross-linking the polymer chains with disulfide bonds. When rubber is stretched, the randomly coiled chains straighten out and orient themselves along the direction of the stretch.

chains are covalently bonded to each other in one giant molecule. Because the chains have double bonds, they have bends and kinks that allow them to stretch. When the rubber is stretched, the chains straighten out in the direction of the pull. The cross-linking prevents rubber from being torn when it is stretched; moreover, the cross-links provide a reference framework for the material to return to when the stretching force is removed.

The physical properties of rubber can be controlled by regulating the amount of sulfur used in vulcanization. Rubber made with 1%–3% sulfur is soft and stretchy and is used to make rubber bands. Rubber made with 3%–10% sulfur is more rigid and is used in the manufacture of tires. Goodyear's name can be found on many tires sold today. The story of rubber is an example of a scientist taking a natural material and finding ways to improve its properties for useful applications.

> **The greater the degree of cross-linking, the more rigid is the polymer.**

PROBLEM 15

Draw a segment of the polymer that would be formed from 1,2-polymerization of 1,3-butadiene.

---

5   COPOLYMERS

Homopolymers are polymers that are formed from only one type of monomer. Often, two or more different monomers are used to form a **copolymer.** Increasing the number of different monomers used to form a copolymer dramatically increases the number of different copolymers that can be formed. Even if only two kinds of monomers are used, copolymers with very different properties can be prepared by varying the amounts of each monomer. Both chain-growth polymers and step-growth polymers can be copolymers. Many of the synthetic polymers used today are copolymers. Table 6 shows some common copolymers and the monomers from which they are synthesized.

There are four types of copolymers. In an **alternating copolymer,** the two monomers alternate. A **block copolymer** consists of blocks of each kind of monomer. A **random**

Table 6   Some Examples of Copolymers and Their Uses

Monomers	Copolymer name	Uses

$CH_2\!=\!CH$ $+$ $CH_2\!=\!CCl$
     $\mid$ $\mid$
     Cl   Cl
 vinyl chloride   **vinylidene chloride**
	Saran	film for wrapping food

$CH_2\!=\!CH$ $+$ $CH_2\!=\!CH$
 (phenyl ring) $\mid$
                   C
                   $\equiv$
                   N
 styrene   **acrylonitrile**
	SAN	dishwasher-safe objects, vacuum cleaner parts

$CH_2\!=\!CH$ $+$ $CH_2\!=\!CH$ $+$ $CH_2\!=\!CH$
    $\mid$ $\mid$
    C $CH\!=\!CH_2$
    $\equiv$ **1,3-butadiene**
    N
 acrylonitrile   **styrene**
	ABS	bumpers, crash helmets, telephones, luggage

$CH_2\!=\!CCH_3$ $+$ $CH_2\!=\!CHC\!=\!CH_2$
      $\mid$ $\mid$
     CH_3 CH_3
 isobutylene   **isoprene**
	butyl rubber	inner tubes, balls, inflatable sporting goods

---

copolymer has a random distribution of monomers. A **graft copolymer** contains branches derived from one monomer grafted onto a backbone derived from another monomer. These structural differences extend the range of physical properties available to the scientist designing the copolymer.

an alternating copolymer	ABABABABABABABABABABABA
a block copolymer	AAAAABBBBBAAAAABBBBBAAA
a random copolymer	AABABABBABAABBABABBAAAB
a graft copolymer	AAAAAAAAAAAAAAAAAAAAAAAAA

```
a graft copolymer    AAAAAAAAAAAAAAAAAAAAAAAAA
                     B         B         B
                     B         B         B
                     B         B         B
                     B         B         B
                     B         B         B
                     B         B         B
```

Nanocontainers

Scientists have synthesized block copolymers that form micelles. These spherical copolymers are currently being investigated for their possible use as nanocontainers (10 to 100 nanometers in diameter) for the delivery of non-water-soluble drugs to target cells. This strategy would allow a higher concentration of a drug to reach a cell than would occur in the natural aqueous milieu. In addition, targeting the drug to the required cells would lower the required dosage of the drug.

6   STEP-GROWTH POLYMERS

Step-growth polymers are formed by the intermolecular reaction of molecules with a functional group at each end. When the functional groups react, in most cases a small molecule such as H_2O, alcohol, or HCl is lost. This is why these polymers are also called *condensation polymers*.

A step-growth polymer can be formed by the reaction of a single bifunctional compound with two different functional groups, A and B. Functional group A of one molecule reacts with functional group B of another molecule to form the monomer (A—X—B) that undergoes polymerization.

$$A—B \quad A—B \quad \longrightarrow \quad A—\boxed{X}—B$$

A step-growth polymer can also be formed by the reaction of two different bifunctional compounds. One contains two A functional groups and the other contains two B functional groups. Functional group A reacts with functional group B to form the monomer (A—X—B) that undergoes polymerization.

$$A—A \quad B—B \quad \longrightarrow \quad A—\boxed{X}—B$$

Because the growing polymer will have the structure A—[polymer]—B, it can cyclize (an intramolecular reaction) instead of adding a new monomer (an intermolecular reaction), thereby terminating polymerization. An intramolecular reaction can be minimized by using a high concentration of the monomer. Because large rings are harder to form than smaller ones, once the polymer chain has more than about 15 atoms, the tendency for cyclization decreases.

The formation of step-growth polymers, unlike the formation of chain-growth polymers, does not occur through chain reactions. Any two monomers (or short chains) can react. The progress of a typical step-growth polymerization is shown schematically in Figure 2. When the reaction is 50% complete (12 bonds have formed between

Step-growth polymers are made by combining molecules with reactive groups at each end.

1295

25 monomers), the reaction products are primarily dimers and trimers. Even at 75% completion, no long chains have been formed. This means that if step-growth polymerization is to lead to long-chain polymers, very high yields must be achieved. The reactions involved in step-growth polymerizations are relatively simple (ester and amide formation). However, polymer chemists expend a great deal of effort to arrive at synthetic and processing methods that will result in high-molecular-weight polymers.

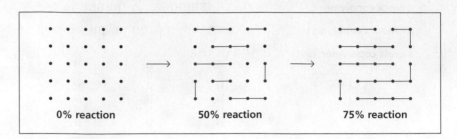

▲ **Figure 2**
Progress of a step-growth polymerization.

7   CLASSES OF STEP-GROWTH POLYMERS

Polyamides

Nylon 6 is an example of a step-growth polymer formed from a monomer with two different functional groups. The carboxylic acid group of one monomer reacts with the amino group of another monomer to form an amide. Thus, nylon is a **polyamide.** This particular nylon is called nylon 6 because it is formed from the polymerization of 6-aminohexanoic acid, a compound that contains six carbons.

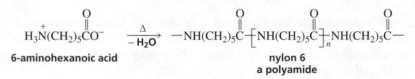

The starting material for the commercial synthesis of nylon 6 is ε-caprolactam. The lactam is opened by a base.

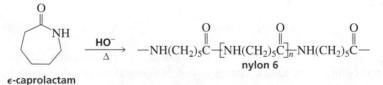

Nylon is pulled from a beaker of adipoyl chloride and 1,6-hexanediamine.

A related polyamide, nylon 66, is an example of a step-growth polymer formed by two different bifunctional monomers—adipoyl chloride and 1,6-hexanediamine. It is called nylon 66 because the two starting materials each have 6 carbons.

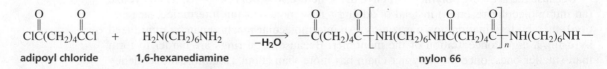

Nylon first found wide use in textiles and carpets. Because it is resistant to stress, it is also used in such applications as mountaineering ropes, tire cords, and fishing lines, and as a substitute for metal in bearings and gears. The usefulness of nylon precipitated a search for more new "super fibers" with super strength and super heat resistance.

PROBLEM 16♦

a. Draw a short segment of nylon 4.
b. Draw a short segment of nylon 44.

PROBLEM 17

Write an equation that explains what will happen if a scientist working in the laboratory spills sulfuric acid on her nylon 66 hose.

PROBLEM 18

Propose a mechanism for the base-catalyzed polymerization of ε-caprolactam.

nylon rope

One super fiber is Kevlar, an aromatic polyamide. It is a polymer of 1,4-benzene-dicarboxylic acid and 1,4-diaminobenzene. Aromatic polyamides are called **aramides.**

The incorporation of aromatic rings into polymers results in polymers with great physical strength. Kevlar is five times stronger than steel on an equal weight basis. Army helmets are made of Kevlar, which is also used for lightweight bullet-resistant vests, automobile parts, high-performance skis, the ropes used on the Mars Pathfinder, and high-performance sails used in the America's Cup. Because it is stable at very high temperatures, it is used in the protective clothing worn by firefighters.

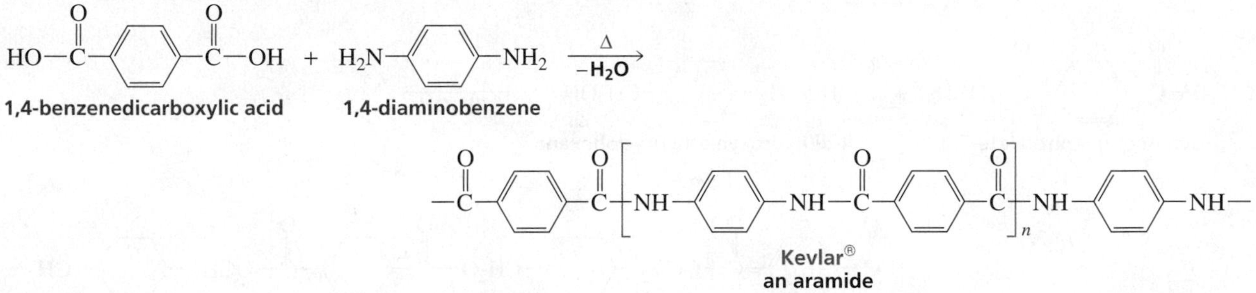

1,4-benzenedicarboxylic acid   **1,4-diaminobenzene**

Kevlar®
an aramide

Kevlar owes its strength to the way in which the individual polymer chains interact with each other. The chains are hydrogen bonded, forming a sheet-like structure.

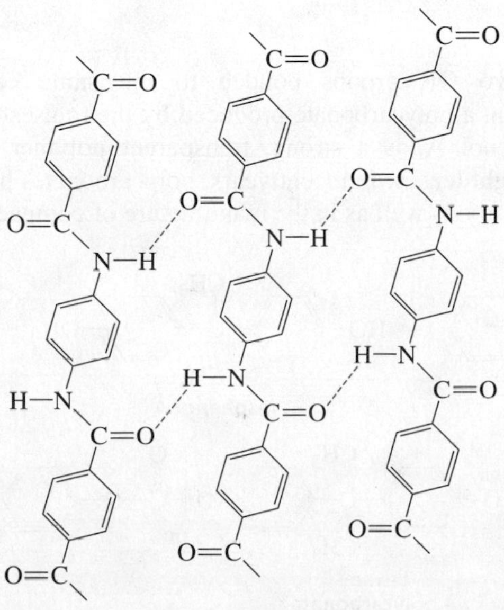

Polyesters

Dacron is the most common of the group of step-growth polymers known as **polyesters**—polymers containing many ester groups. Polyesters are used for clothing and are responsible for the wrinkle-resistant behavior of many fabrics. Dacron is made by the transesterification of dimethyl terephthalate with ethylene glycol. The durability and moisture resistance of this polymer contribute to its "wash-and-wear" characteristics. Because PET is lightweight, it is also used for transparent soft drink bottles.

dimethyl terephthalate + **1,2-ethanediol ethylene glycol** $\xrightarrow[-CH_3OH]{\Delta}$ **poly(ethylene terephthalate) PET Dacron® a polyester**

Mylar balloons

Poly(ethylene terephthlate) can also be processed to form a film known as Mylar. Mylar is tear resistant and, when processed, has a tensile strength nearly as great as that of steel. It is used in the manufacture of magnetic recording tape and sails. Aluminized Mylar was used to make the Echo satellite that was put into orbit around Earth as a giant reflector of microwave signals.

Kodel polyester is formed by the transesterification of dimethyl terephthalate with 1,4-di(hydroxymethyl)cyclohexane. The stiff polyester chain causes the fiber to have a harsh feel that can be softened by blending it with wool or cotton.

dimethyl terephthalate + **1,4-di(hydroxymethyl)cyclohexane (cis and trans)** $\xrightarrow[-CH_3OH]{\Delta}$

Kodel®

PROBLEM 19

What happens to polyester slacks if aqueous NaOH is spilled on them?

Lexan lens in an automobile

Polyesters with two OR groups bonded to the same carbon are known as **polycarbonates.** Lexan, a polycarbonate produced by the transesterification of diphenyl carbonate with bisphenol A, is a strong, transparent polymer used for bullet-proof windows and traffic-light lenses. In recent years, polycarbonates have become important in the automobile industry as well as in the manufacture of compact discs.

diphenyl carbonate + **bisphenol A** $\xrightarrow[-OH]{\Delta}$

Lexan® a polycarbonate

Epoxy Resins

Epoxy resins are the strongest adhesives known; they are extensively cross-linked systems. They can adhere to almost any surface and are resistant to solvents and to extremes of temperature. Epoxy cement is purchased as a kit consisting of a low-molecular-weight *prepolymer* (the most common is a copolymer of bisphenol A and epichlorohydrin) and a *hardener* that react when mixed to form a cross-linked polymer.

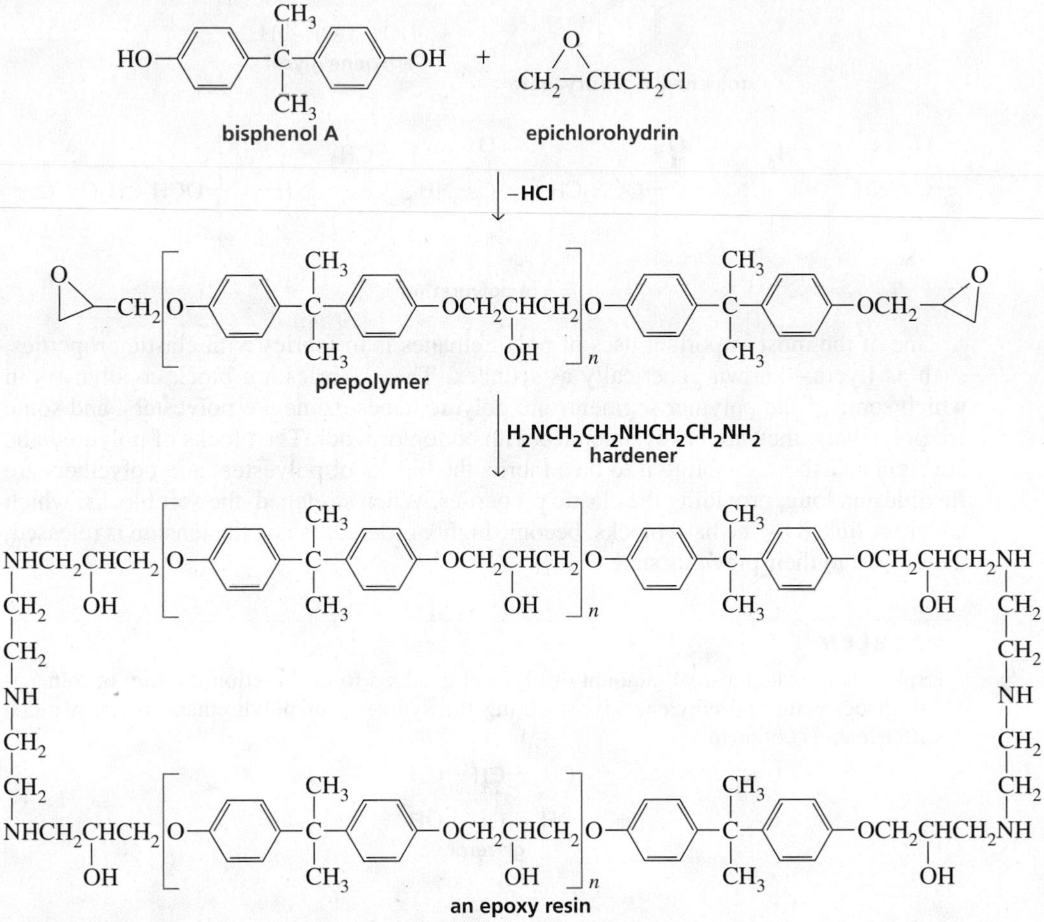

bisphenol A   epichlorohydrin

$-HCl$

prepolymer

$H_2NCH_2CH_2NHCH_2CH_2NH_2$
hardener

an epoxy resin

PROBLEM 20

a. Propose a mechanism for the formation of the prepolymer formed by bisphenol A and epichlorohydrin.

b. Propose a mechanism for the reaction of the prepolymer with the hardener.

Polyurethanes

A **urethane**—also called a carbamate—is a compound that has an OR group and an NHR group bonded to the same carbonyl carbon. Urethanes can be prepared by treating an isocyanate with an alcohol, in the presence of a catalyst such as a tertiary amine.

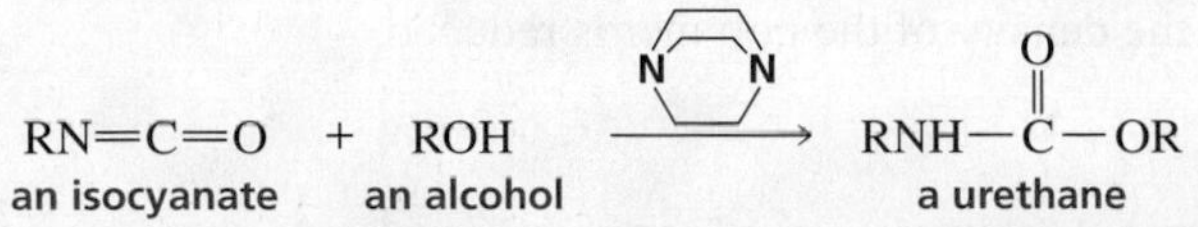

$RN=C=O$ + ROH $\longrightarrow$

an isocyanate   an alcohol   a urethane

One of the most common **polyurethanes**—polymers that contain urethane groups—is prepared by the polymerization of toluene-2,6-diisocyanate and ethylene glycol. If the reaction is carried out in the presence of a blowing agent, the product is a polyurethane foam. Blowing agents are gases such as nitrogen or carbon dioxide. At one time, chlorofluorocarbons—low-boiling liquids that vaporize on heating—were used, but they have been banned for environmental reasons. Polyurethane foams are used for furniture stuffing, bedding, carpet backings, and insulation. Notice that polyurethanes prepared from diisocyanates and diols are the only step-growth polymers we have seen in which a small molecule is *not* lost during polymerization.

$$O=C=N \overset{CH_3}{\bigcirc} N=C=O \quad + \quad HOCH_2CH_2OH \longrightarrow$$

toluene-2,6-diisocyanate   **ethylene glycol**

a polyurethane

One of the most important uses of polyurethanes is in fabrics with elastic properties, such as Lycra—known generically as spandex. These fabrics are block copolymers in which some of the polymer segments are polyurethanes, some are polyesters, and some are polyethers; they are always blended with cotton or wool. The blocks of polyurethane are rigid and short, enabling it to be a fabric; the blocks of polyesters and polyethers are flexible and long, providing the elastic properties. When stretched, the soft blocks, which are cross-linked by the hard blocks, become highly ordered. When the tension is released, they revert to their previous state.

PROBLEM 21

Explain why, when a small amount of glycerol is added to the reaction mixture of toluene-2,6-diisocyanate and ethylene glycol during the synthesis of polyurethane foam, a much stiffer foam is obtained.

$$CH_2-CH-CH_2$$
$$\overset{|}{OH} \quad \overset{|}{OH} \quad \overset{|}{OH}$$
glycerol

---

8   PHYSICAL PROPERTIES OF POLYMERS

Polymers can be classified according to the physical properties they acquire as a result of the way in which their individual chains are arranged. The individual chains of a polymer are held together by van der Waals forces. Because these forces operate only at small distances, they are strongest if the polymer chains can line up in an ordered, closely packed array.

Regions of the polymer in which the chains are highly ordered with respect to one another are called **crystallites** (Figure 3). Between the crystallites are amorphous, noncrystalline regions in which the chains are randomly oriented. The more crystalline (the more ordered) the polymer is, the denser, harder, and more resistant it is to heat (Table 7). If the polymer chains possess substituents or have branches that prevent them from packing closely together, the density of the polymer is reduced.

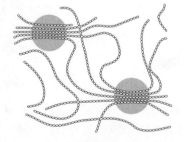

▲ **Figure 3**

The polymer chains are highly ordered in the crystallites (indicated by circles). Between the crystallites are noncrystalline (amorphous) regions in which the polymer chains are randomly oriented.

Table 7 Properties of Polyethylene as a Function of Crystallinity				
Crystallinity (%)	55	62	70	77	85
Density (g/cm^3)	0.92	0.93	0.94	0.95	0.96
Melting point (°C)	109	116	125	130	133

Thermoplastic Polymers

Thermoplastic polymers have both ordered crystalline regions and amorphous noncrystalline regions. These plastics are hard at room temperature but soft enough to be molded when heated, because the individual chains can slip past one another at elevated temperatures. Thermoplastic polymers are the plastics we encounter most often in our daily lives—in combs, toys, and switch plates. They are the plastics that are easily cracked.

Thermosetting Polymers

Very rigid materials can be obtained if the polymer chains are cross-linked. The greater the degree of cross-linking, the more rigid is the polymer. Such cross-linked polymers are called **thermosetting polymers.** After they are hardened, they cannot be remelted by heating because the cross-links are covalent bonds, not van der Waals forces. Cross-linking reduces the mobility of the polymer chains, causing the polymer to be relatively brittle. Because thermosetting polymers do not have the wide range of properties characteristic of thermoplastic polymers, they are less widely used.

Melmac, a highly cross-linked thermosetting polymer of melamine and formaldehyde, is a hard, moisture-resistant material. Because it is colorless, Melmac can be made into materials with pastel colors. It is used to make counter surfaces and lightweight dishes.

Melamine Poisoning

A few years ago, Chinese milk products were found to have been deliberately contaminated with melamine. Although melamine was never approved as a food additive, milk dealers added it to diluted milk to increase its apparent protein concentration. This milk was then used to make such things as powdered milk, infant formulas, cookies, and other foods. The contaminated milk sickened 300,000 Chinese children and killed at least 6 babies. The previous year, about 1000 dogs and cats died in the United Sates as a result of eating pet food with melamine-containing ingredients traced to Chinese suppliers. As a result of these tragedies, the Chinese legislature passed a food safety law and now requires producers to list all additives on food labels.

Designing a Polymer

Today, polymers are being designed to meet ever more exacting and specific needs. A polymer used for making dental impressions, for example, must be soft enough initially to be molded around the teeth but must become hard enough later to maintain a fixed shape. The polymer commonly used for dental impressions contains three-membered aziridine rings that react to form cross-links between the chains. Because aziridine rings are not very reactive, cross-linking occurs relatively slowly, so most of the hardening of the polymer does not occur until the polymer is removed from the patient's mouth.

polymer used to make dental impressions

Elastomers

An **elastomer** is a polymer that stretches and then reverts to its original shape. It is a randomly oriented amorphous polymer, but it must have some cross-linking so that the chains do not slide past one another. When elastomers are pulled, the random chains stretch out and crystallize. The van der Waals forces are not strong enough to keep them in that arrangement, however, so when the stretching force is removed, the chains go back to their random shapes. Rubber and spandex are examples of elastomers.

Oriented Polymers

Polymer chains obtained by conventional polymerization can be stretched out and then packed together again in a more ordered, parallel arrangement than they had originally, resulting in polymers that are stronger than steel or that conduct electricity as well as copper. Such polymers are called **oriented polymers.** Converting conventional polymers into oriented polymers has been compared to "uncooking" spaghetti; the conventional polymer is analogous to disordered, cooked spaghetti, whereas the oriented polymer is like raw spaghetti.

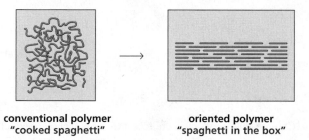

conventional polymer
"cooked spaghetti"

oriented polymer
"spaghetti in the box"

Dyneema, the strongest commercially available fabric, is an oriented polyethylene polymer with a molecular weight 100 times greater than that of high-density polyethylene. It is lighter than Kevlar and at least 40% stronger. A rope made of Dyneema can lift almost 119,000 pounds, whereas a steel rope of similar size fails before the weight reaches 13,000 pounds! It is astounding that a chain of carbons can be stretched and reoriented to produce a material stronger than steel. Dyneema is used to make full-face crash helmets, protective fencing suits, and hang gliders.

Plasticizers

A plasticizer can be dissolved into a polymer to make it more flexible. The **plasticizer** decreases the attractions between the polymer chains, thereby allowing them to slide past one another. Di-2-ethylhexyl phthalate, the most widely used plasticizer, is added to poly(vinyl chloride)—normally a brittle polymer—to make products such as vinyl raincoats, shower curtains, and garden hoses.

An important property to consider in choosing a plasticizer is its permanence—that is, how well the plasticizer remains in the polymer. The "new-car smell" appreciated by car owners is the odor of the plasticizer that has vaporized from the vinyl upholstery. When a significant amount of the plasticizer has evaporated, the upholstery becomes brittle and cracks.

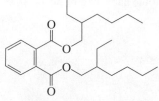

di-2-ethylhexyl phthalate
a plasticizer

Health Concerns: Bisphenol A and Phthalates

Animal studies have raised concerns about human exposure to bisphenol A and phthalates. Pregnant rats exposed to bisphenol A have been found to have a 3 to 4 times higher incidence of precancerous lesions in their mammary ducts. Bisphenol A (BPA) is used in the manufacture of polycarbonates and epoxy resins (Section 7). Although there is no evidence that bisphenol A has an adverse impact on humans, most manufacturers of polycarbonates have stopped using the compound, and BPA-free water bottles are now found in stores.

Phthalates have been found to be endocrine disruptors—that is, they can alter the proper balance of hormones. Therefore, the primary risk they pose is to a developing fetus. It is difficult to avoid phthalates because of the numerous items (for example, the linings of most aluminum food and beverage cans) that contain them.

9   RECYCLING POLYMERS

We saw in Section 2 that polymers are assigned a number from 1–6 that indicates the ease with which that kind of polymer can be recycled—the lower the number, the easier it can be recycled. Unfortunately, only polymers with the two lowest numbers PET (1)—the polymer used to make soft drink bottles—and HDPE (2)—the denser polymer used for juice and milk bottles—are recycled to any significant extent. This amounts to less than 25% of all polymers. The others are found in landfills.

PET is recycled by heating the polymer in an acidic solution of methanol. This transesterification reaction is the reverse of the transesterifcation reaction that formed the polymer. Because the products of recycling PET are the monomers used to make it, the products can be used to make more PET.

poly(ethylene terephthalate)
PET

dimethyl terephthalate   1,2-ethanediol

10   BIODEGRADABLE POLYMERS

Biodegradable polymers are polymers that can be degraded by microorganisms such as bacteria, fungi, or algae. Polylactide (PLA), a biodegradable polymer of lactic acid, has found wide use. When lactic acid is polymerized, a molecule of water is lost that can hydrolyze the new ester bond.

lactic acid

However, if lactic acid is converted to a cyclic dimer, the dimer can form a polymer without the loss of water by ring-opening polymerization. (The red arrows show formation of the tetrahedral intermediate; the blue arrows show the subsequent elimination from the tetrahedral intermediate.)

cyclic dimer of lactic acid

Because lactic acid has an asymmetric center, there are several different forms of the polymer. The extent of the polymer's crystallinity, and therefore many important properties, depends on the ratio of R and S enantiomers used in its synthesis. Polylactides are currently being used in nonwrinkle fabrics, microwavable trays, food packaging, and in several medical applications such as sutures, stents, and drug-delivery devices. They are also used for cold drink glasses. Unfortunately, hot drinks cause the polymer to liquify. Although polylactides are more expensive than nonbiodegradable polymers, their price is falling as their production increases.

Polyhydroxyalkanoates (PHAs) are also biodegradable polymers. These are condensation polymers of 3-hydroxycarboxylic acids. Thus, like PLA, they are polyesters. The most common PHA is PHB, a polymer of 3-hydroxybutyric acid; it can be used for many of the things that polypropylene is now used for. Unlike polypropylene that floats, PHB sinks. PHBV, a PHA marketed under the trade name Biopol, is a copolymer of 3-hydroxybutyric acid and 3-hydroxyvaleric acid. It is being used for such things as wastepaper baskets, toothbrush holders, and soap dispensers. PHAs are degraded by bacteria to CO_2 and H_2O.

glasses made of PLA

3-hydroxybutyric acid   **3-hydroxyvaleric acid**

PROBLEM 24

a. Draw the structure of a short segment of PHB.
b. Draw the structure of a short segment of PHBV with alternating monomers.

SOME IMPORTANT THINGS TO REMEMBER

- A **polymer** is a giant molecule made by covalently linking repeating units of small molecules called **monomers.** The process of linking them is called **polymerization.**

- Polymers can be divided into two groups: **synthetic polymers,** which are synthesized by scientists, and **biopolymers,** which are synthesized by cells.

- Synthetic polymers can be divided into two classes: **chain-growth polymers,** also called **addition polymers;** and **step-growth polymers,** also known as **condensation polymers.**

- Chain-growth polymers are made by **chain reactions,** which add monomers to the end of a growing chain.

- The chain reactions take place by one of three mechanisms: **radical polymerization, cationic polymerization,** or **anionic polymerization.**

- Each mechanism has an initiation step that starts the polymerization, propagation steps that allow the chain to grow at the **propagating site,** and termination steps that stop the growth of the chain.

- The choice of mechanism depends on the structure of the **monomer** and the initiator used to activate the monomer.

- In radical polymerization, the initiator is a radical; in cationic polymerization, it is an electrophile; and in anionic polymerization, it is a nucleophile.

- Chain-growth polymerization exhibits a preference for **head-to-tail addition.**

- Branching affects the physical properties of the polymer because unbranched chains can pack together more closely than branched chains can.

- Nonterminated polymer chains are called **living polymers.**

- The substituents are on the same side of the carbon chain in an **isotactic polymer,** alternate on both sides of the chain in a **syndiotactic polymer,** and are randomly oriented in an **atactic polymer.**

- The structure of a polymer can be controlled with **Ziegler–Natta catalysts.**

- Natural rubber is a polymer of 2-methyl-1,3-butadiene. Synthetic rubbers have been made by polymerizing dienes other than 2-methyl-1,3-butadiene.

- Heating rubber with sulfur to cross-link the chains is called **vulcanization.**

- **Homopolymers** are made of one kind of monomer; **copolymers** are made of more than one kind.

- **Step-growth polymers** are made by combining two molecules with reactive functional groups at each end.

- Nylon is a **polyamide. Aramides** are aromatic polyamides. Dacron is a **polyester.**

- **Polycarbonates** have two alkoxy groups bonded to the same carbonyl carbon. A **urethane** is a compound that has an ester and an amide group bonded to the same carbonyl carbon.

- **Crystallites** are highly ordered regions of a polymer. The more crystalline the polymer is, the denser, harder, and more resistant to heat it is.

- **Thermoplastic polymers** have crystalline and noncrystalline regions. **Thermosetting polymers** have cross-linked polymer chains. The greater the degree of cross-linking, the more rigid the polymer.

- An **elastomer** is a plastic that stretches and then reverts to its original shape.

- A **plasticizer** is an organic compound that dissolves in the polymer and allows the chains to slide past one another.

- **Biodegradable polymers** can be degraded by microorganisms.

PROBLEMS

25. Draw short segments of the polymers obtained from the following monomers. In each case, indicate whether the polymerization is a chain-growth or a step-growth polymerization.

a. $CH_2=CHF$

b. $CH_2=CHCO_2H$

c. $HO(CH_2)_5\overset{O}{\overset{\|}{C}}OH$

d. $Cl\overset{O}{\overset{\|}{C}}(CH_2)_5\overset{O}{\overset{\|}{C}}Cl + H_2N(CH_2)_5NH_2$

e. H_3C — (benzene ring) —NCO + $HOCH_2CH_2OH$, with OCN substituent

26. Draw the repeating unit of the step-growth polymer that will be formed from each of the following pairs of monomers:

a. $ClCH_2CH_2OCH_2CH_2Cl$ + HN (ring) NH $\longrightarrow$

b. H_2N—(benzene ring)—$OCH_2CH_2CH_2O$—(benzene ring)—NH_2 + $H\overset{O}{\overset{\|}{C}}-\overset{O}{\overset{\|}{C}}H$ $\longrightarrow$

c. $O=$(ring)$=O$ + $(C_6H_5)_3P=CH$—(benzene ring)—$CH=P(C_6H_5)_3$ $\longrightarrow$

27. Draw the structure of the monomer or monomers used to synthesize the following polymers, and indicate whether each is a chain-growth polymer or a step-growth polymer.

a. $-CH_2CH-$ with CH_2CH_3 substituent

b. $-CH_2CHO-$ with CH_3 substituent

c. $-CH_2CH-$ with pyridine ring substituent

d. $-SO_2-$(benzene ring)$-SO_2NH(CH_2)_6NH-$

e. $-CH_2\overset{CH_3}{\overset{\|}{C}}=CHCH_2-$

f. $-CH_2CH_2CH_2CH_2CH_2\overset{O}{\overset{\|}{C}}O-$

g. $-CH_2\overset{CH_3}{\underset{|}{C}}-$ with phenyl substituent

h. $-\overset{O}{\overset{\|}{C}}-$(benzene ring)$-\overset{O}{\overset{\|}{C}}OCH_2CH_2O-$

28. Explain why the configuration of a polymer of isobutylene is neither isotactic, syndiotactic, nor atactic.

29. Draw short segments of the polymers obtained from the following compounds under the given reaction conditions:

a. $H_2C\!-\!CHCH_3 \xrightarrow{CH_3O^-}$
(epoxide with O)

c. $CH_2\!=\!CH \xrightarrow{CH_3CH_2CH_2CH_2Li}$ with $\overset{|}{\underset{\overset{\|}{O}}{C}}OCH_3$

b. $CH_2\!=\!\overset{\overset{\displaystyle CH_3}{|}}{C}\!-\!CHCH_3 \xrightarrow{peroxide}$ (with phenyl)

d. $CH_2\!=\!CHOCH_3 \xrightarrow{BF_3,\ H_2O}$

30. Quiana is a synthetic fabric that feels very much like silk.
   a. Is Quiana a nylon or a polyester?   b. What monomers are used to synthesize it?

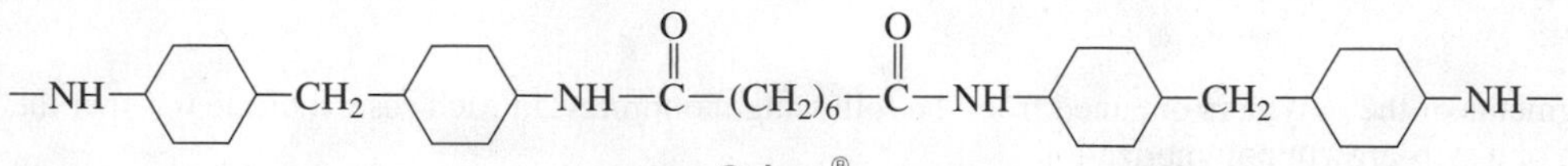

$$-NH-\bigcirc-CH_2-\bigcirc-NH-\overset{\overset{\displaystyle O}{\|}}{C}-(CH_2)_6-\overset{\overset{\displaystyle O}{\|}}{C}-NH-\bigcirc-CH_2-\bigcirc-NH-$$

Quiana®

31. Explain why a random copolymer is obtained when 3,3-dimethyl-1-butene undergoes cationic polymerization.

$$CH_2\!=\!CH\!-\!\overset{\overset{\displaystyle CH_3}{|}}{\underset{\underset{\displaystyle CH_3}{|}}{C}}\!-\!CH_3 \longrightarrow -CH_2\!-\!CH\!-\!CH_2\!-\!CH\!-\!\overset{\overset{\displaystyle CH_3}{|}}{\underset{\underset{\displaystyle CH_3}{|}}{C}}\!-\!CH_2\!-\!CH\!-\!\overset{\overset{\displaystyle CH_3}{|}}{\underset{\underset{\displaystyle CH_3}{|}}{C}}\!-\!CH_2\!-\!CH-$$

(with CH₃CCH₃/CH₃ and CH₃ CH₃ substituents)

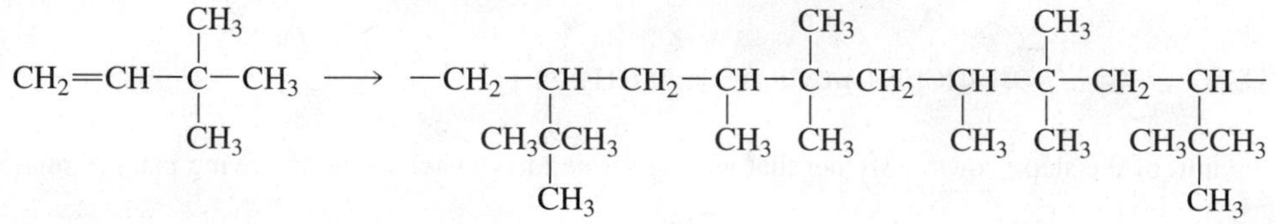

32. A chemist carried out two polymerization reactions. One flask contained a monomer that polymerizes by a chain-growth mechanism, and the other flask contained a monomer that polymerizes by a step-growth mechanism. When the reactions were terminated early in the process and the contents of the flasks analyzed, it was found that one flask contained a high-molecular-weight polymer and very little material of intermediate molecular weight. The other contained mainly material of intermediate molecular weight and very little high-molecular-weight material. Which flask contained which product? Explain.

33. Poly(vinyl alcohol) is a polymer used to make fibers and adhesives. It is synthesized by hydrolysis or alcoholysis of the polymer obtained from polymerization of vinyl acetate as shown here.

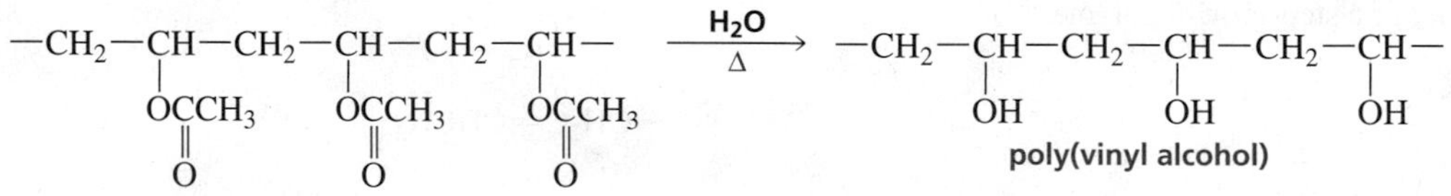

$$-CH_2\!-\!CH\!-\!CH_2\!-\!CH\!-\!CH_2\!-\!CH- \xrightarrow[\Delta]{H_2O} -CH_2\!-\!CH\!-\!CH_2\!-\!CH\!-\!CH_2\!-\!CH-$$

(poly(vinyl acetate) with OCCH₃ / O groups) → (poly(vinyl alcohol) with OH groups)

poly(vinyl acetate)   poly(vinyl alcohol)

   a. Why is poly(vinyl alcohol) not prepared by polymerizing vinyl alcohol?
   b. Is poly(vinyl acetate) a polyester?

34. Five different repeating units are found in the polymer obtained by cationic polymerization of 4-methyl-1-pentene. Identify these repeating units.

35. If a peroxide is added to styrene, the polymer known as polystyrene is formed. If a small amount of 1,4-divinylbenzene is added to the reaction mixture, a stronger and more rigid polymer is formed. Draw a short section of this more rigid polymer.

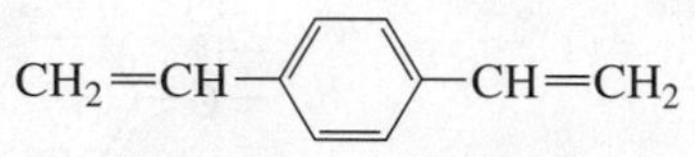

$$CH_2\!=\!CH-\bigcirc-CH\!=\!CH_2$$

1,4-divinylbenzene

36. A particularly strong and rigid polyester used for electronic parts is marketed under the trade name Glyptal. It is a polymer of terephthalic acid and glycerol. Draw a segment of the polymer and explain why it is so strong.

37. The following two compounds form a 1 : 1 alternating copolymer. No initiator is needed for the polymerization. Propose a mechanism for formation of the copolymer.

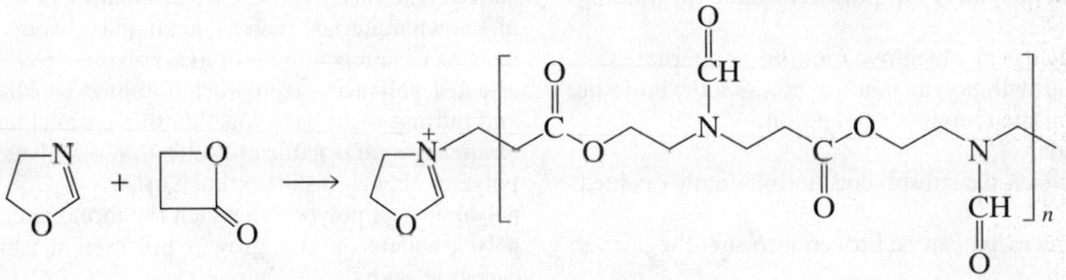

38. Which monomer would give a greater yield of polymer, 5-hydroxypentanoic acid or 6-hydroxyhexanoic acid? Explain your choice.

39. When rubber balls and other objects made of natural rubber are exposed to air for long periods, they turn brittle and crack. Explain why this happens more slowly to objects made of polyethylene.

40. When acrolein undergoes anionic polymerization, a polymer with two types of repeating units is obtained. Draw the structures of the repeating units.

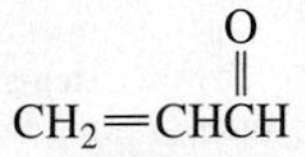

$$CH_2{=}CHCH\overset{O}{\overset{\|}{}}$$

acrolein

41. Why do vinyl raincoats become brittle as they get old, even if they are not exposed to air or to any pollutants?

42. The polymer shown here is synthesized by hydroxide ion-promoted hydrolysis of a copolymer of *para*-nitrophenyl methacrylate and acrylate.
   a. Propose a mechanism for the formation of the copolymer.
   b. Explain why hydrolysis of the copolymer to form the polymer occurs much more rapidly than hydrolysis of *para*-nitrophenyl acetate.

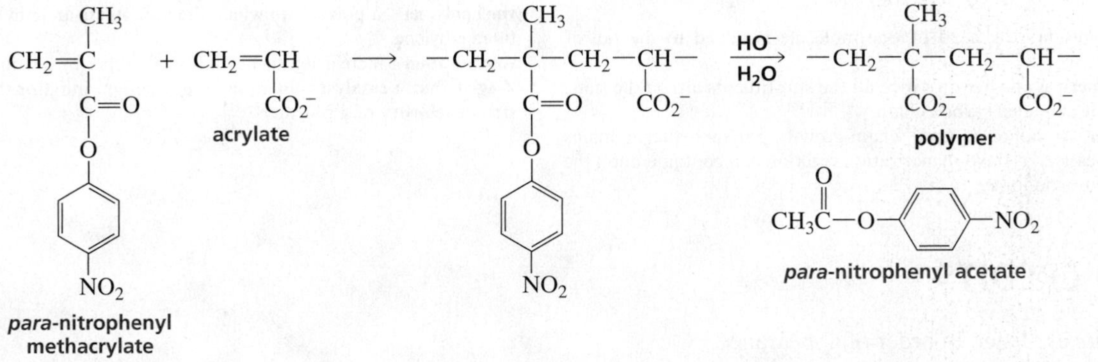

43. An alternating copolymer of styrene and vinyl acetate can be turned into a graft copolymer by hydrolyzing it and then adding ethylene oxide. Draw the structure of the graft copolymer.

44. How could head-to-head poly(vinyl bromide) be synthesized?

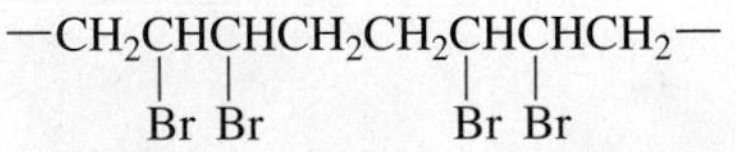

$$-CH_2CHCHCH_2CH_2CHCHCH_2-$$
$$\underset{Br\ Br}{|\ \ |}\qquad\underset{Br\ Br}{|\ \ |}$$

head-to-head poly(vinyl bromide)

45. Delrin (polyoxymethylene) is a tough self-lubricating polymer used in gear wheels. It is made by polymerizing formaldehyde in the presence of an acid catalyst.
   a. Propose a mechanism for formation of a segment of the polymer.
   b. Is Delrin a chain-growth polymer or a step-growth polymer?

GLOSSARY

addition polymer (chain-growth polymer) a polymer made by adding monomers to the growing end of a chain.

alternating copolymer a copolymer in which two monomers alternate.

anionic polymerization chain-growth polymerization in which the initiator is a nucleophile; the propagation site therefore is an anion.

aramide an aromatic polyamide.

atactic polymer a polymer in which the substituents are randomly oriented on the extended carbon chain.

biodegradable polymer a polymer that can be broken into small segments by an enzyme-catalyzed reaction.

biopolymer a polymer that is synthesized in nature.

block copolymer a copolymer in which there are regions (blocks) of each kind of monomer.

cationic polymerization chain-growth polymerization in which the initiator is an electrophile; the propagation site therefore is a cation.

chain-growth polymer (addition polymer) a polymer made by adding monomers to the growing end of a chain.

chain transfer a growing polymer chain reacts with a molecule XY in a manner that allows X to terminate the chain, leaving behind Y to initiate a new chain.

condensation polymer (step-growth polymer) a polymer made by combining two molecules while removing a small molecule (usually water or an alcohol).

conducting polymer a polymer that can conduct electricity.

copolymer a polymer formed from two or more different monomers.

cross-linking connecting polymer chains by intermolecular bond formation.

crystallites regions of a polymer in which the chains are highly ordered.

elastomer a polymer that can stretch and then revert to its original shape.

epoxy resin substance formed by mixing a low-molecular-weight prepolymer with a compound that forms a cross-linked polymer.

graft copolymer a copolymer that contains branches of a polymer of one monomer grafted onto the backbone of a polymer made from another monomer.

head-to-tail addition the head of one molecule is added to the tail of another.

isotactic polymer a polymer in which all the substituents are on the same side of the fully extended carbon chain.

living polymer a nonterminated chain-growth polymer that remains active. This means that the polymerization reaction can continue upon the addition of more monomer.

materials science the science of creating new materials to be used in place of known materials such as metal, glass, wood, cardboard, and paper.

monomer a repeating unit in a polymer.

oriented polymer a polymer obtained by stretching out polymer chains and putting them back together in a parallel fashion.

plasticizer an organic molecule that dissolves in a polymer and allows the polymer chains to slide by each other.

polyamide a polymer in which the monomers are amides.

polycarbonate a step-growth polymer in which the dicarboxylic acid is carbonic acid.

polyester a polymer in which the monomers are esters.

polymer chemistry the field of chemistry that deals with synthetic polymers; part of the larger discipline known as materials science.

polyurethane a polymer in which the monomers are urethanes.

propagating site the reactive end of a chain-growth polymer.

radical polymerization chain-growth polymerization in which the initiator is a radical; the propagation site is therefore a radical.

random copolymer a copolymer with a random distribution of monomers.

ring opening polymerization a chain-growth polymerization that involves opening the ring of the monomer.

step-growth polymer (condensation polymer) a polymer made by combining two molecules while removing a small molecule (usually of water or an alcohol).

syndiotactic polymer a polymer in which the substituents regularly alternate on both sides of the fully extended carbon chain.

synthetic polymer a polymer that is not synthesized in nature.

thermoplastic polymer a polymer that has both ordered crystalline regions and amorphous noncrystalline regions; it can be molded when heated.

thermosetting polymers cross-linked polymers that, after they are hardened, cannot be remelted by heating.

urethane a compound with a carbonyl group that is both an amide and an ester.

vinyl polymer a polymer in which the monomers are ethylene or a substituted ethylene.

vulcanization increasing the flexibility of rubber by heating it with sulfur.

Ziegler–Natta catalyst an aluminum–titanium initiator that controls the stereochemistry of a polymer.

PHOTO CREDITS

Photo credits are listed in order of appearance.

Appendix

pK$_a$ Values

Compound	pK_a	Compound	pK_a	Compound	pK_a
$CH_3C{\equiv}\overset{+}{N}H$	−10.1	O_2N—⬡—$\overset{+}{N}H_3$	1.0	CH_3—⬡—$\overset{O}{\overset{\|}{C}}OH$	4.3
HI	−10	pyrimidine $\overset{+}{N}H$	1.0		
HBr	−9			CH_3O—⬡—$\overset{O}{\overset{\|}{C}}OH$	4.5
$CH_3\overset{+OH}{\overset{\|}{C}}H$	−8	$Cl_2CH\overset{O}{\overset{\|}{C}}OH$	1.3		
		HSO_4^-	2.0	⬡—$\overset{+}{N}H_3$	4.6
$CH_3\overset{+OH}{\overset{\|}{C}}CH_3$	−7.3	H_3PO_4	2.1		
HCl	−7	purine	2.5	$CH_3\overset{O}{\overset{\|}{C}}OH$	4.8
⬡—SO_3H	−6.5	$FCH_2\overset{O}{\overset{\|}{C}}OH$	2.7	quinoline $\overset{+}{N}H$	4.9
$CH_3\overset{+OH}{\overset{\|}{C}}OCH_3$	−6.5	$ClCH_2\overset{O}{\overset{\|}{C}}OH$	2.8	CH_3—⬡—$\overset{+}{N}H_3$	5.1
$CH_3\overset{+OH}{\overset{\|}{C}}OH$	−6.1	$BrCH_2\overset{O}{\overset{\|}{C}}OH$	2.9	pyridine $\overset{+}{N}H$	5.2
H_2SO_4	−5	$ICH_2\overset{O}{\overset{\|}{C}}OH$	3.2	CH_3O—⬡—$\overset{+}{N}H_3$	5.3
pyrrole $\overset{+}{N}H$	−3.8	HF	3.2	$CH_3\overset{CH_3}{\underset{}{C}}{=}\overset{+}{N}HCH_3$	5.5
$CH_3CH_2\overset{H}{\overset{+}{O}}CH_2CH_3$	−3.6	HNO_2	3.4		
$CH_3CH_2\overset{H}{\overset{+}{O}}H$	−2.4	O_2N—⬡—$\overset{O}{\overset{\|}{C}}OH$	3.4	$CH_3\overset{O}{\overset{\|}{C}}CH_2\overset{O}{\overset{\|}{C}}H$	5.9
$CH_3\overset{H}{\overset{+}{O}}H$	−2.5	$H\overset{O}{\overset{\|}{C}}OH$	3.8	$HO\overset{+}{N}H_3$	6.0
H_3O^+	−1.7			H_2CO_3	6.4
HNO_3	−1.3	Br—⬡—$\overset{+}{N}H_3$	3.9	imidazole HN⁺NH	6.8
CH_3SO_3H	−1.2	Br—⬡—$\overset{O}{\overset{\|}{C}}OH$	4.0	H_2S	7.0
$CH_3\overset{+OH}{\overset{\|}{C}}NH_2$	0.0			O_2N—⬡—OH	7.1
$F_3C\overset{O}{\overset{\|}{C}}OH$	0.2	⬡—$\overset{O}{\overset{\|}{C}}OH$	4.2	$H_2PO_4^-$	7.2
$Cl_3C\overset{O}{\overset{\|}{C}}OH$	0.64			⬡—SH	7.8
pyridine $\overset{+}{N}$—OH	0.79				

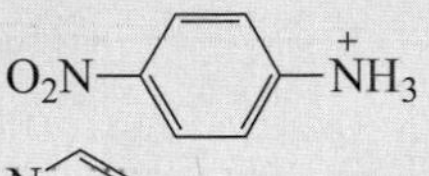

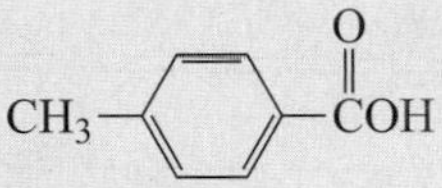

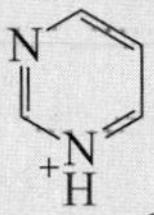

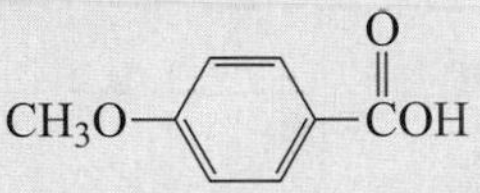

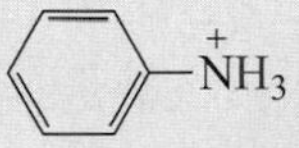

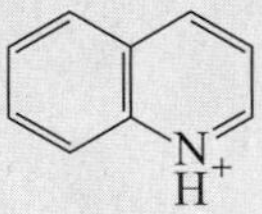

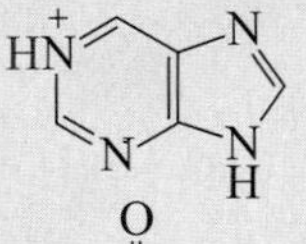

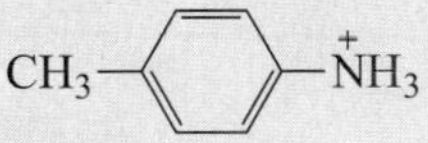

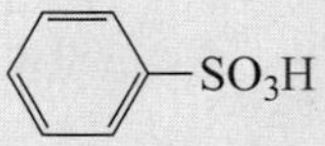

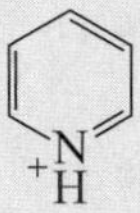

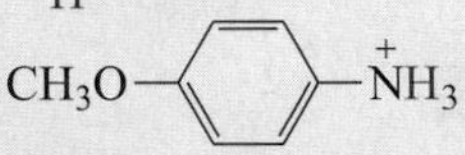

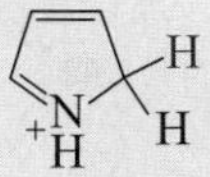

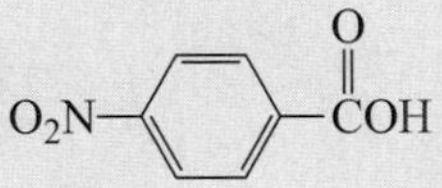

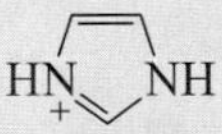

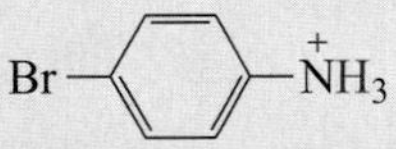

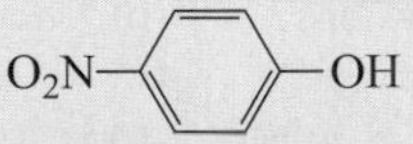

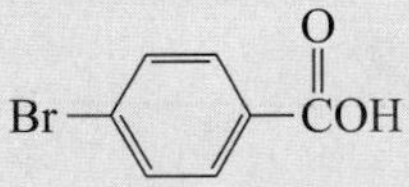

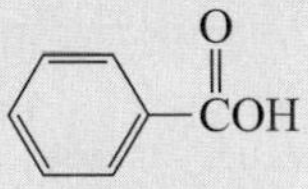

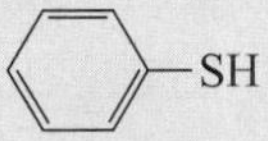

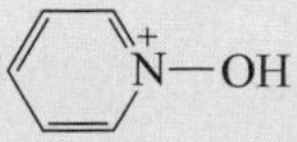

^{a}pK_a values are for the red H in each structure

(continued)

pKₐ Values (continued)

Compound	pKₐ	Compound	pKₐ	Compound	pKₐ
aziridinium (H-N⁺H-H ring)	8.0	$CH_3CCH_2COCH_2CH_3$ (O, O)	10.7	pyrrole (N-H ring)	~17
$H_2NN^+H_3$	8.1	$CH_3N^+H_3$	10.7	CH_3CH (O)	17
CH_3COOH	8.2	cyclohexyl-N^+H_3	10.7	$(CH_3)_3COH$	18
phthalimide (NH)	8.3	$(CH_3)_2N^+H_2$	10.7	CH_3CCH_3 (O)	20
$CH_3CH_2NO_2$	8.6	piperidinium (H-N⁺-H ring)	11.1	$CH_3COCH_2CH_3$ (O)	24.5
$CH_3CCH_2CCH_3$ (O, O)	8.9	$CH_3CH_2N^+H_3$	11.0	$HC\equiv CH$	25
$HC\equiv N$	9.1	pyrrolidinium (H-N⁺-H ring)	11.3	$CH_3C\equiv N$	25
morpholinium (O ring, H-N⁺-H)	9.3	$HOOH$	11.6	$CH_3CN(CH_3)_2$ (O)	30
Cl—⟨ ⟩—OH	9.4	HPO_4^{2-}	12.3	H_2	35
N^+H_4	9.4	CF_3CH_2OH	12.4	NH_3	36
$HOCH_2CH_2N^+H_3$	9.5	$CH_3CH_2OCCH_2COCH_2CH_3$ (O, O)	13.3	pyrrolidine (N-H ring)	36
$H_3N^+CH_2CO^-$ (O)	9.8	$HC\equiv CCH_2OH$	13.5	CH_3NH_2	40
⟨ ⟩—OH	10.0	H_2NCNH_2 (O)	13.7	⟨ ⟩—CH_3	41
CH_3—⟨ ⟩—OH	10.2	$CH_3N^+CH_2CH_2OH$ (CH₃, CH₃)	13.9	⟨ ⟩ (benzene)	43
HCO_3^-	10.2	imidazole (N NH ring)	14.4	$CH_2=CHCH_3$	43
CH_3NO_2	10.2	CH_3OH	15.5	$CH_2=CH_2$	44
H_2N—⟨ ⟩—OH	10.3	H_2O	15.7	cyclopropane (△)	46
CH_3CH_2SH	10.5	CH_3CH_2OH	16.0	CH_4	60
$(CH_3)_3N^+H$	10.6	CH_3CNH_2 (O)	16	CH_3CH_3	>60
		⟨ ⟩—CCH_3 (O)	16.0		

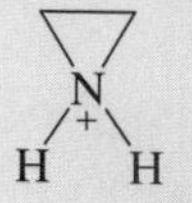

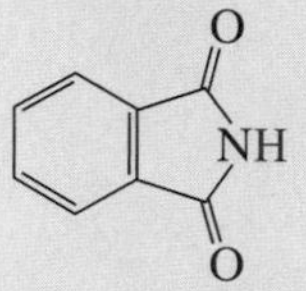

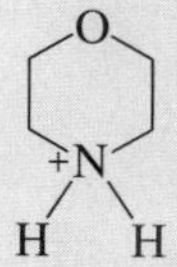

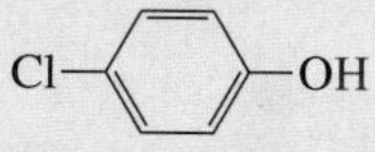

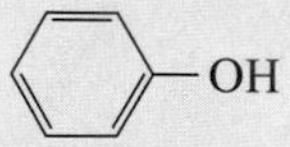

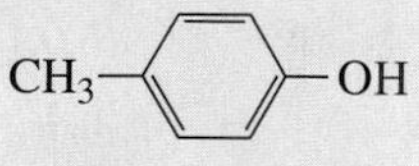

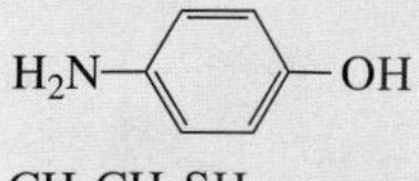

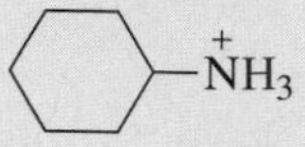

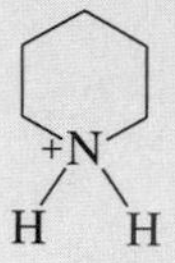

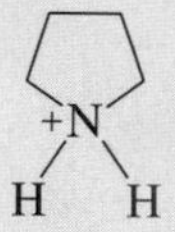

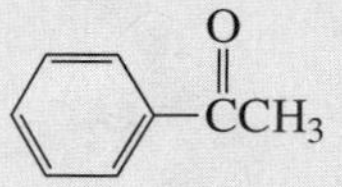

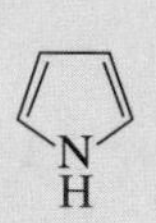

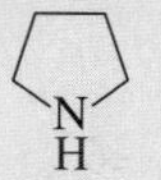

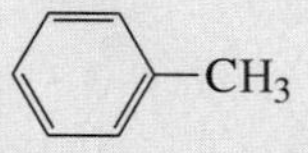

Appendix

Kinetics

HOW TO DETERMINE RATE CONSTANTS

A **reaction mechanism** is a detailed analysis of how the chemical bonds (or the electrons) in the reactants rearrange to form the products. The mechanism for a given reaction must obey the observed rate law for the reaction. A **rate law** tells how the rate of a reaction depends on the concentration of the species involved in the reaction.

FIRST-ORDER REACTION

The rate is proportional to the concentration of one reactant:

$$A \xrightarrow{k_1} \text{products}$$

Rate law: rate $= k_1[A]$

To determine the first-order rate constant (k_1):

Change in the concentration of A with respect to time:

$$\frac{-d[A]}{dt} = k_1[A]$$

Let $a =$ the initial concentration of A;
Let $x =$ concentration of A that has reacted up to time t.
Therefore, the concentration of A left at time t is $(a - x)$.
Substituting into the previous equation gives

$$\frac{-d(a - x)}{dt} = k_1(a - x)$$

$$\frac{-da}{dt} + \frac{dx}{dt} = k_1(a - x)$$

$$0 + \frac{dx}{dt} = k_1(a - x)$$

$$\frac{dx}{(a - x)} = k_1 dt$$

Integrating the previous equation yields

$$-\ln(a - x) = k_1 t + \text{constant}$$

At $t = 0$, $x = 0$; therefore,

$$\text{constant} = -\ln a$$

$$-\ln(a - x) = k_1 t - \ln a$$

$$\ln \frac{a}{a - x} = k_1 t$$

$$\ln \frac{a - x}{a} = -k_1 t$$

HALF-LIFE OF A FIRST-ORDER REACTION

The **half-life** ($t_{1/2}$) of a reaction is the time it takes for half the reactant to react (or for half the product to form). To derive the half-life of a reactant in a first-order reaction, we begin with the equation shown at the bottom of column 1.

$$\ln \frac{a}{(a - x)} = k_1 t$$

At $t_{1/2}$, $x = \dfrac{a}{2}$; therefore,

$$\ln \frac{a}{\left(a - \dfrac{a}{2}\right)} = k_1 t_{1/2}$$

$$\ln \frac{a}{\left(\dfrac{a}{2}\right)} = k_1 t_{1/2}$$

$$\ln 2 = k_1 t_{1/2}$$

$$0.693 = k_1 t_{t/2}$$

$$t_{1/2} = \frac{0.693}{k_1}$$

Notice that the half-life of a first-order reaction is independent of the concentration of the reactant.

SECOND-ORDER REACTION

The rate is proportional to the concentration of two reactants:

$$A + B \xrightarrow{k_2} \text{products}$$

Rate law: rate $= k_2[A][B]$

To determine the second-order rate constant (k_2):

Change in the concentration of A with respect to time:

$$\frac{-d[A]}{dt} = k_2[A][B]$$

Let $a =$ the initial concentration of A;
Let $b =$ the initial concentration of B;
Let $x =$ the concentration of A that has reacted at time t.

Therefore, the concentration of A left at time $t = (a - x)$, and the concentration of B left at time $t = (b - x)$.
Substitution gives

$$\frac{dx}{dt} = k_2(a - x)(b - x)$$

For the case where $a = b$ (this condition can be arranged experimentally),

$$\frac{dx}{dt} = k_2(a - x)^2$$

$$\frac{dx}{(a - x)^2} = k_2 \, dt$$

Integrating the equation gives

$$\frac{1}{(a - x)} = k_2 t + \text{constant}$$

At $t = 0$, $x = 0$; therefore,

$$\text{constant} = \frac{1}{a}$$

$$\frac{1}{(a - x)} - \frac{1}{a} = k_2 t$$

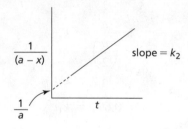

HALF-LIFE OF A SECOND-ORDER REACTION

$$\frac{1}{(a - x)} - \frac{1}{a} = k_2 t$$

At $t_{1/2}$, $x = \frac{a}{2}$; therefore,

$$\frac{1}{a} = k_2 t_{1/2}$$

$$t_{1/2} = \frac{1}{k_2 a}$$

PSEUDO-FIRST-ORDER REACTION

It is easier to determine a first-order rate constant than a second-order rate constant because the kinetic behavior of a first-order reaction is independent of the initial concentration of the reactant. Therefore, a first-order rate constant can be determined without knowing the initial concentration of the reactant. The determination of a second-order rate constant requires not only that the initial concentration of the reactants be known but also that the initial concentrations of the two reactants be identical in order to simplify the kinetic equation.

However, if the concentration of one of the reactants in a second-order reaction is much greater than the concentration of the other, the reaction can be treated as a first-order

reaction. Such a reaction is known as a **pseudo-first-order reaction** and is given by

$$\frac{-d[A]}{dt} = k[A][B]$$

If $[B] \gg [A]$, then

$$\frac{-d[A]}{dt} = k'[A]$$

The rate constant obtained for a pseudo-first-order reaction (k', often called k_{obsd}) includes the concentration of B, but k can be determined by carrying out the reaction at several different concentrations of B and determining the slope of a plot of the observed rate versus [B].

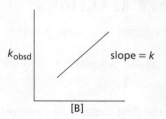

PROBLEMS

1. How long would it take for the reactant of a first-order reaction to decrease to one-half its initial concentration if the rate constant is 4.5×10^{-3} s^{-1} and the initial concentration of the reactant is

   a. 1.0 M? b. 0.50 M?

2. How long would it take for the reactants of a second-order reaction to decrease to one-half their initial concentration if the rate constant is 2.3×10^{-2} M^{-1} s^{-1} and the initial concentration of both reactants is

   a. 1.0 M? b. 0.50 M?

3. How many half-lives are required for a first-order reaction to reach >99% completion?

4. The initial concentration of a reactant undergoing a first-order reaction is 0.40 M. After five minutes, the concentration of the reactant is 0.27 M; after an additional five minutes, the concentration of the reactant is 0.18 M; and after an additional five minutes, the concentration of the reactant is 0.12 M.

   a. What is the average rate of the reaction during each five-minute interval?
   b. What is the rate constant of the reaction?

5. What percentage of a compound, undergoing a first-order reaction with a rate constant of 2.70×10^{-5} s^{-1}, would have reacted at the end of two hours?

6. How long would it take for a first-order reaction with a rate constant of 5.30×10^{-4} s^{-1} to reach 70% completion?

7. The following data were obtained in a study of the rate of inversion of sucrose at 25 °C. The initial concentration of sucrose was 1.00 M.

Time (minutes)	0	30	60	90	130	180
Sucrose inverted (M)	0	0.100	0.195	0.277	0.373	0.468

a. What is the order of the reaction? (Assume that the reaction is either first order or second order.)
b. What is the rate constant of the reaction?

8. Calculate the activation energy of a first-order reaction that is 20% complete in 15 minutes at 40 °C and is 20% complete in 3 minutes at 60 °C.

9. Analysis of a hydrolysis reaction occurring in a dilute aqueous solution of sucrose shows that 80 grams of the original 100 grams of sucrose remain after 10 hours. At this rate, how much sucrose would be left after 24 hours? (Assume a pseudo-first-order reaction.).)

Solutions to Problems

1. a. half-life of a first-order reaction $= t_{1/2} = \dfrac{\ln 2}{k_1}$

$$= \dfrac{0.693}{k_1}$$

$$= \dfrac{0.693}{4.5 \times 10^{-3}\,\text{s}^{-1}}$$

$$= 154 \text{ seconds}$$

$$= 2 \text{ minutes, } 34 \text{ seconds}$$

b. The half-life of a first-order reaction is independent of concentration, so the answer is the same as the answer for **a** (2 minutes, 34 seconds).

2. a. half-life of a second-order reaction $= t_{1/2} = \dfrac{1}{k_2 a}$

$$= \dfrac{1}{2.3 \times 10^{-2}\,\text{M}^{-1}\,\text{s}^{-1}\,(1.00\,\text{M})}$$

$$= 43 \text{ seconds}$$

b. $t_{1/2} = \dfrac{1}{2.3 \times 10^{-2}\,\text{M}^{-1}\,\text{s}^{-1}\,(0.50\,\text{M})}$

$$= \dfrac{1}{1.5 \times 10^{-2}\,\text{s}^{-1}}$$

$$= 87 \text{ seconds}$$

3. One-half of the compound reacts during the first half-life. One half of what is left after the first half-life reacts during the second half-life. One half of what is left after the second half-life reacts during the third half-life, etc. The following table shows that seven half-lives are required to reach >99% completion.

Number of half-lives	Percentage completion	
1	0.50 (100) = 50	50
2	0.5 0 (50) = 25	75
3	0.50 (25) = 12.5	87.5
4	0.50 (12.5) = 6.25	93.8
5	0.50 (6.25) = 3.125	96.9
6	0.50 (3.125) = 1.5625	98.4

Number of half-lives	Percentage completion	
7	0.50 (1.5625) = 0.78125	99.2

4. a. rate $= \dfrac{\text{change in concentration}}{\text{change in time}}$

1st interval:

$$\text{rate} = \dfrac{0.40\,\text{M} - 0.27\,\text{M}}{5\,\text{min}} = \dfrac{0.13\,\text{M}}{5\,\text{min}}$$

$$= 2.60 \times 10^{-2}\,\text{M min}^{-1}$$

2nd interval:

$$\text{rate} = \dfrac{0.27\,\text{M} - 0.18\,\text{M}}{5\,\text{min}} = \dfrac{0.09\,\text{M}}{5\,\text{min}}$$

$$= 1.80 \times 10^{-2}\,\text{M min}^{-1}$$

3rd interval:

$$\text{rate} = \dfrac{0.18\,\text{M} - 0.12\,\text{M}}{5\,\text{min}} = \dfrac{0.06\,\text{M}}{5\,\text{min}}$$

$$= 1.20 \times 10^{-2}\,\text{M min}^{-1}$$

Thus, we see that the **rate** of the reaction is dependent on the concentration of the reactant; it decreases with decreasing concentration. Thus, the rate we have calculated is the average rate for each 5 minute interval.

b. rate $= k$ [reactant], using the average concentration of the reactant during the five-minute interval.

1st interval:

$$2.60 \times 10^{-2}\,\text{M min}^{-1} = k(0.335\,\text{M})$$

$$k = 7.8 \times 10^{-2}\,\text{min}^{-1}$$

2nd interval:

$$1.80 \times 10^{-2}\,\text{M min}^{-1} = k(0.225\,\text{M})$$

$$k = 8.0 \times 10^{-2}\,\text{min}^{-1}$$

3rd interval:

$$1.20 \times 10^{-2}\,\text{M min}^{-1} = k(0.15\,\text{M})$$

$$k = 8.0 \times 10^{-2}\,\text{min}^{-1}$$

The **rate constant**, as its name indicates, is constant during the course of the reaction.

5. $a =$ the initial concentration (set at 100%)
$x =$ the concentration that has reacted at time $= t$

$$\ln \dfrac{a}{a - x} = k_1 t$$

$$\ln \dfrac{100}{100 - x} = 2.70 \times 10^{-5}\,\text{s}^{-1}\,(2\text{ hours})$$

$$\ln \dfrac{100}{100 - x} = 2.70 \times 10^{-5}\,\text{s}^{-1}\,(7200\text{ seconds})$$

$$\ln \dfrac{100}{100 - x} = 0.194$$

$$\frac{100}{100 - x} = 1.21$$

$$100 = 121 - 1.21x$$

$$1.21\,x = 21$$

$$x = 17.4\%$$

6.

$$\ln \frac{a}{a - x} = k_1 t$$

$$\ln \frac{100}{100 - 70} = 5.30 \times 10^{-4}\,\text{s}^{-1}\,t$$

$$\ln 3.33 = 5.30 \times 10^{-4}\,\text{s}^{-1}\,t$$

$$t = \frac{1.20}{5.30 \times 10^{-4}\,\text{s}^{-1}}$$

$$t = 2260 \text{ seconds}$$

$$t = 37.7 \text{ minutes}$$

7. To determine the order of the reaction, we will calculate the first-order and second-order rate constants based on the data provided. Since the rate constant for a reaction should be the same regardless of the set of data we use, the calculations will tell us the order of the reaction.

1st order

$$k_1 = \frac{-\ln \dfrac{a - x}{a}}{t}$$

$$k_1 = \frac{-\ln \dfrac{1.000 - 0.100}{1.000}}{30}$$

$$= \mathbf{3.51 \times 10^{-3}}$$

$$k_2 = \frac{-\ln \dfrac{1.000 - 0.195}{1.000}}{60}$$

$$= \mathbf{3.62 \times 10^{-3}}$$

$$k_1 = \frac{-\ln \dfrac{1.000 - 0.277}{1.000}}{90}$$

$$= \mathbf{3.60 \times 10^{-3}}$$

$$k_1 = \frac{-\ln \dfrac{1.000 - 0.373}{1.000}}{130}$$

$$= \mathbf{3.59 \times 10^{-3}}$$

$$k_1 = \frac{-\ln \dfrac{1.000 - 0.468}{1.000}}{180}$$

$$= \mathbf{3.51 \times 10^{-3}}$$

2nd order

$$k_2 = \frac{\dfrac{1}{a - x} - \dfrac{1}{a}}{t}$$

$$k_2 = \frac{\dfrac{1}{1.000 - 0.100} - \dfrac{1}{1.000}}{30}$$

$$= \frac{1.111 - 1.000}{30}$$

$$= \mathbf{3.70 \times 10^{-3}}$$

$$k_2 = \frac{\dfrac{1}{1.000 - 0.195} - 1.000}{60}$$

$$= \mathbf{4.03 \times 10^{-3}}$$

$$k_2 = \frac{\dfrac{1}{1.000 - 0.277} - 1.000}{90}$$

$$= \mathbf{4.26 \times 10^{-3}}$$

$$k_2 = \frac{\dfrac{1}{1.000 - 0.373} - 1.000}{130}$$

$$= \mathbf{4.57 \times 10^{-3}}$$

$$k_2 = \frac{\dfrac{1}{1.000 - 0.468} - 1.000}{180}$$

$$= \mathbf{4.98 \times 10^{-3}}$$

a. Because the calculated rate constants are relatively constant when the data are plugged into a first-order equation but vary considerably when the data are plugged into a second-order equation, one can conclude that the reaction is first-order.

b. $3.6 \times 10^{-3} \text{ min}^{-1}$

8. Using the equation for calculating kinetic parameters:

$$k \text{ at } 40\,°C = \frac{\ln \dfrac{1.00}{1.00 - 0.20}}{15 \text{ min}} = 1.49 \times 10^{-2} \text{ min}^{-1}$$

$$k \text{ at } 60\,°C = \frac{\ln \dfrac{1.00}{1.00 - 0.20}}{3 \text{ min}} = 7.44 \times 10^{-2} \text{ min}^{-1}$$

$$\ln k_2 - \ln k_1 = \frac{-E_a}{R}\left(\frac{1}{T_2} - \frac{1}{T_1}\right) \quad (40\,°C = 313 \text{ K}; 60\,°C = 333 \text{ K})$$

$$\ln (7.44 \times 10^{-2}) - \ln (1.49 \times 10^{-2}) = \frac{-E_a}{1.986 \times 10^{-3} \text{ kcal}}\left(\frac{1}{333} - \frac{1}{313}\right)$$

$$-2.60 - (-4.21) = \frac{-E_a}{1.986 \times 10^{-3}}(0.00300 - 0.00319)$$

$$1.61 = \frac{-E_a}{1.986 \times 10^{-3}}(-0.00019)$$

$$E_a = \frac{1.61 \times 1.986 \times 10^{-3}}{0.00019}$$

$$E_a = 16.8 \text{ kcal/mol}$$

9. Sucrose is hydrolyzed to form a mixture of glucose and fructose. Because there is excess water (it is the solvent), the reaction is a pseudo-first-order reaction.

First, the rate constant of the reaction must be determined:

$$\ln \frac{a}{a - x} = k_1 t$$

$$\ln \frac{100}{80} = k_1 \times 10 \text{ hours}$$

$$k_1 = 2.23 \times 10^{-2} \text{ hr}^{-1}$$

Now we can calculate the amount of sucrose that has reacted, and therefore, the amount that would be left.

$$\ln \frac{a}{a - x} = k_1 t$$

$$\ln \frac{100}{100 - x} = 2.23 \times 10^{-2} (24 \text{ hr})$$

$$\ln \frac{100}{100 - x} = 0.535$$

$$\frac{100}{100 - x} = 1.71$$

$$100 = 171 - 1.71x$$

$$1.71x = 71$$

$$x = 41.5 \text{ g have reacted}$$

Therefore, 58.5 g would be left.

Appendix:
Summary of Methods Used to Synthesize a Particular Functional Group

Appendix

Summary of Methods Used to Synthesize a Particular Functional Group

SYNTHESIS OF ACETALS

1. Acid-catalyzed reaction of an aldehyde or a ketone with two equivalents of an alcohol.

SYNTHESIS OF ACID ANHYDRIDES

1. Reaction of an acyl halide with a carboxylate ion.
2. Heating a dicarboxylic acid.
3. Heating a dicarboxylic acid in the presence of acetic anhydride.

SYNTHESIS OF ACYL CHLORIDES OR ACYL BROMIDES

1. Reaction of a carboxylic acid with $SOCl_2$, PCl_3, or PBr_3.

SYNTHESIS OF ALCOHOLS

1. Acid-catalyzed hydration of an alkene.
2. Hydroboration–oxidation of an alkene.
3. Reaction of an alkyl halide with HO^-.
4. Reaction of an organocuprate with an epoxide.
5. Epoxidation of an alkene followed by reaction with NaH.
6. Reduction of an aldehyde, a ketone, an acyl chloride, an anhydride, an ester, or a carboxylic acid.
7. Reaction of a Grignard reagent with an aldehyde, a ketone, an acyl chloride, or an ester.
8. Cleavage of an ether with HI or HBr.
9. Reaction of an organozinc reagent with an aldehyde or a ketone.

SYNTHESIS OF ALDEHYDES

1. Hydroboration–oxidation of a terminal alkyne.
2. Oxidation of a primary alcohol with pyridinium chlorochromate or hypochlorous acid.
3. Swern oxidation of a primary alcohol with dimethyl sulfoxide, oxalyl chloride, and triethylamine.
4. Reaction of an acyl chloride with lithium tri(*tert*-butoxy) aluminum hydride.
5. Reaction of an ester with diisobutylaluminum hydride (DIBALH).
6. Ozonolysis of an alkene, followed by reaction with dimethyl sulfide or zinc and acetic acid.

SYNTHESIS OF ALKANES

1. Catalytic hydrogenation of an alkene or an alkyne.
2. Reaction of a Grignard reagent with a source of protons.
3. Wolff–Kishner or Clemmensen reduction of an aldehyde or a ketone.
4. Reduction of a phenone with H_2/Pd.
5. Reduction of a thioacetal or thioketal with H_2 and Raney nickel.
6. Reaction of an organocuprate with an alkyl halide.
7. Preparation of a cyclopropane by the reaction of an alkene with a carbene.

SYNTHESIS OF ALKENES

1. Elimination of hydrogen halide from an alkyl halide.
2. Acid-catalyzed dehydration of an alcohol.
3. Hofmann elimination reaction: elimination of a proton and a tertiary amine from a quaternary ammonium hydroxide.
4. Exhaustive methylation of an amine, followed by a Hofmann elimination reaction.
5. Hydrogenation of an alkyne with Lindlar catalyst to form a cis alkene.
6. Reduction of an alkyne with Na (or Li) and liquid ammonia to form a trans alkene.
7. Formation of a cyclic alkene using a Diels–Alder reaction.
8. Wittig reaction: reaction of an aldehyde or a ketone with a phosphonium ylide.
9. Reaction of an organocuprate with a halogenated alkene.
10. Heck reaction couples a vinyl halide with an alkene in a basic solution in the presence of PdL_2.
11. Suzuki reaction couples a vinyl halide with an organoboron compound in the presence of (PdL_2).
12. Alkene metathesis.

SYNTHESIS OF ALKYL HALIDES

1. Addition of hydrogen halide (HX) to an alkene.
2. Addition of HBr + a peroxide to an alkene.
3. Addition of hydrogen halide to an alkyne.

4. Radical halogenation of an alkane, an alkene, or an alkyl benzene.

5. Reaction of an alcohol with a hydrogen halide, $SOCl_2$, PCl_3, or PBr_3.

6. Reaction of a sulfonate ester with halide ion.

7. Cleavage of an ether with a hydrogen halide.

8. Halogenation of an α-carbon of an aldehyde, a ketone, or a carboxylic acid.

SYNTHESIS OF ALKYNES

1. Elimination of hydrogen halide from a vinyl halide.

2. Two successive eliminations of hydrogen halide from a vicinal dihalide or a geminal dihalide.

3. Reaction of an acetylide ion (formed by removing a proton from a terminal alkyne) with an alkyl halide.

4. Alkyne metathesis.

SYNTHESIS OF AMIDES

1. Reaction of an acyl chloride, an acid anhydride, or an ester with ammonia or with an amine.

2. Heating an ammonium carboxylate salt.

3. Reaction of a carboxylic acid and with dicyclohexylcarbodiimide followed by reaction with an amine.

4. Reaction of a nitrile with a secondary or tertiary alcohol.

SYNTHESIS OF AMINES

1. Reaction of an alkyl halide with NH_3, RNH_2, or R_2NH.

2. Reaction of an alkyl halide with azide ion, followed by reduction of the alkyl azide.

3. Reduction of an imine, a nitrile, or an amide.

4. Reductive amination of an aldehyde or a ketone.

5. Gabriel synthesis of primary amines: reaction of a primary alkyl halide with potassium phthalimide.

6. Reduction of a nitro compound.

7. Condensation of a secondary amine and formaldehyde with a carbon acid.

SYNTHESIS OF AMINO ACIDS

1. Hell–Volhard–Zelinski reaction: halogenation of a carboxylic acid, followed by treatment with excess NH_3.

2. Reductive amination of an α-keto acid.

3. The *N*-phthalimidomalonic ester synthesis.

4. The acetamidomalonic ester synthesis.

5. The Strecker synthesis: reaction of an aldehyde with ammonia, followed by addition of cyanide ion and hydrolysis.

SYNTHESIS OF CARBOXYLIC ACIDS

1. Oxidation of a primary alcohol.

2. Oxidation of an aldehyde.

3. Oxidation of an alkyl benzene.

4. Hydrolysis of an acyl halide, an acid anhydride, an ester, an amide, or a nitrile.

5. Haloform reaction: reaction of a methyl ketone with excess Br_2 (or Cl_2 or I_2) + HO^-.

6. Reaction of a Grignard reagent with CO_2.

7. Malonic ester synthesis.

8. Favorskii reaction: reaction of an α-haloketone with hydroxide ion.

SYNTHESIS OF CYANOHYDRINS

1. Reaction of an aldehyde or a ketone with sodium cyanide and HCl.

SYNTHESIS OF DIHALIDES

1. Addition of Cl_2 or Br_2 to an alkene.

2. Addition of Cl_2 or Br_2 to an alkyne.

SYNTHESIS OF 1,2-DIOLS

1. Reaction of an epoxide with hydroxide ion forms a trans 1,2-diol.

2. Reaction of an alkene with osmium tetroxide followed by hydrolysis with hydrogen peroxide forms a cis 1,2-diol.

SYNTHESIS OF DISULFIDES

1. Mild oxidation of a thiol.

SYNTHESIS OF ENAMINES

1. Reaction of an aldehyde or a ketone with a secondary amine.

SYNTHESIS OF EPOXIDES

1. Reaction of an alkene with a peroxyacid.

2. Reaction of a halohydrin with hydroxide ion.

3. Reaction of an aldehyde or a ketone with a sulfonium ylide.

SYNTHESIS OF ESTERS

1. Reaction of an acyl halide or an acid anhydride with an alcohol.

2. Acid-catalyzed reaction of an ester or a carboxylic acid with an alcohol.

3. Reaction of an alkyl halide with a carboxylate ion.

4. Reaction of a sulfonate ester with a carboxylate ion.

5. Oxidation of a ketone.

6. Preparation of a methyl ester by the reaction of a carboxylate ion with diazomethane.

SYNTHESIS OF ETHERS

1. Acid-catalyzed addition of an alcohol to an alkene.

2. Williamson ether synthesis: reaction of an alkoxide ion with an alkyl halide.

3. Formation of symmetrical ethers by heating an acidic solution of a primary alcohol.

SYNTHESIS OF HALOHYDRINS

1. Reaction of an alkene with Br_2 (or Cl_2) and H_2O.

2. Reaction of an epoxide with a hydrogen halide.

SYNTHESIS OF IMINES

1. Reaction of an aldehyde or a ketone with a primary amine.

SYNTHESIS OF KETONES

1. Addition of water to an alkyne.

2. Hydroboration–oxidation of an internal alkyne.

3. Oxidation of a secondary alcohol.

4. Ozonolysis of an alkene, followed by reaction with dimethyl sulfide or zinc and acetic acid.

5. Friedel–Crafts acylation of an aromatic ring.

6. Preparation of a methyl ketone by the acetoacetic ester synthesis.

7. Preparation of a cyclic ketone by the reaction of the next-size-smaller cyclic ketone with diazomethane.

SYNTHESIS OF α,β-UNSATURATED KETONES

1. Selenenylation of a ketone, followed by oxidative elimination.

SYNTHESIS OF NITRILES

1. Reaction of an alkyl halide with cyanide ion.

2. Reaction of an amide (with an NH_2 group) with $SOCl_2$.

SYNTHESIS OF SUBSTITUTED BENZENES

1. Halogenation with Br_2 or Cl_2 and a Lewis acid.

2. Nitration with HNO_3 + H_2SO_4.

3. Sulfonation: heating with H_2SO_4.

4. Friedel–Crafts acylation.

5. Friedel–Crafts alkylation.

6. Sandmeyer reaction: reaction of an arenediazonium salt with CuBr, CuCl, or CuCN.

7. Formation of a phenol by reaction of an arenediazonium salt with water.

8. Reaction of an organocuprate with an aryl halide.

9. Heck reaction: couples an aryl halide with an alkene in a basic solution in the presence of PdL_2.

10. Suzuki reaction: couples an aryl halide with an organoborane in the presence of PdL_2 and triethylamine.

SYNTHESIS OF SULFIDES

1. Reaction of a thiol with an alkyl halide.

2. Reaction of a thiol with a sulfonate ester.

SYNTHESIS OF THIOLS

1. Reaction of an alkyl halide with hydrogen sulfide.

2. Catalytic hydrogenation of a disulfide.

Appendix

Summary of Methods Employed to Form Carbon–Carbon Bonds

1. Alkene or alkyne metathesis.

2. Reaction of an acetylide ion with an alkyl halide or a sulfonate ester.

3. Diels–Alder and other cycloaddition reactions.

4. Reaction of an organocuprate with an epoxide.

5. Friedel–Crafts alkylation and acylation.

6. Reaction of a cyanide ion with an alkyl halide or a sulfonate ester.

7. Reaction of a cyanide ion with an aldehyde or a ketone.

8. Reaction of a Grignard reagent with an aldehyde, a ketone, an ester, an amide, or CO_2.

9. Reaction of an organozinc reagent with an aldehyde or a ketone.

10. Reaction of an alkene with a carbene.

11. Reaction of an organocuprate with an α,β-unsaturated ketone or an α,β-unsaturated aldehyde.

12. Aldol addition.

13. Claisen condensation.

14. Perkin condensation.

15. Knoevenagel condensation.

16. Reformatsky reaction.

17. The benzoin condensation.

18. Malonic ester synthesis and acetoacetic ester synthesis.

19. Michael addition reaction.

20. Alkylation of an enamine.

21. Alkylation of the α-carbon of a carbonyl compound.

22. Reaction of an organocuprate with an aryl halide or a vinyl halide.

23. Heck reaction: couples a vinyl or an aryl halide with an alkene in a basic solution in the presence of PdL_2.

24. Suzuki reaction: couples a vinyl or aryl halide with an organoborane in the presence of PdL_2.

From Appendix IV of *Organic Chemistry*, Seventh Edition. Paula Yurkanis Bruice. Copyright © 2014 by Pearson Education, Inc. All rights reserved.

Periodic Table of the Elements

1A / 1	2A / 2	3B / 3	4B / 4	5B / 5	6B / 6	7B / 7	8B / 8	8B / 9	8B / 10	1B / 11	2B / 12	3A / 13	4A / 14	5A / 15	6A / 16	7A / 17	8A / 18
1 **H** 1.00794																	2 **He** 4.002602
3 **Li** 6.941	4 **Be** 9.012182											5 **B** 10.811	6 **C** 12.0107	7 **N** 14.0067	8 **O** 15.9994	9 **F** 18.998403	10 **Ne** 20.1797
11 **Na** 22.989770	12 **Mg** 24.3050											13 **Al** 26.981538	14 **Si** 28.0855	15 **P** 30.973761	16 **S** 32.065	17 **Cl** 35.453	18 **Ar** 39.948
19 **K** 39.0983	20 **Ca** 40.078	21 **Sc** 44.955910	22 **Ti** 47.867	23 **V** 50.9415	24 **Cr** 51.9961	25 **Mn** 54.938049	26 **Fe** 55.845	27 **Co** 58.933200	28 **Ni** 58.6934	29 **Cu** 63.546	30 **Zn** 65.39	31 **Ga** 69.723	32 **Ge** 72.64	33 **As** 74.92160	34 **Se** 78.96	35 **Br** 79.904	36 **Kr** 83.80
37 **Rb** 85.4678	38 **Sr** 87.62	39 **Y** 88.90585	40 **Zr** 91.224	41 **Nb** 92.90638	42 **Mo** 95.94	43 **Tc** [98]	44 **Ru** 101.07	45 **Rh** 102.90550	46 **Pd** 106.42	47 **Ag** 107.8682	48 **Cd** 112.411	49 **In** 114.818	50 **Sn** 118.710	51 **Sb** 121.760	52 **Te** 127.60	53 **I** 126.90447	54 **Xe** 131.293
55 **Cs** 132.90545	56 **Ba** 137.327	71 **Lu** 174.967	72 **Hf** 178.49	73 **Ta** 180.9479	74 **W** 183.84	75 **Re** 186.207	76 **Os** 190.23	77 **Ir** 192.217	78 **Pt** 195.078	79 **Au** 196.96655	80 **Hg** 200.59	81 **Tl** 204.3833	82 **Pb** 207.2	83 **Bi** 208.98038	84 **Po** [208.98]	85 **At** [209.99]	86 **Rn** [222.02]
87 **Fr** [223.02]	88 **Ra** [226.03]	103 **Lr** [262.11]	104 **Rf** [261.11]	105 **Db** [262.11]	106 **Sg** [266.12]	107 **Bh** [264.12]	108 **Hs** [269.13]	109 **Mt** [268.14]	110 **Ds** [271.15]	111 **Rg** [272.15]	112 **Cn** [277]	113 [284]	114 **Fl** [289]	115 [288]	116 **Lv** [293]	117 [293]	118 [294]

Main groups — Transition metals — Main groups

*Lanthanide series

57 ***La** 138.9055	58 **Ce** 140.116	59 **Pr** 140.90765	60 **Nd** 144.24	61 **Pm** [145]	62 **Sm** 150.36	63 **Eu** 151.964	64 **Gd** 157.25	65 **Tb** 158.92534	66 **Dy** 162.50	67 **Ho** 164.93032	68 **Er** 167.259	69 **Tm** 168.93421	70 **Yb** 173.04

†Actinide series

89 **†Ac** [227.03]	90 **Th** 232.0381	91 **Pa** 231.03588	92 **U** 238.02891	93 **Np** [237.05]	94 **Pu** [244.06]	95 **Am** [243.06]	96 **Cm** [247.07]	97 **Bk** [247.07]	98 **Cf** [251.08]	99 **Es** [252.08]	100 **Fm** [257.10]	101 **Md** [258.10]	102 **No** [259.10]

aThe labels on top (1A, 2A, etc.) are common American usage. The labels below these (1, 2, etc.) are those recommended by the International Union of Pure and Applied Chemistry.

The names for elements 113, 115, 117, and 118 have not yet been decided.

Atomic weights in brackets are the masses of the longest-lived or most important isotope of radioactive elements.

Further information is available at *http://www.shef.ac.uk/chemistry/web-elements/*

Common Functional Groups,
Approximate pK$_a$ Values,
and Common Symbols and
Abbreviations

Common Functional Groups

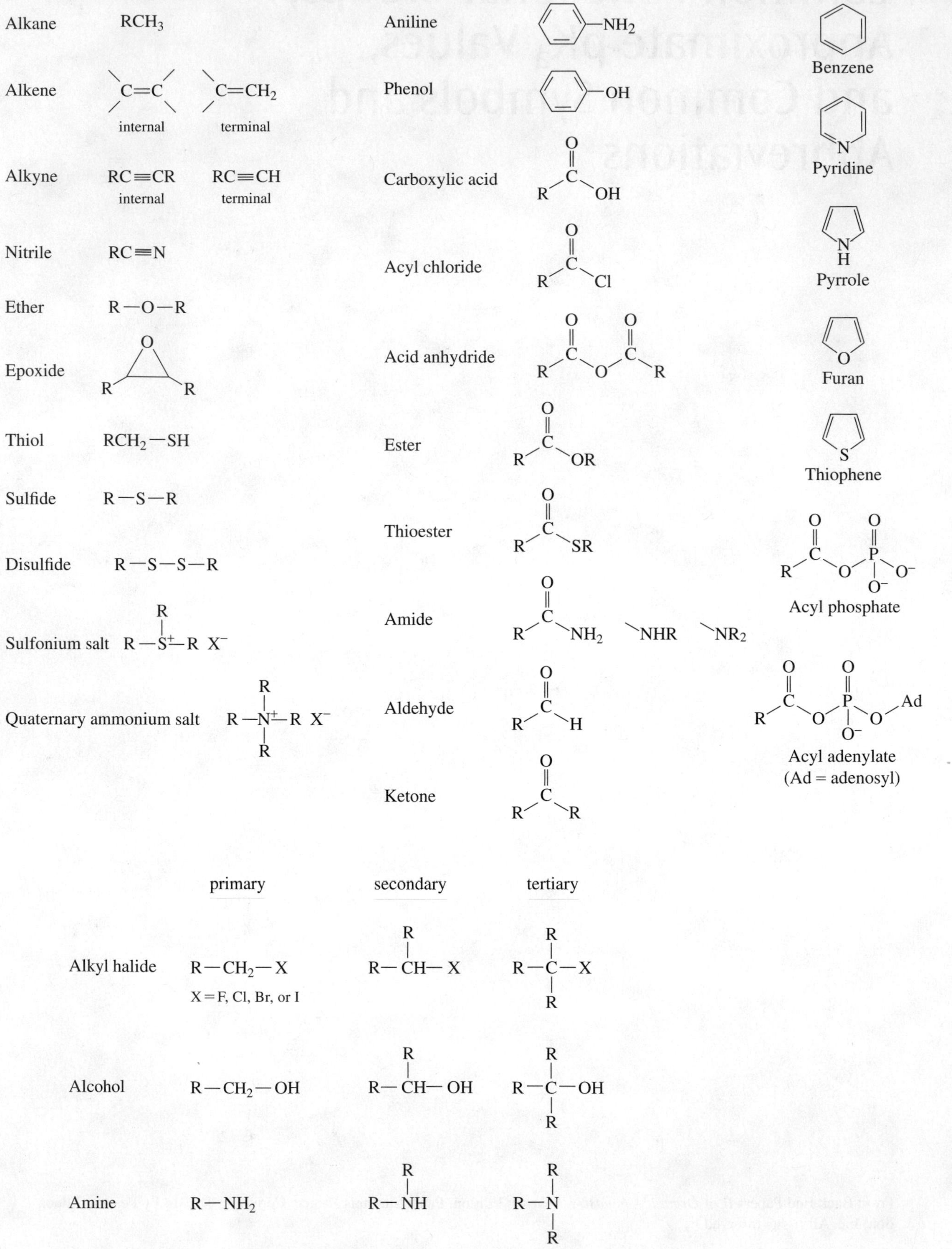

Alkane	RCH_3		Aniline	

Alkane RCH_3

Alkene $C=C$ (internal) $C=CH_2$ (terminal)

Alkyne $RC \equiv CR$ (internal) $RC \equiv CH$ (terminal)

Nitrile $RC \equiv N$

Ether $R-O-R$

Epoxide

Thiol RCH_2-SH

Sulfide $R-S-R$

Disulfide $R-S-S-R$

Sulfonium salt $R-S^+-R \;\; X^-$

Quaternary ammonium salt $R-N^+-R \;\; X^-$

Aniline

Phenol

Carboxylic acid

Acyl chloride

Acid anhydride

Ester

Thioester

Amide NHR NR_2

Aldehyde

Ketone

Benzene

Pyridine

Pyrrole

Furan

Thiophene

Acyl phosphate

Acyl adenylate (Ad = adenosyl)

	primary	secondary	tertiary
Alkyl halide	$R-CH_2-X$ X = F, Cl, Br, or I	$R-CH-X$	$R-C-X$
Alcohol	$R-CH_2-OH$	$R-CH-OH$	$R-C-OH$
Amine	$R-NH_2$	$R-NH$	$R-N$

Approximate pK_a Values

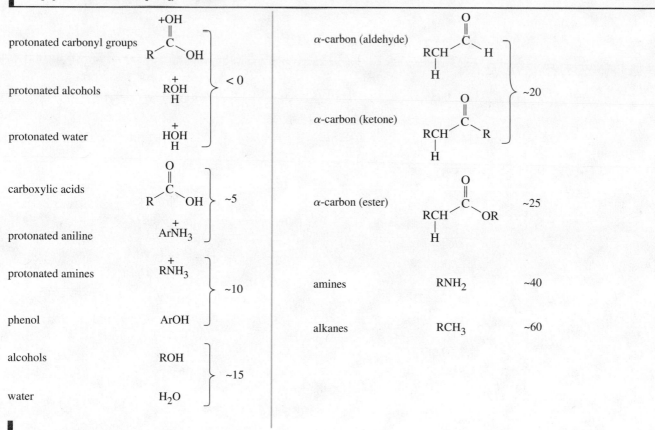

protonated carbonyl groups		
protonated alcohols		< 0
protonated water		
carboxylic acids		~5
protonated aniline	$ArNH_3^+$	
protonated amines	RNH_3^+	~10
phenol	$ArOH$	
alcohols	ROH	~15
water	H_2O	

α-carbon (aldehyde)		~20
α-carbon (ketone)		
α-carbon (ester)		~25
amines	RNH_2	~40
alkanes	RCH_3	~60

Common Symbols and Abbreviations

Å	Angstrom unit (10^{-8} cm)	E	entgegen (opposite sides in E,Z nomenclature)	NaOCl	sodium hypochlorite
Ad	adenosyl			NBS	N-bromosuccinimide
$[\alpha]$	specific rotation	E_a	energy of activation	nm	nanometers
α	observed rotation	ee	enantiomeric excess	NMR	nuclear magnetic resonance
Ar	phenyl group or substituted phenyl group	Et	ethyl	o	ortho
		Et_2O	diethyl ether	p	para
9-BBN (R_2BH)	9-borabicyclo[3.3.1]nonane	H_2CrO_4	chromic acid	PCC	pyridinium chlorochromate
B_0	applied magnetic field	HMPA	hexamethylphosphoric acid triamide	Ph	phenyl
BH_3	borane	HOCl	hypochlorous acid	pH	measure of the acidity of a solution ($= -\log [H^+]$)
t-Boc	$tert$-butyloxycarbonyl	HOMO	highest occupied molecular orbital		
Bu	butyl	IR	infrared	pK_a	measure of the strength of an acid ($= -\log K_a$)
C_6H_5	phenyl	J	coupling constant		
CMR	^{13}C magnetic resonance	k	rate constant	ppm	parts per million (of the applied field)
D	Debye; a measure of dipole moment	K_a	acid dissociation constant		
		K_{eq}	equilibrium constant	R	alkyl group; group derived from a hydrocarbon
DBN	1,5-diazabicyclo[4.3.0]non-5-ene	LDA	lithium diisopropylamide		
DBU	1,8-diazabicyclo[5.4.0]undec-7-ene	$LiAlH_4$	lithium aluminum hydride	R,S	configuration about an asymmetric center
DCC	dicyclohexylcarbodiimide	LUMO	lowest unoccupied molecular orbital		
δ	partial			TBDMS	$tert$-butyldimethylsilyl
Δ	heat	m	meta	THF	tetrahydrofuran
$\Delta G^{\ddagger}$	free energy of activation	MCPBA	$meta$-chloroperbenzoic acid	THF	tetrahydrofolate
$\Delta G°$	Gibbs standard free energy change	Me	methyl	TMS	tetramethylsilane, $(CH_3)_4Si$
$\Delta H°$	change in enthalpy	MO	molecular orbital	Ts	tosyl group (p-$CH_3C_6H_5SO_2-$)
$\Delta S°$	change in entropy	MS	mass spectroscopy	UV/Vis	ultraviolet/visible
DH	bond dissociation energy	μ	dipole moment	X	halogen atom
DIBALH	diisobutylaluminum hydride	$Na_2Cr_2O_7$	sodium dichromate	Z	zusammen (same side in E,Z nomenclature)
DMF	dimethylformamide	$NaBH_3CN$	sodium cyanoborohydride		
DMSO	dimethyl sulfoxide	$NaBH_4$	sodium borohydride		

Index

1330

1336